# RUSSIA AND THE COMMONWEALTH OF INDEPENDENT STATES

# Russia and the Commonwealth of Independent States

## Documents, Data, and Analysis

Zbigniew Brzezinski
Paige Sullivan
Editors

The Center for Strategic and International Studies

M.E. Sharpe
Armonk, New York
London, England

**Library of Congress Cataloging-in-Publication Data**

Russia and the Commonwealth of Independent States : documents, data, and analysis /
edited by Zbigniew Brzezinski and Paige Sullivan.
p. cm.
Includes bibliographical references (p. )
ISBN 1-56324-637-6 (cloth : alk. paper)
1. Commonwealth of Independent States.
I. Brzezinski, Zbigniew K., 1928– .
II. Sullivan, Paige.
DK1.5.R876 1996
947.086—dc20
96-18164
CIP

Printed in the United States of America

The paper used in this publication meets the minimum requirements
of the American National Standard for Information Sciences—
Permanence of Paper for Printed Library Materials, ANSI Z 39.48-1984.

BM (c) 10 9 8 7 6 5 4 3 2 1
BM (p) 10 9 8 7 6 5 4 3 2 1

# Contents

## PART II. The Grand Debates: Whither, Whether, and What?

# A Note on Sources

The documentary materials translated in this collection were drawn from a wide range of sources including both broadcast and print media as well as official publications and releases. Grateful acknowledgment is made for the use of translations from the Foreign Broadcast Information Service (FBIS), the *Current Digest of the Post-Soviet Press* (*CD*), the British Broadcasting Corporation (BBC), United Press International (UPI), and other sources. Translations have been edited to enhance readability and to achieve consistency in the spelling of personal and place names. The chronologies presented in Appendixes A and B are based on an even broader array of press sources. Readers should refer to the Bibliography (Appendix F) for a listing of standard references on the CIS member states.

# Acknowledgments

The editors are pleased to note the valuable contributions made to this volume by several research assistants in the office of Dr. Zbigniew Brzezinski. The original impulse for this documentary volume arose out of a periodic chronology of key developments within the Commonwealth of Independent States (CIS) initiated by Dr. Brzezinski in late 1992. Prepared in his office, the *CIS Update* listed in chronological detail the ongoing efforts both to invigorate and to oppose the consolidation of the CIS as a substitute for the Soviet Union. It was distributed as a public service by the Center for Strategic and International Studies to scholars interested in this subject.

Over the next three years, the *CIS Update*, which was designed and edited by Gudrun Persson, was edited by Melissa Meeker and then by David Riddler respectively, with each author further refining the product. Melissa and David were also involved in collecting the structural documents for this book, after the decision was made in 1994 to take advantage of the assembled materials to produce a single volume that would trace the evolution of the CIS. Others actively involved in producing the book include Marianne Alves, Marek Michalewski, and Patrick Weide.

A special acknowledgment is due to Marek Michalewski and Patrick Weide. Marek deserves thanks for his substantial effort in assembling the statistics for, and preparing the graphs in, Appendix A, along with researching and assembling the maps reproduced there. Patrick merits special thanks for his enormous contributions to every section of the book, including compilation of the charts for Dr. Brzezinski's conclusion, preparation of the "Hot Spot" and annual chronologies, and selection of many original documents. Patrick's long and dedicated hours, as well as his substantive insights, are greatly appreciated.

To all of the above go our sincere thanks for making this final product possible.

# List of Acronyms

| | |
|---|---|
| ASEAN | Association of Southeast Asian Nations |
| ASSR | Autonomous Soviet Socialist Republic |
| BPF | Belarusian Popular Front |
| CBR | Central Bank of Russia |
| CCC | Consultative and Coordinating Committee |
| CD | Current Digest of the Post-Soviet Press |
| CEC | Central Executive Committee |
| CEU | Commission for the Economic Union |
| CFE | Treaty on Conventional Armed Forces in Europe |
| CIA | Central Intelligence Agency |
| CMEA | Council for Mutual Economic Assistance |
| CNDF | Congress of National-Democratic Forces |
| Cocom | Coordinating Committee for Multilateral Export Controls |
| CPRF | Communist Party of the Russian Federation |
| CPSU | Communist Party of the Soviet Union |
| CSCE | Conference on Security and Cooperation in Europe |
| DPR | Democratic Party of Russia |
| EC | European Community |
| EEC | European Economic Community |
| EES | European Economic Space |
| EFTA | European Free Trade Association |
| EU | European Union |
| FIGs | Financial and industrial groups |
| FBIS | Foreign Broadcast Information Service |
| FRG | Federal Republic of Germany |
| GSVG | Group of Soviet Forces in Germany |
| ICBMs | intercontinental ballistic missiles |
| ICC | Interstate Currency Committee |
| IEC | Interstate Economic Committee |
| IGA | Islamic State of Afghanistan |
| IMF | International Monetary Fund |
| INOBIS | Independent Institute of Defense Studies |
| IPA | Interparliamentary Assembly |
| IRP | Islamic Renaissance Party |
| LDPR | Liberal Democratic Party of Russia |
| MID | Ministry of Foreign Affairs |
| MKZhs | International Confederation of Journalist Unions |
| NAFTA | North American Free Trade Agreement |
| NATO | North Atlantic Treaty Organization |
| NIS | newly independent states |
| OPEC | Organization of Petroleum Exporting Countries |
| OSCE | Organization for Security and Cooperation in Europe |
| RDDR | Russian Democratic Reform Movement |
| RFE–RL | Radio Free Europe–Radio Liberty |
| RSFSR | Russian Soviet Federated Socialist Republic |
| SALT | Strategic Arms Limitation treaty |
| SFRYu | Socialist Federated Republic of Yugoslavia |
| START | Strategic Arms Reduction Talks treaty |
| VKP | General Confederation of Trade Unions |
| WEU | Western European Union |

# Russia and the Commonwealth of Independent States

# Introduction: Last Gasp or Renewal?

*Zbigniew Brzezinski*

Will the Russian empire be the first in history to have experienced both dissolution and restoration? This is the key question posed by the formation and the inner dynamics of the Commonwealth of Independent States (CIS), the successor to the Soviet Union. The answer will have vital significance for the geopolitical shape of the world.

The Soviet Union was the successor to the tsarist Russian empire. In both, the inner and dominant core of imperial might was the Russian state, with the Kremlin as its seat of power. With the dissolution of the Soviet Union, the question consequently arises: What in fact is Russia? Most importantly, how does Russia define itself: as a multinational empire or as a national state? That issue perplexes modern Russians as much as it affects their immediate non-Russian neighbors.

A Russia that defines itself as a national state might not always live in peace with its neighbors, but such a self-definition would thereby acknowledge the separate political identity of the non-Russians. A Russia that sees itself as something more than a national state, however, and as the source of a supra-national and quasi-mystical identity, endowed with a special mission in the huge Eurasian geopolitical space formerly occupied by the Soviet Union, is a Russia that claims the right to embrace its neighbors in a relationship that, in effect, denies to them not only genuine sovereignty but even a truly distinctive national identity.

For example, in the view of many Russians today, Ukraine is not a state inhabited by a genuinely authentic and separate nation. Most Russians (according to public opinion polls) perceive Ukrainians as part of a larger Slavic family in which the Ukrainians are a somewhat distinctive but not really different people—"little brothers" who originate from the same cultural roots and thus share a common historical destiny. According to that mindset, it is only natural for the Ukrainian people to be part of a larger multinational state in which effective power is wielded by the Russian majority, that is, "the elder brother." For about a century, the Russian attitude toward Poland was rather similar, and it took several decades of Polish national independence for the Russian attitude eventually to evolve into an acceptance of Poland's separate national status.

It may be helpful to note here some suggestive parallels, both historical and terminological, with the experience of another great imperial nation, Britain. It took several decades for the English to adjust to the notion that the Irish did not willingly belong to the British empire. Moreover, even the difference between being "English" (the inner, dominant ethnic element) and being "British" (the Scots, Welsh, and until not long ago the Irish, as well as the English themselves) has subtle Russian equivalents. "Rossiyskoe gosudarstvo" (which can be translated roughly as the Great Russian state) designates more than a purely Russian state, with non-ethnic Russians also part of that multinational state. It is, in effect, the equivalent of Great Britain. A purely ethnic designation for the Russian state (or equivalent to England) would be "Russkoe gosudarstvo." A "Russkiy" is an ethnic Russian—that is, like an Englishman; a "Rossiyanin" can be a Russian or a non-Russian citizen of the Great Russian empire—that is, like the British.

Thus what happens to the CIS and what happens to Russia are two sides of a single coin. Whatever the CIS becomes—a Moscow-dominated empire, with an increasingly integrated political structure; a zone of protracted ethnic turmoil; or a relatively stable community of cooperative but essentially politically sovereign states—will also determine what Russia becomes. This is an issue that the Russians are debating fiercely, for they know that their future is very much at stake. In many respects, their debate is reminiscent of the choice that Turkey faced earlier in this century: either to be transformed into a more modern, essentially national, and increasingly European post-imperial state, or to remain the center of a multinational empire.

That ongoing grand debate, cast in geopolitical and cultural terms, revolves around a central issue: Is Russia European or Eurasian? Is Russia's destiny to be part of Europe, or is it to play a separate role as the dominant Eurasian power? Given that Russia's demographic center of gravity is in Europe, and that its culture is derived from Byzantine Christendom, Russia can be legitimately seen as the eastern extremity of Europe, in both geopolitical and cultural terms. On the purely personal level, an educated Muscovite or St. Petersburger finds himself more at home in Paris or Berlin, not to speak of Warsaw, than in Beijing, Tokyo, or Singapore. In that respect, Russia is clearly also European.

Those who advocate the European option thus desire to see Russia increasingly move into a closer cooperation with the emerging larger Europe. Abstracting from the specific issues of eventual membership in either the European Union (EU) or the North Atlantic Treaty Organization (NATO), the further implication of a Russia that defines itself as a European state is that Russia evolves into a stable democracy, based essentially on a national state (though a federative one, given that about 20 percent of its population is still non-Russian), while remaining linked in an economically cooperative but politically nondominant relationship with the other members of the CIS. Western-type modernity, social prosperity, and democracy are the central strategic goals of those who advocate the European option.

The Eurasian option is quite different. It posits the proposition that Russia geopolitically and culturally is neither quite Europe nor Asia—but that it has a distinctive Eurasian identity of its own. Instead of focusing on the demographic and cultural bonds with Europe, Eurasianism stresses the special legacy of space-control over the enormous landmass between eastern Europe and the shores of the Pacific Ocean, and of a separate imperial statehood that Moscow shaped through four centuries of eastward expansion. That expansion incorporated and assimilated into Russia a large non-Russian and non-European population, creating thereby also a distinctive Eurasian political and cultural identity. Eurasianism in that sense is both a reality and a mission.

Eurasian doctrines are not new in today's Russia. They started surfacing in the late nineteenth century and became more pervasive in the twentieth. The tsarist empire, until its very last phase, did not need special ideological legitimation. Russian Orthodoxy provided the divine sanction for the imperial mandate. The non-Russians of Eurasia—despite some opposition—were in the main neither politically nor nationally activated and could thus be subsumed in a Russia that was de facto both Eurasian and imperial.

Eurasianism surfaced as a more articulate doctrine in opposition to the communism of the post-revolutionary Soviet Union and in reaction to the alleged decadence of the West.[1] Russian émigrés were especially active in articulating Eurasianism as an antidote to Sovietism, realizing that the gradual awakening of the non-Russians required an overarching doctrine to justify the existence of the Eurasian empire. In the Soviet Union, communism as an internationalist ideology performed that legitimizing function for the non-Russians, and even the very concept of the "Soviet Union" involved the formal recognition of the changing status of the non-Russians. If communism were someday to fade, some other unifying doctrine would be needed to prevent imperial fragmentation.

Eurasianism as a doctrine involved a rejection not only of communism. It was also contemptuous and hostile to the European idea. It repudiated the notion advocated by Russian westernizers that Russia should eventually become part of the West. While communism was seen as a betrayal of Russian Orthodoxy and of the special Russian identity, Europeanism was repudiated because the West was seen as corrupt, anti-Russian culturally, and inclined to deny Russia its right to exclusive control over the Eurasian landmass.

An influential theorist of Eurasianism was Prince Nikolay S. Trubetzkoy. He made the case (in the mid-1920s) that "The Eurasian world represents a self-contained geographical, economic, and ethnic whole distinguishable from both Europe and Asia proper." According to him, the political unification of Eurasia was a historical inevitability, conditioned by its geography and historical experience. The geography is determined by four parallel zones stretching from west to east (the tundra, the forests, the steppes, and the mountains), intersected by north–south rivers, and forming a single unit. The history was determined by the unification of that territory through the Mongolian empire, which subsequently the Muscovite Principality conquered. The Orthodox Russian tsar infused the new imperial and increasingly Eurasian Russia with both a religious and a national content all of its own.

In Trubetzkoy's view, communism was a startling departure from that enduring tradition. "Communism was in fact a disguised version of Europeanism in destroying the spiritual foundations and national uniqueness of Russian life, in propagating there the materialist frame of reference that actually governs both Europe and America, and in nurturing Russia on European theories with deep roots in the soil of European civilization." Thus, if the Russian people "were to reject communism with its European origins, then the connection between it and European civilization would be severed, and the task of strengthening and developing the national historical life of Russia could begin."

He concluded his historical treatise with a ringing appeal to his countrymen: "Our task is to create a completely new culture, our own culture, which will not resemble European civilization. . . . It will be possible only when the attractions of European civilization and of ideologies invented in Europe are exhausted once and for all, when Russia ceases to be a distorted reflection of European civilization and finds her own unique historical nature, when she becomes once again herself: Russia-Eurasia, the conscious heir to and bearer of the great legacy of Genghis Khan."[2]

Small wonder then that in the confused and unsettling post-imperial conditions of the mid-1990s, Eurasianism has been increasingly the vogue among many post-Soviet Russian strategic thinkers. It provides the geopolitical justification, and the cultural legitimacy, for a Russia with a special vocation in the space both of the former Soviet Union and of the former tsarist empire. It projects a Russia that is neither Europe nor Asia, but dominant on the Eurasian continent, with an identity all its own, destined to play (though the words are not used explicitly) an imperial role. Global status as a geopolitical superpower, spatial control, and regional hegemony are the central strategic goals of the Eurasian option.

But is imperial restoration a practical option for Russia? Some Russian nationalists seem to think so. It is not only the extremists—like Vladimir Zhirinovskiy—that do. Aleksandr Rutskoy, the former vice president and a possible presidential candidate, speaks for a very large segment of the Russian people, and reflects a deeply rooted impulse, when he asserts that "it is apparent from looking at our country's geopolitical situation that Russia represents the only bridge between Asia and Europe. Whoever becomes the master of this space will become the master of the entire world. This is why Russia must continue to be a great power."[3] In keeping with this viewpoint, Rutskoy in mid-1994 launched his "great power movement," dedicated explicitly to the idea of restoring Russian dominion within the space once occupied by the Soviet Union (and reported that within only a few weeks he had gained about half a million adherents).

How realistic is that goal? Until this century, great empires endured largely because the subject peoples were politically passive and nationally unaware. Once that passivity and absence of national consciousness started fading, the preservation of empire entailed increasing costs. Democratic imperial home states—such as Britain, France, Holland—were ultimately unwilling to pay the price that increased coercion would have required. Some, like the British, came to that conclusion relatively early, in the face of the growing but essentially peaceful political mass mobilization in India. Some, like the French, learned the lesson more painfully, in the course of the prolonged Algerian liberation struggle.

The Russian empire was exposed to that lesson as well, largely through the Polish uprisings of the nineteenth century. But outside Poland, the level of political self-awareness among the non-Russians was relatively low until well into the twentieth century. Thus the tsarist regime did not pay unbearable costs to sustain itself. Its defeats in the Russo-Japanese war of 1905 and in World War I, and even the subsequent Russian civil war, did not precipitate the empire's general dissolution; only the secession of its more politically and nationally conscious western provinces: Poland, Finland, and the Baltic republics. The Communists thereby inherited an imperial state, though one in which the non-Russians were beginning for the first time to stir and awaken.

The Soviet Union was the new formula for the old tsarist empire. It aimed at the preservation of political power by the center—though that center no longer officially defined itself purely in ethnic, Great Russian terms. (Indeed, in the early phase of Soviet power, the top leadership was composed of an amalgam of Russian, Jewish, Lettish, Polish, Ukrainian, and Caucasian Bolsheviks.) The system at large was to be simultaneously "socialist in content and national in form," which meant that centralized power, which was used to impose an ideologically driven social reconstruction, was to be combined with some respect for non-Russian linguistic, cultural, and even historical diversity. National republics were officially established, their state borders formally (and often very arbitrarily) drawn on the map, and nominal republican governments set up. The pretense was that the Soviet Union was a voluntary association of constituent states.

That diversity, however, was not permitted to cross over into the political realm, and the longer the Soviet system existed, the stronger became the emphasis on the dominant role of the Great Russian "elder brother." Precisely because some degree of politically oriented national awakening among the non-Russians was inevitable, the Kremlin—especially under Stalin—was ruthless in decapitating any non-Russian political elites suspected of harboring national aspirations. In that respect, compulsion as the means of imperial preservation came to be a centrally important instrument. It generated an external image of unanimity that even led many foreign observers to conclude that, indeed, a new formation—a Soviet nation or a Soviet man—had emerged, immunizing the Soviet Union against the fates that befell the British or the French empires.[4]

It was not to be. Contrary to that viewpoint, the Soviet Union did not succeed in immunizing itself against the general consequences of the age of nationalism and more specifically of the contagious effect of post–World War II decolonization. The death of Stalin, moreover, made it more difficult to rely on coercion alone. The post-Stalin leadership felt it had to cultivate the non-Russian Communist elites, thereby indirectly propitiating the gradual spread of nationalist sentiments. For example, Nikita Khrushchev presented Crimea to Ukraine as "a gift" on the occasion of the four hundredth anniversary of Ukraine's incorporation into Russia. In Central Asia, the diversion of large-scale Soviet resources for the upgrading of the republican capitals and greater reliance on ethnically native Communist rulers was also motivated by concerns that global decolonization could prove contagious.

The crisis of survival that the Soviet Union came to face

in the 1980s thus occurred in a setting in which the non-Russian components of the old empire were becoming more assertive and restless. But with the exception of the Baltic republics, and perhaps Georgia, the restlessness was not of the type associated either with Polish nationalism in the tsarist empire or the Irish in the British or the Algerian in the French. It was more a matter of a generalized disaffection on the part of the masses, and especially their intellectuals. This disaffection focused on the demeaning character of the officially imposed worship of everything that was Russian. It also reflected a greater desire for broader autonomy on the part of the national republican, though also still both Soviet and Communist, leaderships.

The disintegration of the Soviet Union in December 1991 was not, therefore, the consequence of an irresistible wave of non-Russian political self-assertion. The old Soviet Union had outlived its day, but the precipitating impetus for its disintegration came from the political conflicts in Moscow, from the struggle between Boris Yeltsin and Mikhail Gorbachev within a demoralized and fractionated Communist Party of the Soviet Union (CPSU)—reinforced by pressures from outside the center for a genuine redistribution of power (especially economic power) between Moscow and the formally "sovereign" capitals in Kyiv, Tbilisi, or Tashkent. The combination of these forces led, to the astonishment of all those who had assumed that the national problem in the Soviet Union no longer existed, to the rapid dissolution of the Soviet Union and its replacement by a quickly improvised new entity called the Commonwealth of Independent States.

However, the majority of the Communist leaders of these independent states, it must be bluntly stated, had their independence bestowed, or even inflicted, upon them. They did not win it either through a national liberation struggle or even by a political contest. To be sure, there were national dissidents, who were willing to defy Soviet coercion and who suffered the personal consequences of their courage and dedication. Nonetheless, the independence that came in late 1991 came largely because the center of Soviet power had collapsed so rapidly. The local, national Communist elites had no choice but to take the lead in a wave of proclamations of national sovereignty and independence, thereby joining in the sudden outburst of national awakenings.

The only exception, but a very important one, was Ukraine. By the summer of 1991, public agitation for national independence infected even the local Communist leadership in Kyiv. Thus both the existing Communist establishment and the long-suppressed nationalist activists joined forces in taking advantage of the Soviet Union's crisis to infuse real substance into Ukraine's formal status as a "sovereign" Soviet republic. That, in the wake of the Soviet Union's dissolution, subsequently provided the basis for Ukraine's firm self-assertion.

Broadly speaking, one can discern the following progression in the process of self-assertion among the non-Russians: The Baltic republics were the first to experience mass demonstrations, with dissatisfaction from below led by a leadership composed of both anti-Communists and non-Communists as well as even some Communist leaders who were swept up by the groundswell of national emotions. The nationalism of the Baltic republics was pervasive, and not surprisingly so, given their previous national independence. Conditions in the Caucasus were in some respects similar, in part because of the deeply entrenched sense of national identity among the Georgians, Azeris, and Armenians. There, too, non-Communists initially took the lead. The above pattern was followed by Ukraine, which as already noted created a surprising coalition of Communists and non-Communists in the progressive assertion of independence, taking advantage of Ukraine's nominal status as a Soviet republic and very deliberately infusing that status with substance even before the formal breakup of the Soviet Union. Last were the Central Asian states as well as Belarus, which were characterized by the absence of bottom–up agitation and by the rapid usurpation of the independence movement by the established nomenklatura.

In any case, as already noted, by 1991 the old formula of the Soviet Union had outlived its day. But its enduring legacy was the highly interdependent economy, which made it more difficult for the newly sovereign republics fully to take charge of their destinies. The collapse of the Union—and of its various institutions of power—spawned a series of states that in the majority of cases lacked not only the ecstatic sense of national emancipation required to sustain the social sacrifice involved in constructing genuinely independent statehood, but even the minimal capacity for economic self-determination.

In that setting, it was unclear whether the newly proclaimed Commonwealth of Independent States was merely a euphemism for a partially restructured Soviet empire or the framework for "a civilized divorce" (to use the apt phrase of the freshly installed Ukrainian president). That very ambiguity dominated the early months of the CIS. But as the documents that follow indicate, the ambiguity endured longer among the astonished non-Russian elites who suddenly found themselves enjoying the fruits and the trappings of national independence than among the Russians themselves.

Within months of the dissolution of the Soviet Union, a series of spontaneous initiatives originated from Moscow, designed to flesh out and to deepen the scope of CIS cooperation, to enhance its status, to create conditions in which the CIS would, at the very outset, be more than the British Commonwealth, then approximate the European Union, then move beyond a confederation or even a federation to eventually become perhaps again just a multinational state dominated by its largest nation.

How effective are these efforts likely to be? Much depends on the strength of the imperial impulse among the Russians and on the tempo of the spread of nationalism within the non-Russian states. In the final analysis, trends within the CIS thus involve a historically significant race between the growing inclination of initially lethargic national elites to value their independence and the Russian determination to exploit their difficulties—especially economic ones—in order to promote a more integrated and centrally controlled entity.

However, how integrated that entity ought to be is also a subject of intense contention among leading Russians. Some Russian politicians, as already noted, believe that an essentially unitary state—in effect, a Eurasian Russian empire—ought to be and will be the final stage. Russian army leaders clearly sympathize, and their efforts to subordinate the nascent non-Russian military establishments under Moscow's control (see Chapter 9) appear to be part of a larger geostrategic design to regain control over the space of the former Soviet Union. (The military determination, manifested immediately after December 1991—to preserve strategic enclaves on the outer fringes of the old Soviet Union, in Kaliningrad, Moldova, Crimea, Abkhazia, Tajikistan, and the Kuriles—was in large measure an instinctive response to the empire's dissolution.)

Other Russian leaders entertain a more enlightened view. In one of the documents cited in Chapter 3 (3.36), the first post-Soviet ambassador of Russia to the United States and currently the chairman of the Foreign Affairs Committee of the Duma, Vladimir Lukin, took on those Russians who, like Rutskoy, have argued explicitly for the restoration of "the Russian-Soviet empire." In Lukin's view, that option "is unthinkable either technically, without prohibitive sacrifices and costs, or politically because Russia would then find itself in a hostile isolation even more dangerous than the one of the Cold War era." And he adds that "all this says nothing about whether this variant is compatible with preserving democracy in Russia."

Lukin suggests instead that "the long-term ultimate task here is to create, by stimulating the natural integration processes rather than by compulsion, close allied relationships of a confederative type. Russia needs to have an internal ring of friends in addition to an outer ring of partners." Though he does not identify which of the former Soviet republics fit into the category of "friends" and which of "partners," he does make it clear that Russia should have a special role as the center of the inner ring: "The level of our relationships with friends should as a rule be qualitatively higher than of our relationships with our partners, and the level of relationships of our partners with our friends should be qualitatively lower than that between us and our friends. . . . There should also be a mechanism in place—ideally a collective one *with a Russian basis*—for resolving extreme situations with the use of force a factor" (emphasis added).

Lukin's emphasis on the special role of Russia highlights the acute dilemma that confronts the moderate Russian nationalists. For the imperial nationalists, the issue is clear: Central control of a unitary Eurasian state by Moscow is the ultimate goal. For the moderates, the situation is more complex: They abjure the goal of imperial restoration, yet they desire closer integration and a special political role in it for Russia. Quite often they point to the European Union as their eventual model, but they overlook several decisive differences between contemporary Europe and the former Soviet Union.

Not only is Europe being integrated by stable democracies—and none of the ex-Soviet republics currently are, nor is Russia itself—but within Europe there is no comparable disproportion in power to that which prevails between Russia and the other former Soviet republics. Germany is militarily integrated into NATO and in that domain is more than balanced by France and the United Kingdom; economically Germany is much stronger than any other European state, but it can also be balanced by a coalition of any two or three major European states while within the EU decision-making process an elaborate voting system of checks and balances effectively precludes domination by any single state. Last but not least, the American presence in Europe, through NATO, also provides welcome reassurance against the domination of Europe by any single power.

The situation in the former Russian empire is dramatically different. Russia is stronger militarily, economically, and politically—and stronger by far—than any of the other former Soviet republics. Even a coalition of all of them could not match Russia's power. Thus any steps by the former republics in the direction of integration—even only economic (including the most rational ones)—have the effect of enhancing Russia's political leverage. The Russians as well as the non-Russians know this—and that is why Ukraine, as the largest and nationally the most ambitious of the newly independent states, and (to a lesser extent) the two potentially economically self-reliant Central Asian states, Turkmenistan and Uzbekistan, have been ambivalent regarding closer economic integration.

The documents selected for Chapter 4 illustrate this ambivalence. On the one hand, the necessity of such "integration" is recognized and its potential benefits much desired, especially given the degree of economic interdependence fostered by the Soviet regime; on the other hand, its political consequences are feared, especially as the new national political elites deepen their attachment to the perquisites of power and as the sense of national awareness becomes socially more pervasive. That is why Ukraine for several years has resisted Russia's entreaties to join the CIS's Economic Union, and that is why the assertive presidents of Kazakhstan and Uzbekistan, though joining the Economic Union, have also been promoting a Central Asian Confederation.

In any case, the asymmetry in power and the painful memories of Russian rule make the non-Russians wary even of the more moderate Russian designs. That injects an enduring element of tension into inner CIS relations. And it gives added salience to the question posed at the outset of this introduction.

It is too early to postulate an answer. A tentative one will be attempted in the conclusion to this volume. The tragedy of Chechnya—the unspeakable slaughter of its population by the Russian Army in late 1994 and early 1995—indicates not only how volatile and explosive that question is but also how difficult it is to provide a categorical answer.

On the one hand, the decision to suppress the Chechen quest for independence—and the manner in which it was executed—bespeaks the worst in the Russian and Soviet imperial tradition. The behavior of the Kremlin seemed almost designed to signal to the non-Russians Moscow's determination to remain the capital of much more than a national state. Viewed in conjunction with other actions taken to reassert Russia's dominant military-political presence for example, in Georgia, Tajikistan, and elsewhere) as well as the energetic pursuit of the efforts to centralize CIS machinery, Chechnya could be viewed as the extreme manifestation of a dominant political momentum within Russia toward an even more assertive imperial restoration.

On the other hand, the above is not the full story. The popular reaction to the war against the Chechens was heavily negative. Not only the atrocities involved but the decision to use force produced large-scale social repugnance. Though it is too early to tell what are the full political ramifications of these social attitudes, they do suggest that there are limits to popular support for policies of overt and coercive imperial restoration. This is not to say that the Russians lack the desire for a revival of the old "Union," but it is to make the point that the public may not be prepared to support imperial policies that would entail excessive costs and burdens.

The chapters that follow document the Russian efforts since 1992 to make the CIS into a more viable instrument of economic and political integration, and the ambivalent—and, in some cases, negative—responses from the newly independent non-Russian states. The materials assembled in this volume should, therefore, help to frame the answer to the central question posed in the beginning of this introduction.

Part I provides the key documents pertaining to the dissolution of the Soviet Union, including the last-minute but aborted attempt to adopt some form of genuine federalism.

Part II traces, in a comprehensive fashion, the debates that ensued in Russia on the sudden discovery that it was no longer the center of an empire. How to explain that; whom to blame; and what to do about it, was the natural, though belated, reaction to the replacement of the USSR by the CIS. It also examines the evolving reaction of the non-Russian elites to the novel post-Soviet context, taking special note of the distinctive Ukrainian posture; reviews several proposed alternatives to the CIS itself; and traces more specifically Russian efforts to revive economic integration and to reinforce CIS political institutions.

Part III then provides an overview of the actual structures, agreements, and protocols that emerged from the foregoing debates and initiatives. On the one hand, it is evident that a major enhancement of the CIS has occurred during the past several years. On the other, it is also clear that opposition to further integration remains strong, either (and most often) through evasion of implementation or through outright refusal to take part. The latter is most obviously the case with Ukraine, especially in regard to the politically sensitive matters of joint security and border controls.

All the foregoing include introductory notes (prepared by Paige Sullivan) designed to give the reader the pertinent historical context, to define more precisely the issues that have dominated the CIS debates, and to draw attention to particularly important aspects of the documents subsequently cited. The Conclusion attempts to delineate the likely prospects for the CIS.

Appendices A and B, the chronologies of key events, are meant to give the reader the opportunity to sense the degree to which what has transpired has been the consequence of a spontaneous reaction by the Russians to salvage as much as possible of the former Soviet Union and the degree to which these efforts, over time, assumed a more deliberate and sustained character. At first, many of the reactions—whether in the five principal areas of ethnic conflict covered in the pages that follow, or more generally in regard to the political, defense, and economic issues that are also addressed in the chronologies—appear largely spontaneous. That, in itself, is impressive testimony to the great-power instinct of many Russians.

However, with the passage of time, one cannot suppress the feeling that a larger design is also being pursued, guided by a historically driven vision. In any case, it is hoped that the chronologies will be useful to the reader not only factually but also as a source of a more dynamic perspective regarding post-Soviet developments. The Western press has often been negligent in noting some of the more significant aspects of the institutional development of the CIS, giving perhaps an excessive impression that the CIS is largely a failure.

Additional appendices present data for the CIS and profiles of the member states as well as directory and bibliographic information.

The editors hope that our efforts will help scholars, analysts, businessmen, and students interested in post-Soviet affairs gain a clearer perspective on the future of a geopolitically vital portion of Eurasia—a future that is still very much subject to conflicting political pressures, national aspirations, economic dynamics, and popular passions.

## Notes

1. For an excellent historical review and summary of Eurasianism's principal tenets, see Françoise Thom, "Eurasianism: A New Russian Foreign Policy?" *Uncaptive Minds*, Summer 1994.

2. N.S. Trubetzkoy, "The Legacy of Genghis Khan: A Perspective on Russian History Not from the West but from the East," *Crosscurrents* (A Yearbook of Central European Culture), no. 9, (1990), pp. 19–20, 58, 60, 68.

3. Interview with *L'Espresso* (Rome, 15 July 1994).

4. This writer encountered that viewpoint while serving as the assistant to the president of the United States for national security affairs in the years 1977 to 1981. Having felt for years that the national problem was the Achilles' heel of the Soviet Union—a viewpoint that he first expressed as a graduate student in his M.A. dissertation—he proposed shortly after entering the White House that an interagency group be established to monitor Soviet nationality problems and to formulate an appropriate U.S. policy toward them. His initiative was opposed (unsuccessfully) by State Department officials on the ground that the Soviet Union did not face any significant national problems that called for a U.S. policy response. The president subsequently approved some specific initiatives in support of non-Russian national aspirations.

# Commonwealth of Independent States

# Russia's Ethnic Republics

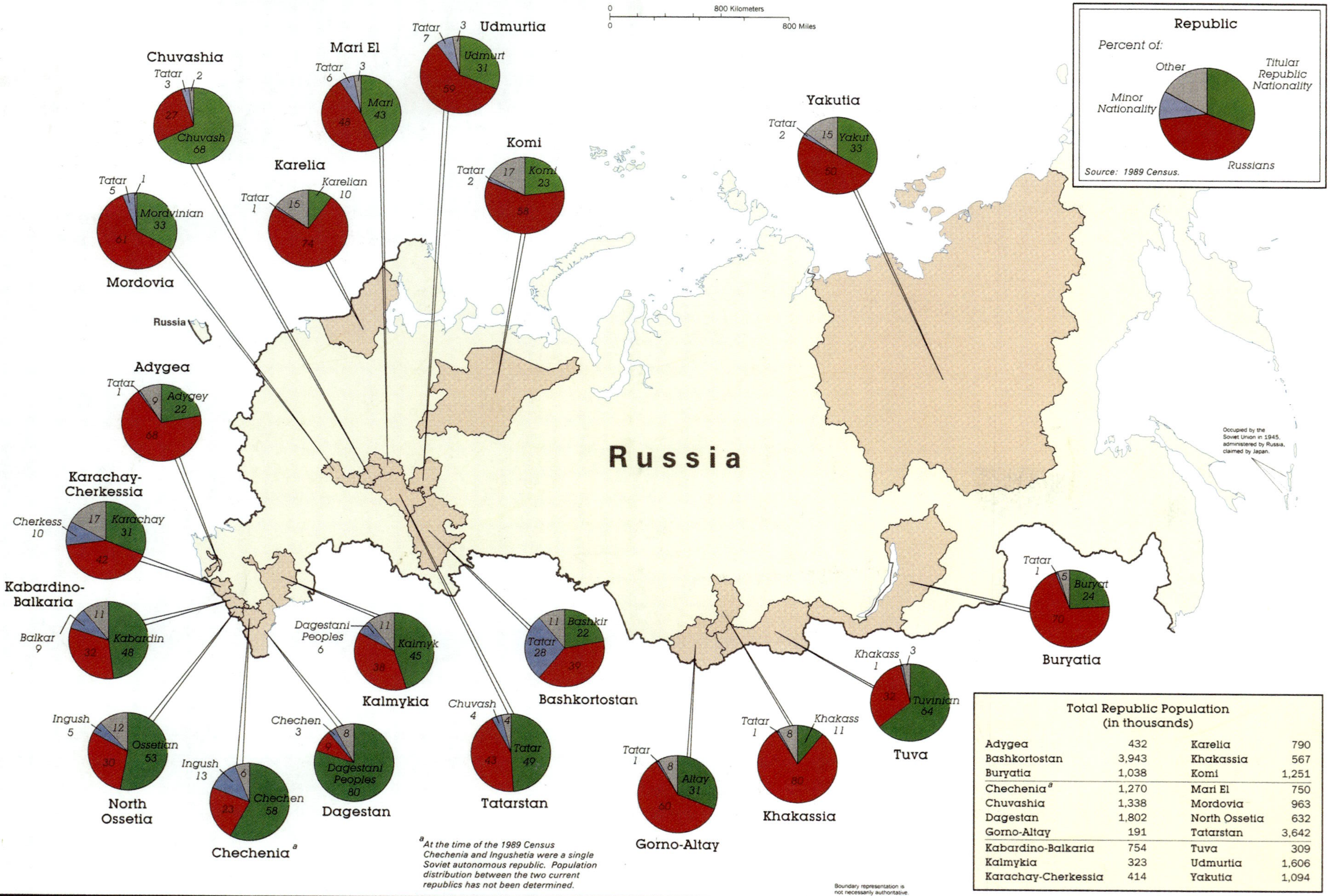

Republic
Percent of:
Titular Republic Nationality
Russians
Minor Nationality
Other
Source: 1989 Census.

0 800 Kilometers
0 800 Miles

Russia

Chuvashia: Chuvash 68, 27, Tatar 3, 2
Mari El: Mari 43, 48, Tatar 6, 3
Udmurtia: Udmurt 31, 59, Tatar 7, 3
Mordovia: Mordvinian 33, 61, Tatar 5, 1
Karelia: Karelian 10, 74, Tatar 1, 15
Komi: Komi 23, 58, Tatar 2, 17
Yakutia: Yakut 33, 50, Tatar 2, 15
Adygea: Adygey 22, 68, Tatar 1, 9
Karachay-Cherkessia: Karachay 31, 42, Cherkess 10, 17
Kabardino-Balkaria: Kabardin 48, 32, Balkar 9, 11
North Ossetia: Ossetian 53, 30, Ingush 5, 12
Chechenia[a]: Chechen 58, 23, Ingush 13, 6
Dagestan: Dagestani Peoples 80, 9, Chechen 3, 8
Kalmykia: Kalmyk 45, 38, Dagestani Peoples 6, 11
Tatarstan: Tatar 49, 43, Chuvash 4, 4
Bashkortostan: Bashkir 22, 39, Tatar 28, 11
Gorno-Altay: Altay 31, 60, Tatar 1, 8
Khakassia: Khakass 11, 80, Tatar 1, 8
Tuva: Tuvinian 64, 32, Khakass 1, 3
Buryatia: Buryat 24, 70, Tatar 1, 5

Russia

Occupied by the Soviet Union in 1945, administered by Russia, claimed by Japan.

Total Republic Population (in thousands)

| Republic | Population | Republic | Population |
|---|---|---|---|
| Adygea | 432 | Karelia | 790 |
| Bashkortostan | 3,943 | Khakassia | 567 |
| Buryatia | 1,038 | Komi | 1,251 |
| Chechenia[a] | 1,270 | Mari El | 750 |
| Chuvashia | 1,338 | Mordovia | 963 |
| Dagestan | 1,802 | North Ossetia | 632 |
| Gorno-Altay | 191 | Tatarstan | 3,642 |
| Kabardino-Balkaria | 754 | Tuva | 309 |
| Kalmykia | 323 | Udmurtia | 1,606 |
| Karachay-Cherkessia | 414 | Yakutia | 1,094 |

[a] At the time of the 1989 Census Chechenia and Ingushetia were a single Soviet autonomous republic. Population distribution between the two current republics has not been determined.

Boundary representation is not necessarily authoritative.

730319 (R00535) 11-93

# The Caucasus and Central Asia

# Major Ethnic Groups in Central Asia

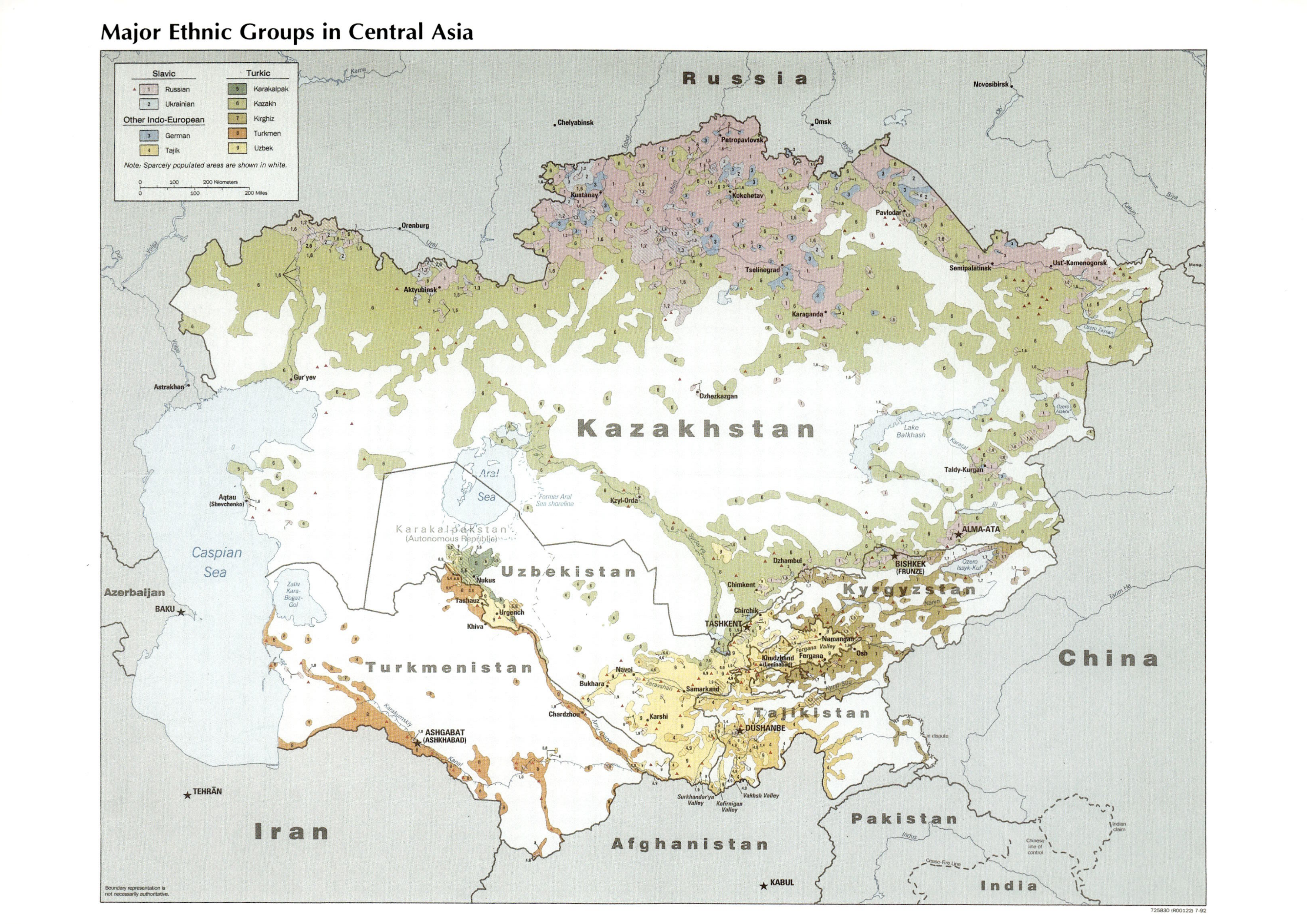

# Major Muslim Ethnic Groups in Armenia, Iran, and the Islamic Commonwealth States

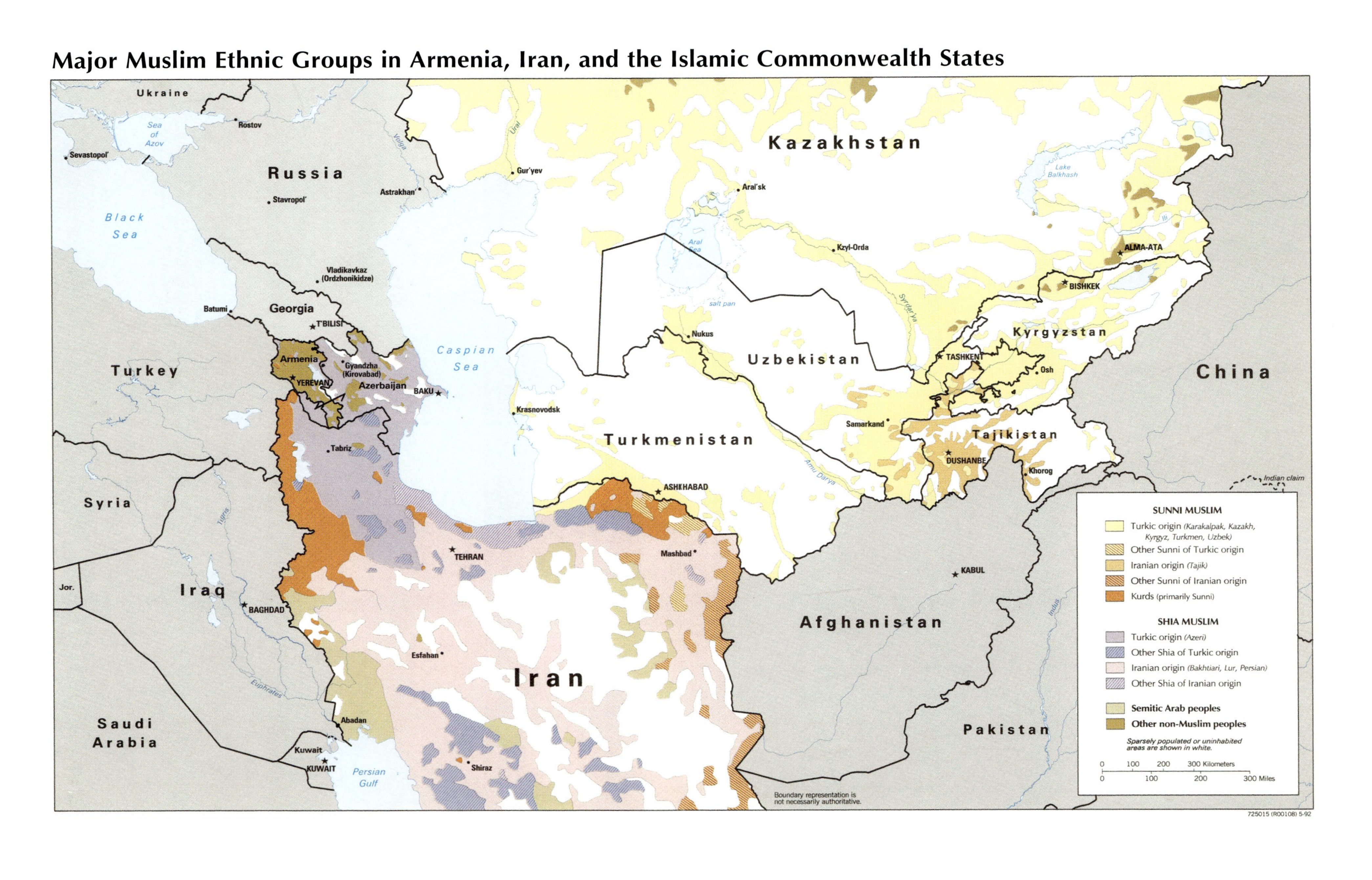

Ukraine
Sea of Azov
Sevastopol'
Rostov
Russia
Stavropol'
Astrakhan'
Volga
Black Sea
Vladikavkaz (Ordzhonikidze)
Batumi
Georgia
T'BILISI
Turkey
Armenia
YEREVAN
Gyandzha (Kirovabad)
Azerbaijan
BAKU
Caspian Sea
Tabriz
Syria
Jor.
Iraq
BAGHDAD
Tigris
Euphrates
Saudi Arabia
Kuwait
KUWAIT
Persian Gulf
Abadan
TEHRAN
Esfahan
Shiraz
Iran
Mashbad
Ural
Gur'yev
Krasnovodsk
Kazakhstan
Aral'sk
Aral Sea
salt pan
Nukus
Kzyl-Orda
Syrdar'ya
Uzbekistan
Turkmenistan
ASHKHABAD
Amu Darya
Samarkand
TASHKENT
Lake Balkhash
Ili
ALMA-ATA
BISHKEK
Kyrgyzstan
Osh
Tajikistan
DUSHANBE
Khorog
China
Indian claim
Afghanistan
KABUL
Indus
Pakistan
Boundary representation is not necessarily authoritative.
SUNNI MUSLIM
Turkic origin (Karakalpak, Kazakh, Kyrgyz, Turkmen, Uzbek)
Other Sunni of Turkic origin
Iranian origin (Tajik)
Other Sunni of Iranian origin
Kurds (primarily Sunni)
SHIA MUSLIM
Turkic origin (Azeri)
Other Shia of Turkic origin
Iranian origin (Bakhtiari, Lur, Persian)
Other Shia of Iranian origin
Semitic Arab peoples
Other non-Muslim peoples
Sparsely populated or uninhabited areas are shown in white.
0 100 200 300 Kilometers
0 100 200 300 Miles
725015 (R00108) 5-92

# Major Defense Facilities in the CIS Territory

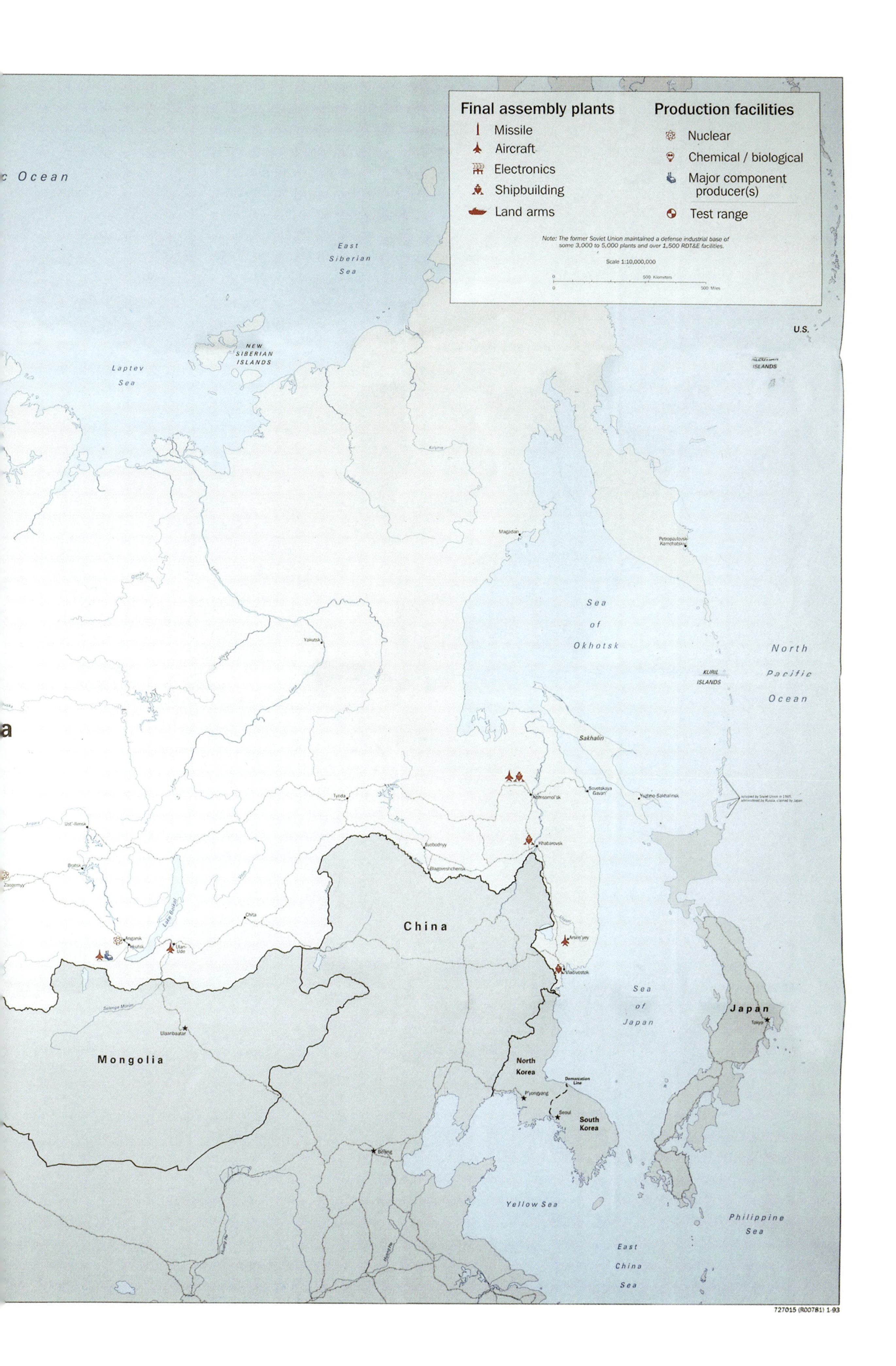
Final assembly plants
Missile
Aircraft
Electronics
Shipbuilding
Land arms
Production facilities
Nuclear
Chemical / biological
Major component producer(s)
Test range
Note: The former Soviet Union maintained a defense industrial base of some 3,000 to 5,000 plants and over 1,500 RDT&E facilities.
Scale 1:10,000,000
c Ocean
East Siberian Sea
Laptev Sea
NEW SIBERIAN ISLANDS
U.S.
Magadan
Sea of Okhotsk
North Pacific Ocean
KURIL ISLANDS
Yakutsk
Sakhalin
Komsomol'sk
Tynda
Khabarovsk
Blagoveshchensk
China
Angarsk
Irkutsk
Ulan-Ude
Chita
Lake Baikal
Bratsk
Ust'-Ilimsk
Arsen'yev
Vladivostok
Sea of Japan
Japan
Tokyo
Mongolia
Ulaanbaatar
North Korea
Pyongyang
Demarcation Line
Seoul
South Korea
Beijing
Yellow Sea
Philippine Sea
East China Sea
a
727015 (R00781) 1-93

# Part I

# The Reorganization of an Empire

# 1
# The Union Treaty Fails

## Introductory Notes

The year 1991 marked the final phase in the long process of the Soviet Union's dissolution. The momentous events which transpired that year included: (a) in April, the drafting of the ill-fated "Nine-Plus-One" treaty or "Treaty of the Union of Sovereign States" at Novo-Ogarevo outside Moscow; (b) in August, an attempt by conservative members of the government to overthrow President Mikhail Gorbachev; (c) declarations of independence by almost every Soviet Socialist Republic between August and December; and (d) between October and December, a failed effort to unify the sovereign republics of the Union under the mantle of an "Economic Community" treaty. The last was an agreement assigned by the State Council to Grigoriy Yavlinskiy, author of the American-influenced "Window of Opportunity" plan for economic reform. Neither the Union treaty, nor the Economic Community treaty, however, was to succeed in halting the fast-moving train of events carrying the Soviet Union to its demise.

A bit of background on the tumultuous political context in which the Commonwealth of Independent States (CIS) was hastily conceived is helpful for understanding the later ambiguities of this organization.

Following the "velvet revolution" of 1989–90 in Central Europe, tensions mounted between Moscow and its republics, which had all declared their "sovereignty" by the end of 1990, meaning they no longer recognized the supremacy of Union laws over republic laws. This did not reflect a radical movement toward national independence in most of the republics. However, many republican leaders desired autonomy over natural resources, finances, and budget matters, as well as the right to engage in foreign trade and other foreign relationships without Moscow's supervision. Gorbachev's answer was to initiate the "Nine Plus One" process on 23 April 1991 at his dacha in Novo-Ogarevo, a rural village outside Moscow.

The "Nine Plus One" agreement, which is included in this collection of documents, conceded considerable autonomy to the leaders of the nine republics which participated (Moldova, Georgia, Armenia, and the Baltic states were absent). The "One" referred to Gorbachev. The treaty stipulated an accelerated program of economic reform, and was intended to be the first step toward a looser form of union between the Center and the republics, but one that retained Moscow as the capital of one enormous state.

Political turmoil intervened, however. President Gorbachev had, on his own without prior consultation with the Supreme Soviet or Congress of People's Deputies, convened the drafting session at Novo-Ogarevo. This omission enraged several of Gorbachev's own government appointees as well as some key members of the Supreme Soviet. A dangerous rift developed between Gorbachev and the anti-reform conservatives. Much was written in the Soviet press about this draft treaty and differing attitudes toward it, samples of which appear in this chapter. The extreme reactions to the treaty (as illustrated by the *Izvestiya* piece, 1.2) figured prominently in catalyzing the August coup attempt against Gorbachev.

Gorbachev was also embroiled in a personal and ideological power struggle with Boris Yeltsin, former Moscow Communist Party chief and ally of Gorbachev in the Politburo. Breaking with Gorbachev in 1987, Yeltsin became an increasingly vocal critic of the Soviet president's frequent compromises with conservative forces and his failure to back the "Five-Hundred Day" economic reform plan, which Yeltsin supported, in 1990. In June 1991, Yeltsin became the first popularly elected president of the Russian Federation. He immediately escalated the struggle for greater Russian autonomy from the Center. An untenable situation developed in which the USSR (represented by the "Center" in Moscow) and the Russian government (also in Moscow) coexisted, but pursued essentially incompatible goals.

The struggle for the military-industrial complex of Russia was at the apex of the political chaos that engulfed the USSR in 1991. Yeltsin attacked the national security priorities of the Soviet government, demanding a reduction in the Soviet defense budget and the conversion of a large group of military enterprises to civilian production. With regard to the "Nine Plus One" process, Yeltsin supported it, but disagreed with Gorbachev over the extent to which the Soviet constitution would supersede those of the sovereign republics. At Novo-Ogarevo, Yeltsin stressed Russia's sovereign rights and talked of "taking a renewed Russia onto the world stage as a sovereign state."

The new Union treaty was scheduled to be signed on 20 August 1991; but between 19 and 21 August, a so-called State Committee on the State of Emergency anointed itself the Soviet leadership, issuing decrees removing Mikhail Gorbachev from power and conferring upon Vice President Gennadiy Yanaev the duties of president of the USSR. As the documents in this chapter illustrate, however, the coup leaders could not cope with the pace or the consequences of the changes occurring in the Soviet Union.

As Gorbachev and his ministers tried to restore some semblance of order following the arrest of the members of the Emergency Committee, a transitional government was erected, the draft law of which appears in this chapter. This law created a "State Council," consisting of the USSR president and the highest officials of the Union republics, to provide coordinated solutions to domestic and foreign policy issues affecting the interests of the republics and the Center. It was this State Council which was put in charge of drafting a treaty on Economic Community. Grigoriy Yavlinskiy, who had previously written an economic reform plan in cooperation with a group from Harvard University, was chosen to consolidate his ideas in a new treaty. A treaty was drafted, but with great difficulty. The republics could not agree on areas of integration or supranational control. Ukraine, as was to become the rule in ensuing months, refused to sign the treaty from the very beginning. The documents in this chapter include Yavlinskiy's draft, his preamble to the draft explaining the politics of drafting such an accord, and some illuminating press reactions to this highly controversial effort.

October and November 1991 were critical months. On 18 October the Economic Community treaty was signed by eight sovereign republics—Armenia, Belarus, Kazakhstan, Kyrgyzstan, Tajikistan, Turkmenistan, and Uzbekistan. (Moldova and Ukraine subsequently initialed the treaty, on 6 November 1991.) This event was heralded as a good omen for the signing of the draft Union treaty in November. Together, these two treaties would provide for a common economic space, common military force, common foreign policy and common borders, but with a "central coordinator" (Gorbachev's term), not a central government per se. The new Union was to be called the "Union of Sovereign States." Nevertheless, as it turned out, these institutions would not be established for several years within the improvised framework of the CIS and even then they would not be unanimously accepted by the former Soviet republics.

An implacable challenge to confederal rule was posed by Russia and Ukraine. Russia could not accept the pace of economic reform or the lack of foreign policy autonomy contemplated by the new Union. Ukraine could accept neither the degree of economic control implicit in the Economic Community treaty nor the limits placed on republican military power by the new Union. In particular, as reported by *Izvestiya*, the Ukrainian Supreme Soviet chairman, Leonid Kravchuk, wanted all the republics with nuclear weapons to share authority over them.

It was Ukraine, in fact, which dealt the final overt blow to the chances for the Union treaty's survival. On 1 December it held a nationwide referendum in which 90.32 percent of the citizens of Ukraine voted for total independence from the Union. Two days later, President Yeltsin issued an official statement on Ukraine, which appears at the end of this chapter. He was the first to unilaterally recognize Ukraine—a bold gesture, which opened the question of de jure recognition of the sovereign republics as independent new states.

The last press commentary in this chapter is from *Izvestiya*. It examines the criteria on which the international community would base its recognition of Ukraine and the other former Soviet republics in days to come. The way was now open for all the former republics to opt out of the Union as totally independent states, a development which was to transpire much more rapidly than expected by the entire rest of the world. Ukraine's recognition, in effect, confirmed the reality of the collapse of the Soviet Union, though it would take thirty days for the complete drama to unfold.

## 1.1
## Draft Novo-Ogarevo Agreement (Union of Sovereign States)

*Pravda*, 27 June 1991
[FBIS Translation]

### I. Basic Principles

1. Each republic which is party to the treaty is a sovereign state. The Union of Soviet Sovereign Republics (USSR) is a sovereign, federative, democratic state, formed as a result of the unification of equal republics and exercising state power within the bounds of the powers with which the parties to the treaty voluntarily invest it.

2. The states comprising the Union retain the right to decide independently all issues of their development, while guaranteeing equal political rights and opportunities for social, economic, and cultural development to all peoples living on their territory. The parties to the treaty will operate on the basis of a combination of values common to all mankind and those belonging to individual nationalities and

will resolutely oppose racism, chauvinism, nationalism, and any attempts to restrict the rights of peoples.

3. The states comprising the Union regard as a most important principle the preeminence of human rights, in accordance with the UN Universal Declaration of Human Rights and other generally recognized norms of international law. All citizens are guaranteed the opportunity to study and use their native language, unimpeded access to information, freedom of religion, and other political, social, economic, and personal rights and liberties.

4. The states comprising the Union see the shaping of a civil society as a most important condition for the liberty and prosperity of the people and each individual. They will seek to satisfy people's needs based on the free choice of forms of ownership and methods of economic management, the development of a Union-wide market, and the realization of the principles of social justice and protection.

5. The states comprising the Union possess the full range of political power and autonomously determine their own national-state and administrative-territorial structure as well as the system of bodies of power and administration. They may delegate some of their powers to other states which participate in the treaty of which they are members.

Those participating in the treaty acknowledge democracy based on popular representation and direct expression of the will of the people as a general, fundamental principle, and strive for the creation of a rule-of-law state which would serve as a guarantor against any tendencies toward totalitarianism and arbitrariness.

6. The states comprising the Union consider one of their very important tasks the preservation and development of national traditions, state support for education, health, science, and culture. They will facilitate the intensive exchange of mutual enrichment of the peoples of the Union and of the whole world with humanist, spiritual values and achievements.

7. The Union of Soviet Sovereign Republics operates in international relations as a sovereign state and a subject of international law—the successor to the Union of Soviet Socialist Republics. Its main aims in the international arena are lasting peace, disarmament, the elimination of nuclear and other weapons of mass destruction, cooperation between states, and solidarity of the peoples in the resolution of the global problems of humanity.

The states comprising the Union are full members of the international community.

They have a right to establish direct diplomatic, consular, trade, and other links with foreign states, to exchange authorized representatives with them, to conclude international treaties, and to participate in the activity of international organizations without encroaching upon the interests of each of the Union states and their common interests, and without violating the international responsibilities of the Union.

## II. The Structure of the Union

Article 1. Membership in the Union

The membership of states in the Union is voluntary.

The states comprising the Union enter it either directly or within the composition of other states. This does not restrict their rights and does not release them from responsibilities under the treaty. They all possess equal rights and bear equal responsibilities.

Relations between states where one state forms part of the other are regulated by treaties between them and by the constitution of the states to which it belongs.

The Union is open for entry into it by other democratic states which recognize the treaty.

The states comprising the Union retain the right to leave it freely in the manner established by the participants in the treaty and set down in the constitution and laws of the Union.

Article 2. Citizenship in the Union

The citizen of a state that belongs to the Union is simultaneously a citizen of the Union.

Citizens of the USSR have equal rights, freedoms, and responsibilities, laid down by the constitution, laws, and international treaties of the Union.

Article 3. The Territory of the Union

The territory of the Union comprises the territories of all the states that form it.

Those participating in the treaty recognize the borders existing between them at the moment of signing the treaty.

The borders between the states comprising the Union may be changed only by an agreement between them that does not violate the interests of others participating in the treaty.

Article 4. Relations Between the States Comprising the Union

Relations between the states comprising the Union are regulated by the current treaty, the constitution of the USSR, and treaties and agreements not contradicting these. Those participating in the treaty build their mutual relations within the Union on the basis of equality, respect of sovereignty, territorial integrity, non-interference in internal affairs, the resolution of disputes by peaceful means, cooperation, mutual help, and the conscientious fulfillment of their obligations under the Union treaty and in interrepublican agreements.

The states comprising the Union are obliged: not to resort to force or the threat of force in relations between themselves; not to encroach on one another's territorial integrity; not to permit the stationing of armed formations or military bases of foreign states on their territory; not to conclude agreements contradicting the aims of the Union or directed against the states comprising it.

The use of the armed forces of the Union within the country is not permitted, except for their participation in eliminating the consequences of natural calamities and ecological disasters and also cases provided for in legislation on conditions under a state of emergency.

Article 5. Jurisdiction of the USSR

The parties to the treaty invest the USSR with the following powers:

—protecting the sovereignty and the territorial integrity of the Union; declaring war and concluding peace; the provision of defense and the leadership of the armed forces and of the border, internal, and railway troops of the Union; the organization and direction of the development and production of armaments and military equipment.

—ensuring the state security of the Union; establishing the conditions of and guarding the state border and the sea and air space of the Union; coordinating the activity of the security organs of the republics.

—implementing the foreign policy of the Union and coordinating the foreign policy activity of the republics; representing the Union in relations with foreign states and international organizations; concluding the international treaties of the Union.

—implementing the foreign economic activity of the Union and coordinating the foreign economic activity of the republics; representing the Union in international economic financial organizations, and concluding the foreign economic agreements of the Union.

—confirming and executing the Union budget, carrying out the issuing of money; holding the gold stock and the diamond and currency reserves of the Union; the direction of space research, the all-Union communications and information systems, geodesy and cartography, metrology and standardization; and the management of nuclear power generation.

—adoption of the Union constitution and the introduction of amendments and supplements to it; adoption of laws within the powers of the Union and establishment of the basic principles of legislation on subjects agreed with the republics; supreme constitutional supervision.

—direction of the activity of the federal law-enforcement bodies and coordination of the activity of the law-enforcement bodies of the Union and the republics in the fight against crime.

Article 6. The Sphere of Joint Authority of the Union and the Republics

The bodies of state power and administration of the Union and the republics exercise jointly the following powers:

—defense of the constitutional system of the Union based on this treaty and the USSR constitution; ensuring the rights and freedoms of USSR citizens;

—determination of the Union's military policy; implementation of measures to organize and make provision for defense; establishment of a common procedure for conscription and the manner in which military service is performed; settlement of matters connected with the activity of troops and the disposition of military facilities on the territory of the republics; organization of mobilization preparation for the national economy; management of enterprises of the defense complex.

—determination of the Union's state security strategy and ensuring the state security of the republics; alteration of the Union's state border with the agreement of the relevant party to the treaty; protection of state secrets; determination of a list of strategic resources and products which may not be exported beyond the Union's borders; establishment of general principles and standards in the sphere of ecological security; establishment of a procedure for obtaining, storing, and using fissile and radioactive materials.

—determination of the foreign policy course of the USSR and monitoring of its implementation; defense of the rights and interests of citizens of the USSR and the rights and interests of the republics in international relations; establishment of the basic principles of foreign economic activity; conclusion of agreements on international loans and credits; regulation of the Union's state foreign debt; unified customs activity; defense and rational use of the natural wealth of the Union's economic zone and continental shelf.

—determination of the strategy for the Union's socioeconomic development and creation of conditions for the formation of an all-Union market; pursuance of a unified financial, credit, monetary, tax, insurance, and price policy based on a common currency; creation and use of the Union's gold reserve and diamond and currency reserves; elaboration and implementation of all-Union programs; monitoring of the implementation of the Union budget and issue of money; creation of all-Union funds for regional development and elimination of the consequences of natural disasters and catastrophes; creation of strategic reserves; the management of unified all-Union statistics.

—elaboration of a unified policy and balance in fuel and energy resources; management of the country's energy system, its gas and oil trunk pipelines and its all-Union rail, air, and sea transport; establishment of basic principles for using natural resources and protection of the environment; coordination of actions in the sphere of the management of water conservancy and resources of interrepublican significance.

—determining the basic principles of social policy on issues of employment, migration, labor safety and conditions, social provision and social security, popular education, and health care; the establishment of unified procedures for pension provision and maintenance of other social guarantees when citizens move from one republic to another; establishment of unified procedures for indexation of incomes and a guaranteed minimum income.

—the organization of basic scientific research and the stimulation of scientific and technological progress; the establishment of common principles and criteria for the training and certification of scientific and teaching cadres; the definition of a common procedure for the use of therapeutic means and methods; promotion of the development and mutual enrichment of national cultures; preservation of the age-old life-style of minority peoples and the conditions for their economic and cultural development.

—monitoring compliance with the constitution and the laws of the Union, the decrees of the president, and decisions adopted within the area of competence of the Union; the formation of an all-Union criminal records and information system; coordination of the fight against crimes committed on the territories of a number of republics; the definition of a unified system for organization of correctional facilities.

Article 7. Procedure for Implementation of Powers of State Bodies of the Union and of Joint Powers of States' Bodies of the Union and Republics

Issues that come under joint jurisdiction are resolved by the authorities and administrative bodies of the Union and the states that comprise it by way of coordination, special agreements, and adoption of the basic principles of legislation of the Union and of the republics and of republican laws in keeping with them. Issues that come under the jurisdiction of Union bodies are resolved directly by them.

Powers which are not placed directly by Articles 5 and 6, either under the exclusive direction of the Union's authorities and administrative bodies or in the sphere of joint jurisdiction of Union and republican bodies remain at the direction of the republics and are implemented by them independently, or on the basis of bilateral or multilateral agreements between them.

After the treaty has been signed the relevant change will be made in the powers of the governing bodies of the Union and the republics.

The participants in the treaty proceed on the basis that as the all-Union market is gradually established there will be a reduction in the scope of direct state management of the economy. The necessary redistribution or alteration of the extent of powers of governing bodies will be carried out with the agreement of the states comprising the Union.

Disputes on questions of implementing the powers of Union bodies or of realizing rights and carrying out duties within the scope of the joint powers of Union and republican bodies are resolved through conciliation. If agreement is not reached, disputes are submitted for examination by the Union's constitutional court.

The states comprising the Union participate in the implementation of the powers of the Union bodies through the joint formation of the latter and also by means of special procedures for coordinating decisions and their fulfillment.

Each republic, by concluding an agreement with the Union, can additionally delegate to it the implementation of certain of its powers, and the Union, with the agreement of all republics, can transfer to one or several of them the implementation on their territory of certain of its powers.

Article 8. Property

The Union and the states that comprise it ensure the free development and protection of all forms of property and create conditions for the functioning of enterprises and economic organizations within the framework of a single all-Union market.

The land, its underground reserves, water, and other natural resources, and plant and animal life are the property of the republics and the inalienable possession of their peoples. The procedure for using, holding, and managing them (right of ownership) is established by the legislation of the republics.

The states comprising the Union make available to the Union those facilities in state ownership that are necessary for the exercise of the powers vested in the Union bodies of authority and administration.

State property that is under Union jurisdiction is the joint property of the states that comprise the Union and is used in their common interests, including the interests of the accelerated development of backward regions.

The states comprising the Union have a right to their share of the Union's gold reserves and diamond and hard currency stocks existing when the present treaty is concluded. Their involvement in the subsequent amassing and use of treasure is defined by special agreements.

Article 9. Union Taxes and Levies

In order to finance the Union's state budget and other expenditures associated with the exercise of its powers, Union taxes and levies are fixed at levels determined by agreement with the republics. Their percentage contributions to all-Union programs are also fixed. The level and designation of the latter are regulated by agreements between the Union and the republics with reference to their socio-economic development indices.

Article 10. The Union Constitution

The present treaty is the basis of the Union constitution.

Article 11. Laws

Union laws and the constitutions and laws of the states comprising it must not contradict the provisions of this treaty.

Union laws on matters within its authority are paramount, and their execution is compulsory on the territory of the republics.

The laws of the republics are paramount on their territory on all matters, with the exception of those falling within the authority of the Union.

A republic is entitled to suspend the operation of a Union law on its territory and to protest against it if it breaches this treaty and contradicts the constitution or laws of the republic adopted within its powers.

The Union is entitled to protest against and to suspend the operation of a republic's law if it breaches this treaty and contradicts the constitution or laws of the Union adopted within its powers.

Disputes in both cases are settled through conciliation or passed on to the constitutional court of the Union.

**III. Union Bodies**

Article 12. Formation of Union Bodies

The Union's bodies of power and administration are formed based on the free expression of the will of the peoples and the representation of the states comprising the Union. They operate in strict accordance with the provisions of this treaty and the Union constitution.

Article 13. The USSR Supreme Soviet

The Union's legislative power is implemented by the USSR Supreme Soviet consisting of two chambers: the Soviet of the Republics and the Soviet of the Union.

The Soviet of the Republics consists of representatives of the republics delegated by their supreme bodies of power. The republics and national-territorial formation in the Soviet of the Republics retain the number of deputies' seats they possess in the Soviet of Nationalities of the USSR Supreme Soviet when the treaty is signed.

All the deputies of this chamber from a republic that is a direct member of the Union have one common vote when issues are being decided. The procedure for electing representatives and their quota are determined in a special agreement of the republics and in the electoral law of the USSR.

The Soviet of the Union is elected by the population of the whole country in constituencies with equal numbers of voters.

The chambers of the Union Supreme Soviet jointly make changes to the constitution of the USSR; accept new states into the USSR; determine the bases of the internal and foreign policy of the Union; confirm the Union budget and the report on its implementation; declare war and conclude peace; confirm changes in the Union's borders.

The Soviet of the Republics adopts laws on the organization of and procedure for the actions of Union bodies; examines questions of relations between republics; ratifies international treaties of the USSR; elects the Constitutional Court of the USSR; endorses the appointment of the Cabinet of Ministers of the USSR.

The Soviet of the Union examines questions of ensuring the rights and freedoms of citizens of the USSR and adopts laws on all questions except for those that are within the competence of the Soviet of the Republics.

The laws adopted by the Soviet of the Union come into force after being approved by the Soviet of the Republics. If a law adopted by the Soviet of the Union is not approved by the Soviet of the Republics, it will be given a second examination by the Soviet of the Union and will come into force on the condition that no less than two-thirds of the chamber's deputies vote for approving it.

Article 14. President of the USSR

The president of the Union is the head of a Union state invested with supreme executive and administrative power.

The president of the Union acts as a guarantor of the observance of the Union treaty, the constitution, and the laws of the Union; he is commander in chief of the armed forces of the Union; he represents the Union in relations with foreign countries; he monitors the implementation of the Union's international obligations.

The president is elected by the Union's citizens on the basis of universal, equal, and direct suffrage by secret ballot for a period of five years and for no longer than two consecutive terms. The candidate who gains more than half the votes cast in the ballot in the Union as a whole and in the majority of the states comprising it is deemed elected.

Article 15. The Vice President of the USSR

The vice president of the USSR is elected together with the president of the USSR. The vice president of the Union performs certain functions of the president of the Union on his authorization and replaces the president when the latter is absent or cannot perform his duties.

Article 16. The USSR Cabinet of Ministers

The Union Cabinet of Ministers is the Union's collegiate body of executive power, subordinate to the Union president and responsible to the Supreme Soviet.

The Cabinet of Ministers is formed by the Union president in agreement with the Union Supreme Soviet.

The heads of government of the republics may participate in the work of the Union Cabinet of Ministers with the right to vote.

Article 17. The USSR Constitutional Court

The Union Constitutional Court examines questions of whether legislative acts of the Union and the republics, decrees of the Union president and the presidents of the republics, and normative acts of the Union Cabinet of Ministers are in accordance with the Union treaty and the Union constitution. It also resolves disputes between the Union and the republics and between republics.

Article 18. Union (Federal) Courts

The Union (federal) courts are the Supreme Court of the USSR, the Supreme Court of Arbitration of the Union, and courts in the Union's armed forces.

The Union Supreme Court and the Union Supreme Court of Arbitration exercise judicial authority within the bounds of the Union's powers. The chairmen of the supreme judicial and arbitration bodies of the republics are ex-officio members of the Union Supreme Court and the Union Supreme Court of Arbitration, respectively.

Article 19. The USSR Procuracy

Supervision of the execution of legislative acts within the Union is performed by the Union procurator-general, the procurators-general (procurators) of the republics, and the procurators subordinate to them.

The Union procurator-general is appointed by the Union Supreme Soviet and is accountable to it.

The procurators-general (procurators) of the republics are appointed by their supreme legislative bodies and are ex-officio members of the collegium of the Union Procuracy. While supervising the execution of Union laws they are accountable to both the supreme legislative bodies of their own states and the Union procurator-general.

**IV. Concluding Provisions**

Article 20. The Official Language of the USSR

The republics determine independently their own state language (languages). The parties to the treaty recognize Russian as the official language of the Union.

Article 21. The Capital City of the Union

The capital city of the USSR is Moscow.

Article 22. The State Symbols of the Union

The USSR has a state coat of arms, flag, and anthem.

Article 23. The Coming into Force of the Treaty

This treaty is approved by the supreme bodies of state power of the states comprising the Union and comes into force after it is signed by authorized delegations.

For the states that signed it, the 1922 treaty on the formation of the USSR is deemed null and void as of the same date.

From the moment the treaty comes into force a most-favored-nation system will take effect for its signatories.

Relations between the Union of Soviet Sovereign Republics and the republics that comprise the Union of Soviet Socialist Republics but that have not signed this treaty should be settled on the basis of the current legislation of the USSR, mutual obligations, and agreements.

Article 24. Responsibility Under the Treaty

The Union and the states comprising it have mutual responsibility for the execution of obligations undertaken by them and compensation for the damage caused by breaches of this treaty.

Article 25. Procedure for Introducing Amendments and Addenda to the Treaty

This treaty or its individual provisions can be canceled, altered, or supplemented only by agreement of all the states comprising the Union.

If need be, upon agreement between the states that have signed the treaty, appendices to it may be adopted.

Article 26. Succession of the Union's Supreme Bodies

To ensure continuity of state power and administration the supreme legislative, executive, and judicial bodies of the Union of Soviet Socialist Republics retain their plenary powers until supreme state bodies of the Union of Soviet Sovereign Republics are formed in accordance with this treaty and the new USSR constitution.

## 1.2

## Reactions to Draft Union Treaty Viewed

Albert Plutnik

*Izvestiya*, 29 June 1991 [FBIS Translation]

The occasion for these notes is the recently published draft "Treaty on the Union of Sovereign States." But what does that have to do with emotion, with our "seething nerves," with political passions roused to such heights that the USSR president recently observed: "We are operating in exceptional conditions, perhaps the most difficult for many years, except for the period of the Great Patriotic War when the enemy was at the gates of the capital or of Stalingrad"?

The point is that this draft has been long awaited, and many people worked on it. And in different ways. Not only in Novo-Ogarevo, of course, but throughout the country. Opposing political forces used every available method, including hysterics (remember, for instance, the events in the Supreme Soviet when the people's deputies "called M.S.

Gorbachev to account"), in the attempt resolutely to define their positions and even to dictate not only particular articles and clauses but even the general thrust of the draft. Because they understood very well that it is a question of the design, the blueprint, according to which a radical reorganization of our entire life is to take place, in all its breadth and multiplicity of economic, political, and social relations.

And now the document has been published. Can it be to everyone's liking? Certainly not. We are too diverse today. We seek things that are too diverse; we see the truth in different lights. Although its supréme, ultimate meaning, you could say, lies precisely in the constructive resolution of many problems that trouble the public mind—solutions designed for the long term and intended to suit everyone. Although in terms of its basic thrust it is intended to provide the essential principle for all fierce political debates on the direction of our future development—in many ways we are offered an attractive outline for redesigning our Union for the sake of society's resolute advance toward progress, toward civil peace, and toward a long overdue acclimatization to the general process of civilization. But the choice has been made, and that means that many alternatives have been discarded. By no means everyone will find his own ideas reflected in this single proposed version; not everyone will accept the rights and wrongs of a different view.

Nonetheless any unbiased observer must, I think, admit that people were not wasting their time at Novo-Ogarevo. The result of joint work by the republics' representatives, the Federation Council, and the members of the preparatory committee formed by the Fourth Congress of USSR People's Deputies is a document that is democratic in content and realistic in character. True, much needs to be clarified and enlarged upon, and many questions will doubtless be prompted by specific provisions relating to the distribution of rights between the Center and the republics, but it is in this area, it seems to me, that there has been a decisive breakthrough. For the first time in the years of Soviet power the draft proclaims what I would call a multiparty economic system. Just as Article 6 of the USSR constitution was revoked some time ago, which meant that the party had renounced its monopoly of power, now the Center is sharing its economic power, largely renouncing its excessive claims. It looks as if at any moment now, on the way to the market, not only prices but the republics are going to be freed. It is not hard to surmise that for our still numerous champions of command-bureaucratic methods of management and staunch supporters of the empire, this whiff—perhaps a mere ghost—of freedom will appear to be the most vulnerable point in the draft.

It is noteworthy that recently, in certain circles, discussions of the draft treaty have usually been linked, so to speak, to the "Window of Opportunity" program drawn up by Soviet and American experts. Whipping up the hysteria, they find a close interconnection between these drafts—on the one hand "the so-called democrats, having grabbed power, are wrecking the Union," and on the other they are nurturing plans for selling off the state to foreign capital. Logical? A shining example of "factional" publicity work is provided by remarks by the well-known Colonel V. Alksnis in the newspaper *Politika*, published by the "Soyuz" deputies' group. Commenting on the long extracts from the speeches at the 17 June Supreme Soviet session by the ministers of defense and internal affairs, and also by the KGB chairman (curiously, the session was closed but "selected" mass media, such as the TV program "600 Seconds" and the newspaper *Politika* were given access to the materials), he seeks to persuade the reader that "the leaders of these government structures, who possess all the information, have stated that unless emergency measures are adopted, our country will cease to exist! And today our people are being taken in by all kinds of Union treaties and similar never-ending stories." So there you have it. In the same discussion, the colonel's worthy interlocutor expresses his view of Yavlinskiy's mission: "It is strange: A man whom nobody has authorized—neither the Cabinet of Ministers nor the Supreme Soviet—suddenly goes off to the West and conducts talks of some kind."

A little earlier, Supreme Soviet member V. Semenov expressed his opinion on the same topic: "I personally do not like the idea that the fate of my state and my people is decided in some office in the White House!" Other people in the Supreme Soviet, as you know, did not like the idea of the fate of the Union treaty being decided in Novo-Ogarevo. . . . Just look at the kind of things that prompt dissatisfaction in our statesmen's minds. What a warped mind you must have to get annoyed about things that you should be thanking people for. Even if Yavlinskiy had indeed been acting without authorization—which is far from being the case—all the same, all joint Soviet-American proposals should be painstakingly examined. If, of course, you are thinking about the people, the country, rather than being consumed by departmental patriotism. . . . Ideas have to be paid for. At least with gratitude. Frankly, I thought gratitude was the only "convertible currency" we still had plenty of. But it seems that here too we have spent it all.

Incidentally, M.S. Gorbachev, speaking in the Supreme Soviet the other day, mentioned by name for the first time, as you know, certain representatives of the right wing who try in all circumstances to impose their own opinion on the country's leadership. Formerly this honor was accorded only to left radical figures. But on this occasion, the president did not get off scot-free, for this "leveling" of the exchange rate. An article promptly appeared in the aforementioned press organ of the "Soyuz" group, openly debating the question of "removing Gorbachev from the political arena." I will quote just one passage. "If the president, totally ignoring the existing constitution, keeps the Congress and Supreme Soviet away from power and, in collusion with the republican authorities, abolishes the state, it would be interesting to know who can decide which is more lawful: these actions on his part, or some kind of coup in defense of the constitution, the "country's fundamental law." Do you see what they are hinting at? "Some kind of coup," however feeble. . . . Taking revenge on the president at a stroke for his Supreme Soviet speech, the "Nine-Plus-One" agreement, and his attentiveness to the "Window of Opportunity" program, the newspaper is basically calling openly for something that is scarcely compatible with the "country's fundamental law."

Departmental patriotism is similar to factional patriotism. It seems to me that the government demonstration, so to speak, that was organized in the Supreme Soviet was the consequence of jealous displeasure in the Cabinet of Ministers at the fact that G. Yavlinskiy seems to be taking a lot on and acting as if he was the Cabinet of Ministers. This is the result of the persistent habit of being monopolists and formulating on every question a decision that is the only right decision, because it is the only one.

A couple of words of warning to conclude. It is very important, in the process of taking opinions into account and improving the draft, to strip it exclusively of its shortcomings, not of its merits. And not to shake the heart out of it in the attempt to perfect it.

## 1.3
## *Pravda* Views Provisions of Union Treaty

Viktor Shirokov
*Pravda,* 29 June 1991 [FBIS Translation]

Last Thursday the draft treaty on the Union of Sovereign States was published in the press. Thus the process of the renewal of the Union has entered the next stage—the time has come for each of us to consider the wording of the articles of the treaty and its individual provisions. This is the most important matter because in the future this lofty treaty is to be the basis of the constitution of the sovereign federal democratic state, its "alpha" and "omega," and we shall all have to live according to this constitution in the years and perhaps decades ahead.

The supreme soviets of certain republics, which have firmly resolved to recreate the Union on new principles, have already had time to discuss and approve this most important draft treaty, which determines the fate of the country. The nationwide discussion of the document and the amendments, comments, and additions submitted during its examination in republic parliaments and the USSR Supreme Soviet will probably be taken into account as well. However, it is already clear right now that the fundamental and extremely responsible stage in the new Union's formation is drawing to a close.

Now let us turn to the text of the treaty. Its preamble sets out in verbal formulas the principles that have prevailed in the political struggle and the passions of the perestroika period and have won the minds of the peoples inhabiting our country. It is here that the sovereignty of the states entering the federal "alliance," the right of nations to self-determination, and the desire and firm intention to create the conditions for the all-around development of each individual and to reliably guarantee individual rights and freedoms are strictly and precisely stipulated.

The republics that are party to the treaty invest the Union of Soviet Sovereign Republics (USSR) with power within the limits of the authority that they freely delegate to it. Henceforth, the Center will exercise its functions of coordination and authority with their knowledge and consent. That is the first of the "fundamental principles" that combine naturally with the next one, which addresses the autonomy of parties to the treaty on all questions of their development and their commitment to guarantee equal political rights and opportunities for socio-economic and cultural development to all the peoples living on their territory.

I would stress this fact as it is the "counterpoint" essential for consolidation of the peoples in a single federation. We all see and feel the acute alarm of those who, as fate dictates, find themselves living outside the borders of their national-territorial entities.

The three basic principles on which the Union is to be built are: (1) free choice of forms of ownership and methods of economic management; (2) development of an all-Union market, that is, a single economic space without which we will wither and die in our national "compartments"; and (3) the realization of principles of social justice and protection.

The section "The Structure of the Union" prescribes the most important features of the organization of the federal state. There is no rigid structure: Entry into the Union can be either direct or as part of another state. It will be open to any democratic state that recognizes it. Parties to the treaty intend to form mutual relations on the basis of equality, respect for sovereignty, non-interference in internal affairs, and the resolution of disputes by peaceful and constitutional means.

The draft treaty strictly defines the jurisdiction of the USSR and the republics and clarifies areas of joint jurisdiction. Powers delegated to the Union relate to protecting the state's territorial integrity, defense and security, implementing foreign policy, coordinating foreign economic activity, approving and implementing the budget, directing space programs, Union-wide communications and information systems, management of nuclear power generation, and sectors where coordinated action is essential to success, such as law enforcement.

The joint jurisdiction of the Union and the republics is being considerably expanded. Essentially, all parties to the treaty will be equals in the solution of questions of strategic importance to the state.

A particularly significant provision says that, on the one hand, the laws of a signatory republic are paramount on its territory while, on the other hand, Union laws on questions under Union jurisdiction shall be binding on the territory of the republics. In a strong and united state there cannot be unlimited sovereignty, and the precise separation of the functions of power must guarantee both the rights and responsibilities of all parties. Disputes in this area will be resolved by the Union constitutional court.

## 1.4
## Coup Leader Gennadiy Yanaev Answers Questions at News Conference

Moscow Central Television, 19 August 1991
[FBIS Translation], Excerpts

[Statement and news conference given by Gennadiy Ivanovich Yanaev, acting president of the USSR, with the participation of Oleg Baklanov, first deputy chairman of the USSR Defense Council under the USSR president; Interior Minister Boris Pugo; Vasiliy Starodubtsev, chairman of the USSR Farmers' Union; Tizyakov, president of the Association of State Enterprises and Industrial, Construction, Transport, and Communications Facilities of the USSR; at the Foreign Ministry Press Center in Moscow on 19 August; video shows participants in news conference asking and responding to questions—recorded]

[Yanaev]: Ladies and gentlemen, friends and comrades: As you already know from media reports, because Mikhail Sergeevich Gorbachev is unable, owing to the state of his health, to discharge the duties of president of the USSR, on the basis of Article 127.7 of the USSR Constitution, the USSR vice president has temporarily taken over the performance of the duties of the president.

I address you today at a moment that is crucial for the destinies of the Soviet Union and the international situation throughout the world. Having embarked on the path of profound reforms and having gone a considerable way in this direction, the Soviet Union has now reached a point at which it finds itself faced with a deep crisis, the further development of which could both place in question the course of reforms themselves as well as lead to serious cataclysms in international relations.

It is of course no secret to you that a sharp fall in the output of the country, which has so far not been replaced by the alternative industrial and agricultural structures, is creating a real threat to the further existence and development of the peoples of the Soviet Union.

A state of ungovernability and multiple authority has arisen in the country. All of this cannot fail to arouse extensive dissatisfaction among the people. A real threat of the country's disintegrating has also arisen with a collapse of the single economic space, of the single space for civil rights, a single defense and single foreign policy.

Under such conditions, normal life is impossible.

In many regions of the USSR, as a result of interethnic clashes, blood is being spilled and the collapse of the USSR would have the most serious consequences, not only internally but also internationally.

Under such conditions, we have no alternative but to take decisive steps to stop the country from sliding into disaster.

As you know, to govern the country and for the efficient implementation of the state of emergency a decision has been adopted to form a State Committee for the State of Emergency in the USSR made up of the following: Comrade Baklanov, first deputy chairman of the USSR Defense Council; Comrade Kriuchkov, chairman of the USSR KGB; Comrade Pavlov, prime minister of the USSR; Comrade Pugo, minister of internal affairs of the USSR; Comrade Starodubtsev, chairman of the Farmers' Union of the USSR; Comrade Yazov, minister of defense of the USSR; and Comrade Gennadiy Yanaev, acting president of the USSR.

I would like to state today that the State Committee for the State of Emergency in the USSR is fully aware of the depth of the crisis that has hit our country. It takes on itself responsibility for the fate of the motherland and is fully resolved to adopt the most serious measures to get the state and society out of the crisis as quickly as possible.

We promise to hold extensive nationwide discussion on the draft of a new Union treaty. Every citizen of the USSR will have the right and the chance to take part in this most important act in a calm atmosphere, and to make up his mind on it, for the fate of the numerous peoples of our great motherland will depend on what the Union becomes.

We intend immediately to restore legality and law and order, to put an end to the bloodshed, to declare merciless war on the criminal world, and to eradicate the shameful occurrences that discredit our society and humiliate Soviet citizens. We will cleanse the streets of our cities of criminal elements and put an end to the arbitrary rule of the plunderers of national property.

We advocate truly genuine, democratic processes and consistent policy of reforms leading to the renewal of our homeland and to its economic and social prosperity, enabling it to take a worthy place in the world community of nations.

The development of the country must not be built on a drop in the living standards of the population. In a healthy society a constant increase in the well-being of all citizens will become the norm. Without relaxing our concern for strengthening the protection of the rights of the individual, we shall concentrate attention on protecting the interests of the broadest sections of the population, those who are most threatened by inflation, the disorganization of production, corruption, and crime.

In developing the mixed nature of the national economy, we will also support private enterprise, offering it the necessary opportunities for developing production and services.

Our priority task will be a solution to the food and housing problems. All existing forces will be mobilized toward satisfying these most vital demands of the people. We call upon the workers, peasants, the labor intelligentsia, and all Soviet people to restore labor discipline and order as quickly as possible and raise the level of production to subsequently move forward in a resolute manner. Our life and the future of our children and grandchildren and the destiny of the motherland depend on this.

We are a peace-loving country, and we shall unfailingly honor all the commitments we have made. We have no claims upon anyone. We want to live in peace and friendship with everyone, but we state firmly that no one will ever be permitted to encroach upon our sovereignty, independence, and territorial integrity. All attempts to speak with our country in the language of diktat, from whatever the source, will be resolutely cut short.

Our multinational people have for centuries lived filled with pride for their homeland. We have not been ashamed of our patriotic feelings, and we consider it natural and regular to bring up the present and future generations of our citizens in this spirit. To stand idly by at this hour, which is critical for the destiny of our motherland, means to take upon oneself the gravest responsibility for the tragic and truly unpredictable consequences.

Everyone who holds our motherland dear, who wants to live and labor in a situation of calm and confidence, who does not accept the continuation of bloody interethnic conflicts, who sees his motherland independent and prosperous in the future, must make the sole correct choice. We call upon all true patriots, people of good will, to put an end to the present Time of Troubles [allusion to the historical period of civil strife in 1605–13]. We call upon all citizens of the Soviet Union to recognize their duty toward the motherland and to render assistance in every possible way to the State Committee for the State of Emergency in the USSR in its efforts to bring the country out of crisis. Constructive suggestions from socio-political organizations, labor collectives, and citizens will be gratefully accepted as a manifestation of patriotic readiness to take an active part in restoring the centuries-old friendship in a single family of fraternal peoples and in the revival of the motherland. Thank you for your attention.

[Correspondent, *Newsweek* magazine]: Where is Mikhail Sergeevich Gorbachev? What is he sick with? Specifically, and concretely, what disease does he have? And against whom are the tanks that we see on the streets of Moscow directed? . . .

[Yanaev]: I have to say that Mikhail Sergeevich Gorbachev is presently on vacation and undergoing treatment in the Crimea. He has, indeed, grown very tired over these past years, and he will need some time to put his health in order. I would like to say that we hope that when he has recovered, Mikhail Sergeevich will return to carry out his duties. At any rate, we will continue to follow the course Mikhail Sergeevich Gorbachev began in 1985 . . . .

[Journalist, *Pravda*]: I have two questions. Perestroika has not brought about tangible results, among other reasons, because there was an absence of a clear-cut tactic and strategy for implementing it. Do you now have a concrete program for regenerating the country's economy? And, secondly: The Russian Information Agency today transmitted an appeal from Yeltsin, Silaev, and Khasbulatov to the citizens of Russia describing the events of the past night as a right-wing, reactionary, anti-constitutional coup. What is your attitude to this statement? That same appeal called for a general indefinite strike. In this connection, will any specific actions be undertaken on this matter by the committee?

[Tizyakov]: Indeed, the perestroika announced in 1985, as you know—this is not a secret to anyone—has not produced the expected results that all of us were waiting for. Our economy today is in a most grave situation. Production is slumping. This has been brought about by a whole series of factors. Among these, of course, is the fact that such a restructuring was being carried out by us for the first time on such a scale. And, of course, there is also the fact that we were seeking ways, and are seeking ways, and in any endeavor it is always possible there might be certain things not properly worked out; errors, let's say.

What lies in store for us? The situation that has evolved served in the main, of course, for the introduction of a state of emergency. You know that links are badly broken in the countryside today—horizontal ones between enterprises. If one considers the preparations for 1991, no doubt many of the people sitting here know that at the end of the year, at the beginning of December, an all-Union assembly of directors was held. There the formation of links between enterprises in all regions was indeed looked at. We managed to get this problem moving in a very meaningful way and at the beginning of January we had already concluded agreements covering 85 percent of movements of output in the technological cycle. This apparently created certain conditions, but later on, in a whole series of so-called sovereignty-related measures that were carried out, this gave rise to the closure of borders recently in republics like Ukraine, Belarus, the Baltic republics, and other republics. This created an extremely difficult situation for the work of enterprises. Production facilities stopped working, and there was uncertainty among the labor collectives.

What will our actions be like? First and foremost, they will be aimed at stabilizing the economy, and naturally we are not giving up and we are not canceling our reforms aimed at moving toward the market. We consider this correct. The only thing is that we are going to have to work it through more precisely and organize things on a higher level in management and in all our actions.

[Yanaev]: I will answer the second question. Today, when the Soviet people were informed about the formation of the Committee for the State of Emergency, I, like several other members of the committee, had contact with the leaders of all nine republics that stated their willingness to enter the renewed Union federation.

We have been in contact with the leadership of many krays and oblasts of the Soviet Union, and I can say that, on the whole, there is support for the formation of the committee and the committee's attempt to extricate the country from the state of crisis it is in.

I had a conversation with Boris Nikolaevich Yeltsin today. I know about the statement by Boris Nikolaevich, Comrade Khasbulatov, and Comrade Silaev. I would like to stress today that the State Committee for the State of Emergency is prepared to cooperate with the leadership of the republics, krays and oblasts, being guided in this by our aspiration to find such adequate forms of developing our democracy and raising the economy. . . .

I think that the appeal to join a general indefinite strike is an irresponsible appeal. Evidently, we cannot afford the luxury of playing some political games when the country is in a state of chaos. Because, ultimately, the majority of these political games turn against our long-suffering people. If we are not indifferent to the fate of our fatherland, if we are not indifferent to Russia's fate, we should find really practical forms of cooperation.

[Journalist, Central Television]: . . . In its message to the people, the Committee for the State of Emergency stated that it will concern itself first and foremost with the interests of the broadest strata of the population and will strive to solve the food and housing problems. Would you say what steps, specific steps, you envisage taking for this, and what resources, what reserves the State Committee has at its disposal?

[Yanaev]: You know, this question is indeed a very interesting, a very important question. The first step that we intend to take is to do the maximum possible to save the harvest. It looks as if we will be adopting the relevant document tomorrow, which will orient us toward adopting emergency measures to save the harvest. Further, we intend to make use of every opportunity, first of all, to carry out a kind of stocktaking of everything there is in the country. As you recall, we said in the statement that once we have done this stocktaking, it will be necessary to tell the people what we have, including material resources that we can mobilize to make progress on the housing problem . . . .

[Journalist, *Argumenty i Fakty*]: Tell us, please, apart from his post as president, Mikhail Sergeevich Gorbachev is also general secretary of the CPSU Central Committee. Who is to carry out the duties of general secretary? And my second question: It has become known that a number of papers, including *Argumenty i Fakty*, *Moskovskie Novosti*, *Kuranty*, *Stolitsa*, and a few others have been closed. For how long have these papers been closed and when will they reopen? Thank you.

[Yanaev]: I think that if we are introducing a state of emergency regime we shall have to reregister some of our mass publications. Reregister. We are not talking about the closure of papers. We are talking about reregistering, because the chaos the country finds itself in is to a large extent the fault of some of the mass media.

As for the post of general secretary, I wouldn't like to comment on it. We have a deputy general secretary. He is capable and hard-working. I think that a party Central Committee plenum or a congress will be able to solve this question. I am now carrying out the duties of the president, and therefore I would not like to use my authority, or lack of it, in order to influence the decision of party bodies.

[Journalist, *Novosti*]: My question is to Comrade Starodubtsev. You have ties with the peasant masses—do you think the peasant masses will support the Committee for the State of Emergency? Thank you.

[Starodubtsev]: In perestroika the heaviest losses have been borne by the peasantry, naturally. This year in particular has dealt the most crushing blow to the peasantry, and the majority of state farms, collective farms, and newly born farm economies are on the verge of catastrophe today. The price parity between the town and the village has been upset. Lack of fuel and spare parts and a sharp decline in deliveries of equipment and other materials to agriculture lead, frankly, to a grave state for the peasantry. I think that, driven to despair, the peasantry hopes today that order will be introduced at last and that the eyes of everyone in our society will be turned to the peasantry and will help it to stand up and see a revival.

[Journalist, *Nezavisimaya Gazeta*]: Could you please say whether or not you understand that last night you carried out a coup d'état? What comparison seems more apt to you, the comparison with 1917 or with 1964? This is the first question. The second concerns newspapers. First, how long will it take to reregister newspapers and what criterion will decide whether or not a particular publication should be reregistered? Who will deal with this? Will political censorship be introduced in the reregistered publications? Thank you.

[Yanaev]: With regard to the reregistration of newspapers, we will try not to drag out this process. I would not like to comment now on the criteria that will be used as the basis for this registration.

As for your allegation that a coup d'état was staged last night, I would beg to disagree with you, inasmuch as we are basing ourselves on constitutional norms. And I assume that confirmation by the USSR Supreme Soviet of the decision we have made will enable us to state that absolutely all the juridical and so-called constitutional norms have been observed. It does not seem to me correct to draw a comparison with either 1917 or 1964. I believe any analogy would be dangerous.

[Unidentified Journalist]: A question for Comrade Baklanov: Please state what measures you plan to take in order to solve, specifically, the housing and food program for servicemen, who have been literally in a disastrous position in some regions since the withdrawal of our troops.

[Baklanov]: I have to say that the meaning of our measures consists in concentrating these existing resources so as now to force through the building of housing for servicemen and their families. You probably know this figure, it is something like 200,000 families in need. This is a manifestly abnormal state of affairs. We have to use our resources and we have to use the resources that, in accordance with the accords, we are to receive from foreign states. That is what the measures mean.

[Yanaev]: I would like to reply to your question regarding President Yeltsin's decree on banning the Communist Party activity. I believe that all decrees and all resolutions to be adopted will be examined from the point of view of the state of emergency we are introducing in the country. But, taking advantage of this opportunity, I would like to stress: What the leadership of the Russian Federation is currently engaged in—building barricades, appealing for disobedience—I think that this is a very dangerous policy, and this policy could lead to the organization of some kind of armed provocation, so as afterward, to lay the blame for the tension, for some kind of excess that might occur, on the leadership of the State Committee for the State of Emergency. We would like to give a very serious warning of this to all Soviet people, especially to the people of Moscow, where a state of emergency has been introduced. We hope that calm and order will be ensured.

[Journalist, *Novosti*]: My question is to Boris Karlovich Pugo. How do you technically see the organization of this regime? . . . Hundreds, maybe thousands of people, will start for the airports today and for places far away from the capital. How is this regime going to deal with this?

[Pugo]: The introduction of military hardware, even including troops, in Moscow, well, it is already completely evident that this is a wholly forced measure. It has been taken only to prevent any disturbance of order in Moscow, to prevent any casualties. As for how things will be controlled, I at least see the development of events in the following way: Provided nobody forces us to extend the arrangement, or to make it long term, we would favor withdrawing all military units and hardware from Moscow as soon as possible.

[Yanaev]: We do not envisage a curfew.

[Pugo]: But that is something that comes within the jurisdiction of the Moscow commander. If it should suddenly come about that this is necessary, well, that is within his jurisdiction. But we did not count on this. We did not consider it necessary to do that today.

[Journalist, Associated Press]: Can you please tell us if your committee is prepared to order the use of force against civilians? And under what circumstances would force be used against civilians?

[Yanaev]: First, I would like to do everything to ensure that the use of force against civilians is not needed. We must do everything to prevent any excesses. And what we are envisaging now, some extraordinary measures, they are not at all linked with any attack on human rights. On the contrary, we want to protect human rights as much as possible. And I would like to hope very much that we will not be compelled, we will not be provoked into using some kind of force against the civilian population.

## 1.5
## Decision of the Supreme Council of the Ukrainian RSR

*Editor's Note*: On 16 July 1990, soon after Russia did so, Ukraine issued its own, even more far-reaching Declaration of State Sovereignty. The essence of this sovereignty was "democracy" and economic autonomy for the Ukrainian Soviet Socialist Republic. In proclaiming its independence on 24 August 1991, the Ukrainian parliament denounced the Union treaty of 1922 that had created the USSR and appointed a Ukrainian government, including a Ministry of Defense and a procurator-general. Ukraine was still, however, at this time considered part of the Union, though it was not clear in what status.

### Concerning the Declaration of Independence of Ukraine
24 August 1991 [Translated by Natalie Gawdiak]

The Supreme Council of the Ukrainian Soviet Socialist Republic resolves:

To declare on 24 August 1991 that Ukraine is an independent, democratic country.

From the moment of the declaration of independence, on the territory of Ukraine only her constitution, laws, and government resolutions and other legislative acts of the republic are valid.

On 1 December 1991 a republic-wide referendum will be carried out to affirm the act of the declaration of independence.

Signed:
Head of the Supreme Rada of the Ukrainian RSR
L. Kravchuk

### Declaration of Independence of Ukraine
Supreme Council of Ukraine
24 August 1991 [FBIS Translation]

Stemming from the mortal danger that hung over Ukraine in connection with the government coup in the USSR on 19 August 1991,

—continuing the thousand-year-old tradition of state-building in Ukraine;

—implementing the Declaration on State Sovereignty of Ukraine, the Supreme Council of the Ukrainian Soviet Socialist Republic solemnly:

DECLARES THE INDEPENDENCE OF UKRAINE
and the creation of the independent Ukrainian state—Ukraine.

The territory of Ukraine is indivisible and inviolate. From this day forth on the territory of Ukraine, only the constitution and laws of Ukraine will be valid.

This act becomes valid from the moment of its approval.

## 1.6
## Referendum on Independence
Interfax, 24 August 1991 [FBIS Translation]

The Ukrainian Supreme Soviet passed a resolution and the text of an act proclaiming the Ukraine an independent and sovereign state within the present-day territory. The Union laws have no effect on the republican territory. According to the ruling, the act takes effect the moment it is approved by the national referendum during the presidential election in the Ukraine on 1 December.

The resolution was passed by 346 votes. The Ukrainian parliament contains 450 deputies, but fewer than 400 were present at the session.

## 1.7
## Yelena Bonner Urges Republics to Observe Human Rights
*Komsomolskaya Pravda,* 3 September 1991
[FBIS Translation], Excerpts

[Interview with Yelena Bonner by G. Vasilyeva: "The Time for Sakharov's Constitution Is Past—As Usual, We Are Late. . . ."]

[G. Vasilyeva]: Yelena Grigorievna, perhaps we are celebrating victory in vain. The putschists have achieved what they wanted. The Union treaty is unsigned.

[Yelena Bonner]: The putsch gave the nudge that resulted in the collapse of the Soviet Union. And this raises an extraordinarily important point, perhaps the most important point that I have to make to the readers of *Komsomolskaya Pravda*. It concerns all the former Union republics, all the autonomous areas, and all the parties. These young sovereign states must come into being without any infringement of human rights. In principle, there are two democratic rights to uphold: the rights of the individual (as set forth in the Universal Declaration of Human Rights) and the rights of the people to self-determination.

On 3 September, I am scheduled to appear before the Danish parliament, which is holding hearings on the subject: "The Helsinki Act: Human Rights—Yesterday, Today, and

Tomorrow." They have invited not just private groups but state delegations from all thirty-five countries represented at the Helsinki Conference (August 1975). This is the main point of my speech: Our August victory is yet to be won. For the republics must have an unconditional right to full independence. But the sole criterion for [giving] economic aid to the republics must be their observance of human rights. Right now there are many alarming reports coming from the republics, signaling attempts to interfere with the rights of certain ethnic groups. I want to remind people of the fact that the Sakharov Committee has considerable influence on public opinion in the West, and this influence will be used to counter these attempts. I say this to the Belorussians, I say this to the Baltic peoples, I say this to the Moldovans, Georgians, Azerbaijanis, Turks, and Uzbeks. I say this to all the republics and their political leaders. Any attempt to carry out repressive policies in dealing with the ethnic minorities will result in these republics being left to stagnate without assistance from the civilized nations and, I hope, without any help from democratic Russia. It behooves a democratic Russia to bear in mind that Russia has already been turned into a prison of peoples—we must not revert to that condition. Right now a new Cabinet of Ministers is being formed, but this is an artificial contrivance of the Center with dangerous implications . . . .

[G. Vasilyeva]: But, Yelena Grigorievna, even Sakharov thought it possible and normal for a union of equal and sovereign states to be placed under the leadership of a parliament in which more than half the seats would be reserved for representatives of Russia.

[Yelena Bonner]: The time for Sakharov's constitution is past. As usual, we are late in trying to jump onto the last car of the train. Today we must consider how to go our separate ways with the fewest losses. It amazes me that the deputies attending the Extraordinary Session of the Supreme Soviet talked for three days about the past when they should have been talking about other things entirely.

[G. Vasilyeva]: Nevertheless, our national territory constitutes a single economic region. In my view it is apparent that if the Union falls apart, everyone will become weaker—and no one will win.

[Yelena Bonner]: The fact that Ukraine has declared its independence does not mean that it has gone off to Australia with all its land and natural wealth. And no one in the West will be buying Estonian sportswear, although I myself am looking for some to buy for my grandchildren. We are all bound together economically, and no sort of national independence will be able to sever these ties.

[G. Vasilyeva]: Within the territory of the RSFSR [Russian Soviet Federated Socialist Republic] there are about thirty autonomous areas—to say nothing of other minority groups. How many pieces can we break up into or is there a limit beyond which a people cannot claim a right to separate statehood?

[Yelena Bonner]: There is no limit, and this very lack of limitation in Europe has led to the existence of Monaco, San Marino, and Luxembourg. As Europe seeks to become increasingly integrated, we are currently coming apart. But the peoples of Europe have known freedom for a good deal more than a decade. A slave who gulps a breath of freedom cannot immediately aspire to integration. He must first cast off his chains.

[G. Vasilyeva]: Right now there are endless explanations being given about who was where during these tragic days—with many sidelong glances and reproaches, often undeserved. How do you feel about this?

[Yelena Bonner]: It is not for people to judge where there is no need to judge. I have neighbors on all ten floors who have known me since 1954. Yet when a campaign was launched against me, only two of these families would speak to me when we met on the stairs. Today I get on with everyone. Yet if the putsch had been successful, I think they would again be looking the other way. It is not for me to judge them. They have been raised in this fashion by our society. When the Supreme Soviet was in session, one person stood up who did not speak Russian very well. He said that he was a Communist, but that he could not comprehend what was happening, and he almost cried. And I almost cried with him. I am always on the side of those who stand up for things.

## 1.8

### Draft Law on the Transition

*Rossiyskaya Gazeta,* 5 September 1991
[FBIS Translation], Excerpt

"Text" of draft USSR law "On the Organs of State Power and Management of the USSR in the Transitional Period," contained in reportage from the Kremlin by *Rossiyskaya Gazeta* parliamentary correspondents Vyacheslav Dolganov and Robert Minasow]

The draft was prepared by the Congress Commission and the signatories to the statement by the USSR president and the leaders of the Union republics "On the Organs of State Power and Management of the USSR During the Transitional Period."

Article 1. The supreme representative organ of Union power during the transitional period is the Supreme Soviet, which consists of two chambers: the Soviet of the Republics and the Soviet of the Union.

The Soviet of the Republics consists of twenty USSR people's deputies and twenty people's deputies from each Union republic delegated by these republics' supreme organs of power. In view of the federative system of the RSFSR [Russian Soviet Federated Socialist Republic], it has forty-five deputies in the Soviet of the Republics. In order to ensure the equal rights of the republics when voting in the Soviet of the Republics, each Union republic has one vote.

The Soviet of the Union is formed by the supreme organs of power of the Union republics from USSR people's deputies in accordance with currently existing quotas.

Article 2. The USSR Supreme Soviet of the Republics and Soviet of the Union by joint decision make changes to the USSR constitution, admit new states to the Union, approve the Union budget and the report on its implementation, and declare war and conclude peace.

The Soviet of the Republics adopts decisions on the organization of and procedure for the activity of Union organs and ratifies the international treaties of the USSR.

The Soviet of the Union examines questions of safeguarding the rights and freedoms of USSR citizens and adopts decisions on all questions within the competence of the Supreme Soviet except those relating to the competence of the Soviet of the Republics. Laws adopted by the Soviet of the Union enter into force after they are approved by the Soviet of the Republics.

Article 3. A State Council is formed to provide coordinated solutions to domestic and foreign policy issues affecting the common interests of the republics. The State Council consists of the USSR president and the highest officials of the Union republics named in the USSR constitution. The USSR president directs the work of the State Council. The State Council defines the procedure for its activity. Decisions of the State Council are binding.

Article 4. The post of vice president of the USSR is abolished.

If the USSR president for various reasons cannot continue to perform his duties (including on health grounds, as confirmed by the findings of the State Medical Commission set up by the USSR Supreme Soviet), the State Council elects a State Council chairman from among its members as acting USSR president. This decision is to be approved by the USSR Supreme Soviet within three days.

Article 5. In order to coordinate the management of the national economy and ensure the coordinated implementation of economic reforms and social policy, the Union republics form the Interrepublic Economic Committee on an equal footing. The committee's chairman is appointed by the USSR president with the agreement of the State Council. The USSR president and the USSR State Council directly exercise leadership of Union-wide organs in charge of questions of defense, security, law and order, and international affairs.

In their activity, the Interrepublican Economic Committee and the leaders of the Union-wide organs are accountable to the USSR president, the USSR State Council, and the USSR Supreme Soviet.

Article 6. All deputies are to keep their status as USSR people's deputies during their term in office, including their right to take part in the work of USSR Supreme Soviet organs.

It is deemed inadvisable to hold sessions of the Congress of USSR People's Deputies during the transitional period.

Article 7. The USSR president is to convene the new USSR Supreme Soviet no later than two weeks after it is formed.

Article 8. The provisions of the USSR constitution are valid insofar as they do not contravene the present law.

Article 9. This law enters into force from the moment it is published.

---

## 1.9
## Yavlinskiy Plan for Economic Union Detailed

I. Demchenko

*Izvestiya,* 11 September 1991 [FBIS Translation]

---

The preamble to the draft treaty on an economic union submitted by G. Yavlinskiy to the State Council says:

"Independent states and current and former components of the USSR, irrespective of their present status, affirming their right to autonomously decide all questions of socio-economic development . . . , wishing to establish normal relations between peoples and protect citizens' economic interests, taking account of the USSR's complex of relations between states that are part of it, believing that solving political, defense, property, and humanitarian questions in relations between states requires time . . . conclude this treaty on an economic union."

The document's text, according to those who drafted it, deliberately did not broach defense and political matters. They believe that this is a problem for further treaties and, perhaps, a question of a more remote time. But this document concentrates only on the economy, which now remains perhaps the only platform on which talks between former republics can be realistically and gainfully held. Incidentally, it is typical that the word "republic" is absent from the document. All participants in the economic union are recognized as sovereign states that are guaranteed full independence.

### A Unified Legal System Operates in the Economic Union Territory

Voluntary participation and equal rights are proclaimed the basic principles of participation in the Union. The draft treaty envisages the regulation of exclusively economic mutual relations between partners, which can be states that are or are not part of the political union of sovereign states. The possibility of associate membership of the economic union is envisaged for states that are part of the USSR and are prepared to take on only some of the commitments specified by the document. Mutual relations with republics that have not joined the economic union are organized in the same way as with foreign states. All these "degrees of economic relationship" between former members of the USSR are very precisely regulated depending on the amount of commitments assumed.

The document says that the economic union's member states agree to pursue a coordinated policy in the spheres of enterprise, movement of goods and services, financial and credit policy, finances and taxes, prices, the labor market and social guarantees, customs regulations and tariffs, foreign economic activities and currency policy, various state programs, standardization, metrology, patents, statistics and accounting, and questions regarding the legal regulation of economic activity. The commitments accepted by Union members make them economically responsible both for fulfilling the treaty and for taking steps that jeopardize it. Throughout the economic union's territory, a unified legal system for economic activity is established for physical and legal persons irrespective of their residence permit, citizenship, place of registration, and so forth. This will obviously stop and prevent the continuation of the war of laws, resolutions, banks, and so forth, the destructive consequences of which have taken such a ruinous toll on our economy. The document's draft says that the economic union's organs are an interstate economic committee, a banking union, and an economic union arbitration service. Key economic policy problems are discussed at a conference of leaders of states' governments, who have the right to veto decisions of the interstate economic committee and its chairman.

### The Economy's Foundation Is Private Ownership and Free Enterprise

Free enterprise is proclaimed the foundation for the growth and development of the economy and is guaranteed in the whole space within the economic union's framework. Priority is given to private ownership. A state's interference in enterprises' economic activities is limited by legislation and will evidently be reduced to minimal economic regulation. Goods and services will be transferred within the unified economic space freely and duty-free, and the restrictions operating in this connection will be lifted. The principle of free movement is also declared for the work force. In any event, participants in the economic union must endeavor to create the conditions for this, which includes the formation of a housing market and evidently—although the document does not say so—the abolition of the institution of the residence permit. Deliveries of the most important types of goods and services at agreed prices in the transitional period will be carried out on the basis of bilateral and multilateral interstate agreements. Generally the document's section on pricing policy is one of the most laconic, and its basic concept is that the economic union's member states conduct a coordinated policy of transition to free price formation.

### A Single Currency and a Coordinated Financial and Credit Policy

It is suggested that states that have signed the document recognize the advantages of retaining the ruble as the common currency of a united monetary system on the economic union's territory and make efforts to strengthen it in 1991–92. The achievement of internal convertibility of the ruble is specified as one of the most immediate tasks. The introduction of national currencies is allowed here, but on the condition that they do not undermine the strength of the ruble. At least two points stem from this. First, evidently the population on the economic union's territory—if it has been concluded with these rules—will have to live on several types of money that are mutually convertible, which is obviously not very convenient. This is especially inconvenient for enterprise, which must be carried out freely in the whole space of the Union. Second, consequently national currencies will right from the start be put at a severe disadvantage compared with the ruble—because, as has already been said above, nothing must restrict the freedom of enterprise.

It is proposed to establish a banking union of central banks of states that have signed the treaty to elaborate and conduct an effective monetary and credit policy on the principles of a reserve system. This will do away with the monopoly of the USSR Gosbank [State Bank] in making key decisions in the area of credit and financial policy and will finally make the banking system really independent of executive authority. The draft treaty specifically makes the banking union responsible for regulating the ruble rate in relation to other currencies, disposing of the gold reserve and foreign currency reserve (should such a thing appear), regulating the general rules regarding commercial banks' activities, fixing the limits of central banks' participation in loans to budget systems, and organizing cash circulation.

This measure, combined with the coordinated budget and tax policy of economic union participants, will evidently enable sovereign states' budget deficits to be limited for some foreseeable time, if only to some extent. Now, as is known, the deficit levels are enormous. For example, around one-third of the whole budget belongs to Russia and almost one-half belongs to Georgia—it is as if republics are competing with one another to see who can succeed in squandering the most. The draft says that in the event that the prescribed limits on the growth of the internal debt are exceeded, the sum is officially registered as debt liabilities to the rest of the economic union's members. The USSR's internal state debt is reregistered and is basically divided among the economic union's members. Commitments made on behalf of the USSR to investors in the USSR Savings Bank and to owners of valuable securities will be honored. Furthermore, the economic union declares itself the assignee of all the USSR's foreign economic liabilities and guarantees that they will be met. It is considered necessary to conclude a special agreement that will specify the share of each component of the former USSR in the total sum of the external debt and in the total debt of other countries to our state.

The document envisages the creation of an economic union budget to finance general expenses. It is proposed that contributions to it are determined in the form of a fixed proportion of produced national income or according to another objective indicator. The document does not allow a situation whereby the Union members refuse to pay their share of it—as republics are now doing in relation to the USSR Union budget. In addition, off-budget funds are set aside for financing targeted programs, one of which is designated as special: the fund for the accelerated development of certain regions.

### Very Rapid Inclusion into the World Economy

Foreign economic activity by Union members will be geared toward achieving the very rapid inclusion of states and entities engaged in economic activity into world economic ties. The consent of all participants in the economic union is required to obtain new foreign loans. Here states participating in the Union have the right to obtain foreign loans and credits autonomously, if they take responsibility for servicing and clearing them.

A large section of the document is taken up by a list of agreements that, aside from the treaty, should regulate economic relations among Union members. The agreements will determine, specifically, the status and powers of the Union's organs, rights of ownership on its territory, the procedure and conditions for the introduction of national currencies, anti-monopoly policy, mutual commitments in the provision of pensions, the procedure for resolving property and other disputes, etc. It is proposed that the economic union be initially concluded for five years.

It is obvious that the text of this treaty—and the authors of the "Five Hundred Day" plan are still talking about the need to conclude it very soon—can serve as a basis for consolidating all republics, society's progressive forces, and leaders who have not lost the ability to interpret events constructively. Its great advantage is its lack of political coloring, which makes it a good basis for talks among authorized representatives of states, irrespective of ideological convictions, nationality, creed, and so on. It is evidently the best chance of agreement; first, because the more time that goes by, the bigger everyone's mutual claims and commitments become, which makes it harder to reach any kind of understanding in principle; and second, because the rapid destruction of the old state structures threatens to bury under the debris not only a unified state but also some of its components—and this is what is now happening in a number of "hot" regions.

As everyone knows, the draft document has been sent out for review by a wide circle of specialists and political figures, including leaders and members of the parliaments and governments of all the republics, leaders of political parties and economic associations and unions, entrepreneurs and economic leaders, ambassadors of leading states, and foreign experts. Evidently discussion of it will very soon be started by the State Council members who were the first to receive the document.

## 1.10

### State Council Views Economic Union

TASS, 16 September 1991
[FBIS Translation], Excerpts

The State Council met in the Kremlin at exactly 10:00 A.M. today under the chairmanship of the USSR president. The leaders of Russia, the Ukraine, Kazakhstan, Belarus, Azerbaijan, Armenia, Kyrgyzstan, Tajikistan, Turkmenistan, and Uzbekistan participated. . . .

G.A. Yavlinskiy, deputy chairman of the USSR Committee for the Management of the Economy, delivered a report entitled "On the Draft Treaty for the Economic Union." He said that the document's starting point is recognition of the independence of all the members of this Union and realization of the extremely grave economic state. The basic points of this concept are entrepreneurship and private ownership and free movement of goods and services throughout the territory of the proposed economic union. As for monetary policy and the banking system, it is proposed to preserve the common currency, while introducing national currencies. A coordinated budget policy and a settlement of all questions of the republics' domestic and foreign debts is also required. Today, an increase in the domestic debt of each of the republics is an increase in the overall debt of the whole union, Yavlinskiy said.

One other vital condition of the functioning of a unified economic union is the speediest possible movement toward freeing prices. A pooling of efforts in working out a unified mechanism for ensuring social guarantees is also seen as inevitable. Here, Grigoriy Yavlinskiy believes, one needs to begin with the question of pensions because, he stressed, "in our country, this is one economic area that requires more attention than others."

In contrast to the food situation issue, which was discussed in an atmosphere of trust and agreement, discussion of the draft economic union treaty clearly did not go smoothly. Here are some views expressed to TASS by leaders of a number of republics immediately afterward.

The Kyrgyz president, Askar Akaev, is convinced that an economic agreement is the most important step toward a future Union treaty. He noted with satisfaction the fact that the State Council members do not differ on this issue and that there are definite opportunities for this matter to be settled at the present sitting using a model similar to the EEC [European Economic Community] structure. The bilateral and multilateral treaties of sovereign republics could form some kind of basis for a future document. However, the Kyrgyz president noted that the alternative economic agreement proposed by G. Yavlinskiy is "unacceptable because of its complexity." Moreover, he spoke highly of the alternative draft agreement put forward by Stanislav Shatalin. "Written lucidly and clearly, his draft suits the present-day situation ideally and ultimately could serve as the core of an economic treaty of sovereign republics," Akaev stated.

I believe that it is too early to sign an economic agreement, said the Turkmen president, Saparmurad Niyazov. He believes that a further two to three months of hard work will be required to eliminate all the uncertainties that have arisen in economic relations among the republics. "Over that period, the sovereign republics, along with the Russian leadership, could clarify their attitudes and work out an acceptable version of an agreement," the president said. He stressed that we will all be faced in the future with trading at world prices and that this will mean inevitable losses, but we are not losing our confidence in the efficiency of the levers of solidarity. For example, if our Caucasian partners are unable to pay for Turkmen gas supplies immediately, we agree to a mutually advantageous agreement envisaging an equivalent exchange. A common interrepublic insurance fund, which should be set up on a voluntary basis, could help during a difficult period.

Despite the clear need for the country's common economic area to be supported and preserved, I am alarmed by the weakening of economic ties between republics, the Azeri president, Ayaz Mutalibov, said. In their present state, the republics will not individually be able to ensure social programs and, thus, gain the support of their peoples. The Azeri leader believes that the process of demarcating republics increased after the August events and that the mistrust that has arisen should be eliminated, primarily by strengthening economic ties. Otherwise, according to Mutalibov, the economic slump will continue, and this could also result in the loss of the democratic gains of recent years. . . .

Grigoriy Yavlinskiy, head of the draft economic treaty, commented: "The draft was adopted as a foundation. The republics will form authorized delegations to seek jointly grounds for the agreements constituting the treaty. A big and very complex task lies ahead."

Unfortunately, we know from experience that it is also possible that this work might be drawn out in time and that principles may be substituted and a kind of "salad" might be created of various programs. However, one thing is of main importance today: Work is beginning, in principle. Judging from the debate, which is coming to an end, it will progress with great difficulty.

For instance, the Kyrgyz president, Askar Akaev, proposed a treaty in a more general form, proceeding as it were from international principles. However, it was decided to adopt the most concrete forms of the document, forms that suit our conditions. I consider that it will prove possible to put the finishing touches on the document in an extremely short period of time. We are at a critical point. The State Council session has just been discussing how to share the existing food around the country. If we do not find new approaches to the problem, we will again be engaging in partition. Work on the main part of the treaty must be finished within one month.

## 1.11
## Kravchuk Disagrees with "One Plus Ten" Statement

Interfax, 3 September 1991
[FBIS Translation]

Ukraine disagrees with certain essential points of the One Plus Ten statement, although it has signed the document, the chairman of the republic's Supreme Soviet, Leonid Kravchuk, said at a press conference yesterday evening. "We disagree, for example, with the idea of some abstract separate center," he said. He believes all central bodies should be interrepublican. Otherwise "we shall have what we have now: the closer you are to the Center the more you get." Kravchuk pointed out that, shortly before the end of the putsch and after it, the Supreme Soviet of the Russian Federation had dealt with matters within the competence of the Union parliament. The Russian prime minister is virtually a co-chairman of the Committee for the Management of the Economy, which is performing the functions of a Union government and includes more members from the Russian Federation than from any other republic. "We see the reasons for it, but we believe at the same time that the wealth accumulated by all peoples of the Soviet Union should be fairly divided among them. We see this as an economic issue rather than a political one," said Kravchuk. He did not share the opinion that the Congress of People's Deputies had started a new coup d'état. "In making suggestions for the agenda, the president and the top leaders of the republics were aware that some of them contradicted the legislation. But the situation in the republics and the Soviet Union as a whole is so grave that emergency measures are perfectly warranted, above all for the sake of preventing economic collapse," he said.

## 1.12

## The Treaty on an Economic Community

*Rossiyskaya Gazeta,* 22 October 1991
[*CD* Translation]

Independent states that are former members of the Union of Soviet Socialist Republics, regardless of their current status, expressing the will of their peoples for the political and economic sovereignty codified in acts adopted by the states' supreme legislative bodies and for the protection of citizens' interests, desiring to establish mutually advantageous economic relations between states, striving for radical economic transformations, and considering the common nature of the problems facing the states in connection with the tasks of getting out of the crisis, changing over to a market economy and entering the world economy, recognizing the advantages of economic integration, a common economic space, and the advisability of preserving economic, trade, scientific-technical, and other relations, conclude this treaty on an economic community.

### Chapter 1. Basic Principles

Article 1. The Economic Community is created by independent states on the basis of voluntary participation by and equal rights for all its members, for the purpose of forming a united market and conducting a coordinated economic policy as an indispensable condition of overcoming the crisis.

Article 2. Membership in the Economic Community entails the making by independent states of the full range of commitments provided for in this treaty and their acquisition of all the rights provided therein.

The member states of the Economic Community are to proceed on the basis of mutual economic responsibility for the implementation of this treaty and are to refrain from any steps that jeopardize the implementation of this treaty as a whole or of its individual provisions.

Article 3. States that have seceded from the USSR and have not joined the Economic Community are to construct relations with the Economic Community on the basis of generally accepted principles and norms of international law. Questions of common interest for the Economic Community that require settlement are to be resolved by special agreements between the Economic Community and the states in question, to be concluded within no more than three months of the date this treaty goes into effect.

Article 4. A member state of the Economic Community has the right to withdraw from the Community. It must notify the other members of the Economic Community of its intention no less than twelve months in advance. Withdrawal from the Economic Community is conditional on the settlement of relations with respect to all commitments associated with membership in the Community, in accordance with the special agreements.

Article 5. The member states of the Economic Community are to reach an agreement on bringing economic legislation closer together, and also on conducting coordinated policy in the following fields:

—entrepreneurship;
—market for goods and services;
—transportation, electrical engineering and information;
—monetary and banking system;
—finances, taxes, and prices;
—the market for capital and securities;
—the labor market;
—customs regulations and tariffs;
—foreign economic relations and foreign currency policy;
—state scientific and technical, investment, ecological, humanitarian, and other programs (including programs to eliminate the consequences of natural disasters and catastrophes) that are of general interest for the Economic Community;
—standardization, patents, metrology, statistics, and accounting.

Article 6. The member states of the Economic Community mutually pledge not to permit unilateral, uncoordinated actions with respect to the division of property that they recognize as joint property. The composition of this joint property is to be determined by a special agreement.

Article 7. In order to carry out a coordinated economic policy and common measures to emerge from the crisis, the member states of the Economic Community will create community institutions endowed with the appropriate powers.

Article 8. This treaty is concluded for a period of three years. No later than twelve months before that time is reached, the member states of the Economic Community are to resolve the question of extending or changing this treaty or concluding a new one.

### Chapter 2. Entrepreneurship

Article 9. The member states of the Economic Community recognize that the foundation for achieving an upswing in the economy is private property, free enterprise, and competition. They are creating conditions to facilitate business activity and are legislatively limiting state interference in the economic activity of enterprises.

Article 10. A member state of the Economic Community pledges to ensure an identical legal system on its territory for physical and juristic persons both of its state and of other member states of the Economic Community. The conditions of the legal system may be spelled out on the basis of bilateral and multilateral agreements.

Article 11. The member states of the Economic Community pledge to conduct a coordinated anti-monopoly policy and to promote the development of competition within the framework of their united market. During the transitional period, the member states of the Economic Community are to carry out coordinated actions with respect to the regulation of prices for goods produced by monopoly producers.

### Chapter 3. The Movement of Goods, Services, and Prices

Article 12. The movement of goods and services on the territory of states belonging to the Economic Community is carried out freely and on a duty-free basis. The importing of goods from third states is subject to the imposition of duties in accordance with the Economic Community's uniform external customs-tariff rates.

Article 13. Striving for the formation of a united market, the member states of the Economic Community recognize the impermissibility of restrictions on the movement of goods and services and pledge to eliminate them within an agreed-upon period.

Article 14. The member states of the Economic Community are to conduct a coordinated policy of changing over to freely determined prices. The member states of the Economic Community are to apply agreed-upon prices to a list of products to be determined jointly.

### Chapter 4. The Monetary and Banking System

Article 15. The member states of the Economic Community recognize that, in order to end the crisis and control inflation, coordinated actions in the field of monetary and credit policy are a major priority.

Article 16. The member states of the Economic Community recognize the need to preserve the ruble as the common currency of a single monetary system, and they have agreed to undertake efforts to strengthen it. The member states of the Economic Community are permitted to introduce national currencies if they fulfill conditions that would prevent detriment to the Economic Community's monetary system. These conditions are to be set by special agreements between individual states and the Economic Community.

Article 17. In order to work out and conduct an effective monetary and credit policy that restrains price increases and maintains the ruble's exchange rate, the member states of the Economic Community are to establish, on the principles of a reserve system, a Banking Union that will include the central (national state) banks of the Economic Community's member states and are to create an Interstate Emissions Bank under the Banking Union.

Article 18. The legal status of the Banking Union and the procedures for forming its administrative agencies are to be determined by a special agreement. The activity of the Banking Union will be regulated by its charter, which is to be confirmed by the supreme bodies of legislative power of the member states of the Economic Community.

Article 19. The following functions are assigned to the Banking Union:

—determining uniform approaches to the implementation of monetary and credit policy, and establishing quantitative parameters (ceilings) for the operations of central banks that are members of the Banking Union;

—setting interest rates on loans granted to commercial banks by central banks that are members of the Banking Union;

—setting reserve requirements for commercial banks;

—organizing interbank settlements;

—regulating the ruble's exchange rate with respect to other currencies;

—managing that part of the gold reserve and the foreign currency reserve that the member states of the Economic Community transfer to the Banking Union;

—organizing the circulation of cash, including the sale of paper money and coins;

—regulating general rules for the activity of commercial banks.

For 1992, the Banking Union is to establish ceilings on the participation of the central banks of the member states of the

Economic Community in providing credits to local budgets.

The directives of administrative agencies of the Banking Union are mandatory for all banks that are members of it.

Article 20. In creating the Banking Union, the member states of the Economic Community consider it necessary to determine, in a special agreement, procedures for the transfer to their ownership and to the ownership of states that do not belong to the Economic Community of appropriate parts of the start-up [*ustavnie fondy*] reserve and other assets of the former USSR state banking system and its equity, limited-liability companies and deposits, as well as of the gold reserve and stocks of diamonds and foreign currency.

Article 21. The member states of the Economic Community are to create, on conditions of equal representation, a supreme bank inspectorate that will be charged with the following:

—monitoring observance of the charters of the Banking Union and the Interstate Emissions Bank;

—considering statements by members of the Economic Community about the infringement of their interests;

—revoking and suspending decisions of the Banking Union that are at variance with its charter.

The Banking Union and the Interstate Emissions Bank must present all necessary documents and materials at the request of the Supreme Bank Inspectorate.

Decisions of the Supreme Bank Inspectorate that are adopted within the bounds of its jurisdiction must be implemented by the Bank Union and the Interstate Emissions Bank.

**Chapter 5. Finances and Taxes**

Article 22. The member states of the Economic Community are to conduct a coordinated budget and tax policy, which means:

—the agreed-upon limitation of the deficits of consolidated state budgets, taking into account off-budget funds, and the establishment of limits on the growth of the internal debt of the member states of the Economic Community. If a state has exhausted possibilities for floating its securities and if one or several member states of the Economic Community exceed the established limits, the excess amount is listed as a debt liability of the other community members, in accordance with agreed-upon rules;

—the standardization of taxation principles, and an agreed-upon policy in the field of taxes that affect the interests of the other member states of the Economic Community.

Article 23. The member states of the Economic Community consider it necessary to conclude an agreement on the division and legal restructuring of the USSR's internal debt, as it stands at an agreed-upon date, between the member states of the Economic Community and states that are not members, and on the procedures for servicing this debt in the future. Part of the debt may be added to the common liability of the member states of the Economic Community. At the same time, an agreement will be concluded on the distribution of centralized credit resources among the banks of the states. The state internal debt is to be determined with consideration for the state budgets' indebtedness to banks for covering price differences for agricultural output as of 1 January 1992.

The member states of the Economic Community affirm the continuity of obligations assumed in the name of the USSR to the holders of deposits in the USSR Savings Bank, USSR state securities, certificates of the USSR Savings Bank, and insurance policies (obligations) of the USSR Ministry of Finance's Chief Administration for State Insurance.

Article 24. In order to finance joint expenditures, including expenditures for the maintenance of institutions of the Economic Community, the member states of the Economic Community are to create a budget for the community.

Within the framework of the Economic Community's budget, the following special funds are to be created;

—a fund to service the part of the USSR internal state debt that has been added to the joint liability of the member states of the Economic Community;

—a fund to service the USSR's foreign debt and the foreign debt of the Economic Community (the portion paid in rubles);

—a fund for emergency situations and for eliminating the consequences of major natural disasters and catastrophes, such as Chernobyl, the Aral Sea, the earthquake in Spitak, and others;

—a fund for special purpose programs;

—and a financial reserve.

The budget of the Economic Community is formed from the dues paid by its members, which are set in fixed amounts. The amounts of these fixed dues and the procedure for forming them are to be determined by a special agreement among the members of the Economic Community. The budget cannot be in deficit.

The Interstate Economic Committee will make quarterly reports to the Council of Heads of Government of the Member States of the Economic Community on the fulfillment of the budget.

Article 25. The member states of the Economic Community may create off-budget funds to finance special-purpose programs. The Economic Community's off-budget funds cannot be in deficit.

Article 26. The member states of the Economic Community recognize the need to conduct special-purpose programs to aid the development of certain regions and provide social support for the population of those regions. To this end, they are to create an appropriate fund by contributing agreed-upon shares of the national income (gross national product) produced on their territories, in current prices. The activity of this fund is to be regulated by a special agreement and the charter.

Article 27. The member states of the Economic Community are to arrange for movements of money and other financial resources and the emission and circulation of securities within the framework of the Economic Community to be regulated in accordance with a special agreement that does not include the free movement of capital.

## Chapter 6. The Labor Market and Social Guarantees

Article 28. The member states are to strive to implement the principle of freedom of movement for the work force on the territories of the member states of the Economic Community and, consequently, are to create the conditions for this, including the formation of a housing market.

Article 29. The member states are not to permit discrimination against citizens on the basis of nationality or on any other basis in questions of job availability, pay or other working conditions, or in providing social guarantees.

The member states are to mutually recognize the educational and skill levels, as confirmed by appropriate documents, that employees obtained in other member states and are not to demand additional confirmation of these levels when hiring people or admitting them to educational institutions, if this condition is not mandatory for everyone.

The member states are to agree on a visa-free system for the movement of their citizens within the territorial bounds of this Community.

Article 30. The member states are to conclude special agreements on regulating migration processes and mutual obligations in the field of social insurance and pensions for citizens of states belonging to the Economic Community.

## Chapter 7. Foreign Economic Relations and Foreign Currency Policy

Article 31. The member states are to reach agreement on the coordination of foreign economic activity and foreign currency policy.

Article 32. The Economic Community, as the legal inheritor of all the foreign economic commitments of the USSR, as well as of the commitments of other countries to the USSR, guarantees their fulfillment. Each member state, when it enters the Economic Community, confirms its participation in the joint fulfillment of these commitments.

The members of the Economic Community are to create a bank that will be the legal successor to the USSR Bank for Foreign Economic Activity, a bank through which settlements related to the repayment of foreign debts and the receipt by the Economic Community of debt payments from other countries will be made.

The members of the Economic Community will assume the settlement of all relations regarding foreign debts with each member of the USSR that does not join the Economic Community.

The members of the Economic Community consider it necessary to conclude a special agreement on determining the share of each member of the former Union of the Soviet Socialist Republics in the overall total of payments on the USSR's foreign debt and in the total of debts owed to the USSR by other countries as of an agreed-upon date.

The indicated shares are to be used for settlements regarding debt repayment with states not wishing to accede to the treaty or withdrawing from the Economic Community. These states are to conduct settlements with the Economic Community up to and including the full payment of all debts falling to them.

If a state that does not wish to accede to the treaty, that withdraws from the Economic Community, or that is a part of it reaches an agreement with all its creditors on restructuring its share of the overall foreign debt, then the independent repayment by that state of the indicated portion of the debt is to be permitted.

Article 33. The Economic Community is to obtain new foreign loans on the basis of an agreed-upon decision by all member states of the Economic Community.

A member state of the Economic Community has the right to obtain foreign loans independently, with all the ensuing obligations regarding the servicing and repayment of such loans.

The granting of credits or other economic assistance to foreign states by the Economic Community or individual member states is to be carried out according to the same procedure.

Article 34. The member states, using a common monetary unit, have agreed that achieving internal convertibility for the ruble is an immediate goal of this Community. They pledge to carry out the necessary preparatory measures on the basis of an agreed-upon program for changing over to internal convertibility for the ruble.

Article 35. Pending the changeover to internal convertibility for the ruble, the member states of the Economic Community recognize the need to establish a uniform procedure for the accumulation of foreign currency receipts to service the foreign debt.

Article 36. The member states are to independently regulate foreign economic activity, the setting of quotas and the issuing of licenses for foreign economic operations within agreed-upon quotas.

Article 37. The member states are to reach agreement on

the preservation of a single customs territory for the Economic Community and are to conduct a coordinated customs policy with regard to third countries.

Questions related to the procedure for establishing customs duties and taxes on exports and imports and levying them in the appropriate budgets are to be resolved by a special agreement.

Article 38. The member states are to reach agreement on common membership in the International Monetary Fund, the International Bank for Reconstruction and Development, the General Agreement on Tariffs and Trade, and other international economic organizations, and they are to confirm the legal validity of applications made to these bodies on behalf of the USSR.

Article 39. The member states are to conduct an independent policy in the field of foreign investments, coordinating this policy when necessary. The member states are also to coordinate their actions in the field of technical, advisory, and other forms of assistance from foreign states and international organizations.

## Chapter 8. The Legal Regulation of Economic Activity

Article 40. The legislative acts of the member states of the Economic Community have supremacy in these states. Physical and juridical persons engaged in economic activity on the territories of these states are guided by their legislation.

Article 41. For the effective period of this treaty, the member states have agreed to bring the norms of their economic legislation closer together, with a view to creating most-favored and equal conditions for entrepreneurship and free trade throughout the space of the Economic Community. They pledge to ensure the conformity of their legislation to the norms of international law and the acts of the Economic Community.

Article 42. If rules other than those contained in the legislation of the member states of the Economic Community exist within the bounds of their jurisdiction, the priority of the treaty and of the agreements and normative acts of the Economic Community is to be recognized.

Article 43. During the period necessary for the settlement of all aspects of economic life by legislation of the member states, acts of USSR legislation, insofar as they are not at variance with this treaty and agreements, are to be used temporarily with respect to unsettled questions, according to an agreed-upon procedure.

## Chapter 9. Institutions of the Economic Community

Article 44. The institutions of the Economic Community are:

—the Council of Heads of Government of the Member States;

—the Interstate Economic Committee;

—the Court of Arbitration of the Economic Community.

The institutions of the Economic Community are formed on a professional basis from representatives of the member states of the Community.

Article 45. The supreme coordinating body of the Economic Community is the Council of Heads of Government of the member states.

Regular conferences of ministers of the member states of the Economic Community will be held in order to resolve questions of cooperation and the working out of a common policy in specific aspects of economic management.

Article 46. The Interstate Economic Committee is the executive body of the Economic Community. Within the bounds of its jurisdiction, it ensures the fulfillment of the tasks of the Economic Community as determined by this treaty and related agreements, as well as by decisions of the Council of Heads of Government of the member states.

Article 47. The work of the Interstate Economic Committee is directed by its chairman.

The chairman of the Interstate Economic Committee is recommended by the Council of Heads of Government of the member states and is appointed and relieved of his duties by the Council of Heads of Government of the member states, by majority vote (more than half).

The chairman of the Interstate Economic Committee is to present the committee's regulations, structure, and estimated expenditures to the Council of Heads of Government of the member states for confirmation.

Article 48. In order to ensure the uniform application of this treaty throughout the member states of the Economic Community, a Court of Arbitration of the Economic Community is to be formed to resolve disputed questions and apply economic sanctions.

## Chapter 10. Agreements

Article 49. Economic relations between the member states of the Economic Community are regulated on the basis of this treaty and related agreements.

Article 50. Within no more than three months from the time this treaty is signed, agreements lasting for the duration of the treaty are to be concluded on the following questions:

—on the status and powers of the institutions of the Economic Community;

—on the formation of the Banking Union, including its character;

—on the Economic Community's Bank for Foreign Economic Activity;

—on bringing closer together the legislation of member states regulating economic activity;

—on the regulation of migration processes;

—on the creation of a Fund for Regional Development and Social Support for the Population, including its charter;

—on the creation of off-budget funds for financing special-purpose programs of the Economic Community and statutes concerning those funds;

—on the settlement of the mutual obligations of states if they withdraw from the Economic Community;

—on the movement of capital and securities;

—on anti-monopoly policy within the framework of the united market of the member states;

—on mutual obligations in the field of pensions and social insurance;

—on scientific and technical cooperation;

—on the patent service;

—on procedures and conditions for the introduction of national currencies by member states;

—on the procedure for changing over to general principles of international economic relations with states that are part of the USSR but have not joined the Economic Community;

—on the division and legal restructuring of the USSR state internal debt and the procedure for servicing it in the future;

—on the principles and mechanisms for servicing the Economic Community's foreign debt.

### Chapter 11. Associate Membership in the Economic Community

Article 52. A state that assumes only part of the commitments specified by this treaty may, with the consent of the Economic Community's members, be granted the status of associate member.

Article 53. The conditions on which states may become affiliated with the Economic Community as associate members are specified by the members.

Article 54. Relations between the Economic Community and associate members are regulated in each individual instance by a special agreement that provides for, in particular, the resolution of questions concerning customs procedures, interstate deliveries and prices, and the Economic Community's participation in determining the associate's budget and off-budget funds.

### Chapter 12. Concluding Provisions

Article 55. The member states of the Economic Community pledge to resolve questions involving the violation of this treaty in the Economic Community's Court of Arbitration in accordance with the jurisdiction assigned to that body, and also through negotiations.

Member states that commit flagrant violations of their commitments stemming from the treaty or its supplemental agreements on specific issues may be subject to sanctions provided for in agreements, up to and including expulsion from the Economic Community.

Article 56. The Economic Community is open to the entry of states recognizing this treaty. The admission of new members will be regulated by a consensus vote among member states.

## 1.13
## President Gorbachev's Appeal to Ukrainian Leadership

*Rossiyskaya Gazeta*, 23 October 1991
[FBIS Translation]

Esteemed Ukrainian people's deputies!

This appeal to you is dictated by a desire to extricate the country as quickly as possible from its grave economic and political crisis and to return people's lives to normal. Today it is a question of the destiny of our multinational homeland. And we are profoundly convinced that only together can our peoples resolve the urgent problems of their development and build a worthy future for themselves and their descendants.

The changes that have occurred in the country since the August events have created a fundamentally new political situation. The obstacles in the way of society's cardinal democratic renewal are being removed. Conditions are evolving so that each people can freely choose its own form of statehood.

Freedom and independence and the conditions for their development are the natural right of every people.

Freedom and independence are also preconditions for the peoples' unification on a truly voluntary and equal basis.

Such a Union is vitally necessary. The entire history of our multinational state and the deep-seated traditions of our peoples' lives speak for themselves. The urgent needs of many millions of people and fundamental international conditions make unification necessary.

For centuries our peoples created a great spiritual culture together. If they were to go separate ways today, this would be an irretrievable loss both for each of them and for world civilization.

We know what state our economy is in now. But it is a fact that it developed for decades within the framework of a unified economic complex. If you were to tear this thick fabric in two or three places, the entire canvas would start to fray and disintegrate. People in all republics and parts of the country are already feeling the pernicious consequences of the rupture of economic ties.

Over the long years of our multinational state's development, the nations and peoples have moved about its vast space. Millions and even tens of millions of people now live outside their own national republics. Demographic upheavals would be the inevitable consequence of their isolation. The dramatic fate of hundreds of thousands of refugees is apparent to everyone.

The military might of our superpower was created by the very hard labor and the sacrifices of several generations. Today, together with other states, we are traveling the path of decisively reducing that might and bringing it into line with the doctrine of defense sufficiency. But the USSR remains one of the two greatest nuclear powers. We bear a responsibility, both to our own people and to the whole world, for ensuring that this terrible force is under reliable control.

Let us be frank. There is still prejudice against the Union in people's minds. Many negative phenomena associated with the rigid unitary model of state structure live on and are still far from being forgotten. But it is now a question of a new Union, in which the very foundations of the economic and political structure, the socially oriented market economy, political and ideological pluralism, and the multiparty system will serve as a guarantee of the free development of every nation and every citizen.

It was for just such a Union that the overwhelming majority of the population of the country as a whole and in each of the republics voted in the referendum on 17 March this year.

The building of the new Union has begun. The signing of the Economic Community treaty is of great significance in this connection. Work on the draft treaty on the Union of Sovereign States has been resumed. In accordance with the joint statement of the USSR president and the leaders of nine republics, the session of the new convocation of the country's Supreme Soviet opened in Moscow on 21 October by decision of the Fifth Congress of USSR People's Deputies.

We address you, esteemed elected representatives of the people, with an expression of hope that Ukraine's representatives will take an active part in this collective work. Ukraine is one of the largest republics in the Union. Its role in our country's development and in everything of which our peoples can be rightly proud is indispensable. We want to believe that it will make an equally worthy contribution to the building of the new, united family of peoples, of which the great Kobzar dreamed.

Let us state frankly: We cannot imagine the Union without Ukraine. We are convinced that the multinational people of Ukraine also cannot conceive of the future without relations of alliance with all our country's peoples, with whom they are linked by a history dating back many centuries.

Accept, esteemed deputies, our wishes for the successful work of the republic's Supreme Soviet.

We send our very kind feelings and wishes to the people of the fraternal Ukraine.

[Signed] M. Gorbachev, B. Yeltsin, S. Shushkevich, N. Nazarbaev, I. Karimov, A. Mutalibov, A. Akaev, S. Niyazov, A. Iskandarov

## 1.14
## Yeltsin's Reactions to a Union Without Ukraine

*Editor's Note*: Ukraine's 1 December referendum on independence forced Boris Yeltsin's hand. Although he supported the Union, he was literally the "co-president" of the Soviet Union, and he wanted to rid the political scene of Mikhail Gorbachev. In the interviews that follow, Yeltsin signals that Russia will probably not sign the Union treaty if Ukraine does not. This, of course, removes any pretext of a need for a Union president and sets the stage for the USSR's collapse. (Yeltsin may have been thinking that he could reunite the republics after Gorbachev's fall.)

**Yeltsin's Pre-Referendum Views**
TASS, 30 November 1991
[FBIS Translation], Excerpts

The Russian president, Boris Yeltsin, stated Monday that he cannot conceive of the Union without Ukraine, whose population will vote tomorrow on whether or not the republic should proclaim its independence.

Addressing journalists following a meeting of the presidents of the USSR and the RSFSR [Russian Soviet Federated Socialist Republic], Yeltsin expressed his apprehension that if Ukraine leaves the Union, "it would rapidly introduce its own national currency and Russia would be left to pick up the pieces."

President Yeltsin said that Mikhail Gorbachev had assured him that he would take every step to see that Ukraine signs the Union treaty.

"As far as I am concerned, I have always said that I am for the Union," the president declared.

[Moscow Central Television, 1900 GMT]

[Interviewer]: In your interview in yesterday's *Izvestiya* you stated that Russia would not sign the Union treaty until Ukraine signs it. Judging from Kravchuk's statements, Ukraine does not intend to participate in the Union treaty. What does this mean? Will there be a Union?

[Yeltsin]: You have paraphrased it somewhat. You have put it inaccurately, just as *Izvestiya* did.

[Interviewer]: You will not—Russia will not—sign?

[Yeltsin]: No, give me the *Izvestiya*.

[Interviewer]: I have it here.

[Yeltsin]: At today's State Council I will be compelled to say that until Ukraine signs the political treaty, Russia will not sign. Well, that is approximately it, although what I said is not entirely accurately reproduced here, but in essence, yes. I cannot envisage a Union without Ukraine. Because if Ukraine were to rapidly introduce its own national currency, the entire ruble monetary mass would collapse on Russia. This cannot be. In general, what kind of Union would it be without Ukraine? The president has told me that he will take all measures necessary for Ukraine to sign the treaty. Well, that is all. I am saying what I have always said—that I am in favor of the Union.

### Yeltsin's Post-Referendum Reaction
TASS, 3 December 1991
[FBIS Translation]

A nationwide referendum on the future of Ukraine took place on 1 December 1991. The Russian leadership declares its recognition of the independence of Ukraine in accordance with the democratic will of its people.

We must now rapidly establish new interstate relations between Russia and Ukraine, preserving traditions, friendship and good-neighborliness, and mutual respect between both peoples, and strictly observe the pledges we have made, including the nonproliferation of nuclear weapons, observance of human rights, and other universally acknowledged international norms.

The basis for a comprehensive and mutually beneficial partnership between Russia and Ukraine was laid down by the treaty of 19 November 1990 and the communiqué of 6 November 1991. We will continue to adhere to this strategic course. We are ready to start work on a draft of a full-scale interstate bilateral treaty, which would meet all the demands of this new stage in Russian–Ukrainian relations.

All-around cooperation between our states, their mutual dependence, and the positive experience of our mutual relations in the past demand a particularly careful attitude.

Mutually beneficial and balanced cooperation between Russia and Ukraine can and must serve as an example of bilateral relations between the republics of the old Union. New possibilities are opening for close cooperation among republics and for forming a genuinely equal community of sovereign states.

The new partnership between Russia and Ukraine—two sovereign states—will make a substantial contribution to security and stability on the continent—in accordance with the CSCE [Conference on Security and Cooperation in Europe] principles—and to the positive processes in the world.

Russian and Ukrainian observation of obligations in disarmament, human rights, and national minorities, including the avoidance of discrimination on the basis of language; openness of borders; freedom of choice of citizenship; and cooperation in the formation of a common economic space and the development of contiguous regions will create a good basis for the rapid establishment of diplomatic relations between our two countries.

## 1.15
## No "Obstacles" to Recognition of Ukraine

*Editor's Note*: Even before formation of the CIS, Ukraine put on the world agenda the issue of whether and how quickly former Soviet republics could expect to receive international recognition as sovereign states. The general referendum ratifying Ukrainian independence on 1 December 1991 caused a great deal of discussion in the world community. Many countries feared the eruption of a territorial dispute between Russia and Ukraine. For some countries, human rights guarantees were a very important criterion for recognizing the statehood of a former republic of the Soviet Union. Many others (including the United States) initially chose to withhold recognition until Ukraine proved it was willing to give up its nuclear weapons and become a "nuclear-free" state. Some Ukrainian politicians, wary of Russian intentions, believed Ukraine should retain at least some of its powerful nuclear missiles and missile production capacity as a bargaining chip in debates with its "big brother." This position was adamantly discouraged by the United States and several European nations, however, becoming in some ways an unfortunate distraction from other issues involved in the stabilization of the region and an obstacle to mutually beneficial relations between Ukraine and the West. As the following article states, however, an important step forward was taken by world organizations in considering what the "universal criteria" for statehood might be. It is noteworthy that the piece that follows mentions Russia's allusion to the status of Russians living in former republics. This was to become a difficult issue for every CIS member and was to figure into Russia's rationale for a more aggressive "peacekeeping" role in the new states. Although old thought patterns would require time to fade, world recognition of Ukraine was an important marker on the path toward the post–Cold War world order, and convincing proof that the Soviet Union was really gone for good.

**Russia and the Community of Nations United in Their Approach**
V. Mikheev
*Izvestiya,* 5 December 1991
[FBIS Translation], Excerpts

The world community of nations is displaying unprecedented unanimity in its approach to the conditions for its diplomatic recognition of the new European power—Ukraine.

In short, in the words of the Australian foreign minister, G. Evans, there are a number of "generally recognized international criteria for statehood." Ukraine must fit the status of a sovereign state in all parameters. Doubt may have been cast on the "generally recognized" concept yesterday. However, with regard to Ukraine the community of nations has displayed rare unanimity, and the criteria for statehood and the terms for recognizing Ukraine de jure in its new capacity coincide in the statements issued by the White House and the Russian president, Boris Yeltsin, in the demands made by NATO and neutral Sweden, and in the wishes expressed by its very close neighbor Germany and far-off Australia.

What are these unified terms? Compliance with the USSR's international disarmament commitments; respect for human rights, including the rights of national minorities (the 12 million Russians, in particular, of course); and implementation of the normative acts adopted in Helsinki and Paris establishing the principles for European communal life. The Russian president's statement adds: "the prevention of linguistic discrimination" and "freedom to choose citizenship."

The second "basket" concerns the subject of the unexpected expansion of the "nuclear club." The United States is insistently calling for Ukraine to implement its desire for a "nuclear-free status" and meanwhile ensure responsible and reliable control over the nuclear weapons on its national territory and prevent the spread of dangerous technologies. The latter is understandable given the increasingly frequent reports that our nuclear physicists are being made alluring offers to work abroad on contract—Iran and Iraq figure among the employers.

The third "basket" holds a reference to one of the precepts of the Helsinki Charter: A change in Europe's existing borders is undesirable, but, if it is unavoidable, it should occur solely by peaceful means and through negotiations. No one denies that the explosive potential of any territorial dispute may rival that of the nuclear arms. Romania's claims to northern Bukovina are an alarm signal. Since up to 85 percent of the borders within the former Union are not marked, the reference in Yeltsin's statement to the idea of Russian and Ukrainian commitment to the principle of "open borders" should be seen as the correct move. Does that settle the problem?

The positions espoused by most states are seen to be consonant on all three "baskets," although there could be some additional emphases in the reactions from the community of nations. Thus Australia is raising the matter of "effective control of its territory" as an important attribute of statehood. How does this apply to Ukraine, where major troop contingents "under the jurisdiction of the Soviet central command" are stationed? On the other hand, the permanent representatives of the North Atlantic Alliance countries unequivocally advocated ensuring reliable control over nuclear weapons and preserving "unified command of these weapons."

The different nuances do not negate the fundamental coincidence of approaches, enabling the following optimistic conclusion to be drawn: With the deideologization of world politics and the fading of Cold War inertia, it has become possible to develop interstate relations on legal foundations that really are universally recognized, on the ethical postulates of the civilized world. It is because of that ideological unity that the voice of the community of nations has become louder and heavier. Everyone who decides, like Ukraine, to become a full-fledged subject of international relations in the future must take that into account. I think that this episode, rather than the united front against Baghdad's tyrannical regime, better displays the features of the "new world order."

People in Kiev are fully aware of that. The Ukrainian foreign minister, Anatoliy Zlenko, confirmed that his state regards itself as the "rightful successor to the former Soviet Union insofar as the agreements signed by the latter are concerned," including the Strategic Arms Reduction Talks (START) treaty, the Helsinki accords, and other human rights agreements. Ukraine, the minister said, intends to eliminate 130 of the 176 ICBMs [intercontinental ballistic missiles] on its territory. In the future Ukraine will become a nuclear-free state. Our involvement in any military blocs is ruled out. In conclusion, Zlenko reassured creditors, Ukraine will abide by the accord of 21 October and pay its share of the USSR's foreign debt.

So, the obstacles have been cleared to formal international recognition of Ukraine. Unlike the Baltic states, however, few are rushing to open their embassies in Kiev. Still, it cannot be ruled out that this step is only a matter of time. The following words are encountered most frequently: "We will consider the matter with neighboring countries" (Norway); "We will make a decision with other EC [European Community] states" (France); "We must wait for an explanation of the position on CSCE [Conference on Security and Cooperation in Europe] principles, including the Paris Charter commitments regarding the observance of human rights" (Sweden); "We will act in accordance with other Asian states" (South Korea); and so on.

The wait-and-see approach does not detract from the main point: There are no formal obstacles to diplomatic recognition, and therefore, according to France's *Le Monde*, it is a matter of months or even weeks.

# 2

# Three Plus Eight: From the USSR to the CIS

## Introductory Notes

As the CIS was being hastily cobbled together (7–8 December 1991) in a meeting held just outside Minsk in the Republic of Belarus, the USSR was "on the brink of explosion." Such terms were used by international reporters because at that moment the Union was bankrupt (every republic had ceased paying taxes months before); all central government bodies had been dissolved in November; 72 percent of the Soviet people no longer believed Mikhail Gorbachev was in control of the country; Ukraine's referendum on independence threw any hopes of its signing the Union treaty into grave doubt; and many people believed that another putsch attempt was in the making. Add to this picture the fact that Presidents Gorbachev and Yeltsin were making it increasingly clear that they could not work with one another—that one or the other would have to go—and you have a complex recipe for dissolution from the top down.

The first two documents in this chapter—press accounts of Ukrainian leaders' pronouncements on foreign and military policy—illustrate that, despite President Gorbachev's vehement statements to the contrary, Ukraine was steadfastly severing itself from the Union. This was being interpreted as the final blow to the Soviet state, making it clear to almost everyone that the old federal way of life was no longer viable.

The Minsk (Belovezh Forest) and Alma-Ata agreements, provided in their entirety in this chapter, are the two formative documents of the CIS. The Minsk agreement was drafted by the three Slav presidents—Yeltsin of Russia, Kravchuk of Ukraine, Shushkevich of Belarus—and their prime ministers at the Belarus government's Viskuli residence in the Belovezhskaya Pushcha (Belovezh Forest) on 8 December. Press reviews said that the three leaders were meeting to discuss certain problems with the current (fifth) version of the Union treaty, which Mikhail Gorbachev had sent to the republics in November for their review. The Kazakh president, Nursultan Nazarbaev, an ardent supporter of the Union treaty, was also supposed to be present (Kazakhstan is the fourth largest nuclear state of the former Soviet Union). For unclear reasons, Nazarbaev did not attend, and the meeting turned out to have quite a different purpose.

The Minsk agreement that resulted from the meeting declared that the USSR had ceased to exist as "a subject of international law" and as a "geopolitical reality." A commonwealth was proclaimed in its place by the three "high contracting parties" (which had been the original founders of the USSR in 1922). Any former republic or third-party state agreeing with the principles established in the Minsk agreement was invited to join the CIS. Fulfillment of the international treaties and agreements signed by the USSR was guaranteed within the agreement. The basic understandings of the three leaders at the time are reflected in the "Declaration on the New Commonwealth" and the "Declaration on Economic Policy," which were attached to the Minsk agreement. These two declarations outline the three leaders' basic understanding of how the commonwealth was to reestablish stability in the region.

What the Minsk declarations did not express, however, was the leaders' fundamentally divergent perceptions concerning the future evolution and role of the CIS. They all saw it as an "amicable divorce" from the Union. However, Boris Yeltsin thought of the CIS as a new type of union, formed to rescue Soviet integration as the Soviet state was falling apart, leading in a few years to a confederal arrangement, similar to the European Union. In addition, Yeltsin saw in the dissolution of the USSR a convenient vehicle for the removal of Gorbachev from the post of president of the Soviet Union. Stanislav Shushkevich regarded the CIS as a vehicle through which Belarus could raise its profile by becoming the new "headquarters" of the CIS. In Leonid Kravchuk's view of post-Soviet life, however, the CIS was a transient arrangement required to ease the transition from Union to independence. He envisioned a loosening of ties over time, as states strengthened their own economies, not the reverse process. None of the three leaders perhaps appreciated fully just how difficult policy decisions would be regarding interstate integration within the CIS.

On 9 December, following the Minsk meeting of the Slav presidents, the Kazakh president, Nursultan Nazarbaev, sat in on the meeting between presidents Mikhail Gorbachev and Boris Yeltsin at which Yeltsin presented the joint resolution of the CIS founding states. Nazarbaev was deeply offended by the exclusion of the other republics from the founding session. Subsequently, on 13 December the Kazakh president and leaders of the other Central Asian states met in Ashkhabad to consider forming their own "Eurasian Union" as a counterbalance to a "Slavic Union," which was what they considered the CIS to be. Ultimately, a senseless division of the former Union into two warring halves (Slavic and Muslim) was avoided. The original text of the "Ashkhabad Declaration" that emerged from the meeting appears here in its entirety. Its purpose was to guarantee Central Asia equal status with the founding members in the CIS. All five Central Asian states (Kazakhstan, Kyrgyzstan, Tajikistan, Turkmenistan, and Uzbekistan) were persuaded to join the CIS when they were assured the status of "high contracting parties" (or founders) in the Alma-Ata agreement, which was concluded at the 21 December meeting of the CIS.

Mikhail Gorbachev retired as the last president of the Soviet Union on 25 December. His resignation in a televised speech to the former Soviet people is contained in this chapter. On 26 December, the USSR Supreme Soviet was dissolved and the Soviet flag over the Kremlin was lowered. This memorable occasion is commemorated in this chapter in an address by the chairman of the Soviet of the Republics before the last extraordinary session of the Supreme Soviet. These two acts confirmed the disintegration of an imperial unit that had lasted for some five hundred years. In a bloodless transition, the second Russian revolution—this time a peaceful, democratic one—had, in effect, ended, or, with the advantage of hindsight, just begun. By its very nature, this transition contained the seeds of a dilemma for the Russian leadership. In some instances, Russia was to assume the duties, obligations, and assets of the Soviet government. In other instances, these things were to be shared among the new member states of the CIS. The CIS was to become the vehicle through which Russia is still today defining itself as a democratic nation.

The remaining documents provided in this chapter reflect the essence of what was to transpire in the first few months of the CIS's existence. Stanislav Shushkevich in his preview of the Minsk Summit of 30 December refers to the "two competing approaches" to organizing the economy—integrationist and minimalist. Stressing the paramount importance of resolving every question legally, he says: "I get the impression that even Russia itself is convinced that in general it cannot occupy a worthy position if it obeys the principle of great-powerism."

The agreement on Councils of Heads of State and Government signed at this first summit established the protocols for meetings and voting on issues. It is interesting to note, as a harbinger of what was to come, how it handles the problem of language. It says: "The official languages of the Council are the state languages of the Commonwealth states." Just below this, it says: "The working language is the Russian language." By the end of the session, it is obvious that the proceedings have been much more difficult than anyone really expected. The press accounts cited at the end of this chapter express apprehension about the relationships evolving within the CIS. Fifteen documents are signed at the summit, most of which concern strategic forces, armed forces, and border troops. The charter, however, which contains the rules and principles by which the CIS will be governed (its "constitution") is not passed. When asked why not, President Shushkevich answers: "It is difficult to elaborate a long-term document during the transitional period." But, in fact, a consensus position on a CIS charter had still not been reached three and a half years later.

## 2.1
## Ukrainian Foreign Minister Views New Political Concept

Radio Kiev, 6 December 1991
[FBIS Translation], Excerpts

At the press conference for Ukrainian foreign journalists, Ukraine's foreign minister, Anatoliy Zlenko, informed all those present about a foreign political conception of the new independent Ukraine.

[Begin Zlenko recording]: I'm glad that the transition to independence is being made by peaceful means, written not in blood but in ink. So the new, independent Ukraine has arrived. New tasks are put forward before the Ministry of Foreign Affairs. The main problem is to develop and implement the whole complex of measures to defend the national interests of

Ukraine. The greater concern is international commitment, and, in accordance with the law on succession of 12 September this year, Ukraine is the successor of the former USSR in treaties that do not contradict its constitution.

The second problem is that of debts. In its statement of 23 October of this year, Verkhovna Rada—the parliament—confirmed its readiness to repay its share of foreign debt and at the same time to obtain its share of the assets of the former USSR. On the question of the armed forces: Ukraine will create its own armed forces, as well as a Republican Guard. This complies with a statement of 22 November 1991, which gradually takes into consideration the existing treaties, agreements, and consultations with other sovereign states and the USSR Ministry of Defense. The number of enlisted men in the army will be established by law on the basis of the principle of reasonable sufficiency. It will also be determined by whatever guarantees of collective defense are agreed upon. We shall also take into consideration the treaty on Conventional Forces in Europe.

The next problem is that of nuclear weapons. Taking into consideration all the factors of national security, Ukraine will carry out its intention to become in the future a neutral nonnuclear state, not participating in military blocs. It will observe three non-nuclear principles. Ukraine will participate in all treaties and agreements on the non-use of nuclear weapons, the START [Strategic Arms Reduction Talks] treaty signed in 1991 between the USSR and the United States, including the section concerning Ukrainian territory. I would like to stress that Ukraine has no control over, and does not wish to control, the nuclear weapons situated on its territory. These weapons must be transferred into the hands of a united command of Ukraine, Russia, Kazakhstan, and Belarus. Nuclear weapons must not be taken from Ukrainian territory, nor transferred for any purpose other than elimination, within the framework of negotiations, and under multilateral international control.

The next bloc of issues concern democracy and human rights. Being a European state, Ukraine has repeatedly talked about its intention to be a full-fledged participant in the all-European process. Accordingly, Ukraine accepts all international legal obligations and commitments resulting from the documents of the CSCE [Conference on Security and Cooperation in Europe]. It will strictly observe the human rights covenants, and will proceed on the principle of the priority of these rights.

The issue of borders: Ukraine has repeatedly stated and states once more that its territory is indivisible and inviolable. It respects the inviolability of existing borders, and it has no territorial claims on any other state. We are categorically against the revision of the Helsinki agreements.

## 2.2

## Ukrainian President Receives U.S. Under Secretary of State

TASS, 7 December 1991
[FBIS Translation]

The Ukrainian delegation is taking a package of important proposals to Minsk, where a meeting will begin today involving the leaders of Russia, Ukraine, and Belarus. The proposals concern the political and economic interrelations between the republics of the former Soviet Union, matters of collective security, and nuclear arms control. This was stated today by the Ukrainian president, Leonid Kravchuk, during a meeting with Thomas Niles, representative of the U.S. president and under secretary of state, who arrived in Kiev to acquaint himself with the principles of Ukraine's domestic and foreign policy after its independence was confirmed by its people. Leonid Kravchuk noted: Ukraine is attaching special significance to relations with Russia and supports Boris Yeltsin's initiative on signing an interstate treaty. It is prepared to sign a similar treaty with Belarus. Naturally, other republics, primarily Kazakhstan, may join them on an interstate basis. In this way, a community of states like the European Community, without any Center as formerly understood, may emerge. This does not rule out the creation of a coordinating body, but an organ similar to the Interstate Economic Committee under Ivan Silaev's leadership.

Ukraine intends to comply with all international pacts, agreements, and treaties and to consistently advocate the destruction of nuclear weapons and the reduction of conventional armaments and armed forces, the president continued. It seeks to develop mutually advantageous cooperation with all countries, including neighboring states and, particularly, wherever Ukrainians live—the United States, Canada, and elsewhere.

## 2.3

## The Minsk [Belovezh Forest] Agreement on Creation of the Commonwealth

8 December 1991 [FBIS Translation]

*Editor's Note:* On 8 December 1991 President Boris Yeltsin, President Leonid Kravchuk, and Supreme Soviet Chairman Stanislav Shushkevich met outside Minsk in a secluded residence in the Belovezh Forest and concluded an agreement establishing a Commonwealth of Independent States. The three newly elected heads of state felt it was time to acknowledge that the Soviet Union no longer existed. In its place they envisaged a free association of former Soviet republics, with common defense forces and a common economic space. Time, however, would reveal the wide gulf in understanding between the "fathers" of the Commonwealth.

Boris Yeltsin would maneuver for Russian supremacy over the organization. Leonid Kravchuk would insist on an amicable separation between equal and sovereign independent states. Stanislav Shushkevich would argue for Belarusian neutrality and a multinational, "rule-of-international-law" organization that would enable Belarus to sow the first seeds of a separate national identity. As "high contracting parties," the three leaders considered their countries the nucleus of the new commonwealth, as they were when the Soviet Union was created. This pivotal assumption, however, was to change in dramatic fashion by the end of the month at the Alma-Ata conference on 21 December.

**Preamble**

We, the Republic of Belarus, the Russian Federation, and the Republic of Ukraine as founder states of the Union of Soviet Socialist Republics (USSR), which signed the 1922 Union treaty, further described as the high contracting parties, conclude that the USSR has ceased to exist as a subject of international law and a geopolitical reality.

—Taking as our basis the historic community of our peoples and the ties that have been established between them;

—Taking into account the bilateral treaties concluded between the high contracting parties;

—Striving to build democratic states governed by law;

—Intending to develop our relations on the basis of mutual recognition and respect for state sovereignty, the inalienable right to self-determination, the principles of equality and non-interference in internal affairs, repudiation of the use of force and of economic or any other means of coercion, settlement of problems by mediation and other generally recognized principles and norms of international law;

—Considering that further development and strengthening of relations of friendship, good-neighborliness, and mutually beneficial cooperation between our states correspond to the vital national interests of our people and serve the cause of peace and security;

—Confirming our adherence to the goals and principles of the United Nations charter, the Helsinki Final Act, and other documents of the Conference on Security and Cooperation in Europe; and

—Committing ourselves to observe the generally recognized internal norms on human rights and the rights of peoples,

We have agreed on the following:

Article 1. The high contracting parties form the Commonwealth of Independent States.

Article 2. The high contracting parties guarantee their citizens equal rights and freedoms regardless of nationality or other distinctions. Each of the high contracting parties guarantees the citizens of the other parties, and also persons without citizenship who live on its territory, civil, political, social, economic, and cultural rights and freedoms in accordance with generally recognized international norms of human rights, regardless of national allegiance or other distinctions.

Article 3. The high contracting parties, desiring to promote the expression, preservation, and development of the ethnic, cultural, linguistic, and religious individuality of the national minorities resident on their territories, and that of the unique ethno-cultural regions that have come into being, take them under their protection.

Article 4. The high contracting parties will develop the equal and mutually beneficial cooperation of their peoples and states in the spheres of politics, the economy, culture, education, public health, protection of the environment, science and trade, and in humanitarian and other spheres will promote the broad exchange of information and will conscientiously and unconditionally observe reciprocal obligations.

The parties consider it necessary to conclude agreements on cooperation in the above spheres.

Article 5. The high contracting parties recognize and respect one another's territorial integrity and the inviolability of existing borders within the Commonwealth.

Article 6. The member states of the Commonwealth will cooperate in safeguarding international peace and security and in implementing effective measures for reducing weapons and military spending. They seek the elimination of all nuclear weapons and universal total disarmament under strict international control.

The parties will respect one another's aspiration to attain the status of a non-nuclear zone and a neutral state.

The member states of the Commonwealth will preserve and maintain under united command a common military-strategic space, including unified control over nuclear weapons, whose implementation is regulated by a special agreement.

They also jointly guarantee the necessary conditions for the stationing and functioning of, and for material and social provision for, the strategic armed forces. The parties contract to pursue a harmonized policy on questions of social welfare and pensions for members of the services and their families.

Article 7. The high contracting parties recognize that within the scope of their activities, implemented on the equal basis through the common coordinating institutions of the Commonwealth, will be the following:

—cooperation in foreign policy;

—cooperation in forming and developing the united economic area, the common European and Eurasian markets, in the area of customs policy;

—cooperation in developing transport and communication systems;

—cooperation in preservation of the environment and participation in creating a comprehensive international system of ecological safety;

—migration policy issues;

—fighting organized crime.

Article 8. The contracting parties recognize the planetary character of the Chernobyl catastrophe and pledge to unite and coordinate their efforts in minimizing and overcoming its consequences. To these ends they have decided to conclude a special agreement that will evaluate the gravity of the consequences of this catastrophe.

Article 9. Disputes over the interpretation and application of the norms of this agreement are to be resolved through negotiations between appropriate bodies and, when necessary, at the level of the heads of government and state.

Article 10. Each of the high contracting parties reserves the right to suspend the validity of the present agreement or individual articles thereof, after informing the parties to the agreement one year in advance of this decision.

The clauses of the agreement may be added to or amended with the consent of each of the high contracting parties.

Article 11. With the signing of this agreement, the norms of third states, including the former USSR, are not permitted to be implemented on the territories of the signatory states.

Article 12. The high contracting parties guarantee the fulfillment of the international obligations binding upon them from the treaties and agreements of the former USSR.

Article 13. The agreement does not affect the obligations of the high contracting parties with regard to third states. The agreement is open to all members of the former USSR, as well as to other states that share the goals and principles of the agreement.

Article 14. The city of Minsk is the official location of the coordinating bodies of the Commonwealth. The activities of bodies of the former USSR are discontinued on the territories of the member states of the Commonwealth.

Executed in the city of Minsk on 8 December 1991 in three copies each in the Belarusian, Russian, and Ukrainian languages, the three texts being of equal validity.

[Signed] For the Republic of Belarus: S. Shushkevich
[Signed] For the Russian Federation: B. Yeltsin, G. Burbulis
[Signed] For Ukraine: L. Kravchuk, V. Fokin

---

## 2.4

**Declaration on the New Commonwealth**
TASS, 8 December 1991
[FBIS Translation]

---

[Declaration by the Heads of State of the Republics of Belarus, the RSFSR (Russian Soviet Federated Socialist Republic), and Ukraine]

We, the leaders of the Republics of Belarus, the RSFSR, and Ukraine

—noting that the negotiations to draw up a new Union treaty are deadlocked, that the objective process of secession by republics from the USSR and the formation of independent states have become a reality;

—affirming that the myopic policy of the Center has led to profound economic and political crisis, to the collapse of industry, a catastrophic decline in the living standards of practically all strata of society;

—taking into account the growth in social tension in many regions of the former USSR, which has led to interethnic conflicts with numerous human victims;

—conscious of our responsibility before our peoples and the world community, and of the urgent practical need for implementing political and economic reforms, declare the creation of a community of independent states, in accordance with which an agreement was signed by the parties on 8 December 1991.

The community of independent states, consisting of the republics of Belarus, RSFSR, and Ukraine, is open for accession by all member states of the USSR, as well as by other states that share the aims and principles of the above agreement.

The member states of the community intend to pursue a policy of strengthening international peace and security. They guarantee to honor international obligations contained in treaties and agreements of the former USSR, and to ensure unified control over nuclear weapons and their non-proliferation.

[Signed] Belarus Supreme Soviet Chairman S. Shushkevich
[Signed] RSFSR President B. Yeltsin
[Signed] Ukraine President L. Kravchuk

8 December 1991

## 2.5
### Declaration on Economic Policy
TASS, 8 December 1991
[FBIS Translation]

[Declaration by the governments of the Republic of Belarus, the Russian Federation, and Ukraine on Coordination of Economic Policy]

It is vitally important to maintain and develop the existing close economic links between our states in order to stabilize the situation in the economy and pave the way for economic revival.

The parties have agreed on the following:

—To conduct coordinated radical economic reforms aimed at creating fully fledged market mechanisms, transforming property relations, and insuring freedom of enterprise;

—To refrain from any actions inflicting economic damage upon one another;

—To base economic relations and settlements on the existing unit of currency, the ruble; to introduce national currencies on the basis of special agreements that guarantee observance of the parties' economic interests;

—To conclude an interbank agreement aimed at restricting monetary issues, insuring effective control of the money in circulation, and forming a system of mutual settlements;

—To pursue a coordinated policy of reducing the deficits of the republican budgets;

—To undertake joint action aimed at ensuring unity of the economic space;

—To coordinate foreign economic activity and customs policy, and to ensure freedom of transit;

—To resolve by way of a special agreement the issue of indebtedness of former Union enterprises;

—To agree within ten days on the amounts of and procedures for financing in 1992 expenditures on defense and on eliminating the consequences of the accident at the Chernobyl nuclear power station;

—To request the supreme soviets of the republics, when formulating taxation policy, to take into account the need for coordinating value-added tax rates;

—To facilitate the establishment of joint ventures and joint-stock companies;

—To draw up during December a mechanism for implementing interrepublican economic agreements.

[Signed] On behalf of the Republic of Belarus, V. Kebich
[Signed] On behalf of the Russian Federation, G. Burbulis
[Signed] On behalf of Ukraine, V. Fokin

## 2.6
### The Ashkhabad Declaration
TASS, 13 December 1991
[FBIS Translation]

Declaration of the heads of states of the Republic of Kazakhstan, the Republic of Kyrgyzstan, the Republic of Tajikistan, Turkmenistan, and the Republic of Uzbekistan: In accordance with the accords reached at the meetings in Alma-Ata (1990) and in Tashkent (1991), heads of states N.A. Nazarbaev, A.A. Akaev, R.N. Nabiev, S.A. Niyazov, and I.A. Karimov gathered for a routine consultative meeting in Ashkhabad. They discussed the situation that has taken shape since the signing in Minsk of the agreement setting up a Commonwealth of Independent States. Following a comprehensive exchange of views and an analysis of the political situation, those attending the meeting declared the following:

We appreciate the desire of the leaders of the Republic of Belarus, the Russian Soviet Federated Socialist Republic, and Ukraine to create, in place of the republics that previously had no rights, independent law-governed states united to form a commonwealth. The Minsk initiative on the creation of a Commonwealth of Independent States, with the participation of Ukraine, is positive. However, this agreement came as a surprise to us.

The participants in the conference agree with the assertion that the process of a new integration of the subjects of the former USSR on the basis of the decisions of the Fifth USSR Congress of People's Deputies has reached a dead end. The Center's shortsighted policy has led to a profound economic and political crisis, the breakdown of production, and a catastrophic decline in the living standards of virtually all strata of society.

The participants in the meeting believe that:

—It is necessary to coordinate efforts to shape the Commonwealth of Independent States.

—The establishment of the Commonwealth of Independent States must be implemented on a lawful basis.

—There must be a guarantee of equal participation by the subjects of the former Union in the process of elaborating decisions and documents on the Commonwealth of Independent States. All the states forming the commonwealth must be recognized as founders and referred to in the text of the agreement as high contracting parties.

—One should take into account in these documents, decisions, and agreements, the historic and socio-economic realities of the republics of Central Asia and Kazakhstan, which, unfortunately, were not considered during the preparation of the agreement on a commonwealth.

—The Commonwealth of Independent States should guarantee the equality of rights of all nations and ethnic groups and the protection of their rights and interests.

—The Commonwealth of Independent States cannot take shape on an ethnic, religious, or any other basis infringing on the rights of individuals or peoples.

—The Commonwealth of Independent States recognizes and respects the territorial integrity and inviolability of presently existing borders.

—In the interests of preserving strategic stability in the world, it is expedient to ensure common control of nuclear weapons and a unified command for strategic defense troops and naval forces.

—It is essential to endorse the treaty concluded earlier on an economic community and to complete work on it.

Proceeding from the above, we declare our readiness to become equal co-founders of the Commonwealth of Independent States, which takes into account the interests of all its subjects. Issues developed within the Commonwealth of Independent States should be examined at a conference of the heads of the sovereign states.

The participants in the consultative meeting acknowledge the fact that Uzbekistan will determine its final position on participating in the Commonwealth after nationwide presidential elections are held on 29 December 1991.

[Signed] N.A. Nazarbaev, president of the Republic of Kazakhstan
[Signed] A.A. Akaev, president of the Republic of Kyrgyzstan
[Signed] S.A. Niyazov, president of Turkmenistan
[Signed] I.S. Karimov, president of the Republic of Uzbekistan

[Dated] Ashkhabad, 13 December 1991

## 2.7
## The Alma-Ata Declaration
21 December 1991
[FBIS Translation]

*Editor's Note:* The Alma-Ata Declaration was signed by eleven heads of state on 21 December 1991. The purpose of this agreement was to incorporate the five Central Asian countries, two of the Transcaucasian states (Azerbaijan and Armenia) and Moldova into the CIS as members of equal standing with Russia, Belarus, and Ukraine. Because these countries had not participated in the meeting in Belovezha Forest, their leaders were justifiably concerned about their rank in the new Commonwealth. The Central Asian leaders demanded and received assurances in the Alma-Ata Declaration that parity would be the guiding principle of the CIS.

### Protocol

The Azerbaijan Republic, the Republic of Armenia, the Republic of Belarus, the Republic of Kazakhstan, the Republic of Kyrgyzstan, the Russian Federation, the Republic of Tajikistan, Turkmenistan, the Republic of Uzbekistan, and Ukraine, on an equal basis and as high contracting parties, are forming a Commonwealth of Independent States.

The agreement on the creation of a Commonwealth of Independent States enters into force for each of the high contracting parties from the moment of its ratification.

Documents regulating cooperation within the framework of the Commonwealth will be worked out on the basis of the agreement on the Creation of a Commonwealth of Independent States and with consideration for the reservations made during its ratification.

This protocol is a component part of the agreement on the Creation of a Commonwealth of Independent States.

Done in the city of Alma-Ata on 21 December 1991, in one copy each in the Azerbaijani, Armenian, Belarusian, Kazakh, Kyrgyz, Moldovan, Russian, Tajik, Turkmen, Uzbek, and Ukrainian languages. All texts have equal force. The original will be kept in the archives of the government of the Republic of Belarus, which will send the high contracting parties certified copies of this protocol.

### Preamble

The independent states:

The Republic of Armenia, the Republic of Azerbaijan, the Republic of Belarus, the Republic of Kazakhstan, the Republic of Kyrgyzstan, the Republic of Moldova, the Russian Federation, the Republic of Tajikistan, the Republic of Turkmenistan, the Republic of Ukraine, and the Republic of Uzbekistan;

—seeking to build democratic, law-governed states, the relations between which will develop on the basis of mutual recognition and respect for state sovereignty and sovereign equality, the inalienable right to self-determination, principles of equality and non-interference in internal affairs, the rejection of the use of force, the threat of force and economic and any other methods of pressure, a peaceful settlement of disputes, respect for human rights and freedoms, including the rights of national minorities, a conscientious fulfillment of commitments and other generally recognized principles and standards of international law;

—recognizing and respecting each other's territorial integrity and the inviolability of the existing borders;

—[knowing that] advantageous cooperation, which has deep historic roots, meets the basic interests of nations and promotes the cause of peace and security;

—being aware of their responsibility for the preservation of civil peace and interethnic accord;

—being loyal to the objectives and principles of the agreement on the creation of the Commonwealth of Independent States;

are making the following statement:

**The Declaration**

Cooperation between members of the Commonwealth will be carried out in accordance with the principle of equality through coordinating institutions formed on the basis of parity and operating in the way established by the agreements between members of the Commonwealth, which is neither a state nor a supra-state structure.

In order to ensure international strategic stability and security, allied command of the military-strategic forces and unified control over nuclear weapons will be preserved, and the sides will respect one another's desire to attain the status of a non-nuclear and (or) neutral state.

The Commonwealth of Independent States is open, with the agreement of all its participants, to states—members of the former USSR, as well as other states—that share the goals and principles of the Commonwealth.

The allegiance to cooperation in the formation and development of the common economic space, and all-European and Eurasian markets, is being confirmed.

With the formation of the Commonwealth of Independent States the USSR ceases to exist. Member states of the Commonwealth guarantee, in accordance with their constitutional procedures, the fulfillment of international obligations, stemming from the treaties and agreements of the former USSR.

Member states of the Commonwealth pledge to observe strictly the principles of this declaration.

**Decision of the Heads of State of the Commonwealth of Independent States**

The member states of the Commonwealth, citing Article 12 of the agreement on the Creation of a Commonwealth of Independent States,

Proceeding from the intention of each state to fulfill the obligations set down by the UN charter and to participate in that organization's work as full-fledged members,

In view of the fact that the Republic of Belarus, the USSR and Ukraine were original members of the United Nations,

Expressing satisfaction with the fact that the Republic of Belarus and Ukraine are continuing to participate in the United Nations as sovereign independent states,

Firmly resolved to promote the strengthening of international peace and security on the basis of the UN charter and in the interests of their peoples and the entire international community,

Have declared that:

1. The states of the Commonwealth support Russia in its continuation of the USSR's membership in the United Nations, including the USSR's membership in the Security Council and other international organizations.

2. The Republic of Belarus, the Russian FSR and Ukraine will support the other states of the commonwealth in resolving the question of their full-fledged membership in the United Nations and other international organizations.

Done in the city of Alma-Ata on 21 December 1991, in one copy each in the Azerbaijani, Armenian, Belarusian, Kazakh, Kyrgyz, Moldovan, Russian, Tajik, Turkmen, Uzbek, and Ukrainian languages. All texts have equal force. The original will be kept in the archives of the government of the Republic of Belarus, which will send the high contracting parties certified copies of the protocol.

## 2.8

**Gorbachev Resigns as USSR President**

Moscow Central Television, 25 December 1991
[FBIS Translation]

Dear compatriots and fellow citizens. Because of the situation that has developed with the formation of the Commonwealth of Independent States, I am ceasing my activity in the post of USSR president. I adopt this decision on principle. I have been firmly in favor of the independence and self-determination of people and for the sovereignty of republics but, at the same time, in favor of preserving the union state and the country's integrity.

Events took a different course. A policy of splitting up the country and disassembling the state—something with which I cannot agree—has prevailed. And after the Alma-Ata meeting and the decisions adopted there, my position on that account has not changed. Besides, I am convinced that decisions of such a magnitude should have been adopted on the basis of a show of will by the people. Nevertheless, I will do everything within my power [to help ensure] that the accords signed there lead to real social harmony and make it easier to emerge from the crisis and to continue the process of reforms.

Addressing you for the last time in my capacity as USSR president, I believe that it is necessary to state my appreciation of the path that has been traveled since 1985, especially as there are quite a few contradictory, superficial, and subjective views on this account. Destiny saw to it that even when I took charge of the state it was clear that all was not well with the country. There was a lot of everything—land, oil, and gas, other natural wealth, to say nothing of the brains and talent with which it was endowed by God—but still our living standards were far behind those of developed countries, and were slipping farther and farther behind them.

The reason was clear even then: Society was being crushed in the jaws of the command and bureaucratic system. Doomed to serve ideology and to bear the terrible burden of the arms race, it was stretched to the limit of its endurance. All attempts at partial reform—and there were many of them—suffered failure one after the other. The country lost its vision of the future. It was impossible to live like that any longer. What was needed was a radical change. That is why I have never, ever regretted not taking advantage of the post of general secretary merely to reign for a few years. I would have viewed that as irresponsible and amoral. I realized that it was an extremely difficult and even risky business to begin reforms on that kind of scale and in our kind of society. Even today I am convinced of the historical correctness of the democratic reforms begun in the spring of 1985.

The process of renovating the country and the key changes in the world community have turned out to be far more difficult than could have been predicted. However, what has been done should be assessed on merit. Society has gained freedom and liberated itself politically and socially. This is the main achievement, which we have not yet fully realized because we have not yet learned how to use freedom.

Still, work of historic significance has been done. A totalitarian system, which has long deprived the country of an opportunity to become wealthy and prosperous, has been liquidated. A breakthrough has been made toward democratic transformation. Free elections, a free press, religious freedoms, representative government, and a multiparty system have become a reality. Human rights have been recognized as the highest principle. A movement toward a mixed economy has begun. Equality of all forms of property is being established. Within the framework of land reform, the peasantry began to revive. Farmers have appeared. Millions of hectares of land have been given to rural inhabitants and city dwellers. Economic freedom for producers has been legalized. Enterprising joint-stock companies and privatization have started to gain force.

As the economy becomes more marketized, it is important to remember that all this is done for the sake of the individual. During this difficult time, everything should be done for his social protection, especially with regard to old people and children.

We are living in a new world. The Cold War is over. The arms race, the insane militarization of the country that disfigured our country, public consciousness, and morality has been halted. The threat of world war has been taken off the agenda.

I want to emphasize again that during the transition period I have done everything possible to retain reliable control of nuclear arms. We have opened to the world and renounced intervention in other people's affairs and the use of troops beyond the borders of our country. In return, we were shown trust, solidarity, and respect.

We have become one of the main strongholds for the transformation of modern civilization by peaceful democratic principles. Peoples and nations have achieved real freedom of choice in seeking their paths toward self-determination. The search for the democratic reform of our multinational state brought us to the threshold of a new Union treaty.

All these changes have demanded a great deal of effort. They came about in a fierce fight in the face of increasing resistance by forces representing what is old, obsolete, and reactionary by the old party and state structures, the economic apparatus, as well as by our habits, ideological prejudices, and the philosophy of leveling down and dependence. They came up against our intolerance, low level of political culture, and fear of change. This is why we have lost so much time. The old system collapsed before the new one could start working. The crisis in society became even more acute.

I know about the dissatisfaction with the present difficult situation, about the sharp criticism of authorities at all levels and of my own activities. But I want to stress once again: Radical changes in such an enormous country, particularly with such a heritage, cannot happen painlessly, without difficulties and disruptions.

The August coup stretched the general crisis to its utmost limit. The most destructive element in this crisis was the collapse of statehood. Today, too, I am alarmed by our people's loss of their citizenship in a great country. The consequences might be very severe for everybody. I feel that it is vitally important to preserve the democratic gains of recent years. They were achieved through historic and tragic suffering. We must not renounce them under any circumstances. Otherwise, all hope for better things will be buried.

I speak about all this honestly and frankly. This is my moral duty. I want to express my gratitude today to all citizens who supported the policy of renewal in the country and joined in the implementation of democratic reforms. I am grateful to the state, political, and public figures, to millions of people abroad, to those who understood our intentions and supported them, who met us halfway in sincere cooperation with us.

I am leaving office with alarm, but also with hope—with faith in you, your wisdom, and strength of spirit. We are heirs to a great civilization. It depends on each and every one of us now—to see that it is reborn and will have a modern and worthy life.

With all my soul, I wish to thank those who stood alongside me during these years in a just and a good cause. Some mistakes could probably have been avoided and much could have been done better, but I am sure that sooner or later our shared efforts will achieve results. Our people will live in a flourishing and democratic society. I wish every one of you all the best.

## 2.9
**Anuar Alimzhanov Addresses the Final Session of the Supreme Soviet**
Radio Mayak, 26 December 1991
[FBIS Translation]

Esteemed people's deputies! As you will have noticed, the flag of the Soviet Union over the Kremlin was lowered today. Last night we all witnessed how the president—the first president of this great country—resigned.

I do not know how the first session of the Supreme Soviet went or how people there felt, but I imagine that they spoke of great things, of world revolution, of social equality, of socialism, of the dream of advancing to communism. Probably there were many good, kind, and fine words there about the future of this vast country.

History has decreed that today I should participate in the last meeting of this body. Let us say plainly that what we dreamed of and spoke of at that first session has not come to pass. There was obviously talk there that by building communism we were skipping entire historical epochs; but it turned out that it is not possible to skip epochs in history. We have again turned toward capitalism; probably not toward a developed form of it, but perhaps toward its least-tamed form.

All this is history. All this is our life. Speaking about the past, of course, we must also give due tribute to the fact that, in fulfillment of the great dream spoken of at that first session, people gave up their lives. The totalitarian system took away the most elite part of our people, our young people. Herein lies its fatal nature.

But there were also victories. There was also unity. There were also exploits. Today we have reached the stage where the old has been destroyed and the new is beginning. We understand perfectly well what we have lost; what will be, we are not yet aware of. However that may be, a new Commonwealth of Independent States has been formed. This is a new phenomenon in world history. What it will show, for the time being, one can only guess at and think about. We would like this new phenomenon to preserve the very best traits of democracy, of a commonwealth of peoples. We would like it to lead people truly along a democratic path, toward social equality, toward an improvement of people's lives, avoiding war and confrontation.

This is our last meeting. Newspapers have written about our previous meetings, that we had missed the boat and that we were dragging out the funeral, but that is the journalists' business. I think that the journalists are also in some kind of crisis now—not all, of course, but there are those who are somehow beginning to dance around the lion that has not yet been slain. In this sense, however it may be, whatever is said, it is precisely our chamber that has held out until the end. I think it has fulfilled its civic duty and the duty of deputies. It has justified the trust of our sovereign republics.

I wanted our president to resign in a human way. Whatever he is, he is a man of history. It is not up to us to judge. Today it is still too early to speak of the past. Nevertheless, as the president resigned yesterday before the entire nation, as the flag of the Soviet Union has been lowered, this means we have the full moral right today to complete our business constitutionally with a clear conscience and to hold our last sitting.

You remember that at the last meeting we had a few questions left concerning comrades being relieved of their posts that had been ratified by the Supreme Soviet. We should discuss this. Prior to this, we should discuss the draft of our declaration, of our decision. It seems to me now that even if anyone has any remarks, addenda, or wishes, we should first discuss our declaration.

I started in a businesslike way, because registering is important, not only because this is our last meeting, but to show that there is a quorum here, if journalists are interested. There is even a quorum from Belarus, which has called back its deputies, over ten people. Present are representatives of Ukraine. Present here are all Central Asian republics and Kazakhstan. Present are deputies, people's deputies of the Soviet Union who have the right to be here at this last final meeting, the final stage of our parliament's work.

Therefore, with a clear conscience we can set about completing our business. That is to draft a declaration ending the USSR and setting up the Commonwealth of Independent States. We have examined the draft. Remarks were voiced. It is necessary to discuss this now. Perhaps someone has an addendum. How shall we adopt it—wholly or in parts?

## 2.10
**Further on the Union Dissolution**
Interfax, 26 December 1991
[FBIS Translation]

On Thursday, 26 December, the House of the Republics of the Soviet parliament passed a declaration on the dissolution of the USSR and the formation of the Commonwealth of Independent States saying that "relying on the will expressed by the top elected bodies of state power of Azerbaijan, Armenia, Belarus, Kazakhstan, Kyrgyzstan, Moldova, the Russian Federation, Tajikistan, Turkmenistan, and Ukraine to form the Commonwealth of Independent States the House of the Republics announces the dissolution of the USSR after the ratification of the agreement on the formation of the Commonwealth by said states."

The House of the Republics urged the top elected bodies and heads of Commonwealth states to do their utmost to guarantee the rights and liberties of people irrespective of their nationality, the peaceful cohabitation of Commonwealth peoples, the democratic development of their states. . . .

The last session of the upper house of parliament also passed a resolution relieving members of the USSR Supreme Court, the USSR Supreme Court of Arbitration, and the Office of the USSR Procurator, and their staffs, of their duties as of 2 January 1992 in connection with the closure of Union government bodies.

For the same reason the chairman of the USSR State Bank, Viktor Gerashchenko, and his first deputy, Valerian Kulikov, have been dismissed.

The House of the Union decided on 23 December not to hold any more sessions.

## 2.11

### Stanislav Shushkevich Previews Minsk Summit Meeting

Oleg Stepanenko and Aleksandr Ulitenok
*Pravda,* 30 December 1991
[FBIS Translation], Excerpts

[*Pravda*]: This is probably the first time since the war that we have greeted the New Year in such an anxious situation.

[Shushkevich]: Yes, the times now are complex and difficult. But they are crucial times. Now it is most important not to plunge into despair. We have everything in order to build a life worthy of man: industrious, skilled hands, a bright intellect, and a strengthening sense of national awareness. In addition, not only destructive processes are under way; creative ones have begun, too. I mean the creation of the CIS. The idea born in Belarus three weeks ago has already found not only broad support but also practical embodiment.

[*Pravda*]: Stanislav Stanislavovich, what place will the upcoming meeting of the leaders of eleven states occupy in this rapidly developing process?

[Shushkevich]: It is hard to answer unequivocally. In Alma-Ata we agreed that eight basic documents would be prepared. Strange though this may seem, the document defining the procedure for reorganizing the armed forces strikes me as the most important in this package. It is necessary to have a clear idea of their present state and of the plans for the phased transfer of military subunits to qualitatively new structures. I also hope that questions of the Commonwealth's charter will also be agreed on, at least conceptually. We might also manage to sign a document on the CIS institutions, because it is on this matter that we still lack full unanimity on certain questions, although, if you take an unbiased look at reality, you cannot particularly count on this in such organs.

[*Pravda*]: Are the materials for the meeting being prepared at a single coordination center?

[Shushkevich]: No, intensive work on the documents is being done in literally all the CIS members. . . .

[*Pravda*]: Stanislav Stanislavovich, are the leaders of the allied states being strengthened in their realization that what is needed is an adequate degree of integration, which will not disrupt the natural relations that have evolved over decades and centuries? Or is what we call the "war of sovereignties" breaking out among some of them?

[Shushkevich]: I will dwell on what is most important: borders. They must be open to all citizens. This question has already been virtually resolved. From the viewpoint of organizing the economy, there are two competing approaches. First, there is an undoubted realization that without integration we will experience exceptionally hard times. Second, the republics cannot find their feet as sovereign states if all the so-called integration structures from the past are preserved. Therefore a fundamentally new approach is needed here. What is the problem? A juridical, legislative vacuum has formed, and an initial normative basis is needed. Even if we work out very few principles and conclude very few treaties, we still need a mechanism and an organ of control with very clear-cut and effective sanctions for violating or failing to fulfill what we outlined together. . . .

[*Pravda*]: Do many people fear the possibility of Russia's diktat? Is this valid?

[Shushkevich]: Since we are building rule-of-law states and a rule-of-law commonwealth, everything must be resolved legally. I get the impression that even Russia itself is convinced that, in general, it cannot occupy a worthy position on the principle of great-powerism. Russia is great in itself. Its greatness must not, in my opinion, be increased by diktat.

Moreover, there is a very easy way out of the commonwealth. The commonwealth, if something were to happen, could turn away from Russia.

[*Pravda*]: The economic situation causes the greatest alarm. Will the problem of coordinating economic programs and reforms arise at the conference?

[Shushkevich]: The first idea is expressed in the old style: to maintain deliveries at least at last year's level of 70 percent, and so forth. To be honest, I do not see any innovations in these premises that would tally with the idea of a political Commonwealth—I cannot see them! I believe that the governments will find a different approach in accordance with the new realities. I do not rule out the possibility that there might not be an overall accord, because there simply must be normal economic relations among sovereign states. But accords must definitely be mutual on the level of inflation, which at once determines spending quotas. Stern measures are needed here. For, if there are none, the Commonwealth could lose its meaning, as then the introduction of national currencies is inevitable. This is virtually a process of disintegration. This is why a fully agreed mechanism of expenditure quotas is needed. It is necessary to rule out legislatively, on the basis of treaties, an improvement in your own position at your neighbor's expense. Questions of your own well-being must be resolved by honest labor. Otherwise there will be no equality. Account should be taken of the fact that there are differences in natural resources, in the republics' levels of development. . . . Perhaps this is why there must be some kind of a correction, for current members of the Commonwealth are sometimes to blame for the state of other members' economies.

[*Pravda*]: But a confidential approach is not an economic category. Economics operates on the basis of self-interest, which is frequently at odds with your neighbor's interests. Clearly, the Commonwealth and its institutions must somehow regulate the balance of interests, after the EEC [European Economic Community] model, for example.

[Shushkevich]: You see, the EC [European Community] includes more reliable autonomous structures, if I may call the states this. They already knew how to manage their domestic affairs, and they have only improved their position thanks to the joint resolution of certain tasks. But we have taken refuge in joint management and built such unreliable, distant economic ties that it is simply terrible.

[*Pravda*]: And yet, will the CIS have a mechanism against monopoly diktat by one of the states?

[Shushkevich] A general prescription is impossible here. For monopoly is sometimes not an economic necessity but a consequence of a past subjective approach. Why, for example, had attention previously focused only on oil extraction in Tyumen? Kazakhstan ranks fifth in the world in terms of known reserves, and oil extraction could have been developed more intensively right there. A redistribution of effort will probably be needed—for we, Ukraine, and Russia all sent our specialists to Tyumen. . . .

But the question is this: Where are investments more profitable? It must be resolved neither out of considerations of the former, central morality nor according to this principle: "We have seen a place, one of the powers-that-be liked it, and so let us invest billions and build it up." Such mobilization of means is impossible now and cannot be squeezed out the way it used to be. We are taking the path of normal management, which is approaching market management. This must be resolved not by new structures, but what is needed is bilateral, trilateral, and quadrilateral treaties, on the basis of which joint economic projects will be implemented. There is no need to create special economic structures for this.

[*Pravda*]: Stanislav Stanislavovich, how do you regard the future—with alarm or with hope?

[Shushkevich]: I am, in general, an optimist by nature, and I know that nothing will come of it if you immediately react with great alarm. I react with hope. I would not have gotten into this business at all if I had not been confident that it was necessary to take precisely this path. . . .

## 2.12
**Agreement on Council of Heads of State and Heads of Government**
30 December 1991 [FBIS Translation]

*Editor's Note*: A provisional agreement on the membership and conduct of Councils of Heads of State and Heads of Government was concluded between the members of the Commonwealth on 30 December 1991. Boris Yeltsin was unanimously elected chairman of the Council of Heads of State. This set a precedent for all subsequent steering committees and councils, which elected Russians as their chairmen. The Russian delegation was also to propose most of the future agreements and protocols to be considered by the Commonwealth member states.

**Preamble**

The member states of this agreement, guided by the aims and principles of the agreement on the creation of a Commonwealth of Independent States of 8 December 1991 and the protocol to the agreement of 21 December 1991, taking into consideration the desire of the Commonwealth states to pursue joint activity through the Commonwealth's common coordinating institutions, and deeming it essential to establish, for the consistent implementation of the provisions of the said agreement, the appropriate interstate and intergovernmental institutions capable of ensuring effective coordination, and of promoting the development of equal and mutually advantageous cooperation, have agreed on the following:

Article 1. The Council of Heads of State is the supreme body, in which all the member states of the Commonwealth are represented at the level of head of state, for the discussion of fundamental issues connected with coordinating the activity of the Commonwealth states in the areas of their common interests.

The Council of Heads of State is empowered to discuss issues provided for by the Minsk agreement on the creation of the Commonwealth of Independent States and other documents for the development of the said agreement, including the problems of legal succession, which have arisen as a result of ending the existence of the USSR and the abolition of Union structures.

The activities of the Council of Heads of State and of the Council of Heads of Government are pursued on the basis of mutual recognition of and respect for the state sovereignty and sovereign equality of the member states party to the agreement, their inalienable right to self-determination, the principles of equality and non-interference in internal affairs, the renunciation of the use of force and the threat of force, territorial integrity and the inviolability of existing borders, the peaceful settlement of disputes, respect for human rights and liberties, including the rights of national minorities, conscientious fulfillment of obligations and other commonly accepted principles and norms of international law.

Article 2. The activities of the Council of Heads of State and the Council of Heads of Government are regulated by the Minsk agreement on setting up the Commonwealth of Independent States, the present agreement and agreements adopted in the development of them, and also by the rules of procedure of these institutions.

Each state in the council has one vote. The decisions of the council are taken by common consent.

The official languages of the Councils are the state languages of the Commonwealth states.

The working language is the Russian language.

Article 3. The Council of Heads of State and the Council of Heads of Government discuss and, where necessary, take decisions on the more important domestic and external issues.

Any state may declare that it has no interest in a particular issue.

Article 4. The Council of Heads of State convenes for meetings no less than twice a year. The decision on the time for holding and the provisional agenda of each successive meeting of the Council is taken at the routine meeting of the Council, unless the Council agrees otherwise. Extraordinary meetings of the Council of Heads of State are convened on the initiative of the majority of Commonwealth heads of state.

The heads of state take turns chairing the meetings of the Council, according to the Russian alphabetical order of the names of the Commonwealth states.

Meetings of the Council of Heads of State are generally to be held in Minsk. A Council session may be held in another part of the Commonwealth states by agreement among those taking part.

Article 5. The Council of Heads of Government convenes for meetings no less frequently than once every three months. The decision concerning the scheduling of and preliminary agenda for each subsequent meeting is to be made at a routine session of the council, unless the Council arranges otherwise.

Extraordinary sessions of the Council of Heads of Government may be convened at the initiative of a majority of heads of government of the Commonwealth states.

The heads of government take turns chairing the meetings of the Council, according to the Russian alphabetical order of the names of the Commonwealth states.

Meetings of the Council of Heads of Government are generally to be held in Minsk. A Council session may be held in another of the Commonwealth states by agreement among the heads of government.

Article 6. The Council of Heads of State and the Council of Heads of Government of the Commonwealth states may hold joint sessions.

Article 7. Working and auxiliary bodies may be set up on both a permanent and an interim basis as decided by the Council of Heads of State and Council of Heads of Government of the Commonwealth states.

These are composed of authorized representatives of the participating states. Experts and consultants may be invited to take part in their sittings.

## 2.13

**Commentary: "Final Attempt" at "Normal Life"**

G. Shipitko

*Izvestiya,* 31 December 1991

[FBIS Translation], Excerpt

*Editor's Note*: From the very beginning, it was realized that economic questions would provide the decisive rationale for integrating the former Soviet republics. Although few leaders of the former republics would agree to any arrangement that entailed a loss of sovereignty, economic integration was a major topic of discussion in the first days of the fragile Commonwealth's existence. In this article, *Izvestiya's* journalist expresses very positive conclusions about the unifying process going on among Commonwealth members.

The meeting of Commonwealth heads of state, which, as is well known, had been preceded by two other meetings, in Minsk and Alma-Ata, began at eleven o'clock on 30 December. Judging by the agenda, the process of the creation of a new Commonwealth is without parallel in the history of state formations and promises to put a stop to the endless question of who will be the legal successor to the collapsed Union. That is a question that vitally concerns not only the millions of people living on our territory of one-sixth of the globe but also, without exaggeration, the entire world community.

So the eleven heads of independent states and the same number of heads of government assembled to answer the question of what basic structure the Commonwealth will adopt. Ten of the thirteen proposed items on the agenda were approved, giving an entirely unambiguous answer—the Commonwealth proposes to have quite firm forms of relations with a set of institutions regulating their relations and common finances to maintain the coordinating institutions. The meeting is discussing agreements on defense questions, on the Commonwealth's television and radio company and its press organ, and on environmental emergencies. The questions of vital interest include an agreement on distributing food purchased with credits from foreign states. The agenda includes questions like temporary measures for the coordinated use of diplomatic and consultative missions, legal succession regarding the former Union's property, and the procedure for signing and implementing agreements on political, economic, scientific and technical, and cultural collaboration among the Commonwealth member states, and also agreements on cooperation in the spheres of culture and space.

. . . But it has already been learned from sources close to the government that it is proposed to create a foreign ministers committee. It is necessary right now to resolve the exceptionally complex and delicate question of interethnic conflicts. A large group of residents of the Dniester region came to meet with the heads of state from Moldova. They are trying to convince the participants of the need to create an autonomous Dniester republic, independent of Moldova.

. . . It is easy to see from the list of questions, given a very approximate assessment of the approaches, that the Commonwealth,while remaining independent in its component parts, is seeking in its actions to create fundamental directions toward unity. That is probably the first encouraging conclusion from the conference that has begun in Minsk. Another equally important conclusion is that the future Commonwealth seems finally to be realizing that without a single economic area . . . it may not prove viable. I think that is an extremely encouraging fact that holds out a promise to all of us moving away from national economic isolation toward the emergence of economic relations in the future. . . .

In the two hours initially allocated for discussion of the ten questions included on the agenda, it was possible to agree on only one of them—the draft agreement on coordinating institutions. . . .

Today it is clear that the Commonwealth of Independent States is the final attempt to switch to a normal life and join the civilized world. Otherwise everyone is threatened by the unpredictability of economic life, which may reduce society to a crisis from which the way out could be long and difficult.

---

## 2.14
### "Commonwealth of Uncivilized States" Observed
Mikhail Mayorov
Interfax, 3 January 1992
[FBIS Translation], Excerpts

---

*Editor's Note:* In this article, the reporter expresses the opposite of the preceding conclusion. Conflicting reports and vantage points at the time showed just how difficult the negotiations, and how grave the disagreements, were.

The relationships taking shape within the Commonwealth have been the source of frustration, rather than inspiration, for people in Europe and worldwide. The Minsk summit of 30 December, according to one of Russia's diplomats appointed to assess its results, vividly showed how dangerous nationalism can be if it is adopted as a dominant form of intergovernmental relations. "So far one can speak only of a 'Commonwealth of Uncivilized States,' considering the imminent political and military disputes, as well as clashes on property-related and other matters," the diplomat said.

According to some observers, nationalism is something that has seriously affected every ex-Soviet republic, big and small. Those blaming the present differences exclusively on Ukraine overlook the fact that Russians regard other nationalities with an undue degree of "ethnic superiority"—an attitude particularly conspicuous among those whom the tide of recent political events has brought right into government cabinets.

The fight for ex-USSR property raging between Russia and Ukraine shows that the two newly independent states longing for access to the international political arena still have to learn a lot about the universally accepted "code of behavior." Ukrainian leaders' desire to have armed forces of their own is understandable, considering the widespread public sentiments in the republic. On the other hand, the West would much rather have the military stay under centralized command within the ex-Soviet borders. Ukraine, for one, is claiming full control over all the non-nuclear vessels of the Black Sea Fleet—Russia's old-time pride—as well as over three military districts with all their hardware, weapon depots, and firing ranges. The shipyards of Nikolaev, where surface ships and submarines for the Northern and Pacific Fleets used to be assembled, will be transferred to Ukrainian jurisdiction, too.

A senior Russian diplomat who asked not to be identified told journalists that "in terms of its territory, Ukraine is another France, and in terms of its ambitions it's another China." However, Russia's ambitions do not seem to be any more moderate, since the same diplomat insisted that Russia should be proclaimed not only the legal successor of the USSR but also a "continuation state" inheriting the full scope of rights from the ex–Soviet Union. "The other independent states may open foreign missions in whatever form they may choose," the diplomat commented.

. . . The most experienced and far-sighted observers, far from trying to find out who's right and who's wrong in the Russo-Ukrainian dispute, have been trying to guess how viable the Commonwealth, which they say is still a far cry from an entity fitting that title, may turn out to be. According to Lithuania's deputy head of parliament, Kazimieras Motieka, the Commonwealth will not last long. The Moldovan parliamentarian Ion Tsurkanu does not see any future for the CIS, either.

No comment has come so far from the politician whom the formation of the Commonwealth of Independent States left without a job and who should regard the present situation from the angle "the worse, the better." But it seems quite possible that Mikhail Gorbachev may one day say something like: "You were unwise enough to prefer Minsk to Novo-Ogarevo; I warned you against it, but you wouldn't listen."

## 2.15

### Former Defense Minister Evgeniy Shaposhnikov Interviewed

Interview by Tatyana Sukhomlinova
*Rossiya,* no. 10, 19–25 April 1995
[FBIS Translation], Excerpts

*Editor's Note:* We have included the following excerpts from an interview with former Defense Minister and Commander General Evgeniy Shaposhnikov because they throw light upon the politics of the collapse of the USSR. Although many republican leaders would have supported a continuation of a Union, Shaposhnikov says they could no longer follow Gorbachev. It can logically be deduced, therefore, that Boris Yeltsin had already become extremely powerful behind the scenes—in typical "Kremlin style." His remarks also seem to indicate full military compliance with Yeltsin's wishes, even though Shaposhnikov adamantly believed that the Soviet Armed Forces should have remained intact.

Evgeniy Shaposhnikov always left top jobs at his own initiative, which is very rare in our country and in our times.

"Every time, my departure was a protest of sorts against the fact that I was powerless to influence a particular situation. . . ."

In August 1991 he took a stand against the GKChP (State Committee on the State of Emergency), believing that "to a certain extent Afghanistan was both an impetus and a catalyst for the disintegration of the Union. It illuminated very dark sides in our life of the Soviet period. . . . For me, the GKChP meant a return to those times and methods. . . .

The same August he became USSR minister of defense. He says that he accepted the appointment without particular elation because he realized that this was too heavy a burden. Generally, he had never thought of it: There had never before in the country's history been any ministers of defense with a background in aviation.

. . . The military is a powerful force. And that always made it attractive, especially for politicians losing their political influence. Whether a minister of defense wants to or not, he acquires the significance of a political figure. This E. Shaposhnikov found out, so to say, on his own hide.

The marshal is still convinced to this day that there was a way to avoid the disintegration of the USSR after the GKChP [formed].

"I saw that the leaders of Union republics did not so much want not to preserve the Union as they no longer accepted Gorbachev. And he, in my opinion, should have responded to this mood and initiated popular elections of a new national president. I think in this case the Union could have been preserved. In a different form, of course, than before. . . ."

But everything went the way it went. And on 21 December 1991 the heads of what were then independent states rather than Union republics gathered in Almaty, where it was said: "The strategic forces across the CIS must remain unified, but the armed forces should be divided. . . ."

E. Shaposhnikov was unequivocally against it. He launched a massive attack and seemed to have won. The leaders decided to return to this subject later.

And they did. On 30 December in Minsk it was decided to divide the armed forces among the CIS components. The ministers of defense, which by then already had been appointed in some republics, and chairmen of defense committees supported their leaders. Only E. Shaposhnikov remained against the idea. And then he wrote the following document.

**"To the Heads of Independent States"— Statement**

"In connection with the fact that the USSR has ceased to exist, and taking into account that after the abolition of the Ministry of Defense there is no unanimous approach to the organizational development of the collective defense and security of the CIS, and taking into account that no transition period is envisaged in dealing with the questions of creating armed forces by some CIS states, which may cause an explosion among the military servicemen and suffering on the part of their family members, I request to be relieved from my responsibilities as the commander-in-chief of the CIS Unified Armed Forces and to be discharged from the armed forces in accordance with proper procedures.

I do not wish to be a part of this. 30 December 1991."

"I fought for the CIS to the last bullet," he says, "because I never thought that we would be infected with this stupidity—be divided by borders, oaths, etc. After all, we have lived together for more than just seventy years. . . ."

Nevertheless, when he was asked to restore the recently dismembered Union by military force, he refused. The request came from, among others, the USSR president.

E. Shaposhnikov reasoned more or less this way then: The leaders of the CIS countries made a decision, albeit an erroneous one. But they had been elected by their people and acted on the people's behalf. . . . They signed the Belovezha agreements and denounced the Treaty of 1922. The USSR president resigned on 25 December. Was it up to the military to become a counterbalance to the people? . . .

"I rejected all these requests because I was aware that preserving the Union by such means would involve bloodshed, human casualties, and, let me put it plainly, unpredictable consequences. . . . At the same time I still defended the idea of the CIS and the common armed forces. And some time at the end of 1991, I began to feel I was becoming not too welcome a figure for the Russian leadership. . . ."

As E. Shaposhnikov tells it, all of 1992 passed in endless argument in favor of integration. In February he appealed to the leaders of the CIS members, asking them to refrain from forming their own armed forces—primarily in those regions where "hot spots" were flaring up or smoldering. He proposed that CIS interstate peacekeeping forces be formed. The reaction to everything he suggested was—silence.

"Then I decided," says E. Shaposhnikov, "that, if we cannot succeed in having a unified armed forces, then at least let Russia have something. At my prompting, President Yeltsin issued edicts transferring to the jurisdiction of Russia the Western Group of Forces in Germany, the Northern Group of Forces in Poland, the Northwestern Group Forces on the territory of the Baltic states, the troops of the Transcaucasus Military District, the 14th Army troops (I did not succeed in transferring all of it, because Moldova by then had passed a law on the subject), and the entire Black Sea Fleet. But the edict on the fleet was subsequently revoked. As to Chechnya . . . because I considered Chechnya a federation component, what was there to divide? (Although I was getting reports that the situation around the garrison was difficult.) That is why we mined approaches to the depots, and sealed the depots. . . .

These days Evgeniy Shaposhnikov is the Russian Federation president's representative at the state company Rosvooruzhenie (Russian armaments). Over the past several years, Russia's influence in the arms market has shrunk somewhat. He does not fully agree with this, however, because usually comparisons are made with the Soviet Period, when the country did not engage in arms trading. "We supplied arms—for slogans, bananas, embraces," says S.E. Shaposhnikov. "It is hard to tell how those $2–3 billion the country did earn were spent. They went to world communism, I think. Rosvooruzhenie was started only in 1993. And Russia immediately achieved a real figure: $2–3 billion. Now we are restoring old ties, entering markets that were not traditionally Russia's—Southeast Asia, southern Africa, Latin America, and the Persian Gulf countries. Moreover, the way the situation is developing, we may become partners with some of our competitors, jointly gaining distribution markets." . . .

. . . "I am not a member of any political party. I like the RDDR (Russian Democratic Reform Movement), Russia's Democratic Choice, S. Glazyev's DPR (Democratic Party of Russia), and G. Yavlinskiy's movement, and the new political entity headed by B. Fedorov."

The marshal laconically rejected a guess that his political ambitions would be satisfied if some political forces close to him saw in him their presidential candidate:

"I am not suitable for this role."

**An Afterword**

Last year, 50,000 published copies of E. Shaposhnikov's book, *The Choice: Memoirs of a Commander-in-Chief*, did not reach the readers. Somebody somehow "excised" from distribution 48,000 copies. Where are they? Who cast "an eye" on them? Perhaps those who recognize his real or potential strength.

# Part II

# The Grand Debates: Whither, Whether, and What?

# 3

# Russia's "Great Power" Debate

## Introductory Notes

The heart of Russia's "Great Power" debate lies in its search for a post-Soviet identity as a fundamentally new state. What is Russia? Who are the Russians? What will Russia's relationship be with the other former Soviet republics? Will the country's leadership choose to develop a modern, Western-style democracy operating from within the existing boundaries of the Russian Federation, or will it revert to an imperial policy? The disintegration of the Soviet Union left many in Russia's upper echelons feeling frustrated, depressed, and powerless—their country now being (from their point of view) devoid of its historical greatness as the former seat of a vast multiethnic empire, evolved from several centuries of aggressive colonization of contiguous territories and their conversion to a religious and dynastic culture that constituted the "Russian myth."[1]

It now appears that nationalism has started to fill the ideological void left by the dissolution of the Soviet empire. Unable to accept the loss of two empires, the nationalists who wish to restore Russia's Great Power status are seeking a "foreign" policy that will compensate for the losses. Many Russian nationalists have become jaded about the promises made in the name of Western democracy and prosperity. They are trying to define just how extensive Russia's real military, political, and economic power ought to be within the fragile CIS (and elsewhere).

The documents in Chapter 3 have been selected to illustrate the debate among the Russian political and economic elite (which are often the same group) on the issues surrounding Russian identity and power projection in the "near abroad."

The Great Power debate contains many sub-issues. This chapter documents six of these—with speeches, interviews, and press commentaries by key members of parliament, the press, the military high command, and the intellectual elite.

The first issue is what constitutes Russia's post-communist geostrategic space. Will borders be redrawn? Instead of being bordered to the west by Poland, Czechoslovakia, Hungary, and Romania as the USSR was, the Russian Federation, as defined by Soviet Party organs, is bordered to the west by Estonia, Latvia, Belarus, and Ukraine and to the south by the Transcaucasus and Central Asia. The problem being raised by several political groups in Russia is essentially whether the government ought to in some way reconstitute the territories governed by the tsars before 1917 and by the Soviet Union after 1922. Given the dubious constitutionality of the new commonwealth and the non-democratic nature of Boris Yeltsin's second presidency, solutions to this question vary greatly from group to group depending on political and economic orientation.

Partially related to the above is the second issue, namely that of Russia's self-proclaimed obligation to maintain peace in "hot spots" throughout the territory of the former USSR. Russian leaders have sought both CSCE and United Nations legitimization of Russia's role as "peacekeeper" in the post-Soviet space. Leaders of the non-Russian CIS states, as shown in this chapter, take an ambiguous position on the presence of Russian troops in their countries. Some feel the need of Russian military assistance, but they fear the loss of political sovereignty this entails. Below the surface lies evidence that Russia has been exacerbating some of these violent conflicts in order to keep the other CIS states weak and receptive to its military and economic influence. Peacekeeping is an important component of the power debate in Russia. It will, in fact, be the most important determinant of whether democratic principles of international relations and engagement will be observed by Russia in the CIS strategic space.

A third Great Power issue is the acrimonious debate over Russian "minority rights" within the other CIS states, and who in the CIS states should receive dual citizenship (Russian in addition to that of the state in which the person is living). This issue addresses a very real dilemma—that of almost 26 million Russians who find themselves living "in foreign countries" following the collapse of the USSR. Still, the issue quickly became part of the Great Power debate when Russia unilaterally declared a policy of "dual citizenship" and began to encourage all Russian-speaking peoples living outside Russia to apply to the Russian consulate in their state for Russian citizenship! Such "citizens" are looked upon as "compatriots" by some participants in Russia's foreign policy debate, as illustrated in the *Izvestiya* article by Valeriy Rudnev (7 September 1992) contained in this chapter (3.22).

A fourth issue simmering beneath the surface of the Great Power debate is: "Should the Commonwealth become a new Soviet Union?" As the economic hardships of the transition to market economics become chronic, many politicians, professionals, and common people in Russia find comfort in lamenting the breakup of the Soviet Union. The former communists of Russia are expert at manipulating these sentiments, and openly espouse a program of reestablishing the Soviet Union. The liberal democrats, who initially accepted the collapse of the USSR and respected the "universal principle" of equal treatment, now find themselves faced with widespread sentiment within Russia for molding the CIS into a second Union.

The question of a renewed Union raises the fifth issue—that of "collective security." If the Commonwealth should be transformed into some sort of Russian-dominated confederation or federation, what will be the new organization's external borders? The new Russian military doctrine, promulgated in November 1993, confirms Russia's "right to export its troops" and contains many vaguely worded, but very aggressive readiness measures. This issue has heated up inside Russia since Yeltsin's invasion of Chechnya, especially in reaction to the "war party" which seemed to heavily influence the Kremlin's policy. (The "War Party" is a name used by the press to refer to military-nationalists; a press account of this group's influence is documented here in 3.30).

In recent years, the Great Power debate has shifted ground from deliberation over Russia's past greatness to the definition of Russia's "special interests" in the CIS states, and whether these interests are legitimate, or merely another way to claim hegemonic power over the Commonwealth member states. This question has been vigorously debated in Russia since December 1993, following the election of large numbers of neo-Communists, Fascists, and Nationalists to the State Duma.

The political factions grappling with these issues inside Russia may be grouped, for the sake of simplicity, under four major headings: (a) the Western-oriented neo-democrats; (b) a large and disparate group of conservative nationalists—dubbed "national-patriots"; (c) the "pragmatic nationalists," who claim to adopt a moderate liberal position on foreign policy and Russian statehood issues; and (d) the neo-Communists and Fascists who adopt extremist, fringe positions, especially on questions of Russia's military role, external borders, and imperial destiny.

### The Neo-Democrats

The neo-democrats assumed positions of power when Boris Yeltsin was elected first president of Russia, in June 1990. Egor Gaydar took control of Yeltsin's economic reform programs, stressing free markets and prices and self-correcting distributional patterns. Andrey Kozyrev became the principal architect of the Russian democrats' foreign policy, which stressed Western humanitarian, non-expansionist values and focused heavily on relations with the United States. This policy endorsed "equal status" among the newly independent states of the Commonwealth, supporting in essence Ukraine's approach to the CIS as the facilitator of a "civilized divorce" among the former Soviet republics. In their speeches and articles, Kozyrev and a small group of Yeltsin advisers and cabinet ministers tried to project a new ideology onto Russian foreign-policy-making—cleansed of Russian chauvinism and rooted in the democratic principles of international relations. They especially counted on Russia's rapid economic transformation and integration into world organizations to restore Russia to the great power status for which many Russians yearned.

Chapter 3 contains several statements that illustrate the political philosophy and policy positions held by prominent democrats surrounding Yeltsin. Stanislav Kondrashov writes (15 January 1992) that "the Commonwealth is vitally necessary—to eradicate the imperial mentality and at the same time to preserve—under different, democratic, conditions—the geopolitical and geostrategic area that ethnic Russians and Russian citizens have grown accustomed to for centuries." A statement by Yeltsin (22 February 1992) defends his government against charges that it is ignoring national interests. The democrats' position on Russia's borders within the CIS is addressed here by Dmitriy Furman. Most importantly, a long article in which Andrey Kozyrev explains his foreign policy philosophy and priorities appears here (20 August 1992) as the centerpiece of the neo-democratic "doctrine." Sergey Shelov-Kovedyaev, Kozyrev's principal deputy in 1992, gives us an insight into working conditions within his ministry and describes the struggle being waged over Russia's use of its armed forces in the CIS.

Kozyrev's address of 22 October 1992 represents a sharp rebuttal to the Supreme Soviet's ever-increasing challenge to his power and authority. This is Kozyrev's last attempt to defend Western liberal democratic principles of international engagement for Russia. The opposition (see "National Patriots" below) has by this time become so entrenched that he is forced to recant not long afterward, hoping perhaps to limit the damage to his former vision. The third piece included here by Kozyrev (22 September 1993) shows a completely different person, who defends the use of Russian soldiers in "hot spots" on former Soviet territory: "Owing to its close historical, political, cultural and other ties with the neighboring states, Russia could not—nor does it have the moral right to—remain indifferent to their requests for help in ensuring peace." Likewise, in a 20 January 1994 paper delivered at a conference on Russia and the Commonwealth, Kozyrev staunchly defends Russia's "special role" in other CIS states. (On 8 February 1993, Russia's ambassador to Ukraine demands dual citizenship for Russians living in Ukraine, and warns that Russia could halt fuel deliveries if Kyiv does not accept Russia's proposals. This ultimatum shows just how hardline the Ministry of Foreign Affairs has become on the issue of Russian dominance.)

Another example of the Russian Foreign Ministry's about-face in policy toward the other CIS states after September 1993 is the letter sent by the Ministry to the British Embassy claiming control over the Caspian Sea's oil and gas resources. That letter asserts Russia's unremitting determination to prevail in "pipeline politics." Finally, the chapter provides two 1995 press accounts of Moscow's new CIS integration policies.

### The National-Patriots

The clash between the Russian White House and the Supreme Soviet in the aftermath of the collapse of the Soviet Union was entirely predictable. The Supreme Soviet deputies had been elected in 1990, when the Communist Party was still the strongest local organization in most Russian towns and oblasts. Most of these legislators were communist nationalists who had fought against Gorbachev's reforms and the Union Treaty drafted at Novo-Ogarevo. They included such people as Vice President Aleksandr Rutskoy (whose views are represented in this chapter) and chairman of the Supreme Soviet Ruslan Khasbulatov, who were both unwilling to come to terms with the breakup of the Soviet Union. Although some of these people, like Khasbulatov, had originally supported Boris Yeltsin, they disagreed with Kozyrev's Western tilt and strongly opposed his lack of any plan for reintegrating the non-Russian CIS states.

As soon as the ink was dry on the Commonwealth Declaration, the national-patriots[2] (those who combined nationalist sentiments with the desire to restore Russia as a Great Power) criticized Yeltsin's foreign policies for being servile to United States influence, irresponsible on the Russian economy, and wrong to allow the ties between the former Soviet republics to lapse. Yuriy Glukhov documents this side of the debate in his *Pravda* article of 24 February 1992.

Another influential member of the national-patriots, Evgeniy Ambartsumov (chairman of the Supreme Soviet Committee for International Affairs and Foreign Economic Relations), directly calls for a hard line with the former Soviet republics in an interview given in April 1992. In his view, Russia cannot afford to be as "civilized" with the turbulent CIS states as it is with Western governments. He demands that the Foreign Ministry defend Russia's interests in the "near abroad," and implies that this cannot be accomplished with "agreements" alone. Like most national patriots, Ambartsumov employs the emotional issue of the rights of Russians living in the other CIS countries to pummel Kozyrev's Foreign Ministry. He even expresses sympathy for the Dniester Republic, where separatist Russians living in Moldova have declared independence with backup from the former Soviet 14th Army under the command of Lt.-Gen. Aleksandr Lebed. Russia has adopted a policy of disclaiming involvement in the Dniester affair, combined with obvious behind-the-scenes support for the breakaway region. (See CIS Hot Spots, Appendix A.)

On 15 June 1992 Ruslan Khasbulatov takes matters into his own hands and declares a Russian foreign policy of intervention into the ethnic conflict between South Ossetia and Georgia (at the time not a member of the CIS). Using the excuse of refugees flowing across Russia's borders, he accuses Georgia of genocide and asks the Supreme Soviet to approve a state of emergency in North Ossetia, thus allowing Russia to send troops to the border. Khasbulatov's statement and the Supreme Soviet's state of emergency declaration are provided here. (For Georgian President Eduard Shevardnadze's response to this bold move, see Chapter 4.)

Such blasts from the Congress were echoed in political groups such as Aleksandr Rutskoy's "Free Russia Peoples' Party." In a piece written for *Rossiyskaya Gazeta* Vasiliy Lipitskiy, who chairs the party, calls for a Russian foreign policy oriented away from the West and toward the ex-Soviet and Eastern European states as well as toward Saudi Arabia, the Persian Gulf states, and Taiwan for economic aid. He castigates Kozyrev for meekly accepting almost as a fait accompli the increasing influence of Turkey, Iran, and Afghanistan in Central Asia. In his diatribe, he outlines the basic tenets of a Eurasianist foreign policy, in which Russia would take on the role of a Eurasian rather than a European "Great Power."

These kinds of critical attacks continued in an ever more serious vein throughout 1992 and 1993. On 30 June 1992 Kozyrev presented his vision of Russia's foreign policy before a closed session of Ambartsumov's Committee. In response, Ambartsumov issued his "recommendations" in the form of a White Paper on Foreign Policy, which opens with the following statement: "The Russian Foreign Ministry does not have an integral concept of its foreign policy in either nearby or distant foreign parts." Konstantin Eggert, in an interesting article provided here from *Izvestiya*, 8 August 1992, characterizes Ambartsumov's foreign policy as Russia's "Monroe Doctrine."

In March 1993, Aleksandr Rutskoy wrote that the CIS was not working. In his article he argues that Russia lost too much in the breakup of the Soviet Union—seaports and critical economic ties. He instructs Russia to take into account the "historically determined interlacing of the (CIS) economies, the problems of their common "Union" and . . . the multinational character of their population." The article is a more modest version of what were to become his later views.

Another article provided here, from *Novaya Ezhednevnaya Gazeta,* illustrates how national patriots often combined their opposition to Yeltsin's liberal economic reform policies with opposition to his government's foreign policy, interchangeably blaming one upon the other. Oleg Bogomolov writes in June 1993 of his disagreements with the Chicago school of economics, in which Egor Gaydar was educated, and about his severe disagreements with Yeltsin's foreign policies.

Yeltsin reaches a point in October 1993 at which he can no longer pretend to work with the parliament. Up to then, Yeltsin and Kozyrev managed to sustain their control over the broad outlines of foreign policy. However, the Supreme Soviet belligerently insists on a definitive role in foreign policy decisions. On October 4, Yeltsin orders parliament to close its doors and decrees that the country will hold new parliamentary elections in December. He wins compliance with this order only with the help of the Russian Army, which fires on almost one hundred deputies who refuse to dissolve the Supreme Soviet, among them Vice President Rutskoy and Ruslan Khasbulatov as leaders of the insubordinate action.

The elections, this time to the State Duma—a new legislature instituted in December 1993 under a new constitution—bring new surprises and challenges. The Russian people elect a large contingent of conservative nationalists, including neo-communists and national-patriots, but most shockingly, the radical right, led by Vladimir Zhirinovskiy (whose party is misleadingly called the "Liberal Democratic Party") wins almost 30 percent of the vote. Yeltsin is forced to compromise with these political forces and evidence points to the fact that not only the "Red-Browns" in parliament, but also the military, begin to play a far greater role from that time on in Russia's foreign policy decisions.

The growing influence of the national-patriots over Russian foreign policy was clearly demonstrated in the new military doctrine promulgated in November 1993. The documents included here describe that doctrine and the view that the "hawks" had won in a decisive political battle over its drafting. The doctrine introduces a new component into Russia's foreign policy. (The contents of this doctrine are documented at some length in Chapter 9.) This is the concept of "peacekeeping"—an idea becoming popular in the West following the dissolution of the Soviet Union. Yuliy Vorontsov, permanent Russian representative to the United Nations, announces on 19 November 1993 that Russia is to become "peacekeeper" in the CIS, which it claims as its "sphere of influence." Russia has tried to obtain the UN's approval of its "peacekeeping" obligation, but has not succeeded in making it an official UN position.

In 1994 and 1995 the national-patriots stepped up their political activities, hoping to win in the next Russian parliamentary elections in late 1995 and in the presidential election scheduled for June 1996. Oleg Bogomolov, in an article contained here from *Kommersant* (15 July 1994), advocates a more rapid economic reintegration of what he calls the "post-Soviet zone." He bases his arguments on the premise that the disruption of interenterprise relations and trade has caused economic crises in the CIS countries. (Other experts believe the crisis was already occurring when the Soviet Union broke up and was deepened in some areas, but dispelled in others by the formation of the CIS.) Bogomolov's views have figured prominently in Russian hard-line economic pressure tactics to accelerate CIS economic integration policies since early 1993. In June 1993, the Russians proposed that the CIS form a supranational "Interstate Economic Committee" (IEC). This was the first suggestion to give any CIS organization supranational powers. Ukraine and Uzbekistan immediately rejected the proposal. Turkmenistan said it preferred not to give CIS organizations such powers. The other states, for the most part, have gone along with the proposal, saying it is time to implement some of the four hundred-odd agreements signed within the Commonwealth.

Upon his release from Lefortovo prison in July 1994, Aleksandr Rutskoy formed a new movement called "Power." He defines the goals of this movement quite explicitly in an interview provided here with the Prague publication, *Lidove Noviny,* as being to restore Russia's innate status as the dominant Eurasian state.

The last three articles in the section provide interesting insights into the anti-American views that have proliferated since Russia's adoption of its imperialist course; the role of the "war party" in Yeltsin's White House (especially over Chechnya policy); and Russian fears of espionage on the part of certain other CIS states.

### The Pragmatic Nationalists

The third group of important players in Russia's foreign policy struggle comprises what might be called the "pragmatic nationalists." Georgiy Arbatov defines this group as follows:

> Although they are quite Westernized in their upbringing and outlook, they are in most cases distinguished by a more realistic, even pragmatic attitude toward Russia, the West, and the world at large. In the Foreign Ministry, such an attitude is shared by the ambassador to the United States, Vladimir Lukin; indeed most of the liberal Foreign Ministry officials who rose during the Shevardnadze years would subscribe to this position.[3]

Arbatov identifies as members of this group: parliamentarians Aleksandr Peskunov, Evgeniy Kozhokhin, and Aleksey Tsarev; older-generation diplomats and intellectuals such as Arbatov himself, Roald Sagdeev, Oleg Bogomolov, Stanislav Shatalin, and Nikolai Petrakov; and younger academics such as Grigoriy Yavlinskiy, Aleksey Arbatov, Sergey Rogov, Aleksandr Konovalov, Sergey Oznobistchev, Emil Payin (Yeltsin's advisor on separatist politics), Leonid Vasilev, and Pavel Baev.

The pragmatic nationalists began to make their influence felt in the foreign policy debate in late 1992 and 1993, and especially after the elections to the State Duma. Recognizing that the impasse between the executive and legislative branches could not be allowed to continue, members of this group intervened to make Russian nationalism more palatable to the West as well as to participants in the Russian debate and in other CIS countries. This relatively small group of intellectuals was able to successfully recast the position of the national-patriots into acceptable, pragmatic terms. The first change, upon which they based their political identification, was to substitute the jargon of "Russian special interests" in the CIS for Russian "Great Power" interests. Arbatov refers to this group as "moderate liberals." They categorically dismiss as "romantic" and "idealistic" the early democratic principles espoused by the liberal democrats. Nevertheless, they leave fairly open the definition of what Russia's "special interests" are, and they make it quite clear that they agree with the national-patriots on the question of whether to use Russian military force when and if necessary to advance those interests. This group is able, by early 1994, to win Andrey Kozyrev's subscription to its views.

One of the pragmatic nationalists' principal spokesmen is the former Russian ambassador to the United States, Vladimir Lukin. In the interview and two articles by Lukin that are presented here, he defines his concept of Russian nationalism and his idea of Russia's role in the Commonwealth. He offers pragmatic concepts, saying Russia should distance itself from the United States and Western liberalism in general, but acknowledges that there is a European side to the Russian soul. He calls for a strong Russian government administrative bureaucracy, led by the intelligentsia, whose mission would be to create a single economic space within the Commonwealth. He does not address what this would mean for the sovereignty of each state, although the implication is that each should willingly sacrifice its sovereignty for access to the protection and resources of Russia. In 1995, Lukin argues the dangers of the international principle of "self-determination" (17 February 1995). Warning of the trouble it can bring to the region for many years to come, Lukin states that Russia should adopt a clear policy of using force in the Commonwealth states, and that "the more convincing it is, the less it will have to be resorted to." Lukin's writings illustrate the dilemma of the moderates—who wish to uphold the sovereignty of the new republics, but foresee the potential for chaos and turbulence in these territories for years to come and realize that Russia's vital interests as a world power remain entangled with all of these new states.

The last document in this section, an interview with the Russian Federation Minister for Cooperation with the CIS states, Valeriy Serov, illustrates the extent to which the official government position on integration within the CIS agrees with the moderate nationalist position. That policy, however, remains a confusing combination of advances and retreats on the imperial front.

### The Right- and Left-Wing Extremists

The "Reds" and the "Browns" comprise the fourth general category of participant in Russia's Great Power debate. These are the extreme-left Communist Party of the Russian Federation and the extreme-right nationalist groups, such as Vladimir Zhirinovskiy's Liberal Democrats and the overt Russian fascists. Russia's right-wing fringe advocates a Russian state that is centralized, authoritarian, and expansionist. Its proponents reject the multinational Russian Federation in favor of a single ethnic state, with one culture and one language. Their vision of a strong Russia, supported by a strong military, is aggressively Eurasianist, supporting a vigorous expansion southward and eastward by military campaign. The "Browns'" supporters include a large segment of the military and large numbers of frustrated nationalists whose support cannot be enlisted by the communists, the democrats, or the moderates. Their greatest threat to Russia's fragile democracy at this time lies in their skillful political game of leading other groups into adopting increasingly conservative policies as they compete for the votes of an increasingly nostalgic populace.

Chapter 3 provides several examples of Russian right-wing extremist positions on issues in the Great Power debate, including Communist Party Chairman Gennadiy Zyuganov's political report to the Third Communist Party Congress.

The documents in Chapter 3 provide key highlights of almost four years of foreign policy toward the "near abroad" in post-Soviet Russia. The documents give points of reference in the struggle that has engulfed Russia's political and intellectual elite, and much of the population at large. Right-wing forces have made obvious gains in the debate, tempered by the ideas of the center-right. For the time being, the former communists have most cleverly manipulated the yearnings of the bulk of the Russian people for stability and the return of social supports offered by the state. Chechnya, however, raises new questions about who controls Russia's foreign policy, as well as about why Chechnya was attacked so ferociously, and how the political premises and precedents set in Chechnya might affect Russian foreign policy in the future. (See CIS Hot Spots, Appendix A.)

**Notes**

1. The Russian "myth" refers to the dynastic ideology of Russian imperialism under early tsars, such as Ivan the Terrible, Peter the Great, and Catherine the Great, who claimed to rule by divine right. These tsars captured large numbers of peasants and lands surrounding Muscovy and "Russified" them—more through religious conversion and dynastic paternalism than through nationalism in its modern sense. Captured lands and peoples were often divided among Russian nobles as rewards for military service and other favors. Through this process, a sense of "Russianness" became rooted in many of the nations surrounding the ancient principalities of Muscovy.

2. Grigory Arbatov, in his article "Russian Foreign Policy Alternatives," *International Security*, Vol. 18, Fall 1993, pp. 5–6, identifies some of the more prominent members of the national-patriot group, including: former Chief of the Yeltsin Security Council Yuriy Skokov, Vice Premiers Oleg Lobov and Mikhail Malei, Sergey Karaganov, Andrey Zubov, Sergey Stankevich, Andranik Migranian, and Aleksandr Tsipko.

3. Ibid., p. 7.

# The Neo-Democrats

## 3.1

**Article Assesses Role of the Commonwealth, Peace**
Stanislav Kondrashov
*Izvestiya*, 15 January 1992 [FBIS Translation]

The television Channel 1 has cited the figures from a poll according to which less than one-third of the people now believe in the stability of the CIS [Commonwealth of Independent States]. The distrust was aggravated by the recent debates between Russia and Ukraine over the Black Sea Fleet. The idea of the CIS is centripetal (albeit without a center) and unifying, yet practice continues to be centrifugal. Currently, this suits some leaders and politicians, but by no means the peoples who are quite reluctant to fight. Yet without the Commonwealth there will be no real peace between the independent states.

To explain this there is no need even to look at the example of Yugoslavia as it unfolds before our eyes or even closer at the example of Azerbaijan and Armenia. You just have to imagine a new—and as yet unpublished—map on which instead of the single Soviet Union there are fifteen independent states. And to take a look around with Russian eyes, remembering history. That of course is difficult—looking with new Russian eyes and not the "common" Soviet eyes, and perception lags behind events but perhaps it should be made to catch up to avoid trouble.

What we can see in Russian on this map is not like what we saw in the USSR. Of the former neighboring states only Norway, Finland, Mongolia, and China remain. The USSR was closer to Western Europe than Russia is now—closer physically. The USSR bordered on Poland, Czechoslovakia, Hungary, and Romania. Russia borders on Estonia, Latvia, Belarus, and Ukraine. It touches on Poland and Lithuania only in Kaliningrad oblast, whose future without the CIS will be complicated. Imagine—without the CIS—the sheer hostility between Ukraine and Russia, in which Belarus inclines toward Ukraine. One option is a railroad blockade of freight from Western and Central Europe to Russia like the blockade that Azerbaijan is implementing against Armenia, only on a far larger and more dangerous scale.

Let us move farther south on the new map. The USSR used to border on Turkey, Iran, and Afghanistan and was but a stone's throw away from Pakistan and India. Russia has different neighbors—Georgia, Azerbaijan, and east of the Caspian Sea it has the large Kazakhstan. The Kazakh republic used to protect Turkmenistan, Uzbekistan, Tajikistan, and Kyrgyzstan from Russia and also protect China over thousands of kilometers of the former Soviet-Chinese border that passed through Central Asia.

What are the options, both extreme and not very extreme? You cannot list them all—from transit visas via Kazakhstan for a citizen of the Russian Federation traveling to, let us say, Tashkent, to the departure of "our" Central Asia to the Muslim world if the historically formed attraction to Russia is weakened and reduced to nothing by the years. At present these are only seeds in the soil of separatism. They will sprout without fail if the centrifugal forces continue to rule and the CIS remains only an empty shell, only a convenient form of bidding farewell to the totalitarian Union.

In the depths of youthful Russian diplomacy has been born the concept of the "near abroad," that is, of yesterday's republics that have become sovereign states. Establishing equitable mutually advantageous ties with them is a priority. That also means filling with content the form of interstate association that is called the CIS. But since diplomatic energy is not infinite and new priorities are fraught with the danger that old ones will be weakened, it is entirely likely that the active foreign policy of Russia and Central Asia will result in passivity in the Near East except, obviously, for Israel, where hundreds of thousands of our fellow citizens are settling.

In general, if we look from Russia, of the former borders

apart from the Norwegian and Finnish borders, only the Far Eastern one remains—with Mongolia and China. Great prospects are opening up there for relations with Japan, a rapidly developing economic superpower, but the underdeveloped, underpopulated Far East is not ready for their accelerated implementation, even in the South Kurils.

If Russia implants democracy and successfully implements radical economic reform, a fruitful involvement in the world economy awaits us. The improvement of the Russian living standard and civic freedom will compensate for the psychological damage caused by the loss of status as a global superpower which millions of people, especially the older generation, cannot help experiencing.

To ensure that this is not a fashionable and implausible new fairy tale about a bright future on the radiant summits of capitalism, the Commonwealth is vitally necessary—to eradicate the imperial mentality and at the same time to preserve—under different, democratic, conditions—the geopolitical and geostrategic area in which ethnic Russians and Russian citizens have grown accustomed to live for centuries. Yes, to nurture one's own national and state dignity, but in no circumstances infringe on the national and state dignity of the Russian giant, is vitally important for all states of the Commonwealth.

Alas, in the story of the Black Sea Fleet, Ukrainian President Leonid Kravchuk failed to display such understanding. He acted too hastily without considering how his "predatory" method would be received in Russia. Boris Yeltsin reacted with obvious delay, not immediately using his position to back up Marshal [Evgeniy] Shaposhnikov and Admirals Chernavin and Kasatonov. And when his position was finally stated—"it has been, is, and will be Russian"—the Ukrainian side suspected it of containing a challenge to its sovereignty. The precondition of the conflict was the general haste in creating the CIS and the obvious incompleteness of its fundamental documents, including those about the fate of the former USSR's armed forces. Now, after the Russian-Ukrainian talks in Kiev, which were belated and, once again, hasty, they seem to have moved toward an agreement, but so far it is hard to regard the agreement as definitive.

Is the present Black Sea Fleet necessary in our present circumstances? What is more important—the Fleet or sausage? These are important questions but they ignore the broader background to the conflict that has suddenly erupted. Behind Ukraine's desire to get the Fleet and, moreover, without prior arrangement, Russia saw the first bold attempt to encroach on its geostrategic area, which it is prepared to share with Ukraine and other CIS members, but does not intend to tear possessions away from itself and hand them over to another state.

Behind the silhouettes of the warships there immediately loomed Sevastopol, the "city of Russian glory," and the entire Crimea and the main Black Sea trading ports, which traditionally belonged to Russia and to which it does not intend to lose free access. Give them an inch and they will take a mile. Especially (let's take another look at the transformed map) insofar as Russia will also be constrained in the Baltic if the settlement with independent Estonia and Latvia is as poorly thought out as with Ukraine and closes the main Baltic ports to it.

The conclusion may be formulated briefly: Live and let live. That simple wisdom is alien to the imperial spirit, but it sees the protection of one's own interest with respect for the interests of others. It also presupposes a consideration of geography and history dictating the compulsory "transparency" of borders. And the utmost balance and circumspection in the difficult construction of the CIS.

---

## 3.2

**Yeltsin Defends His Foreign Policy**
Interview by Nikolay Burbyga
*Izvestiya*, 22 February 1992
[*CD* Translation], Excerpts

---

*Question:* As we know, America and other states, while respecting universal human values and interests, nonetheless never forget their national interests; on the contrary, they even accord the latter paramount importance. But for some reason I haven't heard our politicians say that various steps we are taking serve Russia's state interests. Boris Nikolaevich, does Russia have its own interests? If so, please say what they are, in your opinion.

*Answer:* I have often read and heard reproaches of this kind—to the effect that Russia's leadership has supposedly all but eschewed state interests, and that a retreat is under way on all positions.

I categorically disagree with such assertions.

No, Russia is no longer the main power center of an enormous communist empire. Thoughts of painting the planet red have been discarded. We have rejected the notion that we are surrounded by covert or overt enemies, and that the most important thing in the world is a struggle that we must unfailingly win.

The elevation of these decrepit notions to a principle of foreign policy led to the ultimate collapse of the totalitarian system. They did enormous damage to our people and our national interests. And no less importantly, they did not make a single country or a single people happy.

Domestic and foreign policy are not separated by a wall. And they are based today on the interests of the Russian

Federation's people. I don't think I have to try to persuade anyone that the most important task facing our country is to emerge from the deep quagmire of crisis and to effect a transition to foundations of life that will enable us and other peoples to live and work normally.

It is for this reason that we have an interest in seeing the world truly stable and in affirming the norms of civilized life and strengthening mutually advantageous cooperation in the world community.

We categorically oppose the use of diktat both with respect to states and peoples and with respect to individual citizens. The Russian Federation is firmly committed to strengthening guarantees of human rights and freedoms both within the country and beyond its borders.

These are the most important priorities of Russian foreign policy. We are taking our first steps in foreign policy. Not everyone in our country agrees with them. The influence of imperial thinking is strong. But I think that after a little while we will all understand that a policy of goodwill serves our interests far better than a policy based on force.

## 3.3

## Border Changes, Even if Fair, Are Not Desirable

Dmitriy Furman

*Nezavisimaya Gazeta*, 3 July 1992 [*CD* Translation]

The whole logic of the development of events since the destruction of the USSR is pushing Russia in one direction—toward repeating the path of post-Versailles Germany or today's Serbia. A large country surrounded by weaker countries with national minorities representing the large country, especially if these national minorities are oppressed (and any national minority may feel oppressed, to a certain extent), is virtually unable to withstand the temptation to use its strength. . . .

It is very difficult not to intervene, not to pound the table with one's fist (or pound Kishinev [now Chisinau] with aircraft), when one sees what is happening in Moldova. Rutskoy's threats against Georgia are very understandable—this is a normal human reaction to the nightmare in Ossetia and the nationalistic stupidity of a number of Georgian leaders. The borders that Russia and the other republics inherited are absurd borders, and one simply cannot bring oneself to talk about their unshakability.

But it is not only foreign policy realities that are pushing Russia toward a redrawing of borders. Domestic-policy realities are also pushing it in this direction. The destruction of the Soviet Union, . . . which was a shrewd move by Yeltsin in his struggle for power, could not help but draw reactions from the Russians, who are entangled in their own contradictory aspirations and are increasingly feeling that their leaders have swindled them. If one adds to this the economic reform, the pace of which allows one to hope—as the brightest prospect—for a restoration of 1985's standard of living by the early twenty-first century, one can hardly be surprised at the growing reaction. This growing nationalistic reaction is making the same people who actively destroyed the Union strike Great-Power poses. It is perfectly natural and logical that the democratic wave of 1988–91 is being replaced by a nationalistic wave, which all the careerists who earlier tried to ride the democratic wave are now trying to ride. Stopping it will be just as difficult as it was to stop the first wave.

Everything is drawing us toward a struggle for the redrawing of borders. But this means war, something that, incidentally, the Germans under Hitler were also drawn into gradually. . . . As things stand now, we could unite South Ossetia with North Ossetia without any special effort. The world might even close its eyes to this, and the people would applaud Yeltsin or Rutskoy. Then a war would begin with Ukraine and Kazakhstan. And then what would happen? . . .

The entire logic of the development of events is drawing us onto this path. But nevertheless, this logic must be opposed by everyone who cherishes Russian democracy and, in the final analysis, simply cherishes Russia. . . .

However, the struggle against this trend that is being waged by those who truly seek a democratic Russia is impeded by the chaos of heterogeneous and mutually contradictory principles and values that exists in their minds on the question of borders in the CIS. In fact, although they do not want war and fascism, their moral consciousness protests the unshakability of borders. . . . To recognize the unchangeability of the present borders is tantamount to approving all the injustices and crimes accumulated by history, all the evil of history—from Ivan the Terrible's conquest of the Tatar khanates, after which the Tatars were doomed to live in a state that continually celebrated various anniversaries of the Battle of Kulikovo, to the hiring for jobs in Estonia of Russian workers whom no one warned would someday be required to learn the difficult and strange Estonian language. There are various rights and truths in the world that are mutually contradictory. The right of states to inviolable borders contradicts the right of nations to self-determination, the practical application of which is not at all clear. . . . If you add to this the principle that the rights of indigenous peoples to their historical territory must not be completely denied (although God only knows as of what century a territory can be considered to "historically" belong to a people), and the fact that this introduces even more confusion, because preserving—for example—the Abkhazi's historical rights to Abkhazia, where they have turned out to be a minority as a result of the Stalinist policy encouraging Georgian in-migration, is possible only at the cost of infringing upon the rights

Is it possible to reconcile these various "truths"? I don't think that they can be reconciled, but it is possible to establish a certain hierarchy of these rights and our principles.

No matter how unfair it may be, it is a legal fact that Karabakh is a part of Azerbaijan and Chechnya is a part of Russia. The alternative to recognizing this fact is general and bloody chaos, a war of everyone against everyone in a country . . . that is stuffed with nuclear weapons. This is unfair, but the evil that stems from this unfairness is many times smaller than the evil that would arise if we began a struggle for fairness here, using military force. If the former is an evil, it is the lesser of two evils.

Does this mean that we should put up with all the injustices that are codified by this right of states, that the Ossetians must remain separated forever, and that the Tatars have to continue watching on television and hearing on the radio about the Battle of Kulikovo, which saved Russia from the barbarians? No. . . .

First of all, we must strive to have Russia set an example in resolving conflicts between nationalities. A poor man does not have the right to rob a rich man, but the rich man is morally obligated to help the poor. In comparison to Chechnya and Tataria, we are rich. No one can make us recognize their sovereignty. We have every right (including a moral right) to make sure that Chechnya and Tataria observe our laws, and we may use force in doing so. But we can also simply treat the Tatars and the Chechens humanely by changing our legislation, making it closer to the demands of morality and humaneness, and giving them the opportunity to secede from us legally and in good order. We have the right not to do this, but doing it is also our right, and if we do it, it would be a great and noble act that would change the entire climate in the CIS. Only if we ourselves set such an example will we have the moral right to exert pressure (but not through violence or the threat of violence) on behalf of Russian minorities and peoples like the Gagauz, the Ossetians, or the Abkhaz. We can even strive for border changes in our favor—while absolutely ruling out violence—by appealing to an international court on the Crimean question, for example, and thereby furthering the development of international law. Finally, we can struggle for the creation of a climate and conditions in the CIS in which borders would not have such great significance. . . .

At present, in compensating for our inferiority complex, we have begun to say quite often that Russia is a great power. What this means is not very clear, but it is clear that Russia is a strong country and the strongest one in the CIS. Great demands are made on a great country. We can plunge not only ourselves but half the world into bloody chaos, or we can create a democratic order in the CIS and become a decisive force in creating such an order throughout the world. If Russia's democratic forces halt our slide toward wars and fascism, and if we can get at least ten years of peaceful democratic development, calm down just a little, and come to our senses, the very worst will be behind us. But if this is to happen, the democratically minded people of Russia will have to struggle with all their might against attempts to forcibly change borders, even if they are justified by the loftiest and most obvious principles of fairness.

---

## 3.4

**Shelov-Kovedyaev Comments on Policy Criticism**

Interview by Aleksandr Gagua

*Nezavisimaya Gazeta*, 30 July 1992

[FBIS Translation], Excerpts

---

*Editor's Note:* Shelov-Kovedyaev's description of the Foreign Ministry's problems, and his reaction to parliament's criticism, are especially important because he describes and defends the early policy direction of the neo-democrats. Both Yeltsin and Kozyrev tried to apply the neo-democratic concepts of maximum Russian restraint, and a literal application of international law to the "near abroad," for as long as possible. Soon after the interview recorded here, however, the nationalist viewpoint began to overpower that of the neo-democrats, and the democrats responded by changing their own policy direction.

[Shelov-Kovedyaev]: . . . If in the sphere of policy in the "near abroad," the tilt toward the use of power methods, the lack of coordination of actions, the shift of priorities, the incorrect organization of work, and the unprofessional interference increase, despite the Foreign Ministry's efforts, I do not rule out the fact that I myself will raise the question of my resignation—inasmuch as under such conditions I, really and truly, would not greatly understand in what way I could be of use to the fatherland and the president. . . .

We may speak first and foremost about the fact that the inertia of isolation among states of the Commonwealth engendered by the old Union, and inherited by us after its disintegration, has been overcome. It is significant that even journalists have ceased to concern themselves with when the Commonwealth will cease to exist. It is clear to all now that doubts concerning its viability were premature. Such institutions as the Council of Heads of State and the Council of Heads of Government have shown themselves to be efficient instruments for harmonizing viewpoints and formulating common approaches in policy. Agreement is not being reached on all questions immediately, naturally—the European Community also, for example, developed in just the same way. Many questions to which there were no generally acceptable solutions were on the agenda for years, decades even, and all attempts to force their solution invariably failed. But we have succeeded in coming to terms in such important fields as transport, power engineering, pricing, the functioning of the ruble zone, the environment, and social security; military questions are being decided without serious cataclysms. In respect to the latter it is sufficient to mention the harmonization of the procedure of compliance with the treaty on Strategic Arms Limitation [SALT] and the Conventional Forces in Europe Treaty. A number of states have signed the Collective Security Treaty. A military-political nucleus is thereby taking shape in the Commonwealth. Having devised the appropriate mechanisms and having begun to implement them, the Commonwealth has shown its capacity for responding to the challenge of regional conflicts. Fundamental decisions have been adopted with respect to customs. Finally, the CIS is increasingly taking organizational shape. A working group which prepares meetings of the Council of Heads of State and the Council of Heads of Government has been formed, and the standing orders and rules of procedure of a meeting of the councils have been adopted. The Foreign Ministers Council, the Defense Ministers Council, and a number of sectoral councils or committees—on transport, power engineering, the environment and so forth—are functioning. The Foreign Ministers Council assembles on the eve of a meeting of the Council of Heads of State to put the final touches to the agenda, and the Council of Heads of Government has adopted the decision that the corresponding ministers will assemble on the eve of its meetings for the same purpose. An instrument for the resolution of economic disagreements has been created.

It was clear even before the creation of the Commonwealth that its geometry would be asymmetrical. Actual development proved this forecast correct. Two groups of states have become clearly defined in the CIS: Seven are disposed toward the development of close interaction on a multilateral basis—these are, besides Russia, Belarus, Kazakhstan, Armenia, Kyrgyzstan, Tajikistan, and, of late, increasingly, Uzbekistan; four—Ukraine, Moldova, Azerbaijan, and Turkmenistan—are adopting a more aloof position. It is significant, moreover, that Ukraine and Turkmenistan have as of late been joining increasingly actively in multilateral cooperation and that Moldova and Azerbaijan, which have yet to ratify the Agreement on the Formation of the CIS, are intensively seeking opportunities for association with the Commonwealth.

Has so little happened, and in six months at that?

In addition, a tremendous amount of work is being done in the sphere of bilateral relations with each state of the "near abroad." The process of the conclusion of general political treaties is being completed. I shall not dwell specially on the large-scale bilateral economic negotiations between Russia and Kazakhstan and Russia and Belarus or on the top-level bilateral negotiations in Dagomys and Moscow either. All that the Union formerly made a mess of is now having to be restored on a treaty basis, and this means dozens and dozens of topics for negotiation with each state. The corresponding Russian delegations, which are working very intensively inasmuch as the existing problems require package solutions, have been formed to this end. The most important areas of the negotiations and our positions on each of these issues have been defined. Human rights and humanitarian issues generally are the priorities for us here. Negotiations, however, as distinct from the issuance of instructions, are distinguished by the fact that they take time, and the expectation of immediate results here is for this reason unjustified. The majority of ambassadors have been appointed, and they have taken up their duties, and personnel for the embassies is being selected. The formation of a Commonwealth Affairs Department is being completed in the Foreign Ministry. All this, the selection of people particularly, has required—inasmuch as the extent and the nature of the problems are wholly unprecedented—particular attention and time, if it is considered that we had to begin practically from scratch. But we are completing this organizational period successfully.

The second accusation: All that the Foreign Ministry is doing in the sphere of policy in the "near abroad" it is doing badly, erroneously, and to the detriment of Russia's interests. I will permit myself here to give just a few of the most striking examples of the direct opposite tendency.

The conflicts in the Dniester region and South Ossetia are being settled in accordance with the recommendations and plans, including the involvement of representatives of the Dniester region, which were patiently and painstakingly developed over a period of months by the Foreign Ministry in the course of pursuing bilateral and multilateral initiatives. Special working groups for each conflict situation, Nagorno-Karabakh included, were formed, and permanent representatives to the Mixed Commission were sent to Moldova. Remember how many menacing decisions were adopted by the Supreme Soviet and the congresses of deputies, how many trips for the purpose of saber-rattling were made—they all merely exacerbated the situation. And as a result everyone had to return to the proposals of the Foreign Ministry, which produced results. Furthermore, the signing of a settlement in the Dniester region, for example, not only enabled the supporters of "speaking-straight-from-the shoulder" to get a clearer idea of the alignment of forces in Moldova, but also graphically demonstrated how painful the actual process of settling chronic conflicts can be.

We are all disturbed by the human rights situation in the Baltic countries. But none of the most "decisive" actions, of the Supreme Soviet, for example, were able in any way to influence the international community's attitude toward this question. On the contrary, the appeals of the Foreign Ministry to the CSCE, the Council of Europe, and United Nations led to the Council of Europe's adopting a decision to send to the Baltic a special group of human rights experts.

A third accusation: The Foreign Ministry is operating without a concept of policy in the "near abroad." This, I confess, I absolutely don't understand. Such a policy was created back in March. It was presented to Kozyrev and sent to the president and the two first (at that time) deputy prime ministers—Burbulis and Gaydar—and also to two committees of the Supreme Soviet. The minister and Burbulis and also the committees of the Supreme Soviet made a precise reading of it. But neither the Supreme Soviet nor the government discussed it. Inasmuch as no objections to it were forthcoming, the Foreign Ministry has been guided by it. I would have understood had our concept been deemed inappropriate, but to say that there has been none at all?

Much time has elapsed since then, and for this reason I am now engaged in a collation of both my own thoughts and materials of my colleagues for an updated version of this concept, which I hope to present to the president upon his return from vacation.

[Gagua]: But it is the president who has said repeatedly that the Foreign Ministry is operating without a concept.

[Shelov-Kovedyaev]: I have only one explanation for this. The president has not seen our concept. It has evidently gone astray somewhere in the depths of his staff.

[Gagua]: And what now, from your viewpoint, is most problematic in the sphere of Russia's foreign policy?

[Shelov-Kovedyaev]: The problem of the security of Russia's borders is, as before, quite serious. I am not, of course, talking about a carryover of all the attributes of the Soviet border—barbed wire, machine gun towers, rigorous control, and so forth—to the border between Russia and its neighbors in the "near abroad." We would not want to unilaterally violate the accord on the transparency of internal borders between, for example, states of the Commonwealth. But several states—Ukraine, Moldova, Azerbaijan, and others—did not formerly sign the Agreement on Common Principles for Safeguarding the External Borders of the Commonwealth States. Nor have all signed the agreement on the settlement of customs questions. And some states that have signed these documents, Kazakhstan and Kyrgyzstan, for example, have come to an arrangement concerning reciprocal visits with Turkey that do not require visas. I am by no means questioning these states' sovereign right to adopt such decisions. The issue lies elsewhere—the territory of Russia has come to be open to the penetration of freight or persons from across the border of the former USSR. A similar situation has taken shape on the border with the Baltic countries. It is clear that transparency also works in the reverse direction—for bleeding commodities from Russia, for example. Nor can we lose sight of all that is connected with the border of the former USSR—electronically, the most heavily fortified in the world. It is urgent that all these problems be resolved. The Foreign Ministry has been submitting proposals since April. But insofar as there are many aspects which are outside the jurisdiction of the Foreign Ministry, and require concordance with the border guards, the Ministry of Security, and the MVD, actual steps became possible only recently.

There were many problems with military property prior to the formation of the Russian armed forces. In fact, many weapons were transferred to republics outside of Russia in the final months of the Union. After its disintegration, the Foreign Ministry consistently held that the existence of extra-state armed forces was impossible. But this idea prevailed, as is known. . . .

Things are difficult when it comes to withdrawing forces from the Baltic states. At the negotiations with Latvia, Lithuania, and Estonia the Russian side is upholding the package principle of a solution to the problem, including questions of the future use of former Soviet military property, facilities, bases and such; social problems of redeployed servicemen and their families, military retirees remaining in the Baltic, the status of forces in the period of withdrawal, and so forth.

[Gagua]: At the start of our interview you mentioned the lack of coordination of political actions in the "near abroad."

[Shelov-Kovedyaev]: Two factors may be noted here. First, the unprofessional interference in the sphere of policy in the "near abroad" by persons who dabble in this on an occasional basis, and mainly under the influence of an emotional perception of reality. This may come in the form of political initiatives, or in the appointment of these persons to positions of responsibility. In either case the Foreign Ministry has to make efforts to minimize the consequences. It is sufficient to recall the utter failure of the power approach to a settlement of regional conflicts. Often, super-tough declaratory demarches are undertaken in order to intimidate those to whom they are addressed. Inasmuch, however, as the reaction to them is usually the direct opposite, their authors become totally confused and, unless "insurance has been taken out against them in good time," display a readiness to move considerably further in the direction of a compromise than necessary. It is for this reason that I remain a supporter of diplomatic actions that are, although gradual, consistent and professional. . . .

[Gagua]: But Minister Kozyrev and you have, for all that, an entire ministry. Could these problems not have been resolved more quickly on the basis of self-reliance?

[Shelov-Kovedyaev]: If only things were as you say! Although the Foreign Ministry inherited from the Russian and Union ministries an enormous number of highly skilled specialists, in January, counting your humble servant and his secretary, a few more than ten persons were working in the ministry with countries of the "near abroad"—there was only one special officer in charge of each country! The Commonwealth Affairs Department is still not fully staffed, nor are the departments in charge of the Baltic countries in the Second European Administration. The Foreign Ministry had no trained specialists for work with these countries. People are having to be taught from scratch since, I repeat, the problems for the Foreign Ministry here are entirely unprecedented, beginning with the study of the languages of several states. And this instruction has to be carried out on the march, as they say, while solving current questions.

In addition, the formation of delegations, the selection of ambassadorial candidates, the creation of special regional conflict working groups, the choice of observers for permanent or shift work in areas of the Dniester region, South Ossetia, and Nagorno-Karabakh, the creation of a structure for the prompt evaluation of crisis situations. . . .

We can now say that this work is approaching completion. But it is clear to every unbiased person, I believe, that time has been needed for this, considerable time at that. . . .

The ministry lists of personnel were approved only in April, and there is still no budget! We are still receiving less than 30 percent of the currency resources that we need and 40 percent of the ruble appropriations for the upkeep of the main staff. What fitting support for the new subdivisions this is! The pay scales of our employees for today's times are simply wretched.

The further we go, the worse it gets. We are not being allowed to finance the minimum salary of an ambassador in states of the "near abroad" in dollars, because the majority of states have not as yet left the ruble zone, and it is required that they be paid in rubles. After all, we have already "seen" all this. In the former CEMA countries (and in "socialist countries" generally) our diplomats were paid not in freely convertible currencies, but in the local currency. As a result, these embassies were considered second-rate, and it was frequently the corresponding type of people who ended up there. So what—will we in the new way divide diplomats into "clean" and "unclean"? Who will our employees in the "near abroad" be after this—"priority" or "third-grade"? With whom will we fill the embassies on such terms? Not to mention the fact that even for low pay in dollars some of the necessary engineering personnel could be hired for the embassy locally, but for rubles no local inhabitant would come to work for us. And taking the entire personnel from Moscow would be more expensive, even in rubles. . . .

And even in rubles there is a reluctance to pay at the real exchange rate! Our ambassador in Kiev has calculated that the family of a diplomat of his level of three persons would have to spend R14,000 a month on food alone! This was in May. And now? And the other expenses? No, we are told, let the Russian ambassador be oriented toward state supplies and state prices. And how, after this, will such an ambassador defend the rights of Russians if he himself is wholly dependent on the charity of the authorities of the host state? And I am, naturally, leaving out of the equation such a "trifle" as the organization of obligatory diplomatic receptions. . . .

And what will be the authority of such a "ruble" Russian ambassador in Chisinau when, say, Belarus has determined for its ambassador there, in addition to the ruble allowance, an extra $1,200 a month?

The Foreign Ministry is being continually pestered by the questions: Where are the Russian embassies in the "near abroad"? The Americans are represented everywhere, but we, nowhere. . . .

We are now permanently maintaining our ambassador in Kiev in an unequipped (!) hotel room, paying his expenses "per item." But were we to send only one ambassador without staff on the same terms to the remaining thirteen states, there would be nothing with which to pay the wages of the Foreign Ministry central staff.

In addition, it is not enough just to obtain a building for an embassy. It needs to be renovated, equipped, and so forth. And there is no way that we can obtain the money for this either. Instead, it is once again recommended that the Foreign Ministry halve its overseas staff!

[Gagua]: To what did you refer when you spoke about the danger of a shift of priorities in policy in the "near abroad"?

[Shelov-Kovedyaev]: I have already spoken about the two groups of countries ("7 + 4") within the framework of the Commonwealth. There are the states that signed the Collective Security Treaty together with us. These are our allies. We must be particularly attentive in relations with them.

There are states whose positions are as close as could be to ours on a number of most important issues. In Asia these are Kyrgyzstan, Tajikistan, and Turkmenistan, in Europe, Belarus and Armenia. I believe that in the sphere of bilateral relations with Russia such states should enjoy entirely special advantages. There is no discrimination in this—this is customary political practice.

It is good that such a model has come to be realized in relations with Belarus. A whole package of documents has now been prepared for signing with Turkmenistan, as has a general political treaty with Tajikistan. I believe that they should be signed without delay. Allusions to the situation in Tajikistan, say, would, in my opinion, be inappropriate here. The rivalry of North and South did not arise yesterday and will not disappear tomorrow. In addition, as far as I know, all political forces of Tajikistan would welcome the signing of a treaty with Russia and would recognize its legal validity, regardless of how the situation develops. If, on the other hand, we delay any further, it cannot be ruled out that we would soon lose Tajikistan as a state close to Russia, for which there is no justification. Armenia, on the other hand, signed the Collective Security Treaty, and we, in my opinion, are simply obliged to build our relations with it as our ally in the Transcaucasus.

Georgia has in practice recently been demonstrating the aspiration to develop a close partnership with Russia. Although it is at this time difficult to forecast the results of the fall elections, we should be supporting and developing this trend in every possible way. Otherwise, Georgia will be forced to seek other contracting partners, which would hardly be in Russia's interests.

Russia's relations with Ukraine and Kazakhstan, for example, are an entirely different topic. Our destinies are most amazingly interwoven. We are bound by such a quantity of threads and are united by such a number of common interests and are so interdependent that we should be particularly solicitous and tactful [toward them]. But concurrence does not mean total identity. For this reason it is essential that we adhere precisely to that boundary which in the name of our common interests we cannot transgress, meeting our partner half-way, for damage to Russian interests would inevitably, by virtue of the specifics of our relations, be reflected in damage to Kazakhstan or Ukrainian interests. We must defend these boundaries with the aid of the force of argument, logic, right, and justice inasmuch as we are not in the least bit any less dependent on Kazakhstan or Ukraine than they are on us.

[Gagua]: You have mentioned Ukraine. Russia's policy with respect to this country is criticized most.

[Shelov-Kovedyaev]: Yes, we are, indeed, quite often rebuked for the fact that we have responded "inadequately" to certain actions, including unilateral ones, by Ukraine.

I believe that in the past there were several reasons for such steps by Ukraine. First, considering its history, it was not enough for this country to have gained independence, it was also important for it to prove to itself that it is independent (we are observing the same thing in the Baltics also, incidentally). Second, there was a great temptation, having come face to face with growing economic difficulties, to switch to a struggle against an "external enemy," the more so when there is such a big neighbor as Russia.

I will venture to maintain that it was the absence of a stiff response on the part of Russia that helped Ukraine overcome the syndrome of doubting its own independence, and to convince itself that no one was encroaching on it, and showed its own citizens and the world community the futility of exploitation of the "image of Russia as the enemy."

As a result we have come to witness the start of a change in Ukraine's attitude toward the Commonwealth. And the shouting from Moscow by individual politicians or the parliament of Russia have merely on each occasion made the situation worse, really bringing Russian–Ukrainian relations to a critical point.

Two major problems now remain in our relations, with which all the others are in one way or another connected: the financial problem (including reciprocal payments, commodity turnover, and the fate of the ruble supply following Ukraine's transition to the hryvna) and the problem of the Black Sea Fleet.

We will leave the first to the economists. And it would be quite just to solve the question of the Black Sea Fleet in accordance with the actual Ukrainian concept of a Ukrainian navy, proclaiming that Ukraine needs a coastal defense force, and with it the Naval Symbols Protocol, which Ukraine signed on 16 January 1992.

[Gagua]: Calls for the use of power methods in the solution of Russia's relations with states of the "near abroad" arise with the question of defending the rights of Russians, servicemen included. Usually cited as an example here is America, which, without a second thought, employs force in such cases. . . .

[Shelov-Kovedyaev]: Here we encounter a misunderstanding of the fact that we must deal with the newly independent states, not only with the "old" members of the world community, on the basis of the rules of international law, which excludes the unilateral use of force. Examples of force having been employed with the sanction of the UN Security Council are well known.

It is true that the United States has in some cases acted in violation of this procedure. But, contrary to popular opinion, this was never done on the spur of the moment, without the necessary all-around preparation and without other possible solutions having been exhausted. In addition, all such actions were, as you will recall, very brief and culminated in the evacuation of dozens, and in extreme cases, hundreds, of American citizens, who, in addition, found shelter and every opportunity to continue to work back home, in accordance with their choice.

It is not hard to observe that we are in a fundamentally different situation. Where would we evacuate, for example, hundreds of thousands of people from the Dniester region? What could we offer them—a shameless solution, at their expense, consisting of the problems of our non–Black Earth region? From where would we get the money for this Great

Move from many, many points of the former USSR where conflicts are smoldering or blazing? And, most important, have we asked people whether they want to be evacuated? After all, an absolute majority are demanding observance of their rights on the land that they consider home. See how quickly refugees begin to return to hearth and home just as soon as there is the least glimmer of hope of restoring a peaceful life in the conflict zone! And those who are firmly bent on leaving their old haunts do not, as a rule, wait for an upheaval, and do not request the arrival of troops, but have long since moved to Russia.

Is it, after this, responsible to talk about staging unilateral military actions for the purpose of defending human rights? If anyone believes so, then, please, go ahead, but without me.

After all, it is clear that whereas it is still possible to theorize that the world community would perceive with understanding our unilateral military operations in one location or another, as soon as this diversion started to become protracted, public opinion would undoubtedly turn against us, for there would be only one definition of our military presence outside Russia—occupation. And no one would turn a blind eye to or forgive us our creation, all the more so in our present situation, of new Afghanistans—with all the ensuing catastrophic consequences for Russia.

If we have, indeed, recognized the priority of the law and human rights not just as fine-sounding words, the latter have to be defended by legal methods. Consequently, for the protection of Russians' rights in practice, painstaking work on the realization of the corresponding articles of treaties (which also cannot be achieved immediately and in full) is essential. It is necessary to conclude agreements that are being negotiated at this time. And these, as shown by the situation in the Polish areas of Lithuania following conclusion of the Polish-Lithuanian agreement on the rights of Polish-speaking minorities, are not in themselves a panacea. Enlisting world pressure on one country or another in civilized forms is appropriate, which the Foreign Ministry is doing.

[Gagua]: Andrey Kozyrev recently warned, incidentally, about the possibility of the revenge of the "war party."

[Shelov-Kovedyaev]: In which case the vice-president, as if confirming a report on a meeting of the Security Council, proposed that he resign.

[Gagua]: And the president observed quite sharply that Kozyrev was thereby copying the well-known demarche of Shevardnadze, only without tendering his resignation. What are your thoughts in this connection?

[Shelov-Kovedyaev]: I assume that the minister evidently had his reasons for such a statement.

I can only repeat that I see a grouping of representatives of various parts of the political spectrum that confuse the two concepts "strong policy" (that is, fruitful) and "policy of strength" (that is, coercion) growing increasingly consolidated.

[Gagua]: How do you evaluate the idea of creating a Ministry for Commonwealth Affairs?

[Shelov-Kovedyaev]: It is now possible, as distinct from the practice to date, that foreign policy work with countries of the "near abroad" could be organized with the aid of either a special ministry or the distribution of Commonwealth materials in individual territories. Both paths to me would seem wrong in principle.

The first idea is not new, although credit for its advancement was recently claimed, as unjustifiably as much else, by the speaker of the Russian parliament. A proposal for the creation of a separate Committee (under the auspices of the Foreign Ministry, or independent) or a Ministry for CIS Affairs was advanced last spring. We were able at that time to persuade people of its ineffectiveness.

The formation of such a ministry would cause a new crisis in the Commonwealth since it would be perceived by our partners as Russia ranking them as second-rate countries compared with both the states of the "traditional" abroad and also with the Baltics and Georgia, which are not a part of the Commonwealth.

Further, the formation of a ministry at a time when an analogous intra–Foreign Ministry structure has only just begun to find its feet would be sickeningly reminiscent of the unforgettable Congress of Deputies' administrative urge, which with painful consistency used to force everyone repeatedly with Manilov-like "miracle dreaming" to destroy everything that operated, with only the least bit of efficiency, in the name of a future structure that was unclear to anyone. It is clear that at a time when we may finally in the Foreign Ministry speak about the appearance of efficient departments, such a recarving would indefinitely plunge work with the states of the "near abroad" into a state of disorganization and continuous last-minute rush. Finally, there would simply not be the personnel for such a ministry, and it could not work. It is clear that no self-respecting Foreign Ministry official would go to work in such a ministry since this would mean for him severance from the Foreign Ministry and his formerly chosen profession. Consequently, a new ministry would inevitably be doomed to a long period of amateurism.

The second idea, possibly, has a right to exist in the future, but is clearly premature at this time.

Each state of the "near abroad" is simultaneously, as a rule, a part of several intersecting regional formations taking shape on Russia's borders. For example, the states of Central Asia, Kazakhstan, and Azerbaijan are part of one region with Turkey, Iran, Pakistan, Afghanistan, India, and so forth. Simultaneously, Kazakhstan and Kyrgyzstan, together with Russia, are connected with the Far Eastern region. On the other hand, Azerbaijan together with Armenia, Georgia, Russia, Ukraine, Moldova, Romania, Bulgaria, Turkey, Greece, and a whole number of other countries are part of the extensive Black Sea region. But that same Ukraine together with Belarus and Moldova is part of the Central European region. And Belarus together with the Baltic countries, Russia, Germany, Denmark, Scandinavia, and Finland form the Baltic region. In addition, the formation of a Baltic-Black Sea community with the participation of the states of the Baltic, Belarus, and Ukraine is evidenced by a whole number of indications. A complex system of relationships is taking shape in each of these structures.

On the other hand, rivalry over influence among, for example, Turkey, Iran, and Pakistan and, in the future, Afghanistan is developing in far more than the Central Asia region alone. Russia's concept for maintaining a balance of interests should, accordingly, be imminent.

Consequently, a mechanical assignment of work with states of the "near abroad" by territorial divisions would lead to an artificial and fortuitous disarticulation of the common living fabric of closely interwoven unity into individual, unconnected fragments; the forfeiture of the opportunity to advance initiatives that preserve the interdependence of processes occurring here; and a loss of correct perspective and a vision of the whole, and would increase appreciably the likely emergence of uncoordinated political initiatives. This would confuse work because it simply could not be ruled out that the same executants would be simultaneously receiving separate assignments from separate superiors.

## 3.5

**Andrey Kozyrev Outlines Foreign Policy Priorities**
*Nezavisimaya Gazeta,* 20 August 1992
[FBIS Translation], Excerpts

*Editor's note:* By August, Andrey Kozyrev was fighting for the survival of his liberal, Western-oriented foreign policy, as the following passages clearly illustrate. This article is particularly noteworthy for its honest and critical analysis of the "half-hearted" transition made by the Gorbachev regime, and his portrayal of the Soviet Union's role in the Cold War.

Franz Kafka wrote a horror story called *Metamorphosis.* It describes how one day a man wakes up to find that he has been turned into a horrible arthropod. For this reason the entire world turns away from him, and he himself becomes unbelievably contrary.

Against this we set the Bible story of Jesus Christ before he goes to Golgotha to save mankind and later ascends the mountain and is transformed into God, or, rather, Man-Made-God.

Today, a year after the triumph of democracy in August 1991, Russia and the other CIS states face about the same dilemma: either metamorphosis—a transformation from one kind of monster into another—or a transformation before becoming a true participant in the world process.

Let us not delude ourselves. This question was not resolved once and for all in August. It was only an attempt to preserve the old, to push society back to its former state, which was cut short. And that attempt was doomed. But in the White House, which was defended by tens of thousands of Muscovites, stood the bearers of the new, who were not necessarily devoted to the same alternative for the future.

What happened last August was, rather, that as they drew the final line under the past, the defenders of the White House were opening up a chance for the future. But the question of how to take advantage of that chance is being resolved today. And just over the past year two parties and two trends have emerged quite clearly.

Let us not build up some kind of suspense, as in a detective story. I hope that the thoughts that I put forward will not appear under a garish headline reading "Kozyrev Warns of New Putsch." Although I do understand Andrey Ostalskiy perfectly; it was he who took up precisely this subject as the most attractive for the reasonably good interview he did for *Izvestiya.* I do hope, however, that the conversation will turn out to be deeper and more topical.

Yes, let us hope that the putsch will not be repeated. In fact, progressive-thinking people do stand at the head of the state. The chief of them today is not at all the president who was the head of the Union a year ago. The buttress and the guarantee of democracy is the president of Russia, elected by the people. And it is even more apropos to recall the mandate that he has. It is a mandate for the democratic transformation of Russia together with all the territory of the former USSR, and on the broader plane, the former socialist camp. This is precisely why Russia looks with such hope toward the war-stricken peoples of Yugoslavia, and the peoples of Eastern Europe, painfully feeling their way into the future.

But the resistance offered by opposing forces should not be underestimated. The simplest thing is to be aware of the communist opposition and to pity those who wave red flags because they are associated with decades of heavy labor and hopes, and we should be making a deep obeisance to their exploits, even if they were worshipping at the wrong altar.

. . . It is even simpler to return along with the four-square defenders to the bosom of the CPSU [Communist Party of the Soviet Union]. We can argue until we are hoarse about whether their choice is constitutional, whether the past of the party was constitutional. History itself has pronounced its verdict: Any power that does not come from the people and does not rest on honest and free elections, but on usurpation by a small group of those who claim to possess an absolute truth, is unlawful from the standpoint of contemporary legal and political consciousness. In the twentieth century there is no place for dictatorship either of the fascist or the communist ilk.

It is, however, as if it is precisely the Russians and the other peoples of the USSR, who made a decisive and heroic contribution to the rout of fascism and, in August 1991, the destruction of the communist system, who must face a decisive duel with a third monster that lies in wait for humankind at the close of the present century. This is the plague of national-patriotism or, more accurately, aggressive nationalism.

The great Russian philosopher Berdyaev once said that communism had become the ineluctable fate of Russia. He traced the metamorphosis to totalitarianism from its feudal-tsarist and serflike socialist forms. Today there is a real danger of a new metamorphosis.

And again we clearly sense some kind of inferiority complex in the face of the difficult task of entering the surrounding world. In 1917, convinced that Russia was the sick man of Europe incapable of making it through to the end of the Great War, the Bolsheviks proposed their own primitive but enticing and simple solution. The essence of this solution was that instead of the inevitably prolonged and difficult democratic process of transformation, there would be a cavalry charge to reach (or so it seemed) obvious objectives. Peace for the people, land for the peasants, bread for the hungry—and all of this decisively and boldly, through direct action backed by force. It was in this way that they carried out collectivization and brought order to Czechoslovakia and Hungary, and later to Afghanistan; this was how they strengthened the Union. . . .

And people who moved along a different path evoked a growing hostility, irritation, and abuse. The very fact of the existence of a gradual but steady flowering of liberal, "babbling" democracies was an unacceptable challenge to totalitarianism. Outside it was democracy and the democrats of Nazi Germany that were perceived as Public Enemy No. 1, while inside the communist Soviet Union it was the bearers of these ideas, as a despicable domestic "fifth column" that was interfering in "how things were run."

Even during the decline of the Soviet Union this aggressive rejection of the surrounding world and the desire to force on protective armor against it began to be clothed not in the doomed apparel of Marxism-Leninism but hurrah patriotism apparel. We recall that both in the notorious "Word to the People" and in the appeal issued by the State Committee for the State of Emergency, the emphasis was laid not so much on the defense of socialism against the intrigues of imperialism but, rather, on the loss of superpower status.

Here we have the very essence of a harmful consciousness. Aware of its own inadequacy in today's civilization and its own inability to assert itself in the surrounding world, it seeks a way out through confrontation with that world, thus affirming its own being and ultimately its own right to existence. Properly speaking, this is the deep nature and motive force of any aggressive nationalism, which now threatens not only us but also Eastern Europe and even Western Europe.

It is time to recognize that this threat lies in wait not only for communist and totalitarian, but also democratic societies, especially if the democracy in them has not sufficiently matured. The most fertile ground for the carriers of this virus is the instability of national-state consciousness. The classic example was Germany in 1938, in which, instead of making the difficult entry into the League of Nations, which was being formed at that time, and learning to live according to civilized laws, Hitler chose the shortest route to revival of imperial might and to national catastrophe. . . .

Is this so remote from us in history? The Russian political climate is over-saturated with vituperative discourse about loss of national achievement and status as a great power, and the possibility of becoming one, or several, banana republics. And it is not only the national Communists but also, I would say, the national democrats who are demanding individuality at any price and are looking for justifications for it.

Let us look a little more closely at these arguments. And not just for the sake of polemic, but first and foremost because in many cases it is a question of the real national-state interests, but turned on their heads, and of a neo-confrontational, pseudopatriotic consciousness.

The popular thesis rests on our Eurasian nature and the impermissibility of a pro-Western bias. In fact, if we talk about the West in the geographical sense, then a one-sided orientation toward only our European neighbors would be unjustified. Even though the task remains of joining the European structures and establishing relations with the NATO Euro-Atlantic alliance, which includes the United States and Canada in addition to European countries. And the West as a political concept is the East for at least 70 percent of Russia. The United States, Canada, and Japan are our nearest neighbors in the Asia-Pacific region.

But it is not a matter of geographical discoveries. The attempt to close off America and the entire Western thrust of policy as a priority is transforming the Eurasian nature into an Asiatic one, and Russia as a great power in the heart of Eurasia from a natural connecting link between the highly developed democratic states with market economies (that is what the West really means) into a missing link, the sick man not only of Europe but also Asia, where firm trends toward "pro-Western" development are seen, along with integration on the basis of eagerness for the market economy and the modern organization of society. Whereas the communists are sickened by this direction in foreign policy out of ideological considerations, it evidently frightens the national democrats just as college scares the freshman.

In fact, to use the expression of the Russian historian N.M. Solovyev, Russia has a right to a place alongside the highly developed states and the leading representatives of humankind. However, because of the monstrous zigzag that resulted from the previous attempt to build a home-grown paradise in a single, isolated country, we must make up what was lost in a number of fields and engage, as it were, in external studies to follow the course of civilized behavior here at home and in the world that surrounds us. Neither Peter I nor the thousands of leading minds of Russia in the following centuries were afraid to learn from others. This, however, is humiliating for our new-made democratic state figures, who obviously believe that it is better to be a big fish in a small pond than a small fish in a big sea. Perhaps this explains the desire to make advances to the national-patriots who through their own clamor, particularly within the walls of the Russian parliament, seem to be a much more impressive force than in society as a whole, and to the most backward and reactionary part of the military-industrial complex, and indeed simply the industrial complex, which fears more than anything else in the world the loss of its privileged position in a closed economy and finding itself facing foreign competition.

It is clear, however, that instead of joining the community of democratic states and correspondingly the highly developed world economy, continued autarky would be tantamount to national betrayal and Russia's final slide into the category of a third-rate state, even though by dint of its geographical position and its scientific and technical and resource potential it is called upon directly to make the dash into the club of the elite.

So, dear fellow citizens, shall we walk barefoot into the city for a university education or shall we take delight in the fact that we have never graduated from university, and for that reason "the Soviets have their pride?"

In practice, however, an enormous and powerful country like Russia is simply unable, even when so commanded by parliamentary patriots, to remain on the sidelines of history and idle away its time as some kind of international ignoramus. Willy-nilly it is being drawn into big policy, and on a global scale. And here it turns out that the issue is the direction of that policy: whether it is, as before, opposed to the leading states of the world community or is in alliance with them. It goes without saying that it is a question not of a new military-political super-alliance but of an alliance of states sharing common democratic and market values. . . .

The confusion about the notorious Slavic factor in foreign policy, however, prevents our national democrats from responding precisely and clearly to this simple and apparently convincing question. There is no question that this factor should to some extent be considered in the policy of a state with an enormous Slavic population. But this by no means applies only to the Serbian national Bolsheviks, who incidentally are acting not only against the national interests of the Serbian Slavs themselves, but also against all the other Slavic peoples of former Yugoslavia. We shall not forget that even the Muslims in Bosnia are still Slavs even though they profess a different religion. All the former Yugoslav republics look to Russia with hope and, as I have had the opportunity to be convinced during a trip across former Yugoslavia, with love and expect from it a fair position and protection from the game of war, which, of course, is going on not only in Serbia but also Croatia and Bosnia, and Herzegovina. . . .

Let us consider the scenarios proposed within the framework of the so-called third, special path for Russia, as applied to the political-diplomatic game being played on the very important chess board of the UN Security Council. It is precisely here that Russia has succeeded in inheriting the status and privileges and simultaneously the responsibility as one of the five great powers that are permanent members of the Security Council. Let us not delude ourselves; this by no means happened automatically. There could have been different scenarios. But the concept of the USSR's inheritor state in the United Nations was worked out by Russian diplomats in the closest contact with their English and other Western colleagues. Behind all this could be palpably felt the credit of trust in the democratic leadership in Moscow, and that in the person of the new Russia the international community would not be obtaining a somewhat diminished version of the former Soviet Union, which for many years blocked the work of the Security Council, including by use of its veto, guided as it was by an anti-West and anti-imperialist doctrine.

The civilized world had already welcomed the policy of new thinking that led to the USSR's abandonment of its blocking, confrontational line. By dint of its own halfheartedness, however, this policy did not lead to the emergence in the Security Council of a firm alliance of democratic, highly developed states. As was typical for Soviet diplomacy of recent years, the tactic was employed of slowing the increasing work of the Security Council in standing firm against states that violate the peace or (and this is usually the rule) those same regimes that violate human rights. It is not happenstance but a law-governed pattern that many of these regimes were the ideological allies of the socialist USSR. But the Soviet leadership of the perestroika years was determined that it would not move, as in earlier years, to direct patronage of them through confrontation with the West. However, it lacked the spirit to move not in words but in deeds, in this case through diplomacy, to switch to the other, civilized and democratic side of the barricade. Therefore, as a rule, before a decisive vote the Soviet delegation would remain vague for a long time with respect to its own final vote, sometimes taking time out several times for consultations with Moscow or some other kind of demarche pertaining to its own former clients.

Of course, in particular, specific cases it is appropriate to use old connections and channels of influence to try once again to achieve by peaceful means what must otherwise be extracted from former allies with the aid of sanctions and other harsh measures by the international community. But let us look the truth in the eye: In most cases all these demarches were made with quite clumsy maneuvering that counted not so much on real influence on the "intractable" but rather on fence-sitting. It is not happenstance that this kind of body movement failed to win real respect from either side, even though the West in general, of course, was grateful to Gorbachev for the fact that, in contrast to his predecessors in the post of leader of the Soviet Union, ultimately he did not prevent adoption of the correct decisions. Incidentally, socialist China has also been employing the tactic either of abstaining or voting in favor in the Security Council.

Today, however, more is expected from democratic Russia and the democrats in the Kremlin: real alliance with those who stand guard over international legality and are ready to make use of the most decisive measures for this, not by unilateral military intervention of the type made in the USSR's Afghan and Czechoslovak adventures, but in accordance with the UN Charter. I am convinced that this line (and in the latest votes in the Security Council it has started to take shape in the form of a firm Russian "yes") will evoke great respect both from its new allies and from its former "ideological friends" who, like the former Soviet regime, understand only a sharp tongue.

Really and truly, criticism of the present course from the national Bolsheviks, who are essentially demanding a return to the old position of the Soviet leadership—open opposition to the democratic West and alliance with all forces that oppose it—evokes great sympathy for our role. At least it is clear that it is possible to have if not those, then some other allies. And if we do not stand in the ranks of people of probity, then at least we do not head up a gang of brigands. But obviously neither the one nor the other is enough for the "national democrats," who—as, for example, in the case of sanctions against the Belgrade national patriots—are proposing abstention in the Security Council vote. The logic is in general quite simple: The "democrats" do not want to join hands with the Bolsheviks while the "nationals" are sickened by alliance with the cosmopolitans when the issue goes beyond rebuffing power policy, and with the West. And no references to the "sacred cow" of Serbian-Russian friendship achieve anything here. All we have to do is look at the example of France, which, perhaps, no less than Russia has ancient historical ties with Serbia. It is not by chance that much in the actions by the French is similar to what we are doing. Before voting in the Security Council for the sanctions, the president of Russia sent his own foreign minister to Belgrade and Sarajevo. After the sanctions had been adopted, the president of France himself journeyed to the same places. After becoming convinced that it is precisely Belgrade that has carried and still carries the burden of responsibility for the bloodshed, Russia voted for the sanctions in the Security Council. And France, through the lips of its own president when he returned from his trip to Yugoslavia, spoke in favor of further decisive measures against Belgrade. I think that the vision of both Serbia's and France's long-term interests had an effect here, and also the self-respect of democratic France. I am convinced that out of the same considerations Russia must not go back to a policy of slowing things down, but must go forward so that a Security Council resolution on measures against those who violate the peace and human rights can be put forward not by three powers—the United States, England, and France—as now, but by four great democratic powers. If China with its political abstention in the vote can understand how far in the case, for example, of Yugoslavia, it goes beyond the zone of its traditional interests, then as applied to Russia, this kind of policy was totally at variance with those very special historical, Slavic, if you will, links that we have had and will have with the peoples of former Yugoslavia. . . .

Let us, however, return to the "Slavic factor." Can it, like Orthodoxy, become a real and dominant factor in Russian foreign policy? I think not. First, because today no state is guided by such criteria in its own foreign policy if, of course, we are talking about the civilized, democratic countries. In Europe any attempt to divide ourselves into Slav, Germanic, or French societies threatens to return the situation not to World War II but World War I, and, as applied to the Yugoslav tragedy, a repetition of history in making Serbia the detonator of a global catastrophe. Second, this kind of ethnic-religious foreign policy would be anti-Russian since it would promote a split in Russia itself, where along with the millions of Orthodox Slavs, peoples of other nationalities and religions live, in particular the Muslims. And if hotheads among our so-called patriots are prepared even from the tribune of the parliament to state that under all circumstances Moscow should sympathize with the Serbian militia units in Bosnia and Herzegovina, then why should Russian Muslims not be just as reckless in defense of their fellow believers in that same Bosnia and Herzegovina? Then where would Moscow be? And is this kind of policy not tantamount to shifting the civil war from Yugoslavia to our own motherland? . . .

Nor should we forget that any onesidedness, national patriotism with a red Slavic-Orthodox taint, just like, incidentally a Turkic-Islamic taint, is a scandalous contradiction of the indisputable priority in Russian foreign policy of strengthening good-neighbor relations with "close neighbors" and ensuring the smooth running of the CIS mechanism. At the same time it is precisely in this former Soviet space that no better use can be made of the geopolitical and other advantages of Russia as a great Eurasian power. Both for republics with a predominantly Slavic population and states with an Asian population there is no more natural partner than Russia, which by its very nature cannot be anti-Slav or anti-Muslim, just as it can be neither pro-Slav nor pro-Muslim. What is more, for both the former and latter, Russia will be an attractive and really valuable partner just on the basis of the principles of civilized behavior—those same principles that were worked out and are defended by the CSCE. We earned the gratitude and respect of our neighbors when we actively helped them become full-fledged members of the all-European process. Continuing this line, we shall pursue a truly patriotic policy of democratic transformation both for Russia itself and for the former space of the USSR, and the creation of a community of independent states based on respect for human rights, including national minorities, Russians, Ukrainians, Germans, Uzbeks, Tajiks, Jews, on a common basis of truly democratic institutions of power, equal states, and a market economy. . . .

Look once again at last year's instructive history lesson. If the August putsch finally alienated the republics from the idea of the Union, then the new Russia's adherence to democratic and economic reforms has earned the trust of its neighbors, and it was this that enabled it to move to the Dagomys and Yalta agreements with Ukraine and to a number of coordinating mechanisms for the Commonwealth of Independent States, and to military-political alliance with many of those states, including states in Asia, that just a year ago were apparently ready to look for an alliance with any of their neighbors to the south, but not to the north.

It is also time to remember yet another trend that is capable of reversing Russian foreign policy. This is replacing reckless ideologization for a short-sighted "economization" that is just as reckless. It is not only here that the pressure is felt, and perhaps not even so much from traditional ideological partners but rather from international outcasts—the industrial and military-industrial circles in our country. That same desire can be palpably felt to avoid and cut short the difficult stage in moving to a highly developed society, only in this case not in the political sphere but in the economic sphere. In fact, it turns out that it is not a simple matter to learn to trade with the countries that are more advanced in economic relations, or to penetrate the highly competitive markets that they have chosen. So there is a temptation to accuse them of a reluctance to meet us halfway . . . and in this way to look for even some temporary easy profit from old clients who, under conditions of international sanctions that are damaging to them, are not so fastidious. With feigned naivete we often hear the call: Why is it that sanctions are being introduced against our trade and financial partners and not the West's? Does this not provide proof of some dirty game? Meanwhile, the answer is well known—like master, like man. The fact is that under the communist system, which stifled the Soviet economy in its embrace, there could be no other partners. And today, without changing our benchmarks in domestic and foreign policy and without setting a course toward a change of partners, it is impossible to count on any worthwhile place in the world economy, or on any final elimination of the suspicion among the Western business world, or the complete dismantling of Cocom [Coordinating Committee for Multilateral Export Controls] barriers, and so forth. Otherwise we must again set everything on its head as in these previous seventy years and ascribe the guardedness of the highly developed West as the reason for, rather than the consequence of, our own inability to manage things in a civilized way and show the necessary discernment in the selection of our partners. . . .

In this connection we must speak of the policy of weapons exports. The fact that, even today, for us this is one of the most profitable, unchanging items of foreign trade evokes no doubts, no reproaches. In this field as perhaps in no other, however, there is a great danger of sliding onto the old rails, once again not out of ideological considerations but because of oversimplified economic considerations, when it turns out to be a difficult business to penetrate new weapons markets, where stable pro-Western countries (and these are most often best able to pay) do not acquire weapons to support international terrorism or realize aggressive plans by regimes at war with their neighbors and with all civilized peoples. Whereas the latter only dream of a return to the times when Russia used to supply the weapons of war, more often than not on credit (would that those debts could be repaid now!), the former just as naturally want to convince themselves that the old practice has ended and that the new Russia really does want to build its policy in a new way in particular regions.

And here it is not a question of ideology or the intrigues of the West, but the basic instinct for survival among the new purchasers of Russian weapons who at the same time become the victims of belligerent neighbors armed previously by the Soviet Union. The classic example here is the Persian Gulf, where even today the urgent task is to remove once and for all the threat of a repetition of the Iraqi aggression, and to secure guarantees for the elimination of chemical and other weapons of mass destruction from the Iraqi arsenals. Incidentally, the latter were for us not only military exports but also a geostrategic interest, because destabilization in that region and activation of extremist forces there, whether of the pseudo-revolutionary or Islamic fundamentalist ilk, and even more their possession of very destructive weapons and the means to deliver them over great distances, are capable of spreading with dizzying speed to the vulnerable regions of the Caucasus and Central Asia.

Concern about non-proliferation of missiles and missile technology for military purposes is also a priority for Russian foreign policy in South Asia where considerable potential for conflict exists, and even great and friendly powers, such as India, have still not subscribed to the regime of nuclear non-proliferation. So it happens that any deal, particularly when it is a question of "sensitive" technology on our southern borders, must be weighed on very accurate political scales, not bazaar scales. Only if this is done will it be possible to preserve the military–political links that fully justify themselves, as, for example, with India, and at the same time move on to promising new partners that are not subject to sanctions.

In sum, all of this will be the policy of a great but normal rather than ideologized great power—a power that has respect for itself and enjoys the respect of the international community, a power of which it might be said "tell me who your friends are and I will tell you who you are," without the fear of being assigned to the category of international riffraff. For half a year we have been convinced through positive and negative experience that normal policy, and in general normalcy as a way of behavior abroad, both near and far, has nothing in common with primitive, cavalier, sudden attacks, and improvisation. After seventy years of exclusivity, normalcy is something that we still have to learn. And paradoxical though it may seem, it requires the highest level of professionalism, whether in domestic policy or foreign policy.

When issuing a statement several months ago on behalf of Russia about entry into the Council of Europe, I especially emphasized that whereas in form it was a question of one country's joining a club of other states that already existed, what it was essentially was the beginning of a process of rapprochement and mutual adaptation. The Russian Federation is not simply one of the new members but an entire continent with an enormous and original political, historical-cultural, and economic potential. No one disputed this, and no one will if, of course, we are talking about serious policies for both them and us.

The question lies elsewhere: Will we succeed in not setting our originality in opposition to the democratic community, but in realizing it through cooperation with it? And here it is not a question of playing up to the West. It has learned to live in a civilized and prosperous way without us. For seventy years we also demonstrated our ability to live, but not by the common laws of democracy and the market. And now we can obtain financial stabilization assistance from the International Monetary Fund, while the European Community gains the respect of the Belgrade nationals. This means that once again we have seen demonstrated the adage: Know yourself! It means that once again the "Westerner," who wants a transformed Russia to take its special place not opposite the club of the most advanced states, and not alongside it, but within it, can finally retire with a quality of life that will not make matters worse for Russians.

Of course, instead of national-patriotic political intrigue, true democratic policy faces many objective and subjective difficulties. It would be erroneous to depict the experience gained as only positive. There have been, are, and will be mistakes because only he who does nothing makes no mistakes. Let us emphasize once again that discussion within the framework of this choice and the search for the most effective scenarios are what are most needed both for the government and for society. One problem is the still uninformed nature of public opinion with regard to foreign policy. It is still sometimes difficult to distinguish in journalism between a fondness for various kinds of sensationalism and superficial assessments in the spirit of "rightist" commentators in *Pravda* and *Sovetskaya Rossiya*, the newspaper *Den*, or the "leftists" from *Kuranty*. Nevertheless, there is no alternative to real free speech, which ineluctably brings the dross to the surface. Similarly, there is no alternative to a foreign policy that meets the real national interests.

Ideology is being replaced by a normal vision of geopolitical interests. Today patriotic foreign policy is the persistent realization of four or five main priorities that make it possible to create the most favorable conditions for domestic democracy and the economic transformation of the Russian Federation. First, there is the entry as a great power into the family of the most advanced democratic states with market economies, the so-called Western society, if you will. They are just as much the natural allies of a democratic Russia as they are the sworn enemies of the totalitarian regime, whether it has red, red-and-brown, or simply brown banners. The second priority, which coincides 70 percent with the first in terms of whom it addresses, is the formation of good-neighbor relations with all states along the perimeter of the Russian border. Here, of course, a special place belongs to the CIS members and the major powers of Asia, such as India and China, and neighbors to the south of the CIS. Both these priorities could not better match the third priority—the development of mutually advantageous economic links and attracting investment and support for the domestic economy. The fourth priority is to safeguard human rights, especially for the Russian-speaking populations in neighboring republics. Effectiveness in resolving this task will depend largely on the status that Russia wins for itself in the first two priorities. And finally, the fifth priority is the balanced use in the first four priorities in foreign policy of the advantages of the Russian Federation, such as the unity and diversity of the ethnic groups, religions, and traditions. We have an unprecedented opportunity to be Asians in Asia and Europeans in Europe, and democrats in the world in general. Democrats first and foremost! Loyalty to the democratic choice in domestic and foreign policy means loyalty to the choice for which Russians voted at the first national elections in their history for a president, and which they defended at the barricades raised around the White House on 19 August 1991.

## 3.6
## Address by Andrey Kozyrev Before the Russian Supreme Soviet

Russian Television Network, 22 October 1992
[FBIS Translation]

*Editor's Note:* On 30 June 1992, Andrey Kozyrev warned that a right-wing coup was possible in Russia—by national patriotic forces this time, rather than by communist conservatives. The national patriots were accusing Boris Yeltsin of "over-Westernizing" Russia's foreign policy. An isolationism was emerging that stressed native *(samobytnyi)* traditions and questioned the need for Russia's "return" to Europe as well as reintegration with the world economy. Under pressure from these patriots, Kozyrev drafted a foreign policy even though he basically opposed putting his policy in a single document as too ideological and reminiscent of the Soviet program. In July, however, the Supreme Soviet returned the document to the ministry for revision. Meanwhile, the parliament's Council for Defense and Foreign Policy issued an alternative policy, which emphasized good relations with Western Europe and China, and a steering away from America. It took a pessimistic view of Russia's economic outlook and talked of "encirclement," which could be mitigated by pursuing an "enlightened post-imperial integrationist course" toward the other former Soviet republics, many of which the document said had a "weak historical legitimacy" in terms of borders, ethnicity, and economic development.

By October, feeling his grasp on foreign policy beginning to slip, Kozyrev made a vigorous defense of his ministry's policies to the Russian Supreme Soviet. His speech, which follows, is a strongly worded protest against the adoption of a "Eurasianist," or eastward-leaning, foreign policy. At the same time, however, it tries to mollify the patriots by stressing the importance of Russia's "peacekeeping" role in the "near abroad" and the need for a consistent defense of Russian and Russian-speaking peoples in the surrounding states.

Esteemed People's Deputies! I would like to begin with what worries me most of all. There is the danger that our debate on foreign policy will become something other than the search for the best way of realizing the interests of the country. Under the cover of slogans like "the third way," or "enlightened patriotism" (very similar to socialist patriotism), the choice of a democratic path of development adopted by the First and confirmed by all subsequent Congresses of People's Deputies of the Russian Federation is questioned: the choice of transforming Russia into a democratic country and an equal member of the community of democratic states of the world.

I must say, the government and the president adhere to this choice, and President Yeltsin, speaking from this rostrum, has confirmed this in a recent speech from this rostrum. I perceive it as my duty as Foreign Minister—confirmed by the Supreme Soviet and in the democratic program for reviving Russia as a great power open to the world—to defend the choice your body formulated both here and in the United Nations, and in meetings with the media. A return to confrontation with the world around us, especially in our current state of transition, threatens to lead to a national catastrophe. I am sure that the common sense of the Russian people and their chosen deputies will recognize the threat and will not allow a step backwards, either in domestic or foreign policy.

Esteemed People's Deputies! I do not agree with the panicky and defeatist mood that says that Russia has become a "banana republic," a "third-rate power," and that nobody takes us seriously. My attendance at the UN General Assembly session convinced me of exactly the opposite. Dozens of meetings with foreign ministers and heads of states were imbued with a feeling that they regarded us as a great power and want to do business with us and under no circumstances want us to withdraw from the international community. While my ministry agrees with this, the so-called "near abroad" (*blizhnee zarubezhye*) remains our first priority.

You all probably know the outcome of the Bishkek summit.[1] I will limit myself to only a few remarks about it. On the whole, the summit indicated a turn toward re-integration. On the other hand, this re-integration is occurring in different ways in separate republics. Some republics are prepared to go much further toward integration than others. Probably Armenia, Belarus, Kazakhstan, Kyrgyzstan, Moldova, Tajikistan, and Uzbekistan are ready to move toward integration in the largest number of areas. I do not have to mention Russia.

I must say that our peacekeeping efforts have achieved high respect and are seen as a reserve of stability in the world. You know that some of our partners are at various levels of internal political struggle and conflict and are in the habit of accusing Russia of either lacking involvement or literally the next day accusing her of interfering too much in their affairs.

First they ask for Russian peacekeeping forces and then they accuse them of interference. Of course we are able to carry out this policy partly as a result of our prestige in the world and in the UN Security Council. Once again, this shows how closely countries near and far are linked.

A few words on the painful problem of the Baltic region. We are very glad that the Committee on International Affairs has examined issues linked to Lithuania. But Lithuania is a more or less benign case. Things are much worse in Latvia and Estonia, not so much with respect to the withdrawal of our troops, but with respect to the defense of Russians and the Russian-speaking population. I have provided you with many notes and appeals on this issue.

I would note two aspects of this problem. The first is that our efforts to mobilize international pressure in support of Russian minorities are beginning to show tangible results. Second, the internal struggle in these countries is beginning to recognize the issue of human rights. This is good, and allows us to exert our own pressure not only on foreign media but internally. Of course, this issue will require a much more sophisticated approach in the future and more delicate political work.

**Note**

1. The Kyrgyz capital, Bishkek, was the site of the October 1992 summit of CIS Heads of State and Government. The summit was noteworthy only for its failure to achieve a consensus on what Russia considered the vital question of a unified CIS armed forces and a draft Collective Security Treaty. By October, only six member states had signed the Collective Security Treaty which Russia proposed at the 16 May summit. The Russians also tried to secure an agreement on monetary, credit, and currency policies at the October summit, but only managed to secure commitment to a consultative committee, which Ukraine did not approve.

## 3.7

### Boris Yeltsin Addresses Minsk Summit

Moscow Ostankino Television First Channel Network, 16 April 1993 [FBIS Translation]

[Address by Russian Federation President Boris Yeltsin to representatives at the CIS heads of states summit in Minsk on 16 April—recorded.]

In various political circles, and indeed, in the mass media, they have started speculating of late that the CIS is not capable of resolving any issues at all and almost that the CIS is moving toward disintegration. Of course, this is wrong; therefore, all we heads of state agreed to meet and generally to map out further steps to strengthen the Commonwealth of Independent States. In addition, we have agreed to meet in May in order to make specific decisions relating to the measures already drawn up. The matters facing the CIS now are posed by life itself, and the well-being of our peoples depends to a large degree on these matters being resolved. During the past we gradually realized that it is not practical to regard the CIS merely as a mechanism for a civilized divorce. The euphoria of standing apart and shutting oneself off as states is passing; the striving for restoration and development of economic and political cooperation is becoming increasingly evident. This is not surprising either. Our states are choosing their path of development independently without any force or pressure from outside.

The formation of the Commonwealth did not cause a single conflict between the former Union republics. On the contrary, over the past year, we have acquired our first experience in peacemaking actions on how to put an end to shooting and bloodshed and how to stabilize the situation. Proceeding along the path of strengthening their independence, our peoples are not losing common sense and are maintaining kind feelings toward one another. There is no need to prove today that the CIS countries are united in many more ways than they are divided. The objective community of interests is the basis for this. At our previous meeting, we all unanimously spoke in favor of improving the effectiveness of the work of the CIS. Seven participants of the CIS adopted the Charter of the Commonwealth. Yesterday, the Russian Federation Supreme Soviet ratified the Charter practically unanimously. The main thing today is to give new impetus to the development of the Commonwealth. Its childhood is over; the time has come for it to stand firmly on its own feet.

First, the time has come for the Commonwealth to form its own opinion in world politics. We believe that the formation of a common position with regard to key international problems is both possible and necessary. Why not adopt the practice of putting forward joint initiatives on foreign policy? We can start with issues that are in the interests of all. For example, joint actions on key issues of world politics like preventing weapons of mass destruction and means of delivering them, strengthening international stability, settling regional conflicts, protecting the environment, fighting against terrorism and illegal drug trafficking, as well as putting forward a joint proposal of the Commonwealth members to set up CSCE peacekeeping forces. Another possibility is organizing top-level meetings—for instance, of the Commonwealth foreign ministers with representatives of influential international organizations, such as the European Community, the Council of Europe, ASEAN [Association of Southeast Asian Nations], and some others.

Second, the key to resolving socio-political issues confronting our countries lies in the economic sphere. The time has come—and I think I shall voice a general opinion by saying this, since we have arrived at the same opinion today—to clear away most resolutely all obstacles on the way toward mutually advantageous economic cooperation. The main thing is to establish direct and, more important, free links between enterprises, producers, and consumers of goods, and their direct interaction. I have just returned from the congress of industrialists and entrepreneurs, in which about 5,000 people took part, and they instructed me to tell heads of state here that, without developing further cooperation, the market and the reform will of course progress with great difficulties. A tendency toward forming mutually advantageous unions whose members share the same interests, toward creating various transnational corporations and associations in industry and agriculture, power engineering, transport communications, and in the service sphere is gaining momentum. The task now is to support these trends in an active and concerted manner, and then we can be sure of success.

Third, we need multilateral coordination in the development of individual key industries in the national economy. The Commonwealth already has certain experience of this kind. The CIS electric and power engineering council has been set up. Already operating are a railway transport council, an interstate space council, a regional commonwealth in the field of communication, and a number of others. Agreements on the interstate oil and gas council signed in Surgut have considerable potential. Essentially, we are moving toward industrial cooperation in this sphere, including capital construction, effective use of oil and gas pipelines, scientific and technical matters, etc.

Fourth, the necessity to cooperate on investment is becoming more obvious every day, inter alia to guarantee the development of deposits in Western Siberia, on Sakhalin, on the Barents Sea shelf, and others. We can remark upon these processes or not, we can hinder them or assist them. But we cannot stop them. The development of integration requires free trade, without customs barriers, movement of goods, services, capital, and labor resources, a single monetary system, a more or less uniform legal environment in which enterprises and all sorts of entrepreneurs' structures can work, the formation of a common economic policy or at least its elements. Creating a single economic space and a common market envisages CIS countries setting up an economic union on the basis of a customs and currency union.

The customs union will make it possible to remove customs restrictions of all kinds among signatory countries, and to introduce a single customs rate along the outer CIS perimeter. The difference in price levels will disappear, and the uncontrollable speculative channeling of goods across today's transparent borders will cease.

A few words about a currency union. The ruble zone has effectively disintegrated. As we know, considerable difficulties arise with the introduction of national currencies. That is why we have already put forward our proposals for introducing relevant draft agreements on a single system. Russia is ready to continue with talks on this issue. But in the end we have to decide between either demarcating and creating our own monetary systems, or integrating. There is no third option: preserving the present situation would be dangerous for the economies of the Commonwealth countries.

It is clear that to tackle the range of economic issues we need to create a special permanent committee with appropriate powers. I am sure it is possible to turn the Commonwealth into an effective interstate association capable of coordinating interests and uniting efforts in the cause of economic progress.

Strengthening military-political cooperation seems to be topical. An effective system of collective security is probably not possible without forming a defense union, although not everyone agrees with this. A treaty on collective security can become the core of this system. It is important that the treaty come into force and the organizational structures envisaged by it be created in the shortest possible time.

It is worthwhile to think about establishing a mechanism for its implementation that would take into consideration the possibility of the CIS countries who have not signed the treaty joining it. We should also clearly define the role of the commander-in-chief in the organizational structure of the military-political leadership of the joint armed forces of the treaty member states. The possibility of setting up joint armed forces in the future, perhaps, with the participation of some states under a single coalition command, should not be ruled out.

The development of events inside the CIS and on its borders is more and more urgently demanding joint peacekeeping actions. To regulate and prevent armed conflicts, a set of concrete collective measures should be drawn up within the CIS framework and appropriate mechanisms set up, using wherever appropriate the experiences of the United Nations and the CSCE.

Russia is prepared to take part in peacekeeping actions, but of course it cannot bear the burden of responsibility, both moral and financial, single-handedly. In this connection, I would like to draw your attention to the fact that similar accords are not always fulfilled, putting it mildly.

In order for the Commonwealth fully to live up to its name and fulfill its purpose, it is necessary to ensure strict observance of human rights in all the CIS states. An analysis of local conflicts in the countries of the Commonwealth leads one to conclude that in practically all the primary cause has been violations of human rights or the rights of national minorities. This is of course a very difficult and delicate problem for our Commonwealth. But I am convinced that there would be no friction or conflict if all members of the CIS strictly observed internationally recognized standards of human rights and freedoms. So we call on all our partners to accede to the Universal Declaration of Human Rights and UN pacts on human rights as soon as possible.

We believe that one of the priority areas for the implementation of the CIS Charter is to ensure that a commission on human rights is established and starts work as soon as possible. We also propose that a regional convention to protect persons belonging to national minorities be concluded.

Esteemed council members, it is obvious to Russia that the Commonwealth ought to become an effective interstate association, whose members play a real part in various forms of cooperation and bear their portion of responsibility. Such logic leaves no room for living at others' expense or attempts to derive one-sided advantage at the expense of others. At the same time, it completely rules out the notorious habit of encouraging disciplined partners and punishing the negligent ones. The charter of the Commonwealth of Independent States is essentially aimed at this. The principle of it is simple and comprehensible—equal rights, equal obligations, equal advantages and benefits.

Let's call a spade a spade. He who does not sign the charter will in effect remain outside the main channel of cooperation within the framework of the Commonwealth, with all the consequences that stem from that. It is possible that these countries are placing emphasis on developing bilateral relations. They will find an honest and reliable partner in Russia, with whom they can cooperate on the basis of generally accepted norms of international law. The choice is open to each state.

Participants in the CIS interested in deepening cooperation on joint settlement of economic and other problems could conclude an agreement within the framework of the CIS, of course, after it had been prepared, possibly in May. The signing of agreements of this kind does not mean resuscitating the unified state or returning to the former Union bodies of authority and management. Participants in the conference retain their independence and sovereign rights. We are talking about giving our relations new momentum and multiplying our forces to tackle everyday problems. We have talked about and discussed many of these issues today, when only heads of state were at the first part of the sitting. Thank you all for your attention.

## 3.8

## Andrey Kozyrev on Russia's Peacekeeping Role in the CIS

*Nezavisimaya Gazeta,* 22 September 1993
[*CD* Translation]

One sometimes gets the impression that the world community underestimates or doesn't understand the seriousness of the threat that conflicts in the area of the former USSR pose to the cause of regional and world peace. At the same time, one can see a certain reticence, if not suspicion, regarding Russia's role in settling these conflicts.

There must be complete clarity on these points.

Owing to its close historical, political, cultural, and other ties with the neighboring states, Russia could not—nor did it have the moral right to—remain indifferent to their requests for help in ensuring peace. Back when the world community was just beginning to take a closer look at the "hot spots" in the former USSR, and sometimes simply trying to find them on the map, Russia was the first to provide mediation and its good offices in searching for ways to reach settlements and to aid the populations suffering as a result of hostilities—among them several hundred thousand Russians and Russian-speakers. Russian Federation soldiers and officers have had to stand between the warring sides in order to stop the bloodshed, and in some areas they still are doing so. But in all cases, without exception, Russia's peacemaking and peacekeeping efforts in the former USSR have been made with the consent of and at the request of the parties to the conflicts, and its peacekeeping contingents have been sent into the conflict zones on the basis of appropriate agreements. These agreements are not at variance with the UN Charter. This was the case in South Ossetia and the Dniester region, and operations in Abkhazia and Tajikistan are being readied in accordance with the same principles. The United Nations and the Conference on Security and Cooperation in Europe are regularly briefed on these efforts. Moreover, there is a UN or CSCE presence in virtually all these zones—in the form of special missions, representatives, or observers. They are performing their functions in a worthy manner, and we are grateful for this.

Russia and its neighbors have repeatedly proposed closer cooperation with the United Nations and the CSCE, but so far the response to these proposals has been clearly insufficient. It seemed that things were starting to change when the agreements that were reached between the Georgian and Abkhazian sides this July in Sochi with Russian mediation were reinforced by a Security Council decision to establish a UN observer mission in Abkhazia. However, only 22 of the 88 promised observers have arrived in the conflict zone to date. Moreover, Russia and Georgia, with the Abkhaz side's agreement, have urged the United Nations to send international peacekeeping forces to the contingent stationed in Sukhumi for this purpose. If this contingent had been significantly reinforced and the United Nations had promptly heeded our appeal, it might have been possible to prevent the collapse of the Sochi agreements and the new outbreak of war in Abkhazia. However, the United Nations still hasn't responded, although this does not, of course, justify foot-dragging on our part.

We are today on the threshold of a similar choice in the Nagorno-Karabakh conflict. Whether we like it or not, there is no alternative to a Russian Federation peacekeeping contingent in this conflict either; immediately after a settlement mechanism is set in motion, this contingent should be given the status of a UN force and reinforced with UN units from neutral European CSCE member countries. Here too, we ourselves, as well as the United Nations, must do our historical duty. It would be irresponsible to evade this.

It is time to overcome both our own "internal" foot-dragging and the world community's indecision with regard to Russia's peacekeeping efforts in the former Soviet space. This is not a "neo-imperialistic" space, but a unique geopolitical one in which no one is going to keep the peace for Russia. At the same time, it is an integral part of the global system of international relations that is subordinate to universal legal norms and is undergoing fundamental changes. The "classical" standards with which the United Nations approached peacekeeping operations decades ago are not appropriate today. But we hear doubts: Is it good to have the parties to a conflict and a "neighboring country" participate in peacekeeping contingents? This, it is said, goes against UN practice. But one must proceed from real life, not stereotyped patterns, especially since the new approaches have already proved their effectiveness in the Dniester region and South Ossetia, for example. Of course, as in any endeavor, especially a new one, some snags are unavoidable, but can the answer really be to give up in the face of difficulties? . . .

To put it bluntly, Russia alone has borne the heavy financial burden of real peacekeeping in conflicts along its periphery for two years now. At the same time, the pay that Russian soldiers and officers and their partners in these operations are receiving is hundreds of times less than the pay for UN operations. And this is at a time when we are regularly being charged multimillion-dollar dues to pay for UN peacekeeping operations in distant regions.

Russia recognizes its obligations to the United Nations. But the United Nations is a universal organization, and it too is obliged to recognize that the efforts of Russia and its neighbors to maintain peace and stability in the former Soviet space are at least as important for regional and international security as, for example, extinguishing crises in Central America. We are convinced of the need for genuinely concerned participation and real support—not just sympathetic attention—from the international community.

Needless to say, Russia will not abandon its neighbors and friends, no matter what happens. Unless we find the political will and real resources—troops and hardware, to put it bluntly—for peacekeeping in the former Soviet zone, this vacuum will be filled by others, above all the forces of political extremism, which also threaten Russia itself.

However, cooperation with the world community and the United Nations will make efforts to solve current problems in the new states both easier and less costly for everyone. The UN Secretary General, in a recent article in *Nezavisimaya Gazeta,* rightly stressed repeatedly that any peacekeeping operation always costs less than war, above all in terms of the most precious commodity—human lives. The time has come to turn these ideas into practical actions and to make UN support for peacekeeping efforts by Russia and its neighbors much more substantial.

## 3.9
## Andrey Kozyrev Addresses Russian Ambassadors to CIS States

Aleksandr Krylovich and Georgiy Shemelev
TASS, 20 January 1994 [ITAR-TASS]

*Editor's Note:* By January 1994, Kozyrev's foreign policy approach had rotated one hundred and eighty degrees. It clearly displayed the growing influence of the Russian military on what to do about CIS "hot spots," and on Russia's "peacekeeping" role (which was becoming a euphemism for the use of force). If we compare the following speech with his 1992 address to parliament, the acquiescence to a "national-patriot" viewpoint, probably in order to survive politically, is clearly discernible. His language is that of the moderate nationalists, but the practical results are that Russia has claimed a right to deploy peacekeeping forces to any part of the CIS it deems necessary.

Russia is playing a special role and has a special mission in establishing the Commonwealth of Independent States on the territory of the former Soviet Union, Russian Foreign Minister Andrey Kozyrev said.

The statement was made in his closing speech at a conference on Russia's foreign policy in the Commonwealth and the Baltic states in Moscow today. The conference was attended by the Russian ambassadors to the CIS republics and the Baltic states and representatives of various departments and organizations.

This line lies between two extremist approaches. The first approach is to send tanks and restore some post-imperial space by force. The second one boils down to demands that Russia should leave the former Soviet republics and forget about its historic ties, about what it has achieved over centuries, and its special relations in this space sealed by the common history and culture of the multimillion-strong Russian-speaking population.

"Millions of our brethren, neighbors if you please, do not view and have never viewed Russia as an imperial center. There have been no relations under which one would only take, as a colonial power, while the other would only give. There are lots of examples when the Russian center provided assistance, technologies, culture, and education to other parts of the former 'big country.' That was a process of mutual enrichment. No one will ever be able to escape these historic, humanitarian, economic, and cultural ties.

"I believe that both approaches are anti-historic and unrealistic. The implementation of either one would lead to a repetition of the Yugoslav scenario on the expanses of the former Soviet Union. The sending of tanks would automatically unleash a war of all against all. We have avoided this. And we will never let ourselves be led by pseudo-patriots who are pushing Russia into a fratricidal war with Ukraine and other republics of the former Soviet Union. However, negligence of Russia's 'special role,' and a panicky escape from former Soviet republics, would bring the same result. It is Russian troops that bear the major part of the burden and face bullets. The 'special,' but not imperial, role of Russia is to ward this off.

"Suppose we leave Tajikistan, abandon our peacekeeping role in Abkhazia, Ossetia, Karabakh. Where would millions of refugees go? They will go to Moscow. We cannot and will not allow this to happen. They should stay where they are, but on a new basis," Kozyrev said.

The minister said that he continues to call for creating a legal basis for the presence of Russian troops in the republics that request it. This will protect the troops legally. They will know their rights and duties, and no one will be allowed to say that their presence is illegal. That's what respect for the sovereignty of other states is all about.

The misunderstanding of Russia's "special" role causes hysterical and panicky cries that Kozyrev wants to preserve Russian bases, the minister said. So, if Russian bases give too much trouble to someone in the West, this is a sign that people there have either failed to get rid of Russia's image as an enemy or are trying to revive it.

Kozyrev dwelt upon the foreign reaction to his speech at the opening of the conference yesterday. He said that a statement by the U.S. State Department spokesman in which he argues with him can cause nothing but bewilderment. Kozyrev said he told the U.S. ambassador in Moscow today that he cannot accept such behavior and expects it to be changed.

The reaction of the Baltic neighbors is even stranger, Kozyrev stressed. It seems that someone in Latvia needs an incident. The incident involving Russian army officers in Latvia is over, but someone obviously needed this hysteria about words that in this particular case have nothing to do with Latvia. They even speak about the sovereignty of the Baltic states which they fear Kozyrev may damage. This will be even more comical if one recalls that it was Boris Yeltsin and Andrey Kozyrev who presented representatives of the three republics with a presidential decree recognizing their sovereignty after the August 1991 putsch.

"Speaking about their sovereignty and independence, which they are so carefully protecting from my statements, when they address the United States in a public statement, it seems that they are seeking to get into another dependency," the minister said.

Russia calls for an organized withdrawal of its troops on a contractual and civilized basis, Kozyrev continued. Where they will stay for a long time, a principle for their presence, a legal basis, should be determined. Their withdrawal cannot occur all at once, but should take into account all circumstances that are well known to the Baltic states. This will also strengthen their sovereignty. "So I believe everything is absolutely clear on this issue," the minister noted.

Speaking about the status of the Russian minority in other republics, Kozyrev said there is much speculation and misunderstanding on this issue. This topic cannot be a subject for intrigue because human lives are involved. We will not solve this problem by force, ultimatums, or armed confrontation with the republics as some propose. This would be political adventurism. But neither will we say that there is no such problem in independent states and republics, and that it should not be discussed or even raised because this may be perceived as Russian imperialism or as a concession to red-brown pseudo patriots. We will raise these questions and show responsibility without causing hysteria. We will calmly and seriously work on these problems. However, if, despite endless recommendations of the Conference on Security and Cooperation in Europe (CSCE) and the Council of Europe, and our tactical suggestions, laws (like in Latvia) that lead to deportation of people are adopted, such laws and even their orientation cannot leave us indifferent.

One can violate CSCE recommendations, one can challenge international norms and deport thousands of people because, as it turns out, Moscow should keep silent, Kozyrev should keep silent. Otherwise they will be Zhirinovskiy's agents. We will not yield to blackmailing by our pro-fascist forces, but neither will we give in to that other blackmailing either. We will react in the most decisive way to mass and gross violations of human rights. All these questions can be solved calmly without hysteria. In the event of a reckless line aimed at creating a spot of tension and destabilizing the situation, Russia, of course, will not keep silent and will resort to tough measures. If need be, it will impose economic sanctions. I would like our Western partners to take this seriously, for there would be no double standards in the protection of human rights. Instead of indulging in empty rhetoric about Russian imperialism, we would like the West to take a more serious position on this human rights and therefore international problem," Kozyrev said.

## 3.10
**Russia's "Westernization" Stage Is Seen as Over**
Mikhail Leontyev
*Segodnya*, 24 November 1994 [*CD* Translation]

The time has come to tally some results. One stage of the transformation of post-communist Russia has ended, and a fundamentally new one is beginning. The first period was marked by a course aimed at rapid Westernization and "entry into Europe" and by a completely pro-Western orientation in foreign and domestic policy, with enormous hopes for decisive Western economic assistance and Western solidarity with a country that had thrown off communism; had set everyone free, heedless even of its own losses and interests; had voluntarily and happily capitulated in the Cold War; and had voluntarily taken on the burden of reparations for that defeat.

This period began under Gorbachev and peaked in the first quarter of 1992, when Egor Gaydar said: "We have an opportunity to use for our reform resources that significantly exceed our domestic possibilities." These hopes on the part of political leaders did not just fall from the sky; they were grounded in a very optimistic public mood that was fully commensurate with them. This period ended in defeat and disappointment—a defeat for the West, which for all practical purposes totally missed the opportunity to bring about Russia's "soft" integration into the "Western world," leaving those political forces in Russia that had been counting on a Western future in the position of undoubted political outsiders. Now objectively thinking pro-Western democrats who understand the hopelessness of political mimicry are very much aware of the modesty of their immediate prospects. They hope only for one thing: that they will not pass on the cause of post-communist evolution, which they began, into the worst possible hands. Incidentally, this is what is giving impetus (at least some impetus) to the dreary idea that they themselves might personally nurture some sort of "social democratic future." . . .

The stage of transformation that is beginning today is a national stage. We must get it into our heads that Russia is going to emerge from its very grave and very inevitable crisis on its own, with no support from outside. No one is going to help us, although some can hinder us a great deal. We need foreign investments, and we will get them if all goes well, but investments are not aid. We will get them on very tough market terms.

In this situation, the range of possibilities is much broader than it is under the "pro-Western" vector, which posits a specific ideal model and has patented external inspectors to track and evaluate the parameters of movement toward this model. This new unpredictability gives rise to fear, perhaps fully justified fear, among democrats—both our domestic "Westerners" and the actual foreign ones. National development with an inevitable measure of autarky, first of all a cultural/values autarky, is fraught with the danger of very exotic forms of originality that are not limited by civilization and common sense. At the same time, all national-patriotic constructs imply, in one way or another, the restoration of state management of the economy, something that is hardly possible to accomplish—even at the cost of considerable bloodshed—when the state itself is breaking up. One can console oneself with the thought that no national-socialist prospect for Russia exists, that it is simply a road to disaster, but this is poor consolation, especially for the Russian population.

However, national development creates previously unknown prospects for Russian liberalism and for the building of a truly free economy and an organic societal structure grounded in natural traditions and possibilities and in national cultural values. In building state, public, and economic institutions virtually from scratch and ridding itself of social constructivists who seek to impose on Russia their own "models" (from American to Chinese), Russia will be able to realize advantages that virtually no other country has at present.

There is another undoubted plus in all this, especially as concerns the choice of political solutions: We will no longer have to keep looking slavishly over our shoulder at the West, fearing that we will get a grade of "poor" in "democracy" or in "foreign policy behavior." The period of "training for life" has ended, and everything that can be comprehended has been comprehended. We need partners, but we do not need mentors. In light of the discussions, which have been drawn out to indecent lengths, on admitting Russia even to the Council of Europe, for example, one can assume that when the West finally gets around to admitting us, there simply will be no one left in Russia with whom to discuss the matter. Perhaps that is all to the good.

## 3.11
**Russian Efforts on CIS Integration Assessed**
A. Kolesnikov
*Nezavisimaya Gazeta*, 7 October 1995
[FBIS Translation]

*Editor's Note:* Within Russia, a backlash developed in 1995 against deep economic integration with the CIS states, especially if based on geopolitical and military-political reasoning rather than sound economic rationale. According to writers for Russian economic journals such as *Kommersant Daily*, the only Russian companies that are lobbying for closer CIS economic integration have been those in the energy sector. The economic integration scheme known as "financial-industrial groups" has never taken off (only one serious one exists in Kazakhstan—Interros), and other joint enterprise schemes have also been abandoned. Privatization is only beginning to be developed in Ukraine and Kazakhstan, and exists only on the level of promises in Belarus. Deep integration, therefore, while it should be taken as a serious Russian strategic policy, is not to be rationalized or legitimized on the basis of "restoring essential economic links," as the following article so aptly argues.

The most radical turn in Russian policy has occurred in the area of cooperation with CIS countries. The Ministry for Cooperation with Member States of the Commonwealth of Independent States, originally the State Committee on Economic Cooperation with Member States of the Commonwealth, whose chairman was Vladimir Mashchits, one of the bright stars on Gaydar's team, has renounced its former priorities announced just a year ago.

At the end of 1994, Russia, proceeding from economic and not from geopolitical reasoning, which is the case today, attempted to rid itself of the traditional role of a financial-economic donor of "spiritually close," and not so close, republics. Repayment of credits issued by Russia never did take place. By the way, let us also note it was doubly difficult to repay them because as early as February 1992, massive credits to countries of the so-called ruble zone (the elimination of which took two years) began. According to data of the Institute of Economic Analysis in May 1992, the volume of credit reached 8.4 percent of Russia's GDP (in February it was just 1.2 percent of the GDP). (Simultaneously the import of inflation from other countries began.) Such extensive credits are essentially impossible to repay (between 1992 and 1994, $5.6 billion in state credits had been extended).

At the meeting of Commonwealth Heads of Government in December 1994, Viktor Chernomyrdin threatened his colleagues with a credit cutoff in 1995 if the debt was not settled in 1994.

The priorities changed in less than one year. Unfortunately, the restoration of former ties at any cost, mainly economic cost, is now considered to be of paramount importance. Geopolitics began to determine economics, while, at the same time, Gaydar's version of reform, which was not yet complete but which defined a certain political landscape, was oriented toward the self-sufficient transformation of Russia without the participation of under-reformed or totally unreformed CIS countries. This is what constitutes the complexity of integration in its new variant.

The Ministry for Cooperation with Member States of the CIS does not recognize that the levels of economic development of the CIS countries must be evened out, after which cooperation and integration with Russia will be worthwhile. First, judging by everything, there is no unanimous opinion in the ministry concerning the depth and intensity of integration. Second, the idea of a customs union has received far from enthusiastic endorsement from representatives of other economic agencies: the Ministry of Economics and Ministry of Finance count the losses, while the Ministry of Foreign Economic Relations is concerned with the possible entry of Kazakhstan into the World Trade Organization.

Supporters of a deeper integration have something to say about this. Let reintegration here and now bring certain losses, but later the advantages of closer cooperation, including those of a customs union whose area is expanding, will become self-evident.

The truth, as usual, is apparently to be found somewhere in the middle. It is not worth being hasty only because it is necessary to ensure the security of ethnic Russians against arbitrary action by local authorities, while at the same time, of necessity providing them with jobs. As a start, it would be good to load up the local enterprises and provide work for citizens within Russia. In addition, it is necessary to restrain inflation, where the next regular upswing in imports is inevitable as a result of the increase in the volume of credits. Integration processes really must be synchronized economically and legislatively (work, by the way, is still under way on a model "CIS code," without which civilized economic relations are quite impossible). Synchronization of economic development has not yet reached a level where political and geopolitical motives come into play. A large degree of economic and social prosperity is necessary for that purpose.

It appears that executive power in Russia has been hasty in declaring a geopolitical championship. First-place positions in that tournament will bring no benefits now.

## 3.12
## Kozyrev Responds to Reports of "Impending Dismissal"

Andrey Krasnoshchyokov and Igor Shchyogolev
ITAR-TASS, 20 October 1995 [FBIS Translation]

Russian foreign minister Andrey Kozyrev has voiced his response to the media reports on his impending dismissal.

Yeltsin, departing for his official visit to France together with Kozyrev today, backtracked on his Thursday [19 October] statement at a new conference that he considered replacement of Kozyrev, saying that the foreign minister remains in his post.

"I feel as I did yesterday," Kozyrev told reporters today.

"The president is being exposed to a strong pressure from

dark forces, from people whose minds are closed, who cannot imagine Russia as a normal member of the international community. He needs to be helped to resist this pressure," Kozyrev said.

Kozyrev stressed that he is not pro-Western, but struggles for Russia's interests.

However, a weighed and open approach to this job requires an understanding that it is in Russia's interests to develop ties with the West, and vice versa, Kozyrev continued.

He said that Russian diplomacy should work better to promote Russia's cooperation with international organizations and educate Russians into the notion that Russia may live only with the world around it.

Otherwise, Russia will return behind the Iron Curtain or some sort of fence, with Russian pseudo-patriots sitting behind it and scaring the world with their surprises, Kozyrev said.

## 3.13
### Daily Names Potential Candidates
Maksim Yusin
*Izvestiya,* 21 October 1995 [FBIS Translation]

Before departing for Paris on the morning of 20 October Boris Yeltsin stated at Vnukovo Airport that Andrey Kozyrev will remain foreign minister, admittedly without saying for how long. These words are unlikely to have reassured Andrey Vladimirovich: After the president's news conference Thursday [19 October] the Foreign Ministry chief can hardly be under any illusions as to his future. The fate of post-Soviet Russia's first foreign minister is effectively a foregone conclusion. Who is going to replace him?

The range of candidates is fairly extensive. The list of contenders can be conveniently divided into two groups: Foreign Ministry staffers and "outsiders." As a rule those in the first category are people who do not belong to "Kozyrev's team" and whose promotion to leading positions in Russian diplomacy took place in times when Andrey Vladimirovich was still an unknown official in the USSR Foreign Ministry International Organizations Administration. Younger diplomats who entered the ministry leadership following the Soviet Union's disintegration are closely associated with their boss and his policies.

The figure of a veteran who has never been part of Kozyrev's entourage looks more acceptable in this sense. The names of Russia's ambassador to the United States Yuliy Vorontsov and ambassador to Britain Anatoliy Adamishin are being mentioned as potential Foreign Ministry candidates increasingly frequently. Both have a reputation as fairly conservative politicians (compared to the "early" Kozyrev, at any rate) and should not be rejected by the Duma's communist-nationalist majority.

It would be extremely risky from the viewpoint of their future careers for younger and more ambitious diplomats to accept the offer to head the Foreign Ministry right now. The situation in Russia is too unstable. If Yeltsin loses the upcoming presidential election he could quite easily take his entire team into political oblivion with him, including the Foreign Ministry chief, who will in one way or another be associated with a defeated head of state. This is why few people are mentioning Vitaliy Churkin, Russia's best-known diplomat, who is riding out Moscow's political storms in Brussels, as Kozyrev's possible successor. His time has not yet come, although it cannot be ruled out that Vitaliy Ivanovich's candidacy for the ministerial post will be seriously considered as early as June next year.

The "outsiders" include the ostensible favorite Vladimir Lukin, chairman of the Duma International Affairs Committee, especially as Vladimir Petrovich has never made any particular secret of his desire to hold Kozyrev's job—to hold it not for a few months but for a really long time. Therein lies the difficulty of the choice facing Lukin: Should he accept Yeltsin's possible offer he would, as it were, become a member of the "party of power," heaping on himself all its "sins," which would not be the best pre-election gift for the "Yabloko" bloc. It is hardly worth Lukin's while to jump the gun. In eight months' time, after the presidential election, he will have a good chance of heading the Foreign Ministry without linking his fate with Yeltsin's team.

There is one other possible candidate—Dmitriy Ryurikov, the president's foreign affairs adviser. His nomination seems not so very unlikely, bearing in mind that there have already been precedents for Yeltsin appointing people from his inner circle to key posts in the state (the security minister, for example). In addition, in the last few weeks there have been periodic rumors of Deputy Prime Minister Vitaliy Ignatenko's impending arrival in Smolenskaya Square [location of Foreign Ministry], and the name of Foreign Intelligence Service Director Evgeniy Primakov has also occasionally cropped up.

Be that as it may, there is no doubt that after Kozyrev the Foreign Ministry will pursue the same "presidential foreign policy" as it did under Kozyrev. The only thing for which Yeltsin could never criticize the disgraced minister is for being excessively principled, independent, and unwilling to adapt his views to the Kremlin's constantly changing foreign policy line. Boris Nikolaevich will hardly allow the new head of Russian diplomacy to have shortcomings of which his predecessor, who despite all the forecasts and rumors kept his post for five years, was so fortunately devoid.

## 3.14
**Yeltsin Accepts Kozyrev's Resignation**
ITAR-TASS, 5 January 1996 [FBIS Translation]

Russian president Boris Yeltsin signed an edict today releasing Andrey Kozyrev from the post of foreign minister in connection with his election as deputy in the State Duma, the presidential press service told an ITAR-TASS correspondent.

Forty-four-year-old Andrey Kozyrev has headed the foreign political department since October 1990.

# The National-Patriots

## 3.15
**Russian Foreign Policy Under Fire**
Yuriy Glukhov
*Pravda,* 24 February 1992
[*CD* Translation], Excerpts

*Editor's Note:* The following section looks at developments in the thinking of Russia's "national-patriots," many of whom have influenced Russian policy from their seats in the Russian parliament and on parliamentary committees.

The national-patriot political viewpoint is expressed by members of the military-industrial complex, disillusioned "democrats," some of whom supported President Yeltsin in 1991, and former communists who have chosen radical nationalism over a return to the Communist Party to replace the ideological credo that was lost with the dissolution of the USSR.

As the following article shows so clearly, the nationalists attacked from the outset Yeltsin and Kozyrev's dramatic departure in Russian foreign policy, which, had it worked, would have made the idea of a democratic zone stretching from Vancouver to Tokyo seem less empty.

The new political thinking seemed to promise enormous material benefits for both the USSR and the world as a whole. . . .

Why, then, did we suddenly fall into a deep pit of debt? The loans and credits we have taken total $70 billion already and will top $200 billion by 1995.

First, the disarmament and conversion process itself has proved to be costly and to require special preparations and billions in appropriations, something that was yet another surprise for our politicians and economists. Second, after shifting to payment in freely convertible currency, it turned out that neither we nor our trading partners had any hard currency. The familiar economic relations that had developed over decades were severed, and no new ones have yet been established. The result is losses. Virtually the same thing has happened with our writing off of embarrassing and costly allies and partners (Cuba, Ethiopia, Afghanistan, and others). In one way or another, the scope of cooperation has narrowed, especially with respect to the Third World. Our expanding ties with the West have not compensated for these losses. Our overscrupulousness about arms sales, which resembles a teetotaler's behavior in the company of alcoholics, has also made a hole in the budget.

Unlike us, the practical Americans, for some reason, have not yielded to the euphoria of romanticism and have not cut back either the production or sale of weapons. While the Soviet Army loses its former fighting capability and disintegrates under the influence of the USSR's disintegration, the U.S. Armed Forces are building up their might. While the Warsaw Pact has broken up, NATO continues to grow stronger. And while it was formerly said that there were neither victors nor vanquished in the Cold War and that friendship ultimately triumphed, today the United States has unreservedly declared itself the winner. The Yankees still put their faith in the cult of strength.

Unfortunately, the new political thinking has remained purely our own. Yes, humanity has reached a critical point, threatened by global catastrophe; it is essential to disarm. We ourselves have been exhausted by the mindless arms race. But disarmament and the current unilateral destruction of our military potential are two completely different things. Having torn apart the Union, we have been unable to exploit the enormous positive potential of the USSR's new policy. The Soviet-American disarmament agreements, which were so carefully scrutinized by the experts and based on strategic balance, aren't worth the paper they're written on. We can console ourselves with the thought that the new political thinking and our sacrifice in the name of peace may be duly appreciated in the future: After all, for the second time in the history of the twentieth century, we have burned down our own house in order to light the way for humanity. . . .

The foreign policy of Russia, which has declared itself the Union's successor, is undergoing a transformation: from a policy of peaceful coexistence (Khrushchev and Brezhnev) and partnership (Gorbachev), to allied relations with the West (Yeltsin).

An alliance presupposes an orientation: It is created in the name of something or against someone. However, no such clarity exists. . . . Judging from our leaders' practical actions and pronouncements, the West is being chosen to guide us, since the Cyclops of the Soviet Union preferred to blind itself rather than look at the world with the single eye of the communist vision and has now lost its bearings.

We are on a leash of hunger. Given a desire, the mere manipulation of deliveries of food aid could be used to strangle any regime deemed objectionable.

The Russian leadership says that the fate of economic reforms and democracy itself and the triumph of the incipient renewal process depend on the West's involvement. But what, then, is our own role in this?

We hear constant appeals to the United Nations for help in resolving our conflicts, such as the one in Nagorno-Karabakh. This is tantamount to a declaration of the Commonwealth's inability to manage its own affairs.

The primacy of international law over the laws of individual states is being declared, and the possibility of using force to deal with "disturbers of the peace," as was done with Iraq, is being acknowledged. The Russian Foreign Minister has made his own contribution to the advocacy of "civilized" intervention. I wonder whether he was thinking in abstract or concrete terms. Given the legal chaos that reigns on the territory of the former USSR, such a theory of a "new world order" readily lends itself to intervention in our internal affairs. I wouldn't be surprised if there is soon talk of armed international support for an "independent Crimea," the Kuriles, Kaliningrad, Chechnya, Tatarstan, etc. . . .

It would behoove us to avoid succumbing solely to unrequited love and to consider the extent to which our interests coincide with those of the United States. What role, for example, will be assigned to Russia, which has proposed taking part in the American Strategic Defense Initiative? The role of a trace horse? I think that our immature Russian diplomacy suffers from nocturnal emissions of infertile initiatives that are devoid of either knowledge of the situation or foresight. Hence such failures as the fruitless talks with the Afghan opposition, the futile personal mediation of B. Yeltsin and N. Nazarbaev in the Armenian-Azerbaijani conflict, the scandals surrounding an autonomous area for the Germans and the retargeting of strategic missiles, which frightened Ukraine and Belarus, and much more. . . .

Without belittling the importance of relations with the world's number one power—the United States—or the need to declare Russia's legal succession to the Union, I think that the Russian president should have made his first official visits after the proclamation of the CIS to the Commonwealth capitals. And he should have hurried to Kiev at the first hint of frictions with Ukraine, if only to squelch false rumors that the USSR had been abolished in order to remove Gorbachev, rather than for the sake of preserving Russian-Ukrainian friendship. That did not happen. The moment was lost, and we are seeing the disastrous consequences of the neglect of Russian policy's main priority—the strengthening of the Commonwealth, without which we cannot survive. The Ukrainian chain continues to acquire more and more links of conflict—the army, the Black Sea Fleet, the Crimea, ruble-coupons, resettlement of the Germans, a counteralliance with the FRG, etc. . . .

It is inadmissible to put forward initiatives and assume responsibility for the resolution of issues that affect the interests of the other CIS members without even informing them. And that is precisely what Russian diplomacy is guilty of. Relations within the Commonwealth cannot take second place to relations with the outside world; they must take priority.

No foreign aid can compensate for the losses we have suffered as a result of the disruption of economic ties. . . . The nationwide miners' strike that was supported by the Russian president no doubt cost an amount equal to all the charitable aid we are receiving today. We wouldn't have to go begging had we headed off the decline in the production of industrial goods, metal, and oil, and brought in the harvest. But the decline continues. Meanwhile, our statesmen are demonstrating great skill at borrowing more and more and more from other countries.

Returning to the "peace dividend," it must be said with some sadness that our diplomacy is disastrously unprofitable, both economically and politically. Marx once wrote with delight of the tireless mole of history. I think that mole embodies our zeal today too, although it has changed its Marxist coloring to a democratic one. The trouble is that moles cannot see, and they dig blindly.

---

## 3.16
## Ambartsumov on the Need for Change in Foreign Policy

Interview by Vitaliy Buzuev
*Rossiyskaya Gazeta,* 13 April 1992
[FBIS Translation]

---

[Interview with Evgeniy Ambartsumov, chairman of the Supreme Soviet Committee for International Affairs and Foreign Economic Relations, by Vitaliy Buzuev.]

Until very recently, it looked as if it was going to be far easier building diplomatic relations with the former Soviet republics than with Western countries. But life has overturned that assumption. The Russian Foreign Ministry's initial attempts to take an active stand have left many experts feeling disappointed. Why? I asked Evgeniy Ambartsumov, chairman of the Supreme Soviet Committee for International Affairs and Foreign Economic Relations.

[Ambartsumov]: The young Russian Foreign Ministry appears to have taken the methods of dealing with capitalist countries evolved by the old Union department and projected them into a totally different arena, the turbulent CIS. It is easy to continue the policy of disarmament and easing tension in relations with the West, because there is something to be continued. But totally different methods are needed for working with the CIS countries. Take negotiations, for example, I do not think they can be as "civilized" as our negotiations with Western countries; normal partnerships have yet to evolve within the CIS, and the compromises that Russia seeks to make are interpreted as a sign of weakness. Talks must have more of a "family" atmosphere about them, if you like. This does not mean, however, that dialogue within the "family" (we are not talking about Stalin's "family of peoples," of course) cannot be tough. Yet it is bad when dialogue is reduced to a slinging match. It goes without saying that we have a right to expect greater firmness from our foreign minister when defending Russia's interests in the CIS countries and the interests of Russians who have become foreigners against their own free will. I think many people now realize that the previous foreign policy practice of Russia concluding agreements with the former Soviet republics was detrimental to the cause (the agreements were drawn up by the Foreign Ministry, incidentally). This is mainly because it did not resolve problems connected with the rights of our compatriots and our property. Russia concedes to its neighbors in everything nowadays, receiving nothing in return. The interests of our compatriots are being flouted. They are treated as second-class citizens: They cannot acquire property, and those who do not speak the language cannot get work. Of course, they themselves bear some of the responsibility for the fact that they have not learned Estonian or Lithuanian, for example, but the state in which they live cannot even provide them with textbooks and teachers.

[Buzuev]: Nevertheless, the Russian Foreign Ministry has recently been trying to catch up with events. Take Andrey Kozyrev's recent visit to Moldova and the Dniester region, for example. . . .

[Ambartsumov]: Talking about the Foreign Ministry trying to catch up with events, I am reminded of a recent joke that made the rounds among deputies after the U.S. secretary of state's lightning tour of the CIS countries. Question: Who is the CIS foreign minister? Answer: James Baker. It is a good thing, of course, that Andrey Vladimirovich Kozyrev made this trip. But I am not satisfied with the agreement that our minister signed in Chisinau with three other foreign ministers—Romanian, Ukrainian, and Moldovan. The document lacks balance. It does not consider the interests of another side involved: the Dniester region. . . .

[Buzuev]: Incidentally, the leader of the Dniester Republic, Igor Smirnov, said recently that he cannot understand why Romania is taking part in the talks, while his republic is excluded.

[Ambartsumov]: A perfectly valid complaint. Russia is certainly not obliged to recognize the Dniester region as a sovereign state, but it is nevertheless involved in the conflict. Azerbaijan recognizes that Nagorno-Karabakh is involved in the conflict over there. So why has our minister not secured similar recognition from Moldova? He has had the opportunity to do so, after all. The English political lexicon contains the concept of "bargaining," which translates rather crudely into Russian as "haggling." We really should have engaged in some haggling.

[Buzuev]: Why do you think that "haggling" would have been successful?

[Ambartsumov]: About two weeks ago, some representatives of the Moldovan parliament arrived in Moscow. During talks with us, they agreed that the Dniester region should be given not only special economic status but special political status as well. Admittedly, finding themselves under pressure, they later went back on this agreement. But another group of Moldovan parliamentary deputies visited us recently. We succeeded in drawing up a very balanced document in the course of an evening. I should point out in particular that it was signed by Russian deputies with the most diverse political views. It does not demand recognition for the Dniester region. It proposes special legal status for the region and speaks of the need to separate the hostile sides with the aid of 14th Army units. It looked as if the situation was no longer at a standstill. But then the Moldovan deputies received a telegram of disavowal from Mosanu, chairman of the Moldovan parliament in Chisinau. He said that the deputies did not have the authority to sign such a document. But what has it got to do with authority? The deputies had acted on the basis of their personal beliefs. It is not that Mosanu does not want a peaceful solution to the conflict. Of course not. But he is under a lot of pressure from ultranationalists and the People's Front. These forces are greedy for power, so they are counting on conflict. But it is worth taking a sober view of things: Even if Moldova merges into Romania, it simply will not be to its advantage to drag a "thorn in the flesh" like the Dniester region with it. I am convinced that a solution to the totally irrational Dniester conflict can be found. A "mere trifle" is required: goodwill on the part of both sides involved in the conflict.

[Buzuev]: Another CIS country, Ukraine, is taking part in resolving the conflict in the Dniester region. I know that you have had a dialogue with Ukrainian President Leonid Kravchuk. What is your opinion of this politician?

[Ambartsumov]: Our paths crossed several days ago in Bonn, at a meeting of several European countries' prime ministers, prominent economists, and entrepreneurs. Kravchuk delivered the first report and I gave the second. The main thread of his report was that the formation of the CIS should be welcomed, but that Ukraine must be a completely independent state and not forgo one iota of its sovereignty. The Spanish prime minister, Gonzalez, came up with a good response. He said that Spain cannot control all the processes taking place in the European Community and consciously delegates some of its rights to it. "How do you intend to become part of a united Europe if you refuse to forgo even part of your sovereignty?" Gonzalez asked. Kravchuk had no answer to this. Generally speaking, I must say that he feels he has more freedom abroad than at home, where he is forced to make concessions to national extremists. At the same time, Kravchuk is an intelligent and fairly flexible politician. In my opinion, he is open to dialogue and is prepared to make compromises.

[Buzuev]: Evgeniy Arshakovich, please be frank: Do you believe in the durability of an entity like the CIS?

[Ambartsumov]: No, I do not. Ideally, in my view, we should work toward recreating a confederation of states—albeit not the Soviet Union. Russia should have a leading role. After all, it is something bigger than the Russian Federation. And, moreover, objectively it is a world power.

[Buzuev]: But first the CIS must collapse completely. In your opinion, does that mean further disintegration is inevitable?

[Ambartsumov]: I would like to remind you that all this began with the collapse of a great state—not with the disintegration of the CIS. I am more and more convinced that the breakup of the USSR was a totally ill-considered, irresponsible step. The politicians simply did not foresee the consequences, which will only accumulate. But that is the reality of the situation. Another reality is that it will be far harder getting out of the impasse alone than together. Several states are beginning to realize this, so unifying ideas are not being developed. Georgia and Armenia are drawn toward Russia. This coming together of the CIS countries is also inevitable because it is sometimes impossible to draw geographical or even family boundaries. During the collapse of the British and French empires, the mother countries were clearly separate from the colonies, which subsequently became independent states. But in our case everything is all mixed up together. It is another proof that the process of forming a new union will, unfortunately, not be painless.

## 3.17

### Russian Neighbors Urged to Sign Border Treaties

Aleksey Surkov

*Rossiyskaya Gazeta,* 12 June 1992 [FBIS Translation]

Following the collapse of the Soviet Union, its former constituent parts—which were the sovereign Union republics, drifting along on their autonomous course—rushed around racing each other to visit various Western and Eastern countries, as if they were hurrying to convince the world of the reality of their independence.

At best they started bidding a strained farewell to yesterday's brothers-in-Union. At worst they started drawing national maps, "incorporating" parts of neighboring territory into them.

So much for the provocative and irritating pranks of fledgling statehood. They should not become a substantial part of everyday interstate relations. But the aspirations for a strict policy of cooperation with many republics' "national patriots" (including those in Russia!) are not concealed.

But how on earth can such intentions be allowed to prevail in state policy? The historical roots of our people's fates have become too firmly intertwined, as the Fifth Congress of Russian Federation People's Deputies stressed in its appeal to the Supreme Soviets of the former USSR republics.

The rapid passage of time also helps our historical indissolubility to be recognized.

Having established our "sponger" relationship with the outside world, where everything has long been politically settled and economically "sewn up," many people started to contemplate more soberly the whole ruinous effect of the abrupt dissolution of yesterday's interrepublic contacts and ties. The belief in the need to form our own open, independent Eurasian market for labor, capital, manpower, and services is starting to prevail in the minds of politicians, economists, and legislators. The sooner we grasp this reality and start applying to the building of our own interstate formation modeled on the EEC, the more quickly and easily we will be able to extricate themselves from today's total economic chaos. . . .

The correctness of each independent state's actions to legalize its geopolitical area where it has territorial supremacy, which constitutes an organic part of state sovereignty, is obvious. But are the limits of our territorial independence just as clear and obvious?

Before 1917 Russia had a permanently fixed external border. The domestic administrative-territorial structure comprised the guberniyas and their groups of guberniyas—the governor-generalships. The October 1917 revolution initially destroyed the external border. Russia lost Finland (1917), Estonia (1919), Lithuania (1919), and Latvia (1920).

Immediately afterward an internal redrawing started by means of an arbitrary removal (to the benefit of the revolutionary rulers' immediate interests!) of Russia's state flesh by those who set about zealously establishing Soviet power.

By December 1922 (besides the Baltic peoples, who had been lost) Russia was partitioned into six Soviet republics—the Ukraine, Belarus, and the RSFSR itself, plus the Federation of [three] Transcaucasian Republics that founded the USSR.

However, Russia's partition did not end there. Because of the population's substantial resistance to the insertion of the lumpenized power of the soviets into the specific structure of the Central Asian peoples, the Bolsheviks and Leninists tempted their devotees from surrounding territories by regional autonomization. If only they had secured the retention of power at the local level.

Thus, in October 1924 Turkestan Kray was broken up into the Uzbek and Turkmen Union Republics, the Kirghiz (the future Kazakh) and Kara-Kirghiz Autonomous Republics, and the Karakalpak Autonomous Oblasts, which were part of the RSFSR.

The Moldavian Autonomous Union Republic was formed in October 1924 as part of the Ukraine. Following the return in June 1940 of Bessarabia, seized by Romania from Russia in 1918, Moldavia acquired the status of a Union republic in August 1940.

That is how the truncation of Russia generally, and the birth of new republics in particular, proceeded.

And now, in the conditions created by the existence of autonomous state regimes, we must legislatively enshrine our new lines of republic borders, which in fact have never before enjoyed the status of international borders.

As is known, the establishment of state borders is a political and at least a bilateral process: Contiguous countries take part in it. Initially a treaty on the delimitation of borders is prepared and signed with a neighboring state, that is, the general direction of a border's path between states is determined. A map with the border line marked on it is attached to the treaty.

After this the border is demarcated, that is, the state border is erected on the ground and marked by special and clearly visible border signs (border posts, pyramids, marker signs, and the like). Demarcation is carried out by joint commissions, created by the states concerned on an equal footing.

How will the border process proceed between us? Will there be enough wisdom and diplomacy to prevent the state repartition that goes back to 1917 from turning us into irreconcilable enemies and sowing the seeds of hatred and confrontation in the generations immediately following us?

What can be done to ensure that the permanent state borders between us become transparent borders of peace and good-neighborliness, and not the lines of new "Karabakhs"?

The Alma-Ata agreement on the CIS reaffirmed and enshrined our adherence to the international principle of the inviolability of borders. But this agreement (like others, incidentally) lacks a mechanism to resolve possible territorial disputes. Will we resolve them in a civilized fashion or from a position of strength?

The principle of the inviolability of borders proclaimed by the Helsinki Final Act commits states "now and in the future to refrain from making any encroachments" on these borders and any demands or actions aimed at seizing and usurping part or all of another state's territory.

At the same time this principle, like the principle of territorial inviolability recognized by international law, does not exclude the possibility of changing a state territory's borders by peaceful means through the process of talks.

This mutual understanding is gradually starting to enter the practice of our interstate relations.

In particular, the Agreement on Cooperation Between the Russian Federation Ministry of Ecology and Natural Resources Committee for Geodesy and Cartography and the Lithuanian Republic Department of Geodesy and Cartography was concluded 21 May 1992 in Riga. It was signed by these departments' leaders: by N.D. Zhdanov of Russia and Z. Kumetaitis of Lithuania.

Article 3 of the agreement is noteworthy: "The sides will regularly exchange geographic information (administrative-territorial division, changes in the names of population centers and other facilities, changes in the road and railroad networks, and the suchlike) about their territories.

"The sides will refrain from making on published maps any changes to the position of the state border between the Russian Federation and the Lithuanian Republic before adoption of relevant bilateral governmental agreements on this question."

The same norm in the form of an appendix was also incorporated 21 May 1992 in a similar agreement concluded March 1992 by the aforementioned Russian and Latvian departments.

However, while realizing the importance and delicacy of the problem of territorial delineation between former subjects of the USSR, placing responsibility for resolving it at the level of merely interdepartmental relations would be an unforgivable mistake.

An open, interstate legal act seems necessary, which would regulate the procedure for peacefully resolving (now and in the future!) territorial disputes that arise and simultaneously contain undertakings to adhere to the act on the part of the states that recognize or are signatories to it.

At a recent two-day conference of the Parliamentary Coalition for Reforms in Russia, a memorandum was adopted from conference participants to all parliaments and governments of the former USSR's subject states, containing the proposal to join forces to draw up an open interstate convention on the inviolability of borders and the peaceful resolution of territorial questions. . . .

## 3.18
## Khasbulatov on the South Ossetia Conflict

Moscow Mayak Radio Network, 15 June 1992
[FBIS Translation]

*Editor's Note:* In June 1992, Ruslan Khasbulatov, one of Russia's most prominent national-patriots, as chairman of the holdover Russian Supreme Soviet, boldly asserted Russia's right to intervene in the conflict between Georgia and its South Ossetian province. The following statement asserts Russia's "Great Power" status within the CIS by defining the terms for intervention as "peacekeepers," and by exacerbating the conflict between the two parties. Georgian president Eduard Shevardnadze responded to the speech with shocked surprise and indignation, rightly regarding Khasbulatov's speech as interference in Georgia's internal affairs.

[Statement issued by Ruslan Khasbulatov, chairman of the Russian Supreme Soviet.]

Esteemed citizens of the Russian Federation, as you know from numerous reports in the media, an extremely difficult situation has arisen in a number of regions on Russia's borders. One of the most critical situations that has developed into a bloody confrontation is in South Ossetia. Some political figures in Georgia and armed formations under their control have adopted, it seems, a single course of expelling South Ossetians from their historical homeland in Georgia.

South Ossetian towns, villages, and the city of Tskhinvali are being continuously fired on with all types of weapons, including artillery and missiles. There are numerous casualties. Agreements recently reached between the Russian president and the Georgian leadership with the participation of South Ossetia about an end to the bloodshed and the beginning of peace talks have been cynically ignored by Georgian armed formations; indeed, the attacks have intensified in recent days following a visit to the area by a Supreme Soviet delegation.

Attacks against the civilian population have intensified. Once again there have been numerous casualties in the last few days. At the same time a small contingent of Russian army units stationed on the territory of South Ossetia with peacekeeping aims has come under intense pressure. There have been incidents of fatal attacks on them, of attacks on military warehouses with attempts to seize equipment, or of outright seizure of equipment.

You know that at Russian Congresses of People's Deputies, and during sessions of the Russian Supreme Soviet, the question of the South Ossetian problem has been discussed more than once, because tens of thousands of refugees are fleeing to densely populated regions of North Ossetia, thus creating conditions for a powerful social explosion capable of spreading throughout the North Caucasus. Therefore, this conflict, this war being waged by armed formations in South Ossetia, cannot be regarded as a purely internal affair in Georgia. It directly affects the lives and interests of our people and the state interests of Russia. This is why—and bearing in mind that this problem directly affects the state interests of the Russian Federation—congresses of people's deputies and the Supreme Soviet have adopted many resolutions, in which the Georgian authorities are reminded that the expulsion of the South Ossetian population to Russia, in particular to North Ossetia, is unacceptable.

It must be said frankly that today these actions must be qualified as genocide and a massive expulsion of the South Ossetian ethnic group from its historic motherland. At the same time, striving to resolve this problem on a conflict-free basis with Georgia, which has had friendly relations with Russia for many centuries, the Russian Supreme Soviet has not examined the official appeal of South Ossetia concerning its joining the Russian Federation.

In the present conditions, when it is evidently intended to pursue a de facto policy through the expulsion of an entire people and the abolition of its autonomy, the Russian Supreme Soviet may be forced to study this question immediately, that is to say, with regard to the question of its joining Russia, in accordance with the will and desire of the people and an appeal by the South Ossetian authorities to the Russian Parliament. If the Georgian side does not want the events to head in this direction, we demand that it stop the fighting and find enough courage to sit down at the negotiating table, as was agreed somewhat earlier.

Surely the Georgian leaders must understand that the policy and practice of genocide, and provoking complications with Russia, casts suspicion on the Georgian people themselves, whose courage, kindness, and justice no one has ever doubted. The Georgian side must understand that all exiles without exception who are on the territory of Russia must be returned to their homes, and that it is utterly futile to shed blood over completely unattainable objectives. At the same time, we demand an immediate end to the acts of provocation against Russian soldiers and officers in South Ossetia, who are carrying out their military duty of protecting the civilian population.

We warn that otherwise, Russia is capable of taking immediate measures to defend its citizens from criminal attempts on their lives, and to render harmless those villainous groups that have been shooting to kill, both with regard to the civilian population and Russian soldiers.

## 3.19

**Rutskoy Party Leader Slams Foreign Policy**
Vasiliy Lipitskiy
*Rossiyskaya Gazeta,* 26 June 1992 [*CD* Translation]

*Editor's Note:* Vasiliy Lipitskiy is chairman of the Free Russia Party (Aleksandr Rutskoy's party). He and his party represent a particularly virulent brand of the national-patriotic point of view in their harsh critique of all of Yeltsin's executive branch policies. Lipitskiy expresses the strong feeling of nostalgia for the status and power of tsarist Russia felt by the radical nationalists.

One cannot accuse our government of inactivity in the international arena. But policy, and above all diplomacy, does not consist of actions alone. It entails a system of objectives and priorities united by a common thought or concept. And it is these components that one can follow only with great difficulty in the rush of events.

One gets the impression that the collapse of the USSR has not been followed by any reassessment of values in our foreign policy strategy. Relations with the former republics and now independent states—both those that belong to the Commonwealth and those that do not—have yet to become a priority for our diplomacy, as they should. And this is costing us.

Take the Baltics, for example. Only recently their striving for independence, treated with hostility by the old center, had the full support of the Russian leadership. And so the latter's actions since the center's collapse are all the more puzzling. It seems to have inherited the hard-line policy that it opposed in the past so effectively and with such effect. And, as then, the result of this policy has been directly the opposite of what was intended. The tactic of "turning the screws" (primarily with respect to supplies of energy resources) has indeed caused domestic political complications in these countries and provoked government crises; at the same time, however, it has worsened their relations with Russia.

It is this tactic that—at different times but for the same essential reason—led to the fall of the Prunskiene Cabinet in Lithuania and the Savisaar Cabinet in Estonia, which were optimal partners for us from the standpoint of long-term cooperation. They have been replaced by far more intransigent politicians, and the economic problems designed in Moscow have only hastened those countries' reorientation toward the West.

One can now assess the result conclusively: Russia has lost the Baltics, a region traditionally significant to it, for a long time.

But is a hard-line approach to relations with our former Union partners our Foreign Ministry's fundamental concept? Hardly! All around we can see a striking spinelessness in resolving with other states issues in which Russia's interests are seriously threatened.

Until recently, the Foreign Ministry in effect failed to notice the war in the Dniester region, into which Russia was drawn long ago, whether we like it or not. I note parenthetically that the rejection of A. Rutskoy's plan for using the 14th Army to separate the warring sides has already cost hundreds of human lives. Even greater myopia can be seen where Ossetia is concerned: Talks on transferring weapons to Georgia are pouring oil onto the conflict's flames.

In Central Asia, the Foreign Ministry contemplates with a certain sense of doom the peaceful expansion of Iran, Turkey, and even Afghanistan—an expansion that encompasses the lands that are closest to Russia, as well as their leaders. Commentaries like "east is east and west is west" are viewed only as an admission of the ministry's own impotence.

This is the state of affairs on our immediate frontiers. What is the situation farther away?

Not so long ago, in addition to the Union, we had the "socialist commonwealth," which constituted, if we discard the ideological baggage, a bloc of countries economically oriented toward Moscow. What is left of it today? Severed ties, colossal bilateral losses, growing anti-Russian sentiment, and the total complacency—or rather indifference—of the Foreign Ministry.

A typical example is Bulgaria, whose economy was perhaps the most specialized in terms of serving Russian consumers. We still have an acute need for the products of Bulgaria's agriculture (as well as of Hungarian and Polish agriculture), which produces considerable surplus food. To put it bluntly, that country offers a way—given an effective system of mutually advantageous contacts—to ease our food difficulties.

Bulgarian producers, in turn, are consumers of Russian oil. In 1991–92, President Zh. Zhelev repeatedly sent personal messages to the Russian president proposing that Russia supply energy resources on terms exceptionally advantageous to us. But nothing has happened. Russia isn't meeting its commitments, and we're not receiving Bulgarian products (vegetables, fruit, tobacco, wine, cheese, etc.). Work on a Russian-Bulgarian cooperation treaty has come to a standstill. There is a rapidly growing hostility among Bulgarians toward their erstwhile brothers, and pro-Western political circles have acquired an array of arguments in favor of blaming all the country's economic difficulties on us.

Eastern Europe, along with Vietnam and Mongolia, whose representatives in Moscow are simply ignored, are also becoming inauspicious areas from the standpoint of promoting Russia's national interests.

One is disturbed by an overall desire to follow the beaten

path, without proper regard for the changed situation in the world and Russia's real place in it. To what countries are we turning for help in solving our immense material problems? To the United States, where a presidential election is around the corner and the government is preoccupied with domestic difficulties. To Germany, whose capabilities are being gluttonously swallowed up by its new states—the former German Democratic Republic. To Japan, where aid to Russia will invariably be linked to Japan's territorial claims. It is within this small circle that our country's diplomatic thinking revolves—thinking that is sustained, alas, by deceptive hopes for showers of gold in the form of new credits and investments.

Isn't it time to realize that the club of the strong of this world is burdened with its own problems, and that even if one believes that the club is 100 percent favorably disposed to us, there is little it can do to back up the way if feels? Indeed, chances are that it won't even want to do so, preferring instead to limit its cooperation to the bare minimum that allows Russia to remain afloat, but hardly to embark on any lengthy autonomous voyage.

At the same time, the world does present us with potential partners who have the available resources, are not inhibited by complexes with respect to Russia, and have interests that coincide with Russia's. Among them are Saudi Arabia and the Persian Gulf kingdoms—a path to which Russia seemed to have opened in early June, although this was done not by the Foreign Ministry, but by a Supreme Soviet delegation—as well as Taiwan, which is searching for foreign markets. In terms of their financial capabilities, these countries surpass those from which we are seeking assistance many times over—in some cases by entire orders of magnitude.

However, Russian diplomacy is neglecting their potential. They do not fit into the accustomed westernist stereotype that is unthinkingly embraced by our Foreign Ministry. Moreover, some of the ministry's actions seem deliberately designed to prevent the very possibility of such contacts now and in the future. The Atlantic orientation presupposes close ties with Israel, which deters Arab partners. A new push for rapprochement with the People's Republic of China is afoot, something that could preserve the iron curtain in our relations with Taiwan.

The mentality of an enormous superpower, a nuclear giant whose attention is merited only by those with an equal number of warheads per capita, still hovers over the skyscraper on Smolensk Square. But this picture is far from reality. The new reality demands a different world outlook and different priorities.

Our country is one of the world's major suppliers of raw materials. Objective interests bring it closer to those that are in an analogous situation. For example, Russia could play a prominent role in the Organization of Petroleum Exporting Countries (OPEC)—an entity no less prestigious and influential than the International Monetary Fund, but much more promising from our standpoint. Specifically, oil-producing countries have made concrete proposals for participating in efforts to tap the resources of Western Siberia—they don't want to let European and American energy consumers go in there directly. The ball is in our court. Wouldn't it be better to be a leader of the Third World than an outsider in the First World?

Such a shift in priorities would hardly mean abandonment of our Great-Power status. I already mentioned the leverage afforded by our oil exports, which directly determine the political situation in some countries. Russia has something that is in acute short supply throughout the rest of the world: stocks of mineral resources and untapped sales markets. To industrially developed countries in which shortages of raw materials and energy resources are combined with surplus goods production, we present an invaluable opportunity. We are the masters of the situation here—so why are we on our knees asking for charity?

Needless to say, these opportunities will not realize themselves on their own. Turning them into reality will take a lot of hard work. But most importantly, we need a new concept of Russia's national interests—a concept that can make its actions in the international arena productive.

## 3.20

### Stankevich Calls for More Assertive Diplomacy

Sergey Stankevich

*Izvestiya,* 7 July 1992 [*CD* Translation]

It's good that Russia's foreign policy is finally becoming the subject of a rather impressive public debate: Russia's parliament, its vice-president, and its minister of foreign affairs have all joined in. It's bad, however, that sometimes, instead of engaging in a serious, well-reasoned discussion, participants in the debate get sidetracked into combating caricatures of their own making and using too many stereotyped labels.

The interview with Andrey Kozyrev (*Izvestiya,* no. 151 [June 20—see *Current Digest,* 44, no. 26, pp. 3–5]) left me with the impression of an excessively defensive reaction and a certain nervousness that, in my view, was associated with some annoying inaccuracies.

The minister is hardly correct when he asserts that the Ministry of Foreign Affairs' current "moderate" line, the only thing capable of "stopping Russia's slide into the abyss," is opposed by the "war party"—the solid reactionary mass of "red-browns" and their underlings, who "always have just one response—force."

Allow me the liberty of exclaiming: "There is no such party!" Leaving aside the clinical cases (because only a madman could deliberately drag Russia into a war), the problem can be worded differently: How, by what means, can Russia achieve long-term stability that excludes war in relations with its neighbors?

Even in the view of some quite calm, thoughtful, and well-intentioned people, something is wrong with our foreign policy.

Take the situation in the Dniester region. The crisis there has developed over many months and has taken on some very acute forms. Russia's reaction has invariably been belated and weak, and in my view, it has frequently been mistaken.

For example, it is difficult to find any explanation of the CIS member-countries' statement on the Dniester region from the standpoint of Russia's interests. Without even being legally a CIS member, Moldova obtained the consent of Russia's representatives to all its unilateral demands (the withdrawal of troops, the disarming of the Dniester region guards, and the restoration of "legitimate bodies of power") without giving any simultaneous guarantees of a normal existence for the Dniester region's residents.

This "moderate" line, which unties the hands of one side, promises the Dniester region's residents only the stability of the graveyard.

There is serious doubt as to the advisability of Russia's consent to a quadripartite formula for talks on the Dniester region (with Romania, but without anyone from the Dniester region). In the final analysis, the set of problems with which we are dealing is the collective legacy of the republics of the former USSR. And it is the heirs who must deal with it, in this case Moldova, Ukraine, and Russia. The inclusion of Romania in this process is de facto recognition of its special role and special responsibility for events on Moldovan territory, which clearly plays into the hands of supporters of Moldova's absorption by Romania. If this concession is not used now, then, like Chekhov's gun, it will quite likely be fired in the future.

The enlisting of outside forces in the search for a solution could be useful, but on a broader basis—for example, on the basis of the Conference on Security and Cooperation in Europe.

All accords and declarations concerning a peaceful settlement of the Dniester region's problems are still invariably violated, since Moldova's current leadership has categorically rejected and continues to reject the only sensible solution, which proposes a federal system for the republic.

One can only regret that Russia has not yet insisted on an examination of this main question (either by four sides or by two). Misinterpreting the obvious indecisiveness and inconsistency of Russia's representatives, Moldova's leadership has put its stakes on brute force. The carnage in Bendery was a tragedy of international proportions, and to call it, rather delicately, an "overreaction"(!) and argue that the Moldovan army's full-scale military operation might have been a response to an ordinary exchange of fire is tactless, to put it mildly.

Force will never replace diplomacy and law, needless to say, but, to our great regret, no one has yet succeeded in completely eliminating it from the arsenal of instruments of state policy. If the punitive operation in Bendery had not met with a proper forcible rebuff, diplomats would now have nothing to negotiate about. The Dniester region would have been dismembered and crushed.

That didn't happen, thank God. But is it possible or moral to condemn and stigmatize those who stopped the butchery?

It's difficult to understand what is happening to us. We are told that 450,000 Russians live on the right bank of the Dniester, and right away we are ready to give in to blackmail: let the 150,000 on the left bank be slaughtered; if we make no effort to stand up for them, maybe the rest will be spared. Come on, Andrey Vladimirovich [Kozyrev], is it really appropriate to resort to arithmetic and argumentation as to who should have priority in being spared?

Indeed, the dilemma you present—either "occupy the territory of republics with troops" and "institute a savage reign of terror" or resolve everything in a peaceful and civilized fashion—can only be explained as a polemical excess. Of course, Russia should resolve all problems with all its neighbors in a peaceful and civilized fashion—if our partners agree and are prepared to act in the same way.

But if our partners, while holding nice diplomatic conversations with us, at the same time continue the ruthless butchery, and we agree to such a dialogue, very soon we will no longer be taken seriously, since we will agree to anything. Russia's task is not to "bomb cities" but to stop some very specific murderers and to show convincingly that the problems of ethnic minorities cannot be solved with bayonets and bullets, that Russia will not allow it. Only then will they negotiate with us properly. There is not a grain of an "imperial attitude" here. Looking a little further, this is exactly how war can be avoided, too. But if systematic murders are committed and the diplomatic brakes do not work, Russia has the right—pending intervention by an international court of arbitration—to apply unilateral sanctions.

Back in the 1970s, the world community adopted an important principle: human rights are extraterritorial, the systematic violation of human rights is not an "internal affair," and intervention in such instances is necessary and justified. If, however, there are indications of state terrorism, as there are in the Dniester region, it is even more impossible to conceal it by citing sovereignty. Intervention should be civilized and international, needless to say.

Perhaps I am mistaken, but in my opinion our diplomatic

corps is not doing a vigorous enough job of conveying to the leaders of foreign countries Russia's concern over the situation of Russians in the "near abroad." How else can one explain the lamentable fact that in two months no one has officially responded to the Ministry of Foreign Affairs' memorandum on human rights violations in the Baltic countries? In Estonia, a certain armed grouping that is controlled by the state is declaring war on Russian Federation military men and "colonialists," that is, civilians who are our spiritual countrymen. We, on the other hand, having never received any official written explanations from the Estonian government in response to our note—which was belated, as usual—are continuing bilateral talks, in particular, talks on economic ties that are quite advantageous to Estonia. Why this strange Tolstoyism?

A year and a half has gone by since Russia signed treaties with Estonia and Latvia, in which, among other things, the parties guaranteed full civil rights for all ethnic groups. Flagrantly violating these treaties, the parliaments of Latvia and Estonia have deprived 1.5 million Russian people of their basic rights. Nevertheless, Russia still has not insisted that this question be included on the agenda of bilateral talks with Baltic republic delegations. So where is the priority of "civilized methods"? For how many years will we try to get them to listen to us and to take our pain into consideration?

And how did it happen that the interests of servicemen's families were not taken into account when the agreement on procedures for Estonia's withdrawal from the ruble zone was signed? Why weren't Russian Federation services prepared for Estonia's switch to a strict visa system? These problems were anticipated, after all.

There are quite a few such questions. And they are being asked not by the "war party" but rather by the party of common sense, which is concerned about preserving Russia's dignity. Needless to say, behind all this are the transient working difficulties that are inevitable in a great undertaking. But they must be overcome quickly and effectively, because a buildup of such situations will do irreparable damage to Russia's long-term interests.

Finally, one has to say something about patriotism. What unthinkable shift in consciousness has made us turn this into a swear word and closely associated it with the image of a fascistic degenerate? Isn't it clear that only the selfless patriotic impulse of millions can wrest Russia from the grip of a very grave ailment?

And there is no reason to hastily disown what the state says. Unlike an empire, which presupposes the concentration of a country's manpower and resources on goals of external expansion, a state means that the country turns inward, renounces expansion, and mobilizes its internal forces and resources for an economic and cultural upswing and for a peaceful and civilized breakthrough to the level of the great powers.

I have a clear premonition that Russia will never again be an empire, but that it is bound to become a state. We should all do as much as possible to see that this happens.

## 3.21
## Ambartsumov Foreign Policy Concept Viewed

Konstantin Eggert

*Izvestiya,* 8 August 1992 [FBIS Translation]

*Editor's Note:* The following document offers a particularly poignant critique of the foreign policy recommendations by the chairman of the Russian Supreme Soviet International Affairs and Foreign Economic Relations Committee. *Izvestiya* journalist K. Eggert calls Ambartsumov's policy a "Monroe Doctrine," which would only lead Russia to war. The document is noteworthy for its depiction of the Russian parliament's position on the question of the country's vital sphere of interest. That position only hardened over the next three years. Furthermore, the concept of a Russian "Monroe Doctrine" has been popularized by nationalist politicians seeking an identity for Russia, and embraced by naive Russian citizens, yearning for strong leaders. The following document views Ambartsumov's policy recommendations from the democrats' vantage point, giving its contents the quality of a debate between the two forces shaping Russia's post–Cold War foreign policy.

"Gorbachev's perestroika and 'new thinking' signaled a fundamental move away from the idea of confrontation toward the idea of partnership. It should be pointed out, however, that this move as implemented by Gorbachev and Shevardnadze was seen by many as a disorderly retreat and complete capitulation before the West. A similar impression is formed when analyzing our foreign policy as implemented by Kozyrev."

This is not a quote from *Sovetskaya Rossiya* but an extract from a curious document now at the disposal of our editorial office. It is entitled "Recommendations on the Results of Russian Foreign Minister A.V. Kozyrev's Closed Hearings in the International Affairs Committee on the Foreign Policy Concept of the Russian Foreign Ministry (30 June 1992)." Its author is Evgeniy Ambartsumov, chairman of the Russian Supreme Soviet Committee for International Affairs and Foreign Economic Relations.

The document opens with the following statement: "the Russian Foreign Ministry does not have an integral concept of its foreign policy in either nearby or distant foreign parts." It is well known, however, that Minister A. Kozyrev does have such a concept—or at least its basic principles. It is equally well known that it is not to the liking of the present parliament.

The acknowledgement that the former Soviet republics have the status of equal partners, the reliance on painstaking talks rather than imperial blackmail, and the drawing closer to the developed Western democracies on fundamental issues concerning human rights and the curbing of aggression—these are the basic principles on which the Foreign Ministry's line has hitherto been based. There have been errors and defects, of course, although it would be simply impossible to expect the Russian foreign policy department's actions to be completely faultless. After all, the problem of Russia's relations with the outside world has essentially become a problem of how Russia sees itself. In other words, it has become the subject of domestic political struggle.

Evgeniy Ambartsumov's "recommendations" are interesting, but not because they graphically demonstrate with whose side the parliamentary leadership's sympathies lie—everyone knows that anyway. No, something else deserves attention—the unprofessionalism of the guardians of "powerfulness" [*derzhavnost*] and their persistent desire to see the world as they would like it to be, rather than as it really is.

The following is undoubtedly the central tenet of the recommendations:

"As the internationally recognized legal successor of the USSR, the Russian Federation must base its foreign policy on a doctrine that declares the entire geopolitical space of the former Soviet Union to be a sphere of its vital interests (along the lines of the U.S. 'Monroe Doctrine' in Latin America) and strive to secure the world community's understanding and acknowledgment of its special interests in this space. Russia must also strive to secure from the international community the role of political and military guarantor of stability throughout the territory of the former USSR. It should strive to secure the G-7 [Group of Seven] countries' support for these functions of Russia, including hard currency subsidies for its rapid reaction forces (the Russian 'blue berets')."

The "Monroe Doctrine" is 169 years old. It was directed against interference by the European monarchies of the Holy Alliance, primarily Britain, in the affairs of states in the Western hemisphere. It has undergone considerable changes during its existence. Nowadays, the U.S. leading role on the American continent is determined primarily by natural reasons of a socio-economic nature, rather than by the "big stick" of the Theodore Roosevelt era. Russia should certainly aspire to a similar state of affairs.

But no, Ev. Ambartsumov calls us back to the "gunboat diplomacy" employed at the turn of the century, evidently forgetting that the Moldova and Estonia of today are not Panama in 1903 or Mexico in 1916. A Russian version of the "Monroe Doctrine" is untenable because the price paid for imposing it on the former Soviet republics could only be war, which would mean the end of Russia's democratic development. In these conditions, it is ludicrous to expect the world community—which has only just begun to build a system of genuinely collective security—to voluntarily give Moscow the role of "Eurasian gendarme" and, moreover, finance it from its own pocket.

This does not worry the head of the parliamentary committee, however. He continues:

"The texts of future agreements on the CIS and of bilateral agreements must include the idea of Russia as guarantor and, in particular, give Russia the right to protect the lives and dignity of Russians in CIS countries. There should definitely be a special provision regarding the status of Russian troops in CIS countries, in order to prevent a precedent with the 14th Army in Moldova, which has no right in the eyes of the law."

The stationing of troops in CIS countries is referred to as a long-term option and, moreover, in connection with "protecting the dignity of Russians." Clearly, it is a question of imposing a Russian military presence on our neighbors. This approach can only wreck the timid steps toward military-strategic cooperation and integration that have recently appeared. No one will want to have "human rights guarantors" with submachine guns ready in their territory. Russia is undoubtedly obliged to protect and will protect the interests of its compatriots abroad, using the entire arsenal at its disposal except one—military force.

Russia's policy in so-called "distant foreign parts" is given literally half a page in the document:

"We must be clearly aware that in distant foreign parts, Russia cannot play the USSR's former role, which was based on military might, 'undermining imperialism from within' (a reference to the KGB in the Third World—Africa, Asia, Nicaragua, Cuba, Angola, and Vietnam), and subsidizing 'fraternal communist parties.'

"Hence the need for partnership with the United States and the other G-7 countries, while preserving an independent foreign policy line."

It is hard to find fault with any of this, especially as the Foreign Ministry takes exactly the same stand at present, but the conclusion that follows is surprising:

"China could serve as a national model for Russia's independent foreign policy, as it has equally stable relations with Russia and the CIS, and the United States, and the G-7."

It is not at all clear why China has been chosen as an example for emulation. Beijing's relations with Moscow and Washington are totally undynamic, static, and moreover, very far from ideal. The apparent balance is the result of the present communist Chinese leadership's reluctance and, what is more, inability to develop its relations with the outside world beyond a certain point. We must not forget that the ideological reasoning behind China's foreign policy is radically different (or, at least, should be radically different) from that of the Russian leadership. Proposing that our Foreign Ministry follow the Chinese model in its activity means forgetting this fundamental difference.

Even if you agree that Russia's present foreign policy lacks conceptual depth, Deputy Ambartsumov's proposals are unlikely to improve the situation. Rather, it is the opposite: If this approach prevails, Russia can expect only wars, crises, and international isolation.

---

## 3.22

**Russia Is Again Receiving Compatriots From Abroad**

Valeriy Rudnev

*Izvestiya,* 7 September 1992

[*CD* Translation]

---

*Editor's Note:* The national-patriot movement is partially organized around strong feelings among Russians concerning compatriots "caught" living in foreign countries after the dissolution of the USSR. Part of the imperialistic Great Power movement is dedicated to defending the rights of Russians living in the "near abroad." This defensive mission has been broadened, however, to rationalize the national-patriots' desire to regard all Russian-speaking peoples as compatriots deserving dual citizenship, and to extend Russia's peacekeeping role throughout the CIS. As the following document shows, Russians are aware of the sensitivity connected with interference in other states' internal affairs, but they view "Russianism" as a cultural distinction—deserving a special status within the commonwealth they are trying to build.

The second congress of compatriots opened on 7 September in St. Petersburg. Representatives of more than 10 million people of Russian Federation origin who live outside their homeland [the reference is apparently to those who live outside the former USSR—Trans.] are attending it. One of the main items on the agenda is discussion of the underlying principles for a law on state policy with regard to compatriots. The principles were drawn up by a group of experts from the Independent Institute of International Law under the direction of Professor Igor Blishchenko.

The scholars define compatriots as people who were subjects of the Russian Empire or citizens of the USSR in the past (and their direct descendants), who do not hold Russian citizenship at present, who belong to one of Russia's ethnic groups, and who consider themselves spiritually, culturally, and ethnically linked with Russia. The experts believe that there should be no distinctions in legal status between compatriots who live in the republics of the former USSR (who number nearly 26 million) and those who live in states outside the borders of the former USSR. On applying to a Russian consular institution, all should be issued the same certificate.

However, it's not all that simple, the legal experts from the Independent Institute of International Law warn. One difficult problem is the question of citizenship, the resolution of which will largely determine the situation of Russian compatriots abroad. One solution to the problem is to recognize the institution of dual citizenship.

Another difficult question is, what rights should compatriots be given? It must not be forgotten, the institute's legal scholars point out, that compatriots are usually citizens of another state. And so we should take a cautious approach to granting them rights because an imprudent step could be viewed as interference in [that state's] internal affairs. In such cases, it should be stipulated that in exercising the rights conferred on him by Russia, a compatriot must remain loyal to the state of which he is a citizen.

What does compatriot status do for a person? On the basis of the certificate he receives, the person possessing it can enter Russia at any time on a visa-free basis; is entitled on Russian territory to equal treatment under civil law in the areas of everyday life, entrepreneurial activity, property relations, and freedom of movement; and has the right to use a simplified procedure to obtain Russian citizenship.

As regards privatization, the scholars believe that compatriots should have a preferential right to buy property that they used to own directly or obtained by inheritance. It is suggested that this right would have to be exercised within a certain period of time after the beginning of privatization. Other possible rights include preferential treatment in capital investment and entrepreneurial activity, exemption from some taxes, and other privileges (in all other respects, they would be on equal footing with foreigners).

For its part, Russia is obliged to provide protection for its compatriots. The law should have a special provision on protecting the rights and interests of people of Russian origin, including assistance in finding employment. And here, in addition to general international legal principles in the area of human rights, it is essential to define the rights and responsibilities of Russian officials in providing aid and assistance to people of Russian origin.

The scholars are convinced that a special section of the law should stipulate the right of people of Russian origin in the ex-Soviet republics to automatically obtain Russian Federation citizenship as a second citizenship (citizenship of Russia and of the state in which they live). In the event of repatriation to Russia should a compatriot choose only Russian citizenship, the law should require the Russian government to demand that the state the repatriate is leaving compensate him for all material expenses associated with his resettlement and accommodation in Russia and guarantee the repatriate's right to dispose freely of property he owns on the territory of the state in which he formerly resided.

The experts from the Independent Institute of International Law believe that the adoption of a law on state policy with regard to compatriots will help draw to Russia the intellectual potential of all people of Russian origin and their capital, technical and other knowledge, and ability to manage modern-day production and engage in business operations.

## 3.23

### Rutskoy Article Views Nationalist Trends, Sovereignty

Aleksandr Rutskoy
*New Times International,* 10 March 1993
[FBIS Translation], Excerpt

*Editor's Note:* Former Vice-President Aleksandr Rutskoy was an early leader of the national-patriot faction in parliament. In October 1993 he led the parliament in forcefully resisting President Yeltsin's attempt to dissolve it and call for new parliamentary elections. Rutskoy was imprisoned for his role in the parliament's armed attempt to capture the Moscow radio tower and several other strategic institutions. After leaving prison, Rutskoy established the political party "Power" (Derzhava), which has gained a strong following.

["Abridged" reprint from Russian Academy of Sciences anthology *Forum* of an article by Vice-President Aleksandr Rutskoy, "We Need Each Other."]

Today's reality keeps us permanently on edge due to newspaper publications, TV news, and all that happens around us. The main tensions are connected with "hot spots" because they are a manifestation of troubles in the intra-national relations.

My calculations reveal a total of over 180 actual and potential "hot spots" on the territory of the former USSR, with various degrees of civil unrest. Ethnic tension in the form of open hostilities has already reached Russia's borders and now, having crossed them, is spreading over onto her territory (Ingushetia, the Chechen Republic, North Ossetia).

The system of multilateral agreements, including the CIS Treaty, isn't working. Moreover, relations between the states that we call the "near abroad" were built on mutual distrust from the very outset of their independent existence.

The disintegration of the USSR aggravated the problem of the "Russian-speaking population" and of the status of the army in countries that emerged on the territory of the former USSR. The struggle against "migrants," "occupationists," and others, assuming various forms, shifted from the plane of moral, psychological, and social pressure to the state and legal level and is in many cases raised to the status of official policy.

A dangerous combination of seemingly incompatible tendencies is taking shape. The ex-Union republics carried on the struggle with the former omnipotent central power under the banner of the necessity to broaden their peoples' rights and of securing the right to national self-determination. Having come to power, the same forces not only began to profess the principle of territorial integrity, they started conducting a policy akin to a struggle for "racial purity" coupled with onslaughts on universal human rights. The situation in some Baltic states is undergoing a dangerous metamorphosis: Nationalist passions are whipped up under the disguise of a struggle for democracy, while our Foreign Ministry complacently looks on, trying to "talk" Latvia into having mercy upon her "Russian-speaking" population.

We are facing an increase in the stream of refugees into Russia caused by many factors (the number of refugees has exceeded 1 million people, and it is growing daily). Among them are people of diverse nationalities. The state must pay out vast sums of money to support and accommodate them. What's more, it gives rise to social instability felt in the political atmosphere. Sadly enough, these realities have not yet been fully realized by the country's leadership and the uncontrolled accumulation of explosive material threatens to reach a critical mass. Unfortunately, social and ethnic tensions are growing in regions of ethnic conflict.

#### Enter Mafia

What triggered these conflicts? The reasons for them were common to a number of states in the "near abroad." The most important are: aggravation of the economic and political situation, border problems, traditional ethnic conflicts, violation of human rights, ill-thought-out and uncontrolled migrations that took place in the past, and the environmental crisis.

One more important factor promotes the escalation of conflicts—this is the struggle between local elites and mafia-like groups assuming the form of ethnic conflicts. The circumstance that impairs settling such conflicts is their irrational character, incomprehensible to an uninvolved onlooker, which limits the possibility to influence these developments from the outside.

That people have become accustomed to the tragic nature of events is also very dangerous.

Some symptoms of the oncoming danger are already being observed in Transcaucasia, North Caucasus, Moldova, Georgia, and Tajikistan. I name only the most dangerous and obvious hotbeds of confrontation.

It's also clear that current controversies won't disappear in the near future and that state and public organizations must do their best to settle them by peaceful means. All of us have to learn restraint, mutual understanding, and the art of compromise. A paradoxical situation arises: Many newly born states, ex-Soviet republics, successfully establish friendly relations with many countries of the world, but can't come to terms between themselves. History has made all of us neighbors, we have been neighbors for centuries, and we'll be neighbors in the future. The culture of interdependence is a prime necessity to our states, a sine qua non of survival.

**Not Just a Large Country, but a Great State**

The disintegration of the USSR has left in its territory a number of states, the Russian Federation being the largest among them in terms of territory, size of population, and scientific and technical potential. Russia has lived in the past, and is now living through great hardships, yet she goes along her path as a great power. In her ordeals, she has suffered considerable losses, yet she has largely preserved her colossal potential and her international authority backed by her economic might, cultural traditions, and scientific achievements. Getting rid of stagnation tendencies that degraded our society from above, we must preserve our historical achievements inculcated in popular tradition. If we manage this, we will continue as a great power; if we fail, we'll leave the historical scene. I see nothing wrong in the desire to remain a great power. Being a great power is a result of advancement in the key branches of human activity. It's evidence to the fact that a nation has found its path among civilizations of the world. This has nothing to do with chauvinism. Pride for one's country is a natural and worthy feeling.

Russia is a sovereign state today. Its economic organism has become more integral and its social and cultural landscape smoother and more comprehensible. On the other hand, Russia has "retreated" from its former borders, losing many of its seaports; many of its traditional economic and other ties have been ruptured.

We in Russia must take all this into consideration in searching for the place our country is to occupy among the new sovereign states. We must embrace the whole range of relations, but we must give priority to Russia's own interests, striving wherever possible for their overlap with other countries' interests.

I believe that, in formulating Russia's policy vis-à-vis the countries of the "near abroad," we must proceed from the premise that, apart from separate interests, there are also factors dictating the necessity to consolidate comprehensive relations between these countries, the most important among these factors being the historically determined interdependence of their economies, the problems of their common "Union" heirloom, traditional, cultural, scientific, and other ties, and the multinational character of their population.

Formulating the principles of mutual relations with countries that have emerged instead of the former Union republics, it is necessary to take into account that the steps we make today in this direction will determine the nature of relations with these countries not only in the near future but in the remote future as well. These relations are only being initiated now, but they are going to be of great importance, and therefore they require circumspection, as well as mutual honesty, good will, a principled attitude, and recognition of each other's lawful rights. Unfortunately, this is not happening. Too many accusations are hurled at Russia against the background of unrestrained praise for Turkey, Japan, the United States. . . . I'd say it is immoral to reduce Russia's role in the history of peoples of the former USSR and the role of, I emphasize it, the former Russian empire to the Molotov-Ribbentrop Pact alone. It is noteworthy that Russia, unlike a great many countries of today's "near abroad" and some republics inside her, does not abuse historical arguments in justifying her position. Such attempts must be avoided because the ensuing disputes always assume a heavy and dragged-out nature. Yet Russia has something to counter her popular description as the "prison of peoples." Yes, Russia did wage unjust expansionist wars, but which nation or country has not? However, it will be extremely unjust to reduce our history to them alone.

**Not to Build Dams**

I often consider the question: Is it possible to establish normal interstate relations and even to create a new close-knit alliance on principles differing radically in comparison with the Soviet Union? A rather close one, to the point of creating a unitary confederative state? Certainly, this is not a task for the near future, but it is quite realistic. And, in any event, establishing civilized relations will be an obligatory stage in this process.

This being so, a new integration can start from a rather low level, even ground zero. Already today political practice shows that one cannot intercept the former Union republics halfway on their route to sovereignty. The ensuing process must be carried through and, hence, it would be wiser to support these powerful streams, rather than to build dams in their way.

With those who are prepared for this integration, the establishment of confederate relations can be realized at the interstate level; with those who are not, good-neighborly relations can be developed.

The new type of ties and relations between the countries formerly in the USSR must be built on the principles of mutual economic space. It is not only desirable but necessary for Russia to base her formal relations with these states on the same principles as with any other country of the world.

We must become accustomed to this politically and psychologically new situation in which Russia's relations with former Union republics will henceforth be relations between sovereign states with all the attendant consequences. Trade and diplomatic relations, payments, and border and customs control must be carried out on the basis of these republics' sovereignty. There must not be any "blockades," "sanctions," etc., that could serve as pretexts for blaming Russia. Russia has its own national interests to follow, using all its resources to the advantage of its people. No concessions against the country's interests are admissible.

A few remarks in this connection, concerning the idea of "transparent borders," are being actively discussed at various levels today. On the surface, this idea looks democratic and alluring. Yet the question of to whom and for what our borders will be transparent is not an idle one. We uphold transparency for the sake of freedom of human contacts. But we know that the present openness and uncertainty at our borders daily causes the country enormous damage. They export from Russia whatever they want—suffice it to quote the example of Estonia, which, having no deposits of nonferrous metals, has become one of the greatest suppliers of them on the world market. Many other "undesirable" goods, such as weapons and drugs, are imported to the country through Estonia. Russia is totally unprotected against industrial and any other kind of espionage. We do not want this sort of "transparency."

Sovereign Russia must have completely equipped state borders with all countries, regardless of its current relations with them.

It is too early so far to abolish state borders. The Russian leaders and citizens must unfailingly observe Russia's and its people's interests. We must also secure that these relations take on a normal form and that human rights, including freedom of travel, be observed in all countries.

### To Protect the Rights of Russians

Demanding the observance of a normal frontier, we should remember that 25 million Russians and 3 million representatives of Russia's other peoples live outside the Russian Federation: 43.5 million so-called Russian speakers live in the "near abroad." In short, it is inexpedient to restrict freedom of travel both for "universal humanitarian" reasons and proceeding from the concrete situation in what once was the USSR.

And yet it must be noted that today it is the Russian citizens living abroad who need protection and not the other way around. The neighboring countries must frankly admit that they have no grounds for blaming Russia in this respect. But working out and implementing a conception of protection for the rights of Russian citizens abroad in conformity with the norms of international law is urgently needed. A special concern is protection of economic rights of those ethnic groups whose only historical motherland is Russia and who recognize it as such. One more special concern is introduction of the economic responsibility of foreign countries for the fate of those of their residents who have decided to take Russian citizenship.

The social approach that has been prevalent thus far has suppressed the factor of a person's national identity. The approach from the point of view of human rights implies the unquestionable esteem for it.

### To Avoid a New Yugoslavia

The demand for national self-determination has become one of the most important ones in the present political situation.

However, this just demand plays a destabilizing role in today's political practice. If urgent measures are not taken, this will, in the near future, bring about a total confrontation over the national issue in the country, a second edition of Yugoslav-type developments on a larger scale and with more tragic consequence for the entire world.

In this connection, the urgent task today is to start a struggle with the dangerous stereotype deeply rooted in the mass mentality and according to which the right of nations to self-determination is interpreted exclusively as their right to create national states. This is precisely the nature of many such demands (by Germans of the Volga Region, the Crimean Tatars, the Armenians living in Russia, etc.). The danger of such demands is not duly appreciated.

The principle of creating national administrative formations within Russia has been a historical mistake whose destructive consequences have yet to be fully realized. It has no future for a variety of reasons, including those giving rise to the problems of "indigenous" and "non-indigenous" peoples, border problems, territories, status of such formations, etc. The mounting demands to create national states within Russia result in a growing opposition from the population on whose lands "encroachments" are made; ethnic claims are multiplied and are further whipped up.

The only solution possible in present conditions, promising the possibility to settle similar problems in the future, is to approach this problem from the point of view of individual human rights, to endow the demand of national rights with real weight, that is, to really observe these rights in the cultural, religious, economic, and other spheres. It is also necessary to take into account the psychology and specific economic biases of certain peoples.

Assessing the situation in the country, one should proceed

from the fact that Russia is not an artificial but a historically formed entity. It is a unique state in its size and its multinational character.

In the situation of disintegration of the USSR and declaring state sovereignty by former Union and autonomous republics, we must proceed from the idea of the country's sovereignty recognized by all citizens of Russia, which, in our opinion, must be regarded as a unitary state of all peoples living on its territory.

We believe that building a state not on the principle of national statehood but on that of its division into administrative-territorial units is more fruitful, more in keeping with the worldwide practice, and not contradicting Russia's historical experience. In our opinion, this principle has a future. If we follow it, our country will be able to avoid national conflicts in the future.

At that, the transition to this system should be carried out in stages and must be accompanied by the establishment of new and the restoration of old economic ties and large-scale cultural educational work. This transition can be carried out only by means of consolidating economic, cultural, and administrative independence of both national and national-state formations, and administrative regions and entities, so that they could obtain true rights as parts of Russia.

Some people in the former autonomies fear this prospect without any reasons at all.

If complete and consistent observance of national rights becomes a reality, the problem of national statehood, as it is conceived today, will become less acute. The transition to dividing Russia into administrative-territorial units will become organic and devoid of conflicts.

This is precisely the basis for securing equal rights of peoples, regardless of their numerical strength, and for creating a genuine commonwealth of nations.

We have to admit with regret that the disintegrational processes resulting in the crash of the USSR now also affect the Russian Federation. The state is faced with a real danger of disintegration with all the ensuing political, economic, and social consequences. We have reason to believe that preservation of normal ties and relations within Russia is particularly important for the development of regions as a consequence of historical, economic, and other interregional ties having long-standing traditions.

**Under Damocles's Sword of Disintegration**

On the other hand, destabilization of the situation in any of Russia's regions will immediately affect the whole country. Tension has been growing in other regions as well. We have all reason to affirm that the range of social conflicts based on national factors has dangerously grown, and if the critical line is crossed (the danger of which seems to become more and more real), this may cause more serious consequences than those which accompanied the disintegration of the Soviet Union, primarily because the interests of these regions are much more painfully and closely connected with the age-old Russian territories.

One of the realities of our present life is the obvious weakening of Russian statehood. The role played by the state and its influence on society have diminished; vitally important institutions of statehood have grown weaker. This process was initiated by the struggle of certain social forces against the all-embracing role played by the state apparat, which controlled practically all spheres of social and individual life for decades.

At a certain stage this struggle degenerated into a struggle with the state as such and led to anarchy and disorder in many areas of social life. This tendency is extremely dangerous, all the more so since society is not duly aware of it as yet.

National rights are doubtless one of the most important components of human rights in general. The national factor includes many aspects of human personality: mentality, the ethno-cultural type, religion, social priorities, ideological bearings. Given all that, it is necessary to admit that protection of human rights, in general, unavoidably presupposes protection of human rights of people precisely as representatives of a definite national community, since there are no abstract supranational human beings.

Let us ask ourselves the question: Can former autonomous entities in the Russian Federation claim state sovereignty today, that is, the same rights as Russia has? I do not think so. Russia is sovereign; not so its parts. It is inadmissible to make the Russian statehood and civic concord in the country objects of possible blackmailing actions. Autonomous formations will not be satisfied with sovereignty; they will seek new alliances and will find them.

All peoples inhabiting Russia and the former USSR raise the question of their national peculiarities, traditions, and national self-awareness. The most numerous of the nation, Russians, have until recently remained the only ones unaffected by these processes. The situation is changing now. The rise of Russian national self-awareness is already a fact; it will unavoidably gain momentum and exert a tremendous influence on the situation in the country.

The responsibility for mistakes in the nationality policy and for crimes committed against a number of peoples is not infrequently laid on the Russian people. Complex historical events are far from impartially interpreted to the latter's disadvantage; present generations are blamed for real or ostensible faults by the leaders who ruled their ancestors' state. Of course, I do not mean whole nationalities, but in some regions there are forces professing such views—they play a conspicuous role and affect the social and political situation there in the most negative way. Some politicians have made nationalism their banner, trying to come to power and retain it with its help.

The accusations of Russification and anti-Semitism become louder and louder, which resembles the situation in the 1920s on the eve of such tragic events as the dispossession of peasantry and the mass repressions against the core of the Russian nation. The danger of such accusations is that they may give a second breath to the idea of "the Russian republic," which will mean a new stage of "national squaring-off of accounts" and a direct threat of civil war in the country.

After the disintegration of the USSR, and as Russia's sovereignty was consolidated, the demographic situation in our country radically changed; this applies first of all to the place and the role of the Russian people in society.

In the USSR the latter made up about 40 percent of the state's population. In the present Russian Federation Russians constitute over 80 percent of the population. Today no other people in the country can compare with Russians in numerical strength.

Certainly this does not entitle Russians to a special legal status. The Russian Federation has been and remains a multinational state. The principle of equal rights for all citizens of a state, regardless of their national identity, must be observed unquestionably as one of the basic premises of a democratic society.

Yet the above peculiarity of the country's ethnic composition cannot be disregarded in formulating social, cultural, and political strategies. In particular, no efforts in economics can be successful without it. The policy of reforms can be successful and beneficial to the people, not when it copies a foreign model, but when reforms are worked out and implemented on the basis precisely of this people's historical experience, traditions, business ethics, and mentality. Russia's revival is possible only on this basis. Russia is not a dwarfish state or a banana republic. Unfortunately, the situation is aggravated by the recently initiated scholastic yet dangerous disputes, for instance, whether the Russian nation at all exists. The notion "the Great Russian nation" has come into being and is currently rather popular. These discussions aggravate the already explosive situation against the background of the growth of other people's self-awareness.

The rise of the Russian people's national self-awareness should not be interpreted as an upsurge of jingoism. Today no people in Russia, numerous or otherwise, having a national statehood or not, can be denied the right to national revival. Common approach to each nation and its rights is the obligatory condition for establishing genuinely equal rights.

In conclusion, we note that we have dwelt mainly on what disturbs us today in Russia's relations with the countries of the "near abroad." There are obviously many problems here, yet there are positive changes too. . . .

## 3.24

## Ambartsumov on the Shaping of Foreign Policy

Interview by Sergey Tikhomirov

*Rossiyskaya Gazeta,* 29 May 1993 [FBIS Translation]

*Editor's Note:* As the split between President Yeltsin and the Russian Supreme Soviet became wider, journalists and public alike began to identify Evgeniy Ambartsumov with the parliament's competing foreign policy doctrine. Ambartsumov was a prominent spokesman for the national-patriot foreign policy position prior to the Supreme Soviet uprising on 3–5 October 1993.

[Interview with Evgeniy Ambartsumov, chairman of the Russian Supreme Soviet Committee for International Affairs and Foreign Economic Relations, by Sergey Tikhomirov.]

[Tikhomirov]: Earlier this year the long-awaited concept of our state foreign policy emerged from the depths of the Russian Foreign Ministry. Quite recently we also learned of the concept on this topic elaborated by the Russian president's Security Council. It would be interesting to learn your opinion of these documents and also whether the Supreme Soviet has an analogous concept.

[Ambartsumov]: No normal state can or should have two or three concepts of foreign policy. It is simply irrational, the sign of an anarchy-ridden state. That the government and the Supreme Soviet may, for example, have disagreements over particular provisions of it is another matter. However, the disagreements should be removed, eliminated by coordinating their positions and elaborating the optimal solution.

[Tikhomirov]: What is your attitude to the Security Council's concept?

[Ambartsumov]: A paradoxical situation has come about, you know. The concept confirmed by the president is a classified document. And although our committee and I, in particular, took part in its completion, I have not seen it since it was approved by the president. And I personally do not understand how we can put this concept into effect when we do not know what it consists of.

As far as I am aware, the concept was drawn up as follows. There was Yuriy Skokov as leader of the Security Council Foreign Policy Committee. Its apparatus, with the participation of various departments, above all the Foreign Ministry, and parliamentary committees, particularly the International Affairs Committee, prepared the concept. It then clearly underwent some changes and was signed by the president.

Meanwhile Skokov had left. Whether this might seem to repudiate the document in some way I do not know. I ask myself that question. And I have no answer to it. But the concept has been confirmed and has acquired legal force.

When I asked at the Foreign Ministry where the concept was, it was explained to me that it is a classified document.

Clearly, the document could contain, let us say, some formulations that it is inadvisable to spread around. For example, it might say that some state cannot be trusted, that friendship will not be possible with another, and so forth. Naturally such things should form a classified section of the document.

Incidentally, the foreign policy concept elaborated by the Foreign Ministry does differ with respect to certain questions and in its form from the one prepared by the Security Council, although there are no particular inconsistencies in the approaches. I would specially observe (I express my own standpoint) that I see no point in the existence of two organs—the foreign policy commissions of the Security Council and the Foreign Ministry. There should not be two foreign ministries, as it were. That the president should in current conditions, with the inevitable strengthening of the presidential principle since the referendum, have his own apparatus, something like aides on security, domestic, and foreign policy problems, is another matter. They should pay more attention to foreign policy strategy and defining immediate goals and priorities.

[Tikhomirov]: How, then, can the Foreign Ministry work without having a concept?

[Ambartsumov]: Well, people are born and live their lives. They do not at the beginning work out a concept for themselves. In general they live without one. The postulate that we get nowhere without a program is a purely socialist brainchild.

I observe that the Security Council document is more incisive or, rather, more candid and pragmatic, and is also considerably shorter than the Foreign Ministry one.

[Tikhomirov]: So the latter is not currently operating, then?

[Ambartsumov]: No, why should it be operating? The document says that the Foreign Ministry concept was discussed and approved as a whole at a session of the Foreign Ministry Council for Foreign Policy with the participation of members of the parliamentary Committee for International Affairs. The Foreign Ministry reported on the document to the president and proceeds on the basis of its provisions in its practical activity. Incidentally, there has been no official endorsement of it by the president, although it may not have been required.

Again we are talking about confusion in our state here. Yuriy Skokov strengthened his position, as it were, and set up his own foreign policy commission. When he was organizing it, I asked Kozyrev what sort of commission it was, and he said that he did not know. At the time it seemed that Skokov was gathering strength and influence, but you see how it ended. However, as the saying goes, the man may have departed, but his work lives on. . . .

[Tikhomirov]: Yugoslavia is on everyone's lips today. The events in that unfortunate country are being interpreted in different ways in Russia. Our Foreign Ministry has its own perspective. What is parliament's position?

[Ambartsumov]: We do clearly have differences with our diplomats, above all over how they see the situation in Yugoslavia. Until recently our Foreign Ministry regarded it through the eyes of the West. But that viewpoint is not always sufficiently objective or appropriate. And the story of our attitude concerning the events in Yugoslavia is quite typical in this regard.

We clearly supported premature recognition of Croatia and Bosnia-Herzegovina, although the latter had never existed as a single state. And its emergence as a single state is at variance with its previous constitution. In my view, this premature recognition served to spur on the grievous ethnic war. Incidentally, I warned eighteen months ago—even earlier in fact—that this step should not be taken.

And now, for example, my counterpart from the West German Bundestag (and he is not alone) agrees with me that it was a mistake. Most of our partners in talks support such a stance.

Moreover, take the well-known decision on sanctions made in May last year. The Serbs' responsibility for ethnic cleansing and killing people is indisputable. But all the warring sides—Serbs, Croats, and Muslims—are guilty, albeit to varying degrees. Sanctions, however, were taken only against Serbia.

In this situation the Supreme Soviet proposed modifying our diplomatic service's course and opposed the sanctions. And, after all, they have really come to nothing to date, and have resolved nothing. Despite all this, the Supreme Soviet has never sided unconditionally with Serbia. Some members—yes. And now modification of the Foreign Ministry's course has occurred. Even a short while ago the Foreign Ministry would certainly have supported the proposal on armed Western intervention in the conflict; however, the Kozyrev-Churkin plan, which was drawn up with our participation, is based on the need for a political solution of the problem.

In other words, the differences in approaches have led not to the victory of one or the other but to parliament's view of the problem being taken into consideration to a considerable extent.

I will take yet another aspect. Long before my election as committee chairman, treaties regulating bilateral relations were concluded among the member states of the former Union. It turned out, however, that the concepts of those agreements were not properly studied. Yes, we granted the new states independence—it was not won by them. Without the support of democratic forces and actions in defense of the inhabitants of Tbilisi or Vilnius the acquisition of independence by the new states would have been very problematic.

Thus, even before the disintegration of the USSR, when bilateral treaties began to be concluded they did not take account of one essential element, namely, the rights of our fellow-citizens. The shortsightedness of both the Foreign Ministry and parliament told here. No strategy for the defense of the Russian-speaking population or for its equality in the countries of the "near abroad" was elaborated. Our fellow-citizens ended up second-class citizens against their will.

We all have to work on this now. At the start there was intense pressure on the government and the Foreign Ministry by deputies. And now the Foreign Ministry itself is taking appropriate steps—witness Kozyrev's refusal to attend the last session of the Council of Europe, where Estonia was admitted.

[Tikhomirov]: What specifically are you and your committee proposing?

[Ambartsumov]: Well, given that the Soviet Union was dissolved with such haste and lack of prudence, it is essential to proceed from the new realities. From the very outset I regarded the disintegration of the USSR as a mistake. No, I do not mean that it should have been preserved, but that the process of the "divorce" should have been conducted more smoothly. Incidentally, I recently met with the former German foreign minister, Genscher, and he asked me: "But why did they destroy the Soviet Union?" To which I said: "In my opinion, in order to get rid of Gorbachev." He was surprised: Was not that too high a price? Yes, the cost of the collapse of a great power was indeed too high a price to pay.

Thus we have to take an active part in the political negotiating process with the new states based on the realities, which is being done by our committee members. We have, by the way, played a large role in the talks on the Dniester region, the Moldovan question, Ossetia, and so forth.

We are simultaneously calling for cooperation by the parliaments of the new states. We have begun with the exchange of delegations. We will together search for ways of resolving urgent issues.

[Tikhomirov]: Please say why you did not support Kozyrev when he refused to travel to the Council of Europe in Strasbourg.

[Ambartsumov]: Why did I not give my support? Of course I would have preferred the Council of Europe to have refrained, or deferred the question of Estonia's admission. But I do not think that Kozyrev's demarche was an appropriate method. We have experienced all these boycotts, ostentatious walkouts from sessions, and so forth. I prefer civilized methods in politics. . . . .

## 3.25

### Academics Present Report on CIS Integration

Marat Salimov

*Kommersant-Daily,* 15 July 1994 [FBIS Translation]

*Editor's Note:* The article below is included as evidence of the academic institutions in Russia that have encouraged the growing conviction that the collapse of the Soviet Union was the cause of the economic malaise being experienced by each former republic. Instead of analyzing the inherent weaknesses of the former system of centrally controlled trade and economic linkages, and pointing to that system's responsibility for the post-Soviet economic situation, many statistical reports have been skewed to emphasize the reduction in "deliveries" (not real trade) and to use this as evidence of the need for reintegration.

At yesterday's press conference, academicians Oleg Bogomolov and Stepan Sitaryan presented a report on "Problems of Reintegration and the Establishment of the Economic Union of the CIS Countries." The report suggested that processes of disintegration are still growing in intensity within the territory of the former USSR, and that Russia has to make a much more vigorous attempt to restore trade and economic ties between the Commonwealth countries in this context.

The report was prepared jointly by the "Reform" International Fund and the academy's Foreign Economic Research Institute and International Economic and Political Research Institute. The authors advocate quicker reintegration in the post-Soviet zone. Their arguments are based on the premise that the disruption of cooperative relationships and trade contacts lie at the base of the exacerbation of the economic crisis and the recession in the CIS countries. The following statistics were cited as corroboration. Whereas commodity exchange between the union republics in 1988 constituted (in terms of the GNP) 13 percent for Russia, 27 and 29 percent for Ukraine and Kazakhstan respectively, and from 34 percent to 50 percent for the other republics, the indicators are now less than half as high as they were, and the reduction of reciprocal deliveries is exacerbating the recession in the CIS countries. The authors of the report referred once again to statistics that were popular at the time of the "rebirth of the Russian state," indicating that the complete severance of economic ties between the former republics of the USSR would lead to a situation in which Russia could secure the production of 65 percent of the final product independently, with respective figures of 31 percent for Azerbaijan, 27 percent for Kazakhstan, and 15 percent for Ukraine.

The academicians' conclusions are clearly delayed, because data obtained by *Kommersant-Daily*'s experts suggest that the low point of disintegration has already been reached in the CIS countries. Specific economic policy moves provide evidence of this. The countries are already working on the creation of supranational administrative bodies for the economic and payment union, not to mention the persistent efforts of Belarus and Tajikistan to become part of the ruble zone. The academicians are correct in their assumption that the speed of reintegration and the correspondence of this process to Russia's national interests will depend primarily on Russia's own position on this matter, especially now that there are new political leaders in Kiev and Minsk.

## 3.26

**Rutskoy Discusses New Opposition Movement**
Interview by Petra Prochazkova and Jaromir Stetina
*Lidove Noviny,* 25 July 1994
[FBIS Translation]

We met Aleksandr Rutskoy at the meeting of opposition politicians in Moscow. The general had the seat of honor in the first row.

[*Lidove Noviny*]: Do you recall the Lefortovo prison often?

[Rutskoy]: Sometimes it comes to me. . . . A cell of two by four meters, a small window. . . .

[*Lidove Noviny*]: Are you not afraid that you might be heading back there?

[Rutskoy]: What can I do? I served time in a Pakistani prison. Now I have spent some time in ours, one can survive anything.

[*Lidove Noviny*]: As opposed to Ruslan Khasbulatov, you have returned to political life, appear at the opposition meetings and demonstrations. . . . Is it revenge against the present leadership, or some other motivation?

[Rutskoy]: My role in this process is completely different now. I am working very hard on the establishment of a social-patriotic movement called Derzhava (Power).

[*Lidove Noviny*]: Why the name Power for the movement you are organizing?

[Rutskoy]: The principal point in the program of this movement is the revival of Russia as a power.

[*Lidove Noviny*]: That is not only your goal—Mr. Zhirinovskiy's objective is the same. . . .

[Rutskoy]: His ideas are completely different. I heard, for instance, that he wants to wash his boots in the Indian Ocean [reference to Zhirinovskiy's autobiography—the statement refers to Russian soldiers washing their boots in the Indian Ocean]. The program of my movement will be published in about a week. It discusses the revival of Great Russia in detail. I have been considering it for almost two years.

[*Lidove Noviny*]: You are not alone. Calls for a confederation or some other kind of union of the former Soviet republics are heard in Russia ever more frequently. . . .

[Rutskoy]: No confederation can regain stability in the geopolitical space where the USSR used to stand. Look what happened to the Commonwealth of Independent States—it disintegrated even before it started to exist. There is no place in nature for confederations or some peculiar commonwealths. The only possibility is the unification of nations and nationalities into a single nation. A single economic domain and ruble zone must be created within the 1915 borders.

[*Lidove Noviny*]: What about the states that will not want to enter such a union?

[Rutskoy]: They will have to subject themselves to international norms: that is, trade in world prices, customs, and borders. If they insult us, we will close all their taps, channels, and electric-power network. And we will congratulate them on their complete independence. After just two months, these "sovereign" regimes, which rule there today, will be toppled by their own people, who will join us. Thus, we will gradually create a power that used to exist. It will be done peacefully in a civilized manner.

[*Lidove Noviny*]: So, you want to revive the USSR?

[Rutskoy]: That is an oversimplification. If we look at the map of 1915, we will see that there are no independent states, and even fewer states on which nothing depends. There is only Russia—a great *derzhava*. There is only one possibility—the revival of a state power uniting all nationalities in a single large family. It is the return to the historical truth, to Russia of 1915. Since a lot has changed, however, only a revival within the USSR borders is feasible.

[*Lidove Noviny*]: What do you need to attain these goals?

[Rutskoy]: Power. The main goal of my movement is to take over the power. As long as we do not have power, we cannot achieve a thing. Our minimum program is to obtain a majority in the next election.

[*Lidove Noviny*]: It is no secret that you are also preparing for the presidential elections. What kind of a president do you think Russia needs right now?

[Rutskoy]: Since time immemorial, Russia has been ruled by a strong personality. Everything depended on what kind of a person he or she was. When it was a moral and intellectual person, who loved his or her country, Russia flourished. When it was someone with a crooked character, Russia was debased, just like it is today. Russia again needs a positive figure.

[*Lidove Noviny*]: You believe that Russia has been debased?

[Rutskoy]: Russia today is reminiscent of a one-way street—everything flows out and nothing comes back. Our country is being systematically insulted. They insult the Russians in the former USSR republics, deprive them of the basic human rights, and, at the same time, these states use Russian raw materials, Russian crude oil, Russian natural gas, electric power. And we are just watching. It should be stopped now.

[*Lidove Noviny*]: What do you offer people?

[Rutskoy]: To create a Russian power and society of real socialism—not a society of social egalitarianism, but of social justice. We want to give the people a chance to live in a dignified manner. Dignity rather than humiliation, that is the objective of our movement.

[*Lidove Noviny*]: What is the opposition really?

[Rutskoy]: It is a highly organized association of people with a single leader, a clear program, and mass support. If such an opposition with a single leader had existed in October last year, the White House bloodshed would not have occurred. The opposition must unite. Today, one speaks about nationalism, another about the Russian issue, yet another about the Soviet Union, someone claims that we do not need a president, and someone else calls for a parliamentary republic. It is the same at this session. They spoke about generating some funds, about membership dues. Are we building some kind of a trade union? When I look at this cabal, I think it would be funny if it were not so pathetic.

[*Lidove Noviny*]: So you are in opposition not only to the government and the president but also to the opposition. What intimidates you most about their work?

[Rutskoy]: All their squalor derives from their not respecting the law and not knowing the meaning of responsibility. Nobody is responsible for anything here, no one is subordinated to anyone, no one has to render an account of his actions to anyone. Look at the Duma. What does it do—I myself do not know. To whom is it responsible? Nobody.

[*Lidove Noviny*]: Are you in contact with Ruslan Khasbulatov? Do you intend to cooperate with him in your organization Power in the future?

[Rutskoy]: I meet Ruslan Imranovich [Khasbulatov] relatively often. Neither he nor I, however, have ever planned joint political activity. He helps me a lot. He is a doctor of science, and I have always held him in high esteem. We have prepared the declaration of the Power movement together, and we are working on the program now. If he can contribute to our movement, he will do so.

[*Lidove Noviny*]: Do you have any idea of when the almost disastrous situation in Russian society could turn to the better?

[Rutskoy]: As soon as law is put in the forefront, the economy and politics become stable, and production, work, and business discipline can be revived. Until then, we will float like chaotic molecules in space [sentence as published].

## 3.27

**Text of FIS Report Presented by Primakov**
*Rossiyskaya Gazeta,* 22 September 1994
[FBIS Translation]

[Unattributed account of report delivered to Russian and foreign journalists by Evgeniy Primakov, director of the Russian Foreign Intelligence Service.]

Notwithstanding the clear differences in opinions regarding the future of the CIS and Russia's place in the Commonwealth, two approaches have of late been clearly seen in the

West: one based on cooperation with the Russian Federation as an equal partner and the other based on a "unipolar" world in which Russia is given the role of a country with a strictly limited range of interests and tasks. This conclusion was reached by analysts from the Russian Federation Foreign Intelligence Service [FIS] in an open report entitled "Russia-CIS: Does the West's Position Need Modification?" which was submitted to Russian and foreign journalists at the Foreign Ministry Press Center by FIS Director Evgeniy Primakov.

We are publishing his report.

The main avenue of contemporary development in the international situation is highlighted by a move away from "Cold War values" and a reorientation to civilized interstate relations. But all that does not render the world less complex, does not preclude failures of various countries' national interests to intermesh, and sometimes even conflicts between various countries' national interests. The inertia of thinking and the enduring stereotypes of past practice also take their toll to some extent.

Tracing the development of various processes and trends in the international arena, the Russian Federation FIS could not overlook the fact that influential circles in a number of Western countries interpret the role that Russia may play in uniting the republics of the former Soviet Union as "imperial" and integration as a process aimed at the restoration of the USSR.

Some foreign analysts are promoting with increasing vigor the idea that the irreversibility of the move away from the "Cold War" is directly dependent on keeping the former Soviet Union in a disconnected state. At the same time it is claimed (by Brzezinski and others) that separate CIS states are needed to balance the tendency for Moscow's positions to gain strength. The prospects linked with the results of the recent elections in Ukraine and Belorussia [Belarus] are also being viewed from the viewpoint of the undesirability of centripetal processes on the territory of the former USSR. The conclusion is drawn that the policy of the leading Western countries vis-à-vis the CIS area should be modified with a view to preserving the status quo that took shape following the breakup of the Union. On the one hand, these opinions hid real fears that centripetal processes within the Commonwealth may revive the Union state in its former capacity as the enemy of the West and, on the other, clear pointers regarding the "need" to prevent Russia's growing stronger as a world power.

Needless to say, all Western leaders are not so definite and unequivocal. But even the fact that this topic is being discussed by political circles in the United States and certain European states, along with the calls for a reappraisal of the West's strategy in the sphere of security and for changes to their policy toward Russia and the other CIS countries, is viewed by FIS experts as serious reason for analysis.

## I. Wholesale Generalizations and Reality

The idea of Russia's changing and as yet undefined relations with the former national Soviet republics as "Moscow's imperial ambitions" is not a matter for theoretical dispute. Policy may be—and in a number of cases already is—behind that assessment.

U.S. congressional experts, for instance, rightly point out that in the past year Russia's foreign policy has become more independent, regarding its own vital national interests as being of paramount importance. One can and must concur unequivocally that this change does not revive the "Cold War" era. However, the conclusion that this change in Russia's policy does nevertheless represent a "kind of challenge to the United States" stems from the logic that indeed prevailed during the "Cold War" when one side's defense of its own interests was necessarily regarded as a minus by the other side.

Russia's safeguarding its vital interests is by no means an alternative to its desire for partnership relations with the United States and European and other states. On the contrary, the durability of these relations is ensured by their equitable nature, which manifests itself in the partners' ability to grasp the essence of one another's national interests and uphold them in a non-confrontational climate.

Typically, many foreign experts mainly associate the threat to Russian–Western relations with Russia's stance on the so-called "near abroad."

Some shift of emphasis can be noticed here. The question of whether the centripetal tendencies within the Commonwealth will develop within or outside the democratic process may indeed be worrying both Western politicians and public opinion in the "far abroad." However, often the emphasis is placed on something else: Will the CIS survive at all in its disjointed form or will there be reintegration on its territory? Here the former is seen as beneficial to the West and the latter as contrary to its interests.

Current information indicates that the arguments behind this stance are mainly that:

- reintegration will destroy the sovereignty of the states within the CIS;
- it will at the same time weaken democratic processes throughout the Commonwealth;
- Russia, using its resources, which are incomparable to those of the other CIS countries, will start "flexing its muscles."

These arguments are unfounded.

First, all attitudes toward the breakup of the Union notwithstanding, the tremendous stability of the new states' sovereignty remains an immutable fact. Its attainment is virtually irreversible.

Second, any significant political organizations that condemn the breakup of the USSR do not aim to restore it in its previous form and capacity. Awareness of the irreversibility not only of the CIS countries' state sovereignty but also of the emergence of private ownership throughout the greater part of the former USSR and the development of a mixed economy is growing among these organizations and forces.

Third, the idea of Russia's striving to "take in hand" the other CIS states, using its economic and other advantages to this end, is untenable. The fairly widespread views in the national republics of the USSR that "assets were pumped" from the provinces to Russia and that Moscow "inculcated" excessive centralization in cadre and other decisions initially or historically, so to speak, paved the way for this kind of talk. The latter did indeed take place in the past but the "resumption of Moscow's diktat"—and all serious experts are aware of this—is impossible following the changes in Russia and the former USSR republics' acquisition of sovereignty.

It would be wrong to claim that "integrationist" views and sentiments hold complete sway at present in the CIS countries. There are certain forces in Russia itself and the other CIS states that disregard or underestimate the objective nature of the centripetal tendencies forging their way through various parts of the former Union. In Russia, hypothetically speaking, the "neo-isolationists" rely on "conclusions" that economic integration on the territory of the former USSR would weaken their sovereignty, strengthen Moscow's influence, and complicate the development of relations with other states.

Both groups consider that they express the national idea. And Russian "neo-isolationists" can indeed, for instance, cite the fact that in 1993 alone deliveries to other CIS countries for which no payment was made reached around $10 billion, which is in excess of all the aid that Russia received from the IMF and the International Bank for Reconstruction and Development [World Bank]. (Is that not a counterargument to the big talk regarding the machinations by Moscow, which is allegedly thinking of establishing its own supremacy throughout the Commonwealth!) The "neo-isolationists" believe that Russia should "distance itself" from the other CIS countries because their egotistical credit and monetary policy is dangerously whipping up inflation in Russia.

All that is indeed happening. However, "neo-isolationist" ideas and currents run counter to objective processes. What is more, they harbor considerable potential for conflict. At the same time, according to the Russian FIS's information, a desire is arising in leadership circles in a number of leading Western countries and states in the Muslim world to view "neo-isolationist" currents as a possible mainstay when implementing their policy on the CIS.

### II. Economic Realities

There are a whole series of factors in favor of creating a common economic area in the Commonwealth:

- the traditionally high level of production sharing that has developed over the decades: In the late eighties the RSFSR (now Russia) sold almost twice as much of its output via interrepublic union commodity turnover as it did via trade with foreign countries while the other former Soviet republics sold roughly seven times as much;
- the single technological area that developed over the decades, the unified standards, and the fact that national processing capacities were tied to certain categories and grades of raw materials and semimanufactures;
- the republics' vital need to maintain employment by preserving mutual deliveries and also the very existence of their own industry, whose output, with the exception of cheap natural raw materials, is as yet uncompetitive on the world market;
- the need for investment cooperation in opening up and processing natural resources, and the joint use of important installations, particularly infrastructure;
- the advantages of a coordinated strategy for the conversion of the defense industry;
- the impossibility, given the "transparency" of the borders, of total economic isolation;
- the sizable material losses owing to illegal imports and re-export; the absence of a coordinated financial policy, and difficulties in mutual settlements;
- the unfeasibility in the coming years of a real influx of foreign financial and industrial capital, given the instability of the situation in the CIS countries and the high degree of commercial risk.

Lastly, there is no reason to believe that the CIS will stay aloof from worldwide practice, which demonstrates the advantages of a large-scale economic area for the development of productive forces. It was awareness of these advantages that led to the conception and broadening of economic integration processes in various parts of the world—be it the European Union, ASEAN, NAFTA, or other integrationist groupings.

Naturally in the present conditions, in view of the sovereignty of the CIS states, it is impossible to mechanically restore the economic ties in the forms that existed within the single Union of the past. The movement toward the creation of a common economic area in the CIS is not straightforward and cannot occur without irregularities, digressions, and retreats. Suffice it to say that the formation of a common economic area within the Commonwealth is altogether impossible without in-depth economic reforms in the USSR's former national republics and without squaring their economic mechanisms with the Russian model.

The stage-by-stage resolution of the issue of forming a common economic area is dictated not only by the uneven development of elements for the transition to a market economy in the CIS countries but also by the demands of this process itself. The creation of a common market, including freedom of movement for goods, capital, and the work force, is proposed as the immediate objective followed by the unification of the infrastructure to ensure the normal functioning of the corresponding sectors of the Commonwealth countries' economy.

The creation of a common economic area in the Commonwealth is by no means an easy matter. However, it is hopeless to resist the centripetal tendencies within the CIS, which are particularly manifest at present in the economic sphere.

And counterproductive at the same time.

The creation of a common economic area in the CIS is virtually the only way of reducing tension in interstate relations in connection with the fact that following the breakup of the Union there are around 25 million Russians and so-called Russian speakers who gravitate toward them outside Russia.

### III. Security Realities

A number of factors are prompting the CIS countries to create not only an economic but also a common defense area designed to guarantee their security.

The change in the military-political situation in the world, characterized by a reduction in tension on a global level, the Russian Federation's and other CIS countries' renunciation of the concept of permanent enemies, and the beginning of cooperation with NATO does not mean the elimination of potential threats to their security. The world community is now at a stage where the geopolitical configuration is changing, the militarization of a number of "Third World" countries is continuing, and many nuclear and "threshold" states are situated on or near to the CIS borders, the conflict zone is expanding, encompassing the center of Europe and part of the "outlying areas" of the former USSR.

#### *1. Conflict in the CIS Countries*

The interethnic and interstate conflicts that have broken out in the CIS states directly and in adjacent countries are tending to expand. The situation is aggravated by a number of factors.

First, in postwar history this is the first time that crisis has simultaneously enveloped a host of countries in direct proximity to or bordering on states on whose territory contemporary destructive arms and highly complex technical production units are located. In these conditions the settlement of interethnic and interstate conflicts is a particularly pressing task.

Second, a considerable proportion of the CIS conflict zone is adjacent to Afghanistan, where there is no sign of the situation stabilizing in the near future. In view of the ethnic features (the northern part of Afghanistan is inhabited mainly by Tajiks and Uzbeks) Afghanistan's destabilizing effect on the Central Asian states is intensifying. Moreover it is threatening to the state security of a number of countries, primarily Tajikistan, followed by Uzbekistan. Russia's FIS has information to the effect that there are forces in Afghanistan that want to break the north away and that are striving to create on that basis a Farsi-speaking state incorporating Tajikistan.

Third, the situation in the CIS "hot spots" has been aggravated as a result of states other than Afghanistan, primarily Iran and Turkey, becoming "embroiled" in them. Both these countries are seeking to broaden their influence and are aspiring to the role of regional superpowers. As a result of their "involvement" in the conflicts on CIS territory—and this is of significance not only for Russia alone—there is a "swing to the right" taking place in the alignment of forces in Turkey and Iran.

Fourth, Islamic extremism has a highly negative effect on the crisis situations on CIS territory.

FIS analysts believe that under no circumstances should it be associated with Islamic fundamentalism, which does not presuppose the forcible spread of Islam, much less terrorist methods. However, of late, Islamic extremism has intensified as a movement aiming to spread Islam by force, suppress forces opposed to this, and change the secular nature of the state.

The "effect" of this extremism has manifested itself in both Tajikistan and the Caucasus conflict zone. However, the problem of the spread of Islamic extremism is not locally confined.

Fifth, despite a host of statements, in actual fact an inadequate reaction can be discerned on the part of the world community to the conflict situations that have developed near Russia's borders. For instance, for all the comparable number of victims of the Yugoslav crisis and in the CIS "hot spots," major differences are emerging in peacekeeping diplomacy toward these two crisis zones. The United Nation's sharp reaction involving the use of force to capture several units of combat hardware in Bosnia rubs shoulders with a "polite reference" to the death of Russian border guards when repelling gangs' attempts to infiltrate the Afghan-Tajik border.

#### *2. Peacekeeping Actions on CIS Territory*

All the Commonwealth countries have an interest in their implementation.

Russia's active involvement in settling conflict situations is attributed to its vital interest in a stable situation on its borders and in preventing conflicts having a provocative influence on certain regions of the Russian Federation. People in Moscow cannot close their eyes to the violation of the rights of the Russian-speaking population and the fact that armed operations result in the death of Russian citizens. Refugees are streaming toward the Russian Federation, and huge financial resources are needed to look after them, while their migration is exacerbating the social and crime situation.

Despite the indisputably positive results of the peacekeeping actions in South Ossetia and the Dniester Region, their important role in Tajikistan, and the favorable start to the operations in Abkhazia, they elicit a more or less negative or suspicious reaction in many capitals of the "far abroad." There is criticism of Russia's "special role" in peacekeeping actions on the territory of the Commonwealth countries and the idea that its vital interests are linked with a state of stability in the other CIS countries.

*On the poor international-legal base for peacekeeping operations on CIS territory.* In reality Russia's commitments under the UN Charter, the corresponding UN Security Council decisions, and other international treaties and agreements, including within the CIS framework ("On Collective Peacekeeping Forces in the CIS" of 20 March 1992; "On Collective Peacekeeping Forces and Joint Measures to Provide Them with Material and Technical Support" of 24 September 1993, etc.) constitute the international-legal basis for its participation in peacekeeping activity.

The possibility of the use of Russian peacekeeping contingents abroad in accordance with Russia's international commitments is envisaged in the Russian Federation Law "On Defense" of 24 September 1992 and the Russian Federation presidential edict "On the Basic Provisions of Russian Federation Military Doctrine" of 2 November 1993. The draft law "On the Procedure for the Provision of Russian Personnel for Participation in Peacekeeping Activity" is being examined by the Russian Federation parliament.

It must also be particularly emphasized that no peacekeeping action in the CIS has been conducted without the consent of the conflicting parties although the United States, for instance, has carried out operations in Panama and Grenada without any approval from these countries' authorities.

*On the fact that Russia allegedly pits its efforts against the activity of the United Nations and other international organizations.* By way of confirmation I can cite in particular the words spoken by U.S. Secretary of State Christopher, who said bluntly when addressing the U.S. Senate on 2 March of this year: "We (the United States) do not recognize their (Russia's) right to take any actions in the new independent states save those which are carried out following coordination with the United Nations and other international organs and in accordance with the norms of international law."

*On the lack of neutrality among the Russian forces when implementing individual peacekeeping operations on Commonwealth territory.* There are usually references to the Russian military's "inconsistency" in Abkhazia and Tajikistan. However, the neutrality of the Russian forces involved in resolving conflicts is guaranteed by the pledges made by the Russian Federation when coordinating the terms and framework of the peacekeeping operations with all the interested parties.

*On the predominance of Russian subunits in the Commonwealth's peacekeeping contingents.* This cannot be put forward as an accusation purely because in practice it is not yet possible to ensure full-fledged participation by the other CIS states in peacekeeping operations. The overwhelming majority of states of the "far abroad" are not prepared to send peacekeeping forces here, and the United Nations is not prepared to pay for peacekeeping operations.

*On the inadequacy of international monitoring of Russia's peacekeeping activity.* This is completely refuted, for instance, by the fact that the CIS countries' peacekeeping operations in Abkhazia and Nagorno-Karabakh will be monitored in total by several hundred international observers. UN and CSCE [Conference on Security and Cooperation in Europe] missions have already been operating in the Dniester Region, Georgia, and Tajikistan for a long time. The large Minsk Group of the CSCE on Nagorno-Karabakh has been functioning since 1992. They all have practically unlimited access to the information they require.

*On the inadequacy of the negotiation process in settling conflicts.* Not a single Russian peacekeeping operation has been carried out without preliminary work to organize talks between the parties to the conflict. Moreover, the military phase of the settlement, involving the sending in of disengagement forces (South Ossetia, the Dniester Region, Abkhazia) or the countering of outside aggression (Tajikistan) is always aimed at creating the conditions for intensifying talks with international participation (the United Nations, the CSCE). That talks to find a peaceful resolution to crises should be prolonged is a normal phenomenon in world practice.

Thus, Russia is observing all the internationally recognized conditions of peacekeeping taken together.

### *3. The Problem of the "Transparency" of External and Internal Borders*

Following the USSR's collapse and the formation of the CIS, the question of how to ensure a quality relationship between CIS external borders and the internal borders of Commonwealth countries has sharply arisen. With "transparent" internal borders there is no doubt about the need to protect external borders. At the same time, in the absence of an overall defense area including functions such as a unified system for protecting external borders, there is a need to delimit and demarcate the Commonwealth's internal borders. And this is by no means easy.

The demarcation of Russia's borders with neighboring CIS countries will require huge financial expenditure—which could substantially hamper the reform of the Russian economy and stoke the already tense socio-political situation. The demarcation of Russian borders would be liable to lead to the emergence of new "hot spots" in the CIS—for instance, Ossets, Lezgins, and so forth, would find themselves on both sides of the state border.

Thus, the measures to step up controls of the Russian Federation border with Azerbaijan have shown that in this sector hundred of citizens from Iran, Pakistan, Iraq, and other countries are illegally entering Russia along with contraband, including weapons. As a result of the "holes" in our border protection and visa regulations, and of the lack of coordination in immigration policy between the CIS countries, the number of illegal migrants from various Asian and African countries coming to Russia has increased sharply.

This situation is making it necessary for Russia to stabilize the situation in the "near abroad" through joint efforts with the CIS countries and to restore order on the Commonwealth's external borders, while simultaneously equipping the Russian border—bearing in mind that the approach taken to determining the arrangements in the various different sectors should depend on local circumstances and should rule out the possibility of any damage being done to integration processes on CIS territory.

### *4. Features of Military Organizational Development in the Leading States That Were Formerly "Enemies" of the USSR*

The current phase of the development of international relations has some specific components that Russia and the other CIS members cannot fail to take into account. The United States, Britain, France, and China have currently not only given up their strategic offensive weapons but are also implementing a range of measures to modernize their land-based ICBMs, their submarine-launched ballistic missiles, and their strategic aviation.

Under present-day conditions these states have not missed out on the trend common to all former Cold War participants toward a reduction in military spending. New emphasis is being laid on military doctrines, dictated by the current military-political and military-strategic situation.

The United States's new nuclear strategy, which emerged after the end of the Cold War, preserves the basic principles of the utilization of nuclear weapons—the comprehensive combat use of all components of the "triad"; the provision of conditions for neutralizing enemy defenses; collaboration among different U.S. combat arms; and close coordination of their efforts with the NATO allies.

The single operational plan for destroying the presumed enemy's strategic targets (Strategic Intelligence Operations Plan) envisages supplementing the range of new scenarios and unusual targets. There is a planned transition to "adaptive" planning, which makes it possible to clarify virtually in real time the operational plans for the use of nuclear weapons in response to a changing situation.

While preserving and modernizing their strategic offensive weapons, the United States, China, Britain, and France are emphasizing the development of their national forces. And the United States, with the most powerful strategic offensive weapons systems, is continuing to provide guarantees to a number of non-nuclear powers both within NATO and outside it. For instance, there is a paragraph to this effect in the U.S.-Japanese Security Treaty.

The existing practice, linked to improving strategic offensive weapons and the continued provision of guarantees by the United States, is due to a number of factors:

- the "uncertainty" of the domestic political situation in Russia;
- the continuing presence of nuclear weapons on the territory of Ukraine and Kazakhstan, where the domestic political situation has also not stabilized;
- the need to "restrain" China, particularly in the Asia-Pacific region;
- the acquisition of nuclear weapons by a number of countries, including India and Pakistan;
- the continuing work to develop [*sozdanie*] nuclear weapons in other states (the relevant "indications" of the situation in this sphere in North Korea and Iran are received particularly painfully).

With certain overtones the list of these factors also features in any explanation of political decisions in the sphere of military-strategic organizational development in Britain and France.

China too argues that its policy has an eye to the "need to safeguard its national interests." One way or another, all the measures to improve national strategic offensive weapons are presented as an appropriate reaction to the new dangers of the post–Cold War period.

Irrespective of the weightiness of the reasons behind this practice, it is a reality that Russia—and, clearly, the other CIS countries—cannot ignore. The conclusion has been drawn from this practice that it is necessary at this stage to preserve and develop one's own strategic offensive forces.

### *5. Other Present-Day Security Requirements*

These primarily include the problems of overcoming the environmental crisis which is getting worse throughout the Commonwealth. And the fight against the causes of this crisis requires joint efforts from all the CIS countries for the simple reason that many of the sources of environmental disasters and difficulties are found in several different states (the Aral Sea, for instance).

The same joint efforts are required in the fights against epidemics, which have become particularly dangerous in the context of the deteriorating socio-economic situation in a whole number of areas and the lack of any proper public-health measures appropriate to the "transparency" of our internal borders and the "holes" in the Commonwealth's external borders—particularly in the Central Asia and the Caucasus.

Close coordination of the CIS countries' efforts is also needed in order to successfully combat organized crime. It is necessary to pool efforts in this area because of the "international" nature of organized crime, which came about during the USSR's existence and persists today. We could also conclude that, without effective collaboration between law-enforcement organs and the CIS countries' special services, there will be no chance at all to improve the crime situation on the territory of the former USSR—a situation of increasing alarm not only to the population of the Commonwealth but also to the "far abroad."

## IV. Likely Scenarios for the Development of the Situation in the CIS

### *Scenario A*

Centripetal processes intensify within the CIS. The prerequisites for a common economic area are created and then the area takes shape. General rules of "economic behavior" and unified systems in the sphere of lending, money supply, customs, taxes, arbitration courts, and so forth are elaborated and formalized. Transrepublican companies are set up. While retaining their state sovereignty, the CIS countries "delegate" part of it to the suprarepublican structures required for the functioning of the common economic area.

Along with economic integration (or lagging slightly behind it) there is integration in the military sphere, and a defense area is formed with a unified command and unified subunits designed to protect external borders, undertake peacekeeping missions, and deter potential enemies.

It is not ruled out that in the process of further development the prerequisites will appear for political integration, the most likely form of which could be a confederation.

Events could develop differently under this scenario. Most probably the process would begin with the implementation of agreements on an economic union initially between several CIS members, with the others joining later.

The development of events under this scenario would lead to stabilization, democratization, and the advancement of reform, and could include a transition to a federal system in a number of CIS countries—which would reduce still further the threat of interethnic and interstate conflicts on Commonwealth territory.

The development of events under this scenario would lead to an increase in the power of the CIS, its ability to develop independently, and its competitive strength in international markets. But at a time when democratic processes and economic reforms are developing in the former USSR, this will not result in the clock being turned back to the era of confrontation with the West. On the contrary, this scenario creates the best opportunity for stabilizing the situation throughout the CIS, nullifying the danger of "chaos in a nuclear-weapon state" and producing the necessary conditions for expanding economic cooperation—including by attracting foreign investment.

### *Scenario B*

With direct or indirect outside support, forces advocating "separate development" gain the upper hand in Russia and other Commonwealth countries. This compounds the economic crisis in the former Union national republics and increases the socio-political tension in them. The breakdown of national economic ties and the abandonment of production sharing could become irreversible. The unemployment problem will become acute and the transition to boosting production will be complicated.

The emphasis on nationalism will be accompanied by an intensification in authoritarian and undemocratic trends. The criminalization of society, the infringement of ethnic minorities' rights, and mass violations of human rights will be additional destabilizing factors.

The positions of Islamic extremists in the CIS states with Muslim populations will grow stronger. The intensification of separatist trends will help bring about the collapse of certain states.

Theoretically for Russia the conditions for getting out of the economic crisis could improve in a very short period of time. But the economy does not develop in a vacuum. The question of the need to completely eliminate the "transparency" of borders will arise and will require enormous expenditure. The flow of refugees from certain CIS countries to the Russian Federation will increase. The new geopolitical situation will require considerable additional amounts of defense spending. Finally, Russia will lose its traditional markets, which will be particularly painful just when we are emerging from the crisis and beginning to boost production.

The overall destabilization in the CIS will pose a threat to the world community's security.

***Scenario C***

(Rather, you could call it a "subscenario," since the development of events implied by it would inevitably lead in the final analysis to either Scenario A or Scenario B.)

One of the CIS states (but not Russia) undertakes "unifying" functions. Several republics of the former USSR (without Russia) move closer together. Integration processes begin within the framework of this group of states. One option would be development, whereby definite impetus is given to integration processes on the territory of the entire CIS, and the original group becomes part of a general integrated area. Another option would be that the group turns in on itself [*zamykaetsya v sebe*], which would inevitably push it toward external "centers of influence." [Scenario C ends.]

The influence of leading countries of the "far abroad" on the processes taking place in the CIS is indisputable and, consequently, the scenario that the development of the situation in the former USSR follows will, to a certain extent, depend on those countries.

In recent months there has been a wide divergence of opinions in the West about the future Commonwealth. A great deal will be determined by which approach prevails: reliance on cooperation with Russia as an equal partner (given the irreversibility of democratization and the objective nature of the reintegration processes on CIS territory) or reliance on a "unipolar" world in which the Russian Federation is given the role of a country with a very limited range of interests and tasks. The second approach is unacceptable to Russia and, one way or another, it will reject it.

But, in the main, of course, the prospects for the CIS depends on the Commonwealth countries themselves and, primarily, on the Russian Federation.

## 3.28
## Primakov Reflects on Intelligence Role

*Trud,* 15 October 1994
[FBIS Translation]

[Preface to forthcoming book on Russian intelligence by Academician Evgeniy Primakov, director of the Foreign Intelligence Service, "slightly abridged."]

The Mezhdunarodnye Otnosheniya Publishing House is preparing for publication six volumes of *Sketches on the History of Russian Foreign Intelligence* [Ocherki Istorii Rossiyskoy vneshney Razvedki]. Academician E.M. Primakov, director of the Russian Foreign Intelligence Service [FIS], has written a preface to it. *Trud* was authorized to be first to publish the text of this preface (slightly abridged), which, in our view, is of significance in itself and is of undoubted interest to our readers.

The intelligence service is a necessary mechanism that resolves a whole series of very important state problems. History has proved that. The present day is proving it too.

Is it possible to talk about a distinctive Russian intelligence service with characteristics typical of it alone? Of course there are a whole series of characteristics that distinguish an intelligence service regardless of its national allegiance: its methods and means of operation, a structure enabling it to obtain materials of a political, military, scientific-technical, and economic character, the use of the so-called human factor—in other words, its agents and confidential links—and the use of characteristics of the Russian intelligence service. It may be (I hope that foreign readers will not criticize me for saying so) that they have more patriotism and selflessness than the rest, that (for both subjective and objective reasons) the work of its intelligence agents is less susceptible than the rest to material influences, and that they show a greater tendency toward self-sacrifice for the sake of their people.

At every stage in history, under every system, and in all circumstances an intelligence service has protected the interests of the state. Is that not why the emphasis is being placed on the depoliticization of the intelligence service, a course that we have now consciously moved toward in the post-Soviet era? Of course this depoliticization is relative, because the state has always protected and continues to protect the interests not only of the whole nation but also of the ruling groups. The intelligence service has always been and remains a reflection of the state. As the reader will see, P.I. Rachkovskiy, a Russian intelligence agent at the turn of the century, served the interests not only of the fatherland but also of the tsar's police department by keeping the Russian political émigré movement under surveillance. Nor was the intelligence service insulated from the tragic repressions under Stalin. But it is important to note that because it was not a law enforcement agency it was less to blame than the rest for what happened and, through being very much connected with foreign countries, it absorbed the elite of the armed forces and suffered more than the rest from the bloody repressions. Tens, even hundreds, of intelligence agents were dismissed from their jobs, arrested, and shot.

The questions of depoliticizing and deideologizing the intelligence service are especially important and relevant nowadays. What form do they take? The intelligence service—and this must be said with the utmost firmness—is not involved in Russia's internal political life. Of course, each intelligence officer and employee not only can but is bound to harbor sympathies for this or that political or public force in the Russian Federation. The *mankurt* [the zombie in Aytmatov's novel], the robot, devoid of feelings, emotions, affections, and hostility to things he considers bad, cannot be an intelligence agent. But neither can he be guided by his own political sympathies or antipathies in his daily work. In that work he has to think solely of the country's national interests—paradoxical as it may seem, that is what the depoliticization of the intelligence service means. At the same time, of course, the intelligence officer must be dedicated to democratic values and obey the law—that is extremely important for everyone without exception.

Now a few words on repudiating the ideologization of the foreign intelligence service. This term did not exist at all before 1917. In the USSR's lifetime, however, the intelligence service, like any other state instrument, could not avoid being ideologized. The Cheka, the NKVD, and the KGB, including the foreign intelligence service, were weapons of the party, and it was in that capacity that they conducted their activities under Marxist-Leninist slogans.

The best sources and foreign aides were obtained by Soviet intelligence on ideological grounds. As a deputy director of the USSR Academy of Sciences' Institute of the World Economy and International Relations, I was a good and close friend of Donald MacLean, a senior staffer there. He was one of the most intelligent and well-educated men I have ever met in my life. D. MacLean used to work for Soviet intelligence. As director of a British Foreign Office department he had prime information, sometimes of crucial value to Moscow, which he transmitted to us. As a descendant of seven Scottish lords, D. MacLean was not guided by material considerations. No matter how we view our ideological past today, it is the ideas of that past that brought MacLean and many like him into Soviet intelligence.

Many people predicted that the deideologization of the FIS in present conditions would prevent us from being able to enter into cooperation with foreign information sources. That has not proved to be the case. We are not talking now about the acquisition of foreign assistants. But many people's political interest in cooperation—no longer with Soviet intelligence but with Russian intelligence now—continues to operate. This is because of the dislike of an "unipolar" world, the fear of a possible unilateral reorganization of Europe's postwar borders, and the understanding of Russia's role as a factor of stability in Europe and the world as a whole.

The deideologization of the intelligence service has therefore certainly not eliminated or even undermined it as a major instrument of Russian policy.

Whether the intelligence service is necessary or not is a purely rhetorical question. No reasonably large state, let alone a great state, can or could ever do without one. Clearly it will cease to exist when the state ceases to exist, but such a prospect is purely hypothetical and has nothing to do with reality. Yet there have been two distinct periods in the history of the Russian intelligence service when people either did not realize or doubted the need for its existence. The first period was the formation of the Russian state, and during that period the intelligence service had not yet been formed as an autonomous institution of state power but emerged and developed not only "within the walls" but also as a component of the edifice of diplomacy. We witnessed the second period for ourselves in the recent past, when the euphoria of the end of the Cold War was so mind-boggling that some people in Russia began preaching the abandonment of the intelligence service in the era of "civilized relations" toward which the world was moving.

In 1992 I was interviewed by a British journalist who asked me whether I thought that Russia was ready to stop conducting intelligence work against Britain if Britain did the same toward Russia. In our close, interconnected, and interrelated world, of course, such a question cannot be resolved bilaterally. But what if all the countries jointly decided to abandon intelligence work? Well, then we could talk about it, but to be honest only irremediable daydreamers can believe in such a prospect.

The characteristics of the Russian intelligence service include continuity and a loyalty to the best traditions of the specialist agencies that preceded it. We have never repudiated nor will we ever repudiate all the good, useful, and important work done for the state, society, and the people over many decades and centuries in Russia by the precursors of the FIS including, of course, those in the Soviet Union. But the recognition of this continuity certainly does not mean that we have lost our critical attitude toward our history.

What has the intelligence service been doing since the end of the Cold War, what problems has it been resolving, and on whose altar have its officers been sacrificing their strength, their health, and sometimes their lives?

You may think that at the end of the twentieth century Russia is experiencing one of the most difficult stages in its history. The economic crisis, the unregulated relations between the Center and a part of the periphery, the threat of territorial disintegration or, at any event, the separation of individual parts of it, the resistance to the processes of reintegration on CIS territory, the political need to defend the positions of an independent great power in the international sphere, the fear that the Russian Federation is entering the world economy primarily as a raw materials producer at a time when other great powers have already set course and are successfully heading toward the development of science-intensive, high-technology production—all these are signs of difficult times. But even when things were at their most difficult there has not been a single mature politician who believed that Russia, with its vast human potential, great history, outstanding contribution to world civilization, incalculable natural resources, and great basis in fundamental science, could possibly get stuck at such a stage. It was, is, and will remain a great power. Throughout the history of the Russian state the intelligence service among the other major state mechanisms has helped the state to overcome difficult periods.

Furthermore, the process of emerging from the crisis situation of the 1990s, which is a legacy both of the past and of the mistakes and omissions of the time when these lines are being written, has been happening in an international situation that is far from clear-cut. Yes, the 1980s and early 1990s saw the end of the chapter in international relations when two ideologically different forces stood on opposite sides of the barricades preparing to do battle. The division of the world into two systems ceased to exist; consequently the grounds for Russia to have predetermined permanent enemies also disappeared. But the world did not become less complex because of that. Our state's national interests, which were previously relegated to the background and were often sacrificed to the struggle against the permanent ideological enemies or to support for our permanent ideological allies, became of paramount importance and came to the forefront. This had at least two consequences: a sharp intensification of the search for areas where those interests coincide with the interests of other states and, if they did not coincide, the identification of ways to protect national interests by non-confrontational, political means.

However, was that always possible? The complex character of the answer to that question is reinforced by the fact of the considerable expansion of the zone of regional conflicts after the disintegration of the USSR—a zone that spread to the territory of many national republics of the former Union.

The danger of conflicts expanding has sharply increased because they have developed in conditions of the proliferation of mass destruction weapons and the growth of international terrorism. The long-running and extremely tense situation in Afghanistan has had the most negative impact on these conflicts—the Central Asian conflict zone has now reached as far as this country's borders—and clearly will continue to do so for a long time to come.

The destabilizing influence of Islamic extremism, that is, the violent forms of spreading a militant Islam whose goal is to introduce an Islamic state model, has extended not only to the Central Asian conflict zone but also to the Caucasus conflict zone and will obviously continue to spread for a long time to come. In the post–Cold War period there has been an intensification of interethnic contradictions as well.

Of course, the Russian foreign intelligence service at all stages of its development provided the country's top leadership with reliable and largely advanced information as well as analytical assessments by its experts based on that information. The main area of its work here was and remains the monitoring of all those processes that could damage Russia's interests if things turned out badly for it. As our century draws to a close, the following threats may be named:

- the activity of external forces trying to aggravate or simply exploit the weaknesses of Russia itself and its disagreements with other states that formerly belonged to the USSR; one of the priority tasks of the intelligence service in the mid-1990s is to neutralize the efforts aimed at halting the centripetal tendencies both within Russia itself and within the CIS;
- the prospect of the growth of the "nuclear club" and of a number of Third World countries' gaining access to various systems of mass destruction weapons, and the adoption of political decisions in some of them to begin moving toward the acquisition of nuclear weapons;
- the likelihood of countries previously involved in the Cold War gaining access to new, destabilizing weapons systems;
- intelligence must promptly notify the corresponding Russian state structures about the development and introduction of new types of arms and combat hardware in other countries.

One of the important tasks of the intelligence service here has been and remains to assist the pursuit of an active Russian foreign policy toward both the West and the East (China, India, the Arab states, Asia and the Pacific, and so forth).

The most important task of the FIS in the past, the present, and the future is to help to increase Russia's defense potential and accelerate its socio-economic development.

## 3.29

## United States Trying to "Force Russia and Ukraine Apart"

Boris Filippov

*Rossiyskaya Gazeta,* 28 January 1995

[FBIS Translation]

In November 1994 L. Kuchma made his first foreign visit as head of state not to Russia—as was expected and implied from his pre-election speeches, when he tirelessly spoke of the need for integration with the great neighbor when trying to win the electorate's sympathy in the east of the country—

but to the United States. The point is not that Kuchma and his team would like to play the "American card" in their relations with Russia. It is simply that it is now vitally important for Ukraine to win the favor of the Western world's leader with the aim of obtaining real financial aid to carry out economic reforms, becoming stable in the current multipolar international relations, and solving other important questions. Nevertheless the development of U.S.–Ukrainian relations cannot be considered without taking account of the Russian factor.

The Clinton administration highly rates Kuchma's activity since he assumed the post of president for successfully solving two problems that, in Washington's opinion, will determine Ukraine's future: primarily these are economic reforms and nuclear policy. Ukraine is getting rid of the nuclear warheads left on its territory following the Soviet Union's disintegration and is joining the Nuclear Non-Proliferation treaty, which accelerates the process of cutting nuclear weapons within the framework of the Start I treaty. The United States and Russia are naturally relying on expedited implementation of this treaty.

The United States has now openly started talking about the importance of expanding trade and investment relations with Ukraine, and it has been given trade concessions in accordance with the General System of Preferences.

As of today the total sum of U.S. aid to Ukraine stands at $900 million. Ukraine has become the world's fourth-largest recipient of U.S. aid after Israel, Egypt, and Russia. The United States has already given Ukraine $282 million for economic and humanitarian purposes. It intends to give an additional $200 million in the form of technical assistance for structural reorganization and the development of democratic institutions. Furthermore, the United States has decided to allocate an additional $100 million for settling the balance of payments and clearing the Ukrainian foreign debt, and $25 million is being given in the form of a loan to import food from abroad.

During Kuchma's visit to the United States its commerce secretary, Ronald Brown, and Serhiy Osyka, Ukrainian minister of foreign economic relations, signed a joint statement under which the United States is to start viewing Ukraine as a country with a transitional economy.

The World Bank president, Lewis Preston, stated in turn that if Ukraine continues to move toward the stabilization of its economy and market reforms it could receive credits totaling $500 million to strengthen such sectors as health care, agriculture, and power engineering. At the end of last year Ukraine obtained a $371 million IMF loan.

On the initiative of the United States a special plan concerning Ukrainian energy requirements as a whole as well as a solution to the problem associated with the continued operation of the Chernobyl AES [Nuclear Electric Power Station] was adopted at the July 1994 G-7 meeting in Naples.

The process of developing U.S.–Ukrainian cooperation is continuing even though Republicans, who are in favor of cutting foreign aid levels, have won the majority of seats in Congress. However, this is no threat to Ukraine for the time being, because it is known that the Republicans are much more afraid than are Democrats of Russian intentions to play a leading role in the post-Soviet area and are more favorably disposed toward Ukraine than are Democrats.

When analyzing the state of U.S.–Ukrainian cooperation, especially if it concerns economic, technological, and financial questions, its one-sided nature is striking. But these are serious investments for the future. By giving aid to Ukraine the United States is trying to create a favorable climate for its own wholesale penetration of the region.

When working out its strategy on the political plane, Washington is also trying to drive Ukraine and Russia as far apart as possible in order to reinforce the geopolitical situation that has developed since the USSR's disintegration. Influential lobbies among Americans of Ukrainian extraction are putting pressure on the U.S. government in this direction.

Conversely Moscow, despite the fact that its relations with Kiev are burdened by the problems of non-payments, the Black Sea Fleet, and the Crimea, will do everything in its power to entice Ukraine toward closer cooperation within the CIS framework.

---

## 3.30

**Korzhakov's Role in "Party of War" Viewed**

T. Zamyatina

*Golos*, no. 5, 30 January 1995

[FBIS Translation]

---

*Editor's Note:* The following document criticizes the so-called "War Party" operating from within the Russian White House. This group gained influence during the period between October 1994 and February 1995, when Russia's Chechnya strategy was being deliberated. Its leading figures, General Aleksandr Korzhakov, head of the president's Security Council, and Oleg Soskovets, first deputy prime minister, hold the greatest power. Other key figures are Oleg Lobov (secretary of the Russian Security Council), Nikolai Egorov (deputy prime minister and minister for nationalities and regional affairs), and General Mikhail Barsukov (head, Chief Guards Administration of the Russian Federation and Kremlin Commandant). This group, in close association with the three Russian "power ministers"—Defense Minister Pavel Grachev, Minister of Security Sergei Stepashin, and Minister of Internal Affairs Viktor Yerin—constitute the power behind Yeltsin's throne and would undoubtedly like to set the Russian policy direction for the next twenty years. The "War Party," according to some conservative Russian commentators, sought to replace Viktor Chernomyrdin as prime minister with Oleg Soskovets, and have partially succeeded in altering the political and economic direction of the country. This group has pressed, successfully, for activist policies aimed at restoring Russia's lost "Great Power" status, challenging the Western orientation of democrats such as First Deputy Prime Minister Anatoliy Chubais and Foreign Minister Andrey Kozyrev. (Chubais and Kozyrev were both forced to resign their positions in January 1996.) The adherents to this group's ideology frequently represent the national patriotism of the Soviet period and "Eurasionist" thinking which is often virulently anti-Western and stresses Russia's Asian roots. The group's mood and outlook can be described as fundamentally imperialist.

Throughout the country in January Prime Minister Viktor Chernomyrdin's televised appeal to the citizens of Russia was heard, in which he said that the government's main goal was to stop the bloodshed [in Chechnya]. The message was not heard, however, by the "party of war," that is, the power-wielding ministers and their supporters. It was just as in December, when they did not accept President Boris Yeltsin's instructions to stop the bombings in Grozny.

At the same time, the prime minister spoke of the Russian leadership's preparedness to conduct negotiations on creating neutral zones with all interested sides in the Chechen Republic, but negotiations were broken off, battles continued, and the director of the Federal Service for Counter Intelligence, Sergey Stepashin, issued a directive to deliver a Russian flag to Grozny. The latter was erected over the heavily bombed presidential palace on 19 December.

These "proponents of force" are employing a concept that combines aggression with national dominion to counter the calls by the intelligentsia to use political methods for settling the crisis. It is the hallmark of their "signature." For example, the chief of the Russian Federation Service for Presidential Security, Aleksandr Korzhakov, expressed his opinion during an interview with *Argumenty i Fakty* when he stated that the liberal intelligentsia has "shown a lack of understanding for the necessity of decisive action in critical situations, and the need to consolidate the nation during times of trial" Turning our attention to the general's lexicon, we find that it is identical to statements made by national-patriots, from Zhirinovskiy to Barkashov. Thus, the question over the political sentiments of these representatives is made clear.

As a result of the fact that Boris Yeltsin has become distant from the press, commentators have in recent times sought answers to burning issues more and more often in the book *Memoirs of the President*. Today the section of Yeltsin's candid admissions has become highly relevant in which he describes the vacillation of the army and militia on the eve of the suppression of the revolt in Moscow on 3–4 October 1993. The board of the Ministry of Defense "began to discuss taking the White House," wrote the president. "Chernomyrdin asked, 'So what are the proposals?' In response there was a heavy, somber silence. Unexpectedly for me, the chief of presidential security, Korzhakov asked for the floor. . . . He asked that the floor be given to his officer from the Main Administration for Protection, who had a specific plan for taking the White House. . . . Perhaps, it was precisely at the moment . . . that a sudden ethical transition occurred."

There is no doubt, however, that the decisions to take the White House and to send troops to Chechnya were made by the Russian president. It is important, though, that fellow countrymen know of those "heroes" who are pushing the head of state to adopt decisions bearing such tragic consequences.

Today General Korzhakov is disowning his involvement in the decision to send troops into Chechnya and storm Grozny. To inquire about the latter he recommends asking "presidential advisers, the chief of the president's administration, the former FSK [Federal Counterintelligence Service] deputy director, who has spent a great deal of time in the region, and members of the Security Council. . . ."

We asked presidential advisers. On 11 December, the day troops were sent into Chechnya, the chief specialist among them on conflicts between nations, Emil [Pain], called the decision "suicide for Russian authority," and condemned it. The director of the presidential administration, Sergey Filatov, explained his position to me nearly ten times both before and after the beginning of the military operation; a position that he adopted based on the conclusion of a council of experts and analysts under the chief of state. The council's conclusion essentially consisted of setting up a Provisional Council within Chechnya under the leadership of Umar Avturkhanov and creating acceptable living conditions for people within the part of the republic controlled by Dzhokhar Dudaev's opposition. Such conditions would include providing pensions and measures for social support—thereby weakening the criminal regime of President Dzhokhar Dudaev.

Five months before the beginning of the military operation, on 11 August 1994, Boris Yeltsin, while setting out on a trip along the Volga, stated to us, the journalists accompanying him: "If we use force in regard to Chechnya there will be such turmoil, there will be so much blood, that we will not be forgiven for it."

The president understood everything. He foresaw the tragic consequences of applying force, which the democrats never tired of telling him. In no way did either of his advisers, or Sergey Filatov, have any part in inciting him to pursue the path of war.

Is this not the reason for the blatant hostility demonstrated by those promoting force toward the "so-called analysts," and the "inconsistent liberals," as Aleksandr Korzhakov terms the intelligentsia, for its "one-sided and changing interpretation of protecting human rights?"

When we analyze the "signature" of the "War Party," from my point of view as a graphologist, it is marked by the loss of an ethical position, a lack of humanism, a failure to acknowledge the unconditional value of human life, and a blasphemous gap between words and deeds. How can one perceive any differently the statement made by the secretary of the Security Council, Oleg Lobov, that "our goal is to preserve the lives of Russian citizens in Chechnya"?

Those "peacekeepers" in need of burying "loose ends," that is, evidence of their lack of professionalism both in conducting reforms in the army, and in exposing organized crime, terrorist acts, and ordered killings, had a vital interest in a "small victorious war." It was not all for naught that at the height of the Chechen operation there was an information leak organized by the FSK to the effect that the airborne assault soldier suspected of killing journalist Dmitriy Kholodov was not stationed in Chechnya. As they say, all is written off in war. Also, in accordance with long-standing logic, generals who are victorious are not tried.

However, as all those following the events concur, there will be no victors in the Chechen war. "We will not be forgiven," Boris Yeltsin predicted last August.

It is precisely this, however—forgiveness for the thousands of lives ended in Chechnya—that the "party of war" will not be asking for. It has other methods. It will strive to force the press, Russian society, and the entire world into silence.

There is one gratifying and reassuring side to the anger demonstrated by those promoting force, however: They cannot conquer public opinion. To the cries and demands of the authorities to support the actions of these gallant generals, the people prefer the quiet voice of an intelligentsia member, the bespectacled Sergey Kovalev, the inconsistent liberal, whose conscience has not been broken by either prisons, exiles, or presidential order against "sobbing."

## 3.31
## Press Coverage of CIS Espionage

*Editor's Note:* Although the next two articles do not represent Russian national-patriot thinking per se, they investigate an issue that hard-line Russian nationalists are trying to trump up as an excuse for reinstituting the Soviet-era KGB. That issue is interstate espionage among CIS states, especially the alleged spying of the "near abroad" on the Russian Federation. *Le Figaro* journalist Isabelle Lasserre correctly interviewed several knowledgeable experts on the issue. Her conclusion that many of the spies being caught by Russia may actually be "mafiosi" from other CIS states may be close to the truth. The non-Russian CIS states must be desperate to know what Russian enterprise managers are planning and what Russian economic policies are in the making. Nevertheless, in January 1996 the national-patriots began to make a case for "reconstituting" the old KGB (which was split into two parts, domestic and foreign, in 1991) and for lodging the entire agency within the powerful President's Security Service. The following articles provide insight into Russia's continued existence as a "national security state" and presage what could come in the future if Russian nationalists gain control of security policy within the CIS.

### Espionage Between CIS States
Isabelle Lasserre
*Le Figaro* (Paris), 16 February 1995
[FBIS Translation]

"Certain republics of the former Soviet Union are carrying out operations against us. This year our services have arrested more foreign agents than the KGB and the Interior Ministry combined." Sergey Stepashin, president of the FSK [FCS, Federal Counterintelligence Service]—formerly the KGB—is furious. "We have common roots and the same training, and now these people—our brothers—are working against Russia!" Even the press is indignant: "The unity of the CIS countries is now just an illusion."

These are the new spies, the post–Cold War spies. The experts say that since Russia is no longer felt to be a threat to the West it no longer attracts agents from the West. "Anyone can gain access to information now. Everything is in the press," explained Aleksey Arbatov, director of the Institute for the World Economy and International Relations.

New agents from the "near abroad" have slowly replaced the "traditional enemies from the West." According to the Russian counterespionage service, the number of agents from Eastern Europe, Asia, Islamic states, and especially the countries neighboring the CIS, is growing continually. Ukraine, Uzbekistan, Kazakhstan, and the Baltic countries have now taken the place of the United States and Western Europe in the dock.

Espionage between the former Soviet Republics is a paradoxical consequence of the collapse of the USSR. According to an FSK spokesman it can be explained by "the fact that Russia, which was one of the most secret societies in the world, suddenly opened up. The freedom of speech makes espionage more accessible."

These new spies make the former KGB General Oleg Kalugin smile. "All this is ridiculous. What do you think Estonia—with a population of one and a half million—can do to our country, apart from trying to assess the threat posed by Russia with regard to the sovereignty of the Baltic countries? Moreover, this is not of much interest to us, because Moscow's intentions are unpredictable."

**"Mafiosi"**

Judging by what the experts say, the so-called spies arrested by the Russian services may actually only be "mafiosi." "No CIS republic has the experience or the resources to have secret services worthy of the name," Aleksey Arbatov said. Oleg Kalugin felt that the reason for the publicity in Moscow about CIS spies is the poorly internalized changes in the raison d'être of Russian espionage. "In my time espionage was a war of the intellect. All that is left today is vulgar economic espionage."

Along with many experts he felt that the FSK is now tending "to invent" spies. "Sergey Stephashin wants to preserve the old spirit of espionage. He wants to prove that Russian espionage is still efficient. During the Cold War the KGB was an elite. Today the espionage services feel a bit miserable. I would never go back."

## Spying by Former Republics Claimed

Mikhail Rostovskiy
*Moskovskiy Komsomolets,* 26 November 1995
[FBIS Translation]

The question of whether the former fraternal republics are spying on one another is a very unsavory one, if only because just a year ago most of those who work for and head the secret services of the now-sovereign, ex-Soviet republics operated in a single organization. Tragic incidents like the following are therefore possible in the near future. An ex-cadet at the KGB school in Moscow, now an officer with a new secret service, arrests a former mentor in espionage who has remained to work in Russia. . . .

*Moskovskiy Komsomolets* decided to learn what the relations are among "agencies" of the sovereign states today.

The response to our inquiry, received from Russia's Ministry of Security after exactly a month of stone-walling and procrastination, stated the following: "We collaborate with and have no complaints about our partners. The details are classified information."

An event in which Aron Atabek, a leader of the Kazakh nationalists and editor of the newspaper *HAK*—published, strange as it seems, at the Kolomenskiy Rayon printing office—suffered is apparently among these "classified details." According to the official version, on 29 July of this year Atabek, who resides in Moscow, was attacked by unknown persons, who tied him up and searched his apartment for five hours. "Unknown persons" simultaneously visited the printing office in Kolomna. Atabek was released and accused Kazakhstan's state security service of committing the incident. He went to Baku, where he was granted political asylum. Kazakhstan's National Security Committee declared that it absolutely could not comment on this violation of the law, since it had occurred "on the territory of another state." But at the Russian Ministry of Security I was advised to ask . . . Alma-Ata. An employee at the Kolomenskiy Rayon printing office whom I interviewed, however, told me right away for whom the "unknown" persons worked: "People from our agencies and Kazakhstan's."

The MB [Ministry of Security] appears to have derived a lesson from this incident, however. When a dissident from another of the sunny republics, Uzbekistan, recently received a similar visit, the Russian and Uzbek security agents did not bother to put on masks but arrived completely in the open.

Chechnya is another "classified subject" which the MBRF [Ministry of Security of the Russian Federation] prefers not to discuss. An employee of the ministry's Public Relations Center informed me that the "Ministry of Security is over all of Russia's territory, and Chechnya is a part of the Russian Federation, so there is nothing to discuss." It seemed fairly strange to hear this, since that very day one of the large Moscow newspapers held a briefing at which it referred to the "intelligence agency of the Chechen Republic."

The MBRF's game of silence is absurd also because it is no longer a secret to anyone in the Caucasus today that the Chechen leadership has not only declared "war on Russian and Georgian imperialism" but is even taking action in this direction. It is curious that Chechnya is using Comintern methods in this area and following the guidance of representatives of other Caucasus peoples as a "nation of the victorious dream."

Finally, the most regrettable decision of the Ministry of Security: It labeled as "containing state secrets" a report on the true capabilities of the Chechen leaders, who have repeatedly promised to create a "little Beirut" in Moscow.

Another Russian heir to the KGB, the Foreign Intelligence Service (SVR), has different ways of doing things. Yuriy Kabaladze, chief of the SVR press service, and Tatyana Samolina, its press secretary, refused to answer my questions.

"Is the SVR engaged in operations involving adjacent foreign countries?"

"On the contrary, we collaborate with them. In April the SVR signed an agreement with the secret services of seven CIS countries (Ukraine and Armenia are apparently going to join it later), one of the main points of which is the rejection of operations against one another. In addition, we reached agreement on cooperation in the exchange of information and the joint training of personnel."

At the same time, in a recent appearance on national television, Andris Krastins, a deputy chairman of Latvia's

Supreme Council, offered Western intelligence agencies use of the republic's territory as a base for operations against Russia. The Latvian deputy speaker explained his idea quite candidly: "It is no secret to anyone that serious intelligence forces against Latvia have been established in the Russian intelligence service headed by Primakov."

The Russian Federation's MID [Ministry of Internal Affairs] sent a statement of protest to Latvia's MID. Evgeniy Primakov, chief of Russia's intelligence service, himself spoke out on the subject, however: "We regard those states that are withdrawing as foreign states, and if operations against Russia are initiated somewhere, we shall of course take some sort of counteraction."

The response to these threatening statements was another statement by Krastins, in which he accused Russia of engaging in "psychological warfare against the Latvian Republic."

Russian "competent departments" are allegedly treating Azerbaijan even worse. Azerbaijan's minister of foreign affairs, Iskander Gamidov, was even able to state the exact number of Russian spies apprehended in his republic: "Six during the first five months of 1992." The largest of the alleged operations, conducted by Russian secret services, also involved Azerbaijan.

At the beginning of May of this year the Baku newspaper *Azadlyk* carried the following report (attributed to an unnamed employee of the Ministry of National Security): "With the approval of Ayaz Mutalibov and with the direct participation of workers with Russian military intelligence (the GRU [Main Intelligence Directorate]), a plan has been worked out for a coup d'état in Azerbaijan." Such statements have long since become commonplace in the republic, so that this one too seemed headed for rapid oblivion, if not for . . . a coup d'état did in fact occur in Azerbaijan on 14 May (the new administration managed to hold out for only twenty-four hours, to be sure). Right now, unfortunately, it is impossible reliably to assess the extent of participation by Russian agencies in the change of government. The GRU prefers not to talk to journalists, and all of the Russian Federation's "overt" departments say in unison that all assertions of participation by the GRU are "not consistent with reality."

The only thing reliably known is the fact that when Ayaz Mutalibov, "president for a day," arrived in Moscow, he was put into a special government hospital, and several dozen Russian officers, according to Azerbaijanis serving in the Main Intelligence Directorate of the General Staff of the Russian Federation's Ministry of Defense, were forced hastily to change stations. Russia has now even announced an investigation of Mutalibov.

For now this covers all the well-known incidents involving the Commonwealth's new "cloak and dagger" knights. This does not mean that there will not be more, however. Can anyone say that Azerbaijan's Ministry of National Security is not engaged in operations against sovereign Armenia, that Georgia's Intelligence Service is not interested in knowing how many weapons remain at Russian garrisons in the republic, that Moldova's secret services are not continuing operations against "left-bank separatists," begun so brilliantly (and ending so poorly) with the kidnapping of Igor Smirnov, currently president of the Dniester Region, in 1991? What is more, the Union came apart only a year ago, and many states (particularly in Central Asia) may simply not have had time to set up new, operational secret services.

## 3.32

**Intelligence Chief Views Changed Priorities**

Moscow Television, 19 December 1995
[FBIS Translation], Excerpts

[Studio interview with Evgeniy Primakov, director of Russia's Foreign Intelligence Service, by correspondent Andrey Razbash; from the "Rush Hour" program—live.]

[Razbash]: Hello. Tomorrow, Russia's intelligence officers will mark the seventy-fifth anniversary of their departments. Today, "Rush Hour" welcomes in the studio Evgeniy Primakov, director of Russia's Foreign Intelligence Service. Hello.

[Primakov]: Hello. [Passage omitted.]
[Razbash]: Will you carry out an analysis of the results of the elections among your staff? Do you pay attention to this issue?

[Primakov]: You know we do not vet political beliefs either among the staff subordinated to us, or among our colleagues. Of course, we will do no such thing. Intelligence departments have been depoliticized. Every member of staff can have his beliefs and sympathies, but they must have no influence whatever on his professional work. [Passage omitted.]

[Razbash]: Evgeniy Maksimovich, with the collapse of the USSR, have the priorities of what used to be known as the first main directorate changed?

[Primakov]: The priorities have undoubtedly changed. I can tell you, for instance, that our department never used to be concerned with the task of preserving Russia's territorial integrity as acutely as we are facing it now. We are not an internal intelligence department after all. We are gathering foreign intelligence. But, after all, some external forces wish Russia's disintegration. We need to know their intentions. We must try to neutralize their intentions. This is just one of our directions. I can list many other priorities that we are currently pursuing but that we did not have before. For instance, we are now extremely interested in centrifugal tendencies developing on the territory of the former USSR. Not because we want to eradicate the sovereignty conquered by various republics. But while this sovereignty should be preserved, it is necessary to develop some integrational processes in economic and political spheres. Our intelligence department is not only in favor of it but also working in that direction.

Then, take crisis situations. Never before has Europe been covered by crisis areas to that extent. We thought that the Helsinki agreements guaranteed to a maximum the security of boundaries and so on. And now look: There is a crisis in the south. The intelligence department has shown its good sides there, I think. Generally, we are trying to obtain urgent and accurate information, which ideally is documented, and to report it to the supreme political leadership of this country who need that information to adopt political decisions.

[Razbash]: Evgeniy Maksimovich, several times I came across some shocking articles in the press. They described interaction between intelligence departments that have been known to oppose Russia in a fundamental way. For instance, interaction between Russia and the United States, Russia and Great Britain. And this is described as partnership of intelligence departments. Is this true? Does this partnership exist? If it does, what are its aims?

[Primakov]: You know, there are common spheres of interest among states. Therefore, certain intelligence departments may also have common spheres of interest. For instance, combating organized crime, international terrorism, the drug mafia, as well as exposing dangers that may exist in hot spots and that may lead to destabilization of the situation in large areas. So, we tend to exchange relevant information. We have such partnerships. They mean interaction—not an alternative to intelligence activities.

[Razbash]: I see. Are there any common spheres of interest in the "near abroad"? I know there are agreements banning intelligence activities of one CIS country against another. How do you obtain information, say, about the security of nuclear sites in Ukraine or Belarus, or about the Baykonur space station in Kazakhstan?

[Primakov]: I will tell you unequivocally that if we have such information from some of our sources abroad, we immediately give that information to our Ukrainian or Kazakh partners. But we do not carry out recruiting activity. Nor do we conduct any active intelligence work there.

[Razbash]: I understand that one of your department's priorities is the country's security. The Black Sea Fleet, for instance, is probably an essential factor for this country's security in the south. [Passage omitted.]

[Primakov]: We have certain sources in third countries through which we can obtain information. Naturally, we are in the know of the processes developing in the "near abroad." But we do not use any specific intelligence methods against those countries. There is a complete ban on it. And we have unanimously agreed that we must not work against one another. [Passage omitted.]

[Razbash]: With the disappearance or rather the defeat of communist ideology in the USSR, the ideological basis for recruiting fellow-thinkers, people who believe in communism, has vanished. So, do people want to work for intelligence because of material considerations?

[Primakov]: I think you are wrong there. Even before there was not just an ideological but also a political basis for recruitment. [Passage omitted.] And this political basis still exists. I do not think many people in the world would now want the United States to be the only power dominating the world and imposing its decisions on everybody.

[Razbash]: In other words, a balance of power in the world.

[Primakov]: Yes, a balance of power in the world, and the certainty that Russia is a great power and can therefore play a stabilizing role.

[Razbash]: So, pardon me for asking you, but what is the role of money? Everybody was shocked by the figure of 2.5 million, when the Ames case was reported. What is the role of material incentives?

[Primakov]: Material incentives exist, of course. Many of them have a role to play. But in most cases, I think there are many reasons why people are recruited. I cannot comment on Ames, whom you have mentioned. I cannot comment in any way on his affiliations, I mean his or anybody else's connections with our intelligence department and our intelligence network. This is simply banned by law. But judging by the U.S. press, and judging by Ames's interview, he always stresses that his reasons were not just material, that after meeting Soviet people, after his personal impressions, as he had been here in 1992, he understood that the hostile and aggressively hostile policy of the CIA [Central Intelligence Agency] was wrong. Ames said that this also played a certain role in his decisions.

[Razbash]: Why is the budget of the intelligence bodies not published? After all, they exist on the taxpayers' money. They are funded from the state budget. Should not the taxpayer know how his money is spent?

[Primakov]: He has the right to know, but he can find this out in various ways: from widely disseminated publications or from the members of parliament he has elected. The State Duma has created various committees. The budget committee or the subcommittee for intelligence that existed in the State Duma Foreign Affairs Committee can inform him on the subject. [Passage omitted.]

[Razbash]: At the beginning of our conversation you spoke about the forces, without naming them, that are not interested in having stability in our country, about destabilizing factors and centers. In that respect this coincides with the way the Communist Parties criticize the leadership of this country, which, they say, in dancing to the West's tune, has instituted chaos here, as they call it. Tell me, have you noticed in your work examples showing that our top-ranking functionaries, including those working in the state apparatus, are acting under the influence of somebody or something?

[Primakov]: As for the state's top leadership, I can tell you firmly that they are patriots, who are, without any doubt, taking the country along the path free of any sustained influence from Western forces. But at the same time, many are compelled to take account of the international situation. And you know, those who divide people into those who want confrontation with the West and those who do not want it are wrong. The division is between those who want to settle into normal relations with our former Cold War opponents but to keep these relations on a basis of equality, and those who believe that all methods should be used to move away from the Cold War and to enter European society—all methods, even a ride on the shoulders of one Western country or another—following it as it were—is acceptable. That is the situation. [Passage omitted.]

## 3.33
### Yeltsin Appoints Primakov as New Foreign Minister
ITAR-TASS, 9 January 1996
[FBIS Translation]

President Boris Yeltsin on Tuesday signed a decree appointing Evgeniy Primakov foreign minister of the Russian Federation, the presidential press said.

Before the appointment, academician Primakov headed the Russian Foreign Intelligence Service, the foreign branch of the former KGB.

Primakov will replace Andrey Kozyrev, who was dismissed by President Yeltsin from the ministerial position in connection with his election to the State Duma (lower house of the Russian legislature).

The report of the press service did not say who will replace Primakov as Russia's foreign intelligence chief.

Primakov was born on 29 October 1929 in Kiev into a working-class family and spent his childhood in Tbilisi.

He graduated from the Moscow Institute of Oriental Studies in 1953. After completing a post-graduate course at Moscow State University in 1956, Primakov worked as a correspondent, desk deputy chief editor, and then chief editor at the State Committee for Television and Radio Broadcasting under the USSR Council of Ministers.

In 1962–70, he was a news analyst, deputy editor, and communist daily *Pravda*'s own correspondent in the Middle East.

In 1970, Primakov changed over to active scientific work. He was appointed deputy director of the USSR Academy of Sciences' Institute of the World Economy and International Relations. He was elected a corresponding member of the Academy in 1974 and academician in 1979.

From 1977 to 1985, Primakov worked as the director of the Institute of Oriental Studies and headed the Institute of the World Economy and International Relations in 1985–89.

A Communist Party member since 1959, he was elected an alternate member of the party Central Committee at the 27th party congress and promoted to full membership at the April 1989 Central Committee plenum.

From 1989 to 1991, he chaired the Council of the Union (one of two chambers of the Soviet parliament) and was a member of the USSR Presidential Council and member of the USSR Security Council.

In September 1991, Primakov was appointed the head of the KGB's first main directorate (foreign branch) and the first deputy chief of the KGB. In November 1991, he was appointed director of the USSR Central Intelligence Service and in December the director of Russia's Foreign Intelligence Service.

Primakov is married with a daughter and grandchildren.

# The Pragmatic Nationalists

## 3.34
### Vladimir Lukin on Obstacles to Democracy
Interview by Yuriy Shchekochikhin
*Literaturnaya Gazeta,* 25 November 1992
[FBIS Translation], Excerpts

[Shchekochikhin]: One day, quite recently, at the start of Moscow's summer, a friend, also a journalist, came up to me and said: "I fear that the idea of democracy in Russia is dead. It has not worked out. . . . Everyone is becoming increasingly excited by national patriotism. . . ." Perhaps he is right. There is today no stronger word of abuse than the word "democracy." And people whose names were previously symbols of democracy (Popov, Afanasyev) are today either out of a job or (like Sobchak) being subjected to the most savage criticism. . . .

[Lukin]: You are speaking of symbols, that is, not of real democracy and not of real national patriotism.

[Shchekochikhin]: Symbols they may be. . . . But at approximately the same time some woman or other on the street shouted at me: "Democrats. . . . What have you done to my country. . . ." Of course, I console myself, perhaps she was a little out of sorts, but it made a strong impression, all the same. . . .

[Lukin]: I have a similar story. When I was leaving the Sixth Congress, a woman with eyes bulging approached me and said: "Why are you destroying the Russian people?" I asked: "Who? Me personally?" She became flustered and said: "I do not mean you personally. . . . You generally. . . ."

Still, it is a question of symbols, words. . . . The word "democracy" was in vogue and then went out of style. And in actual fact people are thinking not about democracy or national patriotism but about actual life, existence, prices, crime, and safety on the streets.

[Shchekochikhin]: But remember—and this was quite recently, before and during the First Congress of People's Deputies of the USSR—the enthusiasm of life and the hopes of democracy that simply wafted in the air. . . .

[Lukin]: But the enthusiasm had been implanted not by the structure of the congress itself and not by the principle of elections itself but by the words uttered at that time—and the mere fact that they could be uttered in public. But this was not democracy, this was glasnost: It had been concocted grandiloquently, but quite accurately. Democracy is primarily a form of existence of the state and a certain correlation between the authorities and the civil society, with a particular separation of powers. . . . We did not have such democracy at that time, only hopes for it. And now, when you say that democracy has gone out of style, it is in fact the people who were at that time the symbols of democracy who have gone out of style. Why? Because they made very big advance payments for a rapid solution of problems, and the advance payments have simply remained advance payments. Do you remember Shmelev's article "Advance Payments and Debts"? The advance payments have become debts, and this is the tragedy of the people who made them and of us all.

[Shchekochikhin]: But Boris Yeltsin, when he himself was a member of the opposition, also, in my opinion, made unfulfillable promises. . . .

[Lukin]: This is why Yeltsin also is now less popular. This is part of that same process. . . . Let us, therefore, investigate.

America and the West generally were the example for the people who promised rapid solutions. Remember how they said: "We will now eliminate all the idiotic aspects of our state: the CPSU, the dogma of ideology, the command system in the economy—and all will be in order, as in America. . . ."

This was quite a naive and primitive ideology initiated by certain symbols. One symbol was democracy, which was not true democracy, and for this reason the slogan "Long Live Democracy" was reminiscent of the slogan "All Power to the Soviets." Another symbol was America, which had nothing in common with the real America. It was thought for some reason that as soon as the CPSU had been eliminated, America would automatically be on our side.

But we could not have become a part of the Western world merely by having cleared away the symbols of the old power, about which so much was being said at that time by Yuriy Afanasyev and, even more, Yuriy Vlasov. . . . Vlasov is today with exactly the same passion saying things that are the direct opposite, but passion is, evidently, the principal content of his nature. And were the national socialists to come to power now and good fortune were once again to fail to arrive, Yuriy Vlasov, an honest man after his fashion, would, most likely, just as passionately begin castigating these new authorities also.

[Shchekochikhin]: But it is democracy and the people who were its symbols that are being subjected to the main ordeals as yet. Whence the feeling that things once again have not worked out. . . .

[Lukin]: They have not worked out for the first wave of democrats (I have many friends among them, each of whom I like as an individual), but their problem is that they were ultimately the balalaika players of perestroika. They thought within the framework of their Western-democratic romanticism that it was sufficient to carry out in the country two or three of the simplest social operations, and everything in the country would change for the better. And nothing came of it. Now they are blaming "dark forces" for everything and consoling themselves with the fact that they must be in eternal opposition to the authorities, and let the bureaucrats remain in power. But the question is: Why did they so aspire to power?

[Shchekochikhin]: In case, for that matter, they were accused of only being capable of drawing sketches of a beautiful life and of themselves, having proclaimed the way to them. . . .

[Lukin]: But this is ultimately what has happened! Many of our democrats make it a rule to embark on something and then make a complete mess of it, after which they make a statement to the effect that they need to be in opposition and then to head some party laying claim to head the state. This is the classical condition of the Russian marginal intellectuals who came with the first wave of rejection. I greatly respect them for the role they performed in the destruction of the first stratum of totalitarianism. But then came the next phase, for which, I repeat, they were not prepared. But what is it that is so dreadful that is happening now?

We have a popularly elected president, we have, imperfect, yes, but, for all that, a parliament, we have a multiparty system, albeit very weak (not because anyone is preventing the formation of new parties but for entirely different reasons—purely Russian). We do not have the problem of whether to speak or to hold our tongues, that is, we have freedom of speech. In short, there are all the prerequisites for very serious democratic development. And the talk to the effect that all democrats must go into opposition is simply not serious. And the perception that democracy in Russia has lost and died is connected primarily with the subjective sensation of the stratum of people who have done their work, but who have not themselves come to power, and this power itself has not brought us closer to general happiness as quickly as they expected.

[Shchekochikhin]: Vladimir Petrovich, each of us, most likely, has experienced this feeling of helplessness that comes when one has been approached by people (I refer to my experience as a deputy primarily), and one has not in practice been able to help them. I believe that that has been and is now the case with you at deputies' receptions when you are in Moscow. But what power do the deputies have? And people have gradually begun to grow disenchanted with the fact that their hopes of our electing whom we wanted and of these being our democratic authorities have proved worse than transparent. And public opinion has begun increasingly to incline toward the necessity for some strong, very strong, authoritarian, dictatorial power. . . . This is today's contradiction, possibly. On the one hand there is a desire for democratic authorities, on the other, a strong, strict authority, banging the table with its fist.

[Lukin]: I do not see a contradiction here since I am a supporter of both. I am convinced that no changes for the better are possible in Russia without the participation in these changes of the state. Reforms have never been implemented in Russia such that the state itself has not strictly regulated the reform processes. That the process will proceed of its own accord (I quote a prominent figure) is another illusion. And we need to rid ourselves of this illusion, but not with the aid of hysterical shouts. It is essential to agree to certain compromises.

[Shchekochikhin]: Which, would you say?

[Lukin]: State administration in Russia has to be exercised by bureaucrats, which does not, naturally, rule out the fact that people of a democratic frame of mind should be in the upper echelons of power and should strategically coordinate the changes in society.

But what is the main tragedy of our society? We have no real, sound bureaucracy. When I met with Roh Tae-Woo, president of the Republic of Korea, he gave me several pieces of advice, which I subsequently conveyed to Boris Nikolaevich [Yeltsin]. And his main advice was this: In order for economic reforms to be successful, an effective administration, that is, a strong bureaucracy, is needed. Even if you make a wrong decision, you can, relying on the bureaucracy, quickly revise and adjust it, and, most important, it will be implemented. But if such a bureaucracy is lacking, you could adopt any decision, a brilliant one even, and it would not be fulfilled, all the same.

It is the lack of such a bureaucracy which we are experiencing at this time.

We cannot have an unmanageable state. Russia's historical and genetic background is too ponderous. Otherwise Russia will collapse into tiny parts, to the pleasure of so-called democrats. After this, any person with a mustache would, with the aid of a far-reaching civil war, with tremendous bloodshed, once again combine it into a unified state.

[Shchekochikhin]: But voices calling not only for the preservation of Russia as it is today but also for a restoration of the Soviet Empire are being heard increasingly loudly at this time. And the voices of those who previously were fierce supporters of democracy and the independence of each Soviet republic are being heard in this chorus also.

[Lukin]: People maneuver in politics, of course, and this is a very complex issue. Where is the boundary between natural political maneuvering and political opportunism, when a politician crosses over to the herd, which as of the present moment is the stronger? . . . But I am the judge of no one other than myself. I have always been a supporter of the democratic path of development and have always been an opponent of the country's disintegration. At the First Congress of People's Deputies I voted against having Russian laws take precedence over Union law since I realized full well that there are no first-class laws and second-class laws. I have always been a supporter of unified armed forces and greatly regret that this has not been achieved. I was not a supporter of the country's division, but since things have turned out this way, let us at least keep our state—the Russian Federation—intact. If you consider that even this is an empire, I am for empire. We will forcibly retain nobody, but turning the country into a leopard's skin and making out of the country a further fifteen countries would mean condemning people, both Russians and non-Russians, to bloodshed. . . .

[Shchekochikhin]: But even now Russia is not the great power that earlier figured on maps as the USSR. Nor are we regarded as we were before. . . . Leafing through American papers, I have convinced myself of this yet again. Sometimes for several days running they do not have a single report from Moscow, which means only one thing—we are no longer that great power on which the attention of the whole world was focused, but simply a state. This is both a cause of the explicable nostalgia for past greatness and a pretext for political speculation. . . .

[Lukin]: But this will again depend on what Russia will be. What was Japan when it had an empire and what is it now? And France? The parting with Algeria was a tragedy for it. But can you say that both Japan and France are now weak powers? Could Russia with its boundless territory and 150 million people become a powerful country with exceptional influence in the world? Yes, of course!

[Shchekochikhin]: If we first become an economically powerful country. . . .

[Lukin]: You see, I am not sure that economics is the main thing. The main thing is our souls. And we are now sitting with bowed heads and whining. Democracy is finished, democratic slogans are played out. . . . But we must not just moan! Why, following a devastating defeat in the war, have the Japanese created a powerful state? Because they had the idea of national creation and they wanted to show that even a country without military might can achieve greatness! And de Gaulle reinforced the idea of France's greatness when the empire had really fallen apart.

[Shchekochikhin]: But I fear that under our conditions the idea of national greatness is becoming the idea of nationalism, Russian nationalism, which is being counterposed increasingly to the idea of democracy.

[Lukin]: The national idea has not been formulated with us in practice, and this is why cretinous interpretations of the national idea are appearing. But where are there not cretins?

France's upturn was possible because creative national ideas appeared in the depths of French society. Yes, de Gaulle trampled parliament beneath him and changed governments like gloves, but he won a strong role for France in the world! But de Gaulle did not isolate France from the West, since politics cannot make a civil society isolated from ideas! . . .

[Shchekochikhin]: But, in my opinion, the process of opposing democracy as a symbol and national patriotism as an idea that is taking possession of the masses in just the same way as did the idea of democracy when, earlier, Boris Yeltsin was in opposition has no longer just been discerned, but is under way to the utmost. . . .

[Lukin]: I see another process. . . . First, I do not believe that the idea of extreme nationalism has taken possession of the masses that strongly. . . . Look at the youth! It has American subculture in its nostrils!

[Shchekochikhin]: And are you not frightened by the fact that publications of the extreme right are becoming increasingly popular? That same *Den*, after reading which you want to wash your hands?

[Lukin]: *Den* is a newspaper of civil war. But look at a number of publications on the other side, the extreme left! They also are newspapers of civil war, which are not ascending to dialogue between those who are in the grip of the national idea and the supporters of Western-liberal ideas. . . . The main thing is, after all, hearing one another's arguments!

I am profoundly convinced that Russia has to be Russia. Russia has its own history, its own traditions, and its own national interests. And however many resolutions there are, there will be these interests, just the same, as there were after October 1917, as there were, albeit in distorted form, under Stalin. On the other hand there is also the Russia that has always looked to the West. And it was not, incidentally, the left-wing Chernyshevskiy but Dostoevsky who said that the soul of the Russian man had two sides—Russian and European—that live within and tranquilly get along. So it would be better for us to give some thought to a synthesis of these ideas instead of proposing Marxist extremes.

[Shchekochikhin]: But what for you is the difference between national and imperial interests?

[Lukin]: Very simple! National interests are those that Russia should pursue to the same extent and with the same degree of intensity as any democratic country in the world. What do you think—is America a masochistic country sacrificing its national interests?

[Shchekochikhin]: Of course not.

[Lukin]: Why, then should Russia become a masochist country abandoning its traditional national interests? It by no means follows from this that Russia should be on the side of Saddam Hussein and fight with everyone who disagrees with it! Here is the boundary between imperial ambitions and national interests. There is a balance of interests in democratic countries, serious, tense, sometimes dramatically fine, sometimes not evident, which makes them on the one hand a community, on the other, countries that are each concerned with their own national interests. The problems of the relations of America and Japan and America and Saddam Hussein are at a different level. We also should be a country in which problems between Russia and Japan or Russia and America fit within the framework of world civilization, and support for all dictatorial regimes, like that of Saddam Hussein, does not.

[Shchekochikhin]: You are right, of course, but this is theoretical rightness. But there are, in addition, inner feelings: Many people today feel themselves to be people who have lost, and losers want revenge. When we met with Zbigniew Brzezinski, he said (and his words were quoted in *Literaturnaya Gazeta*): "Russia has lost the Cold War just as Germany or Japan in World War II."

[Lukin]: It was with this issue that I came to America, incidentally. Who had lost and who had won? Was it communism and totalitarianism that were defeated or was it Russia?

[Shchekochikhin]: And you found the answer?

[Lukin]: If you adhere to a purely imperial consciousness, Russia sustained a defeat since it lost many of its territories, some of which were imperial, some that simply could not be deemed imperial. . . . And if there are at all any empires today, it is the republics that have snatched Russian territories only on the grounds that the administrative borders had been carved entirely arbitrarily. Which is why they are today howling that Russia is plundering them, reminiscent of a fellow who has murdered his father and mother and who is demanding of the court clemency on the grounds that he has been left a total orphan. . . . Yes, I repeat, Russia has sustained a defeat, proceeding from the imperial consciousness, but it also paid a very high price, of course, for deliverance from communism and totalitarianism. But Russia has acquired also an opportunity to become a modern, civilized, democratic country. . . . As far as Brzezinski's assertions to the effect that Russia was defeated in the Cold War are concerned, then, to be honest, we were defeated not by the West but by ourselves. Or we conquered ourselves—there is a very complex dilemma here.

[Shchekochikhin]: But I would like to take you back once again to what I have been thinking about continually of late. Owing to a lack of confidence in the democratic outlook, appeals are being made increasingly often—and by an increasingly large number of people—to the idea, as salvation, of a strong authority that is based on the imperial, national-patriotic idea, in which democracy occupies the last place—both democracy as a symbol and democracy as the essence of state rule. I am afraid that were August 1991 to be repeated now, few people would stand in front of the tanks outside the White House. . . .

[Lukin]: For no reason . . . First, there would be no tanks. Second, this is a reflection of those with short memories. . . . Freedom is quite a popular word, but as soon as people want to destroy freedom, they begin themselves to show that this was not the freedom they were dreaming of, that there is another, more ideal. . . .

[Shchekochikhin]: It is easier to gather people in the squares today under national-patriotic slogans than in defense of freedom and democracy. . . .

[Lukin]: There should be no slogans, no mass meetings. We need to learn to govern in such a way as finally to make the streets clean. . . . But this is boring, this is unromantic, and the intelligentsia in Russia is romantic. If there is a slogan, it is unfailingly "Away with!" If there is a speech, it is necessarily, "We are on the edge of the abyss." Whence: "Everything needs unfailingly to be restructured and started from scratch. . . ."

What does this indicate? That we, the intelligentsia, are part of the very flesh of our insufficiently civilized people, by no means that the intelligentsia with us is splendid and that the people are not and need to be replaced. . . .

[Shchekochikhin]: But, Vladimir Petrovich, it was the intelligentsia that both prepared the reforms and supported the reformers. Today, it seems to me, it has come to be in the way of the authorities, the same that it made such, and there is nothing left for it but to return to the kitchen.

[Lukin]: Well, splendid! The main thing in democracy is not (I quote Bulat) that some people burst forth into the governing authorities. Democracy is quite different. It is when there is air so that the intelligentsia might freely express its viewpoint precisely as an intelligentsia. And ideals have collapsed for many people today precisely because the intelligentsia has not become the governing authorities. And what is special about this? Tragedy will come when a totalitarian regime is installed and the intelligentsia is prevented from existing precisely as an intelligentsia, not as the governing authorities from the intelligentsia. . . .

[Shchekochikhin]: I still believe that the disenchantment is not on account of the fact that Russia has not in several years become America. No one seriously entertained this, I believe. No, it is simply that many people have the feeling that they have been deceived. . . . Advantage has been taken of democratic slogans by the old apparatus forces, which have remained in power by making use of these slogans. . . .

[Lukin]: So what? Politics is a cruel business. Yes, the former party nomenklatura is today availing itself of democratic slogans. What, our intelligentsia did not prior to this avail itself of the Marxist-Leninist dictionary in order to live a quiet life, write books, and obtain handouts from the authorities in the form of trips abroad? People should now play the fool or what? So the partocrats, in order to survive, are availing themselves of the new, democratic slogans. After each revolutionary commotion, not only do new people come to power, but the old ones remain! This is the usual historical pattern, and we want something ideal. So we need to work, and the intelligentsia's work is not cleaning the streets but cultivating constructive foundations in society.

[Shchekochikhin]: You now remind me of some figure from the Agitation and Propaganda Department of the Central Committee of the old days. . . . They even had the term: "Criticism should be positive. . . ."

[Lukin]: But I am talking to you about support for democratic institutions, about what the West calls a system of trust, orientation, and a systemic approach to problems. Today even those who know how to work lack criteria for which they are working and earning money. What, merely to amass cash and clear off to America? Or to make their country civilized?

This is how, after all, the Germans and the Japanese and the South Koreans developed. This is what Westernism means to me! Take a look, even the new forms of economics are acquiring some deformed, preposterous nature with us. Why? They lack light and soul. And introducing such is not the business of the Agitation and Propaganda intelligentsia. But this is also the professional work of an intelligentsia cured of the romanticism of democracy.

[Shchekochikhin]: I also am for the intelligentsia being cured of romanticism. But for me this is something a little different. There was a time when the intelligentsia identified itself with power. Our people, intelligent, refined, understanding, had arrived. . . . It seems to me that today it would be better to go back to the kitchens since it was the kitchens that engendered the normal Russian diarchy of strength and soul.

[Lukin]: Let them sit in the kitchen in the evening, but in the daytime involve themselves in normal democratic, civic work.

[Shchekochikhin]: If democratic work means the mass meetings, these are attended less and less often. . . .

[Lukin]: Democratic work begins when the mass meeting ends. The meeting is the incubator of totalitarianism! Have you ever seen a mass meeting at which one person says one thing, and another, the direct opposite? The opposite is hooted down. But after the meeting there are two roads. To take one is to create a civilized society. To take the other means crossing from one mass meeting to another until there finally appears a mustachioed or bearded individual, who speaks at one meeting and breaks up another. This is when totalitarianism begins.

You say, back to the kitchen, that is, once again counterposing the intelligentsia to power. And what is to be done with the beggarly and unsettled old people, with the young people who do not know where to go. . . . Will we then, sitting in our kitchens, counterpose ourselves to the new Kremlin occupants?

[Shchekochikhin]: I hope that the kitchen is not visited by persons with a search warrant. . . .

[Lukin]: If the intelligentsia merely sits in its kitchens, it will necessarily ultimately be gobbled up. . . .

## 3.35
### Bogomolov on Economic Reform, Foreign Policy

Oleg Bogomolov
*Novaya Ezhednevnaya Gazeta,* 11 June 1993
[FBIS Translation], Excerpts

In academician Oleg Bogomolov, as in any normal *homo sapiens*, one can probably find a large number of shortcomings that I am unaware of. But there is one thing of which I am absolutely certain: He was always a very wise and very decent person and was never a conformist. As long ago as 1979, when ardent anti-communist Boris Nikolaevich Yeltsin was heading the party organization in Sverdlovsk Oblast, academician Bogomolov sent the Politburo of the CPSU Central Committee a letter in which he attempted to prove the disastrous nature of the decision to send troops into Afghanistan. At that time he got away with his action and, apparently sensing his own impunity, the academician threw aside all restraint: Thirteen years later he spoke out against the guiding and directing force, this time in the person of the president and prime minister. At that time, in April 1992, he, Georgiy Arbatov, Nikolay Petrakov, and Svyatoslav Fedorov, who had previously been considered true adherents of social reforms, followed one another onto the rostrum of the Supreme Soviet and gave the president and the government a tongue-lashing for their "program of fundamental economic reforms." The democratic public was shocked, and the press immediately reared up, christening what had occurred as an "academicians' mutiny" and branding those academicians as persons with retrograde views and as conservatives. After that, the group of insurgent academicians for a long time disappeared from the television screen and newspaper pages, and Bogomolov, Arbatov, Fedorov, and this time also Nikolay Shmelev (who also was out of favor) were removed from the President's Consultative Council—probably because of their insufficient loyalty.

Currently the voices of these scientists, practically speaking, cannot be heard. But it is not because they are remaining silent. Rather, it is because they are being hushed up. As one can see, even the free press has its blacklists.

In general, this is understandable. The fathers of the nation are completely amenable to an opposition, but not every kind of opposition. Instead, it must be only its extreme forms. If Anpilov or Isakov did not exist, it would be necessary to invent them, in order (a) to have someone to fight with, (b) to have someone to frighten the West with, and (c) to appear, by comparison, to be much more decorous. A moderate, constructive opposition is a completely different matter, since it is difficult to discredit it. It is intelligent, convincing, and, unlike the frenzied opposition, does not terrify the ordinary citizen. Therefore the four disgraced academicians are much more dangerous for the president than Anpilov with his entire army of old ladies.

However, we consider freedom of speech to be unjust only for fools and we open up, by means of the article that follows, a new rubric, "The Constructive Opposition Club." Because the nation has the right to know not only the point of view of M.N. Poltoranyan, but also of those who do not agree with him. At least sometimes. We shall invite as contributors to this rubric intelligent individuals whose opinion "does not necessarily coincide" with the Kremlin's opinion. The first article in this series is by Academician Oleg Bogomolov, director of the Institute of International, Economic, and Political Research.

### How This Began

What was begun by Gaydar and approved by the president was nothing new for economists. Versions of a transition to the market had been discussed long before August 1991. One might recall at least the end of 1990—the conference with the participation of Gorbachev and Ryzhkov, when Abalkin, who was vice premier at that time, reported on the paths for the transition to the market and considered three different alternatives: slow crawling into it; shock therapy; and finally, the so-called moderate-radical alternative. Having agreed in the opinion that the shock alternative would be extremely dangerous under our conditions, most of those present stated that they were in favor of moderate-radical reform. Incidentally, representatives of industry were also present there, and Chernomyrdin made statements.

At that time, just as, incidentally, the current situation is, none of us had any doubt about the need to create market-type economy or about democratization. The measures that were necessary for that purpose were obvious and had even been tested in the practical situation—in the East European countries: privatization; the demonopolization of the economy; liberalization of prices and foreign trade; reform in the tax system; the elimination of the budgetary deficit; the improvement of the monetary system; the development and support of entrepreneurship. The entire question was reduced to the sequence in which to carry out this series of measures. On the eve of the making of these critical decisions—as long ago as the autumn of 1991—both at sessions of the President's Council and in conversations with Gaydar, my colleagues and I expressed our warnings concerning their possible consequences. We asked why it was necessary to begin immediately with the liberalization of prices and wages—two things that would inevitably cause prices to rise sharply and then would cause hyperinflation. It was necessary first to create some kind of competitive environment, so that there would be conviction that the prices would not break loose like an unbroken horse feeling his oats. Unfortunately, no one paid any attention to these warnings. Gaydar's face constantly bore a condescending, skeptical smile, and the president, in my opinion, was under the strong impression exerted by that team and completely trusted it.

But until a certain time we were actually left with the sensation that, at the proper moment, Gaydar would pull some kind of trump card out of his sleeve and everything would change. However, that did not happen.

Then there occurred those same hearings at the Supreme Soviet where we proposed making major adjustments in economic policy. Today everything that we spoke about then is being confirmed in real life and the results of that policy are emerging more visibly. Prices are increasing every month by 20–30 percent, labor is underestimated, to a greater and greater extent the worker is losing any real motivation, and the entrepreneur does not want to invest his funds in the development of production. All this already appears to be very serious. And although we continue to hear reassuring statements that the decrease in production has almost stopped and that price stabilization will occur at any moment, these statements do not agree well with the real facts.

I have never tried to frighten anyone, but if one takes Western criteria, cutting production in half and lowering the standard of living to one-third the previous level over the course of two years can be viewed as a real catastrophe. We are rescued only by the extreme adaptability of the nation, by its ability to survive. "What's good for the Russian is death for the German." And our protest is completely Russian: Inflation is robbing you, so go ahead and rob others if you can. This explains the crime, the thievery, when people steal everything they can from their neighbor and from the state. Thievery always existed in ancient Russia, but today it has become a standard of life.

How do we manage to survive? We sell abroad everything that we can—raw materials, girls, scientists—for a song. . . . We plant vegetable gardens and sometimes even keep chickens on the balconies of city apartments. But go into any rayon social security office and listen to old people who will share their survival experience. Your eyes will fill with tears.

Of course there will be no apocalypse. With our astonishing adaptability and inventiveness, we will come through this somehow. But there will also occur, and is already occurring, a gradual and imperceptible dying out of persons who have not adapted, there will be a moral and physical degradation of society. There is no medicine to prevent this. A person possibly will survive, or he will die. He will starve to death, will slowly expire, even though he might have lived a little longer. Little children are ailing. Unfortunately, from the point of view of anyone in the West, this is already a catastrophe.

People and the nation are attempting to survive. But on the whole this is very humiliating, because we actually are a tremendously rich country and we have a tremendous intellectual potential.

### Concerning Gaydar's Government

In his speech at those memorable parliamentary hearings, Arbatov ended with an old Odessa anecdote: "There were two houses of prostitution across the street from each other. Things were going fine in one of them, but were going poorly in the other one. The madam in the latter one repainted the house and changed the curtains, but still the customers didn't come. Then she asked her more successful neighbor what else she should do. Her neighbor said, 'You have to change the girls!' "

So, let's discuss the question of the "girls. . . ." The government's actions from the very beginning were like the actions of a group of sect members that is convinced of its rectitude and that does not want to listen to others who might interfere with that group's carrying out its plans, because those plans are the correct ones. They also were given conviction concerning this by advice from the IMF. And that sectarian position, that isolation of economic science from practice, caused a large amount of unrest.

We wrote to Yeltsin that it was, of course, a good thing to have a government of people with the same views, when policy is absolutely true and has been worked out in all details. But just imagine that that policy is erroneous, and you have a government of people with the same views, and, moreover, you want to create a parliament of people with the same views!

I think that we are dealing specifically with the situation when they are all linked together in a mutual coverup: they have taken an oath to that policy, it is not working, but they continue to try to convince us that a major breakthrough will occur at any moment. But that breakthrough is not occurring. . . . But the critics are not being allowed to open their mouths. This is the reverse side of the unanimity of views that we already had plenty of at one time. Gaydar was criticized not only by Rutskoy and Khasbulatov. Many scientists expressed extended views on the policy he conducted. But that was simply ignored. No one responded to it.

Gaydar launched myths one after the other. For example, he asserted that, when they began their reforms, the situation in the economy was such that there was no other way out, and if there had been no price liberalization, things would have become even worse. That argument is one of those that might convince a few people, but that are impossible to prove. Serious analysis attests to the fact that the situation would not have been so bad, and that there had been several decisions, but the one that was chosen was the worst one. And certain of Gaydar's people said in general that the main thing was to begin chopping, and it was not important which end to begin. Is that really the right approach?

It is obvious to any specialist: Attempting to stabilize finance without previously defining what the ruble zone is, and without having any control over the republic's banks that are continuing to issue ruble credit, is, in general, a risky business. But it was important for them to display decisiveness. Yeltsin needed obedient followers. Yavlinskiy refused, but Gaydar agreed.

Unfortunately, the press, in my opinion, frequently provides a distorted picture, evaluating everything only in two colors—black and white: Gaydar is a reformer, and those who call for greater control, for increasing the role of the state, are conservatives. The rebirth of the most elementary functions in evaluating and planning the economy, that which previously was done at Gosplan and that which must be done by every state, because without forecast work, without the development of some kind of strategy, a modern state cannot exist—all this is called "a turning back." Well, then, will the invisible hand of the market put everything in its proper place, will it promote the conversion of production and the preservation of some necessary standard of living for the population? That is simply naive. This does not happen anywhere on earth. The state everywhere executes definite and very important functions.

Gaydar and his team introduced the conviction that the market will do everything for us. This is the well-known American concept of the Chicago school—liberalism in its extreme form. But in its pure form the concept has never been embodied anywhere. Reagan was an adherent of it, but only rhetorically, because when he was in office the state kept under its control a very large number of aspects of life. But Clinton currently wants to depart from that and to intensify even more so-called state interventionism. As for the Europeans, for the most part they are adherents of the Keynesian views, according to which the state must introduce an active policy and exert an influence on supply and demand. In Japan and Italy the state has its technical policy and strategy, and strictly monitors the way in which those statewide, national concepts are implemented through private business and a number of drive belts.

But what does the invisible hand of the market mean, if we are talking about social protection? This means the dying out of people who cannot feed or protect themselves. And this also does not happen anywhere else in the world. So the role of the state in the economy is very great. I am not even talking about military production orders or the military industry, or about conversion, where, without state interference, it is completely impossible to achieve any results at all. But in our country the VPK [military-industrial complex] constitutes half the industry.

That is why I consider radical liberalism a very dangerous error. Its adherents represent one extreme, which is perhaps just as dangerous as another one—the national-patriots and orthodox Communists, who want to revive the past. . . .

**About Foreign Policy**

If one takes foreign policy, scientific institutions have also expressed extremely serious criticism in this area. We evaluated negatively those drafts of the foreign policy concept that our government proposed, because of their excessive orientation toward the United States and their underevaluation of the European continent and its role in our priority system, not to mention Eastern Europe and the "near abroad." The relations that took first place were those with America, and those in last place were with Ukraine.

We do not have an intelligent structure or mechanisms for formulating foreign policy. Foreign policy is born in many government offices simultaneously, frequently in an uncoordinated manner. In formulating its principles and in the decision-making process, power structures were involved: the general staff, the ministry of defense, and sometimes also various regional groupings such as GSVG [Group of Soviet Forces in Germany]. In addition, the parliament was a participant in working out international policy, which, in my opinion, was impermissible in such a large analytical job. The parliament should not establish ties, monitor the situation, or issue instructions to embassies. Finally, there exist foreign policy subdivisions within the framework of the president's staff.

However, it seems to me, these links are insufficient if we want to have a foreign policy that is carefully weighted, that is thought out down to the smallest details, and that takes into consideration every kind of long-term consequence. What we need is a serious, scientific base. Previously those functions were fulfilled by the Academy of Sciences, with its international institutes: IMEMO, our institute, and the Far East, Eastern Studies, Africa, and Latin American institutes, which were engaged in studying the state of affairs in individual regions and countries.

Incidentally, I have already mentioned the attitude taken by the authorities to the potential for a scientific approach. They proceed from the assumption that the truth is already known and the only thing left to do is to propagandize it.

About five years ago the newspapers wrote: change your approaches to developing foreign policy; it is necessary to discuss, weigh, and analyze the consequences and to choose from a number of decisions the optimal one. But that has not yet occurred. I do not remember an instance when two different foreign policy approaches collided, when there was any kind of debate. This does not necessarily have to be done publicly. Other forms exist: the holding of various hearings, even closed ones, without broad coverage, at which there could be an analysis of all aspects of a particular foreign policy problem. The results of those hearings should then be used to make the particular decision, resting upon the opinion of the most knowledgeable specialists.

This kind of system is absent in our country, and the existing one operates very weakly. Moreover, it seems to me what is occurring is a certain profanation of the "scientific substantiation" of various decisions. Various centers, foundations, and institutes of "strategic research" are springing up like mushrooms. . . . But my experience tells me that a serious scientific research group can take about ten years to form. Creating in a single year a research center that will operate at the appropriate level is simply unrealistic.

I am very troubled because, when new "centers" of this kind are created alongside serious institutes in the Academy and in counterbalance to them, this destroys the respectable science and favors very dubious sources.

Maybe this is the grumbling of an old scientist who does not understand young people, but it is completely possible that we are present at a process when a serious scientific base is being replaced by pseudoscience, risking the commission of a large number of errors in foreign policy or, even worse, absolute blunders.

Incidentally, certain foreign policy turns taken by the president even today can inspire amazement. Can one really imagine that the president of the United States, having decided suddenly to stop the activities of the U.S. Congress, would send the following request to the FRG [Federal Republic of Germany—West German] chancellor, "In view of the fact that our parliament is very reactionary, I have decided to disband it. Could you please make efforts on our behalf over there in Europe to see that we are supported. . . ." That is simply not done. But that is what happened in our country. The press reported that, on the eve of his "message to the nation," the Russian president requested Helmut Kohl to support the decisive measures being planned by him. And the West, through the mouth of U.S. Secretary of State Warren Christopher and others, was actually given to understand that it supports Yeltsin. But that is completely unprecedented! Maybe some downtrodden developing country can allow itself to suffer that, but Russia. . . .

Although recently it is beginning to appear to me that our leaders' ideas about morality do not agree—not to mention principles—with the most elementary moral principles of Western politicians. French Prime Minister Pierre Bérégovoy shot himself to death because he was persecuted. And he was persecuted because of what we would consider to be a mere bagatelle: In order to buy an apartment, he obtained interest-free credit from an acquaintance. Our people would certainly consider him insane. Because things are generally done differently in our country—if you have climbed your way to power, you have to scrounge, you have to take bribes, you have to use any blessings that you can steal.

**About the West's Position**

We all counted on the West for moral support of the reforms and the process of democratization in the country. But that, first of all, presupposed correct understanding of what was occurring and support for what was working and yielding results. It would have been good if the massive support from the West, including support with money, had actually been a factor in reviving the country. Actually, however, we are rushing farther and farther downward: This policy is not working.

At one time the Marshall Plan was counted on to consolidate Europe and eliminate the enmity between the British and the Germans, and between the French and the Germans. But are we really to believe that at present Western aid is directed at restoring our ties with Ukraine or with Kazakhstan? They have not considered this at all. It is important for them to break us apart. When I begin talking about this, people immediately tell me, "You are repeating the assertions of the national-patriots and the *Den* newspaper concerning agents of influence." But I am only repeating what I repeatedly heard with my own ears from those very influential conversational partners from the European countries. Because the West wants, as rapidly as possible, to make this process irreversible, and they don't give a damn about what effect this has on the nation. It is important for them to have private property, and they are not concerned about who will seize it. Well, they reason, at one time we also had our Rockefellers and Du Ponts. But why should we take that path?

The thing that the West fears most is the bugbear of communism. Therefore it was sufficient to say that the parliament consists of reactionaries who are attempting to restore the communist past. When the West immediately began to regard Khasbulatov as a gangster and the leader of a gangster group, all their favors were given to the president and those who, together with him, are fighting against the parliament. But who is fighting? The same Communists, the same Bolsheviks, but with a reverse sign.

**About What Will Be**

I am not inclined to fall into extreme pessimism. Our society is learning somewhat from what is happening to it and is becoming more mature and more capable of weighing all possibilities. I do not believe that any of the extremes will prevail—leftist-radical liberalism, the national-patriots, or the orthodox Communists. The more sober and more serious approach to reforming society will inevitably triumph because the bulk of the leaders of industry and economic structures, and the scientific intellectuals, have not lost their common sense. And if they are sometimes forced to assent to what is happening today, in the final analysis they will still be able to correct the course of the reforms.

This is especially true since the country's reserves are tremendous. We have already reduced military expenses and shall reduce them even more. For the time being, perhaps, there has been no tangible benefit from this, inasmuch as conversion requires more money than it is capable of giving. But after a certain period of time we shall sense a result. The military-industrial complex is still a glutton that has eaten away a considerable part of the national income.

But the chief prerequisite for rebirth is still linked, in my opinion, not so much with a specific economic policy, with conversion, or with the creation of family farms in agriculture, as it is with state construction, because the chief crisis that we are experiencing today is a crisis of the state system. In essence, we do not have a state, and that which does exist is not monitoring or managing the situation. Laws are promulgated, but not executed. Bureaucratism is monstrous. Bribetakers and criminals remain unpunished. The army is subordinate to no one knows whom, and has already been transformed, in my opinion, into an independent force. Regions are attempting to distance themselves from the Center's unpromising policy.

It is necessary to form a strong state authority and an authoritative state system, which require a re-election of both the parliament and the president. We need political reform, the liquidation of the congress, a two-chamber parliament, the correct distribution of functions between the parliament and the president, that which is called "checks and balances," that is, we need a system of restraints and counterbalances to assure that all the power does not become concentrated in any one of the branches and to assure that there is reciprocal monitoring. It is only on these foundations that we can overcome the scourge of our state system and economy—the crime and corruption that are depriving the entire country of its moral health and making a criminal out of both the authority and the economy. And then it will be necessary to improve the institutions of authority themselves—by means of selection.

After the referendum, this path, if it is not closed, will be moved aside. Once again we are in a stalemate situation, and the president again has the temptation of unconstitutionally implementing his will, of forcing upon the country a constitution that would give him powers that are close to those of a monarch and would reduce the role of parliament to window-dressing. I am extremely depressed at the way that our Supreme Soviet, and even more the congress, are operating. But at the same time I am very troubled by the attacks on the embryonic beginnings of our parliamentarianism. We are not beating up the specific bearers of conservative views. Instead, we are beating up parliamentarianism, freedoms, and the embryonic beginnings of democracy. It is clear that it will be necessary to return to the resolution of these problems, and I would like to hope that they will be resolved in a civilized manner, rather than by having a narrow group of people who are in power force their will on the majority of the population.

Unfortunately, there lives in our genes an orientation toward an idol, a tsar, an orientation toward worshiping some individual, a tendency to be all atwitter because that individual has shaken your hand—we begin to speak breathlessly about him and are even ready to flatter him, as Ryazanov flattered the president. The intelligentsia here is serving as a poor example. All this is so, but, at the same time, time takes its toll, and our chances for authoritarian government are currently very small. Just try to go too far by even the slightest extent and the regions will immediately break off and run in different directions, and this will paralyze the entire central authority. In addition, the army, after all the recent events, will scarcely get involved in the political struggle or take one side or another.

So other times are still coming. True, we do not see any leaders. If only our Russian de Gaulle were to appear, most of the population would trust him—he would be an honest person who would serve the nation, rather than himself and his coterie. That would be much easier. Many people, including myself, would be in favor of that kind of enlightened authoritarian power. Actually, we need the strong hand of a wise politician with a state mind. We don't have anyone like that. We do not even see a state mode of thinking. Gorbachev, incidentally, had that way of thinking. He is a completely different figure, although in his time we were very dissatisfied with many aspects of his activities. We had the feeling that, in his constant maneuvers, he had outfoxed himself. But what can one do if our past leaders did everything to assure that, after they shot their cannon blast, no one would appear to replace them, or if such a figure did appear, he would immediately be stifled.

The natural path is the improvement of democracy. It is completely possible that the pre-election campaign will reveal possible candidates whom we do not yet know, and with the aid of the current mass media we will be able to understand the value of each of them.

And, finally, we have to have a very serious adjustment of economic policy. It is necessary first of all to create those conditions that will assure that the emerging private capital does not go toward trade or into speculative operations, but into production. Today our producers are heavily burdened with taxes. The taxes are such that, if you pay them honestly, you cannot work—it will ruin you. Therefore the evasion of taxes is today the chief concern of any entrepreneur. Even Gaydar himself, who now is giving interviews, says that our tax burden has reached its limit and we have nowhere else to go. But this is nonsense. It is a stupid system and we must change it. Consequently, it is necessary to lower taxes, in order to create an incentive for increasing production, and consequently, as a result of the increased mass of output, you will derive, even with lower tax rates, a greater amount of money.

In addition, we must have serious support of private enterprise, especially in small business—tax holidays and other forms of incentive. This principle is declared constantly, but is never implemented.

We need a sober approach to credit policy: pressuring the bank to make sure it does not issue credit, but, using a selective approach in the area of credit, putting into action a production apparatus.

[We also need] an income policy. If you keep the wages and monetary income of the majority of the population around $30 a month and the retail prices approach the level of worldwide prices, then you must know ahead of time that your industry will not have a market. Because people who get $30 to $40 a month will spend that money only on food. They cannot even think of buying a television set, a car, or a dacha. Consequently, it is necessary to increase people's income in order to create demand. Otherwise your industry will suffocate—those are the ABCs of economic science.

But, first of all, of course, it is necessary to analyze the real reasons why this hyperinflationary increase in prices is occurring, why production is falling, what lies at the bottom of the drop in the standard of living and the impoverishment of the bulk of the population, and how the social differentiation is occurring in society. This requires an objective and public analysis. The nation must know its heroes. Because we still do not know anything about how the government has made decisions, about who is responsible for various errors and failures, or why the promises that were made were not fulfilled. This is not bloodthirstiness. It is the normal practice for any normal state: If you have proposed, insisted upon, and implemented something that failed to yield results or that led to worse results, then you must be held answerable for it. Otherwise we will continue to know nothing, just as we knew nothing during the times of stagnation. Why are ministers being replaced? How has policy changed since the arrival of new people in government? We hear and read that, within the government, there are various points of view, but the thing that this resembles most of all is a fight behind the scenes.

## 3.36
### Vladimir Lukin on Relations with the CIS and with the West

Vladimir Lukin
*Segodnya,* 2 September 1993
[FBIS Translation]

We are perhaps now left with the last chance to extricate ourselves and society from the quagmire of self-torture, reciprocal attacks verging on hysterics, and the sapping of all state offices without exception. We must at long last stop quarreling and destroying. We must summon our strength, rally together, and build—build our home.

When asked what the post-communist ideology is, I usually reply: I think it is an ideology of building a home and running it wisely. A home with a small and capital "h." A home for yourself and your family. A home as your smaller world, the place of your work and your community. A home meaning the region or area when you live. Finally, a home with the capital "h" meaning Russia. For each and every person to concentrate on building such a home is a complex process, a hard and multifaceted experience: You have to find the building material and money, while simultaneously finding the home inside your soul, to overcome the Ivan-with-no-home complex, to stop wandering aimlessly among the absolutes, and return to the concrete soil.

The topic is all the more important because questions of Russia's foreign and security policy are going to occupy a very sizable place in the coming election campaign despite society's being so engrossed in domestic—primarily economic—problems. This has already happened, in fact. The current ruling structures are being bitterly blamed for the "breakup of the Union," for their inability to reliably ensure the security of Russia and its people who find themselves in the newly independent states, and for a "one-sided contest" in relations with the West.

Taking objective stock of Russia's foreign policy over the first "500 days" and summing up experience gained in the meantime are necessary not only to build up intellectual muscle in the election campaign, but also to produce a more effective future strategy—a strategy that is both effective and supported by the Russian people. Let me dwell on two important problems, the main trouble spots of the impending elections—the "breakup of the Union" and "the giveaway game with the West."

The proposition that the "present leaders broke up the Union" is not true. First, it was broken up mostly under the communist leadership, including during their period of agony under Gorbachev. Suffice it to recall the well-known chain of events: the Baltics, Ukraine, Georgia, then Armenia. Second, the Union in its former shape was doomed historically, and the only issue was which path its remaking was going to take—through evolution or explosion. The problem is that the reserves of an evolutionary path were not completely exhausted. For me there is no doubt that they could have been put into effect more skillfully. The possible alternatives like, say, a blend of the British Commonwealth and the EC (in our original packaging, naturally) would have been much safer for making the transition from a unitary state. The stance taken, however, was different—the Belovezh Forest one, which was explosive, established in one fell swoop not only by the well-known decision, but also by the sluggish reaction to its serious consequences, which were absolutely unavoidable and obvious to any professional from the start. Both revolutionary impatience (a temptation to cut a tight knot instead of assiduously undoing it) and the leaders' personal rivalry, as well as political errors, also played a role there. I am going to discuss them separately because we are not completely free from them yet.

All the difficulties and failings (including tragic ones) of the sudden secession affecting borders, the army, the economy and finances, a single legal and democratic space, and a single science and culture seem to have been colossally underrated. Finally, there is the future of Russia itself, which has always been both a territory and a nation. This underrating, in my view, was due to some romantic peculiarities of the radical-democratic mindset that was the hallmark of our initial policy.

The hope that disarming our new neighbors by being kind and magnanimous and counting on their eternal gratitude for having freed them from the yoke of the "imperial center" and the belief that they could only love a repentant Russia led by democrats who broke with the "accursed past"—that all this would lead to a new close-knit family of independent sovereign democratic states that would quickly and easily materialize on the entire post-totalitarian space of the former USSR—ran counter to historical experience and was unrealistic. As should have been expected, reality proved much tougher: self-interest instead of gratitude, militant anti-Moscow sentiments instead of new affection, old suspicions harking back to the Soviet and more distant past, and sharp contradictions and conflicts instead of universal fraternization. And, sometimes, also outright cheating (the latest example is the demarche to Lithuania involving the bilateral agreement). We sobered up in time but wasted much time and many opportunities when you consider that relations with the West at first took the front seat in foreign policy while we neglected our main problem—the problem of our neighbors, the problem of the near, internal circle of Russia.

In particular, no serious attempt was made to coordinate economic reforms in our countries to create—not just in words but in structures—a single economic space (something attempted by G. Yavlinskiy and others and consigned to "criticism by mice"). There was a clear delay in creating Russia's own army, in protecting the Russian-speaking minorities in the adjoining countries—this became the domain of the "national-patriots." There was no mechanism providing for urgent analysis and resolution of conflicts between the countries of the "near abroad" (no room was even provided for this purpose in the Foreign Ministry or the Security Council). Instead, sporadic, one-day summits were held, in whose wake issues that seemed to have been resolved became hopelessly "suspended." Policy increasingly lagged behind events. One of the latest examples is the bloody raid on the Russian border post on the Tajik-Afghan border. Was it really difficult to foresee and to prepare for—especially given our Afghan experience—retaliatory actions by the Islamists after their expulsion from Tajikistan?

As a result of the government's latest efforts, the priorities and style of our policy—primarily as regards economics—have begun gradually to turn toward realism and common sense. Integration sentiments concerning the economy and politics, and even security, have simultaneously started to grow under pressure from the grim reality among our close neighbors after the initial euphoria of boundless autocratic independence. All that is now creating conditions for sharply stepping up our efforts to ensure Russian interests in the "near abroad." True, launching joint actions now would be much harder given greater disorder, economic recession—50 percent of which is attributed to this disorder—and multiple social problems.

Russian strategy in the CIS offers several options that should be decided on as soon as possible. The first one is neo-isolationism, which is essentially a logical conclusion of the very same radical-democratic romanticism whose hopes for an idyll of a new internationalism collapsed ("Ah, if that's the way you want it, we are leaving"). In both cases emotions prevail over reason, and political affectation prevails over an ability for routine statesmen's work. What would be the real significance of Russia's leaving Central Asia and the North Caucasus? First, changing the border involves a multitude of costs. A more important price to pay is political—new borders would cut through the Russian-speaking population (primarily in northern Kazakhstan) with all the unpredictable consequences. Second, removing the Russian presence and influence from there would mean denuding the strategic approaches into the very heart of Russia, creating in this critically important space a strategic vacuum that would be filled by other forces, possibly hostile to Russia.

Third, this would mean leaving to the mercy of fate the weak, fledgling political regimes in the neighboring countries, to virtually doom them to chaos or authoritarian-nationalist degradation. Even forgetting about morals, this would be damaging to the interests of Russia. It can hardly survive as a democracy amid the chronic instability and political turmoil that would poisonously seep into the Russian space through many avenues—border conflicts, streams of refugees, political influence on the provinces, etc.

No less dangerous is, of course, a "national-patriotic" variant to the solution of the CIS problem—an attempt to restore in one form or another the "Russian-Soviet empire." It is unthinkable either technically, without prohibitive sacrifices and costs, or politically because Russia would then find itself in a hostile isolation even more dangerous than that of the Cold War era, if one considers the new alignment of forces and our changed geopolitical situation.

All this says nothing about whether this variant is compatible with preserving democracy in Russia. Happily, Russia does not need this variant: as far as the past is concerned, we need security and economic ties rather than a unitary state. Anyway, the past can be restored only in a slumber or a dream. This does not happen in politics or history.

A real solution to the whole raft of the "near abroad" problems as far as Russia is concerned is striking a complex balance between at least the three factors: Russia's legitimate interests in this space, Russia's realistic changes to realize them, and the interests of our neighbors and partners.

A common formula for this balance can be the system of "neighborliness," long tested everywhere in the world, especially in America. It, however, implies not simply abstract good relations between neighbors—as is often imagined here in this country—but a rather definitive system of mutual obligations between big states and their smaller neighbors, who receive guarantees of security in exchange for their recognition of the "big neighbor's" special interests and influence in proportion to its geographical proximity, and strategic and economic weight. This system of "soft" strategic leadership by Russia in its special interest zone may prove acceptable both to our new neighbors and major partners in the West and East, provided of course that it has been constructed flexibly and ably.

This is a matter requiring many years and involving many things we cannot manage either today or even tomorrow. However, in order to be able to manage them the day after tomorrow, the necessary prerequisites should start to be created today while still retaining these new countries in our area of influence.

The immediate task here is to settle and prevent armed conflicts, to protect the Russian-speaking population, and then to convert this zone, on a collective basis, into a belt of mutual security. The long-term ultimate task here is to create, by stimulating the natural integration processes rather than by compulsion, close allied relationships of a confederative type. Russia needs to have an internal ring of friends in addition to an outer ring of partners.

The level of our relationships with friends should as a rule be qualitatively higher than that of our relationships with our partners, and the level of relationships of our partners with our friends should be qualitatively lower than that between us and our friends.

This hierarchy should not stress form but substance and contents; it should be based on reality and be achieved mainly by positive, that is, political and economic, means. There should also be a mechanism in place—ideally a collective one with a Russian basis—for resolving extreme situations with the use of force as a factor. The more convincing this factor is, the less it would have to be resorted to.

Finally, the formation of such a community of states in a democratic Russia's zone of natural gravity would promote not only security and economic progress for all its members, but also their orderly and joint entry into the world community, as opposed to a humiliating crawl or forcibly dragged into it either as a form of reward, or the opposite—not being dragged into it as a kind of punishment.

As for our relations with the West, the principled line of the top Russian foreign policy officials toward finally overcoming confrontation and achieving rapprochement with the West has undoubtedly been correct. This has helped retain a positive continuity regarding our policy of the late 1980s and given it a serious boost. Unfortunately, inertia and continuity have also been retained with regard to the haste in giving in without bargaining for reciprocal concessions. Apart from a lack of professionalism, this trend has also demonstrated the aforementioned romantic aspects of the radical democrats' mentality. They firmly believed and loudly asserted that once Russia "renounced the old world," knocked communism off its feet, and took on the role of the obedient ally of the United States, the thankful and generous West would at once introduce its profligate daughter to "the Western world," and in any event it would not let us down, but would drag us ashore where the land flows with milk and honey. When it turned out that life is not quite like this, they scrambled to make belated adjustments to this policy, but dual damage has already been done: pampered from the outset by an easy life with a new Russia (and even earlier by Gorbachev and Shevardnadze), our Western partners proved unprepared for an independent Russian policy, and a part of our public, offended by the withholding of the promised Western assistance, has turned away from the West and even started seeing it as the main cause for our woes. Anti-Western and especially anti-American sentiments are in vogue once again and have become a political reality that hinders reasonable policy making. These sentiments have been widely exploited by the forces of the past, which have been calling on the country to start a new "cold war" with the fatally hostile West, which is haunted by the desire to ruin Russia.

This is a war that we are, first, incapable of waging, and, second, do not need whatsoever to pursue our real interests.

The satanic West as presented by our communists and the right wing is an ideological myth just as the ideal West of our romantic democrats; the latter have been lying about it out of ardent love no less than the former did out of animal hatred. The actual West is a complex web of fairly egotistical calculations and apprehensions regarding us. These include:

- not to allow too close a reintegration of the CIS or the restoration of a kind of "Soviet empire" or "Soviet threat" on this basis;
- at the same time, to avert collapse, chaos, and dangerous instability throughout this space;
- to assist in setting up democratic structures on a purely national basis;
- to liquidate nuclear chaos and to prevent the spread of nuclear weapons;
- for domestic policy reasons, to achieve all this at minimum financial and political cost by not doing anything that is burdensome.

This set of interests is far from always being contrary to our own and in many instances corresponds to them. By sticking with the strategic course I outlined earlier, we will not fall out drastically with the West unless we revert to a dictatorship. There will be some periods of chilliness, which is fairly natural. Most importantly, we should not fall into verbose and confrontational hysteria, but rather we need to keep explaining, persuading, and going our own way. If we are strong, everyone will recognize us as such, and some will even pretend to love us.

Another conclusion to be drawn from this is that we have to rely on ourselves above all. Let us finally become realistic—we should not expect any lavish aid (Tokyo appears to have been a culmination of these efforts); moreover, we are unable to make use of what little aid we have. What we need most from the West is the recognition of our legitimate interests in the "near abroad" (preferably with any assistance it can provide in integration processes under way there) and the removal of obstacles preventing Russia from full participation in world trade and economic cooperation. This cooperation will require from us much greater sophistication and consistency in defending our trade and economic interests. Our own firsthand experience has already shown that a political partnership with the United States or other Western countries by no means rules out serious competition in trade and economic relations. One has to be ready for this. In this respect we will not be able to persuade anyone by begging. Hence, we need to act on our own and aggressively.

If, however, Russia is to achieve any serious success in this business that is historically new for our country, it will need something more than just a knowledge of the subtleties of world trade and proficiency in economic diplomacy. What it needs is a national willpower that is set forth in a long-term national strategy and reflected in the clear, thoroughly thought-out actions of an efficient and cohesive leadership. This is the main secret of the startling success stories of postwar Japan and later of the "Asian tigers," which have managed to subordinate everything to a single overriding objective of conquering foreign markets, progressing gradually from low-tech export "niches" to ever higher ones.

Our resources and human capacities will be much greater than those of postwar Japan only, of course, if we maintain the unity of our country and ensure at least some security outside and law-and-order inside. As for national willpower and a single-minded team of governing statesmen, these things are woefully lacking. Without strengthening "the back end" of our foreign policy—its strategic backbone provided by the state—neither ideas, even the best ones, nor tremendous resources will help us achieve success in the international scene. In this respect, Russia's reasonable and strong foreign policy does indeed "begin at home." So, restructuring the Russian house, constructing new good neighborliness, and forging mutually advantageous partnerships with countries of the West and the East—these are the basic priorities of our policy that are closely intertwined with one another.

And one last point—on public support for this policy. We do not have any secrets here. People have enough common sense to figure out where they are being dragged by all sorts of advocates of Russian interests. This is why they are skeptical about the romantic illusions of the radical democrats and have so far not taken the bait of the radical nationalists' alternative. Demand for a reasonable foreign policy, as it is for a domestic one, is immense. If broad support is to be won, the policy has to be aimed substantively at defending the interests of the majority of Russians and be stylistically intelligible and consonant with their sentiments.

This majority is squarely in the center. What they want from their government are simple and understandable things: They are little concerned about human rights in Tibet or the difference between the left wing and the right wing in Peru—although for me these issues have importance—but they are very concerned about the plight of their relatives and friends who have found themselves outside Russia; they are not overly eager "to made the world safe for democracy," but they are very troubled by the security of Russian borders; they give much more thought to domestic disarmament than to international disarmament, wishing that submachine guns would finally cease firing in the streets where their children play. They do not curry favors with the West, nor do they threaten it with revenge; they do not assume a holier-than-thou attitude, nor do they want to be considered worse than any one else; they are not against joining the world community, but they prefer to do this not by crawling on their belly, but by retaining the sense of dignity that behooves a great people with a great history and culture. Let us stick with this policy that is natural for Russia; then it will be accepted both in the country and in the world. And then our people will start respecting us.

## 3.37

### Russia Tells British Embassy It Controls Caspian Sea Resources

Russian Federation Ministry of Foreign Affairs
The British Embassy
No. 120E, 28 April 1994

*Editor's Note:* The letter below is an exact copy of a letter sent by the Russian Foreign Ministry to the Embassy of Great Britain and Northern Ireland with regard to projects dealing in Caspian Sea oil and gas. It is an example of Russia's foreign policy of exerting maximum economic leverage over its CIS neighbors, especially in order to obtain political advantage. Russia demands the right to reject all Caspian oil projects that its former Soviet neighbors are negotiating with Western companies.

The letter was evidently designed to stall a British Petroleum–led consortium that was concluding a $7 billion oil deal to transport oil from Azerbaijan to the West. Russia had been trying for months to gain more control over the oil-rich sea by calling it an "inland lake," owned collectively by all states with borders touching its perimeter. The letter also threatened two other massive projects, both in Kazakhstan—the $20 billion Tengiz oil project and the "Caspishelf" oil and gas venture. Tengiz had been awarded to the American company Chevron, while Caspishelf was being run by a consortium which included Mobil, BP, British Gas, Agip, Statoil, Total, and Shell.

In 1992, the Caspian's oil and natural gas reserves were dispersed among Azerbaijan, Kazakhstan, Turkmenistan, and Russia. But in 1994, Russia started demanding equity stakes in Western energy projects in the region. Lukoil, the largest Russian oil company, secured a 10 percent share in the BP consortium in Azerbaijan. Moscow also achieved in principle a stake in a giant British Gas/Agip venture at Karachaganak in Kazakhstan. Russia was also seeking a stake in the Tengiz project.

The assertion of pre-emptive rights over Caspian Sea resources was the first time Russia provided concrete evidence that it intends to control the flow of oil and natural gas from the ex-Soviet Union to the west. The letter is included here as an example of Russia's potential economic leverage over the other CIS states. Oil and natural gas constitute by far its greatest political-economic source of manipulation and influence over its neighboring states. The politics of energy are certain to be an important determinant of any future organizational form assumed by the Commonwealth.

The Russian Federation Ministry of Foreign Affairs presents its compliments to the Embassy of the United Kingdom of Great Britain and Northern Ireland and has the honor to inform it of the following:

On the 23rd of February 1994 the United Kingdom of Great Britain and Northern Ireland and the Republic of Azerbaijan signed a memorandum on cooperation in the field of power engineering between the two countries, which provides for the holding of talks and the conclusion of an agreement on cooperation in the region of the so-called Azerbaijani sector of the Caspian Sea.

The Russian side would like to draw the attention of the British side to the fact that sectoral demarcation of the sea bed does not exist in the Caspian.

By its very nature the Caspian Sea is an enclosed water reservoir with a single eco-system and represents a object of joint use, within whose boundaries all issues of activities, including resources development, have to be resolved with the participation of all Caspian countries.

Taking the above into account, any steps by whichever Caspian state, aimed at acquiring any kind of advantages with regard to the areas and resources of the Caspian Sea, run counter to the interests of other Caspian states and cannot be recognized.

The Russian Ministry of Foreign Affairs hopes that the above stance will meet with the understanding of the British side, which will bear in mind that any unilateral actions on the Caspian are devoid of legal basis, with all subsequent consequences.

The Ministry takes the opportunity to reassure the Embassy of its great respect.

Moscow
28 April 1994

## 3.38
**Moscow's CIS Policy Changes Assessed**
Valeriy Solovyev
*Nezavisimaya Gazeta,* 9 February 1995
[FBIS Translation]

*Editor's Note:* The following article gains its significance from its insights concerning Russian foreign policy shifts toward the CIS as a result of the Chechnya war. That war, says the author, has brought ambiguous results for Russia and especially for President Yeltsin. The key point here is that a failure with internal policy is very likely to produce a *more* aggressive external policy among Russia's policymakers, even though other CIS member states are likely to be much more wary of Russia's role in the CIS and may try to form "sub-alliances" that might act as counterweights to Russia's potential intervention in internal affairs.

In January the Russian leadership, which seemed to be occupied exclusively with the resolution of the Chechen problem, took a number of important steps in the sphere of integrating the post-Soviet space. Important agreements with Belarus and Kazakhstan are rather obvious. These concern, first, serious economic problems, which for Belarus are multiplied by its total dependence on the deliveries of energy sources from Russia; second, the presidents of these two countries ran into an acute internal political crisis, that in Kazakhstan is further aggravated by increased ethnic tensions.

However, while Aleksandr Lukashenka and Nursultan Nazarbaev are realizing their long-standing and continuously declared aspiration for a closer union with their neighbor, movement by Russia to meet them halfway was rather unexpected. In the course of the last half year, the Kremlin emphatically shunned the Belarusian leader, who continuously emphasized his readiness for the boldest unifying initiatives and urged Russia toward them, and it also repeatedly, in a crude way, rejected the Kazakhstan leader's global integration plan.

What had changed by January 1995? What circumstances obliged Russia sharply to accelerate integration activity? Is it possible to say that Russia is adopting a new strategy in the "near abroad," or is this strictly a tactical maneuver?

### The Political Context

Many analysts have predicted that the military operation in Chechnya will put an end to the prospects for reintegration of the post-Soviet space. But it is exactly the prolongation of the Chechnya "expedition," which almost certainly was planned as a blitzkrieg, that is forcing the Russian leadership to compensate for the serious failure in internal policy with effective steps concerning foreign policy (what is meant is the "near abroad"). Especially since the growing caution of the West with respect to the possible drift of the Yeltsin regime in the direction of an authoritarian government and, accordingly, the prospect of worsening relations between Russia and the West, force the Kremlin to find a counterbalance to this event. It cannot be ruled out that the Russian leadership is sending the West a signal through its reintegration activeness: Continue to support us, or we will start to restore the destroyed empire. The latter is viewed by the West as one of the main threats to its key interests.

The desire to avoid an interpretation of the war as the beginning of a conflict of civilizations and the fear of ethnic and religious ruptures inside the country compel Russia to strive for a "special" relationship, specifically with the formerly Turkic-Islamic Kazakhstan. If it is assumed that the strategic union of Russia, Belarus, and Kazakhstan started to be formed in January 1995, then, according to its socio-cultural and ethnic-religious parameters, it is essentially different from the Belovezh Agreement—this is not a Slavic, but a Eurasian union.

Very important circumstances also exist that are not associated with Chechnya. In taking steps to meet Belarus halfway, the president is playing on getting ahead of parliament, which proposes to conduct a joint meeting of the Russian and Belarusian parliaments in March. In a broader sense, Yeltsin is strengthening his position before an examination in the Duma of the question concerning the denunciation of the Belovezh agreements (the anti-government opposition, after repeated attempts, succeeded in getting it on the agenda).

And, finally, it is possible that this is what is most important. It is likely that the acceleration of the reintegration processes constitutes the start of Yeltsin's preparations for the new presidential elections. Inasmuch as the question of the blame for the destruction of the USSR and the demand for reintegration will inevitably occupy a very important place in the course of the election campaign, then, by correcting the error of December 1991, the president of the Russian Federation knocks out the ground from under his actual critics and potential rivals.

### The Military Strategic Factor

Since [the outbreak of conflict in] Chechnya, not only has the possibility of the expansion of the NATO bloc in an easterly direction increased significantly, but Russia's chances of preventing this have been reduced practically to zero. The threat of the appearance of NATO on the border of the former USSR provides Russia with an incentive to restore an integral defense infrastructure and to form a single military-strategic space. For the time being, only the first step has been taken in this direction—payment for the use of military bases and test ranges has been abolished—but the creation of combined armed forces under the aegis of Russia in the near future can already be seen.

### Economic Motives

Agreements on a customs union and free trade without confiscations and restrictions will greatly promote the revival of the national industry of the three republics, although the effect of the implementation of these agreements for the Russian economy will obviously be ambiguous. On the one hand, Russia almost certainly will have to supply the noncompetitive and energy-intensive Belarusian enterprises with cheap energy sources. On the other hand, Russia has consolidated control over oil and gas pipelines and other communications that cross Belarus and that are strategically important to it, and it retains the right to duty-free transit for energy sources—Russia's main export resource—to Central and Western Europe. As for Kazakhstan, Russia guarantees its participation in the development of the rich oil of the Caspian shelf and gas deposits in Karashaganak. It appears that in this case the greatest gain is achieved by the Russian fuel and energy complex, which is steadily and consistently establishing its strategic control over the oil and gas wealth and the pipeline system in the post-Soviet space. The formation of a single market on the territory of the three republics will put the conclusion of a payments union between them on the agenda (an agreement on the mutual conversion of national currencies has already been signed). In view of the considerable stability and strength of the Russian ruble in comparison with the Belarusian zaichik and the Kazakhstan tenge, this will lead to the domination of the Russian monetary unit and the expansion of the incomparably stronger Russian financial capital. In a strategic perspective, Russia will not only tie the national economies of Belarus and Kazakhstan closer to itself (despite all the efforts of the latter to weaken or disrupt this dependence, they did not succeed), but it will ensure the actual domination of Russian capital in them.

### Russian-Ukrainian Relations

The meeting of the Russian and Ukrainian presidents did not constitute a "breakthrough" in terms of reintegration, and there are weighty reasons for this. Although Ukraine is in a no less disastrous economic situation than Belarus and Kazakhstan, and Kuchma has begun a prolonged political squabble with the Ukrainian parliament, in mutual relations with Moscow, the free hand of the Ukrainian president is seriously restricted by the mood of Ukrainian society. Accelerating a union with the "Muscovites" will undoubtedly cause a tearing away of western Ukraine from authority, split Ukrainian society, and provoke a serious political crisis in the republic.

It is probable that Moscow will also be afraid of including Ukraine in a "trilateral agreement." The economic costs of a close union with the latter look too demanding. It is one thing to supply cheap energy sources to a comparatively "small" Belarusian economy. It is another thing to keep afloat the inefficient and non-competitive Ukrainian industrial giant. The Russian locomotive can easily pull the Belarusian economy behind it, but will it also be able to pull the Ukrainian "coach"? And, indeed, to assist Belarus and Kazakhstan with the relative stabilization of their national currencies is far easier than to assist Ukraine.

However, integration tendencies are also growing in the Russo-Ukrainian relationship. The meeting of the managers of the Russian and Ukrainian border oblasts with the participation of the deputy prime ministers of both countries requires that their countries conclude a customs union. An agreement on the principles of trade and economic cooperation in 1995 proposes the establishment of Russo-Ukrainian oil refineries and timber processing complexes, the creation of joint financial groups and companies and transnational corporations, and also joint-stock facilities that represent a mutual interest in exchange for the delivery of Russian energy sources and the postponement of payment settlements for state credits. Ukraine is opening up somewhat to Russian financial expansion.

**Russia's New Strategy?**

Thus, the reintegration activity of Moscow in January 1995 is the result of a number of factors that are both situational, incited by the war in Chechnya, and strategic. Inasmuch as their combination is of a dynamic and unstable nature, it is difficult to come to an unequivocal conclusion about whether or not the new Kremlin policy will become consolidated.

However, if the "trilateral agreement" begins to be put into practice, then this will signify a cardinal revision of the conceptual principles of Russian policy in the post-Soviet space, and that we were dealing with complete improvisation and situational reactions. But, actually, it is not that important whether it was conscious or spontaneous; Russia maintained a strategy of isolation, although also not without some substantial vacillations and deviations. This policy was cloaked in CIS declarations, and it alternated with sporadic flare-ups of activity in the defense of Russian military-strategic interests.

Russia sent a signal in January about its readiness to shift from an isolation strategy to an integration strategy. This in itself is already an event of principal importance. But it is no less important to understand exactly what type of reintegration Russia intends to implement. Although reintegration undoubtedly carries mutual advantages, it is hardly possible to talk about a "union of equals" with respect to the post-Soviet space. By moving to integration and assuming a significant share of its economic and financial costs, Russia, as the "trilateral agreement" shows, will probably try to consolidate and intensify its objectively dominant position in the economic and military spheres. It will lock in and partially subordinate adjacent economies and national currencies to itself, and it will take control of the system of defense of allies. For the latter, this will mean the infringement of economic and partly political sovereignty. But they objectively have no other way out.

This will not be either a reanimation of the Soviet empire or the revival of the Russian empire. Russia has no need to erase state borders or to consolidate the state-political structures of the new independent states. The Russian ruble and Russian energy sources will successfully replace the general secretary of the Communist Party. This will be an empire of a "new" type, a "velvet" empire, that establishes itself on the financial-economic and military dependence of the post-Soviet republics on Russia and, owing to this, that ensures the political hegemony of Russia in the post-Soviet space. It looks as if it is exactly this kind of objective that the present Russian leadership is setting for itself, and that it is exactly this type of reintegration that it intends to strive for.

---

## 3.39

**Lukin on the Dangers of Self-Determination**

Vladimir Lukin

*Segodnya,* 17 February 1995

[FBIS Translation]

---

*Editor's Note:* Vladimir Lukin's rationale for the dangers of self-determination are convincing and extremely pragmatic. Lukin, however, uses arguments against unleashing the nationalist feelings of every small ethnic minority group to imply that the CIS, without cultural, economic, and political integration administered by the dominant interest group (Russia), is unfeasible. The article ends up as a carefully crafted rationalization of the Great Power point of view in Russia, using modern history rather than Russian imperial tradition to support its line of reasoning. The article is important because it illustrates how widespread the support is within Russia for some version of a restoration of Russian dominance and power over the former Soviet space. By 1994–95 very few, if any, Russian intellectuals were extolling the virtues of a self-contained, democratic Russian state that respects above all the rights and sovereignty of other independent nations, as Andrey Kozyrev and Boris Yeltsin had in 1991.

What took place on the banks of the Sunzha was a terrible tragedy. Who can justify the present methods of resolving the Chechen conflict, the bloodshed on a mass scale, the innocent and the guilty amidst a sea of human suffering? But in this twilight of existence, in the ruins of the beautiful southern city, at least one serious factor arises, one problem on a truly global scale. Or rather—a dilemma: national self-determination or state unity.

Two visual images from my recent diplomatic experience come immediately to my mind. In the center of Philadelphia, among the narrow streets, like museum walls, there is a forgotten bell, erected over two hundred years ago, at the time of American independence (that is, separation from Britain, a stormy, bloody separation). And a grandiose monument, on the pediment of which it is written that the nation presented it to the person who preserved the unity of the country, towers majestically over the nation's capital, its thirty-six columns rising toward the heavens.

The Philadelphia bell and the Lincoln Monument embody the two main principles of today's international life—the principle of a nation's right to self-determination and the principle of preserving territorial wholeness. Both principles have their adherents, martyrs, and heroes. When one speaks of the United States, in the pantheon of national heroes over the Philadelphia insurgents—in the memory of the people and in the opinion of professional historians—one person towers mightily, a person in whose memory there will always be mass demonstrations by adherents of civil rights—a person who razed flourishing Atlanta to the ground for the sake of the country's unity.

In most recent history, both these principles are close neighbors, sometimes even in the same "political credo," for example, in President Wilson's famous "Fourteen Points" (January 1918). The sixth point, dedicated to our country, speaks of the need to observe its territorial wholeness, of a "united and indivisible" Russia. The point dedicated to Austro-Hungary calls for realizing the rights of its peoples to self-determination. Closer to our era, in 1975, the countries of Europe and North America vowed, in the Helsinki Pact, to preserve the territorial status-quo, but fifteen years later they promoted the dividing up of the SFRYu [Socialist Federated Republic of Yugoslavia], the USSR and the ChSFR, into twenty-two new state formations.

True, the West did not immediately change its principles. Let us remember—back in the summer of 1991, President Bush, appearing in Kiev, called on Ukraine to preserve the unity. At that time, the United States declared its support of Yugoslavian unity. It took the Belovezh Forest tragedy and Germany's determined intervention (Slovenia and Croatia were recognized in that same December 1991) to make the West start swimming through the waves of the events and agree, "with hindsight," to recognize the legitimacy of the rights to self-determination. Even today, however, it is best not to talk to the French about Corsica and Normandy, to the Spaniards about Catalonia and the Basque region, or to the English about Scotland and Wales. In conversations on this subject, bombastic Americans usually give a taciturn smile. General Sherman, with his "March to the Sea," which singed the heart of the Southern states, closed this question for the United States.

This silence is understandable. In today's world, in over three-fourths of the 180 existing states, there are minorities numbering over a million people. If all the minorities were to put into effect the principle of self-determination, especially to take it to the level of creating their own state, the world would turn into bedlam. We see reflections of this bedlam on television screens everyday, and in the current stormy decade they have become almost the main ferment of international life.

If we acknowledge, as before, that the latent value of both principles is "equivalent," we risk falling into a minefield, with no end in sight. The question is quite difficult. Was the dismemberment of the former Yugoslavia worth a quarter of a million dead and several million refugees? Superimpose these proportions on our country, make the calculations, and it will turn out that we are just at the beginning of a tragedy that will lead straight to Armageddon!

Let us remember that the superseding of the principle of ethnic self-determination over the inviolability of borders by the last decade of the twentieth century was not restricted to the territories of Europe and the former Soviet Union. The process "proliferated," on an even broader scale. For the first time in many years, the West came out as the initiator of dividing up states that it, strictly speaking, had created and recreated, and whose territorial wholeness it had long guarded. Ethiopia is possibly the most graphic example. After many years of discussing separatism in the world as a whole, and in Africa in particular (look, for example, at the attitude toward the attempts of Katanga to leave Zaire and of Biafra to leave Nigeria), the West, in 1993, sanctioned the division of Ethiopia. For the first time, it openly and even demonstratively violated the principle of territorial wholeness in favor of the principle of "proto-national" self-determination, even when the "self-determining nation" did not have the traditional prerequisites for a state (common history, territory, emotional-psychological community, etc.). Potentially, the appearance on the political world map of a former part of Ethiopia—Eritrea—is an event of extraordinary significance for the future. The world will scarcely survive fragmentation to the level of the sovereign Andorra or Liechtenstein, to the level of the island-state of Nauru (9,300 inhabitants) in the Pacific Ocean. The example of Eritrea is before them, and the flow of arms from the industrial centers of the West and Eastern Europe to the world of ethnic hostility is only increasing. Eritrea is in its own way a symbol, and almost an invitation to the tribalist world to call forth the old gods. This is a very great threat, not only to the half-forgotten "new world order," but also to creating any order in the system of international relations.

We must gather enough intellectual integrity and moral strength to pose in earnest the question of the cost of the process of self-determination for mankind and the main thing, for the individual. One gains the impression that, in 1992–93, the bewitched world, as if wavering from within, kept silent about the real price of self-determination for the inalienable and generally acknowledged rights of the individual. Only last year did the original shock begin to pass, and the problem of self-determination, of the human cost of this self-determination fall into the focus of public discussion. A certain euphoric lift slips away when nationalism has given world history such a particularly and unequivocally progressive turn. From tragic experience comes the realization that the triumph of the principle of self-determination over the constitutional-geopolitical foundation of statehood opens an enormous potential for conflicts. Orientation toward the principle of self-determination is fraught with self-destruction for the world as a whole, for Europe in particular and especially for the unstably modernized Eastern Europe. One must obviously agree with the opinion of Lord Acton, that the principle of the universal right to national self-determination is a step backward in historical development. In today's world, movements that blindly pursue the goal of self-determination at any price undermine the potential for democratic development in the new independent countries and endanger the foundations of sovereignty of the people in democratic states. The time has come to deprive most of them of moral approval and see them for what they really are—a destructive force, drawing the world into the gloomiest eras of the past.

All economic theory calls for market expansion, and the national groups bursting for state self-determination are dooming themselves to progressive lagging behind. No less bleak is the cost of destroying the socio-psychological conditions that constitute the natural living environment of many generations. The destruction of this medium inevitably calls forth a crisis in morals, self-awareness, and culture. Does not true democracy demand variety, not the sacrificial like-mindedness dictated by national intoxication? States that are thought up in someone's head, just like the "cerebral" dismemberment of state expanses into regions, are based on legitimizing the rights of groups and clan interests, to the detriment of the immemorial, basic rights of a person and a citizen. It was not for nothing that communist totalitarianism extolled this troglodytic-prehistoric right. In order to keep the world community from dashing off into the abyss of national frenzy, we should place the rights of the individual, of the citizen, above the rights of a group, a clan, a type, or a family given to fanaticism. At the same time, all the rights of national-cultural autonomy are naturally inviolable.

Integration, not national separation, protection of the rights of the individual, not sacrificially ancestral, ethnic romanticism, should be the basis when determining the political map of the world in a century of dissemination of nuclear weapons, partial paralysis of the United Nations, and the pressure of the local nationalist elite with mercenary interests, whose ideal is to have its own army and its own treasury, and not least, although it is laughable, its own international protocol "at full volume." Not Lincoln's principle: "A government of the people, by the people, and for the people."

## 3.40

**Importance of Russian Interests in "Near Abroad" Stressed**

Sergey Kolchin

*Mirovaya Ekonomika i Mezhdunarodnaya Otnosheniya,* April 1995

[FBIS Translation]

[Article by Candidate of Economic Sciences Sergey Vsevolodovich Kolchin, head of the section on the "CIS States" at the Institute of International Economic and Political Research of the Russian Academy of Sciences.]

With the disintegration of the USSR in the early 1990s a fundamentally new geopolitical reality emerged in the post-Soviet space: Fifteen new independent states appeared. The rise of virtually all, with the exception of Baltic countries (relations with Baltic states are not examined within the framework of this article), and of the successor of the former Union—Russia—was due to the collapse of the USSR, not to the natural logic of their national state development.

The following are grounds for such a conclusion:

- the absence in a number of countries of historical traditions of independent development within the framework of national state formations. In recent history the experience in such a development is limited to several post-revolutionary years and not even in all of them;

- the lack of development of a political system (government institutions, political parties, and so forth). The state system, borders, and internal territorial division, basically, were inherited from former USSR republics, which, essentially, served as decorative formations within the framework of a single Union;
- the extremely high degree of integration of their economies into a single national economic complex and dependence on Russia. In the USSR in 1988 inter-republic trade made up 21 percent of the gross national product (in the European Community, 14 percent). Whereas Russia in isolation from other republics can ensure two-thirds of the production of the final product, for example, Azerbaijan, 31 percent, Kazakhstan, 27 percent, and Ukraine, 15 percent;
- the lack of development of indispensable institutions and functions of an independent state (army, foreign trade, personnel training, and so forth) and, instead of them, dependence on Russian potential.

Furthermore, the specific features of the present situation in the new states lies in the fact that representatives of the old national ruling elite, which was formed during the period of the USSR's existence, are in power in most of them. The brief stay of representatives of nationalist opposition movements in power (Z. Gamsakhurdia in Georgia and A. Elchibey in Azerbaijan) and the rapid collapse of their regimes only confirm this tendency. In reality, out of the present top leaders of the new independent states only Kyrgyzstan's president, A. Akaev, does not belong to the former nomenklatura, but he also experiences a strong pressure from it. A. Lukashenka, Belarus' first president, is also a new figure.

Whereas in these states the internal political situation has not changed greatly, marked shifts have occurred in the positions of the largest member of the commonwealth—Russia. For it the disintegration of the Union and the rise of new independent states signify a new and, more often, unfavorable combination of political, economic, social, and other factors.

On the geopolitical plane Russia has encountered the appearance on its borders of a whole group of states, which are quite unstable and by no means always conduct a pro-Russian policy. Thirteen ports on the Black Sea and five in the Baltic Sea have been lost. Russia is separated from Western Europe over a wide area by a double cordon of countries of the former Warsaw Treaty and former neighbors in the USSR.

In the military-political area the entire security system must be revised fundamentally. Facilities for early detection of enemy missiles remain in Latvia and Azerbaijan. Troop withdrawal from Eastern Europe was followed by a departure from Baltic states. Now the Moscow Military District has become a border district and the question is being raised of establishing a new military district in the west with headquarters in Smolensk.

In the economic sphere Russia, as before, is forced to carry the burden. On 1 June 1994 the total debt to Russia of countries in the "near abroad" totaled 3.241 billion rubles, of which Ukraine accounts for 54 percent, Belarus, 17 percent, and Kazakhstan, 15 percent. While mechanisms of former internal Union economic relations have been dismantled, new ones (of interstate relations) operate poorly.

In the social sphere the problem of Russians who now live in the new independent states has become acute. According to the 1989 census, almost 25.3 million Russians and more than 11 million people of other nationalities, who considered the Russian language their native tongue, lived outside the RSFSR.

There are also a number of other factors determining the need for Russia's special attention to states in the "near abroad" and for an active formation of its policy with respect to these countries.

## Changes in Policy

Evaluating the change in the priorities of Russia's foreign policy, it may be said that the shift in the policy's center of gravity to states in the "near abroad" emerged in 1993–94. In the official line and political debates more and more attention is paid to Russian interests in the space directly adjoining the country's borders.

In our opinion, the basic reasons for such a turn are the following:

1. Russia has begun to realize the change in its status in the geopolitical disposition of forces. "According to the state of socio-economic development," the scientific report of the Center for Foreign Political Research of the Institute of International Economic and Political Research of the Russian Academy of Sciences notes, "Russia cannot be classified among the superpowers and at the present moment it does not need this status, which is beyond its power. It now copes with great difficulty even in the role of a great power, because it does not have the political, economic, and military resources necessary for this. Russia is not in a position to conduct a global foreign policy similar to that conducted by the former Soviet Union and to bear responsibility for global international security."[1]

A more realistic approach predominates in the country's foreign policy now. This is primarily an orientation toward problems directly affecting the situation in Russia, to which, undoubtedly, relations with neighbors in the post-Soviet space belong.

2. The course of events has confirmed that the "civilized divorce" with the former Union republics has not removed from the agenda the need for the development of an integral strategy for Russia's relations with them, and for daily close attention to the situation in these countries. It affects the situation in Russia too severely. Furthermore, relations between Russia and these countries are more profound and vast than ordinary interstate relations. Russia remains the center of attraction in the post-Soviet space and not always in the positive sense. As, for example, RF Prime Minister V. Chernomyrdin notes, "Russia's territory seemingly has become a vacuum cleaner, which absorbs everything that is bad: the drug business and arms smuggling—things that, previously, we never heard about."[2]

3. Matters concerning relations with neighbors and the protection of Russian interests and rights of compatriots in the new independent states have become a trump card played by various opposition powers and forces in Russia itself. Having overlooked these matters at one time, the official leadership (the president, the government, and the Ministry of Foreign Affairs) is forced to make up for the lag, restoring their image of defenders of Russian interests. The population's natural dissatisfaction with the crisis situation in the country is channeled by the opposition into criticism of the "anti-popular" and "treacherous" policy of the authorities.

The concept of "Russia's national or special interests," the zone of which is called the "near abroad," has appeared and has become firmly established in the official political vocabulary. This term itself was borrowed from the Americans, who used it widely, including in the well-known "Monroe Doctrine." During the time of the USSR officials did not use the concept of "national interests," because Marxist dogmas affirmed the principle of internationalism and the primacy of class interests. During the later Gorbachev period the theme of "a return to general human values" was put in the forefront of Soviet policy. Now the time has come for Russia to become aware of and to protect its national state interests.

In February 1993 Russia's President B. Yeltsin put forward the proposal that the United Nations and other international organizations give Russia a mandate for the performance of activity in the sphere of its interests—in the territory of the former USSR. It was a matter of the conduct of peacemaking operations, rendering of humanitarian assistance in zones of conflicts, protection of ethnic Russians, and political settlement through Russian mediation. Russia's military doctrine published in November 1993 stresses that the geopolitical space of the former USSR (with the exception of Baltic countries) is the zone of Russia's vital interests and special responsibility. The program speech by the RF minister of foreign affairs A. Kozyrev at a conference of Russia's ambassadors to CIS and Baltic countries in January 1994 noted the following: "This is where Russia's vital top priority interests are concentrated. The main threats to these interests also originate there."[3]

Finally, the February (1994) message by the RF president "On Strengthening the Russian State" openly points out that a "consistent advancement of national interests through openness and cooperation and ensuring favorable conditions for internal development and the continuation of reforms are the main goals of the policy of the Russian state in 1994." It points out that the CIS is the "sphere of special responsibility and special mutual interests of Russia and its neighbors."[4] Let us pay attention to the fact that the president talks about "mutual interests," softening the sharpness of the statement by the minister of foreign affairs.

Attempts at theoretically substantiating Russia's special role and responsibility in the post-Soviet space have been made. Such a substantiation of the distinctive "Russian Monroe Doctrine" is presented in an especially clear manner in articles and speeches by A. Migranyan, member of the RF Presidential Council.[5] At the same time, whereas the matter of the content and nature of Russian national interests in relations with CIS countries is debated, the very fact of the existence of such interests is essentially not called into question. The authors of the cited report note the following: "States of the so-called 'near abroad' represent a priority zone for safeguarding Russia's national security and, therefore, an all-around development of cooperation with them, a study of the socio-political and economic processes occurring in them, and an effect on them, with due regard for Russia's interests, represent the most important task of Russian foreign policy."[6]

On the whole, Russia's vital interests in the "near abroad" can be reduced to three fundamental concepts: security, stability, and cooperation. The main goals of Russian foreign policy with regard to the new independent states are also determined on the basis of this. In the most concentrated form these goals are indicated in the basic provisions of the concept of Russia's foreign policy approved by the RF Security Council in April 1993. The organization of stable positive mutual relations with countries of the "near abroad" was declared the main content of Russia's foreign policy and the prerequisite for its preservation as a great world power. The actions aimed at undermining integration processes in the CIS were included among the main political threats and challenges to Russia's security. The protection of the rights and interests of national minorities—representatives of RF nations in countries of the former USSR—was considered one of Russia's vitally important interests.[7]

**Reaction in the West**

Russia's efforts to reach these goals in relations with CIS states encountered quite a contradictory reaction in the West.

On the whole, a negative, guarded attitude toward Russia's role in the post-Soviet space predominates there. In the most concentrated form this was expressed by Z. Brzezinski, former U.S. national security adviser.[8] There were also many other similar comments on the part of officials and experts. For example, A. Juppé, France's minister of foreign affairs, stressed that "little by little Russians are beginning to lay their hands on everything in the entire territory of the former Union." In an interview with the newspaper *Croix* he noted the following: "It is possible to understand that Russia is interested in what is happening around it and that, undoubtedly, it must play a special role in what today is called the CIS. However, it does not at all follow from this that it can do this as it wishes. Russia is not destined to play the role of a gendarme in all azimuths."

In the opinion of the newspaper *Segodnya,* "Russia's policy with respect to the 'near abroad' has recently evoked concern on the part of Western countries, including Germany. Recognizing the legitimacy of Russian interests in the CIS, the West is trying to prevent the 'reimperialization' of Moscow's foreign policy."[9] True, other signals are also received from the West. I will mention the well-known statement by President Clinton that the United States would not object to the unification of former USSR republics provided there is voluntary participation and a free expression of the people's will. In our opinion, the West's apprehensions regarding "Russia's imperial aspirations" do not correspond to the real state of affairs.

Having taken the course of restraining Russia in the post-Soviet space, the West should have in mind a number of basic factors. The post-Soviet space is a special kind of space, and such unconventional criteria or clichés as "empire" are hardly applicable to it. Russian national outlying districts nearly always lived better than the "metropolitan country," receiving dividends from the exploitation of Russian resources.

Russia continues to carry the heavy burden of assistance to neighbors in economic and military areas. The West's opposition can lead to the fact that it will have to assume this burden. Furthermore, the regimes in the present CIS republics cannot be called regimes conducting a more democratic and market-based policy than Russia. Russia's displacement from the post-Soviet space will create a vacuum, which countries in no way belonging to Western democracies (for example, Iran and China) will be able to fill quickly.

Thus, if the preservation of stability in the region and support for positive transformations, not the suppression of the former enemy and competitor, are the real goals of the West's policy with regard to Russia and its neighbors from the CIS, this policy should rather promote, not counteract, Russia's efforts in its attempts to somehow cement the disintegrating post-Soviet space.

**Policy Priorities**

Whereas Russia's geopolitical interests and the goals of its policy in the "near abroad" have been defined, each of the spheres of state activity (military, social, economic, and so forth) has its own specific nature requiring a more detailed examination. Defense and security, humanitarian (ethno-social) interests and the protection of citizens' rights, and the economy and interstate economic relations are the basic groups of interests and policy directions.

*Defense and security.* Despite the fact that with the end of the cold war the general international climate around Russia has improved and the danger of large-scale armed conflicts has declined, with the disintegration of the Warsaw Treaty Organization and then of the USSR the status of the security and defense of the Russian state arouses serious apprehensions. NATO continues to exist and even Russia's entry into partnership relations with the bloc does not signify complete harmony of the parties' military interests. In particular, the scenarios of probable conflicts developed in NATO headquarters attest to this.

One of them is presented in the newspaper *Vek*: "An authoritarian government, which immediately begins to intimidate small republics of the former USSR under the pretext of observance of the rights of Russians scattered over all CIS states, comes to power in Russia. Thus, Moscow, with the support of Minsk, demands that Baltic states grant autonomy to Russians. During one and a half to two months, tension increases steadily. Then eighteen Russian divisions with the support of six Belarusian ones will strike a blow along the Polish-Lithuanian border. Lithuania will turn to NATO for help, which at first will deploy 'rapid reaction forces' and then eighteen divisions and sixty-six tactical air squadrons in Poland. In ninety days Western countries will gain the upper hand."[10] Such is scenario No. 7 from the report by the U.S. secretary of defense.

The disintegration of the USSR also signified the breakup of the entire defense system created over decades. Its disorganization and the loss of key infrastructure facilities signify a marked weakening of Russia's military security. Furthermore, the burden of peacemaking operations in the CIS is placed on its armed forces. Under these conditions it is possible to understand Russia's intensified activity regarding the preservation of its military presence in CIS countries, formation of a collective security system, and border protection. . . .

At the same time, the establishment or preservation of Russian military bases must not contradict the norms of international law and must be carried out with the consent of states. The practice of direct or indirect pressure on countries not interested in Russian bases in their territory must be eliminated. The interests of CIS countries in this matter are different. This is exemplified by the Transcaucasus, where in Armenia and Georgia Russian presence receives the support of the republic leadership, and in Azerbaijan, does not. At the same time, Russia is disturbed by the prospects for the appearance of troops of third countries there (for example, the statement by Turkey's minister of defense on readiness to send troops to Azerbaijan indicates such a possibility).

According to press data, the presence of twenty-eight Russian military bases in the new independent states is anticipated.[11] In the course of the visit by the RF defense minister Grachev agreements were reached on this point with Armenia and Georgia. . . .

The problem of safeguarding security in the territory of the former Union is not confined to Russia's military presence. A joint protection of commonwealth borders is an important task. After the disintegration of the USSR Russia received more than 11,000 kilometers of new land borders, while border troops lost up to 40 percent of the military potential, equipped stationing locations, and the necessary infrastructure. Judging by the latest reports, CIS states reached an agreement in the matter of the joint protection of former borders without the introduction of a standard border regime in Russia's new frontiers. The question of the Russian–Azerbaijani border remains open for now (in practice, the border between Azerbaijan and Iran is open and is not protected by Russian border guards).

The adoption of the concept of collective security can solve the entire set of problems connected with safeguarding the interests of CIS members with respect to defense and security. The meeting of CIS ministers of defense in July 1994 resulted in the preparation of the text of the concept for subsequent initialing and examination at the meeting of heads of states. The establishment of regional subsystems was envisaged: East-European (Belarus, Kaliningrad Oblast, and western RF oblasts) with a tentative center in Minsk, Caucasian (Transcaucasian republics and the North Caucasian Region of the RF) with a center in Rostov, Central Asian with a center in Tashkent, and East Asian (Siberia, the Far East of the RF, and eastern oblasts of Kazakhstan). Participation in regional subsystems states that did not sign the 1992 Tashkent Treaty on Collective Security of CIS Countries, that is, Ukraine, Moldova, and Turkmenistan, is anticipated.[12]

*Humanitarian interests and protection of citizens' rights.* The situation of ethnic Russians living in former USSR republics has become one of the most acute problems of Russia's mutual relations with these republics. After the dismemberment of the formerly common Russian ethnic space, a large number of Russians in the "near abroad" have turned into second-class citizens or into stateless persons. Displacement of Russians from former Soviet republics is taking place. In Russia itself the "Russian problem" has become a factor in the internal political struggle.[13]

The acuteness of this problem cannot fail to worry the Russian leadership. The above-mentioned presidential message stresses especially: "Russians' interests in CIS and Baltic countries will be safeguarded only if internationally recognized standards in the area of human and national minority rights are observed throughout the space of the former USSR."[14]

At the same time, negative tendencies in the outflow of ethnic Russians from the new independent states predominate for now. In Russia by now there are more than 2 million refugees, forced resettlers, and so-called economic migrants. Experts at the Center for Economic Conditions and Forecasting under the RF government estimate the minimum influx of those seeking refuge in our country in 1994–96 at 800,000 people and the maximum at 4 to 6 million.[15]

Naturally, such an influx is a heavy burden on the Russian economy and complicates the social situation where there has been the most active influx of migrants. Russia is preparing a program for the protection of the interests and support of ethnic Russians living outside the RF. According to the latest reports, the program is ready and an edict putting it into effect is expected.

In the opinion of the RF Federal Migration Service, negative tendencies in the interrepublic migration within the framework of the CIS are connected with the general socioeconomic crisis, with the lack of legal protection for persons of non-indigenous nationality, with the incompletion of the process of state "demarcation" of countries of the "near abroad" and the weak integration activity in them, with the continuation of the political struggle, in which the "national card" is played actively and the stereotype of enemies as represented by the non-indigenous population, which provokes its immigration, is formed persistently, and with the enhancement of the role of religion in the life of society, which is accompanied by the emergence of interconfessional conflicts.

To this we will add direct armed conflicts in the CIS. More than 500,000 forced resettlers (141,000 from Tajikistan, 86,000 from Georgia, and 82,000 from Azerbaijan) arrived in Russia by the middle of 1994.

In Russia's relations with CIS countries in the area of the population's interstate migration the following basic groups of countries are singled out:

1. Azerbaijan, Tajikistan, Georgia, and Moldova, where military operations are conducted and the outflow of speakers of the Russian language is especially intensive.
2. Kazakhstan and Uzbekistan, where the outflow is going on, despite official statements on the lack of a problem.

3. Kyrgyzstan and Turkmenistan, where migration is growing mainly as the consequence of the deteriorating economic conditions of life for the Russian population and, in part, owing to domestic nationalism and the lack of knowledge of the language. There is readiness on the part of a number of countries to sign an agreement on dual citizenship (which has already been signed with Turkmenistan).
4. Belarus, Ukraine, and Armenia, where there are no signs of a discriminatory policy with respect to Russians and the outflow occurs for purely economic reasons.

Under these conditions among ethnic Russians in the CIS the following alternative behaviors are possible: an attempt at assimilation, despite the sharp decline in the social status; expectation of Russia's support (legal, cultural, social, and so forth); return to the historical homeland or emigration; fight for a reunification with Russia.

Obviously, all four alternatives will be combined in practice and precisely such a sensible combination corresponds to Russia's interests. Every relatively taken alternative is associated with negative consequences.

*Economy and interstate cooperation.* In accordance with the basic gradation of advocates of the economic course into "statists" and "followers of the market" evaluations of the tasks of and prospects for Russia's mutual relations with countries of the "near abroad" also differ. The former focus attention on losses and damages from the rupture of previous economic relations, concluding from this the inevitability of reintegration. They usually cite figures and arguments to the effect that, according to the data of the intersectorial balance, Russia imported products of 102 sectors and exported products of 104 sectors, that it met 23 percent of the needs for machine-building products and one-third of the needs for metallurgical products through import from other Union republics, that the main raw-material base of Russia's nonferrous metallurgy remained in the territory of other CIS countries, and so forth. . . .[16]

Their opponents draw attention to the heavy burden of the economic patronage with respect to the CIS republics for Russia. The figure of the hidden subsidization of these republics in 1992—$17 billion—is well known. Furthermore, they point to the different direction of the courses of economic reforms in CIS countries and Russia, the existing inefficiency of management in neighbors, and non-equivalent conditions of the interstate exchange. As the report of the institute headed by E. Gaydar notes, the "close economic integration among the former Union republics will be directed in large measure toward an exchange of goods not competitive on the world market, reviving the autarkic type of national economies similar to the national economy of the former USSR."[17]

Unfortunately, in both cases only one aspect of the problem of economic interaction within the framework of the CIS is singled out, although quite correctly. On this basis, in accordance with the authors' ideological biases, conclusions are drawn on the correspondence or non-correspondence of the integration with neighbors to Russia's economic interests. There is no overall unbiased analysis or computation of specific economic effects, including indirect consequences. After all, even activities in the post-Soviet space not related directly to economic interests have consequences for Russia's economy. For example, the conclusions and recommendations of the State Duma based on the results of parliamentary hearings "On the Rise of the CIS and Its Present State and Development" (July 1994) note that the establishment of new borders will cost Russia 1 billion rubles per kilometer and in the number of the customs personnel (35,000 people) it has already outstripped the United States twice.[18]

Present interests and opportunities in the sphere of the economy and strategic interests and objectives should be differentiated. For example, the treaty on the economic union with Belarus, sharply criticized by pragmatists, which portends appreciable losses for Russia at the present moment, is capable, from the strategic point of view, of significantly improving its situation (outlet to western borders, access to markets, stability of Russia's western frontiers, and so forth). . . .

Without this, the very observance of Russia's national economic interests is unrealistic. The losses of the Russian treasury alone from non-registered export-import operations carried out through transparent borders with CIS countries amount to billions of dollars.

It is also necessary to decide what industries and commodity markets in neighboring countries are of real interest to Russia and to take specific measures to support and develop them. General talks abut a common market and the restoration of economic ties should be transferred to the plane of specific, economically substantiated decisions.

Russia cannot be interested in the existence of an economic calamity zone as represented by the former Union republics. Therefore, definite assistance for them is unavoidable in the very near future. However, Russia, itself being in the most serious economic crisis, should be most attentive to the efficiency of this assistance and seek maximum opportunities for a reciprocal satisfaction of its economic and social needs.

### Specific Nature of Russian Interests and Policy

*Russia-Ukraine.* Ukraine—the largest of the republics of the former Union—undoubtedly is the most important in the post-Soviet space. It was the one that initiated the disintegration process in the USSR. Within the framework of the CIS,

Ukraine takes a special position on most questions, and Russian–Ukrainian relations, perhaps, have been the most complex in relations among CIS countries. Not long ago Russian experts noted the following: "Kiev's present policy is directed toward creating within the CIS and Eastern Europe a counterweight to Russia's influence in order to isolate it geopolitically."[19]

Problems of nuclear weapons, the Black Sea Fleet, and the Crimea are the most acute problems in bilateral relations. The first two relate to defense and security and seriously complicate the general climate in the CIS. In the opinion of many experts, the problem of the Black Sea Fleet is not so much military as political.

The Crimean question represents a headache for both powers. Objectively, it is difficult to consider Crimea an age-old Ukrainian territory—it became a part of it only in 1954. Russia's support for Crimean separatism also threatens to aggravate similar processes in its territory, and a military conflict with Ukraine. Presidential elections in Ukraine, on the whole, have softened Russian–Ukrainian contradictions. The new leadership triumvirate (Kuchma–Moroz–Mosol) has often declared its aspiration to restore normal relations with Russia, although to interpret L. Kuchma's victory as the success of the pro-Moscow lobby is a simplification. The new president is forced to take the entire range of political forces in the country into consideration in his policy.

Fortunately, serious interethnic contradictions between Ukraine and Russia have not been observed. It is paradoxical, but regions with a large share of the Russian population, which supported L. Kuchma, are by no means the zone of Russian economic interests. The Donets Coal Basin and East and South Ukraine are saturated with non-competitive industries and potential unemployment. Russia is in no way interested in Ukraine's disintegration, or in including these economic crisis regions in its budget.

The specific nature of Russia's and Ukraine's economic interests lies in the fact that oil and gas pipelines to the West pass through Ukraine, which it has often used to its advantage. Ukrainian ports on the Black Sea, which handle large export and import flows, are also of considerable interest to Russia. Recently, Ukraine's role as a source of cheap manpower for Russian enterprises has intensified. Undoubtedly, it is still early to talk about a fundamental change in the climate in Russian–Ukrainian relations, but prospects for their development are more favorable now than under L. Kravchuk.

*Russia-Belarus.* In an environment of acute Russian–Ukrainian contradictions, Belarus played the role of a distinct counterweight to Ukraine. Here the traditions of the union with Russia are quite strong and stable, nationalist forces do not enjoy wide support (Z. Poznyak, leader of the Popular Front, picked up only 13 percent of the votes during presidential elections), and 58 percent of the Belarusians unequivocally came out in favor of an economic union with Russia. After the presidential victories of L. Kuchma in Ukraine and of A. Lukashenka in Belarus, its previous significance for Russia may abate. In any case, Belarus can hardly be viewed in economic terms as a superfluous load for Russia. Objective interests in strengthening relations remain, and specific prospects depend on the coordinated efforts of both sides to strengthen economic interaction. It should be noted that even after the change in leadership, Belarus outweighs other republics in the importance of its bilateral economic relations with Russia.

*Russia-Transcaucasus.* The main principle of Russia's policy in the Caucasus was formulated as far back as tsarist times: "Russia in the Caucasus has no eternal friends or eternal enemies, it has only eternal interests." A. Kozyrev, Russia's minister of foreign affairs, confirmed this old principle to a certain extent in an interview with the magazine *Stern:* "We cannot leave this region so simply. We have historical and geopolitical interests here."

Security is Russia's main interest in the Caucasus. The military conflicts occurring in the territory of all three republics make the situation in the region extremely explosive. A tangle of inveterate problems has emerged here. A stream of refugees, weapons, and criminal activity constantly flows from this region to Russia.

On the economic plane Russia is interested in exporting Azerbaijani oil abroad, and is striving to retain its control. Armenia occupies a special place. Several recently signed agreements (on diamond processing and so forth) attest to good prospects for the joining of Russian raw-material resources and Armenia's skilled manpower.

*Russia–Central Asia.* The Central Asian region, which is quite dissimilar in a number of ways, is of special interest to Russia, first of all, from the standpoint of security. This problem is connected primarily with Tajikistan, where there is a great probability of a repetition of the Afghan events. If Russia withdraws, the country's disintegration, transfer of Islamic fundamentalism to the territory of neighboring republics, and a sharp negative change in the balance of power from Russia's viewpoint are quite possible here.

The problem of displaced ethnic Russians from the former Central Asian republics and Kazakhstan is no less serious. Even in such a relatively calm republic as Kazakhstan the outflow of the population into Russia amounted to 200,000 people in 1993. Russia, which is now not going through the best of times, is vitally interested in stopping this negative process.

On the economic plane Russia pays attention to the problem of exporting energy resources from Kazakhstan and Uzbekistan, which is strategically important to Russia. As E. Pain, member of the RF Presidential Council notes, "attempts by Russia's quite influential political and economic groups to counteract the plans for the transportation of Kazakh and Azerbaijani oil through Turkey and the conclusion of agreements between Kazakhstan, Azerbaijan, and Western companies on the development of oil fields, were a factor in the artificial restriction of the independence of a number of CIS states and in pressure put on them to reintegrate."[20]

During the period following the disintegration of the USSR the special role of the "near abroad" for Russia's state interests and its direct effect on the situation in the country were determined objectively. Despite recognition of this fact, the official political line of Russia's leadership is still being formed, is quite unstable, and is subject to momentary changes. . . .

The solution of the Chechen problem by force introduces new elements into Russia's relations with neighboring republics. On the one hand, it causes natural concern over the aggressive actions of the Russian leadership. Ukraine's ex-president, L. Kravchuk, predicts an intensification of imperial ambitions in Russia's policy in the "near abroad." On the other hand, the position of official authorities in CIS countries on the Chechen problem is very cautious. Similar examples of regional and national separatism exist in many of them. (The Dniester region in Moldova, Abkhazia in Georgia, Crimea in Ukraine, northern oblasts in Kazakhstan, and so forth). Russia's actions in Chechnya intensify the temptation to adopt similar measures in order to suppress these separatists and in any event create an additional precedent for solution by force of internal state problems in the post-Soviet space.

At the same time, the Chechen crisis, obviously, has also illuminated the inefficiency and hopelessness of such military solutions. The lack of public support for them, huge costs for the economy, which is weak as it is, and unprofessional actions by power structures should serve as a warning both for the enthusiasts of the dissemination of the "Russian experience" in CIS countries and for Russia itself in its policy in the "near abroad." One would like to hope that these lessons will cause a transition from political demarches and rash decisions to laborious and very necessary work on the organization of economic interaction and normalization of the social situation in the territory of the former Soviet Union.

**Notes**

1. *Natsional'no-gosudarstvennye interesy Rossii i ee vneshnyaya politika* [Russia's National State Interests and Its Foreign Policy] (Moscow, 1994), p. 1.
2. *Sng: Obshchiy Rynok*, no. 1, 1994.
3. *Rossiyskaya Gazeta*, 19 January 1994.
4. Ibid., 25 February 1994.
5. *Nezavisimaya Gazeta*, 12, 18 January 1994.
6. *Natsional'no-gosudarstvennye interesy Rossii i ee vneshnyaya politika*, p. 5.
7. A. Zagorskiy, "Concept of the 'Near Abroad' in Russian Foreign Policy: Sources, Goals, Tools, and Problems," report from the German-Russian forum (Bonn, June 1994), p. 8.
8. Z. Brzezinski, "Premature Partnership," *Nezavisimaya Gazeta*, 20 May 1994.
9. *Segodnya*, 24 June 1994.
10. *Vek*, no. 24, 1994.
11. *Izvestiya*, 8 April 1994.
12. *Nezavisimaya Gazeta*, 6 July 1994.
13. P. Kandel, "CIS: Two and a Half Years Later," report from the "Big Europe" Commission (Moscow, 1994), p. 7.
14. *Rossiyskaya Gazeta*, 25 February 1994.
15. *Reforma*, no. 5, 1994, p. 8.
16. "Problems in the Reintegration and Formation of the Economic Union of CIS Countries," report under the guidance of academician O. Bogomolov and S. Sitaryan (Moscow, 1994), p. 17.
17. "Russian Economy in 1993, Tendencies and Prospects," report of the Institute of Economic Problems of the Transitional Period (Moscow, 1994), p. 160.
18. *Nezavisimaya Gazeta*, 13 July 1994.
19. *Natsional'no-gosudarstvennye interesy Rossii i ee vneshnyaya politika*, p. 9.
20. Quotation from *Segodnya*, 22 July 1994.

## 3.41

## Chernishev on Caspian Demarcation and the Transcaucasus

Interview by Lusik Ghukasyan
*AZG* (Yerevan), 29 April 1995
[FBIS Translation]

*Editor's Note:* To an overwhelming extent, the strategic economic arguments for controlling energy resources and their export routes underlie the pragmatic viewpoint for a restoration of "Great Power" Russia. Russian nationalist ambitions are uncompromising on the Caspian Sea, exports of Kazakh and Uzbek oil and gas, and Chechen oil. The pursuit of these interests poises Russia for conflict not only with the energy-producing CIS states but also potentially with Western countries that have economic interests in contributing to the development of these rich energy deposits. The "great game," or the contest for the resources of Transcaucasus and Central Asia, has started all over again—and will be a key determinant in the politics of these regions for at least the next decade. Russia has a great many sources of bilateral economic and military leverage over the states of the Transcaucasus and Central Asia, and as pointed out in the following example of pragmatic nationalist thinking, its leaders will resort to any measures they consider necessary to protect their interests.

The following article is also important for its "internal

market" rationale for a reintegration of former Soviet republics. The twenty-first century will be a scramble for markets, and Chernishev correctly points out the advantages to Russia of having ready-made markets for its goods within the former Soviet space, where it will have a competitive edge for some time.

[Interview with Albert Chernishev, deputy foreign minister of the Russian Federation, by AZG correspondent Lusik Ghukasyan in Moscow.]

*According to a report by the IRNA [Islamic Republic News Agency], these days representatives of the five states bordering the Caspian Sea—Russia, Iran, Kazakhstan, Turkmenistan, and Azerbaijan—are discussing issues related to the protection of the environment and the utilization of Caspian resources.*

[Ghukasyan]: Mr. Chernishev, recently, Azeri president (Geydar) Aliev and Turkish prime minister Tansu Ciller signed an agreement whereby Turkey was given an additional share in the consortium for the development of Caspian oil. What is your position on that accord?

[Chernishev]: Our position is that those who own that oil can extract it, and if they do not have sufficient funds to do it then they can form consortia. Clearly, each of the sides has certain rights that it can exercise on its own or that can be transferred to anyone else. Therefore, in principle, we do not object to the fact that Azerbaijan has transferred 6.25 percent of the consortium [to Turkey], especially since the dominant role in the consortium is played not by Turkish but American, British, and French firms. Russia's Lukoil company also has a substantial share: 10 percent. Iran has lost the share given to it by Azerbaijan as a result of Western pressure. However, we reject the contention that the oil will be extracted from Azeri shoals. The truth is that some of these fields are in the deeper portions of the sea and so far from the [Azeri] shore that the Turkmens are saying that it is their territory, not the Azeris'. Therefore, this or that oil-bearing zone cannot be declared the property of one country or another without a prior agreement. We do not object to the main agreement, but such an agreement cannot say that the fields in question belong to a certain state because that immediately becomes a subject of dispute. We have stated this to the Azeris quite candidly and firmly. It must be noted that during the Soviet period, as in all countries, some resources were taken and utilized but some were reserved for the future. Today because of severe economic conditions, the Caspian states have turned their eyes on this oil and are rushing to exploit it.

[Ghukasyan]: In your opinion how should the Caspian Sea be shared? What criteria must be used? Where does the issue of settling the status of the Caspian Sea stand?

[Chernishev]: As you know, there is an international Law of the Sea treaty that defines certain parameters of sea rights. However, there are no standards for sharing the Caspian Sea because, strictly speaking, it is a lake. In this matter much depends not on the international community but on how the five states resolve their differences with one another. There are precedents such as Lake Chad, Lake Victoria, the Great Lakes, and so forth. In all those cases the states concerned have established special conditions. We have still not reached a mutual understanding with the other states in order to do something about the disputes over the status of the sea. There are big differences [between the sides], with proposals ranging from a simple partition to leaving everything as it is. Our starting point is that this is common property and that an agreement is necessary on the joint utilization of these resources so that everyone can benefit. On this point, Iran stands close to our position for the moment. Turkmenistan and Kazakhstan prefer centralization [as published], while Azerbaijan clearly favors partition and taking possession of a portion of the sea. At present the status of the sea is defined by two treaties signed between the Soviet Union and Iran, in 1921 and 1940. It is understandable that a new status is needed now, but as long as a new status does not exist and we have not reached a new accord, the provisions of the old agreement, though imperfect, cannot be disregarded.

[Ghukasyan]: In a recent speech you said that if developments on the issue of Caspian oil take a course that is objectionable to Russia, then Russia would be prepared to use all means of pressure at its disposal. What are those means?

[Chernishev]: Obviously when a country defends its interests it tries to show what it has. No, this is not a threat. When they tread upon our interests we must defend ourselves. However, we must also understand the interests of others. If some parties do not understand our interests or how they are linked to other interests, we can find various means of applying political and economic pressure. For example, how can anyone sail out of the Caspian Sea? One can reach the Black Sea through the Volga River or the Volga-Don canal. We are not saying that we will begin to apply pressure immediately. But if you behave that way and do not take into account our interests or welfare, we will think about how we will behave toward you. We are prepared to take into account your interests and welfare, but let us agree on that. We must think about how everyone can benefit and profit without sacrificing our own interests. Then we can open the gates to the foreigners—the Americans or the British.

[Ghukasyan]: Does Russia have a blueprint for its interests in the Transcaucasus, or is that in its formative stages?

[Chernishev]: When reference is made to a blueprint or a formulated set of interests one should not interpret that as a document that was put together, say, in 1991, and that we act in accordance with that document. As a professional I know that a blueprint may be good for today but not a few days later. Regarding the Caucasus, our interest is the following: We must work on the security of the region jointly with other countries. The issue is not so much our military bases as the need for these countries to feel secure about their defenses. Those countries have inadequate armed forces—that is, for example they may have many soldiers but no air defenses, and that is no security [sentence as published]. However, security also has political and economic dimensions. The military bases constitute only one of many elements. Those bases and the joint defense of borders are essential at some stage. We have sustained enormous political, economic, and other costs because of the transparency of many borders. The reason for the economic cost is that we are so tied to each other that any break in ties may mean the annihilation of entire branches of industry.

[Ghukasyan]: That has already happened.

[Chernishev]: Yes. Perhaps not completely but largely. Had the Soviet Union been a federal state, at least formally, then its economy would have stayed unitary. In my opinion we have gone too far in terms of division. From this perspective, Western nations are setting an example for us by forming unions, the level of which has already reached 50 percent. In the Soviet Union that level stood at 80 percent. It could have dropped to 50 percent. Why was it necessary to go further? We were used to each other. Besides, our not so high-technology goods are needed only in those republics, and their products are needed in Russia. With their goods we do not need to go to outside markets. In any event, even with high-technology goods outside markets would not let us in because the markets are already parceled out.

[Ghukasyan]: One very important question is: What route will the new oil pipeline take to reach the West? Many observers see links between that question and the political developments in the region. What do you think?

[Chernishev]: On the issue of building a pipeline there are two factors that are important for Russia: economic feasibility and political realities.

If there is a war in that region then the pipeline cannot be built. It may be possible to build one only after peace is established. In addition, oil is needed to build a pipeline. At present that oil does not exist, and in that sense perhaps we are only theorizing. According to our information, estimates, and intelligence, even Westerners think that a pipeline cannot be built before 2000 to 2005 because money needs to be obtained first. For the time being it is necessary to use and expand what already exists. Thus Russia has said that it is possible to talk about where the oil will go, but since at present it is not flowing, Russia's existing facilities can be used.

[Ghukasyan]: What facilities did you have in mind?

[Chernishev]: It is possible to connect with our pipeline and deliver the oil to the West. The oil can also be transported by tankers. For the initial period we are proposing that they use partly the pipeline going to the West and partly the pipeline that ends in Novorossiysk. That [port] is relatively small. Certain investments are needed to build the necessary storage facilities there. However, this option can be met in the next two to five years while plans are developed for a new pipeline.

I understand the Turks very well. They want the oil to go through their country. They have excellent modern storage facilities at Yumurtalik, which, in the past, was used for Iraqi oil. Today there is no Iraqi oil flowing, and that facility is idle. Very well, that option can also be used, but how can we get to that point?

[Ghukasyan]: Via Armenia.

[Chernishev]: We are not excluding that option, nor are we saying only via Armenia. The pipeline can pass through Iran, but Westerners do not want that because of political considerations. The oil should pass wherever it is most favorable economically and feasible politically.

[Ghukasyan]: Do you think that a Russian military base in Armenia would contribute to the security of the oil pipeline if it passes through Armenia?

[Chernishev]: That issue was not taken into consideration when the bases were being planned. The objective for the bases was different: to ensure the security of the region and a certain balance of forces. In one sense the base by itself does not solve general security problems. It only signals that if something unexpected happens then certain agreements or provisions take effect. It should not be viewed in an absolute sense. It is rather a symbol, though a very important one.

[Ghukasyan]: Here is what I wanted to clarify. As is known, Russia is interested in having the oil pipeline pass through its territory. On the other hand there is an economically more favorable route that passes through Armenia. However, the situation on that route is unstable. According to your remarks, the oil must not go through that route as long as there is instability. Now, is not Russia interested in the stability of the region by acting as a peacemaker and mediator to settle conflicts in the region? Could you please clarify this situation?

[Chernishev]: The truth is that at the beginning they wanted Russia to withdraw from everywhere, physically as well as politically and psychologically, because we were seen as symbolizing the Soviet Union and imperialist Russia. Although it was insulting, that happened to some extent. I believe that we went too far on this issue.

Let me say candidly that all relatively small states, especially those that face chaos and especially war or crisis, always need the help of someone from outside. Their gazes turned to Europe and the United States, but apparently they did not find anything there. Nobody took this task to heart as much as Russia did—nor will anyone ever do so—because they question what they gain by getting involved in some conflict. As they say, this is a case where "it costs one ruble to get in, two to get out."

Troubled by our initial expulsion, we took the following position. Now you are inviting us in, but we are sorry. Please do not ask for what we could give you in the past through the Soviet format. Now you are independent and distinct, and we are sorry to say that we must comply with certain principles. All these issues are being gradually settled. Relations are being clarified on the basis of mutual understanding and interdependence, and these will become more tangible in the future in a positive sense. That is why we are engaged in mediation missions in conflicts in Abkhazia, Karabakh, and Tajikistan. Unfortunately mediation is, to put it lightly, a very thankless task. It is a noble but thankless task.

[Ghukasyan]: What are the reasons for the OSCE's involvement in the settlement of the Karabakh problem and the active posture it has displayed recently?

[Chernishev]: Initially there was a noticeable intent to drive Russia out of the mediation process. The objective was the following: Russia must not be allowed to establish itself and express its interests in that region. Our position, on the other hand, is this: If you want to help and are genuinely capable to do so then you can participate in a realistic and concrete manner. But if you want just to stand on the sidelines and tell someone else what to do without getting too deeply involved then you should not mediate.

In this case Russia (I do not wish to speak about the other sides) agreed that the OSCE participate on the same basis and contribute to the settlement of the problem. There was a time when there were misunderstandings on their part and suspicions on ours. However, eventually we were able to iron out our differences. In Budapest we came to an agreement that this would be an OSCE operation and that it would have two copresidents. That is very good. Now let us begin to work, and let us work such that we do not step on each other's feet. Now Sweden is being replaced by Finland as one of the copresidents, and we need to work with them well. The Swedes have been there for a long time, while the Finns are just coming in. Newcomers are always unaware of the details of any given moment. However, we are hopeful that with some effort these differences will be settled over time and the work will go forward.

[Ghukasyan]: There is another question that has always intrigued me. I am referring to the roles of Russia and Turkey in the Transcaucasus. Ankara has declared on several occasions that these two countries are the guarantors of the region. It is interesting. On what grounds does Turkey, a relatively less developed country, aspire to have an equal role with a great power like Russia?

[Chernishev]: Of course Turkey has a certain role in this region. That cannot be ignored or denied. On the other hand, it is essential that no one object to the participation of Turkey in the talks on an equal footing with everyone, including Russia. We know that Armenia, and naturally Karabakh, are opposed to such active participation by Turkey. Thus in this case Turkey's role is neither that of an onlooker from the sidelines nor that of a full participant in the process, although it is represented in the Minsk group and the OSCE. Turkey does have a certain, though not so important role, but that is not in any way comparable to the role of Russia or, for that matter, Sweden and now Finland, who are copresidents [of the Minsk group]. On the other hand, Turkey has influence over Azerbaijan and therefore has an impact on the Karabakh problem. That cannot be ignored and must be viewed as a reality that we need to circumvent.

[Ghukasyan]: It is known that as the deputy foreign minister of the Russian Federation you are responsible for Middle Eastern and Transcaucasian countries, including Armenia. For a long time you were the ambassador of the Soviet Union and later Russia in Turkey. Does not that experience have any impact on your current work? Does it not hinder you? Or does it help you?

[Chernishev]: It helps me because Turkey is not an ordinary country. At present it faces a chaotic situation in terms of economic development as well as events related to the Kurdish and other problems. Turkey is surrounded with numerous problems: disputes related to northern Iraq, Cyprus, the Balkans, and Greece. Consequently, my experience in working in all these directions as the ambassador of the Soviet Union and later Russia naturally helps me now. After all, everything was and is of interest to us. Turkey is very close.

Until recently we were neighbors. I can sense quite distinctly what the Turks want on this or that issue. That requires specialization—when all appearances are swept aside, and you see under real light what the Turks think and do.

This has nothing to do with whether Chernishev is pro-Turkish or anti-Turkish. To engage in politics on such issues, one needs to sense and know clearly, without journalistic assumptions or accusations, what can happen really in this or that matter. Consequently, that experience only helps me and does not hinder me.

[Ghukasyan]: I remember the time when the Turkish press published your private conversation with [Turkish president Suleyman] Demirel. In response to one question you spoke in favor of a Kurdish federation. The Turkish president responded immediately and declared that Turkey has been, is, and will always be a unitary state.

[Chernishev]: The exchange was milder than what was presented. There has been talk about the cultural autonomy of the Kurds for many years. Then the Kurds, as they say, expanded their demands, and said that now we must talk about not just administrative and cultural autonomy but a federation. Naturally the journalists were interested in my views, as ambassador. I said that this is generally your own internal affair. We support the principle of the political and territorial integrity of all states in this region, that is, all of Eurasia. You can examine all the alternatives, from cultural autonomy to a federation. Whatever you decide is what will happen. But for them it was painful even to hear the word "federation" from me.

## 3.42

**Russian Minister Previews Summit**
Interview by Yuriy Popov
*Rossiyskie Vesti*, 24 May 1995
[FBIS Translation], Excerpts

*Editor's Note:* In the following interview, Valery Serov, Russian minister for CIS Cooperation, outlines Russia's goals for reintegrating the economies of the former Soviet republics. Serov is a pragmatic nationalist, conveniently arguing that as time passes, all of the CIS countries agree on the need to restore close economic ties. The interview is especially interesting because Serov suggests that his Russian Ministry for CIS Cooperation should be made part of the Interstate Economic Committee, which is the only CIS institution that has been granted supranational powers.

Serov argues that the few existing skilled "specialists" capable of designing an "integration mechanism" for CIS economic cooperation are collected within his ministry. He is probably referring to former "central planning specialists" who controlled the central planning apparatus of the former Soviet Union. In fact, the "integration mechanism" to which Serov refers, and the customs and payments unions which Russia is pressuring the other CIS countries to join, would deeply impinge on the economic sovereignty of other CIS states.

It sounds suspiciously like a variation of the old centralized planning model.

[Interview with Valeriy Serov, Russian Federation minister for cooperation with the CIS states, by Yuriy Popov.]

The next session of the Council of CIS Heads of State takes place on 26 May in Minsk. On the eve of the Minsk summit our correspondent Yuriy Popov asked Valeriy Serov, Russian Federation Minister for Cooperation with the CIS States, to talk abut the current state of integration processes in the CIS.

[Popov]: There are those who believe that the unification process in the CIS countries' economies is in many ways attended by a disproportionate contribution by Russia. It is bearing a major burden to the detriment of its taxpayers. Isn't this process, to a large extent, intended to help Russia?

[Serov]: The entire wealth of world experience of integration convincingly shows that unification processes are bound to fail unless they are based on the partners' mutual commitment.

For the republics of the former USSR, including Russia, three years of sovereign life have been an important stage in the recognition of their nation-state interests. This process has been a pretty agonizing one. There was a time when it seemed to the young states that their fundamental interests lay entirely outside the CIS, while the Commonwealth itself was seen merely as a means of "civilized divorce." Now barely a trace remains of these illusions. Economics and other factors have taken over.

It took only a year to see that the wreckage of mutual links could not be rebuilt overnight, and in many cases it was simply inadvisable. Recognition of this fact is our common achievement, for which we have paid a high price. It is now clear that the CIS members have entirely objective and long-term economic interests vis-à-vis one another.

Russia has them too. Suffice it to recall that as part of the USSR it would import 23 percent of the machine-building products it required from other republics, together with more than one-third of the ferrous and non-ferrous metallurgy products, approximately one-fourth of the chemical and light industry products, and practically 100 percent of many types of non-ferrous metals and rare-earth elements, cotton, locomotives, cars for electric trains, corn- and beet-harvesting combines, and so on. So if Russia is helping anyone by participating in integration it is primarily itself, since it will be very difficult to recreate a powerful Russian state quickly outside the CIS.

Naturally, our partners have their own, entirely material interests vis-à-vis Russia. Incidentally, therein lies an essential prerequisite of effective and mutually advantageous collaboration. We intend to coordinate these interests, in accordance with the accepted procedure in world practice, while endeavoring to prevent one-sided advantages—for one side or another.

[Popov]: Is it possible to create in practice a single economic area on the territory of the former USSR when a genuine market has not yet been established either within the CIS countries or in relations between them? Maybe we should preserve the old system of division of labor and leave it at that?

[Serov]: Your question is entirely natural. In fact, in accordance with economic theory, a single market area presupposes developed market relations both within the cooperating states and among them. It is true also that in both cases market mechanisms are in the process of formation.

But allow me to ask a question in return. How are we to exist today, how are we to implement even traditional trade and cooperative deliveries, without which no national economy in the CIS can function normally? We can no longer cooperate in accordance with the old, "union" principles since states are not interested in resurrecting the old system.

As we see it, the solution is to create a single economic area gradually, as the market foundations in the CIS countries grow stronger. In so doing it is extremely important to ensure the future compatibility of national economic systems. And to achieve this you need to coordinate the transformations being conducted by the countries. This work has already been done in the areas of forming customs and payments unions.

As for resurrecting the old system of division of labor, you are unlikely to find any CIS state that would want this. Of course, a whole range of areas of their specialization will certainly be retained, since they are based on natural conditions or on a high level of production. Elsewhere we are all in for radical restructuring, including with respect to cooperation.

[Popov]: How are the CIS countries fulfilling their commitments under interstate agreements? Is it not the case that the most "disciplined" is Russia?

[Serov]: You have touched a very tender spot in our mutual relations. I would describe the situation as regards the fulfillment of international agreements as unsatisfactory. The actual results of trade and economic cooperation in ensuring interdependent deliveries in 1994 are as follows. According to "Roskontrakt" data, our partners fulfilled by 39 percent (in value terms) their commitments in imports of goods from Russia. And although this figure varies from country to country (Uzbekistan is 74 percent, Belarus 48 percent, Ukraine 30 percent, Kazakhstan 11 percent), the picture cannot be described as rosy.

Russia carried out 43 percent of export deliveries to the CIS states in terms of the value set by contracts, that is, we have nothing particular to shout about either.

It would be unfair to say that this situation is due to the partners' reluctance to fulfill their commitments. There are objective difficulties (crisis and decline in production), but there are also instances of a negligent attitude toward them as well. Clearly, this adverse trend must be eradicated. We need a mechanism that would ensure by economic means the meticulous fulfillment of mutual commitments. We are working on this at the moment.

[Popov]: In Almaty there was serious criticism of the sluggish way decisions already made by the CIS countries with respect to integration are being implemented. This applies in particular to the organization of the Interstate Economic Committee [IEC]. How do things stand at the moment?

[Serov]: In fact, the big hopes the Commonwealth countries are pinning on the creation of the IEC and the entire course of the implementation of the decisions to establish it make a poor match. Suffice it to cite the following example. One IEC presidium session had to be postponed merely because not all the participating states had submitted their plenipotentiary representative candidates by the appointed time. Need I say more. . . .

Now things have gotten moving somewhat, and several sessions of the IEC presidium and collegium have already taken place. At the same time, the committee is not yet fully operating. There are problems with premises and funding. But the main problem is the formation of the apparatus. It turns out that the number of highly skilled specialists capable of tackling the complex tasks of creating the new mechanism of cooperation in the CIS is extremely small. Essentially, they are all concentrated in the Russian Ministry for Cooperation.

This suggests the idea of making the Ministry for Cooperation part of the IEC apparatus, which will enable us to keep cadres together and provide the IEC with quality people. Interestingly enough, there have been similar proposals before from our partners. I think we should heed them. Particularly as our ministry is already effectively performing the IEC's functions.

[Popov]: What about the implementation of the recent summit's other decisions?

[Serov]: The Almaty meeting of CIS heads of state and heads of government made a number of fundamentally important decisions.

I would single out the question of recreating a common scientific area. We are well aware of the role of scientific-technical cooperation in present-day integration processes. For example, in the EC it was the engine that actually powered the participating states to a qualitatively different level of development.

A provisional working group of CIS countries' representatives has now been set up to implement the accords that were reached. It is elaborating such fundamental documents as a blueprint for the recreation of a common scientific-technical area, a program of coordinated actions to implement it, and a draft agreement on it.

The "Principles of the CIS States' Customs Legislation," adopted by the Council of Heads of State, are equally significant, particularly in connection with the formation of the customs union. What we have on the agenda now is the CIS countries' task of bringing their customs codes into line with the "Principles. . . ." Belarus and Kazakhstan have already done this work. According to available information, Uzbekistan, Kyrgyzstan, and Tajikistan have embarked on similar work.

The Almaty meeting also adopted another two important documents aimed at harmonizing national economic systems. I am referring to the blueprint for the mutual legal regulation of economic relations and leveling out of the conditions of economic activity in the states of the Economic Union and the agreement on comparable methodologies and the creation of a common statistical base for the Economic Union.

I don't think there is any need to say a great deal about their significance for the development of integration in its modern form.

The relevant groups and commissions have already embarked on the practical implementation of the adopted decisions.

[Popov]: What practical forms is the integration process currently assuming?

[Serov]: I have just answered that question, in fact. All I can add is that we are actually completing the stage of devising a model for the integration mechanism and are embarking on putting it into practice. The next target is the creation of customs and payments unions, which we see not only as means of removing all manner of artificial barriers in the way of cooperation, but also as a significant step toward a single market area.

[Popov]: What is preventing the independent CIS states from rapidly creating their own "common market"?

[Serov]: There are many factors. First, the extent of the collapse of mutual ties in the period immediately after sovereignties were declared. Restoring them turned out to be a far more complex task. Second (and this derives from the first), the deep crisis in production is hampering this work in a major way. A common market is incompatible with a lack of goods. But the latter is a fact of our life.

Third and last, over the past three years each of the states has carried out reforms according to its own taste, so to speak, without really consulting or considering its partners. As a result our economic systems do not fit together and our task now is to bring them close together, to harmonize them. I have already said how we intend to do this.

The upcoming Minsk summit on Friday I believe will be another step on the road to integration in the CIS. Of course, there are no easy and simple solutions to matters such as these. This is indicated by the agenda, which includes an item on the trouble spot in Tajikistan.

Nonetheless there are grounds for believing that both the general and the specific problems of cooperation among the Commonwealth states will be positively resolved. And that will enable us to forge ahead. All I will mention is the meeting participants' intention to jointly invest in the Yelabuga Truck Plant and also to implement the idea of replacing gasoline with gas in transport.

## The Right- and Left-Wing Extremists

### 3.43

**Zyuganov's Report to Third CPRF Congress**
Address
*Sovetskaya Rossiya,* 24 January 1995
[FBIS Translation], Excerpts

["Political Report of the Communist Party of the Russian Federation Central Executive Committee and the Party's Tasks," delivered by Gennadiy Zyuganov, chairman of the Communist Party or the Russian Federation Central Executive Committee, at the Third Communist Party of the Russian Federation Congress in Moscow on 21 January 1995.]

Comrades!

On opening our third congress today we, Russia's modern Communists, have to reflect and ponder: Who are we, where do we come from, and where are we going?

Our roots are in a party that sacrificed the lives of its best sons and daughters for the people and the fatherland, for the fatherland's independence and its state and cultural grandeur. It is a party whose members crushed fascism and broke through in outer space. It is a party that did not and will not bend under any blows by fate. It is a party whose ranks included Stakhanov and Gagarin, Sholokhov and Leonov, Panfilov and Zhukov, Korolev and Kurchatov, and millions of honest Communists—toilers and patriots.

There is also another party—a bureaucratic party of national betrayal, the party of Trotsky and Vlasov, of Gorbachev and Yeltsin, which always looked upon Russia as its own fiefdom. Now it has cast off its mask, joined an alliance with the criminal bourgeoisie, and destroyed the Soviet state, and is trying to rule the country autocratically and despotically.

Our main historical error and guilt lie in the fact that we cohabited for too long with this party of betrayal in a formally unified organization, we failed to distance ourselves in good time and resist it, we put up with its omnipotence, which inflicted such grave trials on the people and the state.

Now, having reaped the bitter harvest of credulity, carelessness, and lack of political will, the Communists must—in conditions of emergency that do not leave a moment's pause for breath—resolve several tasks at once: to restore the party that essentially means to build it anew, to interpret the lessons of history and map out a program for the future, to counter the actions of the anti-people regime, and to wage a desperate struggle for the survival of Russia and all its peoples.

During the two years since its Second Extraordinary Congress, the Communist Party of the Russian Federation [CPRF] has traversed a long path of struggle and hard trials. Overcoming the difficulties of persecution and proscription, within a brief period of time the CPRF restored the activity of its central and local structures, consolidated itself, and emerged as a statewide political force. It has a sizable representation within organs of legislative power and within local organs of executive power.

The CPRF today comprises 88 republican, kray, oblast, and okrug organizations, over 2,000 rayon and city organizations, and about 20,000 primary party organizations uniting more than half a million Communists.

Considerable theoretical work has been done. A largely innovative draft Party Program has been prepared and submitted for discussion at the Congress today.

The party has preserved its image and firmly upholds the Communists' principled positions on the most important problems of domestic and foreign policy. It is striving to be the unifying element in creating a bloc of state-patriotic forces.

With each passing day, the party's voice is heard increasingly loudly and convincingly, and it is being heeded both by the powers that be and by influential political forces abroad.

But while taking note of all this, can we actually feel complacent? Of course not, because so far we have not achieved the main goal—we have failed to even slow down the country's slide toward the abyss where it is being driven by the incumbent anti-people regime, we have failed to remove this regime from power. Until this is done, we will owe a great debt to the fatherland.

Let us ponder once more the essence and content of the destructive processes now under way in the country.

We are duty bound to examine this question from the broadest possible angle. Today we are talking not only, and not so much about the dismantling of the socialist social system but about the targeted subversion of the foundations of Russian spirituality and statehood in general. Russia today is not being built even on the foundations of capitalist relations. On the contrary, the selfish element that has been let loose is being used to destroy the great Eurasian power whose very existence has hindered the implementation of the "new world order" plans.

The security of any state rests on the "three pillars" of protecting its territory, protecting its people, and protecting the population's way of life. Let us look at today's Russia from this angle.

*Territory.* The country has been pushed four hundred years in the past, within its sixteenth-century borders, and it has been deprived of the results of centuries-long development and its access to the world's trade routes. Vitally important international economic ties have been broken.

Tens of millions of our compatriots have been left outside the Russian Federation's borders as second-class citizens. Checkpoints and customs points have been installed in places where borders never existed and are unnecessary, and they are simply lacking in places where they are needed.

Russia in fact no longer exists as a unified and integral state organism. We are left with its individual elements, with weak legal and organizational links and without any coordination of their activity. We are left with individual links of an economic mechanism that was torn apart while still alive. We are left with territorial formations that increasingly fall under the influence of regional and departmental "elites" pursuing their own selfish interests.

*The people.* Already in 1993 the birth rate had dropped to 1.3 births per woman, while it should be at least 2.1 to achieve the simple replacement of the family. Average life expectancy has dropped to sixty-five years, which is ten to fifteen less than in developed countries.

By the end of last year the minimum pension stood at less than one-half of the subsistence minimum. Prices are going through the roof. Vodka is the only exception—compared with bread, it is now eight times cheaper. The people are being deliberately fed alcoholic and spiritual moonshine. The nation's health has been subverted. Morbidity is rising at a disastrous pace, primarily congenital morbidity. Only 14 percent of children are healthy.

Everything that citizens of the USSR and Russia enjoyed—guaranteed right to work, free education and health care, protection of motherhood and childhood, and much else—all this has been sacrificed so that a handful of compradors can get richer.

In terms of living conditions for the bulk of the population, the results of the past three years fall below the universally accepted definition of genocide.

*Way of life.* Since time immemorial, our compatriots have lived in a communal world, in a spirit of comradeship, mutual assistance, and collectivism. Nowadays everything is being done to destroy this foundation of the people's lifestyle and deprive them of the very opportunity of reciprocal contact between people.

Transportation prices are sky-high, and people cannot afford to attend family funerals.

Crime is rampant—nobody feels safe, and all lock themselves away from the rest of the world behind iron doors and bars.

Vouchers were distributed—sit tight and hold onto Chubays's "two Volgas," keep competitors at bay.

The richness of human contact is being increasingly replaced by the solitary television set, instilling in people the idea that Russia's history is just a senseless chain of mistakes and crimes, that their lives have been lived in vain, that lies and theft are the basis of morality, and that Judas is the ideal human being and citizen.

In order to render the state's collapse irreversible, numerous "mines" with vast destructive power are being planted beneath the state's foundation. Here are just a few of them.

*Social mine.* Society's stratification in terms of the incomes of the strata representing the richest and the poorest 10 percent has reached the proportion of 23:1, which is several times higher than the socially permissible level and threatens to explode at any moment.

*Legal mine.* Today the country lacks even a single federal organ of state power whose legitimacy is beyond all doubt. Total lawlessness is supplemented by total chaos in the system of real power. Governability has declined to such an extent that arbitrary rule and violence are the only remaining methods.

*Geopolitical mine.* Deprived of its natural borders and reliable access to warm seas, the country is losing its state autarky and is doomed to seeing its regions, primarily its raw-material regions. being sooner or later attracted to different systems of economic ties and detached from it.

*Technological mine.* The exportation of industry's fixed assets has exceeded all permissible bounds of security. The degradation of production could at any moment evolve into a terrible catastrophe, even involving planetwide consequences.

*Food mine.* The country has for all intents and purposes lost its own resources for supplying its population with foodstuffs. The destruction of the agro-industrial complex has made it totally dependent on imports.

*Biological mine.* All diseases from the Civil War years—typhus, cholera, diphtheria, and scabies—have reappeared in Russia today. Teachers say that there is not a single school without lice. At the same time, the health-care system is being destroyed, primarily the well-organized system of preventive and hygiene services and our country's pharmacological base.

Virtually all of society's vitally important spheres are in a state of imbalance and threaten to collapse. Bluntly speaking, the situation is unique, unknown, and without even a remote parallel in history.

There is no doubt that the initial push here was given by the bourgeois counterrevolution. But it is also obvious that the matter certainly does not rest here. What is happening is not akin to even primitive capitalism. Here we need some different terms: kleptocracy, universal sellout, pilferage, destatization, dehumanization.

Historical parallels can help us interpret the situation. Let us recall how Russian lands devastated by Teuton and Mongol invaders were gathered bit by bit around Moscow, the land-gathering that was practiced to overcome the evil Time of Troubles in the sixteenth and seventeenth centuries, the opening of a "window to Europe" by Peter the Great, Chancellor Gorchakov's famous words that "Russia is consolidating," the re-creation of state unity and the country's economic might following the collapse of the bourgeois-landowning system in 1917, and finally the headlong revival and attainment of leading positions in the world following victory in the Great Patriotic War.

Despite all the historical and social diversity of these situations, they all have something in common.

First, the extrication from crises was distinguished by special features of the stabilization period. Mechanisms for mobilizing all of society's potential were switched on, priority attention to domestic problems of social recovery and national revival was guaranteed, and the state's external activeness was reduced to a reasonable minimum.

Second, obvious priority was given to nationwide and

statewide interests. A social force, well ahead of its time, was always found and these interests were most fully and most consistently embodied in its activity. Its leading role made it possible to rally the bulk of the people, to awaken their energy and focus it on the most important fulcrums where force had to be applied.

Turning back to the present, we conclude that the country is going through a period when its main tasks are associated with the struggle for national-state self-preservation. At the same time, the objective conditions are such that this struggle fuses with the struggle waged by Communists for real power by the people, for social justice, for socialism. Hence our conception of immediate and long-term tasks.

The first task—in terms of both priority and essence—is to awaken and rally the people and all their social strata, to create an effective alliance of patriotic forces to struggle against the national-state catastrophe.

As our draft program notes, the Communists' real and potential allies in the cause of national salvation are the political parties and movements from the socialist, patriotic, centrist, and consistently democratic spectrum, the trade unions, the workers', peasants', women's, veterans', youth, entrepreneurial, educational, and creative organizations, the Russian Orthodox Church, and religious communities of all confessions.

The common goals that we set are evident from what has already been said. They are simple and obvious:

- to protect Russia's state integrity and the Russian people's unity. This, in turn, dictates the necessity to voluntarily recreate a single union state;
- to achieve not just an economic recovery but a qualitative breakthrough on the basis of the supreme achievements of scientific and technical progress;
- to protect the physical and moral health of the nation and to uphold the traditional values of the people's way of life;
- to restore legality and law and order in the state;
- to ensure civic peace in society, calm and security in each home, prosperity in each family.

The removal of the incumbent anti-people regime from power is a decisive condition for attaining these goals.

Not a single even slightly responsible politician in Russia and abroad can any longer risk describing this regime as "democratic" or as reflecting the opinion and will of the majority of the country's citizens. This is a regime of personal usurpation of power, and it is propped up by bayonets—not so much the army's bayonets, but those of power structures especially created for this purpose. This is a regime of whipping up fear, stepping up political terror, stifling the citizens' democratic rights and freedoms, and destroying society's social gains.

The subjective selfish logic of its behavior is obvious. Being politically weak, it is trying to prop up its power by physically breaking society's backbone and eliminating the most important political institutions in the course of creeping civil war.

The incumbent anti-people regime has no future. If it is forced to exist for a certain period of time in conditions of peaceful and non-violent development and legal stability, it will be immediately swept aside by the opposition. And the entire country is essentially in opposition to it. . . .

As far as Russia's Communists are concerned, the essence of this regime and the ominous consequences of its dominance were obvious from its very inception, and they have always and consistently opposed it.

In the past we said that the proclamation of the supremacy of the Russian Federation's sovereignty over the Union was a dangerous insanity which would inevitably lead to the disintegration of the USSR and then of Russia itself.

We opposed the introduction of the institution of presidential power, adducing proof that in a multinational federal state it would not be a stabilizing factor at all but a source of endless conflicts and arbitrary rule, a factor intensifying the destructive processes.

We warned that the policy of shock therapy would lead to an unprecedented economic collapse.

All this happened long before the party's second congress, during a period when it was only just being reborn and was struggling for the restoration of its right to exist. But the period under review also abounded with struggle against the anti-people policy.

The efforts of party organizations, which were only just getting up on their feet, were aimed at protecting the soviets virtually throughout 1993.

Whereas the March attack by presidential structures against the soviets was successfully repulsed through joint efforts, bloody October and the shoot-out with the Supreme soviet turned into the worse defeat of people's power in Russia's history. It also proved that the Communists and their allies are the most consistent supporters and defenders of democracy.

Incidentally, I think that it is high time to restore this term's true meaning and cleanse it of all the filth that has stuck to it from the dirty paws of the people's butchers and their stooges.

It can be said that we realized in full the meaning of being an opposition party precisely in the wake of October 1993.

Over the past two years I have traveled the length and breadth of virtually the entire country. Wherever I went, I saw the selfless work being done by my party comrades. Without an apparatus, without premises or office equipment, they honorably perform their civic and patriotic duty as Communists. After all, ours is a genuinely people's party, a party of zealots, a party of wholehearted people, a party that is joined not for a career's sake but in answer to a call of the heart. I see many familiar faces in this hall. Many, many thanks to you, comrades, and to the entire detachment of party officials and rank-and-file Communists, to our supporters and friends.

In these difficult conditions the party is learning and is mastering the full arsenal of forms and methods of political struggle. Paramount importance is attached to the very difficult decision to contest the Federal Assembly elections, made by the October 1993 party conference. We were perfectly well aware that the parliament being created within the incumbent regime's framework would be virtually powerless. Its legislative powers are constrained by the executive power's dual censorship and are not backed up by functions to monitor the execution of laws or the government's activity.

Nonetheless, today we can claim that the policy of contesting the elections has been vindicated. The election campaign offered an opportunity to test the combat ability of party organizations, and their ties with the population were strengthened. With just over 10 percent of deputies' seats, the Communist Party exerts noticeable influence on parliament's work.

We entered parliament with a clear-cut program to restore legality and overcome the consequences of the coup d'état. It was further developed into the concept of ensuring civic peace, adopted by the all-Russia party conference in April last year. We said at the time, and still assert today, that it is a real alternative to the so-called Treaty on Social Accord that has been assigned the pitiful role of a fig leaf for presidential dictatorship.

Even within the narrow framework of Yeltsin's constitution, which has been imposed on the country, we are striving to do everything possible to restore and protect people's power. We are talking primarily about amending a series of articles of the constitution with a view to stepping up parliamentary monitoring of executive power. The arbitrary rule by executive authorities and their lack of accountability, which have reached the point of total chaos especially over the past few months, prove that the constitution as a whole must be changed.

Our efforts to ensure immediate adoption of laws on Duma elections, of forming the Federation Council, on presidential elections, and on referendums follow the same track. Three of these laws have already been given their first reading. Obviously, we have to think about adopting yet another law, the need for which may arise at any moment. This is the law on the Constitutional Assembly, empowered to adopt a new constitution.

The party does not look at its parliamentary activity only from the angle of lawmaking. As far as the Communists are concerned, the Federal Assembly is also a rostrum, a means to ensure the cadre reinforcement of party structures, and a major school making it possible to accumulate experience in affairs of state and to constantly keep a finger on the country's pulse. Without such a school, all the talk about coming to power is nothing but bragging.

Growing importance attaches to the activeness of Communist deputies in electoral districts and labor collectives, and their work among the population. In just one year they have visited virtually all regions in the country, telling people the truth abut the authorities' pernicious policies and helping to organize the masses in opposition to the regime.

This means that parliamentary work is most closely linked with extraparliamentary struggle, without which our activity is altogether unimaginable. Just like before, our party is firmly committed to the viewpoint that, given a certain turn of events and the regime's switch to overt dictatorship and repressions, decisive importance will attach to the different forms of actions by the masses.

The experience of the parliamentary elections was also utilized and expanded during the elections to regional organs of power. They proved that the mood of the masses is clearly leaning to the left. Wherever the Communists acted in an organized manner, rallied all patriotic forces, and effectively monitored the observance of electoral legislation, success was almost always forthcoming.

It is typical that, out of the elected deputies who openly declared their party affiliation, 46 percent are Communists. Even the authorities are forced to admit that the opposition forces dominate the majority of local dumas and assemblies today. The "Red Belt" is getting ever tighter around Moscow, reaching up from the south, and extending ever further eastward.

In compliance with the second congress instructions, the Central Executive Committee [CEC] constantly focused its attention on denouncing the Belovezh Forest agreements and re-creating a renewed Union of Soviet Peoples. This was the aim of our repeated initiatives at the Duma, the conferences of fraternal parties, and the First and Second Congresses of the USSR Peoples.

Of course, each and every one of us is aware that the path of re-creating a single country will not be easy. This becomes especially obvious in the light of Russia's impending disintegration. The events in Chechnya, which are fanning the hotbed of a new Caucasian war and a Slav-Muslim confrontation along Bosnian lines, cannot be perceived as anything but the latest step toward the breakdown of Russia-wide unity. You are familiar with the CEC's stance in this regard. We resolutely demand a cessation of combat operations by both sides, the holding of necessary negotiations, and an investigation into the causes of this unprecedented slaughter.

While recognizing each people's legitimate right to determine its own fate, at the same time we believe that any attempts to break down and split the country are contrary to the people's genuine interests. We believe in the common sense of our compatriots. Russia must be a single state, in which the interests of people of all nationalities will be reliably protected.

We perceive the Chechen adventure as yet another provocation against the army and other power structures. The army has been "set up" yet again, and it is being besmirched. On behalf of all Communists, I declare: We believe that people in uniform will remain loyal to their duty to the people. We will do everything to uphold the honor of the fatherland's defenders.

It is an open secret that the imperialist circles are eagerly awaiting the time when they will be dealing with a Russia that has been torn apart and bled dry. In pursuit of this goal, they are prepared to embark on any adventure, up to and including the introduction of foreign troops on Russian territory under the pretext of "peacemaking" or the safeguarding of nuclear and other potentially dangerous facilities. For the time being, people abroad and in the "fifth column" circles are only hinting at that, but these hints may materialize tomorrow. This is why the ideas of Russian patriotism and of the motherland's deliverance from disintegration and devastation today have not only domestic but also foreign political overtones.

In this context I would like to speak at greater length about our assessment of Russia's international position and the world situation as a whole. The development of crisis processes in the socialist countries also produced serious negative consequences for the international situation. It was not just the Soviet Union's state unity that was destroyed, the balance of forces in the world arena, on which international peace was based for over four decades, was also destroyed. The bipolar system of international relations was replaced not by the multipolar system that was the big vision of naive political dreamers, but by a monopolar system meaning global U.S. domination. The results of World War II have been de facto revised. The myth about the pan-European home also burst like a soap bubble.

The Yeltsin–Kozyrev foreign policy course is based not upon a sober appraisal of the realities of contemporary international life but on propaganda utopias in which even our Western "partners" themselves have never believed. Russia has gradually lost all its allies and its international positions. Any timid attempts to bring up Russian interests are cut short with harsh bellows, as happened recently in Budapest. On the other hand, the adventurism and unpredictability of the Russian authorities in their internal affairs are pushing our East and Central European neighbors into NATO's embrace.

In the face of these circumstances, we define our proposals in the sphere of international policy on the basis of two fundamental objectives: To restore the state unity of Soviet peoples and to democratize the global system of international relations.

The country's new foreign policy would mean a return to the protection of national interests, restoration of traditional ties of alliance in all regions of the world, and international solidarity with the countries and peoples struggling for the preservation of their state sovereignty, against the policy of the "new world order."

Wars have recently become again an inalienable factor of international life. But, even in these conditions, the ideal of peace and international cooperation remains immutable as far as the Communists are concerned. We will continue to strive to exclude war forever from the arsenal of means for solving international problems. The principle of peaceful coexistence as a basis of international relations still retains its value today; it has withstood the test of time.

Comrades!

The regime's anti-people policy is largely and personally linked with the individual occupying the president's seat. The CEC has openly declared that Yeltsin's policy and personal behavior denigrate the dignity of our great people and are an insult to the sacrifices made by millions of Soviet people for the sake of mankind's liberation.

This is why the party could not have failed to launch the initiative of collecting citizens' signatures to a petition demanding early presidential elections and non-extension of the powers of the incumbent Federal Assembly. This appeal was heeded by people. Over a period of three months (between 21 September and 21 December 1994), about 3 million citizens signed this petition. This is three times the number required under the still unrepealed RSFSR [Russian Soviet Federated Socialist Republic] Law on Referendums and 1.5 times the number required under the president's draft Law on Referendums.

These are the results: The CEC Presidium and the Communist faction in the Duma issued a statement saying that Yeltsin has no right to ignore the population's opinion and must either resign or call early elections. Otherwise, we reserve the right to demand that a referendum be scheduled in line with the norms existing in all international acts on human rights.

But the pseudo-democrats are also trying to exploit the critical moment of the utmost exacerbation of the crisis of power, especially in connection with the war in Chechnya. This goes a long way toward explaining their strange—at first glance—transformation into fierce critics of their own disgraceful practice of political arbitrary rule.

There are many signs that the stage of the regrouping of forces in the ruling camp is almost over, and that preparations to emerge in the arena are being made by political figures whose main strategic objective is to retain the property and power already acquired by the comprador strata at all costs, including by means of removing the incumbent leadership from power and using strong-arm methods.

The interclan struggle within the ruling grouping forces them to resort to methods of misguiding public opinion like, for example, claims about the coalition nature of the existing government.

In this context we reaffirm the immutable stance of the CEC October Plenum that, for as long as the present anti-people policy is pursued, there cannot be any talk of Communists' participation in a government that is guilty of destroying our great country. It is not a coalition government because coalitions are created as a result of agreement between political forces, not of individual decisions by various people. This is why non-party member V. Kovalev, a member of the CPRF's Duma faction, was expelled from the faction when he agreed to his appointment as minister.

Only if there is a policy change and a government of people's trust is formed will the CPRF leadership, following consultations with party organizations and its closest political allies, be able to make a decision about participating in the creation of such a government.

The systemic crisis raging in the country is called systemic because it has struck all spheres of life without exception and has reached the line beyond which destructive processes become irreversible. Nonetheless, the key problem that a government of the people's trust will have to resolve is economic normalization. For it is essential primarily to make a correct diagnosis.

It is not enough to speak of the pernicious nature of the present economic course—we must understand that this course is logical and realistic in its own cannibalistic way and is pursuing clear potentially attainable goals. What are these goals?

The present economic policy is, as we know, being pursued in response to the dictates of the IMF and that organization has never engaged and is not engaged in creating in the countries under its tutelage an effective Western-style market economy. Its real aim is completely different—forming an economy of a type that cannot exist without sliding into huge foreign debts but that is at the same time capable of paying the interest. The IMF is not interested in the price at which that is achieved.

As proof let us divert ourselves for a minute from our affairs and turn to the experience of countries that have undergone a full course of "treatment" as prescribed by the IMF.

Thus, Venezuela once had the highest per capita income in Latin America. It held important positions among world oil exporters. Since the IMF's intervention Venezuela's foreign debt has increased from $29 billion in 1980 to $35 billion in 1990. Here the country has had to pay $31 billion in interest alone over the ten years. At the same time the export of capital, including the illegal export, was assessed at $35 billion.

One more Latin American country—Peru. It is known that the consumption of calories in food varies for one person from the minimal level of 2,400 to the optimum level of 3,500. In Peru in 1970 this figure was 2,300 while in 1980 it was about 2,000 and in 1990 it was even lower. The poorest strata of the population consumed only 800 calories in 1991, which is less than the amount given to an inmate of Auschwitz.

Michel Camdessus, managing director of the IMF, expressed himself as follows regarding Peru's position: "We believe that the successes achieved in Peru are of extreme importance. This is a program of unusual importance to the whole world. It is a model for the rest of the world. The IMF program cannot be changed on the pretext of fighting poverty."

How can we fail here to recall the words of the well-known U.S. economist and entrepreneur Lyndon Larouche addressed to all those Friedmans, Hayeks, and Sachses and other "pillars" of monetarism: "Since the Nobel Price for Economics was founded it has been received exclusively by those whose 'fundamental works' in this field have turned out to be flagrantly incompetent and only when this incompetence has been confirmed by some national catastrophe resulting from adherence to this doctrine."

The above-mentioned results are painstakingly concealed from the broad public but they cannot fail to be known to the pupils of the IMF figures—to the Gaydars, Fedorovs, Chubayses, and Yavlinskiys who deliberately mislead the people regarding the thrust of their "reforms."

The socio-economic course they have steered since 1992 has led the country to total collapse. The national income has been nearly halved. For virtually all production indicators, for production efficiency, and the living standard of the majority of the population, Russia has been cast back several decades.

The draft budget for 1995 continues the "tough credit and finance policy." Its real expenditure is halved. Peasants, miners, the military-industrial complex, the army, science, education, and the entire social sphere have essentially been left to the tyranny of fate. The indexation of wages and pensions is envisaged only twice a year, with a 40 percent lag behind the rise in prices. Commodity producers are oppressed by taxes and the Russian market is essentially being destroyed.

In this connection the CPRF faction in the Duma has voted unanimously against the draft budget for this year and also expressed its lack of confidence in the government.

They claim "there is no way" other than that dictated to Russia by the Western well-wishers. That is a lie. There are other options that ensure not in words but in reality a way out of the catastrophic situation, but the government is dismissing them.

What do we offer and demand? Absolutely nothing supernatural—merely what is dictated by common sense and elementary calculations.

First, it is essential fundamentally to alter monetary and credit policy by substantially easing it. Despite what is drummed into us, the money supply in the country has been reduced today below any conceivable limit. The shortage of rubles in circulation is compensated by the total dollarization of Russia's economy, that is, it works to service the U.S. internal debt. According to available estimates something like $100 million in cash is circulating on the market and that is 400 trillion rubles [R]—nearly three annual budgets. Some R150 trillion are tied up in non-payments. All this hits the enterprises' working capital. It is the shortage of working capital that lies at the basis of the breakdown of national economic relations.

Second, a fundamental change in taxation policy is required. It is essential to repeal or substantially reduce for real commodity producers the value-added tax, which is the main factor provoking the spiraling of inflation. As a whole, tax pressure on production should be reduced as early as this year by a minimum of 15 to 20 percent and in the next few years it should at least be halved. Otherwise collapse and bankruptcy await virtually all Russian enterprises.

Third, order must be introduced in the state's foreign economic activity. According to the calculations of the Duma committee for economic policy, that will make it possible to attract something like R80 trillion into the budget.

Fourth, we must really ensure a state monopoly over the sale of alcohol, tobacco, and some other products. All these sources of income should be used for the country's urgent needs and not to enrich speculators.

Fifth, we must proceed from the premise that so-called privatization is often implemented counter to public interests and with the most flagrant violations even of existing legal norms. All of Russia's national property has been assessed at a ludicrous sum equivalent to $300 million.

These are, I repeat, elementary measures that would make it possible to lay down the prerequisites for stabilization and to switch to the real reform of the economy in the people's interests. And reforms are essential. Convinced of the need for and the inevitability of Russia's socialist development, the party in no way wants the mechanical reproduction of the path that has been traveled. Our call is not *back* to socialism but *forward* to socialism. And that is, if you like, the leitmotif of our program.

## 3.44
## Vladimir Zhirinovskiy Reviews the Era of Gorbachev and Yeltsin

Vladimir Zhirinovskiy
*Trud*, 1 February 1995
[FBIS Translation]

*Editor's Note:* Vladimir Volfovich Zhirinovskiy emerged as a powerful far-right force in early post-Soviet politics. His Liberal Democratic Party of Russia (LDPR) represents the most extreme form of "Great Power" thinking in Russia, ultra-nationalist, neo-fascist, and anti-Western. Zhirinovskiy was born on April 25, 1946, in Alma-Ata, Kazakhstan. He was educated at Moscow State University. Until 1983 he worked with the USSR Ministry of Defense, with the General Staff of the Transcaucasian command and with the Soviet Society of Friendship and Cultural Relations. From 1983 to 1989 he worked as a legal consultant to Mir Publications. In 1989 he founded the LDRP and built it to such an extent that in the Russian presidential election of 12 June 1991 he came in third out of six candidates, with more than 6 million votes. He is described as an anti-Semitic demagogue, although there have been numerous suggestions concerning his own Jewish background, including one that until the age of eighteen, his surname was Edelstein. Thanks to his clever campaigning prior to the Duma elections of December 1993, the LDPR unexpectedly gained 23 percent of the vote. His political ideas, which include annexation of the Baltic States, Afghanistan, and parts of Alaska, Finland, and Poland, were elaborated upon in his book *The Last Thrust to the South,* published in Moscow in 1993. As a member of the Russian parliament Zhirinovskiy travels widely, openly attacking and vilifying the United States at every opportunity, using phrases such as "American carrion crows," and pledging to fully support rogue nations, such as Libya and Iran. He did poorly in the 1996 Russian presidential race. The following article, full of hackneyed, overdramatic statements and phrases, vividly exemplifies Zhirinovskiy's political views.

[Article by Vladimir Zhirinovskiy, leader of the Liberal Democratic Party of Russia, under the "Viewpoint" rubric: "From Gorbachev to Yeltsin and Beyond"—first three paragraphs are *Trud*'s introduction.]

Judging by the predictions of many Russian politicians, the spring of 1995 will be stormy. But however high the floodwaters of the possible political events may rise, one date will certainly not go unnoticed. It is directly associated with everything that happened here in the last years of the USSR's existence and that is happening now—in Russia and the CIS. April 1995 will see the tenth anniversary of the beginning of perestroika.

To say that this date arouses mixed feelings in our former fellow citizens is putting it mildly. For some people it is a holiday, for others a "jubilee" of mourning. But in any case it was a milestone that marked a sharp historical turning point in our country's history, and indeed in world history—and it cannot be ignored. . . .

Some items in *Trud* have already touched on this theme. And now the LDPR [Liberal Democratic Party of Russia] leader Vladimir Zhirinovskiy, who always wants to be the first to have his say on everything, has offered *Trud* his observations in connection with this special date. Observations that are not, of course, uncontentious. Their style is hard-hitting, emotional, and outwardly paradoxical. True to the principle of openness and seeking to present the widest possible spectrum of socio-political views and sentiments, we offer this contentious article for our readers' judgment.

[Zhirinovskiy begins] The tenth anniversary of perestroika is approaching. Anyone who feels so inclined will soon be writing and speaking on this topic. Or keeping quiet. It all depends on the situation. I will try, as has become traditional, to get in first, and open the season of reflection on the historical role of Mikhail Sergeevich Gorbachev and his perestroika.

Gorbachev is one of the key figures of the century, if not the millennium. He stands at ease alongside Napoleon, Peter [the Great], and Lenin. His reign influenced the entire world and turned that world upside down. It is another matter to ask at what price and with what consequences, who won and who lost, and what did the Russians gain from Gorbachev's global game. And that is what we are going to try to find out.

The system that Gorbachev first shook was absolutely rotten. It was dying. It had sullied its very nature and sunk into bureaucracy. It was a model of a war economy, appropriate to the era of rebuffing outside aggression. But with the appearance of nuclear missile weapons, the real probability of foreign aggression fell sharply. Economic reforms should have been launched. But for thirty years we marked time. Stalin and war were embedded forever in the brains and nervous systems of the then leaders. But now they were gradually dying out. And the young General Secretary Gorbachev, who had never smelled gunpowder, was beginning to shake the system.

People say we should have taken the Chinese path, step by step, with no sudden movements. But that is not for Russia. Russia does not like doing things gradually; here it is either one thing or the other, either hot or cold, either we are all identical atheists or we are all believers to a man, either we are all communists or we are all monarchists, either we all turn out for the *subbotnik* [voluntary Saturday work] or we are all running commercial stalls. No, the gradual approach is not for us. Kosygin wanted phased reforms in 1965, and he came unstuck. So Gorbachev had a choice—go all the way, or turn back; give birth, or abort the embryo. He gave birth.

A professional politician is always particularly interested in tactics. In this respect Gorbachev and his colleagues were pretty skilled. In general the CPSU trained its cadres well in just one art—that of survival in apparatus wars, the art of maneuvers and intrigues. The party did not teach strategy. It did not tolerate strategists in its ranks. That was why it died.

But in 1985 the CPSU was still alive. And how! In 1985 Gorbachev took a sharp ax and started hacking away at the tree. Let us look at how he did it. On the left, he let loose the "Democratic Union," where I began my own political career. The "Democratic Union" rallies in Pushkin Square became famous countrywide. Official propaganda lashed the "Democratic Union," but that only made it more famous. This handful of people did not constitute any kind of social force, but they were spoken of as a serious organization. Even when I realized that this was what someone at the very top wanted.

On the right, Gorbachev let loose "Pamyat," with its patriotic slogans. Inside the party, the detonator, by a whim of fate, was the argumentative Yeltsin. In fact, Gorbachev wanted to use Yeltsin to try to split the party's ruling clique and at the same time to clarify feelings in the Central Committee. The proof? Very simple. Gorbachev, as he himself has said, knew that Yeltsin was planning to deliver a hard-hitting speech. And for some reason, he gave him the floor right at the beginning of the Central Committee plenum. At that time nobody mounted the platform without the general secretary's say-so. So Gorbachev wanted this aggravation.

That Central Committee plenum, where Yeltsin barely muddied the water, showed Mikhail Sergeevich [Gorbachev] a lot. He realized that the Central Committee was a monolith, an assembly of experienced party people who saw perestroika as just another campaign, of which there had been dozens in the years of soviet power. These party people would pin any general secretary to the wall, especially him—young and provincial as he was.

So Yeltsin had to be sacrificed. But, take note, Gorbachev did not send Yeltsin as ambassador or adviser to Mongolia, where he would have been forgotten within the week; he left him in Moscow, where he quickly became the hero of the "new" press and the "Vzglyad" TV program. And so the two of them set off through perestroika together—Gorbachev and Yeltsin, apparently adversaries on the surface, but fatally necessary to each other. Yeltsin became a mighty hammer in the hands of Mikhail Sergeevich, with which he struck at the party system. But according to the laws of dialectics, Gorbachev was raising his own rival and grave-digger. However, to stifle that rival would have meant, for Gorbachev, stifling perestroika itself, tearing up his own political mandate, and being left with those who would never forgive him his "idiosyncrasies." What a dilemma.

Mikhail Sergeevich, convinced that the Central Committee could not be reformed, began to create a parallel structure—the Congress of People's Deputies. The time was ripe for this. A whole brigade of haughty professors, writers, and artists were waiting their moment. For three years they had vied with one another to criticize the CPSU in the pages of *Ogonek* and *Moskovskie Novosti*, in the heavyweight journals, and as guests of the "Vzglyad" TV boys. While the obkom [oblast party committee] members were pursing acceleration and still sweating blood over completely useless party and economic activists, Gorby handed over the hearts and minds of the public to others—to those very same professors and writers. He did this not from weakness, but deliberately. Thanks to glasnost, he secured a parallel power structure—the congress, which set a course of combating the already decrepit Communist Party.

One of the most enigmatic themes from the era of Gorbachev's rule is the story of interethnic conflicts and separatism. In 1988–90 the system was still in control, and any instigator, agitator, or troublemaker could have been put well out of the way. But nobody was touched. People's Fronts were formed and spoke out openly on the air, in the newspapers, and in the parliaments. To make a stronger impression, acts of bloody carnage were started. The turning point was the election of the Russian Supreme Soviet, which immediately began to pull the rug out from under Gorbachev's feet and define Russian sovereignty.

When the echoes of all these events had died down and the Union had broken up, it began to emerge that many active members of these People's Fronts, these all-out nationalists, had apparently been cooperating with the KGB for a long time. Scandals broke out in one place after another. An interesting sign, wasn't it?

And what of the breakup of the Warsaw Pact bloc? Notice that all the East European revolutions took place almost at the same time, as if by order. But why did I say "as if." It was indeed by order. By order of Moscow.

But why did Gorbachev, Shevardnadze, Kryuchkov, and Yakovlev do all this anyway? Why did they hand over Eastern Europe, connive with separatists, and provoke the outlying areas' secession from Russia? There was something objective and very serious behind all this: It was becoming increasingly unsustainable to feed the outlying areas, to supply cheap oil and energy to the "shopwindows" of socialism in Europe. So we were happy to abandon them. Frankly, we "dropped" them, we palmed them off with a "dolly" [bundle of paper cut to look like banknotes] as used to be the practice in the hard-currency "Berezka" stores. They grabbed this "dolly" in the form of sovereignty, fled, opened it, and found just pieces of paper. Look what we are seeing in the "independent" states now. Poverty, down-and-outs. Production is falling, there is no proper business, just wars and forebodings of war. Yet remember how it was ten years ago. They were kings. . . .

There is another aspect too: The kindling of nationalists among small nations inevitably evokes a corresponding upsurge in Russian nationalism, a mighty, Great-Power nationalism. Back in Lenin's day it was Russian nationalism that the communists feared. The USSR was built on playing up to the small peoples' national feelings (look, here's your very own republic, your own government, your own script, your own writers), combined with total suppression of Russian chauvinism. But the communists knew that they had to give the Russians something. And so they lulled them—using the writers—with a gentle, mawkish whispering about Mother Russia, the villages, the forests, the fields. Petr the bayan-player, and Marfa the milkmaid. On television, toothless old crones sang folksy little rhymes. Various bearded types sighed over the loss of the people's traditions. And not a word about Russia's historic might, its influence on world affairs, its amazing wealth. True, in the era of stagnation there was one man who reminded us, to some extent, that Russia is not a land of drunken peasants, but an empire with the sparkling palaces of Petersburg, great historical traditions and achievements, brilliant thinkers, and an advanced culture. That man was called Ilya Glazunov; his chauvinism was combined with no beard, smart foreign suits, and Marlboro cigarettes. They didn't like Glazunov. But I liked him. Many years ago I realized that diehard, "bearded" nationalism puts people off. Russian nationalism should be modern, intellectual, aggressive—fashionable, if you like. Russian nationalism should never be taken to extremes, because extreme nationalism is the road to death. Nationalism is like fire. But you have to know how to handle fire if you don't want to start a conflagration.

In recalling Gorbachev and his perestroika, we clearly perceive a historical thread. Gorbachev destroyed an efficient system, shook the party to pieces, and, by fostering nationalism and separatism in the republics, separated numerous parasites and "younger brothers" from Russia. With the help of anti-Sovietism he separated us from our "allies in the socialist camp." Yeltsin, with one stroke of his bear's paw, finished these jobs off well and truly, and started gradually reviving Russian autocracy. After undergoing serious operations, Russia is thin, emaciated, limping on both feet, but it has come back to the gym to flex its biceps and compete in the world championship again. And here, in the world championships, it needs a deft and quick-witted trainer, because the era now beginning is the era of the division of spheres of influence in the world, where the early bird catches the worm and the rest will have to suck their thumbs for many years to come. Apparently neither our smart economists nor our politicians, worn down by daily cares, understand this. Wake up, boys! Put your glasses on! You can learn to produce the best products in the world, but they'll never let you sell it and they'll find thousands of pretexts. And that cannot be stopped by economic means.

You have to play the world's game—threaten a bit here, cajole a bit there, and in some cases, if you'll pardon the expression, ram it down their throats. And remember the example of Gorbachev again. He did not know how to haggle and make deals. He had an opportunity to extract hundreds of billions of dollars from the outside world for the global concessions that, after all, we would have had to make, because it had become too difficult to feed all that crowd. But Misha was "civilized." He couldn't do that. He wanted to be liked in Paris and in London. The Russians' fate was of less interest to him. Apparently he was not very interested in the economy either, where he made so many hasty mistakes. From 1985 onward, money was not invested in the petrochemical industry, which had always brought in foreign currency for us. Wine and spirit production was destroyed. But that was doubtless part of the overall scenario of perestroika, since, paradoxically enough, reforms can be implemented more easily and more quickly when there is no money, so that there is no alternative to reforms. But that is not the way we are. We are different. And if we withdraw the Russian troops from the Central Asian "states" let us get, in exchange, the money to house officers in proper military encampments, not tents. And let us take the Russians out of these "states." All this talk of the fraternity of peoples in the Soviet period—let us leave that to *Pravda.*

One last thing. Not so long ago, since his resignation, Gorbachev became head of an international ecological organization. An interesting move. I wrote some time ago that it is in the ecological field that internationalism will be revived. An era of nationalism, an era of the right, is beginning in the world today, because a reshaping of spheres of influence is at hand. But at the beginning of the next century green parties will begin to rise throughout the world. An ecological army and ecological police will be created. Countries will be subjected to international sanctions for damaging the environment. Ecological ideas will grip the world just as communist ideas did at the beginning of the twentieth century. But more of this another time. For the moment. . . . Keep an eye on Gorbachev, he'll still spring some surprises.

## 3.45

## Zhirinovskiy Brands Kozyrev, Gaydar "Evil Democrats"

Interview by Wolfgang Briem
*News* (Vienna), 9 February 1995
[FBIS Translation]

*Editor's Note:* The following article provides a valuable insight into Zhirinovskiy's contradictory assessments of Boris Yeltsin. In the preceding article, Zhirinovskiy accused Yeltsin of destroying Russia. In this article, he calls Yeltsin a "patriot." Zhirinovskiy is actually supporting Yeltsin's Security Council, which is viewed by alarmed reformers as Russia's new Politburo, quietly usurping power from a withdrawn Yeltsin. The Chechnya War has clearly enhanced the power of this institution, giving it enormous clout over political and military problems, while Prime Minister Chernomyrdin handles economics. Zhirinovskiy identifies with the council's chairman, Oleg Lobov, and obviously seeks to ingratiate himself with council members, which include the prime minister, the ministers of defense, foreign affairs, and interior, the chief of counterintelligence and the speakers of both houses. The council's decisions do not require parliamentary approval.

["Exclusive" interview with Vladimir Zhirinovsky by Wolfgang Briem in Strasbourg.]

[Zhirinovskiy]: It was a mistake that our tanks and planes did not pulverize Grozny. We should have left nothing but a big crater.

[Briem]: The military leadership grandly announced: Everything will be over in a few weeks.

[Zhirinovskiy]: We said we will capture Grozny quickly. We have captured the presidential palace and driven out Dudaev. Now we will clear Chechnya from the mafia.

[Briem]: Is the army crumbling?

[Zhirinovskiy]: Our army does not have any problems. The armed forces are the best in the world. Only the Moscow democrats and the press are denigrating them.

[Briem]: Soldiers are deserting the "best army in the world."

[Zhirinovskiy]: There is no desertion. Whenever unpatriotic cowards ran away, they were punished. Paratrooper elite units executed these criminal elements according to martial law.

[Briem]: Generals like Aleksandr Lebed have voiced harsh criticisms.

[Zhirinovskiy]: Lebed's statements are wrong. There is only one commander in an army. In the Russian army this is Defense Minister Grachev and certainly not the mad Lebed. He is abused by the democrats and is only boasting.

[Briem]: Does Grachev now have to pay the bill for the failed war?

[Zhirinovskiy]: Never, because he is an excellent commander, a tough guy who stands no nonsense. He will dispose of the gangster Dudaev and his clan. He will completely eradicate the Chechens. But the democrats want Grachev's head. With this they are destroying the entire Russian army.

[Briem]: Does the disaster in Chechnya have any consequences?

[Zhirinovskiy]: We must rearm. Gorbachev and Yeltsin have starved our army. The West demanded disarmament, and we followed like obedient sheep—a perfidious maneuver. Now we have to think back to our real strength. We need new weapons. The armament industry must work at full speed again. As in the past, it must once again be the state's goal to expand our troops strength. We will build new tanks and aircraft. The Russian army is the guarantor of world peace. Where would Europe be today without our soldiers? Vienna, for instance, would certainly be occupied by the Turks today. The Germans would still be in Paris. The West should be grateful to us instead of criticizing our policy. Over the past five to six years, the democrats have tried with great success to destroy our armed forces. But this will change now.

[Briem]: Who are the "evil democrats"?

[Zhirinovskiy]: In particular, [former] Prime Minister Gaydar and Foreign Minister Kozyrev.

[Briem]: Where does Boris Yeltsin stand?

[Zhirinovskiy]: At first, Yeltsin was one of these democrats. But now he is on our side, on the side of the national, patriotic forces.

## 3.46

**Zyuganov on Religion, Russian Idea**

O. Nikolskiy

*Pravda Rossii,* 5 October 1995

[FBIS Translation], Excerpts

[Interview with Gennadiy Andreevich Zyuganov, leader of the Communist Party of the Russian Federation, by O. Nikolskiy; place and date not given. . . .]

[Nikolskiy]: Just a list of the troubles Russia has experienced would probably take up all the columns of *Pravoslavnaya Moskva.* But still, Gennadiy Andreevich, tell us what disturbs you most of all.

[Zyuganov]: All the great times of trouble that befell Russia lasted, as a rule, seven to eight years. Now this period is coming to an end. But I am disturbed by the fact that the enemies of the fatherland could once again lead the country into a web of violence and destruction. The preconditions for such a turn of events are in evidence. There are forces of evil that never wanted anything good for Russia and now are extremely disturbed by the fact that under these incredibly difficult conditions the people are beginning to see the light.

Our fabled Ily Muromets slept on top of the stove for several years and then stood at the intersection of three roads. One road was to transform the entire country into a big Chechnya, the second led to a criminal state, and the third was the path of the good and the just. It is very important to take this last path. After all, during the twentieth century alone we have been at war four times, we have lost almost 100 million of our compatriots, and any internecine warfare will end in tragedy for the Russian people.

[Nikolskiy]: Frankly, believers are cautious about modern communist movements, including the Communist Party of the Russian Federation. And there are quite justifiable reasons for this, including historical ones. Therefore we have a crucial question: How does your party feel about religion and the Orthodox Church now?

[Zyuganov]: I would not say that the attitude of believers toward our party is cautious. Recently my colleague, Deputy Zorkaltsev, who heads up the State Duma Committee on Public Relations and Religious Denominations, and I spent almost two hours talking with His Holiness the Patriarch of Moscow and all Russia Aleksiy II. We discussed questions related to legislation in the area of freedom of conscience and problems pertaining to land laws. We discussed with alarm the penetration of foreign religions into Russia. And his attitude toward us was extremely positive. I have regular contact with the metropolitan of St. Petersburg and Ladozhskiy Ioann, and wherever I am I meet with the local bishops of the Russian Orthodox Church and find understanding and support among them.

A politician who does not understand the colossal and largely unique role played by the Orthodox faith in the establishment and development of our state and culture does not understand Russia itself and cannot lead the country out of its crisis. The history of the fatherland must be considered as a whole, as played out over the millennium. By Christianizing Rus, Saint Vladimir, a prince coequal with the apostles, laid the foundation for internal unity based on the extremely high morality of Orthodoxy. Without this it would have been impossible to live through all the hardships of our history. During the period of 1055–1462 alone the Russian land was subjected to 245 invasions. The decisive power in the difficult life of Russians was not their wealth but the power of the spirit, as perceived and reinforced by Orthodoxy. When I studied history I was struck not only by the vision of Saint Sergey Radonezhskiy, but also by his personal courage. Saint Sergey blessed Peresvet and then he began the Battle of Kulikovo and opened up for his comrades-in-arms the path to a great moral victory accomplished in the name of salvation of the people. I was surprised at the endurance of Patriarch Germogen, who during the Time of Troubles sat in a dungeon surrounded by enemies and called for the people to drive out the Poles. The citizen Minin, Prince Pozharskiy, and thousands of Russian people listened to him. They formed a home guard and drove the enemies out of the capital. And during the Great Patriotic War when the Fascists had seized our country by the throat, the Orthodox Church raised its voice in defense of the fatherland. This played an enormous role in our victory.

Karamzin's "History of the Russian State" would not have existed were it not for the chronicles created in the monasteries. And the architecture, the iconography, the church singing! I have spoken with Vondarchuk about how magnificently the choir sings "War and Peace." He said that it was a church choir from Zagorsk, whose singing was filled with the spirit.

We have respect for Orthodoxy. We recently adopted a new program and regulations where it says that religious convictions are the private affair of a party member. Indeed, in the history of our country the CPSU has gone through various periods in its attitudes toward the church, but we have drawn the appropriate conclusions and will not repeat our mistakes.

[Nikolskiy]: Gennadiy Andreevich, many of our politicians speak about the Russian idea, sometimes including diametrically opposed content in this concept. In your opinion, what should the Russian idea be based on?

[Zyuganov]: I recently edited a large study devoted to this question. And that is what it is called—"The Russian Idea and the State." Historians, legal experts, sociologists, and religious figures participated in its creation. At the basis of the Russian idea lie two fundamental values—Russian spirituality, which is unthinkable without the Orthodox world view, and awareness of our true purpose on earth, and Russian power and statehood. Without moral purification, without spiritual strengthening, stabilization of the situation over the immense expanses from the Baltic to Kamchatka will be impossible.

It is very important to understand the historical peculiarities of Russian civilization. It includes traditions of tolerance and respect for our neighbors. Our expanses were not assimilated as land was in the New World. We proceeded not with the sword but with the cross. We brought literacy and knowledge, and not destruction. Therefore all the peoples living here since ancient times have remained in Russia. What binds us together, what our state was created from, is the Russian people. And without its spiritual rebirth, the rebirth of other peoples of the country will not be possible.

It is very dangerous to use the banner of the Russian idea for unworthy purposes. I see how people who have turned into patriots overnight are trying to use it today. It is not important to them which God you pray to; the main thing is that the money must come in.

[Nikolskiy]: Iniquity, which runs counter to the spirit of Christianity, is increasingly encompassing our society. How can we oppose this?

[Zyuganov]: Russia has successfully driven away those who have come with a sword trying to take over our land and our wealth and to destroy our faith. But we have ended up defenseless before the new types of aggression—the aggression of lies, iniquity, baseness, and greed. The people are very trusting. Now that it is possible through the television screen, the radio, and the mailbox to penetrate into each home, to each family, and deceive the people every hour and every minute, it is difficult to separate the wheat from the chaff.

Look at the lies that surround us, how everything is being misinterpreted! They talk about openness and popular rule, but openness has long been transformed into manipulation of public opinion, and popular rule has degenerated into omnipotence of bureaucrats. They say that we have a democracy, but the president has concentrated in his own hands more power than the Russian tsar and the Soviet general secretary did taken together. And at least the tsar feared God. . . .

We are paying a terrible price for our trustfulness. The population of Russia last year decreased by 940,000. The lifespan is decreasing; for men it was fifty-nine years. They are not living long enough to receive their pension! The people are drinking themselves into oblivion. Last year Russians drank an average of 14–16 liters of alcohol, and if the level of consumption exceeds 8 liters, the nation begins to degenerate. There is a war going on in the south of Russia; there are 6 million refugees in the country who have been driven out of their homes. Some 25 million Russians have ended up outside the state's current borders. Many of them are being subjected to unheard-of persecution and degradation.

The population has been fleeced. First they "deregulated" prices, and savings disappeared. Out of every 1,000 rubles they took 998. Then they deceived the people who, believing the advertising on state television and radio, acquired shares in a multitude of funds and banks that no longer existed.

They took away the social rights of the people, who were actually deprived of freedom to travel. Previously 12 million people went to Crimea alone each year. Now half the health resorts in Yalta are empty at the height of the season. The majority of people cannot go there—a pass costs more than 2 million rubles, and a one-way ticket, 800,000.

But still I think that we will withstand this invasion in which slander and money are the main weapons. The people will begin to figure out what is what. Even fools can guess that you can eat only as much as the stomach can digest, wear only as much as you need to keep warm, and everything else is the devil's work. . . .

[Nikolskiy]: A litmus test with which it is easy to determine the orientation of a state system is its attitude toward the church. Gennadiy Andreevich, do you think the Russian Orthodox Church occupies a proper position in our country?

[Zyuganov]: On the surface, it would seem so. The strong of this world honor the church. But at the same time foreign denominations are making a terribly strong attack on it, and the present authorities are doing practically nothing to protect the country from all kinds of foreign preachers. They are not doing anything to stop the battle against the Orthodox Church, which is now being attacked with greater fury than the CPSU was in its day. I understand quite well why this is being done. If they manage to destroy the spiritual foundations of society, Russia will be defeated. At one time Goebbels, when forming his doctrine regarding our country, said that it is possible to rout the army, to win territory, and so on, but it is impossible to conquer this people without implanting "our own" faith in each village. Now they are trying to implant a faith that does not correspond either to our spirit or to our traditions. From day to day, even on Orthodox holidays, they broadcast foreign preachers on television. There is a dirty invasion by false prophets.

Unfortunately, the State Duma has not yet adopted laws that give priority to traditional religions. Barriers must be placed on the path of penetration of foreign denominations. The country's nation interests demand that the state policy provide support for three world religions—Orthodoxy, Islam, and Buddhism—religions that are traditional for Russia, which have placed spiritual-moral values above mercantile-consumer ones. And this is what disturbs the forces of evil that are trying either to crowd out the traditional faiths or to sow hostility among people who believe in them. The war in Chechnya, incidentally, is an attempt to set Orthodox against Muslims and thus aggravate the situation throughout Russia. We are categorically against such a development of events. Christians and Muslims can and must live together in the world.

[Nikolskiy]: It is no secret that at the present time the material position of the Russian Orthodox Church is difficult. The construction and restoration of temples destroyed or half-destroyed by the state at one time, philanthropic activity—all this costs a colossal amount of money. What possibilities do you see for the state to render financial support to the church?

[Zyuganov]: Many cloisters and temples are national property. There are holy places that are dear and close to each individual, and the state must do everything possible to keep them in good condition, help the church to repair them, etc. On the other hand, many people would be willing to make donations to restore temples but they cannot do so because of their property. And this is a matter for the state, which should implement an economic policy whereby citizens would not be poor.

[Nikolskiy]: One of the most pressing problems of modern Russian life is the millions of abortions. What do you think about this phenomenon?

[Zyuganov]: A woman is meant to be a mother. And when because of bad circumstances, poverty, or a lack of employment or housing she chooses an abortion, this is a very bad thing. It is necessary to support women. We in the Duma are adopting laws on minimum wages and compensations. But if the economy does not work, where will we get the money for social programs? Now every [other] working woman does not even receive her wages promptly.

[Nikolskiy]: The church simply regards abortion as murder. Murder of an unborn child. Do you agree with that?

[Zyuganov]: A person conceived in his mother's womb has the right to see the light of day and live out the time he has allotted to him. But was it not the state that put the woman in a position where she simply cannot clothe or feed her children? I have met with women in Ivanovo oblast. Almost all the textile combines there are standing idle, and in many families nobody has a job. One woman said to me, "I do not wake my children up in the morning—I have nothing to give them for breakfast." So when I see how difficult it is for a woman with even one child, I begin to understand her. She is simply afraid that another one will appear and then she will not even be able to feed the baby.

[Nikolskiy]: Do you not support the idea of a legislative ban on abortions?

[Zyuganov]: There is a time and a place for everything. First let us stop the ruin, the trouble, let us restore the spirituality and culture, let us reach a more or less bearable level of support. I am confident that then our women will make the correct moral decision.

[Nikolskiy]: Gennadiy Andreevich, I would like to know your opinion [about] the restoration of the temple Christ the Savior.

[Zyuganov]: I have thought a lot about why they tore down that temple. How could they. . . . The Temple of Christ was built with public money to mark the great victory of the people, and it had a majestic appearance. It was a symbol of the spiritual force and might of the fatherland. Russia's enemies destroyed precisely these kinds of temples and monasteries.

I am in favor of restoring justice. The temple must be restored. But while the sacred places of Orthodoxy are being restored, I would not want to destroy sanitariums or to close down hospitals or palaces of culture. They must supplement one another.

[Nikolskiy]: And the last thing. Tell us something about yourself, about your family.

[Zyuganov]: I was born in the Orel region, way out in the country. My parents were teachers. My grandfather also taught in a parish school. There were many pedagogues and VUZ [higher educational institution] instructors in our family—their combined tenure was almost three hundred years. My wife works as an engineer at a watch plant. We have children—a son and a daughter. My son graduated from Baumanskiy University with distinction and is now in graduate school, and my daughter has also shown an inclination for scholarship.

I have traveled throughout the country a great deal. I have been almost everywhere—from Kaliningrad to Vladivostok. I know about life in Russia firsthand. I am convinced that if things are bad in the country for the clergyman, the teacher, the policeman, and the doctor, they cannot be good for anyone else.

## 3.47

## Zyuganov Argues for Review of the Left's Tactics

Gennadiy Zyuganov
*Sovetskaya Rossiya,* 14 October 1995
[FBIS Translation]

The autumn session of the State Duma, which opened 4 October, will be dominated, one way or another, by the rapidly approaching parliamentary elections. The Duma hall of sessions will inevitably be used as an arena for the election contest, but you would have to be either very naive or else an inveterate political hypocrite to be upset or indignant about this.

In a state that has I will not say a democratic form, but even a representative form of government, any political force must gain its share of influence on state affairs exclusively by means of free elections. Consequently the elections are not some kind of "diversion," but a central component of the functioning of the state mechanism. Their preparation and holding and the implementation of their results give political life its chief content. Any step in the political arena constitutes a direct or indirect influence on voters, an appeal for their support. That is elementary democracy.

So why all these calls not to turn the Duma platform into a platform for propaganda and agitation? Why the renewed talk of postponing or canceling the elections? The whole point is that elections are by their very nature a peaceful, non-violent way of retaining power or conceding it to one's political opponents. But the events of recent years, particularly the "Black October" of 1993, demonstrate unequivocally that the present ruling regime and the president who heads it *are not capable* of retaining power by non-violent means, but still less do they intend to concede it to anyone, in any circumstances. After all, the next stage in the destruction of the Russian state is in full swing—selling off and placing under foreign control enterprises that play a key role in the functioning of leading sectors of industry, such as "Rybinskie Motory," for instance. A study of Chubays's list of twenty-nine major state enterprises offered for sale leaves not the slightest doubt—the fate of strategically important sectors is being placed in foreign hands. For some reason, figures like Chubays or Kozyrev remain "unsinkable" amid all the reshuffles in court circles. Here interests and forces are involved for which the principles of democracy and the rule-of-law state are mere words.

Hence the attempt to discredit the very idea of elections, the very principle of free expression of the people's will. Hence the feverish search for the slightest pretext or excuse to cancel the elections—from the deterioration of the Chechen crisis to integration with Belarus.

So the main problem for us today is not going to be the question *how* the voter will vote on 17 December, but the question *whether he will be able* to vote at all, and whether the results of the voting will not be trampled underfoot by jackboots and crushed by tank tracks.

*How* people will vote is becoming increasingly clear. It would be wrong to indulge in victory euphoria, but it is a fact that the public mood is patently moving leftward. The elections to the Volgograd City Duma have already been dubbed the "Battle of Stalingrad," and not for nothing. The Communists' victory in twenty-two electoral districts is certainly equivalent, in terms of its moral and psychological significance, to the routing of twenty-two divisions of [German commander] Paulus's army. And as you know, the Battle of Stalingrad was followed by the Battle of Kursk. . . .

But *what* must be done to ensure that a possible victory for left-wing and patriotic forces is not stolen away, that the transitional period of the handover of the levers of state management passes off peacefully? To be honest, any hopes that the regime is prepared even to comply with the constitution that was written by the regime to suit itself are minimal. These gentlemen have grown used to a different language, not the language of law. Therefore it is very important right now to determine whether a force (or forces) exists that is strong enough to paralyze the dictatorial regime's desire to derail the elections or ignore their results. And if such a force exists, how and at what point can it be utilized most effectively, without allowing it to get out of the people's control and establish a new dictatorship? These questions, which go far beyond the bounds of election procedure as such, are nonetheless, in our situation, vital pre-election problems that require careful study and discussion by all interested parties. I believe that the discussion of this group of problems must certainly continue. But for the moment let us return to our first thesis: The sittings of the last Duma session will be decked in the colors of the election contest.

The only question is what methods will prevail in this contest—short-term populist demagoguery, or the accurate examination of the main political issues and the detailed presentation to society of one's own political line for the immediate and longer term.

It is not hard to predict that the temptation to follow the first path will prove very strong. There is, after all, fertile soil for this: the mass non-payment of wages and pensions (which are miserly anyway), the dire situation in the entire government-funded sphere, the unpreparedness of entire regions for the winter, and so forth. All these are urgent problems that must be tackled immediately. But tackled how? That is where the trap lies.

An orgy of socio-economic populism, an auction of meaningless promises, a torrent of impracticable pledges to anyone and everyone, irresponsible draft laws and resolutions costing trillions, even quadrillions, of rubles—*that is exactly what is expected* from the Duma in general, and from all its factions, by the "democratic" and pro-government mass media, which are preparing to enjoy themselves deriding the upcoming comedy, the purpose of which is to discredit parliamentarianism once and for all in the population's eyes, as a futile talking shop. The country is being psychologically prepared for the advent of a "firm hand."

It is a matter of political dignity for the Communists not to allow themselves to be drawn into these games. We say beforehand to the gentlemen from the print and electronic media: Yes, you're absolutely right—it will be a comedy, a vulgar farce. But *why* a farce? Because all its participants know very well that within the framework of the existing regime and the course it is pursuing, it is *impossible* to do anything to alleviate the people's situation or rectify the economic situation. No trillions of rubles, even if by some miracle they could be found, will rectify the situation so long as conditions are being imposed in the country that make productive labor disadvantageous in those key sectors of the national economy that ensure independent and stable economic development in the future; society's cultural and intellectual resources are being destroyed by attrition; and the country's fundamental national-state interests in the international arena are being betrayed.

Therefore the CPRF [Communist Party of the Russian Federation] and its Duma faction are still convinced that the key to overcoming the crisis and avoiding disaster lies in the nature of the authorities and the need for a replacement. And the composition of the government is not the crucial question, in a situation where one person exercises uncontrolled and absolute power. The renunciation of office by the present president—that is the point of intersection at which all problems converge at this juncture. Our faction has not forgotten its own summer initiative on this point and has resumed the collection of deputies' signatures in favor of creation of a commission to bring charges against the president.

People might object that we should not inflame the situation, since it is only a couple of months to the Duma elections, and then the presidential elections will not be far off. Our answer is that even a couple of months could play a decisive role here, since the scientific, technical, cultural, and moral *potential for the future, the potential for development* accumulated by the country in previous decades is being swept away, lost on the wind, with disastrous rapidity. Until such time as it is finally lost we can look forward with hope, but its loss will mean that Russia is struck off the list of countries capable of autonomously tackling the exceptionally complex problems that the twenty-first century has in store for mankind. The key task of preserving and augmenting that potential and creating the conditions to ensure its freest possible exploitation is the basis of the CPRF's election platform.

As for the government, at this Duma session it must without fail report on the results of the first three-quarters of the year, first and foremost on the problems of defaults, the budget deficit, delays to wages and pensions, and the status of food supplies and preparations for winter. The government is hardly likely to be sincere on these points, but all the same a report—and particularly the discussion of it—will help to reveal the true picture and the true cost of so-called stabilization. Then it will become clearer what this year's budget is worth and what should be expected from the 1996 budget.

Today's government financial policy is built on two disingenuous ruses. First, *specious* methods of "combating" inflation, whereby commodity producers and workers in the social sphere are simply not paid for work already done, so that overt inflation is turned into hidden inflation, which is significantly more dangerous, with the risk of an uncontrolled explosion at any moment. Second, the blatant exaggeration of "successes" in this struggle and the formulation of unjustifiably optimistic predictions, which then become the basis of the new draft budget. It is not hard to grasp that attempts to stay within the framework of bogus predictions at all costs will once again generate specious methods of reducing inflation, and so the vicious cycle is repeated. But what do we not know is whether the country's high technologies and defense, science and culture, education and health care, and social protection will survive this next round. The hearings held in the Duma the other day on the draft 1996 budget took a very uncomplimentary view of it. One of the findings is characteristic: No budgetary (read: monetarist) contrivances will help so long as the present economic policy continues, because the budget is only an instrument of policy, not vice versa.

Those who talk about the desirability of having the budget approved this year by the existing Duma are clearly trying to preserve a course that has proved bankrupt and to impose it on the country even in a period in which the correlation of forces in the political arena will tolerate substantial changes, that is to say, they are trying to turn policy into an instrument of the budget as well as making policy hostage to it.

This attempt is hardly likely to culminate in success. But even if it does, nothing can stop the sensible majority in the new Duma, when approving the premier's candidacy, from setting as a condition a substantial change in financial policy and a review of the principles of budget planning.

The changes should emphasize two interconnected points: incentives for the Russian commodity producer and commodity turnover; and social support for those who are not well-off—which means basically expanding the market for those same Russian commodity producers. In the first case, apart from the establishment of elementary payments discipline, this presupposes a reduction in the ruinous taxation, the adoption of protectionist measures in foreign trade, and the fixing of energy prices and transport tariffs; in the second, legislation to set the parameters of the minimum consumer "basket," the linkage of wages, grants, and all forms of social payments to those parameters, and state-guaranteed prices for essentials: bread, milk, baby food, apartment rents, and mail and transport tariffs.

But all this is for the future and will happen only in the event that, not only in the two chambers of the Federal Assembly, but also in the regions' representative bodies, left-wing and patriotic forces have a solid majority capable of markedly influencing the executive power's line, if the executive powers themselves are re-elected centrally and in the regions. Judging by the statements of a number of authoritative leaders, the prerequisites exist for the formation of a coalition of left-wing and patriotic forces in the State Duma and in the regions.

But, I repeat, that is precisely what the present ruling group is most afraid of, and it is trying to derail the normal democratic process of alternation of power. The most characteristic illustrations are the postponement by presidential edict of elections to regional legislative organs and of heads of administrations, and the president's freezing of the law on the procedure for forming the Federation Council, which was approved by both houses of parliament. The motives are absolutely clear. In the first case it was the desire to prevent the formation of local organs of power capable of ensuring honesty in the holding of presidential elections and the summing up of their results. In the second, it was the president's wish to have a Federation Council that is in his pocket, appointed by him personally, as a counterweight to the new State Duma and an obstacle in its path.

It is in this respect that the present Duma could do much in its remaining two months of work. At the opening of the Duma session, our faction proposed that efforts be concentrated on the formation of a legal base for real, not illusory, people's power. Apart from the problems just mentioned, concerning which deputies have already submitted questions to the Constitutional Court, our proposals envisage the creation of a medical commission to investigate the health status of contenders for top state posts, amendments to the concept of parliamentary immunity so that a deputy's mandate cannot be a shield for criminal elements, and the compulsory declaration of officials' income. These are rather modest but realistic proposals, and it is certainly within the powers of the present Duma to implement them. But the more decisive steps—the introduction of amendments to the constitution to widen the control functions of the legislative power and increase the executive's accountability—must be taken by the newly elected deputies.

The CPRF election platform states that the upcoming parliamentary and presidential elections may be the last chance to change by peaceful democratic means the present political and socio-economic course, which has led our fatherland into dreadful tragedy. Our tactics are entirely within the framework of the course of peaceful development. But it is well known that *it is no use begging for peace on your knees. It takes strength, too*. That strength is the strength of the cohesion and organization of all society's sound, patriotic forces, the strength of the combination of parliamentary and various extraparliamentary forms of political struggle—up to and including a general political strike, the prerequisites for which are developing slowly but surely. Finally, it is the strength of flexible tactics innocent of both appeasement of the fatherland's destroyers and adventurism.

We must be fully aware that the left-patriotic opposition could very soon cease to be pure opposition in the people's eyes, at least in parliamentary terms. It will bear an even greater responsibility for the country's fate, and protestations of parliament's impotence are no longer going to impress anyone. In these conditions a major review of the tactics followed hitherto is needed. The past few years' experience has taught us much, but we still have even more to learn.

## 3.48
## Zyuganov Rejects NATO Expansion and Backs USSR Revival

Vera Ivanovicova
*Pravo,* 27 December 1995
[FBIS Translation]

[Interview with CPRF chairman Gennady Zyuganov by Vera Ivanovicova in Moscow; date not given: "You Would Not Talk About Reforms in Your Republic—We Call It Genocide."]

[Ivanovicova]: How does your Communist Party of the Russian Federation [CPRF], which was victorious in the recent elections to the State Duma, differ from the former CPSU?

[Zyuganov]: The CPSU was a system of control, not simply a party. Our party is one of many. It is the strongest and most organized. Its membership exceeds that of all the other parties put together, but it is still a party.

[Ivanovicova]: How does it function?

[Zyuganov]: We have gotten rid of the bureaucratic apparatus; all the secretaries get paid where they work. We have an up-to-date program that accepts as its legacy all the best and proper things: friendship among peoples and respect for those who work. At the same time, it proceeds from a patriotic-state position: It respects the ethnic composition of our multiethnic country and allows for reality both in the West and in the East. It takes the very best not only from the Soviet era but also from the one thousand years of Russian history and takes all the advanced things from the history of the West and other countries. It looks forward and not back, as is sometimes emphasized in the West.

[Ivanovicova]: Do people in the Czech Republic have any reason to fear the return of the Communists in Russia in view of our experiences from the period when we were part of the Soviet bloc?

[Zyuganov]: I would say that you have not learned from these experiences. You are now getting ready to enter a new military bloc. The system of military blocs has become absolutely outdated. It only led to unprecedented hectic armament, confrontation, and the devastation of the planet's resources, to the destruction of all its protection systems—ozone, ecological—and to an energy and demographic crisis and so on. So, it is necessary to take other realities into the new century. I would say good education, elevated culture, and a qualitatively new strategy, in which, on the basis of collective security in Europe, it is possible to balance interests.

[Ivanovicova]: What is your view on NATO enlargement?

[Zyuganov]: It is a bad solution. I will submit to you one of the variants. The decision will be made, and the Baltic republics will enter NATO. But 500,000 Russians, Ukrainians, and Belarusians live together in northeastern Estonia. A referendum will be held there in accordance with international standards, and the people will decide to join—in accordance with international norms—the Russian Federation, because their human rights are being violated systematically and children cannot study in their own language and have no freedom of movement. What will you do then with your NATO? Will you fight Russia over it? No, you will not. There will be no confrontation, however. We do not like it. This is a bad policy.

[Ivanovicova]: What do you see as the main reason for reviving the Soviet Union?

[Zyuganov]: Well . . . let us suppose that someone had taken the Czech nation and divided it into fifteen pieces. How would you have dealt with this?

[Ivanovicova]: But we separated from Slovakia.

[Zyuganov]: The Czechs and the Slovaks separated themselves. But 25 million Russians were left outside the borders of the Russian Federation. So, work it out. We think this is distressing for a nation.

[Ivanovicova]: So, what do you intend to do?

[Zyuganov]: We will resume contacts stage by stage—economic, cultural, and others. At the same time, however, we will not attack anyone's political sovereignty or foist our will on anyone, but we will restore the will of the people. The people will express their opinion. Customs barriers have to be removed. We will live together, but everyone's political institutions will function independently.

[Ivanovicova]: Does your party have plans for solving Russia's economic difficulties?

[Zyuganov]: Yes. We have an economic program, a program to deal with crime, a socially targeted policy. We have already submitted a whole series of laws to the current Duma, but, so far, they have not been passed.

[Ivanovicova]: Which of the implemented reforms do you reject and toward which do you have a positive attitude?

[Zyuganov]: Look—if in your republic, as a consequence of the reforms, output had been reduced by half, everyone had been deprived of all the money they had, and a tide of refugees had been created—and we have 6 million refugees—you would not be talking about reforms. We call it genocide, actual genocide.

One positive outcome is that it is clear to everyone that no one kept anyone in the Union. Everyone worked for the common weal. True, it was not always distributed fairly. Now it is clear that no one will protect himself. Even the until recently wealthy Ukraine, the not always diligent Belarus, and big Russia. They will have to put their potentials together. It is clear that the mass media, which has been monopolized by a narrow group of people, operates as a very destructive force. It is manipulating public opinion and misrepresenting all our history. This is also clear.

[Ivanovicova]: According to the CPRF, how should the situation in Chechnya be resolved?

[Zyuganov]: The most important thing is to resume dialogue and to do so without preliminary conditions. It is important that all the warring sides take part in the talks. Then, the status of the talks should be raised to the level of the first deputy prime minister. It is also necessary to halt the movement of the arms that are continuing to pour into this region. Another of our principles is to interest the neighboring republics in settling the crisis in Chechnya.

[Ivanovicova]: You often talk about the need for changes in government. If you were offered the post of prime minister, would you accept it?

[Zyuganov]: We are discussing this situation, and we will make a decision in a joint forum of our party and with our allies. There are now far too many people in our country who are willing to accept various functions, but they do not want to bear the responsibility. When we carry out a function, we will want to bear the responsibility for our activity. Therefore, we will make a very well-considered decision.

# 4

# Foreign Policy in the Non-Russian States

## Introductory Notes

Parallel with Russia's great power debate documented in Chapter 3, the leaders of the non-Russian CIS states have reacted in diverse ways to Russia's attempts to reconsolidate the former Soviet republics. As their economies have faltered and security issues in the region have become more pressing, some non-Russian CIS leaders have moderated their initial negative reactions to Russia's integrationist policies. In some instances, leaders who initially rejected the CIS as a permanent, supranational organization led by Russia have been replaced by others who have adopted a cooperative, even a subservient, approach to Russia's vision of the CIS. In other instances, leaders of states that do not belong to the CIS have shifted from their original attempts to defend their total sovereignty to a proclivity born of necessity to look for ways to relink their economies with that of the Russian state. These shifts away from primary positions and toward more conciliatory ones can be detected in the documents selected for this chapter. Only Ukraine, considered in the next chapter because of its unique perspective and its sensitive geopolitical position, has so far consistently resisted Russia's pressures to conform to its demands for reintegration.

Although each CIS country's reaction to integration has been colored by its own historical and post-communist circumstances, several trends appear to be taking shape. In some cases, leaders of CIS members have supported Russia's tactics to strengthen the Economic Union and the Collective Security Treaty. In other instances, opposition groups have formed within CIS states and have registered statements of their own against capitulating to Russian pressures. Both these groups' positions are documented throughout this chapter to illustrate the debate in these countries over how to react to Russia.

*Note:* This chapter will give only tangential attention to Ukraine, which is the subject of Chapter 5.

One of the more effective political reactions of non-Russian leaders has been the use of their new constitutions to avoid compliance with CIS agreements and protocols. Kazakhstan, Kyrgyzstan, and Uzbekistan, for instance, have avoided the "dual citizenship" clause for Russians living in their countries (which Russia includes in every Friendship and Cooperation Treaty) by pointing out its unconstitutionality. By making its own constitution supreme over all other laws, each CIS state avoids implementing CIS regulations that would constrain or co-opt its sovereignty. This continuation of the pre-CIS "war of laws" is documented in the Kazakhstan, Kyrgyzstan, and Uzbekistan sections.

Another tactic has been to emphasize bilateral treaties as alternatives and supplements to CIS agreements, along with membership in smaller, regional associations. Non-Russian CIS leaders know these alliances do not necessarily challenge Russia's enormous power and may mean relatively little in the face of Russian imperial pressures, but at this point they do somewhat mitigate Russia's influence within the CIS. Several documents that reflect the emphasis on the bilateral approach to integration appear in this chapter.

Certain leaders have adopted the point of view that CIS integration is inevitable—but over time. They point out the length of time it took the European Union to create its unified customs arrangement and the trouble it is having now reaching agreement on political and even monetary coordination issues. President Mircea Snegur of Moldova, in particular, has used this argument in an attempt to delay the rapid integration being sought by Russian, Kazakh, and Belarusian leaders.

Even when placating Russia's demands for closer economic and military integration, some CIS leaders resist signing an agreement until it has been made essentially toothless—hence the contradictory and often confusing situation within the organization, which cannot consistently implement the more than four hundred agreements it has concluded. As a countermeasure to this tactic, however, Russia has succeeded through its bilateral agreements with CIS states to extract binding economic or military commitments from many leaders. With this bilateral networking approach, Russia obviously hopes to make CIS integration a fait accompli.

Several documents in this chapter reveal another contradiction in goals between Russia and the other former Soviet republics. Several non-Russian CIS leaders would like Russia to provide financial assistance, enter into joint ventures, and extend favorable pricing policies on essential commodities and energy supplies as part of their integration agreements within the framework of the CIS Economic Union. Russia, on the other hand, would like to conclude a CIS-wide customs union agreement, a centralized banking system, a unified currency, and certain other parallel economic systems within the post-Soviet space, which would guarantee its own vital national economic interests. Russian leaders, however, do not want to risk a slowdown in their own economic recovery by tying their budget and resources to the weak and grossly underdeveloped frontier economies of other CIS countries. Moreover, Russian leaders are reluctant to become burdened with the necessity to prop up other state budgets, which would almost automatically transpire under a unified monetary system within the former Soviet space. As a result, Russia has backed away from its original insistence on preservation of the ruble zone.

In spite of the interdependence of CIS enterprises, the contradiction in state economic interests described above imposes a constraint on CIS political and economic integration. Central Asian leaders in particular have sensed Russia's desire to maintain a distance from their economic development needs and have responded with their own regional contingency arrangements, which are documented in Chapter 6.

In general, Russia's integrationist pressures have been most successful in the military sphere. The main issues concerning CIS military integration have been: (1) the ambiguous role of Russian "peacekeeping" and "peacemaking" troops in the CIS; (2) the positioning of Russian military bases and border troops on CIS internal and external borders; (3) the CIS Collective Security Treaty; (4) Russia's option to use force to protect the rights of Russian-speaking peoples in other CIS states. By late 1995, Russia had succeeded in signing military assistance agreements for bases, officer training, border troops, or peacekeeping troops with every CIS country except Ukraine. Tajikistan, Armenia, Azerbaijan, and Georgia have had little choice but to accept the presence of Russian troops on their territories. The one area of resistance has been in the signing of the CIS Collective Security Treaty, which six countries had still refused to do as of September 1995. These countries still balk at defining their national borders as the "external borders" of the CIS.

Chapter 4 provides articles, speeches, and interviews documenting the attitudes and policy reactions of the non-Russian CIS states (except Ukraine) on major integration issues being debated in the CIS. The documents demonstrate how Russia employs a bilateral approach to its advantage in pulling other CIS countries into its orbit, using CIS debates as a forum in which to consolidate its bilateral victories. The documents collected here also give an important sense of the evolution of these debates, in some instances in the words of the leaders themselves, and in others as perceived by a generally free press, CIS-wide. The real debates within the CIS, of course, are occurring behind the scenes at CIS summits and presummit meetings of ministers and other CIS officials. Nevertheless, an attentive press has been able to present a comprehensive picture of these leaders' activism and the results of many of their internal squabbles and discussions.

# Central Asia

The first article in the Central Asian section provides an overview of Central Asian countries' policies toward their Slavic populations.

The documents for Kazakhstan focus on President Nursultan Nazarbaev's support for the reintegration of the former Soviet republics into some kind of confederation. As shown in these documents, Nazarbaev's personal relations with the Yeltsin government have been tense. Nazarbaev considers Yeltsin a weak and ineffective CIS leader, as implied in the 18 February 1994 *Kuranty* article. One of the most contentious early issues between Almaty and Moscow was Russia's recall of all pre-1993 ruble notes, which almost forced Kazakhstan out of the ruble zone. Nazarbaev responded by saying that Russia "does not know what it wants." Another confrontation occurred over Nazarbaev's firm demands for equal treatment and consideration from Russia in the dismantling of its nuclear weapons. In general, Nazarbaev has responded to the abrupt twists and turns in Russian policies with great caution. Kazakhstan's population is almost 50 percent ethnic Russians and Nazarbaev advocates a strong economic, military and political partnership between his country and Russia. Unlike the other former republican leaders who have embraced independence, Nazarbaev has consistently opposed the Beloveh Forest Agreement of 1991 and has supported former President Gorbachev's concept of a more decentralized union. Nevertheless, he has at times taken a moderate position, defending the rights of other CIS members to control their own resources. A noteworthy article by the director of Nazarbaev's Strategic Research Institute, criticizing Russia's energy politics in the Transcaucasus and Central Asia, appears in the Kazakhstan section of this chapter. Because it is written by one of Nazarbaev's strategic advisors, this article carries the president's tacit approval.

President Askar Akaev of Kyrgyzstan has supported deep integration within the CIS and bilateral integration with Russia. Akaev was one of the first leaders to sign the CIS Collective Security agreement. He has fully integrated the Kyrgyz and Russian armed forces, and Russian border guards are positioned in Kyrgyzstan. Akaev, like Nazarbaev, champions the concept of a confederal Eurasian Union, which would include the establishment of a Soviet-type division of economic labor. In March 1995, however, Akaev suggested that Russia was reluctant to enter into extensive economic relations with Kyrgyzstan, presumably fearing that it could end up bearing the burden of Kyrgyzstan's economic recovery. Russia's reluctance to reciprocate for compliance with its wishes by upholding what other leaders perceive as its integrationist responsibilities and "promises" could put Russia in a badly compromised position down the road. Tajikistan's reactions to Russia have been determined largely by the war raging on its territory between political factions fighting for control of Tajikistan's government and future foreign policy positions, particularly toward Uzbekistan. Sergey Kurginyan's revealing article "Islamic Line" explains the thinking on the "Islamic threat" that some Russian political circles are espousing to explain Russia's policy along its southern perimeter.[1] In another article, President Emomali Rakhmonov expresses his hopes that Russia will guarantee Tajikistan's economic recovery after the war. In still another piece the Tajik press documents Russia's backing away from a "special" ruble zone arrangement with Tajikistan, which it had agreed to some months before, illustrating the contradictory role Russia is sometimes assuming in these countries' internal affairs.

Turkmenistan and Uzbekistan are in stronger positions to resist Russian integrationist pressures because of one's enormous oil and gas resources and the other's size and sophisticated military capabilities. As the speeches by Presidents Saparmurad Niyazov and Islam Karimov make clear, both countries have pursued policies of attracting as much Western investment as possible, as a means of both developing their economies and escaping total dependence on Russia. Nevertheless, Niyazov has shifted from delivering strong and outspoken speeches in 1992 and 1993 that defend Turkmenistan's neutrality and economic independence, to statements stressing the need for economic integration. Likewise, President Karimov's reactions have evolved from initial rejection of the CIS to joining the CIS and endorsing the need for accelerated economic integration, albeit without supranational institutions. The documents selected on these countries highlight this evolution in leadership views and pinpoint the stance these two leaders take on the issue of dual citizenship for Russian speakers, which continues to be an irritant in Russia's relations with every Central Asian state, except Tajikistan.

### Note

1. For a thorough analysis of the complex political factors involved in Tajikistan's civil war, see Sergey Gretsky, "Civil War in Tajikistan and its International Repercussions," *Critique*, Spring 1995.

## Kazakhstan

### 4.1
### Nazarbaev Favorably Views Talks with Russia on the Ruble Zone

Evgeniy Dotsuk
*Komsomolskaya Pravda*, 13 August 1993
[FBIS Translation]

Nursultan Nazarbaev considers his latest trip to Moscow to have been successful. Although the fact that Kazakhstan and Uzbekistan will remain in the ruble zone has not yet been stipulated in detail (the government has two weeks to elaborate the mechanism of its functioning), an accord in principle on this does exist. This is, at least, a real step toward the creation of an economic union, in which the Kazakhstani president continues to believe.

Economic uncertainty and the tension in Russian–Kazakhstani relations in recent months caused him once again to undertake a number of energetic actions—a meeting with Karimov, a joint attempt once again to gather together the CIS heads of state, and, finally, the meeting in Moscow. Against a background of disappointment, the Moscow agreement on the preservation of the ruble zone nonetheless looks like a victory.

Kazakhstan felt the obvious danger of being forced out of

the ruble area in June, when the Russian–Kazakhstani governmental talks failed, and Vice Premier Fedorov declared that "for Russia, Kazakhstan is a country just like any other foreign state outside the former Union." This was probably the first time the Russian side had spoken so frankly. It has to be admitted that this came as a surprise both to the Kazakhstani government and its president, since the talks seemed to everyone to be making progress, albeit slowly.

Kazakhstan was clearly faced for the first time with the need to survive on its own. It was then that rumors of the introduction of its own currency began to circulate.

As regards the technical aspect of the matter, according to certain data, the Kazakhstani currency is already almost ready. It could be introduced as a last resort, but this, in the opinion of a number of experts, would still further complicate local problems. Like Ukraine and Kyrgyzstan, Kazakhstan has nothing with which to back up its own money: It does not have in reserve sufficient gold, hard currency, or commodities. After weighing all the "pros" and "cons" of the unpleasant June situation, it was decided to continue the attempts to reach agreement with Russia.

This summer Nazarbaev has repeatedly experienced unpleasant moments in connection with the abrupt turns made by Russian politicians.

As a high-ranking official in his entourage joked, for example, just the statement of the "Slav three" was like the hole of a doughnut sent to Nazarbaev with Shokhin's visiting card—for the statement employed almost all the Kazakhstani side's integration proposals that had been rejected by Ukraine, above all, for approximately eighteen months. Nazarbaev was also stung by the business of the Russian circles' interpretation of the Istanbul meeting—the desire to portray it as an attempt to create a new Asian economic union. This was why the entourage around the president reacted far from unequivocally to the Moscow visit and its results. Some officials close to Nazarbaev were inclined to believe that Russia had shown its disregard for Kazakhstan and that further integration steps were not only humiliating but also useless. It was repeatedly suggested that Nazarbaev think seriously about the prospects for further relations with Russia and take a careful look at Asia. But the sober-minded section of the top power structures in Kazakhstan insists on a more fundamental logic of relations between the two states not dependent on the political sentiments of two to five Russian leaders.

Incidentally, in the situation that had taken shape Nursultan Nazarbaev was most likely relying very much on his good personal relationship with Yeltsin and Chernomyrdin, who, according to him, consider Kazakhstan "a most close and friendly state to Russia and are ready to help it in every possible way." But problems in relations between the two states, as seen from Kazakhstan, start at the level of the deputy prime ministers and ministers.

Also, according to certain data, this time serious tension had again been on the point of arising between the two governments' representatives in Moscow, and the latest dialogue might once again have gotten nowhere. But intervention by the two states' first persons led the talks out of deadlock. As a result, according to unverified information, Kazakhstan has even received some assurances that there will be no unpleasant surprises from the Russian side in the immediate future. Gerashchenko, chairman of the Central Bank of Russia, arrived in Almaty yesterday and met with the president in connection with the recent Moscow accords on employment of the ruble.

Incidentally, for Kazakhstan the ruble agreement will most likely mark the start of a series of complex new compromises with the Russian government. The terms on which Russia will supply its rubles to Kazakhstan are not yet entirely clear. But one thing is for sure: These terms will be quite tough on the republic. Some people at home will again accuse Nazarbaev of waiving state sovereignty by retaining rubles in Kazakhstan. But Nazarbaev has long relied only on his pragmatism. The political dividends, it turns out, are far cheaper.

## 4.2

### Nazarbaev: New Currency Is Not a Break with Russia

Amangeldiy Akhmetalimov and Gennadiy Kulagin
ITAR-TASS, 19 November 1993
[FBIS Translation]

The introduction of the national currency in Kazakhstan "does not mean Kazakhstan's departure from integration in the framework of the CIS economic council and the breakup of its ties with Russia," said the Kazakh president, Nursultan Nazarbaev, at a meeting with leaders of Kazakh northern regions as he was completing his visit to these regions on Friday.

Nazarbaev called on Russia and Kazakhstan, "as eternal neighbors, to live in friendship, unity, and confidence as before."

Speaking on the monetary reform, the Kazakh president declared that "it is necessary to ensure stability of the Kazakh national currency and its convertibility. Therefore, speculation and other shadow transactions with it should be prevented by all means," Nazarbaev said, adding that this was especially important as regards the Russian ruble.

"By organizing the exchange and sale of the Russian currency through currency exchanges we shall create a normal liquidity market for its purchase and sales to carry out organized imports of goods from the Russian Federation that are necessary for Kazakhstan," Nazarbaev said.

## 4.3

**Nazarbaev on CIS Ties, Trade, and Disarmament**
Interview by unidentified Interfax reporter; from the "Presidential Bulletin" feature compiled by Andrey Pershin, Andrey Petrovskiy, and Vladimir Shishlin; and edited by Boris Grishchenko
Interfax, 26 November 1993 [FBIS Translation]

[Interfax (IF)]: Many believe that your statements and recent CIS summits are an indication of a growing split between CIS states on all fronts. To what extent do you hold with such views?

[Nazarbaev]: The ideology of the Belovezh agreement was divisive rather than creative. But I do not yield to pessimism and am still convinced in the need to work toward a CIS economic union now handled by experts.

During last year's summit in Bishkek, I put forward proposals for monetary and customs unions, a CIS bank and a coordinating committee made up of leading CIS states.

In a recent article in the *Izvestiya* newspaper, I wrote that now was the time to examine the experience gained in the creation of the Maastricht union. I also spoke of a supranational currency, calling it by the widely recognized name of altyn. After all, they have ECUs in Europe, don't they? As if by chance, the ruble zone collapsed on the same day, 2 November, as the Maastricht treaty came into effect.

I am very sorry. We have sacrificed our strategic prospects for one-time considerations. To repeat, Kazakhstan and Russia must not be allowed to separate from each other. All talk about the ineffectiveness of our business alliance is nothing but politicking and paraeconomic arguments by the have-beens of the state planning committee who are not in possession of real figures.

Our economic union can evolve along the same lines as the European Community—we must examine carefully the ways in which the European Parliament, Eurogovernment, and commissions operate. They called theirs a European Union. But didn't we fight tooth and nail against the word "union" here? If we cast all ambitions and pipe dreams aside, we will be able to create a common market of former Soviet republics.

The advantages of such a union include common borders, vital supply routes, and great demand for one another's exports. Where on earth can Russia sell its combine harvesters now? Ukraine relied heavily on the prospect of exports to Europe, but there is a glut in the European market just when we are badly in need of Ukrainian products whose shortage led to widespread stoppages in Kazakhstan. We are also in need of imports from Russia, which has now put up trade barriers and forced us out of the ruble zone. Kazakhstan is rich in minerals, which are in great demand on the world market. But our problem is in being a landlocked country, and the problem of transport will always remain.

[IF]: In other words, you are trying to examine potential buyers of Kazakhstan's raw materials far beyond the borders of the former Soviet Union? How many foreign companies are seeking cooperation with your country?

[Nazarbaev]: Quite a lot. Avalanche is the word to describe it best. None of the CIS countries can boast striking such deals as Kazakhstan did. The huge project with Chevron (Tengizchevroil) aims to produce a total of one billion tons of crude oil. Even the first stage of the project spread over twenty-five years will help earn $170 billion in revenues, with 80 percent of this being Kazakhstan's share.

The list of the world's leading companies operating in Kazakhstan include Shell, Mobil Oil, France's Total, British Petroleum and British Gas, and Agip. Chase Manhattan opened its branch here some time ago, and I will meet the Citibank president soon after this interview. Citibank also wants to open a branch here. The Kazakh–Turkish and Arab–Kazakh banks have opened up in Kazakhstan, and a total of 1,500 joint ventures are in operation locally. The world's biggest companies are all represented here.

South Korea's Samsung has been entrenched in this country for some time, and Kazakhstan is self-reliant in the best quality refrigerators now. Other Korean companies have installed assembly plants for color TV sets.

There are enormous opportunities. German and Japanese companies are trying to break into the local market with the best intentions.

Following the introduction of our own currency, we have received substantial support from the IMF, the World Bank, the Asian Development Bank, the European Bank for Reconstruction and Development, and the Bank of Japan totaling $1 billion. This is to help prop up the som [Uzbekistan's currency] in the next two years.

My policy is not to go cap in hand. I always say: come and work here but to mutual advantage. Australian and U.S. companies are involved in gold-mining projects here, and Kazakhstan's annual gold production will reach 50 to 60 tons by 1995–96, which is enough to replenish our gold reserves. We have enough gold. That is why we introduced our currency unit, the tenge. We are capable of producing an annual one thousand tons of silver, which we export to Russia.

[IF]: All of this means greater orientation toward business partners outside the Commonwealth?

[Nazarbaev]: Which way we turn depends entirely on Russia. Should we turn south, we would be met with outstretched arms.

Kazakhstan is ready for wide-ranging cooperation with Russia. But if the vital supply routes including oil and gas pipelines are closed, we will have to turn southward. The southern countries are ready to accept and finance us, but this will deepen the rift with Russia. There is no other option but to try to turn toward the gulf. Central Asian republics are now working on an extensive program to lay down oil and gas pipelines and railway routes toward the Indian Ocean.

However, we want to work together with Russia, including joint ventures in oil and gas production. Why should we try to attract companies from France or other countries instead of Russian ones? We announce a tender and pick the highest bidder. Russia can easily become this. For its part, Kazakhstan is ready for sweeping cooperation and rapprochement of our economies on all parameters. But here again, everything depends on Russian politicians and economists.

Hard as its own plight may be, Russia as a big state should move closer to us rather than fence itself off from other republics. Moreover, we in Kazakhstan and I myself are doing our utmost to strengthen rather than break off the existing contacts.

[IF]: There are some in Russia who accuse Kazakhstan of dilly-dallying with the problem of nuclear disarmament. To what extent do you think such allegations are true?

[Nazarbaev]: Kazakhstan's policy on the issue remains unchanged. It has no intention of becoming a nuclear power, though it has been made to taste all "the delights of nuclear testing." The Lisbon protocol was signed by all states deploying nuclear weapons. Kazakhstan had signed the START I agreement and was the first to ratify it. However, the Lisbon protocol provides for unified strategic forces, which were unilaterally stripped down by the Russian defense ministry in contravention of the protocol. Ukraine can easily argue it was not in breach of the protocol first. I don't really know why Russia behaved that way.

Second, Kazakhstan wants its own share of payment for the enriched uranium that the Americans have pledged to buy.

And finally, we would like to receive assurances from the world community that it will contribute to the elimination of the aftereffects of forty years of nuclear testing in Kazakhstan and provide whatever research or technical assistance is needed.

[IF]: Will you be seeking a treaty on comprehensive military cooperation with Russia?

[Nazarbaev]: Last March, Yeltsin and I signed a paper instructing our defense ministers to prepare such an agreement to be signed at a heads of state level. But this is not ready yet.

[IF]: Has your perpetual optimism ever been frustrated?

[Nazarbaev]: Much too often. And yet, I remain an optimist. I think common sense will prevail in the end. Some of the statements by Russian politicians give me the creeps. I realize Russia is going through an economic and political crisis. I am well aware of how hard this can be. Russia and Kazakhstan are destined to be neighbors, and everything must be done to preserve that.

Our ancestors used to live side by side, and so must those who come after us. We must try to remain good neighbors and sincere friends. In such cases, you must wish your neighbor something you might wish for yourself. So my wish for Russia is: May its democratic forces who aspire for national unity and prosperity win the December election. Kazakhstan will always be close to such a Russia.

## 4.4
## Almaty Views Russian LDP Statement as Provocation

Interfax, 27 November 1993 [FBIS Translation]

Kazakhstan's foreign ministry has described as a provocation a recent statement on Russian television by leader of the Liberal Democratic Party [LDP] Vladimir Zhirinovskiy's attacks on sovereign Kazakhstan and its president.

The ministry could not afford to turn a blind eye to the speech by a candidate taking part in Russia's election marathon, Kazakhstan's deputy foreign minister, Konstantin Zhigalov, told Interfax. He said Zhirinovskiy's statements contained "insulting attacks on President Nursultan Nazarbaev and opened territorial claims to Kazakhstan" and were aimed at fueling ethnic strife both in Russia and Kazakhstan.

The Kazakh official said that no Russian government or foreign ministry officials condemned Mr. Zhirinovskiy's anti-Kazakh statements or his brazen insinuations against President Nazarbaev.

Zhigalov said that a large number of candidates involved in Russia's election campaign were trying to use the status of Russian-speaking communities in Kazakhstan and other countries as a trump card. "Applied to Kazakhstan, this is totally unacceptable," he stressed. The official said Nazarbaev was elected as Kazakh president two years ago after winning 98.6 percent support on the ground including from the local Russian-speaking community.

On the prospect of bilateral relations after Russia had squeezed Almaty out of the ruble zone, Zhigalov said that "no subjective factors will be able to shake our economic or political ties in the future." Kazakhstan depends on Russia for 66 percent of its imports and the same share of its exports. "Our contacts are by far the closest of the rest of the former Soviet Union," he said.

## 4.5

### Nazarbaev on Baykonur and Military Ties with Russia

From the "Presidential Bulletin" column
Interfax, 1 February 1994 [FBIS Translation]

President Nursultan Nazarbaev told Interfax that a military cooperation treaty with Russia would be signed shortly. He had sent a draft of the treaty to Russia's Boris Yeltsin "last March," following which the two governments were going to finalize the document.

The document "tackles the issues of military cooperation and use of testing grounds in Kazakhstan and the Baykonur space launch pad." Nazarbaev explained that the treaty also provided for training of Kazakh officers in Russia, cooperation in manufacture of weapons, and coordination in drafting a common military doctrine.

"If we are to create a common defense space, which is what Kazakhstan wants, we have to cooperate very closely in border protection, personnel training, military doctrine and other areas," the president said.

In his view, the Russian army would have to spend huge sums of money to build new testing grounds if it ceased to use those in Kazakhstan. In the past over 32.5 million hectares were used for testing grounds, but making such large areas available for military needs is out of the question now. Kazakhstan is prepared to lease the installations there on mutually beneficial conditions.

The Kazakh president dismissed the reports that Almaty had demanded that Russia pay $7 billion yearly rent for the Baykonur launch pad. He emphasized, however, that Russia was not the sole country that kept Baykonur in working order. Kazakhstan supplied 1 billion kW hours of energy yearly to Baykonur, which also used underground water, which "is in short supply in the Syr-Daria River." New buildings in Baykonur were constructed of Kazakh materials. "For this reason, the expenses of renting Baykonur for joint operation by Ukrainian, Russian, and Kazakh rocketry are being computed," Nazarbaev said.

(Interfax note: Russian and Kazakh negotiating teams in the talks that started in Moscow Monday are discussing the text of a new agreement on the status and operation of the launch pad and the town of Leninsk. According to a spokesman for the Russian Space Agency, the treaty on the lease of Baykonur and the issue of which legislation would be in existence in the areas rented from Kazakhstan will be discussed shortly.)

The president thinks that discussion of the status of Russian troops in Kazakhstan would be premature at this stage. He said that he had an understanding with Russian Prime Minister Viktor Chernomyrdin—with whom he had met in Davos, Switzerland, during the World Economic Forum—on having experts meet in early February in order to hammer out an agreement on the issues involved.

Nazarbaev rejected accusations of conservative thinking leveled at Chernomyrdin and his new government by the Russian press and recalled that Chernomyrdin's activities during the Moscow events of 3 and 4 October, when he was unambiguously on President Yeltsin's side, had been welcomed by the mass media. Now it is wrong to describe him as a hard-liner even before the government has announced its program, the Kazakh president said.

Currently any Russian government "will have to stay on the course of reform," Nazarbaev thinks.

## 4.6

### Nazarbaev Rejects Dual Kazakh–Russian Citizenship

Interview by Manfred Quiring
*Berliner Zeitung* (Berlin), 11 February 1994
[FBIS Translation]

[Quiring]: You reject dual citizenship. Why?

[Nazarbaev]: I am afraid that this infamous dual citizenship would split our society. With a recently issued ukase I have fundamentally changed some unfortunate passages of the Kazakh Law on Citizenship. I passed on to Boris Yeltsin and Viktor Chernomyrdin the draft treaty on principles associated with the acquisition of citizenship in Kazakhstan and in Russia. We should adopt laws that facilitate this.

[Quiring]: On the eve of elections in Russia statements by Russian Foreign Minister Kozyrev that Russia wants to protect the interests of its fellow citizens in the former Union republics caused excitement. How do you see this today, after Zhirinovskiy's election success?

[Nazarbaev]: The key phrase is "on the eve of elections." An election campaign has its own laws. However, I am also quite sure that such statements have not at all reassured the Russians who live in national republics—in particular not those who never had these problems.

[Quiring]: In Kazakhstan almost 50 percent of the population are Russian. What is your point of view about this fact?

[Nazarbaev]: I am completely calm when I think of the fact that in Kazakhstan there are eight million Kazakhs and five million Russians. Some people in the media want to see a problem where there is none. Often there is talk of migration. However, this process is unavoidable under normal circumstances, and all the more so when an empire collapses.

Kazakhstan, like Russia, is a multinational state. For many centuries Kazakhs here have lived together not only with Russians but also with Ukrainians, Belarusians, Germans, Poles, Uigurs, Dunganes, and—since the fateful time of the Stalinist deportations—also with Koreans, Chechens, Ingushes, and Turko-Mezhets. I am convinced that there will be no ethnic conflict in Kazakhstan.

[Quiring]: But emigration continues to increase.

[Nazarbaev]: I am seriously worried about that. Many Germans, for instance, are leaving us. As regards the Russians, I can give you the following figures: Last year more than 200,000 left the country, not only for Russia. By 1990, 80,000 to 100,000 people left our republic every year. An approximately equal number came to us. However, when listing the current migration figures, one forgets—for whatever reason—to note that last year about 160,000 Russians came to Kazakhstan.

[Quiring]: Among the political personalities who brought about the end of the USSR, you are considered one of the most far-sighted. Most recently, you have distanced yourself from Moscow. How do you see the development of relations with Russia?

[Nazarbaev]: Thank you for the compliment, but I was not invited to "seal the end." In fact, I have always advocated integration in all spheres of our life. Now, also, I am of the view that only together can we find a way out of the crisis.

I cannot agree with the second part of your question. We are not distancing ourselves from Moscow. On the contrary, we are striving for closer contacts within the framework of the CIS. For this purpose, however, we must bring our laws closer together; we must form a common parliament and a common government, which one could base on the Consultative Coordination Committee of the CIS. We must immediately establish customs, banking, payment, and other unions, as they are envisaged in our treaty on the Economic Union.

[Quiring]: For a long time you were one of the most active advocates of the CIS. After the collapse of the ruble zone, do you still believe that the CIS has a future?

[Nazarbaev]: The introduction of an independent currency—as I have often said—was a forced step. Right to the end we hoped to find a joint solution with Russia for the ruble problem. However, the conditions proved unacceptable. However, nothing bad happened for anybody. The difficulties, which are linked with the introduction of national currencies, are unavoidable in view of the temporary delineation. However, it cannot be ruled out that we can achieve a commonly circulated currency in the future, perhaps even to the creation of a common currency, following the model of the European Union.

[Quiring]: In contrast to Russia, the parliament and the local soviets in Kazakhstan have dissolved peacefully. Still, is this a rejection of parliamentary democracy?

[Nazarbaev]: Not at all! Now more than ever the republic needs a professional, constantly working parliament, which is able to adopt laws of a high quality and thus to establish a reliable basis for the reforms, which we are implementing. The special powers given to the president are valid only until a new parliament is convened and apply only to issues linked with the economic reform. The elections for the new parliament will be held on 7 March 1994.

## 4.7
## Nazarbaev Style and Handling of "Russian Question" Examined

Mikhail Shchipanov
*Kuranty,* 18 February 1994 [FBIS Translation]

There is every reason to suppose that our extremely fluid public opinion will clash in the near future with the failure of still another perestroika legend—the legend about the wise reformer and spontaneous democrat, the great friend of free Russia, Nursultan Nazarbaev—nicknamed "the red emir." It is always extremely unpleasant to part with legends, especially against our background of massive disappointments in simple, but radical decisions, the aspiration to the panacea of a firm label. But it is simply necessary to understand from where the legend about Nazarbaev came.

However, to start with it is nevertheless necessary to understand to what extent our prophets, who in spite of the well-known parable have multiplied prolifically in our homeland, are inconsistent. To illustrate, we will take two political figures, who at various times have caused a considerable stir in Russian minds—General Pinochet and the lawyer Zhirinovskiy. Both of them have been treated as fascists without any special reasons, since neither the one nor the other in a strict sense meets the classic national-socialist definitions. But the essence is not to be found in this disappointing misunderstanding.

The embarrassment lies in the fact that the same eloquent prophets, artists at heart, whom from the moment of safe glasnost they invited to the Russian throne of their own Pinochet, called upon to cut with a dexterous stroke, in a fully Chilean manner, all the knots tied by the communists, have fallen into despair, abusing their own people as "not yet mature for democracy," having seen in parliament Zhirinovskiy with his falcons. It would seem that they conjured up a fascist à la Augusto Pinochet—and a fascist, according to their definition. They received instead of gladness, vulgar language. Or are no two fascists alike? There are fascists who are ours and who are not ours.

The problem, of course, lies elsewhere. In the example of V.V.Z. [Zhirinovskiy] it has become quite evident that many of our creators of public political theories and images do not think with arguments and concepts, but with models, so to speak; they put feelings higher than logic and concrete knowledge. Strictly speaking, there would not be anything surprising in this since Bohemians have started to come out into political boundaries in the garbs of prophets, and recent theater critics—have made their debut in the capacity of newspaper political scientists. . . . A whole direction of thought has arisen that can be fully defined as political Bohemianism. The legend of N. Nazarbaev can be explained precisely by the views of this stratum.

In the case with Nazarbaev, the image of the progressive ruler, established during the epoch of the ecstasy of perestroika, completely concealed the real state of affairs in Kazakhstan for many Russian lovers of political literature. To this day, many in Moscow are prepared to close their eyes to the openly authoritarian order of the republic with the blue flag and the absence there of—in the Gaydar sense—real economic reforms, to the national socialism that is gathering force there, and the edge aimed against the Russian community. And Nazarbaev is forgiven a great deal because he is . . . Nazarbaev.

Naturally, Nazarbaev gained a great deal from the fact that, at the moment of the collapse of the Union, he was already a well-known and recognizable politician. He had enough intellect if only to follow in the wake of Gorbachev, but to keep his distance, giving rise to rumors of an independence unknown in the Union. That mask of the independent "perestroika-politician" helped Nazarbaev in the beginning to avoid the Gorbachev trap of the vice presidency, and then not to go to the bottom with the retinue of the last general secretary.

The Belovezh agreements became the first strike at the prestige of the "red emir." On the one hand, he was shown his place, not even having been informed about the plans of the Slavs. But, on the other, it was incredibly comfortable for Nazarbaev behind the cracked walls of the USSR. Nursultan Abishevich, earlier than the other colleagues from Central Asia, understood that an Eastern ruler on his own accord is of little interest without the magical power of the Union looming behind him. And the state of the Union's semi-disintegration made it possible for Nazarbaev once more to "shine," coming out with his own initiatives for the reconciliation of the Center and the republics. Taking into consideration the fact that even President Bush at that moment was an opponent of the liquidation of the empire, the mediator Nazarbaev collected points in the larger world game. He was recognized and well remembered. And the longer the period of semi-disintegration lasted, the more the residues of the might and authority of the Union would be "privatized" by the Kazakh, who had understood the methods of Shevardnadze very well. But Shevardnadze at this moment had virtually departed into non-existence and all the laurels went to Nazarbaev. For this reason, the rapid disintegration of the Union became for N.A.N. [Nazarbaev] the strongest personal defeat. In the economic sense, he already did not add anything to it; in the political sense, the losses were obvious.

Vague and temporizing was Nazarbaev's position during the days of August 1991, but he left the Union with the international reputation of a politician of wise, balanced, and open reforms. The latter circumstance, strictly speaking, in no way assumed concrete form. But the image is really more precious than money. Somehow in the bustle it was forgotten that the liberation of prices, already widely announced, was held back precisely at the insistence of Nazarbaev, and to this day it is unclear what the cost of the ultimatum of the "reformer" of Russia was.

Later the insurance policy of its own natural resources benefited Kazakhstan, in contrast to some of the other post-Soviet republics, the burden of the transition period, and the interest of Western investors in the power of Nazarbaev was hastily proclaimed as the best proof of the effectiveness of the reforms begun under his wise leadership. About the policy that so actively fought for the formation of the CIS nothing bad could be written at all. Either good, or not at all.

Meanwhile Nazarbaev understood one indisputable truth: What concerns Western investors is not his passion for reform or even state democracy (naturally democratic proprieties must be observed), but only political stability.

And as a result, the simple end of stability began to justify any means: the suppression of any free thinking, suppression of the press, the pleasing of Kazakhstan's own nationalists, the forcible change of the demographic situation, and bans even on national-cultural autonomy. Nursultan Abishevich drew conclusions from yesterday's masochism of his Russian colleagues, who for the past two years, as it were, specially tested their own fate, discussing abroad the threats of the red-browns, apparently unaware that in so doing they merely decrease the chances for capital investments from the outside. Nazarbaev, on the contrary, quite in the Soviet manner, likes to talk about the unity of the people of Kazakhstan.

But precisely here the "Russian question"—the fate of the Kazakhstan Slavs, who constitute nearly 50 percent of the republic's population—presents a special danger for the building of the Nazarbaev prosperity. The slightest allusion to the coming aggravation of international relations in the "red emirate," threatening unprecedented shocks—and the ephemeral reformist curtain, put up with such oriental art by Nazarbaev, simply comes tumbling down. The projects of the century will remain unrealized, but the king (more correctly—"emir") risks turning out to be naked. A laughing-stock before everyone.

The realization of such an indisputable fact during the past several months is clearly obscuring Nazarbaev's reason. Meanwhile the crisis in his relations with the Kremlin is slowly ripening. Strictly speaking, Nazarbaev received the first strike from Russia at the moment of the breakdown of the de facto ruble space. Not having received the requested supplies of Russian rubles of the new pattern, Nazarbaev, the "successful reformer," was deprived of one of the bases of his ostentatious economic prosperity—permanent Russian subsidies. The time of prosperity was over.

Since that time, Nazarbaev and Yeltsin have practically not had a single personal meeting, even during the CIS summit meetings. Recently in Ashkhabad Nazarbaev found time to meet and talk tete-à-tete even with Leonid Makarovich [Kravchuk], but by no means with Boris Nikolaevich. Nazarbaev with all his might is demonstrating his dissatisfaction with the policy of Moscow, which has revealed, with some delay, the millions of our own compatriots in the outlying districts of the former empire. A new turn for Nazarbaev in the demonstration of character is the problem of dual citizenship for Russians, who by the will of fate have unexpectedly turned out to be emigrants.

And here Nazarbaev displayed, for a refined and variegated oriental ruler, an unprecedented inflexibility. His unbending "No" clearly troubled Andrey Kozyrev, who with all his might tried to observe diplomatic etiquette. Moreover, Nazarbaev not only does not allow his Slavic subjects to adopt a second Russian citizenship, but he also is among those who are now exerting pressure on A. Akaev, who has already given his assent to the Russian proposal. Why is Nazarbaev so confused by the specter of a second passport among his fellow citizens who are residents of Kazakhstan? Is it not because the queues in front of the Russian embassy better than anything else tell about the real achievements of his regime?

The introduction of dual citizenship clearly does not suit Nazarbaev for two reasons. First of all, the very logic of authoritarianism does not leave the subjects (we are not talking about citizens) room for maneuver. It is no secret that not only Russians are aspiring to Russian citizenship everywhere, but also many "indigenes"—titular nationalities. The classic example is Armenia. Second, dual citizenship, under any propagandistic rubric, is the acknowledgment of national trouble smoldering behind the smiles of Nursultan Abishevich. Indirect support for the "Russian question," an act whose existence can undermine the "Nazarbaev system" the "red emir," who counted on the attractive, painted curtain of complete political and ethnic stability, cannot allow under any circumstances. The comedy of masks must be continued as long as possible. For the time being, large dollar flows are not coming into Kazakhstan, although those in the entourage of the "emir" are evidently counting on this. The West, having become the hostage of its own money, would be forced to support any eccentricities of the ruler.

However, it is not enough for Nazarbaev simply to refuse the Russian proposal for the introduction of the institution of dual citizenship. It is necessary for him to receive directly from the Kremlin the assurance that no "Russian question" of any kind exists in the Kazakhstan subordinated to his will. There cannot be because there can never be one! And this is why, as people in the know assert, Nazarbaev demonstratively procrastinates about the long since announced personal meeting with his colleague Yeltsin, trying to attain, they say, from the Russian president public confirmation of the "international purity" of the policy of Nursultan Abishevich. To cave in to such a concession to Nazarbaev would be a strategic error for Yeltsin. Especially in the present conditions, when even for the defense of the faraway Serbs, the parliamentarians are ready to become a friendly and inviolable wall.

And for this reason, one can expect from Nazarbaev in the near future unexpected unfriendly improvisations with respect to Russia. It is not out of the question that the "red emir" will blame some of his domestic difficulties on the intrigues of Moscow and simply Moscow inflexibility. Variations are possible. But one thing is obvious, many in Moscow will have to part with illusions regarding the wise and just oriental ruler.

P.S. Incidentally, already in America Nazarbaev cautioned against the extension of assistance basically to Russia, whose policy is becoming increasingly more rigid in relation to the so-called near abroad. N.A.N. [Nazarbaev] wisely counseled to support Moscow depending on its faithfulness to democracy. . . .

---

## 4.8

**Nazarbaev: "Nobody Knows What Russia Wants"**
Interview by Enrique Serbeto
Madrid ABC (Spain), 23 March 1994
[FBIS Translation]

---

The 8 March parliamentary elections were regarded by international observers as a departure from the high level of democratic guarantees. This is the first blot on the record of this leader with a reputation of being moderate and sensible, who defends himself by saying that "when you speak of democracy in Central Asia, in Kazakhstan, it cannot be compared with France, Spain, or other countries. If we are speaking of European standards, it must also be said how many years it took to achieve them. It must be borne in mind that women in the United States did not have the right to vote until the beginning of this century. But these states have two hundred years of history, and France, Spain, and Britain have thousands. We have only been independent for two years. We are heading for democracy and the market economy step by step. If we try to turn the democratic cart too quickly, we will risk overturning it."

[Serbeto]: The economic reform and privatization are proving very difficult in Russia. Has Kazakhstan used the lessons of this Russian experience in order to design its own model?

[Nazarbaev]: After many centuries of totalitarian rule, people are accustomed to collective methods of leadership and production, and so, from a psychological viewpoint, it is very difficult suddenly to accept private ownership. But, despite everything, we are prepared to hand over part of Kazakhstan's resources to the country's population. We are conducting not a chaotic privatization but one under the regulation of the government, which is accelerating the privatization process, since it is very difficult from the grass roots. In agriculture, we are not yet considering the total privatization of land, although this is our medium-term aim.

[Serbeto]: Are you in favor of the so-called shock therapy, or of delaying radical decisions to the utmost in order to preserve social peace?

[Nazarbaev]: We started with "shock therapy" when we decided to liberalize all prices, but we must be very vigilant because the process could go wrong at any time. The monetarist prescriptions against inflation are not sufficient in themselves. In the West, they must realize that they reached this situation after many years of evolution. For instance, in Spain Franco left everything ready, but we had no Franco.

[Serbeto]: What is Kazakhstan's nuclear policy?

[Nazarbaev]: Kazakhstan is fulfilling its pledges on the elimination of nuclear weapons. We were the first CIS state to ratify Start I and the Lisbon protocol. The only delays were due to the fact that we were trying to secure guarantees that this is our lawful property and that we will be compensated for the cost of the enriched uranium in our nuclear warheads.

[Serbeto]: The presence in this country of a large population group of Russian origin is viewed as a factor of instability. To what extent do you believe that this is a grave danger to coexistence?

[Nazarbaev]: It would be incorrect to say that this problem does not exist. Some Russian politicians wish to use this for their own ends, but we will not accept this in Kazakhstan, because the Russians of Kazakhstan are our citizens. It cannot be said that human rights are violated here, but if anybody believes this, we invite all UN delegations to check this. In my opinion, people's main concerns are a result of the disintegration of the USSR. The fall of a great empire is always painful, and so it has always been throughout history. I know that Yeltsin and Kozyrev are very experienced people, but these politicians should have thought about the Russians left outside Russia when they signed the Slav treaty dissolving the USSR.

[Serbeto]: Why do you not accept the proposal granting the Russians dual nationality?

[Nazarbaev]: Just imagine if the Russians, who are 52 percent of the population, had two passports in their pockets. They would have one foot in Russia and the other here. That would destabilize the situation in Kazakhstan. Why do they say nothing in Russia about the violation of the rights of the non-Russians living in Russia? A million Kazakhs live in Russia. Have they opened any Kazakh school in Russia? In Kazakhstan, 65 percent of schools operate in Russian, and last year 78 percent of the students starting high school did so in Russian. All the state bodies use Russian, and I myself am speaking to you in Russian. Can any other CIS state say that it respects linguistic minorities as we do? We will resolve all the problems of the Russian-speaking Kazakhs, because this is an internal problem of ours—not that of any other country. Nobody can concern themselves about the Russians better than we can, and we want no assistance from Russia.

[Serbeto]: Russia is trying to attract other CIS states into its military and economic sphere of influence. Have you noted this trend with respect to Kazakhstan?

[Nazarbaev]: Nobody knows what Russia wants. They lack a policy in this respect. It would be better to ask the Russians what they want. We are a sovereign country and are not dependent on Russia for either oil, coal, or food. Since we introduced our currency, we have been living completely independently, and Russia has understood this. We have oil, gas, wolfram, gold, diamonds, and uranium. Russia's relations with Ukraine and Belarus cannot be compared with those it maintains with Kazakhstan. . . .

## 4.9

**Russian–Kazakh Commission Discusses Baykonur Base**
Viktor Gritsenko
ITAR-TASS, 12 August 1994 [FBIS Translation]

A special Russian–Kazakh commission has been working these days at the Baykonur space station on practical implementation of the agreement on "principles and requirements for Baykonur space station utilization" and on the preparation of the space station for winter.

Representatives of eleven Russian ministries and departments had been placed on the commission by the Russian president and the prime minister.

The Baykonur agreement was signed on 28 March of this year and ratified by the two countries' parliaments in July. Now the task is to make the agreement work and to draw an accord on the lease of Baykonur to Russia. The Russian and Kazakh experts have finished alternative drafts of the latter accord. It is scheduled for consideration by official delegations of the two countries in September.

Now, as the dispute over the Baykonur's status is about to be settled, hardships of the space station are coming to the foreground. Last winter, many homes in Leninsk, the urban center of the space facility, suffered from irregular supplies of heat, fuel, water, and power. Locals say the town will be unable to stand another such winter, nor will the space station be served adequately by this town residents.

Another challenge is the irregular financing of Baykonur under Russia's budget scheme. In the future, space station maintenance will be performed on a shared basis by the Russian military space forces and the Russian space agency. The logic behind it is that the Russian space sector has in fact split into civil and military segments.

## 4.10

**Kazhegeldin on Integration with Russia**
Remarks by Akezhan Kazhegeldin recorded by Nail Ishmukhadmerov in Almaty
*Express-K,* 14 December 1994 [FBIS Translation]

In yesterday's issue *Ekspress-K* reported on the press conference held by the country's Prime Minister A. Kazhegeldin jointly with First Deputy Prime Minister N. Isingerin on the results of the conference of heads of CIS governments in Moscow. The subject under discussion was new approaches to integration processes between Commonwealth countries.

Kazakhstan had many issues that require the signing of agreements, including on Baykonur, problems of exploitation of military test sites in Kazakhstan, and the status of Russian servicemen in the republic. We also discussed questions of cooperation in the area of radio engineering, defense, and military technology. In addition, also on behalf of the heads of the two states—Russia and Kazakhstan—and as a follow-up on the March agreements between N. Nazarbaev and B. Yeltsin, the issue of simplified procedure for granting citizenship. Positions have been coordinated by experts; it is hoped that in the next few days the document will be initialed and that it will be signed this year.

In principle, the nineteen documents presented for the heads of governments' consideration (mainly dealing with the new superstructure organ—the IEC [Interstate Economic Committee] of the CIS Economic Council) were adopted without long discussions. "This makes me optimistic; this time I felt especially acutely that integration processes have indeed begun, because the states began to give up on full sovereignty and transfer some of their powers to such interstate organs as the IEC."

### What Does Integration Mean for the Former Union Republics?

"We are facing so much work that we would not have had to do had we not played the fool since 1985. What does integration mean for former USSR republics? It means extra work! It is work that could have been avoided at the time. Had the CPSU together with Mikhail Sergeevich not missed this chance, had there been no crazy escapades in Riga and Tbilisi, after which no minister could tell what exactly had happened. . . . Russia was the first one to ask for sovereignty. From whom? Belarus—that I can understand, from whom. Ukraine also seemed to have reasons to ask for it. From whom did Russia want to become sovereign? That is, the opportunity was lost.

"Essentially, we—our generation, especially economic managers and financiers—will have to fix the mess the politicians have made. Now we are embarking on another part of the job. We have to dig out from the mess inside the country, and on top of that we have this interstate integration work to do.

"We gave the IEC all the powers we could. Russia is very much leaning toward integration; Ukraine is changing right in front of our eyes—the state that all these three years had its own special opinion. But there are still questions: Belarus, Moldova. Turkmenia is still sticking to its special opinion."

### Two Main Principles of Integration: Capital and Property

Integration processes between Russia and Kazakhstan go much deeper. "What were we talking about? About creating between us transnational, industrial-financial, joint-stock, and other companies and legal entities that would agree to merge property. It is my conviction that merging capital and property will always push us farther in the direction we are going."

"We told the RF [Russian Federation] government that we are ready to give up blocks of stock in all radio electronics plants, and require in return only one thing: that the plants not be idle and that taxes be paid on time. Russia appears to be interested in this, and it does not make sense for it to fuss with building new plants.

"We were asked to speed up somewhat the creation of joint banks. Frankly, Kazakhstani banks so far have not been able to enter the Russian market. The tenge is only a year old; a currency must gain strength, and there should be enough of it to conduct intervention in the foreign market. At the same time, we are getting tired of receiving technical loans and credits that sink into the sand and do not return. Therefore we are very interested in strong Russian banks finding their place in our market and providing loans for sectors that need it in RF goods and services for Russian rubles.

"We in the Cabinet of Ministers adopted an appeal to the parliament to speed up the adoption of legislative acts, because if we do not synchronize our market steps and legislation with the RF, we will fall behind it, and subsequently we will face the eternal question—why? Why do they have higher incomes, a better life, and so on."

### Why Did Customs Barriers Arise?

Immediately after signing an official agreement on Baykonur, A. Kazhegeldin had a meeting with Russian Prime Minister Chernomyrdin and the new deputy prime minister, A. Bolshakov, who is responsible in the RF government for liaison with the CIS. "We discussed questions of a mutual strengthening of external borders and eliminating internal customs borders between Russia and Kazakhstan, adapting the legislative base and synchronizing the market steps of the governments so that there would be no need to separate ourselves with customs borders."

Why do customs barriers exist? After the republics acquired sovereignty and were recognized by the world community, they obviously did not have the ability to influence each other's exports and imports. So runaway illegal re-exports began all over the territory of the former Union; the states that suffered most in this were Russia and Kazakhstan. Therefore Russia was the first to start drawing borders and introducing the institution of customs control. Today there is a preliminary agreement. If on 23 December in Almaty the heads of state initial the documents on the foundations of customs legislation and customs statistics, on the basis of this, national laws will be adopted on customs and customs statistics, which in turn will allow us to produce a so-called customs union of CIS countries. Later, if each country that is party to the agreement undertakes protection of external borders of these countries, then the border between these states is removed. "That is, we have to undertake an obligation that we cannot export Russian goods anywhere without Russia having special rights."

### What Will Kazakhstan Get from an Agreement on Karachaganak?

"I am convinced that if capital flows here and we find mutual interests in merging property, the customs questions and others will be resolved automatically. Evidence of this is our agreement on Karachaganak. We deliver the entire output from this deposit for processing at Russian plants. We will receive for the domestic market exactly as much as we need. An agreement has been reached with the Russian Gazprom on construction of two gas pipelines to the east and southeast of Kazakhstan, that is, to the capital. . . . In about two months we will already have the necessary volume we have agreed on. It will be a total of 14 billion cubic meters of gas per 14 million tons of condensate. For the first stage, to the middle of 1995, we undertook an obligation to deliver 4 billion cubic meters and 10 million tons of condensate to the Russian market. The price is great for us! It is not the $55 we have in the south—it is only about $13. Jobs for both sides, although essentially more for us, since the deposit is on our side. The advantages are obvious. We are installing the final link in the gas network, creating a common market of gas consumption—and questions that have been under discussion for two months essentially have been removed.

"Naturally, the Karachaganak agreement will become a catalyst in the struggle between Russian oil and gas companies for Kazakhstan's market. And our problem will be how to make the right choice."

### Where Are We and with Whom?

"I thought that the latest CSCE meeting and B. Clinton's speech at it had a sobering effect on many CIS politicians. I saw a great desire to integrate among those who lately have not particularly supported our integration processes.

"All of us and each republic separately must decide where we stand. Life compels us to do this. Where are we? All of us probably will simply not be admitted into NATO. Because they already are not admitting us into the market. They do not want to give us a lot of money. First, they do not particularly want to part with the money. Second, they have enough of their own problems. Third, we did not spend it wisely. Fourth, generally there is no prospect that it will be repaid. So they want to spend precisely as much as is needed to keep things calm here. We have to understand this. And if we sit complacently in the expectation of this for another three years, we will quietly turn into a banana republic."

P.S. In the course of the press conference A. Kazhegeldin made a remark we liked very much: "There are no leashes—long or short. Only interests!" We gladly subscribe to it.

True, this is not the first time our rulers have talked about integration, while the cart remains where it was. There is, however, a difference today, and a very substantial one. New people are not in the government offices across the commonwealth expanse.

---

## 4.11
### Russian Foreign Policy in Caucasus and Central Asia Viewed

Umirserik Kasenov
*Nezavisimaya Gazeta,* 24 January 1995
[FBIS Translation]

---

Articles about the threat to Russia from the south, referring to the Transcaucasus and Central Asia, are fairly common in the Russian press, but no single state in these regions or all of them together can or will ever pose any kind of threat to Russia.

Then what kind of threat is this? Apparently, it is the danger that Moscow will lose its earlier total control over everything that happens in the former Soviet republics of the Transcaucasus and Central Asia and the danger that other states will become more active in the economy there.

In other words, it is the danger of the loss of spheres of influence, in the language of the "Cold War" era of confrontation, and of the need to view the new independent states in these regions not as autonomous participants in international relations, but as the targets of the acquisitive policies of their neighbors and more distant states.

When the Russian political establishment defines its foreign policy line in the Transcaucasus and Central Asia, it apparently proceeds from the following assumptions:

—the continued sovereignization of the Transcaucasian and Central Asian republics and their establishment and development of foreign policy and foreign economic ties with neighboring southern states will do much to weaken Russia's status as a great power;

—Russia is losing access to sources of raw materials and control over the richest natural resources (primarily energy resources) and gold, non-ferrous, and rare metals of strategic value;

—Russia is facing the real threat of the loss of its monopoly in transportation in connection with the completion of the Trans-Asian Railroad and the plans for oil and gas pipelines and highways to the south;

—while the CIS has been virtually idle, the participation of Azerbaijan and the Central Asian states in the Organization for Economic Cooperation (OEC) has expanded and has taken increasingly concrete forms;

—the vacuum in the Transcaucasus and Central Asia is being filled quickly by the Western states, by virtue of the strength of their oil companies and other firms, and by Turkey and Iran. The economic presence of states in other regions is also growing there.

From the vantage point of Russian politicians and analysts, Russia is losing patience with the constantly growing influence of other states in a territory that Russia views as a sphere of its own vital interests.

What has Russia done in the past, and what can it do now, on its southern flanks to prevent these undesirable geopolitical and geoeconomic changes?

Russia clearly disapproves of the construction of a pipeline from Azerbaijan to the oil terminals near the Turkish city of Ceyhan on the Mediterranean Sea, running through the territory of Iran and Turkey or Armenia and Turkey.

If this pipeline or some other means of transporting oil from Azerbaijan and Kazakhstan to other states, bypassing Russian territory, is built and begins operating, Russia will lose an important source of income and a means of exerting economic pressure on these oil-rich states.

In the opinion of Russian analysts, when Turkey imposed restrictions on the oil tankers crossing the Bosporus on 1 July 1994, it was concerned less about the ecological safety of Istanbul than about the possibility of urging the construction of a pipeline through its own territory. They feel that this was the explicit purpose for the strict limits on tankers crossing the Bosporus.

Thus, Russia is using all available leverage to urge Azerbaijan and Kazakhstan to transport their oil only through Novorossiysk. The work on plans for a pipeline through Greece and Bulgaria—i.e., bypassing the Bosporus—has already begun.

The construction of oil and gas lines stretching hundreds of kilometers will require colossal capital investments by Western investors, and these will be conditional upon their complete confidence in the stability of the states where the pipelines will be located.

The destabilization of the Transcaucasian and Central Asian states could block the flow of new foreign investments, particularly in the oil and gas industry and in the construction of pipelines stretching to the south.

The findings of an expert group from the Moscow State University Independent Institute of Socio-Historical Research and the Soros Fund are interesting in this context: "Through their oil companies, the Western countries are beginning to exert influence in the oil-bearing regions that recently 'opened their doors to the West.' They are penetrating Azerbaijan and Central Asia and setting up 'windows' to the Mediterranean there. Only the possibility of social, economic, or political instability in these regions could stop the process, because Western businessmen are unlikely to invest their money in a high-risk venture" (*Izvestiya,* 19 October 1994).

It is completely obvious that if the situation in the Transcaucasus were stable, there would be no insurmountable obstacles to the construction of a pipeline from Azerbaijan along the northern or southern slopes of the Caucasus, but the present Armenian–Azerbaijani conflict has precluded any major decisions of this kind.

The continuation or escalation of the Armenian–Azerbaijani conflict is objectively helping Russia prevent the construction of a pipeline from Azerbaijan to the Turkish port of Ceyhan on the Mediterranean, because one of the key sections of the future pipeline—the 30 kilometers of the plans to build alternatives to Russian oil and gas pipelines to the south—will have serious and unavoidable geoeconomic consequences.

It is a fact that Kazakhstan's search for alternative pipelines has been fostered by Russia's own actions, such as the limitation of shipments of Kazakhstan's oil to Novorossiysk and Tyumen oil to the Pavlodar Refinery. Turkmenistan's efforts to promote the construction of a gasline to Iran are due in part to the fact that the CIS countries pay only 60 percent of the world price for Turkmen gas or do not pay for it at all, as in the case of Ukraine and some other states.

Ignoring the objective nature of the processes of diversification in the foreign economic relations of the new independent states of Central Asia that were formed as a result of the breakup of the USSR, as well as the stimulation of this process by Russia's own actions, some Russian analysts are wasting their time propounding the myth of some kind of "Turkic belt" that is supposedly meant to separate Russia from the rich raw materials and energy resources of the Central Asian zone and from transportation arteries to the south.

In reality, the Central Asian states have no desire whatsoever for separation from Russia in the economic, military-strategic, or spiritual sphere. This is clear from their relations with Russia throughout the period since the breakup of the USSR. On the contrary, there is more indication that Russia is trying to distance itself from Central Asia in the economic sphere while it is simultaneously striving to preserve its military-strategic presence and external borders there.

Furthermore, this is being done at a time when, according to A. Nikolaev, commander-in-chief of the border troops of the Russian Federation, it costs more than a billion rubles to fortify a single kilometer of the border, and Russia's new borders—along the northern Caucasus and the borders with Kazakhstan, Ukraine, Belarus, and the Baltic states—is 13,500 kilometers long.

As long as Russia lacks fortified borders of its own and the money to pay for this, it should be supporting the economies of other CIS states as much as possible instead of pursuing a tougher economic and financial policy in relations with them. This is not such a high price for the reliable protection of the external borders of the CIS, which are also Russia's own borders at this time.

A harsh decision was recently made on the denial of new Russian credits even to Tajikistan, where the border with Afghanistan is regarded as Russian and where the leadership also complies unconditionally with all of Moscow's demands. The Ministry of CIS Affairs informed the Tajik delegation that "Russia cannot afford to keep the Tajik economy afloat with its own money, and Tajikistan might as well start circulating its national currency" (*Izvestiya,* 30 December 1994).

Clearly, two incompatible goals cannot be pursued simultaneously: Russia cannot withdraw from the Central Asian economy, calling it a "burden," while it secures its military-strategic interests in the same region. Another clearly unattainable goal is the restoration of the situation in which the economies of the Central Asian states were subordinate to Russia's interests, with no consideration for their own—in other words, the perpetuation or augmentation of their role as a "raw material appendage."

Russia, the Transcaucasus, and Central Asia, as well as the neighboring states of Turkey and Iran, are closely interrelated by common and conflicting economic interests.

There is only one way to avoid the exacerbation of the completely understandable conflicts of economic interest. A balance must be struck by securing a prevalence of common interests and reducing the number of contradictory interests. This policy has already been pursued with some success. Some examples are Azerbaijan's inclusion of the Russian Lukoil Company in the Caspian oil contract and Kazakhstan's Karachaganak partnership with Russia's Gazprom.

The Transcaucasian and Central Asian states must share their natural resources with Russia now and in the future, but only on reasonable and mutual beneficial terms, firmly defending their own legal economic interests and their right to diversify their foreign economic relations for the purpose of safeguarding their economic, energy, and ecological security.

The natural course of historical events will lead unavoidably to Russia's renunciation of its claims to a dominant position in the post-Soviet territory and to its "special prerogatives" to maintain order in this zone. If Russia supports, rather than undermines, the independence of the new states that came into being in this zone after the breakup of the

USSR, they can be its strongest partners and allies. This will provide the strongest possible momentum for the reintegration of the post-Soviet territory on the basis of the still unfamiliar, and therefore still unbolstered, democratic principles of relations between states.

This is the only possible way of escaping the "black hole" and the intergovernmental conflicts that could reduce the whole post-Soviet territory to ashes.

## 4.12

**Kazakh Government Hails Russia's CIS Policy**
Interfax, 26 September 1995
[FBIS Translation]

The government of Kazakhstan has given a positive response to Russia's new strategic policy aimed at cementing CIS ties.

"I see no dictate from Russia toward CIS states [in this document—Interfax editor]," Kazakh Prime Minister Akezhan Kazhegeldin told a news conference in Almaty on Tuesday.

He said that at last there were some people in Russia who realized that there were other states surrounding their country with which it was necessary to build good relations.

The prime minister said Russia's earlier focus on the West or America alone appears to be on the way out.

He described as justifiable Russia's drive for taking its own place in what used to be the Soviet Union, though Moscow must work hard to try to make its partners accept this.

Kazhegeldin spoke of a reversal in the views of the Russian political elite. "Things evolve in a way we wish them to," he said.

# Kyrgyzstan

## 4.13

**President Akaev Interviewed on Moscow Visit**
Interview by Aleksandr Peslyak
Moscow Russian Television Network
28 May 1992 [FBIS Translation]

The first and main question is about the purpose of your visit to Moscow. What have you been doing here?

[Akaev]: My purpose is quite prosaic. Last week, or should I say throughout May, nature has been dealing powerful blows to our republic—torrential rains of unprecedented intensity have occurred, accompanied by mudslides many meters deep. Then an earthquake measuring a seven [on the Richter scale] took place in the south of the republic. They have brought the situation in the republic, which was hard enough without them, to a critical point. I will name just a few figures.

More than ten thousand houses have been damaged, 80 percent of them almost totally destroyed; in a word, some eighty thousand people have become homeless. Most of them have no clothes or food because the powerful mudslides not only destroyed the houses but also swept away everything that was inside the houses. In connection with this, naturally we immediately turned our eyes to Russia, because, earlier, when we were part of the USSR, Russia helped us in the event of such natural disasters.

I am very grateful to the Russian government and to the Russian president, Boris Nikolaevich Yeltsin, for the very substantial help they have given us. I hope this will help us to overcome the difficulties facing the republic. That is one side of it. But today we—both the people of Kyrgyzstan and the people of Russia—feel, as Boris Nikolaevich Yeltsin also said at yesterday's meeting, that we need to raise our relations to a new and higher level, and to conclude a treaty of friendship and cooperation between the Russian Federation and Kyrgyzstan.

Those then are the two aims I was pursuing when I met President Boris Nikolaevich Yeltsin, State Secretary Gennadiy Eduardovich Burbulis, and Deputy Prime Minister Egor Gaydar. We agreed on a timetable for the preparation and signing of a friendship and cooperation treaty between Kyrgyzstan and the Russian Federation.

[Peslyak]: Can you reveal the timetable?

[Akaev]: Yes, I can. I hope it will happen in the second half of June.

[Peslyak]: Askar Akadievich, on behalf of the viewers, I would like to express our sincere sympathy in connection with the disaster that has occurred. Could you perhaps explain what forms of aid and interaction were discussed at the talks?

[Akaev]: We raised four main questions. First, we urgently need a favorable long-term loan because we will have to build schools and hospitals. We settled the question of providing a favorable ten-year loan at an annual interest rate of 10 percent.

[Peslyak]: How else will Russia show its active solidarity?

[Akaev]: Russia will help with cash by transferring to Kyrgyzstan during May up to 1.5 billion [currency not specified], which will pay for pensions, student grants, and wages to those who have not been receiving them for two or three months. In addition, there will have to be resowing; for this Russia is allocating up to twenty-five thousand tons of petroleum products. Finally, because of the threat of epidemic, medical supplies are needed, and large deliveries of them are expected from Moscow.

Those then were the main—very specific—things that we need today like we need air. We have appealed to many states of the world with whom we have recently established not only diplomatic but also trade and commercial links, and they are also sending us aid. The Japanese government, in particular, has sent us 20 tons of medicines and transferred $100,000 to us. Now the Turkish government has allocated 1,000 tons of flour for us. We are also expecting help from other countries of the world. But the principal issues have naturally been tackled here in Moscow.

[Peslyak]: In his recent interview the Kazakhstani president, Nursultan Nazarbaev, said that a treaty on friendship, cooperation, and mutual assistance that is being concluded with Russia is the first large-scale interstate treaty. If you are aware of major points of that treaty, and a draft that Kyrgyzstan and Russia have drawn up, are there any similarities and specific points?

[Akaev]: I think that the treaty that we are now preparing together with Russia will be on the same level as the treaty between the Russian Federation and Kazakhstan. The scale is different, of course. The history of mutual relations between Kyrgyzstan and Russia goes back exactly 205 years this year. Two hundred and five years ago farsighted representatives of the Kyrgyz nobility sent their ambassadors to St. Petersburg for the first time in order to come under the protection of the Russian empire. Our friendship and cooperation date back to that time. Of course, I am far from idealizing these relations. Of course there have been bright moments and moments that now give rise to controversy. Nevertheless, something eternal, bright, and kind is characteristic of our mutual relations.

Last August, when the State Committee for the State of Emergency emerged, we the Kyrgyz people, without a moment's hesitation, were among the first to take the side of democratic Russia and President Yeltsin, who courageously fought the forces of reaction who wanted to return us to the era of totalitarianism.

[Peslyak]: Taking into account Kyrgyzstan's serious particularity, which implies possible natural disasters or technogenic catastrophes, could we speak about there being a possibility of retraining and arranging a serious conversion of the army toward its peaceful preventive use in the event of such accidents and disasters? Could this become a solution to the issue of social protection for the army, while retaining its security function?

[Akaev]: First, I would like to say that Kyrgyzstan has adopted a decision not to create its own army. We are, perhaps, the only CIS country to have made a decision not to create an army of its own. We will limit ourselves to setting up a national guard numbering around eight hundred people. We are only just setting it up, and we have laid down the idea that you have just set forth, Aleksandr Mikhailovich. We would like to use this national guard most often since we have frequent earthquakes, mudslides, and avalanches, and we would like this national guard to become a rapid deployment force that could be used to eliminate the consequences of natural disasters. It is known that we are among the six CIS member-countries that have signed the Collective Security Treaty. All this, I believe, provides us with firm guarantees of security.

[Peslyak]: A few words about the economic situation, please.

[Akaev]: Both in industry and agriculture, the rate of production decline amounts to about 12 to 15 percent. But we hoped that this year we would be able to somehow improve the state of affairs in agriculture.

[Peslyak]: We are hoping to maintain industrial and agricultural production at the level of previous years. We are also hoping to continue and successfully develop our reforms.

[Akaev]: First and foremost, we would sincerely like to see the successful advance of reforms in Russia because our economies have common circulatory systems and the heart lies in Russia. So naturally the success of reforms in Russia is of great importance. Then our reforms in Kyrgyzstan will also be successful. I believe that our economy will start to pick up, literally this year or the beginning of next year.

## 4.14

### Military Questions Need for Own Armed Forces

Anatoliy Ladin

*Krasnaya Zvezda,* 5 June 1993 [FBIS Translation]

The army's problems are, as you know, directly dependent on the state of the republic's economy.

In its turn, the army can only expect to receive funds that the budget has at its disposal. In order to appreciate the "wealth" of the budget today, let me cite just a few figures.

In the first quarter of the year Kyrgyzstan's GNP fell 27.5

percent. National income stood at 75.8 percent of the figure for the corresponding period of 1992. Many production units are being axed, and industrial and trading enterprises are "gobbling up" the last remaining stocks of material resources. It is therefore becoming harder and harder to supply the troop units and establishments that switched to the republic's jurisdiction following the breakup of the USSR Armed Forces with all they need.

The worsening problems have forced the republic's military leadership to tell the president, the Supreme Soviet chairman, the prime minister, and the vice president of the difficulties that are retarding the development of the national army. The letter says in particular that an answer must be found in the immediate future to the main question that is primarily worrying officers—does Kyrgyzstan need its own armed forces? The point is that, during an official visit to Japan, Askar Akaev said that Kyrgyzstan is the only CIS member state that will not in the future have an army of its own. At the same time the constitution, which parliament adopted fairly recently, says that "Kyrgyzstan's Armed Forces will be developed in accordance with the principle of self-defense and defense sufficiency." And Article 47 says that "the president of the Kyrgyz Republic is the commander-in-chief of the Armed Forces, and appoints and replaces the supreme command of the Kyrgyz Republic's Armed Forces." All this gives us grounds to think that Askar Akaev is today ready to sacrifice the Armed Forces in order to find a way to break the economic deadlock for his sovereign state and cut spending primarily at the army's expense. It is logical that many units and establishments under the jurisdiction of the State Committee for Defense Affairs are currently short of 30–50 percent of their full complement of officers.

I learned from conversations with officers in various positions in the republic's State Committee for Defense Affairs that most of them see the solution to the army's problems in the further development of integration ties in the defense sphere with Russia and the other CIS signatories to the Collective Security Treaty. Obviously, it is hard for Kyrgyzstan to solve its security problems alone.

## 4.15

### Akaev: Reforms "Impossible" Without Russian Cooperation

Boris Maynaev

ITAR-TASS, 10 June 1993 [FBIS Translation]

"Today we mark the first anniversary of the Treaty on Friendship and Cooperation between the Russian Federation and the Republic of Kyrgyzstan, which is of great significance to us," said President Askar Akaev during his meeting with Russian Ambassador Mikhail Romanov.

According to Akaev, it will be impossible to achieve reforms in Kyrgyzstan without close cooperation with Russia. Old economic ties that had been severed are now being restored on a qualitatively new basis, which is in the interest of both states.

Only culture can strengthen the fraternity of the CIS states, underlined the president. Therefore the republic places great importance on the opening of the Slavonic University and the conservation of a single information space within the CIS.

## 4.16

### Akaev: Russian Aid Needed for Economic Independence

Interview from the "Presidential Bulletin" column

Interfax, 28 January 1994 [FBIS Translation]

Kyrgyz President Askar Akaev believes that his country can truly become economically independent only with assistance from Russia. Speaking at a press conference in Bishkek on Thursday [27 January], the president emphasized that the arrival of Russian Foreign Minister Andrey Kozyrev and the visit of Russian Prime Minister Viktor Chernomyrdin planned for February "would assist with settlement of the issues concerning support of Kyrgyz industrial enterprises that directly depend on economic relations with Russia." "If we break these relations, there is a risk that the Kyrgyz will return to their traditional nomadic life as cattle breeders," the president pointed out.

For the time being, it is impossible to resolve the current economic issues, like supplies of equipment and fuel material and lubricants, without the assistance provided by Russia. This also threatens the future of agriculture.

In connection with this, the president emphasized that dual citizenship would conform with the interest in preserving the economic potential of Kyrgyzstan, since this would help to maintain the working potential of plants where most workers were Russians. He refuted the assertion that this might lead to destabilization of interethnic relations.

Commenting on the changes in the Russian government, the Kyrgyz president said that the situation should not be exaggerated. "The reforms in Russia will continue since now they are irreversible," Akaev pointed out. In his opinion, Chernomyrdin was the man who could carry out the policy of economic reforms in Russia.

Akaev said that if he won the 30 January referendum on confidence in him he would take resolute steps to speed up the reforms. "Kyrgyzstan can't afford to waste any time because it has huge debts, which could go on snowballing," he said.

He said the referendum, called on his own initiative, had three goals: first, to confirm the authority of the head of state, who, under the new constitution, must be the guarantor of stability and of a balance between the three branches of government; second, to confirm the policy of reform; and, third, to stabilize the political situation in the republic.

The president said he would never dissolve parliament, for it had been elected by the people, although under the former regime. "Only the people themselves can determine the future of parliament," he said.

Akaev, during his trips within Kyrgyzstan, said that he had come to the conclusion that the population supported the reform policy but was dissatisfied with the form privatization was taking. For this reason, he was going to take measures to set up bodies for combating corruption and strengthening the republic's economic security, he said.

The same day, Akaev addressed the nation on television and radio. He said the main objective of his policy was to reach ethnic conciliation in the republic and that he would "step up his efforts to establish civil peace in Kyrgyzstan."

He said that, as head of state, he had been doing his best to mitigate the effects of the economic crisis and the social and political shocks experienced by the republic. "My conscience is clear," he said.

He said economic problems that had been taking shape for decades could not be solved within a few months. Today "all citizens of the country, all political parties and movements, both left-wing and right-wing ones, both communists and centrists, stand for market economics, for freedom for the powerful and support for the weak," Akaev said. He said the disagreements were over how rapid the reforms should be and over their social ramifications. There was a need for compromise, he said.

Akaev said the mechanism to control the economic situation was being put into operation "slowly but surely." He said that, by introducing its own currency, Kyrgyzstan had evaded many economic upheavals. The country had also managed to restore its economic ties with other former Soviet republics, in particular Russia, Uzbekistan, and Kazakhstan, the first step toward which was the lifting of customs barriers, the president said.

"We are in the process of forming a single economic area with Uzbekistan and Kazakhstan," he said.

Akaev said that, if he won the referendum, he would take immediate measures to strengthen discipline, give an impulse to the agrarian reform, solve energy and transport problems, support medium-sized and small businesses, create an atmosphere of social partnership between the government, business and trade unions, and improve the position of the pensioners.

## 4.17

## Akaev on Ethnic Issues, Relations with Neighbors

Igor Rotar
*Nezavisimaya Gazeta,* 22 March 1994
[FBIS Translation], Excerpts

During our nearly two-hour conversation, the Kyrgyzstani president managed to shake my skepticism somewhat about the future of Russians in his country. Sadly, my interlocutor asked me not to turn on the tape recorder: "Let it be an unofficial conversation, and you will write an article after you have traveled around our country."

Yet I managed, not without difficulty, to secure Askar Akaevich's consent to publish excerpts from our conversation.

Akaev believes that Kyrgyzstan is vitally interested in the comfortable existence of a Russian ethnic minority, not out of purely humanist considerations. In its ethnic composition, Kyrgyzstan differs appreciably from the other Central Asian states, where Slavs account for a relatively small share (some 10 percent) of the country's population. Tranquility in Kyrgyzstan rests on a precarious ethnic balance between the three most numerous peoples of the state: Kyrgyz (nearly 50 percent), Uzbeks (more than 10 percent), Russians and Ukrainians (nearly 25 percent). In this ethnic "crucible" the Slavic component fulfills the role of a stabilizer that maintains harmony between the peoples. A massive outflow of Slavic people would tip the ethnic balance in the republic, and the probability of interethnic clashes would grow sharply. [Passage omitted.]

The main reason for the Slav emigration is low living standards. The Kyrgyz leader thinks that Western countries have proved in practice totally indifferent to such notions as democracy and freedom of speech and prefer to invest money in countries where they can earn a profit. "We are being helped only by America and Japan: as for Europe, I do not visit it," Askar Akaev added with sadness. As for Kyrgyzstan's neighbors, which have much richer natural resources, they are securing credits all over the world.

The other important specific feature in the life of Kyrgyzstan's Slavs is that, unlike their fellow Slavs in the other Central Asia states, most of them work for defense. The disintegration of the Soviet Union left virtually all workers of the military-industrial complex without a job. The Kyrgyz president pins great hopes on an understanding with Boris Yeltsin on creating an investment fund to turn the enterprises where Slavs work into joint enterprises. "This would finally provide jobs for our Slavs, and they would not be leaving for Russia," Askar Akaev said.

Finally, there is the dual-citizenship problem. In the president's view, it can be granted only to local Slavs and Germans. The granting of dual citizenship to Uzbeks and Tajiks would, Askar Akaev thinks, lead to unpredictable consequences. The Kyrgyz president reminded me that Tajikistan had made territorial claims on his country not so long ago. If one granted dual citizenship to Tajiks, who densely populate areas close to the Tajikistan border, a territorial division would be inevitable.

## 4.18

**Russian Troop Commander Due in Bishkek to Sign Agreements**

Boris Maynaev

ITAR-TASS, 1 April 1994 [FBIS Translation]

General Andrey Nikolaev, commander of the Russian frontier troops, is expected to arrive in Bishkek on Friday to sign a package of agreements on the legal basis for the stay of Russian frontier guards in Kyrgyzstan territory. The documents to be signed include the treaty on the number of Russian frontier guards in the Kyrgyzstan State National Security Committee, the treaty with the Kyrgyzstan Defense Ministry on regulations for Kyrgyz recruiting service in the ranks of Russian frontier guards in Kyrgyzstan. A meeting between the Kyrgyzstani president, Askar Akaev, and General Andrey Nikolaev is planned.

The commander of the Russian frontier troops arrives in Bishkek upon the conclusion of his visit to Tajikistan. He had made a fact-finding trip to the Tajik–Afghan border. General Nikolaev intends to report on the situation in the region to a forthcoming meeting of CIS heads of state and to note the need for the observance of arrangements on joint protection on the border. At a news conference in Dushanbe, Nikolaev said that Russia has strategic interests in the region and noted that Russian frontier guards are not to be withdrawn from Tajikistan's territory in the years ahead.

## 4.19

**Akaev's Views on Democracy and Other Concerns**

Interview by *LG* Editorial Board

*Literaturnaya Gazeta*, no. 43, (26 October 1994) [FBIS Translation], Excerpts

*Editor's Note:* In the following interview, Kyrgyz President Askar Akaev strongly supports Nursultan Nazarbaev's concept of a Eurasian Union, with suprastate structures. Kyrgyzstan and Kazakhstan are part of the firm "integrationist" faction within the CIS. They were even "out in front" of Russian Foreign Minister Andrey Kozyrev on the integration issue. Russia, in fact, may be allowing these two republics to do some of its work in convincing other CIS states of the necessity of closer integration.

On the other hand, the Russian administration is unsure of just how far its wishes to integrate on the economic front. On this issue, Kyrgyzstan and Kazakhstan have been the source of some unwelcome pressure on Russia itself.

Akaev also discusses the interethnic pressures plaguing Kyrgyzstan, Tajikistan, and Uzbekistan. These problems, as he knows, will not disappear and they provide must of the rationale for those republics' acceptance of Russia's military "assistance."

When he was already a mature man, Askar Akaev had the imprudence to write his dissertation on the associative reasoning of computers. That topic was considered super-secret, closed, and Akaev became . . . stuck in his groove. And so he would probably have remained, a strict ivory tower scientist, if life itself had not required his talent on other, political, soil.

And so, upon graduating from the Leningrad Institute of Exact Mechanics and Optics, at which he subsequently worked for a long time (he gave the city on the Neva seventeen years), Akaev returned to his native Kirgizia [Kyrgyzstan], where, having by that time passed through all the stages of a scientific career, he was elected president of the Academy of Sciences. From there he was elected president of Kirgizia . . . twice: in October 1990, as a result of parliamentary voting, and exactly a year later, by the people of the republic.

At night, it is said, the president conducts computer analyses on political topics. But generally speaking, Askar Akaev began his working life as a metalworker. The president intends to mark his fiftieth birthday, which is coming up in November, among his most immediate circle; there will be no celebrations in honor of the occasion in the republic, as they say.

[Akaev]: Perhaps I will in a way be breaking with your tradition, but permit me to have the first word as a guest of your newspaper. I am a reader of *Literaturnaya Gazeta* of long standing, perhaps from the end of the 1960s. I still keep at home many years' worth of clipped articles devoted to the reform of science and conversations with outstanding scientists, "Science Wednesdays." I consider it a great honor to meet you, and I am ready to answer all questions.

[A. Udaltsov, editor-in-chief]: Let us begin with the CIS: I personally do not completely understand—do we have this or not? In practically all the republics, two currents are struggling: one for full independence, and the other for reunification, maybe even in the form of the former USSR.

As a scientist, could you not give us your prognosis, not for the distant future, into which we have all ceased to peer, but for the very nearest. Which forces, in your view, will be victorious—those that are drawing us apart or those that strive to reunify us?

[Akaev]: I think that in the near future, our Commonwealth will be preserved. I personally believe in this and strive to do everything that would promote a closer and more effective integration of our countries. At the same time, one cannot fail to see the truth. We have already adopted more than four hundred documents, but the majority of them are not being implemented at all, or their effectiveness is insignificant. You know that I am in Moscow in connection with a regular meeting of leaders of CIS countries, at which we signed, in addition to other things, the document "On the Intensification of Integrationist Processes in the Commonwealth."

It was at that Moscow summit that President Nazarbaev's idea on the creation of a Eurasian Union was first submitted for free discussion, or, speaking in the language of the documents, "for the first reading." My misgiving was borne out. The majority of CIS members are not ready today to support that idea. As a result, it was tabled until that same distant future into which, as you have said, we have already gotten out of the habit of peering. But that is unfortunate. Peer we must. I was and, as you see, remain, the only one who openly, out loud, supports the idea of President Nazarbaev. Why? Because Kyrgyzstanis, like Kazakhstanis, feel the need for closer integration. And if one of the ways that integration can be strengthened is through the Eurasian Union, whose suprastate structures would be able to guarantee effective realization of joint decisions, and if that Union itself were able to merge with the commonwealth, and possibly even become its nucleus, then would the one really exclude the other? In our view, the direct opposite is true: We are talking about mutually complementary systems here.

Now about the long term: The world of the future, in my opinion, will consist of multiple associations, to speak in the language of mathematics. There is the Eurasian Union; it is not being broadened. There is the CIS, and in the next several decades, all of us who are its current members, undoubtedly, will give priority to collaboration within the framework of our commonwealth. Perhaps that commonwealth will in the future indeed grow into the Eurasian Union.

[Udaltsov]: Soon the Belovezh agreements will be three years old. Since that time, much blood has been spilled; it is being spilled even today. Belorussia [Belarus] is just about the only peaceful one. What do you think—are these miscalculations of policy, or, to be more precise, of politicians—but of which, exactly, and in what, exactly? Or is this an objective natural law, connected with the disintegration of the USSR? It is easier for you to answer since you had the Osh events, after which, it seems, things did seem more or less to calm down in that region. Do you think we could have gotten by without bloodshed?

[Akaev]: It seems to me that ethnically based regional conflicts are a phenomenon that are likelier to arise from the natural order of things than from chance. Throughout the whole world, an ethnic renaissance has taken place and is taking place. It matured in a leisurely fashion, like a volcano, and then the eruptions began. Everywhere, of course, there were their own additional reasons—some in Tajikistan, others in Nagorno-Karabakh, still others in Bosnia and Herzegovina. . . . But to connect this ethnic renaissance only with the boundaries of the former socialist camp is, I believe, incorrect, for its geography is much wider. The disintegration of Yugoslavia or the USSR could, of course, to a certain extent accelerate regional conflicts, but it is hardly worthwhile to seek their main, decisive reasons in this.

The 1990 Osh tragedy could have taken on a scale greater than that in Nagorno-Karabakh: In two days, several hundred people perished there. Thanks to joint efforts with President Karimov, we were able then fairly quickly to neutralize the most dangerous breeding grounds of the conflict. In Kyrgyzstan, practically a front-line state, it is calm today. But for how long? As president, I learned a lesson from the Osh events. I consider internation accord within the republic to be the main, priority task of my policy. And today it looks as though we can be reassured. And a recent sociological poll showed that 80 percent of citizens evaluate interethnic relations in the republic as "normal" or "normal, but with a little tension." For a republic whose people comprise twenty-five major ethnic groups, this is a good sign, you must agree. And still, the danger remains of a new explosion where this already happened once. As a legacy of Soviet power, we have been left with an administrative repartition of the Fergana Valley that gave two of its oblasts—Dzhalal-Abad and Osh—to Kyrgyzstan; one—Leninabad, now Khudzhand—to Tajikistan; and the three remaining—Namangan, Fergana, and Andizhan—to Uzbekistan. Conditional boundaries became state ones, and they by no means always coincide with the ethnic ones. So now three states bear joint responsibility to see that, from a source of heightened danger, Fergana should be transformed into common home for Kyrgyzstanis, Uzbekistanis, and Tajikistanis . . . .

But in answer to your remark, I must say this: there has not been a single refugee from Kyrgyzstan, and I think that there are no dissidents, either. Migrants, yes. Out of Kyrgyzstan's population of more than 4 million, there lived 1.1 million Slavs, including 900,000 Russians, 130,000 Ukrainians, and 13,000 Belarusians. Today 900,000 Slavs remain, of whom 750,000 are Russians. Such a high migration has greatly troubled us, and we have undertaken all possible measures to stop it. And today it is even reversing itself! The number of those leaving Bishkek in eight months of 1994 in comparison with the same period of last year decreased by a factor of five, and during the same time, the number of Russians returning grew by a factor of four. And those people who today are leaving for Russia are no longer selling their houses and apartments, as they were doing before, but are keeping them, leasing them out. That is, they are leaving for Russia or other countries of the CIS to make money, having decided in advance that they would return.

In order to gain an understanding of the reasons for migration and to take the necessary measures to contain it, we held a conference, "Russians in Kyrgyzstan," this year.

Last year we opened the Kyrgyz–Russian Slavic University, with four departments where a course of lectures is also given by professors whom we have invited from Russia.

At the beginning of the 1990s, the high outflow of Russians from the republic was riding the wave of the ethnic renaissance, the affirmation of national statehood, and Kyrgyz as the state language. Sociological polls of the time showed that the language problem was the main reason for migration. Today it is in sixth place; economic problems have emerged in first place. That is, people most of all need work, employment. But here is a concrete illustration: we have approximately thirty enterprises of the military-industrial complex, which employ 70 to 80 percent Russians. These enterprises, which are capable of producing articles that are unique and necessary to Russia, are standing idle. I took Minister Kozyrev, General Gromov, and other representatives of the Russian military-industrial complex there. I did not ask, and do not ask, for assistance. I am proposing to create a joint Kyrgyz–Russian investment fund, which is beyond the power of Kyrgyzstan alone today. We are prepared to give the Russian government or the Russian partners of our enterprises parcels of stock, even controlling ones. Our proposal: Russia will pay R25 billion into the fund and will lend us the same amount; that is our payment, and Kyrgyzstan will return it. President Yeltsin has supported this idea, but there the matter ended. You are worried about the fate of your countrymen in the "near abroad," but after all, these are citizens of Kyrgyzstan and an irreplaceable cadre resource of the republic—do I really have less of a reason to ache for them? So would it not be better for us and you, instead of wrangling with each other, to seek mutually acceptable solutions? . . .

[I. Gamayunov, observer]: It is known that almost all of your Communist Party secretaries occupy important posts even today. To what extent has your nomenklatura been revived, and what are your relations with its old segment?

[Akaev]: I have good relations with them. The thing is, we have no choice. It is Russia that has a choice. It is rich in cadres; it has a remarkable intelligentsia, colossal intellectual potential, and natural resources. Russia is predestined by fate to be reborn into a great power.

But we had no choice. We, perhaps, are the only country in the CIS that did not disperse the Communist Party. The Communist Party of Kyrgyzstan even today is the most organized political force in the country, for the rest of the parties, and we have only ten of them, remind one more of party clubs. All the former Communist Party secretaries even today are in prominent posts in the government and parliament.

What guided me as president, when I made such a compromise? Last year we had a government crisis. I proposed to parliament to nominate everyone it considered necessary to include in the government. Thirty people were nominated, among whom were communists, social-democrats, republicans, and many others. First, the candidates were named by the parties themselves; then they held a straw vote and distributed the portfolios depending on the rating of each candidate. The highest rating was received by Amanbaev, the final first secretary of the Communist Party, who once worked as agriculture secretary. He got the post of first deputy prime minister for agriculture. For five years, before him the first party secretary was Masaliev—he headed the committee on mining. Earlier still, the party in the republic was headed for twenty-five years by Usubaliev—today he is a people's deputy around whom a large deputy group has rallied.

I understand democracy this way: it is obligated to work with those people who are brought forward by legitimate organs of the people's sovereignty . . . .

[Moroz]: Is it true that the Tajik–Afghan border is guarded primarily by Russian border guards? Why do you not participate in this, despite the signed agreements? What consequences for the republics of Central Asia, for Kyrgyzstan in particular, could result from the probable withdrawal of Russian border forces from Tajikistan?

[Akaev]: We are simply, for modesty's sake, not making a show of our participation. But I must say that we are observing the spirit and the letter of these agreements with precision. Quickly upon concluding the agreement, our Kyrgyz battalion was the first to arrive at that border. In winter, it executed a rapid march over snow-covered mountains. It is there to this day. We have reinforced it three times. That is the only strategic road by which the provision of all the Pamirs travels. Jointly with the Russian border forces, we monitor the security of this road, in order to prevent the transport over it of arms and narcotics . . . .

Our som is convertible. Without false modesty, I can say that our currency today is the strongest in the CIS. When we introduced it, a year and a half ago, inflation amounted to 17 percent. Today it is 0.2 percent. On the black and the interbank markets, the currency rates are practically indistinguishable. From the very beginning, we began to conduct a tough monetary-credit policy—and we held out. Of course, we would have had a very hard time without dollar reinforcement. But from the initial financial stabilization, we were able without a break to make the transition to the three-year program of structural transformations in the economy approved by the IMF and the World Bank. Now the som continues to gain strength.

[G. Tsitrinyak, observer]: Your neighbor, Nazarbaev, speaks out categorically against dual citizenship. He says that the people will not understand it. What do you think?

[Akaev]: In 1992, Kyrgyzstan was the first to sign the treaty with Russia on Friendship, Cooperation, and Mutual Assistance, which contains a clause on the granting of dual citizenship on the basis of bilateral agreements. But in our constitution there is no such statute; it did not pass in parliament. So we still need to resolve that question . . . .

All this is all the more vexing in that in the beginning we worked very cordially with our parliament. I note in passing that Russia does not make sufficient use of its parliament. How otherwise to explain the fact that S.S. Alekseev together with his collaborators helped us to write the drafts of many laws, which we passed first and which were then recopied, just about word for word, in other CIS countries? Privatization, land reform, reconstruction of the banking and financial systems—they gave the impetus to all these processes and changed people's outlook. And then in that moment when the greatest mobility was required of the Kyrgyzstan parliament, it became incapable of functioning; its conservative segment began stubbornly to apply the brake to the reforms. What was left for me to do as president? I did not make use of my right to disband parliament, but simply set new parliamentary elections. This is exactly the wish of those deputies whom I call the reform wing. I am sure that the October referendum will become pivotal for Kyrgyzstan, and that the new parliament, which now, per the Russian model, will become bicameral, will continue the course of reforms and the democratization of life in our republic.

## 4.20
## Deputy Foreign Minister on Relations with Russia, CIS

Interview
*V Kontse Nideli,* 4 February 1995 [FBIS Translation]

A whole range of bilateral relations between Kyrgyzstan and Russia will be under consideration during the coming visit of the Russian deputy foreign minister, A. Panov, to Bishkek. Intense preparation of all the necessary documents is currently going on.

Our correspondent met with Kyrgyz Deputy Foreign Minister Alikbek Dzhekshenkulov:

[Correspondent]: How would you describe the present Kyrgyz–Russian relations?

[Dzhekshenkulov]: They are fairly dynamic. Both sides are interested in close friendly relations, which is owed to our history and traditional economic, cultural, and scientific ties as well.

Unfortunately, in recent years there have been losses too. Mostly this is explained by the quick and, for many, unexpected collapse of the Soviet Union. All the citizens of a single huge state woke up one day as citizens of different countries. And the former Soviet republics, having gained their independence, for some time literally relished their freedom. And in this understandable condition of euphoria they somehow forgot about real people. On this wave, some ill-considered and badly prepared laws were adopted, and the outflow from the republics of able-bodied people and highly qualified specialists in many different spheres began. It became clear that this abnormal situation must be changed.

[Correspondent]: Probably all of us now realize that some of the laws that were adopted need correcting.

[Dzhekshenkulov]: That is why we are preparing agreements on various items: cultural and scientific cooperation between Kyrgyzstan and Russia, the main principles for the creation of joint financial and industrial groups, regulations concerning problems of citizenship, the protection of the rights of migrants, the establishment and activities of information and cultural centers, and the establishment of a customs union. Of course, this is not a full list, but all the issues, we hope, will be resolved in a spirit of good will and mutual understanding.

Incidentally, our ministry has good contacts with the Russian embassy. Twice recently Ambassador M. Romanov visited the foreign ministry with all his services. It is easy for us to find a common language.

[Correspondent]: Russia and Kazakhstan recently signed important agreements, thus greatly stimulating the integration process further. And specialists and political scientists have again begun to speak about the idea of the Eurasian Union, put forward by Kazakhstan's president, Nursultan Nazarbaev.

[Dzhekshenkulov]: Of course we are happy with such developments in relations between Russia and Kazakhstan. All of us, the former inhabitants of a single Union, are united by many things and this is what should determine the policy of the leaders. Once again such relations are the cornerstone of our foreign policy, especially toward the CIS countries.

And as for the Eurasian Union, [President] Askar Akaevich [Akaev] was the first CIS president to support Nazarbaev's idea.

[Correspondent]: Some newspapers have reproached Kyrgyzstan's foreign ministry for not issuing a special statement about events in Chechnya.

[Dzhekshenkulov]: Throughout the world there is a definite order. The foreign ministry is not obliged to issue a press statement. In most cases we use diplomatic channels and our offices abroad and foreign embassies in Bishkek to express our position.

Our position from the very beginning has been clear: It is Russia's internal affair. Of course, we cannot remain untroubled by war and deaths, especially of civilians. We expressed our concern and at the same time our hope that the situation will be settled by political means.

[Correspondent]: Since we have started talking about hot spots, we cannot avoid Tajikistan.

[Dzhekshenkulov]: I visited Dushanbe recently as part of a delegation headed by Deputy Prime Minister Ibraimov. We reached agreement on most issues concerning bilateral relations, but some problems remain unsolved. First of all, the problems of land use in border regions. But the negotiations will continue, and we hope that mutually suitable solutions will be found.

The situation in Tajikistan, of course, is very complicated. We are worried about the situation on the Tajik–Afghan border. The acute problems do not pass our country by: the drug business, refugees, the condition of the Kyrgyz diaspora in Tajikistan. The foreign ministry is permanently monitoring the progress of events and informing the country's leadership about them. [Passage omitted.]

## 4.21

**Akaev on the Economy, Policies, and Integration**

Interview by Aleksandr Shinkin
*Rossiyskaya Gazeta,* 25 March 1995
[FBIS Translation]

[Shinkin]: Askar Akaevich, the lives of people today are increasingly dominated by worries about their daily bread. Naturally, my first question is: What is the economic situation in Kyrgyzstan, for which total bankruptcy was being predicted some time ago?

[Akaev]: The situation in our economy is grave, as it is, incidentally, in all former Union republics: major slump in production, enterprises coming to a standstill, unemployment. . . . We are experiencing all this. But the situation is not hopeless.

Over the last few years we have laid down firm foundations for a market economy and have created institutions of private ownership and entrepreneurship. Privatization and destatization of property are actively under way, as is the restructuring of enterprises. . . . To put it briefly, all the prerequisites exist for ensuring the start of the republic's economic recovery. For example, we are expecting noticeable growth of light industrial output already this year.

We are pinning great hopes on the gold extracting sector, although the crisis has not bypassed it either. But gold output started rising already last year. And the quantity of gold obtained from auriferous gravel increased fivefold compared with the previous year. Whereas now we are producing just over 3.5 tons, in some three to five years' time, when the mines are fully operational, in terms of gold production Kyrgyzstan will rank third among CIS countries and twelfth or fifteenth among the world's gold-extracting countries. True enough, for the time being, this is like a bird in the bush. But we are attracting all sorts of foreign investment to ensure that this bird ends up in our hands.

[Shinkin]: And you are thus providing ground for the accusations leveled against you that you are allegedly selling off the country.

[Akaev]: The country is being exported and sold off by criminals and corrupt functionaries. In contrast, the investments and credits are helping us only to consolidate Kyrgyzstan's economic potential. Without them we will be unable to revive our cities and get our industry and countryside back on their feet. After all, the country's economy was founded on intra-Union division of labor—major machine-building enterprises and mining and metallurgical combines that depended entirely on deliveries and orders from outside the republic, on the one hand, and an agro-industrial complex that was primarily geared to raw materials, on the other hand. When the Union disintegrated, our national economy was doomed to come to a standstill.

As a matter of fact, the accusations you mentioned are diminishing. The people are now aware of the truth about the republic's potential and are beginning to realize that nothing is handed out free like manna from heaven, and nobody is going to dish out free meals. But the main lesson lies perhaps elsewhere. People are beginning to realize the simple truth. Our lives are not likely to start improving if we continue to live and work as in the past.

Opening up new deposits using our own resources seemed unimaginable in the past. Things have changed nowadays. The "Solton-Sary" gold deposit, which lies at an altitude of 3,500 meters, was planned, developed, and built using only our own resources, without involving anyone from the "near abroad." We assimilated unique technologies. Furthermore, it took only ten months from the time we drove in the first pegs until we started producing metal. Thus, the help from international financial organizations and leading developed countries is just a backup for the main work we are doing in order to extricate ourselves from the grips of socio-economic crisis.

[Shinkin]: Askar Akaevich, some time ago you drew fire on yourself by introducing your own currency unit. This was done, as the critics claimed, for the sake of receiving the financial aid in question. Is your som not letting you down nowadays?

[Akaev]: The national currency's introduction was certainly not prompted by any desire to receive large credits from international financial organizations. Nor was it done in secret. The problem was discussed with many Russian economists and members of government. The point is that confusion and migration processes within the ruble zone are advantageous for some and not for others.

Kyrgyzstan was in the second category. We added up and saw for ourselves. In 1992 we lost 13 billion [rubles] (I well remember this amount) just through ruble exchange rate fluctuations. This was a vast amount at that time. Furthermore, it increased the budget deficit by almost two-thirds. This calculation convinced me that it was necessary to introduce our own currency.

There was, of course, a lot of confusion to begin with. But we ended last year with inflation at 87 percent. We are planning to bring it down to 30 or 40 percent this year. Let me also note that one out of every two sums is backed either by gold or by Treasury bonds held in U.S. banks. The currency exchange rate is virtually identical on both the black market and the interbank market in our country.

I assume that the collapse of the ruble zone proved advantageous for all. The CIS countries established new relations. And the economic union can now be built on the honest and fair basis of national currencies.

[Shinkin]: In the course of interviews, the leaders of other republics do not fail to mention the problems caused by the disruption of economic ties with Russia. You have kept silent about them. Has it not affected Kyrgyzstan?

[Akaev]: What can I say that is new? Our largest enterprises were geared entirely to consumers who were located mainly in Russia. Now these enterprises have no end of troubles. There are no orders. There are no supplies. The majority of them are standing idle. Workers have not been paid their wages for months . . . .

There is only one way out—the creation of joint production units and financial-industrial groups. I even offered Russia the opportunity to buy the controlling blocks of shares in our twenty-nine largest enterprises. Yeltsin and Chernomyrdin backed this idea. But the president's agreement is to this day "doing the rounds" of bureaucratic offices. A Kyrgyz–Russian joint-stock company for developing the "Sary-Dzhaz" lead and tungsten deposit was recently set up with great difficulties.

You must realize that the point at issue here goes further than just mutual advantage. Russia's prolonged absence from Central Asian markets will only help to squeeze it out of here.

Of course, Kyrgyzstan would not like to see this happen. Any break in ties with Russia would be painful for us. I recently conversed with leaders of Bishkek's Lenin Machine-Building Plant. The shop producing combine harvester chains has reopened there after a lengthy idle period. Its familiar output has been augmented by new products—chains for Zhiguli and Moskvich cars. The main consumer is Russia, which supplies the metal. The plant now urgently needs skilled workers, but they cannot be found.

[Shinkin]: Alas, Askar Akaevich, skilled workers and specialists are now, as a rule, refugees. They were the first among Kyrgyzstan's Slavs, as you said, to leave the republic. The outflow is still going on. According to unconfirmed data, more than two hundred thousand of them have left.

[Akaev]: Let me note that there has not been a single refugee from Kyrgyzstan. Nor, I hope, dissidents. Migrants—yes. But that is a different story.

We have experienced two main stages of migration. The first followed the well-known events in Osh—a tragic conflict. But, thanks to joint efforts with President Karimov of Uzbekistan, we succeeded at the time rather quickly to neutralize the dangerous hotbeds. The second outburst of Slavs' departure was prompted by the euphoria of independence and sovereignty. As president, I learned a lesson even from the Osh events. I consider interethnic accord in the republic the main task of my policy. And I am making all efforts, doing everything possible and impossible, to halt the migration flow.

Here are just some of the measures. Our parliament adopted a law code providing that the land belongs exclusively to the Kyrgyz people. As president, I vetoed it and returned it on three occasions until I succeeded in gaining the support of the majority in parliament. The injustice was eliminated. I did the same with the deputies' amendment to housing legislation, which again was detrimental mainly to the Slavs.

[Shinkin]: But you must agree that your efforts are not sufficient on their own. There is nothing more frightening for the non-indigenous population than local nationalism. The quiet survival of the "foreigner," which cannot be noticed by outsiders.

[Akaev]: I think it was [George] Bernard Shaw who said that nationality is like the spine—a healthy person never notices that he has one. But even if someone in our country does suffer from this plague, all the symptoms of recovery are present. Let us assume that the language problem used to rank first among the factors causing migration. Current sociological polls indicate that it has slipped down to sixth place. At the top of the list, incidentally, the question of cooperation with Russia has been joined by economic difficulties; in other words, people need work most of all.

[Shinkin]: It seems to me, Askar Akaevich, that we have overlooked one of the most acute problems facing Kyrgyzstan's Slavs—dual citizenship. Back in 1992 you signed with Russia a treaty on friendship, cooperation, and mutual assistance, which contains a clause about the granting of dual citizenship. But to this day nothing has actually been done.

[Akaev]: Unfortunately, it is not easy to solve this problem. There is no such provision in our constitution; our parliament did not pass it. There are weighty arguments in favor of doing so. Non-indigenous peoples account for almost half the republic's population of more than 4 million. In addition to Russians, Ukrainians, and Belarusians, we also have large Uzbek, Tajik, German, Dungan, and Uygur communities. . . . There is a total of about twenty-five large ethnic groups and a further fifty smaller ones. Which one should get preferential treatment? All of them? But then there would be no Kyrgyzstan. The Russians? But would such an exception be fair to all the rest? In my view, the best possible solution was reached by Russia and Kyrgyzstan, which recently concluded an agreement on a simplified granting of citizenship.

The adoption of a law on dual citizenship would not be sufficient to maintain Kyrgyzstan's multinational composition. I think that it would be much more important to ensure that the organs of state power provide fair representation not only for the "indigenous" nation but also for the other peoples. In my capacity as president, I will personally take care of this. Otherwise it would be difficult to expect the building of a truly democratic society in the republic.

[Shinkin]: Strange. You talk about market economics and democracy, and in the same breath you speak of the Lenin plant. A pristine monument to Lenin stands in downtown Bishkek as if nothing has happened. Are Lenin and democracy coexisting?

[Akaev]: Let me also add that we are probably the only CIS country that has not disbanded the Communist Party. It remains the best organized political force in the republic. All its former secretaries, even the very last one, held eminent positions in the government and in parliament.

In the wake of the government crisis some eighteen months ago, I suggested to parliament: Nominate all those whom you consider necessary for inclusion in the government. The parties nominated their own candidates to begin with. A total of thirty persons were nominated. The deputies then took a straw poll and distributed the portfolios according to each candidate's rating. Consequently Amanbaev, a former Communist Party Central Committee first secretary, became vice premier, while another—Masaliev—headed the committee for mining affairs. And the skies did not fall in Kyrgyzstan because of this. Democracy makes it incumbent upon me to work with people nominated by the legitimate organs of power.

[Shinkin]: As far as I know, in your capacity as a scientist, you were successful in making light particles obey certain rules and bringing them into a state of accord. But accord in society is evidently difficult to achieve. There is a powerful opposition in the republic. You are perhaps the favorite person of the free press. It tirelessly praises the president's personality, your wife, and your relatives. Are you, Askar Akaevich, not afraid that you might fall victim to your own popularity?

[Akaev]: What can you do—this, as a rule, is the fate of all reformers. But, as a matter of fact, it is not a question of my own personality. When we started building an independent sovereign state, we began with the universal prescriptions of democracy—development of parliamentarianism, separation of powers, freedom of the press. The time has now come to put all this in order and check it against the main criterion of democracy—human freedom. And yet some other CIS countries opted for taking a somewhat different path. But if we want a genuine democracy rather than a cheap imitation of it, we have to drink up all of this cup. If they are to be capable of running their own lives and shaping their common fate, people must first of all learn to freely express their opinion.

The fact that we have a strong opposition is not bad, either. Whether constructive or non-constructive, whether its criticism is to the point or is nothing but propaganda passions—these are the opposition's problems, so to speak.

Ultimately, people can see very well who is sincerely on their side and who is promoting his own ambitions. I regret just one thing: Tolerance and civic spirit are not always present in public debates, even though the authorities and the opposition alike must obey the law's demands.

[Shinkin]: Tell me, have you ever had the idea of following the advice of Griboedov's hero: Get all the troublemakers together and apply the force of authority against them? After all, you would more than likely gain quite a few supporters.

[Akaev]: I am not too sure. I think that there will hardly be anyone likely to force the Kyrgyz people to abandon their freedom of thought. Respect for authority [*chinopochitanie*] was never a trait of our national character even in the past, let alone today.

Here is a typical detail. Whenever I get a free moment, I love browsing through the pages of our national epic poem "Manas," whose millennium will be celebrated this year. Right through the book the great troop leader Manas, who laid the foundations of national statehood, is accompanied by his advisor, who is constantly squabbling: Don't do this, don't listen to them, don't choose those knights. . . . As you can see, we not only honor the freedom of speech but also are accustomed to criticism.

[Shinkin]: But why is there talk about the president's dictatorial manners in the republic?

[Akaev]: Of course, Akaev has been and remains a champion of democracy. It is a different matter that the experience of our statehood has shown that, taking into account our specific economic and social features, culture, and mentality, we have different democratic priorities from the West. In my view, Kyrgyzstan's democratic ideology must combine the West's liberal-democratic ideas with our own people's sound democratic traditions.

I, for example, am convinced that—for the foreseeable future—a developed local self-government is as important as, if not more important than, parliamentarianism. Local communities themselves are capable of most adequately expressing the public interests and will. This is the social area where the political actions of the country's leadership can be really supported or rejected.

Today I put forward this thesis: Strong executive power, local administration, and self-government must be the backbone of the state.

[Shinkin]: Turning back to some more global problems, what is closer to your heart, the CIS or the Eurasian Union?

[Akaev]: Those who are actually striving for the effective integration of our countries will naturally opt for Nazarbaev's idea. But I would not pit the CIS against the Eurasian Union. If CIS integration could be strengthened through the Eurasian Union, whose suprastate structures could be capable of ensuring reliable implementation of joint decisions, and if such a union could be capable of flexibly joining the commonwealth and possibly becoming its nucleus, why should the one rule out the other? After all, in mathematical language, we are talking about complementary systems.

We must take into account realities. The CIS exists. Over the coming decades we, all its members, will give priority to cooperation within its framework.

[Shinkin]: Nonetheless, a special Central Asian bloc does exist in the region, comprising Uzbekistan, Kazakhstan, and Kyrgyzstan. And it is gathering strength.

[Akaev]: We have never striven for isolation. This union is vitally important for us, for all three states. Look at Kyrgyzstan. All the communications linking the country with the outside world—railroads, gas pipelines, and international highways—run across the territories of Kazakhstan and Uzbekistan. We have virtually no oil or gas. Economic integration with the framework of these three states gives us an opportunity to solve these problems. In exchange, Kazakhstan receives our electricity. Nobody is worse off as a result of this. Nor should you forget that our three states are also linked by centuries of shared history and similar mentalities. This is why, and in addition to all else, we have established not only purely business relations but actually human relations, when neighbors are prepared to help their neighbor.

## 4.22

## Bishkek Welcomes Moscow Decision on CIS Relations

Interfax, 26 September 1995 [FBIS Translation]

Kyrgyzstan welcomes Russia's new strategic policy to strengthen interstate relations within the CIS.

Russia's president, Boris Yeltsin, issued a decree on 14 September to enact a document entitled "The Strategic Course of the Russian Federation with the CIS States." The decree asserted the need to "intensify integration within the CIS and to improve coordination of Russian executive bodies' activities in this direction."

The head of the international department of the Kyrgyz presidential administration, Baktybek Abdrisaev, told Interfax on Monday that the decree "reflected the Russian leadership's realization of the importance of restoring ties, established through many decades in the former Soviet Union."

Since proclaiming independence, many CIS states have realized the need to turn to each other, he added.

Abdrisaev said the Russian leadership "builds its relations with former Soviet republics in a well-balanced and sober-thinking fashion."

Abdrisaev emphasized that "not only purely market economic relations should dictate the conditions of our mutual relations, but also those cultural, historical, and human ties that we must preserve at all costs."

## Tajikistan

### 4.23

**Russian Delegation Head on Islamists**
Interview by Dr. Hamdi 'Abd-al-Hafiz
*Al-Sharq,* 30 August 1993 [FBIS Translation]

[Sergey Kurginyan interviewed by Dr. Hamdi 'Abd-al-Hafiz in Moscow: "Has the Confrontation Between Russia and Islam Begun?"]

"I welcome Islam and greatly respect it. I know its fundamentals, its past and its present, and I do not oppose Turkey's involvement for the good of its people, as long as it does not interfere in Russia's internal affairs. But when Turkey puts its hands on Kazakhstan, considering it within its sphere of vital interest, I must speak out and remind it that these lands are in Russia's sphere of influence, and that Turkey has no business there. There are powers aiming to make tense the relations between Russia and Islam; our mission is to thwart that confrontation. If extremism continues and denounces everything Russian in the Islamic republics as blasphemous and reprehensible, then much blood will be shed. Russia must find partners for itself in spirit and faith, in an un-American way, to join the "current" world. Here we might touch on the role of the Arab League or Russian foreign strategy or other matters, but there will be a void as long as Russia keeps to its current path." These were some of the replies of Sergey Kurginyan, president of the Ibda'i Research Center Fund [name as published], who is serving as head of a delegation of Russian academic experts who recently visited Tajikistan to study the situation there in the light of recent events in this republic. The delegation prepared a huge report, examined by *Al-Sharq,* on the situation in Tajikistan and the future of its development. *Al-Sharq* met the man and asked him numerous questions. The text of our conversation follows:

['Abd-al-Hafiz]: What was your delegation's mission in Tajikistan? Which experts comprised it?

[Kurginyan]: Speaker of the Russian Parliament Ruslan Khasbulatov gave us this mission, in light of his unease about the danger of recent events in this republic and the chances of the situation there deteriorating. Of course, no effective steps can be taken in this delicate situation without complete knowledge of it in all its details. Also, I consider Khasbulatov a skillful politician with a clear view of the true situation. That may be because of his familiarity with Islamic matters, because of his background. Also, Ruslan Khasbulatov sought the assistance of the Tajiki leadership as related to this exploratory trip. We went to Central Asia accompanied by a group of political, economic, and party experts, and there we were joined by other experts. The group was able to do much in just four days, to meet with religious and political figures, with military men, sectarian leaders, armed groups, security and interior officials, and all those involved.

We benefited much from the information and study base available to us, such as the study of Islamic movements in the Soviet Union beginning in the early 1970s, and, based on open sources, the Soviet and foreign press. The trip was the crowning achievement of the work that had come before. It is also natural that all news and data remained at the margin of the report, because we are not only trying to cover events, we want to influence them, too.

['Abd-al-Hafiz]: You mentioned in your report that you have some things that may not be published. Who imposed the ban on publishing? What is the subject matter?

[Kurginyan]: No one, but I can prevent them from being published because I am an independent expert. I support the speaker of parliament, because I consider him to be one of the greatest political figures in Russia today, and the most dedicated to the interests of the country. That is why I support him. But I myself decide what I will or will not publish. It is only natural that many things have remained in the dark until now, such as the dimensions of the Isma'ili sect's activities in Central Asia.

['Abd-al-Hafiz]: Why the ban on publishing?

[Kurginyan]: Three things are banned. First, not mentioning names and facts, because researchers do not broadcast secret data. The second relates to the familiarity of the Russian reader, who naturally does not require us to overwhelm him with data about the Isma'ili sect or the Agha Khan and the difference between him and Sayyidi Mansur, or the role of the Isma'ilis in Pakistan, otherwise we would have produced a book. The third thing relates to secret and military data, or data from unnamed sources, since the content at times would reveal precisely who the source was. That could lead to evidence threatening that person's life in Tajikistan's civil war conditions. We cannot reveal sources until after the end of the civil war and the advent of peace. On top of that there is the ethical aspect. For example, we know the practices of the Sufi brotherhoods, but we may not make them public because the Sufis consider them "not for publication." Thus we would not have the right to violate these ethics, even if they were our enemies. And so we may summarize the principles of the ban as follows:

1. No unnecessary examples or details.
2. No revealing of news or information sources.
3. Deference to the level of cognizance of the subject.
4. Adhering to ethics.
5. Not publishing any false or unconfirmed news or information.

['Abd-al-Hafiz]: In your report you level an accusation against Western circles, that they have run a conspiracy to provoke confrontation and conflict between Russia and Islam. Do you have any evidence or proof?

[Kurginyan]: We have proven facts. There are old plans to make Russia turn its attention to the south, after having stripped it of some western territories, so that it will not be watching its northwest. Second, some politicians in Estonia made these plans public in 1988, in the newspapers. They said that the Karabakh issue served to detonate the relations between Armenia and Azerbaijan, to make Russia rely on the Armenians, with the Islamic world getting involved in the conflict, as it was linked to Azerbaijan. Many Estonians, behind whom stand certain interests, exposed this plan. There have also been other statements from Pope John Paul II and former U.S. president Ronald Reagan regarding directing events in Poland. But we have other evidence and statements regarding developments in Afghanistan and the people involved in those events.

Who can fight the Russians today? No one but the Islamic countries, because America would have an all-out political crisis after the first one hundred casualties its soldiers would suffer, whereas the Islamic people are spiritually ready for the sacrifice. The Muslims are ready and so are the Russians. As to the rest of the people living in peace and amity, they are no longer able to fight a war by themselves and mean to win without shedding their children's blood. So now false environmental and population issues are brought up, such as population growth in the South, environmental pollution in the Third World, or its very high birth rates—all meaning that a time bomb is ticking underneath today's world. According to this plan, Russia must become a tool with which to confront the Third World, now that its energies are spent and its strength exhausted.

The United States has said that Turkey is now a regional superpower, though it can be "super" only at Russia's expense. At the same time, we note that things have come to a head in the Crimean peninsula, the northern Caucasus, southern Russia, Bashkortostan, and Kazakhstan—it all looks like a wide-ranging plot. I do not believe that all the powers of this region are anti-Russian. There are in fact some powers loyal to us. It is a question of support and financial, media, and political aid given only to anti-Russian forces, even though they may appear to be anti-American.

Let us consider Azerbaijan. The leader of the "Gray Wolves," Raturkashi [name as published], went to Azerbaijan, and the Azerbaijani interior minister, Iskandarov, received the credentials of the leaders of this organization before representatives of the Western countries. Would it be possible for the citizens of Egypt or any other Arab or foreign country to watch, on their television screens, their interior minister and the deputy head of the ruling party receiving the credentials of an international terrorist organization?

The Gray Wolves group is anti-American—but it is anti-Russian and destructive to Russia. It is likely that the rigid old thinking still dominates the minds of American policymakers, that they are supporting groups opposed to the former Soviet Union, even though the latter has changed directions, as if it had become a criminal group, or cannibals.

I would like to point out here that there are in America individuals personally linked to the present secretary of state, whose importance has been on the rise, and whose methods are never understood, and who has become extremely close to his Russian counterpart as well as to Russian Defense Minister Grachev.

There are of course financial factors, bearing in mind that Minister Grachev's relatives receive not inconsiderable aid from American organizations involved with Russia, though this relationship does not stem only from financial factors—there is a political side.

I welcome Islam and greatly respect it. I know its fundamentals, its past and its present. I do not oppose Turkey's involvement for the good of its people, as long as it does not interfere in Russia's internal affairs. But when Turkey prevents the transport of oil through the straits and puts its hands on Kazakhstan, considering it to be within its sphere of vital interests, I must speak out and remind it that these lands are in Russia's sphere of interests, and that Turkey has no business there. There are powers aiming to make tense the relations between Russia and Islam; our mission is to thwart that confrontation.

['Abd-al-Hafiz]: What is the future of fundamentalism in Russia in light of recent events in Tajikistan?

[Kurginyan]: There is the Islamic Revival Party founded in Bashkortostan and Tatarstan. It is a time bomb in its activity, and if this sort of extremism continues, accusing others and everything Russian of being blasphemous and reprehensible, there will be three phenomena, or three leading forces, the first of which will lead the poor classes of people, described as countercommunism, anti-Russianism, and a false Islamic nature. The second is the nationalistic bourgeois force such as the Islamic Revival Party in Dagestan and Tatarstan, which adopted the slogan of "the (nationalist) community" instead of the slogan of Islam. The third is nationalist democracy; you will hear more talk of "Westernized" Islamic democracy, and all principled people, who will number 70 to 80 percent according to the latest censuses in light of current reforms in the former Soviet Union and the gains made, will oppose these forces. Much blood will be shed in the regions of the Soviet state and similar movements will evolve in Russia itself, as the Russian people cannot be silent for long in the face of these events, the violation of Russian rights, and the march of refugees who today number one million.

In addition, there are in Russia, for example, five million persons who have lived with refugees and say to themselves, they tortured us, expelled us from our homes, and discriminated against us, and their numbers will soon multiply if conditions go from bad to worse.

Because of the clash of these movements and after government pressure through the game of referendums, the constitution, and the like, wars will break out on the borders of these entities, that is, between Bashkortostan and Tatarstan, or between Tatarstan and Russia, or between Russia and Kazakhstan, but after this period of schism there will be a period of unity: the state will become a cauldron in which these conflicts will boil. The wood is now burning under this cauldron, and it is getting hotter. Some of them assassinated Vice-President Victor Bulanishku [name as published], without taking serious countermeasures, and of course he is not the first or the last on the list.

In the wake of the spread of crime and accumulation of weaponry, the future looks very dark. Events cannot be limited to this "sixth of the globe."

Firemen know that forest fires can be fought only with counterfires, started in front of the big fire, and for this very reason a southern Islamic border zone will be set up, and within it the counterfire from Russia will burn. It will suit the federation of the Baltic and Black Sea countries for the same reason. Western Europe is thus planning to build a wall to protect itself from the Russian fire; though this plan is a fantasy, an illusion. It is impossible to fight this fire: it will spread everywhere.

It will lead inevitably to a catastrophe. It appears that some people are looking for quick gains, and don't trouble themselves with the implications of what they are doing and its effect on the future. Every person is acting in his own interest, but the result of these interests will be chaos and conflict.

['Abd-al-Hafiz]: Are you fearful of the collapse of the Russian Federation?

[Kurginyan]: Is Russia's unity holding now? Isn't Tatarstan's position tantamount to secession from the federal entity? Stability is hanging by a thread. All that needs to happen is for confidence in Russia's currency to fail, and here we are sanctioning fiscal reform! The fate of the federal entity is sealed. It will collapse when Russia's provinces forbid the use of the ruble and replace it with local currency. It is likely that the catastrophe will happen within three or four months, because the ruble will not maintain its current price for more than three or four weeks, or a month, and bear in mind that the government's reserves are nearly exhausted. In these circumstances, the Russian Federation will collapse but subsequently reunify, because the exit of the Republic of Yakutia, for example, from Russian administration, would mean the exodus of Russians from it, but they will return if they notice the features of reconciliation between it and Japan, which covets its diamond resources, for example.

I say that whoever plans for the situation in the country to hit bottom, with 100 million Russians fleeing their homes and going back to them angry, is a criminal, whatever he is called: the American secretary of state, the Russian foreign minister, or the Russian president. What we are now witnessing in Russia is nothing but a massive international crime. Sooner or later there will be a Nuremburg trial to consider the personality of those who organized this dangerous, terrible socio-political experiment in a country filled with thousands of nuclear warheads, millions of trained soldiers ready to fight and bristling with arms, and dozens of deteriorating nuclear reactors. Couldn't there have been reform of some other kind that would not require dropping the reins of this headstrong horse? Was it necessary to let the genie out of the bottle without knowing what it would take to get him back in?

['Abd-al-Hafiz]: What is your view, as a researcher and a politician, of the reasons for the emergence of fundamentalism as a political phenomenon?

[Kurginyan]: I consider fundamentalism a normal thing. On the one hand, it expresses a people's resistance to the tribulations being inflicted upon them, and the existence of a certain limit beyond which these tribulations may not go. This limit is called "fundamentalism," but it is not proper to use fundamentalism in normal circumstances as a means of leading a country.

Fundamentalism has the following functions:

1. As a bulwark against the racist policy adopted by the United States for its own interests, which violates the will of other peoples.

2. As a means of leading and controlling the situation (it is an indication of inability).
3. There are those who would use fundamentalism to serve their interests. We may mention in this connection the story of the last shah of Iran, who had huge bank accounts abroad, which were very tempting to some, and when after the revolution Iran asked that they be returned to the country, they were frozen; only a fifth was returned to Iran. The remaining four-fifths remained outside the reach of the people of Iran.

We may say that fundamentalism is a tool for controlling Islamic peoples and preventing their unity. It is a way of keeping them in their current place. There is another cultural aspect of fundamentalism, which at times expresses people's opposition to the march of Western culture and its godless spirit. I warmly welcome this other aspect, because non-Western peoples will perish without their heritage, its spirit and civilization. It is likely that the West has an immunity to godlessness and lack of spirituality, though it may not be conferred on other peoples. I bid "welcome" to fundamentalism as a spiritual revolution and a way of restoring the link to heritage, as a revival of human vision. I consider fundamentalism a vast field for philosophical, social, and cultural study.

I re-emphasize here that it is up to the peoples of the Islamic countries to understand that the events in Tajikistan are very remote from fundamentalism. One might describe them as a *false* fundamentalist phenomenon!

['Abd-al-Hafiz]: You touch on the likelihood of a clash between Russia and Islam. Is it a certainty or not? What can we do to prevent it?

[Kurginyan]: I believe that we are heading toward a multicultural world that will be built on bases unlike those of the Western model and that Islamic culture will regain the greatness it deserves. The Russian Orthodox culture will become great, too. They have a chance to cooperate, and I would like to mention here the Istanbul conference, in which the heads of the Russian Orthodox Church and leaders of the Shiites participated. It is very clear that Turkey is now pushing in a certain direction and that its attention is centered on a Western-style peace. This is a road leading to a new Russian–Turkish war to "absorb" the conflicts threatening the West now. In these circumstances, we must understand the dimensions of the danger and its repercussions on all sides, and conduct a dialogue on the basis of the peace we want—one that is ready for cooperation and dialogue with Russia. I believe that Iran has been Russia's traditional ally since early in the last century. I may say that a large part of the Arab world will, in my opinion, soon be in solidarity with Greece. All this means that Russia must review its foreign strategy. I am now speaking as a person who considers our present foreign policy crazy and meaningless. Its very basis is, necessarily, the evolution of new alliances stemming from a different view of the present world and all the sudden changes in it.

They may be political, economic, military, and diplomatic alliances. Russia must also find itself partners in spirit and religion, but not in the American way, to join the current world. Here we might allude to the role of the League of Arab States or Russia's foreign policy, or other matters, though the gap will remain for as long as Russia follows its present course.

I am not calling for a return to the past—I would consider that a dangerous and terrible step. No, Russia must find its way and orient its policy to the southeast on the St. Petersburg–Tehran and Balkan–Far East axes.

On the Tajikistan front, let me again point out the need to learn the lessons of these events so that the blood of those victims will not have been shed in vain. Otherwise, the outcome will be even more terrible, meaning that our Lord will have turned His face away from us all.

['Abd-al-Hafiz]: Some have accused you of having created the background for the August 1991 coup. In light of that, how do you see Russia's future?

[Kurginyan]: A political struggle is occurring in the country now, and it requires the creation of political beliefs and theories and directing the media to call for certain ideas and their promotion, as well as finding legitimate sources of funding.

Since 1986 I have said that people's morale is more important than technical means. A tank is just iron without a trained crew. I have constantly called on the Communist Party to examine itself and become some kind of Christian–Islamic–Socialist alliance, to focus its efforts on its popularity, because the crew of that tank will not defend the regime as long as it reads the pages of democratic newspapers and can take no organizationally effective measures before the achievement of this basic ideological goal. Neither Mikhail Gorbachev nor anyone else was accused of that at the time.

I once said to Mr. Khrushchev at a conference, "Your plan consists of an attempt to perform an operation with a bad scalpel, which will break and cut the surgeon himself." And on 17 August I raised my voice to say that resorting to force in those circumstances would be a sign of weakness on the part of those using it or resorting to it. I would have welcomed dictatorial rule if I thought it necessary to save Russia, but Yeltsin is not a popular man, nor has he any power or ideology. As to his document on the results of enrichment, it is absolutely without basis. A new coup in these circumstances would lead only to the ongoing destruction of the country. I condemn all attempts to use force as long as it lacks a well-established ideological center. This ideological center will be created in a year or two, and the next situation will require decisive steps against crime, even a declaration of war on it, in the way that Algeria once did.

We must fight crime in every possible way, including mobilizing special forces for that purpose and forming special organizations. No one is planning to begin fighting the opposition in these ways, but the government needs to expand the political arena and keep people informed of what is going on in the corridors of power and politics.

It has been said before, on television, that there is no need for Russian society to fear Yeltsin's authority or the authority of Ruslan Khasbulatov, or the communist opposition. It should worry about the likelihood of the advent of a criminal government regime, with mafia chiefs taking over the reins of power in the country. They would turn the country into a second Lebanon, and the only way to fight them would be with fire and steel. I believe that Russia will enter that phase in approximately 1996. As far as the coup goes, if I had wanted it, I would have accomplished more than the 19 August 1991 coup did.

## 4.24

**Rakhmonov Favors Strengthening Ties with CIS States**

Interfax, 13 April 1994 [FBIS Translation]

Tajik leader Emomali Rakhmonov, at the forthcoming Commonwealth of Independent States (CIS) summit, plans to propose restoring and strengthening economic ties between Tajikistan and other CIS countries, above all, Russia, his aide Saidmurod Pattoev said.

Pattoev told Interfax on Wednesday that, before the summit, to be held in Moscow, Rakhmonov hopes to meet Russia's president, Boris Yeltsin, to discuss the possibility of Tajikistan's joining the ruble zone and of strengthening the Tajik–Afghan border.

The Tajik leadership hopes to receive assistance to restore and stabilize the republic's economy, ruined by the civil war, Pattoev said.

He said Rakhmonov was very interested in the proposal of Kazakhstan's president, Nursultan Nazarbaev, for a Eurasian Union, a community of former Soviet republics closer than the CIS. But the Tajik leader would make no statements on it before he made a detailed study of it, Pattoev said.

## 4.25

**Eurasian Highway a "Historic Chance" for Tajikistan**

Abdugani Mamadazimov

*Nezavisimaya Gazeta,* 3 November 1994

[FBIS Translation]

[Article by Abdugani Mamadazimov, sector head, Institute of World Economy and International Relations, RT (Republic of Tajikistan) Academy of Sciences, under the rubric: "Project": "Construction of a Eurasian Highway Will Consolidate Tajiks: The State Will Have Yet Another Opportunity to Move Ahead in Conducting Reforms."]

The strategic location of Tajikistan, which is situated in the center of Asia, at the crossroads where the basic world cultures and civilizations come into contact, should not convert Tajikistan into a "front line" or "buffer" of one civilization (for example, post-Soviet civilization) with respect to the others. On the contrary, Tajikistan, taking the entire real-life situation into consideration, should play a substantial role both in the West–East (Europe–Asia) dialogue and in intercivilizational interactions of the Western and East Asian cultures.

I might recall that Tajikistan, which, like Uzbekistan and Kyrgyzstan, is part of the CIS, borders on one of the Asian giants—China—and on the south with Afghanistan. In addition, the republic has rather good opportunities for gaining access to influential Asian countries like India, Pakistan, and Iran.

The eastern countries, which have been on the periphery of historical development for several centuries, are gradually coming into the forefront. One hears expressed with increasing conviction the assertion that "the twenty-first century is Asia's century." Even now the eastern part of Asia is contending with the West for economic supremacy over the rest of the world. It is precisely here that one observes the fusion of the Eastern and Western cultures, a fusion that is yielding staggering results and prospects for progress in this region of the world. If one recalls that after the great geographic discoveries Asia gradually yielded its position to Europe, when the world's trade paths shifted from the "Silk Route" on land to maritime, Atlantic paths, then the rebirth of the "Great Silk Route" would symbolize the restoration of Asia's lost positions. The times call for the construction of a Eurasian transcontinental highway. Without even mentioning the global nature of this project, but considering its completely economic side, one can assert that, according to computations made by specialists, the transporting of freight from China's eastern shores to Europe over the existing railroads of the Russian Federation alone will be 20 to 30 percent less than the cost of shipment by sea through the Suez Canal, and the transportation time will be cut in half. But the project for the transcontinental highway in the Central Asian sector must be distinguished from a railroad that encompasses chiefly the northern part of the Central Asian states as it runs across the great steppe of this continent. These adjustments are of a cultural-historical and geostrategic nature.

First, despite the tremendous scope of Central Asia, it is precisely in its central part that one finds the most fertile land, almost all the old cities, the basic architectural and historical monuments, and the bulk of the region's population, and therefore the construction of a highway across that sector will lead to a rapid flare up of business activity on the part of the local population, simultaneously resolving acute problems of a socio-economic nature.

Second, if one recalls that the entire length of the "Great Silk Route" from China to Europe used to be controlled by the Sogdiytsiy-Rakhdonites—ancestors of the Tajiks—then we have the historic right to be a direct participant in this project. (Rakhdonite is the Tajik word *rokh,* "road," and *don,* "knowing," that is, "knowing the caravan road.") We are convinced that the Tajiks, who currently are living in an isolated position, have not lost their historic experience or their ethnic proclivities for establishing economic-trade and cultural contacts with the outside world.

Third, the construction of the road across Tajikistan will lead to Iran—by way of the shortest highways and railroads Dushanbe-Tedzhent-Meshkhed-Tehran—to the Persian Gulf ports or, by way of Turkey, to Europe. Then the transcontinental road through Tajikistan's Badakhshan to the east—across Kulma Pass (Murgabskiy Rayon)—reaches the Karakumskoe Highway. This motor route provides a unique choice: one direction is China and, through it, the countries in the Asia-Pacific region; and the other is to Pakistan and an exit to the Indian Ocean. The dreams that Europeans have had for many centuries will come true—there will be an exit by land to "warm water." In the future the IGA [Islamic State of Afghanistan] has a good opportunity for annexing itself to that highway by constructing a bridge on the Nizhniy Pyandzh-Shcherkhan sector and across Ishkashimskiy Rayon. The inclusion of a greater and greater number of Asian states in the use of the Eurasian highway completely justifies its name and purpose.

Fourth, Russian rubles will function on the territory of Tajikistan. For the time being, the Russian ruble is demonstrating its greatest stability as compared with other new national monetary units with respect to the American dollar and other first-rank currencies. This circumstance is encouraging the establishment of the most effective operations through shipment and the buying-selling transaction, and the rendering of all types of services on the republic's territory. Although Tajikistan is experiencing an acute need for rubles in cash form, the implementation of our project will fundamentally change the situation. With the breakthrough to the east, to the Karakumskoe Highway, Tajikistan will become the connecting link between the countries in the Asia-Pacific region and South Asia, on the one hand, and the CIS and the Middle and Near East, on the other, and through them to Europe.

Fifth, the most important thing, in our view, is that this sector of the transcontinental road across the territory of all of Tajikistan will act as a Tajikistan-wide road of consolidation. Of the three highways linking the capital with the regions, two have a seasonal nature: the passes through the Turkestan, Zeravshan, Gissary, and Darvaz ranges continue to serve as something like demarcation lines between the basic regions of our mountainous republic, reducing to a minimum the contacts among the people inhabiting them. The construction of a Tajikistan-wide road as an inseparable part of the transcontinental one will link all the regions, drawing them into a single economic system.

Since the current level of the republic's financial-economic, material-technical, and intellectual power does not make it possible to implement the entire volume of operations simultaneously, we shall take on as standard equipment the local explosive model of economic development. This model has demonstrated its effectiveness in many countries of the world, but it is especially typical of the rapidly developing countries of the Far East. Therefore we propose implementing our project in two phases:

During the first phase, it is necessary to construct a total of 40 kilometers of road through Kulma Pass . . . and connecting with the Karakumskoe Highway. This 40-kilometer sector connects with the Toktamysh–Murgab highway, and in Murgab we connect with the Osh–Khorog–Dushanbe highway. The rapid growth of that region as a result of the border trade and the through shipments to Dushanbe and farther destinations will promote the gradual weakening of the role played by the state, which, by encouraging the private sector and small- and medium-scale business with the participation of foreign capital, is directing its attention to the second phase in the construction of the road.

During the second phase—the Dushanbe–Khudzhand road, the basic obstacle of which is Anzob Pass, is to be rebuilt. By constructing a tunnel under the pass we will link the capital with Zeravshan Valley, and this will provide a powerful stimulus to the sharp improvement of the road through Shakhristan Pass, with an exit to Leninabad Oblast. As a result we shall emerge from isolation immediately in several directions. To the north, the road connects with the transcontinental railroad; to the west, it will be possible in the future to connect with Iran; to the east, an exit to the Karakumskoe Highway; and to the south, in the future we shall connect with Afghanistan.

Thus, Tajikistan, by standing up for an international cause—the cause of constructing the Eurasian transcontinental highway—has a historic chance of consolidating the entire nation, since a section of this road will unite all its regions.

## 4.26

### Rakhmonov on National Issues and Ties with Russia

Interview by Sergey Ovsienko
*Rossiyskie Vesti,* 26 January 1995 [FBIS Translation]

Emomali Sharipovich, I would like to start our conversation with questions currently of interest to Russia. What is Tajikistan's reaction to the events in Chechnya?

[Rakhmonov]: I have to say immediately that what is happening in Chechnya is Russia's internal affair. As for the Tajikistan people's reaction, it is as follows: All "hot spot" problems must be resolved without fratricidal bloodshed. The Tajik people, who lost 100,000 people during the civil war and saw the republic suffer economic damage to the tune of US$7 billion, reached this conclusion the hard way.

Things are difficult for Tajikistan. But the general political situation is currently tending to improve and gradually normalize. This is also evidenced by the fact that 870,000 refugees have returned to Tajikistan in the past two years, including 90,000 from Afghanistan. Some 7,000 Russians have also returned to the republic.

[Ovsienko]: But the opposition is not abandoning its attempts to seize power in the republic by force. . . .

[Rakhmonov]: I'm not sure it will succeed. Recent years have shown that Tajikistan's problems need to be resolved by peaceful means alone, through negotiation and compromise. Tajikistan's leadership is prepared for this, but the opposition leaders aren't.

Here are just a few examples. Since the agreement "On a Temporary Cease-Fire and the Cessation of Other Hostile Actions on the Tajik–Afghan Border and Within the Country" came into force the opposition has violated it more than fifty times and staged twenty-one armed attacks on border posts. . . .

[Ovsienko]: Mr. President, you are accused of persecuting your political opponents, introducing strict censorship in the national media, and establishing an authoritarian regime. How would you respond to this?

[Rakhmonov]: These accusations against me are nothing new. They appeared literally the day after I was elected head of state at the Sixteenth Supreme Council session in December 1992. My political opponents are still trying to accuse us of things that have never happened in our republic. We are building a rule-of-law, secular state. This means living within the bounds of legal norms prescribed by generally accepted laws and norms of communal living. As for the mass media, they operate within the limits prescribed by the law on the mass media. We simply have no other legal or other norms governing the activity of the media.

[Ovsienko]: What is official Dushanbe's current relationship with the republic's regions—Gorno-Badakhshan, Leninabad Oblast, the Gissar group of rayons? . . .

[Rakhmonov]: Since a constitutional system in Dushanbe was first restored, we have been seeking to maintain equal relations with all the republic's regions. You can see this in both the economic and the personnel policy pursued by the republic's central government. All the republic's regions—the Gorno-Badakhshan Autonomous Oblast, Leninabad, the Karateginskiy Valley, and the Gissar group of rayons—are equally represented in the new government. I am confident that our policy in this respect will not change in the future either, because we realize that a violation of traditional regional policy could lead to the destabilization of the situation in the republic.

[Ovsienko]: You will agree that this is not the only way that the problem of the opposition, including the armed opposition, is being resolved. I am thinking about the process of disarming them. . . .

[Rakhmonov]: The republic's leadership has declared four amnesties and let off more than six thousand people. An edict on the voluntary surrender of weapons was adopted recently. Weapons have been handed in by fifteen hundred gunmen. In one month alone, the population of the Pamir surrendered three hundred firearms.

And in the space of two years government groups have seized around fifteen thousand firearms.

[Ovsienko]: Tajikistan has declared itself to be a democratic, rule-of-law, secular state. In what direction will the republic develop as a result?

[Rakhmonov]: Tajikistan does not intend to copy either a Western or an Eastern model of state system. In the long term we see our republic as part of the CIS.

And the people support the leadership here. The referendum on the new constitution conducted on 6 November last year showed that 95 percent of the population of Tajikistan support this policy of the president. And—an extremely important point—70 percent of the population of the regions where opposition influence is strong came out in favor of the new constitution.

At the same time let me note that the commonwealth itself is still far from perfect. During the years that the CIS has been in existence, more than four hundred different agreements and documents have been adopted. Not all of them are being fulfilled. . . . For example, a number of agreements on the CIS countries' collective security, principally on strengthening the commonwealth's southern borders, which is extremely important for our republic, have not been fulfilled. The Tajik–Afghan border keeps the situation tense not only in our republic but throughout Central Asia.

It is also necessary to take account of the fact that there is no air defense system along the entire length of the border. Violations of Tajikistan's air border are extremely dangerous for CIS countries too. It cannot be ruled out that a border violator could show up in Russian airspace. I have raised the air defense issue with the Russian Ministry of Defense and leadership several times . . . .

[Ovsienko]: Under an agreement concluded with Russia, the funding of all measures to protect the Tajik–Afghan border is based on parity. But to the best of my knowledge, Tajikistan is meeting only 15 percent of its commitments . . . .

[Rakhmonov]: This is indeed the case. But there is an explanation for this that we find persuasive. Tajikistan is the smallest state in Central Asia. But it has the longest southern border, more than two thousand kilometers, mostly with Afghanistan.

The present economic situation in the country, which I talked about at the beginning of our interview, makes it impossible for us to create our own border troops and maintain them. This is why Tajikistan counts on assistance from Russia.

On the other hand, over 80 percent of border post personnel are Tajiks, who are protecting the border along with Russian border guards. We are of course obliged to fulfill our contractual agreements, but at present we find it economically impossible.

[Ovsienko]: Since the conversation has turned to Tajikistan's economy, allow me to ask you—does the republic's leadership plan to introduce its own currency?

[Rakhmonov]: If our parliament gives the go-ahead. As yet only preparatory work is being done.

[Ovsienko]: In building its relations with Russia does Tajikistan intend to cooperate with it within the framework of the CIS or on a bilateral basis?

[Rakhmonov]: One form of cooperation does not preclude the other. After all, around 80 percent of all the republic's economic links are with Russia. Tajikistan benefits from cooperation with Russia, and Russia benefits too, particularly now. One example confirms this: Ivanovo weaving factories are standing idle without cotton. We are prepared to grow cotton for the Ivanovo weavers, but the republic has a fuel shortage . . . .

[Ovsienko]: A few days ago you signed an edict on the formation of a Slavic university in the republic. Does this signify a return to the teaching of Russian in Tajikistan's schools and higher educational establishments?

[Rakhmonov]: This is now the second year that Russian has been taught in the republic's preschool establishments, schools, and higher educational establishments. However, the edict on creating a Slavic university opens up broader prospects for the republic to train the proficient specialists that Tajikistan needs so badly right now.

Construction of the Rogunskaya hydroelectric power station which will earn the republic around US$700 million in income in a single year, is nearing completion. Central Asian countries and neighboring Islamic states are waiting for electricity from the power station . . . .

There are a considerable number of such economic projects in the republic. This is why I made the decision to create the Slavic university, to which we intend to invite teachers from Russia.

## 4.27
### Moscow Prompts Tajikistan to Withdraw from Ruble Zone

Konstantin Levin
*Kommersant-Daily,* 12 April 1995 [FBIS Translation]

Last year the idea of reviving the ruble zone seemed finally to be buried. One only had to uncouple the last ruble car—Tajikistan—but Moscow again promised to prolong for an indefinite period the circulation of the Russian cash ruble on the territory of the Central Asian republic. Having secured the continuation of Russian financial assistance, Dushanbe nonetheless decided in early April to reorient its currency policy and to stop using the Russian currency. Yesterday the group of Russian advisors in Tajikistan also in effect approved the republic parliament's decision to pull out of the ruble zone. [Passage omitted.]

Russia's economic assistance to Tajikistan is, strictly speaking, not a solution to integration problems in the CIS. By rendering Dushanbe not only military-political but also financial assistance, Moscow pursues primarily geopolitical interests in the Central Asian region. Ruble injections into Tajikistan's war-devastated economy are unprecedented. In addition to non-cash credits, the Russian Central Bank has already made 120 billion cash rubles available to the Tajikistan National Bank, and there were plans to provide another 20 billion. The price of political interests, however, has proved too high for Moscow, and therefore it effectively prompted the Tajikistan parliament to adopt the decision to introduce a national currency in the republic.

## Turkmenistan

### 4.28

**Nation's Non-CIS Links Said to Be Favored over CIS Links**

Sergey Tsekhmistrenko
*Kommersant-Daily,* 20 November 1992
[FBIS Translation]

In the last week the growing foreign policy activism of Turkmenistan has attracted attention. On the basis of an analysis of the meetings and visits that took place (and did not take place), the conclusion can be drawn that a concept of the republic's foreign policy, defined by President Saparmurad Niyazov as a policy of "positive neutrality," has begun to take shape.

On 11–12 November the official visit to Ashkhabad of the president of Moldova, Mircea Snegur, was noted. During this visit, a bilateral treaty on friendship and cooperation was to be signed for the first time in the history of relations between Turkmenistan and Moldova. But the visit was postponed until the start of the next year at the initiative of the Turkmen side. As the foreign minister of Turkmenistan, Khalykberdy Ataev, reported, this decision was the result of "matters that appeared right after his visit to Turkey." After this announcement many observers drew the conclusion that Ashkhabad is gradually changing over to a policy of isolationism in relation to its nearby neighbors in order to turn toward distant countries. It is no accident that recently President Niyazov has been limiting his international contacts to discussing potential for investments in the local economy with foreign visitors.

The isolationist trend appeared in the rejection of the proposal of the head of the Russian MID [Ministry of Foreign Affairs], Andrey Kozyrev, who came to Ashkhabad on 6 November, to take part in settling the Tajik crisis. The Turkmen position was formulated in this way: "what is happening in Tajikistan is an internal affair of that state." The refusal to participate in the Russian initiative is related not only to the low effectiveness of the Russian MID's activity in Central Asia. The CIS countries experiencing profound economic crisis are not of interest to Turkmenistan as potential investors. And it is precisely on the basis of this ability that President Niyazov is evaluating his partners at this time.

The Muslim brothers from the Arab countries, including the general secretary of the Organization of the Islamic Conference, Khamid al-Gabid, were the most valued guests of Niyazov in November. After meeting with the secretary on 13 November, the "Turkmen *bashi*" [chieftains] proposed to open a branch of the Conference in Ashkhabad with the status of Committee on Cultural and Social Problems of Central Asia and Kazakhstan. Behind this proposal is the intention to get special aid for Turkmenistan from the Muslim world. It is not impossible that these plans have already received support of rich Arab countries which are influential in the Islamic Conference. According to information from "X," the Saudi King Fahd recently promised Niyazov $3.5 billion in credit. And the influential Kuwaiti Sheikh Abdul al-Baptin, who met with Niyazov on 13 November, reported that under his proposal a number of leaders and businessmen of the Arab countries are creating a Committee to Promote the Development of Turkmenistan.

Turkmenistan's achievements in relation to those of the countries of the West do not look quite as impressive against this background. Western investments have not yet justified the expectations of the Turkmen president. At a meeting with American businessmen held on 16 November, Niyazov accused them of penetrating the Turkmen market too slowly. "You are making a mistake, since Turkmenistan has a favorable position as the gates to Asia and the underbelly of Russia," said the president to Michael Said Ansary, who heads the Mic Corporation, and the former U.S. secretary of state, Alexander Haig, the head of the corporation's "USA-CIS" organization, who had both arrived in Ashkhabad.

Most likely the Western orientation in Turkmen foreign policy will be supplanted even further by the Eastern one after Saparmurad Niyazov's visit to China on 19–23 November. Chinese representatives have repeatedly offered their help to the Turkmen side in the construction and gas- and oil-refining spheres, emphasizing the cheapness and quality of their services. In the opinion of observers, for President Niyazov such reasons are always decisive when determining the foreign policy course. But according to information received by correspondent "X" from some associates of the Chinese embassy to Turkmenistan, Turkmen–Chinese economic ties are experiencing substantial difficulties at this time. They are tied to the fact that the Chinese side is proposing cooperation in the sphere of small and medium-sized business, but the centralized character of the Turkmen economy does not really allow that.

## 4.29

**Russians to Help Establish Turkmen Air Force**
Turkmen Press
*Watan,* 5 August 1993 [FBIS Translation]

Today Turkmenistan's president, Saparmurad Niyazov, met with a military delegation from the Russian Federation headed by Vladimir Zhurbenko, deputy chief of the general staff of Russia's armed forces. Heads of Turkmenistan's military leadership examined questions connected with the further strengthening of Turkmenistan's armed forces. A decree was passed on establishing a national force for air defense on new foundations of command. Orders will be carried out by Turkmenistan's president and the Ministry of Defense.

Other facets of further cooperation greater than that gained in earlier talks about military relations were also discussed by both sides. In the context of these relations questions of joint action connected with the status of Russia's officers serving in Turkmenistan's armed forces were also defined.

"Turkmenistan's military doctrine will remain one of defense," said S.A. Niyazov in the course of the discussions. "Turkmenistan has a policy of positive neutrality, and this has been confirmed in the practice of its own independence. But, as an independent state, it must have a well-equipped army, including an air defense force."

During the talks, representatives of Russia's military delegation said, "In all aspects of building a state, including the military, Turkmenistan has taken an especially honorable and worthy path; a path that considers the interests of its own people."

## 4.30

**Niyazov Reassures Russian-Speaking Population**
Lyudmila Glazovskaya
ITAR-TASS, 17 January 1994 [FBIS Translation]

Whoever we may be by blood, we are all brothers in spirit, brothers in our purpose—it was with these words that President Saparmurad Niyazov of Turkmenistan addressed the Russian-speaking population of Turkmenistan today.

Speaking at the congress of the National Revival Movement, the head of state assured the Russian-speaking population that in the state he heads, there will be no infringement of their civil rights or of the national feelings of all those for whom Turkmenistan has become the motherland. From the lofty tribune of the congress, the president declared that the state and the government guarantee equal democratic rights to all their citizens, regardless of nationality. He pointed out the importance of the Turkmen–Russian agreement on dual nationality, which was signed recently, stressing its role in further strengthening peace, stability, and national accord in Turkmenistan.

## 4.31

**Niyazov Opposes Tough CIS Structures**
Interfax, 14 April 1994 [FBIS Translation]

President Saparmurad Niyazov of Turkmenistan told Interfax Thursday that Ashkhabad was against tough structures within the Commonwealth of Independent States.

"We [CIS countries—IF] have efficient bilateral structures, but the strengthening of independence should not be interfered with by new tough structures. The CIS role should be enhanced, but without those structures, he said. We need to integrate but in a civilized way," he said. Asked about his attitude toward Nazarbaev's proposals on the creation of a Euro-Asian union, he said that "the idea was not new. We should understand the very essence of the idea. One center should not be replaced by another one. If it is a matter of the creation of a new coordinating center, Turkmenistan is opposed. If a new structure will help strengthen independence and our relations, the idea should be analyzed," said Niyazov.

He believes that it is premature to consider the idea of a Euro-Asian union.

Niyazov said that partners should not be sought overseas. CIS partners and Russia are top priority partners for Turkmenistan. "We are grateful to the Russian people that they helped our republic to achieve cultural and other progress," he said.

## 4.32

**President Prefers Bilateral Links with CIS**
From the "Presidential Bulletin" column, compiled by Nikolay Zherebtsov and Andrey Petrovskiy; and edited by Vladimir Shishlin
Interfax, 6 September 1994 [FBIS Translation]

President Saparmurad Niyazov reiterated his republic's preference for bilateral links with CIS member states over multilateral links in the CIS framework.

He believed that agreements that enable CIS centers taking over some of the national sovereignty were non-starters.

[*IF Note:* Niyazov advocated bilateral cooperation and kept his country away from multilateral agreements until Turkmenistan joined, for many observers unexpectedly, the CIS Economic Union treaty at last year's Ashkhabad summit. He is believed to regard the transnational political structures of the Eurasian Union proposed by Kazakh president Nursultan Nazarbaev as centers to which sovereignty would be delegated.]

The Turkmenistan government discussed the agenda of the CIS Advisory and Coordinating Committee [ACC] meeting scheduled to be held in Moscow Wednesday [7 September].

The ACC meeting attended by deputy prime ministers is to lay the groundwork for the meeting of the CIS Council of Heads of Governments, scheduled to be held in Moscow on 9 September. The meeting will be chaired by the Russian deputy prime minister and economic minister, Aleksandr Shokhin, whose mandate of ACC chairmanship expires on 31 December.

Officials in the Russian government's machinery told Interfax that the ACC meeting draft agenda consists of over twenty specific issues in three large fields. Above all, the ACC will discuss joint preparations for celebrations of victory in World War II, in particular, a draft CIS program commemorating the war dead.

Analysis of the implementation of the Economic Union treaty is expected to be the key item on the agenda. The deputy prime ministers will discuss and, possibly, approve the guidelines for the Interstate Economic Committee, which would work for closer integration in the CIS framework and coordination of economic links. [*IF Note:* It was Turkmenistan that vetoed setting up such a committee at the latest CIS summit in Moscow.] Mutual conversion of currencies of CIS member states and a draft of a payment union agreement will also be on the agenda of the meeting.

The ACC will also analyze the implementation of the decisions made by the CIS Councils of Heads of State and of Heads of Government, in particular the Guidelines and Outlook for CIS Integration in 1994–95.

Adoption of legislation in CIS member states on benefits for the families of those who were killed in Afghanistan and other countries where the USSR was engaged in hostilities, an agreement on cooperation in the petrochemical and chemical industries, replacement of locomotives and rolling stock, and the Eurasian Patent Convention will be among the items on the agenda.

The Turkmenistan delegation is expected to discuss with Russian officials the details of Niyazov's visit to Moscow in October.

[*IF Note:* One of the issues in the talks will be an agreement on protection of the Russian-speaking minority in Turkmenistan and a similar document on protection of Turkmens staying in Russia.]

## 4.33

**Russia Seen as "Guarantor of Its Independence"**
Igor Barsukov
ITAR-TASS, 1 October 1994 [FBIS Translation]

The former Soviet Central Asian republic of Turkmenistan sees Russia as a guarantor of its independence, which also promotes its economic development.

"Russia is our key partner. Turkmenistan is a natural ally of Russia and is also absolutely independent in adopting decisions," deputy prime minister of Turkmenistan Boris Shikhmuradov told TASS on Saturday [1 October].

He participated in the 49th UN General Assembly and then visited Washington to meet U.S. officials.

"Cooperation with Russia has become more elaborate and mutually advantageous. We are very grateful to Russia that it not only is the guarantor of our political independence, but also promotes our economic development," Shikhmuradov said.

"Without any reciprocal claims we actively cooperate in the oil and gas industry and in transportation networks," he said, adding that Russia participates in the important Turkmen project of constructing a gas mainland to Europe via Iran and Turkey.

The president of Turkmenistan, Saparmurad Niyazov, is to visit Moscow at the end of the year to sign a whole package of agreements to promote the equal and mutually beneficial relations, according to Shikhmuradov.

He stressed that in the United States he again became convinced that President Boris Yeltsin takes into consideration the interests of CIS partners while protecting the interest of Russia. "All talk about an alleged division of spheres of influence is an invention. Evidently, some people want to aggravate the situation connected with the development of newly independent states after the collapse of the former Soviet Union," Shakmuradov said.

At the same time, he stressed that Turkmenistan believes it is premature to speak about creating any CIS governing bodies.

Shikhmuradov said that while visiting the U.S. State Department on Thursday he handed over the declaration about Turkmenistan's ratification of the Nuclear Non-Proliferation treaty. "On that day the treaty came into force for Turkmenistan," he stressed.

## 4.34

**Niyazov on Political Course and Leadership**
Interview by Yuriy Solomonov
*Literaturnaya Gazeta,* 23 November 1994
[FBIS Translation], Excerpts

[Solomonov]: Mr. President, to recall Kipling's famous phrase: "West is West, and East is East . . . ." Was he not referring to democracy? What do you think, does it incorporate universal values or do we need, for all that, to distinguish British, Russian, and Turkmen democracy?

[Niyazov]: The simplest answer is that democracy is the power of the people. But, you will agree, it explains little in relation to your question. Of course, the discussion should begin with values common to all mankind and the rights and liberties of the individual, regardless of religion, nationality, and country of residence. The right to life. Freedom of movement. Right of ownership. Right to choice of religion. . . . Can it be a question of differences and specifics here? The path toward a democratic society is another matter. Each state has its own. Here Kipling's formula works, so to speak. The East has its own morals, customs, and rules of behavior that have been formulated over centuries. Account has to be taken of this. All attempts to append the European mold right away to the Eastern tenor of life are fraught with serious difficulty. And when the West sometimes tells us that we are lagging behind in the speed of democratic transformations, I always quote the example of the scuba diver: If he comes up from a great depth quickly, wishing to see the sun sooner, he could perish from the well-known bends. Don't rush us. Turkmenistan is moving in the same direction as the world. But you should know the mentality of our people, who are only just beginning to recognize their own statehood, in order to understand why we do not want to and will not blindly copy the experience of others. I do not wish to speak ill of anyone, but you can see what the consequences of haste in this country or the other are.

[Solomonov]: You have one party, no opposition . . . .

[Niyazov]: Yes, one party as yet. The time will come when people have matured for a civilized multiparty and opposition system—all this will come. Some people are, certainly, condemning us at this time, although the numbers of those criticizing us diminish with each political explosion in this former Union republic or the other, it is true. In the West also. I remember that five persons came to see me in 1989, bringing with them the charter of a people's front: Let's form one, they said. The charter was copied letter for letter from the Estonians. Is this serious? I remember that Georgiy Shakhnazarov, an aide to Gorbachev, tried to persuade me: People's fronts need to be created in all the republics, but they will be managed by the state and controlled by the party. There were, after all, people from the Union KGB among those that proposed a people's front with us. Should I call all this the "shoots of democracy"? No, we do not at this time have either a multiparty system or an opposition, but we have no political prisoners either. . . . We are told in addition: You have censorship. Rubbish! We publish everything, but when anti-Russian or anti-Semitic articles appear, what would you advise me to do?

[Solomonov]: Several years ago I would probably have strenuously objected. Now I recall the marches of our Russian fascists, the fence around the Russian parliament building daubed with scurrilous inscriptions, insults against the president, calls for national battles. . . . But, Mr. President, do not be offended with me if I say that such a form of rule has reminded me personally of the power of the CPSU and its general secretary, Politburo, and so forth.

[Niyazov]: I am not offended because you are wrong. In the waning period of the USSR I took part, as you know, in the work of the Politburo. I would sit there, as the representative of a small republic, quietly, detached, remaining silent, and listening to others. And I have to tell you that we could not get anywhere with this "collective intellect." It was clear from these "clever" speeches that the CPSU and the USSR were doomed. This was my first discovery of those times. The second might appear altogether naive, but I have for many years perceived the general secretary as a person who had been popularly elected. Or, at least, elected by all the communists of the Soviet Union. But the choice had been made by about fifteen such persons as himself! The choice that was made by the Turkmen people, entrusting the presidency to me, was surely different from the personnel intrigues in the upper stratum of the CPSU.

[Solomonov]: So you are not one of those that regrets the disintegration of the Union, the demise of a great country?

[Niyazov]: No, I do not. I recall that a year before the Emergency Committee I was vacationing in Foros, at the same time as Gorbachev and Yazov. The general secretary once invited the minister and me to tea. We spoke about many things. I said in some context or other that the first person in the state should know precisely who his associate is, who his secret enemy. A year later, on 9 August 1991, I called Yazov to seek a deferment of army service for young shepherds. I inquired at the same time why he was not vacationing with Mikhail Sergeevich. "I never did vacation with him. Last year this was a coincidence," Yazov said hastily, and I wondered as to why this reaction. The putsch occurred ten days later.

It was clear that it was doomed. It was an attempt to save not the USSR but those that were in power themselves.

I do not regret what happened. The empire was at the end of its working life: temporal, ideological, economic. The reason for the disintegration was not the personalities of individuals. It was a historical inevitability caused by a violation of justice. I have observed repeatedly that our country obtained much as a part of the USSR, primarily in terms of education. But Turkmenistan was never an equal republic. We produced oil and gas, but no one ever knew where it went or at what price. We only knew that we received commands—ship it out, pump it out, send it out. . . . Not one person in Turkmenistan knew where the profit from the sale of 70 billion cubic meters of gas, 15 million tons of oil, and 1 million tons of cotton went. The Center had the nerve to rank us as a backward republic here. And I should regret this?!

But I would like to emphasize here that we are not today casting aspersions on those with whom we lived together. Everyone suffered, it was bad for everyone. Let us build new, equal, bilateral relations. We recently conducted negotiations with the president of Ukraine and reached an understanding on the human level as regards both gas and the debt. We understand Ukraine is not now in a position to immediately pay off its debts running into the billions. We need to seek a way out together, therefore, but to talk as equals. Today, following the disintegration of the USSR, we have states with different economic levels, a different degree of disintegration of the internal infrastructure, and, finally, with different notions of reforms, but this does not mean that we should not cooperate. We cannot avoid one another. We actively support the CIS, therefore. A union of independent states is a sensible mechanism. It is too late now to stigmatize those who thought this up, and to propose endless scenarios of something new is foolish. The idea of the CIS should be welcomed if only because it saved the enormous territory of the former Union from conflagration. The CIS should today be seen, therefore, as a framework mechanism, as a mode of reproduction of civilized relations between equal states of the commonwealth. And the responsibility for participation in this union should be equal. Many people still adhere to the stereotype of Moscow, Russia, being the locomotive that pulls the whole train. It does not! Russia also is experiencing colossal difficulties. If we have conceived of the CIS as a railroad train, each car today needs an engine . . . .

[Solomonov]: Mr. President, how would you comment on the judgments to the effect that Turkmenistan, like other Central Asian republics, has since the disintegration of the Union turned to the East and is moving in the direction of the Islamic world . . . .

[Niyazov]: Listen, I have a Russian wife, my children are Russian! Where am I going? When you hear such talk, try to think: Whom does this benefit? Turkmenistan and Russia are bound by special, uniquely evolved relations, whose severance or winding down is hard, impossible, to imagine. Russia is a most important economic, military, and trading partner and, if you will, a pillar of our independence. Today, and this needs to be acknowledged by both us and the Russians, trade turnover between our countries has declined. And there is immediately increased talk to the effect that Turkmenistan has moved to the East! But let us look at the root of the matter. Here is an example: It is not directly connected with Russia, but it explains much. In Iran we purchased two hundred buses for $19,000 each, fine diesel buses. I could have purchased the same in Lvov and thereby strengthened Turkmen–Ukrainian relations, only I would have had to pay $200,000 per bus. What, pray, has Islam got to do with this? Russia wants to purchase cotton from us at 70 percent of the world price, Turkey and Iran, for 100 percent. . . . Yes, the volume of trade with Russia has fallen 12 percent, but no one in Russia is giving any serious thought as to why this has happened. Using talk about Islamic fundamentalism as a cover is easier than learning to trade in civilized and mutually profitable manner.

When, however, we are asked about fundamentalism, we have, to be honest, a hard time understanding the point at issue. We do not have such a problem. The Turkmen have a very solicitous attitude toward Islam as a religion that essentially saved the nation and helped it to recognize itself and begin to build its life on the basis of the highest spiritual and ethical ideals and principles. But for the Turkmen Islam was never a source for feeling superior to people confessing another faith and it never taught irreconcilability with them. I personally generally believe that Christianity and Islam are religions that are very close to each other. We essentially have the same prophets, the same spiritual and ethical precepts. Both religions, unless ill-intentioned adjustments are introduced to them, presuppose primarily man's intimate conversation with the Almighty. For this reason attempts to somehow politicize Islam are doomed with us. Yes, we are now attempting to revive our religion, but this revival poses no threats to anyone because the purpose of this work is the revival of the culture and history of our people. As far as manifestations of religious or national intolerance are concerned, they will be cut short most strictly as long as I am president. . . .

## 4.35

**Niyazov Favors "Speeding Up" Trade Deal with Russia**

Interfax, 7 April 1995
[FBIS Translation], Excerpts

Turkmenistan's President Saparmurad Niyazov thinks it necessary to speed up the development of a long-term, at least five-year, trade and economic agreement with Russia. He said this on Friday in the course of a five-hour conversation with members of Russia's government delegation headed by Deputy Prime Minister Aleksey Bolshakov.

Bolshakov declared to Interfax that in the course of this meeting the parties discussed a wide spectrum of issues, as the fulfillment of this long-term agreement depends upon their successful solution. In particular, it concerns cooperation in the field of surveying and production of oil and gas, development of transport communication, and military-technical cooperation.

As Bolshakov said, Niyazov appointed Turkmenistan's deputy prime minister Valeriy Otchertsev head of the permanent acting combined commission on developing this long-term agreement and its further implementation.

Bolshakov underlined that "we have reached complete mutual understanding with Niyazov on all issues." In addition, Niyazov expressed great interest in further strengthening and expanding comprehensive cooperation with Russia.

In the course of the conversation, the officials also discussed issues related to military-technical cooperation which, to quote Turkmenistan's deputy prime minister Boris Shikmhuradov, "do not contradict the positive neutrality to which Turkmenistan adheres."

Shikhmuradov underlined that specific agreements on these issues would be reached during Niyazov's official visit to Moscow where he is likely to come to "immediately after the celebration of Victory Day in May. . . ."

## 4.36

### Shikhmuradov Views Niyazov Appeal for Neutrality Status

Interfax, 11 April 1995 [FBIS Translation]

President of Turkmenistan Saparmurad Niyazov has officially appealed to the world community to support the idea of granting his country the status of a neutral state. Ukraine was informed of this by Turkmen Foreign Minister Boris Shikhmuradov.

"We have the support of many Asian countries, the question has also been agreed with the leaders of several European nations," he said. The minister stressed that there is understanding on this matter in Russia.

During the talks between Ukraine, Turkmenistan, and Iran in Tehran on 8–9 April, Ukraine also supported Turkmenistan's desire to be a neutral state.

"We are ready to play the role Switzerland has been playing," Shikhmuradov said. He stressed that such a status would not mean his country would sever ties with the CIS. He also added that at the first stage bilateral consultations would be held with different countries. "Later everything will be registered through appropriate UN structures," the minister said.

Shikhmuradov stressed that Turkmenistan can be given the status of a neutral country on the basis of the 1907 international covenant formulating the key principles of neutrality.

The minister stressed that Turkmenistan has been pursuing a policy of "positive neutrality" for several years.

# Uzbekistan

## 4.37

### Karimov Press Conference on CIS Relations, National Borders, and Other Issues

Interview by Vitaliy Portnikov
*Nezavisimaya Gazeta,* 15 May 1992
[FBIS Translation], Excerpts

[Question]: Mr. President, what position will Uzbekistan take on defense matters at the upcoming meeting of CIS heads of state in Tashkent?

[Answer]: Uzbekistan's position has been an open one from the start. Kazakhstan and other republics agreed with the approach we announced at the meeting in Minsk. Basically, every republic should have its own armed forces—nominal armed forces (in particular, Uzbekistan does not intend to have a force of more than 25,000 to 30,000 men), and they will form part of the OVS [Combined Armed Forces] under the command of Shaposhnikov. At the same time, we believe that the existence of strategic troops is a matter to be decided independently, and we have contributed our bit to that decision. At previous meetings I spoke in favor of the arrangement adopted in NATO. And the fact that Russia, along with Kazakhstan and the other republics, is creating its own armed forces indicates that everybody accepts this arrangement now. Obviously, we will come around to it. It would also be a good idea to conclude a regional agreement on mutual security, a proposal that was made at the meeting in Bishkek. We believe that Russia, as the biggest state, could, by means of such an agreement with the Central Asian republics and Kazakhstan, guarantee stability in the region. The idea is still being worked out. If successful, it would guarantee the inviolability of the borders of the former Union in our region. It would guarantee stability that all the states of the region could endorse . . . .

[Question]: How do you view the prospects of Uzbekistan's cooperation with Turkey and Iran?

[Answer]: When Uzbekistan, as an independent sovereign state, is faced with the dilemma as to which path of development to follow, which path would meet our interests, I say unequivocally that the Turkish path of development is more acceptable to us as a secular, civilized path of society's development. What we need to do is work out our own way, drawing on Turkey's experience. The Iranian model is not right for us. We had occasion to exchange ideas about this in Bishkek and in Ashkhabad, and it seems to me it is the opinion of Central Asia.

[Question]: When you were at the conference of heads of state of the CIS in Kiev, you were in solidarity with Ukrainian approaches to the question of the future of the CIS. Weren't you disappointed that the Ukrainian president, Kravchuk, did not attend the meeting in Tashkent?

[Answer]: Ukraine's position is well known—there's no vacillation there. As for my attitude toward it, I think that every republic, every state in the CIS, must find its own way, and we must respect these paths of development. At the same time, I do recall that when Leonid Makarovich called the CIS a myth I stated, at the same press conference, that CIS is a necessity. Uzbekistan's position on this matter is unequivocal: The CIS must exist. We don't need a return to the old structures that Gorbachev talks about; he has a different understanding. He would like to go back to his Novo-Ogarevo options. Uzbekistan is against that. . . .

[Question]: How do you assess the future of economic cooperation within the CIS?

[Answer]: A lot of people in Uzbekistan think that the CIS should continue to develop, provided that every state remains independent and sovereign. That is the condition for further development of the CIS. Cooperation, interaction, mutual aid—these are all components of its future, as long as the states can build their own independence. But the Russian government under the leadership of Gaydar—I emphasize, under the leadership of Gaydar—has done a lot to speed the transition to market relations and develop reforms. It is by no means confiding its next steps to us. This disturbs us greatly. What will the reform come up with next? What will the Russian government revise tomorrow? And what position will that put us in, since we're functioning in a unified ruble system? If we want to retain the ruble zone we will have to resolve several issues. The banking system, at least, must not be in Russia's hands. That would be wrong; the system should be placed at the CIS level. The printing of money, and monetary circulation generally, should be the province of the CIS rather than any particular government, which could bring any state it pleases to its knees—could do so tomorrow. I will admit that we are getting ready to have our own currency, but when it will happen I can't say. The Baltics, which have been announcing their own currencies for two years now, have yet to make it a reality. The introduction of our own currency is not something ordinary, to be taken lightly. It is tied to all aspects of our life. Unless it is carefully prepared for, it could lead to economic collapse. We will not introduce our own currency without figuring out all the consequences.

[Question]: A lot of people are saying the Russian government is not paying attention to Central Asia, that Russia's influence here has been weakened.

[Answer]: That depends on how you view Russian influence. If it is through the prism of imperial thinking, then we're glad that it is becoming weak here. The principle we advocate and vote for is the full independence and sovereignty of each state, and in this regard I fully endorse Ukraine's position. As for economic, cultural, and human relations, though, I am against weakening those ties. We need to be with Russia in all the complex situations that await us on the path of independence, on the path of Uzbekistan's entry into the world community. These relations must be on the basis of equality in all respects. . . .

## 4.38
**Interview with CIS Leaders**
Ostankino Television, 16 April 1993
[FBIS Translation], Excerpts

[Karimov]: I would like to add something to what Leonid Makarovich has said. I am aware there is currently very great agitation in Moscow and in Russia as a whole among forces suffering from nostalgia for the old Union. Many central publications are, generally speaking, openly attempting to blame all those who in their eyes destroyed the Union and saying that the only way out of the present situation in all our states is to return to the old system and to create a USSR.

As a representative of Uzbekistan, I want to say that we have no such nostalgia. We will never return to the past. For us the only way is forward, only forward. I wish to assure you that in this respect we fully support what is being done by the executive authority under the leadership of President Yeltsin. I am not saying that I am a 100 percent supporter of all reforms. I understand the many mistakes and difficulties involved in effecting reform in Russia. But nonetheless I state categorically that we are equally interested in the victory of reforms, the implementation of reforms, and the implementation of the course being followed in Russia. I am speaking in the name of Uzbekistan.

In this respect, I would like to declare that when it is sometimes said that things are bad everywhere and that there have been no good signs anywhere since the collapse of the Union, I categorically disagree with that.

Many people even forget that, alongside the difficulties we have in the economy, there is a great deal we have worked to achieve, a great deal we have fought for and won since the

Union ceased to exist. It is this aspect to which I should like to draw your attention. And the people who live as sovereign, independent states understand full well where that independence came from. There will be no going back. And so when we speak such a lot of economic difficulties, we must understand that a transition from one system to another is never easy and never happens so well that we live better than we did before.

Therefore we must give this its due and we want to say: independence relates not only to the economy. Independence also means spiritual independence. Independence is above all the independence of nations, the independence of states. In this respect, there is no way back for us. And all those who try to campaign for that quite simply do not know the situation, I think, do not know how people in our regions feel. They do not know the situation or the mood of the people who live in Uzbekistan. . . .

## 4.39

### Karimov Reports on the Economy and Rejects Dual Citizenship

From the "Presidential Bulletin" column compiled by Andrey Pershin, Andrey Petrovskiy, and Vladimir Shishlin; and edited by Boris Grishchenko
Interfax, 29 December 1993 [FBIS Translation]

President Islam Karimov on Tuesday made a report to parliament on reforms in the republic.

The achievements made by Uzbekistan were the result of peace and stability, he said. But "processes in neighboring countries exercise a strong influence on the social and political situation in the republic." He expressed concern over the results of the last parliamentary election in Russia, where "chauvinistic groups are rearing their heads."

Karimov reaffirmed his negative attitude to the idea of dual citizenship. "The introduction of dual citizenship would give rise to inequality between the native population of the republic and the ethnic groups having the citizenship of two countries. In addition, it would weaken their patriotic spirit with regard to the republic where they live," he said.

[*IF Note:* Dual citizenship was a key issue at negotiations between Karimov and Russian Foreign Minister Andrey Kozyrev early in December. Kozyrev insisted that the Uzbek leadership ensure all rights for the "Russian-speaking population," including the right to dual citizenship.

Karimov once again stated his intention to pursue an economic policy coordinated with those of other Central Asian states. He said that he would visit Kyrgyzstan early next year, and that soon after that Uzbekistan would be visited by President Nursultan Nazarbaev of Kazakhstan.

Karimov named privatization, both in industry and agriculture, as Uzbekistan's main economic task.

He paid considerable attention to the introduction by Uzbekistan of its own currency. He rejected "destabilizing" rumors that som coupons would be abolished and that banknotes of five thousand and ten thousand som coupons would be removed from circulation.

He said the som coupons would be replaced by a permanent currency around next July.]

[*IF Note:* The current parliament session approved a budget for next year. It also passed laws on elections to the Oliy Madjlis (the new parliament), on the taxation of individual property, and on income bases for Uzbek and foreign citizens and stateless persons. The elections to the 150–seat Oliy Madjlis will be held next summer, after the term of the present 500–seat Supreme Soviet expires. The new law allows only political parties and governmental bodies to nominate candidates to the new parliament and prohibits members of the armed forces, officers in the Interior Ministry and national security service, clergy and members of religious political parties from running for parliament. Civil servants are allowed to run, but they will have to leave their posts if they are elected.]

## 4.40

### Karimov Discusses the Economic Situation

From the "Presidential Bulletin" column compiled by Nikolay Zherebtsov and Andrey Petrovskiy; and edited by Vladimir Shishlin
Interfax, 12 April 1994 [FBIS Translation]

In a recent exclusive interview with Interfax, President Islam Karimov described the forced transition to a local currency as the biggest problem of his country. He claimed that Uzbekistan resisted the withdrawal from the ruble zone for a long time. He said the worst consequence of the change was the spiritual separation of two million ethnic Russians from their native land. He favored a discussion by interested members of the CIS of the formation of a common monetary system with Russia.

Meanwhile, observers say the president is taking additional steps to stabilize the local currency. Karimov signed a government resolution stopping all transactions with Russian banknotes as of 15 April.

The purpose of the decision is to strengthen the circulation of som coupons and increase their purchasing power. The document permitted all legal entities (public and private) as well as individuals to sell goods and offer services for hard currency on the condition that all transactions are registered.

In keeping with the decision, after 15 April the exchange rates of the som will be determined through regular trading at the Uzbek Currency Exchange. So far exchange rates have been set by the Central Bank of Uzbekistan on the basis of the data supplied by the Central Bank of Russia.

The document provides for the formation of a network of exchange outlets for the public and allows private individuals to exchange currencies provided they have a license from the Central Bank.

Under the decision after 15 April all companies must sell to the Central Bank 30 percent of their foreign currency returns out of which 10 percent will be purchased by the Ministry of Finance for the Uzbek currency reserve and 5 percent by the government of Karakalpakia (an autonomous republic of Uzbekistan) and regional administrations for the formation of local currency reserves.

After 1 May, in keeping with the decision, settlements with companies in other former Soviet republics will be conducted only on a clearing basis or in hard currency.

[*IF Note:* At the moment in Tashkent the Russian ruble sells for 10–11 som and $1 for 200,000–22,000 som.]

## 4.41

**Russian Banknotes Cease to Be Legal Tender**
Shakhnoza Ganieva
"Novosti" newscast, Moscow Ostankino-1 Television
13 April 1994 [FBIS Translation]

A new decision by the cabinet has been published in Uzbekistan. The decision stipulates that banknotes issued by the Central Bank of Russia will no longer be accepted for any form of payment anywhere in the republic from 15 April onward. However, commercial organizations and enterprises, as well as private individuals, will be allowed to sell goods and services for freely convertible currency. Exchange rates between the som coupon, the national currency, and foreign currencies will be fixed in regular auctions at the Uzbek currency exchange, bearing in mind the need to maintain parity between prices on the domestic and foreign markets.

In order to establish a gold and hard currency reserve in the country, all enterprises, irrespective of their form of ownership, will now be obliged to sell the Central Bank 30 percent of their hard currency earnings. Beginning 1 May, all Uzbekistan's settlements with CIS countries will be transacted solely through clearing or in freely convertible currency.

## 4.42

**Karimov Says Russia Has a Great Future**
Interfax 14 April 1994
[FBIS Translation], Excerpts

[IF]: . . . it seems, Uzbekistan, is one of the former USSR republics that has not experienced a substantial industrial recession recently and witnessed even a kind of raising in several fields of industry. Does it mean that you have found a recipe to fight with economic misfortunes?

[Karimov]: . . . We act given the people's wisdom "Having not yet built a new house, do not destroy the old one." Nobody intended to immediately destroy the previous system, toward which the new radical democrats feel hatred; the question is, rather, one of its gradual restructuring because it is impossible to force this process. There is no secret. I am convinced that all reforms should be implemented gradually and that one should not blame the old system for all mortal sins. We followed a creative road but not the path of destruction of everything what was created before.

I believe that the decision of the Russian authorities to widely celebrate the fiftieth anniversary of the victory of the Soviet people over fascism is absolutely just. Now Russia cannot live without the army, without the mighty military-industrial potential that enabled it to support the balance in this bipolar world.

Those republics that primarily shipped raw materials to the former USSR have suffered less since the destruction of economic ties within the collapsed Union than those that produced manufactured goods. For example, Belarus, which has no natural resources, depended entirely upon other Union republics. Such a dependence is extremely dangerous.

Over the past few years by means of barter, in exchange for cotton and other products, Tashkent managed to receive those industrial goods that help to maintain rather a high level of production. But the main thing is that nobody in Uzbekistan intends "to break the old world down to its foundation and then to build a new one."

[IF]: To what do you attribute the relative success of Uzbekistan?

[Karimov]: I believe that everything depends upon those at the helm of this state. For example, "chiefs of laboratories and chief research employees who pretended to be leaders of this nation were ignorantly fulfilling their duties. For example, who could Azerbaijan's former president, Abulfaz Elchibey, a historian, know of economics and policy? This man worked all his life among dusty book shelves."

An analogous situation was found in Georgia as well. In

this republic a philologist, Zviad Gamsakhurdia, was heading the state. For example, I am not going to write poems and to play piano as did, say, the former head of the Lithuanian parliament, V. Landsbergis. When professional politician Algirdas Brazauskas came to power, life in the republic began to return to normal.

There existed a threat that incompetent people would come to power in Uzbekistan in 1989–90, as happened in the Caucasus.

I must say with great regret that such a negative situation obtains in Russia now. The unwillingness of the people to participate in the elections testifies to this fact. And if the people do not trust their ideals, then reforms cannot be implemented. This is a sad fact because Uzbekistan cannot flourish if Russia does not also flourish. I have said this many times.

[IF]: According to available information, about 40 percent of Uzbekistan's citizens do not know the Uzbek language, which is the national language. . . .

[Karimov]: As many as twenty-two million people live in Uzbekistan, of which 77 percent are Uzbeks, Tatars, Kazakhs, Tajiks, Kyrgyzers, Turkmen, and the so-called Russian-speakers representing about 10 percent of the total population comprise other nationalities living in the republic. The overwhelming majority know the Uzbek language; some of them do not speak fluently, but there are some people who do not speak it at all. On the other hand, there are Russians in the republic who speak the Uzbek language brilliantly, for example, Vitaliy Segedin, director of the Akmalyk mining-metallurgical plant.

The Uzbek language decree is a liberal one—the authorities do not demand that all people should freely possess the language, but I believe that they should demonstrate respect for our language because this is sign of respect for the people. I think that the republic's citizens should possess the language at least on the level of everyday association.

By the way, almost 99 percent of Uzbeks know the Russian language, whereas this percentage is much lower in the Transcaucasian and Baltic states. Uzbekistan is an example of respect for the great Russian language. . . .

[IF]: What is the state of the republic's foreign economic relations?

[Karimov]: The interest of foreign capital in Uzbekistan is great, and it seems to me that the Russian leadership perceives this with some jealousy. More than twelve hundred joint ventures are operating in the republic now, including large international corporations. There are branches of six prestigious foreign banks in the republic, and several reputable firms have opened representative offices in the republic.

I believe that Uzbekistan can teach Russia in some areas, for example, in the field of infrastructure construction and in foreign economic policy and transition to market relations. By the way, I have written a book on this problem. At present Tashkent has currency deposits in the world's best banks totaling more than $700 million. What CIS state can boast such a reserve in hard currency created within one to two years? It's my secret how Uzbekistan has managed to achieve this.

Achievement of trade parity with Russia is on today's agenda. Unfortunately, Tashkent buys practically everything from Russia without selling anything to it. At the same time one should not perceive Russia as "a milk cow." It is the richest power, and Russians can be proud of its potential.

[IF]: What is your assessment of the case on corruption in Uzbekistan's supreme echelons of power investigated by the USSR general prosecutor's office and Sh. Rashidov's role (the late first secretary of Uzbekistan's Communist Party—Interfax)? According to the investigation, the greatest number of millionaires [in the USSR] lived in the republic. Where is their money?

[Karimov]: You are right—there was corruption in the republic. The main root of the corruption was the upward distortion of the results achieved in the production of cotton, which were later on transferred into money by Uzbekistan's leaders, who were gaining great profits. However, I believe that the Union authorities were combating these illegal actions by vicious methods, and they did not observe laws themselves.

Nobody took into account that the republic had its own constitution, its own supreme soviet, its own legislature. They arrested deputies of the Union parliament violating their status. Those suspected were hunted with dogs; they were handcuffed without any sanctions of the attorney general.

I believe that the methods used by the investigators of the general prosecutor's office, Telman Gdlian and Nikolay Ivanov, personified complete lawlessness. Such ignorance of the national dignity of the republic's citizens led to the bloodshed in the course of interethnic clashes in the Fergana Valley. In my opinion, all Uzbekistan and, probably, all Central Asia could have exploded as a result of this. The republic could not control Gdlian and Ivanov. They were acting on the order of the CPSU [Communist Party of the Soviet Union] Central Committee.

I believe that an additional investigation of this case should be instituted in the near future. The thing is that funds earned honestly were confiscated from the people accused of corruption at that time. Women were deprived of wedding rings and jewelry received from their ancestors. This was pure humiliation. Where has all the confiscated jewelry disappeared to? How much has remained in Mr. Gdlian's pockets and in the pockets of other investigators? They have not handed over anything to Uzbekistan's Treasury. I think that Tashkent can raise this issue today. Nevertheless, we do not want to trouble people in this difficult time. . . .

[Karimov]: Never in my life did I think that I would be a leader of such a scale and carry a colossal responsibility for the fate of the whole country. For a long time I worked in planning and financial organizations—I was deputy chairman of the government and headed the republic's State Planning Committee. After the discussion of my candidacy in the Politburo of the CPSU Central Committee and at the plenum of the Central Committee, I was confirmed as first secretary of Uzbekistan's Communist Party. It happened just after the tragic events in Fergana. At the same time I believe that I am a member of the party by accident.

I am convinced that any society, even the most democratic one, is not immune to sin. Now, Uzbekistan is accumulating experience while building its statehood. I think that elements of an authoritarian system are needed during the transition period. If the people elected me president, it means that they are confident in my program and see that it will bring stability and peace to this nation. I believe that I must fulfill my duty to the people, that is, I must fulfill my program. I must admit that I have never resorted to independent methods. However, I think that sometimes it is necessary. . . .

[IF]: The mass media often reports about the growth of anti-Russian sentiment in Uzbekistan. . . .

[Karimov]: I do not want to use passionate words, but I'd like to underline that Uzbekistan always treated Russia with great respect and gratitude. This is a sincere feeling.

The expression the "near abroad" is widely used now. Frankly speaking, I do not understand what it means. I think that such terms artificially separate us from the Russian Federation. Authors of this term do not understand themselves what is hidden by this word combination.

Russia has a great future. After the collapse of the USSR, independent young states experienced a kind of euphoria. In some of the republics this independence turned into a fetish. But life has proved that sovereignty is not just loud statements and declarations. Now many people understand that Russia should be the guarantor of peace and stability in Central Asia.

Not long ago in Davos, the prime minister of Pakistan, Mrs. Bhutto, declared that today, when there is no more iron curtain and the republics of Central Asia are open to the outside world, it's high time to start an integration process in this region by creating an association in which Pakistan will be the leader. I categorically oppose this point of view. I believe that the Commonwealth of Independent States is a serious organization and one must take it into consideration.

---

## 4.43

### Karimov Says Russia Is "Losing Uzbekistan's Market"

Vyacheslav Bantin
ITAR-TASS, 17 May 1994 [FBIS Translation]

---

"Russia is losing Uzbekistan's market," Uzbek President Islam Karimov warned in an interview with ITAR-TASS today. The president arrived for a four-day visit to Japan on 16 May.

"It appears that Moscow does not realize that Russia could be squeezed out of Uzbekistan as a result of intensifying international competition for the emergent market in our republic," the president said. "Furthermore, we are eager to win foreign investment in order to develop our economy and to establish close ties with foreign enterprises and attract them to our market." Islam Karimov stressed that Uzbekistan wanted "to retain former links with Russia, but on a new basis."

The president of Uzbekistan called for the establishment of Russo-Uzbek joint ventures based on close ties between industrial enterprises in the two countries dating from the Soviet period. "We are establishing joint ventures with firms in various foreign countries. There are now about fifteen hundred joint ventures in our republic, but there is not one Russo-Uzbek company," the president noted. He pointed out that Russian President Boris Yeltsin also fully agreed that the two republics should set up joint ventures.

Islam Karimov stressed that "the peoples of Russia and Uzbekistan are bound by close spiritual ties." "The Russian people have always stood beside us during those times in our history when things were very difficult for us, and we shall not forget this," the president said.

---

## 4.44

### Moscow Evening News Programs to Be Cut

Moscow Echo Radio Station, 4 June 1994
[FBIS Translation]

---

Uzbekistan plans to stop transmission of all news from Moscow. Beginning next week the main Ostankino evening news programs will be cut. The republic's national television and radio company intends to fill the slots with entertainment programs.

## 4.45

**Karimov Discusses CIS Meeting, Tajik Elections, and Foreign Aid**

News Conference, Uzbekistan Television
29 November 1994 [FBIS Translation], Excerpts

[Karimov]: The third question I want to draw your attention to is the question connected to the meeting of CIS heads of state that took place, as you know, on 21 October 1994.

This CIS meeting examined issues of the further deepening of integration and also agreed on measures to resolve the problems that exist today. I will make use of today's presence of the diplomatic corps and representatives of the press to express once again, to repeat, my attitude to the principle questions, in our view, connected to the prospects for the development of the CIS.

Unlike the supporters of the creation of various kinds of new forms of associations proposing that work begin on forming suprastate political structures and superstructures like the so-called Eurasian Union, in Uzbekistan we believe that today there is no alternative to the CIS, despite the weak points that can be noted in it, and that it meets the national state interests of all its members.

A somewhat paradoxical—I would say incomprehensible—situation is forming: They want to lead the public of the CIS, all CIS countries, into delusion. The fact is that CIS countries that exist today in the post-Soviet area are experiencing various degrees of crisis—politically and economically. It is precisely the disastrous situation of people that sometimes causes nostalgia for those times when the USSR existed. They want to use this to create the impression that at one time everything was good, and now independent princes, as they say, have appeared—independent principalities that are incapable of dealing with their situation. In this sense various forms of association are proposed that will supposedly save the situation; namely, the creation of various suprastate political structures such as a single parliament, a single citizenship, and various suprastate executive structures. But they are given a certain decorative democratic nature, and with this they want to lead everyone who lives in the post-Soviet area into delusion.

Tell me, please, if there is a single parliament, a single citizenship, a single armed forces, what should this state be called? And various, I would say, vagaries around this issue have a single nature—and I will repeat it once more—to bring ordinary people into delusion.

In our view integration has to begin with the economy. As I said at the CIS meeting in Moscow, a house is not built starting with the roof; a house is built starting with the foundation. And the foundation in all times has been the economy. It is economic integration that will create the prerequisites that will allow and ensure integration in all other areas, first and foremost in questions of culture, spirituality, and in questions connected as a whole with cooperation between the states that today make up the CIS.

This tendency comes up against those decisions that were taken at the CIS session, such as on the basic directions of CIS integrational development, the prospective plan for the integration and the development of the CIS, the creation of a payments union, and also the creation of an interstate economic committee. We believe that the formation of these structures is that beginning which will ensure profound integration, primarily on economic questions.

I would like to draw your attention to this fact—everybody, when much is being said about the Eurasian Union, when they speak to their electorates about the need for integration, indeed do little to implement their declarations with specific, practical steps. The creation of an interstate economic committee must ensure economic integration and the implementation of the document adopted eighteen months ago on economic union. More than a month has passed since the decision was made to create the economic committee, but no practical steps on starting the work of this body are being taken. This latest decision may really become the 401st decision made by the CIS that remains on paper. [Word indistinct] to those who say an awful lot and loudly at all forums about integration but in deed do virtually nothing. In the solution of all these problems, the role of Russia as a great power is very significant, but in practice Russia does not go beyond declarations.

# The Transcaucasus

The Transcaucasus is a more ethnically explosive geopolitical region than Central Asia. Leaders such as Armenia's Levon Ter-Petrosyan and Georgia's Eduard Shevardnadze have been forced meekly to submit to Russia's bilateral military pressures and to its basic prescriptions for CIS integration as they try to re-establish order and stability in their war-ravaged countries. For Shevardnadze, this represents a significant shift from his initial refusal even to join the CIS or to accept its role as a supranational organization. The documents selected for this region illustrate the evolution from rejection to acceptance in the attitudes of the three principal Trancaucasian leaders. Nevertheless, the extracts chosen also depict each leader's subtle rejection of Russian domination. Azerbaijan tends to display the strongest opposition to Russia's tactics, being one of the most richly endowed with natural resources. Soon after the collapse of the Soviet Union, Abulfaz Elchibey, a non-communist member of the Azerbaijani Popular Front, was elected president of Azerbaijan. A recurrent theme in Elchibey's speeches, as shown here, was Russia's intervention in the ethnic quarrels brewing in the Transcaucasus. Nevertheless, following the political isolation of the Popular Front, he was driven out of office by the leader of the Azerbaijani Supreme Soviet (who was also the former Communist Party leader and a KGB general), Geydar Aliev. Aliev overrode Elchibey's CIS policies, signing the CIS charter in October 1993. Aliev has positioned Azerbaijan thoroughly within Russia's political and military orbit, although once in power, he too became more independent-minded about exploiting Azerbaijan's rich oil and gas reserves. The Azerbaijan documents depict his attempts to form an energy consortium with Western partners without Russia's approval, which has led to his own troubles with Moscow. Aliev's energy policies tie in directly with Russia's move to control the energy resources of the Caspian Sea, as documented in Chapter 3.

## Armenia

### 4.46

**Steps for Creation of Army Urged**
Mark Harutyunyan
GAMK, 29 October 1992 [FBIS Translation]

Armenia has a problem defending its borders, but it has no army. A qualified general, Norat Ter Grigoryan, has been invited to become the commander of Armenian forces. He will organize the Armenian army gradually with modern concepts relevant to military science. Armenia is implementing a conscription. This means that there is a need for an army, which is a common-sense reality.

The people of Armenia are defending their homes, villages, and farms, especially in the border areas, to the best of their abilities, with the organization of an expeditionary force. The expeditionary force consists of volunteers. It is an auxiliary force, but it is not an army. In this period of getting on our feet, the expeditionary force is playing an important role, but it is not a state solution even if it has the natural backing of the government.

This assessment does not mean that Armenia is completely deprived of armed forces. There are a few regiments, each composed of approximately fifteen hundred soldiers. According to expert opinion, Armenia needs an army of forty thousand soldiers to be able to defend its borders, especially since it does not have the benefit of natural borders with its neighbors. The need to defend the homeland with armies is inevitable until the establishment of peace and cooperation.

The formation of an army is inevitable. Even Switzerland, which has declared its neutrality and which knows that it is not subject to any external threats, has a very modern army. Armenia has insecure borders and cannot be defended with the good will of others. General Norat Ter Grigoryan is expected to express himself as a specialist to provide numbers, an estimate of costs, and the qualitative nature of the army.

The development of the people's civic awareness is of vital importance in this regard. Vazgen Sargsyan faced a negative disposition by the people with regard to conscription during his term as minister of defense. The young men did not respond to their conscription calls, dealing a severe blow to the national aspirations of Armenians. These negative attitudes do not help the reconstruction of our homeland. It is time to leave the narrow circles of power to win the confidence of the masses, by words and deeds, that Armenia and the Armenian nation will be led by basic national values. Such moves will lead the masses to respond positively to conscription calls.

We must contribute to the formation of an Armenian army not out of humanitarian or philanthropic considerations but out of the knowledge that only Armenian platoons, brigades, battalions, and companies can defend the Armenian homeland. Nations have always known to respond positively to the call of patriotism when leaders inspire confidence—as the French did in response to General Charles de Gaulle's call on 18 June 1940. In a democratic system the army is not a political tool. This must also be explained to the people.

There is a large number of Armenian officers and soldiers in the former Soviet and the present Russian army who can return to Armenia and serve in the Armenian army. Because of developments in the Caucasus, Armenian officers and soldiers may not be viewed in a positive light as the descendants of a nation that is destabilizing the region. However, Armenian authorities must offer the families of these soldiers housing—which is not so easy—in order to entice them to return.

Another basic problem is the procurement of military material for the army. The departing Soviet army has transferred arms and ammunition [to Armenia] (although Azerbaijan has benefited more because there were a larger number of military units on its territory). It is evident that military technology is developing very rapidly and that arms are becoming obsolete very quickly. Armenia must eventually produce its own military means and make use of advances made overseas. There is a drive in that direction because the regional powers, especially Azerbaijan, are not standing idle.

This patchwork of thoughts can have meaning only when it is decided in Armenia to form a modern national army, different from a border-guard force. Azerbaijan already has a national army. It does not hide that fact and benefits from the specialized assistance of Turkish officers.

Armenian authorities must recover the time they have lost and hasten the pace of forming a national army. This is not a question of militarism; self-defense and upholding national rights are imperative obligations. This is neither a disposition of aggressionism nor a plot against peace. This is not a easy task, and it will not be easy, but we cannot evade it out of populist considerations.

---

## 4.47

**Ter-Petrosyan on Foreign and Domestic Affairs**
Interview by M. Bchakjian
*Haratch* (Paris) 20–24, 29 October 1992
[FBIS Translation], Excerpts

---

[Question]: Which states are friendly to Armenia?

[Ter-Petrosyan]: I can without any reservation consider Iran a friend. Today Iran is the only state that is truly interested in having to its north a state like Armenia; it would be advantageous for Iran to have a strong and stable state [to its north]. I do not see a similar disposition among our other partners.

Russia has interests in Armenia, or rather in our geographic region. Although Russia is today preoccupied with its internal problems, it seems that it feels subconsciously that it must secure its presence here. That cannot be described as a concrete policy or strategy by Russia. It is more a subconscious feeling because Russia has had aspirations in this region for nearly three hundred years. Therefore it feels subconsciously that perhaps it is too early to terminate its presence in this region. Perhaps in the future Russia will re-evaluate the basis of its foreign policy, and perhaps it will withdraw from here. However, at this time it has not yet formulated such a policy.

[Question]: At present what is the Russian presence in Armenia? What is its level and scope?

[Ter-Petrosyan]: Today our borders are guarded by Russian armies. Russian armies guard the borders of the CIS. Apart from guarding borders, Russian armies are in all republics of the former Soviet Union. That is the only element of Russian presence. Of course, Russia has other means of leverage, such as in finances—Russia is entirely responsible for policies on the ruble. It also has the means to influence the republics through prices—especially the prices of basic materials. It has many such means of leverage. . . .

[Question]: Did you have the same approach when you were a member of the Karabakh Committee?

[Ter-Petrosyan]: No, my views were completely different. At that time we had completely different objectives and we fought, sincerely and steadfastly, for the reunification of Nagorno-Karabakh with Armenia. Why? Because at that time Armenia and Azerbaijan were part of the Soviet Union. Those internal borders were not unalterable from a perspective of international law. They could be changed, and it was possible to change them. The world would very easily accept such an internal change of borders. We believed that through the political pressure we were applying on Armenian and Soviet authorities through protest demonstrations and strikes, we could bring about such changes. However, now that Armenia and Azerbaijan have become independent states, we have to adopt the rules of the new game. Today we cannot talk to the world with similar slogans. Today we have to come to terms with the rules of international law. Obviously our objectives were different then. [Passage omitted.]

---

## 4.48

**Ter-Petrosyan Reviews First Year in Office**
Armen Khanbabyan
*Nezavisimaya Gazeta,* 13 November 1992
[FBIS Translation], Excerpts

---

A year ago, the first president of independent Armenia was sworn in.

The eleventh of November marks one year since the first president of Armenia took his oath of office. In this connection, Levon Ter-Petrosyan reviewed some of the results of his presidency and gave his own evaluation of socio-political events.

During his meeting with journalists, the president remarked that the most serious economic situation in Armenia today prevented him from talking about any achievements. However, the main positive phenomenon, in his view, is socio-political stability, which became possible not so much through the effort of the administration as because of our people, who can foresee the possible catastrophic results of destabilization. The president also emphasized the existence of a civilized opposition in Armenia, whose members do not go to extremes in their rivalry with the government. The situation in the republic is controllable, and its democratic freedoms are developing without being threatened. Work continues on the new constitution.

Our republic managed to overcome the militarist hysteria that, after the summer defeat in Nagorno-Karabakh, could have tempted the republic to become involved in a large-scale war with Azerbaijan. This did not happen because the parliament declined the proposal to give formal recognition to the independent Nagorno-Karabakh republic, but it did not refuse to continue negotiations, as this would also mean a refusal to seek a compromise solution. "The Supreme Soviet decreed that Armenia should not sign any international documents naming Nagorno-Karabakh part of Azerbaijan," said the president, "and this decree deprives us of the possibility of a political maneuver. Meanwhile, I suggested that this idea be phrased differently: Do not sign any documents that contradict the right of the Nagorno-Karabakh people to self-determination. However, the opposition version was accepted because I was not insistent enough."

The president sees Armenia's participation in the CIS structures as an important aspect of its foreign policy. However, Levon Ter-Petrosyan thinks that any further strengthening of the commonwealth may be possible only after the current relations among the former republics are transformed into normal international ties. This is the only chance the CIS has to become a European-type commonwealth. Levon Ter-Petrosyan feels gratified by the level of development in relations with Russia. He has no reason to think that Moscow is conducting either a dishonest or a deceitful policy with respect to Yerevan. "I want to emphasize this circumstance especially," said the president, "because quite contrary statements appear rather frequently in our print media. Meanwhile, the only misunderstanding to cast a shadow on Armenian–Russian relations was related to the violation of military parity between Armenia and Azerbaijan last summer, when the Azerbaijani side captured military warehouses on its territory, and Moscow gave in to pressure from Baku. But later, parity was restored in general, and now Armenian–Russian relations are not tarnished by anything." Levon Ter-Petrosyan feels optimistic in his thinking that the stronger the positions of the Kremlin reformers, the better bilateral relations will be.

The president nevertheless feels optimistic in his expectations that the economic situation will improve with time, especially after Armenia manages to overcome the transportation blockade. Intensive negotiations are going on at present concerning the use of Turkish and Iranian transportation infrastructures by our republic. Trade recovery will lead to qualitative changes in the people's living standards, because today's main problem lies in the fact that Armenia can neither import food and raw materials nor export its own manufacturing goods costing tens of billions of rubles.

As for the current clash between the executive and legislative power, it will disappear gradually after the new constitution is adopted. Nobody is going to try and pull the blanket over to his side by demanding any extraordinary powers. A draft of the new constitution will be presented for public discussion as early as next year. When asked whether he thinks it necessary to introduce direct presidential rule during the transition period, Levon Ter-Petrosyan replied that he was against such measures because "the worst of democracies is still better than the best of dictatorships."

## 4.49

**Armenia Thinks "Caucasian Home" Concept Is Premature**

Karen Topchyan

*Rossiyskaya Gazeta,* 18 June 1993 [FBIS Translation]

Peace, democracy, cooperation—this was the motto of a meeting of political parties from the three Transcaucasian countries held in Tbilisi the other day. The Georgian Justice Party was behind it. The forum was also attended by the Georgian People's Front and Green Party, the Armenian "Dashnaktsutyun" revolutionary federation, the "Constitutional Law" Union, the Republican Party, the "Zharang" national conservative club (all of them from Armenia), and also the Azerbaijani "Yurdash" party. Note that they are all in opposition to their countries' leaderships.

According to the Armenian delegation, the Tbilisi meeting was productive on the whole. As one of the forum's participants, Yervand Gasparyan, said, the Georgian and Azerbaijani sides proposed the signing of a declaration they had prepared on the advisability of creating a "Caucasian Home," which our delegation refused to do. The reason is that the idea of a "Caucasian Home" is clearly premature, so it is still too early to talk about it. Before discussing any specific action to unify the Transcaucasus, the current contradictions and problems among our peoples have to be resolved, in particular, the conflicts in the region have to be stopped. Moreover, the Armenian representative believes, the explanation of the proposal for a Caucasian home had a distinctly anti-Russian flavor. So an effort was being made to blame imperial forces in Russia for the present situation in the Transcaucasus. Disagreeing with this interpretation of the reasons for today's highly dramatic situation in the region, the Armenian delegation observed that the conflicts and unresolved matters in the Transcaucasus have deep historical roots, rather than having been introduced from outside. And if we are talking about interference by other states in the region's affairs, we must be talking about Turkey, which is actively involved in the confrontation in the Transcaucasus.

In the end the sides signed a joint communiqué in which the meeting participants, despite considerable contradictions, "expressed a wish to seek an end to the wars in the Caucasus and contribute to the creation of political mechanisms in the resolution of conflict questions."

## 4.50

**Armenian Premier Proposes Collective CIS Currency**

Sergey Bablunyan

*Izvestiya,* 30 April 1994 [FBIS Translation]

Armenian prime minister Hrant Bagratian has sent a letter to all the CIS prime ministers.

It states, in particular, that financial transactions between the CIS countries have become considerably more complex as a result of the introduction of national currencies. The Russian ruble is used as a single currency for many trade deals between states. The CIS countries look to Russia for loans to execute their transactions, thereby increasing the Russian budget deficit. The prime minister thinks that the way to ease financial difficulties lies in the broader use of national currencies and also proposes setting up a collective currency for the CIS countries, in which it is proposed to carry out all transactions in a CIS interstate bank.

## 4.51

**Bilateral Agreements with the United States Not Ratified**

*Snark* (Yerevan), 29 May 1995 [FBIS Translation]

The Armenian Supreme Council has not ratified the bilateral agreements with the United States concluded in 1992. In his interview with *Snark,* the chairman of the Supreme Council's Committee on Foreign Relations, Davit Vardanyan, said that the Armenian government's delay is a violation of international rules on the introduction of such documents to legislative bodies. According to Vardanyan, the committee has returned the texts of all three documents to the government requiring them to correspond to international norms. The texts had to be in the languages of both parties. But the commission received the texts of the three agreements in English without any explanation. The chairman noted that after returning the texts to the Armenian government, the committee inquired about their fate. The American embassy in Yerevan is concerned about the documents' fate as well. To this day the government has not responded.

## 4.52

**Financing of Russian Military Bases Too Expensive**

*Snark* (Yerevan), 30 May 1995 [FBIS Translation]

Davit Vardanyan, chairman of the Armenian Supreme Council's Committee on Foreign Relations, believes that because Armenia is in a state of war, it is not able to finance 50 percent of the expenses for Russian military bases. Moreover, in his interview with *Snark* he stated that the agreement's clause on 50 percent financing contradicts international standards of foreign base activity, because each country in the world finances its own military bases. Another problem is that weapons of the former USSR were not equally divided among the Soviet republics. After the USSR's disintegration, Azerbaijan received 28 percent of the Caspian fleet's weapons, 130 military planes, and 10,000 vans with ammunition, which by far exceeds Armenia's share.

## 4.53

**Impact of Minsk Summit Analyzed**

Karen Topchyan

*Respublika Armeniya,* 31 May 1995 [FBIS Translation]

The uprooting by the presidents of Russia and Belarus of a post on the Belarusian–Russian border was, seemingly, to have been the symbol of the latest, 17th [as published], meeting of leaders of the CIS, which was held last week in Minsk. But Yeltsin declined to dig up the post, saying that it was not "kingly work," and the post itself was stuck fast in the ground. Chernomyrdin, who substituted for Yeltsin, had, therefore, to confine himself to the cutting of a ribbon symbolizing the state border. Salt was also rubbed into the wounds of Lukashenka, who had been hurt by Yeltsin's refusal to pick up a spade, by Minsk students, who first put up the new-old Belarusian flag with the modified Soviet symbols so beloved by the president at a public restroom and then burned it.

Failing even more to correspond to the planned symbols was the Minsk summit itself. Unlike the Belarusian president, who is literally forcing himself into the guardianship of Russia (at Russian expense, naturally), the other leaders of the CIS displayed no particular zeal on this matter. President Kuchma said that he would wait to break up the border posts, otherwise, he might be doing something "completely stupid." As a result, Ukraine and also Uzbekistan, Moldova, Azerbaijan, and Turkmenistan refused to sign a treaty on joint protection of the external borders of the CIS. A number of Central Asian members of the CIS, and also Azerbaijan, proceeding from their particular view of human rights, rejected the Convention on Human Rights and Basic Liberties offered at the meeting for consideration. Generally, with every meeting it turns out that some country or group of countries has a dissenting opinion on an increasingly large number of questions. The absence from each successive summit of some head of state has become a regular feature. Saparmurad Niyazov and Nursultan Nazarbaev "took ill" and did not go to Minsk on this occasion.

But the very location of the meeting—Minsk—required that the hosts' diligence with respect to land integration be marked somehow. Boris Yeltsin, of course, could not have failed to Lukashenka. "Our integration has gone considerably deeper with Belarus than with other countries of the CIS." But, first, it could just be a question of Lukashenka and, second, of whether the other CIS countries are ready for integration. Any more or less tangible results, on the other hand, may, most likely, be expected only in respect to so-called dire business. Thus, in particular, the leaders of the CIS voted unanimously for extending the stay of the collective peacekeeping force in Tajikistan and Abkhazia. A currency committee also was created without any particular disagreements (only Turkmenistan abstained), but there are still many questions here also. The main one is which currency might be the likely medium of payment in settlements among states of the CIS. The ruble would, in any event, hardly be seriously quoted as a kind of CIS ecu, considering the not particularly respectful attitude toward it in the majority of states of the commonwealth.

The next top-level meeting will be held in Sochi at the start of November. The leaders of the CIS do not, in all probability, greatly miss one another if they are planning their next meeting not for two months hence (as was the case in the first years of the formation of the commonwealth) or three months hence (as was the case at somewhat later rendezvous) but merely after the passage of more than five months. It is a perfectly reasonable assumption, therefore, that several heads of the post-Soviet states invited to the November summit will once again prove to be "unwell."

## Azerbaijan

### 4.54

**Russia's Regional Policy Eyed and Assessed**
Aleksandr Krasulin
ITAR-TASS, 2 April 1992 [FBIS Translation]

"The Slavic Card in Russia's Caucasian Policy" is the topic of a news conference given at the Azerbaijani mission in Moscow on Thursday [2 April]. Scientists, representatives of public organizations uniting various groups of the Azerbaijani population and Azeri reporters took part.

All the states that were included in the USSR and in whose territories conflicts broke out turn to Russia, regarding it as the force that can settle their problems promptly and in a final way, said Akhmed Iskenderov, corresponding member of the Russian Academy of Sciences and editor-in-chief of the journal *Problems of History*. He said Russian leaders are slow to work out the national policy and agree to "incomprehensible compromise entailing sacrifices."

The participants in the news conference reproached Russian news organizations, which, in their opinion, do not give enough coverage to Azerbaijan's problems and do not reflect precisely the situation in Nagorno-Karabakh and the position of various ethnic groups in Azerbaijan. They noted that people of various ethnic groups, among them nearly half a million Russians, live in Azerbaijan and regard themselves as its citizens. Statements about the departure of considerable numbers of Russians from Azerbaijan were described at the news conference as groundless rumors.

### 4.55

**Presidential Adviser Not Optimistic on Future of CIS**
From the "Presidential Bulletin" column compiled by Andrey Pershin, Andrey Petrovskiy, and Vladimir Shishlin; and edited by Boris Grishchenko
Interfax, 26 January 1993 [FBIS Translation]

The president's international policy advisor, Vafa Galuadze, thinks that, in all likelihood, the CIS cannot be preserved in its present form. In his opinion, the commonwealth may only have a future as a consultative organ. Commenting on the meeting of commonwealth leaders in Minsk, at which Azerbaijan participated as an observer, he said that "there is more than one evaluation of this meeting: Certain participants of the summit see it as a success, but the West on the other hand thinks it was without results." The advisor also remarked that several commonwealth states negatively reacted to attempts to interfere in their domestic affairs through various joint documents.

According to Galuadze, however, Baku thinks it would be worthwhile for Azerbaijan to participate at CIS forums as an observer. This, he is convinced, will enable the country "to keep a hand on the pulse of world policy." The republic, however, has principally decided against joining the commonwealth, the advisor stressed.

Azerbaijan's relations with Russia were described by Galuadze as close and friendly.

In the advisor's words, government experts are now developing the question of Azerbaijan's position regarding the treaty, signed in Minsk, on the creation of an interstate bank.

With regard to the results of the visit to Baku by chairman of the Minsk CSCE conference on political settlement of the Armenian–Azerbaijani conflict, Mario Rafaeli, the president's advisor said: Baku is in favor of diplomatic and political settlement of the conflict, but "each step on this path to peace is torpedoed by the Armenian side."

When discussing the outlook for the Minsk meeting and ways to settle the conflict, the Azerbaijani side, the advisor said, [believes] that Yerevan should announce its rejection of any territorial claims on Azerbaijan. Baku leaders think that it is necessary first to annul the resolution to unite Nagorno-Karabakh with Armenia, which was passed by the Supreme Soviet of Armenia in 1989. Galuadze thinks that the Armenian side does not yet appear ready for a Minsk conference. In order for the conference to take place, he thinks, Armenia should end military activities on Azerbaijani territories and remove its troops from Shushi and Lachina.

[*IF Note:* At the meeting last Friday with Mario Rafaeli, Azerbaijani president Abulfaz Elchibey emphasized that the process of political settlement of the conflict is developing slowly, because large states are in no hurry to actively participate in settling the conflict. Elchibey accused Armenia of violating all agreements reached earlier through international mediation.]

## 4.56

### Abulfaz Elchibey Comments on Russia's Role in Caucasian "War"

Report on Interview by Pazit Rabina
*Dvar Hashavua,* 28 May 1993
[FBIS Translation], Excerpts

In Baku it is said that President Elchibey is barely in control and that his coalition is hanging by a thread. There is no room for gentle professors at the top of the Azerbaijani leadership. In an interview at the presidential palace at the beginning of May, however, the president reveals himself to be a man whose sharp tongue is at odds with his immaculate manners and quiet tone of voice.

"To my great regret," Elchibey says, "the war between Armenia and Azerbaijan long ago ceased to be a war between two rivals from the Caucasus. This is a war in which the combating peoples have become the pawns of mightier powers. Russia is stirring up a war in the Caucasus," Elchibey asserts. "Looking back," Elchibey says, "we can enumerate one by one the Russian military officers and KGB personnel who were actively responsible for bringing about the deterioration in the situation. People who once operated within the framework of the Soviet state are now acting as part of the Russian republic and raking in huge profits by selling weapons to the areas of fighting in the Caucasus." Elchibey claims that "the recent Armenian victories in the war can be chalked up to the presence of the Russian Seventh Army, which is stationed in the region and is helping the Armenians in their struggle. Russia has a vested interest in the continuation of the war," the president said, noting that "the conflict between Armenia and Azerbaijan gives the Russians a foothold in the Caucasus."

Elchibey averred that the very week of our talk there had been an indisputable example of Russia's role in fanning the fires of war in the region. Six Russian soldiers, who were taken captive after bloody battles in the Kelbadzhar area, were brought to trial before a military court in Baku. The six Russians were accused of slaughtering thirty-three Azeris with axes and blunt instruments. In addition to murder, they were also accused of planting explosive charges, attempted kidnappings, and terrorist attacks against the local population. From the beginning of the week, rumor had it that the six Russians would be executed. No meaningful attempts to save them were made. Since the soldiers' capture, the Russian government has disclaimed any responsibility toward them, claiming that they were mercenaries, not soldiers. But the six Russians testified that they belonged to Russia's Seventh Army stationed in Armenia. Last week it was officially reported that the six men have been sentenced to death. All the attempts to swap them for six Azeri prisoners have so far been unsuccessful. If the soldiers are executed, it will be the first time in the post-Soviet era that any country has dared to execute Russian soldiers. It is hard to predict the consequences, but anything is possible in the Caucasus.

To counter the existing Russian presence in the Caucasus, Elchibey is planning to form a small, efficient, and modern Azerbaijani army. "Yes," he says, "we are planning a new army that will be based on weapons from Turkey, the United Kingdom, and the United States."

Elchibey is full of praise for Western-style capitalism: "I thank the American oil company, Amoco, and the British oil firm British Petroleum for their extensive investments in the entire region, especially in Azerbaijan."

A group of Pakistani pilots, navigators, and airmen, wearing their flight suits and carrying their instruction manuals, were photographed a day before the interview with the president. In response to an indirect question about Pakistan, a fellow Islamic state and potential ally that could help Azerbaijan build up its Air Force, the president reiterated that he prefers the British, Turks, and Americans. He is right to a great extent. After all, the United States built up Pakistan's Air Force, and the corps structure, not to mention the ranks and uniforms of the Pakistani Army, is a legacy of British colonialism. The initials of the Pakistan Air Force, which are emblazoned on the center of the wings stitched to the pilots' flight suits, are an accurate replica of the British Royal Air Force's logo.

The building of a modern army is the dream of every military industry. The state of war in Azerbaijan offers such an opportunity. The creation of an entire army from scratch would be a veritable gold mine for the economy of a country like Israel, or even the United States, for many years. Therefore, and this is only stating the obvious, many are showing interest, including Israel.

President Elchibey preferred to elegantly avoid the question of why Azerbaijan should not make use of the good services of the State of Israel. However, sources in Azerbaijan and Geneva this week confirmed that nightvision equipment for aircraft had been found in the areas of fighting in Nagorno-Karabakh. The equipment was taken from Turkish helicopters that had crashed in the region. The assessment is that these are Israeli systems or components to enable helicopters to fly over wadis and streams under very difficult flying conditions and to fly close to enemy troop concentrations and carry out commando attacks. Vafa Galuadze does not mince words. Referring to Arkadiy Volskiy, the man appointed by Mikhail Gorbachev to mediate between the warring parties in Nagorno-Karabakh, he says, "He is a filthy man." [Passage omitted.]

When asked about sales of Israeli arms, presidential advisor Galuadze denied the reports saying he never met David Kimche [former Israeli foreign ministry and Mossad official]. But Galuadze has met other Israelis. Last year Galuadze conducted a secret visit to Israel. News of his visit leaked, however, and was published by Armenian journalists in France. In Israel, Galuadze says, he met with Foreign Minister Shimon Peres and proposed cultural and scientific exchanges. "You are leaving a vacuum for Islamic fundamentalism in the region" was Galuadze's friendly warning. "We need more student exchanges between universities," he said. Peres, Galuadze said, agreed with every word. [Passage omitted.]

## 4.57

**President Aliev on the Present Situation and Joining the CIS**

Interview by Aydyn Mekhtiev
*Nezavisimaya Gazeta,* 14 October 1993
[FBIS Translation], Excerpts

[Mektiev]: Mr. Aliev, first of all, allow me to congratulate you on your election to the post of president of the Republic of Azerbaijan. Could you speak about the reasons for holding early presidential elections after Elchibey's election half a year ago?

[Aliev]: When on 17 June this year former president Abulfaz Elchibey left his workplace, in fact ending the execution of his presidential duties, he himself transferred the powers of the head of state to the parliamentary chairman. On the following day the Milli Mejlis [the parliament] confirmed the transfer of these powers. Ways to overcome the crisis were sought. We tried to convince Elchibey of the need to return to his duties, but those efforts were unsuccessful. Delegations of representatives from the intelligentsia went to the village of Kalaki, where Elchibey had taken refuge behind several cordons of security forces, and appeals were sent there in writing, but all was in vain. Therefore we, proceeding from the principles universally recognized in world practice, made a decision to hold a republic-wide referendum on confidence in the president. The results of the poll are known to you: 97.5 percent of the population expressed distrust in Elchibey. As you see, we made this move even though the republic's constitution contains provisions giving grounds for the immediate removal of the president from office for violating the country's Fundamental Law. After the referendum results were announced, a decision was made to hold early presidential elections. The conditions were created for all political forces to participate in the election campaign and to nominate their candidates. Several times I personally appealed from the parliamentary rostrum to the People's Front and its leader—Elchibey himself—to field his candidacy. Yet Elchibey did not do that. On the other hand, representatives of two other political movements declared their wish to run for the post of head of state. Many work collectives and other public organizations also nominated me as a candidate.

[Mekhtiev]: Mr. President, a whole number of opposition movements, and in particular the People's Front, announced their intention to passively boycott the elections and proclaimed their results invalid in advance. How do you assess this?

[Aliev]: I believe that the presidential elections were held under exclusively democratic circumstances. During the election campaign all conditions were provided to ensure freedom of expression. Observers from thirty-five countries and also from such an authoritative organization as the CSCE [Conference on Security and Cooperation in Europe] monitored the conduct of the elections on 3 October as well as the 29 August referendum. Heads of foreign missions also acted as observers. The fact that the People's Front, with all the above conditions, calls the elections illegitimate is very characteristic. The People's Front has been so spoiled by the fact that its members could violate laws and indulge in arbitrariness that it continues to think that it is the opinion of only this organization that is the ultimate truth. However, if the People's Front believes that it has decisive influence on public opinion, why then did more than 98 percent of the electorate vote for the new president?!

[Mekhtiev]: Recently the mass media reported an attempt on your life. Could you elaborate on that?

[Aliev]: My recent observations—and this is confirmed by public opinion polls—show that some extremist political forces, including the People's Front, are not happy with a stable socio-political situation in Azerbaijan. In the past, too, they wanted to destabilize the situation, so the present normalization of the situation in Azerbaijan is obviously not to their liking. This is why they are resorting to the use of force. It was not the first time that they have done so. As you probably remember, while I was working in Nakhichevan last October, they tried to overthrow the legally elected power bodies of the Nakhichevan Autonomous Republic by armed force. For that purpose they mobilized more than three hundred People's Front members armed with automatic rifles and other weapons, but they failed to carry out their plans because voters came out to defend the parliament, and though they were not armed, their number exceeded fifteen thousand. Members of the People's Front threatened me with terrorism. There were reports about acts of terrorism being prepared against me. All these reports were checked. This latest incident is, of course, outrageous, and it well characterizes the People's Front ideologues. It appears that they hired a man—a Turkish citizen, a member of the Gray Wolves organization—and they brought him to Nakhichevan and trained him for more than a month, including target practice and training sessions, and provided him with a sniper's rifle with optical sights and with grenades. They studied the route of my movements and the time of my coming and going to work. As is apparent from the testimony of the suspects, attempts to use a grenade to blow up the car in which I was traveling were not ruled out. I had no doubt that the People's Front, which emerged in its time as a popular movement and which, incidentally, was trusted by many in Azerbaijan and on which many pinned their hopes, has turned into a terrorist organization. No political party or political organization in the world usually has armed units, whereas the People's Front in recent years has formed many armed groups, though I do not know where the funds to support them come from. This puts the People's Front outside the framework of all humane principles and outside the scope of the law and the constitution.

[Mekhtiev]: Is there any intention in light of this to suspend the activity of the People's Front?

[Aliev]: Proposals on the need to terminate the Front's activity have been made, including from parliamentary deputies at recent sessions after the criminal actions by some members of the People's Front came to light. However, I did not agree with this approach and did not even allow this question to be discussed, based on the principles of democracy. At the same time the People's Front can no longer be called "the people's" as it does not represent the people. . . .

[Mekhtiev]: Recently Azerbaijan has become a full-fledged member of the Commonwealth of Independent States. You, as the initiator of its entry into the CIS, were accused by some of an attempt to return Azerbaijan to the sphere of Russia's influence. . . .

[Aliev]: These allegations are groundless and biased. Such statements are intended to influence public opinion. I think that this is already impossible since, before joining the CIS, we repeatedly discussed this question at parliamentary sessions. Incidentally, all Milli Mejlis sessions are broadcast on television and the population can directly oversee the decision-making process. Based on public opinion research results, we have found out that an overwhelming majority of the population supports this move. The opinion of opposition leaders does not express the people's aspirations. They themselves understand that this act has been adopted based on Azerbaijan's supreme interests. Joining the CIS in no way means that Azerbaijan is falling under Russia's influence. I have no doubt, however, that Azerbaijan had to restore normal relations with such a large country as Russia.

[Mekhtiev]: Mr. President, recently there has been some tension in relations between Azerbaijan and the countries of Central Asia. Problems in relations between Baku and the Central Asian capitals emerged after last June when Mr. Elchibey publicly criticized the leadership of the Muslim countries of the former USSR. . . .

[Aliev]: That position of the former president completely contradicted the national interests and historical traditions of the good-neighborly relations between our peoples and did not relate to Azerbaijan's economic interests. Azerbaijan is linked with the regions of Central Asia and Kazakhstan by centuries-old relations. Many Azeris live on the territory of these states, and it is irrational to create artificial barriers in the relations with these states. We have fully moved away from this line. Moreover, I condemn this policy of the former leadership. At present we have reached agreement with all the leaders of the states in the Central Asian region and Kazakhstan on concluding direct treaties and establishing close ties. A package of agreements has been prepared with Uzbekistan, and one is pending with Kazakhstan. I have direct contacts with the presidents of all these republics, and

I have already met with them at the previous CIS session. I think that our relations with these states will develop on all fronts. This will benefit the Azeri people.

[Mekhtiev]: On 8 October you held talks in the Kremlin with the leaders of Russia, Georgia, and Armenia. One of the questions discussed at that meeting was the issue of unblocking transportation routes in the Transcaucasian region. Does the new leadership of Azerbaijan agree with the demand to lift the ban on railway freight transit through Azerbaijan to Armenia?

[Aliev]: This demand cannot be considered outside the context of Resolutions 822 and 853 of the UN Security Council, which call for an immediate withdrawal of Armenian troops from the occupied territories. Neither is it possible to resolve the transportation issue without addressing the entire complex of questions related to the aggression by Armenian armed groups against Azerbaijan. After all, outside the former Nagorno-Karabakh Oblast a large territory of the Republic of Azerbaijan has been captured and hundreds of thousands of people have become refugees. Without resolving these questions, not only is the demand to lift sanctions not serious but it can even be seen as an element of aggression against Azerbaijan. . . .

---

## 4.58

**Aliev Urges "Closer Integration" in the CIS**
Interview by Pavel Alekseev
*Rossiyskaya Gazeta,* 15 June 1994 [FBIS Translation]

---

It will soon be a year since you became president of Azerbaijan. Yet not many years ago most people were sure that G. Aliev's political career was over. . . .

[Aliev]: You could say that I did not come here of my own volition. After leaving the Politburo in 1987 because of serious disagreements with the USSR leadership, including Gorbachev personally, I did not engage in political activity. When, in late 1990, I was not allowed to live in Baku and found refuge in Nakhichevan, again I did not want to get involved in politics. But in 1991 a very difficult situation arose. The autonomous republic [of Nakhichevan] within Azerbaijan is detached from the state's main territory and borders on Iran, Turkey, and Armenia, and therefore, as a result of the war, it found itself blockaded. For several days people held rallies in the squares, demanding that I become head of the republic's Supreme Council, and I was obliged to do so. I started work. It is a very complex region, and our life was not without its difficulties, but at least it is peaceful, unless you count minor border clashes.

There is no war there now either, and there is not the same level of crime as before. Turkey has granted Nakhichevan $100 million in credit. We immediately released $20 million of that to buy food, and that also alleviated the situation. And I thought that this autonomous republic would be my last abode.

But in June last year the power struggle in Azerbaijan intensified, and the republic found itself on the brink of civil war. Then the Azerbaijani leadership, in the person of President Abulfaz Elchibey, appealed to me to come and help lead the republic out of the crisis. I did not want to return, but then I decided that it is wrong to stay on the sidelines when your people are in trouble. Unfortunately, two days after I took office as chairman of the Azerbaijan National Assembly, President Elchibey left Baku in secret and has been in a Nakhichevan mountain village ever since. I was left alone. And for eleven months now I have been trying to sort out the situation in the republic.

[Alekseev]: What are the basic problems facing Azerbaijan at present?

[Aliev]: We are going through a very difficult period. The transition from a situation in which we were all part of a single Soviet state to the achievement of independence and the consolidation of sovereignty is very difficult in itself. The situation is exacerbated by the fact that all the former USSR republics were very closely interlinked, especially economically, and the severance of ties had a very severe effect on Azerbaijan's economy. Then again, the republic is making the transition from one economic system to another, from a socialist economic system, and its laws and principles, to a free market economy. Finally, the third factor complicating the socio-economic situation in Azerbaijan is the war. It has now been going on for six years, some 20 percent of the republic's territory is occupied, and there are more than one million refugees.

[Alekseev]: What can you say about the causes of the conflict, and what must be done to stop it?

[Aliev]: It was not the Azerbaijani side that started this war. There is no point now in analyzing what happened, but I will say one thing. When it all began, Azerbaijan was part of a single centralized state—the USSR—and therefore the Union leadership could have prevented the outbreak of conflict. Or else, if conflict had begun, it could have prevented it from developing. Unfortunately, this did not happen. The

Karabakh conflict was the first on the USSR's territory, and look how many hot spots have developed.

Azerbaijan is now waging a war to preserve its territorial integrity. But all this time I have been trying to put an end to the war through talks. There has been a cease-fire since 10 May. Talks are now under way with the participation of the CSCE Minsk Group and Russia. Maybe it will be possible at last to reach an accord to end the conflict, but on one condition: that the Armenian armed formations free all the occupied Azerbaijani territory, except, of course, Nagorno-Karabakh. Then we can conduct talks on the status of Nagorno-Karabakh.

[Alekseev]: During the Bishkek talks there was talk of disengagement forces. . . .

[Aliev]: Look, whatever happens, in order to stop the war, monitor compliance with the cease-fire accord, and ensure the withdrawal of Armenian armed formations, forces of some kind are certainly needed. The CSCE Minsk Group sees these forces as observers. Russia considers it possible to send in so-called CIS peacekeeping disengagement forces, but this question is still being discussed.

[Alekseev]: How would you assess the situation in the Caucasus? There has been much talk recently of how Russia still has no clear political concept with regard to that region.

[Aliev]: It is hard for me to say whether Russia has a clear stance or not. But it is a fact that the Caucasus is a very complex region and requires a more thoughtful, considered, and intelligent approach. Russia is a great country. And the Caucasus has been part of it for more than two hundred years, be it tsarist Russia or the USSR, in which the Russian Federation was dominant. And Russia should have the same attitude to all the republics of the Caucasus. If there is a slant in one direction or another, it will lead to complications, and, naturally, the loss of its prestige in the region. There should, I repeat, be a very carefully considered approach, irrespective of whether Russia had closer historical links with one people, irrespective of religion. It seems to me that Russia will then have more prestige in the Caucasus, and the Caucasus will be assured of greater tranquility. And greater tranquility in the Caucasus means greater tranquility in Russia.

[Alekseev]: For two years now the former republics have been trying to resolve the problem of the severance of ties in the post-Soviet area within the framework of the CIS. How do you assess that organization's activity and future?

[Aliev]: The organization has a future, I believe. It is needed so that each republic can develop more successfully. That is why, when I became leader of Azerbaijan, I made great efforts to create an atmosphere of trust in the CIS and the right conditions for joining the commonwealth.

To tell the truth, it was not easy. By virtue of various circumstances, a very negative attitude to the CIS had developed in Azerbaijan. I am not saying that everyone felt that way, but the vast majority did. First and foremost, the events of 20 January 1990 had very serious moral effects on the people. On the other hand, the People's Front, while in power, vigorously cultivated anti-Russian sentiments. So it took some months to overcome those sentiments in the public consciousness. And on 24 September last year we joined the CIS. I think it was a necessary and important step, and Azerbaijan should remain in the CIS in the future. But as for the Commonwealth's practical activity, I would not say that the organization has really become established or is functioning properly as yet. It seems somehow that during the meetings of heads of state, which last a day and a half or two days, the CIS exists, but in the intervals between meetings it does not. Therefore, in my view, very energetic measures are needed to enable the CIS to function more successfully.

[Alekseev]: And what does that require? Some kind of supranational structures, perhaps?

[Aliev]: I do not consider that expedient. That implies the resurrection of a unified state, the possibility of impinging on national interests. The former republics are now independent states, and they cannot agree to that. But it is essential to have closer integration.

[Alekseev]: Previously, in the USSR, much was said about the friendship of peoples. In the CIS, they discuss every question under the sun except for this one. Does it not seem to you that now, of all times, this is what is needed?

[Aliev]: The question is indeed very important. At the meeting in Ashkhabad, I stated officially that I consider it abnormal to have a situation wherein two states belonging to one alliance fight among themselves. But I got no response. A second time, in Moscow on 15 April, I again raised the question of how the CIS was created for the purposes of uniting to defend the interests of our alliance, which is why it is unacceptable for two members of the alliance to fight among themselves, and that such a thing had never been seen in world practice. And again this question was not discussed. But you know, alongside the military and economic aspects, the CIS needs a moral climate: What is needed is the strengthening of the friendship of peoples as a means of avoiding future conflicts and hostility toward one another. The CIS should extend its functions to the humanitarian sphere too, to the strengthening of mutual relations between nations and peoples, but in order to do this it is necessary for mutual relations inside the CIS itself to be warm and friendly.

[Alekseev]: In your short time as president, you have shown exceptional diplomatic energy. How would you describe Azerbaijan's political orientation in the international arena?

[Aliev]: I consider a success that I managed to turn mutual relations between the Russian Federation and Azerbaijan toward a positive direction. I met with B. Yeltsin and V. Chernomyrdin and other officials. We signed a number of very important documents, and we joined the CIS. That is, my first steps were aimed at restoring good relations with Russia. And this policy will continue. For Azerbaijan, Russia is both a close neighbor and a large country with which the lives of many of the republic's citizens, living both in Azerbaijan and Russia, are linked.

Naturally, we favor the establishment of friendly relations with all neighboring countries—with Turkey, for instance. These relations were good before, but my visit to Turkey, and the treaty we signed, strengthened our relations still further. Another neighbor is Iran. We have a very long common border and traditional links. Admittedly, the previous leadership did not accord the necessary significance to relations with Iran, and a certain tension arose in our relations. But during a visit to Baku by President Rafsanjani in October, we signed a number of documents.

Georgia is also our neighbor—600,000 Azeris live there—and President Shevardnadze also visited us.

As for other countries, as we move along the road toward building a democratic, rule-of-law state, so we have a great need to develop relations with Western countries with traditions and experience in democracy and a market economy. Therefore, I accepted an invitation from the leaders of Great Britain and France, John Major and François Mitterrand, to visit those states, and we had fairly fruitful talks.

[Alekseev]: You spend a great deal of time on foreign policy, problems within the republic, the Karabakh conflict.... Who helps you in doing so? From the sidelines, just one figure is visible above Azerbaijan's entire political horizon—President Geydar Aliev....

[Aliev]: I have already said that I did not return to Baku of my own volition and I did not plan to get involved in politics. This probably sounds rather high-flown, but the people called me. Therefore, I did not bring any team of my own with me, nor did I prepare one. Admittedly, I know Azerbaijan well, but I left in 1982. Since then, an entire generation of leaders has been replaced. I have no party of my own. I work with the people who are there now.

---

## 4.59

**Aliev on Ties with Russia and Other Topics**
Interview
*Rossiyskaya Gazeta,* 20 August 1994
[FBIS Translation], Excerpts

---

In an endeavor to understand how the republic is living today, our special correspondents have spoken with all sorts of people. People repeatedly mentioned the Soviet Union, many of them experiencing feelings of nostalgia. Tell me, Geydar Alievich, what Azerbaijan nonetheless acquired on the USSR's disintegration. Do you yourself, incidentally, never long for the past?

[Aliev]: Everyone has a longing, some more so than others. I am no exception. This is perfectly natural, but of course, the idea of creating a new Union is nonsense.

You ask what the republic acquired. The answer is simple: independence. Independence is the wish of every people. In the past we spoke of the collapse of the colonial system and said that the countries of Asia and Africa were finding independence. Some people still believe that the Union republics were colonies of Russia. I reject that. It is not true that socialism brought the peoples only misfortune. The economic, scientific, and intellectual potential that Azerbaijan has was created while it was in the Union.

But the movement for states' independence has now been under way in the world for many years. Why should those republics that united in the USSR in 1922 and later not be given the opportunity to set up on their own? Independence had a very high price, but still we achieved it....

[*RG*]: But there do not yet seem to have been any decisive steps in this direction.

[Aliev]: I was transferred to Moscow in December 1982, and I returned in June 1993, eleven years later. During this period the republic's leadership was replaced several times. Therefore there are objective reasons for the slowing of the reforms in Azerbaijan. Our next steps will be privatization and the creation of conditions for private enterprise. We are greatly encouraging foreign investments. This is our path.

[*RG*]: How are Azerbaijan's relations with Russia shaping up?

[Aliev]: This is a very important question, and there are complexities here too. Aggression was committed against Azerbaijan in January 1990. It was committed not by Russia but by the communist leadership of the Soviet Union. I was living in Moscow at the time, and I was in isolation. But I knew perfectly well how many military units of the Soviet army were stationed in Azerbaijan. There were more than enough of them, and there was no need to send in additional forces in order, as Gorbachev said, to instill order.

A huge trauma was inflicted on the people. The people who did not want normal relations between Azerbaijan and Russia and who are now in opposition kept saying the same thing: the Russian army, the Russian army. But there was not a Russian army then, there was the Soviet army, and in it were serving Russians, Ukrainians, Azeris, Uzbeks, Kazakhs. . . . Nevertheless, anti-Russian sentiments were kindled. The People's Front, which came to power in March 1992, took an unequivocal stand: Russia is virtually the enemy. Understandably, relations between Russia and Azerbaijan proved cold. Azerbaijan was not a member of the CIS at the time.

I hope that the Russians know that I made a great effort to change public opinion here. For several months, both in parliament and in labor collectives, I spoke of the need for Azerbaijan to be incorporated in the CIS. This finally happened last September. My first official trip as chairman of the Supreme Soviet [Verkhovnyy Sovet], endowed with presidential powers, was precisely to Moscow. I sensed a reciprocal response from Boris Yeltsin and Viktor Chernomyrdin. Relations between Azerbaijan and Russia now are friendly.

Azerbaijan was part of Russia and the Soviet Union for almost two hundred years. Our ties are of long standing and should not be broken off—this is inadmissible. On the contrary, they must be developed. The absolute majority of our major enterprises are integrated in the Russian economy. The same thing in education, science, technology, and culture. As president of Azerbaijan, I am doing and will do everything to strengthen our ties. But there are forces trying to portray our policy in a different light. Certain Moscow newspapers carry articles that are essentially provocative.

[*RG*]: The state language in your republic is Azeri. At the same time Russian is used not only by representatives of non-indigenous nationalities, but the stratum, so to speak, of Russian-speaking Azeris is huge. People say that if they need to express a simple thought, they express themselves in Azeri, while knowledge [of that tongue] no longer suffices for expressing a complex thought—then they "engage" Russian. Not to mention the fact that the best-known Azeri writers write in Russian. . . . Is there not a problem here, in your view?

[Aliev]: Not at all. You know, a national language is an attribute of statehood. Naturally, our state language is Azeri. This is the full extent of its "privilege." It was the People's Front that shut the mouth of anyone who spoke Russian, even though these very people expressed themselves badly in Azeri.

There was, of course, a complex situation, but I believe that I defused it. It is absurd to think that, by making everyone speak one language, you re-establish this language. Everyone is free to express himself however he deems it necessary. Before [parliamentary] sessions people used to ask me which language they should speak, but now they no longer ask—they themselves choose. Those writers who are accustomed to writing in Russian—let them do so. There are more and more Russian-language programs on our television. The Ministry of Foreign Affairs and the Ministry of Defense present all papers in Russian. When a foreign delegation comes to me, I speak the language that their interpreter is able to translate.

I am convinced that the Azeri people cannot be isolated from the Russian language. Our republic's multinational composition goes hand and hand with the Azerbaijani state's independence. Armenia sought a monoethnic state—and it secured it. Now no one, apart from the Armenians themselves, lives there. I do not consider this an indication of civilized behavior.

[*RG*]: You now represent independent Azerbaijan throughout the world. At the same time you are quite a rare type of leader, who returned to lead the republic a considerable time later. There is, of course, such a thing as diplomatic etiquette. Nevertheless, how, in your opinion, do your partners in talks see you? As a Politburo member who has "changed color"?

[Aliev]: It is hard to say for sure. In those who judge a person mechanically on the basis of his biography and prepared information you sense wariness. But after personal contact has been established and after familiarizing yourself with my practical activity, I believe that you should not be left with any sense of my having "changed color." I left the Communist Party with the same conviction as I served it. I left it, by the way, while the party was still in power.

When I tendered my resignation from the CPSU in Moscow, it organized a commission to check on all my past activity. Foreign correspondents started telephoning: "How do you react to this?" I replied that all this was ridiculous. In the Communist Party the supreme punishment is expulsion from the party. I expelled myself. I do not know what more could have been done to me.

# Georgia

## 4.60

### Shevardnadze to Back Elchibey's Caucasus Idea

Interfax, 22 June 1992 [FBIS Translation]

The chairman of the Georgian State Council, Eduard Shevardnadze, supported Azerbaijani president Abulfaz Elchibey's idea to turn the region into a single Caucasian home.

Georgia will spare no effort to attain this goal, says Shevardnadze's telegram to Elchibey following the latter's election as president of Azerbaijan.

"It's too good to be true," said a Russian expert in this region commenting on this idea. There are many obstacles blocking the advancement to its implementation, he said, including the continuing bloodshed in Nagorno-Karabakh and South Ossetia, the crisis in relations between Armenia and Azerbaijan, and confrontation between Moscow and Tbilisi.

## 4.61

### Shevardnadze Interviewed on Russian Leadership

Interview by Leon Onikova
*Izvestiya,* 22 June 1992
[FBIS Translation], Excerpts

[Onikova]: How do you view the consequences of R. Khasbulatov's statement on the situation in Georgia and the situation as a whole in the Transcaucasus and North Caucasus?

[Shevardnadze]: The consequences did not take long to make themselves felt. The statement has sharply exacerbated the situation.

I have already made a statement on the Russian Air Force's helicopter attack on guards' positions and Georgian villages in the Tskhinvali area on 18 June this year. It was an act of aggression that introduced an entirely different dimension to the conflict. The denials by Russian officials are groundless. The claim that it was a warning strike is in effect an admission that it was an unprovoked attack.

Rather less groundless was the threat made to Georgia by the Russian vice president in a recent phone call to me: "I will send planes up. . . ." The reflexes of the Afghan bombardier prevailed over state wisdom.

In the light of recent events, Khasbulatov's statement constitutes the political and ideological preparatory fire for the infringement of our republic's territorial integrity and sovereignty.

That is precisely how all political parties, the public, and the State Council have described it.

For all my personal devotion to political dialogue and my incessant practical efforts to resolve the conflict by peaceful means, I too cannot describe these threats in any other way.

The 15 June statement sharply reduces the room for our peacemaking efforts and cuts the ground from under the Kazbegi agreements, while excessively expanding the bounds of distrust, aggressiveness, and extremism. Forces that are by no means advocating a political solution to the problem have become more active.

Our task has now become far more complicated. Khasbulatov's statement is of no use to anyone, except perhaps to extremists on both sides. There are leaders in the region who are prepared to give access to the conflict zone to forces operating from an anti-Russian standpoint. If this happens—and things are heading in that direction—then the entire North Caucasus would go up in flames. Together with the conflict already raging in the Transcaucasus, this would create problems for Russia that would make all the current sources of domestic tension fade into insignificance.

The statement, which admits in principle a unilateral revision of borders—in violation of norms of international law that have been ratified by Russia, and of principles accepted by the world community—is helping to revive territorial disputes, which could lead to a wholesale escalation of uncontrollable military actions. All this will have the most negative consequences for the Caucasus, the Transcaucasus, and Russia.

The main thing now—not only for Georgia but for Russia, too—is to take concrete action to restrict and localize the scope of damage done to Georgia and Russia, and then to eliminate it.

[Onikova]: Might Khasbulatov's statement negate B. Yeltsin's statements about the need to accelerate the preparation of a fundamental treaty on the principles of mutual relations between the Russian Federation and Georgia?

[Shevardnadze]: I do not think so. And this is not just a matter of President Yeltsin's incomparably greater influence on the shaping of the Russian state's domestic and foreign policy. Of no less importance is the objective and, one might say, historical need for such a treaty, which the Russian president well understands. I was able to see that for myself during our recent telephone conversation. . . .

[Onikova]: What measures, in your opinion, could help to defuse the new crisis, if only partially?

[Shevardnadze]: You are right, the situation that has arisen over the past few weeks can be described as a new crisis.

It is very important to trace its chronology—it explains a lot. I was in Tskhinvali 13 May and reached agreement on a cease-fire. A few hours later there was a rocket attack on the region where I and my colleagues were. I described it as an attack on our policy of finding a peaceful settlement to the conflict. A more destructive attack followed on 20 May. A group of Ossetian refugees were shot near Kekhvi village. The next day the economic blockade of Georgia began, and the conflict began to spiral even more tightly. On 10 June in the settlement of Kazbegi the leaders of Georgia and North Ossetia signed a protocol on joint action to settle the conflict and immediately embarked on its implementation. On 13 June there was an act of terrorism in Tbilisi and a clash in Gori between guards and servicemen from a tank regiment. Once again people were killed. There was yet another act of terrorism in Tbilisi on 17 June. Destructive forces inside and outside Georgia were clearly becoming more active. It was against this backdrop that Khasbulatov's statement was made. The Tskhinali region is witnessing an escalation of hostilities in which Russian army subunits are participating directly.

This gives some people the excuse to claim that we are not in control of the situation. I will not refer you to the numerous examples of uncontrolled situations within Russia and the CIS. I will not cite the sad experience of certain European countries where, despite all the efforts of their leaders and the world community, conflicts are not abating but merely flaring up with renewed force. Something else is more important to me—namely, explaining why, contrary to all our efforts, attacks are continuing and people are being killed. And nobody should try to prove that people are only suffering and being killed "on the other side."

The losses suffered by the Ossetians have been widely publicized. The information war has been just as immoral as the destruction of lives by force of arms. By emphasizing the ethnic origins of the dead, the writers of such reports are carrying out a second murder—murdering basic morality, human dignity, and justice. As if the dead Ossetians were not our citizens and fellow countrymen, as if their blood was, for us, merely water. It makes me sick. I would note, incidentally, that tens of thousands of Ossetians continue to live peacefully in Georgia—in the Kakhetinskiy region, Kazbegskiy and other rayons, and in Tbilisi—without being subjected to any reprisals. And, with us, they weep for the dead—be they Ossetians or Georgians.

It is amazing but true that propaganda is calling for blood. I prefer the voice of reason. If only to demonstrate how disastrous this senseless war is for both sides, I would mention that since the conflict began more than three hundred Georgians have been killed, around four thousand homes have been burned down, and fifteen thousand inhabitants of Georgian villages have been turned into refugees. Armenians, Russians, Jews, and people of other nationalities have also fallen victim to the conflict.

If I were to make anyone accountable, it would be the politicians who unleashed the conflict. And even if somebody tried to paint me and those who are of the same mind as I am as Ossetian-haters, then, without arguing with them, we would say that we want to save our people from senseless loss of life and suffering. Until the conflict is over, both Georgians and Ossetians will be killed. And an escalation of the conflict could spell doom for Georgia. Nobody but an imbecile could suspect me of wanting that.

The simplistic "black and white" view of the conflict and the popular political cliches are distorting the real picture and leading to irreparable mistakes.

A simple but substantive fact is being lost sight of. For two years people have been living on both sides in an atmosphere of mutual hatred fanned by the deaths of their relatives, the loss of their homes and property, and the impact of propaganda and news attacks. And on both sides of the line people can see no way out other than to destroy the "enemy."

This extremism—I am thinking not of political sentiments but of the state of mind and the consequences of living under extreme conditions—encompasses the fighters too. They are living in the half-burned homes of those they are protecting, eating their bread and seeing their suffering. And it only requires a rumble of gunfire from the "other side" for them to start firing off volleys in retaliation.

Take a look out the window—the State Council building is surrounded by Georgian refugees. They are demanding revenge. The politicians are under constant irrational pressure. The armed formations in the conflict zone are in the same position, only under far worse conditions.

How can we control the situation? We need to do painstaking, delicate, and continuous long-term work. At the political level we need the four-party constructive dialogue that we have already agreed with B.N. Yeltsin. And this should be supplemented by a "lower-level" dialogue between representatives of the sides, be they the commanders of opposed armed groups or just ordinary citizens. We agreed in Kazbegi that the creative intelligentsia, women's and youth organizations, well-known athletes, and journalists from Georgia and North Ossetia would be involved in this work. For a political solution to the problem we need to improve the psychological climate, and we have started moving in this direction.

A four-party observation center has started work in Tskhinvali. We have delegated our own representatives to a joint press center. The other day I talked in Tbilisi with the leaders of the Union of Afghan Servicemen about the formation of an international peacemaking subunit that would keep the sides apart and help bring about a cease-fire. . . .

Nobody has the right to accuse us of twiddling our thumbs. But then Khasbulatov's statement appeared, and everything has virtually come to a halt. I have heard claims from Moscow that "noisy propaganda steps regarding the Georgian side's allegedly persistent desire to solve the problem by peaceful means are being taken" in order to cover up the "strategic line of forcing the South Ossetians out."

This is neither serious nor responsible.

[Onikova]: Given recent events, what ways do you see to settle the situation?

[Shevardnadze]: We have no choice but to seek a path to peace. I have already talked about certain measures we have proposed. We will seek others too, striving for dialogue and conducting it in a spirit of pragmatic compromise. I can clearly foresee two points. The first will be the impossibility of solving all the problems quickly. The second will be the impermissibility of delaying the resolution of the most critical and acute issues. I would include in this category a complete cease-fire, the creation of the most favorable psychological climate for beginning talks, and the return of all refugees—irrespective of their nationality—to their places of permanent residence. This is the first stage, which would allow us to break down the wall of distrust and enmity erected by the conflict. Once that has been removed, there would be scope for action by all our peoples' healthy forces, who have built up many centuries of experience of living together in friendship. By granting these forces a most-favored status, we would reach our goal relatively quickly. If, of course, we are not hampered by people whose aim is to destroy—whatever posts or jobs they may hold.

## 4.62

**Shevardnadze Interviewed on Conflict, CIS**
Moscow Programma Radio Odin Network
21 June 1992 [FBIS Translation]

In an interview to the Turin newspaper *La Stampa*, Georgian State Council Chairman Eduard Shevardnadze expressed the conviction that agreement may be reach on the situation in South Ossetia in a relatively short time if the Russian leadership adopts a correct position and does not support the provocateurs and if President Yeltsin takes part in the talks.

He noted also that Georgia does not intend to join the CIS in the near future but it will strive to strengthen its relations with all member countries of the Commonwealth.

Eduard Shevardnadze reported that a draft treaty between Georgia and Russia is being elaborated now which will establish the legal basis of the stay of Russian troops on Georgian territory.

In reply to a correspondent's question on the fate of perestroika, the former minister of foreign affairs of the former USSR, one of the main architects of perestroika, pointed out that the very fact of the creation of fifteen independent states on the expanse of the Soviet empire has guaranteed perestroika a place in history.

## 4.63

**Abkhazia, South Ossetia Appeal to Yeltsin to Delay Treaty**
Vadim Byrkin
ITAR-TASS, 29 May 1993
[FBIS Translation], Excerpts

The leaders of the Abkhazian and South Ossetian parliaments, Vladislav Ardzinba and Torez Kulumbekov, have appealed to Russian president Boris Yeltsin on Saturday [29 May] to delay signing a treaty on friendship and cooperation with Georgia until a comprehensive settlement of the conflict in Abkhazia and South Ossetia is achieved.

A joint statement to this effect says that "Abkhazia and South Ossetia can no longer remain a part of Georgia, which is unable to act as a guarantor of rights of the people of our republics."

Ardzinba and Kulumbekov expressed concern about "the continuing supplies of weapons, ammunition and military hardware from Russia to Georgia." "We are more concerned about Russo–Georgian talks, at which the Georgian leadership is attempting to decide the fates of Abkhazia and South Ossetia behind our back," the statement said.

## 4.64

**Shevardnadze Discusses Abkhazia and Joining the CIS**
Interview by Daniel Lecomte
ARTE Television Network (Strasbourg)
8 October 1993
[FBIS Translation], Excerpts

Mr. Shevardnadze, you met Boris Yeltsin this morning, and following this meeting, Georgia decided to join the CIS, that is, this commonwealth of the states that made up the former Soviet Union. This, only a few days after you said, "After all that happened in Georgia and mostly after Russia's betrayal, I think Georgia will not join the CIS." Why did you change your mind?

[Shevardnadze]: This is linked to changes within the Russian government. Also, the situation in Georgia compelled me to take this step. Recent difficulties between Russia and Georgia, which became extremely serious with the conflict between Georgia and Abkhazia, were mainly due to these reactionary forces and to their policies. These forces are now isolated in Russia. So, I hope that our decision will help in establishing very friendly relations.

[Lecomte]: We will return to Georgia later, but if I understood your answer properly, last weekend's events [in Moscow] were instrumental in your reversal. [Passage omitted.] Mentioning Georgia, you said recently that the West did not do enough. Do you feel this was true also for Russia? Do you feel that the West is partly to blame? What would need to be done today to reverse the course of events?

[Shevardnadze]: The West's support to Russia is important, but I believe that it is insufficient. In the past six or seven years, since the revolution in the Soviet Union and Russia started, great changes have taken place on our planet, and I would say that this was the greatest revolution this century. International public opinion did not draw conclusions. At the end of the World War II, the United Nations was set up. Current developments are much more radical, and I think that we need something global now. This explains my dissatisfaction.

[Lecomte]: More importantly, you expect Western aid, I assume?

[Shevardnadze]: Aid too, naturally, since the West saved a great deal financially; it saved billions. There is no more confrontation between East and West. There are military, financial, economic, and moral savings. In fact, we, the states on the path to democracy, expect more . . . .

[Lecomte]: Russian troops in the region supported the Abkhazians. Why do you think they supported them, when, in a way, for the Russians this is playing with fire? Why did the Russians and Mr. Yeltsin decide on this support?

[Shevardnadze]: I should differentiate between two sides in Russia. On the one hand is the side headed by Yeltsin, and on the other—I would even use rather crude words—there are the bastards, who did everything they could to raise Abkhaz separatism to the level of fascism. They were the ones who provided the financing, they were the ones who supplied all the modern armament. All the ammunition was supplied by them. Training was carried out with their help, and they even took direct part in the fighting. That is what I would call the policy of reactionary Russia.

I am sure that when the investigation into the putschists has been completed, everything will be clear about how the Abkhaz card was played and how the Georgian cards in general were played. We have been the victims of these games, of this confrontation in Russia. Unfortunately, I could not do anything to prevent this. They were huge forces, which Georgia had no chance of countering. That is why Sukhumi and Abkhazia have been lost and why there is a danger of civil war in Georgia. But I am really counting on the post-revolutionary period in Russia.

[Lecomte]: People have said for the Russians it had to do with access to warm waters, access to the Black Sea, so it was more a reflex from a has-been empire, an imperial reflex, rather than a democratic reflex.

[Shevardnadze]: When I stress the word democratic, I am saying that a democratic state never tries to obtain its aims [as heard]. The events in Abkhazia and Georgia are typical examples of what a totalitarian regime and supporters of totalitarian regimes can do. [Passage omitted: discussion of his role in Sukhumi.]

[Lecomte]: Mr. Shevardnadze, I would like to ask you a question that might seem trivial but is important for us: Who called whom? Did you call Mr. Yeltsin to ask to come to Moscow or did Mr. Yeltsin ask you to come to meet him?

[Shevardnadze]: The initiative was mine, but I also knew that it was also Yeltsin's wish and that he was ready for this meeting. I think that following these events in Moscow, Mr. Yeltsin once more realized that a united and undivided Georgia was more in Russia's interests. [Background video report on current nationalist unrest in the CIS.]

[Lecomte]: I recall that you met Boris Yeltsin this morning and the decision for Georgia to join the CIS was made. First, do you consider this a way to prevent the internal split in Georgia and what specifically did Boris Yeltsin have to offer?

[Shevardnadze]: I would like to say that the decision to join the CIS was not the decision of the parliament—I made this decision. There are certainly difficult times in history when one person has to take all the responsibility; today, I took this responsibility. Parliament might disagree. I must say that it will not be without reason, but I saw in this decision the last chance to rescue my people and my country while preventing its disintegration, preventing civil war, and enabling justice to emerge again in Abkhazia.

[Lecomte]: Thank you for having taken part in this program and for having given us your time to describe the situation in your republic. [Passage omitted.]

## 4.65

### Georgia "Balkanization" Has Perils for Russia

Petr Karapetyan
*Krasnaya Zvezda,* 22 October 1993
[FBIS Translation], Excerpts

Georgia is taking urgent steps to join the CIS. Some 121 parliamentary deputies (the minimum being 110) have already signed a statement in support of Eduard Shevardnadze's proposal on joining the commonwealth. The next stage, apparently, will be the convening of the parliament, at which the decision to join the CIS will be ratified.

This is the second attempt by Georgia's leaders in its recent history to join the CIS. The "pioneer," if you recall, was none other than Zviad Gamsakhurdia. In December 1991, as soon as the artillery of the opposition rebelling against him opened up, he sent telegram after telegram to Moscow requesting Georgia's immediate, instant admission to the CIS. The plan was simple: Having neither the forces nor the capability to crush the opposition, Gamsakhurdia hoped to do this by bringing Georgia into the CIS collective security system. He did not get anywhere at the time.

Nearly two years of history have been repeated. Dismissing the very idea of the CIS following the overthrow of Gamsakhurdia, the new Georgian leaders did all they could to emphasize their independence from Russia and the other CIS countries, while not forgetting to make a big show of their inclination toward the West. Look how much mud was flung at Russia during the Abkhaz war! What humiliation and outrages Russian servicemen and members of their families suffered and indeed continue to suffer in Georgia! One terrible figure: In the past two years, seventy Russian servicemen—think on it, reader—have been killed in Georgia! Only once has the Georgian government deigned to offer condolences to the family of the dead.

And Russia, vilified and slandered, unlike Georgia's beloved West, which fobbed Georgia off with supplies of humanitarian aid and ten UN observers when asked to intervene in the Abkhaz conflict, took on the entire burden of peacemaking in the region.

Tbilisi ran to Russia (for the umpteenth time) when Georgia was at its wit's end. It must be remembered that Eduard Shevardnadze started talking about the CIS when the fate of Sukhumi was practically sealed. Consider the fact that after the fall of Sukhumi Eduard Shevardnadze immediately changed his position and said that there could be no question of joining the CIS. After all this "toing and froing" how can one explain his current burning desire to see Georgia in the CIS, and how come the parliamentarians, immovable until the subject of the CIS was broached, have become so compliant?

The explanation is simple: Tbilisi officials are also in danger because Zviad Gamsakhurdia has snapped up the whole of Western Georgia. Abkhazia is hard at work building fortifications on the borders with Georgia. South Ossetia is making a big show of independence from Tbilisi, and even Ajaria occasionally mentions autonomous status, albeit far more tactfully and diplomatically. . . . Georgia faces the danger of "Balkanization." Things are bad—so the Georgian leaders are prepared to vote for the CIS with both hands.

Are the commonwealth countries supposed to play the part of the collective boy who Georgia has decided will pull the chestnuts out of the fire?

Certainly Russia cannot stand aloof from what is going on in Georgia. If the Balkans are the powder keg of Europe, then Georgia is the powder keg of the Caucasus, and, unfortunately, in Russia's backyard in the south. Given this, political scientists believe, Moscow must first of all come to grips with what is happening in Georgia.

Russia's complex of strategic interests there, from geopolitical and economic to the problems of the Russian-speaking population and the grouping of forces, has taken shape over a long period of time. It is pretty clear that Russian policy in the region—not only in the foreseeable future, as far as that is possible at the moment, but in the very long term—should be shaped by concern for its own interests, rather than by the problems of the latest Georgian political leader, be it Gamsakhurdia, Shevardnadze, or someone else.

It is in this context, I believe, that the Russian foreign ministry decision to respond to Eduard Shevardnadze's request and send a troop contingent to Georgia to guard the Poti–Tbilisi railroad should be viewed. It is particularly emphasized that the contingent's mission is a peacemaking one and, naturally, it will not participate in combat actions on anyone's side. But one should not idealize the situation the Russian contingent could find itself in given that Georgia does not have a military cooperation treaty with Russia, is not a member of the CIS, and, therefore, as the Russian defense minister, Army General Pavel Grachev, said, "is not party to the adoption of joint measures on safeguarding security." On the Georgian front the situation changes at the blink of an eye, but the danger of Russians becoming involved in the conflict exists nonetheless. That is why when introducing the contingent into the conflict zone it is necessary to start devising a whole range of political, diplomatic, and, if need be, other measures to protect Russia's interests in the region.

In turn, Georgia, having invited Russia, as a foreign policy partner, to solve its internal problems, must guarantee observance of its interests.

---

## 4.66

**Shevardnadze Discusses the CIS, Russia, and Other Issues**

Interview by Nato Oniani
Tbilisi Radio Tbilisi Network, 27 December 1993
[FBIS Translation], Excerpts

---

[Oniani]: We present our regular interview with the chairman of the parliament of the republic, Eduard Shevardnadze.

Batono [honorific title] Eduard, the Ashkhabad meeting of the heads of government of the CIS member states was last week's important event. When assessing the results of the meeting in a television interview, you said twice, convincingly: I believe the decision to join the CIS was correct and that its revision is not necessary.

[Shevardnadze]: Good morning. Thank you. I have already expressed my general observations on the Ashkhabad meeting. I shall confirm what I have said: that it was an important stage in the life of the CIS, which has existed only for two years, and from the point of view of unanimity and mutual understanding. Judging from the information I had, this meeting stood out from other meetings, although I was not present at other meetings, I can say so convincingly. Now about some of the conclusions after I got back. Does the CIS have a future? Does it have prospects? Is it viable? I shall touch on only some of the general principles.

In my opinion, the CIS will have a future and become a repository of trust and cooperation if Russia continues to follow the path of democracy and compromise. Even so, the center of the CIS, the key magnetic pole, and the initiator of the CIS is Russia, and the fate of the CIS's future will depend on Russia and the processes under way in Russia. If it does not happen this way, that is, if Russia does not follow the path of democracy, then, of course, the CIS will disintegrate.

I would also like to stress another point, that the CIS in itself becomes one of the factors behind the development of democracy in its member countries as well as in Russia. I have drawn such a conclusion from my own observations and the talks I have had. There are prospects that the CIS could become one of the factors and a major contributor to the development of democracy in these countries.

Third, the CIS will be viable if in practical terms it resists aggressive separatism and becomes in practical terms a solid guarantor of the territorial integrity of the CIS states. . . . Every CIS member desires that this organization become one of the guarantors of territorial integrity and for defeating aggressive separatism.

And one more thing. The CIS must become a necessary organization for everybody if it becomes the guarantor of the integration of its member-states' economy. None of the countries will be able to overcome the deep economic crisis without such integration; I am strongly convinced of this. They will not be able to create a market economy, or the principles of the economy. They will not be able to resolve social problems, and if they cannot resolve social problems then, naturally, social outbursts can always be expected that may become more dangerous than separatist movements if the countries fail to find a solution.

At the present level, social issues are the most palpable and the most acute. I think that within the limits of the CIS, it is quite possible to resolve many social issues on the basis of the development of the process of integration.

There are some other reasons [for joining the CIS], which is why I said that our decision to join the CIS was correct and does not need to be revised and does not require it. By the way, I quoted one of the factors: the document which I signed—that the CIS, and in particular collective security—the principles of collective security and the agreement on collective security, the mechanisms of which we have enacted recently—says that the mechanism is one of the key guarantors of the territorial integrity and stability in these countries.

[Oniani]: You said that unless Russia pursues the path of democratic development, the Commonwealth will disintegrate. Will it disintegrate and the empire be restored? Is this what you mean?

[Shevardnadze]: No, the empire will not be restored. If the Commonwealth disintegrates, each country will go its own way. More emphasis will probably shift toward regional unions, for example, a union of Central Asia, of the Caucasus, or of some other region. And then, the Commonwealth, as we know it now, will disintegrate.

[Oniani]: Thank you. You met with President Yeltsin in Ashkhabad and I would like to quote an excerpt from your interview. You said there: As for Abkhazia, I think that now the Russian president fully understands how acute this problem is for us, the Caucasus and Russia.

A Russian foreign ministry delegation arrived in Tbilisi yesterday. Among other issues, you will probably also discuss with them the issue of Abkhazia. The recent elections in Russia produced both positive and negative results. By positive results I mean that the Ardzinba-Gamsakhurdia-Zhirinovskiy clan has been exposed and both Russia and the world saw the real threat of fascism. And by negative results I mean the heavy vote for Zhirinovskiy. What are the chances for solving the problem of Abkhazia? And also, does the strengthening of relations between Georgia and Russia coincide with Georgia's entry into the CIS?

[Shevardnadze]: I have already spoken about my impressions of my meeting with Yeltsin. A delegation arrived here yesterday. It includes reputable diplomats, as well as the military. My colleagues and I had our first meeting with them yesterday. Of course, we have two problems that are central to relations between Russia and Georgia. These are Abkhazia and Samachablo, [or] the Tskhinvali zone and South Ossetia [word indistinct]. This is not only our problem—but my impression is that this is also, in a certain sense and to some extent, Russia's problem. Therefore, Abkhazia and Samachablo are a touchstone that will test the firmness of our ties.

I do not mean that the issue of Abkhazia should be solved by taking only Georgian interests into account. Both issues, Abkhazia and Samachablo, should be solved by taking into account the interests of all citizens, including, of course, the interests of the Abkhazians. Despite the disaster that unfolded there and the fact that almost all the Georgian population was displaced, I nevertheless believe that there are now better prospects for a settlement than at any other time. This is because, based on my observation, Russia now better understands the danger for Georgia, the Transcaucasus, and the Caucasus region in general, including Russia, of leaving this conflict unresolved.

## 4.67
## Further on Shevardnadze and the CIS

Tbilisi Radio Tbilisi Network, 1 March 1994
[FBIS Translation]

According to a report from the chief of staff to the head of state, information from Head of State Eduard Shevardnadze concerning Georgia's joining the CIS was circulated today among the deputies of the Georgian parliament. The information says: The Georgian parliament must make a significant decision today on whether or not Georgia should join the CIS. The significance of this step is indicated by the choice that our state is faced with today. Today each one of us must decide, as our conscience and intelligence prompt us, whether to follow the path of emotions, illusions, and—proceeding from this—dashed hopes, or to take a hard look at reality and adopt a pragmatic attitude to our everyday issues.

The break-up of the Soviet empire, on the one hand, created fifteen independent states, but, on the other hand, it created a new danger both for the development and existence of these countries and for the whole world. If events developed further according to this scenario, they would end with a complete catastrophe. This is what stimulated the search for ways out of the grave situation, and in my mind, the creation of the CIS was the result of this search.

In the beginning the CIS was not a structure that could effectively resolve the issues of Georgia and other countries, but, in the meantime, in the process of its self-determination, a whole range of significant issues were being worked out and mechanisms for comprehensive cooperation established. Georgia did participate in the work of different organizations of the CIS, but this was chiefly as an observer, which limited our influence on the process of adopting a decision, and our national interests were not fully observed.

During this period, Georgia tried to regulate relations with other former Soviet republics. A fairly strong legislative basis for bilateral political and economic relations was being dynamically created. However, in the new circumstances and against the background of the processes of regional integration, this did not prove to be enough. The development of the processes within the CIS has once again proved the truth that securing political, military, and economic interests can be carried out effectively only if a bilateral cooperation is replaced by the mechanisms of comprehensive cooperation.

The CIS may not represent a definite guarantee for security, but it is one of the factors for legally securing our territorial integrity and the inviolability of our borders.

The regulations of the CIS also envisage the creation of mechanisms for the member-states' collective security, eradicating conflicts, and peacefully resolving disputes.

The realization of the CIS regulations will help restore broken economic links and develop them on a completely new basis, which is a necessary prerequisite for the reforms for transition to a market economy.

Thus, becoming a member of the CIS will help our country to establish its place in the network of international and regional relations, to become a part of the mechanisms of comprehensive cooperation, and to use agreements on collective security and economic cooperation for reinforcing with laws the country's territorial integrity, influencing the development of regional processes, and creating prerequisites for overcoming the economic crisis.

By becoming a CIS member, our country is proving its aspiration for civilized development of international relations and is obtaining additional means for improving the country's social and economic situation; it is becoming a fully fledged member of the CIS with all relevant rights and, of course, responsibilities.

The CIS is in its initial stage of formation. The process of its formation is developing in quite a conflicting manner. Issues are seldom resolved without hot debates. Discussions are often uncompromising and severe.

Will the West react negatively to Georgia's joining the CIS? There are different points of view in Georgia's political circles on this issue. I have the following point of view: Many intergovernmental structures have been created since World War II. Most of them proved viable, and many countries poorer than Georgia have found their way out from crisis and poverty. The CIS must share this experience; there will then be less ground for doubts, and we can show everybody what we can do.

The fact that we have already signed comprehensive agreements on friendship, cooperation, good-neighborliness, and mutual assistance with most of the CIS states gives us the hope that on the basis of work with natural allies we will be able to set up the CIS as a civilized and integrated union, and Georgia can make its honorable contribution in this respect.

I am sure that our active participation in the CIS will assist in overcoming the current crisis, promoting stabilization, and halting the decline in the population's standard of living.

## 4.68

### Shevardnadze Comments on the CIS Summit in Moscow

Interfax, 23 October 1994
[FBIS Translation], Excerpt

Georgian leader Eduard Shevardnadze said he considered the CIS summit in Moscow on 21 October "an event marking the beginning of an active process of interaction between the independent states."

He said on national television on Saturday that he believed that the crisis that had hit all former Soviet republics could be overcome only through economic integration. Shevardnadze said none of the ex-Soviet republics could gain access to the world market on its own, but they could do so by joint efforts.

Shevardnadze said the time had not yet come to carry out the initiative of Kazakh President Nursultan Nazarbaev concerning establishment of the Euro-Asian union. However, Shevardnadze said that such a union could be created in the future and that such countries as Mongolia and unified Korea could join it.

Shevardnadze said that during his meetings with Russian President Boris Yeltsin, Premier Viktor Chernomyrdin, and Foreign Minister Andrey Kozyrev, the issues of bilateral economic cooperation, establishment of the Georgian army, and settlement of the conflicts in Abkhazia and South Ossetia had been discussed.

## 4.69

### Georgian Leaders Appraise Almaty Summit Results

Interviews
*Trud,* 17 February 1995
[FBIS Translation], Excerpts

[Report of interviews with Georgian President Eduard Shevardnadze; Georgian Foreign Minister Aleksandr Chikvaidze; Maj.-Gen. Valeriy Chkheidze, commander of the Georgian Border Forces; and Georgian Prime Minister Otar Patsatsia.]

[Shevardnadze]: Let me begin by saying that, during our meetings with President Yeltsin, Premier Chernomyrdin, Foreign Minister A. Kozyrev, and Defense Minister Grachev, we concluded that the final stage in the settlement of the Georgian–Abkhazian and Georgian–Ossetian conflicts has arrived. Georgia's position on aggressive separatism as voiced in the special addendum to the text of the Memorandum on the Maintenance of Peace and Stability in the CIS, which was signed in Almaty, seems thoroughly justified and fair to me particularly in the context of events in Chechnya. Abkhazia and Chechnya are links in the same chain, and understanding has matured among the CIS heads of state of the danger harbored by aggressive separatism. Our objective assessment of events in Abkhazia and Chechnya elicited the correct response from the Russian leadership. The meeting in Almaty persuaded us that the speediest conclusion of the conflict in Georgia is in the interest of all CIS countries. That manifested itself in unanimous support for Georgia's position on aggressive separatism. . . .

[Chikvaidze]: The Georgian delegation's wording went like this: The states must not support separatist movements in other member states if they arise, must not establish political, economic, and other ties with them, must prevent their use of CIS territories and lines of communications, and must not provide economic, financial, and other aid. This wording was not adopted straightaway—there were votes in favor of a milder version—but we succeeded in getting our way.

The question was unique: why aggressive separatism precisely and not any other kind? This is not merely a theoretical dispute, the problem is more serious. After all aggressive separatism is linked with military force and that is where it differs from Quebec, for instance, where there is also talk of secession but no bloodshed.

[*Trud*]: What methods will be used henceforth to combat aggressive separatism?

[Chikvaidze]: The states pledged not to create a breeding ground for the development of this phenomenon on their territories. Political, economic, and, if necessary, military barriers will be erected in its way. All member countries were unanimous in their agreement. [Chikvaidze ends.]

At our request Maj.-Gen. Valeriy Chkheidze, commander of the Georgian Border Forces, summed up the results of the work done in Almaty:

[Chkheidze]: We discussed the question of the concept for the protection of the CIS state borders. It is gratifying for us that the process of Commonwealth integration is deepening and the quest for a common denominator is successfully under way. That is the main point. However it is necessary to take into consideration the legislation of all member countries in our work so as to secure the sought-after unified legislation without detriment to the republics' sovereignty. . . .

# The States of the Western Region

In the Western Region, Russia has both its strongest ally, Belarus, and its strongest challenger, Ukraine. As these documents illustrate, Belarus's two presidents since the formation of the CIS have held opposing views on CIS integration. Stanislav Shushkevich, several of whose 1992–93 speeches appear here, tried to maneuver an independent course for Belarus within the CIS in an attempt to establish more of a national identity, which Belarus has lacked since tsarist times. Nevertheless, like Azerbaijan's Elchibey, he was politically isolated by his prime minister, Vyacheslav Kebich, who leaned toward close integration with Moscow. In 1993, the Belarusians elected Alekandr Lukashenka, who immediately redirected the country toward full monetary, economic, and military integration with Russia, forming the cornerstone of a new Slavic Union. The Belarus documents contained here have been selected to illustrate the dichotomy between Shushkevich's and Lukashenka's approaches to Russia and the CIS, as well as the extent to which Belarus had forfeited its sovereignty to Russia by July 1995. In his television address on 15 February 1995 (which is contained here), Lukashenka admitted that he had never supported the breakup of the Soviet Union and alleged that Yeltsin had not wanted the Union to dissolve either. In May 1995, Lukashenka signed a bilateral treaty with Russia creating the "Minsk/Moscow Axis" (see Chapter 6).

The third country included in the Western Region is Moldova, whose relations with Russia have been formed primarily by the often violent struggle on its territory over the Trans-Dniester region. The Moldovan documents depict President Mircea Snegur's view that Russia's 14th Army, which has occupied the Trans-Dniester region since Gorbachev's rule, constitutes interference in the country's internal affairs. Press articles confirming Russia's October 1994 agreement to withdraw the 14th Army from Moldova within a period of three years are contained here, along with articles alluding to troubles that have arisen more recently in connection with the withdrawal process. These documents reveal the lack of trust between Russia and Moldova, which is not likely to improve if Russia's generals continue to regard the Trans-Dniester region as vital to Russia's national interests. Another key issue between Moldova and Russia has been the movement to reunite Moldova with Romania.

One article included in this chapter records the signing of a Moldovan-Romanian cooperation agreement. This issue is likely to be a continuous source of antagonism between the two countries. Nevertheless, beset by war and economic disruption, President Snegur changed his decision to remain an observer in the CIS and joined the organization and its Economic Union in September 1993. The documents selected for Moldova also record certain key events in relations between Moldova and Ukraine, a country that is particularly important to Moldova and at one time incorporated part of Moldova. Ukraine announced in 1993 that it would participate in regulating the Trans-Dniestrian conflict. Several articles recording key developments between Russia and Moldova in the issue are contained here. Lastly, to document Moldova's strong orientation toward the West, a press article from 10 July 1995 announcing that Moldova would become the first CIS state to become a full member of the European Council, even before Russia, has been included.

## Belarus

### 4.70
### Shushkevich Reflects on the CIS's Past and Future

Interview by Evgeniy Gorelik and Aleksandr Kryzhanovskiy
*Sovetskaya Belorussiya,* 29 December 1992
[FBIS Translation], Excerpts

"Thirty-Six Hours That Shook the World" was the title of the first interview—published a year ago—that Stanislav Shushkevich gave after Viskuli. Since then there have been many events in the life of the commonwealth, and many interviews. The first one remained the most memorable, though. Now once again, reflecting on the past and contemplating the future, the chairman of the republic's Supreme Soviet speaks to the journalists who met him a year ago right at the steps of the aircraft that came straight from Belovezha.

[Gorelik and Kryzhanovskiy]: Stanislav Stanislavovich, tell us—there, in Viskuli, was your offspring the way you had envisioned, as it now faces the public upon reaching the age of one?

[Shushkevich]: I am not an astrologer or a clairvoyant, but I must say that quite a lot of what had been envisioned is being implemented. However, it is indeed becoming increasingly fashionable to push the negative processes within the CIS to the forefront. We hear from left and right the moans about the worsening state of affairs in the economy, about the complexities in the relationships among the CIS member countries. No matter how much the CIS is criticized, however, no matter how much is said about the negative consequence of its creation, I am convinced that its significance undoubtedly is positive. This is manifest first and foremost in that we have been able, in my opinion, to forestall many negative processes associated with the crushing disintegration of the former Union republics.

I would like to note two factors, the first on a political, or perhaps even moral, plane. I do not know of a single person in the republic today who would be afraid to express his political views. Our society has become more democratic; it is easier to breathe in it now, although so far it is difficult to live. It may be happening slowly, tentatively, but the process of democratic transformation in all the Commonwealth countries is moving forward; the right of nations to self-determination, which was for many years an illusion, is taking real shape.

As regards the economy, alas, the expectations have not been met as yet; production is declining; inflation is stubbornly climbing up; the cost of living is increasing. But who can say that it would have been better without the Commonwealth? I am convinced that it would have been much worse, although sometimes it seems that it cannot be worse than it is. Many parliament members see the existence of the CIS as the sole cause of the failures; they accuse us of bringing about the disintegration of the Union, which—and they know it perfectly well—had practically ceased to exist by the time of the Viskuli summit. For some reason, they forget that they themselves gave the Commonwealth the green light when they unanimously supported the ratification of the Belovezha agreements.

Without question, there could be more positive achievements, but neither can we dismiss entirely the fact that there are two global process going on in parallel today: a complete reorganization of the economy and a new statehood coming into its own. Do I need to tell you how complex and difficult to manage these processes are? Still, by using the Commonwealth levers—weak as they are—we have been able to resolve many issues in a normal, civilized way. Take, for example, transportation, communication, common informational space, and other sectors, where mutually beneficial partnership is gradually being established, and where old threads of economic ties and true market relations are strengthened and new ones emerge.

[Gorelik and Kryzhanovskiy]: Many politicians, including members of former president M. Gorbachev's inner circle, accuse Yeltsin, Shushkevich, and Kravchuk of bringing about the disintegration of the mighty Union. Did not Mikhail Sergeevich himself, though, make the adoption of a new Union treaty impossible by his actions—or rather inaction? Or was what happened in Viskuli a historic inevitability?

[Shushkevich]: Let us look the truth in the face. By the time of our summit in Belovezha, the Union was practically doomed as a single state, despite M. Gorbachev's spasmodic attempts to catch a "second wind" in the Novo-Ogarevo process. I do not think there was a force anymore that could have stopped the fully wound-up flywheel of centrifugal forces.

It was no longer possible to preserve the old "brotherly" family of nations. There was, however, still an opportunity to "divorce" without all-out "court battles," to reach an amicable agreement with respect to the division of both the acquired household possessions and common debts. Unfortunately, the creation of the CIS still did not stem the disintegration processes in the economy and did not save the sovereign states from the flames of local interethnic conflicts. Apparently, it was futile to hope that the extremely complex problems that have been accumulating over decades could be resolved during the first stage of the commonwealth's existence. The time when the old building still could be restored and rebuilt was gone, and could not return. And although the main architect of perestroika kept assuring us that it is this fundamental restructuring that he had been implementing all this time, in fact, everything was inexorably deteriorating.

The scenario of the disintegration of the Soviet empire could have been completely different—much more tragic and bloody—had the leaders of the states that became the commonwealth members not sought a compromise at the negotiating table, had they chosen other methods of resolving numerous conflicts and claims.

[Gorelik and Kryzhanovskiy]: If you do not mind, let us go back to Viskuli for a few minutes. What was the atmosphere of the summit? The famous secrecy that surrounded it begat many rumors. Some say, for instance, that all delegations brought ready drafts of the agreement. Others say that only Burbulis had a clear idea of the future CIS.

Cutting off communications with the outside world was explained by some as the desire to concentrate on the main topic, the task at hand. There are some, though, who interpret it as fear of the reaction by Gorbachev, who could send paratroops in helicopters. What really happened?

[Shushkevich]: A communications blackout, a helicopter, and so on are a figment of someone's imagination. We did not even think about it. The only suspenseful part of the summit, in my opinion, was that up to the last moment we expected N. Nazarbaev to come. We stayed in continuous contact with his airplane; everything was ready—a landing strip and a way to get to Viskuli quickly. He landed in Moscow, however, and got stuck there. Nevertheless, we believed that Nursultan Abishevich would come no matter what, because in February a four-party agreement was reached among Russia, Ukraine, Belarus, and Kazakhstan. On the basis of this agreement, there was only one document prepared in advance—a draft of a communiqué. And for me, for instance, it came as a total surprise that L. Kravchuk suddenly expressed willingness to sign it on the spot, right there in Minsk. Moreover, not only did he not object to closer integration, he even said he intended to go beyond that. For that "beyond," however, to tell you honestly, we did not have anything except some general materials. It was then that the work got under way, as they say, "playing it by ear." I still cannot be absolutely certain whether the Russian side had any draft prepared at home. If so, they must be the greatest conspirators. Or whether it truly was a first-class professional team: All formulae developed quickly; as soon as one of us proffered an idea, it was immediately translated into precise, legally valid stipulations. The night from 7 to 8 December showed an immense capacity for work on the part of the people, who were united by the same desire and goals.

[Gorelik and Kryzhanovskiy]: It is no secret that the Belovezha agreements are now being carefully and exactingly scrutinized by legal experts. Some specialists find faulty wording in them. . . .

[Shushkevich]: It is always hard to argue with authorities in their field, but I want to point out that not a single international association of lawyers, and not a single UN or CSCE organ, has questioned the legality of the Belovezha agreements. I agree that these documents are not a dogma; new documents should be developed based on them. And such work has been done. Its result is the preparation of the draft commonwealth charter, which will be presented for consideration at the upcoming meeting in Minsk.

[Gorelik and Kryzhanovskiy]: We heard that the Belarusian side is not ready to sign this document. . . .

[Shushkevich]: You see, every leader—if he wants to remain such—must be certain that his actions will meet with understanding on the part of the parliament's majority. When we went to Viskuli a year ago, we believed intuitively, deep inside, that the deputies would approve our actions, although we did not have much hope that the documents would be ratified without reservations. Today the situation is different, and there is no certainty that the charter—in the form it has been initialed by our representatives—will then be approved by the parliament. Why? Preliminary consultations with top-level jurists have confirmed our concerns that this document conflicts with some stipulation of the constitutional law—the Declaration of the State Sovereignty of Belarus. Therefore, although I have a constitutional right to do so, I cannot sign it without the approval of the Supreme Soviet. I will gladly do it—if not today, then tomorrow—if the parliament authorizes me. (When this article was being prepared for publication, it was announced that the parliament has authorized S. Shushkevich to sign the document on the CIS charter.—Author.)

[Gorelik and Kryzhanovskiy]: By the way, speaking of the burden of leadership, both our press and the foreign press often promote the idea that the CIS today is holding together only because of the personalities involved. To support this contention, they point to Azerbaijan, which after the resignation of A. Mutalibov—a strong CIS supporter—immediately said goodbye to the Commonwealth. What foundation—one more solid and lasting than the authority of the first person of the state—should the Commonwealth be built on so that its stability would not depend on political storms and cataclysms in any individual country?

[Shushkevich]: The example you quoted does not mean that this is an axiom. No leader today can act only upon his own discretion. As to the Commonwealth's foundation, it should be purely legal, and include both control and sanction mechanisms. It is already clear to everybody that the structures that by now have been created within the Commonwealth framework are inadequate for closer integration and effectiveness of joint actions.

[Gorelik and Kryzhanovskiy]: About 250 documents have been adopted under the CIS banner. Most of them, alas, do not work. Is the main problem only the lack of coordinating organs in the structure? Or is the cause deeper and more profound?

[Shushkevich]: Tell me, did all the stipulations in the Brezhnev Constitution work? And, mind you, that one operated within a rigid, centralized system. I think that our regrettable past continues to have its impact: We produce beautiful agreements and then try endlessly to improve them, instead of adopting smart, effective laws from the start. Unfortunately, most of us so far have limited political experience in running a state; that is why, perhaps, we have not been able to avoid some purely populist decisions within the CIS framework. Look how many initiatives individual state leaders come up with. One gets the impression that for them the most important point is to proclaim an initiative. Instead, we should be taking care to implement concrete problems that cannot wait or be waived aside: for instance, to create an interstate bank, adopt restrictions on printing money, stop the export of goods in exchange for devalued money, and finally get things in order with respect to mutual clearing between subjects of different states.

[Gorelik and Kryzhanovskiy]: Judging by what you said, you are not against the creation of quasi-national structures. Still, we would like to clarify the position of Belarus, which did not support in Bishkek the idea of creating a Coordinating Economic Council [CEC].

[Shushkevich]: There is no contradiction here. We support extensive coordination, first and foremost with respect to the economy. But the original draft on the CEC contained quite a few points borrowed from former Union structures, in particular, the Gosplan [State Planning Committee]. We could not support it in that form. I repeat, unless we coordinate our actions, especially on the world market, we will suffer tremendous losses, which is already happening, with the commonwealth countries elbowing each other in a rush to export as much as possible of raw materials and the few competitive products, thereby bringing down their already relatively low prices. Unless we coordinate our efforts, we will be continuously and shamelessly robbed, now and in the future. Coordination, as you well know, is impossible without appropriate organs, however. In this respect we cannot count on bilateral agreements because there are problems that can be resolved only on the basis of agreements that will include all the former Union republics. If the CIS wants to survive, it will not be able to avoid unpopular steps.

[Gorelik and Kryzhanovskiy]: What do you mean?

[Shushkevich]: May I answer your question with a question: Do you put money aside to save for the future? Of course not, because today it would be simply stupid. Everybody wants to get rid of money as quickly as possible because the banknotes shrink right before our eyes. This means that we have to find a way to boost the value of money, especially considering that the ruble remains the going currency in most states. We simply cannot do without a coordinated tax policy and effective control over the circulation of credit and cash. I would not preclude coordination with respect to a minimum wage, various benefits and subsidies, and other measures that do not add to the authority of governments and parliaments. Still, I think that we will have to accept it if we want in the final analysis to do good for our peoples. Surgeons often have to inflict pain on a patient in order to save him.

[Gorelik and Kryzhanovskiy]: As is known, many citizens of the former Union do not hide their liking for the "great and mighty." It seems to us that the appeal for the creation of a confederation, issued by the congress of Russian deputies, is the result of an undying nostalgia for the past. What do you think of this appeal?

[Shushkevich]: I can only say that the time when a true confederation could be created has been irrevocably lost. Not least the fault for this lies with M. Gorbachev, who, in my opinion, did not delve in depth into these difficult problems. The current processes are irreversible; we cannot turn history back. Its development already is following a different course. The independent countries of the commonwealth already have received international recognition and have taken their place in the world community. Judge for yourself. Belarus is already recognized by more than one hundred states; it conducts its own policy; its voice is listened to by most authoritative international organizations, where it has become an equal and respected member. As to the appeal by our colleagues in the Russian parliament, it is the responsibility of our Supreme Soviet to reply to this.

[Gorelik and Kryzhanovskiy]: Over the past year of the existence of the CIS, the economic situation in most countries has not improved; the standard of living continues to fall. Can we then speak of any gains in the commonwealth as a whole and Belarus in particular?

[Shushkevich]: It is quite clear that an increased national self-awareness, the young republics' entry into the international arena, bringing some order into relations between the countries, and even coming to a realization that some of them live beyond their means—all of this can be put on the commonwealth's credit side. A year of being a part of it gave us an opportunity to adjust our notion of sovereignty, to outline the real shape of statehood, to see the future of our economic system. It has taught us to act in a new way in the changed conditions. I would like to believe that the results of the republic's national economic performance for November are a good sign of these changes. For the first time this year, many industrial sectors not only halted the production decline, but actually increased output. For instance, in light industry, this increase amounted to 6.4 percent; in machine building—2.6 percent; in food processing—3 percent; and in timber and woodworking industry—2 percent. I think that this is not accidental; rather, it is an expected development that became possible thanks to building new bridges of mutually beneficial ties and good partnership with the help of the CIS. The same processes are under way in other commonwealth republics as well; if we manage to preserve calm and mutual understanding, the next year will become for the citizens of the CIS countries a year of many hopes and aspirations to come true.

---

## 4.71

**Shushkevich Addresses Supreme Soviet**
Minsk Radio Minsk Network, 9 April 1993
[FBIS Translation], Excerpts

---

I believe that we should create genuine democracy in domestic policy, ensure principal human rights, and protect citizens from violence by other citizens and the authorities. A law-governed state means a firm balance of the executive and legislative branches both in the center and in localities, based on the constitution and laws.

With regard to the economy, we should create a socially oriented market economy with different forms of ownership, including comprehensive support for individual and private initiative with preservation of the optimum level of social guarantees for the population.

In foreign policy, we should consolidate the independence of our state so that Belarus takes its rightful place in the international community. There can be various models of our state. For example, it might be a strictly national state, based only on the heritage of previous generations. A Slavic option is also possible as a three-path variant: Moscow, Kiev, and Minsk, or the Minsk–Moscow axis. Also, I support the third variant, which is a neutral and politically stable state, which does not have a distinct Eastern or Western orientation. In its political and economic strategy it strives to be a source of stability in the region. Today, we are clearly adhering to an Eastern orientation. We should see this as the reality. Although we should support and develop it, this orientation must not prevent us from establishing other contacts and contacts in the West. This is my understanding of the essence of this problem. . . .

The majority of us in this hall are paying for the military-industrial complex. The percentage of militarists and industrialists among the deputies is higher than among the ordinary population. I would like to stress what I have been talking about in the report. The components of a suprastate have disappeared in Russia. And we are an appendix to those components. We should change the structure of our industry, which was formerly working for the USSR military industry. We do not have to produce missiles, tanks, and to build submarines with such intensity for half the world. We will lose that industry. We must convert it. There is no other way. . . .

Now I will talk about the collective security issue. I believe that the main mistake of the Ministry of Foreign Affairs is that it failed to adopt a scientific approach toward the collective security issue, based on international law. What is the main point here? Since August 1991, when we voluntarily declared to the whole world that we strive for neutrality, your experts and you yourself, Pyotr Kuzmich, should have remembered the Vienna Convention of 1969, which obliges states to abide by signed treaties and fulfill assumed commitments. So why should we not abide by them? There can be a change only if conditions change. Are we supposed to send threatening phone messages? That is the opinion of the military. I understand them. They will dig up anything in order to prove that we are threatened by Poland, Lithuania, Latvia, Russia—not to mention Ukraine. It is so frightening. We should serve as an example of opposition to military escalation and expenditures. We should show that when we now adopt a practically new military doctrine and move away from neutrality, the first reaction of Poland will be: if they act this way we have to reinforce ourselves too. And there goes an escalation. And I think that we should follow the example of other countries, like Poland, the Baltic states, Romania, and Moldova, and assume the status of neutrality [words indistinct] and declare that we have no claims. Assume the status of neutrality, declare that you are the same as us, that you have no claims. I think that we will eventually take this path.

As for NATO, they see that we do not agree to collective security, and I think that we should agree. The NATO generals would be unemployed if we did not have a union here. This is why I believe that the best guarantee of our security is bilateral agreements with Russia. We already have them. It is a more complicated thing to observe these agreements within the CIS and by joint command over our armed forces. Note the approach that we had. What is its superiority over others? We have assumed right away, I think, a very clear line. We said: There are no CIS armed forces on our territory. We consider these forces Russian—we recognize that they are under Russian jurisdiction—and hold negotiations with Russia. It has simplified the whole system. One does not have to meet with all the CIS countries, but only negotiate with the Russian government and the defense ministry. Look how we have straightened everything out. We have concluded all the treaties. We are consistently moving toward withdrawal of the troops. Foreign aid—Clinton has confirmed it now—will allocate houses in Russia for those troops. We are solving that issue. Why do we have to involve someone else in it? . . .

We should only adopt political decisions because we have all the agreements in place. Let me remind you that in 1991, in Kiev, I signed a two-year agreement, which will be valid until 20 March 1994. This means that we already have collective defense. But we intend to take it step by step, because now we cannot help being in a collective defense and we are in one. What is the issue? We are dealing here with a political declaration of the Supreme Soviet about changing our political course. This is how the world will see it. What can one do in this situation when we have found ourselves faced with a stark choice. . . .

## 4.72

### Lukashenka on the CIS and the Belovezh Forest Agreements

Interview by Aleksandr Stupnikov
Moscow NTV, 15 February 1995
[FBIS Translation], Excerpts

*Editor's Note:* In July 1994 Aleksandr Lukashenka burst onto the Belarusian scene. This former collective farm director soon made it clear that he regarded reunification with

Russia as imperative. During his first year in office, he consolidated his control over Belarus—imposing media censorship, ignoring court rulings, and belittling the parliament. Soon after his election, the parliament was suspended due to insufficient electoral turnout. Having captured almost absolute power, Lukashenka proceeded to rule by decree, and to put into action his plan to annex Belarus to the Russian state. Since his ascendancy to power, Russian troops have reappeared in Belarus, the Belarusian language has been discouraged, the old "Soviet republic" Belarusian flag has been readopted, and Belarusian-language schools are being abolished. Lukashenka has accepted more than integration. He has tried to transform Belarus into part of the "Russian myth," culturally, politically, economically, and militarily. The following documents characterizes Lukashenka's "Russificiation" policies.

The agreements and treaties that we have signed with the Russian Federation recently permit us to declare unambiguously that the economic union with Russia has, in effect, been concluded. That's how it is. We sat down and sorted out which options suit them and suit us, too. What more does one need? That is, the main thing is interest. We have an interest in Russia. There is no altruism in the Belarusian president's policy here: it is absolutely in our interest.

Yes, as always, the opposition criticizes me for this—and not just for this, either—saying that in foreign policy I am selling our interest. You know, I'm an economist, not a photographer. I understand perfectly what interest is. Interest is subject to accounting. What is it that can't be understood there? Let's import oil from somewhere, they say. Excuse me, but it's not so simple. Everything has long been divided up in the world. So it's simpler to lose a sales market than to find one. But as for the moral aspect of the matter, I've already spoken about that—it is the unity of these peoples, the fraternal peoples . . . .

[Stupnikov]: You were the first Belarusian Supreme Soviet deputy, the only one in fact, to vote against—it's in the past now—the Belovezh Forest agreements.

[Lukashenka]: Yes, I voted against them. And I'm proud of the fact that I cast my vote against the dismemberment of a great state. Time has shown that we committed a grave error. But those who say that three Belovezh Forest big shots got together to break up the state are taking a very simplified view of things. They had quite different aims. Moscow was at that time without power. Gorbachev did not exercise power. And it is hard to say how I, Lukashenka, would have acted had I been in Yeltsin's place. Yeltsin was fighting for power. And there had to be power in Moscow, so Yeltsin's aims were quite different. . . .

Yeltsin was not in favor of the collapse of a great country, I am convinced. Why should Lukashenka and Yeltsin miss a chance to do a good thing, and thus live on as good leaders in the memory of the Belarusian and Russian peoples? I want to go down in the memory of your Russian and our Belarusian peoples as just such a leader. [End recording.]

## Moldova

---

### 4.73

**Snegur Opposes Centralized Structures in the CIS**
Valeriy Demidetskiy
ITAR-TASS, 27 September 1992
[FBIS Translation]

---

Moldovan president Mircea Snegur spoke out against the restoration of centralized structures in the Commonwealth of Independent States. In an interview with German radio he said he prefers the development of bilateral, political, and economic relations between the former Union republics.

Snegur said that Moldova's recognition by 120 countries, the creation of a national army with a defensive doctrine, and the people's striving to be free are the main proofs of the Moldovan state's viability. The Moldovan president said the Moldovan people favor the economic reforms and realize that the difficulties connected with reforms can be surmounted.

Snegur asserted that the third force had been involved in the conflict in the Dniester region. It often destabilized the situation to please those whom it served. However it was to conclude the agreement with Russia on settling the conflict in order to stop the threat to peace that was established.

---

### 4.74

**Trans-Dniester Representative in Russia Interviewed**
Interview by Vladimir Ryashin
*Pravda,* 16 December 1992 [FBIS Translation]

---

The Trans-Dniester tragedy is only a crisis to many politicians, but for the people who live there it represents hundreds of graves where the grass has not had time to grow yet. At a meeting with journalists yesterday, Igor Mikhaylov, special representative of the Trans-Dniester Moldovan Republic to Russia, reminded them that the people's wounds have not healed yet.

"The only option our republic is willing to accept," I. Mikhaylov said, "is a confederate union with Moldova. Those who question the sovereign status of the Dniester region do not know much about history and do not want to abide by international law. The independence of the Trans-Dniester Moldavian Republic has been confirmed by the will of the people, expressed by them in a referendum. Besides, what kinds of claims can Chisinau have to the Trans-Dniester after the slaughter the armed forces of Moldova committed in Bendery and other cities and rural communities? The UN declaration on ethnic minorities specifically says that ethnic minorities have the right to secede from a state if crimes and acts of genocide are committed against them. Unfortunately, Russian diplomats have excluded our republic from the negotiations that will decide the future of the Dniester region. Furthermore, UN and CSCE structures often take the side of the strongest party in a conflict."

[Ryashin]: What can you tell us about the 14th Russian Army?

[Mikhaylov]: This very day, 15 December, the parliament of the Trans-Dniester Moldavian Republic will be discussing a bill on its status, to legitimize the army's presence in the republic. Of course, the decision to withdraw or not to withdraw the 14th Army must be made by Russia, but the Dniester region will do everything within its power to protect the Russian servicemen—from the resolution of housing problems to the guarantee of pension security.

I sometimes have to explain to foreign diplomats and journalists that there are only a few thousand soldiers in the 14th Army. The commotion over the threat they pose has been artificial. People in the Dniester region remember that units of this army liberated Tiraspol and Bendery from the Fascists in the Great Patriotic War. People do not want the army to leave.

[Ryashin]: But then why did certain forces in Trans-Dniester try to discredit General Lebed, the commander of the army?

[Mikhaylov]: Yes, there was some friction, and it was reported in the press, but today the republic leaders are doing their best to prevent conflicts with the commander of the 14th Army.

## 4.75
**Foreign Ministry Worried About Yeltsin Statement**
Basapress, 6 March 1993
[FBIS Translation]

The Ministry of Foreign Affairs of Moldova expresses its concern regarding the declaration of the president of the Russian Federation, Boris Yeltsin, on the occasion of the "Civic Union" congress on 28 February 1993, where he pointed out that "now is the right time for organizations with international authority, including the United Nations, to offer Russia special power in her role of guaranteeing peace and security on the territory of the former USSR," as is mentioned in a press release of the Ministry of Foreign Affairs. "The desire of a state to embrace the role of peacemaker in this area disagrees with the norms of international rights as long as none of the independent states grants Russia such a mandate. Furthermore, Russia's military units are still present in several conflict areas including the eastern districts of Moldova," the statement underlines. Moldova feels that conflicts must be solved in conformity with the norms of international rights and with the active participation of all international forums, including the United Nations and CSCE, and is rejecting any attempt to interfere in its internal affairs, the statement concludes.

## 4.76
**Parliament Authorizes Snegur to Sign CIS Accord**
Mircea Dascaliuc
Radio Romania Network, 26 October 1993
[FBIS Translation], Excerpts

The presidium of the parliament of the Republic of Moldova, which ensures the current activities of this legislative body until the next parliament is set up, at today's meeting adopted a decision whereby it approves President Mircea Snegur's signing of the accord founding the CIS and the treaty on the creation of the CIS economic union. It also proposed that these two documents be examined in a later ordinary session.

During the discussions, this decision was motivated by the current participation of the government of the Republic of Moldova in more than two hundred multilateral accords and treaties that regulate questions of an economic, social, and legal nature, as well as by its cooperation with more than thirty interstate bodies stipulated in the statutes of the CIS states. It was judged that the postponement, by virtue of certain circumstances, of the ratification of the two documents in question, along with the uncertain judicial status of the Republic of Moldova among the CIS member states, have reduced cooperation possibilities with those states as well as the chances of buying energy sources, fuel, and raw materials needed by the Republic of Moldova.

It was also decided that this situation exerts a negative influence on the conclusion of economic accords for 1994 and on markets for selling the Moldovan products, which are delivered to the Russian Federation. The meeting also emphasized the fact that reason dictates the participation by the Republic of Moldova in the activity of the aforementioned bodies.

Also today in Chisinau, President Mircea Snegur met with leading representatives of the Russian community in the Republic of Moldova. During this meeting, the two sides discussed problems faced by this community, including difficulties in studying the state language—that is, Romanian. Mr. Mircea Snegur assured the representatives that all these shortfalls would be solved if at all possible and welcomed the fact that members of the Russian community were aware of the need to speak the state language, and that they have proved active participants in the process of democratizing society and consolidating the state independence of the Republic of Moldova.

## 4.77
## Kravchuk and Moldova's Snegur Hold a News Conference

Kiev Ukrayinske Telebachennya Network
14 December 1993, [FBIS Translation], Excerpts

*Editor's Note:* The bilateral, almost "fraternal" relationship exhibited in the document below is significant because it offers a potential counterbalance to the Moscow/Minsk axis, if not in military terms, then at least in political ones.

I would like to say that Ukraine supported, supports, and will continue to support the state unity of Moldova. This is a matter of principle. It is the concern of the Moldovan people to resolve their internal affairs regarding the status of their regions, including the Trans-Dniester. However, we are for oneness, for unity of the state, for its sovereignty, and for its independent development. We have declared it previously, and we declare it today—Ukraine will follow such a policy. If some difficulties arise, and Moldova finds it expedient to ask Ukraine for support, we are ready to be a mediator in any questions, following international norms and international documents, and analyzing the extent of accord in interaction with the state. [Passage omitted.]

[Boris Rzhevskiy, Moldovan National Television]: I have two questions for the Moldovan and Ukrainian presidents. First, could you delineate priority provisions of the treaty for 1994? The second question concerns your attitude toward the preliminary results of the elections in Russia.

[Snegur]: I believe that all the aspects of relations between Ukraine and Moldova are of a priority nature, both economically and politically. The economic aspect, undoubtedly, prevailed. Principal import and export figures for 1994 were specified, and the prime ministers signed an agreement on this issue. I think that the issues of Moldova's property on Ukrainian territory and Ukraine's property on Moldovan territory are equally important. We are fully satisfied with the mutual understanding that we achieved on this issue, including the protection of property and the forming of joint-stock enterprises. We have drafted a blueprint for eliminating these problems, and appropriate departments signed an agreement. Ukraine's stance on regulating the armed conflict in the Trans-Dniester region should also be regarded as a priority issue—all the more so since tension in that area is being built up by those who do not agree to draw our stances on this issue closer to one another, regardless of the fact that we have reviewed the standpoint that we took a year ago. We suggest that special legal status be granted to that sector of our republic based on the principles of maximum local self-rule. We are ready to review the language situation that has developed in that region, taking into account existing realities. We have declared our readiness to agree to the activities of some legislative assembly in that region that would be entitled to draft and adopt local laws. Unfortunately, local leaders seem not to understand the changing situation, and one gets the impression that they are ready to reject all these proposals just to be able to call the Trans-Dniester region a sovereign independent state and a member of the artificial Moldovan confederation. Thus, I am going back to my initial postulate that the stance of Ukraine, our great eastern neighbor, is very valuable to us. Many aspects of our lives have intertwined in the course of history, and we have things to discuss. It is enough to mention that Moldovan railroads in many places stretch across Ukrainian soil. Problems do exist. When we lived in the former Soviet Union and were told what to do and carried out orders, there were no problems. Since our states won independence, these issues have emerged.

In fact, I was asked the second question at the airport yesterday when the results of the elections were not exactly known, although even today, we know only preliminary results. Of course, regarding the Russian elections, I believe that they made Russia and the other states think, because of the results. Leonid Makarovich will tell you that we discussed these issues yesterday, but we have not discussed them since, for me, these results were unexpected, to say the least. I want to say that this was a signal for democratic forces, I mean for my republic at any rate. Any turn in domestic policy is possible when dangerous trends are underestimated, and the Russian elections proved this. Perhaps, we should draw conclusions. I do not think that I can give a more detailed analysis at the moment. We will see how events develop in Russia, and we should be careful, especially since we are also on the eve of early parliamentary elections scheduled for 27 February in Moldova and for 27 March in Ukraine. There are forces trying to ban presidents from participation in elections. Some factions are already voicing demands that the president remain neutral and refrain from making statements. How can this be done if there are provocations and insults. The president cannot remain silent because later, he will have to cooperate with the legislature. I want you to understand me correctly, but if these processes are unattended. . . . Our state has the most democratic electoral system, and people should be helped to comprehend it. Voting will be conducted solely through party lists. The law also envisions independent candidates, but this is far from widespread. I think that people should have the difference between the former situation, when there was one candidate on the list, and the current one, when we have perhaps, ten electoral blocs, explained. I think that a more in-depth analysis of the results of the Russian elections could be provided later.

[Kravchuk]: I share President Snegur's standpoint regarding priorities as well as the second problem. I can only add that final assessments and declarations can be made only after the Russian National Assembly convenes its first session and adopts certain declarations and resolutions. We can react to decisions made by a state body. In this case, we are dealing with electoral competition and electoral platforms of political parties and political forces. Thus, I cannot and I do not want to analyze these platforms because every party has a right to voice its political arguments and programs. But frankly speaking, there is anxiety not so much for Ukraine as, forgive my arrogance, for Russia. I told you many times that I was in a very favorable political situation because democracy and sovereignty were initiated by Russia, by Boris Yeltsin, and other democratic forces. You remember that the first declaration on independence was adopted by Russia, and so were democratic laws, including the one on presidential elections. Other newly independent states took up the torch from Russia and adopted Russia's experience. In 1991, when the issue of the further development of states was being resolved, Russia unequivocally decided that it would develop according to democratic principles. Of course, I am worried. When six million electors cast their votes in favor of a great inseparable Russia within its pre-1917 borders during presidential elections, I was worried. Today, this anxiety is greater. The idea of straightforward revanchism was promoted, and this fact should not be concealed. However, I hope that Russia's power, its president, and all forces will stand up to protect democracy and will not allow the ideas of revanchism to develop into a state concept. This can be a political trend of a certain group, but if it develops into a state concept, then I can state with full responsibility that this will be the beginning of great instability in Europe, involving partitions and repartitions. At this stage, I regard this as just populism and a political platform. However, the fact that so many people voted for this idea is alarming, as is the fact that so many people would like to restore the tsarist empire. This is what greatly worries me. I want to emphasize again that I believe in the Russian democratic forces' victory over this temporary political trend, this deviation from the line declared by Russia, a deviation caused by a difficult economic situation. Everybody is using this situation for demagoguery, including political leaders.

## 4.78

**Foreign Minister on Russian Ties and Federation**

Interview by Jerzy Haszczynski

*Rzeczpospolita* (Warsaw), 17 February 1994

[FBIS Translation], Excerpts

When the Soviet Union was collapsing, the Moldovan authorities expressed their criticism of the concept of the CIS. But last year Moldova joined the CIS anyway. How do you explain the change of attitude?

[Botnaru]: Our links with the CIS are of a specific nature. We are a de facto member of the CIS's economic union, but not de jure, since we have not ratified the appropriate documents. My view is that from an economic standpoint, Moldova should be in the CIS's economic union since that is the real situation from which we cannot escape. Our economic relations with the West still leave a lot to be desired, but one must live somehow until that situation changes. Generally speaking, the economic relations of a new country should be multifaceted.

[Haszczynski]: Is Moldova interested in the CIS only from an economic standpoint?

[Botnaru]: We do not participate in the CIS's military and political structures. So far, we have not signed any agreements relating, for instance, to security. What is important is what shape the CIS will take in the future. Experience shows that countries that have traditionally been together can make their lots easier if they act together. But one should take part in such cooperation arrangements only insofar as it does not endanger a country's sovereignty.

[Haszczynski]: Do you not feel threatened by Russia?

[Botnaru]: I regard our relations with Russia to be correct, but there are still many problems inherited from the Soviet Union. First of all the problem of the 14th Russian Army, which is stationed in Moldova. In my view, all problems can be solved if both sides conduct a dialogue. I paid a visit to Moscow recently, where I spoke with Russian Foreign Minister Andrey Kozyrev. We listed the unresolved problems in Russian–Moldovan relations, and we signed a bilateral agreement. Moldova has ratified it but Russia has not yet done so, delaying the resolution of other problems, for instance, the problem of gasoline supplies to former Soviet military personnel residing in Moldova.

A withdrawal of the 14th Army would have an exceptionally positive impact on the development of Russo–Moldovan relations. The problem of the Cis-Dniester region is linked to that.

We think that some elements of Russia's relations with the Cis-Dniester region could be described as interference in Moldova's internal affairs. I mean the issue of Russian aid for the Cis-Dniester region, both in material and political terms.

[Haszczynski]: Was there any specific date set during your Moscow talks for the 14th Army's withdrawal from the usurpatory Cis-Dniester Republic?

[Botnaru]: Moldova's position is that the Russian Army should withdraw by 1 July of this year. I regard that date as realistic. Despite everything, there are fewer [Russian] troops on our territory than used to be the case with Poland or other East European countries. The withdrawal should be unconditional and in accordance with international law and CSCE provisions. Talks are under way. Moldova is absolutely against having foreign troops stationed on its territory. I want to remind you that all countries recognize Moldova within the boundaries in which it proclaimed its independence in 1991, and that includes the Cis-Dniester region.

[Haszczynski]: How do you imagine a calming of the situation in the Cis-Dniester region? After all, fights have already erupted there a few times and hundreds of people have died. Is it possible, for instance, for Moldova to be transformed into a federation?

[Botnaru]: No. I think that the variety of nationalities and their distribution in the various regions make a federal state impossible. Moldova is a small territory (33,700 square km), which would be split into many small units if a federation was to be established. Besides, the base criterion of a federation is not completely clear, either. We used to live in the Soviet Union, and we were told then that it was a federation.

As far as the Cis-Dniester region is concerned, we granted very far-reaching concessions, and we guaranteed that if the issue of Moldova's joining Romania came up, then the people of the Cis-Dniester region could decide the territory's fate through a referendum. Almost all political forces in Moldova think that the Cis-Dniester region should have some kind of special status. The necessity of a decentralization is linked with that. Some amendments to our laws are indispensable, including the law on languages. We must talk about the possibility of using several languages in the country's offices: let us say one of them would be called the state language (Romanian), and the other two would be called official languages (Ukrainian and Russian). The new parliament and the new cabinet should start working on these changes soon.

[Haszczynski]: Do you really believe in a compromise? After all, the authorities of the Trans-Dniester region have announced that the region will take part neither in the parliamentary elections (27 February) nor in the referendum on Moldova's sovereignty and integrity (6 March).

[Botnaru]: We do not seek the authorities' participation, especially given that these are not lawful authorities. We are after the people's participation; the usurpatory authorities' control does not extend to the entire Cis-Dniester region. There are areas and villages that cooperate with us. According to what we have found out, the people living in the areas not under the control of the usurpatory regime will vote.

[Haszczynski]: In official documents, your country is known as the Republic of Moldova. Do the authorities of the republic (which gained independence more than two years ago) think that using the old name of "Moldavia" (which was used in the Soviet period, among others) is inappropriate?

[Botnaru]: "Moldavia" is a word of Russian origin. Right now, Romanian is our official language, and in Romanian our country's name is "Moldova." But we have nothing against someone using the Russian-derived name in newspapers, for instance, as long as the name is used with respect. . . .

---

## 4.79

**Leaders View CIS Ties and Economic Issues**

From the "Presidential Bulletin" column, compiled by Nikolay Zherebtsov and Andrey Petrovskiy; and edited by Vladimir Shushlin
Interfax, 2 August 1994
[FBIS Translation], Excerpts

---

In the opinion of President Mircea Snegur, the main task for former Soviet republics is to form a common economic space with transparent borders and remove customs barriers. The leaders of CIS countries should concentrate their attention on that, Snegur told participants at an international marathon of disabled athletes in Chisinau on Monday.

In an exclusive interview with Interfax, Vice-Premier Valentin Cunev said the Moldovan leadership favors the speedy formation of a mixed economic committee of CIS countries, which in his opinion will help make collectively important decisions concerning the economy, finance, customs rules, and taxation. (The idea of an economic committee was proposed during the latest CIS summit in Moscow last May.)

Cunev remarked that the present circumstances do not permit Moldova to pursue an uncompromising policy with regard to its neighbors while Ukraine is trying to treat Moldova in that way. He said the exorbitant rates for transit across Ukrainian territory have become a serious problem.

In the opinion of the vice-premier, without a joint economic committee the overwhelming majority of documents drafted and approved by the commonwealth will remain unfulfilled.

## 4.80

### Concern About Arrival of New Recruits in Dniester

INFOTAG, 10 July 1995 [FBIS Translation]

Three hundred recruits from the central regions of Russia arrived to join the Russian military contingent, stationed in Dniester.

The arrival of the Russian citizens to the former 14th Army was viewed by the local population as a big contrast to the situation that existed in that army when it was headed by Lt.-Gen. Aleksandr Lebed.

Before his departure for Moscow, Lebed said that during the three years he spent in Tiraspol, the Russian defense ministry had sent only six recruits to that army.

Welcoming the new recruits, the commander of the Russian troops, Maj.-Gen. Valeriy Evnevich, said that not only citizens of Russia but citizens of the former Soviet Union serve in his troops, as many of them had acquired neither Russian nor Moldovan citizenship. The general expressed hope that differences in the citizenship would not result in abuse of position.

The newcomers from Russia have been also greeted by the state secretary of the unrecognized Dniester Republic, Valeriy Litskay, and the military advisor of the Dniester president, Maj.-Gen. Stefan Chitac. The latter said that the soldiers would have "to defend the interests of Russia on Russian soil."

## 4.81

### Dniester and Abkhazia Urge Referendum on Joining Russia

INFOTAG, 12 July 1995 [FBIS Translation]

Delegations of the Dniester Republic and Abkhazia, which are participating in the sessions of the Congress of the Compatriots in Moscow, adopted a joint appeal to the governments and the supreme soviets of these unrecognized republics. The appeal stresses the need to hold referendums on their territories before the end of 1995 on the issue of "restoration of the united country and incorporation in Russia." According to the INFOTAG correspondent in Tiraspol, the Congress of the Compatriots in Moscow was held under the auspices of the State Duma of Russia. The Dniester delegation was headed by Anna Volkova, vice-chairman of the Supreme Soviet.

The branch of the Movement of the Compatriots was established in Dniester only a few weeks before the congress in Moscow. The constituent session of the branch was held on the premises of the Dniester Supreme Soviet. Observers in Tiraspol do not rule out that the current Dniester authorities will hold the referendum on becoming part of Russia at the same time as the elections to the Dniester Supreme Soviet in order to attract more voters. Tiraspol acquired similar experience last spring when, along with the local elections, a referendum on the non-withdrawal of the 14th Army from the region was held.

## 4.82

### Dissatisfaction with Yeltsin Decree on CIS Relations

Interfax, 27 October 1995 [FBIS Translation]

Moldova is unsatisfied with some provisions of Russian President Boris Yeltsin's decree regarding Russia's strategic policy toward the CIS countries, Moldovan Foreign Minister Mihai Popov told a press conference in Chisinau Thursday.

"The decree seems to be only an internal Russian document," Popov said. "Taking into consideration that Moldova is a member of the CIS, this document concerns Moldova's interests," he said.

"In our participation in the CIS we focus on economic cooperation. The military and political aspects of the CIS are likely to be put in place without us. We do not intend to take part in military-political unions," Popov said.

Popov said he thought after reading the Russian president's decree that Russia "will differentiate the CIS countries in terms of their loyalty toward Moscow" and build relations with the CIS countries proceeding from such loyalty. "Whatever documents are adopted in Moscow or in other capitals, Moldova will build its policy proceeding from its national interests," Popov said.

Popov said the foreign ministers of the countries of the Black Sea region were scheduled to meet in Chisinau on 1 November. He said the foreign ministers would in particular discuss the creation of a legal basis for developing economic cooperation between the region's countries.

# 5

# The Ukrainian Perspective

## Introductory Notes

Chapters 1 and 2 documented Ukraine's unique role in triggering the breakup of the Soviet Union. Chapter 5 will further document Ukraine's unique political role following the collapse of the Soviet Union. Ukraine's unwavering national consciousness, combined with its size, as the second largest former Soviet republic, and its national resource endowment, make its survival as a modern, independent European state a plausible prospect. Nevertheless, economic transition has been painful, and internal pressures from Russian-speaking sections of eastern Ukraine have challenged the political abilities of Ukraine's leaders, who must preserve the state's political sovereignty even as they seek long-term friendship and cooperation with Russia in order to soften Russia's imperial ambitions. Despite the pitfalls of such a balancing act, Ukraine has managed to maneuver within the CIS as an equal partner to Russia, insisting on reciprocity in compromise and full respect for its national integrity.

The documents in Chapter 5 reveal the influence of at least three important factors on the attitudes of the Ukrainian leadership. First and foremost, Ukraine was "independence-minded" long before Gorbachev introduced the concept of a "Union treaty." Second, Ukraine's fiercely independent western region has had to vie with a large Russian community in the east. Third, Ukraine's Slavic origins and strong identification with the Russian culture make many Ukrainians receptive to calls for reunification with Russia.

Ukraine's two post-Soviet presidents have had to weigh these three historical factors carefully in all of their political thinking. The two presidents have been strong leaders, with divergent leadership styles, but similar views on the merits of Ukrainian sovereignty. The first, Leonid Kravchuk, assumed the Ukrainian presidency with credentials as ideology chief in the Ukrainian Communist Party, but it was said that after reading materials made available in 1989 he became an unshakable Ukrainian patriot and state builder. As revealed in the speeches in this chapter, President Kravchuk used blunt language in his references to Russian policy from the outset of CIS collaboration. His use of phrases such as "Russian imperial arrogance," "blatant interference," and "Great Power chauvinism," and his emphatic rejection of the CIS as a permanent regional alliance system guaranteed tense relations between Ukraine and Russia.

One of the most important early issues between Moscow and Kiev was the division of Soviet military assets. The fate of the 380-ship Black Sea Fleet was a particularly strong irritant in relations between the two states. In the documents selected here, President Kravchuk challenges Russia's claim that it should own the fleet, declaring that a large portion should be ceded to Ukraine for its independent navy. On the question of ground forces, the Ukrainian government chose to create a 400,000-man army, to be selected from among the nearly one million former Soviet troops stationed in Ukraine. By February 1992, as recorded here, the Ukrainian press announced that this large army was "almost complete." Demonstrating remarkable unanimity on the question of military independence, the Ukrainian parliament refused to enter a joint defense area with the other former Soviet republics or to sign the CIS Collective Security Treaty.

In February 1992, the Russian Supreme Soviet raised an issue which has continued to destabilize Ukrainian-Russian relations ever since—that of the return of the Crimean peninsula to Russia. In a resolution which stated that Nikita Khrushchev's 1954 cession of the Crimea to Ukraine was "unconstitutional" (despite the fact that Russia had declared the Soviet constitution illegal), the Russian parliament challenged Ukraine's territorial integrity. This issue still simmers today under Ukraine's second president and promises to be a continual obstruction to improved Ukraine-Russian relations.

President Kravchuk pulled Ukraine even further away from Russia in September 1992 in reaction to a bilateral treaty, proposed by Russia, which would have positioned Ukraine irrevocably within Russia's security orbit. The treaty was to have included Ukraine in a joint CIS military doctrine and would have awarded dual citizenship to Ukraine's Russian-speaking population of almost twelve million people. The treaty would even have arranged Russia's payment for the transit of its oil and gas through Ukrainian territory in barter rather than hard currency terms. Ukraine's parliament emphatically rejected this treaty, proposing greater independence for both countries, as recorded here.

Several documents in this chapter allude to Ukraine's position on another important issue— that of disposing of the nuclear weapons Ukraine inherited from the Soviet Union. President Kravchuk at first tried to use these weapons as a bargaining lever to gain security commitments from the West and as a deterrent against a Russian attack—an approach which ultimately failed, only isolating Ukraine from Russia and the West.

President Kravchuk maintained his independent course throughout 1993 and the first half of 1994, although his prime minister, Leonid Kuchma, opposed his position against economic cooperation with Russia and the CIS. A speech to the Ukrainian parliament provided here documents Kuchma's contention that "Ukraine's acute economic crisis is a greater national security threat than any posed by a foreign country." In May 1993, Kuchma initialed several CIS economic projects with the prime ministers of Russia and Belarus, aimed at intensifying economic integration among the three countries. (The economic debates within the CIS, including this trilateral economic agreement, are covered in detail in Chapter 7.) Ukrainian nationalist leaders accused Kuchma of violating the state's vital interests by signing these documents. At the time, distrust was so strong of anyone close to the Russian political elite, as Kuchma was to Russian prime minister Viktor Chernomyrdin, that many Ukrainian democrats judged Kuchma to be little more than one of Russia's appointed "governors." Nevertheless, time would prove prime minister Kuchma to be correct about the need to cooperate more closely with Russia on economic matters, and rapidly to resolve the most contentious military and political issues between the two governments.

With the Ukrainian economy near collapse and no real assistance coming from the West, Leonid Kuchma's cooperative approach and focus on economic development began to win general acceptance within Ukraine. By the time the presidential election of July 1994 was held, Prime Minister Kuchma had garnered enough support to win in a very close race with Leonid Kravchuk. Excerpts found here from President Kuchma's inaugural address document his call for reconciliation with Moscow. He initially interpreted his election as a mandate for change, though he knew he would have to balance complex internal forces, some calling for a policy of radical independence from Russia, and others calling for virtual restoration of the Soviet Union. Appealing to the middle in his inaugural address, he said:

> Ukraine stands on the threshold of essential changes in its economic and political course. . . . Ukraine is part of the Euro-Asian economic and cultural space. Ukraine's vitally important interests are concentrated on this territory of the former Soviet Union. . . . Ukraine's self-isolation and voluntary refusal to campaign vigorously for its own interests in the Euro-Asian space was a serious political mistake, which caused great damage, above all to the national economy.

To deal with Kravchuk's legacy of political tension and economic malaise, Kuchma promised to institute serious economic reforms and to cooperate with other countries in the CIS. To initiate this approach, he announced at his inauguration that he was preparing a bilateral Treaty of Friendship and Cooperation with Russia. This tactic at first raised serious doubts among Ukrainian nationalists. Kuchma's later positions on matters involving Russia and the CIS, however, proved him an energetic defender of Ukraine's statehood and international stature as a key nation in the new Europe.

Despite his willingness to cooperate economically with Russia, President Kuchma has had his share of problems in relations with Moscow. In December 1994, Russia delayed indefinitely President Yeltsin's signing of the Treaty on Friendship and Cooperation, which was to have occurred in September. At a public briefing on 11 January 1995 contained here, Kuchma's Chief of Staff, Dmitro Tabachnyk, reveals that Russian Foreign Minister Kozyrev blamed this delay on the participation of Ukrainian "mercenaries" in Chechnya. In February, Russia singled out disagreements over the Black Sea Fleet as a new pretext for holding up the treaty. As documented in the *Komsomolskaya Pravda* article of 10 June 1995 which appears here, President Kuchma made a serious effort to solve this stubborn problem in a meeting at Yeltsin's dacha in Sochi on 9 June 1995. Under the Sochi accords, Ukraine and Russia are each to receive half of the fleet, with Ukraine ceding almost 30 percent of its share in partial payment of its $2.5 billion debt for Russian gas. Under the agreement, Ukraine received assurance that Sevastopol is Ukrainian territory and the fleet's home base will be leased to Russia at an unspecified rate.

The Crimea is an even more threatening source of friction than the Black Sea Fleet. Despite Yeltsin's constant reassurances that the Crimea is an integral part of Ukraine, Russian politicians will not let the issue rest. As the statement written by Viktor Vishnyakov, chairman of the Russian State Duma's Subcommittee on Constitutional Legislation shows, some Duma deputies still strongly favor pushing Russia's territorial claim to the Crimea. His statement is a minutely detailed legal analysis and apologia for the reannexation of the Crimean territory.

To make matters worse, the chairman of the Crimean

Supreme Soviet, Sergey Tsekov, responding to an invitation from the Russian State Duma, delivered an appeal on 14 April 1995 to the Russian government, saying: "We have so far done our best to cope with the crisis ourselves, and we failed. Now we believe there is no way to solve the Crimean problem without Russia." The Ukrainian government immediately charged that the Duma's invitation to Tsekov represented an "unfriendly act." On April 17, Yeltsin once more delayed the Ukrainian-Russian Friendship and Cooperation Treaty, saying: "It will be correct to sign major political documents between Russia and Ukraine only after we are convinced that relations between Simferopol and Kiev do not infringe on the interests of the Crimeans," he announced. Because two-thirds of the Crimea's population claim to be Russians, Yeltsin went on, Russia has "considerable interests there."

On 18 April, a group of fifty Crimean parliament members denounced Tsekov and appealed to Kiev to dissolve the existing parliament. These members promised to form a new Parliament of the Autonomous Republic of Crimea and place it under Kiev's jurisdiction. The crisis subsided on July 5–6, when the Crimean parliament voted for the resignation of its speaker and elected Mr. Yevhen Supruniuk, who is loyal to Kiev and pledges that the Crimean parliament will follow Ukrainian laws. Kuchma prevailed in one of the toughest and most dangerous battles of his presidency.

In addition to problems with Russia, Ukrainian political elites must contend with pressures within the CIS and in their own parliament. Politically, President Kuchma must now stand against Nursultan Nazarbaev's Eurasian Union, Belarus's decision to submit to Soviet-style dependency, and the Kozyrev-Primakov-Chernomyrdin team in CIS institutions. At the May 1995 CIS summit, Russia, Belarus and Kazakhstan created a trilateral customs union. This is to become the "core" of CIS integration. While in Kazakhstan negotiating the agreement, Chernomyrdin publicly urged Ukraine to accede to the protocol. Instead, Ukraine led a group of republics including Moldova, Azerbaijan, and Turkmenistan in rejecting the agreements. Kuchma's firm rejection of a customs union is documented here as one more example of Ukraine's continuing tough defense of its political sovereignty within the CIS.[1]

On 9 January 1995 President Kuchma consolidated his bilateral relations with Georgia in a "Declaration on Prospects for Cooperation Between Ukraine and Georgia and for Joint Approaches to Foreign Relations Issues," the text of which appears here. Georgia and Ukraine have agreed to seek world attention for their countries in the United Nations and other international fora. Both presidents have several times addressed the United Nations, and excerpts from Kuchma's 22 November 1994 UN address are contained here.

Tensions in Ukrainian-Russian relations have continued to mount, pushing Ukraine further toward consolidating its orientation toward Europe. In May 1995 while visiting Riga, Kuchma signed a joint Latvian-Ukrainian statement in which both countries declared: "Threats and political pressure from a bordering state prompt many countries to want to join reliable and stable political alliances to ensure statehood and development."

The documents that follow only begin to capture the complexity of Ukraine's affairs within the CIS. They do, however, elucidate Ukraine's growing determination to build an independent state and to preserve Ukraine's special geopolitical status within the former Soviet space. They also show quite conclusively that Leonid Kuchma, who campaigned on a promise to draw closer to Russia, has upheld the Ukrainian people's vote for sovereignty in the face of Russia's great power ambitions and Ukrainian communists' strong tilt toward Russia and a reconsolidated Soviet Union.

### Note

1. Volodymyr Zviglyanich, "Voting in Belarus and the Independence of Ukraine," *The Ukrainian Weekly*, 18 June 1995.

---

## 5.1

**Kravchuk Comments on Russian Arrogance**

Sergey Tsikora

*Izvestiya*, 16 January 1992

[FBIS Translation], Excerpts

---

*Editor's Note:* The first three documents in this chapter lay out the views of President Leonid Kravchuk on Ukrainian independence, Russian policy toward Ukraine, and the CIS. As President of Ukraine, Kravchuk viewed the Commonwealth as a transitional vehicle for maintaining trade relations among its eleven member states while dismantling the Soviet space.

Kravchuk often delivered impassioned warnings about the potential consequences of Russian interference in Ukraine's sovereign affairs. He understood the signs of resurgent Russian nationalism, and he clearly placed sovereignty above trade opportunities. He insisted that he would "never take a single step in the direction of reducing Ukrainian independence." Unfortunately, Kravchuk was slower to implement economic reforms than political ones, and Ukraine's inflation rate rose to 80 percent annually by the end of 1994. Still, Kravchuk adamantly refused to take Belarus's route and said openly that he considered Belarus's policies a return to old ways. Economic worries did, however, undermine his political support—especially in Crimea and eastern Ukraine. To balance his vulnerability in these regions, he sought a strong relationship with the West. Western sympathies notwithstanding, not even Kravchuk's tilt in that direction could solve Ukraine's geopolitical worries about Russia's power and ingenuity in exerting its leverage. Kravchuk's balancing act is bound to be every Ukrainian president's balancing act until enough time passes to confer legitimacy upon a sovereign Ukraine.

Kiev—Ukrainian President L. Kravchuk devoted his address on republic television, 14 January, to the complex web of economic problems, the creation of Ukraine's own armed forces, and the impermissibility of interference by its neighbors in the internal affairs of an independent state. . . .

The main source of concern was the difficulties the republic has encountered in the transition to the market. Leonid Kravchuk acknowledged that in effect not one component of the machinery necessary for a switch to free pricing is working in Ukraine today—privatization has not been carried out, and Ukraine has not created its own system of credit and finance. It is Russia that has forced us to launch ourselves without preparation into the whirlpool of free prices, the Ukrainian president said.

We are today trying to defend our market by introducing ruble coupons [*kupony-karbovantsy*], Kravchuk continued, but lack of preparation and miscalculations in this matter have today led to prices on products in state trade often exceeding prices in the markets.

"The working people cannot carry incompetent leaders on their backs," the Ukrainian president stated firmly, and added that the guilty men would get their just desserts at the republic Cabinet of Ministers session 16 January.

L. Kravchuk went on to say that, given the current complete indifference of some states of the CIS [Commonwealth of Independent States] to the interests of others, Ukraine is forced to look after itself, and hence intends to introduce its own money. It will be printed by the end of the current year.

It should be noted especially that the theme of current relations with Russia ran through the whole of the Ukrainian president's speech. Whatever issue Kravchuk raised, he without fail mentioned the position of the two republics. It must be noted with regret that mutual understanding has not yet been reached on many acute problems.

The Ukrainian president dwelt in detail on the topic of military building in the republic. He stated that Ukraine has not violated a single clause of the military agreements signed in Minsk and Alma-Ata. But an independent Ukraine will allow no one to dictate to it any conditions whatsoever, especially if they concern its security, the president stressed. We cannot agree to one republic today declaring its right to fleets and other armed forces, and also to embassies abroad. One state cannot be the inheritor and successor of everything to which the peoples of the former Union contributed their labor.

Touching on the currently acute problem of the dividing up of the fleet, L. Kravchuk stated that officially Ukraine has never stated that it is taking over the Black Sea Fleet. We say that we have a right to create our own fleet. And whether it will be formed from the Black Sea Fleet or built at the Ukraine's shipyards is the topic of a separate conversation. But the Ukraine is against the Black Sea being plowed by ships armed with nuclear weapons.

The fact that military-political problems between the Ukraine and Russia are still far from being solved is suggested by the latest news cited by Kravchuk during his television address. He said: Today I have learned that, on the representation of Marshal of Aviation Shaposhnikov, the rank of general has been conferred by Russian President Yeltsin on military men serving in Ukraine. We cannot agree with such actions, Kravchuk stated. This is the jurisdiction of the supreme council, government, and president of Ukraine.

However the most acute issue in the Ukrainian president's address was that of "the blatant interference of some Russian leaders in our internal affairs." Describing this as imperial arrogance, L. Kravchuk cited the latest example from the 14 January issue of *Trud*, which published an interview with Russian Minister Poltoranin. In this article the member of the Russian government allowed himself to use far from diplomatic expressions about the head of Ukraine. Describing his behavior as a display of Great Power chauvinism, "whose carriers in Russia were described as 'small-town police thugs' [*derzhimordy*] by the Bolshevik Lenin," L. Kravchuk said that the Ukrainian government has officially demanded of the Russian parliament, government, and president an end to the whipping up of passions in relations between the two republics and peoples, and to interference in the internal affairs of another sovereign state.

This policy led to extremely serious consequences in the Baltic region, and has provoked conflicts in other regions. There is peace in Ukraine today, and we do not intend to put up with any interference in our life. We will carry out the policy with which the people entrusted us, the Ukrainian president said, and accept no correction from outside.

**Further on Address**
Mikhail Odinets
*Pravda,* 16 January 1992 [FBIS Translation]

Kiev—Ukrainian President Leonid Kravchuk has spoken on Ukrainian television. He dwelt on the problems of strengthening the economy and the monetary and financial system and of the structure of the armed forces, and touched in particular on the Black Sea Fleet.

At the same time a significant part of his address was devoted to instances of interference in Ukraine's internal affairs on the part of certain representatives of the Russian leadership and mass media. "Imperial arrogance is bordering on showing disrespect for Ukraine as a state," he said. Criticizing an interview given to *Trud* by M. Poltoranin, Russian minister of the press and mass media, L. Kravchuk quoted from the newspaper the words that M. Poltoranin is a reliable and longstanding comrade-in-arms of Boris Yeltsin, and stressed: "Even the Bolshevik Lenin was very sensitive to these questions, and called Russian chauvinists small-town police thugs, because he was very much afraid that they would inflict damage on the outposts of the Tsarist Russia of those days."

Addressing the Russian minister directly from the screen and remarking that his message would be conveyed to him, Kravchuk said: "Now you, pardon me, belong to just this category. I would not speak like this if you were the only one, and I would never descend to the level of quarreling with you by proxy. But in this case it has become an illness, a real problem. You consider it your right to interfere in our affairs and interpret our policy; you consider it your right to view Ukraine as a part, an outlying area, of Russia. You liked it when Ukraine was obedient. . . . Then, of course, Ukraine and its leaders followed in the footsteps of such democrat representatives. But now the 'nationalist' Kravchuk is defending the interests of Russians, Jews, and Germans . . . and will continue to defend them, and believes that only here will they find protection and have equal rights. And the democrat Poltoranin and his ilk are taking a different position, and believe they have a right to impose and establish their ways, their laws. . . ."

It was this policy, L. Kravchuk went on to note, which brought extremely serious consequences in its wake in the Baltic, and it could lead to them in other regions. Winding up his speech, L. Kravchuk called the last part of it an emotional outburst of the sort he never allows himself. But everything has its limit, he declared; after all, this concerns not only the Ukrainian president, but all the Ukrainian people. And the "hanba" (disgrace, dishonor, shame, profanation) can be brooked no longer.

It goes without saying that the title of this article should be understood conditionally—the accusations are directed not so much against Poltoranin as against the attitude which, in L. Kravchuk's opinion, has recently been displayed toward Ukraine. And nothing good, as we know, ever comes of conflicts.

## 5.2

**Kravchuk Comments on Relations Within CIS**
Interview with Ukrainian President L.M. Kravchuk
by Vitaliy Portnikov
*Nezavisimaya Gazeta,* 30 January 1992
[FBIS Translation]

[Portnikov]: Leonid Makarovich, the first 100 days of your presidency have passed. I would first like to ask you what you associate with this initial period and how you see the country you have been elected to lead.

[Kravchuk]: If the people elect a president, they above all entrust him with defending their interests. The people must be content and believe that the president will do his duty. And the people's will today is to form an independent state. It is the most terrible thing imaginable for a man to betray a people who have agonized to secure this right throughout their history. That is why I will endeavor to do everything to defend their interests and their future. We can then say—albeit in six months—that we have not merely an independent Ukraine, but an independent, strong Ukraine, which will have its own armed forces, institutions of authority, and laws to meet the people's needs. Then we will say that there is a powerful, strong Ukraine. It will not be economically all that strong or rich—that will take time. But this is the main thing now.

A very interesting situation has taken shape in our country as a whole. The old structures have vanished: However we criticized them, and we did so with good cause, and however we criticized the old unitarian, state-party ideology—they did exist. Now let me look at our state now. We have a Cabinet of Ministers. Who is tackling problems of ideology? No one. We have a Supreme Soviet. Who is tackling problems of ideology? No one. What about a presidential structure. . . . We do not have a philosophy and ideology for the new Ukraine. We want to develop a new, independent, democratic Ukraine, that is true. But we do not have a modern philosophy, one that calls up our roots and our traditions. Rukh [the Ukrainian popular movement] has now taken on the problem of ideology. The congress of Ukrainians, the forum of peoples of the Ukraine in Odessa, and Ukrainians' gathering together and unity are all elements of ideology. Admittedly, they do not have a broad philosophical base because no one has as yet thought about that in earnest. We must create an ideological base for the new Ukraine. Then things will be easier for us, then we will be able to create the appropriate structures, guided by this philosophy.

I think all the time: Well, we have gotten state levers. I can sincerely say that we have no experience behind us. There are no great state structures because Ukraine was in someone else's hands all the time. Our government and ministers worked within the bounds of their own jurisdiction, which extended to 5 percent of the real estate and involved carrying out the Kremlin's orders. The better it did this, the further it advanced. If if did not carry out orders, it went backward. That is why we must learn state leadership of the new Ukraine and state thinking. When someone goes ahead and plows you a furrow, you look for that furrow. But when you yourself have to do it, that is quite another matter. I may be more conscious of this than anyone else because I come to work every day and encounter this every day. I am not saying that everything is fine for Yeltsin, but the structures that he took on—whatever they were—have attained an international level while the former structures in Kiev never went anywhere. Moscow was the highest level that they had to deal with. They did not know a thing about Paris. Yet now we need to deal with other states and we have a totally different range of decisions to make on a totally different scale. Therefore, we need people to whom we can entrust the Ukraine—committed, competent people.

[Portnikov]: One of the problems of Ukrainian statehood may be differences of mentality between the inhabitants of different regions of the republic. . . .

[Kravchuk]: This is rooted in our history. I know western Ukraine: I worked there and I was born in those districts. The constant hints that there is a Ukraine and a "semi-Ukraine" have generated in people's minds such a desire for independence that it is hard to even imagine. I remember my first trip to Lvov. I dropped in at the "Elektron" plant. Things were very hard then, the stores were totally empty. Before that I had driven around several regions and, wherever I went, people spoke to me about that. So I was already geared to thinking that they would ask me about it. I went through one workshop, then another, and no one said anything. Then I myself appealed to people and they replied: That is not the most important thing for us. The most important thing is independence, that is the light at the end of the tunnel. I thought: What strength of will people have—not party leaders who have lived for this and spent time in jail for it, but the working man.

The main point in this was: here we have the Ukraine, here we have its strength, might, and unity, and here we have multiethnic collectives. At first people did not have such a powerful desire for freedom. My wife is from Sumy Oblast and if you go to a village there, their ideas are not quite as crystallized as they are in the west. People say: Certainly we need independence, but we must be together, within the Union. You will not hear this any more in the west. Although I agree that we must not split up, but live on equal terms.

[Portnikov]: Many people are talking about the Ukraine's place in the Commonwealth of Independent States [CIS], but virtually no European policy is being elaborated for the country or its priorities . . .

[Kravchuk]: The Ukraine as a European state has set itself the goal of integrating into European structures. We only say that we do not want to join any blocs. We are a nuclear-free, extrabloc, permanent, neutral state and that is our fundamental policy. We also say that, while integrating, we do not consider it necessary to destroy our ties with the states of the former Union. Our common structures have been built here; our common life is here. But it was based on monopoly. Russia, for instance, has a monopoly on oil, gas, and timber. We do not bear Russia a grudge for that; we just say that it will not be able to provide all the other republics with these products, even if it wanted to do so. The reason why we do not have enough of some things is not that Yeltsin is a bad man. Production has dropped sharply and these regions are already controlled by unknown people. If we continue to pursue the policy of gearing ourselves solely to Russia, at any moment someone—and times here are so revolutionary at the moment that any president may close the pipes at any moment—may be deprived of gas and oil. We will try to obtain these things and we are already working with Iran and other countries now. It is normal for us to seek reserves, and markets are not confined to a single region.

[Portnikov]: You have already spoken of differences between the structures in Moscow and Kiev. Yeltsin has a team effecting economic reforms today. Where is the Ukraine's team? Or will it depend on Russian reforms?

[Kravchuk]: We are not repeating and will never repeat Russian reforms. Our independence from Russia marks our independence from the ruble. The ruble space and the fact that we are tied to monopoly sources of raw materials forces us to follow Russia. But we will take the path that we have chosen. We have people who will be able to do this. We will not say that they can effect another revolution. This will be calm, well-considered, well-thought-out work. We do not want to overextend ourselves. A person in the public eye is not entitled to do that. He is too visible if he does do something.

[Portnikov]: I think that there are many misunderstandings in Russian-Ukrainian relations due to the fact that Moscow does not notice that the Ukraine has indeed become independent. . . .

[Kravchuk]: Yes, undoubtedly.

[Portnikov]: Yet people in Kiev do not take this lack of comprehension into consideration. . . .

[Kravchuk]: No, we are aware of that. I never tire of constantly repeating the same thing at sessions. As soon as Nazarbaev submits a proposal to set up some kind of common committee, I remind him that we represent a state, but our community is not a state. I tell Boris Nikolaevich and he listens. But it is not just a question of one man. There is his whole entourage. The mentality of the Russian leadership does not come from above but from them. I see them sitting at the table, I see them go up to him and say: Let us have the Andreevskiy flag and so on and so forth. . . . I am not accusing anyone, but I think that if people were buzzing around me like flies saying this and that was necessary, I would possibly be the same way because people are telling you eight hours a day that you are a great person and must take charge. Take the problem of the unified armed forces, for instance. That is not a technical but a political problem. Let us think of its possible consequences. Eleven states and unified armed forces. Which state leader will control them? Tell me, who?

[Portnikov]: The president of the largest state?

[Kravchuk]: Why? We will not instruct him to do that. I, for instance, would vote against it.

[Portnikov]: What about a committee then, like the Yugoslav presidency.

[Kravchuk]: Fine, fine. Let me tell you one thing then. If the armed forces are unified and there are 11 civilian bosses, the troops listen to one man. Then—like it or not—a military man, a commander-in-chief, will rise above all 11 republics. The armed forces will rise above politics. That is sure to happen. Armed forces and democracy are incompatible. Not because they are bad. A general cannot be a democrat owing to the logic of his post. Unified armed forces herald the end of democracy, the end of independence. Not because Shaposhnikov is bad. Logic is superior to different people's wishes. And if this military flywheel clashes with the logic of life, we are doomed.

To whom are our troops to swear loyalty? Why spend all your time drumming into a Russian that he is to swear an oath of loyalty to the Ukrainian people? The Ukrainian people include another 12 million Russians. Yet he is told: No, swear loyalty to those Russians living in Rostov. Why should he swear loyalty to the Russians of Rostov and not those in Donetsk? They do not say that; they deliberately egg people on, fueling the question of nationality with comments like: You are a stranger here. . . .

[Portnikov]: You accuse Moscow Television of that kind of propaganda, but can it really be independent not just of its own government, but of the thinking of journalists living in Russia?

[Kravchuk]: I do not accuse it of that, but I do say something else. There are 56 hours of broadcasting per week. Divide 56 by 11 and you get six hours per state. I say: Give the Ukraine four hours a week. We will make the programs ourselves. But they don't want that. They want one independent person, as you say, to carry out one person's will. Yet I say, if this is a commonwealth, I must have my four hours and I will say what I want. That is my right. No Yakovlev [head of Moscow Television] can edit me.

[Portnikov]: Do you see a future for the Ukraine within the CIS?

[Kravchuk]: We have gotten accustomed to living with a center—to some extent Russia was decided as the center. Russians have mentally gotten accustomed to everything else being a part of Russia. Today they cannot surmount that barrier and are still acting as if this center still exists. We must tell them firmly and definitely that an independent Ukraine exists, an equal, full-fledged state that decides its policy itself and can only live in friendship with others on these principles. It cannot do so on other principles. We say this publicly and at the negotiating table. Hard as this may be, we have said that the problems of the Black Sea Fleet must be resolved. They did not want that. They wanted the Andreevskiy flag hoisted right away. We said no. . . . They agreed with us. So, we will gradually teach this great power that everyone is identical—small and great—and that there are no seniors and juniors, large and small. Everyone is equal. When everyone realizes that, we will start to live in harmony. I do not want people to conclude on the basis of any small or even large conflicts that this is the beginning of the end. A very difficult process is under way.

I have said officially and on television that, if we are equal, if we respect one another, if they do not encroach on our independence, and if no one aspires to diktat, no one considers us part of their territory or their people, the Commonwealth will exist. If they do not realize that, the Commonwealth will not be able to be strong. I hope nonetheless that life will make them realize it. If not, life will force us to take a different path. The Ukraine will not renounce its independence, just like Moldova, Belarus, and the other republics, because this constitutes the objective process of mankind's development. Not because some people want it. This is the people's common desire. Anyone who tries to stop it is destined to fail: The people will achieve it sooner or later. I would like this to occur without bloodshed, peacefully, and democratically, so that everyone is aware of this and does not create difficult situations where they do not exist. Ambitions cannot be placed above democracy.

[Portnikov]: I would like finally to diverge from politics. What do you remember when you manage to relax from all these battles, talks, and sessions. . . .

[Kravchuk]: I had a difficult childhood. It was 1939 and reunification was under way. I remember the Soviet troops coming in. I remember collectivization in our village, the livestock being taken away and people crying, not wanting this to happen. Then they were evacuated "to live with the polar bears," as they said then. Then came the war. This also took place before my very eyes. Rovno was bombed, our village was bombed. We hid, first one set of troops came, then another. There were the executions in the church, which I witnessed. I was made to watch them kill Jews—the Hitlerites rounded everyone up to see this to make them afraid and see how they hanged people. I cannot even tell you how many lives were lost before my eyes. I saw the blood and the agony. I cannot convey it. My childhood memories have all this imprinted on them. Now when I am sick and have a fever, I dream of all these dreadful times again and shudder. . . . When the war ended we only learned in June when news reached our village. And I remember my uncle—my father died on the front—flew into the yard on his horse, his pride and joy, and said: The war is over. Everyone wept and I could not understand why. I was so happy then! That is my most vivid childhood memory—the end of the war. I thought that the sun would shine differently and people somehow would be different.

## 5.3

### Background to Conflict over Crimea Viewed

I. Osokin

From the "Novosti" newscast, Ostankino Television

7 February 1992 [FBIS Translation]

*Editor's Note:* The following document characterizes the intensity of the Russian parliament's objection to Ukraine's retention of Crimea. The post-Soviet movement for the return of Crimea to Russia developed with a referendum on 20 January 1991, in which the Crimean population voted for direct links with Moscow. On 23 January 1992 Russia's parliament, catalyzed by Sergei Baburin, leader of the "Rossiya" faction, voted to challenge the legality of Khrushchev's transfer of Crimea to Ukraine. They cited the ancient historical background—Catherine the Great's seizure of Crimea from the Turks in 1772—as support for their decision. In a strict sense Crimea belonged neither to Russia nor to Ukraine, but to the amorphous Russian empire. In 1944 the entire Tatar population (about 200,000) was deported from Crimea in punishment of their alleged collaboration with the Nazis.

Crimea was originally inhabited by the Tatars. Russian troops conquered it at the end of the eighteenth century and the peninsula became part of Russia. In 1954 Khrushchev gave it to the Ukraine and Russia. But there is a new twist to this tale.

The Russian parliament has questioned the legality of the decision to transfer Crimea. Russian parliamentarians proposed that the Ukrainian parliament should also examine this issue. The answer was very harsh. Yesterday's parliamentary resolution says that Crimea is an inalienable part of the Ukraine. The resolution urges that the Helsinki agreements and the principle of the inviolability of existing borders be followed. The Supreme Soviet is reproaching Russian legislators for having committed actions that contravene previously signed treaties. The reproach looks serious; it is true that the agreements on the creation of the Commonwealth speak of the inviolability of existing borders. The Russian Supreme Soviet's decision on Crimea somehow does not really fit into this formula. But on the other hand it must be said that the Ukrainian leaders themselves are to a great extent responsible for the emergence of this issue by laying claim to the Black Sea Fleet. Furthermore, also in violation of agreements that have already been concluded, in the second half of January the Ukrainian General Staff again stated that the entire Black Sea Fleet must be considered as part of the Ukrainian armed forces. Then just after that a document appeared in the Russian Supreme Soviet, which the news agencies called Lukin's memorandum. Lukin is chairman of the parliamentary committee for international affairs. The document proposed that the disputes over Crimea be used to exert pressure on the Ukraine in order to make it renounce its claims to the Black Sea Fleet. And following that Russian legislators did in fact raise the issue of Crimea. . . .

## 5.4

### Ukrainian Leader Opposes Nation Status for CIS

*Xinhua* (Beijing), 11 February 1992

[FBIS Translation]

Ukrainian President Leonid Kravchuk said today that common work in the Commonwealth of Independent States

[CIS] should not deprive Ukraine of independence, because CIS is not a state.

He told *Pravda* that Ukraine opposes attempts to change CIS into a national entity, but leaders of some CIS members seem to have forgotten this when they sit at the negotiation table.

The president stressed that CIS will have no future as long as there are attempts within CIS to reverse life to the empire times.

Kravchuk said the relationship between Ukraine and Russia should be made a priority because there are 11 million Russians living in Ukraine and 5 million Ukrainians in Russia.

The relations between the two countries are of great significance, Kravchuk said, adding that both sides should do everything possible to coordinate themselves well economically and politically instead of entering into quarrels. Every issue between them should be resolved through negotiation, he said.

Kravchuk complained that some persons around Russian President Boris Yeltsin regard Ukraine as part of an empire and not a sovereign state.

He reaffirmed the position that Ukraine has the right to own part of the Black Sea Fleet and that Crimea is part of Ukraine's territory.

Observers here noted that Kravchuk's statements came at a moment when CIS leaders are going to discuss certain tough problems at a meeting scheduled for 14 February in Minsk.

## 5.5
## Supreme Soviet Statement on Crimea Published

*Izvestiya*, no. 27, 12 February 1992
[FBIS Translation], Excerpt

*Editor's Note:* In 1954, Nikita Khrushchev ceded the Crimean peninsula from Russian to Ukrainian suzereignty, as a symbol of unity on the three hundredth anniversary of the Treaty of Pereslavl, which united Russia and Ukraine. Russian patriots now argue that the territory should be returned to Russia because the "unity" has been broken. Russia's policy on Crimea has been to follow a "divide-and-conquer" approach, similar to that exhibited by Russia in Moldova and Georgia. On 23 January 1992 Russia's parliament rejected the legality of the 1954 transfer of Crimea. Despite Ukrainian attempts to resolve the issue peaceably, by May the Crimean parliament in Simferopol had passed an "act of independence," to take effect if passed by public referendum. Since then, Ukraine and Crimea have lurched from crisis to crisis, with the ethnic Russian population of Crimea leading the militant "Russian Association."

On 1 July 1992 Ukraine passed a law giving Crimea the status of an autonomous entity but prohibiting it from transferring to another country without the approval of both the Ukrainian and Crimean parliaments. This eased the momentum toward a referendum for the moment, but tensions have been high between Ukraine and Crimea ever since. This issue is considered to be potentially the most destabilizing for Ukraine, and perhaps for the whole region.

The Ukrainian Supreme Soviet has asked the newspaper to publish its reply statement.

On 23 January 1992 the Russian Federation Supreme Soviet adopted a resolution in which it instructed Russian Federation Supreme Soviet committees together with the Russian Foreign Ministry to examine the question of the constitutionality of the 1954 decisions on the transfer of the Crimean Oblast from the RSFSR to the Ukrainian SSR [Soviet Socialist Republic] and proposed that the Ukrainian Supreme Soviet also examine this question.

The Ukrainian Supreme Soviet believes that these actions could destabilize the socio-political situation in the Ukraine and Russia.

The Russian Federation Supreme Soviet resolution of 23 January runs counter to Article 6 of the Treaty Between the Ukrainian SSR and the RSFSR of 19 November 1990, Article 5 of the communique on the talks between RSFSR and Ukrainian delegations with the participation of a USSR Supreme Soviet delegation of 29 August 1991, Article 5 of the Agreement on the Creation of the Commonwealth of Independent States [CIS] of 8 December 1991, and the Helsinki Final Act.

The history of our countries does not start in 1954, and the search for the truth in events of the past (particularly if it is a question of territorial problems) does not always lead to peace and concord or serve as a constructive road to the future. The Crimea became part of the Ukraine within the framework of the political and legal structures and the procedures and realities that existed in the former USSR at that time.

The Ukrainian Supreme Soviet confirms its adherence to the Helsinki agreements, particularly the principle of the inviolability of existing state borders, and rejects any territorial claims.

The Ukraine has no territorial claims on neighboring countries and hopes for a similar approach and support from all the Commonwealth states, particularly from the Russian Federation, which has traditionally been friendly toward us, in ensuring tranquillity for the population of the Crimea, which is an inalienable part of the Ukraine with the status of an autonomous republic enjoying full rights.

The Ukrainian Supreme Soviet shows its respect for the Russian Federation Supreme Soviet, understands its problems in creating a sovereign state, and declares its readiness for mutually advantageous, equitable cooperation and for constructive talks on questions whose resolution will be to the good of the Ukrainian and Russian peoples and all the peoples of the CIS.

[Signed] Ukrainian Supreme Soviet, Kiev, 6 February 1992

## 5.6

### Efforts to Preserve Joint Army "Useless"

Anders Steivall
*Dagens Nyheter* (Stockholm), 14 February 1992
[FBIS Translation], Excerpt

The Soviet Union no longer exists. But the member states of the Commonwealth of Independent States [CIS], which has replaced the old union, are still far from through with the division of the property.

The most dangerous thing is left—the division of the armed forces. Strong forces want to retain common defenses for the CIS, but disintegration of the 3.5-million-man army now seems unavoidable.

"All attempts to preserve a common army are useless, if you take the political realities into account," Ukrainian Defense Minister Konstantin Morozov said in an interview with *Dagens Nyheter*.

Morozov, an air force general who is himself a Russian, is one of the central figures in a drama which the rest of the world is following with growing concern.

Ukraine, now an independent state with over 50 million inhabitants, has decided on a rapid buildup of its own armed forces.

This has led to increasingly acrimonious relations with Russia, which has 150 million inhabitants and controls the largest part of the old Soviet state's military and economic resources.

The conflict has focused above all on the Black Sea Fleet, to which Ukraine has laid claim but which Russian and the other CIS members want to keep under joint control.

Pessimists see the risk of a devastating civil war between two large nuclear states coming ever closer.

"I am convinced that we will find a compromise. The conflict has been blown up artificially," Morozov said.

Nevertheless, the row over the Black Sea Fleet, which has its home base in the Ukrainian port of Sevastopol on the Crimean Peninsula, and Ukraine's military ambitions have been the flame that has triggered many explosive outbursts from high-ranking officers.

At a stormy meeting in the Kremlin a month ago, the majority of the 5,000 officers present demanded that the armed forces be kept under joint control.

An opinion poll among these officers showed that 71 percent want to restore the Soviet Union which the politicians have abolished.

However, Morozov is not frightened by these threatening noises.

"This is a continuation of the old political line. First people tried to preserve the Communist Party, then they tried to preserve the union. And now, in the face of all sober thinking, they want to preserve the united army."

Together with four other members of the CIS—Belarus, Moldova, Azerbaijan, and Uzbekistan—Ukraine has decided to set up its own armed forces.

It is unclear how large the Ukrainian army will be. There has been talk of everything from 100,000 to 400,000 men.

Ukrainian leaders are trying to reassure the rest of the world by pointing out that what is happening is actually large-scale disarmament. The Soviet forces on Ukrainian territory total around 1.5 million men.

In January the process of transforming the Soviet units into a Ukrainian army began. All soldiers and officers were ordered to swear an oath of allegiance to Ukraine.

The decision caused conflict, because the majority of officers are Russian and the conscripts come from all the former Soviet republics.

However, when it came to the crunch, the drama failed to materialize. So far 90 percent of officers—with the exception of the Black Sea Fleet—have sworn their loyalty to the independent Ukraine.

"The majority of our officers support the political realities with which we are surrounded. Over 350,000 men have already sworn the oath," Morozov said. . . .

## 5.7

### Fleet Upkeep Requires 500 Million Rubles

From the "Vesti" Newscast, 14 February 1992
[FBIS Translation]

A representative of the Ukraine military department has stated that Ukraine is completely financing the upkeep of the Black Sea Fleet. It has allocated 500 million rubles toward this purpose. The money allocated by Russia for these purposes, he said, is not part of Ukraine's budget and cannot therefore be used for this purpose.

## 5.8

### Moves to Join East European "Triangle" Cited

Andrey Kamorin
*Izvestiya,* 19 February 1992
[FBIS Translation]

"Ukraine's path to Europe lies across Poland"—these words uttered in Warsaw by D. Pavlychko, representative of the

Ukrainian Supreme Soviet Commission for Foreign Affairs, attracted the attention of foreign commentators, primarily because of the contradictory role which the Ukrainian delegation played at the recent meeting of CIS [Commonwealth of Independent States] leaders in Minsk.

The special stand that President L. Kravchuk took at the Minsk meeting on questions of cooperation within the framework of the Commonwealth is in clear contrast with the desire for integration shown by Ukrainian diplomacy in the "Central European salient." The most vivid manifestation of this course was D. Pavlychko's statement of Ukraine's desire to join the so-called "triangle" formed by Poland, Czechoslovakia, and Hungary. Let us elaborate that the "triangle" today does not constitute an organizationally drawn up regional formation. It is merely an agreement among three Central European states to coordinate their actions, primarily in respect of the EC, NATO, and the CSCE. There are also plans to extend economic cooperation, although there is still a long way to go to the formation of a small "Common Market" here.

The Polish Foreign Ministry responded promptly to Pavlychko's statement, putting out an explanation that it does not seem possible to include Ukraine in the "triangle" at present because it "is at a different stage of postcommunist changes." So as not to offend the Ukrainians, the Polish foreign policy department did not omit to point to a precedent: Last year's request by Romania, which was also "delicately rejected."

This refusal was perfectly predictable: The "triangle" countries, which recently became associate members of the EEC, have advanced quite a long way along the path of integration in all-European structures, and their being joined by a vast state, one that is still tightly "bound" by the knot of contradictions that exist in the CIS, could set this process back by several stages.

It seems, however, that, although they are keeping Ukraine on the threshold of the "triangle," they are not closing the door in its face. All three member countries have stated their readiness to cooperate with Kiev in implementing specific projects. In particular, Ukraine has been invited to participate together with Poland, Czechoslovakia, and Hungary in the "Jaslo-92" conference, which opens on Saturday [22 February] under the auspices of Lech Walesa.

Obviously, the signing of several quadrilateral Hungarian-Ukrainian-Polish-Czechoslovak agreements this week in the economic and cultural spheres should also be viewed in this context, our Budapest correspondent F. Lukyanov reports. The four countries agreed, in particular, to create free economic zones in regions bordering on each other, to set up a joint bank, and so on. So it cannot be ruled out that the "political geometry" of Central Europe will still change and the involvement of Ukraine here, with its tremendous human and natural potential, could become an important political factor in this part of the continent.

## 5.9

### Retention of Nuclear Weapons Encouraged

Colonel V. Bogdanovskiy
*Krasnaya Zvezda,* 25 February 1992
[FBIS Translation], Excerpts

The mass media in the western region of Ukraine continues to focus on questions concerned with the formation of Ukrainian armed forces. There is nothing surprising in this. Politicians at various levels regard the army as the structure which will enable an independent state to be built. What kind of army is it to be? If you refer to statements by parliamentarians and the president, their line is unambiguous. The Ukrainian armed forces are a state military structure designed for the armed defense of the republic's sovereignty and territorial integrity against external attack.

Both president and parliamentarians have repeatedly stressed that Ukraine has made the decision to remove tactical nuclear weapons by 1994. Everything seems absolutely clear and precise. However, as the mass media indicate, far from everyone is in agreement with this decision. . . .

Some politicians and military men are ready, as we can see, to reject the concept expressed by the parliament and president of Ukraine, while certain others openly describe it as pacifist.

In particular, Lieutenant General V. Stepanov, commander of the Carpathian Military District, insists that nuclear arms reduction be carried out simultaneously rather than unilaterally in every country in the CIS which has them. While Major General of Aviation V. Antonets, commander of Carpathian Military District Aviation, considers that the threat of nuclear war has not been dispelled in the world and that if you renounce nuclear weapons voluntarily you therefore put yourself in a difficult situation.

## 5.10

### President Kravchuk Explains Monetary Reform

Statement by President Leonid Kravchuk
Ukraine Television and Radio Company
12 November 1992 [FBIS Translation], Excerpts

Most esteemed compatriots. I am addressing you on an issue, which is very important to all of us, which is really key in implementing anti-crisis measures at the present stage and in implementing economic reforms. This is the reform of the monetary system, which will doubtless occupy a central place in the transition to a market economy. . . .

Some people maintain that the introduction of a national currency would lead to the severing of economic links with partners in Russia and other states of the CIS. But at the same time they do not want to notice that everything that could be severed was severed a long time ago. And now it is necessary to create new links on new principles, and on principles of mutual advantage and partnership, on principles of protecting the interests of one's own people.

Another important point. The ruble has in fact not existed as a single currency for a long time now. Because in every state a significant difference exists in purchasing power. And this means that other levers for influencing the flow of goods between states do not exist either, only administrative interference. And so that extraordinary situation of a practically complete stopping of payments in interstate accounts, which was caused by delaying the introduction of the Ukrainian karbovanets and establishing its real rate in relation to the currencies of other states, has forced us into this. This was the main reason.

This is our opportunity to renew severed links with CIS countries and rebuild them on civilized principles of partnership. In establishing the rate of our own currency in accordance with its realistic purchasing power, we are preventing a multibillion [currency not specified] turnover of speculative capital, which operated as a parasite in making use of a single monetary unit under conditions of a complete difference in the level of prices in Ukraine and other states of the CIS. This is a very important point.

In this way the very introduction of the coupon into non-cash circulation allows us to develop mutually advantageous production links, necessary for the existence of the economy of Ukraine, with other states of the ruble zone. It is for this very matter that we are introducing a policy of internal conversion of the Ukrainian karbovanets in relation to the ruble.

This means that Ukraine is not leaving the economic space of the CIS, but intends only to introduce a system of economic regulation in its trade and economic relations with other states.

In introducing a national currency, Ukraine is entering a new stage of nation-building. It is acquiring the status of a state, which is in complete control of its own economic space. And it is because of this that I am sure that these steps will be perceived with understanding by the citizens of Ukraine, as steps that correspond to their national interests. Naturally, in my short speech I cannot dwell on all the details of the first stage of the monetary reform. But I would like to report that during these past few days members of the government, executives of the National Bank, and leading experts in financial matters, who will throw light on various aspects of the problem, will appear before you.

As president, elected by the people, I want to assure you all, esteemed citizens, esteemed friends, that the population will not incur any losses from the reforms. This is in fact the main purpose of my appearance. I call on everyone to implement this matter in an organized way, responsibly, in conditions of complete mutual understanding and public calm. . . .

---

## 5.11

**Learn How to Reach an Understanding with Russia . . .**

*Rossiyskie Vesti,* 18 November 1992
[FBIS Translation]

---

*Prime Minister of Ukraine L. Kuchma paid his first visit to Moscow in his new capacity on 22 October. As a result of negotiations with E. Gaydar, acting chairman of the Russian Federation Government, three agreements were signed: on most-favored-nation status with respect to trade; technical assistance to third countries; and trade representations. . . .*

Looking for the proper tone in relations with neighbors—especially Russia—has been proclaimed as another important principle of the Ukrainian prime minister's activities. "We have to reach agreements with Russia on the level of independent states, banks, and enterprises," he believes. "We have to rein in the ambitions of those Ukrainian bureaucrats who yesterday were praying to Moscow and are now spitting on it."

It looks as if this principle will not be easy to apply: Many politicians in Ukraine have reached the pinnacles of power precisely by preaching a break with Moscow. Nevertheless, Leonid Danilovich is set in his intent. "The government will put an end to the 'cold war' with Russia," he says. And, again, he backs it with concrete actions—only nine days after being confirmed in his post he attends negotiations with Egor Gaydar.

Of course, it would have been naive to count on quick progress. Still, there is hope: A Russian-Ukrainian commission has been created; among its priorities will be to discuss the problems of repayment of external debt. There is a visible "light at the end of the tunnel" in other directions as well.

---

## 5.12

**Kravchuk Accuses Russia of "Interference" on Sevastopol**

Aleksey Petrunya
Ukrinform, 9 December 1992 [FBIS Translation]

---

Ukraine was never opposed to the Commonwealth. It is against its low effectiveness, said head of the Ukrainian state

Leonid Kravchuk. "We would like this organization to act in the interests of all its member states," Kravchuk told a news conference on Wednesday after meeting President of Azerbaijan Abulfaz Elchibey. The two presidents signed a treaty on friendship and cooperation. In Ukrainian-Russian relations, according to Kravchuk, "the main political problems stem from the Russian legislature."

"The Supreme Soviet of Russia decided to reconsider the document of 1954 on the transfer of Crimea to Ukraine and the Congress of People's Deputies of Russia instructed the Supreme Soviet to consider the issue of the status of Sevastopol," he recalled. "This amounts to interference by Russia's supreme legislature into the domestic affairs of Ukraine, including such an important sphere as the territorial integrity of our country."

Kravchuk highly assessed the prospects for cooperation between Ukraine and Azerbaijan. During Wednesday's negotiations, he noted, it was agreed in principle that Azerbaijan will continue deliveries of petroleum products to Ukraine in accordance with current agreements.

"Azerbaijan has concluded an economic agreement with Russia and considers it a strong partner," said Elchibey. "However, relations with Ukraine are a different story. There are many problems, on which our views coincide with those of Ukraine. It is primarily secession from the ruble zone, solution of economic and ecological problems. We also have similar views on issues of territorial integrity of the states. We also have a possibility to exchange experience in agriculture and oil production."

## 5.13

## Political Bloc Considers Signing CIS Rules "Treason"

Interfax, 26 December 1992 [FBIS Translation]

*Editor's Note:* Right-wing democratic groups in Ukraine have vociferously protested Russia's actions and pronouncements within the CIS. The Congress of National Democratic Forces consisted of eleven democratically oriented political parties in Ukraine, which lobbied Kravchuk not to sign the CIS Charter as written.

To sign the CIS (Commonwealth of Independent States) Rules in the proposed version would be treason to Ukraine—goes a statement issued Friday by the Congress of National-Democratic Forces (CNDF).

"Russia's reactionary forces aim at turning the CIS into a new world empire," insists the Congress' council. "The proposed rules are no less treacherous than the 1922 treaty and mean a new colonial yoke we'd never be able to throw off."

The CNDF calls on all the parties and organizations to unite into "an anti-imperialistic democratic front" and urges the president, parliament, and government to refrain from signing the rules and thus protect the act of Ukraine's independence.

The CNDF is a bloc of over ten political parties and movements, including Ukraine's Republican and Democratic parties.

## 5.14

## Europe Orientation Seen as Only Realistic Course

Dmytro Tuzov

*Narodna Hazeta*, no. 1, January 1993

[FBIS Translation], Excerpts

### "We Are Europeans"

. . . We find the beginnings of a European identity as far back as the third century B.C. The Hellenic philosopher, Isocrates, called upon the Europeans to unite against Persia to defend the continent against Asiatic expansion. Isocrates was not heeded, and for many more centuries Europe remained torn between various blocs, alliances, and religious beliefs. The French historian Besancon describes the stratification of Europe as follows: "Historically, there were three Europes: the first, wealthy Europe stretches from Madrid to Vienna through London, Paris, Rome and Berlin. Between the sixteenth and seventeenth centuries, there existed a second Europe—less brilliant, poorer, but nonetheless European—which included Poland, the Baltics, Ukraine, and Belarus. At times, Sweden and Hungary were part of this 'second Europe.' Muscovy comprises the third Europe—poorer still, remote, and 'barbarian.' " (*Le Figaro* [Paris], 6 September 1991).

The eastern portion of the continent, especially Ukraine and Belarus, is subject to significant influences from this Muscovite "Europe": the eastern orientation of countries in the so-called Byzantine belt is traditional. Even the ancient Achaeans established ties mostly with Asia Minor, because the dangerous Adriatic lay across the route to the West. This is how Mykhaylo Hrushevskyy defines the strategic error made by Volodymyr the Great: "Volodymyr consciously and energetically pushed Rus' in this direction [toward Byzantium] . . . at that time it was not possible to foresee that Western culture was destined to flourish and Byzantine culture was fated to decline." Ultimately, Moscow did no more than imitate Babylon, which in its time applied as an instrument of state-building the theory of empire proposed by Zeno the Stoic: "to live not as separate states governed by different laws, but together, under a single system, like a herd that grazes on a common pasture."

### "Spiritual Aggression"

All attempts to interbreed West with East produce only negative results. If the Norman influence on the formation of the English, French, or Ukrainian state is an internal phenomenon, as it were, today the hot eastern wind more often than not carries with it not the wisdom of the Koran or the cosmogony of Ramakrishna, which educated Europe has long known, but the acrid smoke of local and world conflicts and interethnic hostilities. The expansion of the East into geographic Europe has long since passed beyond the benchmark of mutually enriching cultural influences and, instead, recalls spiritual aggression. The export of violence to the continent, especially to Ukraine, is linked to the dividing up of spheres of influence by Asiatic mafias, which, as they set up their legal and illegal businesses, are preparing the infrastructure for new masters and binding the state to their interests with their capital, thereby pushing the state into the opposite direction from Europe—precisely at the time when various dreamers are claiming that we are drawing nearer to the West.

Those Ukrainian politicians who seek "free oil" must realize that Ukraine's leading position in Asia is illusory. To consolidate such a position, it is necessary to wage wars for centuries with all neighboring states like the wars between Iran and Iraq, in which the participants have long since forgotten what originally caused the outbreak of hostilities. Clearly, we will not be satisfied with the colonial status of a "bamboo democracy," in which the Asiatics thrive in business, while the indigenous population, most of which still fails to understand the rules of three-card monte, especially in light of privatization by auction, engages exclusively in circulating newspapers. In any case, there exists evidence that cannot be faked that we are part of Great Europe: anthropological, historical, cultural, linguistic, tribal, etc. Someone has even said that every one of our genes shouts Europe. But we also understand something else. There is a limit to how long we can keep relying solely on this evidence. After all, it is not only a case of the Soviets having traumatized the psyche of the population and destroyed its sense of private property, but the depth to which these deformations have penetrated social consciousness. A broken street light is evidence of the mental aberration of one specific person, but the inability to at least stabilize our own economy raises doubts regarding the normalcy of our society as a whole. Being transformed into another Turkey, which is not being permitted to join Europe as its applications to become a member of the EC are methodically rejected, will not only cause moral damage to Ukraine, but will make it more difficult for her to get out from under Moscow's influence. . . . Russia, whose influence on the European process must certainly not be ignored, is herself confused on the West-East issue.

Incredibly, of all the post-Soviet states, it is Russia that enjoys the highest degree of trust in the West—and not only because of the West's sympathies toward Gorbachev, the reformer of the "evil empire," and Yeltsin, the victor over the State Committee for the State of Emergency, but also as a result of historical factors associated with the Europeanization of the Muscovite empire that began under Peter I and was continued by his successors (especially the German, Catherine II). Not only have the Muscovites skillfully rewritten history, copied the Dutch flag, and refashioned a religious song into a hymn, but they farsightedly invested large sums of money into education, science, and culture. Small wonder that in Europe Russia is not identified with the Lenin epic and with wars of occupation but with the works of Tolstoy, Dostoevsky, and Solovyev. It is this kind of associative thinking that prompted international recognition of Russia as the legal heir of the defunct Union. But even though the elite European club demonstrates in every possible way its openness to new members, in all probability the Russian question will hang in the air for a long time or will be limited to associate membership. It is hardly likely that the EC will risk taking on the burden of the Eurasian giant.

### There Is No Alternative

Of course, all this is a matter for the distant future. Today the Community's commissions are determining how much benefit will be derived from membership in the EC of Sweden, Austria, Switzerland, Finland, Ireland, Liechtenstein, and Norway—members of the European Free Trade Association, countries with developed market economies and established democratic systems. The members of the "Vysehrad triangle" [as of 1 January, undoubtedly a quadrangle—Ed.] are knocking with growing urgency at the door of Great Europe: Poland, Hungary, and Czechoslovakia have already become associate members of the Community. The time has come for signing similar agreements with Rumania and Bulgaria. To be sure, the founders of the European union have not yet reached an understanding with the opponents of this union, of whom there are many. Proud old Great Britain is seriously troubled by the likelihood of losing its sovereignty, Denmark rejected Maastricht in a referendum, and all of Europe fears the prospect of waking and falling asleep to the sound of the rousing marches of Germany, which is not likely to abandon its economic hegemony over the continent. Thus when Prime Minister John Major of Great Britain hopes that after a period of time Western Europe will see "strong democracies, moving toward growth and full membership in the EC" in the East, he is not merely being polite. For the first time in many years, "brilliant Europe" has turned its eyes eastward in search of answers to questions about the continent's future. Will a force be born there that will be able to extend the

European vector from London, through Paris, Rome, and Berlin, eastward, and will it help to find the balance that is so badly needed in the whole world today?

Kiev, too, finally has an opportunity to gain an important voice in world politics, but only if it succeeds in ridding itself of its complex of "Hellenistic effeminacy," owing to which the building of statehood was replaced by dreaming and nostalgia for the past "when we were Cossacks." Unique economic reforms must become the foundation of the future Ukraine. New laws must stimulate the growth of production, the introduction of the latest technologies, and the influx of capital. This is not a new path. It is possible to avoid the fate of Third-World countries, the common pasture fought over by external forces, only by accomplishing an economic miracle. We simply have no other alternative.

## 5.15

**Nationalists Reject Russian Cooperation Accord**
*Molody Ukrainy,* 6 January 1993
[FBIS Translation]

*Editor's Note:* The Ukrainian National Assembly strongly supported President Kravchuk's rejection of a Russian-dominated CIS in 1992 and 1993. At that time, the assembly was still predominantly composed of nationalist parties and movements, and former communist groups loyal to Ukrainian nationalist ambitions.

["Statement by the Ukrainian National Assembly in connection with the intentions to sign the CIS Charter and the Russian-Ukrainian Agreement on Good-Neighborly Relations, Cooperation, and Partnership"; issued in Kiev on 1 January.]

The Ukrainian National Assembly fully supports the policy toward full establishment of Ukraine's statehood and its non-participation in blocs.

Following the decision adopted by Russia's Supreme Council on the illegal aspect of the 1954 act on the transfer of the Crimea to Ukraine and the discussion of the status of Sevastopol at the recent Congress of Russia's People's Deputies, the Ukrainian National Assembly rejects, in principle, the signing of the CIS statute and the agreement "on good-neighborly relations," until the position on these questions of Russia's supreme legislative organs is changed.

In international practice, diplomatic relations are broken following such statements, and agreements on friendship or joining common blocs are not signed.

Russia needs the signing of the CIS Charter and the conclusion of a bilateral agreement to continue (using the agreement as cover) interference into Ukraine's internal affairs.

Ukraine should, on the contrary, take advantage of the still-available openness of the Russian leadership to Western influences and try to attract attention to this matter by demonstratively refusing to sign not only the CIS Charter, but also the Agreement on Cooperation and Partnership between Ukraine and Russia.

[Dated] Kiev, 1 January 1993

## 5.16

**Kravchuk Supports CIS Customs Union**
Interview with Ostankino Television
16 April 1993 [FBIS Translation], Excerpts

*Editor's Note:* By April 1993, President Kravchuk's position on introducing a separate Ukrainian currency had changed. Although he still resisted joining a common CIS "ruble zone," he argued in favor of a "supranational currency." Ukraine's inability to pay its interenterprise debts and its difficulty establishing supplier and producer relationships with other CIS states, especially Russia, were already beginning to seriously threaten the viability of the sovereign state.

[Correspondent]: A question for the president of Ukraine. Leonid Makarovich, at today's meeting it was proposed that some kind of supranational structures be created for improving the coordination of economic cooperation among the CIS states. How do you envisage economic cooperation in the CIS?

[Kravchuk]: Today we did not discuss the question of the creation of supranational structures, but I would support a supranational formula by which a supranational currency would be created. A currency union should be created, for then there would be a currency capable of determining life within the framework of the CIS, of improving the settlement of accounts, of improving all economic ties in mutual payments, regulating financing, banking, and other structures. I would support that, although in questions of economic interaction, consultative, coordinating structures could be created within the framework of economic interaction.

I would like to say a few words since everybody knows my position with regard to the CIS and often link the prospects of the development of the CIS, its strengthening, or as journalists say and write, the disintegration of the CIS, with Ukraine's position.

Today I can state categorically that whether the CIS exists and what sort of entity the CIS is will depend on 25 April.

If Russia goes along the path of reforms, democracy, and strengthening the processes begun in Russia back in 1990, having confirmed this course in August 1991 before the whole world, if the people of Russia and Russia support this course on 25 April, the CIS will exist, will be a democratic formation, and will be a structure which will bring new life to us and to all. I believe that the people of Russia and Russia will stand up and support this course with all their might. If something else should happen—which I don't believe will be the case—then the CIS will not exist. I state this with full responsibility because the forces that are operating there today do not want the CIS but the USSR, and a return to the Union is not possible. Second, if somebody embarks upon that path, then that path will be watered with lots of blood. That can be said with full responsibility.

Therefore, the future of the CIS, its content and prospects, and the democratic future of all our states within the CIS, and not only within the CIS, today depend on 25 April. I am profoundly convinced of this, and I should like to state my conviction—the conviction of a man who has a particular position, and you know what my attitude is to the heads of state: I often criticize them sharply. But today I can say with firm conviction that 25 April will decide the fate of the CIS and the future of our peoples and of our states. . . .

## 5.17
## Ukrainian Cabinet Statement Defends Economic Union with Russia, Belarus

*Uryadovyy Kuryer,* 12 August 1993
[FBIS Translation]

[Undated "Statement by the Press Secretary of Ukraine's Cabinet of Ministers"]

Lately, quite often, deliberately or inadvertently, in speeches of some political figures and in the mass media, actions by the Ukrainian Cabinet of Ministers have been represented in a false light, and the position and intentions of Ukrainian Prime Minister Leonid Kuchma regarding the recent signing in Moscow by the heads of the Ukrainian, Russian, and Belarusian governments of the statement on the intention to create an economic commonwealth have been misrepresented. Here, the age-old underhanded machinations and the "telephone right" [decisions made by putting pressure upon officials through telephone calls] are resorted to. Particularly critical passions around this question flared up after the newspapers *Uryadovyy Kuryer* and *Golos Ukrainy* published one of the drafts of the "Agreement on the Creation of an Economic Union (Commonwealth)." This created grounds for holding, on 29 July 1993, a "roundtable" of political parties and public organizations where some of the participants in the discussion brutally slandered members of the Ukrainian government and its head. Besides, the publication was not the initiative of the editorial boards or of the Cabinet of Ministers Press Service. Moreover, the published material was a badly formulated draft variant of the document, which is just a Russian Federation proposal. The published text had not been worked on by specialists and, the main thing, it had not been discussed at a Cabinet of Ministers meeting. It did not incorporate the numerous remarks, additions, or corrections without which the Ukrainian government cannot accept the text of the agreement.

In view of the fact that problems with present economic ties and their effectiveness, as well as questions of economic integration, are extremely important and truly strategic for the young Ukrainian state, some politicians and journalists are trying to use them in a biased or openly mendacious manner in their dirty political maneuvers. They depict the actions by the Ukrainian prime minister as his personal position, which contradicts Ukraine's general state policy and interests and is at variance with documents that have been signed by the president of our state.

In this connection, I have been authorized to state this:

1. The actions by Ukrainian Prime Minister Leonid Kuchma that are aimed at intensifying mutual economic ties and economic integration constitute direct fulfillment of instructions of the Ukrainian government that are contained in the declaration adopted by the heads of those states that are members of the Commonwealth of Independent States and in other documents that were signed on 14 May 1993 by presidents of the states, including Ukrainian President Leonid Kravchuk. It was, in particular, written in these documents: "*To authorize heads of governments* of those states that are members of the Commonwealth to organize work on preparing corresponding draft documents . . . ," and also "Heads of states that are participants in the Commonwealth find it necessary for this purpose *to focus the efforts of their governmental structures and organs of the commonwealth on preparing, in the near future, corresponding documents and on making practical steps. . . .*"

2. Ukrainian Prime Minister Leonid Kuchma repeatedly stated and reiterates now his position to the effect that he will never sign a document directed against the Ukrainian people's interests or a document that in any way defies the Ukrainian state's sovereignty.

However, the policy aimed at the restoration and further development of mutually advantageous economic ties objectively conforms to the interests of the young Ukrainian state and to the interests of every family and of every citizen of our country, no matter what politicians of various levels and denominations who often know nothing about the economy or true human needs would say.

Besides, and this must be stressed in principle, Leonid Kuchma only signed a statement by the heads of the governments, more specifically, the declaration on intentions—a document that has no legal consequences, does not need to be coordinated with the parliament, but only testifies to the orientation of the government's intention regarding the restoration of precisely economic ties. The draft documents—treaties and accords—are presently being further specified and supplemented taking into account ideas, positions, and points of view or representatives of all branches of power in Ukraine, Russia, and Belarus.

3. The discussion and criticism of practical steps made by the Ukrainian government, as also by governments of any country of the world, must be freely conducted by all political movements, parties, and citizens, but without misrepresentation or deliberate falsification. The press service of the Cabinet of Ministers will continue to react accordingly to all attempts at elucidating the policy of the government and the position of its head in a non-objective or misleading manner.

[Signed] Dmytro Tabachnyk, press secretary of the Ukrainian Cabinet of Ministers

## 5.18
**Further Russian Pressure Expected**

Volodymyr Chemerys

*Ukrayina Moloda,* 29 October 1993

[FBIS Translation]

The results of the elections to Russia's Federal Assemblies will affect the political situation in Ukraine no less than the future composition of our own Supreme Council. For us the dramatic quality of the situation lies in the fact that—even without knowing ahead of time how many seats in the State Duma [Council] the party known as "Russia's Choice" will have, and how many the party of Shakhray and Volskiy will have—we can already make the following statement: the "war party" will prevail in Moscow. And this will be not merely war in Tajikistan and Georgia, but also economic and (entirely real) military pressure on Ukraine.

The smashing of the White House in Moscow could become a Pyrrhic victory for Yeltsin unless he takes measures to "gain ground" around the Kremlin and assumes the ideological leadership of Russian chauvinism. This is being demanded of him not only by the "strong men" in the government, but also by the Russian Army, whose passivity in October signified a final warning to its own president. Yeltsin's advisors in their most recent broadcasts have been publicly demanding that he answer the following question: Will Russia's federal structure be a model for a future confederation of the former USSR republics?

Whereas victory for the Russian president's partisans in the parliamentary elections is a virtual certainty, a win for Yeltsin himself in the 1994 presidential election is still questionable. The inhabitants of Russia's outlying areas are alien to Gaydar's liberal phraseology and have started to lean toward a revival of the Union. They would welcome a "change of direction" in the Kremlin administration and would suddenly take back their trust in Yeltsin. At the present time, therefore, when Russia has expanded its influence in the Caucasus and Central Asia, we must anticipate that the Kremlin will step up its attacks on Ukraine during the interval of time between 12 December and 12 June. And taking into consideration the fact that the Moscow authorities have been using tough and even fiery language with their internal Russia opponents, we cannot reject the idea that they will now begin to use the language of tanks with Ukraine.

No matter how paradoxical it may seem, the United States is an ally of Russian imperialism. Clinton's statements and the recent visit by Secretary of State Christopher to the CIS countries have demonstrated very clearly that for official Washington the problems of democracy and human rights (among them—the right of a nation to create its own state) are subordinate to interests of state power. These interests demand that the bet be placed—first and foremost—on the force which is dominant in this region. Such a force within the CIS is, of course, Yeltsin's Russia. Therefore, the priority of Russia in American policy is understandable. It is also important to understand something else: why in the trade between the economically strong United States and the economically weak Russia it is the weaker country which is winning. In addition to granting a great deal of financial assistance, American government officials are also engaging in some shadowy dealings to Moscow's advantage. For example, Christopher's stopover in Riga and unequivocal warning to the governments of the Baltic countries; rejecting the Eastern European countries' bid to join NATO; the West's winking at Russia's military intervention in the Caucasus; and—finally—pressure on Ukraine. One gets the impression that the West is permitting Moscow to have its traditional sphere of influence—Eastern Europe and Eurasia, and—in addition to this—is untying Moscow's hands in this part of the world. A consequence of such a policy could be the revival of a state or quasi-state creation similar in form to the Russian Empire or the USSR, which—despite its dependence on Western aid—would be neither pro-Western nor pro-American. After gathering its strength, Russia would emerge from under the West's control. Whether Washington has an alternative to this course of events is not clear.

However, there are already tendencies which could be detrimental to the above-mentioned development of events. One of the persons expressing such tendencies is U.S. Secretary of Defense Les Aspin, who has called for a slowdown in granting Russia membership in NATO. Perhaps he is genuinely amazed at Ukraine's nuclear weapons, or it is possible that his eyes were opened by the many speeches made by the former Ukrainian defense minister, Morozov, during the latter's visit to America. A further obstacle to Moscow's expanding influence is Tajikistan, where the Russian Army could be bogged down, just as it formerly was in Afghanistan. Another interesting variant is likewise possible—a split in the ranks by some of the independent players in Western Europe, primarily Germany, whose interests are far from identical with America's interests.

One way or another, Russia's expansion is already proceeding in all directions, and none of the above-named variants is capable of stopping it without Ukraine's participation. The latter is the only one of the Eastern European countries which has the potential capacity to oppose Russia. Moreover, there is no longer any choice for Ukraine: If Russia continues to follow its presently clear channel, sooner or later Ukraine will lose its sovereignty.

Nowadays the possibility of foreign-policy maneuvering has been narrowed down for Ukraine in comparison with 1992. At that time there was a possibility of creating a bloc of certain former USSR countries around Kiev; and—as a result of this—Russia would have lost the mechanism for influencing its neighbors through the CIS. Today, however, Georgia and Azerbaijan have been driven to their knees—and they are no longer Ukraine's allies. The non-intervention by official Kiev and the intervention by the UNSO [Ukrainian People's Self-Defense Forces] in the Dniester regional conflict have led to a strain in the relations between Ukraine and Moldova, along with the appearance of the Dniester Republic as a Russian staging area for penetrating Ukraine. Our government's absolute passivity with regard to establishing contacts with Eastern Europe (the Baltic states, Poland, and Hungary) has deprived Ukraine of further natural allies; the strengthening of these countries would not have been to Russia's liking at all. If we add to all this the demoralizing influence of the Masandrivka Protocols, the following conclusion may be drawn: The Ukrainian leadership is not prepared to protect and defend our national interests.

It cannot be said, however, that nothing is being done in Kiev to counteract the pressure from the East and West. These days Ukraine's nuclear arms constitute a kind of card game which is played by the president and the Supreme Council every time we have visitors from across the ocean: Kravchuk assures them of his readiness to get rid of the nuclear warheads, whereas Plyushch performs the thankless task of putting the brakes on the president's intentions. During Christopher's visit Speaker Plyushch, by strongly adopting the ideological stance abandoned by Konstantin Morozov, assumed the function of the principal opponent of the American emissary.

It is understandable that Ukraine intends to do everything possible to retain its nuclear weapons—practically its sole guarantee of the existence of an independent state and its stability. But the method of retaining them—playing a game around the ratification of SALT-1, unsupported by political or economic steps—is a strategical loser. Only future elections in Ukraine will be able to change the state of affairs. To vote for those forces which are ready to protect the national interests and capable of proposing a concept of an independent foreign policy, and, at the same time, advocates of democracy and economic reforms: No other choice remains for the citizens of Ukraine.

## 5.19

**Further Kravchuk, Plyushch Comments [on Russian Military Doctrine]**

Interfax, 5 November 1993

[FBIS Translation], Excerpts

"There has never been a precedent set in the world whereby a state would defend people of any nationality if they are citizens of another state," Ukrainian President Leonid Kravchuk said while commenting on the statement, at the request of journalists, of the Russian Defense Minister Pavel Grachev on the need to defend the Russian-speaking population in other countries. The Ukrainian president pointed out, referring to Grachev's words, that the recently adopted Russian military doctrine contains this statement.

Kravchuk also said that, as he believes, "nationality has a subordinate meaning with respect to citizenship." If one adheres to this principle, he added, "there is no need to defend anybody." "Ukraine will defend Russians living in Ukraine," he emphasized.

At the same time the Ukrainian president recognized that he had not seen the official text of the Russian military doctrine. "I don't think that Russians have overstepped all norms of civilized behavior and humanism by writing in their doctrine that they have the right to use nuclear weapons first," he declared.

Chairman of Ukraine's Supreme Soviet Ivan Plyushch who was present at this conversation indicated that "one can understand when somebody asks for security but not when somebody proposes it." In his words, "there is no need to defend the Russian-speaking population in Ukraine." "I do not know of any case where a Russian addressed us with a request to defend him," he underlined. . . .

## 5.20
## Kravchuk on Intent to Possess High-Precision Arms

Ustin Gritsenko
*Kommersant-Daily,* 10 June 1994 [FBIS Translation]

It has become possible in the course of the ongoing talks on "the division of the Black Sea Fleet" to produce a draft agreement under which Russian ships would after all be deployed in Sevastopol. Kiev, according to Russian Ambassador to Ukraine Leonid Smolyakov, agrees to this in principle. But Leonid Kravchuk, forced to make concessions that do not add luster to his image as a "patriot," put forward a new slogan in the course of his Zhitomir tour. In his words, "Instead of nuclear missiles Ukraine intends to have high-precision arms."

The latest round of Black Sea Fleet talks discussed the division of not only the bases but also the shore-based infrastructure. By the start of the discussions, which moved from Sevastopol to Kiev, the experts, according to Admiral Feliks Gromov, commander-in-chief of the Russian Navy, had "basically determined" the parameters in accordance with which the division of ships will take place. Members of the delegations were briefed on the results of this work. Some, of course, had more ideas "up their sleeve," hearing which Admiral Gromov thinks, "Different points of view on separate questions cannot be ruled out." Agreement on Sevastopol, though, was on the whole sewn up. The programs for further talks include the division of the remaining naval bases along the Crimean coast.

Considering that the talks on the Black Sea Fleet would hardly be over by 26 June (election date), Kravchuk avoided serious statements about dividing the fleet. But on the other hand, he told the officers and cadets of the Zhitomir Air Defense Radio Electronics Higher Military School about new prospects of boosting the republic's military might. He pointed out that in Ukraine (as also "in Russia, and in the United States") "scientific exploration" is under way to develop some sort of "high-precision weapon with assigned parameters." Leonid Makarovich [Kravchuk] did not go into details, hinting vaguely that this weapon "is capable of engaging installations that carry danger and of protecting from weapons of mass destruction, nuclear weapons included." It is this secret weapon that Ukraine intends to add to its arsenal "instead of the nuclear missiles that are being destroyed." Interestingly, Kravchuk divulged this "military secret," attesting to the president's tireless care about the army and the military-industrial complex, precisely in the homestretch of the election race and precisely before the grateful audience of a military school. In the course of the Zhitomir tour, however, he also spoke to the workers of a woodworking combine (prospects of the sector's growth), and to those who work in agriculture (also prospects, but those concerning agriculture). Should Kravchuk continue to concentrate on similar problems (avoiding the less-than-advantageous subject of the Black Sea Fleet), there may appear a chance for Russia and Ukraine finally to agree about the fleet.

## 5.21
## Kuchma Gives Inauguration Speech

Kiev Radio Ukraine World Service
19 July 1994 [FBIS Translation], Excerpts

[Inauguration speech by President Leonid Kuchma at the Supreme Council in Kiev—live relay]

Dear compatriots, esteemed foreign guests. Today's event is proof that the Ukrainian state lives, is getting on its feet, and is consolidating itself in the world. I would first like to thank you, the electorate, for placing such a high trust in me and electing me as the president of Ukraine. I am deeply grateful to the people of Ukraine for demonstrating such a high civic consciousness, for not being indifferent to the fate of the motherland, and for carrying out a civilized and democratic restoration of political power at the center and in the localities.

I would like to express thanks to the Supreme Council of Ukraine which, through legislation, guaranteed the process of political transformation on the principles of democracy and steadfastly abided by human rights, peace, and progress. I give due regard to Leonid Makarovych Kravchuk, who took the first difficult steps on the path of establishing Ukrainian statehood. During the elections the people of Ukraine expressed their desire to live in their own state. However, they would like Ukraine to be prosperous, democratic, and powerful.

Ukrainian statehood cannot be an end in itself. A state is for the people and not a people for the state. It is precisely this faithful thesis which should become our conviction and should be filled with real content. A state that is incapable of defending its citizens from spiritual and material impoverishment is worth nothing. The Ukrainian state is not an icon to which one should pray. It is an exceptionally important institution which should work effectively in the interests of people and serve them. I am convinced that the entire state building should be subordinate to the task of economic and spiritual revival. As president of Ukraine, I see my main tasks as providing the citizens of Ukraine with safety, social protection, stability, and creating the conditions for cultural development and a dignified life on one's native land. First, though, it is necessary to renew the legal system. The torrent of irresponsibility and crime that has taken over Ukraine should be stopped. We should not allow crime, while fusing with part of the administrative apparatus, to seize power in Ukraine one step at a time. Measures are necessary, and I will take particularly tough ones in the near future, which the legislative base correspondingly allows. There will also be an end to the plundering of Ukraine.

Ukraine stands on the threshold of essential changes in its economic and political course. The need for such changes was confirmed by the people's active participation in the elections. People have placed the responsibility for conducting reforms on the single entire system of executive power headed by the president. I would like to assure you that I intend to persist with this aspiration and to use fully the powers given to me to conduct such reforms. A single realistic path has been created to overcome the socio-economic crisis in which the Ukrainian economy finds itself.

The time for reform came long ago. Its directions were determined. The main ingredient, the political will to implement it, was lacking. It will be necessary to demonstrate this will now. Marking time means death for the Ukrainian economy and the Ukrainian state. Only immediate and resolute actions by the authorities, implementation of budgetary and monetary reform, liberalization of taxation policy, bringing order to the currency regulation, and foreign economic activity can correct the situation. Reform should ease the people's lives, so it should be well thought out.

Ukraine can only exist as a social state, in which the strong are given the possibility of fulfilling themselves, and the weak are given social protection. If today, during the difficult transition period, the state does not give people more substantial support, then science, health care, culture and the life of the nation itself will end up under threat of final destruction. Under these conditions, ensuring the stability of the state should be a fundamental issue for Ukraine's political leadership. The policy of the president and government should be comprehensible to people and it should be consistent. People should again have the opportunity to look to the future with hope.

As president of Ukraine I will do everything possible to consolidate the nation and to overcome everything that violated the Ukrainian state's internal stability. This is the situation in Ukraine today. Therefore, one has to act quickly but carefully. The state does not have time for experiments, and the people do not have the strength or patience for them. I have always said, and repeat now: Our aim is not revolution, but systematic, resolute, and consistent renewal of the economic and socio-political system in Ukraine. The main means to achieve this are realism and common sense in both domestic and foreign policy.

When determining domestic political strategy today, what should first be understood is that Ukraine is a multiethnic state. Any attempts to disregard this fact threatens a deep split in our society and the failure of the idea of Ukrainian statehood. Ukraine is mother to all of its citizens, regardless of nationality or religion, regardless of what they consider to be their native language. In the near future I intend to propose a change to the current legislation with the aim of granting official status to the Russian language, while the Ukrainian language retains its state status. [Applause]

We should repay our debts to Ukrainian culture, while simultaneously creating the best conditions for the free development of the national cultures of all people who live on the territory of Ukraine. We should consolidate, not break up, society at this critical moment in Ukrainian history. To work honestly for the good of Ukraine, for the glory of Ukraine: This is the main demand, which should be put to all of the citizens of our state.

Without a doubt, the conditions, without which reforms or any movement forward are impossible, are the formation of a strong and effective state power. This envisages the strengthening of a single executive vertical structure as the fundamental instrument for implementing statewide policy. At the same time, relations between all branches of power should be stabilized. The president, Supreme Council, and judicial authority should have one aim so that nothing hinders our common activity on behalf of the people.

The absolute priority of the constitution should be acknowledged in relations between bodies of power. Its clauses cannot be changed unilaterally. I repeat, its clauses cannot be changed unilaterally without mutual agreement. Taking into account the main political realities and the dynamics of socio-economic and political processes in our society, I, as president, consider it necessary to resume the constitutional process, having first elaborated in detail the mechanism for implementing it.

The other objective reality, in relation to which Ukraine's domestic policy will be determined, is the need for further development of local self-government and the extension of regions' rights and responsibilities, above all in the socio-economic sphere. Market reforms are incompatible with tough administration, especially when it comes from one center. The competence and responsibility of all bodies of state power and the strengthening of various levels of administration should be clearly consolidated legally.

Ukraine will keep the continuity of its foreign policy in that part relating to its international obligations. It will steadfastly fulfill agreements and demand the same attitude from its partners. At the same time adjustments will be made to foreign policy so as to make Ukraine's international policy more dynamic and effective. Its efficiency should be determined not by the number of visits and agreements, but by concrete political and economic results.

The priorities of Ukraine's foreign policy should not be determined by ideology, but by the nationwide interests of our state. In concrete terms this means that realism, specificity, and pragmatism should replace political romanticism and euphoria and a certain vagueness, characteristics of the initial period of state-building.

Historically Ukraine is part of the Euro-Asian economic and cultural space. Ukraine's vitally important national interests are concentrated on this territory of the former Soviet Union now. These are also sources of necessary goods, raw materials, and energy-carriers. It is the most realistically accessible market for the products of Ukrainian producers. We are also linked with those countries, former republics of the Soviet Union, by traditional scientific, cultural, informational, and even family ties. Millions of Ukrainians permanently reside there. Ukraine's self-isolation and its voluntary refusal to campaign vigorously for its own interests in the Euro-Asian space was a serious political mistake, which caused great damage, above all to the national economy.

We should not simply be present in the Commonwealth of Independent States but should learn actively to influence policies within the Commonwealth and resolutely defend our own interests, while clearly not forgetting about our partners. If we will not take part in the established rules of the game then those rules will nevertheless be set but without us and to the detriment of our interests. I am convinced that Ukraine can assume the role of one of the leaders of Euro-Asian economic integration and establish civilized, mutually favorable relations between interested parties. In this context the normalization of relations with Russia, our strategic partner, is of principal significance. The signing of a comprehensive and broad treaty on economic cooperation with the Russian Federation, whose preparation is practically complete, could become the first step in this direction.

This treaty should also become a good basis on which to resolve those political and economic absurdities that have accumulated in mutual relations between Ukraine and Russia. Relations with Western countries should be filled with a new realistic content. We should be sincere and ready for fruitful cooperation with all countries and with every one of them individually. It is necessary to move from exchanging declarations to full economic cooperation as soon as possible and to abolish decisively the obstacles that have been devised. Particular attention will be paid to cooperation with those countries and international economic organizations that display true business interests in Ukraine as a developed industrial state. Our aim is to integrate with the international economic system, not as a deindustrialized raw material appendage but as a partner with full rights.

Dear compatriots, I do not have answers to every question because that is impossible. It was not in vain that Winston Churchill said that people who have all of the answers are not to be trusted with the leadership of the state. I have never promised and do not promise now that all our misfortunes will come to an end tomorrow. The economic and social situation has greatly deteriorated during the last six months. The country is almost on the verge of an economic catastrophe. In order to prevent it, it will be necessary to adopt difficult and unpopular measures. On no account must we succumb to the temptation of implementing pseudo-stabilization by means of cosmetic measures, close our eyes to the existing crisis, and deceive the people. At any rate, we are obliged to stop the slide into the abyss and the march to nowhere. It is only when we stop this fateful process of degradation, the dying out of our economy and society, that we will begin to stand on our feet and only then will we have a future.

I would like all citizens to know that there will be difficult trials and that not all of our misfortunes and our shortages have come to an end. But I assure you that I will do everything to change the situation. I am deeply convinced that we will attain this. The personal commitment and responsibility of the president of Ukraine will guarantee this. We will also survive the difficult autumn and winter. Ukraine will undergo the last and the most difficult period of trials, and we will have a country worthy of our industrious people, of our glorious past, and of our children and grandchildren. Only joint work and respect for laboring people will give abundant fruit. . . .

## 5.22

**Dubinin-led Delegation Arrives in Kiev for Talks**
Interfax, 11 August 1994
[FBIS Translation]

A Russian government delegation led by special envoy Yuriy Dubinin arrived in Kiev on 11 August. The delegation will hold talks on the drafting of a treaty of friendship and cooperation between Ukraine and Russia.

Dubinin said he had arrived in Kiev in a good mood and hoping for constructive work. "We need to put on paper what has developed over many decades," he said.

According to Interfax-Ukraine, the two sides intend to discuss concrete provisions for the treaty between the two countries, including the provision on dual citizenship.

Ukrainian legislation does not provide for granting dual citizenship to the republic's inhabitants. But some Russian politicians insist on including the provision in the treaty.

It is expected that Dubinin will make a statement to journalists on 12 August on the results of the talks.

## 5.23

**"No Change" in Stance on CIS Security System**
Interfax, 17 August 1994 [FBIS Translation]

"There have been no changes in Ukraine's position on the collective security system of the Commonwealth of Independent States (CIS), nor are any changes expected in it," the Ukrainian president's official spokesman, Mikhail Doroshenko, said, commenting on media reports that the Ukrainian Armed Forces Chief of Staff Anatoliy Lopata had suggested stronger military ties with other CIS countries.

At a news briefing in Kiev on Wednesday, Doroshenko cited a statement made by Ukrainian President Leonid Kuchma at a recent news conference he held jointly with his Kazakh counterpart, Nursultan Nazarbaev, to the effect that Ukraine would closely consider Kazakhstan's proposals for more intensive integration within the Commonwealth. Kuchma, however, is not in a hurry to make up his mind about these proposals, the spokesman said.

Kuchma, at the same news conference, suggested giving priority to CIS economic integration.

## 5.24

**Parliament Ratifies NPT with "Reservations"**
Viktor Demidenko and Mikhail Melnik
ITAR-TASS, 16 November 1994 [FBIS Translation]

The Ukrainian legislators ratified the Nuclear Non-Proliferation Treaty [NPT] by a majority vote on Wednesday. Ukraine joins the treaty with reservations, which were adopted in a law. The treaty "does not cover in full the unique situation which emerged after the USSR collapsed," Ukraine's amendment states. The lawmakers also issued a supplementary statement to the law.

Although Ukraine owns nuclear weapons it acquired from the USSR, these weapons will be dismantled and disposed of under state control to prevent reuse of nuclear materials. "Ukraine will use nuclear materials exclusively for peaceful purposes," the statement said.

However, the presence of nuclear weapons on Ukrainian territory as well as their proper maintenance until their complete elimination does not run counter to Articles 1 and 2 of the treaty.

Ukraine will deem any attempt at violating her borders or territorial integrity by any nuclear power "as emergency circumstances which jeopardize her top priority interests."

The law on Ukraine's joining the Nuclear Non-Proliferation Treaty will come into effect upon all the party states' signing an international legal document to guarantee her security.

## 5.25

**Text of Resolution [on Nuclear Non-Proliferation Treaty]**
Kiev Radio Ukraine World Service
16 November 1994 [FBIS Translation]

[Text Resolution by Supreme Council, detailing reservations to Nuclear Non-Proliferation Treaty, adopted in principle, at Supreme Council session in Kiev—live]

The law of Ukraine on Ukraine's accession to the Nuclear Non-Proliferation Treaty dated 1 July 1968.

On the basis of the provisions in the Declaration on State Sovereignty of Ukraine of 16 July 1990:

statements by the Supreme Council of Ukraine on additional measures with regard to guaranteeing Ukraine's acquisition of non-nuclear status of 9 April 1992;

resolutions of the Supreme Council of Ukraine on ratifying the agreement between the Union of Socialist Soviet Republics and the United States on the reduction and limitation of strategic offensive weapons, signed on 31 July 1991 in Moscow, and the protocols attached to it, signed on behalf of Ukraine on 23 May 1992 in Lisbon, of 18 November 1993;

and resolutions of the Supreme Council of Ukraine on the implementation by the president of Ukraine and the government of Ukraine of the recommendations contained in Point 11 of the resolution of the Ukrainian Supreme Council on the ratification of the agreement between the USSR and the United States on the reduction and limitation of strategic offensive weapons, signed on 31 July 1991 in Moscow, and the protocols attached to it, signed on behalf of Ukraine on 23 May 1992 in Lisbon, of 3 February 1994; the Supreme Council resolves:

To accede to the Nuclear Non-Proliferation Treaty of 1 July 1968, with the following reservations:

1. The provisions of the treaty do not completely embrace the unique situation which has arisen as a result of the collapse of the nuclear state of the USSR;

2. Ukraine is the owner of the nuclear weapons inherited by it from the former USSR. After the dismantling and destruction of the given weapons under its control, and according to the procedures which will exclude the possibility of reusing nuclear materials which are components of these weapons, according to their primary purpose, Ukraine intends to use the aforementioned materials exclusively for peaceful purposes;

3. Until the nuclear weapons are completely liquidated, and also until corresponding work on maintaining, servicing, and liquidating the nuclear weapons is carried out, the presence of nuclear weapons on Ukrainian territory does not contravene Articles 1 and 2 of the treaty;

4. The threat or use of force against the territorial integrity and inviolability of the borders or the political independence of Ukraine on the part of any nuclear state, and equally, the use of economic pressure aimed at subjugating to its own interests the exercising by Ukraine of rights inherent in its sovereignty, will be regarded by Ukraine as exceptional circumstances that have threatened its highest interests;

5. The documents on Ukraine's accession to the treaty will be passed to the depositary countries after it has come into effect by means of this law;

6. This law takes effect once Ukraine receives security guarantees from the nuclear states formulated by means of signing a corresponding international legal document with Ukraine.

## Kuchma Addresses Parliament on Accession

Kiev Radio Ukraine World Service
16 November 1994 [FBIS Translation], Excerpts

[Address by Ukrainian president Leonid Kuchma on Ukraine's accession to the Nuclear Non-Proliferation Treaty (NPT) to a Supreme Council session in Kiev]

In the USSR on the wave of perestroika we were called an African country, with the one difference that we had nuclear weapons. I do not know what they will call us because we have missiles with nuclear warheads that are beyond their expiration date. For this reason, I would like to turn to you, esteemed deputies, with a request to give me a little more time for my speech.

[Supreme Council Speaker Moroz]: Does anyone disagree with this?

[Deputies]: No.

[Kuchma]: Esteemed chairman, deputies. In addressing this problem I would like to remind you all that world experience has proven that people will forgive politicians for everything apart from untruths. The truth is that Ukraine today does not have the choice of being nuclear or non-nuclear. The choice has been made by the activity of the previous and present parliament, and by virtue of Ukraine's international obligations, and finally by the actual situation. The situation is as follows: The course of the world process of nuclear disarmament depends on our decision today. If we have decided to trade with the world community then we have chosen the most unfavorable moment and the most unsuccessful item of trade. We will surprise the civilized world, since the decision on the accession to the treaty is the concluding stage that logically is conditioned by all the previous political acts approved by the Supreme Council of Ukraine in 1994. This is the declaration on the state sovereignty of Ukraine that was mentioned today, the statements of the Supreme Council on the non-nuclear status, the resolution of the Supreme Council on the ratification of START I, and the Lisbon protocol that envisages Ukraine's accession to the NPT and the trilateral statement of 14 January 1994 that was also approved by the Supreme Council, in accordance with which the withdrawal of warheads to Russia will actually be carried out in two and a half years. This prevents them being used by Ukraine. Any further delay in resolving this issue would, in my view, cause the whole world community to stop dealing with us as a country that does not know how to honor its obligations.

So what are the gains that those politicians of ours who are effectively obstructing the process of accession or, what is more, coming out for Ukraine's nuclear status, are hoping for? Are they national security or the political and economic advantage and benefit?

I will remind those who, in a flight of artificial patriotism, have forgotten that Ukraine not only has no respective production, cannot use nuclear weapons in defense terms, and cannot even make use of the nuclear warheads inherited by it for power engineering purposes, because the dismantling of warheads, in accordance with nuclear safety requirements, is possible exclusively at the enterprises that manufactured them, i.e., in Russia, which was also recorded in the trilateral statement by the three states. Experts estimate that the creation and improvement alone of a safekeeping system for the nuclear weapons, which are present on our territory, can cost between $10–30 billion. That is as far as the economic benefit is concerned, as it were.

As for production of our own, to complete the nuclear warheads production cycle requires the investment of at least $160–200 billion over 10 years, which was also something that was recalled here today and was set down in the notice handed to you. So have we any choice, as we are trying to make out to the people? Who from the advocates of nuclear games will stand up now and say who needs to sell and mortgage all of Ukraine's belongings in order instead to bring it the happiness of a nuclear arsenal of its own? . . .

## 5.26

**Kuchma Writes to Shevardnadze on Cooperation, CIS**
Sakinform
29 December 1994 [FBIS Translation]

Eduard Shevardnadze has received a letter from Ukrainian President Leonid Kuchma. In particular, the letter says that seeking efficient ways to overcome the economic crisis, stabilize home policy, and increase the standards of living of the population is a major task for Ukraine, Georgia, and other CIS countries. The letter says that in its striving to implement economic reforms, Ukraine considers developing the comprehensive cooperation with Georgia and other CIS countries as a matter of particular importance. "With regard to this, I support your idea within the United Nations, other international organizations, and the CIS," Leonid Kuchma writes. He emphasizes that he regards regular Ukrainian-Georgian consultations at the highest level as useful and therefore gratefully accepts Eduard Shevardnadze's invitation to visit Georgia.

## 5.27

**Presidential Assistant States Political Priorities**
Rayisa Stetsyura
*Uryadovyy Kuryer,* 7 January 1995
[FBIS Translation]

"Ukraine's external political course will be oriented toward cooperation with CIS countries, first and foremost with Russia." This was stated by the Ukrainian president's first assistant at a briefing on 4 January at the Ukrainian Presidential Administration building. He believes that, this year, "the Russian factor in mutual relations will be a determining one for our country." Oleksandr Razumkov is convinced that Ukraine will sign a large-scale agreement on friendship and cooperation with Russia in the near future and that all problems regarding the Black Sea Fleet will be resolved.

In his opinion, Ukraine must exert maximum effort to raise Ukrainian-Russian relations to a qualitatively new level. This not only concerns the political aspect. Oleksandr Razumkov stressed that the idea of creating transnational corporations, with the participation of CIS countries, is very promising. It will also make it possible to avoid many political problems.

The first assistant to the Ukrainian president believes that such problems must also be urgently resolved within the state. A reorganization of state power must become the first step in this direction, as it will have a beneficial effect on economic processes. In Oleksandr Razumkov's opinion, they will be successful, provided that the vertical structure of executive power is clearly defined and a legal venue is created to rule out contradictions between the legislative and executive powers. This must be taken care of by the constitutional bill "On State Power and Local Self-Government in Ukraine."

## 5.28

**Kuchma Warns 1995 Will Be Toughest Yet for Economy**
Interfax, 12 January 1995 [FBIS Translation]

*Editor's Note:* During late 1993 and 1994 confusion and delay characterized the economic reform program in Ukraine. Most of the communist-dominated parliament objected to private property and distrusted far-reaching economic reform, preferring to continue state subsidy policies that could bankrupt the economy. When Kuchma entered office in June 1994 he began to tackle the economy and to enforce sound fiscal and financial practices. The short press account below characterizes briefly the Ukrainian economic crisis.

"The year of 1995 will be the most difficult for the Ukrainian economy," Ukrainian President Leonid Kuchma told the editors-in-chief of the regional media in Zhitomir today.

Kuchma said that "either the situation will be changed for the better or we will face a complete crash." "We have no more reserves," he added.

Commenting on the draft state budget for 1995, Kuchma said it would be "the toughest one compared to all past years." He said "allocatios will be directed only to vital needs, including social security, law enforcement, health care, the army, and some others." Kuchma said he did not intend to boost money supply and resolve economic problems by raising inflation.

Kuchma said the Ukrainian leaders planned to cancel "numerous tax privileges." Furthermore, he said that in 1994 the total tax waivers amounted to 268 trillion karbovantsy, decreasing budget revenue by 38 percent.

## 5.29

**Kuchma Adviser: Russia "Apparently" Halts Pact Talks**
INFOBANK (Lvov), 13 January 1995
[FBIS Translation]

*Editor's Note:* By January 1995, it had become obvious to observers of the CIS countries that both the "war party," which had prevailed in Russia's policy toward Chechnya, and the national patriots, who essentially reject the

Belovezh process in its entirety, were prevailing over Russian policy toward the CIS. These groups, which form the core of the Russian great power advocates, stalled the final signing of a bilateral agreement between Russia and Ukraine designed to improve relations between the two states by resolving the outstanding debt, territorial, and naval forces issues which continue to be a source of intense subterannean pressure. Were the agreement to be signed, it would go far to legitimize the Ukrainian state, despite its troubles, but the Russian patriots of Russia's supreme legislature—the State Duma—had not yet accepted this change in the post-Soviet geopolitical order. It could, judging from the insights and the attitude expressed in the following article, take a very long time for Russia to accept the realities of Ukrainian sovereignty.

At a briefing on 11 January, Dmytro Tabachnyk, chief of the presidential administration, stated that as of today, 90 percent of the text of a wide-ranging agreement between Ukraine and Russia has been accepted by the two sides. Nonetheless, the agreement is not likely to be signed in the very near future. Mr. Tabachnyk further stated that Russia has apparently decided to temporarily halt the negotiation process.

No consultations regarding this agreement have been held between Kiev and Moscow since the beginning of the year. Oleg Soskovets, a deputy prime minister in the Russian government, postponed his trip to Kiev, scheduled for late December, citing as a reason the complex situation in Chechnya. Mr. Soskovets was due to discuss the final wording of the text with Yevhen Marchuk, the Ukrainian deputy prime minister. Afterwards, the Russian Foreign Ministry issued a statement, in which it denounced the presence of Ukrainian mercenaries in Chechnya. Although it is well known that mercenaries from various countries are actively fighting in the Chechen war zone, the Russian Foreign Ministry decided to single out Ukrainian mercenaries. Perhaps Russia is, indeed, looking for a pretext to cancel any further talks on the subject of the nearly completed agreement as beneficial to Ukraine, since it will become yet another guarantee of Ukraine's security interests.

Once Yeltsin and Kuchma sign this document, the negotiations regarding the Crimea and the Black Sea Fleet immediately acquire a different character. In accordance with this agreement, Russia will only be able to lease the naval ports and facilities on the Crimean coast. Moreover, any further discussion on issuing Russian passports to the residents of the Crimean peninsula will be simply meaningless. At a time when Yeltsin is facing growing opposition at home to Russia's military invasion in Chechnya, the chances of concluding a Russian-Ukrainian agreement seem even more remote. Besides, Ukraine has yet to conclude an agreement on friendship and cooperation with Romania, whose territorial claims to Ukrainian soil are somewhat rhetorical in nature and, in any event, by far less intimidating than Russia's.

## 5.30
## Declaration on Prospects for Cooperation Between Ukraine, Georgia

*Uryadovyy Kurier*, 19 January 1995
[FBIS Translation]

*Editor's Note:* As the war in Chechnya heated up and tensions between Russia and Ukraine smoldered, especially those connected with Ukraine's attitude toward genuine CIS integration along the lines espoused by President Yeltsin and other Russian politicians, Ukraine continued to seek out alliances with former Soviet republics and Western states. Such alliances, it is hoped, will buttress Ukraine's claims to independence and help Ukrainian leaders find support for their position in CIS debates.

[Text of declaration signed by Leonid Kuchma of Ukraine and Eduard Shevardnadze of Georgia: "Declaration on Prospects for Cooperation Between Ukraine and the Republic of Georgia and for a Joint Approach to Foreign Relations Issues, Tbilisi, 9 January 1995."]

Ukraine and the Republic of Georgia, striving to ensure the positive development of bilateral relations and to define the priorities of future cooperation, in order to endow this cooperation with the necessary dynamism on the basis of the Agreement on Friendship, Cooperation, and Mutual Assistance Between Ukraine and Republic of Georgia,

confirming the historical and spiritual closeness of the peoples of Ukraine and the Republic of Georgia,

recognizing that the existence of the democratic, sovereign, and territorially integral states of Ukraine and the Republic of Georgia is of fundamental importance to international security and to the strengthening of interstate cooperation and [effecting] historical changes in the system of international relations,

- recognizing in particular the importance of wide-ranging friendly relations between Ukraine and Georgia for strengthening international security in the Black Sea region and in the Caucasus, as well as in the Eurasian corridor,
- based on the shared conviction and resolve of Ukraine and the Republic of Georgia to extend democratic processes and steadfastly proceed along the path of

economic market transformations and civil accord in society,

- upholding the principles of equal and mutually beneficial interstate economic cooperation within the framework of the CIS and to the same degree with European and other international economic organizations and groupings,
- reaffirming their dedication to the universally recognized objectives and principles of the UN Charter, the Helsinki Final Act, and other fundamental documents of the CSCE,
- have signed this Declaration in order to build up friendly bilateral relations between them and to coordinate joint approaches to important issues in international relations, such as:

1. Ukraine and the Republic of Georgia welcome the historic prospects that the restoration of state independence and sovereignty has afforded each of them with respect to their national rebirth and democratic development. They each steadfastly support the efforts of the other to create a society founded on the supremacy of law and full respect for the rights and fundamental freedoms of the individual.
2. Ukraine and the Republic of Georgia attach great importance to the Agreement on Friendship, Cooperation, and Mutual Assistance that they concluded, the value of which has been demonstrated since the time that it was signed and ratified by the parliaments of both states.
3. Ukraine and the Republic of Georgia will continue to expand the legal basis of bilateral cooperation for the purpose of comprehensively promoting the development of traditional economic, cultural, scientific-technical, and other relations.
4. Ukraine and the Republic of Georgia believe that the principles of territorial integrity, inviolability of borders, peaceful settlement, and non-use of force or threats of force—together and in integral conjunction with other universally recognized principles of international law—are fundamental to regional and European peace and security.
5. In connection with this, they stress the extreme danger of any manifestations of separatism that resort to force within the boundaries of internationally recognized territorially integral states, taking into consideration the fact that such manifestations pose a real threat to stability and peace not only in the states in question but also in adjoining regions.
6. In accordance with the UN Charter, Ukraine and the Republic of Georgia recognize the right of any state to apply the necessary constitutional measures to safeguard its own political, economic, and territorial integrity.
7. At the same time, Ukraine and the Republic of Georgia believe that the effective settlement of conflicts can be guaranteed only through persistent and steadfast attempts by all parties to find peaceful negotiatory means to regularize such situations.
8. Ukraine and the Republic of Georgia hold that the settlement of the Georgian-Abkhazian and Georgian-Ossetian conflicts on the territory of Georgia are the internal affair of the Republic of Georgia. The decisions of the Budapest summit with respect to the regularization of the situation in the zone of the Georgian-Abkhazian conflict and the condemnation of ethnic cleansing were reaffirmed.
9. Recognizing the conflict in the Chechen Republic to be the internal affair of the Russian Federation, Ukraine and the Republic of Georgia nonetheless express their concern over the reports of numerous victims, especially among the civilian population. They regret the fact that this conflict is growing in scale and resulting in mass bloodshed and call upon the parties to exhibit restraint and make an urgent effort to find peaceful means of resolving the conflict and to observe human rights.
10. Ukraine and the Republic of Georgia call upon the UN, the CSCE, the European Union, other international organizations, the world's leading powers, as well as the CIS countries to devote more attention to the problems in the Caucasus and make additional political and economic efforts to attain an all-encompassing peace in this region. They express the hope that the world community will substantially expand the process of developing and implementing specific programs and measures designed to create the necessary conditions for ensuring stable and enduring development in this explosively dangerous region.
11. Ukraine and the Republic of Georgia believe that the most important factors in strengthening stability, peace, and civil accord in their countries are the implementation of profound democratic transformations accompanied by the unconditional observance of the rights and fundamental freedoms of the individual, as well as the protection of the rights of ethnic and religious minorities.
12. Viewing bilateral relations as an important priority, Ukraine and the Republic of Georgia regard it as necessary to make effective joint efforts designed to perfect the methods of realizing the agreements they have reached in various fields of cooperation.

    At the same time, they stress the prospects of developing economic relations between regions in both countries.

13. Ukraine and the Republic of Georgia will expand bilateral trade and economic cooperation. In doing so, they will devote particular attention to the development of new forms of industrial cooperation—production cooperation, joint business enterprise, etc.
14. Ukraine and the Republic of Georgia believe that an important prerequisite of the development of broad economic cooperation between them is the creation of reliable transportation links. In conjunction with this, the parties believe it useful, jointly with other interested countries, to begin developing and realizing energy transit projects, as well as establishing a ferry service between the Black Sea ports of Ukraine and Georgia.
15. Ukraine and the Republic of Georgia will coordinate their efforts more closely and cooperation in international organizations and structures, including the CIS organs. In connection with this, they will begin bilateral consultations on urgent issues in international relations and hold them on a regular basis.
16. Ukraine and the Republic of Georgia agree that the principal tendencies in the member countries of the CIS confirm the priority nature of economic cooperation and the diversity of forms of participation by these states in the Commonwealth, including in CIS organs. They underscore that the CIS organs should be open to all member states and should develop as interstate rather than supranational bodies.
17. Ukraine and the Republic of Georgia assess positively the prospects of developing the Black Sea Economic Cooperation [ChES], which unites countries that vary in size and potential. They attach special significance to multilateral cooperation in investments within the framework of the ChES, the potential of which has grown with the creation of the Black Sea Bank of Trade and Development, and urge all potential donors to support the establishment of this bank.
18. Ukraine and the Republic of Georgia will actively cooperate in promoting the transborder dissemination of information about national attainments in the spheres of culture, science, and education, expand cultural and information exchanges, and support the development of international contacts among young people.

[signed] *President of Ukraine*
*Leonid KUCHMA*

[signed] *Chairman of the Parliament of Georgia*
*—Head of State*
*Eduard SHEVARDNADZE*

## 5.31

**Kuchma to Sign Trade, Economic Pact in Moscow**
UNIAR, 23 January 1995
[FBIS Translation], Excerpt

The signing of the major treaty between Ukraine and Russia has been postponed so far. Ukrainian President Leonid Kuchma will travel to Moscow on 24 January, accompanied by high government officials, to sign a trade and economic agreement for 1995. They will also discuss the provision of critical imports of fuels and the granting of components for Ukrainian enterprises that are cooperating with Russia.

Observers believe that Russia is seriously concerned by Ukraine's ever increasing activity in establishing economic ties with republics of the CIS and by its growing political influence in the Caucasus. Russia recently signed several agreements with Belarus and Kazakhstan, which in effect signifies the creation of an economic union between these three republics. Russia can afford to delay signing of the major political and economic treaty with Ukraine, which has been in a state of preparation for such a long time.

In the view of the experts, Russia aspires to include Ukraine in those decisions that would be favorable to Russia in terms of transporting and fitting out future oil and gas pipelines from Azerbaijan and Central Asia. This could be viewed as Russia's reaction to the quite successful Ukrainian-Turkmen agreements, which disturb Russia both from the economic and political points of view. However, because Ukraine can secure certain influence in Central Asia and the Caucasus, it is able to defend its positions and achieve those compromise options where its political and economic interests will be taken into account. . . .

## 5.32

**Ukraine Attends Interstate Economic Committee Session**
Kiev Radio Ukraine World Service
11 March 1995 [FBIS Translation]

A session of the presidium of the Interstate Economic Committee [IEC], which was set up by a decision of the CIS heads of state, got under way in Moscow on 10 March. A government delegation of Ukraine, led by Deputy Prime Minister Serhiy Osyka, is taking part in its work. Aleksey Bolshakov, chairman of the IEC Presidium and deputy prime minister of the government of the Russian Federation, greeted the participants in the session on behalf of the government of Russia. He wished them success in achieving the main goal of the Interstate Economic Committee—development and interaction of economies of the CIS member states.

Consideration of a draft agreement on setting up an Interstate Currency Committee [ICC] was among the priority issues at the session of the IEC Presidium. The ICC is supposed to promote multilateral cooperation in the field of currency and credit relations, development of forms and methods of coordinating the monetary, credit, and currency policies of the signatories to the agreement on setting up a payments union among CIS countries. As is known, Ukraine will join the payments union only after its national currency is put into circulation. However, at the current session the delegation of Ukraine submitted a number of fundamental proposals concerning the draft agreement on setting up an Interstate Currency Committee. In particular, these are proposals concerning the procedure of seceding from the ICC, which should be based on internationally recognized principles.

The activities of the Interstate Currency Committee will undoubtedly contribute to developing economic cooperation between the CIS countries, bringing their currency procedures closer to each other, speeding up mutual settlements, and regulating debt issues. The desire of all the participating sides to build their economic relations on the basis of mutual respect and equality should be regarded as the main factor in this process. The draft agreement on setting up an Interstate Currency Committee will be submitted for the consideration of the Council of the CIS Heads of State.

The session of the IEC presidium also addressed other important issues.

## 5.33

### Ukraine Abstains from Joining CIS Customs Union

UNIAN, 11 March 1995
[FBIS Translation], Excerpts

. . . At UNIAN's request, Ukraine's Deputy Prime Minister Serhiy Osyka commented on this fact as follows:

"Ukraine supports the idea of setting up the Customs Union in general, but the distance to it should be covered not at rocket-escape velocity, but at a normal pace." According to him, the agreement on free trade, concluded between Ukraine and the Russian Federation, will contribute to coordinating the fiscal, taxation, and economic functions of the customs borders between the two countries. "The law enforcement and political functions of the customs borders will remain unchanged," Serhiy Osyka pointed out.

In his opinion, the issue of Ukraine's accession to the Customs Union will depend on the effectiveness of the free trade procedure between Ukraine and the Russian Federation.

## 5.34

### Yeltsin Considers "Tough Approach" Toward Ukraine

Article by Mikhail Berger; Report by Yanin Sokolovskaya
*Izvestiya,* 21 April 1995 [FBIS Translation]

The Russian-Ukrainian talks on the Black Sea Fleet are once again deadlocked. As distinct from previous similar situations, however, Russia, to judge from information disseminated by ITAR-TASS, is prepared to be tough with Ukraine. Boris Yeltsin's words in this regard, as reported by presidential aide Dmitriy Ryurikov, appeared as follows: Unless Ukraine moves toward us over the question of the Black Sea Fleet, Russia will take this as a sign to reconsider its economic and financial agreements with Ukraine.

A tough approach to foreign policy in the near abroad seems to be becoming increasingly popular in the top echelons of power. Politicians and journalists had not had time to react to Andrey Kozyrev's statement on the possibility of using armed forces to defend the Russian-speaking population in the CIS, when Boris Yeltsin threatened to revise the economic agreements with our closest neighbor—Ukraine.

True, things are going no further than declarations at present. But this is the very reason why there is still time to try to assess the possible consequences of such a policy if Russia were to make the transition from words to deeds. Let us consider the possible results of taking a tough approach economically, as something more specific and predictable than the use of military force.

Which agreements, in fact, can we be talking of? It is possible to assume with a high degree of probability that President Yeltsin had in mind the recently signed agreement on restructuring (deferring) the Ukrainian debt. This agreement elicited a highly negative reaction from many parliamentarians and is regarded by its critics as another "sale of the motherland." Meanwhile, in the opinion of a high-ranking staffer in the Russian government, this agreement is no less advantageous to Russia than to Ukraine. Until recently, the problem of mounting Ukrainian debts had gotten no further than the talks stage. But now Russia has already started getting real, live money from Ukraine. The terms of the treaty are so tough, the staffer emphasized, that it can be torn up on the very first delay in payments.

In other words, a certain ultimatum-like approach is already enshrined in the agreement with Ukraine, and an extremely tough approach can be taken regardless of the development of the talks on the Black Sea Fleet. So the introduction of an additional political excuse to revise the economic agreement seems superfluous.

Of course, it would create a mass of problems for Ukraine

to declare itself bankrupt and incapable of repaying debts. The IMF, above all, might suspend the granting of reserve credits to Ukraine. Thus, the threat of "revision" is fraught with very grave economic consequences for Ukraine. But Russia will not get off lightly either. The point is that the agreement on restructuring the Ukrainian debt was part of Russia's complex, multimove game with the IMF, which resulted in reserve credits of $6.8 billion granted to Russia. It cannot be ruled out that, in order to "punish" Ukraine, Russia itself will have to accept punishment.

On the contrary, a solicitous attitude to the agreement on debts promises Russia big economic advantages. One of the leaders of the Ministry for Cooperation with CIS Member States told *Izvestiya* that Russia did not receive a single cent of the $600 million due from Ukraine last year. Without the agreement the result could be the same this year too. But Ukraine made the first payment three days early—27 March—and will pay $406 million before 1 September. The agreement on restructuring the debt of $2.7 billion, apart from securing a chance of actual repayment with interest, enables Russia to substantially increase gas exports to Europe, to resolve the problem of transit payments, and to participate in the privatization of Ukrainian enterprises.

Of course, the treaty also benefits Ukraine, and at first sight it seems possible to use it to exert pressure at the talks on the Black Sea Fleet. Tearing it up, however, could place Ukraine in such a desperate economic position that there will be no question of any "movement" toward us over the problem of the fleet.

Boris Yeltsin's words, made public by Dmitriy Ryurikov, about the Ukrainian side's supposedly unconstructive approach to the division of the Black Sea Fleet cannot be said to have alarmed Kiev politicians. They evidently had a premonition that these problems would arise. Kiev was far more troubled by the statement which the Russian president made four days earlier about his concern at the fate of the people of the Crimea. Ukrainian politicians perceived it as "disgraceful interference in Ukraine's internal affairs."

To date no statement by Ukrainian parties or authorities concerning Yeltsin's "fleet" statements has been made public. *Izvestiya* was told at the country's Ministry of Foreign Affairs that the information about Yeltsin's position came through unofficial channels. It is now being checked out and elucidated by the Ukrainian Embassy in Russia. "If this information is confirmed, Kiev's reaction will be consonant with it."

The president's staff has also decided not to react officially to the Russian presidential aide's unofficial statement. Reliable sources have told us that this question was not discussed at all during the meeting which Leonid Kuchma and Yevhen Marchuk had with heads of deputies' factions.

Ukrainian Acting Prime Minister Yevhen Marchuk himself says that he was satisfied with the results of his meeting with Yeltsin and Chernomyrdin. The contentious point of borders, included in the large-scale treaty, has at last been resolved with the Russian side. Henceforth it sounds as follows in the version proposed by Ukraine: "The sides respect each other's territorial sovereignty and confirm the inviolability of the borders that exist between them."

---

## 5.35
## Kuchma News Conference Discusses Current Issues

Kiev Radio Ukraine World Service, 29 May 1995
[FBIS Translation], Excerpts

---

[Moderator]: Today's news conference . . . deals with the results of [Kuchma's] visits to the Latvian and Estonian republics and participation in the conference of the CIS heads of state in Minsk . . . .

[Kyselyov]: Esteemed Leonid Danylovych, I am Serhiy Kyselyov, Ukrainian office of the *Literaturnaya Gazeta*. During your visit to Latvia you said Ukraine could possibly change its non-aligned status. Several days after, Ukraine's Defense Minister Valeriy Shmarov, while summing up the results of the Ukrainian-U.S. exercises at the Yavorov training ground, said that the peacekeeping exercises did not mean at all that our state will change its non-aligned status. Therefore, my first question is: How can you explain the difference between the above statements? And the second question in that vein is: In a week's time you intend to pay an official visit to NATO headquarters in Brussels—what will be discussed there? What have you prepared for that meeting? Thank you.

[Kuchma]: Thank you. First and foremost, I am going there at the invitation of the European Union. So, I would ask you not to confuse things, as they say. If there is enough time—and I will take advantage of this—I will definitely visit NATO headquarters. . . .

True, I did say that [changes thought], as a matter of fact, I reiterated something I have been saying all the time and everywhere. That is that no country has the right of veto, that NATO's doors should be open to any country, Ukraine included, and that we must really try to expand cooperation, including with NATO. The peacekeeping exercises in Lvov Region were held for this reason, so that we could understand each other better. . . .

[Levytska]: URP-Inform and *Za Vilnu Ukrayinu* newspaper, Lviv, Levytska. Esteemed Mr. President, in Minsk a number of CIS states signed agreements on the joint protection of borders and on a common customs union. Does this mean that on 26 May the CIS split into two parts: the countries that became fully dependent on Russia and the countries that will cooperate on a bilateral basis? And the second question is: You visited the Baltic countries and met with Ukrainian communities. Were they asking for any protection, like the Russian-speaking population was doing, according to repeated statements by the Russian leadership? Thanks.

[Kuchma]: I believe you are politicians and political scientists, so draw your own conclusion on whether the CIS split into two halves. However, I think the fact that Ukraine, for example, did not join the customs union or the agreement on the protection of borders is our own business, first and foremost, because this is our policy. We understand that if there are external borders and no internal borders, there will be no Ukrainian state, there will be one state only. The same refers to the customs union because this is as lengthy a task as joining NATO, as they say, because it is necessary to update . . . .

. . . I had an official, so to speak, meeting with the Ukrainian community in Latvia and they met me unofficially in Tallinn on a square in front of the municipal council. And everywhere there was a warm welcome and open conversations. True, there are problems today, including those facing our diaspora. The main problem, as you know, is that of citizenship. I believe that we must show understanding of the situation that has taken shape in the Baltic countries. If there are fewer Latvians in Latvia, for example, than people of other nationalities, this is a big problem. We must understand their measures to resolve this problem. But when taking these measures it is necessary to ensure that the non-indigenous population, or national minorities, as they say, do not feel like guests, as a matter of fact. It is necessary to move in the direction of adopting positive decisions to deal with the issue. And there were conversations both with the president of Latvia and the president of Estonia. I do not want to mention the steps which have already been taken, but they have been taken and they give me grounds to say this.

By the way, in Tallinn we were met by people who were carrying posters saying that our main task, after all, is to preserve the Ukrainian state. And I have never heard a single word on their part—although there were many people in the square—that pressure was exerted on them, that they were not allowed to work and live normally, although these problems do exist, as you understand. And the leaders of those states acknowledge this. So, the task now is to solve these problems at the government level.

[Zarya, . . . ]: Sergey Zarya, Interethnic Information Bureau. Mr. President, excuse me for speaking in Russian. In Minsk you said that the question of borders is a question of sovereignty. Then why did Mr. Horbulin agree to sign a joint Air Defense Treaty with Russia? Our military objected, but Mr. Horbulin went ahead and signed it.

[Kuchma, . . . ]: I don't know whose gossip you are repeating because Horbulin did not sign any treaties and had nothing to do with them. So ask your source about it. We did not sign anything like that in Minsk, and Horbulin was not there, and he was not in Almaty, either. . . .

## 5.36

**Moroz Says Ukraine Will Never Join NATO**
UNIAN, 29 May 1995
[FBIS Translation]

"Ukraine will not join NATO, and neither will Russia," said Ukraine's Supreme Council Chairman Oleksandr Moroz in an exclusive interview for UNIAN. "The Partnership for Peace formula makes it clear: Render unto Caesar what is Caesar's," the speaker said.

Moroz told UNIAN that he had held that view since 1993 when the Supreme Council approved the country's foreign policy guidelines. Taking account of the objective circumstances, such as the existence of common Ukrainian-Russian air defense and strategic forces, as well as the existence of the Black Sea Fleet on the Ukrainian territory, these relations must be formalized in legal terms, he said. "I am not talking about military blocs or alliances. We need a legal framework for cooperation, including military cooperation, between neighboring countries. Perhaps we need a similar framework with NATO, if you consider the interests of Ukraine and her neighbors who may join the organization in the future."

The parliament speaker said Ukraine had to seek other ways of protecting its interests, by promoting the transformation of the Black Sea region into a zone of peace and reducing confrontation between the military organizations of its neighbor states.

## 5.37

**Duma Deputy Favors Return of Crimea**
RIZ, 5 June 1995
[FBIS Translation], Excerpts

[Press conference held by Viktor Grigoryavich Vishnyakov, deputy chairman of the Committee on Legislation and Judicial-Legal Reform, chairman of the Subcommittee on Constitutional Legislation of the State Duma, Russian Federation Federal Assembly.]

## 1. On the Russian Federal Status of Crimea

The situation surrounding the Republic of Crimea is approaching a dead end.

In January 1991, an all-Crimea referendum was held, at which 93 percent of the population of Crimea spoke out in favor of "restoration of the Crimean ASSR [Autonomous Soviet Socialist Republic] as a subject of the Federation of the USSR." In February of 1991, the Supreme Soviet of Ukraine ratified the Law on Restoration of the Crimean ASSR as a Component of the UkSSR, thereby distorting the results of the referendum.

In September of 1991, the Supreme Soviet of Ukraine adopted the Law on Delineation of Powers and Authorities Between Ukraine and the Republic of Crimea, which secures the status of the Republic of Crimea as a component of the Ukraine. The Republic Movement of Crimea collected 247,000 signatures in favor of holding a referendum on the independence of Crimea (according to the Law on Referendum of the Republic of Crimea, 180,000 were needed).

In May 1992, the Supreme Soviet of the Republic of Crimea adopted the Statute on Proclaiming State Independence of the Republic, and ratified and implemented the Constitution of the Republic of Crimea. The Supreme Soviet of Ukraine adopted a resolution instructing the Supreme Soviet of the Republic of Crimea to repeal the Statute Proclaiming State Independence of the Republic and the statute on holding an all-Crimea referendum, because they contradicted the Constitution of Ukraine.

But the "war of laws" went even further. In November 1994, the Supreme Soviet of Ukraine repealed en masse over a hundred legal statutes adopted by the Supreme Soviet of the Republic of Crimea on grounds that they supposedly contradicted the Constitution and the laws of Ukraine. As a result, the struggle surrounding the legal status of Crimea became even more exacerbated.

In this struggle, the Ukrainian authorities are taking an ever harsher position. On 14 April 1995, the Supreme Soviet of Ukraine sent an appeal to the State Duma of the Russian Federation, stating that the questions concerning Crimea are "exclusively the internal affair" of Ukraine. The appeal concluded that the decisions of the State Duma regarding Crimea, and especailly the invitation from Moscow to the chairman of the Supreme Soviet of the Republic of Crimea to address the State Duma, contradict generally accepted international-legal standards and principles.

By what means are we to emerge from the dead end into which the Crimean problem has been pushed by Russian politicians, whose actions are dictated not by the state interests of Russia, but by immediate, personal ambitions?

Historically, Crimea never belonged to Ukraine. Such a state did not even exist until 1918. Prior to 1783, Crimea was a khanate dependent on Turkey. When the Khan Girey voluntarily went under the scepter of Catherine II, she issued the manifest of 8 January 1783, in accordance with which Crimea was annexed by Russia in February 1784, as part of Tavrich Oblast.

We know that the fate of Crimea was determined in 1954 by the independent decision of N.S. Khrushchev. The question of the transfer of Crimea was prepared and decided in surroundings of strictest secrecy. On 1 February 1954, a secret note was sent in response to N.S. Khrushchev's statement, and was signed by Suslov and Pegov. It spoke of the USSR Supreme Soviet Presidium on reviewing the "joint presentation" of the RSFSR Supreme Soviet Presidium and the UkSSR Supreme Soviet Presidium on handing over Crimea Oblast from the jurisdiction of the RSFSR to the jurisdiction of the Ukrainian SSR. Provision was made for everything—the date of the meeting (19 February 1954), the identify of those invited, who would call the meeting to order, who would be given the floor for speeches, and which of the members of the USSR Supreme Soviet Presidium would speak with "approval and support" of the party leader's initiative.

At that time, there was no "separation of powers," and the "matter" moved along at a rapid pace. Already on 5 February 1954, the RSFSR Supreme Soviet Presidium, at a closed session, adopted the resolution, "On the Order of Transfer of Crimea Oblast from the Jurisdiction of the RSFSR to the Jurisdiction of the Ukrainian SSR." It contained a point calling for ratification by the USSR Supreme Soviet Presidium. On 13 February 1954, the chairman of the Ukrainian SSR Supreme Soviet Presidium, D. Korotchenko, asked the USSR Supreme Soviet Presidium to hand over Crimea Oblast from RSFSR Jurisdiction to that of the Ukrainian SSR.

On 19 February 1954, the USSR Supreme Soviet Presidium ratified the "joint presentation" of the two presidiums of republican supreme soviets. The "joint presentation" was not published, however. Moreover, it was not found in the state archives. The question thus arises, did it ever exist?

On 26 April 1954, the USSR Supreme Soviet ratified the USSR Supreme Soviet Presidium Edict of 19 February 1954 and resolved to retroactively introduce the appropriate amendments to Articles 22 and 23 of the USSR Constitution. After that, the RSFSR Supreme Soviet, by its Law of 2 June 1954, also retroactively, introduced amendments to the RSFSR Constitution, excluding Crimea Oblast from the jurisdiction of the RSFSR.

In the process of all these actions, "along the way," fundamental principles secured in Articles 15 and 18 of the USSR Constitution were violated. According to these articles, the USSR was called upon to protect the sovereign rights of the union republics, and primarily their right to territorial integrity. Article 18 directly stated that the territory of a union republic could not be changed without its consent.

The decision of the USSR Supreme Soviet Presidium directly violated the sovereign rights of the RSFSR, since the "consent" for change in the jurisdiction of the RSFSR was given not by the RSFSR Supreme Soviet, but by the RSFSR Supreme Soviet Presidium.

Also violated were Articles 6, 13, 14, 16, 19, 23, and 33 of the RSFSR Constitution. Article 6 of the RSFSR Constitution, for example, stated that the land, its mineral resources, water, forests, factories, plants, and other objects of state property are all-people's property. We might ask, why then were these objects, which are located on the territory of Crimean Oblast, "handed over" to Ukraine on an unlawful basis—by the decree of the USSR Supreme Soviet Presidium?

While Article 14 listed the oblasts and krays of which the RSFSR consists, at the same time Article 16 emphasized that "the territory of the RSFSR cannot be altered without the consent of the RSFSR." According to Articles 19 and 23 of the Constitution, the question of the territorial make-up of the RSFSR is under the exclusive administration of the RSFSR Supreme Soviet.

All these violations of the principles of the constitutional order of the USSR and RSFSR received proper legal evaluation, but already in an entirely different historical situation. On 21 May 1992, the Russian Federation Supreme Soviet adopted the decree, "On a Legal Assessment of the Decisions of Supreme Organs of the RSFSR State Power on Change in the Status of Crimea, Adopted in 1954." In it, the RSFSR Supreme Soviet Presidium Decree of 5 February 1954 was denied any legal validity from the moment of its adoption "by reason of violation of the RSFSR Constitution and legislative procedure."

However, the decree of 21 May 1992 contained a certain inconsistency. It noted that, in connection with the conclusion of a bilateral treaty between Ukraine and Russia on 19 November 1990, in which the parties rejected territorial claims, the regulation of the question of Crimea must proceed by means of interstate negotiations between Russia and Ukraine with the participation of Crimea, and on the basis of a popular referendum.

But if the decree of 5 February 1954 was deemed not to have any legal force, then why at the same time proclaim Russia's renunciation of all territorial claims on Ukraine? Once again, as countless times before, political concessions and compromises achieved at the price of violating constitutional principles have given rise to a new round of contradictions surrounding the Crimean problem.

The position of the jurists sharply deteriorated, as they were now forced to prove that the legality of the decisions of the RSFSR organs of state power, as decreed by the Supreme Soviet of the Russian Federation on 21 May 1992, retains its force.

Naturally, the question may also be resolved by means of interstate negotiations between Russia and Ukraine with participation of Crimea and on the basis of popular will, but under one condition: Strict adherence to the legal principles proclaimed by the Constitutions of the USSR and RSFSR. Specifically, the Treaty of 19 November 1990 spoke of the respect of the territorial integrity of the RSFSR and the UkSSR "within the currently existing boundaries within the framework of the USSR" (Article 6). In connection with the fact that the USSR still existed in 1990, the question arises: Why, in violation of the USSR Constitution, did the union republics independently resolve questions of recognizing boundaries between them, sanction their change, etc.?

Based on the fact that drastic changes had occurred in the legal situation surrounding Crimea since 1954, it is necessary to place the concept of "newly discovered circumstances" at the basis of this evaluation. Who could foresee in 1954 the disintegration of the Union, the division of unified state power, unified territories, property, citizenship, laws? Even in the most delirious dream, one could not foresee the coming to power in Ukraine of extreme nationalists, who implemented a hostile policy toward Russia and who refused the very idea of negotiations on handing over Crimea to Russia.

There are at least two possible options if the Ukrainian Supreme Soviet continues on a unilateral path, from a position of force, to repeal those statutes adopted by the Supreme Soviet of the Republic of Crimea which it does not like. Based on the situation in Crimea, such actions will continue to be regarded by the citizens of Crimea as unlawful, forceful, directed at the Ukrainization of the population of Crimea, and an infringement on the rights and freedoms of the Russian population, which comprises the majority in the Republic of Crimea, created by an 83 percent majority on 20 January 1991.

The first option is the presentation of the Crimean problem for discussion by the heads of state of the Commonwealth of Independent States.

The second option is an appeal to the International Court in The Hague.

Crimea, in strict accordance with the law, has been and remains Russian. And the recognition of this immutable legal fact will remove unnecessary tension in relations between Ukraine and Russia.

### 2. On the Russian Federal Status of the City of Sevastopol

By its Decree of 9 July 1993, "On the Status of the City of Sevastopol," the Supreme Soviet of the Russian Federation confirmed the Russian federal status of the city of Sevastopol within the administrative-territorial boundaries of the city okrug as of December 1991. The Russian Federal Supreme Soviet Committee on Constitutional Legislation was assigned the task of preparing a Russian Federation draft law on securing the federal status of the city of Sevastopol in the Constitution of the Russian Federation.

However, almost two years have passed, and the problem of the status of the city of Sevastopol is growing more acute. The Ukrainian side is rather skillfully implementing the principle of "divide and conquer," under which the legislative and executive branches in the city of Sevastopol are not interacting as they should, on the basis of various "checks and balances," but are being used as a devilish mechanism in the hands of the authorities, located in Kiev, for setting against one another and mutually weakening the organs of power and administration in Crimea.

Indicative in this connection is the Edict of the President of the Republic of Crimea dated 9 September 1994, "On Organization of State Administration in the Republic of Crimea." Annoyed by the endless efforts of the Kiev authorities to bribe the corps of deputies, President of the Republic of Crimea Yu. Meshkov is taking rather decisive actions. In connection with the "encroachment of the Supreme Soviet of Crimea on the extent of lawful powers and authorities," states the edict, "and the serious consequences stemming from this, associated with the loss of manageability, the activity of the Supreme Soviet of Crimea, and the rayon and city Soviets of People's Deputies is hereby suspended." Their powers and authorities are handed over to the President of the Republic of Crimea and the heads of local administrations. The prosecutor's office, in conjunction with the security service and the MVD [Ministry of Internal Affairs] of Crimea, are instructed to verify the "merging" between part of the deputy corps and the criminal structures.

At the same time, Ukrainian authorities are using "legal" arguments:

1. "The city of Sevastopol is an integral part of Ukraine, and therefore its status may be determined only by Ukraine."

Sevastopol was founded in 1784, and from that moment sovereignty over it was exercised by the Russian Empire. Administration of the city was performed by a military general-governor, who by his duties was the commander of the Black Sea Fleet. The city of Sevastopol was a specific administrative okrug, managed by the military-naval administration, which was appointed directly in St. Petersburg.

The legal principle upon which the Russian federal status of the city of Sevastopol is based is, as we know, the decree of the RSFSR Supreme Soviet Presidium dated 29 October 1948, which separated the city of Sevastopol from the Crimea Oblast, making it into an independent administrative-economic center with its own specific budget, and relegating it to the category of a city under republic subordination.

The city of Sevastopol remains to the present time a part of the Russian Federation. The decree of 1948 was not repealed by anyone. The imposition of Ukraine's jurisdiction over the city of Sevastopol in 1977 was accomplished unilaterally, without the appropriate decision by the constitutional organs of the RSFSR.

Since the formation of the independent states of Russia and Ukraine within the scope of the CIS, the legal status of the city of Sevastopol has not undergone any changes.

2. "The city of Sevastopol is a city of Ukrainian republic subordination."

The decree of the USSR Council of Ministers dated 25 October 1948, "On Measures for Accelerating the Restoration of Sevastopol," signed by I. Stalin, stated: "Sevastopol shall be categorized among the cities of republic subordination." At that time, republic subordination clearly meant the subordination of Sevastopol under the RSFSR Council of Ministers, and certainly not under the Ukrainian SSR Council of Ministers. It was specifically under the RSFSR Council of Ministers that this same decree established a special administration for the restoration of the city of Sevastopol.

It follows from the RSFSR Supreme Soviet Presidium Decree of 29 October 1948, relegating Sevastopol to the category of cities of republic subordination, that Sevastopol was removed from the jurisdiction of Crimea Oblast, which at that time was part of the RSFSR.

However, relegation of Sevastopol to the category of cities of republic subordination was not specified in the RSFSR Constitution in 1948. The RSFSR Constitution did not contain any list of cities of republic subordination.

The RSFSR Supreme Soviet Presidium Decree of 5 February 1954, "On Transferring Crimea Oblast from the Jurisdiction of the RSFSR to the Jurisdiction of the UkSSR," and the USSR Supreme Soviet Presidium Edict of 19 February on this question also contained no mention of Sevastopol.

The subsequent edict of 29 October 1948 was neither amended nor repealed. Legally, it retains its validity to the present time, and the Russian Federation as the legal successor of the RSFSR even today exercises its sovereignty over the city of Sevastopol.

Therefore, all unilateral statutes of Ukraine declaring the city to be under Ukrainian jurisdiction cannot be recognized as legal.

State sovereignty over Sevastopol, in accordance with international law, was never handed over to anyone. International law demands, and this is confirmed by the practice of the International Court, that any transfer of state sovereignty to a territory be in the form of an international treaty. The Vienna Convention on the right of agreements defines a treaty as an international agreement concluded between states in written form and regulated by international law. Neither the edict of the RSFSR Supreme Soviet Presidium nor the decree of the USSR Supreme Soviet Presidium are international treaties. These are domestic documents, proscribing certain actions by the administrative organs of the USSR and RSFSR.

3. The edict of 29 October 1948 speaks not about administrative-territorial, but about the administrative-economic status of Sevastopol.

First of all, in the above-mentioned edict the discussion is not about the administrative-economic status, but about the administrative economic center, which is relegated to the category of cities of republic subordination. Sevastopol represents a specific administrative-economic center primarily because the decisive administrative authority belonged to the military administration.

The special regimen for administering the city was defined by its special purpose—to be a military-naval base for the Black Sea Fleet.

A special legal regimen was established for the city which concerned, specifically, the residences and activities of its residents, the procedures for entry and exit, the financing and supply of the city, and other issues. The position of the city as a military-naval base of union subordination stemmed from its importance to the defense and security of the country.

Even after its legal formulation as a city of republic subordination, the special-purpose designation of the city continued to play a decisive role as a city of union subordination. According to the USSR Constitution of 1936, all military affairs were relegated to the USSR, with the appropriate subordination to the Narkomat [People's Commissariat] of Defense in the city of Moscow.

4. "Since 1954, Sevastopol has been financed from the budget of Ukraine, and not the Russian Federation." These unfounded statements of Ukrainian nationalists have no basis in fact.

The financing of Sevastopol between of 1954 and 1958 was conducted out of the union budget. Aside from special-purpose financing, at the end of the program for restoring Sevastopol in 1954, the USSR Council of Ministers adopted the decree of 26 July 1954, No. 1508, "On Measures for Continued Development of the Agriculture, Cities, and Resorts of Crimea Oblast." Within the scope of this decree, at the expense of the union budget, in which the relative share of Russia was 85–88 percent, an extensive list of raw materials, equipment, transport, and monetary funds, including currency, were allocated specially for the development of Sevastopol and Crimea Oblast for the period 1955 to 1958. By 1958, and through 1990, the special-purpose capital investments continued to be made from the union budget. They were directed toward creation and development of the resort network, communications, transport means (including pipelines), improvement of power supply and development of culture.

Practically until 1991, not only financial, but also organizational functions, including the passport regimen on the territory of Sevastopol, were implemented under the direct management of the USSR Council of Ministers, without the participation of the Ukrainian SSR with Council of Ministers.

The most complex legal collisions arise in connection with the signing of the Treaty of 19 November 1990 between Russia and Ukraine, according to Article 6 of which the parties recognize and respect the territorial integrity of the RSFSR and the Ukrainian SSR within the "boundaries currently existing within the scope of the USSR."

Russia is the legal successor of the USSR with regard to its right to use the port at the military-naval base of Sevastopol. This does not mean the submission of any territorial claims to Ukraine. The discussion here is not about claims to territory belonging to Ukraine, but under the administrative-territorial boundaries of Russia when the USSR existed. The fact is that the Treaty of 19 November 1990, in which the "High Agreeing Parties" (represented by the presidents of Russia and Ukraine) recognize each other as sovereign states and confirm the existing boundaries within the scope of the former Union, signifies that Ukraine, by Article 6 of this treaty, has confirmed that Sevastopol remains a city of the Russian Federation.

We will note that, in seeking out various legal hitches, the Ukrainian side is assiduously evading the main point.

Thus, the Declaration of State Sovereignty, adopted by the Supreme Soviet of the Ukrainian SSR on 16 July 1990, in proclaiming the state sovereignty of Ukraine, does not say a word about the USSR, about the USSR Constitution as the main legal basis for interrelations of the Union with the union republics, and the initial legal base for determining the competency of the union republics. Thus, the declaration states that the resolution of questions of "all-union property (common property of all republics)" is performed on the basis of agreement between the republics which are subjects of this property. However, the USSR Constitution did not contain any concepts of "all-union" property, or "common property of the union republics." There was a singular state (all-people's) property, and the union republics did not have the right of ownership to it.

Transformation of all-people's property into common property of all republics was needed by the authors of the declaration to justify the anti-constitutional actions of Ukraine in seizing the all-people's property located on the territory of Ukraine.

The State Duma of the Federal Assembly of the Russian Federation must, as soon as possible, confirm once again the Russian federal status of Sevastopol; decisively declare that Ukraine's encroachment on a part of Russian territory contradicts the Constitution of the Russian Federation and all other legal statutes, including the Treaty of 1990; and confirm the jurisdiction of Russia over Sevastopol as the main base of the Russian Black Sea Fleet.

For the purpose of accelerating the stabilization of the situation surrounding the status of Sevastopol and the Black Sea Fleet and retaining good neighborly relations with Ukraine, the State Duma of the Russian Federation must instruct the government of the Russian Federation to take all possible measures to conclude negotiations with Ukraine on the status of Sevastopol and the Black Sea Fleet no later than July of 1995 (UP TO THEIR STRICT TIE-IN WITH FULFILLMENT OF ALL OTHER AGREEMENTS).

## 5.38
**Talks End; Documents Readied**
Moscow Television, 9 June 1995
[FBIS Translation]

The working meeting between the presidents of Russia and Ukraine has just ended in Sochi.

The main issue in today's dialogue between the two presidents concerns the Russian Ukrainian fleets' bases. Moscow wants to have Sevastopol as its main navy base as well as two Crimean airports—Gvardeyskiy and Oktyabrskiy—and two communication points. All other places of vessel deployment should be given to Ukraine in line with previous agreements. Unfortunately, the stance of the Ukrainian side was not always consistent. Kiev has departed from the previously announced principle of separate deployment. During recent meetings with Yeltsin in Moscow and Minsk, Leonid Kuchma said the two fleets should be deployed in Sevastopol together. Russia believes this is not acceptable and does not intend to return two of the four Sevastopol bays to Ukraine. All in all there are 220 places for mooring vessels in the city's bays. Ukraine is offering Russia only 28 percent of them, and this is despite the fact that the Russian part of the fleet is four times bigger than that of Ukraine.

Apart from the fleet deployment problem, the presidents of the two countries are likely to discuss economic problems, confidentially at first, and then at an expanded sitting with participation of foreign and defense ministers and experts.

## 5.39
**Further on Agreement**
ITAR-TASS, 9 June 1995
[FBIS Translation]

The Russian Black Sea Fleet will be based in Sevastopol under an agreement signed here today as a result of the Russian-Ukrainian summit.

The four-hour talks between Presidents Yeltsin and Kuchma resulted in the signing of two documents—the agreement on the Black Sea Fleet and a joint communique.

Ukrainian President Leonid Kuchma told journalists after the signing that Russian-Ukrainian summits will henceforth be held every month, and the prime ministers of the two countries will also meet on the same regular basis.

Yeltsin and Kuchma will next meet in Crimea in July.

## 5.40
**Yeltsin Adviser: Agreement Not "Ideal"**
Interfax, 9 June 1995
[FBIS Translation]

The Russian president's adviser on national security, Yuriy Baturin, has said he does not consider the Russian-Ukrainian agreement on the Black Sea Fleet as "ideal."

However, he told an Interfax correspondent in Sochi that he believes that the agreement has "rich potential opportunities for resolving the Black Sea Fleet issue."

Baturin said that many diplomatic concessions were made at the talks "without which given the time shortage the agreement would not have been achieved." "We have been trying to resolve the problem for three years already. Will another three years be needed?" he went on to say.

Baturin said that "if large steps cannot be made, smaller ones should be made more often." He said that is why the presidents, the prime ministers, their advisors and ministers have decided to meet quite frequently. He said this means that Russian-Ukrainian contacts have been intensified.

Baturin said the agreement for the first time mentions that the interests of Russia and Ukraine in the Black Sea's basin coincide. He said this wording is very important and it means even more than the term "strategic partnership" which was also introduced into Russian-Ukrainian relations for the first time.

Baturin said that the presidents agreed on the site for the Ukrainian navy's headquarters while the decision was not fixed in the document. He said it would be inappropriate for a sovereign Ukraine to fix the stationing of its navy's headquarters in a bilateral agreement.

## 5.41

### "Text" of Black Sea Fleet Agreement

*Pravda,* 10 June 1995
[FBIS Translation]

The Russian Federation and Ukraine, hereafter known as the "Sides," fully resolved to strengthen friendship and cooperation between the Russian Federation and Ukraine, note the coincidence of the two states' interests in the Black Sea basin and, based on Russian-Ukrainian documents signed previously in this sphere, have agreed on the following:

Article 1. The Russian Federation Black Sea Fleet and the Ukrainian navy are to be formed on the basis of the Black Sea Fleet. The Russian Federation Black Sea Fleet and the Ukrainian navy are to be based separately.

Article 2. The main base of the Russian Federation Black Sea Fleet together with the headquarters of the Russian Federation Black Sea Fleet are to be in the city of Sevastopol.

The Russian Federation Black Sea Fleet will use installations of the Black Sea Fleet in the city of Sevastopol and will also use other basing and deployment locations for ships, aviation, shore-based troops, and operational, combat, technical, and rear support installations in Crimea.

Article 3. The Sides' governments will settle questions pertaining to the property of the Black Sea Fleet and will sign a separate agreement on that matter, mindful of the previous accord on the division of the aforementioned property on a 50/50 basis.

Article 4. The Russian Federation is to receive 81.7 percent of the ships and vessels of the Black Sea Fleet, Ukraine—18.3 percent.

Article 5. When dividing up the armaments, military hardware, and support facilities of the shore-based defense forces, marines, and land-based naval aviation of the Black Sea Fleet, the Sides will work from the situation existing as of 3 August 1992.

Article 6. If one Side is interested in using installations which under the terms of this agreement are designated for the use of the other Side, questions will be resolved by the conclusion of special agreements in each specific instance.

Article 7. Each officer, warrant officer, and petty officer of the Black Sea Fleet has the right to freely determine his future service.

Article 8. The Russian Federation will participate in developing the socio-economic sphere of Sevastopol and other population centers where the Russian Federation Black Sea Fleet is to be based.

Article 9. To preserve stability in the Black Sea region and ensure safety at sea, the Sides will pool their efforts in interaction and cooperation in the naval sphere. The organization of and procedure for cooperation in this sphere will be determined by the Agreement on Cooperation between the Russian Federation Fleet and the Ukrainian navy.

Article 10. The Sides will continue talks on the Black Sea Fleet and, in particular, the elaboration of the legal status and conditions governing the presence of the Russian Federation Black Sea Fleet on Ukrainian territory, the procedure for mutual settlements connected with the resolution of the problem of the Black Sea Fleet, and other questions.

Article 11. A Russian-Ukrainian Joint Commission consisting of the state delegations of the Russian Federation and Ukraine at the talks on the Black Sea Fleet is to be formed to monitor the fulfillment of the accords on the Black Sea Fleet.

The commission is instructed to draw up specific parameters for the division of Black Sea Fleet installations.

## 5.42

### Russian-Ukrainian Accords Prompts U.S. "Dismay"

Stanislav Menshikov
*Pravda,* 17 June 1995 [FBIS Translation]

The Sochi accords between the Russian and Ukrainian leaders caused concern across the ocean. Outwardly, Washington wanted to emphasize its neutrality and even expressed ostentatious satisfaction at the settlement of the "bitter infighting" over the Black Sea Fleet. But it is clear from the U.S. press that behind this outwardly calm reaction lies something very different. Some commentators claim that the Sochi agreements were achieved "against a background of deteriorating relations between Moscow and Kiev." Others note the vagueness of the documents signed, which makes it possible for the sides to depart from the accords, as has happened more than once in the past. It is stressed that the domestic political situation in Ukraine and the stance of the nationalists there will hardly help President Kuchma to follow the agreements to the letter.

These predictions may not be far from the truth, but they also indicate that by no means everyone in Washington found the signs of a rapprochement between Russia and Ukraine to his liking. As one commentator noted frankly, although the "specter of the IMF hung over Sochi," the Russian-Ukrainian summit testified to the diminishing role and influence of the United States in the region. It was also recalled that at one time George Bush's mediation was apparently needed in order to resolve the Russian-Ukrainian dispute on tank exports, and later Bill Clinton helped to settle disagreements between Kiev and Moscow over nuclear weapons. But now, as *The Washington Post* puts it, the only sign of a U.S. role is that Yeltsin and Kuchma met in the Sochi Radisson-Lazurnaya Hotel, which is owned by Americans.

Such claims may seem unsubstantiated to the reader. Surely the share-out of the Black Sea Fleet and other questions of Russian-Ukrainian relations come exclusively within the sphere of authority of these two sovereign states themselves. What business is it of Washington's, you may wonder, how Moscow and Kiev decide contentious issues between themselves? In this situation three is a crowd, so to speak. But it seems that people across the ocean look at things differently. Both Russia and Ukraine, thanks to the ill-fated reforms, have found themselves in the position of poor relations with the West. And the "high road of civilization" where the liberals are taking us exacts more political and economic dues than the legendary bandit Solovey.

As for Ukraine, Washington now counts it firmly among its own, and deems it part of the NATO sphere of interests. Listen, for instance, to the reasoning of prominent figures in the U.S. administration. U.S. Secretary of State Christopher called Ukraine "one of the pivots of European security." His deputy Talbott added that "because of its geographical location it plays a key role in the new Europe that has sprung up since the end of the cold war." Defense Secretary Perry explained that "Ukraine is of vital significance for U.S. national security interests." All these statements were made not just anywhere, but during visits by the U.S. ministers to the Ukrainian capital. And President Clinton himself, when he was in Kiev in May, did not omit to emphasize Ukraine's place in the U.S. plans.

Washington began to display a heightened interest in the fate of that country immediately after the breakup of the Soviet Union, but especially after last year's elections, which brought President Kuchma to power. Initially Washington was concerned about his pre-election statements in favor of a rapprochement with Russia and the support he won from the Russian-speaking population. However, as *The Washington Post* writes, these misgivings were dispelled when it became clear that the gap between pre-election statements and real policy is considerable. There was approval for the new president's leaning in the direction of market reforms, his tough confrontation with the Crimean autonomous formation and his own parliament, and his recent statement that Ukraine cannot remain outside military blocs. (There is really only one military bloc in existence in Europe now—NATO.)

All this won praise across the ocean. *The Wall Street Journal* expressed the hope that "Ukraine could return to the fold of European countries much more quickly than its northern neighbor, on condition that it is allowed to do so and is not included in Russia's sphere of interests. Ukraine is a powerful factor for stability in Central Europe and the Black Sea region. Whether as a buffer or a bridge (as Kuchma sees his country's role), an independent Ukraine could play an important role as a counterweight to Russia's influence in Europe." So whether Ukraine is admitted to NATO or not, the unenviable role of anti-Russian pawn in the West's game is in store for it in America's geostrategic plans.

The newspaper went on to note that during his presidency Kuchma has achieved greater rapprochement with the West than with Russia. This, it says, is entirely the fault of Russia, because of its "great-power syndrome."

These thoughts were published not long before the Sochi meeting. And then there was dismay: The sides managed without an intermediary from across the ocean. Apparently there is no need to pay yet more dues to the gentlemen of fortune on the "high road of civilization." The gentlemen are offended. But they will be even more offended if things go even further and hopes for the early admission of Ukraine to the NATO system prove vain, and instead—who knows what may happen—it "follows the example of Belarus" (as the same *Wall Street Journal* wrote in horror). That was what Washington liked least about the Sochi meeting. But it is too soon to rejoice. Western strategists do not abandon their plans and objectives so easily. Too tasty a morsel is in danger of being wrested from their voracious maw, and it is one, moreover, that they think they have already paid for. So we can expect more dirty tricks from our "civilized" partners.

---

## 5.43
## Concern over Nuclear Fuel Deliveries from Russia

Interfax, 17 July 1995 [FBIS Translation]

---

There are violations in the schedule of delivery of nuclear rods to Ukraine under the trilateral agreement on nuclear disarmament, Ukraine's Deputy Foreign Minister Konstantin Grischenko told Interfax. . . .

"We cannot but be worried about this," he added.

According to Grischenko, a Ukrainian delegation headed by Defense Minister Valeriy Shmarov raised the issue during its visit to the U.S. Grischenko was also a member of the delegation.

"A nuclear power plant should have an emergency stock of fuel for 18 months," he added, saying the violations made the stations use the emergency stocks.

According to the trilateral statement of the U.S., Russian,

and Ukrainian presidents, Ukraine will pull out nuclear warheads in exchange for nuclear fuel. For its part, the U.S. will fund the dismantling and recycling the nuclear charges under the Nunn-Lugar program.

---

## 5.44

**Kuchma: Too Early for Customs Union**
UNIAN, 17 July 1995
[FBIS Translation]

---

Ukrainian President Leonid Kuchma, who arrived in the Belarusian capital today on an official visit, told journalists at the airport there were no border problems between the two countries.

He said: "Ukraine is in favor of a customs union, but it is too early yet. This is not an issue for the present. We would be pleased to establish a border-crossing system so that our citizens would not be inconvenienced, but we are not going to remove the customs barriers."

He recalled that the two countries' European neighbors, Poland, the Czech Republic, and Hungary, have allowed themselves 10 years to form a customs union. Kuchma stressed the need to create a legislative framework and sign a free trade treaty before such a union is established.

---

## 5.45

**Analyst on Russia's "Big Brother" Aspirations**
Natalya Filipchuk
*Golos Ukrainy,* 22 September 1995
[FBIS Translation]

---

If one accepts the idea that Russia's reassumption of the "big brother" role in the "CIS family" will gain popularity during the election campaign, it may be inferred that the election campaign has started. Russian Federation President Boris Yeltsin has put the first brick under his chair (so that it will not shake too much) by issuing an edict that will determine Russia's strategic course toward the development of relations with CIS countries.

The text of the document was published in full by quite a number of the mass media. There has still been no official reaction from Kiev, although Boris Tarasyuk, first deputy foreign minister, pointed out that "It is difficult to imagine integration processes that are regulated by edicts of the head of just a single member state."

"While developing relations with CIS partners, it is necessary to be firmly guided by the principle of doing no harm to Russian interests," reads the document. Before one grasps the full meaning of the text, one catches the continuation: One of Russia's main political goals is "consolidation of Russia as a leading force in the formation of a new system of interstate political and economic relations on the post-Union territory, acceleration of the integration processes within the CIS . . . and gradual expansion of the Customs Union that will encompass states affiliated with Russia by a profoundly integrated economic and strategic political partnership."

Incidentally, with regard to the latter, the "profound economic integration and strategic partnership" apply, first and foremost, to Ukraine. We recently signed corresponding documents in Sochi. At the same time, our sober-minded attitude toward the Customs Union does not seem to bother our northern neighbor, as from now on, Russia intends to seek recognition on the part of the CIS countries "of the fact that the CIS is, primarily, a zone of Russian interests."

To put it another way, the cited statements of the edict indicate that Russia is sick and tired of complying with the opinions of its partners, and has decided to exercise the rights of the "big brother" and to take the initiative and, concurrently, forbid the other CIS members to pursue their independent external policies. This is explained as follows: "To demand that the CIS states fulfill their commitments and refrain from participation in unions or blocs aimed against any of the states." Over time, the Council of Europe may also be listed among such blocs. Just recall that recently, after the question of Ukraine's admission to the Council of Europe was raised at a parliamentary assembly session, some Russian politicians stated that Russia wanted nothing to do with that organization, because its attitude toward Ukraine was more "amiable." How then should Ukraine treat the Council of Europe and NATO, which has recently been a matter of annoyance for the Russian leadership, and so on?

With regard to the pursuit of an independent external policy, which is one of the conditions for a state's existence, it also appears that precisely this edict is directed against the CIS states. If one adheres to the clause on the "non-participation in blocs and unions," then all CIS countries, with the exception of Russia (or the other way around), must immediately withdraw from the Commonwealth.

Therefore, Ukraine's Foreign Ministry and also President Kuchma have something to ponder. So far, our state has refrained from an official reaction. However, something has already become obvious: An election campaign has started in Russia, and Yeltsin dreams of getting additional votes by taking advantage of the moods prevailing among the citizens—nostalgia for the great and indivisible [Russia]. For Ukraine, the role of small-fry has been prepared.

## 5.46
### Yeltsin's Edict on Links with CIS Assessed

Interview with Borys Oliynyk
Elena Myloserdova
*Kievskie Vedomosti,* 26 September 1995
[FBIS Translation]

[Interview with Borys Ilich Oliynyk, chairman of the Ukraine Supreme Council Commission for Foreign Affairs and Relations with the CIS.]

From time to time, official Moscow "issues" state documents that involuntarily cause some bewilderment. Given our independence and sovereignty, some decisions by the Russian Duma—for example, on the Black Sea Fleet or the status of Sevastopol—are obviously not in step with the fact that Ukraine is an independent state and stopped being a Soviet republic some time ago. In this regard, Russian President Yeltsin's edict "Russia's Strategic Course on the Development of Relations With the CIS States" was not an exception.

Last week, Russian Federation Council Chairman Vladimir Shumeyko declared that Ukraine will soon become a full-fledged member of the CIS Interparliamentary Assembly (MPA), which he also heads. If a hasty and unexpected visit by Belarusian President Mr. Lukashenka (read: Yeltsin's emissary) is added to these developments, an impression of a mass attack on Ukraine is created.

But let us not make guesses. Let us give a word to Boris Oliynyk, chairman of the Supreme Council Commission for Foreign Affairs and Relations with the CIS.

[Myloserdova]: Borys Ilich, what is your opinion about Yeltsin's recent edict?

[Oliynyk]: My opinion about the personality of Boris Nikolaevich has been known since 1990 and, contrary to our democrats, I warned: He is a changeable and impulsive man. It is necessary to deal with him with particular caution and with witnesses present, because it is possible that Yeltsin—given all his objective qualities—does not at all know what he will do tomorrow. Precisely this has been confirmed, although such an edict did not surprise me. When I spoke about this before, our democrats attacked me: How could I oppose "the father of universal democracy"? But the events that followed confirmed that I, to my great regret, was right. This also concerns the full-scale agreement which we still cannot sign, the Black Sea Fleet problem, and the bombardment of the legally elected parliament. The left forces, by the way, have assessed this event, while democrats have not done so. We only heard that all leaders of contiguous countries have justified Yeltsin's activities. When such actions are forgiven, they lead to Chechnya. All this has been done by "the father of democracy." Hence, this edict did not surprise me.

With regard to its essence, each state must advocate its national strategic interests, but must not forget that its neighbors also have their national strategic interests. At least our foreign policy is based on the American model—live and make it possible for other people to live. Therefore, one must not shape his behavior as though he were alone, but one must consider the fact that similar people continue to live near him.

[Myloserdova]: Could you provide more detail?

[Oliynyk]: Let us take the first postulate. I am citing: "Russia's primary vital interests in the spheres of the economy, defense, security, and protection of Russians' rights, on the provision of which national security is based, are concentrated on the territory of the CIS states." This is a tenet from the American strategy: U.S. vital interests are those which the United States has the right to specify. In Zanzibar? Yes. In Somalia? Yes. In Yugoslavia? Yes. Even bombing can be applied there. Therefore, there is no difference between Yeltsin's statement and the American concept.

[Myloserdova]: What do you think about the tenet "strengthening Russia as a leading force in the formation of a new system of interstate political and economic relations on the territory of what used to be the Soviet Union?"

[Oliynyk]: This is precisely the U.S. imperial model with its "vital interests." They also have the phrase "specific interests." But each state has its own "specific interests" and one should remember them. I do want to say: Whoever says it and whatever it is that is said, I have my own position and I do not change it. The Moscow team must not be confused with Russia and the Russians. Teams come and go. The Russian people suffer; they, like the Ukrainian people, have great problems. No one has the right to separate [the peoples] and claim loudly that someone is enemy no. 1 or 2.

If the Moscow team, however, thinks it is possible to specify Russia's vital interests wherever it likes, we, as a sovereign state, must do everything to ensure our national security.

[Myloserdova]: At the same time, good relations with Russia as the no. 1 partner should be preserved.

[Oliynyk]: Undoubtedly. Therefore, I say that one should not mix up the [government] team with the country. One way or another, we are neighbors and we will have good relations with the Russian people, because we must not be divided by barbed wire.

[Myloserdova]: With regard to the country, I agree with you. But political issues have to be resolved with the team.

[Oliynyk]: Yes, and therefore one should not be afraid. Our [politicians] are afraid when the point is relations with the CIS, the MPA, and so on. They are afraid that we will get into a force field. Who, however, is pushed into a force field? It is up to us not to enter it. At least economic relations should be developed if it is profitable for the entire state. We must guarantee our independence politically.

[Myloserdova]: Let us return to the MPA. Why is Mr. Shumeyko so confident when he is drawing conclusions?

[Oliynyk]: He, as a rule, confuses reality with desires. Since the issue on joining the MPA has been raised, it is under consideration in our commission, and its members differ on whether Ukraine should join the MPA. At the demand of the members of the commission, a package of documents on the MPA—which we do not have—should be submitted to us. Then the Foreign Ministry should submit its expert assessment to us. Our commission will draw its conclusions (will vote for or against the Foreign Ministry's option). Deputies with the right to a legislative initiative, however, have the right to submit this issue for consideration at the session of the Supreme Council.

[Myloserdova]: Are you personally a supporter or opponent of joining the MPA?

[Oliynyk]: I support the friendship of all the peoples of the world and I favor the concept that we are living together according to God's will. Our relations should be built in such a way that our grandchildren will be able to peacefully look into each other's eyes. But we should determine our specific rights. Based on our will, we could delegate only part of our power [to the MPA], but not as the eight CIS countries have done. Every president has the right to specify his status. If we do not express a will to join the MPA nothing could be made of it . . . Economic relations, however, must exist, because capillaries and vessels cannot be cut; one will be unable to discuss sovereignty and independence later because he will be bleeding to death.

With regard to the political aspect, if we consider Yeltsin's last edict, of course, the imperial mood is being felt there. If one state, however, claims "specific interests," I do not intend to join the field of these specific interests.

## 5.47

**Udovenko Seen Rallying Other CIS States Against Moscow**

Viktor Zamyatin

*Kommersant-Daily,* 7 October 1995

[FBIS Translation]

*Nezavisimost,* a Kiev-based newspaper, has published a letter by Foreign Minister Hennadiy Udovenko to President Leonid Kuchma. This sort of correspondence is, of course, confidential, particularly so this letter: In effect, it is a "reply to Chamberlain" [a strong negative response to an ultimatum], for it contains an assessment of Boris Yeltsin's recently announced "strategic line" in relations with CIS countries.

Generally speaking, there is nothing particularly unexpected about the letter—Kiev has been repeating at all levels that Russia does not see Ukraine as an equal partner and will not do so in the near future. The same is said in the minister's letter: "Russia does not intend to develop its relations with CIS countries on the basis of international law." Meanwhile, Udovenko believes that further integration within the Commonwealth, the need for which Moscow has been emphasizing on every occasion (and which was one of Leonid Kuchma's key electoral slogans), is leading to the watering down of CIS countries' sovereignty, subordination of their interests to those of Russia, and the recreation of a centralized superpower. The Foreign Ministry also warns the president about the realistic nature of Moscow's recent threat to call Ukraine a bankrupt country and to demand that all its debts be paid off in Ukrainian assets. Since all this runs counter to Ukrainian interests, the minister proposes a number of emergency measures.

What is needed first of all is collaboration with the states whose views on the CIS are similar to those of Ukraine in order to devise a joint approach to various problems of the Commonwealth. These states are not mentioned in the letter, but they are well known: Azerbaijan, Turkmenistan, and Moldova. The head of the Ukrainian foreign service also thinks it necessary to calculate the upper limit of Ukraine's economic dependence on Russia and to spot the enterprises and industries that are actually linked to the Russian economy and will be unable to function on their own. The letter asserts that, as a matter of fact, Russia may well demand Ukrainian gas-industry assets (pipelines and storage facilities) and shares in key Ukrainian enterprises as payment for Ukrainian debts. Yet, Leonid Kuchma has repeatedly stated that he will not allow anything like that to happen.

*Nezavisimost* did Udovenko a disservice—a scandal arose. Meanwhile, Udovenko's meeting with Andrey Kozyrev, on which the two ministers could agree only in New York, is scheduled for 13 October. Kiev expects much from the meeting: It should clarify, for instance, what line Moscow will take after Boris Yeltsin's words about Ukraine's penchant for bending agreements in its favor and Kuchma's resolute disagreement with this statement.

Now, after the publication of Udovenko's letter, Moscow appears to have obtained yet another moral trump card, for it can say that the game against it is not being played honestly. A moral trump alone may prove to be insufficient, however. But then one will have to admit that Udovenko was right to some extent after all.

# 6
# Alternative Confederal Concepts

## Introductory Notes

In previous chapters in this volume, we have documented key aspects of the policy debate over whether the CIS mission is to facilitate a "civilized divorce" among the former Soviet republics, or to furnish the mechanism for reintegration into some kind of new union through which the new Russian state will invariably attempt to carry out its great power ambitions. The viewpoints of the political elites in the CIS states have been so divergent on this overarching issue, and on the detailed subissues within each sphere of integration—economic, political, and military—that little progress has been made toward operative integration of any kind. There has, however, been a tendency for some leaders to take the question of integration more seriously and to admit that some kind of economic integration, as a first step, would hold advantages over what exists. Parallel with the major debate over the CIS's direction, certain strong political leaders within several CIS countries have been taking steps toward the formation of new potential "unions" or confederations. All of these potential unions claim to be compatible with the CIS, at levels below the umbrella commonwealth apparatus. It is still difficult to ascertain the truth of this claim, or to determine whether these subassociations could end up replacing the CIS. For the present, these confederations provide a balance to Russia's great power ambitions.

Chapter 6 documents the key events in the evolution of the concepts for four of these potential new unions: (a) a Russia/Belarus confederation (or the Moscow/Minsk axis); (b) a Slavic Union; (c) a Eurasian Union; and (d) a Central Asian Union. Each of these represents an association between two or more states which is to be built on stronger mechanisms than bilateral treaties or CIS agreements. An approximate analogy might be drawn with organizations like the European Free Trade Association (EFTA), or the European Coal and Steel Community, which preceded or operated in parallel with the European Community for some time, until the Community absorbed them.

Each of the potential new confederal concepts contains extensive references to eventual political and military integration, but starts out with economic integration mechanisms. As a result, most of the policy emphasis in these potential unions has so far been put on economic standardization. Several factors have influenced the decision to seek economic integration first. For one thing, economic modernization is difficult to achieve through bilateral contacts alone; often an entire region coordinating its economic development goals can achieve those goals faster and more efficiently. Furthermore, production relations in the Soviet empire were often regionally oriented, and it is tempting to try to reestablish these relationships. If, however, these new associations or unions do partially reinstate Soviet economic structures, they could exclude much of the restructuring being contemplated in the economic reform plans of several of the former republics. Still another factor influencing CIS leaders is the social cost associated with economic transformation. Leaders have weighed the economic benefits of reinstituting old economic patterns with the costs of restructuring and reform, and have endorsed a restoration of old ties in the interests of achieving social stability.

Strong political leaders have so far been the driving force behind these potential confederal concepts. In the case of the Moscow/Minsk axis, President Aleksandr Lukashenka has been the strongest proponent. After his election in July 1994 he carried out plans that had been in place since 1992, but which had never materialized under Stanislav Shushkevich. The Slavic Union has also been championed by Lukashenka, who would like to draw Ukraine into the nearly complete union between Russia and Belarus. The Slavic Union has strong adherents among Russian right-wing and left-wing political leaders. Vladimir Zhirinovskiy and several of the neo-communist groups, as well as leaders of the national-patriot groups (Sergey Baburin, for example) have often called for a Slavic Union as the nucleus of a new Russian empire. The Eurasian Union, which is a loose confederation, approaching a new Soviet Union, is Kazakh President Nursultan Nazarbaev's creation. A formal draft of this concept was presented to CIS leaders on 12 April 1994. So far, the current Russian administration has sidelined the plan, but Nazarbaev has continued to advocate it. The Central Asian Union is led jointly by the presidents of Uzbekistan, Kyrgyzstan, and Kazakhstan. Formed partly in reaction to the unification tactics of the Slavic states, and partly in order to break out of the isolation of the colonial era and open up new channels of trade and communications, these leaders have structured a union almost identical to the CIS, but organized along ethnic and regional lines within the post-Soviet space. It is clear that Moscow regards the Central Asian Union as a challenge to its zone of influence in Central Asia.

In each case, these quasi confederal unions have been initiated to meet the perceived inadequacies of the CIS, and to integrate its members more closely and efficiently. Although striving toward a "confederal concept" in each instance, the proponents of the new confederal concepts have actually steered toward reestablishing the old federal Soviet economic, military, and political relationships. (For example, the Eurasian Union Treaty is worded almost identically to Mikhail Gorbachev's nine-plus-one Union Treaty, with phrases added that purport to respect individual nations' sovereignty and territorial integrity.) In actual fact, it is difficult for CIS leaders to distinguish between "federation" and "confederation," not yet having strong national identities or legal systems to bring to a confederal association of states. Furthermore, it is difficult to conceive of a true confederation of equal states which would include Russia, which dwarfs the other former republics economically, militarily, and politically.

In the documents which follow, the leaders advocating each confederal concept voice their arguments and opinions of what is needed in the post-Soviet space. These leaders' statements are interspersed with positive and negative reactions to the potential new unions. A special introductory note for each section fills in the background to the documents.

# Russia/Belarus Axis

Despite an active political debate in which the opposition to a union with Russia swelled almost at the "last minute," and continued to grow with hindsight, Belarus's leadership took many of the steps required to unify its newly independent state with Russia in February and May 1995. Belarus's tilt toward Moscow has been based primarily on pragmatic and strategic considerations: 90 percent of its enterprises produce for the Russian market; it is dependent on Russia for almost all of its energy supplies; it shares its eastern Slav ethnic roots with Russia; and Belarus's geostrategic position constrains its foreign policy choices.

Actions to form a Russian/Belarusian confederation began early. In July 1992, Belarusian Prime Minister Vyacheslav Kebich and Russian acting Prime Minister Egor Gaydar signed a new "Comprehensive Union" package of bilateral treaties. This package embraced cooperation in military and economic matters and looked toward significant political alignment in the near future. Egor Gaydar called the agreement "a step in the direction of a confederation within the CIS framework."

Nevertheless, a confederation was not to be concluded between the two countries so fast. The issues of Belarusian military neutrality and confederation with Russia were on a direct collision course and spawned a fierce battle in parliament, as well as in the public domain for nearly two more years. The main subject of debate was over whether Belarus ought to join the CIS Collective Security Treaty, which was seen by Stanislav Shushkevich as a grave threat to Belarusian sovereignty. Shushkevich was thoroughly committed to the principle of neutrality, saying it should define Belarusian foreign policy for centuries to come, especially given Belarus's desire to maintain positive relations with its Baltic and Central European neighbors.

Moscow, however, had made it pretty clear that economic cooperation with Belarus was out of the question as long as it claimed strict neutrality. Prime Minister Vyacheslav Kebich, who supported the concept of confederation, became Moscow's ally on the issue and argued that Belarus ought to sign the treaty, with certain reservations. The strong Belarus military-industrial complex supported this position, as did pan-Slav factions in parliament and other organizations. The rest of the populace was slow to become embroiled in political questions, and was pro-Russian to a much greater extent than, for example, Ukraine or Kazakhstan, even though they appeared in surveys to be no less committed to "democratic values."

The first documents in this collection highlight the sharp policy differences between Prime Minister Kebich and parliament head Shushkevich. The neutrality debate between government and parliament lasted until the election of the first Belarusian president on 10 July 1994, with the consequence that there was neither an abrupt break with Moscow, nor was the confederation consummated.

In 1993, however, several key economic battles emerged to accompany the struggles over Belarusian neutrality and to cause trouble spots between Moscow and Minsk. One was the question of Belarus's continued membership in the ruble zone; another was about a proposed "monetary union" between Russia and Belarus; and another was over Moscow's use of energy as an economic weapon against Belarus. Moscow may have unnecessarily created a pretext for later popular hesitancy about the merits of a confederation by using its energy weapon, although at the time it probably considered Belarusians too politically inactive to worry about such things. Moscow raised energy prices in March 1993, cut down on its energy supplies in response to Belarus's gas and oil debt, and maintained this pressure on Belarus in order to obtain its signature on the CIS Collective Security Treaty and the CIS Economic Union.

Also in March 1993 Belarus took policy steps which de facto amounted to a split from the ruble zone. The National Bank of Belarus (NBB) was the force behind this decision, floating a Belarusian currency parallel to the Russian ruble, and establishing a Belarusian foreign currency exchange, on which the Russian ruble was listed as a foreign currency. One year later, the NBB began to seriously question the treaty on a merger between the monetary systems of Belarus and Russia, which was signed by the prime ministers in April 1994. Under the treaty, the NBB was to have been abolished, and the Russian Central Bank was to control all emissions, as well as the monetary, credit and currency policies of the two countries. Moreover, the treaty would have gone so far as to consolidate the two country's budgets. The documents that follow provide insights into the NBB's principle concerns and into the political positions of Belarusian opposition groups and Russia on these issues.

In July 1994, with the election of Aleksandr Lukashenka, all of these disputes were settled after two visits to Moscow, and Belarus moved squarely into Russia's orbit. In the documents to follow, Lukashenka voices his pragmatic but abiding loyalty to Moscow, stressing that Belarus has no other options. These documents make clear Lukashenka's opinion that the Belovezh Forest meeting was a tragedy. The leader of the Belarusian Popular Front, Zenon Poznyak, on the other hand, strenuously objected, saying: "We need to break from Russia . . . a criminal, bureaucratic state." Nevertheless, the process moved inexorably forward with Lukashenka's enthusiastic embrace of all of Russia's overtures. Only in December 1994, when the Russian State Duma suggested that Belarus and Russia hold a joint session of the two parliaments to discuss unification into one state, did Belarusian Parliament Speaker Myacheslav Hryb protest that such a joint session was unconstitutional—the standard reply whereby a CIS state gracefully declines a "suggestion" from Moscow.

This move by the Russian parliament activated the opposition still further and began to raise doubts among the Belarusian people. These included low, but audible, rumbling in the parliament and in the oppositionist Popular Front against Lukashenka's stridently pro-Russian bias. Ironically, the press accounts collected here would make it appear that in the last hour, the Belarusian people might be becoming more politically active and nationally conscious. Speaker of the Belarusian parliament Myacheslav Hryb mounted a wave of complaints against Russia's policies, especially after the Russian parliament made its bid for the joint session of the two parliaments. Hryb attacked Moscow, saying it "does not always base its relations with Minsk on the equality principle." Russia was forced to accelerate its diplomatic actions for implementing a confederation with Belarus, anticipating that the election of a new Belarusian Supreme Council could halt or seriously delay these plans, which included restoration of Belarus's Soviet-era flag.

Consequently, in January 1995, a comprehensive Friendship and Cooperation Treaty was signed between the two nations following a three-day visit to Minsk by Russian Foreign Minister Kozyrev. This was to be followed by the signing of a comprehensive package of bilateral agreements which would go far toward unifying the two governments in economic and military relations. On 14 May, Belarus held parliamentary elections and a popular referendum on Lukashenka's policies of economic integration with Russia. The referendum gave Lukashenka surprisingly strong backing for his policies, thereby causing a political setback for the Belarusian nationalists. Nevertheless, the nationalists kept up their loud protests. According to Popular Front leaders, the customs agreement and others which accompanied the Friendship and Cooperation Treaty would result in "the liquidation of the independence of the state, the complete poverty of the Belarusian nation, huge losses. . . ." They also charged that these were political, not economic, agreements in that they provided for "free transit and maintenance services for military establishments of the Russian Federation on the territory of Belarus," and made several other concessions, such as full-scale free trade. The free trade measure threatened Belarusian enterprises that might try to compete with much larger and more efficient Russian enterprises. In Belarus, for example, average per capita income comes to just $9.20 per month, whereas in Russia it is $120.00. The nationalists also accused Lukashenka of isolating Belarus from the West in favor of a reannexation to Russia.

To complicate the situation further, Belarusian parliamentary elections had failed to attract enough voters to elect a new Supreme Council in February 1995. The old Supreme Soviet, still under the leadership of Myacheslav Hryb, was kept alive, but Lukashenka made it clear he considered it illegal and rejected several pieces of its legislation. On Belarus's five-year independence day, 27 July 1995, the Popular Front held a public demonstration which grew to nearly one thousand people and erupted into violence near the end. This gave Lukashenka serious cause for reconsideration. Since then, he has increased his contacts with the West, established closer ties with Belarusian economic organizations, instituted an emergency "Economic Crisis Program" of his own, and ceased to emphasize Russia in his public speeches, even saying that "several mistakes have been made by the presidency." It is still unclear where plans to unify the two countries are headed, but they can certainly not be ratified until Lukashenka arranges for the election of a new Supreme Soviet.

Since his election, Lukashenka has also actively campaigned for the unification of the three Slavic states. In his television address on 29 August 1994 he called Russia "elder brother" of the Belarusian people and said he stood "not only for unification of Belarus and Russia, but for all Slavic nations." This declaration puts Lukashenka in the political camp which supports Russia as the great power leader of a separate Slavic bloc within the CIS, although he usually refers to the three eastern Slavic states and does not include Slovakia, the Czech Republic, or Serbia (as, for example, Zhirinovskiy does). Lukashenka's pro-Slav union stance aligns him with the left-wing communist and pan-Slav parties in Belarus. He is also like these groups in other ways. On market issues, Lukashenka flatly opposes making land private property and supports the continuation of collective farming.

In sum, Belarus has experienced its own version of the tumultuous and bewildering search for identity occurring in every CIS country. Although the Moscow/Minsk Axis is still a distinct entity, and could become the nucleus for reintegrating the CIS states, Belarus is going through a maturation process. If Lukashenka can establish absolute power over weaker nationalist elements, Belarus is quite likely to enter into a confederation with Russia, becoming an appendage, which might worry some Central European states, such as Poland.

## 6.1
## Economic Treaty with Belarus Issued

Agreement

*Izvestiya*, 8 February 1992 [FBIS Translation]

*Editor's Note:* Early in 1992, Belarus entered into a bilateral economic treaty with Russia. Parliamentary Chairman Shushkevich accepted the fact that Belarus was economically dependent on Russia's markets, although he tried to open avenues that would reorient Belarusian trade more to the West and tried to maintain Belarus's neutrality on defense matters. Despite his efforts, it was clear from the beginning of the CIS that Belarus was sovereign and independent only in words.

For the purpose of creating the most favorable economic and legal conditions for the entrepreneurial activity of all kinds of economic structures and the development of market relations and broad economic ties between the Russian Federation and the Republic of Belarus, the Contracting Parties have agreed on the following:

Article 1. Trading and economic relations between the Russian Federation and the Republic of Belarus are carried out within a single economic area on the basis of mutual advantage.

Article 2. The organs of power and management of the parties to this agreement, both at their centers and local level, take all necessary coordinated economic, legal, and organizational measures ensuring freedom of enterprise and the sale, acquisition, and movement of goods, work, services, and investments on their own territories and between the Russian Federation and the Republic of Belarus.

Article 3. A single monetary unit (the ruble) is used as the means of financial settlement, money circulation, granting of credit, and other kinds of financial transactions on the territories of the Russian Federation and the Republic of Belarus.

The governments of Russia and Belarus and their central banks provide arrangements for the unhindered movement of all kinds of monetary resources.

The central banks systematically inform each other about the volumes of credit and money issue and the state of money circulation.

Article 4. Enterprises and organizations of all legal organizational forms registered on the territory of the Contracting States are entitled to set up subsidiary enterprises, branches, departments, offices, and other individual subunits on the territory of the other state without let or hindrance provided the appropriate legislation is observed.

Article 5. In their activity and economic transactions on the territory of the other Contracting State, enterprises and organizations of all kinds of ownership comply with the latter's tax system.

Article 6. The organs of power and management of the

Russian Federation and the Republic of Belarus pursue an active anti-monopoly policy, support the new economic relations, and do not allow the adoption of normative acts which restrict the free development of market relations.

Article 7. For the purpose of priority implementation of bilateral treaties on deliveries of goods, the Russian Federation and the Republic of Belarus conduct a coordinated export-import policy with respect to goods and services of substantial significance to the exchange of commodities between the parties to this agreement.

The governments of the Contracting Parties pledge not to introduce additional payments for transportation across their territory (transit) by any means of transportation, including pipelines. This procedure applies to all kinds of freight, including export and import freight.

Article 8. A bilateral intergovernmental commission is created for the elaboration and pursuit of specific measures in execution of this agreement and the monitoring of its observance by the government of the Russian Federation and the Republic of Belarus.

## 6.2

## New "Comprehensive Union" Created with Belarus

Vitaliy Portnikov

*Nezavisimaya Gazeta,* 22 July 1992

[FBIS Translation]

*Editor's Note:* Prime Minister Vyacheslav Kebich, former Belarusian Communist Party chairman, succeeded in deepening the union between Belarus and the Russian Federation in July, cementing Russia's military role in Belarus and consolidating the quasi-confederation between them. The border between Belarus and the Baltic states represents a critical issue in relations between the Baltic republics and Russia. Kebich's allusion to the "closing" of this border has an ominous ring to it, which sounds as though calculated to warn the Baltic states against too independent a line. By the end of 1992, Kebich had won his battle with Stanislav Shushkevich over the question of a military alliance with Russia. Though Shushkevich continued to prevent Belarus from signing the CIS Collective Security Treaty, the bilateral treaty soon became a more powerful short-term tool in Russian leaders' hands than the Collective CIS Treaty.

After six hours of tiring coordination Belarusian Prime Minister Vyacheslav Kebich, Russian Federation Acting Prime Minister Egor Gaydar, and leaders of both republics' ministries signed a package of documents which the Belarusian premier described as a "comprehensive union." "Neither I myself nor my predecessors ever succeeded in signing such an agreement as the one we are signing today," Kebich emphasized. He urged journalists to draw special attention to the document on military cooperation as being most important. The Belarusian premier recalled that this aspect was absent from the interstate treaties concluded by the Russian Federation with Kazakhstan and Uzbekistan earlier.

In addition to military union, the Russian-Belarusian accords embrace very broad cooperation in the economy—even including setting up an interrepublic economic coordination council. Egor Gaydar concurred that the agreement is a step in the direction of a confederation within the CIS framework (Kebich said "tiny little steps") and expressed hope verging on confidence that some CIS countries—Kazakhstan and Kyrgyzstan, above all—will join this union.

This statement by the head of the Russian government takes the accords reached between the Russian Federation and Belarus beyond bilateral contracts, as it were. Answering a question from *Nezavisimaya Gazeta*'s correspondent as to whether the Russian-Ukrainian "reconciliation" at Dagomys had influenced the conclusion of the Russian-Belarusian union, Premier Kebich "unequivocally declared"—yes. It obviously became clear to the Belarusian leaders after Dagomys that there simply is no alternative to rapprochement with Russia, but Kebich refused to answer a question about the possibility of Ukraine's subscribing to the Russian-Belarusian accords.

Russia and Belarus also agreed, according to V. Kebich, "on the joint closure of certain borders." To all appearances, it may be a question, above all, of the borders with the Baltic countries, which will now have to regard Belarus as a definite ally of Russia (although the Belarusian premier remarked to journalists that the signing of agreements with Russia does not mean that Belarus has abandoned its policy of equidistance from its neighbors).

## 6.3

## Shushkevich, Kebich Differ on Confederation

Igor Sinyakevich

*Nezavisimaya Gazeta,* 19 September 1992

[FBIS Translation]

During a visit to Germany, Belarusian leader Stanislav Shushkevich gave a news conference at which he expressed his attitude toward plans to create a confederation within the framework of the CIS. According to ITAR-TASS, he stated: "My attitude toward this question is not what the newspapers have been writing recently. The issue of creating a confederation was missed in 1990, and after 1990 it is very difficult to return to it. Unfortunately, later too, in 1991, the USSR president did not heed advice on this issue. I believe a confederation is now hardly likely. Anyway, I do not have authorization from parliament to sign documents on closer interaction beyond that within the CIS. It is necessary to achieve truly legal relations on the basis of the documents already signed and not go further."

References to the absence of authorization are the usual device used by the Belarusian Supreme Soviet chairman to avoid souring relations with Russia and at the same time not to follow in the wake of its "imperial policy." Like any "centrist," Stanislav Shushkevich merely reflects the real correlation of political forces.

Belarusian Premier Vyacheslav Kebich is more pro-Russian than the head of parliament. This is explained by economic pragmatism, on the one hand, and, on the other, by the need to strengthen his position on the eve of the referendum on early elections to parliament. The word "confederation" was heard for the first time in Moscow on 20 July during the signing of a package of Belarusian-Russian bilateral agreements. It may be assumed that this was a sort of "trial balloon" with the aim of eliciting public reaction to the said innovation. As was to be expected, the opposition, in the shape of the Belarusian People's Front, mounted a campaign of criticism of the Moscow accords and designated them the result of "the capitulatory position of the comprador-style government of Belarus." Understandably, the premier will now be more cautious regarding the fraternal embraces of the "great eastern neighbor." All the more so inasmuch as Russia has already broken the 20 July agreements. In accordance with the latter, the base price of the Russian petroleum delivered to the republic was not to exceed 2,800 rubles per ton before the end of the year. But it was precisely for the cheap petroleum so needed by the Belarusian ex-communists on the eve of the referendum that Russia obtained through the Moscow agreements a whole range of military-political and geoeconomic concessions.

The correlation of forces in the republic is such now that neither the abrupt disintegration of ties with Russia nor the voluntary return of its sovereign rights as the result of Belarus's entry into a confederation will be welcomed by the public.

## 6.4

**Security Commission Chairman on Treaty**

Interview by Igor Sinyakevich

*Nezavisimaya Gazeta,* 17 March 1993

[FBIS Translation]

[Sinyakevich]: Many Belarusian politicians today speak of the need for the republic to accede to the CIS Collective Security Treaty. There are also opponents who think it necessary for Belarus to maintain its neutrality. What do you think?

[Hryb]: If you look at questions of the security of the Republic of Belarus from the legislative viewpoint, we have determined our stand. At the Tashkent meeting of the CIS heads of state and heads of government, Stanislav Shushkevich, the head of our delegation, did not sign the treaty whereby the republic would have joined the collective security system. Proceeding from this fact, we are working out our own stand regarding the creation of the armed forces of the Republic of Belarus.

[Sinyakevich]: In a recent interview, however, Nikolay Shorynin, the Belarus prime minister's advisor, declared that the republic would sooner or later accede to the Collective Security Treaty. Are any changes expected in the republic's official stand on this matter?

[Hryb]: It is impossible to give a clear answer on whether we have gained anything from not acceding to the Collective Security Treaty, or whether we have lost by doing so. Many deputies and representatives of the executive branch think that, had we been among the states that signed the Collective Security Treaty, we would be finding it easier now to resolve the questions of securing arms and material for our armed forces. Belarus has concluded bilateral treaties on military and technical cooperation with Russia and Ukraine. But these agreements are not always fulfilled. There are various forces hindering it. We cannot simply ignore this. Of course, our sovereignty and our neutrality are very important and valuable notions. World experience shows, however, that neutrality is not to be interpreted in so lopsided a manner. Indeed, the neutral European states are members of a European security system. So there are different opinions, and everything will depend on our parliament's deciding whether we remain in a position of pure neutrality, or we accede to the Collective Security Treaty with certain reservations. Such as, for example, a reservation to the effect that we will not be supplying our armed forces for actions on the territories of other countries. Provided, of course, that this is accepted by the other parties to the treaty.

[Sinyakevich]: Does the Belarusian leadership regard military cooperation with Russia as enforced and temporary, or is this a strategy?

[Hryb]: As I see it, we should orient ourselves toward cooperation with Russia over a long period. There are also economic factors prompting us to cooperate: Russia has raw materials while we have a concentration of manufacturing sectors. It is not so easy to break into Western markets filled with goods, whereas Russia is a fine market for our products. I believe that we should be interested in a very close, well-disposed, and intimate cooperation with Russia. I see this as giving us advantages and adding appeal to our plan.

## 6.5

**Communique Notes Importance of Ties**
ITAR-TASS
18 March 1993 [FBIS Translation]

*Editor's Note:* By March 1993 a turning point had been reached in Belarusian–Russian relations. Russia had agreed to the creation of a joint monetary and banking system, based on the ruble. These plans ran into difficulties, originating in the different paces of economic reform espoused by the two governments.

Minsk reported today on the results of the visit to Belarus by a Russian parliamentary delegation led by Ruslan Khasbulatov, the parliamentary speaker. "The sides have expressed their resolution to strengthen their treaty and legal basis" the document notes.

The sides spoke in favor of bringing closer together their legislation on socio-economic questions, and also creating in the near future a concrete mechanism for the functioning of a common credit and monetary system based on a common monetary unit—the ruble—the currency of the Russian Federation.

The parliamentary heads agreed to begin work immediately on drawing up proposals to create the legal basis for a single monetary system for the two countries, to implement coordinated budgetary, tax, credit, customs, and price policies, and to form a single economic and customs union. Both sides confirmed their interest in interacting during the implementation of economic reforms. They noted the constructive approach of both sides in tackling problems in the military political sphere.

The chairman of the Russian Federation Supreme Soviet expressed his satisfaction at the ratification by the Belarus Supreme Soviet of the START I treaty.

## 6.6

**Kebich Endorses CIS Collective Security Treaty**
From the "Presidential Bulletin" feature compiled by Andrey Pershin, Andrey Petrovskiy, and Vladimir Shishlin; and edited by Boris Grishchenko
Interfax, 18 March 1993 [FBIS Translation]

*Editor's Note:* In the first half of 1993, Vyacheslav Kebich persistently lobbied for his pro-Russian, integrationist views, which included ceding supranational powers to CIS institutions. His position was that of the non-dogmatic, socialist wing of the former communists who thought of Russia as the "center" of a new confederation.

Prime Minister Vyacheslav Kebich told heads of local executive power bodies Thursday [18 March] that Belarus's position with regard to the CIS Collective Security Treaty should be reviewed. He considered that the republic should join the treaty, but on the following two conditions: that citizens of Belarus carry out their military service only in the republic itself; and that the Belarusian armed forces do not take part in the quelling of regional conflicts on the territory of other states. Kebich was convinced that the Belarusian army would soon prove incapable of defending the republic if Belarus did not sign the treaty. "The situation which is developing outside Belarus compels us to strengthen our borders," he said. The prime minister was of the opinion that a number of CIS states posed a considerable threat to the republic. These were states in which armed conflicts were taking place and political tension existed.

Kebich doubted the practicality and wisdom of the republic's achieving its declared goal of becoming a neutral state. He declared in the name of the government the necessity to back down from that goal made by parliament.

The prime minister considered it essential to set up an Economic Union of Commonwealth States. He said that this should incorporate agreed fiscal, credit and monetary, hard currency, and external customs policies. Economic trade relations would be built on the rigorous observation of agreements with the imposition of "harsh sanctions" on any state which ignored these. Kebich considered that the Interstate CIS Economic Court which was under formation must be given the appropriate powers. After stressing that the above was an official Belarusian governmental statement Kebich announced the results of his consultations on the formation of the Economic Union with the prime ministers of Russia, Kazakhstan, and Ukraine. He said that the Belarusian proposal had been backed unanimously.

The prime minister once again confirmed the government's policy on developing a "socialized economy." After speaking in support of the market and private ownership Kebich said that privatization in the republic should take place on a stage-by-stage basis. First of all, service sector enterprises would be privatized, then consumer goods manufacturing, agricultural enterprises, and, finally, steps would be taken to privatize large plants, the Belarusian prime minister announced.

## 6.7

**Kebich Favors Creation of CIS Economic Union**
Interfax, 18 March 1993
[FBIS Translation]

*Editor's Note:* As illustrated in the press note below, Prime Minister Kebich continued to cite two conditions on the deployment of Belarusian troops under the CIS Collective Security Treaty. These conditions, however, meant little in terms of Belarus's political position with respect to the CIS. The Collective Security Treaty determines "external borders" of the CIS (implying a single confederated state), and legitimizes Russia's CIS-wide "peacekeeping" role.

At a conference of heads of local executive organs on Thursday [18 March] in Minsk, Kebich said that such a union envisages implementation of a coordinated fiscal, monetary-credit, currency, and customs foreign policy, establishment of trade-economic relations based on unconditional observance of treaties, and usage of "tough sanctions" with respect to the parties which violate these treaties. The premier believes that the currently created CIS Interstate Economic Court should have appropriate authorities.

Kebich made the point that the premiers of Russia, Kazakhstan, and Ukraine have unanimously supported this initiative put forward by the chairman of the Belarusian Council of Ministers in the course of consultations devoted to the creation of the Economic Union.

At the conference Kebich also proposed to revise Belarus's attitude toward the CIS Treaty on Collective Security. The head of the Belarusian government believes that the republic should enter the system of the Commonwealth's collective security under two obligatory provisions, namely, the citizens of Belarus will serve only on the territory of their homeland and the Belarusian armed forces will not participate in settlement of interethnic conflicts on territories of other states.

## 6.8

**Country "Quietly" Leaving Ruble Zone**
Egor Glukharev
*Kommersant-Daily,* 20 March 1993
[FBIS Translation]

*Editor's Note:* It was becoming apparent by the end of March 1993 that Belarus's slow pace of reform was incompatible with Russia's comparatively lively one. This was the primary reason for Russia's reluctance to become the primary creditor and monetary regulator in a CIS ruble zone. Belarus, after realizing that it would not receive special Russian subsidies or discounts on the oil it imported from Russia, decided to introduce its own ruble as a separate and freely floating currency on the Interbank Hard Currency Exchange of Belarus. By floating the Russian ruble and forcing 100 percent of earned rubles to be sold to the Central Bank, Belarus transformed the Russian currency into a foreign one on the Belarusian market.

Representatives of the Belarusian leadership continue to state that their republic does not intend to leave the ruble zone. Recent developments, however, attest to the contrary. Yesterday a joint decree by the Council of Ministers and the National Bank of Belarus "On Regulating Settlement with Republics of the Former USSR" was officially circulated. Its content makes it possible to conclude that a national currency is, in fact, being introduced. *Kommersant-Daily* correspondent Egor Glukharev comments on the situation that has taken shape.

Despite the understanding reached in Bishkek at the end of last year by heads of states in the ruble zone a republic leaving the ruble zone is supposed to notify parties to the agreement at least two months in advance of the introduction of national currency. Belarus is putting its own bank notes into parallel circulation and is adopting decisions that amount virtually to quitting the ruble zone. In particular, the joint decree adopted "without much ado" by the Council of Ministers and the National Bank of Belarus, and officially circulated yesterday, envisions a transition to floating official quotations for the monetary units of former USSR republics. Quotations will be fixed on the results of trading at the Interbank Hard Currency Exchange of the Republic of Belarus. This is a de facto recognition of the fact that the Belarusian and Russian rubles are two different currencies, and the decree draws a very clear distinction between the two.

The occasion for the signing of the new decree is innocent enough—to solve the problem of settlements with Russia, namely, to create a reserve of Russian rubles through the obligatory sale of all their rubles by Belarusian commercial structures to the National Bank of Belarus at the official current rate. Subsequently, these rubles are to be sold on the domestic currency market at the exchange-quoted rate. The purchased currency must be used within twenty days to pay for contracts with foreign suppliers, and, failing this, the monies are to be sold back to the National Bank.

According to specialists, the decision to officially float the Russian ruble actually amounts to the introduction of a Belarusian monetary unit. As it tries to solve the problem of settlements with Russia, however, and given the deficit of Russian rubles and the lower purchasing power of the Belarusian ruble, Belarus will soon come up against the problem of devaluation of its national currency, like what happened to Ukraine at the end of last year.

Increasingly, the present decree envisions 100 percent sales of Russian rubles whereas the normal quota for the mandatory sale of hard currency, in particular the dollar, is 20 percent of export takings. Thus, a curious situation arises where Russian rubles are valued in Belarus higher than dollars, and have in fact become a foreign currency.

## 6.9

**Popular Front Warns Against Pro-Russian Orientation**

Vasiliy Romanovskiy
*Vecherniy Minsk,* 24 December 1993
[FBIS Translation]

The Board of the Belarusian Popular Front [BPF] has issued a statement deeming the results of the Russian elections as a victory of fascist and communist extremists.

"Political strife in Moscow has been aggravated," reads the document. "The events in Russia have testified to the danger of being in this country's neighborhood. Union or concord with Russia means being hostage to unforeseen and aggressive Russian policies, as well as the unstable situation in that country.

"The BPF believes that Belarus should immediately leave the CIS. The Belarusian-Russian border should become a fully established state border.... A 'common ruble zone' with the country that permanently stands on the brink of civil war is out of the question. Union with Russia, which is being supported by the government, will lead us into the deadlock of murderous Russian conflicts and lead to an economic and political breakdown. Any further orientation toward Russia will bring destruction to Belarus."

## 6.10

**Discussion of Economic Integration**

Gennadiy Ezhov
ITAR-TASS, 2 July 1994
[FBIS Translation]

*Editor's Note:* A year passed, allowing Russian leaders to assess the political costs of allowing Belarus to opt out of the ruble zone. These were judged to be too high. In April 1994 a bilateral treaty providing for the merging of the Russian and Belarusian banking and monetary systems was drafted and submitted to the two governments for deliberation. The treaty would cede Belarusian economic sovereignty to the Russian Federation. Negotiations were to require a full year, but by mid-1995 Russia and Belarus were ready to join their economies through separate agreements on banking and payments, customs, and other trade regulations.

Russian Prime Minister Viktor Chernomyrdin met his Belarusian counterpart Vyacheslav Kebich for negotiations today to discuss the implementation of the bilateral treaty on monetary merging.

Russia is ready to change some treaty articles and sign an agreement on amendments, the Russian premier told correspondents upon the meeting's completion. In his words, the new agreement will be concluded with Vyacheslav Kebich to "stop political games around the close economic integration of Russia and Belarus."

The treaty of April 1994 provides for the transfer of all powers of the Belarusian Central Bank to the Central Bank of the Russian Federation which runs counter to the current Belarusian constitution. It seems that the Belarusian Central Bank will have to pass over its regulating functions to the Russian bank, chairman of the Russian Central Bank and member of the Russian delegation, Viktor Gerashchenko said here today.

Asked about the issuance of a Russian state credit to Belarus, Kebich said that Russia is to issue 150 billion rubles to Belarus in the first half of 1994 under the agreement of 24 May. Only 20 billion have been issued, he added.

Belarus hopes to receive an additional credit in the second half of 1994 but this will depend on who controls the money emission in the single system of the two states. Russia is now ready to allocate 42 billion rubles to eliminate the consequences of the Chernobyl accident on Belarusian territory.

The negotiations did not discuss the second stage of Belarusian presidential elections, Kebich told correspondents. He stressed it is an internal affair of the state. At the same time, the Russian prime minister did not exclude a meeting with leader of the Belarusian presidential race and parliament deputy Aleksandr Lukashenka.

## 6.11

**Signs Monetary Merger Progress Protocol**

Gennadiy Ezhov
ITAR-TASS, 3 July 1994 [FBIS Translation]

Russia and Belarus today signed a protocol on the progress in the implementation of the April agreements on uniting their monetary systems. Russian Prime Minister Viktor Chernomyrdin and Vyacheslav Kebich, the head of the Belarusian government, put their signatures under the document. The document reflects, on the one hand, the progress in the unification process, and on the other, it outlines very tough (in terms of time), objectives to work out ways and mechanisms of carrying out the monetary emission and exchange of monetary notes.

Russian Deputy Prime Minister Aleksandr Shokhin said in an ITAR-TASS interview that concrete mechanisms could be drawn up by August.

The treaty on merging the monetary systems of Russia and Belarus, signed in April 1994, contained a provision that the Russian Central Bank should have the power to control and regulate the activities of the National Bank of Belarus.

The protocol says that, if a need arises to introduce certain adjustments into the documents signed earlier, this can be done. However, in doing this Russia will be guided not by political, but by purely professional, banking, and finance considerations, Aleksandr Shokhin said.

According to the deputy prime minister, this means that if the National Bank of Belarus acquires a different status from that recorded in earlier agreements, a number of points in those agreements will have to be adjusted accordingly. The two sides agreed to outline in full detail by August 1994 how emission is to be regulated. For instance, if the Russian Central Bank carried out the emission through the National Bank of Belarus by using the corresponding control mechanisms, this could be practiced instead of the Russian Central Bank taking over all the controlling functions, and it would not contradict the existing Constitution of the Republic of Belarus.

## 6.12

## Shumeyko Predicts Increased Integration with CIS

Interfax, 14 July 1994
[FBIS Translation]

Chairman of the Federation Council Vladimir Shumeyko is not inclined to overly exaggerate the consequences of the new Belarusian leaders' possible refusal to create a monetary union with Russia. He told journalists during his visit to Interfax agency on Wednesday that the Belarusian parliament had opposed such a union, because if the state loses control over monetary emission, it partly loses its sovereignty.

He said, however, that the Russian and Belarusian leaders would find a new form of integration, because integration as such reflects the vital interests of the two states. In his opinion, in the near future Russia's economic contacts with the CIS countries would become stronger, which will make their political ties stronger as well. Eventually, the Commonwealth of Independent States may develop into a confederation, he said.

In the meantime, the Russian Federation must establish borders with the former Soviet republics to protect Russia from the uncontrolled inflow of goods, drugs, and weapons from foreign states, said Shumeyko. The border regime will be different in different sections, he went on to say. It must be extremely stringent in the sections bordering on the Baltic states, while relations with those countries must be friendly.

He announced that as before he favored the idea of extending the mandates of the president and parliament. He also said that many public figures and statesmen shared this view.

## 6.13

## National Bank Seeks to Amend Monetary Union with Moscow

Interfax, 14 July 1994 [FBIS Translation]

The Bank of Belarus is insisting on amending the treaty of unification of its monetary system with Russia, signed in Moscow on 12 April. A source at the National Bank of Belarus (NBB) told the Financial Information Agency on Thursday that the bank's board had drafted amendments to be sent to experts at the Central Bank of the Russian Federation in the near future.

Among other things the amendments provide for changing article five of the treaty, because its current wording implies the abolition of the NBB. The Belarusian National Bank also suggests reserving in the treaty the right of the Bank of Belarus and the Central Bank of Russia to the joint control of emission, monetary, credit and currency policies. Besides, it proposes the coordination of credit emission and refunding rates.

The NBB insists on the exclusive right to independently carry out monetary emission in Belarus and control the activity of banks and creditor institutions. It is suggested opening correspondent accounts of the NBB at the CBR and of the CBR at the NBB.

The National Bank of Belarus is for setting the exchange rate of the Belarusian and Russian currencies on a parity basis proceeding from the real buying power of either currency computed by the national boards of statistics and irrespective of the free-market cross-rate of the Belarusian and Russian rubles against the U.S. dollar.

Instead of consolidating the state budgets of the two countries, the NBB suggests jointly coordinating the budgets of consolidated budgets in percent of their gross domestic products and jointly choosing methods and sources of covering the deficits, the minimum state debt, and terms of servicing it.

The proposed amendments provide for an inquiry by experts into the equivalency of incomes of either side after the abolition of customs duties.

Russia and Belarus early this month signed a protocol to pledge to amend the treaty, if need be. A Russian-Belarusian summit meeting is expected to take place in the first days of

August where negotiations on the unification of monetary systems would continue.

## 6.14

**Lukashenka Favors Pragmatic Relations with Russia**

Editorial from the "Vzglyad" program
Ostankino Television, 29 July 1994
[FBIS Translation], Excerpt

[Aleksandr Lukashenka] says that in the past his opponents had accused him of being anti-Russian, but this was not the case. He says: "In Belarus there is no politician who is more favorably inclined toward Russia than Lukashenka. You will not find such a politician in Belarus. Therefore, for me the concepts of independence and sovereignty are specific actions, specific actions by politicians. Up to now they have been just a routine uproar, particularly in parliament, in the Supreme Soviet. This was played as a political card. I would like to stop playing this political card.

"I am a pragmatist, a man from industry. When your Aleksandr Shokhin visited us—Aleksandr Shokhin has a pragmatic view of relations between Russia and Belarus—we reached a common understanding remarkably quickly on how to tackle this problem. We reached the understanding that today it is necessary to tackle urgent human problems, economic problems. We will tackle economic problems. Then we will tackle such things as confederation, federation, or former Soviet Union. Unfortunately, at present we tried to tackle this problem from the top downward, whether to be a confederation or a federation. In my opinion, if we go along this road we will not achieve any confederation or federation; we will not achieve any drawing together but more likely we will soon not even be speaking to each other. That is, I am in favor of the problem of Russian-Belarus relations being tackled from the bottom upward."

## 6.15

**Parliamentary Speaker Views Integration with Russia**

Viktor Kuklov
ITAR-TASS, 29 July 1994 [FBIS Translation]

*Editor's Note:* Below, Myacheslav Hryb reacts to the economic pressures partially responsible for Belarus's preparedness to sign a treaty unifying the monetary systems of Belarus and Russia. All supplies bought from Russian factories would be purchased at the same price paid by Russian consumers, which would mean reductions of 50 percent, or more.

"The Russian armed forces' presence in Belarus is a historical necessity, an unavoidable legacy. We cannot ask our neighbor to withdraw all its troops immediately because they need to be stationed somewhere. A bilateral agreement has been signed, which recognizes that the situation is temporary," chairman of the Belarus Supreme Soviet Myacheslav Hryb said in an interview in the Zvyazda newspaper, which is published in Minsk. He also noted that Russia is not to blame for this state of affairs and that "the Commonwealth of States can be compared to a family that owes its existence to skillful compromise."

Belarus agreed with the timetable that Russia put forward for the withdrawal of its troops and will help with transit. Russia, in turn, will give the republic discounts on the purchase of fuel.

The speaker of the Belarus parliament stressed again that the signing of the treaty on the unification of the Belarusian and Russian monetary systems did not mean that they had actually been unified. The Supreme Soviet needs to ratify a number of articles in the treaty before this can happen. Nine articles are currently valid, involving taxation and customs tariffs among others.

"I personally support integration, but not by huge steps," Myacheslav Hryb said, "although I must admit that it would be very much to our advantage to buy fuel, for example, at the prices for which Russians buy it. Russia currently supplies us with oil at $80 a ton, whereas it supplies its own consumers for $40. This, of course, affects the price of our own output. In addition to this, Russia is still the main market for our goods. We cannot sell them in Poland or in the West."

## 6.16

**President Pins Hopes for Security on Russia**

ITAR-TASS, 10 August 1994
[FBIS Translation]

*Editor's Note:* Aleksandr Lukashenka was elected president of Belarus in June 1994. He has been described by critics as a megalomaniac, a dictator, and a paranoid personality who has done much to turn the clock backward in the CIS, playing into the hands of Russian radical nationalists and empire-builders. Lukashenka was thirty-eight years old at the time of his election, having formerly achieved the sole distinction of being a collective farm director who had an understanding for his electorate. The document which follows is merely the precursor to Lukashenka's subsequent actions to join Belarus with Russia, and while doing so to create a deep constitutional crisis in Belarus over presidential and parliamentary rule.

"Who are we to ensure our security with? My answer is—Russia. Our economy and our army are most closely linked with Russia," Belarus President Aleksandr Lukashenka told ITAR-TASS today. He added that "many Belarus enterprises were part and parcel of the common military-industrial complex and are tightly 'linked with Russia.' What is more, they manufacture goods of world quality standard which are supplied both to the Belarus and the Russian armies. There is a double link between the Belarus and Russian armed forces: The Belarus army cannot be ensured with everything it needs without close economic ties with Russia. There are also some purely military aspects, too. I do not intend to spoil the good relations existing between out two armies."

"We have agreed to conclude a treaty of friendship and good-neighborly relations," Lukashenka stated. "It is strange that no such agreement was signed so far. We have completed the formation of all the articles of the treaty to merge our monetary systems. But we have to catch up at least a bit with Russia in order to carry out this unification."

The Belarus president stressed that he attached priority importance to the republic's relations with Russia. "Moscow is not trying to dictate the pace of the rapprochement between our two states," he added. "There is no sign of any diktat in our relations. On the contrary, I was somewhat surprised during the negotiations by the understanding, tact and even, I may say, emotion with which the Russian leadership treated Belarus and its new leadership."

Speaking about his view on the problems of the Commonwealth of Independent States (CIS), the president said: "Today it is our duty to safeguard the slightest form of unity left among the states of the former union. The CIS is the only thing left and the only thing that holds us together to a certain extent. This is why we must strengthen and expand the Commonwealth. And I shall do everything in my power to help raise the extent of the Commonwealth's integrity to an increasingly higher level."

## 6.17

### Lukashenka Urges Unification of Slavic States

From the "Presidential Bulletin" feature compiled by Nikolay Zherebtsov and Andrey Petrovskiy; and edited by Vladimir Shishlin, Interfax, 29 August 1994
[FBIS Translation]

*Editor's Note:* Aleksandr Lukashenka does not believe in the Belovezh Forest agreement. He would have preferred a confederated Soviet Union, with planned and coordinated production managed by state enterprises. He was to give many television addresses urging the reunification of the three eastern Slavic states, and even threatening dire consequences for Ukraine if it did not give up its sovereignty and become part of the Slavic Union.

President Aleksandr Lukashenka, speaking on Sunday night over the national TV, doubted the expediency of the construction in the Russian city of Lipetsk of a tractor-building plant. He said that the plan is being developed by the Russian government.

Lukashenka believes that there is no sense for Russia to spend money for the construction of a plant similar to the Minsk Tractor Plant and proposed to use better the available capacities for the construction of tractors in Belarus.

Calling the Russian people an "elder brother" of the Belarusian people, Lukashenka said that he stands not only for the unification of Russia and Belarus, but for all Slavic nations. "The Belarusian people is sincerely for the unification with Russia," he believes.

*Interfax Note:* Lukashenka, as is known, was the only deputy in the Belarusian parliament who voted in 1991 against the ratification of the Belovezh Forest agreements on the creation of the Commonwealth of Independent States and the elimination of the Soviet Union. During his election campaign, the future president, addressing deputies of the Russian parliament, spoke in favor of the union of three Slavic nations, Russia, Belarus, and Ukraine.

He said that he is flatly against the idea of making land private property. He said that there are many individual farms in the republic, whose land is used with less efficiency than the land of collective and state farms.

He doubted the expediency of auctioning production enterprises and buying them by their work collectives. The auctioned enterprises in Belarus cannot elevate labor productivity, said Lukashenka. Tomorrow, 30 August, President Lukashenka will be forty. Interfax congratulates the Belarusian president.

## 6.18

### "Dynamics" Forces Lukashenka to Face Many Issues

Yuras Karmanov
*Nezavisimaya Gazeta,* 20 December 1994
[FBIS Translation]

The "Eastern vector" in Minsk's policy again made itself felt during the Budapest summit of the CSCE member states. In his statement Aleksandr Lukashenka expressed concern about NATO's moving its borders to Belarus. In view of the Brussels failure of the Russian "Russia-NATO" program and Belarus's subsequent refusal to join the NATO Partnership for Peace program, the official Minsk position should be seen as having been made clear once and for all. The dynamics of geopolitical changes in Eastern Europe, conditioned by the drift by the Baltic states and the Visegrad group toward NATO, is forcing Belarus to define its own place. The memorandum on security guarantees, signed in Budapest by Great Britain, the United States, and Russia, to a large extent eases confrontation between countries orientated toward Russia and those oriented toward NATO, without, however, removing the cause of the confrontation. In practice this will mean that Russia's aspiration to keep Belarus within the sphere of its political influence will be growing in direct proportion to the growth of confrontation in the East European region. In this process, its energy dependence on Russian oil and gas is becoming the main lever to pressure Minsk. Belarus's debt to Russia for gas supplies has exceeded one trillion Russian rubles. The conflict between Minsk and Ankara, provoked by a mild wave of spy mania, with a subsequent rupture of diplomatic relations, in the runup to a switchover to world prices in mutual settlements for energy carriers between Minsk and Moscow deprives the republic of an alternative to Russian oil. This alternative could have been the Baltic-Black Sea terminal [kollektor] envisioning the participation of Turkey, the Baltic countries, and the Visegrad group. The pending agreement on a customs union, with a leading role for Russia, creates all prerequisites for Belarus's economic incorporation.

The emerging military-political integration of Russia and Belarus within the framework of the Collective Security Treaty, signed by ex-speaker Stanislav Shushkevich but not ratified by the Belarusian parliament, can become a starting point toward the formalization of two confrontation blocs in Eastern Europe. The so-called "cordon sanitaire" option, as a possible buffer between Russia and Europe with the participation of Ukraine, Belarus, and the Baltics, can be counted out automatically. The official moves by Belarusian parliament Speaker Myacheslav Hryb can be seen as an attempt to create a counterweight to the Belarusian Foreign Ministry's official position. Recently he has stepped up his activities on the world arena, by visiting the United States, Poland, and Britain. Hryb does not consider it necessary to coordinate his moves with the republic's Foreign Ministry. Attending the fortieth session of the North Atlantic Assembly, Hryb spoke in support of Belarus's earliest possible accession to Partnership for Peace. The head of parliament did not read in the Visegrad group's wish to join NATO any negative consequences for Belarus. The Belarusian parliament speaker described their position as quite reasonable and justified, so saying at a press conference on the results of his visit. Moreover, when in Warsaw, Hryb began to discuss the question of Belarus joining the Visegrad group, in which Lithuania and Ukraine are also seeking membership, and even obtained a promise of support from Hungary and Poland.

In all probability, Myacheslav Hryb's attempts to create an alternative to Lukashenka's one-sided orientation toward the Kremlin are dictated by the awareness of the fallout within the country from open confrontation between Moscow and the West. Given that the Supreme Soviet has not as yet defined the country's foreign policy doctrine, disputes about Belarus's geopolitical affiliation are already taking on the character of open polemics. The mounting pro-Russian movement, which has risen on the wave of the "Eastern vector" proclaimed by Lukashenka, is running the risk at a certain point of colliding with the "Western vector" of the democratic opposition, which is seeking support in the opposing camp.

The attitudes in Belarus to the events in Chechnya demonstrate the remoteness of the two poles. Addressing a press conference by the democratic opposition, Belarusian National Front leader Zenon Poznyak described the Chechnya developments as the beginning of the restoration of an imperial Russia. He said that the Belarusian National Front Seym has decided to send Dzhokhar Dudaev a letter expressing admiration for the fortitude of the Chechen people in the struggle for the right of nations to self-determination. The Belarusian National Front leader assessed the Belarusian president's silence as full support by Aleksandr Lukashenka for the Kremlin's position.

---

## 6.19

### Hryb Sees Russian Unification as Unconstitutional

From the "Presidential Bulletin" feature, compiled by Nikolay Zherebtsov and Andrey Petrovskiy; and edited by Vladimir Shishlin
Interfax, 29 December 1994 [FBIS Translation]

---

Parliamentary Chairman Myacheslav Hryb said neither he nor the parliament presidium had been asked to approve a recent proposal to convene a joint session of the Belarusian parliament and the Russian parliament's lower house to discuss unification between Belarus and Russia. The proposal came from the State Duma, Russia's lower house, which suggested convening such a session next March.

"Unification between Belarus and Russia would contradict our constitution," Hryb told Interfax in Kiev, where he led a parliamentary delegation on an official visit. "I see this as an initiative of some separate group of members of the Russian State Duma."

But he said the matter could be put on the Belarusian parliament's agenda if a group of its members proposed this.

Moscow "does not always" base its relations with Minsk on the equality principle, Hryb complained.

Unification was also the subject of an international congress in Moscow, which was held several days ago and decided to set up grassroots action committees to collect signatures under petitions to call referendums in Russia, Kazakhstan, Belarus, and Ukraine on the reunification of those countries.

"Only immediate steps to reintegrate the republics, bringing them together into an economic, political, and defense union, would enable each of them to surmount the total and deep crisis," the head of the International Congress of Industrialists and Entrepreneurs, Arkadiy Volskiy, told the forum, whose motto was "Toward reintegration through consensus."

He said the reintegration of Russia, Ukraine, Belarus, and Kazakhstan should go through four stages: creating an open alliance of the four post-Soviet states; lifting all restrictions on the movement of goods, capital, and labor between them; making the four countries' laws as similar as possible, and forming supranational administrative bodies. "We must appeal to the peoples of the republics and via referendums force their governments to start concrete reintegration processes," Volskiy said in an interview with Interfax. "I don't have the slightest doubt that the people would accept the idea of unification referendums."

However, another congress delegate, Nikolay Gonchar, a member of the Federation Council, the upper house of the Russian parliament, told Interfax: "At this point we can only talk of an economic union. As for a political union, all the forces who hold power in the member countries of the Commonwealth of Independent States would rise up against that and, as far as a defense union goes, some forces in the West as well would be against it."

## 6.20

**Press Release: Embassy of the Republic of Belarus to the United States of America**
4 January 1995

At a press conference held 29 December 1994 the Speaker of the Parliament of the Republic of Belarus Myacheslav Hryb informed journalists that parliamentarians of the Russian State Duma suggested to hold a joint session of the Belarusian and Russian Parliaments to discuss unification of Russia and Belarus into one state. Commenting upon this fact the Speaker stated that the issue was consulted neither with the president nor with the Supreme Council of the Republic of Belarus.

"The most important thing," M. Hryb noted, "is that this proposal contradicts our Constitution. We cannot support such a decision."

## 6.21

**Friendship, Cooperation Treaty Initialed**
Interfax, 24 January 1995 [FBIS Translation]

*Editor's Note:* Andrey Kozyrev's visit to Minsk in January 1995, as sherpa for President Yeltsin, marked the end of the negotiation phase between the two "brother" states. A treaty was prepared for signature during Yeltsin's upcoming visit; speeches were written; and announcements for the West and other third countries were drafted. Kozyrev took full advantage of the occasion, pointing to the broader implications of the Agreements, to say that the Moscow/Minsk access could even "become a new model," obviously referring to a framework for confederation which other CIS members could emulate.

Culminating Russian Foreign Minister Andrey Kozyrev's three-day visit to Minsk in January 1995, Belarus and Russia initialed a treaty on friendship and cooperation, and concluded a consular convention and a protocol on consultations between the Russian and Belarusian foreign ministries.

Kozyrev told a press conference in Minsk on Tuesday [24 January] that "a good foundation has been laid for President Yeltsin's visit to Belarus."

Yeltsin is expected to visit Belarus at the end of January or the beginning of February.

Kozyrev stated that "a tendency toward the two countries' economic integration has made itself felt." He also noted that the restoration of economic contacts does not mean an automatic roll-back to the past. Economic contacts must be resumed on a mutually advantageous basis, Kozyrev stressed.

"The year 1994 made it clearer what the CIS actually is—a form of civilized divorce of former Soviet republics, or a model of new association," he said. He noted that "the form of civilized divorce is of great importance," adding that "there are clear integration processes in relations between Russia and Belarus."

## 6.22

**Press Briefed on Results of Visit**
Minsk Radio
24 January 1995 [FBIS Translation]

The visit of Russian Foreign Minister Andrey Kozyrev to Minsk has ended. A news conference was held today in the national press center for Belarusian and foreign journalists during which the head of the Russian foreign political department summed up the results of the visit and answered questions from the journalists.

[Begin Kozyrev recording.] Today's conversation with Mikhail Chyhir gave me hope. He confirmed the aspiration, which is also present on our part, to put into effect and realize an agreement on a customs union, on payments, and so forth, i.e., those agreements that have recently been reached and enable us to speak about Belarus and Russia going over to a new quality of economic relations. I think that it is even setting a certain example for the CIS. You know that agreements of approximately the same character are taking shape between us and Kazakhstan. In a number of aspects, our two countries are setting new patterns, new levels of cooperation. Yesterday's meeting in the Supreme Soviet gave me hope and joy. During that meeting, two members of our delegation, representatives of the Duma and the Council of Federation, participated actively. They are both present here now at the news conference; therefore, if you have any questions, you can put them directly to our legislators. The main thing is that both our legislators and the Supreme Soviet [of Belarus], as I understand it, have aspirations to support those agreements, both political and economic. There is a good chance for ratification of the political agreement after their signing by the presidents—at least that was my impression after yesterday's meeting. And what is not less significant, there is a good chance that the legislators will try to work in unison so that the laws they adopt will reflect, of course, internal specifics and, first of all, the interests of each state, as well as the foreign policy and other aspects of activities, but at the same time making sure that they comply with each other. Then economic cooperation will develop actively too.

In short, the meetings that have been held leave a very good and optimistic feeling, including the meeting yesterday with representatives of creative intellectual workers, which show that, although with difficulties, a new model of equal mutually beneficial, good-neighborly partners' relations between the two brotherly peoples and states is taking shape. [End recording.]

## 6.23

**Article Criticizes Customs Pact with Russia**

Uladzimir Kulazhenka

BELAPAN, 26 January 1995 [FBIS Translation]

[Article by Uladzimir Kulazhenka: "If the Belarusian-Russian Agreement on a Customs Alliance Is Ratified, Belarus Will Lose Its Economic Independence," from the "Analysis, Commentaries, Forecasts" feature No. 15, 18–24 January 1995.]

During the last two weeks passions have raged in the Supreme Soviet of Belarus around the detailed redistribution of the expenses sector of the budget, for the purpose of increasing expenditure on health protection. Conditions became even more tense when the MPs [Members of Parliament] received a package of documents signed by the governments of Russia and Belarus which, according to the Prime Minister Mikhail Chyhir, will become the basis for a conclusion to the Treaty on Friendship, Cooperation, and Good-Neighborly Relations between the two states. MP Uladzimir Zablotski announced that it was exactly the tactless action of signing the agreements that will determine the shortage of resources for health protection, education, and culture. In the newspaper *Literatura I Mastatstva* the signed agreements were called "the deal of the week." Members of the "shady cabinet" of the Belarusian Popular Front consider that the agreements signed about the establishment of a customs alliance between Russia and Belarus, about the main principles of the establishment of financial-industrial groups, and others, will result in "the liquidation of the independence of the state, the complete poverty of the Belarusian nation, huge losses, and the irreversible desecration and destruction of national, material, and cultural possessions."

A particularly great deal of criticism was directed to the customs alliance, on the main principles for the creation of financial-industrial groups, and on the ensuring of mutual convertibility and the stabilization of the exchange rates of the Belarusian ruble and the Russian ruble. The extent to which this criticism is justified can be judged by an analysis of the Agreement on a Customs Alliance between the Republic of Belarus and the Russian Federation. This is not only an economic, but a political act. So, in the Protocol on the Introduction of a Free Trade Regime Without Exceptions or Limitations Between the Republic of Belarus and the Russian Federation it was recorded, "the Parties, confirming the earlier bilateral agreement concerning free transit and maintenance services for military establishments of the Russian Federation on the territory of the Republic of Belarus, have agreed that from 1 January 1995 a free trade regime will be introduced between the Republic of Belarus and the Russian Federation on full scale with no exceptions or limitations. . . ." Commentary is unnecessary. Apart from anything else, Belarus will keep the troops of a foreign country on its land for nothing. The conclusion of a customs alliance may look like an insignificant event far from the needs of the people (goods are conveyed somewhere or other across the border, they are sent somewhere, some sort of stock taking is done, etc.). At first glance, this has no meaning for the people at all. In fact, this is far from true. It was not by chance that the Russian side accelerated the official signing of the agreement.

Sensing the unpredictable atmosphere and the real possibility that the political situation in Belarus may change after the forthcoming elections to the Supreme Soviet, the Russian side demanded that the "Protocol for the Introduction of a Free Trade Regime . . . " should contain a note about the necessity of internal state procedures for the legal implementation of the bilateral agreements about the preservation of the Russian Federation's military establishments on the territory of Belarus, and the agreement on a customs alliance, dated 6 January 1995, of which Moscow is to be notified through diplomatic channels.

At first the Belarusian government kept the agreements signed with Russia secret, hoping it would succeed in not submitting them for ratification. But at the demand of the opposition it had to distribute the texts of the agreements among the parliamentarians. And this was not mere chance because the content of the customs agreement provokes a lot of questions.

For instance, Point 2 of Article 1 reads: "The Contracting Parties define the customs alliance as an economic amalgamation of the states" though it will be based on certain principles. When and where was a customs alliance the economic amalgamation of states? Undoubtedly, they pursued political ends. Article 2, "Mechanism and Stages of the Creation of the Customs Alliance," is disadvantageous for Belarus. It specifies two stages of making the alliance. In the first the Parties are to carry out "the unification of the foreign economic, customs, financial, tax, and other laws concerning the foreign economic activities within four months from the date of signing the Present Agreement" (it means before elections to the Belarusian parliament). It is clear why one of the Contracting Parties is in such a hurry. It is not clear how the present composition of the Supreme Soviet, which is near its expiration and overloaded with lawmaking work as it is, can accept this. The following note in Article 2 is particularly interesting: "the power to regulate relations between the customs alliance and other countries and international organizations shall be delegated to one of the Contracting Parties." It is evident that Russia will hardly delegate power to Belarus. One can see that here the agreement defines Russia's aspiration to take upon itself the regulation of relations with third countries on behalf of our republic. If the customs agreement is ratified, then in accordance with it common customs tariffs and a common customs policy will be established toward third countries. This means that Belarus's independence in the international economic arena will be liquidated. And we will be deprived of the opportunity to enjoy our unique geopolitical situation. In a disguised form it is Russia's aspiration to establish a supernational organ which will control customs services. Article 6 reads: "The Contracting Parties shall ensure the unity of control of their customs services . . . ." But this will be later, by order of the governments, without consent from the parliaments. And further: the Contracting Parties "shall arrange joint control over the circulation of goods and vehicles. The procedure and organization of official control shall be stipulated by special agreements (protocols) between the customs departments of the Contracting Parties." It is clear what the "joint control" means. Belarus will lose the right to independent actions in the sphere of foreign economic activity, the more so, as this will be carried out "on condition of reliable customs control at the external borders." This means that our external borders will be "our own" for Russia as well.

In the "Memorandum on the Expanding and Deepening of Belarusian-Russian Cooperation" this was spelt out even more candidly. Here the task was determined as the security of "effective combined defense" of external borders of the customs alliance. Following the Agreement on a Customs Alliance, Belarus will lose hundreds of millions of dollars, as customs duties will be revoked not only on those goods which Russia sells to Belarus or the other way round, but also on those which Russia sells to a third country and shifts across the territory of Belarus. The scale of external export is incomparable between Belarus and Russia. Nevertheless, the Russian Federation as the deciding side, concerned for its interests, took the opportunity to include a note in the "Memorandum on the Expanding and Deepening of Belarusian-Russian Cooperation" which read, "supplying fuel to the Republic of Belarus will become more comfortable if it takes active steps to liquidate its debts for fuel supplies from the Russian Federation." As they say, friendship is friendship, but money is something else.

The agreement on the customs alliance is so beneficial for Russia and so damaging for Belarus that Russia agreed to the following: "the current agreement will be temporarily applied from the moment of signing and will come into force on the date both sides exchange notifications that all the necessary internal state procedures have been fulfilled." As apparent, Belarus will lose all sovereignty and independence if the customs agreement is ratified.

## 6.24

### Upbeat on Russian-Belarusian Relations

by BELINFORM for TASS
ITAR-TASS, 3 February 1995 [FBIS Translation]

*Editor's Note:* This article summarizes the scope of the package of articles signed by Moscow and Minsk, whose leaders now talk openly of using the agreements as a model for the other CIS countries.

Russian Duma Speaker Ivan Rybkin completed the first day of his Belarusian visit today with a meeting with Executive

Secretary of the Commonwealth of Independent States, Ivan Korotchenya.

The one-hour-long meeting of the Duma speaker and Belarusian President Aleksandr Lukashenka focused on aspects of bilateral relations, including the official Belarusian visit of Russian President Boris Yeltsin scheduled for late February.

"Belarus and Russia cannot live without each other. We have always been brothers and must remain such. I think that a politician who has different opinion is a short-sighted person," said Lukashenka.

In his opinion, Ivan Rybkin is both the Duma leader and "a most prominent and promising Russian politician."

Ivan Rybkin has also met Chairman of the Belarusian Supreme Soviet Myacheslav Hryb. He noted that the Russian State Duma is positive about the idea of Russian-Belarusian economic integration. The latter will help both economies to solve numerous problems.

Rybkin told a news conference today that the Russian-Belarusian relations are a model of cooperation for the whole Commonwealth. In his words, many sovereign republics have come to understand the need for more active participation in integration.

As for the military-political union between Belarus and Russia, Rybkin thinks that all the prerequisites are there. Not a single Commonwealth sovereign state is now capable of independent protection of its borders, which means they should have more understanding than doubts in the sphere of military cooperation.

The Duma speaker has been received by Belarusian Prime Minister Mikhail Chyhir and Belarusian Foreign Minister Vladimir Senko.

Estimating the results of his official Minsk visit, Rybkin noted he is satisfied.

Speaking of his meetings with the Belarusian leaders, he noted they dwelt on boosted economic integration. The sides discussed the package of documents they have signed and the treaty to be signed during Yeltsin's visit to Belarus.

## 6.25
## Opposition Leader Promises to "Sever Moscow Ties"

Eve-Ann Prentice
*The Times,* 4 February 1995
[FBIS Translation]

*Editor's Note:* Zenon Poznyak is leader of the Belarus Popular Front. The front is a weak, but vocal nationalist opposition group in Belarus. President Lukashenka has cut this group off from all access to the Belarusian media, banning it from TV and radio. In April 1996, Poznyak began broadcasting his criticism of Lukashenka's "fascist policies" from a sympathetic Ukraine, whose president also opposes Belarus's Slavic absolutism, and refuses to sacrifice Ukrainian sovereignty to Russia, as Belarus has. Ukraine's harboring of Poznyak has become another highly sensitive political issue in the triangular politics among Belarus, Ukraine, and Russia.

Belarus will risk Russia's wrath and sever ties with Moscow if the opposition Popular Front wins elections on 14 May, Zenon Poznyak, the front's leader, said.

"We need to break from Russia . . . a criminal, bureaucratic state," said Mr. Poznyak, whose party is believed to be gaining popularity and who is in London meeting MPs, police, and education officials at the government's invitation. Minsk, the Belarus capital, is capital of the Commonwealth of Independent States—the loose economic and security group comprising all the former Soviet republics except the Baltic states—and Mr. Poznyak promised to take his country out of the CIS if he wins.

"We have to prepare our own people for a siege by Russia," he said. "I know for sure that Russia would consider invading Belarus. All telecommunications, roads, and railroad networks from Russia to the West go via Belarus. We also have an important nuclear early warning radar system," he said.

The elections will be the first parliamentary poll in Belarus since the country declared independence from the Soviet Union in August 1991. They will be a key in helping to decide whether the country develops a market-reform economy, or reinforces cultural ties with Moscow. Parliament has been dominated by communists, but some observers believe that the Popular Front is gaining ground in traditional communist strongholds in the countryside.

In what may be a sign that President Yeltsin is concerned about the prospect of a Popular Front victory, the Russian leader is to lead a delegation to Belarus this month to sign a new friendship agreement with President Lukashenka, who was elected last July. Mr. Lukashenka this week called for a referendum, to be held at the same time as the parliamentary elections, to decide whether the old Soviet symbols should be reinstated on the nation's flag and whether Russian should rank alongside Belarus as a national language.

Mr. Poznyak said: "The chances of Russia becoming a democratic state are bleak. The impoverishment of the people makes the government look for an external enemy, such as Chechnya. It is like Afghanistan again; they have imperialistic intentions. The Popular Front hopes to build a democratic state, regarded as mature and along the lines of those in Western Europe."

The opposition leader believes, however, that Minsk will

not become another Grozny. "I know our nation and our people, we are not violent by nature.

"Yeltsin got where he is by slogans," Mr. Poznyak said. "He is an old party communist and has the old mentality. He is good at leading from the top of a tank."

## 6.26

**Russia Reintegration Support Seen Falling Away**
Aleksandr Starikevich
*Izvestiya,* 8 February 1995
[FBIS Translation]

*Editor's Note:* In February 1995, the Belarusian center was still seen as a viable political force, even by the Russian newspaper *Izvestiya.* Opposition to unification with Russia was still strong enough in parliament to entirely squelch the idea of a joint parliamentary session with Russia's lower house—the Duma. Despite the *Izvestiya* journalist's pessimism, however, events three months later were to open another opportunity for Russian expansionist aims and were to push Belarus into the Russian parliament's open arms.

One after another Russian and Belarusian politicians have been announcing the latest steps down the road of economic and political integration. But with each new spiral fewer and fewer people believe in or welcome the coming unification. [Passage omitted.]

The Belarusian People's Front—one of the real contenders for victory in the parliamentary elections—is traditionally negative about such an alliance. Gennadiy Karpenka, leader of the centrist party bloc, also believes that the policy chosen by the president in the sphere of Russo-Belarusian ties is "hampering the establishment of normal economic relations between the two states." Admittedly, there are still the left-wingers—who will settle for nothing less than the restoration of the USSR—but they are now unable to even organize a more or less noisy street rally.

Consequently, during his stay in Minsk Ivan Rybkin clearly felt as if he was being hit by contrasting showers. The exuberance of the presidential team gave way to a scandal in parliament, where the speaker wanted to make a short speech. He was allowed to speak, but with great difficulty. The only result of Rybkin's visit was his attempt to smooth out the effect of the, to be blunt, not exactly brilliant idea of holding a joint session of the Russian and Belarusian parliaments.

It is not only the Belarusian political elite that is cool about its neighbors. Many ordinary people are coming to realize that, for all its good intentions, Russia can never solve Belarusian problems. The political cynicism shown by certain Russian representatives who stated during the visit that "the Belarusians should not sit on the fence behind Russia's back" (with regard to the Chechnya conflict) did not go down well.

It is clear that the Russian side is not mistaken about the possibility of strengthening integration periods at this moment in time. The current Belarusian parliament has, at most, three months left, and it is no longer entitled to make changes to the Constitution. It is very hard to predict today what the new Supreme Council will be like. Since this is the case, the next wave of Moscow's efforts on the Belarusian front will subside after a certain time, as has happened in the past.

## 6.27

**Yeltsin Comments on Minsk Talks**
From the "Vesti" newscast
Russian Television, 21 February 1995
[FBIS Translation]

*Editor's Note:* Boris Yeltsin positioned himself squarely in the unification camp in 1995, as the following document illustrates. Responding to the nationalist patriots, the former communists, and millions of Russians who have been unable to accept the collapse of the Soviet Union, he pressed forward, despite the economic objections, with a plan to become an "empire builder," starting with the weakest CIS link. In February 1996, exactly one year after the meeting referenced below, Aleksandr Lukashenka made his first official visit to Moscow, where the two presidents announced that a "union" would be formed in March.

President Yeltsin and the officials accompanying him arrived in Minsk today at mid-afternoon. Today is the first day of President Boris Yeltsin's official visit to Belarus.

[Begin recording.] [Correspondent V. Skvortsov]: Today's meeting in Minsk between the presidents of Russia and Belarus was a logical continuation of the active process of agreements from last year. Aleksandr Lukashenka declared on more than one or two occasions that comprehensive cooperation with Russia was a priority thrust of his policies.

There is just no other solution. The economic crisis continues to deepen. An indirect consequence of this was an unusual surge in crime in once quiet Belarus. Incidentally, the presence of the border and the weak national currencies suit Belarusian profiteers. Russia accounted for 84 percent of Belarus trade turnover last year. Close integration with Russia and the restoration of the sundered links seem to be the only salvation for both civilian and military industry.

A considerable step in that direction was taken in Minsk today. Apart from an interstate treaty on friendship, goodneighborliness, and cooperation, signed by Yeltsin and

Lukashenka, a number of agreements on economic and military cooperation were adopted.

[Lukashenka]: We have dotted the "i"s in all areas. We have concluded a major treaty. We have signed a major customs agreement, a specific one, on the administration of the customs system, and we have signed a model—for others, I mean—treaty on border issues, on cooperation in protecting the state border of the Republic of Belarus, and on building up our western border, working together jointly.

[Yeltsin]: In all of Russia's foreign policy, foreign policy with the CIS countries is paramount. In the whole of the CIS's foreign policy, foreign policy with the CIS, in the CIS with Belarus is paramount.

[Skvortsov]: The undoubted success of the Minsk meeting is likely to make Moscow's talks with Kiev easier, where people think of sovereignty more often than the economy. [End recording.]

## 6.28

## Grachev: "Early" to Talk About "Coalition"

Interfax, 21 February 1995
[FBIS Translation]

Russia's Defense Minister Pavel Grachev stated in Minsk on Tuesday . . . after talks with his Belarusian counterpart Anatol Kastenka, that "it's early" yet to talk about Russian-Belarusian "coalition" armed forces. He said that "in principle he is not rejecting this idea, but there is no coordination mechanism so far between the Russian and Belarusian defense ministries."

Pavel Grachev, who is accompanying President Yeltsin during his official visit to Belarus, told Interfax correspondent Boris Grishchenko that two documents were signed on Tuesday between the Russian and Belarusian armed forces: the agreement on the technical maintenance of airfields and the reception of planes, and the agreement on state acceptance of military equipment which implies that Belarusian military-technical experts will be delegated to Russian defense enterprises and Russian experts to Belarusian defense enterprises.

"We shall go further afterward, taking up questions of training, and material and technical supplies, and then raise the question of coalition forces," said Grachev.

He said that "well-equipped and modern troops are needed to create joint armed forces." "Therefore, the time has not come yet for such a military coalition," he went on to say. "It is early yet to take up such global issues," he added.

He expressed satisfaction that many problems, including the basing of Russian planes on Belarusian airfields, had been solved. He also noted that "we have no disagreements on the removal of our strategic missile forces from Belarus."

He said that "the two early warning facilities in Baranovichiy and Vileika will be under Russia's jurisdiction" and that there is an agreement on paying the work of the Russian personnel servicing these facilities with due account taken of the economic possibilities of both countries.

## 6.29

## Yeltsin on Protecting CIS Border

ITAR-TASS, 22 February 1995
[FBIS Translation]

Russian President Boris Yeltsin's second day in Belarus opened today with a meeting with CIS Executive Secretary Ivan Korotchenya for talks on implementing decisions made at the recent CIS summit in Almaty. The summit elected Yeltsin for a second term as chairman of the CIS Council of the Heads of State.

Yeltsin's schedule was very busy on Tuesday [21 February]. He held talks with President Aleksandr Lukashenka of Belarus and signed the treaty of friendship, good-neighborliness and cooperation, as well as two other agreements—on joint protection of the Belarus state border and on joint efforts in upgrading customs services.

After the signing, the Russian president recalled that the two nations shared a common historical experience over many centuries. "This created the basis for signing the treaty and other documents on deeper integration of our two countries. Among all CIS countries, Belarus has the greatest rights to such a relationship due to its geographical location, its contacts with Russia, our friendship, and the progress of its reforms," the Russian president said. [As received.]

As concerns the military aspect, Boris Yeltsin singled out border questions. He noted that "throughout the years of existence of the Commonwealth of Independent States we were unable to find a solution to the problem of protecting the Commonwealth's external border. Only this time, we managed to sign a good agreement thanks to the attitude of the president of Belarus and the leadership of the republic. All borders will be well locked and at the same time no one stands to lose anything."

Asked if Belarus was likely to lose its sovereignty as a result of rapprochement with Russia, the Russian president replied: "All European countries tend to keep together. The European Union has many countries which are sovereign, independent, and pursuing their own policy. But they coordinate all their activities among themselves. This results in smaller costs to every country."

Lukashenka replied to the same question using the words of Russian Foreign Minister Andrey Kozyrev: "May God forbid any encroachments on Belarus's sovereignty. Russia would be the first to lose much."

## 6.30
## Reportage on Russian President Yeltsin's Two-Day Visit

### Yeltsin's Speech at Academy of Sciences
Radio Minsk, 22 February 1995
[FBIS Translation], Excerpts

*Editor's Note:* Boris Yeltsin's assurances that Russia had no intention of absorbing Belarus were made long before the presidential election campaign started in Russia. As the election drew nearer, Yeltsin's rhetoric grew stronger and more openly nostalgic about the old Russian imperial reach. This document is noteworthy because it shows Yeltsin's "old school" style of appealing to a foreign audience by making the promises of a "guardian and a protector," all the while disguising the political and economic significance of what he is saying.

[Speech by Russian President Boris Yeltsin at a meeting with Belarusian President Aleksandr Lukashenka and Belarusian intellectuals at the Belarus Academy of Sciences in Minsk.]

Dear friends,

This is the last meeting on Belarusian soil before the end of my visit. So, perhaps at the end of my speech I shall give a brief account of what President Lukashenka and I did.

Dear friends, our visit to Belarus has been very important for me. Both in Russia and Belarus today many people are seriously concerned as to whether the thread that has linked our peoples for centuries will be broken, whether we shall forget that the Russian and Belarusian peoples have the same roots. This must not be allowed to happen in any circumstances. It would deal irreparable damage to our states. It would be mocking history, an act of violence against the fates of our peoples. . . .

I think you know that in Russian society today there are various viewpoints coming up against each other on the issue of the nature of the national interests of the Russian Federation. There are forces in our country that are aspiring to restore the political domination of Moscow over the whole territory of the former Soviet Union, to restore the Russian empire in its previous dimensions. I think that such views are not only incompatible with international law, they are dangerous primarily for Russia itself. They threaten to turn into innumerable misfortunes both for it and for the other peoples. History cannot be turned back and attempts to restore the former Soviet Union under a new name are doomed. Russia's adherence to respect for territorial integrity of independent states cannot be doubted by anybody. This equally applies to both our western and southern neighbors.

However, we expect that everybody will adhere to this principle. We will not go for any redrawing of borders. This is the road to nowhere. This is the road to new artificial crises and conflicts.

Genuine interests of our country lie elsewhere: consolidation of stability of genuinely democratic states situated near its borders and the development of equal and mutually beneficial relations with them. This is our stance and we will not shift from it.

Development of our mutual relations has nothing to do with some kind of absorption of Belarus by Russia, as some people say in Belarus and in Russia. We do not have such intentions and, of course, we won't have them in the future. In-depth integration, more in-depth integration than it is today, close interaction, this is the choice and this choice [pauses] has been reinforced in the documents, in the treaty, and in several agreements which we signed yesterday together with Belarusian President Lukashenka. . . .

Integration became a prevailing tendency in the CIS last year, moreover, Belarus and Russia went for the maximum speed [as heard]. I would like to confirm once again that we stand for the closest, the most intensive Russo-Belarusian relations in all spheres. And indeed, Belarus was the first country among all the CIS countries, with which we decided to sign ten packages of agreements on in-depth integration. After Belarus and Russia, it is possible that Kazakhstan may join the process. Thus in this way, a certain nucleus will emerge in the Commonwealth and states will be bolder and more resolute in coming together . . . .

Our states had firmly decided to create a common market for goods, services, capital resources, and work force, and to cooperate closely in the spheres of investments, finance, and industry. In particular, our states agreed to unify the principles for investment activities and acquisition of real estate as well as their protection. Concrete measures will be taken in creating industrial and financial groups.

We have agreed today—and we will draw up a specific document to this effect—that our enterprises and organizations will enjoy equal conditions in terms of price, customs duty, and taxes. [Applause.]

Agreements on a customs union and on joint guarding of state borders have been signed—just imagine, what a saving it is: We protect the external border of the CIS; we protect the border of Belarus; we protect the border of Russia, and we do it together, and we do it all in Belarus. Thereby, of course, the expenditure is smaller and the quality and efficiency of border protection is higher.

Just recently, a program has been adopted for actions of

Russia, Belarus, and Kazakhstan in implementation of accords on further expansion and intensification of mutual cooperation. This program includes specific measures that are designed to help all of us in overcoming the economic crisis more quickly, so that people can, at long last, feel positive changes . . . .

A long-term plan for the integrational development of the CIS was approved last October. It includes such questions as preparation of draft agreements on joint innovation activity, developing joint scientific projects in the fields of ecology, health protection, space utilization, training scientific cadres, restoration of a single information support system for scientific research.

By the way, Aleksandr Lukashenka, when I put forward my proposal regarding cadres, I did not only mean science. I meant cadres in general: industrial cadres, social cadres, cadres involved in health protection, culture, education, and so on and so forth. I meant cadres at different levels: managers, middle management, and others. In essence, this would be an attachment and training at an enterprise or organization involving a person who already has higher education and who will then become a well-known director or manager in Belarus. That is what we are proposing.

I think that this system is more effective because we are using it with Western countries. Two thousand of our people study in Japan, 1,000 in the United States, 1,000 in Great Britain, 1,000 in Germany, and 1,000 in France. The duration of study is one year in each case.

We managed to negotiate an agreement under which we are only paying for their food, while the cost of the entire process of study is covered by the host country. I will tell you about our initial experience relating to specialists who have already returned. They are indeed gaining modern experience, especially in advanced technologies, which can then be immediately invested . . . .

It is necessary to agree on the mutual recognition of doctors' and candidates' certificates. We are neighbors who lived together for such a long time, and now we do not recognize each others' certificates.

At the present time, almost all exchanges of scientific journals and books have stopped. There are virtually no joint publications. Our science has taken a unilateral step and decided to send, at its own expense, journals that are of interest for your academy to the Belarus academic library. This is just the beginning.

In addition to that the Russian Academy of Sciences has fully preserved the membership of the Belarus scientists in the editorial collegium of its journals, and its scientific councils, and in all the commissions. A number of the Russian scientific societies are including the teams and separate scientists of Belarus as their members. However, these initiatives and measures of the Russian scientists are not sufficient. Our nations sharply need an intergovernmental agreement on scientific and technical cooperation, and it should be signed as soon as possible. The work is being finished on this agreement, and I think that the agreement will regulate many issues, which are currently hampering the cooperation between the scientists of our nations . . . .

The integration from the bottom between Russian and Belarus peoples is under way, and we should help this natural process. I mean financial assistance as well. Of course, we will find some solution to this matter, let us suppose that first, it should not be said that I have brought something in my pocket, but at least we have decided that we will allocate or, in other words, give the first R150 billion as a long-term credit. Later, we shall find some other kind of financial support.

The people felt by themselves the consequences of the disruption of the ties. As far as we know, nowadays the absolute majority of Russian and Belarusian people—and I was convinced at the factory now about that as well as everybody speaks loudly about this—want the rapprochement of our states, and closer rapprochement means deeper integration.

Dear friends, there is such a close degree of relationship of trust and mutual understanding between Russian and Belarusian peoples, which rarely exists between two close neighbors. This is a reliable guarantee that all the difficulties in the way of our close cooperation could be overcome. In this field, our intentions and those of the Belarus president are the most serious, and I think that you will support us in these efforts.

Thank you. [Applause.] . . .

**Yeltsin Looking Forward to "Belo-Rus"**
From the "Vesti" newscast
Russian Television, 22 February 1995
[FBIS Translation]

---

Today is Boris Yeltsin's second day in Belarus and it started with a meeting with CIS Executive Secretary, Ivan Korotchenya. During the conversation, ways to implement the decisions adopted at the recent Almaty summit were discussed.

The second day of Boris Yeltsin's stay in Belarus started with a visit to the Minsk Car Factory (MAZ). The directors of Belarus's large industrial enterprises have to call for immediate reintegration with Russia. Supplies of components have been discontinued and Belarus tractors, trucks, and refrigerators are losing the vast Russian market which is their only market. A joint project with a German company cannot save the factory. But the results of yesterday's talks between the two presidents hold a real chance to improve the situation for the production managers and engineers and workers of Belarus state enterprises who, together with the republic's entire population, are coping with yet another 50–100 percent rise in food prices.

[Begin recording.] [Yeltsin]: We will move, first, toward a deeper integration and the setting up of individual large corporations and then we will simply unite and there will be a Belo-Rus.

[Lukashenka, laughing]: Yes, Belo-Rus! [End recording.]

**Yeltsin, Lukashenka Sum Up Results**
Moscow Radio, 22 February 1995
[FBIS Translation], Excerpt

*Editor's Note:* The following document, a conversational recording, is enlightening in that it sounds like two dictators discussing in detail their plans for their countries' joint activities in every sphere.

Now, some more details about one of today's main events. By that I mean the Russian president's visit to Belarus, which has just ended, and an appraisal of this visit from the mouths of the two presidents, Boris Yeltsin, and Aleksandr Lukashenka.

[Begin Yeltsin recording.] Further, everything that we have signed must be put into practice. A number of issues have not found their way into the agreement. We must also bring them together and discuss them among the representatives of both states, and include them in one document—a protocol—so that absolutely all issues are included in them. At the same time, both presidents must have control over them, so that not a single issue is forgotten. We have problems in the motor vehicle construction industry, where there are a number of factories. Now the world has begun a policy of deepening [cooperation]; let us set up a common corporation in motor vehicle construction between Belarus and Russia. It must be advantageous to everyone and in everyone's interests. Of course, everything must be considered, and some kind of model of this idea must be made, and then we can begin to implement it.

I liked the president; and just now at this conference and in the academy we also said that we must restore the system that existed before in education, in the exchange of postgraduates and students on special courses, and so on and so forth. We decided that we will do this. We also spoke about issuing diplomas of various kinds, as many as we could. So, we have now immediately issued instructions on this, and we are to immediately issue a general edict. We are going to sign it together, and the diplomas will be jointly recognized on the territory of Belarus and Russia. [End recording.]

[Lukashenka]: For me personally, a young politician, Boris Nikolaevich's coming here was important to me. You will probably recall how people were saying that nothing could be decided between Lukashenka and Yeltsin, they could not talk to each other or come to any agreement. I will say to you frankly, and I can, of course, say this publicly, that we have become friends. It could not be any other way. I think that this is a tradition of our fraternal relations and our fraternal states.

Boris Nikolaevich feels somewhat—well, if not shy, then embarrassed when we say that we need not simply to integrate but to join together in certain areas. We spoke about this one-to-one. I am happy to speak about this—I am not afraid that when people talk about Russia devouring Belarus or, God forbid, Belarus devouring up Russia! I am not afraid of this. I promised the people one thing at the elections—complete integration with the fraternal Russian state, and integration between the peoples. Therefore, I am moving toward this directly, including by way of a referendum. And we will be holding this referendum. So we are implementing a concept of the two presidents—to move on behalf of the economy and the lives of the people. And this will without doubt lead to a solution to political questions. The lives of the people will lead toward this. I am not afraid of that either. I would simply like to thank the Russian delegation and the president. They have responded to our request in a very positive manner. Practically all the questions have been solved, and the main thing is that we are placing equal conditions on all our enterprises—Russian and Belarusian—as of today. These are our common enterprises, and the MAZ [Minsk motor-vehicle works] is not just a Belarusian but a Russian enterprise, too, and we must together show concern for their development and the benefit of our peoples. [End recording.] [Passage omitted.]

## 6.31

**Moscow-Minsk Border, Military Agreements Hailed**
Vasiliy Kononenko and Aleksandr Starikevich
*Izvestiya*, 23 February 1995 [FBIS Translation]

The series of documents signed during President Boris Yeltsin's visit to Minsk gave material form to the achievements of Russian diplomacy on the Belarusian front. Russia's leader assessed relations between Moscow and Minsk as the best within the CIS. They are now consolidated by the Treaty on Friendship, Good-Neighborliness, and Cooperation, as well as other agreements.

Boris Yeltsin was greeted at Minsk-2 airport with flowers and kisses. The flowers were presented by young girls, and the kisses were from his Belarusian colleague Aleksandr Lukashenka. In the Belarusian capital itself, the highest level of readiness was displayed beforehand by the services ensuring the visit's security. Even before B. Yeltsin took off from Moscow, certain streets in Minsk were closed, and militiamen were placed along the entire route to be followed by the presidential motorcade. Viktor Sheyman, secretary of state of the Security Council, tightened up security to such an extent that the militiamen surrounding the government residence on Voyskovyy Pereulok would not even allow Sheyman himself through for a long time.

The services responsible for supporting the journalists in their professional activity worked with their traditional inefficiency, as was the case during Bill Clinton's visit. Many representatives of the press simply did not reach the official event, and many had all kinds of communications problems. But otherwise everything was fine.

"In the European Union, coordination of economic and foreign policy activity costs less," Boris Yeltsin said. Well, now the two states—and first and foremost, Russia—can save on this. The discussion of Belarus's debt for energy sources proved difficult, but the difficulties were overcome thanks to an accord on a barter basis: Russian enterprises (first and foremost "Gazprom") received the right to participate in share ownership of Belarusian enterprises. Specific projects for the creation of joint production facilities will be implemented by finance and industry groups, while the governments of the two countries have pledged to assist them. It became clear that the presidents also signed treaties on coordinating operations to protect the borders and the control of customs networks.

"For a long time we have been working toward an understanding with Belarus as to what border security is," Col.-Gen. Andrey Nikolaev, commander of the Russian border forces, said in an exclusive interview for *Izvestiya.* "Within the framework of the commonwealth, this problem has not yet been resolved and is not likely to be in the near future. In relations with Belarus we have reached a remarkable and unexpected stage of mutual understanding. For instance, the treaty that was signed contains a definition of what constitutes security on the borders with Lithuania, Latvia, and Poland. This means that Russia is shifting its border interests 615 km to the west of the administrative borders. Under the bilateral agreement we have pledged to set up an operational group of border troops (headed on the Russian side by the deputy commander of border forces). The Belarusian side, in turn, is creating a group of representatives to the Russian Main Border Forces Directorate with a view to upholding their state's interests on the actual borders. And in the agreement we avoid the contentious issue of representation of army subunits on the territory of Belarus. The border troops will be brought in under a special statute (which was signed during the visit—[author's note]) only to protect the border, and nothing more."

Several documents were signed by the foreign ministers. Among these documents, the agreement on cooperation in the military sphere stands out. Russian Defense Minister Pavel Grachev said that talks on this subject "went remarkably smoothly and calmly." "We have already resolved the problem of basing our long-range aviation at the Baranovici airfield. We have now reached an agreement on material and technical services for the two countries' armies, with the introduction of state quality control [gospriemka] at specialized enterprises in Russia and Belarus." In addition, the question of creating an integrated ABM defense system was discussed.

Summing up the results of the visit, it can be noted that the Russian leadership has achieved successes of some importance, meeting its interests in the economic, military, and other spheres. This must be formulated in documentation, since whereas at present the Belarusian political elite is putty in Moscow's hands, in the foreseeable future there could be major domestic political changes in Minsk. Then Russia would encounter undesirable complications.

## 6.32

## Parliamentary Elections and Referendum in Belarus

Press Release
Embassy of the Republic of Belarus to the United States of America, 16 May 1995

*Editor's Note:* The May elections to the Belarusian Supreme Council, and the referendum, were the turning points for Belarus in deciding what direction its domestic and foreign policies would take. The results of the election were disappointing for Belarusian democrats and Popular Front oppositionists. Even though there was good voter turnout throughout the country, its distribution failed to meet electoral requirements in most provinces, and only 19 deputies out of 260 were elected. The referendum provided the final blow to Belarusian representative democracy. Lukashenka won a mandate for extensive presidential powers, including that of dissolving the Supreme Soviet. These results meant that Lukashenka's reintegration policy would prevail in an environment of public confusion and lack of political will.

On 14 May, Belarus held parliamentary elections and a referendum. At a press conference on 15 May the Chairman of the Central Electoral Commission of Belarus A. Abramovich announced preliminary results of the first round of elections to the Supreme Council and of the referendum.

Despite predictions of widespread apathy, overall voter turnout was high—64.5 percent. But only 19 of 260 deputies (7.3 percent of the seats) were elected in the first round of voting. Elections in 27 electoral districts of 260 (in Minsk—in 11 of 42) failed to attract the requisite number of voters.

Runoff elections will be held on 28 May.

According to the preliminary results of the referendum, voters gave President Lukashenka a strong show of backing (82.4 percent) for his activities, aimed at furthering economic integration with the Russian Federation and surprisingly high approval (83 percent) for the equal status for the Russian language alongside Belarusian. Seventy-five percent of voters supported a proposal to establish new national symbols—state flag and state emblem. In addition, 77.6 percent of voters agreed with the need of making amendments into the Constitution for empowering the president to dissolve the Supreme Council if it violated the Constitution.

As reported by press agencies, a team of observers from the OSCE came to the conclusion that the voting "had peaceful and adequate character."

## 6.33

**Foreign Minister of Belarus on the Results of the Referendum**

Press Release
Embassy of the Republic of Belarus to the United States of America, 18 May 1995

Commenting on the results of parliamentary elections and a referendum, the Foreign Minister of Belarus Uladzimir Syanko in an interview with Reuters said, "While we move closer to Russia economically, we cannot give up what's most important—our sovereignty, independence, and statehood."

"Belarus sovereignty is not for sale. These are not mere words. We have to look at reality. Belarus has strong ties with Russia . . . but I think national consciousness will grow quickly."

## 6.34

**Russians Eye Union with Belarus**

19 May 1995
UPI


The Russian Parliament launched a bid to bring Belarus back under the direct rule of Moscow Friday, overwhelmingly supporting a motion aimed at uniting the neighboring nations into a single state.

Lawmakers in the lower house of parliament voted 249 to 0, with one abstention, to charge a State Duma committee with the task of drawing up legislation that would lead to the unification of Russia and Belarus.

The move to meld the two Slavic states into one was led by nationalist lawmaker Sergey Baburin, who called for a referendum in December asking Russians whether they favor the reunification.

Baburin blasted the idea of a long-discussed monetary union with Belarus that many believe would hurt Russia's more stable economy, urging fellow legislators to push instead for "full socio-political unity."

The bid to bring Belarus back under Russia's wing was buoyed by voters in the former Soviet republic of more than 10 million, who gave strong support to closer ties with their giant neighbor in a referendum Sunday.

A significant majority of voters in Belarus gave their backing to closer economic links with Russia, official status for the Russian language and a return to state symbols similar to those scrapped after the Soviet breakup. The referendum marked a defeat for narrowly supported nationalists in Belarus and a sign it will continue to gravitate toward Moscow as it seeks to solve serious economic troubles many see as a result of breakup of the centralized Soviet system.

The Russian legislators' move reflects a widespread Russian desire to see a revival of something like the Soviet Union—or at least a Slavic Union of Russia, Belarus, and Ukraine—with Moscow as its center.

In a separate State Duma vote Friday, lawmakers passed an official statement commending Belarus President Aleksandr Lukashenka for holding the referendum and saying they hope his initiative "will find support and understanding among the presidents of several Commonwealth of Independent States nations."

The Russian government's courtship of Belarus has been a quieter affair, as President Boris Yeltsin and his men worry that tight economic ties could drag Russia down toward the level of Belarus and hurt its bid to crush inflation.

Ivan Korotchenya, the Belarusian secretary of the Commonwealth of Independent States, said Thursday that Yeltsin and Lukashenka would take advantage of a CIS summit in the Belarusian capital Minsk next week to sign a broad-based economic agreement that would also be open to other members of the twelve-nation group.

A summit meeting in Minsk last year produced a customs agreement among the neighboring nations as well as pacts ensuring Moscow's continued military presence in Belarus, a buffer state separating it from Europe and NATO's possible expansion into Poland.

## 6.35
## Caution Advised on Unification Moves with Belarus

Maksim Sokolov
*Kommersant,* 6 June 1995 [FBIS Translation]

The last weekend in May enriched the political-economic situation in Belarus with two important innovations. On 27 May, after the symbolic removal of the border gate and its replacement with a birch tree planted by Chernomyrdin and Lukashenka, the Russian-Belarusian border columns were removed, and on 28 May, after the second round of parliamentary elections, the RB [Republic of Belarus] was without a parliament as well: the term of the old one had run out but a new one failed to be elected.

The symbolic ceremony to remove the border gate was accompanied by no less symbolic statements of the two presidents. President of the RB Lukashenka summed up and said: "The road is now open. Those who impede traffic will be put in the appropriate places," alluding to the idea that during integration he is prepared to plant more than just birches [Russian verb also means "imprison"]. Ukrainian President Kuchma, commenting from afar on the birch-border gate procedure, noted sardonically: "The posts can be broken. The important thing is not to break the wood [make a lot of mistakes.]"

Despite how appropriate that desire is, its practical realization may not prove to be so simple—that became obvious the very next day when reports from the voting precincts showed that the elections were not accomplished and there is no parliament in Belarus.

Undoubtedly, the history of post-communist parliamentarism on the territory of the former USSR cannot be called glorious—suffice it to recall the RF Supreme Soviet and its deplorable end, and then too the Duma's activity can hardly be a subject of admiration. Nonetheless, the role of parliament as a whistle mechanism for releasing steam is difficult to deny, since where the safety valve is absolutely riveted shut, a tranquil life can only be achieved in two ways: either by direct terror (the USSR in the 1930s–1950s), or by using food to tame subjects who in addition still remember the past terror very well (the Brezhnev oil dollar prosperity). In the sense of using these methods, the capabilities of the present Belarusian regime seem questionable.

Direct terror presupposes not only a smooth-running repressive mechanism, but also the country's blind isolation from the outside world, and that contradicts not only the emotional procedure with the border gates and the birches, but also the very geopolitical position of the small republic, which is at a busy Eastern European crossroads. Given the impossibility of properly shutting a country off from relatively liberal neighbors, any half-formed terrorist intentions will produce not so much a frightening as an irritating effect, as happened in 1989 at the decline of the CMEA [Council for Mutual Economic Assistance] "people's democracy." An unwritten social contract with subjects, that is to say, steadily increasing the standard of living of working people in exchange for their abandoning any social activism at all, requires the appropriate economic resources—no wonder there was perestroika; meaning the actual denunciation of the social contract a la Brezhnev coincided with the end of the oil dollar miracle. But in Belarus, the economic situation is such that there simply are no resources with which to buy people off. The only real resource may be ignoring the de facto failure to fulfill tariff-customs agreements with Russia—without saying so, letting subjects live on contraband at the expense of the Russian budget. Blaming the Belarusian leadership in this case would not be completely fair: they are not trying to get rich, just survive.

The political options which are now being developed amount either to changing legislation in hindsight and considering the parliament empowered when 40 percent of the deputies are elected, or merging the old and the new Supreme Soviets, or having direct presidential rule. In any case, this means only that social contradictions (which given the average monthly wage of U.S.$5 may be extremely critical) will begin to be channeled not to the parliamentary hall but to the street. Given such prospects, it is quite difficult to plan Russian-Belarusian relations for the long-term: while as applied to one partner (Russia), the word "stability" has a certain sense, this word no longer applies at all to the Belarusian partner. An altogether unpredictable political mishmash may appear beyond the birch on the border at any time.

The strangest thing here is that although the specific nature of the Belarusian leadership was no secret to anyone and the degree of democratism of the election procedure was obvious from the very start, Russian politicians amicably showed unprecedented enthusiasm right after 14 May, when the first round of elections was combined with a referendum on integration with Russia. The fact that all the fears and doubts were rejected by the Russian communists was more or less understandable, since it would be strange for a communist not to rejoice in a communist renaissance on his neighbor's land. However, the president of the RF rejoiced as well, immediately congratulating the integrator Lukashenka and recognized democrats like the chairman of the Duma committee on foreign affairs, Vladimir Lukin, who was expected to understand that "there are secret, strategically fateful decisions which overrule all accounting calculations." "Fateful decisions" meant the reunification of the RF with the RB, while "accounting calculations" meant the economic price which Russia must pay for it. Accounting calculations can, of course, be overruled—the novel by Vasiliy Aksenov discusses the similar problem of the reunification of the USSR with the Crimean Peninsula, and one of the Politburo "little portraits" notes: "We do not economize on ideology." But it follows from Lukin's reasoning that "accounting calculations" are the only obstacle to the Anschluss, while the political prerequisites are in fine shape.

However, it would be best to refrain from the second assumption at least until the results of the second round of elections to the Belarusian Supreme Soviet are summarized. On the one hand, it would at least clarify what is happening on the Belarusian political scene; on the other, collating statistical data on the plebiscite and both rounds of elections to the Supreme Soviet would enable us to establish more precisely whether there were improprieties, and if so, then to what extent, and exactly what the value is of this plebiscite overall. Ultimately, if the natural mission of the Belarusian leadership is to push up the price for themselves in every possible way (including by means of manipulating the elections), then the mission of Russian politicians and diplomats which is just as natural is to understand what really is behind all the plebiscite manipulations. It has now become clear that nothing is behind them except the complete political and economic impasse in which the Lukashenka administration finds itself—but that was certainly almost obvious back on 15 May. Waiting another two weeks to talk with Lukashenka, taking into account the changed conditions is elementary diplomacy. But this elementary procedure proved to be unavailable to the Russian leadership, and they preferred, based on the principle of "ringing the bell without looking at the church calendar," to make "fateful decisions" and break the border gates, which produced the biting comment by their neighbor Kuchma.

# The Slavic Union

The Slavic Union is another confederal concept which has evolved within the CIS framework. Its supporters come from both the far-left and the far-right political factions in Russia, Belarus, and Ukraine. It is based partly on the reinvigoration of the pre-Bolshevik Slavophile tradition and doctrine, but in more pragmatic terms also on the "close natural ties" between the Slavic peoples. There are fewer language difficulties; the peoples share a strong ethnic identity; and the countries were integrated under the former Soviet economic, military, and political systems. The "Slav Card" is being played by several political parties (for example, Vladimir Zhirinovskiy's Liberal Democratic Party) for its populist appeal and power to capture votes.

In the documents collected in this section, we see the Slavic Union concept being embraced by the Russian and Belarusian Slav Assembly—an umbrella party which desires a return to Slavic roots. The Belarusian component of the Slav Assembly, "Belaya Rus" (White Russia), wants to reinstate the "purity of the Slav home." The Slav Assembly is well organized, with regional organizations in all the oblasts and major rayons of Belarus and Russia. It therefore considers itself a political structure rather than a civic association or club. The ideology of the Slav Assembly is that the Slav peoples possess the greatest intellectual resources, energy reserves, and traditions in the world, but they are split politically. The unity of these strong peoples must be restored. These goals are voiced here by Mikalay Syargeev, co-chairman of Belaya Rus. The economic program of the Slav Assembly is socialistic, with individual "freedom to work," but "various forms of ownership" and no mention of private property.

Without Ukraine, obviously, the idea of a Slavic Union would be dead. On 26 June 1994 Ukrainian President Leonid Kravchuk spoke out urgently against the concept of a Slavic Union, calling it a "'dangerous idea' because it would split in two those states where Slavs and non-Slavs live," and this would be followed by "horrible confrontation with the Asian states," which would rush to form their own union. Kravchuk also made the point that no confederation should be based on ethnic considerations alone. Following Kravchuk's removal in the 10 July 1994 Ukrainian presidential election, commentator Aleksandr Boroday wrote that the elections of Aleksandr Lukashenka (on the same day) and Leonid Kuchma would exert strong influence for the Slavic ideal and would advance the concept of a Slavic Union. Nevertheless, President Kuchma has not met these expectations and has never mentioned support for this movement or the concept. Lukashenka, however, as noted in the last section, has become a strong ally and spokesman for Slavic integration.

The conservative right-wing Russian national patriots also support the restoration of a Slavic Union. Several members of the Committee for the State of Emergency, which led the attempted coup against Gorbachev (Vasily Starodubtsev, for example), are in the twenty-one-member organizing committee for the work of the Slav Assembly. The leader of the organizing committee is Boris Mironov, who was fired by Yeltsin for his "fascist-like statements," and former KGB general and head of the Russian National Assembly Aleksandr Sterligov. Their involvement is documented here.

The ultranationalist, proto-fascist Liberal Democratic Party (LDP) has also embraced the political goal of forming a Slavic Union (but supports the idea of this Slavic community reaching out to embrace the Eurasian heartland). In April 1994, the Congress of Slav, Orthodox, and Christian Peoples elected Vladimir Zhirinovskiy, leader of the LDP, as their president. Zhirinovskiy's concept of a Slavic Union includes the Eastern European Slavic states—the Czech Republic, Slovakia, and Serbia—with which he wants Russia to form a military-political alliance. In his speech to the Congress, which is documented here, Zhirinovskiy stressed that he believes the Balkan war was the catalyst of the idea of a Slavic Union.

A center-right group which supports a variation of the Slavic Union, but would include Kazakhstan (in a kind of "Slavic Home Plus One") is the Russian coalition consisting of: the Union of Renewal, the Party of the Majority, the Civic Union (with branches in Belarus and Ukraine), and the Agro-Industrial Union. On 28 December 1994, this group held a conference in which it supported the full unification of Russia, Belarus, Ukraine and Kazakhstan. At the conference, Civic Union leader Arkadiy Volskiy voiced his full support for the political, economic, and military reunification of these states, with supranational structures, including a unified parliament, constitution, and combined armed forces. He posits four stages for creating such a unified state: (1) form an open union based on treaties and quadrilateral agreements; (2) form a joint customs union which lifts all restrictions on the free flow of goods, labor, and capital across members' borders; (3) align the legislatures; and (4) build supranational structures.

The following documents highlight the key organizing activities of these left, center-right, and right-wing groups which support some form of Slavic Union. How influential this concept becomes will partially depend on the future electoral successes of its adherents in both presidential and legislative contests.

## 6.36

**Russian Economists Do Not Approve of Slavic Union**
KAZTAG-Almaty, 22 July 1993
[FBIS Translation]

A group of prominent Russian economists and statesmen released a statement in which they characterized the economic policy reflected in the decision to form an economic union adopted by the government heads of Belarus, Russia, and the Ukraine as "shortsighted."

In the statement that was signed in particular by Stanislav Shatalin, Leonid Abalkin, Vadim Bakatin, and other leaders of the Reforma International Fund, it is noted that the agreement of the heads of three Slavic states could be welcomed if not for the fact that it "can be viewed as a political reaction to steps taken by the countries of Central Asia, Kazakhstan, and Azerbaijan toward a closer cooperation with Turkey, Iran, Pakistan, and Afghanistan."

The leaders of the Reforma Fund believe that the decision adopted by "the three" without having previously consulted Kazakhstan and the Central Asian states and clarified their relationship with the union members, "considerably restricts the possibility of participation in the Union." "Moreover, such a decision can alienate these countries from participation in the economic union. It was Kazakhstan that has always been one of the major initiators of a much closer integration and a system of coordinating agencies."

The authors of the document underline that such a foreign economic policy toward Kazakhstan and other CIS countries is "shortsighted and does not suit Russian or other countries' interests either in a short- or a long-term prospect."

"The main way of deepening economic integration with the former USSR countries is seen in the development and implementation of the declaration on setting up an economic union within the CIS frameworks. The efforts of all states should be focused on this issue. Limitations and obstacles should not be introduced into this essentially complicated process because they can undermine the fundamentals of such a fragile, and so badly needed, consensus reached in Moscow," the statement says.

## 6.37

**Slavic Economic Union Membership Assessed**
Fedor Burlatskiy
*Nezavisimaya Gazeta,* 22 July 1993
[FBIS Translation]

*Editor's Note:* Fedor Burlatskiy expresses the views of the neo-democrats who perceive great problems involved in creating an exclusive economic union of Slavic states.

Burlatskiy makes the case for an even-handed policy with all former Soviet republics and the rejection of a reactionary policy against perceived Central Asian "collusion." The Slavic Union movement is shown here to be an evasion by Russia of the insecurity it feels in dealing with the Transcaucasus and Central Asia.

In Brezhnev's times our politicians advertised extensively and strenuously the common European home. In Gorbachev's times, preference came to be given the Euro-Atlantic home, where relations between the USSR and the United States took precedence. There is no exception to be taken to this in principle, although it may be judged variously whether our country derived many dividends from this strategy. But what is obvious to all is that in our enthusiasm for the building of new homes lacking as yet not only a roof but even the semblance of a foundation, we have increasingly demolished the Eurasian home, which Russian policy, diplomacy, and culture have been erecting for centuries.

Two events of a single week are graphic confirmation of this. The first was the meeting of a group of countries in Istanbul. This is the list of them: Turkey, Pakistan, Iran, Uzbekistan, Kazakhstan, Azerbaijan, Kyrgyzstan, Turkmenistan, Tajikistan. These countries agreed in principle on the formation of a common economic zone. The second was the signing of the agreement between Belarus, Russia, and Ukraine on close economic cooperation.

These events reflect new vectors of a geopolicy which could in the future change the character of the Eurasian continent.

Let us begin with the meeting near Moscow. Why did President N. Nazarbaev or his representative not take part? Why was the Belovezh Forest model repeated once again? The explanations given by the leaders of the three governments withstand no criticism. Nazarbaev, who prior to the putsch, during the intoxication after the failure of the putsch and in the period of sobering up following the intoxication right up to the present has been an enthusiast of economic integration, was once again not invited, shoved aside, and insulted. The attempt to portray this as fortuitous is simply ridiculous. No, deliberate policy is behind this. The same questions arise concerning the leaders of Uzbekistan and the other states of Central Asia displaying an interest in preservation of the common economic space with Russia.

It may be said that the meeting near Moscow was revenge for the formation of the Muslim bloc. But this is wrong. If Russia is dreaming of a European home, why can Central Asian states not build their homes on their borders with peoples of kindred civilizations? Besides, at the Kiev meeting the participation only of the three Slav states was planned in advance. So we are faced with a design. And it consists of the exclusion from the economic union of the Caucasian and Central Asian states of the former USSR.

I was recently told of a staggering fact. Prior to the meeting in Belovezh Forest, a trio of the president of Russia's advisors had prepared a paper on three pages for B.N. Yeltsin. It proposed an "uncoupling of the cars from the train," that is, Russia withdrawing from the USSR and getting rid of the Union republics. The argument for this was that Russia could then derive enormous revenue from the sale at world prices of energy resources to the former republics of the Union. Another argument: These republics were, like leg-irons, preventing Russia from accomplishing a rapid transition to the market.

If this is the case—and it can be seen from the entire subsequent course of events that, indeed, it is—everlasting shame will descend on the names of the authors of the treacherous paper. They will occupy a place alongside the two false Dmitriys, and, further, for company, to make up a trio, some furtive Byron will be added to them.

What has Russia gained from the great disintegration? Economically it has gained nothing. The increased revenue for oil and gas has not compensated for the decline in production on account of the destruction of the evolved business relations. And Russia has undertaken the reform—the leap into a financial hole, more like—together with the other states of the ruble zone all the same. And we have not as yet been able to do anything particularly good on our locomotive without the cars: However it has puffed and blown off steam, it has not rolled out of the impasse, despite the efforts of the switchmen replacing one another.

It can now be seen that the creators of this policy are not content. They are continuing to cut off from the economic union the majority of countries of the CIS. Even Kazakhstan, whose Russian-speaking population constitutes half, is no exception. Such a foolish policy of self-destruction is historically unparalleled.

This policy needs to be rectified as quickly and energetically as possible. It is necessary to apologize to Nazarbaev and the presidents of the other republics concerned. It is necessary urgently to convene a meeting of the leaders of the CIS to form an economic community along the lines of the common market in Europe and, most importantly, it is necessary, finally, to recognize the simple truth that without preservation of the Eurasian home, we will not be needed in any other homes.

Europe, America, and Japan are already laughing at our claims, as before, to play the part of great power. Our might is being squandered in deference to the momentary interests of petty politics and narrow-minded politicians.

Here also, for that matter, we see some inexplicable duality. Displaying no concern for economic and cultural integration with the Central Asian countries, we are once again prepared to become bogged down in military conflicts in this region. I refer to the events on the border of Tajikistan and Afghanistan. We should think ten times before allowing ourselves to become involved in a war which—the experience of the Afghan adventure showed this—we obviously cannot win. Here is imperial thinking in physical form: weapons, the blood of our soldiers, certainly, economic support, no, get out of it as best you can.

How many sweet-sounding speeches have been delivered as of late on the Russian idea! It is conceived of this way and turned around that way. But when the most acute question of the preservation of Russia as the center of Eurasian civilization has arisen, all the verbal trumpery has turned to ashes.

Take a listen to the speeches of the leaders of the new independent states. They all speak excellent Russian. And this applies to the political and cultural elite and to the majority of people in these countries. And this is most valuable capital for Russia. The United States detached itself from Britain two hundred years ago. But, having preserved English, it has remained the closest ally of the British.

The dilemma of a union of Slavic states or a Eurasian society is contrived. More precisely, it is an ideological fabrication. Dividing such a multinational country as the former USSR along racial or religious lines is impossible and dangerous. This is fraught with the danger of the disintegration of Russia itself, in which there are Tatarstan and Kalmykia and Chechnya and Sakha (Yakutia). In addition, can it be forgotten that the main disintegrating factor thus far has been the position of Slavic Ukraine?

If we behave in intelligent fashion, the peoples which negotiated together with us innumerable historical ordeals and which have made a tremendous contribution to Eurasian civilization will preserve what has been accumulated, regardless of the nature of the political unions with Russia. If, on the other hand, we take the path of further pan-Slavism or a one-sided Europeism, our descendants will in several decades find themselves in a completely different world. And it will, believe me, be the worst of worlds for a Russia scorned by its natural allies. Central Asia will be condemned to incorporation in the zone of influence of Turkey, the Far East, Japan, Russia, Ukraine, and Belarus, Germany, and West Europe.

We need to return to our common Eurasian home. Not in order to attempt to reconstitute from scratch the empire or the Union: In order to build new democratic relations jointly, equally and in the common interests of all participants. Then more heed will be paid to us in both the European and the Euro-Atlantic homes.

## 6.38

### Slav Assembly Meets, Adopts Party Platform

Alyaksandr Zhuk
Radio Minsk, 13 December 1993
[FBIS Translation]

*Editor's Note:* The Slav Assembly's platform is reviewed in the next document. This group combines the democratic ideal of market development with a reintegration of those states which founded the first Soviet Union in 1921. Ignoring the clear pro-independence position of Ukraine, this group expounds its racist views quite openly. It is well organized at grass-roots levels in Russia and Belarus and therefore offers a credible challenge to other contenders in the 1995 parliamentary elections and the presidential election of 1996.

Over 100 delegates gathered for the Slav Assembly congress. They adopted the party's program and outlined an action plan for the near future. What are the aims of this organization? Here are some features of the program. Belaya Rus is in favor of the unity of the Belarusian, Ukrainian, and Russian peoples, social justice, moral, and ecological purity of the Slav home. I will ask Mikalay Maksimavich Syargeev, the co-chairman of the assembly duma, to comment on the results of the congress:

[Syargeev]: The results of the congress pleased us. We summed up the outcome of the activity of our organization in the period since the founding congress, i.e., we have now made it clear that we have regional organizations in all oblast centers and in many rayon towns. That is the first thing. Second, we have gone through the stage of establishment, i.e., from a small group of enthusiasts who were organized as a club, we are now approaching the creation of a real political structure. Most importantly, the new version of our organization's rules and the program were passed.

The main aim is this: We have an analysis section in which we say that the strength of the Slav world is potentially gigantic, for it possesses the very greatest intellectual resources and energy reserves in the world. But its weakness is that the Slav world is politically and ethnically split. Therefore, the strategic task of our program is the restoration of unity, first and foremost, of the Slav-Russian peoples, i.e., the Great Russian, Belarusian, and Ukrainian peoples.

The second very important aspect is that we aspire to have a market economy socially oriented toward the person, to have freedom of work and creation, and various forms of ownership. At the same time, we shall aspire to carry out a strong social policy, i.e., our people must be fed, clothed and, so to speak, provided for.

[Zhuk]: Will the assembly be able to exert an influence on the political situation in the republic?

[Syargeev]: I think that it is already exerting some influence, the more so in that we are the initiators and now active participants in the biggest political association in the republic, the People's Movement of Belarus, and we act within the framework of that movement, which now already has its MP's association in the Supreme Soviet. So we shall obviously influence the situation in the Republic. [Passage omitted.]

[Zhuk]: The Slav Assembly program also includes a provision of the official status of the two languages—Belarusian and Russian. [Passage omitted] [End recording.]

## 6.39

**Slavic Union: Union with Russia "Urgent"**
BELAPAN-Minsk, 13 December 1993
[FBIS Translation], Excerpt

The First Congress of the Slavic Union White Rus [Belaya Rus] was held in Minsk on 11–12 December. The main issue on the agenda was the party's platform. According to the platform, the Slavic Union Belaya Rus is a part of an international Slavic movement that was established following the collapse of world stability, which had been ensured by two world powers—the USSR and the United States. The program emphasizes that the Slavic civilization is currently in a critical situation, and the only way of overcoming this crisis would be establishing a Slavic Union, whose nucleus would be a union of Slavic-Rus nations—the United Rus [Sobornaya Rus]. Unification of Rus, in the opinion of the Slavic Union, is the only possible way to provide military, political, and economic security for its historical components, including Belarus. In this regard, measures for a more rapid integration with the Russian Federation are inevitable and urgent for Belarus. According to the participants in the congress, the unification of Rus is a path to economic development and prosperity. Among the main objectives of the "nationally oriented economic policy" are fundamental changes in the tax policy, which should stimulate domestic production and the development of an infrastructure, the introduction of a state monopoly in external economic relations until the end of the period of economic stabilization, and the cancellation of the "destructive conversion of the defense industry complex."

## 6.40

**Slav Congress Elects Zhirinovskiy President**
Moscow Television, 3 April 1994
[FBIS Translation]

Unity, national dignity, and independence have been proclaimed the main political principles of the World Congress of Slav, Orthodox, and Christian Peoples which took place in Moscow today. The congress proclaimed a union of Slav peoples. Vladimir Volfovich Zhirinovskiy, chairman of the Liberal Democratic Party of Russia, was elected president. A Slav parliament and government was also formed.

## 6.41

**Zhirinovskiy Seeks Union of Slavic States**
Interfax, 3 April 1994
[FBIS Translation]

*Editor's Note:* Vladimir Zhirinovskiy represents the militant nationalist wing of the Slavic Union movement. Zhirinovskiy, however, changes his message to suit the crowd and the moment. He also identifies with the "Eurasianists" and the right-wing national patriots.

Leader of the Liberal Democratic Party Vladimir Zhirinovskiy stated in Moscow during the opening of the World Congress of Slavic Orthodox and Christian Peoples that a union of Eastern European Slavic states might make the twenty-first century an age of Slavic civilization.

He said that Western Europe had always feared the unification of Eastern European Slavic states and made efforts to prevent them from creating a union. "The time has come to stop taking instructions from Paris and London," he said.

He believes that the Balkan war was the catalyst of the idea of a Slavic union and that Russia must form up a military-political alliance in Eastern Europe.

"If the Russian troops come to the Balkans, they will never tolerate humiliation of the Slavic peoples. We shall send as many divisions to the Balkans as our Slavic brothers will ask," he said.

He emphasized at the same time, that Slavs would never become aggressors.

Leader of the Russian National Congress Aleksandr Sterligov said that the economic crisis in Russia was evolving into a catastrophe. This can be judged, he continued, from the unprecedented industrial recession and from the incompetence and inefficiency of the authorities.

Sterligov demanded that the events in Moscow last October be investigated and that their actual organizers not be allowed to evade punishment.

He said that the democrats were preparing for a civil war and were forming up armed units.

He urged the opposition to do everything they can to prevent Russian-American military exercises in the Urals.

In his opinion, the United States wanted to obtain information about the Russian army's combat capability so as to become better prepared for the occupation of Russia.

He said that public committees should be set up to organize actions of protest against Russian-American military exercises.

The World Congress of Slavic Orthodox and Christian Peoples is being held on the initiative of the Liberal Democratic Party which claims that the event is being attended by delegations from Poland, the Czech Republic, Slovakia, Serbia, Bulgaria, Ukraine, and Belarus.

## 6.42

### Three Parties Sign Appeal for Reunification of Slav Lands

Gleb Cherkasov

*Segodnya,* 17 June 1994 [FBIS Translation]

Aleksandr Tikhonov, chairman of the "Consolidation" party; Igor Karpenko, leader of the "Slavic Unity of Ukraine"; and Nikolay Sergeev, chairman of the "Belaya Rus' Slavic Assembly" yesterday circulated an "Appeal to the Heads of State and Parliaments of the Republic of Belarus, the Russian Federation, and Republic of Ukraine." The appeal contains a call to pass "legislative acts on the unification of the sovereign states to form a confederation or some other mutually acceptable state entity." In the opinion of its authors, "the creation of a confederation will make it possible to prevent disintegration of the unified community and to solve political, economic, and defense tasks."

Speaking at a press conference devoted to the promulgation of the appeal, the leaders of the parties stated that the pro-integration feeling prevalent among the peoples of the three countries does not always meet with adequate understanding among members of the political elites of Russia, Ukraine, and Belarus. At the same time, according to the assurances of the "Consolidation" chairman and State Duma deputy Aleksandr Tikhonov, Speaker Ivan Rybkin intends to meet with the authors of the appeal as early as 21 June. Mr. Tikhonov thinks the speaker's stance is drawing closer to proposals expounded in the appeal.

As for practical steps to create a unified state entity, it is suggested that referendums be held in all three republics following which the supreme organs of representative power could take specific actions to carry into effect the peoples' will. In the view of the authors of the appeal, the new Slavic statehood ought to have unified armed forces, a unified "ruble" system, and equality of state languages.

It is worth pointing out that the very fact of signing such an appeal is, to a certain extent, a new method in the political struggle both in Russia and in the Ukraine, as well as in Belarus. Until recently pro-integration and unification sentiments were mainly exploited by radical-communist and national-patriotic organizations, so the same sentiments expressed by "centrist" associations may have greater repercussions.

## 6.43

### Kravchuk Terms Slav Union Idea "Dangerous"

Viktor Demodenko and Mikhail Melnik

ITAR-TASS, 26 June 1994 [FBIS Translation]

"Alliances based on ethnic identity are dangerous per se, for they inevitably engender nationalism, pan-Slavism, and all that ensues from it," Ukraine's President Leonid Kravchuk told ITAR-TASS today, when asked about a possibility of creating a Slav union. "Both politicians and historians condemn it, knowing that that is where aggravated nationalism is rooted," he said.

Leonid Kravchuk also believes that this question is not quite specific in practical terms. "As for a union, one should then speak in terms of East Slavs, rather than Slavs. Otherwise, what should one do with Poland, Slovakia, Yugoslavia, or Bulgaria? Thus, the Slavs will be divided into Western Slavs and Eastern Slavs," he summed up. One should bear in mind that about thirty million Russians, Ukrainians, and Belarusians live beyond the borders of Russia, Ukraine, and Belarus. "The Slav Union will inevitably create problems for them." "By having such a union, we will split in two those states where Slavs and non-Slavs live." Kravchuk is convinced that "by this step we will begin dividing states of the former Soviet Union." He expressed the belief that this would be followed by creating Asian and other unions, and "then horrible confrontation will begin."

"One should not raise questions that are ready to be resolved today, neither theoretically nor practically. One should live and follow the chosen path: build independent states but not on the basis of ethnic identity," he said. "We are building a civil society in Ukraine and Russia is building the same society in its country. If we pursue a different policy a social explosion is possible. Russia is also not 100 percent a Slav state."

"I advocate economic relations, integration, and the use of the CIS in resolving issues that can be resolved by good will," Kravchuk said. In his opinion, one should not try to establish new structures and organizations immediately, he said. One should do everything to avoid confrontation between states and peoples, he said.

## 6.44

**Shakhray Sees Three-to-Four-State CIS Confederative "Nucleus"**
Interview
*Novaya Ezhednevnaya Gazeta,* 6 July 1994
[FBIS Translation]

Sergey Mikhaylovich, I have two questions for you: What are the prospects for the reintegration of the CIS states and how do you see its stages?

[Shakhray]: There is no alternative to reintegration. The only question is how it will happen. Either on a voluntary and equitable basis with the preservation of the political sovereignty of the states—and then, legally speaking, it may be only a confederation—or by an uncivilized, imperial method, with people washing their boots in some ocean or other [allusion to expansionist Zhirinovskiy remark].

The difficult situation in the economy, the impossibility of defining borders and establishing customs control to protect the country's economy, and the undeveloped state of the budget, taxation, and banking systems means that the factors of economic reintegration are being turned into a platform for internal political struggle both in Russia and in other countries of the CIS.

Reintegration will be complex and will take place in several stages: First, it is absolutely essential to preserve the CIS by creating a confederative (Eurasian) union not in place of but within the CIS; in other words, it is necessary to form a nucleus consisting of three or four states.

I believe that the main conditions for reintegration can and will be as follows: the fight against crime and economic union.

The interim, transitional form of reintegration will be through bilateral agreements such as the Russia-Belarus or Kazakhstan-Belarus type, and so forth. In this way the network of bilateral treaties will create the model for the future Eurasian union.

Subsequently, when the political and economic forms of reintegration have been perfected, we will surely proceed along the lines of the Maastricht accords in which sets of draft agreements will be submitted to referenda.

That will take several years—on condition that political stability is preserved in the CIS countries, of course.

## 6.45

**Integration with Ukraine, Belarus Viewed**
Aleksandr Golts
*Krasnaya Zvezda,* 16 July 1994
[FBIS Translation]

[Article by Aleksandr Golts: "Ukrainian and Belarusian Presidents Win Mandate of Confidence. Exercising It Will Take Strength and Willpower"]

Comparing the post-Soviet area to a unified organism that has broken down into different parts and lamenting this fact (sometimes sincerely, but sometimes speciously) have become frequent occurrences of late, when many parts of the former USSR are suffering from the inevitable hangover induced by a distorted sense of independence. Now that the peoples of Ukraine and Belarus have given those they have elected president an obvious and clear mandate for reintegration, there is hope that we will all be able to create some kind of new association which will allow all its participants to develop.

But the experience of the post-Soviet years shows that the best way to destroy an undoubtedly positive idea—be it the idea of democracy or of independence—is to decide that if it is acceptable to both peoples, it will come about by itself. That is the case with integration too. After all, just putting a torn fabric back together by no means always results in an integrated whole. Let us be frank—the period of almost three years since the Belovezhskaya Forest meeting has left its mark. And if the question of closer ties than those which exist in the CIS is really put on the agenda, their organization will by no means be painful.

In my view, the main thing today for Moscow, Minsk, and Kiev is to decide just what specifically they mean when they talk about reintegration. After all, it is different ideas of the aims and tasks of the CIS on the part of our countries' leaders (some seeing it as a chance to preserve everything positive accumulated during the Soviet years, others seeing it merely as a device for a civilized divorce) that have made this a not particularly effective association.

The same thing could happen to the new ideas about integration. If we proceed on the basis of the election programs of Leonid Kuchma and Aleksandr Lukashenka, we can be said to be dealing with more or less definite intentions. They could in principle be used to organize any policy, including the policy followed by their predecessors. And that is no accident. Our countries, which are going through economic crises—caused, among other things, by the breakdown in economic links and the loss of markets—are simply preordained to restore their ties. And this should be done irrespective of who occupies the presidential residence. The entire question is just how it should be done.

Let us be frank, restoring mutually advantageous economic ties is currently just a political slogan. In practice it is a highly complex balance of priorities and interests. For instance, the paramount vital interest of Ukraine and Belarus is obvious. This is to obtain from Russia sources of energy cheaper than those available on the world market. But Russia's interest is not so obvious. Our production slump has brought about a crisis in sales of enterprise output. Quite simply, our industry produces costlier and lower-quality output than is available abroad. In an attempt to protect its own producers, Moscow is creating concessions for them (which is causing a lot of controversy). If we are talking about simple production-sharing or expanding commodity turnover, then Russia—insofar as I understand it—will have to indirectly subsidize inefficient production in Ukraine and Belarus as well.

And this despite the fact that the two republics' leaders have views of reform that are very different from those of Moscow. So if we are talking seriously about economic union, we need to agree on the overall concept for reforming our economies. After all, talking about the advantages of integration, the EU, which people in our country love to refer to, was able to be established when the economic situation in the West European states had evened out somewhat. But so far the economies of our three republics are functioning on the basis of fundamentally different rules.

It is this fact rather than nationalistic idiocy that has led to the emergence of all these customs posts, entry and exit regulations, and all the other things that so annoy the peoples of all our states. But if we continue to make economic progress not only at different speeds but in different directions—as is the case at the moment—the customs posts will never disappear since they are an attempt by the state to regulate inequalities in development and distribution between neighboring states.

It is another matter if we manage to reach agreement on the overall concept of reform. Within this framework programs would be elaborated in which interstate cooperation would make goods produced in the three republics competitive on the domestic and world markets. And it is not just a question of supplying components. The contribution of Ukraine and Belarus could be special privileges in the use of Black Sea or Baltic ports, preferential transit regulations, and much else besides.

But there is one "but"—and one which, incidentally, the EU encountered as well. Any integration leads to a certain amount of sovereignty being transferred to suprastate organs. And this annoys those who are accustomed to thinking in terms of absolute independence. Yet the more effective integration is, the less room there is for notorious separatism. Let us not forget that optimizing the economy will unambiguously require not only the elaboration of unified approaches in the sphere of financial activity and production relations, but also in defense, security, and foreign policy. After all, deciding each time what weapons to buy where, where to deploy troops and to what end, what maneuvers to participate in, and whom to apply sanctions against should be agreed with those with which we have a common economic interest.

But we are talking about areas where until recently Kiev and Minsk had sought and found symbols of their own sovereignty (we would recall merely the unfinished saga of the division of the Black Sea Fleet). But whether the new representatives of the Belarusian and Ukrainian elites, who have only just gained access to power, will agree to coordinate not only the level of interest rates and the size of the money supply, but also military programs and participation in certain international programs remains an open question.

And this question—of striking a balance between sovereignty and commitments within a union—has not been unambiguously resolved either by France, Britain, or a number of other West European states. As for our three states, the real potential for integration will not be evident before all its conditions are clearly and precisely spelled out. And that is work that needs to be expedited, not least by Russia.

---

## 6.46
## Creation of Slav Union Viewed

A. Boroday
*Krymskie Izvestiya,* 11 August 1994
[FBIS Translation]

---

Policy in the East European region, which has thus far depended on the three "Belovezh diehards"—Leonid Kravchuk, Stanislav Shushkevich, and Boris Yeltsin—could change appreciably and influence the situation in the Slav countries with the appearance of the new presidents of Ukraine and Belarus—Leonid Kuchma and Aleksandr Lukashenka.

The newly elected heads of the Ukrainian and Belarusian states emphasized in their campaign programs the desirability of their countries' close cooperation with Russia. In addition, Aleksandr Lukashenka insisted on political and economic integration with his northeast neighbor. Leonid Kuchma was considerably more guarded in his statements that concerned cooperation with Russia. But we will see the presidents' true colors some time hence inasmuch as it cannot be ruled out that the campaign promises of one of them were merely a bluff.

The Belarusian head of state has now found himself in a very difficult situation. First, he is the first president of Belarus, and this will require of Aleksandr Lukashenka sufficient discretion and level-headedness. The first two years will be difficult for Belarusians, I believe, since the institution of the presidency has yet to be perfected legislatively, and conflicts between the executive and the legislature are therefore possible. Especially since Aleksandr Lukashenka has, compared with Leonid Kuchma, no experience of work in the top echelons of power. His very modest position on a commission of the Supreme Soviet of Belarus has hardly given him substantial political equipment. Second, Russia, understanding that it could soon lose not only Ukraine but Belarus also, will try in any way to keep it near it. It cannot be ruled out that the northeast neighbor will begin to "bear down" economically. Although it has to be said that such a development of events is possible only provided that Aleksandr Lukashenka moves to confront Boris Yeltsin. Currently, however, the president of Belarus is disposed toward a close alliance with Russia, realizing that the voice of the Belarusian state in the world community will not be heard any time soon and that, consequently, it should not "gamble" entirely on assistance from the West. It is not surprising, therefore, that Aleksandr Lukashenka declared in the course of the campaign battles that Belarus's attitude toward the Partnership for Peace program would be coordinated with Russia.

Relations between Ukraine and Belarus will shape up pretty well since neither party has yet any claims on the other. Although the unification of Belarus, Ukraine, and Russia in a single Slavic union can be ruled out at this time. Even were the Belarusians to agree to it, Leonid Kuchma would not, I believe, want to be under the thumb of Russia, especially since Ukraine has already undergone a certain evolution since the times of the proclamation of independence. In the immediate future, therefore, Belarus and Ukraine and Russia will be finding one another's range, and the creation of a unified bloc and union among them is possible only after a certain length of time, when all three states have their own voice in the world.

## 6.47

**National Patriots Urge Restoration of Slav Union**

Oleg Artyushin

ITAR-TASS, 1 October 1994

[FBIS Translation]

Editor's Note: The following document briefly characterizes the views of the national patriots on Slavophilism. The patriots distinguish themselves from nationalist by pointing to their traditional Slavophile origins, literally the "soil-bound" (*pochvennik*) tradition which focuses on the historic existence of a supranational community on the central Eurasian land mass. During the Russian empire, these peoples were basically free to pursue their own paths, while their "elder brother" provided rich cultural inputs but tolerated the parallel development of other cultures. The "imperial" approach stressed the rights of individuals and communities, rather than nations. In contrast, the Russian nationalists stress the development of state structures in which the ethnic Russian nation would be seen and treated as superior, and "protective" of lesser nations. The nationalists endorse Russification techniques to a far greater extent than the national patriots do. Consequently, the nationalists have been more willing to make common cause with the overtly pro-Soviet neo-communists. The article names among the patriots who support the Slavic Union Vasiliy Starodubstev, a member of the Emergency Committee which initiated the aborted coup attempt against M. Gorbachev. Aleksandr Solzhenitsyn is a patriot in the Slavophile tradition who condones a voluntary reconstitution of some sort of Slavic union, and views Soviet adventurism as the kind of colonialist-type imperialism that ruined Russia's true spirit.

National Patriots of Russia, Ukraine, and Belarus called for the possible restoration of the union of three Slav nations.

They held a conference in the city of Bryansk on Saturday, which was attended by 200 representatives of twenty-four political parties and movements from thirty regions of Russia, Ukraine, and Belarus. The conference was also attended by a Serbian delegation.

The participants set up a twenty-one-member organizing committee to carry out the idea of restoration. The committee will be headed by former KGB general, head of the Russian National Assembly, Aleksandr Sterligov.

Apart from Sterligov, other committee members include General Valentin Varennikov, Agrarian Union Leader Vasiliy Starodubstev, Orlov Region Administration Head Egor Stroev, former Soviet Coal Industry Minister Mikhail Shchadov, former leader of the Russian State Committee for the Press Boris Mironov, and leader of the Workers' Union Dmitriy Igoshin.

They also agreed to establish a party of national and patriotic forces of Russia in the run-up to parliamentary and presidential elections to be held in 1996. An organizing committee set up to coordinate work on the establishment of the party will be headed by Boris Mironov, recently sacked by President Boris Yeltsin for "fascist-like" statements.

The party congress is expected to be held by the end of the year.

## 6.48

**Zhirinovskiy Proposes Slav Economic Union**
23 October 1994
UPI


*Editor's Note:* In the case of the Economic Union, the distinctions between nationalists and patriots become blurred. Both agree that the collapse of the centrally controlled Soviet economy was a disaster and seek the reinstitution of a supranational economic "bloc," dominated and controlled by Russia. For nationalists, however, economic union represents only the first stage of unification.

Russia's ultranationalist Liberal Democratic Party is going to propose that Slav nations form an economic union, LDP leader Vladimir Zhirinovskiy said.

In a letter to lower parliament house chairman Ivan Rybkin, Zhirinovskiy says the proposal is to be considered by an academic conference in April or May 1995 and by a "congress of Slav peoples held at the governmental and legislative levels" two or three months later. The congress would draw up a legal basis for the future Slav community, the LDP leader says. He says his party has already formed an organizing committee for the conference and congress, headed by Aleksey Zvyagin, a parliament member and professor at the Moscow Commercial University.

Zhirinovskiy asks the lower house to back the proposal and says he has made the same request to upper house chairman Vladimir Shumeyko.

## 6.49

**New Electoral Association Supports Economic Union with Three Former USSR States**
Interfax, 1 December 1994


*Editor's Note:* The so-called Third Force coalition was led by Arkadiy Volskiy, leader of the Union of Industrialists and Entrepreneurs, or "Civic Union." Volskiy is spokesman for the coalition, and probably one of its founders. His support for a Slavic Union coincides with his organization's call for a strong center in Russia, and a harsh form of federalism. For this group of industrialists, as for the Zhirinovskiy ilk nationalists, an economic union with supranational powers is merely the first step toward inevitable political unification. A majority of the executives in Russia's privatized enterprises are veterans of the old party and state hierarchies, and Volskiy's group finds common cause with the former, or (depending on one's viewpoint) "neo"-communists.

The leaders of a new electoral association, the Third Force, have announced the convocation of a congress of citizens of Russia, Ukraine, Belarus, and Kazakhstan in Moscow between 28 and 30 December to discuss practical steps needed to bring about a common economic union of these countries.

Arkadiy Volskiy, the leader of the Civic Union, which is a member of the new alliance, told a news conference on 1 December that the congress would discuss holding referendums in the four countries, to bring about their speedy unification.

Volskiy said that, in his opinion, it was the disintegration of economic ties between the republics rather than the reforms which had been the main cause of their decline.

Volskiy added that the first stage of unification at an economic level would inevitably be followed by the amendment of laws and later probably by political unification. He thought that other countries could join later, on a voluntary basis.

"It is time countries got over the sovereignty mania and realized that we will not survive without each other," Volskiy said.

According to him, economic and political recovery is possible only with the reintegration of former Soviet republics into a new economic and military strategic entity.

The Party of the Majority, the Union of Renewal and the Agroindustrial Union are among the parties which make up the Third Force, together with the Civic Union.

## 6.50

**Lukashenka Favors Tripartite Economic Unity**
BELAPAN-Minsk, 14 December 1994
[FBIS Translation]

*Editor's Note:* The following report on a meeting between A. Volskiy and A. Lukashenka epitomizes the degree of authoritarian control desired by supporters of an economic union. Lukashenka's willingness to issue a decree establishing the industrial structure of the country verges on blatant dictatorship, if not megalomania.

President of the Republic of Belarus Aleksandr Lukashenka voiced resolutely in favor of unification of the economies of Belarus, Russia, and the Ukraine, and approved the steps toward this, which were made by the participants of a meeting of industrialists of Belarus, Russia, and Ukraine the other day in Minsk, Arkadiy Volskiy, chairman-coordinator of the Council of the International Congress of Industrialists and Enterprises, president of the Russian Union of Industrialists and Entrepreneurs, informed the press. The leaders of the national organizations of industrialists of the three Slavic countries acquainted the Belarusian president with the agreements that had been reached at the meeting: on the most-favored regime for the countries and on the establishment of transnational companies and industrial-financial groups common for the countries, as well as a statement for the heads of states, governments, and parliaments, which contained a call for creating a real alliance of the economies of Belarus, Russia, and Ukraine with some subsequent steps in the political field. Mr. Volskiy stressed that since the very beginning of their cooperation the industrialists and enterprises of the three countries called upon their colleagues in Kazakhstan to back up the cooperation on a mutually beneficial basis.

Answering a question on the prospects of integration, Mr. Volskiy said that it was inevitable. The same is as regards the other CIS member states. "We had been constructed as one large factory. We have so many specialized enterprises that it is absurd to develop these productions in every newly independent state, and it requires a lot of money that Belarus, Ukraine, and Russia lack." But having started establishing financial-industrial groups and transnational companies in practice without the participation of the governments, the industrialists came across the impossibility of solving concrete problems because of the difference in the national legislations. These problems can be solved within the framework of regulation of interstate relations in the political and legal fields. According to Arkadiy Volskiy, Aleksandr Lukashenka expressed his willingness to support the establishment of the first transnational companies by means of a presidential decree authorizing the unification of the values of basic funds of the enterprises under amalgamation into transnational companies, corporations, and groups.

**6.51**

**Industrialists Support Slav Union Plus Kazakhstan**

Interfax, 28 December 1994



An international congress "Toward Integration Through Accord" opened in Moscow on Wednesday. Participants in it decided to form pressure groups to collect signatures in favor of holding referendums on reunification in Russia, Kazakhstan, Belarus, and Ukraine.

"Immediate efforts to reintegrate the republics into an economic, political, and defense union will make it possible for each of them to overcome their profound crisis," the chairman of the International Congress of Industrialists and Entrepreneurs Arkadiy Volskiy told the forum.

According to him, the process of reintegration must go through four stages: form an open union between Russia, Ukraine, Belarus, and Kazakhstan; lift all restrictions on the free movement of goods, capitals and labor; adjust legislations; build supranational management structures.

"We must appeal to the peoples of our republics and use referendums to force the authorities to start a real reintegration processes," Volskiy told Interfax. "I have no doubt that the people will give a positive answer to the question concerning reunification."

In the meantime a member of Russia's Federation Council, Nikolai Gonchar, who also attended the congress, told Interfax that currently "only an economic union can be discussed; a political union will be opposed by the forces now in power in the CIS countries, and some forces in the West will object to a defense union."

Convened at the initiative of the Civic Union-Third Force association, the forum gathered spokesmen for 128 political parties and organizations from the former Soviet republics.

# The Eurasian Union

The Eurasian confederal concept originates partially from pre-revolutionary Russian political theorists who saw Russia's power and might emanating from its position between Europe and Asia. In brief, these theorists embraced an anti-European, anti-Atlanticist doctrine, which held that Russia is not European, but a combination of European and Asiatic. They proposed the construction of a great Eurasian state which could eclipse the culturally bankrupt societies of the West, and become a great superpower and leader of the Eurasian heartland.

Within the context of the CIS, however, Kazakhstani president Nursultan Nazarbaev is the author of the Eurasian Union concept. It is not clear that Nazarbaev adheres to the pre-Soviet doctrine of Eurasian power, or that he is a "neo-Eurasianist" (although he may be). It is clear that he envisions the restoration of something similar to the USSR, but professes that it must incorporate the ideals of sovereignty, territorial integrity, and independence for its member states.

In the documents collected here, speaking before the USSR Supreme Soviet on 26 August 1991, Nazarbaev proposes the formation of a confederation, which he would call "Free Union of Soviet Republics." He associates his proposal with the ideas of Andrey Sakharov. (The prize-winning scientist and human-rights activist had pressed for a Confederation of Free and Independent States at the first USSR Congress of People's Deputies in May 1989, comparing his concept with the British Commonwealth.) Nazarbaev's organization would contain all former Soviet republics that wished to join and would be managed by supranational councils for defense, transportation, and communications. He specifically excludes a council for foreign affairs, conceding autonomy in this realm to the individual states. He concentrates economics in an interim economic council and community, to which any former republic may belong even if it decides not to join the Union. Nazarbaev's proposal lacked many specifics. Consequently, coming in the wake of the attempted coup and Ukraine's declaration of independence, as well as the unsuccessful discussions associated with Gorbachev's Novo-Ogarevo Union Treaty, he was unable to reassure the republican deputies that his confederation would be truly democratic.

Since the Belovezh Forest declaration of 8 December 1991 and the formation of the CIS, Nazarbaev has continued to advocate a new confederation of free and equal republics. At times, his proposal has seemed to represent the only clear formulation of what CIS integration would entail. None of the presidents or legislatures has definite ideas on how to integrate a commonwealth, or how such an integrated organization would operate. Nazarbaev's concept also has the advantage of coming from a man who is a known entity, has consistently advocated the same course, and who offers his own state as a supportive, but impressive counterweight to Moscow's authority. His "Eurasian Union" appeals to some of the Central Asian states, but is still rejected for various reasons by the other CIS leaders.

During 1992–93 CIS politicians were in no mood to contemplate seriously the persistent arguments of a man who had never accepted the breakup of the Union. These were years of frenetic foreign contacts, writing of new laws and constitutions, elections, introduction of economic reforms, and political organizing. Nevertheless, the winters of 1992 and 1993 were particularly harsh. Fuel was in short supply. Several violent interethnic conflicts were raging and many people in the afflicted regions were starving. In many places, the flow of food and other products through the economic pipelines of the former republics had slowed to a trickle, or stopped altogether, causing high unemployment and sending prices into the stratosphere.

During this time, the Commonwealth was faltering. It had utterly failed to create a common defense force and had stalled on every economic issue it took up—currencies, monetary policy, customs union, interenterprise debt, etc. In June 1993, the three Slavic leaders created a trilateral economic union within the CIS framework without consulting Kazakhstan or any other CIS state. Kazakhstan had already been conspicuously left out of the Belovezh Forest meeting, and this second exclusion was glaring. In July, Russian president Boris Yeltsin arranged an extraordinary summit between Russia and the Central Asian leaders in Almaty. The documents collected here record the meeting. This meeting demonstrated that Russia understood it could not leave out Kazakhstan or the other Central Asian nations in its integration plans. This was the first meeting Moscow had held with the Central Asian leaders since forming the Commonwealth. It gave the Central Asian states new status within the Commonwealth and Nazarbaev's proposal for a Eurasian Union a new lease on life.

Nazarbaev has continued to refine and defend his concept. Other CIS leaders have registered lukewarm reactions, although Kyrgyz president Askar Akaev has consistently supported it. The documents here focus on the 1994–95 period in which the Eurasian Union was being taken more seriously, but still receiving negative reactions, especially from Russia and Belarus, whose leaders would prefer to integrate the entire CIS, or the Slavic states and Kazakhstan under the CIS umbrella.

In the documents that follow, Nazarbaev draws attention to the fact that his proposed union would provide the apparatus currently lacking in the CIS to ensure that its four-hundred-plus documents and protocols (each one signed by varying numbers of CIS members and ratified by even fewer of the parliaments) would be observed by each member state. Nazarbaev also makes particularly strong reference to the need to align the legislation of the CIS states through the creation of a Eurasian Union parliament, which would become the organization's supreme consultative and deliberative body. Modernization within the post-Soviet space is impossible without a supranational legislative mechanism. (In Chapter 8 the reader may judge how well the CIS Interparliamentary Assembly, created for this purpose, is working.) An alignment of CIS members' legislation would of course also mean a sacrifice of essential sovereignty, and this particular feature is not likely to be acceptable to most CIS leaders for some time to come. This and a number of other binding articles in Nazarbaev's draft, aimed at guaranteeing the ability to implement Union decisions, actually resemble a federation more than a confederation, although Nazarbaev contends that his concept is the latter.

One surprising aspect of the draft is that it assumes side-by-side existence with the CIS, which Nazarbaev insists would be compatible with the Eurasian Union. The Eurasian Union is only being offered as an option, he says, which CIS states may or may not join. The mere existence of such an integrated, supranational organization (if it contained Russia) would, however, create pressures to join, because it could be harmfully exclusive of other states in its policies and laws.

The Eurasian Union draft of April 1994, which Nazarbaev presented during a CIS summit, evoked a landslide of press reaction, not only in the CIS but in the Western industrial states. There was much conjecture among Western reporters that the proposal would carry far-reaching repercussions within the CIS. Some European commentators surmised that the supranational Economic Council would improve the success of the new states in their foreign economic activity. Several Western correspondents saw the Eurasian Union as the possible basis of a new association of sovereign states, replacing the CIS for the purpose of strengthening the security, and social and economic modernization process in the post-Soviet space. On one issue, that of citizenship, the Eurasian Union improves matters by eliminating the "dual citizenship" concept and establishing automatic citizenship for any Union resident moving to another Union state.

The article included here by Kazakh writer Asylbek Bisenbaev puts a positive slant on one clause in the proposal which would distinctly curtail freedom of the press. This measure would create an Executive Committee Information Bureau which would have the power to censure any articles which it judges "unfriendly" to parties of the treaty or that could be "harmful to relations between them." Presumably this clause exists in order to rid the Union of the slanderous press campaigns sometimes launched against a leader or a particular event in another state, but it removes the sanctity of free speech and freedom of the press.

Only Belarus and Kyrgyzstan immediately supported the Eurasian Union proposal. In Russia, opinions ranged across the gamut. Aleksandr Solzhenitsyn (*Izvestiya*, 4 May 1994) fiercely attacked the proposal and suggested a union of Belarus, Russia, Ukraine, and Kazakhstan as the only feasible variant of the Eurasian Union. Others, confusing Nazarbaev's concept with neo-Eurasianism, which has more than one hundred models, interpreted it as an attempt to advance Central Asia's dominance in the region. Some high Russian officials (for example, Sergey Shakhray) expressed support for the proposal, saying that such integration was inevitable. Askar Akaev, on the other hand, blamed Russia's coolness on its imperial ambitions and its jealousy of Kazakhstan. It is worthwhile noting that in Ukraine, when the proposal was first put forward, Leonid Kravchuk rejected it out of hand, but Leonid Kuchma called it "the optimum solution to the crisis situation in which the sovereign states of the former USSR have found themselves."

At the September 1994 CIS summit of heads of state the issue of the Eurasian Union draft was once again brought up, but deep-sixed. The proposal appeared dangerous to some and confusing to others. Yeltsin, Kuchma, and Shevardnadze all said the concept "needed reflection," though Yeltsin said it contained valuable ideas which could be absorbed into the CIS integration process. The *Izvestiya* article recording the proposal's reception at the summit drew a picture of total lack of enthusiasm. Russia retained control of the summit session and the direction in which the CIS was going. Reactions to the concept have continued in the same vein since then. Each leader views the CIS as preferable to the Eurasian Union's restoration of a structure reminiscent of the USSR.

The following documents provide a thorough discussion of Nazarbaev's proposal and internal CIS reactions. It is clear that Nazarbaev does not intend to give up his struggle. Time will tell whether his ideas will help to construct a new USSR.

## 6.52
**Nazarbaev Proposes Confederative Treaty**
Moscow Central Television, 26 August 1991
[FBIS Translation]

Esteemed people's deputies! The past week has been not just days, not just time. It is an era separating us from our entire past life. We celebrate the victory of democratic forces over reaction. We have deep respect for the Muscovites who stood in the way of those who were prepared to chase us back into the stable of a totalitarian regime. We cordially thank the Supreme Soviet of the RSFSR [Russian Soviet Federated Socialist Republic] and Yeltsin, president of Russia, who took a decisive and uncompromising position in a serious and tragic situation. We deeply grieve for those who perished on the streets of Moscow, for those who, at the cost of their lives, defended the young democracy. At this time, we cannot but also remember all of those who joined so selflessly in the struggle against reactionary forces in practically all regions of the country—the students of Alma-Ata, who, as far back as 1986, went out with their bare hands against military hardware, the victims of Tbilisi, Yerevan and Baku, Vilnius and Kishinev.

However, there is acute concern for the future of the country. We need to take on the broad problems facing us, which, as events of the past few days have shown, are far from being solved. We now all have a different awareness, and this obliges us to look at the future of the Union and the Union Treaty in a new way.

For me it is obvious that the renewed Union can no longer be a federation. We have spent too long chasing after the past. The declaration of independence by the Ukraine and the similar decision being prepared by Belarus are evidence of new historical realities which we must not turn away from, or pretend that they do not exist. The situation has changed significantly.

You know that I was an active supporter of the speedy signing of the Union Treaty. I remain one. But recent events have shown how explosive and fraught with the danger of bloodshed the old scheme was.

In our time, we failed to listen to Andrey Dmitrievich Sakharov who proposed the Union as a community of equal republics. He said before we did that the republics had matured for that. But the time has now come for us to understand the correctness of what he said.

How do I envisage that future union? Having entered into contractual economic agreements among ourselves, we republics have in mind broad economic relations with everyone who agrees to it. It is not to the benefit of any of the fifteen republics in economic terms to go their separate ways now.

Common Union bodies retain certain functions. In my opinion this should be the protection of common borders, a supreme council for control [*kontrolnyy komitet*] over nuclear arms under the Ministry of Defense, in which all members of the Union would take part, controlling those who have their finger on the missile launch button.

The Union Ministry of Defense is the ministry for the defense of our borders. As for the army in each republic—and this has already been decided by the RSFSR and Ukraine, and others have also decided this, and Belarus also intends this—we send the requisite number of servicemen for the Union Ministry of Defense. If an external threat arises, the ministry combines all our armed forces to defend the union.

We should have a joint Union transport network, as well as a republican transport system, including an air transport system. There should be interrepublican service lines [*kommunikatsii*], and our own service lines should be under our own control. The same applies to communications: there should be interrepublican communications and our own.

International relations are shared only in working out general trends, including the problems of disarmament, general world problems, and so on. But each republic must have its own ministry of foreign affairs with full rights, regulating its foreign political and economic activity and foreign economic and political treaty relations with other states. Each republic must also have completely autonomous consular administrations, so our citizens do not have to depend on the Ministry of Foreign Affairs of the USSR to be allowed to travel abroad or to invite guests from abroad.

We are used to the abbreviation USSR. I propose leaving it and expanding it into the Free Union of Sovereign Republics [*svobodnyy soyuz suverennykh respublik*]. By republics I have in mind all republics, including the autonomous ones that have declared and want to declare themselves sovereign. We must give the right to representatives of any nationality to speak in their national language at all Union and republican forums and give every person the right and opportunity to hear their deputy in their mother tongue. In other words, we are proposing that a confederative treaty be concluded.

I am convinced that only then shall we attain genuine equality for the republics. Perhaps only through this shall we achieve a true federation: I am convinced of it. I should like to note, by the way, that the Novo-Ogarevo treaty is in effect practically a confederative treaty. We are being cagey in regarding it as a federal treaty. So let us admit this honestly, and finally put everything in its rightful place. The conclusion of a non-economic treaty can only be done in a stable and tranquil atmosphere, and only with republics that fully guarantee human rights and trust each other. Guarantees of human rights for people of all nationalities, religious faiths, and party allegiances, apart from fascist parties, of course, must be assured without fail on our territories in line with the Helsinki accord. To this end, Kazakhstan is prepared to be the first to invite an international commission to operate permanently in our multiethnic republic. We want to completely end violations of human rights that we have inherited from the old system.

We are also prepared to invite an international commission of experts to examine all patients in all hospitals, maybe in our republic, too. We are going to immediately launch a review of all so-called economic crimes, most of which were a punishment for ordinary entrepreneurial activity. We also regard all this as an assurance that we shall build a law-governed state and as a guarantee of republican sovereignty.

I call for an immediate solution to the question of giving full freedom to the Baltic republics, Moldova, and Georgia, and all who have expressed their aspiration for independence and autonomy by legal, democratic means, without any settling of accounts or the presentation of the notorious claims for debts amounting to billions, because even one human life is invaluable, and the people of our country have paid with so many for the love of freedom which they have displayed. And are these victims the last?

So, the ten-plus-one formula remains a thing of the past. Then, I must say straight out that in the new Union there should not be any Union cabinet of ministers, no Union parliament, apart from the treaty relations entered into by the republics. The moment of truth has arrived today. I cannot imagine any other basis on which Kazakhstan will enter a union with the other republics.

That, so to speak, is strategy. But the tactics for the immediate future must consist of the urgent formation not of a cabinet of ministers, but of a transitional interrepublic economic council. In it representatives of all the republics will work on an equal footing, joining in an economic community. But before that, we must give those who have decided to leave the opportunity to realize that right. All may join that interim economic council, including those who decide to secede from the Union. But economic cooperation will still continue, for the time being, for we are aware that, while they will leave politically, they will continue to cooperate economically with the others.

We have a legally elected president and parliament, which continue their activity. This should be mainly directed at ensuring that people should survive the winter and not die of hunger.

That is my view, in general terms, of the future Union. In conclusion I want to stress—so that no one should harbor any illusions—that Kazakhstan will never be any region's underbelly and will never be anyone's little brother. We shall enter the Union only with equal rights and equal possibilities. Thank you for your attention.

## 6.53

**Nazarbaev Statement at News Conference**
Moscow Ostankino Television
16 April 1993
[FBIS Translation], Excerpts

Esteemed ladies and gentlemen, esteemed representatives of the mass media: I imagine how impatiently and tensely you are awaiting the results of our meeting. It is really very extraordinary that there were few who suspected it a week ago. Although I would say that it is very timely—a meeting that was achieved through long suffering, yet was quite usual. It has been waiting its turn for a long time, and the peoples and the leaders of the countries of the Commonwealth have long been asking themselves why is it that there are more and more documents and agreements after CIS summits, but our life is becoming worse and worse?

Just think what a paradox we have here. The more documents we adopt on coordination and integration, the greater the lack of coordination there is in our actions, the more quickly and fruitlessly the previously unified economic space slides apart, thereby undermining not only the ties that have evolved, but also the very fate of ordinary people. Thus, the idea of the meeting was born. The awareness of this paradox grew with every day, with each of our steps toward integration.

Finally, the moment came when concern over the growing destabilization in the CIS led us to realize a very important thing. We realized that such a paper integration could not continue any longer, that it gave rise to endless papers and agreements. Our people are expecting us to take real, responsible steps.

It is concern for the state of affairs in the CIS that gave rise to the renowned appeal by the president of Russia at the beginning of March to the heads of our states. It is concern over the lack of results of that integration that did not allow us to remain silent. In reply, Alma-Ata gave its support to Moscow, and the idea was born of holding a meeting of the heads of states in the nearest future at which one question would be discussed, a question that was approved on the agenda today. It is the question of the further strengthening of the CIS. No other questions were examined. No documents were signed apart from the protocol on instructions that I, as chairman, have been instructed to sign and that I will speak about further. Despite the rather complex situation in each of our states, all the states of the CIS viewed that initiative sympathetically and responded to it with great interest. None of us remained indifferent, because this question worries us and is of concern to us all.

I would like to draw your particular attention to this. Despite the difficult conditions and the rather difficult situation in Russia—in spite of certain opinions—the Russian president, just as we all are, is concerned with the problems of the CIS. At the beginning of March, he appealed to all of us. I emphasize the role played by Russian President Yeltsin in today's meeting. This is important because, at the present state, precise and correct actions by the Russian leadership and by all power structures in Russia determine more than the political stability in Russia itself. In a sense, the development of events in Russia could seriously affect the progress of democratic reforms and market transformations in all CIS states today. I think it is in the interest of all of us that the process of reforms begun in Russia should continue. A president elected by popular vote and a legitimate parliament are guarantors of this . . . .

I am well aware that the question of whether or not to move toward closer union in the CIS is very difficult and complicated for a politician and a head of state. It is not every man who can find in himself the courage to answer this question. The result is destabilization, collapse, suffering, and torment. I wonder, how many more victims and how much more suffering must there be before we appreciate the need for closer coordination? How many more losses must we sustain before we stop dithering and doubting in this regard? Or perhaps the heads of state of our Commonwealth simply need the assistance of our parliaments? Take a look at the interesting experience being gained in this respect here, on Belarusian soil.

The question came up regarding the need for Belarus to join the Treaty on Collective Security. The Belarus parliament examined it and adopted a decision. Perhaps other Commonwealth countries should adopt this experience. After all, the question of the CIS's future is not a political question. It is simply a human question—the most everyday question—the question of people's futures, of the possibility for us all to live normal human lives.

Before this meeting, all of us were moving, as it were, in a crowd, but each of us had his own thoughts. Each of us had his own ideas about the intermediate stages along the way, and the speed of travel. We all had confused and individual ideas about which way we were moving toward integration, what we wanted from the Commonwealth, what we were joining together for, and what we intended to use the Commonwealth for as an instrument. Each of us had his own vision of a desirable near and distant future and his own views on our relations with the other participants in this difficult and unexplored voluntary integration.

Now this extraordinary meeting has taken place. At the meeting, we stopped, as it were, and determined the direction in which we should proceed. Which of us is already prepared to go in that direction? How do we organize an effective joint movement? How do we join our resources? How do we employ them collectively in the interests of our peoples?

At the next meeting, which we planned by mutual accord to take place in mid-May, we will decide how to organize effective joint advancement, how to unite our resources, and how to apply them collectively in the interests of our peoples. . . . We will be examining substantial work carried out by working groups on documents that we made rough plans for today.

For its part, Kazakhstan proposed a whole range of questions of principle and strategy to be expanded at the next meeting. These questions concern key issues: closer integration in the economic, military, and foreign policy areas.

We have suggested setting up a union of agricultural workers. A meeting has already been held in Akmola, and it has been set up by all ministers in the former USSR states. A council of oil workers has held a session in Surgut. We have invited all industry ministers to convene in Karaganda at the end of May. Seven states have signed documents on the Collective Security Council, the CIS Charter, the Economic Zone, and the Single Currency Zone. They were ready for close cooperation, hoping that our subsequent summits would really become important milestones on the road to a new integration in real terms. We cannot help but take into consideration the hopes and expectations of our peoples. Thank you for your attention.

---

## 6.54
## Russian, Central Asian Leaders Hold Summit—Ties Viewed

Vitaliy Portnikov
*Nezavisimaya Gazeta,* 6 August 1993
[FBIS Translation]

---

### The Kremlin Has Finally Noticed Central Asia

If we forget for a moment that the leaders of Russia and the Central Asian countries are gathering in Moscow to stop the escalation of civil war in Tajikistan, it can be acknowledged that we are witnessing a truly unprecedented event. Indeed, every cloud has a silver lining. Boris Yeltsin held separate meetings with the leaders of the Baltic countries, he met with the leaders of Ukraine and Belarus—and each time these meetings turned out to be historic—but he has not yet held separate meetings with the heads of the Central Asian states. Obviously, there were many collective gatherings after the meeting of the leaders of Russia, Ukraine, and Belarus in Belovezh Forest that led to the dismantlement of the USSR and creation of the CIS, but apart from Yeltsin and his Central Asian partners, they were invariably attended by Kravchuk, Snegur, Shushkevich, and Ter-Petrosyan, for whom Central Asia, judging from their countries' geographical position, holds much less interest than for Russia. As a matter of fact, leaders of the Central Asian countries have also met with each other more than once. But Yeltsin did not attend. Now, dispatching his special representative, Anatoliy Adamishin, to Central Asia, the Russian leader has effectively put forward a new concept of collective responsibility of Russia and local countries for stability in the region. What is effectively at issue is the creation of a new geopolitical union—this time without European republics of the former USSR, which are not particularly interested in such an association. It will be recalled that this is occurring literally a few weeks after the demarche of the prime ministers of the three Slavic states, who announced—almost like their heads of state in Belovezh Forest—the creation of an economic union of Russia, Ukraine, and Belarus; and also after sharply critical statements by some high-ranking Russian politicians accusing the Central Asian countries of desiring to set up their own economic union without Russia, but including

Turkey, Iran, and Pakistan. But the goings-on in Tajikistan have pushed Russia and the Central Asian countries toward each other.

**Nazarbaev as Guarantor of Russian Stability**

Kazakhstan's unexpected absence from the economic union of Russia, Ukraine, and Belarus looks as inexplicable and conspicuous at this former union republic's non-participation in setting up the CIS. Yes, Kazakhstan is not the most immediate neighbor of Ukraine and Belarus. For Russia, however, it is not simply a country next door, but a guarantor of the secure existence (if not survival) of the Russian state. It is Kazakhstan with its half-Kazakh, half-non-Kazakh population that makes it possible for Moscow to feel itself to be part of Europe, by giving it a bridge into Asia; a Kazakhstan that turned into a sort of buffer between two different civilizations not only for Russia, but for the whole world. Kazakhstan's significance for Russia appears even greater when, looking at the map, you realize how narrow a strip of land separates this country from the predominantly Muslim republics of the Russian Urals. One can imagine the consequences of Almaty's switch from a favorable policy vis-à-vis Moscow to a confrontational one. This is why one cannot stop wondering how deep-seated the Kremlin's desire is not to pay heed to Nursultan Nazarbaev—not only an outstanding politician, but also a real "political gift" to the current Russian authorities. At a time when it is becoming increasingly evident that "an Islamic revolution" in Central Asia in one form or another is inevitable, Russia would be wise to hold onto Kazakhstan with both hands. This is its only chance not to bog down in Central Asia, but to remain a respected and welcome partner for the region's countries.

**Karimov Rethinks Viewpoints**

On the eve of the Moscow meeting, the Uzbekistan president has made a whole range of political statements on Tajikistan that would confound observers. The Uzbek president, who had actively supported the forces that defeated the current Tajik opposition, now calls on Dushanbe to start talks with opposition leaders. Evidently, Karimov does not need instability on the Tajik-Afghan border because it can at any moment turn into a factor for instability on Uzbekistan's borders with Afghanistan or even with Tajikistan. Besides, the Uzbek president is unlikely to be satisfied with the fact that all the key positions in Dushanbe have been seized by people from Kulyab and not those from Leninabad, who have traditionally oriented themselves toward Tashkent.

Taking part in the Tajik events, Islam Karimov has shown who can be the region's master. Now it is essential for him to add to this a reputation as Tajikistan's "pacifier." Sooner or later, the instability in Tajikistan may backfire against its neighbors. For Uzbekistan, for its fledgling statehood, the blow may prove fatal. This is why Islam Karimov has to come to grips with the most complex problem in his political biography. Not only will his political future depend on it, but also the future of the republic itself. Inclined to make pragmatic decisions, Islam Karimov is set to confound, and repeatedly so, those who are inclined to view him as a marginalized post-communist leader.

**Niyazov as Observer**

Turkmenistan intends to distance itself from the rest of the Central Asian countries trying to resolve the Tajik conflict jointly with Russia. President Saparmurad Niyazov said in advance that a Turkmen military contingent could not be sent to Tajikistan and that he would attend the Moscow summit as an observer. The very concept of the society that Saparmurad Niyazov is building in Turkmenistan requires such a distance. There is no denying, however, that this viewpoint evokes understanding and approval among Turkmenistan's neighbors from among foreign countries that are "far" from Russia, but close to this country. Saparmurad Niyazov has more than once met with his regional neighbors without Yeltsin—and did not look like an observer at all. It is Niyazov, moreover, who can be seen as the initiator of the process of independent communication between heads of the former Soviet republics of Central Asia.

**Akaev at the Helm of a European Kyrgyzstan**

Kyrgyzstan tried to play the role of a peacekeeper in the civil war in Tajikistan until military units were sent there (which resulted in a conflict between the president and the parliament). Although Bishkek's political clout is not very strong, this gesture by Akaev (as well as some others, including the introduction of the som) was designed to produce a favorable external effect—to demonstrate that Kyrgyzstan, being a CSCE member like the other Central Asian countries, is willing to play according to European rules. Perhaps it is precisely from the Kyrgyzstan president that Russia can expect support for its new concept of collective responsibility. This support may be brought to nil in the event of a lack of understanding between Russia and Kazakhstan, albeit in a veiled form.

**Rakhmonov as a Transitional Figure**

The current Tajik leadership's stubborn unwillingness to start talks with the opposition only emphasizes its provisional nature. Still, there is a big difference between the refusals to do so that are coming from Prime Minister Abdumalik Abdullodzhanov, who is from Leninabad, and Kulyab-born Supreme Soviet Chairman Emomali Rakhmonov. The struggle for power is not over yet, and the Kulyab people, who have seized the commanding positions from the Leninabad group by force, are unlikely to feel secure. The Leninabad clique may obtain the highest posts even through compromise. For the Kulyab people, however, war means life and power. The leaders of Russia and the Central Asian countries are unlikely to persuade Rakhmonov. Anyway, decisions by the Moscow summit may help the current Tajik leadership in its castling process and open the way for a negotiating process.

### Stability Will Suit All

Even a superficial analysis of the approaches taken by the Central Asian leaders shows that any durable union between these countries and Russia is thus far out of the question. The only thing that can actually unify the interests of Moscow, Almaty, Tashkent, Ashkhabad, Bishkek, and, sometime in the future, Dushanbe is stability in a region where any heavy-handed movement is fraught with civil wars, massacres, bloodshed, and even the collapse of a state. This understanding alone could be a factor contributing toward a successful summit in Moscow.

## 6.55

## Karimov Comments on CIS, Euro-Asian Union

Mikhail Kalmikov and Vilor Niyazmatov
ITAR-TASS, 27 April 1994
[FBIS Translation]

"I have no doubt that we needed the Commonwealth of Independent States. Otherwise we would have had a Yugoslavian version here after the collapse of the Soviet Union. History shows that such apprehensions are grounded," Uzbek President Islam Karimov told a joint press conference here, devoted to the results of French President François Mitterrand's state visit to Uzbekistan.

Replying to an ITAR-TASS question, the Uzbek leader noted that he believed the future of the CIS lay in economic integration, including the formation of interstate transcontinental corporations, which could unite Russian and Uzbek enterprises in joint-stock societies.

"We believe that the future of the CIS depends precisely on the formation of such associations, uniting the enterprises of Russia, Uzbekistan, and other republics. It will be even better if such associations are created not on a bilateral, but on a multilateral basis," Karimov stressed.

Asked by ITAR-TASS to comment on the idea of the Euro-Asian union, recently voiced by Kazakh President Nursultan Nazarbaev, Karimov said: "Much is left unsaid when the term Euro-Asian is used. If this implies a single parliament, single suprastate structures and even a single citizenship and a single constitution, it means the restoration of the old union, no matter what it is called. I am vehemently opposed to any comeback of the past. It is not for this that we have achieved sovereignty and are now engaged in national revival. There is no return to the past. History moves only ahead."

"To put it in a nutshell," Karimov stated, "the Euro-Asian idea smacks very much of populism, no matter who voices it."

## 6.56

## Karimov: Eurasian Union Dead; No Hurry to Join PFP

Vyacheslav Bantin
ITAR-TASS, 18 May 1994
[FBIS Translation]

"Agreement has been reached with Kazakhstan's President Nursultan Nazarbaev on scrapping the proposal to set up the Eurasian Union," Uzbekistan's President Islam Karimov, on his first visit to Japan, told a news conference in Tokyo today.

The president expressed the view that the idea to set up the Eurasian Union was no more than a "populist slogan." "If we are unable to enforce real economic integration between CIS countries, how can one talk about setting up the Eurasian Union?" he said. Islam Karimov stressed the importance of the creation of the CIS, noting that "without the Commonwealth, the breakup of the former USSR could have resulted in the start of civil war" in several republics.

Speaking about NATO's Partnership for Peace (PFP) program, the president made it clear that Tashkent does not intend to hurry in deciding whether to join it. "We are now studying the issue," Islam Karimov said. He noted that Uzbekistan is consulting with Russia on the issue of joining the NATO program. "We have good understanding with Russia on the matter," the president said.

## 6.57

## Nazarbaev Proposal for Eurasian Union

*Kazakhstanskaya Pravda,* 7 June 1994
[FBIS Translation]

[Draft proposal signed "President of the Republic of Kazakhstan": "Formation of a Eurasian Union of States."]

*Editor's Note*: In the following proposal, Nursultan Nazarbaev makes it clear that he considers the CIS a spent force. He offers a detailed, written manuscript which proposes extensive voluntary reintegration of the former Soviet republics into a Eurasian Union.

At the present time all CIS countries are continuing to experience a profound crisis in all spheres of social life—the economy, policy, ideology, and international relations—and socio-economic tension is on the rise. This is happening under conditions where the development of the CIS is being determined by two trends. On the one hand the continued formation of national statehood is taking place, on the other, a trend toward the integration of the CIS countries is being observed.

The CIS, as an interstate association, is performing a positive role in the legal structuring of the interstate relations of its constituent countries. CIS potential is not exhausted. Nonetheless, the structure of CIS organs that exists at the present time is preventing the realization to the full extent of the available integration potential. Not only the leaders of the CIS countries, but also a large part of the population of these states are calling attention to this.

The experience of CIS functioning in past years shows the need for a transition to a new level of integration that would guarantee observance of the jointly adopted commitments by all the participating states.

International practice shows that any interstate association goes through different stages in its development and is supplemented by new forms of cooperation. The CIS has appreciable advantages—the high degree of integration of the economy and similar socio-political structures and mentality of the population and also the multinational composition of the majority of the republics and common historical traditions.

All this testifies to the need for a combination of the process of national-state building and the preservation and development on this basis of interstate integration processes. The logic of history is such that integration in the world community is possible only by the joint efforts of all CIS countries taking advantage of the powerful integration potential that has taken shape over decades.

The actual conditions show that while perfecting the mechanisms of the CIS, we should not regard it as the sole form of association. As practice has shown, the further development of the CIS countries is being held back by the insufficiency of the intrinsic potential of each of them. The development of the latter is possible only with the economic integration of the countries of the post-Soviet space on a new, market basis.

The inherited structures of the single national economic complex are continuing to corrode. Outmoded forms of economic relations are objectively dying away. At the same time perfected production engineering relations corresponding to the economic interests of our countries in the near and distant future are being disrupted.

Market reformations have universal regularities. No country can ignore them without succumbing here to economic romanticism. It would be expedient to combine the efforts geared to market reform of the economies of the countries of the former USSR on the basis of the close business relations that have already taken shape over the course of decades.

As world practice shows, only with collective efforts are transitional societies in a position to undertake successful modernization. At the same time, on the other hand, we see that the continuing attempts to tackle this task by individual countries of the CIS alone are, as before, proving unsuccessful. They will remain such until the realization of economic integration on new conditions. On the other hand, the unrealistic nature of the attempts at a reorientation toward some regional economic associations in the far abroad is obvious.

The lack of coordination in price policy with respect to exported raw material, which is having a negative effect on the states' economic position, has become a serious problem for them. On the other hand, this is introducing an element of instability to the settled world economic relations and forcing third countries to adopt strict sanctions. Exports of raw materials and energy resources are our states' most important revenue item. There is an urgent need in this connection for a unified export policy within the CIS countries in the interest of all the participants, with the adoption of serious measures in the event of some country's failure to observe the quotas and prices agreed upon.

An important component of the successful implementation of market reforms is an improvement in the national legislative bodies of the CIS countries. Further modernization is impossible without the rapprochement of the legislative foundations of economic activity, inasmuch as the current differences between them are becoming a serious obstacle to integrative processes in the economy.

Considering the differences among the countries in levels of development of the market economy and the democratization of political processes, we propose the formation of an additional integration structure—a Eurasian Union—in harmony with the activity of the CIS. Account is taken here of the polyvariant nature of the integration and the different pace and the heterogeneousness and varying vectors in the development of the CIS states. This is reason to speak of the urgent need for the formation of a new economic order in the CIS. The purpose is the harmonization of economic policy and the adoption of joint programs of economic reforms binding on the participating states.

The socio-economic and political crisis is occurring

against the background of the multinational composition of the population of practically all the CIS states. As a consequence, interethnic tension leading not only to intrastate tension but also, in a number of cases, growing into interstate conflicts is on the rise. This situation is undermining the very institution of the CIS. Consequently, it is essential by joint efforts to devise mechanisms for deterring, localizing, and extinguishing various types of conflicts.

At the present time all CIS countries are searching for forms of constitutional arrangement appropriate for the internal conditions. But, as practice shows, neither the unitary nor the federative CIS states can be considered completely stable.

A solution to the questions of economic integration dictates the need for the creation of political institutions possessing a sufficient volume of authority. They must incorporate within themselves the regulatory functions of states' mutual relations in the economic, political, legal, environmental, cultural, and educational spheres proper.

Thus the time has come for the removal of the obstacles to interaction at the highest level and the simultaneous creation of the instruments for this.

A process of disintegration in the sphere of science, culture, and education is taking place at this time. The once unified cultural and educational space is becoming disconnected. The claim under these conditions that "science has no borders" is proving to be simply unfounded. Against the background of the intensifying socio-economic difficulties, the drain of specialists from the sphere of science, culture, and education, the decline in intellectual potential, and the fall in the standard and quality of education are increasing. These processes are leading not only to the rupture of the once unified system, but also to isolation from the cultural and scientific achievements of world civilization.

At the same time, on the other hand, the aspiration to intensify integration processes in the economy and policy should be based on the preservation and development of a concerted policy in the sphere of culture, education, and science. It is essential to preserve and augment the internationalization of the processes of the acquisition and practical use of new knowledge. The integration of R&D in the S&T sphere has become an inalienable part of the globalization of industrial activity in general.

The isolation of the post-Soviet space from the world cultural and scientific community is fraught with the danger of a new stage in the lag of the technological sphere.

A key task for the new states is the assurance of territorial integrity and security. At the present time the post-Soviet space is a zone of instability and the combination of various types of conflicts and is also experiencing the impact of centers of tension outside of the CIS. The guarding of the exterior borders and the stabilization of the situation in the conflict regions may be accomplished only by the joint efforts of all the states concerned and requires the concerted approach of the participants to a range of questions of a defense nature.

The problem of environmental safety remains one of the most painful and unresolved problems in the CIS countries. The environmental tension has been caused by a number of factors. The consequences of the testing of nuclear weapons and the activity of the nuclear power stations, the contamination of the environment with industrial waste, and the degradation of the natural environment as a result of man's economic activity (the running dry of water basin, deforestation, soil erosion, and so forth) may be attributed to these.

These problems are urgent for practically all CIS countries, especially since the main zones of ecological disaster are located in border areas, as a rule. They have been caused by the community of the evolved production engineering base and the methods of economic activity, the basis of which was the extensive focus. Not one of these problems can be resolved today by the independent efforts of one, even the biggest, state. Environmental preservation is a global task requiring large-scale capital investments and a combination of the efforts of all states.

The draft of the creation of a new integrative association, provisionally entitled the Eurasian Union, is offered for discussion.

### Eurasian Union

The Eurasian Union is a union of equal independent states that is geared to the realization of the national-state interests of each participant and that has an aggregate integration potential. The Eurasian Union is a form of the integration of sovereign states aimed at a strengthening of stability and security and socio-economic modernization in the post-Soviet space.

Economic interests determine the bases of the convergence of the independent states. The political institutions of the Eurasian Union should adequately reflect these interests and contribute to economic integration.

#### *Principles of Association*

The following principles and mechanism for the formation of the Eurasian Union are proposed:

- —national referenda or parliamentary decisions on states' membership in the Eurasian Union;
- —participant signing of a treaty on creation of the Eurasian Union based on principles of equality, non-interference in one another's internal affairs, and respect for sovereignty, territorial integrity, and the sanctity of borders. The treaty shall contain the legal and organi-

zational prerequisites for an extension of integration in the direction of the formation of an economic, currency, and political union;
—associate membership in the Eurasian Union is not permitted;
—decisions shall be adopted in the Eurasian Union on the basis of the principle of a necessary majority of four-fifths of the total number of participants.

Independent states shall be members of the Eurasian Union given fulfillment of the following preliminary conditions:
—obligatory observance of adopted interstate agreements;
—mutual recognition of the evolved state and political institutions of Eurasian Union participants;
—recognition of territorial integrity and sanctity of borders;
—renunciation of economic, political, and other forms of pressure in interstate relations;
—termination of mutual hostilities.

New countries shall be admitted to the Eurasian Union following the presentation of expert findings concerning their readiness to join the Eurasian Union by the unanimous vote of all members of the Eurasian Union. The expert findings shall be produced by a body formed on a parity basis by the states that have expressed consent to become members of the Eurasian Union.

Eurasian Union states may participate in other integration associations, including the CIS, on the basis of associate or permanent membership, or they may have observer status.

Each participant may withdraw from the Eurasian Union, having given the other states advance notice no later than six months prior to the adoption of the decision.

The formation of the following supranational bodies is proposed:

A Council of Heads of State and Heads of Government of the Eurasian Union—the supreme organ of political leadership of the Eurasian Union. Each participant shall preside in the Eurasian Union for six months at a time in Russian alphabetical order.

The supreme consultative-deliberative body is the Eurasian Union Parliament. The parliament shall be formed by way of the delegation of deputies of the parliaments of the participants on the basis of equal representation from each participant or by way of direct elections. Decisions by the Eurasian Union Parliament shall take effect following their ratification by the Eurasian Union state parliaments. The question of ratification must be considered within one month.

The main area of Eurasian Union parliamentary activity shall be to coordinate the legislating bodies of the participants to ensure the development of a single economic space and the accomplishment of tasks pertaining to the protection of the social rights and interests of individuals and mutual respect for state sovereignty and the rights of citizens in Eurasian Union states.

A common legal base regulating the mutual relations of the economic transactors of the participants shall be created via the Eurasian Union Parliament.

A Eurasian Union Foreign Ministers Council for the purpose of the coordination of foreign policy activity.

The Eurasian Union Interstate Executive Committee—the permanent executive and control body. The leader of the Executive Committee shall be appointed in turn from representatives of the participants by the heads of state of the Eurasian Union for a time that they determine. The structures of the Executive Committee shall be formed from representatives of all the participants.

The Eurasian Union, in the shape of its Executive Committee, should acquire the status of observer in a number of important international organizations.

The Eurasian Union Executive Committee Information Bureau. The adoption of a special commitment or law by the participants to prevent unfriendly words about the parties to the treaty that could damage relations between them.

An Education, Culture, and Science Council. The shaping of a concerted educational policy and cultural and scientific cooperation and exchange and joint activity in the creation of textbooks and aids.

For the more in-depth coordination and efficiency of the activity of the Eurasian Union countries, the creation in each of them of a state committee (ministry) for Eurasian Union affairs would be expedient.

At the level of ministers of Eurasian Union countries the holding of regular meetings and consultations on questions of health care, education, labor and employment, the environment, culture, the fight against crime, and so forth.

Encouragement of the activity of non-governmental organizations in various spheres of cooperation in accordance with the national legislation of the Eurasian Union participants.

The official language of the Eurasian Union, together with the functioning of the national bodies of language legislation, shall be Russian.

Citizenship. The unrestricted movement of citizens within the Eurasian Union will require the coordination of external—in relation to third countries—visa policy. Upon a change in the country of residence within the Eurasian Union, an individual shall, if he so wishes, automatically acquire the citizenship of the other country.

A city at the intersection of Europe and Asia, Kazan or Samara, for example, could be proposed as the capital of the Eurasian Union.

### *The Economy*

For the purpose of creating a single economic space, it is proposed to form within the framework of the Eurasian Union a number of supranational coordinating structures:

—a commission for economics under the auspices of the Eurasian Union Council of the Heads of State elaborating the main directions of economic reforms within the framework of the Eurasian Union with regard to the interests of the national states and presenting them for confirmation to the Eurasian Union Council of Heads of State;

—a commission for raw material resources of Eurasian Union exporting countries for the harmonization and confirmation of prices and quotas for exported raw material resources and energy and the signing of the corresponding interstate agreement. Coordination of policy in the sphere of the mining and sale of gold and other precious metals;

—a fund for economic and technical cooperation formed from contributions of the Eurasian Union countries. The fund would finance promising, research-intensive economic and S&T programs and assist in the accomplishment of a wide range of tasks, including legal, tax, financial, environmental, and so forth;

—a commission for interstate financial and industrial groups and joint ventures;

—a Eurasian Union international investment bank;

—a Eurasian Union interstate arbitration tribunal for economic matters legally ruling on contentious issues and imposing penal sanctions;

—a commission for the introduction of a monetary unit of settlement (transferable ruble).

***Science, Culture, Education***

It is proposed to implement a number of measures to preserve the potential achieved in previous decades and increase integration in this sphere:

—to create Eurasian Union common research centers for basic research in the sphere of modern learning;

—to create a Eurasian Union fund for the development of scientific research uniting the research outfits of various countries;

—to create a committee for relations in the sphere of culture, science, and education under the auspices of the Eurasian Union Council of Heads of Government;

***Defense***

The conclusion within the Eurasian Union framework of the following accords is proposed:

—a treaty on joint actions to strengthen the national armed forces of Eurasian Union members and to protect the exterior borders of the Eurasian Union.

The Eurasian Union proposes the creation of a common defense space for the coordination of defense activity:

—the formation of a collective Eurasian Union peacekeeping force for maintaining stability and extinguishing conflicts in the participating countries and between them. With the consent of Eurasian Union participants, in accordance with international rules of law, the dispatch of the peacekeeping force to conflict zones on the territory of the Eurasian Union;

—the presentation of a collective appeal of Eurasian Union countries to international organizations, including the UN Security Council, for conferring the status of peacekeeping force on the joint contingent;

—the creation of an interstate center for problems of nuclear disarmament with the participation of representatives of international organizations.

All states of the Eurasian Union, aside from Russia, shall maintain their nuclear-free status.

***The Environment***

The formation as soon as possible of the following mechanisms is essential:

—an environment fund under the auspices of the Eurasian Union Council of Heads of State realizing within the framework of the Eurasian Union environmental programs and financed by all the participating states;

—the coordination of operations with international organizations to reduce the degree of environmental pollution;

—the adoption of short-term and long-term programs on major problems of environmental restoration and the elimination of the consequences of ecological catastrophes (the Aral, Chernobyl, the Semipalatinsk Nuclear Test Range);

—the adoption of a Eurasian Union interstate treaty on the storage of nuclear waste.

History is affording us an opportunity to enter the twenty-first century by the civilized path. One method, in our view, is the realization of the integration potential of the initiative pertaining to the creation of a Eurasian Union reflecting the objective logic of the development of the post-Soviet space and the will of the peoples of the former USSR to integration.

[Signature and date illegible]
President of the Republic of Kazakhstan

---

## 6.58

**Foreign Minister Rejects Eurasian Union Proposal**
Interfax, 17 June 1994
[FBIS Translation]

---

Uzbekistan's Foreign Minister Saydmukhtar Saydkasymov said that the Kazakh President Nursultan Nazarbaev's proposal of a Eurasian Union is an attempt to restore the former USSR. He was speaking in Tashkent on Friday [17 June] at a meeting with the heads of the diplomatic corps accredited in the republic.

Saydkasymov said the CIS has not exhausted its opportunities yet and a Eurasian Union would mean a return to the past. "Uzbekistan links its future with the sovereignty obtained by the republic," he said.

Kazakhstan's Ambassador to Uzbekistan Nazhameden Iskaliev told Interfax that "every state has the right to express its opinion on issues in which it has an interest." Iskaliev believes that the idea of an Eurasian Union would be discussed at the September summit of the heads of CIS states.

"Of course, certain corrections will be introduced into the document, nevertheless it should be treated with great attention," Iskaliev said.

## 6.59

**Eurasian Union: For and Against**
Asylbek Bisenbaev
*Kazakhstanskaya Pravda*, 27, 30, 31 August 1994
[FBIS Translation], Excerpts

*Editor's Note:* The following article reveals the large number of supporters Nazarbaev's Eurasian Union plan claimed when it first appeared. This was perhaps because so many Russians regarded the CIS as an ineffective creation; they felt that Russia lacked the firmness shown by Nazarbaev. Of particular note is the support reported on behalf of the international community. However, by introducing a new version of the union treaty, Nazarbaev was appropriating Russia's role and gaining an upper hand, psychologically if not politically, within a reconstituted union. Yeltsin, therefore, warily distanced himself from the concept, calling it "premature." The development of the Eurasion option also entailed the danger of Russia's being drawn into regional conflicts and even assuming the burdens of the old union. Yeltsin knew Russia had little capacity for assuming such a responsibility. Russia, furthermore, was still hopeful that it could pull Ukraine back to its fold, and preferred to avoid the creation of a Russian-Muslim bloc.

[27 August]

The draft "Formation of a Eurasian Union of States" is an initiative that has evoked the most extensive press in the CIS in the spring and summer of 1994 and the close attention of the news media of the far abroad. The majority of journalists has evaluated the draft as "sensational" and as having "had powerful repercussions." It may be said that the idea has fallen on fertile ground and has been taken up seriously by the press, part of which, even while according it a guarded or negative evaluation, has involuntarily attracted public interest in the new integration idea and the personality of the president of Kazakhstan.

Many observers link N.A. Nazarbaev's idea concerning the creation of a Eurasian Union (EAU) with his lecture at the M.V. Lomonosov Moscow State University. But he launched the first trial balloon on the need for the formation of a new integration formation during his visit to Great Britain. Speaking in the Royal International Relations Institute, N. Nazarbaev declared: "The development of the post-Soviet space is being determined currently by two trends: On the one hand, the formation of national statehood is under way; on the other, the trend toward integration of the countries of the CIS. There is an urgent need for a reform of the Commonwealth of Independent States itself that would ensure the creation in this region of a belt of stability and security and would enhance the degree of predictability of political evolution." Despite this, the idea of the creation of an EAU was for many politicians unexpected, although thinking on the transformation of the CIS had been in the air, so to speak. N.A. Nazarbaev himself and many analysts and politicians have expressed repeatedly the need for the conversion of the amorphous CIS into a really operative union. "More than 300 documents have been adopted in the time of the CIS's existence," N.A. Nazarbaev explained in his interview with the Japanese paper *Mainichi.* "But no results are in sight because a number of states partially rejects some of them, interpreting the content of the agreements in their own interests."

The international community always awaits the overseas trips of the president of Kazakhstan with interest since, during the meetings with the leaders and the public of other countries, he frequently presents enterprising proposals for a solution of global and regional problems.

N.A. Nazarbaev's speech at the Moscow State University, in which he set forth his basic vision of the EAU, evoked an active response on the part of the international community. Many representatives of diplomatic circles of the United Nations recognize that the president of Kazakhstan is in the CIS "the most sober-minded and attractive politician of high international standing who has a realistic evaluation of the situation both in the republic and in the Commonwealth as a whole." They believe that the basic principle of his proposal concerning the creation of an EAU is the "equality, respect for the sovereignty, and the independence of each state, the rights of the individual, and the individuality of each state." Realization of this initiative, in the opinion of foreign diplomats, would enable the EAU members, while preserving the states' economic independence, to coordinate more precisely questions of foreign policy and foreign economic activity. In the view of UN representatives, the military aspect of an

EAU and the principles of its structure would seem the most complex. Forecasting the possible response on the part of countries of the CIS to Kazakhstan's initiative, VOA observed that "the proposal to create on the basis of the CIS a union similar to the European Union will, most likely, be supported in the states of Central Asia. Ukraine, which is actively defending its independence, will most likely reject the idea of membership of any union that specifies a central parliament, common citizenship, and a common currency."

A number of news media have perceived President N. Nazarbaev's new initiative as a "forced piece of impromptu politics" (ABV, 22 April 1994, for example). It is hard to agree with such an approach. You could make an analysis of all N. Nazarbaev's speeches and see for yourself that, behind the initiative, which appeared somewhat of a surprise, there is lengthy work on the Kazakhstan leader's part on the shaping of integration on the basis of new principles. Let us recall some of them. At the UN General Assembly's Forty-seventh Session, the president of Kazakhstan expressed concern that "the fragile structure of our Commonwealth is not yet fully taking account of the age-old traditions of the interaction of the countries and peoples of this part of Eurasia. As a result, the processes of transition to an economy of free markets and democracy in the CIS are being accompanied by a growth of socio-economic and political instability and an exacerbation of existing conflicts and the emergence of new ones."

Back in the perestroika period, N. Nazarbaev was expressing his idea of an economic union, and his position in the period of the signing of the Belovezh Agreement was essentially pivotal in the preservation of the unity of the new states. Returning to the events of that period, Ya. Plyays in the article "Present and Future of the CIS" (*Nezavisimaya Gazeta,* 3 March 1994) believes that the most significant mistake of the politicians, both former and present, was the signing by the leaders of the three republics on 8 December 1991 of the Belovezh Agreement "without the consent and in defiance, even, of the wishes of the peoples expressed in March of that same year for the preservation of the Union of SSR.

"The Belovezh Act was a mistake because it interrupted the process of reform of the USSR and its conversion from a unitary state, if not into a genuinely federative, then, at least, into a confederative state." Analyzing the state of the integration processes in the post-Soviet space, Ya. Plyays concludes: "The present CIS is an ineffective formation and, generally, not all that viable. We need to honestly acknowledge this and, having boldly abandoned it, switch emphatically to the creation of a confederative union of Eurasian states, with the intention that it could in time develop into a stronger national-state association."

Speaking of the invariability of President N.A. Nazarbaev's policy of a strengthening of ties between the republics, A. Gurskiy also notes the negative impact of the Belovezh Agreement. "The October arrangement of thirteen heads of republics was suddenly muffled by the forest echo of the decision, which flew around the world, adopted in December's Belovezh Forest. And, in what might this forest itself have resulted for our peoples and what guns of a regional, interethnic, and fratricidal caliber might it have begun to fire had it not been for the wisdom of Kazakhstan and its Uzbek neighbor. Meeting on the insistence of Nursultan Nazarbaev, the leaders of ten union republics signed in the evening of 21 December 1991 the Alma-Ata Declaration on the appearance on the political map of the world of the Commonwealth of Independent States and on the termination of the existence of the USSR and a number of documents for easing this transitional period" (*Vestnik Kazakhstana,* 6 May 1994).

The draft "Formation of a Eurasian Union of States," which has been published in the press and which has been sent out to the heads of states of the Commonwealth, could be the basis of the formation of a new association of states. I would like to call attention to several fundamental points of this draft. The idea of an EAU does not in the least signify the demolition of the CIS, as some politicians believe. N. Nazarbaev's draft says that "the CIS, as an interstate association, performs a positive role in the legal structuring of the interstate relations of the countries that are a part of it, and the potential for action of the CIS is not exhausted." In addition, the EAU states could participate together with the CIS in other integration associations, with associate- or permanent-member status or observer status. On the other hand, some politicians and journalists have perceived the EAU project as a resuscitation of the former USSR. But they have overlooked such propositions as "non-interference in one another's internal affairs and respect for sovereignty, territorial integrity, and the sanctity of borders." The EAU is thus a form of integration of sovereign states for the purpose of a strengthening of stability and security and social and economic modernization in the post-Soviet space.

Considering the experience of the activity of the CIS, the imposition of preliminary conditions for membership of the EAU is proposed: obligatory observance of interstate agreements that have been adopted, mutual recognition of the evolved state and political institutions of the participating countries, recognition of territorial integrity and the sanctity of borders, the renunciation of all forms of pressure in interstate relations, and an end to mutual hostilities.

A "golden rule" suggested by Jean Monnet at the time of the unification of Europe was the formula: "go as far as possible in the mechanisms of union. But never touch the sovereignty of the members." As we can see, the draft of the EAU, abiding by this formula, has proposed much that was in the mechanisms of the Union, but has not touched the foundations of the sovereignty of the new states. Moreover, the concept of the EAU strengthens this sovereignty.

This aspect of the problem is, truly, extraordinarily important. Many people see as a threat the possibility of the restoration of the USSR. But when it is a question of integration on a voluntary, democratic basis whose purpose is not military-political supremacy but economic cooperation and a strengthening of stability, the attitude toward such an association changes. U.S. President B. Clinton's speech, for example, which coincided with the debate on the discussion of the draft of the EAU, was an important event. The American president declared that he did not believe that the possibility of the democratic and legal reunification of the republics of the former Soviet Union would represent a threat to the countries of the Baltic, Europe, or the interests of the United States. "This will depend on whether such decisions are really adopted voluntarily and are in keeping with the wishes of the majority or not," he emphasized in an interview published on 4 July 1994.

For the accomplishment of the tasks set in the draft of the EAU, the formation of a number of supranational bodies is proposed. Including a Council of Heads of State and Government with the rotating chairmanship of each participant, and EAU Interstate Executive Committee—a standing executive-control body—and so forth. An EAU Parliament is to be an important body providing for the coordination of sets of legislation. A number of observers have noted also the great flexibility of the problem of citizenship, which is considerably superior to the idea of the introduction of dual citizenship and corresponds more precisely not only to the national interests of the countries and the citizens but also to international standards—"upon a change of country of residence within the framework of the EAU, an individual may, as he wishes, automatically acquire the citizenship of the other country."

In addition, the draft provides for the creation of an EAU Executive Committee Information Bureau with the adoption of a special undertaking concerning the inadmissibility of unfriendly expressions directed at parties to the treaty that could be harmful to relations between them. A multitude of examples of the news media being used for such unseemly purposes could be adduced. The question of the northern oblasts, oppression of the Russian-speaking population, and such is regularly raised in respect to Kazakhstan, for example. Similar statements are being made, what is more, by people that are far from having a knowledge of the actual situation in the republic and also by those that would like to enhance their political popularity thanks to a "winning theme." When the bulk of the information space of the CIS is concentrated in single hands, such statements seriously undermine stability. The adoption of the corresponding law or commitment could do away with this problem.

We could quote the example of the situation in Kazakhstan being evaluated via the established stereotype concerning the growth of Islamic fundamentalism and interdominion and clan contradictions. Thus, S. Kurginyan has quite a lot to say in his material ("Content of the New Integrationism," *Nezavisimaya Gazeta,* 7 July 1994) about "dominion-clan," tribalist tendencies. Granted all the outward attractiveness and the mass of "predigested facts," the substantial quantity of material can only evoke in the reader, who really comprehends the situation in depth, regret or consternation. Speaking of the various models of Eurasianism, and there are, indeed, quite a lot of them, and they are frequently of a directly opposite content, S. Kurginyan has failed to notice that the president of Kazakhstan invests in this concept a purely geographical meaning. If a union of states that are located in Europe and Asia could be called not "Eurasian" but something else, does this need to be underpinned by an ideological base? Unfoundedly contriving for N. Nazarbaev's draft farfetched models "in the spirit of apologetics for the Golden Horde," and "reconstruction of the USSR," and, even more, "neo-fascist Eurasianism," as the author of the article does, is, at least, inappropriate. If we attempt to take as a basis S. Kurginyan's procedure, which is distinguished by a one-sided view, it may be said that his article is a reflection of the interests of the "republic party-nomenklatura elites," which aspire on the basis of a common idea (communist once again, perhaps) and a "balanced attitude toward the values and experience of the Soviet period" to restore the "indestructibility of a new union state."

The draft of the EAU has, naturally, evoked a serious response primarily in Kazakhstan. Different political forces and individual authors have evaluated the idea and, subsequently, the draft itself differently. The majority of deputies of the Supreme Council of the republic has supported the idea of a Eurasian union, judging by polling results (93.9 percent).

O. Suleymenov, leader of the People's Congress of Kazakhstan, defined his position almost immediately. It is necessary right now to create an integration "think tank," O. Suleymenov believes, which would before the end of the century have worked up the theory of the Eurasian Union and the paths and methods of its interaction with the European and North American confederations (*Biznes-Klub,* 28 June 1994). In addition, the leader of the People's Congress of Kazakhstan Party believes that the draft of the EAU presented by President N.A. Nazarbaev has been useful on a purely practical level also—"the structures of the CIS have stepped up their work sharply. Important documents were signed in record time at the meeting of the heads of 16 April" (*Kazakhstanskiye Novosti,* 18 June 1994). This approach shows O. Suleymenov's consistent position on questions of the integration of the Eurasian space. We may, thus, remind the readers of an interview in the newspaper *Aziya* that was published in July 1993, in which the poet spoke of the need for the formulation of a theory of Eurasian integration. Other party members supported their leader also. *Khalyk Kongresi* published on 12 April an article by Professor B. Alikenov in the Topical Problems column, which makes a positive assessment of the idea of a Eurasian union. The author emphasizes that, although its creation is not for Kazakhstan the sole way out of the crisis, it is the most efficient as of this time.

The international independent newspaper *Turkistan* (12 April 1994) expresses the fear that the idea of an EAU could be a step en route to the revival of the empire. At the same time, on the other hand, it recognizes that both in Russia and within Kazakhstan itself this initiative has knocked the ground from beneath the feet of separatists and reactionary forces. The difference in evaluations of the idea of the EAU, in the paper's opinion, is an objective point, but this initiative has, on the whole, captured the people's mood.

[30 August]

A. Solzhenitsyn commented in sharply negative words on President N. Nazarbaev's initiative and proposed the unification in a single state of the Slav republics and Kazakhstan (*Izvestiya,* 4 May 1994). He kindly consented here to "release" the other peoples from this state. The famous Kazakhstani writer I. Shchegolikhin expressed his attitude toward A. Solzhenitsyn's interview, noting that this position represented the "customary Soviet, Bolshevik style: pin a topical label on, and that's all there is to it." As far as North Kazakhstan is concerned, "it is sufficient to glance at the topography to see that Southern Siberia is North Kazakhstan. But we are making claims on neither Russia nor Siberia, and neither politicians nor even writers are talking about this, and there are among them such daredevils as are not a bit inferior to the Nobel Prize winner either in national ambitions or rashness of generalizations" (*Kazakhstanskaya Pravda,* 13 May 1994).

Kh. Kozha-Akhmet, chairman of the Azat Movement, even issued an appeal in which he expressed a protest at the "separatist adventure of President Nazarbaev regarding the creation of a Eurasian union." In his television interview on a news program of the Tan Television and Radio Company of 23 June of this year, he called for resistance to the EAU project and the formation of a popular front and various headquarters for resistance to this idea. But, as subsequent events have shown, there are fewer and fewer of those that wish to flock to the banners of Kh. Kozha-Akhmet. They are made leery, evidently, not only by the absence of any constructive ideas but also by the Azat GDK's [further expansion unidentified] borrowing of the experience of the Communist Party, which formerly was continually in the business of resolving all problems by the creation of so-called "three-man teams," "operational groups," and headquarters for combating the latest phenomenon. Extremism in policy has never been supported by Kazakhstanis, and individual figures still do not understand or are unwilling to understand this. It is not surprising that the popularity of such parties and movements has sunk to the lowest level. . . .

Certain distortions of the EAU project were reflected in a number of publications of the Russian press also. Thus *Obshchaya Gazeta* (17–23 June 1994) published the article "Our Response to Nazarbaev" based on sociological data of the Vilchek Service. It reflected, unfortunately, the old habit of criticizing without knowing the subject of the discussion. The article says, for example, that "Boris Yeltsin's comments on Nazarbaev's plan for the creation of a Eurasian Confederation contained nothing other than an ironical intonation. . . . In any event, the Russian president made it understood that the referendum proposed by the president of Kazakhstan will not take place." I would like to recall that the draft deals with the formation of a Eurasian union of states and that the republics would join it here in the wake of national referendums or decisions by the parliaments. One has the impression that the Vilchek Service, which placed in the newspaper the results of a sociological survey of 2,000 Muscovites, familiarized itself with the plan for the formation of an EAU insufficiently seriously. Nonetheless, the results of this survey show that from 33 to 38 percent of Muscovites supported N. Nazarbaev's idea, whereas 24.5 percent took a negative view of it (18 percent were sharply opposed). As the newspaper observes, "even among the people that in principle take a positive view of the idea, almost one-third does not believe that this idea can be implemented." But one further figure, on which the sociologists did not comment, is indicative—only 15.5 percent of Muscovites believe that it will be easier for Russia to extricate itself from the crisis alone. . . .

As far as Russia, without which the Eurasian idea is simply dead, is concerned, "total disarray reigns there as yet." The opponents of the idea in Moscow, in V. Verk's opinion, proceeding from the danger of Russia's conversion into a Turkic power in the first twenty years of the twenty-first century even, consider possible rapprochement on "horse and rider" principles. The seriousness of the assertion concerning Russia's "Turkization" in the next twenty years evokes puzzlement, to say the least. Various excesses are possible in polemical ardor, and a journalist is free to employ various methods of calling attention to his publication, but in this case they have assumed a cyclopedic nature.

If, however, we lend an ear to the opinions of real scholars, several publications may be distinguished. Thus, V. Tishkov, director of the Russian Academy of Sciences Ethnology and Anthropology Institute, writes in his article: "Nazarbaev's recent proposal concerning the creation of a

Eurasian union would seem exceptionally important since a powerful economic and social and cultural symbiosis between Kazakhstan, Russia, Kyrgyzstan, and a number of other post-Soviet states is preserved, and it is more rational to use it for people's common reconstitution, not to break things down by strict borders, which, among these three states at least, the present generation of politicians will not succeed in imposing on the population. This would not damage either the sovereignty of the peoples or the culture of the peoples, and it could spare us tragic mistakes. It is gratifying that Nazarbaev's initiative has been supported by many top politicians, Sergey Shakhray, for example" (*Nezavisimaya Gazeta,* 8 April 1994).

The newspaper *Aziya* (No. 22) offers the reader an interview with A. Brudnyy, corresponding member of the Academy of Sciences of the Kyrgyz Republic. Making a very high assessment of the EAU idea, the scholar observes that "whether it is realized or not, it will go down in history. A material point: I do not believe that it should be counterposed to the CIS since other unions are possible within the framework of the Commonwealth also. The Eurasian Union is stronger by design. It does not infringe on the CIS nor does it infringe on the various sovereignties. Such a union is possible, I believe, between Russia, Kazakhstan, and Kyrgyzstan . . . ." A. Brudnyy believes that imperial ambitious are the cause of Russia's cool attitude toward the president of Kazakhstan's initiative. In his opinion, "this idea is not exclusively the personal invention of Nursultan Nazarbaev, it reflects a trend that has supporters in both countries. He has caught it and expressed it. As president. And it is President Yeltsin's duty to 'speak for Russia.' He has not, however, expressed a precise position. Nor has his team either."

A. Sergeev ("EAU: Reason Against Chaos," *Rik,* No. 23, 16–22 June 1994) believes that publication of the draft of an EAU put an end to the speculation of certain news media and politicians concerning the impromptu, improvised nature of this initiative of the Kazakhstani president. In his opinion, "even the most 'thick-skinned' politicians are beginning to understand that without a new commonwealth of equal and independent states filled with creative meaning there remains just one path—into the abyss."

B. Aubakirov writes in an article entitled "Back to the Future" (*Argumenty i Fakty,* No. 24, June 1994): "The proposal concerning the creation of a Eurasian union of states, seemingly advanced spontaneously by N. Nazarbaev at a meeting with students of Moscow State University, has acquired a real edifice. In the form of a published draft—with a detailed study of both the general idea of such a union and of its individual 'components.' This is eloquent testimony to the fact that a true politician does not make incidental statements." The author notes that President N. Nazarbaev's proposals, which are geared to a strengthening of integration processes, "have either been glossed over in silence or have been supported in order to mollify public opinion." The main obstacle in the way of integration, in his opinion, are the leaders of the new sovereign states, "who did not even dream that they would end up with enormous power. They have gotten a taste for it and will hold on to it until the end. The examples of Ukraine, Tajikistan, and Azerbaijan are confirmation of this. It is immaterial who is at the helm—communists or democrats—they will not give up the 'riches' that have fallen into their laps. Even if they have to cover the country with blood, as we have seen in Georgia." In this connection the directive elite of the new states will see the idea of an EAU as a "hankering to limit their power." And it is not important that the draft speaks of entirely new principles of association. They will interpret them in a way that is beneficial for themselves, juggle them, or simply gloss over them in silence. We need to be prepared for this lest we once again suffer a major disappointment." Yu. Kirinitsiyanov (correspondent of *Rabochaya Tribuna*) also notes in the newspaper *Novoe Pokolenie* (23 June 1994) the cool attitude toward the EAU idea of neighboring republics. . . .

In his subsequent interview with *Nezavisimaya Gazeta* (24 June 1994), S. Shakhray said that the statement of Andrey Kozyrev, Minister of Foreign Affairs of Russia, that Russia was undoubtedly prepared to conduct negotiations on a confederation, but on the Eurasian Union format, only following "the specific decisions of parliaments and the political leadership," seemed unwarranted.

The EAU idea was actively supported by V. Shumeyko and many other top political figures, who took account of the mood of the bulk of the population of Russia and the CIS and who are attempting to construct the image of consistent supporters of integration. Specifically, S. Shakhray deemed it possible even to publish the draft Confederative Agreement that he had put forward in 1992 and that he had announced in the PRES's election platform. Nonetheless, a comparative analysis of the drafts of the EAU and the Confederative Agreement points to the more thorough development of the Kazakhstan document. V. Shumeyko considered even that the advancement of the EAU idea had "prompted some people in the direction of more active integration steps," and the success of the meeting of heads of state of the Commonwealth in Moscow was "largely due to President Nursultan Nazarbaev" (*Kazakhstanskaya Pravda,* 6 May 1994).

Various organizations adopted a positive attitude toward the EAU idea. The International Democratic Reform Movement issued an appeal to the Moscow conference of heads of state of the CIS "Yes to the Eurasian Union." It speaks on behalf of more than sixty group participants operating in the independent states of the need for support for President N. Nazarbaev's initiative.

The "Toward a New Accord" forum, which was founded on the initiative of the International Democratic Reform Movement, fully supports the initiative concerning the creation of a Eurasian union. The forum staged on 18 June 1994 was the conference "The Eurasian Community: The Common in Diversity," in which representatives of thirty parties and sixty grassroots movements of the post-Soviet states took part. The conferees adopted an appeal to the people, members of parliament, and heads of state: "We support the draft of a Eurasian union devised by N.A. Nazarbaev, president of Kazakhstan, and call on the heads of state to view it from the standpoints of the interests of their peoples."

The General Confederation of Trade Unions (VKP) and the International Confederation of Journalist Unions (MKZhs) called for support for N. Nazarbaev's initiative. The VKP is confident that the association would "give the peoples of the Commonwealth countries a real opportunity for a better future, the speediest way out of the global impasse, and the creation of living conditions worthy of man" (*Panorama,* 7 May 1994). On behalf of the members of fifteen journalists unions, the MKZhs declared that "the embodiment of the president of Kazakhstan's initiative would afford an opportunity for removing the grounds for any kind of conflicts" (*Kazakhstanskaya Pravda,* 16 April 1994).

Shaykh R. Gaynutdin, chairman of the Religious Administration of Muslims of the Central European Region of Russia, mufti, and chief imam-khatyb of the Moscow Central Mosque, evaluated highly and supported N.A. Nazarbaev's draft. A meeting was held in the Embassy of the Republic of Kazakhstan in the Russian Federation with Professor E. Bagramov, chief editor of the journal *Evraziya, Narody, Kultura, Religiya,* and Professor G.E. Trapeznikov, president of the Russian-Hellenic Religious Unity International Foundation, who consider timely and fruitful the appearance of this document, which points to a prospect of a revival of the relations of the Eurasian peoples with regard to the new historical realities (*Kazakhstanskie Novosti,* 2 July 1994).

I. Rybkin, chairman of the State Duma of the Federal Assembly of the Russian Federation, observed during a meeting with K. Saudabaev, minister of foreign affairs of the Republic of Kazakhstan, that the idea of the creation of a Eurasian union was being received very well in Russia. "And despite the fact that not all states of the CIS are yet prepared for association on such a basis," he said, "we are open to such a union" (*Kazakhstanskaya Pravda,* 3 June 1994).

I. Korotchenya, executive secretary of the CIS, fully supported the idea of an EAU. As he told journalists, the realization of this idea would not harm the CIS in the least.

L. Kravchuk, the former president of Ukraine, commented negatively on the proposal concerning the creation of an EAU. "I do not believe that we should today disband the CIS and create a Eurasian union," he declared. "This idea is premature. In addition, this could not be done without regard to the opinion of all citizens. We would be forced to conduct a referendum, as was the case at the time of the creation of the European Union" (Reuters). "President Nursultan Nazarbaev's initiative," Leonid Kuchma, the newly elected leader of Ukraine, believes, "represents just about the optimum solution of the crisis situation in which the economy of the sovereign states, former republics of the USSR, has found itself. . . . The great danger for all countries of the CIS, Ukraine included, would be to remain outside such a formation."

"The initiative of Kazakh President Nazarbaev on the creation of a Eurasian union should be seen as positive," V. Zametalin, press spokesman for the chairman of the Council of Ministers of Belarus, said. "That its realization will be just as complex as answering the question of why such a structure as the Commonwealth of Independent States is inoperative is another matter." The recent elections in Belarus showed that the electorate had a positive attitude toward integration projects. Account was taken of this by A. Lukashenka, the new president of Belarus, also.

The position of the leadership of Uzbekistan is ambivalent. For several months the press has been earnestly exaggerating the differences between the two leaders of Central Asia. But the conclusive position of president of Uzbekistan I. Karimov was determined only recently. In his letter to N. Nazarbaev, he writes: "The idea of the creation of a Eurasian union instead of completing the economic integration of the CIS countries offers a solution of the crisis situation that has taken shape in certain states that gained independence following the disintegration of the Union of SSR by political means and, essentially, if we investigate more deeply, a renunciation of sovereignty and statehood and a return to the previous formats and norms of the unitary associated state." A definite position, in any event. It is regrettable that the draft of the EAU, which contains not a whiff of encroachment on the sovereignty of the republics and which contemplates voluntary, equal cooperation in all spheres, has come in for such a distorted interpretation. But I. Karimov's practical actions are, for all that, directed into an integration channel, which was shown by Uzbekistan's membership of the Central Asian integration zone.

In the past, at the time of the formation of the European Community, Jean Monnet, who is called the father of a united Europe, advanced the following formula in respect to Great Britain, which could pertain to Uzbekistan also: "Do not wear out your nerves on Great Britain. It will always start by saying 'no.' It will affiliate with the Community because it will have no choice." This forecast in respect to Great Britain was fully justified.

Kyrgyzstan President A. Akaev observed in his letter to

N.A. Nazarbaev: "I consider it necessary to emphasize once again that I have from the very outset unreservedly supported your initiative concerning the creation of an EAU (the sole president among those of the CIS countries, possibly), I support it now, and I will remain a convinced supporter of it." A. Akaev submitted a proposal concerning the creation of a joint commission of experts of interested states for the more in-depth and high-quality study of the mechanisms and principles of realization of this initiative.

K. Zatulin, chairman of the Committee of the State Duma for CIS Affairs, acknowledged that N. Nazarbaev's idea concerning the creation of a Eurasian union has a future, but needs detailed study. It is essential that this work be performed thoroughly, the Russian deputy emphasized, first among a group of experts and only then at the "political responsibility" level.

At a news conference on the results of his visit to Pakistan, E. Rakhmonov, chairman of the National Assembly of the Republic of Tajikistan, speaking about the initiative concerning the creation of an EAU, observed that this idea deserved attention.

Commenting on President N. Nazarbaev's initiative, E.A. Shevardnadze termed it "very interesting and deserving of attention."

L. Meri, president of the Republic of Estonia, observed during his official visit to Almaty that "such associations and such integration are the future" (*Panorama,* 18 June 1994).

*Deutsche Allgemeine* (No. 175, 18 June 1994) carried undêr the heading "Eurasian Union: Ways Out of the Crisis" the opinion of a number of representatives of diplomatic departments of the near abroad in Kazakhstan. V. Sokolov, minister-counselor of the Russian Federation in Kazakhstan, writes: "This idea is not contrary to the interests of Russia and has, on the whole, obtained a benevolent response. Although not all share it. The idea of Eurasianism is not new, it has been around for several dozen years. But it acquires in President Nazarbaev's initiative a new, opportune resonance, thereby attracting attention." G. Bezhuashvili, minister extraordinary and plenipotentiary of the Embassy of the Republic of Georgia in Kazakhstan: "The EAU idea is acceptable to Georgia because this union could create guarantees of stability within the state—political and economic—which is what we currently lack. Mechanisms of the elimination of interethnic conflicts not only in Georgia but in all countries of the former USSR could be found." A. Grigoryan, first secretary of the Embassy of Armenia in Kazakhstan: "The attitude toward the idea of a Eurasian union of states is favorable in the leadership of Armenia as a whole."

N. Johns, ambassador of Great Britain in Kazakhstan, declared his country's readiness to send to Kazakhstan British experts experienced in work in the structures of the European Union and the CSCE to study President N.A. Nazarbaev's initiative concerning the EAU.

A number of opinions on the position of the Russian leadership in respect to the draft of the EAU has been expressed in the Russian press. We shall cite two of them. V. Portnikov in the article "The Dream as an Instrument of Real Policy. There is No Eurasian Union, But It Is Already Getting in Many People's Way" (*Nezavisimaya Gazeta,* 7 May 1994) described the idea of an EAU as a dream and saw it as a successful test "of the true nature of Russian integration intentions—a test that has demonstrated to the former Soviet republics once again that they should not count on Moscow's effective support, unless they agree to such a semicolonial scenario of 'integration,' with unpredictable consequences, as the Belarusian scenario or show up on the Kremlin porch following a bloody civil war, like the Tajik leaders."

Aleksandr Vladislavlev, president of the Independent Foundation for Realism in Policy, writes, stressing the priority in Russia's foreign policy toward the CIS countries (*Nezavisimaya Gazeta,* 6 May 1994): "The priority of this direction is recognized by all, but, unfortunately, nothing has been done or is being done here as yet. . . ."

"It is astonishing, therefore, when we not only lack our own initiative but do not even have the time to respond to the initiative of others." A. Vladislavlev believes that Nursultan Nazarbaev has put forward "an idea that is undoubtedly interesting and that moves in the direction of the development of the integration process. Of course, many aspects, which require clarification and further study, arise here. But reacting to this important initiative with total official silence is not simply incomprehensible, it is impolite also. Such an approach is enormously damaging to us and casts a shadow on the readiness and desire of Russian diplomacy to do business respectfully and properly with our most important partners."

In his letter to Kazakh President N.A. Nazarbaev, B.N. Yeltsin observed that "we need to move toward unification, while preserving, of course, the sovereignty of the states, and on a purely voluntary basis. This is an imperative of life itself. Integration should be of a profound nature and should extend to all spheres: the economy, policy, science, military matters, the environment, and the social sphere. The creation of supranational control mechanisms will most likely be needed also." Having made specific mention of certain issues, Russian Federation President B. Yeltsin believes that "the most in-depth study of the issue is needed to ensure that, at the upcoming meeting of heads of state of the Commonwealth in September, all determine together how best to advance the cause of integration."

The broad palette of opinions on President N. Nazarbaev's new initiative reflects the need for the stabilization of the post-Soviet space and an intensification of the integration processes, primarily in the sphere of the economy. No CIS state is in a position to accomplish the transformations in isolation from its neighbors. The draft of the EAU determines a sufficiently extensive range of zones of cooperation. Professor V.V. Kiyanskiy, who was elected at the Interparliamentary Assembly of CIS States chairman of the Environmental Protection Commission, believes: "Ecologists have to the greatest extent felt the consequences of the centrifugal forces, having been left on their own to confront problems requiring a collective solution, in respect, for example, to the Caspian, the Aral, the Urals, Kapustin Yar. Through the ecological prism, the Eurasian Union is a command of the times" (*Kazakhstanskaya Pravda,* 2 July 1994).

In his article "EAU: Stalemate" (*Karavan,* 12 August 1994), S. Mekebaev says that "the idea of a Eurasian union is closer and more comprehensible to political scientists and theoreticians than practical politicians." If the responses of politicians and statesmen of the CIS countries are analyzed, the majority of practical politicians have expressed agreement in principle with the basic propositions of our president's initiative. It is sufficient to cite the example of the parliamentary hearings organized by the Committee of the State Duma of the Russian Federation for CIS Affairs, at which representatives of the parliaments of Ukraine and Belarus and diplomats were present. It was observed during the hearings that the EAU was the most developed and substantiated of the integration ideas that have been proposed as of this time and that it corresponds entirely to the cherished aspirations of the majority of the population of the Commonwealth countries. Leonid Kuchma commented positively on Nursultan Nazarbaev's initiative, noting that not political, but economic, interests would prevail in this association. N. Nazarbaev's draft has caused a sharp upsurge of both theoretical studies on problems of the integration of the post-Soviet space and of practical measures pertaining to a resolution of this problem. Thus, the Council of Foreign Ministers of the CIS states has adopted the decision to hold a meeting of experts of the Commonwealth countries for a detailed discussion of N.A. Nazarbaev's proposal. A.L. Adamishin, First Deputy Minister of Foreign Affairs of the Russian Federation, declared: "An integration plan, which encompasses a whole number of aspects of the activity of the CIS, is being prepared at this time, and the Russian position is that the president of Kazakhstan's ideas could be incorporated in this common integration plan of development of the Commonwealth."

It is today clear that the draft of the formation of this integration association is already performing a practical role. As N. Nazarbaev observed during his visit to St. Petersburg: "Even for ensuring that people may go to see each other in the now independent states without hindrance, we should if only for this reason settle the question of association in a community or, if you will, a union. We need to alleviate the processes occurring in our countries, approach people in civil, human fashion. . . . The times themselves demand that the issue of the greatest convergence and integration of our economies and peoples be part of the agenda" (*Sovety Kazakhstana,* 9 August 1994).

The creation of the union of three Central Asian republics is a practical result of N. Nazarbaev's initiative also. The first meeting of the Council of Prime Ministers of Kazakhstan, Uzbekistan, and Kyrgyzstan in Bishkek initiated real work within the framework of the new union of the three states. These trilateral accords are even today a model for the further expansion of integration processes.

The EAU has ceased to be an idea to whose implementation people aspire and has become a realistic project, which will undoubtedly be supplemented and revised in accordance with the new conditions, a project that is already being realized in practice. This is a long path. But we are being prompted toward this path not only by national interests but also the objective logic of history consisting of the world trend toward integration.

Making a full survey of all the responses to President N.A. Nazarbaev's initiative today is, naturally, quite difficult. Each new day is bringing information concerning the positions of this politician, statesman, and scholar or the other of various states. And this flow will increase by the day. Especially since there is an opportunity today even to discuss not only the idea of integration but also a prepared document.

Little over a month remains until the next meeting of heads of state of the CIS—sufficient time for a thorough study and analysis of the Eurasian Union project. Let us hope that the leaders do not repeat the old mistakes and that the possibility of uniting efforts for a resolution of common problems is realized.

---

## 6.60

### Nazarbaev Struggles to Push Eurasian Union

Boris Vinograd
*Izvestiya,* 29 October 1994
[FBIS Translation]

---

[Report by Boris Vinogradov: "Another Union—But This Time Eurasian? Why Moscow Refutes Nazarbaev's Idea"]

*Editor's Note:* Despite the initial attention the EAU received, it did not ultimately succeed. There were many complex psychological and political reasons why its first adherents pulled back from their original enthusiasm and rescinded their support. After 1994, the EAU continued to be mentioned in speeches, but only in a token, lukewarm manner.

Belarus's President Lukashenka adamantly opposed an EAU because he was devoted, in a fanatical way, to a Slavophile "center" for the CIS. The other Central Asian nations seemed to harbor suspicions of Nazarbaev's motives, not wishing to see a Kazakh/Russian coalition become dominant in their region of the world. There has definitely been political competition between Nazarbaev and Uzbekistan's Islam Karimov for influence over Central Asia. The Transcaucasian leaders—Aliev, Ter-Petrosyan, and Shevardnadze—somewhat listlessly continued to share an interest in a reconstituted economic union and an alternative to Russian hegemony within the CIS, but could do little for Nazarbaev on their own. The Moldovan president, Mircea Snegur, simply rejected any supranational institutions and clung to the basic, but workable, approach of strong bilateral ties with as many former republics as possible. As the following article so aptly notes, however, the idea of some kind of confederal community lingered in many leaders' minds because it had struck a definite chord in the public consciousness. The reader might expect to see several confederal options appear, and perhaps disappear, over the next few years as state leaders determine just how much or how little integration they and their citizens can tolerate.

Ever since Nursultan Nazarbaev thought of setting up a Eurasian Union on the expanses of the CIS—and this was back in March this year—the attitude both of the masses and individuals toward this idea has remained almost unaltered. Perception fluctuates between impenetrable incomprehension and a reserved "acknowledgment of the general picture."

Efforts to test the water and promote this idea at the recent summit in Moscow also proved unsuccessful. Kazakhstan's leader barely managed to keep the Eurasian Union question on the agenda, and then only in the "any other business section." Aleksandr Lukashenka—a newcomer to the presidents' club—demanded that it be taken out of the discussion altogether.

Nursultan Abishevich might have expected such blatant ostracism from anybody else but the man who stood alone in his own parliament when he ventured to vote against the 1991 Belovezh Forest Agreements. According to logic, a kindred spirit had to be found in him at least. The lack of logic in the Belarusian leader's move only reflects all the confusion and chaos surrounding the Eurasian Union.

Few people will remember now that the abbreviation EAU appeared in our midst before CIS as a product of the dream of the great humanist Andrey Sakharov for a union of the peoples of Europe and Asia. During the collapse of the superpower, it was proposed as the new name most in line with the geopolitical essence of the former fraternal, now independent, republics. It most probably failed to gain a foothold for lexical considerations. The word "union" warmed the hearts of some, and shocked others. So closely linked is it with the phrase " . . . inviolable union of free republics" [from the Soviet national anthem].

Obviously, Nazarbaev took account of this rather important circumstance when he circulated his draft to all CIS aspirants "for perusal and recommendations." I do not know how things progressed with the perusal, but no recommendations followed. According to the results of the Moscow summit, the following configuration of forces emerges.

Boris Yeltsin sees in the EAU many positive points which could be used in the course of integration, but he believes that the peoples "do not want this." Leonid Kuchma has yet to clarify his stance: "It requires reflection." At the end of the conference, Lukashenka declared that he "had been incorrectly understood" when the agenda was being confirmed, and he had thereby completely obscured the position of Belarus. Mircea Snegur remained enigmatically silent. Eduard Shevardnadze supported it "in principle" as an innovation with which it is possible to step with dignity into the twenty-first century. Geydar Aliev also expressed support, but he hedged it around with a mass of provisos and comments. Levon Ter-Petrosyan reiterated the Russian sentiments. Islam Karimov gave an unequivocal "no." Saparmurad Niyazov, who attended the conference only in the physical sense, uttering not a single word at it, kept his thoughts to himself. Askar Akaev, who considers himself a close friend of Nazarbaev, approved. Emomali Rakhmonov seemed also to be "in favor," but he showed no particular enthusiasm.

In other words, the participants in the discussion formed up into a defensive circle, clearly not wishing to make too much of the issue. Everyone seems to understand that it is a matter of some kind of an alternative to the CIS, into the formation of which Moscow is putting so much effort. To vote for the Almaty project would mean opposing Russia, to whom the other members of the Commonwealth are in debt.

Everyday economic reality determines cognizance. One way or another, integration within the framework of the CIS is progressing according to a plan dictated from Moscow. The creation of the Interstate Economic Committee [IEC], where Russia has 50 percent of the votes, formalized its leading role for the first time. Experts believe that this figure is artificially low and does not reflect the real weight of a state possessing such mighty resources. However, Boris Yeltsin managed to obtain a rejection of the principle of consensus—that is, unanimity when taking decisions—which gives him a chance to act more authoritatively.

Behind the fine rhetoric and good intentions, perspicacious Russians detected in the EAU a desire to return to the old system of the distribution of raw material assets. So it is that Konstantin Zatulin, chairman of the Duma Committee for CIS Affairs, suggests: The majority of republics have already realized that they have no future without Russia, but as yet they are not all prepared to pay the real political price for a chance to hitch themselves to its apron strings once again.

Nazarbaev's position is special. His influence abroad rivals that of Gorbachev during the era of perestroika. His laurels on the international arena give rise to jealousy in the Kremlin, and half-hearted utterances regarding Russia's interests prompt suspicion. Nursultan Abishevich is finding it hard to convince Boris Nikolaevich that he does not intend to restore what the latter demolished. (The draft contains an item on the inviolability of borders and the renunciation of strong-arm pressure.)

The domestic situation is another matter. Moscow observes somewhat maliciously that Nazarbaev is between the devil and the deep blue sea. Namely, the nationalists, who curse him for his pro-Russian inclinations, and the Russian-speaking population, demanding dual citizenship and compliance with human rights. It is no accident that the aspiration to transfer the capital from Almaty to northern Kazakhstan was viewed as a political maneuver aimed at lessening the discontent of the Russians [in Kazakhstan].

The very same force of inertia is operating in the case of the EAU. Nazarbaev thumps his chest, trying to demonstrate the sincerity of his kindly feelings toward Russia and the purity of his intentions. Official Moscow is in no hurry to believe him, discerning other "motives" between the lines of the precocious draft. Almaty's proposals to introduce the free movement of citizens and uniform visa regulations, to make the ruble the pan-national currency, to create a single defense system, parliament, and legislation, to set up one executive committee as a suprastate organ, and so on, met with a positive response only from an insignificant proportion of Russian politicians.

At the beginning, Sergey Shakhray, Arkadiy Volskiy, Gennadiy Zyuganov, Mikhail Gorbachev applauded enthusiastically. But the applause faded. Now, each of them has furnished himself with his own plan for the reintegration of the post-Soviet area, having been persuaded that the idea of reviving the union has proved more tenacious in people's minds and maybe highly popular at the next elections.

In these conditions, the Almaty initiative has two chances. It will either assume central place in future discussion on the given theme, having undergone adjustments from the sidelines, or else it will be buried by the joint efforts of those who, from the outset, saw it more as a piece of craftiness than as an acknowledged necessity. As well as those who bank on exploiting the nostalgic feelings of the voters, and who are not averse themselves to assuming the role of the new messiah.

Incidentally, Nursultan Abishevich says that he is ready to share copyright, just so long as the "matter goes through."

## 6.61
**Skepticism Toward Eurasian Union Idea Noted**
Marat Salimov
*Kommersant-Daily,* 25 October 1994
[FBIS Translation]

A press conference with Nursultan Nazarbaev, president of Kazakhstan, that was held on Saturday was devoted entirely to problems of the realization of Nazarbaev's idea of the creation of a Eurasian union. Despite the fact that this idea aroused no enthusiasm in the majority of heads of state at the Moscow summit, Nazarbaev is sure that the real integration of the new independent states may occur only on the principles stated in the concept of creation of the union.

At the news conference the president did not conceal his disappointment with the results of the meeting of heads of state of the CIS. This disappointment was connected not only with the rejection of the idea of a union but also with the fact that the economic integration of the Commonwealth countries has in practice come to a halt. In Nazarbaev's opinion, the main reason preventing the Commonwealth countries from restoring those that have been lost and developing new ties in the economic and humanitarian spheres is the absence of supranational bodies capable of monitoring and directing the integration processes. Nazarbaev observed that a barrier blocking the creation of such structures is the negative attitude of the leaders of some states of the Commonwealth toward the integration process as a whole. Specifically, the president of Kazakhstan mentioned the presidents of Belarus and Uzbekistan, Aleksandr Lukashenka and Islam Karimov, who are definitely opposed to the creation of supranational bodies and who at the meeting proposed that the item concerning the creation of a Eurasian Union be removed from the agenda. Despite the fact that this matter was nonetheless discussed, it was decided merely to note the proposal concerning the creation of a union and to use the ideas contained in the concept of its creation to stimulate integration processes in the CIS.

*Kommersant-Daily: The Nazarbaev draft concerning the creation of a Eurasian Union was presented for consideration by the heads of government of the Commonwealth countries in the spring. It provides for the creation of an economic and political union of the former republics of the USSR, including supranational executive authorities and an interstate parliament, which would coordinate the legislative activity of the participants.*

It is significant that Belarusian President Aleksandr Lukashenka also expressed dissatisfaction with the results of the Moscow meeting following his return to Minsk. That is, he stuck to the opinion that he had expressed at a news conference on the threshold of the summit—the signing of interstate agreements could not get integration moving.

Despite the absolute skepticism of Lukashenka and Karimov's rejection of the principle of integration itself, the first supranational body was created in Moscow. But the position of the presidents needs to be given its due. Of the more than 450 documents signed by the heads of state and government of the Commonwealth countries, no more than one-fourth are actually working. It is entirely likely in this connection that the Russian delegation will propose new initiatives at the December meeting of heads of state of the Commonwealth in Almaty.

## 6.62

**Ministry Official Rejects Nazarbaev Proposal on Union**
Interfax, 1 November 1994
[FBIS Translation]

Leonid Drachevskiy, director of the CIS Department of the Russian Foreign Ministry, told Interfax in Moscow on Tuesday that Russia and other CIS countries were not yet ready to create a Euro-Asian union, which was initiated by Kazakhstan's President Nursultan Nazarbaev.

He addressed a meeting in the Russian Foreign Ministry, involving representatives of public organizations, members of the Russian delegations at the talks with CIS countries, as well as staff members of the ministries and departments concerned.

Drachevskiy said "the Commonwealth countries are advancing toward a sort of a community, be it a union or something else." "But haste is out of place here," he said.

## 6.63

**Nazarbaev Urges "Vigorous" Integration Push**
Interview by Vladimir Ardaev
*Izvestiya,* 30 December 1994
[FBIS Translation], Excerpt

Just before New Year, your *Izvestiya* correspondent interviewed Nursultan Nazarbaev, president of the Republic of Kazakhstan.

[Ardaev]: Nursultan Abishevich, you are a consistent defender of the idea of integration among the peoples and states of the former Union. How would you describe the outgoing year in that regard?

[Nazarbaev]: When I put forward the idea of creating the Eurasian Union, I set myself several tasks. First, to break the deadlock over the CIS. I believe that some movement has been made in this regard. We have finally formed the Interstate Economic Committee and signed an agreement setting up a free-trade zone, which is the first stage in organizing a customs union. Parliamentary collaboration has become more active. The experts have approved the pact on peace and stability in the CIS, which will clearly be discussed at the presidents' meeting in Almaty at the start of 1995.

All these things are based to a greater or lesser extent on the ideas of the Eurasian Union. There is also understanding for our approach to citizenship, particularly with regard to making the procedure for acquiring and changing citizenship as simple as possible. More constructive approaches are emerging on the question of protecting external borders. Second, my task was to formulate a general concept for the integration process, a kind of model or norm to work toward. This makes it possible to see clearly the main outlines of integration in the long term. Finally, one of the main schemes was to arouse the mass consciousness of people whose lives are affected by the success or failure of integration.

Of course, many things were not done, but I am not going to reduce the problem to the conduct of the political elites. There are objective factors creating barriers to integration. I respect the attitude of many of my colleagues who put forward perfectly sensible arguments. But the concept of the pace of play operates in politics too. I do not accept the argument that integration will occur of its own accord sooner or later. Time can be wasted here. That is why vigorous efforts are needed.

But we must not resort to extremes. Of course it is possible to restore the rigid centralized state. But that would certainly not do people a favor. The second stereotype is a faith in an instant integration into the European or, say, Asian home. It is time to get over these romantic illusions.

I see the future of the independent states being decided on a purely voluntary basis.

[Ardaev]: In your opinion, what will 1995 be like? What will we be able to do away with and what will we build in the new home, and under what conditions?

[Nazarbaev]: It has already become customary to cite the great Russian economist Kondratyev's idea on the model of long-term economic cycles. The analogy is weak, of course, but we are clearly standing on the threshold of a wave of integration. The first to sense this, by the way, were the businessmen, entrepreneurs, and directors. Probably this has to do with the fact that feedback in the business sphere is more effective than in politics. I hope that the will of the political elites will be channeled in this direction.

To be specific: I would like to see the results of the activity of the economic structures which have been created—in particular the Interstate Economic Committee.

Clearly the problem of simplifying the acquisition or change of citizenship will be resolved. It is to be hoped that economic legislation will be harmonized. The implementation of major economic projects between Kazakhstan and Russia will begin.

The republics are increasingly beginning to understand that the joint solution of the majority of common problems is also the constructive path which will enable us to find our place in the world.

The main thing, in my view, is that a similar process has also begun in Russia. I have met with many scientists and politicians and I felt, especially in the second half of 1994, that Russia's role and place in the new geopolitical area have begun to be reassessed. In other words, its three-year policy of "repelling everyone" is producing an understanding that this will cause political and economic harm, most of all to Russia itself. . . .

---

## 6.64

**Nazarbaev on Eurasian Union, CIS Future**
Interview by Viktor Kiyanitsa
*Moskovskie Novosti,* 15–22 January 1995
[FBIS Translation]

---

*Editor's Note:* The following document was chosen for Nazarbaev's interesting explanation of his belief in regional blocs. He offers this explanation as a reason why the former Soviet republics must not think they can simply attach themselves to other regional blocs in the West, and should instead form their own bloc.

Three years ago you became the first popularly elected president of Kazakhstan. And something else happened almost simultaneously—the signing of the Belovezh agreements and the disintegration of the Union. How did you perceive it then and how do you evaluate it now?

[Nazarbaev]: As distinct from many of my colleagues, I have been and remain a consistent supporter of integration. Mine has always been a clear position—it is essential not only to recognize and declare but also to engage the colossal potential of economic, cultural, and human relations in the Eurasian space. But history does not know the subjunctive mood. It is only in analytical scenarios that it is permissible to argue in the style: "What if the USSR had not been demolished?" In practical politics people operate mainly in the categories of real, not conditional, time. The disintegration of the USSR means a complex set of objective historical and personality factors. The great German sociologist Max Weber once seriously asked whether there would have been World War I had Archduke Ferdinand not been assassinated in Sarajevo. The disintegration of the USSR undoubtedly has its inner logic. But the problem is now in a different key. First, everyone recognizes the complexities and difficulties of an autonomous surmounting of crisis. Second, we are talking about the relations of politically independent states.

[Kiyanitsa]: The Commonwealth of Independent States was created at that time in Almaty, with your active participation. You described it at that time as a "form of civilized divorce." Are you not now disappointed in the sense that "the mountain has given birth to a mouse?"

[Nazarbaev]: The Commonwealth of Independent States was created, for all that, in Belovezh. As far as the situation that has taken shape since the Belovezh agreements is concerned, it has been very complex. In December 1991 we faced the potential threat of a breakup of the entire post-Soviet space along the lines of the Slav and Turkic republics, Christians and Muslims. It was on that basis that actions were pursued to preserve the manageability of the processes in this vast space. We took account of the global aspects of this decision also, incidentally. A colossal arc of instability could have had critical consequences for all of Asia and Europe.

[Kiyanitsa]: Nonetheless, you have all these years voiced dissatisfaction with the amorphousness of the CIS. And the meeting planned for December was canceled. What was the reason for this?

[Nazarbaev]: Concerning postponements—if there are valid reasons, why not accommodate them. And on this occasion there are valid reasons. The documents had not been prepared, and this was, besides, connected with the tragic events in Chechnya. I will say frankly that Kazakhstanis cannot fail to be concerned that blood is being spilled on the territory of a friendly state. As before, the impact of events in Russia on its neighbors is great. Consider also the fact that many Chechens live among us and that they are all anxious for the fate of near ones and dear ones. People wish for the speediest end to the bloodshed.

While viewing the Chechen problem in the context of Russia's territorial integrity, I consider a search for peaceful paths of a settlement essential and possible, nonetheless. I spoke about this on New Year's Eve with Boris Nikolaevich Yeltsin. I permitted myself to express concern at the protracted conflict on Russian soil. Taking as a basis the opinion of the community and the mood of the thousands of Chechens living in Kazakhstan included. And I recently had a lengthy telephone conversation with Viktor Stepanovich Chernomyrdin. I offered to send humanitarian assistance for the civilian population of Chechnya. Food and medicine have now been dispatched from Kazakhstan. I plan to meet with the Russian leaders, and this difficult subject—the Chechen crisis—will be discussed in our conversation. I am prepared personally to make every effort for the speediest settlement of the conflict.

[Kiyanitsa]: All these years you have been a consistent sponsor of integration, which has gained you the reputation of a supporter of a "new Union," just about. Was this required by your political image or is it a question of some inner requirement?

[Nazarbaev]: It is not a problem of my image or my inner requirement. There are in politics clear imperatives which you may accept or reject, but they continue to exist, regardless of personal likes or dislikes. There is such an imperative in the modern world—regional integration. The whole world today is first and foremost an aggregate of integrated blocs. No one, other than utterly archaic isolationists, denies this. It is wholly a question of how to integrate and with whom. I am convinced that Eurasian integration is of great importance for the post-Soviet states. And it is not so much a question of historical and cultural realities even as of exclusively pragmatic factors: economic geography, transport accessibility, unity of infrastructural system, relatively similar production engineering base, and so forth. It is time to rid ourselves of "romantic integration" into the developed integrated blocs of East and West. It is time to understand that no one needs us there.

[Kiyanitsa]: At the last meeting of heads of the CIS you presented the latest initiative—the EAU [Eurasian Union]. Many people deemed it untimely. Some saw it as your personal failure, when you were left among your colleagues practically on your own. Do you intend to return to this idea at the coming meeting or will you put if off "until better times," when public opinion and leaders have "matured"?

[Nazarbaev]: I would remind you that the documents of the Moscow summit point to the need for use of the valuable ideas of the EAU project for a further intensification of integration processes in the CIS. It is not the case that this project or the other is accepted immediately 100 percent in politics. I do not believe that such a large-scale project could have been accepted immediately, but its positive significance is already great enough. First, the CIS has finally gotten moving—an Interstate Economic Committee (one of the pivotal ideas of the EAU) has finally been formed. Questions of interparliamentary cooperation are coming to be analyzed more actively. In accordance with the EAU project, this is for the future a most important issue: the convergence of legislation primarily in the economic sphere. A set of questions of specific military-political cooperation has been discussed at expert level. A political document in development of the ideas of the EAS—a peace and stability pact—will be prepared for the Almaty meeting.

Second, a number of elements of the EAS are being realized at the bilateral relations level also. Specifically, we are hoping in the very near future for a concerted decision on a simplified procedure of the change and acquisition of citizenship between Kazakhstan and Russia. Third, the EAS project has exerted great influence on the mass consciousness on the scale of the entire CIS. No other foreign policy document in the CIS has, perhaps, throughout its existence evoked such extensive discussion.

[Kiyanitsa]: In the years of your presidency Kazakhstan has acquired all the signs of a sovereign state and the final attribute of statehood—a national currency being introduced a year ago. What are the results of the first year of "full and final" sovereignty?

[Nazarbaev]: It has already been said that Kazakhstan never previously controlled its economy in full. And this assertion was correct up to this year. Now we have this experience. And although it has begun not entirely successfully—the inflation indicators have deteriorated, and the decline in production has continued—signs of positive change in the direction of financial stabilization are already visible, nonetheless. Having signed the standard bank agreement with the International Monetary Fund, we are adhering to all the necessary conditions. This obligation is, naturally, stimulating foreign investment. As of today agreements have already been signed on official financial assistance with the IMF, the World Bank, the European Bank for Reconstruction and Development, the Asian Development Bank, and the government of Japan on the allocation of long-term loans on favorable terms (totaling $910 million altogether). Eighty-eight individual credit agreements for a sum in excess of $1.5 billion have been signed.

Of course, all transformation involves certain hardships for the population. Nonetheless, the level of the average wage, in terms of which we have established ourselves in the CIS in second place behind Russia, has been rising steadily here since September. By the start of 1995 this indicator will be over $70. I myself recently appealed to parliament for the immediate discussion and adoption of most urgent legislation determining the legal basis of economic relationships and regulating questions of ownership, investments, bankruptcy, mortgages, and so forth.

## 6.65

**Lukashenka Criticizes "Eurasian Union" Plan**

Moscow Radio Rossii Network, 25 January 1995
[FBIS Translation]

*Editor's Note:* Lukashenka's rejection of Kazakhstan in the following document could pose problems for Russia, which can afford neither to ignore the Muslim populations to its south, nor to risk a Slavic/Muslim split within the CIS. To complicate matters for President Yeltsin, the Slavophile national patriots (of whom Solzhenitsyn is a member) are gaining a modestly large following in Russia. A split between Slavophiles and radical Eurasionists (like Zhirinovskiy) is a dangerous political possibility within the larger political picture.

Aleksandr Lukashenka, president of Belarus, has categorically distanced himself from Russian-Kazakhstan unification, and has sharply criticized Nursultan Nazarbaev's idea of setting up a Eurasian Union instead of the Commonwealth of Independent States. In an interview to a group of journalists, which included our Minsk journalist Yuriy Svirko, Aleksandr Lukashenka said:

[Begin Lukashenka recording.] I have nothing to do with that so-called Eurasian Union, which began taking shape with the signing of Russian-Kazakhstan agreements in Moscow. Moreover, I am also convinced that Boris Yeltsin does not consider these signed agreements as a Eurasian Union. We clearly decided at the last meeting of heads of state that we had the Commonwealth of Independent States, and that we wanted to cooperate at various levels. It has been demonstrated by Ukraine, Belarus, and Russia. Here it is, and let us get on with it. And any attempts by certain politicians to declare that Belarus, Russia, and Kazakhstan have set up some new Eurasian Union, or have started setting it up, believe me, I can speak for myself, and for Yeltsin, and I can tell you that there is nothing like that. We have decided that we shall continue to develop the trends which are being shaped in the CIS. I am ready for this. I will not depart from this policy. [End recording.]

## 6.66
**Meshkov Supports Eurasian Union**
Governmental Telegram
*Krymskie Izvestiya,* 11 February 1995
[FBIS Translation]

*Editor's Note:* A very interesting letter was sent to CIS leaders by Yuriy Meshkov, then president of separatist Crimea. In it, Meshkov requests designation as "staff headquarters" of a Eurasian Union. The Crimean nationalists' rejection of Ukraine and the Slavic confederal concept led them to opt for a Eurasianist course. In this letter, the Crimean government head is also clearly enunciating the direction in which he wishes to see Russia go.

Esteemed Nursultan Abishevich! I request that you promulgate our appeal in the high forum of Commonwealth countries in Almaty.

*Appeal to the CIS heads of state participating in the Almaty meeting.*

Esteemed Gentlemen!

Ardently supporting your efforts to strengthen cooperation in the sphere of defense, the economy, safeguarding of borders, customs control, and humanitarian contacts, I am hoping for further development of this cooperation within a framework which implements the concept of the Eurasian Union.

The Crimean Peninsula has always received all guests of every nationality, from any corner of the former USSR, with cordiality and hospitality. We are deeply convinced that henceforth as well, for all of time, Crimea must not only be the health and resort center for all people from the CIS countries, but must become a peninsula of peace, friendship, and unification of the citizens of Commonwealth countries.

In the event that the Eurasian Union is established, we propose to make Crimea the location of its staff headquarters. Objective circumstances exist in favor of this:

(a) Proximity to all the CIS countries and convenient transportation links;

(b) The kindly predisposition of Crimean Republic residents to representatives of all nationalities, and excellent environmental and climatic conditions for fruitful work;

(c) Our readiness to afford high-quality premises and everything else necessary to accommodate representatives of the Commonwealth countries on short notice.

I am relying on the wisdom of Commonwealth heads of state in this most important matter—the unification awaited by all our peoples.

With sincere respect,

[Signed] President of the Crimean Republic Yu. Meshkov
City of Simferopol, 9 February 1995
No. 1262

## 6.67
**Nazarbaev: People Will Support the Eurasian Union**
ITAR-TASS, 19 November 1995
[FBIS Translation]

"The idea of a Eurasian Union is a very serious concept of the post-Soviet era. It has its logical sources in the world practice," president of Kazakhstan Nursultan Nazarbaev said in an interview with ITAR-TASS. Nazarbaev is [in Paris to participate] in celebrations to mark the fiftieth anniversary of UNESCO.

"Integration processes are forging their way ahead in Western Europe, Southeast Asia, and the Middle East," the Kazakh head of state said. "The twenty-first century is the century of integration. We lived together for several centuries in the territory of the former Soviet Union, developed a common culture and have a common history. Hence the people are greatly inclined toward one another, and the idea of a Eurasian union will in the end win over minds of the people," he said.

Speaking today at a solemn session in the Paris-based headquarters of UNESCO, Nazarbaev noted that Kazakhstan strives for developing equal and mutually beneficial relations with all states of the world on the basis of fundamental principles of the UN charter, the Helsinki act on cooperation and security in Europe, and other documents. He highly appreciated the selfless activities of UNESCO to protect ideals of humanism, [including] human rights and liberties and the spiritual wealth of peoples.

# The Central Asian Union

The Central Asian Union is a recently formed association of Central Asian states (excluding Tajikistan and Turkmenistan), but it also refers to a process of gradually increasing cooperation among the Central Asian states, which started at a 4 January 1993 meeting of Central Asian leaders in Tashkent. Or perhaps the starting point should be pushed back to the December 1991 Belovezh Forest meeting of Slavic leaders, from which Kazakhstan was excluded, thereby offending the Central Asian republics by not initially including them in the ranks of the founding members of the CIS. Whatever is taken as the beginning point, the Central Asian countries began a process of regional integration among themselves at a level which they explain comes under the CIS umbrella and is compatible with CIS integration goals. It would, however, be difficult to say that this process is creating a regional unit, because each country is entirely unique in its level of political and economic development and in its national composition and aspirations. Nevertheless, the process had progressed from a concentration on coordinating transportation and environmental projects in 1993 to instituting an organizational basis for military and political cooperation by 1995.

The documents collected in this section depict an ongoing debate over the true intentions of the leaders of the Central Asian Union. On 13 May 1993 an article appeared in *Nezavisimaya Gazeta* authored within the Russian Ministry of Foreign Economic Relations, which charges Central Asian leaders with "undermining the interests of Russia," stressing the "Muslim component" as the main base for cooperation, and forming a "Central Asian Regional Alliance." A rebuttal of the accusations made in this article appeared in an open letter from the press office of the Embassy of the Republic of Uzbekistan published by *Nezavisimaya Gazeta* on 23 May 1993. This debate is documented here.

In July 1994 the process of Central Asian integration took a more serious conceptual direction. At a press conference following a meeting of the presidents of Kazakhstan, Kyrgyzstan, and Uzbekistan it was announced by the director of the Kyrgyz president's Institute for Strategic Studies that "it is not enough to form a unified economic space." Councils for the three countries' heads of government and state and foreign and defense ministers were established at this meeting and planned to meet regularly thereafter. Later in July, President Askar Akaev stated at a press conference that he expected the Slavic and Transcaucasian republics to follow the Central Asian states' example, saying he was convinced that the integration of Russia, Belarus, and Ukraine would take place. Such statements suggested that the Central Asian leaders still may feel excluded by the Slavic CIS countries, and are reacting to this situation by launching a strong regional initiative.

The concept of the Central Asian Union raises many questions concerning the participation of Tajikistan, Turkmenistan, and even Uzbekistan's adherence to any supranational decisions made by the organization's councils. Turkmenistan's President Niyazov is opposed to any collective structures in general. Tajikistan, at this time, is almost in the status of a Russian protectorate. Uzbekistan's President Islam Karimov also maintains an extremely independent political and military stance. Some of these questions are addressed in this collection by Kyrgyz President Askar Akaev.

As the following highlights of the Central Asian Union's activities show, progress toward real coordination, even on economic issues, has been slow. It is also questionable whether these countries have the resources to make cooperation produce effective results. One article in this section contains a Kazakh economist's pessimistic analysis of the ability of the Central Asian states to form any kind of union without Russia. The three presidents, in fact, often point to their joint goal of simulating the CIS to take more resolute steps toward operative integration, rather than to present the CIS with any competitive structures. In April 1995 the presidents of Kyrgyzstan and Uzbekistan announced their readiness to join the Russian/Belarus/Kazakhstan customs union as soon as possible. Another goal is to draw international attention to security problems in the region.

In conclusion, the Central Asian Union is still more of a consultative body than a true confederation. Its regular meetings have continued, and a plan for economic integration through the year 2000 has been approved. Nevertheless, there is little evidence yet that this trilateral union presents any threat to Russia's interests in the region or to the CIS.

## 6.68

## Agreement on Economic Cooperation in the Implementation of Projects of Mutual Interest

Interfax, 13 May 1992
[FBIS Translation]

Guided by the principles of friendship and solidarity, Iran, Kazakhstan, Kyrgyzstan, Pakistan, Turkey, Turkmenistan, and Uzbekistan hereby agree upon the following:

1. To accelerate the construction of a trans-Asian railway. To complete the construction of the Sarakhs-Meshkhed section by 1995 and to speed up the development of the Druzhba (Friendship) station of the border railway section linking the Republic of Kazakhstan and the People's Republic of China.

2. To design and build a gas pipeline that would supply natural gas from Turkmenistan to the Islamic Republic of Iran, to the Turkish Republic and to Europe in the following volumes: 15 billion cubic meters a year at the first stage and 28 billion cubic meters a year at the second stage. The states concerned shall ensure the transit of gas from Turkmenistan through their territories during at least thirty years.

3. The signatories to the agreement have found it expedient to design, build, and reconstruct sections of the highway linking Istanbul, Teheran, Islamabad, Ashkhabad, Tashkent, Bishkek, and Almaty, pending the opening of international motor communication. To implement the said projects the signatories have found it expedient to set up corresponding coordinating councils consisting of the leaders of corresponding bodies.

**Joint Communiqué**

"The presidents and prime ministers of Iran, Kazakhstan, Kyrgyzstan, Tajikistan, Turkmenistan, and Uzbekistan hereby state the following:

1. Proceeding from the belief that economic and trade contacts constitute a considerable part of their relations, the parties will make efforts to broaden trade and develop joint investment projects.

2. The parties will exchange delegations, organize exhibitions and conferences, and other events facilitating the sale of goods and services.

3. The parties have expressed readiness to exchange goods on a mutually advantageous parity basis, to grant each other the most favored nation status and to set up joint banks and small and medium enterprises.

4. Attaching great importance to the establishment and development of motor, railway, air and sea transportation and lines of communication, the parties have agreed to develop and implement bilateral and multilateral projects in the said areas.

5. The parties note that the construction of gas and oil pipelines, the production of oil and gas, and the building of refineries, will allow them to ensure normal economic development of the given region and to transport natural gas and oil to other countries.

6. The parties have agreed to consider the possibility of introducing soft customs dues and a simpler procedure of crossing the border to ensure free movement of people and goods through their territories in accordance with international provisions and agreements.

7. With the purpose of promoting tourism and improving mutual understanding in the region, the sides will jointly study each other's culture and history, and restore historical and cultural monuments.

8. The parties note that the Economic Cooperation Organization plays an important role in broadening mutual cooperation, and agree that the members of this organization should make every effort to attain its goals.

9. The parties emphasize that by broadening regional cooperation they pursue no plans for establishing any blocks threatening the interests of other states.

10. The parties have agreed to instruct their foreign ministers to explore the possibility of convening a conference on interaction and trust-building measures in Asia.

11. The parties find it expedient to hold regular consultations on problems of mutual interest. [No closing quotation marks. As received.]

## 6.69

## Central Asian Leaders Conclude Summit in Tashkent

*Editor's Note:* The next five entries give alternative accounts of the 1 January 1993 meeting of Central Asian leaders who came together in order to increase their chance to negotiate from strength with Russia. Each report contains a slightly different emphasis. Each one however, exhibits the important motive of economic integration behind each new report.

### News Conference

Interfax, 3 January 1993
[FBIS Translation]

Uzbekistan, Turkmenistan, Tajikistan, Kyrgyzstan, and Kazakhstan have confirmed their allegiance to the Commonwealth of Independent States, which, in their view, is still relevant. President Islam Karimov, of Uzbekistan, has said that at the same time each CIS member state wants to have guarantees it is free to develop the way it deems right. President Karimov was speaking at a news conference in Tashkent following Monday's meeting of the leaders of the five republics in the Uzbek capital.

President Karimov said the meeting was a "complete success." Among other things the conference decided to specify the term "Central Asia." From now on it should be applied to the region comprising Kazakhstan, Kyrgyzstan, Tajikistan, Turkmenistan, and Uzbekistan.

President Nursultan Nazarbaev, of Kazakhstan, told the news conference the Tashkent summit had taken "appropriate steps" to bring about a common market of Central Asian countries with common taxation, customs, pricing, investment, and export policies. President Nazarbaev said the term "common market" would incorporate the principles of cooperation and integration of all economies of the region.

At the same time he warned against locking up oneself within the bounds of the region and called for broader cooperation with other countries, in the first place, Russia. President Nazarbaev said that the five republics would like to preserve the ruble zone on the principles of equality. He said there should be a banking union to control monetary policies and emission, where one member state would have one vote. President Nazarbaev and other participants in the summit believe that this would guarantee a lasting ruble zone.

The presidents of Kazakhstan, Kyrgyzstan, Turkmenistan, and Uzbekistan said they were prepared to step up their efforts to help eliminate the effects of the conflict in Tajikistan, for instance, to provide additional food and medical supplies.

President Saparmurad Niyazov, of Turkmenistan, suggested holding the next summit in Ashkhabad.

### Karimov, Nazarbaev Comment

From the "Presidential Bulletin" feature compiled by Andrey Pershin, Andrey Petrovskiy, and Vladimir Shishlin; and edited by Boris Grishchenko
Interfax, 4 January 1993
[FBIS Translation]

Leaders of the five [Central Asian] states held a joint press conference on the results of the Tashkent meeting. At the press conference they proposed that the geographic understanding of "Central Asia" should henceforth be understood as the region encompassing the territories of Kazakhstan, Kyrgyzstan, Tajikistan, Turkmenistan, and Uzbekistan. It is their belief that the new term will require changing the cumbersome phrase "republics of Central Asia and Kazakhstan."

The president of Uzbekistan, Islam Karimov, ascertained "the complete success of the meeting." He spoke against the assertion that the CIS has exhausted itself. At the same time I. Karimov did not rule out the possibility of forming a general market of countries of Central Asia, which will envisage a general customs duties, pricing and export policy, and also mutual harmonization of the economies of all states. I. Karimov noted in connection with this that "appropriate steps (toward a general Central Asian market—IF [Interfax]) have already been taken" at the summit in Tashkent. The president of Uzbekistan called for the maintenance of one equal ruble zone in which the administration of financial policy and monetary emission would be realized on the principle of "one state, one vote." Karimov believes that this approach might guarantee the "durability of the ruble zone." He assessed the possibility of withdrawing from the zone as "undesirable."

At the press conference, the establishment of a number of joint programs of the states of the Central Asian region was announced. This concerns, in particular, the processing of surplus crude oil, extracted in Kazakhstan, at enterprises of Turkmenistan and Uzbekistan which have available free capacities. It was decided to create a unified information space. With this aim, the Tashkent TV and radio center will be used as a base to begin transmission on the whole region and the publication of a general Central Asian newspaper.

States of the region will form expert committees on energy (with headquarters in Bishkek), on oil (Ashkhabad), on cotton (Tashkent), on grains (Almaty).

The president of Kazakhstan, Nursultan Nazarbaev, proposed that the general symbol of the states of Central Asia be the plane—a tree which has one root and five branches.

### Leaders Adopt Joint Communiqué

ITAR-TASS, 4 January 1993, [FBIS Translation]

Five Asian republics of the former Soviet Union ended their summit meeting in Tashkent on Monday by agreeing to pursue a concerted policy aimed at creating a common market in the region.

The participants also agreed to set up an Interrepublican Coordinating Council and exchange governmental representatives to coordinate joint activities and oversee the implementation of agreements.

The meeting was attended by Kazakh President Nursultan Nazarbaev, Kyrgyz President Askar Akaev, Turkmen President Saparmurad Niyazov, Uzbek President Islam Karimov, and Tajik Parliament Speaker Emomali Rakhmonov.

They discussed the economic and political situation in the region and in the CIS as a whole and called for stronger economic and financial ties, stressing the need to preserve the ruble as a common currency. They also decided to set up an international fund for the salvation of the Aral Sea, the joint communiqué adopted at the summit said.

In addition, the five states agreed to create a regional information network, including television with the headquarters in Tashkent and a regional newspaper.

The leaders at the summit said their republics will continue humanitarian aid to conflict-torn Tajikistan which signed a treaty of friendship, cooperation, and good-neighborly relations with Uzbekistan. Uzbekistan also signed a cooperation agreement with Kazakhstan.

They suggested holding such meetings on a regular basis. The next meeting will be held in the Turkmen capital of Ashkhabad in April.

---

## Agreements, Discussion Outlined

Turkmen Press-TASS, 5 January 1993
KAZTAG Press-TASS, 5 January 1993
[FBIS Translation]

---

[Turkmen Press-TASS, 5 January 1993]

Ashkhabad 5 January TASS—The need to establish closer economic cooperation among countries in the region was emphasized by participants in the recently concluded meeting of regional heads of state in Tashkent.

They reached agreement on the elaboration of a specific mechanism for permanent monitoring of the implementation of interstate and intergovernmental treaties and agreements.

Turkmenistan's newspapers today widely published accounts of the meeting and a joint communiqué, adopted as a result of a discussion of the political and economic situation in countries of the region and of an exchange of views on matters aimed at further strengthening equal and mutual beneficial contacts between the sovereign states.

The heads of state instructed the respective governments to work out issues connected with the problems of the Aral and Caspian seas in the interests of the states of the region and decided to set up an international foundation for the salvation of the Aral.

Participants in the meeting [agreed] to hold such meetings of the heads of state and government regularly. Next meeting is to be held in Ashkhabad in April.

Alma-Ata 5 January TASS—The following heads of state had a meeting in Tashkent on 4 January: President Nursultan Nazarbaev of the Republic of Kazakhstan, President Askar Akaev of the Republic of Kyrgyzstan, Tajikistan Parliament Speaker Emomali Rakhmonov, President Saparmurad Niyazov of Turkmenistan, and President Islam Karimov of the Republic of Uzbekistan.

Participants in the meeting discussed in detail the political and economic situation in the states of the region, had a useful exchange of views on matters aimed at further strengthening equal and mutually beneficial economic and humanitarian relations between the sovereign states, considered the state of affairs in the Commonwealth of Independent States, and discussed other matters of mutual interest.

During the discussion, participants made proposals on the pursuance of a coordinated policy in economic, financial, and other spheres in the light of a stage-by-stage transition to market relations.

The heads of state reviewed the implementation of bilateral and multilateral treaties and trade-and-economic agreements between countries of the region, pointing out that these became a basis for constructive solutions to political, economic, social, and humanitarian problems, including that for a certain softening of the consequences of production slump, inflation, and decline in the living standards of the population.

Participants in the meeting emphasized the importance of closer economic cooperation among countries in the region and reached agreement to work out a specific mechanism for constant monitoring of the implementation of interstate and intergovernmental treaties and agreements.

The heads of state reaffirmed readiness to render every kind of assistance to the fraternal Tajik people and supported measures being taken by the leadership of the Republic of Tajikistan to take it out of the critical political and economic situation as soon as possible, and declared for increasing moral support for the constitutional authority of Tajikistan and humanitarian aid with food, medicines, and clothes.

Within the framework of the meeting, there was an exchange of views on matters concerning international cooperation and the coordination of action in this respect, including those in strengthening regional security and peace.

The heads of state decided to exchange ambassadors and instructed their ministries of foreign affairs to decide the matter before 1 February 1993, according to an established procedure.

As a result of the exchange of views, the heads of state instructed their governments to work out matters connected with the pricing policy, the development of communications, the provisions of energy resources, the problems of the Aral and Caspian seas in the interests of the states of the region.

The sides decided to set up an international foundation for the salvation of the Aral and deemed it necessary to hold foundation meetings in Kzyl Orda, Nukus, and Tashauz.

The participants in the meeting were unanimous about the advisability of holding meetings of the heads of state and government on a regular basis. Next meeting is to be held in Ashkhabad in April.

## 6.70
**Uzbekistan Repudiates Charges of Undermining Russia's Interests**

Open Letter to Editor
*Nezavisimaya Gazeta,* 25 May 1993
[FBIS Translation]

Dear Sir: 30 May marks the first anniversary of the treaty on fundamentals of interstate relations, friendship, and cooperation between Russia and Uzbekistan. Collaboration between the two countries is strengthening and developing. Regular meetings of presidents B. Yeltsin with I. Karimov make it possible to find forms of political, economic, and cultural ties, optimal to both states, which suit the interests of Russian and Uzbek people. The visit of V. Chernomyrdin, chairman of the Council of Ministers of the Russian Federation, to Uzbekistan in March of this year provided fresh impetus in bilateral relations.

Analyzing the current level of collaboration of Uzbekistan and Russia at a session of the Supreme Soviet of the republic in May of this year, President I. Karimov stressed that "with regard to the conduct of profound democratic transformations in society, formation of a market economy, and the establishment of bilateral, mutually advantageous relations between two independent states, Uzbekistan and Russia have identical views."

However, the fact that certain forces do not like the consolidation and development of mutually advantageous collaboration of the two states is becoming increasingly clear. The article by V. Yurtaev and A. Shestkov "Asiatic Gas to Flow West. New Alliance Harms Russia's Interests," published on 13 May of this year in *Nezavisimaya Gazeta,* provides graphic proof of that. What is significant is that its authors are officials of MVES [Ministry of Foreign Economic Relations] of Russia. That is evidently designed to convince readers of the substantiated nature of conclusions reached in the article and be accepted by them as a presentation of the position of an important government organ of Russia on questions raised in it.

The principal goal of this article is to prove that centralized cooperation, which formed as a result of a meeting of the heads of state of the region on 4 January at Tashkent, bears a character that is hostile to Russia, detrimental to its political and economic interests in Central Asia and the Middle East, violates previous agreements within the framework of the CIS, and intensifies the position of those countries which are interested in the creation of a "Turkish belt" along the southern borders of Russia, separating it from a broad region rich in raw material and energy resources.

Uzbekistan is especially condemned by the authors for, they assert, occupying Tajikistan and striving to establish its hegemony in the southern part of Central Asia as well as expanding its territory at the expense of neighboring countries.

The danger of such reasoning is obvious. Left without attention, it is capable of sowing seeds of distrust in relations between states of Central Asia and Russia and among themselves as well. That, in turn, may complicate the development of regional cooperation, disrupting the efforts of countries aimed at resolution of economic difficulties, leaving them one-on-one with numerous political and economic problems.

In that connection we consider it necessary to provide an appropriate explanation of the true motives behind the Tashkent meeting of the heads of five Central Asian states on 4 January and the significance of the decisions adopted in the course of it.

Even though the January meeting in Tashkent was called "unexpected" by journalists, in reality, preparations for regional cooperation befitting the new realities which had formed as a result of the formation of independent states in Central Asia were being made gradually by the countries in that region. In 1991–92 there were several meetings of the leaders of Kazakhstan, Kyrgyzstan, Tajikistan, Turkmenistan, and Uzbekistan in the course of which promising forms of multilateral political and economic relations between the countries were explored.

Action by the governments was determined by objective factors—the aspiration of jointly surviving with the least possible losses the breaking up of economic ties following the disintegration of the USSR and preventing a sharp drop in the living standards of the people while laying the foundation of future economic development.

It is specifically an understanding of the fact that singlehandedly it would be impossible to survive under difficult conditions of the period involving transition to a market economy that led the heads of state of Central Asia to think about new forms of cooperation, the bases of which were created on 4 January in Tashkent.

Governments in the region are unified by their fate, cultural similarities, and traditional economic links which were determined by historically formed division of labor among the peoples of the region. The goal of such cooperation is to live in peace, helping each other work to mutual advantage, while consolidating efforts and raising friendly ties to a qualitatively new level and utilizing the mighty economic potential of the region for entry into the civilized market with minimal losses.

Heads of state of Central Asia have repeatedly stressed that the creation of the new commonwealth does not signal a break with the CIS as a whole or with certain individual states forming a part of it.

They unanimously agree that great hopes are being attached to cooperation within the framework of the CIS, its preservation and strengthening. They stress that they unified not for the purpose of seeking a handout from their CIS partners. The new collaboration is primarily called upon to resolve economic problems and there is not even any mention of infringement on the interests or sovereignty of others. As indicated by the time that has passed since January, all assertions concerning "the Muslim component" as the main feature of cooperation among Central Asian countries have lost their validity. There are no grounds to speak of an aspiration to isolate those states on the basis of their ethnic or congregational features. Moreover they are actively seeking ways for establishing bilateral ties with CIS states and other countries, striving to establish themselves as full members of international organizations: the United Nations, CSCE, IBRD [International Bank for Reconstruction and Development], UNESCO, and others.

Independent countries of Central Asia strictly recognize the existing boundaries and are prepared to cooperate closely with other states of the Commonwealth, particularly with the Russian Federation.

In the article, however, the actions of independent states aimed at a resolution of their domestic economic problems are declared as undermining the interests of Russia. The absurdity of such allegations is obvious.

Neither the documents adopted based on the results of the meeting of 4 January in Tashkent nor the declarations and actions by the heads of state of Central Asia provide any grounds whatsoever for accusations of that nature. Consequently, allowing themselves to voice them, the authors must come to grips with the fact that they are producing idle fabrications, juggling with the facts, interpreting events loosely, and ascribing plans to government figures of independent states of which there was never even any mention.

The article contains an entire series of factual inaccuracies and fabrications by the authors. In none of the documents is it possible to read about the creation of the Central Asian Regional Alliance (TsARS). Not a single speech by the heads of state of the Central Asian commonwealth contains even a hint of their adherence to the idea of creating a "Greater Turkestan," a "Turkic Belt," especially inclusion in it of the Muslim territories of Russia.

What violation of an agreement by the Central Asian states concerning the unified strategic defense space within the framework of the CIS can there be if none of the documents even mention a defense alliance of the new commonwealth? It does not exist in practice. More than that, I. Karimov, president of Uzbekistan, repeatedly focused attention on the key role of Russia in the maintenance of peace and stability in the region.

It is unfortunate that the unseemly distortion of facts was engaged in by personnel of the MVES of Russia, a state organization called upon to develop foreign economic relations and not create the image of an enemy of Russia consisting of the peoples of Central Asia who are traditionally friendly toward the Russian people.

Publication of the article one day before the meeting of heads of state and government of the CIS in Moscow, in the course of which the principal question under discussion was that of the creation of an economic alliance, bears a provocative character, and is nothing other than an attempt to heat up the situation, to cause a clash between leaders of the CIS countries.

Unfortunately it is necessary to surmise that certain forces once again utilized *Nezavisimaya Gazeta* for the purpose of undermining Russian-Uzbek relations.

In the situation that has formed, reference to freedom of speech in a democratic society appears not only unconvincing but rather a mockery, since a democracy cannot provoke interstate conflicts, interethnic tension, or manifestations of religious extremism. The indicated article, however, is aimed specifically at that.

[Signed] Press Service of the Embassy of the Republic of Uzbekistan in the Russian Federation.

*From the editorial office.* The reader who directed attention to the given article published in *Nezavisimaya Gazeta* in the "Ekonomika" section could not help but be surprised by the fact that in the last few paragraphs the authors take a sharp turn away from the "petroleum and gas" theme to a political prediction of a highly sensational nature. They predict (pay attention!) a swift occupation of Tajikistan by Uzbekistan with the involvement of Russia in such an act. The scenario of these events is also described in considerable detail involving the participation of Turkish intelligence services, certain forces in Afghanistan, etc. In other words, the end of that article by the two Russian MVES staffers has a distinctly non-economic and an exceptionally extravagant character.

In this connection *Nezavisimaya Gazeta* declares that the given viewpoint is that of the authors of the article which was published due to an oversight by the editorial office in a specialized section of the newspaper. If *Nezavisimaya Gazeta* had at its disposal information on such sensational plans, it naturally would immediately publicize them on the front page of the newspaper and not in the "Ekonomika" section. Thus far we have no reason for doing so, and we regret the publication of the article with such serious predictions of the future and interpretation of past events, without having information in accordance with which it would have been possible to determine the correctness of the authors' position.

[Signed] *Nezavisimaya Gazeta*

## 6.71

**Integration Among Central Asian States at "New Level"**

Interfax, 11 July 1994
[FBIS Translation]

The meeting between the presidents of Kazakhstan, Uzbekistan, and Kyrgyzstan held in Almaty last Friday "showed that integration between the three Central Asian states reached a qualitatively new level," the director of the Institute for Strategic Studies under the president of Kyrgyzstan, Asylbek Saliev, told Interfax on Monday [11 July].

According to him, President Nazarbaev's statement concerning the need to form a unified economic, political, and military space also testified to this.

As Saliev said, the union does not conflict with the CIS; on the contrary, in practice it is prepared to effectuate the ideas that failed to be reflected in the framework of the Commonwealth. He believes that the forming of coordinating councils involving the heads of government and state, foreign ministers, and defense ministers, and the establishment of the joint Central Asian Bank for Cooperation and Development show how serious the three countries' intentions are.

The Kazakh, Kyrgyz, and Uzbek leaders realize that "it's not enough to form a unified economic space" Saliev said. "Industrial development must also be coordinated. The union of three Central Asian countries builds mechanisms for preemptive solutions to numerous regional problems both in the socio-economic and in the political and military areas," he said.

## 6.72

**Akaev on Creation of Central Asian Union**

From the "Presidential Bulletin" feature compiled by Nikolay Zherebtsov and Andrey Petrovskiy; and edited by Vladimir Shishlin
Interfax, 13 July 1994
[FBIS Translation]

President Askar Akaev believes that creation of regional associations like the union of three Central Asian countries (Kazakhstan, Kyrgyzstan, and Uzbekistan) will increase efficiency of the CIS as a whole. Speaking at a press conference in Bishkek, Akaev voiced a supposition that the Slavic and Transcaucasian republics will follow example of the above-mentioned republics. "I am convinced that integration of Russian, Ukraine, and Belarus will take place, the issue concerns only the time of the conclusion of this alliance between these countries," Akaev declared.

In his opinion, the results of three Middle Asian presidents in Almaty lay the ground for a hope that the next CIS summit scheduled for this September will make a resolute step to strengthen integration trends.

Mentioning possibility of Tajikistan's joining the Central Asian Union, Akaev noted that it will depend upon establishing firm civil peace and stability in this country. At the same time Akaev said that Kyrgyzstan, Kazakhstan, and Uzbekistan, whose peacekeeping contingents are participating in the defense of the Tajik-Afghan border will assist in stabilizing the situation in Tajikistan by creating premises for its possible joining the single economic and political Central Asian space.

Speaking about probability of Turkmenistan's joining the Central Asian countries union, Akaev reminded that the leadership of this country has a traditionally restrained attitude toward participation in any collective structures within the CIS—the republic has not signed the Collective Security Treaty and the agreement on creation of an economic union. "Turkmenistan pursues this course in the sphere of its relations with its neighbors in Central Asia; however, one should not rule out that in the near future the situation can change," Akaev said.

He underlined that despite several contradictions in the opinions between the leaders of the Central Asian union member-countries they have no principal differences. In particular, he declared about the absence of any contradictions with Uzbek President Islam Karimov and called the latter "a deep, constructive, and pragmatic politician."

Akaev believes that on the whole the course of the Uzbek leader coincides with the aims and tasks of the neighboring states and the CIS as a whole. Akaev agrees with Karimov, who deems it necessary to restrict freedom of speech "during the transition period." In Akaev's words, in this case one should speak exclusively about restriction of a moral-ethical nature but not about "suppression of critics." Akaev added that freedom of press in Kyrgyzstan, which is not limited by anything, is a destabilizing and destructive factor. Explaining his opinion, Akaev accused the Kyrgyz press of irresponsible staining of political leaders, artificial stirring of passions, and provoking conflicts between different power branches.

---

## 6.73

**Central Asian Economic Zone Viewed**
Boris Plyshevskiy
*Rossiyskie Vesti,* 6 September 1994
[FBIS Translation]

---

In early January 1994 Kazakhstan and Uzbekistan signed an agreement on the formation of a common market, the repeal of customs borders, and the free movement of citizens. In the same month Kyrgyzstan acceded to the agreement. In April at Lake Issyk-Kul the leaders of the three countries reached a fundamental agreement on the creation of a regional economic union and the general coordinating organs. This decision was explained by the slowness of the formation of CIS structures and the failure to fulfill the interstate agreements and decisions adopted by the heads of state and government.

At the second meeting of the three countries' presidents in Almaty, documents were adopted which essentially formed a Central Asian alliance. And although many of the declared decisions are for show and will require a long time to be implemented, it would be wrong to consider them to be merely a declaration of intent. Some of the decisions made are already being implemented in practice.

An interstate council consisting of the presidents and prime ministers has been created. Its structure includes a council of prime ministers, which resolves economic questions, a council of foreign ministers, which should coordinate foreign policy issues, and a council of defense ministers, for questions of preserving stability in the region.

The structure of the interstate council provides for the institution of a permanent executive committee with coordination, consultation, analytic, and information functions, which will be located in Almaty. The executive committee chairman has already been appointed—a representative of Kazakhstan. The Interstate Council session in Bishkek 6 August reached agreement on the conditions of the executive committee's activity and approved the statute on the creation of the Central Asian Bank.

The creation of this coordination organ anticipates the solution of the same question within the framework of the CIS. Proposals for the formation of an interrepublican economic committee with some controlling functions and the right to make binding decisions, prepared back in September last year, will again be examined by the heads of state only at the forthcoming meeting. The emergence, alongside the secretariat in Minsk, of similar regional organs in Almaty is becoming a reality. Of course it is appropriate here to recall N. Nazarbaev's proposal on the creation of a Eurasian Union.

Initially N. Nazarbaev's proposals were seen only within the framework of the CIS. The states prepared for closer forms of economic and political integration were to create a Eurasian Union. In August it was stated that this integration was also possible within the framework of the Economic Cooperation Organization in which countries of the Near and Middle East (Turkey, Iran, Pakistan, Afghanistan) take part alongside a number of CIS states. But such plans are in the future. Today the reality is the Central Asian Union.

What are its prospects? Of course the Central Asian Union under certain conditions could merge with CIS organs. But its independent existence is also possible.

Primarily this is promoted by the presence of rich mineral and fuel and energy resources in the region and favorable climatic conditions for food production. Previously the region's economy was specialized in meeting the requirements of the union national economic complex for raw material and fuel. The processing sectors were insufficiently developed and the republics' requirements for foodstuffs and industrial goods were met mainly by imports from Russia. Actually it was the disruption of economic ties and the reduction of trade turnover with Russia which obliged the leadership of the Central Asian states to seek joint ways of resolving economic problems.

The priority avenues of regional cooperation include the buildup of the extraction and export of fuel, the production of export commodities like cotton and non-ferrous metals, the processing of agricultural raw material, the joint construction of a gas pipeline and an oil pipeline, railroads, and communications lines. . . . All this should help the three states to reach the markets of the Near and Middle East, Europe, and Asia. In two or three years the three could become independent of Russian supplies of petroleum products, ferrous metals, and construction materials. . . .

The formation of a common Central Asian market is also promoted by the active penetration of foreign capital into the economy of the region's states and by the restoration of traditional ties with Asia's Muslim countries. Foreign businessmen are attracted by the rich resources, the cheap manpower, preferential conditions for investments and the export of profits, enormous markets for the sale of commodities, and political stability.

The expansion of this alliance through the entry of other

states of the region is so far rather problematic. Turkmenistan objects to the formation of any coordinating organs in the CIS. Tajikistan is in the grips of a keen political conflict and so far intends to remain in the ruble zone.

As for Russia, its leadership advocates the consolidation of the CIS as a whole. At the same time, taking into account the reality of the emergence of the Central Asian Union, it is expedient to set up collaboration with the coordination institution of regional cooperation. Otherwise the Asian union will never become a Eurasian one.

## 6.74

### President of New Central Asian Bank Interviewed

Interview by Tleuzhan Esilbaev
*Pravda,* 27 September 1994
[FBIS Translation]

As people will know, the agreement founding the Central Asian Bank for Cooperation and Development was signed by the presidents of Kazakhstan, Kyrgyzstan, and Uzbekistan in Almaty at the start of July this year. It could become the prototype for a CIS international bank. Satybaldy Sazanov was recently confirmed as president of the bank by decision of the heads of those states.

[Esilbaev]: The idea of setting up a Central Asian Bank has been in the air for several years now. What, in your view, speeded up the practical implementation of this idea?

[Sazanov]: First, it is important to organize normal, multilateral, interstate settlements for trade and other transactions envisaged by government decisions. Second, it is important to provide credit for the three republics' strategic programs and finance facilities of regional significance.

There are a host of other questions concerning, for example, the convertibility of the national currencies—the Kazakh tenge, the Kyrgyz som, and the Uzbek sum. After all, the way things are now if someone from Almaty say, decides to go to Bishkek but doesn't have any Kyrgyz som, it's a problem to buy that currency in Kazakhstan. You have exactly the same picture in Tashkent, and with the tenge in Bishkek. Our bank intends to set up the exchange of national currencies. We will be opening branches in Tashkent and Bishkek in the very near future.

[Esilbaev]: Satybaldy Sazanovich, clearly the bank which you direct did not appear in a vacuum, did it?

[Sazanov]: Alas, we are starting from scratch. Our initial incorporation capital of $9 million is made up of contributions of equal amounts by the three founders—$3 million each from Kazakhstan, Kyrgyzstan, and Uzbekistan.

[Esilbaev]: It cannot be ruled out that Russia, a large part of whose territory is on the Asian continent, will become a founder member; the same applies to Turkey, Iran, Pakistan, Mongolia, and China.

## 6.75

### Central Asian Economic Integration Plan Complete

KAZTAG, 10 October 1994
[FBIS Translation]

A session of experts of Kazakhstan, Kyrgyzstan, and Uzbekistan has been held in Bishkek with the participation of representatives of the executive committee of the Interstate Council. It considered and discussed items on the agenda of the next session of the council of prime ministers of the three states to be held in Tashkent.

The economic integration program between Kazakhstan, Kyrgyzstan, and Uzbekistan was finalized. A new section has been included—"Geological Structure"—which sets out the creation of a coordination council with the aim of integrating the efforts of the geological services to study the Central Asia region.

Agreement was also reached on creating an interstate joint-stock company, "Ak-su," on the basis of the Aksu corn-processing combine and the Kyrgyz starch syrup works.

Kyrgyzstan will consider the question of its share of participation in financing construction of the second segment of the gas pipeline passing through its territory and the West Kazakhstan-Kumkol oil pipeline.

A mechanism and principles have been worked out for implementing joint plans in the area of the pharmaceutical industry and health care, and approval was given to a draft agreement on cooperation and interaction in the sphere of research into earthquakes and the forecasting of seismic danger.

The experts drew up proposals to supply the three republics with sugar as well as control and measuring devices for gas, water, and electricity. The execution of the measures outlined will provide the possibility to improve supplies of medicines to the population at more accessible prices, to reduce the cost of sugar production, ensure greater satisfaction of the demand for it, and reduce imports of this product.

Apart from that, draft documents were considered on measures to bring economic legislation more into line on concepts of use of energy sources and electricity, and on other issues.

## 6.76

**Akaev: Trilateral Central Asian Union Active**
Boris Mainaev
ITAR-TASS, 4 November 1994
[FBIS Translation]

"At long last, we came to have a mechanism for real economic unity. We have a united bank, customs association and, most importantly, we agreed on (joint) use of arterial roads," Akaev said.

"We should live and surmount the crisis together. There is no alternative to it. Nobody is waiting for us on the world market, all of its sectors are occupied," he added.

Akaev regretted the fact that other CIS member states "have not yet come to realize this." "One year ago we signed documents on founding a united bank, but it did not begin working until now," he complained and added that the idea of a united economic space was not realized so far, either.

The Kyrgyz president referred to the Interstate Economic Committee, established by the Commonwealth, as "extremely necessary and vitally important agency of economic integration" and emphasized that its activities would largely depend on the personality of its leader.

"If (the leader) is an active and respectable person, we shall get the business moving. In the opposite case, everything will remain at dead center," he emphasized.

## 6.77

**Prospect of Central Asian Integration Explored**
Kenes Akhmetov
*Ekspress-K,* 9 February 1995
[FBIS Translation]

*Editor's Note*: The transformation of the former Soviet Union into a new arrangement of interdependent, modernizing countries is bound to be a prolonged process. Whether the Central Asian Union "prevails" over other confederal concepts is almost irrelevant. What is to the point is that the Central Asian countries are experimenting with layers of bilateral and multilateral integration and planning frameworks, which will all contribute to the process of transformation within the region and the broader CIS. The following discusses economic progress in Central Asia, drawing the conclusion that little can happen without Russia. While this may be true in the current stage of transformation, it would be shortsighted to simply dismiss the regional formations and institutions being established, which might at some later date play an important role in CIS events.

Can the Central Asian Union become, with the passage of time, an alternative to the relatively incapable Commonwealth? Analysis shows that this version of events is impossible. Despite the integrative efforts of the leaders of the former Central Asian republics of the USSR, centripetal processes are building up in the region, and economic reality encourages unification, which is inconceivable without Russia.

In the autumn of last year, RK [Republic of Kazakhstan] Prime Minister Akezhan Kazhegeldin made the following statement: "We have our interests in Russia. And if there are forces in Russia that will not let us in the door, we must climb through the *fortochka* [small hinged window pane for ventilation]. It is our market."

That statement was made soon after President N. Nazarbaev spoke optimistically about Central Asian economic space, but a week earlier had signed in Istanbul a joint declaration with the leaders of five other Turkic states, in which there is mention of the reinforcement of "pan-Turkic" integration.

But Kazhegeldin failed to agree with N. Nazarbaev's views. He simply formulated a reality without taking consideration of which Central Asian countries might cease their existence without even having begun to integrate. Only Russia is today the guarantor of the region's security in the face of the threats of religious fundamentalism, tribalism, and the military and economic expansion of China. Also, Russia is the largest creditor and trade partner of Central Asia. And its influence is constantly growing.

For Kazakhstan its Central Asian partners play a considerably lesser role than Russia does. In 1994, Kazakhstan's exports to Russia constituted, according to intergovernmental agreements among the CIS member states, more than 60 percent of the total, and to the Central Asian countries, 20 percent.

Moreover, Russia's share in Kazakhstan noticeably increased precisely after the proclamation of independence, and Kazakhstan's exports to Central Asia during recent years has noticeably decreased.

The leaders of Central Asian integration, in addition, are not distinguished by punctuality in fulfilling reciprocal pledges. Kazakhstan fulfilled its shipments in accordance with intergovernmental agreements to Uzbekistan by 37 percent, and to Kyrgyzstan by 48 percent. In turn, Uzbekistan and Kyrgyzstan realized their pledges to Kazakhstan by 60–65 percent. For purposes of comparison: Turkmenistan, which stands all alone, fulfilled its pledges for shipments to Kazakhstan by 85 percent.

Kazakhstan, as the militarily strongest state in Central Asia, also rejected the idea of creating a defensive union on parity principles. That became obvious after the ratification by its parliament of the agreement on deploying Russia's strategic nuclear forces on the territory of Kazakhstan and

Table 6.1

**Export and Import Shipments** (in percent)

| | Exports | | | | Imports | | | |
|---|---|---|---|---|---|---|---|---|
| | 1991 | 1992 | 1993 | 1994 | 1991 | 1992 | 1993 | 1994 |
| Russia | 62.0 | 72.1 | 69.7 | 73.7 | 66.0 | 74.3 | 70.9 | 80.3 |
| Central Asia, total | 16.7 | 12.3 | 12.7 | 12.0 | 14.0 | 9.6 | 15.5 | 4.73 |
| including: | | | | | | | | |
| Kyrgyzstan | 3.5 | 2.4 | 2.3 | 3.3 | 4.1 | 2.7 | 1.3 | 0.2 |
| Tajikistan | 2.5 | 1.1 | 1.4 | 0.6 | 1.3 | 0.4 | 0.5 | 0.6 |
| Uzbekistan | 9.2 | 6.0 | 6.9 | 6.5 | 6.6 | 3.7 | 9.0 | 3.5 |
| Turkmenistan | 1.5 | 2.8 | 2.1 | 1.6 | 2.0 | 2.8 | 4.7 | 0.4 |

the signing of an intergovernmental agreement on the use of test ranges on our territory by the Russian military.

Kazakhstan has actually revealed its cards in the Central Asian game. Balancing on the junction of the leading geopolitical forces in the region, it is playing a very complicated game in attempting to coordinate its interests and resources with other people's. For Russia, Kazakhstan is becoming a reliable ally in the region: for China, a partner in preserving the calm in Sinkiang; and for Iran and the other Islamic countries, an outpost of Islam. For all of them simultaneously, Kazakhstan is at the same time a promising and capacious sales market. Uzbekistan and Kyrgyzstan are acting for Kazakhstan, to a greater degree, as competitors in the region.

The realities make the plan for creating a Central Asian Union, more than anything else, an object of theoretical analysis. This kind of attempt was undertaken by experts from Kazakhstan, Turkmenistan, and Kyrgyzstan.

The conclusions that were drawn by the scientists are not comforting. They establish the fractionalism among national markets in the Central Asian countries. Competition for the world market of goods, credit, and investments has intensified. The limitation of financial and technological resources has become obvious. The countries in the region are not yet capable of opposing the economic pressure exerted by the stronger states.

## 6.78

**Uzbek, Kyrgyz, Kazakh Presidents Sign Accords**

A. Kondrashov from the "Vesti" newscast
Moscow Television, 10 February 1995
[FBIS Translation]

A few hours after the end of the Almaty summit of Commonwealth heads of state, three Central Asian leaders decided to continue the meeting on a smaller scale.

Nursultan Nazarbaev, Askar Akaev, and Islam Karimov renewed their efforts to deepen the integration between the Asian states that was started back in 1991 immediately after the breakup of the Soviet Union. As a result of today's negotiations, the presidents of Kazakhstan, Kyrgyzstan, and Uzbekistan signed an agreement on an Interstate Council in which they themselves feature in the role of chief coordinators. A Council of Heads of Government and a Council of Foreign Ministers of the three republics were also set up. In conclusion, the leaders of Kazakhstan, Kyrgyzstan, and Uzbekistan set up the Central Asian Bank of Cooperation and Development.

In response to the question as to whether the Asian republics fear all this might be taken to separate action, bypassing Russia, the President of Kazakhstan Nursultan Nazarbaev, answered:

[Begin Nazarbaev recording; shown addressing news conference alongside other presidents] Russia and Belarus have signed accords, Russia and Kazakhstan have just done so, should everyone take offense at that? No one is becoming offended at all: On the contrary, I reckon that within the CIS, as the CIS Charter states, bilateral and regional alliances are not excluded. I believe, the three of us believe, that this strengthens the CIS. [End recording.]

## 6.79

**Central Asian Presidents Issue Communiqué on Cooperation**

*Narodnoe Slovo,* 15 April 1995
[FBIS Translation]

["Communiqué" on "regular" meeting of Kazakh President Nursultan Nazarbaev, Kyrgyz President Askar Akaev, and Uzbek President Islam Karimov on 14 April in Chimkent, Kazakhstan]

On 14 April, a regular meeting of the heads of three Central Asian states—Kazakh President Nursultan Nazarbaev, Kyrgyz President Askar Akaev, and Uzbek President Islam Karimov—took place in the city of Chimkent.

The heads of state exchanged opinions on a number of issues relating to the situation in the Central Asian region and relations between the three countries.

The presidents' negotiations again confirmed the unity of the positions of these states in further developing mutually beneficial relations. In the course of the meeting, concrete approaches were defined for the fuller use of the potential available for cooperation between the three countries and for the acceleration of the process of integration of the Central Asian states.

The presidents of Kazakhstan, Kyrgyzstan, and Uzbekistan considered the key aspects of the process of implementing the Treaty on Creating a Single Economic Space.

Concrete recommendations for further strengthening cooperation within the treaty framework were made.

The program for economic integration between the Republic of Kazakhstan, the Kyrgyz Republic, and the Republic of Uzbekistan up to the year 2000 was considered in detail and approved. Information from the executive committee of the Interstate Council of the Republic of Kazakhstan, the Kyrgyz Republic, and the Republic of Uzbekistan, and the Central Asian Bank for Cooperation and Development was heard.

The presidents of Kyrgyzstan and Uzbekistan emphasized their adherence to a single customs space and declared their readiness to join the customs union of the Russian Federation, the Republic of Belarus, and the Republic of Kazakhstan, which will become a new testimony to the Central Asian republics' aspirations to strengthen the processes of integration.

The heads of state considered the political processes under way in their countries in their broad historical context.

President Askar Akaev of Kyrgyzstan and President Islam Karimov of Uzbekistan believe that, in the prevailing socio-political conditions, the decision to hold a referendum on extending the term in office of President Nursultan Nazarbaev is the most optimal, and will serve to secure social stability, and ensure the consequent and dynamic implementation of economic transformations in the name of the prosperity and well-being of Kazakhstan's people.

The meeting of the heads of state confirmed once again the need to continue dialogue between the leaders of the three countries, and the long-term prospects for extending cooperation between Kazakhstan, Kyrgyzstan, and Uzbekistan in the context of CIS integration processes.

The heads of Kazakhstan, Kyrgyzstan, and Uzbekistan, having analyzed the development of processes in the region in the economic, political, and humanitarian spheres, stated that a definite legal basis and the political preconditions have been created for the formation of an Association of Central Asian States.

In order to secure stability in the region, the heads of the three states deem it expedient to establish a joint battalion of peacekeeping forces under the UN aegis, and noted with satisfaction that this initiative had received the backing of UN Secretary General Boutros-Ghali.

They ordered their heads of government and ministries of defense to introduce concrete proposals for its formation.

The presidents of Kazakhstan, Kyrgyzstan, and Uzbekistan decided to support and undertake the necessary efforts in the international arena to promote the Kazakh initiative to organize a permanent UN seminar in Tashkent on the problems of security and cooperation in Central Asia.

The parties expressed their intention in the future to carry out an active policy within the CIS framework, to encourage the strengthening of the Commonwealth with the aim of preserving regional stability in the region, and achieving mutual understanding and further developing international cooperation.

[Dated] Chimkent City, 14 April 1995

## 6.80

### Official Outlines Functions of Interstate Council

Interview by Tleuzhan Esilbaev
*Pravda,* 10 August 1995
[FBIS Translation]

The Executive Committee of the Interstate Council of the three countries is a fundamentally new interstate formation in the Central Asian region formed a year ago on the initiative of Presidents Nursultan Nazarbaev, Askar Akaev, and Islam Karimov. What does it represent and what goals does it pursue?

[Serik Primbetov]: The appearance of the Interstate Council and its Executive Committee was preceded by the decision of the leaders of the three republics to create a single economic space within the confines of Kazakhstan, Kyrgyzstan, and Uzbekistan. We are integrating on the basis of centuries-old ties of neighborliness, preserving national distinctiveness and sovereignty here. But our rapprochement under the new historical conditions does not in the least mean regional exclusiveness. The treaty on the creation of the single economic space is open to other countries of the CIS.

If, however, we are speaking about the principal functions of the Executive Committee, they are manifold. We organize meetings of the presidents and the heads of government and the foreign policy departments and experts on this matter or the other. In addition, we have been entrusted with the elaboration of substantiated drafts and decisions of a conceptual nature for consideration by the Interstate Council and also joint proposals on the rational interstate division of labor.

The staff of the Executive Committee, manned by experienced, highly professional employees, is performing a great deal of work on the creation of practical economic conditions for the mutual investment of capital, including the attraction of foreign investments and credit, in spheres of the economy of mutual interest.

[Esilbaev]: The leaders of the three neighbor republics have jointly signed a number of important documents. How do they work for integration?

[Primbetov]: Their basic propositions are geared to the long term inasmuch as integration is a lengthy process, not one that can be accomplished overnight or in a year. As the experience of the states which are members of the European Union shows, it took them forty to fifty years to get there.

The main thing today is that the leaders and the community of a majority of republics of the former USSR realize that before moving onto the world market, it is necessary to create a regional market. In my view, many people mistakenly understand the market as freedom of each and everything. Let the doors be opened wide to all who can make the effort, they say, and let everything come in and go out. But the experience of the developed countries testifies that a market may take shape only if there is strict state regulation.

The Executive Committee has published the digest "Brief Survey of the State of the Economy of the Participants." It presents the main socio-economic indicators of the development of our republics in 1991–94. It is not difficult upon familiarization with them to conclude that it is unrealistic to hope that we can manage without the support of and firm ties with our traditional partners. This is why each republic that is a part of the Interstate Council must with great perseverance seek the full-fledged work of the Economic Union as an integrated formation with permanent and stable ties making it possible to ensure the dependable cooperation of production and the development of trade and other forms of cooperation.

We have held approximately twenty meetings of experts. A draft program of economic integration up to the year 2000 has been drawn up on the basis of their results. This program, which has been approved by the premiers, includes more than sixty-nine projects in all the main branches of the economy—fuel and energy, iron and steel, mechanical engineering, the agro-industrial complexes, and chemical and light industry.

In addition, there is the task of examining priority investment projects. There are ten of them altogether. Primarily the stabilization and development of the raw material base of the Karatau phosphorite basin. After all, our trade and economic cooperation with the two neighbor countries is built on the fact that Uzbekistan supplies us with gas, and Kyrgyzstan, with electricity. It is these types of products that account for the lion's share of the commodity turnover. For this reason the phosphorite basin performs a big role both for Kazakhstan and for Uzbekistan and Kyrgyzstan, which use a large quantity of the Karatau raw material.

Unfortunately, the Karatau phosphorite basin has thus far been developed extremely unsatisfactorily. This problem requires the speediest solution by the efforts of both Kazakhstan and Uzbekistan, Kyrgyzstan, and Russia, more than forty of whose enterprises work on Karatau raw material.

Economies in energy resources could produce rapid returns. And meters and gauges are needed for this first and foremost. This is why the development of capacity for the production of instruments for recording the consumption of gas, water, and electric power is the second main investment project.

[Esilbaev]: What, in your view, is holding up the entry of Turkmenistan and Tajikistan into the Interstate Council?

[Primbetov]: Turkmenistan, as far as I know, has yet to express its intentions with regard to the treaty on a single economic space. Nor have we yet made it an official offer. I believe that Turkmenistan's position may at this time be regarded as being wait-and-see. It obviously wants to see what the results are. If Turkmenistan sees that the treaty is working, its affiliation to the single economic space will not be far off. As far as Tajikistan is concerned, it will become a full member of the Interstate Council, I believe, as soon as the domestic political situation in this republic normalizes. Especially since the emblem of the alliance of our states is a plane tree under the sun. And plane trees, as we know, have five branches.

We regard Central Asia as a single whole. And the whole civilized world understands Central Asia as being five states together. Were they all together, they could, naturally, perform a more appreciable role in the world community and in the world economy, since there is within the confines of our states everything necessary for their successful development. As long only as there is peace and harmony.

# 7

# The Debate on Economic Integration

## Introductory Notes

Chapter 7 documents key issues in the CIS debates on economic integration. In many ways, these economic issues represent the continuation, within a transformed political and economic context, of the 1990–91 debates (documented in Chapter 1) over the USSR's Economic Union. During that debate, the former republics could not agree on how far they wanted economic independence to go. With Ukraine as a catalyst, most of the former republics decided they wanted complete economic independence following the collapse of the Soviet Union. They wanted to sever all economic bonds with the former system and they did not want to replace former economic linkages with a Russian version of the same. Nevertheless, socio-economic problems associated with reform and significant troubles in each country's bilateral economic relations with Russia forced most CIS leaders to accept at least the concept of economic integration by the spring of 1993. Now the debate concerns not whether, but how far these leaders would like economic integration to go. As the documents collected here reveal, while the non-Russian CIS leaders support the principle of forming a single economic space, they still hesitate when it comes to assigning supranational authority to any CIS economic executive agency.

The first big economic issue on the CIS agenda was to redefine the status of the ruble zone. Although CIS leaders were intent upon establishing independent economies, an agreement had been signed at the 21 December 1991 Alma-Ata meeting to preserve the ruble as the single monetary unit on the territory of the CIS. Most CIS states (especially Belarus and the Central Asian states) were committed to retaining the ruble as the common CIS currency. Nevertheless, the Russian administration soon began pressing for supranational monetary policies (which it said were required for the ruble to remain the common unit of exchange), which convinced non-Russian leaders that their sovereignty was at stake.

Those monetary policies included a demand from Russia that a CIS Interstate Bank be created, which would be located in Moscow and would control all currency emissions as well as regulate all bank credits to enterprises or government budgets. An attempt was made in October 1992 to draft an Interstate Bank agreement, but only six states signed and it was doubtful whether even those six would be able to obtain ratification by their parliaments. Had the Interstate Bank agreement been implemented, it could easily have brought each CIS country financially to its knees. For one thing, it would have obligated each parliament to give up its authority over its Central Bank. For another, decision making in the CIS Bank would be assigned on the basis of each member's capitalization and economic potential, which meant that Russia would dominate the institution.

Guaranteed Russian domination was therefore the main issue that split the countries apart in this important area, where policy coordination could have actually aided with the early economic transformation process. When Ukraine introduced the karbovanets in November 1992, acting Russian Prime Minister Egor Gaydar reassured the other CIS countries that Ukraine's dropping out of the ruble zone would have little effect on their ruble rate. Both Russia and Ukraine agreed not to interfere with the exchange rate between the ruble and the karbovanets, but to allow the market to determine the rate. This encouraged other CIS leaders to consider introducing national currencies, and even Belarusian Prime Minister Kebich, in one of the documents in this collection, voiced his opinion that each state should make its own decision whether or not to introduce its own currency.

By the end of 1992, most CIS leaders had instituted steps to introduce their own national currencies. Three countries—Belarus, Kazakhstan and Uzbekistan—retained the ruble as their national currencies until late fall 1993, but Russia's recall of all pre-1993 rubles in July, and its reluctance to issue credits to the ruble-strapped CIS republics, forced even these stalwart supporters of economic integration out of the single economic space. It may have been that Russian economic officials decided that they also did not want to risk bankrupting their own state coffers by becoming the official commonwealth "treasury."

The second major issue to appear on the economic front

was that of Russia's use of its energy weapon to extort political and military concessions from the near abroad. After Ukraine introduced the karbovanets, President Yeltsin and Egor Gaydar both declared in December 1992 that Russia would deliver oil to non-ruble states only at world prices, and only in return for freely convertible hard currencies. This and the new payments settlement policy adopted by the Russian Central Bank for all state-owned enterprises caused Ukrainian enterprise debts to skyrocket to about 25 billion rubles. Russia convinced many in the West that it could not afford to "carry" the other former republics' energy debt, but Russian officials failed to put that debt in the context of the ruble shortage problem which Russia itself had caused. Whether economically justified or not, Russia's energy payments policy led to a catastrophic fall in oil deliveries to Ukraine, ending up in December at only about 5 percent of deliveries expected in 1992. After that, Russia did not hesitate to establish linkage between the other republics' energy debts and its own agenda at CIS summits.

Throughout the economic debate, articles have appeared in CIS media which have pointed out Russia's use of its economic prowess and leverage to blackmail the countries of the "near abroad" into accepting its policies and vision for the CIS. One such article appears here in the Russian periodical *Rabochaya Tribuna*, disclosing the way in which Russia uses the debts of its partners to extort concessions from them in economic and foreign policy (and other matters). The article also points to Russia's own responsibility in creating the ruble shortage in the "near abroad," and confirms Russia's hand in the development of interenterprise payments problems.

By May 1993 every CIS country was experiencing, for a number of reasons, almost complete economic collapse. The karbovanets collapsed, partly because Ukraine failed to institute effective economic reforms. Other states were caught in the whirlwind of civil war, or stagnation associated with the economic patterns established under Soviet colonial rule. The interenterprise payments problem had spiraled out of sight. When CIS heads of state met on May 14, the non-Russian leaders made an abrupt collective about-face in their stance on independence and signed a Declaration of Intent to form a CIS Economic Union. In an *Izvestiya* article contained here, Andrey Illarionov, leader of Egor Gaydar's analytical team, attributes this change of heart to the CIS states' need for Russian ruble credits. Indeed, the Russian administration had warned a few weeks before that it was planning to impose tough restrictions on the granting of state credits to the former republics' budgets, in accordance with intergovernmental agreements and within the Russian Federation's budget limits. Any outstanding debts of state-owned enterprises in the former republics were to be transformed into state debts owed by the governments to the Russian government. Without Russian subsidies, many of the other CIS enterprises would fall into bankruptcy.

By mid-1993 economic integration was seen as the only way out of the crisis in which each state found itself. In accordance with the 14 May 1993 Declaration of Intent, the CIS executive body, the Consultative and Coordinating Committee (CCC), was instructed to prepare the legal basis for a deepening of economic integration among the CIS states. Twenty-five legal documents were to be prepared by the end of July 1993, all to be negotiated and adopted on the "principle of consensus." The framers of the CIS Economic Union used the European Community as its model. The problem was that the CIS countries were rarely able to arrive at a compromise position in its debates which would facilitate a consensus.

Before the CCC could complete its work, the prime ministers of Belarus, Russia, and Ukraine met in Moscow on 3 July 1993 to adopt a "Resolution on Urgent Measures to Deepen Economic Integration," forming a preliminary economic alliance among the three Slav countries. This alliance became known as the "Tripartite Economic Alliance," although it never went very far and was not implemented. At the time, economists like the ones writing in *Nezavisimaya Gazeta* in the series of documents collected here (for example, S. Shatalin, L. Abalkin, and V. Bakatin) interpreted the resolution as a political reaction to the 4 January 1993 Tashkent meeting of Central Asian states (documented in Chapter 6) at which plans were drawn up for closer economic cooperation with Turkey, Iran, Pakistan, and Afghanistan. All of the points contained in the tripartite resolution were contained in the provisions of the CIS Treaty for the Creation of an Economic Union. However, the tripartite agreement resolved a number of issues that had been the subject of heated debate within the full CIS. The emotional chain of events put in motion by this joint declaration is well documented in this chapter. Many principal actors in the CIS administrative framework were astonished by the joint resolution. In the documents to follow, the CIS Executive Secretary, Ivan Korotchenya, who knew nothing until the agreement was announced in the press, voices his displeasure over the "joint Slav agreement."

An article from *Segodnya* (30 October 1993) discusses the problems with the formation of joint or "transnational" enterprises. This issue is significant because in 1995 Ukraine and the other CIS governments began to enter into Russian-inspired "financial and industrial groups" (FIGs). The article contained here reveals the problems that can be encountered under such plans, which are open to extensive manipulation by the partner who contributes the greatest share of capital.

In 1994, the trend toward melding the region's economies continued—in principle. At the Moscow CIS summit in April, the need was discussed for an Interstate Economic Committee (IEC), which would possess supranational managerial powers over the CIS national economies. An agreement on forming such a committee was drawn up, but ultimately watered down into a Commission for Economic Union, which would have mainly analytical and advisory responsibilities. This was probably because Ukraine and Turkmenistan objected to a body to which each republic would assign much of its authority over its own economic affairs.

The Belarus and Ukrainian elections of 10 July 1994, however, rejuvenated the idea of forming the IEC. Yeltsin was sure that the new leaders of these two Slavic nations would make commitments to the goals of economic integration, and decided to add more pressure to move forward. Meeting in mid-July, the CCC responded to a request from Boris Yeltsin to lay the legal foundations for an IEC, whose role would be to administer a customs union, a payments union, and perhaps even manage the fifty intersectoral cooperative bodies which had grown up over the short life of the CIS. Resolving the question of how the IEC and the CCC would work together, the CCC was transformed into the Board of the IEC.

Chapter 7 documents in explicit detail the evolution of the IEC debates, the fiercest ones being over the decision-making structure within the body and whether it should have the power to impose sanctions. In one of the documents in this series, the new CCC Executive Secretary, Aleksandr Shokhin, expresses his view that the future of CIS integration depends on the IEC. At the October 1994 CIS summit, which every CIS head of state attended, further plans for the IEC were laid. Despite Belarus's protests, it was decided that its head office would be set up in Moscow. One ominous note sounded during the summit, however, is discussed in the *Segodnya* article appearing here, and that was Yeltsin's suggestion that former Soviet Gosplan chief Baybakov be brought back to help run the agency.

In November, the IEC presidium met for the first time. In his message to the gathering, Yeltsin asked the body to draw up plans for CIS customs and payments unions, which would both create the conditions for making the transition to a CIS common market. Though the momentum toward a regional economic union seemed to be gathering, Russia may have overplayed its hand in dismissing the disagreements that were still being heard in other CIS capitals. Moreover, the new Ukrainian leader turned out to be a disappointment to Moscow. When Belarus and Russia triumphantly initialed a customs union agreement in December 1994 and asked Ukraine to join, Leonid Kuchma rejected the idea. Kuchma had seen how Russian customs officials stood side by side with Belarusian officials on Belarus's borders with Poland, Latvia, Lithuania, and Ukraine (though no Belarusian customs officials appeared on *Russia's* borders with other countries) and helped determine Belarus's import and export policies with third countries. Obviously, Kuchma had second thoughts when it came time for Ukraine to sign such an agreement.

In January 1995, realizing they stood little chance of drawing in Ukraine, a customs pact was signed among Belarus, Russia and Kazakhstan which envisaged unifying their customs and other trade regulations by mid-1995. According to economic analysts in Russia's Economics Ministry, however, this agreement is expected to be just another facade due to the wide disparities in these countries' economic interests, levels of political and economic reform, and disagreements over trade policy with third countries. Significant problems in bilateral economic relations between Russia and Kazakhstan are reviewed in the documents found in this chapter.

Notwithstanding his rejection of the customs union, Kuchma has sent delegations to the IEC presidium meetings. In March 1995, as documented in a Kiev Radio broadcast, the presidium's agenda was to draft an agreement on establishing an Interstate Currency Committee (ICC). Ukraine submitted a number of proposals on the draft agreement, mostly concerning methods of seceding from the ICC, which it said should be based on internationally recognized principles. At the IEC meeting, Tajikistan and Kyrgyzstan submitted applications to join the customs union formed by Russia, Belarus, and Kazakhstan. Ukraine conspicuously did not ask to join. In a press interview documented here, Ukrainian delegate Serhiy Osyka says: "Ukraine supports the idea of setting up the customs union in general but the distance to it should be covered not at a rocket-escape velocity, but at a normal pace." President Kuchma, in a 17 July 1995 visit to Minsk, confirmed Ukraine's position on this seminal issue, adding that a customs union should follow the signing of a free trade treaty between Russia and Ukraine.

Of particular note is a recent Russian initiative in economic integration at local levels within the CIS. A significant piece from *Kommersant-Daily* reports on a meeting in July of the local governors of twenty-seven oblasts in Russia, Belarus, and Ukraine. This first "intergovernors" conference was held under the direct auspices of the Foreign Minister Andrey Kozyrev in Novgorod; the second, reported on here, was held in Minsk. The participants attacked the Interstate Economic Committee as totally ineffective. They also called for joint investments in their oblasts and direct agreements between micro-economic entities to stimulate the process of integration at local levels. The significance of these meetings lies in their grass-roots approach to economic integration, and in their potential for forming a local groundswell in support of rapid CIS integration.

The economic integration movement within the CIS continues to ebb and swell. As can be observed from the press accounts selected for this chapter, the view now exists that integration is moving forward; at least most CIS leaders accept the principle of melding their economies into some form of economic union. There is also the impression that Moscow's influence over the pace of integration and the economic policies of the near abroad has increased. However, the Interstate Economic Committee has not yet become operative, and none of the twenty-five agreements comprising the legal foundations of an economic union has been fully accepted or implemented. The picture is complex, with activity at several levels and on many issues at once. In general, bilateral economic agreements are still preferred by the non-Russian CIS leaders as the means for socio-economic improvement. Other multilateral economic groupings, and banks, are also being formed within the former Soviet space, such as the Central Asian Development Bank.

It might be said that much of the recent integration momentum has been motivated by popular nostalgia for the Soviet Union and by the pragmatic desire of the "Red Directors" in Russia to shore up their new wealth and power. Essentially, notwithstanding the idle agreements, however, an intricate framework has been laid and the issue of CIS economic integration is still very much alive.

## 7.1 Further on Gosbank

Ivan Zhagel
*Izvestiya*, 27 December 1991
[FBIS Translation]

*Editor's Note:* The following document describes the situation immediately following the creation of the CIS, in which each new state suddenly found itself forced to apply to the Russian Central Bank (CBR) to obtain credits with which to meet its budget requirements. This meant trouble for both Russia and the other republics. Inflationary pressures on Russia in 1992 mounted to the point where they threatened its stabilization program. The CBR issued credits to other republics amounting to 10 percent of GDP in 1992, but total credits amounted to a staggering 40 percent of GDP. Banking and access to international credits within the CIS could be called the primary economic issue for former communist states in their social transformation to modern industrial, pluralist states.

It has become known that V. Gerashchenko, chairman of the country's Gosbank [State Bank], has resigned. His resignation has not yet been accepted, but this is obviously a matter of a few days or even hours. In any case, the liquidation commission authorized by the Russian parliament and the Central Bank is already working at the Gosbank.

Not many people know that the day before the meeting of the independent states' leaders in Alma-Ata, where, among other documents, an agreement to preserve the ruble as the single monetary unit throughout the territory of the CIS [Commonwealth of Independent States] was signed, the Russian Federation Supreme Soviet Presidium adopted a resolution on the practically immediate liquidation of the Gosbank and the transfer of its premises, documents, and specialists to the control of the Russian Central Bank.

Meanwhile, if the Commonwealth is to have a unified currency, then there must also be a unified supranational organ to regulate all questions of monetary policy. This alone can be the USSR Gosbank's legal heir. Other options could be regarded as an attempt on the rights of other CIS members.

Incidentally, I telephoned V. Matvienko, chairman of the board of the Ukrainian National Bank, and inquired whether he knew that the liquidation commission authorized by the Russian parliament had been working at the USSR Gosbank since 23 December. V. Matvienko replied that no one had agreed to such a step with the Ukraine. In his opinion, this could lead to a whole series of "piquant" situations. For example, there is now a catastrophic lack of ready money in Ukraine. Hitherto such questions have been resolved at the USSR Gosbank. If it is liquidated and not replaced by a jointly created supranational banking organ, then Ukraine and all the other republics will have to turn . . . to the Russian Central Bank for additional bank notes.

However, we should not be in a hurry to accuse Russia and its parliament of creating a "piquant situation." The question of creating a unified banking system and an interstate banking organ as the legal heir to the USSR Gosbank has already been placed on the agenda repeatedly. The Russian Central Bank has, as a rule, been the initiator of this. True, its proposals have not always seemed attractive to the others, but, instead of seeking ways to a compromise, the former republics have sent representatives not endowed with powers to discuss the draft. No wonder the question of the Interstate Bank has still not been resolved.

The absence of even a mention of an interstate banking organ in the documents signed in Alma-Ata seems a mistake which can and must be eliminated at the next meeting in Minsk. Otherwise the collapse of the economy cannot be halted.

All the independent states are now preparing their

budgets for next year. It is natural that under the present complex conditions these budgets will, as a rule, end up with a deficit. It will be possible to stop the financial holes only by means of credit emission. But who will be able to give permission for this and to decide the size of the increase in the money supply?

In principle there can be two options here. In the absence of an interstate organ the Russian Federation Central Bank will regulate credit emission. At the same time, however, all the other members of the CIS must virtually get their budgets approved . . . in the Russian parliament.

With the second option the central banks of each state will independently carry out credit emission. But this is nonsense! Such a path will lead not only to a rapid devaluation of the ruble but also to the erection of barriers on the borders which will cancel out all the Alma-Ata accords in the economic sphere.

In short, the question of the Interstate Bank is overdue. And it cannot be resolved unilaterally, even by liquidating the USSR Gosbank.

## 7.2

## Kazakhstan's Nazarbaev Favors Ruble Zone

Arkadiy Rotmistrovskiy
ITAR-TASS, 11 September 1992
[FBIS Translation]

Kazakhstan is for maintaining the ruble zone, a unified economic, credit, and monetary space, Kazakhstan's President Nursultan Nazarbaev said during a meeting with heads of a number of CIS countries' agrarian academies and other participants of a symposium on problems of the development of international scientific and economic relations in the system of agricultural industrial complex, now under way in Alma-Ata, the capital of the republic.

The prevailing situation in republics of the former Soviet Union, the continuing fall in production in each of them, and the economic crisis increasingly convince us about the necessity of restoring disrupted national economic ties, the president said.

Nursultan Nazarbaev supported the idea of the formation of a Council of Presidents of the Commonwealth's Academies of Agrarian Sciences. The president is of the view that the integration of scientific research will enable government bodies to find more rational forms and methods for a transition toward market relations, especially in such important spheres of reformation as decentralization of property, improvement in investment policy, and establishment of effective contacts with entrepreneurs of other countries.

## 7.3

## Central Banks Open Talks on Forming Ruble Zone

Interfax, 17 September 1992
[FBIS Translation]

During a two-day meeting which opened in Alma-Ata on Thursday [17 September], representatives of the governments and central banks of the former Soviet republics will formulate the basic provisions of the former currency union—"ruble zone."

One of the managers from the National Bank of Kazakhstan told "IF" [Interfax] that the documents prepared at the meeting will most likely be submitted for review by the heads of states at the Bishkek summit in early October.

The representatives of ten republics initially agreed to attend the meeting, including Russia, Belarus, Kazakhstan, Tajikistan, Kyrgyzstan, Uzbekistan, Georgia, Armenia, Lithuania, and Moldova. However, Aleksey Dudkin, deputy chairman of the Kazakh bank, said that Ukraine and Turkmenistan joined in the talks at the last minute. Representatives from Azerbaijan, Estonia, and Latvia are not participating in the meeting.

Managers from the Central Bank of Russia [CBR] said the bank will call for the creation of a single currency zone with the ruble as the main or reserve currency, if the nations decide not to form a banking union.

Deputy chairman of the CBR Aleksandr Khandruev told "IF" that the bank "advises setting up an interrepublic payment union, regardless of whether a banking union is organized." The leaders of Belarus and Kazakhstan said they plan to remain in the "ruble zone."

## 7.4

## Shokhin Cited on Interstate Bank, Foreign Debt

Interfax, 12 October 1992
[FBIS Translation]

Aleksandr Shokhin, vice premier of the Russian government, called the decision of the Council of the Commonwealth's heads of state in Bishkek on creation of an Interstate Bank of CIS a compromise. In his interview with Interfax he pointed out that at the first stage the bank "would be an accounting chamber." The need for such body is great, Shokhin said, as mutual payment accounts are not balanced in the economy of the CIS states and the mechanisms for their regulation are not established.

According to the vice premier, the issue of when the

Interstate Bank will become "a real bank" is "to be solved not today and even not tomorrow." Shokhin noted that the main task for today is to define the principles for bank's functioning. He believes that "it will be based not on democratic principles (one country—one vote) but on principles of a joint stock company—according to the capital."

Shokhin reported that at the recent negotiations in Kiev between the governmental delegations of Russia and Ukraine devoted to the problems of the former USSR's foreign debt, the parties stood on the positions that "the old system of common responsibility does not work." Russia proposed two variants for solving the debt problem.

According to Shokhin, the first variant envisages handing over to Russia all assets and liabilities on the foreign debt of the former USSR. The second foresees separate servicing of the foreign debt with a single subject controlling it, that is, the Russian Foreign Economic Bank.

Shokhin said that the Ukrainian party promised to work out its own position with respect to foreign debt before the session of the Paris Club scheduled for late October in the French capital.

## 7.5

**Ruble Zone Agreement Problems Viewed**

Ivan Zhagel

*Izvestiya,* 13 October 1992 [FBIS Translation]

The agreement signed in Bishkek on preserving the ruble zone on the territory of six CIS states—Russia, Kazakhstan, Uzbekistan, Belarus, Kyrgyzstan, and Armenia—inspires hope that at long last a solution to the poser of uncontrolled credit emission, at least by the twelve republic central banks, will be found.

Just prior to the meeting in Bishkek, V. Gerashchenko, acting chairman of the Central Bank of Russia, stated to an *Izvestiya* correspondent that he saw no possibility of stabilizing the ruble and the Russian economy as a whole without a coordinated credit and monetary and currency policy and without setting up an interstate bank, which would regulate all these questions. However, despite the fact that fundamental agreement between the six CIS countries already exists, the enormous complexities that will certainly be encountered en route to creating interstate banking structures must not be underestimated.

The most serious problem is that as of today all republic central banks are subordinate, or more precisely accountable, to their national parliaments, which must have the final say on this question. But this will not be so easy for them. At any rate, not one of them has made such a decision.

The prospects that an interstate bank will be set up mean that the parliaments will be forced to harmonize their budgets, or at least their budget deficits, with this suprastate structure. This means that the parliaments are being deprived of the chance to make up for their failings in implementing economic policy by credit emission. In other words, they will no longer have such an inexhaustible source of financial resources as their own tame central banks.

This problem will certainly surface during ratification of the Bishkek agreement by the republic parliaments. It can be expected that the magnificent six will be reduced to five, four, and so on. At any rate, the parliamentary debates could be quite prolonged, which means that the period of the ruble's instability will be dragged out.

There is one more serious problem that could affect the time taken to create unified banking structures. When the idea of an interstate bank was discussed a year ago, Russia stated that representation in this organ and the share of votes on its board must correspond to each republic's economic potential. And this, in general, is fair. That is the principle on whose basis most international financial organizations—for instance the IMF itself, in which the United States plays a leading role—were built.

However, at the time this position was sharply criticized by Ukraine. The point is that Russia's economic potential exceeds that of all the other former Union republics put together. Which means that they would be obliged to remain simple extras in the interstate bank leadership, deciding nothing and influencing nothing. That is, all the states in the ruble zone would effectively have to get their budget deficits approved in the Russian parliament.

An even bigger imbalance on the interstate bank's board could occur now, when Russia's economic potential is opposed by only five states. True, the Russian leadership and Russian bankers have repeatedly stated that they voluntarily agree to limit their influence on the interstate bank. But this also has its limits, about which lengthy debates could develop in the Russian parliament.

## 7.6

**Shokhin: Review of Foreign Debt Payments Needed**

Ivan Ivanov

ITAR-TASS, 27 October 1992

[FBIS Translation]

Russian wants to review the mechanism of foreign debt payments since it is the only former Soviet republic that is paying the debt now, according to Russian Deputy Prime Minister Aleksandr Shokhin.

"The mechanism of joint responsibility (for debt payments) is not working," he told reporters on Tuesday [27 October], adding that Russia wants to take over either the debt and assets of other former Soviet republics or the right to manage former Soviet debt and assets so that it becomes the only partner for foreign creditors.

Shokhin said [while] several CIS states have already transferred their foreign debts to Russia, others have not yet adopted the final decision. Ukraine wants to pay its share of the foreign debt itself, he added.

Russia remains interested in getting humanitarian aid from the West. A group of Russian envoys left for Tokyo on Tuesday where they would attend the conference on humanitarian and technical help to the Commonwealth. Shokhin said that their main task is to work out mechanisms of rendering aid to Russia and the rest of the Commonwealth.

The forum will be used by Russian envoys also for meetings with representatives of the World Bank and the European Bank for Reconstruction and Development at which they will discuss legal and technical issues of granting credits to Russia, according to Shokhin.

## 7.7

**Interstate Bank Accord Unlikely to Be Ready for CIS Summit**

From the "Presidential Bulletin" feature compiled by Andrey Pershin, Andrey Petrovskiy, and Vladimir Shishlin; and edited by Boris Grishchenko
Interfax, 29 October 1992
[FBIS Translation]

Head of the economic affairs department of the Russian State Committee for economic collaboration with the Commonwealth member states Adrian Budyanu considers that there is little likelihood that the draft agreement on the Interstate Bank will be ready in time for the next meeting of the heads of CIS states.

*IF* [Interfax] *Note*: A resolution was passed at the recent Bishkek summit by which all nations bound by the agreement on the ruble zone would draft, by 1 December, their proposals on the form that this bank would take and on a multilateral accounts system. Experts from the Commonwealth nations will meet in Minsk from 3–5 November to discuss these matters.

Budyanu told Interfax correspondent Boris Zvyeriyev that the Interstate Bank is seen in Russia as nothing other than an international accounts bank. Here he pointed out that Russia was in favor of the number of votes on the bank council being proportional to the amount of capital at the disposal of its members, but several states, he continued, were in favor of a "one-country-one-vote" system.

*IF Note:* Interfax has been informed that the Russian leadership is currently discussing sanctions against any of the CIS nations who failed to observe recent agreements concerning the coordination of the activities and protection of the interests of those states which belong to the ruble zone in the event a national currency is introduced by them. Such sanctions include halting the shipment of commodities, refusing to prop up the cash circulation of any one nation, prohibiting the opening of ruble accounts in Russian commercial banks by residents of any particular state, and halting energy supplies for rubles.

## 7.8

**Gaydar, Kebich Comment on New Ukrainian Currency**

Interfax, 13 November 1992
[FBIS Translation]

Commenting on Ukraine's introduction of its own monetary unit, the karbovanets, acting Russian Prime Minister Egor Gaydar said that Russia and Ukraine now need an urgent agreement on new mutual credits in new currency. Neither the Russian nor the Ukrainian government, he said, will interfere with the market exchange rate of the ruble to the karbovanets. He said the introduction of the karbovanets will have no disastrous consequences for any of the ruble zone states.

Belarusian Prime Minister Vyacheslav Kebich said, "We have practically no ruble zone; what we do have is a bilateral ruble crisis which has put an end to the ruble zone." He thinks it is each state's internal matter whether to introduce its own currency or not.

Ukrainian Prime Minister Leonid Kuchma said this is a measure all CIS countries will ultimately have to take.

## 7.9

**Impact on Interstate Bank Noted**

Ivan Zhagel
*Izvestiya,* 14 November 1992
[FBIS Translation]

It would be a mistake to think that the breakup of the single ruble area is beginning specifically with the Ukrainian president's decision. The real start of this process came on 1 July this year, when Russia imposed tough reciprocal payment conditions on all former Union republics, which restricted the flow of money across borders. As a result of uncoordinated credit issuances by national banks, the ruble in Russia began to differ in weight, so to speak, from the ruble in Ukraine, Belarus, Kazakhstan, and so on. So by introducing the karbovanets, Ukraine merely legitimized the fact that a number of national currencies have emerged.

The recent intergovernmental agreement in Bishkek, under which an interstate bank should be set up, does not guarantee the preservation of the single ruble area either. Incidentally, a conference of experts opens next Monday in Moscow; the experts will work on the practical implementation of the agreement. They conceive of an interstate bank not as an organ that unites the republics' banking systems, but merely as a structure that will enable economic ties to be preserved under conditions in which a number of national currencies exist.

This is the view of A. Khandruev, deputy chairman of the Central Bank of Russia: "It would be senseless to impose unrealistic tasks on the interstate bank right away, to establish it as a common issuance center controlling the monetary systems of all the republics. This is impossible, primarily for political reasons. Therefore, at the initial stage its main business should be organizing multilateral settlements between former Union republics."

Well, the mechanism for settlements between republics and their enterprises remains the same as before—via correspondent accounts [*korrespondentskiy schet*]. The only difference is that these accounts will now be opened not in national banks but in the interstate bank that is being set up. With time it is planned to extend this practice to commercial banks as well, which will open correspondent accounts with each other directly and thus link together enterprises from various republics. It is clear that this system envisages the existence of a number of national currencies.

It is true that the interstate bank's draft charter contains a point that envisages the function of managing the cash and credit issuance by central banks. This will only be possible, however, after appropriate decisions have been made by national parliaments, which will voluntarily renounce their right to control the activity of national banks and delegate this right to the interstate bank.

Incidentally, it would be good if one national parliament were to show some initiative on this issue and thus set an example to the rest; otherwise, the process of separating the monetary systems will continue. In the first place, this will prompt Russia to introduce its own Russian ruble, as the continuing chaos in the money supply impedes the implementation of further economic reforms. According to available information, these decisions are already being discussed in the Russian echelons of power.

## 7.10

**Kiev, Moscow Agree on New Rules for Banking, Economic Relations**

Interfax, 18 November 1992 [FBIS Translation]

In his exclusive interview with Interfax, Vadim Getman, the chairman of the National Bank of Ukraine, declared that at the negotiations in Moscow on 17–18 November, Russia and Ukraine worked out a joint position concerning the establishing of new correspondent relations and principles of accounts between the banks and economic subjects of two states. In Getman's words, the necessity of introduction of new principles of accounts of Ukraine with Russia and other countries of the ruble zone was raised after the introduction of the Ukrainian national currency, the karbovanets. Now a Ukrainian enterprise can procure goods in Russia only with the consideration of the available rubles (or hard currency) and vice versa. The head of the National Bank added that an analogous scheme would be used at the level of interstate accounts.

According to Getman, the key issue of the negotiations is the fact that Russia agreed to allow to its commercial banks to open correspondent accounts directly in the Ukrainian banks, by-passing the Central Bank of Russia.

Vadim Getman also reported that Russia and Ukraine agreed to complete the solution of all issues on interstate accounts and accounts between the enterprises before 1 March 1993.

Getman said that in the near future Ukraine and Russia agreed to make an inventory of payments of both countries' enterprises to each other. Getman pointed out that at present the debt of Russia to Ukrainian exporters constituted "several hundred billion rubles." Getman reported that at the Moscow negotiations the parties discussed an issue concerning the division of assets and liabilities of the former USSR State Bank. According to the chairman of the National Bank of Ukraine, the parties should receive appropriate shares of the assets and liabilities.

## 7.11

**Karbovanets Is a Lifesaver for Ukraine's Power Industry**

Sergey Leskov

*Izvestiya,* 2 December 1992 [*CD* Translation]

Talks with a Ukrainian government delegation headed by Vice Premier Yu. Ioffe were held at the Russian Ministry of Fuel and Power. The amounts and terms of oil deliveries from Russia were discussed at the talks.

After Ukraine left the ruble zone, first B. Yeltsin and then E. Gaydar declared that Russia would deliver oil to that state (as well as to others that take the ruble out of internal circulation) at world prices, with accounts to be settled solely in freely convertible currency. The situation was aggravated by the position of the Central Bank of Russia, which on 1 July changed the procedures for settling accounts with the former Union republics. As a result, Ukrainian enterprises' debts to Russian suppliers began to rise catastrophically and are now estimated at 25 billion rubles.

This whole tangle of problems led to the result that in November, oil deliveries from Russia to Ukraine dwindled to a thin trickle. At the end of the month the Lisichansk Oil Refinery, Europe's largest, was shut down. The Kremenchug and Odessa refineries are on the verge of shutting down. According to experts' estimates, Ukraine has enough reserves of fuel oil to last only a week. The impending energy famine is being felt especially keenly by motor transport enterprises, which have reduced the number of trips they make to a minimum, while gasoline is hardly being released at all to private individuals.

At the same time, specialists believe that Ukraine itself has helped aggravate the difficulties. The Ukrneftekhim [Ukrainian Petrochemical] state company, which was created at the beginning of the year and was supposed to set up joint enterprises with the Russian oil industry, has been accused of machinations involving fuel and has been liquidated. In the same decision, Yu. Ioffe banned deliveries of petroleum products from Ukraine to commercial structures in the CIS.

According to unofficial information, during the Moscow talks the Ukrainian delegation had only one weighty trump card in its hands—Russia's need of the new Ukrainian currency, the karbovanets, for upcoming trade operations. This trump card did not bring any particular gains, however, and the talks did not end with the signing of any documents.

However, as our staff correspondent N. Lisovenko reports, the news has spread in Kiev that Ukraine has been given assurances that two million tons of crude oil will be delivered before the end of the year. This has caused rejoicing at oil refineries, although the amount is only 5 percent of the deliveries that were expected in 1992. But in any case, the question of deliveries of petroleum products from Russia to Ukraine next year has not been studied at all and remains a big question mark.

## 7.12

### Republics Fail to Meet Export Commitments to Russia

I. Kirilina and E. Lyadeeva, Administration of Statistics and Material Resources
*Rossiyskaya Gazeta,* 10 January 1993
[FBIS Translation]

In 1992, despite the fall in the levels of production and the disintegration of the market, the geography of Russia's economic relations with countries of the former Union on the whole experienced no serious changes.

At the same time the reduction in levels of production changed the nature of the distribution of output between the states. The limited resources of the Commonwealth countries were primarily used for internal republic consumption. As a result exports began to decline, which made it impossible to fulfill mutual intergovernmental agreements. During the first nine months of 1992, Belarus, for example, delivered to Russia only 6 percent of the automobile fuel stipulated by annual agreements, only 2 percent of diesel oils, 28 percent of diesel fuel, 39 percent of truck tires, and 32–33 percent of bulldozers and scrapers. Kazakhstan shipped to us 55 percent of the oil, 65 percent of the gas, 65 percent of the coal, and only 5 percent of the grain; Kyrgyzstan delivered 63 percent of the trucks; Armenia 20–22 percent of the synthetic rubber and AC motors and 16 percent of the metal cutting machines; Moldova sent 47 percent of the wood-working machines; and Turkmenistan delivered 25 percent of the wool and less than 10 percent of the petroleum products stipulated by the annual agreements. Latvia delivered only 66 percent of anticipated minibuses.

## 7.13

### Ukrainian Deputy Prime Minister Discusses USSR Debt

Kiev Radio Ukraine World Service, 20 January 1993
[FBIS Translation]

Ihor Yukhnovskyy, Ukraine's first deputy prime minister, gave a news conference today at the Ukrainian Cabinet of Ministers. He informed journalists about issues regarding the foreign debts and assets of the former Soviet Union, and replied to the questions of those present.

He said Russia has thus far allocated to our state a sum to the value of $81 billion (for the debts). Ihor Yukhnovskyy reported that repayment of this sum may begin only after it has been analyzed in detail. According to Ihor Yukhnovskyy, Ukraine has so far checked the justification regarding $38 billion. As to the money the Soviet Union once lent to other countries, it has now been established that the debtor states have a fairly low payment capability.

The first deputy prime minister emphasized the importance of dividing fairly the foreign assets and liabilities between the former republics of the USSR and stressed the complexity of this process.

He also set forth his point of view on a number of other topical economic problems.

## 7.14

**Leaders to Sign Economic Union Declaration**
Ivan Ivanov
ITAR-TASS, 13 May 1993 [FBIS Translation]

Intensive efforts by a number of CIS countries, primarily Russia and Kazakhstan, toward deeper economic integration of CIS countries are expected to lead to the signing of a declaration on the establishment of an Economic Union at the CIS summit here on Friday [14 May].

The document will recognize the need to establish an Economic Union of CIS countries and indicate approaches for experts to be guided by when drawing up specific agreements about a mechanism for the creation of an economic community on a larger part of the territory of the former Soviet Union. Highly informed sources believe that "all heads of state will apparently put their signatures to such a declaration."

It is so far difficult to say for certain which specific provisions the declaration will contain: experts are yet to work on it, and the presidents themselves are, most likely, to introduce some adjustments to it at their meeting on Friday.

One can only suppose that, since Russia as of now is actually the locomotive of reforms among CIS countries, a Russian economic reform will be adopted as a basis for CIS countries' common strategy in an economic union, considering Russia's economic potential.

Russian experts on the economic cooperation of CIS countries believe that the formation of a customs union should become the first step toward the establishment of an economic one. This presupposes the lifting of customs duties and quantitative restrictions—quotas and licenses in trade among CIS states—and at the same time the fixing of a common tariff with regard to third countries. With this end in view, customs union member countries will have to agree to follow a common strategy and a common pace of the realization of economic reforms.

This also presupposes a consistent introduction of market-determined forms of economic management, the pursuance of a coordinated pricing policy, the harmonization of taxation systems (having in mind the creation of a single taxation system subsequently), the joint elaboration of an effective anti-monopoly regulation and a common foreign economic policy with regard to third world countries.

It must be taken into account that both a strategy to achieve the adequacy of economic reforms and the carrying out of foreign economic activities would require that the customs union member states to a certain extent give up part of their national sovereignty both in foreign economic relations and in purely national approaches to reforms.

Therefore, it can be supposed that if no effective political accords are reached among CIS states before these economic measures, the customs union idea may burst as a soap bubble.

The formation of an economic union which would effectively function in the interests of all member states would require a number of other serious stages.

The recreation of a single monetary system (the formation of a currency union) and the pursuance of a rigidly coordinated currency and financial policy are an indispensable condition for the creation of a common economic space.

This may be connected with CIS states' consent to create an interbank union and transfer the right of casting the deciding vote to the Central Bank of Russia on such matters as emission, amounts of credits, and the rate of interest on them.

For the Economic Union to function effectively, CIS countries are also to make efforts consistently to bring national legislations closer together and, most importantly, create supranational agencies for coordination and management and refer to them part of the functions of national managerial bodies.

Incidentally, the latter condition is expected to be implemented as early as Friday when the leaders of CIS states will sign documents on the formation of a CIS Consultative and Coordinating Committee and an Executive Secretariat.

Apart from these two items and the signing of a declaration on the establishment of an economic union, the tentative agenda of the Moscow summit also includes items concerning a program for the drafting of documents on further strengthening of the CIS and on permanent plenipotentiary representatives of CIS members at the CIS charter-provided agencies.

## 7.15

**Union Seen to Threaten Russia's Interests**
Mikhail Berger
*Izvestiya,* 18 May 1993 [FBIS Translation]

The conference of CIS heads of state that took place last week in Moscow culminated, as already reported, in the signing of a declaration on the formation of an economic union.

The fact that an agreement was reached at the highest political level to set up an economic union and general CIS coordinating organs such as the Consultative and Coordinating Committee can be considered the indisputable achievement of both Russian economic diplomacy and that of the former Union republics. From December 1991 until very recently, the leaders of the former Soviet republics were in no hurry to tie themselves to any kind of specific commitments in relation to their Commonwealth partners, seeing any supranational organs as a threat to their own economic independence.

So it is possible that, being practically sentenced to economic cooperation, the former fraternal republics have indeed come to understand the need to integrate.

However, even if the coordination mechanism does not have serious shortcomings, in no way does it mean that all sides will benefit equally from the formation of an economic union. Andrey Illarionov, leader of the Russian prime minister's analysis and forecasting group, surmised that what lies at the bottom of the sharply changed sentiments of the CIS states who have displayed their readiness to form an economic union is primarily Russia's first timid attempts to impose order in mutual settlements with its Commonwealth partners. It is above all a question of the so-called technical credits which are allocated to purchasers of Russian products in nearby foreign countries without any preconditions. In practice this means the virtually free transfer of Russian resources to neighboring republics' citizens and enterprises.

Last year technical credits reached 10 percent of Russia's gross domestic product [GDP]. The proportion of free aid given by the United States to other countries does not exceed 1 percent of GDP.

It is the danger that Russia will sharply cut subsidies to neighboring economies by reducing the amount of technical credits that has, in Illarionov's opinion, forced the move to very rapidly set up the economic union, which, he believes, could turn into an instrument reproducing Russia's donor role in relation to the Commonwealth states.

Illarionov, the Russian prime minister's chief economic analyst, says that what is worrisome is the fact that so far the Russian side has not, for all intents and purposes, made the Economic Union's formation conditional on any demands favoring its own interests.

## 7.16

**Karaganda Conference Aims at "Economic Commonwealth"**

Oleg Stefashin

*Izvestiya,* 1 June 1993 [FBIS Translation]

A meeting of heads of government and plenipotentiary representatives of the governments of CIS countries convened on the initiative of Kazakhstan President Nursultan Nazarbaev ended in Karaganda with the creation of an intergovernmental council for industry and also an interstate joint-stock company and corporation.

The participants in the conference, which was attended by delegations from ten states—Russia, Azerbaijan, Armenia, Belarus, Georgia, Kazakhstan, Kyrgyzstan, Moldova, Ukraine, and Uzbekistan—noted that the reason for holding it was the disintegrating processes that have become more marked of late and which are bleeding the sovereign states' economies dry as well as giving rise to additional difficulties in their mutual relations. Attempts to establish an economic commonwealth on a fundamentally new basis have certainly been made before, but these have not led to the desired result. In the opinion of Kazakhstan President N. Nazarbaev, who spoke at the conference, this is primarily due to the lack of fully functioning CIS coordinating organs, and also to the lack of proven mechanisms for implementing the reforms that have been adopted and effectively monitoring their execution.

It is intended that the establishment of the intergovernmental council for industry will go some way toward finding a solution to this problem.

The conference also saw the adoption of a decision to establish the "Karatau" interstate joint-stock company and an interstate coal and steel corporation and the signing of agreements on basic principles governing cooperation in the military-industrial complex and machine building.

## 7.17

**CIS Economic Union Faces "Serious Test"**

Artur Vardanyan

*Kommersant-Daily,* 17 June 1993

[FBIS Translation]

The idea of an Economic Union of CIS countries is facing its first serious test. Yesterday, in Moscow, the leaders of the Russian government held bilateral talks with representatives of Kazakhstan and Belarus, which focused on credit and payment settlements. No specific solutions were found, but the sides are intent on continuing the search. Their efforts may result in the preservation of the ruble zone or may trigger the launch of national currencies in Belarus and Kazakhstan.

If yesterday's talks between Russian Deputy Prime Minister Aleksandr Shokhin and Kazakhstani Prime Minister Sergey Tereshchenko were preplanned, the visit to the Russian capital by Vladimir Zalomay, Belarusian state secretary for relations with the CIS, was prompted by a telegram in which Moscow asked Minsk to make it clear by 1 July whether Belarus will stay in the ruble zone or whether it will introduce its own currency. In effect, both talks were held to discuss one subject: credits and their settlements. This issue is of importance for Russia's relations with all the CIS countries, but it holds particular interest with regard to Belarus and Kazakhstan because these are the countries that, judging from the pronouncements by their top leaders, are to make up the "core" of the CIS economic union on the basis of the ruble zone.

The talks dwelled on matters of the conversion of technical credits into government ones and Russia's granting fresh credits to Belarus and Kazakhstan in 1993. Meanwhile, Russia estimates Kazakhstan's debt for these credits at 300 billion rubles [R] and that of Belarus at $267 billion. The debtors, for their part, offer estimates that are 10–30 percent lower. An even greater gap is evident in the approach to new credits: Almaty has requested an astronomical amount of R4 trillion, and Minsk is willing to put up with R300 to 500 billion. At the same time, both Belarus and Kazakhstan, while technically remaining in the ruble zone, insist that they be given a ten-year grace period for repayment of the credits.

New ruble credits are the centerpiece of the debate. Yesterday, this issue remained unresolved. Probably during the next round of talks a radical solution may be found: Russia would meet its partners' credit requests to such an extent that they would be faced with the need to introduce their own currencies. Russia pursues similar policies with regard to the former republics of the USSR—which are not regarded as the "core" of the Economic Union—the basic characteristic of which is the retention of the ruble zone. A compromise solution, however, seems more probable, which, strictly speaking, serves Russia's interests to a lesser extent: The credits would be given in the requested amounts on condition that the republics stick with the ruble zone, and the decision on the introduction of national currencies in Belarus and Kazakhstan would be postponed.

While the Russian-Belarusian talks were limited to the subject of credits, the Russian-Kazakhstani meeting got around to considering the oil issue as well. As a result, the fuel and energy ministers, Yuriy Shafranik and Kadyr Baykenov, initialed an agreement setting up a Russian-Kazakhstani company to deal with the prospecting, production, and transportation of Kazakhstani oil. It was also tentatively agreed that the agreement would set quotas for Russian oil exports to Kazakhstan for the second half of 1993.

## 7.18

**Importance of Economic Cooperation Viewed**

Aleksandr Golts

*Krasnaya Zvezda,* 19 June 1993

[FBIS Translation]

People grow accustomed to anything. Thus, we, the inhabitants of the former Soviet Union, seem to have gotten accustomed to living in conditions of crisis, be they political or economic crises. Who could be amazed by multiple price hikes, huge demonstrations, or mass strikes? But the events in Ukraine this week do, in my view, merit very close attention. Not only owing to their scale, although the strikes and protest actions have encompassed virtually the whole southeast of the republic. And not even owing to the political consequences, although the strikes have forced the people's representatives to agree to hold a referendum on confidence in the president and the parliament.

In my view, this crisis has highlighted an extremely curious trend, which may, in the future, be crucial in the so-called post-Soviet area. For all that the economic position of all the states in this area leaves a great deal to be desired, Kiev, in the Soviet tradition, has managed to cause itself major additional difficulties. I am referring to the attempt to put the idea of total and absolute independence into practice. It is probably possible, although far from always rational, to achieve this in politics or defense. But in the current circumstances the desire to secure economic independence is comparable to trying to be independent of the law of gravity, with all the attendant consequences or, more precisely, the consequences that fall on the heads of the supporters of "independence."

Remember that it is Kiev that has opposed—resolutely and extremely consistently—any attempts to coordinate economic efforts within the CIS framework. Ukraine blocked the creation of common organs, which it constantly saw as the revival of the notorious "center." It was Kiev that carried on an utterly pointless dispute over its share of the USSR's foreign assets, preventing our foreign creditors from adopting the decision to reschedule debt repayments. And Ukraine was more zealous than the others in erecting customs and other barriers in the way of normal commodity turnover between the republics. Trade and economic links that had taken not years or decades but centuries to create were blocked. The result is obvious: economic collapse.

I am by no means writing this in order to reproach and accuse anyone in particular. Russia, too, has plenty of people willing to destroy and smash what we have in the name of their own, often speculative, idea. I am writing this to, once again, emphasize that the idea of a "sovereign economy," an idea which is being implemented most consistently in Ukraine but which has affected all the new states (and is still exciting certain Russian regional leaders, who are now campaigning for oblast sovereignty), has not stood the test of practice. It turns out that you cannot escape from the crisis on your own.

A whole series of objective factors precludes this. To wit, the approximately identical level of the CIS states' economic development, the fact that they are tied to specific sources of raw materials and subassemblies, and their obvious inability to sell competitive goods on the oversaturated world market in the near future all make it necessary to seek means of integration rather than try to run as far as possible from one another.

For the time being the fear of once again ending up in a totalitarian empire has prevented us from being aware of this. The euphoria over total sovereignty has also gotten in the way. But now that the republics have laid the foundations of their own statehood, something else is becoming increasingly obvious. Namely that it is not collaboration but the lack of it that poses a threat to this statehood.

Significantly, this has been realized by those who encounter the daily hardships of life nowadays. They were more perspicacious and rational than some politicians. Events in Ukraine have shown that jingoistic statements to the effect that Russia is mainly to blame for the crisis do not wash. On the contrary, demands to join the CIS Economic Union and the banking, customs, and payments unions have taken an important place among the political demands. Now may be the first time that there has been such a clear call for reintegration (without, of course, encroaching on state independence).

Leaders can no longer ignore these calls. It is indicative that at the moment the situation reached critical level, the Ukrainian president did not run the risk of postponing the meeting with his Russian counterpart. What is more, if this meeting has not brought concrete solutions to contentious problems, it has at least indicated the ways by which a solution to the deadlock must be sought. And the free trade agreement may be called a step in the right direction.

It goes without saying that it is too early today to say that this approach has won over the Ukraine. But let us not forget that it has been Kiev's policy throughout the CIS's existence that has been the main factor delaying the pooling of Commonwealth economic efforts. And if the development of the political process causes this policy to be reviewed, the CIS may find itself with fresh prospects.

I would remind you that one of the most rational leaders, Kazakhstani President Nursultan Nazarbaev, is drafting a very interesting integration concept. He is proposing, in particular, to develop collaboration as West European countries did in the past by cooperating in certain sectors and formulating a common policy for the production and sale of certain kinds of output.

At any rate the CIS states seem to be starting to emerge from the period of estrangement. I would hope that this trend will grow stronger and gain momentum.

## 7.19
## [Republics] Sign Statement on Economic Integration

ITAR-TASS, 10 July 1993 [FBIS Translation]

The text of the statement of the governments of the Republic of Belarus, the Russian Federation, and Ukraine on urgent measures to deepen economic integration is as follows:

Taking account of the historic closeness of the peoples of the Republic of Belarus, the Russian Federation, and Ukraine, the contiguity of their territories, the closeness of the level of development, and the mutual links between their economies, the governments of the three states have decided to implement urgent measures for closer economic integration.

The interests of the states dictate the necessity of preserving a single economic area where in conditions of market relations production will effectively develop; goods, services, and capital will move freely; and efforts will be united for implementing joint economic projects. While expressing the firm intention of participating in the development and implementation of the treaty on the creation of a economic union within the framework of the CIS, the governments expect that the measures for a closer trilateral integration of their economies will aid in speeding up the implementation of the goals and principles of this treaty.

The governments believe immediate measures with respect to economic integration must cover the process of production, investments, foreign trade, and also financial credit, currency, and social relations. These measures are aimed at halting the breakdown in traditional production links and aiding the dynamic and harmonic development of the economies of the three states and the carrying out of economic reforms in the interests of raising the population's standard of living.

Such integration is conditional on the compatibility of their economic systems. To ensure this, close cooperation will be organized in the implementation of economic reforms and there will be coordination of their aims and of the stages and speed of their implementation. It is planned to draw their national economic legislation closer together, primarily in order to create the most favorable conditions for the work of enterprises.

The governments agreed to make it their priority to remove tariff and non-tariff restrictions in trade-economic relations between the three states, to set up a customs union and common customs territory, and also to form a common market of commodities, services, and capital on the basis of a single policy in the field of price-making and investments and in harmonizing the taxation systems.

Special importance is attached to the functioning of the monetary system on the basis of a rigid policy consensus in the field of credit and cash emission. It is planned to coordinate the budget policy by establishing a single limit for deficits of consolidated budgets. An active joint policy will be pursued in the social sphere. At the same time, citizens will receive an opportunity to freely possess property, move around, and live and work on the territory of any of the three states. At the same time, the governments particularly stress that they will strictly protect the legal interests of their fellow countrymen who live outside their states.

The governments note that close integration of their economies will demand joint management of many areas of economic activity. The essential coordinating bodies will be set up for this purpose and allocated the appropriate management functions. It is planned that their activity will include decision-making processes as practiced by international organizations.

The governments proceed from the fact that economic integration cannot be achieved in isolation, without a wider interaction in all aspects of politics, defense, and legislation.

In this regard, they call upon the parliamentarians of the three states to consolidate their efforts in pursuing a coordinated foreign policy, based on adherence to international legal norms, the goals and principles of the UN Charter, and the obligations to maintain territorial integrity of states, which they have undertaken within the framework of the Conference for Security and Cooperation in Europe, as well as their efforts in bringing national legislations as close together as possible and in this way creating a firm legal foundation for deeper mutual economic integration.

The governments are convinced that close cooperation on the basis of the principles set out in the current statement will be able to settle the fundamental issues of our mutual relations and allow us to leave behind the practice of putting forward unilateral complaints.

The governments have charged V.A. Zalomay, deputy chairman of the council of ministers of the Republic of Belarus; A.N. Shokhin, deputy chairman of the council of ministers and government of the Russian Federation; and V.I. Landyk, deputy prime minister of Ukraine, with organizing practical work on preparing by 1 September a draft treaty between the Republic of Belarus, the Russian Federation, and Ukraine on deepening economic integration and bringing into effect the tenets of the current statement.

The governments declare that states sharing the aims and principles of economic integration set forth in the present statement—and not belonging to any other economic unions or associations, if such participation could violate the integrity of the customs borders of the contracting sides, hamper the pursuance of a coordinated foreign economic policy with regard to third states, or in another way infringe the legitimate economic interests of the sides—may accede to this treaty.

[Signed] Chairman of the Council of Ministers of the Republic of Belarus: V. Kebich
Chairman of the Council of Ministers of the Government of the Russian Federation: V. Chernomyrdin
Prime Minister of Ukraine: L. Kuchma

## 7.20

**Kebich Says Union Open to "All the CIS"**
Ivan Ivanov
ITAR-TASS, 10 July 1993
[FBIS Translation]

Moscow, 3 July—[date as received] Today's meeting at the residence of the Russian Government near Moscow of Russian Prime Minister Viktor Chernomyrdin, Belarusian Prime Minister Vyacheslav Kebich, and Ukrainian Prime Minister Leonid Kuchma ended with the adoption of a resolution on the formation of an economic alliance by the three countries. They signed the statement drafted by the governments of the three countries on urgent measures aimed at deepening economic integration.

Responding to questions from journalists after the signing ceremony, Chernomyrdin said that "the signing of the treaty on deepening economic integration will be a history-making event for our states. We are moving forward and making our respective economies into an integrated whole in all spheres." Kebich stressed the importance of the decisions taken today and said: "Common sense has finally got the upper hand." The Belarusian prime minister especially mentioned the fact that today's meeting and its results will not lead to some separatism within the CIS framework. He said that "the economic alliance is open to all the CIS member states who wish to join it, but on definite terms." The Russian Deputy Prime Minister Aleksandr Shokhin explained that the economic alliance would be open to all the CIS member states, which do not intend to join other economic alliances.

Leonid Kuchma described today's meeting of the three prime ministers as a "natural process."

After the meeting, Russian-Ukrainian agreements on cooperation in oil and gas production were signed. According to Russian Minister of Fuel and Energy Yuriy Shafranik, the agreements include a mechanism of smooth transition to trade in energy carried out at world prices.

## 7.21

**Hastens Central Asian "Isolation"**
Artur Vardanyan
*Kommersant-Daily,* 13 July 1993
[FBIS Translation], Excerpts

The most notable thing about the meeting near Moscow [of Belarus, Russia, and Ukraine heads of government in Gorky near Moscow on 10 July] was the absence of representatives of Kazakhstan. According to unofficial information from Russian and Kazakhstani government circles, the reason was economic differences between the two states. The stumbling block in relations between the republics is property rights in the Baykonur space complex. The crisis of credit and financial relations has also worsened. It consists of the clash between Russia and Kazakhstan on debt repayment questions. The essence of the differences is that Russia is demanding that Kazakhstan should return the credits granted in 1992 and 1993.

The approved 1992 debt amounts to 240 billion rubles. As for 1993 debt, which is nearly the same size, Kazakhstan refuses to pay it, proposing a mutual retirement of debts instead. [Passage omitted.]

One way or another, the statement "on urgent measures to deepen economic integration" is a statement by three countries. This document has gone through a curious transformation in comparison with the Declaration of CIS Heads of State on an Economic Union adopted in Moscow on 14 May. Whereas in May the "main issue" consisted of ensuring "mutual interests of the Commonwealth member states," two months later the governments of the three republics have come to "believe that measures to ensure a closer trilateral integration of their economies will promote a faster implementation of the goals and principles of the treaty on the establishment of an economic union." The return to the idea of a three-component nucleus of an economic union formed by a different set of republics may be assessed in various ways. The good thing about it is Ukraine's changed stance. This gain, however, is more of a tactical nature given the fact that relations between Moscow and Kiev are hard to predict. Clearly, the bad thing about it is that it is once again (like in Belovezh Forest) damaging to the integration interests of the Asian republics, primarily those of Kazakhstan. This loss may very well prove to be strategic, and an economic union would become—even before having been born—a bargaining chip between a union of Slavic republics and a union of Central Asia whose isolation is being hastened by the statement of the three.

## 7.22

### CIS Executive Secretary on Economic Union

Ivan Korotchenya
*Nezavisimaya Gazeta,* 14 July 1993
[FBIS Translation]

Two reports that appeared in the mass media within a period of a couple of days could not but rivet the attention of the public. One report was from Istanbul and the other from Moscow. Both dealt with burning issues of economic integration among the former USSR republics. Both questioned the results of CIS agreements on the creation of an economic union, signed by nine leaders of Commonwealth states.

Let me briefly recall the results of the Istanbul conference of the leaders of the states that are members of the Organization for Economic Cooperation [OEC]. As it was reported, along with Turkey, Iran, Pakistan, and Afghanistan, former Soviet republics of Azerbaijan, Kazakhstan, Kyrgyzstan, Uzbekistan, Tajikistan, and Turkmenistan attended. A unanimous declaration was adopted at the conference, in which the OEC's main goal was depicted as safeguarding the social and economic well being of its members. The final document noted that the participants at the conference agreed to enhance the role of free enterprise in order to boost the process of integrating the OEC states with the world economy.

Projects approved in Istanbul include those on the development of transport infrastructure and telephone communications, cooperation in science and technology, and the removal of customs barriers hampering trade. Construction of roads was called a priority task aimed at granting access to the sea for countries that have no sea borders. Other priorities include the creation of a bank for development and trade, an institute of culture, the introduction of most favored nation status for OEC members, and granting free flow of capital and workers.

The participants at the conference agreed to open joint OEC coordinative organs—an airline and a shipping company in Iran, an insurance company in Pakistan, and a bank for trade and development in Turkey.

While assessing the results of the discussion in Istanbul and the documents signed there as "a firm foundation for the creation of another large economic union," Kazakh President Nazarbaev and Kyrgyz President Akaev pointed out the importance of "mutually advantageous cooperation between the OEC and the CIS for the benefit of all nations" and that the experience of interstate relations within the CIS may be useful for the OEC.

What happened in Moscow? The prime ministers of three Slav states gathered there, conferred for a while, as they did a year and a half ago in the Belovezh Forest, and decided to adopt a joint declaration. The declaration does not offer any new ideas, unusual approaches, or unexpected proposals. The declaration contains all the provisions of the Treaty on the Creation of an Economic Union. While reading the text of this document one may notice that the prime ministers have managed to resolve a number of problems that were the subject of heated debates, including the issues in which participants have advocated their specific stance, in particular the issue of creating coordinating organs.

It is very relevant that the prime ministers acknowledged that the adequacy of economic systems is a necessary condition for integration. This means lifting tariff and non-tariff barriers in trade and economic relations, the creation of a customs union with a single customs space, and in the future, the creation of a common market of goods, services, and capital. It is symptomatic that an epithet "strict" has been used for the first time in a CIS document with reference to the most important and delicate sphere—that of financial relations. The agreement envisions strict coordination of policy in the spheres of issuing loans and printing money, including the establishment of a standardized budget deficit ceiling.

Numerous adversaries of Belarus joining the collective security treaty and advocates of Ukrainian sovereignty were surprised at the unanimous acknowledgement by the prime ministers that "economic integration cannot be achieve in isolation, without broad comprehensive cooperation in the fields of politics, defense, and legislation."

Does this mean that the Economic Union of nine states has been rejected and that the idea on its creation has repeated the fate of many useful but unimplemented CIS projects? Is the existence of the CIS not threatened by the fact that its members overtly grouped in accordance with their Muslim or Slav origin? I am not going to make any forecasts, because forecasts are an ungrateful thing in big politics. However, as a person involved in many events in the Commonwealth, I am not indifferent to its fate and I do not quite understand the haste demonstrated by the prime ministers and their unmasked striving to counter the Istanbul move of the Asian republics with an equally strong move of the Slav states. Would it not be better to gather at a common table and resolve all the problems of the day, like normal families do? Today, it is hard to say what has prevailed in Moscow—concern over the economy paralyzed by the crisis or regular political calculations. I have no doubt that time will bring answers to these issues.

Arguments used by the signatories to the "tripartite union" in order to justify this determined step seem unconvincing to me. They said that there was no chance to create an economic union of nine states, because CIS members have been lacking in accord and mutual understanding and there was no hope left for the situation to change in the near future. However, the prime ministers must have known that virtually all the disputable issues had been resolved at the 1 July Consultative and Coordinating Committee session. The document was sent to CIS governments for final signature before the presidential summit.

Now, let us see what is the authors' vision of the legal basis of a broad union, its "pillars" and "roof," compared to the construction suggested by the three prime ministers.

Speaking in terms of official documents, both the small and the big unions see their aim as creating conditions for the stable development of the economies of their members and, in the future, creating a single economic space on the basis of market relations. This means the free flow of goods, services, capital, and workers, a single customs space, and a coordinated policy with regard to loans, budget, prices, and foreign economic relations. At first glance, this means a return to the previous state of things. However, this is a purely superficial resemblance—the new consolidation is supposed to be made on a strictly voluntary and mutually advantageous market basis.

The authors of the project distinguish several stages of building a new union. Each stage will move their states from simple forms of cooperation to more elaborate forms. First, through interstate association of free trade, the abolition of quotas, taxes, and licenses, through the introduction of single tariffs, the gradual formation of a customs union is envisioned. Lifting the barriers hampering the development of interaction among the producers will lead to the formation of a common market. Finally, when a full-scale union is formed with a single market of goods, services, and capital, a single monetary and banking system will be created and facilitate implementation of a common economic policy.

Each stage depends on a number of factors. For example, a free trade zone cannot be introduced without coordinating and standardizing the prices, because otherwise, goods from "cheap" areas will flow to more "expensive" areas regardless of all the borders.

A customs union will not be possible without a coordinated policy toward third states, and a common market will not exist without a joint investment and social policy. However, these issues cannot be resolved by even the most organized voting and by the adoption of separate legal acts. These things demand time and patient and diligent work. Only these conditions can grant vitality to any union.

The draft "treaty of nine" envisages a package of measures whose implementation will be provided for in separate agreements. In the experts' opinion, the package of accompanying documents should consist of fifteen agreements regulating relations between the states in the sphere of mutual interests. These agreements will, in fact, be the legal mechanism for the fulfillment of the honorable ideas and principles of the union. Such a package cannot be omitted also during the realization of the "treaty of three."

Some documents are inevitable in order to build the structure of an economic union. I will name some of them. These are agreements on common conditions and mechanisms of supporting cooperation in the sphere of production, development of direct ties between enterprises, standardizing tariff and non-tariff measures in the framework of free trade, and coordinating the policy of prices and taxes. It is not a secret that one of the most complicated problems of the CIS is that of regulating credit, financial, and convertible currency relations. No matter what variant of a common economic space will be chosen, these problems will have to be resolved.

Both the Istanbul meeting and the "Moscow weekend" of the prime ministers have demonstrated all over again that the independent states resemble boats that once hurried away from an old big ship and then, having enjoyed all the pleasures of lonesome sailing, are now trying to get together again. The recent meetings have demonstrated that even the most ardent advocates of "virgin independence" are now ready to give it up and pass certain functions to transnational structures. Life has proved that independence and sovereignty have little in common with political and economic isolation. This is even more true with regard to former union republics that have shared a common fate for decades. Their attempts to reach the desired shore of plenty and progress all by themselves, in most cases, have proved to be futile. Judge for yourself. According to official statistics, the volume of industrial production in the CIS has decreased by 18 percent in 1992, while in the sphere of capital construction the figure is 45 percent. This process was more painful in Armenia, Kyrgyzstan, and Tajikistan. The first half of 1993 has brought another 16 percent decline in industrial production. The volume of foreign trade between the CIS members and the third states has dropped by 22 percent, consumer goods production has gone down by 10 percent. At the same time, retail prices have increased ninefold and the number of unemployed has reached some one million people.

In experts' opinion, one of the main reasons behind the developing paralysis of the economy of newly independent states is the destruction of former economic ties. The restoration of these ties on a new mutually advantageous basis has become the number one task for every state. As Ukrainian Prime Minister Kuchma once said: "A year of our states' independence has demonstrated their total mutual dependence."

In this light, I would like to draw your attention to one aspect. The opponents of unions and broad integration are placing an emphasis on bilateral relations. Afraid of the revival of the former command center, they consider bilateral relations a sufficient and the only possible form of cooperation among the former USSR republics. However, the two years that have passed after the divorce have persuasively proved that criticizing the drawbacks of the former system is much easier than the practical establishment of new economic relations. Independent states have found themselves in a paradoxical situation, when the single center ceased to exist and, without the conductor, even duets started to play out of tune. The reason is plain to see—bilateral relations cannot solve the problems of our closely intertwined and mutually dependent economic systems. We need full-scale integration that would stipulate not only trade on favorable terms but also broad cooperation and even division of labor within the framework of a single economic space. This space can be restored on a market basis by means of a multipartite union, no matter whether it consolidates three or nine states. There has been enough material evidence in support of this idea.

For example, Ukraine can reach an agreement with Belarus on the direct purchase of tractors and MAZ trucks. However, Belarus cannot fulfill the agreement without Russian enterprises that produce 70 percent of the components of these machines. Thus, a whole chain of contracts would have to be signed with Russian enterprises. Yet, Uzbek, Kazakh, and Tajik enterprises also contribute to the production of trucks. The coordination of this system of agreements without common favorable conditions in the spheres of customs, taxes, transport, and the like has turned out to be practically impossible. If superficial political considerations are rejected, the only alternative to the economic union is transition to an isolated subsistence economy. The modern world does not know any examples of efficient functioning of this kind of economy—be it Albania or North Korea. Attempts by some leaders of the newly independent states to abolish objective economic laws, as was once done by Bolsheviks, and limit relations among our states to bilateral contracts, are doomed to failure.

This is why the idea of creating a firm framework in order to safeguard equal conditions for entrepreneurship and business within the CIS seems logical and timely.

There exists one more aspect—both the treaty and the declaration view the unification of legislation regulating economic relations as one of the conditions for success.

First of all, these are legal acts facilitating the free flow of goods, services, workers, and capital; standardizing taxes and prices, regulating civil, transport, and anti-monopoly legislations. Discrepancies and drawbacks in existing documents must be corrected if they hamper the development of a single economic space. National legislatures will have to perform this strenuous task.

In my opinion, committees in charge of economy should immediately draft a list of legal acts that would facilitate the fruitful participation of Belarus in the economic union. If we are truly concerned with market reforms, if we genuinely want Belarus to recover and overcome the crisis, we should put aside our political ambitions and devote the rest of our "deputies' life" to diligent work on drafting laws. The fate of the program of the creation of a single economic space to a great extent depends on real active support from the legislatures of the Commonwealth.

## 7.23

### Russian Decree on Credits to Former Soviet Republics

*Federatsiya,* 15 July 1993 [FBIS Translation]

For the purpose of regulating the issuance of state credits in the currency of the Russian Federation to state governments of the former republics of the USSR and for their more efficient utilization, the Supreme Council of the Russian Federation decrees as follows:

1. Establish that state credits are to be issued by the Council of Ministers and the government of the Russian Federation to state governments of former republics of the USSR in accordance with bilateral intergovernmental agreements at the expense of and within limits of funds allocated for these purposes in the republic budget of the Russian Federation.
2. The Council of Ministers and the government of the Russian Federation is to establish the order in which state credits are to be issued, repaid, and serviced;

   Stipulate the maximum overall amount of allocations in drafts of the republic budget of the Russian Federation (including the one for 1993) for credits to state governments of former republics of the USSR and for corresponding expenses associated with their servicing.
3. Establish that the granting of technical (payment) credits to governments and to central (national) banks of former republics of the USSR will be discontinued as of 1 July 1993.
4. The Council of Ministers and government of the Russian Federation, together with the Central Bank of the Russian Federation, are to:

   Conduct negotiations within one month with governments and central (national) banks of former republics of the USSR regarding conversion of indebtedness for technical (payment) credits granted in 1992–93 into state debts owed by these governments to the Russian Federation;

   Conduct negotiations with interested states regarding the granting of state credits to them in 1993, bearing in mind that these credits may be granted only after the indicated conversion by these states of all debts for technical (payment) credits obtained by them from the Russian Federation in 1992–93;

   Conduct of negotiations to be governed by the fact that the balance of unpaid state credits granted governments of former republics of the USSR are indexed in accordance with changes in the ruble to U.S. dollar exchange rate at the current rate of the Central Bank of the Russian Federation. The indexed amount of state credit is subject to repayment.
5. Adopt the proposal of the Council of Ministers and the government of the Russian Federation concerning inclusion of appropriations from the republic budget of the Russian Federation for issuance of indicated state credits as part of the national debt of the Russian Federation and the corresponding increase, in the established sequence, of the maximum amount of national debt of the Russian Federation by 2.3 million rubles [R].

   Recommend that the Central Bank of the Russian Federation record the indebtedness within limits of the indicated amount of national debt for a period up to the year 2000 with its repayment starting in 1995 in equal annual increments without the accrual of interest.
6. The Council of Ministers and the government of the Russian Federation are to introduce proposals before 1 July 1993 concerning the size of allocation from the republic budget of the Russian Federation for the granting of credits to governments of former republics of the USSR in 1993.

   Henceforth, until the adoption of the maximum allocations for the issuance of state credits by the Supreme Council of the Russian Federation in the republic budget for 1993, the Council of Ministers and the government of the Russian Federation are authorized to issue state credits up to a total sum of R800 billion to governments of former republics of the USSR, with a corresponding increase in expenditures from the republic budget of the Russian Federation.
7. The Council of Ministers and the government of the Russian Federation, following coordination with the Central Bank of the Russian Federation, will submit proposals to the Supreme Council of the Russian Federation pertaining to conditions involved in servicing the part of the national debt of the Russian Federation associated with the issuance of state credits to governments of the former republics of the USSR in 1993.
8. Establish that state credits may be repaid pursuant to the agreement for the delivery of goods, by making other payments with property (transfer of property, packages of shares in key production facilities), as well as with payments in rubles or a hard currency.
9. Grant the Central Bank of the Russian Federation the right to place at the disposal of governments and central (national) banks of states which preserved the ruble as legal tender ruble banknotes of the 1961–93 series (state treasury bills of the USSR and bank notes of the State Bank of the USSR, as well as bank notes of the Central Bank of the Russian Federation) on the condition that appropriate intergovernmental and interbank agreements are signed as stipulated for states with which agreements were concluded on the terms and conditions governing utilization of the currency of the Russian Federation on their territory as legal tender, assuring the unconditional return of the cash to the Central Bank of the Russian Federation if national currencies are introduced by these states or upon the withdrawal of the corresponding bank notes from circulation.

Recommend that the Council of Ministers and the government of the Russian Federation study whether there is a need to form currency reserves on the basis of deductions by states signing the indicated agreements in order to ensure the withdrawal of ruble currency.

In the case of states not signing such agreements, the allocation of rubles may be made at the expense of and within the limits of state credit granted to governments of these states. The total amount of cash cannot exceed 25 percent of the total amount of monetary emission of the Central Bank of the Russian Federation.

10. The Central Bank of the Russian Federation will quote on a regular basis the ruble exchange rate for national currencies of former republics of the USSR with which agreements were not concluded regarding the terms and conditions for utilizing the currency of the Russian Federation on their territory as legal tender, with the systematic publication of that information in the press.
11. The Council of Ministers, the government of the Russian Federation, and the Commission of the Council of the Republic of the Supreme Council of the Russian Federation for the Budget, Plan, Taxes, and Prices are to provide for a rise in the deficit ceiling of the republic budget of the Russian Federation in accordance with this decree, when working out indexes of the republic budget of the Russian Federation for 1993, in accordance with the Supreme Council of the Russian Federation decree of 14 May 1993 on introducing a law on the republic budget of the Russian Federation for 1993.

[Signed] R.I. Khasbulatov, chairman of the Supreme Council of the Russian Federation
Moscow, House of Soviets of Russia, 30 June 1993
No. 5301–1

## 7.24

**On the Statement by the Governments of Belarus, the Russian Federation, and Ukraine on Urgent Measures to Deepen Economic Integration**

*Nezavisimaya Gazeta*, 21 July 1993
[*CD* translation]

*Editor's Note:* The following is an editorial that finds irony and surprise in the signing of a trilateral Slavic economic agreement so soon after the signatures were dry on an Economic Union for the CIS. At this point in time, Russian politics were probably pushing the decision to form a trilateral economic space more than any strategy for catalyzing greater CIS cooperation. Russian national patriots, who adhered to the Slavophile tradition of Russian greatness, were becoming highly vocal in the parliament, and Russia was headed for a constitutional crisis over the powers of the presidency.

The heads of government of Belarus, Russia, and Ukraine recently issued a statement on the need to deepen economic integration through the development of market relations and closer trilateral cooperation. . . .

This statement cannot fail to raise a number of questions. . . . How does it mesh with the decision by the CIS Heads of State on creating an economic union that would include all the states that signed the well-known Declaration of 14 May 1993 in Moscow? Does this statement nullify the declaration or propose a new way to realize the idea of creating an economic alliance according to the plan worked out in Belovezh Forest? One cannot avoid the impression that this statement by the heads of government of the three republics was dictated by more than just a recognition of the need for closer economic integration. Regardless of the intentions of its initiators, it can be seen as a political reaction to the actions being taken by the states of Central Asia, Kazakhstan, and Azerbaijan toward closer economic cooperation with Turkey, Iran, Pakistan, and Afghanistan. . . .

The last paragraph of the statement says that the treaty can be joined by states that "are not involved in any other economic unions or associations." And although provisos are made there that participation in other unions is possible if it does not harm the interests of this union, the point of this wording in the context of the whole statement is unambiguous.

This is an extremely important issue. It cannot be resolved by impulsive statements. . . .

This initiative by the three states, for all its progressiveness, cannot be regarded as a way to accelerate the implementation of the treaty on establishing an economic union within the framework of the CIS. The adoption of this statement without preliminary consultations and without ascertaining the positions of Kazakhstan and the Central Asian states puts significant limitations on the possibility of their participating in the union. Furthermore, such a statement could push those states away from participating in the economic alliance. And this is despite the fact that, throughout the entire existence of the CIS, Kazakhstan in particular has acted as one of the chief initiators of closer economic integration and the establishment of a system of coordinating bodies.

It seems to us that this kind of foreign economic policy with respect to Kazakhstan and other states in the CIS is shortsighted and is not in the interests of Russia and the other states in either the short or the long term. In our view, the main way of deepening economic integration with the near foreign countries [other former Soviet republics] lies in developing and implementing the Declaration on creating an Economic Union within the CIS framework. It is on this that all states should concentrate their efforts, without bringing into this inherently difficult process restrictions and obstacles that could undermine the foundations of the accord, so fragile and so necessary that was achieved in Moscow.

[Signed] International Reform Foundation: S. Shatalin, L. Abalkin, S. Assekritov, V. Bakatin, V. Kiryushin, S. Sitaryan, M. Shakkum

## 7.25
## Barriers to Russian-Kazakh Economic Integration Assessed

Revmira Voshchenko
*Rossiya*, 21–27 July 1993 [FBIS Translation]

*Editor's Note*: Economic differences and indecision between Alma-Ata and Moscow also contributed to the motivation for the three Slavic states to signal that they were ready to deepen their own economic cooperation. The *Rossiya* article below offers insight into the issues that bedeviled Russian Eurasianists.

Alma-Ata—Top-level negotiations in Uralsk and in both capitals, the Omsk meetings of government delegations and the heads of administration of contiguous regions, meetings of leaders of enterprises of the iron and steel complex, oil workers, agrarian specialists, and industrialists in Surgut, Akmola, Karaganda . . . A mass of documents on the restoration and further development of economic relations between Russia and Kazakhstan has been signed. The parties have scrupulously specified the terms of duty-free trade and questions of the elimination of customs barriers and the establishment of direct contacts between the commodity manufacturers and suppliers.

But . . . The lines of communication are still overloaded with wordy telegrams and hours-long negotiations: non-payments, absence of quotas and licenses, plants coming to a halt in Kazakhstan. . . .

It is paradoxical, but, following all these numerous intergovernment signings and agreements, the chaos has only increased. And the speeches heard at a recent republic meeting of Kazakhstan President Nursultan Nazarbaev and Prime Minister Sergey Tereshchenko on the economic situation in the republic and the state of credit and monetary relations with countries of the CIS were not at all comforting.

Ties to Russia, the prime minister emphasized, are of decisive significance for Kazakhstan, but it is they which are having particular difficulty taking shape. The series of top-level meetings has not produced an effective result. Nor was the last one in Moscow in June any exception. Russia refused to count as payment of Kazakhstan's debt its own enterprises' actual debts to Kazakhstan. Nor was an agreement for 1993 reached in the credit sphere either. The winding down of interstate economic relations could bring about the total paralysis of many enterprises and increase social tension in both states.

So why are the intergovernmental negotiations not producing the anticipated results? I put this question to Sauk Takezhanov, co-chairman of the Interparliamentary Commission for the Cooperation of the Russian Federation and the Republic of Kazakhstan.

Each side, my partner believes, is defending primarily the interests of its own state. And only recently the latter were part of a common economic organism and had as a common goal the building of a socialist economy. But the problem is that this term implied by no means equal partnership but the pumping of the raw material resources of our republic in the name of the prosperity of the so-called all-union economy. Nonetheless, business relations, abandoning which is inexpedient, did take shape. They need to be modified, but such that highly marketable production be developed at home on the basis of our own raw material, the necessary components for this being obtained from the former suppliers.

Another problem is that the documents which are adopted have no legal basis. Each state en route to the market formulated many laws, but they differ markedly. It is for this reason that our commission intends to look for conceptual approaches bringing the legislative practice in the two countries closer together. An agreement was reached at the first session of the commission, for example, on the economic and legal conditions for the development of business relations between Russia and Kazakhstan. It emphasized that the principal figure in cooperation is the commodity producer and industrial manager contributing to the fulfillment of bilateral trade and economic agreements.

A harmonious duet is not resulting as yet, in S. Takezhanov's opinion, although the difficult dialogue in a search for a convergence of the positions of federal Russia and unitary Kazakhstan continues. Thus it has been decided to establish a Russian-Kazakhstan bank to regularize reciprocal payments and monetary circulation, obviating the need for government intervention. So is an economic union taking shape? Wait a while. The Russians are proposing to their Kazakhstani partners the creation of a business cooperation association, bringing together all enterprises, associations, and stock companies which have received state support. The priority areas of activity would be the harmonization of supplies of most important product types, assurance of the conditions for efficient capital investments, the creation of a variety of joint ventures, and the establishment of a permanent arbitration body for the examination of industrial disputes. Kazakhstani members of parliament, on the other hand, are sure that reforms in Kazakhstan and Russia being implemented independently of one another will not contribute to an intensification and expansion of bilateral relations. Nor does the contemplated association promise this—it pretends to nothing more than the preservation or restoration of the former overly centralized contacts.

In order to remove the barriers obstructing relations of the commodity manufacturers, it is necessary, Almaty believes, to legislatively confirm the procedure of duty-free trade in industrial-engineering products, and of equal importance, of the use in Kazakhstan of the Russian ruble, to streamline transport flows, etc.

It is thus a question of imparting some legal basis to the rules of the game, which have not become firmly established as yet, without infringing the interests of either party.

## 7.26

### CIS Economic Union Concept Advocated

Interview by N. Zhelnorova
*Argumenty i Fakty*, no. 30, July 1993
[FBIS Translation]

*Editor's Note:* Nikolai Gonchar, who is interviewed below, is a prominent member of the Russian new left. Chairman of the Moscow Soviet since 1989, Gonchar endorsed Boris Kagarlitsky's August 1991 initiative to establish the Party of Labor, which is committed to a form of socialist self-management of society and the economy. This and other new democratic socialist groups have allied themselves with the trade unions. The party's platform argued that "Thatcherites" had come to power in Russia, and vowed to combat what they termed "wild-cat privatization," and opposed "market Stalinism." Their goal was to defend labor interests in the transition to the market. Kagarlitsky warned that capitalism was taking the route familiar in the Third World, where state interests engaged in an exclusive form of capitalist modernization. Yeltsin, he maintained, could easily become an authoritarian president, because he represented sections of the old nomenklatura and state managers. Gonchar espouses a distinctly Marxist viewpoint and seems to identify the Economic Union with a rehabilitation of Trotsky's Fourth International and working-class power, a concept with which he sympathizes.

Interview with N. Gonchar, chairman of the Moscow City Soviet and leader of the Moscow Civic Union branch, by correspondent N. Zhelnorova; . . . ]

[Zhelnorova]: Nikolay Nikolaevich, something, it seems, is on your mind?

[Gonchar]: It is the idea, making its way in politicians' minds, that whoever is not with us is against us. This is a sign of political impotence. Reaching agreement is difficult, brains are needed here. But to counterpose yourself to someone else you need a cast-iron throat in order to orate at a demonstration that there should be "no more compromises with this scum." Of course, he is scum since you lack the brains. . . . But our history shows that the winner on each occasion is a previous loser.

[Zhelnorova]: You have begun to actively advocate the creation of an economic union within the framework of the CIS because the situation has reached crisis point?

[Gonchar]: It is a cri de coeur! For Moscow, for its industry, for its population—it is the sole salvation.

We cannot go on like this. See here, in 1991 industrial production declined 2 percent, in 1992, 25 percent, and in just five months of this year, 17 percent! The main reason is the severance of economic links. And the strain in the economy is immediately echoed in civil conflicts.

[Zhelnorova]: But Russia, Ukraine, and Belarus recently agreed to conclude an economic agreement. What else?

[Gonchar]: A "mere trifle" is needed—that it be fulfilled. Many such decisions were adopted in 1992, but after each of them the economy was torn apart increasingly.

What is the problem? Or is it a question of those who sign the decisions and do not fulfill them? And one further pertinent question—where is Kazakhstan?

I believe that there can only be one guarantee of the fulfillment of such agreements—the will of the peoples. This is why I propose that we conduct a referendum and learn the opinion of the peoples as to whether they need an economic union or not.

[Zhelnorova]: But the Union also fell apart so easily for many reasons: It was a colossus with feet of clay, dozens of people wished to become presidents, premiers. . . .

[Gonchar]: And? The American system is strengthened by such desires. Truly, some politicians conceived a desire for both personal aircraft and carpeted walkways in front of them, and in the aircraft, their own pilots. And others wanted to be the personal pilots.

When do people take to the streets? When they have no other way of changing their fate. So they need to be given an opportunity to express themselves legally—at a referendum.

It would be beneficial, incidentally, to those who signed the documents in Belovezh Forest also. If people respond "no" to an economic union, the formation of the CIS was the right step, not the voluntarism of Yeltsin, Kravchuk, and Shushkevich. And if they say "yes," well, then, "the people's voice is the voice of God." After all, the presidents were not, I hope, dissembling in talking constantly about economic integration. They would have something to lean on.

Even under these conditions we should not be pulling someone or other apart from someone else but seeking what unites them. Some people are reluctant to part with the idea of sovereignty. But why is it not possible to create an economic union with regard for the current state independence?

[Zhelnorova]: There is no harm in wanting, but in recent years the peoples have somehow adapted to living in isolation and would hardly decide to return to the recent past.

[Gonchar]: Why to the past? When, after World War II, people began to create a united Europe, there was an entrepreneur by the name of Monnet who said: "Economic solidarity needs to be substituted for politics." A wise thought! Now no one in Europe is shouting about the fact that a Greater German Empire is being created. Why is this not possible with us? True, there are banks which earn money exchanging coupons and rubles. But in Europe there is the ECU and the Deutsche mark and the franc. And one does not prevent the other. A solution could, therefore, be found with us also. If there is the desire . . .

So we need to choose: either the first version—civil war: "the armor's strong, and our tanks are fast"—or the European version, when the Deutsche mark has come to resolve economic problems better than the Wehrmacht and the Schmeisser submachine gun.

[Zhelnorova]: But will politicians have a desire to hold a referendum aimed at the creation of an economic union?

[Gonchar]: Politicians should understand that the economy influences everything. Remember that the uprising on the Battleship Potemkin began with rotten meat. And, further, today's national elites need to be given an opportunity to preserve what they have. Otherwise they will fight to the last. You want a banner, coat of arms, army? By all means. Your own monetary unit? Go right ahead.

[Zhelnorova]: A good idea, although reaching agreement will not be easy.

[Gonchar]: It is better to take a long time reaching agreement than to quarrel quickly. Account has to be taken today of people's serious dissatisfaction. Someone said recently at the meeting of foreign ministers of Russia and Ukraine with officers in Sevastopol: "If the politicians cannot find a solution, we will suggest it to them—restoration of the Union." And the room erupted in applause.

Or remember the recent decision on the division of the Black Sea Fleet. This is in fact a sentence to its virtual destruction! Truly, how can you divide the indivisible? You know the parable? Two women with one child came to King Solomon, and each tried to prove that it was her child. The king then said: "Very well, let us cut him in two." To this one replied: "Very well," and the other: "No, you must not!" And the king gave the child to the one who had not wanted such a cruel sharing for she was the true mother. One has the impression that for those who wanted to divide up the fleet it is someone else's. . . .

Referenda should be held in the fall. We should say what kind of union we are talking about: a common unit of currency, a common bank, not subordinate to any government. No borders or customs. And we should obtain from the people an answer without any ideological motives, on the basis of the reality of life.

[Zhelnorova]: Interesting, but what about the local currency? In the old days it was said: "God gave the money, the devil, the hole. And God's money runs into the devil's hole." Will this be the case with us?

[Gonchar]: When a start is made on integration of the economy, much will become clear. The Ukrainian government may, for example, print as many coupons as it likes, but with a common money this item would not wash. A common bank would be responsible for this. If the politicians need to fulfill some of their wholly unsupported promises and cover intrastate problems with local money, print as much as you wish. But then your coupon would slide downward in relation to the common currency. In a word, there would no longer then be any idle talk about who was getting rich at the expense of whom. . . . But transrepublic financial-industrial structures should operate in accordance with common legislation in order not to have to hide from taxes in, say, Kazakhstan or Ukraine.

But the main thing is that capital would flow in the direction in which the level of political stability were higher. So compete to ensure that you have this stability. And if hysteria is spurred somewhere or other, capital . . . will float into another state.

There is the term *real politics.* An economic union with Ukraine and with Kazakhstan and with Belarus is not simply advantageous to Russia today, where the situation is better than in Ukraine, but essential. . . . How much demagogy we have heard about the fact that they are, allegedly, robbing us, that we are feeding them. . . . This is, you know, as though the hands had decided that they were constantly feeding the mouth for nothing and had held out the feet.

At the start of 1933 two ministers—Russian and Ukrainian—wrote their premiers that to compensate for the severed relations in iron and steel industry alone Ukraine would require $8 to $10 billion, and Russia, $28 to $30 billion. And such a price has to be paid merely for the fact that politicians cannot reach an agreement! Perhaps it would be cheaper to change the politicians?

## 7.27

**Nazarbaev Views Results**
Oleg Velichko
ITAR-TASS, 7 August 1993
[FBIS Translation]

The leaders of Russia, Kazakhstan, and Central Asian Republics called on all CIS states to hold a summit meeting in Moscow on 7 September to sign documents establishing an economic union of CIS states.

Commenting on the results of the meeting, Kazakhstan's President Nursultan Nazarbaev said that Russia, Kazakhstan, and Uzbekistan signed an agreement committing them to stay in the ruble zone. "This is what we need today. It is necessary to create a core and there is no need to force anyone into the Commonwealth," Nazarbaev said.

The Kazakh leader said the participants in the summit meeting also signed three documents on Tajikistan. Additionally, Russia and Kazakhstan issued a joint statement and instructed the ministers of defense and foreign affairs to work out within a month a document on military cooperation which would dot all "i"s and cross all "t"s in determining the future of the Baykonur space center in Kazakhstan. "In this connection I suggested establishing an international company with the participation of Russia, Ukraine, and Kazakhstan," Nazarbaev said.

## 7.28

**Analyst Warns Against Undue Haste in Implementing Ruble Zone**
Grigoriy Selyaninov
*Kommersant-Daily,* 19 October 1993
[FBIS Translation], Excerpt

Construction of the new ruble zone has entered upon the phase of actually coordinating and bringing together the economic legislation of the participating countries. This process will be supported by bilateral consultations. The decree signed by Viktor Chernomyrdin "On Measures to Implement the Intergovernmental Agreements on Unifying Monetary Systems," concerning which "B" reported on Saturday, has confirmed the personnel composition and departmental structure of the Russian government delegation. Meanwhile, however, an "opposition" has emerged within the Russian government with regard to the model of the ruble zone to be implemented; and this allows us to assume that the rapid pace undertaken in creating the new ruble zone will hardly be sustained. It's possible that a corrective adjustment will ensue—one aimed at protecting and defending Russia's economic interests. The situation which has evolved is analyzed below by Grigoriy Selyaninov.

The pace set in creating a ruble zone based on the Russian ruble has been impressive. On 7 September the agreement of the six countries "On Practical Measures for Creating a New Type of Ruble Zone" was signed. On 23 September signing was completed on bilateral agreements providing for Russia's partners joining in its monetary system. By mid-October Russia and the other countries of the new ruble zone had reached agreement on the need to standardize economic legislation in accordance with the Russian model. Some of the already adopted documents have been ratified by the parliaments of the ruble-zone countries.

The following reasonable question arises: Why are Armenia, Belarus, Kazakhstan, Tajikistan, and Uzbekistan with such readiness and haste essentially abandoning their economic sovereignty in favor of Russia? In Minsk and Almaty they reply that restoring economic ties with Russia is the only way to keep their economies afloat and thereby retain their national states. This answer corresponds to reality. Economic ties within the CIS [Commonwealth of Independent States] have developed in accordance with the following specific cycle: at first, euphoria with national independence, now—the recreation of economic ties as the only realistic method of smoothing over the economic crisis to a certain extent, and—in the future—it is entirely probable that there will be a resumption of the centrifugal tendencies, accompanied by the slogan of restoring sovereignty and creating a national currency.

However, there is also a more tangible price to be paid for abandoning economic sovereignty (most likely, temporarily). Membership in the ruble zone allows the partners to count upon an expansion of Russian deliveries of supplies within an increase of their deficit in trade with Russia. Eliminating this deficit under the conditions of a unified monetary system is more than problematical. Furthermore, the targeted convergence of the currency exchange rate of the ruble in the ruble-zone countries (with all the difficulties of achieving this goal under the conditions of different levels of development in the currency markets concerned) will reduce the economic motivation for exporting goods to Russia. And this—in turn—will expand the trade deficit.

Russian officials (with the exception of Boris Fedorov—a traditional opponent of the Central Bank, in accordance with whose prescriptions the ruble zone was created) until recently were loyal with regard to the matter of the ruble zone. A crack appeared only after the IMF [International Monetary Fund] came out with a criticism of the ruble zone in a confidential letter to Viktor Chernomyrdin. The Fund bases its argument on the concept that a model in which one country determines and specifies the monetary-credit policy of other countries is not a very viable one. The IMF's recommendations boil down to advising against any attempt to set a speed record; it would be better to retain "individual and separate" currencies in the various countries concerned during the period of preparation for a currency union.

The letter from the IMF revealed that within the Russian government there is an opposition to an overly fast formation of a ruble zone, which could lead to a premature transfer of rubles to the republics involved—something that would not be at all to the ruble's advantage. Both Egor Gaydar and Aleksandr Shokhin turned out to be among the oppositionists. As a result, there are grounds for assuming that the "unification of the monetary systems" will be slowed down. The decree "On Measures to Implement the Agreements on Unifying the Monetary Systems" has set the task of establishing and coordinating a system of indicators—one which will show the budget deficit, price increases, the currency exchange rate, as well as the pace of privatization, etc. The closer Russia's partners come to approximate these indicators, the more they can count on obtaining Russian rubles (they have the most need for rubles in the form of cash). If this system is really put into operation, then the new ruble zone will be presented with its first test of strength. However, even if this test is successfully passed, we cannot rule out the possibility that—as time goes by—the phase of disintegration will ensue.

---

## 7.29

### Convertible Local Currencies Could Settle Russia–CIS Debts

Aleksandr Krotkov
*Rabochaya Tribuna,* 22 October 1993
[FBIS Translation]

---

Help! Russia intends to blackmail "near abroad" countries using their growing debt to Moscow as a tool of political pressure on its neighbors. This is the way some newspapers commented on the results of the recent meeting of the Russian Federation Council of Ministers Current Affairs Commission headed by Oleg Soskovets, at which they discussed the issue of fulfilling treaties and agreements on production ties between CIS member states.

Well, we have to admit that a proposal was indeed put forward at the meeting to use the economic debts owed to us by "near abroad" countries to resolve foreign policy problems (such as, for instance, the rights of the Russian-speaking population or the ownership of the Black Sea Fleet). It came, however, not from the commission chairman or even the head of the foreign policy department, but from Russia's chief privatizer—Anatoliy Chubays—whose responsibilities, it should be noted, are limited to strictly domestic matters.

I agree that Anatoliy Borisovich's influence on the government is extremely great, and it would not hurt him to weigh more carefully the words he utters. But in this particular instance it is more likely that he stepped into "unfamiliar territory" out of desperation.

Seriously speaking, the economic crisis that has struck the Commonwealth countries has resulted in a situation whereby our neighbors have not delivered to us $3.03 billion worth of goods. To be fair, our indebtedness to them also is in excess of $2.319 billion. And now this $700 million positive balance has turned into a mortally dangerous clot in the way of developing foreign economic relations between Russia and the former brotherly republics.

This clot cannot dissolve on its own: the CIS countries do not have in their treasuries Russian rubles they could use to pay for our goods, while the Central Bank of Russia refuses to accept USSR banknotes that are no longer in use here. This is logical. The whole idea of introducing the Russian ruble was, I would like to remind, precisely this: to shield the country from the tidal wave of Soviet banknotes from the "near abroad," whose uncontrolled influx boosts inflation.

Our new national currency, however, in its current shape proved to be a double-edged sword, one side of which has dealt a painful blow to domestic enterprises. They now are frequently unable to get payment for the goods they deliver to CIS countries. And the Central Bank is forced to play hide-and-seek with itself, having promised to issue to neighboring countries 800 billion rubles [R] in so-called interstate technical credits so that they could pay us back.

And since this also boosts inflation, Russia is trying by hook or by crook to withhold the promised handout: by the middle of October it had allocated only R113 billion in credits to the near abroad. By the end of the year their total amount will reach a maximum of R400 billion. In other words, in the economic battle with its neighbors, Russia also has tied itself into such a knot that now, no matter how it tries to "bite" its partners, it only bites, excuse me, its own tail.

How can it untie this knot? Vladimir Mashchits, head of

the Russian state Committee for Economic Cooperation with CIS Member States, proposed at the Current Affairs Commission's meeting to first unload this mind-boggling problem onto the producers (they are now independent, so let them figure it out with their foreign partners on their own); and second, to reach an agreement with CIS countries on converting our mutual debts into promissory notes secured by real estate (if the plants cannot pay their bills, let them become the property of the note holders).

In principle, this is not a bad idea. A similar mechanism exists in all civilized countries of the world. However, as Anatoliy Chubays correctly warned the meeting participants, Russia will lose more than it will gain as a result of such operations. It will acquire ownership of hundreds of backward "bedridden" enterprises. To make them work will require enormous investment, for which we do not even have enough money for ourselves.

Actually, even if acquisition of enterprises in "near abroad" countries were indeed a super-lucrative deal, it would still be naive to count on getting rich soon. In order for such operations to take place, it is necessary to bring to a common denominator the respective codes of law of the CIS countries. If we recall how difficult and hopeless the process of reaching consensus between the branches of power was even in Russia proper, one can imagine what debates the settlement of this issue will spawn between the sovereign CIS republics. Because at issue, I want to emphasize, is not just signing some non-binding official papers—it is partial acceptance by our neighbors of our model of economic reforms, permitting mortgaging and selling enterprises.

Certain hope in this respect stems from the recent agreement on the creation of a new ruble zone on CIS territory. At least, that is the way it may look to the uninitiated. In the general opinion of experts, however, the aforementioned agreement, alas, would not facilitate in any way economic ties between neighboring countries. Because in the final analysis the real weight of the ruble is determined in each of them by the seriousness of its own *internal* problems.

So it turns out that in Kazakhstan, for example, the ruble has one-third of the dollar content of that in Russia. Which means that if we accept rubles from Kazakhstan in payment for Russian goods at the rate of one to one, we devalue our exports there by a factor of three.

The situation, in short, is dismal. The absolute hopelessness and helplessness in the voices of most Russian statesmen speaking at the Current Affairs Commission meeting would depress the most irrepressible optimists. Had it not been for the commission chairman, who resolutely dismissed this universal moaning and demanded from the procession of sad speakers at the rostrum an answer to the question "What to do?," the discussion almost certainly would not have produced any results whatsoever.

In the end, however, the discussion did get onto a constructive track: the only salvation for economic ties with the near abroad is the soonest possible introduction of convertibility of local currencies. You are offered soms in payment for goods—sure! You take them, exchange them at the bank for rubles, and no headache. If you are offered karbovantsy, lits, or Kazakh or Belarusian rubles—the procedure is the same.

True, implementation of this idea will require considerable effort. But why not make an effort if you know it will not be wasted?

---

## 7.30
## Problems of Transnational Enterprises in CIS Discussed

Evgeniy Spiridonov
*Segodnya,* 30 October 1993
[FBIS Translation]

---

The state's degree of participation in transnational companies has not been decided yet.

The Collegium of the Russian Federation State Committee on Economic Cooperation with the CIS States [Goskomsotrudnichestvo] discussed the draft of an intergovernmental agreement on cooperation in the development of transnational enterprises yesterday. The draft proposes cooperation in the support of joint financial and industrial groups, production corporations, associations, unions, and joint ventures, as well as commercial establishments, and financing and credit institutions.

Goskomsotrudnichestvo Deputy Chairman Marat Khusnutdinov said that the work on this program should restore the economic ties that existed earlier and worked so effectively among enterprises of the countries belonging to the Economic Union. The program was drawn up in conjunction with the Ministry of Economics, the Ministry of Foreign Economic Relations, the State Committee on the Administration of State Property, the State Committee on Industrial Policy, the Committee on Machine Building, the State Committee on the Defense Industry, and the Committee on Contracts. It is the first of a group of documents which will have to be finalized before the end of this year for the restoration of the economic zone in the territory of the former USSR. Pyotr Kormilitsin, the head of Goskomsotrudnichestvo's recently established Main Administration for the Coordination of Foreign Economic Operations, told the *Segodnya* correspondent that the main article of the agreement that is still in question concerns the degree of participation by budget-carried organizations in the transnational economic structures. The current plan proposes that 51 percent of the capital stock be retained as the state property of countries participating in the establishment of the joint production units. This stock would be turned over in trust to the government agencies whose representatives would be members of the supervisory councils of the transnational economic structures.

The very same government agencies that worked together to draft the agreement—i.e., Goskomsotrudnichestvo, the Ministry of Economics, the Committee on Machine Building, and so forth—would serve, according to Pyotr Kormilitsin, as the Russian co-founders of the joint financial-industrial groups and other economic and commercial structures. This participation in the management of corporations, in his opinion, would be warranted by the privileges and licenses they would grant the economic entities for the fulfillment of intergovernmental obligations in reciprocal deliveries of goods for state needs and the cooperative network.

The final decision on the degree and forms of participation by the state in international associations in industry, trade, insurance, and banking is being complicated, in Mr. Kormilitsin's opinion, by the varying speed of reforms in the Commonwealth countries and by the fact that the economic language of intergovernmental communication has not lost its political dialect. This is why economic integration without expressly stipulated participation by the state in the operations of the joint production structures has not worked. Machine assembly plants in Russia, for example, hoped to rely on Ukrainian suppliers of components, but customs officials said they could not.

According to Mr. Kormilitsin, the draft agreement has been approved on its merits and will now be discussed by the Russian Council of Ministers. If it grants its approval, the document will be forwarded to the governments of the other states.

---

## 7.31

**Moscow Blamed for Kazakhstan's Decision Not to Join Ruble Zone**

Aleksandr Krotkov

*Rabochaya Tribuna,* 5 November 1993

[FBIS Translation]

---

The current Russian leadership's harsh treatment of its opponents (or competitors—call them whatever you wish) is moving from the sphere of domestic policy to foreign economic operations, at least in relations with nearby foreign states. The interests of the great nation, which associate the Kremlin with macroeconomic stabilization, have caused Moscow to encourage its most loyal Kazakh and Uzbek neighbors to institute national currencies of their own.

I will remind you that both of these Central Asian republics were willing until recently to keep the ruble as their main form of legal tender, but normal trade with Russian enterprises and payment for imports from the north required not simply the ruble, but the new Russian rubles. Many rubles—hundreds of billions. Meanwhile, Moscow was increasingly reluctant to lend money to its southern neighbors, feeling that this kind of generosity could only escalate galloping inflation.

On the one hand, the rising prices in Russia were partly a result of the extension of colossal credits to former associates in the indestructible Union. On the other, cutting off the supply of rubles to Kazakhstan could have closed the hundreds of enterprises living on exports to that republic. After all, it would have had no way of paying for our goods, and we would still be overcome by the inflation caused by these production cuts.

There were only two ways of breaking out of this vicious cycle: Our colleagues in the ruble zone could have been asked to acknowledge the Central Bank of Russia as the only official body issuing currency (but that would have brought them to their knees before Moscow in the financial sphere) or forced to institute their own currencies, which could then be exchanged for rubles at an acceptable rate.

Then Russia made a stipulation a week ago: It was willing to give its two southern neighbors rubles in cash, *BUT(!)* only for half a year and only at the regular Central Bank rate of interest (which businessmen in the country have called usurious). Furthermore, they would have to put up 50 percent of the credit in hard currency as collateral. If these neighbors would then refrain from introducing their own currencies in the next five years, Russia would return the collateral and would not charge interest on the credit.

It is understandable that Kazakhstan and Uzbekistan felt that these terms (especially the five years of ruble "bondage") were unacceptable in these troubled times.

One thing is clear: There will be no increase in the quantities of Central Asian musk melons, watermelons, and other exotic fruits in Moscow stores in the near future. The Russians will manage to survive somehow without these luxuries, but can they live without cotton? . . .

---

## 7.32

**Case Argued for Economic Union with Ukraine**

Interview by Vladimir Kuzmishchev

*Rossiyskaya Gazeta,* 23 February 1994

[FBIS Translation], Excerpts

---

[Interview with Nikolay Gonchar (chairman of the Federation Council Committee for the Budget, Financial, Currency, and Credit Regulation, Monetary Emission, Tax Policy, and Customs Regulation) by Vladimir Kuzmishchev. ]

[Kuzmishchev]: Nikolay Nikolaevich, I recently heard that you are calling for economic reunification with Ukraine in connection with the situation in the Crimea. To be honest, my first thought is God forbid. Loss-making mines in the Donbass, exhausted mines in the Krivoy Rog area, the mighty defense industry in Dnepropetrovsk in need of restructuring. We just won't be able to cope. . . .

[Gonchar]: Imagine that a third of the modules are removed from your computer here and you are asked: How are we to get it to work? Simple. You either replace the parts that have been removed with some new components, which is a costly, awkward, and exacting business. Or you put back the old ones. Then you can use the computer.

[Kuzmishchev]: That is debatable. It transpires that Ukraine produced 18 percent of the former Union's GNP. Some went on reproduction and on intrarepublic deliveries, some to Central Asia, Georgia, and the Baltic. And, no matter how you look at it, Russia's dependency on Ukraine was minimal. But Ukraine experienced maximum dependency on Russia, as the last few years have shown.

[Gonchar]: Now let us give the people of Tyumen a say from these positions. They say that the whole Russian economy has been shored up and still is being shored up with their petro dollars and gas dollars, and you can wash your hands of and forget the rest. Let them survive on their own. It will end up with neither one thing nor the other. The system can be restructured or, to continue the computer analogy, a new program loaded. But why destroy things?

[Kuzmishchev]: But political realities must be taken into consideration. . . .

[Gonchar]: The former political system turned out to be fiction. If it could collapse because of a conflict between two leaders, it did not exist. I suggest that we evaluate reality. The reality is that borders which were established in an extremely arbitrary manner in the past (everyone was after all convinced that they were a formality) suddenly proved real. So far only economically, but the logic of confrontation is not always the logic of common sense. It dictates its own laws. Border guards carrying assault rifles have already appeared. The absurd thing is that a person's membership of a particular nation is only decided by geographical latitude and longitude. . . .

[Kuzmishchev]: You have been able to say things to me a journalist that you cannot say as a politician. Having traveled the entire length and breadth of Left Bank Ukraine I was never able to understand how the thinking, lifestyle, psychology, or, as people are now saying, the mentality of a peasant, a miner, an engineer in those areas differs from the mentality of someone from Tula, Saratov, Omsk, or Minsk, say. There has never been a language barrier. It was harder for me to understand, say, what a coastal dweller from Arkhangelsk was saying than a peasant from Kharkov. According to Dal [nineteenth-century Russian ethnographer] a person belongs to the nationality in whose language he thinks. Second, what have the millions of Russians who came from Russia to Ukraine or to Kazakhstan on an institute placement or in response to the authorities' appeal done to hurt us? By political realities I meant the situation today.

[Gonchar]: The saying that politics is economics in a concentrated form is more appropriate today than ever. Economic integration represents the means of survival for us. There is a cheap work force in Ukraine. Tell me please how we will be able to stem the flow of people from there who are desperate to come to our country to work. They are even willing to take half pay, but what are we to do with our own people? This threatens a social explosion. Now let us take the situation in the Crimea. The recently elected president Yu. Meshkov was the leader of the "Russia" bloc. A fairly eloquent name. Their objective is to leave the economic wreckage of Ukraine and form a single economic area with Russia and other states. Russia now seems to be accommodating that aspiration. But there are many problems here. We do not after all even share borders. Electricity comes from Ukraine as does the water for agricultural irrigation, and so on. The Crimea is greatly indebted to Ukraine. What will happen to Left Bank Ukraine where there are similar pro-Russian sentiments? But no one there is advocating the revival of the Soviet and Socialist Union. It is a question of economic integration.

[Kuzmishchev]: The chain reaction will hit Kazakhstan next. Whereas the breakup of the Union caused a series of wars in outlying regions, the breakup of Ukraine and Kazakhstan would mean. . . .

[Gonchar]: There are plenty of politicians in Ukraine and our country who would not be averse to playing that card. Both in order to remain in big-league politics and in order to reach new heights. Suppose someone in Ukraine wanted to postpone parliamentary and presidential elections. Is the possible deterioration of the situation in the Crimea not an excuse for those politicians? And aren't our radicals even better?

[Kuzmishchev]: But doesn't integration involve a partial loss of sovereignty for states with a weaker economy?

[Gonchar]: Sovereignty is not the aim, but the means of attaining contemporary civilization and a high standard of living. People in Ukraine were convinced that autonomy is the way to success. Now the result is clear. No one on the long-established Western market is waiting for us or for them. Look how quick they were to reach agreement with us and drive down prices of aluminum and oil. We thought that we would inundate the West with these goods and take off like the Arab Emirates did in the past. That is not the case. Besides, it is offensive if you have high technologies to place the emphasis on raw materials when all countries are trying to sell finished output. But for the time being we are prepared to trade finished output mainly among ourselves. . . .

[Kuzmishchev]: But how specifically are the first steps to be taken? How long has there been talk of creating a ruble zone at the highest level and it has all been to no avail. . . .

[Gonchar]: The key issue here is the number of money supply centers. Needless to say, Russia will never agree to there being several points deciding the money supply. There must only be one of these points. Europe has already reached that conclusion—a currency institution has been set up there and it will have a single currency. Why are our politicians opposing proposals dictated by life itself? Remember the advertisement that begins with the words: "Ladies and Gentlemen, you have thirty seconds to think about it. . . ." There are no deals that are any easier or grander than those involved in playing with exchange rate differences and brokering currencies transfers. This is a colossal network and can produce billions out of thin air. On the one hand, these billions affect prime cost and thus bring soaring prices and inflation; on the other, they require political protection and lobbying for their interests. I therefore propose that, if the politicians do not reach agreement, let the people reach agreement. The politicians must be presented with the fait accompli of the people having reached agreement. Let us hold a referendum. With one sole question: Is it necessary to create a single economic area?

[Kuzmishchev]: The population of Ukraine will probably vote "for," but Russia is another matter. What a hullabaloo there has been over the creation of a single monetary system with Belarus. But here, after all, the scale is far greater. And the political forces of those opposed are far more aggressive. It will be enough for someone in Russia say that this will whip up inflation, and then. . . .

[Gonchar]: If you so desire, shortcomings can be found, exaggerated, and dreamed up in any plan, any idea. Let us first grasp what is meant by "unification" [*ob''edinenie*] or the creation of a "single economic area," the specific steps involved.

There are of course plenty of people in our country, in Ukraine, and in Kazakhstan whose political career is based on the policy of separation [*razyedininie*]. Many people are only interested in nationalist hysteria so that they can come to power. Some individuals have already succeeded, but the parade of sovereignty indisputably had adverse consequences for most of the population. At the same time we have enough sober politicians who are capable of telling people that economic integration by no means signifies that Russia will include Ukraine in its budget and exchange rubles for coupons on a one-to-one basis.

[Kuzmishchev]: What does it mean then?

[Gonchar]: Primarily that the economic and legal barriers should be taken down and the duties removed. And private capital given the freedom to invest money in the Ukrainian economy.

[Kuzmishchev]: But, as Marx said, capital is not sentimental, it goes where there is the greatest profit. It is hard in Russia to force someone to invest in production in conditions of inflation, never mind in Ukraine.

[Gonchar]: A contentious issue. Even from the purely economic viewpoint. You only have to remember Ukraine's mighty oil refining industry, which is driving down our monopolists' prices. Anyone with a reasonable knowledge of the Ukrainian economy can name straight off several sectors in which it is profitable to invest capital. That is not the point. Entrepreneurs are primarily interested in political stability and legal protection. Russian firms, say, are willing to refine oil in Ukraine, but what guarantee is there that Ukraine won't appropriate the oil tomorrow? There must be seamless legislation on property.

[Kuzmishchev]: Common legislation.

[Gonchar]: No. Ukrainian and Russian legislation. The main thing is that the laws should not contradict one another. In a word, there must be an incentive for a change of legislation. If you want to employ an incentive, then think it through. Capital, I repeat, goes where there is greater social and political stability. If we squabble, there will be no capital. That is, the economic motor is cranked into action via politics. And, on the contrary, every stage of political rapprochement must have legal and political backup. Let us give Russian and Ukrainian capital a chance to work together. If it is just a question of Russian capital, there is one system of taxation, if it is just Ukrainian, then another applies. But if it is a joint project, there must be lenient taxes to provide incentives. It is not important what language people speak. Or what they wear. They will be united by profit and a common cause. When people come along and undermine the common cause out of personal interest, Russian businessmen will tell their Ukrainian colleagues: The plan will collapse if this idiot forces the situation. Would you be so kind as to have it out with him? Give money to his election rival, give money to the party that backs integrationist processes and we will deal with our Russian loonies. People of the same blood must be united by common interests.

[Kuzmishchev]: Do you think that this is the most pressing problem now?

[Gonchar]: Why has the rich United States suddenly taken the initiative and started strengthening the economy of Mexico? Mexicans were dying to get across the border, so economic potential had to be created there. The situation is far more acute in our country. Things are so bad in Ukraine that an explosion may be about to occur. We are hoping in vain that the debris will only fly inward. There is no alternative to economic integration. People in the Crimea have realized that and are clearing the way. The others are only hampered by the fact that some politicians in both capitals are afraid to acknowledge their mistakes, which have brought national tragedy.

## 7.33
## Committee to Explore Creation of CIS Monetary System

Konstantin Smirnov
*Kommersant-Daily,* 15 July 1994
[FBIS Translation]

There are two main tasks involved in the construction of the Economic Union of the CIS countries at this time: the creation of a supranational administrative body and the establishment of an effective settlement system for the Commonwealth countries. These tasks will not be completed within the near future, but a decision was made at yesterday's meeting of the CIS Consultative and Coordinating Committee (CCC) to return to the idea of the Interstate Economic Committee (MEK) and appoint an international team of experts to plan a settlement system for the CIS countries.

The short history of the Commonwealth of Independent States already has some predictable patterns of its own: Decisions requiring no commitments are approved with relative ease, but as soon as there is any mention of real limits on national sovereignty (or at least the need to give up part of the state budget), the pattern of action in CIS agencies is one step forward and two steps back. The history of the long-suffering Interstate Economic Committee is the most indicative example of this. The initial response to Nursultan Nazarbaev's idea of establishing a standing body of the Economic Union in Moscow, with specific supranational functions, was more than favorable in the majority of CIS capitals, but during the Moscow summit in April 1994 the as yet unborn MEK was turned into nothing more than the CCC Commission of the Economic Union (KES), with no directive or supervisory functions whatsoever. A statute on the KES was prepared for the last meeting of the CCC in Moscow, but by then everything had reverted to the old pattern.

First Alma-Ata, which had secured approval of the decision to establish supranational institutions of the Central Asian Economic Union at the last meeting of the three presidents, revived the idea of the MEK. Then President Yeltsin asked Deputy Premier Aleksandr Shokhin, the chairman of the CCC, to make all of the arrangements for the replacement of the stillborn KES with the MEK by the time of the September summit in Sochi. Besides this, after Leonid Kuchma won the presidential election, his support for the MEK could also be expected.

The CIS settlement system has not taken shape yet either. In April this year, the heads of state approved a plan for a payment union, proposing the eventual creation of a monetary union of the members of the Economic Union. The idea of the payment union was supported by commercial banks in Russia and the other CIS countries. At yesterday's meeting of the CCC, the Russian Ministry of Cooperation with the CIS Countries and the Association of Russian Banks gained acceptance of the idea of appointing an international team of experts to plan a settlement system for the CIS countries. President Sergey Yegorov of the association and Deputy Minister of Cooperation with the CIS Countries German Kuznetsov became the co-chairmen of the international team. According to German Kuznetsov, proposals on the creation of the payment union will be prepared for the CCC by 1 September. The first step in the normalization of settlements between CIS countries was the approval of the standard draft agreement on measures to secure the mutual convertibility of national currencies on a bilateral basis at the last meeting of the CCC.

**Proposed Status of Interstate Economic Committee of CIS Economic Union**

The committee will be a standing executive and coordinating body of the Economic Union; it will be accountable to the Council of Heads of State and the Council of Heads of the Government of the CIS.

The decisions the committee makes within the confines of its jurisdiction will be binding for the states of the union whose representatives voted in favor of them.

**Standard Draft of Bilateral Agreements on Measures to Secure Mutual Convertibility of National Currencies**

- mutual alignment of the national monetary systems in operation within the territory of the republics party to the agreements;
- gradual institution of the mutual convertibility of national currencies—the stabilization of national currency exchange rates in relation to one another and to hard currency during the first stage, and complete mutual convertibility during subsequent stages;
- establishment of liquid currency markets in the republics party to the agreements;
- regular (at least weekly) adjustment of the official exchange rates of national currencies in relation to one another and to the dollar on the basis of supply and demand in domestic currency markets by the central (or national) banks of the republics.

## 7.34

## Supranational Body to Coordinate CIS Economic Policies

Konstantin Smirnov
*Kommersant-Daily,* 23 July 1994
[FBIS Translation]

Analysis of economic integration in the former USSR gives rise to the following conclusion: In the CIS, they harness quickly, but ride slowly. Lots of agreements and treaties have already been signed, but their implementation leaves much to be desired. An attempt to mend the situation will be made in September 1994 in Sochi. Yesterday, a group of experts representing virtually all CIS countries was set up to prepare constituent documents of an Interstate Economic Committee [IEC] for the Sochi summit.

The history of setting up the committee, as *Kommersant-Daily* already reported in its issue of 15 July, has a certain detective-story tinge. Before it had time to come to life, the committee was replaced at the Moscow summit this past April with the Commission of the Economic Union as part of the CIS Coordinating Consultative Committee. In the view of Vladimir Pokrovskiy, deputy minister for cooperation with CIS countries and head of the Russian expert group in talks on setting up the IEC, last spring those "on high" in the Commonwealth were not mature enough to form such a committee.

In addition, Mr. Pokrovskiy emphasized in an exclusive interview for *Kommersant-Daily,* the first draft statute on the IEC was palliative in nature and for that very reason the committee, designed to be the CIS Economic Union's working body, did not have real supranational functions. Moreover, according to *Kommersant-Daily* experts, the creation of an Interstate Economic Committee would have called into question the status of the Consultative and Coordinating Committee (CCC). Now, a Solomonic decision has been found. The IEC, according to the concept of its architects from Russia, Kazakhstan, and Belarus, will be given a supranational status and will be directly subordinate to the Council of CIS Heads of Government. The CCC together with its chairman will be converted into the IEC's board. Aleksandr Shokhin, current CCC chairman, liked this solution, so the committee decided to elevate the commission's status to that of an interstate committee. The shift occurred without a snag at the latest CCC meeting (see *Kommersant-Daily* of 15 April).

Thus, with one month to go before the Sochi meeting of the CIS prime ministers, the IEC "detective story" appears to have come to an end. The sides have begun to finalize the draft agreement on setting up the committee and its draft statute. According to its architects, the new executive and coordinating body is to assume supranational functions in the areas of anti-monopoly and customs regulations, the standardization and certification of mutual deliveries, energy supply, transportation, and communications.

Only the problem of settlements and payments between CIS countries will be left outside the committee's terms of reference. This area of relations between the former USSR republics will be regulated by the Interstate Bank—as far as the servicing of government credits and payments by state-run enterprises is concerned—and by commercial banks of CIS countries, which even at this stage service up to 70 percent of all mutual settlements. Another international group of experts tasked with devising a concept of the CIS countries' payment system must draw up by 1 September its proposals on setting up a single center for the quotation of national currencies against each other; this could be done, for instance, by the Moscow Interbank Currency Exchange.

## 7.35

## Russian Official on Interstate Economic Committee

Evgeniy Spiridonov
*Segodnya,* 26 July 1994
[FBIS Translation]

An international group of CIS experts, which was tasked by decision of the CIS Consultative and Coordinating Committee [CCC] on 14 July 1994 with drawing up a statute on the

Interstate Economic Committee [IEC], started its work today so as to submit the document to the meeting of Commonwealth heads of government scheduled for September 1994. This was said to your ADS correspondent by Vladimir Pokrovskiy, deputy Russian Federation Minister for Cooperation with CIS Member States, who joined the expert group from the Russian side.

While the recently created Commission for the Economic Union (CEU), which operates as part of the CCC, performs only expert and analytical functions, he said, the IEC, according to its authors' concept, must be the Economic Union's management body endowed with operational and executive functions in the areas that will be turned over to it by the Commonwealth governments. That is to say, as a supranational management structure, it will be able to make decisions binding on all CIS members. According to Mr. Pokrovskiy, existing interstate sectoral cooperation bodies (a total of around fifty), whose operation has been plagued by discord, may be "dragged" into the IEC structure in the capacity of its departments, he said.

If a universally acceptable document that would legally formalize a higher level of integration between the economies of CIS members is prepared by September, at the next meeting of Heads of Government Russia will put forward specific initiatives on free trade—an agreement on which has already been signed—customs tariffs, and mutual payments.

Mr. Pokrovskiy thinks that the fiercest arguments in the process of drawing up the IEC statute may be triggered by the debate on the committee's decision-making procedure. At present, most CIS members advocate the one-state-one-vote principle, regardless of their varying economic potentials. Still, one can draw on the EC's method of equalizing "weight categories" between member countries in the voting procedure, taking into account their economic potentials. In this case, for instance, Russia would be given enough votes to block any IEC decision that would put it at a disadvantage economically.

In Mr. Pokrovskiy's view, the states should show the political will to entrust the IEC with the power to impose economic sanctions should any state disregard its decisions and thereby cause damage to other members of the Economic Union. The IEC's weight will be boosted, he thinks, if the Consultative and Coordinating Committee, made up of deputy prime ministers, becomes the new committee's board.

---

## 7.36

### Outlook for CIS Economic Union

Yuriy Shishkov

*Rossiyskiy Ekonomicheskiy Zhurnal,* 15 August 1994

[FBIS Translation], Excerpts

---

From the very beginning, we must note that for many years, the Soviet economic and political space was not merely unified, but also maintained a certain functional equilibrium. Yet it has been over three years now since the formerly unified space broke up into fifteen separate parts and entered a state of wandering and searching for means to a new equilibrium. Along this path it has encountered everything—territorial claims, military conflicts, civil wars, deep economic crisis which gripped the newborn independent states, and the desire of some to get rich at the expense of others. The impression is of some confused conglomeration of mutual claims and suspicions, assurances of friendship and compromises. At the same time, *two stages* are clearly apparent here.

The point of departure of the *first* of these stems from the spring of 1990, when there was an increasing tendency toward sovereignization. The course toward independence was proclaimed by Estonia (30 March), Lithuania (4 May), Russia (12 June), Ukraine (16 July), and Armenia (28 August). After August of 1991, the centrifugal tendency reached a critical mass and took on avalanche-like proportions, fed by the euphoria of independence. There were hopes that the very fact of economic independence would allow every former union republic to more effectively utilize its natural and other resources, to quickly implement economic reform, and to embark upon a successful autonomous course in the open sea of the world economy. The leaders of the Baltic states, Ukraine, Turkmenistan, Georgia, and Azerbaijan spoke most confidently about this. Life, however, did not confirm these optimistic suppositions. In practice, it turned out that independence is not only a blessing, but also that it involves many problems and difficulties, often entirely unforeseen. The initial hopes for rapidly overcoming the painful disintegration of former ties, the forced integration into the world economy, and the mastery of new foreign trade horizons, was replaced by the understanding that even after the divorce it would still be necessary to live in the "same apartment" for some time to come.

The *second* stage began in spring 1993. An even deeper strategic shift became evident. Earlier, beginning with the Treaty on Economic Cooperation signed on 18 October 1991 and with subsequent agreements of a multilateral character, it was quietly assumed that the search was on for organizational forms of *civilized divorce.* Today, the problems of *real reintegration* of the twelve national economies have appeared on the agenda. This new tendency strengthened when the Russian reserve of "charity" toward the other CIS states was exhausted. In 1992, Russia granted them so-called "technical credits" toward payment for the power resources and other goods supplied by Russia in an amount equivalent to \$15–\$17 billion. (With high rates of inflation, such interest-free credits are a direct loss for Russia.) In the first seven months of 1993, such aid totaled \$1.2 billion more than the previous year. Moreover, the CIS states received considerable benefit in the form of hidden subsidies, determined by the varying degree of increased prices on goods which they imported from Russia, on the one hand, and those which they supplied to Russia on the other hand. Finally, Russia was forced to put an end to the flooding of its domestic market with devalued Soviet rubles, which were emitted in large volumes by the national banks of other CIS states to pay for imports from Russia.

As a result, it became apparent that the traditional system of "feeding" the former Soviet republics at the expense of the center had finally faded into the past, and that now each of the new states must itself earn money "to live on." At the same time, it became clear that for many of them, the resolution of such a task was beyond their means. There was no basis to depend on their own means and, consequently, they had to opt for uniting seriously, and for a long time. Even the leader of Turkmenistan, a country rich in natural resources, who intended to rapidly turn this country into a new Kuwait, sharply altered his negative attitude toward the restoration of a unified economic space within the boundaries of the CIS.

A period of hopes for the *transformation of the old,* integral command-distributive economic space inherited from the USSR into an equally integral market space had come to an end. A period of the *formation of a new* common market comprised of the domestic markets of twelve countries, which were somewhat autonomous and somewhat dissimilar, had begun. It was necessary to hastily undertake a search for suitable mechanisms for the reintegration of the national economies.

### Model of Reintegration Borrowed from the European Community

From the very beginning, the model developed and used in the European Community was viewed as the standard for the future economic interaction of the CIS states. However, its competent emulation was hindered by two circumstances. The *first* of these was the fact that, for some time after the disintegration of the USSR, the economic "transparency" of the inter-republic boundaries was retained, and up until the fall of 1993 there was also a unified ruble space within the CIS. It seemed that the decisive elements of the economic union toward which the EC countries had stubbornly been moving for three and a half decades, already existed within the confines of the CIS. Therefore, the problem, it seemed, was reduced merely to implementing a task which was not too difficult: to finish building the "lower stories" of the building of integration—to create a customs union, a unified market of capital, and so forth, all the while retaining and strengthening its finished "roof." The *second* circumstance was the following. The conviction prevailed that on the whole the entire situation in the CIS was much more favorable than in the EC, whose participants had in the past been entirely independent states, with their own economic, legal, and institutional specifics. The situation in the CIS was believed to be principally different. Here all the countries were as similar to each other as twins, since their national economies had quite recently been a single whole and, it would seem, were well adapted to each other. Moreover, the states of Western Europe had moved forward by feeling their way, by the method of trial and error, while the CIS countries already knew the trail which they had blazed, making it possible for them to follow it confidently and quickly.

As a result of the indicated circumstances, the understanding of the need for consistent passage by the CIS states through all the stages of integration from a zone of free trade to a currency and economic union did not come all at once. At first it was considered enough to retain the transparency of the borders and the unity of the customs territory, and the rest seemed to be a simple task of secondary importance. In the spring of 1992, there was intensive work on the agreement on a customs union, and in April of that same year it was signed. At that time, it was still not difficult to implement. The boundaries between the CIS member states remained rather transparent, and a set of trade and political tools for relations with third countries, which still rested on a unified system inherited from the USSR, could be rather easily unified. However, due the prevalent orientation toward "divorce" at that time, the CIS states were in no hurry to recognize the multilateral agreements which had been signed.

At the same time, events followed their own course, and for objective reasons the "transparency" of the interrepublic borders quickly evaporated. Differences in the rates of economic reorganization caused different levels of price liberalization in the CIS countries, which led to significant differences in levels of economic development. And this, naturally, gave rise to the rush of goods, including vitally important resources, from the countries with relatively low domestic prices to the others, where they were higher. Just as strong was the drain of material goods, caused by the repeal of the state monopoly on foreign trade. Raw materials, fuel, and metal began to flow in a broad current to the "far abroad," i.e., to the world market, where a higher price could be obtained for them (sometimes two to three times), with payment in hard currency. Often such a drain occurred through the customs territory of other CIS states. All this forced Russia and the other Commonwealth states to introduce strict quantitative and tariff limitations on export, and to create customs services at the borders. As of 1 January 1993, Russia expanded a unified regimen of setting quotas and licensing not only on exports to the Baltic states, but also on exports to all the CIS states.

Parallel with this trend was the unchecked erosion of the

Table 7.1

**Relative Share of Workers Employed at State Enterprises in 1992** (% of overall number)

| Country | % | Country | % |
|---|---|---|---|
| Armenia | 68.1 | Moldova | 60.9 |
| Azerbaijan | 68.4 | Russia | 77.5 |
| Belarus | 75.7 | Tajikistan | 54.5 |
| Georgia | 75.1 | Turkmenistan | 56.1 |
| Kazakhstan | 75.4 | Uzbekistan | 61.1 |
| Kyrgyzstan | 68.5 | Ukraine | 81.8 |

*Source:* The World Bank, *Statistical Handbook 1993, States of the Former USSR* (Washington, The World Bank, 1993), pp. 6–7.

unified monetary space. When the old Soviet ruble was a unified means of payment in the CIS zone, independent national banks of the member states were able to implement uncontrolled emission of credits and to use them to settle accounts with each other, especially with Russia, which accounts for up to 60 to 80 percent of the foreign trade of most of these states. This formed the means for flooding Russia with a devalued mass of money, which spurred on the already galloping inflation. It was necessary for Russia to introduce its own currency and to change over to a strict credit policy with regard to the other CIS states. Each of them (except for Tajikistan) in turn introduced its own national currency.

Thus, *the initial situation changed significantly.* Almost nothing was left of the ready "stockpiles" for the future economic union, and it became quite clear that it would be necessary to pass through the entire "building cycle," from foundation to roof. The decision to prepare the appropriate multilateral treaty on a step-by-step movement toward an economic union was adopted by the heads of the CIS states on 14 May 1993. Its text was prepared in only one-and-one-half months. On 1 July it was coordinated at the level of vice premiers, and on 24 September of the same year it was signed in Moscow by the heads of nine states: Armenia, Azerbaijan, Belarus, Kazakhstan, Kyrgyzstan, Moldova, Russia, Tajikistan, and Uzbekistan. Turkmenistan joined in the Treaty on the Economic Union on 24 December 1993, as an equal member, and Ukraine on 15 April 1994, as an associate member.

Many principles of this treaty testify to the fact that its participants relied on West European integration theory and practice. The goals of the union and the means of attaining them were rather similar to those which were fixed in the treaty on Instituting the EES [European Economic Space] (1957), in the agreement on the EFTA [European Free Trade Association] (1960), and in the Unified European Act (1987). Thus, according to Article 3 of the treaty, "The Economic Union presupposes the unhindered movement of goods, services, capital, and manpower; coordinated credits, budget, tax, price, foreign economic, customs, and currency policy; harmonized economic legislation of the agreement parties."

Article 4 of the treaty exactly reproduces the stages of integration laid down in the Rome Treaty: "The Economic Union is created through a step-by-step intensification of integration and coordination of action in the implementation of economic reform. An interstate (multilateral) association of free trade; a customs union; a common market of goods, services, capital, and manpower and a currency (monetary) union will facilitate this coordination process."

Thus, after several unsuccessful efforts, having skipped several stages of development, in order to get rapidly to the finish line, the CIS states had to reconcile themselves with the fact that they would have to move ahead in the same way as Western Europe did, i.e., step by step. . . .

The political disintegration of the USSR occurred at the very beginning of the transition from such a command-distributive model of economic management to the market model. Therefore, the economic space of the former USSR divided between the fifteen sovereigns began to be managed not through one centralized mechanism, but through fifteen mechanisms—smaller, more decentralized, but of the same type as the former one. A large part of industry, transport and other infrastructure, and even a certain portion of agriculture in the CIS, still remains under state ownership (see table). Also, the newborn independent countries cannot ensure the functioning of their economies in any other way but through the centralized redistribution of the resources under their management. And since the latter are rather limited, there is no other way out but to establish strict governmental control over all these resources, limitation of their loss to neighboring republics, and concentration of the national monetary and fiscal sphere in their hands.

Thus, despite the declaration of intent to preserve the "transparency" of the borders, immediately after the division of all-union property there occurred an unstoppable spontaneous formation of fifteen closed economic organisms, genetically repeating the basic traits of the Soviet economic model. This barrier to the path of reintegration will remain until denationalization of the economy reaches a level characteristic for the majority of the developed countries which have a market economy. . . .

The *third* barrier arises from the fact that, for a number of historical and political reasons, the transition to the market model of economic management in the various countries of the CIS is taking place at different rates. This means that, for a certain period, the post-Soviet economic space turns into a mosaic of different variants of transitional models of economic management. This circumstance erects invisible barriers between the CIS states. However, aside from this, such asymmetric transition to a market economy gives rise, as we have already said, to a flow of goods to countries where domestic prices are higher. And with the introduction of national currencies or quasi-currencies, this stimulus is coupled with factors associated with uncomplementary market exchange rates. All this forces the young states to shut themselves off from one another.

The *fourth* medium-range obstacle is associated with the changes in price structure after elimination of state control. Bringing domestic prices closer to world prices exposes the irrationality of certain former interrepublic movements of goods. These trading patterns have been disrupted, and often it is necessary to seek new partners beyond the boundaries of the CIS, which leads to a relative reduction in trade for each former republic, and the CIS as a whole. In the future, a reorientation of trade will be supported by the need for resource- and energy-saving technologies required to deal with increased costs of power and raw materials, whose prices were artificially low in the USSR. The need for buying such technologies will demand a partial reorientation of the foreign trade relations of the CIS countries to the West, and this factor will indirectly hinder the process of reintegration for a prolonged period of time. This will last until a new balanced structure of foreign trade relations between the CIS countries is established, and until a new model of international export specialization for each of them has been formed.

Along with the above-named medium-range factors, the process of reintegration will be opposed by one other circumstance *of a deeper, structural character.* The insufficiently developed and weakly diversified economy of the Central Asian and Transcaucasus countries of the CIS, as well as Moldova will hinder the process. The sectoral structure of the economy as a whole, by the level of industrial production, the level of labor productivity, and the social well-being of the majority of the population, position these countries closer to the developing states than to the developed industrial countries of the West. . . .

The division of the post-Soviet economic space into a relatively developed industrial center, and a developing periphery, becomes even more valid when examining the industrial structure of the CIS countries. High technology sectors of machine building, electronics, electrical technology, and the chemical industry, which are usually associated with the development and production of arms, are concentrated in Russia, Belarus, and Ukraine. Certain types of machine building are present also in the Transcaucasus, Uzbekistan, and other countries of the periphery. However, the main industries on the periphery are the mining and primary processing of mineral resources, food, and light industrial products.

At the same time, at that stage of economic development when agriculture, the mining industry, and the "lower stages" of processing dominate, national economies are not so much complementary as they are competing partners. There is little basis for any broad and effective division of labor between them. Therefore, from a trade perspective, the developing and even the moderately developed countries may remain indifferent to one another for a long time, while between the highly developed countries there is actively increasing mutual attraction.

In 1990, for example, the volume of per capita intraregional exports among the countries of Western Europe comprised $3,121, while the per capita volume of West European exports to third countries was $1,250. For Latin America, the amounts were $412 and $262, respectively, and for Africa—$6.50 and $110. In Western Europe, the ratio of intra- and extra-regional trade is approximately 2.5:1; in Latin America, 1:6.4; and in Africa, 1:17.9. It is quite obvious that the less developed the national economies, the less their economic interest in one another, and the more actively they seek partners from among the number of industrially developed states.

For the CIS, this means that its least developed member states are not inclined toward broad economic cooperation with each other. They may be "attracted" either by a more developed center of the former USSR (the European part of Russia, the Urals, Ukraine, Belarus), or by the even more developed West. And in fact, statistics show that almost all the CIS countries are oriented in their exports toward Russia, to a lesser degree toward Ukraine, and (with the exception of Azerbaijan and Tajikistan) export only an insignificant portion of their products outside the boundaries of the CIS. Aside from the higher technical-economic level of development of Russia, the huge volume of its domestic market is, of course, also well-known here. Interstate trade among the Central Asian countries is not great. For four of them it does not exceed 20 to 25 percent of the total volume of exports, and only for Kyrgyzstan does it reach 32 percent. Moreover, interstate trade among the three Transcaucasus states is quite negligible—not more than 4 percent (in some measure this is determined by the military conflicts in the region; however, even in peacetime the relative share of intraregional exchange was very small here).

Under such conditions, the tangible integrational processes among the CIS countries referred to here as "the periphery" are improbable in the foreseeable future. Even if the coordination of their economic policy outlined after the Tashkent meeting of 4 January 1993 by leaders of the five countries of Central Asia were to grow, say, into the creation of a Central Asian free trade zone, the real effect would be rather insignificant. The experience of most of the "free trade zones," "customs unions" and "common markets" in the developing world speaks in favor of such a conclusion. The interstate trade among the ASEAN [Association of Southeast Asian Nations] countries, which in 1960 comprised 21.7 percent of their total exports, by 1970 had declined to 14.7 percent; in 1980 it did not exceed 17.8 percent, and in 1990—18.5 percent.

Thus, it is reasonable to conclude that the integrational process on a *multilateral* basis in the CIS zone still does not meet any favorable objective prerequisites. No matter how important the political decisions adopted in this regard may be, it is unlikely that they will go beyond the realm of good intentions. Meanwhile, however, real multilateral integration of the national economies of the CIS will progress at an extremely slow rate. It is true that fairly adequate material prerequisites exist here for *one-sided* integration according to the "heliocentric" model, when each of the less developed countries of the Commonwealth will be "drawn" ever more strongly to its developed nucleus, and primarily to Russia. The tendency toward such development was clearly outlined already in the first years after the disintegration of the USSR. Here, however, much will depend on how greatly Russia itself and its economic subjects are interested in strengthening and intensifying economic ties with the countries of the post-Union periphery. There is no unanimous answer to this question. There are a number of economic and political conditions in support of such interest, among which, in particular, is the threat for Russia of losing prospective markets in the CIS countries. In evaluating this circumstance, we must not, however, forget that Russia itself is a less developed country in comparison with the leading states of the West. And therefore, due to this objective reality, Russia is more interested in economic rapprochement with the West, and not with the peripheral partners of the CIS.

The practical experience of recent years shows that the tendency toward a reorientation of Russia's trade relations with the "far abroad" is intensifying. While the specific weight of the latter in the overall volume of Russian exports in 1989 comprised 33.3 percent, and in 1990—35.5 percent, in 1991 it reached 41.2 percent, and in 1992—52.5 percent, in 1993—60.7 percent, and in the first quarter of 1994—75.7 percent. At the same time, within the CIS itself Russian exports are ever more greatly concentrated on the countries of its developed nucleus, i.e., on Ukraine and Belarus. In 1990, their share of Russian exports to the current Commonwealth was 57.2 percent, in 1991—61.3 percent, and in the first quarter of 1994—already 70.8 percent. An analogous tendency is observed in the changes of the geographical structure of Russian imports. The peripheral countries of the CIS are becoming ever more marginal partners of Russia. Therefore, their propensity toward integration with it risks being a case of unrequited love. And this would mean a slowing of the process of reintegration within the CIS economic space, even in its one-sided "heliocentric" variant....

Many multilateral government agencies of a sectoral character have been formed within the CIS, including the Eurasian Coal and Metal Association. Thus, as is customarily said, "much work has been done."

Nevertheless, many leaders of the Commonwealth countries admit that the institutional system of the CIS operates most ineffectively. By the summer of 1994, around 500 different multilateral documents had been signed within the scope of the CIS, but most of them are not being realized in practical application. And there is nothing sensational about this. The same situation is characteristic for most international organizations of an integrational character which operate in the developing regions. The main reason for such ineffectiveness is also clear: *The economic conditions for real merging of the national economies have not yet matured.*

Not understanding this, or not giving this factor proper attention, certain leaders of the CIS countries are striving to correct the situation by means of intensifying the supranational institutions of the Commonwealth. The repeated initiatives by Kazakhstan President N. Nazarbaev have become particularly notable in this regard. Back in September of 1992, he proposed creating administrative rather than coordinating structures of the CIS, which would in fact turn it into a confederation. In March 1994, he spoke out with the idea of transforming the Commonwealth into a Eurasian Union, and at the beginning of June he publicized a detailed plan for such a union. It proposes the creation of supranational power structures (a council of heads of state and governments, and EAC [Eurasian Union] Parliament, a Council of Ministers of Foreign Affairs, and an Interstate Executive Committee), which would make decisions by a qualified 4/5 majority of votes, under the condition that each of the countries would have one vote. These decisions would be binding for the member countries. It is presumed that this would make it possible to create the legal and organizational prerequisites for intensifying integration toward economic, currency, and political union, and for the creation of a unified defense space.

N. Nazarbaev's initiative was met with suspicion on the part of most of the leaders of the CIS countries, although it did find support among certain Russian politicians. It seems, however, that in spite of all the good intentions which lie at the basis of such an initiative, the idea of a Eurasian confederation, at least for the immediate period, does not have any chance for success—primarily due to the above-named objective economic reasons. These are supplemented by important political reasons—subjective as well as objective.

Among the subjective ones is the "anti-imperialist syndrome," which has been deeply imprinted in the national memory of the people. Any steps in the direction of confederation are perceived by a significant part of the population of the peripheral CIS countries as a tendency toward restoration of Russian hegemony. Even if the current leaders of these countries understand the ephemeral nature of such suspicions, they can still not ignore them, if for no other reason but that the nationalist opposition will use such attitudes for strengthening its influence and for seizing power. In such a political atmosphere, it is difficult to expect success for the idea of the Eurasian confederation.

There is also an important objective political obstacle in the path of strengthening the supranational power structures of the CIS. This is the huge imbalance of forces in favor of Russia. It accounts for 59 percent of the GNP, 91 percent of oil drilling, 77 percent of the natural gas, 58 percent of the smelted steel, two-thirds of the machine building production, and the overwhelming portion of the scientific-technical and military potential of the Commonwealth. Already because of this, in the CIS, as at one time in CMEA [Council for Mutual Economic Assistance], there objectively exists the danger that the interests of the "younger brothers" will be subordinated to the interests of "big brother." After all, what is beneficial to Russia is ultimately also beneficial to the CIS as a whole. But not everything that is necessary for Russia or even for the Commonwealth as a whole corresponds to the interests of individual member states at a given moment. Therefore, for them it is extremely important that decisions of the organs of the Economic Union be adopted unanimously. In other words, each state must have the right of veto. Only in this way can they guarantee for themselves their own individual key to the start-up mechanism of this union, and guarantee their own economic security. However, it is specifically this principle that is fixed in Article 23 of the CIS Charter, and repeated word for word in Article 27 of the Treaty on Development of an Economic Union: "Decisions of the Council of the Heads of State and the Council of the Heads of Governments are adopted through general agreement—consensus. Any state may declare its lack of interest in one question or another, which should not be viewed as an obstacle to adoption of the decision."

It is characteristic that this principle is copied almost word for word from the Charter of CMEA, whose members also strived to safeguard themselves from the dictate of the USSR. However, such a procedure, which is suitable to the members of the association, is extremely ineffective for the common cause because any collective decision may be torpedoed by any one country which is dissatisfied with it. CMEA spent years "hammering out" one or another form of cooperation, but suffered from the same problem.

The European Community also spent a long time languishing in a similar procedure. It took three decades to almost entirely edge out the principle of consensus from the practice of the EC Council. However, we cannot forget that direct analogies are incorrect here. After all, a relative equilibrium of forces existed within the EC from the very beginning. In 1958, the share of the FRG [Federal Republic of Germany] comprised 36.2 percent of the total GNP of the six EC countries, of France—32.8 percent, and of Italy—18.5 percent. Under these conditions, the small participant countries could maneuver within this triangle, supporting that "angle" of it which was closest to them in each specific case. With the entry of Great Britain, the balance of powers in the EC was strengthened even more. Today, however, the situation has changed, and the unified Germany has become clearly dominant in the Community. However, the integration here had already gone so far that this circumstance cannot stop the process.

If Germany had been unified already in the late 1950s, it is unlikely that the institutional system of the EC would have taken on its present form, and the integration process could hardly have gone as far as it has by the present time. In the CIS, however, the imbalance between economic and political forces is a huge obstacle in the path not only of confederation, but also of an economic union of fully sovereign states. And this creates significant additional complications for reintegration of the post-Soviet economic space.

Thus, a new stage in post-Soviet history has begun. The pendulum of history has swung in the direction of unification of the economic and political efforts of the CIS states, which had initially been intoxicated with independence. Both in Moscow and in Kiev, as well as in the other capitals of the states entering into the Commonwealth, the voices of sober pragmatists are sounding ever louder, becoming capable of rising above the nationalistic emotions. Under the pressure of the problems which have crashed down upon them, the leaders of the peripheral Commonwealth countries are even ready in some measure to ignore the "anti-imperialist syndrome." Most of the state leaders here obviously believe that the problem of reintegration of the post-Soviet space lies in the sphere of political freedom, and that everything here comes down to a search for the correct decisions, which take into consideration the attitude of the broad popular masses. After all, it is no accident that N. Nazarbaev proposes to launch the construction of the Eurasian Union with the passage of referenda in the CIS countries.

However, we believe that such logic is based on deep delusion. It loses sight of a most important circumstance. In the three to four years which have elapsed, powerful forces which determine the subsequent course of events have gone into action—*economic regularities.* Today the decisive role is played not by leaders and political parties, not by conceptions and programs, but by inherent economic processes taking place in each country which is a successor of the USSR. . . .

## 7.37

### Guidelines for CIS Economic Committee Agreed

Interfax, 8 September 1994
[FBIS Translation]

*Editor's Note:* Based on what the author of the preceding article said, the reader can better understand just how impotent the Interstate Economic Committee probably will be. The fact that some decisions will inevitably be based on "economic weight" of the Member States means that Russia will dominate the Committee, and that most other Member States will ignore its decisions, due to the "principle of consensus."

The delegations of twelve CIS member nations agreed on the text for guidelines for the Interstate Economic Committee. Russian Deputy Prime Minister Aleksandr Shokhin told journalists that this was the main result of the CIS Consultative and Coordinating Committee (CCC) meeting held in Moscow Wednesday.

Shokhin, who chairs the CCC, said that his committee had come to the conclusion that the CIS Economic Union's commission, an auxiliary mechanism rather than a body of authority, had to be replaced by an agency to which CIS member nations could delegate some of their powers.

Decisions will be made in the Interstate Economic Committee through straightforward democratic procedures such as consensus and "qualified majority" and by the economic weight of the member states, Shokhin said.

Shokhin is convinced that if the ISEC is set up soon, it will handle numerous issues that now encumber the agenda of the Council of Heads of State, the Council of Heads of Government and CCC which will focus on the current issues in bilateral and multilateral relations.

He said that the CIS has now reached the stage when the Commonwealth is capable not only of adopting documents but also of tackling specific issues.

## 7.38

### Sangheli Supports Interstate Economic Committee

Interfax, 8 September 1994 [FBIS Translation]

Moldova supports the creation of an Interstate Economic Committee as a coordinating and executive body of the CIS economic union, and the adoption of an agreement on a single payment space, Moldovan Prime Minister Andrey Sangheli told Interfax before his departure to Moscow to attend the meeting of the Council of the CIS Prime Ministers, which is to take place 9 September.

He also intends to speak on the consequences of the natural disasters which hit Moldova in summer.

Reliable sources told Interfax that at their meeting, CIS prime ministers are expected to adopt an appeal to all CIS countries to render aid to Moldova to help it overcome the aftermath of the drought, hurricane, and torrential rains in the summer. The total damage is estimated at 1,200 million leu (4 leu = $1).

## 7.39

### Chernomyrdin Views Proposed CIS Economic Body

Vladimir Taranov and Gennadiy Ezhov
ITAR-TASS, 9 September 1994
[FBIS Translation], Excerpt

The Council of the Heads of Government of the CIS countries today agreed to set up an Interstate Economic Committee [IEC] [Mezhgosudarstvennyy Ekonomicheskiy Komitet—MEK] within the framework of a CIS Economic Union—the first body within the Commonwealth with distributive functions. Russian Prime Minister Viktor Chernomyrdin said this at a news conference at the President Hotel. He stressed that a final decision on setting up the IEC would require the approval of the Council of CIS Heads of State, which is scheduled for the beginning of October this year. So far, the agreement to set up the IEC at government level has been signed by all those participating in the council with the exception of Turkmenistan and Azerbaijan, which will work on their positions before the meeting of the Heads of State.

"The appearance of an IEC," the Russian prime minister emphasized, "will make it possible to work on the adoption of decisions within the CIS framework before they are carried out, which has not, by and large, been the case in the Commonwealth thus far." At the initial stage, the IEC's basic functions will be performed by the CIS Consultative and Coordinating Committee. It is proposed to create an IEC presidium from the CIS governments operating in the Coordinating and Consultative Committee and to create the collegium on a permanent basis.

It is proposed to locate the IEC headquarters in Moscow. Over fifty Commonwealth bodies operating today will be considerably reduced once the IEC has been created, so that the CIS structure does not give at the seams, Viktor

Chernomyrdin observed. This will be done, basically, in order to save money. Important decisions will be taken in the IEC by consensus; the others by a majority of 80 percent of the votes, where Russia is allocated 50 percent of their total number.

Turning to the question of the creation at today's sitting of CIS Heads of Government of a Payments Union within the framework of the Commonwealth, Viktor Chernomyrdin said Russia will not in the immediate future choose to amalgamate its monetary system with that of the Republic of Belorussia [Belarus]. The republic will have to bring its economy up to Russia's level. Russia will also be cautious, Viktor Chernomyrdin said, in its approach to the question of removing customs barriers with Commonwealth countries. "We were not the first to introduce them here, but today the situation has changed too significantly. A great deal now goes not only from Russia, but also through Russia, for example, drugs and weapons," the Russian prime minister stressed.

Viktor Chernomyrdin said the discussion of the other items on the agenda of the Council of Heads of State session (there were more than twenty of them in all) was constructive, but not very easy. The only package of agreements giving rise to no objections was, he emphasized, the documents on the joint marking of the fiftieth anniversary of victory in the Great Patriotic War. . . .

## 7.40

### Kiev Against CIS Common Strategic Space

UNIAR, 9 September 1994
[FBIS Translation]

As a UNIAR correspondent reported, Russia's Prime Minister Viktor Chernomyrdin, while opening a regular session of the Council of CIS Heads of Government, first and foremost reminded that "the European Economic Community has been taking shape for thirty-five years. We do not have so much time, therefore, there should be a different pace," he pointed out.

One of the main issues considered in Moscow is the issue on setting up an Interstate Economic Committee (MEK). This structure is endowed with suprastate functions, although, according to Mr. Chernomyrdin, countries should themselves determine exactly which issues will be within MEK's competence.

Ukraine's delegation expressed certain reservations concerning the system of voting within the framework of the MEK. Depending on the importance of the issue, the authors of the draft proposed four systems of voting. To adopt a decision on the most important issues, 80–85 percent of votes at the MEK will be required, other issues are to be resolved through a consensus. Ukraine will have 14 percent of the votes, Russia—50 percent. In the opinion of some Ukrainian experts, Ukraine can refuse to join this or that proposal by the leadership of the MEK, if it wishes to do so. Others believe that the mechanism of defending [Ukraine's] own interests is not clear enough.

This is why the Ukrainian delegation agrees to signing the agreement on setting up the Interstate Economic Committee with the following reservation: The term of the validity of the proposed mechanism of voting should be limited by the end of 1996.

The Ukrainian delegation refused to sign in any form the agreement "On the payments union" because no system of payments is available as yet in the countries—members of the MEK, it believes.

Among the issues considered at the meeting there were issues of military and technical cooperation. The Ukrainian delegation came out against the formation of a single strategic space and spoke out for establishing bilateral relations. As Col.-Gen. Ivan Bizhan pointed out in his speech at the meeting, "we believe that Ukraine's state borders have already been outlined and there is no need to raise the question on common CIS borders."

The participants in the meeting discussed over twenty issues on the agenda of the session of the heads of government of the CIS member states.

## 7.41

### Special Opinions Voiced on Economic Issues

Vladimir Taranov and Gennadiy Ezhov
ITAR-TASS, 9 September 1994
[FBIS Translation]

Azerbaijan, Turkmenistan, and Ukraine expressed special opinions during the debating of two fundamental documents in the economic package of documents at a meeting of the Council of Heads of the Government of the Commonwealth of Independent States (CIS) on Friday. These are the agreements on setting up an Interstate Economic Committee and a Payments Union in the framework of the CIS Economic Union.

Azerbaijan's Prime Minister Suret Guseynov told ITAR-TASS after the first stage of the discussion that he will not sign either of these agreements on Friday as the Azeri parliament is not yet ready to approve of the delegation of some supranational functions of economic management to an Interstate Economic Committee, the proposed executive body of the Economic Union of the CIS.

Turkmenistan, while not rejecting the idea of an Interstate

Economic Committee, asks for a month of deferment, until CIS Heads of State meet in October, for additional debating of the matter in the republic. Nevertheless Turkmenistan's delegation is prepared to sign the agreement on forming a Payments Union today.

Ukraine's attitude is more complicated. Ukrainian Prime Minister Vitaliy Masol told ITAR-TASS that, in his opinion, a Payments Union should be formed gradually because it provides free conversion of national currencies and their market quotation. It would be expedient to form a Payments Union by concluding bilateral agreements between the republics. "For free quotation of the Ukrainian coupon, Ukraine should have a stabilization fund, a gold reserve in the amount it does not have at present," Masol said.

Ukraine's special stand on forming an Interstate Economic Committee is determined by its associate membership in the Economic Union, Igor Mityukov, Ukrainian Deputy Premier for the economy, told ITAR-TASS. He said Ukraine does not intend to alter its status in the Economic Union. Mityukov said that if Ukraine signs the agreement on forming an Interstate Economic Committee it will keep associate membership. It will participate in debating only whatever matters it deems necessary. "If the documents on the Interstate Economic Committee and the Payments Union are finally coordinated, taking into account Ukrainian amendments, on Friday, we shall be ready to sign them," the Ukrainian deputy premier said.

The memorandum on the basic trends of integration among the CIS countries was initialed during the first part of the meeting. There have been no special objections. Foreign ministers of Commonwealth countries who held a separate meeting on this matter and defense ministers agreed upon all aspects of the document which is expected to be signed on Friday. Amendments and supplements to the document that have not been submitted to it within a month will be agreed upon and attached to the signed text of the document as supplements.

## 7.42

## Economic Union Augurs Reintegration of CIS States

Anna Ostapchuk
*Nezavisimaya Gazeta,* 7 October 1994
[FBIS Translation]

*Editor's Note:* The following document was chosen to illustrate the Russian parliament's intense involvement in the CIS reintegration question. The Belarusian and Ukrainian parliaments have joined in on the issue of formulating "financial-industrial groups" (FIGs). If these factions of their respective parliaments were able jointly to influence their state's legislation, reintegration could become more than wishful thinking.

The creation of an interfactional deputies group, the Economic Union, including representatives of both houses of the Russian parliament, was announced yesterday evening. It includes nineteen people representing various groupings of the Federation Council united by one objective: to begin work on the economic reintegration of the former USSR countries "from below," by forming transnational financial-industrial groups. By 6 October similar groups had been created in Belorussia [Belarus], Ukraine, and Kazakhstan. The minimum program of all these four entities is to work out and get through their legislatures corresponding laws on financial-industrial groups.

The creation of a structure that tasks itself with coordinating the reintegration efforts of the legislative branches of the four CIS republics is unprecedented. No fewer than several hundred documents have been adopted on the subject of economic integration within the CIS framework. Unfortunately, however, it has to be admitted that so far they are not working. An economic union on Commonwealth territory has not materialized.

"It has not happened because it was envisioned as an association of bosses, not producers," Nikolay Gonchar, initiator of the new integration model and chairman of the Federation Council's Budget and Finances Committee, believes. According to him, for real reintegration to begin, a common legal field is needed. Quite obviously, it can be created in two ways: either by adopting uniform laws (while there is no common legislature) or by working out coordinated laws—analyzing and adjusting current laws and trying to bring new laws as close together as possible in the drafting stage. It is not hard to see that the second path appears to be the most realistic.

According to Gonchar, the interfactional Economic Union group includes deputies who share the following theses: preservation of an industry-based economy is the major prerequisite for Russia's survival; it can be helped only by massive nonstate investments; investment will begin only when production is integrated, and only integration will restore it; the model of integration through financial-industrial groups is recognized as quite effective.

The interfactional group does not encroach on the structure of factions. Nikolay Gonchar said that "if a faction is categorically against integration, its representatives will not join us." The Economic Union is essentially a working group for coordinating different interests represented in parliament and for promoting the ideas of economic integration. In Ukraine, the statement on the Economic Union was signed by twenty-one deputies; in Belarus, by sixteen; in Kazakhstan, twenty-one; and in Russia, nineteen. Russian parliamentarians sharing this position include Aleksandr Pochinok, Yuriy Yakovlev, Vladimir Semago, Yuriy Boldyrev, Aleksandr Belyakov, Oleg Ochin, and Sergey Shakhray (in all, nine deputies of the State Duma and ten members of the Federation Council).

The first law drafted by the Economic Union with the aim of forming a common legal space is the Law On Financial-Industrial Groups. The draft has been sent to the Supreme Soviets of Ukraine, Belarus, and Kazakhstan.

## 7.43

**Moscow Seen Regaining Former Authority in CIS**
Sergey Parkhomenko
*Segodnya,* 22 October 1994
[FBIS Translation], Excerpt

*Editor's Note:* The following brief article indicates the authoritative tone and manner adopted by President Boris Yeltsin at the sixteenth CIS summit.

[Passage omitted.] The sixteenth summit since the CIS's inception was attended by all heads of state without a single exception. It was organized in Moscow without frills, in a modest and businesslike manner. Everything has returned once again according to cycle: in the absence of more serious business to attend to, the big boss summoned lower-level clerks for yet another briefing to give them guidelines and chart the prospects—well, of course, also to hear their views as behooves a democrat and to turn down quietly their objections and proposals.

Summing up the talks that lasted nearly three hours, Russian President Boris Yeltsin said at a news conference yesterday that participants in the meeting unanimously approved an agreement on setting up the Interstate Economic Committee (IEC). "Not all participants in the meeting demonstrated a full unanimity of views. Not all states are ready to be equally involved in the Economic Union," the Russian Federation leader conceded, stressing in the same breath that the draft of the IEC constituent document was "passed without amendments, right in its initial version." No one voted against it. Leonid Kuchma, who for the first time attended the talks as Ukrainian head of state, deserved particular praise: "The situation was entirely different from what it was under the former president. Today we did not have arguments with Ukraine on a single question." [Passage omitted]

Despite the Belarusian president's protests, it was decided to set up the IEC head office in Moscow. In Boris Yeltsin's opinion, the key argument was the availability of "experienced specialists" on interstate cooperation in the Russian capital. It seems unlikely, however, that leaders of the sovereign countries will be delighted by the possible comeback of Comrade Baybakov's [a Soviet Gosplan chief] time- and battle-tested old guard (which, as is known from the ancient wheeze, has "an immense destructive capacity") to full-time decision-making activity. It seems rather that another argument of the Russian president held sway: in his opening address he proposed that if any complications or delays arise over the creation of the IEC, its powers be temporarily turned over to the Russian Economic Cooperation Ministry. Apparently, the prospect that national economies will be simply subordinated to a Russian ministry did produce the desired impression.

Journalists who attended the news conference also were given the opportunity to form their opinions about the tone and style of the talks that had just finished. When asked by a correspondent of a Tajik newspaper about the prospects for recreating the ruble zone. Boris Yeltsin responded that he himself has "always supported the idea," but locally there are—so far—some officials who oppose the introduction of a single currency "with whom we will fight and simply replace them." [passage omitted]

What is it that comes to mind in this regard? Ah, yes: "Pluralism is essential for us—there cannot be any disagreement."

## 7.44

**Russian Newspapers Comment on CIS Summit**
ITAR-TASS, 22 October 1994
[FBIS Translation]

The creation of the Interstate Economic Committee became a feature event of the regular session of the Council of CIS Heads of State, *Rossiyskaya Gazeta* reported today. For the first time in the history of the new commonwealth an institution with controlling, distributing, and executive functions has been created. Actually, it is the firstling of the supranational structure, which will take decisions on some matters, mandatory for all the CIS member states.

*Delovoy Mir* carries an article by its political observer entitled "Economic Integration Speeded Up in CIS." The newspaper stressed that problems of the integrated development of the new commonwealth was in the focus of attention of the CIS leaders. Their solution is connected with pursuing an active policy, aimed gradually turning the CIS into an effective alliance of sovereign states. The session discussed and adopted the memorandum entitled "guidelines of integration development of the Commonwealth of Independent States." The development of effective economic cooperation, which is the main condition for overcoming the crisis and ensuring the economic direction of integration. According to the memorandum, it has become evident that it is necessary to improve the efficiency of the CIS working bodies and to create a mechanism for cooperation inside the new commonwealth. The necessity emerged to give supranational powers to some of those bodies. It is for this reason that the council decided to create the Interstate Economic Committee—a permanent working body of the economic alliance, whose task is to ensure the effective work of alliance and the rational development of integration in the CIS.

The session discussed a number of military matters. The priority task in this respect is the formation of the collective security system, the observer wrote.

The session of the council may become a breakthrough in the process of economic integration of the CIS member states, *Kommersant-Daily* wrote. The results of the session are unprecedented: The agreements on the creation of the payment alliance and on the setting up of the Interstate Economic Committee were adopted unanimously. Participants in the session reserved the right to interpret those agreements in their own way. President Yeltsin said, however, that "a stop has been put in the discussion of the need for economic integration." Progress was achieved in the settlement of some political problems. By its structure the economic integration of the CIS member states is approaching the highest world standards. The same question remains: In what way will the signatories to the agreements behave during their implementation? Now the agreements reached by the presidents leave much room for one's own interpretation of the commitments assumed. The difference between relations inside the group of independent states and show-off programs is that in the first case the quest for solutions is impossible without taking into consideration conflicting interests. Anyway, integration has been given a start.

Yeltsin said at a final press conference that the creation of the Interstate Economic Committee, the need for which was discussed on more than one occasion at previous meetings of leaders of former USSR republics, was the main result of the session, reported *Nezavisimaya Gazeta.* The creation of the committee has become a reality. Besides, a decision was made on the future functioning of the payments and customs unions of the CIS member states. The Interstate Economic Committee was created in line with the best traditions of "commonwealth cooperation." All the CIS member states agreed that the committee was a necessity. It was one of the few unanimous decisions in the short history of the CIS. However, it is the presidents and governments of the CIS member states who will decide in what way each particular country will take part in its work and what powers it will delegate to the committee. Thus, *Nezavismaya Gazeta* continues, the Interstate Economic Committee has little chance to become a truly working structure. It is more probable that it will be an auxiliary body for those who wish to avail themselves of its services.

A high-flown and elaborate diplomatic protocol, which was typical of the initial period of CIS history, has become a thing of the past, according to *Segodnya.* The Moscow session was unemotional, modest, and businesslike.

The problem of the creation of an Euroasian alliance, included in the agenda on the initiative of Nursultan Nazarbaev, again sparked a lively discussion. The idea was discussed from a theoretical point of view, because most of the CIS leaders who attended the session believe that the CIS has not yet exhausted its potentialities. In the opinion of *Krasnaya Zvezda,* the creation of the Interstate Economic Committee may give a new impetus to the development of the CIS.

---

## 7.45

**Nazarbaev on Setting Up Interstate Economic Committee**

From the "Presidential Bulletin" feature compiled by Nikolay Zherebtsov and Andrey Petrovskiy; and edited by Vladimir Shishlin
Interfax, 24 October 1994 [FBIS Translation]

---

Kazakh President Nursultan Nazarbaev believes that the setting up of an Interstate Economic Committee [IEC], the first transnational body in the CIS, is evidence that the Commonwealth is very much alive. He said that this step fit the idea of a Eurasian Union. In Nazarbaev's view, the IEC could be chaired by somebody not coming from the old structures but a well-known personality in the CIS, such as Grigoriy Yavlinskiy (*IF Note:* a leader of the Yavlinskiy-Boldyrev-Lukin group in the Russian State Duma, a well-known economist and a coauthor of the Treaty on Economic Community of the former USSR's republics in 1991).

Even though the CIS summit only took note of the Eurasian Union project, its ideas have played and will play a significant role in the integration of the Commonwealth member nations, Nazarbaev thought.

---

## 7.46

**Chyhir on Results of Moscow CIS Summit**

From the "Presidential Bulletin" feature compiled by Nikolay Zherebtsov and Andrey Petrovskiy; and edited by Vladimir Shishlin
Interfax, 25 October 1994
[FBIS Translation], Excerpt

---

At a press conference on Monday evening in Minsk, head of the Cabinet of Ministers Mikhail Chyhir expressed his opinion saying that "one should not demand immediate results from the agreement on creation of the Interstate Economic Committee (IEC) signed by heads of CIS states." "This supranational body has no leadership yet but it must work on a permanent basis," he said. "However, having signed this document, the presidents have given their private guarantee that the agreement will function," Chyhir noted.

He expressed his conviction that the IEC would not infringe upon national interests of CIS states. In his opinion, this body will assist each republic "in testing the integration process on a scale which it deems appropriate."

Chyhir reassured listeners that the IEC is an organization which controls and manages, it will primarily be busy with transnational sectoral issues. Among them Chyhir mentioned power engineering, communications, and transport.

Chyhir noted that in the course of his meeting with Russian Premier Viktor Chernomyrdin in Moscow they reached an agreement that economic cooperation between Russia and Belarus would deepen irrespective of the results of integration processes taking place within the Commonwealth.

Commenting on the proposal of Belarusian President Aleksandr Lukashenka to station the IEC headquarters in Minsk, Chyhir expressed his doubt that the republic "will be able to tow this burden." He pointed out that there would inevitably appear problems related to transport, communication, accommodation of the committee's employees if the IEC were located in Minsk. "Even if we allocate apartments, highly qualified specialists would hardly agree to come to Minsk," he said. . . .

The agreement on a Payments Union signed by the heads of CIS states is necessary for mutual accounts, Chyhir said. He called the creation of an interstate bank, the work on which was suspended due to the absence of large mutual projects, a practical move for implementing this agreement. Chyhir said that the information stating that Belarus has withdrawn its share of capital from the bank under creation is invalid.

"Of course, the bank will not work within the next half a year. But there is a need for its creation in the future," he indicated.

## 7.47

**Yeltsin Message to Interstate Economic Committee Presidium**
Interfax, 18 November 1994
[FBIS Translation]

Russian President Boris Yeltsin has sent a message of greetings to the participants in the first session of the Presidium of the Interstate Economic Committee which is to open in Moscow on 19 November, the presidential press service announced on Friday.

As the chairman of the Council of CIS Heads of State, Yeltsin expressed the conviction that it was a "remarkable event in the life of the CIS." The beginning has been laid for the work of a body which possesses great powers and which is expected to set the mechanism of the Economic Union into motion, says Yeltsin's message. The immediate task, the message goes on to say, is to establish a Customs and Payments Unions, and to create conditions for the transition to a common market of goods, services, capital and labor resources.

Yeltsin believes that "favorable conditions have been created for the integration of the CIS countries." He also expressed confidence that the Interstate Economic Committee would become an efficient instrument for strengthening the CIS.

## 7.48

**Russian Minister on Economic Union Aim**
Interview by Aleksandr Zolotarev
*Rossiyskie Vesti,* 8 December 1994
[FBIS Translation]

[Interview with Marat Khusnutdinov, Russian Federation Minister for Cooperation with CIS Member States, by Aleksandr Zolotarev. ]

[Zolotarev] Marat Khayrutdinovich, what does the CIS integration program adopted in October represent?

[Khusnutdinov]: It consists of not one but several documents. They are the memorandum "The Main Directions of the Development of Integration in the Commonwealth of Independent States" and the "Long-Term Plan for Development of CIS Integration." They were adopted as "one package," as they complement one another. The memorandum contains basic agreements on the aims of integration and ways of moving toward it. The Long-Term Plan sets out for perhaps the first time concrete measures, indicating specific time limits and the people responsible for their implementation.

The plan did not arise out of nothing. It is built on the fairly solid legal contractual basis created earlier. You only have to recall the treaty on an Economic Union, the agreements on the formation of a free trade zone as the first stage of a customs union, the documents on a payments union, cooperation in the field of investments activities, and efforts to facilitate the development of coproduction, and so on. Several of those adopted earlier are now no longer in force. For example, the introduction of national currencies in the CIS states "repealed" the agreements on forming a single ruble zone.

I would like to express my satisfaction that, for the first time in the CIS republics' as yet short life, favorable conditions have appeared not only for normalizing but for expanding multilateral ties among them. So we have every reason for talking about the beginning of a new phase in the development of the CIS states' integration.

[Zolotarev]: How long will it take to implement the program?

[Khusnutdinov]: It is likely to be difficult to give a concrete schedule. You see, we have formulated a plan to organize effective economic cooperation and the gradual formation of an Economic Union, as well as cooperation in the humanitarian and social spheres and in the sphere of human rights. Add to that the formation of a security system and the implementation of the idea collective peacekeeping for the Commonwealth states, as well as joint measures to guarantee the security of their external borders.

As you see, there are various tasks which will be carried out as we become ready to do so. By the end of this year a draft agreement should have been produced on a Customs Union and in the first half of 1995—a draft blueprint and plan of action to form a common market for goods, services, capital, and manpower. Draft agreements for the formation of a currency union will be ready by mid-1998.

[Zolotarev]: Are any steps being taken to that end, or is everything as yet limited to discussions?

[Khusnutdinov]: Practical work is, of course, being carried out for the purpose of implementing the agreements signed. Here are some examples. You know that there is an agreement on the formation of a Payments Union. The Ministry for Cooperation with the CIS States and the Bank of Russia are holding talks on concluding the necessary bilateral agreements on direct action with those states interested in bringing the union into being. Models for these agreements have been produced.

A start has been made on forming the future common market's production infrastructure: Work is being carried out on the formation and development of transnational associations. At the end of March 1994 Russia and Kazakhstan came to an agreement on the principles for creating Russian-Kazakh financial-industrial groups. Similar agreements with Ukraine and Belarus are being prepared. The formation of such groups is being organized on the basis of intergovernmental agreements.

In particular, schemes are being examined to form a financial-industrial group from the "Gazprom" Russian joint-stock company and the "Kazakhgaz" holding company, to set up the "Mezhdunarodnyye Aviamotory" financial-industrial group with Ukraine, and to organize a Russian-Kirghiz [Kyrgyz] financial-industrial group to be called "Roskyrgyzinvest." From May of this year a Coordinating Council for Problems of Transnational Structures has been operating on a permanent footing at the Ministry for Cooperation with the CIS States. What also should be pointed out are the joint projects in the areas of investment, scientific and technical cooperation, and the deepening of cooperative and direct ties between enterprises.

[Zolotarev]: But who is monitoring all these extensive and multifaceted activities?

[Khusnutdinov]: There is no single coordinating body. All the Commonwealth's statutory bodies are working in this area. As far as the future is concerned, executors for each of the measures planned have been identified. They include the Council of Foreign Ministers, the Council of Defense Ministers, the Collective Security Council, the Permanent Consultative Commission for Peacekeeping Activities, the Councils of the Leaders of Foreign Economic Departments and the Customs Services, the interstate and national banks, the Interstate Industrial Council, the Interstate Scientific and Technical Council, and others.

[Zolotarev]: But how, nevertheless, can the implementation of decisions be monitored?

[Khusnutdinov]: As I have already said, the integration plan envisages specific measures to monitor the progress of the agreements' implementation. But both the Council of Heads of Government and other CIS bodies will deal with this in their sessions.

Clearly, sanctions can be applied to those who do not carry out the decisions adopted. It is, however, necessary to recognize that work on this question has not yet been completed in the Commonwealth. The memorandum notes that a mechanism for cooperation which ensures the implementation of the multilateral agreements is essential, that Commonwealth bodies can make executive decisions including mandatory ones, and that an effective monitoring system is necessary.

[Zolotarev]: What role does the Interstate Economic Committee [IEC] have in the adopted program of integration?

[Khusnutdinov]: The Economic Union has a permanent coordinating and executive body in the shape of the IEC, which performs monitorial and administrative functions. Now the whole range of questions pertaining to the socio-economic sphere of cooperation are passing into its jurisdiction. The IEC will also examine questions prepared by the Council of Foreign Ministers, the Council of Defense Ministers, and other Commonwealth bodies, if their resolution entails significant expenditures.

The Committee has been given wide-ranging rights. On a basis of consensus it can adopt decisions on matters of principle concerning the Economic Union's development and on specific economic questions, taking the states' economic potential into consideration. The IEC prepares the most important questions for scrutiny at sessions of the Commonwealth's Council of Heads of Government and Council of Heads of State.

All IEC integration measures directed at forming a common market will one way or another serve to maintain and deepen cooperation between technologically interconnected works. It is expected to play a particularly large role in ensuring the normal functioning of energy systems, transport, communications, gas and oil pipelines, and other facilities and spheres of a transnational nature.

## 7.49

**End to Concessions in Trade Sphere Warned**
ITAR-TASS, 9 December 1994
[FBIS Translation]

*Editor's Note:* As the following article shows, Russia started taking immediate steps to bring problems to a head for the other CIS states, which would create strong pressures for economic reintegration. While the non-Russian states should be held accountable for making interenterprise payments, their economic structuring problems and dependence on Russia (except for energy) were not entirely of their own making. The question of bilateral debts to Russia creates the greatest long-term imbalance within the CIS.

Russia will be forced to introduce the principle of advance payment in trade and economic agreements and contracts for the supply of commodities. In order to resolve debt problems, financial penalties for failure to pay on time will become a mandatory condition of such agreements. This was announced today by Viktor Chernomyrdin, head of the Russian government, when he addressed the meeting of the Council of CIS Heads of Government at the President Hotel in Moscow.

We have gone as far as we can on payments for supplies and cannot make any further concessions, Chernomyrdin notes. This is the umpteenth time we have had to bring up the question of the debts owed to the Russian fuel and energy sector by the CIS countries. The Russian prime minister disclosed that on 1 November 1994 these debts amounted to 7.5 billion rubles [R], and almost $6.6 billion of the total is for gas. Gazprom, Chernomyrdin said, has been forced to raise the question of reducing gas supplies for countries that do not pay.

Chernomyrdin proposed that the Commonwealth give priority to resolving the problem of nonpayments. Let us agree, the Russian prime minister said, that any new questions that come up in 1995 will only receive joint examination if the problems of 1994, connected with debts and nonpayments, are resolved in a positive way.

Referring to credit and monetary relations between the independent states of the Commonwealth, Chernomyrdin said that since the establishment of the CIS, the Russian government had granted state credits to these countries worth the equivalent of $5.6 billion, in order to make it easier for the partners to adjust to the new conditions of economic cooperation.

However, Chernomyrdin went on, we now have to acknowledge that these credits have not only not resolved the problem of nonpayments, but have also exacerbated the debt situation. The settlement of these debts is proceeding extremely unsatisfactorily. We have to bring about a qualitative change to our relationship and put into practice the principle of mutual benefit and the idea that our relations with CIS countries take priority over those with other countries.

## 7.50

**Objectives of Interstate Economic Committee Viewed**
Andrey Storch and Gennadiy Ezhov
ITAR-TASS, 9 December 1994
[FBIS Translation], Excerpt

The main objective of the CIS Interstate Economic Committee (IEC) is to make the CIS's plans and agreements practical, Russian Prime Minister Viktor Chernomyrdin said in an opening address.

The IEC should avoid becoming another speechifying organ thrusting away practical challenges it is expected to handle, Chernomyrdin stressed. There is no point in debating on whether national sovereignty is dented by the IEC, an authority with certain executive functions, the prime minister said.

Chernomyrdin said the IEC is expected to be joined by CIS states which want solutions to issues rather than mere discussions. And this requires the IEC to become a think-tank and an executive center of the CIS.

"We have founded an organization from which we expect much. And its range of responsibilities is quite broad: From rapid management of specific painful problems, such as energy supplies and compliance with agreements, to strategic integration problems," Chernomyrdin continued.

The prime minister said CIS members expected the IEC to have just these roles when they proposed both coordinating and executive functions for the committee. It is obvious that rights should be jointly delegated to the IEC to provide for its efficient work, Chernomyrdin said.

## 7.51

### Russia, Kazakhstan, Belarus Sign Accord

*The Moscow Times,* 31 January 1995
[UPI, © 1995 by Independent Press]

The agreement, signed Saturday, envisages unifying trade regimes and customs rules by the second quarter of this year, lifting all mutual trade barriers and eventually creating a Customs Union, which could also include additional former Soviet republics, ITAR-TASS reported.

"We created a strong nucleus for economic union," Prime Minister Viktor Chernomyrdin said after the signing, which came just after Russia and Kazakhstan pledged to unite their armies.

But economists interviewed Monday said the three nations' different interests and economic situations, and their disagreements over trade policy with other countries, would derail any attempt to put the pact into practice.

"So far I have a feeling that once again this will just be a piece of paper," said Yelena Ishenko, an economist in the foreign trade department of the Russian Economics Ministry's Center for Economic Analysis and Forecasting. "First we need more political coordination; then there can be concrete economic agreements. Now, union with these republics, where who-knows-what is going on, would be suicide," she said, referring especially to the porous borders with other neighbors.

The pact requires members to unite their economic, monetary, and foreign trade policies. Previous integration agreements, including an attempt to create a ruble zone, have run aground on concerns that member countries' inflationary monetary policies could undermine the Russian economy.

Another Russian economist, who asked not to be identified, said the three countries would be unlikely to unify their import policies. "Russia has high tariffs on imported cars to protect its car industry. In Kazakhstan, they don't make cars," so tariffs are kept low to benefit the consumer.

He said Kazakh President Nursultan Nazarbaev had refused a request by Russia last year to raise tariffs. He said, "We are not going to pay for the welfare of Russian manufacturers to the detriment of Kazakh consumers."

Russia too does not sound ready to compromise. Russian Customs Committee spokesman Anatoliy Kruglov said recently that Kazakhstan and Belarus would face Russian customs barriers until they adopt Russian customs policies, *Kommersant-Daily* reported. A top Western economist, who asked to remain anonymous, said that potential for integration today is limited because the former republics "are not at the same stage of political and economic reform."

He said he suspected most recent integration agreements had been motivated by "nonmarket" concerns, such as popular nostalgia for the Soviet Union.

Some leaders have argued that reintegration would repair economies hit by a sharp decline in interrepublic trade since the Soviet Union collapsed.

A 1994 World Bank study blamed the "precipitous decline" in interrepublic trade from 1990 to 1993 partly on export restraints "at least as severe as those that impede trade with third countries."

The Russian economist said removing barriers would increase Russia's exports, but not significantly.

"It would legalize the trade that is going on already," he said. "The goods that are now going across the steppe would start going down the highway."

Ishenko said increasing trade would not necessarily increase income for Russia. She said 50 percent of Russian deliveries to former republics were not paid for in the third quarter of 1994—up from 40 percent in the second quarter.

## 7.52

### Ukraine Attends Interstate Economic Committee Session

Radio Ukraine, 11 March 1995
[FBIS Translation], Excerpt

A session of the presidium of the Interstate Economic Committee [IEC], which was set up by a decision of the CIS heads of state, got under way in Moscow on 10 March. A government delegation from Ukraine, led by Deputy Prime Minister Serhiy Osyka, is taking part in its work. Aleksey Bolshakov, chairman of the IEC presidium and Deputy Prime Minister of the government of the Russian Federation, greeted the participants in the session on behalf of the Government of Russia. He wished them success in achieving the main goal of the Interstate Economic Committee—development and interaction of economies of the CIS member states.

Consideration of a draft agreement on setting up an Interstate Currency Committee [ICC] was among the priority issues at the session of the IEC presidium. The ICC is supposed to promote multilateral cooperation in the field of currency and credit relations, development of forms and methods of coordinating the monetary, credit and currency policies of the signatories to the agreement on setting up a

Payments Union of the CIS countries. As is known, Ukraine will join the Payments Union only after its national currency is put into circulation. However, at the current session the delegation of Ukraine submitted a number of fundamental proposals concerning the draft agreement on setting up an Interstate Currency Committee. In particular, these are proposals concerning the procedure of seceding from the ICC, which should be based on internationally recognized principles.

The activities of the International [as heard] Currency Committee will undoubtedly contribute to developing economic cooperation between the CIS countries, bringing their currency procedures closer to each other, speeding up mutual settlements, and regulating the issue of debts. The desire of all the participating sides to build their economic relations on the basis of mutual respect and equality should be regarded as the main factor in this process. The draft agreement on setting up an Interstate Currency Committee will be submitted for the consideration of the Council of the CIS Heads of State. . . .

## 7.53

### Ukraine Abstains From Joining CIS Customs Union

UNIAN, 11 March 1995
[FBIS Translation]

Tajikistan and Kyrgyzstan have submitted applications to the Interstate Economic Committee seeking admission to the Customs Union. Ukraine did not submit an application to join the Customs Union. Aleksey Bolshakov, deputy head of the government of the Russian Federation and chairman of the Presidium of the Interstate Economic Committee, relayed this news to journalists after the Presidium and the board of the Interstate Economic Committee ended their 10 March session.

At UNIAN's request, Ukraine's Deputy Prime Minister Serhiy Osyka commented on this fact as follows:

"Ukraine supports the idea of setting up the Customs Union in general, but the distance to it should be covered not at a rocket-escape velocity, but at a normal pace." According to him, the agreement on free trade, concluded between Ukraine and the Russian Federation, will contribute to coordinating the fiscal, taxation, and economic functions of the customs borders between the two countries. "The law enforcement and political functions of the customs borders will remain unchanged," Serhiy Osyka pointed out.

In his opinion, the issue of Ukraine's accession to the Customs Union will depend on the effectiveness of free trade between Ukraine and the Russian Federation.

## 7.54

### CIS Governors Criticize "Integration from Above"

Olga Kolotnecha
*Kommersant-Daily,* 14 July 1995
[FBIS Translation]

It seems that phenomenal activity is characteristic not only of some of Russia's regional leaders but also of their colleagues from the former Union republics. Often in their desire to set up direct economic links, the governors even outdo the government. Recent confirmation of this has been provided by a conference of the heads of twenty-seven oblasts of Ukraine, Russia, and Belarus that just ended in Minsk. Conference participants signed a cooperation agreement envisioning the provision of the most favorable conditions for effective economic integration.

Russia can be considered the initiator of the intergovernor movement: The first conference was held in January in Novgorod under the personal patronage of Andrey Kozyrev and the subsequent ones—in spring in Novgorod and this time in Minsk—were attended by representatives not only from the Ministry of Foreign Affairs but also from the Cooperation Ministry [Minsotrudnichestva]. The Minsk meeting gained in its official weight from a presentation by Belarus President Aleksandr Lukashenka. True, the authorities' attention did not prevent the governors from subjecting "integration from above" to serious criticism. A special target of attack was the Interstate Economic Committee, which, in the region's opinion, has failed to become an effective structure, and with it all heads of the Commonwealth states for the non-viability of most of their decisions. A real alternative to the cumbersome bureaucratic machine of the CIS, in the participants' view, can be direct interregional cooperation, conditions for which, however, should be created by the Commonwealth governments.

So as not to drag out this process, the governors immediately drafted an inter-government agreement "On Cooperation of Regions of Russia, Belarus, and Ukraine." The fourteen articles of this document, its authors believe, should bring closer the bright economic future of the CIS. The agreement in particular provides for mutual investments, the conclusion of agreements directly between regional economic entities, the creation of joint ventures and interregional shareholding companies, and also a free exchange of labor and intellectual resources. Separate treaties should regulate freight, cooperation of law enforcement bodies, environmental protection schemes, and other spheres of mutual interests.

The main condition for the implementation of the agreement, according to the heads of the regions, is the conclusion of a customs union. The request to speed up this process is contained in a conference address to the governments of the three countries. The governors' aspirations quite coincide with the hopes of the Russian government, which somewhat clarifies the participation of the Russian Foreign Ministry as a midwife. The criticism of the authorities heard at the meeting only strengthens the impression of an accurate calculation on the part of the latter: Even by criticizing the government, the governors implement its will. It is noteworthy that even the delegates from Ukraine, whose government cannot be suspected of particular affection for customs unions, were well disposed toward integration. A representative from Kiev oblast states that the next meeting in September will necessarily be held in Kiev even if this is not to the liking of the Ukraine authorities. And although it is not quite clear yet who has earned the main laurels for the successful operation "in the enemy rear," it is obvious that a powerful "fifth column" has appeared on the territory of one of Russia's main adversaries in the customs war.

## 7.55

**Kuchma: Too Early for Customs Union**
UNIAN, 17 July 1995,
[FBIS Translation]

Ukrainian President Leonid Kuchma, who arrived in the Belarusian capital today on an official visit, told journalists at the airport there were no border problems between the two countries.

He said: "Ukraine is in favor of a Customs Union, but it is too early yet. This is not an issue for the present. We would be pleased to establish a border-crossing system so that our citizens would not be inconvenienced, but we are not going to remove the customs barriers."

He recalled that the two countries' European neighbors, Poland, the Czech Republic, and Hungary, have allowed themselves ten years to form a Customs Union. Kuchma stressed the need to create a legislative framework and sign a free trade treaty before such a union is established.

# 8

# On Constitutional and Parliamentary Processes

## Introductory Notes

Two instruments have been created within the CIS for guiding and coordinating the process of political integration. Russian politicians have relied more on these two institutions than any others within the CIS infrastructure to carry out its attempts to advance a restorationist agenda. The CIS Charter and the Interparliamentary Assembly (IPA) are to become, at least in line with Russia's agenda, the unitary constitution and parliament of a new confederation among the CIS states. Despite the determination of the strong Russian leaders who have headed these two institutions, however, their progress has been very slow, forcing Russia to resort more and more overtly since the spring of 1995 to a unilateral military program to rebuild its strength and influence in the region.

Those who supported a new Union wrote the draft CIS Charter in early 1992. Notwithstanding Russia's protestations that the charter in no way diminished the member-states' sovereignty, it was obvious to most CIS leaders that the language in the document was integrationist. The language which set the long-term direction and goals of the organization left little doubt that the final destination was a renewed Union. The goals, which were directly contradictory to those of equal partnership and loose consultation sought by most of the other CIS leaders (with the notable exception of Kazakhstan's president, Nursultan Nazarbaev), spawned a contentious battle, reminiscent of one between the Soviet center and the periphery. The two-year process of holding out for a diluted document more in line with perspectives of the non-Russian CIS leaders is documented in this chapter, although at the time the specifics on arguments over the charter were kept secret.

After much debate, the final charter was brought to its present form, which does little to directly threaten any CIS member's sovereignty or territorial integrity. Nevertheless, as indicated in one Ukrainian observer's analysis in this chapter, many of the terms (for example, "external borders"), which have become catchwords for a restored Union, are contained in the charter, and the organs referred to as CIS "coordinating institutions" are obviously slated to become supranational bodies. Russia quite clearly needs the charter to legitimize its Great Power policy goals in the "near abroad." As a result, those CIS member states which rejected integrationist goals were forced to react to Russia's imperial ambitions more than to the charter itself.

The IPA, established in September 1992, was envisaged as the vehicle for coordinating and aligning CIS member-states' legislation. As its two successive hard-line chairmen (Ruslan Khasbulatov and Vladimir Shumeyko) from the Russian Federation have made clear, however, it was looked upon by Moscow as the pivotal instrument for transforming the Commonwealth into a close confederation of states. The documents selected here follow the evolution of the IPA's plenary sessions as disclosed in articles by Russian spokesmen (often parliamentarians), who confidently predict eventual success in integrating the CIS, and speeches by its chairmen.

# The CIS Charter

The draft CIS Charter was first debated at the 25 December 1992 CIS heads of state summit. From the reports, it is obvious that the battle over provisions in the charter was fierce. Ukraine and Turkmenistan rejected the entire document, reporting during subsequent press interviews that it needed "further study." Kravchuk made an open statement endorsing a bilateral approach over the collective one, in keeping with his position that the CIS was a transitory organization. Belarus's Shushkevich steadfastly maintained that the charter was incompatible with his country's commitment to military neutrality. Only five of the ten CIS members (Georgia and Azerbaijan had still not joined) initialed the draft.

The draft was scheduled to be the main subject of discussion at the next CIS summit on 22 January 1993. On 5 January, *Golos Ukrainy* published a detailed critique of the articles contained in the twenty-three-page draft by V. Skachko. His analysis, contained here, brings out the blatantly re-integrationist character of this new "constitution." Article 2, for example, states that the purpose of the CIS is: "Comprehensive and balanced economic and social development of member states within the framework of a joint economic space, along with interstate cooperation and integration." Article 9 states that: "obligations that arise during the period of participation in this charter link the corresponding states until their total fulfillment." These statements actually imply that the states must assign certain supranational authorities to the CIS.

Russian leaders were quite aware of the controversial language in the draft and used every method they could think of to ease the way for its acceptance. Another article in this collection quotes Russian Deputy Prime Minister Aleksandr Shokhin, advising Russia and the other Commonwealth members *not to sign* the draft. His reasoning was specious; because the draft gave each state one full year in which to sign, he advocated allowing the parliaments to take their time to ponder it and make suggestions for amendments. He also conspicuously mentioned that "Ukraine would not be able to initiate any dramatic steps." Ukraine, of course, is an essential member of the charter. The Russians were worried that President Kravchuk would resign from the CIS over the charter issue. If its parliament had enough time to consider the draft, however, Russia calculated that Ukraine would go along, and it would have time to "work on" other member states. Under the rules the council had adopted, any state not signing the charter within one year would automatically become an observer, and its intra-CIS relations would be conducted at the bilateral level only.

Ukraine was not the only worry, however. In an interview recorded here, a few days prior to the January summit, Turkmen Deputy Prime Minister Boris Shikhmuradov announced that Turkmenistan opposed any "rigid structures" and rejected the entire concept "of the creation of a center." He said his country considered the charter "unnecessary" given the other structural documents which had already been signed. He stressed the feeling of his administration that the CIS should develop along "horizontal," not "vertical" lines. Likewise, in a pre-summit statement included here, Stanislav Shushkevich emphasized that the many different goals expressed in the charter "should be reached in stages . . . ," implying that they should not all be contained in one overarching document.

Russia was also concerned that it had troubles with the Central Asian states. Yeltsin's statement as he boarded the plane for the summit, which appears in this series, was that all the CIS states "had to sign" (even though he did not expect Ukraine to do so) because "we must not allow the Central Asian republics and Kazakhstan to break away."

At the January summit, all the CIS leaders were deeply concerned about their countries' economies, and especially about payments to Russian enterprises. Some accepted the charter as a way to meet Russia's wishes and to perhaps persuade Russia to be more generous with its interstate ruble credits. The outcome was that seven of the ten members signed the draft. They all had one year in which to obtain ratification by their parliaments.

As documented in this collection, the Kazakh, Russian, and Kyrgyz parliaments ratified the charter in April. Armenia, Tajikistan, and Uzbekistan ratified the charter in December 1993. Georgia ratified in March 1994, forced by its need for Russian assistance in its civil war in Abkhazia. The Belarusians waged a long battle between the prime minister, Vyacheslav Kebich, and Speaker Shushkevich (Belarus still had no president), but finally ratified the charter in April 1994 with the reservation that its troops would not fight outside its own borders. Ukraine and Turkmenistan never signed or ratified the charter.

---

## 8.1

**Working Group Coordinator on CIS Charter**
Interview
*Narodnaya Gazeta,* 5 December 1992
[FBIS Translation], Excerpt

---

[Interview with I.M. Korotchenya, coordinator of the CIS Working Group and people's deputy of the Republic of Belarus, by unidentified *Narodnaya Gazeta* reporter. ]

Two events—the anniversary of the signing in Belovezh of the Agreement on the Formation of the Commonwealth of Independent States and the completion of the lengthy and strenuous work on the draft CIS Charter—are deserving of a balanced analysis and an exacting appraisal. It was this which led to *Narodnaya Gazeta*'s interview of I.M. Korotchenya, Coordinator of the CIS Working Group and People's Deputy of the Republic of Belarus.

[*Narodnaya Gazeta*]: Ivan Mikhailovich, you do not deny

your direct involvement in the first anniversary of the Commonwealth?

[Korotchenya]: No, I do not. I have participated directly in the preparation and realization of the vast majority of meetings of leaders of the CIS and in the elaboration of many important documents. Since the Council of Heads of State appointed me coordinator, I have been ex officio, as they say, at the source of many initiatives and practical steps of the Commonwealth and, naturally, bear my share of responsibility for all that has been unsuccessful and failed to come to fruition.

[*Narodnaya Gazeta*]: The CIS is commemorating its anniversary, although it is not a question, of course, of the date. What is important is that the Commonwealth, for which breakup was predicted from its very first days, virtually, is strengthening and developing. . . .

[Korotchenya]: Politicians have observed repeatedly that the CIS is a unique formation. Unique if only because communities are usually created for the sake of an intensification of processes of integration. Take, for example, the European Common Market, the Organization of American States, the British Commonwealth. . . . Despite the pronounced differences, they are all connected by a single purpose—strengthening and expanding all-around cooperation.

The states of the CIS, however, while having proclaimed approximately the same ideas, have in actual policy more often moved, unfortunately, in a different direction. If they have not directly prompted disintegration processes, neither have they decisively prevented them. The statements of the leaders of certain countries to the effect that they consider the CIS a transitional formation and see it merely as a mechanism of the civilized divorce of recently inseparable sister republics are widely known.

The unflattering assessments of the Commonwealth expressed in domestic and foreign media are largely justified, alas. Although it makes sense for objectivity's sake to quote other opinions also, of the *New York Times,* for example: "Had the leaders of the three Slav peoples not managed to reach agreement in Belovezh Forest, the disintegration of the Soviet empire could have had far grimmer consequences both for the peoples inhabiting it and for the world."

Many attempts were made in the year of the Commonwealth's existence to convert it from an amorphous formation into a structure capable not only of declaring its good intentions but of consistently realizing them also. We may put in this category the institution of the Economic Court and the idea of the creation of an economic body endowed with coordinating functions. Finally, we cannot disregard the activity of the numerous, already more than fifty, interstate councils and commissions. These are the bodies which are involved in applied, specific, and, frequently, urgent business.

But, I repeat, I am not about to exaggerate the positive undertakings in the activity of the CIS. The time allotted for its formation and people's reserve of trust in it are drying up. The draft Commonwealth charter to be submitted to the December meeting of the heads should, we believe, be the test precisely determining each state's attitude toward the fate of the Commonwealth.

It is in the present version oriented toward the achievement of two principal goals. On the one hand the charter should lay a sufficient legal foundation for the states' cooperation within the CIS framework. And the second goal, which is dominant today, perhaps, is the preservation by all the states of full, wholly unlimited sovereignty.

[*Narodnaya Gazeta*]: Could you not describe the actual preparation of the draft in more detail? We realize what a complex business this is.

[Korotchenya]: The Tashkent meeting of the Council of CIS Heads of State in May of this year may be considered the start of the elaboration of the CIS Charter. It was there that the decision on the preparation of the draft was adopted in accordance with the proposal of B. Yeltsin, president of Russia. Shortly after, legal experts, economists, political scientists, and diplomats were sitting at a common desk. Authorized representatives of the states in the working group joined in actively. In a word, a very authoritative joint team was assembled.

We were rendered great assistance by scholars from various Commonwealth countries, and, as we now say, the remote outside. Their studies and scholarly publications were frequently "support structures" of the draft. I recall, for example, meetings and talks with Professor Emile Noel, who had come to Minsk at our invitation. President of the European University in Florence—one of the world's leading centers of learning—he was formerly present at the start of the European Community, and elaborated the theoretical foundations of its composition and functioning. He was then for many years head of the EC Executive Council. And, of course, his unobtrusive recommendations and counsel were for us a considerable support at the initial stage.

[*Narodnaya Gazeta*]: And what is the fate of the idea of the creation of a flexible, multilevel system of participation in the CIS—popular and supported by the leaders of the majority of states?

[Korotchenya]: As we all know, during the preparation of the charter there were many arguments and a clash of mutually exclusive viewpoints, when finding wording for this article or the other and this section or the other acceptable to all was incredibly difficult. But right from the outset there was among us complete unity on one point: Everyone wanted to see the Commonwealth as open as possible and accessible to cooperation in any form and for any state. It was here that the ideas of the leadership of Russia on "different speeds" of entry into the CIS and "different levels" of participation came in useful. I will go further, a formation with such diverse aims and uniting such dissimilar subjects cannot be rigid and carved out by a single yardstick.

This is why, according to the present version of the charter, associated members may join the CIS also with full rights and obligations of members. These are the states which would like to cooperate not in all but merely in some forms of activity. For example, they may wish to tackle economic problems together, but pursue an absolutely independent policy in the sphere of military organizational development. In addition, in accordance with a decision of the Council of Heads of State, representatives of all other countries could be admitted to meetings of bodies of the Commonwealth as observers.

Nor is the reverse path forbidden anyone either. The sole condition for a state wishing to quit the CIS is that it announce this twelve months prior to its withdrawal.

[*Narodnaya Gazeta*]: It is said that appreciable adjustments have been made to the main goals and principles of the Commonwealth which were proclaimed formerly. . . .

[Korotchenya]: Nothing of the sort. As we all know, these goals and principles were originally formulated in the Belovezh agreements and later developed and amplified in the Alma-Ata Declaration and other subsequent documents and have in the draft charter acquired, if we may put it this way, a certain consummate and concentrated form. But the main goals of the voluntary and equal association of sovereign states remain as before. These are all-around cooperation in the economic, political, defense, humanitarian, and other spheres of life; realization of the rights of each country to freedom of choice in its development path; the creation of a common economic space; and the coordination of economic policy. Cooperation in the assurance of peace and security and joint protection of human rights and basic liberties in accordance with the standards of international law is another goal. And, finally, concordance of action contributing to the formation on the territory of the states of the Commonwealth of a single legal space is ultimately envisaged.

The principles of mutual relations within the framework of the Commonwealth were also declared in the first, initial documents. The most important of them are good-neighborliness, mutual understanding and cooperation, recognition of and respect for the integrity of the territories and the inviolability of the borders, non-interference in one another's internal affairs, conscientious discharge of the commitments assumed, the solution of disputes only by peaceful means, and the rejection in international relations of the use of force or even the threat of force.

[*Narodnaya Gazeta*]: Ivan Mikhailovich, as far as may be judged from the scant press reports, the main stumbling block, virtually, during preparation of the charter was the Commonwealth's coordinating bodies. The president of Kazakhstan, our prime minister, and several other leaders are propounding the idea of the creation within the framework of the Commonwealth of structures endowed with command authority to a certain extent. The president of Ukraine and his Moldovan and Turkmen colleagues are championing the principle of full sovereignty of their states. What has the result been, whose ideas have acquired "charter recognition"?

[Korotchenya]: The charter stipulates precisely that the highest bodies of the Commonwealth are the Council of Heads of State and the Council of Heads of Government. They alone are endowed with the right to adopt decisions, exclusively by way of consensus. What is more, any state may declare its interest in this problem or the other and decline to participate in its discussion. There are, generally, in the charter practically no imperative, mandatory points. For example, the formation of a common legal space is not proclaimed as a duty of states of the Commonwealth; mention is made merely of the desirability of such actions. Each country retains the full, wholly unabridged right to determine actual policy.

Nonetheless, it would be wrong to maintain that the states which sign the charter would have in respect to one another no duties other than those which are recorded in, for example, the Helsinki agreements. While confirming these obligations (and it is superfluous to mention how important they are in respect, say, to the inviolability of borders and the observance of human rights), the charter at the same time provides a general outline of the mechanism which under our specific conditions would make it possible to put these principles into practice.

If, say, we are speaking of a single economic space within the framework of the CIS, it has to be borne in mind that controlling or coordinating institutions are, naturally, necessary for its normal functioning. Specialists in all countries of the Commonwealth without exception are agreed that without such structures, extricating ourselves from the vise of today's brutal crisis and reattaching the completely severed threads of long-standing ties will be very difficult.

It cannot be ruled out that these circumstances will be

taken into consideration by the heads of state when they examine the section of the charter on the Consultative Economic Commission and the working group.

[*Narodnaya Gazeta*]: Does the charter record the birth within the framework of the Commonwealth of new bodies?

[Korotchenya]: To some extent, yes. Aside from the main, "legislative" bodies—the Councils of the Heads of State and Heads of Government—it incorporates Councils of Ministers of Foreign Affairs and Defense and the above-mentioned Consultative Economic Working Commission, to which is entrusted the elaboration of proposals pertaining to the development of socio-economic cooperation and the realization of multilateral decisions, and also the Interparliamentary Assembly, the Economic Court, the Human Rights Commission, and sectoral cooperation bodies.

The new executive body could be called the Committee of Permanent Representatives, perhaps. It should incorporate authorized representatives of the members of the Commonwealth at, it is anticipated, ambassador extraordinary and plenipotentiary level.

The committee would be entrusted with the organization of sessions of the councils of the heads of state and heads of government, fulfillment of their instructions, the preparation of information and proposals concerning the exercise of cooperation in various spheres, the organization of meetings of experts, and certain other functions. Organizational-support work would be assumed by the secretariat of the Committee of Permanent Representatives.

The states should also, undoubtedly, be represented in the Committee and the Economic Commission, which also will work on a standing basis, by top-class professionals with experience of both international work and economic management. In addition, and this is particularly critical for the host country, it is essential to create for these people conditions both for their work and their residency befitting their rank. They should not be huddled together in hotels, and their work should not be paid according to some standards other than diplomatic.

I am sure that the heads of state understand this and are already taking practical steps. I refer to the decision adopted at the July meeting in Moscow on the construction of a building for CIS headquarters in Minsk and certain other documents.

[*Narodnaya Gazeta*]: Does all this mean that appreciable transformations in the present working group are inevitable?

[Korotchenya]: The decision, as I have already said, will always be up to those who have the right to sign the document. I can only say that, in the event of the adoption of the charter, the functions of both the Committee of Permanent Representatives and the secretariat will be somewhat different. . . .

---

## 8.2
**Proposals for Resolving Constitutional Status of CIS**

B. Pugachev
*Rossiyskaya Gazeta,* 9 December 1992
[FBIS Translation]

---

8 December 1991— S. Shushkevich, chairman of the Supreme Council of the Republic of Belarus, B. Yeltsin, president of Russia, and L. Kravchuk, president of Ukraine, sign an agreement in Minsk creating the Commonwealth of Independent States. The agreement stipulates "that the USSR is ceasing its existence as a subject of international law and a geopolitical reality."

Article 11 of the agreement proclaims that "from the moment of the signing of this agreement the application of third-country norms, including those of the former USSR, is prohibited on the territory of the signatory states." In other words from the moment of signing of the agreement three union republics considered themselves as having withdrawn from the USSR.

12 December 1991—The Supreme Council of the RSFSR with its resolution ratifies this agreement. On the same day the Supreme Council of the RSFSR denounces the agreement creating the Union of Soviet Socialist Republics. The withdrawal of Russia from the USSR was thereby granted legal status.

There is no doubt that the procedure followed in the dissolution of the USSR and the withdrawal of Russia from it contradicted norms of the USSR constitution which was then in force.

Neither the president of the USSR nor the Supreme Soviet of the USSR provided an appropriate legal evaluation of such a procedure involving "liquidation" of the USSR. Moreover, the law of the USSR "On organs of state power and administration of the USSR in the transition period," adopted on 5 September 1991 at the Congress of People's Deputies of the USSR, contained an anti-constitutional Article 8—"Provisions of the constitution of the USSR are effective in the part that does not contradict current legislation." The diluted nature of such a formulation opened up broad possibilities for arbitrary interpretation of the union constitution. At the same time the Congress of People's Deputies delegated its rights to amend the constitution of the USSR to the Supreme Soviet of the USSR.

In that manner the Congress of People's Deputies of the USSR itself committed a crude violation of the union constitution for the sake of political expediency. Continuing on that course, the Supreme Soviet of the USSR and the president of the USSR failed to make an appropriate legal evaluation of the procedure itself involved in the dissolution of the USSR, which was undoubtedly a very crude violation of their constitutional duties.

The Minsk agreements of 8 December 1992 [sic] were supplemented with a number of other legal enactments. The most important one of which was "The protocol on the agreement on creation of the Commonwealth of Independent States" signed on 21 December 1991 in Alma-Ata. In accordance with that protocol the CIS itself was changed in a radical manner—their intent to join it was declared by the Republic of Azerbaijan, Republic of Armenia, Republic of Kazakhstan, Republic of Kyrgyzstan, Republic of Moldova, Republic of Tajikistan, Turkmenistan, and the Republic of Uzbekistan. At the same time it was coordinated that the agreement on creation of the CIS goes into effect for each of the sides from the moment of its ratification. The fact that corrections and amendments to the agreement, signed at Minsk, were applied to the protocol is of a paramount importance. Such corrections were also made during ratification of the agreement on the formation of the CIS in the republics and especially in Ukraine. There is no doubt that all of these corrections, just as the indicated protocol of 21 December 1991, had to be ratified by the Congress of People's Deputies of the Russian Federation, which, however, was not done. All the more since the Alma-Ata declaration, adopted the same day by representatives of Azerbaijan, Armenia, Belarus, Kazakhstan, Kyrgyzstan, Moldova, Russia, Tajikistan, Turkmenistan, Uzbekistan, and Ukraine, proceeded from the premise that "with the formation of the Commonwealth of Independent States the Union of Soviet Socialist Republics ceases to exist." In this manner the fact that the USSR ceased to exist was recorded not in an interrepublic agreement but in a declaration which did not have the status of a law-establishing document. In a strictly legal sense the USSR continues to exist to this day, inasmuch as an international law agreement was never concluded concerning its partitioning among the former union republics.

All of the subsequent treaties within the framework of the CIS bore the stamp of legal dualism—without dividing the population of the USSR by means of an international law agreement and not determining its successors, the sides participating in the CIS were likewise unable to establish reliable legal foundations for cooperation.

Not having resolved the issue of USSR succession in the form of an agreement, CIS leaders signed a temporary agreement on the Council of the Heads of State and the Council of Heads of CIS Governments on 31 December 1991. In accordance with that agreement the Council of the Heads of State is empowered to discuss numerous issues, "including problems of succession occurring in connection with the dissolution of the USSR and the elimination of union structures." These problems of succession, including those regarding union property, common union infrastructures of the state (as well as army infrastructures), and citizenship, however, were not resolved on the basis of an agreement.

The above makes it possible to question the constitutionality of the agreement on creation of the CIS and the protocol on that agreement of 21 December 1991 just as that of the decrees of the Supreme Soviet of the RSFSR on ratification of the agreement on formation of the CIS and the 12 December 1991 denunciation of the agreement on formation of the USSR.

Constitutionally the resolution of this issue may be restored by two methods.

First method. The Sixth Congress of People's Deputies of Russia, neglecting to ratify the agreement on formation of the CIS, failed to fulfill its direct constitutional obligations. The next regular Congress of People's Deputies should return to the examination of that issue. Ratification of the agreement on formation of the CIS at the congress, changes in the constitution of the Russian Federation in connection with that agreement, and requirements rigidly formulated by executive power concerning legal resolution of the issue of succession to the "abolished" USSR through an agreement could restore constitutionality with regard to this issue.

The second method involves a study of the case by the Constitutional Court of the Russian Federation. By virtue of Article 2, Paragraph 1 of the law on "The Constitutional Court of the RSFSR," the Constitutional Court cannot evade the study of as important an issue as the correspondence of the agreement on creation of the Commonwealth of Independent States with the provisions of the constitution of the Russian Federation. Constitutional law must be observed in all countries.

## 8.3

## Snegur: Moldova Will Not Subscribe to CIS Charter

A. Pasechnik

*Pravda,* 11 December 1992 [FBIS Translation]

In an interview on national radio Mircea Snegur, president of Moldova, declared that he would not sign the CIS Charter, even if that creates severe economic consequences for it. He considers bilateral ties to be the best variant of mutual relations among countries of the former USSR. Commenting on certain statements by responsible officials of the Romanian MFA, which stipulate deadlines for the unification of Moldova with Romania, he indicated that he strongly favors independence for the republic.

## 8.4

**Disagreements May Prevent Signing of CIS Draft Charter**

From the "Presidential Bulletin" feature compiled by Andrey Pershin, Andrey Petrovskiy, and Vladimir Shishlin; and edited by Boris Grishchenko
Interfax, 18 December 1992 [FBIS Translation]

On Thursday [17 December], the Belarusian parliament gathered for a closed session to discuss the draft charter of the CIS. According to parliamentary sources, the need to consider this issue arose during the discussion of the Belarusian military doctrine, whose individual provisions have to do with relations between the CIS republics and with the CIS as a whole. Chairman of the Belarusian parliament Stanislav Shushkevich is reported to have backed the initiative of a group of deputies to continue the discussion of the CIS Charter. Local observers believe that the parliament's opinion may influence the position of the Belarusian leader during the CIS summit in Minsk.

Having noted that the parliamentary session was closed, Ivan Korotchenya, the coordinator of the working group of the Council of the Heads of State and of the Council of the Heads of Government, refused to tell Interfax [IF] about the situation, but, as a people's deputy of the Republic of Belarus, he said that he had a cautious approach to the spontaneous discussion of the charter.

IF Note: Prior to his appointment to the post of coordinator of the working group, Ivan Korotchenya was a member of the presidium of the Belarusian parliament and chairman of the Commission for Openness, the Media and Human Rights.

"If the deputies start changing the draft charter, coordinated and signed by CIS experts and representatives, this document may not be signed on 25 December," he told Interfax in an exclusive interview. Korotchenya explains this possibility by the fact that "not all of the changes made by the deputies may be accepted by the CIS leaders."

As we reported earlier, the draft charter has a provision on the ratification of this document by the parliaments of all of the CIS signatory states.

IF Note: Earlier, President of Kazakhstan Nursultan Nazarbaev said, speaking on Channel 1 of the Ostankino TV company, that a great controversy is going on on the eve of the Minsk summit. "Reports say that Ukraine, Moldova, and other states disagree with the draft charter. If this is so, the CIS that was established last year, no longer exists," he said. On Thursday, it was announced that Stanislav Shushkevich had canceled his visit to Chisinau. Local observers explain this decision by the Belarusian leader's reluctance to give rise to speculations that he was going to Chisinau to persuade Snegur to join the CIS Charter.

## 8.5

**Shushkevich Urges Moldovans to Collaborate in CIS**

Svetlana Gamova and Eduard Kondratov
*Izvestiya,* 23 December 1992 [FBIS Translation]

Relations with Russia and the CIS were the focus of attention during a visit to Chisinau by a Belarusian delegation headed by S. Shushkevich.

Voicing concern at the relations taking shape between Moldova and Russia, the Belarusian Supreme Soviet chairman attempted to persuade Moldovan parliamentarians that the conference of CIS Heads of State in Minsk is a lever that the Moldovan side can use to resolve its problems. The fact that more than 80 percent of Belarus's commodity turnover is with Russia points to the need to further develop relations, above all with the east, Shushkevich declared. The same is also evidently characteristic of Moldova's economic ties. According to the Belarusian Supreme Soviet chairman, "there will be no miracle with world markets." "If they dismember us, we will lose everything," he emphasized. "They are dividing us and cleaning us all out one by one. Therefore we must unite in our poverty to oppose the well-developed system for collecting tribute from the republics of the former Union."

Moldova prefers at present to be just an observer within the CIS framework, afraid of being crushed by the structures that are being created within it. But it also expects Russia to take steps which will confirm its respect for Moldova's constitutional foundations.

Two points were emphasized at the meeting that took place, and Moldova does not consider it possible to trust Russia and the Commonwealth as a whole unless they are eliminated. The first is the presence of the Russian 14th Army on Moldovan territory. The second, according to Moldovan Supreme Soviet Chairman A. Mosanu, is Russia's support for the Trans-Dniester region. Making a choice in favor of the CIS now would mean "being misunderstood by the majority of the population." To restore trust, the presidium of the Moldovan parliament believes, "Russia must publicly declare that it defends constitutional Moldova and does not support the separatist movement."

At a news conference after signing a treaty of friendship and cooperation it was announced that Moldovan President M. Snegur and Prime Minister A. Sangheli will participate in the Minsk meeting of CIS Heads of State. Shushkevich called this a "courageous action."

## 8.6

## National Democratic Forces Critical of New CIS Charter

Radio Ukraine, 27 December 1992
[FBIS Translation]

The signing of any version of the CIS Charter by Ukraine representatives would be a betrayal of the Ukrainian people. This was stated by the Council of the Congress of National Democratic Forces of Ukraine after discussing and becoming familiar with the draft of that document. In the opinion of members of the council, the new designers of the single and indivisible state [Russia] want to do with the help of the CIS Charter what they failed to do with the help of the new union treaty. The Council of the Congress believes that signing this document would impose new bondage on our people. The leading body of the congress called on all democratic forces of Ukraine to unite into an anti-imperial front.

## 8.7

## CIS Working Group Coordinator Interviewed

Interview by Aleksey Eroshenko
*Rossiyskaya Gazeta,* 31 December 1992
[FBIS Translation]

At the intersection of two alleys in Minsk there is a gloomy and chilly looking building of unremarkable appearance whose offices for a year now have been linked by invisible capillaries to the circulation system of the giant organism called the CIS. Before the New Year our correspondent in Belarus, Aleksey Eroshenko, met here with I.M. Korotchenya, the coordinator of the CIS working group. He speaks:

Over the year the CIS has existed there have been many attempts to transform it from an amorphous formation into a community capable not only of declaring good intentions but also of implementing them consistently. The Economic Court, the Coordination Economic Council conceived but not yet born, and much else deserves the best possible fate on this plane. Nor can we fail to value the activity of over fifty interstate councils and commissions which have made their contribution to the CIS life support system. But these are all just "units" and not an integral structure of the CIS locomotive. The charter submitted for approval by the next summit meeting should be the universal mechanism which defines in principle each subject's attitude toward the fate of the Belovezh Forest formation.

A formation which unites so many dissimilar "characters" cannot be strictly cut according to a single template. None of the document's forty-five paragraphs contains a single mandatory, arbitrary point. Each country retains the right to determine its own policy. Thus, on the one hand, the legal base is formed for cooperation among states and on the other their sovereignty is preserved with nothing to encroach upon it. That is what we cudgeled our brains over. Here Russia's ideas on "different speeds" for joining the CIS and different levels of participation in it were very useful indeed.

Mutual accounts between the states should be just as trouble-free as they were in the recent past between the republics of the former Union. This optimistic forecast is connected with the creation of an interstate bank, the draft agreement on whose formation has already been signed by experts of eleven countries, although for some details complete unanimity has not yet been reached. In this case the search for a single international means of accounting has been tricky since many republics have their own currency in circulation. Everyone agreed beforehand on the Russian ruble. Although the most attractive proposal was the one for the introduction of a special inter-nation currency like the European Economic Community's Ecu [European currency unit].

## 8.8

## Kravchuk, Conference Delegates Discuss CIS Charter

Interfax, 4 January 1993 [FBIS Translation]

Ukrainian President Leonid Kravchuk has said that before the CIS Charter can be signed, Ukraine must conclude large-scale economic agreements and interstate treaties with Russia and other CIS member states. He was speaking at a consultative meeting of delegates from political parties, socio-political associations, trade unions, cultural workers, women's and youth organizations in Kiev on Monday [4 January]. The consultative meeting discussed whether Ukraine should put its signature to the Charter of the Commonwealth of Independent States at the forthcoming CIS summit.

Kravchuk argues that the emphasis must be placed on wider bilateral and multilateral cooperation within the CIS. The Ukrainian president says that any integration process must stem from the idea of resolving economic problems first and political problems second.

Participants in the conference have said most delegates believe the charter in its present form should not be signed. They argue that if signed after all, the charter would upset the principle of free membership of the CIS and run counter to the agreements concluded in Belarus and Alma-Ata.

Most speakers on behalf of their organizations asked the president, the cabinet, and parliament to consistently protect Ukraine's sovereignty in keeping with the Ukrainian Independence Declaration Act of 24 August 1991 and the will of the people expressed in the referendum of December 1991.

The consultative meeting was attended by the Ukrainian Prime Minister Leonid Kuchma and first deputy speaker of the Ukrainian parliament Vasiliy Durdinets.

## 8.9
## Ukrainian Government Declines to Sign

Interfax, 5 January 1993
[FBIS Translation]

The CIS Charter offered for signing now coincides neither with Ukrainian potentialities and legislative norms, nor with its national interests. It answers to the interests of those forces which want to return to a united state, said Ukraine's president, Leonid Kravchuk, on Monday [4 January] at a consultative conference with spokesmen for twenty-six Ukrainian political parties, associations, and movements. Participants in the conference discussed Ukraine's attitude to the plans to sign the CIS Charter.

According to L. Kravchuk, at first it is essential to resolve common problems, finalize the necessary documents, and consider the consequences of signing the charter for all members of the Commonwealth.

Ukraine, said its president, will not consider the prospects for participating in the CIS in the capacity of an associated member or observer. According to him, the republic might consider the entire package of documents with addenda and explanations to it, not the suggested CIS Charter. However, it might take decades to work out such documents.

World public opinion, said L. Kravchuk, is interested in having the CIS Charter signed, because it fears that the development of events in the former USSR might take the Yugoslav road, and that nuclear arms will get divided uncontrollably. "I don't think that our internal problems interest them," the president said.

Ukraine also has forces that want it to be a member of the CIS, said L. Kravchuk. "I held a post in the CPSU when the USSR existed as an empire. Now I'll spare no efforts in opposing the attempts to return to an empire," he said. "A different attitude would be a step away from my people and from my country." According to L. Kravchuk, Russia has always been beset by the idea of a superpower and a leading force in the CIS and beyond it. "We are not criticizing it for this. Russia pursues a policy that answers to its interests," he said. Ukraine must not interfere in the internal affairs of any country. However, the Ukrainian president said, as a member of the United Nations and of the CSCE, Russia should be guided in its policy by international norms and principles; it has no right to prevent the Ukrainian people from expressing their will. Besides, said L. Kravchuk, if there are wise politicians in Russia, they'll do their best to avoid confrontation between the two countries.

The president dismissed the assertions that Ukraine's failure to sign the CIS Charter will bring about economic pressure on it. "No nation will part with its wealth in favor of Ukraine simply for its signature to the charter. President Yeltsin told me in a cable that from now on Russia will determine the general quota of oil exports, not quotas for each consumer state, as it did previously," said L. Kravchuk.

## 8.10
## Proposed CIS Charter Criticized

V. Skachko
*Ukrainy,* 5 January 1993
[FBIS Translation]

This question inevitably arises if one carefully reads to the end this twenty-three-page document entitled "Charter of the Commonwealth of Independent States" (Draft). Along with the question of "Who needs it?" there appears another no less important one: "What for?"

In order to answer these questions, in my opinion, it is necessary to return to the sources—to 8 December 1991 when the CIS was born below Minsk, and 21 December of the same year, when ten of the fifteen former union republics unified in the Commonwealth. The CIS was regarded differently in the various countries, but they all noted one factor: the Minsk–Alma-Ata Act signaled the final demise of the USSR of the old type, but also the start of a new process for those ten states which did not find sufficient national fortitude to say "farewell" to the largest and newest empire and embark on an independent voyage. They all lived on hopes and anticipations, the different nature of which comprised a tight new complex of contradictory problems, in part with directly opposite content and directions for their resolution. The CIS, as the final death knell of the Soviet Union, was necessary for all, but with respect to their answer to the question of "What for?" the countries diverged. Some of the new states (including Ukraine) viewed the CIS as a mechanism for a civilized divorce, so that it would really not be necessary to break off primarily economic ties. Some other states were willing to continue performing the role of the newest "underbelly" (A. Solzhenitsyn) of Russia and have that powerful ally and (military, economic, and finally, political) patron protecting against solicitations of other states of the world. Finally, Russia occupied a separate position within the CIS. It was recognized by the world community as the sole legal successor to the Soviet Union and which, according to the statement of its president in Kiev in March of last year, in fact declared: "We never have ceded from the USSR." It was then that the contradictory nature of the CIS was revealed in a clear manner: to be equal partners and to strive to dominate at the same time is impossible. In the year of its existence the CIS irrevocably proved this new truth which, at the same time, is as old as the world. The CIS failed to resolve a single serious problem in any spheres of activity of the young independent states. Still, the achievements within the framework of the Commonwealth may be attributed only to the bilateral interstate agreements, including the ones between Russia and Ukraine. For Russia, however, that is apparently not enough.

Does the "CIS Charter" reflect these realities? Yes, fully and entirely. For while proclaiming absolute de jure equality and independence of members of the Commonwealth, it also de facto preserves the impossibility of realizing these principles. At the same time if the charter would simply establish another incontrovertible truth of the day, that it is impossible to be absolutely independent in a world which is moving toward planetary integration, there would be nothing to be afraid of. But Article 2 of the document already states that the purpose of the CIS is "Comprehensive and balanced economic and social development of member states within the framework of a joint economic space, along with interstate cooperation and integration."

Further on it gets "even better": Article 9 proclaims that "obligations that arise during the period of participation in this charter link the corresponding states until their total fulfillment." In other words, if you signed the charter, and its fulfillment is detrimental to your national interests, you must still fulfill its provisions. Otherwise (Article 10) with regard to the violating state "measures may be initiated which are allowed under international law." What kind of measures these are is not indicated. Why?

There exists a certain principle: a state is totally independent only when it fully controls primarily its economic space. Not some "general" one, but its own, and establishes contacts with any state on that basis. "The CIS Charter," however, in Article 19, offers "the formation of a common economic space based on market relations and free movement of goods, services, capital, and manpower." That same article speaks about "development of transportation and communication systems, coordination of credit and financial policy, cooperation in standardization and certification of industrial production and goods," etc. It is not difficult to imagine the reaction, for instance, of American Ford to a proposal by Japanese "Toyota" to standardize production. We, however, are being offered just that under the guise of "formation of a common economic space, a European and Eurasian market, and customs policy" (Article 4).

The term "external borders" of the CIS is incorporated in "the charter," which calls for joint protection by combined means, including military forces, stipulating the creation of common forces which could be used to resolve disputes, naturally, upon agreement of the sides (Articles 11–13). What if there is no such agreement? See Article 10 above.

The charter also talks about "joint coordinating institutions" (Article 4). I attempted counting the number of such "institutions" just in the economic sphere: Council of Heads of State, Council of Heads of Government, Consultative Economic Working Commission attached to two councils, Economic Court, organs (councils, committees) of branch cooperation, and also the Committee of Permanent Representatives headed by a coordinator with his own secretariat.

Some of these "institutions," of course, engage in dealing with political problems as well. In politics, however, they are assisted by the Council of Ministers of Foreign Affairs, the Council of Ministers of Defense, the Main Command of Unified Armed Forces, the Council of Commanders of Border Forces, the Commission on Human Rights, the Interparliamentary Assembly with its Council. Forgive me, of course, for such a long list, but all of these organs also have their rights and their powers. Instead of a comment I will allow myself to raise another question: What will remain within the competence of national leaders of the so-called independent states?

"The charter" also contains the rather strange Article 20: "Member states cooperate in the sphere of law, particularly through the conclusion of multilateral and bilateral treaties on the extension of legal assistance and promotion of rapprochement in national legislation." That is exactly what it states, no more and no less. Even if one hypothetically imagines that, for example, democrats and conservatives come to power in two individual states, not just economic cooperation will be needed (which is quite possible) but it will also be necessary "to conduct consultations and negotiations for the purpose of working out proposals to eliminate such contradictions." In what manner? What is more important is why is it necessary to bring closer together something that cannot be brought closer together either by treaty or through economic sanctions or even through aggression. Since in either case those to suffer will be the people who, in picking their leadership, may have possibly made a mistake, but they alone have the power to correct it. Legislative alignment under pressure merely preserves non-democratic regimes in individual states. That is confirmed by world experience.

We are talking, however, not about universal problems but about the fact that the Ukrainian delegation in Minsk will be asked to sign this"CIS Charter." There will probably be some economic pressure applied as well in accordance with the principle "either food or independence," coupled with the demand to continue the existence of the CIS. But no longer even in the form of a mechanism for a civilized divorce and conclusion of mutually advantageous bilateral or multilateral agreements, but in another form—that of a peculiar suprastate formation, which is not too far removed from a new "unified and indivisible" state. For even such interstate agreements (according to Article 5 of "The Charter"), "concluded within the framework of the Commonwealth, must suit the goals and principles of the Commonwealth, the obligations of the member states according to the present charter." Authors of the document let it be clearly understood that the CIS will act only with a voluntary rejection of independence, which means it will no longer be what it was conceived as, which was an association of truly independent states. That is why "the CIS Charter" is a hidden trap in the manner of the Soviet Union.

Almost all of the serious political forces have arrived at this conclusion in Ukraine in recent months. State leadership is expressing approximately the same thoughts. Then what is preventing the final expression of opinion regarding the CIS? After all, everyone, I repeat, everyone knows that.

## 8.11
## CIS Charter Seen as Rejected for Political Reasons

Lyubov Khazan
*Nezavisimost,* 6 January 1993
[FBIS Translation], Excerpts

Three diehards of the Eastern Slavic policy prevailed over the fourth just over a year ago in the celebrated Belovezh Forest.

People call it a year of lost opportunities, a year of lost illusions. The illusions have given way to the same old stereotypes, and it turns out that 52 percent of our fellow citizens already regret the demise of the Soviet Union.

The hope of sovereign prosperity was vanishing with each hopeless day while the unforgettable Vitold Pavlovich closed his eyes to the corruption and dealt the final blow to the almost lifeless economy. His high-level patrons and the political leaders who claimed the exclusive right to love our common motherland should know that they were instrumental in Vitold Pavlovich's success. The 52 percent who miss the Union are also partly the result of your efforts, ladies and gentlemen.

According to the Oriental calendar, the year of the monkey, which is not quite over yet, promised us the sneers of fate. One of them, perhaps the most fascinating, is the CIS Charter that was drafted by a certain working group.

It was supposed to consist of ten states, and representatives from most of them took part in coordinating the draft. Ukraine did not participate in these efforts, and judging by our president's fairly harsh words of 4 January, the new charter looks like an attempt to restore the old imperial USSR. . . .

The president cleared up all of the details. When he was asked who needed the charter, he replied that the charter appealed to the forces that want to turn the CIS into the Union. When he was asked where these forces are located, he replied that, first of all, they are abroad, because people there are frightened by the prospect of another Yugoslavia and the spread of nuclear weapons. Second, they are in Russia, because Russia wants to reclaim the glory of a superpower. Third, they are in Ukraine, because those who are campaigning in favor of the charter (primarily pro-communist forces) expect the Union to give them a share of the power.

The president concluded by proposing a compromise: He would consider the draft charter only in a package with other documents—supplements, clarifications, etc. According to his calculations, their compilation will take at least ten years. Therefore, there is no need to bother with the charter now. As they say in the East, by that time either the donkey will drop dead or the shah will expire. The only indisputable thing at this point is our president's ability to make points where no one expected him to. There is no question that the venerable leaders of the political parties and movements gave him the highest marks. This was clear when the irrepressible patriot Stepan Khmara and the equally ardent patriot Larisa Skorik nodded their approval: The president made them happy.

## 8.12
## Shokhin: Russia "Should Not Sign" CIS Charter

Ivan Ivanov
ITAR-TASS, 19 January 1993
[FBIS Translation]

"Russia, as well as other CIS states, should not sign the charter of the Commonwealth at the CIS summit in Minsk on 22 January," Russian Deputy Prime Minister Aleksandr Shokhin told reporters on Tuesday.

The draft charter gives member states a year to sign it, including ratification in parliament. Thus, the approval of the charter in Minsk will only signal the beginning of the process of joining it in the legal framework existing in every CIS state, he explained.

Shokhin does not expect any dramatic developments at the summit. "Neither a new organization will be created, nor the Commonwealth will collapse," he said, adding that "Ukraine will not be able to initiate any dramatic steps."

The Minsk summit will begin the formation of a new legal structure of the Commonwealth which will last for a year.

Even if the Russian delegation does not go to Minsk, it would not impede Russia from joining the charter the next day, according to the deputy prime minister.

Shokhin believes such a "soft" scheme of joining the charter is the correct one as it allows parliaments to participate in the process. Among the most serious Commonwealth problems, Shokhin listed trade and economic relations, as well as credit and monetary issues in 1993. He spoke in favor of creating an interstate bank of the Commonwealth "which is to act as a clearing chamber at the first stage."

## 8.13

## Deputy Prime Minister Rejects Draft CIS Charter

Interfax, 21 January 1993
[FBIS Translation]

In his interview with Interfax Boris Shikhmuradov, the republic's deputy prime minister, declared that Turkmenistan will not sign the CIS Charter. In Shikhmuradov's words, the leadership of Ashkhabad "does not perceive the draft of the charter conceptually." "We oppose rigid structures and deny the idea of creation of any center. Life demonstrates that any issue within the so-called single space can be solved on the basis of bilateral relations. Direct contacts do work and they do not work in the CIS," Shikhmuradov said.

"We welcome a flexible approach with respect to the draft of the charter and believe that the charter is not necessary; there are already basic documents about the creation of the CIS which have been signed. It is necessary to develop horizontally that fragile sovereignty which we have now," Shikhmuradov believes.

## 8.14

## CIS Heads of State, Government Work on Draft Charter

### Russia Favors Signing

Moscow Mayak Radio, 21 January 1993
[FBIS Translation], Excerpt

[Announcer]: It seems that almost all our news items today are in some way connected with the latest meeting of the leaders of the countries in the Commonwealth of Independent States, which opens tomorrow in Minsk. . . .

[Korshunov]: Leonid Grigorievich, beginning, I think, almost from the very first meetings of CIS leaders, there have been claims that the CIS structure will not be long-lived, that it is on the verge of ceasing to exist. However, the most recent meetings would seem to have refuted such claims. Even so, the situation does not permit complete optimism. What might be the consequences if the CIS were to disappear for the territory that used to be the Soviet Union in the event of the current participants in the CIS not signing documents on the further existence of this structure?

[Ivashov]: Well, first I would like to point out that the draft charter will bring the problems of the CIS to a head. In the event of its not being signed by some of the Commonwealth members, they will automatically become observers of the Commonwealth, as it were. Then, the bases for mutual relations will be decided at a rather different level, that of bilateral relations, maybe of multilateral relations, but such close multilateral cooperation will, no doubt, be complicated.

[Korshunov]: When people talk of a possible danger for the CIS or the probability of a particular state not signing the CIS Charter, Russia seems (to be above) suspicion. Is there any possibility or danger that Russia, too, might not sign such a document?

[Ivashov]: At all the meetings, whether of experts, ministers of foreign affairs, deputy ministers of foreign affairs [words indistinct] in Minsk, Russia has been in favor of signing and has merely made constructive contributions to drawing up a draft that would suit everyone. If something quite extraordinary happens and Russia does not sign the CIS Charter, the consequences could, to be blunt, be disastrous. And, in such circumstances, the situation for Russia would be most unfavorable. It would be cut off from close cooperation with the Central Asian region, and bearing in mind that Russia is a Eurasian state and that the greater part of its territory is in fact in Asia, the priority trend of cooperation in the military and in other spheres must be in this region and, all the more so, when the leaders of these states are preparing for closer cooperation and are ready for profound integration. It is not fortuitous that of the six states that signed the Treaty on Collective Security in Tashkent, four of them were the states of Central Asia and Kazakhstan. Thus, it is not in Russia's interest not to sign the Commonwealth Charter.

### Yeltsin Says Ukraine Must Sign

Interfax, 22 January 1993
[FBIS Translation]

President Yeltsin told journalists before flying off to the Belarusian capital Minsk on Friday [22 January] that it was necessary to sign a CIS Charter at the Minsk summit, and all the countries but Ukraine were prepared to do so. However, he noted that "to sign the CIS Charter without Ukraine would be undesirable."

In Mr. Yeltsin's words, participants at the summit are in for hard work and will have to make some hard decisions. He said there would be three main issues on the summit's agenda. The first one concerns the signing of the charter. The president emphasized that the alliance of five Central Asian states should be prevented from sudden secession.

Mr. Yeltsin regards nuclear arms as the second issue to settle with Ukraine and Kazakhstan. As for Belarus, an agreement with it has already been reached. He wished the parliaments of Ukraine, Kazakhstan, and Belarus would ratify the START I Treaty.

The third "very hard" issue in Mr. Yeltsin's view is that of the combined CIS armed forces, the status of which, in his words, is declining. The president urged reaching final agreement on these forces.

He said he was going to focus on the situation in Tajikistan in Minsk. In his opinion, to stop the bloodshed in that republic, Russia's 201st Division should be backed up by peacekeeping troops from other CIS states.

**Three States Refuse to Sign**
Interfax, 22 January 1993
[FBIS Translation]

Discussion of the draft CIS Charter at the summit in Minsk on Friday has shown a divergence of views by various sides. At the joint assembly of the Council of the Heads of State and the Council of the Heads of Commonwealth Governments, the majority were in favor of signing the draft charter.

Belarus announced its readiness to sign the draft, without part 3 regarding collective security, and the provision on dividing border troops. Russian President Boris Yeltsin remarked that "absolutely all provisions of the charter" suit Russia and they correspond with the country's national interests. Kazakhstan was in favor of approving the draft charter.

Until the recess, however, Turkmenistan's position was still unclear; the country does not categorically refuse to sign the charter, but has several reservations. The Moldovan delegation announced that due to the difficult political situation in the republic, and the fact that the Moldovan parliament has not yet ratified the founding documents of the CIS, Chisinau cannot sign the draft charter at this time.

Ukrainian President Leonid Kravchuk confirmed that Kiev did not find it necessary to sign the draft charter in its present form.

The heads of Commonwealth states, taking into account the Ukrainian president's announcements, decided to discuss possible compromise versions during the recess in order to find a formula which Ukraine would find acceptable to join the charter.

Experts think it possible that for instance, a memorandum could be signed or a protocol to the CIS Charter, which would reflect Ukraine's interests.

In addition to this, according to information from Minsk, separate meetings of the Council of Heads of State and Heads of Commonwealth Governments have begun. The presidents have returned to the question of the CIS Charter.

In the first half of the day on 22 January, at the joint meeting the heads of states and Commonwealth governments in principle approved an agreement on a CIS interstate bank and a document on regulating an interstate securities market.

## 8.15

**Seven States Sign Charter**
Leonid Timofeev
ITAR-TASS, 22 January 1993
[FBIS Translation]

Heads of seven CIS states signed the CIS Charter today, although with some reservations. Discussions on its content have lasted the last several months. Moldova, Turkmenistan, and Ukraine appeared to be unprepared for this step.

Explaining the Kiev stand, Ukrainian President Leonid Kravchuk told the final news conference "the adopted wording corresponds to the current situation in the Commonwealth." However, the economic situation is now the most important for Ukraine. He said the time will come for Ukraine to determine its stand on the issue.

The Ukrainian leader also said the Commonwealth as a structure has potential to upgrade its work in all spheres, both political and economic.

The meeting of the [Council of Heads of States] reached an agreement to sign a common statement on the charter. In the opinion of Russian President Boris Yeltsin, the most important thing about the document is that the charter is declared open for signing by any state head any time.

The meeting focused on a set of economic issues. In particular, it reach an agreement on creation of an interstate bank and principles of its work. Government heads considered fourteen economic issues and prepared four issues for discussion together with state heads. All of them were signed, with the exception of ones sent back for improvement. In the opinion of Russian Premier Viktor Chernomyrdin, the meeting, held at a high professional level, was a success.

As for political aspects of the meeting, one should note

the discussion of the Tajik situation, which resulted in a decision to close the Tajik-Afghan border that is, in the words of Uzbek President Islam Karimov, the gates for penetration of arms and drugs on CIS territory. Five states, which signed the agreement on collective security, decided to send a battalion each to reinforce the border.

At the same time, state heads voiced an opinion that Tajikistan and its new leaders, after stabilizing the situation in the republic, must be able to ensure the inviolability of borders by themselves.

Commenting on correspondents' request for the results of the first year of CIS existence, Boris Yeltsin said the Commonwealth "underwent normal and natural processes of self-awareness in the world for young, newly established states." The Commonwealth was living, growing experienced, and has done much, said the Russian leader.

He stressed a "chaotic disintegration" of the nuclear state was prevented. Understanding of the need to preserve the whole set of relations, especially under conditions of an economic crisis, has been reached. "We understood we cannot live without each other," noted the Russian leader.

In the opinion of head of the Belarusian Supreme Soviet Stanislav Shushkevich, results of the CIS existence are evident. "State and government heads have never before met each other with such respect and understanding of each other's dignity. This is the main result of the Commonwealth," he said.

## 8.16

### Kuchma Interviewed by Japanese Papers on CIS Charter

From the "Diplomatic Panorama" feature, compiled by Andrey Pershin, Andrey Petrovskiy, and Vladimir Shishlin; and edited by Boris Grishchenko
Interfax, 4 March 1993
[FBIS Translation], Excerpts

In an interview to Japanese *Tokyo Shimbun* and *Chunichi Shimbun* daily papers, Ukrainian Prime Minister Leonid Kuchma answered their correspondents' questions concerning the current economic and political situation in the republic, as well as Ukraine's relations with its CIS partners. . . .

Q: Will Ukraine be supportive of the policy of greater integration in the framework of the Commonwealth of Independent States and sign the CIS Charter?

A: I don't know a single person who said that the CIS policy was sensible after he has analyzed it. I regard the CIS merely as a screen, and nothing else. In the meantime, I wouldn't take it as a tragedy if Ukraine signed the CIS Charter. In fact, by doing so we would cut the ground from under some of the politicians who seek to set apart Russia and Ukraine. On the other hand, under the present-day situation, I can also see the reason for not signing the charter. In his decision making, the president, certainly, was proceeding from the interests of political conciliation in this country. With Ukraine being divided into two rival camps, it is evident that our most important mission today lies in preserving equilibrium, hence conciliation, in Ukraine.

## 8.17

### Kazakh Parliament Ratifies CIS Charter on 14 April

Fyodor Ignatov
ITAR-TASS, 16 April 1993
[FBIS Translation]

The parliament of Kazakhstan ratified the CIS charter late on Thursday [14 April].

Parliament Speaker Serikbolsyn Abdildin expressed hope that the document, which was signed on 22 January in Minsk, will promote the creation of an economic union among former Soviet republics which may be later joined by former Comecon members.

## 8.18

### Russian Parliament Ratifies CIS Charter on 15 April

Ivan Novikov
ITAR-TASS, 15 April 1993
[FBIS Translation]

The Supreme Soviet (Parliament) of Russia on Thursday ratified the Charter of the Commonwealth of Independent States (CIS) adopted in Minsk on 22 January 1993. The decision won a majority at a joint meeting of the parliamentary chambers.

Anatoliy Adamishin, first deputy foreign minister, who presented the document the charter meets the national state interests of Russia, emphasized [sentence as received].

Adamishin believes that the charter is an important instrument for equitable operation between Russia and CIS partners.

The charter is aimed at enhancing centripetal tendencies in the CIS and at consolidating it, Adamishin said. "This signifies that a new tendency aimed at establishing a zone of stability and active cooperation throughout the former Soviet Union has emerged and is growing stronger," he added. The charter is yet another proof that the CIS makes headway toward civilized work, not a civilized divorce as some predicted, Adamishin said.

The diplomat stated that the adoption at the same time reflected growing awareness of the fact that political independence of their states does not rule out close interconnection between them and moreover that such interconnection is of benefit to all.

He recalled that the charter sets out better guarantees for human rights. "To us this is a factor of exceptional importance since about 25 million of our compatriots live outside Russia," Adamishin stressed.

## 8.19

### Further on CIS Charter Ratification

From the "Diplomatic Panorama" feature by Andrey Borodin, Dmitriy Voskoboynikov, and Igor Porshnev
Interfax, 15 April 1993
[FBIS Translation]

Russia's parliament ratified the CIS charter on Thursday [15 April].

In an appeal before the vote, Speaker Ruslan Khasbulatov asked lawmakers to ratify the charter despite all its flaws. He called the document the core of cooperation between the Commonwealth countries, especially in the run up to the 25 April referendum.

Khasbulatov rebuked the executive structures for their attempt to claim foreign policy as its own sphere of activity. He declared that the holding of a CIS summit in Minsk on 16 April and a meeting of the leaders of national parliaments on the same day testified to attempts to become fully separate.

Reporting on the CIS charter at the session, the first deputy foreign minister Anatoliy Adamishin noted that the statute consolidated a new trend aimed at creating on the territory of the former USSR an environment for cooperation in the economic, humanitarian, and other fields.

In his words, the charter is yet more proof of the Commonwealth's advance toward integration rather than toward a civilized divorce, as was earlier expected.

This international document has already been ratified by Kazakhstan's parliament without amendments or reservations. Kyrgyzstan's parliament is to ratify it on Friday.

"Russia should not lag behind other CIS countries, all the more because it has ratified fewer CIS documents than the other Commonwealth states," Adamishin said.

## 8.20

### Minsk Approves CIS Charter "With Amendments"

ITAR-TASS, 15 November 1993 [FBIS Translation]

The government of Belarus has approved with amendments the charter of the Commonwealth of Independent States which was adopted in Moscow on 22 January 1993.

The charter has now to be ratified by the republican parliament.

An amendment to the section "Collective Security and Military Political Cooperation" says that the use of the Belarusian armed forces in other CIS states, as well as the deployment or use of the armed forces of other CIS states in Belarus, should be sanctioned by the Belarusian parliament.

In Article 16 of the section "Prevention of Conflicts and Settlement of Disputes" the phrase "all possible measures" is understood by Belarus as all possible non-military measures. The republic proceeds from the fact that disputes indicated in Articles 17 and 18 of the CIS Charter should be resolved in accordance with generally accepted principles and norms of international law, the government said.

Belarus believes that the provisions of Articles 28 and 29 concerning the Coordinating and Consultative Committee and the secretariat and Article 30 concerning the main command of the unified armed forces do not fully correspond to the present-day realities and need correction. The latter should be made in the form of amendments to the CIS Charter as envisaged by its Article 42.

## 8.21

### Belarus May Have to Leave CIS if Charter Not Ratified

Interfax, 14 January 1994
[FBIS Translation]

Unless Belarus's parliament ratifies the CIS Charter before 23 January, the republic will be automatically removed from the composition of the Commonwealth, CIS Executive Secretary Ivan Korotchenya declared in his interview with Interfax.

He recalled one thesis from the CIS Charter confirmed by heads of states 22 January 1993 in Minsk. The document stipulates that "those sponsor states which adopt commitments according to the present charter within a year after its adoption by the Council of Heads of States are member states of the Commonwealth." At the same time Korotchenya stressed that "the charter is brought into force for all sponsor states, from the moment they hand over instruments of ratification for storage or for sponsor states which will hand over their instruments of ratification in a year following adoption of this charter."

According to Korotchenya, as of today only six states have ratified the CIS Charter, namely Armenia, Kazakhstan, Kyrgyzstan, Russia, Tajikistan, and Uzbekistan. Korotchenya said that they expected ratification from Azerbaijan and Georgia in the near future. Russia, Kazakhstan, and Tajikistan have already forwarded their instruments of ratification to Belarus's governmental archives, to the country-depository of the CIS.

At the Ashkhabad summit, leaders of the Commonwealth countries have prolonged the term of ratification of the CIS Charter for Moldova for three months more, said Korotchenya.

The session of the Belarusian parliament is to resume its work in Minsk on 18 January. Korotchenya, a people's deputy of this representative body, believes that the issue on ratification of the Commonwealth charter "must be considered without any delay during the first day of the session's work."

### 8.22

**Georgian Parliament Votes to Join CIS Charter**
ITAR-TASS, 1 March 1994
[FBIS Translation]

At its evening session today the Georgian parliament adopted a resolution that the Republic of Georgia should join the CIS Charter. There were 125 parliamentarians voting for this decision, 64 against, 4 abstained, and 3 deputies did not take part in the voting.

Eduard Shevardnadze, head of the Georgian state, who made a speech during the stormy debate on the question of ratification, stressed that Georgia "has no other choice and that this has already become obvious during the events in Abkhazia."

# The Interparliamentary Assembly

The intra-CIS debates concerning the formation of an Interparliamentary Assembly were tense, but perfunctory. An IPA was established in early 1992, with seven states as signatories, primarily because a majority of the CIS leaders considered it necessary to facilitate a smooth separation from the previous economic planning system and interrepublican economic and legal obligations. They also realized that some coordination in the re-drafting of national laws and regulations might be useful or necessary. Nevertheless, the IPA rapidly became the target of intense Russian pressure for integration into a unitary confederal state, with a joint parliament. This collection of documents highlights the outcomes of the IPA Council's plenary sessions, as it tried to define its role within the CIS.

The first plenary session was held on 14 September 1992 in Bishkek. Ruslan Khasbulatov, speaker of the Russian parliament, was elected the Council's chairman. Following the meeting, Khasbulatov announced here that "this is the beginning of the creation of a new confederation." In the interview included here, he compares the IPA to the European Parliament, and denies that his plans for it have anything in common with reinstituting the USSR Supreme Soviet. It is noteworthy, however, that despite the momentous importance Khasbulatov attached to the organization, its agenda in January was restricted to coordinating regulations on social security, conditions for entrepreneurs, and some trade activities, as well as establishing a joint publication on legal issues.

On 29 September 1992, Khasbulatov issued his own "Guidelines" (signed only by himself) expressing the goal of the IPA as being to "provide legal support for a common economic space within the framework of the Commonwealth and to prevent and eliminate juridical conflicts in the process of regulating social and other processes." (The phrasing raises the question of why the IPA should try to eliminate all disagreement, which is bound to occur among independent national legislatures.) The guidelines, which are included here, list ten areas in which Khasbulatov hopes to "eliminate contradictions between national legislation."

At the second plenary session of the IPA Council on 28 December, Khasbulatov proves himself to be an extremely ambitious and controlling chairman. In his opening statement, he says: "Given the chaotic and unpredictable nature of the talks between the heads of the CIS member states, the Interparliamentary Assembly should take charge of the integration processes."

Although Chairman Khasbulatov attempts to lead a discussion on coordinating the constitutions of CIS member states, the representatives confine themselves to sharing experiences and to a discussion of values and principles they hold in common. The disparate views of the agency's role in the Commonwealth are evident from post-session interviews included in this selection of documents. Only the Kazakh representative draws attention to the "former unified state, which existed for hundreds of years," as being a model for creating a "life-support system" in the CIS. The Russian Supreme Soviet representative, Sergey Sirotkin, who is also quoted here, says the IPA is responsible for recreating "the Union's complex legacy at a qualitatively new level."

Nothing much has been written about the IPA's activities during 1993. Membership remained stagnant at six. Leadership attention shifted to the draft Economic Treaty, introduced during the spring CIS summit and to the extraordinary developments in Russia's internal politics. Khasbulatov was removed from the chairmanship following the Russian Supreme Soviet's October uprising against the Yeltsin government. It appears that the IPA's influence over the economic integrationist goals harbored by Russia and Kazakhstan became eclipsed by the Economic Union treaties being drafted within the full CIS and separately among its members (see Chapter 6).

During a special meeting of CIS parliament heads in St. Petersburg in February 1994, Vladimir Shumeyko is elected by secret ballot as the new IPA chairman. A *Segodnya* article in this collection refers to a list of laws which the Russian delegation announces it will submit "under Russian sponsorship" for enactment at the upcoming March IPA plenary session. This list, somewhat ominously, includes procedures for imposing martial law on CIS territory—"On Mobilization," "On the Procedure for the Development, Production, Testing, and Delivery of Armaments and Military Hardware," among others.

Vladimir Shumeyko is another strong sponsor of the Russian restorationist agenda. A series of 1994 press accounts in this collection reviews his goals for the IPA's role in strengthening the CIS. In November, Yuriy Lepskiy, writing here in *Trud*, notes the lack of attention being accorded to the IPA's activities, but predicts that it will nevertheless be the determining force in shaping future CIS integration. It is interesting to note that, in analyzing the IPA's potential impact, he foresees the 1996 Russian presidency going to those who are now "quietly and professionally creating the legal basis" for the renewed Union.

The sixth plenary IPA Council session was held on 20 February 1995. The main issue on the agenda, which is reviewed here, was how to strengthen the IPA, which had not progressed far despite all the dedication of its Russian chairmen. The explanation given for its weak performance in *Rossiyskaya Gazeta* is that the association's initiatives are "not effective due to the disparity between national legislation"—a rather obvious statement making it appear that the Russian delegation might have been deluding itself. Speaking on the eve of the February plenary session, Shumeyko moves far to the right, openly supporting Nursultan Nazarbaev's initiative to create a "Eurasian Parliament." (Shumeyko is re-elected chairman at the session, by secret ballot.) Following the meeting of parliament heads, Shumeyko says a new draft IPA Convention to be discussed in May will raise the institution to a level equal in authority with the CIS Councils of Heads of State and Government. His optimism seems to correspond more with Russian policies than with a consensus of CIS members, of which there were seven as of mid-1995. The series of documents on the IPA included here ends with an interview with Vladimir Shumeyko on 16 May 1995 in which he confidently discusses the outcome of the council's sixth plenary session, going into some detail on what he calls the "model Civil Codes" for the CIS's participating members. The interviewer questions him at some length on why these are considered "model" codes.

## 8.23

**Khasbulatov on New Confederation**

Interfax, 14 September 1992

[FBIS Translation]

Chairman of the Russian Supreme Soviet Ruslan Khasbulatov believes the establishment of an interparliamentary assembly of CIS nations is "the beginning of the creation of a new confederation." He announced this Monday [14 September] in Bishkek, where he arrived to participate in the assembly, which starts 15 September.

According to Khasbulatov, Ukraine's absence in no way affects the work of the interparliamentary assembly, since Ukraine plans to participate in the assembly in the future.

Chairman of Kyrgyzstan's parliament Medetkan Sherimkulov pointed out that he views the Interparliamentary Assembly as similar to the European parliament and doesn't believe it is a repetition of the USSR Supreme Soviet. He said the assembly will become an independently operating organ which will have the right to protest decisions made by heads of Commonwealth states.

## 8.24

**IPA Agenda Described**

Interfax, 15 September 1992

[FBIS Translation]

A parliamentary assembly of CIS countries is opening in Bishkek today. In the morning the council of the assembly consisting of the speakers of parliaments held a session.

On the first day the role of parliaments in expanding relations between community countries is to be discussed, as well as the draft regulations of the assembly. The working bodies are also expected to be elected.

There will be a debate on the efforts of parliaments to guarantee social security, the coordination of legislations in the economy, the foundation of a joint weekly publication, the formation of a parliamentary club of political movements and parties.

According to the press service of the assembly, the Kazakh delegation does not think it useful to form a club because the assembly does not have the purpose of coordinating the actions of political movements and parties.

A draft resolution on cooperation between parliaments in economic legislation has been circulated. The aim is to create favorable legal conditions for entrepreneurship and free trade throughout the CIS.

The draft resolution says that economic laws, including laws on enterprise, finance, banking, prices, and taxes, should be based on common concepts and their adoption should be coordinated. It is necessary to exchange information about legislation that has been passed or is in the makings and it is desirable to remove contradictions between the laws of different member countries.

## 8.25

### Document on Aligning Legislation in CIS States

*Rossiyskaya Gazeta,* 19 September 1992
[FBIS Translation]

"Basic Guidelines for Rapprochement Between the National Legislation of the Commonwealth Member States," signed by CIS Assembly Council Chairman R. Khasbulatov in Bishkek on 15 September.

The Interparliamentary Assembly of the CIS member states recognizes the need to provide legal support for a common economic space within the framework of the Commonwealth and prevent and eliminate juridical conflicts in the process of regulating social and other processes. The Interparliamentary Assembly sets itself the goal of assisting by every means the creation of the most favorable and equal legal conditions for the development of enterprise, economic activity, and free trade throughout the Commonwealth, economic security, and also the creation of common guarantees of citizens' rights.

1. The Interparliamentary Assembly deems it advisable to bring about a rapprochement between national legislation in the following areas:

—the status and general conditions of the activity of enterprises and other economic entities;
—legal support for common energy networks and nuclear power engineering;
—legal regulation of interstate transport networks;
—conditions for the movement of goods and financial resources between states and common procedures for mutual settlements;
—customs rules and tariffs;
—basic conditions for the movement of the work force and guarantees of the labor and social rights of employees;
—conditions and regulations for the exchange of information between states;
—basic principles for the rational exploitation of national resources and for ecological security;
—foreign economic activity, including foreign investments and currency regulation;
—inventions, discoveries, industrial models, and trademarks.

2. It is recommended that activity to bring about rapprochement between national legislation be based on common concepts and be carried out in a coordinated fashion, as far as possible ensuring that the corresponding acts are adopted at the same time.

3. With the aim of eliminating contradictions between national legislation, the Interparliamentary Assembly and the Supreme Soviets (parliaments) will use such forms of work as: recommended (model) legislative acts of the Interparliamentary Assembly, mutual exchange of information on legislative acts after adoption or, where necessary, during preparation, joint discussion of legislation issues, examination of questions of rapprochement between legislation in Commonwealth coordination institutions, scientific conferences and recommendations from academics on ways and means of bringing about rapprochement between legislation.

4. In the event of the discovery of substantive differences in legislative regulation which hinder the formation and functioning of a unified market, a consultative conference of the chairmen of Supreme Soviets (parliaments) is to determine the timetable and procedure for preparing proposals to overcome these differences.

5. In view of the important role of legally binding acts in the legal regulation of relations in the member states of the CIS, the Council of Commonwealth Heads of State is to be instructed to discuss ways and means of coordinating and aligning the legally binding acts as far as their jurisdiction applies.

[Signed] Assembly Council Chairman R. Khasbulatov
[Issued in] Bishkek, 15 September 1992

## 8.26

### Khasbulatov Addresses IPA Session

Interfax, 28 December 1992
[FBIS Translation]

Given the chaotic and unpredictable nature of talks between the heads of the CIS member states, the Interparliamentary

Assembly should take charge of the integration processes, chairman of the Russian parliament Ruslan Khasbulatov, who also chairs the Interparliamentary Assembly, said at the opening of the second plenary meeting of the Interparliamentary Assembly in St. Petersburg today.

The plenary meeting is attended by representatives of Armenia, Belarus, Kazakhstan, Russia, Tajikistan, and Kyrgyzstan, and by observers from Azerbaijan. Among the guests are delegates of the Parliamentary Assembly of Northern Europe, the Northern Council, Finland, and Bulgaria. The agenda includes various aspects of the constitutional reform in the CIS, the implementation of the agreements on coordinating the national legislative systems and principles regulating citizenship matters in the Interparliamentary Assembly's member states.

Having referred to the opinion of many observers, Khasbulatov said that the disintegration of the USSR was not a fatal phenomenon, but the result of serious political errors. He also expressed the opinion that the Interparliamentary Assembly should establish the exact cause of the Soviet Union's breakup.

To quote Khasbulatov, the interest in the assembly's activity has increased lately, largely due to a splash of integration moods, which can be seen from various opinion polls and from the activity of political parties and movements.

He said that the assembly should draft a common policy, reflecting the equality of the CIS member states and excluding any pressure or roll-back to the past.

The Russian Movement for Democratic Reform believes that the plenary meeting of the Interparliamentary Assembly should attach priority importance to the search for solutions that would stop disintegration and create conditions for restoring normal economic contacts between the newly independent states.

On the eve of the plenary meeting, the Movement for Democratic Reform circulated a statement in Moscow, which says that due account should be taken of the bilateral and interstate interests of the CIS states and that the Assembly might discuss the possibility of setting up special state bodies in charge of the entire complex of relations between the CIS member states.

## 8.27

**Participants Cited on Petersburg Interparliamentary Assembly**

Unattributed report

*Rossiyskie Vesti,* 31 December 1992

[FBIS Translation]

*Editor's Note:* The following articles document the discussions which have transpired within the CIS Interparliamentary Assembly, (IPA), an institution slated to bring constitutions and legislation into line with one another, especially in the economic realm. The IPA has assumed responsibility for coordinating, above all, the human rights legislation and control mechanisms of the CIS states. Russian participants, as seen in the following, dominate in these discussions.

St. Petersburg—Beksultan Ishimov, chairman of the Kyrgyzstan Supreme Soviet Commission for Legislation Issues, Human Rights, Defense Matters, National Security, and the Struggle Against Crime:

"Constitutional reform cannot be confined to repairing the existing fundamental laws but is called upon to seriously update their content, principles, and forms. While in St. Petersburg we have seen for ourselves that work on draft constitutions is in full swing not only in our country but also in other Commonwealth countries. We have also seen for ourselves that many of us are trying out national constitutions to combine problems of today and a future which is still hard to imagine. This means that we need to exchange experience and information."

Babken Ararktsyan, chairman of the Armenian Supreme Soviet:

"In speaking of the creation of their own constitutions, my colleagues who have spoken have frequently referred to the experience of other states. But the CIS countries were missing from these lists. I believe that it is possible to create within the framework of the Interparliamentary Assembly a structure which could make an expert appraisal of the constitutions which are still being drawn up. Being the basis of the state's legislation, the constitution must always be the focus of attention for countries which intend to cooperate closely in the future.

Nikolay Ryabov, deputy chairman of the Russian Supreme Soviet:

"We very much hope not only for an exchange of experience in the course of constitutional reforms but also that we will confirm and enshrine in the new constitutions our adherence to common principles. People's lives depend on this, as does, in the final analysis, something of which all people in our states dream: of returning to the principles of friendship, cooperation, and mutual respect which we always had.

Rakhmet Mukashev, member of Kazakhstan's delegation:

"The new independent countries have inherited from the former unified state, which existed for many hundreds of years, a common infrastructure, a unified life-support system, and numerous interregional economic ties. Their disruption in recent years has been one of the most important causes of the crisis. In addition to purely political causes, the splintering of once unified legislation and the emergence of gaps and contradictions in it have been a specific cause of this breakup. It would be useful to complete the specific accords on legislation by creating mutual plans for eliminating groundless legal disagreements and creating legally binding acts under the auspices of the Council of Heads of State and Heads of Government and other structures. Of course, there can be no question of drawing the Commonwealth countries' entire legislation closer together, merely of eliminating contradictions with regard to the common infrastructure, the common 'life-support systems,' and interregional deliveries."

Sergey Sirotkin, chairman of a Russian Supreme Soviet subcommittee:

"Control mechanisms are essential in the sphere of human rights. Even the very best constitution with a very broad catalogue of rights is never an adequate guarantee unless rights are backed up by control. A complex legacy of the Union's breakup is the fact that the population of the now sovereign states has been wrested from the unified humanitarian area. We are responsible for recreating it at a qualitatively new level. With any degree of our integration in the 'European' or 'Asian' home, we remain, all the same, a unified geopolitical area."

Evgeniy Shaposhnikov, commander-in-chief of the CIS Joint Armed Forces:

"The breaking up of the armed forces of the former USSR into constituent parts has been far from painless. The second serious loss has been the lowering of the armed forces' combat ability and of the CIS states' level of security. Unless we wish to dissemble, we should admit that the combat readiness of the Commonwealth's army is a cause for very serious alarm. Certain states are reckoning on building strong and effective modern defenses on their own. The high command seeks to coordinate the Defense Ministries' efforts on questions of military and military-technical policy in the Commonwealth. In this we receive support and understanding from Armenia, Kazakhstan, Kyrgyzstan, Tajikistan, Uzbekistan, and Russia. Ukraine and Turkmenistan participate in military cooperation only on certain questions, and the Belarusian Defense Ministry constantly vacillates. Moldova is moving farther and farther away from the CIS military organization. We need a clear-cut and definite approach in the positions of parliaments and their assistance. The high command proposes the idea of creating a defensive union which will constitute the military basis of the CIS countries' collective security system, and its doctrine must be defensive and be a generalized reflection of national military doctrines."

## 8.28

**Russia to Sponsor CIS Laws at Interparliamentary Assembly**

Elena Tregubova

*Segodnya,* 2 February 1994

[FBIS Translation]

*Editor's Note:* The following article is significant because it documents Russia's dominant sponsorship of CIS-wide laws.

[Elena Tregubova reports under the "CIS" rubric: "Federation Council Delegation to Interparliamentary Assembly Named. 'Russia Remains Main Sponsor of Legislation'."]

The Federation Council has prepared a list of representatives to attend the upcoming meeting of the CIS Interparliamentary Assembly [IPA]. The delegation will apparently include Federation Council Speaker Vladimir Shumeyko; Vadim Gustov, chairman of the Federation Council's Committee for CIS Affairs; Evgeniy Pavlov, his deputy; and Viktor Stepanov, Federation Council member and chairman of the Karelian Supreme Soviet.

Vadim Gustov recalled that the meeting of CIS parliamentary heads at St. Petersburg on 8–9 February will focus on the election of an Interparliamentary Assembly chairman "to replace the departed Ruslan Khasbulatov." According to Gustov, reports that Vladimir Shumeyko has recently offered the post of "CIS speaker" to the "fired" Stanislav Shushkevich "do not square with reality, because this post can only be held by an acting head of parliament." He said that "talk about Shushkevich took place before recent events."

The chairman of the Federation Council's Committee for CIS Affairs listed legislative enactments that would be worked out under "Russia's sponsorship." Some of them are based on the CIS "Treaty on Collective Security" (many provisions of which Stanislav Shushkevich opposed as speaker of the Belarusian Supreme Soviet); they include enactments on the procedure for imposing martial law on CIS territory, "On Mobilization," "On the Procedure for the Development, Production, Testing, and Delivery of Armaments and Military Hardware," and so forth.

The IPA plenary session scheduled for the beginning of March will set up a number of commissions which, the CIS committee chairman assured, "Russia will join, bringing all its ideas along, because in the light of its experience it has always been and will be the main source and drafter of legislation within the CIS."

## 8.29
### CIS Interparliamentary Assembly Meets in St. Petersburg: Shumeyko Views Tasks
Svetlana Ivanova
ITAR-TASS, 8 February 1994
[FBIS Translation]

*Editor's Note:* It is not entirely clear what Vladimir Shumeyko thinks he can do to get CIS states' parliaments to ratify pending CIS treaties, but he clearly hopes to apply pressure on state parliaments through their IPA delegates.

St. Petersburg, 8 February, TASS—Speakers of the two chambers of the Russian parliament, Vladimir Shumeyko and Ivan Rybkin, have arrived here today to attend a session of the Council of the CIS Interparliamentary Assembly.

Shumeyko told ITAR-TASS at the airport here that even in dealing with international matters some leaders had the habit of "looking at what the neighbor will do" and waiting for somebody to make the first step. "This is why," the speaker of the Council of the Federation (the upper chamber of the Russian parliament) believes, "the Economic Union treaty, signed on 24 September 1993, has still not been ratified by CIS parliaments."

Shumeyko believes that the main task confronting the current session of the Interparliamentary Assembly is to create a mechanism to settle problems linked with the ratification of interstate treaties. Shumeyko thinks the leaders of the delegations, attending the current session, are able to reach an agreement on the simultaneous ratification of the Economic Union treaty by all the CIS parliaments.

## 8.30
### Shumeyko Says CIS Strengthened
Gennadiy Talalaev
ITAR-TASS, 8 February 1994
[FBIS Translation]

St. Petersburg 8 February TASS—Vladimir Shumeyko, elected chairman of the Interparliamentary Assembly (IPA) of the Commonwealth of Independent States by secret ballot here today, told ITAR-TASS that the assembly has a great role to play in strengthening the Commonwealth. "No matter what economic and other agreements may be signed by the executive, they are difficult to translate into life without a legislative basis which provides a firm foundation," Shumeyko said in an exclusive interview with ITAR-TASS.

"This foundation can be created only by the parliaments of the Commonwealth member states. If we learn to consider coordination of legislation of all CIS states at IPA sessions, we shall be able to remove unnecessary barriers hampering the development of market economies and combine the national economies of the Commonwealth member states quite soon. I mean legislation concerning taxes, export/import duties, and the customs union," Vladimir Shumeyko said and added that much work was lying ahead.

As concerns IPA's political significance, it is also great, according to the assembly chairman. Interparliamentary assemblies have long existed in Europe and Asia, he said. Now an Interparliamentary Assembly has also appeared on the territory of the former Soviet Union. "It must have a significant weight in the world because it has brought together countries which have embarked on the path of democratic reforms," Shumeyko said. He stressed the need of establishing contacts with other, older interparliamentary assemblies.

## 8.31
### Shumeyko Says Democratic Reform Depends on Integration
Lyudmila Ermakova
ITAR-TASS, 29 October 1994
[FBIS Translation]

"Time and new circumstances serve us, because they produce an increased awareness of the fact that our prevalence over the economic crisis and the fate of democratic reforms in the CIS member states depend on the results of integration," said Vladimir Shumeyko, chairman of the Council of the CIS Interparliamentary Assembly, when addressing the latter's fifth session here on Saturday.

The integration has reached the stage of concrete decisions based on stage-by-stage shaping of an Economic Union, Shumeyko pointed out.

Heads of CIS member states passed a memorandum setting out guidelines for integration at their recent summit. To encourage integration, it is necessary to work out its legislative basis, he said.

The way Shumeyko sees it, the main task of parliaments of the CIS member states consists of "working out a thoroughly considered legislative basis for shaping the Economic Union, intensifying efforts aimed at drawing the national legislations closer together, and coordinating (parliaments') law-making activities for the purpose of forming a unified law-governed space."

We should propose that the CIS Heads of State adopt a convention on the CIS Interparliamentary Assembly, which would enhance its legitimacy both on interstate and interparliamentary levels and correspond to the international practice of parliamentary organizations, he said.

The final documents adopted at the Moscow summit of

the CIS Heads of State this month "expanded the horizons for coordinating the law-making activities" in particular, as far as entrepreneurial activities, taxation, and investments are concerned. Shumeyko described the models of a civic code and a criminal-procedural code as major contributions toward the integration.

The peacemaking function of the assembly is ever growing in importance, Shumeyko said.

The tangible result of such efforts may be seen in the Bishkek Protocol, which brought about cessation of hostilities in Nagorno-Karabakh last spring, he recalled.

The Council of the CIS Interparliamentary Assembly formed a peacemaking task force to tackle the Georgian-Abkhazian crisis at its session late in the evening on Friday. Another peacemaking task force, which is to tackle the crisis in the Trans-Dniester region, will soon be formed.

In conclusion, the chairman of the upper house of the Russian parliament thanked Azerbaijan, Turkmenistan, Kazakhstan, and Uzbekistan for the support they gave to the inhabitants of the Sakhalin region who suffered badly from recent natural disasters.

## 8.32

### Shumeyko Argues CIS Court Not Necessary

Interfax, 30 October 1994
[FBIS Translation]

Chairman of the Russian Federation Council and the CIS Interparliamentary Assembly Vladimir Shumeyko believes that the Commonwealth of Independent States does not need a single court.

Answering the question put by Interfax he told the news conference in St. Petersburg on Saturday that the only thing to be discussed is the economic court of arbitration within the CIS framework.

Shumeyko said that the CIS countries are trying to draw closer their legislation systems and as soon as this is achieved court decisions of CIS member states will be similar.

Meanwhile, chairman of the Belarusian parliament, Myacheslav Hryb said that the economic court of arbitration had been created in Minsk. It will resolve financial disagreements between CIS member states.

## 8.33

### CIS Assembly Seen Shaping Integration

Yuriy Lepskiy
*Trud*, 4 November 1994 [FBIS Translation]

Unfortunately we write little and know even less about what happens at sessions of the Interparliamentary Assembly. To us it seems to be a particularly dull and uninteresting event. I also thought that way but I was wrong.

Now I believe that it is at the Tauride Palace in St. Petersburg that the economic, political, and social models of our life in the twenty-first century are being quietly and professionally shaped (let me remind you that the end of the twentieth century is just six years away). It is not impossible, furthermore, that we will find out about the "Tauride" deliberations far earlier than that, for example, during the 1996 summer and fall political season in Russia.

What does the Interparliamentary Assembly actually do? Briefly, the official version is that it is actively preparing the legislative base for a new integration of the republics of the former USSR. Unofficially, however, the Interparliamentary Assembly has begun building the new home that we will possibly move into in a few years' time. The foundations of that building are being laid now; model legislation (or acceptable legislation common to all the sovereign republics of the CIS) is being written and adopted. The assembly's present session, for example, has approved part one of the CIS states' Civil Code. To give you an idea of this document's content, let me mention just a few of its 447 articles: "Protection of honor, dignity, and professional reputation," "Right to protection of personal privacy," "Right to inviolability of the home," "Enterprise," "Official and commercial secrecy," "Money," "Basic provisions on the joint-stock company," "Public funds," "Right of ownership and other property rights. . . ."

The important thing is that this document, which does not contain a single declaration, is entirely practical, adapted to our actual daily lives, and has the form, essence, and force of a law which is recommendatory for the time being. I would remind you that the Civil Code is a legislative model common to all CIS states. I stress the word common.

What else does the assembly plan to adopt in common for the whole CIS? Next year it will be discussing a model Criminal Code, Labor Code, and Housing Code. At the Tauride Palace they will be examining the following common laws: "On the Principles of Settlements Among Economic Components of the Interstate Economic Committee Member Countries," "On Securities," "On Joint-Stock Companies," "On Limited-Liability Companies," "On Agreed Common Standards for the Production of Output in the CIS States," "On the Basic Principles of Taxation," "On the Concept of Collective Security of the States Party to the Collective Security Treaty," "On the Protection of Consumers' Rights." . . . This is far from a complete list of laws common to the CIS republics, but it is enough for us to understand that we are going to be living together in earnest for a long time to come. Whereas in the past the foundations of our common home were built on the shifting sands of the political declarations written by S. Mikhalkov and G. El-Registan, but in reality held together by the steel bonds of the empire, now they are being created on a quite firm legal foundation and written by the full-fledged representatives of sovereign states and their strength will be determined by how much the partners need one another.

Around four years ago, if you recall, the most widely used words in our political lexicon were "renewed Union." This meant that if the republics of the USSR were to get greater autonomy and Moscow fewer administrative functions, everyone would be content and happy in the Union thus renewed. Now it is clear that these ideas were quite naive. Moscow would never actually surrender its administrative privileges to the republics voluntarily and the opposite of that would mean the suicide of the system. It sometimes seems to me that the unrestrained sovereignty bandwagon, the rapid formation of republic bureaucracies, the bloody regional conflicts, the explosions of nationalism, and the ethnic cleansing are quite simply the real price of the renewal of the Union. Undoubtedly the price is too high. However, that is the price that is being paid for the structural alteration of a gigantic country and the attempt to transform it into what really is a union of sovereign and equal members. It seems to me that the best guarantees of sovereignty and equality are the existence of ethnic property owners in the republics and a local bureaucratic administrative apparatus which will never simply surrender its powers to Moscow just like that. Whether all this is worth even a single human life is another matter. . . .

One way or another the Interparliamentary Assembly, in my view, is already laying the foundations today of a really new Union of Sovereign States and not simply a commonwealth of republics of the former USSR. I wonder what the supporters of the restoration of the USSR will do when everything that we have talked about becomes a political reality? Either they will publicly thank the Interparliamentary Assembly or they will choose to fight for ideological purity (or rather for the second and third "s" in the abbreviation—denoting "Soviet" and "Socialist"). In any event, it is the Interparliamentary Assembly which has done quite a lot to ensure that in 1996 the electorate who previously voted for the Umalatovs, Zhirinovskiys, Konstantinovs, and Anpilovs give their vote to those who made the renewed Union a reality not at rallies but by quietly and professionally creating its legal basis.

However, it is not that easy. I only had to ask the heads of the republic delegations a question about the "renewed Union" to immediately receive in reply a series of moral lectures on what a journalist should and should not ask. Answering my question on the renewed Union, Abish Kekilbaev, chairman of the Kazakhstan Supreme Council, Pyotr Luchinskiy, speaker of the Moldovan parliament, Yashar Aliev, deputy chairman of the Azerbaijan National Assembly, and Ara Saakyan, first deputy chairman of the Armenian Supreme Council, on the contrary, did everything they could to stress the sovereignty of their republics. Only Abish Kekilbaev cautiously used the term "integration." Essentially I was taught a lesson in political good manners and it served me right! Of course, I should have remembered that in the house of a hanged man you do not talk about ropes, so among the heads of the CIS parliaments you should scarcely utter the word "Union," especially not in the sense of "indestructible."

The movement toward one another, or integration, is now taking place with great caution on the thin fall ice which has finally formed on the surface of the recently stormy sea of national ambitions and interrepublic conflicts. Any injudicious movement and the ice could break and everyone would drift away on their own ice floe. The important thing here is to do nothing which would cause panic, obstruction, or even endanger this cautious movement.

Vladimir Shumeyko, chairman of the Interparliamentary Assembly, believes that the process of political integration should not be hurried but should occur strictly after the strengthening of economic ties. The economy must dictate to the politicians both the forms of unification, the timetable, and the intensity of that process. I think he is right. For too long everything has been the other way around in our country: Politics forced the economy and indeed people's lives to fit into the framework of a single ideological doctrine. We know what the result of that was. Now the solution of political problems—Karabakh, Abkhazia, the Trans-Dniester region—will have to wait for the results of economic integration. And as for the "renewed Union," Shumeyko answered that question using the words of Ukrainian President Kuchma, who commented: "Anyone who does not regret the disintegration of the USSR has no heart; anyone who hopes to revive it has no head." Well, you cannot bring back the past, do not make the same mistake twice. That is all there is to it.

However, not everything that existed in the past has gone forever. Prior to the Interparliamentary Assembly session in the Tauride Palace an international conference opened on the legal problems of the social protection of veterans of the last draft of the war. It seems that they, drafted as seventeen-year-olds in the last year of the war, still do not even have the status of Great Patriotic War Veterans. Furthermore, many of those who do have that status still do not have decent housing, high-quality medical services, or a worthwhile pension even to this day. . . . There is just one thing that no one can take away from them: their common memory of the war, of their dead comrades, of the difficult days and nights at the front and in the rear, and of the hard postwar life given wholeheartedly to their common motherland—the Soviet Union.

These people are still alive, they have not disappeared into the past along with the name of the country, its cities, streets, and lanes. And everything that subsequent generations consider past has not gone, it all continues to live on with them and therefore with us. If you do not understand that, you have neither a heart nor a head.

## 8.34

### Report on CIS Interparliamentary Assembly Council Session

Sergey Alekhin
*Rossiyskaya Gazeta,* 22 February 1995
[FBIS Translation]

[Report by Sergey Alekhin: "Members of Parliament, Unite! Single Information Area Essential to CIS"]

More than twenty questions concerning economic, social, and cultural cooperation between Commonwealth states were placed on the agenda of the CIS countries' Interparliamentary Assembly [IPA] Council session which took place in St. Petersburg.

Questions of strengthening the IPA and of its cooperation with the parliaments of CIS member states occupied a central position in the work of the nine countries' delegations.

But from a practical point of view, there is the question of changing the status of the IPA, which does not at present have any direct influence on the political and social processes in Russia and the CIS countries, although quite a few very necessary and appropriate documents have been produced within the framework of this interstate formation. But they are not effective due to the disparity between national legislation. Speaking at the meeting of CIS Heads of State and Government in Alma-Ata on the eve of the IPA Council session, Vladimir Shumeyko, speaker of the Russian parliament's upper chamber,. supported an initiative by Kazakh President Nursultan Nazarbaev to create a so-called Eurasian parliament.

At the final press conference, Vladimir Shumeyko, who has again been elected to the same post, emphasized the extreme importance of the draft convention's first paragraph confirming the assembly as an organ of interstate cooperation. In his opinion this should raise the assembly to the level of the Council of Heads of State and the Council of Heads of Government. Shumeyko considers that the role of the CIS member countries' IPA will without doubt be strengthened in the future and it is possible that it will be renamed and transformed into a parliamentary assembly with the right to adopt laws which are binding throughout the Commonwealth's territory.

However, it is pointless at present to think about the IPA's development and the enhancement of its role when we do not possess a common information area. People in the states, which not so long ago comprised a single country, now know more about the situation in Germany or life in the United States than about the joys and misfortunes of their nearest neighbors across customs checkpoints. Therefore, a decree was adopted at the current assembly on the formation of a special organ under the IPA Council: the Council of Commonwealth Countries' News Agency Leaders.

Yuriy Sizov, executive director of the Council of CIS State News Agency Leaders, emphasized in a conversation with your *Rossiyskaya Gazeta* correspondent that the CIS states' aspiration to more efficiently provide broader and more diverse information about the political, economic, and social processes taking place in the CIS should be at the heart of a coordinated information policy.

Questions of strengthening peace and security on the territory of Commonwealth states were examined at the IPA Council's February session. It was proposed that IPA peacekeeping groups continue the quest for ways to achieve a political settlement in the Georgian-Abkhaz and Trans-Dniester Region conflicts. Particular attention was paid at the current meeting to the preparation of recommendatory legislative acts, aimed at stepping up the struggle against organized crime in Commonwealth countries. A program committee and working groups have been formed to produce a model Criminal Code and Code of Criminal Procedure.

Other questions of interparliamentary cooperation were also examined at the current assembly.

## 8.35

### Interparliamentary Chief Discusses CIS Integration

Interview by Yuriy Lepskiy
*Trud,* 16 May 1995
[FBIS Translation]

The Sixth Plenary Meeting of the Interparliamentary Assembly of the CIS Countries took place in St. Petersburg. *Trud*'s political commentator talked with its chairman, Vladimir Shumeyko.

[Lepskiy]: There is a persistent suspicion with respect to the CIS Interparliamentary Assembly: with the help of this organization, the Soviet Union is being quietly reconstructed. . . .

[Shumeyko]: Well, it is 1925 again! It is impossible to reconstruct the former USSR, either quietly or noisily. What we are actually doing goes by another name—the strengthening of integration within the framework of the CIS.

[Lepskiy]: But, perhaps, this is basically no different?

[Shumeyko]: Here is the current agenda of the IPA [Interparliamentary Assembly]. We intend to examine and adopt the second part of the model Civil Code for the CIS's participating members. I would remind you that the Civil Code is the law which regulates the daily mutual relations of a person with respect to property and with respect to the organizations and structures which represent in our countries the new market relations. This time, we are also discussing a model Criminal Code and Criminal Procedure Code for the Commonwealth's countries. . . .

[Lepskiy]: Why are these codes called "model" codes?

[Shumeyko]: Because the CIS countries' parliaments view them as models for the development of their own laws. During this process, the essential strategic ideas of the model are preserved, while the details may well be modified or adapted to the needs of a specific state. But, as a result, within the CIS's framework, we obtain a unified legal domain without infringing the sovereignty of the independent states. Just imagine how important the adoption of unified criminal procedure legislation will be for us. This will make it possible to find, to neutralize, and to punish criminals rapidly, efficiently and without delays, regardless of the state in which they committed the crimes. For the time being, however, our disunity is creating splendid opportunities for criminal structures to evade accountability.

And here is one more example for you. We are once again examining a draft of model legislation on financial-industrial groups in the CIS. Such groups will establish their own banks and insurance companies. It is very important that the mutual relations of the various states' enterprises "within" a financial-industrial group will not be international in status. Within the framework of this association, the partners maintain domestic prices and economic ties have been simplified to the maximum extent. This will make it possible for us to preserve the high-tech industries which existed within the bounds of the former USSR and which fell apart together with it. For example, the production of civilian aircraft. And there are plenty of examples! Just recently, I was at Baykonur. Today, Russia is paying Kazakhstan $20 million a year to lease the cosmodrome. As for the cosmodrome itself, there is only enough money at least to maintain this unique installation in operating condition somehow and to preserve its infrastructure. It turned out that there, just in the Buran [shuttle] program, there were 650 absolutely new technologies. But both Baykonur and Buran were developed through the efforts of the former union's many republics, since Russia alone was not able to do this then, and it is not able to do so now. This is why we consider it essentially important to maintain a unified legal and economic domain on the territory of the former union. I hope you agree that this is scarcely reminiscent of a process for the reconstruction of the USSR in the form that the leftist radicals picture it.

[Lepskiy]: But are you not afraid that, when you finally restore a unified economic domain and a unified market within its former borders, when all these subcontractors and suppliers again turn to one another, then the citizens of the CIS countries will one day, to their consternation, discover in the shops the Rubin [Ruby] and Izumrud [Emerald] television sets, even though they will have barely begun to get used to the Sony and Philips receivers so standard for the rest of the world? To put it briefly, do you not believe that all this economic integration is beneficial only to those who are used to living within the bounds of a closed, non-competitive market?

[Shumeyko]: First of all, if a market is closed and non-competitive, then this is no market at all, rather, it is called something else. But, if this is a real market, then people will always buy in it what is of the better quality and what is cheaper. You know that, in Kaliningrad oblast, there is a city—Sovetsk—and there is a pulp and paper combine. So, this combine produced cardboard of disgracefully poor quality, but the production was sold under those Soviet-era conditions in one of the USSR's republics. The plant did not sell its own cardboard, it supplied it. The [Soviet] Union collapsed and market relations of one sort or another arose and it became very clear very quickly that no one needed the "Soviet" cardboard. Now they are producing remarkable wallpaper and selling it with enormous success. Thus, I want to tell you that the enterprises which are being integrated today into a unified economic domain are no longer the ones which existed previously and that the republics are not the ones which made up the USSR at one time. Much has changed. And what we want to create will not be a closed market and it will contain all the best things produced in other countries, as well as those things at which the USSR was good at one time. I am talking about the high-quality space technologies, the achievements of aircraft building, and many other things about which *Trud*'s readers know as well as I do. So, Sony and Philips will not disappear from our shops, but we will also not be building a new Russian economy for the sake of increasing employment in Japan or Germany.

[Lepskiy]: As of 12 May, a year had passed since the day that a real armistice ensued in Nagorno-Karabakh. It is well known that achieving this was the successful result of the efforts of the peacemaking group under the auspices of the Interparliamentary Assembly, which finally led to the signing of the Bishkek Protocol. Attempts to stop the bloodshed in Karabakh had been undertaken previously as well. Why, in your opinion, did the Interparliamentary Assembly's peacemaking activities prove to be more successful here than those of others?

[Shumeyko]: Because these peacemaking activities were carried out by deputies who knew the problem very well from the inside. And also because the deputies from the various sides had one thing in common: they had all been elected by their own peoples and were not seeking in the negotiating process either personal profit or political dividends. They were sincerely striving for one thing—for peace. By the way, we understood that it is precisely on the basis of the Karabakh example that the work of the peacemaking deputy groups may be effective. Now a peacemaking deputy group under the auspices of the Interparliamentary Assembly is also at work in Moldova. Recently, we met with Eduard Shevardnadze and discussed for a quite lengthy period the Interparliamentary Assembly's participation in settling matters in Abkhazia. I think that the leaders of many CIS states have sensed the Interparliamentary Assembly's peacemaking capabilities. And this is good.

[Lepskiy]: Is it not possible to apply these capabilities in Chechnya?

[Shumeyko]: I think not. You see, the Interparliamentary Assembly is, all the same, an international organization, while the conflict in Chechnya is an internal Russian matter, inasmuch as Chechnya is a component of the Russian Federation.

[Lepskiy]: Do you want to say that Chechnya cannot be viewed as an independent party in the conflict?

[Shumeyko]: Exactly so.

[Lepskiy]: How would you regard the idea of establishing, under the auspices of the Interparliamentary Assembly, a summer camp or holiday resort, where children from Armenia, Azerbaijan, and Nagorno-Karabakh, from Ukraine and Russia, and from Georgia and Abkhazia would live and vacation together?

[Shumeyko]: In my opinion, it is an excellent idea. It works for the future. It would then be a sin for fathers whose children are friends to shoot one another. I support it. And what is more. At the end of May, the heads of state of the CIS are supposed to gather in Minsk for a routine meeting. It is anticipated that the Convention on the CIS Interparliamentary Assembly will be adopted at this meeting. The convention about which I am talking provides for the establishment of a special Interparliamentary Assembly humanitarian fund. We could allocate money from this fund for the realization of this idea. It is something worthwhile.

## 8.36

## Future UN Role as Main Peacekeeper in CIS Considered

Boris Sitnikov
ITAR-TASS, 31 August 1995
[FBIS Translation]

United Nations, 31 August (Itar-Tass)—"In conditions where the United Nations' strained resources do not permit it to undertake the conduct of its own full-scale peacekeeping operations in conflict zones in the territory of CIS member countries, it should delegate its powers to the CIS peacekeeping forces and provide financial, technical, and economic assistance to the efforts of the Commonwealth member states aiming to maintain peace and settle conflicts in the hot spots of the Commonwealth of Independent States," Deputy Chairman of the Federation Council of the Federal Assembly of the Russian Federation Valerian Viktorov said at a special session of the Interparliamentary Council which is being held here on the occasion of the fiftieth anniversary of the United Nations.

"A pressing problem for the United Nations is to ensure regional approaches to maintaining peace and security. It must be admitted that the United Nations and the Organization for Security and Cooperation in Europe are unfortunately unable as yet to provide adequate assistance in preventing and settling regional conflicts, including in Nagorno-Karabakh, Trans-Dniester, and Abkhazia.

"In these conditions, the Council of the Heads of State, the Interparliamentary Assembly, and the Council of the Heads of Government of the Commonwealth member states, seeking to observe the generally accepted international legal norms, standards, and principles, have to prevent threats to peace and ensure security in the CIS region through their own efforts," said Viktorov, who represents the CIS Interparliamentary Assembly at the forum.

"A collective security system is now being formed within the framework of the Commonwealth of Independent States on a regional basis, which enhances the role played by parliamentarians in terminating armed conflicts.

"Pursuant to a decision of the Council of the Interparliamentary Assembly, parliamentarians have formed peacekeeping groups to deal with the problems of Nagorno-Karabakh, facilitate the settlement of the Georgian–Abkhazian conflict, and bring an end to confrontation in Trans-Dniester," Viktorov said.

"The Interparliamentary Assembly, jointly with the Ministry of Foreign Affairs of the Russian Federation, has succeeded in initiating the signing of the so-called Bishkek protocol by the leaders of the parliaments of the conflicting countries. On the basis of this document, a bloody war was halted in the zone of the Nagorno-Karabakh conflict, and ceasefire was ensured in the conflict zones in Abkhazia and Trans-Dniester with the help of Commonwealth member countries, the Russian Federation above all.

"It appears that the regional approach to the solution of security problems is an important means of strengthening stability in the whole world," Viktorov said.

"It is advisable for the Interparliamentary Assembly to summarize experience gained by parliamentarians in peacekeeping actions and arrange, jointly with the United Nations and other international organizations, a seminar on "Parliamentarianism and Peacemaking."

"A seminar of this kind could be held in St. Petersburg where the headquarters of the Interparliamentary Assembly of the Commonwealth of Independent States is located," Viktorov said.

# 9
# The Military Evolution

## Introductory Notes

The dissolution of the Soviet empire left its prodigious armed forces in an ambiguous position. Did they "belong" collectively to the Commonwealth, or to the newly independent Commonwealth states on whose territory they were deployed? The founders of the Commonwealth had very different perceptions of what the answers to this question should be. Russian President Boris Yeltsin, who had presumably discussed this issue with the Soviet defense minister, Evgeniy Shaposhnikov, before the seminal Belovezh Forest meeting, supported the concept of a single, combined CIS armed forces (or a "single CIS defense space"), with Russian officers in full control. For Yeltsin and Shaposhnikov, there was no need for debate. Splitting up the post-Soviet forces would create a security vacuum into which surrounding countries would move, eager for influence in the European and Asian regions of the former Soviet empire. Although both Yeltsin and Shaposhnikov had supported greater autonomy for the former republics during the Novo-Ogarevo talks in 1990, they saw no reason why the new states should want to try to build their own national armies.

A plan for unified Commonwealth armed forces, drafted by Shaposhnikov and the former Ministry of Defense, was discussed at the first summit of CIS Heads of State. This plan was debated at the summit alongside an alternative submitted by the former Soviet chairman of the Committee for Military Reform, General of the Army Konstantin Kobets. The Kobets plan was less centralized and provided only for general coordination among the separate member-state defense ministries. It, ultimately, was the scheme adopted by the CIS Heads of State, reflecting the fears of the former republics that Russia was showing its hegemonic intentions in its defense plans for the Commonwealth. Both plans are analyzed in this chapter by noted Russian military commentator Pavel Felgengauer.

For the other members of the Commonwealth, the military questions weighed heavily on their minds. The Ukrainian president and CIS co-founder, Leonid Kravchuk, in particular, viewed any talk of a unified army as tantamount to the restoration of the USSR and an insupportable breach of sovereignty of the newly independent states. Ukraine therefore shocked Moscow in January 1992 by immediately declaring full control of the former Soviet forces stationed on its territory, and announcing its intention to build a large national army of some 400,000 soldiers. Moreover, Kravchuk and his government declared that the Black Sea Fleet, stationed in Sevastopol on the Crimean peninsula (transferred to Ukrainian suzereignty by Khrushchev in 1954), would become the new Ukrainian navy. Ukraine, in fact, quickly assumed leadership of a small bloc of CIS states that insisted on building their own armies (including Moldova and Azerbaijan). Within a year this bloc had persuaded other non-Russian CIS leaders to take the same course of action. By May 1992, most former republics had established a national Ministry of Defense and were building national armies. The debates over how to divide the former Soviet forces had become acrid disputes, raising much speculation in the press that the CIS was nothing but a fig leaf.

Two events in May demonstrated conclusively that Russia was in for a long and difficult struggle and that the non-Russian CIS heads of state had rejected the CIS as a unified security alliance. On 7 May, President Boris Yeltsin announced that Russia would create its own independent Russian armed forces and Ministry of Defense, indicating that he had decided to pursue a unilateral approach to unifying CIS military capabilities; on 15 May, the CIS Heads of State debated a collective security treaty, with only six countries becoming signatories by the end of the day.

Significantly, Yeltsin chose to build his defense structure on the foundations of the former Soviet military-industrial complex, rather than create a whole new system, and he put a military officer, rather than a civilian, in charge. (Army General Pavel Grachev was appointed Russian defense minister, a position he continued to hold until mid-1996, when Yeltsin established an alliance of convenience with Grachev's rival, Aleksandr Lebed.). How Russia would define its military mission in the CIS and elsewhere immediately became a new security issue for the former Soviet republics.

These leaders fear for the security of their own states in view of a Russian military run in accordance with conservative Soviet military beliefs. This includes the long-standing Soviet military tradition that the formulation of an adequate defense is the sole province of military leaders. Non-Russian CIS leaders worry that there are no effective civilian controls over the Russian military now that Communist Party control has ended and a weakened Russian president appears to rule largely as a result of the personal backing of his defense minister. (By mid-1994, these apprehensions were intensified by the alliance between the Russian military and the increasingly powerful KGB, now split into several new agencies.) Leaders of the Soviet successor states are well aware that military reforms planned during the Gorbachev era had never been implemented. Moreover, the Russian military has suffered the most in the post-Soviet transformation and shown itself to be radically nationalist in its political makeup. The big difference between the Soviet and Russian armies is that now the army will serve *Russian* Great Power goals and interests rather than those formulated in terms of a socialist internationalist ideology.

Of even greater concern is the size and economic preponderance of the Russian military complex. The Soviet military structure, with all its attendant socio-economic problems, still operates in Russia. Essentially, the economy is still a mobilization economy with a great many ministries engaged in research and production for the armed forces, albeit with reduced resources. Although Russian military spending has been reduced from Soviet days, it is still an unknown quantity, and the military is agitating for far bigger budgets. (Revelations about real Soviet defense spending continue to astonish Western and Russian citizens alike. Some estimates state that the defense sector in the USSR employed 10 million people and consumed more than 25 percent of the Soviet gross national product. *Business Week* [29 July 1991] quotes 52 percent.) All these facts make non-Russian CIS leaders extremely nervous.

On 15 May 1992 the focus in the Council of CIS Defense Ministers shifted from the reapportionment of former Soviet troops and bases to the drafting and signing of a CIS collective security treaty. At the 15 May Tashkent summit of CIS Heads of State, a Russian draft was signed by six of the eleven Commonwealth members (Georgia had still not joined the CIS). The paucity of support for the treaty was a telling sign of CIS resistance to Russian defense policy. The treaty, often referred to in the CIS as the Tashkent agreement or Tashkent alliance, was rejected by Ukraine and Turkmenistan, whose governments had still not acceded to the treaty when this account went to press in January 1996. The treaty itself is included in Chapter 10. To date it is more of a political symbol than a binding alliance or security doctrine.

Three key unresolved issues have been on the CIS military agenda since the signing of the Tashkent agreement: (1) How should the collective security interests of the CIS be defined? (2) Should the CIS refer to its "external borders" or to "internal borders" of member states? and (3) What kind of coordination should occur among the defense ministries of the twelve CIS member states? For non-Russian CIS leaders, another inescapable issue has been: How will Russia choose to utilize its total forces (about 2.1 million) and soldiers deployed on army bases in other CIS states, and how will Russia relate to Ukraine—the only other real military power in the CIS (with about 400,000 forces)?

### Abandonment of the Unified CIS Armed Forces

In 1993, the Soviet military machine continued to devolve into separate, tiny armies (with the exception of Ukraine's army). Without unanimity, the CIS finally abandoned all pretense of creating unified conventional forces on 15 June 1993. Strategic forces were, however, maintained under CIS collective control (under Russia), as decided early in 1992. Nevertheless, even after the collapse of the program for a single defense space, military issues remained the primary bone of contention in CIS relations. Several pieces in this chapter document the collective security treaty debate and the mid-June events that forced Russia to switch to a bilateral and increasingly unilateral approach to military "integration."

A new CIS general staff was created and subordinated to the Council of Ministers of Defense, with General Viktor Samsonov as chief of staff. Samsonov (fifty-two years old at the time) had been chief of staff of the CIS Unified Armed Forces, and before that general staff head of the USSR armed forces as well as USSR first deputy defense minister. Supreme Commander Evgeniy Shaposhnikov was relieved of duty and appointed secretary of the Russian Security Council when the combined armed forces were abolished. Several documents in this chapter describe these events and their subsequent impact on the CIS military debate.

The creation of twelve separate armies was fraught with difficulties. Most forces of the former Soviet republics were made up of mixed ethnic groups, but usually a majority were Russians, and most officers were Russian. Consequently, loyalties to Russia within these forces has in many cases lingered. Since President Yeltsin's face-off with the extreme nationalists in the Russian parliament in October 1993, and the electoral success of nationalists and former communists in the December 1993 parliamentary elections, Russia's relations with the "near abroad" (the former Soviet republics) have evolved into a tougher, more aggressive policy, based on the overt declaration that Russia possesses a "sphere of influence" in the Soviet successor states.

Russian leaders often define their "special interests" in CIS states in terms of their sacred duty to protect the minority rights of the Russian diaspora now living in these countries. Even Andrey Kozyrev defended and encouraged a Russian-dominated economic and political/parliamentary integration among CIS states. Paradoxically, although the CIS states admit that they need closer economic association, most do not want a restoration of the USSR, which they fear is the dominant agenda for most Russian politicians.

On 2 November 1993 the Russian military promulgated its new military doctrine. Several journalist accounts of the doctrine's orientation and significance are contained in this chapter, along with two interpretative articles on the subject. One of these is by Foreign Minister Andrey Kozyrev, writing in *Krasnaya Zvezda* for the neo-democratic position on the doctrine. The other is by Army General Pavel Grachev, writing in *Nezavisimaya Gazeta* on 4 June 1994, representing Russia's great power military posture. In the West, the consensus was that the "hawks" in the Russian parliament, Foreign Ministry, and military emerged as victors in the struggle over the drafting of this doctrine.

With its promulgation, Russia launched a campaign to obtain official sanction from the United Nations and individual Western states as the principle "peacekeeper" in the CIS. For Russia, the role implies that the CIS states are its special sphere of influence. No one can deny the need for some type of operational peacekeeping presence in Soviet successor states embroiled in civil war, such as in Tajikistan, Moldova and Georgia, Armenia and Azerbaijan. Yet the extent to which Russia has manipulated these confrontations to its own advantage (especially in Moldova, Georgia, and Azerbaijan) has made Western countries distrustful of its "sphere of influence" and "peacekeeping" rhetoric. Ukraine, in particular, views Russia's campaign as an attempt to win the right to intervene militarily at will in the CIS states.

During 1994, Russia adopted a bilateral approach to its military relations with the CIS states. Part of the approach included the establishment of military bases and deployment of border troops in the Transcaucasus and Central Asia. This policy has been pursued aggressively and, in light of Russia's twin campaigns in Chechnya and in the Caspian Sea oil pipeline war, is a cause of high anxiety among all the post-Soviet, non-Russian states. Despite the deplorable social conditions in Russia's army, it is the strongest tie that binds the CIS states at this time.

Partly in reaction to Russia's bilateral and regional military pressure, the non-Russian states began to seek separate accommodations with NATO in 1994. By the middle of the year, all the CIS states (including Russia) had joined NATO's Partnership for Peace (PFP) except Azerbaijan, which signed the accession documents in May 1995. Each state has adopted a neutral position on NATO's expansion into Central Europe, using publicly veiled statements such as Moldovan President Mircea Snegur's in an interview with the German newspaper *Handelsblatt* in November 1995: "We do not think that NATO's possible broadening towards the East will bring any danger to Moldova's security," adding "when NATO begins the expansion, it should necessarily consider all factors, all possible consequences, first and foremost—the Russian factor."

Because all the non-Russian states, except Ukraine, have tiny armies, their activities in the PFP are necessarily limited. Nonetheless, the PFP has been integrating the PFP countries into an "interactive program" including such activities as training CIS armies and guard units to conduct peacekeeping operations and humanitarian assistance programs, as well as how to control air flights and establish cooperation between civil authorities and military commands. Assisting CIS soldiers to learn English is also a part of some PFP programs in the CIS states. Although Russian military leaders are presumably not pleased about so much interaction, there has been little said publicly about the PFP. The principle taboo is on NATO membership per se.

As usual, Ukraine is a special case. Because of its size and geopolitical importance in the region, its consultations with NATO have gone farther than with the other non-Russian successor states. In September 1995, as documented here, Ukraine and NATO held "sixteen plus one" talks within the framework of the PFP program. Before this, only Russia had been granted this privilege. In October, President Kuchma felt it necessary to oppose the eastward expansion of NATO, but couched his opposition in terms of not wanting a division of Europe, but favoring an "evolutionary" expansion of NATO after the Russian presidential election. In December, Ukraine announced that it would expand its relations beyond the PFP, but not to the point of joining either NATO or the Tashkent alliance.

Several documents describing the types of programs NATO is establishing in the CIS through the PFP, and political statements on the part of CIS leaders (meant to reassure Russia that they do not intend to join NATO), are contained in Chapter 9. These documents give the reader a sense of the delicate but firm way in which non-Russian leaders are treating the issue of CIS external ties as well as the difficulties Russia is having creating any kind of voluntary single defense space within the CIS.

By October 1995, hard-line Eurasianists, who believe that Russia's strategic interests lie to its south and stress the union of Slavic, Transcaucasian, and Central Asian states in a strong alliance, were drafting a second Russian military doctrine that is reputedly far more aggressive and hegemonic than the 1993 doctrine. The influence on the Russian Ministry of Defense of extremists such as Anton Surikov, general director of a mysterious Independent Institute of Defense Studies (INOBIS), has been widely conjectured upon in the Russian press. Mr. Surikov has written a long paper cataloguing external threats facing Russia and putting the United States and its policies in the CIS at the top of the list. Subsequent articles by Surikov in right-wing journals have advanced the thesis of a Western conspiracy against Russia and given long explanations of hostile intentions toward Russia emanating from the United States and its fellow members of NATO. The institute and its personnel, who possess sophisticated computer design equipment and a large, modernized office building, were purportedly consulted on a frequent basis by the Russian defense minister, Pavel Grachev. It may be that Surikov's articles constitute nothing more than "paper rhetoric" designed to placate nationalist forces. However, those who want to restore Russia's great power status in the world are using nationalism and external "adversaries" to unite the country and to restore Russia's identity. The views expressed in these articles are not new, but harken back to Eurasianist national themes tolerated by the Soviet Communist Party in order to give the population something around which to rally in difficult economic times. Surikov's paper and articles are provided here to demonstrate how volatile Russian politics had become by late 1995, and leading up to the 1996 Russian presidential election.

The following documents capture the evolution in the CIS military debates among CIS heads of state and government and other officials from January 1992 to December 1995. In certain respects, this debate follows the progressive politicization of the Russian military and the increasingly aggressive direction in the development of Russia's military doctrine after the fateful storming of the Russian parliament in 1993. The debate has become something like a dance at which Russia chooses its partners, while they carefully try to end the dance without stepping on their very large and demanding partner's toes.

## 9.1
## Comparison of Military Reform Plans Charted

Pavel Felgengauer
*Nezavisimaya Gazeta,* 31 December 1991
[FBIS Translation], Excerpts

When Yeltsin, Kravchuk, and Shushkevich agreed at Belovezhskaya Pushcha on 8 December on how to first disband the USSR and then take apart the vestiges of the inheritance, they did not begin to consult either with President Gorbachev, Nazarbaev, or moreover, with Bush. The only person they called was USSR defense minister Marshal of Aviation Shaposhnikov. Republic leaders could never have acted so decisively without a mutual understanding with the highest military leadership. But then again, the preliminary agreement was, we need to assume, reached beforehand, most likely during Marshal Shaposhnikov's meeting with President Yeltsin on 3 December 1991.

We have every reason to believe that it was the minister of defense who finally coordinated with the Russian president personnel changes in key leading posts in the army and, first of all, the retirement of General of the Army Lobov who was removed from his post as chief of the general staff on the morning of 7 December. Other arrangements were also made in expectation of inevitable changes: the elimination of the USSR Ministry of Defense in its present form and the creation of the new Commonwealth military structure where, instead of the previous, purely formal supreme commander—USSR President Gorbachev—a quite real commander-in-chief of the "Combined Armed Forces" from the military will appear (so far, the only real contender for this post is Marshal of Aviation Evgeniy Shaposhnikov).

On 11 December, during an expanded session at the Ministry of Defense, President Gorbachev called on the military leaders to "save the Union" but he encountered little understanding or sympathy among them. In the corridors, they joked maliciously: He should have issued those appeals, they said, on 19 August—then things would have been worked out. In December 1991, the military leadership had quite different problems: how to adapt to the new state (more accurately, non-state) structure and to the newly emerged CIS [Commonwealth of Independent States].

By the time of the conference at Alma-Ata on 21 December, when the republics of Central Asia, Armenia, Azerbaijan, and Moldova had joined the CIS, the Ministry of Defense had prepared and submitted a package of documents on the new armed forces structure (see Diagram 9.1.1). However, the republics did not adopt this proposal (although they anticipated some sort of independent armed forces for Ukraine). Besides, an alternative draft was submitted from the Committee for Military Reform (chairman—General of the Army Konstantin Kobets), which received serious support in the republic delegations (see Diagram 9.1.2).

After Alma-Ata, the agreements were continued and a compromise plan for the transition period was drawn up during the conference with the republic ministers of defense in Moscow on 26–27 December: An Agreement on Defense Issues Among the CIS Member States. (Elements from the

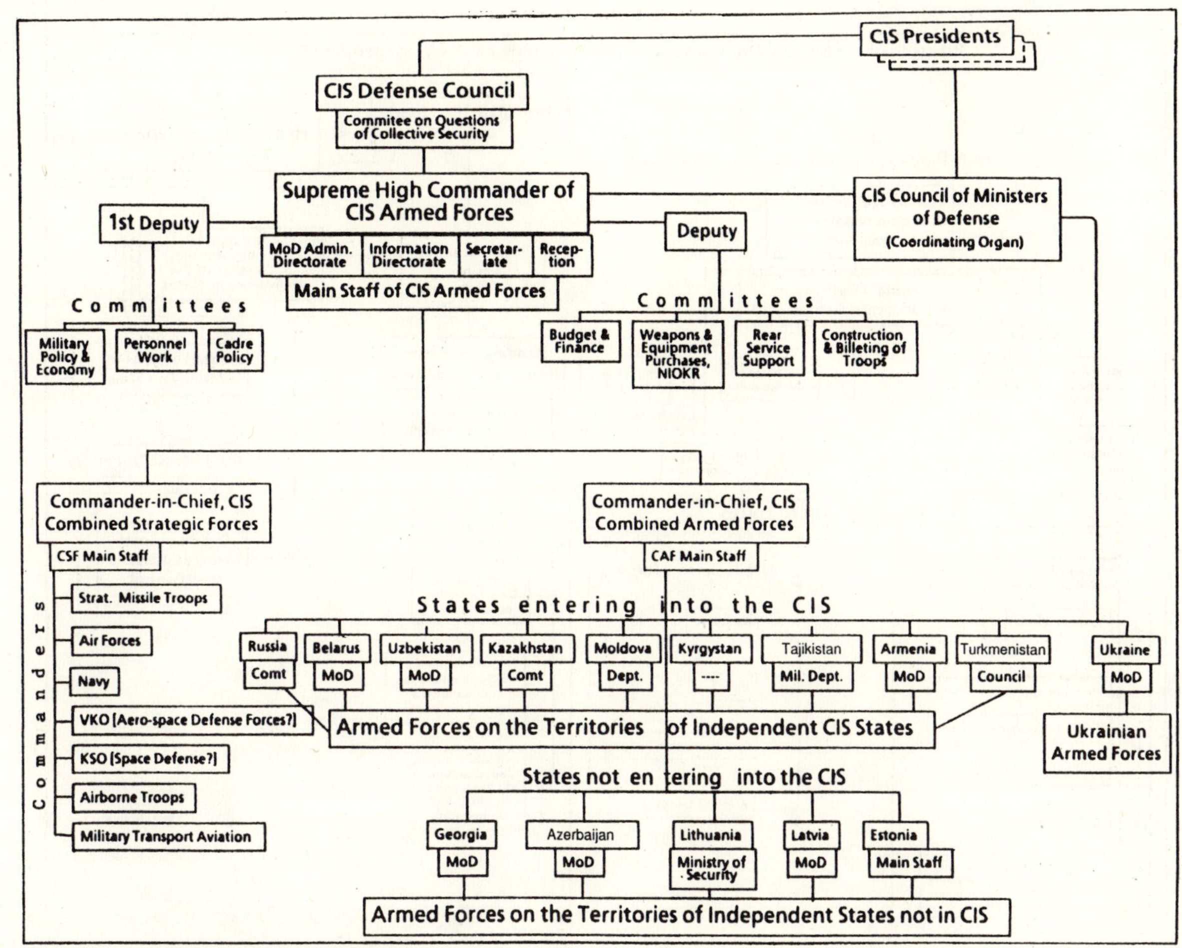


Diagram 9.1.1 **Ministry of Defense Plan**

drafts of both the Committee on Military Reform and the Ministry of Defense were utilized in the agreement reached.) Representatives from all CIS member states agreed that an adequately lengthy transition period was needed for a relatively peaceful transformation of the armed forces of the former USSR.

The Agreement on the Status of the Allied Armed Forces on the territory of the Commonwealth Member States was approved (without any special friction) (troops will not interfere in internal matters and servicemen have been directed not to participate in any political activity; separate agreements will be concluded on all remaining issues in each specific case). An interim provision on the CIS Council of Ministers of Defense was also approved. The CIS Council of Ministers of Defense will consist of the republic ministers of defense (the chairmen of the appropriate state committees) and also the combined armed forces commander-in-chief and the combined armed forces chief of the general staff. The Council of Ministers of Defense must coordinate the policies of the individual independent states on issues of military policy and also ensure civilian control over the army that was left and still exists when the state that created it no longer exists. The Council of Ministers of Defense will convene in turn in the capitals of all of the independent states and a representative of the state that hosts the Council of Ministers of Defense at that time will be the chairman. Any decisions will be adopted by the Council of Ministers of Defense only unanimously and at the same time they will be recommendatory in nature.

They have not yet managed to find a streamlined formula for an interim compromise on more substantive issues.

During the session, it was ascertained that although all "Commonwealth member states confirm their legal right to establish their own armed forces" (Article 6 of the agreement), so far only three states are actually insisting on the immediate establishment of their own army: Ukraine, Moldova, and Azerbaijan. The remaining republics have agreed to transfer the share in the common inheritance owed to them to the allied command since they understand very well that they (with the exception of Russia) have neither the economic nor financial resources even to simply maintain those modern armed forces that are deployed on their terri-

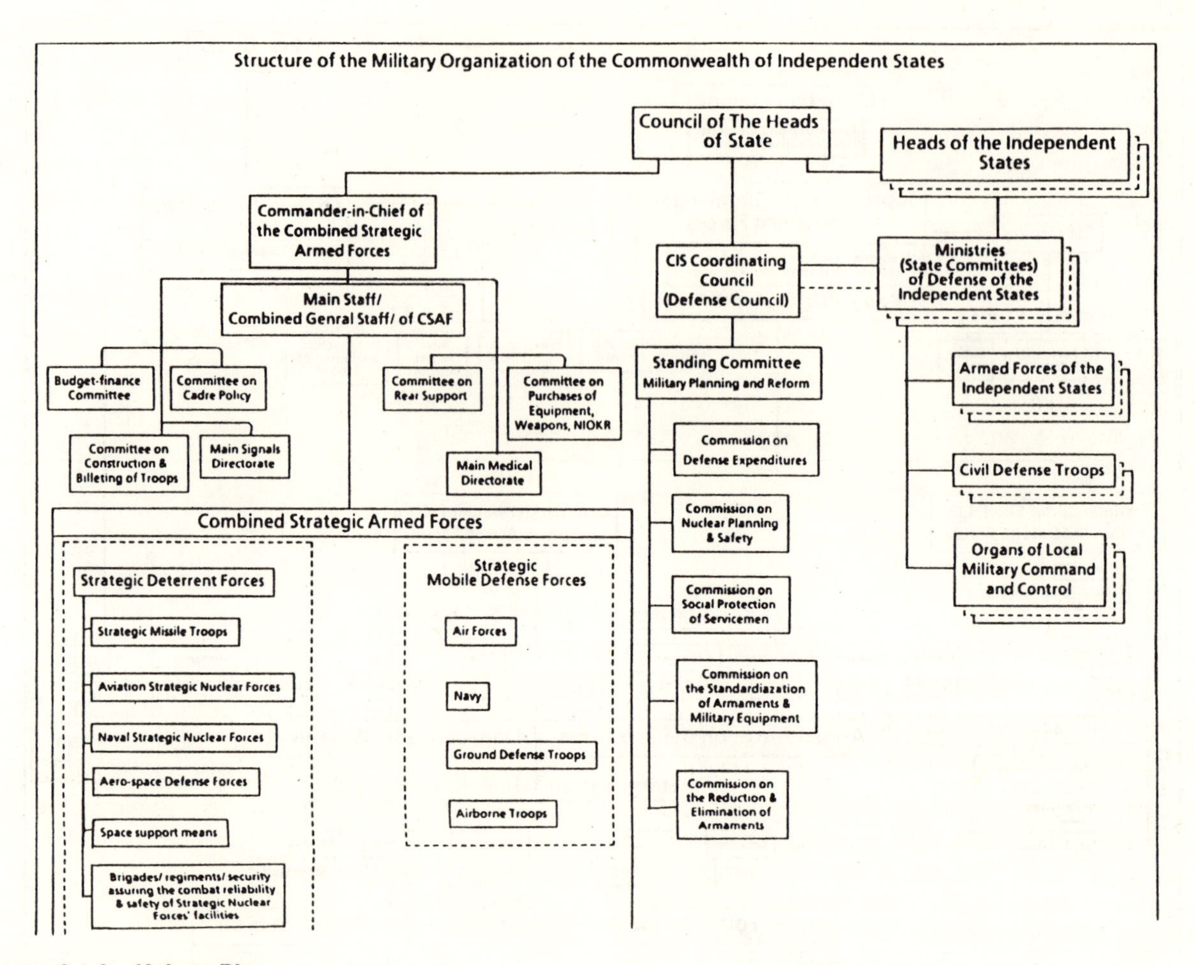


Diagram 9.1.2 **Kobets Plan**

tory today. Most are talking about conducting further rational military structural development.

Thus, the army will still be divided, but into quite unequal shares:—into the combined armed forces which in turn consist of the "Strategic Forces"—"armies, division-sized and smaller units, institutions, institutes, strategic and operational reconnaissance, the RVSN (Strategic Missile Troops), the Air Forces, Navy, Air Defense Troops, Airborne Troops," and so forth; and also "Combined General Purpose Armed Forces"—"armies, division-sized and smaller units, institutions, institutes, and other military facilities that are located on the territories of the Commonwealth member states concerned that are not part of the strategic forces of the independent state (the list for each Commonwealth participant is defined by a separate protocol)"; or into some independent states' own armed forces.

If a large number of the states that have proclaimed their independence see the need and advisability of concluding a defensive alliance (they actually were ready to sign a treaty on the CIS at Novo-Ogarevo), the current Ukrainian leadership does not intend to participate in military alliances and the Agreement on Defense Issues must cease to be in force with regard to Ukraine by the end of 1994 (the Ukrainian delegation's amendments to Articles 4 and 9).

There is every reason to assume that verbal formulas for an interim compromise on defense issues will be found in Minsk on 30 December. In any case, the majority of members feel sufficient optimism. General of the Army Kobets, while answering a *Nezavisimaia Gazeta* correspondent's questions, stated: "There are no fundamental differences between Russia and Ukraine on defense issues. I think that the agreement will be adopted in Minsk." . . .

This political burlesque will certainly disrupt to some degree the campaign planned by the Ukrainian authorities to compel servicemen deployed in Ukraine to take the new oath "for loyalty to the Ukrainian people." At the same time, it is easy to imagine how difficult it will be for Marshal Shaposhnikov to command the remaining army (combined armed forces), now officially totally composed of different nationalities, when he achieves what he is seeking and has been confirmed as its commander-in-chief.

## 9.2

**Armed Forces "Keystone"**
From the "Novosti" Newscast
Ostankino Television, 12 January 1992
[FBIS Translation], Excerpt

Kebich, the prime minister of Belarus, has declared that the question of the armed forces is the keystone of our fragile Commonwealth. In his opinion, if the leaders of the republics fail at their conference in Minsk on Friday, 14 February, to find acceptable solutions on financing the army and on the future structure of the armed forces, then the question of the Commonwealth's existence in the future will be extremely problematic. Today, however, work was under way in Minsk to prepare for the forthcoming meeting . . . .

[I.M. Korotchenya, coordinator of Commonwealth working group]: Painstaking work is under way in order somehow to unite all these sorts of nuances which have been drawn up locally by the governments of the republics into a unified whole, what can be unified, that is; and to draw up a single document for the heads of states to discuss and sign.

[Zhuk]: All the heads of state and heads of government of the Commonwealth countries are preparing for the Minsk meeting in the most painstaking way. Incidentally, the prime minister of the Republic of Belarus, Vyacheslav Kebich, has visited one of the military units of the Belarus Military District.

[Kebich]: I have come in order to draw up my own approaches, the government's approaches to the further fate of the army.

## 9.3

**Attitudes Toward Military Noted**
From the "Novosti" Newscast
Ostankino Television, 12 February 1992
[FBIS Translation]

In Minsk, about one hundred top army command staff experts are preparing documents for the upcoming meeting of Commonwealth presidents. Marshal Shaposhnikov himself is supervising the drafting of the military documents.

As you know, the focal issue today is the maintenance of the army and its very prospects for existence. The army today is effectively paid for by Russia. Even after cuts in weapons purchases, the costs are awesome. Some republics are prepared to pay only for their own army; others think Russia should bear any extra costs because the army in the republics defends the strategic interests primarily of Russia; and others still would prefer to instill some decisive clarity into the money issue, proposing bilateral defense treaties instead of a CIS army. Under this kind of approach, Russia should pay if it wants to defend itself, its allies think. We hope to get a report with details of what is happening in Minsk by this evening.

## 9.4

**Armed Forces "Might Collapse"**
Interfax, 12 February 1992
[FBIS Translation]

Political scientist Maj.-Gen. Vladimir Dudnik has acknowledged the existence of considerable differences on military matters among the Commonwealth member states. He warned that, unless these differences were narrowed at next Friday's [14 February] CIS summit in Minsk, the "artificially united CIS armed forces" might collapse.

Asked by Interfax to predict the possible outcome of the pending top-level meeting in Minsk, Dudnik observed that continuing differences on military development might cause the Russian Federation to follow the example set by other ex-Soviet republics and speed up the formation of its own armed forces and defense ministry.

Commenting on the future of the Black Sea Fleet, a major stumbling block in Russo-Ukrainian relations, the general said that the problem could only be settled by reviewing the existing strategic concepts and radically restructuring the fleet.

## 9.5

**Belarus: "Independent Army Concept" Backed**
Postfactum, 12 February 1992
[FBIS Translation], Excerpt

On 12 February the group for preparation of the meeting of CIS state and government leaders managed to discuss only several drafts from the twelve-page document package at its session at Kolodishchi Army Base near Minsk. The discussed documents, in particular, included the regulations on social insurance and civil rights of retired military servicemen, regulations on the authorities of state leaders in defense issues, regulations on the council of defense ministers of the Commonwealth, and several other documents.

Virtually at once the draft regulation on the authorities of the Commonwealth armed forces supreme military bodies was revoked from further discussion. Leonid Privalov, deputy chairman of the Belarus Supreme Soviet's standing commission on national security, defense, and crime, told Postfactum that none of the representatives of the CIS armed forces command was able to explain to him what the CIS united armed forces are.

For this reason Privalov concluded that Moscow does not have any clear concepts of this issue itself. He also believes that the documents submitted for discussion do not express the existent situation as they do not clarify the positions of the states urging for the formation of their own armies and others currently overlooking this problem. In the words of Privalov, the session has been rather poorly organized. He assumed that over such a short period it would be difficult to discuss the whole package of documents in details. It would make far more sense to draft a single fundamental law and equip it with the basic solution of all military issues.

## 9.6
## "Transitional Period" Needed

Colonel P. Chernenko and Captain O. Odnokolenko
*Krasnaya Zvezda,* 13 February 1992
[FBIS Translation]

The conference of the Commonwealth of Independent States working group which is preparing the ground for the meeting of the CIS heads of state in Minsk on 14 February began work at Kolodishchi near Minsk on 11 February. Marshal of Aviation E. Shaposhnikov, commander-in-chief of the CIS armed forces, and the defense ministers, chairmen of defense committees, and military experts from the CIS countries are taking part in the conference. The only absentees at the moment are the representatives of Uzbekistan, who have been delayed by bad weather.

Unfortunately, the conference is being held behind closed doors and journalists have not been allowed access there at the moment. It is presumed that the agenda of the CIS Heads of State meeting will include around twenty questions, the majority of them military. There will be a discussion of draft laws and agreements on defense, military doctrine, the social protection of servicemen and their families, the reduction of the armed forces, and so forth.

Clearly it will not be easy to resolve these questions since on many problems the Heads of States have their own, sometimes diametrically opposed, opinions.

As Marshal of Aviation Evgeniy Shaposhnikov stated, he was and remains a supporter of the balanced approach to military problems and considers that a two-to-three-year transitional period is needed to resolve all the questions connected with the armed forces. However, the commander-in-chief refrained from making a prediction on the results of the CIS heads of state meeting.

Meanwhile, addressing officers and non-commissioned officers at the Rogachev Motor-Rifle Division on 11 February at Uruchye, where they are stationed, V. Kebich, chairman of the Republic of Belarus Council of Ministers, promised officers reliable social protection and pension provision. According to his statement, the republic must have its own army, and all the property of the Belarusian military district belongs to Belarus, with the exception of strategic weapons, of course.

Expressing the government's viewpoint, Vyacheslav Kebich stated that after reduction the Belarus Army will number no more than 80,000 men. It is planned to reduce one division a year since a larger reduction would be too costly. He also expressed the intention to halt wholesale conversion and to support defense plants in every way. The military hardware which they produce is planned to be sold. Otherwise, the prime minister stated, we will not be able to survive.

It has become clear that recently Belarus's position on the creation of a nuclear-free zone within the republic's borders has been changing radically. No one is going to talk with a non-nuclear power, Kebich stated at the meeting with servicemen. There was no comment on that statement.

## 9.7
## Deepening Crisis "Papered Over"

From the "Novosti" Newscast
Ostankino Television, 14 February 1992
[FBIS Translation]

Without an army there cannot be a strong state. It is precisely the army that currently determines to a large extent how peaceful and stable our life will be in the near future. Therefore, what was being decided in Minsk today was not simply the question of a united army—to a large extent, the future of the Commonwealth was being decided.

[Begin recording.] [Correspondent A. Gerasimov]: The third Minsk meeting began, at the initiative of the long-serving airman Shaposhnikov, with a display of combat air equipment for the republican leaders at one of the Minsk garrisons. One can only guess at the political motives for such a start. But the meeting itself developed quite logically. It was almost totally devoted to military questions.

First of all, it soothed the world community. Not one of the meeting's participants rejected the idea of united strategic deterrent forces. The nuclear button will remain in the same hands as before. In the opinion of many experts who accompanied the leading officials to the conference of the heads of state, the meeting has, to say the least, not brought the former Union republics close together. At the same time, the optimism of the documents signed increases from one meeting to the next.

The deepening crisis is papered over in fine style by various declarations of cooperation. Officially, none of the participants has rejected the Commonwealth of Independent States, but the clear split with regard to the united general purpose armed forces may considerably speed up this process. At one extreme are the eight states that are putting their forces under a united command, and at the other extreme are Ukraine, Moldova, and Azerbaijan, which want to take command of the Soviet Army property that is in their territory. Evidently, this course of events motivated the former members of the Union, in one of the documents that was adopted, to conclude an agreement on renouncing the use, or threat of use, of weapons.

[Kazakh President Nursultan Nazarbaev]: The Commonwealth, the sort of Commonwealth we want, does not yet cause satisfaction because we do not have any sanctions or strict monitoring of the implementation of our decisions. And when we do actually prepare specific documents, when they are subsequently discussed, the specific elements in them somehow disappear and I personally am not satisfied with the way we postpone some specific questions from one meeting to the next.

[Gerasimov]: According to the facts that we have, only a few of the fourteen [as heard] planned documents were signed today. The CIS heads of states will hold their next meeting on 20 March in Kiev. Tomorrow in Minsk, there will be talks between Yeltsin and Kravchuk that will be of no less importance than the meeting that has just ended. [End recording.]

## 9.8
## Defense Officials View Meeting

Postfactum, 14 February 1992
[FBIS Translation]

Ukrainian Defense Minister Konstantin Morozov: The Ukrainian Army already exists. As many as 352,000 servicemen have sworn allegiance to Ukraine to date. Before the opening of the Commonwealth summit in Minsk, which is devoted to the military issue, Col.-Gen. Konstantin Morozov told Postfactum Ukraine would build its own defense system as befits a sovereign European state. He believes the time is past when Moscow could dictate its will to others. Morozov observed that this pressure is felt also at the present talks. But this is not the path that can be crowned with success, he said. The basis for agreement is mutual respect for each other's interests. The interests of Ukraine are known. They are bolstered by the strong enough economy, the will of the republic's government, and the will of the Ukrainian people. Observers believe it is these factors that played not the least role in shaping the Ukrainian position. It is likewise obvious that not all the republics which were recently part of the Soviet Union are prepared to take on themselves all cares for the maintenance of their own army. Observers think this may be manifest in several hours after the beginning of the summit and many of the republics may strike a certain defense alliance with Russia.

For instance, Eduard Simonyan, deputy chief advisor on national security issues to the Armenian president, emphasized that the Armenian delegation is against ill-considered decisions and comes out for a transition period that would enable the parties to specify their positions and elaborate mutually coordinated decisions. Simonyan thinks if the Union no longer exists and the armed forces cannot be maintained in their old form, they should be reformed both in structure and the form of command. But this work requires time and, what is all-important, coordination and mutual understanding. Certainly, nothing can be simpler than to nationalize the units of the Seventh Army stationed in Armenia, but this is not a way out of the situation, Simonyan observed. It is high time to learn to solve problems in a civilized way. He said the Armenian delegation has exactly this sentiment when it sits down at the negotiating table.

Representatives of the Central Asian republics and Kazakhstan were the first to support Russia's course of creating the uniform armed forces. For instance, Lt.-Gen. Sagadat Nurmagambetov, chairman of Kazakhstan's State Committee for Defense, noted: Kazakhstan stands for the single armed forces, for their single supply and the single personnel training. It wants the army to be staffed under both territorial and exterritorial principles, specifically the strategic forces. Col.-Gen. Petr Chaus, acting defense minister of the Republic of Belarus, expressed the supposition that this meeting would fail to reach an agreement on the integral armed forces. He believes some military union is possible but it is already obvious that Ukraine, Azerbaijan, and Moldova will pursue an independent path. Already today they have de facto their army as, for example, Ukraine. Others envisage a transition period of two to three years for this purpose as, say, Belarus. These are the points on which agreement should be reached, Chaus thinks.

Marshal Evgeniy Shaposhnikov, commander-in-chief of the Commonwealth's armed forces, asked if there is any hope for an accord about the single armed forces, answered: Yes, and there is a basis for this accord. The Commonwealth defense ministers have already reached some agreement in previous days. At least, we have reached mutual understanding on thirteen documents. This is the foundation on which my hope rests.

## 9.9
### Kravchuk Interviewed Upon Arrival
Interfax, 14 February 1992
[FBIS Translation]

Speaking in the airport upon his arrival in Minsk today (14 February), Ukrainian President Leonid Kravchuk said he firmly stuck to the view that the CIS should not have joint armed forces, for this was only possible in a single state. "The CIS is not a state," President Kravchuk stressed. Apart from this, he said, it was clear that having joint armed forces "is dangerous for all of us," for they could be used to strangle the democratic movements in all Commonwealth states.

As regards the Black Sea Fleet, the Ukrainian president doubted the issue could be resolved in the course of the Minsk summits, though he expressed the hope that at least the basic principles would be worked out in the Belarusian capital for the solution of the naval problem.

President Kravchuk said he would like to have a bilateral meeting with Russian President Yeltsin in Minsk and emphasized that they needed no mediators.

Speaking about economic problems to be dealt with in the course of the summit, Mr. Kravchuk expressed the hope that the meeting would work out an agreement on economic relations within the CIS. "It would be a mistake to assume that Russia can strangle Ukraine economically by refusing to sign such an agreement," he said. "Yes, Russia is able to do this, but thereby it will kill itself as well—it'll simply die a day later."

## 9.10
### Nazarbaev Proposes Initiatives
Leonid Sviridov
Radio Rossii, 14 February 1992 [FBIS Translation]

The meeting of the heads of the Commonwealth of Independent States member states is continuing in Minsk. An interval was announced at 1400. [Seytkazi Mataev], press secretary of Kazakhstan, spoke to journalists, who are thirsty for any piece of news. He told the journalists that a definitive agenda has not been approved yet. [Mataev] confirmed reports that Ukrainian President Leonid Kravchuk is chairing the meeting on the proposal of Boris Yeltsin, and that Ukraine and other republics have proposed discussing economic issues.

At the start of discussion on military issues, differences emerged among the republics. Kazakh President Nursultan Nazarbaev proposed new initiatives. He had four proposals:

The first proposal is to declare a transitional period, but Nazarbaev himself disowned that proposal immediately because Moldova, Ukraine, and Azerbaijan already have set up national armed forces. The proposal has been withdrawn.

The second proposal is to set up separate armed forces in each republic, and then to establish a defense alliance.

The third proposal is to allow CIS member states to join the defense alliance at their discretion.

The fourth proposal is to declare all armed forces on the territory of the republics their own national armies, and then to conclude a defense alliance.

According to [Mataev], there was an exchange of views on the wording about what should be considered strategic forces and what should be regarded as conventional armed forces. [Mataev] believes that there are no differences of opinion here, just variant interpretations.

A news conference is being postponed to 1900. There will be a break of one hour at 1530, when the presidents will have lunch. It looks as though the journalists are ready to descend on the presidents with numerous questions. There are 150 CIS journalists and 180 foreign reporters accredited to the meeting of the heads of the CIS member states.

## 9.11
### Shaposhnikov Address on Nagorno-Karabakh Issue
ITAR-TASS, 19 February 1992 [FBIS Translation]

Marshal of Aviation Evgeniy Shaposhnikov, commander-in-chief of the Commonwealth of Independent States Joint Armed Forces, has addressed the Council of CIS Heads of State in connection with the events in Nagorno-Karabakh. The address said:

The events around Nagorno-Karabakh are taking on an ever more tense and menacing character. The scale of combat operations is widening. The number of casualties is rising. Towns and villages are being destroyed. There are continued attempts to draw servicemen into the conflict, and continuing encroachments on the arms, military hardware, and stocks of material resources belonging to the Transcaucasian Military District.

All these actions run completely counter to the statement of the CIS heads of state of 14 February 1992 on the impermissibility of using or threatening to use force and on the need to resolve contentious problems by exclusively peaceful means, through negotiations.

The situation is being made worse by the persistent attempts of the leadership of the Azerbaijani Republic to set up its own armed forces on the basis of the Fourth Army of the Transcaucasian Military District which, in turn, prompts the Armenian side to create its own army on the basis of the Seventh Army of the Transcaucasian Military District.

The implementation of these intentions will result in

drawing regular units and formations of the Transcaucasian Military District into combat operations and will inevitably turn the conflict into a large-scale fratricidal war.

The development of events in the Transcaucasus is causing serious concern among states bordering the Azerbaijani Republic and the Republic of Armenia.

The command of the CIS Joint Armed Forces deems it necessary to undertake urgent measures in order to prevent Transcaucasian Military District troops from being drawn into combat operations and to prevent the conflict around Nagorno-Karabakh from escalating. For these purposes it is essential to solve the following priority issues:

1. To employ the entire political authority and capabilities of leaders of independent states to solve the conflict around Nagorno-Karabakh by political means.
2. Until a political solution to the problem is found, the Transcaucasian states should refrain from creating their own armed forces.
3. To examine the issue of forming a contingent of interstate armed forces of the CIS, which will be subordinate to the Council of CIS Heads of State, for the purpose of maintaining order, stabilizing the situation, and insuring a solution, and insuring a solution to interethnic and internal political issues by political means.

## 9.12

### Shaposhnikov Answers Questions

Interview by Irina Lapina on "How Are We Going to Live?" Program
Moscow Television, 22 February 1992
[FBIS Translation], Excerpts

[Question and answer session with Air Marshal Evgeniy Shaposhnikov, commander-in-chief of the Commonwealth of Independent States Armed Forces, in the Russian TV Studio in Moscow; "How Are We Going To Live?" program presented by Irina Lapina, also reading questions telephoned in to Shaposhnikov—live.]

[Lapina]: Hello, here is our live program entitled "How Are We Going To Live?"

Our guest today is Marshal Evgeniy Shaposhnikov, recently the minister of defense, and nowadays commander-in-chief of the CIS Unified Armed Forces. Evgeniy Ivanovich, you are the first military man in our program, but of course this is not the reason why I am welcoming you here especially warmly. I remember your answers when you were still a deputy and were being vetted for the ministerial post. Your answers were precise, military style, but also humanely wise. So let us start our program—and now we have a couple of minutes while the questions are being prepared so help me to clarify one point.

The change of your title is, naturally, connected with the breakup of the Union and with the formation of the CIS. So, what about the Unified Armed Forces? Have there been the same changes as with the Union? I ask because the process could not fail to influence the army. So, do we have an army as a single whole at the moment, or is what we call Unified Armed Forces as different from the old army as the CIS is different from the USSR? Let us start from here.

[Shaposhnikov]: Thank you. Dear television viewers, many greetings, I will certainly try to answer all your questions, and as far as possible, will do so during the program. Well, on the first issue, I would like to say the following: Indeed, with the formation of the CIS—and the CIS as we know is neither a state nor a superstate superstructure. Therefore, there are naturally no ministers in charge of any area within the CIS. Therefore, in connection with changes that have taken place in our states, a change in the armed forces has taken place, since the army is part of society. We have actually been fighting for a long time for the armed forces to be unified. In my opinion, this is the most acceptable idea for a transitional stage since so many visible as well as invisible links connected all our states—take the common economy, the common military-strategic space, the common procurement system of the troops, the common system of orders for equipment and weapons. Groups of armed forces were located on the territory of the former Union, regardless of the administrative borders of the states, and so on and so forth. Therefore, we have been repeatedly putting forward this idea of unified armed forces.

However, at the last Minsk meeting of the heads of independent states a decision was made—at last—and I am actually pleased that there is some clarity now. Because, you know, some people had been talking about unified, others about single, yet others about independent [armed forces], and so on and so forth. The eight states that decided to stick to the position of unified armed forces, that is, unified by the common aims and tasks that I have just been talking about, they decided to create unified armed forces.

Well, three states did not join this agreement. They are on their way toward the creation of their own armed forces. One of the issues was that of the command of the Unified Armed Forces. Therefore, you know, there has to be a commander-in-chief of the Unified Armed Forces. The Council of Heads of State decided to appoint me. There are also two command authorities that have been set up—the strategic forces and the general purpose forces, which are organizationally included into this chief command authority. All other candidates for the positions of commanders of specific services will be confirmed at the next meeting in Kiev on the 20th [of March]. . . .

[Shaposhnikov]: Yes, we have been talking a lot about military reform for a long time. This discussion began before the August events. Two concepts were elaborated. One concept was an unofficial one. It was advocated by a group of officers and people's deputies of the USSR—as it then was—and by some Russian deputies. Another concept was worked out in the Defense Ministry, in the former Defense Ministry. I would like to say that the more radical reform, or reform concept, was elaborated by the officers. The reform elaborated by the Defense Ministry was milder and covered a longer period, and to a large extent was a cosmetic reform.

Therefore, as soon as I took over the post of defense minister I would like to say that I, as it were, combined these two reforms into one. We began implementing it. But at that time the Union still existed. The first steps that we managed to take are at present being implemented. But then there began this confusion over defining the armed forces, defining them as independent for some states, and so on. The actual reform of the armed forces themselves was slowed down somewhat, I admit this. This was because I was devoting all my strength and energy to somehow trying, in the broad sense, to preserve the armed forces as a whole. We are at present working on the internal reforms.

Now we have made the decision to hold a meeting, literally sometime after 23 February—actually on the 27th and 28th—with the defense ministers of the Commonwealth countries at which we will begin elaborating the documents that will guide the reform. Therefore, it can be said that the reform is progressing in two directions. In the large and broad sense it is a reform of the armed forces on the scale of the Commonwealth countries. The second part in the narrow sense is the part that directly concerns the armed forces themselves. . . .

[Lapina]: . . . What do you think about the redrawing of borders? If one looks at history, then territorial claims have been the most frequent cause for the outbreak of wars. Well, the independent states really do have a lot of claims. What is going to happen, Evgeniy Ivanovich?

[Shaposhnikov]: I am categorically opposed to redrawing all the borders that exist between the CIS states since this could lead to very severe consequences. In general there are no precedents. At the initial stage, before the formation of the CIS, there was some mention of this. But at the present time the political leaders of our states have shown sufficient responsibility and wisdom; on 14 February in Minsk an agreement was signed to the effect that the countries of the Commonwealth would never use armed force against each other. In the same way it has been declared that all questions that arise can be resolved only by political methods, peacefully, by means of talks.

[Lapina]: Thank you. There are some questions here which I simply cannot overlook. Teryukhina of Saratov asks: Evgeniy Ivanovich, on behalf of all mothers I ask you not to send Russian lads to places where wars are going on. Let them sort things out themselves. Our boys are dying. It is even hard to read this: there are very many questions like this. Soldiers on compulsory service are sometimes really just boys, and how many tragedies there are. What is your opinion on the army's role in so-called interethnic disputes? They do need to be quelled and contained somehow, and the heat needs to be taken out of them, but on the other hand there is no need to send young boys into that furnace.

[Shaposhnikov]: Yes, I agree with you. My position is that the army should never—and in no circumstances—be used in interethnic conflicts, or to solve political tasks, or to back particular aspirations of individual leaders within the CIS. Therefore, taking advantage of the agreement reached by the heads of state in Minsk on the non-use of the armed forces in dealings with each other, I sent all the heads of state an appeal on the 19th [not further specified]. That appeal contained three points. The first was that they should apply all their political sense and all their political authority to settling the Karabakh problem peacefully. The second was that national armed forces should not be set up in the CIS countries in the Transcaucasian region until there is a political settlement of the Karabakh problem, since it is clear for what or how those armed forces would be used. The third was that they ought probably, perhaps, to think about establishing within the CIS, in addition to the armed forces, some kind of body or corps along the lines of the United Nations, which could be used to stabilize the situation in a particular region, but only volunteers could be sent to that corps. We are now finalizing proposals to this effect, and we shall put them forward in specific form in Kiev on 20 March . . . .

## 9.13

### Shaposhnikov Views Future Role of the Armed Forces

Interview
Radio Moscow, 24 February 1992
[FBIS Translation], Excerpts

. . . . The first question was what are the prospects for reform of the ex-Soviet Army?

[Shaposhnikov]: I believe that in the end all the armed forces can be divided up according to national or state principles—the CIS is not a state, but a group of states which does not have a single center, so the final objective of any state is complete independence and everything which follows from this. [Passage omitted.]

[Announcer]: It is often said that the CIS Armed Forces are placed above the state. What is your opinion?

[Shaposhnikov]: At present, eleven chiefs of strategic forces and eight chiefs of conventional forces are subordinate to me, and that is difficult. I cannot alone make any decision, particularly if it is a serious and important one. It is a little easier in the area of current business—we have established a Council of Defense Ministers of the CIS to complete the documents which were not able to be signed in Minsk, as well as the mechanisms for the implementation of decisions. The defense ministers will work in close contact with their respective presidents. . . .

## 9.14

**Kravchuk on Disputes with Russia, CIS Armies**
Helsinki Radio, 26 February 1992
[FBIS Translation]

Ukrainian President Leonid Kravchuk, who arrived in Finland this morning, said during an interview at the airport that he was very confident that disputes between Russia and Ukraine will be solved. Kravchuk believes that the dispute on the Black Sea Fleet will be solved, possibly next month.

The president of Ukraine said that he believes that the dispute will be solved at a meeting in Kiev around the 20th of next month. On the one hand the dispute depends on the committee working on the matter, while on the other it depends on what the two states, Russia and Ukraine, want. The president of Ukraine also believes that every member country of the Commonwealth of Independent States will form its own independent army. He did not view this as a special threat to the member countries of the CIS.

Leonid Kravchuk, together with four other CIS presidents, will sign the CSCE Final Act at the Finlandia Hall this afternoon.

## 9.15

**Armed Forces Being Financed Mainly by Russia**
Interview
*Krasnaya Zvezda,* 6 May 1992
[FBIS Translation]

[Interview with Col.-Gen. V. Samsonov, chief of the CIS Joint Armed Forces General Staff, by unnamed correspondent; date and place not given: "Plenty of Agreements, But They Are Working Poorly So Far. CIS Joint Armed Forces Chief of General Staff Answers Our Correspondent's Topical Questions"]

*Krasnaya Zvezda* published an interview on 18 March with Col.-Gen. V. Samsonov, chief of the CIS Joint Armed Forces General Staff. Judging by the letters reaching the editorial office, it generated a great deal of interest among readers—many of whom have proposed making such meetings a regular feature of the newspaper and having the chief of the General Staff give regular assessments of current events, talk about the problems being resolved in the Joint Armed Forces, and answer the questions posed by the readers' letters. We put this request to Viktor Nikolaevich and, having obtained his consent, asked several currently topical questions.

[*Krasnaya Zvezda*]: It is well known that the State Commission for the Creation of a Russian Federation Defense Ministry, Army, and Navy has announced that the Russian Armed Forces' administrative structures will be set up on the basis of the former USSR Defense Ministry and the USSR Armed Forces General Staff. How is the General Staff participating in this work, have mutual understanding and coordination been achieved with the State Commission on everything, and what problems and difficulties can be identified in this area?

[Samsonov]: The main work is, of course, being implemented by the State Commission. At its request, representatives of the General Staff are being involved in drawing up materials for reference, background information, and analysis purposes, as well as in holding various kinds of consultations and operational calculations.

The main difficulty is caused by the tough time limits.

[*Krasnaya Zvezda*]: A month ago you stated that the forward-based strategic echelon in the former USSR's defense system had been completely destroyed. How can such significant damage be offset during the creation of the Russian Army? Is some kind of work under way to restore the strategic echelon of the armed forces under the new conditions?

[Samsonov]: In the West the basis of the USSR armed forces' forward-based strategic echelon was provided by troops (forces) under the Western, Northern, Central, and Southern Groups of forces. With the troop withdrawal from the East European countries a situation has come about whereby the forward-based echelon has to all intents and purposes started disappearing. And, naturally, that would have meant considerable damage for a unified state such as the Soviet Union. But today we can no longer assess the situation from these positions. The CIS has been set up and the geopolitical situation has completely changed. We no longer treat the West as our enemy. So there is no need to restore the forward-based strategic echelon in the same form or give it the same tasks as before.

Although work is undoubtedly under way to set up new CIS Joint Armed Forces troop groupings and arrange for their strategic arrangement. But this work arises from entirely different assessments of the existing military danger, is based on the concept of "mobile defense," and presupposes the creation of structures which will be primarily aimed at preventing local conflicts.

As far as Russia is concerned, we need to take account above all of its security interests and its realistic potential. During the elaboration of proposals on the creation of Russian armed forces, the questions of their organizational development and alignment will naturally be examined.

[*Krasnaya Zvezda*]: Is the General Staff currently exerting full control [*kontrol*] over the military units and combined units under the jurisdiction of the CIS Joint Armed Forces? What problems are there in this area?

[Samsonov]: Controlling the troops' activities is one of the tasks carried out by the General Staff. But, with the formation of the CIS and the announcement by some of its member states that they are to set up their own armed forces, this task has been made considerably more difficult. The "privatization" of the troops located on these states' territories, the ambitions of certain politicians, the unilateral resolution of questions concerning the army, and the reluctance to reach a compromise sometimes considerably complicate the General Staff's ability to control the troops' lives and activities. This in turn could lead to a destabilization of the overall situation.

The main task today, in my view, is to ensure reliable monitoring [*kontrol*] of nuclear weapons. I can confidently assert that the General Staff, along with specialized forces and means, is confidently carrying out this task.

Another topical problem requiring constant General Staff control is staffing. Agreement was reached at the Minsk summit of heads of state this February that the CIS Joint Armed Forces should be manned by units of troops (forces) allocated out of the CIS states' own armed forces. But this question has not been finally worked out as yet.

As for troops (forces) which are not part of the sovereign states' armies but are located on their territories, as well as the groups of forces in the East European countries, the Baltic, and Mongolia, they come under the direct control of the General Staff, which continues to administer all of their activity, including the withdrawal.

[*Krasnaya Zvezda*]: How are the Commonwealth states participating in financing and providing backup for the CIS Joint Armed Forces?

[Samsonov]: Under the agreement on the formation of a unified defense budget signed between them in Minsk on 14 February 1992, the Commonwealth states pledge to participate in the formation of a unified defense budget by paying fixed contributions into it. The same agreement stipulates that the size of these fixed contributions and the procedure for their payment will be determined in the Commonwealth Heads of Government Council. But, given that no such council decision has yet been made, there is no Commonwealth defense budget. Therefore the Joint armed forces are still being financed mainly out of the Russian Federation budget.

[*Krasnaya Zvezda*]: Attacks on military units are continuing in various regions. What statistics are currently available on these attacks? Do you have any generalized information on the human and material losses suffered by the Armed Forces as a result of such actions? What measures are being taken to rule out losses in the future?

[Samsonov]: The CIS Joint Armed Forces High Command and the General Staff have accurate and exhaustive information about human and material losses. A clear-cut system of operational information about all illegal acts committed either by or against the troops has been worked out.

This year alone there have already been 305 attacks on military installations, including 34 on weapons and supply dumps, and 90 on guards (sentries). As a result of these attacks, 31 servicemen have been killed, 72 have been wounded, and more than 7,500 firearms and 500 units of combat and other equipment have been stolen. Most of the attacks have been in the Transcaucasus—265—with 16 in the North Caucasus. There have been four attacks in Central Asia, four in Siberia and the Far East, and eight in the central part of Russia.

Servicemen often have to show considerable restraint, persistence, and sometimes even courage in negotiating with local and state organs of power and administration with regard to each attack on military installations and seizure of people, weapons, or military hardware.

Close contacts have been established on all military and defense matters with the sovereign states' defense ministries (committees). The legal basis for the functioning and collaboration of strategic and general-purpose troops (forces) with the republics' military departments has been worked out and continues to be developed and improved. Measures are being taken to step up security around military installations and to provide them with technical means and fortifications. The training and selection of the personnel and the tasks of the sentry and internal services are being implemented in a higher-quality fashion. But, because of low manning levels at a number of combined and other units, it has proved harder to solve this problem. I wish to make no secret of the fact that we have even been forced to use officers to protect certain installations.

I am profoundly convinced that whatever measures are taken by the troops to protect and defend installations, this problem can only be completely resolved by reducing socio-political tension and ending armed conflicts in the regions.

[*Krasnaya Zvezda*]: The last summit in Kiev is often said to have succeeded in resolving military issues. At the same time, there has been concern about the fact that no mechanism has been developed to implement the adopted agreements and accords. What can you say on that score?

[Samsonov]: Yes, the Kiev summit should have provided the foundation for the planned and civilized reform of the former USSR's armed forces. All the more so as thirty-four documents were adopted on military matters at the Minsk and, particularly, the Kiev summits, whereas just five were passed at the first three summits.

What were these documents? I would recall that the Commonwealth heads of state signed agreements defining the powers of the top CIS organs on questions of defense, the position and functions of the Joint Armed Forces during the transitional period, the legal basis for their activity, and the principles underlying their manning and the provision of military hardware and other material means. Decisions were passed on the Joint Armed Forces High Command, and appointments made to leading positions in their structures. As you can see, a sizable chunk of the military issues which constituted a stumbling block at previous summits was successfully resolved. But in practice all these agreements and accords are working poorly so far.

The negotiations with Ukraine about clarifying the list of combined and other units to be part of the Strategic Forces have been difficult. There have also been a lot of contradictions over the General-Purpose Forces.

Here we need to understand that the new political and economic relations between the Commonwealth states require not a cosmetic reorganization of the armed forces, but a profound revision of the principles of military organizational development and a search for original approaches to the resolution of the questions of guaranteeing the Commonwealth's defense capability.

[*Krasnaya Zvezda*]: It is well known that Russia, as a subject of international law and a UN Security Council member, is considered the successor to the former Union when it comes to carrying out the commitments under the START Treaty and the Paris CFE Treaty. How are these issues being resolved at the moment?

[Samsonov]: May I briefly remind readers that the Commonwealth's stance on these issues is that the START Treaty—signed 31 July last year—should be ratified without delay. Russia advocates its strictest fulfillment. Moreover, it advocates in principle the complete elimination of nuclear weapons and other mass destruction weapons.

The following measures have currently been taken to limit and reduce strategic offensive weapons. I will dwell on just a few points. Thus, around 600 land-based and sea-launched ballistic missiles (almost 1,250 nuclear warheads) have been stood down from combat alert. Of this total, 130 ICBMs and 49 launch silos have been destroyed, and equipment has been dismantled at 81 silos.

Furthermore, six nuclear missile submarines have been decommissioned. Since last October all existing rail-mobile ICBMs have been at their permanent base sites. Tu-95MS and Tu-160 heavy bombers have been stood down from combat alert. Production of existing types of long-range air- and sea-launched nuclear cruise missiles is ending.

And one last point to which I would like to draw attention—the number of strategic offensive weapons on combat alert will be reduced to the level agreed in the treaty within a three-year period, instead of in seven years.

Now to turn to the CFE Treaty. Russia views it as one of the cornerstones of the future European and world security system. It is making efforts to ensure that the treaty comes into force and begins operating in full in the very near future.

The question of implementing the treaty is a subject for painstaking work between the independent Commonwealth states. The complexity of the problem lies in the search for decisions acceptable to the CIS members regarding the devolution of the former USSR's treaty commitments to the newly formed states. Several consultation sessions at expert level are currently under way. A number of difficulties have emerged in the course this work. One significant difficulty is the problem of the distribution of the levels of arms held by the USSR and due to be limited under the treaty. The emergence of this problem is entirely natural, since it is bound up with the division of the former USSR's property, and each successor state is relying on its right to a share of this property, not least as regards defense potential.

But the generally constructive mood which, despite the presence of certain differences, prevails in the work on these problems, and the understanding of the treaty's importance shown by all interested parties gives every reason to believe that mutually acceptable solutions will be found.

## 9.16

### Committee Debates CIS Collective Security Treaty

Interfax, 8 April 1993 [FBIS Translation]

The Russian parliament's Committee on International Affairs and External Economic Ties debated the CIS Charter and collective security treaty on Thursday [8 April] following the Russian president's request to speed up the review of these documents.

In committee chairman Evgeniy Ambartsumov's view, the ratification of the treaty gives at least some opportunity to keep the CIS intact. The pact has been ratified by five of its six signatories. The Belarusian parliament is now debating its ratification.

Although the committee recommended parliamentary approval of the document, it noted that "given some asymmetry on defense between Russia and other signatories, the pact will obviously mean in practical terms Russia's military assistance to these countries rather than the other way round." The committee believes that with current military and political instability in the former USSR, that "would be tantamount to attempts to involve Russia in various military conflicts with undesirable consequences."

It viewed as unacceptable for Russia Article 4 of the treaty, which makes the granting of such aid an automatic decision, and voiced displeasure with the treaty's clause saying that any decision to use armed forces would rest with the presidents of the participant states in disregard of what parliament thinks. The committee recommended government talks on amendments to the treaty.

Regarding the CIS Charter, the committee argues that the principle of recognizing the existing borders (Article 3) might cause certain problems for Russia. It claimed for one that since Russia's frontier with neighboring states had not been demarcated, Russia's state border was not yet clearly defined. It also said that such issues as the status of the Crimea and Sevastopol remained unsettled.

## 9.17

### Shaposhnikov Urges Integration of CIS Armed Forces

From the "Novosti" newscast
Ostankino Television, 13 May 1993
[FBIS Translation]

The preparations for the meeting of CIS leaders, which is known to be taking place in Moscow tomorrow, are continuing. Today, the CIS ministers of defense met in the capital of Russia.

[Begin correspondent Sergey Omelchenko recording]: Improving and strengthening defense capacity and military cooperation is on the agenda of today's session of the Council of Ministers of Defense of the Commonwealth countries in Moscow. Exactly a year has passed since the treaty on collective security was signed, which six states are party to today, but up to the present time, it has not been filled with specific content. No mechanism has been worked out for implementing it.

The ministers of defense gathered today to put this state of affairs right and make specific decisions.

[E. Shaposhnikov, commander-in-chief of the CIS Joint Armed Forces]: The time has come when one has to move toward integration, and precisely according to the formula quality rather than quantity. Let there be a smaller number of states, but real ones. I perceive the Commonwealth as a qualitatively new type of cooperation among our fraternal peoples.

[Omelchenko]: The doors of the Commonwealth's defense alliance are open to all those who are interested in collective security. It was announced at the session that one could become a member of the alliance on a bilateral basis. Thus, the ministers of defense of the CIS countries are counting on coming out at a qualitatively new level of military cooperation. [Video shows scenes at meeting and interview with Shaposhnikov.]

## 9.18

### Shaposhnikov Interviewed on Military Reform Plan

Interview by Aleksandr Zhilin
*Moskovskie Novosti,* 30 May 1993 [FBIS Translation]

[Interview with Marshal Evgeniy Shaposhnikov, commander-in-chief of the CIS Joint Armed Forces, by Aleksandr Zhilin, *Moskovskie Novosti* military observer; place and date not given: " 'Shaposhnikov's Plan' Not Yet Shelved"]

The Russian Defense Ministry did not support Shaposhnikov's proposals aimed at reforming the Joint Armed Forces at the latest meeting of the Council of CIS Defense Ministers. Nonetheless, the Joint Armed Forces commander-in-chief still hopes that his plan will be adopted.

[Zhilin]: Why was it that your concept drew so vehement a protest from the Russian Defense Ministry? Wasn't it because you suggested that the Joint Armed Forces be set up in the absence of normal national armies? No one in the world, it seems, has ever embarked on this path.

[Shaposhnikov]: Indeed, up to the present day joint armed forces have been set up on the basis of full-blooded national armies. This is a classical scheme, but should we blindly emulate it? Our situation is not classical, either. The collapse of the USSR resulted in a process leading in the opposite direction: Joint armed forces are turning into national ones. What is better in this situation, to wait until the breakup is complete or to set up joint structures along parallel lines preserving, as much as possible and utilizing in a rational fashion, the most important components of the systems that had been designed to operate as a single whole and are capable of operating only as such.

[Zhilin]: Have you not been late with your proposals?

[Shaposhnikov]: I have always argued for maintaining the unity of the main pillars of the joint defense space for some time. During the current transition period, we could have ensured a civilized transition from Joint Armed Forces to national armies.

Alas, this idea has been rejected. General-purpose troops of the Joint Armed Forces that have been mechanically, often arbitrarily, turned into national units and have encountered a large number of new problems in addition to old ones, ranging from social to logistical, operational, legal, and international matters. In the final analysis, the republics themselves have realized that, in order to overcome this extremely difficult situation, we have to join forces.

[Zhilin]: *Moskovskie Novosti* extensively reported about your plan in its previous issue. But here is a new objection from Pavel Grachev: "Russia is not against the Joint Armed Forces, but setting them up now would be premature, because it will require enormous extra financing," which, he said, will have to be fully paid by the Russian Defense Ministry.

[Shaposhnikov]: Let us consider an elementary example. Let us imagine that, under a national defense doctrine, the Russian Army consists of three divisions for which the requisite money has been appropriated. One of the divisions is placed under the operational command of the Joint Armed Forces commander-in-chief. Evidently, this will not require additional investment. This is precisely what the proposed principle on the formation of the CIS Joint Armed Forces boils down to, but if Russia is to form its armed forces exclusively on its own, the most primitive measures to protect its borders will translate into about 20 trillion rubles—to say nothing of saturating the border areas with weaponry and equipment. The same applies to the other states of the CIS.

We are not intent on creating some artificial structures. Moreover, while arguing for the maintenance of some of the global defense systems of the former Soviet Union, we are in effect saving Russia and the other CIS states the hair-raising and absolutely unjustifiable expenses both in the short run and in the more distant future. This is why we believe it is necessary to create, on the basis of available units, the following operation commands: the "South" Air Defense Command, including Russia, Kazakhstan, and the Central Asian states; and the "Caucasus" Air Defense Command, which would combine the forces and equipment of Russia, Armenia, Azerbaijan, and Georgia. It is worthwhile to consider creating joint ABM and Air Defense Commands tentatively called "West." I think that decisions that, hopefully, will be made by the Council of Heads of State of the CIS in July would open up the prospect for moving in precisely this direction.

[Zhilin]: But if these plans still have to be shelved until full-blooded national armies are formed . . .

[Shaposhnikov]: It would be a pity. What does it mean, for instance, for the Central Asian countries to create "full-blooded national armies" from scratch? It is perfectly clear that some of them are objectively unable to cope with this task without resorting to help from other countries outside the CIS. This assistance will have to be paid for, if not with money, then with their independence.

There is another side to the problem. If the republics of the former USSR start looking for help outside the CIS, the East Europeans will begin drifting toward NATO. Even at present we feel the desire on the part of Turkey, Iran, and Pakistan to exploit what they perceive as an emerging "geopolitical vacuum" to strengthen and broaden their influence on the Transcaucasus and Central Asia. Given that each of these states has completely different interests, we will in no time find ourselves among new military-political blocs, alliances, and groupings within the CIS. Zones of rivalry are bound to emerge, which will give a powerful boost to instability. As a result, the states of the former USSR will join an arms race against each other. Furthermore, if this insane marathon is dangerous for the world community, it will be economically fatal for its participants.

[Zhilin]: The plans are that the CIS Joint Armed Forces are to include peacekeeping forces. With what functions will they be entrusted?

[Shaposhnikov]: Yes, we have envisioned the creation of a force that will deal with the settlement of conflicts within the CIS. The emphasis is put on the political solutions to the problems, but an effective mechanism is also essential to regulate conditions on which peacekeeping forces could be brought in: the withdrawal of the combatants from combat positions, control over ceasefires, measures to deter the extremist forces trying to thwart the implementation of agreements, etc. My dream is to set up an interstate military and political body within the CIS that would symbolize the indivisibility of the historical fortunes of the CIS peoples, the unity of combat traditions of their defenders, and the reliability of their joint defense and collective security system.

## 9.19

**Expert Sees Hawks' Win on Russian Military Doctrine**

Aleksandr Zhilin

*Moskovskie Novosti,* 16 November 1993

[FBIS Translation], Excerpts

The military doctrine, recently approved by the Security Council and the president of Russia, is remarkable if only in that such a document has appeared for the first time in the state's history. Specialists point out, however, that it carries a stamp of rivalry between various political interests. The doctrine has reflected a battle between the "moderates" and the "hawks" outside the walls of the General Staff.

According to Russian Federation Defense Minister Pavel Grachev, the doctrine's authors have worked on the assumption that as of now "Russia has no probable adversaries" or potential enemies. Nonetheless, commentators did not ignore the fact that the document, which rose on the yeast of political declarations first by the Soviet Union and then Russia about the renunciation of first use of nuclear weapons, does not contain such an obligation in its text. On the other hand, it contains references to the possibility of "exporting" military force. Presumably, the world will not draw a conclusion from this that Russia is making preparations for an offensive war, but it will hardly add to its neighbors' calm . . . .

As for the former union republics, "the more Russian military bases are located on their territory," [Maj.-Gen. (ret.) Gennadiy] Dmitriev believes, "the quicker a single economic and military union will be restored." Therefore he is not worried over the fact that observers read into the military doctrine Russia's possible expansion over the entire CIS. This expansion does not mean that tomorrow new wars will begin. These may also be "cold," where the state maintains its interests from a position of force. Which is what we are seeing in Russia's relations with Georgia, Armenia, Azerbaijan, Tajikistan, Moldova . . .

And if Russia itself does not see "probable adversaries" around itself, the foreign policy section of its military doctrine will hardly pacify the neighboring and the more distant countries, which are accustomed to looking back at it as a potential threat.

Another dubious aspect of the military doctrine is the provision about the possibility of using army subdivisions "to assist the law enforcement bodies." Because the document does not specify in what instances this "assistance" is allowed, there is a high likelihood of this provision's being interpreted loosely. In effect, it sanctions the use of the army in internal political conflicts.

Some military analysts therefore suspect that what they regard as an overhasty adoption of the doctrine was designed to provide some justification for the authorities' actions on 3 and 4 October. The president, who received support from the army in those days, apparently hopes for it in the future as well. Few will know that not all the troops brought into Moscow have been withdrawn from the capital. A special purpose airborne regiment is being formed with the facilities of a communications division in Sokolniki. The 27th Motorized Brigade of the Moscow Military District is being attached to paratroop forces. The stationing of commando units in and around the capital in the absence of facilities to airlift them to zones of potential military conflict suggests that these troops are designed for "internal use."

The paradox consists in that at another spiral of political confrontation, military force may turn also against the president.

At any rate, the army is increasingly inclined to think that it alone is capable of putting things in the country in order. "Pavel Grachev has made the most of the 4 October victory. At present no one has any doubts that it is the army that controls the situation in the country. I think that as of now the period of endless compromises has finished and an era of order begins, and it will be enforced by us, the military"—this conclusion from the military doctrine was made by military expert Colonel (ret.) Dmitriy Kharitonov. He is not alone in his appraisal. It was not accidental that at one of the recent sessions of the military collegium Pavel Grachev, responding to a remark about the likelihood of the doctrine's being amended by the new parliament, noted: "We shall amend the parliament."

In the opinion of Colonel Konstantin Ivanov, an officer at the General Staff, "never before have the power ministers moved so close to the helm of political power." And should the incumbent president become unamenable to them for some reason or other, "Boris Yeltsin's illusions about his power functions will be dispelled at once."

In presenting the military doctrine to journalists, the defense minister said that it will not be considered by the Federal Assembly. It seems that society is again being deprived of the right of oversight in a sphere that affects both its well-being (as a taxpayer) and its security.

## 9.20

### Vorontsov: Military Doctrine Allows Peacekeeping

Boris Sitnikov
ITAR-TASS, 19 November 1993 [FBIS Translation]

United Nations Organization 19 November TASS—A memorandum of Russian Foreign Minister Andrey Kozyrev to UN Secretary-General Boutros Boutros-Ghali about a new military doctrine of Russia evoked a lively interest here. Journalists, accredited at the UN headquarters, asked Russia's permanent representative to the United Nations Yuliy Vorontsov to explain the meaning of the message, handed by Russian Deputy Foreign Minister Sergey Lavrov during his meeting with Boutros-Ghali on Wednesday.

"The deputy foreign minister spoke about a new strategic military doctrine of the Russian armed forces, which includes a special chapter about UN peacekeeping operations. It assesses peacekeeping operations most positively. At the same time, it mentions operations on the overall territory of the former Soviet Union, on which CIS member states will reach an agreement. It will become an additional element for the peacekeeping process in general," Vorontsov said.

"Thus, along with the participation in UN peacekeeping operations, the doctrine provides for the beginning of so-called regional peacekeeping operations within the CIS boundaries," the Russian representative added.

## 9.21

### CIS Security White Paper Extracts Published

Viktor Litvinov
*Izvestiya,* 20 November 1993
[FBIS Translation], Excerpts

The paper that we are presenting to readers today was only issued at the end of this year. *Izvestiya* has obtained the exclusive right to quote excerpts from it—it is entitled "Toward Security Through Cooperation." It is a report from the commander-in-chief of the CIS Joint Armed Forces to the heads of the CIS countries and the leaders of their military departments on the state of the CIS's national armies, the organization of collective defense, their collaboration in this sphere, and the development of the strategic situation in the world and on Commonwealth territory.

Giving *Izvestiya* a reprint copy prepared for publication, Maj.-Gen. Georgiy Klimchuk, chief of the information and intelligence sector [*napravlenie*] at the CIS Joint Armed Forces Staff, said:

"This is the first time in world practice that we are unprecedentedly offering for free distribution full official analytical materials on the CIS Armed Forces based on seventy government documents. We are revealing the aims, principles, content, structure, and practical activity of our armies. Stripped of unnecessary secrets, this information can and should serve as a source of confidence among the governments and peoples of our countries.

"The paper has been drawn up by the CIS states' Center for Information and Analytical Support in conjunction with the Commonwealth Joint Armed Forces High Command. Leading military scientists, including specialists from the Russian Academy of Sciences, the Russian Academy of Natural Sciences, and other academies, took part in writing it. It is to be published in Russian and English in three versions—a concise edition in the form of a 150–page booklet; a regular, 600–page edition; and an expanded edition including all the acts of legislation adopted in the Commonwealth states on the question of armed forces organizational development, with photographs and biographical information on the leaders of the CIS military departments.

"All the factual information is current as of May 1993. Amendments will be made a year later in the next edition, if one is published," General Klimchuk stated.

He also said that it is hard as yet to say how big the print run will be—specialists are looking at the demand for the paper and the number of requests for it, but a decision has already been made to send a folio free of charge to the leaders of states, including major world powers, to the United Nations and NATO, and to CIS states' power ministries; other citizens and organizations will be able to acquire the "White Paper" for hard currency or its equivalent.

It will comprise the following rubrics—the state's geopolitical position, military doctrine, the organization of defense, armed forces organizational development, the system of manpower acquisition, and social and legal protection for servicemen. We do not have room to quote from each of these—we will select what are, in our view, the most substantive points.

#### The System of Organizing Military Cooperation in the Commonwealth

The system underlying the organization of military cooperation in the Commonwealth can be tracked on the following flowchart. [No flowchart given.] Incidentally, it was first published in *Izvestiya* no. 158 in July 1992. There have been a few changes to it over the past eighteen months. [Passage omitted.]

#### Structure and Composition of CIS Strategic Forces

The structure and composition of the CIS strategic forces comprises the Strategic Rocket Forces, naval strategic nuclear forces, and airborne strategic nuclear forces.

Table 9.21.1

| Missiles | Number of missiles deployed | Number of warheads | Deployed at (number of missiles) |
|---|---|---|---|
| RS-10 (SS-11) | 326 | 326 | Bershet (60), Teykovo (26), Krasnoyarsk (40), Drovyanaya (50), Yasnaya (90), Svobodnyy (60) |
| RS-12 (SS-13) | 40 | 40 | Yoshkar-Ola (40) |
| RS-16 (SS-17) | 47 | 188 | Vypolzovo (47) |
| RS-20 (SS-18) | 308 | 3,080 | Dombarovskiy (64), Kartaly (46), Derzhavinsk (52), Aleysk (30), Zhangiz-Tobe (52), Uzhur (64) |
| RS-18 (SS-19) | 300 | 1,800 | Khmelnytskiy (90), Kozelsk (60), Pervomaysk (40), Tatishchevo (110) |
| RS-12M (SS-25) | 288 | 288 | Lida (27), Mozyr (27), Teykovo (36), Yoshkar-Ola (18), Yurya (45), Nizhniy Tagil (45), Novosibirsk (27), Kansk (27), Irkutsk (36) |
| RS-22 (SS-24) rail-mobile | 33 | 330 | Kostroma (12), Bershet (9), Krasnoyarsk (12) |
| RS-22 (SS-24) silo-based | 56 | 560 | Pervomaysk (46), Tatishchevo (10) |
| Total | 1,398 | 6,612 | |

Note: More than 20 percent of the intercontinental strategic missiles deployed on the territory of the CIS countries are mobile, and 90 percent of their warheads are multiple warheads.

The memorandum of understanding establishing initial figures in connection with the START Treaty between the USSR and the United States lays down that the grouping of intercontinental strategic missiles will comprise 1,398 missiles.

Where are the missiles actually deployed and how many nuclear warheads do they carry? This is covered by Table 9.21.1.

More than 20 percent of the intercontinental strategic missiles deployed on the territory of the CIS countries are mobile, and 90 percent of their warheads are multiple warheads. Incidentally, the prestigious annual British publication *The Military Balance 1992–1993*, published by the International Institute for Strategic Studies, confirms the figures on strategic missiles published in the "White Paper."

*Sea-launched strategic nuclear forces* are located: in the Northern Fleet—at the Nerpichya base (six Typhoon-class nuclear submarines with 120 RSM-52 missiles, each missile equipped with ten nuclear warheads. *Izvestiya* also discussed these boats in Nos. 50–52 in February 1992): at the Yagelnaya base (six Navaga-class nuclear submarines with 96 RSM-25 missiles, four Delta II nuclear submarines with 64 RSM-40s, one Navaga-M nuclear submarine with 12 RSM-45s, and three Delta III, IV submarines with 48 RSM-50s); at the Olenya base (two Navaga III, IV nuclear submarines with 32 RSM-50s and seven Dolphin nuclear submarines with 112 RSM-54s); and at the Ostrovnoy base (nine Delta I nuclear submarines with 108 RSM-40s).

*Pacific Fleet:* at the Rybachiy base there are three Navaga-class nuclear submarines with 48 RSM-25s, three Delta Is with 36 RSM-40s, and nine Delta III, IVs with 144 RSM-50s; at the Pavlovskoe base there are three Navaga nuclear submarines with 48 RSM-25 missiles and six Delta I nuclear submarines with 72 RSM-40s.

In all, the Russian Navy has 62 nuclear submarines with 940 strategic missiles.

The ICBM stockpiles for the submarines, the plants where the missiles are produced, test ranges, and the sites where missiles are loaded onto the submarines are depicted on the following map. [Map indicates Olenya, Nerpichnya, Yagelnaya, Ostrovnoy, Pavlovskoe, and Rybachiy as submarine bases; Okolnaya as an SLBM loading and storage site; Revda as an SLBM storage site; Nenoksa as a test range and SLBM storage site; Severodvinsk as an SLBM loading and silo refitting or elimination site; Zlatoust as an SLBM production site; Pashino as an SLBM elimination site; Krasnoyarsk as an SLBM production site; and Bolshoy Kamen as an SLBM silo refitting or elimination site.]

*Heavy strategic bombers* are part of the Air Force's Long-Range Aviation. This means bombers with a range of more than 8,000 km and missile-carrying aircraft equipped with long-range nuclear air-launched cruise missiles (ALCMs).

At the time of the signing of the START Treaty the Soviet Union had 147 Tu-95 and 15 Tu-160 heavy bombers, of which 84 and 15 (respectively) carried long-range cruise missiles. These accounted for 735 and 120 nuclear warheads.

All heavy bombers were deployed on the territories of three former Union republics. At Mozdok and Engels in Russia, at Priluki, Uzin, and Ukrainka in Ukraine, and at Semipalatinsk in Kazakhstan. There are now 79 bombers on Russian territory (2 Tu-160s, 25 Tu-95MSs, 45 Tu-95K22s, and 7 Tu-95Ks), and 43 aircraft on Ukrainian territory, including 16 Tu-160s. There are 40 Tu-95MSs in Kazakhstan.

Long-Range Aviation also includes 34 tanker aircraft. These were all based at the Engels Airfield in Saratov Oblast, but, just before the collapse of the USSR, some of them were redeployed to Ukrainian territory. They are still there today.

The breakdown of the Commonwealth's strategic forces

Table 9.21.2

| Country | ICBMs A | ICBMs B | SLBMs A | SLBMs B | Heavy bombers A | Heavy bombers B | Total A | Total B |
|---|---|---|---|---|---|---|---|---|
| Belarus | 81 | 81 | — | — | — | — | 81 | 81 |
| Kazakhstan | 98 | 980 | — | — | 40 | 240 | 138 | 1,220 |
| Russia | 912 | 3,970 | 788 | 2,652 | 79 | 271 | 1,779 | 6,893 |
| Ukraine | 176 | 1,240 | — | — | 43 | 372 | 219 | 1,612 |
| Total | 1,267 | 6,271 | 788 | 2,652 | 162 | 883 | 2,217 | 9,806 |

*Note:* The column head letters denote: A—platforms; B—warheads [*zaryady*].

Table 9.21.3

| Type of armament | Azerbaijan | Armenia | Georgia | Moldova |
|---|---|---|---|---|
| Tanks | 220 | 220 | 220 | 210 |
| Armored fighting vehicles | 220 | 220 | 220 | 210 |
| of which, infantry fighting vehicles and armored fighting vehicles with TV | 135 | 135 | 135 | 130 |
| of which, armored fighting vehicles with TV | 11 | 11 | 11 | — |
| Artillery systems | 285 | 285 | 285 | 250 |
| Combat aircraft | 100 | 100 | 100 | 50 |
| Attack helicopters | 50 | 50 | 50 | 50 |

*Note:* Explanation of TV is unknown.

at the time the "White Paper" was drawn up was as shown in Table 9.21.2.

Belarus has stated its readiness to abide by previously signed international agreements, has ratified the START I Treaty and the Lisbon Protocol, has signed the Non-Proliferation Treaty, and has concluded a treaty with Russia on the status of the strategic nuclear forces on its territory. These include not only the strategic missiles and nuclear warheads due to be withdrawn to Russian territory by the end of 1997, but also servicing and maintenance, intelligence-gathering system, space communications, ABM defense, and missile early-warning units.

Kazakhstan too has ratified the START I Treaty and the Lisbon Protocol, but has not yet acceded to the Non-Proliferation Treaty, although it has stated its desire to do so. It is leaning toward recognition of Russia's right to have ballistic missiles, nuclear warheads, and strategic aircraft; approaches to other complexes and systems are being worked out. (Many experts claim that the Kazakhstani leadership's decision in this area will largely depend on Ukraine's stance on strategic nuclear weapons. Kiev's example could also influence Almaty's non-nuclear status—V.L.)

Ukraine has officially stated its adherence to non-nuclear status and its readiness to abide by the international commitments stemming from the Lisbon Protocol. But the current Ukrainian parliament has not set any deadline for carrying out these commitments, the START I Treaty has been ratified with major reservations, and the state itself intends to accede to the Non-Proliferation Treaty as a country possessing nuclear weapons.

### Conventional Arms

On 15 May 1992 Georgia and seven CIS states—Azerbaijan, Armenia, Belarus, Kazakhstan, Moldova, Russia, and Ukraine—signed a joint statement reaffirming their adherence to the Treaty on Conventional Armed Forces in Europe [CFE]. Kazakhstan signed the agreement as a state, part of whose territory is covered by the treaty. It has no troops or armaments on this territory.

Under these accords the quotas for armaments and hardware were divided up between the Transcaucasian states and Moldova as follows in Table 9.21.3.

In actual fact the Azerbaijani Armed Forces have four motorized rifle, one tank, and two artillery brigades which organizationally make up two separate army corps, as well as a detachment of warships, one air squadron, and a separate Spetsnaz battalion.

The "White Paper" does not quote a precise figure for the number of troops in the republic or for the number of armaments. It only indicates that, following the Russian Army's departure, there are 20 Mi-24 and Mi-8 helicopters left behind, along with 70 Czech-made L-29 aircraft and 16 Su-24 and MiG-25 reconnaissance aircraft.

*The Military Balance* ascribes 400 tanks, 470 armored vehicles, 120 aircraft, and 14 helicopters to the country. But there is probably nobody who could quote an accurate figure

Table 9.21.4

| Type of armament | A | B | C |
|---|---|---|---|
| Tanks | 1,525 | 275 | 1,800 |
| Armored fighting vehicles | 2,175 | 425 | 2,600 |
| of which, infantry fighting vehicles and armored fighting vehicles with TV | — | — | 1,590 |
| of which, armored fighting vehicles with TV | — | — | 130 |
| Artillery systems | 1,375 | 240 | 1,615 |
| Combat aircraft | 260 | — | — |
| Attack helicopters | 80 | — | — |

*Note:* The column letters denote: A—regular units; B—in storage; C—total. The explanation of TV is unknown.

today. Some of this equipment has been "consumed" in the flames of war, other equipment has broken down owing to a shortage of spares and poor servicing.

What is known is that the republic's Defense Ministry was allocated 3 billion rubles in 1992 and that Azerbaijan plans to have a 50,000–strong army by the year 2000.

According to the "White Paper" the Armenian Armed Forces comprise 10,000 men, or 15,000 together with other military formations. The law allows the army to have up to 30,000–35,000 men.

The CIS Joint Armed Forces High Command report claims that Armenia does not have the amount of combat hardware which the republic is permitted to have under the CFE Treaty. The two divisions which Russia transferred to it had 180 tanks, 180 infantry fighting vehicles [IFV], 60 armored personnel carriers [APC], 130 artillery systems, and several dozen "Osa," "Strela," and "Igla" surface-to-air missile systems, as well as the "Shilka" self-propelled anti-aircraft gun.

Armenia's greatest shortage is its lack of aircraft. It only has a squadron of Mi-24 and Mi-8 helicopters. Admittedly, the Institute for Strategic Studies gives the following figures—the republic has 250 tanks, 350 fighting vehicles, and 7 helicopters.

The Moldovan Armed Forces consist of regular troops and a reserve. The regular troops are based on ground units, air defense forces, and army aviation. The army comprises four groupings, each with a motorized rifle brigade and an helicopter squadron. The total strength of the armed forces is 10,000 men. They have been equipped at the expense of the 14th Army units which were located on the south bank of the Dniester. Their aviation comprises 38 MiG-29s. Efforts are being made to exchange them for helicopters. Combat hardware, including T-55 tanks and 2,000 assault rifles, is also reaching the republic from Romania.

According to *The Military Balance* figures, the republic has no tanks or armored vehicles.

### Belarus

Armament and combat hardware quotas for Belarus are as shown in Table 9.21.4.

The Belarusian Armed Forces are based on the Belarusian Military District, which comprise 130,000 servicemen—not counting the 40,000 men in the strategic forces. There was one serviceman for virtually every 43 inhabitants of the republic. The figures for Russia, for instance, were one serviceman for every 634 members of the population; for Ukraine, one for every 98 people; and for Kazakhstan, one for every 118 people.

By the end of this year the Belarusian Army should comprise 87,000 men—10,000 officers and generals are to be cut. Stress is being laid on the creation of airborne assault and assault landing units and mechanized brigades. The air force consists of seven to eight air regiments. Some 1,600 tanks, more than 1,200 IFVs and APCs, and 130 aircraft are due to be scrapped. According to the British annual publication, there are currently 1,850 tanks, 1,390 armored vehicles, 617 aircraft, and 80 helicopters in the republic.

### Russia

Armament and combat quotas for Russia are as follows in Table 9.21.5.

Russia's Armed Forces were set up by Boris Yeltsin's edict of 7 May 1992. They are based on the Strategic Rocket Forces, the Ground Forces, the Air Defense Forces, the Air Force, and the Navy. Apart from these, the armed forces include the Airborne Forces, the Military Space Forces, rear service units, and military construction and troop-billeting units.

In terms of military administrative organization Russia has eight military districts (the Leningrad, Moscow, North Caucasus, Volga, Ural, Siberian, Transbaykal, and Far East Military Districts) as well as, currently, three groups of forces (the Western, Northwestern, and Transcaucasus Groups of Forces). At the time when its own armed forces were set up the Russian Army had 2.8 million men; there should be 2 million by 1995.

The "White Paper" devotes around 55 pages to the Russian Army. . . .

Table 9.21.5

| Type of armament | A | B | C |
|---|---|---|---|
| Tanks | 1,300 | 5,100 | 6,400 |
| of which: | | | |
| in regular units | 700 | 4,275 | 4,975 |
| in storage | 600 | 825 | 1,425 |
| Armored fighting vehicles | 1,380 | 10,100 | 11,480 |
| of which: | | | |
| in regular units | 580 | 9,945 | 10,525 |
| in storage | 800 | 155 | 955 |
| of which, infantry fighting vehicles and armored fighting vehicles with TV | — | — | 7,030 |
| of which, armored fighting vehicles with TV | — | — | 574 |
| Artillery systems | 1,680 | 4,735 | 6,415 |
| of which: | | | |
| in regular units | 1,280 | 3,825 | 5,105 |
| in storage | 400 | 910 | 1,310 |
| Combat aircraft | — | — | 3,450 |
| Attack helicopters | — | — | 890 |

*Note:* The column letters denote: A—flanks: Leningrad and North Caucasus Military Districts; B—Moscow, Volga, and Ural Military Districts and Kaliningrad oblast; C—total. The explanation of TV is unknown.

Table 9.21.6

| Type of armament | A | B | C |
|---|---|---|---|
| Tanks | 680 | 3,400 | 4,080 |
| of which: | | | |
| in regular units | 280 | 2,850 | 3,130 |
| in storage | 400 | 550 | 950 |
| Armored fighting vehicles | 350 | 4,700 | 5,050 |
| of which: | | | |
| in regular units | 350 | 4,000 | 4,350 |
| in storage | — | 700 | 700 |
| of which, infantry fighting vehicles and armored fighting vehicles with TV | — | — | 3,095 |
| of which, armored fighting vehicles with TV | — | — | 253 |
| Artillery systems | 890 | 3,150 | 4,040 |
| of which: | | | |
| in regular units | 390 | 2,850 | 3,240 |
| in storage | 500 | 300 | 800 |
| Combat aircraft | — | — | 1,900 |
| Attack helicopters | — | — | 330 |

*Note:* The column letters denote: A—Odessa Military District; B—Carpathian and Kiev Military Districts; C—total on Ukrainian territory. The explanation of TV is unknown.

## Ukraine

The Ukrainian Armed Forces were set up on the basis of the Kiev, Carpathian, and Odessa Military Districts, part of the Black Sea Fleet, and other military formations located in the republic. They number 700,000 men in total.

The concept developed in Ukraine for the armed forces' organization development envisages the creation of two operational commands—the Western and Southern Commands—as well as airspace defense troops based on the air force and the air defense forces. Armed forces strength is planned to be 420,000 by 1995 and 230,000 by the year 2000.

It is planned that the armed forces will comprise three branches of service—ground, air, and naval forces. It has also been stated that missile and space troops will begin to be set up. It is planned that the organizational and manpower structure of combined units will include three motorized rifle regiments and tank, self-propelled artillery, surface-to-air missile, and separate anti-tank battalions. Logistical subunits will include a reconnaissance battalion, a communications battalion, a combat-engineer battalion, a chemical warfare defense battalion, and others.

It is planned that the Ukrainian Navy will have 100 ships and vessels by 1998, as well as around 40,000 men. We would recall that the Black Sea Fleet currently comprises 45 large surface combatants, 28 submarines, more than 300 medium-sized and smaller ships and vessels, 151 aircraft, and 85 deck-borne helicopters.

The most modern combat vehicles and aircraft are in service with the Ukrainian Army; these include the T-72 and T-64 tanks, the BMP-2, the Su-27 and MiG-29, the "Smerch" and "Uragan" multiple rocket launcher systems, "Tochka" tactical missile systems, and S-300 surface-to-air missile systems. Enterprises from nine defense-sector [*oboronyy*] and seven machine-building ministries work for the country's military-industrial complex. The state can independently produce strategic missiles and other kinds of military output.

Armament and combat hardware quotes for Ukraine are as follows in Table 9.21.6 (see page 463).

## 9.22

**CIS Armed Forces Command Abolished**
Interfax, 22 December 1993
[FBIS Translation]

The CIS Council of Defense Ministers meeting Wednesday in Ashkhabad has decided to annul the CIS Joint Armed Forces Command. Instead, a staff will be formed, subordinate to the CIS Council of Defense Ministers, which will be in charge of coordinating military-technical cooperation among the CIS countries.

Those present agreed that they recommend General Viktor Samsonov to be approved at the summit as the chief of staff. Earlier the general was chief of staff of the CIS Armed Forces.

## 9.23

**Defense Agreements Signed**
Interfax, 23 December 1993
[FBIS Translation]

At a session of the CIS Defense Ministers Council in Ashkhabad on Thursday [23 December], Russia signed bilateral treaties on military and technical cooperation with Tajikistan, Turkmenistan, Azerbaijan, Kyrgyzstan, Kazakhstan, Belarus, and Armenia.

As a Russian negotiator has told Interfax, the documents allow not only promotion of military cooperation bilaterally but also reinforcement of the CIS collective security system.

The session also finally resolved the issue on funding headquarters to be set up to coordinate military cooperation among the CIS states. According to an agreement, Russia is to foot 50 percent of the headquarter's maintenance bill. The remaining costs are equally shared among the other collective security treaty's participant states—Kazakhstan, Belarus, Tajikistan, Armenia, Kyrgyzstan, and Uzbekistan. The headquarters will be housed in the former building of the CIS Combined Armed Forces Command in Moscow. The agreement will be submitted for endorsement at the CIS summit on Friday.

## 9.24

**Kozyrev Writes on Military Doctrine**
Andrey Kozyrev
*Krasnaya Zvezda,* 14 January 1994
[FBIS Translation], Excerpts

The results of the December elections to the Federal Assembly have generated a debate abroad about possible changes to Russia's foreign and military policy. Citing the irresponsible, profascist statements made during the election campaign, certain observers have been predicting a strengthening of Moscow's "imperial aspirations."

I have already stated my attitude to such statements. I would like to reemphasize that Russia's line in the foreign policy and security sphere is determined by the Russian Federation president. This situation, which is enshrined in the new democratic Constitution, reliably protects that line from the influence of national extremism. The policy basically remains the same—both in terms of being geared to the development of good-neighborly relations of partnership with the outside world and in terms of being open and predictable.

Another guarantee of the stability and consistency of the country's foreign and military policy is the Russian Federation military doctrine approved by the Russian president in early November. Together with the foreign policy concept ratified in April, it is an integral part of the emergent concept for the country's security.

The distinctive features of our military doctrine are its defensive thrust and the strict account it takes of the country's actual requirements as regards ensuring national security. In terms of its content it is entirely comparable with the doctrinal provisions of most democratic states. I am referring above all to the similarities in the assessment of the military-political situation and the prospects for its development both globally and regionally. A key factor is the absence of any potential military enemy for Russia. According to the doctrine, the main threat to the country's security is posed by local conflicts and the spread of nuclear and other weapons of mass destruction.

The Russian military doctrine is an obstacle in the way of

any attempts to abuse military force. While using the full panoply of means available to it to safeguard its security, Russia will at the same time give priority to peaceful—above all, political and diplomatic—means. Military force can only be used for self-defense purposes, in order to rebuff aggression.

The Russian armed forces will be shaped to take account of the major changes in the world and the country's actual economic potential. By the end of this century, after the transition to the mixed principle of manpower acquisition, they will be substantially smaller, but more mobile and equipped with effective weapons systems and military hardware in sufficient quantities to ensure the defense of the country and its allies.

This provision of the doctrine can be illustrated by the navy. It will fully retain its role as a factor determining Russia's might as a great power. Moreover, given the cuts envisaged by the Russian-U.S. START II Treaty, the naval component of our strategic "triad" will have increased. The decommissioning of obsolete ships will be combined with efforts to equip the navy with modern hardware. The ships flying the St. Andrew's ensign should embody the most advanced achievements of Russian science and technology.

The fact that all this is not only necessary but possible was something I saw for myself at Severomorsk in Murmansk Oblast, where the electoral district, which I represent in the State Duma, is located. An immense potential is concentrated there—the flower of Russia's navy. This was also attested by my visit to Russia's naval outposts in the Far East and to the ships of the Black Sea Fleet. The navy is traditionally one of the most flexible foreign policy tools at the disposal of the Russian state, whose interests naval seamen and diplomats have defended and will continue to defend together. I am therefore all the more aware of the social need—requiring state support—of those who serve the motherland on its ocean and maritime borders. . . .

It would be a mistake to make out that our new approach to nuclear weapons increases the risk of nuclear war. The fact that we do not intend to be the first to use any weapons, and see nuclear weapons as a last resort, is of fundamental importance.

The military doctrine's provisions take full account of CIS collective security interests and are geared to strengthening defense cooperation among the CIS states on a multilateral and bilateral basis. The prerequisites for this are shared national security interests, a standard armed forces structure and standardized weapons systems, and an interest in maintaining the former Union's defense infrastructure within your own territory. . . .

The new view on the nature of a possible war, which recognizes the minimal likelihood of the unleashing of world nuclear and conventional war along with the increased risk of local wars, including those on the territory of the former USSR, orients our diplomacy toward more active cooperation with other states with regard to peacemaking. The CSCE declaration condemning aggressive nationalism, which underlies many local conflicts—a declaration adopted on Russia's initiative—is of great significance.

In accordance with the doctrine the Russian Federation Armed Forces acquire a new function: participation in carrying out peacemaking operations. These are implemented by decision of the UN Security Council and other collective security organs or in accordance with international agreements, primarily within the CIS context. It is a matter of Russia's fulfilling not some sort of gendarme functions, but a mission fully consonant with its status and duties as a permanent member of the UN Security Council. The Russian troops taking part in such operations strictly respect the sovereignty of the sides involved in the conflict and act with their consent and in accordance with a clearly defined mandate.

In fulfilling peacemaking functions on the territory of the former USSR, the Russian Federation is prepared to cooperate closely with all interested countries and international organizations, primarily the United Nations and the CSCE. We are entitled to expect the world community to provide material support for this activity. We are raising the question of creating an international voluntary fund for these purposes.

The military doctrine envisages completing before 1996 the withdrawal to Russian territory of the Russian units stationed outside Russia. We will seek to ensure that the agreements being elaborated with a number of countries regarding this properly take into account Russia's security interests and also the interests of the Russian-speaking population in these countries.

At the same time the doctrine realistically proceeds from the premise that the interests of the security of the Russian Federation and other CIS members may necessitate the stationing of Russian Federation forces outside its territory. But, of course, this will be done only on the basis of appropriate international legal documents and with the consent of the states on whose territory our armed forces will be stationed.

In outlining the main directions for the development of Russia's cooperation with other countries in the military-technical sphere, the doctrine officially sets the relevant departments, the Foreign Ministry included, tasks which we effectively began to work on from the first months of 1992. On the one hand, there is a need to maintain the country's export potential in the sphere of conventional arms at the proper level, partly so the hard currency proceeds can be used for the needs of military production and for purposes of conversion. On the other hand, we cannot allow deliveries of arms and military equipment to lead to the spread of technology making it possible to create weapons of mass destruction and delivery systems for them, to the undermining of regional stability, or to the breaking of embargoes or other international accords to which Russia is a party. . . .

The adoption of the basic principles of the foreign policy concept and military doctrine marks an important milestone in the development of democratic Russia's political and military thought. They graphically reflect the profound changes in the life of our state and society. Free from ideological rhetoric and claims to superpower status, these documents are called upon to help form in the world an image of democratic Russia as a peace-loving state with natural national interests. Not regarding any state as its enemies, Russia is prepared to see as partners all states whose policy is not detrimental to its interests and does not contravene the UN Charter.

Only thus will we be able to reliably ensure the security of the motherland.

---

## 9.25

**Shikhmuradov Signs NATO Partnership for Peace Program**

Moscow Radio Ekho, 10 May 1994 [FBIS Translation], Excerpts

---

Turkmenistan today became the first Central Asian country to join the NATO Partnership for Peace program. Turkmen Deputy Prime Minister Boris Shikhmuradov signed the document in Brussels.

---

## 9.26

**Comments on Partnership with NATO**

Aleksandr Mineev

ITAR-TASS, 10 May 1994 [FBIS Translation]

---

"Turkmenistan is not seeking NATO's help, but is striving to develop partnership with the alliance on a mutually advantageous basis," Boris Shikhmuradov, deputy prime minister of this Central Asian republic, told an ITAR-TASS correspondent. At NATO headquarters in Brussels today, he signed the outline document of the "Partnership for Peace" program. So, eighteen countries have joined the program, including eight former Soviet republics.

Turkmenistan, the deputy prime minister said, has possibilities of broad cooperation with NATO member-countries in all spheres, including the military one. He said talks were under way on joint programs for training officers of the Turkmen Armed Forces and on cooperation in other areas of military construction.

At the signing ceremony, B. Shikhmuradov said Turkmenistan sees the main purpose of its participation in the program set out by NATO member-countries in objectively promoting a solution to the principal task, which is strengthening Turkmenistan's sovereignty and turning it into a fully independent member of the international community.

Along with developing partnership relations with new democratic Russia, the deputy prime minister noted, his republic is developing cooperation with West European countries and the United States, Asian states, and CIS neighbors. He specifically noted Turkmenistan's potential role as a bridge between Europe and south and Southeast Asia. [Passage omitted.]

---

## 9.27

**Almaty Signs NATO Partnership for Peace**

Aleksandr Mineev

ITAR-TASS, 27 May 1994 [FBIS Translation]

---

Kazakhstan has become the nineteenth state to join the Partnership for Peace program proposed by the NATO leadership. At the alliance's headquarters in Brussels on Friday the republic's foreign minister, Kanat Saudabaev, put his signature to the program's framework document. Kazakhstan's embassy in Brussels noted that the republic was the second nuclear state, after Ukraine, to give official support to the initiative put forward by the NATO countries' heads of state and government. This has taken place soon after Kazakhstan's accession to the Nuclear Non-Proliferation Treaty and this, Almaty believes, is clear confirmation of Kazakhstan's firm political desire to strengthen international security and stability.

Speaking at the signing ceremony, K. Saudabaev expressed the hope that the implementation of the Partnership for Peace concept would make it possible to rule out the emergence of any grounds for new confrontation after the ending of the Cold War, would create real possibilities for a smooth transition to an atmosphere of stability, mutual trust, and cooperation throughout the world, and would ensure firm guarantees for the security and territorial integrity of Kazakhstan.

Acceding to this initiative, he stressed, opens up for the republic broad prospects for military-political, economic, and technical cooperation with NATO in terms of setting up a modern army in accord with democratic principles and world standards, for conversion of the defense industry and for participation in peacekeeping operations under the UN aegis. The minister said that the importance of Kazakhstan's accession to this program is that it contains a stimulus for the further development of economic reforms, transition to a market economy, and the construction of a law-based democratic state.

Kazakhstan's foreign minister had a conversation with Sergio Balanzino, NATO deputy secretary general.

## 9.28

### Grachev Gives Overview of Military Doctrine

Army General Pavel Grachev
*Nezavisimaya Gazeta,* 9 June 1994
[FBIS Translation], Excerpts

As we know, following the collapse of the Soviet Union and the formation of a number of independent states on its territory, Russia's state-legal status has changed radically. It has become an autonomous subject of international relations, the USSR's successor as permanent member of the UN Security Council, and one of the five world nuclear powers.

The cardinal changes in the geopolitical and military strategic situation in the world, the new nature and system of international relations, and the creation of Russian armed forces have put before Russia the task of elaborating and adopting a fundamentally new national military doctrine.

On the one hand, the end of the confrontation which had proceeded under the sign of the struggle between two systems, with its projection onto all aspects of international life, has not only reduced the threat of global war but has laid down the preconditions for establishing constructive cooperation between states which were previously confronting each other and for a fundamentally new relationship between Russia and the world around it.

However, it must be clearly understood that these new relations do not by any means rule out the emergence of differences and contradictions, which at times are quite sharp.

The main thing, in our view, is that they should be resolved and settled within the framework of normal collaboration between states, with the specific nature of national interests taken into account.

On the other hand, attention must be paid to the relatively high level of tension in individual regions of the world. The probability of local and regional wars continues to exist, as well as of armed conflicts within individual states on the basis of national-ethnic, territorial, religious, and other contradictions. A particularly dangerous challenge to regional and international stability is presented by the growth of aggressive nationalism in various regions of the world.

From the very beginning of work on the military doctrine it was absolutely clear that what had to be done was not simply to amplify military-doctrinal views which existed in the Soviet period but to elaborate fundamentally new stances and approaches.

Essentially for the first time we stated that we would protect not the ideology but the vitally important interests of the country, which do not affect the security of other states in any way and are ensured within the framework of mutually advantageous interstate relations based on equal rights.

The cornerstone of our military doctrine is the provision that Russia does not regard any state as its enemy. This has brought about the need for a radical review of approaches to the entire spectrum of problems of military-organizational development. . . .

The "Basic Provisions of the Russian Federation Military Doctrine," which are of a most general and conceptual nature, have made it possible to unite all these legislative acts and departmental documents in a single whole, a kind of "military constitution." For the first time in our country's history, this document was thoroughly examined at a number of sittings of the Russian Federation Security Council and ratified by Russian President B.N. Yeltsin. Thus the "Basic Provisions of Military Doctrine" are an official normative-legal document, a constituent component of the concept of Russian Federation security, and represent a "document of the transitional period—the period of the formation of Russian statehood, the implementation of democratic reforms, and the formation of a new system of international relations." In other words, the adopted "Basic Provisions of Military Doctrine" constitute not dogma but a document open to relevant revision as the military-political situation develops. It is intended to ensure the implementation of a minimum of two basic functions. . . .

Another function, the informative function, is also highly significant. The military doctrine enables the peoples of our country and the world community to correctly understand Russia's aims and tasks in the sphere of the struggle to avert possible armed conflicts and wars and in preparations for repulsing possible aggression against Russia and protecting its vital interests.

Structurally speaking, the "Basic Provisions of Military Doctrine" are made up of a preface, political, military, military-technical, and economic sections, and also a concluding section.

The political principles of the military doctrine reflect two interconnected tasks—the prevention of war, and readiness to repulse an aggressor.

Russia rejects any war, use of military force, or threat of force as means of achieving political, economic, and other aims. It advocates the adoption of commitments by all states not to use military force first and adheres to the principles of the inviolability of state borders and non-interference in the internal affairs of other states. All questions under dispute should be resolved only by political and diplomatic means.

As I have already noted, the political principles of the military doctrine indicate for the first time that the Russian

Federation sees no state as its enemy. I will say frankly that for us the professional military, who have gotten used to having the most specific of guidelines, this turn of events has created a lot of additional difficulties. Primarily we had to retune ourselves psychologically, but there were also difficulties of what I would call a technical nature. This primarily concerns questions of strategic planning, the instruction and training of cadres, and finally, the determination of the defense budget. Today we are building our defense policy, gearing ourselves to taking existing and potential threats into account. These are expounded fairly fully in the doctrine. While Russia has no enemies, like any state, it has its vitally important interests. We are capable and prepared to protect them . . . .

I believe that Russia is entitled to expect a different attitude to Paragraph 5 of the CFE Treaty. However, the lack of a solution to the problem of flank restrictions can be viewed as evidence that by no means everyone has yet jettisoned the bloc-style thinking of the Cold War era.

Russia believes that the main principles for resolving its military security problems are the maintenance of stability in the regions adjoining its borders and compliance with international commitments. At the same time the doctrine indicates that Russia's security should be protected without prejudice to other countries' security or the security system as a whole.

Regarding the matter of maintaining international peace and security and preventing wars and armed conflicts, the Russian Federation views as partners all states whose policies do not prejudice its interests or contradict the UN Charter. A priority for us is to establish cooperation within the CIS—cooperation with its members in resolving problems of collective defense and security and coordinating military policy and defense building.

At the present time we are formulating a blueprint for collective security within the CIS framework. The basis for this is the Collective Security Treaty. It is becoming increasingly evident that no state in the Commonwealth can develop normally in various spheres without close integration. And there is no sense in looking for the notorious "hand of Moscow" or a new "imperialist thinking" on Russia's part. This is an objective assessment of present realities. We favor an equal pooling of efforts by all CIS member states in order to resolve problems that arise.

On a regional level, Russia is implementing cooperation with CSCE member countries and other states and military-political structures in adjacent regions which have an existing or nascent collective security system. But here too there is a major peculiarity. Russia is not just a European but also an Asian state. Consequently, in order to protect our security and vital interests we cannot retire into Europe alone; we must make efforts to establish collective security systems in other regions too, including the Asia-Pacific region . . . .

In accordance with this, the main objective of Russia's peacekeeping activity is to protect its national security interests through encouraging the establishment and maintenance of peace and stability both on the planet as a whole and in various regions.

In realizing this objective, Russia firmly adheres to the generally accepted norms and principles of international law as well as the specific norms and rules for carrying out peacekeeping activity. The latter category includes:

—matching the form and content of peacekeeping steps to the situation to be resolved;

—giving priority to political over military ways and means of settling a situation;

—strictly observing and protecting human rights;

—using military contingents of peacemaking forces only with the consent of the opposing sides and with the approval of the international community.

In the context of the said norms and principles the leading areas of the Russian Federation's peacemaking activity are:

—mediating in settling crisis situations and preventing the escalation of conflicts;

—influencing the conflicting sides by diplomatic means above all but also, in extraordinary circumstances, by military means with a view to ensuring the fair and peaceful resolution of disputed issues;

—taking measures to preserve peace by disengaging the conflicting sides by using military contingents of the Russian Federation Armed Forces operating under UN auspices or sent into the crisis zone with the consent of the conflicting sides within the framework of the Collective Security Treaty;

—determining and supporting structures which seek to strengthen peace and to prevent the emergence or the continuation of a conflict.

Thus it is possible to maintain that there is nothing in Russia's peacemaking activity or in the tasks entrusted to its armed forces that runs counter to the fundamental, universally recognized norms and principles of international relations.

All this confirms once again Russia's desire to be a full and equal member of the world community in the resolution of international peacekeeping questions.

The second fundamentally new task for us is to assist internal affairs organs and internal troops of the Russian Federation Ministry of Internal Affairs in localizing and blocking regions of conflict, ending armed clashes, and disengaging opposing sides and also to defend strategically important installations according to the procedure laid down in existing legislation.

The possibility of fulfilling such tasks is provided for in the legislation of many countries. It is not a question of the armed forces' assuming any special internal functions but of the possibility of additionally involving forces and equipment in halting bloodshed. It must be realized here that it is just as legitimate to come between, let us say, Ossetians and Ingush who have taken up arms as between Serbs and Muslims in Bosnia.

What is new is the provision that "the security interests of the Russian Federation and other CIS states might require the stationing of troops (forces) and means outside the territory of the Russian Federation."

This provision has elicited a highly ambivalent response abroad, particularly in certain "near" countries and in Eastern Europe. Such a reaction, however, is, rather, nothing but an attempt at political speculation because the same paragraph unequivocally states that "the conditions for such stationing and manning are defined in the appropriate international legal documents." In other words, these questions are in full accordance with the universally recognized norms of international law.

The need of military bases outside Russia's territory is dictated, above all, by the interests of maintaining stability in individual regions. Their deployment is initiated primarily by those states that are in need of additional stability factors.

As regards military installations, the need for these outside Russian territory is determined in some cases by their purpose (for example, missile attack early warning system radar stations) and in other cases by their technical uniqueness and by the impossibility of creating a replacement of equal worth on Russian territory in the very short term.

In all cases military bases and installations are deployed outside Russia by mutual agreement between the sides.

Here I would like to dwell on the thesis, which is being greatly exaggerated by certain people, that Russia is supposedly ready to interfere in its neighbors' internal affairs. I will cite from the doctrine: "The Russian Federation abides by the principles of the peaceful settlement of international disputes, respect for a state's sovereignty and territorial integrity, non-interference in its internal affairs, inviolability of state borders, and other universally recognized principles of international law." This is set forth in the military doctrine's political fundamentals, which take priority over its other constituent parts. . . .

At the same time the doctrine defines for the first time that the needs of the armed forces and other troops for armament, military hardware, and property will be met with due regard for the country's scientific, technical, and economic potential.

A few words about military-technical cooperation with foreign countries.

Whereas military-technical cooperation in the former Soviet Union was aimed at supporting welcome regimes and, as is known, was more ideological in nature, today Russia regards it as, above all, a constituent part of the balanced support for its own economic interests. Therefore continued trade in arms and military hardware will be one of the elements of activity in this sphere. But, at the same time, what is new is the emergence in the military doctrine, as one of the basic principles of Russia's policy, of the following: "The inadmissibility of deliveries of arms and military hardware which might exacerbate a crisis situation, undermine regional stability, or violate embargoes or other corresponding international accords to which the Russian Federation is a party."

In addition to sales of weapons, it is also planned to expand contracts for training foreign servicemen, above all from CIS countries, in Russian educational institutions.

The final section of the "Basic Provisions of Military Doctrine" states that the Russian Federation guarantees the fulfillment of the provisions of this document. At the same time it will strictly abide by the UN Charter and by universally recognized international legal norms and principles.

This is a brief resume of the crux of the main doctrinal provisions aimed at ensuring Russia's security and the defense of its vital interests.

Like the whole world, Russia is now going through an extremely difficult and crucial historical period. It is a period of the creation of a new world order. To all appearances, its character will predetermine the destiny of mankind in all countries and peoples for many decades. It is in precisely this world-historical context that the search must be made for the role and place of the Russian state in the still evolving and largely contradictory and impermanent geopolitical balance of forces and interests.

## 9.29

### Samsonov on CIS Collective Security Concept

Moscow Radio, 22 July 1994
[FBIS Translation]

A regular conference of the CIS defense ministers took place in Moscow this week. Problems of setting up CIS joint armed forces were discussed. Speaking about the results of the work, Col.-Gen. Viktor Samsonov, chief of staff for coordination of military cooperation of the CIS countries, noted:

The concept of collective security is the totality of what has been agreed by the treaty participant states on averting and removing threats, on joint protection against aggression, and on guaranteeing sovereignty and territorial integrity.

I would like to draw attention, in particular, to the clause of the concept that says the strategic nuclear forces of Russia fulfill the function of restraint from possible aggressive intentions against all CIS participant states.

I will note that the Commonwealth's collective security is based upon the principles of the indivisibility of security, equal responsibility, the collectiveness of defense, of a consensus in the adoption of fundamental decisions in the sphere of defense.

The concept also notes that in the long term participant states may take the decision to create joint armed forces. Therefore, collective security may be created by collective peacekeeping forces. The main directions for creating effective collective security have been outlined: the drawing closer together of defense legislation, consultations on problems of military organizational development, the development of common approaches for training troops, coordinating questions of operations organization of territories, carrying out joint operational measures and combat training, coordinating both operational and other plans and programs, training cadres, coordinating questions on manufacturing, and repairing equipment and others.

The concept provides for three stages in the formation of a collective security system. At the first stage, the creation of national armed forces must be completed, a system of military and military-technical cooperation must be worked out, and a legal base for the operation of the collective security must be created.

At the second stage, a coalition grouping of troops and a joint air defense system must be created and the questions of the formation of joint armed forces must be examined.

At the third stage, the creation of a collective security system must be completed in a practical form.

## 9.30

### Further on Samsonov Comments

Gennadiy Meranovich

*Krasnaya Zvezda,* 22 July 1994 [FBIS Translation]

At the press conference held here yesterday, Col.-Gen. Viktor Samsonov, chief of staff [for the coordinating of military cooperation], commented on the results of the session of the CIS Council of Defense Ministers on 18–19 July. He noted that the greatest disagreements among participants in the session were evoked by the draft decisions regulating the performance of peacekeeping operations in the Georgian-Abkhazian conflict zone and material-technical, financial, and cadre support for peacemaking operations in the CIS framework. According to the decision of a joint session of the Council of Defense Ministers and the Council of Foreign Ministers, General Samsonov reported, the question of the form of the CIS countries' participating in this operation is to be examined promptly.

The adoption of the draft concept of collective security of the states party to the Collective Security Treaty was described as the session's most important result. In this regard Col.-Gen. Viktor Samsonov drew the journalists' special attention to the provision of the concept in which the Russian Strategic Nuclear Forces are assigned the functions of providing a deterrent to potential aggressive intentions directed against states party to the treaty.

## 9.31

### General Volkov Proposes Changes in CIS Collective Security System

Major General Vasiliy Volkov

*Nezavisimaya Gazeta,* 20 August 1994

[FBIS Translation]

[Article by Maj.-Gen. Vasiliy Petrovich Volkov, candidate of legal sciences, and representative of the CIS Executive Secretariat Council of Defense Ministers.]

More than two years have elapsed since the Collective Security Treaty was signed in Tashkent. The peoples of the former Soviet Union, and not they alone, breathed somewhat easier. Some, because there was now hope that the focal points of the interethnic and other conflicts that had erupted full force by that time on the territories of certain states of the Commonwealth would be extinguished, others, because they understood that the new regional community professing the principles of non-aggression, non-interference, and good-neighborliness in mutual relations with other states was prepared to use force in the sole instance of it, the community, or any of its members being subjected to some aggression on the part of third countries. These declarative postulates of the treaty were subsequently bolstered by real actions. The nuclear weapons were concentrated mainly on the territory of Russia and the strategic missiles were no longer targeted at facilities of the former probable enemies.

So the world community could sleep easy. But can the mother, wife, or family of the soldier and officer of the Commonwealth peacekeeping forces (Russian servicemen, mainly), who are literally separating with their bare hands the inhabitants of one and the same state who are avid to exterminate each other, sleep easy? The peacekeepers themselves quite often find themselves caught up in this fight.

It may now be affirmed that everything with regard to our commitments to the world community is being fulfilled unswervingly. Yet our attempts to create an effective regional system of collective security, within whose framework not only problems of military security but also, perhaps, at this stage, problems of greater urgency for the Commonwealth of Independent States could be tackled successfully, remain in many instances merely good intentions, with which the road leading whither is well known.

Only the efforts of Russia, perhaps, are somehow as yet holding back the points of tension that are at times slowly dying, at times flaring up. One has the impression here that it is losing far more than it is gaining. Not to mention the charges of "imperial ambitions," Russia is sustaining considerable economic losses. It is losing its sons. Is it not too high a price for, as some people maintain, "the defense of Russia's interests in the near abroad"? I believe that what we have there are the interests of the whole Commonwealth, not just of Russia.

God forbid, but what is at this time someone else's business could come to be the epicenter of tragic events like those that have occurred in the Dniester region, Karabakh, Abkhazia, and Tajikistan. It makes no difference here what goals were being pursued or who ignited these conflicts. The main thing is that people, the absolute majority of whom are totally innocent, are dying.

In order for such things to have been ruled out in the future, real political will should have been displayed yesterday, even. Although even today this is not too late.

But, as the almost three-year experience of the CIS shows, political will alone is frequently insufficient for ensuring that the documents adopted in the Commonwealth at the highest level operated flawlessly. If we go back to that cart in which the Collective Security Treaty is peacefully slumbering, the political will of its participants does, it would seem, exist, and all want to pull this cart in the same direction (as distinct from the participants in the operation from Krylov's well-known fable), but there have been no real, tangible results as yet.

Evaluating this situation, you involuntarily make a comparison with the activity of the NATO bloc. You may take a varying view of this organization and question the very need for its existence under modern conditions, but the fact that no armed conflicts are permitted within the NATO framework and that, if they do arise, they are quickly cut short with the use of all possible means is indisputable. NATO has a mechanism for the realization of adopted decisions. With us, however, this mechanism is far from perfect.

In expressing my position on this issue, I would like to mention that, as a representative of the Joint Armed Forces Main Command and, subsequently, a representative of the Commonwealth of Independent States Executive Secretariat Defense Ministers Council and simultaneously, for almost a year now, having been acting chief of the Department of Interstate Political-Legal and Military Cooperation of this Secretariat (may there be no wincing, as they read these words, among employees of the Finance Office of the CIS Military Cooperation Coordination Staff, where I am down for all the types of allowance—I hold the latter position on a voluntary basis, that is, on an unpaid basis, strictly in accordance with clause 7 of Article 10 of the law of the Russian Federation "Status of the Serviceman") and participating in practically all sessions of the Commonwealth Council of Heads of State, the Council of Heads of Government, and the Defense Ministers Council since the moment of their formation and in certain sessions of the Foreign Ministers Council, I have concluded that the attempts to transfer international experience of the adoption and realization of decisions onto our reality are not always justified. There are many reasons of both an objective nature for this: The absence thus far of the necessary legal base and supranational bodies whose decisions would be binding on all states and the readiness of all bodies, organizations, officials, and citizens to abide unswervingly by the decisions adopted by arms of the Commonwealth, and here the reason is that same legal nihilism to which all of us became inured over decades, and much, much else, including a lack of interest in and, in some cases, active resistance to the integration processes occurring in the Commonwealth on the part of individual states, organizations, and officials. This is natural; each has his own interest here.

Considering this, it will take a considerable amount of time to create the conditions where the decisions adopted within the framework of the Commonwealth are fulfilled unswervingly.

And in addition, while there is an absence in our states of a law-based, civil, democratic society, and while all processes are controlled only by individuals or groups of individuals, not rules of law, we need a chief who can lead us to that same law-based, democratic society. This is the mentality of the majority of our society. And we simply cannot escape this.

Considering what has been said, a number of organizational-legal measures are necessary in the immediate future, in my view, to ensure that the Collective Security Treaty operates for the good of the Commonwealth of Independent States, each participant, and, yes, the entire world community as a whole.

The signing of the Collective Security Treaty in May 1992 pursued military-political goals, primarily. This meant preservation of the single defense space, joint armed forces of the Commonwealth, the unified command of the strategic nuclear forces, and certain other points, which corresponded in full to the constituting documents of the Commonwealth of Independent States. The adoption in states of the CIS of legislative instruments on the creation of their own armed forces and on neutrality and such is evoking in the Commonwealth a cool attitude, to put it mildly, both toward the very idea of the treaty and toward all other documents adopted in its development.

In addition, collective security, it is customary to believe, means mainly joint defense against a military threat, and this, let us be realistic, has been pushed back considerably at the present time compared with the Cold War period.

Yet the very concept of security incorporates several subspecies. According to some criteria, security is subdivided into political, economic, environmental, military, and so forth; according to other, most general, criteria, into security of progress, social security, and so forth; according to yet others, into the security of the person, society, the state, and a system of states and planetary security.

To ensure that the involvement of CIS states in the Collective Security Treaty is more compelling and, most important, necessary and useful for all members of the Commonwealth, it is essential to determine the priority threats to each state. For the Republic of Belarus, possibly, these could be environmental and economic threats, for the Republic of Tajikistan, political and military threats, and so forth. Having all united together within the framework of the Collective Security Treaty, each state could participate in the areas that it needs most.

And now concerning the role of the chief in this process. The entire system of collective security at the first stage could operate under the direct leadership of the head of the state that is the chairman in the statutory bodies of the Commonwealth—the chairman of a Collective Security Council. To ensure continuity in leadership and the fulfillment of the adopted decisions, it is essential to have on a permanent basis one first deputy chairman of the Collective Security Council and three deputies, who could be responsible for the entire set of questions on problems of security in one of four regions of the CIS: Eastern Europe, the Caucasus, Central Asia, East Asia.

There could under the first deputy be a small staff consisting mainly of citizens of the state whose representative is the first deputy himself. The basic preparatory work (preparation of draft documents, their concordance and substantiation, and so forth) should be performed in the regional structures of the system of collective security of the CIS and also in the Defense Ministers Council, the Foreign Ministers Council, and the Military Cooperation Coordination Staff, and also, if necessary, in other arms of the Commonwealth.

The entire organizational-support work on the final polishing of the documents and the preparation and realization of sessions of the Collective Security Council could be assumed by executive secretariat.

Why do I speak about this in such detail? Recent sessions of the Commonwealth Defense Ministers Council and the Foreign Ministers Councils have confirmed once again that the majority of subscribers to the Collective Security Treaty are reluctant to have a structure of the system of collective security that presupposes its strict centralization and the creation of costly new interstate bodies. But it is not even the economic difficulties that frighten some states, although some of them are allocating up to 4 percent of their budget, which is sparse by today's standards, for the upkeep of the interstate bodies that already exist. The main thing is that such a system of collective security and the assignments that it is proposed tackling within its framework are not to the states' liking. Consequently, it is necessary to change the system and give it different assignments. I believe that at this stage, whatever degree of integration we achieve in the immediate future, it is essential to shift the brunt of the work on problems of security to the states and the regions and to reserve for the center the solution of organizational-legal questions and the elaboration of the conceptual propositions of the collective security of the Commonwealth.

The fact that the main intellectual potential in all areas of the proposed activity of the Collective Security Council is concentrated in the states and that it needs to be utilized to the maximum extent in the interests of the entire Commonwealth also speaks in favor of the creation of just such a system of collective security.

The framework of an article does not allow the proposal of a solution of all the complex problems of the collective security of the Commonwealth. But it would be better, in my view, to discuss this at a conference of leaders of the staffs of the security councils of the participants in the Commonwealth of Independent States, that is, the people who feel in their bones and who know better than anyone all the problems of the security of their states, the regions, and the CIS as a whole. It is simply amazing that we have been attempting thus far to decide these most important questions for the Commonwealth without their participation.

It may be assumed with a great degree of probability that some of the proposals that have been expressed in this article will not suit some people—particularly those who have no interest in a strengthening of the Commonwealth in all areas, specifically in the creation of a dependable defensive alliance. But the integration processes in the Commonwealth have already begun, and no one can stop them.

## 9.32

## CIS General on Military Cooperation

Interview

*Rossiyskie Vesti,* 20 September 1994

[FBIS Translation], Excerpt

Following the breakup of the USSR, there has been less and less talk in the Western press about possible aggression by our country.

Recently, however, certain Western politicians have suddenly started once again fearing a threat from the East. Why? Their concern is due to the fact that the former Soviet republics have decided to deepen their military cooperation and have started thinking about creating a defense union. The West sees in the CIS collective forces the renascent military might of the USSR. How valid are these fears, when Russia's military doctrine is of a highly peace-loving nature? . . .

[Kozyreva]: Boris Evgenevich, is the creation of a "defense union" as a counterweight to NATO a myth or reality?

[Pyankov]: Recently this question has been put to me quite frequently both by former Warsaw Pact allies and by representatives of Western military departments. I am profoundly convinced that the question of creating a defense union cannot be resolved right away. The conditions for forming it do not yet exist. The young sovereign states are now creating national armies. I would even say this: The fragments of the former Soviet Army are disintegrating. In other words, not one of these states is tackling questions of any close military union or seems to have any intention of tackling them in the immediate future. All the CIS members are faced with other problems: completing the dividing up of the army, reducing it, and strengthening their national armies.

[Kozyreva]: The conditions for creating the defense union do not exist today, but might they still emerge tomorrow?

[Pyankov]: You see, both politicians and military people understand that today it is still necessary somehow to guarantee the states' security with the minimum expenditure on national armed forces. The former Soviet Army has melted away like ice. Today the independent states cannot maintain even those units of that former army that remain on their territory. Ukraine is a vivid example. It has inherited three mighty military districts: Carpathian, Kiev, and Odessa. If you also include units under central jurisdiction—all brigades and divisions—you can even talk about five districts. Ukraine is not in a position to maintain such an army. Today it is being reduced. It will not be increased again even in the future. Today we speak of collective security, but the defense union question is not mentioned directly at all. At conferences with the heads of the CIS states' defense departments we resolve questions of the moment, as the saying goes. We have just been discussing problems of repairing military hardware. This is the crux of the matter: Hardware goes out of commission, but repair plants are scattered across the territory of the former Union. Some states are able to repair only armored vehicles, others only aircraft. . . . Henceforth we will repair all hardware together. But does this really mean that the defense union is being created? The USSR war machine will not be revived either today or tomorrow. . . .

[Kozyreva]: How do you see the problem of relations with the NATO bloc? A number of politicians in the West believe that it is time to disband NATO. What is your viewpoint on this?

[Pyankov]: I believe that the disbandment of NATO would accord with the interests of all the world's peoples. The Warsaw Pact does not exist, and Russia's president and defense minister have repeatedly declared that they do not see NATO as their enemy. Of course, we are not yet allies like Britain and France, let us say. But it is a lengthy process to develop relations of alliance. The mistrust between our peoples was sown over long years. People in both Europe and in the world must understand that Russia is not the USSR, which saw as its chief task to defeat NATO.

[Kozyreva]: Probably the Russian-U.S. joint exercises on the Totsk Range contributed precisely to greater mutual trust, did they not?

[Pyankov]: Undoubtedly. These exercises were the first step on the path of mutual rapprochement. We are not enemies, and we can be allies, but for the time being we are partners. Joint exercises help us to understand one another. I am an old general who has served in the army for forty years. Throughout these years the politicians told us that the United States, West Germany, and the other NATO countries were our enemies. We studied the potential of the enemy armies, their weak and strong points. . . . Today we must change. We are slowly freeing ourselves from old stereotypes. I believe that no military danger threatens us today. The fact that we might be conquered economically is another matter. We should be concerned about that.

[Kozyreva]: And yet, how prepared are the CIS joint forces to defend their territory, and will they be able to repulse possible aggression?

[Pyankov]: But who is going to attack? A specific image of the enemy has not been formed in our military doctrine. Military matters are a specific science. The military must know against whom they are to defend themselves. If there is a specific enemy, it will at once be clear whether we are weak or strong. If we are weak, we will create groupings, stocks, reserves. . . . But our task today is not to have a strong fist to repulse aggression but to be friendly with everyone. It is up to politicians and diplomats to have their say here. By 1995 Russia will have an army of just 1.5 million men.

[Kozyreva]: How do you rate the participation of CIS national armies in internal conflicts?

[Pyankov]: Unfortunately Russia bears the whole burden. There are many undesirable points for it in doing so. People are once again calling it an occupier and saying that it is once again demonstrating its imperial ambitions to the world. When the problem of the war in Tajikistan was discussed at the conference of leaders of the CIS states' military departments, I personally said: "Let us all provide one regiment each and give joint assistance." But what is the position today? Only Uzbekistan and Kyrgyzstan have each provided a battalion. Fortunately, the situation is being normalized in other hot spots. But it must be pointed out that there are still no joint peacemaking forces.

[Kozyreva]: But new hot spots are appearing. For example, Chechnya. . . .

[Pyankov]: Chechnya is Russian territory, and I would put the question like this: Let neutral forces go there, for example, joint forces of Ukraine and Kazakhstan. Because, if we are going to talk about collective security, then we must act together and help one another. But if Russia sends its own troops into Chechnya, then the situation can only worsen: Russian troops might be viewed as occupation troops, and both government troops and opposition detachments would fire at the soldiers. But our CIS allies are in no hurry to help us at present. I hope, however, that we will still succeed in creating genuine peacemaking forces of all the CIS states, and then our task of ensuring the security of the CIS with minimal armies and with minimal material expenditure will have been resolved.

## 9.33

### Report Views CIS Defense Issues

ITAR-TASS, 21 September 1994
[FBIS Translation]

"Changes in the military-political situation in the world characterized by decreasing tensions on the global level, by the rejection by Russia and other CIS states of the concept of permanent adversaries, by the beginning of interaction with NATO does not mean that potential threats to security have been eliminated completely," according to a report of the Foreign Intelligence Service headlined "Russia-CIS: Does the Western Position Need Correction?" made public on Wednesday.

"Interethnic and interstate conflicts in the CIS and neighboring states tend to expand." Thus, the intelligence service possesses information that Afghanistan, which borders on the CIS, "has forces striving to separate the north of the country, populated mostly by Tajiks and Uzbeks, and create on its basis a Farsi-speaking state which would include Tajikistan."

Iran and Turkey also work to expand their influence claiming the role of "regional superpowers" which does not allow them to stay away from conflicts on the CIS territory, according to the report.

Islamic extremism—"a movement aimed at forced spreading of Islam, suppression of forces resisting it, changes in the secular character of states"—negatively tells on the situation in hot spots in the CIS and especially in Tajikistan, the document added.

Its authors note that the "export of the ideology of militant Islam acquires the character of a serious threat also outside the CIS and this means that contraposition to it meets the interests of the whole world community."

The Russian intelligence also noted "inadequate reaction of the West to conflicts close to Russian borders." The number of victims in hot spots of the CIS is comparable to that in former Yugoslavia; however, "there exist serious differences in peacekeeping diplomacy regarding the two crisis areas," the report says. Moreover, many foreign countries criticize the "special Russian role" in peacekeeping operations in the CIS and the thesis that Russian vital interests depend on stability in other CIS countries. The intelligence service denounces "the double standards in assessing the rights and obligations of the West and Russia" and believes it to be groundless to claim that "Russia contrapositions its efforts to the actions of the United Nations and other international organizations."

Despite all post-"cold war" changes in military doctrines of leading Western powers, Russia and other CIS countries have to take into account that the countries are not going to refrain from modernizing and developing their offensive armaments. "This practice results in the conclusion that, at the present stage, the CIS countries have to preserve and develop their own strategic offensive forces. However, in conditions when nuclear forces in the CIS, according to international agreements, can exist only in Russia, the necessity of creating a single defense space of the Commonwealth is growing," the report says.

Other security requirements of today include "problems of overcoming the ecological crisis," joint efforts in the fight against organized crime, drug trafficking, and the smuggling of radioactive materials and weapons, according to the document.

"Thus the realities of security confirm that economic, political, and military integration in the Commonwealth . . . meets the demands of the time and is a natural and objective process," the report says.

## 9.34

### [Kazakh] Agreement with Russian Federation on Military Cooperation

#### Kazakh-Russian Agreement on Military Affairs

*Sovety Kazakhstana,* 19 October 1994
[FBIS Translation]

[Treaty between Republic of Kazakhstan and Russian Federation on Military Cooperation, signed in Moscow on 28 March 1994]

The Republic of Kazakhstan and the Russian Federation, hereafter referred to as the contracting parties,

Guided by the Treaty on Friendship, Cooperation, and Mutual Assistance Between the Republic of Kazakhstan and the Russian Federation of 25 May 1992.

Mindful of earlier agreements on cooperation in the sphere of defense within the confines of the Commonwealth of Independent States and on a bilateral basis in the interest of guaranteed collective security,

Aware of the need for the precise and consistent fulfillment of the obligations assumed by the contracting parties in connection with the Treaty on the Reduction and Limitation of Strategic Offense Arms of 31 July 1991 and the protocol signed in Lisbon on 23 May 1992, hereafter referred to respectively as the START I Treaty and the Lisbon Protocol,

Acknowledging the need for united effort and concerted action for reliable joint defense within the confines of the common military-strategic territory,

And expressing the wish to give the military cooperation between the contracting parties a new quality and to provide it with a legal foundation, have agreed as follows:

*Article 1.* For the purposes of this treaty, the following terms will be defined in this way:

"Strategic nuclear forces" (SNF)—military elements, including large and small units, institutions, organizations, and facilities, armed with or storing strategic nuclear weapons, and the units securing their operations.

"Integrated military units"—large and small units of the Armed Forces of the Republic of Kazakhstan and the Armed Forces of the Russian Federation assigned by the contracting parties for joint defensive missions.

"Facilities used for defensive purposes"—test sites, military facilities, the facilities of industrial representatives, and battlefields, located on parcels of land within the territory of the contracting parties and capable of being used by the parties jointly or transferred by one of the parties to the other, including lease transfers, for use for military purposes in the interest of strengthening the defensive capabilities of both parties.

"Delivery system"—the intercontinental ballistic missile (ICBM), the heavy bomber (HB), and the air-launched cruise missile (ALCM).

"Nuclear munitions"—the ICBM or ALCM warheads containing a nuclear charge.

*Article 2.* The contracting parties reaffirm their commitment to friendly interstate relations based on the principles of mutual respect for state sovereignty and territorial integrity, the inviolability of borders, the peaceful settlement of disputes and the refusal to use force or threats of force, and the conscientious fulfillment of treaty commitments in accordance with the Treaty on Friendship, Cooperation, and Mutual Assistance between the Republic of Kazakhstan and the Russian Federation of 25 May 1992, as well as the observance of other common standards of international law.

In the event of a situation threatening the security, independence, or territorial integrity of one of the contracting parties, the Republic of Kazakhstan and the Russian Federation will hold consultations without delay and undertake specific actions to give one another the necessary assistance, including military assistance, in accordance with international law, the bilateral Treaty on Friendship, Cooperation, and Mutual Assistance of 25 May 1992, and the Treaty on Collective Security of 15 May 1992.

*Article 3.* Strategic nuclear forces within the territory of the Republic of Kazakhstan and the Russian Federation will perform missions in the security interests of the contracting parties.

The Republic of Kazakhstan, with a view to the existing system for the functioning of the strategic nuclear forces located within its territory, will assign these military units of the strategic nuclear forces the status of strategic nuclear forces of the Russian Federation—Russian military elements temporarily deployed within the territory of the Republic of Kazakhstan.

Until all of the strategic nuclear weapons temporarily located within the territory of the Republic of Kazakhstan have been eliminated or withdrawn to the territory of the Russian Federation, the decision on the need to use these weapons will be made by the president of the Russian Federation with the approval of the president of the Republic of Kazakhstan.

In these cases, the Russian Federation will guarantee the institution of organizational and technical measures to preclude the unauthorized use of strategic nuclear weapons located within the territory of the Republic of Kazakhstan.

The terms of the presence of strategic nuclear forces within the territory of the Republic of Kazakhstan, corresponding to the standards of international law, will be defined in a separate agreement.

*Article 4.* All movable and immovable military property will belong to the contracting party within whose territory it was located on 31 August 1991.

The Russian Federation acknowledges the right of the Republic of Kazakhstan to receive compensation (in monetary form or some other form) equivalent to the value of the materials, agreed upon by the contracting parties, in nuclear munitions and delivery systems, as well as equipment and other property of the strategic nuclear forces located within the territory of the Republic of Kazakhstan on 31 August 1991 before their withdrawal to the territory of the Russian Federation.

Appraisals of the value of the materials and equipment and of the Russian Federation's expenditures on their maintenance, transport, and recycling, as well as the proportional share of compensation to be granted to the Republic of Kazakhstan, will be conducted according to a procedure agreed upon by the contracting parties.

Property rights to facilities, buildings, and installations erected after 31 August 1991 or weapons, vehicles, equipment, and property brought into the territory after this date will be exercised by the contracting party financing these operations. In the event of shared financing, property rights will be defined in separate agreements with consideration for proportional contributions.

The contracting parties reaffirm the possibility of the use of facilities and installations located within the territory of one contracting party by the armed forces of the other. The list of military facilities and installations and the procedures and terms of their use will be established in separate agreements.

Proceeding from the need for improvement in joint defense and the consolidation of national security, each of the contracting parties may turn the property of its own armed forces over to the other party for possession and use on mutually beneficial terms, including the terms of lease, in accordance with its own legislation.

One contracting party will not be obligated to compensate the other, unless other agreements stipulate otherwise, for improvements made by the other party in military facilities or on parcels of land located within the territory of the first party and used for military purposes, or for buildings or installations remaining on these grounds at the time this treaty expires, or for the early surrender of facilities and parcels of land.

*Article 5.* The status of facilities used jointly by the contracting parties for defensive purposes will be defined in line with the legal authority of the Republic of Kazakhstan and the Russian Federation as states responsible for the management of these facilities and their operation and material and technical support, as well as the joint authority of the contracting parties in the supervision of the activity and use of these facilities for the enhancement of the defensive capabilities of the parties.

During the performance of functions connected with the management, operation, and material and technical support of the SNF and of defense facilities leased from one another, the contracting parties will be fully responsible for their safe operation and the maintenance of the necessary levels of nuclear safety and other types of security.

During these processes, each of the contracting parties pledges to refrain from actions that might in any way prevent the other party from fulfilling its obligations, including those stemming from the START I Treaty and Lisbon Protocol, and prevent the functioning of its government agencies and/or damage state and/or private property.

The Russian Federation will take measures agreed upon with the Republic of Kazakhstan to eliminate the aftereffects of the operations of strategic nuclear forces located within the territory of the Republic of Kazakhstan, as well as facilities used for defensive purposes and turned over to the Republic of Kazakhstan by the Russian Federation. In the event of emergencies, the contracting parties will take immediate measures to eliminate the causes and will notify one another of this without delay. . . .

*Article 6.* With a view to the importance of the strict observance of the provisions of the USSR-U.S. Treaty on the Limitation of Anti-Ballistic Missile Defense of 26 May 1972, and with a view to the mutual interests of the Republic of Kazakhstan and the Russian Federation, the contracting parties will proceed from the knowledge that the Sary-Shagan test site will be used for the purpose of developing and improving ABM systems or components deployed within the region as specified in Article III of that treaty. The conditions of the use of the Sary-Shagan test site by the contracting parties will be defined in a separate agreement.

*Article 7.* The contracting parties will give one another mutual assistance in the implementation of multilateral international treaties and political commitments for the reduction and limitation of strategic offensive and conventional arms.

Each of the contracting parties must consider the interests of the other party during the conclusion of treaties and agreements with third states on military cooperation and deliveries of equipment and weapons.

*Article 8.* The defense ministries of the contracting parties will draft and conclude separate agreements on matters pertaining to the joint planning and use of troops in the interest of the mutual security of the parties and will plan and conduct joint operations in the preparation of command and control agencies and the training of troops within the territory of either of the parties by mutual agreement.

The contracting parties may form integrated military units under a joint command. . . .

*Article 9.* The management, personnel hiring, and material and technical supply procedures for facilities used jointly by the contracting parties for defensive purposes, and their integrated military units and joint command, will be defined in separate agreements.

*Article 10.* The contracting parties will cooperate in the sphere of military intelligence.

Each of the contracting parties pledges not to conduct military intelligence activities directed against the other party.

*Article 11.* The contracting parties will conclude an agreement on the use of the forces and resources of the Navy of the Republic of Kazakhstan and the Navy of the Russian

Federation in the Caspian Sea basin for joint operations to safeguard the security of the parties.

*Article 12.* Questions connected with the legal status of the servicemen of the armed forces of one of the contracting parties serving within the territory of the other party and of members of their families, their pension security, and other matters pertaining to the social and legal protection of these individuals, will be addressed in a separate agreement.

The contracting parties will extend the guarantees of the application and exercise of social and civil rights envisaged in their legislation to their citizens in military service outside the boundaries of their state.

The contracting parties will acknowledge the validity of military titles, state honors, and educational and pension documents of servicemen, the privileges granted to servicemen, individuals with military discharges, and members of their families, in accordance with the existing legislation of the contracting parties, with a view to their term of service in the armed forces of the former USSR and their subsequent service in the armed forces of the contracting parties, including contracted military service.

The contracting parties will guarantee civilian personnel equal rights, irrespective of their citizenship, to employment in military units and the enterprises and institutions of their armed forces and will include this period of their employment in their total term of service for pension eligibility.

When one of the contracting parties is inactivating its military units, establishments, and institutions located within the territory of the other party, the former party will compensate civilian personnel in accordance with its own labor legislation.

The contracting parties will consult one another on ways of improving and coordinating their national legislation, including laws on the financial and social security of the servicemen and civilian personnel of the armed forces and on the privileges granted to servicemen and individuals with military discharges and the members of their families.

*Article 13.* Members of the staff of military units, establishments, and institutions will not require visas to cross the state border of the contracting parties or require travel passports or special notations in passports if they carry identification (military service cards or passports) and travel authorization papers from their commanding officers (furlough passes or travel orders), and their minor children will not have to meet these requirements if their names are listed on the documents. When they are sent to a new service location or their permanent place of residence, they will transport their personal belongings across the state border between the contracting parties without the payment of duties, taxes, and other fees.

Subunits, units, and teams of more than fifty servicemen of the armed forces of one of the contracting parties may cross the state border of the other party after advance notification and by agreement of the defense ministries of the parties.

*Article 14.* Material and technical supply operations for military elements will be conducted by the defense ministries of the contracting parties on mutually beneficial terms, guaranteeing the maintenance of their armed forces and integrated military units at a high level of combat readiness and combat effectiveness, and will be regulated by separate agreements.

*Article 15.* The activities of the military elements of one of the contracting parties located within the territory of the other party will be financed by the party with jurisdiction over them.

Questions connected with the circulation of the national currencies of the contracting parties for the daily needs of the servicemen and military elements of the parties located within their territory will be regulated in accordance with the agreement between the National Bank of the Republic of Kazakhstan and the Central Bank of the Russian Federation.

*Article 16.* Each of the contracting parties pledges not to violate the state and public security of the other party and the personal safety of its citizens during its activity in facilities and on parcels of land belonging to the other party.

*Article 17.* The contracting parties will agree on policy in the sphere of joint development, production, repair, and shipment of arms, military vehicles, and material and technical resources in the interest of the comprehensive support of the armed forces, facilities used for defensive purposes, and integrated military units, and will coordinate aspects of military-technical cooperation, securing the preservation and development of existing cooperative relationships between enterprises developing and producing weapons and military hardware. Deliveries and services will be performed on a duty-free basis at prices set by each of the contracting parties for their own needs. Prices and rates will be agreed upon by the parties and will be defined in a separate agreement in each case. Questions connected with the coordination of policy in the sphere of arms and military hardware and reciprocal deliveries of goods (and work or services) will be addressed in special agreements on the basis of joint weapons programs.

The contracting parties will create an intergovernmental commission on industry and scientific research and experimental design projects for the pursuit of the policy agreed upon in the military-technical sphere, with the preservation and development of existing patterns of specialization and cooperation.

The contracting parties will create an intergovernmental commission on military-technical cooperation by the Republic of Kazakhstan and the Russian Federation for the pursuit of the policy agreed upon in the military-technical sphere.

*Article 18.* The contracting parties will retain their existing procedures for the education and training of officers and junior military specialists for the armed forces of the parties on the basis of the corresponding agreements.

*Article 19.* The contracting parties will retain the existing network of all types of communications, air defense, anti-ballistic missile defense, and early warning systems and supply lines and will agree on measures for their development.

The contracting parties will cooperate in the sphere of military transport movements. The procedures of this cooperation will be defined in a separate agreement.

The contracting parties will retain the common air space for flights by military and civilian aircraft and the joint flight control system on the basis of the corresponding agreements.

*Article 20.* For the purpose of reinforcing discipline and order in the armed forces, in facilities used for defensive purposes by the contracting parties either jointly or on the terms of a lease, and in integrated military units, the contracting parties will coordinate operations in the law enforcement sphere.

*Article 21.* The contracting parties will plan measures jointly and render mutual assistance in the resolution of ecological problems connected with the aftereffects of military operations.

*Article 22.* This treaty is not directed against any other states and will not affect the contracting parties' rights and obligations stemming from other international treaties to which they are party.

*Article 23.* The contracting parties will not allow the use of their territory by a third state for activity directed against the other contracting party.

*Article 24.* For the purpose of implementing the provisions of this treaty, and in the interest of broader and more intensive cooperation in the sphere of defense, the contracting parties will form a joint committee, which will act in accordance with a statute approved by the parties.

*Article 25.* This treaty may be amended and supplemented by mutual agreement of the parties.

The treaty must be ratified and will go into force on the date of the exchange of instruments of ratification.

The treaty will be concluded for a term of ten years. It will be renewed automatically for the next ten years unless one of the parties notifies the other, in writing and at least six months before the expiration of this term, of its wish to withdraw from the treaty.

This treaty will be in force as an interim agreement on the date it is signed.

Done in Moscow on 28 March 1994 in two copies, one in the Kazakh language and one in the Russian language, with each version being equally authentic.

Republic of Kazakhstan
[Signed] [Signature illegible]

Russian Federation
[Signed] [Signature illegible]

### Protocol Contents Concerning Article 4

*Sovety Kazakhstana,* 19 October 1994
[FBIS Translation]

[Protocol Memo of Agreement on Meaning of Article 4 of Treaty Between the Republic of Kazakhstan and the Russian Federation on Military Cooperation of 28 March 1994.]

The Republic of Kazakhstan and the Russian Federation will proceed from the understanding that the reference in the first paragraphs of Article 4 of the Treaty Between the Republic of Kazakhstan and the Russian Federation on Military Cooperation of 28 March 1994 to the property rights of the Republic of Kazakhstan to the movable military property located within its territory on 31 August 1991, specifically mentioning nuclear munitions, will apply to the material of these munitions and not to the munitions in assembled form.

Republic of Kazakhstan
[Signed] [Signature illegible]

Russian Federation
[Signed] [Signature illegible]

## 9.35

### CIS Military Integration Prospects Viewed

Dmitriy Trenin
*Nezavisimaya Gazeta,* 4 November 1994
[FBIS Translation]

Within the framework of the revitalized debate concerning the reintegration of the post-Soviet states, a notable place is occupied by questions of convergence in the military-political and military spheres. Things have not been confined to debate. It may be affirmed that for the first time in three years the trend toward the drawing together of the defense space is starting to be the prevailing trend. A minimum of three most important circumstances are contributing to this: an awareness by the political elites of many new independent states (NIS) of the enormous difficulties that attend the independent building of national defense systems and, consequently, the gravitation toward a military alliance with Russia; the gradual formation in Russia itself of the political will in support of the military-political union of the countries of the CIS as a means of stabilizing the situation on the periphery of the Russian Federation and the creation of a "good-neighbor" zone around Russia; finally, as a derivative of the first two, the far greater realism in the vision of the actual ways to realize military integration plans.

The purpose of the present Russian defense policy in the "near abroad" is, evidently, the restoration—on a new basis and on a different scale—of the unity of the military-strategic space of the former Union (minus the Baltic), which was torn apart or seriously undermined as a result of the disintegration of the USSR. Designed to achieve this purpose is a strategy whose main components are the creation of a military-political alliance out of the CIS, headed by Russia; the close coordination of the efforts of the Russian Federation and the NIS in guarding, and if necessary, defending the external borders of the Commonwealth; the restoration of military-economic relations within the former Union military-industrial complex; the conversion of the CIS into a regional organization that is recognized by the world community as bearing (with Russia having the leading role) the main responsibility for the settlement of armed conflicts on the territory of the post-Soviet states. It is thus obvious that the former approach based on the theory of collective security and the practice of the division of the once united Soviet Army is giving way to a new approach aimed at realization of the principle of collective defense and the creation under the aegis of Russia of joint, and in the future, united, armed forces of the CIS.

It might at first glance seem that those that in the fall and winter of 1991–92 were defeated in the argument over the fate of the Soviet military legacy have been compensated by history: The course of events has confirmed that they were right and, on the other hand, set an inordinately high price for the success of the recent "nationalizers" of the army. Nonetheless, it has to be seen that irreversible changes, which rule out a direct return to primary unification ideas, have occurred in the past three years in all the former Soviet republics.

We need first and foremost to scrutinize the concept of a common Eurasian strategic space as the cornerstone of the majority of integration constructs. The unity of this space is seen as natural and stabilizing, and its rupture as, correspondingly, unnatural and destabilizing. The weakening of Russia's position in the world, the conflicts in the post-Soviet states, the expansionist aspirations of certain contiguous countries—these are the main arguments adduced in support of this proposition. The conclusion—unite before it is too late—sounds logical.

What is the basis of this unification? The imperial interests of a vast multinational state served as this basis in past times. Today there is no such state, and it will hardly emerge tomorrow. It has to be a question of a community of security interests among a number of independent states, a number of which are located in Eastern Europe, others in the Transcaucasus, and yet others, in Central Asia. How great can this community be?

There are two instances wherein it could be sufficient for the formation of a military-political alliance. The first, traditional instance: a perception by the community of a common threat. It was such a perception that formerly united Norway and Italy, Portugal and Turkey around the United States and beneath the flag of NATO. The second instance is a natural consequence of multilateral integration, in which it is essential that the building of a federative structure entails the "federalization" of defense. This process is occurring currently within the framework of the European Union. How do matters stand in the CIS in this connection?

For the majority of NIS a common external threat (and its perception) is absent. The attention of Armenia is concentrated on Azerbaijan and Turkey, and of Uzbekistan, on Tajikistan and Afghanistan, and Belarus is, possibly, free of the perception of any threat at all. The total resistance of the CIS countries to Russia's repeated attempts to enlist them in joint peacekeeping operations in conflict zones on the territory of the former USSR testifies to the degree of real, not sham, community of perceived threats.

As far as broad reintegration, the result of which could be a unified defense system, is concerned, all attempts at "reunification" have thus far proven unsuccessful mainly owing to the fact that they have conceptually been addressed to the past and have been based on material and ideological structures whose positions have been incessantly eroded. In this sense there should be even fewer hopes for a Eurasian union than for the CIS.

A number of regions are actually forming in the space of the "subworld of the USSR": a new Eastern Europe, Transcaucasus, and Central Asia. The differences in the geostrategic position of the NIS and their security requirements, threat levels, and so forth are extraordinarily great and are continuing to increase. They are, of course, drawn together by one circumstance: All three new regions are situated on the periphery of Russia, which has interests in each of them. But even in this case the relations among the new East Europeans, Caucasians, and Central Asians are indirect. Thus the unity of the strategic space—in the sense of a buffer separating it from the traditional abroad—really exists only for Russia and is a fiction for the NIS.

Consequently, it cannot be expected—even less demanded—of Belarus that it will perceive the situation in Gornyy Badakhshan as directly affecting its fundamental interests or that Uzbekistan will display concern for a strengthening of Russia's Far Eastern borders. On the other hand, Russia and some of the other states have an undoubted interest in each of the newly formed regions. Under these conditions, what is more beneficial from Russia's viewpoint—relying on an illusory community of interests of all the former republics and heaping onto itself the burden of building and maintaining a new Warsaw Pact, whose history could prove shorter than, and its fate similar to the fate of the prototype? Or seeking less all-embracing, but more efficient, longer-lasting, and cheaper options? This, it would seem, is an important question, the answer to which will help impart the optimum parameters to the planned military-political integration.

The author sees the following version of an answer. Instead of a single, but predominantly formal military-political alliance within the framework of the entire CIS, Russia could adopt a policy of building a system of regional agreements for deterring and warding off possible power challenges to itself and its new neighbors. In each region Moscow would rely on the countries whose long-term security interests are so consonant with its regional interests that no change of leaders or governments could rapidly alter Russia's national strategic priorities.

Obviously, in Eastern Europe this means Belarus, which covers a most important strategic axis, secures our ground line of communication with the rest of Europe, and brings Russia as close as possible to its Kaliningrad enclave. A union with Georgia, which is interested, like Russia, in preventing the regional expansion of neighboring states, is natural in the Transcaucasus. In Central Asia our strategic ally on the southern and eastern axes is Kazakhstan.

Close Russian-Belarusian relations in the military sphere ensuing from the organic convergence of the two countries could be developed on the basis of a bilateral security treaty, which would not cause apprehension among neighbors: Ukraine, Poland, and the Baltic. The corresponding supreme political and military authorities would be formed and joint armed forces would be created within the framework of regional treaties on the collective defense of Central Asia and the Transcaucasus. Even though the allies of the new Russia would not be that many, they would be key, supporting states bound to Russia by strong ties. Russia could accord its allies dependable security guarantees.

The separation from the ranks of CIS countries or a group of allies poses the question of Russia's relations with the other, "non-allied" states. Would Russia not thereby repel them and hurl them into neighbors' embraces? Would regional balances not thus be to Russia's detriment?

If we are speaking of Eastern Europe, then, despite the good prospects of economic integration and certain coincidental security interests of Russia and Ukraine, their military alliance is hardly possible politically as a consequence of the manifest domination therein of Russia. In addition, an alliance concluded despite the manifest absence of a military threat in Eastern Europe could itself give rise to the apprehensions of neighboring Central European states and provoke a process whose result might be the reconstitution of the enemy image. Under such conditions Russian-Ukrainian cooperation in the defense sphere could include the basing in Crimea on the corresponding terms of the Russian Black Sea Fleet, the integration of air defense and ABM systems, the coordination of border activity, and of course, the military-technical cooperation of the two countries.

In the new situation the strategic significance of Moldova for Russia is peripheral. A military threat to our country from this direction is absent. In addition, Moldova is split, and under these conditions a military alliance between Moscow and Chisinau could be of significance only to Tiraspol. For its part, despite the economic attachment to the countries of the CIS, Chisinau is firmly attuned toward a neutrality that excludes a foreign military presence. Russia's interest consists of promoting a settlement of the conflict between the Left Bank region and the rest of Moldova, not of acquiring bases "with a view of the Balkans."

There can hardly be any expectation of an alliance with Azerbaijan with the intention of deterring potential challenges on the part of Turkey and Iran (and where else?). Account should be taken also of the steadfast anti-Russian mood of part of the Azerbaijani political elite. Cooperation with Baku could include military-technical and border issues and lease of the radar station. Armenia, on the other hand, is definitely oriented toward an alliance with Russia, for which this country also is of obvious strategic interest. Nonetheless, despite the presence of Russian bases and border guards and also Moscow's assistance in the formation of Armenia's armed forces, an official military alliance with Yerevan—prior to the settlement of the Karabakh conflict, in any event—could be detrimental to Russia's position in Azerbaijan and in the region as a whole.

Finally, in Central Asia the military-political line that Russia is actually in a position to hold extends, probably, along the southern and eastern borders of Kazakhstan, which should be its strategic ally. The other countries are either extraordinarily unstable or are pursuing goals barely consonant with Russia's or demonstratively prefer neutrality. In addition, as experience shows, threats emanating from the Central Asia-Middle East region cannot be countered with traditional bloc building. Bilateral and multilateral military cooperation in specific areas (air defense, borders, military-technical cooperation) could produce greater benefits.

So a system of Russia's collective defense alliances

would include a security treaty with Belarus, a defense treaty with Georgia a special agreement with Armenia, and a defense treaty with Kazakhstan. Affiliated with this system would be a package of agreements with Ukraine and agreements on border protection, the unification of air defense systems, the leasing of facilities, and military-technical cooperation with other countries of the CIS.

Contrary to the widespread notion, collective security is not a lower form of military-political integration compared with collective defense. In actual fact, these two constructs differ in principle. Whereas the first is designed to ensure security against encroachments emanating from within some community ("collective") of states, the second is aimed at repulsing threats directed from outside. Collective defense could, of course, also imply the collective security of the allies in respect to one another, but this aspect is undoubtedly subordinate. Finally, "security" does not require indication of the potential aggressor since all would obtain equal guarantees against one another but "defense" could not be built without a sufficiently certain vision of a probable enemy. In practice this means that it would be expedient, together with the creation of regional defense systems, to preserve the system of security of the CIS, which could be entrusted with the mission of practical peacekeeping on the territory of the Commonwealth.

## 9.36

### PFP Viewed as "Waiting Room" Before Joining NATO

From the "Diplomatic Panorama" feature
Interfax, 21 December 1994 [FBIS Translation]

Kyrgyzstan regards the NATO Partnership for Peace [PFP] program as a waiting room before joining NATO.

Kyrgyz Foreign Minister Rosa Otunbaeva said in Bishkek Tuesday [20 December] that Kyrgyzstan was drafting a document on joining the partnership and a Kyrgyzstan-NATO program.

According to Otunbaeva, Kyrgyzstan's joining Partnership for Peace would contribute to regional security and help Kyrgyzstan deal with the effects of natural disasters and in carrying out technological and training programs.

She believes that Central Asian countries should not trail behind the others. They have to form their own peacekeeping forces and keep them prepared for various operations, notably peacekeeping in Tajikistan.

Otunbaeva emphasized that the conflict in Chechnya is a Russian internal affair. She called for avoiding bloodshed and looking for resolution of the conflict through political dialogue.

## 9.37

### CIS Divided on Treaty on Defense of CIS External Borders

Interfax, 9 February 1995 [FBIS Translation]

Azerbaijan categorically disagrees with the draft treaty on cooperation for defense of CIS external borders submitted for consideration by the CIS summit opening in Almaty on 10 February, Azer; Foreign Minister Hasan Hasanov told the joint meeting of the CIS defense and foreign ministers and the commanders of the CIS border guard troops in Almaty this evening.

The Azerbaijan delegation said the draft treaty did not conform with the national interests of the country. Baku insists on preserving both internal and external borders in the CIS. Hasanov said Armenia had "occupied" more than 130 kilometers of the Azerbaijani border with Iran and Armenia and "seized" 20 percent of Azerbaijani territory. Baku proposed that the draft treaty be amended and discussed at the next CIS summit.

The Ukrainian delegations also stated a special position on the draft treaty. Kiev said national border guard troops must defend the Ukrainian state borders.

However, Russian Foreign Minister Andrey Kozyrev, chairing the meeting, said that the document had been recommended for consideration by the CIS summit in Almaty and could be discussed in detail at the next meeting of the CIS leaders if the appropriate instructions were given.

Kozyrev said all CIS members had backed the initiative of Kazakh President Nursultan Nazarbaev on signing a memorandum on peace and stability in the CIS in Almaty. The memorandum will confirm inviolability of borders and will be aimed at preventing any actions undermining inviolability of borders.

The signatories of the memorandum will agree to refrain from military, political, economic, and other forms of pressure as well as from participating in unions and blocs aimed against any of them, and to prevent on their territory actions of organizations and individuals aimed at undermining territorial integrity and inviolability of borders of the signatories.

## 9.38

### CIS Plans Four Regional Collective Security Zones

14 February 1995
[BBC] © Copyright 1995
The British Broadcasting Corporation

Moscow: CIS military integration will start with the formation of a chiefs of staff committee and four regional collective security zones in the Commonwealth, said Lt.-Gen. Leonid Ivashov, the secretary of the CIS Defense Ministers Council.

He told an Interfax reporter Tuesday [14 February] that in all, four military regions were planned to be set up on CIS territory. Western, Eastern, Central Asian, and Caucasus zones will unite nine CIS member states, along with cooperation on an irregular basis by Ukraine, Moldova, and Turkmenistan.

According to Ivashov, the Western zone is reserved for Belarus "as the key element," and for the Kaliningrad and Smolensk regions of Russia. "Ukraine and Moldova will be in touch with them, if needed," Ivashov explained. The Caucasus zone will include Azerbaijan, Armenia, Georgia, and the North Caucasus republics within the Russian Federation. The Central Asian zone is reserved for Kyrgyzstan, Tajikistan, and Uzbekistan. The three republics will cooperate "with Turkmenistan on some elements." Kazakhstan, Russia, and part of Kyrgyzstan will belong to the Eastern zone.

Ivashov stressed that no special groupings would be set up within these zones, nor would army units be transferred there. "Everything will be done depending on what is available in these zones," he said. It is assumed that if one of the states belonging to a zone is attacked, the rest will help it to repulse the aggression. A planning body will determine what force and means are necessary to curb the attack.

"The coalition defense forces will train jointly and will have common combat-readiness standards for troops and headquarters," Ivashov reported, adding that joint exercises are planned to be carried out as well. He announced that these proposals would be submitted to CIS heads of state at the end of 1995.

## 9.39
## NATO, Bishkek to Make Cooperation More Active

Interfax, 16 February 1995 [FBIS Translation]

NATO is interested in developing and activating cooperation with Kyrgyzstan within the framework of the Partnership for Peace program. This statement was voiced at a press conference in the republic's Foreign Ministry Thursday [16 February] that was devoted to the arrival of a NATO delegation to this country headed by Colonel Dan Kwist.

It was declared that the Foreign Ministry and Defense Ministry have prepared a document on the republic's participation in Partnership for Peace program that will be considered at NATO headquarters in Brussels; an individual program of actions will be developed on the basis of this document.

Participants in the press conferences underlined NATO's interest in rendering assistance in projects on environmental protection, research programs, military production conversion, and technical assistance to fight natural disasters.

Dan Kwist expressed interest in Kyrgyzstan's active involvement in programs, seminars, and conferences on regional economic and environmental security.

A representative of Kyrgyzstan's Foreign Ministry expressed interest in Kyrgyz servicemen's participation in exercises of the U.S. Armed Forces as supervisors.

## 9.40
## Prudnikov Views Unified Air Defense System for CIS

Sergey Ostanin

ITAR-TASS, 17 February 1995 [FBIS Translation]

Commonwealth defense departments have started to implement an agreement on an allied air defense system. The agreement was signed by Commonwealth leaders on 10 February in Almaty, commander-in-chief of the Russian air defense troops and commander-in-chief of the Commonwealth allied air defense system Col.-Gen. Viktor Prudnikov told a news conference here today.

In his words, the departments are elaborating basic legal documents on military-technical cooperation in air defense.

After disintegration of the former Soviet Union, air defense means and forces were divided by the former Soviet republics and, as a result, became less efficient. "Many systems are losing combat readiness and the personnel is losing skills," said the general.

Commonwealth leaders with the exception of Moldova and Azerbaijan have decided to [pool] efforts in protection of the Commonwealth air space and assign means and forces from each state to the allied air defense system. The latter is expected to have a coordinating committee to include air defense commanders of each member state, their deputies, and other high-ranking officials.

The military-technical cooperation provides for the delivery of material, repairs of armaments, and training of the personnel, said Prudnikov. Cooperation details have not been specified. This will be done later during meetings with air defense commanders.

In the opinion of the commander-in-chief, "the creation of the allied air defense system will help settle many problems associated with the stabilization of national air defense forces and with consolidation of defensibility and sovereignty of states."

## 9.41
### Further on Prudnikov Comments
Interfax, 17 February 1995
[FBIS Translation]

The joint air defense force of the Commonwealth of Independent States will concentrate on air surveillance and the exchange of information, the chairman of the force's coordinating committee and the commander-in-chief of the Russian air defense forces General Viktor Prudnikov told newsmen Friday at his headquarters in Balashikha near Moscow.

Neither aircraft rocket launchers nor fighter jets belong with the joint air defense force, Prudnikov said, adding that under "a plan for interaction" it is up to the Commonwealth republics to decide what units and hardware are to be detailed for service on the joint force.

In Prudnikov's view, the Commonwealth republics are not likely to be able to build their own armed forces without help from Russia. Today, Russian air defense men serve in Latvia, Azerbaijan, Belarus, Kazakhstan, Tajikistan, Turkmenistan, and Uzbekistan, he said.

According to Prudnikov, air defense installations are made and undergo repairs mostly in Russia. Other Commonwealth republics will be supplied with air defense hardware in keeping with bilateral agreements.

Addressing the results of the Commonwealth's Almaty summit, Prudnikov said that all the republics, except Azerbaijan and Moldova, signed an agreement on the creation of a joint air defense force. Problems of air defense will be settled in the context of bilateral relations between Russia and Azerbaijan, and Russia and Moldova, he told newsmen at Balashikha.

Russia's Foreign Ministry had been instructed to discuss prospects for the joint protection of air space with the Baltic republics.

In response to journalists' questions, Prudnikov said that "on the whole" the Russian air defense force showed "a good fighting potential and was capable of meeting its objectives." Ninety percent of its active components—rocket launchers and fighter planes—were of the newest makes, he said.

## 9.42
### Collective Security Council Chief Interviewed
Interview
*Krasnaya Zvezda,* 15 June 1995
[FBIS Translation], Excerpt

[Interview with Gennadiy Shabannikov, general secretary of the Collective Security Council, by an unnamed questioner.]

[Shabannikov]: The Collective Security Council is the supreme political organ of the states adhering to the Collective Security Treaty signed 15 May 1992 in Tashkent. The tasks which the Collective Security Council must resolve include examining questions connected with ensuring the fulfillment of this treaty: holding consultations to coordinate positions of the member states if a threat arises to commonwealth security, territorial integrity, or sovereignty of one or several signatory states or if there is a threat to international peace and security; elaborating measures to improve defense management in signatory states; examining questions of aid, including military aid, to a signatory state subject to aggression from any state or group of states; taking steps deemed necessary to maintain or restore peace and security (reports of such measures are immediately sent to the UN Security Council); coordinating the signatory states' activity on the main avenues of mobilization training of the armed forces and economic structures; the elaboration of recommendations for the main avenues of military-technical policy, and the provision of weapons and military equipment for the armed forces of the signatory states.

[*Krasnaya Zvezda*]: Who are the signatories to the treaty and who belongs to the council?

[Shabannikov]: The initial members of the Collective Security Treaty were the six states which signed the treaty: the Republic of Armenia, the Republic of Kazakhstan, the Republic of Kyrgyzstan, the Russian Federation, the Republic of Tajikistan, and the Republic of Uzbekistan. In 1993 the Republic of Azerbaijan, the Republic of Georgia, and the Republic of Belarus acceded to the treaty. In accordance with the Collective Security Treaty the Collective Security Council has been joined by the Collective Security Treaty signatory states and the commander-in-chief of the CIS Joint Armed Forces. Since the formation of the Joint Armed Forces Main Command in 1993 and the staff for the coordination of the CIS states' military cooperation, the Collective Security Council composition has been revised. In accordance with the decision of the heads of the Collective Security Treaty signatory states adopted 24 December 1993, the Collective Security Council includes the heads of state, foreign ministers, and defense ministers of the signatory states and the general secretary of the Collective Security Council. The Collective Security Council's supreme consultative organs are the council of foreign ministers for questions of coordinating domestic and foreign policy and the council of defense ministers for questions of coordinating military policy and military organizational development which include, respectively, the foreign ministers and defense ministers of the Collective Security Treaty states.

[*Krasnaya Zvezda*]: And what is the role of the Collective Security Council general secretary?

[Shabannikov]: The Collective Security Council general secretary is appointed from the ranks of civilians. He is assigned the holding of political consultations, the coordination of the Collective Security Treaty states' positions on military questions, the preparation of council sessions, and the generalization of questions for discussion and decision; on instructions from the Collective Security Council he represents it in relations with heads of state, international organizations, and the mass media, and represents the signatory states' common interests on questions of collaboration with NATO and other military-political groupings, blocs, and alliances, and he leads the work of the Collective Security Council secretariat.

## 9.43

**Kozyrev Addresses Ambassadors, Federation Council on CIS**

6 July 1995



*Editor's Note:* In mid-1995 the Russian Foreign Ministry stepped up its advocacy on behalf of CIS military and economic integration. A concerted effort was made by Foreign Minister Kozyrev and his staff to get the United Nations member states to condone Russia's legitimate role as "peacemaker" and "peacekeeper" within the CIS, which Russia claims is "one security space." The United Nations has never sanctified Russia's role as peacekeeper in the region, however, preferring to express its approval of "combined CIS peacekeeping troops," and the efforts of other international bodies, such as the OSCE, to monitor "hot spots" in the region. Most of the non-Russian leaders view "peacekeeping" as a mask for Russian intervention in the internal affairs of the CIS states.

A meeting of Russia's ambassadors to the CIS is being held at the Foreign Ministry. Foreign Minister Andrey Kozyrev told the ambassadors that relations with the CIS are central to Russia's foreign policy. He also stressed that message in an address to the Federation Council. Kozyrev passed on to the ambassadors the greetings of President Yeltsin, who also emphasized the central importance of relations with the CIS. Kozyrev went on to discuss the continuing economic reintegration of the CIS states and their crucial role in peacemaking. He noted that the organization had played a very important role in maintaining stability in the former Soviet Union since its collapse. The meeting is continuing. . . .

[Text of report by Radio Russia]

[Presenter]: A two-day meeting of Russian ambassadors to the CIS countries opened in Moscow today. President Boris Yeltsin greeted its participants. Prime Minister Viktor Chernomyrdin said he intended to meet the ambassadors. Foreign Minister Andrey Kozyrev opened the meeting. . . .

[Kozyrev, voice]: Not only has it been found possible to generally ensure a civilized divorce, but in what is in historical terms a very short time, for we are only speaking of two to three years, it has been found possible to ensure that integration trends and trends toward cooperation in a number of areas have reappeared.

[Presenter]: On the same day, Andrey Kozyrev addressed the Federation Council, and our parliamentary correspondent Vyacheslav Osipov has the details.

[Osipov]: He stated straightaway that relations with the CIS were a priority for the Russian Foreign Ministry. The main problem besetting relations within the Commonwealth, noted Kozyrev, was the implementation of accords that were reached.

Russia will give sympathetic consideration to specific credits, but it will, at the same time, take account of strategic interests where relations with the CIS are concerned.

The level of Russia's relations with Belorussia [Belarus] was termed as being "unprecedented"—but the key issue does remain: political accords should not be suspended in midair. Kozyrev stressed that the problem of ensuring security in the former union republics needed to be met with understanding and support on the part of the parliament and international organizations. Kozyrev said the United Nations had adopted a resolution to support Russia's peace efforts in Tajikistan. He also said it was impossible to resolve the problem of ethnic minorities by diplomatic methods alone. Kozyrev welcomed the establishment of a commission dealing with expatriate affairs, headed by Sergey Shakhray. . . .

* * *

[Text of report by RIA news agency]

[Moscow, 6 July, RIA Novosti correspondent Viktor Bezbrezhnyy] For Russian diplomacy there is no task more important than the strengthening of the Commonwealth of Independent States. Such is the injunction of Russian President Boris Yeltsin, conveyed on his behalf by Foreign Minister Andrey Kozyrev to the Russian ambassadors in the CIS countries.

The minister spoke at the working meeting which has opened at the Foreign Ministry today. He stressed that "the CIS is a zone of important vital interests for Russia." But

Russia too, according to Kozyrev, "is an object of vital interests for the CIS states." Characterizing the situation in the Commonwealth, the head of Russian diplomacy noted that in the recent period "there have occurred serious changes in the desired direction of the CIS—a period has set in of digesting the experience of economic ties and political interests which weakened after the breakup of the USSR." Kozyrev believes that in carrying out integrationary processes the Commonwealth has found "a successful, flexible scheme of multivariants and multi-rate development." Calling the promotion of economic cooperation the "basis of integration," the minister identified as a major aim "the creation of a common market." Among other main objectives of Russian diplomacy in the CIS, Kozyrev names the turning of the Commonwealth into an influential regional organization, the ensuring of security along the perimeter of the boundaries of the CIS countries with countries which are not members of the Commonwealth, and the fight against terrorism and contraband. . . .

* * *

[Excerpt from report by Interfax news agency]

[Moscow, 6 July] Russia is ready to integrate in the CIS as far and in such forms as its partners are ready, Russian Foreign Minister Andrey Kozyrev said in Moscow on Thursday, opening a conference of Russian ambassadors to the CIS states. According to Kozyrev, integration within the CIS was still based on the principle of multispeed and multioption development. "It was prompted by life itself," he added. . . . According to Kozyrev, "cooperation ensured not only a civilized divorce, but a successful start of pulling together most countries of the former Soviet Union." Kozyrev said the Commonwealth's functional basis had been and will be economic cooperation. "An important step in this field has been made as Russia, Belarus, and Kazakhstan form a customs union and a mechanism of trilateral interaction, creating landmarks to move toward and toward which other CIS states may move," he added. Many CIS states already show interest in such forms of cooperation, Kozyrev remarked. He stressed the customs union was open for all CIS states sharing its goals and principles. In this context, economic integration will require "ever increasing transparency of internal borders," Kozyrev said. At the same time, unfounded aid and unrecoverable loans should be abandoned, he said. "Both in politics and economics there must be equality of rights and duties for all partners," he emphasized. Nevertheless, Kozyrev was convinced that "everything should not be brought to pure mathematics." "Of course, we should account for all loans and aid, but we should also remember our plans in a longer term. We should see the forest of our long-term interests behind the trees of concrete figures. All this, maybe later, will justify itself," he said.

**Kozyrev Discusses CIS role in Peacemaking**

[Excerpts from report by ITAR-TASS news agency]

[Moscow, 6 July, ITAR-TASS correspondent Dmitriy Gorokhov and Yuriy Kozlov] The CIS is an important instrument of maintaining stability in the post-Soviet expanse and thus contributes consistently to the consolidation of regional and global security. Russian Foreign Minister Andrey Kozyrev said this at a meeting of Russian ambassadors to CIS countries that opened here on Thursday. Kozyrev said it is necessary to emphasize this role of the Commonwealth and to ensure that it is mentioned in all documents of international forums.

The peacemaking activity of the CIS spreads, to Nagorno-Karabakh, to the zone of the Georgian-Abkhaz conflict, and to Tajikistan, Kozyrev said. He said the world community, including the United Nations and OSCE, should share the considerable load borne by the Commonwealth in settling regional conflicts.

There must be "no room for double standards" in the area of peacemaking as far as sovereign states and the United Nation's members are concerned, the minister said. As a result of persistent efforts of Russian diplomats, the Security Council registered a few days ago the collective appeal of CIS countries for launching a full-scale operation of the United Nations in Tajikistan, Kozyrev said.

The interaction of independent states in peacemaking and in the military-political area should be ensured by the collective security system of the CIS, the minister said. "It is not meant to form a closed bloc opposed to the rest of the world, but a mechanism that will be playing the stabilizing role" in the huge expanse of the former USSR, Kozyrev said. . . .

It is the second time that a conference of Russian ambassadors to CIS countries has been held by the Russian Foreign Ministry with the participation of representatives of other ministries and agencies concerned. The debates have a working character and are held behind closed doors. Such meetings are planned to be held on a regular basis in the future.

## 9.44
**"Sixteen Plus One" Talks Open in Brussels**

Aleksandr Mineev
ITAR-TASS, 14 September 1995
[FBIS Translation]

The North Atlantic Alliance and Ukraine today began their "Sixteen plus One" political dialogue. Earlier out of twenty-six countries participating in the NATO program "Partnership for Peace" only Russia had been granted this privilege.

Ukrainian Foreign Minister Gennadiy Udovenko arrived in Brussels for the meeting with NATO Secretary-General Willy Claes and then took part in a meeting behind closed doors of the NATO Council together with the ambassadors of sixteen member-countries of the alliance.

Before the opening of the NATO Council meeting, a ceremony was held to adopt Ukraine's individual program within the framework of the "Partnership for Peace" program.

The program provides for joint army exercises, personnel training, and exchange of peacekeeping experience, as well as assistance in the development of democratic control over the Ukrainian armed forces.

Ukraine has been one of the most active participants in the NATO program since the beginning of its implementation. It has participated in most military exercises conducted within the framework of the program and even permitted the exercises to be held on its territory, whereas Russia has taken part only once, in a naval exercise.

During the "Sixteen plus One" meeting, the members of the NATO council and the Ukrainian foreign minister conducted "an in-depth analysis of problems relating to the strengthening of European security," and stressed their desire to "develop relations both within the framework of the "Partnership for Peace" program and beyond, the parties said in a joint communiqué.

The NATO secretary-general said in his speech that the alliance and Ukraine would exchange visits of their high officials and hold consultations, including under the "Sixteen plus One" format on the most important issues of mutual interest, which include European security architecture, prevention and settlement of crises, nuclear security, non-proliferation of weapons of mass destruction, and disarmament.

Claes stressed that NATO attached "key importance to its relations with sovereign, independent, and democratic Ukraine and its place in a new European architecture of security and cooperation."

Ukraine made a special note of NATO's contribution to the creation of an atmosphere of confidence in the Euro-Atlantic region with its "respect for territorial integrity, existing borders, and the rights of national minorities."

Udovenko recalled that Ukraine was the first CIS country to sign the "Partnership for Peace" program. "We are open to cooperation with NATO and Europe in building a single and indivisible system of European security," he said. He then stressed that the development of good relations between NATO and Russia was in Ukraine's interests.

Kiev expressed the desire to receive a delegation from NATO to inform Ukraine about the results of NATO's "internal discussion" of procedures and terms of its eastward enlargement.

Asked about Kiev's view on the NATO enlargement plan, Udovenko said Ukraine itself had no plans for joining the alliance in the foreseeable future but it had no right to veto its neighbors' ascension to the North Atlantic alliance. "It is important for Ukrainian national interests to be taken into consideration in the process of expanding NATO. We would not like to become a buffer state between an enlarged NATO and signatories to the Tashkent security treaty," he said.

---

## 9.45

**Agreement with NATO to Set Up Contacts**

From the "Diplomatic Panorama" feature
by Aleksandr Korzun, Igor Porshnev,
Evgeniy Terekhov, and others
Interfax-Ukraine, 14 September 1995
[FBIS Translation]

---

Ukraine and NATO have agreed to step up their contacts, NATO headquarters reports.

An agreement to this effect was reached between Ukrainian Foreign Minister Gennadiy Udovenko and NATO Secretary-General Willy Claes in Brussels.

Udovenko attended the North Atlantic Council meeting on 14 September and presented a draft of the Ukraine-NATO cooperation program.

The NATO headquarters says in an official report that the cooperation will include exchange of top-level visits, in particular a visit to Kiev by the NATO secretary-general.

Consultations will be held continuously between Ukraine and NATO at various levels on the European security prevention of conflicts, peace missions, nuclear safety, non-proliferation of weapons of mass destruction, disarmament, and arms control.

Ukraine will step up its activities in the framework of the Partnership for Peace program and the North Atlantic Council. Ukraine invited a team of NATO experts to visit Kiev to present their findings on the issue of NATO enlargement.

In the discussion held in Brussels on 14 September, NATO expressed support for Ukraine's sovereignty, independence, territorial unity, economic reforms, and democratic changes. It was said that an independent, democratic, and stable Ukraine is a key factor of stability in Europe.

The North Atlantic Council also took note of the significant contribution made by Ukraine to the UN peacekeeping operations, notably in the former Yugoslavia. NATO also welcomes Ukraine's efforts to resolve peacefully all the issues in its relations with its neighbors, the NATO headquarters report says.

To reach the political aims of the NATO Partnership for Peace program, Ukraine is working for openness in defense planning and budget-making, securing democratic control over the armed forces, and expanding military relations with

NATO with the purpose of joint planning and training. Thus reads part of the presentation document of Ukraine's cooperation program with NATO submitted to a special session of the North Atlantic Council in Brussels Thursday.

The Ukrainian delegation is led by Foreign Minister Gennadiy Udovenko.

In keeping with the document, the Partnership for Peace program involves on the Ukrainian side an operational group of generals and officers, two peacekeeping battalions, and an airborne landing battalion, a squadron of Ilyushin-76 transport planes and a squadron of Mi-8 helicopters with crews, two ships, and a motorized civil defense brigade. Ukraine will also offer two test sites for exercises.

Ukraine signed a framework cooperation agreement on 8 February 1994 and handed the presentation document to the NATO leadership on 25 May. A senior official told Interfax-Ukraine that Kiev finds the Partnership for Peace program a timely and promising step that should not draw new dividing lines in Europe. It will allow all interested sides to take practical political and military steps.

"Ukraine also believes that the Partnership for Peace program is a key mechanism for maintaining peace in Europe," the diplomat added.

## 9.46

### Marchuk: Partnership Does Not Mean Joining NATO

From the "Presidential Bulletin" feature compiled by Nikolay Zherebtsov and Andrey Petrovskiy; and edited by Vladimir Shishlin
Interfax, 14 September 1995 [FBIS Translation]

Ukraine's individual partnership program with NATO does not mean that the country is joining the military alliance, Prime Minister Yevhen Marchuk declared during a visit to an international health protection exhibition in Kiev Wednesday (13 September).

He confirmed Ukraine's status as a non-aligned country, saying that the main thing is for Ukraine not to become a buffer between NATO and what is not NATO.

(Interfax Note: The North Atlantic Council held a special session in Brussels Thursday at the ambassadorial level attended by a Ukrainian delegation led by Foreign Minister Gennadiy Udovenko. The individual partnership program was presented before the session. On 8 February 1994 Ukraine signed a framework Partnership for Peace agreement with NATO and on 25 May 1994 submitted the presentation document to the NATO leadership. Immediately thereafter Kiev started working on the individual partnership program.)

On the situation in the former Yugoslavia, NATO air attacks on Bosnian Serbs in particular, Marchuk said that, whatever the reasons, bombing is not the best policy.

He also said that Ukraine must carefully analyze everything happening in the former Yugoslavia, NATO, and Russia before changing its attitude toward NATO.

The prime minister also said that his meeting with his Russian counterpart regarding the division of the Black Sea Fleet will not take place before the end of the month.

According to Marchuk, the Ukrainian side suggested that the experts of the two countries meet to update the draft agreements on the principles of mutual settlements and the city of Sevastopol. "One should not waste time on the discussion of questions undecided even by experts," the prime minister believes.

## 9.47

### Aliev on Readiness to Develop NATO Relations

Interfax, 20 September 1995
[FBIS Translation]

Azeri President Geydar Aliev confirmed Baku's readiness to develop relations with NATO within the framework of the Partnership for Peace program Tuesday, at a meeting with a NATO delegation headed by a representative of the NATO allied command, Italian Brig.-Gen. Angelo Areno.

The president told the delegation that the Azeri presentation document necessary to join the program will be submitted in the near future.

According to Aliev, the Partnership for Peace program aims at "consolidation of global peace."

The president hopes that NATO will render political support to Azerbaijan in finding a solution to its territorial integrity problems. "The ongoing Armenian aggression against Azerbaijan must be denounced," he said.

Brig.-Gen. Areno said that NATO is ready to expand cooperation with Azerbaijan.

According to him, an official NATO representative in Azerbaijan will soon be nominated to coordinate cooperation between NATO and Azeri Ministry of Defense.

## 9.48

### Strategic Policy Toward CIS Published

*Rossiyskaya Gazeta*, 23 September 1995
[FBIS Translation]

*Editor's Note:* (January 1996) The following edict appeared in the Russian media suddenly, with no preparatory debate, announcements of pending doctrinal materials, or allusions as to who the author or group of authors might be. The edict marks a fundamental departure in Russia's policy toward the CIS, in that it makes close integration among CIS members a top-priority "strategic goal." The edict is strongly worded, and has understandably raised the suspicions of other CIS leaders about Russia's intentions. This suspicion may be well-founded, because were a Russian leader to appear who is dedicated to the reinstatement of the Soviet Union, this "strategic plan" would be used as the policy guideline. At the 19 January 1996, CIS summit, President Yeltsin continuously referred to Russia's "strategic plan" for the CIS, indicating that the September edict is being used as the "blueprint" or "working model" for guiding Russian CIS policy. Likewise, the 10 January 1996 appointment of Evgeniy Primikov as foreign minister, and Primakov's subsequent allusions to the primacy of CIS reintegration among his foreign policy goals, points to the possibility that he and the staff of Russia's Foreign Intelligence Service (SUR), where Primakov was appointed director in 1991, may have authored the edict. It is of interest to note that the edict's provision for forming a "government commission" for implementing the strategic policy toward the CIS had not been carried out as of 15 March 1996. The Commission would have pushed the Russian Foreign Minister and the Ministry of Foreign Economic Affairs into the background on CIS affairs. Instead, Evgeniy Primakov, as the new foreign minister, appears to have claimed for himself the key decision-making role on CIS matters, eclipsing even the so-called Ministry for Cooperation with the CIS member states, headed by Valeriy Serov.

[Russian Federation president's edict No. 940, dated Moscow, the Kremlin, 14 September, and signed by Russian Federation President Boris Yeltsin plus appended document on "Strategic Policy of Russia Toward CIS Member States."]

**On Approval of the Strategic Policy of the Russian Federation Toward CIS Member States**

With a view to deepening the integration processes within the CIS and enhancing the coordination of activity by Russian Federation organs of executive power along this avenue, I decree:

1. The appended Strategic Policy of the Russian Federation Toward CIS Member States is approved.

Organs of executive power at all levels are ordered to be rigorously guided by it in all their practical work.

2. A government commission for CIS questions is to be created for the implementation of the Strategic Policy of the Russian Federation Toward CIS Member States.

The Russian Federation Government is to approve a statute on the aforementioned government commission and its composition.

3. This edict comes into force on the day it is signed.

[Signed] Russian Federation President B. Yeltsin
[Dated] Moscow, The Kremlin, 14 September 1995

**Strategic Policy of the Russian Federation Toward CIS Member States**

The development of the CIS is in line with the Russian Federation's vital interests, and relations with CIS states are an important factor for Russia's inclusion in the world's political and economic structures.

***I. Objectives and Main Tasks of the Strategic Policy of the Russian Federation Toward CIS Member States***

1. The priority given to relations with CIS member states in the Russian Federation's policy is primarily determined by the fact that: our main vital interests in the spheres of the economy, defense, security, and protection of the rights of Russians are concentrated on the territory of the CIS, and the safeguarding of these interests constitutes the basis of the country's national security; effective cooperation with CIS states is a factor that counteracts centrifugal tendencies in Russia itself.

2. The main objective of Russia's policy toward the CIS is to create an economically and politically integrated association of states capable of claiming its proper place in the world community.

3. When developing relations with our CIS partners, it is important to be firmly guided by the principle of not inflicting any damage to Russia's interests. The partners' diverse interests must be coordinated on the basis of balanced mutual compromises.

4. The main tasks of Russia's policy toward the CIS states are: to ensure reliable stability in all its aspects: political, military, economic, humanitarian, and legal; to promote the establishment of CIS states as politically and economically stable states pursuing a friendly policy toward Russia; to consolidate Russia as the leading force in the formation of a new system of interstate political and economic relations on the territory of the post-Union space; to boost integration processes within the CIS.

***II. Economic Cooperation***

5. Mutually advantageous economic cooperation should be seen as the fundamental prerequisite for solving the entire package of questions concerning mutual relations with CIS member states. This task must be accomplished by means of

the best possible combination of multilateral relations within the Economic Union framework and of bilateral forms of relations.

6. The development of trade links must proceed from the premise that they constitute an important instrument for stabilization of the economic situation and extrication from the crisis both in Russia and in the CIS member states.

7. One of the most important ways to organizationally strengthen the CIS is the gradual expansion of the Customs Union, which is made up of states linked with Russia by a deeply integrated economy and a strategic political partnership.

Matters must be driven toward the gradual accession to the Customs Union of the remaining members of the Economic Union as conditions for this ripening.

It is necessary to aim for a convergence of national economic systems by elaborating, jointly with our partners, ways to improve standard model acts within the framework of the CIS member-states' Interparliamentary Assembly.

The model for variable-speed integration proposed by Russia within the CIS framework is not mandatory. But our partners' attitude toward this model will be an important factor determining the scale of economic, political, and military support by Russia.

With a view to consolidating and expanding the Customs Union, bilateral agreements should not give the parties to them equal (let alone greater) advantages than those enjoyed by members of the aforementioned Union.

8. The normalization of payments and settlements relations must be perceived as the most important condition for developing and improving cooperation. In this context, it is necessary to accelerate the practical work on forming a Payments Union, with a view to introducing common rules for the organization of currency markets in CIS member states, adopting a currency exchange rate determined in line with supply and demand, attaining mutual convertibility of national currencies, and using the Russian ruble as the reserve currency for the foreseeable future.

9. The formation of a viable integrated association presupposes effective cooperation in the development of production, science, and technology.

The most important avenues for this work must be: to identify common scientific and technological priorities and pursue a joint strategy for the purpose of forming a unified scientific and technological space; to coordinate the processes of restructuring national economies through the elaboration of common interstate investment programs; to ensure conditions for joint work by scientists and production specialists; to pool efforts for solving ecological and environmental protection problems.

10. It is necessary to consistently resolve the task concerning the convergence of legal and economic conditions so as to create joint property on the basis of free entrepreneurial activity and create, in the long term, a common capital market. Comprehensive state support must be given to the creation of financial-industrial groups consisting of enterprises and banks from CIS member states, and to the formation of transitional production, scientific, technical, and other structures.

11. The necessary conditions must be provided for the functioning of the Interstate Economic Committee of the CIS member-states' Economic Union as an international organization pursuing the goal of boosting the effectiveness of cooperation within the CIS framework.

### *III. National Security*

12. Matters must be driven toward the creation of a collective security system on the basis of the Collective Security Treaty [CST] of 15 May 1992 and of bilateral treaties between CIS states. States parties to the CST should be encouraged to unite within a defensive alliance on the basis of shared interests and military-political objectives.

Consistent work must be done to fine-tune the mechanism for implementing agreements already achieved within the CIS member framework in the defense sphere and to preserve military infrastructure facilities on the basis of mutual agreement. There should be a switch to the principle of military basing in the event of reciprocal interest, with clear-cut regulations governing the legal position of Russian military bases and the status of servicemen and members of their families living in these states.

These questions should be resolved in line with the military-political situation prevailing in each CIS member state.

It is necessary to ensure that CIS member states honor their pledges to refrain from participating in alliances and blocs aimed against any of these states.

13. Cooperation between CIS member states must be deepened in the sphere of state border security.

It is necessary to proceed from the premise that the reliable protection of borders along the CIS perimeter, which is their common cause and should be effected through joint efforts, corresponds with Russia's national interests and the common interests of CIS member states.

Work must be completed on settling the package of border questions with CIS member states and creating a treaty-legal basis for the stationing of Russian Federation Border Guard Troops in these countries. There should be a study of the possibility of creating in the future regional command units of CIS member states border guard troops, and efforts should be made to create a uniform system for the protection of their borders.

At the same time, and with due consideration for the foreign political situation, the Russian Federation's state border with all adjacent states should be defined in line with the Russian Federation Law "On the Russian Federation

State Border," while bearing in mind that the retention of the principle of open borders within the CIS corresponds with Russia's long-term plans. Practical work should be guided by differentiated approaches toward individual sectors of the Russian Federation state border, depending on bilateral relations with different states.

14. It is necessary to proceed from the premise that peacekeeping activity aimed at settling and preventing conflicts in CIS member states, primarily by peaceful political and diplomatic means, is an important component of efforts to ensure stability on CIS territory.

Matters must be directed toward ensuring that peacekeeping activity becomes the fruit of collective efforts by CIS member states via their more active participation at all stages of activity to set up a mechanism for effective solution of peacekeeping questions within the CIS framework. All work along this avenue must be done in collaboration with the United Nations and the OSCE, and effective participation by these international organizations must be sought in settling conflicts on CIS territory.

In this process, and when collaborating with third countries and international organizations, it is necessary to seek their agreement that this region is primarily a zone of Russian interests.

Cooperation along the line of security services ought to be developed, primarily with a view to preventing the use of CIS territory by special services from third countries for purposes hostile to Russia. Cooperation with CIS countries' security services should be organized in strict compliance with Russian Federation legislation and respect for Russia's international commitments in the area of human rights.

### *IV. Humanitarian Cooperation and Human Rights*

15. It is necessary to step up Russia's cultural exchanges with states in the "near abroad" and cooperation with them in the areas of science, education, and sports. Russian television and radio broadcasting in the "near abroad" should be guaranteed, the dissemination of Russian press in this region should be supported, and Russia should train national cadres for CIS member states.

Special attention should be placed on restoring Russia's position as the main educational center on the territory of the post-Soviet space, bearing in mind the need to educate the younger generation in CIS member states in a spirit of friendly relations with Russia.

Reciprocal exchange of information and cooperation should be organized between CIS member states' national commissions for UNESCO.

16. It is necessary to aim for real guarantees of equal rights and freedom for all citizens of CIS states regardless of nationality and other differences. The expression, preservation, and development of the ethnic, cultural, linguistic, and religious uniqueness of all peoples inhabiting these countries should be encouraged.

There should be active promotion of Russians' adaptation to the new political and socio-economic realities in the countries, former USSR republics, of which they are permanent residents.

In the event of violation of Russians' rights in CIS countries, the solution of questions concerning financial, economic, military-political, and other cooperation between Russia and the specific state could be used as a possible means of influence depending on that state's real stance in terms of observing the rights and interests of Russians on its territory.

Russians should be helped in exercising production, economic, entrepreneurial, commercial, cultural-enlightenment, and educational activity.

Work should be completed on elaborating within the CIS framework a treaty basis regulating mutual relations between CIS member states in the terms of observing the fundamental rights and freedoms of man and of ethnic minorities, bearing in mind the need to fundamentally ease the acuteness of nationalities problem.

### *V. Coordination of Activity for Solving International Problems and Collaboration with the World Community*

17. Our CIS partners should be persistently and consistently guided toward the elaboration of joint positions on international problems and the coordination of activity in the world arena. Joint efforts should be applied to achieve the further affirmation of the CIS as an influential regional organization and to establish broad and mutually advantageous cooperation with authoritative international forums and organizations at the CIS level.

18. It is necessary to proceed from the premise that the growing efficiency of mutual cooperation must become the most important factor capable of preventing centrifugal tendencies within the CIS.

Efforts should be concentrated mainly on coordinating the CIS member states' positions in the United Nations and the OSCE and their approaches toward relations with NATO, the EU, and the Council of Europe.

### *VI. Basic Tasks of Bilateral Relations*

19. Bilateral relations with CIS member states should be formed with due consideration for the specific features of each of these states and their readiness for different forms of deeper integration. In the elaboration of Russian policy toward CIS member states on a bilateral basis, it is necessary to proceed from the premise that the following are in line with Russia's interests: assistance in stabilizing the political, economic, and social situation in these countries; their far-

reaching and stage-by-stage gradual voluntary involvement in integration processes and in the collective solution of questions concerning military security, border protection, and peacekeeping activity; prevention of confrontational aspects in Russia's relations with CIS member states.

20. The following must become subjects of permanent concern in bilateral economic relations: maintenance of balance-of-payments equilibrium on the basis of long-term agreements; preservation of Russia's leading position in CIS markets, especially with respect to markets for finished products; performance of obligations in production-sharing relations, prevention of debts resulting from commodity deliveries, and normalization of payments and credit relations; favorable conditions for the transit of Russian goods across the territory of these states.

### *VII. On the Domestic Russian Mechanism for Implementing This Strategic Policy Toward the CIS*

21. The Russian Federation government fine-tunes the domestic Russian mechanism for the implementation of strategy toward the CIS, specifies the powers of ministries and departments performing certain functions concerning mutual relations with CIS member states, and systematically consolidates the legal basis of cooperation with these states.

22. The Russian Federation Ministry of Foreign Affairs, exercising state management over the Russian Federation's relations with foreign states, ensures the implementation of common strategy and the coordination of foreign policy with CIS member states.

23. The Russian Federation Ministry for Cooperation with Member States of the CIS is assigned responsibility for implementing the Russian Federation's economic and social policy in the area of mutual relations with CIS member states.

The ministry in question organizes the work and coordinates the activity of CIS executive organs and Russian Federation executive organs on the development of cooperation with CIS member states in the economic and social spheres; organizes the negotiating process and drafting of bilateral and multilateral treaties and agreements with CIS member states on economic and social questions; assists the foreign economic activity of Russian Federation economic components with CIS member states; monitors the performance of agreements reached with CIS member states and the honoring of commitments under Russian Federation treaties and agreements with these states.

24. A government commission for CIS questions is to be set up as an organ ensuring the implementation of Russian Federation state policy as regards cooperation with these states.

Advancement along all these avenues will create a real opportunity in the long term to build a regional international organization capable of ensuring political and socio-economic stability on the territory of the CIS.

## 9.49

## Conceptual Provisions of a Strategy for Countering the Main External Threats to Russian Federation National Security

Anton Viktorovich Surikov, General Director
INOBIS (Institute of Defense Studies), Moscow
October 1995

There are a significant number of external threats to Russian Federation (RF) national security at present that are difficult to classify. At the same time, several of the main threats of this kind are rather obvious.

An analysis shows that above all the United States is the main external force potentially capable of creating a threat to RF military security and to Russia's economic and political interests abroad and of exerting substantial influence on the economic and political situation within Russia and on Russia's mutual relations with the former Soviet republics. As a rule, the United States implements its policy toward Russia in coordination with other Western countries, Israel, and Japan. "Assistance to processes of democratization and of transition to a market economy with the help of the West and in close, equal partnership and cooperation with it" is declared to be the West's official policy with respect to Russia. At the same time, recent experience demonstrates that the West places its own interpretation on all the above terms. In particular, the term "equal partnership" is understood to mean unconditional movement in the direction of U.S. and Western policy in the international arena. And the West's "help" for Russia is extremely limited in nature and determined by the fulfillment of a whole series of preconditions. On the whole, it appears the principal mission of U.S. and Western policy with respect to Russia is to keep it from turning into an economically, politically, and militarily influential force and to transform post-Soviet space into an economic and political appendage and raw materials colony of the West. Because of this, it is the United States and its allies that are the sources of main external threats to Russia's national security, and they should be considered the principal potential enemies of the Russian Federation.

### 1. The Nature of the Main Threats to Russia's National Security Caused by External Factors

*A. The line of the United States and its allies toward intervening in Russia's internal affairs to impose on it paths of development in a direction favorable to the West represents the greatest danger.* The comprador model of building the economy suggested to Russia by the International Monetary Fund (IMF) and the World Bank consists of orienting Russia toward exporting raw materials and importing everything else, encouraging default on the debt to Russia by CIS countries, as well as encouraging outflows of capital from Russia to the West and, at the same time, stifling national industry, science, and agriculture. Attempts at destroying the high-tech potential of national industry and, above all, of our military-industrial complex by not admitting Russia to world markets for arms and for space, missile, aviation, and nuclear technologies and nuclear materials are most obvious at present. Protectionist measures against Russian exports of fuel for nuclear power stations, opposition to the Russian-Iranian nuclear deal, and the hysteria over cryogenic engine deliveries from Russia to India are examples of such attempts.

On the whole, the economic model being realized threatens to degrade the country's economic potential and eliminate the unified domestic market, which can in turn become a basis for regional separatism and for raw materials regions and maritime regions to fall away from Moscow, thereby reducing the country to fourteenth- and fifteenth-century borders. It should be noted that, in following that line, the West finds support among part of the Russian elite and relies on Russian comprador business, which has become especially entrenched in speculative-finance banking structure and export-oriented raw materials sectors of the economy. The West is least interested in the growth of internal accumulations of any kind whatsoever in Russia, otherwise such accumulations could be used for modernizing the national processing industry and agriculture, for conversion and rescue of high technologies of the military-industrial complex, and for maintaining the armed forces' combat power and solving their social problems. A so-called strategy of black holes, through which material and financial resources are being pumped out of the country, is being implemented to prevent this. And although Dudaev's Chechnya has been the best-known "black hole," it is far from the only one. The comprador-oriented export policy being effected within the framework of the described "black hole strategy" in the oil and natural gas industry, non-ferrous and in part ferrous metallurgy, the timber industry and production of mineral fertilizers is leading to colossal plundering of the country's national wealth in favor of a narrow group of people—the so-called new Russians.

The outflow of capital from Russia, with the TEK [fuel-energy complex] accounting for the lion's share, is estimated to be from one billion to two billion dollars monthly. The bulk of these funds settle in foreign bank accounts of "new Russians" or are invested abroad in real estate, stocks, and bonds. But even the export receipts from that return to Russia often go for comprador imports and for importing expensive consumer goods into the country for the "new Russians." Or they go, via the budget, to commercial and above all "authorized" banks for "investment." Or in the final account they are directed toward realization of various expensive projects of dubious importance, above all in the construction area. It should be recalled here, for example, that the Mafia in Sicily traditionally makes its main money specifically on construction contracts by inflating by many times the estimate for work done.

It is also important to note that all major operations of pumping resources and funds out of the country are being carried out with the involvement of foreign partners. In terms of petroleum exports alone there are approximately twenty various joint ventures operating in Russia today. And according to up-to-date information from Russian special services, a considerable number of the foreign associates of such organizations are persons connected with intelligence agencies of Western countries. It follows from the Federal Security Service press release "On Federal Security Service Activity in 1994" that "around ten identified agents and around ninety specialists and advisers whose affiliation with foreign special agencies is undoubted have been exposed just within the system of a number of major RF economic departments. Over a hundred foreign firms and organizations, including those in banking, are being used as cover by special services." Special services of Western countries have full access today to all documentation of joint ventures and other partners of Russian exporters. They have original financial documents, they are knowledgeable about the movement of commodity resources and financial flows, they have information on bank account numbers of the "new Russians" and on their real estate and securities transactions abroad. And it should be understood that the way the outflow of resources and capital from Russia is taking place today has become criminalized to the highest degree and represents not only a violation of domestic laws but also the grossest violation of laws of the Western countries themselves. Consequently, foreign intelligence agencies have in their hands compromising criminal information on many Russian parties to foreign economic activity and on the politicians and state officials connected with them. As a result, these representatives of the Russian business and political elite are not completely independent in their actions and are extraordinarily vulnerable to the pressure of outside forces who possess compromising information. It follows from what has been said that they are by definition incapable of following a consistent policy conforming to Russia's national interests. The most striking example of that situation lately appears to be Russian government policy on the question of Caspian oil. Another notable example is the "peacemaking process" in Chechnya.

*B. Turkey's political penetrations and U.S., U.K., and FRG economic and intelligence penetration into Azerbaijan is in full swing.* This is the most suitable base of operations for subsequent Turkish and Western expansion in the direction of Central Asia, the Volga region, and the North Caucasus using the "Turkic" and "Islamic" factors. And Turkey is acting here as an instrument by which U.S. policy is being pursued.

The principal goal of such a policy is to establish Western control over energy resources and, above all, petroleum resources of the Caspian Sea. Caspian oil reserves are commensurate with the oil reserves of the North Sea and Alaska when their exploitation began, but oil resources of the latter two will be exhausted in the foreseeable future. Therefore to avoid the Persian Gulf region's total monopoly of oil exports, the West is very interested in Caspian oil. It should be noted that the problems of Caspian Sea oil arose immediately with the USSR's disintegration. It was then that the regimes of Caspian states cast doubt on the 1940 Soviet-Iranian treaty on the division of the Caspian and began its repartition without prior arrangement. And this is being done illegally—the Caspian is not a sea, but a lake. Since that is so, the rules of international maritime law do not extend to it. Any decisions relating to the use of Caspian resources must be made based on a consensus of all Caspian countries of the CIS. And the 1940 Soviet-Iranian treaty, ratified at one time by the USSR Supreme Soviet, can be revoked only by parliamentary vote, not by a decision of particular RF government representatives. It follows from this that the so-called contract of the century concluded by the Azerbaijani government with an international consortium of eleven oil companies is illegal from the start, but the West openly ignores this fact.

With the actual assent of a number of highly placed Russian government officials and businessmen, the question of the rightful ownership of Caspian shelf resources is being replaced by a discussion of pipeline routes over which it is proposed to pump "early" oil.

Another aspect of the West's Caspian policy is the attempt to cut Russia off from the Transcaucasus by encouraging separatism in the North Caucasus, above all in Chechnya. In particular, there are projects for establishing an anti-Russian "Confederate of Mountain Peoples" made up of Dagestan, Chechnya, Ingushetia, Kabardino-Balkaria, Karachay-Cherkessia, and Adygea. It is presumed that this entity will gain direct egress to the Black Sea and to Turkey through the territory of Abkhazia. Plans for assisting separatism in Tatarstan and Bashkiria for the purpose of actually cutting off from central Russia regions of the Urals, Siberia, and the Far East rich in energy resources are being considered in the longer term.

*C. Western policy with respect to NATO's future is seen as an attempt to isolate Russian and ultimately oust it from Europe.* An eastward expansion of the NATO bloc obviously is inevitable and is planned in several stages. In the first stage (over two to three years), Poland, the Czech Republic, Slovakia, and Hungary will be accepted as NATO members. In the second stage (tentatively by the year 2000), it is planned to accept Slovenia, Romania, Bulgaria, and, if possible, Lithuania, Latvia, and Estonia in NATO. The inclusion of Finland and Austria in NATO is likely at this same stage. Finally, the acceptance of Ukraine in NATO is possible at the third stage (approximately in 2005). But Russian will not be accepted in NATO under any circumstances.

West Germany is the main instigator of NATO's eastward expansion (the decision on expanding NATO was ultimately made after the withdrawal of the Western Group of Forces from the former East Germany had been completed). In fact, we are dealing with a resumption of German expansion in the eastern and southeastern directions twice interrupted in this century and being accomplished this time primarily by political and economic methods under the American "nuclear umbrella." The United States is another instigator of NATO's eastward expansion. In the opinion of a number of influential representatives of the American elite, such a line will permit reinforcing the U.S. leading position on the European continent and thereby compensate somewhat for America's obvious economic weakness compared with the European Union headed by the West Germany.

Many Western politicians now express assurances that they do not plan a NATO expansion by means of the Baltics and Ukraine or the stationing of Western troops and nuclear weapons in Eastern Europe, but there are no grounds for believing this. Just two years ago Russia was assured that they did not plan to expand NATO at all, even by means of Poland, Hungary, and the Czech Republic. Moreover, there generally was uncertainty in NATO during 1991–92 with respect to the future of this military-political alliance. Today, however, proposals for stationing tactical nuclear weapons in the Czech Republic and Poland are being discussed, operational plans for the movement of NATO mobile forces to the Baltics in case of its conflict with Russia are being drawn up, and the idea of establishing a 60,000-member "Baltic Corps" from troops of Poland, Denmark, and West Germany is being discussed.

*D. The line toward Russia's unilateral disarmament, which threatens strategic stability in the world, also should be examined in the very same way.* With respect to strategic nuclear weapons, this line is being fulfilled today in two main directions. First, in the absence of financing, a rapid degradation is occurring in strategic systems presently in the inventory, and much Research Development in Technology and Engineering (RDT&E) in this area has been slowed or entirely halted. Second, international agreements clearly unfavorable to Russia are being imposed on it: the START II treaty and proposed changes to the 1972 Anti-Ballistic Missile treaty.

With respect to the START II treaty, there are two groups of arguments, against its ratification, military-strategic and political. In speaking of the first group of arguments, one should single out the problem in inequality in so-called rapid uploading potentials of the U.S. and Russian strategic forces (expected ratio 5:1 in favor of the United States) and U.S. attempts to change the regime of the ABM treaty under the pretext of a need to create a "tactical ABM defense system." It should be emphasized that, in practice, if the U.S. delegation's proposals at the Geneva talks are accepted, they legalize the U.S. right to create a strategic system for defending its own territory against ABMs. And the Republican majority in the U.S. Congress is stepping up pressure on the U.S. administration to persuade it to take such a step even without coordinating with Russia and in spite of international restrictions in force.

In considering the second group of arguments, one should direct attention to the fact that, as of the moment, the START II treaty was signed in January 1993, there were illusions in Russia regarding the possibility of friendship and partnership between it and the United States. Because of this, skews in the treaty favoring the United States did not seem so important, but today START II treaty shortcomings appear quite differently under conditions of the approaching "cold peace" caused by the NATO bloc's planned eastward expansion. At the same time, the practice being followed of implementing treaty provisions without prior arrangement, without its ratification, and on a unilateral basis may lead to irreversible consequences in the very near future.

Above all, observance of the principle of quantitative equality with the United States in strategic arms may become practically unattainable. The importance of observing this principle is explained by the fact that the majority of Western politicians are not military specialists and are capable of grasping only the simplest quantitative parameters characterizing the ratio of both sides' nuclear forces. Under these conditions, the widening gap between Russian and U.S. strategic nuclear forces, in terms of the number of operational nuclear warheads, obviously will be perceived in the West as grounds for regarding Russia as a second-rate nuclear power, which the United States, the only remaining nuclear superpower, will be able to subject to nuclear blackmail for the purpose of dictating its will. (The expected result of START II implementation in a version curtailed for financial reasons is no more than 500–600 Topol-M missiles in the Strategic Missile Troops by 2003–5, and new nuclear-powered missile-armed submarines have not been built at all since 1990.)

The situation also is largely similar with respect to the Conventional Forces in Europe (CFE) treaty. Russia's partners in the CFE treaty refused to accommodate Russia until recently on the question of so-called flank quotas; that is, to agree that Russia can have as many heavy weapons in the inventory in North Caucasus and Leningrad [St. Petersburg] military districts as necessary for the country's defense, and not as many as specified by the treaty concluded in November 1990 under quite different military-political conditions, when the Soviet Union and Warsaw Treaty Organization still existed. Only when it became clear that Russia would unilaterally refuse to fulfill this part of the CFE treaty if that line was continued did NATO begin to show readiness to take into account Russia's interests. But in exchange they demand that Russia remove objections to the NATO bloc's eastward expansion.

*E. Western attempts to counteract integration trends operating within the CIS framework are obvious.* This is manifested most openly with respect to Belorussia [Belarus], which is more ready than the other former Soviet republics to undertake close integration with Russia. On the whole, however, this opposition as well as NATO's eastward expansion, the activeness of Turkey and Western oil companies in the Caucasus and the Caspian, attempts to coerce Russia into unilateral disarmament, barring it from world markets of high-tech products, and, finally, the economic model being imposed on Russia by the IMF and World Bank all are links in the same chain: creeping expansion of the United States and its allies, with an ultimate goal of eliminating Russia as a state and turning its territory into a raw materials colony of Western countries.

**2. Strategy for Neutralization of External Threats and for National Survival of the Russian Federation**

*A. A most rapid, fundamental change in the country's economic course appears to be fundamentally important to the Russian state's survival.* The general outlines of such a change are presented in detail in programs of a number of political parties and blocs that plan to take part in the 17 December 1995 parliamentary elections. They include, in particular, rejecting cooperation with the IMF and World Bank; reversing the privatization of state property; imposing elementary order in foreign trade, in the banking system, and in exporting sectors of the economy (even within limits of existing legislation, which will permit reducing the outflow of capital and thereby increasing state investments in converting the military-industrial complex and in modernizing and restructuring national industry); increasing import tariffs for fifteen to twenty-five years; that is, until national industry and the national agrocomplex can withstand the competition of imported goods relatively painlessly; extraordinary measures in fighting organized crime and corruption, and expropriation of money and property acquired through criminal means; and economic integration within the CIS framework.

*B. Preventing illegal exploitation of Caspian Sea shelf resources by Western oil companies is a vitally important goal for Russia, and the problem of deterring the "Turkic" and "Islamic" factors should be considered above all in this light.* The main task here in the short term is a rapid end to the war in Chechnya by imposing constitutional order throughout its territory, providing for the elimination of Dudaev's armed units, and disarming the population. It is obvious that this is possible only by force; therefore, it is advisable to stop the so-called peacemaking process and renewing operations of federal forces to disarm and eliminate illegal armed units.

Another very important task is to prevent fulfillment of the "Caspian oil contract" in its present form. In this case it is advisable to carry out measures such as officially refusing to recognize that part of the Caspian stipulated in the contract as a zone of Azerbaijan's jurisdiction; taking practical steps, including also forceful steps, should it be necessary, to stop any oil production activity of foreign companies in the former Soviet part of the Caspian until its legal status is determined; preventing Turkey's territorial tie with the main part of Azerbaijani territory; and exerting pressure on the regime in Baku; for example, by creating threats of a fragmentation of Azerbaijan and of an Armenian military offensive on Gyanzha [sic] and Yevlakh.

*C. Opposition by force to the NATO bloc's eastward expansion seems an extremely urgent task.* At the same time, in the case of Poland and other East European countries, it is obvious that Russia has no real opportunities to hinder this by way of force, and threats not backed up by corresponding actions only discredit a state. An example of such discrediting was Russia's reaction to the NATO military operation in the Balkans in September 1995. But creating a military bloc of CIS countries, particularly the involvement of Central Asian countries in confronting the NATO bloc, is also obviously needed. With respect to Ukraine, it obviously will refuse to participate in such a military alliance in the foreseeable future.

The situation with respect to Belorussia [Belarus] is different. Close military cooperation on a bilateral basis should be developed here, and a key element of such cooperation should be the deployment of tactical nuclear weapons on the territory of Belorussia, in Kaliningrad, and on naval vessels of the Baltic fleet. Such a step is needed because, for economic reasons, Russia cannot permit itself to have as large an army today as the USSR did ten years ago. NATO now surpasses Russia by two to three times in the number of troops and conventional weapons in Europe. This gap will grow even more after Poland, Hungary, and the former Czechoslovakia join the alliance. Under such conditions the only possible and economically realizable way to deter NATO is by relying on tactical nuclear weapons capable of overcoming enemy superiority in conventional weapons, that is, to adopt the strategy to which the NATO bloc itself adhered during the "Cold War." And it is a question not only of the Western Theater of Military Operations (TVD), including the former Soviet-Polish border and Baltic Sea, but also of the Northern Theater of Military Operations, encompassing Russia's border with Norway and the Barents Sea, and the Southern Theater of Military Operations, encompassing the Black Sea and bases of Russian troops in Crimea, Abkhazia, Georgia, and Armenia—tactical nuclear weapons must become the basis of Russia's defense in all three European theaters of military operations.

The situation is completely different with respect to the Baltic as compared with Eastern Europe. In general, a neutral status of the Baltic republics similar to that of Finland during the "Cold War" probably would meet Russian interests, but if NATO accepts the Baltic republics as members, RF armed forces should be sent to the territory of Lithuania, Latvia, and Estonia.

It should be noted that Russia has the legal and moral grounds for introducing troops to the Baltics. First, an extension of the NATO military infrastructure to this region would present extreme danger to RF national security interests. During the "Caribbean crisis" [the Cuban missile crisis], when the USSR began deploying nuclear weapons on Cuban territory at Cuba's request, a similar situation from the U.S. standpoint provoked a naval blockade of the island by the United States accompanied by direct threats of military invasion and led to the most acute crisis of the "Cold War." Inclusion of the Baltic states in NATO would present no less a threat from Russia's standpoint than did the deployment of Soviet nuclear weapons in Cuba at one time from the U.S. standpoint.

Second, there are illegal, anti-democratic regimes in Estonia and Latvia similar to those that previously existed in South Africa and Southern Rhodesia [Zimbabwe]. In these biethnic republics, one ethnic group arbitrarily deprived people of the other nationality of their civil rights and usurped all power. Under these conditions, the part of the population being discriminated against (so-called non-citizens) are entitled to create their own parallel structures of authority and power. If force is used against them, they have the right to turn to Russia for armed support. Lithuania, does not recognize the "Molotov-Ribbentrop Pact." Consequently, Russia and Belorussia have the right to take back Klaipeda and Vilna kray.

Third, the Baltics is a criminal zone living chiefly on smuggling and controlled by mafia structures. Considering the precedent of the U.S. invasion of Panama and the arrest of General Noriega, Russia also has the right to arrest and indict a large number of Baltic figures in Russian courts. It is obvious that a Russian return to the Baltics must be accompanied by the deportation to the West of persons who sullied themselves by complicity in discriminating against people of a different nationality and who do not wish to live in republics where the scope of civil rights does not depend on nationality.

With respect to presumed Western reaction to the probable introduction of Russian armed forces into the Baltics, an analysis shows that no one in the West plans to fight Russia over the Baltics. Economic sanctions are possible, but they most likely will not be in the nature of a total embargo. Above all this concerns the export of Russian energy resources. In particular, it is expected that in the foreseeable future Europe will experience a natural gas consumption deficit of 100 billion cubic meters per year. At the same time, Russian natural gas reserves make up one-third of world reserves. The experience of the conflict over the "gas/pipes" deal in the 1980s persuades us that West Germany, France, Italy, Finland, Greece, and Eastern Europe will continue to buy raw material resources from Russia as before, which will provide funds for conversion of the domestic military-industrial complex and for the country's reindustrialization.

Finally, in case of a total break in relations with the United States, Russia has such convincing arguments for it as the nuclear-missile potential and the threat of proliferation of weapons of mass destruction around the world, which with skillful tactics can play the role of a kind of trading card. And if Russia is persistently driven into a corner, then it will be possible to undertake to sell military nuclear and missile technologies to such countries as Iran and Iraq, and to Algeria after Islamic forces arrive in power there. Moreover, Russia's direct military alliance with some of the countries mentioned should not be excluded, above all with Iran, within the framework of which a contingent of Russian troops and tactical nuclear weapons could be stationed on the shores of the Persian Gulf and the Strait of Hormuz.

*D. Concerning the question of the attitude toward strategic nuclear forces, it should be noted that the Russian nuclear potential is one of the few arguments convincing to the West that Russia inherited from the USSR and that is not yet fully destroyed.* There needs to be a most rapid formation of a program for developing the strategic nuclear forces based on the fact that they must develop within the framework of the START I treaty over the next fifteen years. The RF Ministry of Defense must develop such a program in a short time, and parliament must provide financing for its realization. Funds for these purposes could be found, for example, if financing is terminated for recovery work in the Chechen Republic and a large number of other programs, the need for which is not obvious.

An analysis shows that if the strategic nuclear forces develop within the framework of quantitative limits of the START I treaty, then this is a technically and economically fully realizable option, even considering Russia's loss of production capacities of the former USSR Ministry of General Machine-Building on Ukrainian territory. And in the first stage the warranty operating life of part of the multiple independently targetable reentry vehicle (MIRV) ICBMs in the inventory today—UR-100N, R-36 M UTTKh, [possible translation—"Upgraded specifications and performance characteristics"], R-36 M2, AND RT-23 UTTKh—obviously should be extended to twenty years. In the second stage (by the beginning of 2009), ballistic missiles equipped with six medium-power warheads and now being developed within the scope of RDT&E for creating the D-31 naval missile complex obviously should be deployed (400 to 500 of them) in silo launchers of the above MIRVed ICBMs as well as in certain UR-100K ICBM silo launchers. After 2008 (when the effect of the START I treaty expires) it is advisable to begin deploying approximately another one hundred such missiles mounted on railroad flatcars.

It appears extremely important to offer opposition to U.S. plans for creating a "tactical anti-ballistic missile defense system" and in this connection changing the terms of the 1972 ABM treaty. These plans essentially are another attempt at dragging the Strategic Denfense Initiative idea in through the back door, and they present a significant threat to strategic stability in the world and provoke China and other "small nuclear countries" to build up in their nuclear missile forces. In China's case, for example, its nuclear forces, which are heavily inferior to the strategic forces of Russia and the United States, can be completely depreciated by the deployment even of a very limited U.S. ABM defense system. In view of this, a sharp quantitative increase in Chinese nuclear-missile forces, above all in the MIRVed ICBM grouping, should be expected if a U.S. ABM defense system is deployed. This in turn obviously would provoke India, which in that case will follow China. Pakistan would then also undoubtedly will join in the nuclear race.

Russia must not consent to any changes in the text of the treaty that would contradict the part of it that prohibits giving tactical ABM defense systems characteristics permitting their use for strategic ABM defense purposes. Arguments according to which Russia and the United States should cooperate in the area of creating a "tactical ABM defense system" in view of the fact that they allegedly have common enemies sound altogether unconvincing. It is obvious that countries such as Iran, Iraq, and North Korea are not Russia's enemies. Second, any kind of cooperation between Russia and the United States will hardly be possible under conditions of the approaching "cold peace" connected with the upcoming NATO bloc expansion. Finally, by virtue of a policy of "double standards" being followed by the United

States relative to the Israeli nuclear program, which is aimed against Russia among others, any U.S. argument regarding the question of non-proliferation of nuclear weapons should be viewed with suspicion. In view of the power of the pro-Israel lobby in the United States, one should not expect any serious steps by the United States to force Israel to give up its nuclear potential.

On the whole, we should take into account the fact that, as an analysis shows, the regime of the Nuclear Non-Proliferation Treaty (NPT) most likely cannot be preserved over the long term and the number of nuclear powers will grow steadily. Israel already has approximately two hundred nuclear devices in the inventory. The range of Israel's nuclear weapon delivery vehicles is up to 2,500 km (i.e., Moscow is within reach of Israel's nuclear forces). It is obvious that Israel will not give up its nuclear potential and accede to the NPT under any circumstances. It should be understood that Israel's nuclear potential was created not just for deterring a non-nuclear attack by Arab countries, but also for blackmailing the USSR to compel it to exert a deterring influence on the Arabs in case of their conflict with Israel and the latter's military failures.

A final decision on creating nuclear-missile forces of small size to deter China and Pakistan was also made in India not long ago by the country's political leadership. And this decision is the product of national consent of all of India's political forces, and in all likelihood no arguments by the world community about the inadvisability of turning India into a nuclear power will be accepted by India's leadership and no threats of sanctions against India will influence its resolve to create its own nuclear-missile forces. At the same time, it is obvious that equipping the Indian armed forces with nuclear weapons will deprive the world community of arguments against Pakistan turning into a nuclear power, already the eighth in count (after Russia, the United States, U.K., France, China, Israel, and India). Along with this, Algeria, in which there is a great probability of Islamic forces coming to power, also has everything necessary for creating this kind of potential in a short time if its leadership makes the corresponding political decision.

Under the conditions now forming, Russia has two possible options for behavior. The first option presumes continuation of the present RF Ministry of Foreign Affairs line toward cooperating with the United States in order to pressure potential Third World possessors of nuclear weapons to give up realization of their nuclear programs. It appears that such a line will suffer total failure in the short term. The second option presumes Russia's refusal to follow the U.S. line in the question of non-proliferation of nuclear weapons and missile technologies and an unfolding of its cooperation in the nuclear-missile area above all with India. India is one of the few countries whose national interests do not run counter to Russia's. Russian and Indian cooperation in the nuclear-missile area modeled after U.S.-U.K. relations, where the United States supplies the U.K. with delivery vehicles—Polaris and Trident missiles—on which the U.K. installs its own nuclear warheads, is seen as most rational. And it is important to remember that India is also capable of creating its own nuclear-missile forces independently and in the extreme case undoubtedly will do this. But it would be extremely advantageous for it to cooperate with Russia in this matter. The economic advantage to Russia in the event of such cooperation also is obvious.

In the more distant future Russia could also develop cooperation in this area with Iran and a number of the Arab countries. It appears that such cooperation not only would bring Russia appreciable commercial advantages and political influence in Southwest Africa and North Africa, but also could deter Russia's Third World partners with respect to the content and direction of their work in this area.

*E. The course toward integration within the CIS framework, above all with Ukraine, Belarus, and Kazakhstan, must become a very important direction of Russian policy.* With respect to Belorussia and in part to Kazakhstan, it can be said that these republics probably will welcome integration with Russia. The situation appears much more uncertain with respect to Ukraine. At the same time, an analysis shows that, judging from everything, the results of Kuchma's "reforms" will be even worse than those of Gaydar's "reforms" in Russia. In contrast to Russia, Ukraine has no oil or natural gas, and the West will not be able to put Ukraine on full support. This fundamentally distinguishes the prospect of Ukraine's development from that of the Baltics, for example. Because of the small size of its population, the latter can be subsidized by the West in the extreme case within the range of $3–5 billion per year—this is little for Ukraine but enough for the Baltics. On the whole, it should be expected that in three to five years Ukraine's economy will approach a final collapse and the republic quite probably will go to pieces. Under these conditions its eastern and southern parts obviously will express a desire to reunite voluntarily with Russia. Realizing this, the West and western Ukraine's nationalist forces may try to provoke a conflict between Russia and Ukraine. Crimea might be the cause, and the goal would be to start the two peoples quarreling and sow hatred between them, as the West succeeded in doing in the former Yugoslavia with respect to the Serbs, Muslims, and Croats, and thereby make any future reunion of Russian and Ukraine impossible.

In this connection it should be emphasized that the West's goal is to provoke a sharp deterioration of Russia's relations not just with Ukraine but also with such countries as China and Iran, and to create powerful, constant pressure on the periphery of post-Soviet space in the Tajik-Afghan and Asia Minor zones. It appears that, on the one hand, Russia should be decisive in following its line concerning NATO expansion, the Baltics, Chechnya, the Caspian shelf, and the situation on the Tajik-Afghan border. But, on the other hand, it should react with extreme caution to provocations of west Ukrainian nationalist forces in Crimea and in eastern and southern Ukraine, especially as Crimea's present political elite is comprador-oriented and totally corrupt.

On the whole, it appears that if a judicious policy is followed, there are grounds for counting on restoration of a union in five to ten years made up of Russia, Belorussia, Kazakhstan, the majority of Ukraine, as well as the Dniester region, Abkhazia, and South Ossetia. And Russia's relations with the Transcaucasus and Central Asia could develop according to the model of relations that existed earlier within the framework of the Council on Mutual Economic Assistance, and with Moldavia, the Baltics, and western Ukraine according to the model of Soviet-Finnish relations between 1944 and 1991.

This document may be used in developing a new Russian Federation military doctrine.

Document prepared by
Advisor, Institute of Defense Studies,
Candidate of Technical Sciences, Anton Surikov

## 9.50

### New Defense Doctrine Would Stem NATO Expansion

Aleksandr Lyasko
*Komsomolskaya Pravda,* 29 September–6 October 1995 [FBIS Translation]

[Report by Aleksandr Lyasko: "Doctrine Is New, But Looks Old."]

I learned from reliable sources that some time ago the General Staff completed its formulation of a version of the basic principles of Russia's new defense doctrine.

The first element is a demand for amendments to the Conventional Forces in Europe (CFE) treaty on the so-called flank quotas. The changes boil down to the idea that Russia should be allowed to have as many heavy arms as it considers necessary to its defense in the arsenal of the North Caucasus and Leningrad [St. Petersburg] military districts. Hitherto the arms ceilings have been laid down by the 1990 treaty, which the participating countries signed back in the time of the USSR and the Warsaw Pact. With the collapse of both structures Russia is refusing to accommodate the Turkish and Norwegian restriction demands.

Another detail of the new defense strategy is Russia's stance with regard to NATO's eastward expansion. The Defense Ministry believes that the president was slightly overdoing it when he spoke of a CIS military alliance in response to the East European states' admission to NATO. The president's idea that Uzbekistan and Kyrgyzstan will help us in a confrontation with NATO cannot be attributed to anything but emotion, especially since there is no point in counting on Ukraine's involvement in a new military alliance. Belarus is another matter. People on the Arbat [square] (i.e., the Defense Ministry) see the deployment of tactical nuclear weapons on Belarusian territory (as well as in Kaliningrad oblast and on Baltic fleet ships) as a key element in military cooperation with that republic.

According to our interlocutor at the General Staff, this is the only means that we have at our disposal for curbing NATO expansion. The country just does not have the money to maintain the army that we had ten years ago. That is why national strategists consider it necessary to eliminate the enemy's edge everywhere: the western theater of operations (the Polish border and Baltic waters) along with the northern theater (the border with Norway and the Barents Sea) and the southern theater (the Black Sea and Russia's bases in Crimea, Abkhazia, Georgia, and Armenia). Tactical nuclear weapons will be deployed everywhere.

The military department's next sensational idea involves dramatic action in connection with NATO's expansion. As regards Poland and the other countries of Eastern Europe, Russia is currently unable to stop this process by force. However, the plan presupposes that if NATO agrees to admit the Baltic republics, Russian Federation armed forces will immediately be moved into Estonia, Latvia, and Lithuania. Any attempt by NATO to stop this will be viewed by Russia as the prelude to a nuclear world catastrophe.

The draft new military doctrine casts doubt on the international system for monitoring the nonproliferation of weapons of mass destruction. Russia will decide the question of sales of military missile technologies to India, Iraq, Iran, and Algeria under certain conditions. A direct military alliance with Iran is not ruled out.

According to the high-ranking General Staff officer, the preliminary outlines of the defense doctrine formulated by General Grachev's department have been cautiously approved by the minister himself and his first deputy and constitute the military's response to the lack of any consistent policy by the Foreign Ministry and presidential structures on questions of military security. According to some of Grachev's statements following his talks with Yeltsin in Sochi, the army is ready to begin erecting a nuclear shield over the besieged fortress, which is how it sees Russia.

It is of course quite possible that the actual draft or, to be more precise, its publication is just a means of putting pressure on the West: Do not toy with Russia or it will return to the pre-perestroika positions, start to bristle, and instead of a civilized partner will be forced to become a bogeyman as before for Western democracy, a friend of dark Islamist forces, and a spawning ground for Zhirinovskiy and others. Regrettably, this is no empty threat, given the fact that the authors of the draft by no means lack allies in the Duma and within the Kremlin Walls.

## 9.51

### "Provisions" May Underlay New Military Doctrine

Aleksandr Lyasko

*Komsomolskaya Pravda,* 27 October–2 November 1995 [FBIS Translation]

[Article by Aleksandr Lyasko under the "*Komsomolskaya Pravda* Investigation" rubric: "Will Grachev Invade the Baltics? A New Russian Military Doctrine Is Being Drawn Up at a Secret Defense Institute."]

Three weeks ago *Komsomolskaya Pravda* reported, citing a senior officer of the General Staff, that the formulation of a new Russian military doctrine had begun at the Defense Ministry. Later we were able to establish that the trail of the document in question, which is being eagerly cited by military men, leads back to a man by the name of Anton Surikov, military expert at a rather closed institution called the "Independent Institute of Defense Studies" (INOBIS).

What is INOBIS? Some 30 km northeast of Moscow, if you follow the Yaroslavl highway, you will come to the city of Kaliningrad. Hidden deep within one of the city's pleasant parks is an eight-story building in pale pink brick with no signboard. If you try to get into the building the security guard will examine your documents meticulously and summon his superior. Then the superior will call the security chief, and together they will ask you to explain in detail the purpose of your visit. If they are satisfied with your answers, they will assign you to an escort in a severe suit that barely meets across the chest—also with no identification marks.

On the eighth floor, in an office under a portrait of Lenin, you will most likely be greeted by Mr. Anton Surikov, who introduces himself as an "independent scientist." He works surrounded by powerful computers, half a dozen telephones, and a poster giving instructions on how to handle a U.S. M16 rifle. Don't be surprised if your conversation is interrupted from time to time by telephone calls from Washington, London, or Delhi. The staffer will answer the callers in good English, but with an accent betraying the training of a Soviet military instructor of the mid-1980s.

It was thanks to the assistance of this inquisitive man of science that we were able to obtain the document with the ominous title "Conceptual Provisions of a Strategy for Countering the Main External Threats to Russian Federation National Security." This work was prepared "by way of initial data for the pursuit of military-technical studies."

Let us note particularly that INOBIS is a complex organization originally founded by respected private individuals. They include: Nikolay Alekseevich Sham, former chief of the USSR KGB Main Economic Counterintelligence Directorate; Vitaliy Khusseynovich Doguzhiev, USSR deputy prime minister and minister of general machine building (that was the Soviet euphemism for the missile and space industry); Yuriy Dmitrievich Maslyukov, formerly chairman of the USSR Gosplan (State Planning Committee).

Every year, 1995 included, INOBIS carries out a number of studies financed by the Defense Ministry on research regarding strategic nuclear weapons, the military use of space, and anti-ballistic missile defense. To cap it all, certain fragments of the "Conceptual Provisions" bear a suspicious resemblance to remarks made by Defense Minister Grachev after his meeting with the president in Sochi. These interesting similarities and coincidences prompt the obvious thought that the text of the "Provisions" contains what Grachev cannot say by virtue of his office, but which the Defense Ministry would like to make known urgently to probable allies and, equally, adversaries.

Above all, the document confirms what we have said before: It examines the question of a close military alliance with Belarus involving the deployment of "tactical nuclear weapons on the territory of Belarus, in Kaliningrad, and on the naval vessels of the Baltic fleet." It also proposes the siting of nuclear weapons and delivery vehicles in both the Northern (the border of the Kola Peninsula and Barents Sea) and Southern (the Black Sea and Russian Federation bases in Crimea, Abkhazia, Georgia, and Armenia) theaters.

The theme of the admission of the Baltic republics to NATO is also developed in an interesting way in the "Provisions." It is recognized that Russia's interests would be met by a neutral status for the Baltic republics—along the lines of Finland during the "Cold War." However, the document says, "if NATO admits the Baltic republics as members, Russian Federation armed forces should be sent to the territory of Lithuania, Latvia, and Estonia."

The "moral and legal justification" whereby Russia is to be absolved of its sins in connection with a possible occupation of the Baltics is particularly curious. First, the "Provisions" relate from Russia's viewpoint, the inclusion of the Baltic republics in NATO would constitute no less a threat than, from the U.S. viewpoint, did "the deployment of Soviet nuclear weapons in Cuba." Second, the defense researchers believe that one ethnic group (in Estonia and Latvia) has deprived people of another nationality of their rights and has usurped power. Therefore the noncitizens who are discriminated against are "entitled to create their own parallel structures of authority and power" and, in response to the use of coercion against them, to appeal to Russia for armed support. The Russian population of the Baltics is thus assigned the unenviable role of hostages whom the Russian armed forces may save by once again depriving the Baltic republics of their independence.

The authors of the "Provisions" believe that the West is not likely to go to war with Russia over the Baltics. And there will not be a total blockade, if only because they value our gas exports and the possibility of thereby covering 100 billion cubic meters of the gas shortfall in Germany, Italy, and Eastern Europe. Or if the enemy takes a very strong line, we could intimidate them by selling nuclear missile technologies to Iran, Iraq, or Algeria. We could even form a military alliance with Iran, "within the framework of which a contingent of Russian troops and tactical nuclear weapons could be stationed on the shores of the Persian Gulf and the Strait of Hormuz."

What do the authors propose that Russia should do after it has washed its warheads in the waters of the Persian Gulf? Extend the operating life of MIRVed (multiple independently targetable reentry vehicle) ICBMs to twenty years. After the year 2008, start deploying one hundred ballistic missiles with six medium-power warheads each, in the version mounted on railroad flatcars. Finally, in extremis, bury the Treaty on the Nonproliferation of Nuclear Weapons—because according to the "Theses," the number of nuclear powers is going to increase steadily in any case.

That point is particularly noteworthy. The text contains references to two hundred nuclear weapons belonging to Israel with a range that would reach Moscow. China is also likely to build up its nuclear missile forces. At the moment, Russia, following official Foreign Ministry policy, is cooperating with America to contain the possessors of nuclear weapons in the "Third World." The document's authors believe this line will fail completely in the near future. Instead, we should sell delivery vehicles first to India (just as the Americans sold Britain the Polaris and Trident missiles for British warheads) and in the longer term to Iran and certain Arab countries.

This could mean the prospect of unfurling a nuclear umbrella over such fine men as Saddam Hussein—in the event, of course, that the "Provisions" really do get incorporated in the Defense Ministry's future work plan. And they will—at the end of the text, it says in black and white: "This document may be used in developing a new Russian Federation military doctrine."

What might the immediate future of the "Conceptual Provisions" be? They are expected to be used in the introductory volume to the future Russian military doctrine. In the course of twelve to eighteen months General Staff officers will be able to come up with a whole host of appendices to them, such as an operational plan for sending troops into the Baltic republics. By 1997, this whole collection of folios could be approved by the next defense minister. After that it will become clear that local conflicts in the Caucasus and tactical nuclear strikes against NATO troop accumulations in Eastern Europe are not the last surprise being prepared for us by the military strategists and "independent researchers."

## 9.52

### NATO Delegation Examines PFP Participation Option

Basapress, 31 October 1995 [FBIS Translation]

A NATO delegation comprising representatives of European armies led by Italian Army Colonel Carlo Greco is visiting Moldova to assess the possibilities of Moldova's participation in the "Partnership for Peace" (PFP) program. During their meetings with Moldovan officials, the delegation will discuss possibilities of conducting peacekeeping operations and humanitarian assistance programs, controlling flights, and establishing cooperation between civil authorities and military commands. They will be familiarized with the activity of Moldovan peacekeepers, systems of civil defense, and military instruction and training. Aspects of assisting Moldovan soldiers to study English will be also examined.

Viorel Cibotaru, Defense Ministry spokesman, told Basapress that the visit will conclude on 3 November. Tiraspol leader Igor Smirnov declared in his speech on 29 October during the Dniester All-Deputy Congress that NATO officers would participate in the military maneuvers of the Moldovan army. Defense Ministry representatives denied this in an interview with Basapress.

## 9.53

### President Confirms Plans Not to Join NATO

INFOTAG, 16 November 1995
[FBIS Translation]

Our participation in the Partnership for Peace program does not mean that Moldova is planning to join the North Atlantic Alliance, as claimed by separatist Dniester leaders, President Mircea Snegur stated in the Ministry of Defense today. Mr. Snegur, who is the supreme commander-in-chief of the Moldovan armed forces, remarked that the national army also contributed to the country's integration into international structures. It has been developing cooperation with the armed forces of Bulgaria, Turkey, Romania, Ukraine, United States, and the Russian Federation. Moldova's participation in the Partnership for Peace program includes, first and foremost, assistance in personnel training, consultancies, exchange of experience, and peacekeeping operations, the president stressed. As stipulated by the constitution, Moldova is a neutral state, shall allow no foreign troops to be deployed on its territory and cannot enter any military blocs or unions, Snegur said. The top ministry officials, who took the floor at today's collegium, said that the supreme commander-in-chief should act as co-author in drafting the laws related to military issues. The collegium confirmed some progress in national army development, as exemplified by the first graduation of officers from the Alexandru cel Bun Military College. At the same time, Snegur emphasized the importance of eliminating such drawbacks as low discipline, lack of officers' responsibility for their subordinates, and poor patriotic upbringing of the younger generation.

## 9.54

**Snegur: Country "Cannot Become" NATO Member**
INFOTAG, 30 November 1995
[FBIS Translation]

Seeking integration into European Union structures, Moldova nevertheless cannot become a NATO member, as its constitution proclaimed Moldova a neutral country that shall not join any military blocs, President Mircea Snegur stated in his interview with the *Neue Zuricher Zeitung* newspaper of Switzerland and the *Handelsblatt* of Germany.

"We do not think that NATO's possible broadening toward the East will bring any danger to Moldova's security," he said, adding that "when NATO begins its expansion, it should necessarily consider all factors, all possible consequences, first and foremost—the Russian factor."

The president expressed concern that the Russian State Duma still had ratified neither the basic Moldovan-Russian treaty, signed as long ago as 1990, nor the agreement of 21 October 1994, on the withdrawal of Russian troops from Moldovan territory. "We are looking forward to the forthcoming elections to the State Duma with interest, for the would-be forum will largely influence official Russia's stance on the Dniester issue," Mr. Snegur said.

Moldovan President Mircea Snegur also stated in the aforementioned interview that he is "dreaming of organizing, as soon as possible, a team of like-minded allies so that to conduct the second stage of reforms in the next five years with the fewest possible errors," the presidential press service said.

Observers have regarded this statement as Snegur's intention to struggle for a second term as president of Moldova.

"I have been Moldovan president since 1990, and now I must report to the nation what I have done, and to outline new priorities. I believe we have something to tell the people," Snegur said. He believes that it is a priority to make 1996 a turning point in economic reform. This implies economic growth, making the Bankruptcy Law effective, exports and wages grow, successful privatization, revision of budget policy in favor protecting the people.

In the political field, Snegur places priority on bringing up Moldovan youth in the spirit of patriotism.

## 9.55

**Kiev Wants to Expand Links with NATO, WEU**
UNIAN (in Ukrainian), 26 December 1995
[FBIS Translation]

Ukraine wishes to extend cooperation with NATO and the Western European Union (WEU) and will continue to develop cooperation with European and trans-Atlantic organizations, Ukraine's Foreign Minister Gennadiy Udovenko told UNIAN on 26 December.

He stressed, however: "Ukraine as a non-bloc national is worried about its national security, as it may end up between an expanded NATO and the Tashkent alliance."

According to the minister, Ukraine has no intention of joining either NATO or the Tashkent bloc, but will not limit its cooperation with NATO to an individual program within the Partnership for Peace. "NATO too, has expressed an interest in cooperating with Ukraine," he added.

## 9.56

**Defense Minister Shmarov Considers Joining NATO "Absurd"**
Interfax, 31 January 1996
[FBIS Translation]

The claims that Ukraine is striving for NATO membership are absurd, Ukrainian Defense Minister Valeriy Shmarov told students at the Belarusian Military Academy in Minsk. The idea of NATO's eastward expansion is "premature," said the minister. Ukrainian legislation provides for the "off-bloc and neutral position of this country. Bilateral relations with CIS and other countries are a different thing," said Shmarov.

Ukraine has changed its stand on military threats, said the minister. In his opinion, a serious enemy will not appear in the next few years or decades.

"Local conflicts are possible, but there will hardly be several of them at one time. It is also unlikely that neighboring countries and even more unlikely that faraway countries may be united against Ukraine as a common enemy," he said.

Commenting on relations with Russia, the minister said these countries and their armed forces "are most eligible for cooperation. They simply will not survive without each other."

The minister also named Belarus as a strategic partner of Ukraine.

# Part III

# The Formal Structure of the CIS

# 10

# Major CIS Structural Agreements and Protocols

## Introductory Notes

The structural documents contained in Chapter 10 form the legal framework of the Commonwealth. Agreements and treaties are signed by heads of state and submitted to the state parliaments for ratification. The Council of Heads of State has to date considered more than 450 agreements, most submitted in packages by the councils of Heads of Government, Ministers of Defense and Foreign Affairs, and other working commissions. These agreements and protocols are attached to the core conceptual framework, which consists of the Commonwealth Charter, the Economic Union Treaty and the Collective Security Treaty. Most of the agreements, and even the core framework treaties, have lacked the political will or machinery to carry them out, making the CIS structure an impotent structural shell so far, despite Russia's determined efforts to consolidate the organization into a substitute for the Soviet Union.

The principal political structural instrument, the Commonwealth Charter, is a loosely worded conceptual document, hastily drafted by the Belovezh Forest trio. It has still not been signed or ratified by Ukraine or Turkmenistan (whose presidents have only signed the Commonwealth Declaration). Most interstate political, economic, and military activity is still conducted on bilateral terms. The Interparliamentary Assembly (IPA), under strong Russian leadership, was supposed to coordinate CIS laws, thereby paving the way for the practical implementation of CIS agreements. So far, however, the IPA has been ineffective because its signatories insist on subordinating CIS laws to their own constitutions, and on the sovereignty of their national governments and parliaments.

The CIS Economic Union Treaty has enjoyed little more success than the charter. The treaty has still not been signed by Ukraine or Turkmenistan, and most of the agreements attached to it have fewer than ten signatories. Whereas a great many economic agreements signed in early 1992 played a vital role in implementing the "divorce" among former Soviet republics, the framework for the reintegration of the republics has so far failed to achieve its purpose. In 1995, the thrust toward economic integration was shifted toward a multilevel variant. This approach has shown signs of bearing greater fruit, with a trilateral Customs Union Agreement joined by Russia, Kazakhstan, and Belarus being trumpeted as the first giant step. Even this agreement, however, faces severe difficulty in the practical implementation stage because of the diversity in political and economic development of its signatories.

The Collective Security Treaty has also been rejected by Ukraine and Turkmenistan. This treaty has also failed in its mission to form the framework for building a unified CIS armed forces. Only the strategic command, which has been brought under total Russian control, has a unified command structure. As a consequence, a new regional approach to CIS security is being discussed within the CIS Council for Defense Ministers, as documented in Chapter 9. It remains to be seen whether a legal structure and enforcement mechanisms can be be designed to carry out this approach. The other military agreements in this section are basically only statements of principle, lacking the political will or the unified command structure to carry out its military aims. In sum, the entire structural framework of the CIS so far lacks the political support required for successful implementation of unification principles. The only political impetus so far emanates from Russia, in concert with Belarus, Kazakhstan, and sometimes with one or two other CIS leaders. This is insufficient to create the kind of integrated interstate organism implied by the word "commonwealth," and certainly an insufficient basis from which to move toward a confederation. So far, at least, any pragmatic integrationist measures which have been implemented have been the result of unilateral pressure or force by the Russian Federation.

# The Political Framework Documents

## 10.1

**Charter of the Commonwealth of Independent States (22 January 1993)**
*Rossiyskaya Gazeta*, 12 February 1993
[FBIS Translation]

The states which have voluntarily united within the Commonwealth of Independent States (hereinafter the Commonwealth), basing themselves on the historic community of their peoples and the ties established between them, acting in accordance with the generally recognized principles and norms of international law, the provisions of the UN Charter and the Helsinki Final Act, and other documents of the CSCE,

—seeking to ensure by joint efforts the economic and social progress of their peoples, fully resolved to implement the provisions of the Agreement on the Creation of the Commonwealth of Independent States, the protocol to that Agreement, and the provisions of the Alma-Ata Declaration,

—developing cooperation among themselves to ensure international peace and security and equally for the purpose of maintaining civil peace and interethnic accord,

—desiring to create conditions for the preservation and development of the cultures of all the peoples of the member states,

—seeking to improve the mechanisms of cooperation in the Commonwealth and enhance their effectiveness,

—have decided to adopt the Charter of the Commonwealth and have agreed on the following:

**Section I. Aims and Principles**

Article 1. The Commonwealth is founded on the principles of the sovereign equality of all its members. The member states are autonomous and equal subjects of international law.

The Commonwealth promotes the further development and strengthening of the relations of friendship, good neighborliness, interethnic accord, trust, mutual understanding, and mutually advantageous cooperation among the member states.

The Commonwealth is not a state and does not possess supranational powers.

Article 2. The aims of the Commonwealth are:

—cooperation in the political, economic, ecological, humanitarian, cultural, and other spheres;

—the all-around and balanced economic and social development of member states within the framework of the common economic space, and interstate cooperation and integration;

—the guaranteeing of human rights and basic freedoms in accordance with the generally recognized principles and norms of international law and documents of the CSCE;

—cooperation among member states to ensure international peace and security, the implementation of effective measures to reduce arms and military expenditure, the elimination of nuclear and other types of weapons of mass destruction, and the achievement of general and complete disarmament;

—assistance to citizens of member states in free association, contacts, and movement within the Commonwealth;

—reciprocal legal assistance and cooperation in other spheres of legal relations;

—peaceful solution of disputes and conflicts between Commonwealth states.

Article 3. In order to achieve the Commonwealth's aims member states, proceeding on the basis of the generally recognized norms of international law and the Helsinki Final Act, build their relations in accordance with the following interconnected principles of equal value:

—respect for the sovereignty of member states, for the inalienable right of peoples to self-determination, and for the right to determine their future without external interference;

—inviolability of state borders, recognition of existing borders, and rejection of unlawful territorial acquisitions;

—territorial integrity of states and rejection of any actions aimed at dismembering another state's territory;

—non-use of force or the threat of force against the political independence of a member state;

—resolution of disputes by peaceful means in such a way as to avoid jeopardizing international peace, security, and justice;

—supremacy of international law in interstate relations;

—non-interference in one another's internal and external affairs;

—the guaranteeing of human rights and basic freedoms for all regardless of race, ethnic origin, language, religion, and political or other convictions;

—conscientious discharge of commitments assumed under Commonwealth documents, including this charter;

—consideration of one another's interests and of the

interests of the Commonwealth as a whole, provision of assistance in all spheres of their mutual relations on the basis of mutual assent;

—the pooling of efforts and provision of support to one another for the purpose of creating peaceful conditions of life for the peoples of Commonwealth member states, and the ensuring of their political, economic, and social progress;

—development of mutually advantageous economic, scientific, and technical cooperation, and expansion of the processes of integration;

—spiritual unity of their peoples, based upon respect for their identity, and close cooperation in the preservation of cultural values and cultural exchange.

Article 4. The following belong to the spheres of joint activity of member states and are carried out on an equitable basis via common coordinating institutions in accordance with the commitments assumed by member states within the framework of the Commonwealth:

—guaranteeing human rights and basic freedoms;
—coordinating foreign policy activity;
—cooperating in forming and developing a common economic space, the pan-European and Eurasian markets, and customs policy;
—cooperating in developing systems of transport and communications;
—protecting health and the environment;
—questions of social and migration policy;
—struggle against organized crime;
—cooperating in the sphere of defense policy and the protection of external borders;

This list may be extended by the mutual assent of member states.

Article 5. Multilateral and bilateral agreements in various spheres of member states' mutual relations form the fundamental legal basis of interstate relations within the framework of the Commonwealth.

Agreements concluded within the framework of the Commonwealth should correspond to the aims and principles of the Commonwealth and the commitments of member states under this charter.

Article 6. Member states promote the cooperation and development of ties among state organs, public associations, and economic structures.

**Section II. Membership**

Article 7. The states which have signed and ratified the Agreement on the Creation of the Commonwealth of Independent States of 8 December 1991 and the Protocol to that Agreement of 21 December 1991 by the time of the adoption of this charter constitute the founding states of the Commonwealth.

The founding states which assume commitments under this charter within one year of its adoption by the Council of Heads of State constitute the member states of the Commonwealth.

A state which shares the aims and principles of the Commonwealth and assumes the commitments contained in this charter can also become a member of the Commonwealth by joining to it subject to the assent of all member states.

Article 8. On the basis of a decision by the Council of Heads of State, a state desiring to participate in certain categories of Commonwealth activity may join the Commonwealth as an associate member on the terms set by the agreement on associate membership.

By decision of the Council of Heads of State, representatives of other states may attend sessions of Commonwealth organs as observers.

Questions concerning the participation of associate members and observers in the work of Commonwealth organs are regulated by the rules of procedure of such organs.

Article 9. A member state has the right to secede from the Commonwealth. A member state shall inform the depositary of the charter of such intention in writing twelve months prior to secession.

The commitments which arise during the period of participation in this charter are binding upon the relevant states until they are discharged in full.

Article 10. Violations of this charter by a member state or the systematic non-fulfillment by a state of its commitments under agreements concluded within the framework of the Commonwealth or of decisions by Commonwealth organs shall be examined by the Council of Heads of State.

Measures permitted under international law may be applied to such a state.

**Section III. Collective Security and Military–Political Cooperation**

Article 11. Member states shall pursue a coordinated policy in the sphere of international security, disarmament, arms control, and the organizational development of the armed forces and shall maintain security in the Commonwealth by, inter alia, groups of military observers and collective peacekeeping forces.

Article 12. In the event of a threat to the sovereignty, security, and territorial integrity of one or several member states or to international peace and security, member states shall immediately activate the mechanism of mutual consultations with the aim of coordinating positions and adopting measures to eliminate the threat, including peacemaking operations and the use, where need be, of Armed Forces in exercise of the right to individual or collective self-defense in accordance with Article 51 of the UN Charter.

The decision on the joint use of the armed forces shall be made by the Council of Heads of State of the Commonwealth or the interested member states of the Commonwealth in line with their national legislation.

Article 13. Each member state shall take appropriate measures to ensure a stable situation along the Commonwealth member states' external borders. On the basis of mutual assent member states shall coordinate the activity of border troops and other competent services which exercise control and bear responsibility for ensuring observance of the prescribed procedure governing the crossing of member states' external borders.

Article 14. The Council of Heads of State is the Commonwealth's supreme organ on questions concerning the defense and protection of member states' external borders. Coordination of Commonwealth's military–economic activity shall be exercised by the Council of Heads of Government.

Collaboration by member states in the implementation of international agreements and the resolution of other questions in the sphere of security and disarmament shall be organized by joint consultations.

Article 15. Specific questions of member states' military–political cooperation shall be regulated by special agreements.

### Section IV. Preventing Conflicts and Resolving Disputes

Article 16. Member states shall take all possible measures to prevent conflicts, primarily along interethnic and interreligious lines, which could entail violation of human rights.

They shall, on the basis of mutual agreement, render assistance to one another in settling such conflicts, including within the framework of international organizations.

Article 17. Commonwealth member states shall refrain from actions which could be detrimental to other member states and result in the exacerbation of potential disputes.

Member states shall conscientiously and in a spirit of cooperation make efforts toward a fair and peaceful resolution of their disputes through talks of the achievement of accord on an appropriate alternative procedure for settling the dispute.

If member states do not resolve the dispute via the means indicated in the second part of this article, they may refer it to the Council of Heads of State.

Article 18. The Council of Heads of State is empowered, at any stage of a dispute whose continuation could threaten the maintenance of peace or security in the Commonwealth, to recommend to the parties an appropriate procedure or method for settling the dispute.

### Section V. Cooperation in the Economic, Social, and Legal Spheres

Article 19. Member states shall cooperate in the economic and social spheres along the following avenues:

—forming a common economic space on the basis of market relations and the free movement of goods, services, capital, and labor;
—coordinating social policy, elaborating joint social programs and measures to ease social tension in connection with the implementation of economic reforms;
—developing systems of transport and communications and energy systems;
—coordinating credit and financial policy;
—promoting the development of member states' trade and economic ties;
—encouraging and mutually protecting investments;
—promoting the standardization and certification of industrial goods and commodities;
—legally protecting intellectual property;
—promoting the development of a common information space;
—implementing joint environmental protection measures, providing mutual assistance in eliminating the consequences of environmental disasters and other emergency situations;
—implementing joint projects and programs in the spheres of science and technology, education, health care, culture, and sports.

Article 20. Member states shall cooperate in the sphere of the law, in particular by concluding multilateral and bilateral treaties on affording legal assistance, and shall promote the alignment of national legislation.

In the event of conflicts between the norms of member states' national legislation regulating relations in spheres of joint activity, member states shall hold consultations and talks for the purpose of elaborating proposals to remove those conflicts.

### Section VI. Organs of the Commonwealth

#### *The Council of Heads of State and the Council of Heads of Government*

Article 21. The Council of Heads of State is the supreme organ of the Commonwealth.

The Council of Heads of State, in which all member states

are represented at top level, shall discuss and resolve fundamental questions connected with member states' activity in the sphere of their common interests.

The Council of Heads of State shall assemble for sessions twice a year. Extraordinary sessions of the council may be convened on the initiative of any one of the member states.

Article 22. The Council of Heads of Government shall coordinate the cooperation of member states' organs of executive power in the economic, social, and other spheres of common interest.

The Council of Heads of Government shall assemble for sessions four times a year. Extraordinary sessions of the council may be convened on the initiative of any one of the member states' governments.

Article 23. Decisions of Council of Heads of State and the Council of Heads of Government shall be made by common assent—consensus. Any state may declare it has no interest in a particular question, which should not be regarded as an obstacle to the adoption of a decision.

The Council of Heads of State and the Council of Heads of Government may hold joint sessions.

The working procedure of the Council of Heads of State and the Council of Heads of Government shall be regulated by the Rules of Procedure.

Article 24. The heads of states and heads of governments shall take the chair at sessions of the Council of Heads of State and Council of Heads of Government by rotation in the Russian alphabetical order of the names of the Commonwealth member states.

Sessions of the Council of Heads of State and Council of Heads of Government shall be held as a rule in the city of Minsk.

Article 25. The Council of Heads of State and the Council of Heads of Government shall create working and auxiliary organs on both a standing basis and a temporary basis.

These organs shall be formed from representatives of the member states vested with the corresponding powers.

Experts and consultants may be invited to participate in their sessions.

Article 26. Conferences of leaders of the corresponding state organs shall be convened to resolve questions of cooperation in individual spheres and to elaborate recommendations for the Council of Heads of State and the Council of Heads of Government.

### *The Foreign Ministers Council*

Article 27. The Foreign Ministers Council, on the basis of decisions of the Council of Heads of State and the Council of Heads of Government, shall coordinate the foreign policy activity of the member states, including their activity in international organizations, and shall organize consultations on questions of world policy of mutual interest.

The Foreign Ministers Council shall perform its activity in accordance with the statute approved by the Council of Heads of State.

### *Coordinating and Consultative Committee*

Article 28. The Coordinating and Consultative Committee is the standing executive and coordinating organ of the Commonwealth.

In execution of the decisions of the Council of Heads of State and the Council of Heads of Government, the committee:

—elaborates and submits proposals on questions of cooperation within the framework of the Commonwealth and of the development of socio-economic ties;

—promotes the implementation of accords in specific areas of economic mutual relations;

—organizes conferences of representatives and experts for the preparation of draft documents to be submitted to sessions of the Council of Heads of State and Council of Heads of Government;

—ensures the holding of sessions of the Council of Heads of State and the Council of Heads of Government;

—assists the work of other Commonwealth organs.

Article 29. The Coordinating and Consultative Committee is made up of permanent plenipotentiary representatives, two from each Commonwealth member state, and a committee coordinator, appointed by the Council of Heads of State.

For the purposes of providing organizational and technical backup for the work of the Council of Heads of State, the Council of Heads of Government, and other Commonwealth organs, the Coordinating and Consultative Committee has a Secretariat, headed by the committee coordinator—deputy chairman of the Coordinating and Consultative Committee.

The committee operates in accordance with the statute ratified by the Council of Heads of State.

The committee's seat is in the city of Minsk.

### *The Defense Ministers Council; The Joint Armed Forces High Command*

Article 30. The Defense Ministers Council is an organ of the Council of Heads of State on questions of military policy and the military organizational development of the member states.

The Joint Armed Forces High Command exercises leadership of the Joint Armed Forces, as well as of groups of military observers and collective forces maintaining peace in the Commonwealth.

The Defense Ministers Council and the Joint Armed Forces High Command perform their activity on the basis of the relevant statutes ratified by the Council of Heads of State.

***The Council of Commanders of Border Troops***

Article 31. The Council of Commanders of Border Troops is an organ of the Council of Heads of State on questions of the protection of the external borders of member states and the provision of a stable situation along them.

The Council of Commanders of Border Troops performs its activity on the basis of the relevant statute ratified by the Council of Heads of State.

***The Economic Court***

Article 32. The Economic Court operates for the purpose of ensuring the fulfillment of economic commitments within the framework of the Commonwealth.

The jurisdiction of the Economic Court includes the settlement of disputes arising in the fulfillment of economic commitments. The court may also settle other disputes referred to its jurisdiction by agreements of member states.

The Economic Court is empowered to interpret the provisions of agreements and other Commonwealth acts on economic issues.

The Economic Court performs its activity in accordance with the Agreement on the Status of the Economic Court and the statute on it ratified by the Council of Heads of State.

The seat of the Economic Court is in the city of Minsk.

***Commission on Human Rights***

Article 33. The Commission on Human Rights is a consultative organ of the Commonwealth and monitors the fulfillment of commitments on human rights which the member states have assumed within the framework of the Commonwealth.

The commission comprises representatives of Commonwealth member states and operates on the basis of the statute ratified by the Council of Heads of State.

The seat of the Commission on Human Rights is in the city of Minsk.

***Organs of Sectoral Cooperation***

Article 34. On the basis of agreements between the member states on cooperation in the economic, social, and other spheres, organs of sectoral cooperation may be established to carry out the elaboration of agreed principles and rules of such cooperation to promote their practical implementation.

The organs of sectoral cooperation (councils, committees) perform the functions envisaged in the present charter and in the provisions on them, ensuring the examination and settlement on a multilateral basis of questions of cooperation in the relevant spheres.

Leaders of the corresponding organs of executive power of the member states are members of the organs of sectoral cooperation.

The organs of sectoral cooperation adopt recommendations within the limits of their competence and, whenever necessary, also submit proposals for consideration by the Council of Heads of Government.

***The Working Language of the Commonwealth***

Article 35. The working language of the Commonwealth is Russian.

**Section VII. Interparliamentary Cooperation**

Article 36. The Interparliamentary Assembly holds interparliamentary consultations, discusses questions of cooperation within the framework of the Commonwealth, and elaborates joint proposals in the sphere of the activity of national parliaments.

Article 37. The Interparliamentary Assembly comprises parliamentary delegations.

The organization of the activity of the Interparliamentary Assembly is carried out by the Council of the Assembly, which comprises leaders of the parliamentary delegations.

Procedural questions of the activity of the Interparliamentary Assembly are regulated by its standing orders.

The seat of the Interparliamentary Assembly is the city of St. Petersburg.

**Section VIII. Finances**

Article 38. Expenditure on financing the activity of the Commonwealth organs is apportioned on the basis of proportional participation by the member states and is set in accordance with special agreements on the budgets of the Commonwealth organs.

The budgets of the Commonwealth organs are ratified by the Council of Heads of State on the basis of proposals by the Council of Heads of Government.

Article 39. Questions of the financial and economic activity of the Commonwealth organs are examined according to the procedure established by the Council of Heads of Government.

Article 40. Member states are autonomously responsible for the expenditure associated with the participation of their representatives, experts, and consultants in the work of Commonwealth conferences and organs.

**Section IX. Final Provisions**

Article 41. The present charter is subject to ratification by the founding states in accordance with their constitutional procedures.

The instruments of ratification are handed over to the government of the Republic of Belarus, which will notify the other founding states of the submission of each instrument for safekeeping.

The present charter enters into force for all founding states from the moment instruments of ratification have been handed over for safekeeping by all founding states, or, for the founding states which have handed over their instruments of ratification, one year following the adoption of the present charter.

Article 42. Amendments to the present charter may be proposed by any member state. Proposed amendments are examined in accordance with the rules of procedure of the Council of Heads of State.

Amendments to the present charter are adopted by the Council of Heads of State. They enter into force following ratification by all member states in accordance with their constitutional procedure, from the date on which the government of the Republic of Belarus receives the last instrument of ratification.

Article 43. In the process of ratifying the present charter, the Commonwealth's founding states may make reservations and appeals* on Sections III, IV, and VII and on Articles 28, 30, 31, 32, and 33.

Article 44. The present charter will be registered in accordance with Article 102 of the UN Charter.

Article 45. The present charter has been done in one copy in the state languages of the Commonwealth founding states. The original copy is stored in the archives of the government of the Republic of Belarus, which will send certified copies thereof to all founding states.

The present charter was adopted 22 January 1993 at the session of the Council of Heads of State in the city of Minsk.

[Signed]
For the Republic of Armenia, L. Ter-Petrosyan
For the Republic of Belarus, S. Shushkevich
For the Republic of Kazakhstan, N. Nazarbaev
For the Republic of Kyrgyzstan, A. Akaev
For the Republic of Moldova
For the Russian Federation, B. Yeltsin
For the Republic of Tajikistan, E. Rakhmonov
For Turkmenistan
For the Republic of Uzbekistan, I. Karimov
For Ukraine

## 10.2
## Statement by the Council of Heads of States Belonging to the CIS

Translated by David Ridler
22 January 1993

The agreements adopted and the mechanisms elaborated within the Commonwealth make it possible to regulate problems of political, economic, humanitarian, military, and other cooperation by international legal means.

The heads of CIS member states are united in the view that the Commonwealth possesses the requisite potential to improve its activity on the basis of existing agreements. However, all the participants in the conference of CIS heads of states in Minsk state their resolve to continue efforts to improve the effectiveness of CIS activity in the economic and political spheres.

The states which have signed and those which have not signed the Decision on the CIS Charter will concentrate their efforts first and foremost on finding ways to overcome the economic crisis and on forming effective ties among the components engaged in economic activity during the transition to market relations.

The heads of state deem it essential to remove persisting obstacles to the development of mutually beneficial economic cooperation.

The heads of state consider that the CIS countries' relations, above all their economic relations, will serve to ensure conditions for normal cooperation among those countries.

The Decision on the CIS Charter is open for signature by those states which are prepared to sign it.

Completed in the city of Minsk on 22 January 1993 in one authentic copy in the Russian language. The authentic copy is lodged in the archives of the government of the Republic of Belarus, which will send an attested copy to the states which have signed the statement.

[Signed]
For the Republic of Armenia, Levon Ter-Petrosyan
For the Republic of Belarus, Stanislav Shushkevich
For the Republic of Kazakhstan, Nursultan Nazarbaev
For the Republic of Kyrgyzstan, Askar Akaev
For the Republic of Moldova, excluding political sphere, Mircea Snegur
For the Russian Federation, Boris Yeltsin
For the Republic of Tajikistan, E. Rakhmonov
For Turkmenistan, S. Niyazov
For the Republic of Uzbekistan, Islam Karimov
For Ukraine, Leonid Kravchuk

# The Economic Accords/Protocols and Treaties

## 10.3

**Protocol on Economic Relations Published**
*Rossiyskaya Gazeta,* 3 January 1992
[FBIS Translation]

[Text published under the rubric "Official Section: Protocol for the Coordination of Measures Being Taken to Preserve Economic Ties in the First Quarter of 1992 Between Enterprises, Associations, and Organizations Located on the Territories of the Commonwealth of Independent States."]

The governments of the Commonwealth's independent states, bearing in mind the need for coordinated actions to create favorable working conditions for enterprises, associations, and organizations located on their territories and the need to implement the interstate agreements on the principles of trade and economic cooperation for 1992, have agreed:

1. In order to prevent the shutdown of production facilities enterprises, associations, and organizations—irrespective of their forms of ownership—which are located on the territories of the Commonwealth's member states and to ensure that the economic relations operating in 1991 are preserved in the first quarter of 1992 and that contracts for delivering raw and other materials, components, and finished products are concluded and met on a scale of not less than 70 percent of the volumes of deliveries in the first quarter of 1991.

2. Before 1 February 1992 enterprises, associations, and organizations are to submit data to the state statistics organs on the volumes and destinations of deliveries under contracts concluded for the first quarter of 1992. Administrative organs of the Commonwealth's member states are to summarize the data submitted for consideration in determining quotas and issuing notices of pairing arrangements for deliveries made under the interstate agreements for 1992.

3. In the period up to 15 January 1992 the administrative organs of the Commonwealth's member states are to furnish suppliers and consumers with plan documents specifying pairing arrangements for the delivery in 1992 of goods according to quotas (attachments No. 1 and No. 2 to the agreements), and in the period up to 15 February 1992 they are to make the necessary adjustments taking account of the concluded contracts for the first quarter of 1992 under established economic ties.

4. Licenses for the export of output above the prescribed quotas are issued only by decision of the states' governments or by organs authorized by them.

The organs regulating deliveries of goods for state needs are to be made responsible for determining the specific suppliers and consumers in accordance with the quotas and for monitoring the conclusion and fulfillment of goods delivery contracts between enterprises.

The sides undertake to inform each other of the license-issuing procedure operating in each state.

5. Economic entities are to be assisted in maintaining and developing reciprocal goods deliveries, including by intersector and intrasector cooperative arrangements, taking account of prevailing production volumes without making additional demands for reciprocal deliveries of other goods.

6. A mutually coordinated procedure for making settlement transactions for licensed goods allowing financial control over their export to be exercised is to be elaborated and approved. Manufacturing enterprises and brokerage organizations make settlement transactions for licensed goods in accordance with the approved procedure.

When output is dispatched without licenses 90 percent of the money received by the supplier from the purchaser is transferred to state budget revenue and 10 percent to the relevant central (national) bank.

7. It is to be agreed that all settlements by the states' enterprises and organizations for reciprocal goods deliveries are made in rubles at free or regulated prices taking account of the states' existing normative acts.

8. Incentives and sanctions are to be applied to enterprises and organizations in order to realize goods deliveries under interstate agreements in accordance with the states' existing legislation. The specific terms of delivery and responsibility for meeting them are stipulated in the contracts concluded by enterprises.

9. It is to be ensured that a coordinated statistical accounting methodology is utilized in the states in order to monitor the progress of the fulfillment of reciprocal deliveries of the output stipulated by interstate agreements.

Questions which arise in the process of reciprocal deliveries are to be periodically reviewed.

10. The organs for managing transport (rail, water, air, road, and other) of the Commonwealth's member states are to ensure that transport resources are allocated in the first quarter of 1992 on presentation of the freight by senders.

11. The leaders of the governments undertake to issue the necessary normative acts ensuring that the commitments ensuing from this protocol are met.

[Signed]
For the Azerbaijan Republic, Azizbekov
For the Republic of Armenia, Arutyunyan
For the Republic of Belarus, Kebich
For the Republic of Kazakhstan, Tereshchenko
For the Republic of Kyrgyzstan, Iordan
For the Republic of Moldova, Chebuk
For the Russian Federation, Gaydar
For the Republic of Tajikistan, Khaeev
For the Republic of Turkmenistan, Charyev
For the Republic of Uzbekistan, Sharipov
For the Ukraine, Masik

## 10.4
## Trade, Economic Cooperation Accord

TASS, 15 February 1992
[FBIS Translation]

**Agreement on the Regulation of Relations Between the States of the Commonwealth in the Sphere of Trade and Economic Cooperation in 1992**

In order to create favorable conditions for the development of trade and economic cooperation between enterprises and organizations located on the territory of the Commonwealth states, the heads of state and heads of government of the Commonwealth states, hereinafter termed the "parties," have agreed on the following:

Article 1. Trade and economic relations between the states are to be carried out in the framework of the Commonwealth on the basis of mutual advantage, and actions causing economic damage to one another are to be refrained from.

Article 2. The parties use a single monetary unit (the ruble) for reciprocal payments between economic subjects, for credit, and other financial operations within the Commonwealth.

Where individual states of the Commonwealth introduce their own national currency, the procedure for payments will be determined by separate agreements.

Payments between enterprises and organizations of the Commonwealth states are implemented at free market prices. It is possible to set a price ceiling on individual, very important types of products, mutually agreed by the sides, which are supplied within the framework of intergovernmental agreements.

The parties agree to regulate prices on the products of monopoly enterprises on the basis of a special agreement.

Article 3. The sides undertake to ensure unhindered transit across the territory of their states for goods and services being supplied to other states, and to coordinate ceilings for tariffs for the transit of goods.

Article 4. The parties undertake to adopt measures to eliminate double or multiple taxation of income from trade between enterprises of the Commonwealth states.

Article 5. The parties agree not to impose quotas, licenses, or other forms of non-tariff restrictions on supplies of output, with the exception of a list of goods according to a catalogue included in intergovernmental agreements.

Until this list is agreed, quotas and licenses shall be imposed in accordance with the procedure customary in each state.

The parties agree to notify the organizations responsible for implementing the interstate agreements of the quotas for exportation of output to states of the community, in accordance with the volumes fixed by the intergovernmental agreements. The said quotas are the basis for issuing licenses for exportation of output to the relevant states of the Commonwealth. The notifications matching buyers to suppliers and issued by the regulating bodies of the parties within the limits of the fixed quotas are considered as licenses for exportation of output from their territories.

The parties agree, until the adoption of national normative acts regulating the exportation and importation of goods, to ensure unhampered transit across their borders of goods being moved on the basis of matching warrants issued by the authorized bodies of the parties.

Article 6. The parties agree to adopt the necessary normative acts guaranteeing fulfillment of interstate agreements on deliveries of goods, including the conclusion within one month of economic contracts between the parties' enterprises for delivery of output in 1992, in implementation of the said agreements and their implementation.

The parties agree within one week to adopt the necessary normative acts, ensuring in February of the current year the conclusion by the enterprises, associations, and organizations located on the states' territories, of economic contracts for delivery in the year 1992 and their implementation, according to the volumes of deliveries notified by the government bodies in keeping with the intergovernmental agreements. The parties will ensure the acceptance of orders for supply of output by all enterprises, associations, and organizations as per the intergovernmental agreements within the limits of the fixed quotas, and also responsibility for nonconclusion of contracts for the supply of output, in accordance with the volumes indicated, in the form of a fine and forfeits for the non-fulfillment of delivery commitments. The scale of the fine and forfeit is to be agreed within one week.

The parties agree to instruct the enterprises, associations, and organizations located on the parties' territories, when concluding contracts for delivery of output under intergovernmental agreements, not to unilaterally set consumers' additional requirements for reciprocal deliveries of output. By consent of the parties, direct contracts may be permitted with individual regions of the parties as part of agreed volumes of delivery for the year 1992.

Article 7. The parties pledge to take measures in order to maintain and develop coproduction links within and between industries for deliveries of raw and other materials, semi-finished goods, and component items, used in technological conversions, and other output not envisaged by intergovernmental agreements, as a rule at a level of not less than 70 percent of the volume of deliveries in the year 1991.

Article 8. The parties consider it impermissible for enterprises located on the parties' territories to unilaterally make additional demands for reciprocal deliveries of output when concluding contracts for delivery of output in implementation of intergovernmental agreements.

For monitoring the implementation of intergovernmental agreements, for agreeing times of bilateral deliveries, and for examining disputed issues regarding their implementation, the parties are setting up bilateral commissions made up of representatives of the state departments concerned.

Article 9. The parties are organizing records of reciprocal payments and accounts regarding commercial and non-commercial operations on the basis of payment balances, and are endeavoring to make them balance. The parties recognize the right of each of the parties, for purposes of ensuring payments balance, to introduce restrictions with regard to payments and accounts with other parties, and to demand the undertaking of state commitments for clearing the balance of payments.

Article 10. The parties agree not to permit the re-export of goods subject to quotas and licenses without the consent of the authorized body of the state on whose territory the particular goods were produced. In the event of the reexport of such goods, the state on whose territory they were produced is entitled to introduce measures additional to those in intergovernmental agreements to restrict the export of goods to the territory of the state which permitted the re-export, and to demand the full hard currency earnings from the re-export as compensation for the loss.

Article 11. The parties agree that in order to establish a common legal base regulating relations between economic subjects in the various Commonwealth states, to adopt within one month an agreement on common conditions for deliveries of goods between organizations in the Commonwealth states.

Article 12. The parties are to form a Consultative Customs Council, comprising representatives of the parties, to elaborate and implement a joint customs policy, to coordinate cooperation among the customs services, and to record customs statistics, and they instruct the relevant bodies to draw up within one month a statute governing this Council and its membership.

Article 13. The parties agree to conclude agreements within one month on the procedure for settling disputes that arise between enterprises and organizations in the Commonwealth states as they engage in trade and economic cooperation, and to establish within the same period the bodies necessary to consider them.

Article 14. The agreement comes into force from the moment it is signed.

Done in Minsk on 14 February 1992 in one original copy in the Russian language. The original copy is preserved in the archives of the Government of the Republic of Belarus, which will send a certified copy of the agreement to the signatory states.

The agreement has been signed by the representatives of Azerbaijan, Armenia, Belarus, Kazakhstan, Kyrgyzstan, Moldova, Russia, Tajikistan, Turkmenistan, Uzbekistan, and Ukraine.

The representatives of Azerbaijan added the footnote: "From this year, taking account of the world proportions of prices."

The representatives of Turkmenistan added the footnote: "Taking account of the gradual transition over two years to world prices in mutual payments for raw materials and output."

## 10.5

**Foreign Debt Agreement**
TASS, 14 March 1992
[FBIS Translation]

### Agreement on Additions to the Treaty on Succession Concerning the Foreign State Debt and Assets of the USSR

Confirming their commitment to the fulfillment of their obligations which proceed from the treaty on legal sucsion concerning the foreign state debt and assets of the USSR adopted on 4 December 1991, the states joined in this treaty have agreed on the following:

Article 1. As the main members of the Interstate Council for monitoring the servicing of the debt and the use of the USSR's assets, the Russian Federation and Ukraine are cochairmen of the Interstate Council. Other members of the Interstate Council appoint a third co-chairman on the basis of rotation.

Article 2. The authorized bank provides full information on its operations to the Interstate Council and its members.

The Interstate Council appoints an independent auditor on a competitive basis who is provided with any documentation of the authorized bank which he may need for drawing up a report submitted to the Interstate Council at least once a year.

Article 3. The Foreign Economic Bank retains its rights as authorized bank. The relevant alterations are being made to the Foreign Economic Bank's charter. They are listed in the supplement to the present agreement to lend the authorized bank international status. The Interstate Council, at which the sides are represented with plenipotentiary powers, remains the highest body for the authorized bank. The Interstate Council sets up a standing observer body at the level of experts, to be located in Moscow.

The votes of plenipotentiary representatives on the Interstate Council are distributed in accordance with the proportionate burden borne by the sides in contributions to pay off the state debt of the former USSR. Decisions of the Interstate Council are adopted by no less than 80 percent of the votes of council members taking part in sessions. Sessions of the Interstate Council have valid authority when plenipotentiary representatives with at least 80 percent of the votes are present at them.

Article 4. The powers and functions of the interstate commission to work out criteria and principles in relation to the distribution of all the property of the former USSR abroad are transferred to the Interstate Council and the activity of the commission is terminated.

Article 5. The memorandum on mutual understanding of 28 October 1991 will be open for signing on behalf of the governments of the states joined in the present agreement who have not signed the aforementioned memorandum.

Article 6. The present agreement comes into force from the moment of its signing. It was completed in the city of Moscow on 13 March 1992 in one original copy in Russian. The original copy is being kept in the archives of the government of the Republic of Belarus, which will send a certified copy of it to the states which have signed this agreement.

The agreement has been signed by representatives of the governments of Armenia, Belarus, Georgia, Kazakhstan, Kyrgyzstan, Moldova, Russia, Tajikistan, and Ukraine.

The following document is appended to the agreement:

**The Position of the Republic of Uzbekistan's Delegation Regarding the Draft Agreement on Additions to the Treaty on Legal Succession in Respect of the USSR's Foreign State Debt and Its Assets**

1. The agreement's preamble confirms the adherence of the agreement's participants to fulfillment of the obligations stemming from the 4 December 1991 treaty on legal succession in respect of the USSR's foreign state debt and its assets. Uzbekistan did not sign that treaty in connection with differences over methodology for apportioning the foreign debt (disagreement with the overall figure).
2. The Republic of Uzbekistan is unable to agree with the proposed procedure for the adoption of decisions by the Interstate Council (the decisions are adopted under Article 3 of the agreement by no less than 80 percent of the votes of the council members taking part in a session). Russia with 61.34 percent of the votes has the potential to block any decision.
3. The provisions of this agreement are at variance with the provisions of the joint communiqué on the results of the interstate consultative conference on questions of creating the necessary conditions for the timely fulfillment of the debt liabilities of the former USSR, which was signed on 25 February 1991 in Kiev. It is envisaged to remove the USSR Foreign Economic Bank from the jurisdiction of the Russian Federation. The present agreement is restricted only in the sense that the co-chairmen will propose the candidature of the bank's deputy chairman on questions of servicing the foreign debt.

Arising from this, we consider it impossible to sign the agreement, and also the protocol on foreign debt and assets and the charter of the Interstate Council.

At the same time, this does not signify a refusal on the part of the Republic of Uzbekistan to take part in clearing off the foreign debt.

**Rules of the Interstate Council on Supervising the Servicing of the Debt and Use of the Assets of the USSR**

Article 1. The Interstate Council to supervise the servicing of the debt and the use of the assets of the USSR is set up in accordance with the treaty on legal succession with regard to the external state debt and assets of the USSR (hereinafter referred to as the treaty) with the aim of resolving all questions connected with servicing and repaying the USSR's external debt, and also with managing the foreign currency debts owed to the USSR and other assets of the USSR and their sale.

The Interstate Council is created from plenipotentiary representatives of the parties to the treaty. Any party which unilaterally halts debt-servicing payments without the agreement of the other parties which are members of the Interstate Council loses the right to a deciding vote in the Interstate Council and, until it resumes payments and makes up the incurred indebtedness, acquires the status of observer.

The Interstate Council is the supreme body of administration of the USSR Foreign Economic Bank (the authorized bank), which fulfills the functions of an agent of the parties which have signed the treaty, in questions of administering the debt and assets of the former USSR. The Interstate Council, at the level of experts, is creating a standing supervisory body located in Moscow. (A special unit will be set aside at the Foreign Economic Bank which is independent of any state and which is responsible exclusively for the servicing of the foreign debt, with clearly defined powers.)

Article 2. The members of the Interstate Council are appointed by the supreme executive bodies of the sovereign states.

The cochairmen of the Interstate Council are the authorized representatives of the Russian Federation and Ukraine, and a third cochairman is appointed on the basis of rotation from among the authorized representatives of other parties.

Article 3. The parties grant the Interstate Council powers on the following questions, connected with the implementation of the treaty:

—monitoring the fulfillment of commitments regarding the debt of the USSR and the distribution of the assets of the USSR in accordance with the treaty;
—holding an inventory and an appraisal of the debts and assets of the USSR;
—adopting decisions on the operative regulation of the commitments regarding the debt of the USSR and the distribution of assets of the USSR;
—carrying out the reorganization of the USSR Foreign Economic Bank without prejudicing credit agreements which have been concluded;
—implementing the monitoring of the activity of the standing supervisory body;
—attracting independent financial and juridical consultants;
—examining other issues which have been tabled for discussion by the Interstate Council, the initiative of the parties, the supervisory body, or the board of the authorized bank.

Article 4. Votes in the Interstate Council are distributed between its members in proportion to the shares specified for each of the sides under Article 4 of the treaty, and by additional agreements between the sides.

The Interstate Council is empowered to adopt decisions on issues submitted for its consideration, if Interstate Council members representing in total no less than 80 percent of the votes of the sides which signed the treaty are taking part in a sitting.

Decisions of the Interstate Council are adopted by a qualified majority of 80 percent of the votes represented by the Interstate Council members taking part in a sitting.

The sides which signed the treaty have the right to participate in the work of the Interstate Council as observers.

Article 5. A working apparatus is being set up to organize the current activity of the Interstate Council. The activity of the Interstate Council and its working apparatus is financed by the parties on a shared basis in proportion to the payments to pay off the external debt.

Article 6. The Interstate Council ceases its activities after the fulfillment of the treaty.

* * *

The Rules were signed by representatives of the governments of Armenia, Belarus, Kazakhstan, Kyrgyzstan, Moldova, Russia, Tajikistan, Ukraine.

**Protocol on Foreign Debt and Assets**

The governments of the Azerbaijan Republic, the Republic of Armenia, the Republic of Belarus, the Republic of Georgia, the Republic of Kazakhstan, the Republic of Kyrgyzstan, the Republic of Moldova, the Russian Federation, the Republic of Tajikistan, Turkmenistan, the Republic of Uzbekistan, and Ukraine have agreed to the following:

1. To entrust the Interstate Council with supervising the servicing of the USSR's debt and the use of its assets with regard to submitting proposals directed at enhancing the efficiency of the work of the Interstate Council, including the allocation to it of essential funds, the setting up of the working apparatus, and other matters, for consideration by the regular sitting of the Council of Commonwealth Heads of Government.

2. To instruct the Interstate Council to present within one month to the Council of Heads of Government of the Com-

monwealth information on the existing assets of the former USSR abroad.

Completed in the city of Moscow on 13 March 1992 in a single original copy in Russian. The original copy is being kept in the archives of the government of the Republic of Belarus, which will send an authorized copy to the states who have signed the said protocol.

## 10.6

## Agreement on the Former USSR's Internal Debt

TASS, 14 March 1992
[FBIS Translation]

### Agreement on the Principles and Mechanism for Servicing the Former USSR's Internal Debt

Recognizing the need for continuity in meeting the former USSR's obligations toward citizens, proceeding from the need to provide financial backing for measures aimed at economic recovery and transition to the market, and to pursue coordinated financial, credit, and social policies, the governments of the states participating in this agreement have agreed on the following:

Article 1. For the purposes of the present agreement, the state's internal debt consists of money owed by the former USSR government to the population in respect of deposits still held in savings banks. It also consists of the state loans floated in 1982 and 1990 but not yet repaid, including additional compensation, income, and interest deriving from them. Also included are amounts of compensation for citizens' contributions to long-term insurance contracts and other debts owed to the USSR State Bank, the USSR State Insurance, and the USSR Savings Bank arising from funds used to finance the expenditure side of the union budget.

Article 2. The parties accept liability for repaying the former USSR's state debt to the public in sums proportional to the residual debt outstanding on 1 January 1991 on balances held with branches of the USSR Savings Bank on the territory of each of the parties.

The remaining part of the debt—that owed to the USSR State Bank, the USSR State Insurance, and other components of the internal debt—will be apportioned among the parties on an integrated scale, based on the respective contribution of each party to produce national income and the utilization of centralized capital investment from the union budget on the territory of each party, as averaged out over 1986–1991.

Article 3. All expenditures involving the servicing of the former USSR's internal debt for the period starting 1 January 1992 are made by the sides at the expense of their state budget.

Article 4. To convert state domestic premium loan bonds issued in 1982 quickly and to convert the former USSR's state treasury obligations regarding the loan issued in 1990 by entering the relevant sums into bank accounts and by exchanging state loan bonds and other forms of attracting the funds of the population of the states party to this agreement;

Not to conduct future prize draws on the USSR state domestic premium bond issued in 1982.

Article 5. The parties' finance ministries are charged with taking appropriate amounts of the former USSR's state debt to the population onto their balances from the USSR State Bank and converting it into the debt of these states in the manner agreed with the central (national) banks. The procedure for the division and mutual settlements of the residual part of the former USSR's state domestic debt is determined by a separate agreement, taking into account the provision emanating from Article 2 of the present agreement.

Article 6. The agreement comes into force from the moment it is signed.

Completed in Moscow on 13 March 1992 in one original copy in Russian. The original copy is kept in the archives of the Republic of Belarus, which will send a certified copy of it to the states which have signed this agreement.

The agreements have been signed by representatives of the governments of Armenia, Belarus, Georgia, Kazakhstan, Kyrgyzstan, Moldova, Russia, Tajikistan, Turkmenistan, Uzbekistan, and Ukraine. Moreover, the representatives of Belarus, Moldova, Turkmenistan, and Ukraine have made individual notes.

### Appeal of the CIS Heads of Government and the Republic of Georgia on Adopting the 13 March 1992 Agreement on the Principles and Mechanism for Servicing the Former USSR's Internal Debt

Considering the social implications involved in servicing the USSR's internal debt and considering it is essential to safeguard the rights of every citizen of the former USSR, irrespective of the present place of residence within the former USSR's territory, the CIS heads of government and the Republic of Georgia appeal to the heads of former USSR states that have not joined the CIS to accede to the aforementioned 13 March 1992 agreement on the principles and mechanism for dividing and servicing the former USSR's internal debt and to instruct the appropriate bodies to ensure that it is implemented.

The appeal was signed by representatives of the governments of Armenia, Belarus, Georgia, Kazakhstan, Kyrgyzstan, Moldova, Russia, Tajikistan, Turkmenistan, Uzbekistan, and Ukraine. Moreover, the representatives of Belarus, Turkmenistan, and Ukraine have made individual notes. Appended to the document are the following proposals by Belarus, to which Ukraine and Moldova have affiliated themselves:

### Proposal by the Republic of Belarus on the Agreement on the Principles and Mechanism for Servicing the Former USSR's Internal Debt

Article 2 in the aforementioned agreement envisages that the Commonwealth states accept liability for repaying the former USSR's state debt to the population of their states in sums corresponding to the residual debt outstanding on 1 January 1991 on balances held with branches of the USSR Savings Bank on the territory of each of the Commonwealth states.

This debt arose as a result of the borrowing of savings bank funds by union bodies. It was distributed unevenly around the union republics.

In light of this, we believe that the agreement on the partition of the internal debt should be signed at the same time as the agreement on the partition of the assets and liabilities of the former USSR State Bank.

## 10.7
## CIS Treaty on the Formation of an Economic Union

TASS International, 24 September 1993
[Translation by the Council of Advisors to the Parliament of Ukraine]

The states—participants of the current treaty, hereinafter referred to as agreeing parties,

—based on the historic closeness of their peoples and realizing the importance of expansion and intensification of versatile and mutually advantageous economic relations;
—respecting the sovereignty of every nation and confirming their commitment to the goals and principles expressed in the constituent documents on the formation of the Commonwealth of Independent States;
—in order to create favorable conditions for the dynamic and harmonious development of the economies and for the implementation of economic reforms aimed at the improvement of the living standards of their peoples;
—realizing the objective necessity of forming and developing a single economic space, based on the free movement of goods, services, labor, and capital, as well as the necessity to strengthen the direct ties between the economic subjects of the agreeing parties;
—realizing also the importance of technological interconnections of highly integrated scientific and technical, as well as industrial potentials of the states;
—in order to create favorable conditions for organic integration of their economies into the world economy;
—emphasizing the equality of all the nations of the former USSR and the urgent necessity to resolve the problems connected with the inheritance of the property of the former USSR, in order to secure the development and intensification of economic cooperation;
—based on the generally accepted norms of international law;

have agreed on the formation of an Economic Union.

### Chapter 1. Goals and Principles of Economic Union

Article 1. The Economic Union shall be established on the basis of voluntary participation, respect for sovereignty and territorial integrity, equal rights and mutual responsibilities among the agreeing parties for the implementation of the principles of this Treaty.

In their activities within the borders of the Economic Union the agreeing parties shall adhere to international legal principles, including:

—non-interference in the internal affairs of others, respect for human rights and liberties;
—peaceful solution of conflicts and non-use of any kind of economic pressure in interstate relations;
—responsibility for assumed commitments;
—elimination of race or of any other kind of discrimination in relation to any legal entities or individuals of the agreeing parties;
—mutual consultations with the purpose of coordinating positions and in order to take appropriate measures in the event of economic encroachment against any of the agreeing parties, on the part of any state or group of states, which are not the participants of this treaty.

Article 2. The goals of the Economic Union are as follows:

—creation of conditions for stable development of the economies of the agreeing parties with the purpose of improving the living standards of their populations;

—gradual formation of the single economic space on the basis of market relations;
—provision of equal possibilities and guarantees for all economic subjects;
—joint implementation of economic projects, which are of interest to all agreeing parties;
—joint efforts to resolve ecological problems and liquidation of consequences of natural disasters and catastrophes.

Article 3. The Economic Union envisages:
—free movement of goods and services, capital and labor;
—coordination of monetary, credit, budgetary, taxation, foreign economic, customs, currency, and price policies;
—harmonization of the economic legislation of the agreeing parties;
—availability of a common statistical base.

Article 4. The agreeing parties agree that the Economic Union shall be formed through gradual deepening of integration and coordination of actions in the process of implementation of economic reforms, through:
—an international (multilateral) free trade association;
—a single market of goods and services, capital and labor;
—a currency (monetary) union.

Each form of integration shall be realized through a complex of special interconnected measures, which shall be adopted and implemented in accordance with other agreements to this one.

**Chapter 2. Trade and Economic Relations**

Article 5. In accordance with Article 4 of this treaty, in order to create an international free trade association, the agreeing parties concurred on the following principles:

—a gradual decrease in and final abolition of customs duties, taxes and collections, quantitative and other restrictions and limitations;
—harmonization of customs legislation and mechanisms of tariff and non-tariff regulations;
—simplification of customs procedures;
—unification of statistical customs forms and documentation;
—gradual equalization of tariffs for cargo and passenger transportation, as well as transit tariffs, preserving the principle of free transit;
—prohibition of non-legitimate reexport to third countries.

Article 6. In the process of establishing a customs union, the agreeing parties agreed to abolish tariff and non-tariff regulation of the movement of goods, services, and labor, as well as to:

—establish common tariffs with states which are not participants in this treaty;
—coordinate foreign trade policies with states which are not participants in this treaty.

Article 7. In the process of making the transition to a single market the agreeing parties shall:

—create the necessary legal, economic, and organizational conditions for free movement of capital and labor;
—create conditions for fair competition, including the elaboration of anti-monopoly regulations;
—conduct coordinated policy in the fields of transport and communications, aimed at realizing effective cargo and passenger transportation;
—ensure equal economic conditions for capital investment in the development of the economies, and elaborate effective mechanisms for protection of rights and interests of investors.

Article 8. Trade relations will be based on free market prices, which will be set through the integration of internal markets of the agreeing parties. The agreeing parties pledge not to use discriminatory price policies toward their economic subjects, regardless of nationality.

Article 9. Agreeing parties shall not conduct without coordination any unilateral activities of a non-economic character with the aim of limiting access to their markets.

**Chapter 3. Entrepreneurship and Investments**

Article 10. Agreeing parties shall ensure legal national regulation of economic activity by residents on the territories of member states of this Treaty.

Article 11. Agreeing parties shall promote the development of direct economic relations between economic subjects, creating favorable conditions for strengthening productive cooperation.

Article 12. Agreeing parties shall promote the creation of joint ventures, transnational corporations, networks of commercial and financial credit institutions and organizations.

Article 13. Agreeing parties shall coordinate their investment policies, including the attraction of foreign investment and credits in branches of mutual interest, and shall conduct joint capital investments, including those made on a barter basis.

### Chapter 4. Relations in the Sphere of Money, Credits, Finance, and Currencies

Article 14. Agreeing parties shall coordinate their policy in the sphere of money, credits, currency, and financial regulations.

Article 15. In the functioning of interstate free trade, the agreeing parties shall use in their financial relations:

—a multicurrency system, which shall include national currencies functioning in separate states;
—a system based on the Russian Federation ruble.

The transition to a single currency system for mutual payments will be ensured whenever a currency union has been created. This system should be based on a common (reserve) currency, which shall be the most utilized and stable currency.

Article 16. The creation of a monetary and currency system based on national currencies shall be implemented in stages through the creation of a Payments Union based on the following principles:

—mutual recognition of national currencies and recognition of their quoted values;
—the realization of payments in national currencies with the use of multilateral clearing mechanisms through the Interstate Bank and other payment centers;
—the introduction of a mechanism for coordinating deficits in balance of payments;
—establishing a standard conversion rate for national currencies in current operations;

As the integration process deepens, the Payments Union shall be transformed into a unified currency system which shall stipulate:

—the use of floating exchange rates and the coordination of limits on their standard fluctuations;
—the introduction of a banking mechanism for controlling exchange fluctuation rates;
—the achievement of full currency conversion between national currencies.

Article 17. The agreeing parties will join a unified ruble zone based on the use as legal tender of the Russian ruble in accordance with the undermentioned:

—until the activity of the Interstate Bank commences as the emissions institute of the states using the ruble as a national currency, the authorities of the central (national) banks which make credit and monetary emissions shall be delegated to the Central Bank of the Russian Federation. Central (national) banks of these states commit themselves to coordinate their credit emission with the Russian Federation Central Bank;
—relations between central (national) banks shall be established through a special interbank treaty;
—limitations on the use of rubles in the interstate payments of states which use the Russian ruble shall be removed;
—[a balance of payments deficit between states will not be treated as a mutual debt which is subject to regulation].*

**Note:* Proposal by Ukraine, supported by all experts except those of the Russian Federation.

Article 18. The states of the common ruble zone commit themselves to common principles in the implementation of monetary and credits policies:

—on deposits of jointly owned money on their territories in keeping with anticipated price indices;
—on standards governing obligatory reserve demands on commercial banks acting on the territories of the agreeing parties;
—on the maximum volume of credits to be issued to government and local power organs by the commercial banks of the agreeing parties;
—on the level of the discount rate on loans granted by central (national) banks to commercial banks;
—on the rules of payment between economic subjects, and also between commercial banks, including rules of opening an account by a bank and by a non-resident economic subject from a third country;
—on regulation of commercial banking activities;
—on a regular basis provide each other with the balances of central (national) banks and of the banking system as a whole, as well as with other required information.

These states will apply a coordinated ruble exchange rate to hard currency and to the currency of third countries including participants in the Economic Union who use their own currencies.

Article 19. Agreeing parties which enter into the ruble zone will carry out a coordinated budget policy, which stipulates:

—coordinated limitations on the consolidated budget deficit, as compared with the gross national product;

—methods of financing states' budget deficits;
—conditions, kinds, and magnitude of non-budget fund formation;
—limits on increases in the amount of internal debt on the emission base.

The agreeing parties commit themselves to the implementation of coordinated measures aimed at the consecutive decrease of consolidated budget deficits and non-budget funds.

Agreeing parties which exceed the coordinated deficit of their consolidated budgets, should, during a stipulated term, take measures for their normalization.

Article 20. The agreeing parties shall implement a harmonization of their taxation systems. They shall unify basic types of taxes, and the legal regulations which govern the procedures of taxation and taxation rates.

The harmonization of the agreeing parties' taxation systems will be implemented through a special treaty on taxation policy within the framework of the Economic Union, and through a unified methodology of cost accounting.

## Chapter 5. Social Policy

Article 21. The agreeing parties will provide their citizens with "non-visa" regulations within the territory of the Economic Union.

Article 22. The agreeing parties recognize the need to coordinate their policies in the sphere of labor relations, taking into account statements, conventions, and recommendations of the International Labor Organization; to regulate household incomes on the basis of the condition of the manufacturing and consumer markets; and to maintain the standard of living for the handicapped and low-income families.

Article 23. The agreeing parties will extend mutual recognition to documents of education and worker qualifications without additional confirmation, where conditions and the character of the work permit it.

Article 24. The agreeing parties will not permit discrimination against citizens on the basis of nationality or any other basis in the provision of jobs, payment for labor, determination of work conditions, and extension of social guarantees.

Article 25. The agreeing parties shall coordinate policy in the sphere of labor conditions and labor protection, taking into consideration the generally accepted international rules, developing general requirements for norms and rules of labor protection, state supervision, and inspections of working conditions.

Article 26. The agreeing parties shall agree specially on rules for labor force migration, commitments dealing with social security, pension guarantees, and other issues requiring mutual consent among the states forming the Economic Union.

## Chapter 6. Legal Regulation of Economic Relations

Article 27. Economic relations among the agreeing parties and the subjects of their economies shall be regulated by this treaty, bilateral and multilateral agreements, international legal norms, and national legislation. When this treaty's norms and rules are at variance with national legislation, the rules and norms of international law and of this treaty shall prevail.

Article 28. The agreeing parties recognizing the need to standardize the regulation of economic relations, agree to align themselves with the norms contained in this treaty and under international law.

In light of the foregoing, the parties agree:

—to develop model legal statements regulating economic relations;
—to coordinate and closely align national legislation with model projects and international legal norms in order to eliminate contradictions among them;
—to agree on the adoption of national legislation on economic issues;
—to examine normative statements in advance in order to ensure their correspondence with international legal norms, this treaty, and bilateral and multilateral agreements.

## Chapter 7. Institutions of the Economic Union

Article 29. In order to sustain the activity of the Economic Union the agreeing parties shall use existing and establish new joint executive and coordination bodies.

Article 30. Procedures for establishing, operating, and financing the institutions of the Economic Union, together with their coordination with other CIS economic bodies, shall be regulated by special agreements.

## Chapter 8. Concluding Statements

Article 31. Membership in the Economic Union requires acceptance of the full scale of commitments and the extension of all rights stipulated by this Treaty; it does not prohibit economic ties between states which are not members or with other economic groups and societies, as long as these do not run counter to the interests of the Economic Union.

The agreeing parties bear mutual responsibility for implementing this treaty, and abstain from any steps which would threaten its non-fulfillment.

Article 32. Any state wishing to assume only a portion of the commitments contained in this treaty may receive the status of associate member by consent of the members of the Economic Union. Conditions for joining the Economic Union as an associate are determined by its members.

Article 33. The agreeing parties are committed to discussions of issues connected with the interpretation and implementation of the current treaty through negotiation or through an appeal to the Commonwealth of Independent States Economic Court.

If the Economic Court declares that a member state has not fulfilled one of the commitments imposed by the current treaty, the state shall comply with the judgments of the court.

The agreeing parties shall agree separately on procedures for considering issues connected with the economic relations of the subjects of member states of the Economic Union, as well as on a system of sanctions to be used in cases of non-compliance with the tenets of this treaty.

If it is impossible to settle issues through negotiations or through the Commonwealth of Independent States Economic Court, the agreeing parties agree to discuss them in other international legal bodies in accordance with rights and procedures.

Article 34. No reservations with respect to this treaty shall be permitted.

Article 35. This treaty shall be concluded for ten years, and shall be automatically extended for five years if none of the agreeing parties declares its intention to withdraw from it.

Every agreeing party may declare its secession from this treaty, having informed the other agreeing parties no less than twelve months in advance. Secession procedures are discussed in a special agreement.

Article 36. This treaty shall be ratified by the agreeing parties in accordance with their constitutional procedures.

Any state which recognizes this treaty may join by consent of the treaty members.

This treaty comes into force after the third instrument of ratification has been submitted to the state depositor.

For every state which ratifies this treaty or joins it after the third instrument of ratification has been submitted to the state depositor, the treaty shall come into force on the thirtieth day after the state has submitted its instrument of ratification or a statement on joining the treaty.

The state depositor of this treaty is the Republic of Belarus.

Accomplished in the city of Minsk, 24 September 1993, in a single original copy in the Russian language. The original copy of this treaty shall be kept in the archives of the government of the Republic of Belarus, which will send a certified true copy to the states which have signed this treaty.

[Signed]
For the Republic of Armenia
For the Republic of Belarus
For the Republic of Kazakhstan
For the Republic of Kyrgyzstan
For the Republic of Moldova
For the Russian Federation
For the Republic of Tajikistan
For Turkmenistan
For the Republic of Uzbekistan
For Ukraine

## 10.8

## Russia, Belarus, Kazakhstan Sign Customs Union Agreement

*Rossiyskaya Gazeta,* 28 January 1995
[FBIS Translation]

The government of the Republic of Belarus and the government of the Russian Federation, on the one hand, and the government of the Republic of Kazakhstan, on the other hand, referred to hereinafter as the "contracting parties," seeking to further develop balanced and mutually advantageous economic relations, expressing the intention to continue to implement the Treaty on the Creation of an Economic Union of 24 September 1993, and desiring to initiate the establishment of a Customs Union between them, have agreed as follows:

Article 1. The contracting parties shall establish a single Customs Union, whose objectives, establishing principles, mechanisms and phases of creation, and operating procedures, as well as the distribution of customs duties, taxes, and fees, and the terms governing the imposition of temporary restrictions and the implementation of customs control, are set forth in the Agreement on a Customs Union Between the Russian Federation and the Republic of Belarus of 6 January 1995.

Article 2. The contracting parties shall assume the full rights and responsibilities arising from the Agreement on a Customs Union Between the Russian Federation and the Republic of Belarus of 6 January 1995, with respect to the objectives, operating principles, and mechanism and phases of creation of the Customs Union, the distribution of customs duties, taxes, and fees, and the terms governing the imposition of temporary restrictions and the implementation of customs control. At the same time, regulation of the foreign economic activity of the Republic of Kazakhstan shall be effected in accordance with the Agreement Between the Government of the Russian Federation and the Government of the Republic of Kazakhstan on a Standardized Procedure for the Regulation of Foreign Economic Activity of 20 January 1995.

Article 3. With a view to implementing this agreement, the contracting parties, on the basis of a separate agreement, shall establish an executive agency of the Customs Union.

Prior to the establishment of an executive agency of the Customs Union, the contracting parties shall abide by the provisions of Agreement Between the Government of the Russian Federation and the Government of the Republic of Belarus of 6 January 1995.

Article 4. The contracting parties have agreed that the Agreement Between the Government of the Russian Federation and the Government of the Republic of Kazakhstan on a Standardized Procedure for the Regulation of Foreign Economic Activity of 20 January 1995, and the Protocol on the Introduction of a Free Trade Regime, without exclusions and limitations, between the Russian Federation and the Republic of Kazakhstan of 20 January 1995, shall constitute inalienable parts of this agreement.

Article 5. This agreement shall not affect the right of any contracting party to adopt, in accordance with international law and its domestic legislation, any measures necessary to safeguard national security, public order, public health and morals, the cultural and historical heritage of their peoples, and rare animals and plants.

Article 6. Disputes and disagreements between the contracting parties with respect to the interpretation and/or application of the provisions of this agreement shall be resolved through consultation.

Article 7. This agreement shall not affect the validity of other international treaties of the Republic of Belarus, the Russian Federation, and the Republic of Kazakhstan that do not conflict with this agreement.

Article 8. Each contracting party may cease its participation in the agreement by furnishing official notification in writing to the other contracting parties of its intention to withdraw from the agreement twelve months prior to said withdrawal.

Article 9. This agreement shall be applied provisionally as of the date of signing and shall enter into force on the date of final notification of completion by the Republic of Belarus, the Russian Federation, and the Republic of Kazakhstan of internal state procedures required for its entry into force.

## 10.9

## Commonwealth Countries Sign Payment Union Agreement

*Rossiyskaya Gazeta,* 28 January 1995
[FBIS Translation]

### Agreement on the Creation of a Payment Union of the Member States of the Commonwealth of Independent States

The member states of the Commonwealth, referred to hereinafter as the parties, believing the creation of a Payment Union and of an effective payment system to be a necessary condition for the normal functioning of an Economic Union, fostering conditions for the free movement of goods and services, seeking to ensure balanced trade and economic relations, promoting the growth of the economic potential of their national economies on the basis of mutually beneficial cooperative and other economic relations, mindful of the close historical, cultural, and ethnic ties of their peoples, and promoting conditions for the gradual establishment of an optimal system of mutual settlements, have agreed as follows:

Article 1. To create a Payment Union through the voluntary drawing together of the parties for the purpose of ensuring uninterrupted settlements through the use of mutual convertibility of their national currencies and the formation of a payment system on this basis.

The parties view the creation of a Payment Union as a gradual process and shall set about its implementation through the conclusion of bilateral and multilateral agreements.

In the subsequent stage the parties may adopt measures to establish a multilateral system for effecting settlements in a collective currency.

Article 2. The establishment of the Payment Union shall be based on the following principles:

- recognition of national currency sovereignty and the role of the central (national) banks of the parties as emission centers and agencies of cash-credit and currency regulation on the territory of each party;

- the inadmissibility of any restrictions on the territory of each party on the acceptance and use of the national currencies as payment instruments in trade and non-trade transactions set forth in contracts;
- the establishment of state foreign trade and currency regulations that promote the development on the territory of each party of a full-fledged (liquid) currency market for transactions relating to the buying and selling of its national currency for the national currencies of the other parties and for other currencies;
- the ensuring of guaranteed conversion of a national currency into the currencies of the other parties with regard to socially significant payments: remittances of pensions, alimony, state entitlements, disbursements, and compensation payments, including payments to employees to compensate for damages arising from injury, occupational illness, or other loss of health sustained on the job; sums paid on the basis of sentences, rulings, determinations, and resolutions of judicial and investigative agencies; payments relating to the death of citizens; monetary compensation to victims of political repression and members of their families and their heirs; and compensation for expenses incurred by legal, investigative, arbitration, notary, and other law-enforcement agencies;
- the granting of permission to authorized commercial banks to effect settlements relating to foreign economic transactions and to extend credits to correspondent banks and other non-residents that are parties to foreign economic transactions;
- the inadmissibility of administrative restrictions with respect to determination of the currency of payment in the concluding of contracts between economic entities of the parties;
- the permitting of non-residents to hold a national currency and to use it to pay for goods and services and to effect payments of a non-trade character within the framework of the parties.

Article 3. The payment system to be established on the basis of the Payment Union shall service settlements relating to: trade turnover in the non-state and entrepreneurial spheres; non-trade transactions; services in the transportation, communications, and other spheres; state, bank, and commercial credits; currency exchange (conversion) transactions; and the buying and selling of currency in cash through the banking systems of the parties.

The participants in the payment system shall be:

- the governments of the parties, with respect to bilateral and multilateral agreements concluded between or among them;
- the central (national) banks that ensure the functioning of the payment system and the narrowing of differences between currency norms and regulations;
- commercial banks authorized by the central (national) banks to process transactions in foreign currencies;
- an Interstate Bank, which shall constitute a specialized institution of the Payment Union and shall effect settlements between the central (national) and other banks of the parties, and perform other functions in the interests of the Payment Union;
- economic entities—legal entities, entrepreneurs operating without the formation of legal entities, and physical persons—that are residents of the parties, with respect to the performance of non-trade transactions.

Article 4. The participants in the payment system may, at their discretion, use the national currencies of the parties or other currencies as the currency of payment.

International settlements shall be effected through correspondent accounts in the central (national) and authorized commercial and other banks.

The participants in the payment system may freely place funds in the national currency of another party on the latter's domestic money market in the procedure established by the legislation of that party.

The parties shall use a standardized currency exchange rate for all types of foreign economic transactions.

The participants in the payment system shall use an exchange rate determined by supply and demand on the parties' currency markets.

The parties shall seek to take coordinated steps to maintain the stability of the exchange rates of their national currencies.

With a view to regulating the exchange rates, the central (national) banks shall create and utilize stabilization funds in freely convertible currencies and precious metals.

The parties shall ensure access to their domestic currency markets for non-resident banks, as well as for the Interstate Bank, in accordance with their national legislation.

The exchange of the parties' national currencies by physical persons shall be effected by commercial banks and other banking institutions at the exchange rate in effect on the currency market.

The parties shall seek to liberalize regulations governing the exchange of balances in the accounts of non-residents in their national currencies for the national currencies of the other parties.

The central (national) banks shall not be responsible for the performance of obligations incurred by the parties' commercial banks.

Article 5. With a view to promoting multilateral cooperation in the sphere of currency-payment and credit relations, narrowing differences between currency regulations relating to the multilateral settlement-credit relations of the Parties and improving such regulations, and developing forms and methods of coordinating cash-credit and currency policies, the Parties shall create an Interstate Currency Committee.

Article 6. In order to provide information and technical support for the payment system, the Interstate Bank and associations of commercial banks shall establish data banks. When necessary, a specialized international organization to provide information services for the payment system shall be created in the form of a closed-type joint-stock company, a controlling bloc of shares in which shall be held by the central (national) banks, with the remaining shares to be distributed among commercial banks.

Article 7. The parties shall cooperate and take concerted actions in currency control matters on both a multilateral and bilateral basis, in connection with which they shall direct their respective executive bodies to conclude the necessary agreements.

Article 8. The parties shall take measures to foster favorable credit-banking and currency-financial conditions for the establishment and development of the operation of transnational financial–industrial and banking groups and the broadening of international production specialization and cooperative production, scientific–technical and investment cooperation, and reciprocal trade turnover.

The parties shall direct the Interstate Bank to facilitate the provision of credits and financing for interstate projects, in conjunction with the commercial banks of the parties and international credit–financial organizations.

Article 9. The parties shall intensify reciprocal contacts in order to solve the non-payments problem, as well as to reduce the volume of barter deals, by taking measures to introduce interest-bearing notes in international economic turnover, as well as a multilateral clearing mechanism through the Interstate Bank and other specialized institutions.

Article 10. Any disputes or disagreements between the parties regarding the interpretation and/or application of the provisions of this agreement, as well as other disputes relating to the rights and responsibilities of the parties under this agreement or in connection with it, shall be resolved in the following procedure:

- through direct consultations between the parties concerned;
- within the framework of a special reconciliation procedure involving the creation of working groups to study the dispute and to issue recommendations;
- on the basis of other procedures established by international law.

A shift to a subsequent procedure shall be made possible by the mutual consent of the parties between which the dispute or disagreement arose, or at the request of one of the parties, provided efforts to reach agreement have been unsuccessful for a period of six months from the date of commencement of the procedure.

Article 11. This agreement may be altered or amended by the mutual consent of the Parties.

No qualifications with respect to this agreement shall be permitted.

Article 12. This agreement shall be open for accession to any member state of the Commonwealth of Independent States that recognizes the provisions of this agreement in effect at the time of accession and that expresses a readiness to abide by the agreement in its entirety.

Accession shall be effected on the terms and in the procedure set forth in a separate agreement with the acceding state; said agreement shall be reconciled with the parties in advance and shall be subject to their approval in accordance with their internal state procedures.

Article 13. Each Party shall have the right to freely withdraw from the agreement by providing written notification to the depositary at least six months prior to withdrawal. The depositary shall inform all the parties of said withdrawal within 30 days. In the event of a breach by any party of the provisions of this agreement that causes substantial damage to the achievement of its objectives, the other parties shall be empowered to adopt a decision to suspend the agreement or individual provisions of it with respect to that party or to adopt a decision to expel that party from the agreement participants.

In order to settle potential disputes and claims, including property-related disputes and claims, the provisions of this agreement shall remain in force with respect to a party that has ceased its participation until such time as all claims have been settled in full.

Article 14. This agreement shall enter into force as of the date of placement with the depositary of a third notification that the parties that have signed it have completed all the internal state procedures required for its entry into force.

The depositary of this agreement shall be the Republic of Belarus.

Executed in the city of Moscow on 21 October 1994, in one

original copy in the Russian language. The original copy shall be placed in the archives of the government of the Republic of Belarus, which shall provide a notarized copy to each state that has signed the Agreement.

## 10.10

### Commonwealth States Sign Agreement on Aid to Refugees

*Rossiyskaya Gazeta,* 28 January 1995
[FBIS Translation]

**Agreement on Aid to Refugees and Forced Resettlers**

The states signatories to this agreement, henceforth referred to as the parties, on the basis of commonly accepted principles of international law and humanitarianism, confirming their obligations in accordance with international agreements aimed at the protection of human rights, taking into consideration the critical situation that has formed in connection with the rise in the number of resettlers and refugees on the territory of the former USSR, recognizing their responsibility for the fate of people experiencing hardships and deprivations, and recognizing the need to extend aid to the refugees and forced resettlers, have agreed as follows:

Article 1. For the purposes of this agreement a refugee is recognized as an individual who is not a national of the party granting asylum, who was forced to abandon the place of his permanent residence on the territory of another party as a result of violence or persecution in various forms against himself or members of his family, or because of a real threat of persecution because of his racial or national origin, religious faith, language, political convictions, or affiliation with a certain social group in connection with armed and international conflicts.

An individual who has committed an offense against peace or humanity or some other premeditated criminal act cannot be recognized as a refugee.

Article 2. For the purpose of this agreement an individual who, being a national of the party which granted asylum, was forced to abandon the place of his permanent residence on the territory of another party as a result of violence or persecution in various forms against himself or members of his family, or because of a real threat of persecution because of his racial or national origin, religious faith, language, political convictions, or affiliation with a certain social group in connection with armed and international conflicts is recognized as a forced resettler.

Article 3. The status of refugee or forced resettler is determined in accordance with this agreement, generally recognized norms of international law, and the legislation of the party which granted asylum and is confirmed by the issuance of appropriate documentation.

Article 4. The state of departure, with the cooperation of interested parties:

- Carries out evacuation of the population from zones of armed and interethnic conflict, granting the opportunity for its unhindered and voluntary movement to the territory of one of the parties on the basis of the provisions of Article 1 and Article 2 of this agreement;
- Ensures the personal and property security of evacuees, striving for a cease-fire of public order during such evacuation;
- Matters concerning financial, material–technical, food supply, medical, and transportation support of evacuees will be resolved among the interested parties.

Article 5. The parties providing asylum assume the following obligations:

- To ensure the availability of necessary social and household services for refugees and forced resettlers in the areas of their temporary deployment;
- To assist refugees and forced resettlers with job placement in accordance with legislation adopted by each of the parties on employment.

Article 6. The parties assume the following obligations:

- To assist refugees and forced resettlers with their demands and the acquisition of documents necessary for resolution of questions connected with citizenship;
- To assist refugees and forced resettlers with the acquisition of certificates of marriage, birth, labor books, and other documents at their place of former residence that are needed to resolve issues connected with pensions, confirmation of work background, travel abroad, etc.;
- To assist with acquisition of certificates on relatives living on the territory of the state abandoned by the refugee or forced resettler, as well as on his property left there.

Article 7. The state of departure reimburses refugees and forced resettlers the value of housing and other property left or lost by them on its territory, and compensates for damage to health and loss of earnings. The amount of material compensation is determined on the basis of evaluation by the state of departure.

The order of accounting for lost personal and real property belonging to refugees and forced resettlers and determination of material damage and payments of compensation is determined jointly with the interested parties.

Article 8. The parties will create an Interstate Fund for Aid to Refugees and Forced Resettlers.

The conditions, order of formation, and utilization of fund assets will be determined by statute, which will become an intrinsic party of this agreement.

Article 9. The Consultative Council for Labor, Migration, and Social Protection of the Population of the Commonwealth of Independent States will provide practical assistance with realization of the provisions included in this agreement.

Article 10. Every refugee or forced resettler has the right to appeal to courts on the territories of the parties.

Article 11. The parties shall take measures to ensure their participation in international agreements on problems of refugees and forced resettlers.

The parties shall bring national legislation into accord with international legal norms in that sphere.

Article 12. This agreement is subject to ratification.

The agreement goes into effect after deposit of the third instrument of ratification with the depository. In the case of parties that ratify it later, it goes into effect on the date they deposit their instruments of ratification.

Article 13. Any one of the parties may denounce this agreement by means of written notification submitted to the depository. The agreement terminates for that party six months after the day of receipt of such notification by the depository.

Article 14. With the consent of all parties this agreement is open to other states sharing its goals and principles, which may join it by submitting documents on such affiliation to the depository. Affiliation is considered to be effective on the day the depository receives notification concerning agreement to such affiliation.

Done in the city of Moscow in one original copy in the Russian language. The original copy is kept in the archives of the government of the Republic of Belarus, which shall forward a certified copy of it to the states that are signatories to this agreement.

The agreement has been signed by the heads of state of Azerbaijan, Armenia, Belarus, Kazakhstan, Kyrgyzstan, Russia, Tajikistan, Turkmenistan, and Uzbekistan.

# The Military Accords

## 10.11

**Agreement on the Powers of the Highest Bodies of the Commonwealth of Independent States on Questions of Defense**

30 December 1991

Translated by David Ridler

The member states of the Commonwealth, hereafter referred to as "the member states,"

—guided by the principle of the provisional agreement on a Council of Heads of State and a Council of Heads of Government of the Commonwealth of Independent States of 30 December 1991,

—recognizing the need to ensure the security of each member state,

—taking into consideration their interest in coordinating the activity of the member states of the Commonwealth in solving questions of strengthening defense capability,

—proceeding from the understanding that the Joint Armed Forces include the strategic forces and the armed forces of the member states of the Commonwealth—on the decisions of these states—as well as special purpose forces (apart from the armed forces of states who are not part of the Joint Armed Forces), have agreed to the following:

Article 1. The Council of Heads of State is the highest body of the Commonwealth on questions of defense.

The Council of Heads of Government carries out the coordination of the military–economic activity of the Commonwealth.

Article 2. The Council of Heads of State:

—works out and implements the military policy of the Commonwealth;

—determines the concept of collective defense and the basic directions of military construction;
—adopts the military doctrine and nuclear strategy of the Commonwealth;
—establishes the procedures for adopting decisions for the use of nuclear weapons and the system of measures for excluding unsanctioned use, and also the procedure for carrying out the unified control of nuclear weapons and other weapons of mass destruction;
—determines by referral to the heads of government of the Commonwealth the volume of allocations and material–technical resources for defense and the upkeep of the Joint Armed Forces of the Commonwealth;
—establishes the composition and structure of the Joint Armed Forces of the Commonwealth, creates the Main Command of the Joint Armed Forces of the Commonwealth, and determines its powers;
—establishes the procedure for the performance of military service in the Joint Armed Forces of the Commonwealth;
—reviews the plan for the development of the Joint Armed Forces of the Commonwealth, the mobilization plan of the Joint Armed Forces of the Commonwealth, and the plan for their adaptation in wartime;
—appoints on recommendation to the Council of Ministers of Defense, which in the future will be the Council of the Ministers of Defense, a commander-in-chief, chief of the General Staff, and deputy commanders-in-chief of the Joint Armed Forces of the Commonwealth, and also a commander of the Strategic Forces;
—confers the rank of Army General and its equivalent;
—adopts decisions:
—on the introduction of martial law throughout the territory of the Commonwealth in the event of aggression or the threat of aggression against the Commonwealth, against several of its member states or one of them, on a declaration of war, on the conduct of military operations, on the lifting of martial law, on the cessation of the state of war and on the conclusion of a peace treaty;
—on the procedure for bringing into force normative acts in wartime and of the termination of their validity;
—on the use of contingents of the Joint Armed Forces of the Commonwealth through the necessity to fulfill international treaties and obligations;
—and other decisions on the more important questions of defense.

The decisions of the Council of Heads of State are taken on the basis of a consensus.

Article 3. The Council of Heads of Government:

—submits to the Council of Heads of State the draft unified defense budget of the Commonwealth;
—works out, together with the High Command of the Joint Armed Forces of the Commonwealth, an agreed program for developing armaments and military hardware for the Joint Armed Forces of the Commonwealth for an appropriate period, the finances for this program within the limits of allocations for defense and the upkeep of the Joint Armed Forces of the Commonwealth, and the priorities in meeting orders for military goods;
—establishes the procedure for adopting armaments, military hardware, and other military property for the Joint Armed Forces of the Commonwealth and the procedure for their material and technical provisioning;
—determines the procedure for conducting scientific research and developmental work in the defense field, and ensures via the appropriate bodies of the independent states that the Joint Armed Forces of the Commonwealth are equipped with armaments, military hardware, and other material requirements and that the Joint Armed Forces of the Commonwealth are provided with essential services;
—agrees on the annual contingent of citizens who are liable for conscription for military service in the Joint Armed Forces of the Commonwealth and the annual number of forces-trained specialists liable for training or retraining;
—draws up mobilization plans for the national economy and plans for accumulating material resources for the mobilization reserve;
—establishes tasks for preparing and handing over to the Joint Armed Forces of the Commonwealth means of transport and communications and other material and technical facilities when mobilization is declared as well as other mobilization tasks in war time;
—takes decisions on the creation, development, and procedure for the use of defense facilities and lines of communication between the communications and transport networks on the territory of member states in the interests of collective defense, and the system of management within the Joint Armed Forces of the Commonwealth;
—takes decisions regarding social and legal guarantees and financial, material, housing, domestic, and pension provisions for servicemen of the Joint Armed Forces of the Commonwealth, people released from military service and their families, and the families of servicemen who have been killed (or died) in carrying out the obligations of their service.

Article 4. A Council of Defense Ministers from the member states is formed to coordinate military construction. A High Command of the Joint Armed Forces of the Commonwealth is created to carry out decisions taken by the highest bodies of the Commonwealth on defense matters. The provisions on the Council of Ministers of Defense and on the High Command of the Joint Armed Forces of the Commonwealth are endorsed by the Council of Heads of State.

Article 5. This agreement comes into force on signing and applies to the states-signatory to it. Done in the City of Kiev on 20 March 1992, in one original copy in the Russian language. The original copy is to be kept in the archives of the government of the Republic of Belarus, which will send the certified copies to the states-signatory to this agreement.

This agreement is signed by representatives of Armenia, Belarus (with the addition of the words: "with transitional period for the Republic of Belarus-years"), Kazakhstan (with the addition of the words: "It is necessary that measures are taken for creating an effective technical measure for controlling and blocking the use of nuclear missile systems within the shortest possible time"), Kyrgyzstan, the Russian Federation, Tajikistan, and Uzbekistan.

## 10.12
## CIS Agreement on Guarding of Borders
30 December 1991
Translated by David Ridler

**Agreement:**

On the guarding of the state borders and maritime economic zones of the member states of the Commonwealth of Independent States,

—the states that are party to this agreement, henceforth called "the Commonwealth member states," in accordance with the agreement of the Council of Heads of State of the Commonwealth of Independent States on armed forces and border troops of 30 December 1991,

—proceeding from the need to implement mutually acceptable decisions in the interests of guarding the state borders and maritime economic zones of the Commonwealth member states, and taking into account the system and the principles for ensuring inviolability of those borders which have come into being, have agreed the following:

Article 1. In this agreement, the terms below mean:

1. "State borders of the member states of the Commonwealth of Independent States"—the sections of the state borders of independent Commonwealth member states with states that are not in the Commonwealth.
2. "Border troops"—formations of border troops of the Commonwealth and of states' own border troops.
3. "Own border troops"—formations of border troops belonging to a Commonwealth member state.
4. "Commonwealth border troops"—formation of border troops which are not own border troops.

Article 2. The guarding of the state borders and maritime economic zones of the Commonwealth member states is implemented by Commonwealth border troops or by states' own border troops.

Article 3. With the aim of ensuring their security, the Commonwealth member states pledge not to undertake actions on state borders and in maritime economic zones that are to the detriment of the political, economic, or other interests of other Commonwealth member states. The establishment of state borders and changes to their system of operation are implemented by mutual accord with neighboring states and with regard for the interests of the Commonwealth member states.

Article 4. The Council of Heads of State is the supreme coordinating body of the Commonwealth of Independent States in the sphere of guarding of state borders and maritime economic zones of the Commonwealth member states. The Council of Heads of Government carries out the coordination of measures to ensure the guarding of state borders and maritime economic zones.

Implementation of decisions of the Council of Heads of State and the Council of Heads of Government on matters of the guarding of state borders and of maritime economic zones of the Commonwealth member states is effected by the joint command of border troops, which coordinates the activities of border troops.

The regulations on the joint command are confirmed by the Council of Heads of State.

The leadership of the Commonwealth border troops is provided by a commander-in-chief for border troops appointed by the Council of Heads of State.

Article 5. Prior to conclusion by Commonwealth member states of interstate agreements on borders and maritime economic zones and their system of operation, the organization and activities of border troops are regulated by the Acts of the Commonwealth, national legislation of states, and law-making Acts of the former USSR that do not contradict the latter.

Matters of manning, finance, and material–technical supply for Commonwealth border troops and of the social and legal status of those serving in them are regulated by special agreements of the Commonwealth member states.

Article 6. This agreement is open for states which are not member states of the Commonwealth of Independent States to subscribe to it.

Done in the City of Kiev on 20 March 1992 in one original document in the Russian language. The original document is kept in the archives of the government of the Republic of Belarus, which will send a certified copy to the states that have signed this agreement.

Article 1 (Ukraine's wording). The states' borders of the Commonwealth member states are of identical status for the whole of their length, then as given.

Article 4 (Azerbaijan's and Ukraine's wording). The implementation of decisions of the Council of Heads of State and Council of Heads of Government on matters of the guarding of the state border and of maritime economic zones is effected by the competent bodies of the Commonwealth member states.

Collaboration by the Commonwealth border troops and Commonwealth member states' own border troops is implemented on the basis of separate agreements.

The agreement was signed by representatives of Armenia, Belarus, Kazakhstan, Kyrgyzstan, the Russian Federation, Tajikistan, and Uzbekistan.

The representative of Moldova made a note: "Issues of the guarding of the state borders of the republic of Moldova are resolved on the basis of bilateral agreements with the main command of the CIS border troops."

Article 1 (in Ukraine's wording) and Article 4 (in the wording of Azerbaijan and Ukraine) quoted above were signed by the representative of Ukraine.

## 10.13
## CIS Agreement on Strategic Forces

30 December 1991
Translated by David Ridler

*The Agreement on Strategic Forces was concluded between the eleven members of the Commonwealth of Independent States on 30 December 1991.*

### Preamble

Guided by the necessity for coordinated and organized solutions to issues in the sphere of the control of the strategic forces and the single control over nuclear weapons, the Republic of Armenia, the Republic of Azerbaijan, the Republic of Belarus, the Republic of Kazakhstan, the Republic of Kyrgyzstan, the Republic of Moldova, the Russian Federation, the Republic of Tajikistan, the Republic of Turkmenistan, the Republic of Ukraine, and the Republic of Uzbekistan, subsequently referred to as "the member states of the Commonwealth," have agreed on the following:

Article 1. The term "strategic forces" means: groupings, formations, units, institutions, the military training institutes for the strategic missile troops, for the air forces, for the navy, and for the air defenses; the directorates of the Space Command and of the airborne troops, and of strategic and operational intelligence, and the nuclear technical units and also the forces, equipment, and other military facilities designed for the control and maintenance of the strategic forces of the former USSR (the schedule is to be determined for each state participating in the Commonwealth in a separate protocol).

Article 2. The member states of the Commonwealth undertake to observe the international treaties of the former USSR, to pursue a coordinated policy in the area of international security, disarmament, and arms control, and to participate in the preparation and implementation of programs for reduction in arms and armed forces. The member states of the Commonwealth are immediately entering into negotiations with one another and also with other states which were formerly part of the USSR, but which have not joined the Commonwealth, with the aim of ensuring guarantees and developing mechanisms for implementing the aforementioned treaties.

Article 3. The member states of the Commonwealth recognize the need for joint command of strategic forces and for maintaining unified control of nuclear weapons, and other types of weapons of mass destruction, of the armed forces of the former USSR.

Article 4. Until the complete elimination of nuclear weapons, the decision on the need for their use is taken by the president of the Russian Federation in agreement with the heads of the Republic of Belarus, the Republic of Kazakhstan, and the Republic of Ukraine, and in consultation with the heads of the other member states of the Commonwealth.

Until their destruction in full, nuclear weapons located on the territory of the Republic of Ukraine shall be under the control of the Combined Strategic Forces Command with the aim that they not be used and be dismantled by the end of 1994, including tactical nuclear weapons by 1 July 1992.

The process of destruction of nuclear weapons located on the territory of the Republic of Belarus and the Republic of Ukraine shall take place with the participation of the Republic of Belarus, the Russian Federation, and the Republic of

Ukraine under the joint command of the Commonwealth states.

Article 5. The status of strategic forces and the procedure for service in them shall be defined in a special agreement.

Article 6. This agreement shall enter into force from the moment of its signing and shall be terminated by decision of the signatory states or the Council of Heads of State of the Commonwealth.

This agreement shall cease to apply to a signatory state from whose territory strategic forces or nuclear weapons are withdrawn.

---

## 10.14
## CIS Agreement on the Status of the Border Troops

TASS, 30 January 1991
[FBIS Translation]

---

The member states of the Commonwealth, fulfilling the agreement of the heads of the CIS member states on armed forces and border troops of 30 December 1991, have agreed on the following:

Article 1. The protection of the states' borders and the maritime economic zones of the member states of the Commonwealth is carried out by the border troops of the Commonwealth or by the states' own border troops.

Until new normative documents regulating the activity of the border troops are adopted, they are guided by the documents of the Commonwealth, by the national legislation of the states, and by normative documents of the former USSR that do not contravene it.

The border troops of the Commonwealth are not involved in the fulfillment of other tasks, with the exception of the elimination of the consequences of natural calamities, accidents, and disasters.

Article 2. The border troops of the Commonwealth are formed from personnel on the basis of the principles defined by a separate agreement.

Article 3. The participating states in the present agreement recognize the necessity for the existing system for training and raising the qualifications of cadres for the border troops of the Commonwealth to be used and developed.

The training of cadres is carried out to orders from the commander-in-chief of the border troops and commanders of the states' own border troops.

Article 4. The social and legal guarantees of servicemen in the border troops of the Commonwealth, of individuals discharged from military service, and of members of their families is regulated by the 14 February 1992 agreement between the CIS member states on the social and legal guarantees of servicemen, individuals discharged from military service, and members of their families, and by the national legislation of the member states of the Commonwealth.

Article 5. Controlling bodies of Commonwealth border troops undertake their activities in conjunction with appropriate state bodies, enterprises, and organizations of the Commonwealth member states.

Commonwealth border troops conclude contracts with enterprises and organizations of the Commonwealth member states to carry out work for developing and modernizing weaponry and military equipment and for other matters concerned with facilitating the activities of the Commonwealth border troops. Commonwealth member states assist in the conclusion of such contracts.

Movements of formations and units of Commonwealth border troops, and other actions undertaken outside their permanent (base) areas, are carried out in accordance with decisions of the Border Troops Joint Command by arrangement with the government of the states where they are located or with bodies empowered by that government.

Article 6. The member states of this agreement, in the interests of protecting the state borders and maritime economic zones, provide for the military formations of the border troops of the Commonwealth the land, air, and sea space and internal navigation routes necessary for this movement and assist in the movement of these formations.

The provision of air flights and of navigational and hydrographical supplies for ships of the Commonwealth's border troops, as well as the use of mooring and port facilities, airfields and airports, railways and roads and their construction on the territory of the member states of the Commonwealth, connected with the protection of the borders and maritime economic zones, is carried out gratis.

The member states of this agreement provide the border troops of the Commonwealth with plots of land for the siting of engineering and technical buildings and monitoring facilities along the border, for use without time limit free of charge.

The construction of new roads, bridges, buildings, and other facilities in the interests of protecting the borders on the territory of the member states of the Commonwealth is carried out with the consent of the competent bodies.

Article 7. The member states of this agreement take a proportional part in financing the border troops of the Commonwealth and providing them with material and technical supplies. The volume of expenditure for these purposes and the procedure for their financing is determined by separate agreements.

Article 8. The states participating in this agreement retain for the Commonwealth border troops the land which they had before this agreement was signed and also provide them with electric power and communal and other services. The procedure and conditions of use of the land plots that had been allocated (with the exception of cases provided for by Article 6 of this agreement) and of the provision for the Commonwealth border troops of all manner of services are determined in accordance with the laws of the host state and agreements on the mechanism of the activities of the Commonwealth border troops on the territories of the Commonwealth states.

The movable property of the Commonwealth border troops is in their ownership and use. The procedure of what can be done with it is determined by the Council of CIS Heads of Government.

The motor transport resources of the Commonwealth border troops have registration numbers and distinguishing marks. The registration numbers and marks are established in agreement with the High Command of the Commonwealth Joint Armed Forces.

Article 9. In cases where laws have been violated by persons who are members of the Commonwealth border troops or members of their families, the laws in force on the territory of the Commonwealth member state where these violations of the law occurred will apply.

Article 10. The Commonwealth border troops have a border flag, naval and air border flags, and identification symbols, the description of which and the procedure for the use of which are approved by the Council of Heads of Government.

Article 11. The Commonwealth border troops servicemen wear the prescribed military uniforms and are allowed to keep and carry standard issue firearms in accordance with the established procedure.

Article 12. The procedure for the stay and the status of the Commonwealth border troops on the territories of states which are setting up border troops of their own and which are not members of the Commonwealth are determined by separate agreements.

Article 13. Each state signatory to this agreement has the right to withdraw from this agreement. A state intending to withdraw from the agreement notifies the depository state and all other participant states in writing about its decision to do so. Such notification is given not less than six months in advance of the proposed date of withdrawal from this agreement. The border troops of the states which are not members of the Commonwealth could be incorporated in the Commonwealth border troops on the basis of a special agreement.

Article 14. The present agreement comes into force from the moment of its signing. Done in the city of Kiev on 20 March 1992 in one original copy in the Russian language. The original copy is to be kept in the archives of the government of the Republic of Belarus, which will send the certified copies to the participant states-signatories to the agreement.

This agreement is signed by representatives of Armenia, Kazakhstan, Kyrgyzstan, the Russian Federation, and Tajikistan.

## 10.15

**Strategic Forces' Status Agreed**

TASS, 15 February 1992
[FBIS Translation]

["Agreement Among the Member States of the Commonwealth of Independent States on the Status of the Strategic Forces."]

The Republic of Azerbaijan, Republic of Armenia, Republic of Belarus, Republic of Kazakhstan, Republic of Kyrgyzstan, the Russian Federation, Republic of Tajikistan, Turkmenistan, the Republic of Uzbekistan, and Ukraine, hereinafter referred to as the "Commonwealth member states," guided by the agreement on the strategic forces among the member states of the Commonwealth of Independent States,

—taking into account the role of the Commonwealth's strategic forces in ensuring the security of the Commonwealth member states,

—confirming their commitment to the principles and norms of international law,

Have agreed to the following:

*Article 1. For the purposes of the present agreement,* the following terms mean:

1. "Strategic Forces"—military formations and installations which are under united command, the list of which is determined by each state in agreement with the command of the strategic forces and is confirmed by the Council of Heads of State.

2. "Military Formations and Installations"—military units, establishments, military training establishments, enterprises, organizations, military representations, air bases, test sites, command points, and other installations of the strategic forces.
3. "Host State"—a state which is a member of the Commonwealth on whose territory strategic forces are deployed.
4. Place of Deployment (Basing)—the territory allocated for use by the strategic forces.
5. Immovable Property of the Strategic Forces—the military settlements, airfields, naval bases, ports, access rail tracks, structures at combat positions, training areas, firing ranges, fixed site command and control facilities, communications equipment, residential buildings, and other structures in use by the strategic forces and guaranteeing their ability to function, which are located on the areas of land granted to them for temporary use.
6. Movable Property of the Strategic Forces—all forms of weaponry, ammunition, and military equipment, including the necessary means of transport and other material and technical resources in use by the strategic forces.
7. Persons Belonging to the Strategic Forces—servicemen and civilians performing service or working in formations of the strategic forces.
8. Members of the Families of Persons Belonging to the Strategic Forces—spouses, children, and other dependent relatives.

*Article 2. General Provisions*

1. Any state which is a member of the Commonwealth may be a party to the present agreement regardless of whether military formations and installations of the strategic forces are deployed on its territory.
2. The purpose of the strategic forces is to ensure the security of all states which are party to the agreement and they are maintained at the expense of fixed payments from these states. (The property of the strategic forces is the joint property of all the states which are party to the agreement—Armenia dissenting.)
3. Each of the Commonwealth member states gives its consent to the permanent or temporary deployment and functioning of military formations and installations of the strategic forces in their places of deployment (bases), in which they were deployed and were functioning at the moment of the signing of the present agreement. Alterations to the places of deployment are implemented by agreement between the parties to the present agreement.
4. The deployment of military units and facilities of the strategic forces on the territory of a Commonwealth member state in no way affects the sovereignty of that state. The strategic forces do not interfere in the internal affairs of the host state.

   Military units and facilities of the strategic forces on the territory of a Commonwealth member state and persons belonging to them are obliged to respect and observe the laws of that state.
5. The Commonwealth member states do not commit actions obstructing the strategic forces in the fulfillment of their functions, unless they contradict the legislation of a sovereign state.

*Article 3. Acquisition of Manpower by the Strategic Forces*

The strategic forces acquire their manpower on the basis of the principles defined in a separate agreement.

*Article 4. Command of Strategic Forces*

1. The strategic forces function as an independent strategic grouping.
2. Command of the Commonwealth strategic forces is exercised by the commander of strategic forces, subordinated to the Council of Heads of State and the commander-in-chief of the Joint Armed Forces of the Commonwealth of Independent States.
3. The Strategic Forces Command

   - draws up plans for the combat application of the groupings, formations, and units of strategic forces;
   - organizes combat duty (operational service) (*boevoe dezhurstvo* [*boevaya sluzhba*]), plans and carries out other measures to maintain the strategic forces in the requisite state of combat readiness;
   - exercises direct operational control of groupings, formations, and units of strategic forces; organizes and carries out measures to maintain the safety of the civilian population and measures to protect the environment;
   - carries out the functions prescribed to it in the general system of measures to prevent unauthorized actions involving nuclear weapons;
   - places orders with scientific and industrial organizations of the member states of the Commonwealth on a contractual basis for the development and supply to the strategic forces of armaments and combat hardware in line with approved armaments programs, and finances the works that are carried out within the limits of the budget allocations earmarked for these purposes;
   - and carries out measures to observe international

treaties on nuclear weapons and other forms of weapons of mass destruction.

4. A decision on the need to use nuclear weapons is taken under the procedure laid down in Article 4 of the Agreement on Strategic Forces between the member states of the Commonwealth of Independent States of 30 December 1991.

*Article 5. Legal Position of Persons Who Are Members of the Strategic Forces and the Members of Their Families*

The legal position of persons who are members of the strategic forces and the members of their families is regulated by the agreement between the member states of the Commonwealth of Independent States on Social and Legal Guarantees for servicemen, persons discharged from military service, and members of their families.

*Article 6. Relations Between the Strategic Forces and the State Bodies, Enterprises, and Organizations of the Commonwealth Member States*

1. Military command bodies of the Commonwealth's strategic forces carry out their activity in cooperation with the state bodies, enterprises, and organizations of the Commonwealth member states.
2. The Commonwealth's strategic forces conclude contracts with enterprises and organizations of the Commonwealth member states to carry out work to create, modernize, and destroy weapons and military equipment and work on other matters of providing back-up for the activity of the strategic forces. The Commonwealth member states assist in concluding such contracts.
3. Movements by units of the strategic forces, exercises, maneuvers, and other activities organized for the operational and combat training of the strategic forces outside the confines of their places of permanent deployment (base) are conducted in accordance with plans agreed with the body authorized by the government of the Commonwealth member state on whose territory it is intended to conduct those activities, or with the consent in each case of that government or of the body authorized by it.

   The Commonwealth member states grant the military units and facilities of the strategic forces the necessary transportation facilities and land, air, and sea space for their movement, in accordance with the above plans.
4. The construction of new roads, bridges, buildings, and permanent radio and radio-electronic structures with defined frequencies and capacity in the places of deployment of the strategic forces and the construction of other immovable facilities of the strategic forces is carried out with the consent of the responsible bodies of the Commonwealth member state on whose territory it is proposed to construct the new facilities.
5. When areas of land in use by the strategic forces are left, they revert to the host state. The subject of immovable facilities built there with the funds of the strategic forces is settled in accordance with the legislation of the host state or with a relevant agreement.

*Article 7. Financing the Strategic Forces and Providing Them With Material and Technical Back-Up*

1. The Commonwealth member states participate proportionately in financing the strategic forces and in fulfilling international obligations to reduce and destroy them. The volumes of expenditures on the above aims and the financing procedure are determined by a separate agreement.
2. The procedure for providing material and technical back-up for the strategic forces and payments in the currency of the host state is determined by the Council of Heads of Government of the members of the Commonwealth.

*Article 8. Property of the Strategic Forces*

1. The host states reserve to the strategic forces the immovable property which they had at the time of the signing of the present agreement; they provide them with electricity and communal and other services. The procedure and conditions under which the strategic forces use the areas of land allocated to them and also for providing the strategic forces with all types of services are determined in accordance with the legislation of the host state.
2. The movable property of the strategic forces is at their disposal and for their use. The procedure for dealing with it is determined by the Council of Heads of Government of the Commonwealth.

   The Commonwealth member states undertake not to hinder the transfer of movable property of the strategic forces outside the state by agreement with the host party.
3. The road transportation vehicles of the military formations of the strategic forces have registration numbers and distinguishing signs. Unified registration numbers and signs are established by the main command of the Commonwealth's unified armed forces.

*Article 9. Matters of Jurisdiction*

In cases of crimes and misdemeanors committed by persons belonging to the strategic forces or by members of their families, the legislation in force on the territory of the Commonwealth member state where the crimes or misdemeanors were committed shall apply.

*Article 10. Procedure for Withdrawal from the Agreement*

In exercise of its sovereignty, each member state of the Commonwealth has the right to withdraw from this agreement. A member state of the Commonwealth which intends to withdraw from the agreement informs the depositary state and all other member states in writing of its decision to act in this way. Such notification is to be given at least one year before the proposed withdrawal from the agreement.

*Article 11. Coming into Force of the Agreement*

1. The present agreement is subject to ratification by each member state of the Commonwealth in accordance with its constitutional procedures. The instruments of ratification are handed over for safekeeping to the government of the Republic of Belarus, which by this agreement is appointed as the depositary.
2. The present agreement comes into force ten days after the instruments of ratification have been handed over for safekeeping by all member states of the Commonwealth.
3. The depositary immediately informs all member states of the Commonwealth:
   (a) of the handing over for safekeeping of each instrument of ratification;
   (b) of the coming into force of the present agreement;
   (c) of any notification of the member states of the Commonwealth to withdraw from the present agreement in accordance with Article 10, and of the date of their withdrawal;
   (d) of any matter requiring the provisions of the present agreement to be revised or defined more precisely.

The agreement comes into force from the moment that it is signed.

Completed in Minsk on 14 February 1992 in one original copy in Russian. The original is kept in the archives of the government of the Republic of Belarus, which will send a certified copy to the states which have signed this agreement.

The agreement was signed by representatives of Azerbaijan, Armenia, Belarus, Kazakhstan, Kyrgyzstan, Russia, Tajikistan, Turkmenistan, Uzbekistan, and Ukraine.

The Azerbaijan representative added the footnote: "On condition that the strategic units located only on the territory of Azerbaijan are financed, and that those units are withdrawn by the end of 1994."

The representative of Armenia added the note: "With a dissenting opinion."

The Kazakhstan representative noted that "an agreement on the test sites will be concluded with the Republic of Kazakhstan."

The Kyrgyzstan representative specified: "Inclusion of Article 2, Point 2."

The Ukraine representative made the annotation: "With the exception of Article 2, Point 2, and Article 10 (for Ukraine, in accordance with the Minsk agreement, the term for withdrawal is the end of 1994)."

## 10.16

## General Purpose Forces Agreement

TASS, 18 February 1992
[FBIS Translation]

["Agreement Between the Republic of Armenia, Republic of Belarus, Republic of Kazakhstan, Republic of Kyrgyzstan, the Russian Federation, Republic of Tajikistan, Turkmenistan, and the Republic of Uzbekistan on General Purposes for the Transitional Period."]

The states participating in the present agreement, referred to as the "participant states," proceeding from the need for a mutually acceptable and organized settlement of matters in the area of directing the general purpose forces,

bearing in mind the role of general purpose forces in ensuring the security of the participant states,

guided by the Agreement of the Council of Heads of Member States of the Commonwealth of Independent States on Armed Forces and Border Troops of 30 December 1991,

have agreed on the following:

Article 1. The participant states shall form joint general purpose forces.

The term "general purpose forces" means: groupings, formations, units, institutions, military training establishments, other military formations, and military facilities which do not form part of the strategic forces of the Commonwealth of Independent States, as well as the armed forces of the participant states themselves, operationally subordinated, with their consent, to the commander-in-chief of the Joint Armed Forces.

The list of military formations subject to inclusion in the general purpose forces and their deployment is determined by separate protocols for each participant state.

Article 2. The directing of the Commonwealth participant states' own armed forces is carried out by the ministries of defense (defense committees) of those states.

Article 3. The state of the general purpose forces is determined by a separate agreement between the participant states taking into account their national legislation.

Article 4. The material, technical, and financial backing for the general purpose forces is regulated by separate agreements.

Article 5. Each participant state has the right to withdraw from the present agreement, having given the other participant states at least six months notification of this.

Article 6. The agreement comes into effect when it is signed, and for the Republic of Belarus from the moment it is ratified by the Supreme Soviet of the Republic of Belarus.

Concluded in Minsk of 14 February 1992 in one original copy in Russian. The original is kept at the archives of the government of the Republic of Belarus which will send the states which have signed the present agreement a certified copy.

The document has been signed by representatives of Armenia, Belarus, Kazakhstan, Kyrgyzstan, Russia, Tajikistan, Turkmenistan, and Uzbekistan.

The president of Armenia made the following note: To add to Article 1, after the word "protocols," the following: "and is approved by the Council of Heads of State."

## 10.17

## Agreement on Defense Budget

TASS, 18 February 1992
[FBIS Translation]

["Agreement Between the Member States of the Commonwealth of Independent States on Formation of a Single Defense Budget and the Procedure for Financing the Armed Forces of the Commonwealth States."]

The Azerbaijan Republic, the Republic of Armenia, the Republic of Belarus, the Republic of Kazakhstan, the Republic of Kyrgyzstan, the Republic of Moldova, the Russian Federation, the Republic of Tajikistan, Turkmenistan, the Republic of Uzbekistan, and Ukraine, henceforth described as "the Commonwealth member states," proceeding from the need to make financial provision for the armed forces of the Commonwealth states, have agreed on the following:

Article 1. A single defense budget being formed to finance the armed forces of the Commonwealth states includes the following forms of expenditure on an annual basis:

—maintenance of the army and navy (money allowance for servicemen, salaries of permanent members of staff and office workers, food supplies, payment for clothing and related gear, payment for and storage of special fuels, repair and preparation of weapons, military equipment and property, transportation costs, lease of telecommunications facilities, maintenance of cosmodromes, special testing grounds, bases and depots, covering in rubles any foreign currency spent on maintaining troops on the border, operational, supply and other expenditure connected with providing necessary supplies to the troops);
—payment for weapons, military equipment, and property, including purchase of nuclear munitions;
—payment for scientific and technical production;
—capital construction and capital repairs, including specialized construction and housing construction;
—provision of pensions for servicemen and members of their families.

Article 2. Expenditure on the upkeep of the army and navy will be determined on the basis of the de facto numerical strength of servicemen, workmen, and support personnel of the armed forces of the Commonwealth states; of established levels of and maintenance standards for weaponry, military hardware, and assets; of existing prices and tariffs; of combat training plans; of production and commercial activities; and of other factors.

Expenditure on armaments, military hardware, and facilities will be determined on the basis of their planned supply volumes to maintain the armed forces of the Commonwealth states within the limits of the funds budgeted.

Expenditure on scientific and technological products will be determined on the basis of planned volumes of scientific research and development for military purposes, with account taken of their results and relevance, within the limits of the monies budgeted.

Capital investment in and expenditure on construction of new facilities and reconstruction and expansion of existing fixed assets will be determined as the sum total of spending on creation of proper conditions for combat and special training, on living standards, on storage, maintenance, and repair of armaments, military hardware, and assets, on command bodies, on medical services and recreation for personnel, and also on provision of servicemen and the members of their families with accommodation, social amenities, and consumer facilities.

Expenditure on pensions for servicemen in the armed forces of the Commonwealth states and the members of their families will be determined on the basis of the number of pensioners and the pension sums due to them.

Article 3. The member states of the Commonwealth will recognize the desirability of defining the single defense budget according to the prices of the base year, with allowance for the forecast price index and for social safeguards for servicemen and the members of their families.

Article 4. The draft single defense budget will be drawn up by the high command of the Commonwealth states' armed forces, considered by the Council of Defense Ministers (chairmen of defense committees) of the Commonwealth member states, and submitted to the Council of Heads of Government of the Commonwealth. The single defense budget will be endorsed by the Council of Heads of State of the Commonwealth upon representation by the Council of Heads of Government of the Commonwealth.

Article 5. The member states of the Commonwealth undertake to join in formation of the single defense budget by way of payment of fixed contributions. The extent and procedure for payment of the fixed contribution of each Commonwealth member state will be determined by the Council of Heads of Government of the Commonwealth.

Article 6. If, in the course of a year, the need arises for additional expenditure not envisaged in the single defense budget as endorsed, a procedure for payment into the single defense budget and for financing of the armed forces of the Commonwealth states will be determined in accordance with this agreement.

Article 7. An annual account on implementation of the single defense budget will be submitted by the commander-in-chief of the armed forces of the Commonwealth states to the Council of Heads of State of the Commonwealth.

This agreement will come into effect when it is signed. Done in the city of Minsk on 14 February 1992 in one original copy in Russian. The original will be held in the archives of the government of the Republic of Belarus, which will forward a certified copy to the signatory states.

The document is signed by the representatives of Armenia, Belarus, Kazakhstan, Kyrgyzstan, Russia, Tajikistan, Turkmenistan, and Uzbekistan, some of whom have made personal notes.

The document contains the following special option, signed by the presidents of Azerbaijan and Ukraine: The Republics of Azerbaijan and Ukraine will not join in the formation of a single defense budget for the upkeep of the joint strategic forces, but will assume a share of the financing solely of the upkeep of the strategic forces on their territory for a period of time determined for the Republic of Azerbaijan and Ukraine in accordance with the Minsk agreement between the members of the Commonwealth of Independent States on strategic forces, dated 30 December 1991.

## 10.18
## CIS Agreement on Joint Armed Forces for the Transitional Period

20 March 1992
[FBIS Translation]

The member states of the Commonwealth, henceforth called "the member states,"

Proceeding from the need for the mutually acceptable and organized solution of the questions of reforming the armed forces of the former USSR,

Have agreed on the following:

Article 1. The CIS Joint Armed Forces are not directed against states which are not participants in this agreement. The Joint Armed Forces are being created for the transitional period with the aim of providing security for the member states, preserving the army command, preventing conflicts, and coordinating the reform of the former USSR armed forces.

Article 2. The CIS Joint Armed Forces include the Strategic Forces of the Commonwealth created in accordance with the agreement reached by the CIS member states on the strategic armed forces of 30 December 1991, and also the General Purpose Forces created in accordance with the 14 February 1992 agreement on the General Purpose Forces between the Republic of Armenia, the Republic of Belarus, the Republic of Kazakhstan, the Republic of Kyrgyzstan, the Russian Federation, the Republic of Tajikistan, Turkmenistan, and the Republic of Uzbekistan.

Article 3. The Strategic Forces of the Commonwealth are subordinated directly to the command of the CIS strategic forces. The military formations and facilities of the General Purpose Forces, with the exception of the member states' own armed forces, are subordinated directly to the commander of the General Purpose Forces.

Article 4. The authority of the supreme bodies of the CIS and the high command is applied to the member states' own armed forces which have been transferred with their consent to operational subordination to the High Command of the Joint Armed Forces, only in respect of questions of operational subordination.

Article 5. Each member has the right to withdraw from this agreement by notifying the other member states not less than six months in advance.

Done in the city of Kiev on 20 March 1992, as one original copy in the Russian language. The original copy is to be kept in the archives of the government of the Republic of Belarus, which will send a certified copy to the member states which are signatory to this agreement.

The agreement was signed by representatives of Armenia, Belarus (with the addition of the words: "With the stipulation of a transitional period of two years for the Republic of Belarus"), Kazakhstan, Kyrgyzstan, the Russian Federation, Tajikistan, and Turkmenistan.

---

## 10.19
**CIS Statute on Joint Command of Border Troops**
20 March 1992
[FBIS Translation]

---

### I. General Points

1. The Joint Command of Border Troops is a permanent collegial body of the border troops of the Commonwealth and its states' own border troops, which effects the implementation of the decisions of the Council of Heads of State and the Council of Heads of Government of the Commonwealth on matters of the protection of the state borders and maritime economic zones of the CIS member states, and the coordination of the activity of the border troops.

   In its activity, the Joint Command is guided by international legal acts, acts of the CIS, the legislation of the CIS member states, and the enforceable enactments of the former USSR that do not contradict it, and also by this statute.
2. The staff of the Joint Command of Border Troops comprises: The commander-in-chief of the border troops, the chief of general staff of the border troops, the commanders of the states' own border troops, and authorized representatives of the CIS member states, but not exceeding a total of two persons from each member state. The commander-in-chief of the border troops and each CIS member state possess one vote in the Joint Command.

The Joint Command invites specialists and experts, as required, to meetings of officials of the border troops.

Representatives of states, formerly republics of the USSR, which are not members of the Commonwealth can take part in the work of the Joint Command as observers.

### II. Powers of the Joint Command of Border Troops

1. The Joint Command of Border Troops:
   —coordinates the activity of the border troops of the Commonwealth and the states' own border troops;
   —submits proposals to the Council of Heads of State and Council of Heads of Government and to the top bodies of state authority and administration of the CIS member states on matters of the protection of state borders and maritime economic zones, and on maintaining the activity of the border troops of the Commonwealth and the states' own border troops;
   —works out general principles and proposals on developing the system for protecting state borders and maritime economic zones and allowing persons and means of transport across state borders;
   —organizes and ensures the implementation of acts of the Commonwealth and the legislation of CIS member states by the border troops of the Commonwealth and the states' own border troop;
   —Works out coordinated approaches on matters of manning, training of personnel, and maintaining combat readiness of the border troops of the Commonwealth and the states' own border troops.

   The Joint Command can also examine other matters which relate to the protection of state borders and maritime economic zones and to the activity of the border troops of the Commonwealth and of the states' own border troops.
2. Members of the Join Command have the right:
   —To represent the Joint Command, at its instruction, in state and public organizations, including international, on matters within the competence of the Joint Command;
   —To receive the necessary information on the activity of the border troops through the governing bodies of the border troops.

### III. The Organization of the Work of the Joint Command of Border Troops

1. Conferences, which are held as required under the chairmanship of the commander-in-chief of the border troops, are the main form of work of the Joint Command. Conferences are convened on the decision of the commander-in-chief or on the proposal of the permanent members of the Joint Command.
2. Decisions of the Joint Command are passed on the basis of consensus and are announced by orders of the commander-in-chief of the border troops and the commanders' of the states' own border troops.

   When no consensus is reached, an issue is submitted to the Council of Heads of State or the Council of Heads of Government.
3. Other matters of the Joint Command's work are determined by regulations worked out by this command.

## 10.20
**Declaration on Rejecting Use of Force**
TASS, 23 March 1992
[FBIS Translation]

["Declaration on the Non-Use of Force or the Threat of Force in Relations Between the CIS Member States," issued by the CIS summit in Kiev on 20 March.]

The Commonwealth member states:

—noting that rejection of using force or threatening its use, which is enshrined in the UN Charter and the CSCE Final Act, is an obligation which all states must observe,
—confirming their obligations to tackle all contentious problems by exclusively peaceful means, as was stated by the CIS heads of state in Alma-Ata on 21 December 1991 and in Minsk on 14 February 1992,
—with the aim of ending bloodshed and localizing and averting tension,
—and guided by the desire of the peoples of the Commonwealth for peace, security, and good-neighborliness, declare as follows:

Member states

1. Do not permit the use of force or the threat of its use.
2. Do not deliver arms to zones of conflict.
3. Prevent attacks on troop units aimed at seizing weapons.
4. Refrain from organization or encouraging the organization of irregular forces or armed bands, including mercenaries.
5. Refrain from fanning tension in relations between the Commonwealth states.
6. In the event of disputes arising between them, they will apply their efforts conscientiously and in a spirit of cooperation to reach a fair decision based on international law, within a short period of time.

   To these ends, they will use such means for the peaceful settlement of disputes as talks, investigation, mediation, reconciliation, arbitration, court examination, or other peaceful means of their own choosing, including any procedure for settlement that was agreed before the disputes occurred and to which they may have been party, and apply the principles, tenets, and norms for the peaceful settlement of disputes that have been developed within the framework of the UN Organization and the CSCE.
7. Encourage the use of various forms of people's diplomacy and public initiative with the aim of averting the threat of inter-state conflicts.
8. Support the efforts of the international community and its instruments to settle conflicts on the territory of the Commonwealth states.

Done in Kiev on 20 March 1992 in a single original copy. The original is kept in the archives of the government of Belarus, which will forward a certified copy of this convention to the signatory states.

* * *

The declaration was signed by representatives of Azerbaijan, Armenia, Belarus, Kazakhstan, Kyrgyzstan, Moldova, the Russian Federation, Tajikistan, Uzbekistan, and Ukraine.

## 10.21
**CIS Treaty on Collective Security**
15 May 1992
[FBIS Translation]

The states participating in the present treaty, henceforth referred to as "the participating states,"

—guided by the independent states' declarations of sovereignty,
mindful of the creation by the participating states of their own armed forces,
—taking coordinated actions in the interests of ensuring collective security,
—recognizing the need to strictly fulfill the treaties that have been concluded with regard to the reduction of arms and armed forces and the strengthening of confidence-building measures, have agreed to the following:

Article 1. The participating states confirm their commitment to refrain from the use or threat of force in interstate relations. They pledge to resolve all disagreements among themselves and with other states by peaceful means.

The participating states will not enter into military alliances or participate in any groupings of states, nor in actions directed against another participating state.

In the event of the creation of a system of collective security in Europe and Asia and the conclusion of treaties on collective security to the end—for which the contracting parties will strive unswervingly—the participating states will enter into immediate consultations with each other for the purpose of incorporating the necessary intentions in the present treaty.

Article 2. The participating states will consult with each other on all important questions of international security affecting their interests and will coordinate their positions on these questions.

In the event of a threat to the security, territorial integrity, and sovereignty of one or several participating states or of a threat to international peace and security, the participating states will immediately activate the mechanism of joint consultations for the purpose of coordinating their positions and taking measures to eliminate the threat that has emerged.

Article 3. The participating states will form a Collective Security Council consisting of the heads of participating states and the commander-in-chief of the CIS Joint Armed Forces.

Article 4. If one of the participating states is subjected to aggression by any state or groups of states, this will be perceived as aggression against all participating states, to this treaty.

In the event of an act of aggression being committed against any of the participating states it will give it the necessary assistance, including military assistance, and will also give support with the means at their disposal by way of exercising the right to collective defense in accordance with Article 51 of the UN Charter.

The participating states will immediately inform the UN Security Council of any measures taken on the basis of this article. When implementing these measures, the participating states will abide by the corresponding provisions of the UN Charter.

Article 5. The Collective Security Council of the participating states and the organs to be created by it undertake the coordination and ensuring of joint activities by the participating states in accordance with this treaty. Until the aforementioned organs have been created, the activities of the armed forces of the participating states will be coordinated by the High Command of the Commonwealth Joint Armed Forces.

Article 6. Any decision to use armed forces for the purpose of repulsing aggression in accordance with Article 3 of the present treaty is adopted by the heads of the participating states.

The use of armed forces outside the territory of the participating states can be effected exclusively in the interests of international security in strict compliance with the UN Charter and the legislation of participating states in the present treaty.

Article 7. The location and functioning of installations in the collective security system on the territory of participating states is regulated by special agreements.

Article 8. The present treaty does not affect any rights and commitments under other bilateral and multilateral treaties and agreements currently in force and concluded by the participating states with other states, and it is not directed against third countries.

The present treaty does not affect the participating states' right to individual and collective defense against aggression in accordance with the UN Charter.

The participating states pledge not to conclude international agreements that are incompatible with the present treaty.

Article 9. Any questions arising between the participating states in connection with the interpretation or application of any provision of the present treaty will be resolved jointly in the spirit of friendship, mutual respect, and mutual understanding.

Amendments to the present treaty may be incorporated on the initiative of one or several of the participating states and adopted on the basis of mutual agreement.

Article 10. The present treaty is open to all interested states which share its aims and principles.

Article 11. The present treaty is concluded after five years with a subsequent renewal.

Any of the participating states has the right to withdraw from the present treaty if it notifies the other parties of its intention at least six months in advance and fulfills all the commitments resulting from withdrawal from the present treaty.

The present treaty is subject to ratification by each of the signatory states in accordance with its constitutional procedures. The ratification documents will be handed to the government of the Republic of Belarus, which is thus being appointed depository.

The present treaty comes into force immediately after the ratification documents have been submitted for safekeeping by the participating states which have signed it.

Done in the city of Tashkent on 15 May 1992 in one original copy in the Russian language. The original copy is kept in the archives of the government of the Republic of Belarus, which will send certified copies to the states which have signed the present treaty.

---

## 10.22

**Tashkent Statement on Armed Forces Cutbacks**
*Rossiyskaya Gazeta,* 23 May 1992
[FBIS Translation]

---

["Statement by the Heads of CIS Member States on the Reduction of the Armed Forces of the Former USSR," adopted at 15 May 1992 session of the Council of CIS Heads of State in Tashkent.]

Confirming the adherence of the states belonging to the CIS as the legal heir of the former USSR to the international commitments of the former USSR in the field of disarmament and arms control, being persuaded of the need for the further development of the process of the consolidation of security and cooperation in various regions and in the world as a whole on the basis of the radical reduction of existing military potentials,

The heads of the CIS states state:

I

The CIS states confirm their intentions to reduce the armed forces of the former USSR within deadlines agreed among themselves which do not contradict international treaties.

II

The CIS states undertake not to implement unilateral decisions and actions that could harm the fulfillment of the international commitments of the former USSR for the reduction and limitation of armed forces and armaments.

Done at the city of Tashkent 15 May 1992 in a single authentic copy in the Russian language. The authentic copy is stored in the archives of the government of the Republic of Belarus, which will send a certified copy to the states that have signed this statement.

The document has been signed by representatives of Azerbaijan, Armenia, Belarus, Kazakhstan, Kyrgyzstan, Moldova, the Russian Federation, Tajikistan, Turkmenistan, Uzbekistan, and Ukraine.

## 10.23

## Treaty on CIS Collective Security Published

*Rossiyskaya Gazeta,* 23 May 1992
[FBIS Translation]

[Signed by representatives of Armenia, Kazakhstan, Kyrgyzstan, the Russian Federation, Tajikistan, and Uzbekistan, adopted at 15 May session of the Council of CIS Heads of State in Tashkent.]

The states participating in the present treaty, hereinafter referred to as the "participating states,"

—guided by the independent states' declarations of sovereignty,
—mindful of the creation by the participating states of their own armed forces,
—taking coordinated actions in the interests of ensuring collective security,
—recognizing the need to strictly fulfill the treaties that have been concluded with regard to the reduction of arms and armed forces and the strengthening of confidence-building measures, have agreed as follows:

Article 1. The participating states confirm their commitment to refrain from the use or threat of force in interstate relations. They pledge to resolve all disagreements among themselves and with other states by peaceful means.

The participating states will not enter into military alliances or participate in any groupings of states, nor in actions directed against another participating state.

In the event of the creation of a system of collective security in Europe and Asia and the conclusion of treaties on collective security to that end—for which the contracting parties will strive unswervingly—the participating states will enter into immediate consultations with each other for the purpose of incorporating the necessary intentions in the present treaty.

Article 2. The participating states will consult with each other on all important questions of international security affecting their interests and will coordinate their positions on these questions.

In the event of the emergence of a threat to the security, territorial integrity, and sovereignty of one or several participating states or of a threat to international peace and security, the participating states will immediately activate the mechanism of joint consultations for the purpose of coordinating their positions and taking measures to eliminate the threat that has emerged.

Article 3. The participating states will form a Collective Security Council consisting of the heads of participating states and the commander-in-chief of the CIS Joint Armed Forces.

Article 4. If one of the participating states is subjected to aggression by any state or group of states, this will be perceived as aggression against all participating states to this treaty.

In the event of an act of aggression being committed against any of the participating states, all the other participating states will give it the necessary assistance, including military assistance, and will also give support with the means at their disposal by way of exercising the right to collective defense in accordance with Article 51 of the UN Charter.

The participating states will immediately inform the UN Security Council of any measures taken on the basis of this article. When implementing these measures, the participating states will abide by the corresponding provisions of the UN Charter.

Article 5. The Collective Security Council of the participating states and the organs to be created by it undertake the coordination and ensuring of joint activities by the participating states in accordance with this treaty. Until the aforesaid organs have been created, the activities of the armed forces of the participating states will be coordinated by the High Command of the Commonwealth Joint Armed Forces.

Article 6. Any decision to use armed forces for the purpose of repulsing aggression in accordance with Article 3 of the present treaty is adopted by the heads of the participating states.

The use of armed forces outside the territory of the participating states can be effected exclusively in the interests of international security in strict compliance with the UN Charter and the legislation of participating states in the present treaty.

Article 7. The location and functioning of installations in the collective security system on the territory of participating states is regulated by special agreements.

Article 8. The present treaty does not affect any rights and commitments under other bilateral and multilateral treaties and agreements currently in force and concluded by the participating states with other states, and it is not directed against third countries.

The present treaty does not affect the participating states' right to individual and collective defense against aggression in accordance with the UN Charter.

The participating states pledge not to conclude international agreements that are incompatible with the present treaty.

Article 9. Any questions arising between the participating states in connection with the interpretation or application of any provision of the present treaty will be resolved jointly in the spirit of friendship, mutual respect, and mutual understanding.

Amendments to the present treaty may be incorporated on the initiative of one or several of the participating states and adopted on the basis of mutual agreement.

Article 10. The present treaty is open to all interested states which share its aims and principles.

Article 11. The present treaty is concluded after five years with a subsequent renewal.

Any of the participating states has the right to withdraw from the present treaty if it notifies the other parties of its intention at least six months in advance and fulfills all the commitments resulting from withdrawal from the present treaty.

The present treaty is subject to ratification by each of the signatory states in accordance with its constitutional procedures. The ratification documents will be handed to the government of the Republic of Belarus, which is thus being appointed depositary.

The present treaty comes into force immediately after the ratification documents have been submitted for safekeeping by the participating states which have signed it.

Done in the city of Tashkent 15 May 1992 in one authentic copy in the Russian language. The authentic copy is kept in the archives of the government of the Republic of Belarus, which will send certified copies to the states which have signed the present treaty.

## 10.24
**"Text" of Declaration on State Borders**
ITAR-TASS, 7 August 1993
[FBIS Translation]

**Declaration on Inviolability of the State Borders**

The states which have signed this declaration, reaffirming their adherence to the UN Charter and the principles of the CSCE, as well as the basic documents of the Commonwealth of Independent States;

—stressing that the inviolability of the borders and territorial integrity of states are fundamental principles of international relations, and their observance presents the most important condition toward the preservation of international peace, security, and stability;
—recalling that the territory of a state is inviolable and cannot be an object of the use of force in violation of the UN Charter;
—expressing their concern in connection with incidents of serious violation of their borders by countries which are not members of the CIS;
—proceeding from the inalienable right of the states for individual and collective self-defense in accordance with Article 51 of the UN Charter, state the following:

1. Ensuring the inviolability of the borders of the states which have signed this declaration is seen by them as a sphere of joint, vitally important interests, is their common cause, and is being implemented on a multilateral or bilateral basis.

2. The states-signatory to this declaration will consider any encroachment on their borders to be unlawful actions, which give grounds for the adoption of retaliatory and commensurate measures, in accordance with international law, including the use of armed force, and in the form of individual or collective self-defense. They will jointly avert and thwart any attempts at incursion from outside on the territory of any of the states signatory to this declaration.
3. The states signatory to this declaration will, on their territory, thwart the activities of individuals, groups, or organizations which are aimed at violating the inviolability of the borders of these states.
4. The states signatory to this declaration bear collective responsibility for the inviolability of their borders with third states. At the same time, none of the states signatory to this declaration bears any obligation to unilaterally ensure the security of the borders of another state.
5. The states signatory to this declaration reaffirm their readiness, by way of talks with the participation of all the interested parties, to seek ways to end and avert armed conflicts on the borders.
6. Internal stability within the states signatory to this declaration is a necessary condition of security on their borders. For this purpose each of them will take the necessary steps to strengthen democratic institutions and to achieve national consensus on the basis of respect for human rights and basic freedoms.
7. The states signatory to this declaration count on support and understanding of its articles on the part of neighboring states and the international community as a whole.

[Signed]

For the Republic of Kazakhstan—N. Nazarbaev
For the Russian Federation—B. Yeltsin
For the Kyrgyz Republic—A. Akaev
For the Republic of Tajikistan—I. Rakhmanov
For the Republic of Uzbekistan—I. Karimov

## 10.25

**CIS Unified Air Defense Agreement**
*Rossiyskaya Gazeta,* 25 February 1995
[FBIS Translation]

### Agreement on the Creation of a Unified Air Defense System for the CIS States

The CIS states, having signed this agreement and referred to hereinafter as the "member states," guided by the CIS security principles, basing themselves on the Decision on the CIS Council of Heads of State's Memorandum "Basic Guidelines for the Integrated Development of the CIS" and the long-term plan for the integrated development of the CIS of 21 October 1994, and proceeding on the basis of the need to pool their efforts on the member states' air defense and the protection of their airspace borders, have agreed as follows:

Article 1. The member states shall create a unified air defense system for the CIS member states.

The unified air defense shall include member state air defense forces and means (some forces and means) operating on the basis of a coordinated plan and carrying out the tasks envisaged under Article 2 of this agreement.

The principles underlying the formulation and implementation of the unified air defense system's missions shall be determined by the Statute on the CIS Member States' Unified Air Defense System, which is part and parcel of this agreement.

Article 2. The unified air defense shall be set up to resolve the following tasks:

- Ensuring the protection of the airspace of the member states' state borders;
- Implementing joint monitoring of the procedure governing the utilization of the member states' airspace;
- Informing member states about the aerospace situation and providing warnings of missile or airborne attack;
- Conducting coordinated actions by the member states' air defense troops to repulse airborne or missile attack.

Article 3. With a view to coordinating efforts to create and improve a unified air defense system and to coordinate the actions of the unified air defense system troops and forces, a Coordinating Committee for Air Defense Issues shall be set up under the CIS member states' Defense Ministers Council (hereinafter, the Coordinating Committee).

The commander-in-chief of the Russian Federation Air Defense Troops shall be chairman of the Coordinating Committee.

Members of the Coordinating Committee shall include the commanders of the member states' air defense troops (air defense and air force) as well as, at the decision of the CIS member states' Defense Ministers Council, the deputy chairman of the Coordinating Committee and other officials.

The statute on the Coordinating Committee shall be ratified by the CIS member states' Defense Ministers Council.

Article 4. The member states shall ensure the constant combat readiness of the troops and forces allocated to the unified air defense system, and their actions in carrying out their joint air defense missions.

The procedure for cooperation between unified air defense system forces and means shall be laid down by the plan elaborated by the Coordinating Committee in conjunction with the CIS member states' Military Cooperation Coordination Headquarters, taking account of the plans for the utilization of member states' air defense (air defense and air force) forces and means; the plan for cooperation between unified air defense system forces and means shall be ratified by the CIS member states' Defense Ministers Council.

Direct command and control of the air defense (air defense and air force) troops and forces of each member state shall be exercised by the commanders of those states' air defense (air defense and air force) troops taking account of the plan for cooperation between unified air defense system forces and means.

The coordination of the actions of unified air defense system forces and means shall be implemented from the Russian Federation Air Defense Troops Central Command Post.

Article 5. Some of the unified air defense system forces and means shall be on permanent combat standby protecting the airborne borders of the member states on the basis of the plan for cooperation between unified air defense system forces and means.

Command and control of the unified air defense system's standby forces and means shall be exercised from member states' air defense (air defense and air force) command posts, while their actions shall be coordinated from the Russian Federation Air Defense Troops Central Command Post.

Article 6. Air defense armaments and military hardware shall be supplied on the basis of bilateral agreements between the member states' governments, while repairs of air defense armaments and military hardware shall be effected under the procedure laid down by the CIS Council of Heads of Government.

The member states pledge not to sell and not to transfer to other states which are not party to this agreement air defense armaments and military hardware defined on the list ratified by the CIS Council of Heads of Government on the basis of the proposal from the CIS member states' Defense Ministers Council, nor to divulge information constituting member state military secrets.

Article 7. The training of military specialists for unified air defense system forces and means shall be effected on the basis of bilateral agreements between the member state governments.

Article 8. This agreement is open to other states to join with the consent of all the member states.

Article 9. Each member state is entitled to withdraw from this agreement once the relevant written notification has been sent to the depositary. As far as the member state is concerned, the agreement lapses one year from the receipt of notification by the depositary.

Article 10. The agreement of 6 July 1992 on the air defense system shall be deemed to have lapsed.

Article 11. This agreement comes into force from the day it is signed.

Done in the city of Almaty on 10 February 1995 in a single copy in Russian. The original [*podlinny ekzemplyar*] shall be held in the archives of the government of the Republic of Belarus, which will send a certified copy to each signatory state.

**Appendix to the Agreement**

**Statutes on the CIS Member States' Unified Air Defense System**

1. The CIS member states' unified air defense system shall be set up with a view to protecting the member states' air borders, monitoring the procedure governing the utilization of airspace, providing warning of a threatened or incipient airborne or space attack, and protecting the member states' most important installations from air strikes.
2. The unified air defense system unites the member states' forces and means for the joint resolution of missions in the interests of all member states.

   The air defense forces and means of each member state resolve missions to provide air defense for their territories independently and in conjunction with the air defense troops of adjoining member states.
3. The coordinated resolution of tasks in the interests of the air defense of all member states is based on the following principles:

- The use of troops and forces under the member states' plans and the coordinated plan for cooperation in the interests of all the member states' air defenses;
- The equipping of the member states' air defense troops with armaments and military hardware on the basis of the agreed military–technical policy;
- The collaboration of unified air defense system forces and means where state borders intersect, and also in the member states' airspace;
- The centralization of notification and command and

control of unified air defense system forces and means by the member states' air defense (air defense and air force) troops and the Russian Federation Air Defense Troops Central Command Post;

- The unity of the main requirements for combat readiness and the combat and operational training of the member states' air defense (air defense and air force) command and control organs and troops.

4. The joint actions of the member states' air defense forces and means are coordinated by the chairman of the Coordinating Committee for Air Defense Issues through the commanders of the member states' air defense (air defense and air force) troops.
5. In order to organize cooperation, representatives of the chairman of the Coordinating Committee and the CIS member states' Defense Ministers Council of the CIS members, as well as by the agreement of the creation of the CIS states' unified air defense system.
7. The chairman of the Coordinating Committee has a deputy ratified by the CIS member states' Defense Ministers Council on the basis of representations from the Coordinating Committee.
8. The chairman of the Coordinating Committee is obliged to:

- Coordinate joint operations by unified air defense system forces and means;
- Organize in conjunction with the commanders of the member states' air defense (air defense and air force) troops the elaboration of the plan for cooperation between unified air defense system forces and means and submit it for ratification to the CIS member states' Defense Ministers Council;
- Inform the commanders of the member states' air defense (air defense and air force) troops about the aerospace situation, the integrated unified air defense system network, and unified cooperation signals;
- Elaborate in conjunction with the commanders of the member states' air defense (air defense and air force) troops proposals and recommendations for the further development and enhancement of the combat readiness of unified air defense system forces and means, equipping them with armaments and military hardware;
- Elaborate in conjunction with the commanders of the member states' air defense (air defense and air force) troops a plan for the holding of joint operational exercises (training sessions for the member states' air defenses, as well as training and methodology sessions, and submit it for ratification by the CIS member states' Defense Ministers Council (funding for the measures envisaged under the plan shall be provided by the states taking part in it).

9. The chairman of the Coordinating Committee is responsible for organizing cooperation between the member states' air defense forces and means.
10. The chairman of the Coordinating Committee is entitled:

- To inform the member states' heads of state and defense ministers about the progress of operational and combat training, and the combat standby condition of the forces and means allocated to the unified air defense system;
- To take part, by agreement with the member states' defense ministers, in monitoring the progress of operational and combat training and the state of readiness of standby forces and means allocated to the unified air defense system, including those involving the use of performance-grading intercept targets;
- To make recommendations to the commanders of the member states' air defense (air defense and air force) troops on questions of the combat use of the troops in the event of a threatened air attack or the threat of intrusion to the member states' airspace, as well as on questions of the organization of combat standby duty;
- To make annual recommendations on the questions of improving combat standby duty, cooperation, combat and operational training, and the assimilation of new military equipment by the troops allocated to the unified air defense system;
- To coordinate the training of air defense specialists in the interests of the unified air defense system;
- To establish a procedure for mutual information on the actions of unified air defense system standby forces and means.

11. Leadership of the unified air defense system and cooperation is carried out in Russian.

**Note**

The Republic of Azerbaijan did not sign the Concept for the Collective Security of the States Party to the Collective Security Treaty, or the Agreement on the Creation of a Unified Air Defense System for the CIS Member States.

The Republic of Moldova did not sign the Agreement on the Creation of a Unified Air Defense System.

## 10.26
**Treaty on Border Protection Between CIS, Non-CIS States**
*Rossiyskaya Gazeta,* 7 July 1995
[FBIS Translation]

["Treaty on Cooperation in the Protection of the Borders of the Participants in the Commonwealth of Independent States with States That Are Not Members of the Commonwealth."]

The participants in the Commonwealth of Independent States that signed this treaty, hereinafter called the parties,

—guided by the generally accepted principles and rules of international law and a desire to develop friendly, neighborly relations, and to contribute to mutual support for security on the borders of the states of the Commonwealth,
—recognizing the need for cooperation in the protection of the borders of the participants in the Commonwealth of Independent States with states that are not members of the Commonwealth, and
—confirming their commitment to the provisions of the UN Charter, the principles of the OSCE, the provisions of the Helsinki Final Act, the Charter of the Commonwealth of Independent States, and other documents on border issues that have been adopted by the parties and have legal force for them, have agreed as hereunder:

Article 1. For the purposes of this treaty the terms listed below shall mean the following:

Borders—sections of the state borders of participants in the Commonwealth of Independent States with states that are not members of the Commonwealth.

Council of Commanders—Council of Commanders of the Border Troops.

Border Troops—border troops of the parties.

Article 2. The purposes of the parties' cooperation in protection of the borders are:

—protection of the borders with regard to the interests of the parties;
—effective struggle against international and domestic terrorism, all manifestations of separatism and nationalism, drug trafficking, illegal immigration, and the illegal movement of weapons, ammunition, and radioactive, toxic, and psychotropic substances and also other objects and freight whose importation and exportation are banned by national legislation of the parties and international agreements;
—the development of a treaty-legal base for the cooperation and rapprochement of the legislation of the parties on border issues.

Article 3. The parties shall establish and develop between themselves equal partner relations aimed at the effective accomplishment of the tasks for strengthening peace on the borders.

The parties shall be mutually responsible for the protection of their section of the border with regard to the interests of the security of the parties.

The parties shall recognize the priority of the decisions (adopted on the basis of consensus) of the supreme authorities of the Commonwealth on questions of support for the security of the borders.

Article 4. The parties shall protect the borders in accordance with national legislation by concerted or joint efforts with regard to the interests of the parties on conditions that shall be determined by the corresponding bilateral or multilateral agreements and arrangements.

In the event of a threat to the security of the borders arising, the parties shall immediately hold mutual consultations for the adoption of the appropriate measures to eliminate the threat that has arisen.

Article 5. The parties shall be entitled to adopt measures to ensure the protection of their borders with the aid of a contingent necessary for this or another participant in the Commonwealth of Independent States on the basis of international agreements.

Article 6. The parties may, with regard to national legislation, create regional joint commands (operational groups, joint staffs, or coordination councils) for the coordination of joint efforts in the protection of the borders and the realization of border policy.

Article 7. The parties shall cooperate in scientific studies, in the creation of joint programs, planning, and the manufacture and introduction of new technical facilities for protection of the borders included.

Article 8. The parties shall cooperate in raising the level of the border troops' provision with equipment.

The procedure and the conditions of preferential supplies to the border troops of special equipment for the border troops and material assets shall be determined by a separate agreement.

Article 9. The parties shall, where necessary, in the interests of protection of the borders, conclude bilateral or multilateral agreements regulating the use by the border troops of sections of territory, the air space, water, and land, airports, airfields, ports, moorings, railroad approach tracks, and motor highways and also obtain the necessary information pertaining to meteorological support for flights of border aviation and navigation and hydrographic support for boats of the border troops.

Article 10. The parties shall for the timely adoption of decisions on protection of the borders provide for the continuous operation of the process of the collection and processing of data and the forecasting of the situation on the borders and the constant mutual exchange of information and also for the preparation of proposals concerning measures of a preventive nature adopted in international practice.

The parties shall maintain in working order the existing special communications channels of the border troops and other competent services and simultaneously adopt joint measures to create new channels (systems) of communications and information within the framework of programs drawn up by the participants in the Commonwealth.

Article 11. The parties shall not transmit to anyone material and information of an official or secret nature obtained from one another without the written consent of the party from which this material and information were obtained.

Article 12. The parties shall cooperate in questions of operational-search activity in the interests of protection of the borders conducted by arms of the border troops authorized for this in accordance with the legislation of each party.

Article 13. The parties shall harmonize (coordinate) their border policy in relations with contiguous countries that are not participants in the Commonwealth.

This treaty shall not affect the rights and obligations of the parties with respect to other current bilateral and multilateral treaties and agreements and is not aimed against third countries.

Article 14. The parties shall adopt measures to harmonize their legislative and other law-making instruments regulating questions of the protection of the borders and their procedures.

The parties shall for the creation of the most favorable and equal conditions in support of the activity of the border troops exchange information on the adoption of new national legislative instruments affecting questions of the protection of the borders and their procedures.

Article 15. The parties shall on a contract basis render one another assistance in the training of officer personnel and junior specialists for the border troops.

Diplomas and certificates and also other documents attesting that the corresponding education or specialty has been obtained, including those issued prior to this treaty taking effect, shall be recognized on the territory of the parties.

Article 16. The parties shall, if necessary, conclude separate agreements on questions of the establishment for servicemen and their families and other citizens participating in the protection of the borders outside of their states sums of compensation and privileges with regard to the regional singularities of service, state and interstate insurance, and material, financial, and other forms of support.

Article 17. The parties shall make provision on contractual terms for the admittance to their general health institutions for observation and treatment (hospitalization) servicemen and retirees of the border troops and members of their families and also for the allocation of passes to departmental sanatoriums and recreation centers at the expense of the party that has sent them.

Article 18. Disputes arising in the interpretation and application of this treaty shall be resolved by way of consultations and negotiations between the parties.

Addenda and revisions to this treaty may be made on the initiative of one or several of its signatories and adopted on the basis of consensus.

The parties shall entrust to the Council of Commanders the coordination of measures for the fulfillment of this treaty. To this end the Council of Commanders may, following coordination with the interested parties, create temporary working bodies of representatives of the Ministries of Foreign Affairs, border troops, and other interested ministries and departments of the parties. The procedure of realization of these measures shall be determined by a separate agreement with the host party.

Article 19. This treaty shall take effect the day of the presentation to the depositary of notice in triplicate of compliance with the intra-state procedures necessary for it to take effect. For a party that notifies the depositary of compliance with such procedures after the treaty has taken effect, it shall take effect the day such notice is presented to the depositary.

Article 20. The treaty is concluded after five years and will automatically be extended on each occasion for the subsequent five-year period. Each party may withdraw from this treaty by way of written notification of the depositary of this no fewer than six months prior to the expiration of the corresponding period.

Other participants in the Commonwealth of Independent

States may accede to this treaty after it has taken effect.

Article 21. This treaty is to be registered with the UN Secretariat in accordance with Article 102 of the UN Charter.

Done in the city of Minsk on 26 May 1995 in one authentic copy in Russian. The authentic copy shall be kept in the Executive Secretariat of the Commonwealth of Independent States, which shall send each state that signed this treaty its certified copy.

The treaty was not signed by Azerbaijan, Kazakhstan, Moldova, Turkmenistan, Uzbekistan, and Ukraine.

Minsk, 26 May 1995.

# Conclusion

# The Russian Bloc: Hegemony, Cooperation, and Conflict

*Zbigniew Brzezinski*

The CIS is still in its formative stage. Since 1992 a great deal of structural growth has taken place, but in a setting of continuing confusion over means as well as ends, and amid much conflict. The four charts below summarize in graphic form the institutional growth of the CIS as a binding network.

The charts, however, cannot convey by themselves the degree to which that process has been moved forward by deliberate pressure, including political manipulation, economic leverage, and even military intervention. That reality emerges more sharply in the preceding chapters. It is testimony to the degree of commitment and initiative involved in the Russian efforts to reverse, at least in part, what transpired in December of 1991.

The charts also cannot convey the degree to which this process has been derived from the reality of a genuinely shared interest in some reconstitution of the economic ties that previously existed in the Soviet Union. Even the most nationally ambitious elites in the newly independent states recognize that a renewal and a consolidation of economic cooperation with Russia is essential to their well being. That basic reality has reinforced the impetus for more formal institutionalization of the CIS.

Still another aspect that the charts cannot convey is the extent of continued evasion of the implementation of the many CIS arrangements. In part, this is due to the inevitable confusion inherent in the post-Soviet conditions, including the disruption produced by the very uneven shift away from a centrally controlled economy in the former Soviet space. The resulting massive chaos is simply not susceptible to management by an orderly process, on the manner, say, of the European Union's deliberate step-by-step integration.

But it must also be noted that some of the evasion is also quite intentional. It is a form of indirect opposition on the part of the new political elites to pressures from Moscow which they cannot resist directly. These elites not only sense the long-range political implications of growing economic integration, but they also follow and fear the debates in Moscow regarding Russia's special role in the geopolitical space of the former Soviet Union. Much of that debate must have an ominous ring for the newly self-governing and increasingly nationally self-conscious non-Russian leaders.

The present situation in the CIS is in some ways strikingly suggestive of the condition of the former Soviet bloc in the 1970s and 1980s. (It is to be noted that similarities do not imply identity; they note, however, some important parallels, significant differences notwithstanding.) The CIS, much like the former Soviet bloc, is a combination of Russian hegemony, many elements of cooperation, and of open as well as often hidden conflict. Russia, just like the old Soviet Union in the former Soviet bloc, is the central player, but its power is not unlimited. Coercion may be the last resort, intimidation may lurk in the background, but bargaining, pressure, and cooperation are also omnipresent.

As in the former Soviet bloc, Moscow's current influence in the various capitals of the CIS differs greatly. In the 1970s and 1980s, the prevailing relationship between the Soviet Union and Yugoslavia was friendly but very wary, with the Yugoslavs utterly determined to maintain their complete independence. The relationship between the Soviet Union and Poland, with the latter's communist regime very dependent on Moscow but also motivated by sensitive nationalist feelings, was more complex, but it certainly was not one of total subordination. On the other hand, Czechoslovakia, especially after 1968, and also the regimes of East Germany

**CIS Major Agreements and Treaties** (December 1992)

| | Political | | Economic | Military/Security | |
|---|---|---|---|---|---|
| | Alma-Ata[1] Declaration | IPA[2] | Ruble Zone | Collective Security[3] Treaty | Peacekeeping[4] |
| Russia | x | x | x | x | x |
| Armenia | x | x | x | x | x |
| Azerbaijan | x | | x | | |
| Belarus | x | x | | | |
| Georgia | | | x | | |
| Kazakhstan | x | x | x | x | x |
| Kyrgyzstan | x | x | x | x | x |
| Moldova | x | | x | | x |
| Tajikistan | x | x | x | x | x |
| Turkmenistan | x | | x | | |
| Ukraine | x | | | | |
| Uzbekistan | x | x | x | x | x |

[1]The Alma-Ata Declaration, signed on December 21, 1991 by eleven republics, ensured the former Soviet republics not a party to the Minsk Agreement, signed on December 8, 1991 by the Russian Federation, Belarus, and Ukraine creating the CIS, could enter the new organization on a parity basis.

[2]The Interparliamentary Assembly, formed in early 1992, was devised to better coordinate the legislation dockets of the republics. The IPA seeks to ensure that CIS agreements, declarations, and protocols are properly implemented.

[3]Signed on May 15, 1992.

[4]Formed on March 20, 1992, the CIS peacekeeping agreement provides for CIS troops to intervene in the various conflict zones in the former Soviet land space and for prevention of future ethnic and political conflicts in the region.

**CIS Major Agreements and Treaties** (December 1993)

| | Political | | | Economic | | | Military/Security | |
|---|---|---|---|---|---|---|---|---|
| | Alma-Ata Declaration | Ratified[1] Charter | IPA | Ruble Zone | IEC[2] | Economic Union[3] Treaty | Collective Security Treaty | Peacekeeping |
| Russia | x | x | x | x | x | x | x | x |
| Armenia | x | x | x | | x | x | x | x |
| Azerbaijan | x | | | | x | x | | |
| Belarus | x | | x | | x | x | x | |
| Georgia | | | | | | | | |
| Kazakhstan | x | x | x | | x | x | x | x |
| Kyrgystan | x | x | x | | x | x | x | x |
| Moldova | x | | | | x | x | | x |
| Tajikistan | x | x | x | x | x | x | x | x |
| Turkmenistan | x | | | | | | | |
| Ukraine | x | | | | | | | |
| Uzbekistan | x | x | x | | x | x | x | x |

[1]This column indicates if the republic's parliament ratified the CIS Charter, thus making membership in the organization official and legal. The Charter outlines the organizational framework of the CIS, both its functional duties and its guiding principles.

[2]The Interstate Economic Committee, formed on December 9, 1993, is the coordinating body of the Economic Union. Like the IPA, it serves to harmonize the legislation of the republics in the sphere of economics.

[3]Formed on September 24, 1993.

and Bulgaria (with the latter even desiring at one point to join the Soviet Union) were far more directly subordinated. And all, except for Yugoslavia, were militarily integrated into the Moscow-run Warsaw Pact.

In the current CIS, a great deal of diversity similarly prevails in the relationship between Russia and the other states. At one extreme are Ukraine and Uzbekistan. The former has made considerable progress in internal democratization—it is the only CIS member to have had an orderly and democratic presidential transition—and it is also the only one to have created a significant national army of its own. Moreover, Kyiv has refused to integrate that army into the CIS system. It has also asserted itself (as the documents cited show) on a variety of other critically important economic, political, and security issues. It is, in effect, managing to maintain political independence despite continued economic dependence.

Uzbekistan likewise has jealously protected its new status

**CIS Major Agreements and Treaties** (December 1994)

| | Political | | | Economic | | | Military/Security | |
|---|---|---|---|---|---|---|---|---|
| | Alma-Ata Declaration | Ratified Charter | IPA | Ruble Zone | IEC | Economic Union Treaty | Collective Security Treaty | Peacekeeping |
| Russia | x | x | x | x | x | x | x | x |
| Armenia | x | x | x | | x | x | x | x |
| Azerbaijan | x | x | | | x | x | | x |
| Belarus | x | x | x | | x | x | x | |
| Georgia | x | x | | | x | x | x | |
| Kazakhstan | x | x | x | | x | x | x | x |
| Kyrgystan | x | x | x | | x | x | x | x |
| Moldova | x | x | | | x | x | | x |
| Tajikistan | x | x | x | x | x | x | x | x |
| Turkmenistan | x | | | | | | | |
| Ukraine | x | | | | | | | |
| Uzbekistan | x | x | x | | x | x | x | x |

**CIS Major Agreements and Treaties** (June 1995)

| | Political | | | Economic | | | Military/Security | | |
|---|---|---|---|---|---|---|---|---|---|
| | Alma-Ata Declaration | Ratified Charter | IPA | Ruble Zone | IEC | Economic Union Treaty | Collective Security Treaty | Peace-keeping | Air-Defense Agreement[1] |
| Russia | x | x | x | x | x | x | x | x | x |
| Armenia | x | x | x | | x | x | x | x | x |
| Azerbaijan | x | x | | | x | x | | | |
| Belarus | x | x | x | | x | x | x | | x |
| Georgia | x | x | | | x | x | x | | x |
| Kazakhstan | x | x | x | | x | x | x | x | x |
| Kyrgystan | x | x | x | | x | x | x | x | x |
| Moldova | x | x | | | x | x | | x | |
| Tajikistan | x | x | x | x | x | x | x | x | x |
| Turkmenistan | x | | | | | | | | x |
| Ukraine | x | | | | | | | | x |
| Uzbekistan | x | x | x | | x | x | x | x | x |

Note: The Unified Air-Defense Agreement, signed on February 10, 1995, provides for the joint protection, monitoring, and use of air space. An early warning system for all members is another element of the agreement.

as a sovereign state. The most populous of the Central Asian states and more homogeneous religiously and ethnically, it is less vulnerable to the external exploitation ("*divide et impera*") of internal ethnic strife—a weakness from which a majority of the other CIS members suffer. It is governed autocratically—in effect, through a personal dictatorship—but with a longer-range design in mind, patterned in part on the predemocratic experience of some of the economically successful Asian countries.

Most important of all, the Uzbek elite and population appear to be motivated by a deeper sense of historical identity than is the case with the other Central Asian states, and they even entertain some regional hegemonic aspirations. The Uzbeks identify themselves with—some would say usurp—the grand legacy of Tamerlane's empire and favor the creation of large Turkestan, embracing the Central Asian region. Much like the Ukrainians, the overall orientation of the Uzbeks is to safeguard and strengthen their suddenly obtained national independence.

At the other extreme are Belarus, Georgia, and Tajikistan. The Belarusians have been largely russified. To all effect, the native language is gone, the economy is merely an extension of the Russian one (with essentially no self-sustaining components), and a large majority of the Belarusian public even voted in late spring of 1995 to adopt Russian as their state language, to restore Soviet-era official (largely ideological) holidays, to return to the Soviet-Belarusian flag and the Soviet-Belarusian national coat of arms, and even to register their desire for a closer union with Russia. (It is noteworthy that this vote was effusively praised by Boris Yeltsin on 24 May in a rare television address to the Russian people.) Though a formal incorporation of Belarus into Russia has been delayed—in part, because of Russian reluctance to assume immediately the economic costs of subsidizing Belarus—politically and militarily Belarus is becoming an extension of Russia.

Of course, one cannot exclude a later reaction against the current trend. It is possible that in time Belarusian nationalism could ferment and revive. That could happen if developments in Russia, especially socio-economically, were to turn very sour. But for the near future and probably beyond, the sovereignty of Belarus is again nominal, largely as it was in the Soviet time.

Georgia and Tajikistan are also examples of political dependence and subordination. In the Georgian case, that is especially tragic because one would have expected the cultured and historic Georgian people to be more successful in establishing their independence. Georgia was an ancient kingdom, independent until 1801 and again, though only very briefly, in the early 1920s. The Georgians have a genuine sense of national identity and they appear to desire independence. Indeed, after the dissolution of the Soviet Union, the Georgians refused to join the CIS at all.

However, the lack of internal unity, ineffective national leadership after independence, clanish conflicts, and also Russian manipulation of the Abkhazian and Ossetian minorities produced such destructive internal violence that ultimately Georgia had to request Russian assistance. It was "granted" in return for Georgia's full adhesion to the CIS, the acceptance of three major Russian military bases on Georgian soil, and the inclusion of Georgia in other Russian-controlled CIS security arrangements and border controls. Though certainly still more autonomous, and clearly more nationally distinctive, than Belarus, Georgia's sovereignty is now narrowly circumscribed.

Similarly, the social and ethnic conflicts in Tajikistan created an opening for even more direct Russian military intervention. This effectively stripped Tajikistan of any genuine independence. That intervention, initially also supported by Uzbekistan because of the latter's regional ambitions, has so badly damaged the social structure of the country and has inflicted so many casualties that its destructive impact can only be compared to the war on Chechnya.

In effect, one can delineate two broad constellations within the current CIS. The dominant one by far, led by Moscow, is composed of Russia itself (with almost one-half of the total population of the CIS and commanding about two-thirds of its resources), Belarus, Georgia, Armenia (fearful of Turkey and hostile to neighboring Azerbaijan, hence supportive of Russia), Tajikistan, Kyrgyzstan, and (more ambivalently) Kazakhstan.

The other constellation, much weaker and largely defensive, is based on Kyiv, with Ukraine's independent stance supported, sometimes overtly and sometimes tacitly, by Moldova (which is geographically sheltered by Ukraine), by Uzbekistan as well as by Turkmenistan (both sheltered geographically by Kazakhstan), and by oil-rich Azerbaijan. This cluster is clearly resistant to Moscow's efforts to transform the CIS into a politically as well as economically more integrated system. Accordingly, these states have tended to support Ukrainian reluctance to give the CIS a supranational status. In private bilateral meetings, the top Ukrainian leadership has been urged by them to maintain its stand in favor of a loose CIS.

Kazakhstan, while generally supportive of Moscow, has played a deliberately ambiguous role in these internal CIS contests.[1] With almost as many Russians in its population as native Kazakhs and hence extremely vulnerable, Kazakhstan has to be especially careful not to alienate Russia. Its skillful and impressive president, Nursultan Nazarbaev, in propitiating Moscow, has carefully maneuvered for greater international recognition for Kazakhstan, and has sought to promote some regional Central Asian cooperation, while also advocating a new Eurasian union, built on closer Russian-Kazakh cooperation.

The collapse of the old Soviet bloc was the consequence of the weakness and failing of the Soviet Union itself. The future of the CIS similarly is dependent, to the greatest degree, on how Russia, and its aspirations, evolves. The latter appear to be increasingly driven by a strategy to create a cohesive power structure in most of the space previously occupied by the Soviet Union. This goal is now expressed quite openly, and is well summarized in an edict entitled "Strategic Policy of Russia Toward CIS Member States," dated 14 September 1995 and signed by Boris Yeltsin. This policy edict (which appears in this collection as item 9.48) conveys bluntly the prevailing spirit of those who dominate the Kremlin's current thinking on Russia's special interests in the former Soviet republics. A commission of the Russian government has been created to see that the policy is implemented.

Some of the policy objectives listed in the edict which most obviously coincide with the goals and objectives of "great power" thinking in Russia are:[2]

> 1. —our main vital interests in the spheres of the economy, defense, security, and protection of the rights of Russians are concentrated on the territory of the CIS, and the safeguarding of these interests constitutes the basis of the country's national security; effective cooperation with CIS states is a factor which counteracts centrifugal tendencies in Russia itself. . . .
>
> 4. —to ensure reliable stability in all its aspects: political, military, economic, humanitarian, and legal;
>
> —to promote the establishment of CIS states as politically and economically stable states pursuing a friendly policy toward Russia;
>
> —to consolidate Russia as the leading force in the formation of a new system of interstate political and economic relations on the territory of the post-Union space; . . .
>
> 12. —Matters must be driven toward the creation of a collective security system on the basis of the Collective Security Treaty (CST) of 15 May 1992 and of bilateral treaties between CIS states. There should be encouragement of the intention of states parties to the CST to unite within a

defensive alliance on the basis of shared interests and military-political objectives. . . .

—It is necessary that CIS states honor their pledges to refrain from participating in alliances and blocs aimed against any of these states. . . .

13. —Work must be completed on settling the package of border questions with CIS states and creating a treaty-legal basis for the stationing of Russian Federation Border Guard Troops in these countries. There should be a study of the possibility of creating in the future regional command units of CIS states border guard troops, and efforts should be made to create a uniform system for the protection of their borders. . . .

—At the same time, and with due consideration for the foreign political situation, the Russian Federation's state border with all adjacent states should be defined in line with the Russian Federation Law "On the Russian Federation State Border," while bearing in mind that the retention of the principle of open borders within the CIS corresponds with Russia's long-term plans. . . .

14. —Cooperation along the line of security services ought to be developed, primarily with a view to preventing the use of CIS territory by special services from third countries for purposes hostile to Russia. . . .

17. —Our CIS partners should be persistently and consistently guided toward the elaboration of joint positions on international problems and the coordination of activity in the world arena. Joint efforts should be applied to achieve the further affirmation of the CIS as an influential regional organization and to establish broad and mutually advantageous cooperation with authoritative international forums and organizations at CIS levels. . . .

18. —Efforts should be concentrated mainly on coordinating the CIS states' positions in the United Nations and the OSCE and their approaches toward relations with NATO, the EU, and the Council of Europe. . . .

The above passage succinctly defines Russia's goals. But how these broad objectives are pursued, and how they are defined in practice—i.e., whether through voluntary cooperation or through compulsion—depends in the first instance on what transpires in Russia itself over the next ten to twenty years. In broad terms one can envisage at least five different Russian futures:

i. a truly democratic Russia that is also impressively successful in its economic recovery;
ii. a Pinochet-type authoritarian and highly nationalist Russia that also attains impressive economic recovery and growth;
iii. an economically successful Russia, but beset by continued political instability, perhaps with alternating cycles of semi-democracy and authoritarianism;
iv. a democratizing, perhaps even somewhat anarchistic, Russia beset by continued economic crises;
v. a politically authoritarian and nationalist Russia, unable to cope effectively with its economic dilemmas.

Obviously, many more variants could be listed, but the five above delineate the basic options. And of these five, only the first two would make likely the full attainment in the near future of the more ambitious Russian goals regarding the CIS—though quite obviously they would be defined and pursued in strikingly different ways. A truly democratic and prosperous Russia inevitably would be less likely to attempt to coerce other CIS members into an involuntary union. It would be more likely to heed the advice of those liberal Russians who have warned against the view "that the full independence of the former republics would mean greater economic, political, humanitarian and strategic losses for Russia; [that view] also misjudged the outside world's tolerance of Russian neo-imperialistic policy and the regular use of force, a probable outcome if this course were adopted."[3]

Instead, Russia's power of attraction would be derived from the combination of democracy and prosperity. That attraction would be especially powerful if Russia's prosperity were to be visibly higher and socially more pervasive than that of the neighboring states. Popular pressures within them on behalf of socio-economic integration with Russia would be likely to overwhelm residual nationalist sentiment.

Under such extraordinarily favorable circumstances, the CIS could eventually become a cross between the European Union and NAFTA. And though NAFTA involves an enormous disproportion in power between the United States and its partners (as is the case with Russia in the CIS), the democratic character of the American political system respects and protects the latter's political sovereignty. In brief, in seeking a more integrated CIS a democratic and prosperous Russia would be more likely to follow the course advocated by Vladimir Lukin, as cited on p. 7 of our introduction.

The second option for Russia's future would involve both a different concept of integration and different means of seeking it. A Pinochet-type Russia, driven by nationalist zeal, led by an energetic military figure (say, Lt.-Gen. Lebed) and sustained by a thriving economy, would be much more likely to become overtly imperialistic. Such a Russia would most probably be imbued with a special sense of a Eurasian mission and subscribe to the Eurasian doctrines. It would also be very tempted to exploit the presence in the "near abroad" of the Russian minorities in order to disrupt and undermine its neighbors, especially in Ukraine and Kazakhstan. In brief, it would define itself in a manner consistent with the views expressed by Aleksandr Rutskoy, cited on page 5 of our introduction.

Since such a Russia would lack a supernational ideology to justify imperial unity, a function that communism performed in the Soviet Union for the non-Russian Soviet political elites, under the second option Moscow would have to rely on Great Russian nationalism to generate among the Russian people the needed enthusiasm and the willingness to use force. That, in turn, would stimulate stronger opposition from the non-Russians, enhancing Moscow's reliance on coercion as the major instrument for the transformation of the CIS into a more traditional hegemonic structure.

For reasons that do not require prolonged elaboration, neither of the above options is very probable—at least not in the near future. The current Russian combination of pluralism, authoritarianism, and anarchy is not likely soon to be transfigured into a stable democracy and genuine prosperity. The requisite political culture of transnational compromise is also lacking. It is, therefore, unlikely that the CIS can evolve voluntarily in the course of a decade or so into a truly integrated structure, centered on Moscow.

Similarly, the recreation of an imperial structure, dominated by an economically powerful and very nationalistic Pinochet-type Russia, seems unlikely. Chile's Pinochet enjoyed the backing of disciplined and unified armed forces. He imposed martial law on a homogeneous nation. And he could tap the talent and competitive energies of a large entrepreneurial class, grounded in a well established free market tradition. None of these conditions currently prevail in Russia. Moreover, since it is very likely that a blatant effort to reestablish the empire would produce violent resistance from the "near abroad," the costs involved could become prohibitive, even for an economically successful Russia.

Short of these two extremes, the more likely future involves one of the remaining three options. But none of them entails a Russia healthy enough to attain in the next decade or so the more ambitious goal of a politically subordinated and economically integrated CIS. Without both political stability and impressive economic recovery, Russia will simply lack the means either truly to attract or effectively to dictate.

At the same time, it is also important to note that even a weakened or troubled Russia is still bound to be much stronger than any of the newly independent states. Hence continued progress in enhancing the CIS as a Russian-dominated bloc is to be expected. Moreover, since Russia has already made more strides in its geopolitical than in its socio-economic recovery, it is not surprising that by mid-1995 statements by top Russian officials, who previously had emphasized the importance of economic interdependence, were increasingly stressing the need to create within the entire space of the former Soviet Union (short of the Baltic Republics) "a collective security system," including central control over the external borders of the CIS.[4]

Though Russia's own evolution, as well as strength, is likely to be decisive insofar as the future is concerned, the degree to which the new states achieve internal viability, and the extent to which they are drawn into wider international cooperation, will also influence the evolving geopolitical landscape. The political consolidation and economic recovery of the new states can be enhanced by closer contacts with the outside world. Increased international recognition as well as economic ties can strengthen the sense of national confidence of the new elites, making it somewhat easier for them to try to evade the more stringent Russian aspirations for the CIS.

However, the central reality is that all of the new states are very vulnerable to Russia's leverage. All of them face serious, in some cases even threatening, domestic problems. Some of these problems have already been exploited by Moscow, and it is safe to assume that it will happen in the future as well. Almost all of the newly independent states have either large Russian minorities and/or are very vulnerable to internal ethnic conflicts. In several cases, particularly in Central Asia, their borders were defined quite arbitrarily by Moscow, thereby creating the preconditions for territorial conflict. The war in the Caucasus between Armenia and Azerbaijan has already greatly weakened the viability of both states, and a replication of such conflicts elsewhere is certainly not to be excluded.

The future of Kazakhstan is especially uncertain. To survive as an independent state, Kazakhstan will have to navigate with extraordinary skill. To avoid a collision with Russia, it will have to support CIS integration even while seeking to dilute it or to favor alternatives to it—while at home the Kazakhstani regime will have to bend over backward to avoid an internal collision between its Kazakh and its Russian citizens. But tensions are rising—and the proposed redesignation in mid-1995 of the country as the Kazakh Republic (instead of, as heretofore, the Republic of Kazakhstan) seems to indicate a more nationalistically assertive Kazakh attitude. It is almost inevitable that internal Kazakhstani-Russian tensions will intensify, as the Kazakh people become more nationalistic and as the Russians become more openly resentful.

Geopolitical and economic support from the outside, while not decisive, could still be of some consequence. For example, one may assume that China would probably prefer an independent Kazakhstan to one that is reintegrated into Russia. Thus the Kazakh regime may be able to take advantage of the Chinese interest to encourage its powerful neighbor to treat Kazakhstan as a useful buffer state. Large-scale international participation in the exploitation of Kazakhstan's energy resources would also contribute to Kazakhstan's economic consolidation, especially if new pipelines to world markets were to be constructed outside of territories under Russian control. Western support for a series of new pipelines would diminish Kazakhstan's vulnerability to Russian leverage.

As already briefly noted, Kazakhstan provides a geographical shelter for the other Central Asian countries. Without an independent Kazakhstan, they will be far more vulnerable to Russian pressures. Their ability to resist such pressures is already weakened by rivalries among them, potential border conflicts, and ongoing ethnic hostility. That ethnic hostility is not only directed at Russian colonists, with the latter often contemptuously hostile toward the new ruling national elites, but it involves intense and often violent antagonisms among the different Moslem tribes and clans, who live within countries that are not yet quite fully homogeneous nation-states.

That internal vulnerability heightens the importance of international support for the new states. The new countries of Central Asia already enjoy the sympathy of Turkey, Pakistan, and Iran, all of whom have a geopolitical as well as a religious interest in Central Asian independence. Turkey has been especially supportive of the region's efforts to give a more modern definition to its national renaissance. Moreover, like Kazakhstan, both Uzbekistan and Turkmenistan are anxious to gain direct access to Western markets. Hence they have also been actively seeking external support for new pipelines as an alternative to the existing ones that run partially through Russian territory. The same is true of oil-rich Azerbaijan, whose geopolitical isolation in the Caucasus can only be overcome if direct pipelines to the West, through non-Russian territory, are internationally financed.[5] In fact, the pipeline issue is becoming in many respects the critical test of the international community's capacity to sustain Central Asia's and Azerbaijan's political independence.

Among the newly independent states, Ukraine towers over all others. Its future, therefore, is going to be of the greatest significance, both for the other CIS members as well as for Russia itself. Ukraine's survival as an independent state, and especially its ability to draw a clear distinction between economic cooperation and political independence, will directly influence not only the future character of the CIS but perhaps even the self-definition of Russia. Indeed, the Eurasian option for Russia would be either meaningless or threatening to Russia without Ukraine. It would really cease to be a "Eurasian" option and become in fact a predominantly Asian option, thereby further widening the traditional gap between Russia and the West. Only with Ukraine reintegrated into Russia might the Eurasian option have some geopolitical and economic validity for Russia.

Without explicitly debating the issue, Ukraine—like Russia—thus also faces the grand option of either Europe or Eurasia. But in Kyiv the choices are not cast in the form of a debate over the country's "soul" or "special mission," nor is the debate driven by a quest for great-power status. Rather, the choice is likely to be made pragmatic, dictated by what eventually happens in Ukraine itself. Hence a prediction here is somewhat easier: If Ukraine succeeds politically and economically, it will gravitate toward the European Union and even some form of a relationship with an expanded NATO. Thus its choice will be Europe. If Ukraine fails, it will be sucked into a politically more integrated CIS, and hence Eurasia will also be its destiny.

So far, Ukraine's first five years have been a mixture of surprising political success and persisting economic difficulties. The political success has been twofold: Ukraine has made significant progress in political democratization and it has avoided any major ethnic collision with its large Russian minority. Political democratization has involved a successful presidential transition as well as the absence of the kind of political violence that occurred in Moscow itself. The Ukrainian political elite has demonstrated a considerable ability to compromise when necessary, thereby avoiding head-on collisions.

At the same time, Kyiv's liberal policy regarding ethnic diversity and a general mood in Ukraine of ethnic and linguistic accommodation has avoided any serious ethnic collisions, outside of the rather specific Crimean problem. In fact, many of the 12 million Russian-speaking citizens of Ukraine actually consider themselves to be Ukrainian—not unlike the Irish in Dublin who speak mainly English but feel Irish. Last but not least, a sense of pride in Ukrainian statehood is beginning to permeate the Ukrainian elite, with even those political leaders (like President Leonid Kuchma himself) who had earlier advocated closer economic union with Russia taking an uncompromising stand on the question of national sovereignty.

The economic difficulties remain Ukraine's primary vulnerability. They are also the source of the greatest Russian leverage. However, if it is applied in a heavy-handed fashion, it is likely to reinforce Ukrainian nationalism. Thus both Moscow and Kyiv know that their relations must be guided by a very careful and subtle balancing of political sensitivities and economic realities. The foregoing limits Ukraine's vulnerability to Russian economic leverage while also giving Ukraine some time both to implement economic reforms and to consolidate wider ties with the outside world.

It is noteworthy that international support for Ukrainian independence has been growing. The United States and Western Europe, notably Germany, have come to recognize Ukraine's geopolitical importance and the influence its future is likely to have on Russia's own evolution. As a result, Ukraine has been gaining in external geopolitical and economic support, all of which enhances its long-term prospects. Indeed, *if* economic reforms are implemented and *if* political stability continues, five years from now Ukraine, given its resources, population, size, and favorable geographic location, could outstrip Russia in economic recovery.

Paradoxically, Ukraine's political and economic success may be the best hope for Russia itself and for the CIS more generally. A politically independent Ukraine, economically cooperating with Russia—and, over several decades, accustoming Russia to the reality of Ukraine's independence—would also help Russia to make the ultimate choice in favor of the European option over the Eurasian one. The historical dilemma posed in our introduction would then be answered in a positive fashion. In turn, that would permit the evolution of the CIS into a more stable and cooperative system, minimizing some of the unavoidably enduring elements of conflict.

That hopeful prospect is, admittedly, at best a long-term one. Though it is not to be excluded, in the shorter run the dominant trend is likely to remain the Russian effort to infuse the Russian bloc with more overtly hegemonic characteristics. Given the relative weakness and vulnerability of most of the newly independent states, that effort in the foreseeable future is likely to continue to make progress—as this volume demonstrates. Nonetheless, in the long run, it cannot resolve the fundamental historical reality that neither the old Soviet Union, based on ideological cohesion, nor the Tsarist Russian Empire, superimposed on politically dormant non-Russians, can be recreated. If the CIS is to avoid becoming a geopolitical space dominated by destructive conflicts—highly debilitating to Russia itself—the elements of voluntary cooperation must be enhanced by the deliberate forsaking of hegemonic aspirations. That is the historical challenge which the CIS poses to Russia.

## Notes

1. By way of example, one may cite the Moscow-sponsored CIS treaty on joint control of external CIS borders (i.e., borders with non-CIS members). The official text, announcing the signing of the treaty in Minsk on 26 May 1995, concluded as follows: "The treaty was not signed by Azerbaijan, Kazakhstan, Moldova, Turkmenistan, Uzbekistan, and Ukraine."

2. "Strategic Policy Toward CIS Published," *Rossiyskaya Gazeta,* 23 September 1995.

3. Alexei Arbatov, "Russian National Interests," in Robert Blackwill and Sergei Karaganov, eds., *Damage Limitation or Crisis?* Washington-London, 1994, p. 61.

4. See, for example, a speech by Foreign Minister Andrey Kozyrev to the Russian Ambassadors to the CIS, 6 July 1995.

5. Despite American reluctance, Turkey, Germany, Britain, and Austria have become actively engaged in supporting a pipeline from Turkmenistan which would bring Turkmen gas to international markets via Iran. Turkey is also very actively engaged in having both Turkmen natural gas and Azerbaijani oil flow through Turkish territory to an outlet on the Mediterranean.

# Appendices

# APPENDIX A

# CIS "Hot Spots"

## *Chronologies of Key Events*

### Introductory Notes

Since the formation of the Commonwealth of Independent States, several civil wars have erupted on former Soviet territory, each one of massive proportions in terms of the violence it has perpetrated and the destruction that has rained down on the new states. The following chronologies address five zones of violent conflict, or "hot spots," which have erupted on the former territory of the Soviet Union: (a) Chechnya; (b) Abkhazia; (c) Nagorno-Karabakh; (d) the Trans-Dniester Republic; and (e) Tajikistan. The purpose of these chronologies is not simply to give a factual account of developments, but to provide abstracts that depict the political thinking, tactics, strategic reasoning, goals, and interests of the main players in hot spot zones. Their aim is also to provide a sense of the deep problems surrounding the geopolitical self-determination of the post-Soviet states.

Each Hot Spot Chronology offers an abstract of the month-by-month state of play as these conflicts grew, became internationalized, and ebbed into tension-filled holding patterns while players calculated their next moves. Each conflict has provided a convenient rationale for the creation of a Russian power base, either for reasons of conflict resolution, the protection of the minority rights of Russian people living in another CIS state, or border surveillance. Russian military contingents dispatched to sites of conflict are labeled "peacekeeping forces." These forces play an important role in the constantly evolving interaction between diplomacy and coercion in the CIS hot spots. The chronologies attempt to convey a sense of what the troops being sent to conflict zones have been doing there, of the political conditions that have surrounded them, and of their strategic import, not only for Russia but also for other CIS states.

In several cases, Russia has competed with the United Nations or the Organization for Security and Cooperation in Europe (OSCE) in trying to provide a plan or deploy international forces in the conflict zone. For example, Russian commentators have called the OSCE plan for a cessation of hostilities in Nagorno-Karabakh "anti-Russian" because it would prevent the deployment of Russian peacekeepers in the Karabakh enclave. Only by monitoring the exhaustive political bargaining and negotiations between Russia, the combatants, and participating mediators, does the larger picture of the role of Russian troops come into clearer focus.

# The Chechnya Conflict

### Background Note

The Chechen-Ingush Autonomous Republic, a Soviet republic of 1.3 million people situated in the North Caucasus, proclaimed its independence from the Russian Federation in October 1991. This tiny mountainous republic, bordered by Georgia in the south and by the Russian republics of North Ossetia in the west and Dagestan in the east, is the home of a fiercely proud and independent-minded people whose enmity with Russia reaches far back in history. Its oil-rich territory is considered to be of special importance by Russia, because pipelines from Russia's Caspian oil fields wind through Grozny, Chechnya's capital, to the Black Sea, where the oil is pumped aboard tankers for export.

On 27 October 1991 the Chechen separatist leader, General Dzhokhar Dudaev, and his Chechen National Congress forcibly deposed the republic's communist leaders, seized power in Grozny, and declared their intent to break away from Russia. Two weeks later, presidential elections of dubious democratic merit were held among five candidates, resulting in a landslide victory for Dudaev, who subsequently declared the independence of a new state to be known as Chechnya. To date, no other state has recognized Chechnya's independence.

Immediately following the Chechen declaration of independence, Boris Yeltsin declared a state of emergency and commanded Russian troops and tanks to head toward Grozny. Dudaev answered with the creation of the Chechen national guard, appealing to all Muslims to "turn Moscow into a disaster zone." On 10 November, however, the Russian parliament rejected Yeltsin's decision and called for negotiations. Russian troops were commanded to retreat. This dramatic episode was followed on 31 March 1992 by eighteen federative administrative units of the Russian Federation signing a Federation Treaty drafted by Moscow. The Treaty lays out the basic division of powers between Russia and its federative unit governments. Chechnya and Tatarstan, however, refused to initial the document.

For almost three years, Dudaev ruled quasi-independently—free to run his affairs more or less as he wished. Misreading his popularity, Russia's behind-the-scenes assumption seems to have been that Dudaev would be toppled by internal opposition forces. Dudaev faced opposition from within his own power base and from other groups, such as that of Umar Avturkhanov, head of the Chechen Provisional Council, which controlled several districts in the northwest. Russia's support for Avturkhanov, by his own reports, was substantial. In July 1995, on return from a visit to Moscow, he said he had received 2 billion rubles from Russia over the 1992–95 period. Moscow was unable, however, to affect internal political change despite Dudaev's weak hold on the country and its increasingly desperate economic situation.

To the Chechen people, Dzhokhar Dudaev had been something of a hero—the first Chechen in Soviet history to become a general commanding a strategic bomber wing at the age of thirty-six. His family had been deported to Kazakhstan in 1944, the year he was born, along with almost the entire population of the Chechen-Ingush Autonomous District. Stalin had feared the Chechens might collaborate with the Nazis in order to win independence, given the long history of animosity between the Russians and the Chechens. In Kazakhstan, General Dudaev attended elite Soviet military schools and married a Russian woman, Alevtina. He worked his way up to major-general, commanding a division of Soviet strategic bombers based near Tartu, Estonia, from 1987 to 1990. While there, he learned Estonian and showed remarkable tolerance for Estonian nationalism. He even refused to carry out the orders of the Soviet government to shut down the Estonian television and the parliament. He returned to Grozny from Estonia in 1990.

In December 1994, Russia poured out its full strength on the mutinous leader and his republic. The timing of Russia's escalation to use of force, three years after Chechnya's claim of independence, coincided with a dramatic event in the political economy of the region—Azerbaijan's signing of a consortium agreement with Western companies to develop and pipe oil to the West, bypassing Russia. This put Russia's vital interests in the region into clearer perspective for Yeltsin, who apparently could no longer tolerate Dudaev's recalcitrance. Russia apparently viewed the war as necessary in order to assure its role in decisions involving future routes for Azerbaijani energy exports. These exports will take one of two principal routes, either northward through existing pipelines via Grozny, or westward through Georgia and Turkey, thereby boosting the Turkish economy and Turkish influence in Central Asia.

By asserting full control in Chechnya, Russia is also meeting the challenge Turkey poses to its power in the region. With Chechnya pacified, Russia may well be in a good position to argue that the northern route through Grozny is more secure than the western route through Turkey. As this strategy unfolds, Moscow has also put increasing pressure on Azerbaijan's president, Geydar Aliev, and Georgia's president, Eduard Shevardnadze, who both have consistently supported genuine power-sharing arrangements within the CIS. So far, Aliev has not succumbed to Russian demands for either stationing Russian troops along the Azerbaijani-Irani border, or participating in a unified air defense system. Georgia, on the other hand, has agreed to the establishment of Russian military bases on its territory.

On 1 November 1994, which is where this chronology begins its evolutionary account, the situation in Chechnya began its tragic descent into violence. The official date of Russia's military attack on Chechnya is 12 December, but fighting is reported to have flared up in early November. The battle reached its highest pitch in mid-December, when Russian tanks, bombers, and ground troops were sent in full force to Grozny after its refusal to heed Yeltsin's ultimatum to lay down arms. Although few Russians would disagree that Moscow has a legitimate claim to Chechnya, which has been part of Russia for more than a century, Yeltsin's decision to use force appears to have won little support among Russian politicians or the public. Even charismatic military leaders have spoken out against the invasion—most notably Aleksandr Lebed, who would play a key role in ratcheting down the conflict in mid-1996, after the reported death of Dudaev and Yeltsin's electoral-season break with the "war party."

Reactions within the other CIS countries have varied. Most leaders supported Moscow, due to its legitimate fears that Chechnya could pose a threat to its territorial stability by encouraging other republics to break away in similar fashion. However, groups of sympathizers in several CIS countries, including Ukraine, Azerbaijan, and Georgia, have surreptitiously participated in the conflict on the side of the Chechens. From the CIS point of view, of great import was Russia's willingness to use overwhelming force in the face of humiliating military reverses.[1] These events have dampened the hopes of many CIS states that a democratic Russia which respects the sovereignty of the newly formed states on its perimeter will emerge from the ruins of the USSR.

In sum, the Russian-Chechen crisis is the sequel to a long history of enmity between the two peoples. The policies of both Boris Yeltsin and Dzhokhar Dudaev were driven by deeply ingrained perceptions of national and ethnic interests. Russia's policies in the war have revealed a side of its new statehood, however, which is being critically questioned in the West as well as by some of its own citizens. The conflict is also being cautiously evaluated by other CIS members who must decide how to structure their own national policies toward Russia. Some leaders, like Eduard Shevardnadze, support Russia's choice of military intervention. Others, perhaps, privately question what it may mean for them.

### Note

1. One way to put Russian actions into perspective is to compare them with Serb misconduct in the Bosnian war. At the height of the Bosnian war, Bosnian Serb forces were firing some 3,500 artillery shells per day into Sarajevo. On the other hand, at one point during the Chechnya conflict, Russian aircraft dropped 4,500 bombs (not artillery shells) on Grozny *in one hour*, producing a vast number of civilian casualties, many of whom were Russians who had stayed in Grozny. Unlike many Chechens, they had no place to go and no rural relatives to take them in—Ed.

## Chronology of Key Events

The following chronology starts as the Russian military buildup on the borders of Chechnya begins, in early November 1994. Up to that time, relations between Russia and Chechnya had become highly volatile, according to Russia because of Dudaev's "criminal activities" and the unconstitutionality of his government.

### 1994

***November 1994***

**1** • Sergey Filatov, Yeltsin's head of administration, suggests the use of force in Chechnya in order to establish a "state of emergency," which would "normalize the situation for ordinary people."

**11** • Russian Vice-Premier Sergey Shakhray calls for the "voluntary resignation of Dzhokhar Dudaev and free democratic elections in Chechnya."

**13** • The Provisional Council, main opposition group to Dudaev, in which Ruslan Khasbulatov is active, says it is likely to receive large contributions of military equipment from "certain CIS countries." Commander of the Provisional Council's armed forces is Beslan Gantemirov, who is using armed forces to internally dislodge Dudaev's government.

**15** • Dudaev tells Interfax that he believes Russia and the United States will confront one another in the Middle East, but that Russia will be defeated. Dudaev also calls China a real threat for Russia, citing a "quiet occupation of Russia by China, to which the patient mentality of the Chinese contributes." He says the interests of Russia and China will clash in Kazakhstan in particular.

**17** • Ruslan Khasbulatov says there are still opportunities to resolve the Chechen situation, calling on as many people as possible to rise up against Dudaev in order to solve the problem peacefully. He predicts a Russian invasion—"to teach people a lesson."

• The Russian Defense Ministry denies that Russian troops are approaching Chechnya, but many tanks manned by Russian crews are reported breaking into Chechen areas from Northern Ossetia.

**18** • Dudaev argues in *Die Zeit* that Russia is forcing Chechnya into war, saying "we are being provoked into launching a counter offensive." He notes extensive bombing carried out by Russia in efforts to destroy the Chechen economy.

**21** • The Chechen General Staff notes the concentration of some 50 armored vehicles and 5 to 6 infantry battalions in the Naura region and troops and equipment building up in neighboring Dagestan. Pavel Grachev continues to deny reports of a Russian build up.

**23** • The opposition Chechen Provisional Council, Dudaev

oppositionists, launch gunship attacks on Shalinskiy Tank Regiment—Dudaev's main striking force. Press secretary for the Provisional Council Ruslan Martagov says Dudaev forces suffered considerable losses and he makes clear the damage was inflicted by the Provisional Council forces.

**24** • Dudaev introduces martial law and mobilizes all Chechen males between the ages of seventeen and sixty.

• The head of the Provisional Council's armed units, Zaimdi Choltaev, says his forces have blockaded Grozny's main routes and approaches.

**25** • Provisional Council forces strike down a helicopter and two transport aircraft headed toward the Naura region. Choltaev claims the destroyed aircraft were delivering munitions and weapons to Dudaev forces.

**26** • Press reports confirm that Chechen opposition forces have seized the Interior Ministry and the State Security Committee. Fierce battles are under way and the Presidential Palace in set ablaze. Dudaev forces, despite logistical inferiority, resist successfully and launch counteroffensives. They seize 20 armored vehicles, 3 tanks, and a total of 120 people, 58 of whom are Russians, thus proving Russia's involvement in the opposition.

• Dudaev accuses the Russians of "open military intervention" and "open sanctioning of barbaric air raids on civilian and economic installations." But, he says, "there is no weapon in the world capable of breaking the spirit of a people that has risen to fight for its freedom and independence. The true sons of Chechnya and the Caucasus will stand like a solid cliff in the path of the aggressors and will defend the young Chechen state."

**27** • Dudaev asserts that opposition forces operating in Grozny are composed of Russian troops and equipment. Responding to press questions regarding the opposition, Dudaev asks "what internal opposition in the world has ever had assault aircraft? The opposition is Russia."

• Dudaev captures seventy Russian servicemen and holds them hostage.

**28** • Several North Caucasian leaders appeal to Yeltsin to halt the "bloody conflict in Chechnya" and to "ensure the constitutional order and protection of the rights and legal interests of the people."

• Dudaev holds Russian authorities responsible for the heightened level of conflict. He repeats his request for an objective UN assessment of the situation, threatening to deal with the 200 Russian hostages he now has as "mercenaries and thugs."

• Duma Chairman Ivan Rybkin says military intervention is well within Yeltsin's legal domain as guarantor of the Russian Constitution.

• The Democratic Union passes resolution warning of a "Second Caucasian War." It notes that the last such war lasted forty-seven years (1817–64) and was the longest of all wars waged by the Russian Empire. It notes Russian crimes against the Chechens, including the banishment of all Chechens to Central Asia in 1944. It questions whether the Chechen people can now voluntarily submit themselves once again to Moscow's authority. It calls Moscow's policy to subordinate Chechnya by force not only a crime, but a mistake, just as the Afghan war was. It warns that Russia will never win a final victory and that thousands of Chechens and Russians will perish.

**29** • Yeltsin appeals to Chechen combatants to lay down their arms, and issues a forty-eight-hour ultimatum, after which Moscow will declare a "state of emergency" in Chechnya (although he does not say what this means). Musa Merzhurev, Dudaev's spokesman, says the ultimatum will cause all Chechens to unite.

• Dudaev holds an emergency meeting of his Congress to discuss Yeltsin's appeal. Members are split almost evenly between signing the Federation Treaty and continuing the resistance.

• Russian Su-27 combat planes raid Grozny airport. Three hundred people are reported dead and all planes stationed at the airport burned. Dudaev forces strike down Su-27 with a stinger guided missile—allegedly supplied by Islamic leaders in a neighboring state.

***December 1994***

**1** • Moscow secures Georgian leader Eduard Shevardnadze's guarantee that the Chechen-Georgian border will be completely sealed to prevent Georgian sympathizers from joining Chechen separatists.

• Provisional Council leader Umar Avturkhanov calls for a Russian invasion of Chechnya.

• Several hundred volunteers from the Transcaucasus and the Baltic states arrive to join Dudaev's forces.

**2** • General Leonid Ivashov, secretary of the CIS Defense Ministers Council, does not rule out sending military observers from the Commonwealth states into Chechnya, but protests that "the conflict in Chechnya is an internal affair, and CIS participation will only be possible with a request from Moscow's top political leadership."

**3** • Russian forces continue to build up on Chechnya's borders. General staff team headquarters are established north of Grozny.

**4** • Dudaev supports talks with appropriate leaders of the

Russian Federation, but demands an immediate end to the military aggression.

• Maj.-Gen. Polyakov resigns over Chechnya.

**5** • The former Russian vice-president, Aleksandr Rutskoy, says the country is on the brink of a large-scale war in the North Caucasus. He warns of a tragedy for Russia if it does not close all borders with Chechnya, barring its relations with any Federation republic.

**6** • Ruslan Khasbulatov supports Russian federal authorities and calls for a surrender of arms. He criticizes the Provisional Council's military attack on Grozny of 26 November, noting that it failed because of "the absolute lack of preparedness for the operation from a military and political point of view, not Dudaev's strength."

**7** • Chechnya forms a government commission for talks with Russia, headed by Finance Minister Taymaz Abubakarov.

**8** • Grozny hands over several Russian POWs, captured during the 26 November abortive offensive against Grozny, but says it will continue to hold others.

• Grachev warns of "toughest option" against Chechnya if it does not cease its resistance. Says units of the Armed Forces, Internal Troops, and Spetnatz are concentrated in the North Caucasus to implement this option.

• The Karabakh Intelligence service reports that a special battalion of the Azeri Army has been dispatched to Chechnya to help Dudaev's government.

**13** • The Russian Social Democratic Union issues a statement opposing the use of force in Chechnya.

**14** • Aide to President Levon Ter-Petrosyan of Armenia says the events in Chechnya are Russia's internal affair, and that events in Chechnya "cannot affect Armenian-Russian relations in any way."

• Georgian President Eduard Shevardnadze tells Radio Tbilisi that the policy of genocide will rebound like a boomerang on everyone responsible. Nevertheless, he then tells journalists that no other path of action is available to Russia, and that preservation of territorial integrity is of vital importance to the Russian state.

• Estonians picket the Russian Embassy in Tallinn to protest the invasion of Chechnya.

**15** • The Russian ultimatum to surrender arms expires: five aircraft bomb Grozny, killing twenty, mainly Russians. The village of Pervomaiskiy is bombed all night. Russian troops continue to tighten the blockade around Grozny.

• Shevardnadze's office declares that Chechnya has had repercussions in Abkhazia, causing it to declare a full combat alert.

• Latvian Congress condemns Russian military aggression in Chechnya, citing the right of self-determination of peoples. It calls on the Russian government to withdraw its armies.

**16** • The Caucasian Confederation of Russia issues a statement saying that war has already begun in Chechnya, which means that war could grip the whole North Caucasus tomorrow. Its position is that Russian troops must be withdrawn and the sides brought to the negotiating table.

• The Lithuanian parliament offers to send a "mission of good will" to Chechnya in order to help restore peace. The offer of mediation would be carried out with a delegation of representatives from all factions of the parliament.

• A discussion between Grigoriy Yavlinskiy and Egor Yakovlev on the events in Chechnya is recorded in *Obshchaya Gazeta*:

> [Yakovlev]: The past week will become the same kind of tragic page in our history as 25 December 1979, the day of our invasion of Afghanistan, or 13 January 1991—the day the bloodshed began in Vilnius.
>
> [Yavlinskiy]: The executive branch has completely cut off whatever was left that connected the Russian administration to society. . . . As a result of the military gamble all preconditions are in place for the disintegration of the military and the disintegration of the state. . . . It has become clear for everyone how dangerous the current constitution is. It ensures complete unaccountability of the executive authority. Neither the parliament nor the public can influence the president's actions. . . . As to the lawmakers' reaction, we should not forget that the Duma is in constant danger of being disbanded. ... They, together with Zhirinovskiy, are willing to take any unscrupulous steps to cover up any action of the president.
>
> [Yakovlev]: The diagnosis of the current event that many agree on is that we are moving in one direction: from a state of emergency in Chechnya to a state of emergency in Russia.

**18** • Georgian President Shevardnadze again stresses that Russia must defend its state interests and territorial integrity in Chechnya. He adds that if aggressive separatism is not curbed in Chechnya, it could cause the Russian state to disintegrate. "Georgia is interested in the preservation of an integral and stable Russia," he says.

**19** • Russia completely seals off its border with Azerbaijan with border guards. All transport and movement of people will be forbidden.

• ITAR-TASS announces that no volunteer detachments to aid Dudaev's regime have been organized in any North

Caucasus republic. The ITAR-TASS correspondent held talks with leaders of Dagestan, Kabardino-Balkaria, Adygeya, Ingushetia, and Karachaevo-Cherkessia. They all stressed that statements made by the nationalist Confederation of Caucasian Peoples were untrue. The leaders did express their intent to offer humanitarian aid, however.

**20** • Ukrainian Ministry advises Ukrainian citizens temporarily staying in Chechnya and caught in the conflict zone not to take part in the confrontation.

**21** • Chechen Vice-President Zelimkhan Yandarbiev urges international organizations not to send humanitarian aid to Grozny, "as the city is blocked by Russians and it is they who get everything." He says the Russian Army is using chemical weapons which cause damage to the skin.

• Fierce fighting continues in northern Grozny, which is bombed many times during the night. Dudaev calls again for peace talks with Russia, but on equal terms for both parties. He says there is no response from Russia.

• Dudaev blames the West's policy of non-interference for Russia's actions. He is also angry that a summit of Muslim countries in the Moroccan city of Casablanca did not produce a statement of support for Chechnya. Following the conference, however, Libya, Iran, and Saudi Arabia all issued statements on the Chechen problem.

**22** • Dudaev says on republican television that it is better to die with honor in accordance with the laws of a holy war than to enter into slavery. He calls on his fellow believers to fight against the "satanic methods of Russia."

**24** • As of 24 December, nearly 24,000 forced emigrants from Chechnya have been registered in various Caucasian republics.

**25** • Ruslan Khasbulatov, reiterating what Russian officials have said, says that an all-out assault has not yet begun. He estimates that between 150,000 and 180,000 residents remain in the city and that about 1,000 residents have been killed so far.

**27** • Yeltsin orders an end to the bombing of Grozny and stresses that a political solution is still possible. He also appeals to Russian servicemen in Chechnya to disarm the bandit formations.

• Vladimir Smirnov, chief of the Russian Federation President's Expert Council and chairman of the "Military for Democracy" movement, resigns saying: "As chairman of the coordinating council of the 'Military for Democracy' movement, I cannot allow my name to be linked directly or indirectly to the military adventurists in state structures. The military adventure in Chechnya is just the last straw, and it is to all intents and purposes the consequence of the policy pursued by our military leadership. From my standpoint, military reform is entirely lacking and the defense minister has no idea of what needs to be done or how to go about it."

• Foreign minister of Azerbaijan announces that "although it wants a peaceful solution in Chechnya, his country wants no split with Russia over the crisis in the rebel Muslim republic."

• The Moscow Helsinki Group and a number of other human rights groups picket in front of the administration building of the Russian president in protest against the bloodshed in Chechnya.

**28** • Valentin Sergeev, head of the Russian government press service, says that Russian troops are performing operations aimed at getting closer to the outskirts of Grozny, but that no assault has begun.

• Speaking about the Russian blockade of its borders with Azerbaijan, Azeri state advisor for foreign policy says: "Events in Chechnya are harmful for Russia since they exert influence on the international rating of the Russian state, and on its image as peacekeeper of regional and ethnic conflicts settlement."

**30** • Chairman of the Dagestan State Council, Magomed-Ali Magomedov, says that not a single soldier has gone to Chechnya, but that Dagestan "welcomes refugees from Chechnya and has rendered medical and humanitarian assistance to them."

• Dudaev's proposal for a New Year's ceasefire goes unheeded, and heavy bombing in Grozny's suburbs continues. In an interview, Dudaev says he will accept any condition for talks, except disarmament of the Chechen people.

**31** • Dudaev, in his New Year's address to the people, says, "Despite the great might of the Russian army which is attacking Grozny, we have won a moral victory. Nineteen-ninety-five will become the year of a powerful moral upsurge for the people of Chechnya."

## 1995

### *January 1995*

**1** • Yeltsin issues an edict creating a Temporary Oversight Commission for Observance of Citizens' Constitutional Rights and Freedoms, to be formed from representatives of the Russian Federation Presidential Staff and the Federal Assembly Federation Council and State Duma. V.A. Kovalev, deputy chairman of the Federal Assembly State Duma, is appointed chairman of the Commission. The

Commission's duty is to monitor compliance with the Russian Constitution of actions taken by subunits of the Russian Armed Forces and Russian Ministry of Internal Affairs Troops. The Russian Federation Presidential staff is to provide financial, informational, and material support for the Commission's activity. The Commission is to operate until "constitutional legality" is restored in Chechnya.

**2** • Viktor Sheynis, Russian State Duma deputy, after visiting Grozny, reports to the Duma: "This operation is monstrous in the way it was planned and in the way it is being carried out. Houses are being destroyed. The center of Grozny is on fire. The refinery is also on fire. Information that Grozny has been taken under control by the government forces is a blatant lie."

**3** • Georgian president, Eduard Shevardnadze, announces that camps have been set up in Abkhazia where militants are being trained to be sent to Grozny. Shevardnadze proposes that Georgian and Russian border troops should erect strict joint controls on the border.

**4** • Through the German ambassador, the European Union countries protest to Russia over the bloodbath in Chechnya. Internal disagreements among the EU countries have prevented any statement until now. Danish Foreign Minister Helveg Petersen denies that the EU will recognize Chechen independence, and says Chechnya cannot be compared with the Baltic states, which the EU never recognized as Soviet republics.

• The OSCE proposes to send a group of experts to Chechnya to observe the human rights situation there.

• Lt.-Gen. Aleksandr Lebed says "the Chechen problem cannot be resolved by military means."

• Human rights activist Sergey Kovalev says he expects that Russian troops will storm Grozny the next day. He sharply attacks the tactics of the Russian Defense Ministry and other power ministries, saying Grozny is already nothing but ruins which remind him of pictures of Stalingrad during the Great Fatherland War. He will go to Moscow and insist on talking with Boris Yeltsin, whom he calls responsible for events in Chechnya.

• ITAR-TASS reports seventy-eight incidents involving journalists in the zone of conflict in Chechnya since 1 December, in which the rights of the press were violated. Reports of Interior troops opening fire on journalists, and detention of foreign journalists have been corroborated by reliable sources in the area.

• Grigoriy Yavlinskiy, leader of the Yabloko parliamentary group, calls on Yeltsin to volunteer his resignation as a result of the events in the Chechen Republic. He notes: "We think it is indisputable that neither Boris Yeltsin nor his ministers in Army uniform can now settle the conflict in Chechnya. There is one way out; they should go for the sake of our country's future." The President's dismissal is impossible both formally and juridically, which is why Yabloko calls for voluntary resignation.

**5** • The Russian Federal Migration Service reports that more than 130,000 people have left Chechnya since the start of the military action.

• The president's palace in Grozny is bombed. (The bunker below the palace, however, reportedly remains intact.)

**6** • Downtown Grozny is reported still under the control of Chechen volunteers. The Russians do not know where Dzhokhar Dudaev is at this time. A military source in Mozdok emphasizes the fact that the Russian forces have started a new tactic—the use of small mobile groups to "cleanse" Grozny of "bandit forces."

• Russian Interior Forces report that the Russian-appointed "government of Chechnya's national revival," headed by Salambek Khadzhiev, former USSR minister of the oil industry, is expected to come to Grozny that day. This government is composed of Dudaev's adversaries.

• The head of the Chechen National Revival Government, Salambek Khadzhiev, begins his work. In an interview, he is asked how long he expects to work in this capacity. He responds that it depends on how long it takes for the situation to return to "normal." He thinks this will take five or six months in the short time-frame, or eighteen months in the long time-frame.

**10** • An interview with Dudaev is recorded in *Zavtra*: [Dudaev]: "About 100 nationalities live here in Chechnya—Russians, Chechens, Jews, Armenians. They are living, as the saying goes, in cramped quarters. De jure they are citizens of one country. Whom are they disturbing? Why are they being bombed, and why are missile strikes being delivered on them? Why even be surprised that the bombing starts and stops on command—that yesterday a Duma delegation of Yushenkov was here, and today a new air strike will be delivered on Grozny!"

• An interview with the president of North Ossetia is recorded in *Severnaya Osetiya*. President A.Zh. Galazov says: "I do not think that the Chechnya problem can be separated from the overall chain of events linked to the collapse of the former Soviet Union. Chechnya is the only one of the links in this chain that is associated with the tragedies of Nagorno-Karabakh, South Ossetia, Abkhazia, Ingushetia, and North Ossetia."

• Salambek Khadzhiev, prime minister of the Government for National Revival of the Chechen Republic, says he intends to set up a system of aid for the refugees, then gather together a number of managers, primarily of plants and industrial facilities, to think about restoring the refineries and other plants. Asked what are the government's levers of control, he replies: "the personal influence of the leaders," and "the economic influence of Russia." He says the program of economic recovery will cost in the neighborhood of 1 trillion rubles.

• A military journalist writes in *Krasnaya Zvezda* about the Chechen forces: "We were initially told that there were only a few bandit formations operating in the city, and that the internal troops would be able to disarm them. We now realize that everything has turned out to be far more complex. There are well-trained and superbly armed and equipped military formations of mercenaries and professionals operating in Chechnya. . . . And they have no shortage of patrons or instructors. Russian troops are finding the going even harder in Grozny itself. The streets have been cleverly closed off with concrete blocks in such a way that combat hardware has fallen into specially primed 'mousetraps' and is then fired upon from overlooking apartment windows."

• According to Grozny sources more than thirty chemical bombs were dropped on Grozny on 9 January. The Russians bombed the city the whole day. Before the dropping of the chemical bombs, units of Russian troops left the positions from which they bombarded the city. Grozny sources also report that the Russian military leadership is sending soldiers to Chechnya belonging to ethnic minorities, mostly Chuvashs, Mordvins, Tatars, Bashkirs, Komi, Kalmyks, and Marii.

• The Russian Ministry of Economics estimates the cost of rebuilding the housing stock, engineering installations, and other facilities this year could run between 2.3 trillion and 2.7 trillion rubles. The likely cost of restoring the oil and gas complex will run to at least 800 billion rubles (although the oil stock could yield an immediate financial return). The most cautious estimates of the economic losses in 1995 from the Chechnya war range between 3.1 trillion and 3.5 trillion rubles.

• In an interview in Baku, special representative of the Russian president to Chechnya cannot answer the questions: "Why didn't Moscow react to Dzhokhar Dudaev's proposal to begin political negotiations without preliminary terms?" or "Why doesn't Russia use the very methods and means that it recommends to its neighbors, in particular Azerbaijan, for settlement of its internal problems?"

• ITAR-TASS reports on Foreign Minister Kozyrev's statement to representatives of the OSCE leadership and the Hungarian ambassador on 10 January, namely that "restoration of constitutional order in Chechnya is Russia's aim," and "it is an internal affair of the Russian Federation." He says the OSCE delegation is concentrating solely on humanitarian rights. Their task, he says, is to restore the validity of all rights and freedoms that were infringed by the Dudaev regime for three years on Chechen territory.

• A forty-eight-hour truce is declared in Chechnya on 10 January. The truce saves the lives of hundreds of city residents, who gather in the streets after weeks of cold and starvation.

**12** • At 8:00 A.M., the morning of the expiration of the forty-eight-hour truce, gunfire opens on Grozny. Russian aircraft is also heard overflying the capital. The president's palace remains the main target of the Russian forces.

• Viktor Chernomyrdin meets with members of the Chechnya diaspora in Moscow, and says that political methods should be used to resolve the Chechen conflict. These methods include the creation of peace zones and amnesty for members of illegal armed units who voluntarily lay down heavy arms, and a broad dialogue on political settlement in Chechnya among the Chechen people and their representatives.

• *Izvestiya* reports that Chernomyrdin has taken over command of the whole Chechen operation. He has approved a plan for the recreation of structures of power in Chechnya, including a transitional government and restoration of the economic life. A. Volskiy, as a member of the conciliation commission, is contacting Chechen entrepreneurs. R. Abdulatipov is preparing a meeting with the republic's religious leaders. Shumeyko is gathering together leaders of the republics, krays, and oblasts of the North Caucasus in order that they can provide assistance in restoring organs of power and government in Chechnya. Yeltsin has begun consultations with North Caucasus leaders. Still, according to the article, the disarmament of Dudaev's guerrillas is the sine qua non for the Kremlin. The Army will remain in Grozny under arms. This means danger of further bloodshed. But Filatov has given assurances that a shift has been made toward non-violence.

**17** • Military expert Sergey Surozhtsev, writing in *Novoya Vremya,* says: "It is now absolutely clear that long before the official decision of the Security Council on the introduction of forces into Chechnya, the Russian forces and special services had taken the unequivocal course of using force to resolve the conflict with Grozny, although political methods were by no means exhausted." The work was done crudely and soon both the Russian and foreign press began to shed light on the "activity" of the bodies of the counterintelligence service and the Main Intelligence Directorate of the Defense Ministry, whose representatives did everything they could to avoid secret participation in the "establishment of constitutional order in Chechnya." One of the representatives of the opposition once appeared on Russian television and openly demanded that Moscow cease delivery of old combat equipment and send the most up-to-date equipment. In conclusion, Surozhtsev says that Grachev sent in ten regiments (almost eight times what the "enemy" had), but still does not have complete control of Chechnya. The Russian Army has suffered huge losses of manpower and combat equipment. He says the armed forces must be completely reformed.

**18** • Konstantin Zatulin, chairman of the State Duma's committee for CIS affairs, tells press correspondents that the Chechen conflict has seriously complicated perspectives of integration processes within the Commonwealth. "One can speak of integration seriously only with Belarus," Zatulin said. In light of the Chechen conflict, certain CIS member states have attempted to distance themselves from Russia, he said. In particular, he mentioned Ukraine. But, he added, the war in Chechnya will not stop after Grozny is seized. In his opinion, Chechnya will play the part of Nagorno-Karabakh in Russia. He criticized the executive branch for "incompetence."

• Vladimir Shumeyko, speaker of the Russian parliament's upper house, does not rule out the imposition of a "state of emergency" in Chechnya, which he defines as: "a special legal regime which is to help restore order by using interior troops only, without resorting to heavy weapons."

• Dudaev makes the statement that: "This conflict cannot be solved by military means. Even if you bring all the armies together it is impossible to crush the nation's spirit; it is a gift of nature. The Almighty created the Chechen people so that they could not be slaves. This struggle has been going on for 300 years."

**28** • *Izvestiya* reports that since 25 January, the Central Bank of Russia has started to manage the country's banking system with a consideration for national identity. Dagestan, North Ossetia, Kabardino-Balkaria, and Karachaevo-Cherkassia were sent the following telegram: "In the future, until special instructions, the following procedure is established for the clients of commercial banks. Cash operations by commercial banks are forbidden. Cash may only be paid out by offices of the Central Bank of Russia, strictly for wages, pensions, and social needs. Other payments are banned." Thus, the Central Bank is being mobilized for operations being conducted in Chechnya. The reporter goes on to remark that the Central Bank's policy is not only offensive to bankers, it is "simply stupid." Money for Dudaev's supporters can be brought in from any part of the country—whether it is brought in from Dagestan or not makes no difference. He also points out that: "The Central Bank suspects the Southern republics of complicity with Dudaev although this contradicts the Russian authorities' story that the conflict is of a local nature and the people who fueled it do not enjoy support from their neighbors."

**29** • Writing in *Moskovskie Novosti*, Sanobar Shermatova considers the impact of the Chechen war on two former republics—Tajikistan and Azerbaijan. In Tajikistan, she says, "President Emomali Rakhmonov has spoken for the first time about the need for an armistice with the opposition as a decisive factor in stabilizing the situation in the country . . ." She says that "some explain the president's statement by his shaken faith in the might of his military ally and by the necessity of independently settling relations with the opposition." In Azerbaijan, she says, the Russian Ministry of Foreign Affairs has an account to settle with President Geydar Aliev. "Throughout the past year and a half, Aliev step by step has been outplaying his opponents, by making concessions but invariably striving to achieve his goals. With the arrival of Aliev, Azerbaijan joined the CIS, but refused to let the Russians establish military bases on its territory or border forces on the border with Iran." But she says that Azerbaijan is being punished for its wins. The blockade of Chechnya has drawn a noose around its economy, which has deprived it of foreign shipments. The economy is expected to plummet further in coming months. She cites that a million refugees are in Azerbaijan, and almost 20 percent of the territory has been occupied. A major new offensive by Armenian forces in Karabakh could promote the next overthrow.

***February 1995***

**1** • The Russian government press reports that they have data which shows that the militants are getting ready to start sabotage activities, such as blowing up the dam of the reservoir.

• *Segodnya* correspondent writes an article about Russian cabinet fears that Chechnya might become the economy's "black hole," sucking in huge sums of money. It is likely, they also fear, that criminal structures will rally around the flow of capital into Chechnya, which will render it unstoppable. This would collapse the policy of financial stabilization, the ruble's crash, hyperinflation, and an increase in regional separatism. Moreover, cabinet members fear that regional elites might draw on the Chechen precedent as a means of pressuring Moscow to receive substantial funding with no strings attached. The article goes on to say that many government experts believe that "not a ruble from the federal budget should be invested in the reconstruction of the Chechen economy. Even the oil processing and oil industries are unattractive." He points out that experts are saying that the population is leaving Chechnya anyway, and is unlikely to return. For this reason, some in the government believe it more advisable to "help them settle down elsewhere." The most convenient regions, he says, are said to be Krasnodar kray and Rostov oblast. In this way, the cost of rebuilding Chechnya's social infrastructure can be cut fifteenfold. Government officials dealing with the problem believe that all funds being channeled to Chechnya should be handed to a federal body that will run the republic until elections can be held. In the specialists' view, giving Chechnya any tax breaks or export benefits would be impermissible.

**2** • Salambek Khadzhiev, head of the Government of National Revival, accuses the Russian mafia of having run Chechnya for the past three years. He gives the following figures: "In the period 1991–1994 Russia pumped 20–30 million tons of oil to Dudaev. There flowed back—after refining—at best 50 percent. But here there are also the subtle details of payment to be taken into account; you could delay payment or salt money away in convenient banks. In general, there were many ruses here capable of producing big money."

• Anatoliy Kulikov, deputy Interior Minister and commander of Russia's Interior Ministry's troops in Chechnya, is appointed head of the joint command of the Russian group of forces on the Chechen territory. He tells reporters Chechnya is a "completely criminal zone and a completely militarized zone." He says further that the West is not supporting Dudaev because it understands what is behind him. He quotes a passage from Dudaev's book *The Thorny Path to Freedom*, which he says proves that Dudaev is an international terrorist.

**7** • Russian combined forces establish control over strategic facilities in the center of Grozny. Dudaev's supporters move to the southern and eastern districts of Grozny, where one-story private houses are located. A Russian government spokesman tells the press that "Russian Interior Ministry troops will perform mopping-up operations in these districts."

**8** • As before in history, the Sundzha River becomes the major confrontation line between the federation forces and Chechen formations. A spokesman for Dudaev says Dudaev's formations still possess "a substantial amount" of heavy military equipment, in particular tanks. According to him, Dudaev's home guards have captured seventy-two units of Russian armored vehicles during the first two weeks of this operation.

**9** • Dudaev orders the Chechen armed forces to immediately evacuate Grozny. The city will be surrendered to Russian forces at midnight. The Chechen defense committee says this does not mean, however, that the Russians will be in control of the situation. Dudaev orders his mobile assault groups to remain in the city and to conduct guerrilla warfare behind the lines of the Russian troops.

• Large parts of Grozny are ablaze as Russian troops move to cut off the last remaining road leading south from Grozny. A senior Chechen military commander, Aslambek Abdulkhadzhiev, says that the city will "blow sky high." "There are a lot of secrets in Grozny," he adds, referring to unnamed tools with which to destroy the city. "We'll build a new city."

• Russian Chief of Staff Col.-Gen. Mikhail Kolesnikov tells reporters that Dudaev "has gone mad and he must be destroyed." . . . "There is no need for a court trial, investigation, and collection of evidence," he stresses. He says Federation forces know about the possible places where Dudaev can stay.

• Konstantin Borovoy, leader of the Economic Freedom Party, gives an interview to *Obshchaya Gazeta*, in which he says he has spoken with Dzhokhar Dudaev in person. He says he was escorted to Dudaev by his people. He says the main purpose of his trip was "to convince Dudaev that the citizens of Russia, the mass of them, do not support the war in Chechnya and that terrorist acts against Russia's civilian population are unfair." He reports that Dudaev said that the image of the Chechen terrorist is being created by the Federal Counterintelligence Service (FCS). Following the meeting, Borovoy met with Col.-Gen. Kulikov, who asked whether Dudaev would not agree to surrender voluntarily. Borovoy says he explained that there is no way one can hope for that. He tells Kulikov that Russia "has effectively lost the war, so why resist?" He says Kulikov responded: "Oh come on, in three days we shall encircle Grozny and then everything will be over." Borovoy says: "My impression from these talks and from our military is that the sense of realism has been fully lost."

• *Moscow News* publishes a survey of Russian public attitudes toward the war in Chechnya. The data show that 42 percent would definitely "vote for opponents of the war" in the next election, while 25 percent were undecided if it would affect their choice of a candidate.

**14** • Professor Sayed Abdullo Nuri, leader of the Islamic Revival Movement in Tajikistan, tells the press that "Today we cannot say that the Muslim people of Chechnya have been defeated or that they have failed. This is no defeat; it is the light of jihad and struggle. These people have our enemy, our colonialist, by the throat. In the near future, we will see that this phenomenon will arise in Dagestan and then come to the Ingush. This movement will continue in the whole of the area."

**16** • Yeltsin issues a decree containing measures to restore the economy and social sphere of the Chechen Republic. He sets up a State Commission for the Restoration of the Economy and Social Sphere, to be formed from Federation executive agencies.

• Russian Security Council Secretary Oleg Lobov says federal authorities are ready to negotiate "practically at any level." He puts great emphasis on current talks between commander of the Russian interior troops Anatoliy Kulikov and chief of the Chechen staff Aslan Maskhadov. The Security Council has decided that the talks, now mediated by Ingushetia, should possibly involve other parties, he says.

• A report obtained from allegedly "reliable sources" says the Security Council orders all structures engaged in restoring Chechnya order should give priority to the Russian population of Chechnya in all cases—in the distribution of humanitarian aid and in allocating posts in the reestablished local administrations.

**17** • Dudaev's elder brother is reported arrested in Grozny. There is no immediate confirmation of the arrest.

• Chechen mufti leader, Muhamed Alsabekov, says the clergymen are looking for peaceful ways out of the present situation. He says that in talks with Viktor Chernomyrdin he was promised that Moscow would not send its representative to rule Chechnya. He remarks that "the Muslim leaders will support neither Ruslan Khasbulatov, nor Salambek Khadzhiev or Umar Avturkhanov while forming new power bodies. . . . New leaders will appear during the peace settlement."

**20** • Minister Grachev tells the press that the only viable scenario for resolution of the conflict is an ultimatum from federal troops demanding a total arms surrender by the militants. Defending the Russian operation thus far, he says: "We came up against a situation in which the correlation of forces along the main thrusts of the operation was one to three in favor of the opponent."

• A spokesman for the Russian Ministry of Defense confirms the arrest of Dudaev's elder brother, Bekmurza Dudaev, who is being questioned by counterintelligence officers.

**23** • Yeltsin creates an office in the presidential staff to represent the Chechen Republic. A document is issued which says this step was motivated by the need to ensure representation of the republic in the federal bodies of power.

**28** • Director of the Institute of Economic Analysis Andrey Illarionov tells the press that expenditures for the operation of the federal forces in Chechnya are about $5 billion, or 2.5 percent of the Russian GNP. He says these estimates do not include assets needed to restore the republic's economy. Illarionov adds that expenditures for defense amounted to 4.1 percent of the GNP over eleven months of 1994, and jumped to 6.6 percent in December, resulting in a growth of the budget deficit to 10.4 percent of the GNP.

• Baltic groups for parliamentary relations with Chechnya maintain that the West does not analyze, and evades the main problem of Chechnya—the nation's right to self-determination—and by supporting Russia carries out an amoral policy.

• The UN human rights commission calls for immediate termination of the combat operations and of violations of human rights in Chechnya. It calls upon the UN Supreme Human Rights Commissioner to continue his dialogue with the Russian government.

## *March 1995*

**6** • The European Union refuses to sign an interim agreement on trade and partnership with Russia until an effective ceasefire in Chechnya is reached. Sources clarify that the active military operations in Chechnya over the weekend made signing the document impossible.

• The Western press reports that Dudaev formations are still getting into Grozny at night and killing Russian soldiers. One pocket of militants is still fighting in Grozny despite Moscow's claim that it had wiped out the last bastion of Chechen resistance.

**9** • The OSCE announces its intention to establish a permanent mission in the Chechen republic.

• Dudaev relays a message to the Lithuanian press, telling reporters that the war in Chechnya may last for fifty years.

• Chechen television is reestablished with the help of Krasnodar and Stavropol territory. Telecasting is based at Znamenskoe, carries 1.5 hours per day, and covers 500 km. Subjects of programming include disarmament of Chechen units, settlement initiatives of the Russian command, and economic problems of Chechnya.

**10** • Dudaev announces a general mobilization in the country and calls for implementation of the Muslim laws of Shari'ah in Chechnya.

• Russians shell the Argun area for three hours, then send in T-80 and T-72 tanks to break the defense line of Chechen volunteers.

**11** • German Foreign Minister Klaus Kinkel says that Russia's handling of the Chechen crisis has serious repercussions on the network of international relations in which Russia moves. He refers to the IMF which will delay negotiations on a $6.3 billion credit program.

• The foreign minister of France says the EU is determined as never before to promote economic and democratic reforms in Russia. He says the Chechen crisis undermines the partnership between the EU and Russia. At the same time, he says the EU does not favor isolating Russia.

• Moscow radio reports that a congress of the Chechen people is held in Shali under the chairmanship of Dudaev.

The congress adopts a resolution that, henceforth, the country will live according to military law and that the town of Shali has been declared the new capital of Chechnya.

**14** • Head of the presidential administration, Sergey Filatov, rules out the possibility that Dzhokhar Dudaev could take part in talks on a peace settlement of the Chechen conflict. He accuses Dudaev of "global-scale human rights violations."

• A Dudaev aide, Khamad Kurbanov, tells reporters that negotiations on principles of Chechnya's integration in Russia are possible. He made the statement after a brief walk-out by Dudaev's supporters in protest of a speech made by Salambek Khadzhiev.

• Participants of the "Peace Initiative in the Caucasus" conference held in Moscow approve a plan for gradual settlement of the Chechen crisis. The plan outlines several stages, ending with a treaty between the Russian Federation and the Chechen Republic on the division of powers, similar to the one signed between Russia and Tatarstan.

**18** • Commander of the Chechen troops, Aslan Maskhadov, tells "Vesti" news that the Russians now say they will capture Shali. But he asks "So what?" Even if they do, he says, the war will continue until the last Chechen is killed. (Other reports suggest that not all Chechens are willing to die for General Dudaev.)

• Nikolay Semenov, head of the territorial administration, reports that Russia has contributed 12.5 billion rubles to Chechnya to pay overdue salaries to employees of state organizations.

**19** • Russian warplanes bomb Shali, killing at least four. A major offensive is launched against Dudaev's strongholds—Argun, Gudermes, and Shali—bringing them under constant air and missile attack.

**23** • Boris Yeltsin issues a decree "On Interim Bodies of State Authority in the Chechen Republic" creating a Committee of National Reconciliation, which is to find ways to achieve reconciliation, and to draw up a draft constitution of the Chechen Republic, and stage free elections in Chechnya. The committee is to assume that executive power in the republic will be exercised by the national Revival Government of the Chechen Republic (under Prime Minister Khadzhiev) until a body of executive authority is set up in compliance with its new constitution in accordance with the decision of the Provisional Council of the Chechen Republic.

• News agencies announce that Dudaev has moved his headquarters from Shali, but the whereabouts of the new headquarters is unknown.

**24** • A Chechen radio station, "Svobodnyy Kavkaz" (Free Caucasus), is set up in Krakow. A representative of the Chechen information center in Krakow, which was established on 13 January, said the radio would go on the air this year. The Russian embassy in Poland protested against the programs of the Chechen Information Center. Its protest was dismissed on grounds that the Center was engaged in humanitarian activities.

• OSCE representative Istvan Gyarmati returns from Chechnya to Moscow and tells the press that he hopes the OSCE can help create the framework for real talks.

**28** • The OSCE says it plans to open a permanent mission in Grozny as early as mid-April, fearing the war in Chechnya could spread to neighboring republics. The mission's role would be to mediate between the Russian authorities and the international community over problems concerning the delivery of humanitarian aid, citing French medical aid group whose doctors were banned by the Russians over allegations they were smuggling weapons to rebel Chechens.

**29** • State Duma Deputy Vladimir Lysenko publishes an article in *Trud*, describing the Chechnya war in stark terms. He accuses the war party in Moscow of trying to prosecute the Chechen war to its "victorious conclusion," that is to say full control over the whole territory of the republic by force. He charges this group with being afraid to face the grim responsibility for their deeds, thus prolonging the judgment day. He predicts that under this scenario, Russia will end up with another Ulster in the North Caucasus. Another scenario would be immediate withdrawal of federal troops and granting self-determination to the Chechen Republic. But he says this "recognition of one's own defeat" is unacceptable to the Russian leadership. He ends by condemning both the military scenario and the political policy of forming a federal government on Chechen territory. He defends the right of the Chechen people to select their own leaders and calls for Moscow to put its political ambitions to one side.

**30** • The issue of the Caspian oil deal, which has become more complex due to Chechnya, is examined by radio Turan commentators in Baku. The Kurdish separatists want to lay the pipeline from Azerbaijan through the eastern regions of Turkey, which they call Kurdistan. Turkey wishes to lay the oil pipeline through its eastern regions (after eliminating the Kurds) to the Mediterranean coast. Several Russian sources have confirmed, the radio commentators say, that Moscow and Ankara agree that Turkey will not interfere in Chechnya and Russia will turn a blind eye to the suppression of the Kurdish movements by Ankara. Russia's Foreign Ministry has indirectly confirmed this understanding by expressing its sympathy with Turkey's problems on the border with Iraq, and Ankara has announced its understanding in the matter of genocide against the Chechens by the Russian army.

*April 1995*

**4** • *Kommersant-Daily* carries an article in which costs of economic reconstruction in Chechnya are examined. The total figure is estimated at 5.3 trillion rubles. This includes replacement of the housing stock on which 450 billion rubles must be spent and reconstruction of the fuel and energy complex, which will cost about 539 billion rubles. Until the plan drawn up by the State Commission on Rebuilding Chechnya, the banks will grant a credit against a government guarantee. The credit will be paid out of Chechen oil exports. It has been proposed to exempt the republic's oil exports from all duties and taxes, which will earn 360 billion rubles in 1995 alone.

• Federation forces took Shali and Gudermes, says *Kommersant-Daily* writer D. Kamyshev. But this does not mean disarmament. Dudaev is ready for new battles, presumably from mountainous areas which can be cut off from the rest of Chechnya, which will be "peaceful."

• The press carries reports of refugees continuing to pour into Dagestan. Officially registered so far are 75,544 refugees. Some 30,000 have yet to register.

• Nikolay Starodymov writes in *Krasnaya Zvezda* that with the taking of Shali and Gudermes it may confidently be said that there no longer exists a unified front of struggle by Dudaev's supporters against federation troops. A turning point in the military confrontation has been reached—at least where heavy artillery is concerned.

**5** • Rumors have spread that Dudaev's militants are trying to relocate gradually to Dagestan, taking positions in the mountain regions. Dagestan stresses that the rumors are unfounded.

• Russian aircraft resume shelling in the southwest portions of Chechnya.

**6** • Emil Payin, leader of interethnic relations in the President's Analysis Center and member of the Presidential Council, says guerrilla warfare is likely to develop. He says that Dudaev was never a universal national hero in Chechnya, but a popular military leader. With his military defeat, his popularity will decrease.

**11** • The Chechen Defense Council, headed by Dzhokhar Dudaev, states that it is "still committed to a political settlement of the conflict between Chechens and Moscow," attempting to include itself in the political settlement process despite the odds. Dudaev continues to say the war will cease if Russia will withdraw its troops, but if not the war will continue.

**24** • A "Peace and Accord" Charter is signed by in Grozny detailing the aspects for settlement. Key provisions include:

1. immediate cessation of hostilities;
2. amnesty to be offered by the Russian parliament for those engaged in "illegal activities" in the separatists' movement;
3. disbanding armed formations and confiscation of their equipment and weaponry;
4. preparations for general elections to legislative bodies of power in Chechnya;
5. establishment of national accord committees in every town and district.

Signatories to the charter, according to the Russian press service, were Chechen leaders Umar Avturkhanov, Salambek Khadzhiev, and Mufti Muhamad Arsanukaev and federal representatives Nikolay Semenov and Aleksandr Babak. All of the Chechen signatures, however, were those of leaders installed by a Russian presidential decree, not those of the Dudaev regime.

**28** • Separatists forces agree to a ceasefire after talks with Russian Army force commander Gennadiy Troshev. Troshev tells reporters that Chechen military commander Maskhadov "agreed to a ceasefire, did not object to proposals to allow the militants [to] go home and not to persecute those who fought against the federal troops." Maskhadov did not, however, agree to disarm and said the fighters will do so only when Russian troops fully leave Chechnya.

**29** • Interfax quotes Dudaev as saying that the Chechens "do not need interim truce, moratorium, or amnesty from Russia" and that such attempts are merely Moscow's efforts to demonstrate its "peaceable disposition in the face of the world and at deepening collapse in the Chechen territory."

**30** • Despite the moratorium on continued fighting, skirmishes continue in many villages, most of which occur at night. OSCE delegate to Chechnya Sander Meszaros says both sides are in violation as shooting and cannon fire is widespread near the Bamut village, Vedeno, and around Grozny. He says that should the sides observe the moratorium, there is a good chance for successful negotiations, based on his group's assessment of the situation.

*May 1995*

**1** • Eight hundred to two thousand separatist fighters are reported in and around Grozny covertly continuing combat operations, according to Russian press services. Four hundred operate as snipers and several Russian servicemen were killed by sniper fire. Most federal troop units in Grozny enter into combat activity daily.

**4** • Russian First Deputy Prime Minister Oleg Soskovets announces a government of national revival has been formed in Chechnya. He notes the body "faces a lot of difficulties" and includes some members "unloyal to the federal government." He urges continued dialogue and blames Dudaev for violating the moratorium on combat actions.

• An article in Moscow's *Segodnya* discusses morale and psychological problems emerging within Russian troops of the Ministry of Internal Affairs (MVD). Author Leonid Kostrov talks of "Chechnya syndrome" in which MVD servicemen are tormented by their role in Chechnya as the media characterizes their activities as murderous. Kostrov reports that several military and civilian psychiatrists hold sessions with troops in the combat zones and report that the troops suffer from low morale and psychological problems. Says one Russian general, "it is extremely difficult to perform one's duty if amid a mass of accusations one begins to feel a murderer." Also contributing to the "Chechnya syndrome," according to Kostrov, is the absence of substantial social protection for servicemen and their families. Traveling allowances, compensation for injuries, and aid to injured servicemen's families all influence this low troop morale and psychological deterioration.

**8** • Basaev charges that Yeltsin's moratorium on combat activities is a ploy to gain favor in the press and the international arena as Russia prepares for VE Day. He indicates that Chechen forces will force federal troops to leave through a guerrilla war, tactics "sure to make the Russian leadership sit down at the negotiating table on conditions of the withdrawal of Russian troops." Basaev says to defeat the separatist forces, federal troops would have to "cover the mountains with dead bodies." He resolves that there will be "no peace in Chechnya until the last Russian soldier leaves its territory."

**11** • A press report indicates that Dudaev has issued an elimination order of Grachev for violating accords the two reached in December 1994. The report notes an attempt will be made to execute the Russian defense minister when he visits Chechnya soon. A group of eight specially trained terrorists are reportedly training for the move.

**17** • Chechen Chief of Staff Maskhadov announces his readiness to enter into talks with Grachev, saying "I am deeply convinced that this meeting could exert a radical influence in order to stop further bloodshed." The two are scheduled to meet on 22 May.

**18** • Responding to reporters' questions in Beijing regarding possible talks with Chechen commander Maskhadov, Grachev says preconditions must be met before negotiations can begin. He lists a full ceasefire, full Chechen capitulation and surrender of weapons, and that the Chechen side recognize Chechnya as a constituent entity in the Russian Federation.

• Russian Deputy Finance Minister Oleg Velichko says Russia has earmarked 410 billion rubles to restore the Chechen economy, 365 billion by way of investments and the rest financed by the government. Velichko also notes the government will lobby commercial banks to invest in the Chechen region.

**19** • Dudaev describes a peace proposal offered by Chernomyrdin as resulting from talks with "officials from a regime propped up by Russian bayonets." Chernomyrdin proposed a round-table conference involving the National Accord Committee, the Territorial Board of Federal Authorities, the OSCE mission delegates, and representatives of Dudaev's entourage. He says the Russian leadership have exhausted their political credit but does not rule out resolution by peaceful means.

• Battlefield update: Fighting continues throughout most of the Republic. Russian aircraft have been reportedly bombing Dudaev positions in south, south-east, and west Chechnya. Russian General Mikhail Egorov, commander of Russian troops in the region, has vowed not to bomb villages but, according to press releases, such is not the case. In Grozny, Russian units fire powerful salvos from "Grad" and "Uragan" missile complexes set up in the area. The Russian military leadership believe Chechen militants hide in the mountains south of Grozny to orchestrate terrorist attacks and guerrilla war. Bamut, Vedeno, Mesket, Alleroi, Shali, and Merzoi-gala remain intense fighting centers and have been the scene of massive Russian air bombardments.

**22** • Dudaev says talks will never take place that involve elements of the National Revival Government in Chechnya (Umar Avturkhanov and Salambek Khadzhiev) as he regards this body a puppet government, secure only because of Russian troops. Dudaev also insisted that federal troops must withdrawal for talks to take place. (Talks slated for today obviously did not take place and Dudaev announces that neither he nor his representatives will participate in 25 May talks announced by Russian delegate to Chechnya Nikolay Semenov.)

**23** • Press reports note a Dudaev and associates conference taking place in the southern mountain region of Chechnya. Sources close to Dudaev confirm the meeting in which a peace plan was discussed. Elements of the plan include:

—an immediate ceasefire controlled by international observers;

—full withdrawal by the Russian military, political operatives;
—measure to broaden the security zones;
—storing heavy weapons under international control;
—return of refugees;
—presidential and parliamentary elections.

The conference participants also stipulate talks will not include Avturkhanov.

• OSCE head in Chechnya Sander Meszaros announces talks will likely take place in the OSCE mission in Grozny on 25 May. The mission has sent appeal to both sides' leadership asking to halt combat activities by 00:00 on 25 May and enter into talks at 10:00 the same day. Press reports note Usman Imaev will head the Chechen delegation though Dudaev is supposed to be present.

**25** • Talks are held in Grozny but break down after three hours. The Russian delegation, headed by Semenov, reportedly storms out of the OSCE building making no comment to journalists. According to Chechen delegation head Imaev, "Russia was not prepared for the talks." A Russian spokesman says the talks did not break down but were suspended. Each side, according to him, confirmed its original positions but expressed interest in settling the matter by political means. OSCE officials believe talks will resume in the next few days once each side consults its respective field commanders.

• Several Chechens, mostly women, hold a rally after the delegations break up demanding "an immediate stop to the injustice inflicted on the Chechen people." Many note an increase in air strike and combat operations in their respective villages carried out by Russian troops.

**30** • Imaev announces a decision made by the council of Chechen field commanders to extend combat hostilities into other regions of the Russian Federation. He says the decision was "provoked by the acts of wanton cruelty and marauding" committed by Russian troops against Chechen civilians. The council will give their proposal to Dudaev and ask for a new policy toward fighting Russia.

**31** • Imaev announces that Dudaev has rejected the proposal noted above, opting instead to focus on eliminating the Russians from Chechen territory.

***June 1995***

**3** • Russian troops seize Vedeno, a Dudaev stronghold. A Russian spokesman says the operation resulted in minimal losses for federal troops while the militants lost many men and equipment. The fall of Vedeno alters the balance of forces according to several commentators, changing significantly the correlation of forces in the foothills. Dudaev has longed prepared the village as the main stronghold for defense.

**4** • Dudaev appeals to Clinton to intervene in the Chechen conflict, according to a Russian MVD spokesman.

**9** • Press reports indicate that federal forces have encircled all of Dudaev's forces, which split into four groups, according to Russian military sources, after the fall of Vedeno. An official with the Russian military says they have "blocked the militants, making it impossible for them to move out of the mountains and into the plain in Chechnya's south."

**11** • Troshev reports that all escape routes from Chechnya's mountains, where most Dudaev militants operate, have been sealed off, including routes to Georgia, Dagestan, and Ingushetia. He believes all bandit formations will be disarmed within the next two weeks. Shatoy, the last Dudaev stronghold, is in danger of falling to federal forces as well.

**12** • In a press interview, Dudaev discusses the war in Chechnya and Russian aims in the Caucasus. Asked why his forces can resist so successfully attacks by the Russian military, he says it is because the Chechens have been doing so for 300 years. He elaborates on the notion of "Russism"—a ploy by the Russian military and official leadership to recruit Chechen (or Abkhazian, Georgian, Azeri) criminals and join them with Russian operatives to create an artificial "opposition" group. The group then stirs up ethnic or civil strife and tensions flare, promoting discord between government officials and the domestic population. This is the so-called opposition in Chechnya, Dudaev says, and they have implemented similar operations in Georgia, Nagorno-Karabakh, and Lezhastan (in Azerbaijan). According to Dudaev, these tactics will soon yield complete control in the Caucasus to Russia, who will then move north to recapture the Baltics. In his words, "If the world . . . does not stop this devil or if it does not make Russia a constitutional state (there will never be democracy there), it will be confronted with severe shocks."

**14** • Sixty to eighty Chechen gunmen enter the Russian town of Budennovsk, taking hostages and seizing administrative buildings. The group is armed with automatic weapons and grenade launchers, entering the town in coffins by bus. They reportedly ravage the town, killing several town workers, women, and children. Stavropol Territory authorities notify the federal security services (FSB) and Yeltsin dispatches FSB director Sergey Stepashin. Russian Prime Minister Chernomyrdin is also informed of the attack, believed to be a Dudaev ploy to continue the separatist war despite several recent setbacks.

• Russian Deputy Prime Minister Soskovets chairs an emergency session, composed mostly of officials from the MVD, FSB, and the Federal Border Service, to discuss events in Budennovsk. He informs the participants of Yeltsin's personal involvement in the matter and that special "Alpha," "Vega," and OMON units have been dispatched to the area. As of 1730 GMT, the militants keep hostage about 500 (varying figures reported) hospital patients and staffers, threatening to kill all should one of their fighters be killed or shot at.

• 1900 GMT: Chief doctor of the hospital seized by the militants contacts Russian officials with two demands made by the militants:

1. Urgent meeting between Yeltsin, Chernomyrdin, and Dudaev;
2. Cessation of combat and the withdrawal of federal troops.

The group threatens to blow up the hospital if the demands go unmet.

• 1930 GMT: Russian Government issues statement urging citizens to "stay clam, show restraint, and be vigilant." They assure the Russian people that "law enforcement bodies have taken all measures to render harmless and eliminate the terrorists." The statement expresses condolences for those whose families members were injured and that aid will be forthcoming.

**15** • Dudaev denies involvement in the Budennovsk events through a telephone exchange with Moscow's ITAR-TASS. "Not a single armed structure subordinate to me had ever had any orders or instructions to carry out terrorist actions on the territory of Russia," he claims. Such operations "only bring the national liberation struggle of the Chechen people into disrepute." Imaev, also in on the conversation, acknowledges the possibility that the invasion was carried out by desperate Dudaev supporters frustrated with recent losses.

• A FSB spokesman says the federal authorities have "absolutely reliable proofs" that the Budennovsk events were carried out the Chechen militant Shamil Basaev, an experienced commander who served in Abkhazia and Nagorno-Karabakh. Talks between the militants and Russian officials have made no progress.

**16** • Grachev announces that force may be needed to "liquidate these bandits as fast as possible." He denies the possibility that the group can simply leave through negotiations and that they have nowhere to retreat to.

• Chernomyrdin arrives in Budennovsk to conduct talks with the militants. He issues statement indicating he will now command talks with the militants and reminds citizens to remain calm and to display restraint.

• Dudaev officials continue distancing themselves from the events, saying the government of the Chechen republic initiated no action, signed no document, nor made any preparations for the terrorist attack in Budennovsk. Chechen Minister for Foreign and Economic Relations Ruslan Madiev says the "Chechen people should not be identified with the criminals who committed the gangster crime. . . ." He expresses deep condolences to those injured and the families losing loved ones.

**18** • Chernomyrdin holds telephone talks with Basaev, who spells out the demands of the militants once again. He calls for an immediate end to combat operations, full withdrawal of Russian troops from Chechnya, and (a new demand) a referendum in Russia on the status of Chechnya. Chernomyrdin made clear that he possesses no power to initiate a referendum but agreed to make all others possible. Basaev indicates he needed to consult his group and plans were made to continue talks at 1000 tomorrow. A preliminary agreement was reached to free all hostages in return for an immediate halt to combat operations followed by peace talks to solve all outstanding problems.

• Basaev gives news conference, saying he is not a terrorist but a leader of a "diversionary action." He lists several atrocities committed by the Russian military over the last few years similar to the events in Budennovsk and that no one has labeled them terrorist acts. He claims responsibility for the some of the deaths but denies any were "hostages." He notes he has no reason to believe Chernomyrdin is bluffing when assuring him that the demands will be met and names Imaev as the chief delegate to talks when negotiations take place.

• Chernomyrdin arranges to meet the demands set forth by Basaev. He provides eight buses for transporting the militants back to Chechnya (Vedeno) and announces Russian Commander in Chechnya General Kulikov will halt combat activities as of 2000, 18 June. Police are to escort Basaev and his men safely into Chechnya. Chernomyrdin's only demand was that all the hostages be set free. Volunteers have been recruited (local officials and State Duma members) to ride with the militants to ensure they will not be fired on or ambushed, a demand made by Basaev.

**19** • Chernomyrdin informs Basaev all buses are in place, combat actions have been halted, and a Russian negotiating team (lead by First Deputy Minister for Nationalities Affairs Vyacheslav Mikhaylov) is ready and waiting at the OSCE mission building in Grozny.

• Talks open at the OSCE Grozny mission. Imaev and Akhmed Zakaev, field commander, lead the Chechen delegation and Mikhaylov, Semenov, and Kulikov head the Russian team. Talks focus on technical needs for a ceasefire and on the release of hostages. No major progress is recorded.

• 1620 GMT: The bus convoy carrying the militants

leaves the Stavropol region for Chechnya. They will be escorted roughly 30 km outside of the region and then proceed on their own. Approximately 150 people travel with the militants as "human shields," mostly local officials and State Duma members. (Thirteen journalists also are part of the volunteer force.)

**20** • Press reports reveal that Dudaev and the Russians have agreed to keep individual federal units in Chechnya to "prevent possible acts of terrorism by militants in the republic." Russian representatives assure the Chechens in the talks that all those surrendering their arms will be guaranteed safety and freedom to travel. Talks continue on military matters and on implementing and maintaining a ceasefire. The moratorium on combat operations is extended for three days.

• Yeltsin expresses his satisfaction with Chernomyrdin's efforts in the Budennovsk events, particularly for the safe release of the hostages. The two decide to investigate the incident further. Yeltsin also held talks on the same topic with Federation Council chairman Shumeyko. The two discuss responsibility for the tragedy and focus on strengthening national security.

**21** • In a press interview, Kulikov discusses the progress of negotiations. He announces that he informed the Chechen team they must publicly denounce terrorism and demands they hand over the terrorists within three days. Should the Chechen side not comply, Kulikov warns combat operations would again resume. Kulikov continues discussion of military issues concerning the conflict's settlement and how to go about disarmament, arguing that Dudaev militants should stockpile their weapons under international supervision and in small groups 2–3 km apart. Arrangements for other armed fighters to surrender their weapons have been made. They will be given vouchers upon giving up their arms that protect them from criminal prosecution. A protocol is signed by Kulikov and the Chechen delegation, now headed by Maskhadov, who claims to have full authority to negotiate and make agreements on Dudaev's behalf.

**22** • Yeltsin announces his government will work toward a peaceful resolution of the Chechen conflict, saying too much emphasis was placed on the military route. Yeltsin says "a strict line at political settlement of the Chechen problem is the strategic position of the federal authorities."

**30** • Talks in Grozny are held to discuss the political problems associated with the Chechen conflict. Six issues were agreed to by the sides. The main issue was agreeing not to discuss the political status of Chechnya until after democratic elections are held. The other five outline the ways in which the Chechen people participate in political deliberation and how they can best freely express their opinion. Russian delegation head Mikhaylov says he believes the six agreements provide a constructive approach to the conflict's settlement.

• Two members of the Chechen leadership agree to resign. Umar Avturkhanov, head of the National Accord Committee, and Salambek Khadzhiev, leader of the National Rebirth government, announce they would step down "for the sake of peace in Chechnya." Both leaders insist, however, that the "unconditional resignation of Dudaev and his government" is necessary for further compromise. They contend that Dudaev's "extremism . . . has thrown many years back not only the Chechen Republic but also Russia with its fledgling democracy."

***July 1995***

**1** • A communiqué is issued on the progress of the talks in Grozny. The statement says the sides continue to work out political issues for holding elections and for ensuring their legitimacy. Agreements concluded thus far include guarantees by the state bodies of Chechnya and the Russian Federation not to pursue conflict participants, and measures to monitor public speeches that may damage ethnic relations and on working out arrangements for the safe return of those forced out of the Chechen Republic.

**2** • Commander of Russian ground troops in Chechnya Vladimir Bulgakov says Dudaev is using the Grozny talks to resupply his fighters with food, matériel, and weapons and for moving its troops in the mountains. As for allegations by the Chechens that they are strategically located in Russia to carry out terrorist acts, Bulgakov asserts this is a bluff to keep the Russians in "permanent psychological tension."

**3** • Dudaev holds conversation with deputy head of the Russian delegation Arkadiy Volskiy in which he states he will step down "with my cabinet if the sovereignty of Chechnya is recognized." Dudaev has rejected the so-called zero-sum plan proposed by Russia which calls for the resignation of all current government officials and their replacement by representatives from all political groups in the republic in November elections.

**7** • Yeltsin signs presidential edict setting up the mechanisms for a permanent deployment of the Russian military in Chechnya.

**10** • OSCE chief mediator in Chechnya Sander Meszaros believes that the negotiators are making significant progress. He notes talks have narrowed differences on reaching agreement on the status of Chechnya. While the two still differ as to what that status will be, agreements on a number of political issues have paved the way for addressing this point, the most controversial in all of the talks. He praised the fact that military consultations are continuing, the exchange of POWs is being worked out, and the two sides have devised a system of communication that is being tested. Military commanders of the Russian Federation, however, assert that Dudaev is still re-arming.

**16** • Talks are suspended due to the serious illness of Chechen negotiator Imaev. A technical break was also announced because, according to Dudaev delegate Maskhadov, "the sides have come to the point of deciding the most fundamental issues and need to consult." The issue is deciding on the fate of the Chechen Republic. Russian delegation head Mikhaylov says all the main provisions for settlement have been basically agreed on.

**19** • In press interview, Basaev says the talks in Grozny should achieve peace first and foremost. After arriving at that, he argues independence and freedom for the Chechen people should be achieved. He recognizes that the republic will always have to deal with the Russian Federation but as long as his regime is in power, "we do not want to be under its heel." Asked what he would do should a Chechen referendum opt for the republic to remain in the Russian Federation, Basaev says he will not oppose the people's will and keep fighting. He will continue the struggle, though, "using other weapons and other methods."

**26** • Press secretary for the Dudaev regime Ramzan Muzaaev says Chechnya is ready to enter a power sharing relationship with Russia in which certain functions will be carried out by the Chechens and others by the Russian Federation.

**30** • Russian and Chechen officials sign a military agreement that calls for the immediate cessation of military actions, provides for the exchange of POWs, halts all terrorist activities, and outlines a staged pullout of Russian troops. Special monitoring commissions have been formed. Maskhadov and federal troop commander Anatoliy Shirokov are the guarantors of its fulfillment.

**31** • The Russian Constitutional Court rules that Yeltsin's edict on 9 December 1994 ordering the invasion of Chechnya to stop the "activities of illegal armed formations" was legal, 16 to 3. Five of the judges expressed reservations about certain provisions of the edict, including a secret order giving Defense Minister Grachev "sweeping powers" to quell the three-year-old independence bid.

***August 1995***

**1** • A Chechen separatist leader, Abu Movsaev, proposes the partition of Chechnya into two states. The Dudaev regime would assume control of the mountain districts southeast of Chechnya, to include villages such as Argun, Gudermes, and Nozhai-Yurt. The plan notes the regime would not lay claim to the rest of the republic, which would remain in the Russian Federation. Movsaev was active in the hostage incident in Budennovsk.

**2** • Press reports note widescale infighting among the Chechen separatists. Information coming from Vedeno, Argun, and Shali says that the Chechen field commanders regularly ignore orders from Dudaev and are voicing their intention of taking power by force. At a meeting of commanders organized by Maskhadov, clashes erupted killing one commander and injuring two. According to the reports, the commanders use the "most foul language toward one another" and there is obvious discord among them.

**4** • Russian Interior Ministry troops replace the army in Chechnya. The troops will transit through Ingushetia and the number of troops entering will be much smaller than those leaving. The troops are to assist Chechen police in implementing the military agreement concluded by the sides.

**5** • POW exchanges remain troubled as neither side is releasing nor acknowledging the true number of POWs it holds. Russian Lt.-Gen. Anatoliy Romanov says he received word from the Chechen field commanders that they would proceed on an all-for-all exchange but, according to him, they have not been forthcoming. Talks will not continue, says Romanov, until the military provisions of the latest agreement are met and "the Chechen side is aware of this."

**10** • Leader of the Chechen talks in Grozny Khozh-Ahmed Yarikhanov, who replaced the seriously ill Imaev in late July, suspends talks because of Russian non-compliance with the military accords reached thus far. In particular, he points to the POW issue where Russians have not been forthcoming in the "all-for-all" principle. Additionally, Yarikhanov asserts the Russians continue military operations despite a moratorium on fighting and Russian check points have not been allowing the free return of those forced out of the republic. Tensions are again on the rise and Russian military officials have prepared for renewed offensives.

• Russian Commander of the North Caucasian Military District Anatoliy Kvashnin proposes the Russians unilaterally withdraw all troops from Chechnya and disembark all guard posts. The move, says the commander, would provide a trust factor to the talks and allow the military agreements to be implemented more successfully. The main factor hindering progress on the accords is that the Chechens are afraid to disarm. Russian withdrawal would eliminate this fear and smooth the way for implementation of the accords and for the talks to continue.

**12** • Talks resume in Grozny between military experts and individual members of the official delegations.

**14** • Agreement is reached on an approach to disarmament between Romanov and Maskhadov. The two decide to commission four groups of representatives comprised from the federal forces and Chechen formations which will be dispatched to Chechen settlements to handle disarmament procedures.

• The Russian government issues statement condemning Chechen separatists for not faithfully abiding by the terms of the military accords reached on 30 July. The statement notes the strong potential the agreement holds for arriving at a peaceful settlement to the conflict and notes the Chechens continually thwart the tenets of the agreement. It announces all measures will be taken to end the activity of the illegal armed formations engaging in violence, sabotage, and terrorism.

**18** • Construction material and other forms of material aid sent to the Chechen Republic for reconstruction are being pilfered along the way, according to press reports. Officials note barely one-third of equipment sent for reconstruction purposes reaches its intended destination. The rest is pilfered by, according to the reports, transportation agents, suppliers, and those in charge of storage and distribution. Russia has earmarked over 5 trillion rubles for reconstruction of the Chechen economy but due to the forgoing, not much has been accomplished.

**21** • Chechen militants enter Argun and seize the internal affairs department. The assault, led by field commander Alaudi Khamzatov, with alleged connections to organized crime, numbers 200–250 in strength. The group was well-armed. Federal forces surrounded the town and Russian officials said force would not be excluded to extract the rebels should they refuse to lay down their arms and surrender. While the Russians charge the move violates the military agreement, Khamzatov asserts Argun is "my native town and I have been appointed its military commander." OSCE official Meszaros, after arriving in Argun, could not comment on whether or not the Chechen actions represent a violation of the agreement.

**31** • In a press interview, Dudaev representative Khamad Kurbanov asserts that any move to isolate Dudaev from settlement proceedings is futile. He indicates that Dudaev still controls all the processes, from negotiations to implementation of agreements. Any decision taken without his participation, knowledge, or approval is "doomed to failure and can only damage both Russian interests in the Caucasus and the peace process," Kurbanov says.

***September 1995***

**1** • A voluntary surrender of weapons program begins in Grozny. Individuals are given five days to surrender their weapons. After that, weapons being carried without a legal right to do so by their carriers will be forcibly taken.

**3–5** • Disarmament is taking place in many Chechen villages, according to press reports. Fighters in Pervmayskaya, Shatoy, and several other villages are reportedly complying with federal authorities to turn in their weapons. Some still resist and Maskhadov informs the Russian command his representatives are traveling to Chechen settlements to organize steps for disarmament. Romanov announces disarmament is evolving extremely slowly, saying only 1,000 pieces have been thus far recovered. The commander also notes the difficulties associated with organized arms traders, who are penetrating the villages and procuring most of the weapons themselves. Some are not at all involved in the Chechen conflict but are simply "enterprising individuals."

**5** • Grachev announces his serious displeasure with the progress of implementing the military accords. He expresses dismay at the fact that Russian servicemen are being killed while negotiations move on. Grachev reminds officials that the Federation Council has passed a resolution which allows Russia to use "political and force measures" against Chechens failing to comply with the accords.

**6** • Over 3,000 people gather in Grozny to celebrate the republic's fourth anniversary of independence. Dudaev representatives addressed the rally with chants supportive of the Dudaev regime and quite critical of the Russians, whose "gangs provoke the sons of Chechnya to terror!"

**7** • Russian Security Council Secretary Oleg Lobov announces an all-Russia search for Dudaev has been launched and a criminal case has been opened. He says it is nonsense to include him in Chechen elections, for criminals have "no right to contest elections." Should his views change and he enter rehabilitation, Lobov says things would be looked at differently.

**11** • After Yeltsin urges his participation in Chechen settlement proceeding, former Russian parliamentary speaker

Ruslan Khasbulatov announces he will indeed enter the peace talks. A native Chechen, Khasbulatov could be instrumental in achieving some Russian-Chechen accord. In a press interview, he says he sees his role as a preparer "of elections in Chechnya in taking the socio-political niche which exists between the federal authorities and the Chechen people." One of the first steps Khasbulatov will take is to remove the Moscow-imposed regime of Khadzhiev and Avturkhanov, as he sees their participation in the talks as a major obstacle to any settlement.

**13** • A deadline for widespread disarmament is reached. The agreement, signed by Russian and Chechen officials at the OSCE mission in Grozny, obligates Chechen militants to hand over weapons by 20 September. An issue raised by the Chechen leaders was the establishment of local defense units (SDU). The Russians argue such forces be created only where there exist no law enforcement bodies, while the Chechens insist on their right to establish SDUs in all towns and villages.

• Romanov announces that in the last five days, over 8,000 federal troops have returned to their permanent stations. Romanov says that Dudaev was quick to take advantage of the situation by increasing his forces in the south of the republic, "tipping the military balance in his favor again."

**14** • Volskiy announces that some 60 percent of the Chechen military formations are still controlled by Dudaev. The rest, he believes, are under the command of Khadzhiev, Avturkhanov, and Labazanov. In addition, there are scattered Islamic battalions which have no intention of handing over their weapons. All of this, he says, complicates immensely disarmament procedures and for controlling the various factions.

### Summary at the End of 1995

The military agreements concluded on 30 July are not fully implemented. Disarmament activities are met with some resistance as splinter factions emerge among the Chechen formations, and no agreement is reached on the establishment of local self-defense units. Sporadic fighting occurs in some regions where the Russians have been active in "disarming illegal armed formations," which damages mutual confidence and trust between the sides. Reconstruction efforts, despite significant funding and material allocations from Moscow, proceed slowly and are hampered by pilfering and graft. As to the status of Chechnya, several plans are announced but the sides will not agree to discuss the matter until military operations are completely suspended, widespread disarmament has occurred, and the return of refugees has been adequately achieved. As for Dudaev, he still maintains that Chechnya, in whole or in part, will achieve true independence and his mission will not be completed until this happens. The Russians insist elections must be held before discussions on Chechnya's status. Election dates are canceled and rescheduled because of the refugee problem and the fact that combat operations have not altogether halted. In October, Khadzhiev resigns and is replaced by Doku Zavgaev, who in early December signs an agreement with Chernomyrdin that grants Chechnya certain spheres of autonomy within the Russian Federation. In mid-December elections, Zavgaev is declared the winner. At year's end, Russian and pro-Dudaev forces battle over the city of Gudermes.

# The Georgia-Abkhazia Conflict

## Background Note

The first freely elected Georgian parliament bears some responsibility for inciting the separatist movement in Abkhazia. Immediately following his election as head of the new parliament in January 1990, Zviad Gamsakhurdia introduced a system of republican prefects to monitor local officials' political activities. This system, which was being used in almost all the former Soviet republics, was only supposed to give prefects the power to report local abuses of republican laws. However, Gamsakhurdia went far beyond this, expanding the center's powers at the expense of local authorities and ignoring calls for local self-rule. In an upsurge of national assertiveness, several of Gamsakhurdia's candidates for post of prefect were refused by ethnic Abkhazian minority leaders, who held most of the positions of power in the former Abkhazian Autonomous Soviet Socialist Republic. The self-rule issue and the politically naive and highly sectarian rhetoric employed by Gamsakhurdia heaped fuel on the nationalist Abkhazians' struggle for autonomy and strengthened the local appeal of secessionist leaders.

As early as 1989, the smaller nations of the North Caucasus seem to have realized the advantages of pulling together to recreate a large "North Caucasian Mountain Republic" that would include the Abkhaz Republic, Checheno-Ingushetia, North Ossetia, Kabardino-Balkar, and

Karachai-Cherkess. A large republic by this name had existed from 1921 to 1924. The Georgian government, however, would hear nothing of it. The loss of Abkhazia was anathema to Georgian national leaders. Instead, they preferred the concept of a Caucasian "Commonwealth" or "Home"—a confederation which would have included Georgia and given it a dominant role in the region.

Going back much further, Abkhazia gained its independence in 1921 following the disintegration of the first Russian empire; subsequently it formed a confederative state with Georgia in 1922. This state became part of the USSR. Since 1978, the Abkhazians have been campaigning for the secession of their autonomous republic from Georgia, and its incorporation into the RSFSR. The Soviet government continually rejected these demands, but awarded the republic large-scale economic aid and cultural concessions. These concessions obviously did not satisfy Abkhazian separatists.

After much political infighting, the Abkhazian Supreme Soviet restored its 1925 constitution in July 1992, declaring Abkhazia to be a sovereign state. The Georgian parliament immediately annulled the declaration, and President Eduard Shevardnadze sent in Georgian National Guard detachments to restore Georgian national authority, thereby igniting the long and gruesome conflict which was to continue for that year and most of 1993.

During the fighting, the Abkhazians received covert Russian military support with the presumed purpose of bringing the independent-minded Shevardnadze to his political knees. In 1993, General Pavel Grachev entered Georgia at will, ignoring Georgian sovereignty, giving vent to the Russian military's open disdain for the man who had "helped Gorbachev destroy the Soviet Union." Russian soldiers helped the Abkhazians defeat Georgian forces during July 1993, after which Russia tried to impose debilitating terms on Shevardnadze. He initially resisted, but by the fall of 1994 he was forced to join the CIS and to agree to the humiliation of Russian military bases on Georgian territory and Russian Border Guards to patrol Georgia's border with Turkey. The fractious Georgians were unable to consolidate their state, even in self-defense.

The Abkhazians have continued to appeal to Boris Yeltsin for unification with Russia, but after obtaining what it wanted from Shevardnadze, Russia shifted its policy to that of support for Abkhazia's semi-autonomous status within Georgia. Shevardnadze had made Defense Minister Grachev swear that he would control the separatist movement in Abkhazia and keep the republic within Georgia. Exhibiting its displeasure with these arrangements, the Abkhazian government, on 6 December 1994, issued a demarche aimed at acquiring its independence by announcing the inauguration of parliament head Ardzinba as president of Abkhazia, in accordance with a new constitution adopted on 26 November. The United Nations, Russia, and the CSCE refused to recognize Sukhumi's action and proclaimed their support for the territorial integrity of Georgia. Ardzinba's bold move raised concerns for the stability of the peace process, and opened the possibility of new violent confrontation. In the meantime, President Shevardnadze voiced his support for the deployment of CSCE troops in the Georgian-Abkhazian conflict zone, but said that, so far, few countries of the "Minsk Group" were interested.

President Shevardnadze meanwhile attended CIS summits where he supported the Russian government's actions in Chechnya. The only small gesture of independence he made in August 1995 was to agree to a pipeline to carry Azerbaijani oil through Georgia, to Turkey and to points west. Four days after he attended meetings on this pipeline arrangement, he was almost killed by a bomb carefully placed near his car.

The following chronology documents key events as they unfolded during the tragic, destabilizing civil war between Abkhazia and the Georgian government, and Russia's role in manipulating the conflict. Georgia's strategic location in the North Caucasus and its refusal for more than two years to join the CIS made Russian military leaders and many national-patriots in the Supreme Soviet eager to seize the opportunity to intervene in Abkhazia. Moreover, Georgian independence rhetoric, and talk of a "Caucasian Home" during the 1991–92 period were perceived as serious threats to Russia's national interests in the region and its integrationist policies for the CIS. From the chronology emerges a picture of a Georgian defeat and a Russian victory—built on the ancient destabilizing strategy of "divide and rule."

## Chronology of Key Events 1992

### *July 1992*

**23** • The Supreme Soviet of the autonomous republic of Abkhazia restores the 1925 constitution, proclaiming control of the republic's "state sovereignty." Georgian parliament annuls the declaration. Abkhazian Supreme Soviet Chairman Vladislav Ardzinba denies that Abkhazia is seceding from Georgia.

### *August 1992*

**11** • Supporters of the ex-Georgian president, Zviad Gamsakhurdia, take hostage Interior Minister Roman Gventsadze and other Georgian officials in Abkhazia.

14 • Georgian National Guard detachments are deployed in Abkhazia. Georgian leader Eduard Shevardnadze announces all measures will be employed to secure the release of Interior Minister Roman Gventsadze and other officials taken hostage. Abkhaz authorities call the deployment an "occupation," violating the April 1992 bilateral agreement, which permitted Georgian National Guard troops to enter Abkhazia only with permission from the Abkhaz government.

15 • Following the first day of talks between Georgian and Abkhazian officials, a ceasefire agreement is reached, allowing a phased withdrawal of Georgian National Guard units. The following day, fighting erupts in the Abkhazian capital, Sukhumi, prompting a wide-scale crackdown by National Guard units in which more than 200 Abkhazians are arrested. Thirty-nine are killed. Soon after, a Russian airborne regiment is flown to the area to safeguard strategic military installations and to evacuate the 1,700 Russians in the republic.

17 • Shevardnadze declares that Georgian forces have fully restored their authority in Abkhazia and have removed the Abkhazian government. The Georgian State Council guarantees Abkhazia's right to self-determination within Georgia, but refuses to recognize Abkhazian independence.

18 • Despite Shevardnadze's declaration, National Guard troops shell Sukhumi in an attempt to force Vladislav Ardzinba to resign. Ardzinba and a group of Abkhaz parliament members flee north to Gudauta, allowing Georgian troops control of major installations in Sukhumi.

19 • With a curfew and martial law imposed, Georgian National Guard units move to establish a military government replacing the pro-independence Abkhazian parliament. Clashes continue throughout the city as Abkhazian separatists launch "a campaign of armed resistance to the Georgian occupation."

21 • Efforts to peacefully contain the situation in Abkhazia fail as Georgian National Guard units violate an agreement to withdraw and renew clashes with Abkhazian forces. The Georgian State Council issues a statement diminishing hopes for a peaceful resolution to the conflict and announcing Georgian units will not withdraw from the region. The situation worsens as the Confederation of Mountain Peoples of the Caucasus (a loose association of ethnic groups of the northern Caucasus within Russia) announce intentions to assist Abkhazian separatism.

24 • Shevardnadze demands immediate withdrawal of the 1,000 volunteers of the Confederation of Mountain Peoples of the Caucasus assisting the independence cause. The volunteers were incorporated into local Abkhazian forces in the coastal town of Gudauta.

• Boris Yeltsin says he will take steps to end the Georgian-Abkhazian conflict and dispatches Russian State Secretary Gennadiy Burbulis to Sukhumi.

25 • The Georgian State Council's military commander, Col. Giorgi Karkarashvili, issues ultimatum to Ardzinba, demanding that he resign within twenty-four hours or incur a major offensive by Georgian National troops on Gudauta. Ardzinba agrees to Moscow talks between Yeltsin and Shevardnadze on condition that representatives from Abkhazia and the North Caucasus are allowed to participate. The Moscow talks are slated for 3 September.

***September 1992***

3 • Negotiations between Russia, Georgia, and Abkhazia in Moscow result in a ceasefire, effective 5 September, establishment of a tripartite monitoring committee, and agreement that Georgian troops will remain in Abkhazia. The agreement includes a clause forbidding the North Caucasian republics in the Russian Federation from participating in the conflict.

8 • Shevardnadze accuses the Abkhazian leadership of "gross violations" of the 3 September ceasefire.

12 • In an series of meetings in Sukhumi, the tripartite monitoring commission revises the ceasefire agreement amid continuing clashes. The Abkhazian parliament cables President Yeltsin, appealing for a Russian security guarantee for the Abkhazian government.

21 • The Russian Foreign Ministry issues a statement which says Russia "is most profoundly concerned" that Abkhazia and Georgia are failing to comply with the terms of the 3 September ceasefire agreement. Georgian deputies in the Abkhazian parliament charge that the Abkhazians are "totally ignoring" the ceasefire.

23 • A new ceasefire agreement is reached calling for the withdrawal of all armed groups from the region and the formation of a tripartite commission empowered to enforce the ceasefire process.

25 • Russian Supreme Soviet passes resolution criticizing the Georgian leadership, accusing Shevardnadze of using "violence to solve complex problems of interethnic rela-

tions." A second resolution suspends the transfer of Russian arms and equipment to Georgia, including those already promised. Georgia accuses Russia of interference in its internal political affairs.

**28** • President Yeltsin meets with Shevardnadze in Moscow to review the situation in Abkhazia. The first group of fighters from the Confederation of Mountain Peoples of the Upper Caucasus departs Abkhazia, returning to Grozny.

**31** • The situation is confused by a split in the Abkhaz leadership along ethnic lines, with the leader of Georgian Abkhazians in the Abkhaz parliament, Deputy Chair Tamaz Nadareyshvili, calling for the resignation of Chairman Vladislav Ardzinba, still in hiding. Georgian Justice Minister Dzhoni Khetsuriani demands that the Abkhaz parliament dissolve, charging that the electoral law of July 1991 was faulty. The 1989 census shows Abkhazia with only 17.8 percent Abkhazians and 45.7 percent Georgians. The remaining major ethnic groups are Russians and Armenians.

*October 1992*

**2** • Violating the September Russian mediated ceasefire, Abkhazian forces overrun Gagra, on the Black Sea coast. Accusations of Russian involvement escalate as Shevardnadze blames "reactionary forces in the Russian parliament" and conservatives in the Russian military for supplying arms to separatist troops. Russian Foreign and Defense Ministry spokesmen respond, insisting that Russian soldiers remain neutral. However, Boris Yeltsin warned Russia would take "appropriate measures" if Russian lives are threatened.

**6** • Abkhazian troops take the villages of Gantiadi and Leselidze, establishing firm control of northern Abkhazia.

**12** • Georgian-Russian relations deteriorate as Georgian State Council announce Georgia will assume control of Russian weapons and military equipment on its territory. Russian Defense Minister Pavel Grachev describes the move as a "flagrant violation of all earlier accords," warning that attempts to seize Russian military hardware will lead to "armed clashes."

**13** • Talks between Yeltsin, Shevardnadze, and Ardzinba are postponed indefinitely when no agreements could be reached. While the Georgian delegation called for a return to military positions occupied prior to 1 October, the Abkhazians insist on complete withdrawal of Georgian troops.

*November 1992*

**2** • Georgian forces seize a Russian arms depot in southern Georgia. Shevardnadze condemns the incident but Russian army officials accuse Georgia of pursuing a deliberate anti-Russian policy. Georgian Defense Minister Tengiz Kitovani announces he personally ordered the seizure.

**19** • Georgian and Abkhaz forces reach a ten-day ceasefire agreement to allow Russian troops to leave Sukhumi.

*December 1992*

**1** • Clashes in the Ochamchira district of Abkhazia erupt following the deployment of Georgian combat aircraft. Shevardnadze declares a state of emergency in the Abkhazian capital Sukhumi and the Ochamchira district.

**3** • Shevardnadze imposes martial law throughout Abkhazia and vows to militarily crush the Abkhazian drive for independence. The hard-line announcement comes in the wake of an artillery attack, blamed on Abkhazian forces, on the Black Sea port of Sukhumi. Fourteen are killed and more than twenty wounded.

**7** • Russia pushes the United Nations for a UN Security Council resolution demanding all parties to the Abkhazian conflict honor the 3 September ceasefire agreement. Ardzinba writes Yeltsin protesting the exclusion of Abkhazian representation in the mediation process and accuses the Russian military of arming the Georgian army in its fight against the independence-seeking Abkhazian forces.

**8** • According to Russian news reports, a brigade of Turkish volunteer soldiers are fighting alongside Abkhazian forces against the Georgian military.

**13** • Following a meeting between Georgian and Abkhazian officials, a new ceasefire agreement is reached calling for the withdrawal of all heavy weaponry by 18 December. The agreement also stipulates withdrawing all military hardware from both sides of the Gumista River area, north of Sukhumi, and from the Ochamchira district.

**21** • Georgian and Abkhazian forces clash in heated battles near the Gumista River outside Sukhumi. Statements by the Abkhaz leadership note an increase in Georgian forces near Gudauta in preparation for a large-scale offensive.

**27** • As clashes escalate, Abkhazian forces shoot down a Georgian Mi-8 military helicopter over the Abkhaz district of Ochamchira.

## 1993

*January 1993*

**4** • Shevardnadze appeals to UN Secretary-General Boutros Boutros-Ghali for an immediate dispatch of United Nations peacekeeping forces to the Georgian-Abkhazian conflict zone. Two months ago, Shevardnadze issued a critical statement accusing the United Nations of inaction in the region.

**5** • In response to an Abkhazian offensive, the Georgian army launches a counterattack near the Gumista River in Abkhazia. Heavy fighting also erupts in the northwestern region of Sukhumi, which is currently held by the Georgian military. Over 200 are reported to have been killed. A Russian Su-25 combat aircraft is shot down over the village of Eshera near Sukhumi. It is not yet clear whether the plane was shot down by Georgian or Abkhazian forces.

**10** • Georgian forces seize forty-five Russian troops taking over a Russian garrison in Lagodekhi to prevent Russian control of the garrison's weaponry. The hostages are released two days later following extensive negotiations between Russian and Georgian officials.

**14** • Russian Deputy Defense Minister Georgiy Kondratev calls for concluding a bilateral treaty with Georgia specifying the status of Russian troops in Georgia. The Russian military refuses to withdraw until the volatile Abkhazian conflict is resolved.

**15** • Georgian national security advisor Tedo Japaridze says joining the CIS is out of question for Georgia, and "it is about time someone explained the meaning of this commonwealth." Asked if Georgia misses Western aid going to CIS, he replies, "I doubt that any of the CIS republics besides Russia get anything. That is why the West also realizes now that it has to make its allocations separately, to each country so all the aid is not sucked into the Russian black hole."

**17** • Shevardnadze says Iran, in view of its political, economic, and military strength, can be an important guarantor of regional security in Central Asia and the Caucasus. He regrets that relations between Tbilisi and Tehran have not properly developed due to two centuries of Russian and Soviet imperialism.

**18** • Shevardnadze alleges Abkhaz separatists are connected with groups pressing for reviving the erstwhile Soviet Union in order to stop formation of an independent Georgia.

**21** • Head of the Russian delegation to talks with Georgia, Feliks Kovalyov, reveals that Russia is considering withdrawing its troops from Georgia, citing financial concerns. Russian troops deployed in South Ossetia and Abkhazia as peacekeepers are continually threatened, adding to Russian concerns.

**29** • Russian Lt.-Gen. Sufian Beppaev, former commander of Transcaucasus Military District, says 10 percent of the former Russian troop strength remains in Georgia—15,000 troops. He claims they are incapable of performing in combat and should be withdrawn.

• Aleksandr Chikvaidze describes a bilateral "friendship and cooperation" agreement signed between Abkhazia and the "Dniester Republic" as illegal. He says the agreement can only lead to confusion and more conflict and in no way contributes to a peaceful settlement of the Georgian/Abkhazian conflict. Further, he claims the agreement infringes on the territorial integrity of Moldova and Georgia.

• Georgian radio report more than 15,000 volunteers from the North Caucasus stand ready to fight against Georgians. This was stated on Gudauta TV by an influential Abkhazian member of parliament.

*February 1993*

**5** • Georgian Defense Ministry accuses Abkhazian forces of shelling a reservoir in the Ochamchira region, disrupting regional water supplies. The Georgian military command in Sukhumi issues a statement blaming Russian aircraft for an assault on Georgian positions. During the assault, a Russian Su-25 combat aircraft was downed by Georgian anti-aircraft units. Russian officials contend the plane was escorting humanitarian aid shipments to Abkhazia.

**13** • As sporadic clashes between Georgian and Abkhazian forces continue along the Gumista River, a detachment of 700 Georgian National Guard reinforcements arrive in the Georgian-held Abkhazian capital Sukhumi. Fighting continues in the Ochamchira district, with Abkhazian forces downing two Georgian Mi-8 military helicopters.

**15** • Georgian Defense Minister Tengiz Kitovani says the only real threat to Georgian forces is the presence of Russian troops. Kitovani warns Russia not to interfere in the conflict.

**23** • Shevardnadze accuses the Russian leadership of seeking an armed conflict with Georgia and demands their withdrawal from Abkhazia and Adzharia. Grachev orders Russian troops to "shoot to kill" to defend against Georgian units seeking weapons and munitions.

**24** • Georgian parliament passes resolution attributing full responsibility to the Russian government for "aggressive action" against Georgian forces by Russian troops stationed in Abkhazia. Shevardnadze threatens a total mobilization if Russian troops do not withdraw. The resolution follows the killing of one civilian and the wounding of eight others in the bombing of Sukhumi, the Georgian-controlled Abkhazian capital, on 20 February, which the Georgian authorities claimed was carried out by a Russian aircraft. Russia denied the charge and countered by claiming that Georgian artillery was firing on its positions.

27 • Grachev arrives in Abkhazia unannounced to inspect Russian troops. Shevardnadze criticizes him, claiming the visit was a diplomatic impropriety reflective of Russia's "ultimatum-oriented approach" to protect its strategic interests in the North Caucasus and the Black Sea region. He maintains, however, his commitment to further dialogue with Russia.

*March 1993*

2 • Shevardnadze states "the beginning of the disintegration of Georgia will become the beginning of the disintegration of Russia, because an unstable Georgia will stimulate the most negative processes in the northern Caucasus."

18 • In a letter to Shevardnadze, the ethnic-Azeri "Dayag" organization offers its military and political support.

19 • Georgian anti-aircraft units shoot down a Russian Su-27 combat aircraft over the village of Esheri, north of Sukhumi. The incident aggravates already tense relations between Moscow and Tbilisi over the role of the Georgian-based Russian military units stationed in Abkhazia.

26 • The Abkhazian parliament condemns the recent transfer of Russian weaponry and munitions to the Georgian military.

27 • Meeting with a visiting CSCE delegation, Shevardnadze states that new opportunities for rapprochement between the conflicting sides result from the military stalemate. The CSCE delegation has been touring the country as part of a three-month human rights fact-finding project.

*April 1993*

1 • Heavy fighting erupts between Georgian units, stationed in their stronghold of Sukhumi, and attacking Abkhazian forces. Dozens of civilians are wounded in the artillery exchanges.

4 • Ardzinba appeals for international assistance, demanding a UN condemnation of Georgian human rights violations against the Abkhazian population. Ardzinba further asserts his "willingness for peace" only when Georgian units withdraw.

6 • Grachev says Moscow is standing by its position that Georgia must remain an entity with effective autonomy guaranteed to the Abkhazian, South Ossetian, and Adzharian regions. He confirms that the Russian troops stationed in the war zone remain neutral unless attacked.

7 • On the eve of scheduled Georgian-Russian meetings in Sochi, the Georgian parliament votes against a resolution withdrawing Russian troops in Abkhazia. The Sochi talks aim at reaching a durable ceasefire agreement.

• Georgia and Russia agree to station Russian troops in Abkhazia and elsewhere in Georgia until the end of 1995, as stipulated in a February draft agreement. They also establish a 3 km demilitarized zone between Georgian and Abkhaz forces. No Abkhaz representatives were present at the talks.

8 • Russian and Georgian delegations stumble over settlement specifics. Grachev objects to the Georgian variant because "Georgia puts all obligation in implementing the agreement on Russia, leaves nothing for itself, and the third party—Abkhazia—is not mentioned at all." No crucial military issues are resolved and the developing rapprochement diminishes.

12 • Shevardnadze meets with Ukrainian officials to forge a "joint political front against Russia." The Georgian and Ukrainian presidents should conclude a twenty-part series of bilateral agreements, including a treaty on friendship and cooperation.

17 • Georgian Foreign Ministry protests "aggression by the Russian Federation against Georgia." They charge that Russian military facilities are being used by Abkhazian forces as bases for their military operations.

*May 1993*

13 • A mudslide in North Ossetia in Russia damages natural gas pipelines transporting Russian energy supplies to Georgia and Armenia. Meanwhile, secret negotiations on fuel supplies continue between Georgian and Azeri officials. According to Azerbaijan's terms, Georgia will be allowed to purchase Azeri fuel only if none of it is transferred to Armenia, which remains under Azeri blockade.

14 • Shevardnadze and Yeltsin arrange a ceasefire in Abkhazia effective 20 May. The leaders call for withdrawing

heavy military equipment and artillery from the combat area and ban flights over Abkhazia beginning on 25 May. In response to these talks, Ardzinba offers an Abkhaz unilateral ceasefire conditional on a Georgian military withdrawal.

**21** • Following a meeting with Ardzinba, Yeltsin special envoy to Abkhazia Boris Pastukhov states deployment of a tripartite Russian-Abkhazian-Georgian peacekeeping force is "unrealistic," recommending instead a CSCE or United Nations force.

**29** • Abkhazian and South Ossetian parliaments appeal to Yeltsin to delay signing a treaty on friendship and cooperation with Georgia until a comprehensive settlement of the conflict in Abkhazia and South Ossetia is achieved. A joint statement says that "Abkhazia and South Ossetia can no longer remain a part of Georgia, which is unable to act as a guarantor of rights of the people of our republics." Both sides express concern about "the continuing supplies of weapons, ammunition, and military hardware from Russia to Georgia."

**30** • Georgian intelligence reports about 500 armed Cossacks transferring from Krasnodar Kray to territory controlled by the Abkhazian forces.

***June 1993***

**2** • Soldiers from "Dniester" Russian forces in eastern Moldova arrive to support Abkhazian independence. The cooperation is based on a January 1993 Cooperation Treaty between Abkhazia and the self-declared "Dniester" Republic.

**5** • Shevardnadze's chief military advisor Col. Vladimir Chikovani accuses Abkhazians of using the Moscow-mediated truce to build up their military might and prepare a new large-scale assault on the city of Sukhumi. He says the Abkhazian side violates the ceasefire agreement consistently, confirming the Moscow agreements on a ceasefire and political solution will scarcely work at this stage.

**6** • Reacting to severe shelling of Sukhumi, Shevardnadze warns of a Georgian military offensive response. The delivery of Russian humanitarian aid is continuously endangered by sporadic fighting and by repeated attempts at downing the Russian helicopters transporting the aid shipments. Russian Foreign Minister Andrey Kozyrev announces his intention to visit Tbilisi, Sukhumi, and Gudauta in a new mediation attempt.

**10** • Kozyrev characterizes talks as deadlocked, where Abkhazians are demanding unconditional withdrawal of Georgian Army units and Georgians insisting on restoring the Russian-Georgian border on the river Psou.

***July 1993***

**1** • Abkhazian forces launch a large-scale offensive to retake the Georgian-controlled Abkhazian capital Sukhumi. Abkhazian forces shoot down a Georgian Su-25 combat aircraft and move rapidly against Georgian positions near the town of Ochamchira, 80 km south of Sukhumi.

**6** • Responding to the escalation of fighting with Abkhazian forces, Shevardnadze declares martial law throughout the area. In Tbilisi, Georgian Prime Minister Tengiz Sigua threatens to break all relations with Russia if Moscow continues to arm and supply Abkhazian forces.

**7** • Georgian Prime Minister Tengiz Sigua says Georgia will recall its ambassador from Moscow if Russia continues with arms, equipment, and ammunitions supplies to Abkhazia.

**8** • Abkhazian forces continue their advance on Sukhumi, seizing a key hydroelectric power plant and several strategic heights overlooking the city. The UN Security Council responds by unanimously adopting a resolution for deploying UN peacekeeping forces in Abkhazia once a durable ceasefire is reached.

**14** • Russian Deputy Foreign Minister Boris Pastukhov arrives in Tbilisi to review the situation in Abkhazia with Georgian officials. He announces the Moscow-based peace talks suspend indefinitely, stalemated by Georgia's refusal to meet Abkhazian demands for full restoration of Abkhazian governmental institutions and complete Georgian military withdrawal.

**21** • A ceasefire proposal is drafted as a compromise between Russia and Georgia. The move follows Georgian rejection of a Russian-mediated Abkhazian proposal calling for an immediate withdrawal of Georgian forces. The compromise plan calls for the trilateral monitoring of a ceasefire, the coordinated deployment of peacekeepers, and the withdrawal of Georgian forces.

**26** • The Abkhaz parliament debates and approves a plan to end the conflict with Georgia.

**28** • Meeting in Sochi, Russian, Georgian, and Abkhazian officials formally sign the recent ceasefire agreement scheduled to enter into force within ten days.

*August 1993*

**1** • Russian troops, part of a tripartite ceasefire monitoring group, arrive in Abkhazia.

**2** • Shevardnadze welcomes UN involvement in the Abkhazian conflict, saying the UN role serves as a good model for other conflicts in the Transcaucasus region. Based on an agreement reached the previous week, a force of fifty UN peacekeepers is to join a trilateral (Russian, Georgian, Abkhazian) "control group" empowered to monitor and enforce the ceasefire.

**10** • In accordance with the provisions of the 27 July Abkhazian ceasefire agreement, Russian, Abkhazian, and Georgian officials agree to specific terms for the staged disengagement of all armed formations from the conflict zone.

**16** • A tripartite meeting in Sochi of Georgian, Russian, and Abkhazian officials results in an agreement to withdraw all troops from the conflict zone in accordance with the Russian-mediated disengagement plan reached in late July.

**24** • Meeting in Moscow, Russian and Abkhazian officials agree to further negotiations on the Abkhazian conflict to be held under UN auspices with Russian mediation. The UN Security Council adopts a resolution (No. 858) calling for a six-month deployment of an additional eighty-eight ceasefire observers for the Abkhazian accord.

*September 1993*

**6** • The Abkhazian parliament in Gudauta accuses Georgian forces of failing to meet its scheduled withdrawal of military forces and equipment in accordance with the tripartite ceasefire agreement. The United Nations announces it will hold talks on the Abkhazian conflict on 13 September in Geneva.

**12** • The planned mediation talks to be convened in Geneva by the United Nations are postponed due to the Abkhazian protest over Georgia's repeated failure to comply with the scheduled withdrawal of its military equipment from Abkhazia.

**16** • Abkhazian forces launch an offensive violating the seven-week-old ceasefire agreement. Abkhazian forces break the Georgian blockade of Tkvarcheli, advancing to Georgian-held cities of Sukhumi and Ochamchira. As Abkhazian forces move against Sukhumi and initiate severe artillery attacks, Shevardnadze hastily flies there to personally direct its defense. Responding to the ceasefire violation, strong condemnations are issued by the Russian Foreign Ministry and the UN Security Council calling on the Abkhazian forces to withdraw immediately.

**18** • As the Abkhazian offensive continues, Russian Defense Minister Pavel Grachev meets with Abkhazian parliamentary chairman Vladislav Ardzinba in Gudauta and with Shevardnadze in Sochi. Following the meetings, Russia formally condemns the Abkhazians for the offensive and criticizes Georgia for refusing to negotiate with the Abkhazians and calls for economic sanctions to be imposed on both sides.

**20** • Abkhazian forces reach the outskirts of Sukhumi, establishing a nearly complete circle around the city by assuming positions in the hills overlooking the area. The attack intensifies after Georgian troops refuse the Abkhazian offer of withdrawal through an Abkhazian-established corridor from the city. With heavy artillery bombardment and growing casualties, Shevardnadze issues international appeals for assistance.

**22** • Reflecting the intense situation in Sukhumi, the third civilian passenger aircraft is shot down by Abkhazian forces advancing on the airport.

**24** • As Abkhazian forces enter Sukhumi and street-by-street fighting erupts, nearly 4,000 civilian residents are evacuated from the city. With Georgian reinforcements stalled 15 km south of Sukhumi, the Georgian defenders are bolstered by a mere 300 troops, making the fall of Sukhumi inevitable. Ousted President Gamsakhurdia, exploiting the chaos of the battle for Sukhumi, returns from exile to the central Georgian Mingrelian district capital Zugdidi. It is reported, however, that Gamsakhurdia has ordered his forces to advance on Sukhumi to assist Georgian troops against the Abkhazian offensive.

**26** • After eleven days of intense fighting, Abkhazian forces fully seize the Abkhazian capital Sukhumi, forcing the retreat of Georgian troops and causing Shevardnadze to flee south. The Abkhazian victory in Sukhumi represents the last significant military challenge to Abkhazian separatists, as Sukhumi was the last remaining Georgian stronghold in the region. Russian forces based in the Abkhazian city of Gudauta evacuate 10,000 civilian refugees prior to the fall of Sukhumi.

**29** • Kozyrev urges the Abkhaz leadership to lay down their arms. He says that Russia imposed economic sanctions on the Abkhaz.

**29** • The deputy chairman of the Abkhaz Supreme Soviet

says that Abkhazia would hold a referendum on secession from Georgia and requests protection from Moscow.

*October 1993*

**1** • Independence-seeking Abkhazian forces continue their advance through Georgian-held areas of Abkhazia, causing thousands of Georgian refugees to flee. They seize the coastal town of Ochamchira and push on to attack the town of Gali, the last remaining town in Abkhazia under the control of Georgian forces. In Tbilisi, Shevardnadze appoints his former Defense Minister Tengiz Kitovani as commander of all Georgian army units.

**4** • Georgian troops retake the western town of Khoni.

**6** • The Georgian government issues an urgent appeal to the UN Security Council warning of the plight of over 20,000 Georgian refugees who fled Abkhazia eastward through the mountains to Svanetia. The appeal identifies Svanetia as a disaster area and asks for immediate humanitarian assistance to evacuate the refugees. In response to the growing crisis in the area, the United States dispatches a plane-load of food and emergency medical supplies. The UN Special Representative for Georgia, Eduard Brunner, announces that the Geneva talks on Abkhazia are encouraging and suggests the future involvement of Russia, which he says has "a legitimate interest" in the stability of the Transcaucasus. Ardzinba appeals to the international community for aid and affirms that Abkhazia will soon adopt a new democratic constitution with guaranteed rights for all national and ethnic groups.

**8** • After a meeting with the leaders of Armenia, Azerbaijan, and Russia, Shevardnadze announces that Georgia would seek membership in the CIS as their last hope in the civil war. He cites the use of economic blackmail by Russia to force Georgia into the organization: "if you want gas, oil, raw materials . . . then join the CIS." Reaction in Tbilisi is characterized by anger and shock. Shevardnadze justifies joining of the CIS by saying that none of the CIS agreements contain language that implies a loss of sovereignty and that the threatening elements in Russian national politics have been isolated and do not therefore pose a threat to those who do join the CIS. The participants of the meeting agree on the need to coordinate efforts in conflict mediation with the help of international organizations and on deployment of Russian troops to restore rail links from the Black Sea to Tbilisi and to Armenia.

**9** • Russia and Georgia sign an agreement legalizing the status of Russian troops currently stationed in Georgia. They also sign a protocol on the joint use of all ports and airfields. The agreement does not specify a date for the final withdrawal of troops from Georgia.

**13** • Georgian Foreign Minister Aleksandr Chikvaidze states that Georgia may cede control of some of its military bases to Russia in exchange for military aid.

**14** • Ukrainian air crews evacuate 7,000 Georgian refuges from Abkhazia.

• The Russian government issues a statement condemning Abkhazian forces for looting property belonging to refugees from Abkhazia, for human rights violations, and mass-scale ethnic cleansing.

**19** • Grachev states Russia could not offer military aid to Shevardnadze because Georgia is not a member of the CIS or its Collective Security Treaty. He says that although Russian, Azeri, and Armenian troops might participate in the securing of the road from Poti to Tbilisi, any other participation in the conflict would be viewed as Russian interference in the domestic affairs of Georgia.

**20** • Georgian forces launch a major counteroffensive against Gamsakhurdia's forces and eject them from Poti, Khoni, and Lanchkhuta. The rail junction of Samtredia is also recaptured. Russian forces are rumored to be assisting the Georgian military in neutralizing opposition.

**25** • Georgia signs the CIS Collective Security Treaty, clearing the way for Russian, Armenian, and Azeri troops to defend Georgian transport routes.

**28** • Forces loyal to ousted Georgian President Zviad Gamsakhurdia launch a counterattack, retaking Khobi. However, Georgian forces recapture the town the following day.

*November 1993*

**9** • Russian Foreign Ministry warns the Abkhaz will face sanctions should their troops or weapons cross the Inguri River, separating Abkhazia from the rest of Georgia. He states also that Georgia should not commit aggression against the Abkhaz and that Moscow would not remain indifferent to such aggression. There is presently a concentration of Georgian troops on the Abkhaz border.

**11** • Shevardnadze states reunifying Georgia is the immediate goal. He says that he is not ashamed at having needed Russia's help in defending his country, although prefers Georgia to have been able to defend itself.

*December 1993*

**2** • The first round of negotiations to resolve the conflict under the auspices of the United Nations, with Russian assistance and the CSCE's participation, ends in Geneva. The memorandum signed by the Georgian and Abkhaz delegations says the two sides undertake the responsibility not to use force or threaten to use force against each other during the period of time of negotiations aimed at a full-scale political settlement. The sides assume responsibility for creating conditions for the prompt return of the refugees.

**20** • Heated clashes erupt between Russian forces and Georgian paramilitary units of the "Mkhedrioni" militia in the eastern Georgian town of Telavi. Further destabilizing the internal situation are reports of civil disorder and criminal acts in the Azeri-populated districts of Bolnisi and Marneuli in Georgia. Following meetings with the local Azeri community, Shevardnadze forms a special detachment of elite troops to restore order in the two districts.

## 1994

*January 1994*

**1** • Shevardnadze rejects a statement made by parliament speaker Goguadze claiming the Georgian situation is so dire that the economic and military aid of Russia is the only solution. Shevardnadze concedes adopting the Georgian coupon as the country's currency last year was a mistake, but that Georgia would need many more months before it could rejoin the ruble zone. Within a year, the Georgian coupon fell from a ratio of 1:1 with the Russian ruble to 100:1. On the issue of Russian troops being used to enforce law and order in Georgia, Shevardnadze comments that "a country which could not instill law and order on its own does not deserve independence."

**4** • Reports from Russian peacekeeping forces stationed in Georgia indicate that Abkhazian forces are engaged in sporadic attacks targeting Georgian refugees as they are repatriated to their homes in Abkhazia. These attacks, although minor and infrequent, continue to delay the return of the estimated 250,000 Georgian refugees displaced from their homes in Abkhazia.

**12** • At Georgian-Abkhaz peace talks in Geneva, Abkhaz delegation leader Sokrat Dzhindzholia proposes deploying Russian troops as peacekeepers in the region. The political status of Abkhazia remains far from settled and experts are scheduled to convene in Moscow in February to address this question. Dzhindzholia later states that "life itself has . . . meant [Russia must] take part in the destiny of the peoples that were joined by the Soviet Union. There is simply no other way out. No other countries nor the United Nations have the strength or the means to do this. What is more, no one knows better than Russia the situation on the territory of the former USSR."

**18** • Continued clashes between Georgian military units and Abkhazian forces in the Gali region of Abkhazia threaten to delay the return of Georgian refugees to their homes in the region.

**28** • Abkhaz Prime Minister Sokrat Dzhindzholia and other high level Abkhaz and South Ossetian officials voice concern over the imminent signing by Russia and Georgia of a treaty on friendship and cooperation. They fear that the treaty would threaten their autonomous status by giving Georgia access to Russian arsenals, and have appealed to the Russian leadership to postpone signing until after settlements have been reached in the Abkhaz and South Ossetian conflicts.

*February 1994*

**2** • A group of Russian parliamentary deputies cautions Yeltsin that signing a treaty with Georgia is premature and could destabilize the entire Transcaucasus.

**3** • In Tbilisi, Yeltsin and Shevardnadze sign a bilateral treaty on friendship and cooperation as well as twenty-four other agreements on trade and economic ties, scientific and cultural cooperation, the status of Russian border guards in Georgia, and military basing rights. They also discuss the conflict in Abkhazia and South Ossetia and Georgia's possibility of rejoining the ruble zone. Shevardnadze characterized the agreements and the visit as important for establishing peace and stability throughout the Caucasus.

**9** • Yeltsin and Shevardnadze appeal to Boutros-Ghali to send UN peacekeeping troops to Abkhazia. ITAR-TASS reports that Abkhaz Prime Minister Dzhindzholia stipulates that if troops are deployed, they only do so on the border between Abkhazia and the rest of Georgia.

• Georgian officials state that Georgian civilians are fleeing Abkhazia to escape ethnic cleansing. Abkhazian parliamentary speaker accuses Georgia of shelling a hospital in Sukhumi.

**10** • As fighting between Georgian troops and Abkhazian forces escalates, the Abkhazian parliament formally adopts a declaration of independence from Georgia. This move repudiates the earlier UN- and Russian-brokered agreement to determine the status of Abkhazia through a national referendum.

**14** • Shevardnadze appeals for Russian peacekeeping troops under UN coordination to Abkhazia. Abkhazians agree in principle but want the deployment limited to the Georgian border with Abkhazia.

**18** • The Council of Ministers of Abkhazia appeals for Russian intervention to end the Georgian-Abkhaz war, lest "tragedy will be unavoidable and it may be followed by a new, more large-scale and bloody Caucasian war, which may also involve Russia."

**22** • In Geneva, negotiations under UN auspices between Georgian and Abkhazian representatives resume. A primary element of the talks is the return of 200,000 to 300,000 refugees, mostly ethnic Georgians, displaced from the heavy fighting in the region.

**25** • The third round of UN-sponsored negotiations between Abkhaz and Georgian leaders ends in stalemate. Unresolved issues include conditions for the return of 200,000 Georgian refugees to Abkhazia and the future status of Abkhazia within Georgia.

***March 1994***

**1** • After a stormy debate, the Georgian parliament ratifies Georgia's membership in the CIS, 121–47 (4 abstentions).

**10** • Georgian parliament resolves to disband the Abkhazian parliament, further straining relations between the two.

**13** • Before leaving for his meeting with Warren Christopher in Vladivostok, Kozyrev tells reporters that he saw "no opportunity for the use of peacekeeping forces from Western countries for operations in Georgia." This is apparently in response to Clinton's statement of 7 March: "The United States would be inclined to support a UN peacekeeping operation in Georgia, an operation that would not involve U.S. military units." Following the meeting with Christopher, there were no reports as to whether this issue was addressed.

**18** • Georgian parliamentary deputy Irina Sarishvili delivers a statement to the parliament asserting Georgian membership in the CIS is illegal. As a result, international experts have been called to Georgia to determine the validity of Georgian membership. He equates joining the CIS with joining Russia.

**25** • Countering the Georgian offensive, Abkhazian forces retake the village of Nizhnaya Lata, east of the Abkhazian capital Sukhumi, and advance to seize two villages in the Svaneti district, outside Abkhazia. The Abkhazians initiate artillery attacks against Georgian military positions in the Gali district and issue an ultimatum demanding the withdrawal of all remaining Georgian troops from Abkhazia. The United Nations condemns the escalation of violence and calls on both parties to resume talks.

• Georgians and Abkhazians agree to resume negotiations in Moscow with Russian mediation.

**26** • Georgian government reports 95 percent of its population is in favor of joining the CIS.

***April 1994***

**3** • In Moscow, Georgian and Abkhaz representatives sign an agreement ending the latest round of violence and providing for the return of 250,000 Georgian refugees to their homes. Both sides call for UN peacekeeping forces to be deployed in Abkhazia.

**15** • CIS heads of state adopt resolution to send Russian peacekeepers to the Georgian-Abkhaz conflict area. CSCE and UN observers already are active in the region.

**22** • Concluding four days of UN–brokered talks on the Abkhazian conflict, the Georgian and Abkhazian delegations agree to reconvene in Moscow on 10 May to discuss a UN proposal of a federal agreement that would maintain Georgian territorial integrity while attempting to meet the demands of the Abkhazian side. The Abkhazian parliament has threatened to issue its own declaration of sovereignty, however.

**24** • An unnamed participant in negotiations between Georgia and Russia on the settlement of the conflict states that any peacekeeping forces deployed in Abkhazia will be Russian, but will be under the control of the UN, the CSCE, Georgia, and Abkhazia.

**25** • Shevardnadze states that if the 5 May UN-sponsored talks do not record significant progress, CIS peacekeepers will be requested.

***May 1994***

**13** • Moscow negotiations continue to falter as the Georgian parliament adopts a resolution requiring its delegation to refuse a draft agreement on the military withdrawal of troops and deployment of CIS peacekeepers. The Georgian delegation signs the agreement anyway and assents to the deployment of several thousand CIS peacekeeping troops along the Abkhazian border with Georgia proper.

**14** • Contravening the resolution passed the previous day, Georgian delegation signs a draft agreement to disengage the warring forces and deploy CIS peacekeepers along the Inguri River, the de facto border between Abkhazia and Georgia. The Georgian parliament objects, claiming it does not provide adequate safety for the repatriation of Georgian refugees to Abkhazia.

• An Abkhaz delegate to Moscow talks reports the CIS peacekeeping force, primarily Russians, will number 2,500–3,000 and will be deployed before the end of May.

**18** • Georgian parliamentary deputies call on the country's leadership to disavow the ceasefire agreement signed in Moscow. Radical spokesman Irakli Tsereteli proposes a no-confidence vote in Shevardnadze, who defends the agreement as "the most realistic solution."

**23** • Moscow announces Russian troops from the Group of Russian Forces in the Transcaucasia will form the backbone of the 3,000–man peacekeeping force to be deployed in Abkhazia.

**31** • Talks on the repatriation of Georgian refugees to Abkhazia deadlock over Abkhazia's demand that all returnees sign a declaration to abide by the laws of the Republic of Abkhazia. Abkhazia additionally wants to bar Georgians participating in last fall's fighting from returning. The two sides also disagree on the total number of refugees: Georgia estimates 290,000 displaced; Abkhazia calculates 185,000.

***June 1994***

**2** • Russia's Federation Council rejects a presidential request for deploying Russian troops to Abkhazia as part of a joint CIS peacekeeping force. Chairman of the Federation Council's Committee on Defense and Security claims that there is no legal basis for deploying Russian troops beyond its borders.

• Russia's Foreign Ministry lifts economic sanctions imposed on Abkhazia last September.

**3** • The Federation Council's decision rejecting Russian peacekeepers to Abkhazia is met with disappointment from the Georgian and Abkhaz sides. Spokesman for Yeltsin indicates Yeltsin is "likely to exert pressure" on the Council to reverse its decision. CIS Executive Secretary Ivan Korotchenya, recently completing a tour of the Central Asian and Caucasian states, says most CIS heads support CIS peacekeepers in Abkhazia.

**8** • Boutros-Ghali recommends increasing UN observers to Abkhazia to the UN Security Council. He argues for an independent UN operation in close coordination with the CIS contingent.

**9** • Yeltsin signs a decree ordering the creation of a peacekeeping force for deployment in Abkhazia. The force will be comprised of three battalions from the Group of Russian Forces already stationed in the Transcaucasus. The actual date of the deployment must still be decided by the Federation Council, which previously rejected Yeltsin's request to send peacekeepers to Abkhazia. The Federation Council is scheduled to meet on 21 June.

**14** • According to Russian Deputy Defense Minister Georgiy Kondratev, Russian paratroopers stationed in Gudauta would soon be deployed along the Inguri River, which marks the de facto Abkhaz-Georgian border. He indicates that the approval of the Federation Council is not necessary because the troops were already in the region. Georgian opposition groups protest the move on the grounds that the troops in Gudauta fought on the Abkhaz side in the civil war. At an extraordinary session of the Georgian parliament opposition deputy Nodar Notadze calls for Shevardnadze's resignation, while others demand the suspension of the Abkhaz peacekeeping operation. No votes are taken on the issues.

**15** • Russian field engineers clear mines in the Gali region in preparation for Russian peacekeepers. Commander-in-chief of the Russian ground forces announces that Russian peacekeepers will return fire if attacked, but "it is not their task to disarm or eliminate" armed formations.

**21** • Following an address by Grachev, the Federation Council overwhelmingly votes in favor of deploying peacekeepers to Abkhazia, reversing the 3 June rejection of the proposal.

**22** • Deputy Foreign Minister Boris Patukhov invites the United Nations to send more observers to Abkhazia and expresses his hope that the United Nations will "formally endorse" (i.e., finance) the operation, which will cost an estimated 10–11 billion rubles.

**24** • Two battalions of Russian peacekeeping troops take up positions along the 12–kilometer security zone separating the warring parties. In all, between 2,500 and 3,000 Russian troops will monitor the area along the Inguri River. Russian troops will oversee the return of Georgian refugees to Abkhazia and enforce a curfew within the security zone.

**28** • Grachev says Russia welcomes an international peace force in Abkhazia "on any terms . . . even under any command."

• Abkhazian officials accuse Russian peacekeepers of violating the agreement on the repatriation of Georgian refugees. The Abkhaz claim that Russian soldiers transit masses of Georgian refugees without regard to whether or not those returning participated in the war against Abkhazia. The repatriation agreement stipulates that only those Georgians who did not fight against Abkhazia would be permitted to return.

*July 1994*

**1** • The UN-brokered peace plan for Abkhazia continues with the return of 5,000 Georgian villagers to their homes in the region. Although Russian peacekeeping troops are in place between the Abkhazian and Georgian military forces, sporadic fighting erupts in the Kodori Gorge. In meetings with Shevardnadze in Tbilisi, UN representative Eduard Brunner announces the UN established a fund to aid Georgian refugees.

**6** • Russian Deputy Defense Minister Georgiy Kondratev dismisses the charge that Moscow is acting unilaterally in Abkhazia, citing other CIS states—Uzbekistan, Kazakhstan, and Belarus in particular—offered troops to monitor the security zone along the Inguri River. A decision on this matter will be made at the 18 July meeting of the CIS Council of Defense Ministers.

**12** • An agreement on the disengagement of hostile forces from the Kodor Gorge stalled when Georgian Defense Ministry officials claim Abkhazian troops violated agreement provisions. Georgian and Abkhazian forces exchanged fire repeatedly in the 50 km strip of land located between the Kodor and Inguri rivers in Abkhazia.

**21** • The UN Security Council votes unanimously to endorse the deployment of Russian peacekeeping forces in Abkhazia. The UN Security Council, however, also increases the number of UN observers monitoring the ceasefire implementation in Abkhazia.

**23** • Shevardnadze considers the UN Security Council resolution "the first and very important stage in settling the conflict." He praises Russia, which "took upon itself huge responsibility by deploying its peacekeepers in the zone of conflict." Due to Russian and UN efforts, Shevardnadze says, the refugees will return safely to their homes.

**24** • Abkhazia issues an ultimatum demanding that Georgia withdraw all its forces from the Kodor Gorge by 30 July and that those Georgians residing near the Gorge be disarmed. Abkhaz deputy defense minister says if the demands are not met, Abkhazian armed forces will "undertake specific measures" to force Georgian troops out. The Abkhaz also have refused to discuss the refugee problem until Georgian forces withdraw.

• Ardzinba states "only the closest ties with Russia" can ensure the development of Abkhazia. The comments demonstrate the Abkhazian strategy to forge political ties with Russia in their quest to secure independence from Georgia.

**25** • Shevardnadze, speaking on Russian television, blames Defense Minister Tengiz Kitovani for starting the war in Abkhazia. Shevardnadze states that Kitovani had disobeyed orders when he sent troops into Sukhumi, the capital of Abkhazia.

*August 1994*

**1** • Vadim Gustov, chairman of the Federation Council's CIS Affairs Committee, says that Russia's diplomatic and peacekeeping initiatives "are helping to stabilize the situation in the Georgian-Abkhazian conflict zone." Gustov, who recently led a Russian parliamentary delegation to the area, reports that fighting had stopped and that refugee repatriation is proceeding. Nevertheless, Gustov says that Georgian troops refuse to disengage from the Kodor Gorge, while the Abkhaz side actively works to delay the return of the Georgian refugees. In addition, negotiations on a political settlement for Abkhazia remain deadlocked.

**5** • Quadpartite negotiations were scheduled to resume in Sochi to discuss procedures for the safe return of refugees to Abkhazia, the withdrawal of Georgian forces from the Kodor Gorge, and on reaching a political settlement for Abkhazia. However, the Abkhaz delegation plans to boycott the session until the Georgian troops are withdrawn.

**6** • Georgian troops, monitored by UN and Russian observers, began withdrawing from the Kodor Gorge. A Georgian Defense Ministry spokesman says that 250 soldiers left the canyon. Two platoons of Russian peacekeepers will patrol the Kodor region.

**10** • A meeting between Abkhazian and Georgian officials is convened by Russian Foreign Ministry officials to review the current ceasefire agreement. Following the meeting, Georgian Defense Minister Vadriko Nadibaidze states his support for Russian peacekeeping forces in the area and hopes Georgian refugees return shortly.

**16** • Tens of thousands of Georgian refugees gathered in Zugdidi threaten to march back to Abkhazia en masse if the repatriation process is not sped up.

**17** • In a television address, Ardzinba states that the mass return of Georgian refugees threatens Abkhazia's state system. He says that his government would consider a selective return of refugees only after all armed Georgian formations withdraw from Abkhazia. He dismissed Tbilisi's claims that all Georgian troops have indeed left.

• Ardzinba meets with Tatar President Mintimer Shaimiev. They sign a friendship and cooperation treaty, recognizing each other as "subjects of international law."

**19** • The aforementioned treaty raises protest from the Georgian and the Russian foreign ministries. In Tbilisi, a Foreign Ministry spokesman argues the treaty constitutes a threat to the territorial integrity of both Georgia and the Russian Federation. In Moscow, Foreign Ministry officials contend signing, which violates provisions of the Russian-Georgian Friendship Treaty signed in February 1994. An Abkhaz spokesman refuted the Russian argument, pointing out that Moscow has not yet ratified the Russian-Georgian Treaty, thus it lacks legal force and cannot apply to the Abkhaz-Tatar Treaty.

**22** • Russia sends an additional battalion of troops to Abkhazia, bringing to 2,200 the number of Russian peacekeepers deployed in the conflict zone.

**31** • The nominal leaders of both the Russian and Georgian delegations failed to appear in Geneva at UN-sponsored talks on a political settlement in Abkhazia. Georgia's Dzhaba Ioseliani refused "on principle" to attend, stating that Abkhaz officials continued to delay the repatriation of Georgian refugees to their homes in Abkhazia. Russia's Deputy Foreign Minister Boris Pastukhov's absence was not explained.

***September 1994***

**1** • UN–sponsored talks on the Abkhazian conflict between Russian, Georgian, and Abkhazian representatives resume in Geneva. Although a preliminary agreement is reached for the safe return of thousands of Georgian refugees displaced from their homes in Abkhazia, the Georgian delegation is harshly critical of the Abkhazian government's delay in allowing for this return. Meeting in Tbilisi with Shevardnadze, U.S. Ambassador to the United Nations, Madeline Albright, reports that the United States insists on the speedy return of the Georgian refugees as the first step in reaching a settlement. Additionally, she states that the deployment of Russian peacekeeping troops must be temporary and will be "under international scrutiny."

**14** • The Abkhaz government accuses Russian Deputy Foreign Minister Georgiy Kondratev of initiating a mass repatriation of Georgian refugees to Abkhazia's Gali rayon. The Abkhaz government announces that it has put its troops on alert to prevent the expected influx. A Russian Foreign Ministry spokesman denied that Moscow and Tbilisi had reached a separate agreement on the refugee situation.

• Grachev says that the deployment of Russian peacekeepers in Abkhazia may be extended beyond the six months originally agreed upon. He argues that the Abkhaz and Georgians have yet to show that they are capable of living in peaceful harmony.

**16** • Another round of Russian-mediated talks on the repatriation of refugees and the political settlement of the Abkhaz question opens in Sukhumi. The two sides reportedly reach agreement linking the return of the refugees to the withdrawal of all Georgian military equipment from the hotly contested Kodor Gorge.

**20** • Speaking before parliament, Shevardnadze attempts to link basing rights for the Russian army to a satisfactory solution of the Abkhaz conflict. Unimpressed, the vocal opposition faction in parliament denounces Shevardnadze's policy, accusing him of capitulating to the Russians.

**23** • Ardzinba states that refugees would not, as previously agreed, be allowed to return to the Gali region on 1 October and that only 200–300 Georgians would be allowed to return to Abkhazia in October. The statements spark a mass demonstration of Georgian refugees in the western Georgian city of Zugdidi.

• Abkhaz troops are laying mines along the Inguri River, which serves as the de facto border between Abkhazia and Georgia. In July, Russian minesweepers cleared the area along the Inguri as a precondition of the deployment of Russian peacekeeping troops into Abkhazia.

**28** • Russia's special envoy to Georgia, Feliks Kovalev, says that Russia "has no intention of considering ratification" of the Russian-Georgian friendship treaty (signed on 3 February 1994) "until the resolution of the Georgian-Abkhaz and Georgian-Ossetian conflicts." The treaty would provide for Georgia's territorial integrity. In response, Georgia's ambassador to Russia, Valerian Avdadze, says that "Georgia's independence depends to a great extent on Russia's position. Georgia will be independent if Russia wants it [to be]. . . ."

***October 1994***

**12** • Three months after deployment of Russian peacekeepers to the area, six Georgian families return to Abkhazia's Gali rayon in accordance with the established procedures.

The UNHCR representative overseeing the operation complained about barriers to repatriation created by the Abkhaz government and states that as many as 33,000 refugees could have been returned by now had the Abkhaz authorities cooperated.

**28** • In Tskhinvali, South Ossetia and Trans-Dniester Republic sign an interstate treaty on friendship and cooperation.

**31** • A draft agreement on cooperation between Russia and Georgia is completed. It provides for training Georgian officers in various Russian military schools and Russia's active participation in supplying the Georgian army with military equipment in exchange for Russian military bases in Georgia.

***November 1994***

**2** • In Georgia, UN Secretary-General Boutros Boutros-Ghali confirms the UN position on the territorial integrity and inviolability of Georgia's borders. He emphasizes the good relations between UN observers and Russian peacekeepers in the zone of Abkhazian conflict. He says that soon the UN will build up an observer contingent in Abkhazia which will consist of 136 emissaries from 26 countries.

• Following talks with Georgian Prime Minister Otar Patsatsia, Boutros-Ghali declares that further progress on the repatriation of over 250,000 Georgian refugees to their homes in Abkhazia depends on the course of political talks being mediated in Geneva by UN officials. According to Boutros-Ghali, a conference is being considered for next month between Russian, Georgian, and Abkhazian officials.

• Commenting on the results of Boutros-Ghali's visit to Georgia, Chairman of Georgia's Parliamentary Commission on Security and Defense Nodar Natadze says that "at present, the UN is unable to render any practical influence to settle the conflict in Abkhazia."

**12** • A bilateral agreement on border enforcement and cooperation is signed between the visiting head of the Russian border troops and Georgian officials. The agreement calls for joint efforts to enforce border security, including cooperation in customs, enforcement, and immigration operations.

**15** • Georgia and Abkhazia fail to reach agreement in Geneva talks. The Georgian delegation insists that it would continue the talks only if Abkhazia recognized its territorial integrity. However, the latter says that Georgia and Abkhazia have equal rights and therefore Abkhazia could choose its future on its own—to be part of Georgia or exist as a separate state. The impasse hinders solutions to the refugee problem: only 300 are now returning monthly to Abkhazia under a formal procedure. Proposals from UN and Russian intermediaries to increase the number to 3,000 are rejected by Abkhazia.

• According to officials of the UN High Commission for Refugees stationed in Georgia, Abkhazian forces are engaged in a campaign of harassment and intimidation against Georgian refugees returning to Abkhazia.

**26** • The Abkhazian parliament ratifies a new constitution proclaiming the Republic of Abkhazia as a sovereign, law-based state historically established according to its right to self-determination. Ardzinba is elected Abkhazia's first president.

**27** • Shevardnadze assesses the Abkhazian Supreme Soviet's declaration of the republic's sovereignty as an attempt by the separatists to set up an independent state. He says that Georgia will not conduct talks with the Abkhaz side as a representative of an independent state.

**28** • Commenting on the constitution adopted by Abkhazia's parliament, Russian government declares that it will continue to duly respect Georgia's territorial integrity and characterizes the adoption of the constitution as an act contradicting obligations assumed by the Abkhaz side toward reaching a peaceful settlement of the Georgian-Abkhaz conflict.

**29** • Ardzinba says that the new constitution would make little change in the status of Abkhazia which was a sovereign state under the 1925 constitution. The Supreme Soviet of Abkhazia announces it is necessary to continue the traditions of more than 1,000 years of Abkhaz statehood and that "Abkhazia is not breaking off the negotiating process with Georgia and is prepared to continue with the aim of creating a union state with two equal components."

***December 1994***

**9** • The Azerbaijani Foreign Ministry announces that the unilateral decision of the Abkhazian Supreme Soviet is a flagrant violation of Georgia's sovereignty and that it is in contrast with the peace negotiations and creates new serious threats in the region.

## 1995

***January 1995***

**4** • Diplomats and officials of humanitarian organizations in Tbilisi concur that incompetence is the only quality of most of the men in power in Georgia. Shevardnadze still considers the situation in Abkhazia a "personal tragedy," stating that "if Sukhumi had fallen, I would have resigned." Given that Sukhumi did fall and he did not resign, Shevardnadze had to alter his statement by saying that he "felt like it," but the people wanted him to stay in power. The chairman seems to feel that some hope for resolution remains.

His comments on Russia: "I do not believe Russia is interested in a weakened Transcaucasus because this weakness could lead to a major destabilization of the region and will have consequences reaching to Russia. . . . The majority of the independent states support Russia's desire to preserve its territorial integrity."

**6** • Deputy Chairman of the Russian parliament's upper house Ramazan Abdulatipov visits Ardzinba and commander of Russian peacekeeping forces in Abkhazia, Lt.-Gen. Vasiliy Yakushev. Abdulatipov first sought to check on reports that there are bases on Abkhaz territory that train militants for fighting alongside Dudaev in Chechnya, and second, to discuss the extension of the six-month mandate given earlier to Russian peacekeepers in the zone of the Georgia-Abkhazia conflict.

While concluding that the peacekeepers had played a positive role in stabilizing the situation, Abdulatipov says Russia "had undertaken a burden that was too heavy" and that no country or international organization could simultaneously wage an internal conflict and conduct five or six peacekeeping operations.

**10** • The Abkhazian government warns illegal (non-Abkhazi) militants to withdraw from the Gali district (which borders on Georgia's Zugdili district). Otherwise, it will destroy bandit formations and terrorist groups.

**12** • Georgian Defense Minister Vardiko Nadibaidze turns down a Ukrainian offer to sell modern combat equipment to Georgia for its armed forces because Russia will hand over similar equipment to Georgian forces free of charge.

**13** • Kitovani leads a group of 200–250 partially armed veterans of the Georgian-Abkhaz conflict (in small buses) to western Georgia intent on "restoring the territorial integrity of the state." Kitovani says his actions were agreed upon with the Russian military. Shevardnadze issued orders for his interior, defense, and security ministers to make every effort to stop and disarm the group.

**14** • Kitovani and his group are detained and held in the village of Simoneti in western Georgia.

***March 1995***

**3** • Georgian and Abkhaz experts end recent talks by defining the devolution of power between Tbilisi and Sukhumi. The two sides agree that the future treaty will give Tbilisi peacemaking powers on crucial issues (foreign policy, economics, military policies). The location of the new legislature is still to be determined

**10** • Russia and Georgia sign intergovernmental accords on increasing joint protection of the country's external border. Shevardnadze believes the accords will be a stability factor for the entire Transcaucasus region.

**16** • Georgian Deputy Foreign Minister Mikhail Ukleba reveals that a Georgian delegation in Moscow sent a letter to the Russian Foreign Ministry protesting the presence of an "Abkhazian working team" within the Russian Ministry for Cooperation with the CIS. Ukleba says the Russians were understanding of the Georgian position and vowed to monitor more closely the activities in which Abkhazian separatists are allowed to participate.

**21** • Georgian Prosecutor General's office sanctions the arrest of Ardzinba for war crimes and genocide against the Georgian people. Prosecutor Anzor Latsuzbaya says his office possess sufficient evidence to prove Ardzinba's guilt in provoking the military conflict in Abkhazia. Latsuzbaya also says a suit must be brought against all those involved in acts of genocide against the Georgian people.

**23** • At a Paris conference on European stability, Georgian Foreign Minister Aleskandr Chikvaidze says events in the Caucasus region endangers European stability. He urges conference participants, particularly European Community members, to join efforts to achieve stability in the region.

**26** • Russian Defense Minister Pavel Grachev holds "confidential" meeting with Ardzinba. The two discuss settlement possibilities for the conflict and Grachev defends the presence of the Russian peacekeeping forces in the republic.

***April 1995***

**3** • In an interview with Tbilisi radio, Shevardnadze says Abkhazia is on the road to chaos and anarchy. He comments how the recent Georgian-Russian military cooperation treaty and talks with the Russian leadership and with UN officials—all new developments—have worked to bring the conflicting parties closer to agreement but notes the extreme positions held by separatist forces make negotiation difficult.

**26** • In an interview with Interfax, Shevardnadze expresses dismay with Moscow's Abkhaz policy regarding the role of Russian peacekeeping forces in the region. He laments the passive behavior of the peacekeeping forces during a recent attack by Abkhazian separatists on returning refugees. He points out relations between Georgia and Russia would be firmer if "we felt support of Russia's counterefforts confirming reality of Moscow's break with Abkhazia's separatist regime and if we felt support of the peacekeeping forces."

**28** • The Supreme Council and the Council of Ministers of Abkhazia announce they cannot support extending the Russian mandate for peacekeeping forces in the conflict "if no positive developments result." The mandate is set to expire 15 May and sporadic fighting continues as refugees returning to their homes are harassed and sometimes beaten. Both sides say their peoples are victims of such abuse and claim peacekeeping forces do nothing to mitigate the clashes. The Georgian side says the mandate will be extended only if the forces assume police functions and "provide real protection for the Georgian population."

***May 1995***

**9** • After a series of meetings, the UN Security Council votes to extend the UN observer mission in the Georgia-Abkhazia conflict zone until 12 January 1996. The Council praised the efforts of the CIS peacekeeping mission, singling out Russia's role in particular.

**13** • Commander of Russian peacekeeping forces in Georgia, Vasiliy Yakushev, announces his forces stand ready to guarantee the security of the Georgian population in Abkhazia. Georgian Defense Minister Nadibaidze tells journalists this marks a clear change in the Russian position and welcomes the new stance.

• After talks in Moscow, Russian chief delegate to the United Nations Sergey Lavrov says Georgian and Abkhazian officials "confirmed their obligations to prevent resumption of the armed confrontation" and agreed they should "live in a single state." Lavrov only regrets the insignificant progress in negotiations over the political status of Abkhazia and on the slow pace of returning refugees.

**19** • Abkhaz representatives reject signing a document that includes Abkhazia in a Georgian federative state with broad autonomy. Georgian officials see the rejection as a ploy to prevent a comprehensive political solution. According to Russian mediator Deputy Foreign Minister Boris Patukhov, Abkhazia suffers from the "victor's syndrome."

**22** • Ardzinba tells journalists that Georgian refugees can begin returning to their homes in Abkhazia with "full security" on 25 May. Some 200 refugees should be received weekly—a figure Ardzinba declares will allow Abkhaz authorities to keep "irregulars and war criminals away."

**23** • In a press interview, Russian Lt.-Gen. Yakushev describes the security zone as "extremely complex" and rife with banditry. He laments the fact that some 700,000 mines still exist and blow up cattle and people frequently. He also says that refugees allowed to return to Abkhazia should be increased to 1,000 a week.

**27** • Shevardnadze meets in Tbilisi with Russian Deputy Prime Minister Nikolay Yegorov. No details are disclosed but reports indicate they discuss settlement of the ethnic conflicts in Abkhazia and South Ossetia. In speaking to journalists after the meeting, Yegorov declares Russia's "unequivocal support for the territorial integrity of Georgia" and condemns separatist activity in Abkhazia.

***June 1995***

**5** • The former Georgian parliament member and foreign minister, Murman Omanidze, warns that Georgia should not trust Russia's mediation efforts given that, as he points out, quite a few Russian structures own major enterprises in Abkhazia. He declares that the separatist element in Abkhazia must "be blown up from within" and Georgia can achieve this by supporting opposition in Abkhazia to the separatist regime (Ardzinba).

**7** • Russian peacekeeping troops will oversee the return of Georgian refugees to their homes in the Ghali and other Abkhazian districts. Georgian Deputy Prime Minister Tamaz Nadareishvili says the refugees will not have to undergo the "preliminary questioning" previously employed by Abkhaz authorities to ensure militants were not allowed into the republic.

**9** • Shevardnadze expresses readiness to meet with Ardzinba on the condition that the refugee return process begin immediately. He laments the fact that despite widespread international recognition of Georgia's territorial integrity, Ardzinba continues with referendums in Abkhazia to determine attitudes on independence or for rejoining Russia or Georgia. Given the prevailing international position, Ardzinba "should show some respect to the rest of the world," says Shevardnadze.

**14** • In a press interview, Shevardnadze discusses how separatist events in the Transcaucasus and the Russian Federation could evolve into a Balkan-like situation. He says the fragmentation of any country's territory is inadmissible and that the West could help in this regard by expressly denouncing separatism, extremism, and terrorism. Should separatist activity spread in the former Soviet Union space and in the Russian Federation, Shevardnadze says the "Yugoslav tragedy would be nothing by comparison."

**16** • In talks with OSCE Sec.-Gen. Wilhelm Hoeynck, Shevardnadze praises the OSCE role in settling ethnic conflicts. Pressed as to what degree of autonomy Abkhazia can expect, Shevardnadze says they will enjoy all the attributes of statehood within the state of Georgia. Hoeynck declares he will personally participate in settlement activities.

• In a press interview, Nadareishvili discusses the dire situation in many Abkhazian cities, where "Abkhazians face a real threat of extinction." He points out that no one can oppose the Ardzinba regime, that they control all sources of power and that many people have been executed by firing squads. He believes that the return of refugees, given this context, will provoke more violence as Abkhazian police units and administrative structures take defensive positions. Nadareishvili says "we will not be able to regain Abkhazia by peaceful means. I believe it is necessary to use military force."

**24** • In a meeting of the Georgian Constitution Commission, participants discuss two variants to a new constitution. One envisages a federal arrangement where Abkhazia, Ajaria, and Tskhinvali become self-governing regions but are subjects of the federal state, Georgia. The other proposes a strict unitary Republic of Georgia, necessary in many opinions to prevent confrontations between regions and separatism. Shevardnadze favors the federative model and declares a constitution reflective of this is in the "final stage."

**27** • Ardzinba issues statement of dissatisfaction concerning talks between Georgian and Russian officials on deployment of Russian peacekeepers throughout Abkhazia. Ardzinba says participation of the Abkhazian leadership in all talks is necessary for complete settlement of the conflict.

***July 1995***

**2** • The State Constitution Commission approves a draft constitution, 60 to 4. It provides for a federative arrangement where Abkhazia will have the attributes of statehood. Defense and border issues, infrastructure maintenance, and monetary policy will be the responsibility of the "center." The draft will be sent to parliament for debate.

**6** • Tbilisi radio reports that major agreements on national repatriation were achieved, allowing for the safe return of refugees. Representatives from both the Georgian and Abkhazian sides met in the Zugdidi district village of Anaklia.

**11** • In Georgia for talks with Shevardnadze, CIS Interparliamentary Assembly Chairman Vladimir Shumeyko pledges 10 billion rubles for humanitarian and refugee use, stressing the aid would go through the appropriate Georgian channels. Shumeyko emphasizes the protection of Georgia's territorial integrity reinforces Russia's strategic interest in the region. The Abkhaz leadership was "bewildered" that humanitarian aid intended for Abkhazia should go through Georgian state bodies. In their opinion, they will never receive the money given the disbursement arrangement.

**16** • UN special envoy Eduard Brunner arrives in Tbilisi. In talks with Georgian officials, Brunner hears the pessimism of Nadareishvili, who believes settlement negotiations are futile. "There is only the military way to return Abkhazia," Nadareishvili says in discussions. "Ardzinba is a state criminal" and "we have come to the conclusion that we do not need either the United Nations, the [OSCE] in Europe, or Russian mediation." For his part, Shevardnadze urges stepped up measures, saying "further delays in the process could cast doubts on the part of the international community to find a political settlement to the problem." Brunner declares the UN is working to get back Abkhazia, but that a political solution involving Ardzinba must be found. According to the envoy, no "perceptible progress" has been made.

**21** • Disagreements over Georgia's approach to the Abkhazian conflict surfaces between Shevardnadze and Georgian parliament member Jaba Ioseliani, leader of the Mkhedriono opposition faction. Ioseliani favors full-scale military activities without Russian help while Shevardnadze has been urging resolution through continued talks and negotiations with Russian mediation. Ioseliani vows to transform his group into a political party to fight against what he sees as rising despot, referring to the Shevardnadze regime.

**25** • In an address to parliament, Shevardnadze urges the body to adopt the draft constitution submitted for their review creating a federated Georgian state. Shevardnadze says it is the right course for the country at the moment and regulates fairly the territorial organization of the state. Parliament members agree to debate the document for eight hours daily until 3 August.

***August 1995***

**1** • Tamaz Nadareishvili and Zurab Erkvania, head of the Tbilisi-based Supreme Council and the Council of Ministers of the Abkhazia express their satisfaction with the Russian proposal for conflict's settlement. The proposal calls for the safe return of refugees, confirms the territorial integrity of Georgia, and provides for wide autonomy for Abkhazia. The plan is to be used as the framework for the next round of talks.

**2** • Patsatsia holds talks with head of the OSCE mission in Georgia Dieter Boden. Boden spoke of the possibilities of rendering humanitarian aid to the refugees from Abkhazia and urged the Georgian prime minister to allow increased Russian participation in the settlement proceedings, believing that Russia can play the decisive role.

**7** • Reports surface that the Abkhaz separatists are being supplied weaponry and supplies by Tatarstan. The Republic of Tatarstan, part of the Russian Federation, and Abkhazia have signed a friendship and cooperation agreement.

**14** • The conflict enters its third full year, with the death toll reaching 10,000 lives. Tbilisi estimates that over 300,000 refugees have been created as a result of the fighting. Despite the peace agreement signed in May 1994, sporadic fighting continues as no political resolution has yet been achieved.

**18** • In a press interview, Shevardnadze denies responsibility for "unleashing war" in Abkhazia. He states the intention of sending into Abkhazia Georgian National Guards in 1992 was to "restore the control" over the main rail and road routes with the aim of averting robberies of food trains passing through Abkhazia and western Georgia on their way to Tbilisi.

• The Supreme Council of Abkhazia announces that Ardzinba has prepared a special decree on an amnesty for the participants in the "liberation war." The move is predicted to upset many ethnic Georgians living in territories controlled by the secessionists.

***September 1995***

**5** • In a press interview, Shevardnadze says the Abkhaz leadership has once again "destroyed yet another chance [for settlement] through its irresponsible conduct," referring to Abkhazia's walkout on talks being held in Moscow. He believes the Abkhaz leadership is procrastinating until Moscow holds elections and politicians with favorable views toward Abkhazia will assume power. As to settlement possibilities, Shevardnadze says "I still adhere to peaceful settlement, but everything has not only a beginning but also an end."

**6** • CIS Interparliamentary Assembly chairman Shumeyko contends the policies of Ardzinba were criminal from the outset. These policies, he argues, led to the war, the human casualties, and the destruction brought to the home, villages, and towns in Abkhazia. Says Shumeyko, "Ardzinba [will] finally face God's judgment for his deeds."

**7**• UN envoy to the Georgian-Abkhazian peace talks Eduard Brunner addresses a UN Security Council session outlining the "destructive policies" he concludes were orchestrated by Ardzinba. Brunner further charges it was Ardzinba and the Abkhaz delegation who broke the Moscow talks, which, in his opinion, is sure to lead to a deterioration of the situation in the Caucasus region.

**11** • Abkhaz ideologue Yuriy Voronov is found murdered in his home in Sukhumi. Voronov was an ethnic Russian who served as vice-premier in the Abkhazian government and was a key part of the Abkhaz negotiating team. Abkhazian ambassador to Moscow Igor Akhba reveals that he has no doubts the killing was contract work and political. Voronov favored independence for Abkhazia and argued for equal union of the two states.

**14** • Abkhazian Foreign Minister Leonid Lakerbaya says their intelligence sources have identified increased Georgian troop buildup near the Gali district. Lakerbaya sent letters to the CIS peacekeeping command and to the UN military observers' command urging them to take action. He believes the buildup (due to receive reinforcement from the Georgian army) is in preparation for cutting off the region from Abkhazia and rejoining it to Georgia. In talks with Russian Deputy Prime Minister Aleksey Bolshakov, Shevardnadze denies the troop buildup, saying "one should be very imaginative to think of such a thing."

**17** • The Abkhaz Security Service believes Russia and Georgia will launch a joint mission to "restore Georgia's territorial integrity." Security Service sources say Russia has given warships to Georgia for a landing operation to be launched simultaneously with military operations in the Gali district. The service further asserts that a Russian contingent may take part in the combat operation.

**18** • Ardzinba requests that Russian Foreign Minister Kozyrev provide him with a copy of a supposed document signed by Chernomyrdin and Shevardnadze which, according to Abkhazian special services, provides for the "forcible methods of resolution of the Georgian-Abkhazian conflict." Ardzinba says the Abkhaz leadership desires the document so they can "make an objective assessment of the situation."

## Summary at the End of 1995

The Georgian-Abkhazia conflict has yet to be resolved. The sides still do not agree on what the official status of the Republic of Abkhazia should be in relation to Georgia. The Abkhaz insist on a federative structure while the Georgians are intent on a confederative structure which maintains Georgian territorial integrity. Georgia has said the Abkhaz would enjoy the same attributes of statehood and retain wide autonomy.

The Abkhaz desire complete independence. No major fighting has occurred and the rumored joint Russian-Georgian military operation to retake the republic has not developed. All sides confirm their commitment to arrive at a political settlement through peaceful means. The UN, OSCE, the U.S. and NATO, the EU, and the CIS peacekeeping command are the multilateral organizations involved in talks and the settlement proceedings.

# The Nagorno-Karabakh Conflict

## Background Note

Starting in late 1988, a wrenching war of tragic proportions raged itself into a tension-filled lull, which now hangs over the separatist enclave, punctuated intermittently by low-intensity outbursts of guerrilla warfare. Ostensibly about reclaiming "historical lands," the Nagorno-Karabakh conflict is also very much about Russia's oil interests and external competition for control over the region. In this instance, Moscow has brilliantly played each combatant against the others, accomplishing in the process several of its imperial goals in the Caucasus. To fulfill these goals, it has made use of its by now familiar policy of sending in "peacekeeping" troops, albeit with some muted resistance from Western organizations.

As with most of the CIS "hot spots," the history of the Nagorno-Karabakh conflict precedes the formation of the CIS. Ethnic Armenians and Azerbaijanis have been fighting for hundreds of years. The most recent feud began in December 1988, when Moscow once again refused an Armenian request to incorporate Karabakh within the Armenian Soviet Socialist Republic. Fighting had broken out between Armenia and Azerbaijan following a wave of anti-Armenian pogroms in Azerbaijan and the subsequent erection of a physical and economic blockade between Armenia and Nagorno-Karabakh. The intensity of the hostilities increased in 1989 when the Armenian parliament passed a "Resolution on the Reunification of Armenia with Nagorno-Karabakh." This open insult to Azerbaijan's territorial integrity was followed by Nagorno-Karabakh's 1990 election of twelve deputies to the Armenian parliament, which Azerbaijan, of course, regarded as illegal. Following this challenge, the "Republic of Mountainous Karabakh" announced its existence and its secession from Azerbaijani jurisdiction in September 1991.

Each state has its own stubborn reasons for continuing the long and gruesome battle. In its own defense, Armenia claims it is liberating its historical lands from the illegal apportionments made during the Stalin era. (In 1926, the population in Nagorno-Karabakh was 117,000 Armenians and 13,600 Azeris.) Azerbaijan, on the other hand, regards arbitrarily established Soviet boundaries of the 1920s as legal, and refuses to recognize Armenia's historical arguments.

Russia's historical interest in the Nagorno-Karabakh region dates back to its eighteenth-century competition for the lucrative trade routes leading to Iran and Asiatic Turkey. Russia's increasing appetite for raw materials, notably silk, cotton, and copper, fueled its colonization of Azerbaijan in the early nineteenth century (oil had not yet been discovered in the Caspian Sea shelf). Over everything else, however, the area's attraction stemmed from its strategic value as the Transcaucasian corridor penetrating deeply into Iran, and for its vantage point on Turkey's eastern flank. In the nineteenth century, therefore, Russia became the first European power to establish direct rule over the Transcaucasian slice of the Middle East, far ahead of Britain's colonization of Egypt and France's mandate over Syria and Lebanon.

Returning to today, the chronology which follows begins in January 1992, after the hastily convened Belovezh Forest meeting from which the rather ill-conceived and unplanned CIS sprang. To reestablish its long-standing dominance in the Transcaucasus, the Russian government began to work partly through CIS institutions to assert its "peacekeeping" role in the Nagorno-Karabakh war. For this purpose, it used the Soviet 7th Army, which was deployed in Armenia before the USSR dissolved. During the war, Moscow has employed the 7th Army to support both the Armenians and the Azerbaijanis in the struggle. Moscow's zigzag diplomacy in the region has sometimes met with suspicion and resistance both from countries within the CIS and on the part of international organizations like the United Nations and the Organization for Security and Cooperation in Europe (OSCE) (formerly the Conference on Security and Cooperation in Europe [CSCE]). Nevertheless, the Western powers have been weak in their response, and fearing a major conservative Islamic backlash in Azerbaijan, largely extended their support for Moscow's ceasefire plan in the Karabakh war, as laid out in the May 1994 Bishkek Protocol.

Moscow's rivalry with the CSCE/OSCE over resolving the Karabakh conflict is elucidated in this chronology, as are relations between Russia and the principal parties in the negotiating process, including Turkey, Nagorno-Karabakh, Iran, Armenia, Azerbaijan, the OSCE, the United Nations, and the CIS leaders. Karabakh has, in fact, become something of a test of whether Russia will cooperate with international bodies in the future in resolving CIS hot spots, or whether it will insist on dominating the "peacekeeping" process, thereby forming its own secure sphere of influence. The chronology reflects a sense of Russia's intentions, tactics, and strategy over the course of the gruesome war. The endpoint of its strategy has not, however, been reached, for President Geydar Aliev remains a thorn in Russia's side in his refusal to countenance Russian incursions against Azerbaijani independence.

## Chronology of Key Events

### 1991

*December 1991*

**28** • Armenians of Nagorno-Karabakh (NKR) hold parliamentary elections and prepare to establish statehood. The Azerbaijani government refuses to recognize the republic and cancels its "autonomous" status.

**1992**

*January*

**2** • Presidential rule is decreed in Azerbaijan by President Ayaz Mutalibov.

• Nagorno-Karabakh's leadership ignores Mutalibov's decree of presidential rule.

• President Ayaz Mutalibov announces that "war is on in Karabakh. Previously a local conflict, it is escalating into a full-fledged war." Reports indicate that in the past week, 6,000 Azerbaijanis have been driven from their homes.

**6** • NKR Supreme Soviet appeals to the United Nations and CIS states for recognition, declaring the republic ready for diplomatic relations with all states.

**8** • Director of the geological museum in Gadrut Artur Mkrtchyan is elected chairman of the NKR Supreme Soviet. The parliament is made up of eighty-two seats, eighteen of which are reserved for Azerbaijanis.

**13** • Chairman of the NKR Supreme Soviet announces that NKR and Azerbaijan are in a state of war.

**25** • In a press interview, Armenian Foreign Minister Raffi Ovanesyan warns that the Nagorno-Karabakh conflict can become another Yugoslavia. He reveals also that Armenia has concluded friendship and cooperation treaties with Russia and applied for membership in the United Nations and the CSCE.

**27** • NKR Prime Minister Oleg Yesayan visits Armenia seeking food aid. During the visit, he informs Armenia's prime minister about parliamentary elections and efforts to form a government.

**29** • Azerbaijan cuts off its gas pipeline to Armenia. Fuel now reaches Armenia through the Georgian branch line.

**31** • Armenian Foreign Minister spells out the Armenian position on NKR to a session of the Helsinki Council of Foreign Ministers. He claims the conflict "is not a territorial issue, and Armenia does not have any territorial claims." He states the NKR problem is one of human rights and the right of citizens to self-determination.

• Mutalibov describes the conflict as an "internal affair and a matter of principle for Azerbaijan," vowing not to internationalize the conflict by appealing to international organizations.

*February 1992*

**2** • TURAN news agency reports intensive Armenian armed formations along the Azeri border, transported there by Soviet army aircraft. The Armenian forces reportedly are equipped with arms and matériel of the Soviet motorized rifle regiment stationed in the Khankendi region (Stepanakert).

**6** • In an address to Azeri political leaders, Azeri Prime Minister Hasan Hasanov notes Russia will soon become the enemy in view of its increasingly pro-Armenian stance. He believes a recent Russian-Armenian security treaty will serve to reinforce Russian support for Armenia in the NKR conflict.

**8** • Azeri presidential advisor R. Musabekov states that Russia is not viewed as a neutral, objective force in the conflict by Azeris but decidedly pro-Armenian. While realizing the difficulty of resolving the conflict without a Russian role, he laments the "obtrusive form" taken of late by the Russian leadership.

**13** • NKR's Cabinet of Ministers warn CIS heads that should the Russian leadership resubordinate military units on Azeri territory to Azerbaijan, the regional situation will deteriorate, lead to considerable casualties, and damage Karabakh and Armenian relations with Russia. The Cabinet also reveals that over 1,500 missiles have been fired on Stepanakert since November 1991, killing 49 and wounding 159.

**20** • In a public address, Mutalibov says instability in Azerbaijan directly relates to the conflict in the NKR and appeals to all political currents, movements, and parties—regardless of ideology—to join in overcoming the difficulties. The same day, Azerbaijani Popular Front (APF) leader Abulfaz Elchibey demands Mutalibov's resignation for the dire state of affairs he has caused.

**28** • Mutalibov agrees to combine the general forces in Azeri territory with the CIS army. The coalition will be subordinate to the Azeri president and CIS Commander Shaposhnikov. The Azeri leadership has previously insisted on incorporating the general forces units stationed in Azerbaijan into an Azeri army. The move should ease tensions between the general forces and the Azeri republic.

***March 1992***

**2** • The Azeri town of Shusha incurs heavy shelling, damaging energy, water, and electricity resources. Both the CIS 366th regiment based in Stepanakert and Armenian forces deny a part in the shelling.

**6** • Under strong pressure, Mutalibov resigns as Azeri president. Presidential powers are passed over the Supreme Soviet Chairman Yaqub Mamedov. The APF, under Elchibey, further demands the resignation of the whole republican leadership, charging the regime has performed poorly in the NKR conflict.

• The CIS 366th regiment, stationed in Stepanakert, is withdrawn. NKR leaders lobbied against the move as they believe the withdrawal provides the Azeri leadership the opportunity to wage full-scale warfare against NKR.

**11** • Turkey and Iran announce interest in forming a peacekeeping division to resolve the conflict. Iranian Foreign Minister Ali Akbar Velayati proposes a multinational force be dispatched to the region while Tehran negotiates the conflict's solution. The Turkish plan, revealed in a memo from the Turkish Embassy in Moscow, envisages both Armenia and Azerbaijan inviting CSCE experts to examine the situation and assign a special ceasefire monitoring mission. Turkey asserts that both sides should pronounce intentions to settle the conflict peacefully.

**19** • A ceasefire, brokered in Tehran talks between representatives of the warring sides, is announced. The agreement, effective 20 March, encompasses monitoring committees comprised of Iranians to be established in Baku, Yerevan, Stepanakert, and Shusha, provides for exchanging the dead, and lifts economic sanctions on the region. Iran's interest in the conflict's resolution is linked in part to Iranian efforts to become a major player in developing Caspian Sea oil reserves.

**22** • Armenian-populated villages of Karachinar, Kharkhaput, and Manashit in the Shaumyan district come under heavy fire. Reports note Azeri troops amassing near other Armenian villages and residents begin to flee. Armenian officials say the shelling violates the recent ceasefire in effect since 20 March.

**30** • An Iranian delegation arrives in Baku to implement stage two of the Islamic nation's efforts to arrive at a peace settlement. In meetings with Azeri leader Mamedov, Iranian Deputy Foreign Minister Mahmud Va'ezi emphasizes the importance of a durable ceasefire, pledging additional Iranian units if needed. Va'ezi also says Iran is prepared to intermediate the exchange of hostages.

• In a press interview, Turkish Ambassador to Russia Volkan Vural states Turkey considers the Karabakh region as an inseparable part of Azerbaijan and that the province must remain part of the Azerbaijani Republic. He notes Ankara's support, however, for a peaceful solution to the problem. Turkey has consistently supported Azerbaijan in lieu of its centuries old trading links and common Turkic origins. Like Iran, Turkey also seeks a prominent role in the Caspian Sea oil basin's development.

***April 1992***

**1** • A CSCE delegation arrives in the NKR to meet with its leadership. Delegation head Jiri Dienstbier announces plans to hold an international conference on the NKR conflict within the CSCE framework. NKR leadership make clear that "Nagorno-Karabakh is a state formation, and its presence is obligatory in any process concerning the republic's interests."

**7** • Azeri units upset relative calm in the NKR by opening massive rocket and gun fire on Stepanakert, violating the Iranian-brokered ceasefire. Attacks are also opened in the Armenian villages of Norachen, Berdashen, and Nurishen. The activities halt when Armenian defense forces fire a salvo into Azeri positions.

• Russian special envoy Vladimir Kazimirov meets with Va'ezi in Baku. The two discuss Russia's role in conjunction with Iranian mediation efforts and express support for UN and CSCE participation in maintaining the ceasefire.

**12** • Azeri units launch massive missile and artillery bombardment on Stepanakert and Skhnakh in the Askeran district. NKR defense committee reports transportation of ammunition for Grad rocket launchers; anti-aircraft and artillery pieces are delivered to Shusha, controlled presently by Azeri units.

**14** • Artur Mkrtchyan, NKR Supreme Soviet chairman, is shot dead in his Stepanakert apartment. Acting NKR leader Georgiy Petrosyan reports the death was accidental, not a murder. The Azeri leadership deny any part in Mkrtchyan's death.

**16** • Accusing the Russian military of supporting Armenian combat activity against Azeris in the NKR, Azerbaijan Defense Ministry calls for withdrawing of all Russian troops from its territory.

**18** • Azerbaijan Defense Ministry issues statement to CIS member states charging Russia with inciting the conflict in Nagorno-Karabakh and continuing as the aggressor. The statement charges that Russia actively supplies arms and equipment to both sides and demands criminal proceedings be held against Russians serving in the CIS 4th Army. Extensive commentary purports that Russia supports Armenia to better its own chances of regaining a hegemonic role in the Caucasus.

**25** • CSCE representatives arrive in Yerevan in preparation for arrival of a CSCE observer team in the NKR. The team's mandate should be approved in Helsinki by the CSCE Committee of Senior Officials. Jiri Dienstbier of the CSCE Council of Foreign Ministers dispatched a similar preparation mission to Baku.

**30** • Russia and Azerbaijan agree on transferring all general-purpose troops stationed on Azeri territory to Azeri jurisdiction. The agreement nullifies previous arrangements to have joint Azeri-CIS control over the troops.

***May 1992***

**1** • Agreement is reached between representatives of the Russian and Azerbaijani Defense Ministries to withdraw all CIS Joint Armed Forces from Azeri territory by the end of 1993. Some of the unit's troops will be transferred to Azeri jurisdiction and others will join regiments in Russia.

**2** • Turkish Prime Minister Suleyman Demirel addresses the Azerbaijani parliament saying Azerbaijan must do better to bring their case regarding the NKR to the world community and offered Turkish diplomacy in doing so. Demirel also recognized the dissolution of the USSR affords ethnic Turks new opportunities to restore their fraternal links.

• Intensive artillery shelling continues unabated on Stepanakert. Armenian defense forces have repelled Azeri troop advances but the shelling brings the capital to a critical situation with foodstuffs emptied, communication systems down, and city water and energy resources badly damaged.

**4** • Armenian government organizes military call-up in line with the presidential decree signed last spring. The republic's Defense Ministry, military commissariat, executive committees of rayons and city soviets, the Interior Ministry, and the Ministry of Education are all involved in the conscription process which according to military commissar Col. Levon Stepanyan underlines the fact that the conscription is a national endeavor. Those born between 1970 and 1973 who have not yet served, and those born in 1974, are covered in the draft.

**6** • Armenian Vice-President Gagik Arutyunyan announces all defense equipment belonging to the ex-USSR will be controlled by the Defense Ministry. The decision evolved out of a defense ministry meeting that sought a comprehensive solution to defend Armenia's border.

**7** • Tripartite talks in Iran begin. The Armenian, Azerbaijani, and Iranian presidents sign a statement for a phased solution involving a permanent ceasefire, deployment of international observers, freeing transportation routes, exchanging prisoners, and ways to improve Armenian-Azeri bilateral relations. Commentators believe Iran's mediation, because of its clear neutrality, provides the best settlement opportunity.

**8** • Tripartite participants in the Iran talks sign a ceasefire agreement and pledge lifting the embargo on Karabakh. Implementation is doubtful as several political organizations in Baku express skepticism over the agreement.

**15** • A state of emergency is introduced in Baku as a result of internal power struggles between Mutalibov—who resigned in March and is being considered for reinstatement by parliament—and APF supporters. Clashes between these factions have threatened Azeri unity in defense of the NKR. As a result of the power struggle, Armenia has stepped up its campaign and CIS forces, at Russia's behest, have become active on the Armenian side.

**18** • Azerbaijan Defense Ministry reports that CIS troops invaded the rayon center of Sadarak in the NKR, shelling the rayon and making way for Armenian defense formations to follow. Two CIS soldiers who deserted their units entered into Baku and spoke to journalists regarding CIS involvement, saying groups of thirty to forty men are being assembled from Georgia and sent to fight in Armenia against Azerbaijan.

**21** • Armenian government urges the Nakhichevan leadership to request Iranian military observers to the autonomous republic to secure the region as a zone of stability. Increased fighting threatens stability on the Nakhichevan-Armenian border as CIS troops become more involved with the Armenian effort and as Turkish troops assemble on the region's border. Iran agrees in principle to the deployment, pending approval by appropriate leaders.

**23** • A group of Iranian observers arrives in Nakhichevan.

***June 1992***

**1** • CSCE-sponsored peace talks begin in Rome with eleven nations participating. Armenia and Azerbaijan send delegates but representatives from Karabakh refuse to participate without full diplomatic recognition. No agreements are reached.

**3** • Azerbaijan blockades Armenia, crippling the Armenian economy. The natural gas pipeline to Armenia through Georgia is also shut down due to violence in South Ossetia. Industrial production in Armenia falls 51 percent for the first five months of 1992.

**4** • President Ter-Petrosyan states Armenia will accept a peace settlement that has first been accepted by the NKR government.

**7** • APF leader Abulfaz Elchibey wins presidential election. He vows to increase Azerbaijan's effort in prosecuting the war, to re-establish strong links with Turkey, and to resist Russian attempts to reinvigorate their control in the post-Soviet space. In connection with this, Russia steps up its support to Armenia and begins talks with Nakhichevan parliamentary leader Geydar Aliev, a former Politburo member who remains pro-Russian. Rift between Aliev and Elchibey increases as the two compete for support from Nakhichevan's people and over control of the autonomous republic.

**7–12** • Azerbaijan launches counteroffensives to retake Armenian control of Karabakh, capturing fifteen villages in the Mardakert. Armenia alleges that Azerbaijan is trying to open a corridor to the Azerbaijani-populated Autonomous Region of Nakhichevan through Armenian territory. Artillery bombing is recorded on virtually the entire Armenian-Azerbaijani border.

**14** • Armenian self-defense forces counterattack. The Armenian defenders retake the village of Kichan. Azerbaijan maintains air superiority and by day's end controls the Shahumyan district. Armenian villages, including Karachinar, Buzlukh, Erkech, Manashid, and Kharkhaput, are completely destroyed, forcing 10,000 people to flee southward.

**16** • Armenian government threatens withdrawing from the CIS if member states fail to provide military assistance under the terms of mutual defense pact signed in May 1992.

**18** • Armenia and Azerbaijan reach a preliminary agreement for a ceasefire at CSCE-sponsored talks in Rome.

• Armenia calls reservists under age thirty-five and declares a one-month state of emergency in Nagorno-Karabakh.

**27** • Azerbaijani President Elchibey vows to "recapture within two months all territory in the NKR lost to Armenian forces" and rejects political autonomy for Armenians in Nagorno-Karabakh.

***July 1992***

**5** • A large-scale Azerbaijani offensive is launched against the northern Mardakert district of Karabakh. The offensive seizes most of the district, forcing upward of 70,000 Armenian villagers to flee their homes.

• The Armenian delegation at the eleven-nation CSCE Rome peace talks walk out in protest of Azerbaijani aggression.

**6** • The Rome CSCE session adjourns with no significant results.

**7** • After several Azerbaijani victories against Armenian villages, the Azerbaijani delegate to the Rome CSCE talks, Araz Azimov, agrees to a July ceasefire.

**8** • The Armenian Supreme Soviet resolves to support the rights of NKR and declares unacceptable any documents "referring to the Nagorno-Karabakh Republic as being within the Azerbaijani state structure."

**15** • The Armenian government announces that the delegations from Karabakh and Armenia will not be attending the scheduled CSCE Rome talks to protest recent Azeri attacks against the Armenians of Karabakh. The Armenian border district Goris is being shelled.

***August 1992***

**2–4** • Azerbaijani and Turkish delegations to the CSCE talks in Rome walk out in protest against a reference recognizing the delegation from the Republic of Karabakh. To date, the Karabakh delegation has been limited to "interested party" status.

**5** • Following talks in Moscow with the Armenian foreign minister, Russian Foreign Minister Andrey Kozyrev proposes that new Karabakh peace talks be held in southern Russia aimed at reaching a preliminary ceasefire agreement. The Russian proposal also suggests a second stage involving the deployment of a UN- or CIS-sponsored peacekeeping detachment in Karabakh.

**10** • An Azeri attack on Armenian territory around the village of Artsvashen in western Azerbaijan prompts President Levon Ter-Petrosyan to appeal to the signatories of the CIS Collective Security Treaty to "carry out their obligations" in coming to Armenia's aid. Russian spokesman emphasizes that the treaty envisages negotiations before any military involvement.

**21** • Armenian and Russian leadership agree in Moscow to Russian military support for Armenia and a legal status for Russian forces in Armenia. The Russian decision to side with Armenia reflects its displeasure with anti-Russian Azerbaijani President Elchibey.

**24–27** • In a letter to the UN Security Council, Armenia requests the Security Council convene immediately to discuss the Karabakh situation. It calls for UN involvement to counter renewed Azeri offensives. CSCE mediator for the Karabakh conflict Mario Rafaeli presents his proposal for establishing a sixty-day ceasefire and deploying military observers to prevent escalation.

**27** • Kazakh President Nazarbaev offers to mediate the conflict. Trilateral negotiations at Foreign Ministry level begin in the Kazakh capital, Alma-Ata, and produce a ceasefire agreement to begin 1 September. Azeri attacks continue, however.

***September 1992***

**1–3** • Following Kazakh mediation attempts, Armenian and Azeri representatives meet in the Idzhevan border region, sign a protocol calling for the introduction of a ceasefire, and promise Armenian-Azeri dialogue. Despite the protocol, Azeri attacks continue against Armenian population centers, in violation of the ceasefire.

**10** • The fifth round of the CSCE's Rome peace talks on Karabakh ends in stalemate as Azerbaijan continues to refuse to abide by the CSCE ceasefire proposal. American and French delegates to the talks condemn Azerbaijani aggression.

**18** • The Azeri leadership calls on all Armenian forces to withdraw in return for a guarantee of safe passage from Karabakh through the Lochin corridor.

**19** • Armenian, Azerbaijani, Georgian, and Russian defense ministers sign an agreement in Sochi calling for a ceasefire in Karabakh. The agreement also calls for the formation of a peacekeeping force composed of various CIS states and establishes a two-month moratorium on military activity, allowing for a phased withdrawal of military units in the region.

**25** • The Russian-brokered ceasefire comes into effect. It is accompanied by protocols on the withdrawal of arms, prisoner exchanges, and on stationing observers from CSCE.

***October 1992***

**1–30** • Armenian and Azerbaijan fighting over Nagorno-Karabakh persists during October, despite the ceasefire agreements.

**12** • Russia and Azerbaijan sign a treaty of "Friendship, Cooperation and Mutual Security."

**31** • Moscow talks are scheduled for next week between the defense ministers of Russia, Azerbaijan, Armenia, and CIS Joint Armed Forces Commander-in-chief Shaposhnikov. Observers believe Russia and Armenia will press Azerbaijan to consent to dispatching Russian Armed Forces to Karabakh. Observers also believe no consent will be forthcoming as many Azeris believe a Russian presence aggravates the situation.

***November 1992***

**6** • Ter-Petrosyan and NKR parliament chairman Georgiy Petrosyan announce their willingness to enter into a ceasefire without any preconditions. Azerbaijan, however, demands Armenia withdraw its forces from Shusha and Lochin before any agreements are discussed.

**13** • Azerbaijani Defense Ministry reports renewed fighting after some calm in various rayons and towns. They claim Armenian units launched attacks against Azeri positions in Khodzhalinskiy, Gyunkyslakh, Seidlyar, and Kazanchi. The report indicates the offensives were repelled by Azeri forces, who made the units retreat after loosing thirty men.

**2** • As the Republic of Nagorno-Karabakh's parliamentary president Georgiy Petrosyan arrives in Yerevan to meet with Armenian officials, Azeri forces wage artillery attacks on the Karabakh capital of Stepanakert and the district of Askeran. Azeri bombing attacks by Su-25 combat aircraft then follow against the positions around the Lochin corridor connecting Nagorno-Karabakh to Armenia.

5 • Azeri combat aircraft conduct bombing attacks against Armenian villages and the regional center in Martuni district. Republic of Nagorno-Karabakh forces retake several strategic positions near the village of Marzilli.

6 • Armenia border villages in the Askeran and Martuni districts, as well as the capital Stepanakert, are subjected to heavy shelling by Azeri artillery units. Following bombing attacks by Azeri combat aircraft over Martuni, Republic of Nagorno-Karabakh forces shoot an Azeri Su-25 jet fighter and an Mi-24 assault helicopter. A later Azeri offensive against the Askeran district is repelled by Nagorno-Karabakh forces.

8 • Sporadic Azeri attacks against Republic of Nagorno-Karabakh positions around the vital Lochin corridor probe the defenses of the area in preparation for a possible wider attack.

11 • Azeri forces initiate bombing and GRAD multiple-missile artillery attacks on the Armenian districts of Goris and Gapan, killing six and wounding sixteen residents. Armenian Foreign Minister Arman Kirakosyan warns that due to the Turkish cancellation of the recent energy agreement coupled with the continuing Azeri-imposed blockade of Armenia, the Armenian government will be forced to consider reopening its 800–megawatt Medzamor nuclear power station. In Tiflis, an Armenian delegation signs an economic and energy transport agreement with Georgian officials.

13 • Azeri combat aircraft conduct bombing attacks over the Armenian district of Gapan and Azeri artillery units attack the Goris district, resulting in twenty-five residents killed and forty-six wounded.

15 • Azeri armored forces launch an offensive into Armenia, penetrating the border to attack the villages in Armenia's Gapan district. Soon thereafter, artillery attacks begin against Armenian villages in the Megri and Idzhevan districts.

16 • Azeri artillery units shell the districts of Askeran and Martuni. The regional center and villages of Vardashen and Charpaz in Martuni suffer particularly heavy damage. Reportedly, the Azeri attack on Askeran utilizes a new type of rocket, a ten-meter-long S-200 unguided surface-to-air missile, modified for longer-range artillery attacks and not believed to have been in the Azeri arsenal, indicating a recent acquisition. Clashes erupt in the Mardakert district and around the Lochin corridor.

21 • Armenian villages in the Goris and Megri districts are subjected to heavy artillery shelling by Azeri forces. Armenian Vice-President Gagik Arutyunyan heads a delegation meeting with Iranian officials in Tehran. Armenia formally requests that Russia, Ukraine, and Belarus dismiss all Armenian officers serving in their national armies. It is estimated that there are nearly 5,000 Armenian officers serving in these republics.

22 • Following heavy shelling, an Azeri force of tanks, armored vehicles, and combat aircraft launch an offensive in Armenia's Krasnoselsk district. The Azeri thrust penetrates 3.5 kilometers into Armenian territory before being repulsed by Armenian defenders.

23 • Azeri forces mass on the border of the Idzhevan district in preparation for attack. Azeri artillery shelling is resumed in the Gapan district, with heavy damage in the villages of Yekhvard and Agarak.

27 • The Armenian Foreign Ministry issues a statement protesting the Azeri "aggression against the Republic of Armenia and the encroachment on its sovereignty and territorial integrity."

29 • Armenian officials conclude an agreement with Turkmenistan calling for the daily shipment of 11 million cubic meters of natural gas. Armenia's daily basic minimum need is nearly 12 million cubic meters of natural gas. Responding to Armenia's appeal for energy, the Russian government promises to dispatch another 7 million cubic meters of natural gas once the pipeline through Georgia is repaired and operational.

29 • An Azeri offensive of armored vehicles and tanks is launched against the Idzhevan district of Armenia. Azerbaijan and Iran conclude a barter-type agreement calling for the export of 250 million cubic meters of Iranian natural gas to Azerbaijan in exchange for an unspecified amount of Azeri petroleum-related products. The Azeris refuse an Iranian offer to mediate the Karabakh conflict.

30 • The Russian military command, following an earlier accord with Azerbaijan, cedes nineteen border posts to Azeri forces. The frontier posts include eleven in Lenkoran and eight in Prishib.

## 1993

### *January 1993*

3 • Following a general Azeri mobilization of forces and the Azeri imposition of a state of emergency in its Fizuli district, clashes erupt between Armenian forces in the Martuni district and Azeri positions in the bordering Fizuli

district. Armenian villages of Gevorgavan and Yenokavan are heavily damaged. After halting the Azeri attacks, Armenian forces are able to hold their positions in the Martuni district. Armenian positions around the Lochin corridor connecting Karabakh to Armenia are also subject to Azeri attacks.

7 • Closed, unofficial "consultative" talks on Karabakh are convened in Moscow with Russian, Armenian, Azerbaijani, Turkish, and U.S. participation. No representation from Karabakh was present.

9 • Commenting on the joint Bush-Yeltsin statement on Karabakh, Parliamentary Chairman Georgiy Petrosyan welcomes the international involvement in the "aggression of Azerbaijan against Karabakh." The Karabakh leader stresses that international recognition of the NKR will help the conflict's settlement.

11 • Armenia's "Snark" news agency publishes what it asserts is a top secret governmental report on Karabakh. The six-point document states Armenia is ready to negotiate without preconditions except for a refusal to enter agreements recognizing Karabakh as a part of Azerbaijan. The Armenian government's position, according to the document, proposes "recognizing Karabakh as a territory without status" and calls for "negotiations to determine Karabakh's status."

15 • As the blockade intensifies the energy crisis in Armenia, Armenian Industry Minister orders all but seven essential industrial enterprises to close by 1 February. Public transportation shuts down, homes remain subject to severe heating and electricity rationing, and virtually all of the country's industries stand idle.

15 • Azeri government criticizes France, United States, and other Western countries for sending aid to Armenia, claiming it will "enable Armenia to continue its aggression against Azerbaijan."

21 • Azeri government blockades garrisons of Russian Lenkorani border detachments and prohibits Russian flights over Azerbaijan.

18 • NKR State Defense Committee Chairman Robert Kocharyan concludes meetings with Armenian officials in Yerevan, where participants reviewed the situation in Karabakh and discussed upcoming CSCE talks.

• Azeri artillery attacks escalate against positions around the Lochin corridor connecting Karabakh to Armenia. Azeri attacks are also launched against villages in the Martuni and Askeran districts.

25 • Nagorno-Karabakh forces retake the strategic village in the Mardakert district. As NKR forces push deeper into the Azeri-occupied district, a total of eleven Armenian villages are retaken.

*February 1993*

2 • NKR government announces it will introduce a citizenship program for all Karabakh residents including a plan to issue passports and identification papers.

22 • Nagorno-Karabakh forces liberate the village of Aterk, one of the largest in Mardakert, from Azeri units. Karabakh defense forces, following the consolidation of their positions in Mardakert, move to retake the strategic Sarsang hydroelectric and reservoir complex. Fighting spreads to the Azeri stronghold of Agdam with sporadic but intense battles.

25 • In a joint statement coinciding with the CSCE talks on Karabakh, Russia and the United States urge all parties in the Karabakh conflict to cease military action and to enter into peaceful negotiations. A NKR delegation participates in the eleven-nation CSCE talks on Karabakh, convened in Rome after a five-month break.

*March 1993*

2 • Azeri forces continue to retreat from positions in the Mardakert district of Karabakh. NKR self-defense forces, advancing almost daily on Azeri strongholds in the area, are seeking to regain the Armenian villages in the Mardakert district which were seized by Azeri forces in a major offensive last year.

• In Rome, Nagorno-Karabakh representatives participate in the CSCE talks. The participants reach an agreement to send an Advance Group of Observers to the area.

3 • To date, NKR forces have succeeded in recapturing nearly twenty-five Armenian villages in the Mardakert district of Nagorno-Karabakh.

17 • NKR units retake the Armenian villages of Tonashen, Gyulatag, and Dzhanyatag in the Mardakert district. Azeri units renew artillery attacks against the capital, Stepanakert.

18 • CSCE-sponsored talks resume in Geneva seeking a durable ceasefire agreement. The talks also encompass a new bilateral initiative by the Turkish and Russian foreign ministries but are stalemated due to non-inclusion of NKR representatives. The Rome talks included the democratically elected NKR government in a limited but participatory role.

*April 1993*

**1–2** • Nagorno-Karabakh defense forces drive back Azeri troop attacks on the Lochin corridor to the outskirts of Kelbajar. The Kelbajar region for months has been a staging ground and munitions stockpile for attacks against the corridor. The Azeri president introduces a two-month state of emergency, effective for all of Azerbaijan, and initiates a mass mobilization of Azeri citizenry.

**3–4** • Nagorno-Karabakh self-defense forces seize the district center of Kelbajar. More than five thousand Azeris are allowed safe passage from the area. Karabakh forces also advance on the Azeri stronghold of Fizuli, just southeast of Gadrut, and fifteen miles from the Iranian border. Sporadic Azeri attacks continue against Armenian positions defending the Lochin corridor.

**5–6** • Turkish Prime Minister Suleyman Demirel states that Turkey "would not abandon" Azerbaijan and compared the situation to a chess match—"we must play our men carefully to ensure that both Turkey and Azerbaijan emerge victorious." Fighting continues in the mountain passes north of Kelbajar and around Kubatly.

• Armenian forces launch a new offensive in Nagorno-Karabakh, capturing a large part of Azerbaijan territory. The Russian 7th Army, located in Armenia, actively supports Armenians in the battle.

**7** • Azeri President Elchibey formally appeals for direct Turkish military assistance. In a telephone conversation with Israeli Prime Minister Yitzhak Rabin, Elchibey appeals for Israeli diplomatic support in condemning Armenia. To date, there has been no official Israeli position on the Karabakh conflict.

**8** • Armenia aligns itself with foreign countries calling for cessation of combat activities in Karabakh conflict. The government emphasizes that all parties should realize NKR authorities control the situation and organize their self-defense independently, thus any talks require their participation.

• Russian Defense Minister Pavel Grachev brokers a ceasefire agreement, halting military operations by noon on Friday, 9 April. Further talks between the Armenian and Azeri prime ministers are scheduled for 13 April. Turkey deploys additional combat forces along its 156–mile (250–km) border with Armenia and continues daily Turkish Air Force flights through Armenian airspace. Turkey dispatches 10.5 tons of aid to Azerbaijan and issues a strong protest that the United Nations Security Council's statement on the situation in Karabakh was too weak.

• Speaking at a press conference in Switzerland, Nagorno-Karabakh parliamentarian Zori Balayan defends recent actions by Nagorno-Karabakh forces in the Lochin corridor, arguing they were necessitated by continued Azeri military attacks. Baroness Caroline Cox, deputy speaker of Britain's House of Lords, states that Karabakh's action "was prompted by Azeri attempts to cut off the Lochin humanitarian corridor."

**9** • Armenian forces seize the strategically important settlement of Govshatly, cutting off three neighboring regions of Azerbaijan (Kubatly, Zangelan, Jebrail) from the rest of the republic. Armenian forces concentrate large numbers of military hardware in the region.

• Armenian Defense Minister Vazgen Manukyan argues that the latest success of the Karabakh defense forces may bring Azerbaijan to negotiations. As for Karabakh's status, he does not imagine the Republic will remain part of Azerbaijan. Its own army is the only real guarantee of Karabakh's independence. As for the principle of inviolability of borders, the minister believes that the former borders inside the Union did not arise from natural processes but were drawn arbitrarily, hence they cannot be regarded as sacred. Touching upon statements by Baku on Armenia's aggression, he says the republic is not waging war and calls the Azerbaijani side the aggressor because of the blockade and shelling of Armenian border areas.

**10–11** • Fighting continues around Fizuli. Nagorno-Karabakh self-defense forces consolidate their hold over eighteen villages and strategic mountain passes in southwestern Azerbaijan bordering Iran. Azeri armored forces kill thirteen and wound eight as they seize two Armenian villages in the Gapan district. Azeri Ambassador to Iran Nasib Nasibzade formally appeals to Iran for assistance and Azeri State Secretary Panakh Husseinov arrives in Tehran for meetings. A CSCE delegation arrives in the region, meeting with leaders in Baku and Yerevan.

**12** • Yeltsin and Nazarbaev agree to coordinate their activities in connection with the conflict. They will also cooperate with the CSCE Minsk group.

**13** • In Baku, Turkish President Turgut Ozal calls for complete and immediate withdrawal of Armenian forces from Azerbaijan, warning "no one should doubt Turkey's support for Azerbaijan." In the same trip, he proposes a formal Turkish-Azeri military alliance to fully ensure Azeri control over Nagorno-Karabakh.

**14** • The Turkish Foreign Ministry reports Turkey is set to ratify the Turkish-Azeri "solidarity and cooperation" accord signed last November, said to contain a mutual defense clause giving "Turkey a legal basis for helping Azerbaijan." This follows several recent reports in the Turkish press on the transfer of rocket launchers, ammunition, and assorted light weaponry from Turkey to Azerbaijan. The presence of Lt.-Gen. Erdogan Oznal, head of the Turkish Chief of Staff's Special Warfare Unit, in the Ozal delegation fuels further speculation on Turkish military assistance to Azerbaijan.

**16** • Armenian Defense Ministry charges Turkey with sending arms and troops to Azerbaijan via Nakhichevan, warning Armenia might shoot down Turkish transport planes crossing its air space.

**17** • The Organization of the Islamic Conference assures Azerbaijani President Elchibey that its member states are ready to render "moral and material" help to Azerbaijan.

**19** • A two-day ceasefire begins, allowing a visiting CSCE delegation to tour Nagorno-Karabakh and the Kelbajar district. The ten-member delegation will prepare for a 600–member CSCE monitoring mission. NKR parliament confirms it will cooperate with all mediation attempts and urged a new framework of bilateral negotiations to overcome the deadlock.

**21** • While in Ankara for Turkish President Turgut Ozal's funeral, the presidents of Armenia and Azerbaijan meet. An agreement pledging the parties to fully participate in the CSCE mediation effort results. In Istanbul, the Turkish Export-Import Bank reports that it will grant $250 million in aid to Azerbaijan, $50 million dollars in trade credits and the remainder to support Turkish investment in Azeri industry. The Turkish parliament formally ratifies a previously negotiated consular and military agreement with Azerbaijan.

**26** • The CSCE Committee of Senior Officials, responding to an Azeri request, convene an emergency session in Vienna to examine the situation in Karabakh. The meeting, chaired by Swedish diplomat Anders Bjurner, establishes a working group consisting of Armenia, Azerbaijan, Turkey, Russia, and the United States to work toward a durable ceasefire framework. U.S. delegate to the talks John Kornblum condemns the "seizure of Kelbajar by Armenian forces" and rejects the official Armenian position denying any official involvement in support of NKR defense forces. Azeri Ambassador to the United Nations Hasan Hasanov appeals to the UN Security Council to impose "all appropriate sanctions" against Armenia.

**26** • In Moscow, Russian mediators arrange secret negotiations between Azeri officials and representatives of the NKR's State Defense Committee. Although limited and secret in nature, the talks are the first direct Azeri-Karabakh talks and, as such, are a departure from the previous Azeri position of denying any recognition of the NKR. The legitimacy of the Nagorno-Karabakh delegation is in question, however, as it has no official authorization from NKR's parliament.

• Speaking at a Yerevan news conference, Armenian Vice-President Gagik Arutyunyan urges international recognition of the NKR as an effective means toward the "suspension of hostilities."

**27–28** • NKR self-defense forces initiate a partial withdrawal from their positions in the recently secured Kelbajar district between Armenia and Nagorno-Karabakh.

**29** • United States, Turkey, and Russia draw up a new peace plan for Nagorno-Karabakh intended to restart the stalled CSCE-sponsored negotiations.

**30** • UN Security Council unanimously passes resolution calling for immediate halting of hostilities and demands full withdrawal of "all occupying forces," including "local Armenian forces," from the Kelbajar district between Armenia and Nagorno-Karabakh.

***May 1993***

**4** • In a press interview, Azeri presidential advisor Vafa Guluzade characterizes the UN resolution as a great victory for Azeri diplomacy. He hails the fact that for the first time, Armenia is finally recognized as an occupational force and should it refuse to withdraw from the captured regions of Azerbaijan, it is sure to do so under pressure from the United Nations.

**7–15** • A series of gunfire exchanges take place, concentrated mostly around Nakhichevan and in the NKR rayons of Agderinskiy, Fizuliniskiy, and Kubatlinskiy. Armenia says Azerbaijan initiated the fighting in seeking to safeguard its positions while Azeri officials declare Armenia wants to expand the zone of combat to better position itself for bargaining in negotiations.

***June 1993***

**2** • Armenia and Azerbaijan agree to a mediatory initiative of Russia, United States, and Turkey.

**7** • A newly revised CSCE plan on Karabakh is submitted to the Armenian, Azerbaijani, and Nagorno-Karabakh governments. The plan, based on UN Resolution 822, calls for an immediate ceasefire, a phased withdrawal of military forces from Kelbajar, and eventual lifting of the Azeri blockade.

**10** • The Armenian and Azeri foreign ministries announce their acceptance of a CSCE-sponsored peace plan. The CSCE initiative is based on the 30 April UN Security Council resolution calling for a ceasefire, a withdrawal of forces, and the deployment of CSCE observers.

**15** • Azerbaijan National Assembly elects Geydar Aliev as chairman of the country's Supreme Soviet (suspended since May 1992 after the Azerbaijani Popular Front came to power). Yeltsin approves, characterizing Aliev as "an experienced and authoritative politician" who would help improve Russia-Azerbaijani relations.

**16** • Armenian forces in Nagorno-Karabakh take advantage of political chaos and unrest in Azerbaijan resulting from political infighting and launch a new attack.

**22** • Newly installed acting parliamentary Chairman of NKR Karen Baburyan meets with the Armenian president and vice-president to review the situation and the recently endorsed CSCE plan. Speaking at a press conference, Baburyan states that Karabakh is following the foreign policy course set by its December 1991 national referendum, whereby the Karabakh population voted overwhelmingly to seek independence.

**25–27** • Karabakh self-defense forces regain control of the regional center of Mardakert, seized by Azeri forces in a major offensive last year. Azeri artillery units respond by launching rocket attacks on Stepanakert.

**30** • Azerbaijan National Assembly votes to appoint rebel Colonel Surat Husseinov as prime minister. Parliament Chairman Aliev says one of his principle tasks would be "to recapture Azerbaijan's lands," thus calling into question Azerbaijan's commitment to the latest CSCE Karabakh peace plan.
• Advancing on Azeri positions around their stronghold of Agdam, Karabakh forces liberate all but a four-mile strip of land along Karabakh's eastern border. The advance solidifies control over the strategic district seized by Azeri forces in last year's offensive.

*July 1993*

**2** • The Armenian Communist Party Central Committee calls on the Armenian government to extend official diplomatic recognition to the Nagorno-Karabakh Republic and for the simultaneous signing of a bilateral treaty on friendship, cooperation, and mutual security with that republic.

**5** • Kocharyan states his forces have no intention of occupying the Azeri stronghold of Agdam. Nagorno-Karabakh forces down an Azeri combat aircraft on a bombing raid of the Lochin humanitarian corridor connecting Nagorno-Karabakh to Armenia. The pilot is a Ukrainian mercenary. CSCE Special Envoy to Nagorno-Karabakh Mario Rafaeli postpones his scheduled tour of the region due to the escalation of fighting.
• Azeri Prime Minister Surat Husseinov arrives in Agdam with reinforcements consisting of over 2,500 infantry forces and several armored vehicles and tanks.

**10** • CSCE envoy to Karabakh Mario Rafaeli arrives in Baku to advance the CSCE negotiated settlement. In a sign of growing internal discord, over five thousand people demonstrate in Baku in support of ousted President Elchibey. A strike of over fifty municipal government officials is also staged. The first direct flight from Baku to Tabriz (Iran) is inaugurated.

**14–15** • Azeri artillery units violate an informal ceasefire arranged for a visit by CSCE envoy to Nagorno-Karabakh. The barrage includes attacks with S-200 surface-to-air long-range missiles.

**23** • Karabakh forces backed by tanks take the strategic town of Agdam, advancing to within one kilometer of Fizuli.

**24** • Aliev calls on the United Nations to "restrain" Armenian aggression, arguing that "any references by the Armenian side to the fact that the armed forces fighting in Nagorno-Karabakh are not subordinate to the republic of Armenia are without foundation."
• In a significant development, officials from the Republic of Nagorno-Karabakh and Azerbaijan speak by telephone and announce a three-day ceasefire effective at midnight. The discussion is the first high-level recognition of the NKR government by Azeri officials.

**25** • The Turkish Foreign Ministry issues a statement that it will bring pressure to bear on the United Nations and other international bodies to secure a complete withdrawal of Armenian forces from all occupied Azerbaijani territory.

**28** • Based on the successful discussions of 24 July, a meeting between Azeri and NKR officials is held along the border of the Mardakert region of NKR. The three-day ceasefire agreement is extended for another five to seven days and an agreement is reached to pursue further high-level talks between Nagorno-Karabakh Foreign Minister Gukasyan and Azeri Defense Ministry officials.

*August 1993*

**3** • Azeri artillery units launch several attacks against population centers in Armenia's Noemberyan district and in the

Askeran, Gadrut, and Martuni districts of the NKR. Kocharyan arrives in Yerevan for an official two-day visit and is scheduled to meet with officials of the Armenian government.

**5** • Talks between Azeri Defense Ministry officials and NKR representatives lead to another extension of the current ceasefire agreement. Sporadic artillery attacks by Azeri forces, however, continue to threaten the durability of the ceasefire agreement.

**11** • In an official statement by the government of the Republic of Nagorno-Karabakh, Stepanakert leaders call for the United Nations and the CSCE to dispatch independent observers and peacekeeping detachments to Karabakh in an effort to guarantee the safety of the population of the Republic. The statement further calls for direct bilateral talks with Azerbaijan as the proper avenue for initiating a durable mediation process.

**18** • Attacks by Azeri combat aircraft over Armenian's Gapan district leave seven residents dead and thirty-four wounded. Further aircraft attacks are reported in Nagorno-Karabakh, with Azeri bombers dropping cluster bombs on Armenian villages in the Martuni district. Azeri shelling also intensifies in Armenia's Megri, Gapan, and Goris districts.

• Responding to the Turkish and Azeri requests for an urgent meeting of the UN Security Council, the Security Council issues a statement calling for an immediate, complete, and unconditional withdrawal of Nagorno-Karabakh forces from the recently seized areas of Azerbaijan. The UN Security Council further urges Armenia to "use its unique influence to this end."

**19** • Nagorno-Karabakh forces continue their advance on Azeri military positions, seizing the southern Azeri stronghold of Dzhebrail, 14 km from the Azeri-Iranian border. UN observers report the advance forced 30,000–50,000 Azeris to flee eastward into Azerbaijan and south toward Iran. Following a meeting in Moscow with his Armenian counterpart, Turkish Foreign Minister Hikmet Cetin threatens to push for UN sanctions against Armenia if Nagorno-Karabakh forces fail to withdraw from the recently seized areas of southwestern Azerbaijan.

• Republic of Nagorno-Karabakh and Azerbaijani officials agree to a new five-day ceasefire. The two sides also agree to hold a summit to discuss a long-term ceasefire as well as continue their peace talks.

**21** • Russian special envoy on Nagorno-Karabakh Vladimir Kazimirov issues a statement criticizing the current CSCE Karabakh negotiation framework, labeling the plan ineffective and inadequate, and claiming that the CSCE lacks experience in large-scale peacekeeping operations.

**24** • Tens of thousands of civilians continue to flee toward Iran to escape the ongoing fighting in the region of Dzhebrail.

**26** • Negotiations between Azerbaijan Deputy Parliament Chairman Afiaddin Dzhalilov and Nagorno-Karabakh Foreign Ministry fail to arrive at a ceasefire agreement.

• Kazimirov travels to Nagorno-Karabakh at the suggestion of Aliev. After touring southwestern Azerbaijan, Kazimirov meets with officials in Stepanakert to discuss possibilities of resolving the conflict. On 27 August, he flies to Yerevan to meet with Armenian President Ter-Petrosyan and returns to Baku the following day.

***September 1993***

**6** • Aliev and Yeltsin hold talks in Moscow on Nagorno-Karabakh situation, discussing basic principles for settling the conflict. Aliev also states that he gained permission from his parliament to explore the idea of joining the CIS. Grachev says Aliev requested that he use "the authority of the Russian Army, the Defense Ministry, and personal relations with the Armenian leadership" in solving the conflict.

**9** • In a formal letter of protest, Armenia accuses Ukraine of supplying arms to Azerbaijan and asks for an explanation from Kiev. The note stresses that this action "runs counter to the efforts undertaken by the international community to resolve the Karabakh confrontation peacefully and violates Article 10 of UN Security Council Resolution 853." Officials in Kiev deny the charge, saying Ukraine only assists Azerbaijan in equipment restoration.

**15** • Ter-Petrosyan and Yeltsin meet in Moscow to review the regional ramifications of the Nagorno-Karabakh conflict and to discuss Armenian-Russian economic and trade relations. Sporadic artillery attacks by Azeri forces are reported in Armenia's Noemberyan, Idzhevan, and Tavush districts along the border with Azerbaijan.

**17** • A five-nation Commonwealth of Independent States (CIS) interparliamentary delegation, empowered to mediate the Karabakh conflict, arrives in Yerevan to begin a tour of the region. The mediation group consists of parliamentarians from Belarus, Kazakhstan, Kyrgyzstan, Russia, and Tajikistan and was formally constituted under the CIS Interparliamentary Assembly. After meeting in Yerevan, they will fly to Baku to continue their peacemaking mission.

**20** • After several postponements, the Azerbaijani National Assembly votes 31 to 13 for the renewal of Azerbaijan's membership of the CIS.

**23** • Direct bilateral talks between Karabakh and Azerbaijan officials continues with a meeting in the Askeran border district. The two sides review economic needs of Karabakh and the bordering districts of Azerbaijan as well as lifting economic and transport blockades of Karabakh by Azeri forces. The current ceasefire agreement is scheduled to hold until 5 October.

**25** • In Moscow for the CIS summit, Armenian President Ter-Petrosyan and Aliev hold a meeting to discuss the situation in Karabakh and the continuing blockade. The meeting, arranged by the Russian Foreign Ministry, is coordinated by Kazimirov.

***October 1993***

**1** • The ceasefire agreement in effect since July between Karabakh and Azerbaijan is extended for another month following a meeting in Moscow between Aliev and Kocharyan.

**2** • According to a report in a Moscow newspaper, Azerbaijan's agreement to become a member of the CIS will not have a significant impact on the Karabakh conflict. The article states that, "since Armenia conducts no military actions in Karabakh, the military situation there will, as it did before, depend on further progress in the direct talks opened between Stepanakert and Baku rather than on directives from Yerevan."

**11** • Azeri officials announce their rejection of the latest CSCE-proposed timetable for the settlement of the Karabakh conflict. Their main reason for objecting to this schedule is the exclusion of the Lochin humanitarian corridor from the agreement. The deadline for responding to the schedule is extended beyond the previous 7 October date. Both Armenia and Nagorno-Karabakh had previously accepted the CSCE settlement by the original deadline.

**12** • Addressing the UN General Assembly, Aliev states that complete and unconditional withdrawal of Armenian forces from all occupied territory in Azerbaijan is an essential precondition for convening the CSCE Minsk conference on a settlement of the conflict.

**14** • UN Security Council unanimously adopt Resolution 874, supporting the continuation of peaceful negotiations by the Minsk Group within the framework of the CSCE. According to a spokesperson for the Armenian Foreign Ministry, Armenia finds the resolution "acceptable," but is concerned by certain references within the language of the text, specifically that Karabakh is mentioned as part of Azerbaijan, and references to preserving Azerbaijan's territorial integrity.

**18** • In response to a CSCE deadline for a formal reply to its proposed mediation timetable proposal, Azerbaijan rejects the CSCE plan. Armenia and Nagorno-Karabakh previously ratified the plan before the original 7 October deadline.

• Baburyan issues a statement to Boutros-Ghali expressing approval of Resolution 874. The message applauds UN support for direct negotiations between Karabakh and Azerbaijan as initiated by Russia, but expresses dismay over the specific language found in the text, including the use of the definition "Mountainous Karabakh district of Azerbaijan Republic."

**20** • Azerbaijan condemns UN Resolution 874, passed on 14 October, as "unacceptable," offering no explanation.

**23** • The ceasefire agreement in Karabakh is broken by Azerbaijani attacks against the southern Karabakh district of Gadrut and by fighting around Horadiz, Zangelan, and Kubatly. Kocharyan states Karabakh authorities had contacted Azeri and Russian officials demanding observance of the ceasefire agreement following Azerbaijan's seizure of five defensive positions between Fizuli and Dzhebrail. Officials in Baku claim they have no control over various formations in southwestern Azerbaijan.

**25** • Kozyrev blames Karabakh Armenians for recent ceasefire violations, urging them to cease their offensive and withdraw their forces.

**26** • Russian negotiator Kazimirov, accompanying the CSCE delegation to Baku, seeks to counter local disillusionment with the CSCE mediation effort and warns both sides against attempting to scare each other with predictions of an internationalization of the conflict.

**28** • A ceasefire is reestablished due in part to the mediation efforts of Iranian President Ali Akbar Rafsanjani. The Iranian Foreign Ministry previously stated its desire to avoid the conflict, but with Rafsanjani in Baku it once again sought the role of mediator. The negotiated truce is scheduled to commence at 7:00 P.M. local time.

***November 1993***

**2** • Western sources report that Azerbaijan enlisted a force of Afghan special forces, or mujahidin, to bolster its army.

**5** • Karabakh Foreign Minister Arkadiy Gukasyan issues a statement on conditions for a true peace process listing three concerns: the political status of Karabakh, lifting the Azeri-imposed blockade of Armenia, and withdrawing Karabakh forces from Azerbaijan. Gukasyan further states that negotiations between Azeri and Karabakh officials remains the main instrument for peace, citing the bilateral talks as the only means to date of having brokered effective ceasefire agreements.

**10** • President Yeltsin sends envoy Kazimirov to Stepanakert, Baku, and Yerevan to negotiate an end to the hostilities between Armenia and Azerbaijan. The Foreign Ministry also calls for the withdrawal of all Karabakh forces to their positions held before the violation of the 21 October ceasefire.

**22** • Armenia accepts the latest version of the CSCE devised timetable for the Karabakh conflict. Azerbaijan rejects the terms of the mediation plan.

**29** • Armed formations deployed in Azerbaijan's Beylagan region attack Armenian positions in Fizulinskiy rayon. NKR armed forces repel the attack and inflict a strong counterstrike. Despite relative calm in the conflict zone, the Azerbaijani side continues aggravating the situation daily, provoking in planned manner the Karabakh army to take countermeasures.

***December 1993***

**1** • Rakhman Mustafa-Zadeh of the Azerbaijan president's press service says that the non-constructive posture of Armenia is the largest obstacle to the CSCE peace process. He makes it clear that Moscow can and must play a more active part in resolving the Armenian-Azerbaijani conflict. He says that Baku recognizes Russia as a great power having special interests in the Transcaucasian region.

**3** • Kocharyan states the NKR leadership agreed to the CSCE peace plan, hoping that Azerbaijan would reply in kind. He says Azerbaijan rejects not only the schedule endorsed by the United Nations and the CSCE, but also the very idea of the Minsk conference, the aim of which is the peaceful settlement of the Karabakh conflict. Despite the UN provision prohibiting arms supplies to the conflicting sides, Azerbaijan is purchasing tanks, military aircraft, and mercenaries.

**14** • As Azeri attacks intensify against Karabakh forces in the Beylagan district, northeast of Fizuli, sporadic artillery attacks are launched against Armenian population centers in the Idzhevan and Kazakh districts of Armenia. Later Azeri artillery attacks are initiated from positions within Nakhichevan.

**16** • Russian officials broker a ten-day ceasefire between Karabakh and Azeri forces. The ceasefire, following days of intense clashes in the district of Beylagan, allows for retrieving the dead and wounded. Violations occur the next day, however, when Azeri troops attack the Martuni region of Nagorno-Karabakh.

**28** • Karabakh forces are forced to retreat from mountain positions surrounding Agdam, following fierce fighting that resulted in heavy casualties on both sides. The battle is attributed to the presence of Afghan mujahidin and Azeri orders to shoot deserters fleeing from the front line. Fighting is also continuing in Mardakert rayon.

**30** • As the Azeri offensive continues, attacks erupt in Karabakh's Askeran, Martuni, and Mardakert districts. Azeri artillery attacks are further initiated against population centers in Armenia's districts of Idzhevan, Tavush, and Kazakh.

## 1994

***January 1994***

**4** • A relatively successful ceasefire along the Azerbaijani-Armenian border for the past several weeks is upset when Azeri artillery units launch attacks against the villages of Movses and Aygepar in the Tavush district.

**12** • Nagorno-Karabakh military sources report a large buildup of Azerbaijani military forces along the southeastern Karabakh borders. Additional concentrations of Azerbaijani military units are reported in the northern district of Shahumyan and along the vulnerable district of Mardakert, a longtime target of Azeri artillery and armored attacks.

• The CSCE's working group on the Nagorno-Karabakh conflict, the "Minsk Group," convenes in Vienna without official representation from parties to the conflict. They review the increasingly tense situation in and prepare for a tour of the area by CSCE Minsk Group Chairman Anders Bjurner later in the month.

**18** • In meeting with advisors, Aliev reveals that the large-scale military buildup of Azeri forces along its borders with Nagorno-Karabakh is aimed at exploiting the internal discord erupting in Armenia. Aliev reportedly plans a military offensive to regain the political initiative from his opposition within Azerbaijan.

**31** • Aliev rejects the latest Russian peace plan for Nagorno-Karabakh one day after Armenia accepted the plan, which provides for Russian troops to enforce a two- to three-week ceasefire starting 1 February. Aliev tells Kazimirov, plan architect, that he would not talk about a ceasefire while "the aggressor continues to occupy Azerbaijani territory." Speculation is that Aliev is hesitant to halt hostilities while Azerbaijan is slowly regaining lost territory.

***February 1994***

**2** • Newly released population statistics indicate the Karabakh population is steadily returning to its preconflict level. Chairman of the Karabakh Commission on Refugees and Humanitarian Aid Lenston Ghoulian reports that nearly 25,000 refugees returned to their homes in Nagorno-Karabakh for 1994. The birthrate also increased as housing is being reconstructed, living conditions improving, and the economic situation somewhat stabilizing.

**2–9** • Continuing their massive offensive launched in mid-December, Azeri forces maintain coordinated attacks against the Mardakert and Martuni districts of Karabakh as well as the strategic Kelbajar district between Armenia and Karabakh. With the continued aid and training from Turkish military units, the Azeri army is increasingly mounting impressive tactical gains against Armenian positions defending Karabakh. The significant role of a detachment of 1,000 Afghan mujahidin greatly bolsters the Azeri military capabilities.

**10** • A survey of Karabakh citizens reveals that 55.2 percent still favor independence for Karabakh. Some respondents opted for Karabakh's reunification with Armenia and some suggested Karabakh's entry into the Russian Federation.

**18** • Organized by the Russian Foreign Ministry, defense ministers of Armenia and Azerbaijan meet in Moscow to negotiate a durable ceasefire agreement. Talks center on the Russian proposal for an immediate ceasefire, redeploying of all armed formations from the conflict zone, returning foreign mercenaries, and the eventual monitoring of the agreement by Russian forces.

• Karabakh forces repel Azeri attacks on the Armenian villages of Tonashen and Mataghis in the northeastern part of the Mardakert district in Karabakh. The Azeri offensive continues their advance against the Kelbajar district between Armenia and Karabakh. Heavy fighting is reported at the strategic Omar Pass, a vital mountain pass linking Kelbajar with Khanlar, which returns to full Armenian control following the complete consolidation of its mountainous heights.

***March 1994***

**3** • A large-scale Azeri offensive is launched against Karabakh's southeastern border. The attacks, coinciding with the scheduled arrival of a CSCE delegation to the Karabakh capital Stepanakert, allow the Azeri troops to advance partially around the town of Fizuli. Artillery bombardments target Armenian population centers in other districts of Karabakh.

**4** • A Moscow newspaper reports that the government of Azerbaijan has officially rejected the latest Russian-proposed peace plan. The plan would have called for the creation of a demilitarized zone between the front lines of the warring parties and for the introduction of military observers and international peacekeeping forces. The plan called for a simultaneous retreat of Armenian and Azerbaijani troops to establish this zone. Baku refused to pull back its troops, citing the occupation of Azeri territory by Karabakh troops.

**9** • An unattributed radio commentary broadcast on Radio Baku International Service editorialized that Russia is eager to establish a military base in Azerbaijan and to have a hand in the resolution of the Karabakh conflict. Aliev, however, opposes the idea and demands Moscow to comply with international agreements and laws prohibiting Moscow to act as an internationally sanctioned peacekeeper in the former Soviet Union.

**15** • Kocharyan declares before the Ministers' Council that "The Republic of Mountainous Karabakh is still supporting immediate cessation of the bloodshed and peaceful negotiations on regulation of the Karabakh conflict. Based on this, the Karabakh side supported the recent peacemaking efforts of the CSCE and Russia, though not all of the topics were acceptable to the RMK." Kocharyan stresses a durable ceasefire can only be attained with international observers and peacekeeping forces.

• Baburyan condemns all peace initiatives to the conflict, stressing that only representatives of the democratically elected government of Karabakh are empowered to conduct negotiations. The statement clarifies Karabakh's hesitation over the Armenian-Azeri meeting recently held between the parliamentary chairmen of the two countries.

**17** • Kocharyan briefs Armenian officials in Yerevan on the latest Azeri attacks in Karabakh. Meanwhile, over Karabakh, an Iranian C-130 "Hercules" military-transport plane crashes in mountainous terrain north of Stepanakert, killing all aboard. An Iranian delegation is immediately dispatched to investigate the circumstances of the crash. While Azeri officials charge that the plane, although seriously off course, may have been shot down by Karabakh forces, a number of contradictory reports emerge.

18 • Following a tour of Karabakh, Russian parliamentarian and leader of the Russian National-Republican Party Nikolay Lysenko urges Russia to recognize the independence of Karabakh and cites the common geopolitical interests of Armenia, Russia, and Karabakh.

25 • Talks in Moscow between representatives of Armenia, Azerbaijan, and Karabakh, under the auspices of the Russian Foreign Ministry, conclude with a review of a Russian draft ceasefire proposal. The Russian plan, based on the 18 February ceasefire agreement between the parties, is to be further revised and then submitted to each of the three parties' governments for ratification. The CSCE is also actively involved in this Russian mediation effort.

26 • Azeri forces escalate their attacks against Armenian villages around the southern town of Fizuli. Azeri artillery units initiate bombardments of the Armenian-held eastern town of Agdam, followed by Azeri air assaults on the area.

31 • A CIS interparliamentary delegation begins meetings with Azeri, Armenian, and Karabakh government officials. Previous CIS efforts have included a ceasefire agreement last September, which although violated by Azerbaijan, effectively established a durable peace for several weeks.

*April 1994*

1 • The Republic of Nagorno-Karabakh and Azerbaijan State Commissions on Hostages and Prisoners of War organize and initiate an exchange of prisoners. Six Azeris are freed in exchange for one Armenian prisoner.

1–3 • A delegation of the Interparliamentary Assembly of the CIS begins touring the region in "search of a mechanism" to mediate the Nagorno-Karabakh conflict. The CIS delegation is to meet with various officials in Stepanakert, Yerevan, and Baku during their tour.

11 • A four-day meeting of the CSCE working group on Nagorno-Karabakh, chaired by negotiator Jan Eliasson, convenes in Prague. The meeting, with a Karabakh delegation participating, produces a twenty-three-point "confidence-strengthening" document of humanitarian measures including a call for the free passage of humanitarian aid to Armenia through neighboring Turkey.
• An Azeri attack against the village of Tonashen in the northern part of Karabakh's Mardakert district results in several casualties. The Karabakh forces advance in their counterattack, retaking the village of Talysh in the Mardakert district and securing the village of Gulistan in the northern Shahumyan district.

19 • Armenian Parliamentary Chairman Babken Ararktsyan arrives in Stepanakert to meet with Nagorno-Karabakh officials. They review the latest Russian peace proposal and coordinate next month's CIS interparliamentary meeting of Armenian and Azeri delegates.

23 • In a speech to the CIS heads of state, Aliev urges CIS states to condemn Armenian aggression against Azerbaijan, and apply sanctions if it does not end. At the meeting CIS heads of state issue a statement asking the international community, including the CSCE and the United Nations, to support actions taken by the CIS to end the conflict.

25 • Meeting in Moscow to discuss the Russian mediation effort for Nagorno-Karabakh, Azeri Parliamentary Chairman Rasul Guliev and Kozyrev announce an agreement of "unspecified modifications" to the latest Russian peace plan. Kazimirov criticizes the Azeri position for its refusal to recognize Nagorno-Karabakh as an equal party in the peace plan. A further round of meetings between the Russian defense minister and the Azeri delegation seeks to reach some consensus on a preliminary agreement.

26 • Despite diplomatic progress, Azeri armored forces attack the southern Karabakh district of Martuni and nearby Fizuli and the northeastern Mardakert district. Karabakh self-defense forces hold their positions.

28 • Responding to the latest proposal of the Russian peace initiative, NKR authorities accept the plan, based on the 18 February accord reached among the defense ministers of Armenia, Azerbaijan, and Nagorno-Karabakh under Russian coordination. Karabakh's acceptance follows yesterday's announcement of Azerbaijani assent to the plan. According to the terms of the Russian proposal, a ceasefire is to enter into force on 29 April, followed by a staged withdrawal of armed combatants 20 km from the region, and the eventual creation of a "buffer zone" between the sides, to be enforced by a contingent of CIS peacekeepers. The next stage of the plan mandates a lifting of Azerbaijan's blockade of Armenia and Karabakh and calls for the return of refugees to their homes. Negotiations over Karabakh's future political status is to take place during this third stage. This delay in addressing Karabakh's political standing, however, is a fundamental obstacle to achieving a durable resolution to the Karabakh problem, according to NKR representatives.

*June 1994*

1 • Shelling and sporadic attacks by Azeri artillery units over the border in Nakhichevan follow a previous incident

in late May which broke the relative calm along the Armenia-Nakhichevan border. Armenian border guards retaliate and destroy the newly formed Azeri base immediately across the border from the village of Gyounoupi.

**7** • Grachev's ceasefire plan, entailing a predominantly Russian peacekeeping and monitoring force, is amended. According to reports, Russian troops will now constitute only about one-third of the CIS force. The troops will remain in Azerbaijan no longer than six months. Meanwhile, Britain offers to send a detachment of troops under CSCE auspices to monitor a ceasefire in Nagorno-Karabakh.

• Kazimirov completes a tour of Baku, Stepanakert, and Yerevan, where he negotiates with all three sides a peaceful settlement to the six-year war. Kazimirov blames the lack of progress on the inability of the parties to compromise and urges all sides to dispense with "maximalistic aspirations." As regards peacekeeping troops, Kazimirov states that the forces would be "observers with corresponding control and security bodies" and they "would not have disengagement functions."

**8** • Guliev signs the Bishkek protocol at the IPA session, which formalizes the ceasefire and disengagement of forces. Azerbaijan originally refused to sign the Bishkek protocol in early May, but has abided by its conditions.

**10** • Aliev acknowledges that peacekeeping forces are necessary to achieve a lasting peace settlement of the Nagorno-Karabakh conflict. He insists, however, that all troops introduced as peacekeepers operate under the CSCE mandate.

**14** • A meeting is held along the Armenia-Nakhichevan border in an attempt to resolve the recent border skirmishes and reach a peace agreement. Armenia proposes establishing a trade center along the border. The Azeri side refuses, arguing it is not possible while the two sides are at war.

**18** • With meetings between the CSCE negotiator, Jan Eliasson, and Armenian officials under way in Yerevan, Azeri forces launch an artillery attack against Armenia's Noemberyan border district, violating the 15 May ceasefire agreement between the two countries. The Azeri forces follow with an attack against an Armenian border post near the village of Barekamavan, killing six border guards, injuring one, and kidnapping another.

**24** • Six years of war, blockade, and economic collapse force many Armenians to seek relief. Russia's Ambassador to Armenia Vladimir Stupishin reports about 20 percent of Armenia's 3 million people have left the country in the last two years. Some 10,000 Armenians have registered with the Russian embassy signaling their desire to obtain Russian citizenship and emigrate to the Russian Federation.

**25** • An Armenian-Greek center opens in Nagorno-Karabakh to unite and help the Greek residents of Karabakh. The center was founded by a Greek inhabitant of the Mardakert village of Mehmania, where Greeks have been living since the nineteenth century.

• Kazimirov holds talks with a senior Iranian official regarding Russia's peacekeeping initiative in the contested territory. The Iranian side reportedly approves Russia's plan as "close to the requirements of reality." Kazimirov also meets with Azerbaijani President Aliev in Baku. Aliev reaffirms his desire that the conflict be solved on a multilateral basis, combining the efforts of the CSCE, Russia, Iran, and Turkey.

**28** • Kazimirov announces that the basis for the political settlement of the Nagorno-Karabakh conflict is practically ready, pending agreement on complicated wording necessary to assuage the conflicting parties. He reaffirms that the 12 May ceasefire in the conflict zone is holding.

***July 1994***

**3** • Azeri forces initiate an attack against the Armenian border village of Dara in the Vardenis district. The eight-hour attack is eventually repulsed by Armenian border guards.

**5** • Azeri artillery units resume with sporadic shelling against the Seisoulan and Talish villages in Karabakh's Mardakert district. Artillery attacks then target the villages of Shourabad and Shotlani in the Agdam district. The attacks violate the standing ceasefire agreement between both sides.

**7** • Chief-of-staff of Turkey's armed forces Dogan Gures travels to Baku for talks with Aliev. Gures reaffirms his country's willingness to participate in a multinational peacekeeping force in Nagorno-Karabakh under the auspices of the United Nations or CSCE and denies Armenian press reports that Turkey is supplying arms to Azerbaijan.

**8** • In response to Gures's offer, Russian Defense Minister Grachev says that Russia would oppose any unilateral move undertaken by Turkey in the Transcaucasus—a region of vital interest to Russia. A Russian Foreign Ministry official says that the deployment of Turkish peacekeepers would unnerve Armenia and thus serve to exacerbate tensions. Armenian President Ter-Petrosyan concurs, stating that only Russian peacekeeping troops could ensure stability in Karabakh. Turkey's ambassador to Azerbaijan, however, says that Baku retains the prerogative to decide which country to invite to send peacekeepers.

**9** • NKR officially rejects Turkish troops from participating in the peacekeeping contingent since Turkey is not a neutral player in the conflict. Evidence of Turkey's active participation includes its blockade of Armenia, supply of weaponry and military training to Azerbaijan, and the existence of Turkish prisoners of war.

**12** • Guliev says an understanding was reached on a potential settlement of the Karabakh conflict and a peace agreement could be signed by the end of July. The announcement came following a CSCE parliamentary session in Vienna where Guliev met with U.S. and Russian representatives and with the chairman of the CSCE Minsk Group.

**15** • Ter-Petrosyan press secretary Levon Zurabyan says Karabakh Armenians offered to withdraw troops from all occupied lands, except the Lochin corridor in exchange for security guarantees for its population. Zurabyan also says that Armenia will reject Turkish participation in any peacekeeping or observation force as long as it remains a "party to the conflict by blockading Armenia."

**19** • Kazimirov will visit Baku, Yerevan, and Stepanakert to consult on draft agreements for settling the conflict. Russia has prepared a fundamental draft agreement containing sixteen articles, a diversified set of addenda, and a time frame for troop withdrawal. According to Kazimirov, there is no alternative to the Russian draft, and attempts to create one (i.e., the CSCE Minsk Group) serve only to impede a resolution to the conflict.

**26** • In Baku, Kazimirov states the latest Russian peace plan offers a more realistic settlement. He says the sixteen-point Russian plan, unlike the CSCE proposal, envisages establishing a secure demilitarized buffer zone and deployment of international observers and CIS peacekeeping forces to monitor a durable ceasefire.

**27** • In a news conference, Vann Ovaisyan, member of the bureau of the opposition union of the Armenian Revolutionary Federation, commented positively on the creation of Russian military bases in Armenia. He rejects the possibility of Armenia's becoming a part of the USSR (in the event of its restoration) or any other such empire.

• Aliev staff member says Azerbaijan supports Russia's deployment of peacekeeping forces in Nagorno-Karabakh on the condition that Armenian forces withdraw from Lochin and Shusha and that a political settlement on the status of Nagorno-Karabakh is reached. The staff member envisioned a force of 3,000–6,000 troops, of which between 60 and 90 percent would be Russians, operating under a CSCE mandate.

• Defense ministers of Armenia and Azerbaijan and NKR military commander sign agreement legalizing the ceasefire accord reached in May. The three sides also pledge to work toward an agreement on the deployment of peacekeepers. In Nagorno-Karabakh officials reacted with lukewarm enthusiasm to the agreement, acknowledging that the document "creates effective preconditions for stopping the armed conflict," but warning of the difficulty of maintaining peace without the presence of an international disengagement force.

***August 1994***

**1** • Kazimirov attributes the relative success of the ceasefire to the realization by all parties of the rising human and economic costs of conducting war. He warns of certain forces in Armenia and Azerbaijan seeking to exploit the conflict for their own political gain, alluding to the power struggles being waged in Armenia and Azerbaijan. He also expresses his disgust for those "criminal Mafia structures who are getting fat on this war, making money from the sale of fuel, oil, and guns."

**3** • Azerbaijan receives deliveries of ammunition, including artillery shells and anti-tank mines, from Turkey. In addition, a new group of Turkish military instructors arrives in Azerbaijan to train recent Azeri recruits.

**4** • Russia and Armenia are close to agreement on permanent Russian military bases in Armenia. A Russian Defense Ministry spokesman proclaims "it will be first Russian base on the territory of another state." Russia's 127th Motorized Infantry Division is currently stationed in Armenia. This group is expected to be supplemented with air combat forces, anti-aircraft defenses, and radar stations.

**6** • A month-long Human Rights Watch mission to Armenia, Azerbaijan, and Nagorno-Karabakh found that regular military units subordinate to the government of Armenia played a significant role in the Karabakh Armenians' offensive launched last spring. According to HRW, the units in question often served to secure communication lines, thus freeing up Karabakh Armenian forces to continue the offensive.

• In a press interview, Hasan Hasanov highlights many obstacles to settling the conflict. The first is that many countries are unwilling to recognize Armenia as the aggressor. Armenia's true goals are the second obstacle. For example, Armenia has no intention of liberating all the territory occupied outside of Nagorno-Karabakh. It is refusing to liberate the Lochinskiy and Kelbadzharskiy districts. Armenia is unwilling to give back Shushinskiy district, which, although on the territory of Nagorno-Karabakh, was populated in the vast majority by Azeris.

**9** • The major task of Russian foreign policy in the CIS countries—to form a common defense space—has met with stubborn Azerbaijani resistance as Azerbaijan is the only member of the CIS that does not have Russian soldiers on its territory. However, Russia will not leave Azerbaijan as easily as it did the Baltics. Since Aliev will not sign the Russian peace plan for Karabakh, the ascension to power of people loyal to Moscow cannot be ruled out. Neither can the "Georgian version" be excluded, under which Russian troops would be greeted in Baku as saviors.

**13** • Negotiations on a formal settlement end without significant progress made. Azerbaijan reportedly retracts an earlier pledge to recognize the Nagorno-Karabakh Republic as a party to the conflict. The Baku delegation continues to link negotiations on the future political status of Nagorno-Karabakh to the withdrawal of Armenian troops from Azerbaijan. Karabakh Armenians refuse to retreat from the strategic town of Lochin.

**16** • Surat Husseinov says Azerbaijan will discuss political settlement for Nagorno-Karabakh if Armenia guarantees the safe return of the 200,000 Azerbaijani refugees to their native lands, currently occupied by Karabakh Armenians. He also called on the Armenian forces to vacate Shusha and Lochin. If these conditions are not met, "Azerbaijan will be ready to recover its own territory."

**20** • Negotiations between Russian and Azeri Defense Ministry officials end without an agreement on border security and the placement of frontier troops. Russia seeks to secure a joint presence of troops to enforce Azerbaijan's increasingly vulnerable borders.

**24** • Husseinov says in interview that the ceasefire will not last forever and that Azerbaijan should use this time to strengthen its army and improve military preparedness. He continues that Azerbaijan must first liberate its seized territory in order to achieve peace and that any concessions by Azerbaijan over the status of Nagorno-Karabakh would be "an extremely dangerous step."

**30** • Armenia and Karabakh will unite their monetary systems and the Armenian drum will be the sole legal tender in Nagorno-Karabakh. When asked if the agreement would provoke any negative political repercussions, the chairman of the Armenian Central Bank replied that it would not likely "affect politics." However, an Azerbaijan foreign policy advisor denounced the agreement as "another provocation aimed at thwarting a peace settlement."

***September 1994***

**6** • Azeri units based in Nakhichevan launch artillery attacks against Armenian positions across the border. The shelling targets Armenian defensive positions in the area of Yeraskhavan and precedes another Azeri attack from within Azerbaijan against the village of Vahan in the Krasnoselsk district. The previous night, the Tavush district along the Armenian-Azerbaijani border was shelled by Azeri units.

**8** • Armenia's Ministry of Foreign Affairs issues statement condemning the bombing of civilian targets in Armenia and places all responsibility on the Azerbaijani government. The ministry also links the bombing to Azerbaijan's recent hard-line stance on Nagorno-Karabakh negotiations.

**9** • Regional government officials announce the return of 250 refugee families to their homes in Mardakert district villages. NKR government is providing these families with free transportation to the area, as well as food provisions for an initial period of their return.

**12** • A meeting of the eleven-nation CSCE Minsk Group on Karabakh convenes in Vienna to review the situation and prepare for a larger summit in Prague. The CSCE criticizes the latest Russian mediation effort for failing to include a prominent role for the CSCE and accuses Russia of adopting a dangerous unilateral approach. The Russian delegation responds by boycotting the remainder of the session.

**13** • Azeri artillery attacks are launched against the villages of Levonarkh and Shourabad in the Mardakert and Agdam districts. Units of the Turkish paramilitary "Grey Wolves," deployed in the region as part of Turkey's military aid to Azerbaijan, advance to Nagorno-Karabakh's eastern border.

**14** • The CSCE's Committee of Senior Officials convenes in Prague to discuss the situation in Nagorno-Karabakh, criticizing the Russian mediation plan for contradicting the CSCE's own peace initiative. The Turkish delegation also criticizes the Russian plan and calls for the deployment of its own troops as part of an international peacekeeping contingent. Although pro-Azerbaijan, the Turkish delegation refuses to support any peace plan without Turkish military involvement.

**22** • Ter-Petrosyan blames a power struggle between Russian and CSCE mediators for the delay in reaching a solution.

The mediators "seem less interested in the conflict than in the distribution of power after it is settled."

**26** • Azeri artillery units launch attacks against the villages of Aghabekalendg, Levonarkh, and Talish in Karabakh's Mardakert district. It is also revealed that a new force of 200 armed members of the Grey Wolves organization has been dispatched from Turkey in preparation for a new Azeri offensive and to train units of the Azeri army. Nagorno-Karabakh officials warn this development, as well as Turkish assistance in continuing Azeri attacks, may greatly damage the current ceasefire in effect and hinder ongoing mediation efforts by Moscow and the CSCE.

**28** • Ter-Petrosyan says Armenia neither occupies any Azerbaijani territory nor harbors any ambitions to do so. Rather, Ter-Petrosyan contends it is the Nagorno-Karabakh Armenians who, in the name of self-determination, prosecute the war with Azerbaijan.

**29** • Aliev addresses UN General Assembly, calling on the world community to pressure Armenia into withdrawing from Azerbaijani territory. Aliev guaranteed the safety of the Armenian population of Nagorno-Karabakh and mentioned (unspecified) concessions on their status within Azerbaijan.

***October 1994***

**6** • Kazimirov announces Russia's frustration with the CSCE Minsk Group's mediation efforts for Karabakh, claiming the CSCE is manipulated by some countries to legitimize their geopolitical interests in the region, rather than as a proper vehicle for true conflict resolution. The CSCE effort is also competing with a Russian mediation effort which is more in line with Russia's increasingly activist strategy for the states of the former Soviet Union, or "near abroad."

**12** • A CSCE delegation led by Deputy Chairman of the Minsk Group Pierre Anderman arrives in Yerevan after holding a series of meetings in Baku. The group meets with Foreign Minister Vahan Papazyan to discuss the next steps in the peace process. Talks center on maintaining the current ceasefire, increasing the dialogue and trust between representatives of the conflicting parties, improving cooperation between Russian and CSCE negotiators, and addressing the logistics and make-up of a peacekeeping contingent.

**19** • Following earlier meetings between the Russian and Armenian defense ministers, Russia announces deployment of a squadron of MiG-23 combat fighter-interceptor aircraft to Armenia. The deployment is part of the CIS collective defense treaty and seeks to solidify the regional air defense system. Russian officials are seeking to establish five military bases in the Transcaucasus: two bases in Armenia and three in Georgia.

**24** • During the closed CSCE session in Vienna, Russian delegation head Vladimir Shustov demands the CSCE agree that Russia lead in handling the Nagorno-Karabakh conflict. Russia holds that the eleven-nation CSCE Minsk Group's performance has been dismal to date and the 1992 formation of the CSCE group empowered to handle the Karabakh question is flawed by its lack of legal standing.

**31** • Aliev says Azerbaijan will negotiate the political status of Nagorno-Karabakh only after Armenian troops withdraw from Azeri territories. Aliev favors restoring the demographic situation in Nagorno-Karabakh that had existed in 1988. According to the official census reports there were about 70 percent Armenians in Nagorno-Karabakh. At present there is not a single Azerbaijani on the territory of the former autonomous region.

***November 1994***

**7** • Armenia expresses willingness to open its railways for transportation loads from Azerbaijan to Nakhichevan and agrees to traffic humanitarian cargoes from Turkey destined for Azerbaijan, via Armenia. Azerbaijan has been blockading Armenia for over two years while Turkey has refused transportation of even humanitarian aid for Armenia for the past year and a half.

**16** • In a meeting with CSCE Minsk Group Chairman Anders Bjurner, Aliev agrees to the CSCE proposal for the deployment of international peacekeepers to the region, but remains adamant about the necessity for retaking the Lochin humanitarian corridor connecting Karabakh to Armenia as well as the city of Shusha. To date, Karabakh officials have agreed to the deployment of CSCE peacekeepers to a limited area along the border between Karabakh and Azerbaijan.

**17** • Yeltsin and Aliev discuss the Azeri position regarding Karabakh, as well as the complicated issue of the oil deal signed between Baku and a consortium of Western oil companies. Yeltsin stresses the urgent need for a political solution to the Karabakh conflict and informs Aliev that the NATO Partnership for Peace security group was seriously considering offering a peacekeeping force for the region, a proposal reported to be ready for the upcoming CSCE summit in Budapest.

***December 1994***

**6** • According to the Armenian president, the conflict cannot be resolved until Karabakh is recognized as a sovereign country and before the disengagement forces are deployed in the conflict zone.

## 1995

### *January 1995*

**1** • Azerbaijan's Democratic Congress views Russian control of Azeri borders as "unjustified economic sanctions against Azerbaijan." The Congress views such actions aimed at CIS member states as indicative that Russia views the CIS as a "puppet organization" to pursue its neo-imperialist policy.

**2** • Armenian Defense Minister Vazgen Sarkisyan expresses satisfaction over the republic's "close" military cooperation with Russia. Responding to comparisons of Chechnya with the Nagorno-Karabakh conflict, Sarkisyan said that it is "Russia's internal affair" though "Russia's actions toward Chechnya are similar to the steps once taken by Azerbaijan toward Karabakh and by Georgia toward Abkhazia."

**3** • In Baku, Turan Abbas Ali Novrusov, commander of border troops in Azerbaijan, says "we are a sovereign state and there is no necessity for our borders to be guarded by the military men of other states" (referring to a Russian guard on the Azeri-Iranian border). He emphasizes "there can be only technical help by Russia for reliable guard of Azerbaijan's border."

• State Advisor for Foreign Politics Vafa Guluzade notes the "stumbling stone" of negotiations is Armenian liberation of Shusha and Lochin. The issue prevents signing of the Major Political Agreement (MPA), providing the schedule and mechanisms of liberation of the occupied Azeri territories. Armenia has refused to liberate Shusha and Lochin. OSCE's peacekeeping forces can be sent to the region only after signing the MPA.

**4** • Azerbaijan's Democratic Congress states "Russia's position toward the Muslim conflicts in Nagorno-Karabakh, Bosnia, and now on the Northern Caucasus testifies to its religious intolerance and intention to turn these conflicts into religious confrontations."

**6** • Kazimirov reiterates that for the beginning of the OSCE peacekeeping operation in the conflict zone, the MPA must be concluded, appropriate resolutions of the UN Security Council adopted, and the operation's planning, financing, and duration determined.

**9** • Kocharyan signs a decree on 9 January creating an NKR Defense Ministry. Maj.-Gen. Samuel Papayan is appointed Defense Minister.

### *March 1995*

**14** • An article in Moscow's *Segodnya* says an oil pipeline crossing through the Nagorno-Karabakh region could ease the conflict in Azerbaijan. Author Armen Valesyan notes the prospects for energy supplies and dividends paid for transit provides both Armenia and Nagorno-Karabakh with ample incentives for compromise. Valesyan also argues the pipeline could become a guarantee for future peace in the region, as Western companies with oil operations in the Caucasus would have a vested interest in political and social stability in the conflict zones and in the republics themselves.

**15** • Azerbaijan's internal affairs minister orders disbanding of OMON (special purpose police unit), noting the unit has become uncontrollable. Personnel have been ordered to hand in their weapons to the authorities. The situation remains calm and unit members do not believe force will be used against them.*

• In Baku, Azerbaijani government troops surround OMON special police headquarters. Fighting between the two factions also breaks out elsewhere in the country as Deputy Interior Minister Rovshan Dzhavadov calls for the resignation of President Aliev and the Azerbaijani government. Dzhavadov says Aliev and the government have brought the country to the "verge of ruin."

**20** • Government troops and Interior Ministry forces hold off a coup attempt, but dozens are killed and wounded. Aliev implicates former President Ayaz Mutalibov and former Prime Minister Surat Husseinov as masterminds of the operation in which Aliev was to be assassinated. Some two hundred arrests are made.

**21** • A new organization forms in the NKR, "Democratia." The group will work to increase voter participation in elections and to ensure democratic elections. Through mass media and holding political dialogues with other organizations and political leaders, the group hopes to create a legal state in the NKR.

---

*The Azerbaijani special police apparently caused a deterioration in the republic when it seized the building of the Prosecutor-General's Office on 3 October 1994 and demanded the resignation of Prosecutor-General Ali Omariv, who was subsequently dismissed. The special police also demanded the resignation of Parliamentary Chairman Guliev and Interior Minister Usubov. As a result of the Baku events and an abortive attempt by OMON to remove the legitimate authorities in Gyandzhe, Prime Minister Surat Husseinov is removed from his post for charges of maintaining contacts with the rebels and aiding their efforts. Aliev labels the events an attempted coup.

**27** • Aliev says to OSCE officials that Azerbaijan will uphold the ceasefire. "Status-quo does not suit us, we are going to obtain full peace." OSCE delegates Vladimir Kazimirov and Anders Bjurner, returning from meetings with Nagorno-Karabakh representative Arkadiy Gukasyan, tell Aliev it is necessary to keep the ceasefire and urge him to speed up the exchange of prisoners and hostages.

**30** • Demonstrators in Yerevan seek immediate resignation of Armenian President Levon Ter-Petrosyan and the Armenian cabinet. The opposition proposes that a coalition government of national accord be formed until the Constitution is adopted and a government established following 5 July 1995 parliamentary elections.

***April 1995***

**6** • In a press interview, NKR President Robert Kocharyan states his government is still pursuing international recognition for the NKR and doubts OSCE negotiations will be enough to solve the conflict, given Azerbaijan's present stance. He also states the internal power struggle in Azerbaijan will do little to decrease its resolve to "preserve its territorial integrity" given it has changed leaders eight times since 1988 and its position on the NKR remains unaltered.

**8** • An article in Baku's *Zerkalo* editorializes that the recent Russian-Armenian treaties on friendship, economic cooperation, and Russian military bases in Armenia signals Russia is trying more than ever to regain its control over the Transcaucasus and that the military-political union the treaties forged between the two eliminates Russia as a neutral party in negotiations over the conflict's settlement.

**11** • Russia's State Duma conducts hearings on the NKR conflict, held on the eve of the "Eightieth anniversary of the Genocide of Armenians." Armenian Ambassador to Russia Yuri Mkrtumian is present but his Azeri counterpart telegraphed the State Duma two days ago indicating his belief that the hearings would have a pro-Armenian tendency that Azerbaijan thus could not take part.

**17** • In a press interview, Karabakh Supreme Council deputy Maksim Mirzoyan argues the oil contract "is the main obstacle" preventing NKR recognition by Azerbaijan. He holds that despite the fact that the contract is still "just a paper," it is a paper giving millions of dollars to the Dudaev, Gamsakhurdia, and Aliev regimes. Thus, Aliev continues fighting knowing Armenia will have to give if it expects concessions from the oil deal. Fellow Karabakh Supreme Council deputy Vasiliy Aghajanyan notes that "no Azeri oil tricks" will weaken the NKR's resolve to reach independence, noting it is not the Republic of Armenia who will determine the NKR fate but the people of Nagorno-Karabakh.

**18** • Armenian government offers to withdraw its troops from the disputed territory in exchange for ending the Azeri blockade of Armenia. Azerbaijan Foreign Ministry welcomes the proposal but suggests agreements to exchange prisoners should solidify before withdrawal. Baku also announces its willingness to grant the NKR self-rule within Azerbaijan but NKR forces continue demanding full independence.

**21** • Hasanov charges present OSCE Chairmen Vladimir Kazimirov and Anders Bjurner with being ineffective in resuming stalled talks. Beginning 21 April, Finnish diplomat Heikki Talvitie will assume the chairmanship of the OSCE Minsk Group, a move supported by Hasanov.

**27** • NKR First Deputy Prime Minister and Minister of Economy Spartak Tevosyan calculates $2.5 billion in damage to the NKR as a result of "Azeri aggression." About 8,000 industrial, agricultural, and construction objects, 7,000 buildings, 172 educational, and 47 medical institutions suffered as a result of enemy shellings. In addition, several cultural monuments were destroyed, according to the minister.

***May 1995***

**3** • A Russian Federation delegation lead by Federation Council CIS Affairs Committee Chairman Vadim Gustov arrives in Yerevan for an eventual tour of the NKR conflict zone. Gustov says the delegation is to help the Russian parliament and the CIS Interparliamentary Assembly "define the roles they can play" in the conflict's settlement, insisting that any Russian role will not favor a particular side.

• In meetings with Turkish Foreign Minister Erdal Inonu, Aliev says Azerbaijan is leaning on Turkey to held bring the NKR conflict to an end. He hopes that Turkey will support Azerbaijan within the framework of the OSCE and in overcoming economic difficulties as a result of the war. Inonu responds favorably, saying Turkey advocates the withdrawal of Armenian troops, restoration of Azerbaijani territorial integrity and that Turkey will assume more responsibility for ensuring peace and security in the region.

• The CIS Affairs Committee delegation meet with Aliev. Aliev commends the group for their efforts and encourages them to promote continued talks on a settlement. The only issues remaining, according to both the NKR and Azerbaijan regarded the exchanging of prisoners, Armenian troop withdrawal from Shusha and Lochin, and the political status of the NKR. Aliev says mediation efforts should focus on bringing each side's positions closer.

**12** • Marking the first anniversary of Karabakh-Azeri truce, Armenia releases twenty-seven Azeri prisoners who returned to Baku on an Armenian plane accompanied by an International Red Cross delegation. Aliev responds by announcing Azerbaijan will unilaterally release all POWs in Azerbaijan and other grounds.

**15** • Moscow-based OSCE negotiations fail to produce an ultimate resolution to the conflict. OSCE representative Rene Nyberg comments that the talks passed in a "business-like and open atmosphere" but no radical progress was made. Key problems discussed during the talks remain an Armenian presence in Shusha and Lochin, unaccounted-for POWs, and the political status of the NKR. Armenian, NKR, and Azeri representatives fail to agree on the order in which these issues should be tackled.

**25** • Armenian Foreign Ministry announces suspension of its activity in the OSCE Minsk Group in connection with an Armenian gas pipeline bombed a few days ago. Azeri State Advisor for Foreign Affairs Vafa Guluzade denies Azerbaijani involvement in the pipeline's destruction, saying the suspension is an Armenian and Russian attempt to sabotage OSCE efforts that do not conform to Russian interests in the Caucasus. Another round of talks is scheduled for 15 June in Helsinki.

***June 1995***

**2** • Armenia announces it is reentering the OSCE-sponsored talks. In Yerevan, Ter-Petrosyan holds talks with Iranian Deputy Foreign Minister Mahmud Va'ezi. Va'ezi expresses pleasure that talks will resume and underlines his country's opposition to resuming the war. Va'ezi pledges his government will work to ensure a durable peace and return stability to the region.

• Azeri parliament lifts the state of emergency invoked 4 October 1994.

**15** • OSCE-sponsored talks begin in Helsinki with delegations from Armenia, Azerbaijan, and the NKR present. Two draft approaches to settlement are introduced, one proposing a stage-by-stage process of withdrawal and the other envisaging an all-encompassing solution to be effected once all sides narrow key differences spelled out earlier. (See entry for 15–20 May.) Once agreement is reached on a settlement approach, peacekeeping operations will be launched.

**23** • Armenian Deputy Foreign Minister Vardan Oscayan labels the latest Helsinki talks as "constructive," saying that despite continued discord between Azerbaijan and Armenia over Karabakh security, the key problems were discussed in detail and the sides appear closer to agreement.

**28** • Gulizade announces a major step was taken in the Helsinki talks when Azerbaijan and Armenia agreed in principle to previously disputed issues. "A major political agreement is not far off," hailed Gulizade. He highlights the fact that the NKR will assume special status within Azerbaijan and that the ultimate resolution agreement will be arranged so that no side will feel like the winner or the loser.

**29** • Armenian press agencies report Azeri attacks in northeast Armenia, violating the ceasefire. The Armenian Defense Ministry say Armenian defensive positions were fired on around Verin-Champarak and Nerkin-Champarak in Krasnoselsk district by Azeri units hoping to draw Armenia into a wide-scale war. The NKR Public Information Office reported no incidents and said all sides were observing the ceasefire.

***July 1995***

**5** • In a press briefing, Armenian Deputy Foreign Minister Vardab Oskanyan states the NKR has attained "de-facto self-determination" and is in "full control of the external and internal situation." He says the main goal of the Minsk Group is to tackle the problem of NKR's political status.

**13** • NKR Press Secretary Semyon Afian charges Azerbaijan with making unrealistic demands before the Minsk Conference begins—demands unacceptable to the NKR and Armenia and thus preventing the talks from being held. Specifically, Afian says, Baku's demands that peacekeeping forces be deployed between Armenia and Azerbaijan isolate the NKR from Armenia, its only source of food and other supplies. Baku also demands NKR forces withdraw from their positions before the conference begins. Afian asserts this would position Azeri units in "choice front lines" from which they could resume military operations against Karabakh forces. Azeri requests that Turkish troops take part in the international peacekeeping forces is also termed unacceptable by Afian because Turkey is decidedly pro-Azerbaijan.

• OSCE Minsk Group Co-chairmen Heikki Talvitie and Valentin Lozinskiy arrive in Baku for discussions with Aliev. They focus on the key problems noted in entry above, but according to both co-chairmen, no progress is made. Azeri Foreign Minister Hasanov says Azerbaijan will be ready to "sign a political agreement after Armenian forces withdraw and indigenous people return to their homes."

• From Baku, the delegation went to the NKR for talks with its leadership. Delegation head Talvitie emphasized to Kocharyan that the conflicting sides "have no other way but to agree to compromises." Talvitie says these compromises can be worked out in talks in Austria set for 24 July.

**20** • Newly elected Chairman of the NKR Supreme Council Karen Baburyan announces the NKR will proceed on developing its statehood. The immediate goals of the new parliament include forming a strong army, creating a developed economy, and cultivating positive foreign relations with neighbors and abroad.

**24** • Azeri-Armenian talks open in Baden, Austria. Aliev restricts the topics of the Azeri representative (FM Hasanov) to only the liberation of Lochin and Shusha and the return of refugees displaced from these regions.

***August 1995***

**2** • In a press conference, NKR Foreign Minister Gukasyan discusses results of Baden talks, saying they revolved around two concerns. First, NKR security was discussed and the NKR delegation proposed an involved Armenian role by either giving Armenia a mandate to interfere in further Karabakh-Azeri conflicts or by stationing Armenian soldiers in the NKR. The second concern was raised by the Azeris and again dealt with liberating Shusha and Lochin. Gukasyan laments the talks produced little because the Azeris will not address any other issue until liberation is achieved. As to OSCE arrangements for a peacekeeping deployment, this will occur when the sides reach political consensus. More talks are scheduled in Moscow in late September, according to Kazimirov.

**7** • Turkey appoints Kemal Ayhan—ex-Turkish ambassador to the USSR—as Turkish special envoy to the Karabakh talks. In discussions with Aliev, he vows to maintain Turkey's support for Azerbaijan in the conflict's settlement, saying "occupied Azeri territories must be liberated. . . . This is our principal position and we will not change it."

**14** • Fighting continues along Armenia's northeast border with Azerbaijan in the Noemberyan region. Relative calm is reported on all other fronts.

**20** • Hasanov claims Russian OSCE Minsk Group Cochairman Kazimirov "does not understand the tasks of the mediators." He criticizes him for again trying to "dodge the basic issues of the settlement, notably the liberation of occupied Azerbaijani lands."

**31** • Armenian Deputy Foreign Minister Vardan Oskanyan says tensions are again rising as another round of talks is scheduled for 4 September in Moscow. The deputy minister says tensions always flare when talks are due to resume and indicates Azerbaijan is most unhappy with recent developments since settlement of the conflict is of "vital importance" for them. He predicts no groundbreaking results to occur given the sides' positions on Lochin and Shusha. For the Armenians, the future of Lochin is the "fundamental guarantee of the security in Nagorno-Karabakh," according to Oskanyan.

***September 1995***

**4** • Moscow-based talks are held. Participants continue to discuss resolution of the Lochin and Shusha regions, still "occupied" by Armenian forces. The OSCE, the United States, Russia, and delegations from Armenia, Azerbaijan, and the NKR take part in the talks. Possible security alternatives to the NKR without Armenian control of Lochin were proposed.

**5** • The OSCE announces plans to establish permanent representation in the NKR settlement proceedings. The decision comes after a visit to both Baku and Yerevan by OSCE special envoy Andre Erdesh. In talks with Ter-Petrosyan, Erdesh believes OSCE participation at this new level will make the overall peace initiative more effective.

**12** • The U.S. Embassy in Azerbaijan clarifies the U.S. position regarding the conflict. In a statement sent to the Azeri leadership, the United States confirmed:

—the United States respects the territorial integrity of Azerbaijan;
—the United States does not recognize the self-proclaimed NKR;
—the United States does not recognize Kocharyan as the president of any institution;
—the U.S. mission is the restoration of peace in the region and will continue talks with those involved in the issue to further this process.

**14** • Oskanyan holds a news conference to discuss results of the talks, which ended on 9 September. He indicates Armenia achieved a number of favorable provisions in agreements signed with Azerbaijan dealing with the geography of the confrontation zone, the disengagement of fighting units, and on disarmament procedures. He argues that the ceasefire has been holding up relatively well over the last year and a half, allowing for political conditions "quite conducive for the peaceful resolution of the problem."

## Summary at the End of 1995

The Azeri side persists in its refusal to discuss any political agreement until Armenian troops leave the Lochin and Shusha regions while the Armenians contend their presence in the territories is necessary to maintain food and material supplies to the NKR, isolated as a result of the Azeri blockade still in effect. Moreover, Armenia believes that Azerbaijan will resume intense military operations should these troops withdraw. Fighting still continues in many areas along the Azeri-Armenian border. Analysts predict some concessions may be forthcoming on the part of Azerbaijan as oil negotiations come closer to completion.

# The Moldovan/Trans-Dniester Conflict

## Background Note

Moldova is the only CIS state that has confronted a Russian army stationed on its territory and supporting a separatist Russian insurgency against its state. This insurgency has been in progress since the autumn of 1989, when the pro-Soviet, hard-line communist community of Russian patriots living on the left bank of the Dniester River in eastern Moldova proclaimed a "Trans-Dniester Soviet Socialist Republic." In 1991, the enclave renamed itself the "Trans-Dniester Democratic Republic" (TDR). In claiming its independence, the self-proclaimed republic appropriated 15 percent of Moldova's territory, and formed its own separate government and national guard. Following Moldova's declaration of independence on 27 August 1991, the TDR rejected Moldovan sovereignty, refusing to participate in the political processes of the country. This disagreement evolved into armed conflict in the late fall of 1991, and entered a critical stage in January 1992 over the control of Bendery, the TDR's largest industrial center. In July 1992, the Moldovan and Russian presidents formed a Joint Supervision Commission in the Dniester area to oversee an official ceasefire and negotiations on the issue of the withdrawal of Russian troops from Moldova. Today, the fighting has slowed to a trickle, but the political status of the TDR is still being contested by Moscow, Moldova, and the president of the TDR, Igor Smirnov.

Moldavia was annexed by Russia from Romania in 1812 and again in 1940. Many Moldovans still speak Romanian and Moldovan nationalists, including the pro-Romanian Popular Front, sympathize with the political goal of reunification with Bucharest. Following the demise of the Soviet Union, Trans-Dniester secessionists feared that Moldova might be swallowed up by a "greater Romania." To forestall this possibility, they imposed linguistic and cultural Russification policies on the region under their control, and proclaimed the political goal of becoming part of a restored Soviet Union or a Greater Russia.

The postcommunist Russian government preserved the Gorbachev regime's policy of privately supporting the separatist ambitions of the TDR, mainly because it considered the region a useful base for the 14th Army. Beyond this, however, many deputies in Russia's Supreme Soviet, including Vice-President Aleksandr Rutskoy, openly extended political and economic support to the secessionists in 1991, identifying TDR inhabitants as a Russian minority whose rights required defending. Notwithstanding its behind-the-scenes assistance and support, however, the Russian government stopped short of officially recognizing the independence of the self-declared republic.

The former Soviet Union's 14th Army in Moldova, made up of more than 10,000 troops, has played an active role in the TDR's insurgency movement. Created for possible action in the Balkan peninsula in 1945, under the command of the Southwestern Military Theater, the 14th Army was once a formidable combined force composed of armored, artillery, tactical, missile, air defense, Spetsnaz, chemical, air reconnaissance, and engineering units.[1] Under the former Soviet Union's command, the army was stationed partly in Ukraine, partly in Moldova, and was a component of the USSR's Odessa Military District. When the USSR fell, Ukraine was able to assume control of its portion of the 14th Army. Moldova, on the other hand, was not as forceful. In February 1992, Russia unilaterally announced its control of the Moldovan units.

With the help of the 14th Army, the Trans-Dniester insurgents also formed and trained separate paid forces of their own, calling them a "Republican Guard." Numbering about 8,000 men, this unit is actually an army of experienced, skilled veterans of the USSR armed forces. Its officers come mostly from the 14th Army under "loan" or "transfer" arrangements. (It is easy to imagine how difficult it could be to arrange a genuine "withdrawal" of Russia's 14th Army when many soldiers could conveniently transfer into the Trans-Dniester "Republican Guard.")

Over and above the guard units, Russia has posted a Spetsnaz (special forces) unit, several internal security troops, and "border troops" in eastern Moldova. Added to these are several thousand Cossacks and other veterans of the USSR armed forces who came to Moldova in 1992 to oppose the new government and to make sure that Moldova did not join Romania. These recruits have been provided with residences and are being called "local inhabitants" by the TDR authorities.

The TDR has an economy designed to support military activities. It is an important producer of defense-related products, once forming an important sector of Moldova's economy. Working in the defense factories are highly skilled military reservists, numbering in the thousands, many of whom have combat experience gained in Afghanistan.

The commander of the 14th Army in Moldova for two years was General Aleksandr Lebed. Lebed established firm control over the Trans-Dniester Republic, in effect reestablishing Russia's influence over the entire country. Though Moscow frequently denies that Lebed was following Moscow's orders, Lebed maintained control while Moscow sought to legalize the troops' presence in the TDR by pressuring Chisinau to "host" the troops on its territory. In June 1993, Moldova was first on Moscow's list of newly independent states in which Russia sought basing rights, a demand which President Mircea Snegur rejected out of hand.

From 1992 to 1995, while Russia did not reject outright Moldova's demands for withdrawal of its troops from the TDR, it stalled the withdrawal by stipulating conditions to which Chisinau could not agree. Principally, it called for granting the TDR "special" political status, which Chisinau accepts, but the two have not been able to agree on how to define this status. Russia refuses to withdraw its troops until a political solution broadly acceptable to the TDR is found.

A "Russian-speaking" enclave on the left bank of the Dniester, which remained pro-Soviet and pro-Union after the collapse of the USSR and which is supported by Russian ultranationalists, is of particular concern to Ukraine. Like Bucharest, Kiev has every reason to try to prevent the formation of an independent Trans-Dniester state. In September 1995, when Presidents Snegur and Smirnov met to discuss the political status of the TDR, the Ukrainians participated as mediators. Romania, however, was excluded due to sharp objections on the part of the TDR.

In sum, the protracted Trans-Dniester conflict is about much more than the protection of minority Russian rights in Moldova. The 14th Army also has a much longer and more complex history than merely that of a unit of "peacekeeping troops" in a troubled area of the newly independent states. Russia's perceived national security interests are partly at stake, as are those of the new states of Moldova and Ukraine. The following chronology explores, month by month, the remarkable political interplay between the political leaders and officials of Chisinau, Tiraspol, Moscow, Bucharest, and Kiev.

### Note

1. Vladimir Socor, "Russia's Army in Moldova: There To Stay?" *RFE-RL Research Report,* 18 June 1993, p. 43.

## Chronology of Key Events

## 1992

### *January 1992*

**3** • Slavic separatist forces in Trans-Dniester seize the Ministry of National Security in Bendery.

**8** • "Dniester Republic" paramilitary forces fire on the headquarters of the Bendery police. The Bendery city soviet (now under the control of communist forces) announces the formation of its own military to oppose the Moldovan police.

**9** • "Dniester Republic" Supreme Soviet resolves to subordinate all ex-USSR troops stationed there (about 20,000) to "Dniester Republic" authority, to triple officers' salaries, to institute an oath of allegiance to the "Dniester Republic" and introduce military conscription.

• "Dniester Republic" paramilitary forces attack a Moldovan convoy, seizing equipment and armaments of a former Soviet MVD battalion which Moldova was transporting back to Chisinau. Moldovan soldiers are beaten and tied up, under orders not to return fire.

**10** • In a meeting with FSU troop commanders of FSU troops in Moldova, Snegur reasserts his country's aim to form a 12,000–man republican army. He emphasizes that all officers and non-commissioned officers, regardless of nationality, are welcome to join, and offers full social guarantees. Only those wishing to serve in the Moldovan army will take an oath of loyalty.

**12** • At a demonstration in Moscow sponsored by Russian nationalist and communists, several hundred marchers break down the gate of Moldova's mission, overrun the compound and chant slogans in favor of the "Dniester Republic." Snegur cables Yeltsin complaining of a "violation of diplomatic immunity . . . by adversaries of the Alma-Ata accords." Despite the incident, Snegur pledges to promote CIS relations and closer Moldovan-Russian relations.

**13** • According to the Hungarian Foreign Ministry, Moldova is in full compliance with international standards in observing the rights of ethnic minorities.

• Snegur cables Yeltsin again to complain about Russian mercenaries "taking part in armed raids" with the "Dniester Republic" armed guard. Snegur notes that the "Dniester Republic" is a "bridgehead of the reactionary forces of the former USSR which oppose the Alma-Ata accords, try to preserve the empire, and seek revenge for their defeat in the August 1991 putsch."

**15** • In his inaugural address, Snegur announces that the cornerstone of his policies will be "the observance of human rights." He pledges to build a government "capable of ensuring cooperation of all political forces" in the country.

**16** • The "Dniester Republic" issues several laws on defense, which serve to maintain the existing force structure of the army units on its territory, financing them from its budget and employ the officers and NCOs on contract to the "Dniester Republic," with significant pay increases.

• Deputy Commander-in-Chief of the CIS Armed Forces Col.-Gen. Boris Pyankov meets with Moldovan officials in Chisinau to discuss the status of the 14th Army in Moldova. He rejects Moldova's demands that the 14th Army command be disciplined for its participation in the bloody events in the "Dniester Republic." He also asserts that Moldova has no juridical rights over left-bank FSU units (i.e., the 14th Army), only over those on the right bank.

**20** • During a meeting of the All-Army Officers Assembly, CIS Commander-in-Chief Shaposhnikov announces that the 14th Army has been subordinated to the Ground Forces and that Col.-Gen. Boris Gromov, perceived as a reactionary leader in the armed forces, has been named its commander.

• Minister of Internal Affairs Ion Costas issues an appeal to the International Helsinki Federation accusing the "Dniester Republic" leaders of "trying to artificially create a confrontation between the Moldovan and Russian-speaking population," through "armed provocations" against Moldovan authorities and their families. The appeal claims the "Dniester Republic" leaders "are talking in the language of force and ultimatums," and are supported and armed by "reactionary circles in Moscow." The minister pleads for international attention to the events on the left bank.

• "Dniester Republic" guards once again seize CIS military equipment due to be transferred to Moldovan control. CIS officers and soldiers make no attempt to interfere with the seizure.

**22** • Snegur reports to a visiting international delegation that Yeltsin has failed to respond to both cable requests that the Russian President intervene to curb the Russian-sponsored insurgency in eastern Moldova. Snegur said that Chisinau would appeal to the United Nations and other international NGOs to intervene in Moldova if Yeltsin refuses to help.

**23** • Moldova's parliament votes to introduce the country's own currency, the Moldovan leu, to replace the ruble.

**25** • One day after a joint Moldovan-Romanian "Council of Union" issued a proclamation calling for the restoration of the "Romanian national unity state," Snegur meets with his Romanian counterpart, Ion Iliescu, on the Moldovan side of the border. The topic of reunification is not broached; rather the two announce that a "treaty of friendship and cooperation" will be signed in March.

**26** • In an interview, Snegur reasserts that a "large majority of Moldovans prefer independence after centuries of control by Russians, Turks, and even the Romanians." Snegur said that the thought of reunification remains "a very long time away."

**31** • Moldovan police in Bendery are beat by Russian crowds and "Dniester Republic" guards.

***February 1992***

**1** • A Moldovan kolkhoz chairman from the left bank is shot and killed by "Dniester Republic" "border guards" near Dubasari. Two days later a Moldovan police officer is killed by these same border forces in Dubasari.

**11** • Prime Minister Muravschi announces that Moldova will not contribute financially to the CIS general-purpose forces, rather only to those forces on Moldovan territory that have pledged support for the republic. He also states that the future status of the 14th Army will be decided in talks between Moldova and CIS military.

**12** • The chairman of the "Dniester Republic" Defense Committee states that in the event of an attack by Chisinau, the "Dniester Republic" would commandeer the large arsenal in the area (i.e., from the 14th Army) in order to "arm the entire adult population . . . and prepare an Afghanistan here."

**18** • The United States and Moldova formally establish diplomatic relations.

**21** • According to Moldova's defense minister, Ion Costas, the Moldovan army is to be created from the remnants of the 14th Army located on the right bank of the Dniester. Units located on the left bank will be given the status of an army on the territory of an independent state, and will be withdrawn to Russian territory in eighteen to twenty-four months. According to Costas, this was a preliminary agreement of talks with CIS Deputy Defense Minister Boris Pyankov in Chisinau.

**24** • Snegur signs a decree stating that all residents of Moldova would be offered Moldovan citizenship regardless of ethnicity or Romanian language ability.

**25** • The Moldovan MFA sends a note to the Russian MFA protesting the presence of Russian Cossacks in the "Dniester Republic."

***March 1992***

**2** • Dniester guardsmen and Russian Cossacks surround the Dubasari district Police Department demanding the personnel to abandon or face destruction. The district police is staffed mainly by Moldovans who would be asked to assist the Moldovan government in resisting separatist fighting. The guardsmen and Cossacks, backed by armored personnel carriers, capture the building and take positions on all bridges and roads on the left bank of the Dniester. Late in the day, Smirnov imposes a state of emergency in Dubasari. Snegur describes the event as a "provocation aimed at disrupting Moldova's admission to the United Nations and the CSCE."

**3–15** • Fighting intensifies as TDR guardsmen and Russian Cossacks seek to solidify their positions in and around Dubasari and in controlling all roads and bridges to the left bank of Dniester. Moldovan government forces and security units sought to repel the Dniester advances and retain control of lost areas. Heavy fighting occurred in Dubasari and attacks on legitimate power structures, terror against civilians, and repeated bombing incidents blowing up bridges occur throughout Dniester. Both sides, TDR forces and Moldovan government forces, repeatedly try to capture 14th Army weapon and equipment cashes near Tiraspol but are mostly held in check by CIS forces.

**6** • Snegur telegrams Yeltsin to again firmly protest against Russian Cossacks' support of TDR separatists. Another telegram is sent to Ruslan Khasbulatov, Supreme Soviet of the Russian Federation chairman, challenging Khasbulatov's proposals that Russian committees and commissions be formed to discuss the TDR situation. Snegur says such deliberations violate the Alma-Ata Declaration pledging signatories to abide by the principles of observing sovereignty, equality, and non-interference in internal affairs.

**13** • In a Moscow meeting with CIS government heads, Moldovan Prime Minister Muravschi denounces TDR leaders for having isolated the left bank of the Dniester from the rest of the republic and for "refusing to have contacts with the Moldovan leadership." He proposes a resolution of the conflict by forming a single province of all left-bank rayons in Dniester and awarding the new province the status of free economic zone.

**15** • Moldovan government demands all armed formations in the Dniester left bank "voluntarily surrender their arms to the authorities." Should armed groups not comply, Moldova will take the required measures "to protect citizens . . . from banditry." The groups are given forty-eight hours to comply.

**20** • Snegur says Moldova will appeal to Romania for support should the danger of civil war not drastically diminish. TDR separatists, fighting especially against greater integration between Romania and Moldova, believe the move would further polarize the situation and fighting would grow more intense.

**23** • A quadripartite group is formed in Helsinki talks. The group is composed of experts from the Romanian, Russian, Ukrainian, and Moldovan foreign ministries established to coordinate efforts to seek a settlement to the conflict.

**31** • In remarks to the Moldovan parliament, Snegur laments the fact that TDR leaders are unwilling to resolve the conflict by peaceful means. He says that in rejecting Moldovan proposals for meetings, a ceasefire, and for free economic zone status, the TDR leadership has "chosen the road of bloodshed."

***April 1992***

**1–3** • As fighting continues, the quadripartite group holds talks, labeled by one Romanian official as "productive." The group issues a communiqué whereby the opposing sides vow to settle the conflict exclusively by political means, stress the importance of human rights, and recommend steps to prevent the escalation of bloodshed.

**1** • TDR Supreme Soviet declares illegal Snegur's imposition of a state of emergency that includes Dniester. The deputies also "instruct" Smirnov and other government leaders to appeal to the officers of the 14th Army that they act as "guarantor of the Dniester region." Moldovan officials increasingly question the 14th Army's neutrality as detachments from the army repeatedly violate Moldovan airspace, preparing for CIS aircraft landing in Tiraspol military fields. Frequent visits to Tiraspol by Russian generals without notifying the Moldovan Defense Ministry further alarm the Moldovan government.

**4** • Snegur warns that 14th Army involvement in the Dniester conflict "would be a catastrophe on a scale hard to predict." Army Assistant Commander Vladimir Baranov says the army is not defending left-bank separatists but assumes the role of safeguarding the weapons dumps and army installations, necessary because of the massive supply of arms and equipments left over from Soviet rule. Baranov stresses, however, that many officers,' NCOs', and servicemen's families are being attacked and "when their fathers, mothers, brothers, and sisters . . . beg them to defend their native land, it is hard to restrain people."

**6** • Romanian Minister of National Defense says that although Romania is not involved in the Dniester conflict and is not supplying arms and matériel to Snegur's government, "if Moldova puts the request to us, we will undoubtedly meet it." He emphasizes such a request would be perfectly legal since Moldova is a UN member and takes part in the Helsinki process and thus could buy supplies to "defend its territorial integrity."

• At the urging of Russian Vice-President Aleksandr Rutskoy, the Sixth Congress of Russian People's Deputies adopts resolution "on assistance to ensure human rights in the Dniester area." In his remarks, Rutskoy stresses that Russia must protect Russians wherever they live and alluded to the Dniester conflict as an area where ethnic Russians were suffering from discrimination. He urged Russia to guard against these abuses and suggests employing Russian military resolve to such end.

**7** • Snegur and Moldovan Parliament Chairman Aleksandr Mosanu individually telegram the Russian leadership protesting Rutskoy's message and charging him with "slander." Both believe the speech was aimed to assure Dniester separatists that they would enjoy support from Russian communist parliamentarians and also signals renewed Russia's intentions to rebuild its former empire which "should put the sovereign and independent states on the alert." The telegrams were sent to Yeltsin, Khasbulatov, the Sixth Congress of People's Deputies, and CIS heads of states and parliaments.

**8** • In a press interview, former Moldovan Prime Minister and Christian-Democratic Front of Moldova Chairman Mircea Druc speak favorably about Romanian-Moldovan unification, saying "we can do anything to make Romania whole again, on the condition it is something constructive and creative." He also discusses the simple aims of the National Council for the Union (which he chairs): "Our role is to convince as many people as possible to demand the union." Druc laments the fact that the "great powers" may not want to see a united Romania with 300,000 inhabitants "in the center of Europe and at the mouth of the Danube."

**9** • Moldovan protests against the Russian resolution continue. Moldovan Vice-President Victor Puskas charges the Russian position is a "gross interference in the domestic affairs of an independent and sovereign state, . . . a fruitless attempt to continue the old policy of the imperial center vis-à-vis the formerly rightless national fringelands." Many note that Rutskoy's statements encouraged Russophobia and in fact make the 14th Army an occupational force. Massive meetings and demonstrations take place, escalating tensions and hostility.

**10–15** • Russian Foreign Minister Kozyrev holds a series of meetings with Ukrainian, Moldovan, and Dniester leaders. Talks centered mostly on a Russian peacekeeping role in the Dniester conflict. Kozyrev defends the Russian position that only Russian peacekeeping forces alone can level the conflict and that the 14th Army already in Moldova has served this function. Moldovan and Ukrainian representatives espouse different views than Kozyrev in talks, mainly challenging Russian neutrality and viewing the process in Moldova as "orchestrated by conservative forces from Moscow." Snegur asserts that the Russians intend "to create a reliable outpost in Ukraine's rear" for use should Ukrainian and Russian relations enter serious conflict.

**20** • Smirnov forms a Dniester army by presidential decree, inviting in particular ex-officers of the 14th Army to join the 12,000–man-strong force.

**21** • In a press interview, TDR Vice-President Aleksandr Karaman argues that events in the Dniester region are preventing the unification of Moldova and Romania and thus serve to preserve Dniester's autonomous status. He clarifies that the Dniester region would agree to federation status in the republic of Moldova with equal rights, but since key governmental posts in Moldova are held by representatives of the People's Front—which advocates Romanian-Moldovan unification—the TDR must remain on guard.

**23** • In a press interview, Ukrainian National Assembly member Dmitriy Korchinskiy reveals that a small paramilitary organization under National Assembly auspices is fighting in the Dniester conflict on the Dniester side. Korchinskiy says the group supports the "ideological and moral motives" of Ukrainians in Dniester.

**28** • Conflict settlement talks deadlock in Chisinau as Moldova rejects a Dniester proposal to make Dubasari a peace zone and to withdraw armed formations there. No date is set to resume talks. The quadripartite commission begins work on establishing ceasefire monitoring units to be dispatched along the conflict zone.

*May 1992*

**2** • Heavy fighting continues and four Cossacks are killed in Cosnita from a mine explosion. Smirnov declares the TDR will observe the ceasefire but seeks the status of an independent state for Dniester. The quadripartite commission (Moldova, Romania, Ukraine, and Russia) also pieces together a conciliatory commission composed of representatives from Moldova's and Dniester's parliaments and governments with talks set for 5 May.

• Pressures on Snegur to hold a referendum on unification with Romania continue as 3,000 Moldovan soldiers appeal to Snegur to hold such vote. The appeal, published by various media sources, says not holding the referendum contradicts Moldovan national interests.

**5** • The conciliatory commission holds talks in Bendery. The Moldovan side is represented by First Deputy Prime Minister Konstantin Oborok, who maintains Moldova simply seeks to preserve its independence and territorial integrity. The Dniester side, led by TDR Vice-President Aleksandr Karaman, holds that nothing can be discussed until all armed formations are withdrawn from the Dniester region. Some third-party participants believed both sides should first focus on solidifying the ceasefire and in separating the combatants. Both of these goals have been achieved in Bendery and in Dubasari, where fighting has indeed been minimized and seven ceasefire monitoring posts have been established.

**8** • A new ceasefire agreement is signed in Bendery by conciliatory commission representatives. The document imposes an all encompassing ceasefire in all of Dniester, provides guaranteed control of the armistice, phases disengagement of armed forces, and provides observers to establish more monitoring posts in heavy combat zones.

**12** • Responding to new evidence that the 14th Army continues its support for Dniester separatists, Snegur again telegrams Yeltsin in protest. Dniester paramilitary formations are heavily equipped with 14th Army weapons, equipment, ammunition, telecommunications, and other military hardware. Snegur believes the Russian decision to unilaterally place the 14th Army in the Dniester region under Russian jurisdiction negatively affected the Dniester situation as separatists continue fighting because they know support is forthcoming from Moscow.

**13** • The TDR plans to open missions in Moscow and Kiev, despite a proclamation by the Moldovan parliament that this is unconstitutional. TDR officials moved forward with the plan after recent trips to both capitals by Dniester parliamentarians. The Russian Foreign Ministry says no such thing "would happen at the official level."

**15** • The first wave of eighty military observers arrives in Bendery. The group is composed of twenty observers from each commission member (Russian, Romania, Ukraine, and Moldova) and will focus on meeting the goals of the 8 May ceasefire agreement. Fighting heats up in Bendery and Dubasari, two areas where tension had subsided.

**16** • In a press conference after visiting the conflict area, Moldovan parliament president Alexandru Mosanu argues the conflict "could be settled peacefully if chauvinistic and pro-imperial forces. . . . did not interfere in our domestic affairs," referring to Russian hard-liners who consistently argue for a Russian presence in Dniester and are believed to support Dniester separatists through the 14th Army. Moldovan government continues its calls for withdrawing the army.

**20** • In a press interview, Dniester Supreme Soviet Chairman Marakutsa argues the Dniester maintains its guard to ensure the region does not become part of a "Greater Romania." He notes his understanding of those Moldovans wanting unification but that a referendum in Dniester clearly reveals opposition to unification and to remain autonomous. Marakutsa laments the fact that all the options offered to Dniester—cultural autonomy, free economic zone, special status within Moldova—do not afford Dniester (or Gagauz) the right not to join Romania.

**23** • Snegur telegrams Boutros-Ghali saying the Republic of Moldova faces "serious danger and its security, independence, and territorial integrity is challenged by the brutal and violent involvement . . . of the 14th Russian Army." He notes the whole territory on the left bank of the Dniester is "occupied" by 14th Army forces who have been blatantly part of combat operations. Snegur terms this involvement as "an act of military aggression, an undeclared war," and that the Russian government bears all responsibility for its consequences. Snegur urges Boutros-Ghali to distribute copies of the telegram to UN Security Council members in order that they be briefed on the situation.

**25** • Addressing a closed-door parliamentary session, Snegur asserts parliamentarians have two options: to stop military operations in Dniester or to declare a state of war with Russia. In his address, Snegur opposes the first option, revealing his frustration with the Russian leadership and what course he favors.

*June 1992*

**2** • Moldova's foreign minister claims: "Russia is attempting to exclude Ukraine and Romania from the process of regulating the conflict in the Trans-Dniester region and to conduct the negotiations with Moldova on a bilateral basis."

• In a national address, Snegur denounces "those who are trying to make Moldova into a base for continuing the imperial games of the former Union and its legal successor, Russia." Moldova and other former republics "regarded the CIS as a means for discarding the former Soviet empire in a peaceful and civilized way; Moscow seeks to use the CIS as a new form of the USSR and to install pro-Moscow governments. . ., resenting our independent course . . . Moscow uses separatist movements as a means for pressing Moldova into changing its course and for installing a pro-Moscow regime in Chisinau."

**3** • A meeting of the newly formed Russian Security Council convenes to discuss the problems facing ethnic Russians in the "near abroad," and particularly in "Dniester Republic." The Security Council calls for the protection of Russian interests in the "Dniester Republic."

• Grachev issues another warning to Chisinau, stating that should military action against the "Dniester Republic" be undertaken, he could not guarantee the restraint of the 14th Army.

**4** • Moldovan Foreign Minister Tiu meets his Ukrainian counterpart, Zlenko. Both sides agree that Ukrainian-Moldovan relations represent an important factor in maintaining regional stability. Tiu stresses that his "government appreciates the neutral stance taken by Ukraine and that distrust for Russia remained high."

• Moldovan Defense Minister Costas releases intelligence information documenting transfer of arms and ammunition to the "Dniester Republic" forces by the 14th Army.

• Snegur says that he would resign and lead a "national liberation movement" if a federation settlement is imposed on Moldova.

**7** • In an interview with *Le Monde*, Kozyrev is asked whether Moldova's Dniester region "would some day become part of Russia." Kozyrev responded that he "would not rule that out." He suggests that Moldova create a region with a special status and "very close links, privileged links, with Russia."

**8** • A majority of the Moldovan government resigns.

**11** • Left-bank deputies, most of whom are Russian or Ukrainian, interrupt their parliamentary boycott to join their right-bank counterparts and vote for a list of six principles for a political settlement of the conflict in eastern Moldova. The points represent Chisinau's views more closely than they do Tiraspol's. In Tiraspol, leaders of the "Dniester Republic" reject any settlement short of "republican" status for the region. Meanwhile on an unofficial visit to Moscow, Smirnov calls on Russia to guarantee the would-be republic's independence.

**12** • In a report drafted by the quadripartite Joint Group of Military Observers, practically all ceasefire violations in eastern Moldova are blamed on "Dniester Republic" forces. The joint group is comprised of twenty-five officers each from Ukraine, Russia, Romania, and Moldova with each side rotating the chairmanship of the group.

**18** • The "Dniester Republic" announces that former Deputy Chief of Staff of the 14th Army, Col. Stefan Kitsak, has been promoted to major-general, appointed "defense minister," and formally been instructed to "form a Dniester army." Kitsak is an ethnic Romanian, native to Northern Bukovina, and participated in the invasions of Hungary in 1956, Czechoslovakia in 1968, and Afghanistan in 1979.

**19** • Dniester forces using APCs attack the lone remaining Moldovan police station in Bendery. In response, Moldovan police reinforcements and units of Moldova's nascent army counterattack, seizing control of much of Bendery at the cost of five dead.

**20** • In a cable to the Russian government, Snegur implores that it stop the 14th Army from intervening "against the defenders of the Republic of Moldova." Snegur warns that any intervention "would mean that the Russian Federation is starting a war against Moldova."

• In a surprise attack, spearheaded by "tanks and soldiers of the 14th Army," Russian forces drive out Moldovan defenders from Bendery.

**22** • Kravchuk shifts gears and calls for the breakaway "Dniester Republic" to be given the status of an autonomous republic within Moldova. Previously Kravchuk did not openly support the "Dniester Republic" separatist aspirations. But Snegur and Shevardnadze issue a joint communiqué that reads: "the NIS are faced with a recurrence of Russian imperial thinking" and that "conflicts develop precisely in the area where Russian troops are located." "Russia supports authoritarian and neo-communist forces [within the republics], a course of action that endangers Russia itself."

**23** • Moldovan Foreign Ministry official warns instability in Trans-Dniester and lack of international support may force Chisinau to shift toward Romania, marking "the collapse of its policy of independence."

• In an article in the Russian government's *Rossiisskaya*

*gazeta*, Presidential Counselor Sergey Stankevich calls on the 14th Army to defend the Slavic minority, thus protecting "a thousand years of history [and] legitimate interests."

• Snegur vows to resist the destruction of "our state . . . by the Russian military, Cossacks, and mercenaries. . . . I will not be frightened and will not bow my head to the Russian leadership's threats." This last statement is in response to statements by Yeltsin and Rutskoy denouncing "Moldovan aggression" in the "Dniester Republic" region.

• Issuing an appeal to "governments, parliaments and peoples of the world," the Moldovan parliament's Presidium implores all states to help arrest "the armed aggression against Moldova." The UN secretary-general responds by dispatching a three-member team to Moldova to ascertain the situation of human and ethnic rights in the republic.

**24** • Russian military and government officials admit that soldiers of the 14th Army did participate in military operations against Moldovan forces on 20–21 June. Furthermore, officials report that whole battalions of 14th Army troops have been transferred to the "Dniester guard." At this time however, the question of who gave the orders remains unclear.

• Senior Moldovan government and parliament officials refute "the myth concerning Moldova's alleged desire to become part of Romania. That is completely untrue . . . Moldovans want complete independence and freedom. At most 5 percent of our population have an interest in an association with Romania. . . . Moldova struggled for its independence and will never give it away to anyone." The officials claim that "unification propaganda" originates from imperialistic Romanians and from Moscow's "disinformation campaign designed to mislead Russians and justify outside intervention in Moldova."

**25** • The presidents of Russia, Moldova, Ukraine, and Romania meet in Istanbul and issue a communiqué on the conflict in Moldova. The communiqué calls for: (1) an immediate and unconditional ceasefire; (2) the neutrality of Russia's 14th Army; (3) Russian-Moldovan talks on the status of the 14th Army and the terms of its withdrawal; and (4) a "political status" to be granted by Moldova to its eastern area.

**27** • The Russian Foreign Ministry delivers Moldova a protest note over "the actions undertaken by Moldova's forces in the Dniester area." The note accuses Moldova of "using the most modern armaments," "placing the 14th Army in a difficult situation," and showing "a persistent unwillingness to look for a peaceful settlement of the problem." Citing potential "highly adverse consequences for regional peace," the Russian MFA warns that "the leadership of the Russian Federation cannot stand idly by" in the conflict. It demands that Moldova "not permit further bloodshed" and "enter into constructive negotiations for settling the situation around the Dniester region."

**28** • Moldova's Foreign Ministry replies that the escalation of the conflict stems from the "Dniester Republic" use of tanks and artillery supplied by Russia's 14th Army and that "Dniester Republic" forces were using those weapons to bombard localities remote from the area of conflict.

• Maj.-Gen. Aleksandr Lebed replaces Nekachev as commander of the 14th Army in Moldova. Lebed announces that the 14th Army will maintain a stance of "armed neutrality."

• Russian State Secretary Gennadiy Burbulis says that Russia is prepared to apply "economic sanctions" to force Moldova to agree to the creation of a "Dniester Republic." Indeed, Snegur expresses concern about Russia's "incipient economic blockade" of Moldova, including a slowdown in the delivery of goods as well as a toleration of the "Dniester Republic's" recent closure of a Russian gas pipeline to Moldova. Snegur states, however, that "we are not going to put up our hands in surrender."

**29** • The UN fact-finding mission dispatched to Moldova is prevented from inspecting Bendery by Russian insurgents who open fire on the delegation, forcing the UN group along with accompanying Moldovan government officials and the U.S. envoy to Moldova to seek refuge in the last Moldovan-controlled police station in Bendery. The group remains under fire for three hours.

**30** • Snegur nominates Andrey Sangheli as prime minister of Moldova, replacing Muravschi who resigned along with most of the cabinet on 8 June. Between late 1990 and early 1992, Sangheli chaired the conciliation commission which negotiated with the breakaway "Dniester Republic." In his acceptance speech, Sangheli pledges to actively seek a political settlement with the left-bank Russians.

***July 1992***

**1** • Lebed, in his first news conference, terms the right-bank city of Bendery "an inalienable part of the Dniester Republic" and "the Dniester Republic itself a small part of Russia." He calls the CIS "an assemblage of abnormal states," and considers that his appointment in Moldova "is connected to a massive shift in the policy of the Russian government." One observer notes that the "Dniester Republic" could become "the ideal testing ground for the new policy."

• In response to alleged Moldovan attacks, Russia's acting Prime Minister Gaydar issues an order authorizing Russian forces to open fire in response to attacks.

**3** • Yeltsin and Snegur meet in Moscow and agree in principle on the following sequence for settling the conflict: a ceasefire; creation of a demarcated corridor between the forces; introduction of "neutral" peacekeeping forces; granting of "political status" to the "Dniester Republic"; and finally, bilateral negotiations on withdrawing the 14th Army. Yeltsin agrees to resume normal shipments of goods to Moldova. The two presidents also establish a "hotline."

**4** • Lebed criticizes Russia's policy of "going with an outstretched hand to the world's cabinets, instead of building up a great power capable of imposing its will." He calls for an end to "political blathering and begging for aid around the world." Lebed saves some venom for the Moldovan government, which he accuses of "committing genocide on the border of Moldova and the Dniester Republic."

**6** • The CIS heads of state agree in principle to create joint peacekeeping forces and deploy them within the next few weeks to eastern Moldova. The forces, which may number between 2,000 and 10,000 soldiers, will be comprised of troops from Russia, Ukraine, Belarus, and Romania and Bulgaria—two non-CIS states. The Russian contingent will be over and above the 14th Army. Yeltsin states that Snegur requested the deployment. Moldova will cover the expenses borne by the peacekeeping troops.

**7** • A formal ceasefire is signed by Moldova's defense minister, the "Dniester Republic" military commander, and the commander of the Russian Federation's land forces. The agreement provides for an immediate and unconditional ceasefire, the redeployment of all heavy weaponry, and the creation of trilateral monitoring groups.

**9** • Russia's Supreme Soviet adopts a resolution advocating the use of the Russian army "to separate the parties in conflict before the CIS peacekeeping forces go into action." The resolution also proposes that Moldova's CSCE membership be challenged on the grounds that Moldova "has committed genocide."

• Just two days after the signing of the ceasefire agreement, the commission of Russian, Moldovan, and Dniester military observers found "Dniester Republic" forces in violation of the accord. Moldovan forces were found in full compliance.

• In his speech to the CSCE summit in Helsinki, Snegur calls for the use of CSCE peacekeeping mechanisms in Moldova. Moldovan officials signal that this appeal supersedes the earlier consent to the CIS plan.

**11** • The CIS plan for deployment of CIS peacekeepers to Moldova collapses as Belarus, Romania, and Bulgaria all rescind pledges to participate and call for the use of CSCE mechanisms.

Meanwhile, Russia fails to obtain a CSCE mandate for the peacekeeping operation in Moldova.

**14** • The "Dniester Republic" Supreme Soviet refuses an offer from Chisinau to include four TDR representatives in the Moldovan government. Furthermore, left-bank deputies decide against returning to the Moldovan parliament. The Supreme Soviet calls on Russia and Ukraine to assume the functions of "protecting powers" and to represent the interests of the "Dniester Republic." Marakutsa further states that "the continuation of the war is the only real course in relations with Moldova."

**16** • Snegur appeals to all democratic forces in the CIS, warning against "nationalist-chauvinist forces still hoping to restore the empire. These forces threaten the independence of all newly independent states."

**21** • Yeltsin and Snegur sign an armistice agreement ending hostilities in eastern Moldova. Smirnov is present but does not sign. The agreement reiterates the provision that should Moldova's statehood status change, the Dniester area would have the right to decide its own fate.

**24** • The number of refugees fleeing from the left-bank conflict is estimated to be over 50,000.

**29** • Peacekeeping forces, authorized by the 21 July armistice, begin taking up positions in the conflict zone. Russia contributes six battalions while Moldova and "Dniester Republic" forces provide three battalions each.

**31** • Snegur appeals to UN Sec.-Gen. Boutros Boutros-Ghali to send a team of UN observers to oversee the implementation of the Russian-Moldovan convention of 3 July on the principles for peaceful settlement of the conflict in eastern Moldova. Snegur writes that "destructive forces" are violating the terms of the convention, "causing profound concern and raising doubts about the other sides' sincerity."

***August 1992***

**4** • In an article critical of the Russian government for being too hesitant in aiding the insurgent "Dniester Republic," the journal *Sobesednik* describes the aid so far rendered: "Aid is being given behind the scenes. Employees of Russia's Ministry of Internal Affairs are serving in the Dniester Battalion, the OMON's equivalent in the Dniester Republic. The Russian government pretends not to see when Russian factories sell firearms and military vehicles to

Tiraspol written off as part of conversion. The Dniester banks are connected to the outside through accounts in the Russian Central Bank. Volunteers from Russia—and not only Cossacks—are fighting in various armed formations."

• Describing the 14th Army as "local," Lebed claims that "the Dniester people have a right to the army." He also reiterates his recognition of "the Dniester Republic and the legal organs of power of this republic on whose territory the 14th Army is based." Lebed sees three possibilities for the future of the "Dniester Republic": either its accession to Russia (akin to Kaliningrad); in the event of Ukraine reuniting with Russia, the "Dniester Republic" could unite with this new state; or it should become independent but with close economic links to Ukraine and Russia. Lebed extends the scope of the "Dniester Republic" to also include any right-bank peoples who wish to secede from Moldova.

**7** • In meetings with new Prime Minister Sangheli, Gaydar states that Russia is "satisfied with the new Moldovan government's constructive approach to the whole range of issues in Russian-Moldovan relations, its moderation and realism, and the evident wish to work toward stability of the situation in the region."

**10** • Moldova observes a national day of mourning for the victims of the war. Snegur condemns "the senseless and barbaric war which has been imposed upon this peaceful land." Meanwhile, the ceasefire holds and no casualties have been reported since 1 August.

**11** • A Moldovan delegation led by Sangheli leaves for the United States. Moldovan officials will sign documents associated with Moldova's admission to the IMF and World Bank, and will seek U.S. private investment for the agricultural and food-processing sectors of the Moldovan economy.

**12** • Moldovan and Russian delegates begin the first round of talks on "the status of Russia's troops [in Moldova] and a time-table for their withdrawal." The talks, mandated by the Sangheli-Gaydar communiqué of 7 August, will deal with both the 14th Army, based in Tiraspol, and the 300th paratroop regiment in Chisinau.

**14** • Troop withdrawal talks adjourn. The Russian Foreign Ministry describes them as "preliminary and exploratory." The MFA announces that the talks covered "not only the status of Russian troops and their stage-by-stage withdrawal," but also "the prospects of military cooperation among the two states," which is necessitated by the "fundamental changes to the defense system of the former USSR that would result from the withdrawal of troops in that sector."

• Concomitant with the staging of a military parade in Tiraspol, Smirnov announces that the "Dniester Republic" intends to form its own army. Lebed further states that the 14th Army would assist in the undertaking.

**19** • Russian ultranationalist Vladimir Zhirinovskiy calls for reducing Moldova and the Baltic states "to the size of Liechtenstein" as part of a general change of borders in favor of Russia.

**24** • The "Dniester Republic" Supreme Soviet debates a draft "law on languages" that would reimpose the Russian alphabet for the "Moldovan language" (i.e., Romanian). Moldovan schools in the left bank reinstated the Latin alphabet in 1989 and have resisted "Dniester Republic" authorities' orders to switch to Cyrillic.

**26** • Moldovan Ambassador to Russia, Petru Lucinschi, warns of "a situation on the Yugoslav model in which Moldova's eastern area would be cut off" from the rest of the country. According to Lucinschi, the TDR forces are using the current ceasefire, coupled with the presence of the Russian peacekeepers, to consolidate their gains and strengthen their state.

***September 1992***

**1** • Grachev issues written instructions to Lebed on "the impermissibility of [making] political statements." This marks the second such request to Lebed. Ambassador Lucinschi complained to Grachev about Lebed's recent statements calling the Moldovan government "fascist" and Lebed's pledge of continued Russian support, including military assistance, to the TDR. According to Lucinschi, this type of rhetoric undermines Yeltsin's policy, contravenes CIS commitments, fuels anti-Russian sentiment, and exacerbates "an explosive situation" in Moldova.

• Yeltsin and Snegur meet for the fourth time in two months. Yeltsin offers to serve as intermediary between Chisinau and Tiraspol concerning the political fate of the TDR. Snegur welcomes Yeltsin's offer and confirms that Moldova is prepared to grant the right-bank city of Bendery the status of a free economic zone and to negotiate the political status of the left bank "provided that the territorial integrity of our state is maintained."

**2** • The TDR marks the second anniversary of the proclamation of the "Dniester Soviet Socialist Republic." Smirnov tells the gathered crowd that "the republic has survived only thanks to Russia and the 14th Army." Smirnov also announces that Russia agreed to provide 2 billion rubles of credit to Tiraspol for the purchase of foodstuffs.

• The aforementioned fear of a "Yugoslav scenario" in Moldova may have prompted officials there to assume a soft policy on the status of Russian troops. Lucinschi states that "unlike the Baltic states, Moldova does not insist on an immediate withdrawal of Russian troops and is prepared to reach agreement on the conditions of their temporary presence on Moldovan territory."

• Snegur political adviser predicts that should Moldova not grant a political status to the TDR, "there will be a renewed conflict followed by a Russian economic blockade of Moldova, the deterioration of the Moldovan economy, and the threat of a political coup." He admits "many of Moldova's problems can only be resolved through Moscow's mediation."

**8** • The TDR Supreme Soviet approves a language law requiring the use of the Russian alphabet for "all situations in which the Moldovan language is used." The legislative body also officially approved the formation of the TDR air force consisting of "airplanes and helicopters based on its territory." Some aircraft have already been turned over to the TDR by the 14th Army.

**9** • A team of Moscow-based human rights observers from "Memorial" return from a fact-finding trip to Moldova and refute allegations that the Moldovan side had committed widespread abuses during the fighting.

**11** • Tiraspol continues to develop state structures of its own. Smirnov signs a decree establishing "Dniester border guards," subordinated to the newly formed "Dniester Ministry of National Security." Tiraspol also boasts of creating the TDR's own banking system, fully separate from Moldova's.

• A CSCE fact-finding delegation assesses that the conflict in eastern Moldova is "a political, not an interethnic conflict." This assessment directly contradicts the claims of the Tiraspol and Moscow authorities who maintain that the conflict arose from ethnic discrimination against Russians in eastern Moldova.

• The TDR decides to establish a customs service, another in a series of steps, facilitated by the ceasefire and presence of Russian troops, toward establishing a fully functioning state apparatus in Tiraspol.

**13** • Romania's Foreign Minister Adrian Nastase states that the question of unification with Moldova should be settled in and by Moldova, but regrets "we do not receive signals from [Chisinau]." He adds that Moldova must remain in Romania's "sphere of influence."

**14** • TDR authorities announce their intention to introduce TDR citizenship. Not surprisingly, officials in Moldova express concern that the "presence of the peacekeeping forces is being used by the Tiraspol leaders to consolidate illegal state structures in the Dniester area."

**16** • Marakutsa states that "Russia's support for the Dniester region is not only moral and political, but also material and military." He reports that he was recently received in Moscow by Yeltsin and Gaydar, where he learned that the Russian leadership planned to condition the 14th Army's withdrawal on Moldova's acceptance of a political status for the TDR. According to Marakutsa, the Russian government also pledged credits to Tiraspol and will soon conclude an economic agreement with the separatist region.

**17** • The second round of bilateral troop withdrawal talks ends "without any results." Lucinschi told the Russian delegation that Moldova would accept a 1994 target date for the withdrawal but the Russian delegation refused to discuss specific dates. Snegur assured the Russians that all 14th Army officers and NCOs would be welcome to join the Moldovan army. Snegur also urged that the Dniester leadership be involved in the talks since their acceptance of any deal is de facto required.

**18** • Yeltsin promotes Lebed to Lieutenant-General.

**30** • Snegur states that "Moldova's independence is the choice of its people and no one has the right to conduct a policy opposing that choice. . . . The existence of a Moldovan independent state is in the interest of all its neighbors, including Romania."

***October 1992***

**1** • In an address to the United Nations' General Assembly, Moldovan Foreign Minister Nicolae Tiu urges the United Nations to send ceasefire and human rights observers to the TDR, where "pro-communist imperial forces, the military-industrial complex, and the upper ranks of the ex-Soviet army have launched a veritable war." Tiu supports a proposed UN resolution on the Russian troop withdrawal from the Baltic states and asks that the issue of Russian troops in Moldova be added to the debate.

**2** • Colonel Stanislau Khazheev is appointed the TDR's first "Minister of Defense." Khazheev inherits 35,000 troops to comprise his forces, according to TDR estimates. These troops are being supplied by continuing arms procurement, according to Smirnov.

• On the occasion of the 200th anniversary of the founding of Tiraspol (ironically as a military settlement of the Russian empire) Kozyrev and Rutskoy both send congratulatory messages to the leaders of the TDR.

**15** • Grachev states that the withdrawal of the 14th Army will only be possible when the conflict in the region is settled. He claims that 14th Army units are staffed by personnel from the region and that they would refuse to withdraw unless the conflict was over.

**16** • Snegur outlines Moldova's policy on the "Dniester" question to a group of Russian journalists. Snegur maintains that Moldova will continue to resist its transformation into a "federation of republics" and the creation of a Dniester Republic with its own army and security apparatus. Chisinau is ready, Snegur continues, to grant Tiraspol "self-government" with political, economic, and cultural autonomy, but only within an "integral and indivisible" Moldova.

**22** • Smirnov announces new conditions on negotiations with Moldova toward a settlement of the conflict, namely that Moldova must adhere to the CIS and the ruble zone. Smirnov reiterates Tiraspol's demand that Moldova become a confederation in which the TDR would have full-fledged state structures and its own armed forces.

**25** • Moldova's Foreign Minister Tiu issues another appeal to UN Secretary-General Boutros-Ghali protesting Russia's "interference in the internal affairs" of Moldova "on the pretext of defending the rights of ethnic Russians." The message claims that Russia's policy exacerbates destabilizing trends in Moldova.

***November 1992***

**3** • Lebed denounces the Dniester leadership's proposals to Chisinau on the delimitation of powers which would in effect "confederalize" Moldova. Lebed terms this stance as "servile." He further charges that the TDR leadership is becoming bureaucratized and was allowing its military force to "die a slow death."

**7** • The TDR celebrates the anniversary of the Bolshevik revolution with rallies and demonstrations. Smirnov praises Soviet achievements and chastises those who renounce them. The Dniester press agency comments that the "TDR's very existence strengthens the political forces in Moscow that seek to restore a 'Greater Russia.' " Leaflets distributed at the rallies proclaim that "Dniester's struggle against Snegur" reflects a "determination to restore the USSR."

**9** • Snegur military adviser Nicolae Chirtoaca briefs NATO's Political and Military Committees on the current situation in the TDR. Chirtoaca asks NATO to send observers to the region to monitor the implementation of the ceasefire accords and the troop withdrawal negotiations. Since Moldova agreed to accept a bilateral negotiating framework, under Russian economic and military pressures, Chisinau has sought to enlarge the framework and has made repeated appeals to the CSCE and the United Nations to involve themselves in the process.

**14** • According to Moldovan presidential adviser Vaslu Malakhov, Kiev is concerned over the possible secession of the TDR from Moldova. Ukraine views the TDR as "a forward base for Russian aggression" and has concluded that its own interests require it to support Moldova's territorial integrity.

**16** • *Novoe Vremya* concludes, based on interviews with Moldovan officials, that Chisinau sees Austria's relationship with Germany—two separate states despite the identity of the language—as the model for Moldova's own relationship with Romania. The latest Moldovan opinion survey shows that only 9 percent of Moldovans desire reunification with Romania. The multiethnic composition of Moldova can help shape a Moldovan national consciousness distinct from the Romanian.

**17** • On the occasion of the Russian army's "Conscript Day," Lebed addresses a gathering of local draftees who were conscripted into the 14th Army. Lebed instructs them to defend the Russian homeland and peace on the Dniester. It is reported that of the 1,500 new conscripts, 80 percent will serve in the 14th Army and 20 percent in the Dniester guard.

**20** • The third round of negotiations on the withdrawal of the 14th Army ends at an impasse. The Russian side reportedly proposes disbanding the army and transferring its assets to the "local authorities" (i.e., the TDR). Chisinau rejects this proposal and protests against the continued transfer of 14th Army equipment and personnel to the TDR forces. Russia's chief delegate to the talks, Col.-Gen. Eduard Vorobev, causes a stir by warning that disputed issues will ultimately be "settled by Russia on its own." The Moldovans term this statement "yet another demonstration of disrespect for Moldova's independence and territorial integrity."

**29** • The TDR security minister, hitherto known as Vadim Shevstsov, acknowledges that he is in fact Vladimir Antyufeev, a former high-ranking official of Soviet Latvia's KGB and OMON in Riga. He states that he and others of his ilk have been assigned "by Russian democratic forces" to beef up the TDR security forces. Antyufeev refutes allegations of corruption leveled against him by Col. Mikhail Bergman, military commander for Tiraspol. He also denies Bergman's charge that the State Secretary of the TDR, Valeriy Litskay, is a former KGB officer involved in corruption. The next day, however, Lebed appears at a press conference and produces Litskay's KGB personnel file. These events highlight the increasingly hostile relations between 14th Army officials and the TDR authorities.

*December 1992*

**3** • Noting continued and intensified equipment and personnel transfers from the 14th Army to the TDR guard, Snegur's military adviser Chirtoaca states that the transfers may make any withdrawal agreement "symbolic and purely formal" because the army would simply undergo a new change from "Russian" to "Dniester." Chirtoaca reiterates that only participation by international organizations can lend credibility to the troop withdrawal negotiations.

• Moldova's delegate to the UN General Assembly's Third Committee, Vitalu Snegur, lambastes what he terms the "screen of silence" that has been drawn over the situation in the TDR. Snegur outlines the large-scale violations of human rights of the indigenous population by the TDR authorities. He lists: the banning of the Latin script; the reintroduction of Russian communist textbooks in place of Moldovan ones; the closure of Moldovan schools and universities; the elimination of all Moldovan language newspapers; the jamming of Chisinau radio and TV broadcasts; the introduction of conscription into the 14th Army; and the widespread purges and arrests of Moldovans who oppose the TDR. Snegur reiterates Moldova's plea for UN observers in the area.

**7** • Snegur states that "with every passing day, from one meeting of the heads of [CIS member] states to the next, the desire of certain states' leaders to return to the organization of the former USSR is becoming increasingly apparent."

## 1993

*January 1993*

**3** • Mihai Gonta, chairman of Chisinau State University, writes "pro-imperial forces try by all means to keep their influence and even restore former political geography of the USSR under a new name: the CIS, confederation, joint economic space, and so on." He says a referendum on independence would serve Russian interests, cooling relations between Moldova and Romania.

**5** • Russian 14th Army Commander Lebed says Russia will open consular office in TDR, which will grant Russian citizenship to anyone who desires it.

• Lebed accuses United States of conducting "an imperialist policy vis-à-vis Russia and the TDR."

**7** • Snegur offers to grant TDR status of "self-governing territory" with a free economic zone. Smirnov insists on recognition of the TDR, that it have its own government and army, and enter into confederation with Moldova.

**14** • Mircea Druc, chairman of the rump Popular Front and former prime minister, accuses some Moldovan politicians of aspiring to form a "Greater Moldova" (a union of Moldova with northern Romanian provinces). This area forms the historic Moldovan principality. Druc, who supports unification with all of Romania, moves his base of operations to Romania.

**15** • Public opinion poll shows 46 percent feel Moldova is headed in wrong direction. A majority reject joining the CIS or Romania; 65 percent accept idea of referendum on Moldova's future and would support the "Republic of Moldova."

**19** • Moldovan deputy minister of foreign trade outlines reasons why Moldova should not reunify with Romania: (a) new "economic center" would control investments, credits, and technical assistance; (b) Moldovans would be second-class citizens; (c) "Trans-Dniestrian Republic" (TDR) probably would dissolve, alienating Russian citizens; (c) tensions would rise in Transylvania, making Romania a major source of primitive tension and chaos in Europe. He calls for consolidation of Moldovan statehood.

**25** • Abkhaz delegation signs Treaty of Friendship and Cooperation with "TDR"—the first known agreement between insurgent movements in CIS states, both reportedly supported by the Russian military.

• Commander of 14th Army, Lebed, advocates accession of TDR to Russia in arrangement similar to Finland and Tsarist Russia. This strongly implies that Moldova's independence is fleeting and he predicts the current Moldovan leadership will face criminal prosecution.

**27** • Parliamentary presidium approves Snegur's proposal to call a referendum confirming the country's independence.

*February 1993*

**4** • Petru Lucinschi is elected president of the Moldovan parliament. He stresses his immediate concerns are the settlement of the Dniester conflict and adopting a new constitution. Dniester parliamentary speaker Marakutsa describes Lucinschi as "bright, flexible, and cunning politician who can look for compromises."

**9** • Talks between Yeltsin and Snegur fail to produce agreement on withdrawal of 14th Army from Moldova.

Yeltsin insists on linking the withdrawal with the political status of Dniester area.

**17** • In meetings with Dniester parliamentarians, Lucinschi reveals his government is prepared to make "major concessions and compromises" on the TDR's status. He contends, however, that despite changes in Moldova's parliament sure to increase settlement possibilities, meaningful dialogue between Chisinau and Tiraspol is prevented by Dniester leaders insisting on making Moldova a federal state.

***March 1993***

**6** • Moldovan Foreign Ministry issues a press release expressing concern over statements made by Yeltsin in February where he said the time is right for organizations with "international authority, including the United Nations, to offer Russia special power in her quality of guaranteeing the peace and security on the territory of the former USSR." The statement highlights the fact that any state's desire to act as peacemaker when not given a mandate by the state(s) concerned is inconsistent with international norms and rejects any attempt of interference in its domestic affairs.

**9** • The Democratic Party of the Dniester region issues a statement condemning Russian "nationalists and pro-communists" for their role in the Dniester conflict. The statement claims such forces are to blame for the impasse in settlement talks and their continued activity is sure to prevent a peaceful and dignified resolution.

**10** • Negotiations in Moscow between Chisinau and Tiraspol representatives agree only on the inviolability of Moldovan borders. Serious differences remain over the administrative-territorial division as Dniester insists on having republic status within a federated Moldova and that existing TDR power structures be kept largely intact.

**20** • Settlement talks take a new level as Moldovan Parliamentary Chairman Petru Lucinschi and Prime Minister Andre Sangheli meet with Smirnov and Marakutsa in Chisinau. They continue discussions of the Dniester region's status and began to address socio-political and economic problems resulting from the conflict. The men informed journalists they agreed only to hold talks regularly and to prevent another escalation of fighting.

***April 1993***

**8** • In meetings with Swiss ambassador to Moldova, Snegur indicates his willingness to demilitarize the Dniester region and pledges his firm support to resolve the conflict "only by political means." He later discounts assertions that Moldova is equipping its armed forces with modern weaponry and is seeking close relations with NATO because they plan a new offensive, instead saying 1993 is the year for peacefully settling the conflict.

**9** • A fourth round of talks takes place in Chisinau on the withdrawal of the 14th Army. Two main issues are discussed, one focusing on the legal status of the 14th Army troops and the other focusing on a timetable and conditions for withdrawal. The first is hindered by the fact that should the 14th Army's legal status be agreed to, it would constitute foreign troops on Moldova's territory against its will. This would have grave international implications. The second issue involves Russian insistence that a political settlement be reached before withdrawal, rejected by Moldova because such a plan leads to "political pressure on the Moldovan leadership for the adoption of decisions in the interests of separatists, anti-constitutional forces." Fourteenth Army Commander Lebed voices his belief that troops should not withdraw until the socio-political situation levels and until the logistics of withdrawal are clarified. Says Lebed, the biggest problem "is that there is no place to go."

**15** • Moscow offers citizenship to all TDR inhabitants. Representatives of the Russian Embassy's Consultating Section in Chisinau will begin providing the necessary services to all interested persons on 22 April.

**17** • In a press conference, Moldovan Deputy Foreign Minister Ion Botnaru explains that Russia's insistence on arriving at a political settlement before withdrawal of the 14th Army "provoke riddles and anxiety" and that such a condition was never envisioned by the quadripartite group. He declares Russia's interest in settling the conflict through this mechanism has declined and that it is apparent Russia wants unilateral control over settlement talks and peacekeeping forces. Such being the case, Botnaru declares "we will endeavor to settle the situation in Dniester by three states—Romania, Ukraine, and Moldova."

***May 1993***

**12** • Moldova signs a memorandum with CSCE Moldovan mission head Timothy Williams outlining the mission's activity in settling the conflict. Involving the CSCE gives Moldova the opportunity to internationalize the conflict and to mitigate against Russian pressures to dominate the settlement process.

**17** • Snegur endorses the Russian-dominated peacekeeping mechanism and the "linking issue," both major concessions necessitated by the lack of international attention and support for Moldova.

*June 1993*

**3** • TDR leaders demand that Moldova rescind parts of the 1991 declaration of independence, join the CIS, and renounce its national army.

**8** • Moldovan Defense Minister Creanga charges "the Russian side's intention to resolve the whole range of military issues unilaterally is a serious flaw, reflecting contempt toward a small state and an erroneous assessment of the real situation in Moldova. Russia is of course a great state, but this does not entitle it to abuse its power and unilaterally resolve issues affecting this or that other state."

**10** • Marakutsa declares that Dniester has good relations with representatives of executive and legislative powers from Russia, and that it will permit the acquisition of fuel and cereals from Russia. He also states that Russia helps TDR by providing hard cash.

**16** • Moldova's Foreign Ministry rejects as "unacceptable under any circumstances" the proposals made by Yeltsin for establishing military bases in Moldova and other newly independent states under agreements with their governments."

**22** • Snegur states Moldova's interest in creating a European collective security system under CSCE auspices based on NATO structures.

**25** • Troop talks deadlock for the sixth time. Russia rejects Moldova's proposal to invite CSCE observers to the session. Moldova "expressed more clearly its position" in refusing political conditions, called for troops to withdraw by the second half of 1994, rejected basing rights, and offered to build housing in Russia for 14th Army officers.

*July 1993*

**5** • TDR leaders claim the Crimea, Donbass, and Odessa regions of Ukraine support them and believe their cooperation will lead to the formation of "Novorossiya."

• Lucinschi observes that if Russia recognizes Moldova as a state, it ought to abide by "appropriate norms." He criticizes Russia's continued political and military involvement, arguing such a policy "violates all international rules." He asks whether Russian policy seeks to help Moldova's Russian minority through cultural and educational programs or "with the gun." "We do not ask for much," he says. "Only that we be respected as a state."

**9** • Moldova redoubles efforts to internationalize negotiations with Russia on the Dniester conflict and the 14th Army. Lucinschi says Moldova's leadership wants UN and CSCE participation in resolving the conflict.

*August 1993*

**3** • Yeltsin sends message to Snegur reemphasizing Russia's commitment to the 21 July 1992 agreement. He indicates he wants to help negotiate a settlement "through Russia's mediation," something unacceptable to Chisinau because it places the TDR on equal footing with Moldova and lets Russia claim the role of arbitrator in Moldovan affairs.

**5** • Moldovan parliament fails to ratify the CIS economic treaty.

**9** • A walk-out by TDR delegates to the parliament results in the rump Popular Front—whose members comprise only 10 percent of the parliament—attaining veto power over legislation.

**18** • For the second time in two months, a NATO delegation is visiting Moldova. The Moldovans asked for political support from NATO and its member states in securing the withdrawal of the 14th Army and for securing an independent and neutral Moldova. According to alliance spokesman, NATO is interested in stability in the region and seeks closer cooperation with Moldova, adding that Moldova needs a military to uphold its neutrality.

**24** • Snegur complains to Yeltsin of activities by "reactionary forces in Russia working against Moldova's independence and territorial integrity" by supporting the TDR. He urges Yeltsin to help "increase the level of confidence" in Russian-Moldovan relations and build "bilateral relations up to contemporary standards and free from prejudice."

• Lebed says Moscow authorized him to execute certain political functions. "I have taken on, and strive to vigorously fulfill, the roles of peacekeeper, diplomat and neutral party. I have asked for some kind of diplomatic confirmation since I am simultaneously the commander, ambassador, adviser-delegate and military attache. This request has now been granted."

**27** • Lebed accepts to run for the TDR Supreme Soviet.

*September 1993*

**3** • Hard-line Russians, including Viktor Alksnis, attend anniversary rallies for the TDR. They call for the restoration of the USSR and the prosecution of Yeltsin and other NIS leaders.

**7** • Lebed says his army is based in "Dniester territory" and expressed confidence that he "will serve there for a long time to come."

**9** • A session of the Gagauz Supreme Soviet appeals to the CIS states to recognize the "Gagauz Republic" as an independent state and member of the CIS. They also resolve to hold a joint session of the Gagauz and Dniester Supreme Soviets to call for recognition and accession to the CIS political and economic agreements, adding more pressure for Moldova accede to the CIS.

**10** • Moldovan government issues a statement accusing Russian deputies of encouraging secessionist tendencies in the Gagauz and Dniester areas. Such incitement is seen as an effort "to blackmail Moldova and keep it within the former empire's zone."

***October 1993***

**8** • In a UN speech, Moldovan Foreign Minister Nicolae Tiu criticizes the presence of Russian troops on Moldovan territory, and accuses them of encouraging separatists of the Dniester Republic. He declares "we consider unacceptable Russia's insistent proposals to be entrusted with a UN mandate for peacekeeping operations in conflict zones in the former USSR. It is clear that these efforts are aimed at justifying a continuing military presence on the territories of independent states. The final goal is obviously the revival of the former imperial structures with the blessing of the international community."

**13** • Moldova appeals to United Nations for support in troop removal.

**15** • Lebed resigns his parliamentary seat in the Dniester Republic Supreme Soviet after receiving 88 percent of votes.

**22** • Taking exception to a speech made at the UN by Moldova's foreign minister, Russia issues a statement which claims that the troop talks are proceeding "constructively" and that no outside help is required or desired.

**26** • Nicolae Andronica, chairman of the Moldovan parliament's commission for legal affairs and chief delegate to talks with Tiraspol, indicates that an "autonomous formation" in TDR is "acceptable" and points to the Social Democrats' proposals for a new constitution that "would not rule out a federal structure" for Moldova. Vice-chairman of the Agrarian Party (Moldova's largest) says TDR should be granted "economic independence and local self-government" and urges that "we should not be afraid of words" such as "federation."

**27** • CSCE commends Moldovan leaders for their efforts to settle the armed conflict in the TDR. The CSCE promises to step up the organization's activity in Moldova and will monitor the February parliamentary elections in which left-bank Moldovans may be prevented from participating.

***November 1993***

**4** • Moldova accuses Russia of breeches in the ceasefire agreement. Specific complaints include (1) allowing penetration of additional forces into the security zone; (2) blocking inspection of suspected illegal arms stockpiles in Bendery; (3) tolerating aggressive picketing of the last Moldovan police station by TDR Russian communist groups.

**8** • Moldovan President Snegur visits the left-bank area still belonging to Moldova. He vows the Trans-Dniester will never be given up and denounces "those who played the game of Russian reactionary circles and, through the force of the former Soviet army, created a phantom republic here."

**10** • The CSCE mission in Moldova is extended another six months. Snegur assures the mission chief Timothy Williams that Chisinau agrees on granting the TDR a special legal status "entailing a high degree of local self-government." The CSCE mission has thus far been unable to facilitate the withdrawal of Russian troops and the brokering of a political settlement of the TDR conflict.

***December 1993***

**16** • Russian Ambassador to Moldova Vladimir Plechko announces an agreement on the 14th Army's withdrawal has been reached. The only issue remaining is the exact timing of the withdrawal. He says the army's presence in the region has been a stabilizing factor.

**1994**

***January 1994***

**14** • Moldova's acting Foreign Minister Ion Botnaru rejects Russian proposals for dual citizenship for Russians in Moldova, stating that while it may be a stabilizing factor in Turkmenistan, it could "boomerang" elsewhere. He also urges that official Russian statements about Russians living abroad be "specific, not inflammatory."

**15** • Leaders of the Dniester Republic accuse Russian 14th Army commander General Lebed of starting a "civil war" in order to install a new leadership.

**19** • The Dniester Republic's President and Supreme Soviet announce a ban on participation of its residents in the upcoming Moldovan parliamentary elections, and declare a state of emergency. Meanwhile, Gen. Lebed continues his anti-corruption crusade against the "hard-line communist" Dniester leadership, who in turn asked for sanctions against Lebed. It is unclear whether Lebed has Moscow's support.

**20** • Smirnov and the Supreme Soviet on 19 January forbade the holding of Moldova's parliamentary election in the territory under their control and declared a state of emergency until 1 March, banning public gatherings, imposing restrictions on the media, and stipulating criminal prosecution of persons engaged in electoral activities.

**21** • Moldova's Foreign Ministry requests in a diplomatic note that Moscow denounce a statement by Vladimir Zhirinovskiy calling for the 14th Army to remain in Moldova, and stating that the Dniester Republic would soon become part of Russia through political, diplomatic, economic, and military means. The note expresses concern that Zhirinovskiy, through such propaganda, could jeopardize negotiations concerning Russian troop withdrawal, as well as exacerbate the Dniester conflict and make a political settlement more difficult.

**26** • Round eight of negotiations between the Russian and Moldovan governments on the 14th Army is postponed, allegedly because of Russian insistence on special political status for the region. Supreme Soviet Chairman of the "Dniester Republic" Marakutsa announces the same day that Moldova ought to be divided into a confederation of three states: rump-Moldova, Dniester, and Gagauz.

**26** • In a press interview, Zosim Bodiu, chairman of the Executive Committee of the Agrarian Democratic Party, says his party will work toward turning Moldova into a "demilitarized zone." The idea, contemplated until now only privately among the leadership of the Agrarian Democrats and some other political parties entails disbanding Moldova's still nascent army and pledging the country's neutrality.

**28** • Ignoring objections of the pro-Romanian opposition, the Moldovan parliament's Presidium approves Snegur's proposal to call a referendum on the country's independence. Labeled a "popular consultation" in an apparent reference to the 1991 referendums held in the Baltics, the poll will be conducted simultaneously with the anticipated legislative elections on 27 February. Snegur and the parliamentary majority had urged such a referendum since 1991 but earlier the pro-Romanian minority used its veto power.

***February 1994***

**1** • Moldova's acting foreign minister, Ion Botnaru, says that the CSCE plan for settling the Trans-Dniester conflict "has been examined and accepted at the highest level as the basis for negotiation." The CSCE plan, endorsed by the CSCE meeting in Rome, proposes considerable autonomy for the "Dniester Republic."

**2** • *Rossiyskie Vesti* dismisses the notion that Gen. Lebed acted independent of Moscow when he unleashed the 14th Army on Moldova in 1992. This corroborates earlier statements to that effect by Yeltsin advisor Sergey Stankevich and other Russian officials. The article stated: "Only now, summing up all the facts, we have come to understand every step of that Army's commander was authorized by the hierarchy of Russia's Ministry of Defense." Had the 14th Army pulled out of Moldova, it "would have meant incurring the anger of millions of compatriots and losing a valuable strategic outpost oriented toward the Balkans."

• Russian-Moldovan troops remain in stalemate. However, General Lebed reportedly tells an officers' assembly that the 14th Army is to be reorganized this year, either as a force based abroad or as an "operational group," with Lebed still in command.

**11** • Smirnov announces all travelers between the republic and rump-Moldova (as well as any other states) will have to pass through customs and checkpoints at their border. Dniestrian border troops will man the points.

**16–17** • Russian Deputy Defense Minister Col.-Gen. Georgiy Kondratev, responsible for Russian peacekeeping troops, visits Russian troops in Moldova "to determine their needs." He meets with the Russian-dominated armistice commission (composed of representatives from Trans-Dniester, Moldova, and Russia), and political and military leaders of the "Dniester Republic." He concedes that peacekeepers cannot remain there forever, but he calls for Russian-Moldovan "military cooperation on a bilateral basis" and for giving Russia's 14th Army the status of an "operational army group" with basing rights at Tiraspol and two other Moldovan cities.

**20** • Senior Dniester Republic official Aleksandr Porozhan reveals that thirty new graduates of the Russian military academies in St. Petersburg and Moscow will be "returning" to Dubasari (in the Dniester Republic) soon. Moldovan Defense Ministry officials dismiss the idea that they were from Dubasari as a cover for an attempt to transfer the men to the Dniestrian military forces. Security officials in Tiraspol were reported to have said that forty officers from Russia's revamped Ministry of National Security had joined the Dniestrian ministries of State Security and Internal Affairs, both of which already have on staff former KGB and MVD officers from Russia.

**25** • President Snegur reiterates his acceptance of the CSCE plan for the conflict's resolution. Plans are formulating which allow for maximum administrative autonomy for the "Dniester Republic," but require a single constitution and single army for the whole country. Secessionists, however, demand full-scale statehood.

***March 1994***

**1** • *Zavtra* publishes a dialogue between leading Russian ultra-nationalists Aleksandr Nevzorov and Aleksandr Prokhanov. The two men describe the Dniester Republic as "our favorite child" and praise Lebed for having led the 14th Army in "masterful operations which threw out Romanians [i.e., Moldovans] from all their positions."

• Yeltsin sends a letter to Smirnov outlining a political settlement. Particulars include: "a wide autonomy for the Dniester region as part of the Republic of Moldova"; a single army, security service, and financial system; a "role for Russia together with other countries" in guaranteeing the eventual settlement.

**4** • In a press interview, Lebed states the 14th Army's manpower level is at 85 percent of statutory level, high for the Russian military today. This is possible because of recruiting soldiers on contract both in Russia and locally. Local recruits become citizens of Russia and take the Russian military oath and pledge to obey Russian law, an act in blatant violation of international law.

**17** • Yeltsin special envoy to Moldova Vladlen Vasev arrives in Chisinau to confer with Snegur and Smirnov. Vasev proposes a settlement guaranteeing Moldovan territorial integrity, but also granting the TDR considerable autonomy. Moscow has viewed the division of Moldova into two federated republics as consistent with preserving its territorial integrity.

• Yeltsin states that Russia will accept the CSCE settlement proposal, which Moldova already accepted. Chisinau remains concerned, however, that Russian troops encouraging Dniesterian separatism could delay settlement of the conflict and undermine the CSCE plan.

**21** • Vasev says that "Russia has geostrategic interests in Moldova and also means to defend the Russian-speaking population." This is the first such statement by a Russian official, and it is expected to be followed by increased pressure for military bases in Moldova. Moldova has rejected Russia's use of the term "Russian-speakers" because it comprises 35 percent of Moldova's population, rather than the 13 percent who are ethnic Russians.

***April 1994***

**8** • The Moldovan parliament ratifies Moldova's membership in the CIS, with reservations, as well as accession to the economic union. The reservations were that Moldova would not participate in any military pacts or the ruble zone. The Trans-Dniestrian state secretary says that this would enable talks on the resolution of the conflict between Tiraspol and Chisinau to resume at a new level.

**9** • Moldovan President Snegur and Trans-Dniestrian leader Igor Smirnov hold renewed talks on determining the legal status of Trans-Dniestria. The talks were moderated by the CSCE mission and Russia, and the basis of a proposed agreement is one prepared by the CSCE which clearly delineates the division of powers between Chisinau and Tiraspol, giving considerable autonomy to the latter.

**28** • Snegur, Smirnov, Russian mediator Vladlen Vasev, and CSCE chief of mission in Chisinau sign political document providing for future negotiations on defining the future status of "Trans-Dniester Republic's statehood under law."

***May 1994***

**14** • Moldova's new foreign minister, Mihai Popov, says Moldova seeks to "cooperate with a democratic Russia which should see in Moldova a responsible and correct partner. . . . Any political and economic pressure should be excluded from this relationship. We do not accept statements suggesting that near abroad states must become satellites of Russia."

**18** • Press reports reveal the Russian side of troop withdrawal talks insists on securing basing rights for the 14th Army in Moldova. Lebed corroborates the report, indicating that to "persuade" Moldova to grant basing rights, it is not necessary to openly use force; economic measures are enough.

**23** • The power struggle between Russia's 14th Army and the leadership of the self-styled republic continues as the army issued a statement revealing that the top leadership of the "Dniester Republic" applied for and received Russian Federation citizenship.

**25** • Russian government blocks the shipment of "Dniester" rubles (printed in Russia's mint) despite a ruling by Russia's State Arbitration board authorizing release of the "Dniester Republic's" new currency. Smirnov indicates the delay causes a destabilizing economic situation in the republic. He lambastes the decision, charging that "certain circles in Russia and Moldova seek to blackmail the Dniester Republic into concessions on matters related to the withdrawal . . . of the 14th Army."

*June 1994*

**8** • In a press interview, "Dniester Republic" Supreme Soviet chairman Grigoriy Marakutsa says Trans-Dniestria was "an inalienable part of the Russian state's southern region, [which] also includes Crimea, Odessa oblast, and a number of other [Ukrainian] oblasts, [and is] known as Novorossiya."

**13** • Moldovan Foreign Minister Mihai Popov dispels notion that Chisinau resigned to a permanent 14th Army presence on Moldova's territory. He reasserts the 14th Army withdrawal from Moldova "without any linkage to any other issues."

• The commanding officers of the "Dniester" guards announce that if Russia and Moldova reach an accord on the withdrawal of the 14th Army, the "Dniester Republic" will claim all of the 14th Army's equipment. The "Dniester Republic's" Supreme Soviet passed an edict to this effect.

**14** • Kozyrev says that "Russia clearly understands that there must be no foreign troops in a sovereign state," but that a withdrawal of the 14th Army is "complicated" and will take time. Moldova's foreign minister says that Russian troops were a "Soviet legacy" for which "Moldova does not blame Russia."

**21** • An international conference on peacekeeping in CIS states, organized by Russia's State Duma and various Russian government ministries, was addressed by "Dniester Republic" President Igor Smirnov, who told a large Western audience that Russian peacekeeping in Moldova responded to the wishes of the conflicting parties and did not entail any political interference or pressure upon the parties.

**23** • "Dniester Republic" President Igor Smirnov says the 14th Army's reluctance to withdraw from Moldova rests on monetary, not patriotic considerations. Inordinately high pay and the relative low prices in the area combine to make 14th Army soldiers economically better off than their compatriots in Russia.

**24** • Lebed says the "Dniester Republic" has no future with its current leadership and calls for early, comprehensive, and internationally monitored elections. According to Lebed, without significant political change, "the Trans-Dniestrian people will only be facing more disasters and finally Trans-Dniestria as a republic will decay from its roots with no external interference."

*July 1994*

**12** • Troop withdrawal talks deadlock as the two sides disagree on linkage between the 14th Army's withdrawal and a political settlement for the "Dniester Republic." Russian Deputy Foreign Minister Sergey Krylov says withdrawal is being "impeded by the undefined nature of Trans-Dniester's status." The two sides did agree to initiate negotiations with Ukraine regarding the transit of 14th Army convoys through Ukraine in the event of a withdrawal settlement.

• Basapress reports "Dniester Republic" forces illegally occupy a military outpost in the right-bank (western Moldova) town of Bendery, the scene of major fighting in July 1992.

**13** • The Joint Control Commission meets to address violations of the 1992 armistice. The Commission concludes that "Dniester Republic" forces repeatedly entered the security zone entrusted to Russian peacekeepers, set up border posts there, illegally stopped and checked Moldovan peacekeepers and vehicles on patrol, and blocked access to military observers on inspection.

**16** • At an international conference on security in Central Europe held in Romania, Russian State Duma First Deputy Speaker Mikhail Mityukov, a member of Russia's Choice, says a settlement of the "Dniester" conflict must respect Moldova's sovereignty and integrity, while "granting to the Dniester region . . . state status." He says that the 14th Army guarantees peace in Moldova and that its withdrawal is a bilateral matter between Moldova and Russia.

**27** • "Dniester" forces set up a "customs station" outside the right-bank city of Bendery to establish a "border" that would separate the "Dniester Republic" from the rest of Moldova. Representative of the Russian side of the Joint Control Commission says the Russian peacemaking command authorized the decision. An armed standoff ensued, but the Moldovan side withdrew. The Bendery "customs station" is the third of its kind established by "Dniester" forces on the right bank, a violation of the July 1992 armistice convention.

• Vladimir Solonar, the leader of Moldova's Socialist Unity Bloc, the second largest parliamentary grouping, advocated the "unification [of the NIS] at least in the form of a confederation," and the accession of Moldova to a "military-political union with the other republics of the CIS." He defined the bloc as "pro-Union . . . and pro-Russian because Russia would form the [confederation's] center. . . . Russia is more than just the former RSFSR. . . . One way or another, our entire state was Russian."

### *August 1994*

**3** • The 14th Army will undergo restructuring. Certain command structures would be dissolved, some personnel will be demobilized, and the reduced force will become an "operational group" of the 59th Motorized Infantry Division. The restructuring is to be completed by 1 September.

**8** • Moscow's announced intention to recall its peacekeeping forces from Moldova, placing Chisinau in the unenviable position of requesting that Russian troops remain in Moldova, temporarily. Unwilling to face the superior "Dniester" forces without a disengagement force, Moldova's co-chairman of the Joint Control Commission, Maj.-Gen. Victor Catana, says that under the terms of the armistice, withdrawal of the peacekeeping force requires the consent of both antagonists.

• An assembly of 14th Army officers drafts a message to Grachev strongly urging Lebed be retained as commander of the 14th Army. According to the officers' assembly, "only the authority of Aleksandr Lebed can induce the officers and servicemen to fulfill their mission."

**10** • Moldova and Russia initial an "agreement on the legal status, procedure, and timetable of the withdrawal of Russian military units temporarily located on the territory of Moldova." The announcement came during the tenth round of negotiations in Chisinau. The withdrawal is to be completed within three years and "will be synchronized with the political settlement of the Dniester conflict and the determination of the special status of the Dniester region of Moldova."

**14** • Lebed attacks the 10 August troop withdrawal agreement and the decision to restructure the 14th Army. He warns that "Dniester Republic" authorities sit poised to commandeer arms and ammunition from army stockpiles, a move according to Lebed "guaranteed to destabilize the situation."

**15** • Yeltsin effusively praises the 14th Army, Lebed in particular, for "defusing the conflict and halting violence in 1992" and for "controlling the situation" since the armistice was signed in July 1992. Yeltsin warned against "any hasty actions or decisions" that could serve to renew tensions in the region.

**16** • According to a commentary in *Krasnaya Zvezda*, the "number one" advantage to the recently signed troop withdrawal agreement is that "it removes the poisonous accusations about Russia's alleged 'imperial ways' in keeping its army on the territory of sovereign Moldova without a legal status and allegedly refusing to withdraw it."

**18** • Russian Defense Ministry officials say they do not doubt the 14th Army will be kept in Trans-Dniester but under a different name. Russian Federation Council Deputy Chairman, Valerian Viktorov, cautions it will take at least six months for the Russian government to "draft and sign" the agreement to withdraw the troops from Moldova.

**23** • Khazeev says the breakaway region maintains an armed force of "no less than that of the 14th Army" in aggregate manpower (around 10,000). "Dniester" military doctrine and legislation, organization, and uniforms mirror Russia's, "otherwise it would be difficult to reunite with it."

• Lebed says that the cost of transporting the 14th Army's massive weapons stockpiles from Moldova to Russia would be prohibitive and predicts Russia would not pay for the withdrawal. Lebed believes that since Moldova also cannot pay for the withdrawal, "we should come to an agreement on the conditions for keeping the troops here."

**26** • Grachev meets with Lebed, saying the 14th Army's structure will remain as is, less a 20 percent cut in the command staff. Lebed seemingly saved his job as commander of the 14th Army.

### *September 1994*

**7** • Russian officials continue to distance themselves from the 10 August troop withdrawal agreement. The deal, termed "idiotic" by Lebed, has met with near-constant criticism. Deputy Foreign Minister Sergey Krylov argues that because the issue of withdrawal was agreed upon and linked to a political solution in the Trans-Dniester, "the 14th Army's presence in Moldova is therefore legitimate." He praised Lebed and his minions for "stabilizing" the region.

**17** • Moldova's Helsinki Committee issues appeal to international organizations and human rights watch groups, decrying the state of human rights in the separatist Trans-Dniester region. The appeals note that Dniester authorities have embarked on a campaign of linguistic apartheid. Although Trans-Dniester's population is 40 percent Moldovan, 28 percent Ukrainian, and 25.5 percent Russian, only 20 percent of schools teach in Moldovan (and of those only three schools teach in Latin script). The state of Ukrainian language education is even more appalling—only 0.5 percent of the schools teach in Ukrainian, while 77 percent teach in Russian.

**20** • Lebed's continued service as commander of the 14th Army gains unexpected support from Snegur. Snegur says Lebed "knows how to maintain order in the army. This is very important as the army has huge amounts of armaments which must not fall into the hands of separatists." Snegur welcomed Lebed's efforts to fight corruption in the breakaway region.

**28** • Talks open in Tiraspol on establishing a "special status" for the separatist region within Moldova. Russian special envoy Vadim Vlasev predicts long and contentious discussions.

**29** • Addressing the UN General Assembly, Snegur calls for international support to secure the removal of Russian troops from Moldova. He highlights Moldova's "geopolitical and cultural affiliation to the European democratic space" and voices concern that "isolation from that space would face us with the reemergence of those influences from which we have suffered in the recent past."

***October 1994***

**7** • Moldovan officials accuse Russia of violating the 1992 armistice convention by unilaterally withdrawing a peacekeeping battalion from the security zone on the left bank, allowing TDR units to infiltrate other areas of the security zone. Chisinau fears that further reductions in the peacekeeping force would permit the heavily armed Dniester forces to launch "provocations" against Moldova, a pretext to keep the Russian 14th Army in Moldova as "stabilizers."

**12** • The status of the 10 August troop withdrawal agreement remains unclear following three days of talks between Moldovan and Russian officials in Chisinau. Editorial changes to the agreement appear to weaken the linkages sought by Russia between the troop withdrawal and a satisfactory political solution in the TDR.

***November 1994***

**3** • Snegur reconfirms his position that the Army of Moldova and armed units of the Trans-Dniester should be subordinate to a single Moldovan command.

**7** • Marakutsa states the Dniester region can coexist with Moldova as a single state within Moldova's current borders if the eastern district receives a special juridical status.

**11** • The Moldovan delegation within the joint control commission on resolving the conflict in Trans-Dniester says that it is against reducing the peacekeeping contingent in the region without coordinating the issue beforehand. Apparently, Russia is going to reduce its peacekeeping forces from four to two battalions before 19–28 November.

**19** • Nicolae Chirtoaca, former presidential military counselor in Moldova, says Russia's decision to reduce peacekeeping forces in the Dniester conflict area might have unpredictable consequences. After signing the 14th Army's withdrawal agreement, a fragile balance was set in Dniester.

**22** • Russia begins downsizing TDR peacekeeping forces. Four battalions of the 27th Motorized Infantry Division are to be replaced by two other battalions of the same division. The decision is explained by a relative stabilization in the region.

**28** • Russian Deputy Defense Minister Col.-Gen. Georgiy Kondratev urges the 14th Army personnel to continue service in Trans-Dniester. He says that since no mechanism for withdrawal exists, the soldiers should engage in their planned activities.

## 1995

***January 1995***

**6** • Lebed tells a delegation of CIS MPs that, on the question of the 14th Army's withdrawal, "this would be General Lebed's departure, a briefcase in hand," while the rest of the troops would stay in Moldova because the army is mostly staffed with locals. Lt.-Gen. Lebed, the most popular officer in the Russian Army, has been an outspoken critic of the war in Chechnya. Lebed is regarded as a hero to soldiers because his actions give voice to the feelings of bitterness and betrayal in an army fallen on hard times since the Soviet empire's collapse. Although Lebed claims to have no presidential aspirations, "Russia's long history shows clearly that the ones who come out on top in political disputes are the ones who have the army on their sides," the newspaper *Kommersant Daily* observed.

• Valeriyan Viktorov, deputy speaker of the Russian Federation Council, stated that the Russian-Moldovan agreement providing for the withdrawal of the 14th Army from the territory of Moldova within three years "has so far been producing more questions than answers." The deputy speaker noted that a key obstacle to finalizing the agreement is the problem of transporting arms and equipment across the territory of Ukraine. Viktorov admitted that Lt.-Gen. Lebed has a very high prestige among the population of Trans-Dniester and is also respected by the Moldovan leadership.

• In Deputy Speaker Viktorov's opinion, Moldova's adherence to all CIS structures might facilitate the Dniester conflict settlement. He also said that the 14th Army's transformation into a Russian military base "in principle, is not rejected by the Moldovan leadership, though it invokes Moldovan Constitutional provisions according to which on Moldovan territory no foreign military formation should be deployed."

**9** • According to Chisinau Radio, a special commission of the Russian Ministry of Defense will soon arrive in Tiraspol vested with the authority to dismiss Lt.-Gen. Lebed from his post as commander of the 14th Army. The dismissal would supposedly be in reaction to Lebed's stern criticism of President Yeltsin's policies in Chechnya. Asked what he will do if the commission dismisses him, Lebed said "Let them sack me and you will see what I will do." Lt.-Gen. Lebed recently met with Ernest Muehlemann, Council of Europe rapporteur on Russia, where allegedly they spoke about the contingency of Russia's acceptance into the Council of Europe on the withdrawal of the 14th Army from Moldova. Lebed's comment was that the scheduling of such a withdrawal was the key, and that the key factor is that 90 percent of the army's forces consist of local inhabitants. The fate of these people must be "guaranteed."

**12** • Moldovan Parliamentary Chairman Petru Lucinschi commented on the political integration of the CIS republics. He feels that the lawmaking synchronizing process within the Commonwealth, the proposal of a common CIS social charter, the strengthening of the Interparliamentary Assembly (IPA) to that of the European Parliament (in terms of dealing with the economy, budget, legal codes, and a resolution for the Dniester and Gagauz problems) would assure Moldova's integrity.

• Smirnov wants Russia and the OSCE to act as the guarantors of the non-use of force in a demilitarized zone in Moldova.

• The next Russian/Moldovan summit on the status of Trans-Dniester is scheduled for 15 February 1995.

**13** • At the invitation of Moldova's parliament, a peacekeeping group of the CIS IPA, headed by Valeriyan Viktorov, had a number of meetings connected with the problem of settling the Dniester conflict. The IPA's peacekeeping efforts have encompassed other "hot spots," including the Tajik-Afghan border, Abkhazia, and Karabakh. All this attests to the strengthening of the IPA's role within the CIS.

***March 1995***

**14** • Moldova's Parliament Chairman Petr Lucinschi calls for removing arms and equipment belonging to Russia's 14th Army from the Dniester region before troops pull out. Lucinschi says that Tiraspol leaders' strong views on the region's special status conform with Moldova's constitution. Kishinev intends to ask the public to decide whether its peace plan is acceptable.

**18** • In an interview with *Sovetskaya Rossiya*, Dniester's Security Minister Vadim Shevtsov states that Dniester is the object of a geopolitical struggle which includes Russia, the Baltics, and the Western states. He says the Dniester Republic complicates the plan for establishing the Black Sea–Baltic Confederation started by Poland's resident CIA officer in 1990. "Dniester leaders consider themselves Russians," says Shevtsov, "and we are defending the southern borders of the Slavic state. The separation of Ukrainians and Russians is nonsense and will end sometime soon." Shevtsov claims his republic's formation prevented Romania from swallowing up Moldova in 1992. He also claims that Moldova would have committed "genocide" against its 500,000 Russian-speaking people had the republic not been formed.

• Regarding the 14th Army, Shevtsov says Lt.-Gen. Lebed shrank the army from 10,500 to 3,500 and demoted it to a unit which guards military property yet to be sold. He goes on to defend the Dniester region's right to this property. He criticizes Lebed as an enemy of the current Dniester "president," Smirnov, who is preparing to dissolve the republic from within, doing Russia's bidding. He accuses Lebed of coming to Dniester in order to keep the 14th Army from joining with the people of the Dniester region, which he says was not Russia's plan. He denounces Lebed as a Russian government puppet, on a mission to preserve Russia's interests in the region while ignoring the interests of the Dniester people.

***April 1995***

**4** • Russian Federation Council Speaker Vladimir Shumeyko calls for withdrawing Russia's 14th Army from the Dniester region, saying "the experience of the Chechen conflict shows that weapons should be urgently withdrawn from hot spots." Shumeyko says plans by Russian and

Moldovan defense authorities already have a draft plan for the withdrawal.

**17** • A European parliament delegation headed by Elisabeth Schrodter visits Moldova and the Trans-Dniester region. Schrodter says the 14th Army's withdrawal should be internationally monitored but, after meeting with Lt.-Gen. Lebed, receives the impression the withdrawal will not be a rapid one. Trans-Dniester leaders make clear to Schrodter their insistence on preserving the region as an independent state but would agree to a confederation with Moldova.

**19** • Russian Lt.-Gen. Aleksandr Lebed says given the current economic and political situation, withdrawing the 14th Army from the Dniester region within three years is "impossible." In addition to the economic and political concerns, Lebed also notes the Army Withdrawal Agreement's provision to simultaneously settle the conflict and remove the army adds to the difficulties.

**26** • Russia's State Duma adopts resolution opposing plans to remove the 14th Army from the Dniester region in Moldova, 267 to 0 with two abstentions. The resolution, although non-binding, says the troop pullout could increase tension in the region, noting the 14th Army was a strong stability factor.

**28** • TDR parliamentary Speaker Marakutsa warns the Snegur initiative to constitutionally change the official Moldovan language to Romanian is sure to further complicate relations between Chisinau and Tiraspol. Snegur recently ruled out closer integration with Romania but his initiative reveals that pro-Romanian forces are influencing his policy decisions.

***May 1995***

**4** • Commentary continues over the consequences of the 14th Army reorganization and Lebed's imminent dismissal. Many believe tensions will increase should Lebed depart as army commander because he was instrumental in safeguarding the weapons and equipment garrisoned in the region. TDR activists under Smirnov, who claim the equipment belongs to them, are expected to take to the streets in support of their claims. Moldovan parliamentarian Nikolay Andronik says the army's withdrawal is a Russian matter damaging Russia's prestige, not Moldova's.

**10** • In Victory Day rallies in Tiraspol, Smirnov urges Moldova and the TDR "to build their relationship the way equal-right states usually do. . . . The sooner Chisinau recognizes Dniester as a separate state," he says, "the sooner we restore relations in all spheres."

**22** • Talks between Snegur and Smirnov are slated for 7 June. In meetings with State Duma CIS affairs committee chairman Konstantin Zatulin, Moldovan Prime Minister Andrey Sangheli says he believes the stage is set for the conflict's resolution and that a mechanism for the 14th Army's withdrawal will soon be found. He praises Lebed for his contribution in maintaining stability in the region and asks that he not be removed and commander.

**23** • OSCE representatives arrive in Chisinau for settlement talks. Negotiations have been stalled since the parties disagree on two important elements: withdrawal of the 14th Army, which Tiraspol rejects because of possible destabilization; and subordination of TDR power structures to the TDR government, rejected by Chisinau as it implies statehood. According to Lebed, negotiations are futile and the only "machinery capable of solving any task is the 14th Army command."

**25** • Lebed reportedly tenders his resignation, which awaits Yeltsin's signature, to Russian Defense Minister Grachev.

***June 1995***

**2** • Fourteenth Army officers appeal in writing to Yeltsin not to dismiss Lebed and to impose a moratorium on further reductions in the army's personnel. The note indicates rising pro-Romanian and anti-Tiraspol activities in Chisinau that may reignite armed conflict, especially in Lebed's absence. This renewed fighting, the appeal warns, may spill into neighboring regions.

**7** • Snegur and Smirnov meet in Snegur's residence to discuss the TDR situation and a draft decree on special status for the TDR region. OSCE Chairman of the permanent mission in Moldova Mikheal Whitgent and Russian special envoy Laslo Kova attend the meeting. Snegur and Smirnov agree that talks thus far have served to eliminate a new outbreak of war but have done little for an actual settlement. Snegur also advances Moldova's proposal, endorsed by the OSCE, for a special autonomous status of the TDR within a united Moldova. Smirnov counters with a proposal to first delineate responsibilities between Tiraspol and Chisinau and from there talk about the region's legal status. Snegur rejects the proposal as unacceptable since it proceeds to build relations on an interstate basis. The two decide to continue talks in early July.

**13** • Basapress in Chisinau reveals that a ranking Russian official in Moldova informs them that the current deployment of Russian peacekeepers in Trans-Dniester "would not be able to guard the 14th Army depositories should the need arise." The informant says the peacekeeping force simply does not have the technical facilities for such a task.

**16** • New Russian General of the 14th Army Valeriy Yevenich arrives in Trans-Dniester joined by Deputy Commander of Russian Land Forces Col.-Gen. Anatoliy. The two are forced to land 40 km outside Tiraspol as demonstrators, organized by the Union of Trans-Dniester Women, blocked the runway. (The group protested Lebed's dismissal, and the disturbance was calmed only when Lebed arrived on the scene.) Yevenich is an ethnic Russian who has served in the armed forces since 1968. He graduated from the Ryazan Higher Airborne Command College in 1972 and from the General Staff Military Academy in 1992.

• Asked what type of advice he has for his successor, Lebed answers "No pieces of advice." He does express anxiety over the massive arms and equipment cache garrisoned in the TDR, saying no political solution has yet been agreed to and that they are sure to be plundered by central Moldovan government forces and local TDR forces. He labels the region "a delayed time bomb" and praises the 14th Army's efforts for saving thousands of lives. Regarding settlement, Lebed stands "for a legitimate, civilized solution of the problem," vowing to continue his effort in the settlement process. Fourteenth Army officers lodge an official protest to the Russian Supreme Command over Lebed's removal.

**20** • Russian Defense Ministry officials tour the TDR region, meeting with junior and senior Russian troop officers. The ministry officials focused on determining who will protect Russian Federation citizens in the TDR and Moldova and on plans to prevent a Baltic-like situation where people divided up according to nationality after Russian troops withdrew.

**21** • The Russian Duma passes a document imposing a moratorium on the Defense Ministry's decision to downgrade the 14th Army and withdraw heavily from Dniester until the conflict has been solved politically. Moldovan Foreign Affairs Ministry issues statement that the Duma's stance interferes in the internal affairs of a sovereign nation, accusing the body of hindering "the positive peace process backed by the international community, in particular the OSCE."

**23** • In a press interview, Smirnov announces a document is now being prepared on governing interrelations between Tiraspol and Chisinau, predicting the document will be signed by Snegur and himself and that both republics' parliaments will approve the plan. He says the recent Duma decision (noted above) was "useless and did not change the situation."

**26** • Grachev arrives in Tiraspol to examine the situation, meeting with Russian servicemen and their families regarding their fate. He meets later with Snegur in Chisinau to discuss the issue of Russian troops on Moldovan territory and what to do with the weapons and equipment. Grachev reminds Snegur of the emerging European security structure and Russia's role as a "stabilizing factor" in the region. He also stresses no staff reductions are envisioned before a political agreement is reached.

***July 1995***

**3** • In Chisinau, Snegur informs U.S. diplomats that OSCE efforts were successful in scheduling a new round of settlement talks "to continue in a constructive fashion." Regarding Russian proposals for military bases in the republic, Snegur reiterates his republic's constitutional banning of foreign military deployment on its territory. Further talks are scheduled for 5 July.

**5** • Discussions between Snegur and Smirnov produce no political agreement. Nevertheless, the sides do approve a number of proposals regarding economic cooperation and pledges not to use force against one another or enter into alliances against one another. More complete settlement hinges on the political status of the TDR. TDR leaders demand the region be accorded statehood while Moldova agrees only to extensive autonomy within the framework of the Moldovan constitution, viewing the TDR proposal as counter to their constitution and to international practice. Both sides, however, encouraged Ukrainian participation in the settlement process, in negotiations and as part of international peacekeepers.

**18** • In a conference with TDR leaders, public organization heads, industry managers, and media representatives, Smirnov again emphasizes talks between the TDR and Moldovan officials aimed at strengthening TDR statehood and securing recognition as an independent state by Chisinau.

**20** • Yevnevich confirms destruction of 14th Army's weaponry is being carried out. The destruction is taking place 25 km from the town of Rybnitsiy, where no disturbances to neighboring Ukraine would occur. Yevnevich also says that rumors regarding preparations for renewed fighting in the Trans-Dniester are unfounded. "The situation is calm," he says. "Those who come here on a visit say it is a resort here. This is indeed true."

**31** • A Dniester representative announces the two sides have agreed on the legal status of the TDR. No details were released but the representative said the groundwork is being laid for a meeting between Snegur and Smirnov set for 13 September.

*August 1995*

**2** • Smirnov addresses a meeting of TDR officials, saying the fact that the Russian peacekeepers are still here and have been for over three years "proves that the military clash was an open aggression on the behalf of Moldova." He praises the agreement between Tiraspol and Chisinau on non-aggression and believes the TDR increased defense capabilities ensure the conflict will not reignite. As to a political solution, Smirnov remains adamant that "one way or another, Moldova will have to recognize our statehood."

**10** • In a press interview, Marakutsa warns that the internal political developments in Moldova may lead to a deterioration of the talks between the two sides. He refers to the collapse of the Moldovan Agrarian Party when Snegur and eleven parliamentary deputies left the party to create a new presidential party, Moldova's Revival and Accord. The new group seems intent on continuing Moldovan-Romanian integration and further "Romanization" of Moldova. Marakutsa made clear the Tiraspol leadership categorically rejects such a move and this "will not promote settling the conflict in Trans-Dniester."

**17** • The Tiraspol City Council calls for the resignation of Marakutsa and Vyacheslav Zagryadskiy, chairman of the Dniester Republican Bank, for "being under the Moldovan leadership's thumbs" by supporting the introduction of the Moldovan leu. TDR officials argue such a move would "in the final analysis, [mean] the absorption of Dniester by Moldova."

*September 1995*

**1** • In a press interview, Smirnov discusses the status of the Dniester Republic upon reaching its fifth year of independence. He says the remarkable thing is that the republic has been formed and is assuming the attribute of statehood. He acknowledges that the Dniester people are experiencing difficulty, just as all people in the FSU are, and to speak of achievements would be an exaggeration. He praises the fact that the republic was formed as a counterbalance to nationalism and that the TDR "guarantees everyone who lives here an opportunity to speak in their native language and rules out any interethnic conflict." As to whether the TDR will survive in the face of economic and political isolation, Smirnov answers with an emphatic yes.

**7** • Snegur continues to express optimism that a political solution will be found to the conflict which maintains Moldova's territorial unity. In a press interview, he indicates the Moldovan leadership will be patient and persistent in talks with Tiraspol for as long as "necessary for them to comprehend that their intentions to found a confederation are fruitless."

### Summary at the End of 1995

Both leaders remain steadfast in their positions and no political solution seems possible in the near future. While fighting has been quelled for quite some time by Russian peacekeeping forces and the 14th Army armaments depots are effectively controlled, tensions remain. Ukraine has increasingly become involved in the settlement process (nearly 40 percent of the Dniester citizens are Ukrainian). Moreover, relations between the Chisinau and Tiraspol are governed by a series of bilateral agreements and protocols documented in the lines above and the Dniester Republic has not been recognized by any international institution or individual country. Snegur has repeatedly expressed the intent to grant the Dniester region the status of an autonomous republic within Moldova. Dniester leaders reject the idea and will settle for nothing less than the republic's complete sovereignty.

# The Tajikistan Conflict

## Background Note

The conflict in Tajikistan is more than an interethnic struggle between Slavs and Muslims, although a Muslim awakening did occur in the late 1980s, influenced by Gorbachev's policy of glasnost. Politically, the new awareness of the Muslims gave rise to the Islamic Renaissance Party (IRP), which holds broad appeal in rural parts of the country. The party's leadership claims not to be fundamentalist, although it has received financial and other assistance from fundamentalist Islamic groups in Afghanistan and Iran. As evidence of its secular persuasion, the IRP has called for a secular, democratic Tajikistani government, and its leaders formed an oppositionist coalition with leaders of the much smaller "Democratic Party"—composed mainly of a weak but vociferous Tajikistani intelligentsia. However, the real, underlying source of conflict in Tajikistan is a complex web of interclan rivalries, Muslim pretensions, and the external efforts of Uzbekistan and Russia to frustrate Tajikistan's emergence as a viable sovereign state.

Internally, the IRP–DP coalition opposes the ruling communist majority in Dushanbe, which has continued to hold power since the USSR's demise. The Tajikistani Communist Party, however, was never the ruling majority's main power base. The communists never managed to overshadow entirely the country's clan structure. The strongest clans, in fact, provided Tajikistan's communist leaders, often fighting among themselves for supremacy within the Communist Party. The 1990–91 movement by the Soviet republics toward greater political and economic autonomy led to the rise of one clan, the Khodjenti of Leninabad, to parliament and the presidency in the October 1991 elections.

In contrast, however, with the easy hold on power maintained by the communists of other post-Soviet Central Asian republics, Rakhmon Nabiev, one of the Khodjenti clan's leaders, faced strong opposition demonstrations by the IRP–DP anti-communist opposition coalition. To counter this coalition, Nabiev formed his own coalition government, which included democrats, Islamists, and others. Instead of producing peace, however, this volatile mixture led to the outbreak of fighting in May 1992 between Nabiev and the opposition.

In the ensuing unstable situation, the southern Kulyabi clan began to vie for more political power, ostensibly out of resentment of concessions Nabiev made to the Islamists. Kulyabi leader Sangak Safarov created a military organization with which to fight both Nabiev's government and the IRP–DP coalition forces, gaining the upper hand in late 1992. To complicate matters, external parties were involved. Saudi Arabia, Iran, and Pakistan were said to be sending aid to the Islamic Renaissance Party and Uzbekistan came to the aid of the Kulyabi clan. Further, because the government made a Muslim leader of the Pamiri clan acting President in September (in an attempt to mollify the IRP), the Kulyabis were able to accuse it (probably erroneously) of trying to install an Islamic regime. On gaining power in the 1992, however, the victorious Kulyabis abolished the office of President and appointed one of their leaders, Emomali Rakhmonov (who remains in power today), as head of both parliament and state. Thus, the Kulyabi clan prevailed over the Khodjenti, the Pamiris, and the opposition coalition.

Subsequently, other Muslim clans in the eastern and southeastern regions of Tajikistan began to fight the government. In self-defense, Dushanbe, joined by other Central Asian states, Uzbekistan in particular (and later Russia), pointed to the Islamic component of the opposition as justification for their intervention to support the communist elite.

Where does Russia come into all this? In 1992, Russia confined its interest to Russian border troops, which remained stationed in Tajikistan patrolling the Afghan border, and the 201st Motorized Infantry Division in Dushanbe afforded Russia quick access to Tajikistan's internal political affairs. Moreover, the CIS, under Marshal Shaposhnikov, attempted to tie Tajikistan into a "single CIS strategic space," conveniently neglecting to acknowledge the political sovereignty of the country.

Ironically, the Russians at first missed the Islamic content of the struggle, with democrat Evgeniy Ambartsumov, former Chairman of the State Duma's Foreign Affairs Committee, calling for Russian support of the IRP and its allies. In the summer of 1993, the situation underwent an abrupt change. At that time, fighters from Afghanistan crossed the border and killed more than twenty Russian border guards, which catalyzed a far more active Russian interest in Tajikistani affairs. Defense Minister Pavel Grachev flew to the country and Boris Yeltsin told Foreign Minister Andrey Kozyrev to coordinate a Russian policy on Tajikistan. A Russian-Tajik treaty was hastily drafted, signed, and ratified in a two-week period. Russia henceforth asserted the right to defend both the border and the entire region. The chronology that follows provides a clear sense of Russia's rapid takeover of Tajikistan's government, army, and territory. Boris Yeltsin was quite easily able to obtain approval for its hegemonic actions from other Central Asian countries by presenting the problem as the need to contain Muslim radicalism.

Uzbekistan, especially, favored Russia's intervention in Tajikistan and support for the pro-communist Tajikistani regime. Uzbekistani President Karimov understandably feared the spillover effect which could result if a democratic and Islamic-allied government were installed in Tajikistan. Karimov, in fact, was issuing warnings to international organizations such as the United Nations and the Conference on Security and Cooperation in Europe (CSCE) about "Islamic fundamentalism" in Tajikistan as early as March 1992. Karimov's position can partially be explained by the fact that Uzbekistan has a large Tajik miniority, both republics having been formed from territory of Turkestan and Bukhara in the 1920s. For its part, Tajikistan has harbored strong resentment toward Uzbekistan for keeping the administratively and culturally important cities of Samarkand and Bukhara. Karimov hopes to silence the Islamic and secular opposition in both Tajikistan and Uzbekistan and to promote his agenda for a greater Turkestan. He made this agenda known when he invited the leaders of Kazakhstan, Kyrgyzstan, Tajikistan, and Turkmenistan to a summit in Tashkent in January 1993, which resulted in the creation of a Commonwealth of Asian States.

It was partially in order to thwart Karimov's ambitions that Russia reasserted its interests in Central Asia in 1993. When Karimov tried to form a Central Asian Union in 1994, only Kazakhstan and Kyrgyzstan responded. Even then, Nazarbaev and Akaev sided clearly with the integration movement within the CIS, which was the opposite of Karimov's intentions.

In sum, the key developments set forth in the following chronology illustrate the timing and increasing intensity of Russia's peacekeeping operations in Tajikistan, which afford Russia a strategic regional outpost near both Uzbekistan and Kazakhstan in addition to giving it a favorable vantage point from which to involve itself in the affairs of Afghanistan, Pakistan, and Iran.

## Chronology of Key Events

### 1992

*February 1992*

**7** • Tajik President Rakhmon Nabiev calls on all political parties and movements to unite and promote national concord, saying such action is necessary on the "eve of the anniversary of the February tragedy." Here, he refers to February 1990 riots in Dushanbe, sparked by rumors that Armenian refugees would receive housing before Tajiks. The incident triggered a wave of political factionalism and opposition activity.

• Nabiev meets with CIS border troop Commander-in-chief Kalinichenko to discuss securing Tajik border. Afghan mujahidin fighters continually attempt to cross into Tajik lands to support religiously-based opposition parties.

**28** • Tajik Supreme Soviet prepares a constitutional amendment precluding the establishment of an Islamic theocracy in Tajikistan by parliamentary means. The Islamic Renaissance Party, legalized in 1991, has gained considerable influence in Tajik political and social life, especially in the south.

*March 1992*

**18** • CSCE representatives arrive in Tajikistan to review its ability to uphold CSCE obligations. In meetings with opposition leaders, the CSCE delegates are told that human rights and international law standards are repeatedly being violated by the current regime, which is establishing anti-democratic rule.

**22** • Demonstrations take place in Dushanbe against the "pro-communist dictatorship," organized by the Democratic Party of Tajikistan and the Rastokhez Popular Movement; 1,500 to 2,000 participants are involved.

*April 1992*

**2** • Demonstrations continue in Dushanbe. Opposition leaders demand the resignation of parliament leaders including chairman Safarali Kendzhaev, a new constitution, and elections for new parliamentarians on a multiparty basis. In talks with Nabiev, the opposition says that should "softened demands" not be met and political persecution continue, a National Congress of the Tajik people will form to challenge official bodies. Hundreds of thousands of Muslims are reportedly en route to Dushanbe in support of the opposition.

**6** • The government of Tajikistan appeals to the demonstrators to return to their families and work. They note that all "economic and social aspects of life . . . are in a severe crisis. . . , held hostage to political games." The appeal says all issues should be resolved through peaceful means. Despite the call, opposition leaders only increase their demands on the government, adding a call for the resignation of Nabiev and his government. Rallies continue around the clock.

**12** • Tajik Supreme Soviet decides to meet opposition demands. Presidium leaders and Parliament Chairman Kendzhaev will resign, drafting of a new constitution will be accelerated, and parliamentary elections will be held once the new constitution passes. Talks resume between opposition and government leaders, with many political observers noting that the conflict is near its end.

**14** • As rallies continue, the opposition insists the Supreme Soviet convene to address its demands. Leaders of the Democratic Party, the Popular Movement Rastokhez, and the Islamic Renaissance Party send an appeal to the president and the parliament, declaring that if demands are not met swiftly, they will not be responsible for actions taken by rally participants. During the transaction, a column of armored personnel carriers passes through main streets, signaling impending armed conflict.

**20** • Deputy head of the Islamic Renaissance Party issues ultimatum to Tajik parliament, giving the body until noon 21 April to meet its demands. Deputies refuse to accept the ultimatum but agree to hold talks with opposition leaders.

**21–24** • Opposition leaders organize armed detachments and surround the Supreme Soviet building. Some deputies are taken hostage and armed protesters make their way to the front of the parliamentary building. Kendzhaev resigns and Nabiev says no protesters taking part in the hostage-taking will be prosecuted until after the 24th. Opposition leaders promise to end the rallies on the 24th and they do end because opposition leaders agree that the demands have been met.

**23** • Nabiev appoints Kendzhaev Chairman of the National Security Committee. The Supreme Soviet announces no prosecution actions will be taken against rally participants.

**25** • Tensions flare as new demonstrations in Dushanbe take place. Supporters of the government arrive from the Kulyab region, demanding Kendzhaev be reinstated and that the Supreme Soviet resolutions meeting opposition demands be abrogated. Opposition supporters again start to amass in the city, threatening serious conflict.

**29** • Clashes break out in the southern region of Kulyab, a communist stronghold where government supporters dominate. Tajik Islamic leader Akbar Turadzhonzoda says communist supporters launched a campaign of open terror against his people and against Islamic Renaissance Party members. In Dushanbe, riots threaten as government supporters chant slogans such as "Down with Islam! Down with Democracy which split the Soviet Union!" and "Long live Safarali Kendzhaev!" Despite talks between opposition and government leaders, the country is on the brink of civil war.

***May 1992***

**1** • Tajik parliament imposes presidential rule, giving Nabiev powers to control the legislative, judicial, and executive branches of the republic, powers to suspend all political parties and movements, and to impose a moratorium on rallies and demonstrations.

• Tajik parliament forms trilateral commission composed of government officials, opposition spokesmen and representatives of government support groups. The commission aims to prevent further deterioration of the situation and to work out a plan of national reconciliation.

**2** • Nabiev issues a decree to form battalions and incorporate them in a special brigade. The Tajik Defense Committee is charged with organizing the battalion which, according to battalion commander Col. Burikhon Dzhabirov, will act as a national guard corps.

**7** • Nabiev announces establishment of a National Reconciliation Government (NRG) involving current government leaders, opposition parties, and religious movements. The body will have broad powers and authority and pledges to ensure equal rights and political coexistence, denouncing further armed conflict. In a press interview, Nabiev says the main political figures of the NRG would form a "coalition council," but declares "not to renounce the powers given to me by the people." Government supporters and oppositionists continue their rallies, many becoming heavily armed. Clashes escalate.

**14** • In a press interview, Tajik defense officials, noting mujahidin formations on the Tajik-Afghan border, reveal that CIS troops stationed in Tajikistan will assist in controlling the border situation. Intelligence reports indicate Tajik opposition fighters have appeared in Afghanistan in search of weapons and support.

***June 1992***

**28** • Fighting continues in the Kurgan-Tyube oblast between supporters and opponents of the new coalition government. Rallies in Dushanbe, organized by the "united bloc of democratic forces" (the opposition parties included in the government coalition), demand Nabiev's resignation, charging him with ineptitude. Opposition parties accuse Nabiev of holding talks in Khodzhent and Kulyab with former Vice-President Narzullo Dustov and former Supreme Soviet Chair Kendzhaev, both deposed in May and opposed to cooperating with the opposition. Nabiev denies the charges.

***July 1992***

**8** • Amnesty is granted to participants in the April and May rallies that foreshadowed the conflict in Tajikistan.

**27** • A peace agreement is signed by representatives of political parties, government, religious leaders, and security service representatives. The agreement commits signatories to exchange hostages, lift roadblocks, disband armed formations, and hand over illegal weapons by 3 August. Nabiev does not attend the signing ceremony.

***August 1992***

**2** • In a press conference, Tajik democratic leader Shodmon Yusup charges Russia with domestic interference and calls for the immediate withdrawal of Russian troops stationed in Tajikistan. Russia's 201st Motorized Infantry Division was involved in July fighting incidents in the Kurgan-Tyube oblast, but, according to Lt.-Col. Nikolay Surkin, the division maintains a neutral position. Says Surkin, CIS troops are the only real troops in the republic and have acted to separate the opposing sides and to confiscate weapons in accordance with the 27 July Khorog agreement.

**4** • Tajik Deputy Prime Minister Usmon Davlat laments the fact that few firearms were surrendered when the date set to do so came on 3 August. Only 19 out of an estimated 17,000 firearms were turned into surrendering stations set up by officials. Despite this, Davlat says implementation of the Khorog agreement is "gaining momentum" since the arrival of commission members and opposition representatives to the conflict zones.

**10** • The executive committee of the Kurgan-Tyube oblast, where fighting has been most intense, reports that over 300 people were killed and over 350 are missing as a result of June–July fighting.

**19** • In televised address, Tajik Internal Affairs Minister Mamadaez Navzhuvanov says the Khorog agreement has failed, factions are amassing weapons, and deaths continue mounting—over 1,000 thus far. He says no paramilitary group intends to surrender weapons and that Commonwealth countries and Afghanistan illegally arm the warring sides. Tens of thousands of refugees have resulted.

**26–28** • In a meeting with Tajik military experts, CIS Joint Armed Forces Commander-in-chief Shaposhnikov proposes introducing CIS "Blue Helmets" into Tajikistan. He says such troops could be successful at disengaging combatants after a ceasefire agreement. Shaposhnikov will tour the conflict zones over the next few days. On the 28th, he signed agreements with Nabiev to deploy military forces where the political situation is most tense.

***September 1992***

**1** • Military observers from the CIS states of Russia, Kazakhstan, and Kyrgyzstan arrive in Dushanbe, headed by Russian General Shamilov. Speaking to journalists, Shamilov says the group will take ten days to assess the situation in the various conflict zones, stressing the group will not take sides.

**7** • Cornered by opposition forces at the Dushanbe airport, Nabiev resigns as president "with aim of further stabilizing the situation and ending the fratricide." Presidential powers were transferred to Supreme Soviet Chairman Akbarsho Iskandarov.

**8** • In a press interview, Uzbek President Islam Karimov speaks of the dangers for Uzbekistan caused by the Tajik civil war. He laments that the Afghan-Tajik border is virtually non-existent, that weapons and drugs are constantly transported across it, and that the country is entirely in the hands of fundamentalists—an alarming development. He expresses deep concern for the 1.3 million Uzbeks living in Tajikistan.

**10** • One thousand troops arrive from the republics of Uzbekistan, Kazakhstan, Kyrgyzstan, and Russia to assist in controlling the Tajik-Afghan 2,000–kilometer border zone. Border violations have significantly increased over the last few months as combatants search for weapons, supplies, and training grounds.

**9** • The Bishkek CIS summit fails to produce an agreement on sending peacekeeping forces to Tajikistan.

**14** • Preparations to send Kyrgyz peacekeeping troops into Tajikistan are suspended as the Kyrgyz Supreme Soviet votes to cancel the contingent, fearing for the safety of Kyrgyz servicemen.

**25** • Rebel forces control Dushanbe briefly but are driven out by government forces. While the rebels control broadcasting facilities, former Supreme Soviet Chairman Safarali Kendzhaev, an ally of Nabiev's, claims himself President, and appeals to non-Tajiks not to leave, denouncing the "Islamic fundamentalism" of the current leadership. During the attack, Russian troops in Dushanbe remain neutral, but retain control of the airport and main railway stations.

***November 1992***

**10** • Acting President Akbarsho Iskandarov and his coalition government resign, calling for a ceasefire as pro-communist militias lay siege to Dushanbe and occupy much of Tajikistan. Iskandarov said that he was resigning to save the country from destruction.

**19** • The Supreme Soviet elects Emomali Rakhmonov, a former communist and chairman of the executive committee in the southern Kulyab region, as speaker of parliament (the de facto president).

**26** • Parliament exempts conflict participants from criminal prosecution and decides to mark 26 November as an annual Day of Peace. Communist militias lift the blockade of

Dushanbe and open the road between the capital and Kurgan-Tyube region.

**8** • Deputy Commander-in-chief of CIS armed forces Col.-Gen. Boris Pyankov says up to 5,000 CIS peacekeeping troops will be deployed to Tajikistan as heavy fighting continues despite the ceasefire.

***December 1992***

**10** • After five days of fighting, pro-government forces take control of the capital.

**12** • The former communist Emomali Rakhmonov, chairman of the Tajik Supreme Soviet since the collapse of a pro-Islamic coalition government (in November), opens peace talks with Islamic opposition forces. However, continuing heavy fighting is reported near Kafarnihan, the Islamic forces headquarters 25 km east of the capital, and fighting also continues in Dushanbe.

**15** • More than 100,000 refugees are stranded on the banks of the Oxus River, between Tajikistan and Afghanistan. People are dying of cold in the freezing conditions. Some 5,000 people a day are entering Afghanistan, braving an icy river crossing. The fighting displaces 10 percent of the republic's 5,000,000 people.

**20** • Government forces take Kafarnihan after three days of heavy fighting. Clashes continue when opposition forces retreat.

## 1993

***January 1993***

**4** • Tajik and Uzbek leaders meet for the first time to discuss Tajik conflict.

**8** • A state of emergency is declared. Russia's 201st Motorized Infantry Division is activated, and curfews imposed.

**9** • Relative calm results from the introduction of a curfew.

• CIS states send food aid to mountainous areas of Tajikistan, where people are dying of malnutrition (Pamir, Garm).

**12** • Tajik leader Emomali Rakhmonov meets with Russian Border Troops Command staff to discuss provision of Russian troops and matériel. He says he sees Russian troops in Tajikistan as a "stabilizing factor" for CIS southern border.

**21** • Unofficial talks between Yeltsin and Rakhmonov end with Russian promises to provide food, medicine, and fuel and 200 billion rubles in technical credits.

**22** • The United Nations announces it will provide $20 million in humanitarian aid to Tajikistan; $1 million is earmarked for housing the 40,000 refugees in the region.

• CIS Minsk summit resolves to send four battalions of Commonwealth peacekeeping forces (mostly Russian) to Tajikistan in order to patrol the Afghan border. Rakhmonov welcomes the decision, saying it would demonstrate "Tajikistan is not alone and all CIS countries are working together to protect the borders."

**25** • Tajik government troops attack last centers of armed opposition (Romit Gorge—70 km from Dushanbe). The government hopes to quell all opposition before CIS peacekeeping forces move in.

**26** • UN representatives visit Tajikistan, expressing support for CIS Minsk summit decision to send peacekeeping forces to end the conflict.

• Russian border guard units go on alert due to planned oppositionist attacks from Afghanistan.

**27** • Foreign Minister Rashid Alimov receives U.S. Ambassador to Tajikistan Stanley Escudero to discuss bilateral relations. Escudero reports that aid will arrive in next few days.

• High-ranking officers from Supreme Command of the CIS Joint Armed Forces, headed by First Deputy Chief Vladimir Krivonogikh, visit Tajikistan to assist with organization of the Tajikistan Ministry of Defense and creation of a national Tajik Army.

**29** • A state of emergency is declared along the Afghan border.

***February 1993***

**4** • In Dushanbe, Grachev holds a talk with Rakhmonov, saying thirty-one representatives of the Russian Defense Ministry will remain in Tajikistan to help form a local army. He emphasizes that in keeping with the CIS collective security treaty, Tajikistan should be included in the joint air defense system. The Russian 201st Division will also remain in Tajikistan and will be under Russia's command. In an address to the Russian troops, Grachev declares that the region is strategically important to Moscow, and that the Russian troops serve as a bulwark against Islamic fundamentalism.

**21** • Government forces retake Romit Gorge and open up

access to the Pamir region by gaining control of Garm, Komsomolabad, Novabad, and Tajikabad. Government forces also capture the strategic rebel stronghold of Tavildara, 200 km east of Dushanbe.

*March 1993*

**3** • CIS peacekeeping forces from Kyrgyzstan arrive in Tajikistan and are stationed on the Tajik-Afghan border. Uzbekistan has already delivered a battalion, and troops from Russia and Kazakhstan are expected soon. Each battalion consists of 500 soldiers.

**12** • Three hundred thousand persons who became refugees during the Tajik civil war are reported to have returned to their homes. The conservative government in Dushanbe is using the refugees' return as a measure of the normalization achieved in Tajikistan.

**30** • Kyrgyzstan withdraws its battalion of 500 border troops from Tajikistan.

*April 1993*

**2** • According to the National Security Council, there are about 500 Afghan mujahidin on Tajik territory waiting for orders from the opposition to begin combat operations.

**4** • Opposition militants constantly try to invade Tajikistan from Afghanistan but are stopped by border guards. The Tajikistan Foreign Ministry protests to Afghanistan, saying that this is "giving reasons to doubt the sincerity of the repeated assurances by the leadership of Afghanistan that it would take all necessary measures to prevent such acts."

**15** • To slow migration of the Russian-speaking population out of Tajikistan, the leadership considers revising the law on the state language to give Russian equal status with Tajik.

**27** • Kozyrev describes Tajikistan as a region of particular interest for the Russian Federation because of the presence of a Russian-speaking population of more than 200,000. (Many of the Russian speakers have fled to escape the civil war, but large-scale out-migration of the Russian-speaking population had begun already in 1990.)

*May 1993*

**18** • UN experts are prepared to help Tajikistan draw up a new constitution and to help draft laws on elections and other legal documents.

**25** • Yeltsin and Rakhmonov sign a treaty on friendship, cooperation, and military assistance plus seven other agreements on economic, scientific, and technical cooperation and on the status of Russian frontier troops in Tajikistan.

**28** • Talks between the official Tajik authorities and leaders of military formations of the Gorno-Badakhshan autonomous region in the Pamirs take place. The Pamir delegation advances the terms on which the opposition is ready to end resistance. They include: strict observance of the Tajik Supreme Soviet's amnesty decree, which, in their opinion, is not always honored; the organized return of refugees from the Pamirs and their guaranteed personal safety in places of permanent residence; the reorganization of so-called Pamir self-defense detachments into government battalions of the national army and the interior ministry; and the quick opening of the Osh-Khorog humanitarian corridor.

**29** • Afghan and Tajik rebels attack a Russian border post.

*June 1993*

**3** • Units of the Tajik Ministry of Internal Affairs reinforce the Tajik-Afghan border due to renewed threats of military provocations by Tajik opposition formations based along the border.

**8** • The Afghan army has been actively organizing and arming Tajik opposition forces. Tajikistan warns that such actions are fraught with serious consequences, and reserves the right to involve CIS collective security treaty members in combating the Afghanis.

**21** • CIS troops in Tajikistan shell the border areas in Afghanistan's Qunduz province. The attack reportedly causes heavy loss of life. CIS forces have been frustrated in their attempts to stop the insurgents without attacking the rebels' bases in Afghanistan.

*July 1993*

**10** • Rakhmonov announces creation of a multilateral commission with representatives from Tajikistan, Uzbekistan, Russia, and Afghanistan to deal with the Tajik-Afghan border situation.

**15** • The Russian parliament approves a Foreign Ministry order to take "all necessary measures to protect and ensure the safety" of Russian border guards in Tajikistan. It also ratifies a treaty with Tajikistan signed on 5 May 1993 creating a legal basis for Russian troops to engage in combat operations. Citing "a threat to the national security of the

Russian Federation," Yeltsin threatens unilateral action if parliament does not act quickly. Deputy Defense Minister Konstantin Kobets implies that Russian forces, now authorized to use aircraft and missiles, might attack bases in Afghanistan; he says that Russian troops would take action on the entire territory of the conflict, acknowledging that "most of the weapon stockpiles and terrorist training camps are in Afghanistan."

**16** • Grachev and a high military delegation tour Tajikistan and the border area where twenty-four Russian soldiers were killed on 13 July. Calling the cross-border attacks "an undeclared war against Russia," he promises to punish those who led the attack and to strengthen Russian forces in Tajikistan. Tajik and Russian military officials criticize other CIS leaders, especially in Central Asia, for not contributing enough.

**21** • Saber-rattling over the Tajik-Afghan border crisis continues. The Chief-of-Staff of the Russian Security Ministry Viktor Barannikov states that "the border guards have received an order to open fire . . . even across the border," and underlines that Russian troops have "the moral right to raid Afghan territory if violations of the border do not stop." Emotional debates are going on in the Russian and Kazakh parliaments over their respective roles in Tajikistan, with some deputies in both states expressing fears of repeating the Soviet experience in Afghanistan.

**31** • Kazakh President Nazarbaev states his support for proposals which call for the use of UN observers, and appeals to the leaders of Pakistan, Iran, Afghanistan, Turkey, and Saudi Arabia to formulate a solution for peace.

***August 1993***

**1** • The Russian government issues a threat to launch a preemptive strike on the Tajik-Afghan border if Tajik opposition forces and their Afghan supporters continue to concentrate along the Afghan side of the border.

• Evgeniy Primakov, acting as special envoy for Yeltsin, meets with Iranian leaders in Tehran to discuss the ongoing crisis on the Tajik-Afghan border. The two nations agree the crisis should be resolved by negotiations between the government and the opposition. In Tajikistan, Primakov emphasizes the importance of Iran's role in negotiations.

**4** • According to Kozyrev, Russia's national interests dictate that it play a forceful role in Tajikistan. Russia's presence there is aimed at preventing the renewal of civil war, providing security for the multinational population there, developing Tajik democracy, and ensuring that the entire region does not fall prey to political extremism.

**5** • Tajik Deputy Prime Minister Abdujalil Samadov states that Tajikistan's government refuses to accede to Kozyrev's appeal that it negotiate with opposition leaders in order to end the fighting in Tajikistan.

**7** • The Kazakh, Kyrgyz, and Tajik heads of state meet with Yeltsin to seek a solution to the Tajik-Afghan border conflict. The participants decide to increase the number of troops guarding the border; to increase humanitarian, military, and economic aid to Tajikistan; and to retaliate if outside attacks on the border continue. The declaration also calls for a political solution to the conflict.

**15** • The foreign ministers of Tajikistan and Afghanistan conclude several days of negotiations with a communiqué in which the Afghans say they will try to stop attacks on Tajikistan from Afghani territory, and the Tajiks say they would not allow attacks on Afghanistan except in cases of self-defense.

***September 1993***

**9** • Kozyrev meets with Tajik leaders in Dushanbe after telling the press that Russia is to send several thousand more troops to the area to reinforce the approximately 15,000 that are already there. He urges the leadership to begin a dialogue with the armed opposition which has set up a government-in-exile just across the border in Afghanistan. The Tajik foreign minister says that he would have nothing to do with those "criminals."

***October 1993***

**5** • Kazakh Foreign Minister Tuleitai Suleymenov asks the United Nations to formally recognize the CIS peacekeeping troops guarding the Tajik-Afghan border as UN peacekeeping forces. These troops currently compromise units from Kazakhstan, Kyrgyzstan, Russia, and Uzbekistan.

**22** • The commander-in-chief of the CIS peacekeeping forces, Boris Pyankov, meets with leaders of the self-ruled Gorno-Badakhshan region in the Pamir mountains, urging them to cooperate with the CIS forces' leadership.

**25** • CIS coalition forces in Tajikistan inflict their first assault on what they refer to as the "Islamic opposition." It is a combined air and artillery strike on guerrillas as they attempt to cross the border from Afghanistan.

**26** • The government of Tajikistan regains control of Dushanbe from opposition fighters from Kulyab.

***November 1993***

**5** • The Tajik-Afghan border is cited as the most difficult hot spot in the CIS. The opposition is reportedly regrouping on the Afghan side for operations in the winter.

**9** • The heads of the security agencies of Tajikistan, Russia, Kazakhstan, Uzbekistan, and Kyrgyzstan meet to discuss setting up a joint force to protect the Tajik-Afghan border and normalize the situation in Afghanistan. The agreement differs from an earlier one signed by them in that it refers to interference in Tajikistan's internal affairs.

**16** • CIS joint peacekeeping forces launch war games in Tajikistan to test their combat capabilities in mountain and desert terrain. Military sources say that the training is in preparation for a potential full-scale invasion of Tajikistan by the opposition rebels training in Afghanistan.

**18** • Kozyrev, spending two days in Dushanbe, signs a treaty of friendship and cooperation with Tajikistan. Kozyrev announces his satisfaction with the talks with the Tajik government on the stabilization of the conflict and the securing of the Tajik-Afghan border.

***December 1993***

**6** • Tajik First Deputy Supreme Soviet Chairman Abdumadzhid Dostiev states that Tajikistan's leadership has not and does not intend to negotiate with the armed Tajik opposition because the opposition has no support within the country.

**15** • The head of Russia's border troops states that CIS peacekeeping forces in Tajikistan may need strengthening before the spring, when they expect renewed attacks. He also suggests that it may be in Russia's interest to extend the CIS agreement, to provide military support to the Tajik government, which is due to expire at the end of December.

**22** • The commander of CIS forces in Tajikistan, Russian General Boris Pyankov, states that troop reinforcements will not be needed, but that earlier CIS resolutions calling for troops from Kazakhstan, Kyrgyzstan, and Uzbekistan have not yet been met and are needed. The question of whether to put these troops under the control of the CSCE or the United Nations has not been resolved, partly because the United States raises the question of Russia's ability to remain neutral as a peacekeeping force.

**21** • Tajik Islamic Renaissance Party chairman Muhammadsharif Himmatzoda says his party is now willing to share power with the present "neo-communist" government. He says continued fighting only plays into the hands of Russian national extremists.

## 1994

***January 1994***

**4** • A Russian Orthodox priest is killed shortly after conducting a service for members of Russia's 201st Motorized Infantry Division, the core of the CIS peacekeeping troops in Tajikistan. Tajik authorities report that the killing is in connection with a robbery.

**12** • The Supreme Council Presidium of Tajikistan extends the state of emergency for six months.

***March 1994***

**14** • The foreign ministers of Kazakhstan, Kyrgyzstan, Tajikistan, Uzbekistan, and Russia meet in Dushanbe to discuss the situation in Tajikistan, where they have all committed troops.

• Tajik Foreign Minister Rashid Alimov reports that the Tajik government has requested the UN Security Council to recognize CIS troops on its soil as a UN peacekeeping force.

**24** • The CIS military command holds its first set of joint exercises in Tajikistan not far from the border with Afghanistan. Press reports that the exercises included mock attack aircraft, tanks, and helicopters, which is unusual because the primary threat in the area comes from Tajik rebels who do not have air and armor support.

***May 1994***

**6** • Rakhmonov concludes two days of talks with top Russian leaders in Moscow. According to the communiqué issued at the end of the visit, the alleviation of Tajikistan's severe economic crisis and a resolution of the Tajik civil war dominated the agenda. Rakhmonov urges accelerating the unification of the Russian and Tajik monetary systems.

**7** • Russia's Ministry of Foreign Affairs issues a warning that Russia will use any necessary measures to secure the Tajik-Afghan border. Russia accuses the Tajik Islamic opposition of planning a spring offensive.

**31** • A lieutenant-colonel of the Russian border guards in Tajikistan is killed in Dushanbe. Not long afterward, three more Russian officers are found dead. The rash of killing may be related to the forthcoming peace talks in Tehran. While certain Tajik officials may seek to derail the talks, others in the Tajik government want Russia to take a harder line against the Tajik opposition, which, not surprisingly was blamed for the deaths.

*June 1994*

**6** • The number of Russian officers killed in Tajikistan in the last ten days rises to seven. Russian officials say the attacks are an attempt to undermine Russia's commitment to stay in Tajikistan. The Tajik opposition in exile rejects charges that it was responsible for the killings, blaming instead those "forces which are not interested in resolving the conflict peacefully." The opposition also calls on Russia to adopt a more neutral stance, otherwise its legitimacy as mediator in the conflict would be diminished.

**9** • Russian forces along the Tajik-Afghan border shell suspected Tajik rebel bases in Afghanistan in response to rocket attacks by rebel forces. Commander of Russian border guards in Tajikistan Anatoliy Chechulin reiterates that Russia will use any means necessary to protect Russian servicemen deployed in Tajikistan. He states that increased rebel activities across the border signals a prelude to a larger summer offensive being planned by the rebels.

**15** • Tajik Deputy Defense Minister Ramazan Radzhabov and his five bodyguards are killed near the town of Garm, 150 km from Dushanbe. No group claims responsibility for the killings, but the region is considered an opposition stronghold. Despite the killings, UN-sponsored peace talks are still scheduled to begin on 18 June in Tehran. Representatives of the Tajik government, the Tajik opposition, and the Iranian, Russian, and Pakistani governments will participate along with UN mediators.

**16** • Tajik government and opposition representatives to the Tehran peace talks outline their positions in interviews with Interfax. According to the government negotiator, the Tajik government seeks a ceasefire, the repatriation of refugees, and the creation of a constitutional system. The opposition for its part also stresses the need for peaceful settlement and issues a warning (intended for the Russians) that failure to reach an accord could lead to Afghan-like conditions. He emphasizes that the Tajik opposition must be consulted before any new constitution can be ratified.

**17** • Approximately 2,000 Russians and other Slavs reportedly migrate from Tajikistan every month. Overall, only 120,000 Slavs remain in the republic, a substantial decline from the 600,000 that lived there in the Soviet era. (The report did not indicate whether the new figure includes the 20,000 Russian troops currently stationed in Tajikistan). The emigration continues apace despite far-reaching concessions to the Russian-speaking population, including the right to hold dual citizenship. The report cites bleak economic prospects, in addition to the obvious security concerns, as causes for the exodus.

**18** • UN-sponsored peace talks begin in Tehran. UN special envoy for Tajikistan Ramiro Piriz-Ballon states that a ceasefire is a prerequisite for the difficult task of finding a lasting peace settlement. Piriz-Ballon also announces that the United Nations stands prepared to "immediately send its representatives to Tajikistan" to verify compliance with any agreement. Despite a status quo that favors the government's forces, the Tajik opposition announces its willingness to agree to a ceasefire along the Tajik-Afghan border.

**21** • In a move that could derail the ongoing peace talks, Tajikistan's armed forces launch attacks against suspected rebel strongholds within Tajikistan, killing sixteen people and confiscating a large cache of arms.

**23** • The UN-sponsored peace talks in Tehran yield no positive steps toward ending the civil war in Tajikistan. Tajik opposition demands that the Tajik government release political prisoners, legalize all political parties, and grant the right to publish newspapers. On the question of disarmament, the opposition proposes a bilateral disarmament under the supervision of the United Nations, CSCE, Russia, Iran, Afghanistan, and Kazakhstan. The government counters with a demand that the opposition disarm unilaterally.

**26** • The Tehran talks fail to produce even a temporary ceasefire agreement as each side blames the other for lack of progress in the talks. However, a Russian observer indicates that the opposition may drop its demands linking any ceasefire agreement with a release of political prisoners, the restoration of a free press, and the legalization of political parties.

**27** • Talks in Tehran conclude without agreement on a ceasefire, according to Russian news agencies. Despite this failure, both sides express optimism that the next round of talks, scheduled for the end of July in Islamabad, will lead to an end to the conflict and a normalization of the political situation in the beleaguered country. The opposition proposes a ceasefire, effective 1 August, conditioned on the release of political prisoners held by the Tajik government. Government representatives countered that a minimum of four months was needed to prepare the release and amnesty of the opposition forces currently incarcerated.

**28** • Representatives of the Tajik government and Tajik opposition forces issue a joint declaration supporting a ceasefire and national reconciliation, although no timetable for a cessation of hostilities is set. UN mediator Ramiro Piriz-Ballon says that the two sides made considerable progress and that the Tajik government committed itself to prepare to free political prisoners and grant amnesty to opposition leaders.

*July 1994*

**1** • Underscoring the sense of optimism that emerged from the Tehran peace talks, Tajik Labor Minister Shukhurdzon Zukhorov, who led the government delegation, announces that the majority of ministers and the Tajik head of state, Rakhmonov, believed that the opposition's demands could be met, but not immediately.

**14** • Tajik refugees have begun to return home from Afghanistan. According to the Tajik Foreign Ministry, there are about 20,000 Tajik refugees in Afghanistan. The Tajik opposition claims 100,000.

**19** • Tajik First Deputy Interior Minister Gennadiy Blinov says that a state of emergency in Dushanbe and regions adjacent to the Afghan border has been prolonged in order to secure these areas against incursions by "uncontrolled armed opposition groupings."

• Despite evidence indicating a decrease in border incidents by two-thirds from previous years, commander of the CIS peacekeeping forces Col.-Gen. Valeriy Patrikeev says that the current deployment of 7,500 soldiers is inadequate, calling for a twofold increase in troop strength. He appeals to the governments of Kazakhstan, Uzbekistan, and Kyrgyzstan to provide considerably more peacekeeping troops for the area.

**22** • According to chief-of-staff of the Collective Peacemaking forces of the CIS, Col.-Gen. Bessmertniy, the situation is deteriorating in the republic and no improvement has been recorded in the past month. He also notes that the second round of intra-Tajik negotiations in Tehran has no constructive effect and there are more frequent cases of crossing the state border both from Afghanistan into Tajikistan and in the opposite direction.

• Opposition forces attack Tajik government forces east of Dushanbe, capturing fifty-six soldiers and major equipment. The attack represents the largest operation within the country since the end of the civil war.

**26** • CIS and Tajik government forces launch attacks at rebel positions. Su-25 planes destroy equipment commandeered by the rebel forces, but prisoners' whereabouts remain unknown. General Patrikeev says his forces were conducting searches and were negotiating with local rebel leaders—activities in clear violation of Russia's oft-stated mission only to protect the Tajik-Afghan border, not to interfere in the internal imbroglio in Tajikistan.

**28** • Rakhmonov meets with defense, interior, and security ministers, commanding them to destroy the rebels. The command of Russia's 201st Motorized Infantry Division stationed in southern Tajikistan had refused "for the time being" requests from the Tajik government to intervene on its behalf against the rebels. Moreover, Russian media reports that CIS border forces turned back attempts by armed opposition groups to cross the Tajik-Afghan border. Russian Defense Minister Grachev uses the degradation of the Tajik situation to lobby for effective, centralized command of border guards and armed forces in the CIS. Meanwhile, thirty of the fifty-six missing soldiers did defect to the rebels, according to a Tajik Defense Ministry spokesman.

*August 1994*

**3** • General Patrikeev is convinced no alternative to the Russian presence exists. If Russian troops leave Tajikistan, the country will turn into a "burning torch." The only way to restore peace in the republic is to strengthen Russia's role.

**9** • Tajik parliament prepares an amnesty decree releasing political prisoners, meeting one of the opposition demands.

**11** • Russia's First Deputy Foreign Minister Anatoliy Adamishin blames the Tajik opposition for driving the country to the brink of its second civil war in two years. Adamishin says that the opposition, having failed to achieve its goals of democratization and power-sharing, is resorting to violence to advance its agenda. He urges the opposition to cooperate with the United Nations, the West, and Russia as an alternative to bloodshed. Despite this plea, Adamishin criticized the UN's apparent bias in favor of the Tajik opposition.

**17** • Despite the conflict's escalation, the Dushanbe government decides not to renew the state of emergency imposed at the end of 1992. Tajik Deputy Minister of National Security Anatoliy Kuptsov states that the state of emergency was allowed to elapse due to the 25 September presidential election and referendum on a new constitution.

**19** • Heavy fighting resumes on the Tajik-Afghan border, resulting in the deaths of at least fifty Tajik opposition fighters and seven Russian border guards. The attack is jointly coordinated by opposition forces on the border as well as opposition forces within Tajikistan. Russian troops were forced to call in helicopter gunships and Su-27 bombers to repel the attack.

**24** • Tajik opposition leader Akbar Turadzhonzoda* says that the opposition considers the upcoming presidential elections illegal and is calling for an election-day boycott. He also says that the opposition forces would resume their

---

*The Islamic Renaissance Party and other religiously-based opposition groups joined under the umbrella of the Movement for Islamic Revival in Tajikistan (MIRT), in which Turadzhonzoda is a dominant figure. In coalition with the Democratic Party and the Rastokhez Movement they formed the United Tajik Opposition.—Ed.

offensive on the border, claiming that it is the only way to unseat the Dushanbe government.

**31** • Interfax reports that nearly all the Tajik refugees who wanted to return home have done so. The UN High Commissioner for Refugees in Central Asia, Philippe Labreveux, estimates that over one million people left the country as a result of the civil war. About 350,000, mostly ethnic Russians, have not returned and it is highly unlikely that they will in the future.

***September 1994***

**1** • Rakhmonov announces that an amnesty decreed by the Presidium of Tajikistan's Supreme Soviet would include some members of the Tajik opposition who are accused of inciting the 1992 civil war.

**2** • In a statement circulated at the UN Security Council, Rakhmonov protests the use of mercenaries by the Tajik opposition and warns that these "soldiers of fortune" are contributing to an escalation of the conflict threatening instability in all of Central Asia. The complaint names Algeria, Afghanistan, and Sudan as states supplying fighters to the Tajik opposition. Rakhmonov appeals to those states and others to support the peaceful resolution of the conflict and cease the flow of mercenaries to the region.

**7** • Tajikistan's Supreme Soviet votes to postpone the presidential election and referendum on a new constitution until 6 November. The decision is made in hopes that the Tajik opposition could be persuaded to participate. Rakhmonov comments, "The opposition must also be given a chance to nominate its own candidates."

**10** • Tajik government troops incur additional losses in two days of fighting east of Dushanbe. The opposition forces captured the town of Tavil-Dara, effectively cutting off the capital from the southern and eastern parts of the country. Government losses are reported to be high. Opposition sources said that government troops are offering little resistance due to shortages of fuel and ammunition. Russian troops reportedly have stayed "in their barracks" and do not appear willing to involve themselves in the internal fighting in Tajikistan: they maintain that their sole function is monitoring the Afghan-Tajik border.

**12** • In the wake of reported military setbacks, a Tajik government delegation travels to Tehran to hold talks with the Tajik opposition in exile and representatives of the Russian Foreign Ministry. The talks were characterized as preparatory discussions for the resumption of formal negotiations later this month.

**13** • Tajik government forces launch a counteroffensive in an attempt to retake the Tavil-Dara rayon east of Dushanbe. The Tajik air force, which days earlier was grounded due to fuel shortages, strikes against opposition positions, forcing residents of several villages to flee.

• Grachev advocates tougher measures to deal with the situation in Tajikistan. He calls for a stronger group of forces and for the CIS Border Forces along the Tajik-Afghan border to be integrated with the army to form a "powerful fist." He predicts the situation in the war-torn country will become "more acute."

• A key opposition faction, the Democratic Party, considers participating in upcoming presidential elections. The party, along with the nationalist Rastokhez movement and the Islamic Renaissance Party, was blamed for instigating Tajikistan's civil war in 1992 and was outlawed in 1993. Reports from the talks in Tehran also indicate that the Rastokhez Movement may be willing to take part in the elections. The Islamic Renaissance Party refuses to recognize the legitimacy of current election results.

**17** • UN-Russian-Iranian sponsored negotiations between the Tajik government and opposition result in the agreement on a temporary ceasefire. It will become effective as soon as UN monitors can be deployed to the area. It is scheduled to last until 5 November. Both sides also agreed to release political and war prisoners.

**21** • Citing Tajikistan's inability to monitor its own border with Afghanistan, Patrikeev says that Russian border and peacekeeping troops would remain in the republic for another year or two.

**23** • Russian border and peacekeeping troops braced for stepped-up combat operations by the Tajik opposition along the Tajik-Afghan border as well as inside the Tajik republic. The Tajik opposition wants to gain advantageous positions before the arrival of UN observers and with them the implementation of a ceasefire. The observers are expected by 28 September.

***October 1994***

**19** • Talks between the Tajik government and the Tajik opposition resume in Islamabad, Pakistan. The head of the Tajik government's delegation, First Deputy Chairman of the Tajik parliament Abdumajid Dostiev, says that the talks would cover a permanent ceasefire, repatriation of refugees from Afghanistan, and the future system of government in Tajikistan. Dostiev also said that he would encourage the participation of the opposition in the 6 November presidential elections and referendum on a new constitution. Meanwhile, eleven UN monitors arrive in Dushanbe to observe compliance of the temporary ceasefire.

**20** • The talks hit a potential snag when the Tajik Islamic opposition leader Turadzhonzoda accuses the government of murdering two political prisoners. The accusation followed a Helsinki Watch report which stated that the Tajik government reneged on its promise to release twenty-seven political prisoners, one of the opposition's preconditions for resumption of the talks.

*November 1994*

**8** • The Council of CIS Heads of State extends the mandate of the CIS joint peacekeeping forces in Tajikistan until 30 June 1995 and broadens the powers of the commander of these forces, Maj.-Gen. V. Yakushev.

**12** • The Tajik armed opposition ignores the agreement on ceasefire along the Tajik-Afghan border, trying to cross into Tajikistan from Afghanistan. Concentrations of these forces are spotted near the border in the Afghan provinces of Tahor and Badakhshan. Armaments are being delivered there. Instructors from several Arab states are known to have arrived in the area.

**17** • Opposition leader Akbar Turadzhonzoda charges the Russians with violating the ceasefire agreement. Under the Islamabad truce agreement, all the parties in the conflict have no right to regroup, let alone engage in combat. He says that the CIS troops fired heavy artillery shells and helicopter-borne missiles at opposition troops near the village of Olur. Lt.-Gen. Chechulin says that there has been no ceasefire on the Tajik-Afghan border and that since the ceasefire agreement came into force on 20 October, Russian border guards have stopped twenty-nine attempts by armed groups to cross the border and were subjected to seventeen attacks by opposition militants.

*December 1994*

**5** • According to the Tajik Foreign Ministry, 38,000 Tajik refugees returned home from Afghanistan and 21,000 remain there. According to the Tajik opposition, about 80,000 refugees from Tajikistan remain in Afghanistan.

**6** • Patrikeev reports that the situation at the Tajik-Afghan border took a slight turn for the better compared with August–September. He praises the inter-Tajik ceasefire agreement reached in Islamabad and control over its implementation on the part of international bodies, particularly the United Nations, which sent a group of military experts to Tajikistan. He also describes first steps aimed at disarming illegal militant units in the Gorno-Badakhshan region as "encouraging." He says that inhabitants of several districts began to surrender arms. On the other hand, the general situation in Gorno-Badakhshan region is complicated because there are several scores of pro-opposition groupings in the region and they posses about 2,000 firearms.

## 1995

*January 1995*

**4** • Press reports reveal the Foreign Ministry of Tajikistan sent a letter to CIS leaders containing an urgent request to introduce additional measures to reinforce and protect the Tajik-Afghan border in line with the May 1992 CIS collective security treaty signed in Tashkent.

**6** • According to the commander of the armed wing of MIRT (the Movement for Islamic Revival of Tajikistan), Rizvon Sadirov, now that Russia has "launched an aggression against Muslim Chechnya," he has declared a holy war (*jihad*) on the Russian servicemen in Tajikistan.

**9** • In *Novaya Ezhednevnaya Gazeta*, Dodo Atovullo, a Tajik refugee, condemns the fact that the war of over two years has gone virtually "unnoticed" despite the fact that the number of victims is greater than in the Yugoslavian, Abkhazian, or Karabakh wars.

• Rakhmonov declares that the government of Tajikistan needs Russian border guard services, a clear appeal to the border guards to safeguard Tajikistan's border. Otherwise, the regime feels it could not remain in power for even a day.

**11** • Tajik Deputy Foreign Minister Rakhmatollahov calls on the Islamic Republic of Iran to use its influence to facilitate the convening of the fourth round of Tajik peace talks. Moreover, he expresses Tajikistan's readiness to participate in these talks.

**12** • UN military observers meet with Rakhmatollahov to deal with issues relating to the implication of the Tehran and Islamabad inter-Tajik agreements and the preparations for the regular round of the inter-Tajik talks in Moscow.

**14** • Tajikistan's foreign policy priority is still to strengthen ties with Russia and other CIS countries, according to Talbak Nazarov, Tajikistan's new foreign minister. He notes that his department will continue the dialogue with the opposition.

*March 1995*

**13** • In a press interview, First Deputy Director of the Russian Federal Frontier Force Col.-Gen. Aleksandr Tymko says fighters from MIRT launched a spring offensive, with between 6,000 and 8,000 members standing poised across the Tajik border ready to challenge the frontier guards. Although Tymko does not rule out a major breakthrough attempt, he notes that the Islamic fighters are employing a tactic of "quiet infiltration." Tymko claims that the frontier guards possess sufficient manpower and equipment to repel a major attempted breakthrough, saying "I don't think the fighters across the border have enough strength to oust Russia's frontier guards."

**15** • MIRT leader Abdullo Nuri, a major component in the Tajik opposition, expresses willingness to talk "at any time and at any place" to UN military observer head Gen. Hassan Abasa. Nuri says the opposition will not initiate a spring offensive, but border skirmishes between Russian-Tajik border guards and Tajik opposition groups continue.

**28** • Russian Federal Border Service Director Andrey Nikolaev reports that the armed Tajik opposition is increasing attacks on the Tajik-Afghan border under MIRT leadership. Nikolaev notes that the opposition has obtained massive military supplies including small arms, mortars, multiple-rocket launchers, and choppers in preparation for a May offensive. Nikolaev says that while the opposition has ample weaponry and forces, the likelihood of rebel-induced destabilization in Tajikistan is receding because the opposition "has lost influence and initiative."

***April 1995***

**1** • Grachev announces his opposition to reinforcement of Russian forces in Tajikistan, saying he believes the Tajik leadership is not "fully using its own means of settling the conflict." Grachev also comments that to stake the settlement of the conflict solely on force in "unpromising."

**3** • Uzbek President Karimov meets with Tajik opposition leaders in Tashkent. The participants exchange views on ensuring regional security and establishing peace in Tajikistan. Karimov calls on the leaders to renounce military means of attaining their goals and encourages them to accept CIS, United Nations, and CSCE roles in the conflict's settlement. While hailing the meeting as constructive, the Tajik leadership is "troubled" over the fact that they were not informed of the meeting.

**8** • A major offensive is launched against Russian border guards from the Afghan border in the Pamir mountains. Russian Commander Chechulin describes the incident as the beginning of a "large-scale spring-summer combat operation by the irreconcilable Tajik opposition." Chechulin further asserts that, given MIRT's commitment to a political settlement through talks, they are either "phony words or they do not control their field commanders." Because of the casualties suffered by the border guards (ten), the Tajik government should take a more active role in the Pamir section of the border.

**10** • Russian aircraft fire missiles on the border post in Dashti-Yangul settlement to quell Islamic militant activity in the area. Russian border service reports that its border guards now completely control the situation.

**17** • Emerging from talks in Moscow, Tajik Prime Minister Dzhamshed Karimov declares "we are fully in favor of a close alliance with the Russian Federation" and announces Russia will provide the republic with "concrete military aid." Karimov also notes plans for other types of cooperation but that special emphasis was placed on deepening economic and military integration.

**19–22** • In Moscow talks with Tajik opposition leaders and government officials, Kozyrev stresses Russia will not "tolerate the death of its servicemen on the Tajik-Afghan border" and "will use all means available to ensure a political and democratic process." Kozyrev also indicates Russia's support for a greater UN role as well as other international organization involvement in the conflict's settlement. His comments draw strong protest from opposition leaders, saying he used "individual estimation" in characterizing events in the Pamir region. Despite the protest, talks went as scheduled, with both sides agreeing to hold a fourth round of direct, inter-Tajik talks. Rakhmonov and MIRT leader Nuri pledge to meet again.

**29** • Russian Duma deputies arrive in Dushanbe to exchange views on the situation with Rakhmonov, and to establish further cooperation between Tajikistan and Russia.

***May 1995***

**5** • Russian frontier guard troop commander Anatoliy Chechulin reports a concentration of over 300 militants on the Tajik-Afghan border. He says the frontier forces have enough resources to repel an armed invasion but their presence is troubling.

**11** • Lt.-Gen. Pavel Tarasenko replaces Chechulin as commander of Russian border troops in Tajikistan.

**12** • In a press interview, Rakhmonov discusses settlement prospects at the upcoming talks in Kabul. He says his true aim is to achieve a truce with the opposition and he believes they want this as well. He expresses hope that "common sense and reason will prevail over political ambitions," indicating no military solution is viable. Rakhmonov also notes his appreciation of the UN and Russian roles in the conflict, saying both sides agree on a continued Russian presence to monitor the Tajik-Afghan border.

**16–19** • Rakhmonov and Nuri hold talks in Kabul, discussing implementation of ceasefire agreements, confirm the inviolability of the border, and agree to repatriate refugees to their homes. MIRT's deputy chairman Turadzhonzoda agrees that progress was made and the fact "that the recent political antagonists have reached out to each other . . . inspires hope for a peaceful future in Tajikistan." In a joint statement, Nuri and Rakhmonov describe the meeting as beneficial and say talks will continue between their representatives in Alma-Ata on 22 May to hammer out differences. The statement also expresses their gratitude to the United Nations, neighboring countries, and to Afghan President Rabbani for the role of each in the conflict's settlement.

**22** • The fourth round of inter-Tajik talks, stalled for three months, open in Alma-Ata. UN mediator Piriz-Ballon praises the Kabul meeting between Nuri and Rakhmonov as providing the positive context for current discussions. A major topic to be discussed involves introducing roughly 2,000 blue helmets to the conflict zone—half of which will be Russian. Says deputy head of the opposition delegation Otakhon Latifi, such a plan will "kill two birds with one stone. First, it will restore peace in Tajikistan . . . and second, it will remove from Russia suspicion of imperial tactics."

**24** • Piriz-Ballon expresses pessimism over the outcome of the inter-Tajik Alma-Ata talks. Opposition leaders demand another revision of the constitution and call for a coalition government with new presidential and parliamentary elections. The new demands further polarize the positions of the two sides. In a press interview, Rakhmonov categorically rejects the opposition's demands, saying "I have sworn an oath that I will use all my knowledge and skill, experience and power to defend the constitution, the fundamental law, and will keep it sacred." Opposition delegation head to the talks, Turadzhonzoda, says he will take the opposition's case to Kyrgyz President Akaev and will press him to bring up the matter at the CIS Minsk summit. Talks are suspended with agreements to extend the ceasefire to another three months, to exchange POWs, and to build relations on the basis of mutual trust. Talks are to resume in Alma-Ata on 25 July.

***June 1995***

**11** • In a press interview, Nuri discusses his pleasure with Rakhmonov and in the meeting they had in mid-May, saying he considers Rakhmonov's "approach and his meeting with officials representing the refugees to be a good omen." He laments the violence caused by the war and speaks of the need to provide stability for the children's future, their need to "study sciences, be artisans, and prepare themselves for the future of the country." He urges Tajiks to forgive one another, to give up claims of "office, regionalism, egotism, revenge, and hostility, and to unite" in creating a country able to meet all the republic's needs.

**12** • UN Secretary General Boutros Boutros-Ghali terms the recent talks in Alma-Ata "insignificant" and proposes extending the UN observer mission mandate and to exert UN control over north Afghanistan. His comments appear in a report to the UN Security Council on the situation in Tajikistan.

**19** • Tajik government draws up a list of political war prisoners to be exchanged on a "one-to-one" basis spelled out in the Alma-Ata agreements signed in mid-May. The list includes opposition supporters active since 1992. Tensions between the opposition and government officials mount as the Tajik government announces that those involved in killing civilians, policemen, and Tajik and Russian soldiers will be tried as criminals, not as fighters defending an ideal. Their fate rests in the Tajik Supreme Court. The opposition declares the charges were concocted to discredit the opposition.

**28** • Border clashes erupt as a group of twenty-two militants tries to cross the border from Afghanistan into Tajikistan. Russian border guards issue warnings and the group opens fire. The border troops return fire, preventing the group from succeeding.

• In a press interview, Tajik Foreign Minister Talbak Nazarov describes Russia as Tajikistan's "chief partner and ally." He acknowledges Russia's geopolitical interests in the region, Tajikistan in particular.

***July 1995***

**2–5** • Nuri holds an opposition conference in Talukan, Afghanistan, where a resolution to increase attacks on Tajik government officials and military positions is passed. A 200–man rebel force, under Commander Salam Mukhabbatov, is reportedly infiltrating the Garm region from which it will launch a terrorist campaign.

**3** • Nuri issues a statement adopted by MIRT declaring that the opposition forms an inseparable part of Tajikistan. He says opposition policy is to promote friendly relations and notes the important favorable change in Uzbek policy toward the opposition and its leadership.

**4–6** • Rakhmonov visits the southern Khatlon region, scene of the most intense fighting. Local clans engage in constant gun battles as they try to re-carve their spheres of influence. Rakhmonov gives a series of addresses criticizing the fact that "all posts at enterprises are occupied by people put there through force or nepotism." As regards the dire economic conditions besetting the region, Rakhmonov says governmental funds have been plentiful and while much progress has been made in rebuilding businesses, industries, learning centers, and hospitals, he points the finger at the local Kulyab when asking where most of the money went.

**11** • In a news conference, First Deputy Prime Minister Makhmadsaid Ubaydulloev discusses settlement issues and the status of opposition activities. He says much more focus has been placed on refugees, POWs, and disengagement matters and immediate provision of secure living conditions. He promises his government's resources to achieve these aims and invites international support in resolving the conflict and providing humanitarian aid. Ubaydulloev speaks harshly of recent opposition activity, noting in particular a more radicalized Nuri faction forging stronger ties with the Afghan elite and waging a campaign of terror. He announces another amnesty decree issued by Rakhmonov, but the environment for talks, in his view, is worsening.

**13** • Opposition groups charge that in certain districts outside Dushanbe, Tajik security police are "rounding up scores of Muslims" for interrogation, killing many in the process. Opposition representative Zafar Rakhmonov says opposition members have been pulled out of their homes by masked security officers.

**19** • Talks are held in Tehran, hosted by Iranian President Rafsanjani at the presidential palace. Rakhmonov, his foreign minister, labor minister, and vice-president make up the Tajik delegation and Nuri, MIRT first deputy Turadzhonzoda, and other deputies represent the opposition. Opening the talks, Iranian Foreign Minister Velayati underlines his country's desire to see the conflict settled and pledges Iran's commitment to finding common ground. The minister cautions participants on the danger of the Tajik conflict evolving into a Cambodia or Vietnam and calls on both Nuri and Rakhmonov to see that this does not happen. Some press reports believe Iran has brought pressure to bear on the opposition and this meeting is a result. During the visit, the Tajik (Rakhmonov) delegation signs twelve intergovernmental agreements covering economic and cultural relations with Iran.

**22** • No major progress is reported as talks in Tehran end. Observers note the important fact that Rakhmonov even agreed to hold discussions during his state visit to the Iranian capital, indicating that the president seriously desires resolution. The next (fifth) round of talks, according to Tajik presidential spokesman Davlatali Davlatov, will focus on the composition of the Congress of Peoples of Tajikistan.

**24** • MIRT spokesman Akbar Turadzhonzoda announces in an Iranian newspaper interview that the Tajik opposition does not want control over executive power but wants 40 percent representation in parliament. "We believe that in order to achieve a sustainable peace in Tajikistan, we should declare a two-year transition stage with Rakhmonov's leadership and with the participation of a National Reconciliation Council."

***August 1995***

**1** • Rakhmonov receives Russian Deputy Foreign Minister Albert Chernyshev in Dushanbe for consultation on the conflict. Both agree that the inter-Tajik talks should proceed in the same direction since no alternative exists. Rakhmonov confirms his desire to settle the conflict peacefully and says that the immediate aims of talks remain: establishing a truce, return of all refugees, and the elimination of tensions in the Central Asian region. Mechanisms are now in place for displaced persons to return safely and the process is already under way in the Gorno-Badakhshan region and in some districts of Dushanbe.

**2** • In a news conference, commander of Russian border troops in Tajikistan Tarasenko says the situation on the Tajik-Afghan border is alarming. He says the Russian command has identified a large buildup of militants in many border districts who they believe seek to cross into Tajik territory for operations against the Tajik government. He notes certain forces in Afghanistan and in certain circles in Tajikistan do not wish to see the situation stabilized, in particular black market operatives and drug dealers.

**10** • Iranian Foreign Minister Velayati meets with UN representative in Tajikistan, Ramiro Piriz-Ballon. The two discuss Iran's positive role in bringing the sides closer to agreement and providing the impetus for continuation of talks. The officials also discuss the role of the United Nations and other international agencies capable of affecting some resolution, with Piriz-Ballon highlighting the necessity of continued Iranian participation in settlement proceedings.

**14** • Piriz-Ballon meets with Tajik opposition leader Nuri. He tells Nuri that the United Nations does not wish the fifth round of talks to take place unless the leaders are sure significant progress will be made. Nuri says the opposition will persist in its demands for increased participation in political affairs and for more representation in the political machinery throughout Tajikistan.

**17** • Nuri signs a protocol on achieving peace in Tajikistan. Piriz-Ballon is present at the signing, held at the Afghan Foreign Ministry. The protocol, already approved by Tajik authorities, extends the current ceasefire for another six months and provides for talks to continue in a peaceful atmosphere. Nuri complements Afghan president Rabbani for his work in the Tajik peace process.

**18** • The Russian border command and Tajik rebels operating on the Afghan-Tajik border agree to dismantle firing positions in the Gorno-Badakhshan region.

**23** • Despite the above agreement, clashes and mortar fire continue on the Tajik-Afghan border. The exchange occurs primarily in Khorog region. A Russian border command spokesman terms the incident a unilateral violation of the ceasefire extension and charges that the Tajik opposition leaders are "either reluctant or unable to keep their militants under control." In a related statement, Nuri charges the Russian border troops with blame for the incident. Nuri says that, because the opposition forces are not required to disengage until a "full political settlement is achieved," the Russians try to provoke the opposition and are thus met with "fierce resistance."

**30** • CIS Commander of the joint peacekeeping forces in Tajikistan Valentin Bobryshev reports that the Tajik opposition has been strengthening its forces in the Garm region and the Obikhingou valley. Over the last three months, the commander says their military presence has been complemented by 2,000 to 2,500 more troops and mercenaries. He believes the purpose of buildup is to continue assaults from the Afghan border in order to "keep the country's leadership tense" and to force concessions at the next round of talks.

***September 1995***

**18** • Inter-Tajik talks are canceled because of conflicting views over where the talks should be held. The Tajik leadership prefers Ashkhabad while the opposition pushes for Tehran or even the Afghan capital.

**19** • The Russian foreign minister is in Dushanbe for talks with the Tajik leadership. The minister indicates that the purpose of his trip is to ascertain the steps needed for implementing agreements reached in Moscow between Russia and Rakhmonov.

## Summary at the End of 1995

Fighting continues on the Afghan-Tajik border as militants continually launch attacks. The opposition refuses to change its demands and the Tajik leadership refuses to heed them, especially demands for changing the constitution and for overturning the last elections. Plans for forming a National Reconciliation government composed of representatives from all factions guide the talks when they are held. The Movement for Islamic Revival continues to accuse Russia of hegemonic ambitions in Tajikistan, and to pursue its secular aims for a more balanced representation of the political, ethnic, and clan elements of the country's society. The United (IRP–DP) Opposition continues to oppose the Kulyabi clan's hold on power through Rakhmonov, the former communist party, and Russia.

# Appendix B

# Chronologies of Key CIS Developments, 1992–1995

## *Political, Economic, Security*

### Introductory Notes

When the Soviet empire collapsed in December 1991, the twelve former republics that became signatories of the Belovezh Forest agreement creating the Commonwealth of Independent States (CIS) were unprepared and ill equipped for statehood. They lacked experienced legislatures, constitutions, and independent social institutions. More important, they were each tied (purposefully) into a highly centralized and controlled Soviet budget, monetary system, transportation and communications system, electricity grid, energy supply system, defense system, and industrial production and distribution system. No republic was economically self-sufficient. Nor could any republic even begin to create a banking or monetary system without first creating a national currency and divorcing itself from the Russian ruble. Each one was dependent on monetary emissions from the Central Bank of Russia for budgetary activities and state sector salaries, which made up the lion's share of each republic's economy.

Carving out statehood from the remnants of empire will mean first of all disengagement from the Soviet economic, political, and military networks over which Russia assumed control, followed by a diversification of the links binding these states to the imperial "center" (Moscow). New bilateral and multilateral linkages will have to be formed in and outside each one's own region of the huge former Soviet space. The ultimate success of this state-building process will depend to a large extent on how well each former republic can balance its own vulnerabilities to Russia's bilateral demands and pressures with its autonomous national political, economic, and security goals. Diversification is made all the more difficult, however, by the fact that the national-patriotic elements of Russia's political elite have taken control of Russia's foreign policy agenda—with respect to both the CIS and countries outside the CIS.

As discussed in Chapter 3, the Russian national-patriots comprise several distinct sectors of Russian society, including the former Communist Party nomenklatura, the Slavicists, the KGB and other security services, the Eurasianists, and many of the pragmatic "industrial managers" and "financiers." Many members of these groups simply refused to accept the collapse of the Soviet Union and the validity of the Commonwealth of Independent States. In mid-1992, in fact, the Communist Party of the Russian Federation declared the Belovezh agreement illegal, in deference to the March 1991 Soviet referendum upholding the retention of a Union government. The nationalists, national-patriots, and communists elected to the Russian parliament in 1993, as seen in previous chapters, virulently opposed and spoke out against the Yeltsin/Kozyrev foreign policy orientation toward Western democracies from the very beginning of the CIS. These groups have been entirely effective in enlisting the attention and the sympathies of the Russian population.

The following chronologies detail who the players formulating and implementing Russia's CIS policy are, how they think and interact with players from the non-Russian republics, and how they articulate their professed goals and strategies for the CIS. The reader can judge how sincere Russian politicians and governments officials are when they address Russia's overall socio-political, economic, and collective security agendas. The chronologies begin with Andrey Kozyrev and end with Kozyrev's dismissal and the appointment of Evgeniy Primakov, former chief of the Russian Foreign Intelligence Agency, as Russian foreign minister on 10 January 1996. Primakov's appointment could well mark the end of the foreign policy debate inside Russia and the beginning of an era of attempted restoration of the Russian imperial state. The extracts in the chronology provide liberal quotations, which epitomize the "rights of ethnic Russians" and "Eurasian space" rhetoric used by Russian government officials, and record the carefully guarded, but unmistakably anti-imperial remarks of the non-Russian CIS elites when talking with the press or writing in political journals.

The future of the CIS will depend on how the non-Russian states react, how Russian youth reacts, and how the West reacts to Russian imperial-oriented tendencies. Under Primakov, the KGB will undoubtedly become more active in holding the CIS together, and Russia will try to use the CIS to counteract the West's "unipolar" world plan—Primakov's shorthand for the emergence of the United States as the world's sole superpower. Under Primakov, Russia's foreign policy will be in the hands of a master Eurasionist, whose ultimate goal will be to spread Russia's influence throughout the Caucasus, Central Asia, and the Middle East. In such a case, the Commonwealth will cease to be the vehicle for independent state-building in the former Soviet territory and will become the vehicle for carrying out Russia's post–cold war hegemonic ambitions.

# 1992

## January 1992

### *Economic*

**5** • Vneshnekonombank Vice-Chairman Tomas Alibegov announces that the CIS hopes to derive some $30 billion from selling the debts owed to the former USSR by socialist partners and Third World countries at a discount on secondary credit markets.

**30** • Italy makes available $2.3 billion in credit for the non-Russian republics.

### *Political*

**31** • In a press interview, Kazakh President Nursultan Nazarbaev speaks favorably on CIS prospects. He says coordinating bodies have been established to assure the "Commonwealth will eventually operate smoothly and relations between the member states will be settled." Asked about Western aid, Nazarbaev urges the CIS states to rely only on their own strengths as "genuine aid cannot be ensured to 300 million."

### *Security*

**5** • CIS Commander-in-chief Evgeniy Shaposhnikov discloses that only Russia, Kazakhstan, Armenia, Kyrgyzstan, and Tajikistan are interested in maintaining a unified command over former Soviet general purpose forces.

**8** • According to Ukrainian President Kravchuk, telecommunications between himself, Yeltsin, Shushkevich, and Nazarbaev are reliable enough to prevent a unilateral nuclear launch.

**10** • Ukraine cuts defense communications with the CIS command due to its dispute with Russia over nuclear weapons.

**16** • CIS leaders hold a summit in Moscow to discuss military affairs, coordinate CIS foreign policy, and establish an organizational group to prepare future meetings of CIS heads of state and government. The presidents of Uzbekistan, Moldova, and Turkmenistan are absent, but delegations from these states participate.

• Transcaucasus Military District Commander Patrikeev states that the district should be placed under Russian command.

**17** • Over 5,000 ex-Soviet officers meet in Moscow to discuss the future of the armed forces. The officers urge retention of unified armed forces and improved social guarantees for servicemen.

**21** • Commander of the Northwestern Group of Forces Col.-Gen. Valeriy Mironov says his forces have been transferred from Soviet to Russian jurisdiction.

**22** • CIS military affairs spokesman Lt.-Col. Leonid Ivashov announces that Russia will assume 62.3 percent of the CIS military budget in 1992. Ukraine will pay 17.3 percent and Kazakhstan 5.1 percent. The nine other members will contribute the remaining 15.3 percent.

**24** • Russian sources claim that 286,000 CIS servicemen and their families are homeless.

**28** • In Russia, Head of the General Staff Academy Col.-Gen. Igor Rodionov calls for his academy to become the training center for CIS officers. He argues that the academy should be under Russian supervision.

**29** • Yeltsin announces CIS and Russian arms cuts. He states that 600 strategic land and sea missiles have been taken off alert status and proposes eliminating 130 land-based missile silos and halting production of heavy bombers and cruise missiles.

## February 1992

### *Economic*

**14** • At the CIS Heads of State summit in Minsk, leaders sign an economic document which seeks to regulate trade and economic cooperation, and calls for the ruble to remain the sole monetary unit of the CIS.

### *Political*

**3** • The National Public Opinion Studies Center in Moscow conducts a poll in Russia, Ukraine, and Kazakhstan on whether individuals in these nations believe the CIS has a future. The results:

| | Kazakhs | Russians | Ukrainians |
|---|---|---|---|
| predict deeper integration | 28% | 8% | 7% |
| predict worsening conflict | 10% | 20% | 20% |
| predict difficulties in reaching accord | 28% | 39% | 29% |
| predict breakup of union | 16% | 16% | 22% |

The study polled 1,597 Russians, 517 Ukrainians, and 240 Kazakhs and reports a 3 percent margin of error. The Center concludes that ordinary Russians and Kazakhis are more optimistic about the future of the Commonwealth than Ukrainians are.

• Ukrainian President Leonid Kravchuk discusses Ukraine's independence and relations with Russia with the Ukrainian press. He states first of all that Ukraine will be a non-nuclear state and will abide by all international agreements regarding disarmament signed by the former USSR, adding that all ex-Soviet republics should honor such agreements. He argues against the popular notion in some Russian circles that Ukraine is simply a "small Russia" and says that Ukraine will adamantly work toward consolidating its independence. As for an impending Ukrainian-Russian conflict, Kravchuk says the prophets are wrong, citing several referenda conducted in Ukraine revealing the extent to which the Ukrainian people support independent Ukraine, including a number of ethnic Russians in the country.

• In Moldova, President Mircea Snegur argues that the CIS will last into the "near future" and will work to ease the economic strain on the former Soviet republics, but believes the republics will leave the body once they are integrated into Western economic structures. Asked about possible reunification with Romania, an issue gaining in popularity since independence, Snegur warns against implementing such a union without consulting the people first.

**6** • In Azerbaijan, Prime Minister Hasan Hasanov Azeri says Russia could become Azerbaijan's political enemy. Hasanov argues that Russia's recent support to Armenia will mobilize political forces in Azerbaijan against Russia.

**8** • A CIS Heads of Government summit opens in Moscow. Every republican prime minister participates except for the Ukrainian prime minister, Vitold Fokin. They discuss ecological cooperation, coordination of interstate power engineering and hydrometeorological services, food and agricultural materials, and economic interrelations. Economic issues dominate because many factories throughout the CIS stand idle and receive no deliveries of production materials and energy supplies. The delegates draft an eleven-point plan addressing economic concerns and agree to name Russia as legal "successor to and guarantor of the foreign credit agreements of the Commonwealth member states." Ukraine argues that each republic should maintain separate and direct communications with creditors, and Russian Deputy Prime Minister Gaydar does not object. Five additional documents are signed addressing economic interaction, but leave many issues unsettled. The Russian delegation accuses Ukraine of hindering many agreements. The fact that Ukraine is represented by a deputy economic minister indicates it is retreating from the community.

**15** • A CIS Heads of State summit is held in Minsk. All leaders attend (though a few leave unexpectedly). Security issues dominate discussions, in particular creation of Joint CIS Forces (see chronology under Security, 15 February 1992). Commentators note the meeting's political significance in that the leaders have learned they have made progress and communicate effectively despite widespread differences. Twenty documents are signed, focusing on military and economic issues.

**18** • In Ukraine, Kravchuk responds to allegations that Ukraine seeks association in a "triangle" union among Poland, Hungary, and Czechoslovakia, saying Ukraine "will not abandon one treaty for another." He says, however, that while Ukraine is not searching for such an alliance at the state level, the country does want to pursue cooperation with all countries. He argues that, in the end, Ukraine must cooperate with the CIS states to "strengthen its independence."

• In Russia, President Yeltsin meets with U.S. President Bush requesting Western assistance to help Russia's and the former Soviet republics' economies. Some Western states, principally Germany and Britain, support establishing a fund to aid the Soviet successor states and Russia but the United States has objected, arguing the new states should first join international economic bodies such as the IMF to help strengthen their economic discipline. Experts believe \$5-\$12 billion will be needed for the stabilization fund to be effective.

• A struggle is building for control over the Central Asian republics, according to Interfax. Middle Eastern countries have been increasing their humanitarian aid to the Central Asian states and Central Asian officials have participated recently in a Tehran summit of the Organization for Economic Cooperation, where the main topic was feasibility of an Islamic Common Market. While most of the Central Asian officials at the summit recognize some merit in playing the "Islamic card," they realize the dangers inherent in doing so. Should Islamic fundamentalism take root in Central Asia, reformist forces would recede, turning the "region into a source of global danger."

**19** • Belarusian parliamentary head Shushkevich tells Interfax that the CIS is viable, citing as evidence the recent Heads of State meeting in Minsk. He says the CIS permanent working center in Minsk will sustain CIS existence and that the "fears regarding the instability of the Commonwealth and its coordinating organs are in my view unfounded."

**25** • In an address to the Council of Europe in Strasbourg, Russian parliamentary speaker Ruslan Khasbulatov argues that the CIS is not, contrary to prevailing opinion, in a crisis. He defends the slow progress of the organization by saying "no one should have expected the new Commonwealth to start to live in full-blooded life as soon as it was proclaimed. . . . [But] to speak of a crisis . . . is frivolous and superficial. It is simply being born in conditions of crisis."

**27** • In Russia, republican Supreme Soviet chairmen hold a conference to discuss documents adopted in Minsk. Members debate creation of a Council of Heads of Parliament, but the idea is rejected by the Ukrainian delegation, which argues that the two existing councils, the Heads of State and the Heads of Government, are sufficient to debate and carry out CIS initiatives. All participants (with the exceptions of Moldova and Uzbekistan) sign the following documents:

—Agreement on Interparliamentary Cooperation in the Legal Sphere;

—Agreement on Consultative Conferences of Representatives of Supreme Soviets;

—Protocol on Interparliamentary Information and Reference Service;

—Protocol on the Elaboration of Multilateral Acts.

### *Security*

**3** • During a visit to the United States, Russian Maj.-Gen. Nikolay Stolyarov states that tension in CIS armed forces is growing, particularly in Ukraine. He cites the pressures to swear more than one military oath as a major centrifugal force splitting the armed forces.

**4** • Lt.-Gen. Vasiliy Vorobev announces that Russia is the only CIS member to allocate funding for the first quarter of the 1992 CIS budget.

**10** • Ukrainian President Kravchuk asserts that the gravest danger to the CIS is the maintenance of a unified armed force. Nazarbaev dissents and urges a unified CIS army.

**11** • CIS defense ministers meet in Minsk to discuss the army's future. Meanwhile, Moldovan Prime Minister Valeriy Muravschy announces that his country will not contribute to the CIS armed forces budget.

**12** • Ukraine and Belarus refuse to sign military preparatory documents for the CIS summit in Minsk due to disagreement on the composition and redistribution of CIS forces among the commonwealth states.

**14** • The CIS Heads of State meet in Minsk. The military accords are minor and few members sign all documents. Moldova does not sign an agreement on strategic forces and Belarus does not sign an agreement on general purpose forces.

• Russian air crews "defect," with six Su-24 bombers, from Ukraine to Russia, having refused to swear an oath of allegiance to Ukraine.

**15** • Shaposhnikov is confirmed as commander-in-chief of the CIS Joint Armed Forces.

**18** • Shaposhnikov predicts that the armed forces of the USSR will split into national armies.

**28** • Azeri President Ayaz Mutalibov announces that Baku has agreed to subordinate general purpose forces to both the Azeri president and the CIS military command.

## March 1992

### *Economic*

**13** • The CIS Council of Heads of Government agrees to accept joint responsibility for repaying the USSR's debt. Russia will repay about 61 percent, Ukraine 16 percent, and the remaining six members split the rest.

### *Political*

**3** • Azerbaijan, Tajikistan, Moldova, Uzbekistan, Kazakhstan, Kyrgyzstan, Armenia, and Turkmenistan are admitted to the United Nations. The General Assembly vote is unanimous.

• Sazhi Umalatova, head of an organizing committee to convene an emergency Sixth Congress of USSR People's Deputies, argues in a press interview that the disbandment of the USSR Supreme Soviet is illegal. Umalatova's activities spark widespread debate, with some arguing for a body "capable of building a powerful and prosperous multinational state" and others believing a renewed Congress of USSR Deputies will lead to a "war of governments." Says Umalatova, "there is no reviving the old Union, and the objective of the Congress is to set up a body around which the republics, regions, and districts of the former Soviet Union will unite on a new basis."

**6** • In talks with Turkish Foreign Minister Hikmet Cetin, Ukrainian President Kravchuk criticizes Russia's recent actions regarding CIS relations. He identifies dangers for the CIS should Russia continue to try to impose its will on other CIS states without consultation and compromise. Kravchuk is especially critical of Russia's decision to cut off gas supplies to Ukraine and raise energy prices for certain CIS states. In such an environment, says Kravchuk, "it is very difficult to establish friendly ties."

**8** • In a press interview, Gorbachev discusses CIS prospects. While objecting to any association short of preserving the Union, he says he supports the efforts of republican leaders to make the newly independent states viable. Given the degree of republican interdependence, he notes the need for coordinating mechanisms and increased political interactions. However, he criticizes the CIS for not fulfilling these needs. "Policy must be coordinated. Corresponding bodies must be established."

**10** • In a press interview, Kazakh president Nazarbaev asserts that supranational coordinating structures are needed within the CIS. Nazarbaev especially believes economic coordinating bodies are necessary to "shape a common market of a new type" and ensure newly independent states "gradually enter the mainstream of contemporary civilization." In the absence of such bodies, Nazarbaev believes the CIS "will be an empty sound, a beautiful form that is filled with no real content." As to Western involvement in the CIS, Nazarbaev believes aid should be directed toward the individual republics making up the CIS, as it "is naive to try to reform the entire social giant at once. It would be better to form islets of developed economy and democracy which would have a tendency to expand."

**11** • Azerbaijan's main political opposition party—the Popular Front of Azerbaijan (PFA)—declares the republic is not officially a CIS member because the republic's Supreme Soviet and national council have not yet ratified the Commonwealth Chater. Political infighting over membership in the CIS is intense, with the PFA lobbying against and former communists (former President Mutalibov, Supreme Soviet chairman Mamedov, and Prime Minister Hasan Hasanov) supporting membership in the body.

**13** • In Moscow, the Council of Heads of CIS Governments convenes. Armenia, Kazakhstan, Kyrgyzstan, Moldova, Uzbekistan, and Ukraine are each represented by a prime minister, and Russia, Belarus, Turkmenistan, and Tajikistan send plenipotentiary representatives. Azerbaijan sends an official not empowered to sign documents. The attendees sign documents on sixteen of the eighteen issues on the agenda. Most entail economic affairs such as payment of pensions due, tax policies, principles of a customs union, and a host of questions dealing with the organization of economic ties. Of major significance is the settlement of the USSR's lingering external debt and foreign holdings issue. Ukraine favors collective responsibility. Two major items on the agenda dealing with commerce and railway transport are not resolved. Commentary notes the calm atmosphere in which the meeting takes place, as distinct from the volatile aura of past discussions. The following agreements are adopted:

—Agreement on Scientific Ties;
—Agreement on Standardization;
—Agreement on Taxation Policy;
—Agreement on Pension Provision;
—Agreement on Plenipotentiary Missions;
—Protocol on Price Setting;
—Scientific-Technical Facilities Joint Use;
—Interstate Protocol on Railways;
—Agreement on Scientific Personnel;
—Food Import Commission Protocol;
—Foreign Debt Agreement;
—Protocol on Internationalist Soldiers;
—Scientific-Technical Cooperation;
—Agreement on Customs Policy Principles;
—Agreement on USSR's Internal Debt;
—Interstate Protocol on Banking.

**17** • In Russia, a Sixth Congress of USSR Deputies is held in Podolsk. Umalatova is elected Presidium head; a fourteen-person committee and the participants approve documents declaring the Commonwealth of Independent States illegal and recognizing the Soviet Union as a "geographical reality." Participants also plan a future meeting to chart a campaign for restoration of the Union, revival of communist ideology, and resignation of the Russian government. Most former republican leaders denounce the work of the Congress and believe the body has no legal or legitimate basis for existing. The Supreme Soviets of many republics also claim the Congress's actions were illegal.

**20** • In Kiev, CIS heads of state convene for the second CIS summit. Uzbek President Islam Karimov chairs the meeting. The leaders are to select key CIS military posts. In his remarks, Ukrainian President Kravchuk criticizes the organization, saying the body has failed resolutely as "not a single military, political, or economic issue has been resolved within the framework of the CIS." Kravchuk also notes the tense situations in Moldova and Azerbaijan, which further complicate CIS relations. Despite the pessimism, the following agreements are signed:

- Appeal to the United Nations for help with Chernobyl;
- Declaration of the Non-Use of Force or Threat of Force in Relations between CIS Members;
- Agreement of Groups of Military Observers and Collective Peacekeeping Forces in the CIS;

- Protocol on Rail Transport Council;
- Decision of the Joint Command of the Border Troops;
- Agreement on the Status of Border Troops of the CIS;
- Agreement on Army Manpower Acquisition.

**26** • Writing in *La Stampa*, ex-Soviet President Gorbachev warns that the CIS is headed for disaster should political leaders fail to erect coordinating mechanisms. He criticizes Yeltsin for failing to reform the multiethnic USSR and hints that its demise was largely Yeltsin's fault. Relations among the former Soviet republics will continue to disintegrate, he says, because the CIS is not designed to encompass the former republics in a sincere organization.

**27** • A CIS interparliamentary conference ends in Alma-Ata. Seven CIS states sign an agreement on the creation of an Interparliamentary Assembly (Armenia, Belarus, Kazakhstan, Kyrgyzstan, the Russian Federation, Tajikistan, and Uzbekistan). Ukraine opposes the new body, believing it will create a "third tier" of suprastate structures.

• In Ukraine, economic advisor to President Kravchuk Lionel Stoleru indicates Ukraine sees no future in the CIS and wishes to integrate into the European Community. Stoleru says the whole of Ukraine's "economic program is oriented toward leaving Russia, to cutting the links, including the link of the currency, without cutting trade."

***Security***

**5** • Shaposhnikov dismisses air force Maj.-Gen Mikhail Bashkirov for swearing allegiance to Ukraine. Ukrainian Defense Minister Morozov promptly reinstates Bashkirov.

**14** • CIS Deputy Commander-in-chief General Boris Pyankov claims that all nuclear weapons have been removed from the Transcaucasus.

**20** • CIS heads of state meet in Kiev. Opinions differ over military issues. An agreement on a CIS peacekeeping force is among seventeen documents signed, rejected only by Armenia and Azerbaijan. Kravchuk and Yeltsin exchange harsh words over the question of Russian domination of CIS military and political structures. The Council, however, confirms the appointments of CIS Armed Forces Chief of the General Staff Col.-Gen. Viktor Samsonov, CIS Strategic Forces Commander-in-chief Yuriy Maksimov, and CIS General Purpose Armed Forces Commander-in-chief Vladimir Semyonov.

**April 1992**

***Economic***

**12** • The World Bank announces plans to lend $12–15 billion to the former USSR.

***Political***

**2** • Kyrgyz President Askar Akaev tells the press that the recent Kiev summit of CIS heads of state produced radical results only in the military field, but failed in the most important area, economics. Akaev believes that until a "strong foundation for economic cooperation" is achieved, the CIS will remain a fragile body. "Free trade, free provision of services, free movement of people without customs or barriers, and the free movement of capital around the countries of the Commonwealth," according the Akaev, are the requirements for an effective organization. Akaev also believes no additional suprastate structures are needed and that existing councils are sufficient coordinating bodies to achieve the above tasks. Concluding his interview, Akaev says the most important element for a successful Commonwealth "is to find more mutual confidence with each other."

• In Azerbaijan, a news conference is held with the theme "The Slavic Card in Russia's Caucasian Policy." Public organizations, Azeri journalists, and some political officials attend the conference, which highlights the fact that all former Soviet republics experiencing ethnic strife and political conflict turn to Russia for assistance, only to have Russia manipulate the situation for its own interests. Many participants note this phenomenon in Russia's handling of the conflict in Nagorno-Karabakh, and that the Russian press fails to report accurately on developments in the conflict.

• In Georgia, Russian Foreign Minister Kozyrev holds talks with Georgian President Eduard Shevardnadze. The two lay the groundwork for a "big treaty" which will provide a contractual and legal basis for relations between the two countries. Kozyrev tells journalists the main element of his trip is to ensure successful relations under the new conditions.

**6** • CIS intelligence service representatives agree to coordinate their efforts. The agreement, according to Kazakh KGB Chairman Bulat Baekenov, more clearly defines the tactics and strategies of the intelligence services.

**8** • Eduard Shevardnadze states that he is "not optimistic about the future of the CIS." He believes that since the body originated with little preparation, circumstances will deteriorate and the body will be unable to function because of disagreement and political infighting.

**9** • In Ukraine, the National Assembly adopts a statement viewing a Yeltsin decree that transfers the Black Sea Fleet (BSF) to Russia's jurisdiction as a "declaration of war." The statement gives the Assembly's full support to President Kravchuk's decree on building a Ukrainian armed forces that envisions the Black Sea Fleet as part of the military.

**10** • Belarusian parliament head Stanislav Shushkevich defends the CIS during a visit to France. He tells French journalists: "Nowadays, it has become unpopular to defend the concept of the CIS, but, for as long as the Commonwealth has existed, my position has not altered. I am still an ardent supporter of a just Commonwealth of truly independent states." Along these lines, Shushkevich argues that a deteriorated CIS will mean the economic collapse of most of the republics "that is not only detrimental to them, but it is very unprofitable for the West." He also touches on organizational issues critical for an effective CIS, the consequences of inaction by outsiders in many of the conflict zones in the former Soviet Union, and on the ways in which the global powers could assist the CIS and its members.

**13** • In Bishkek, a meeting of the Central Asian republics is held. Kazakh President Nazarbaev speaks of the negative consequences of ethnic division and the conflicts now raging in some republics. He also warns of the friction developing in Ukrainian-Russian relations and points out that these two countries "bear too much responsibility for the destiny of all [CIS] peoples," something he asks all the leaders to consider. He asks the Central Asian republics to take steps to combat this trend.

**14** • Georgian President Shevardnadze rules out joining the CIS, saying Georgia "will not have anything in common with the CIS." He indicates, however, that the republic needs help from Russia and advocates equal relations with it.

**15** • CIS representatives meet in Kiev to discuss USSR property issues. Experts focus on shares of property due to each republic. A draft agreement is drawn up that will be presented to the Council of CIS Heads of State.

**22** • Moscow columnist Anatoliy Karpychev, writing in *Kuranty*, warns that Russia should not engage in hard-line politics when crafting CIS policy. He points out the negative mindset of many CIS republican leaders created by Russia's "imperial policies" and the disintegrative effects such a policy has even if it is only in rhetoric. He writes that a hard-line policy will not solidify the CIS but ensure its demise and, contrary to the view of many Russian politicians, "would not be Russia's strength but its weakness."

**23** • In Moldova, President Snegur meets with Moldovan Democratic Party (MDP) representatives who question joining the CIS. In particular, the MDP leadership argues that "independent Moldova cannot reconcile itself with the pressure of another independent state," referring to the secessionist movement in Trans-Dniester.

**24** • In Alma-Ata, CIS interior ministers meet to discuss crime prevention. The participants all note an increase in organized crime that "recognizes no borders," especially when many Commonwealth borders are poorly patrolled. They note the destabilizing effects such crime has on the political, social, and economic well-being of their countries and people and agree to establish specific cooperation measures to combat such activity.

• Central Asian leaders (less Tajik President Nabiev) again meet in Bishkek. Participants focus on creation of a Turkic-Asiatic economic and political space, with Uzbekistan's Karimov arguing that the national wealth of the Central Asian states should first and foremost benefit the people in the region and secondly be used for its neighbors (hinting at Russia). Kyrgyzstan's Akaev reminds all that Russia remains a very reliable partner for the republic and cautions against ignoring Russia. According to some commentators, the meeting consolidated the region's desire for forming a Turkic-Asian community and for using such a policy to sever the traditional Russian role in Central Asia, but without negative economic consequences for the republics.

**29** • In Ukraine, Russia and Ukraine begin talks on Black Sea Fleet (BSF) issues. Reports highlight the secrecy of the talks. The Ukrainian Defense Ministry's chief naval expert Anatoliy Katalov says the talks will focus on reaching a common approach to dividing BSF equipment, personnel, and bases. Ukraine views the BSF as the fleet of the former USSR, while Russia considers it part of the CIS.

**30** • Ukrainian and Russian negotiators sign a "communiqué on progress" pledging that both sides will "adhere to a moratorium on any unilateral actions complicating the situation around the Black Sea Fleet."

• Representatives of ten oblasts from Russia, Belarus, and Ukraine meet in Chernigov. They sign an agreement at the local oblast level outlining the main directions for cooperation during their transition to market economies. Also stipulated in the agreement are provisions for cooperation in science, culture, and public health.

### *Security*

**2** • Former commander of the Northern Fleet Admiral Feliks Gromov is promoted to CIS Naval Forces first deputy commander-in-chief, replacing the retired Admiral Ivan Kapitanets.

**7** • CIS defense ministers sign ten out of eleven draft CIS military agreements. Ukraine and Azerbaijan do not initial documents concerning the creation of CIS common defense structures.

**15** • Azerbaijan and the CIS main naval staff reach agreement on the division of the Caspian Sea Flotilla. The two sides agree that Azerbaijan and Russia will retain control of one-quarter of the ships and facilities of the flotilla. The fate of the remaining 50 percent is to be determined during negotiations with Kazakhstan and Turkmenistan.

**29** • Military Intelligence Directorate (GRU) Chairman Col.-Gen. Evgeniy Timokhin announces that his organization will be subordinated to the Russian government, and not the CIS.

## May 1992

### *Economic*

**9** • CIS economic experts reach agreement on a working document establishing national currencies in the former republics. Those creating their own monies will function outside the "ruble zone."

**29** • The Ukrainian deputy prime minister and economics minister, Vladimir Lanovoy, states that the CIS has no future "on the economic level."

### *Political*

**1** • In Kiev, the Ukrainian Republican Party (URP)—an influential faction of 12,000 members—holds a third congress. The body votes for Ukraine to leave the CIS, considering that the "main mission of the CIS—the civilized self-liquidation of the empire—has been completed."

**6** • In Moldova, the Moldovan Christian-Democratic Popular Front urges the republic to completely secede from the CIS, consolidate its independence from Russia, and reorient its economy toward cooperation with the West. The chairman of the party's Executive Committee, Yuriy Roshka, says Moldova must take these steps to strengthen its independence and sovereignty. Other measures advocated by the faction include introduction of a national currency, secession from the ruble zone, and a friendship and cooperation agreement with Romania. Roshka also calls for labeling the Russian 14th Army in Moldova an "occupation force."

**10–12** • In Ashkhabad, seven Central Asian leaders hold talks (Turkmenistan, Iran, Kazakhstan, Kyrgyzstan, Pakistan, Uzbekistan, and Turkey). The participants focus on promotion of economic cooperation and deepening ties among them. A number of documents are signed, of which Turkmen President Niyazov assures that none collide with CIS member states' interests and in fact may prove beneficial to some republics. The agreements signed are left open for other interested parties.

**14** • In Moscow, Kazakh President Nursultan Nazarbaev says the future of the CIS is "very obscure," believing that "mutual threats will lead to a deadlock not only in relations between neighboring states but the [CIS] as a whole." He stresses that Kazakhstan will continue to build "equal, good-neighborly" relations with Russia, a policy "vital for Kazakhstan."

• The EC Commission holds a seminar in Maastricht attended by CIS officials. The seminar is organized to provide information on European integration they might find useful in CIS development. Armenian Vice Premier G. Aresyan says the EC experience can be helpful in this regard. He notes that the impulse for European integration stemmed from the "need to prevent wars, . . . to halt Europe's economic decline."

**17** • Yeltsin tells press he is satisfied with the Tashkent summit, describing it as "useful and fruitful." He assures listeners that the CIS is "a live and growing organism and I believe in its productivity." Of the thirteen documents adopted, Yeltsin hails the Treaty on Collective Security, signed by Russia, Kazakhstan, Uzbekistan, Turkmenistan, Armenia, and Tajikistan as the most important. The treaty is left open for other CIS members.

### *Security*

**13** • In Tashkent, CIS defense ministers meet to draft an appeal to the heads of state urging them to preserve the joint armed forces.

**15** • The CIS heads of state meet in Tashkent. Only Russia, Kazakhstan, Uzbekistan, Turkmenistan, Belarus, and Armenia are represented by heads of state; the other five by heads of government. Ukraine, Belarus, Azerbaijan, and Turkmenistan reject the Collective Security Treaty. Other agreements include the manning and financing of frontier troops, the formation of CIS peacekeeping forces, and fulfilling the USSR's obligations on chemical weapons control.

**18** • While traveling to Washington, Nazarbaev announces that his government will allow Russia to base its nuclear missiles on Kazakh territory.

**25** • Armenia, Azerbaijan, Belarus, Georgia, Moldova, Russia, and Ukraine inform NATO of their plan for complying with the treaty on Conventional Forces in Europe. Russia will maintain 54 percent of the allotted troops and hardware, Ukraine 27 percent, and Belarus 12 percent. Armenia, Georgia, Moldova, and Azerbaijan split the remaining 7 percent.

**26** • CIS defense ministers meet in Moscow and agree on the composition of the CIS Strategic Forces. They will consist of the Strategic Rocket Forces, nuclear components from the air force and the navy, the ballistic-missile warning system, anti-missile defense systems, and space forces.

## June 1992

### *Economic*

**17** • On 26 June CIS prime ministers will meet to discuss a wide range of economic issues. Coordinator of CIS's working group Ivan Korotchenya tells reporters "the economic agenda will be vast as usual." The prime ministers will discuss, in particular, ways to protect the interests of states belonging to the ruble zone in the event of new national currencies being introduced in some other states.

• In Bishkek, all fifteen former Soviet republics sign a

protocol stipulating that by 1 September states will agree on volumes of trade for 1993 based on free-market prices.

**18** • The exchange rate of the Ukrainian coupon is falling rapidly against the ruble and is now worth less than one ruble. Early in July Ukraine expects to withdraw from the ruble zone and later introduce its own currency, the hryvna.

**19** • Russia's Central Bank proposes to the Russian parliament that it declare insolvent several central banks of other CIS members, including Ukraine's. The action would represent a "penalty" for failing to fulfill agreements on the coordination of credit and monetary policies within the ruble zone. With many billions in debt to Russia, Ukraine has still decided to issue 300 billion rubles in credits to Ukrainian enterprises. The enterprises will use this money to settle accounts with Russian enterprises, dealing a heavy inflationary blow to Russia's economy. The Central Bank also recommends that the government set strict limits on supplies for Ukraine from Russian enterprises.

**20** • Ukraine cuts the flow of Russian oil through the Druzhba pipeline to Hungary and the Czech Republic by 25 percent. The decision seeks to force Russia to pay $5.5 million in past-due transit fees.

**25** • In Minsk, CIS representatives talk inconclusively on currency issues. (Moldova and Azerbaijan send only observers.) Russia, planning to standardize its internal ruble exchange rates from 1 July, needs specific agreements on which states would stay within the ruble zone, and on procedures for the introduction of separate currencies by certain republics. Russia, which controls ruble printing, is currently forced to extend credit to other republics. Armenia, Azerbaijan, Belarus, Moldova, Kazakhstan, Russia, Turkmenistan, and Uzbekistan agree to introduce standardized customs tariffs and categorize foreign trade lists.

• The leaders of eleven Black Sea states meet in Istanbul to sign a Declaration of Black Sea Economic Cooperation. The states include: Russia, Turkey, Bulgaria, Greece, Azerbaijan, Armenia, Ukraine, Moldova, Georgia, Romania, and Albania. The document states that the Black Sea region must become a sea of peace, stability, and prosperity.

• Presidents Kravchuk and Shevardnadze confirm intentions to sign major interstate economic and political agreements.

### *Political*

**2** • The Ukrainian parliament rejects Russia's 21 May Supreme Soviet vote annulling the 1954 decree transferring the Crimea from Russia to Ukraine.

**10** • Russia and Kyrgyzstan sign a friendship treaty.

**23** • Boris Yeltsin and Leonid Kravchuk meet in Dagomys to work on a comprehensive political treaty "reflecting the new quality of their relationship." They discuss transferring ex-Soviet property abroad to Ukraine and methods to introduce a Ukrainian currency.

**30** • The Ukrainian parliament passes constitutional amendments on the status of the Crimea. These include Crimean autonomy and new citizenship provisions which read: "each citizen of the Republic of Crimea is simultaneously a citizen of Ukraine."

### *Security*

**2** • Boris Yeltsin announces an immediate strengthening of surveillance along Russia's border with other CIS states. Formal border controls with Azerbaijan, Estonia, Latvia, Lithuania, and Ukraine and a customs border with Georgia will be established. According to an *Izvestiya* article, visa-free regimes recently introduced in some CIS states have been allowing "criminals and spies" to enter Russia through transparent ex-Soviet internal borders.

**11** • Kazakh soldiers mutiny in Arkhangelsk oblast, Russia, objecting to "serving in a foreign state."

**13** • The former Soviet Turkestan military district will be disbanded by 30 June 1992. Kazakhstan, Kyrgyzstan, Turkmenistan, Tajikistan, and Uzbekistan fall within the district's jurisdiction.

## July 1992

### *Economic*

**6** • The CIS heads of state meet in Moscow. Agreement is reached on creating a CIS Economic Court, dividing USSR property, and coordinating new currencies when member states depart from the ruble zone.

**9** • In Ukraine, Vice-President and Minister of Economics Vladimir Lanovoy declares that he will resign because "intimidation has set in." He tells *Nezavisimaya Gazeta* that he is "sick and tired of playing the role of mimic. . . . I would like to engage in reforms, not support their mimicry." Asked whether he received the Ukrainian president's support in his independent economic reform positions, he replies: "I did not receive the president's support." Lanovoy calls the program for privatizating state property a restoration of neo-socialism through collective forms.

**10** • Armenia, Belarus, Kazakhstan, Russia, Tajikistan, and Uzbekistan establish the Interstate Economic Committee and the Interstate Ecological Fund. Ukraine attends the meeting but does not sign. Azerbaijan, Moldova, and Turkmenistan do not attend.

**14** • Vice-President Aleksandr Rutskoy visits Moldova to engage in "civilized dialogue." State advisor to the Moldovan president Cheslav Chobanu emphasizes the need to shift attention from the political to the economic sphere." "Russia is our largest partner and we always maintain traditionally beneficial relations," he notes.

**28** • A detailed article in *Rossiyskie Vesti* analyzes Russia's "special economic relations" with former Soviet republics. These represent a combination of economic and geopolitical interests plus a concern for Russians living abroad. The author says it is "impossible to separate these three groups of interests from one another. . . . We would like to retain deep integration with all the former republics of the USSR making no exception for anyone. . . ." The model of integration lies in "a single economic area, including a common monetary and credit system; non-restrictions on the shipment of goods, services, and capital; free movement of peoples; a coordinated socio-economic policy; and a common line *toward third countries.*"

### *Political*

**6** • The Council of CIS Heads of State meets in Moscow. (Azerbaijan does not participate.) The main result is an agreement to establish joint peacemaking forces to intervene in CIS conflicts. Also, an economic court is established, with headquarters in Minsk. Kyrgyz President Akaev describes it as "a mechanism for the resolution of disputes concerning economic cooperation and . . . for exclusive discipline over economic obligations." A proposal by Kazakh President Nursultan Nazarbaev to establish a consultative economic coordination council and a military coordinating council is met with general approval. Nazarbaev points to OPEC and NATO as models for the putative councils. Three documents are signed on issues of legal succession to the USSR regarding treaties, property, archives, debts, and assets. A second group of issues, on which a protocol is signed, concerns collective security, including missile early-warning systems, space control, anti-aircraft defense, a collective security council, the composition of CIS strategic forces, leadership of the CIS joint armed forces and the protection of state borders. The four states with nuclear weapons on their territory agree to meet separately to discuss transferring these weapons to Russia. The question of interim control over nuclear weapons in Ukraine had been removed from the agenda, as it was felt that there was no likelihood of an agreement at the summit. Ukraine has insisted on retaining operational control over long-range nuclear weapons deployed on its territory. The participants warn that by next summit, states which have not ratified their membership in the CIS (Azerbaijan and Moldova) should do so. The concept of a statute for the CIS was mooted and met with general agreement, even from Ukrainian President Leonid Kravchuk, who had previously opposed the creation of CIS bodies.

**9** • The Crimean Supreme Soviet votes against holding the referendum on independence planned for 2 August.

**11** • Kasatonov and Ukrainian Naval Commander-in-chief Boris Khozhin meet in Sevastopol to sign an agreement that neither side will take unilateral actions with regard to the Black Sea Fleet (BSF). Nevertheless, a BSF naval infantry unit seizes the military commandant's office of the Sevastopol garrison and forcibly removes the Ukrainian commander.

**17** • Armenia and Georgia establish diplomatic relations.

**26** • Except in the Black Sea Fleet and the Caspian Sea Flotilla, CIS naval ships begin to fly the Russian naval flag.

### *Security*

**2–3** • CIS defense ministers meet in Moscow. Although the meeting fails to remove tensions between Ukraine and Russia over administrative controls of nuclear weapons in Ukraine, CIS agreements are signed on anti-missile defense and control over space projects, air defense, and on a council of collective security.

**6** • The CIS heads of state meet in Moscow. They agree in principle to create peacekeeping forces numbering between 2,000 and 10,000 with the first deployment to monitor the ceasefire in Moldova. The question of the administrative control of Ukraine's nuclear weapons remains unsolved. The participants agree to restructure the CIS Border Forces by establishing a CIS Council of Commanders of Border Forces. Finally, Col.-Gen. Boris Pyankov is confirmed as CIS deputy commander.

• Shaposhnikov announces that the main functions of the CIS command are centralized control over strategic nuclear arms, coordination of military doctrines and military reforms of CIS member states, and the settling of armed conflicts both inside and along the periphery of the CIS. The CIS command will have 300 military and 100 civilian employees and will be subordinate to the CIS heads of state. The CIS command will also manage the meetings of the CIS Defense Ministers' Council. That council will have a committee for coordinating nuclear strategy and a secretariat.

**10** • Director of the CIS Naval Press Service Novikov warns that CIS naval personnel are upset over delays in receiving their pay. R1.5 billion have yet to be distributed.

**16** • CIS defense and foreign ministers meet in Tashkent to discuss deployment of peacekeeping forces within the CIS.

**27** • CIS Naval Commander-in-chief Admiral Vladimir Chernavin becomes Russian naval commander-in-chief and a member of the Russian Ministry of Defense Collegium. He asserts that the Baltic Fleet will remain in the Baltic, and that the fleet command hopes to secure bases in Tallinn, Estonia, and in Liepaja, Latvia.

### August 1992

#### *Economic*

**15** • Azerbaijan announces its intention to assume its $1.63 million share of the former USSR's foreign debt.

**28** • Russia is able to pay only $2 billion of its external debt in 1992, according to Acting Prime Minister Egor Gaydar. Russia and the CIS were to have paid $9.8 billion of the former USSR's debts, but as of 27 July Russia had paid approximately $1 billion.

**28** • In Russia, Aleksei Mamonton, chief of the Currency Marketing Department at the Moscow Interbank Currency Exchange, calls the fall in the dollar rate of the ruble the result of a true "panic" in currency trading. Inflationary expectation is so high that speculation is running rampant. (The rate stood at $1 for 205 rubles on 27 August.) The Gaydar government is surrendering one position after another and monetarism is falling. According to one reporter, "the Central Bank's new leadership has, for all intents and purposes, rejected the stabilization of the ruble and embarked on the tried and true path of printing money. Rubles will become more plentiful and goods more scarce."

**31** • Russia and Belarus sign an agreement under which Russia will pay the Belarusian share of the total foreign debt of the former Soviet Union, which amounts to US$3.5 billion, or 4.13 percent of the total.

#### *Political*

**3** • At a meeting in Yalta, the Ukrainian and Russian Federation presidents sign an agreement on the division of the Black Sea Fleet, providing for a transitional period extending to 1995 which allows the Black Sea Fleet to be removed from the CIS joint armed forces and "placed under the direct jurisdiction of the Russian Federation and Ukraine." It is expected that the fleet will be divided at the end of the transitional period. The agreement stipulates that during the transitional period, Russia and Ukraine would exercise joint command over the Fleet, enjoy equal use of its bases and facilities, recruit to it equally by conscription, and allow servicemen of the fleet to swear allegiance to the state of which they are citizens.

#### *Security*

**12** • CIS High Command Representative Colonel Vasiliy Volkov announces that a protocol on procedures for using CIS peacekeeping forces is signed. Azerbaijan, Belarus, Ukraine, and Turkmenistan reportedly do not sign.

**18** • The CIS Armed Forces Committee of the Chiefs of Staff meets in Moscow. Chiefs of staff and deputy defense ministers for member states less Moldova participate.

**31** • The CIS Joint Armed Forces Command is formally established.

### September 1992

#### *Economic*

**1** • A senior banker from the Central Bank of Russia (CBR) admits to an ITAR-TASS reporter that the CBR exceeded its authority when it instructed Russian enterprises to curb shipments to Commonwealth states because of their debts to Russia. "We did not take into account the commitments of the Commonwealth members and tried to protect only our own interests," Chief Administrator Sergey Panov told TASS. The new leadership of the bank canceled the order.

**8** • Russia concludes agreements with Belarus, Turkmenistan, and Uzbekistan according to which Russia will assume these republics' share of the former Soviet Union's foreign debt in return for their portion of the Soviet Union's assets.

• Georgia issues a decree which restricts imports and exports of goods on Georgian territory, irrespective of the form of ownership. The decree sets quotas and authorizes rules and regulations for Georgian trade.

**14** • Moscow television reports that as the first congress of industrialists opens in Kazakhstan, a list of economic woes is compiled. Since the beginning of 1992, 80,000 people have lost their jobs. More than 300 factories have ceased part or all of their production activities. Prices are skyrocketing, but there is nothing to buy. Laws and economic decrees are issued, but ignored, and people are becoming catastrophically impoverished.

**20** • In Belarus, the government bans free sales of hard currency, and suggests assessing exporters a tax on hard currency profits of 10 percent of the contract price. If the tax is not paid within thirty days of receipt of profits the penalty will be 100 percent of the unpaid amount.

**22** • In Russia, the Central Bank suspends indefinitely financial transactions between Russia and Ukraine, on orders from the Russian government. Transactions will resume when the two governments settle mutual payments for goods supplied. Russia contends that Ukraine is exporting inflation to it when Ukrainian enterprises purchase Russian goods with credits issued by Ukraine. Although Russia has the monopoly on printing money, all former Soviet republics can issue credits to their industries.

**24** • In an exclusive interview with Interfax, Chairman of the Belarusian parliament Stanislav Shushkevich speaks in favor of reintroducing elements of economic planning. The prime minister's chief economic advisor, Yevsey Makhlin, shares the view that a general and detailed plan for economic development in Belarus "for several years ahead, taking into account political and economic shifts eventually taking place" is needed. Makhlin says that economic plans can be and are always amended to take account of "unexpected realities."

**27** • Kazakhstan and Japan agree to form a Kazakh-Japanese coordinating structure for economic cooperation.

• President Sapamurad Niyazov says that Turkmenistan plans to build a democratic and secular state with diverse forms of ownership. He expresses confidence that the country can ease the pain of economic transition through sales of oil, gas, cotton, and related processed products.

***Political***

**16** • The first Interparliamentary Assembly of CIS member states is held in Bishkek. Participating are delegations from Armenia, Belarus, Kazakhstan, Kyrgyzstan, Russia, and Tajikistan—six of the seven states which had agreed to an interparliamentary assembly in April. Uzbekistan sends no delegation. Ruslan Khasbulatov, chairman of the Russian Supreme Soviet, is elected for a one-year term as chairman of the Assembly, whose seat would be in St. Petersburg. He speaks strongly in favor of increased cooperation among CIS states, mentioning a "supranational parliament" and common citizenship.

• The CIS summit meeting scheduled for 25 September is postponed until 9 October to allow more time for finalizing the working documents for discussion at the meeting. The delay apparently stems from disagreement between Russia and Ukraine over control of the nuclear weapons stationed in Ukraine.

**25** • Crimean Republic's Supreme Soviet adopts amendments to rectify the "defects" in the constitution adopted in May. It confirms that the Crimean Republic is "a legal, democratic, secular state within the structure of Ukraine," whose jurisdiction "is determined by its constitution and the law of Ukraine on the delimitation of powers between the bodies of power of Ukraine and the Crimean Republic."

***Security***

**3** • CIS Council of Defense Ministers meets in Moscow. Russia and Ukraine continue to argue over administrative control of nuclear weapons. Ukraine insists that the strategic forces on its territory should be under the jurisdiction and control of the Ukrainian Defense Ministry.

**29** • Shaposhnikov warns European countries against interfering in talks between CIS states over nuclear weapons control.

**30** • In *Krasnaya Zvezda*, Shaposhnikov calls for Russian supervision over CIS nuclear weapons. In addition, the CIS commander-in-chief asserts that the United States wants to create a "unipolar world" and calls for CIS to play a role between the North and the South.

**October 1992**

***Economic***

**8–9** • A CIS summit is held in Bishkek where a joint session of the Heads of State and Heads of Government convenes. Agreements include coordinating economic legislation, the creation of an international TV company, and the mutual recognition of property rights.

***Political***

**1** • Uzbekistan and Kyrgyzstan sign a treaty on friendship and cooperation and conclude a trade and economic treaty for 1993. Included in the friendship treaty is a defense cooperation agreement, covering material provisions for the armed forces and officer training.

**9** • The CIS Council of Heads of State meets in Bishkek. (No delegation from Azerbaijan is present.) Agreement is reached on a number of economic issues. A "consultative working commission" will be set up "under the Councils of Heads of State and Government"—a watered-down version of a proposed permanent Consultative and Coordinating Economic Council—on convergence of the economic laws of commonwealth states. The establishment of an interstate bank to coordinate fiscal policy in the ruble zone was agreed to in principle by six member states—Belarus, Russia, Kyrgyzstan, Uzbekistan, Kazakhstan, and Armenia. Military matters discussed at the summit include the situation in Tajikistan, the general concept of military security in the CIS, and the status of strategic and nuclear weapons in CIS member states. Agreement is reached on sending humanitarian assistance to Tajikistan and on reinforcing military units present in the republics, but not on sending peacekeeping units. A statute on the CIS joint armed forces high command is also signed, which according to Commander-in-chief of the CIS forces Evgeniy Shaposhnikov will "allow the basis of military policy and the collective defense of CIS states to be determined" and "put an end to uncertainty." Regarding strategic and nuclear forces, Shaposhnikov claims that the situation remains essentially unchanged, with Russia, Belarus, Kazakhstan, and Ukraine in the process of negotiating bilateral agreements. Ukraine, however, remains reluctant to relinquish control over the nuclear missiles on its territory.

*Security*

**6–7** • CIS defense ministers meet in Bishkek.

**8–9** • A joint session of the Councils of Heads of State and Government convenes in Bishkek. All ten members participate and Georgia and Azerbaijan send observers. The parties agree to cooperate in ensuring the stability of the Commonwealth's external frontiers. On the issue of nuclear weapons control, Belarus and Kazakhstan agree to turn over their launch codes to Russia and to dismantle their weapons within three years. Ukraine refuses to go along. A CIS draft treaty on defense and collective security is signed except for Ukraine and Moldova. It calls for members to defend each other from external threats. Yuriy Maksimov is removed from the position of CIS commander-in-chief for strategic rocket forces. No replacement is named.

**16** • The CIS High Command expresses concern over the disintegration of the former Soviet air defense system in Central Asia and the Caucasus. Radar stations continue to close as military specialists flee ethnic fighting.

**November 1992**

*Economic*

**1** • In Ankara, Turkey hosts a summit bringing together the leaders of Azerbaijan, Kazakhstan, Turkmenistan, Uzbekistan, and Turkey to create a "Turkic Common Market." Participants sign a declaration on economic cooperation and agree to reconvene annually.

*Political*

**13** • CIS heads of government meet in Moscow. They are unable to reach agreement on the Commonwealth charter, with Ukrainian Prime Minister Leonid Kuchma declaring his government unwilling to sign the document in its present form.

*Security*

**5** • CIS military officials fail to agree on a document establishing the composition of the CIS strategic forces.

**12** • CIS Armed Forces Commander-in-chief Boris Pyankov states that the enduring potential for hostilities breaking out in the CIS, particularly in Central Asia and the Caucasus, makes the creation of CIS peacekeeping forces a necessity.

• Shaposhnikov calls for a Russian-Ukrainian summit to discuss the disposition of nuclear weapons still deployed in Ukraine.

**18** • Shaposhnikov states that he favors a NATO-style arrangement for the CIS in which each of the members would provide a specified number of troops to unified forces, which would carry out agreed assignments.

**21** • The CIS Council of Defense Minister's secretary, Lt.-Gen. Leonid Ivashov, warns that the armed forces of the former Soviet space were divided among the successor states in an "irrational" manner.

**December 1992**

*Economic*

**6** • In Armenia, a dramatic energy crisis evolves. The republic is receiving less than 4,000,000 cubic meters of gas per day, as compared to requirement of 7,500,000. The shortages are primarily a consequence of a blockade by Azerbaijan—itself a result of the dispute over Nagorno-Karabakh—which has forced the closure of pipelines bringing Turkmen gas to Armenia via Azerbaijan. Supplies from Russia via Georgia are down by one-third, due to a combination of pressure from Azerbaijan, pilfering, and local disruption as a result of fighting in the area. Azeri pressure is also said to be behind Turkey's decision not to go ahead with scheduled daily deliveries of 30–35 Kw of electricity to Armenia, due to begin on 1 December.

**7** • CIS agricultural figures are released. Total grain production rises 18 percent primarily from Kazakhstan's record harvest. State procurement of potatoes and processed sugar has dropped all around. Meat production falls 14 percent and milk, 12 percent.

**17** • The World Bank estimates the former Soviet external debt at $75.4 billion in mid-1992. Repayment arrears at that time amounted to $9.4 billion. Nearly 52 percent of the total debt is due for repayment within the next three years.

*Security*

**11** • At the 7th Russian Congress of People's Deputies, Shaposhnikov calls for "strong coordinating bodies" between Kazakhstan, Uzbekistan, Kyrgyzstan, Tajikistan, and Armenia. He states that Turkmenistan and Belarus favor closer integration into a NATO-type security system.

**16** • The Belarusian parliament votes by an overwhelming majority not to join the CIS Collective Security Treaty concluded in Tashkent on 15 May.

**21** • Shaposhnikov meets with CIS representatives. He complains that the CIS Collective Security Treaty and arrangements on peacekeeping forces have failed to work. Moreover, he states that a recent inspection of the strategic nuclear forces revealed security violations.

**28** • Shaposhnikov expresses concern about the combat readiness of the CIS forces.

# 1993

## January 1993

### *Economic*

**1** • Ivan Korotchenya, coordinator of the CIS staff in Minsk, says the CIS is just a system of "units," not an integral structure. He characterizes "The Economic Court, the Coordination Economic Council, as conceived but not yet born. . . ."

• The Moldovan first deputy minister of economics, Deakonu, says it is better for Moldova to be part of a single economic space (buying petrol at 1.2 times the domestic price) than to face the plight of the Baltics, buying resources at 3–3.5 times domestic prices. He notes the economic crisis in Moldova during the first eleven months of 1991 (industrial production fell 25 percent from same period in 1990; trade shrank by 50 percent; investment by 30 percent).

**4** • The Central Asian states hold a major regional summit where the governments declare a Central Asian Economic Union, but claim it does not compete with the CIS Economic Union. Moscow regards the summit as a definite challenge to its authority and leadership within the CIS.

**5** • Prime Minister Kuchma tells Ukrainian reporters that "First and foremost, Ukraine is dependent on Russia." He points to fascination with sovereignty as the cause of the breakdown in former Soviet economic supply networks. "Every state wants to make everything for itself," he says.

**6** • In an interview preceding a bilateral summit with Yeltsin, President Kravchuk says, "I'm in bed with an elephant. This is a nightmare—I'm afraid that I might wake him up." He says that Ukraine is primarily interested in an energy agreement, and is dependent on a single "Russian pipe." He is sure that Russia will have an energy quota in mind, but Ukraine will present its own "quota list." (Ukraine produces 90 percent of aeroengines in the CIS, and most sugar and sunflower oil.) He adds that Ukraine and Russia must start to reform absurd "unbreakable ties," and initiate new and mutually advantageous ones.

• In Belarus, the independent paper *Belarusian Businessman* forecasts that as a result of disrupted CIS inter-enterprise payments and high taxes for social protection, the GNP will not grow in 1993.

**7** • Russia and Kazakhstan conclude an agreement under which Kazakhstan will supply 14 million tons of oil to Russian refineries while Russia will supply more than 12 million tons of oil from western Siberia to Kazakhstan, in order to save on transportation costs.

• Ukrainian Deputy Prime Minister Viktor Penzenyk discusses Ukraine's share of the former Soviet debt to foreign banks and governments, which is 16.4 percent. Ukraine and Russia were to sign an agreement on the debt before the end of 1992, but failed.

**10** • Several Ukrainian political parties call for an end to the country's isolationist policy, which they say is "leading Ukraine into a blind alley." They protest Ukraine's withdrawing from the CIS, claiming that separation will lead to mass poverty and destruction.

**11** • The Ukrainian government decides not to give Russia authority over administering its portion of the former Soviet debt.

**12** • Ukrainian Prime Minister Kuchma threatens to resign because of Russia's new policy to raise energy prices, and the Russian Central Bank's refusal to pay Russian state enterprise debts to Ukraine. (Gerashchenko says the Russian state is not responsible for its enterprises' actions.) In response, Kuchma tells Ukrainian enterprises to require prepayment from Russian customers.

• In Turkmenistan, President Niyazov states that he believes the country's new currency, the manat, will be stable because the economic potential of the country is strong. Turkmenistan possesses indigenous supplies of oil and gas, ores, and mineral salts (including iodine).

• In Russia, Arkadiy Volskiy, president of the Russian Union of Industrialists and Entrepreneurs, says an agreement to step up economic links between the CIS republics is essential. He says priority should be given to agro-industrial and pharmaceutical sectors, because agricultural processing, storage, and transport are in such lamentable condition that only 50 percent of the harvest reaches the shops. He adds that $5 billion worth of grain and $2 billion of other foodstuffs are procured abroad. If only half of these funds were directed toward the infrastructure, the situation would improve dramatically. Russia has only 60 percent of the medicines it needs, he continues. "Economics are a far greater danger to democracy than the issue of a referendum on the constitution." On reforms, he says "price liberalization is not a panacea. It should be the final stage of reform, whereas we are still in the initial stage."

• CIS-wide data from the State Committee for Cooperation shows that the volume of industrial production in the Commonwealth countries dropped 18–20 percent in 1992 in comparison with 1991, while the GNP declined by 20 percent. Real monetary incomes of the population decreased twofold. Prices rose fourteen times, wages increased approximately seven times. Trade turnover among CIS member states decreased 50 percent.

**14** • In Tajikistan, the minister of economics says regions of country have lost their workforce due to fierce fighting. (The exodus of Russian citizens from Tajikistan has been very heavy—50,000 in the last few months of 1992.)

• In Georgia, Deputy Prime Minister Roman Gotsiridze, returning from Moscow, supports economic cooperation with Russia in exchange for Russian subsidies in the amount of "tens of billions of rubles." The Georgian parliament issues a strong rebuke of Gotsiridze's recommendations and demands withdrawal from the CIS.

**15** • The Georgian government signs an agreement to join the single monetary "ruble zone." Russian pressure in the form of withholding 15 billion rubles from the Georgian budget in December 1991, combined with total denial of transfers in January 1992, is a critical factor in the decision.

**18** • In Kyrgyzstan, President Akaev says the Russian changeover to world pricing with CIS states bears witness to the impossibility of achieving constructive economic relations within the CIS framework. He calls for use of Kyrgyz "trump cards" at the 22 January Minsk summit.

• The CIS working group announces that it will consider forming a consortium to build an automobile factory in Elabuga, Tatarstan. Tatarstan would hold a 25 percent share; Ukraine—15 percent; Belarus—10 percent; Uzbekistan—10 percent; and Kyrgyzstan—5 percent. Components would be manufactured in the republics entering the concern. Russian Prime Minister Chernomyrdin is the originator of the concept. The factory in Elabuga was started in the 1970s, but was suspended for lack of funds.

• In Ukraine, President Kravchuk holds a news conference on the upcoming Moscow meeting. He says an economic policy must be built on links between states that are needed, but some links are not needed by any state. He notes that meeting international standards of production, quality, and accounting would provide an alternative to tight integration among CIS states.

**19** • Also in Ukraine, several Donbass mining associations say the economic survival of all CIS states is possible only under conditions of restoring the ties established by the former USSR.

• In Kazakhstan, visiting President Ter-Petrosyan arranges natural gas deliveries to Armenia. All industry has stopped in Armenia due to the energy shortage.

• Kazakh President Nazarbaev tells journalists that economic liberalization rules dictated by Russia "have not justified themselves"—but then calls for a revival of state planning.

**21** • In Ukraine, the government announces that it will pay its share of the former Soviet debt (16.4 percent), but will also retain its share of former Soviet assets.

• Ukrainian Prime Minister Kuchma says the Ukrainian economic crisis is a result of "disintegration" and the "decolonization" of the economy. He says that under conditions of political stability, Ukraine and Russia could sign mutually advantageous treaties of economic "partnership."

**22** • The second CIS summit is held in Minsk. A document creating a CIS Interstate Bank is signed by seven members. Russian CIS delegates describe this as a major achievement. (The Bank is never implemented.)

• Igor Shichanin, Russian chief of CIS affairs, says the CIS was formed to ensure a unified national economy. He says quite a few sectoral coordinating organs have been set up—in railroad transport, fuel, energy—and that the level of integration in the CIS is much higher than in the EC, although the comparison is not justified.

**25** • In Belarus, President Shushkevich discusses the CIS summit of 22 January, saying that the economic agreements are the most important. He wants the CIS to adopt strict financial and credit rules. He also desires a strong Interstate Bank in which Russia has 50 percent of votes.

**26** • In Russia, the Customs Committee toughens customs regulations on borders with other CIS states, citing too many exports of staples under the guise of "barter" deals.

• The three largest commercial banks of Russia, Ukraine, and Belarus (Promstroybank, Prominvestbank, and Belpromstroybank, respectively) sign an agreement to coordinate banking activity.

**27** • At a World Economic Forum in Davos, Belarusian Shushkevich says: "the two major European economic groups will be the EEC and the East European Economic Community (EEEC), which means that the CIS, and the two will begin drawing closer—perhaps as early as 2000—to form a "common European home."

### *Political*

**1** • In Belarus, President Shushkevich calls the CIS a "historic inevitability," and says "every other variant would have been worse." He places his faith in the Interparliamentary Assembly for determining the approach to integration. He thinks it is too early for a confederation, as Russia wishes, but not for a "real common market, currency, and common charter."

**3–4** • The summit of Central Asian states in Tashkent erects a system of bilateral and multilateral treaties, which they will call the "Central Asian Commonwealth." (Attendees: Kazakhstan, Kyrgyzstan, Tajikistan, Turkestan, Uzbekistan.) Uzbek president Karimov says each state wants guarantees that it is free to develop the way it deems correct. Kazakh president Nazarbaev says steps are being taken to start a Central Asian common market, with common taxation rates, customs, pricing, investment, and export policies. He adds that the five would like to preserve the ruble zone—but on a principle of equality with Russia. The participants reconfirm their commitment to the CIS in general, but complain of being seen by some as "younger brothers" within the CIS. Awareness of the shortcomings of the CIS is widely seen as the motivating force behind the decision to set up some form of regional association.

**4** • In Ukraine, President Kravchuk says that before his government will sign the draft CIS Charter, it must conclude large-scale economic agreements and interstate treaties with Russia and other CIS member states. He argues that the emphasis must lie on wider bilateral and multilateral cooperation within the CIS. In his opinion, any integration process must first resolve economic problems, then address political problems. Most Ukrainian political leaders oppose the charter—arguing that it would upset the principle of voluntary membership in the CIS, violating the declarations signed in Minsk and Alma-Ata.

**5** • In Turkmenistan, President Niyazov says the CIS is developing on the basis of bilateral agreements. He says every CIS state needs to develop its own sovereignty.

• In Ukraine, President Kravchuk says the CIS draft charter does not serve Ukraine's national interests. He says Ukraine will not participate in the CIS as an associate or observer. In his view, Ukraine might consider some documents on the agenda, but these could take decades to work out. "I held a post in the CPSU when the USSR existed as an empire. Now I'll spare no efforts to oppose attempts to return to an empire." "Russia pursues a policy which meets its interests, but as a member of the UN and the CSCE, Russia should be guided by international norms and principles."

• In Kyrgyzstan, President Akaev says the CIS does not meet the needs of its members in its present form. He calls for a three-tier arrangement consisting of: (a) a "corps" of member states; (b) associate members; and (c) observers.

• In Ukraine, commenting on the CIS Charter, President Kravchuk says: "who needs a statute? . . . those who want to return to a unified state, and to turn the Commonwealth into a union. . . . as for why Russia wants the CIS, it is because Russia has never abandoned its intentions to be a superstate and a leading force in the CIS and beyond the boundaries of the CIS."

**6** • On the question of dual citizenship (11 million Ukrainian citizens are of Russian origin and 5 million of Ukrainian origin), President Kravchuk says: "It is difficult for me to explain why a Russian in Ukraine is free to choose his language, school, theater, cultural environment, . . . when a Ukrainian in Russia is not. . . . Nowhere in Russia will you see a single newspaper nor hear a single broadcast in Ukrainian—even in Yakutia."

• In Belarus, the state radio proposes that all future CIS sessions be held in Minsk. CIS presidents will vote on this proposal on 22 January, where the CIS is expected to be "given a second wind." The headquarters of the CIS Coordinating Office are already in Minsk, but summits are moved from capital to capital.

• In Russia, *Izvestiya* calls the Central Asian summit a complete surprise. However, it denies the existence of a "Slavic bloc" and says Russia has problems similar to those of its southern neighbors. It adds that Russia and Central Asia have very major long-term interests in common—economic and geopolitical.

• In Ukraine, the Ukrainian National Assembly publicly alleges that "pro-imperial forces" are pushing Ukraine to sign the CIS Charter.

• Mikhail Horyn, Ivan Drach, and Dmitriy Pavlychko say that Ukraine must not sign the CIS Charter on 22 January. Vyacheslav Chernovil says Rukh advocates Ukraine's secession from the CIS. Stepan Khmara concurs with Rukh.

• The Ukrainian National Assembly rejects the Russian-Ukrainian Agreement on "Good Neighborly Relations, Cooperation, and Partnership" until the Russian Supreme Soviet rescinds its decision calling the 1954 transfer of Crimea to Ukraine illegal, and also the December 1992 resolution to give Russia control of Sevastopol.

**7** • In Ukraine, President Kravchuk advises the authors of the CIS Charter to "get down to work and spend 10 years on it."

**10** • In Ukraine, Naval Commander Borys Kozhyn registers opposition to the Ukrainian-Russian Yalta agreement on the Black Sea Fleet. He says his task is to transfer the disputed navy to the Ukrainian Deputy of Defense's jurisdiction without waiting for the political solution envisaged by the Yalta agreement.

**11** • The Ukrainian deputy minister of defense, Ivan Bizhan, says Ukraine is not taking control of its strategic nuclear forces. He notes that the administration of strategic weapons is still unified under the CIS Alma-Ata Agreement of 1991. However, he says the Ukrainian president wants the power to block the launching of strategic missiles.

**14** • The Central Asian nations sign an international relations treaty in Bishkek, which Kyrgyz President Akaev says is needed to ensure the inviolability of existing borders. Territorial disputes among Central Asian states, which began in 1989 when the Soviet Union collapsed, have not yet been resolved. The treaty establishes diplomatic relations and embassies within the region.

• In Russia, First Deputy Prime Minister Vladimir Shumeyko says the future of the Commonwealth will depend to a great extent on the results of the upcoming summit in Minsk. He refuses to make forecasts, but assesses the effectiveness of the CIS economic mechanism as "something between bad and unsatisfactory." He says bilateral cooperation within the CIS framework is more promising.

• An article in *Rossiyskaya Gazeta* investigates the possibility of the rise of a united "Turkestan" in the wake of the Tashkent summit of 4 January. It mentions a book by Lev Gumilev—"Millennium Around the Caspian," which has appeared in Baku this year.

• In Ukraine, disaffected Donetsk political representatives appeal to President Kravchuk not to withdraw from the CIS, expressing the dominant views of miners in eastern Ukraine. (One day later they call for Kravchuk's resignation.)

**17** • In Crimea, the All-Crimea Movement organizes a rally in Sevastopol—calling for recreation of the USSR and shouting slogans of "Back to Russia." O. Kruglov, chairman of the Sevastopol Committee of the National Salvation Front, says the city must become Russian even if a Slavic confederation is formed.

**20** • In Russia, Andrey Kozyrev says Yeltsin will put forward a "solid package of principled ideas to strengthen the Commonwealth." "Maybe this would be a CIS Doctrine," he continues.

• Following a pre-summit meeting of CIS Council of Ministers of Foreign Affairs, it is announced that Ukraine and Turkmenistan are opposed to the CIS draft charter.

**22** • On the day of the summit, in Tashkent, Kravchuk announces that "We oppose creation of any rigid suprastate structures with a center of any kind. Any problem within the so-called single space can be solved on the basis of bilateral relations."

• Yeltsin tells reporters before flying to Minsk that a charter must be signed, noting that "to sign the charter without Ukraine would be undesirable." He emphasizes that the five Central Asian nations should be prevented from sudden secession.

• In Minsk, after a day of debate, there is only partial agreement on a charter for closer political and economic integration. The charter is signed by seven of the ten attending states. (Georgia and Azerbaijan do not attend.) Only Russia and Kazakhstan express full support for the original draft charter, with Belarus expressing doubts about collective security arrangements, and Uzbekistan objecting to clauses covering human rights. Ukraine, Moldova, and Turkmenistan refuse to sign; all three will remain members of the CIS while they consider their positions. The ten republics reach an agreement on the creation of an interstate bank, to act as a clearinghouse for trade among them and to coordinate monetary, credit, and budgetary policy for those republics which remain in the ruble zone. The summit rejects Russia's proposal to take control of nuclear weapons in Ukraine, Belarus, and Kazakhstan. Russian President Boris Yeltsin says a door had been left open, giving the dissenting republics three months to sign the charter. Ukrainian President Leonid Kravchuk says that the charter framework is too binding, but acknowledges that refusal to sign the charter does not mean that his republic is leaving the CIS.

• The CIS summit ends. Members deeply disagree over the outcome, especially on the form of relations (i.e., bilateral or suprastate authority given to coordinating bodies).

• In Ukraine, on the day of the summit, the speaker of the Ukrainian parliament, Ivan Plyushch, sends a letter to Ruslan Khasbulatov, head of the Russian parliament, accusing the Russian Supreme Soviet, in its latest decree on Sevastopol, of "a relapse into the past, an attempt to subdue Ukraine, and an attempt to set the two nations against one another and lead them to bloodshed."

**23** • President Kravchuk says an all-Ukrainian referendum would be required to change the administrative status of Sevastopol. Russia will not resolve the problem without Ukraine, regardless of decrees or decisions adopted. He warns people not to yield to provocations.

**24** • In Belarus, President Shushkevich is asked whether some former Soviet republics were forced to sign the CIS charter. He answers that: "Nobody signed the charter. Nobody, not a single side. A memorandum on *the attitude* to the charter was signed. The charter is very flexible because it can hardly be otherwise." He concludes: "There is no pressure here. There is a door which is kindly opened: Come in, please. What pressure is exerted on states which did not join the charter? Or what pressure is exerted on Belarus because we do not participate in the collective defense system? No pressure at all. There is no big brother here."

**25** • In Georgia, a third round of Russian-Georgian talks on a treaty of friendship and cooperation begins in Tbilisi. Talks will address a trade and consultative agreement. According to the Russian ambassador to Georgia, Vladimir Zemskiy, Georgian officials intend to sign an agreement drawn up by the Russian side. Opposition voices in parliament accuse Shevardnadze of "pro-Russian political orientation"—to which Shevardnadze replies that his priority is to orient Georgia toward Germany and the United States.

• In Ukraine, Defense Minister Morozov says a large number of officers who took loyalty oaths to Ukraine did so expecting that the CIS armed forces would be united, and that housing and wages would be provided. He calls for their resignation saying: "It is impossible to serve in the army of a sovereign state while submitting to ideas about violating its sovereignty."

**26** • A *Nezavisimaya Gazeta* article says a gradual but steady disintegration of the integrational core is occurring in the CIS.

• In Georgia, International Policy Advisor to President Shevardnadze Vafa Guluzade thinks the CIS cannot be preserved in its present form, and may only become a consultative organ. He says the Minsk summit was "without results," and West thinks this also. He adds that Baku thinks Azerbaijan should only participate in CIS as an observer, and has decided not to join.

**27** • In Crimea, the new commander-in-chief of the Black Sea Fleet, Eduard Baltin (appointed 15 January as a result of an agreement between Yeltsin and Kravchuk), arrives in Sevastopol.

**28** • In Ukraine, Leonid Smolyakov, Russian ambassador to Ukraine, says Russia plans to open consulates-general in Odessa, Lvov, and Sevastopol. He says Ukraine may open consulates in St. Petersburg, Tyumen, and Vladivostok.

***Security***

**1** • In Belarus, the government establishes its own National Armed Forces, a Defense Ministry, and a law on the "Status of Servicemen," which includes privatization of housing for servicemen.

**5** • Russian General Leonid Ivashov of CIS Defense Ministers' Council says a CIS unified armed forces would consist of nuclear weapons, collective peacekeeping forces, and a contingent for the prevention of conflicts. He says the CIS states are divided on the numbers of troops to commit to any future unified force. Ukraine is at particularly sharp odds on strategic force strength. (Ukraine's unilateral decision to take administrative control of the nuclear arms deployed on its territory will be on the agenda of the 22 January summit in Minsk.) The CIS commander-in-chief insists he alone must be in charge of CIS nuclear forces.

• Russian Rear Admiral Ivan Semenov discusses the fate of the Baltic and Black Sea fleets. He says that the navy's shore infrastructures, built up over centuries, suddenly turn out to be situated on the shores of various countries. He says more than 90 percent of all base facilities and combat personnel of the Black Sea Fleet are concentrated in Odessa, Nikolaev, and Kherson oblasts. Ukraine claims the entire system of bases as its own without a word about compensating Russia for real estate, bases, and technical and logistical facilities. If Russia concedes, the fleet will be deprived of bases, not only for ships, but for aircraft and shore forces. A new system of bases would have to be built for Russia's share of the fleet. Likewise, four of five former USSR naval bases on the Baltic Sea are in Estonia, Latvia, or Lithuania. The Russian navy has kept Baltiysk, but Semenov says that 50 percent of the surface forces, all of the Baltic Fleet submarines, and 15 percent of all naval aircraft were based in the Baltic and will remain there temporarily. It would cost R22 billion to build a new Baltic Fleet base and take five to six years. To rebuild the Black Sea Fleet would cost R1 trillion, and take ten to twelve years.

**9** • In Russia, Aviation Marshal Evgeniy Shaposhnikov discusses the failure of the Treaty on Collective Security, signed by only six states in Tashkent in May 1992. In a little more than six months, the Treaty has been ratified by only three parliaments—those of Kazakhstan, Tajikistan, and Kyrgyzstan. Shaposhnikov says he will work more with the parliaments of the CIS states in 1993. He adds: "For some time I thought, and even now I would like to think, that the Commonwealth will be more than just a word. But I can see how the situation is also changing. The sovereignization syndrome is creeping into the CIS countries' military departments."

**10** • In Belarus, the armed forces give oath of allegiance to the new Belarusian state. Col.-Gen. Pavel Kozlovskiy says officers wishing to return to any other republic of the CIS are free to do so.

**15** • In Georgia, National Security Advisor Tedo Japaridze says joining the CIS is out of the question for Georgia, and "it is about time someone explained the meaning of this Commonwealth." Asked if Georgia misses Western aid going to the CIS, he replies, "I doubt that any of the CIS republics besides Russia gets anything. That is why the West also realizes now that it has to make its allocations separately to each country so all the aid is not sucked into the Russian black hole."

**20** • The CIS Council of Defense Ministers meets in Minsk on the eve of the summit: discusses national armies, position of career servicemen, future of nuclear arms of former USSR. Council Secretary Leonid Ivashov says, "The politicians have managed to divide the manpower and hardware of the former USSR. . . . So far, the combat readiness of the armies of independent countries . . . is rather doubtful. A standard, well-oiled machine of army control has been ruined. Research into new types of military hardware, material supply of ground forces and navy, manning policies, everything has been ruined."

**22** • In Belarus, at the Minsk summit, Ukraine refuses to sign a nuclear arms accord. Kravchuk announces that Ukraine will form its own nuclear arms control center. Marshal Shaposhnikov responds that Ukraine lacks appropriate mechanisms. Russia and Ukraine agree to hold bilateral talks on terms of dismantlement and transportation of nuclear ammunition. Belarus agrees that its nuclear weapons should be controlled by Russia. Kazakhstan supports the supreme command of the combined forces of the CIS on what is to be included in the strategic forces.

**25** • CIS Supreme Commander Marshal Shaposhnikov says Russia is the undisputed heir to Soviet nuclear potential and that all her demands are justified. Belarus and Kazakhstan agree to this without material compensation. Ukraine demands compensation, which the marshal says should be resolved in one month.

## February 1993

### *Economic*

**2** • In Ukraine, President Leonid Kravchuk says the reason Ukraine left the ruble zone is partly because the ruble is inflationary. He adds: "If Russia had the dollar, everyone would stay in that zone."

• In Uzbekistan, President Karimov states that although Uzbekistan has an economic base of cotton and gold to support its own currency, it will stay in the ruble zone unless Russia switches to its own currency.

**6** • In Ukraine, Prime Minister Leonid Kuchma says that Ukraine has embarked on a course that will end the economic war with Russia. He attributes Ukraine's leaving the ruble zone to Russia's actions to block Ukrainian financial and banking transactions. His conclusion is that "there can be no single ruble area" and that every former union republic will introduce its own currency in the near future.

### *Political*

**3** • In Baku, Georgian President Eduard Shevardnadze arrives for a meeting with Azeri President Abulfaz Elchibey. Detailed discussions are held on a prospective "treaty of friendship, good neighborliness and mutual security" and a bilateral treaty is signed which envisages an expansion in trade and economic relations. Both leaders express the hope that the treaty will improve the political climate in the Transcaucasus and will act as a stepping stone to a summit between Azerbaijan, Georgia, and Armenia.

**6** • An argument breaks out between Russia and Ukraine on border troops. The press service of Ukraine's border troops refutes a statement by Lt.-Gen. Anatoliy Parakhin, deputy commander of the Russian Federation border troops, saying that the initiative to establish a Ukrainian-Russian boundary comes from Ukraine.

### *Security*

**3** • Belarus plans to organize two border customs posts and a subunit of patrol boats on the border with Ukraine in response to Ukraine's decision to open a customs post at the border.

## March 1993

### *Economic*

**2** • In Moldova, economic experts say energy resources for Moldova are dependent on the unpredictable Russian price policy and the energy war between Russia and Ukraine. That war started as a result of an ultimatum delivered by the Russian gas concern Gazprom. Price increases on gas from Ukraine to Europe and from Russia to Ukraine have occurred because Ukraine refuses to pay for lubricants at world prices. Observers state that the energy crisis could spell catastrophe for Moldova. A 16–20 percent reduction in required deliveries to Moldova is expected. Experts reveal that only 40–50 percent of lubricants needed for the economy are being received. Gasoline usage for the last few years has been reduced by two to five times, diesel-oil and coal by two times, fuel oil by 1.7 times. Russia will also introduce customs duties on the export of natural gas. The Moldovan government intends to build a port at Giurgiulesti which will open a door to energy resources from the West. Moldova will also participate in building an atomic power generator in Cernavoda/Romania.

• Representatives from twelve former Soviet republics meet and sign an agreement to form an intergovernmental council on oil and gas. Latvia and Turkmenistan are not represented and the Estonian delegation has only observer status. The agreement seeks to ensure adequate production and supplies among its members, but will not seek to influence world markets. It provides for three institutional levels of cooperation: a council of heads of government on oil and gas; a similar ministerial structure; and a permanent secretariat to be located in Tyumen. No multilateral agreement is reached on prices, nor do the Russian and Ukrainian delegations resolve their dispute over the price of Russian gas supplies. The Russian Fuel and Energy Minister Yuriy Shafranik believes that the agreement will lay the foundation to restore Russia's coordinating role in the fuel and energy complex of the former USSR.

**4** • Russia and Ukraine fail to agree on fuel prices and debt payments. According to Viktor Yushchenko, chairman of the board of the Ukrainian National Bank, the debt owed by Russian economic structures to their Ukrainian partners amounts to 600–650 billion rubles, as compared to 100 billion owed by Ukraine.

**8** • The Belarusian National Bank stops paying for goods delivered by Ukraine, Tajikistan, Turkmenistan, Kyrgyzstan, and Georgia, calling these goods repayment of debts which those states owe to Belarus.

**9** • Georgian President Eduard Shevardnadze appeals to Russia to honor an agreement made in Bishkek in October 1992, when Russia agreed to provide Georgia's cash needs. Georgia has not received funds for four months.

• The Belarusian Association of Independent Trade Unions calls for the restoration of a single economic space within the CIS.

**18** • Belarus, Russia, Kazakhstan, and Ukraine discuss formation of an Economic Union. Belarusian Prime Minister

Vyacheslav Kebich calls it essential to set up a CIS Economic Union. It would incorporate fiscal, credit, and monetary policies as well as a joint external customs policy. "Harsh sanctions" would be imposed on any state which ignored the agreements. He emphasizes that the Interstate Economic Court, which is being created, must be given adequate power to take action.

**31** • In Kazakhstan, President Nazarbaev initiates a conference during which Russia, Belarus, Kazakhstan, Kyrgyzstan, Ukraine, Moldova, Georgia, Azerbaijan, Armenia, Tajikistan, and Uzbekistan sign a draft agreement on interstate cooperation in agrarian-industrial relations. The agreement will also stipulate unification of investment, interbank, and scientific-technical policies.

***Political***

**3** • The Russian delegation to the UN formally requests that international organizations, including the United Nations, grant Russia special powers as a guarantor of peace and stability in the regions of the former USSR. To achieve its goal, the delegation presents the UN with a document: "Participation of Russia in Peacemaking Operations in Ex-Soviet Countries." The paper argues that Russian participation in peacemaking operations is required to ensure Russia's security. It also makes the claim that Russia's awareness of its special responsibility for stability and human rights in the former Soviet Union calls for a response to appeals from other CIS countries. The paper says that the legal basis for Commonwealth peacekeeping operations is laid out in the "Agreement on Military Observers and Collective Peacekeeping Forces in the CIS," signed at the Kiev summit on March 20, 1992. According to that agreement, decisions on sending in peacekeeping forces are made by the Council of the Heads of State when a request is submitted by each conflicting side, or if a ceasefire is arranged before the peacekeeping force can be sent in. The main tasks of the peacekeeping forces resemble functions of UN peacemaking forces. They include: "supervision over the fulfillment of terms of ceasefire, disengagement of the warring parties, creation of demilitarized zones and humanitarian corridors, creation of conditions for negotiations on the peaceful settlement of conflicts, suppressing mass disorders, and assisting in ensuring human rights."

• In Ukraine, the Foreign Ministry protests against Russia's demand in the United Nations for special peacekeeping powers, stating that Russia was not authorized by the CIS to make such a request. The delegation of such powers would be a blatant violation of all existing international legal norms, including principles of the UN Charter and the main CSCE documents. Russia's claim carries the unconcealed threat of conveying "police" functions to one CIS state. This would inevitably lead to dictat and interference in internal affairs of other CIS members, constituting a threat to Ukraine's sovereignty and territorial integrity. Ukraine vows never to agree with Russia's position.

• In Ukraine, Prime Minister Leonid Kuchma charges that: "the CIS is simply some kind of cover and nothing more." At the same time he disagrees with President Kravchuk, saying he does not consider signing the CIS Charter "a great tragedy." He believes that signing the charter may take the advantage away from "all the politicians trying to set Russia and Ukraine at odds."

• Ukraine's Supreme Council attacks the questionnaire circulated on 25 December 1992 among the people's deputies of the Republic of Crimea at the initiative of the Russian parliament's provisional working commission on the status of the City of Sevastopol. The Council states that the Russian parliament has no grounds for questioning the current status of the Republic of Crimea and the City of Sevastopol. The Council's statement reads: "The fact that the highest legislative bodies of the Russian Federation would raise the issue of the status of the City of Sevastopol before the Russian Supreme Soviet, and would consider establishing confederative relations among Russia, Ukraine, and the Republic of Crimea, and holding a referendum on Crimean independence constitutes open interference in Ukraine's internal affairs, and encroachment on the territorial integrity of a sovereign state which is a UN member and a CSCE participant." The Presidium also calls it a violation of the CIS Minsk Agreement, which contains obligations to respect the territorial integrity and inviolability of existing state borders.

**12** • In Moscow, delegates from all member states, plus Azerbaijan, with observer status, participate in the CIS Heads of Government summit. They sign agreements on forming an interstate council for the protection of industrial property, forming a coordination council for the interstate broadcasting company, use of military satellite communication systems, forming an interstate commission for military and technological cooperation within the CIS. Commenting on the meeting, Ukrainian Prime Minister Leonid Kuchma says he favors "one economic union and the sooner it emerges, the better."

**17** • Boris Yeltsin appeals to CIS leaders for closer cooperation among the states, particularly in creating collective peacekeeping forces; coordinating foreign policy; joint diplomacy and defense; joint protection of human rights (i.e., for Russians outside Russia); setting up transnational associations in branches of industry, agriculture, electric power, transport, and services; joint activity in production and investment; formation of joint custom systems; shaping a single economic space; and moving toward a common market.

**18** • President Nursultan Nazarbaev supports Yeltsin's call for strengthening the CIS. He proposes joint actions on several fronts, including a collective security treaty, a single currency, an interstate economic committee, which would be responsible for all economic policy in the member countries. He sends a letter to all CIS states that have signed the charter.

• Presidents Mircea Snegur of Moldova and Saparmurad Niyazov of Turkmenistan meet briefly and express a common view on the CIS. In their estimation the CIS mechanism does not provide for stable cooperation among its members. Efforts to create a rigid structure within the CIS only lead to further fragmentation. Snegur describes Moldova's membership in the CIS as that of an "associate member, non-signatory of the charter." He notes that Moldova does not want the CIS transformed into a political and military organization.

**30** • Representatives of CIS governments meet to discuss the statute of the Commonwealth Coordinating and Consultative Committee (CCC), which is to be the CIS executive body. This committee is expected to replace the existing Working Group which consists of Heads of State and Heads of Government. According to Alexander Shokhin, the Russian draft proposal for a CCC is based on the European Community model. Sessions would be presided over in turn by each CIS state. He says that "such a model would encourage Ukraine, Moldova, and Turkmenistan to approach the CIS in a more pragmatic fashion."

**31** • The Kazakh parliament ratifies the CIS Charter. President Nursultan Nazarbaev notes that the seven states which have already signed the Charter have four major tasks: to create a unified bank with a coordinated currency-credit policy; to create an economic committee to supervise signed agreements; to create a single defense area; to define CIS policy toward former Soviet states which have not signed the Charter.

### *Security*

**2** • Despite claims of neutrality, Belarus appears willing to cooperate on the military side within the CIS. Myacheslav Hryb, chief counselor to the head of parliament on defense issues, states that Belarus may join the CIS Collective Security Agreement as an associate member. It would sign on three conditions: (1) Belarus would not be obliged to send troops to foreign countries; (2) the Belarusian army would not be taking part in any combat operation beyond its borders; (3) arms and military equipment from Belarus would never be used in any armed conflicts between other parties.

**17** • CIS heads of state discuss the Collective Security Pact, focusing mainly on the need to create a joint security zone as a counterbalance to the nearby regional powers of China, Japan, Iran, and Western Europe.

**20** • Russian Defense Minister Pavel Grachev stresses the importance of the Northern Fleet, given Russia's loss of operational positions in the Baltic and the Black Sea. Russia's sea strategy will be based on two concepts: a peacetime concept, based on naval containment of potential adversaries; and a wartime concept, providing for the organization of an adequate defense of the coast and sea communications and stopping potential aggression. Priority will be given to the construction of a new generation of submarines. Within the next two years, the fleet will be cut by 20 percent. Quality is to prevail over quantity. Grachev emphasizes the necessity of arming the navy with aircraft carriers for strategic tasks. Each of the four Russian fleets should have at least three such ships. At present, there are only four of them.

**24** • There is speculation on why Belarus changes its position on the Collective Security Treaty. Non-official sources say that Commander-in-chief of the CIS Joint Armed Forces Evgeniy Shaposhnikov gave Belarus an ultimatum: join the collective security system, or the military and technical supplies to the Armed Forces will stop. Belarusian army has no means to manage without Russian supplies. Supreme Soviet Chairman Stanislav Shushkevich disagrees with Prime Minister Vyacheslav Kebich and reaffirms the republic's need for neutrality.

## April 1993

### *Economic*

**28** • At the meeting of CIS Council of Heads of Government in Minsk, the Statute of the Coordination and Consultative Committee is adopted. This committee will be made up of vice-premiers of all CIS states, and will be entrusted with supervising the implementation of decisions adopted within the CIS, especially economic ones. The purpose of the committee is to strengthen economic ties within the CIS. Russian Premier Viktor Chernomyrdin comments: "We are finally starting to really move toward a close economic alliance."

### *Political*

**8** • Kazakh President Nursultan Nazarbaev on relations with Russia: "Today close relations between Russia and Kazakhstan create a stabilizing factor throughout the entire Eurasian area. . . . Our republics are objectively tied to each other by economic and military-political relations, which have roots in the deep past and are directed into the future." He advocates the strengthening of the Russian state as the guarantor of stability in Eurasia, and links the prosperity of Kazakhstan with the prosperity of Russia. He states that the strategic task of the Armed Forces of Kazakhstan is to cooperate with the Russian Army.

**10** • A regional summit of the Central Asian heads of state scheduled for 22 April is canceled. Turkmen President Saparmurad Niyazov expresses doubts about the importance or usefulness of a regional common market, preferring bilateral relations as a basis for foreign policy.

**13** • Ukraine and Georgia formalize a treaty of friendship, cooperation, and mutual assistance. Both sides sign twenty agreements covering all types of bilateral cooperation. Both parties observe that Georgia could have long since solved the problem of Abkhazia through peaceful negotiation but a "third force" constantly builds up tension in the region (referring to Russia).

**14** • The parliament of Kazakhstan ratifies the CIS Charter.

**15** • The parliament of Russia ratifies the CIS Charter.

**16** • In Minsk, an emergency CIS summit takes place. The main result is general support for democratic reforms in Russia, and for Yeltsin. No documents are signed at the summit, which is devoted to general discussion on strengthening the Commonwealth. Yeltsin and Nursultan Nazarbaev call for CIS states that have not signed the CIS Charter to decide finally whether they wish to be members or not. Yeltsin states that non-signatories "would in effect remain outside the main channel of cooperation within the framework of the CIS, with all the consequences that stem from that."

***Security***

**6** • The CIS military command proposes to standardize defense-related legislation throughout the CIS, especially with regard to social protection for servicemen and their families, and coordinate activities among the security organs of CIS member states. Commander-in-chief Evgeniy Shaposhnikov says it is time to put an end to "unlimited sovereignty" in the CIS, and to promote greater economic and military integration.

**9** • Russia's Defense Ministry rejects Ukrainian accusations that Russia has taken unilateral actions with respect to the Black Sea Fleet in violation of the Yalta agreement. The ministry makes the counter-accusation that the Ukrainians are trying to avoid responsibility for delaying negotiations on the question of the Black Sea Fleet as well as for the policy adopted by Ukraine's Defense Ministry aimed at breaking up the fleet and changing its structure. The Russian Defense Ministry accuses the Ukrainians of systematic violation of the Yalta agreement by trying to take command of the fleet and bypassing the Russian and Ukrainian presidents.

**27** • The Russian parliament ratifies a bilateral security treaty with Belarus.

**May 1993**

***Economic***

**18** • Uzbekistan retaliates against Kyrgyz currency reform by threatening to cut off its entire energy supply. The Uzbeks fear that Kyrgyzstan will not be able to retire its outstanding debts.

• Russian Deputy Prime Minister Shokhin states that the CIS economic union should benefit Russia because it is the largest and wealthiest economic partner in the CIS. He also acknowledges that an economic union would entail a partial loss in political, as well as economic, sovereignty for all of the successor states of the Soviet Union.

**19** • Uzbekistan reconsiders the previous day's threat against Kyrgyzstan and restores the energy supplies that had been partially cut off.

• A "high-speed clearinghouse" is created by the CIS and joined by more than thirty commercial banks from major industrial regions in the CIS and Baltic states. Its purpose is to speed up interrepublican payments which may take anywhere from a week to a month to clear.

**30** • The second regular session of the CIS Council of Leaders of Foreign Economic Departments begins in Alma-Ata. Participants include: Belarus, Kazakhstan, Kyrgyzstan, Moldova, Russia, Tajikistan, Uzbekistan, and Ukraine. Nazarbaev expresses his discontent with the situation in CIS markets, calling for more integration and a more professional approach to business. The session results in a protocol on general principles for improving tariff and non-tariff regulations within the framework of a free trade regime, and agreements on setting up a CIS Council for Export Control and agreements on information exchange in the field of foreign economic activities.

***Political***

**23** • The third plenary sitting of the Interparliamentary Assembly takes place in St. Petersburg. The following states are represented by members of their Supreme Soviets: Belarus, Armenia, Kazakhstan, Kyrgyzstan, Russia, and Tajikistan. Azerbaijan, Moldova, and Ukraine send observers. The Assembly adopts resolutions on harmonizing the legislation of member states in the fields of information, foreign investment, and military affairs. Chairman Ruslan Khasbulatov stresses that the path of isolated development and the hope of painlessly overcoming political, economic, military, and other difficulties are becoming increasingly illusory and unrealistic for CIS states. Lyudmila Fomicheva of ITAR-TASS says that the "main result of today's session is perhaps the obvious integrational processes that are now visible among not only CIS member countries, but also [other] republics that were part of the FSU."

***Security***

**14** • CIS defense ministers meet and most agree on the urgent need for an integrated defense system.

**18** • CIS Joint Armed Forces Commander Shaposhnikov states that Russia and Uzbekistan support the creation of a collective security system for the CIS, but have "their own approaches to the problem." Uzbekistan prefers a joint command, with every CIS defense minister having a veto over defense decisions.

**26** • The first deputy chairman of the Ukrainian parliament reiterates that it is opposed to allowing Russia to maintain a naval base at Sevastopol.

**28** • Evgeniy Shaposhnikov recommends the creation of three new integrated military commands: (1) "Yug" [South] air defense system, which would include assets from Russia, Kazakhstan, and southern Central Asia; (2) "Kavkaz" [Caucasus] air defense system, which would include Russia, Armenia, Azerbaijan, and Georgia; (3) "Zapad" [West] unified command of anti-missile and air defense assets.

• Shaposhnikov warns that failure to increase defense cooperation in the CIS could lead to instability and the creation of security blocs in the former Soviet space, as countries on Russia's periphery gravitate toward their non-CIS neighbor states.

**June 1993**

***Economic***

**1** • The first CIS Coordinating and Consultative Committee session is held to discuss cooperation in a wide variety of areas between CIS states. Russian Deputy Prime Minister Aleksandr Shokhin is elected to a six-month term as chairman of this committee. The CCC sets 30 June as the date for consideration of the first draft of the Treaty on Economic Union. Georgia's Deputy Prime Minister Avtandil Margiani confirms his country's intention to participate in the Economic Union, even though Georgia is not a formal member of the CIS.

**10** • The five Central Asian states and Azerbaijan reportedly will become members of the Asian Development Bank, providing another source for loans apart from the IMF, EBRD, and World Bank. It is thought that the ADB will be more lenient on requiring clean human rights records than the other lending institutions.

**14** • Azerbaijan decides to leave the ruble zone, following Kyrgyzstan's example. As of 15 June, the manat will be the only legal tender. The ruble will be removed from circulation on 20 June. Until then, Azerbaijani citizens may exchange their currency at a rate of 10 rubles per manat.

**15** • Russia sets stringent conditions for Belarus to continue to receive rubles. Acceptance of the conditions would allow Belarus to continue to receive rubles until 1 October 1993. A spokesman for the Belarusian National Bank says that the conditions are unacceptable.

**17** • CIS member states are urged by CCC Chairman Shokhin to expedite the introduction of national currencies so that IMF standby loans will be released (the loans are tied to inflation rates, interest rates, and the issuance of credits). He also states that Russian credits to former Soviet republics will soon be reduced, and that Russia intends to double natural gas prices to the republics as well as to domestic customers.

**21** • The Crimean parliament issues a resolution introducing a free economic regime on the peninsula. Although this is portrayed by the mass media as an attempt to leave Ukraine, parliamentary chairman Mykol Bagrov says Crimea and Ukraine would overcome their economic crisis together. It is important to note, however, that on the peninsula there is significant support for separation from Ukraine.

**22** • Kazakh Prime Minister Tereshchenko says trade relations with Russia have sharply deteriorated. He says that in the latest round of trade negotiations, Russia has insisted that Kazakhstan's total trade deficit with Russia should be transformed into "sovereign debt" with conditions analogous to Western loans. He claims that Russia's tough stance is intended to push Kazakhstan out of the ruble zone. He states that Russian negotiators have even gone so far as to suggest that Kazakhstan introduce its own currency.

**23** • The CIS Intergovernmental Council for oil and gas suggests to the governments and parliaments of CIS countries that they introduce order to the system of mutual accounting between suppliers and consumers of energy. "Essentially it is a case of the restoration of the union of sovereign states within the framework of the fuel and energy complex" (*Rossiiskaya Gazeta*). There is reason to be pessimistic about such an agreement, though, because most republics are behind in payments as well as deliveries.

**24** • A session of the CIS Interstate Committee on Agro-Industrial Complex finishes today. The representatives of the CIS states sign a number of agreements on economic cooperation, such as creating an interstate scientific-technical center. Essentially this is yet another meeting in which CIS representatives express their desire to reintegrate their economies after two years of devastation caused by the dissolution of the Union, but will probably fail to implement the agreement.

• A St. Petersburg conference on legislative and economic activities of CIS enterprises opens today. It is held under the auspices of the CIS Interparliamentary Assembly and is being attended by economists, political leaders, bankers, and engineers from all CIS states except Uzbekistan and Turkmenistan. Even representatives of Azerbaijan attend. Rector of St. Petersburg University Tarasevich tells ITAR-TASS that he thinks the high level of representation should ensure progress in solving the problems of paralysis of many enterprises caused by the collapse of the USSR.

• A Deutsche Bank spokesman says interest payments on the estimated US$25 billion commercial debt owed by the Soviet successor states will for the seventh time be delayed by ninety days. Russia has assumed responsibility for the FSU's government and commercial debts which currently total approximately US$88 billion.

• An official for the CCC says member states must decide by 1 July whether they wish to remain in the ruble zone. Those former republics that have issued their own currencies have been asked to return all rubles to Russia.

### *Political*

**17** • In addition to the Black Sea Fleet agreement, Yeltsin and Kravchuk agree to: "intensify work on a comprehensive political treaty between Russia and Ukraine; accelerate the process of reaching an agreement on dual citizenship; cooperatively solve fuel and energy problems; and draft an agreement on jointly selling shares in Ukrainian and Russian enterprises." Yeltsin also confirms Russia's readiness to offer Ukraine security guarantees that would come into force after the Ukrainian parliament ratifies START I and adheres to NPT.

**26** • The Tajik parliament ratifies a treaty of friendship, cooperation, and mutual assistance between Tajikistan and Russia, which was signed by Rakhmonov and Yeltsin on 25 May 1993.

• The next CIS summit meeting scheduled for 16 July in Yerevan is postponed because the Treaty on Economic Union is not ready, and it is the most important item on the agenda. An agreement is expected by the end of the summer.

• The acting president of Azerbaijan, Geydar Aliev, is granted power by parliament at the end of June. He is a former KGB general, "ruthless" former Communist Party chairman of the republic, and former member of the Soviet Politburo. Some in Russia think that the rise to power of a former comrade might soon be followed with Azerbaijan's accession to the CIS Charter.

### *Security*

**1** • Over 200 Black Sea Fleet ships raise the Russian naval ensign (St. Andrew's flag), although reports confirm that no combat ships have done so. The support ships are manned and controlled mostly by Ukrainians. The protest is motivated by economic concerns: the men want higher wages and feel that Russia can provide them.

• Ukrainian Defense Minister Konstantin Morozov accuses anti-Ukrainian groups in the Crimea of instigating the strike for higher wages on the Black Sea Fleet support ships.

**2** • Ukrainian Prime Minister Kuchma suggests that Ukraine lease the base at Sevastopol to Russia, reasoning that the Russians will be there a long time anyway, so they may as well pay for the base. Ukrainian nationalists oppose this idea as a violation of Ukraine's sovereignty, but Rukh leader Vyacheslav Chernovil says that it could be workable.

**3** • In a letter dated 28 May and sent to the seven current signatories of the Collective Security Treaty, Belarus Supreme Soviet Chairman Shushkevich clarifies the republic's position on the pact. The letter states that the use of Belarusian troops beyond the country's borders is only permissible through a decision by the republic's Supreme Soviet. Furthermore, Belarus has the right to discontinue its participation in the collective security system the moment all Russian military and strategic forces have been removed from its territory (RFE/RL).

**10** • Yeltsin says Russia should adopt military basing practices similar to those of the United States in order to maintain a military presence in Georgia, Moldova, Armenia, and Central Asia. He calls for the conclusion of intergovernmental agreements to formalize the Russian military presence in those areas.

**14** • Yeltsin unexpectedly names CIS Commander-in-chief Shaposhnikov to the post of secretary of the Russian Security Council. Yeltsin suggests that Shaposhnikov has become available because as the CIS states build their own armies, Shaposhnikov's responsibilities are shrinking.

**15** • The Council of CIS Defense Ministers disbands the unified armed forces command, replacing it with a "united headquarters for coordinating military cooperation."

• Shaposhnikov is replaced by a less influential officer, Col.-Gen. Viktor Samsonov. The collapse of the unified forces leaves unresolved the pressing issue of strategic nuclear forces in Ukraine and Kazakhstan.

**16** • "There are no CIS combined forces today, and their creation in the future is problematic. . . . We are doomed to cooperation, but that will be in the future," says ex–CIS Joint Armed Forces Commander-in-chief Evgeniy Shaposhnikov.

**18** • CIS Commander-in-chief Shaposhnikov tells *Izvestiya* that although he is no longer head of the CIS Joint Armed Forces, he still holds the codes to the nuclear arms on FSU territory.

**28** • *Izvestiya* reports that Ukraine and Russia are ready to sign an agreement on the dismantling of Ukraine's warheads and missiles. This agreement was apparently drafted while Chernomyrdin was visiting Ukraine, but not signed before his departure on 28 June. The agreement would state that both sides are to share responsibility for the dismantlement and share the proceeds from the sale of fissile nuclear materials in them. According to this report, the dismantlement must be complete by 1995, which represents a big concession by Ukraine.

**30** • Ukrainian Deputy Foreign Minister Boris Tarasiuk contradicts the Russian press, saying that Russian sources are exaggerating the progress of the negotiations. It is, however, likely that an agreement on ICBM maintenance is ready, although not yet signed.

• Evgeniy Shaposhnikov is appointed secretary to the Russian Federation Security Council in the wake of the dissolution of the CIS Joint Armed Forces and therefore as their commander-in-chief. The "nuclear button" is still in Shaposhnikov's possession, although he says it will probably soon be transferred to someone else, most likely the country's defense minister, Pavel Grachev.

• Ostankino television reports that Shaposhnikov's appointment to the position of secretary to the Security Council is rejected by the Russian parliament. Khasbulatov declares the vote invalid, so a second one will take place, but the initial vote suggests that Shaposhnikov will have a hard time with the confirmation.

**July 1993**

***Economic***

**10** • The prime ministers of Russia, Ukraine, and Belarus meet outside Moscow to implement increased economic integration. They pledge to prepare a draft agreement before 1 September, and announce that the agreement is "likely" to be open to other countries, under certain, unspecified conditions. The proposed agreement goes further than the Economic Union Treaty signed in mid-May. Kuchma is the most circumspect among the participating prime ministers: he emphasizes that the agreement will not lead to the restoration of the Union. The statement addresses urgent measures to deepen economic integration and the intention of the three countries to participate in the development of the treaty on the creation of the CIS Economic Union. They propose lifting tariffs and non-tariff barriers to trade, and creating a customs union. The prime ministers stress that the document does not presuppose the breakup of the CIS and should not introduce separatism within the CIS. Nonetheless, Kazakh Prime Minister Tereshchenko expresses dismay at the proposed "Slavic Monetary Union" because President Nazarbaev had been one of the most vocal of the CIS member state leaders in support of further economic integration. He questions the deliberate exclusion of Kazakhstan from the proposed union.

**10** • Relations between Russia and Kazakhstan worsen markedly following the formation of a pan-Slavic economic union. Although the three Slavic states say the union is open to newcomers, they reserve the right to accept or reject applicants.

• The Russian media accuses Kazakhstan of not paying its bills and not honoring delivery agreements. Kazakhstan, meanwhile, accuses Russia of trying to push it out of the ruble zone. The Russian radio notes that "it has become standard practice in Kazakhstan to blame every shortcoming in the republic's economy on Russia."

• Uzbekistan's first deputy prime minister states that the country is being forced to introduce its own currency (the "som") due to Russia's uncompromising monetary policies. Apparently Russia has stopped supplying banknotes and has begun to replace old rubles with new ones meant for use only inside Russia. Kazakhstan, too, has complained about the withdrawal of old rubles, and many enterprises are reported to be in pre-strike position due to cash shortages. There are reports of the imminent introduction of a Kazakh national currency.

• Russia and Ukraine agree on oil prices: Russia is to sell 20 million tons to Ukraine in 1993 at a price of $80 per ton, and as of December the price will rise to $100 per ton. (The current world price for oil is $120 per ton.)

**13** • Russian Deputy Prime Minister Shokhin tells the newly independent states that have joined the Economic Cooperation Organization (ECO) that they must choose to either stay in the organization or join the newly formed union between Russia, Ukraine, and Belarus. The Central Asian republics and Azerbaijan have recently participated in an ECO conference in Istanbul, at which they agreed on further measures of economic integration. The pan-Slavic agreement is seen in Central Asia as an effort to push them out of the ruble zone and as a possible first step toward the dissolution of the CIS, to which the ECO is seen as a possible alternative.

• The Erk Democratic Party of Uzbekistan spokesman declares that the three Slavic states want to push Central Asia out of the CIS, and "whatever happens, the funeral of the CIS is being held."

**14** • *Narodnaya Gazeta* carries an article by CIS Executive Secretary Ivan Korotchenya on the Central Asian and Slavic economic unions, in which he queries: "Does this mean that the economic union of nine states has been rejected and that the idea behind its creation has mimicked the fate of many useful but unimplemented CIS projects? Is the existence of the CIS not threatened by the fact that its members are overtly grouped in accordance with their Muslim or Slav origins? . . . Would it not be better to gather at a common table and resolve all the problems of the day? . . . Arguments used by the signatories to the 'tripartite union' in order to justify this determined step seem unconvincing to me. They said there was no chance to create an economic union of nine states, because CIS members have been lacking accord . . . and there was no hope left for the situation to change in the near future. However, the prime ministers must have known that virtually all the disputable issues had been resolved at the 1 July Coordinating and Consultative Council session."

**20** • Russia and Tajikistan sign three economic agreements stipulating: (1) that Tajikistan's R49 billion trade debt will be converted into sovereign debt, to be repaid over a five-year period beginning in 1996; (2) that Russia will extend to Tajikistan a R60 billion loan, with which it may purchase Russian imports; and (3) that Tajikistan agrees to the conditions required by Russia for remaining in the ruble zone, which presumably means that the Tajiks have agreed to concede some of their sovereignty in monetary policy to Russia.

**24** • The Russian Central Bank releases a statement saying that ruble notes issued before 1993 will no longer be valid in Russia. This is apparently an attempt to curb inflation both domestically and in the "near abroad," by invalidating the ruble overhang. Another goal may be to eliminate the membership of the former Soviet republics that remain in the ruble zone. The reactions in the ruble zone are not consistent with the Bank's plan, however, to say nothing of the domestic turmoil the action caused. Belarus, Kazakhstan, and Uzbekistan are generally supportive, saying that they would stay in the ruble zone for now, and would phase out old rubles less quickly than Russia. Armenia reminds Russia that it has an agreement calling for a six-month warning on any changes in currency, and subsequently states that it will continue to use the old rubles indefinitely. Azerbaijan states that it will hasten the replacement of its rubles with the manat. Georgia gives its citizens one week to exchange their rubles for the national coupon. Moldova plans to accelerate the introduction of the leu, withdrawing large-denomination ruble notes on 26 July, but retaining notes of 200 rubles or less to circulate beside the national coupons.

**26** • Yeltsin modifies the currency reform on 26 July, allowing Russian citizens to exchange up to 100,000 rubles in old banknotes, and extending the timeframe in which they are allowed to do this to 31 August. This is in response to widespread criticism both at home and from foreign ministers of the other CIS states.

• The Russian Finance Ministry sharply opposes the recall of old rubles, and releases a statement which includes the following: "The exchange will have an extremely adverse impact on mutual relations with former USSR republics, which still to varying extents use rubles as cash. Old banknotes are circulating in these countries, many of which are not yet ready to introduce their own currencies. The exchange has caused chaos in these countries, and worsens their attitude toward Russia as a trustworthy partner. Moreover, questions over the future of the ruble zone have not been resolved, as a result of which a new spiral of inflation in Russia is inevitable as gratis deliveries of Russian cash continue to flow into the former Soviet republics because banks have ceased to issue money equivalent to Russian money since July 1992."

• The U.S. *Journal of Commerce* reports that in 1993 Russian oil exports (for the first six months of the year) to other members of the CIS had fallen by 40 percent of what they were for the same period in 1992. Natural gas deliveries have fallen by some 50 percent.

**30** • Russia has cut off oil deliveries to Ukraine because of outstanding payments due. Apparently Ukraine owes Russia about 250 billion rubles for oil delivered since 1992. On 9 August, Russian TV reported that deliveries had resumed, and that an agreement had been reached on back payments and unit prices. On 7 August, Russia suspended shipments to Azerbaijan because of a failure to deliver a promised $10 million in barter goods.

### *Political*

**18** • In Russia, political scientist Igor Klyamkin states that reintegration of the post-Soviet republics will soon be put on the top of the political agenda; he cites the Civic Union as the instigator of this move.

**22** • CIS Executive Secretary Korotchenya and chairman of the Secretariat of the Council of the CIS Inter-Parliamentary Assembly Mikhail Krotaw have signed an agreement on the interaction of the organs of which they are heads. The document "establishes the legal basis for the already existing good relations between the working collectives of these CIS organs."

**28** • The presidents of Kazakhstan and Uzbekistan issue an appeal to other CIS heads of state to hold an emergency summit in the beginning of August. The two cite the following reasons for their urgent appeal: "regional separatism, the isolationism of individual states, and their desire to get out of the crisis on their own or at the expense of the economic interests of neighboring states," and that these concerns are taking the place of economic integration. Their concerns have been sparked by the currency fiasco and the Slavic Economic Union.

***Security***

**2** • The Ukrainian parliament votes to claim jurisdiction over the nuclear weapons on its territory. The amendment adopted does not assert Ukraine's right to operational control over the weapons.

**7** • Parliamentary Chairman Shushkevich agrees at a Parliamentary hearing to postpone until the fall a referendum on Belarus adhering to the CIS Collective Security Treaty.

• Russian Defense Minister Grachev and the chairman of Kyrgyzstan's State Committee on Defense, Umetaliev, sign two agreements on military cooperation. One provides for either country to lease land on the territory of the other, in exchange for equipment and training. The other maintains supply systems for military units and for equipment with local industries. The agreements also cover the issue of protection for military personnel situated on either party's territory, as well as the provision of housing.

**9** • In Ukraine, the Foreign Ministry tries to reassure its international partners by declaring an ambiguous status with regard to its nuclear weapons: "Ukraine has a unique status: the republic is not a nuclear state, but it has nuclear weapons," said Yuriy Serheev, leader of the Foreign Ministry's Press Service. He adds that "Parliament should ratify START I along with the Lisbon Protocol to the Nuclear Non-Proliferation Treaty." (The Ukrainian Supreme Soviet takes a different view and its position usually prevails.—Ed.)

**14** • The Russian government has approved a draft agreement on nuclear weapons elimination that will be proposed to Ukraine: Russia would reprocess the highly enriched uranium from weapons into reactor fuel for Ukrainian power plants, and Russia would assume responsibility for the long-term storage of the other weapons components. The agreement does not appear to contain any security guarantees or recognition of Ukrainian ownership of the weapons, two conditions Ukraine is likely to attach to any agreement.

**15** • The removal of five ICBMs from Pervomaysk base in Ukraine had begun, in accordance with an agreement between Russia and Ukraine to deactivate some nuclear weapons.

**22** • ITAR-TASS reports that Russian Defense Minister Grachev accuses Ukraine of moving to acquire operational control of the nuclear weapons located on its territory by issuing orders canceling all Russian directives concerning the weapons, and subordinating the special units guarding the weapons to the Ukrainian military establishment. The loyalties of the troops guarding the weapons has been ambiguous until this point, and Ukraine's move to take operational control appears designed to clarify the issue. Grachev supposedly possesses the launching codes for the nuclear weapons on former Soviet territory, which *Komsomolskaya Pravda* reported were given to him by Shaposhnikov shortly after he stepped down from his position as CIS commander-in-chief.

**23** • Ukrainian Defense Minister Morozov apparently suggests that Ukraine may try to join the NPT with the special status of a "transition country" with nuclear weapons. This would assume that Ukraine has weapons on its territory, but that they are in the process of being destroyed, after which their special status will be removed. This conciliatory move can be seen as an attempt to placate hardliners who do not want to sign the treaty, and the international community who wants them to accede. The parliament has the power to accept or reject the treaty; in the case of a rejection, relations with Russia could be worsened to a point at which prospects for implementation of START could be hurt.

• Belarus formally accedes to the NPT, and receives the promise of $59 million in U.S. aid. The fifty SS-25s on Belarusian territory are to be removed by the end of 1994.

**27** • Defense Minister Morozov claims that Ukraine has begun the process of dismantling ten SS-19s located at Pervomaysk. U.S. officials respond by saying they will begin to provide some of the Nunn-Lugar assistance that has been promised for nuclear dismantling. Morozov announces that Ukraine will eventually dismantle all of its SS-19s, but refuses to discuss a schedule for dismantling SS-24s. Subsequently, Ukrainian officials note that elimination of the SS-24s is not necessary in order to comply with START I, implying that they are not planning to get rid of them in the near future.

**August 1993**

***Economic***

**1** • Nazarbaev issues a decree increasing state control over financial transactions with other successor states of the former Soviet Union. All Kazakh enterprises are to close bank accounts in other countries of the region and henceforth conduct business only within the CIS through Kazakh banks. The chairman of the Supreme Soviet of Kazakhstan announces that Kazakhstan will introduce its national currency, the tenge, by the end of the year.

**2** • In Uzbekistan, the government announces that it will indefinitely delay the exchange of old ruble notes for 1993 ruble notes due to a shortage of the new notes. A Russian

Central Bank official also notes that only 5 percent of Uzbekistan's currency requirements will be met with new notes, and those transferred will be given to Uzbeks making trips to Russia.

• In Georgia, the government officially withdraws from the ruble zone in response to the Russian Central Bank's invalidation of old rubles. Georgia's national currency, the "lari," will be introduced once the economic situation has stabilized in the country, until which time the only legal tender will be the interim coupon, introduced in April, the value of which is plummeting.

**4** • In Ukraine, Prime Minister Kuchma announces that the Ukrainian government's past fears of being pushed out of the ruble zone by Russia have been vindicated by the Russian Central Bank's ruble exchange. He also repeats calls for an economic union with Russia, as long as it will not compromise Ukrainian sovereignty.

**6** • Kazakhstan, Russia, and Uzbekistan agree in principle to form a ruble monetary union. A formal pact is to be presented to the three national parliaments within about a month. The arrangement is open to other states as well. This does not preclude the eventual introduction of their own national currencies.

**11** • Prime ministers Chernomyrdin and Kuchma sign a draft agreement in Moscow on 11 August on the supply of Russian oil and gas to Ukraine. Ukraine is to pay in dollars for the fuel, the price of which will rise to world levels by early 1994. Amounts to be delivered are to remain at the current level of 30 million tons of oil and 60 billion cubic meters of gas. Any reexport of the imports will require Moscow's permission.

• Belarus is to formally introduce its new currency, the zaichik, by 15 August, and it will reportedly be the only legal tender by the end of the week. This is apparently because Russia is not providing Belarus with enough rubles to meet demand or to replace purchasing power.

**15** • Russia cuts gas deliveries to Belarus because of payment arrears. Belarus owes Russia 100 billion rubles. Gas is apparently being supplied only to homes, bakeries, dairies, and some other important enterprises, from the republic's reserves.

**16** • Russia increases gas deliveries a small amount when it is shown that Belarus has paid off 10 percent of its debt.

• In Ukraine, the government declares the value of the karbovanets will no longer be determined by the Central Bank, but rather by trading against other currencies in the Kiev currency exchange. Within a week its value plummets from 6,000 to the dollar to 19,000 to the dollar (on 19 August). The government announces that the introduction of the hryvna is not yet possible because the government cannot set up a hard currency fund to stabilize it.

**18** • The Asian Development Bank (ADB) admits Kazakhstan, Kyrgyzstan, and Uzbekistan, bringing its total membership to fifty-six. Member states are already members of the EBRD, but membership in the ADB gives them access to more concessional development loans. Last year Azerbaijan, Tajikistan, and Turkmenistan indicated interest in joining the bank, and it was reported in the 10 June 1993 issue of the *Guardian* that all five Central Asian states and Azerbaijan would soon join it.

**19** • In Belarus, Shushkevich and Yeltsin meet and appear to agree on the transformation of technical credits issued by the Russian Central Bank into sovereign debt owed to Russia by Belarus. This agreement, it is hoped, will stabilize the value of the Belarusian ruble next to the Russian ruble, and open new credit lines for purchasing gas and oil from Russia. Russia has reached similar bilateral agreements with the governments of Armenia, Kyrgyzstan, Moldova, Tajikistan, Turkmenistan, Ukraine, and Uzbekistan within the last few days.

• The vice-president of the Moldovan Parliament holds a meeting with a delegation from Ukraine in which he notes that 22 percent of Moldovan foreign trade is conducted with Ukraine. The head of the IMF mission to Moldova says Moldova must establish close economic links with other CIS member states if it is to emerge from its economic crisis.

**20** • In Turkmenistan, President Niyazov announces that his country will fully replace the ruble with its national currency, the manat, by 1 November. The initial rate of exchange is to be one manat to the dollar. The exchange of rubles for manats is to begin on 1 September, at a rate of 1,000 rubles per manat, and citizens are allowed to change up to 30,000 rubles. Niyazov contends that Russia's introduction of new rubles hastened the decision to introduce the manat.

• Ukrainian parliamentary speaker Ivan Plyusch expresses his opposition to any economic union between Russia and Ukraine, stating that the republic is opposed to the creation of "a new superstate." He views the economic union as the first step in this direction: "an attempt to restore not only a single economic space, but also a single citizenship, and a single state administration," which would be "absolutely unacceptable" in this country. On the same day, a round table of communist and pro-communist parties is held in Kiev. Members urge that Ukraine form an economic union with its Slavic neighbors.

**22** • Reuters reports that Japan wants to build a massive oil and gas pipeline from Central Asia through China which would deliver fuel to the Pacific Rim countries. Apparently, Japan will ask Kazakhstan, Turkmenistan, and Uzbekistan to participate in the project, which envisions a 6,000-km pipeline. The Japanese government will soon conduct a feasibility study, which will be followed by the formation of a promotional organization with leading trading establishments and oil companies. Japan is already helping in the construction of a refinery in Bukhara.

**23** • "A new type of ruble zone" is agreed upon by representatives of Armenia, Kazakhstan, Russia, and Uzbekistan. Although no details are provided, the agreement seems to imply the eventual merging of monetary, fiscal, and trade policies.

**24** • A report in *Moskovskie Novosti* discusses Russian monetary policy: "the Russian Central Bank has bluntly demanded that the newly independent countries make up their minds as to how they want to use the Russian ruble: by converting their national banks into the Russian Central Bank's branches, or by introducing their own currencies." The article goes on to state that the Central Asian countries will use their bargaining chips (nuclear test grounds, the Tajik problem, and "the threat of Islamic fundamentalism") in order to obtain better monetary conditions from Russia.

**26** • A session of the Coordinating and Consultative Committee (CCC) is held in Minsk, during which the principles of the CIS economic union are finalized. CIS economic integration is to be based on the principles of the European Community, and to take place in stages, the first of which will be the creation of a free trade association. The second will be the creation of a customs union, and the third a single market of goods, services, and capital. This draft treaty will be ready for signing at the 7 September Heads of State meeting.

### *Political*

**3–5** • The Moldovan parliament holds a special session to consider the ratification of Moldova's membership in the CIS, and other CIS commitments which President Snegur has signed, mostly economic. Moldova has consistently denied joining CIS political and military agreements. Both the president and the prime minister urge ratification of the CIS Charter, citing the economic situation as requiring continued union with the CIS states.

**4** • The Moldovan parliament is five votes short of ratifying participation in the CIS. Majority leaders then call for dissolution of the legislature and rule by presidential decree. The opposition, which is pro-Romanian, campaigns against ratification, saying that the CIS is a tool being used by Russia to reestablish domination over former Soviet republics.

**18** • A CIS Human Rights Commission is discussed at a meeting of experts in Minsk, as envisaged in Article 33 of the CIS Charter. A draft statute on the commission is presented by Russia, which has been the driving force behind the commission. A Russian Foreign Ministry official says the talks are difficult because some countries view human rights as a purely internal matter.

**23** • The meeting of the Council of Heads of Government scheduled for 27 August in Minsk is postponed until 7 September.

**24** • During talks in Ashkhabad, Armenian President Ter-Petrosyan and Turkmen President Niyazov sign bilateral agreements on economic and cultural cooperation, including one ensuring the restoration of crucial gas deliveries to Armenia. A friendship treaty is also signed.

**27** • Representatives of the communist parties of Belarus, Russia, and Ukraine meet in Minsk and bemoan the fate of the USSR, saying that at least Belarus, Kazakhstan, Russia, and Ukraine ought to be restored as Soviet republics.

### *Security*

**3** • Russian and Ukrainian officials criticize U.S. plans for airstrikes on the Bosnian Serbs. The Ukrainians fear for the UN peacekeeping troops which would be left hostage to the warring sides, and contend that airstrikes would be accompanied by large-scale ground troop deployments. Russian Deputy Foreign Minister Boris Kolokolov tells parliament that Russia has been committed to resolving the crisis by political means, and refers to military intervention as inadmissible under current UN resolutions.

• The Russian Foreign Ministry criticizes Ukraine's assertion that Kiev will hold onto the forty-six SS-24s on its territory, even after the ratification of START I, and says that Russia cannot accept Ukraine's status as a temporary nuclear power.

**4** • The Russian government releases a statement criticizing Ukraine's handling of the nuclear weapons on its territory, charging that recent Ukrainian actions violate a number of international agreements, especially the NPT, and threaten world stability. The statement also holds that Ukraine's inability independently to maintain the weapons on its territory increases the likelihood of a nuclear accident.

**5** • The Clinton administration is reportedly close to adopting a policy which allows the United States to intervene diplomatically in regional and ethnic conflicts on the territory of the former USSR. The plan's aims are to broker settlements to disputes before they become armed conflicts which could prove destabilizing for Yeltsin, and for the CIS's already conflict-ridden southern tier, or provide a context for Russian troop involvement.

• In Russia, it is noted in *Krasnaya Zvezda* (4 August) that the accession of Russia to the CIS Collective Security Treaty was prompted by the escalation of the Tajik-Afghan conflict. With its accession, Russia's participation in conflicts along its southern tier becomes legitimized.

• A Russian Foreign Ministry spokesman tries to allay what he refers to as Washington's exaggerated fears of becoming involved in conflicts in the former Soviet Union.

**10** • The head of the directorate for arms control and disarmament of the Ukrainian Ministry of Foreign Affairs, Konstantin Hryshchenko, denounces Russian assumption of control over the weapons in Ukraine, and notes that under CIS agreements and international law, Ukraine is entitled to ownership of the weapons. Moreover, Ukraine has ceded control of the weapons to the CIS command, and although it was dissolved in June, any decision adopted by the CIS can be repealed only by those who adopted it, not by one party.

**11** • Yeltsin reportedly promises Kuchma that Russia will compensate Ukraine for nuclear weapons transferred to Russia. Kuchma denies that the Ukrainian government is trying to obtain operational control of the weapons, but says the Defense Ministry might be doing so.

**12** • Ukraine's Foreign Ministry reasserts its contention that the SS-24s on its territory are not covered by START I, but apparently Defense Minister Morozov, while in Washington, said that the dismantling of SS-24s would not begin until after the ratification of START I, implying that they are covered by the treaty.

**17** • German Defense Minister Volker Ruehe, during his visit to Ukraine, confirms that Germany will offer financial assistance to Ukraine for the dismantlement of nuclear arms. The offer is contingent upon Ukraine's adhering to START I and NPT.

**19** • The Ukrainian government releases a statement that asserts that the fate of the nuclear weapons in Ukraine will be decided by parliament. The statement also outlines the government's interpretation of START I and the Lisbon protocol as requiring only a 36 percent reduction in the nuclear forces of Ukraine. The Lisbon protocol, meanwhile, holds that Ukraine will join the NPT as soon as possible.

**20** • Col.-Gen. German Burutin discusses the mission and structure of the CIS defense forces in an article in *Rossiyskaya Gazeta,* in which he states that a "coalition rapid response force" is needed. He goes on to outline what he thinks ought to be the three basic components of the CIS Joint Armed Forces as they are outlined in an agreement adopted at Bishkek on 9 October 1992:

—groups of military observers and a collective force to maintain peace in the CIS;

—a force to prevent local conflicts on the external borders of the CIS;

—contingents of forces to wage combat actions in large-scale armed conflicts.

**22** • The Council of CIS Defense Ministers is scheduled to meet for two days in Moscow starting 23 August. Key issues to be discussed are said to be the Tajik-Afghan border situation and reorganization of the CIS command structure. Defense Minister Grachev is to chair the meeting.

• The Council appears to have made little progress in its search for a mechanism that will coordinate CIS members' defense policies. Russia's unwillingness to underwrite the costs of maintaining a permanent joint command appears to be the major obstacle to reaching an agreement. ITAR-TASS reports that Grachev has urged the Council of Defense Ministers to reorganize into a working body that would coordinate military cooperation. The chief command during the transitional period will be headed by Col.-Gen. Viktor Samsonov, chief of staff of the CIS Joint Armed Forces.

• Before the meeting, Grachev says that until such a joint military body could be set up, CIS states ought to establish temporary "coalition" groups of forces to deal with specific security problems, such as the Tajik-Afghan border conflict. Tajik Defense Minister Shishlyannikov concurs, and suggests that Kazakh, Kyrgyz, Russian, and Uzbek troops take part. Grachev repeats his position that Russia prefers a political solution to the crisis, and notes that the Russian 201st division, already in Tajikistan, is trying to set up coalition forces.

## September 1993

### *Economic*

**3** • While in Massandra with Kravchuk, Yeltsin announces that the CIS summit scheduled for 7 September will be postponed until 24 September. Apparently the delay was at the request of Kravchuk, who wants to allow the Ukrainian parliament to discuss the proposed economic union. RFE/RL reports that interest in some form of economic reconciliation is so important to some of the republics that even Moldovan President Snegur and Azeri parliament leader Aliev have decided that they will attend, even though their countries "are no longer members of the CIS." Shevardnadze, as of 3 September, had not yet responded to an invitation to attend.

**6** • In Kyrgyzstan, Prime Minister Tursunbek says that his country will "fearlessly join the CIS Economic Union," and that Kyrgyz independence is not in danger, thanks to the "timely introduction of their own national currency."

**7** • Representatives of the governments and national banks of Armenia, Belarus, Kazakhstan, Russia, Tajikistan, and Uzbekistan sign an agreement detailing measures for the creation of a new ruble zone. The agreement is to be accompanied by bilateral documents between Russia and the others outlining measures by which the countries would unify monetary, fiscal, banking, and customs policies. The strictly regulated circulation of national currencies (other than the ruble) will be permitted during a transitional period. The agreement is open to other countries.

• Russia and Kazakhstan have agreed that Russia will assume responsibility for Kazakhstan's share of the FSU's foreign debt in exchange for transferring FSU assets on Kazakh territory to Russia. Yeltsin and Nazarbaev meet on 7 September to prepare for the signing of the new "ruble zone" agreement among several CIS states.

• Russia also agrees to assume Armenia's share of the FSU's foreign debt in exchange for assets on Armenian territory. Russian Deputy Prime Minister Shokhin signs a similar arrangement with Azerbaijan.

**8** • Prime Ministers Chernomyrdin and Kuchma reach an energy deal in which a Swiss holding company will purchase Russian oil and sell it to Ukraine, then sell Ukrainian goods throughout the FSU as its repayment. Ukraine will end up 8 million tons short of the 40 million it needs for the winter, but at least this way, according to Kuchma, Ukraine will be able to complete its harvest.

• Prime ministers Chernomyrdin and Kebich sign an agreement uniting the economic systems of Russia and Belarus. The agreement resolves the problem of gas prices and transport costs, and places the Belarusian gas transport system under the Russian state gas monopoly, Gazprom. Belarus also agrees to remain in the ruble zone as one of six former Soviet republics who agreed to do so the day before. A representative of the Belarusian National Bank hails what he refers to as not just a monetary union, but also an economic union.

**11** • *Segodnya* hails the equity-for-debt arrangement between Russia and Belarus as a model for former Soviet states with mounting debts to Russia. The Belarusian agreement raises none of the political outcry raised by Ukraine against the fleet-for-debt swap proposal.

**16** • Kuchma apparently tells Ukrainian TV viewers that because Ukrainian-Russian ties are so interwoven, the only way out of Ukraine's current economic crisis is through economic union with Russia. He argues that Ukrainian politicians must devise a way for economic union not to impinge on political sovereignty.

**17** • At the fiftieth anniversary celebration of the liberation of Bryansk from the Germans, Prime Minister Chernomyrdin observes that: "All borders must be eliminated between Russia, the Ukraine, and Belarus. All customs and customs posts must be eliminated as soon as possible" (ITAR-TASS).

**24** • The prime ministers of Armenia, Azerbaijan, Belarus, Kazakhstan, Kyrgyzstan, Moldova, Russia, Tajikistan, and Uzbekistan initial an Economic Union Treaty during a meeting of the Council of CIS Heads of Government. The new agreement seems to supersede the tripartite/Slavic economic union concluded earlier. Turkmenistan and Ukraine announce that they will probably only take out associate membership in the union. Chernomyrdin acknowledges Ukraine's position and says that an effort will be made to alleviate Ukraine's concerns in the hopes that it will eventually join. The Georgian prime minister attends as an observer. The Economic Union Treaty is still only a framework document, however, which provides for the free-flow of commodities, services, capital, and work force; coordinated fiscal, budget, tax, price, and foreign economic policies.

**25** • At a news conference in Moscow following the signing of the Economic Union agreement, senior Moldovan officials explain why Moldova, which is not a member of the CIS, signed the agreement. Most of the journalists present view the signing of the agreements as de facto adherence to the CIS. The officials justify it because of Moldova's need to regain its former position in CIS markets, especially Russia. Ceslav Ciobanu, President Snegur's economic advisor, stresses that the document is strictly economic, and will be of help in economic reform.

**26** • The leaders of Rukh and Ukrainian nationalists adamantly oppose the CIS Economic Union. The leader of the Nationwide Movement of Ukraine, Larysa Skoryk, says that although the treaty contains a number of provisions that are not in Ukraine's interests, Kiev has the right not to fulfill those articles as an associate member.

• Belarusian Prime Minister Kebich expresses pessimism about prospects for implementing the Economic Union Treaty, noting that: "Now it is necessary to work out about thirty-five individual agreements so that this union works. I am more than convinced that it will be very difficult to work them out. . . ."

**28** • An article in *Pravda* on the significance of the Economic Union states that its formation is "undoubtedly the most significant event in the CIS's entire brief, chaotic history. . . . The new independent states, barely having broken away from mother Russia and feeling the awkwardness of their position, have once again rushed to cling to its breast, where there are preferential credits, understated prices well below those on the world market, and energy sources delivered without set dates of payment. . . . It is quite possible that the Economic Union will be a prototype for a "Union of the Peoples."

**30** • The Crimean parliament emphatically disagrees with Ukraine's associate membership in the new Economic Union. It says that it is in Ukraine's national interest to have full membership. It accuses Kiev of conceding to forces bent on isolationism.

*Political*

**6** • Aliev and Yeltsin hold talks in Moscow on Nagorno-Karabakh. Aliev agrees on principles for settling the conflict. It is unclear whether these principles call for the deployment of Russian "peacekeeping" troops in the area. Aliev also states that he will attend the 24 September CIS summit in Moscow, and has permission from his parliament to explore the idea of joining the CIS.

**7** • Russia suspends taxes and customs duties on Moldovan imports from 1 August until 1 November, giving Moldova more time to decide whether it will join the CIS, in which case these measures will be permanently lifted. Russian markets are by far Moldova's largest, and the taxes raise prices of Moldovan goods in Russia to an uncompetitive level. Moldovan leadership favors accession to the CIS and its economic community, but the parliamentary minority is opposed and holds a de facto veto. The leadership therefore wants to extend this deadline to 1 January, in order to allow for multiparty elections to parliament, hoping the new parliament will vote for accession to the CIS.

• Lt.-Gen. Aleksandr Lebed, commander of Russia's 14th Army in Moldova, is running as candidate for the Supreme Soviet of the Dniester Republic, which he claims will join Russia on a federative or a confederative basis in the foreseeable future. This is not the first time Lebed has called for the republic's accession to Russia, but it is the first time he has done so as part of a political platform.

**8** • Both the chairman of the Belarusian Supreme Soviet, Shushkevich, and the prime minister of Belarus, Kebich, call for closer economic ties with Russia as the only way to maintain independence. With no reform program of its own, even Belarusian nationalists have been calling for closer cooperation with Russia. Belarus has been hard hit by Russia's raising of its export prices to near–world market levels in an effort to force the republics that benefit from its subsidies out of the ruble zone.

**8–9** • A session of the Gagauz Supreme Soviet appeals to the CIS states to recognize the "Gagauz Republic" as an independent state and member of the CIS. They also resolve to hold a joint session of the Gagauz and Dniester Supreme Soviets to call for recognition and accession to the CIS political and economic agreements. This adds to already strong demands by Russia and the Dniester Republic that Moldova accede to the CIS.

**10** • The Moldovan government issues a statement accusing Russian deputies who recently visited Moldovan regions of Gagauz and Dniester of encouraging their secessionist tendencies in an effort to recreate the Soviet Union. The fueling of these secessionist demands is seen as an effort "to blackmail Moldova and keep it within the former empire's zone."

**14** • Russian parliamentary chairman Khasbulatov, who is also chairman of the CIS Interparliamentary Assembly (IPA), proposes that the IPA hold joint elections by CIS member states. He adds that the joint parliament could form a common CIS government. The IPA has been "one of the more effective CIS institutions to date and could have an important role to play in harmonizing the legislation of CIS member states if a CIS Economic Union gets off the ground."

**15** • Tajikistan and Russia sign an agreement merging their monetary systems. This bilateral agreement follows the 7 September agreement creating a new ruble zone. "These steps embrace the banking, budget, and customs spheres, and the pricing policy, in other words, the entire set of measures necessary to make the ruble's worth in Tajikistan on a par with that in Russia."

**17** • Khasbulatov writes to the parliaments of the CIS nations suggesting that they delegate certain unspecified powers to a supranational parliament that would coordinate their political, economic, and military activities. Khasbulatov is expected to raise this proposal with the IPA at its next scheduled meeting, on 25 September.

• The Moldovan government appeals to the Council of Europe to help fend off Russian deputies whom they hold responsible for inciting the Gagauz and Dniester republics to attempt to secede and join a Russian Confederation.

**18** • ITAR-TASS reports that Vice-President Rutskoy, who has been at least temporarily stripped of his duties, has called for the recreation of the Soviet empire. This he said while speaking at a regional conference convened by Khasbulatov, apparently in an effort to counter the first meeting of the Council of the Federation, called by Yeltsin. Rutskoy claims that Russian foreign policy is dictated by the United States, and the people must resist Yeltsin's policies.

**20** • The Azerbaijan National Assembly, or its rump parliament, votes to renew its membership in the CIS. Chairman Aliev reassures parliament members that the vote will in no way compromise Azerbaijan's independence. Besides becoming party to the CIS charter, Azerbaijan decides to accede to the Treaty on Collective Security and the National Assembly will join the IPA. "It is also envisaged that at the forthcoming September Commonwealth summit meeting Azerbaijan will sign a treaty on an Economic Union" (ITAR-TASS).

**21** • As a result of Yeltsin's decree stating that Russian deputies are allowed to take part in the IPA only with his explicit permission, Khasbulatov will no longer head the Assembly. The IPA's press center says that the consultative meeting of the CIS heads of parliament is unlikely to take place as scheduled on 25 September.

**24** • Thousands of members of the Azeri Popular Front

party in Nakhichevan, an Azeri region separated from Azerbaijan proper by a swath of Armenian territory, protests the signing by parliamentary chairman Aliev of the papers committing Azerbaijan to CIS membership. Former Azeri President Elchibey, who was ousted in July after his Popular Front's support experienced a steady decline, has taken refuge in Nakhichevan.

• President Yeltsin says at the CIS summit in Moscow that "the Commonwealth has a realistic prospect of becoming one of the most powerful integrated unions in the world." This is broadcast on a Moscow Ostankino television program entitled "The CIS: From Strife to Accord." The program also airs images of two major oil refineries in Ukraine operating at 10–15 percent of their capacity because of the lack of oil . . . the mayor of Donetsk is then shown stating that "we will not survive in isolation." Kravchuk, filmed for the program, says, "I am convinced that the most vehement opponents of Economic Union will see that it is good for us," in reference to the Economic Union.

### *Security*

**9** • Belarusian Defense Minister Kazlouski reiterates terms under which Belarusian troops would participate in CIS peacekeeping operations, saying as well that he agrees in principle with the Collective Security Treaty's goals. Belarusian troops would not be permitted to participate in operations in FSU "hotspots," but rather in conflicts on the republic's perimeter. Shushkevich, however, will still not sign the treaty which parliament voted to join last fall, ostensibly because it is a betrayal of national sovereignty.

**29** • Col.-Gen. Boris Pyankov, commander of Collective Peacekeeping Forces of the CIS, is interviewed on the new forces by a *Krasnaya Zvezda* correspondent, in which he states that although there has been a lot of talk about the creation of Joint Armed Forces, they "have simply not been created." In light of that, and of the rising extremism in the CIS, it was seen as necessary to take on the task of creating the Collective Peacekeeping Forces. They are to number approximately 25,000, and their mission will be to "stabilize the situation in Tajikistan, and maintain peace."

## October 1993

### *Economic*

**8** • Gazprom is apparently planning to forgive Ukraine its oil-delivery debts in exchange for long-term leases on Ukrainian gas distribution facilities at Kiev, Uzhhorod, and Mykolaiv; storage facilities in western Ukraine; and some port facilities at Illichivsk and Odessa. Gazprom reports that Ukraine's debt now totals 710.5 billion rubles. Gazprom also accuses Ukraine of siphoning off oil destined for Western Europe from pipelines that travel through Ukraine. Gazprom has apparently had to pay $10 million in breach-of-contract fees for this.

**11** • Belarus is hoping that Russia's decision to transport gas through Belarus rather than Ukraine will provide Belarus with an opportunity to buy Russian gas more cheaply. Yeltsin and Walesa signed an agreement in August on gas transport that would divert 80 billion cubic meters of gas through Belarus (approximately four times the amount that now goes through Belarus). Ukrainian officials label the agreement as anti-Ukrainian, as Belarus's increase would come directly from gas now transported through Ukraine.

**14** • The karbovanets loses 25 percent of its value vis-à-vis the U.S. dollar in one day on the Ukrainian Interbank Currency Exchange. Speculation as to why points to the persistence of the low fixed exchange rate for the mandatory sale of 50 percent of hard currency earnings in Ukraine. The new prime minister, Zvyahilsky, as well as the new finance minister, Pyatachenko, apparently oppose the removal of the law. Former Economics Minister Pynzenyk called the law "an illegal, absurd, and criminal decision." The same day, the Belarusian zaichik loses 50 percent of its value vis-à-vis the dollar on Belarusian exchange markets. A spokesman for the Belarusian National Bank cites the halt in the Bank's selling of rubles to Belarusian enterprises at a discounted rate as the reason.

**19** • Vladimir Mashchits, chairman of the State Committee for Economic Cooperation with Commonwealth States, says that the overall debt of the CIS states to Russia amounts to about 3 trillion rubles. He adds that the CIS states have no way of paying this back to Russia in 1994, and that although Russia, too, owes money to some CIS states, its debt to them is significantly smaller.

• The Russian government's Commission on Operative Questions expresses dissatisfaction over the trade problems afflicting suppliers in the Russian Federation. Apparently the suppliers are owed the equivalent of $2.5 billion by other members of the CIS. The Commission proposes speeding up the creation of the CIS Economic Court, as well as increasing penalties against delinquent payments and simplifying customs procedures, as ways to help in alleviating this problem.

• Moldovan President Snegur and Russian Prime Minister Chernomyrdin sign an agreement which stipulates that Russia will take over Moldova's share of the FSU's foreign debt in exchange for Moldova's share of the FSU's property outside Moldova. Russia also agrees to continue beyond 1 November to exempt Moldovan imports from duties and quotas to which other non-CIS countries' imports are subject. This concession by the Russians is made in the hope that Moldova will soon join the CIS, and that Snegur just needs a little time to convince the parliament of the economic benefits associated with joining the CIS.

• Moldovan Prime Minister Sangheli tells reporters that Moldova will introduce its national currency, the leu, in November. The arrival of the leu has been long anticipated by the international lending institutions; it will be pegged to the Russian ruble.

**22** • Belarusian Prime Minister Vyacheslav Kebich states that Belarus cannot not survive economically, or politically, without close cooperation with Russia. An article in *Rossiyskaya Gazeta* stressed that the states of the CIS need to integrate through the strengthening of the Economic Union and the implementation of many other agreements on "finance, duties, and the movement of goods." This would, according to the article, stave off further declines in production and unfulfilled delivery contracts. The article also notes that to reduce their relationships to solely commercial matters would only make matters worse, because none of them has enough hard currency reserves to sustain such a relationship.

**27** • Russian Deputy Prime Minister Shokhin expresses reservations over the feasibility of the ruble zone at a press conference. He is particularly concerned that the legal system the zone will require is very complex and will take time to formulate. He advises the other five republics who signed the 7 September agreement to clarify what it is that they expect from the zone.

### *Political*

**22** • Commonwealth experts are said to be preparing a draft decision to start a CIS newspaper, which will be entitled *Sodruzhestvo Nezavisimykh Gosudarstv—Obshchiy Rynok* (Commonwealth of Independent States—Common Market). Its initial print run is expected to be 100,000 copies, run in early 1994, and appearing as a supplement to Egor Yakovlev's *Obshchaya Gazeta.* Nine hundred fifty million rubles and $200,000 from the Commonwealth budget are expected to be allocated to its publication.

### *Security*

**1** • Belarusian TV reports that Russian troops are to be completely withdrawn from Belarus by 1996, when the remainder of the nuclear arms located on Belarusian territory are scheduled to be removed. Originally, Russian troops, whose numbers in Belarus are between 35,000 and 40,000, were not scheduled to be out until 1998. The Russian and Belarusian parliaments still have to ratify the new timetable. Problems could arise in Russia due to the housing shortage for returning troops, the reason many of the troops are in Belarus in the first place; most of them were on route from Eastern Europe to Russia when told to remain in Belarus until they could be accommodated in Russia.

**8** • Foreign Minister Kozyrev expands on Russia's assumed role in the "near abroad" during an interview with *Izvestiya*: Russia will strive to create effective peacekeeping forces for the conflict areas of the FSU, as well as in other areas of the world. He highlights Russia's danger of "losing geopolitical positions that took centuries to conquer." He says that Russians should worry about Asians meddling in the area, not Americans: "we have plenty of neighbors in Asia who are prepared to send soldiers and weapons into the former Soviet republics, even under the guise of peacekeeping forces." Kozyrev therefore ominously establishes a connection between peacekeeping in the "near abroad" and maintaining spheres of influence.

**12** • Anatoliy Zlenko, foreign minister of Ukraine, says that Russia's seeking of special UN status for its troops participating in the "near abroad" is unacceptable. Zlenko says that Ukraine favors peacekeeping in the territory of the FSU only on the basis of a decision by the UN Security Council. A Ukrainian TV broadcast characterizes Russian involvement in the Georgian/Abkhaz conflict as the manifestation of Russia's old "imperial" ambitions.

**13** • The defense minister of Turkmenistan, Danatar Kopekov, denies Russian press reports that planes from the Afghan airforce have bombed Turkmen territory on 11 October. Kopekov refers to the reports as "disinformation of a provocative nature," and the Turkmen press publishes a statement by Afghanistan's ambassador to Turkmenistan declaring the Turkmen-Afghan border is a "line of friendship and accord."

**20** • Deputy parliamentary speaker of Ukraine Vasil Durdynets apparently tells Western reporters that the debate over accession to START I will finally take place in November. Following their closed debate on Ukraine's military doctrine, however, various officials appear to view Ukraine's obligations under START I and the Lisbon protocol differently: According to Kravchuk, after ratification all Ukrainian SS-19s would be scrapped; according to Dmtyro Pavlychko, chair of the parliamentary committee on foreign relations, only 36 percent of Ukraine's delivery vehicles would be dismantled (this is based on the assumption that the limits required under START I will be applied equally and proportionately in all the nuclear republics of the FSU).

• Kravchuk's statements following the meeting also raise the question of who controls the weapons. He states that they will be taken off alert and de-targeted. The former was already thought to have been done, and the latter (targeting) was previously controlled by the CIS Joint Armed Forces, and is now thought to be in Russia's control.

Thus Kravchuk's statements imply that Ukraine has much more control over the weapons than mere veto power over a launch.

• A train carrying potentially leaky nuclear weapons from Ukraine to Russia has apparently been stopped at the border because Ukraine wants guarantees that it will be compensated for the material inside the warheads. The Russian Federation's Defense Ministry issues a statement accusing Ukraine of "misconduct," of not realizing the imminent danger posed by the damaged weapons, and of being distracted by "disorder in supreme power echelons."

**25** • Ivan Plyushch, chairman of the Ukrainian parliament, states that if Russia, Ukraine, and the United States were to sign a collective security agreement, parliament might be persuaded to ratify START I. He notes that Ukraine's relations with Russia were damaged by the Russian parliament's claim to Sevastopol and the proposal by the Shakhray/Stankevich bloc, Party of Russian Unity and Concord, to make Russia the center of a new confederation. These statements, Plyushch says, "may influence Ukraine's position on its nuclear-free status."

**27** • A military delegation from Belarus visits Ukraine in order to discuss military relations in 1994. They meet with Ukrainian defense minister Radetskiy, and deputy defense ministers Bizhan and Oliinyk.

• In response to a question asking that he describe Russia's relationship with the former Soviet republics, Yeltsin aide Sergey Karaganov says that: "Most of the former Soviet republics, apart from the Baltic ones, start to understand that it is extremely hard for them to survive politically and economically, especially economically, without the might of Russia. So basically all of them are, in one way or another . . . returning. That doesn't mean, however, that Russia wants to get them all back, and to take over politically."

## November 1993

### *Economic*

**2** • Russia's requirement that all members of the ruble zone have enough gold or hard currency with which to buy Russian rubles virtually precludes Kazakhstan from joining the zone. This new condition ruins the basis for the agreement unifying the Russian and Kazakh monetary systems. Russia has said it will provide Kazakhstan with the technical assistance necessary to develop its own monetary system and to introduce its own currency.

• Russian Deputy Finance Minister Andrey Kazmin notes that the debate over the new ruble zone has gone from one of a technical nature to that of a political nature. The key issue is Russia's willingness to give members of the zone new rubles. Kazmin now says Russia is willing to help other states set up their own interim currencies instead of the immediate creation of a ruble zone. Russian policy calls for alignment of states' monetary policies before it will supply new rubles because of the inflationary effects of last year's transfers to the former Soviet republics.

• The Uzbek government announces it is literally being pushed out of the ruble zone by the stringent terms of Moscow's Government and Finance Ministry. Officials say that even if Uzbekestan was able to meet the conditions, Russia would simply introduce new ones. The ruble zone is, at least for the time being, not an option for Uzbekistan either. Uzbekistan and Kazakhstan plan to introduce a common currency for economic interaction between the two states, which now use the ruble.

**4** • Russia may again cut off, or dramatically reduce, natural gas and oil shipments to Belarus in retaliation for payment arrears. Russia delivered only 430,000 tons of oil to Belarus in October, in contrast to the usual 1 million tons. Rozkontrakt said it would supply only 200,000 tons in the month of November.

• Moldova has lost as much income due to Russia's imposition of duties on imports from Moldova as it did because of last year's drought, which forced Moldova to seek international aid. The duties have the further effect of pressuring Moldova to join the CIS Economic Union, and to sign the CIS Charter. As a result of lost exports, Moldova will most likely have to slow down its pace of economic reform.

**5** • Uzbek Deputy Prime Minister Khamidov says Russia's new conditions for membership in the ruble zone place in doubt its sincerity about creating a unified economic space. The new conditions were not previously agreed upon, and none of the republics have the hard currency or gold deposits required to meet them. Some have gone as far as to say that a crisis in the CIS has resulted from Moscow's requirements (*Pravda*, 4 November). For example, huge amounts of old ruble notes have been flooding Kazakhstan and Uzbekistan, causing shortages and hoarding, and creating destabilizing situations in those countries.

• In an attempt to monitor the inflow of rubles from the other former Soviet republics, President Nazarbaev of Kazakhstan will require that banks set up special accounts in which pre-1993 rubles are to be deposited. The Council of Ministers passes a subsequent decree stipulating that the deposit of old rubles received in payment for consumer goods and some basic services will not be subject to the deposit requirement.

**8** • Belarusian Chairman of the Supreme Soviet Shushkevich says the economic relationship with Russia is one of the most important issues facing Belarus today. He stresses that it is crucial for Belarus to remain in the ruble zone and to form a customs union with Russia in order to remain competitive on Russian markets. The Belarusian government advises the parliament to ratify the treaty creating a monetary union with Russia.

**10** • In Alma-Ata, the Kazakh and Uzbek presidents sign an agreement outlining the simultaneous introduction of new national currencies in the republics, although no date is specified. The move is designed to prevent the flooding of one republic with old rubles in the event one of them changes currencies first. The coordination of policy will delay currency reform in Kazakhstan, however, because Uzbekistan is not prepared for immediate change.

**11** • Russia and Tajikistan will discuss the creation of a new type of ruble zone. Although Tajikistan may introduce its own currency, it will remain in the ruble zone, as will Kazakhstan and Uzbekistan.

**12** • Kazakhstan and Uzbekistan introduce their new currencies: the tenge and the som. Despite the coordination of policies, Uzbekistan's slower approach, involving the continued use of rubles, will probably invite an inflationary inflow of old rubles from neighboring republics. Kazakhstan, on the other hand, plans a brief stage during which old rubles will be phased out, and by 25 November the tenge is to be the exclusive legal tender in Kazakhstan.

**15** • Russia may require that the nations of the former Soviet Union pay off all accumulated debt and pay in advance for new oil shipments before any oil/gas will be delivered for the rest of the year. The Russian Committee for Economic Cooperation with CIS Countries reported last month that payments in arrears amount to 386 billion rubles owed Russia. Apparently the Russian government also plans to investigate whether these countries have been reexporting oil received from Russia, and if so, this will be taken into account in future negotiations.

**16** • Russian Deputy Prime Minister Shokhin publishes an article in *Izvestiya* in which he notes that some former Soviet republics unhappy with Russia's new ruble policies have threatened that ethnic Russians living in these republics would be in for a hard time. He says this type of blackmail cannot be taken seriously, but should not be overlooked. The article continues: "First, it is absurd to use ultimatums when talking to Russia; second, Russia has adequate levers and means to oppose the methods of marauding 'people's diplomacy,' to ensure that the architects of this policy regret it. And the old ideological litany about 'imperial practice' should not be hyped up—Russia is able to defend its current interests."

**17** • Belarus, Moldova, and Armenia prepare to leave the ruble zone by the end of November. Only Tajikistan remains committed to the ruble zone agreement, which all six, plus Russia, signed in September.

**18** • The introduction of the tenge and som has not gone smoothly: the new national currencies are difficult to come by in Alma-Ata and Tashkent. As a result, in Alma-Ata, panic buying is reported and many shops were closed. In Tashkent, many are scrambling to exchange new high-denomination ruble notes for old and low-denomination ruble notes (up to and including 1,000 ruble notes), which are to circulate beside the som, at least until the end of the year.

• Russian Deputy Prime Minister Shokhin meets with Ukrainian officials to discuss mounting Ukrainian debt and future oil deliveries from Russia. Although a previous debt-equity swap arrangement reached at Massandra was rejected by the Ukrainian parliament, Shokhin expresses optimism after this first round of talks. He says the Ukrainians appear close to accepting the exchange of Ukrainian assets (presumably pipelines through which Russian gas and oil travel to Western Europe) for debt relief. The Ukrainian debt is now estimated to be near 500 billion rubles. It appears the Ukrainian government will have to make a deal, because Shokhin indicates that if Russian credits are to continue, the Ukrainian debt will have to be significantly lowered.

• The Belarusian parliament ratifies the CIS Economic Union and a monetary union agreement with Russia. Belarus must now subordinate its monetary and fiscal policies to the Russian Central Bank. The chairman of the Belarusian Central Bank says that Russia still refuses to subsidize Belarus. Many of the deputies who are in favor of the agreements think the result of acceding to the Economic Union will be Russia's continued subsidization of Belarus. However, Shushkevich cautions against ratification of the economic agreements.

**20** • The Ukrainian government, in an attempt to mitigate the effects of rising costs of Russian oil and gas, relieves Ukrainian enterprises of export duties imposed on goods destined for Russia's gas and oil ventures, including Gazprom and Rosneft. According to Interfax, Ukraine's debt to Russia amounted to 714.6 billion rubles on 15 November, which includes 490.7 billion for gas, 206.3 billion for oil, and 17.6 billion for other petroleum products. Ukraine's gas debt to Turkmenistan is also significant: 85.8 billion rubles.

**22** • Tajik citizens are reacting to the flood of old Soviet rubles from neighboring republics who have introduced their own currencies by hoarding goods, which causes shortages and fuels inflation. Tajikistan is the only post-Soviet republic which still uses the old rubles, and has stated that it will remain in the ruble zone with Russia, and will not introduce a national currency to function beside the ruble.

• Uzbekistan has threatened to cut off vital oil supplies to Kyrgyzstan if the Kyrgyz government does not pay its $9 million debt within three days. Kyrgyzstan has asked the IMF to help.

• Armenia introduces its national currency, the dram, because of its inability to meet Russia's conditions for remaining in the ruble zone. In Yerevan, only U.S. dollars and 1993 rubles are being accepted in stores, and the dram has lost half of its value vis-à-vis the ruble, although it was only found to be officially exchangeable with U.S. dollars. Nagorno-Karabakh also decided to switch to the dram.

• Russian Deputy Prime Minister Shokhin expects the 3 trillion rubles owed by other republics to Russia for energy debts to be transferred into promissory notes and floated on the CIS financial markets. Shokhin calls for the republics to start paying in advance for gas and oil deliveries, for the opening of the CIS Interstate Bank to be hastened, and for the introduction of convertible national currencies on the territories of the former republics. Shokhin questions the feasibility of debt-equity swaps in the CIS, citing as a hindrance the relative inexperience with property commercialization.

**23** • Russia and Tajikistan sign an intergovernment agreement whereby Russia is to grant Tajikistan a cash credit worth 1 billion Russian rubles. Tajikistan plans to exchange pre-1992 bank notes currently circulating in Tajikistan for the credits. The prime minister of Tajikistan notes that with this first step, Tajikistan is now a member of the ruble zone.

• Belarus exhausts its fuel supplies and starts running on reserves, which can be expected to last only one week, according to the energy minister. Belarus is at least 140 billion rubles in debt for energy supplies, 28 billion of which is owed to Russia.

**25** • Moldovan President Snegur introduces the leu as the national currency of Moldova. His decree states that all rubles will have to be exchanged by 2 December 1993, at a rate of 1 leu per 1,000 rubles. After 2 December, the leu will be the only legal means of payment.

**26** • Ukraine cuts off electricity exports, and ceases to transport Russian energy supplies through its territory to Eastern Europe, due to its domestic energy crisis.

### *Political*

**11** • Russian Foreign Minister Kozyrev states that he favors linking other CIS states' acceptance of Russia's proposals on dual citizenship with Russia's economic aid to the former republics and with progress on CIS re-integration.

**15** • The head of the Russian presidential Commission for Citizenship says all Russians living in the "near abroad" should be guaranteed identical rights to the citizens of those countries, and that in fact they ought to be granted dual citizenship. He says that if dual citizenship were to be institutionalized, it would be a path toward mutual trust among the states.

**22** • A member of a Ukrainian delegation to the CIS (of which Ukraine is not a member), who wishes to remain anonymous, states that: "more than two hundred agreements adopted during the CIS period of existence have not made it an efficient organization because it lacked appropriate structures that would guarantee implementation of law-abiding acts. A real delimitation on the basis of national independence, ownership rights, and a national currency had not taken place in order to later unite on a new basis" (UNIAN).

**26** • TASS reports that Kazakh President Nazarbaev says he would never want to jeopardize relations with Russia, but he had to note that some of the campaign rhetoric being used in Russia is disturbing even to him. Especially disturbing are statements about the need to "protect" Russians in the "near abroad." An earlier statement by Nazarbaev provoked a protest from the Russian government: he said the policy is analogous to Hitler's policy protecting Sudeten Germans in Czechoslovakia.

• Chairman of the Moldovan parliament Petr Lucinschi states that "no one, not even the representatives of the opposition, feels any doubt now that Moldova must be a part of the CIS and maintain 'good, warm' relations with the Russian Federation" (*Rossiyskie Vesti*).

**30** • The missions and embassies of nine countries of the former Soviet Union lodge a complaint with the Moscow mayor for his introduction of a "special order of residence" in the city, whereby citizens of all former Soviet countries have to register to remain in the city, calling it humiliating to citizens of their countries. (Armenia, Azerbaijan, Belarus, Georgia, Kazakhstan, Kyrgyzstan, Moldova, Ukraine, and Uzbekistan lodge the complaint.)

### *Security*

**2** • Russia has suggested that the UN create a voluntary fund which would support peacemaking operations on the territory of the former USSR. At this time, according to a Ministry of Defense official, Russia has been the only former Soviet state involved in all of the peacekeeping operations, and largely finances them on its own.

**3** • Ukrainian Deputy Foreign Minister Boris Tarasyuk announces that Ukraine will not turn over any deactivated SS-19s to Russia until Ukraine, Russia, and the United States resolve the question of Ukraine's compensation for surrendering the weapons.

**5** • The council of CIS commanders of border troops meets in Minsk to discuss operations, cooperation, equipment provision, finances, and the training of personnel. Only Georgia, Moldova, and Tajikistan neglect to send representatives.

• Upon returning from a trip to Washington, Belarusian Defense Minister Pavel Kazlouski says that Belarus is entitled to half of a fund of $400 million set aside by the U.S. government to help former republics of the Soviet Union disarm. He justifies his claim by citing Belarus's surrender of its nuclear arms. To date, Belarus has received $59 million from the United States to help pay for the dismantlement.

• The Japanese government agrees to aid Belarus with the disposal of its nuclear arms. Earlier in the year, Japan pledged $100 million to all the former Soviet republics for this purpose. They scheduled a first joint Japanese/Belarusian disarmament committee meeting for 9 November.

**10** • The Ukrainian Defense Ministry repudiates claims by *Izvestiya* that it is planning troop cuts to 250,000 troops by the end of 1995. The Ukrainian Defense Ministry says that after reductions, the army is planned to be 450,000 strong.

**16** • CIS joint peacekeeping forces launch wargames in Tajikistan to test their combat capabilities in mountain and desert terrain. Military sources say the training will prepare the troops for a potential full-scale invasion of Tajikistan by opposition rebels training in Afghanistan.

**18** • A 600–page report entitled "Security Through Cooperation," on the state of CIS defense capabilities, and particularly its Collective Security Treaty, is published in Moscow. The document is supposed to circulate without restriction.

• *Sevodnya* reports that three versions of the document are available for purchase with hard currency: an 80-page short version, US$299; a 150-page abridged version, US$299; and the full-length document, US$1,675. Parts of the paper were published in *Izvestiya* on 20 November. These extracts mostly comprised data on the composition of the armed forces of each of the member-states of the CIS; little of a doctrinal nature was in these excerpts.

• The Ukrainian parliament votes on 17 November to ratify the START I Treaty, but attaches so many conditions that prospects for its implementation in the near future are very dim: Parliament reserves the right not to adhere to Article 5, which would require that Ukraine accede to the Nuclear Non-Proliferation Treaty; it demands compensation for the tactical nuclear weapons removed in 1992, foreign financial assistance for the dismantlement, and security guarantees recognizing the inviolability of Ukraine's borders. Parliament also reaffirms Ukraine's ownership and administrative control of the weapons. The resolution, which stipulates dismantlement of only 36 percent of the launchers, and 42 percent of the warheads, will be implemented only if financial assistance is received.

**19** • Ukraine's permanent envoy to the United Nations, Viktor Batyuk, states that the ratification of START I by Ukraine is a historic achievement, but goes on to say that unfortunately, a solution has not been found to the problems associated with disarmament, namely security guarantees and compensation. He indicates that only when these problems are solved will the treaty be implemented. Batyuk estimates that it will take about seven years for START to be honored.

**19–20** • The Russian and U.S. governments react negatively to Ukraine's conditional ratification of the START I Treaty, accusing the Ukrainian parliament of reneging on previous statements. However, they leave open the possibility of future negotiations, saying that the resolution does not necessarily reflect the positions of the Ukrainian government or President Kravchuk. Kravchuk is also critical of the resolution, saying that he would resubmit the treaty for ratification in March, after the new parliament has been elected. However, Kravchuk's sincerity is questioned by many analysts. Ukrainian Deputy Prime Minister Shmarov releases figures on financing the START I treaty accession [the Ukrainian interpretation] at $1.6–1.8 billion. He says that to disarm completely would cost $3.8 billion (if accurate, $1 billion higher than previous estimates).

**23** • Russian officials allegedly hint they will use economic pressure in order to get Ukraine to give up its nuclear weapons.

**30** • At CSCE meetings in Rome, Russia's proposal to lead peacekeeping operations in the former Soviet space are debated. None of the Western powers, the EU, or NATO are willing to take on the role themselves, but are also unwilling to let Russia have the job. The Baltic states and Ukraine express strong reservations about the proposal. Whereas most Western delegates strongly criticize Ukraine, Canada's foreign minister offers to mediate between Russia and Ukraine. Foreign Minister Zlenko accuses Russia of mobilizing the criticism, adding that: "Russia has the nuclear button and that's where the threat comes from." Kozyrev asks for the support of the CSCE for Russian peacekeeping in the former Soviet Union, and suggests that the political coordination of all peacekeeping operations undertaken by NATO, NACC, WEU, and CIS be taken over by the CSCE. Later, Kozyrev acknowledges that Russia's peacekeeping proposal stands scant chance of acceptance.

• At a conference on Black Sea cooperation, President Kravchuk refers to Ukraine's nuclear weapons as "material wealth," and demands that Ukraine be compensated for them when they are given up.

• *The Washington Times* reports that Ukraine is trying to take operational control of its weapons, and may succeed by 1994. The allegation is later denied by Kravchuk.

## December 1993

### *Economic*

**1** • An updated chart on levels of radiation contamination in Belarus, Russia, and Ukraine is released by specialists in Russia: percentages of land in which contamination levels are one curie per square kilometer or higher are as follows:

| | |
|---|---|
| Belarus: | 23 percent |
| Russia: | 0.6 percent |
| Ukraine: | 5.2 percent |

• Belarusian Prime Minister Kebich meets with Russian Prime Minister Chernomyrdin to discuss economic ties, including Belarus's entry into the ruble zone, its debt to Russia, and its energy crisis. Deputy Russian Prime Minister Gaydar has set new conditions on the monetary union with Belarus, which include legislation allowing the creation of a common market for land, other property, and securities.

**7** • A *Kommersant-Daily* article says oil debts in the former Soviet Union serve to promote the integration of CIS as well as non-CIS countries, such as Ukraine. It explains that the necessity to repay debts lays the groundwork for setting up multilateral financial institutions in the former Soviet space.

**9** • The eleven member states of the Black Sea Economic Cooperation Parliamentary Assembly (BSEC) agree to create a joint bank to facilitate regional trade and industrial projects, to be called the Bank for Black Sea Trade and Development, based in Thessaloniki, Greece. Bulgarian President Zhelev issues a statement which calls for the countries of the area to overcome ethnic and religious hatreds, so as to facilitate the development of a prosperous region. A Russian, Evgeniy Kotovoy, is elected to chair the BSEC secretariat.

**17** • A joint Belarusian-Russian commission, meeting in Minsk, orders the Belarusian Finance Ministry and National Bank to prepare the monetary and fiscal conditions of the country for the imminent economic union with Russia.

• Armenia, Moldova, and Ukraine apply for membership in the GATT, and are formally accepted. Working groups are established to accomplish their accession.

**18** • The CIS Interstate Bank is officially established and ten CIS member states commit five billion rubles as start-up capital, with each member's contribution based on its proportion of inter-CIS trade. The purpose of the Bank is to facilitate multilateral clearing of trade transactions within the CIS. Viktor Gerashchenko, head of Russia's Central Bank, is elected chairman.

**23** • A *Segodnya* article suggests that Ukraine's share of the Black Sea Fleet be used as collateral to back Ukraine's energy debt to Russia.

**29** • The Russian Information Agency reports that Uzbekistan plans to reduce oil and petroleum product imports from CIS countries by at least 15 percent next year in an apparent effort to become more economically independent.

### *Political*

**3** • Georgia becomes a full member of the CIS, the last of the ex-Soviet republics (excluding the Baltic states) to do so. (Moldova and Ukraine are "Associate Members.")

**4** • The Lviv city council reacts to Moscow's requirement for all citizens of former Soviet republics to register with Russian authorities by introducing a tax on Russian residents of Lviv. The Russian Foreign Ministry is apparently negotiating to change the law.

**7** • The secretary general of the CIS, Ivan Korotchenya, speaks before Belinform on the second anniversary of the formation of the CIS. Korotchenya cites the failure to implement multilateral agreements as the greatest obstacle to the integration of the CIS, but predicts that the creation of the Coordinating and Consultative Committee will alleviate this problem.

• On the same subject, *Nezavisimaya Gazeta* editorializes that the CIS has expanded within 1993 to coincide with the boundaries of the former Soviet Union. However, "if Russia fails to convince the other ex-Soviet republics that its intentions have a logical boundary and that its pressure will not spread beyond mutually acceptable limits, next year may well witness the conversion of the CIS into an association of integrated foes."

**10** • Interfax reports that Moldova plans to become a full member of the CIS, and will apply following its parliamentary elections. The charter stipulates that for an ex-republic to become a full member of the CIS, the charter must be ratified by 22 January 1994, but Moldova requests an exception to the rule because, it argues, ratification will not be possible until after elections scheduled for 28 December.

**13** • Ukrainians express alarm at the gains made by the Liberal Democratic Party and the Communists in the Russian parliamentary elections of 12 December. The opposition party, Rukh, intensifies its campaign against Ukraine's unconditionally giving up all of its nuclear weapons. The leader of the Democratic Party of Ukraine compares Zhirinovskiy to Hitler. Volodymyr Yavorivskiy, another democratic leader, warns Ukrainian democrats of the need to form a united front in order to prevent the same thing from happening in Ukraine. Belarusian Parliamentary Chairman Shushkevich also expresses concern over the Russian election results, noting that if the democratic parties do not form a unified front with which to counter Zhirinovskiy's rising power, he will begin his campaign to recreate the Russian empire. Some in the liberal minority in Belarus welcome the destabilizing results, which, they point out, may make Russia concentrate more on its own problems than on dominating the affairs of Belarus. Moldovan Prime Minister Snegur is also wary, noting that during the campaign Zhirinovskiy referred to Moldova as a province of Russia, whose governor-general is 14th Army Commander Aleksandr Lebed.

• Presidents Kravchuk and Snegur sign bilateral cooperation accords in which they pledged to respect each other's territorial sovereignty and increase cooperation in the legal and economic spheres.

• President Nazarbaev, a staunch supporter of reintegrating the ex-Soviet republics, modifies his views. In an interview with *Profil* (Vienna), Nazarbaev states that: "There was a time, after the Soviet Union disintegrated, when we were very romantic. We thought we could reunite. I, too, was wrong in that respect. Now I know that every state needs a chance to live and balance its own independence. Unfortunately, certain Russian politicians still have a Soviet way of thinking, which prevents such unification."

**15** • Iranian President Rafsanjani calls for increasing his country's ties with all countries of the CIS, especially economic ones.

**25** • CIS summit: Although many agreements are signed at the Ashkhabad summit on 24 December, there is little evidence of the CIS's increased viability. It was agreed that implementation of the Economic Union should begin even though some of the members' parliaments have not yet ratified the agreement. A Russian draft accord on minority human rights is not signed because of stiff opposition from a number of CIS states. Apparently, a number of bilateral meetings which do not include Russia take place in Ashkhabad. Boris Yeltsin is re-elected chairman of the CIS for a six-month period, and Russian diplomat Gennadiy Shabannikov is elected General Secretary of the CIS Collective Security Council, even though the Treaty on Collective Security has not been implemented. ITAR-TASS reports that the CIS members have decided to request that the UN grant the CIS status as an international organization. Support from the international community is perceived by Russia as legitimization.

**27** • Georgian President Shevardnadze, in a radio interview, expresses confidence in the longevity and viability of the CIS, provided Russia follows a democratic course. Conceding Russia's predominant influence and importance, he says, "And still, the center of the CIS, the magnetic pole and the initiator is Russia, and the fate of the CIS's future will depend on Russia and the processes under way in Russia . . . if Russia does not follow the path of democracy, then, of course, the CIS will disintegrate."

**28** • Ukrainian television broadcasts a discussion among Ukrainian officials in the wake of the Ashkhabad summit. Deputy Foreign Minister Boris Tarasyuk notes that Ukraine had signed 21 agreements, including a political declaration which stipulates that members will honor each other's territorial integrity and refrain from using force against one another. Political observer Valeriy Zhyhun notes that the CIS charter is to come into effect on 22 January, and that only those states which have ratified it at that time will be members. Thus, Ukraine has to decide whether it wants to integrate more closely with the CIS, or "seek ways to abandon it." Tarasyuk states that a two-tiered CIS may emerge, one tier of signatories to the charter, and a second tier of associates.

**31** • Kazakh Foreign Minister Suleymenov praises the pledge made by CIS member states not to interfere in each others' internal affairs and to respect one another's territorial sovereignty. He stresses the importance of the Economic Union, saying, "we should follow the example set by Western countries," and eventually establish a transnational currency.

### *Security*

**1** • The Black Sea Economic Cooperation Parliamentary Assembly opens its second conference in Kiev. Participating delegations are from Albania, Armenia, Azerbaijan, Georgia, Moldova, Romania, Turkey, and Ukraine (Russia was not mentioned). Greece sends an observation team. Kravchuk puts forward a number of proposals on security cooperation in the Black Sea area, including detailed naval training activities, the adoption of a declaration on the inviolability of naval borders, and a declaration stipulating that members would not use their naval forces against one another.

**2** • A *Moscow News* article carries a warning from the deputy chief of staff of the Russian Defense Ministry's Main Directorate for Nuclear Weapons, Vitaliy Yakovlev, explaining that if Ukraine's nuclear weapons have exceeded their six-year service life, attempts to dismantle them might be dangerous, if not impossible. Yakovlev also warns that the weapons could be made into radiological weapons, in which the radioactive material could be dispersed over a target area. These statements contribute to what appears to be a new Russian campaign to pressure Ukraine into giving up its nuclear weapons.

**7** • A Russian Foreign Ministry spokesman says the nuclear arms issue in Ukraine has reached a new level of urgency, and that Ukraine should heed the mounting international consensus that it must relinquish the weapons.

• A majority of CSCE members advocate a peaceful solution to the Dniester conflict, and the unconditional withdrawal of the 14th Russian Army from Moldova.

**15** • The head of Russia's border troops states that CIS peacekeeping forces in Tajikistan may need reinforcements before spring, when they expect renewed attacks from the opposition based in Afghanistan. He suggests that it may be in Russia's interests to extend the CIS agreement, which is due to expire at the end of the month, providing for military support to the Tajik government.

• Representatives of the U.S. State Department and National Security Council who visit Dushanbe are reportedly studying the legality of CIS peacekeeping forces in the country. Russia and Tajikistan have applied to have these forces converted to the status of UN peacekeeping forces.

• Following the success of the Liberal Democratic Party in Russia's parliamentary elections, Belarusian first Deputy State Secretary for National Security says that the republic will continue to send its weapons to Russia. He says that he believes the president and prime minister to be guarantors of Belarusian security and that even if the LDP gains control of the parliament, he believes the "button" would remain with the executive. Opposition leaders asserted that Belarus should follow Ukraine's example and halt the transfer of weapons to Russia because of the instability there.

• Yeltsin's first public statement on the Ukrainian parliament's ratification of START is carried by Interfax. Yeltsin accuses Ukraine of cheating and deception in its nuclear weapons policies, calling its approach "evil." Ukrainian officials dismiss these charges.

**16** • The head of Rukh, Vyacheslav Chernovil, calls Yeltsin's comments diversionary tactics, an attempt to shift world attention away from the "crashing defeat of democracy" in the Russian elections and onto Ukraine. The leader of the Ukrainian Social Democratic Party observes that Yeltsin's statements signify he has "begun to respond to the new balance of power in Russia."

**17** • Premature reports appear in the Western press claiming that an agreement has been reached regarding compensation to Ukraine for dangerous nuclear fissile material removed from its warheads. Despite substantial discussions between U.S. Deputy Defense Secretary William Perry, U.S. Ambassador-at-Large Strobe Talbott, Russian Deputy Defense Minister Georgiy Mamedov, Ukrainian presidential advisor Anatoliy Buteiko, and Ukrainian Deputy Prime Minister Valeriy Shmarov, no final agreement appears to have been reached.

**21** • Trilateral negotiations among the United States, Russia, and Ukraine appear to be leading to a debt-equity swap (nuclear weapons for energy debt), along the lines of the earlier Massandra agreement.

**22** • The commander of CIS forces in Tajikistan, Russian General Boris Pyankov, states that troop reinforcements will not be needed, but that earlier CIS resolutions calling for troops from Kazakhstan, Kyrgyzstan, and Uzbekistan have not yet been met, and are needed. The question of whether to put these troops under the control of the CSCE or the United Nations has not been resolved, partly because the United States raises the question of Russia's ability to remain neutral as a peacekeeping force.

• According to the Belarusian Defense Ministry, twenty-seven of its eighty-one strategic SS-25s have been sent to Russia, with the rest to follow by January 1996.

• Ukrainian Defense Minister Vitaliy Radetskiy announces that Ukraine will form a working relationship with CIS military bodies, and particularly with the Russian military. The Ukrainian parliament refutes Radetskiy's words, saying they express his own opinion only.

**23** • The CIS defense ministers meet in Ashkhabad, Turkmenistan (which is not a member of the CIS), and decide to transform the Joint Military Command into the CIS Joint Staff Committee, which will be smaller and will execute orders issued by the CIS Council of Defense Ministers. Col.-Gen. Samsonov will continue to head the committee. The formation of a joint staff has reportedly been agreed upon, with Russia to pay for half the costs. Col.-Gen. Pyankov urges Kazakhstan, Kyrgyzstan, and Uzbekistan to fulfill their commitments to the operation. Russian and Ukrainian Defense Ministers Grachev and Radetskiy apparently hold a productive meeting in which they agree on many issues, reflecting improved relations between the two defense ministries after extensive personnel changes in Ukraine's Defense Ministry. "Our views coincide almost entirely," Radetskiy said. Prior to the meeting, Turkmen President Niyazov states that "if the interests of independence are not infringed in any respect, and if Turkmenistan's proposals for an economic alliance are taken into account, then Turkmenistan does not rule out the possibility of becoming a full member."

**28** • Kazakh Defense Minister Nurmagametov advocates closer military ties with Russia, calling the treaty of friendship and cooperation between Kazakhstan and Russia the legal basis for such ties.

• First Deputy Defense Minister of Ukraine Ivan Bizhan denies that Ukraine has agreed to join the CIS Council of Defense Ministers. He reminds the press that Ukraine did not sign a single document at the Ashkhabad summit, and that any decision to join any arm of the CIS could only be decided by the president and the parliament. These two parties are deciding whether to send their representatives to the CIS Joint Armed Forces and the Council of Defense Ministers. Bizhan's comments appear to be calculated to discredit Radetskiy, who is considered by many to be too pro-Russia and pro-CIS.

**29** • Deputy Foreign Minister Boris Tarasyuk announces that Ukraine will deactivate twenty SS-19s and twenty SS-24s, although the timing of the deactivation remains unclear.

# 1994

## January 1994

### *Economic*

**1** • In Azerbaijan, the manat becomes the only legal tender as of 1 January. The exchange rate is to be 1 manat per 10 rubles.

**5** • President Yeltsin meets with CIS Executive Secretary Korotchenya to discuss strengthening the CIS. They focus on the problems associated with the Economic Union, especially the fact that there is still no legal basis for the union. Korotchenya asks Yeltsin to set forth measures which would create supranational interstate structures such as a payments union, and a free trade area.

• Russia and Belarus decide to merge their monetary systems by mid-January, and to make the Russian ruble the only legal tender in the two republics.

**8** • Tajikistan replaces all pre-1993 Soviet and Russian rubles with new Russian rubles, which are to be the country's only legal tender. The move is in keeping with the November 1993 agreement between Tajikistan and Russia, which subordinated Tajik economic policy to Russian authority in exchange for Russian support of the failing Tajik economy.

**11** • Reuters reports that Kazakhstan and Uzbekistan have agreed to create a common market by the year 2000, by abolishing all tariffs, quotas, licenses, and other trade barriers between the two countries. Kazakh President Nazarbaev says the agreement abolishes borders between his country and Uzbekistan.

• A representative of Russia's Ministry of Fuel and Energy, Vladimir Trofimov, announces that Russia has been strictly observing the schedule of oil deliveries to Ukraine. He also notes that Ukraine and Russia have settled on the price of oil at $80/ton rather than the previously announced $100/ton (the world-market price).

• Russia extends to Georgia a long-term credit of 10 billion rubles in order to revive its faltering economy. According to a Georgian embassy official in Moscow, the Georgian power supply is to be reintegrated into the southern Russian power grid.

• *Izvestiya* reports that Goskomsotrudnichestvo (the State Committee for Cooperation with Commonwealth Members) confirmed that Russia has transferred 30 billion rubles to Tajikistan in order to help it reenter the ruble zone. Apparently a total of 60 billion rubles has been earmarked for Tajikistan's reconstruction. Russian officials give Tajikistan six months to fulfill the requirements for membership in the "ruble zone," including subordination of its central bank to Russia's, coordination of tax, budget, and customs policies, and the implementation of economic reforms.

• Russia's Finance Ministry subsequently contradicts the report that Tajikistan will join the ruble zone, but confirms the extension of Russian credits.

**12** • Belarusian parliamentary chairman Shushkevich and Boris Yeltsin issue a joint communiqué calling for a new treaty on friendship and cooperation. Ostankino TV reports that the talks removed barriers between "what is essentially a single people."

• Two remaining disagreements are the rate of exchange between Belarusian and Russian rubles, and the price of oil in Belarus, which Moscow wants to raise above Russia's.

**13** • The Russian government outlines requirements Georgia must meet in order to rejoin the ruble zone. They include adopting all the economic laws of Russia, merging Georgian customs laws with Russia's, refraining from all independent credit emissions, and appointing a Russian to one of the vice-presidencies of the National Bank of Georgia.

• The Russian Foreign Ministry requests consultations between the Azeri and Russian governments concerning the creation of a free economic zone in Azerbaijan and Dagestan. Such an arrangement would in effect eliminate the borders between Azerbaijan and the Russian Federation republic of Dagestan. An advisor at Russia's Baku Embassy observes that "The creation of a free economic zone serves to maintain historical and cultural relations between the two people. . . ."

• An agreement on economic trade and cooperation between Ukraine and Georgia takes effect, with a list of goods that will be traded duty-free under the framework of the agreement.

**16** • Kyrgyzstan and Uzbekistan sign a number of agreements defining prospects for the further development of bilateral economic relations, most importantly an agreement which will permit the entry of Kyrgyzstan into the nascent Kazakh-Uzbek economic union.

**17** • In apparent response to former Russian First Deputy Prime Minister Gaydar's resignation, Belarusian Prime Minister Kebich and First Deputy Prime Minister Myasnikovich defend the economic merger between Russia and Belarus, which Gaydar opposed. Kebich asserts that Belarus's entry into the ruble zone will not cause the ruble to decline in value; rather, it will have the long-term effect of stabilizing it. Myasnikovich praises the final agreement, stating that Gaydar's view was not that of the Russian government.

**20** • After meeting in Kiev with President Kravchuk, Kazakh President Nazarbaev proposes the creation of a single currency unit in the CIS, the "altyn." No reactions to the proposal are noted in the press.

**25** • Deputy Prime Minister Valentin Landyk states that Ukraine will accede to the CIS Economic Union at the CIS summit in March. The Ukrainian leader stresses that his republic should rely primarily on relations with the states of the former Soviet Union and Eastern Europe in order to improve the economic situation in the country.

**26** • The Russian Ministry for Cooperation with CIS States reports that shortfalls in energy product deliveries to CIS states amount to 16.8 million tons of crude oil and oil products and 16.4 billion cubic meters of natural gas.

• Uzbek President Karimov addresses an international congress of industrialists and entrepreneurs in Tashkent, where he states that Uzbekistan's future is inconceivable without Russia. He unequivocally favors restoring economic ties among the CIS states and Eastern Europe, but emphasizes not all interstate ties should be resurrected, pointing out that Russia "continues to play the role of dictator" in the former Soviet states.

**27** • Belarusian Prime Minister Kebich reassures Russia that the ouster of Parliament Speaker Shushkevich will not affect the planned signing of agreements on monetary union between the two countries.

• A UNIAN correspondent reports research results on the likely impact of Ukraine's not signing the economic union. The Cabinet of Ministers estimates annual costs to Ukraine would be $4.6 billion, and therefore recommends Ukraine's cooperation with the Economic Union as an associate member.

***Political***

**1** • In Russia delivering his New Year's Eve broadcast, President Yeltsin pledges to defend the interests of Russians living abroad, implying those in states of the former Soviet Union. He says that "on the basis of law and solidarity, we defend and will defend our common interests."

**4** • In Tajikistan, authorities do not release information about the killings of a Russian Orthodox priest and nine Russian Baptists, and reports that the killings are linked with a robbery are unconvincing. They attribute the reluctance to explain the killings to a fear that it will scare Russians living in Tajikistan and result in greater Russian interference in Tajik internal affairs. The priest was killed shortly after conducting a service for members of the Russian 201st Motor-Rifle Division, the core of the CIS peacekeeping troops in Tajikistan.

**7** • In Kyrgyzstan, President Akaev proposes that Russians in Kyrgyzstan be given at least temporary dual citizenship as a way to rally their support behind him. A concession was reportedly necessary because the Russian-speaking industrial workers have been hardest hit by Kyrgyzstan's economic decline. He also hints that the state language law might be reconsidered. The Kyrgyz nationalist party Asaba immediately attacked the proposals.

**10** • In Georgia, President Shevardnadze rejects parliament speaker Goguadze's statement claiming the situation in Georgia is so dire that Russian economic and military aid is the only solution. Shevardnadze concedes that the adoption of the coupon last year had been a mistake, but he said that Georgia needed many more months before it could be decided to rejoin the ruble zone. Within a year, the Georgian coupon fell from a ratio of 1:1 with the Russian ruble to 100:1.

• On the issue of Russian troops being used to enforce law and order in Georgia, Shevardnadze states that "a country which could not instill law and order on its own does not deserve independence."

**12** • At a Moscow news conference, representatives of several human rights organizations report that Russian troops in the NIS committed brutal acts on civilians. Helsinki Watch, present at the press conference, places much blame on the Russian government for not preventing the escalation of human rights abuses, particularly in Moldova. Apparently, rather than those responsible for such abuses being disciplined, some were even promoted or awarded medals.

• The Bureau to Coordinate the Fight against Organized Crime and other Dangerous Forms of Crime on CIS Territory starts work in Moscow. The participants agree that criminal formations have "spread out" across the whole of the former union. Criminal activities include "illegal" businesses which export raw materials, large-scale bank fraud, automobile theft, drug and firearm import and distribution, and foreign currency dealings.

• A Moldovan newspaper prints a document submitted by Russia at the Ashkhabad CIS summit last month which calls for Russians living in the "near abroad" to be granted dual citizenship. While Moldova avoided the issue at Ashkhabad, officials there appear to be weakening in their resolve to oppose such a move.

**13** • In Minsk, negotiators from ten CIS states concluded talks on migrant labor. They are to sign an intergovernmental agreement in March. Specific issues which arose due to the collapse of the USSR are pension rights, medical coverage, and safety standards for migrant workers. Currently, Russia has a labor agreement only with Ukraine, but that is not preventing the emigration of thousands of workers from Ukraine to Russia searching for higher wages.

**14** • Moldova's acting Foreign Minister Ion Botnaru rejects Russian proposals for dual citizenship for Russians in

Moldova, stating that while it may be a stabilizing factor in Turkmenistan, it could "boomerang" elsewhere. He also urges that official Russian statements about Russians abroad be "specific, not inflammatory."

• Botnaru also rejects Russia's claim that the railroads from Moldova to Russia would not be able to handle the evacuation of Russia's 14th Army stationed there. He says that Moldova would internationalize the troop withdrawal and peace settlement negotiations if the deadlock is not broken.

**19** • During a meeting with U.S. Ambassador to Russia Thomas Pickering, Foreign Minister Kozyrev states that if Russia does not maintain a role in the "near abroad," the result will be chaos and Russia will be flooded with refugees. He also says that Russian troops will remain in neighboring countries only under the terms of bilateral agreements, as in Tajikistan. These statements are to clarify remarks made on 18 January, which caused alarm in many former Soviet republics, especially the Baltics.

• Apparently, Kozyrev also rejects the use of force or ultimatums against other states in the "near abroad" over Russian minorities living there, but, he says, the question of Russian minority rights cannot be ignored.

• Kozyrev responds sternly to continued criticisms of his statements, saying that Russia would not sit idly by if the rights of its minorities in the near abroad were violated. He says that Russia would use stronger means to protect these minorities if international legal norms were violated. (No evidence of such violations has been found by international human rights organizations.)

**20** • In Kiev, Kazakh President Nazarbaev meets with Kravchuk, where they sign eight agreements, including one on mutual indebtedness, as well as a Treaty on Friendship and Cooperation. They assert that the NATO Partnerships for Peace program has great positive potential, and express concern over attempts to destabilize the NIS from the outside and the rising tide of ethnic conflict. Moreover, the two leaders express their negative attitudes to the idea of dual citizenship. The two reiterate intentions to develop stronger bilateral relations.

**21** • Moldova's Foreign Ministry asks Russia, in a diplomatic note, to distance itself from a statement by Vladimir Zhirinovskiy in which he states the 14th Army ought to remain in the Moldova, and that the Dniester Republic would soon become part of Russia through political, diplomatic, economic, and military means. The remarks are broadcast on Dniester Republic TV. The note expresses concern that Zhirinovskiy, through such propaganda, could jeopardize the delicate negotiations under way concerning Russian troop withdrawal, as well as exacerbate the Dniester conflict, making it impossible to arrive at a political settlement.

**23** • In Kyrgyzstan, Russian Foreign Minister Kozyrev discusses the Russian-speaking minority issue with republican leaders. Kyrgyzstan has been under pressure for the past year to offer dual citizenship.

**27** • In an interview with Rome RAI Due Television, Russian Deputy Foreign Minister Anatoliy Adamishin states the following about Russia's evolving foreign policy: "In what direction is it changing? It is becoming more pragmatic—let us say, more interested in the defense of our national interests. . . . Sometimes it is said that in these [former Soviet] republics we are going back to imperialist ways, that we want to rebuild the Soviet Union. It is not true. We have seen, we have established, that unless Russia plays an active role, it is very difficult to have stability, security, and tranquility on our border. The former republics want Russia's active role; they want it."

***Security***

**3** • In Belarus, Belarusian parliament speaker Stanislav Shushkevich signs the CIS Collective Security Treaty. Even after the conservative Belarusian parliament voted to sign the treaty in April last year, Shushkevich declined to sign it, on the grounds that it compromised Belarusian sovereignty. The report implies that Belarus's harsh economic reality is forcing Shushkevich to allow for more conciliatory relations with Moscow and the CIS, and that Shushkevich may even be planning to meet with Yeltsin soon.

**5** • Shushkevich sends a letter to the CIS secretariat confirming his signing of the Collective Security Treaty. He notes, however, conditions that no Belarusian troops will be used on the territory of other CIS states, and that no other state's troops shall be placed or used on Belarusian territory without the express permission of the Belarusian parliament.

**8** • In Ukraine, Ukrainian parliamentary member and leader of the Rukh party Les Tanyuk accuses President Kravchuk of withholding documents apparently signed at the CIS Ashkhabad summit. The allegations heighten suspicion regarding what he actually did sign there.

**9** • Western press agencies report that the trilateral agreement between Ukraine, the United States, and Russia on the disarmament of Ukraine was imminent. The complicated agreement, crucial parts of which were kept secret, involves the sale of highly enriched uranium from the weapons of the former Soviet states to the U.S. government and U.S. energy agencies. Some of the uranium will be reprocessed and returned to Ukraine for use in its nuclear energy program, and Russia will cancel some or all of Ukraine's energy debt to Russia.

• The deal also reportedly includes formal recognition by Russia and the United States of Ukraine's borders and sovereignty. However, the issue of security guarantees, required by the parliament before they ratify any disarmament agreement, has not yet been clarified, although reports hold that no security guarantees will be given until NPT is ratified.

• The agreement is designed in such a way as not to require parliamentary approval in Ukraine, a process that would most likely hold up the process indefinitely. It is not certain, however, that Kravchuk will be able to proceed with such a sensitive issue exclusively by executive order.

• In Ukraine, reports surface that the republic will be a non-nuclear state within three years. Other reports stipulate that no timetable was set because it might anger the parliament, in which case a seven-year period is envisioned, in keeping with the START I deadline.

**10** • In Ashkhabad, the Council of the CIS Border Troop Commanders meet to discuss illegal migration and drug trafficking across external borders of the Commonwealth. Also on the agenda are principles of cooperation in logistic supplies to troops.

**14** • Presidents Clinton, Kravchuk, and Yeltsin sign the Trilateral Agreement in Moscow. Clinton says at a press conference that Ukraine's SS-24s will be deactivated within ten months. Radio Mayak, on the other hand, reports that the agreement does not include a timetable, but does indicate that they will be dismantled within "the shortest time possible." According to RFE/RL, the timetable will not be made public, even when completed.

• Russia will deliver 100 tones of nuclear fuel to Ukraine in exchange for approximately 200 warheads.

• In Ukraine, Kravchuk is not quick to announce the above agreement. His parliamentary opponents have been voicing their opposition to a deal arrived at largely behind their backs; Rukh leader Chernovil charges him with "shameful capitulation."

**15** • In Baku, Russian Border Troops Commander-in-chief and Federal Border Service head Andrey Nikolaev meets with President Aliev. Nikolaev states that Russia is not concerned with the border between Azerbaijan and Armenia, but rather with Azerbaijan's "external" borders.

• Ukrainian Foreign Minister Zlenko and Defense Minister Radetskiy state the Trilateral Agreement does not need ratification by parliament because it is an agreement, not a treaty; moreover, it does not introduce new issues, but rather addresses parliament's reservations which were manifested in the START I conditional ratification.

**17** • Nikolaev states that in 1994, Russian border troops would continue to guard the "external" borders of several CIS states including Armenia, Georgia, Tajikistan, and Turkmenistan. Nikolaev argues, "I agree that cooperation in defending the border, when sovereign independent states cooperate in defending the border, is a completely new thing. . . . You understand, two years have gone by, and we finally realized that it is very important to have not only rights in the CIS, but probably obligations as well."

**18** • Foreign Minister Kozyrev, at a meeting of Russian ambassadors to the NIS, states that it is necessary for Russia to keep its military presence in the former Soviet republics in order to prevent forces hostile to Russia from filling the "security vacuum."

• In Minsk, military delegations from Belarus and Ukraine sign an agreement on military cooperation. Belarusian Deputy Defense Minister Maj.-Gen. Vasiliy Dzyamidzik and Ukrainian Deputy Defense Minister of Armaments Col.-Gen. Ivan Olinyk fashion the agreement for military cooperation in 1994 similar to one the two countries signed in December 1992.

**19** • In Ukraine, Deputy Foreign Minister Tarasyuk says a detailed agreement concerning the disassembly and transport of missiles has not been concluded, and he dismisses Russian assertions that all missiles could be transferred to Russia within one or two years. Tarasyuk says such a move would create unsafe conditions in Russian nuclear arms storage areas.

**20** • Ostankino reported that a political battle is brewing in Ukraine over the Trilateral Agreement. Parliament, which opened today, is scheduled to debate the treaty at this session.

• In Kiev, the Ukrainian and Uzbek defense ministers sign military cooperation agreements, signaling the deepening of ties.

**21** • Belarusian Foreign Minister Aleksandr Stytchev states at a meeting with the West European Union that Belarus has now transferred 34 of its 81 strategic missiles to Russia, in addition to the short- and medium-range missiles transferred earlier. He continues to say that Belarus hopes to be a nuclear-free state by the end of 1996, and that the West ought to grant more aid to dismantle its conventional weapons.

**22** • *Izvestiya* reports that Kazakh President Nazarbaev sent requests for US$1 billion to dismantle its nuclear arsenals. Apparently, after Kazakhstan's missiles have been dismantled and the warheads transferred to Russia, Russia will sell the uranium contents to a U.S. firm.

**25** • The Ukrainian parliament puts off debating the Trilateral Agreement, as well as Kravchuk's proposal to ratify NPT. The influential speaker of the parliament, Ivan Plyushch, has come out backing Kravchuk on both deals, which may bode well for their passage.

**28** • The defense ministers of Kazakhstan, Kyrgyzstan, and Uzbekistan spend two days meeting in Bishkek, where they discuss the coordination of air defense systems, technology exchanges, and scientific research work. As reported by

ITAR-TASS, a protocol on military cooperation is signed by Uzbekistan and Kyrgyzstan.

• Belarusian parliament elects a new leader, Myacheslav Hryb, who has said that his foreign policy will not differ much from that of his predecessor, Stanislav Shushkevich. However, while Shushkevich was against Belarus's signing of the CIS Collective Security Treaty, Hryb has always supported it.

**29** • *Krasnaya Zvezda* editorializes that recent moves by officers of the Strategic Rocket Forces to declare allegiance to Ukraine represent steps to complete Ukraine's takeover of nuclear weapons on its territory.

• The article also continues the barrage of accusations that the nuclear weapons on Ukraine's territory are not in safe hands: it claims that 60 percent of rocket force units in Ukraine are not combat ready because of personnel shortages and that there are approximately 500 warheads located in the Pervomaysk storage facility which is six to eight times the amount previously estimated. Because only up to 320 warheads have been removed in 1993, the figure of 500 suggests that either more missiles have been deactivated, or that there are some "spare" warheads, previously unaccounted for in Ukraine.

### February 1994

#### *Economic*

**1** • Kazakhstan, Kyrgyzstan, and Uzbekistan remove customs on their common borders in the first step toward their previously agreed-upon economic union.

**4** • Gazprom reduces gas shipments to Ukraine because of Kiev's failure to pay its gas debt to Russia, which is estimated at over 1 trillion rubles. Ukraine has made no payments on the debt for the past few months.

**8** • The prime ministers of Moldova and Russia sign an economic cooperation and trade agreement. Chernomyrdin says the balance of trade between the two republics is "shaping up fine," with Moldova paying for all imports from Russia except fuel.

**9** • The Russian government demands 20 to 40 percent of revenues from Kazakh oil exports transited through Russian pipelines.

**14** • Belarusian Prime Minister Kebich states that the imminent monetary union between Russia and Belarus is not a political-economic union, and that Belarus will forge its own path to a market economy. In Russia, the union has come under attack as being too costly for the Federation. Documents for the union are being prepared in anticipation of Russian Prime Minister Chernomyrdin's upcoming visit to Belarus.

**16** • Russia may increase gas supplies to Ukraine and Belarus in the first quarter of 1994. Accumulations of surplus gas and oil in pipelines due to withholding forces Russia to ship it. Storage facilities are insufficient to allow a complete cut-off of energy supplies to Russia's debtors.

• A Russian parliamentary group, called the "Union of 12 December," opposes monetary union with Belarus, saying that it would be too costly for Russia and would result in a 5 percent increase in inflation. They do, however, support close political and military ties.

• Russian Prime Minister Chernomyrdin and his Armenian counterpart Bagratyan sign a trade and economic cooperation agreement, according to which Russia will annually supply Armenia with US$141.3 million worth of goods and a continuing supply of energy. The agreement does not provide for Russian credits to Armenia.

• Russian First Deputy Prime Minister Soskovets and Minister Bagratyan also agree that Russia will supply equipment and specialists to help reactivate the Armenian nuclear power plant at Medzamor. Armenia will reportedly finance the operation.

**22** • Turkmen officials claim they will halt gas supplies to Ukraine if its debt is not paid. Ukraine has paid only $3.4 million of a $154.1 million debt to Turkmenistan for gas supplied in 1994. Bills for 1993 remain unpaid.

**23** • Nikolay Gonchar, chairman of the Council of the Russian Federation Budget Committee, calls in *Rossiyskaya Gazeta* for an economic union of Ukraine and Russia. "In Crimea they have already understood," he says, that there is "no alternative" to such a union, and that the "parade of sovereignties has had negative consequences." Gonchar warns that "in Ukraine, things are so bad" economically that "at any moment there could be an explosion," which would reverberate beyond Ukraine's borders.

**26** • Turkmenistan says it will resume gas shipments to Ukraine in exchange for shipments of food and other consumer goods. The Turkmen government says that Ukraine has already begun to pay off some of its debt in convertible currency.

#### *Political*

**3** • President Boris Yeltsin and Eduard Shevardnadze sign a bilateral treaty on friendship and cooperation and twenty-four other agreements on trade and economic ties, scientific and cultural cooperation, the status of Russian border guards in Georgia, and military basing rights. They also discuss Georgia's possible reentry into the ruble zone. Shevardnadze characterizes the agreements and the visit as vital to establishing peace and stability throughout the Caucasus.

**8** • The Russian Federation Council chairman, Vladimir

Shumeyko, is elected chairman of the CIS Interparliamentary Assembly. He is the second Russian to hold this position; Ruslan Khasbulatov was the first.

**9** • Russian Foreign Minister Kozyrev and President Nazarbaev meet to discuss tensions between their two countries, over dual citizenship, military cooperation, and the fate of the Baykonur space complex. Although Kozyrev avers concurrence on all issues, Nazarbaev later reiterates his opposition to dual citizenship in an interview with *Komsomolskaya Pravda*.

**12** • General Mikhail Dmitriev, chief of the Analytical Administration of Russia's Foreign Intelligence Service (SVR), tells journalists that he is responsible for preparing psychological profiles for the Russian leadership on foreign leaders, including those of the CIS states. Dmitriev points out that all of the work is done from Russia, and there is no need therefore to engage in spying.

**15** • The Moldovan ambassador to Ukraine, Ion Borsevici, one of President Snegur's right-hand men, says: "Ukraine's wish to defend its independence plays an exceptionally important role for the defense of our own country's interests. . . . Ukrainian-Moldovan good neighborly relations transcend the framework of state-to-state relations. . . . Ukraine provides an umbrella against those forces which want to bring us into a neo-Soviet brotherhood."

• Uzbekistan interrupts the broadcast of Russian TV for one day due to unpaid debts to the Uzbek Ministry of Communications.

**17** • The CIS statistics committee reports that CIS populations are declining, due to decreasing birth rates. It also reports that Belarus, Russia, and Ukraine are experiencing intensive aging of their populations.

**20** • Ukraine's deputy foreign minister, Boris Tarasyuk, says the violation of Ukrainian rights in Crimea should be a matter for concern. He states that despite cries from Moscow that ethnic Russians' rights are being violated, the opposite is actually happening in Crimea, where Ukrainians make up only 25 percent of the population. The Ukrainians there complain that Russian authorities have opposed opening Ukrainian schools and newspapers, have censored Ukrainian radio and TV broadcasts from Kiev, and have issued "anti-Ukrainian propaganda."

**22** • The Ukrainian minister for nationalities and migration, Oleksandr Yemets, accuses Russia of ignoring the rights of its largest minority, the Ukrainians. He says he has received complaints from Ukrainians in Russia about authorities who are openly hostile about establishing Ukrainian schools, even where Ukrainians live compactly. The 1989 census puts the number of Ukrainians in Russia at 4.4 million, but Yemets claims there are between 6 and 10 million Ukrainians in Russia.

• Kyrgyz President Akaev meets with Yeltsin in Moscow to discuss Russian-Kyrgyz relations. The press writes that Akaev has been under pressure to allow for dual Russian-Kyrgyz citizenship, but has said he will not allow it without parliamentary approval. He is eager to stop the flow of Russians leaving Kyrgyzstan (reports estimate that 100,000 Russian-speakers have left Kyrgyzstan since the collapse of the Union), and while in Moscow he allegedly unveils a series of proposals designed to make Kyrgyzstan more appealing for Russians. He is susceptible to Russian pressure because of the economy's dependence on Russia.

• Akaev recommends his own code for interethnic relations. The provisions are as follows: (i) adequate representation of national minorities in the state administration system; (ii) refusal to countenance discrimination against persons who do not speak the national languages of the NIS; (iii) a solution to the problem of dual citizenship.

• Uzbek President Karimov meets with the chairman of the Tajik Supreme Soviet Rakhmonov in Tashkent, where they sign a trade and economic agreement on economic aid which Uzbekistan will provide its neighbor, including: fuel, agricultural machinery, fertilizers, and seeds which will save spring planting. Karimov announces that Tajikistan will not be allowed to join the Uzbek-Kyrgyz-Kazakh economic union until its border with Afghanistan is calm.

***Security***

**1** • The Ukrainian Foreign Ministry, in its weekly press briefing, condemns the "hardening" of Russia's policy toward the newly independent states of the former USSR. The head of the ministry's information department, Yuriy Serheev, says that Ukraine categorically opposes linking the protection of minority rights with the presence of foreign troops in CIS states.

• Serheev notes that at the December CSCE conference in Rome, Russia is told to evacuate all troops from Moldova and the Baltics. Russia has been exacerbating tensions in these states by provoking intolerance among ethnic Russians there, and "in essence the ideology of great-state chauvinism is being revived."

**2** • Russian Defense Minister Pavel Grachev, on a visit to Tbilisi, announces that Russia would like to open three military bases in Georgia and two more in other parts of the Caucasus, within the framework of the CIS Collective Security Treaty. The current agreement between Russia and Georgia allows Russian troops to stay until 1995. The proposed bases would give the Russians approximately 23,000 men in the Caucasus.

• A group of Russian parliamentary deputies, including State Duma Chairman Ivan Rybkin and Egor Gaydar, send a letter to Yeltsin cautioning that signing a treaty with

Georgia prematurely could destabilize the entire Transcaucasus.

**3** • The Ukrainian parliament votes overwhelmingly to ratify, with reservations, the START I Treaty and the Lisbon protocol following a speech by President Kravchuk which stresses that Ukraine must denuclearize in order to avoid international isolation. U.S. President Clinton pledged to double economic aid to Ukraine if it ratified the treaty.

• The Trilateral Agreement can allegedly be put into effect, under which 200 weapons will be transferred from Ukraine to Russia in ten months. However, crucial issues must be resolved, including compensation of Ukraine for the earlier removal of tactical nuclear weapons, and a timetable for the Trilateral Agreement.

**8** • Ukrainian parliamentary deputy Sergey Holovaty charges that the official text of the Trilateral Agreement which was published publicly is not, in fact, what was passed in parliament. He claims that an amendment to the treaty stipulated that no warheads would be transferred until compensation and security guarantees were signed. This partially explains the confusion expressed by some deputies after passage of the resolution.

• During talks in Kiev on the Trilateral Agreement, Yuriy Dubinin, head of Russia's delegation, asks for clarification of whether parliament voted for complete or partial denuclearization.

**10** • President Clinton announces that he expects U.S. aid to Ukraine for the dismantlement of nuclear weapons to be doubled, and that he will convey the message to President Kravchuk in March. He also says that upon accession to the NPT, the United States will offer Ukraine security guarantees, as provided for in the Trilateral Agreement (RFE/RL).

• Turan (Baku) reports that the signing of a military pact between Kazakhstan, Kyrgyzstan, and Uzbekistan comes as a surprise to Moscow. The Russian ambassador to Kyrgyzstan meets with the Kyrgyz defense minister, who expresses his readiness to sign a bilateral treaty with Russia. Preparation of treaty documents will begin immediately, and will be ready when Chernomyrdin visits Bishkek later in the month.

**14** • The Russian Foreign Ministry objects to Ukraine's "misinterpretation" of the CSCE foreign ministers' statement calling for the withdrawal of the 14th Army from Moldova. The Ukrainians were responding to Foreign Minister Kozyrev's claim of Russia's right to station troops in the "near abroad."

**23** • Grachev claims that Russia now has approximately 16,000 peacekeeping troops in the "near" and "far" abroad, and that Russia is the only state in the FSU to have constructed a viable, if not yet fully effective, armed forces.

• Kozyrev states that Russia does not wish to act alone in peacekeeping operations in Bosnia or other former Soviet states, then expresses Russia's desire for "serious help." In an article in *New Times*, No. 4, Kozyrev writes that no international "legislation" of Russian peacekeeping operations is necessary because Russia's deployments are already fully legal. He stresses that "a UN mandate and the presence of UN observers confirms the impartial character of Russian peacemaking missions."

**24** • CIS defense ministers meet in Moscow and discuss the following issues: transforming the CIS joint armed forces command into a headquarters for the coordination of military activities; defining the extent of interaction among signatories of the Collective Security Treaty; the Tajik-Afghan border conflict; and peacekeeping on the territory of the post-Soviet states. Of the twelve former Soviet republics, only Moldova fails to send a delegation. ITAR-TASS reports suggest that centrifugal tendencies in CIS military cooperation have been arrested.

**25** • *Moskovskiy Komsomolets* reports that the armed forces of CIS countries lost 3,000 officers to the Russian military in 1993. The report estimates that over 1,000 officers have left the Kazakh armed forces, 800 left Belarus, and 1,000 left Ukraine, for higher wages offered to officers in Russia. The report predicts increased military migrations in 1994.

**28** • The chief of the Russian General Staff, Mikhail Kolesnikov, states that Russia expects to establish about thirty Russian military bases in states of the former Soviet Union under bilateral treaties. He says bases in Armenia, Azerbaijan, and Tajikistan will be set up with troops already stationed in these countries, and in all of the other former Soviet republics, except Ukraine and the Baltics, Russia will build bases with its own forces.

• Russia's chief of staff for coordinating military action among CIS member states, Col.-Gen. Viktor Samsonov, says CIS forces now guarding the Tajik-Afghan border are inadequate to hold back stepped-up strikes in spring and summer. Every CIS state, except Ukraine, Belarus, and Moldova, has signed a treaty pledging to send peacekeeping troops to the region.

### March 1994

#### *Economic*

**1** • The CIS Coordinating and Consultative Committee (CCC) meets in Moscow to discuss the Economic Union, and specifically mechanisms for Ukraine's participation in the union.

• The Russian minister of economics, Shokhin, says the CCC has approved the draft agreement for Ukraine to join the Treaty on Economic Union, and that Ukraine might change its policy and become a full member of the economic union.

• Shokhin notes that: "Ukraine's associate status is a temporary situation."

• Ukrainian Deputy Prime Minister Valentin Landyk tells journalists in Moscow that Ukraine will remain an associate member of the CIS Economic Union. He says Ukraine wishes to remain free to cultivate relations with non-CIS states.

**2** • Gazprom, in another attempt to extract payment from Belarus, Moldova, and Ukraine, threatens to cut off gas supplies. The company accuses Ukraine of siphoning off gas destined for Europe, a charge Kiev denies. A senior official in the Ukrainian gas company Ukrazprom warns, however, that Ukraine may be forced to tap gas transit lines in the future if its own supplies are cut.

• Uzbek President Islam Karimov and Boris Yeltsin conclude an agreement on economic integration. The agreement provides for coordination on economic reform and fiscal policies, and encourages interstate ties between enterprises. Karimov says the introduction of dual citizenship in Uzbekistan might be detrimental for Russia, implying that a huge outward migration could take place from Uzbekistan to Russia.

**4** • Belarus guarantees the transit of gas exports from Russia to Western Europe across its territory, despite Russian cuts in gas deliveries to Belarus itself.

**4–6** • Russian and Western news agencies report further cuts in natural gas supplies to Ukraine, bringing deliveries down to about one-fifth the normal level by March 6. Supplies are not terminated, as warned, despite Ukraine's continued failure to repay its 1.5 trillion ruble debt.

**5** • *Trud* states that Ukraine's admission to the CIS Economic Union as associate member is troublesome. While Ukraine's embassy in Moscow implies that the issue has been resolved, Ivan Korotchenya, executive secretary of the Coordinating and Consultative Committee, says: "At the moment it is too complex to admit Ukraine. The republic has not signed that Economic Union Charter."

• Nevertheless, Committee Chairman Aleksandr Shokhin states that the basic idea of Ukraine's joining has been accepted and the timing is up to Ukraine.

**7** • Russia says it will continue to provide Ukraine with 40 million cubic meters of natural gas daily, at least until talks on Ukraine's debt resume in Moscow on 10 March. Gazprom charges that Ukraine is daily siphoning off another 40 million cubic meters of gas on its way to Western Europe.

—Belarus is receiving 40 million cubic meters of gas from Russia daily since it initiated payment on some of its arrears.

—Moldova is receiving partial shipment, in the amount of 11 million cubic meters daily, because it has started to pay off its debt, mostly with agricultural goods.

**8** • The Russian Ministry of Economics may ask Ukraine to compensate Russia for its $500,000-a-day loss, allegedly due to the gas Ukraine has been siphoning off. Russia says it will accept payment for Ukraine's arrears in the form of property rights to Ukrainian pipelines and oil installations.

• Gas deliveries to Belarus increase when Belarus makes a 30 billion ruble payment on its 400 billion ruble debt.

**9** • Ukraine makes a $9 million payment on its $900 million debt to Russia on the eve of scheduled talks in Moscow. The deputy economics minister of Ukraine again hints that a debt-equity swap may be in the works.

• Belarusian National Bank Chairman Stanislav Bahdankevich says that Belarus will not enter into a monetary union with Russia under the terms thus far proposed. Belarus hopes for the exchange of its weaker currency on a one-to-one basis with the Russian ruble and domestic Russian prices for oil and gas. Russia is apparently demanding that the National Bank of Belarus be replaced by a branch of the Russian Central Bank in Minsk, and that a single government budget be formed for Russia and Belarus. Russia is also unwilling to lower energy prices for Belarus. Bahdankevich says that under these terms the economic situation in Belarus could only worsen.

**10** • Gazprom decides that Ukraine's debt for 1993 will be paid half in cash, half in equipment, and payments for 1994 will be made in rubles or hard currency. The two parties agree in principle on joint management of Ukrainian pipelines (to be privatized soon), and of underground storage facilities, in exchange for partial debt forgiveness.

• An Uzbek delegation signs an agreement with Belarus in Minsk on economic cooperation and trade. Uzbek officials acknowledge their country's $13.98 million debt to Belarus, which they will start to repay in January 1995.

**11** • Moldova agrees to pay promptly its 126 billion ruble debt to Russia. Ninety-six billion will allegedly be borrowed from international financial institutions, and another 30 billion worth of high-quality agricultural goods will be shipped to Russia to cover the remainder.

• Moldova's arrears are partly the result of prohibitive customs duties and excise taxes imposed on Moldova for failing to ratify CIS documents in August 1993. Moldovan President Snegur tells Gazprom that in order to prevent this from happening in the future, all trade restrictions between CIS countries should be removed.

**14** • Kazakh President Nazarbaev and Uzbek President Islam Karimov create a "common economic zone." Trade between Kazakhstan and Uzbekistan has risen dramatically since January.

• Gazprom reduces gas deliveries to Belarus by 10 million cubic meters daily. A Gazprom official puts the Belarusian debt at 400 billion rubles, or $230 million, and says that Belarus is in no hurry to pay it back.

**16** • According to Reuters, the Russian Fuel and Energy Ministry indicates that debtor nations that cannot pay back energy bills to Russia may be asked to transfer property rights instead.

**21** • Belarusian Prime Minister Vyacheslav Kebich says that Moscow has reduced the price of gas for Belarus from $80 per 1,000 cubic meters to $50, dated retroactively to January 1994. As a result, Belarus's debt to its Russian suppliers decreases by 150 billion rubles. The opposition in Minsk is skeptical: "the only thing for free is cheese in the mousetrap," said People's Deputy Sergey Naumchuk.

**22** • Gazprom warns Ukraine that it will stop delivery of natural gas if Ukraine does not live up to its end of the 10 March bilateral agreement. The agreement stipulates that Ukraine must pay $100 million of its $900 million debt by 1 April, and it must turn over an unspecified amount of shares in Ukrhazprom, the Ukrainian gas concern. Officials from Gazprom states that officials in Kiev have yet to draw up a list of enterprises in which Gazprom would receive shares, and to determine the interest rate Gazprom will receive.

**23** • The press secretary to Belarusian Prime Minister Kebich, Ulyadzimir Zamyatalin, states that arrangements for the monetary union between Belarus and Russia are nearly complete. Zamyatalin states (contrary to earlier reports) that officials from the two countries have agreed on identical prices for fuel; a 1:1 exchange rate for the two countries' currencies; and making the National Bank of Belarus the main bank of a sovereign Belarus. Each country will maintain its own budget. The Russian Duma is scheduled to begin debating the monetary union on 24 March.

• Terms outlined by Zamyatalin are not favorably received in the Russian Duma, where some say that an official currency can have only one master. They feel Belarus must cede sovereignty over its monetary and credit policy to Russia. Russian Deputy Minister of Economics Sergey Ignyatyev says that Belarus will have to relinquish some of its political sovereignty.

**25** • The chairman of the Russian State Duma's Committee for CIS Affairs, Konstantin Zatulin, blames the delay in Russian-Belarusian talks on monetary union for raising tension in other spheres of Russian/Belarusian relations. Serious differences between Belarus and Russia are becoming glaringly apparent. Zatulin suggests resolving the issue by public referendum. The leader of Russia's Choice, Egor Gaydar, says a customs union is much more important than a monetary union.

**30** • A private corporation called Respublika, in Kiev, has bought the portion of Ukraine's gas debt which is owed to Turkmenistan. Apparently Respublika will take over the 1993 debt of $671,900,000 and the 1994 debt of up to $500,000,000. The deal stipulates that 35 percent of the debt will be paid in cash, and the rest in Ukrainian goods.

### *Political*

1 • After a stormy debate, the Georgian parliament votes by 121 in favor, 47 against, and 4 abstentions, to ratify Georgia's membership in the CIS. When Shevardnadze approved accession to the CIS in early October last year, the parliament was not in session.

• Yeltsin sends a letter to the leader of the "Dniester Republic," Igor Smirnov, in which he outlines, in a tone highly deferential to Smirnov, a framework for the political settlement of the region's conflict. He calls for "broad autonomy for the Dniester region as part of the Republic of Moldova; a unified army, security service, and financial system; and a role for Russia together with other countries" in guaranteeing the eventual settlement.

• President Snegur says that the newly elected Moldovan parliament must approve a new constitution and join the Economic Union of the CIS. He denies advising that any CIS agreements other than economic be joined. "We take part in integration on the issues that do not infringe upon our independence," he says.

**3** • Yeltsin and Uzbek President Karimov conclude a Russian-Uzbek Agreement which notes a mutual interest in developing bilateral links. Conspicuously absent from the document, according to *Izvestiya* of 3 March, is reference to the issue of dual citizenship. Karimov assures Yeltsin that he will not tolerate any human rights infringements on ethnic grounds.

**4** • The Russian Foreign Ministry holds a seminar on Russians and Russian speakers in the CIS and the Baltics. A top official at the Ministry for Foreign Affairs, Leonid Drachevskiy, says that integration within the CIS is the only way to defuse the controversy surrounding the issue of Russian speakers in the former Soviet republics. Gennadiy Mozhaev, who represents the Association for Ties with Foreign Compatriots, states that Russia's strategic task is to "keep all Eurasian territory of the former Soviet Union if not under control, then under strong influence. . . . From this point of view it's an advantage for us to have a big number of Russians in the near abroad."

• Interfax reports that Aleksandr Denisov, identified as Russia's senior advisor to the United Nations, says: "The Russians must stay wherever they are historically strong [and not migrate back to Russia] and need only economic support."

• Opposition in Georgia to President Shevardnadze and his decision to accede to the CIS is increasing. *Komsomolskaya Pravda* and the Georgian Iberia news agency report that at least sixteen parties have united to oppose Shevardnadze and joining the Commonwealth in a powerful statement to that effect published on 4 March.

**9** • An unattributed radio commentary broadcast on Radio Baku International Service editorializes that Russia is eager to establish a military base in Azerbaijan and to have a hand in the resolution of the Karabakh conflict. President Aliev, however, opposes the idea, and Baku has requested that Moscow comply with international agreements and laws that do not allow Moscow to act as internationally sanctioned peacekeeper in the former USSR.

**10** • The CIS Heads of State meeting scheduled to be held at the end of March is postponed until 15 April at the request of Kazakh President Nazarbaev.

• Dumitriu Motpan, leader of the winning party in Moldova's legislative elections, says he will form a single-party government. His platform includes the following: full independence from Romania and Russia; rejection of Russia's demands for military bases on Moldovan territory; insistence on the withdrawal of Russian troops; turning Moldova into a demilitarized zone; Moldovan participation in economic, but not political or military, agreements of the CIS.

**16** • The Council of CIS Foreign Ministers meets in Moscow to discuss CIS coordination of major foreign policy issues. The foreign ministers of Azerbaijan, Georgia, Kazakhstan, Kyrgyzstan, Tajikistan, and Russia, as well as ministry officials from Armenia, Belarus, Moldova, Turkmenistan, and Ukraine, attend the session.

• After the meeting, Kozyrev says the Council discussed ways to protect CIS external borders. They decided to form a consultative commission in which they would peaceably solve interstate disputes. Kozyrev says a consensus was reached among the participants to apply for the status of an international organization at the United Nations.

• During a joint meeting of CIS defense and foreign ministers, Kozyrev says that Russia has a historic duty to guard the Tajik-Afghan border because "it is a frontier of the CIS . . . and . . . it is clear that except for us, no one can solve these issues." He also says that peacekeeping activities do not require international approval.

• The Ukrainian Ministry of Foreign Affairs releases a statement prior to the meeting which reads: "[Ukraine] does not join military and political unions, blocs, or groupings. The Ukrainian Ministry of Foreign Affairs told participants that the creation of a security system parallel to NATO is dangerous and inexpedient. Ukraine is in favor of global collective security on the territory of Europe."

**18** • A Georgian parliamentary deputy, Irina Sarishvili, delivers a statement to parliament asserting that Georgian membership in the CIS is illegal. As a result, international experts have been called to Georgia to determine the validity of Georgian membership. Sarishvili equates joining the Commonwealth with joining Russia.

**21** • Russian Deputy Foreign Minister Anatoliy Adamishin tells a group of Russian and French journalists that Russia has neither the will nor the power to conduct an imperialist foreign policy, and it supports the political independence of the other states of the former Soviet Union, especially Ukraine. He says that if Ukraine as a state breaks apart, it would have grave consequences for Russia, and Russia is trying to prevent such a development "by all means." He asserts that Crimea should remain part of Ukraine, and Russia has no territorial claims against Ukraine.

**23** • On the second day of his official visit to London, Kazakh President Nazarbaev rules out the possibility of Kazakhstan rejoining the ruble zone. He also advocates modifying the CIS to form a smaller grouping named the "Eurasian Union" which would be a "belt of stability and security" and from which countries engaged in military hostilities would be excluded.

• In Moscow less than one week later, Nazarbaev elaborated on this proposal, saying that the new "Eurasian Union" would coordinate the economic, military, and foreign policies of its members. Nazarbaev also suggests that Russia and Kazakhstan could form the union on their own if other states are not interested.

**26** • Turkmen President Niyazov sends a message to Yeltsin supporting the agenda of the next CIS summit, but also stating that Turkmenistan will not join any of the new CIS structures. Turkmenistan is party to some CIS agreements, but has not acceded to the charter, and therefore is not a full member of the Commonwealth.

• The Georgian Security Ministry reports that in accordance with information recently obtained, some 95 percent of the Georgian population is in favor of joining the CIS.

**29** • The Russian chairman of the CIS Interparliamentary Assembly, Vladimir Shumeyko, says the time has come to raise the issue of closer integration within the CIS. Shumeyko says that "supranational bodies" may be needed to turn the CIS into "a kind of union." He adds that the union should not be similar to the Soviet Union, and that the sovereignty of each of the additional states should be preserved.

• In a meeting with CSCE Commissioner for Ethnic Minorities Max van der Stoel, Russian Minister for Nationalities Sergey Shakhray says that over the past few years 356,000 ethnic Russians have emigrated from Central Asia to Russia. Shakray cites statistics showing that 28 percent of those who returned did so "with the real threat of violence or persecution," and 19 percent with "insult and humiliation to national dignity."

• In Moscow, head of the Russian Fund for Constitutional Reforms Oleg Rumyantsev suggests that Crimean citizens be given dual Russian-Ukrainian citizenship, which would be used as a test case for the idea across the entire ex-Soviet Union.

• The same day, Konstantin Zatulin, chairman of the committee in charge of relations with the CIS states, says that in order to avoid a conflict with Crimea, Ukraine must give it considerable economic autonomy. "To put pressure on Ukraine could cause an explosion. For Ukraine there is only one outcome, that is to offer Crimea the widest possible scope in the economic domain. . . . Zatulin said, "Russia has never played and will never play the Crimean card, [but] it is impossible not to take into account Russia's special interests in Crimea."

**30** • On his last day in Moscow, Nazarbaev tells journalists that his "Eurasian Union" proposal was inspired by the European Union, and was not meant as a formal proposal, but rather was meant to provoke discussion.

• Nazarbaev's proposal is, nevertheless, very well received by some in Moscow. An article in *Krasnaya Zvezda* praises and expands the idea: "The union must have a common parliament. This is to make it possible to bring together the laws of the states in the union. The Economic Union needs an interstate executive secretariat. . . . Even the currency could be common and supranational. . . . It is perfectly within our power to create a union on totally new principles, without the former imperial aroma. . . ."

• Yeltsin receives the unexpected idea with cautious optimism, and says that he will take time out to consider it. He also says that the CIS already embodies many of the EU ideas, and he sees no point in replacing one good idea with another.

***Security***

**3** • Moscow Russian Television Network runs a program that states that the CIS Defense Ministers' Council has been reassigned as the main consultative body on military matters of the Collective Security Council. It is proposed that the Council be composed of 250 people, half of whom are Russian.

**4** • Russian Foreign Minister Kozyrev sends a letter to every CIS foreign minister emphasizing the importance of gaining observer status as an international organization at the United Nations. He urges that efforts be made for the CIS to gain recognition as a regional organization by the EU and CSCE.

• These efforts appear to be part of new Russian campaign to obtain international organization status for the CIS, which will impart greater legitimacy to CIS peacekeeping efforts.

• The head of Russia's delegation to the CSCE, Vladimir Shustov, argues that Article 52 of the UN Charter allows regional organizations to take action to maintain peace and security, and that therefore there is no need for Russia to seek approval for its peacekeeping efforts in the CIS.

• Russian and Kyrgyz officials conclude an agreement which regulates the service of Russians in the Kyrgyz armed forces. Russian citizens apparently make up a large portion of Kyrgyzstan's officer corps. The agreement provides for Russians to have individual contracts with the Kyrgyz Ministry of Defense. After the contracts expire, the Kyrgyz ministry will have to buy the Russians housing anywhere in Russia, except Moscow or St. Petersburg, or, if they choose not to leave Kyrgyzstan, give to the Russians the housing they currently occupy at the time their service ends.

• The Chief of General Staff for military cooperation among CIS states, Col.-Gen. Viktor Samsonov, meets with Georgian Prime Minister Otar Patsatsia in Moscow. Samsonov says that the collective security of the Commonwealth states is being strengthened, especially in view of the Russian-Georgian Treaty on Friendship, Neighborliness, and Cooperation signed in February. ITAR-TASS reports that this treaty "reflects the intentions of each side to facilitate the temporary stay of Russian military formations on the territory of Georgia, and assistance with the creation of a Georgian Armed Forces. . . ."

• Press agencies report that President Kravchuk has announced that a train carrying sixty warheads from ICBMs has departed for a dismantling site in Russia. Delays in shipping the warheads allegedly resulted from Russia's delinquency in making nuclear fuel deliveries. *The New York Times* publishes Kravchuk's warning that further warhead shipments are contingent on Russia's actions, particularly natural gas deliveries to Ukraine.

**10** • *Krasnaya Zvezda*'s Sergey Prokopenko reports that the newly independent states are clearly unable to afford independent armies. He writes that "national armies are becoming a national disaster for the independent states of the post-Union space," and that "one cannot expect economic progress and a corresponding easing of the crisis in these armed formations in the foreseeable future."

**11** • ITAR-TASS reports that a five-year treaty on coordinating military activities has been signed by Belarus and Russia. The treaty includes the following provisions: they will cooperate in the military sphere, consult each other in the event of an attack, maintain links in military production spheres, and jointly train military cadres. The treaty is signed by the Belarusian First Deputy Foreign Minister Grigoriy Tarazevich and the Russian ambassador to Belarus.

**13** • Before leaving for his meeting with Warren Christopher in Vladivostok, Foreign Minister Kozyrev tells reporters that he sees "no opportunity for the use of peacekeeping forces from Western countries for operations in Georgia" (RFE/RL). He is apparently responding to Clinton's statement of 7 March: "The United States would be inclined to support a UN peacekeeping operation in Georgia, an operation that would not involve U.S. military units."

**14** • The foreign ministers of Kazakhstan, Kyrgyzstan, Tajikistan, Uzbekistan, and Russia met in Dushanbe to confer on the situation in Tajikistan, where they have all committed troops to defend the Tajik-Afghan border.

• Tajik Foreign Minister Rashid Alimov reports that the Tajik government has requested the UN Security Council to recognize CIS troops on its soil as a UN peacekeeping force.

• Reuters and Interfax report that Russian Border Troops Commander Andrey Nikolaev has signed an agreement in Tbilisi which provides for Georgian troops to serve in Russian units guarding the Georgian-Turkish border. Nikolaev signs a similar agreement in Yerevan the next day which provides for Armenian troops to participate in Russian border units.

**19** • After Ukraine ratifies START I, Russian parliamentarians begin informally to debate START II. Already some have voiced the concern that the treaty leaves Russia at a disadvantage, a signal that a contentious debate is in the works. Apparently the Duma has not yet set a date for formal discussion of the treaty.

• U.S. Defense Secretary William Perry receives assurances from Kazakh President Nazarbaev that all of the SS-18s in Kazakhstan would be shipped to Russia for dismantling. RFE/RL reports, however, that the warheads were not sent with the missiles of strategic bombers already removed to Russia. Nazarbaev states that an agreement will soon be reached on compensation for the highly enriched uranium contained therein, which would facilitate transfer of the warheads.

**20** • Defense Secretary Perry signs an agreement in Kiev which provides for an additional $100 million of Nunn-Lugar aid to be given to Ukraine, in addition to $135 million already pledged, to aid in its nuclear disarmament program.

**21** • Ukrainian officials confirm that a second shipment of sixty nuclear warheads is en route to Russia even though, as Deputy Prime Minister Valeriy Shmarov notes, Ukraine has not yet received any reactor fuel in exchange. Shmarov also says that the United States has not paid Russia the $60 million it is owed under the terms of the Trilateral Agreement. The *Washington Post* reports that U.S. officials have downplayed this issue, stating that the Russians are having no problem implementing the agreement.

• The United States and Ukraine sign another agreement whereby American nuclear weapons will no longer be targeted at Ukraine. Ukraine reportedly cannot do likewise because control of its weapons remains in Russian hands.

• *Krasnaya Zvezda* reports that two missile regiments have started to withdraw from Belarus to Russia. They are part of a grouping of strategic military forces which comprises eight missile regiments in possession of road-mobile missile complexes containing a total of seventy-two missile launchers. Under this year's jointly agreed-to timetable, half of these forces will be withdrawn this year, and the rest by the end of next year.

**22** • Col.-Gen. Georgiy Kondratiev, Russian military commander in charge of peacekeeping operations, announces that Russia will increase recruits for peacekeeping operations because Russians are the only ones capable of "separating warring factions" in the CIS. Reuters reports that Kondratiev says Russia will still seek an international mandate and financing for its peacekeeping operations in the former Soviet Union.

• Kondratiev notes that Russia has over 9,000 peacekeeping troops in the region, stationed as follows: 6,000 in Tajikistan, 2,000 in the Trans-Dniester region, and 1,500 in Abkhazia and South Ossetia. Kondratiev says that Russia has spent 26 billion rubles out of its own budget in 1993 to finance the operations.

**24** • The CIS military command holds its first set of joint exercises in Tajikistan not far from the border with Afghanistan. RFE/RL reports that the exercises include mock attack aircraft, tanks, and helicopters, which is unusual because the primary threat in the area comes from Tajik rebels who do not have air and armored support. Ostankino reports that the exercises were in no way carried out to intimidate Afghanistan.

**25** • Vladislav Petrov, a spokesman for the Russian Ministry of Atomic Energy, criticizes the United States for not giving Russia a $60 million advance payment for the nuclear fuel it will ship to Ukraine, as provided for in the Trilateral Agreement.

**26** • At the 48th session of the UN General Assembly, the Ukrainian representative to the United Nations, Vladimir Khandogiy, resolutely rejects the notion of the CIS becoming a subject of international law and a regional organization which could conduct peacekeeping operations in the region. His words take Russian representative Yuriy Vorontsov, who is actively campaigning to obtain such status for the CIS, by surprise. Vorontsov attributes Ukraine's stance to "its being insufficiently informed on the character of the

Commonwealth and the agreements between its neighbors." He says that the treaty, which is the legal basis of the CIS's peacekeeping activity, was signed by CIS members, including Ukraine, on 20 March 1992, in Kiev.

• *Izvestiya* notes that the CIS obtained observer status in the UN by a resolution of the General Assembly, without a vote.

## April 1994

### *Economic*

**1** • Ukrainian Deputy Prime Minister Valentin Landyk tells Interfax that Ukraine has paid Gazprom $100 million, and that a joint-stock company will be set up through which Gazprom will receive its promised shares in the Ukrainian gas industry.

• Talks are scheduled to resume on 10 April.

**8** • Western and Russian intelligence sources say that Turkmenistan has cut off its natural gas supply to Azerbaijan, which makes up about half of Azerbaijan's supply, due to delinquent payments. Baku TV reports that Azerbaijan is unable to keep its end of a deal under which it is to pay for the gas with oil industry equipment and other products.

**12** • Russian Prime Minister Chernomyrdin and Belarusian Prime Minister Kebich finally signed their agreement on monetary union. The agreement outlines two stages in forging the union: the first will begin on 1 May with the lifting of trade and customs barriers, at which time Russia will also be permitted to use strategic arms installations in Belarus; the second stage, which has yet to be approved by the Russian and Belarusian parliaments, envisions setting the exchange rate of the Belarusian ruble to the Russian ruble at a 1:1 rate, after which the Russian central bank will be the sole issuer of rubles, and will conduct the monetary policy, in both countries. The chairman of the National Bank of Belarus, Stanislav Bahdankevich, who has been opposed to this last condition for some time, declined to sign the agreement.

• Ukraine and Russia reportedly reach an agreement which will guarantee a Russian supply of gas through the summer. Ukraine has paid Russia $87 million, not quite the $100 million they agreed to pay by 10 April. They have now agreed to pay another $600 million in cash and shares in Ukhrazprom by June. Ukraine's total debt has been estimated at $900 million. The next round of meetings is scheduled for 10 May in Kiev.

• Moldovan Prime Minister Sangheli categorically rejects Moldovan entry into the ruble zone, calling it "a bluff." He says, "today it is unrealistic to talk about the unification of monetary systems. On the contrary, it is necessary to accelerate the process of stabilization of national currencies in all former Soviet republics."

**13** • Tajikistan's Prime Minister Abduzhalil Samadov announced that Tajikistan intends to join the CIS Economic Union and the ruble zone, so far composed only of Belarus and Russia. After meeting with Chernomyrdin, they announce that they have reached a preliminary agreement on the unification of Tajik and Russian monetary systems, Interfax reports. They also sign an agreement providing for a Russian loan of 80 billion rubles to Tajikistan.

• Belarusian National Bank chairman Bahdankevich has called the monetary union between Belarus and Russia unconstitutional. He says that because the union gives Russia the sole right to issue rubles in Belarus, it gives a foreign power the right to impose its rule on Belarus. Some Russians still fear the inflationary effect on Russia that could result from the 1:1 exchange rate of the weaker Belarusian ruble to the Russian ruble. The Belarusian ruble currently stands at 18,600 to one Russian ruble. Interfax reports, however, that since individuals could exchange only up to 200,000 Belarusian rubles, the inflationary effect will be limited. To date, neither parliament has ratified the agreement, and all of the details have not been worked out.

**19** • At an EBRD meeting in St. Petersburg, Russian and Belarusian representatives disagree over the terms of the monetary union signed a week earlier. Specifically, Belarus seems to be rethinking the issue of subordinating its monetary policy and national bank to those of Russia.

**20** • Turkmen officials agree with representatives of Georgia on a settlement of the latter's gas debt, which is estimated at $200 million. Georgia will pay Turkmenistan in the form of consumer goods. Turkmenistan had threatened to cut off supplies on 1 May if the arrears are not paid.

**21** • CIS countries owe Russia "hundreds of billions of rubles for telephone and postal services," an ITAR-TASS correspondent reported. The problem stems from the fact that while Russia pays other CIS countries for such services, payments by the latter, when they have the money, are held up at the border while their national currencies are changed into rubles.

• Moldovan Prime Minister Andrey Sangheli says that the republic will pay off its gas debt to Gazprom by the end of May. ITAR-TASS reports that last week Gazprom cut off gas supplies to Moldova by a third and threatened to cut them off completely if payment was not made. The report says that Moldova has already begun to make payments but the 108-million-ruble outstanding debt is largely owed by consumers in the Trans-Dniester region, which is suffering severe economic difficulties.

**26** • Ukrainian President Kravchuk says that although there are no real differences between "full" and "associate" membership in the CIS Economic Union, Ukraine must join as associate member because the parliament made some "comments and stipulations" when it signed the Treaty on the Formation of the CIS which he cannot ignore. He says that Ukraine will nonetheless play a full and active role in the union.

**28** • Interfax reports that Armenian Prime Minister Grant Bagratyan has reiterated an earlier proposal to create a CIS currency to be used beside CIS members' national currencies in order to facilitate interstate financial transactions. Bagratyan argues that such a currency would help to stabilize local currencies and slow the growth of debts owed to Russia by other CIS states. He says the idea is based on the European Union's ECU.

**29** • The presidents of Kazakhstan, Kyrgyzstan, and Uzbekistan hold a closed-door summit in Cholpon-Ata near Lake Issyk-Kul, during which Kyrgyzstan became a formal member of the Economic Union which was set up by Kazakhstan and Uzbekistan in January. The communiqué issued after the meeting stresses the need for increased cooperation in the spheres of politics, culture, and especially economics. They offer the new Central Asian Economic Union as a basis for a CIS Economic Union, and invite all CIS states to join.

### *Political*

**8** • ITAR-TASS reports that the Moldovan parliament has ratified Moldova's membership in the CIS, with reservations, as well as accession to the Economic Union. The reservations are that Moldova will not participate in any military pacts or the ruble zone. The state secretary of "Trans-Dniester" says that this would enable talks on the resolution of the conflict between Tiraspol and Chisinau to resume at a new level.

• Moldovan Prime Minister Andrey Sangheli reiterates that his country has not joined any CIS security or military-political cooperation agreements, and will only participate in economic agreements.

**9** • Moldovan President Snegur and Trans-Dniester leader Igor Smirnov hold renewed talks on determining the legal status of Trans-Dniester. The talks are moderated by the CSCE mission and Russia, and the basis of a proposed agreement is one prepared by the CSCE that clearly delineates the division of powers between Chisinau and Tiraspol, and gives considerable autonomy to the latter.

**12** • The pilot issue of the first official CIS weekly publication *SNG: Obshchiy Rynok* (CIS: Common Market) appears. It is financed by contributions from all the CIS member states, and contains articles from all of them, including policy statements.

**13** • A Russian journalist who has been reporting on the situation in the Taldy-Kurgan oblast in northern Kazakhstan is arrested by Kazakhstani authorities. They had charged him with inciting interethnic discord in July 1993, but he ignored the charge and continued to act as spokesperson for Russians in Kazakhstan. Siberian Cossack leader Viktor Ochkasov predicts that the arrest will further heighten tensions in the region.

**14** • Reuters reports that the Foreign Ministry of Romania has criticized Moldova for joining the CIS: ". . . the natural place of the republic of Moldova as an independent and sovereign state is in the big family of European nations, and by no means in Euro-Asian structures. . . . The foreign policy acts by the parliament of Moldova seem to show a tendency toward the latter."

• President Saparmurad Niyazov of Turkmenistan says that his country is against tough structures within the Commonwealth. He says the countries of the CIS "have efficient bilateral structures, but the strengthening of independence should not be interfered with by tough new structures."

• The Council of CIS Foreign Ministers meets in Moscow to discuss signing a treaty on members' sovereignty, territorial integrity, and the inviolability of their borders. Russian news agencies says that Armenia has written and proposed an alternative draft.

• Russian Foreign Minister Kozyrev says that the declaration is "a useful document giving lie to speculations that the CIS is some kind of neo-imperialist club."

• The foreign ministers fail to reach agreement on the rights of national minorities. Significant controversy arises while discussing a document to that effect, some CIS members voicing fear that the document would give Russia carte blanche to use force for the "protection" of Russian minorities in their states.

• The foreign ministers agreed on what Kozyrev describe as a UN-type flag, which will be proposed to the heads of state when they meet on 15 April. The flag is a yellow circle on a blue background.

**15** • The heads of the twelve CIS states meet in Moscow to discuss a wide array of issues. Significantly absent from the summit is Kazakh President Nazarbaev, who claims he has the flu. He is also apparently upset that his idea of creating a Eurasian Union to succeed the CIS was rejected by Moscow. The Russian media suggests that Nazarbaev had a "diplomatic disease." Also absent is Sergey Shakhray, Russian deputy prime minister, who has taken to Nazarbaev's idea and even drafted a confederacy agreement.

• The Moldovan delegation for the first time includes a representative from the "Trans-Dniester Republic," Deputy Prime Minister Viktor Sinev, which resulted from the recent talks between Trans-Dniester leader Smirnov and Moldovan President Snegur.

• The leaders sign a document on territorial integrity, sovereignty, and inviolability of borders proposed by the foreign ministers, which Ukrainian Deputy Foreign Minister Tarasyuk says will prevent any of the economic agreements from being threatening to Ukrainian independence.

• The meetings are somewhat overshadowed by renewed tensions between Russia and Ukraine over the Black Sea Fleet.

**19** • An interview with Georgian President Shevardnadze on the results of the CIS summit is published in *Izvestiya*. He says: "Our meeting in Moscow has undoubtedly revealed an increasingly deepening trend toward the formation of new ties between the independent states within the framework of the Commonwealth, accompanied by strict consideration of and respect for their national and state interests."

**26** • ITAR-TASS reports that the Moldovan parliament has ratified the CIS Charter, which President Snegur had signed at the Heads of State meeting in Moscow on 15 April.

• Kazakhstan's ambassador to Russia, Tair Mansurov, explains his president's idea of a Eurasian Union to replace the CIS, proposed weeks ago, but jettisoned by Russia.

• It is hard to call the CIS a Commonwealth. In the three years of its existence, over 400 joint documents have been adopted, but more than half of them have not been ratified by the national parliaments, and those that have been ratified are not operating. In addition, some members of the CIS are in a state of open war and many are waging economic battles against one another. In these conditions it is hard to talk about the mutual aid and support characteristic of a truly integrated association.

**27** • Moscow headquarters of ITAR-TASS hold a meeting of CIS states' news agencies, which is attended by representatives from Armenia, Georgia, Kazakhstan, Kyrgyzstan, Moldova, Russia, Tajikistan, Turkmenistan, Ukraine, and Uzbekistan. The participants are invited to participate more actively in cooperation and to make full use of the technical potential of ITAR-TASS. ITAR-TASS director-general Vitaliy Ignatienko says, "The task of preserving and developing the unified information expanse is equally important for all republics of the former USSR in the conditions of the establishment of their political and economic independence."

**28** • Moldovan President Mircea Snegur, Trans-Dniester leader Igor Smirnov, Russian mediator Vladlen Vasev, and CSCE chief of mission in Chisinau meet near Tiraspol and sign a political document providing for future negotiations on defining the future status of "Trans-Dniester Republic's statehood under law" under Russian mediation and taking into account Russian and CSCE views, neither of which have advocated statehood for the republic. RFE/RL reports that Moldovan officials have stated that the concession was made in an effort to facilitate an accord.

**29** • President Yeltsin, in a speech to leaders of Russia's Foreign Intelligence Service, emphasizes his intention to pursue a more assertive foreign policy vis-à-vis both the "near abroad" and the West. He dismisses concerns voiced in the West about Russian neo-imperialism, stressing that Russia views the states of the former Soviet Union as areas of vital interest, and Russia will step up its efforts to integrate them with Russia. Yeltsin also says that forces in the West and the "near abroad" who charge Russia with a neo-imperialist policy are attempting to exacerbate relations between Russia and these states.

***Security***

**1** • Russian Minister of Atomic Energy Viktor Mikhailov states that 60 nuclear fuel rods had been sent to Ukraine in exchange for the 120 warheads received from Ukraine. He also complains that the United States still has not lived up to its side of the Trilateral Agreement which, at this point, requires the United States to transfer $60 million to Russia. Mikhailov says that future shipments are contingent upon U.S. transfers of money and Ukrainian shipments of more warheads.

**5** • The Russian ministries of defense and foreign affairs issue a joint statement expressing Moscow's frustration with what it considers unfair criticism of its "near-abroad" peacekeeping operations. This comes after the United Nations fails to extend support to Russian peacekeeping operations in the former Soviet space. The statement is carried in *Rossiyskie Vesti*.

**6** • ITAR-TASS reports that President Yeltsin has signed a directive endorsing a Defense Ministry proposal to establish thirty military bases on the territory of CIS states and Latvia which would serve the purpose of bolstering Russian security and serve as grounds for weapons testing.

• Officials from the foreign affairs and defense ministries attribute the misguided directive to "technical errors," suggesting that it might have simply been an outdated document that Yeltsin signed. Others contend that Yeltsin may have intended to confirm a plan outlined on 28 February by Russia's General Staff Chief stipulating that Russia would establish thirty military bases in the "near abroad" on the basis of bilateral agreements with those states. Foreign Minister Kozyrev says that he does not know where the document came from, nor with whom it was agreed. The abundance of explanations and confusion point to the disarray within Moscow's government circles, especially where foreign policy is concerned. Radio Rossii reports that Kozyrev has said it was an attempt to set the president, government, Foreign Ministry, and Defense Ministry at loggerheads.

• Boris Miroshnikov, chief of counterintelligence operations at the Federal Counterintelligence Service, reportedly told ITAR-TASS that the states of the FSU have developed their own intelligence centers and have started active intelligence and subversive operations against the Russian Federation. Miroshnikov says that they are exacerbating instability in the hot spots of the "near abroad" and using Islam and other forms of aggressive nationalism to do this.

**10** • Interfax reports that spokesmen from Armenia and Georgia say their countries' leaders have no objection to the continued presence of Russian military bases on their territory under the guidelines of existing agreements. An aide to Azeri President Aliev says that the issue no longer concerns Azerbaijan because there are no more Russian troops stationed on its territory. However, the aide also says that negotiations are continuing about the leasing to Russia of Gaballa radar station in northern Azerbaijan, which, he says, is a military installation and not an army base.

• Moldovan Deputy Defense Minister Brig.-Gen. Tudor Dabija tells Interfax that Russia will never be allowed to have army bases in Moldova, and that the only thing the two countries could negotiate is the full withdrawal of the 14th Army from the Dniester region.

• Kazakh Deputy Foreign Minister Bulat Nurgaliev also speaks with Interfax on this issue, saying that the Kazakhstani-Russian treaty on friendship and cooperation, signed on 23 May 1992, provides for the "joint use, in mutual interests and in mutual agreement, of army bases, shooting ranges, and other defensive facilities on their territories."

**14** • Representatives from the defense ministries of eleven CIS states (Moldova is not present) meet in Moscow, where they adopt documents establishing the "Common Security Council." Delegations whose states are signatories to the Collective Security Treaty (CST) also adopt a declaration that provides for increased defense cooperation and affirms their participation in NATO's Partnership for Peace program (PFP). The declaration allegedly describes PFP as a real alternative to the enlargement of NATO, and describes the CST as a permanent part of what will eventually be an integrated European security system, and a similar security system in Asia.

**15** • The CIS heads of state adopt a resolution whereby Russia will send its peacekeepers to the Georgian-Abkhaz conflict area, where there are already CSCE and UN observers, according to ITAR-TASS.

**19** • ITAR-TASS reports that Moldovan and Russian representatives sign an agreement coordinating their efforts to combat terrorism and organized crime and drug trafficking through their foreign and national intelligence services. Russia has already signed such agreements with Armenia, Belarus, Georgia, Kazakhstan, Ukraine, and Uzbekistan.

**20** • At a signing ceremony on military and technological cooperation with Turkey, Russian Defense Minister Pavel Grachev says that Russia will set up military bases in CIS countries with their consent. He specifies that Russia will have three bases in Georgia and two in Armenia as well as a missile warning facility in Azerbaijan at Gaballa.

**28** • A UNIAN correspondent reports that Russian Defense Minister Pavel Grachev has issued instructions for all officers of Ukrainian descent to be removed from the Russian army. The order allegedly follows the failed talks on the Black Sea Fleet on 22 April, when Grachev is reported to have stormed out of the negotiations and flown back to Moscow without informing the Ukrainians or even his own entourage.

**29** • *Segodnya* reports that Yeltsin and Nazarbaev have signed an agreement on the disposition of the nuclear arms in Kazakhstan under which Russia would assume jurisdiction over the strategic forces, and all warheads would be removed within fourteen months. Kazakhstan's silos and missiles are to be destroyed within three years. No more information is made available on the dismantling process or the issue of compensation for highly enriched uranium contained within the warheads.

• *Rossiyskaya Gazeta* publishes the text of a directive signed by Yeltsin on 10 April calling for the creation of Russian military bases in countries of the former Soviet Union on the basis of bilateral agreements. Unnamed Russian Foreign Ministry personnel suggest that this is part of a plan to create a "zone of stability" on Russia's borders and a way of protecting Russia's interests in its "near abroad." No Baltic states are mentioned in this directive, and Russian officials claim that an earlier directive's mentioning of Latvia as a possible host for a Russian military installation was a mistake, and maybe even a deliberate provocation.

**30** • Ukrainian Strategic Forces commander Col.-Gen. Igor Sergeev says that Ukraine is fulfilling its responsibility under the Tripartite Agreement and transferring warheads to Russia, in exchange for which Ukraine is receiving deliveries of fuel for its nuclear power stations. Sergeev reported that Ukraine has delivered 180 warheads to Russia so far (Kiev Radio Ukraine).

## May 1994

### *Economic*

**5** • In an interview with *Komsomolskaya Pravda*, Grigoriy Yavlinskiy, leader of the moderate-reform bloc Yabloko, says that he regards the collapse of interrepublican economic ties as the main culprit in the overall economic decline of the former Soviet Union. He indicates that the wholesale replacement of the political elite is necessary to begin the process of reintegration.

**16** • Belarusian Prime Minister Vyacheslav Kebich announces that the Russian ruble may replace the Belarusian currency in two months. Kebich reiterates the need for economic union with Russia to save Belarus's flagging economy. He is confident that the economic situation will improve significantly shortly after the implementation of an Russian-Belarusian economic union (ITAR-TASS).

**17** • During a state visit to Japan, Uzbek president Islam Karimov calls for genuine integration among the republics of the former Soviet Union. He called the CIS "a screen" behind which certain politicians "carry on their affairs." Karimov says that the agreement signed by Kazakhstan, Kyrgyzstan, and Uzbekistan on the creation of a single economic space is open to Russia and the other republics of the ex-Soviet Union.

• Karimov issues a warning to Moscow that "Russia is losing Uzbekistan's market... [and] does not realize that [it] could be squeezed out of Uzbekistan as a result of intensifying international competition for the emergent market in our republic." Karimov stresses that his country wants "to retain former links with Russia, but on a new basis." He points to the fact that not one Russian-Uzbek joint-venture exists, as proof of a lack of cooperation.

• Interfax reports that Russia and Ukraine have reached agreement on a payment schedule for Kiev's 1.4 trillion ruble energy debt. The negotiations, characterized as "businesslike and friendly," end with Ukraine agreeing to pay Russia $125 million by the end of May.

**23** • In a meeting with Britain's energy minister, Timothy Eggar, Azeri President Aliev says that his country "intends to look into the problem of expanding cooperation with foreign companies in the development of oil" reserves. Aliev, who recently completed a visit to Great Britain, praises the level of cooperation between Baku and London, and expresses optimism that the cooperation will be deepened. Eggar, visiting Baku in connection with an international exhibition devoted to oil and gas extraction in the Caspian Sea, said that his country fully realizes the potential benefits of cooperation with Azerbaijan.

**24** • Interfax reports that as of 1 June the Belarusian ruble will be the only legal tender in Belarus until the monetary union, signed 12 April, goes into effect. Realization of the monetary union (which the Belarusian National Bank chairman described as "stillborn") hangs in the balance as both sides continue to flesh out the exact terms of the agreement. Nevertheless, Belarusian radio reported that Russia will extend to Belarus a R150 billion credit for the first half of 1994.

**31** • The *Financial Times* reports that the Russian government has informed the British embassy in Moscow that Russia is demanding the right to veto Caspian Sea oil projects, jeopardizing foreign activities in Kazakhstan and Azerbaijan.

***Political***

**4** • Kyrgyz President Askar Akaev meets with Russian Federation Council Chairman Vladimir Shumeyko to speak about strengthening the Russian population in Kyrgyzstan in order to stem the tide of Russian emigration. They discuss Russian aid to Kyrgyzstan, especially to its regions, which have "suffered from many disasters," according to Shumeyko. Akaev said that Russia is Kyrgyzstan's major partner and ally, and needs Russia's aid.

**5** • *Rossiyskaya Gazeta* reports that parliamentarians of Russia and Kyrgyzstan have met in Bishkek and signed an agreement which calls for the promotion of all forms of relations between the two countries, essentially supporting what Akaev and Yeltsin called for the day before.

• *Nezavisimaya Gazeta* publishes an interview with the chairman of the Duma's Committee for CIS Affairs, Konstantin Zatulin. Zatulin, who professes to be "an admirer of empire," said that "policy toward CIS is Russia's internal policy" and called for "special treaties that would codify the special relationship" between the "near-abroad" states and Russia. He added that "the special status of regions inhabited by ethnic minorities [in the "near abroad"] must be backed by Moscow's guarantees." Specifically, he mentioned the "Dniester Republic" in Moldova, eastern Ukraine, Crimea, and northern Kazakhstan. Zatulin advocated the signing of bilateral treaties along the lines of the recently signed Russian-Georgian pact.

**18** • Russian Foreign Ministry spokesman Grigoriy Karasin announces that the Russian Ministry of Foreign Affairs has produced a draft set of guidelines on supporting Russians in the other Soviet successor states. The stated aim of the new guidelines is to facilitate the integration of Russians into the life of the host state while preserving the Russians' cultural identity. The guidelines call for greater cooperation among government ministries, parliamentary committees, and public organizations concerned with ethnic Russians in the former USSR.

• In a speech to the Russian Duma, Aleksandr Lukashenka, one of the front runners in the presidential race in Belarus, calls for the reunification of Russia, Ukraine, and Belarus into a single state, ITAR-TASS reported. Lukashenka also calls the creation of the CIS "a mistake." Chairman of the Duma's Committee for CIS Affairs Zatulin attempts to distance the Duma from Lukashenka's views, and asks that the record indicate that Lukashenka's speech "in no way reflects the State Duma's position."

**19** • Interfax reports the creation of a new coordinating body within Russia's Ministry for Cooperation with the CIS (created in January 1994). The new organ will be tasked with implementing Russia's economic strategy vis-à-vis the other CIS states. The body will also serve in an advisory capacity to the Russian government. Its members will include, among others, representatives of the Central Bank of Russia and the directors of large Russian state-owned and private companies.

**23** • During a three-day visit to Russia, Britain's foreign secretary, Douglas Hurd, says that Russia has the right to carry out peacekeeping operations in the former USSR, granted that these operations are requested by the governments involved in the conflicts and that the missions are conducted in the spirit of United Nations and CSCE documents.

**24** • In remarks to his NATO counterparts, Russian Defense Minister Pavel Grachev attempts to reassure his audience that Russia's mediation and peacekeeping efforts in the "near abroad" are not fueled by imperialist aspirations. He stressed the need to find political, not military, solutions to the problems at hand in the former Soviet republics. Reuters reports that NATO continues to object to Moscow's request that the CFE Treaty be amended to allow Russia a larger force in the Caucasus than originally permitted.

**27** • The Council for Foreign and Defense Policy (SVOP), an ex officio foreign policy advisory board comprised of leading academics, publishes its second set of theses on Russian foreign policy in *Nezavisimaya Gazeta*. The report calls for the creation of a system of political, military, and economic cooperation, in the form of a new union on the territory of the FSU, that would support the interests of Russia in the "near abroad."

**31** • In a meeting with CIS Executive Secretary Ivan Korotchenya, Kazakhstan's President Nursultan Nazarbaev reportedly agrees to send troops to Abkhazia as part of a CIS peacekeeping force. Korotchenya, who completed a tour of the Central Asian states (except Turkmenistan), indicated that Nazarbaev's concept of a Eurasian Union would be the main item on the agenda at the CIS summit scheduled for September. Korotchenya also meets with UN Deputy Secretary-General Goulding in Dushanbe. The talks focus on coordinating United Nations and CIS efforts to end the civil war in Tajikistan.

### *Security*

**4** • Kazakh Defense Minister Sagadat Nurmagambetov says that military cooperation with Russia is the cornerstone of Kazakhstani strategic policy, and he sees increasingly close cooperation between the two countries' military establishments. Interfax reports that these statements are made in conjunction with the anniversary of the creation of Kazakhstan's own armed forces. The minister also says that the strategic forces in Kazakhstan are to be removed by the year 2000.

• Vladimir Shumeyko, chairman of the CIS Interparliamentary Assembly, urges the CIS to create a contingent of peacekeeping forces from CIS countries, a Postfaktum correspondent reports. This contingent, he said, "should solve the internal problems of the CIS."

**5** • RFE/RL reports that the secretary of the CIS Council of Defense Ministers, Lt.-Gen. Leonid Ivashov, says that the CIS military command plans to propose that cooperation between the CIS and NATO as two defense alliances be part of the Partnership for Peace plan (PFP). Ivashov and other CIS military leaders just returned from meetings with NATO leaders in Brussels, where they discussed Bosnia and PFP. ITAR-TASS reports that Ivashov expressed concern that the CIS countries are not adequately protected by NATO's PFP, and it does not solve the security problems in the CIS, especially the issue of peacekeeping. These remarks of Ivashov can be regarded as another in a series of attempts by Russian officials to obtain for the CIS recognition on an international level as a legitimate regional organization.

**14** • According to President Leonid Kravchuk, 180 nuclear warheads have been removed from Ukraine to Russia, fulfilling Ukraine's obligations under the Trilateral Agreement signed in January. He says that all 49 SS-24 strategic missiles in Ukraine have been deactivated and that none of the 176 missiles remaining in Ukraine is targeted at the United States.

**16** • According to Defense Minister Radetskiy, all nuclear warheads will be withdrawn from Ukraine in three to four years. Since the Trilateral Agreement in January, Ukraine has transferred approximately 60 warheads per month to Russia. Radetskiy indicated that this pace would be maintained. Ukraine is believed to possess about 1,800 nuclear warheads.

## June 1994

### *Economic*

**1** • Belarusians will go to the polls on 23 June to elect that country's first president. Seven candidates have been registered, including Vasil Novikau, the leader of the Party of Communists of Belarus (PCB). Novikau's platform, according to Belarusian radio, includes social protection and consumer subsidies. He reportedly favors a restoration of the former USSR. The PCB says it will cooperate with two other leftist candidates, parliamentary speaker Kebich and Aleksandr Lukashenka (who last week, in an address before the Russian Duma, called for the creation of a union of Slavic states, if either becomes president).

**6** • Russia's deputy prime minister, Aleksandr Shokhin, indicates that the outcome of Belarus's presidential election will significantly affect the pace of monetary union between the two countries. According to Shokhin, the new president, armed with a fresh mandate, will have to initiate the constitutional changes necessary for the agreement to be implemented. Presumably, a victory by one of the leftist candidates (Kebich, Lukashenka, Novikau) would accelerate the pace of implementation, whereas an electoral victory by one of the more nationalistic candidates, Shushkevich or Paznyak, may necessitate further negotiations on the provisions of the agreement.

**7** • At a press briefing in Moscow, Russian Ministry of Foreign Affairs spokesman Grigoriy Karasin states that Russia has nothing against cooperation between Caspian littoral states (Azerbaijan, Kazakhstan, Turkmenistan, Russia, and Iran) and interested third parties to undertake exploratory work and resource development on Caspian territory. Karasin later qualifies his statement, saying that there are serious limitations on unilateral action because of the vulnerability of the Caspian's ecosystem and due to the fact that the territorial demarcation of the Caspian seabed has not been resolved. Thus, Russia claims a de facto veto over any deal made between a Caspian state and a third party. Not surprisingly, the Russian foreign minister sent a letter of protest to the British embassy in Moscow over the signing of a British-Azeri memorandum on cooperation in prospecting for oil and gas in the Caspian Sea, according to RFE/RL and the *Financial Times*.

• Interfax reports that the World Bank will extend a $1.5 billion loan to Kyrgyzstan for the implementation of a three-year program for economic restructuring. In 1994–95 Kyrgyzstan will also receive $550 million to cover a balance-of-payments deficit and social programs.

**8** • The Tajik economics and planning minister, Rustam Mirzoem, stated that a "full merger of the Russian and Tajik monetary systems can be achieved no sooner than by the end of this year," ITAR-TASS reports. Mirzoem stresses, however, that the causes for the delay stem from technical and organizational, and not political, problems. He says that Tajikistan faces severe cash shortages and that the remaining tranche of a 120-billion-ruble credit from Russia will not even cover pensions and social outlays in the war-torn country.

**9** • According to *Izvestiya*, Ukraine has agreed to construct 145,000 square meters of housing in the gas-producing region of northern Russia as partial payment for Kiev's reported $800 million gas debt. Increasingly, Ukraine is relying on non-monetary payments, including ownership transfer of domestic energy infrastructure, to reduce its expanding energy debt to Russia.

**22** • Interfax, quoting Ukraine's deputy finance minister, Boris Sobolev, reports that Ukraine's debt to the countries of the former Soviet Union exceeds $4.3 billion. These debts, Sobolev says, are not being regularly repaid due to continuing disputes over the division of the former USSR's assets and liabilities. Overall, Sobolev estimates Ukraine's debt to be about $6 billion, but stresses that Kiev promptly repays its debts to Western creditors.

**24** • The *Financial Times* reports that Russia has signed a $10 billion oil and gas deal with a Western consortium to develop energy reserves off Russia's Sakhalin Island.

**28** • According to Interfax, Uzbekistan's President Islam Karimov issues a decree on 27 June making Uzbekistan's new currency, the som, the only legal tender as of 1 July. The som will replace the temporary coupons issued in November 1993 when Uzbekistan withdrew from the ruble zone citing Russia's excessive demands for membership. Uzbek citizens can begin to exchange their coupons for soms on 28 June.

• According to Kazakhstan's new minister of the oil industry, Ravil Shardabaev, Russia began cutting off almost all of Kazakhstan's oil exports in May, causing a near-complete shutdown of Kazakhstan's production and refining facilities, the *Financial Times* reports. Currently, the only export pipelines available for use by Kazakhstan traverse Russia, and many observers believe that Russia is using this leverage to gain shares in the abundant gas and oil fields of Kazakhstan. Russian officials refute this interpretation, complaining rather that Russia loses about $300 million per year by transporting other countries' products through its pipelines.

**29** • World Bank officials announce that the bank will extend an additional $500 million loan to Russia to upgrade production in the oil industry, where production continues to fall (12 percent in 1993) due to outdated equipment and neglected facilities. The new funds are reportedly earmarked for pipeline improvement and environmental protection measures. Over the last two years Russia has received over $1 billion from the World Bank to upgrade its oil sector.

### *Political*

**8** • *Nezavisimaya Gazeta* publishes the full text of Kazakh President Nursultan Nazarbaev's project for the formation of a Eurasian Union. Nazarbaev initially proposed the idea for a new union last spring and was met with ambivalence from other CIS leaders. In recent weeks, however, the concept has resurfaced amid talk of the need for greater integration of the states of the former Soviet Union. The plan calls for the drafting of a new treaty forming the legal and organizational prerequisites "for deepening integration with the purpose of forming an economic, currency, and political union." Associate membership in the Eurasian Union would not be permitted under Nazarbaev's plan. A cessation of hostilities between CIS member states would be a prerequisite for membership. The preamble to the plan states that the existing structure of CIS bodies has not exploited fully the potential for integration. Nazarbaev contends that the Eurasian Union would act only as an additional integration structure, and not an alternative to the CIS.

• In a 11 June interview with *Nezavisimaya Gazeta*, Nazarbaev claims that "it is generally believed that common sovereignty is superior to each separate sovereignty of a particular state, and that it is more useful." Nazarbaev also said that "no one can be content with the CIS that we have today. Some 400 documents are not being implemented. I am not saying that the CIS is not serving its purposes and that it must be disbanded. It is good to have something rather than nothing. But the CIS is moving in one direction, and life in another."

**9** • Federation Council speaker Vladimir Shumeyko believes that the processes of integration in the former Soviet Union will lead to the creation of a federative, if not confederative, state. He says that the restoration of the Soviet Union is impossible because the USSR rested on the ideological pillar of the CPSU. Shumeyko's comments came at a session of the CIS Interparliamentary Assembly (IPA) in St. Petersburg.

**14** • Kyrgyz President Askar Akaev laments the continuing emigration of ethnic Russians from his republic, calling it a potential "ethnic Chernobyl" for Kyrgyzstan and Russia. According to the president, 170,000 of the 918,000 Russians living in the republic in 1990 have emigrated; Russians now account for 17 percent of the population, compared with 21 percent in 1990. Akaev blames general economic collapse for the exodus, but also specifically cites the declaration of Kyrgyz as the official language of the republic and the preferential treatment accorded to ethnic Kyrgyz. Akaev calls for the adoption of Russian as a parallel state language and for the granting of dual citizenship to ethnic Russians. Both of these measures would require amending the Kyrgyz constitution, an action the Kyrgyz parliament would be reluctant to do.

• RFE/RL reports that Russian representatives at a CSCE conference in Prague have rejected conditions proposed by the CSCE for international recognition of Russian peacekeeping operations in the CIS. The draft proposal stipulates that all sides in a conflict must agree to the introduction of peacekeepers and specifies that peacekeeping troops should not be deployed indefinitely. In spite of the disagreement, CSCE observers report favorably on the conduct of Russian peacekeepers in South Ossetia.

**15** • Speaking at a meeting of high-level CIS functionaries, CIS executive secretary Ivan Korotchenya says that "Regarding the reasons why many agreements within the CIS framework do not work or work rather feebly, they are of an objective nature. First, every country lives according to its own laws. Second, many agreements have been signed only by some CIS members. Finally, there are difficulties with the ratification of adopted documents by the parliaments of the CIS countries."

• In an interview with *Rossiyskaya Gazeta*, Azeri President Aliev says that "energetic measures" are needed to enable the CIS to function properly. Aliev observes that "during the meetings of heads of state, which last a day and a half or two days, the CIS exists, but in the intervals between meetings it does not." He defends his decision to bring Azerbaijan into the CIS in September 1993, saying that "it was a necessary and important step, and Azerbaijan should remain in the CIS in the future."

• *Segodnya* reports that, according to an observer of ethnic Russian migration, as many as six million "forced migrants" may arrive in Russia in the next two years. Dmitriy Rogozin, leader of the Congress of Russian Communities organization, further points out that, according to a government directive of 1 May 1994, each refugee should be issued 1 million rubles from state budget funds for resettlement (6 trillion rubles or $30 billion).

**16** • Interfax reports that at the end of June Russian President Yeltsin intends to announce a comprehensive program of aid to ethnic Russians living abroad in the former Soviet republics. The program reaffirms Russia's adherence to the notion of dual citizenship and its desire that Russian be made a parallel official language in the ex-Soviet space.

• On the same day Interfax reports on Andrey Kozyrev's comments to a session of the Council on Foreign Policy in Moscow. Kozyrev says that Russia will unequivocally support the desire of any former Soviet republic to set up a confederation, or even a union. He praises Russia's "strength to avoid confrontation and bloodshed." He says that it is "necessary to advance toward setting up a real commonwealth rather than a neutral community."

• Kozyrev deals a subtle blow to Nazarbaev's Eurasian Union concept when he says that although Russia is as "ready for integration as our partners are. . . . we would like to avoid a situation in which the proposal and discussion of such promising ideas [for closer integration] hinders us in resolving concrete pressing problems, or distracts our attention away from them," ITAR-TASS reports. Russia, Kozyrev says, should remain the "locomotive of reform," and the stronger Russia is, economically and politically, the greater the benefits for the rest of the CIS.

**17** • According to Interfax, Uzbekistan's Foreign Minister Saydmukhtar Saydkasymov says that Nazarbaev's Eurasian Union proposal is an attempt to return to the past and restore the former USSR. He believes that the CIS has not exhausted its possibilities, but that Uzbekistan "links its future with the sovereignty" acquired upon the dissolution of the Soviet Union.

**21** • An international conference on peacekeeping in CIS states, organized by Russia's State Duma and various Russian government ministries, is addressed by "Dniester Republic" President Igor Smirnov, who tells a large Western audience that Russian peacekeeping in Moldova responds to the wishes of the conflicting parties and does not entail any political interference or pressure upon the parties. He indicates that the conflicting parties appreciate Russia's role "far more than that of the United Nations."

• Russia's State Duma chairman Ivan Rybkin holds talks with a Belarusian parliamentary delegation headed by First Deputy Chairman Vyacheslav Kuznyatsov. Rybkin states that, in order to facilitate greater integration, the CIS member states must work harder for unified, standardized legislation. He also says that the Russian State Duma supports the idea of a parliamentary meeting of Russia, Ukraine, and Belarus to discuss the results of the 1991 agreement forming the CIS.

**28** • An editorial in *Nezavisimaya Gazeta* characterizes Lukashenka as a Belarusian Zhirinovskiy and warns that an electoral victory by Lukashenka will bolster the right-wing forces in Russia who favor a dictatorship and the restoration of the borders of the former USSR.

• According to Interfax, the CIS Committee in the Russian Duma will conduct hearings on the future development of the CIS on 5 July. Committee chairman Konstantin Zatulin expresses his hope that these hearings will dispel any notion of rescinding the Minsk accords, the death knell of the USSR.

### *Security*

**2** • Anton Buteiko, a senior foreign policy advisor to Kravchuk, says that accession to the nuclear non-proliferation treaty is "not of great urgency" and that economic reform retains a much higher priority on the legislative agenda. Buteiko downplays the importance of formal accession to NPT, pointing out that the denuclearization of Ukraine is proceeding according to schedule. RFE/RL points out, however, that many of the security assurances provided to Ukraine under the Trilateral Agreement signed in January depend on Ukraine's accession to the NPT.

**28** • In an address to graduates of the Russian military academies, Russian President Boris Yeltsin states that "nobody and nothing can free Russia from the political and moral responsibility for the fate of countries and peoples, which for centuries have moved forward together with the Russian state."

• The outcome of Ukraine's presidential election on 10 July could have an impact on how quickly, if at all, that country accedes to the NPT. According to ITAR-TASS, President Kravchuk indicates that he will call on parliament to immediately accede to the NPT "with certain provisos" if elected. Interfax and Reuters quote Leonid Kuchma as saying that NPT accession is a low priority and that $1 billion for denuclearization from the United States is required before he will lobby parliament to join the NPT.

• On the same day, Ukraine's parliament votes to reopen discussion on the question of continued transference of nuclear warheads to Russia as part of the Trilateral Agreement signed by Ukraine, Russia, and the United States, Intelnews reports. The item was proposed by the leader of the nationalist UNA/UNSO faction, Oleg Vitovich, who believes that Ukraine's national security is jeopardized by the agreement. The issue will be considered by the defense and national security committees, which will produce recommendations for the full chamber of parliament.

## July 1994

### *Economic*

**2** • Russian Prime Minister Viktor Chernomyrdin meets with his Belarusian counterpart, Vyacheslav Kebich, to discuss implementation of the agreement on monetary union between Russia and Belarus. Chernomyrdin indicates that several provisions of the agreement need to be amended before the Russian State Duma will approve the accord. RFE/RL speculates that Russia wants the National Bank of Belarus to be subordinated to the Russian Central Bank before the agreement is adopted.

• Reuters reports that the two parties agree to iron out all the remaining details of the agreement within a month, but little progress is observed on the crucial issues of the future of the Belarusian central bank and the rate of exchange of Belarusian coupons to Russian rubles. Belarusian head of state Myacheslav Hryb indicates his skepticism that progress on these and other issues could be worked out in a month. He contends that more than thirty agreements need to be finalized before monetary union becomes a reality.

**8** • According to *Izvestiya*, Gazprom, Russia's state gas concern, is acquiring shares in its Moldovan counterpart, Moldgas, as payment for a 168 billion rubles ($84 million) debt for Russian gas. However, according to RFE/RL, nearly half of the 168 billion is owed by industrial consumers in the "Dniester Republic," currently outside the control of Chisinau, where local authorities have already begun to transfer assets to Russian state companies. Similar asset transfers have been used in Ukraine, which is reported to owe Russia as much as 1.5 trillion rubles ($750 million).

*Political*

**6** • In an interview with *Novosti*, Kuchma accuses the government of lacking a realistic foreign policy. He argues that the Ukrainian Foreign Ministry relies too heavily on relations with the United States to the detriment of relations with Germany and other European partners. Kuchma also says that Ukraine's policy in the field of nuclear disarmament had been "insufficiently tough."

• Konstantin Zatulin, chairman of the State Duma's CIS Committee, says that the committee has debated four possible procedures for reintegrating the former Soviet republics: an economic union; bilateral integration with Russia (e.g., Belarus); a Slavic Union; and a Eurasian Union, as proposed by Kazakhstan's President Nazarbaev. Zatulin denies that his committee unanimously endorses Nazarbaev's plan, noting several shortcomings that need to be addressed.

• According to Russian news agencies, Kazakhstan's Supreme Soviet votes to move the country's capital to Akmola (formally Tselinograd) in the northern part of Kazakhstan. President Nursultan Nazarbaev has long advocated the move, claiming that the current capital, Almaty, is overcrowded and unable to expand. Other observers cite Almaty's proximity to China as the principle reason behind the move. Kazakh intellectuals have suggested that the move will help reinforce Kazakhstan's hold on the largely Russian northern oblasts.

**7** • In the final debate between the two candidates, Kravchuk says that he has no intention of dissolving parliament but indicates that he will not allow that body to trim his powers if reelected. As regards the economy, Kravchuk says that the failure to proceed with reforms was due to a lack of consensus on the general concepts of economic change. Kuchma uses the opportunity as an attempt to clarify his position on relations with Russia. "Economic union with Russia means equal, good neighborly cooperation. In the Soviet era we developed together," he says.

**8** • The leaders of Kazakhstan, Kyrgyzstan, and Uzbekistan agree to form a defense and economic union, UPI and Interfax report. Under the terms of the agreement, an interstate council of heads of state will be created to help coordinate the standardization of legislation. In addition, there will be councils for foreign affairs and defense, and a Central Asian Bank for Cooperation and Development (with $3 million in start-up capital from each state). The three leaders emphasize that the new union is not meant to become a replacement for the CIS, but most observers believe that it resembles a micro version of Kazakh President Nazarbaev's Eurasian Union concept. Uzbek president Karimov says, "We are acting within the framework of the CIS, but we want to call on all countries of the commonwealth to move from declaration and inaction toward concrete deeds, like our Central Asian Union."

• In a televised debate, Lukashenka accuses Kebich of contributing to the breakup of the USSR and of lethargically implementing monetary union with Russia. Lukashenka provides viewers with a basic outline of his platform: stop inflation, fight corruption, crush crime, and restore ties to the former Soviet republics. Kebich denies any complicity in destruction of the USSR, and stresses that he was working diligently to restore close ties with Russia. He promises to lower prices on staple goods and increase social protection.

**11** • International press agencies report that Aleksandr Lukashenka wins a resounding victory over Prime Minister Kebich to become Belarus's first president. Preliminary reports give Lukashenka nearly 80 percent of the vote, while Kebich receives only 14 percent. Seventy percent of the eligible voters participate in the election, thereby validating the results.

• Not unexpectedly, Prime Minister Kebich tenders his resignation immediately following announcement of the election results.

• Russian and Western press agencies report that former prime minister Leonid Kuchma was pronounced the winner of Ukraine's presidential election by the Central Election Commission. Almost 70 percent of eligible voters took part in Sunday's run-off election, far above the 50 percent necessary to validate the results. Mr. Kuchma won 52 percent of the vote compared with 45 percent for the incumbent President Kravchuk. Regionally, the election followed the pattern established in the 26 June elections, with Kuchma capturing huge majorities in the east and Crimea, and Kravchuk dominating in the more nationalistic west. The swing oblasts appear to have been in the central regions (e.g., in Poltava, Kuchma outpolled Kravchuk 59 percent to 37 percent, whereas in the first round Kravchuk edged out Kuchma 29 percent to 27 percent).

**26** • According to UNIAN, an international conference will be held in Kiev to establish a League of Parties of the Baltic–Black Sea Region. The organizers, who include representatives of Ukraine's Republican, Democratic, and Green parties, expect parties from Ukraine, Lithuania, Latvia, Estonia, Poland, Belarus, Moldova, Romania, and Bulgaria to attend. Organizers envision the formation of a "Baltic–Black Sea Bloc of free peoples" as an alternative to integration through the CIS or "Eurasian Union."

• (Postscript: The conference closed on 30 July. A statement on the creation of "the League of Parties of Countries Between the Seas" was signed by representatives from every country except Romania and Bulgaria, whose delegates claimed a lack of authority to sign. However, with many other delegates professing a similar lack of authority to adopt concrete measures, the conference failed to produce a declaration of basic principles and program of action for the League.)

• In an interview with *Kievskie Vedomosti*, Vyacheslav Igrunov, deputy chairman of the Russian State Duma Committee for CIS Affairs, says that Russia "would welcome a slow and gradual correction in Ukraine's course" with regard to integration within the CIS. Igrunov warns that civil unrest in Ukraine will invariably spill over into Russia. He also says that, "for Russia, quicker integration means quicker deterioration of its economic situation. . . . Russia will integrate with Ukraine very carefully." When asked to what degree Ukraine should trust Russia, given the failed "partnerships" of 1654 and 1922, Igrunov replies that times are different and "Russia is simply not able to restore the empire."

• Kuchma, in a speech to strategic missile forces, says that the erroneous policies of the former leadership have resulted in self-isolation of Ukraine from Russia and other CIS member states. Kuchma expresses confidence that relations with Russia will improve and numerous problems will be resolved.

**27** • In an interview with *Rossiyskie Vesti*, Azeri President Aliev says that after having attended three CIS summits, "I cannot say that the CIS is an organization which fully meets its purpose. . . . I cannot regard the Commonwealth's work as satisfactory, although I am convinced that the organization is necessary."

**28** • Vladimir Shumeyko, speaker of Russia's Federation Council and chairman of the CIS Interparliamentary Assembly, calls for closer integration among former Soviet republics and the transformation of the CIS into a "confederation . . . a single community with a single goal and a single program . . . on the territory of the CIS." The path to "political unity" is marked by economic integration, he says. He rejects the notion that such a system would "violate the sovereignty of independent states."

**29** • CIS Executive Secretary Ivan Korotchenya tells Interfax he will propose to UN Secretary General Boutros Boutros-Ghali that a fund be established to support peacekeeping operations and provide assistance for the repatriation of refugees in the CIS. Korotchenya says that the world community should fund a portion of the CIS peacekeeping operations "because all states are vitally interested in peace and security in this region."

### *Security*

**4** • The joint staff for coordinating military cooperation among the states of the CIS meets in Moscow to discuss prospects for forming a collective security system and a coalition force within the CIS. Lt.-Gen. Leonid Ivashov, the secretary of the CIS Council of Defense Ministers, says that the participants of the meeting recognize the existence of a military threat to individual CIS countries and to the CIS as a whole. Ivashov reportedly introduces a proposal outlining the creation of a military-political union of CIS countries under a supranational joint command which would be subordinated to the Collective Security Council. Ivashov admits that there are obstacles to the formation of such a military union, but reportedly suggests that closer military integration might be achieved by signatories to the CIS Collective Security Treaty, with non-signatories joining solely on matters of air defense and defense production.

**26** • Kazakhstan signs an agreement with the International Atomic Energy Agency (IAEA), which will allow IAEA inspectors to monitor Almaty's adherence to the nuclear non-proliferation treaty (NPT) which it signed in December 1993. IAEA Director Hans Blix says that the agreement hopefully decreases the likelihood that plutonium or other substances from Kazakhstan's 104 SS-18s could be smuggled. Blix also says that preliminary investigations find no unusual levels of radiation around the former Soviet nuclear test site at Semipalatinsk. Blix acknowledges that higher levels could have been recorded earlier, and that underground radiation levels could be higher as well.

**28** • According to a Belarusian Defense Ministry spokesman, Belarus is keeping to the timetable for the withdrawal of Russian troops from the country, despite requests by Moscow to maintain some Russian military installations in Belarus. Ostensibly, Russian forces will be withdrawn from Belarus when the last long-range nuclear missiles are removed in 1998. Russia, however, maintains that certain installations in Belarus are of vital strategic interest to Russia, including a radar station in Hantsevich and a communications center for Baltic Fleet warships in Vileika. The issue will be discussed by the two countries' leaders at the summit on 3 August.

**29** • At a meeting of the Council of the CIS Border Troop Commanders in St. Petersburg, representatives of eleven CIS countries sign agreements aimed at strengthening the border regime on the external borders of the former Soviet Union. Ukraine, officially not a member of the Council, does not sign the agreements, but indicates that the documents would be forwarded to Kiev for further study.

• Uzbekistan, Tajikistan, Turkmenistan, and Russia also announce the signing of a "three plus one" agreement on cooperation to protect the CIS external borders in the area of the Amu Darya.

## August 1994

### *Economic*

**2** • The economies of the CIS member states continue to decline in the first half of 1994, according to the CIS statistical committee. Belarus experiences the sharpest decline in GDP, a 31 percent drop compared to the same period in 1993. Kazakhstan's GDP declines by 27 percent, Ukraine's by 26.5 percent, and Russia's by 17 percent. Overall CIS gross domestic product declines 20 percent over the same period in 1993.

• Inflation rates for June show a great disparity within the CIS. According to the CIS statistical committee, inflation remains low in Ukraine (1 percent), Russia, Kyrgyzstan, and Moldova (3–5 percent), but is high in Uzbekistan and Turkmenistan (20 percent), Azerbaijan and Belarus (30 percent), and Kazakhstan and Armenia (40 percent).

• In the energy sector, all the oil-producing republics except Uzbekistan (where production rose 37 percent) witness a decline in production in the first half of 1994 (Russia down 14 percent, Kazakhstan 16 percent, and Turkmenistan 9 percent). The four states register increases in the production of natural gas, however. In Russia production rises 2 percent, in Kazakhstan 37 percent, in Turkmenistan 45 percent, and in Uzbekistan 9 percent.

• Official unemployment figures remain surprisingly low throughout the CIS. A total of 2.1 million people in the CIS have applied for unemployment compensation in the first half of 1994. Over half of the 2.1 million unemployed reside in Russia (1.26 million). Many economists dismiss the "official" unemployment rate, and claim that the rate of "hidden" or structural employment is much greater.

• Most CIS countries report a positive balance of trade with partners outside the former Soviet Union during the first half of 1994, raising exports 9 percent compared with the same period last year. According to the CIS Statistical Committee, the bulk of the CIS's exports consisted of raw materials (40 percent), while machinery and industrial goods accounted for only 10 percent.

• Moldovan President Snegur tells Interfax that he favors a common economic space, with transparent economic borders and without customs barriers, on the territory of the CIS. According to Snegur, the creation of such a system should be the singular focus of the CIS heads of state.

**3** • Belarusian President Aleksandr Lukashenka announces following a summit meeting with Russian President Yeltsin that the two sides have agreed to forestall monetary union until the two economies are better suited to meld together. Lukashenka cites the inflation rate disparity as one example of the incompatibility of the two economies. Russian Prime Minister Chernomyrdin offers Belarus a 150 billion ruble credit in addition to advice on implementing economic reform: "We know our mistakes and tell you what must not be done."

• Lukashenka reports that Yeltsin and Chernomyrdin will travel to Minsk in mid-autumn to sign a Treaty of Friendship and Cooperation with Belarus.

• On the functioning of the CIS, Lukashenka says: "Our countries have no alternative. But we must fill CIS declarations with more concrete contents. The time of simple letters of intent is over."

**5** • ITAR-TASS reports that the prime ministers of Kazakhstan, Kyrgyzstan, and Uzbekistan met to discuss the further economic integration of their countries. An interstate council will coordinate the work of the economic ministries, including better alignment of laws on economic matters in the three republics.

**9** • Gazprom warns Belarus and Ukraine that natural gas supplies will be cut off in September if payments are not made on their debts. According to Gazprom, Ukraine owes about 2 trillion rubles (about $1 billion), and Belarus owes 700 billion rubles (about $350 million). Gazprom also complains that Ukraine has failed to privatize pipeline facilities promised to Gazprom in exchange for debt relief.

• According to a Basapress report of 29 July, Moldova's energy debt to other CIS countries equals 1,170 lei ($286 million), about 60 percent of which is owed to Russia. However, according to the Economics Ministry in Chisinau, Moldova is being charged up to 20 percent above the world price for its natural gas by CIS suppliers. A ministry spokesman says that Moldova hopes to secure a 200-billion-ruble loan from Russia to help pay its debts.

• Gazprom claims that overall the former Soviet republics' gas debt equals 3 trillion rubles ($3 billion).

**13** • According to *Kommersant*, Russia's state oil concern, Lukoil, secured a 10 percent share in the oil field development project to be signed by a consortium of Western oil companies and the Republic of Azerbaijan.

**16** • On the heels of concluding a troop withdrawal agreement on 10 August, Russian Prime Minister Viktor Chernomyrdin meets with his Moldovan counterpart, Andrey Sangheli. The Russian side agrees to extend a 70-billion-ruble credit to Moldova to help that country pay its energy debts to Gazprom for past deliveries. Sangheli calls on the Russian parliament to ratify the Russian-Moldovan treaty signed in 1990. He indicates that Moldova's parliament has nearly completed a draft law on the special status for Moldova's Trans-Dniester region. The law would provide the separatist region with broad economic and cultural autonomy while preserving Moldova's territorial integrity.

• Despite the credit, Moldgaz, Moldova's state gas concern, will conduct negotiations later in the month regarding a transfer of assets to Gazprom. Under the preliminary agreement reached in early July, Gazprom would acquire approximately 35 percent of the shares in Moldgaz.

**17** • Gazprom's windfall grows as Ukraine agrees to pay 20–25 percent of its reported $1.5 billion debt between August and October (Ukraine's net energy debt is actually smaller because Gazprom owes Ukraine close to $400 million in transit fees). The two sides also agree on means to improve the payment mechanism for Russian gas, and also to allow Gazprom to contract directly with individual gas consumers in Ukraine.

**18** • Gazprom completes negotiations with Belarus on repayment of Belarus's gas debt, which reportedly stands at $425 million. The agreement signed by Belarusian Prime Minister Mikhail Chyhir calls for total repayment of the debt by 1 April 1995. Despite the aforementioned agreements, Gazprom reiterates its threat to cut off gas supplies to Russia's western littoral should those countries delay the repayment of their debts.

**19** • Due in part to the growing debt problem, Russia's energy exports to CIS countries drop significantly during the first seven months of 1994 compared to the same period last year. Oil exports fall 33 percent, while natural gas deliveries drop 10 percent. Not surprisingly, exports to "far abroad" countries witness an increase of 15 percent for natural gas and 12 percent for oil.

• Ukrainian officials meet with a delegation from Turkmenistan to discuss the resumption of natural gas supplies to Ukraine. Turkmenistan cut off gas shipments to Ukraine in February due to Kiev's $700 million debt. Since then Ukraine has managed to reduce that figure to a reported $279 million. The two sides reach a tentative agreement on the price of natural gas and the means of payment. The relevant documents will be submitted to the two governments for approval.

**25** • *Izvestiya* publishes details of proposals drafted by Russian officials to be submitted to the CIS Council of Heads of Government meeting in September. One proposal calls for the creation of a Payments Union, with mutually convertible national currencies. Such a system could "become the basis for establishing a Currency Union." Another proposal would turn the existing Consultative and Coordinating Committee (CCC) into the Intergovernmental Economic Committee (IEC) as a "controlling and decision-making" body, issuing "binding decisions in transnational spheres: transport, communications, oil and gas pipelines." Participant-states would be expected to "renounce some of their national prerogatives and turn them over to the supranational body." According to *Izvestiya*, the authors of the proposals expect only four or five states to join the IEC initially.

**30** • According to Interfax, Kazakhstan, apparently succumbing to months of pressure, has agreed to establish a working group with Russia to draft agreements on the development of the Karachaganak natural gas field in northwestern Kazakhstan. In 1992 an international competition for the right to develop the deposits was held and British Petrol (BP) and Italy's AGIP were chosen. However, in tactics reminiscent of similar deals in Azerbaijan, the Russian government has lobbied intensively for Russian firms, in particular Gazprom, to have a share in the project.

***Political***

**1** • Negotiations at CIS headquarters in Minsk are under way to discuss amendments to the CIS Charter, which was signed in the Belarusian capital on 22 January 1993. The CIS Executive Secretariat press service attributes the amendments to new realities that have emerged in the Commonwealth.

• Belarusian Foreign Minister Vladimir Senko says that the development of relations with Russia, Belarus's primary economic partner, is the priority of Minsk's foreign policy. Senko points out that over 70 percent of Belarusian exports go to Russia.

**3** • A commission of Russian, Belarusian, Ukrainian, and Kazakhstani intellectuals has been established to promote closer integration of the four republics. The commission's major objective, according to the chairman of the State Duma's CIS Committee and one of the commission's initiators, Konstantin Zatulin, is to convince people of the need for closer integration. Zatulin believes that apprehension about integration stems from misunderstandings that it will be based on "elder brother–younger partner" mentality.

**4** • President Yeltsin issues a statement condemning Latvia's newly adopted citizenship law, and adds an explicit warning to other former Soviet republics to take "responsibility for the provision of the civil, political, economic, social, and cultural rights of [ethnic Russians]." The statement continues: "Russia cannot be reconciled to the fact that hundreds of thousands of ethnic Russians in a neighboring country are put in what is essentially a humiliating situation. We are proceeding from the fact that they have a right to remain in those states where they were born or where they have lived for many years. . . . I, as president of the Russian Federation, state that the Russian Federation will give all the support it can to our fellow countrymen living abroad and confirm the readiness of our state to grant Russian citizenship without impediment to our fellow countrymen who so desire . . . and also to defend and offer protection to all ethnic Russians in Latvia and outside Russia generally."

• The leaders of Kazakhstan, Uzbekistan, and Kyrgyzstan meet in Almaty and sign an agreement on the creation of a Central Asian Union. According to Kyrgyz president Akaev, the agreement "is a breakthrough on the integration front politically, economically, and culturally. The tripartite agreement . . . should be a good incentive for other CIS countries [who] realize that it is far easier to solve complex problems together than in isolation. . . . This is probably a true step toward a confederation, a step toward the creation of a Eurasian alliance" proposed by Nazarbaev. All three presidents stress that membership in the Union is open to any interested country.

**10** • On an official visit to Ukraine, Kazakh President Nazarbaev urges Kiev to support his idea for a Eurasian Union. Nazarbaev calls the CIS ineffective because it lacks a common policy on anything. Nazarbaev stresses that the Eurasian Union would not represent a resurrection of the USSR. Kuchma assured Nazarbaev that his administration would thoroughly study the Eurasian Union plan in preparation for the October CIS Heads of State summit.

16 • According to UNIAN, a Ukrainian Foreign Ministry delegation visits the Central Asian states of Uzbekistan, Kazakhstan, and Kyrgyzstan to discuss official positions on cooperation in the CIS. Ukraine will reportedly focus its discussions on issues of inter-CIS trade and economic relations, and mechanisms for implementation of interstate treaties and agreements.

• According to Interfax, 80 percent of the text of the Russian-Ukrainian Partnership, Friendship, and Cooperation Treaty is now complete. The final text of the treaty is to be coordinated with relevant ministries and parliament committees. Further negotiations are scheduled for mid-September. The Ukrainian side apparently has dropped its insistence that the issue of the Black Sea Fleet be resolved before a comprehensive bilateral treaty could be signed. Rather, a Ukrainian delegation spokesman says, "this treaty provides solutions for the general political and economic problems between Ukraine and Russia and if every separate political issue is to be included in the treaty, it will never be signed." Concerning the sensitive issue of dual citizenship, Interfax quotes unnamed sources as saying that the two sides have reached a compromise whereby dual Ukrainian-Russian citizenship could be granted on a case-by-case system by an act of the Ukrainian parliament.

**18** • According to Valerian Viktorov, deputy chairman of Russia's Federation Council, CIS member states are "fed up with their sovereignty" and predicted the formation within one or two years of a "tough" CIS in which "the republics will resolve jointly all economic and social problems." He says that currently Russia is "the locomotive capable of pulling [the republics] out of crisis." Finally, Viktorov advocates the creation of "a legislative body to determine the rules of the game for the CIS countries." Significantly, Viktorov did not mention the need to "resolve jointly" military and security problems.

**24** • According to Kazakhstan's President Nursultan Nazarbaev, the proposal to create a Eurasian Union has been registered for consideration by CIS leaders at the upcoming October summit, Interfax reports. The Eurasian Union would entail a single customs and defense space, single borders, and coordination in foreign policy and trade for all its members. The official language of the Union, Nazarbaev says, would be Russian. Nazarbaev reiterates that states currently engaged in armed conflict would be permitted to join. He points out that since 1989 "ten times more people than the Afghan war—almost 150,000—have been killed" in armed conflicts on the territory of the former USSR.

**28** • According to Oleg Rumyantsev, chairman of Russia's Fund for Constitutional Reforms and a long-time democrat, the integration processes within the CIS "should result in the formation of a new single state, [which would be] neither the Soviet Union nor the Russian empire [but] a union state," which might be called the "Russian Union." The new Union, Rumyantsev tells Interfax, would have both federal and confederal features and would meld together "Russian territories and other former Soviet republics" including "most of Ukraine."

• Calling the Russian people "the older brother" of Belarusians, Belarusian President Aleksandr Lukashenka appears on national television and calls for the unification of Russia and Belarus, and indeed of all Slavic countries, Interfax reports. "The Belarusian people sincerely stand for reunification with Russia," he says. This marks the first occasion since the election that Lukashenka, who ran on a unification platform, supports the notion publicly. Previously, he cautioned that any attempt to recreate the USSR was unrealistic and that each republic's sovereignty must be respected. Lukashenka's statement, coupled with the remarks by Sergey Karaganov, a close Yeltsin aide (see below), casts doubt on the longevity of an independent Republic of Belarus.

### *Security*

**1** • Russia and Ukraine sign an agreement on the joint protection of their 1,500-km border. According to the commander of the Russian Border Guards, Andrey Nikolaev, the Russian-Ukrainian border will be divided into sections which will be guarded by either Russian or Ukrainian guards, but not both. The agreement, which has "no precedent in world practice," reportedly represents a reduction of tensions between the two countries, and is viewed as a cost-saving mechanism for both. Currently, 2,500 Ukrainian and 2,000 Russian border guards patrol the frontier.

**3** • According to Clinton administration officials, Vice-President Al Gore secured "no specific promises" from Ukraine's President Kuchma that Kiev would accede to the nuclear non-proliferation treaty. Nevertheless, Gore remains optimistic that Ukraine will agree to sign the NPT.

• One Ukrainian member of parliament does not share Gore's rosy outlook. Nikolay Porovskiy, a member of the Defense and State Security Committee, tells Interfax that Ukraine's accession to the NPT will be a lengthy process. He argues that Ukraine needs $6 billion to dismantle the nuclear weapons on its territory, and that aid from the United States has been woefully delinquent. He advocates the cessation of missile dismantlement provided for under the terms of the Trilateral Accords signed in January. Ukraine's nuclear systems, he says, need to serve as a deterrent while the country is establishing its statehood. He notes that his views are shared by numerous members of the Ukrainian parliament.

**4** • According to the director of the Russian space agency, Yuriy Koptev, at least 28,000 Russian troops are needed to maintain the Baykonur space station in Kazakhstan. Koptev argues that any fewer soldiers could lead to an "irreversible collapse" of the facilities; four years ago 70,000 troops were based at the facility.

**8** • According to Interfax, the Kazakhstani and Turkish defense ministers sign a "preliminary" agreement on bilateral military cooperation. Initially, cooperation will be developed in the realm of arms production. Kazakhstan's defense minister, Sagadat Nurmagambetov, rejects the notion that the Kazakhstani-Turkish agreement conflicts with Kazakhstan's CIS military obligations.

**19** • Ukraine's President Kuchma eschews his past ambiguity concerning the NPT and emerges firmly in favor of joining the non-nuclear regime. Kuchma tells *Moloda Ukraina* that he will submit the pact to parliament for approval in October. He expresses confidence that they will pass it. Kuchma's move may be seen as an attempt to pave the way for a successful visit to the United States in late November. In a transparent criticism of the West's policy of linking economic aid to Ukraine's accession to the NPT Kuchma says: "A single problem remains between us and the United States—the Nuclear Non-Proliferation Treaty. It must be signed. . . . Then we'll see whether the West again will put conditions on aid to Ukraine."

• In response, however, parliamentary speaker Oleksandr Moroz says Ukraine's nuclear dismantlement was a mistake, and that parliament will assiduously study the document to determine whether Ukraine's interests would be served by acceding. Moroz also shows his disdain, not only for the U.S. policy of linking aid to the NPT, but also for the slow pace of aid disbursements to Ukraine, stating that "we will get rid of the nuclear weapons, as soon as those countries which are interested in it make corresponding steps . . . measured in real money, for the reduction in our nuclear arsenal."

**22** • Labeling the present CIS collective security system a "failure," Maj.-Gen. Vasiliy Volkov, representing the staff of the Council of CIS Defense Ministers, tells Interfax that so far only Russia has been able to "control" regional tensions and conflicts within the CIS. He advocates a restructuring of the present system with one promoting greater coordination and cooperation. One step Volkov mentions is to make the Russian president chairman of the CIS Security Council. Volkov also advocates an "interested party" system by which states could participate in certain areas of special interest but withhold from areas deemed not vital to that particular state; Volkov's proposal closely resembles the current state of CIS collective security—a system he condemns as ineffectual.

**24** • According to Ukraine's deputy prime minister, Valeriy Shmarov, plans are nearly complete to transfer strategic bombers from Ukraine to Russia. Shmarov says that the long-range Tu-160 and Tu-95 bombers have no role in Ukraine's military doctrine. Russia and Ukraine are likely to reach an agreement on compensation for the forty-one planes shortly, Interfax reports.

**26** • Ukraine's President Kuchma continues a government housecleaning by removing Vitaliy Radetskiy from the post of defense minister and replacing him with Shmarov, who will become Ukraine's first civilian defense minister. Mr. Shmarov indicates that he does not plan any changes to Ukraine's military doctrine. He says that he does support Ukraine's accession to the Nuclear Non-Proliferation regime.

• In an interview with *Segodnya*, foreign policy expert and member of the Presidential Council Sergey Karaganov argues against NATO's eastward expansion and advocates a 1,500 km "buffer zone" between Western Europe and Russia. But Karaganov cautions against hastily reintegrating Ukraine into a military alliance with Russia as it would alarm the West. Karaganov, however, states that "Belarus is an absolutely crucial country for us. . . . Once we are in close alliance with Belarus, the problem of Ukraine will virtually be solved."

**31** • Russian Foreign Intelligence (SVR) spokesman Yuriy Kobaladze told ITAR-TASS that his agency is not "working against" the former Soviet republics, but is closely monitoring events there.

## September 1994

### *Economic*

**7** • The CIS Coordinating and Consultative Committee meets in Moscow and agrees in principle to Russia's proposals to create an Interstate Economic Committee (IEC) as the first supranational body of the CIS. The IEC will be endowed with executive and managerial powers and participating states will transfer certain national prerogatives. The IEC will assume jurisdiction over certain "transnational systems" (such as power grids, gas and oil pipelines, transport, and communications) and over "CIS property" or "jointly owned assets of member states" (such as industrial and financial corporations). The IEC will also be responsible for developing a payments union (which could be transformed into a monetary union), and a CIS free trade zone and eventual customs union which would theoretically guarantee a free flow of goods, capital, and labor within the CIS. Russia will provide 50 percent of the financing for the IEC and will obtain 50 percent of the votes in the IEC; the remaining 50 percent are still to be apportioned among the remaining eleven CIS member states, presuming, of course, that all 11 decide to participate. (According to Interfax, should all the CIS member states participate, the vote distribution would be as follows: Ukraine, 14 percent; Belarus, Kazakhstan, and Uzbekistan, 5 percent each; and the remaining seven states, 3 percent each.)

**8** • Ukraine's Finance Ministry releases figures on that country's debts to other CIS states. In total, Ukraine's debt to the former republics is $3.4 billion. The bulk of the debt—$2.7 billion—is owed to Russia. Ukraine also owes Turkmenistan $671 million, Moldova $28 million, and Kazakhstan $1.3 million. According to Ukraine's head of the foreign debts department in the Ministry of Finance, Aleksiy Berezhnoy, Kiev has signed bilateral debt repayment agreements with all of its creditor nations. Ukraine will pay Russia back in 3–5 years, Moldova in 3 years, Turkmenistan in 2 years, and Kazakhstan in 1 year.

**9** • The Council of CIS Heads of Government approves in principle the aforementioned creation of the Interstate Economic Committee (IEC). The agreement seems to include certain safeguards to prevent Russian dominance over the IEC. Decisions made by the IEC that involve "substantial" expenditures will require a consensus, while lesser issues require an 80 percent majority (of which Russia is entitled to 50 percent). The Presidium and Collegium of the IEC will be based in Moscow. Further discussions on the extent of the IEC's power will be held at the upcoming CIS Heads of State summit. Azerbaijan and Turkmenistan abstain from signing the document on the creation of the IEC, reserving judgment until the summit. Ukraine voices reservations regarding the notion of creating supranational bodies in the CIS.

• Interfax reports that, on written instructions from President Leonid Kuchma, the Ukrainian delegation makes it clear that Ukraine's participation in the IEC will be limited to matters of interest to Ukraine. The Kiev representatives also object to institutions which "limit our sovereignty."

• According to RFE/RL, Belarus, Armenia, and Georgia emerge as the main proponents of economic integration. Yerevan and Minsk also reportedly advocate the creation of a CIS citizenship.

• Russia's apparent disappointment with the outcome of the session manifests itself in Moscow's decision to maintain certain tariff barriers on goods imported from CIS countries. Chernomyrdin says that the barriers will remain until "the necessary preconditions" are met.

**10** • Stating that Belarus must reach Russia's economic level before a monetary union between the two is feasible, Russian Prime Minister Viktor Chernomyrdin rules out any such union for the near future, Interfax and ITAR-TASS reports. Belarusian income and inflation rates, for example, are disproportionate to corresponding rates in Russia. Although economic disparity is given as the primary reason to postpone monetary union, the real reason may lie in Belarus's unwillingness to cede its entire economic sovereignty to Russia. Newly elected President Aleksandr Lukashenka, an ardent supporter of union during his campaign, has objected to handing over control of monetary policy and emission to Russia.

**19** • The prime ministers of Uzbekistan, Kazakhstan, and Kyrgyzstan, the three countries which form the Central Asian Union, will meet in Tashkent in mid-October to discuss further economic integration. The leaders will discuss mechanisms to accelerate the implementation of agreements and means to make them compulsory for the signatories.

**20** • The government of Azerbaijan and a consortium of oil companies from the United States, Britain, Norway, Turkey, and Russia sign an agreement to develop three oil fields on Azerbaijan's Caspian Sea shelf. According to RFE/RL, the Western companies will own the majority share in the project (the Russian firm, Lukoil, will own 10 percent), but about 80 percent of the profits will go the Azerbaijan—a potential windfall, according to Azeri President Aliev, of $34 billion over the next thirty years. Negotiations on transport of the oil were postponed. Currently no secure overland pipeline exists to export the product.

• In line with previous statements (including a letter to the British embassy), the Russian Foreign Ministry announces that the Russian government will not recognize the legitimacy of the deal. Russia's objections are ostensibly rooted in concern over the ecological impact of any development project in the Caspian.

• The next day, Russia's Foreign Minister Kozyrev reiterates that the Russian MFA had "expressed . . . Russia's official view" on the matter. Kozyrev's statement is precipitated by reactions of disbelief from officials in Russia's oil industry that Russia would object to the deal.

• Russia will soon request a $500 million loan from the World Bank to rehabilitate and modernize West Siberian oil fields, Interfax reports. In 1993 Russia obtained a $610 million loan from the Bank.

**21** • The Azeri government indicates that it will ignore Russia's objections to the recently signed Caspian Sea development agreement. An unnamed Azeri source tells Interfax that the deposits covered by the agreement are located on the Azeri section of the Caspian shelf "which we have been using for decades." The source also claims that Russia's ambassador to Azerbaijan has joined Russia's Fuel and Energy Ministry and Lukoil in approving the contract.

**26** • According to ITAR-TASS, Russia's energy minister, Yuriy Shafranik, proposes the creation of a multinational coordinating committee to oversee the development of offshore oil deposits in the Caspian Sea. In addition to Russia, participants will include Turkmenistan, Kazakhstan, and Azerbaijan. Shafranik says that Turkmenistan and Kazakhstan have already given their assent to such a committee. Eventual Iranian membership in the committee is also proposed. According to the draft proposal on creating the committee, each state will be given sovereign rights over territorial waters up to ten miles from its coastline. The committee would supervise the exploitation of mineral resources on the Caspian shelf.

### *Political*

**4** • Negotiations on a Russian-Ukrainian Friendship and Cooperation Treaty stall when Russia objects to Ukraine's wording on the question of inviolability of borders. According to Intelnews, the Ukrainian version of Article 2 of the treaty reads: "the parties accept existing borders between the two countries and pledge that they do not have now, nor will have, any territorial claims against each other." Russian negotiators reject this wording, arguing that nationalists in the parliament (which must ratify any treaty) would object to such an article. Rather, Russia proposes the more ambiguous wording: "the parties respect their territorial integrity and the unchangeable nature of their borders in accordance with the Conference on Security and Cooperation in Europe's [CSCE] 'Final Act.' " The treaty will reportedly also contain a proviso, insisted upon by Russia, that forbids Ukraine from joining any military blocs or unions without Moscow's approval. Belarus concludes a similar agreement with Russia.

**6** • Reuters reports that U.S. Ambassador to the United Nations Madeleine Albright having concluded her tour of CIS conflict zones (Moldova, Georgia, and Nagorno-Karabakh), travels to Moscow where she gives her tacit approval of unilateral Russian peacekeeping in the former Soviet republics, but cautions Moscow that the international community would pay very close attention to Russia's actions: "The burden of proof is on Russia to prove its commitment to accepted international principles, to the sovereignty of the newly independent states, and to adopting a neutral stance in ethnic conflicts." Albright adds that UN observers in Abkhazia have determined that the Russians "have now become a neutral force."

**7** • Russian media reports the creation of the Assembly of Russian Compatriots in the "near abroad." The organization, whose founders include the ministries of foreign affairs and finance, representatives of the presidential administration, centrist State Duma deputies, and prelates of the Russian Orthodox Church, targets "all those who regard Russian as their *native* language" (emphasis added). Russia has often claimed its rights to protect ethnic Russians in the so-called "near abroad." However, the use of "Russian-speakers" as a criteria greatly expands the purview of Russian protection. According to reports, the Assembly will provide emergency economic relief to Russians in the former Soviet space. The Assembly justifies its activities on the stated assumption that Russians in the former Soviet Union are treated as "second-class people." The founding congress of the Assembly will be held this autumn.

**13** • Interfax reports that the Standing Commission of the CIS Interparliamentary Assembly (IPA) issued a statement criticizing the new rules for the transit of the Turkish Bosporus Straits. The rules, enacted in January and enforced as of 1 July, call for twenty-four hours' notice to be given before the passage of oil tankers, and special permits for the transit of nuclear waste or chemicals. Russia has long protested that the new rules contravene the 1936 Montreaux Convention governing the straits, but the protest by the IPA may signal a more united front by the CIS oil-producing states.

• The IPA statement is preceded by an agreement, signed between Gazprom, the Russian state gas concern, and a Greek consortium, that seeks to circumvent the transit of the Turkish straits. The *Financial Times* of 12 September reports that a 350 km pipeline is to be built through Bulgaria to the northeastern coast of Greece. The pipeline, which will reportedly carry Russian, Kazakh, and Azeri oil, could move 20–40 million tons of oil annually, and will take three years to construct, at a cost of $600 million.

**14** • At a meeting of the CSCE's Committee of Senior Officials in Prague, Russia offers a new proposal for a CSCE framework for Russia's peacekeeping operations in the CIS. The previous proposal allows for significant CSCE monitoring and mediation in the settlement of conflicts in exchange for political endorsement and CSCE financial support for the Russian operations. The new proposal, according to NATO and non-NATO observers, dilutes the CSCE's role and broadens Russia's opportunities to act unilaterally in the CIS, but with CSCE financing. The new proposal is tabled by the meeting's delegates.

• Interfax reports, however, that Yeltsin and Kozyrev will raise the issue of peacekeeping in the CIS at the forthcoming session of the UN General Assembly. A Russian Foreign Ministry spokesman says that Russia seeks both UN acceptance of the CIS as a framework for regional peacekeeping and "an optimum balance of political and financial responsibility."

**15** • Kazakhstan's ambassador to Russia, Tair Mansurov, reignites debate on the creation of a "Eurasian Union." Mansurov labels the proposal a more advanced form of integration among CIS states. The Union's functions would include: coordination of economic policy; adoption and mandatory implementation of joint reform programs; creation of common political institutions including a central "consultative-deliberative" parliament; and "an agreed approach" to military and border tasks.

**17** • Nazarbaev further elaborates his proposal, saying that it is linked to the Russian ideology of "Eurasianism."

**19** • Russia's ambassador to the Court of St. James, Anatoliy Adamishin, says that one of Yeltsin's main objectives during the upcoming London summit will be to secure London's understanding of "Russia's foreign policy priorities, such as the organization of the post-Soviet space, the settlement of conflicts on former Soviet borders, and Russian peacekeeping." Adamishin uses the "rights of ethnic Russians" rhetoric by complaining that "a new iron curtain" is being raised, this time to keep out "an inflow of people from Russia. . . . The difference between the rights of Britons in Russia and of Russians in Britain cannot be tolerated much longer; no country can tolerate such discrimination," he complains.

**20** • The secretary general of the Council of Europe, Daniel Tarschys, announces that Ukraine and Belarus need to overcome "legal difficulties" before joining the thirty-two-member Council. For Ukraine, Tarschys says, the adoption of a new constitution is necessary for membership. For Belarus, the issue is parliamentary elections. Nine former Soviet bloc countries have already been admitted to the European Council and seven more await acceptance—Albania, Belarus, Croatia, Latvia, Moldova, Russia, and Ukraine.

• Tarschys meets with Moldovan President Mircea Snegur and expresses satisfaction with Moldova's recent parliamentary elections and new constitution. These two factors, he says, should accelerate Moldova's accession to the Council of Europe. On the issue of Moldova's separatist problem, Tarschys labels the "Dniester Republic" as "absolutely illegitimate."

• (Brief Analysis): The Russian Foreign Intelligence Service (SVR) has issued a report entitled "Russia-CIS: Does the Western Position Need Correction?" which states that all major political forces in Russia and the CIS support reintegration of the post-Soviet "Eurasian space," and that this "natural, historical trend" is not a manifestation of Russian neo-imperialism. Rather, the document contends, the process of reintegration serves to hinder the rise of nationalist forces in Russia. The spokesperson for SVR Director Evgeniy Primakov says the report intends to clarify the Russian position on this issue on the eve of the U.S.-Russian summit.

The report envisions CIS development along three possible paths. The first holds that centripetal tendencies will create conditions for the formation of a single economic and security space, and eventually forming a voluntary confederation. Because of the "democratic" nature of this scenario of reintegration, the new and improved CIS would not result in a return to confrontation with the West, the SVR concludes.

The second scenario assumes those forces within the CIS who advocate "isolated" development will emerge victorious, with direct or indirect "external" support. In such a scenario, "a lurch toward nationalism is likely to be accompanied by growing authoritarian and anti-democratic tendencies," the report says. The report blames "isolated" (i.e., independent) development for Islamic extremism in Central Asia, and for the reversal of reforms in Russia.

The third scenario envisions one of the other CIS countries (not Russia) assuming the "unification" function. The resultant non-Russian bloc of CIS countries could follow either of the first two paths.

• Russia and Ukraine continue their dispute over language in a draft Friendship, Cooperation, and Partnership Treaty. Russia continues to propose language from the CSCE Final Act. Kiev considers that language too ambiguous (it permits peaceful, negotiated border changes) and wants Russia to renounce territorial claims on Ukraine.

• Kozyrev calls Russia and Ukraine "twin brothers" (but "brothers" just the same). The commander of the Black Sea Fleet offers as a solution to tensions between the RF and Ukraine the "economic, political, and military drawing together of the two Slavic states." He calls on Ukraine's new leadership "to subdue the ambitions of 'national-patriots.' "

• President Leonid Kuchma offers reassurances that

Ukraine's national interests remain his top priority in signing a treaty with Russia. He maintains, however, that "the treaty must be signed, as Russia remains our main partner."

• Belarusian President Aleksandr Lukashenka tells *Frankfurter Allgemeine Zeitung* he has no intention of ceding Belarus's political or economic sovereignty to Russia. He says he will attempt to cooperate with Russia on the same basis as with other Western countries.

• The following day, Lukashenka blames Russia's energy prices for unpopular economic measures undertaken by the Belarusian government (e.g., lifting of some food subsidies).

**21** • Russian Foreign Minister Kozyrev says the United States should expand bilateral economic relations with Russia in order to help create "stability and prosperity" in the CIS. Kozyrev says Russia envisions the CIS as a European Union–like structure, and "not a centralized state or a new USSR." He claims a "lack of any imperial ambitions in Moscow."

• Nazarbaev's Eurasian Union proposal receives support from several prominent Russian politicians at a conference in Alma-Ata. The Russian State Duma's CIS affairs committee chairman Konstantin Zatulin and Russian Communist Party leader Gennadiy Zyuganov endorse the proposal, but disclaim any intent to reconstruct the USSR. The conference, attended by representatives of eight former republics, resolves to establish a non-governmental organization to promote the Eurasian Union "and other integration projects."

**26** • In his address to the UN General Assembly, Boris Yeltsin says the CIS states constitute "Russia's foreign policy priority." Ties with these states "exceed those of mere neighborliness. This is rather a blood relationship," Yeltsin says.

**28** • Dmitriy Rurikov, who is in Kiev for talks on the draft state treaty, insists that the document must provide for dual citizenship in order to "facilitate the legal defense" of ethnic Russians in Ukraine and to "offer an additional guarantee to persons who don't want to break links with Russia." Rurikov indicates that dual citizenship "is a basic position of Russia's policy not only toward Ukraine, but also toward the other CIS states."

### *Security*

**9** • An opinion poll measuring political views of Russian military officers highlights trends of growing political restlessness within Russia's armed forces. The results, published in *Der Spiegel*, show that 62 percent of the respondents believe that Russia requires "authoritarian rule" to solve its problems; less than half support market reforms and privatization; 80 percent seek "Russia's restoration as a great power respected in the entire world;" and absolute majorities "do not want Yeltsin as president" or Grachev as defense minister. The survey finds Lt.-Gen. Aleksandr Lebed is the clear choice among the officers to institute "a strong-arm policy" or "a Bonapartist solution."

**12** • Russia lobbies for changes to the 1990 Conventional Forces in Europe (CFE) treaty. The Russian Defense Ministry would allow Russia to reinforce its flank regions—the Leningrad and North Caucasus Military Districts—to levels greater than the treaty permits. Russian Defense Minister Pavel Grachev announces Russia would be prepared to withdraw a considerable number of troops and arms from the Kaliningrad oblast should CFE flank limitations be readjusted.

**15** • Ukrainian parliamentary speaker Oleksandr Moroz proposes an international conference in Kiev in 1995 to discuss Ukraine's accession to the NPT. Action on the treaty is unlikely before the end of the year. The proposal deals a blow to President Kuchma who had hoped to present a ratified NPT as a "gift" to President Clinton when the two meet in Washington in November. The United States considers Ukraine's accession to the non-proliferation regime an issue of paramount importance in relations between the two countries.

**20** • *Nezavisimaya Gazeta* reports that, according to Pavel Grachev, servicemen are humiliated and some are in "a strongly revanchist mood."

## October 1994

### *Economic*

**6** • Deputy Prime Minister Aleksandr Shokhin, Russia's representative to the IMF, World Bank, and EBRD meetings in Madrid, proposes that Western credits to other CIS states be used to pay the recipient country's debt to Russia. Shokhin also implicitly links repayment of CIS debt to Russia with Moscow's ability to repay Western creditors.

• Citing the failure of the CIS to promote economic integration within the FSU, parliament deputies from Russia, Ukraine, Belarus, and Kazakhstan form four factional deputies' groups into an "Economic Union," *Nezavisimaya Gazeta* reports. Rather than wait for an effective CIS legislature to develop, the four country groups will introduce legislation in their respective legislatures that has been coordinated at the drafting stage. Members of the "Economic Union" faction in the Russian parliament include Deputy Prime Minister Sergey Shakray and Yabloko co-chairman Yuriy Boldyrev.

**9** • Ukraine's President Leonid Kuchma admits that his country is incapable of paying its share of the former USSR debt ($13.6 billion with annual servicing of $700–800 million) and indicates that Ukraine would have been better off had it joined the other republics in ceding its share of the debt to Russia in exchange for former Soviet assets abroad (the so-called "zero-option"). Russian Prime Minister Viktor Chernomyrdin insists that Ukraine pay its energy debt to Russia ($1.5 billion) and presents a plan to Kuchma that would transfer control of strategic Ukrainian energy facilities to Ukrainian-Russian joint-stock companies and joint energy ventures in exchange for partial debt relief.

**12** • The collapse of the Russian ruble on 10 October sends economic shock waves throughout most former Soviet republics, where the relatively stable ruble (relative to the local currency) is considered almost a hard currency. In Moldova, Finance Minister Leonid Talmaci reports that companies dealing with Russia lost up to 40 percent of expected revenues as a result of "Black Tuesday." In Uzbekistan, Azerbaijan, and Armenia the value of local currencies immediately plunges 20 percent, suggesting strong links to the ruble. In Georgia, Finance Minister David Yakobitze simply states "we are completely dependent on Russia as concerns this question. . . . The masses will suffer."

***Political***

**7** • The fiftieth anniversary of Ukraine's liberation from Nazi forces provides the setting for talks between Chernomyrdin and Kuchma. The two discuss progress on the comprehensive bilateral treaty and the Black Sea Fleet negotiations.

• Belarusian President Lukashenka and Moldovan President Snegur also attend the liberation commemoration and meet briefly with Kuchma. Afterward, Lukashenka says that should the next CIS summit fail (as the previous ones have) to address the major issues haunting the post-Soviet space, Belarus "will draw more attention to establishing bilateral relations."

• A reorganization in Azerbaijan following an abortive coup attempt against President Aliev (on 5 October) continues with First Deputy Prime Minister Fuad Guliev named as interim prime minister, replacing Surat Huseinov, who was implicated as one of the coup plotters.

**9** • Following several days of Russian-Iranian consultations concerning the recent oil deal between Azerbaijan and a Western consortium, Iran opposes the deal, "fearing a strengthening of Western influence in the Caspian region." The Iranian delegation flies to Moscow for a meeting of Caspian littoral states to discuss future development of the Caspian basin.

• Russia circulates a document at the United Nations outlining its position on the Azeri oil project and future development of Caspian basin resources. The document reasserts Russia's claim that any "unilateral steps" by Caspian states in developing maritime resources violates the Soviet-Iranian agreements of 1921 and 1940, which called for joint use of the sea's resources by states bordering the sea. The document repeats the Russian position that the norms of international maritime law do not apply to the Caspian, and calls on Caspian littoral states to adopt new documents (prepared by Russia and Iran) updating the pre-war agreements. The document concludes that Russia "reserves the right to take such measures as needed, at a suitable time, to restore the legal order and eliminate the consequences of unilateral steps. . . . All responsibility for possible material damage rests on those who take the unilateral steps."

**10** • Preliminary discussions in preparation for the CSCE meeting in Budapest illustrate a growing rift between Russia and most other members over security issues in the former Soviet Union. Russia continues to seek political and financial support for its peacekeeping operations in the "near abroad." In addition, Russia's proposals to renegotiate CFE flank sublimits in the Caucasus and northwestern Russia have drawn negative reaction from other CSCE states. Finally, Russia proposes at the last minute that the CSCE suspension of rump Yugoslavia be lifted. For its part, the CSCE has insisted that before Russian peacekeeping operations receive a mandate, CSCE monitors must be in place and specific time constraints be placed on the deployment of Russian troops. Additionally, CSCE officials insist political negotiations in CIS "hot spots" be placed solely under CSCE auspices and reject Russia's undermining of the CSCE peace settlement in Nagorno-Karabakh.

• Despite acknowledgement that it is "clinically dead," Russian officials submit a proposal for reorganizing the CSCE into a "pan-European security body" with a broadly defined "coordinating" role over regional associations such as NATO and the CIS. The United States and the EU reaffirm their opposition to the proposal. Germany, which holds the rotating EU presidency, counters that the CSCE should be used as the "first resort" in regional crisis management.

• Aliev imposes a sixty-day state of emergency in Ganje where supporters of the former prime minister reportedly tried to seize power. Baku radio reports names of nine other government ministers slated for sacking in the wake of the attempted coup. In addition, fifty-six officials of the National Security Ministry are also fired for alleged involvement in the events.

• Yuriy Ushakov, Russia's chief delegate to CSCE, says Russia would welcome the support, but not "supervision," of the CSCE for its peacekeeping operations. By support, Ushakov indicates that Russia seeks logistical, financial, and other "material" assistance for operations, in addition to CSCE support for Russian mediation efforts in these areas.

**11** • The third round of Ukrainian-Russian negotiations on the draft Treaty on Friendship, Cooperation, and Partnership ends without resolution of two key sticking points: dual citizenship and inviolability of borders. Russia continues to insist that dual citizenship be available for Ukraine's 11 million-strong ethnic Russian minority as a means to improve their lives in Ukraine (no mention of dual citizenship for the 4–6 million Ukrainians living in Russia, however). On the border question, Ukrainian officials seek a "recognition" of state borders, while Russia is willing to concede mere "respect" for the present frontier.

• A conference of the Caspian littoral states opens in Moscow, with Russian Foreign Ministry spokesman Grigoriy Karasin calling for an agreement on a regional cooperation organization, modeled after the Caspian Sea Cooperation Council (created in February 1992).

• Kazakhstan's President Nursultan Nazarbaev requests resignation of his government, led by Prime Minister Sergey Tereshchenko, citing its inability to implement economic reform. Nazarbaev's request reverses an earlier pledge to give Tereshchenko until the end of 1995 to demonstrate substantial economic progress. Nazarbaev nominates (and the parliament subsequently confirms) Akezhan Kazhegeldin for the post of prime minister. Kazhegeldin, who was deputy prime minister in the Tereshchenko government, drafted the most recent economic reform program.

**12** • Russian Foreign Minister Andrey Kozyrev tells the Moscow conference of Caspian littoral states that he has spoken with Azeri President Aliev, who agrees that the Caspian is an asset to be shared by all littoral states.

• Azeri parliamentary speaker Rasul Guliev promises at least a one-month discussion of the ratification of the oil contract with the Western consortium.

**13** • Two Russian newspaper articles seek to debunk the myth that all 25 million ethnic Russians in the "near abroad" are potential refugees. In *Segodnya*, ethnographer Vladimir Kozlov condemns "alarmist" trends in the Russian media concerning the refugee issue. According to Kozlov, nearly 70 percent of the current "Russian refugees" fled war-torn countries like Georgia or Tajikistan, while only 1.5 percent came from Estonia and Latvia, which are accused of institutional discrimination in the Russian press. Furthermore, Kozlov maintains that most refugees are highly educated, urbanized Russian-speakers capable of making significant contributions to Russia's economic transition. In the other article, appearing in the Russian *Obshchaya Gazeta*, Andrey Fadin concludes that the plight of ethnic Russians in the "near abroad" is being used as an instrument, a "fifth column," by the authorities to further their goal of restoring the empire.

• Azeri parliament votes to arrest former Prime Minister Huseinov on charges of treason. At the time of the decision, Huseinov's whereabouts are unknown. The agriculture and security ministers are also placed under arrest.

**18** • A meeting of heads of state of the Turkic-speaking countries (Azerbaijan, Kazakhstan, Kyrgyzstan, Turkey, Turkmenistan, and Uzbekistan) commences in Istanbul. The gathering focuses on strengthening and deepening economic and cultural integration of the Turkic-speaking world.

At the conclusion of the summit the following day, leaders sign a document calling for closer political, cultural, and economic cooperation. The heads of state also signal their support for plans to construct pipelines to transport oil and gas from Central Asia to Europe via Turkey. In the post-summit press conference, Turkey's President Demirel objects to the Russian Foreign Ministry's warning to participants not to engage in "pan-Turkic rhetoric."

**19** • Belarus and Russia complete a draft Treaty on Friendship and Cooperation meant to prevent future misunderstandings and to deepen cultural and economic relations. Belarusian Deputy Foreign Minister Pyotr Byalyaeu stresses that the document does not diminish Belarusian sovereignty, stating that close ties to Russia strengthen Belarus.

**20** • According to Dmytro Tabachnyk, head of Ukraine's presidential staff, Russia agrees to remove references to the dual citizenship question from the text of the Ukrainian-Russian friendship treaty. The issue of language regarding the inviolability of borders remains a key sticking point.

**21** • The whereabouts of Azerbaijan's former prime minister, and current Enemy Number One, Surat Huseinov, remain a mystery even as he issues a statement to the international community asserting his innocence and accusing Aliev of conducting a Soviet-like purge of his opponents. Huseinov claims Aliev has appropriated money from the oil deal for his family's use.

### *Security*

**5** • Ukraine's President Kuchma submits a letter to parliament urging it to accede to the Nuclear Non-Proliferation Treaty (NPT) before the treaty comes up for renewal next year. Presidential spokesman Mykhailo Doroshenko says the letter makes "good on his promise to the international community . . . that Ukraine should join the NPT." Ukraine's foreign minister, Gennadiy Udovenko, says Ukraine still wants "guarantees of national security," but indicates that Kiev would not link NPT accession to such guarantees.

**6** • The dire financial straits of the Russian army are highlighted when several Russian oil companies refuse to supply fuel to the army due to unpaid bills. Russian Defense Minister Grachev informs the Cabinet that only 38 percent of the year's defense budget has been dispersed; barely enough to cover salaries, Interfax reports. Grachev says the army's debt for weapons, food, and other supplies exceeds $1 billion, a fifth of which is owed for energy supplies alone.

**7** • Russian and Belarusian border guard commanders sign a protocol on expanding bilateral cooperation on border defense. Under protocol, the two states agree to "coordinate actions" of the border troops on Russia's and Belarus's borders with Ukraine. The two sides also agree to develop by 30 October a concept for a "single regime for the defense of Russia's and Belarus's borders with" Poland, Latvia, and Lithuania.

**11** • Following meetings with Russian Federation Council speaker Vladimir Shumeyko, Armenian President Levon Ter-Petrosyan announces that an agreement on the status of Russian military bases in Armenia will be signed soon in Moscow. For his part Shumeyko characterizes Russia's relations with Armenia as "closer than with any other CIS country." The two states "have practically the same view" on the future development of the CIS and on bilateral economic, trade, and military relations. Earlier in the visit, Shumeyko pledges 110 billion rubles in credits as the first tranche of a comprehensive aid package to Armenia. Sixty billion rubles is earmarked for reconstruction of Armenia's idle nuclear power station. Shumeyko, however, denies that a Russian-Armenian confederation has developed.

• The IMF agrees to extend Armenia a $500 million loan on the condition that the Armenian government implement tough economic reform measures. Previously, the IMF refused to grant credit to Armenia, claiming that the funds could be used to purchase arms for the Karabakh conflict.

**15** • A Russian Defense Ministry source informs Interfax that lack of funds for weapons disposal could lead to serious accidents in Russia. For example, eighty-five decommissioned nuclear submarines remain at their bases because there is no money to dismantle them. According to the source, the Finance Ministry "owes" the military 9 trillion rubles.

**16** • Ukraine's Deputy Foreign Minister Boris Tarasyuk says his country's accession to NPT still depends on security guarantees from other nuclear powers.

**17** • The draft proposal for CIS Collective Security is removed from the upcoming CIS summit agenda, at Russian request. The draft security framework was initialed on 20 June by eight of nine CIS signatory countries to the CIS Collective Security Treaty. Ukraine, which is not party to CST, continues to object to tightened military cooperation among CIS states.

**19** • Russian Defense Minister Grachev announces that Russia will dispatch a squadron of fighter aircraft to Armenia to provide air defense for Armenia and to protect Russian troops stationed there.

## November 1994

### *Economic*

**8** • At talks between Turkmen and Ukrainian presidents, Turkmenistan gives Ukraine a seven-year deferral in payment of the $700 million debt for gas. This sum is to be transformed into a credit at 8 percent annually.

**10** • Kazakhstan and Russia agree to establish an interstate industrial finance consortium (Sotrudnichestvo) on a parity basis.

**12** • Ukraine and Uzbekistan sign the Declaration on the Main Trends in Economic Cooperation, Agreement on Avoiding Double Taxation of Incomes and Property, Agreement on the Trade and Economic Cooperation for the Year 1995. Uzbek President Islam Karimov says that he is in favor of intensifying economic ties within CIS, but against the creation of single parliament and government of the CIS and against the Eurasian Union. Leonid Kuchma supports his view.

**15** • A contract between the Western consortium and the Azeri State Oil company last September on joint development of three Caspian oil deposits is ratified by the Azeri parliament. The document provokes a negative reaction from the Russian Foreign Ministry, which refuses to recognize the agreement provisions. RF Foreign Ministry spokesman Grigoriy Karasin threatens political consequences if the 1921 and 1940 Soviet-Iranian treaties are ignored. Baku observers note a possible Russian-Iranian alliance to protect Transcaucasian geopolitical interests.

**16** • Russia cuts the gas supply to Belarus by 50 percent due to the republic's failure to repay debts to Gazprom. Vladimir Kurenkov, Belarus's gas industry minister, says the reduced gas shipment could prove truly disastrous for industry.

**18** • Gazprom resumes full-scale natural gas supplies to Belarus, after Belarus pledges regular payments.

**19** • Russian state oil company, Rosneft, agrees to exchange oil with Kazakhstan oil company to keep refineries of both countries fully occupied. Russia will supply hydrocarbon raw materials to Chimkent and Pavlodar plants and Kazakhstan will supply the Samara plant as was done during the USSR period. Annually each side will supply around 4 million tons of oil.

The possible cooperation is not limited to this. The Russians expressed an interest in buying Kazakh coal, corn, and non-ferrous metals. Rosneft can pay for these goods in oil and petroleum products. This proposal will arouse the interest of the Kazakh side. Rosneft is also looking for a new market in Kazakhstan which is less saturated than the Russian market. The company intends to sell about 4 million tons of various petroleum products in Kazakhstan annually. The Russian oil industry workers who have free technical and personnel resources also offered the Kazakh side their assistance in the capital repair of oil wells and in the exploitation of one to two oil deposits. Kazakhstan is ready to propose that the Russians participate in joint oil prospecting on the Caspian Sea shelf.

***Political***

**1** • Moldova, like other CIS states, is not ready to accept the political integration concepts of the Eurasian Union project proposed by Kazakhstan's President Nursultan Nazarbaev, says Moldova's permanent envoy to the CIS, Jacob Mogoryanu. He says "this option could be of tragic consequence under present conditions, when most CIS countries are forming their own statehood and democratic institutions."

**3** • The Supreme Council of Crimea debates the situation afterthe Supreme Council of Ukraine's resolution "On the political and legal position of the Crimean Autonomous Republic" of 22 October is passed. The Crimean Supreme Council Presidium proposes to return to the September 1992 constitution. Parliament Chairman Sergey Tsekov believes that the main problem in relations between Crimea and Ukraine is Ukraine's unitary structure which, in his view, gives rise to an abnormal situation in which Crimea's parliament must follow Ukraine and change the peninsula's constitution for the fourth time in two and a half years. Tsekov believes that "the key to settling problems between Ukraine and Russia is in Simferopol," hoping that Kiev will renounce its policy of ultimatums and threat.

Deputy Anushevan Danelyan feels that any resolution should be coordinated with a Ukrainian delegation. He proposes to continue the process of negotiation.

• Crimea's Russia-Unity group puts forward a motion to hold a Crimea-wide referendum to determine whether the Crimean constitution should be brought into line with that of Ukraine.

**5** • According to the Ukrainian Military Department chief, Moscow's stance at the bilateral negotiations on the Black Sea Fleet remains rather tough. According to the official, neither president will agree to radical concessions in the near future, because "such steps could be suicidal for both Kuchma and Yeltsin."

• Crimean Supreme Soviet Presidium adopts an appeal to the presidium and parliament of Ukraine expressing its inability to eliminate contradictions between the constitutions of Ukraine and Crimea because of the exceptionally difficult political and socio-economic situation in Crimea. Deputies include in the appeal a call for wisdom, restraint, and political perspicacity and ask Ukraine to refrain from taking any radical measures to resolve the problem.

• Andrey Senchenko, deputy prime minister of Crimea, points out that "The only justification for Crimea's statehood is its desire to reform its economy."

**16** • Ukrainian parliament ratifies the Nuclear Non-Proliferation Treaty, with reservation.

**24** • According to a top-ranking official from the Russian Foreign Ministry, Russia, the United States, and Great Britain will grant security guarantees to Ukraine only if ratification documents on Ukraine's joining the NPT do not contain "reservations or ambiguities." Ukraine's Supreme Soviet resolution on joining the NPT gives rise to many questions. It is not clear in what status, as a non-nuclear or as a kind of semi-nuclear power, Ukraine will join the NPT. Russia will not be satisfied with Ukraine's "semi-nuclear" status.

**25** • Kazakhstan's Foreign Minister Kasymzhomat Tokaev says that dual citizenship in Kazakhstan may be a serious source of instability and entail heavy consequences.

***Security***

**2** • The Tajiks, Uzbeks, and Russians sign agreement on strengthening the southern borders of the CIS. The border along the Pyanj and Amu Darya rivers will be strengthened, not only by the border guard but also by the river navy. This should decrease the border provocations and prevent the smuggling of narcotics and weapons.

**12** • Following meetings with the visiting head of the Russian Border Guard Service, Azeri officials agree to a bilateral treaty with Russia on border protection and enforcement, including cooperation in immigration and customs.

**17** • The council of border troop commanders of CIS members agree jointly to defend CIS external borders. Thus a single system of defending CIS external borders is created. Ukraine does not sign the agreement, because it is not a full member of the Council, taking part in it only as an observer.

**December 1994**

***Economic***

**1** • The CIS Intergovernmental Council sets up an operating working group on the agro-industrial sector and an intergovernmental science and technology center, as a move to promote economic cooperation among the CIS states.

**2** • Turkmenistan reaches an agreement with Armenia, Azerbaijan, and Georgia, according to which it will sell gas to these states in 1995. Natural gas is Turkmenistan's main source of income.

**7** • Moldovan government begins to restore economic ties within the CIS. Three joint economic commissions—Moldovan-Russian, Moldovan-Ukrainian, and Moldovan-Belarusia—have been set up in order to strengthen the integration process. Moldova intends to set up a joint venture with Russia to service the main gas pipeline along which "Rosgaz" exports gas to Romania, Bulgaria, and Turkey. According to the government, Moldova "gets 95 percent of its energy sources from CIS countries—mainly Russia . . . [and] 70 percent of Moldovan output is exported eastward as well." Moldova is aware that it is more advantageous to restore its lost positions in the Eastern market than to work its fingers to the bone to make a breakthrough into the Western market. Consequently, while striving to raise Western loans, the republic is looking Eastward with hope.

**9** • Viktor Chernomyrdin, addressing the meeting of the Council of CIS Heads of Government in Moscow, announces that Russia will be forced to introduce the principle of advance payment in trade and economic agreements and contracts for the supply of commodities. Financial penalties for failure to pay on time will become a mandatory condition of such agreements. He says that Russia has gone as far as it can and cannot make any further concessions. This is the umpteenth time that the question of the debts owed to the Russian fuel and energy sector by CIS countries has been brought up. On 1 November 1994, these debts amounted to 7.5 billion rubles, with almost 6.6 billion rubles of the total for gas. Since establishment of the CIS, the Russian government has granted state credits to those countries worth $5.6 billion, in order to make it easier for partners to adjust to the new conditions of economic cooperation. Credits, however, have only exacerbated the debt situation. The settlement of these debts is proceeding extremely unsatisfactorily.

**16** • Kazakh Prime Minister Akezhan Kazhegeldin meets with the chairman of the State Committee of Kazakhstan, Sarybay Kalmurzaev, in Moscow to consult on Kazakhstan's plans for privatizing several of its defense enterprises, including the Khimvolokno Association in Qostanay, the enterprise in Stepnyak in Qoqshetau oblast, the Stepnogorsk plant in Aqmoly oblast, and many others. Russian officials have indicated their willingness to discuss the list through diplomatic channels. The Russian State Property Committee will recommend investors after "the appropriate appraisals have been made." During meetings between the State Property Committees of Russia and Kazakhstan, an agreement is reached to exchange documents on the practice and main problems of privatization in the two countries.

• Addressing the nature of economic reform in each state, Kazakh Prime Minister Kazhegeldin observes: "We were out of synch in implementation of economic reforms. This is the only reason, I believe, that prompted Russia to push its former allies out of the ruble zone." In Kazhegeldin's opinion, the Kazakh model of economic reform is in principle no different from the Russian model—but lacks the law on private ownership of land, with only an edict permitting the purchase, transfer, or mortgage of a plot of land for farm centers or plants. The legal basis of this edict needs to be adjusted, and this needs to be done quickly, because people are afraid of waiting. He continues, "Although the Russian government failed last fall to refrain from certain steps that led in October to madness on the currency market, we need, on the whole, to catch up with Russia and this will, perhaps, facilitate our task of reform of the economy—the use of Russia's example."

**20** • Russian and Ukrainian representatives of the "Interstate Euro-Asiatic Coal and Metal Association" meet in Donetsk to discuss pricing of railroad services to the coal and metal industries of the two countries. It is disclosed at the meeting that a 1-ruble increase in the price of a unit of raw material increases the price of a consumer good by 4–6 rubles. Therefore, there is concern by CIS companies about the monopolist position of the Russian Railway Ministry, and concern by the Railway Ministry about non-payment for rail service. Many CIS countries have common problems in this area. A decision is adopted at the meeting to send the problem of railway tariffs and payments for coal and metal transport to the CIS Interstate Economic Committee (which was formed in October 1994) for resolution.

***Political***

**1** • Supreme Council of Crimea adopts an appeal to the State Duma of the Russian Federation and Russia's Federation Council, noting that they should guarantee Crimea's statehood.

**5** • An agreement on percentage staff of the CSCE peacekeeping forces on CIS territory is adopted by the heads of the CSCE member states. Russian forces are to make up 30 percent of total forces.

**8** • An intergovernmental Russia-Crimea protocol is signed on trade and economic cooperation in 1995; it does not include Russia's obligations for delivering Russian goods to Crimea. The situation for Crimea could change for the better if Sevastopol becomes Russia's naval base. Whether the Russian navy will receive the whole of Sevastopol as a naval base or only its harbor depends on political decisions in Kiev and Moscow.

**9** • Russian Prime Minister Viktor Chernomyrdin, outlining priority directions in the economic development of CIS countries, says an agreement on the creation of a free trade zone in the CIS and the implementation of joint investment programs should be given priority in CIS activities. Ivan

Korotchenya, executive secretary of the CIS council, says that the future of CIS depends on the Interstate Economic Committee. The summit of CIS Heads of Government meeting in Moscow approves twenty documents predominantly connected with the creation of an economic union. Ukraine refuses to sign agreements on military affairs because it is not a member of the CIS Defense Council. Ukraine signs a document on the CIS accident and disaster quick response corps. Under the agreement, the force will incorporate national accident relief subunits subordinated to a single intergovernmental command which will be on call to fly to emergency areas.

• Russia and Ukraine conclude an agreement on the "zero option," regarding the division of the former USSR's assets and liabilities. Russia will retain all the assets and in return it will pay the interest charges on the Ukrainian share of the former USSR's debts to foreign creditors.

**14** • Andrey Kortunov, head of Russia's Academy of Sciences Department of Foreign Policy of the United States, predicts that electoral victory for the U.S. Republican Party will worsen U.S.-Russian relations. He complains that more Republicans see Russia's foreign policy as a manifestation of neo-imperialism and an aspiration to revive the Soviet Union in some oblique form. In addressing Russia's policies toward the "near abroad," he writes:

"This does not mean, of course, that Russian policy does not at the level of specific foreign policy actions in particular regions suffer from neo-imperial ambitions. I would like to add, in addition, that with respect to its so-called near abroad, our problem is further exacerbated by the fact that each department has its own 'policy.' Mr. Kozyrev has 'his,' which does not coincide entirely with the policy of Chernomyrdin. We saw this for ourselves when the treaty with Azerbaijan was signed; Mr. Grachev has 'his,' which, in turn, differs appreciably from the policy of Mr. Lebed. And so forth. Speaking of some unified strategy is out of the question. And this, of course, is causing perfectly justified concern both in the CIS and beyond. America included."

***Security***

**1** • In Moscow during the meeting of the Council of CIS Defense Ministers, Pavel Grachev explains the issue of the increasing the powers of the commander of the CIS collective peacekeeping forces in Tajikistan. The main purpose of additional powers is that if an emergency situation arises in a particular section in Tajikistan, Lt.-Gen. Valeriy Patrikeev may decide autonomously to use peacekeeping forces and then submit a report to the defense ministers of the states whose peacekeeping forces are used. The second objective is to introduce organizational and manning level changes to the management structure of the peacekeeping forces and to allow the commander to recommend the appointment or dismissal of officers and generals. The ministers support the proposal.

• Grachev calls the meeting one of the most productive and points to the increasing military and technical integration among CIS states. Speaking on the peacekeeping operation in the Georgian-Abkhaz conflict area, he says that all participants of the meeting express their intention to assist in its implementation. He says Ukraine does not rule out the possibility of sending its peacekeeping unit to the conflict area, as well.

**5** • Ukraine signs the NPT in exchange for guarantees on its security from the United States, Russia, and Britain. Under the assurances, the parties recognize the territorial integrity of Ukraine and the nuclear parties commit not to attack a non-nuclear state.

# 1995

## January 1995

***Economic***

**5** • Evgeniy Marchuk, Ukrainian deputy prime minister, announces that Russia and Ukraine want to conclude a political treaty, containing more than twenty agreements on cooperation in various fields. Marchuk emphasizes the weakness of Ukraine's position in the treaty talks, given its debt to Russia of more than $5 billion.

**8** • Yuriy Yarov, Russian deputy prime minister, is appointed chairman of the Collegium of the CIS Interstate Economic Committee by CIS presidents.

**9** • Since May 1994, a group of twenty Russian banking experts from major Russian banks has been working with the Russian "Ministry for Cooperation with Member States of the CIS" (MCMS) to design a blueprint for a common accounting system. In October 1994 an Agreement on Creation of a Payments Union was adopted. According to the evolving blueprint, the Payments Union will consist of a set of rules instead of an organization, which will define the mutual convertibility of national currencies. The rules will adhere to the model Belarus/Russian bilateral agreement, entitled: "On Measures to Guarantee Mutual Convertibility and the Stabilization of the Exchange Rate of the Russian Ruble and the Belarusian Ruble."

• According to Adrian Budyanu, chief of the Russian Ministry for Cooperation with Member States of the CIS, Russia is now owed about R8 trillion by the other CIS member states. Much of this is energy debt. Russia proposes partial payment in property shares belonging to the indebted enterprises.

**10** • Uzbekistan takes a leap forward in its reform plans for 1995. *Pravda Vostoka* outlines a new mandate from the state committee for managing state property and supporting enterprise. The priorities for the new year will be:

1. Moving the center of gravity for implementing privatization to the territorial branches of the State Property Committee.
2. Demonopolizing existing management structures.
3. Denationalizing the mining, fuel, and energy complex and cotton processing industry.
4. Faster privatizing of large enterprises.
5. Changing priorities to denationalization, transferring property to private ownership, regardless of the sphere of activity and area of industry the enterprise is in.
6. Encouraging foreign investment. There are already more than 1,300 joint-stock ventures registered in Uzbekistan. Among them are such well-known firms as Newmont, Coca-Cola, Pepsi-Cola, Rank Xerox, Mercedes Benz, Daewoo, Bursel, and Buller.

**11** • The 19–25 December (1994) issue of *Delovoy Mir* summarizes a report issued by the newly formed (1994) "Institute of Economic Analysis," whose board is represented by well-known Russians including Abel Aganbegyan, Aleksandr Granberg, Stanislav Shatalin, Egor Gaydar, and Boris Fedorov, and foreigners Richard Layard, Anders Aslund, and Jeffrey Sacks. The report analyzes Russia's 1994 economic policies, concluding that it was a "lost year." Contrary to Viktor Chernomyrdin's beginning-of-the-year commitment to "combat inflation through non-monetary methods," and "not to allow shock therapy in the future," the prime minister drastically reversed course in October (following the "Black Tuesday" collapse of the Russian ruble)—instituting a new shock "therapy" based on monetary principles.

• The 19–25 December (1994) issue of *Delovoy Mir* also contains an analysis of the economic impact of migrant workers entering and leaving the Russian Federation. The article concludes that between 1995 and 2000 an approximate parity will be reached between incoming migrants from the "far" and "near" abroad. Workers entering Russia from the "far abroad" (non-CIS) are led by the Chinese, followed by Turks, Bulgarians, North Koreans, Macedonians, and Germans. Their numbers are expected to make up 40 percent of the workers entering Russia between now and the year 2000, with the CIS accounting for the remaining 60 percent. The article predicts that the number of illegal immigrant workers will grow, due in part to what it calls "super-liberal migration legislation" in Russia's near abroad. The report estimates there were 1.63 million illegal immigrant workers in Russia by mid-1994, concentrated within major cities and the Far Eastern region of the Russian Federation. The future, the article concludes, will depend on migration legislation which is currently being formulated, alluding to possible Russian intervention on CIS border policies.

### *Political*

**4** • Writing in Rotterdam's *Algemeen Dagblad*, Kazakh President Nazarbaev views the Eurasian Union as a "partnership among equal, independent states that must defend the national interests of individual member states." Nazarbaev believes his plan for a Eurasian Union will succeed after general agreement with its principles at the October 1994 Heads of State summit. He supports the CIS, but says it has no means to implement its decisions and agreements. Nazarbaev adamantly supports supranational bodies, saying they form the crux of his Eurasian Union proposal.

**10** • Uzbek advisor to the Russian Foreign Ministry speaks of the Uzbek notion of creating a "greater Uzbekistan." He says the idea, bandied about by Central Asian politicians, "does not reflect real tendencies of the Central Asian region" and even the Uzbek "leadership does not share the idea." Competition among the prospective countries—Turkey, Azerbaijan, Turkmenistan, Uzbekistan, Kazakhstan, and Kyrgyzstan—over land and water resources, export markets, and foreign investments, in the expert's opinion, will continue to divide the Turkic community, preventing unity, at least for now.

**11** • Prime Minister Aleksey Bolshakov indicates that Kazakhstan may join the Customs and Payments Union among Slavic states. Kazakhstan had sought to forge closer links with Turkey but, while not abandoning such links, does not feel this course will ultimately succeed.

**13** • The CIS Interparliamentary Assembly adopts a declaration on self-government, which is supposed to ensure local democratic control by citizens over their own affairs. (In other East European governments self-government has referred to government control through appointed, subsidized membership organizations in the private sector.

**14** • Talbak Nazarov, Tajik foreign minister, says Tajikistan's foreign policy will focus on Russia, Uzbekistan, and the CIS. Nazarov believes strengthening ties with these countries will aid with reconstruction following the civil war.

**18** • Konstantin Zatulin, chairman of Russia's State Committee for Relations with CIS Member States, says that since

Russia's invasion of Chechnya, certain CIS members have tried to distance themselves from Russia, complicating integration prospects. According to Zatulin, "One can speak of integration seriously only with reference to Belarus."

**23** • Kazakh President Nazarbaev previews the upcoming CIS Heads of State summit scheduled for mid-February in Alma-Ata. He says closer integration among CIS members is "imminent." He speaks of increased ties among Russia, Belarus, and Kazakhstan—forged as a result of the new customs union, and how this union will serve as a model for CIS members. Nazarbaev argues that bilateral CIS agreements will grow into multilateral structures. He notes that the February summit participants will consider a draft Peace and Accord Act, something that "would be a blessing for all peoples in the post-Soviet area."

**28** • CIS Executive Secretary Ivan Korotchenya, writing in *Rossiyskaya Gazeta,* points to recent progress in laying the groundwork for long-term CIS integration. He praises the development of the Economic Union Treaty, which provides for various levels of membership, and notes that most of the committees associated with the Union are ready to function. He doggedly expresses optimism for the survival of the CIS.

### *Security*

**1** • The chief of the Main Department on Military Budget and Financing, Col.-Gen. Vasiliy Vorobev, announces that the defense budget of 44 trillion rubles proposed for 1995 fails to meet even the minimum needs of the Russian army and fleet.

**6** • President Yeltsin sends New Year's greetings to servicemen in Chechnya: "I know how important it is to know that you are remembered and believed and people's hopes are vested in you when you are so far away from home, performing a responsible and dangerous combat duty." He goes on to say that "your relatives and friends will live in a more tranquil and safer country thanks to your heroic military labor. . . . I am convinced that the people of Chechnya will understand the importance of what you are doing in the name of preserving the homeland, the united and indivisible Russia."

**7** • Interfax reports that the Russian defense complex suffers from a decrease in production, which has seriously undermined defense capabilities. Cuts in allocations for research and design and for the purchase of arms and military equipment have led to cancellation of the production of 175 types of military equipment. The country's defense complex will soon be incapable of building new generations of arms.

• If the space industry declines any further, Interfax reports, all defense programs in space for 1996–97 will be canceled. The share of modern weapons in Russia's armed forces could fall to 10 percent by the year 2000, compared with 60–70 percent in the armies of industrialized Western countries.

**10** • In a letter to Belarusian President Aleksandr Lukashenka, Russian President Yeltsin expresses a wish for greater cooperation between their two countries. Belarus has agreed to lease two military bases—the phased-array ballistic-missile early warning radar under construction at Baranavichy and the Russian submarine communications center in Vileika, near Minsk—to Russia for twenty-five years for a nominal sum covering the maintenance of the bases. In return, Belarus will be given free access to the military information Russia will collect at Baranavichy and to Russian air defense missile ranges.

## February 1995

### *Political*

**2** • The Crimean Supreme Council urges Ukraine to join the Russian/Belarus Customs Union, believing such a move would raise prospects for both Ukraine and Crimea to emerge from their economic crises. In a statement issued by the Supreme Council, it is argued a Customs Union will "restore the severed links, and remove the artificial barriers between the republics."

**7** • In Ukraine, the director of the Foreign Ministry department for CIS problems, Aleksandr Danilchenko, says Ukraine favors CIS integration. In a press interview, he forecasts that the upcoming Heads of State meeting will consider forming a scientific information union, something President Kuchma has been pushing for. This indicates, says Danilchenko, "that the CIS leaders want to design mechanisms to activate cooperation within the CIS." He did note, however, that increased ties should proceed horizontally and that Russia remains a priority for Ukraine.

• After several months of tense relations, Russian First Deputy Prime Minister Oleg Soskovets says, "Ukrainian-Russian relations have become civilized, and it is now possible to expand and consolidate economic and political ties between the two states." A Russian delegation is now in Ukraine seeking to resolve outstanding issues; Black Sea Fleet issues, payment of debts, and energy deliveries are a few. Soskovets and his Ukrainian counterpart, Evgeniy Marchuk, both speak of positive developments in their relations and tell journalists that a number of documents are near completion that would resolve some of these contentious issues.

• Moldova's Prime Minister Andrey Sangheli clarifies that his government is "vitally interested in the Economic Union . . . working at full capacity." Sangheli says both he and Moldovan President Snegur agree that measures to increase and make more efficient economic activity among the republics are needed and it is right that this topic be a high priority on the 10 February summit agenda.

**8** • Russian Foreign Minister Kozyrev denies allegations that Russian policy is geared toward "swallowing up" Belarus. He says the large, interstate treaty promoting integration between the two "has nothing to do with imperial policy" and indicated Russia would like to see relations with Ukraine develop similar to those of Russia and Belarus.

• Soskovets and Marchuk initial a friendship, cooperation, and partnership treaty in Kiev. Other documents are signed relating to the Black Sea Fleet base, which will be in Sevastopol, and on deliveries of military hardware.

**9** • In an appeal to President Kuchma, the Council of Democratic Parties of Ukraine urges the republic's withdrawal from the CIS. The appeal notes that the CIS is "acquiring an ever more immoral character and threatening Ukrainian sovereignty." The Council particularly criticizes the events in Chechnya as grounds for Ukraine's leaving the Commonwealth.

• Central Asian political analyst Kenes Akhmetov addresses the issues of Central Asian integration, concluding that a Central Asian Union as an alternative to the CIS is "impossible." Akhmetov argues that, despite the joint declaration in Istanbul of five "Turkic" leaders indicating their intentions to form a "pan-Turkic" union, developments in the region and the economic realities of the individual republics show that such a desire is "more than anything else, an object of theoretical analysis." The critical factor is Russia. Akhmetov argues that Russia is the region's only security guarantor given the "threats of religious fundamentalism, tribalism, and the military and economic expansion of China." Furthermore, as trade statistics show, none of the republics could survive without strong ties to the Russian economy. Kazakhstan also poses difficulties as it increasingly has shown its desire to augment, at a minimum, trade relations with Iran and China, creating competition in its markets with Uzbekistan and Kyrgyzstan. For these reasons, Akhmetov believes "pan-Turkic" integration is a long way from reality.

**10** • The CIS Heads of State summit opens in Alma-Ata. In his opening remarks, Yeltsin expresses dissatisfaction with developments in 1994. In a tone suggesting Russia's frustration with the non-implementation of agreements, Yeltsin says, "I deem it necessary to state most frankly that the results of the past year were unsatisfactory. We have agreements on cooperation but they are not being implemented." Yeltsin argues that there is no alternative "to a policy of integration and consolidation of the CIS." His chairmanship of the Council of CIS Heads of State is extended for another year, as Azeri President Aliev's candidacy is rejected due to the war in Nagorno-Karabakh.

• The summit produces ten documents. Of major significance is a memorandum "On the Maintenance of Peace and Stability in the Commonwealth of Independent States," pledging that the republics will not use political, economic, or military pressure on other CIS members to achieve foreign or domestic goals.

**11** • Belarusian President Lukashenka, acknowledging the achievement of major goals, nevertheless expresses pessimism about ever implementing the agreements concluded. "I am not quite assured that these documents will be implemented, because all of them were signed either with reservations or with some other considerations," he warns.

• Russia's State Duma Committee on CIS Affairs argues that the CIS Interparliamentary Assembly needs reform. Committee chairman Zatulin notes that the numerical composition of the body gives some republics the advantage over others. Zatulin points to the European parliament as an example the IPA may follow to ensure objectivity and fairness in debating resolutions and in adopting CIS-wide decisions.

**13** • In Moldova, Foreign Minister Mikhail Popov reveals that the central thesis of Moldovan foreign policy is the development of bilateral relations with CIS countries, in particular Russia, Ukraine, and Belarus. Popov indicates that active cooperation with these countries offers Chisinau the best hopes of emerging from the present economic crisis. Of great importance also, notes Popov, is developing its relations with Romania, the United States, Germany, France, Britain, Italy, and Canada. Engaging in multilateral, international institutions, while a separate concept, is also a priority in Moldova's diplomacy.

• Armenian President Ter-Petrosyan comments on the Alma-Ata summit. He criticizes the peace and stability memorandum, which, in his view, conflicts seriously "with the norms of international law, particularly as regards human rights and the rights of national minorities."

**14** • Vladimir Shumeyko is reelected IPA chairman, having received unanimous support at a closed door session in St. Petersburg. Shumeyko comments on IPA development. He says the body has now been recognized as an interstate cooperative body. In this regard, Shumeyko places the IPA on equal footing with the Council of CIS Heads of State and the Council of CIS Heads of Government. Shumeyko believes the next stage for the IPA will be to transform the body into a parliamentary rather than interparliamentary assembly, thus giving it more authority to make binding decisions. Eventually, he argues, the CIS will see an independent parliament with representatives elected along the lines of the European parliament.

**15** • Kyrgyzstan, Kazakhstan, and Uzbekistan form an Interstate Council to continue integration trends among their republics.

**16** • Belarusian Supreme Soviet chairman Myacheslav Hryb believes the proposals by Shumeyko to expand the powers of the IPA are "ill-timed." Hryb says the body should *not* be turned into a "supranational representative body of the Commonwealth similar to the USSR Supreme Soviet."

**21** • Yeltsin and a Russian delegation arrive in Minsk for talks. Yeltsin and Belarusian President Lukashenka sign a treaty on friendship, good-neighborliness, and cooperation and other agreements on the joint Customs Union and on joint protection of borders. In a joint news conference, the two presidents speak of their relationship as a model worthy of emulation. One political observer notes that the "Slav brothers are showing other partners in the CIS quite a realistic model of two countries' cooperative integration."

**22** • In an interview with the German radio station Deutsche Welle, Lukashenka says Belarus will not seek NATO membership but will continue to remain under the political influence of Russia.

**23** • Officials of five agrarian parties from Russia, Ukraine, Moldova, and Armenia meet in Moscow to discuss restoring ties. Ukrainian Agrarian Party Chairman Sergey Dovgan emphasizes that if a referendum on the attitude to the USSR were held in the CIS countries today, it would produce the same results as on 17 March 1991, when the overwhelming majority of the voters favored retaining the Union.

**25** • Nazarbaev predicts a Eurasian Union by the year 2000, by which he means a new version of the Soviet Union. Formation of the new Union will start with an Economic and Customs Union and plans to settle citizenship and language questions.

**27** • Agreement is struck between the State Duma's LDPR (Liberal-Democratic Party of Russia) faction (whose leader is V. Zhirinovskiy) and the Crimean Supreme Soviet. The parties claim they will combine efforts to work out and implement joint political programs that will draw Crimea closer to Russia.

***Security***

**1** • Addressing a veterans group in the House of Officers, Belarusian President Lukashenka says military security is "one of the most important sectors of our state system." While noting the republic's efforts to constitutionally make Belarus a non-nuclear state and assume a neutral state status, he highlights the difficulties of these goals given the geopolitics of the region.

• Improved relations with Russia have made Georgia's population more receptive to Russian troops in the Transcaucasus, says a joint statement by the Georgian Procurator's Office, the Security Service, the Ministry of Defense, and the Ministry of Internal Affairs. Despite terrorist acts carried out in Georgia and abroad by destructive forces in the republic that do not want to see improved Georgia-Russian relations, the report says, Russian servicemen and the Georgian people should "not yield to provocations."

**9** • Armenians oppose the CIS peace pact and will not sign the document in Alma-Ata if their remarks are not taken into account. Foreign Minister Vagan Papazyan says the document ignores human rights principles and the right of national self-determination.

**14** • During the Alma-Ata summit Ukraine adopts the rigid stance that protection of external borders means former Soviet borders. Azeri Foreign Minister Hasanov objects that using such a definition, Azerbaijan would lack a mechanism for safeguarding its frontiers with Armenia. Hasanov confirms Azerbaijan's intention to press for amendments to the CIS Treaty on Collective Security.

**17** • Colonel-General Viktor Prudnikov will assume command of the CIS allied air defense system. The joint system is agreed to by CIS members (excluding Moldova and Azerbaijan) at the Alma-Ata summit. Prudnikov tells journalists that the air defense force will concentrate on air surveillance and exchange of information. He stresses that the agreement excludes anti-aircraft rocket launchers and fighter jets from the joint force, and makes clear that the agreement is strictly circumscribed to broad cooperation. Most CIS states will be supplied the hardware necessary to implement the agreement through bilateral agreements with Russia.

**21** • Russian Defense Minister Pavel Grachev indicates that it is too early to speak of a "coalition" of armed forces with Belarus, but does not reject the idea in principle. He stipulates that "well-equipped and modern troops are needed to create joint armed forces and that therefore the time has not come for such a military coalition." He confirms that the Belarusian early warning facilities in Baranovichiy and Veleika will be under Russian jurisdiction.

**23** • CIS land forces Commander Semenov says that a joint CIS or Eurasian Union armed forces will eventually be set up. He reiterates that both a Commonwealth defense area and a Commonwealth economic area are vital needs, arguing that the CIS member states must adopt the same defense strategy, tactics, and structure. Semenov asserts that joint peacekeeping forces will significantly contribute to military integration, but that Russia is a lone figure in carrying out peace missions at this time. He doubts that a truly joint CIS air defense system will be set up any time soon, mainly because "everything is in shambles" in Transcaucasia where the air defense force used to be stationed. Tremendous expenditure will be needed to restore the system, he says.

• A joint Russian/Belarusian draft agreement containing thirty clauses—one of which elaborates a customs union, another of which establishes joint financial and production groups—is ratified by the Belarusian parliament. In a somewhat defensive statement to his Cabinet of Ministers, President Lukashenka says: "The constitution is not being violated today in any respect. It would be good if we had a chance in these difficult times to work more or less evenly and to move the economy. This is what concerns me most of all."

**28** • Russian Defense Minister Pavel Grachev says: "Russia will have no professional Army under current economic conditions. It is necessary to raise the salaries of contractual servicemen by five to seven times in order to have professional troops." He uses events in Chechnya to illustrate that servicemen "cannot show their worth," and urges the revival of the army and more attention to the armed forces on the part of society and all power structures. "The army is sick," he says, "and must be cured by common efforts."

## March 1995

### *Economic*

**2** • Financial expert for Moscow's *Finansovye Izvestiya* Evgeniy Vasilchuk warns of CIS disintegration should the economic crisis throughout the region continue. He attributes the economic crisis to the "absence of an effective mechanism [for] multilateral cooperation." Despite plans to create an economic union, Vasilchuk says economic integration of the CIS as a whole is prevented by the following factors:

1. The Interstate Economic Committee and the joint official accounting-payment institutes are not in full operation.
2. Certain Commonwealth states block the unified conception and organization for protection of the Commonwealth boundaries, hindering efforts to create an effective customs union.
3. Western refusal to help Russia regain the lead role in post-Soviet geopolitical/geoeconomic space, giving rise to nationalistic forces destructive to integration.

Vasilchuk believes that "the indecisiveness of politicians and their disregard for the strategic benefits of multilateral economic cooperation threatens the Commonwealth with historical non-existence."

**16** • The Turkmen company Turkmenelektro says it may resume delivery of oil supplies to Ukraine in exchange for manufactured goods. Kiev still owes Ashkhabad over $1 billion for previous gas and oil deliveries.

• A draft copy of a report entitled "Guidelines for International Economic Activity" prepared by the Ukrainian Ministry of Foreign Economic Activity gives cooperation with CIS and Baltic countries top priority. Enhancing relations with the European Union is priority number two, with a goal to achieve associated member status in the organization within the next six years. Other notable aspects of the draft include:

—increased investment in mining industries within the CIS;
—creation of international/multinational corporations;
—increased bilateral relations in regions within Russian Federation;
—continued participation in CIS economic structure.

**20** • In Tajikistan, businessmen call for accelerating integration programs within the CIS and for building an economic union of CIS countries. Appealing to CIS heads of state, prime ministers, and parliaments, the business groups believe the new political situation "holds out a hope for equitable and comprehensive cooperation" between Tajikistan and CIS countries.

**27** • The World Bank considers four separate loans totaling $150 million to Azerbaijan. Sixty-five million dollars is to redress balance of payments problems; $45 million to modernize Baku's water supply system; $20 million to develop market infrastructure; and $20 million to update the oil industry.

### *Political*

**3** • In Ukraine, the Council of the Interregional Reform Bloc (MBR) issues a statement on Ukrainian-Russian relations, which categorically rejects the neo-communist call to restore the USSR and declares that the Reform Bloc intends to cooperate with Russian democratic forces.

**6** • According to Deputy Prime Minister Sergey Shakhray, the Russian government, backed by a presidential decision, will develop a state program to support the Russian community abroad in political, economic, and cultural spheres. Shakhray admits, however, that the lack of funds may halt the project. Meanwhile, a related commission debates the guidelines for higher education policy in the CIS and the Baltics, the urgency of creating a single educational space in the former Soviet Union, and the need to preserve the Russian language.

**9** • Responding to the British press on questions about NATO's eastward expansion, Belarus President Lukashenka asks: "How would Great Britain feel if Russian tanks were deployed in France along the Channel coast? But in my country's case, NATO tanks would only be a few meters away. Belarus is very concerned." He asserts Russia knows quite well that NATO is trying to advance, not only to Belarus's frontier but to Russia's as well. He also mentions that Russia is "being provoked," and that "Belarus does not want her people to be trapped between two opposing blocs. Why should anybody want to crush a small state like ours? The West's policy has already led to instability in Russia."

**10** • In Georgia, the Supreme Soviet and Council of Ministers of the Abkhaz Autonomous Republic issue a protest statement against the Russian State Duma's recommendation that the Russian-Georgian border along the river Psou be opened in order to alleviate the acute shortage of foodstuffs and vital necessities in Abkhazia. Opening the border would cause even greater socio-political destabilization in the region and make it easy for illegal armed formations to enter the autonomous republic from the North Caucasus and vice versa. The statement notes the Russian State Duma's contradiction of a CIS Heads of State decision which stresses the permanence of borders and the territorial integrity of Georgia.

• In Uzbekistan, President Islam Karimov works to increase Uzbekistan's international status. He hopes to attract world attention to the Aral Sea problem and the need for peace and tranquility in Uzbekistan so that he may garner international aid.

**11** • Russia's Counterintelligence Service signs a cooperation agreement with the Georgia Security Service, ostensibly aimed at combating crime and terrorism. The agreement provides for building a joint "data base on criminal gangs."

**13** • The Russian Foreign Ministry views Kazakhstan's parliamentary crisis as a phase of "internal development," and voices hope that the crisis will be overcome by constitutional means and that the next legislative elections will adhere strictly to law and recognized democratic norms.

**15** • In Moscow, CIS secret service chiefs meet to discuss joint crime fighting measures and agree to establish a formal organization to coordinate their activities. The head of the Federal Counterintelligence Service department, Yuri Demin, hopes the conference will produce an alliance among the secret services of CIS countries, dismissing the notion that such an alliance is a resurrection of the KGB. Demin says joint activities would focus solely on crimes threatening national security such as terrorism, money laundering, corruption, smuggling, and sabotage of strategic installations, but assured that each service would retain its independence.

• In Belarus, Social Democratic leader Hramada Aleh Trusau opposes Russian-Belarusian agreements on a customs union and military bases, saying he hopes parliament has the "common sense" not to ratify the agreements. Russian Deputy Prime Minister Aleksey Bolshakov threatens to reimpose customs duties if the agreements are not ratified but Trusau says it's a bluff, noting Russia would be the loser if customs duties were introduced.

**16** • In Russia, President Yeltsin meets with Foreign Ministry officials to tell them they will remain the country's main coordinators of foreign policy, thus preserving the role of Andrey Kozyrev. Yeltsin's decision neutralizes his threat to establish a special organ within the presidential staff devoted to foreign policy, something he proposed earlier to the Federal Assembly.

**20** • Russian First Deputy Prime Minister Oleg Soskovets says the main goals of a Russian delegation in Kiev are to determine Russia's share of the Black Sea Fleet and to secure a debt agreement covering money owed by Ukraine to Russia for energy deliveries. Soskovets notes that "considerable progress" was made on both issues, which smooths the way for speedy resolution.

**21** • In Belarus, President Aleksandr Lukashenka proposes new legislative elections and a republic-wide referendum on economic union with Russia on 14 May. He includes in the referendum some questions regarding the status of the Russian language in Belarus and some privatization in the republic.

**22** • An article in Moscow's *Literaturnaya Gazeta* highlights the rising trend of Kazakh nationalism. Author Aleksandr Samoylenko cites as evidence a "secret letter" sent to President Nazarbaev by ninety-nine deputies of Kazakhstan's parliament threatening a three-year moratorium on amendments to the constitution, a move initiated by a wing in Kazakhstan's Communist Party. Samoylenko argues that this wing, now organized under the banner of "national-patriotism," fears that other nationalities will rule Kazakhstan. National-patriotism leader Kuanshalin says introducing state bilingualism, and private ownership, equals "cutting off the Kazakhs' tongue and giving the land to foreigners." Commercial firms and banks, fearful of national-patriotism's views, call for a nationwide referendum on private ownership and the nature of the state—which Samoylenko offers as proof of the growing concern over nationalist sentiments.

**23** • In Georgia, Shevardnadze speaks with Tbilisi radio on ties with Russia and defends the recent Treaty on Military Cooperation with Russia. He acknowledges that while the presence of foreign troops on one's own territory is never desired, Georgia needs Russia for political and territorial security —a vital component of Georgia's long-term independence. Shevardnadze clarifies the fact that he has approached several other capitals seeking military support for Georgia, but "not a single country offered help."

**27** • In Russia, several Duma deputies criticize First Deputy Prime Minister Oleg Soskovets, charging he does not understand the political situation in Ukraine and thus betrays

Russia's interests in negotiations with Ukraine. Led by Konstantin Zatulin, chairman of the State Duma Committee on CIS Affairs, several Russian deputies call for a reassessment of Russian-Ukrainian relations, urging a tougher Russian line. Soskovets says Russia must acknowledge Ukraine's independence and its right to run its own internal affairs, regardless of who occupies the Kremlin.

**29** • Mikhail Shchipanov writes in *Rossiyskaya Gazeta* about a rising trend toward the formation of interest groups within the CIS. He notes an increase in the number of bilateral agreements which may lead to greater cohesion in CIS relations. Russia and Ukraine continue to narrow their differences but political and military disputes (Black Sea Fleet issues, Crimea, dual citizenship) prevent conclusion of an interstate Treaty on Friendship and Cooperation. Shchipanov also notes the dangerous nature of CIS plans to build an oil and gas pipeline from Northern Kazakhstan to a Russian port on the Black Sea coast. The plan could destroy much of the general transportation system within the CIS—a system which "undoubtedly" makes the post-Soviet area more cohesive.

**31** • In Uzbekistan, President Islam Karimov tells diplomats and foreign journalists that any alternative to the CIS is "absolutely untenable." Karimov points to Russia as an "important strategic partner and great power."

• Moldovan President Mircea Snegur and Uzbek President Islam Karimov sign a bilateral friendship and cooperation treaty. The document includes seventeen agreements on specific trade, economic, and cultural issues. In a news conference, Snegur says he hopes that the treaty will become a "firm basis for joint activities in the United Nations, OSCE, CIS, and other international organizations." Karimov says he favors "mutually beneficial" economic cooperation, but objects strongly to establishing transnational political structures in the CIS framework.

### *Security*

**2** • More than 3,000 Black Sea Fleet workers appeal to the Russian State Duma to alleviate the wretched conditions of fleet facilities. The appeal asserts that Russia is intentionally withholding funds to ruin the fleet and calls for immediate funding for materials through Russia's Defense Ministry and for Russia to repay debts as soon as possible.

**3** • In Armenia, President Levon Ter-Petrosyan discusses implementation of the CIS air defense system with CIS Air Defense Committee chairman Col.-Gen. Viktor Prudnikov.

**4** • In Georgia, Shevardnadze and Prudnikov discuss the CIS air defense system. They agree to establish a group of experts to examine Georgia's current system and a coordinating committee to work out a plan of action.

• Russian sailors remain hostile as they depart from their posts in the Black Sea Fleet facilities, now occupied solely by the Ukrainian navy. Following the wartime credo "nothing should be left for the enemy," the sailors remove all portable objects, destroy communication links, and dismantle surveillance systems.

• Kazakhstan pushes actively for renewed cooperation in defense industries with Russia. Kazakhstan still has delivery obligations to the Russian army and navy but Russian finances prevent payments to Kazakhstan, thus jeopardizing the Kazakh defense industry.

**7** • The former Riga OMON (Special Purpose Militia Detachment) in the Dniester region is redeployed to Moscow. The military commander of the "Dniester Republic," Mikhail Bergman, says Russian generals will manage the redeployment, but does not know what new duties the forces have been assigned.

**8** • Ukraine makes official inquiry into the nature of remarks made by Russian navy commander Feliks Gromov in a meeting with Ukrainian officials. Gromov apparently voiced opposition to the joint deployment of the Ukrainian and Russian fleets in Sevastopol.

**10** • *Izvestiya* carries an interview with Russian Air Defense commander-in-chief Viktor Prudnikov, who says a CIS joint air defense system will preserve existing CIS countries' systems and integrate them into a single complex. He says the task remains difficult because many CIS countries' systems (such as Georgia's and Armenia's) are destroyed. Other difficulties include funding for rendering all state systems combat-ready. Prudnikov also stresses that, as stipulated in the Almaty agreement, the joint system's mission is only to enable participants to monitor CIS airspace. Reactions to air space violations or attacks will still be controlled at the state level. Prudnikov will travel to CIS capitals to assess each state's requirements for hardware and specialists.

• In Kazakhstan, *Kaztag* interviews Chairman of the Committee for the Defense Industry Kadyr Baekenov. Baekenov describes economic, scientific, and technological agreements which the two nations have signed in the defense sector. The Kazkontrakt joint-stock company and Oboronresurs, a state-owned enterprise, will ensure that Russian and Kazakh defense industries provide each other with supplies, material resources, and components. Baekenov suggests creating an interstate structure for cooperation among all CIS defense industries.

**12** • In Belarus, members of the Popular Front continue to collect signatures from Belarusian citizens to be submitted to the president and the Supreme Soviet. The appeal, now with over 40,000 signatures, protests against "the participation of Belarusian servicemen in any military conflicts of other states, against locating Russian bases or troops on Belarusian territory, and against the participation of Belarus in the CIS collective security system."

**13** • Ukrainian President Kuchma and Russian Deputy Foreign Minister Yuriy Dibinin discuss Black Seas Fleet issues in Kiev. Kuchma says the Black Sea Fleet is the "only stumbling block in Ukrainian-Russian relations" and that the issue should be resolved before Yeltsin's upcoming trip to Kiev.

**16** • Russian Strategic Missile Forces chief Vladimir Krivomazov says the remaining SS-25 "Topel" ICBMs will be withdrawn from Belarus by 25 July.

**20** • In Moldova, parliament approves the first version of the republic's military doctrine. The doctrine declares Moldova a neutral state that will build its security through preventive diplomacy, and building good neighborly relations with other defense departments in the region. The doctrine also foresees creating a military force "adequate for the protection of its territory and citizens."

**21** • In Russia, Commander-in-chief of Russian ground forces Vladimir Semenov says troops should be brought up to full strength and "manned with the best soldiers." He notes that of the 300,000 new recruits needed to meet full strength status, only 10 percent have been recruited, leaving many military districts grossly understaffed. Semenov also calls for the ground forces portion of the armed forces to be brought up to 50 percent, saying the unreasonably small share of the ground forces at present "is very dangerous in the current situation."

• Russian Defense Minister Pavel Grachev arrives in Tbilisi for talks with Shevardnadze and Defense Minister Vardinko Nadibaidze on military and technical cooperation, a CIS Collective Security Treaty, and a proposal for a treaty on military bases for Russia.

**22** • Addressing CIS heads and parliamentary leaders, Black Sea Fleet Air Force commanders warn that the fleet is threatened by a reduction in units and logistical support bases and by political rivalry between nationalist-separatists and reformers in Ukraine and Russia. The commanders express dismay with the United States, who they claim "is spreading its interests eastward," so that Ukraine and Russia become weaker militarily. The group urges CIS lawmakers to lay aside grievances and mistrust concerning collective defense to rest, strengthen the unity of the Red Banner Black Sea navy, and form a single center to coordinate national air force actions.

**24** • In Russia, navy Commander Felix Gromov says a "single Black Sea Fleet under erratic [dual] command will be a strange structure." Speaking before the Duma, Gromov argues that the fleet must be based in Sevastopol and "must fly the flag of one state, whose interests it represents and whose political tool it is."

**25** • Russian army General Pavel Grachev and Georgian leader Shevardnadze initial a Treaty on Military Cooperation, granting Russia four military bases in Georgia for a twenty-five-year period. Shevardnadze views the accords as an important factor in Georgia's stability.

**27** • Shevardnadze says he appreciates the Russian military presence in Georgia but warns that Georgia's parliament is free to declare the military base agreement "null and void" should relations with Russia deteriorate.

• Georgia's parliament remains split over the defense accords. While a majority supports Shevardnadze's policy of strengthening relations with Russia, the "Republicans" in parliament regard the accords as interference in domestic affairs, and call for Defense Minister Vardiko Nadibaidze's release, and for terminating the Russian peacekeeping presence in Abkhazia after 15 May, when its mandate expires.

**30** • In Armenia, more than 1,500 soldiers take part in joint Armenian-Russian military exercises. Deputy Commander of Russian troops in the Transcaucasus Pavel Labutin credits the exercises with exhibiting a high degree of training and level of understanding between commanders and personnel of both armies.

### April 1995

#### *Economic*

**3** • The Central Asian Bank for Cooperation and Development grants its first loan—to the joint-stock company Sayam. Serik Primbetov, chairman of the executive committee of the intergovernmental council, notes the importance for the region of the electrical project Sayam will undertake. Primbetov comments that funds for other projects in Kazakhstan, Uzbekistan, and Kyrgyzstan are under review and that the bank will soon prove its worth.

**10** • The CIS Interstate Economic Committee (IEC) presidium meets to discuss a CIS customs union. Ukraine's Deputy Prime Minister refuses to sign, commenting that it is too early for such a broad agreement on free trade. He says the bilateral Russian-Ukrainian economic cooperation agreement will suffice and will be a test of such cooperation. At the same meeting, representatives adopt a resolution to advise CIS Heads of State to form an Interstate Currency Committee (ICC), which would work to promote cooperation in currency exchange policies within the CIS, and promote currency emission control. According to press accounts, the structure of the ICC is being "discussed behind closed doors."

• The Russian Ministry for Cooperation with Commonwealth Member States reports that in the first two months of 1995, Russia reduced its oil exports to other CIS members to 3.2 million tons, which is 42 percent of its deliveries in the same two months of 1994.

**13** • The agreement between Russia and Kazakhstan to create a free flow of goods across borders through uniform customs controls has failed. The Kazakh Ministry of Finance has been the principal spoiler. Belarus and Russia are progressing toward agreement on the removal of border controls, but the tripartite idea has come to an impasse.

**14** • Article by Georgiy Bovot says that without peace between Armenia and Azerbaijan, there will be no multi-billion-dollar contract to develop the Caspian oil fields. Bovot examines the geopolitical game swirling around the two countries—a game that is much larger than the struggle for Nagorno-Karabakh. Namely, he analyzes the strategic economic contest among Russia, Iran, Turkey, and the West for oil. The article summarizes Azerbaijan's position, asserting that by signing a contract with a Western consortium, it has established the precedent of distributing Caspian Sea resources unilaterally. Now, the issue is the route the Caspian oil will take to Western markets. Moscow continues to insist that the oil cross Russia, more precisely Dagestan, and the ill-fated Chechnya, to terminals in Novorossisk. The article also discusses the positions of all major players in the game, including Washington, DC, the European Union, Iran, and Turkey, concluding that there is much less economics than politics involved.

• In Ukraine, Shevardnadze and Kuchma sign a treaty which provides for Ukrainian assistance to Georgia with developing petroleum plants and coal mining. Ukraine will also construct hydroelectric power stations and sell electricity to Georgia at "reasonable rates." Press account says the plan for a "Euro-Asiatic" corridor for transporting Azerbaijani oil products across Georgia to Ukraine is gaining visibility. The plan will allow Ukraine to "straighten out its energy transport route" and Georgia to "earn transport taxes for its budget."

**25** • At a meeting in Bishkek, the prime ministers of Kazakhstan, Kyrgyzstan, and Uzbekistan approve a five-year economic integration program. The program gives priority over the next two years to cooperative production of small electrical engines, gas meters, medicines, and fertilizers.

**27** • In Russia, Finance Minister Vladimir Panskov denies charges that Russia underwrites over half of the other CIS countries' economies, saying the Russian government provides credits only to Belarus, Tajikistan, and Armenia. Russian economic and financial aid to other CIS countries, he says, is "very insubstantial."

**28** • Azerbaijani oil from the Caspian is to be transported to Europe through one of four variant pipelines. One provides for transportation to Turkey via Iran, which is unacceptable to the international consortium that signed an agreement on the joint development of three Caspian oil deposits in September 1994. Another would build a pipeline via Armenia to a Turkish Black Sea port, but can only be fulfilled if Armenia and Azerbaijan reach a political agreement on the Nagorno-Karabakh conflict. A third variant is to build a line through the Northern Caucasus to Novorossisk in Russia, in which Russia is interested. The fourth would cross Georgia and wind up in Batumi. In this case, the transportation mode might be railroad. The consortium will meet in London on 3–4 May to discuss construction of the pipeline. Oil extraction is scheduled to begin in eighteen months.

**29** • Sergei Primbetov, chairman of the Intergovernmental Council of the Republics of Kazakhstan, Uzbekistan, and Kyrgyzstan, states that the Central Asian countries should form a regional economic union like the GATT. The region is economically interdependent; Kazakhstan depends on Uzbekistan for gas and on Kyrgyzstan for electricity. Uzbekistan is pursuing a strategy toward a self-sufficient economy, but is simultaneously integrating its economy on a regional basis.

### *Political*

**3** • Leaders from the agrarian parties of Russia, Belarus, Moldova, Armenia, and Ukraine's Peasant Party issue a declaration calling for the reintegration of the people of the former Soviet Union. Noting the suffering and deprivation caused by the USSR's break-up, the members call for recognition of the common historical roots and traditions shared by the people in the post-Soviet republics and claim that only through unity can the populations of the republics enjoy social and economic well-being.

• In a related story, representatives of the agrarian parties agree at a meeting in Moscow to create an International Association of Agrarian Parties to influence governments and work to cultivate the integration they desire.

**7** • In talks with U.S. Defense Secretary William Perry, Islam Karimov urges increased U.S. participation in Central Asia, a presence he terms a "guarantee of security." Karimov lists three factors threatening Uzbekistan's independence: (1) imperialist ambitions in Russia, which intensify daily; (2) the fundamentalist threat from the south; and (3) the problem of irreversibility of certain systems from the past.

**18** • Russian Foreign Minister Kozyrev announces that Moscow reserves the right to intervene militarily to protect the rights of ethnic Russians living in former Soviet republics. Kozyrev notes that some 240,000 Russians have emigrated from CIS states to Russia, calling it evidence of the harsh treatment they are subject to.

**23** • In Belarus, President Lukashenka rejects a Russian proposal, signed by Russian Prime Minister Chernomyrdin, calling for stationing 100 Russian customs officers on the Belarusian-Ukrainian border. (Ukraine has refused to join the customs union with Kazakstan.) Lukashenka says Belarus can control its Ukrainian border by itself and asks Russia to "trust Belarus."

**28** • Armenian Prime Minister Grant Bagratyan denounces recent calls, made chiefly by communists, for an Armenian-Russian confederation. Bagratyan argues for Armenia to preserve its sovereignty and independence and criticizes those who "try to cash in on today's hardship and spread defeatism in order to win political capital and seize power." He notes the country's reforms have now begun to yield "positive results."

**29** • The Council of CIS Heads of State appeals to peoples of the former Soviet Union to pool their efforts to prevent war. Striving "to consolidate world peace," the leaders urge all peoples "to resist new threats to peaceful life . . . and to eliminate existing and avert potential conflicts."

### *Security*

**12** • In Ukraine, Deputy Prime Minister and Defense Minister Valeriy Shmarov argues against swift entry into NATO, calling instead for a gradual entry into European political structures and participation in the Partnership for Peace. He comments that quick admission into NATO would be a "hasty step disturbing the European balance of power, resulting in a new confrontation."

**19** • In Russia, Defense Minister Pavel Grachev receives support from the Foreign Ministry for his statements indicating that Russia might not fulfill certain CFE treaty provisions. Foreign Ministry spokesman Gregory Karasin says that although his ministry tended to support full compliance with CFE in the past, he hopes the United States and other nations "will understand our position and take account of the Russian Federation's interests."

**21** • CIS foreign and defense ministers agree to extend the mandate for peacekeeping operations in Tajikistan and Abkhazia until the end of 1995. The extension plan still awaits approval by CIS Heads of State who are due to convene on 26 May.

• In Tajikistan, Lt.-Gen. Valentin Bobryshev is named the new commander of Russian border forces, replacing Colonel-General Patrikeev.

**22** • In Ukraine, acting Prime Minister Evgeniy Marchuk says Ukraine "will never allow the Black Sea Fleet to be headquartered de facto—much less de jure—in Sevastopol," saying the presidential agreements on the Black Sea Fleet "make no sense."

**26** • Russian Strategic Missile Forces Chief-of-staff Col.-Gen. Viktor Yesin reports that all Soviet-era nuclear warheads have been transferred from Kazakhstan to Russia. (Kazakhstan possessed 104 giant SS-18 ICBMs, each loaded with ten nuclear weapons.)

• Secretary of the CIS Council of Defense Ministers Lt.-Gen. Leonid Ivashov says that the CIS defense ministers plan to establish a Joint Chiefs of Staff Committee of the CIS armed forces. The committee will include chiefs of the General Staffs of the CIS national forces.

**27** • In Ukraine, President Leonid Kuchma tells citizens that Ukraine must adopt a new military doctrine as soon as possible. "After a new doctrine has been adopted, we should organize our armed forces in accordance with it . . . and not leave our army to deal with its problems on its own."

**30** • President Boris Yeltsin signs into law the Duma bill that increases service in the army by two years.

## May 1995

### *Economic*

**3** • In a quarterly report on the CIS economies, *Kommersant-Daily* says that in comparison with 1989, the last year in which some growth was observed in the Russian economy, the gross domestic product and national income will decline 54 and 57 percent respectively in 1995. It predicts similar declines for 1996. According to current forecasts, inflation will not fall below 200 percent. The structural and technical degradation of Russian industry will also continue.

• In Russia, head of the Ministry for Cooperation with CIS Member States Valeriy Serov says Russia will continue to conduct talks on restructuring CIS states' debts on a bilateral basis. Only one bilateral agreement has been signed to date—with Ukraine on 20 March 1995. Kiev is allowed to defer its debts, which were in default in 1994. Ukraine will give Russia promissory notes, denominated in U.S. dollars, for the remainder of the debt. Because the Russian federal budget does not provide for granting state credits to the CIS countries in 1995. Several other CIS countries have asked Russia to restructure their debts, including Kyrgyzstan, Georgia, Moldova, Uzbekistan, and Turkmenistan. In preparation, the International Economic Committee has prepared proposals for creating an International Currency Committee, which will be considered when the Heads of Government meet at the end of May.

• *Kommersant-Daily* reports that the agreement on rescheduling CIS countries' debts sounds good in principle, but in detail has a long way to go. The CIS ruble zone died in March 1995, the article's author states, when Tajikistan decided to introduce its own national currency. Thus, the hope that a ruble zone would annul all debts evaporated also. Now, the members are coming to grips with a payments agreement, but so far progress has only been pursued at bilateral levels. Not surprisingly, at the CIS level, Russia's Ministry for Cooperation with CIS Member States is playing the lead role.

**4** • Phase one of the Belarus-Russia customs union ends on 9 May. During this phase, Belarus has simplified regulations for rail and air transport across the Belarusian-Russian border, but all cargoes and the routes they will take must still be declared. The Belarusian State Customs Committee announces that "it will take some time to create a customs union."

• An article in *Vechirniy Kyyiy* by Mykola Mykhalchenko, president of the Ukrainian Academy of Political Sciences, warns against continuing the current line in Ukrainian-Russian relations. The author is pessimistic about democratization in Russia and says Russia is following a "neo-imperial foreign policy." He criticizes Russia's economic policy toward the newly independent states, saying that Russia has appropriated all former Soviet reserves of precious metals and stones, and other material treasures of the former empire, leaving Ukraine with the rundown mines of Donbass, which would have ruined Russia financially had it tried to modernize them. He is firmly critical of the outlook in Ukraine that Ukraine's only economic course is to reintegrate its economy with Russia. He says "this viewpoint has nothing in common with scientific studies." He calls economic "integration" a policy of strengthening economic dependence on Russia, which is the opposite of what Ukraine needs. Russia's attitude toward Ukraine, he says, is already a traditional one of an imperial center toward a province. For this to change, Russian government must institute the following principles:

- an unconditional acknowledgement of Ukraine's territorial integrity and decisive restraint from entering pro-Russian separatists' activities;
- rejection of using economic levers to exert pressure on Ukraine;
- termination of psychological and informational warfare with Ukraine;
- conducting balanced policies to resolve the old problems (in particular, that of the Black Sea Fleet).

**5** • In Belarus, leaders from nineteen Russian, Ukrainian, and Belarusian oblasts hold a three-day meeting in the city of Homel devoted to issues of economic integration, including direct business and trade ties. The leaders issue an appeal to their governments to step up the process of economic integration among their countries.

• *Rossiyskaya Gazeta* publishes contents of the tripartite Customs Union Agreement among Russia, Belarus, and Kazakhstan. The agreement supplements the Customs Union Agreement between the Russian Federation and Belarus of 6 January 1995.

**6** • In Russia, the Ministry of Foreign Economic Affairs introduces a bill to restore restrictions on exports and imports. The Duma passes its own bill, which is very similar to the ministerial draft. A State Duma committee on economic policy is also formed, which is dominated by "etatists" who favor a return to nontariff regulations on trade as soon as the IMF loosens its hold over Russian economic policy.

• *Kazakhstanskaya Pravda* carries an article that warns of the dangerous state of Kazakhstan's nuclear plants—which are all plants of the Chernobyl type—"Aleksandrov boilers" (named after academician Aleksandrov who designed them). Kazakhstan contains 40 percent of the world's known uranium reserves—but the expense of developing safe nuclear plants and burying the waste is far too great for the new state at this time, the article argues.

**12** • In order to pay Ukraine's gas debt to Russia, the Ukrainian government announces that it will cut off gas supplies to those Ukrainian companies and residents that do not pay the government during the summer.

**13** • In Belarus, *Respublika* reports President Lukashenka's statement that debts to Russia hang over the republic like a sword of Damocles. The debts for oil and gas, valued at world prices, are especially onerous. By the end of 1994, Belarus's debt to Russia for energy amounted to $460 million; $420 million for gas and $40 million for oil. The debts have been partially caused by the unwillingness of consumers to pay for energy supplies, as well as an attitude that "we can live off of Russia." The Russian government has demanded repayment in the form of shares in Belarusian companies, which if adopted as a solution could put Belarus's best companies in Russian hands, according to the article. The article goes on to criticize Russia for unjustly distributing the former Soviet Union's property, keeping all oil and gas complexes located in Russia as its own monopolistic property. The author suggests that each new state should have received a share in the oil and gas resources equal to its share in the former USSR's national income. In conclusion, however, the author states that Belarus must unify its economy with Russia's and receive energy for below-world-market prices.

**15** • In Kyrgyzstan, President Askar Akaev receives a Consultative Council on Labor, Migration, and Social Security of the CIS States. He says that all CIS states should adopt this council's recommendations, giving priority to solving social problems from which all former Soviet republics suffer.

**16** • Azeri and Georgian representatives hold a joint press conference discussing prospects and benefits of transporting Caspian Sea oil through Georgia.

• In Uzbekistan, Savdogar (Tashkent) reports on President Karimov's speech at a conference of Central Asian states, calling for a single "moral, economic, and political zone in Central Asia." The prospects for future cooperation were outlined at a meeting on 14 April 1995 in Shymkent, which noted that the foundations for a union of Central Asian states now exist. The author of the article notes that the public movement "Turkestan—Our Common Home" is growing.

**17** • In Kazakhstan, a conference on the Caspian Sea in Alma-Ata ends. The participants were Russia, Azerbaijan, Iran, Turkmenistan, and Kazakhstan.

**18** • The CIS summit in Minsk on 25–26 May will consider the accession of other countries to the customs union agreement. Yuriy Kravchenko, chairman of the State Customs Committee of Ukraine, announces that Ukraine will not join the union because of the lack of economic and political parity among the CIS states.

**23** • *Vo Slavu Rodinu* reports that customs control at the "Kozlovichy" and "Mokrany" border passes will be conducted jointly by Belarus and Russia in the near future. Russian customs officers arrived in Brest on 16 May 1995.

**25** • The third assembly of the International Congress of Industrialists and Entrepreneurs is held in Chisinau on 25 May. President Mircea Snegur hosts the congress, which includes CIS states and businessmen and managers from Central European countries and China.

• In Belarus, CIS heads of government and heads of state meet to discuss economic issues, including restructuring of interstate accounts from 1992 to 1993.

• In Ukraine, a Supreme Council deputy, Yu. Boldyrev, writing in *Donetskiy Kryazh* (19–25 May), reviews the terms of a $1.5 billion IMF credit to Ukraine. The credit will allow Ukraine to pay for gas imports from Russia and Turkmenistan and to support the value of the karbovanets. According to the agreement, Ukraine must pay for gas in hard currency. If one payment is missed, the agreement on debt restructuring becomes void. Boldyrev criticizes the IMF for acting on behalf of creditor countries, not recipient countries. The author states that $1.5 billion paid to the Russians will have a negligible impact on the Ukrainian economy, which needs $20 billion in one lump payment. In his opinion, the former Soviet states should reunite on a new basis. Without Russia, Ukraine is a "banana republic," he says. His opinions echo those of many deputies in the Supreme Council who support the formation of a new Slavic union.

**30** • Boris Yeltsin and Viktor Chernomyrdin point to the "conclusion of the first stage" of forming a customs union as the main achievement of the CIS summit of 25–26 May. On other questions, such as "CIS borders," Ukraine's Leonid Kuchma argues that CIS member states would not accept the idea of external borders, only borders of independent states.

• Russian border troops suspend railway traffic between Azerbaijan and Russia, alleging that Azeri cargoes contain arms and drugs.

***Political***

**3** • In Belarus, President Lukashenka says he believes that Belarus "is prepared for the closest possible integration with Russia." He predicts that in the 14 May referendum, 80–90 percent of Belarusians will support increased integration and making Russian the state language. Asked whether the world will recognize the merging of the three Slavic states—Ukraine, Belarus, and Russia—Lukashenka says he believes this will depend on the peoples of these countries.

**6** • In Ukraine, Kuchma updates journalists on Black Sea Fleet talks, saying Ukraine is prepared "to hand over practically the whole infrastructure of Sevastopol for stationing the Black Sea Fleet. [However], Sevastopol cannot be handed over to Russia as a naval base." Kuchma highlights the problem as the only remaining one in Ukrainian-Russian relations. He places much hope in his proposal to establish joint customs and frontier posts on the Russian-Ukrainian border as a means to resolve the Black Sea Fleet issue but says Russia has not responded.

**13** • The CIS IPA holds its sixth plenary session. All members less Uzbekistan send delegations to discuss coordination of lawmaking in the CIS states. The Assembly approves eleven documents covering such areas a migration of labor, consumer rights, protection of citizens, and the rights of prisoners of war. Most delegates support continued work on unifying the civil legislations of the CIS countries and believe that the IPA should receive a higher stature in governing CIS affairs. In his address, Chairman Shumeyko laments that the body's potential "is not fully used."

• The Council of the IPA decides to turn the body into an interstate organization. The new body would parallel in stature the Council of CIS Heads of State and the Council of CIS Heads of Government. The draft document outlining the IPA's new role will be submitted for approval at the 26 May heads of state meeting in Minsk and, once accepted, forwarded to the individual republics for parliamentary ratification.

**17** • In Russia, Foreign Minister Kozyrev says Russia favors an energetic "foreign policy which will be firm and even aggressive but not in a military sense." Kozyrev was quick to point out that Russia will not pursue a policy that is "nationalist in the sense of being xenophobic, imperial, or chauvinist," saying Russia's foreign policy priorities will not allow these elements. Overall, he says Russia will transform its ideologically driven policy into a policy that meets national needs.

**18** • While in Russia for talks on bilateral relations, Turkmen President Niyazov signs accords with Yeltsin. Both agree to protect the Russian ethnic minority in Turkmenistan and Turkmen nationals residing in Russia. A number of other agreements are reached in the defense sphere, including a strategic partnership agreement that will last until the year 2000, the first of its kind in CIS relations.

**20** • In Kazakhstan, President Nazarbaev forms a working committee to promote CIS cooperation. Khaliq Abdullaev will chair the new body, named the State Committee of the Republic of Kazakhstan for Cooperation with CIS Countries.

**23** • The CIS executive office in Minsk releases the agenda for the 26 May CIS summit. The Council of Heads of State is expected to review eight issues, the main ones being creation of an interstate currency committee, an extension of the peacekeeping mandate in Tajikistan and Abkhazia, and procedural issues connected with the Collective Security Treaty. Another fourteen issues are to be tackled by the Council of Heads of Government. Observers forecast fierce debate over drafting of a convention on human rights and freedoms.

**25** • Uzbek President Karimov talks about the CIS in a press interview, explaining that Uzbekistan regards the CIS's main role as economic, along the lines of the EU. He expresses his hope that the CIS—like the EU—will soon have free movement of goods, capital, and people.

**26** • At the CIS summit in Minsk, as expected, there was disagreement on a number of issues. Only four republics (Russia, Belarus, Kazakhstan, and Kyrgyzstan) approve a unified border security system and most republics judge the Convention on Human Rights as premature. Agreement is reached on Russia's proposal to establish a regional currency committee and on the shipment of special and military-related cargoes. At Georgian leader Shevardnadze's urging, the CIS leaders extend the peacekeeping mandate in Abkhazia and Tajikistan. While Yeltsin and Russian Prime Minister Chernomyrdin view the meeting as success, others are not so enthusiastic. Belarusian President Lukashenka believes the summits are increasingly addressing questions of secondary importance instead of dealing with economic matters which will "unravel all the knots." Uzbek's Karimov agrees.

**29** • Tajik President Rakhmonov applauds developments between Russia and Belarus, saying that in the future, "all obstacles in relations between Russia and Tajikistan will also be removed."

### *Security*

**1** • In Uzbekistan, President Karimov creates a National Security Council tasked with drafting a law, "On National Security," in coordination with the Ministry of Defense. The council reports directly to the president, in a fashion similar to the Russian president's Security Council.

**4** • In Belarus, Defense Minister Anatoliy Kostenko writes that the republic will follow a policy of "minimum defense sufficiency," thus abandoning the neutrality doctrine. The policy includes rearming the troops with state-of-the-art weapons systems and creating a superior training program at all levels. In 1995, he says Belarus plans to disband 54 formations, units, and institutions, and reform 228, as a result of which the armed forces will be reduced by 16,000 men, and by 1 January 1996 total numerical strength will be 100,000 (*Krasnaya Zvezda,* 4 May).

• In Azerbaijan, President Geydar Aliev announces on television that Azerbaijan will cooperate with NATO under the auspices of the Partnership for Peace by providing information on compliance with the CFE (Conventional Forces in Europe) Treaty in the Caucasus. Azerbaijan has new ambassadors in Brussels and Vienna, which Aliev hopes will help Azerbaijan "consolidate its ties with NATO." Aliev notes that Russian armed forces have been based in Armenia and Georgia, while Azerbaijan has only its own armed forces to defend its borders.

• Turkish Foreign Minister Erdal Inonu says Turkey will continue to provide "friendly" military assistance to Azerbaijan. In the same statement, he mentions Turkey's desire to have a pipeline for the export of Azerbaijani oil through Turkey, but Azerbaijani Foreign Minister Hasan Hasanov says the possibility of having the pipeline run through Armenia on its way to Turkey is simply "not on the agenda."

**7** • In Russia, Defense Minister Pavel Grachev reviews prospects for the Russian army on the occasion of its third anniversary. He reminds officers that 700,000 servicemen and 45,000 pieces of combat hardware have been redeployed to Russia from "Eastern" Europe and the Commonwealth countries over the past few years, the largest any country has experienced. He says Russia's strategic nuclear forces remain Russia's chief deterrent, but that the armed forces are equally important.

**17** • Presidents Boris Yeltsin and Sapamurad Niyazov tell journalists that Turkmenistan and Russia now have a "strategic partnership" through the year 2000. Turkmenistan is the first CIS member nation with which Russia has established this relationship, Yeltsin asserts. He adds that the two presidents exchanged views on counterbalancing "Islamization" of Central Asia with CIS coordination, emphasizing that Islamization could have dire consequences and pointing out Russia's problems with Afghanistan.

• In Kyrgyzstan, Supreme Council deputies ratify a military cooperation treaty between the Kyrgyz Republic and

Russia and another on how and when the Kyrgyz government may use Russian military facilities on its territory and one on the status of servicemen.

**15** • In Ukraine, the secretary of the National Security Council, Vladimir Gorbulin, tells journalists that a thoroughly balanced approach must be adopted toward NATO, stating that: "it cannot be only NATO and Russia, or NATO and Ukraine, or Russia and Ukraine." He adds that Ukraine has proposed a new model for NATO's relations with other states.

**19** • In Georgia, the chairman of the Supreme Council of the Autonomous Republic of Abkhazia and deputy prime minister of Georgia, Tamaz Nadareishvili, addresses the Georgian parliament. Nadareishvili harshly criticizes Russian peacekeeping troops in the Georgian-Abkhaz conflict zone, noting that they have merely demarcated the border between Georgia and Abkhazia. In his words, refugees from the zone will be unable to return to Abkhazia as long as Abkhaz leader Ardzinba stays in power, implying that Russian troops are aiding the leader. He says that 700,000 mines have been laid in the Gali district alone. Moreover, he notes that some Russian troops have no right to be part of a peacekeeping action (referring to forces of the airborne assault regiment No. 345 which participated in the storming of Sukhumi). Another deputy reads aloud a petition signed by 70,000 Abkhazian refugees calling on the Georgian government to refuse Russian participation in peacekeeping operations and demanding that Russia withdraw its contingent from the Georgian-Abkhaz conflict zone.

**24** • In Ukraine, Deputy Defense Minister Vasiliy Sobkov says Ukraine will offer two military training ranges (Privko near Lvov and Shirokiy Lan in the Odessa military district) for lease to Western troops. Germany, France, and Britain currently send their troops to Canada for training.

**25** • Director of Russia's Federal Border Service Andrey Nikolaev says that a single system for defending the borders of the Commonwealth is the next step after signing the Treaty on Joint Guarding of the External Borders of CIS. Seven CIS heads of state have initialed the treaty. Ukraine, Turkmenistan, Azerbaijan, and Moldova are refusing to sign. This will make it possible to establish a "single border regime" on CIS territory. The situation now, he remarks, is that each state has its own vision of collective security and guarding of borders. Touching on the most complicated "common borders" issue, Nikolaev says CIS states need to coordinate their "positions" toward powers near the border, using the example of China, with which Russia, Kazakhstan, Kyrgyzstan and Uzbekistan each has its own relations. "On the one hand, this is correct," he says, "but on the other hand, it would be sensible to link the foreign policy lines of all the CIS countries."

• At the Minsk session of the Council of CIS Heads of State (26 May 1995) military issues will dominate the agenda, while, atypically, economic ones take a back seat. Uzbek President Islam Karimov remarks on Uzbek television, "I do not like the way this agenda is formed. I think that the CIS was above all created to resolve economic issues." Karimov explains that most of the heads of state are against the CIS's becoming a subject of international law, or turning into the former Soviet Union. In his view, the European Union is held together by the resolution of economic matters, and that is what ought to form the basis of the CIS. He would like the free movement of all goods and people across member-state lines.

**26** • In Russia, Boris Yeltsin submits to the lower chamber of the Russian parliament several agreements with Kazakhstan on the status of army units temporarily deployed on Kazakh territory, and on the Russian citizens under contract in the armed forces of Kazakhstan. He tells parliament that the agreements meet the interests of the Russian Federation, "creating a legal basis for the deployment of Russian army units on the territory of Kazakhstan. . . ." President Nazarbaev is also working for ratification of the agreements on Russian citizens under contract in the armed forces of Kazakhstan, on which he issued a decree on 24 May.

**30** • Russian special envoy and chief negotiator in talks with Georgia Feliks Kovalev tells reporters that at the Minsk CIS summit of heads of state, the mandate of CIS peacekeepers in the Georgian-Abkhaz conflict zone was "clarified." "The CIS leaders signed a document which says that CIS peacekeepers will more actively carry out their peacekeeping job," said Kovalev.

### June 1995

#### *Economic*

**3** • In Armenia, transportation ministers of Armenia, Iran, and Turkmenistan sign a cooperation agreement in Yerevan for Armenia to treat the use of Turkmen transport facilities as repayment for gas supplied to Yerevan. Armenia has experienced serious energy shortages due to the Azerbaijani and Turkish blockades. The agreement facilitates Turkmenistan's ties with Armenia through Iran.

• Moldova's foreign trade turnover increases 21 percent during the first four months of 1995 over the same period in 1994, with exports to the West up by 72 percent. Western countries, unlike CIS members, do not apply excise and other taxes on Moldovan exports. Moldovan goods are not competitive in the CIS due to their high prices, which result from the high share of energy costs in production. In 1994, prices in Moldova averaged 2.4 times those in Russia, due to the incompatibility between the leu and the ruble.

**5** • In Russia, news agencies report the surprising strengthening of the ruble against the dollar, which is beginning to crowd dollars out of circulation and create ruble "shortages." Russia's currency reserves are growing (now topping $8 billion according to some reports) due to the reductions in credits issued by the Central Bank and the policies of its forty-five-year-old leader, Tatyana Paramonova.

• Ukrinform reports on the creation of a Russian-Ukrainian joint-stock gas-transporting society—Gaz Transit. The Russian monopoly gas firm, Gazprom, is to contribute its share of the company in the form of "retired" penalty payments on Ukraine's delayed payments in 1994 for gas supplies. Ukraine is to supply two gas storage facilities—Bilche-Volytsko-Uherske and Bohorodchanske. However, Russia has valued these facilities at $23.6 million and $82.2 million, respectively, while Arthur D. Little says they are worth $9.7 billion and $958 million, respectively.

**8** • In Moldova, parliament ratifies the CIS agreement on the formation of the Interstate Economic Committee (IEC) within the CIS Economic Union and the Agreement on Establishment of the CIS Payment Union.

**13** • Ukraine will receive $1.5 billion from the IMF in five tranches over a period of twelve months. It has already received the first tranche of $70 million. Each subsequent tranche will be granted after a review of Ukraine's 1995 economic reform. (In October 1994, Ukraine received an STF credit of about $700 million from the IMF. On 1 June 1995, Ukraine received a credit of 85 million ECU from the European Union.) The Ukrainian government has pledged to privatize within the first six months of 1995 about 1,000 medium and large businesses, as well as to liquidate five bankrupt enterprises and to lift grain export quotas.

**14** • *Kommersant-Daily* reports that the Russian firm Lukoil has formed a new oil consortium in Azerbaijan, in which it will take the largest share (35 percent), while the Italian firm Agip and the U.S. firm Penzoil take 30 percent each. The Azerbaijani state oil company is given a 5 percent share in the consortium. The oil will come from the "Karabakh" oil deposit in the Azerbaijani sector of the Caspian shelf.

• In Ukraine, the government imposes a 15 percent import duty on imports of gasoline, coal, and diesel oil. In addition, VAT exemptions, excise tax exemptions, import duty exemptions, and customs duty exemptions are all being eliminated. Proceeds from the new duties will be used to create a stabilization fund for the fuel and energy complex.

**15** • Georgia owes Turkmenistan $394 million for gas delivered during 1993 and 1994. As a consequence, Turkmenistan has reduced gas deliveries from Turkmenistan to Georgia from 4.3 million cu. meters to 1 million cu. meters per day.

**16** • In Georgia, Abkhazia protests Russian peacekeeping forces in Abkhazia, in connection with the take-over of the Inguri hydroelectric station. Abkhaz separatists can no longer pressure the Georgians by limits on electricity supplies.

• The steady fall of the dollar against the ruble continues—as $500 million are sold and the Central Bank of Russia buys $300 million. A major panic seems to be setting in. Some politicians delight in the "de-dollarization" of the Russian economy. Others say the end result will have no lasting effect on the economy and that measures should be taken to stabilize the dollar's value and halt the decline.

• In Russia, *Aragil Electronic News Bulletin* (in English) writes that Russian armed forces gain control of the Shatoy and Nozhay-Yurt settlements in Chechnya, which removes the last obstacles to building a Baku-Groznyy oil pipeline. Moscow and Baku are to build this line to the port of Novorossiysk. *Aragil* speculates that the Russian government presented Aliev with an ultimatum on the pipeline, probably offering something in return linked with the Karabakh war. *Aragil* says that it is hard to predict what sort of pressure Moscow will apply on Armenia, but Russia will probably keep the Republic of Mountainous Karabakh army in place as long as possible to keep the pressure on Azerbaijan.

**19** • Russia and Kazakhstan complete the first phase of accords to create a customs union between Belarus, Kazakhstan, and Russia, to regulate and structure payments between economic entities in the three republics, and to govern the leasing of Baykonur space launch site.

**20** • *Izvestiya* reports that in the first six months of 1995, Russia paid the London Club (of Western bankers) $350 million, enough to cover the interest on the 1992–93 debt. Total Russian debt to the London Club amounts to $25 billion, which Russia intends to pay in the last half of 1995, along with interest on the 1994 debt. This will bring total Russian payments in 1995 to $1 billion. Total payments in 1995 on Russia's overall debt, which now stands at $120 billion, will amount to $6 billion.

• In Uzbekistan, Ukrainian President Kuchma and President Karimov discuss economic cooperation—in gold, cotton, gas, aviation, and other areas. The two countries have increased trade to about $90 million—not very high, given the compatibility of political goals held by both presidents in the CIS arena. An obstacle to greater trade is that the two parliaments have not ratified a free-trade agreement signed during Islam Karimov's visit to Ukraine in 1994. Ten documents exist, including one on economic integration, and others in the fields of health care, tourism, and information exchange between foreign ministries. The increase in numbers of agreements, comments *Ukrayina Moloda* (23 June 1995), is beginning to look like a "strategic partnership." Two days prior to Kuchma's visit, President Karimov repeatedly spoke on the future of the CIS, each time more radically asserting the inviolability of Uzbek borders and not agreeing to a unified CIS command.

**26** • Russia's State Duma once again fails to adopt a Land Code. (On 14 June, 105 deputies voted against the code.) The opponents include a large faction of communists (particularly the Agrarian Party) who insist that private land ownership be removed from the draft code altogether. But Valentin Kornilov, aide to State Duma deputy Mikhail Lapshin, comments: "Today millions of people possess documents affirming their right to own land, and taking that land away from them would be the same as declaring a civil war."

**27** • In Georgia, President Shevardnadze announces that Tbilisi is ready to eliminate all obstacles to routing Azerbaijani oil across Georgia to the Black Sea port of Batumi.

**30** • A *Rossiyskaya Gazeta* article examines IMF policies in the CIS states. It criticizes the division of CIS states into groups, saying that these groups create payments problems among the states. The article claims that Russia is owed $6 billion by former Soviet republics, but that the IMF will not allow it to repay outstanding debts to non-CIS states by calling in these debts. The IMF has asked Russia to restructure and even in some cases to "suspend" these debts. The author calls for a CIS strategy for dealing with the IMF, including the creation of a clearinghouse payments union based on the country's own economic structures.

***Political***

**2** • In Azerbaijan, President Aliev delivers his usual double message concerning the CIS. He says the Azeri state is convinced the Commonwealth must develop and be strengthened, but stresses that the organization must not develop into a new center of a single state similar to the Soviet Union. He explains that all republics base their participation in the CIS to the extent that it helps consolidate their independence and sovereignty.

• IPA General Secretary Andrey Vermishev praises the results of the CIS Minsk Summit. He believes the meeting "demonstrates that the time for the formation of the Commonwealth as a single organism is nearly over," emphasizing he has no doubts that the body "is moving toward a grand Eurasian alliance like that proposed by [Nazarbaev]." Like IPA Chairman Shumeyko, Vermishev believes the increased powers of the IPA will greatly enhance implementation of approved CIS agreements and documents.

**3** • Turkmenistan and Armenia enter into transportation cooperation agreements with Iran. The transportation, energy, and trade ministers of each republic, after meeting in Yerevan, announce the agreement, which establishes a joint venture for transportation services to be headquartered in Ashkhabad.

**5** • In Armenia, Foreign Minister Vagan Papazyan indicates that unless the CIS develops equal economic links, there will be no future for the body. Papazyan believes the organization was necessary in the aftermath of the USSR's dissolution but notes that relations must develop in a positive way, not to any one member's or bloc's advantage.

**6** • Speaking to an international conference on regional cooperation, Kyrgyz President Akaev believes "cardinal reform" to establish a working Central Asian cooperation system is needed. He speaks of creating a special regional administration that draws from political, economic, and social institutions. Such a body, once established, he believes, will provide a framework for preventing economic, political and military conflicts.

**7** • In Moldova, parliament ratifies two CIS agreements, the Agreement on the Establishment of the Interstate Economic Committee of the CIS Economic Union and the Agreement on Establishment of the CIS Payment Union. (President Snegur signed these documents on 21 October 1994.)

**10** • Yeltsin and Kuchma sign an agreement on the Black Sea Fleet (BSF), both believing it meets the interests of Russia and Ukraine. The agreement acknowledges that the Russian BSF and the Ukrainian navy are to be formed out of the former Soviet BSF. The Russians will base their share in Sevastopol and will use the Crimea for additional basing of ship, troops, and material supplies. Russia is to take ownership of 81.7 percent of the fleet's ships and vessels and Ukraine the remainder. An important part of the agreement stipulates that Russia will invest in the socio-economic development of Sevastopol and other areas used by the Russian part of the BSF. Talks will continue to narrow remaining differences and the agreement provides the Ukrainian and Russian naval officers the power to make ad hoc decisions. Russian officials, in particular BSF commander Admiral Eduard Baltin, voiced displeasure with the agreement, which they say reduces the fleet's personnel by one-third and hands over to Ukraine equipment and material they are not capable of managing. In Ukraine, views were mixed, with more nationalist-minded politicians denouncing the agreement, saying Ukraine "sold out." Others, like Foreign Minister Udovenko, believe the agreement provides the framework for increased cooperation regarding the BSF and lays the groundwork for improved Russian-Ukrainian relations, hampered for the last few years because of BSF disagreements.

**14** • In Russia, the State Duma rejects a resolution submitted by Russian communists charging that the CIS has produced "extremely negative political and economic effects." The communists call for reintegrating the republics, and specifically demand all Duma committees to develop a conceptual framework for economic, political, financial, and military union between the Russian Federation and Belarus that other CIS states can join. The resolution receives 197 positive votes, needing 226 for full passage.

• In Georgia, Shevardnadze meets with CIS interior ministers. He speaks of the dangers associated with increased organized crime, noting that the problem is *not* international in scope.

**20** • In meetings in Tashkent, Ukrainian President Kuchma and Uzbek President Karimov sign bilateral documents pledging expanded cooperation. A treaty on economic integrations, specifying cooperation in science, technology, education, public health, tourism, and sports, is also signed.

**30** • An article in Moscow's *Kommersant-Daily* discusses the effects of increased nationalism on Russians living in Kazakhstan. Kazakhs are irate that they are left looking for work while Russians, Tajiks, and Uzbeks fill many state and private employment positions. Additionally, state coffers are being drawn down by social subsidies that provide the minimum necessities of life to incoming refugees from neighboring "hot spots," like Tajikistan and Nagorno-Karabakh. This, concludes the article, is why Kazakh movements to drive many Russians and other foreigners out are increasing in numbers and strength.

***Security***

**1** • Eleven heads of CIS state security bodies report that as a result of a two-day meeting they have agreed to join forces to combat organized crime in their states. Specifics of the agreement include creating a joint database, engaging in technical cooperation, and providing joint training and exchange of personnel.

**2** • Boris Yeltsin sends an official letter to the chairman of the Russian Federation Council renewing the mandate of CIS peacekeeping forces in Abkhazia until the end of 1995. CIS peacekeeping forces in Abkhazia, Nagorno-Karabakh, and Tajikistan so far consist entirely of Russian troops, because no other CIS state has "volunteered" to deploy troops to the conflict zones.

**5** • An article in *Demokratychna Ukrayina* analyzes Russia's geopolitical goals in its so-called "near abroad." It emphasizes Russia's inability to resolve serious disputes other than by force, economic coercion, and political manipulation, which exacerbates tensions and instability in the former Soviet republics. One particularly insightful passage from the article says: "In addition to well-tested means for assembling the three Slavic republics around it as the foundation for its future as a renewed superpower—interference in events in the Crimea, the foot-dragging in dividing the Black Sea Fleet, and the use of oil and gas dependence—there exist other ways, such as persistent attempts to draw Ukraine into military-political formations, pressure to establish a joint "external" CIS border, make "internal borders porous," change customs regulations to the Russian Federation's advantage, renew economic cooperation, including in the sphere of manufacturing weapons, and so on." The paper offers an articulate analysis of the important strategic tasks of NATO in banning the proliferation of weapons of mass destruction in this region, and in helping to create a new collective security structure for the region.

**8** • Addressing the UN disarmament conference, President Nursultan Nazarbaev says that a vertical, "geopolitical" zone of nations has emerged in Eurasia, which belongs to "neither the West nor the East" in terms of security interests. He defines the zone as containing Russia, Central Asia, Iran, Pakistan, and India. Despite their differences, Nazarbaev says, these states form an "integral group in terms of their potential influence on the balance of forces in Eurasia." He says security in Europe and Asia largely depends on the orientation of these states. And insofar as Russia is the largest and most powerful, its actions "will determine the stability and changes in Eurasia."

**16** • Ivan Zayets, the deputy chairman of Ukraine's parliamentary Commission on Foreign Affairs and Relations with CIS Countries, tells an interviewer that the question of Ukraine joining NATO rests on three choices. The first is for Ukraine to become a member of the collective security system of the CIS, that is, the Tashkent military bloc. The second is for Ukraine to agree to special terms with NATO, and remain between the two blocs. The third is for Ukraine to become an equal, full member of NATO. He calls the Tashkent Treaty a "purely military bloc structure," which would contravene Ukrainian legislation. He notes that this bloc unites politically and militarily unstable countries. "Ukraine's membership in this bloc would distance it from Europe and from the Ukrainian peoples' choice on 1 December 1991" (*Kievskie Vedomosti*, 16 June 1995).

**21** • In Uzbekistan, visiting Ukrainian President Leonid Kuchma stresses Uzbekistan's important geopolitical position in Central Asia, and says Ukraine looks upon it as a reliable strategic partner. He says he is prepared to raise Ukrainian-Uzbek relations to a higher level. President Karimov tells reporters that "political experts believe that developments in the post-Soviet space will largely depend on the development of relations between Ukraine and Uzbekistan." He adds that he fully shares that opinion.

**22** • General Vasiliy Yakushev, commander of Russian peacekeeping troops in the Georgia-Abkhaz conflict zone, tells reporters that Russia has fulfilled all its commitments under the agreement signed on 14 May 1994 for sending Russian troops into the zone. He says that peace has been established (although he admits it is a rather fragile peace); all armed formations of both warring sides have been withdrawn and peacekeepers have created conditions for the organized return of refugees to their homes. He says that though Russian troops have been criticized for inactivity, he will not use "his boys" as policemen, asserting that "bandits" and "gangs" (not terrorists or combatants) are carrying out "raids" on the peaceful population. His position is that Russian soldiers cannot guarantee the safety of Georgian-Abkhazian citizens because they are there to "stop armed conflict," not settle "disputes" among "citizens," which he says must be carried out by political means.

**July 1995**

***Economic***

**1** • *Trud* describes the harsh effects of the Russian economic blockade against Abkhazia, which was supposed to be temporary—six months ago. The article concludes that Abkhazia has only one hope—Turkey—but even the sea routes between Sukhumi, Trebizond, and Istanbul could be cut off, it says. Moscow closed the border with Abkhazia because Chechen gunmen were penetrating it to obtain medical and food supplies. Meanwhile, some deputies in the Russian Duma have initiated discussion on annexing Abkhazia.

**5** • An economic integration plan—"Turkestan Is Our Home"— designed to join the five Central Asian states into an economic system has been completed in line with President Islam Karimov's May initiative. Central Asia accounts for 20 percent of the coal, 18 percent of the gold, and 92 percent of the cotton produced in the former Soviet Union. The plan will attempt to spur economic reforms and speed up the regional industrial and agricultural development of the most competitive areas of the five states' economies. Great attention has been paid in the plan to unemployment, which is the highest in the CIS states. The total population of the Central Asian nations is about 56 million, which grows at an average annual rate of 1.7 percent.

• Russia reduces its deliveries of gold to Armenia by half and deliveries of silver to less than a third. Platinum and a number of other metals will be reduced due to late payments by Armenian enterprises. The precious and semi-precious metals are used in Armenia's jewelry industry (*Kommersant-Daily*, 5 July 1995, p. 4).

**6** • Andrey Kozyrev addresses a conference of Russian ambassadors to the CIS states in Moscow saying that integration is based on the principles of "multi-speed" and "multi-option" development. During the speech, he said the CIS customs union was open to any state wishing to join, but reminded the ambassadors that economic integration would require "ever increasing transparency of internal borders."

**7** • The Council of CIS Heads of Government issues a decision recommending the settlement of CIS interstate payments on a bilateral basis, in dollar equivalent terms, employing the value of the ruble in relation to the U.S. dollar on the date payments are made, or as agreed by the parties. It recommends sending disputes to the Economic Court of the Commonwealth of Independent States.

**11** • *Finansovye Izvestiya*, No. 48, p. 3, analyzes the financial status of Armenia and Azerbaijan. In 1995, the budget deficit for Armenia is set for 12 percent of GDP. The Bank of Armenia has announced that it intends to be guided by Western currencies, and that "the Russian ruble is something which we must fight and which must be ousted from domestic monetary turnover." Currency circulation is highly saturated with dollars. According to some estimates, total dollar capital is three times greater than the volume of money supply in the national currency. The Azerbaijani manat underwent drastic devaluation in 1994, while prices rose by 2,831 percent. The Bank envisions reducing the budget deficit to 4.7 percent of GDP in 1995. Most of the deficit will be covered by bonuses from foreign oil companies on oil contracts and international finance organizations.

**12** • Russia and Belarus sign an accord on currency and export regulation. One of the goals is to detect criminal activity in the economic sphere.

**13** • Ukraine and Uzbekistan conclude economic exchange agreements (Uzbek gas for Ukrainian machinery and pipes) intended to further reduce both countries' dependence on Russia. An article in *Kommersant Daily* (23 June 1995) describes Uzbekistan's "Turkestan" model of development, oriented toward Kazakhstan and Kyrgyzstan. In this relationship, Uzbekistan dominates in terms of population and real economic might (it is one of the world's leading producers of gold, cotton, and fuels). The article adds that ethnic Uzbek communities in southern Kazakhstan, Kyrgyzstan, and Tajikistan exert a real socio-political and economic influence.

• Turkish private companies have invested about $1.5 billion in Turkmenistan, and Turkey is very interested in building a Turkmenistan-Iran-Turkey-Europe gas pipeline. An interstate council session on prospects for cooperation with Turkmenistan will be held in September.

**14** • *Rossiyskaya Gazeta* publishes the provisions of the CIS Interstate Currency Committee (MVK) Statute, which was signed by the Council of CIS Heads of Government on 26 May. The Committee is a standing body of the CIS, with its own statute under the Agreement on Creation of a Payments Union. The stated goal of the Committee is to develop a system of mutual quotas for soft CIS currencies and the principle for their conversion. To date, the Committee has overseen the signing of bilateral agreements between Russia and Belarus, and Kazakhstan and Turkmenistan.

• An *Izvestiya* article (20 June 1995) describes the costs of joining the CIS customs union for former USSR republics. Armenia, for instance, calculates that its entry into this union would incur losses of $50 million a year. Other Caucasus states also react coolly to the idea. Georgia, for instance, is considering pegging its currency to the U.S. dollar. However, Uzbekistan and Turkmenistan support the customs union. Trade is hampered by huge indebtedness to Russia. To pay off these debts, the non-Russian states are attracted to markets in the "far abroad" and not to potentially risky institutions such as a payments union or a customs union, which would be dominated by Russia, say the authors.

**17** • In Ukraine, President Kuchma announces his support for joining the customs union between Belarus and Kazakhstan. However, he says a free trade agreement needs to be signed first so the two countries may "get accustomed to each other and see how to develop their relations." He says trade between Belarus and Ukraine has declined, with the main reason being the breakup of relations following the collapse of the USSR. However, he says, "to restore relations is much more difficult than to break them."

**18** • CIS interstate committees have been established on statistical standardization, anti-monopoly laws, aviation, and a number of other economic questions, totaling fifty in all. The Interstate Economic Committee (MEK) has the right to decide how these committees will operate. However, to date they do not work and the Russian deputy minister for CIS Cooperation says "real integration in the CIS is not yet taking place."

**19** • The Georgian journal *Mimomkhilveli* reports that the "Secret War for Caspian Oil Enters a New Phase." The war between Russia and Turkey for acquisition of the transport route for Caspian oil has taken on strategic meaning for the United States, which must balance Russia's influence in the "Eurasian corridor."

**25** • Yerevan hosts a CIS agricultural council session among eight CIS countries. The council will study the Armenian experience of land privatization. Armenia is the only former republic which has fully carried out an extensive land privatization program.

• Russia and Ukraine proceed with plans for re-integrating their economies. Experts agree to create transnational corporations in several industries, including: prospecting, mining, energy resource processing, gas pipeline construction, pipe production, and aluminum production. Four agreements will be signed, the goal of which is to "restore the earlier lost economic ties between the two republics of the former USSR." Ties will be reforged in machine-building, non-ferrous metals, extraction of oil and gas, and cooperation in the space industry. Ukrainian producers, who have capacity but few raw materials, will receive stable channels of supply while Russia will be spared from building expensive processing enterprises.

• The U.S.-Ukrainian Joint Commission on Trade and Investment meets, in accordance with an agreement signed in May by President Bill Clinton and Leonid Kuchma. Commission experts say the United States supports Ukraine's efforts to join GATT/WTO, and will increase aid to Ukraine for these purposes.

**27** • In Russia, First Deputy Foreign Minister Igor Ivanov denies reports of Russia's intentions to apply economic pressure on Azerbaijan to stop its planned development of Caspian Sea oil reserves.

• Ukrainian Prime Minister Evgeniy Marchuk announces that Russia has for the third year in a row restructured its gas debt. Ukraine's gas debt for 1995 totals $1.014 billion, with $100 million owed for natural gas for the winter of 1995. Debts for 1994 total $1.5 billion and for 1993, $2.5 billion.

**28** • Kiev Radio reports on Evgeniy Marchuk's trip to Moscow on 25–26 July. Prime Minister Marchuk says four agreements were signed. One provides for joint operation and upkeep of Ukraine's 1,100-kilometer pipeline across Ukraine, which he says is "almost not functioning." He says "this is not a transfer of ownership, but mutual use of the pipeline by Ukraine and Russia." A framework agreement was also signed on principles for establishing Russian-Ukrainian financial-industrial groups. He says that in the near future it is planned to launch at least ten financial-industrial groups, transnational corporations, and joint ventures. One financial-industrial group is already producing aircraft engines.

• *Kommersant Daily* reports on an authoritative article in *Ukrayina Moloda* charging Russia with circulating the outline for a plan to discourage the British firm JKX from investing in Ukrainian liquid gas fields located in the Crimea and the adjoining shelf of the Black Sea. The plans are categorically denied by Russian officials. The author is impressed by the expertise and plausibility of the charges, given Russia's potential loss of political influence over Ukraine if significant amounts of British capital were to be invested in Ukrainian gas.

***Political***

**10** • In Strasburg, the Committee of Ministers of the Council of Europe vote to make Moldova and Albania the thirty-fifth and thirty-sixth members of the Council of Europe. Official recognition of the two will be made at a ceremony on 13 July.

**11** • Vladimir Shumeyko, chairman of Russia's Federation Council, visits Georgia in order to sign an interparliamentary cooperation agreement. He stresses that protection of Georgia's territorial integrity is part of Russia's strategic interests. He also promises 10 billion rubles in Russian aid to Georgia to be used for delivery of foodstuffs to Abkhazian citizens. In delivering the aid to Georgia, he stresses that Russia is abiding by international law and does not intend to bypass Georgian authorities.

• The Council of CIS Foreign Ministers meets in Bishkek. Delegations from Russia, Armenia, Ukraine, Kazakhstan, Tajikistan, Moldova, and Kyrgyzstan discuss European security, relations with NATO, and strengthening OSCE effectiveness. In a press conference after the meeting, Kyrgyz Foreign Minister Otunbaeva says that although all the ministers desire closer relations with the OSCE, not all agree on CIS foreign policy objectives. The needs of Ukraine, Belarus, and Moldova are different from those of the Central Asian republics and hence require different strategies.

**12** • First Deputy Commander-in-chief of the Russian navy Igor Kasatonov criticizes Ukraine for its nonobservance of the Sochi Agreement, which supposedly resolved the Black Sea Fleet issue. President Yeltsin sends a letter to Kuchma urging him to "attentively consider" Russia's interests, reminding him that the Sochi Agreement calls for a settlement of the issue "as quickly as possible."

• Nazarbaev and Yeltsin sign a treaty on joint border protection. The treaty provides for a united command for the frontier troops which will guard the external borders of Russia and Kazakhstan.

**13** • In Azerbaijan, Russian Federation Council Chairman Vladimir Shumeyko discusses events in the Caucasus while visiting Azeri President Aliev. He says President Yeltsin has urged him to organize a conference on problems in the Caucasus.

**18** • A *Kommersant Daily* article (24 June 1995) by Professor Anatoliy Kolodkin, president of the Russian International Law Association, discusses the legal status of the Caspian Sea. Kolodkin writes that the 20 November 1993 agreement between Azerbaijan and Russia "On Cooperation in Prospecting and Developing Oil and Gas on the Territory of the Azerbaijan Republic" erroneously implies that the Caspian Sea can be cut up into territorial sectors—a practice he associates with medieval times when a monarch extended his land into bodies of water.

**21** • In talks with Austrian Foreign Minister Wolfgang Schalenberg, Moldovan President Snegur reveals that Chisinau's foreign policy will be directed toward full integration into European structures. He unveils plans for executing this policy and recaps some of the agreements Moldova has already signed working toward this end. Snegur has increasingly expressed his dissatisfaction with Russia and CIS, viewing the body as a vehicle for Russian imperialism.

**25** • In a meeting with EU officials in Tashkent, Uzbek President Karimov says his republic desires closer relations with Europe in order to achieve quick integration in the world economy. The parties discuss the geopolitical situation in Central Asia, requirements for transition to a market economy, and ways the republic can lay the groundwork for eventual EU membership.

**27** • In Uzbekistan, Russian Prime Minister Chernomyrdin discusses debt problems with President Karimov, the treatment of the Russian diaspora in Uzbekistan, and the situation in Tajikistan and Afghanistan. Both express the need for a political settlement to the Tajik civil war and Chernomyrdin assures Karimov that Yeltsin will use Russian resolve to aid in this matter. Both express a desire to strengthen bilateral relations.

**28** • Following his meeting with Chernomyrdin, President Karimov says Uzbekistan needs Russia, but a "democratic Russia which accepts us as an equal in an all-round way, which welcomes our successes and is ready to hold out its hand and help us. . . . This is the kind of Russia we consider close to us." He praises Chernomyrdin for his efforts in helping to consolidate relations between the two, saying if leaders of the Russian Federation "are like [him], then in my personal opinion our cooperation will be consolidated and we will have a good future."

***Security***

**4** • In Russia, Yeltsin decrees the formation of the "Russian Military Brotherhood." The goal of the Brotherhood is to strengthen Russian statehood and defense capability, and to foster "military-patriotic education and extrabudgetary funding for military social programs." Article one of Yeltsin's edict reads: "1. The Russian Federation Government will examine the Statute on the Russian Military Brotherhood and provide it with comprehensive assistance in its activity" (*Rossiyskaya Gazeta*, 4 July 1995, p. 5).

**6** • In a speech to Russian ambassadors to CIS member states, Foreign Minister Andrey Kozyrev calls the CIS an important "instrument of stability in the post-Soviet space." He says the peacemaking burden in the CIS has spread to Georgia, Tajikistan, and Nagorno-Karabakh, and calls on the world community, including the OSCE and the United Nations, to share the load.

• In Russia, Foreign Minister Kozyrev speaks of the need for Russian military bases throughout the CIS, arguing that many CIS states are threatened from the abroad which in turn threatens the security of the Russian Federation. Echoing the thoughts of President Yeltsin, Kozyrev says there is no more important element in Russian foreign policy than to continue integration within the CIS. This, says Kozyrev, "is prompted by life itself." Kozyrev makes clear that this is not an imperial policy and, indeed, such Russian initiatives are desired in many of the former Soviet republics.

**14** • In Georgia, Abkhaz leader Vladislav Ardzinba sends a letter to President Boris Yeltsin protesting a change made at the 26 May summit in the CIS mandate for peacekeeping troops in the Georgian-Abkhaz conflict zone. He especially objects to new items in the mandate which stipulate: the "creation of conditions for the secure and dignified return of (Georgian) refugees and the disbandment and withdrawal of all voluntary formations." He calls both of these items "interference" in the affairs of local bodies, and that the item on the safe return of refugees is "questionable."

**15** • In Ukraine, Leonid Osavolyuk of the State Committee for the Defense of Ukrainian Borders tells reporters that Russia has still not consented to begin negotiations on the legal delimitation of joint borders. He adds that if Russia continues to drag its feet, Ukraine may unilaterally begin the delimitation of the Ukrainian-Russian border.

• A meeting of the Ukrainian-Moldovan border commission produces a protocol on border delimitation. As an example of problems that must be overcome, Aurelian Danila, head of the Moldovan delegation, reports that Moldovan officials are less ready than their Ukrainian colleagues to discuss the transfer to Ukraine of Moldovan railway assets and infrastructure, based on Ukrainian territorial rights. These problems will be discussed again in October.

**18** • Col.-Gen. Dmitriy Volkogonov, prominent historian, says in *Krasnaya Zvezda* (18 July 1995) that "combat operations in Chechnya have undoubtedly complicated the entire reform process in Russia. The recent conflict between the legislative and executive branches is an added proof of that. The most regrettable thing is that many forces from the Right and Left quite often make use of Chechnya for their own narrow political ends."

**26** • A platoon and a group of officers from the Kyrgyz Republic will take part in NATO exercises in the United States 6–28 August. Kyrgyzstan joined the Partnership for Peace in June 1994. The leadership considers expanding direct contacts between Kyrgyz army units and NATO troops an "imperative of the times." However, the republic ultimately prefers non-alignment, and its contacts with NATO are only of a "peacekeeping nature," say *Krasnaya Zvezda* editors.

**28** • A *Moskovskie Novosti* article examines the potential for a regional security system in Central Asia, led by Russia. The two main problems for Russia are instability in the region resulting from the Tajik conflict and Central Asian leaders' desire to diversify their security contacts. Although all five Central Asian states recognize Russia's special role and interests in the region, not all agree on the real intentions behind Russia's military presence in these states. The article notes that discussions are taking place on the idea of convening a conference in Tashkent on security and confidence measures, with participation of the five Central Asian states, Russia, Afghanistan, Pakistan, India, Iran, and China, under the aegis of the United Nations and with participation by the major international organizations (*Moskovskie Novosti*, 18–25 June 1995).

• Lt.-Col. V. Shevchenko, a Ukrainian Supreme Council deputy and member of the Commission on Issues of Defense and State Security, analyzes in *Molody Ukrainy* the risks for Ukraine of signing the CIS Treaty on a Unified Air Defense System. His concerns are heightened by Belarus's virtual annexation by Russia. He points to the November 1993 Russian Military Doctrine and its influence on Russia's interpretation of the Treaty as another serious concern.

## August 1995

### *Economic*

**1** • Russia and Kazakhstan sign an agreement on basic principles of trade and economic cooperation, emphasizing the principle of "equality" in the form of direct contracts between companies. Intergovernment agreements will control how payments are to be made, however. The agreements forbid use of non-tariff barriers to trade, except for products subject to international obligations. According to Russia's Ministry on Cooperation with CIS Member States, Russia has now signed bilateral trade agreements with every CIS country except Azerbaijan.

**2** • CIS countries announce they intend to "restore Soviet construction might" through collective construction agreements reached on 2 August entitled "Agreement on Creation of a Common Research and Technological Zone" and "On the Main Principles of Organizing Cooperation Between CIS States." The CIS countries have reduced their construction outside the former USSR by a factor of four since 1992. Earnings from construction have dropped overall from $4 billion to $1 billion. Experts estimate there is huge potential to recapture these earnings, which number in the thousands, by maintaining and reconstructing facilities already built abroad.

**8** • An article in *Problemy Teorii i Praktiki Upravleniya* provides a detailed analysis of the problems and trends of foreign investment in the CIS. It criticizes foreign investors who concentrate on building assembly plants, and import the parts from their own countries.

**15** • Ukraine announces it will develop oil and gas found in the Black Sea shelf, in what it terms "its own economic zone." It estimates that 1,240 billion cubic meters of gas and 215 million tons of oil are available off Ukraine's Black Sea coastline. However, current technology owned by Ukraine is incapable of boring deep enough to recover the fuels. Ukraine has therefore announced an international tender to the right to prospect for and process oil and gas on the shelf.

**17** • In Russia, an agreement on commercial, economic, and trade cooperation is signed between Moscow and the Gagauz Autonomous Republic.

**18** • In Ukraine, the poor condition of the electricity-generating industry may force the country to limit consumption to between four and six hours of power per day. The crisis is due to a lack of fossil fuels for non-nuclear electrical plants. Only two out of eight power units are working, and even that is at reduced capacity. Only with Russia's help has Ukraine maintained a 49.5 power grid frequency. "If Russia pulls out," says Deputy Energy Minister Yuriy Nasedkin, "the automatic system controls will start switching off consumers."

**23** • The Intergovernment Bank (IB) set up to service settlements within the CIS may be granted the right to invest its assets in securities of the countries that invested in the bank's incorporation capital. IB Council Chairman Tatyana Paramonova (also acting chairman of the RF Central Bank) has pointed out that the IB's charter should be amended so the bank can widen its powers.

**25** • In Russia, an interbank crisis is caused by the insolvency of several banks. The Moscow interbank market stops all operations on 24 August for the first time since the beginning of Russia's market reforms. The Central Bank is being called upon by unnamed Russian banking representatives to use available financial reserves to prevent a bank crash. Exchange rate restrictions have been extended until the new year in an effort to prevent the dollar's fall to below 4,400 rubles to the dollar.

• Konstantin Borovoy, leader of the Economic Liberty Party, blames the Russian government's credit and monetary policy for triggering the banking crisis. In Borovoy's opinion, the CBR should have lowered its refinancing rate (currently 180 percent per year) long ago and taken measures to bring down the ruble rate. In addition, artificial restrictions on the ruble exchange rate caused economic instability.

**26** • An article in *Segodnya* (26 August 1995, p. 4) reviews the economic crisis in the Dniester region, blaming its lack of statehood status for its poor economic condition.

**28** • Foreign investment in Ukraine for the first half-year (1995) increases by $116.3 million over the same period in 1994, totaling $566.4 million. This includes $35.9 million in investment by CIS and Baltic states (6.3 percent of the total).

• Ukraine begins to procure fuel supplies for the winter. A tough winter is predicted, and the state is unprepared. Natural gas supplies are particularly low. UkrGazProm's accumulated debts to Russia's GazProm stand at US$228 million. Ukrainian economist Volodymyr Chernyak said miscalculations on fuel and energy demand and supply could lead to catastrophic consequences, culminating in a number of oblasts and regional enterprises being disconnected from the energy supply.

• A Belarusian delegation to the IMF blames President Lukashenka's economic policies for the IMF's delay in providing promised loans. Points made by the delegation include: (1) Three years ago Belarus received $100 million; the money disappeared while economic conditions continued to deteriorate; (2) One year ago, the newly established government drafted a reform program, yet almost none of the program's points were accomplished.

**29** • In Turkmenistan, the Ministry of Gas and Oil is authorized to accept construction work performed by Ukrainian building companies as payment for gas arrears. Total value of the construction work is US$23.7 million, almost half of Ukraine's gas debt to Turkmenistan.

• In Ukraine, Deputy Prime Minister Viktor Penzenik makes the sensational announcement that "the unique moment" for introducing the national currency—the hryvna—has arrived. The opportunity is made possible by a $2 billion reserve stabilization fund and the fact that the country has virtually paid for the gas it uses. He claims the hryvna will be exchanged at a rate of 100,000 karbovantsy to 1 hryvna. The dollar will be hypothetically worth 1.5 hryvna.

• Ukraine will pay GazProm in bonds for past natural gas debts. According to the Russian-Ukrainian agreement on restructuring Ukraine's debts for natural gas signed 18 March 1995, the bonds may be cashed in twelve years, although Ukraine should begin buying them back in 1997. The Russian bank Natsionalnyy Kredit has been selected as the authorized depositor and payments agent in dealing with loans to Ukraine.

**30** • Uzbekistan's inflation rate is eight times lower than in 1992, and the budget deficit is below the limit coordinated with the IMF. He says the country is capable of becoming self-sufficient in grain and oil production and can eliminate large grain imports.

*Political*

**2** • In Moldova, the chairman of the parliamentary Commission on Human Rights and National Minorities, Vladimir Solonari, argues against joining NATO or the CIS defense system. He says Moldova "cannot go around Kiev," referring to Ukraine's position of balanced openness and neutrality, and confirming that Moldova will follow its lead.

**3** • In talks between Ukraine and Russia on the Black Sea Fleet, negotiators seek to resolve issues of basing rights and command posts. In the latest Russian proposal, Russia would take Sevastopol as its main base and command post, with minor bases in Feodosiya and Kerch, while Ukraine would utilize bases in Balaklava (outside Sevastopol) and Donuzlav, 140 km northwest of Sevastopol.

**6** • In a *Lesnaya Gazeta* article, Russian commentator Valeriy Begishev questions Russia's role in FSU republics and the CIS. He observes that Yeltsin's recent speech to military graduates, in which he says Russia "feels a political and moral responsibility for the fate of the countries and peoples who marched alongside the Russian State for hundreds of years," is sure to alarm CIS leaders.

**7** • Ukraine announces its view that Russian naval ships and forces based in Sevastopol should be categorized as a foreign military force on the territory of another country. Under international law, the foreign country would be required to pay rent for the facilities and compensate the host country for environmental damage. Instead, the Russians want Sevastopol to be given Russian territorial status.

• In Belarus, Deputy Chief of Staff Vladimir Zametalin accuses the Russian media of campaigning against Lukashenka and his policies. He says he is convinced such activity is orchestrated by certain Russian circles "opposed to integration in the CIS."

**10** • Belarusian First Deputy Foreign Minister Valery Tsapkala discusses Belarus's foreign policy. He says Belarus needs to forge close relations with its neighbors, Russia, Poland, Ukraine, and the Baltics and says relations with these nations should focus on economic cooperation and contacts. Furthermore, Belarus maintains its neutral military position. The minister calls for a pan-European collective security arrangement, saying that NATO is obsolete because its purpose was to contain communism. Instead, he affirms the post–Cold War importance of the OSCE and the UN.

**15** • In Tbilisi, the Abkhaz Council of Ministers expresses alarm about U.S. "activism" in the Transcaucasus, especially its contribution to constitution drafting in the region.

**31** • During a special session of the CIS Interparliamentary Assembly, Russian Deputy Chairman of the Federation Council Valerian Viktorov argues that the United Nations should endorse the role of CIS peacekeeping forces in CIS hot spots where UN resources are insufficient to conduct full-scale peacekeeping operations. Viktorov argues that one of the main problems for the United Nations is to ensure a regional approach to peace and security. To date, Viktorov says neither the United Nations nor the OSCE have provided adequate assistance in preventing and settling regional conflicts, including Nagorno-Karabakh, Transdnistria, and Abkhazia.

*Security*

**4** • In Ukraine, Deputy Foreign Minister Konstantin Krishchenko discusses defense cooperation with Kazakh Defense Minister Saghadat Nurmaghambetov. Kazakhstan is interested in training its personnel in Ukraine, exchanging teaching manuals and training methods, and sending some of its weapon systems to Ukraine for repair.

**5** • Belarus delays the transfer of its strategic nuclear missiles to Russia. A Belarusian official says the Russian Defense Ministry is having problems in preparing for the transfer because of Chechnya.

• *Izvestiya* publishes an article that alludes to the growing unpopularity of Turkmen President Sapamurad Niyazov. The regime has built shrines to Niyazov, instituted a leadership cult around him, and indoctrinates youth with loyalty "oaths" and other forms of dogma. A large resistance movement seems to be forming. One youth is quoted as saying "No one in our school intends to learn the oath of the Turkmen." The catastrophic state of the economy, in which many are half-starving, adds fuel to the fires of resentment and alienation.

**8** • Writing in *Rossiyskie Vesti,* Gennadiy Voronin, deputy chairman of the Russian Federation State Committee for Defense Industries, details the perilous state of Russia's shipbuilding industry. The number of warship expeditions has been cut by three-fourths and duty calls to foreign ports by "several tens." Less than one-quarter of scheduled repair work is being carried out. Ships which have not served half of their established service lives are being decommissioned from the fleet. Technical facilities and installations are below standard. The bulk of the warships are laid up, especially large ones such as the heavy nuclear-powered missile cruisers *Lazarev* and *Nakhimov.* Four of five aircraft-carrying cruisers have been decommissioned. The average age of Russian ships is substantially greater than that of U.S. ships. The author recommends that a "law on the state defense order" be enacted immediately.

**10** • Belarusian Defense Minister Valery Tsapkala refers to the CIS Collective Security Pact as "purely a consultative mechanism," which does not abrogate Belarusian neutrality.

It has no supranational control over Belarusian armed forces. Belarus could not deploy forces without express authorization by its Supreme Council.

• In Russia, Andrey Kokoshin, first deputy defense minister, and Army General Mikhail Kolesnikov, chief of the Russian Federation Army General Staff, meet with administration heads to discuss the army's urgent winter food situation. The Defense Ministry tells the government that defense spending must be "brought into line with real prices" and "regularized," meaning that it must be raised.

• Russian naval command sends orders to headquarters of the Black Sea Fleet to dismantle the flight control center at the aerodrome of Gvardeiskoe (Simferopol, Crimea) before 16 October. Fifty percent of the aerodrome facilities, equipment, arms, property, and material resources are to be transferred to Ukraine's navy by that date. One Russian military expert says that if carried out, the dismantlement would mean the loss of Russian control over Crimean air space for the Black Sea Fleet command.

**11** • *Karavan* (11 August 1995) describes the Kazakh army's many serious problems. Seventy percent of all former officers have been discharged over the past three years—and not only Russian speakers, as some claim. Accordingly, there is a shortage of officers in the army and in the Ministry of Defense. Another problem is the embezzlement of army property and arms, which is rampant. Add to this starvation, homelessness, and no pay and you have a thoroughly demoralized and incompetent army.

**17** • Belarus suspends the transfer to Russia of its last eighteen SS-25s. Lukashenka is quoted (without full corroboration) saying: "since we are going to unite with Russia anyway, then why should we aimlessly ship military equipment to and fro?" *Belorusskaya Delovaya Gazeta* suggests that Belarus did not make the decisions about which weapons to keep. He points out that Boris Yeltsin left Minsk on 22 February 1995 after very successful meetings with President Lukashenka. The next day, Lukashenka announced that Belarus would not scrap its tanks. Perhaps, the author writes, Russia is testing out a trump card, which is the presence of a nuclear sword located directly at NATO's possible future borders.

**22** • Lt.-Gen. Evgeniy Malashenko, former chief of staff of the Carpathian Military District, reviews Russian "military reform" at length in *Krasnaya Zvezda* (22 August 1995). It recommends that Russia reinforce its troops in the West and North Caucasus because in the West there used to be troops from four groups and four military districts, but now there is just one—the Moscow District, which has become a "border district."

**24** • Russian Maj.-Gen. Vladimir Osadchiy reports that expenditures on the Chechnya war have outstripped the military budget for fiscal year 1994–95 by 1.9 trillion rubles. He says the desperate situation in the army and the navy result from the fighting in Chechnya and from "financial errors" by the Ministry of Finance, noting in particular that the Defense Ministry tops the list of Russia's debtors and has no funds to pay servicemen or civilian personnel.

**28** • Russian Atomic Energy Minister Viktor Mikhaylov says on television that Russia is destroying chemical weapons using underground nuclear explosions. Even though such explosions contravene the nuclear test ban treaty, currently drafted, the method is "approximately 100 times cheaper than any other means," says the minister. He says he does not take seriously France and China's rationale for carrying out additional nuclear tests.

• Defense Minister Pavel Grachev announces that "nuclear weapons will remain Russia's main war deterrent." At a meeting marking the fiftieth anniversary of the Russian nuclear industry, Grachev argues "nuclear force must be adequate to the likelihood of a possible threat," and warns that Russia should not lag behind other nuclear countries in security matters.

**29** • A new S-300 missile complex is established in southern Ukraine, near Kherson. Military authorities witness a powerful missile salvo "destroy" sixteen Mi-8 helicopters and Su-23 aircraft in less than a minute. The new missile complex was developed by Ukrainian specialists with Russian participation.

**30** • In Uzbekistan, parliament approves the country's military doctrine of neutrality. The doctrine prohibits Uzbekistan from producing, purchasing, or deploying nuclear weapons and favors the indefinite extension of the Nuclear Non-Proliferation Treaty. Uzbekistan "considers it inadmissible to replace the previous ideological confrontation between East and West with confrontation based on ethnic and religious grounds."

### September 1995

#### *Economic*

**1** • Yeltsin delivers a strong speech to Russia's commercial bankers and entrepreneurs. He emphasizes that the Russian government intends to hold the line on reducing inflation. He says, "People tell me that high inflation is to the advantage of some banks. When inflation is high, dubious debts rapidly depreciate. . . . The leadership of such banks does not pay a high price for miscalculations. . . . But if you (want financial stabilization), then support the state's efforts to solve the problem of inflation." He emphasizes the necessity for banks to finance the industrial sectors, and not to let Russia become a country "that develops only its raw material sectors."

**4** • Aleksandr Lukashenka announces that Belarus will build "market socialism." He defines this as "a society with a high degree of social protection, a strong material and moral motivation to work, without shortages or cues. . . . " He calls for a "new pricing and financial-credit policy" in order to keep the country self-sufficient in foodstuffs. This policy, he says, should "keep the incomes of the farmers at a level which encourages them to increase production, making the social standing of the urban and rural population more or less equal. . . ."

**11** • The Ukrainian State Oil, Gas, and Oil-Processing Industries Committee chairman, Yevhen Covzhok, asks Russia to pay more for its natural gas shipments to Europe through Ukrainian territory. Russia now pays \$0.55 per 1,000 cubic meters of gas per kilometer, but Ukraine wants to increase this rate to US\$1.6. Ukraine also asks Russia to pay for the gas to keep transit machines operating, which brings the total price to US\$1.75 per 1,000 cubic meters. Ukraine plans to sign three separate treaties covering purchasing, transit, and storage of natural gas.

**12** • Ukrainian President Kuchma and St. Petersburg Mayor Anatoliy Sobchak sign a trade and economic cooperation agreement. Kuchma laments that the Black Sea Fleet remains a significant problem, and that Ukraine gets "all the blame for this problem. . . . " Sobchak says he believes the fleet should belong to all the former Soviet republics.

**14** • The Russian Fuel and Energy Ministry is instructed to coordinate with the Kazakhstani government the question of guaranteed deliveries via the Caspian Pipeline Consortium network of 4.5 million tons of Kazakhstani oil per year. The American company Chevron, which is conducting oil drilling operations at the Tengiz deposit, is left out of the deal. Oman has promised to provide most of the necessary financing, including its own investment and loans from Western banks. However, many large banks have turned the project down.

***Political***

**1** • The head of Kazakhstan's State Committee for Cooperation with CIS Countries, Kalyk Abdullaev, says the committee's work is distinguished from that of the Foreign Ministry in that the committee works with all state agencies in implementing agreed CIS policies, whereas the Foreign Ministry assists in the determination and deliberation of those policies. Put succinctly, the committee's work is to achieve the "utmost promotion of integration processes and the restoration of economic ties."

**3** • The UK announces it will support Ukraine's admission to the Council of Europe despite the outcome of Russia's bid for entry. Rifkind assured Ukrainian Supreme Council Chairman Oleksandr Moroz that British representatives were already advancing Ukraine's admission at a Paris meeting of the Political Affairs Committee of the Council of Europe. In their talks, Rifkind also encouraged Ukraine to take full advantage of NATO's Partnership for Peace program, ensuring Moroz that the organization's enlargement threatens no one.

**6** • In Russia, the CIS parliamentary speakers from Armenia, Azerbaijan, Georgia, and Russia meet at IPA headquarters in St. Petersburg to assess the conflicts in the Caucasus. While they differ on settlement approaches, the leaders agree that resolution is paramount for stability in the entire Transcaucasus region. The parliamentarians, according to IPA chairman Shumeyko, "did everything within their status to influence developments in the region."

**11** • Aleksandr Lukashenka demands the creation of a political-military bloc of CIS states, asserting that such a body is necessary to confront NATO enlargement. Lukashenka believes a military-political bloc within the CIS structure would prove invaluable for world security: "What is happening in the Balkans today could not have happened ten years ago when the Soviet Union acted as the guarantor of stability in the world," he claims.

• Sergey Naumchik, leader of the Belarusian Popular Front, denounces Lukashenka for supporting the idea of building a CIS military bloc. He feels that such a bloc would threaten Belarus's neutrality doctrine, as stipulated in the constitution. Moreover, he argues, Belarusian participation would "forever block integration into European institutions, at the same time aggravating the domestic situation and increasing tensions within Belarusian society."

**12** • Russian, Ukrainian, Belarusian, and Kazakh officials declare in a jointly published statement that most CIS states would be unable to fulfill their CFE treaty obligations by the November 17 deadline due to economic hardships. All four claim that funds are not available to dismantle military equipment and weapons falling under the CFE treaty.

**14** • Chairman of Moldova's United Democratic Congress Party Valeriu Matei states that most Moldovan politicians and academics oppose the military-political bloc proposed by Belarusian President Lukashenka. Matei argues that such a bloc would lead to restoration of the Soviet Union and "smother the former Soviet states' independence and sovereignty."

• Kyrghyzstan indicates that it will soon join the customs union treaty currently participated in by Russia, Belarus, and Kazakhstan.

**15** • Russian Prime Minister Chernomyrdin flies to Tbilisi for talks with Shevardnadze. The two discuss a number of bilateral agreements covering railway transportation, legal codes covering criminal, civil, and family cases, pensions, currency, and export control.

• Belarusian Foreign Minister Syanko denies President Lukashenka ever proposed establishment of a military bloc, but says he merely spoke of the past when the Warsaw Pact could have prevented the Yugoslav crisis. Syanko says journalists misreported what the president actually said. Russian Foreign Minister Kozyrev says such a formation is proceeding within the CIS Collective Security framework, but that the treaty seeks to "strengthen the common security of our states and realize our common interests." In fact, says Kozyrev, "Russia is against a new division of blocs."

**16** • The presidential decree "Russia's Strategic Policy Toward the CIS" is published. Tenets of the policy include:

- ensuring overall stability in political, military, economic, humanitarian, and legal areas;
- helping ensure political and economic stability in the CIS states that will pursue friendly relations with Russia;
- positioning Russia as the leading force in a new system of economic and political relations;
- solidifying the CIS integration process; and
- positioning Russia as the leading educational center and ensuring generations to come are educated with a favorable Russian view.

Much of the document envisages stepped-up efforts toward military integration, consolidation of the Collective Security Treaty, and implementation of previous CIS defense agreements. The new policy asserts that concerted efforts must be taken to position the CIS as a leading regional organization "which can enter into sweeping cooperation with leading international forums and organizations."

**26** • Prime Minister Akezhan Kazhygeldin says Kazakhstan supports Russia's new strategic policy toward the CIS. He sees no "dictatorial" mandate in it. He agrees with Russia's placing heightened importance on CIS relations, and less on those with the West.

• Baktybek Abdrisaev, head of the Kyrgyz president's commission on international relations, says Kyrgyzstan welcomes the new Russian policy edict toward the CIS. He praises the Russian leadership for rebuilding relations with the former Soviet republics "in a well-balanced and sober-thinking fashion."

• Borys Oliynyk, chairman of the Ukrainian Commission on CIS relations, says he is not surprised by the September 15 Yeltsin edict and says Ukraine will not support the document. He says it contains the seeds of potential conflict among the CIS member states because each has its own strategic goals. This policy, he feels, leaves little room for flexibility and Russia clarifies Russia's hegemonic aims in the CIS. He particularly objects to the article that tries to forbid CIS states from entering into international fora and institutions that might "compete with the CIS."

***Security***

**6** • Russian Defense Minister Pavel Grachev visits several new military schools which have been established in Russia's Far East. In one of his speeches, he credits the Russian army with saving Russia from disintegration. He gives a speech to military officials on military reform.

**8** • Tajik President Rakhmonov visits President Boris Yeltsin in Moscow to sign several bilateral agreements. He says Russia is Central Asia's security guarantor. "Russia's readiness to render our republic material and technical aid, financial and food assistance, preparation for a customs union, and establishment of Russian-Tajik financial and industrial groups provides graphic evidence that Russia is Tajikistan's principal strategic partner," he says.

**13** • Leonid Kravchuk, former president of Ukraine, says Ukraine's only foreign policy option is to balance between Russia and the West. "The main difficulty (in Ukraine's relations with Russia) is the mentality of the Russian leadership, which still does not see Ukraine as an equal partner in the international arena."

• President Kuchma denies that Ukraine's overtures toward NATO will intensify tensions in its relations with Russia. He strongly criticizes President Boris Yeltsin's recently expressed intention to form an opposing military bloc should the Western alliance choose to expand.

**14** • The super-modern antisubmarine vessel, the *Admiral Chabanenko* (named after the legendary Northern Fleet Commander Andrey Trofimovich Chabanenko), is returned to the Russian navy by the Yantar shipyard on the Baltic Sea where it has been assigned to the Northern Fleet.

• Ukraine and NATO launch a series of political dialogues on the "sixteen plus one" formula. Until now, Russia has been the only member of the Partnership for Peace engaged in a privileged "special relationship" with NATO. The "sixteen plus one" program envisages holding joint exercises, training assistance, exchanges of experiences with peacekeeping operations, and assistance with establishing democratic means of controlling national armed forces.

• President Kuchma meets with Israeli Prime Minister Yitzhak Rabin in Kiev on 13 September to speak about Ukraine's determination to maintain its independence. He clarifies his intention to adopt a pro-American policy, but suggests a tripartite Ukrainian-Israeli-U.S. cooperation program, saying that Israel has a vested interest in maintaining "a strong and stable Ukraine." Kuchma explains that Ukraine purchases oil from Iran because Russia has raised its prices, but that Ukrainian-Iranian ties are strictly civilian, and there will be no military cooperation. Earlier in the day President Kuchma's staff tells Rabin that Russia has asked Ukraine to join its nuclear deal with Iran, but Ukraine turns down the offer.

• CIS Foreign Ministry representatives meet at the Headquarters for Coordination of CIS Military Cooperation. CIS Executive Secretary for Military Cooperation and Security Oleg Putintsev says the purpose of the meeting is to define peacekeeping in the CIS member states. Putintsev reports that each state supports a "peacekeeping" role for the CIS, but refers to constitutional limitations on its participation. The Russian delegation proposes to wait for a response from the United Nations and the OSCE on its request for recognition as "peacekeeper" in the region. Belarus strongly objects. It is decided to resume the discussion on 27 September, after which a definition will be submitted to the Council of CIS Heads of State.

**15** • President Boris Yeltsin issues an edict announcing a new Russian "Strategic Policy Toward the Commonwealth of Independent States" (see Chapter 3).

## October 1995

### *Economic*

**6** • Ukrainian Foreign Minister Gennadiy Udovenko sends a confidential letter to President Kuchma, which is published by the Ukrainian *Independence* paper. Udovenko negatively assesses Russia's strategic policy toward the CIS which has been approved by President Boris Yeltsin. He says the Ukrainian Foreign Ministry deems it necessary to assess the admissible level of Ukraine's economic dependence on Russia and determine the number of enterprises which are "tied" to the Russian economy and could not function if economic relations with Russia were broken. Udovenko suggests that the countries whose views on the CIS are close to Ukraine's should work out a joint approach to resolving problems related to the CIS. "Russia has no intention to work out its relations with the CIS in line with international law, respecting their territorial integrity, sovereignty, and the principle of non-interference into domestic affairs," Udovenko says. He says the "integration, whose necessity is proclaimed in Yeltsin's corresponding decree" in fact means undermining the CIS countries' sovereignty, subordinating their activity to Russia's interests and restoring the centralized superpower." Udovenko continues that: "It should be considered possible that following its earlier proclaimed attempt, Russia will declare Ukraine a bankrupt state and demand that Ukraine compensate its debt by its property." He says he believes Moscow might take several steps. It "may put forward property claims and demand that Kiev transfer to Russia the Ukrainian network of main pipelines and gas storage facilities, control blocs of shares in enterprises of the key industries, etc." Udovenko expresses regret over the publication of his confidential letter to Kuchma and tells the press the letter is not an official document.

**19** • Ukrainian Prime Minister Evgeniy Marchuk says the core Russian-Ukrainian economic agreements for 1996 will be signed before the end of November. He is referring to agreements on trade and economic cooperation between Ukraine and Russia in 1996, and on gas supplies to Ukraine and transit of Russian and Turkmen gas through Ukrainian territory. Marchuk says he believes that the situation concerning free trade within the CIS "is getting more complicated rather than easier" following the creation of the customs union between Russia, Belarus, and Kazakhstan. He says that after the treaty on the customs union comes into effect, Ukraine and Kazakhstan will have to coordinate their bilateral trade relations with "supranational structures."

**23** • A high-ranking Russian diplomat tells Interfax that Moscow has received no official explanation from either the Azerbaijani government or the international oil consortium concerning the decision to accept two routes for exporting crude oil from the southern part of the Caspian Sea—through Russia and through Georgia. The decision was made at the beginning of October. He says Moscow does not object to the parallel transportation of crude oil on the condition that the Russian pipeline is given priority. "Up to Novorossiysk, the Russian oil pipeline is in need of minor repair, estimated at about $60 million. The construction of the Georgian pipeline is still to be completed, which will require an investment of $200 million to $300 million, and will take about eighteen months," says the diplomat.

**31** • The deputy speaker of the Georgian parliament calls the international oil consortium's decision to transport Azeri oil across Georgia "comparable only with the republic's entry into the UN." Referring to Russia's role, he says that the Russian side was essential, because Georgia would not have been able to transport oil across its territory on its own." In the deputy speaker's opinion, "the decision to transport Azeri oil across Georgia is just the start of an international program to build a Eurasian oil pipeline across Georgia and Azerbaijan."

### *Political*

**4** • A suggestion to convert schools in eastern Ukraine to the Ukrainian language causes an outpouring of political passions, especially in Russian-speaking Kharkov. Some candidates for deputies in the elections make names for themselves by promising not to permit the forcible Ukrainianization of the schools. Most of the schools remain Russian-speaking after the uproar, while volunteer classes in Ukrainian language are instituted.

**5** • Viktor Timoshenko, writing in *Nezavisimaya Gazeta*, says that unlike some other CIS participants, the Russian Foreign Ministry has its own plans for "closer integration." A meeting of CIS foreign ministers in Moscow—six ministers from Azerbaijan, Armenia, Georgia, Russia, Tajikistan, and Uzbekistan and six first deputies from Belarus, Kazakhstan, Kyrgyzstan, Turkmenistan, Moldova, and Ukraine—ends in failure to sign any of the seventeen documents they discuss. Russia was interested in creating a concept of CIS peacekeeping activities, so it put forward several ideas on forming working agencies—a Council and a Secretariat of CIS Collective Security—as well as a draft procedure for funding peacekeeping functions. Ukraine took greatest exception to these proposals, saying "the legitimacy of peacekeeping activity is determined by already existing authoritative international organizations." The second package put forward by the Russians was coordination of the CIS countries' actions on the international scene. The Russian delegation believes united CIS working agencies should be set up to coordinate CIS relations with ASEAN, UNESCO, and other international organizations. These proposals found no support either. The conference was supposed to pave the way for the CIS summit planned for late October 1995, but no decision on the summit was made. The date will likely be moved further down the road.

• Ukrainian Supreme Council Chairman Oleksandr Moroz and Polish Sejm Speaker Jozef Zych announce at their 5 October meeting that Ukraine and Poland should coordinate their actions in the Council of Europe by working out common decisions and defending them. The two speakers also agree to coordinate their countries' legislation. Moroz stresses that following the Council of Europe's decision to admit Ukraine, both countries have "equal responsibility for the situation in Europe."

• Ukrainian President Leonid Kuchma, speaking at a meeting with Polish Sejm Speaker Jozef Zych in Kiev, says that Ukraine and Poland are strategic partners not only in rhetoric but in action. Kuchma adds that Ukrainian Defense Minister Valeriy Shmarov is visiting Poland at that moment, and that a consultative meeting of the Ukrainian and Polish presidents has been held recently.

**10** • An article in a Lvov publication, *Post-Postup,* entitled "In the East Ukraine Borders Russia, in the West, America," says Kiev must balance between two axes—Washington and Moscow—in its geopolitics. The article appraises Prime Minister Evgeniy Marchuk's October visit to the United States. On the economic level, things did not go well. The IMF has established the condition on U.S. credits to the Ukrainian economy that mandatory sales of hard currency by enterprises must be abolished, yet the Ukrainian parliament has prepared a draft resolution calling for 100 percent mandatory sale of hard currency! Illustrating Ukraine's method of playing Washington off against Moscow, Marchuk stated that, "in the event of Moscow's unconstructive policy with regard to the Black Sea Fleet base, Kiev does not rule out a possibility to appeal to the United States for arbitration."

**12** • An article by Tetyana Sylina in *Silske Zhyttya* (Ukrainian) entitled "The CIS Train Has Been Offered a 'Strategic' Course, Ukraine Is in No Hurry to Stand in Line for a Ticket," reveals Ukraine's position on the recently promulgated Russian edict (14 September) proclaiming a new strategic policy toward the CIS. The following excerpts are taken from the article:

> "To all appearances the already notorious 'Strategic Course of Russia Toward Developing Relations with CIS Countries' was born without assistance from the Russian Foreign Ministry. This conclusion is obvious even after a superficial glance over the document's blunt wording, which makes it more like a manifesto. . . . As for who concocted the manifesto, society was denied that knowledge. There are many institutions and officials who consider themselves experts on foreign policy issues, including the Commonwealth Affairs Ministry and the Duma Committee for CIS Affairs and Relations with Compatriots, as well as 'analysts' of the Russian Defense Ministry and KGB in the [Yeltsin advisor Aleksandr] Korzhakov mold. . . . Although not particularly enthusiastic about its implementation, the CIS countries have regained the gift of speech. Upon returning from Moscow, the Ukrainian delegation (led by Deputy Foreign Minister Kostyantyn Hryshchenko) noted with pleasure: " 'For the first time ever, the Ukrainian stance was consistently supported by many states. In the past, on the eve of every summit, they whispered into our ears, 'You guys are our last hope.' During the summit, however, they preferred to keep quiet and watch Ukraine 'throwing itself at the tank.' Outspoken about its refusal to dance in the crowd, Ukraine appeared to outside observers as a lonely heretic facing isolation. . . . This time, we have chosen a different path, thinking not only about whether, and why, proposals made by integrators from Moscow are not suitable for us, but also why they are unacceptable to others. And it turned out that if you convincingly point out the unfounded and premature nature of an idea, they start to listen to you. If a Moscow proposal is unacceptable, you may come up with your own. . . . Our delegation went to Moscow with two proposals: to enforce the notion of 'state' rather than 'internal' or 'external' CIS borders, and . . . to examine the issue of CIS support for returning deported citizens. . . . When signatories of the Tashkent Treaty (Collective Security Treaty) approved a set of documents about the financing and activities of the Col-

lective Security Council, Ukraine demurred. As of today, there is no clear concept of the collective security system in the CIS. Nor has it been decided against 'whom' or 'what' the Commonwealth should defend itself. Besides, who is supposed to defend it? After the collapse of the Soviet Union, only Russia and Ukraine were left with big armies. The rest of the states have 'toy' armies that are short of equipment, qualified officers, and recruits. One can understand their intention to hide under someone's 'security umbrella.' It is unlikely, however, that everyone will feel comfortable under such an 'umbrella' because everyone has different geopolitical interests. Therefore, one can state that the CIS collective security system is presently nothing more than a political slogan."

**13** • Moldovan President Mircea Snegur says that relations with Russia have priority for Moldova over those with Romania and Ukraine. As an "economic member of the CIS," Moldova is at the same time orienting itself toward "Western values in politics and economy." The partnership agreement with the EU and the "Partnership for Peace" with NATO should ensure Moldova's stability, says the president. Moldova does not want to become a member of any military bloc, nor agree to a stationing of troops.

**20** • In an interview with *Kievskie Vedomosti*, Russian State Duma Chairman Ivan Rybkin is questioned about Russian-Ukrainian relations. One year has passed since the Ukrainian parliament's speaker, Oleksandr Moroz, visited Russia. No reciprocal visit by the Russian speaker has been made. Rybkin indicates that relations are cool. His answers are evasive and repeat the "formula" provided by the Foreign Ministry. He urges Ukraine to join the CIS Interparliamentary Assembly (IPA). He says that the Russian delegations go to meetings of the IPA with copies of adopted Russian laws and bills; every CIS parliament is given the "opportunity to become familiar with these bills." Asked whether he is satisfied with the Russian parliament's Committee for CIS Affairs and Ties with Compatriots, which many Ukrainians believe engages in activities that violate their direct functions and serve to weaken ties between Ukraine and Russia, Rybkin replies evasively, recounting the difficulties encountered by the 25 million Russians living outside their "homeland." He says the problem of Crimea is a "tough problem," referring to Crimea as a "component of Ukraine," but then saying: "All decisions regarding Crimea should be well-considered and take into account the Crimean people's will." He says the Black Sea Fleet is still a problem, especially in the context of the Yugoslav civil war, implying that Ukraine is trying to join the NATO military bloc by participating in NATO peacekeeping units in Bosnia. When asked whether Russia and Ukraine are in a "state of cold war," he answers: "I am sure our relations could be closer, and I wish they were. We should not stick to any narrow selfish nationalistic interests, many of which are fictitious."

**26** • Ukrainian President Kuchma, touring Latin America, tells journalists in Brasilia that "Russia bears in mind only its own interests in relations with other CIS countries." He feels "neither respect nor common equal partnership" with Russia, he says. The problem of the Black Sea Fleet could have been solved much sooner, he believes. "Ukraine is making great concessions on this issue," he adds.

**27** • During President Leonid Kuchma's visit to Buenos Aires, Ukraine and Argentina are to sign a declaration on the principles of political relations respected by both states.

**28** • Konstantin Zatulin, chairman of the Russian State Duma Committee for CIS Issues and Contacts with the Diaspora, talks about the 20 October resolution on the development of integration processes between the Russian Federation and Belarus. This is a resolution "To discuss the need to hold a Russia-wide referendum on relations with the Republic of Belarus after the program for integration is drafted." The first clause of the resolution reads: "To recommend that the president of the Russian Federation submit for discussion by the Russian Federal Assembly the program for integration between Russia and Belarus." Asked what prompted the Duma to pass the resolution, Zatulin says the Duma considered the 14 May referendum in Belarus which confirmed that the Belarusian nation is striving to reunite with Russia. In light of this referendum the Duma wanted to express its absolute support for profound integration between the two states. State Duma deputies "expect that further rapprochement between our two countries . . . will contribute to the increase of their wealth, . . . facilitate the development of Slavic culture, and strengthen Russia's and Belarus's authority in the international arena."

### *Security*

**6** • Col.-Gen. Viktor Samsonov, chief of staff for Coordinating Military Cooperation Among CIS States, writes in *Krasnaya Zvezda* that events in the Balkans are assuming a dangerous character, necessitating the immediate formation of a CIS mechanism for joint consultations so as to coordinate their positions (on Bosnia). He writes that "such actions by the CIS should not be viewed as the formation of a bloc and return to confrontation. In the current international atmosphere this is above all an attempt to ensure their own security by collective efforts within the framework of a regional collective security system." He goes on to say that "Military security is just one of the many forms of security that the majority of countries want to achieve. The practice of world development shows that this goal can be achieved in practice only on the basis of the pooling of states' efforts.

. . . The objective conditions for the formation of such an alliance based on the community of interests and military-political aims already exist in the CIS states. The corresponding interstate documents reflecting the countries' desire for closer cooperation in the military security sphere also exist." He writes further that implementation of CIS security agreements would "increase CIS influence on the development of world events directly affecting their collective interests." All of this implies that Samsonov is arguing for collective intervention by CIS states in the Yugoslav conflict, or at least coordination of the CIS position with Russia's perceived interests.

**11** • Within the framework of the interstate agreement between the Republic of Kazakhstan and Russia "On Joint Efforts to Protect the CIS Countries' External Borders," Russian and Kazakh border guards will perform joint service in Kazakh outposts. Maj.-Gen. Vladislav Prokhoda has been appointed commander-in-chief of these troops.

**13** • Pavel Tarasenko, commander of the Russian Border Troops in Tajikistan, says in a news conference that the Tajik-Afghan border is subject to violent attacks by the Tajik opposition forces and that more than 1,500 opposition fighters are poised on the border, prepared to attack. Opposition authorities describe these words as "lies and slander." They say the Tajik united opposition's military forces are not preparing to attack, but are awaiting the fifth round of negotiations, and want peace in the region.

**19** • Ukrainian Defense Minister Valeriy Shmarov says Ukraine will not join with Moscow in any future military alliance aimed at countering NATO, but neither does Ukraine intend to seek full membership in the Western alliance. Reacting to a recent veiled threat by Russia to create a new military bloc to counter an eastward expansion of NATO, he says: "Ukraine is a neutral state, is not a member of any bloc, and will not take part in any military bloc." The minister says Ukraine has joined NATO's Partnership for Peace scheme and "is currently looking at ways of cooperating with the Western alliance," but will under no circumstances join NATO's management structure. On 25 September Defense Minister Pavel Grachev had warned that if any of the three former Soviet Baltic republics were to join NATO, Russia would "create its own alliance." Shmarov's reply, made during a visit to the autonomous republic of Crimea, comes on the same day that Russian President Boris Yeltsin says he will sack his foreign minister, the strongly pro-Western Andrey Kozyrev. Kozyrev's fall is seen as symptomatic of a hardening of Russian attitudes against the West in general and NATO in particular.

**23** • Eduard Baltin, commander of the Black Sea Fleet, does not intend to carry out a single decision of the commander-in-chief of Russia's navy and the Ukrainian defense minister regarding the transfer of the Black Sea Fleet infrastructure until a corresponding decision is adopted by the presidents of Russia and Ukraine.

**27** • The Russian State Duma adopts the Russian Federation Law "On Defense" in its third reading. The new law makes a number of organizational changes, such as providing for involvement in defense of the border troops, the internal troops, the railway troops, FAPSI (Federation Government Communications and Information Agency) troops, and civil defense troops. Article 10 says: "The Russian Federation president can use the Russian Federation Armed Forces to perform tasks that fall outside their remit in accordance with federal laws." The law remains to be brought into force (or not), which will depend on the Federation Council and the Russian Federation president. One of the drafters does not expect the law to have an easy passage or to be quickly approved by presidential edict. Debates are inevitable.

**29** • In the Russian parliament the Russia's Choice bloc criticizes the recent increase in the military budget in connection with an increase of the army by 200,000 men, which will cost the budget 2.7 trillion rubles.

**31** • Abkhazian Premier Gennadiy Gagulia announces that Russia has lifted its sea blockade of Abkhazia and the port of Sukhumi is reopening.

**November 1995**

***Economic***

**8** • Ukrainian officials call for a new free trade agreement between bordering oblasts of Ukraine, Russia, and Belarus, which is supposed to boost local trade. A previous agreement signed in 1993 is judged ineffective because 200 commodities, including fuels, non-ferrous metals, timber, and cellulose, were not put on the list and are still subject to non-tariff regulations.

**14** • Ukrainian President Kuchma holds talks with top Chinese leadership on military-technical cooperation. Ukrainian-Chinese trade reached $900 million in 1994. Most of this was Ukrainian exports ($850 million). China is interested in the Nikolaev Shipbuilding Yards, capable of building large warships and auxiliary craft.

**15** • Writing in the Kazakh journal *Karavan-Blits,* Viktor Verk expresses his view that if the communists come to power in Russia and if Viktor Chernomyrdin is toppled, Almaty will be forced to give up several planned joint programs, particularly the establishment of joint financial and industrial groups. He explains that left-of-center parties in Russia are calling for restrictions on assistance from Moscow to former "little brothers." The Zyuganov faction receives its mandate from populist promises it makes to voters. Furthermore, as communists move to fulfill their slogans of "restoring the USSR," relations with Kazakhstan will become "frosty." Verk believes that attempts to play the Russophone card will intensify because Zyuganov is closer to nationalist patriotism than to orthodox communism. "Things have never worked out well when we have had Russian nationalist patriots in charge of things," he explains. Verk says people who believe that once "reds" take over in Russia, the same will happen in Kazakhstan, are wrong because the Kazakh communists have no program. He conjectures further that Russian communists might grant one of the petitions that claim the electoral law is unconstitutional, which could delay the emergence of a new Duma indefinitely.

**21** • Russian Fuel and Energy Minister Yuriy Shafrannik says Russian-Ukrainian trade turnover has grown by 1 percent in 1995. He believes that next year the trade turnover will grow primarily through more Russian energy supplies to Ukraine. Ukraine's total energy debt to Russia for 1995 is $186 million compared to $133 million for 1994. Ukraine plans to cover $60 million of the debt by supplying nine fishing vessels to Russia.

**25** • An article in *Kommersant-Daily* assesses prospects for the CIS Customs Union, which has three signatories: Kazakhstan, Russia, and Belarus. At a regular trilateral commission meeting held in Moscow under the chairmanship of Vice Premier Aleksey Bolshakov, the members decide to eliminate the accelerated acceptance of new members in the Customs Union. Countries will be admitted only after they meet minimum procedures and fulfill all requirements established by the founding states. These requirements include adopting a list of complex tariff regulations and non-tariff barrier agreements. Uzbekistan, Tajikistan, and Kyrgyzstan have applied to join the union. In response, the procedures were worked out for accepting new members.

**28** • Turkmenistan and Russia form a joint-stock company, Turkmenrosgas, which will control 100 percent of Turkmen gas supplies to CIS markets. President Niyazov emphasizes that the joint venture does not limit the interests of potential investors from third countries: "If they have mutually advantageous proposals, let them come."

**29** • Turkmen President Saparmurad Niyazov and Slovak Republic Prime Minister Vladimir Meciar approve the idea of creating a new trilateral grouping with Ukraine under which the Slovak Republic will buy Turkmen gas with payment in commodities and services. The deal depends, however, on Russia, which must approve the transit of gas through the Russian pipeline. No alternative exists at this time. Turkmenistan has made efforts to build a new pipeline with Iran and Turkey to the west, as an alternative to the Russian pipeline, but Western investors lose enthusiasm when Iran is included. Meanwhile, Moscow has plans for Turkmen oil and gas, expressed by Viktor Chernomyrdin during his spring visit to Turkmenistan. Chernomyrdin proposed exporting by Russian gas pipeline to Europe as much as 125 billion cubic meters of natural gas from Turkmenistan, a deal which would be much more profitable for Russia than exporting Siberian gas abroad. However, Niyazov has made no response to Chernomyrdin's proposal to date, at no surprise to local observers.

• Viktor Hladush, Ukraine's First Deputy Minister of Trade and Foreign Economic Relations, says the Ukrainian government believes it is inexpedient to sign any more interstate agreements on trade and economic cooperation within the CIS and plans to shift to direct contracts with economic agents of CIS states, full-fledged free trade, and formation of financial and industrial groups. The decision comes following a sharp drop in volume of goods supplied under interstate agreements within the CIS. Total trade turnover within the CIS for the first nine months of 1995 amounted to 85.4 percent of the level for the same period in 1994. The share of trade under interstate agreements in total turnover fell from 8.7 percent to 4.5 percent. The CIS rule of reconciling mutual supplies at world prices is unprofitable, says Hladush, inasmuch as many Ukrainian prices are higher than world prices. Ukraine will continue to purchase fuel and energy supplies under interstate agreements, but only to cover the minimum needs of the communal and production sectors. Ukraine owes CIS states $3.46 billion, including due and overdue interest.

• Moldovan presidential spokesman Vasiliy Grozavu tells the press that Moldova does not intend to join the CIS military-political union or any other military-political blocs. This would contradict the Moldovan Constitution which reads that the country is neutral, Grozavu says. "Moldova's membership in the CIS consists of participation in comprehensive mutually beneficial economic cooperation with all CIS member states," he says.

• Ukrainian Foreign Minister Hennadiy Udovenko tells the press that the success of Ukraine's economic reform depends to a large extent on assistance from the European Union (EU). Only three EU member-countries out of fifteen (France, Spain, and Italy) have so far ratified a partnership and cooperation agreement between Ukraine and the European Union. Udovenko says these three share a common view with their Ukrainian counterpart on the processes taking place in the CIS in general and Russia in particular. He says that the 14 September decree on Russia's strategic interests prompted concern and various reactions in EU member-countries.

***Political***

**2** • Moldovan oppositionist Nicolae Dabija, deputy chairman of the Party of Democratic Forces, writes in the weekly *Literatura şi Arta* that Moldova has reached a dead end in its new constitution. He says the Moldovan Constitution "in fact only ratifies the Ribbentrop-Molotov pact and looks like an annex to the pact. It federalizes Bessarabia. It makes collective farms eternal. It tells lies (e.g. [by referring to] the Moldovan language, Moldovan people, Moldovan nation). It states that we are not a nation, but a population. Private property is not protected. Teaching of religion in schools is prohibited," Dabija says. In fact, he concludes, Moldova as a state with an "independent" appearance is really a Russian "creature," with its "sovereignty declared and never achieved, and which will estrange us from our origins."

**3** • After a five-month pause in the activity of CIS supreme collective organs, the CIS Heads of State meet in Moscow for a summit. In the period under review, no breakthrough toward closer integration in the political or economic spheres has been achieved. Prime Minister Viktor Chernomyrdin announces that forty-seven agreements have been adopted, and an Interstate Economic Committee and customs and payments unions have been created. Nevertheless, CIS member-states are in arrears to the Russian fuel and energy complex to the tune of 14 trillion rubles, including 12 trillion rubles for gas. The sum of Russian state credits to its partners in recent years is $5.8 billion. The Interstate Economic Committee (IEC) proposes delegation to itself of additional powers which would allow it to adopt decisions on the formation of the Economic Union and the activity of its organs. This proposal has been debated before, but the Armenian delegation objects to any point which would give the IEC the right to approve statutes and rules of collective organs. Azerbaijan and Ukraine have an even tougher stance. They are against the IEC's powers to make operational decisions and see no need to create supranational organs. A summit of the CIS Heads of State is due to take place in December.

**11** • Vladimir Zhirinovskiy and Ukrainian businessman Volodymyr Bezymyanyy hold a meeting which they describe as an encounter between the future presidents of Ukraine and Russia. They tell journalists that they will take all measures to unite Ukraine and Russia.

• The Supreme Soviet chairman of the so-called "Dniester Republic," Grigore Maracuta, says Dniester will adhere to all CIS economic, customs, and security structures. "Our integration with all CIS institutions will prevent Moldova's integration with Romania and will contribute to its remaining an independent state," he says. Dniester's constitution stipulates that Dniester is an independent, sovereign country, parliamentary-presidential, with independent branches of power. The "republic" plans to adopt the constitution by referendum.

**14** • Russian and Estonian negotiators launch another round of talks on their border dispute. In 1991, when Estonia regained independence, it claimed some 2,500 square kilometers of Russian land, referring to the 1920 Tartu peace treaty. Tallinn is ready to drop its territorial claims in exchange for Russia's recognition of the treaty. Russia is willing to recognize the historical value of the document only.

**21** • Russian Foreign Ministry spokesman Grigoriy Karasin tells the press that Russia and Ukraine disagree considerably on the "joint citizenship" issue. Ukraine would prefer to end dual citizenship entirely, he says. This would mean that if a treaty is signed, people who are now citizens of both Ukraine and Russia would have to give up one of their citizenships. As for Russians, they believe that "in the situation that emerged following the disintegration of the USSR, it is necessary to recognize people's right to be citizens of not only the state they live in at present, but also of the state they were born in or lived in for a long time, and not force them to choose the citizenship of one country," Karasin explains.

**22** • The Russian State Duma votes down a motion put forward by two factions to ask Russian President Boris Yeltsin to dismiss Defense Minister Pavel Grachev. In their appeal, the Russia's Choice and Women of Russia factions urged lawmakers to "rid the army of such a minister." This move follows Grachev's statement that he will order the army to vote for the Our Home Is Russia [party] in upcoming parliamentary elections scheduled for 17 December. Faction members testify that Grachev has violated the law "On Defense," which bans campaigning in the army.

**28** • Ukrainian Prime Minister Evgeniy Marchuk tells *Le Soir* that the Ukrainian government was never sent an advance copy of the 14 September decree issued by President Boris Yeltsin on Russia's relations with the other CIS member-states. Even in Russia, this decree was interpreted as a sign that Moscow was returning to ideas close to Soviet imperialism. Marchuk says he does not know whether other countries received copies of the decree. Interviewer says Belarus obviously did, because it is clearly acting on it.

**29** • Dmitriy Markov, Ukrainian President Kuchma's press secretary, tells the press that "The Ukrainian president will continue to follow principles of balanced and predictable relations, the more so with Russia, while pursuing his foreign policy."

• Moldovan presidential spokesman Vasiliy Grozavu says President Mircea Snegur emphasizes the OSCE's priority role in mediating the settlement in the Trans-Dniester region.

*Security*

**2** • Pavel Grachev addresses a meeting of the CIS council of Defense Ministers held prior to the summit of Heads of State. He says that so far no common approach has been formed with regard to the creation of a collective security system and a unified armed forces. Grachev suggests removing from the agenda the issue of a committee of chiefs of staff which would plan the use of troops and coordinate plans for building CIS armed forces, saying it is too early considering the absence of a common approach to a collective security system for the Commonwealth.

**8** • In a move which strengthens Russian-Kazakh military cooperation, Russia announces it will supply military aircraft to Kazakhstan in November and December. The Russian spokesman does not specify what kind of aircraft will be supplied.

**11** • The CIS Council of Defense Ministers meets and discusses training of CIS peacekeeping troops. Participants (except Moldova, Ukraine, and Azerbaijan) approve an agreement on the preparation of training programs. The joint air defense system is also discussed, but financing poses difficulty. Russia wants all members to contribute, but this is unacceptable to some because each has an air defense system which is more or less developed. The necessity of a single approach is questioned.

**21** • Ukrainian Defense Minister Valeriy Shmarov and Austrian Defense Minister Werner Fasslabend sign an agreement on military cooperation, under which the two countries will cooperate in peacekeeping, European security, military personnel, military medicine, economics, and other aspects of security.

**23** • On 23 November 1995, Lt.-Gen. Aleksandr Sokolov, deputy commander of the land troops of the Russian Federation, proposes a function transfer from Russian peacekeeping troops deployed in the eastern part of the Republic of Moldova to the Operative Group of Russian Troops, the successor to the 14th Army. The Moldovan Joint Control Commission for peaceful settlement of the situation in Dniester has issued a communiqué rejecting the Russian proposal. The communiqué cautions that Article 4 of the Moldovan-Russian convention of July 1992 stipulates that the Russian Army units deployed in Moldova are to observe strict neutrality. A transfer of functions would also be tantamount to a violation of the 1994 accord between Moldova and Russia regarding the juridical status, manner of operation, and deadlines of Russian military units temporarily deployed on the territory of the Republic of Moldova. The 1994 accord stipulates the full withdrawal of Russian troops from Moldova.

**25** • The Moldovan Ministry of National Security captures a Russian agent infiltrated into the leadership of the national Defense Ministry.

**27** • Russia and Ukraine reach agreement to hand over the Black Sea fleet based in Crimea to Ukraine's Navy beginning 1 December 1995. Agreement is based on Sochi talks between Ukrainian Defense Minister Valeriy Shmarov and his Russian counterpart, Pavel Grachev. Under another agreement with Russia, Ukraine will receive a military garrison in Kerch, where aircraft, missile, naval, and airborne testing ranges, airfields, shipyards, and ship maintenance shops are located. Ukraine will also take over arsenals of two ammunition depots, including an aviation ammunition depot belonging to the Black Sea fleet.

**29** • Commanders of the Baltic countries' armed forces concluding a joint meeting in Riga announce an agreement to create a joint Baltic headquarters which will conduct training programs within the Partnership for Peace program. A permanent Baltic battalion staff will also be created. Estonian peacekeeping forces will be incorporated in the Nordic unit and Lithuanians and Latvians within the Danish battalion.

**December 1995**

*Economic*

**13** • The European Bank for Reconstruction and Development (EBRD) announces that it will provide Ukraine with a loan worth more than US$75 million to promote economic reform in the republic.

• Ukraine signs an agreement on cooperation with the Konrad Adenauer Foundation in Kiev. Deputy Prime Minister Ivan Kuras stresses that this action is additional convincing evidence of a rising trend in the development of German-Ukrainian ties.

**16** • In an interview with *Vysokyy Zamok*, the Russian consul in Lvov, Anatoliy Kovalev, explains that he has made an economic study of western Ukraine and sent a report to Moscow, which is being studied at the highest levels. The study analyzes the economic potential of factories in the west, the kinds of investment they require and what channels for selling products. Kovalev says western Ukraine is the most stable region politically and that Russia and other countries could safely invest their capital and technologies there. He holds out prospects of such economic activity creating many new jobs. "Our concept is thus to return Russia (with Ukraine's consent) to these eight Ukrainian oblasts at the state level," he says. "This requires a long-term program and partnership, and Russia should be represented in this process by the most serious organizations. Kiev has also supported this program."

**18** • The frequency of current in Ukraine's power grid grows following its reconnection to the Russian power grid. After disconnection of the two power grids on 4 December the frequency dropped to a close to critical level as far as nuclear power plants are concerned.

**26** • Viktor Chernomyrdin announces at a meeting of the CIS Interstate Economic Committee that the CIS countries owe Russia 15 trillion rubles for energy resources. Over eleven months of 1995, Russia exported 61 billion cubic meters of natural gas, 25 million tons of oil, and 10 billion kW of electrical energy to CIS countries. Chernomyrdin says that in 1996 the Russian budget earmarks only 200 billion rubles as credits to CIS countries. These credits will be set aside for the purchase of ready-made products, not energy supplies, says Chernomyrdin.

*Political*

**7** • The Tajik Communist Party leader, Shodi Shabkolov, says his goal is the re-creation of the USSR as a voluntary union of fraternal peoples. "The communists of Tajikistan are now very actively participating in the movement for unity, friendship, and fraternity of the peoples of the Union," he explains.

• Ukrainian Parliamentary Speaker Oleksandr Moroz stresses that "mistakes were made when the CIS was set up because the disintegration of the USSR should have been based on the specific condition that the future subjects of the Commonwealth would continue to exist even after the Soviet Union was gone." He says that as things are, one member has ended up with everything and the others with nothing. He says, however, that Ukraine could defend its interests better if it were a full, rather than an associate, member of the organization. He says that joining the CIS Interparliamentary Assembly would pose no threat to Ukraine's sovereignty. Leonid Kravchuk, the former Ukrainian president, notes that "setting up the CIS was not a simple process and is still continuing." He emphasizes that the European Union was set up over a thirty-five-year period and that in the initial stages the CIS failed to resolve the problem of a "civilized divorce."

**8** • Ukrainian Foreign Minister Gennadiy Udovenko says he is "astonished" by Romanian Foreign Minister Teodor Melescanu's statement openly implying a territorial claim against Ukraine. Ukraine and Romania have territorial disputes going back to before World War II. The Molotov-Ribbentrop pact, which divided much of Eastern Europe between Nazi Germany and the Soviet Union in 1939, ceded vast tracts of Romanian land to the Soviets. Since the end of the Cold War, nationalists in Romania have been calling for an abrogation of the pact, despite the 1947 Paris Peace Treaty which recognized Soviet and subsequently Ukrainian gains under the pact. Melescanu said in the Romanian parliament that Ukraine "refused to give back Zmeyiniy Island, situated in the Black Sea."

**13** • The Ukrainian president's chief of staff, Dmitriy Tabachnik, tells the press that Kiev and London have developed completely new relations, and want closer cooperation. He says London now treats Ukraine as an entity quite outside the former USSR. He recalls that the Ukrainian and British leaders have met on numerous occasions in 1995 and that the British parliament is one of the first to ratify the Ukrainian-European Union partnership and cooperation agreement. The two countries plan to sign a joint declaration on 15 December which will express a more active rapprochement between the two countries and thus enhance Ukraine's international standing.

• Azerbaijani President Geydar Aliev says the slowing in Azerbaijani-Iranian relations must be overcome. "One cannot allow cooling in our bilateral relations," he declares during a meeting with Iranian Deputy Foreign Minister Mahmud Va'ezi in Baku. "Azerbaijan will not allow any other country to influence the Azerbaijani-Iranian relationship."

**14** • Russian Prime Minister Viktor Chernomyrdin states categorically that there will be no unification of Russia and Belarus.

**18** • The presidents of Kazakhstan, Uzbekistan, and Kyrgyzstan meet in southern Kazakhstan to discuss the economic integration plan they adopted in April. They decide to increase the Central Asia Bank's nominal capital from $5 million to $9 million. They also set up a Council of Defense Ministers in order to stiffen regional security. During the meeting they appeal to the UN secretary general to form a UN-sponsored peacekeeping battalion as a reserve force to be deployed on Kazakhstan's southern border with Uzbekistan.

• Moldovan and Belarusian Defense Ministers General Pavel Creanga and Lt.-Gen. Leonid Maltsaw tell the press in Chisinau that the parliamentary elections in Russia should not lead to the restoration of the ex-USSR. "The elections and their results are Russia's domestic affair and they should not influence interstate political relations," the Moldovan minister says. "The idea of restoring the USSR should be renounced, because this might result in confrontation, including armed confrontation, between the countries of the former USSR," he concludes.

• A foreign policy advisor to Moldovan President Mircea Snegur, Petru Dascal, says that the Russian parliamentary election results cannot change Moldova's external political course. "This course is determined by the constitution and the national Foreign Policy Concept, and is being consistently implemented," he explains. "We are building civilized, mutually beneficial relations with all states, including the Russian Federation."

**20** • Gennadiy Zyuganov says there has been a complete debacle of anti-communism in Russia and a complete downfall of radical democrats who were unanimous in their disdain for Russia. He says recent Duma elections in Russia show that the government's course is backed by only one out of ten who went to the polls. He says the elections also show the defeat of the centrists who do not enjoy support from many social segments of the population.

**21** • Evgeniy Primakov, in an address to mark the seventy-fifth anniversary of Russian Foreign Intelligence Service, says the service's aim of preventing Russia's territorial disintegration has acquired a new importance. He says the Foreign Intelligence Service has done much to strengthen Russia's peacemaking role in the southern crisis and in the Armenia-Azerbaijan, Georgia-Abkhazia, Tajik-Tajik, and other conflicts. The service supports the recent emergence within the former USSR of centripetal trends and processes, he says. "The future of the CIS lies in such processes," he stresses, "as long as there is a sensible approach to reality and an understanding of the need to inject several radically new elements into the policy of economic, scientific, cultural, information, and military-political integration."

• "Moscow does not recognize division of the Caspian Sea into national sectors and rejects any unilateral actions in this direction," according to the deputy director of the Foreign Ministry's Asia Department, Maksim Peshkov. Russia's position is calculated to prevent any of the five countries bordering on the Caspian from unilaterally developing any part of its resources or profiting from them.

**26** • The presidents of the CIS states, who are planning to hold a summit in mid-January 1996, will probably not respond to a petition from the "Dniester Republic" to join the CIS, says *Nezavisimaya Gazeta* reporter Alan Kasaev. Mircea Snegur will be adamantly opposed to the petition, and Eduard Shevardnadze and Geydar Aliev are likely to support him.

**28** • Results of the Russian State Duma elections show that four electoral associations were the "big winners." The CPRF (Communist Party of the Russian Federation) for the left-wing forces; "Yabloko" for the right-wing forces; the LDPR (Liberal Democratic Party of Russia) for the national patriots; and the "party of power" in the shape of "Russia Is Our Home" (NDR).

• *Moskovskie Novosti* reporter Akakiy Mikadze writes that Igor Georgadze, former head of the security service in Georgia, is living in Pavel Grachev's dacha in Moscow. Mikadze conjectures that Georgadze may be used as a tool by Moscow authorities if Georgian policy (especially concerning the transportation of Caspian oil) does not suit Moscow.

***Security***

**15** • Ukrainian President Kuchma says that "a strong and sovereign Ukraine is a major factor in European stability," before returning home after a three-day stay in London. He stresses that Ukraine is not seeking membership in NATO. "For Kiev even to apply for admission to the North Atlantic alliance would have a negative role from all perspectives, but especially politically, by acting as a destabilization factor on the European continent," Kuchma says. "Ukraine should keep clear of any alliances."

• Ukrainian President Kuchma and the British secretary of state for defense, Michael Portillo, meet to discuss bilateral relations. Kuchma calls for more active relations between the two defense ministries. Portillo tells the press that the possibility is being considered of holding joint maneuvers of British and Ukrainian servicemen on the territory of Ukraine by the end of next year.

**22** • The CIS Council of Border Guard Commanders meets in Dushanbe and signs twenty-six documents coordinating border policy throughout the Commonwealth, especially along the Tajik-Afghan border. The meeting is chaired by the director of the Russian Federal Border Service, Andrey Nikolaev. Among the documents signed is a Russian-Turkmen agreement on the presence of up to 1,000 Russian border guards and military advisers in Turkmenistan. Nikolaev notes progress toward the creation of a single information space for CIS border troops through coordination of research activities. The term of the Uzbek, Kazakh, and Kyrgyz military units stationed in Tajikistan is due to expire on 31 December, but Nikolaev favors extending their stay.

• The presidents of Uzbekistan, Kyrgyzstan, and Kazakhstan appeal to the UN Secretary General to consider the possibility of forming a regional peacekeeping force under UN aegis to be deployed in the south of Kazakhstan. They stress that the new force can be used in Bosnia and various trouble spots rather than to maintain stability in Tajikistan where CIS peacekeepers are now deployed. Andrey Nikolaev, the Russian director of the Federal Border Service, says the three-nation initiative should not be seen as an attempt to contrast regional forces to CIS collective security forces. Nikolaev says that he hopes for further Dushanbe-Moscow cooperation because Tajikistan is finding itself in a key role in ensuring the stability and security of Central Asia and the CIS as a whole. "By solving the problem of Tajikistan, we solve those of each CIS state," he says.

**27** • Russia completes removal of the last strategic missiles from Ukraine scheduled for this year. It has bought thirty-two strategic SS-19 missiles from Ukraine; this does not violate international agreements, as Moscow may maintain 105 missile complexes after ratification of START-II.

**28** • President Boris Yeltsin creates a Foreign Policy Council, which purportedly will include leaders of the Foreign Ministry, Defense Ministry, Ministry for Foreign Economic Affairs, Ministry of Finance, Federal Security Service, Foreign Intelligence Service, and Federal Border Service as well as Dmitriy Ryurkov, the president's aide for foreign policy. The council will engage in analysis and forecasting of the international situation and will work directly under the coordination of the head of state.

# 1996

### Postscript

On 5 January 1996, Russian Foreign Minister Andrey Kozyrev submits his resignation to President Yeltsin. Kozyrev has served in the post since October 1990. On 9 January 1996, President Yeltsin appoints Evgeniy Primakov as the new foreign minister. Primakov formerly headed the Russian Foreign Intelligence Service, a branch of the former Soviet KGB. Many analysts predict that the appointment signals a change in Russian foreign policy to a more hard-line approach toward the CIS and the West.

# APPENDIX C

# CIS Data

## Political Chronology of CIS Member States

| | Established as Soviet Socialist Republic | Declares Sovereignty | Declares Independence | Joins CIS |
|---|---|---|---|---|
| Armenia | 29 Nov 1920 | 23 Aug 1990 | 23 Aug 1990 | 21 Dec 1991 |
| Azerbaijan | 28 Apr 1920 | 23 Sep 1989 | 30 Aug 1990 | 21 Dec 1991 |
| Belarus | 1 Jan 1919 | 27 Jul 1990 | 26 Aug 1991 | 8 Dec 1991 |
| Georgia | 25 Feb 1921 | 9 Mar 1990 | 9 Apr 1991 | 4 Oct 1993 |
| Kazakhstan | 26 Aug 1920 | 25 Oct 1990 | 16 Dec 1991 | 21 Dec 1991 |
| Kyrgyzstan | 1 Feb 1926 | 15 Dec 1990 | 31 Aug 1991 | 21 Dec 1991 |
| Moldova | 15 Oct 1924 | 23 Jun 1990 | 27 Aug 1991 | 21 Dec 1991 |
| Russia | — | 12 Jun 1990 | 12 Jun 1990 | 8 Dec 1991 |
| Tajikistan | 5 Dec 1929 | 24 Aug 1990 | 9 Sep 1991 | 21 Dec 1991 |
| Turkmenistan | 27 Oct 1924 | 22 Aug 1990 | 27 Oct 1991 | 21 Dec 1991 |
| Ukraine | 20 Aug 1920 | 16 Jul 1991 | 24 Aug 1991 | 8 Dec 1991 |
| Uzbekistan | 27 Oct 1924 | 20 Jun 1990 | 31 Aug 1991 | 21 Dec 1991 |

*Source*: Embassy of Republics, Washington, DC; *Keesing's Record of World Events*.

## National Currencies

| | Currency Name | Date Introduced |
|---|---|---|
| Armenia | dram | 24 Nov 1993 |
| Azerbaijan | manat | 15 Jun 1993 |
| Belarus | Belarusian ruble | 25 May 1992 |
| Georgia | Georgian coupon | 2 Aug 1993 |
| Kazakhstan | tenge | 27 Dec 1993 |
| Kyrgyzstan | som | 10 May 1993 |
| Moldova | leu | 29 Nov 1993 |
| Russia | Russian ruble | — |
| Tajikistan | Russian ruble | — |
| Turkmenistan | Turkmen manat | 1 Nov 1993 |
| Ukraine | hryvnia* | 2 Sep 1996 |
| Uzbekistan | Uzbek som | 15 Jun 1994 |

*Source*: Embassy of Republics, Washington, DC.
*Replacing karbovanets.

## Energy Dependency in the CIS

| | Energy imported from Russia as a share of total state consumption | |
|---|---|---|
| | Natural Gas | Crude Oil |
| Armenia | 0 | * |
| Azerbaijan | 0 | 14 |
| Belarus | 100 | 91 |
| Georgia | 27 | 82 |
| Kazakhstan | 0 | 0 |
| Kyrgyzstan | 0 | * |
| Moldova | 100 | * |
| Tajikistan | 0 | * |
| Turkmenistan | 0 | 16 |
| Ukraine | 56 | 89 |
| Uzbekistan | 0 | 55 |

*Sources*: FBIS reports; RFE/RL Research Reports.
*Republics that do not have oil refineries.

# List of Key CIS Officials

## Political

*Executive Secretary of the CIS:* Ivan M. Korotchenya, Republic of Belarus

### *Interparliamentary Assembly (IPA)*

*Chairman:* Vladimir Shumeyko, Russian Federation
*Members*
Russia
Armenia
Azerbaijan
Belarus
Georgia
Kazakhstan
Kyrgyzstan
Moldova
Tajikistan
Uzbekistan

### *The Council of Heads of State*

*Members*
Russia: Boris Yeltsin
Armenia: Levon Ter-Petrosyan
Azerbaijan: Geydar Aliev
Belarus: Aleksandr Lukashenka
Georgia: Eduard Shevardnadze
Moldova: Mircea Snegur
Kazakhstan: Nursultan Nazarbaev
Kyrgyzstan: Askar Akaev
Tajikistan: Emomali Rakhmonov
Turkmenistan: Saparmurad Niyazov
Ukraine: Leonid Kuchma
Uzbekistan: Islam Karimov

### *The Council of Heads of Government*

*Members*
Russia: Viktor Chernomyrdin
Armenia: Grant Bagratyan
Azerbaijan: Fuad Guliev
Belarus: Mikhail M. Chyhir
Georgia: Otar Patsatsia
Moldova: Andrey Sangheli
Kazakhstan: Sergey Tereshchenko
Kyrgyzstan: Apas Jumgulov
Tajikistan: Jamshed Karimov
Turkmenistan: Saparmurad Niyazov
Ukraine: Pavlo Lazarenko
Uzbekistan: Utkar Sultanov

### *The Council of Foreign Ministers*

*Members*
Russia: Evgeniy Primakov
Armenia: Vagan Papazyan
Azerbaijan: Hasan Hasanov
Belarus: Uladzimir Syanko
Georgia: Aleksandr Shikvadze
Molodova: Mihai Popov
Kazakhstan: K. Tokaev
Kyrgyzstan: Roza Otunbaeva
Tajikistan: Talbak Nazarov
Turkmenistan: Boris Shikhmuradov
Ukraine: Hennadiy Udovenko
Uzbekistan: Abdulaziz Komilov

## Economic

### *Council of Leaders of Foreign Economic Departments*

*Members*
Russia: Evgeniy Yasin
Armenia: Armen Egiazaryan
Azerbaijan: Saleh Mammadov
Belarus: Vladimir Radkevich
Georgia: Lado Papava
Kazakhstan: Altai Tleuberden
Kyrgyzstan: Andrey Iordan
Moldova: Valeriu Bobutac
Tajikistan: Izatullo Khayoev
Turkmenistan: Chary Kuliev
Ukraine: Serhiy Osyka
Uzbekistan: Utkar Sultanov

### *Interstate Economic Committee (IEC)*

*Chairman:* Aleksey Bolshakov, Russian Federation

*Members*
Russia
Armenia
Azerbaijan
Belarus
Georgia
Kazakhstan
Kyrgyzstan
Moldova
Tajikistan
Uzbekistan

### *Economic Union Treaty*

*Members*
- Russia
- Armenia
- Azerbaijan
- Belarus
- Georgia
- Kazakhstan
- Kyrgyzstan
- Moldova
- Tajikistan
- Uzbekistan

### *Economic Court*

*General Secretary:* Viktor Gonchar, Russian Federation

*Members*
All CIS states may appeal to the Economic Court for arbitration of interstate disputes.

## Military

Council of Heads of State. *See* **Political** heading.

Council of Foreign Ministers. *See* **Political** heading.

Council of Defense Ministers

*Chairman:* Leonid Ivashov, Russian Federation

*Members*
- Russia: Igor Rodinov
- Armenia: Vazgen Sarkissyan
- Azerbaijan: Safar Abiev
- Belarus: Leonid Maltsev
- Georgia: Vardiko Nadbaidze
- Kazakhstan: S. Nurmagambetov
- Kyrgyzstan: Murzakan Subanov
- Moldova: Pavel Crenja
- Tajikistan: Sherali Khayrullaev
- Turkmenistan: Danatar Kopekov
- Ukraine: Valeriy Shmarov
- Uzbekistan: Rustan Akhmandov

### *Council of Commanders of CIS Border Troops*

*Chairman:* Andrey Nikolaev, Russian Federation

*Members*
- Russia
- Armenia
- Azerbaijan
- Belarus
- Georgia
- Kazakhstan
- Kyrgyzstan
- Moldova
- Tajikistan
- Turkmenistan
- Ukraine
- Uzbekistan

### *Joint Staff*

Coordinates Military Cooperation Between CIS States

*Officers' Chief of Staff:* Viktor Samsonov
*First Deputy:* Boris Gromov
*First Deputy:* Bronislav Moelichev
*Deputy Chief:* Vladimir Krivonogikh
*Spokesman:* Viktor Koltunov

### *CIS Air Defense Committee*

*Chairman:* Viktor Prudnikov, Russian Federation

*Members*
- Russia
- Armenia
- Belarus
- Georgia
- Kazakhstan
- Kyrgyzstan
- Moldova
- Tajikistan
- Turkmenistan
- Ukraine
- Uzbekistan

# Appendix D

# Country Profiles

## Republic of Armenia: Political Profile

**Capital: Yerevan**
**Official language: Armenian**
**Currency: Dram (introduced 24 November 1993)**

### Head of State:

***President Levon Akopovich Ter-Petrosyan,*** b. 1945. Ter-Petrosyan worked at the Armenian Institute of Literature in the mid-1970s and at the Matenadazan Archive intermittently from 1978 to 1990. A radical nationalist, he was a member of the Karabakh Committee and was in prison December 1988–May 1989. He was leader of the Armenian Pan-national Movement and in August 1990 was elected Chairman of the Supreme Soviet (de facto head of state). He was confirmed in office by an overwhelming majority in direct elections for the post of Executive President of the Republic in October 1991. The popularity of his regime decreased gradually as a result of the protracted civil war in Armenia and consequent economic conditions. One major problem in 1993–94 was the conflict between pragmatic Armenians (like Ter-Petrosyan) favoring closer links with Turkey and the more extreme, nationalist groups—the Union for National Self-Determination and the Armenian Revolutionary Federation—opposing such links. In 1996 Ter-Petrosyan declared victory in a closely contested presidential election in which he was opposed by a former Karabakh Committee ally, Vazgen Manukyan.

### Main political parties:

*Armenian Democratic Party,* f. 1993 formerly the Communist Party of Armenia, it was dissolved in September 1991, re-legalized in 1992, renamed in 1993.
*Armenian Pan-national Movement,* f. 1989.
*Armenian Revolutionary Federation—Dashnaktsyutun,* f. 1890 by the ruling party in independent Armenia; 1918–20, prohibited under Soviet rule but continued its activities in other countries; permitted to operate legally in 1991 then banned again in 1994–95.
*National Democratic Union,* f. 1991 as a splinter party for the Armenian Pan-national Movement.
*Party of Democratic Freedom,* f. 1905.
*Republican Party of Armenia,* f. 1990 following a split in the Union of National Self-Determination.
*Union of National Self-Determination,* f. 1989.

### National legislature and distribution of seats:

Unicameral assembly with 190 members

In July 1995 elections, the Armenian pan-National Movement and its allies won 70 percent of the seats in a new parliament. The Communists came in third.

## Republic of Armenia

Total Area: 11,506 sq. miles (29,800 sq. km.)

*Source: States of the Former Soviet Union,* CIA 1992.

## Republic of Armenia: Economic Statistics

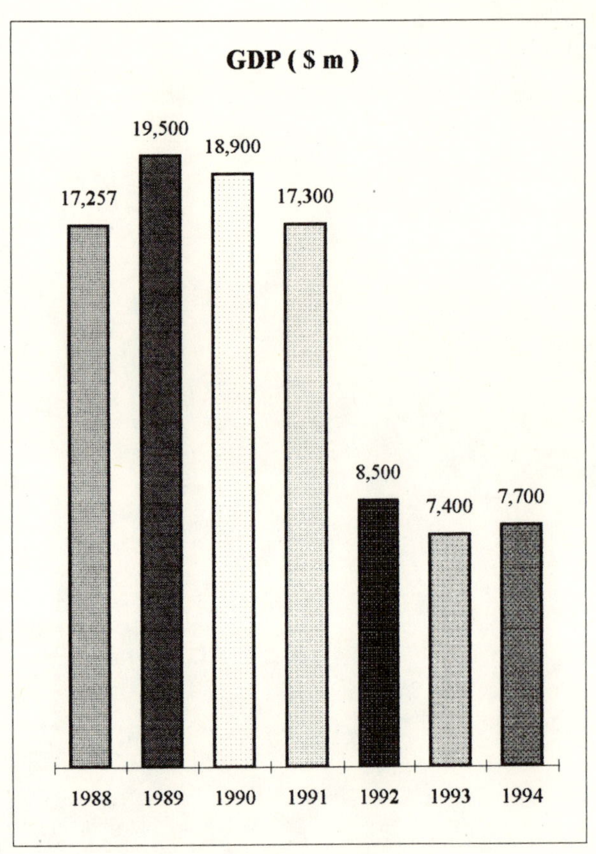


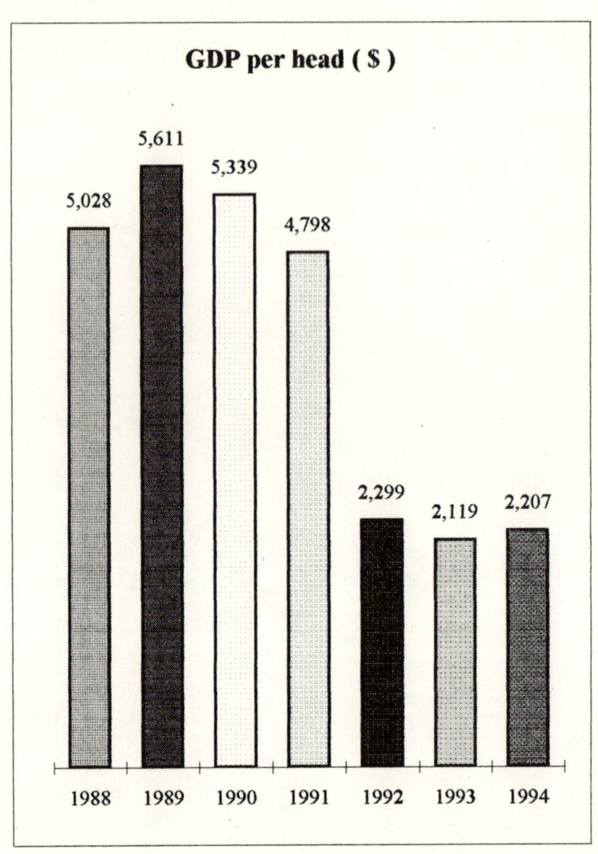


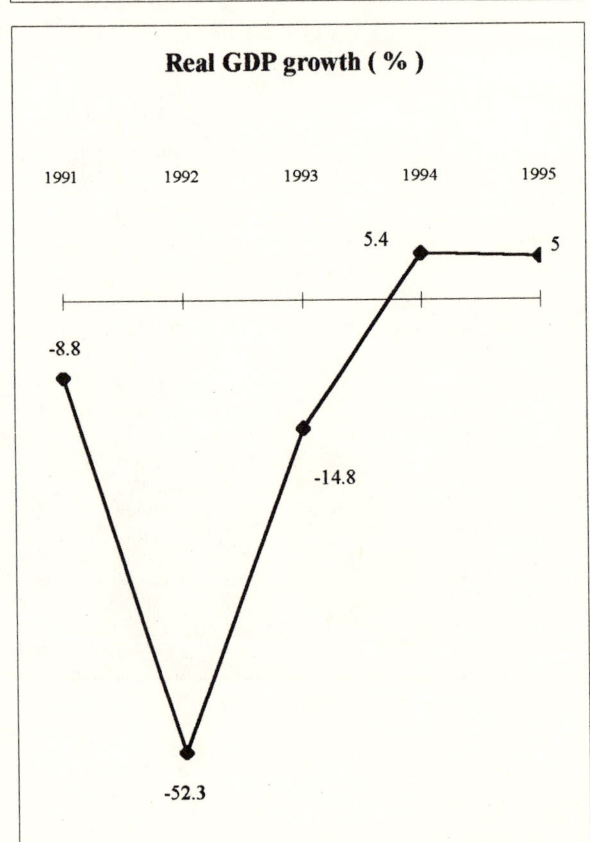


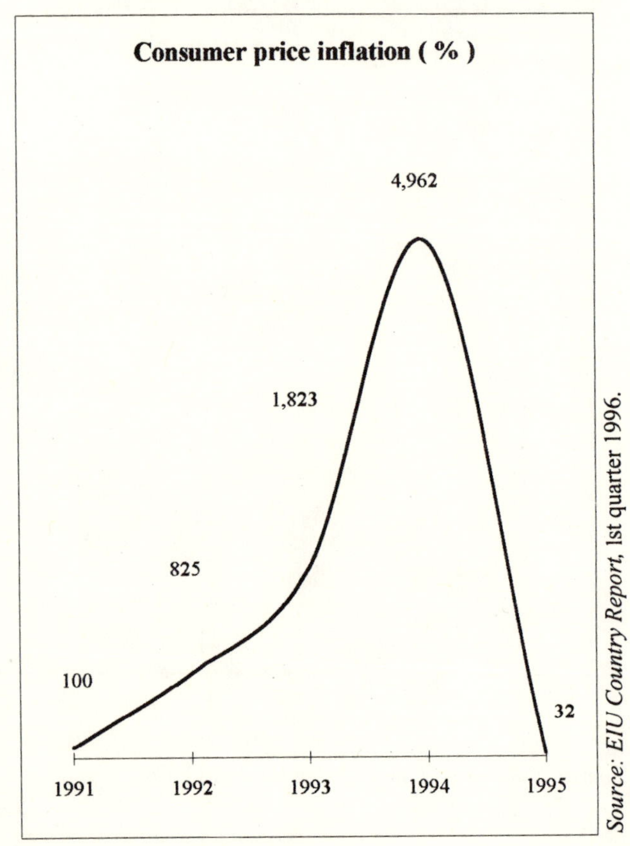


*Source: EIU Country Report*, 1st quarter 1996.

## Republic of Armenia: Demographic Statistics (1989)

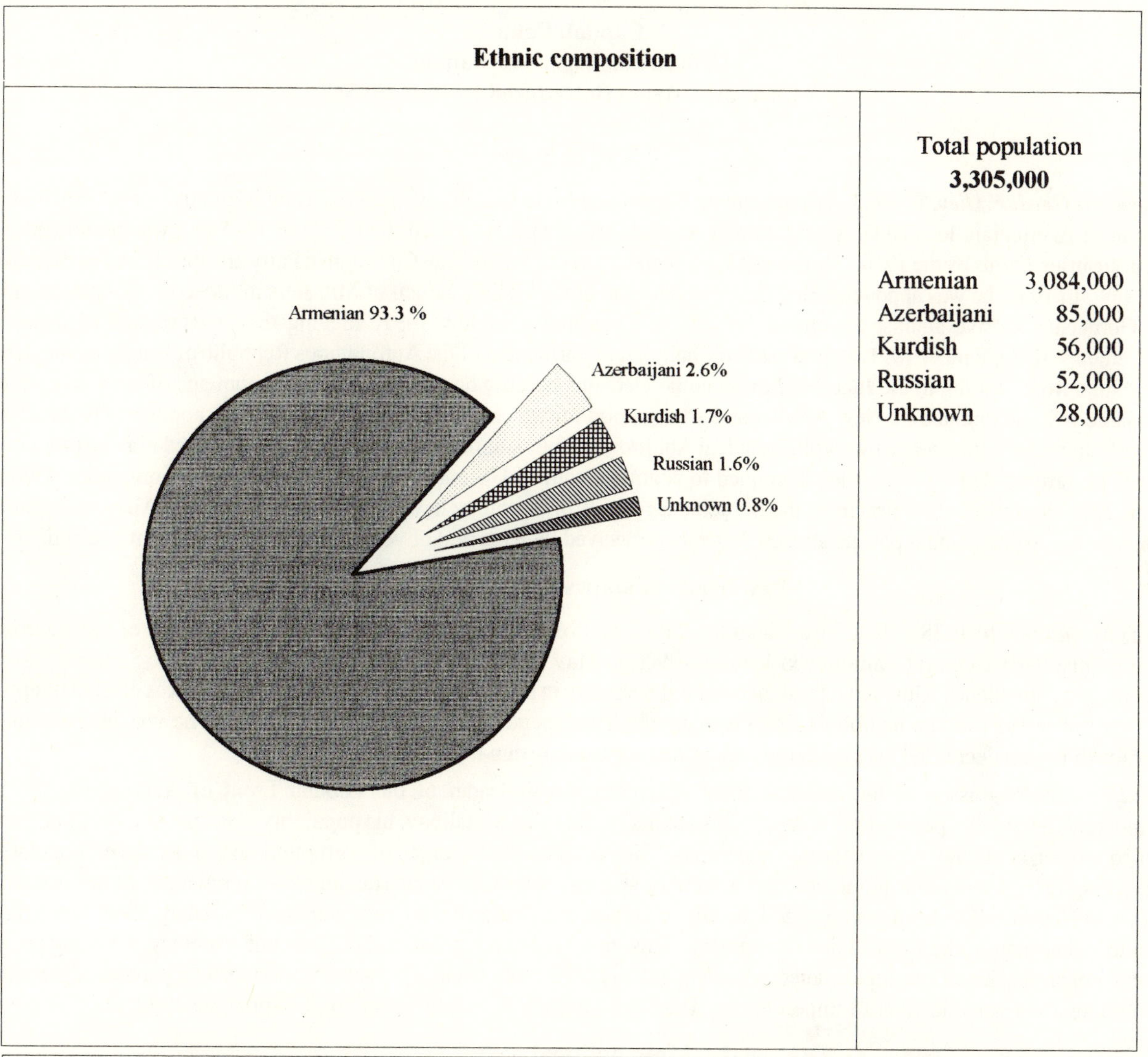


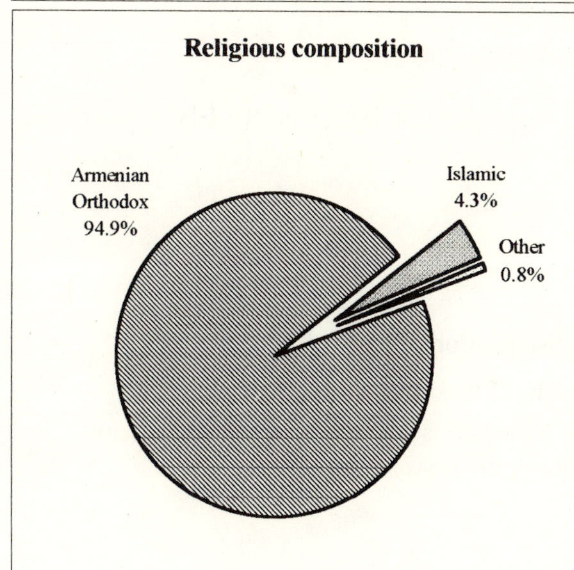


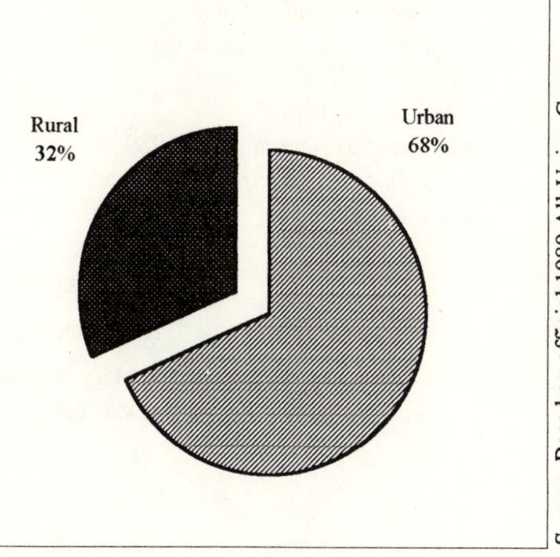


*Source:* Based on official 1989 All-Union Census.

# Azerbaijani Republic: Political Profile

**Capital: Baku**
**Official language: Azerbaijani**
**Currency: Manat (introduced 15 June 1993)**

## Head of State:

***President Geydar Aliev,*** b. 1923. A graduate of Azerbaijani State University in 1957, Aliev is a native of the Nakhichevan region of Azerbaijan, located between Iran and Armenia (see map). He joined the CPSU in 1945 and was prominent in the republican apparatus by the 1960s. He became First Secretary of the Azerbaijani Communist Party and thus leader of the republic in 1969. In 1982, he was appointed First Deputy Chairman of the USSR Council of Ministers in Moscow. Dismissed in 1987 by Gorbachev's drive against corruption, he left the Communists in July 1991, alleging their suppression of democratic movements. In September 1991, he was elected Chairman of Parliament of the Autonomous Republic of Nakhichevan. He was prevented from running in the June 1992 presidential elections because he exceeded the maximum age of sixty-five years. In the same year, he founded the New Azerbaijan Party, support for which demonstrated his continuing popularity in the country. In early June 1993, threatened by revolt, President Abulfaz Elchibey summoned Aliev to Baku and offered him the premiership, which he refused. On 13 June Aliev attempted to negotiate with Col. Surat Husseinov, a rebel army commander. Two days later, Aliev was elected Chairman of the Supreme Soviet. With Elchibey having fled to Nakhichevan, Aliev was granted a majority of the presidential powers and, on 3 October, received 98.8 percent of votes cast in direct presidential elections.

## Past heads of state since independence:

***Ayaz Mutalibov,*** b. 1938. Mutalibov joined the CPSU in 1963 and climbed his way to First Secretary of the Azerbaijani Communist Party Central Committee in January 1990. In May 1990, he was appointed Chairman of the Supreme Soviet of Azerbaijan (President). Although he resigned as First Secretary in the aftermath of the August coup, he was elected unopposed in September 1991 as the republic's first President since independence from the Soviet Union. The war in the Nagorno-Karabakh region decreased his popularity and he was forced to resign in March 1992.

***Abulfaz Elchibey*** assumed the presidency by direct election as leader of the Popular Front of Azerbaijan—the main democratic-nationalist party—on 7 June 1992. Like his predecessor Mutalibov, his popularity was hurt as a result of the war, which worsened already poor economic conditions. This, coupled with charges of corruption, led to decreased popularity of his regime. His demise was imminent after a military skirmish with Col. Surat Husseinov—Commander of the Azerbaijani forces in Nagorno-Karabakh—who seized the city of Gyanja and demanded the resignations of Elchibey, the Prime Minister, and the Chairman of the Milli Majlis (parliament). Husseinov ordered his forces into Baku and Elchibey, knowing the army would not protect him, fled. Impeachment proceedings were set in motion and, after some wavering, 97.5 percent of participants in a referendum voted in favor of impeachment. The Milli Majlis endorsed this result on 1 September 1992.

## Main political parties:

*Communist Party of Azerbaijan* disbanded September 1991, re-established November 1993.
*Independent Azerbaijan Party.*
*Musavat—"Equality" Muslim Democratic* existed in 1992; re-established in 1992, promotes Islamic values and the unity of Turkic peoples.
*New Azerbaijan,* f. 1992.
Popular Front of Azerbaijan, f. 1989.
*Social Democratic Group,* f. 1989.
*United Azerbaijan.*

## National legislature and distribution of seats:

Unicameral assembly with 125 members

In November 1995 parliamentary elections, Aliev's New Azerbaijan party won a clear majority. Participation of other parties was banned or (as in the case of Musavat) restricted.

# Azerbaijani Republic

Total Area: 33,436 sq. miles (86,600 sq. km.)

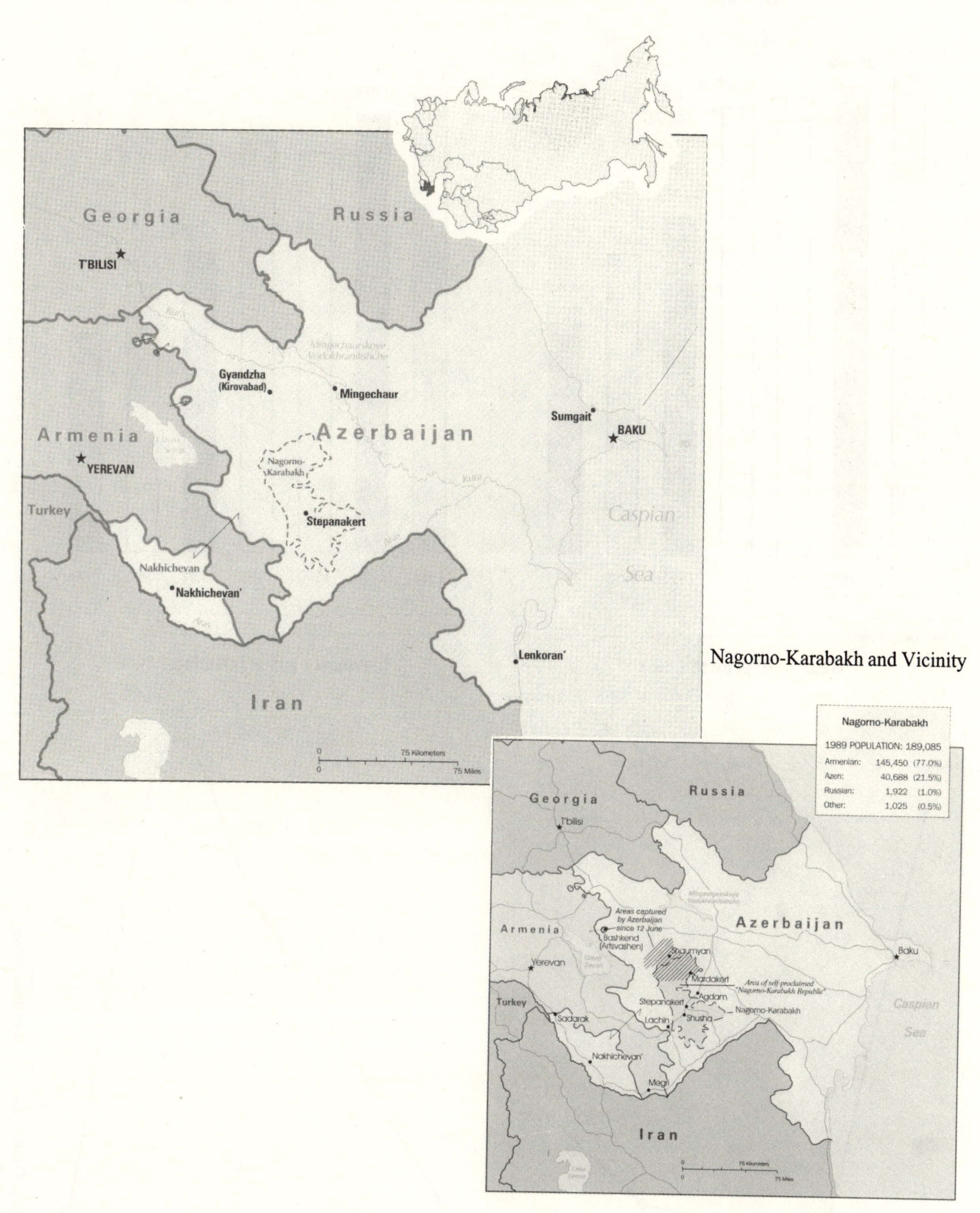


Nagorno-Karabakh and Vicinity

*Source: States of the Former Soviet Union*, CIA 1992.

## Azerbaijani Republic: Economic Statistics

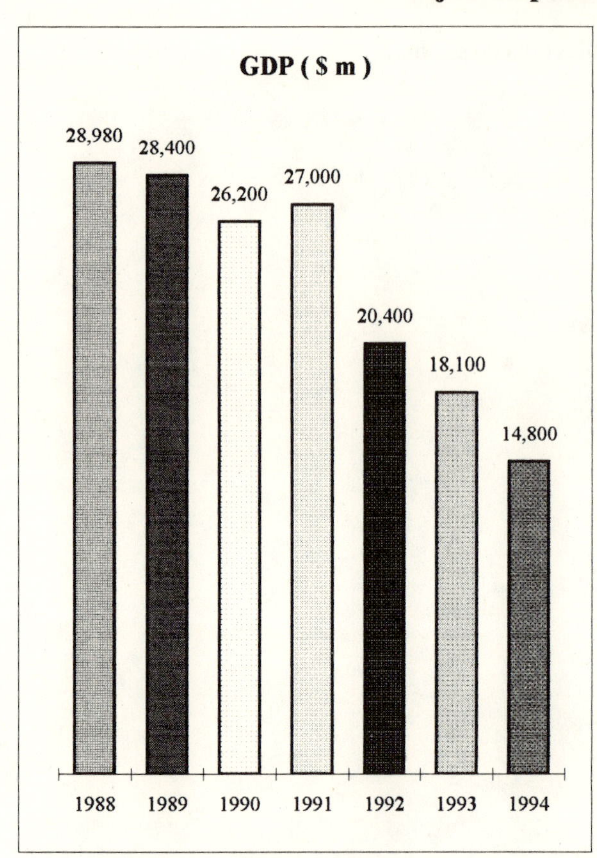


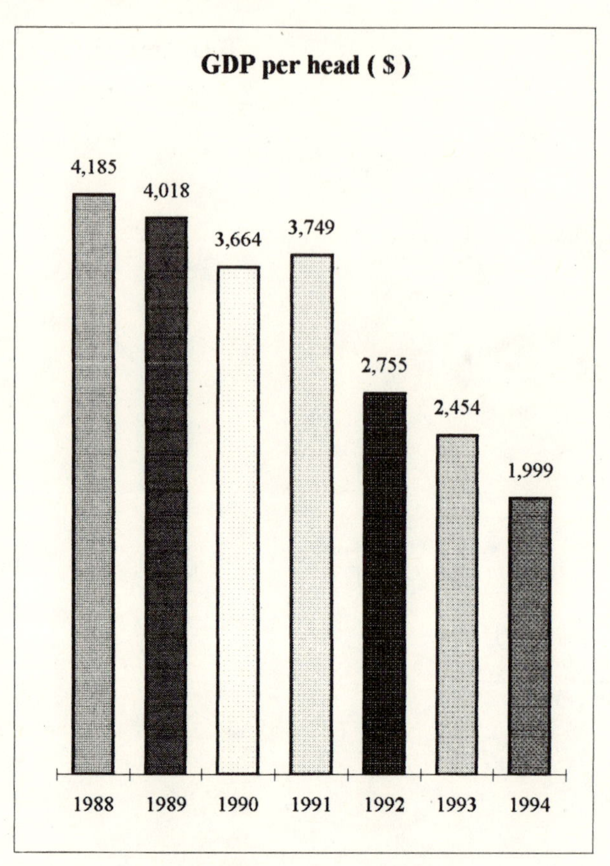


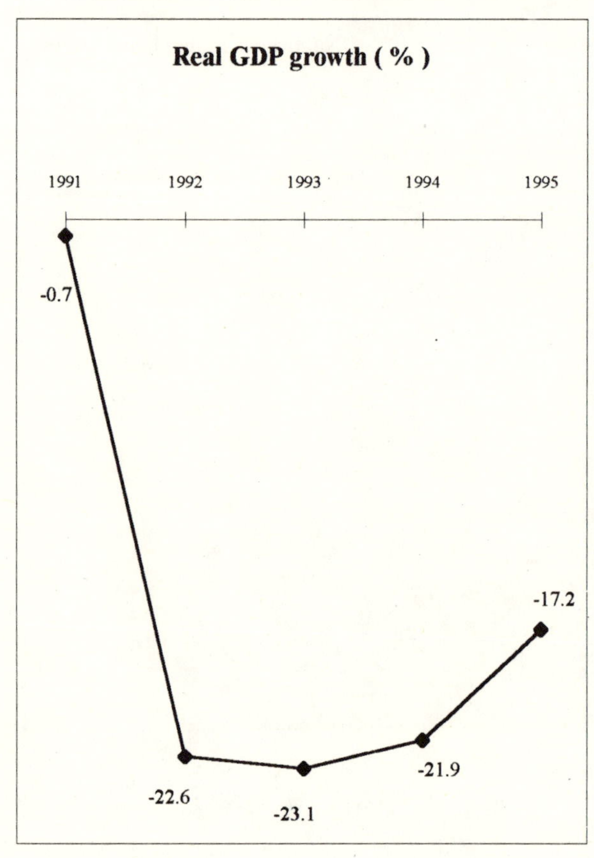


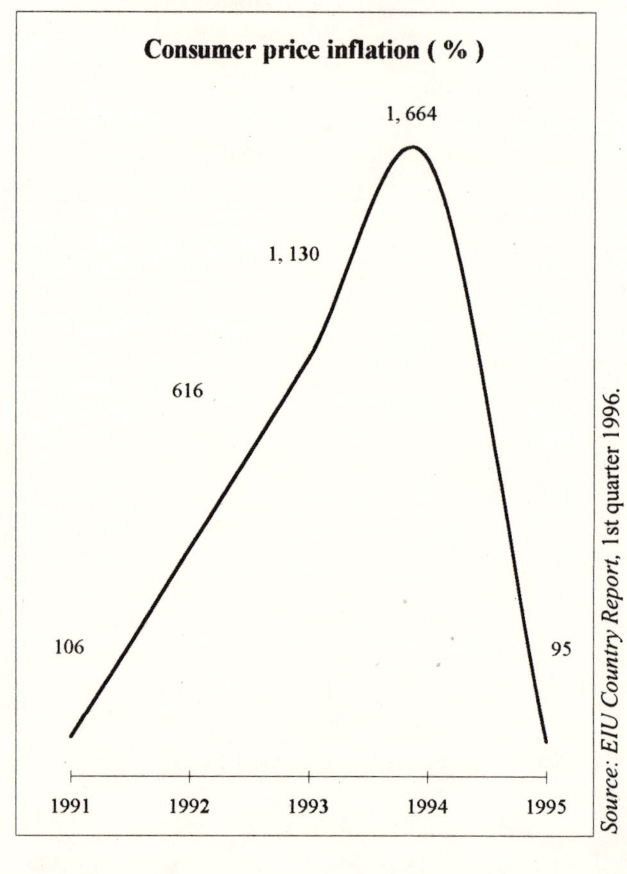


*Source: EIU Country Report*, 1st quarter 1996.

# Azerbaijani Republic: Demographic Statistics (1989)

## Ethnic composition

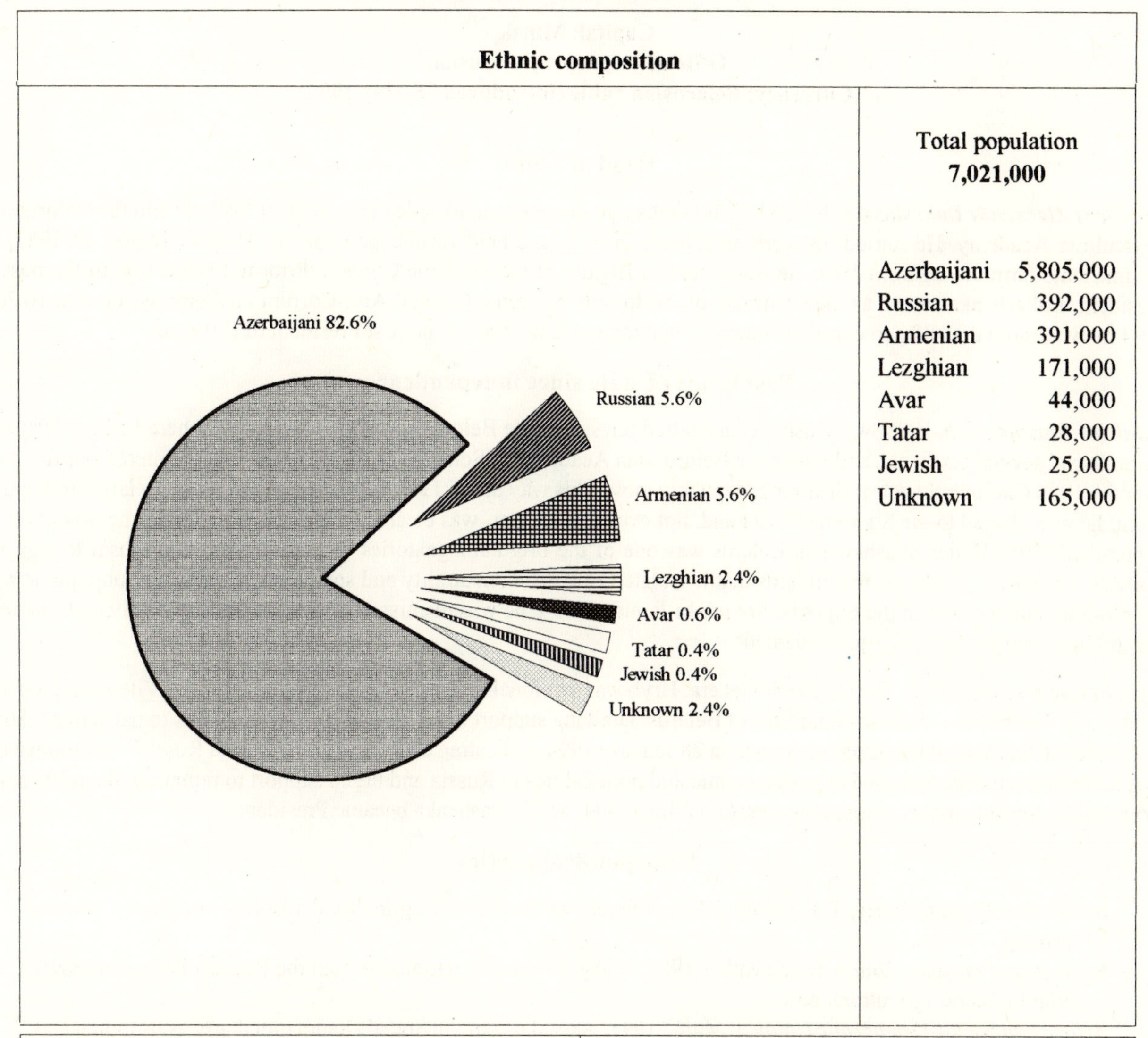


Total population
**7,021,000**

| | |
|---|---|
| Azerbaijani | 5,805,000 |
| Russian | 392,000 |
| Armenian | 391,000 |
| Lezghian | 171,000 |
| Avar | 44,000 |
| Tatar | 28,000 |
| Jewish | 25,000 |
| Unknown | 165,000 |

## Religious composition

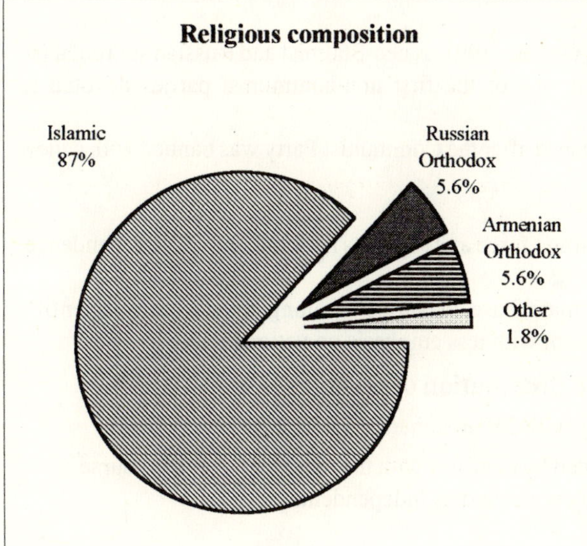


## Rural / Urban population split

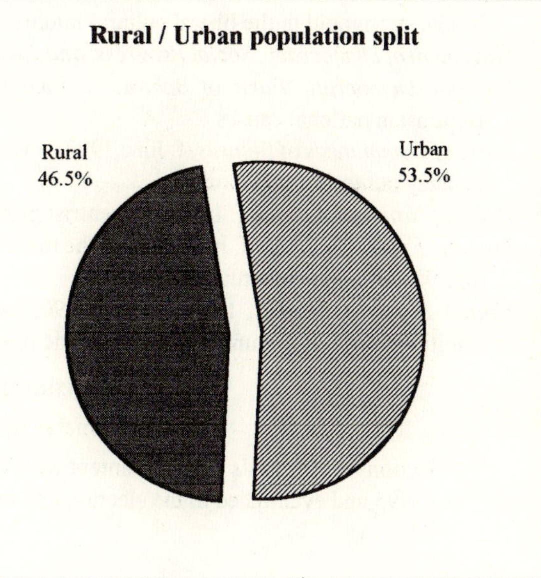


*Source:* Based on official 1989 All-Union Census.

# Republic of Belarus: Political Profile

**Capital: Minsk**
**Official language: Belarusian**
**Currency: Belarusian ruble (introduced 25 May 1992)**

## Head of State:

***President Aleksandr Lukashenka,*** b. 1954. Lukashenka graduated from Mogilev Pedagogical Institute and the Belorussian Agriculture Academy. He started his working activity in 1975 and held various positions in Mogilev region. In 1987, he became state farm manager. In 1990, he was elected a Deputy of the Supreme Council. Prior to his election to the post of President, Lukashenka served as the Chairman of the Interim Supreme Council Anti-Corruption Commission. On 10 July 1994 he won more than 80 percent of presidential election votes to become the first President of Belarus.

## Past heads of state since independence:

***Stanislav Shushkevich,*** b. 1934. Shushkevich studied physics at the Belorussian State University, where he later lectured. He authored several textbooks and joined the Belorussian Academy of Sciences in the early 1970s. He entered politics only after the 1986 Chernobyl nuclear disaster and quickly moved his way up. In 1990, with the support of the Belarusian Popular Front, he was elected to the Supreme Soviet and, not even a year later, was elected Chairman of the Supreme Soviet on 19 September 1991. Under Shushkevich, Belarus was one of the original signatories of the CIS, but his refusal to sign the Collective Security Treaty, on the grounds that it violated Belarusian neutrality and sovereignty, made him unpopular with the pro-Russian majority in the Supreme Soviet. In January 1994, he was dismissed by the parliament for alleged financial misconduct, charges he categorically denied.

***Myacheslav Hryb,*** b. 1937. During the Soviet era, Hryb was a colonel in the police force in the northern Vitebsk region and head of the security and defense committee of Belarus. Drawing support from the conservative factions in parliament, Hryb was elected Chairman of the Supreme Soviet on 28 January 1994, indicating a victory for the "Great Russia" communist old guard. He immediately proposed closer economic and political ties to Russia and urged support to remain in the ruble zone. Hryb lost the presidential election, however, on 10 July 1994, and Lukashenka became President.

## Main political parties:

*Belarusian Peasant Party,* f. February 1991, defends the interests of agricultural workers and favors private farming.

*Belarusian Peasants' Union,* f. September 1989, is less Belarusian nationalist than the Peasant Party but draws from the same agricultural base.

*Belarusian Social-Democratic Union,* f. March 1991, takes the German Social Democratic Party as its model and has its stronghold in the liberal cultural intelligentsia.

*Movement of Democracy, Social Progress, and Justice,* f. October 1991, is neo-Stalinist and Russian supremacist.

*National-Democratic Party of Belarus,* f. June 1990, is one of the first non-communist parties devoted to Belarusian national causes.

*Party of Communists of Belarus,* f. June 1992, was established after the Communist Party was banned with a view to carry out the Party's work.

*Party of National Accord,* f. 1991, is a centrist grouping.

*Popular Front of Belarus,* f. June 1989, is the main opposition force and supports national revival, independence, and Western-style institutions.

*United Party of Belarus,* f. November 1990, has its main base of support among the technical/scientific intelligentsia. It is committed to democratic development with less emphasis on nationalism.

## National legislature and distribution of seats:

Unicameral assembly with 260 members

Elections to Belarus's new parliament were disrupted by conflicts with the president over the course of 1995 and eventuated in the election of a majority registered as Independents.

## Republic of Belarus

Total Area: 80,155 sq. miles (207,600 sq. km.)

*Source: States of the Former Soviet Union,* CIA 1992.

## Republic of Belarus: Economic Statistics

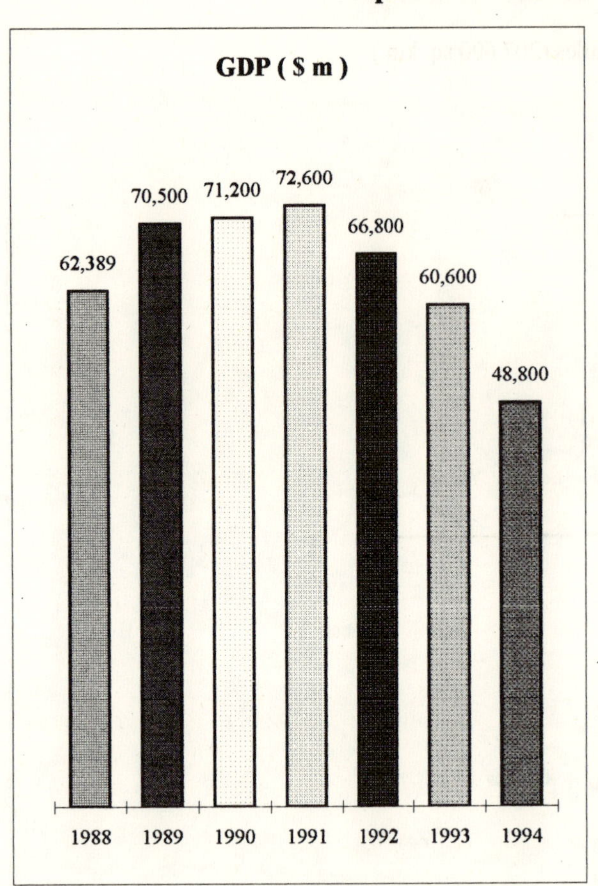


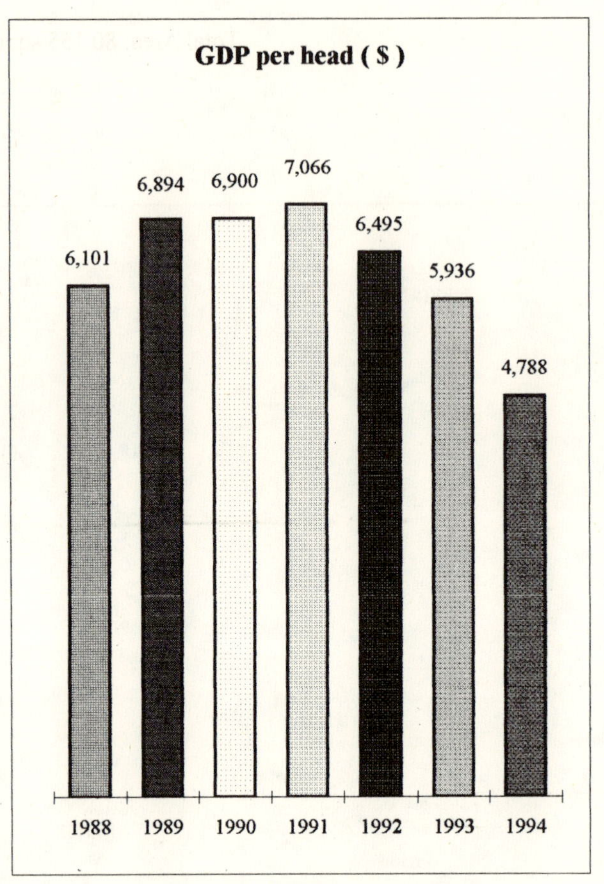


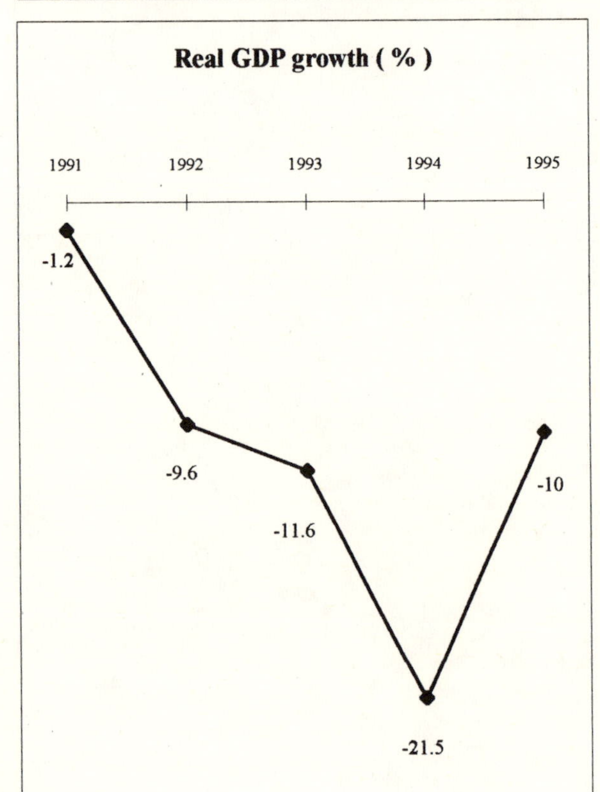


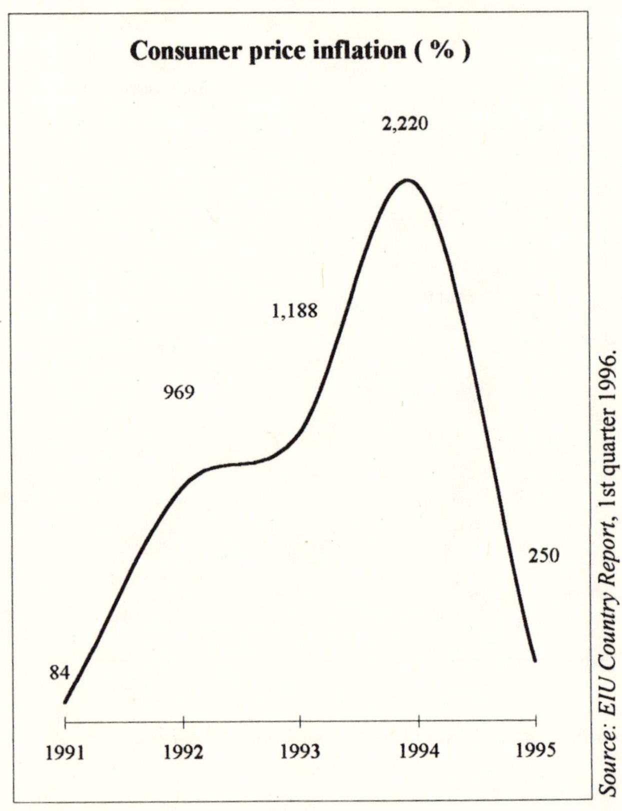


*Source: EIU Country Report*, 1st quarter 1996.

## Republic of Belarus: Demographic Statistics (1989)

### Ethnic composition

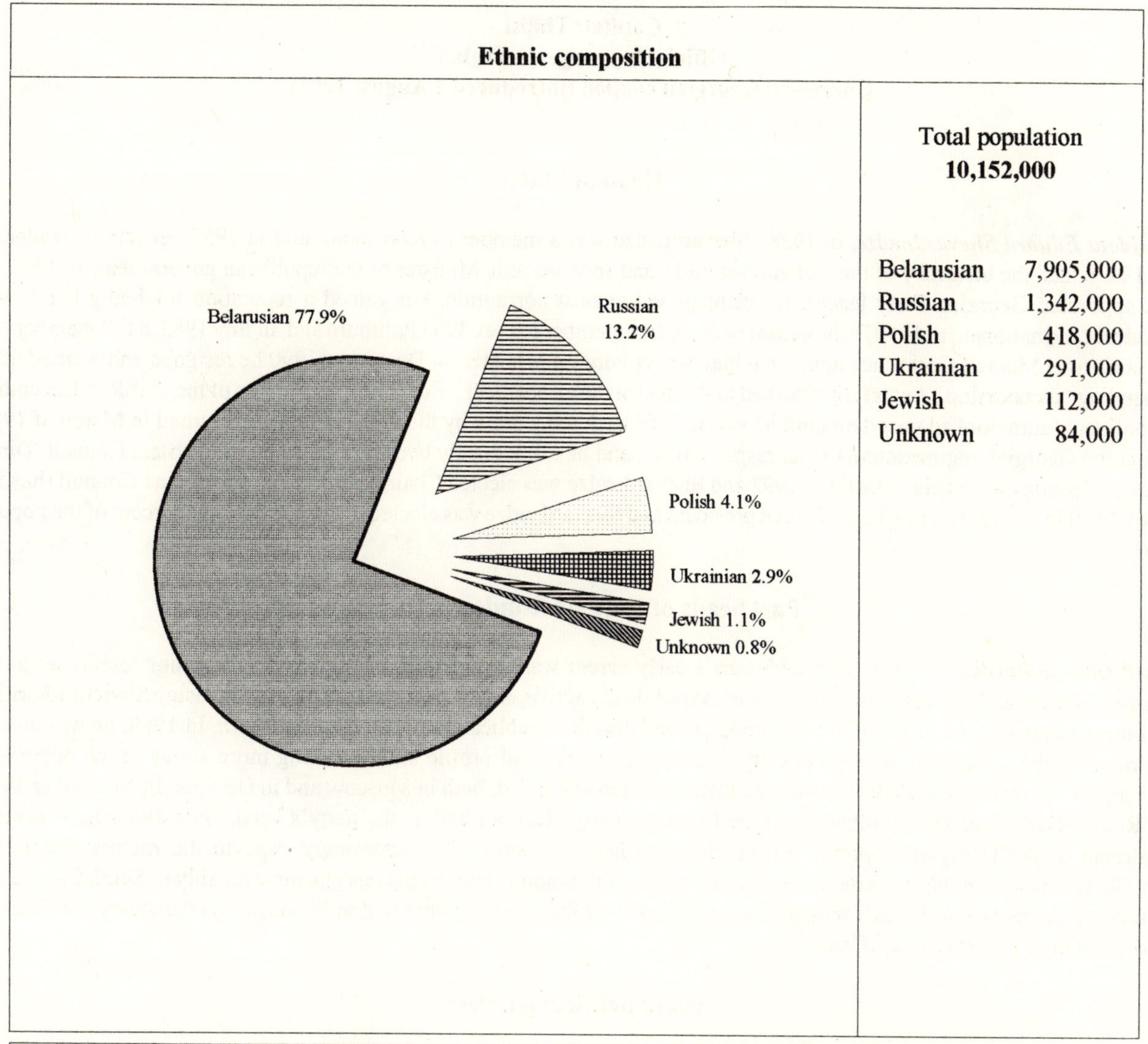


Total population
**10,152,000**

| | |
|---|---|
| Belarusian | 7,905,000 |
| Russian | 1,342,000 |
| Polish | 418,000 |
| Ukrainian | 291,000 |
| Jewish | 112,000 |
| Unknown | 84,000 |

### Religious composition

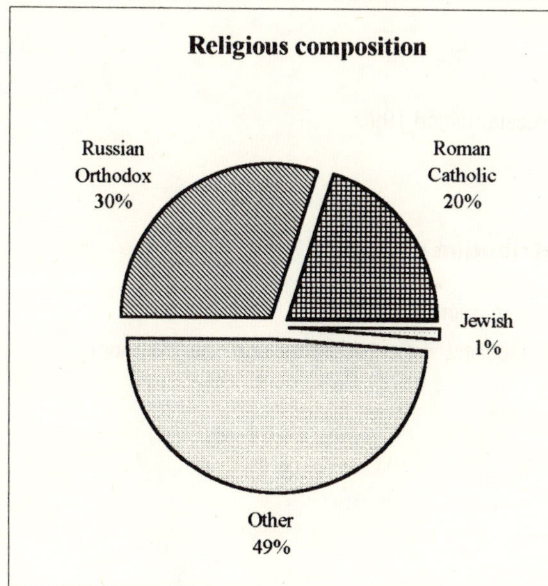


### Rural / Urban population split

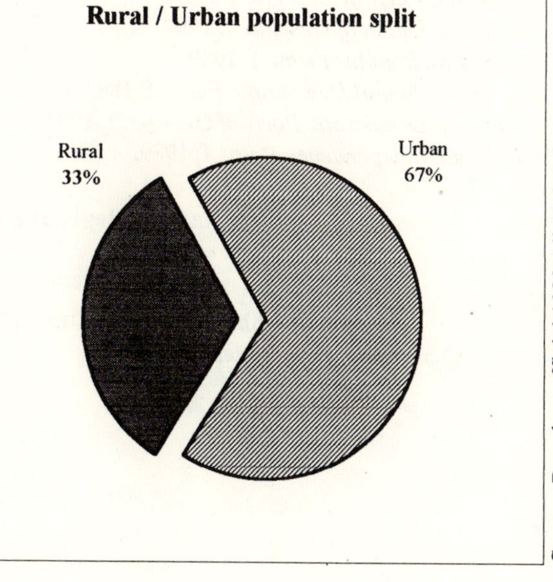


*Source:* Based on official 1989 All-Union Census.

# Republic of Georgia: Political Profile

**Capital: Tbilisi**
**Official language: Georgian**
**Currency: Georgian coupon (introduced 2 August 1992)**

## Head of State:

***President Eduard Shevardnadze,*** b. 1928. Shevardnadze was a member of Komsomol and in 1957 became its leader. In 1961 he joined the hierarchy of the Communist Party and soon became Minister of the republican government. In 1971, he was appointed Georgian Party leader. He campaigned against corruption, but gained a reputation for being harsh with dissidents and nationalists. In 1978, he became a candidate member of the CPSU Politburo and, in July 1985, a full member—at the same time Mikhail Gorbachev appointed him Soviet Foreign Minister. In December 1990 he resigned and warned of an impending "dictatorship," but briefly returned to the post at the end of 1991. Following the demise of the USSR in December, his political future looked uncertain until he was invited back to Georgia by the new regime. He returned in March of 1992, giving the Georgian regime international respectability, and he immediately became Chairman of the State Council. Direct popular elections were held in October 1992 and Shevardnadze was elected Chairman of the new Supreme Council (head of state). In 1995 Georgia restored the office of president and Shevardnadze was elected to the post with 73 percent of the popular vote.

## Past heads of state since independence:

***Zviad Gamsakhurdia,*** b. 1939. Gamsakhurdia's early career was in literature. He worked as a senior researcher at the Rustaveli Georgian Literature Institute. He became politically active, engaging in several projects opposing Soviet authorities, becoming publisher and editor of the first underground dissident publication in Georgia, *Samizdat.* In 1989, he was elected chairman of the nationalist St. Iliya Society, increasing his political profile and becoming more active in the opposition movement. As such, Gamsakhurdia was repeatedly arrested and jailed, both in Moscow and in Georgia. In November 1990, the Round Table Free Georgia Party won the first multiparty elections and as the party's head, Gamsakhurdia was made Chairman of the Georgian Supreme Soviet (de facto head of state). His increasingly chauvinistic rhetoric threatened non-Georgians from holding political positions in various autonomous regions in Georgia, most notably in South Ossetia and Abkhazia. Opposition to his rule mounted and a series of military riots culminated in his ouster on 6 January 1992. He is reportedly in hiding in Grozny, Chechnya.

## Main political parties:

*Agrarian Party of Georgia,* f. 1994.
*Citizens' Union of Georgia,* f. 1993.
*Georgian Popular Front,* f. 1989.
*Georgian Social Democratic Party,* f. 1893; dissolved 1921; reestablished 1990.
*National Democratic Party of Georgia,* f. 1981.
*National Independence Party,* f. 1989.

## National legislature and distribution of seats:

Unicameral assembly with 235 members

In 1995 elections Shevardnadze's Citizens' Union of Georgia won 106 seats and the National Democratic Party 34 seats.

## Republic of Georgia

Total Area: 26,911 sq. miles (69,700 sq. km.)

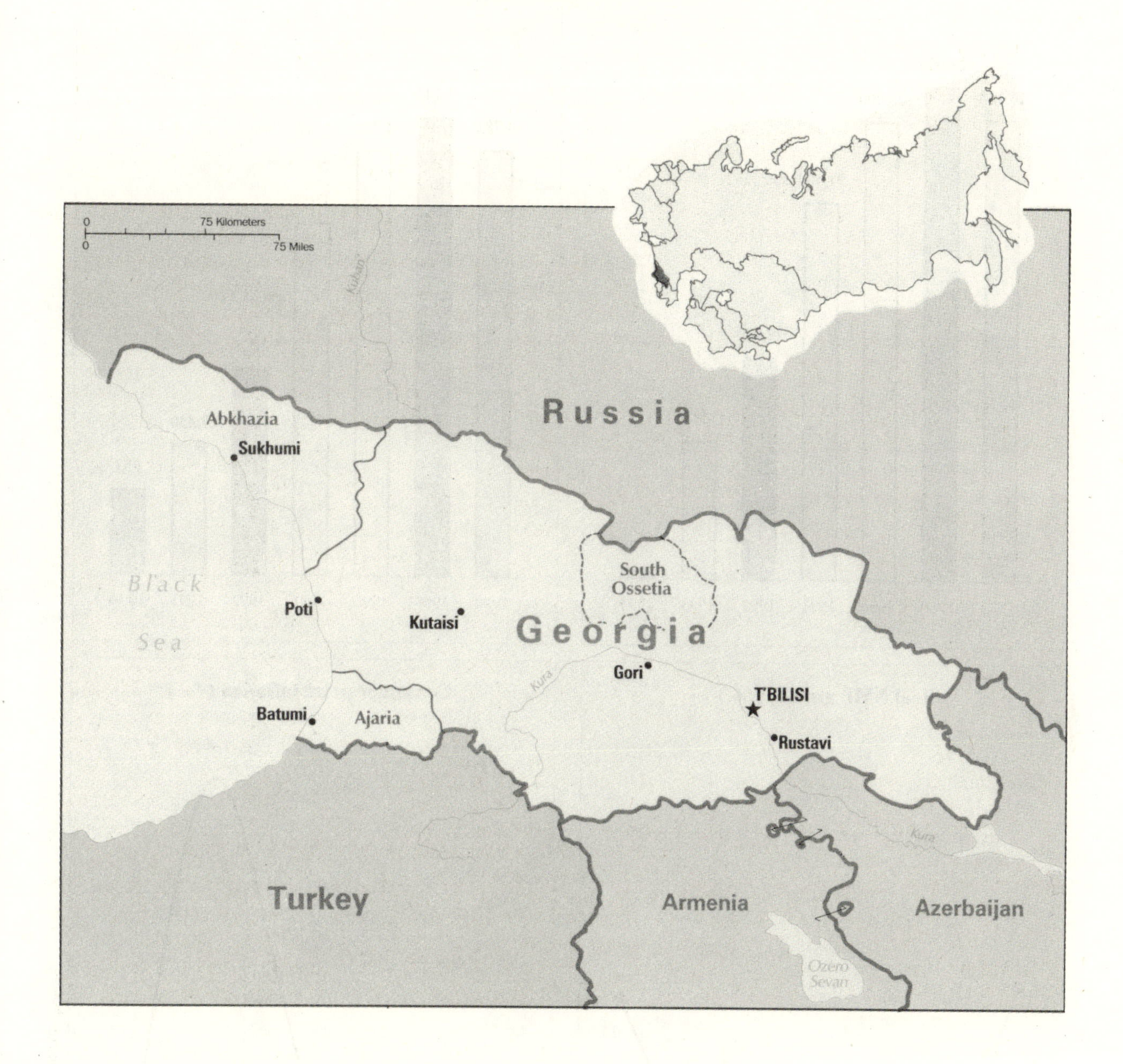


*Source: States of the Former Soviet Union,* CIA 1992.

## Republic of Georgia: Economic Statistics

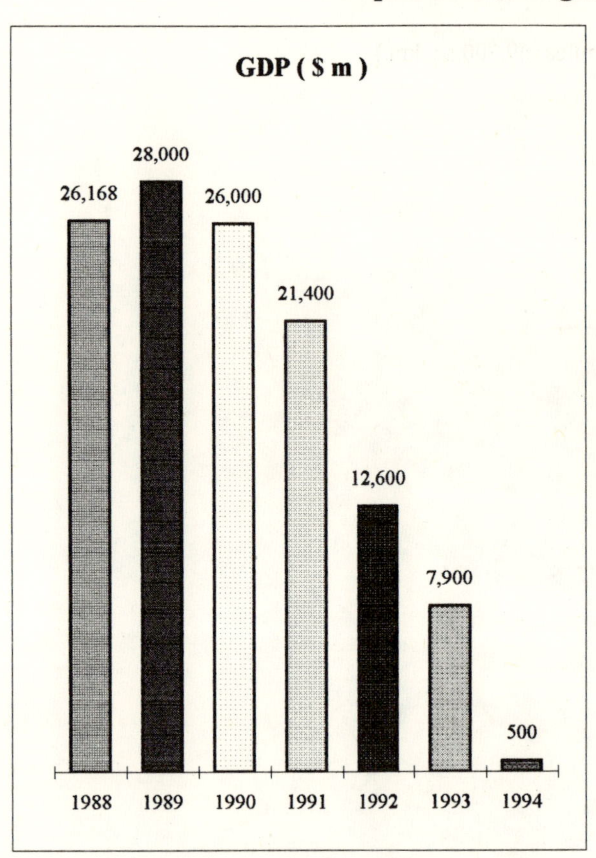


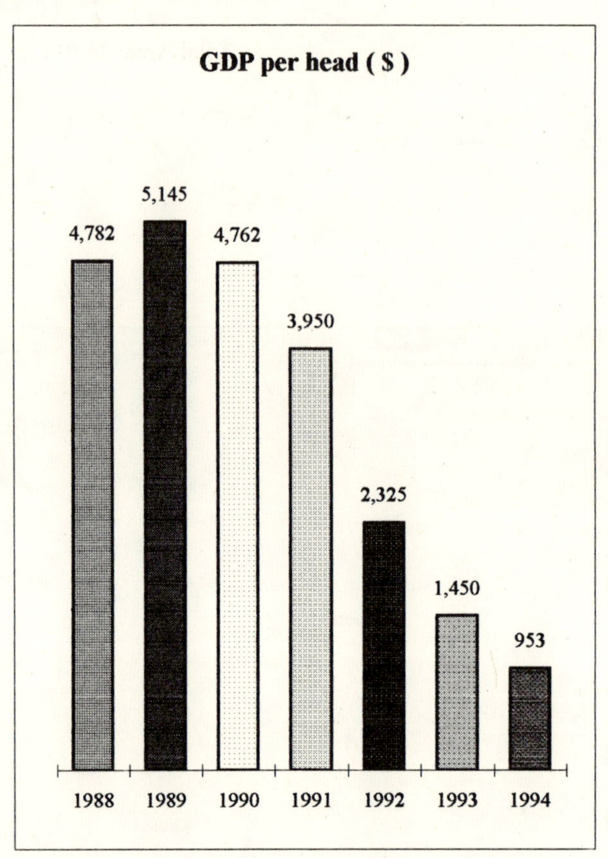


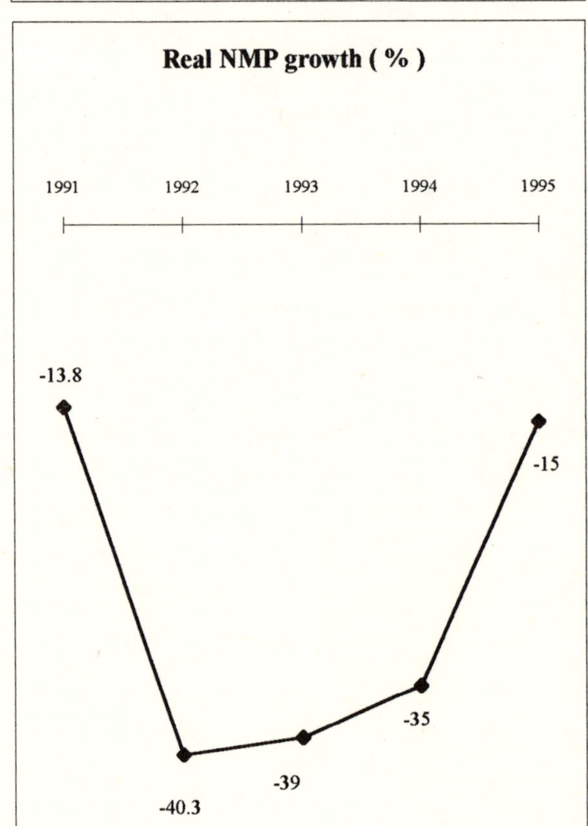


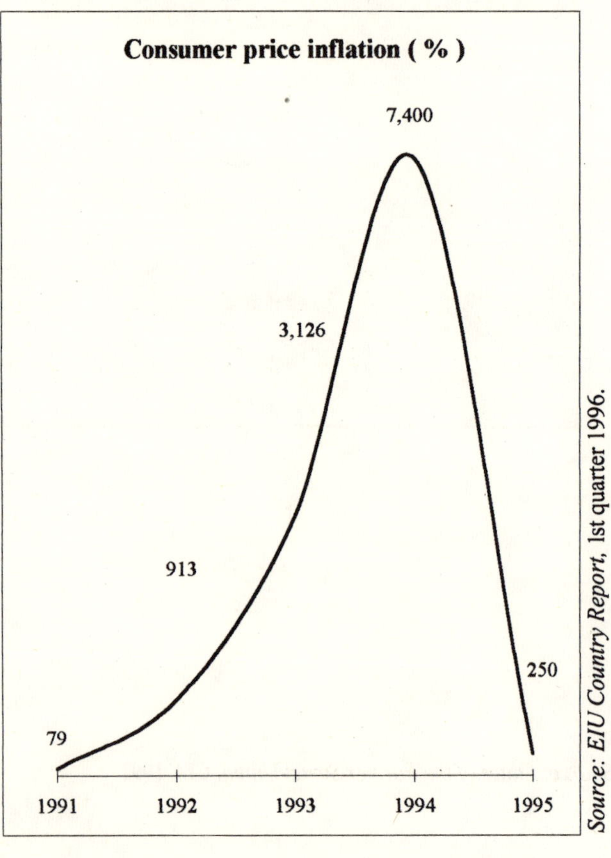


*Source: EIU Country Report*, 1st quarter 1996.

## Republic of Georgia: Demographic Statistics (1989)

### Ethnic composition

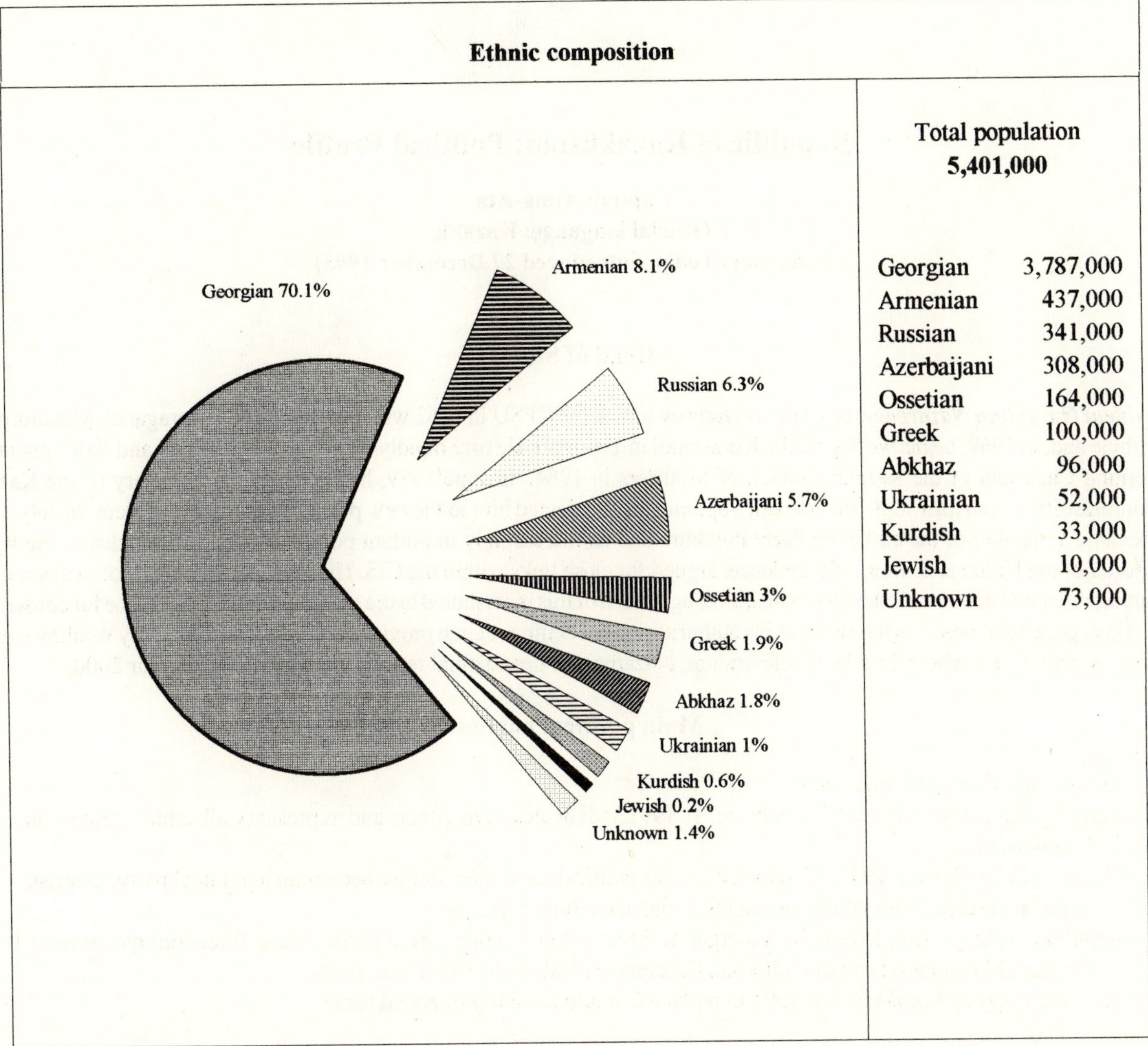


**Total population 5,401,000**

| | |
|---|---|
| Georgian | 3,787,000 |
| Armenian | 437,000 |
| Russian | 341,000 |
| Azerbaijani | 308,000 |
| Ossetian | 164,000 |
| Greek | 100,000 |
| Abkhaz | 96,000 |
| Ukrainian | 52,000 |
| Kurdish | 33,000 |
| Jewish | 10,000 |
| Unknown | 73,000 |

### Religious composition

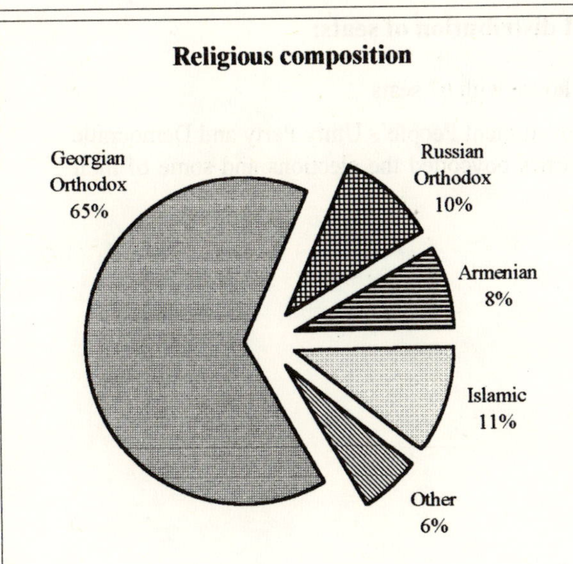


### Rural / Urban population split

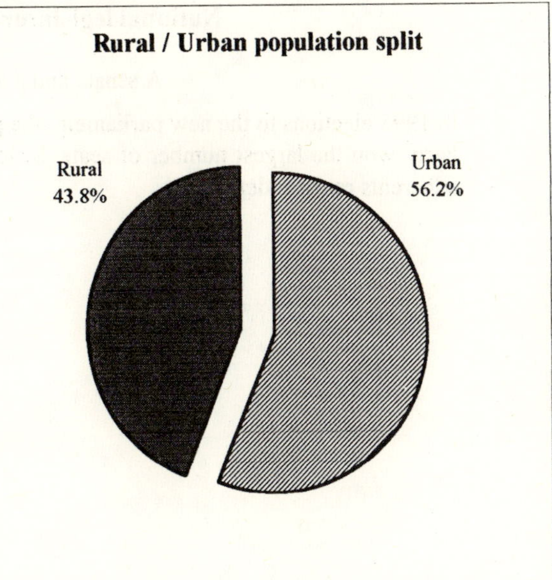


*Source:* Based on official 1989 All-Union Census.

## Republic of Kazakhstan: Political Profile

**Capital: Alma-Ata**
**Official language: Kazakh**
**Currency: Tenge (introduced 27 December 1993)**

### Head of State:

***President Nursultan Nazarbaev,*** b. 1940. Nazarbaev joined the CPSU in 1962 while working at the Karaganda Metallurgical Combine and, in 1969, began work with the Komsomol in Temirtau. He rose rapidly in the republican Party and state apparatus, becoming Chairman of the Kazakh Council of Ministers in 1984. In June 1989, he became First Secretary of the Kazakh Communist Party. In April 1990, the Kazakh Supreme Soviet elected him to the new post of Executive President. In July 1990 he became a member of the all-Union Party Politburo and an increasingly important politician outside Kazakhstan. He was a supporter of the Union and, after independence, argued for close links within the CIS. His main concern was political stability, resulting in limited democratic developments, although much of this is attributed to the general cultural preference for consensus. He enjoys good relations with the West, as his authoritarian but benign regime provides stability in a potentially wealthy county and in the region as a whole. In a 1995 referendum, Nazarbaev's presidential term was extended to the year 2000.

### Main political parties:

*Democratic Party of Kazakhstan.*
*People's Congress Party of Kazakhstan,* f. 1991, advocates civil peace and represents all ethnic groups in Kazakhstan.
*People's Unity Party,* f. 1993. Originally a socio-political movement before becoming a political party; centrist, opposing radical nationalism; promotes social and ethnic harmony.
*Republican Party–Azat,* f. 1992 by a merger of three nationalist opposition parties: Azat (Freedom movement), the Republican Party, and the Jeltoqsan (December) National-Democratic Party.
*Socialist Party of Kazakhstan,* f. 1991 to replace Communist Party of Kazakhstan.

### National legislature and distribution of seats:

A senate and a lower house with 67 seats

In 1995 elections to the new parliament, the pro-government People's Unity Party and Democratic Party won the largest number of seats. Several parties boycotted the elections and some of their adherents ran as independents.

# Republic of Kazakhstan

Total Area: 1,049,155 sq. miles (2,717,300 sq. km.)

*Source: States of the Former Soviet Union,* CIA 1992.

## Republic of Kazakhstan: Economic Statistics

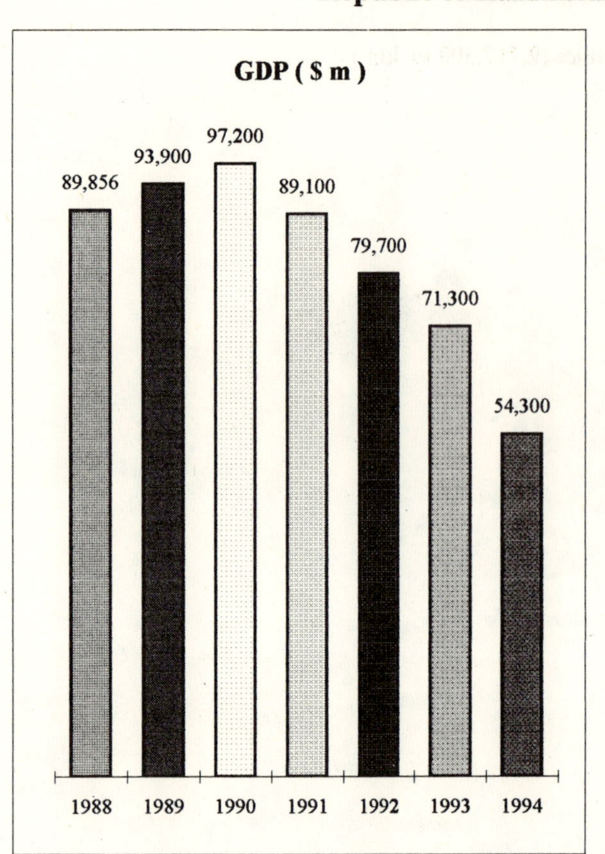


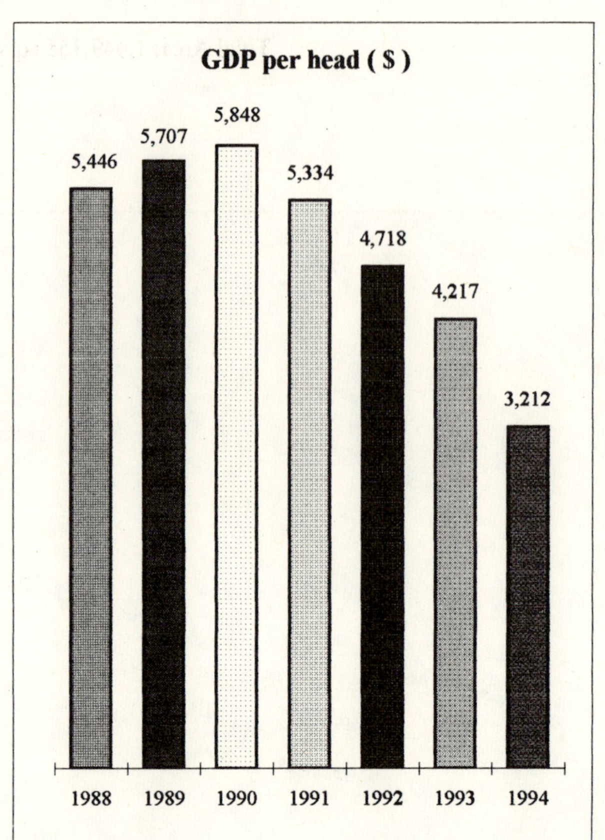


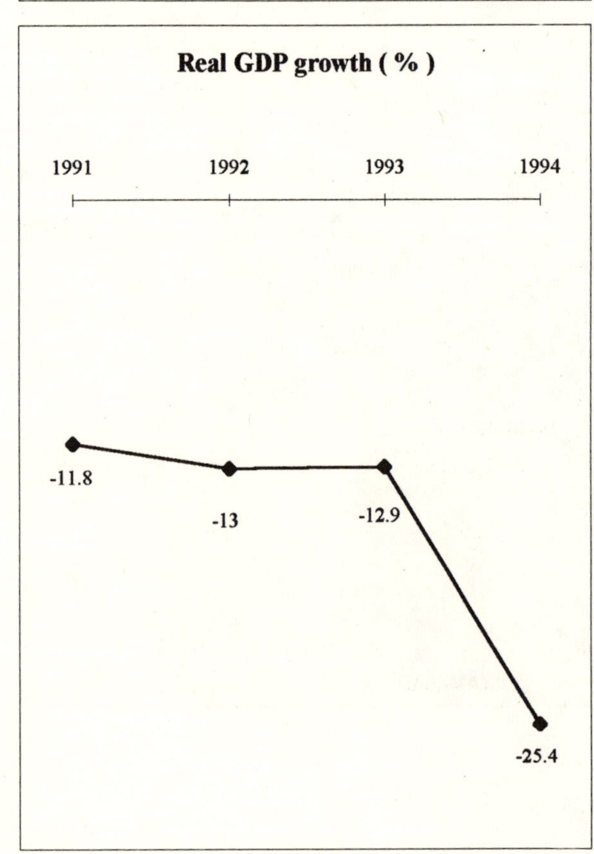


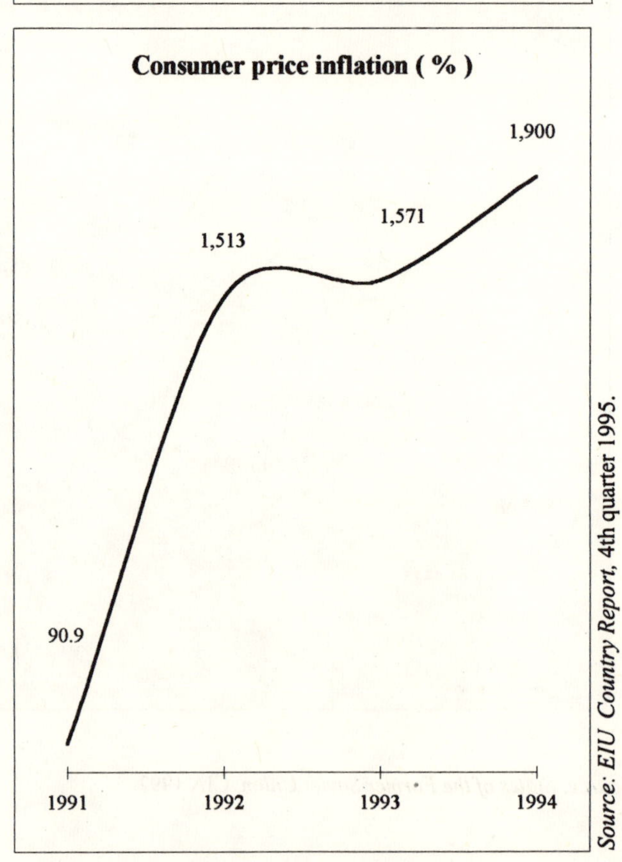


*Source: EIU Country Report*, 4th quarter 1995.

# Republic of Kazakhstan: Demographic Statistics (1989)

## Ethnic composition

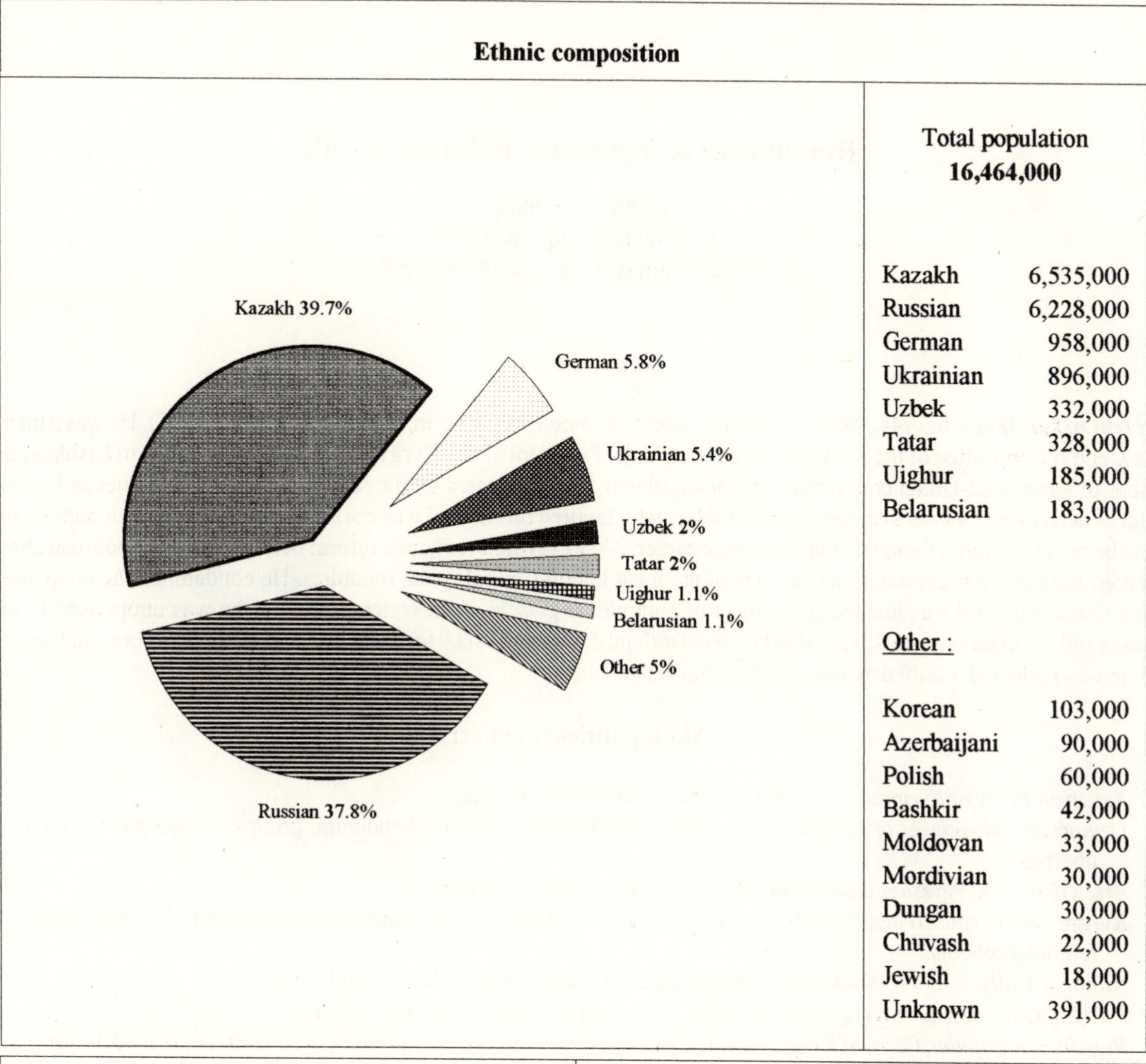


**Total population**
**16,464,000**

| | |
|---|---|
| Kazakh | 6,535,000 |
| Russian | 6,228,000 |
| German | 958,000 |
| Ukrainian | 896,000 |
| Uzbek | 332,000 |
| Tatar | 328,000 |
| Uighur | 185,000 |
| Belarusian | 183,000 |
| Other : | |
| Korean | 103,000 |
| Azerbaijani | 90,000 |
| Polish | 60,000 |
| Bashkir | 42,000 |
| Moldovan | 33,000 |
| Mordivian | 30,000 |
| Dungan | 30,000 |
| Chuvash | 22,000 |
| Jewish | 18,000 |
| Unknown | 391,000 |

## Religious composition

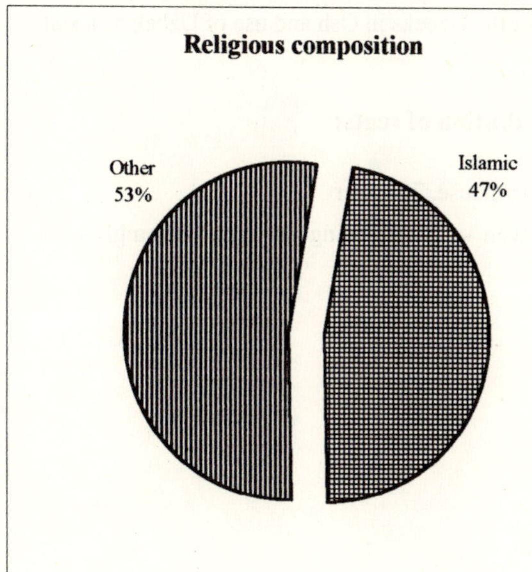


## Rural / Urban population split

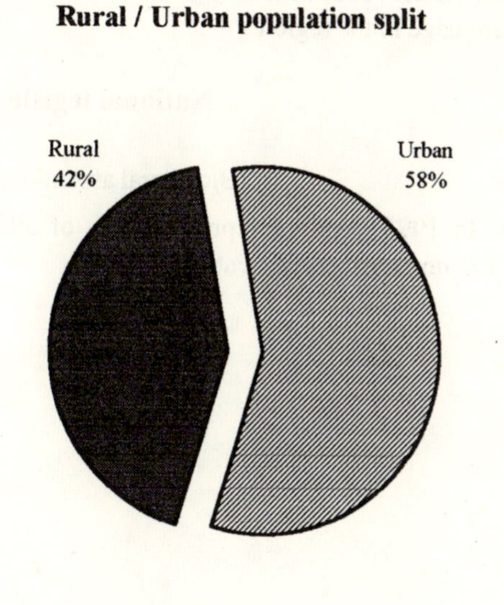


*Source:* Based on official 1989 All-Union Census.

## Republic of Kyrgyzstan: Political Profile

**Capital: Bishkek**
**Official language: Kyrgyz**
**Currency: Som (introduced 10 May 1993)**

### Head of State:

***President Askar Akaev,*** b. 1944. Akaev was a professor who joined the CPSU in 1981 (he resigned in 1991). He was a member of the Central Committee of the Kyrgyz Communist Party, President of the Kyrgyz Academy of Sciences in Bishkek, and a member of various all-Union committees. A known liberal, he was elected by the republican Supreme Soviet as Executive President in October 1990 as a compromise candidate who favored reform and was not connected with the dominant factions, having been in Leningrad for most of his academic career. Akaev favored economic reform before instituting political changes and he ensured that Kyrgyzstan remained one of the most liberal Central Asian republics. He condemned the coup attempt against Gorbachev and was himself the subject of communist putschists. In October of 1991 he was unopposed in direct elections and resolved to make Kyrgyzstan the "Switzerland of Central Asia," focusing also on economic reform. In December 1995 he was reelected, easily defeating a Communist challenger.

### Main political parties:

*Agrarian Party of Kyrgyzstan,* f. 1993, represents farmers' interests.
*Democratic Movement of Kyrgyzstan,* f. 1990. Coordinating body for democratic groups; campaigns for civil liberties.
*Erkin (Free)–Kyrgyzstan Democratic Party,* f. 1991, a leading opposition.
*Kyrgyz Democratic Wing,* f. 1990, works for greater religious tolerance and the construction of mosques and religious schools.
*National Unity,* f. 1991. Moderate democratic party seeking to unite different ethnic groups.
*Party of Communists of Kyrgyzstan,* f. 1992. Successor to Kyrgyz Communist Party.
*Republican Popular Party of Kyrgyzstan,* f. 1993. Centrists, founded by prominent scientists and academics.
*Uzbek Adalet (Uzbek Justice),* f. 1989. Advocates autonomy for the Uzbeks in Osh and use of Uzbek as a state language in the region.

### National legislature and distribution of seats:

Bicameral assembly: 35 seats in upper house, 70 lower

In 1995 elections representatives of all major parties won seats, replacing an overwhelmingly Communist parliament.

# Republic of Kyrgyzstan

Total Area: 76,641 sq. miles (198,500 sq. km.)

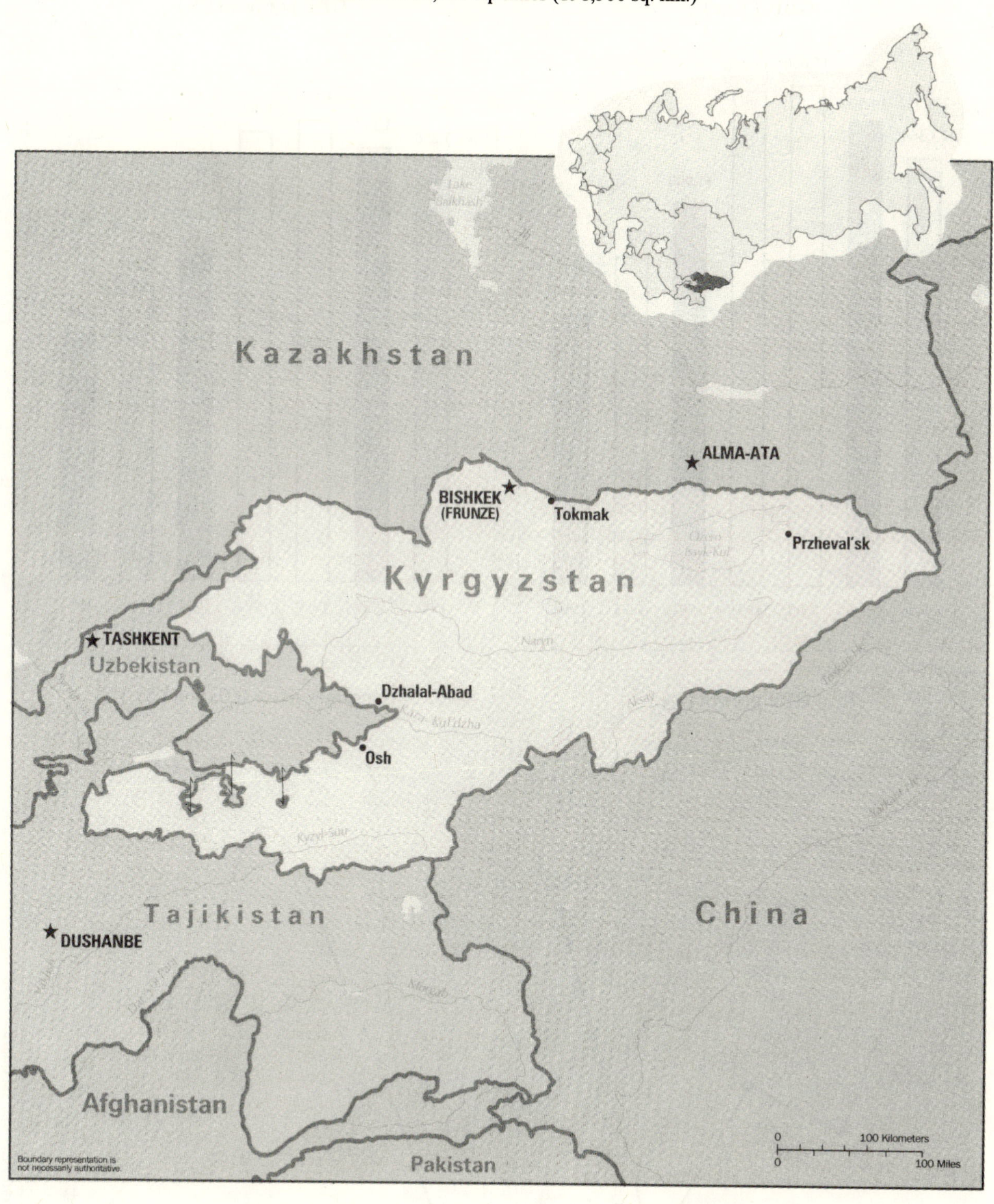


*Source: States of the Former Soviet Union,* CIA 1992.

## Republic of Kyrgyzstan: Economic Statistics

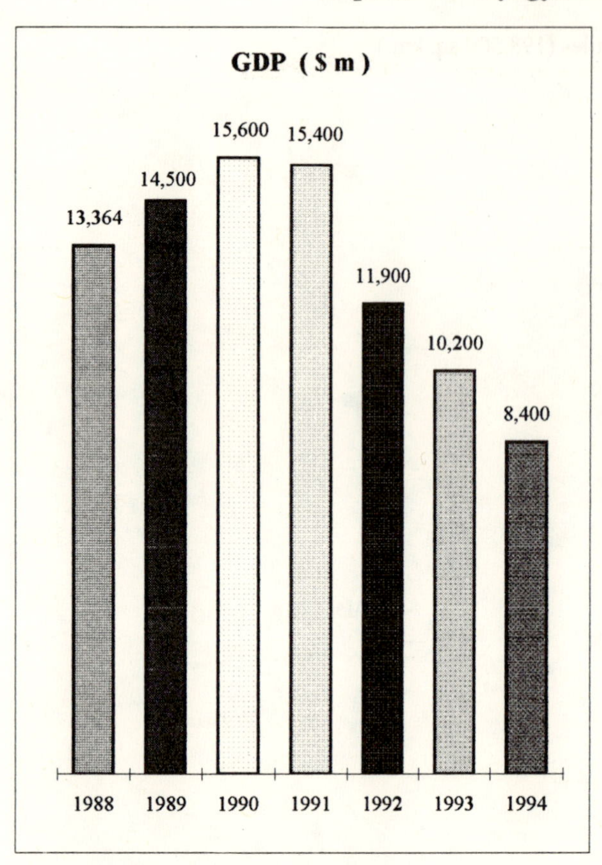


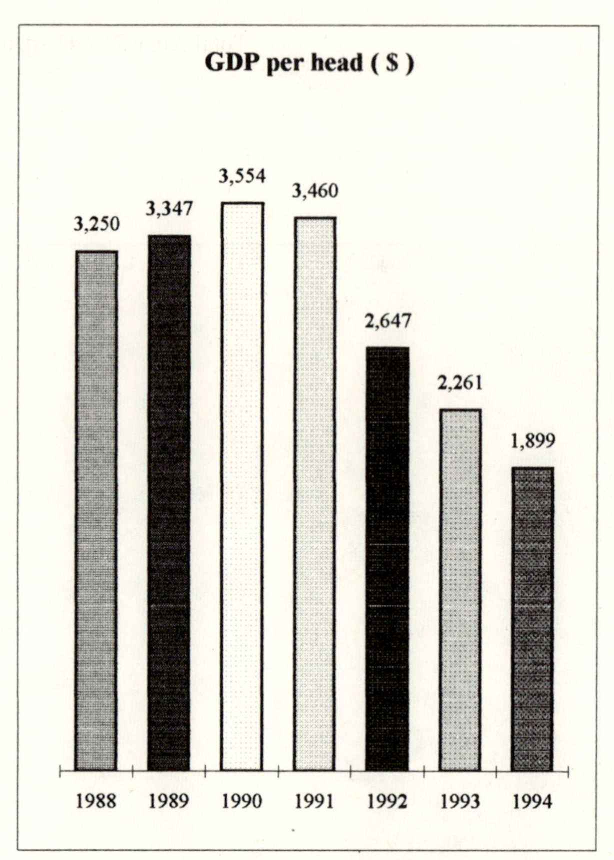


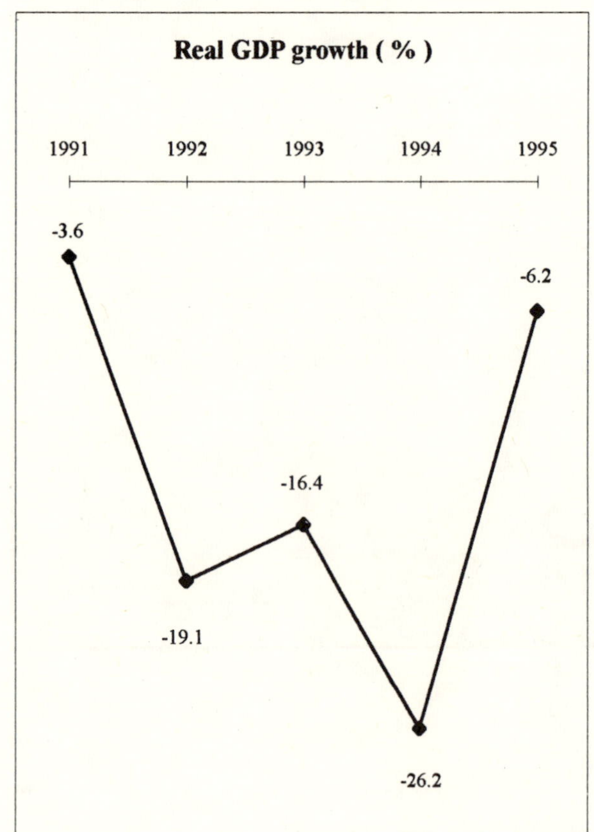


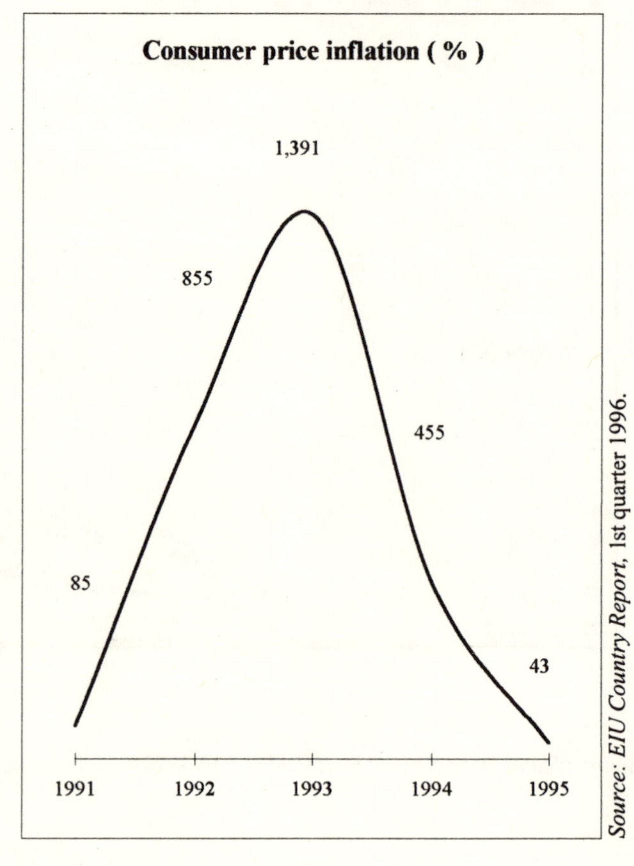


*Source: EIU Country Report,* 1st quarter 1996.

# Republic of Kyrgyzstan: Demographic Statistics (1989)

## Ethnic composition

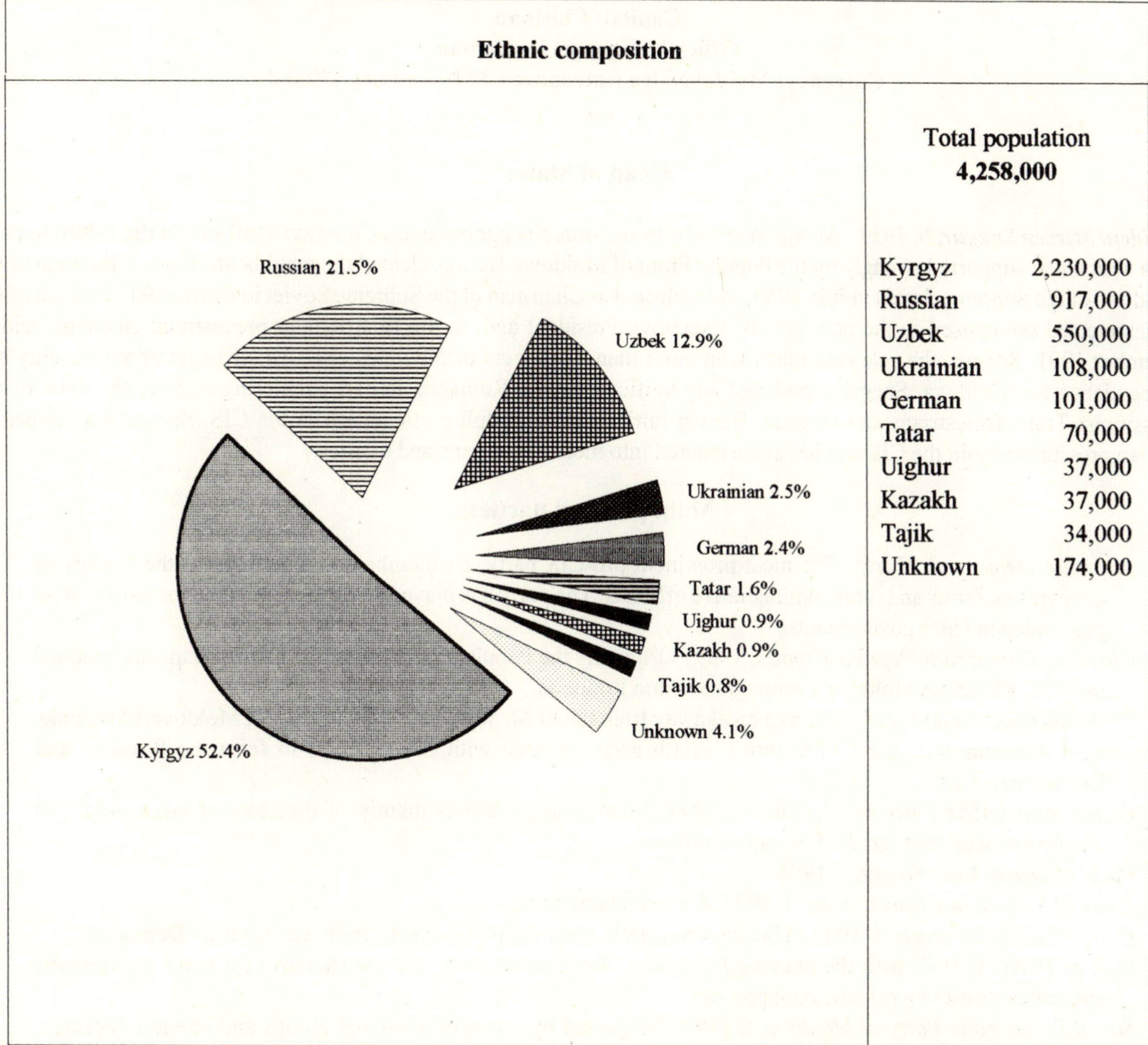


**Total population**
**4,258,000**

| | |
|---|---|
| Kyrgyz | 2,230,000 |
| Russian | 917,000 |
| Uzbek | 550,000 |
| Ukrainian | 108,000 |
| German | 101,000 |
| Tatar | 70,000 |
| Uighur | 37,000 |
| Kazakh | 37,000 |
| Tajik | 34,000 |
| Unknown | 174,000 |

## Religious composition

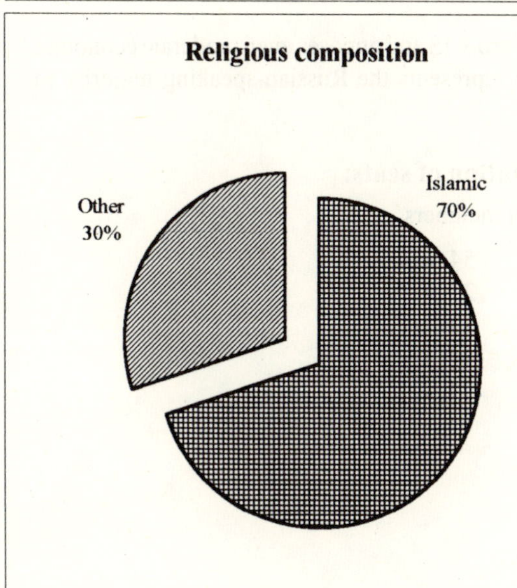


## Rural / Urban population split

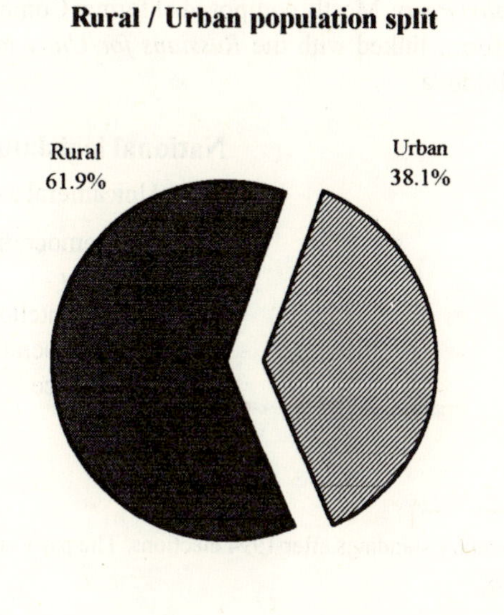


*Source:* Based on official 1989 All-Union Census.

## Republic of Moldova: Political Profile

**Capital: Chisinau**
**Official language: Romanian**
**Currency: Moldovan leu (introduced 29 November 1993)**

### Head of State:

***President Mircea Snegur,*** b. 1939. An agronomist by profession, Snegur became a Communist official in the 1980s, but was also a nationalist supported strongly by the Popular Front of Moldova. He was elected the republican leader, Chairman of the Presidium of the Supreme Soviet in July 1989, and reelected as Chairman of the Supreme Soviet in April 1990. In September, he was elected unopposed to the new post of Executive President and, in the first popular presidential elections held in December 1991, Snegur, the sole candidate, won more than 98 percent of the votes cast. An advocate of sovereignty and independence for Moldova, Snegur repudiated any unification with Romania, in part attempting to allay the fears of the secessionist Trans-Dniestrians and Gagauz. Having initially rejected full participation in the CIS, Snegur succumbed to Russian pressures to join the CIS and has since entered into most CIS organs and treaties.

### Main political parties:

*Agrarian Democratic Party.* The most prominent pro-CIS party, its membership comes from the leaders of cooperative farms and rural administrative officials. The party has played a critical role in the formation of all post-independence governments.

*Christian Democratic Popular Front,* f. 1989. Originally the Popular Front of Moldova, the group was renamed in 1992; advocates Moldova's reintegration into Romania.

*Congress of Intelligentsia,* f. 1993 as a breakaway from the PFM; advocates an independent Moldova, close links with Romania with the CIS at arm's length away; teamed with other groups to form the Peasants and Intellectuals List.

*Democratic Labor Party of Moldova,* f. 1993. Membership consists mainly of directors of large industrial enterprises who want gradual economic reform.

*Party of Democratic Forces,* f. 1995.

*Party of Revival and Conciliation,* f. 1995. A presidential party.

*Party of Social Progress,* f. 1995. A social-democratic party created in a split from the Agrarian Democrats.

*Reform Party,* f. 1993 with the backing from large financial interests and the support of a major newspaper; advocates restricting privatization process.

*Social Democratic Party of Moldova,* f. 1990. Supported by urban professional groups and ethnic minorities; supports independence and economic reform.

*Socialist Party.* Mostly composed of former Communists, strongly pro-CIS and opposes even moderate economic reform; linked with the *Russians for Unity* movement which represents the Russian-speaking majority in Moldova.

### National legislature and distribution of seats:

Unicameral assembly with 104 members

| | |
|---|---|
| Agrarian Democratic Party | 54 percent* |
| Socialist Party | 27 percent |
| Peasants and Intellectuals bloc | 10 percent |
| Christian Democratic Popular Front Alliance | 9 percent |

*Parliamentary standings after 1994 elections. The party subsequently split.

## Republic of Moldova

Total Area: 13,012 sq. miles (33,700 sq. km.)

*Source: States of the Former Soviet Union,* CIA 1992.

## Republic of Moldova: Economic Statistics

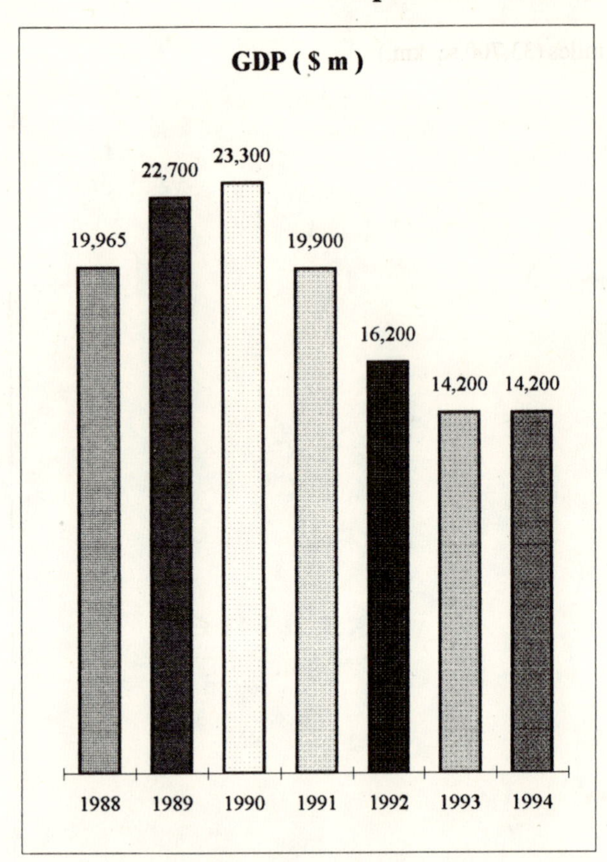


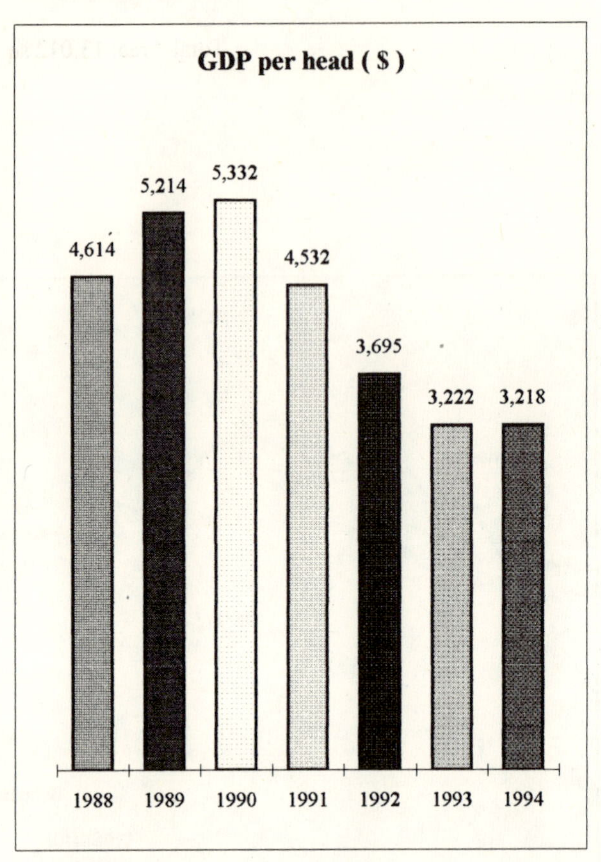


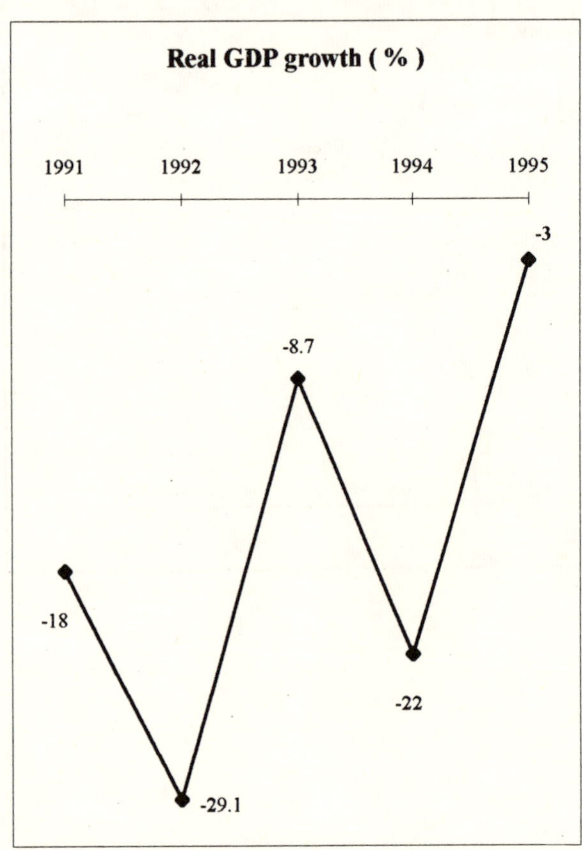


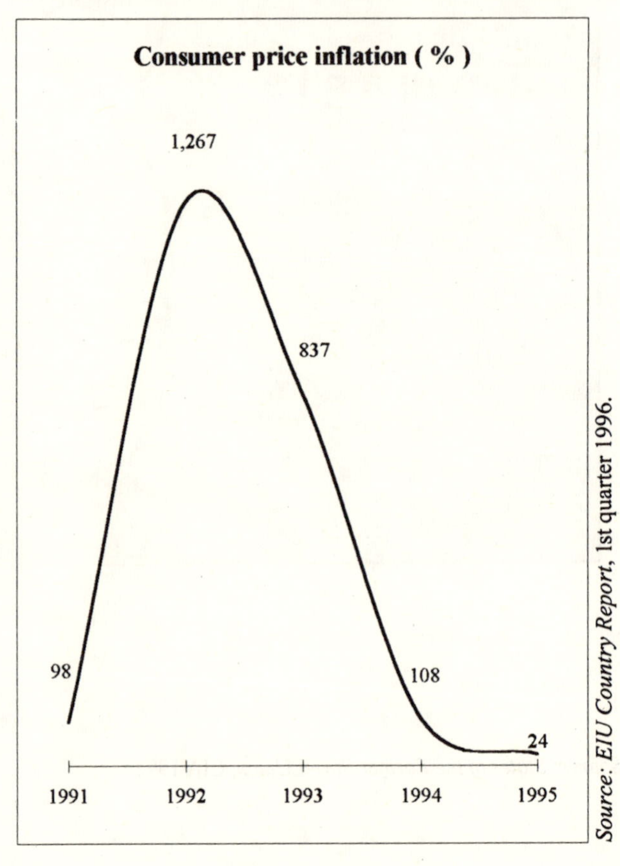


*Source: EIU Country Report*, 1st quarter 1996.

## Republic of Moldova: Demographic Statistics (1989)

### Ethnic composition

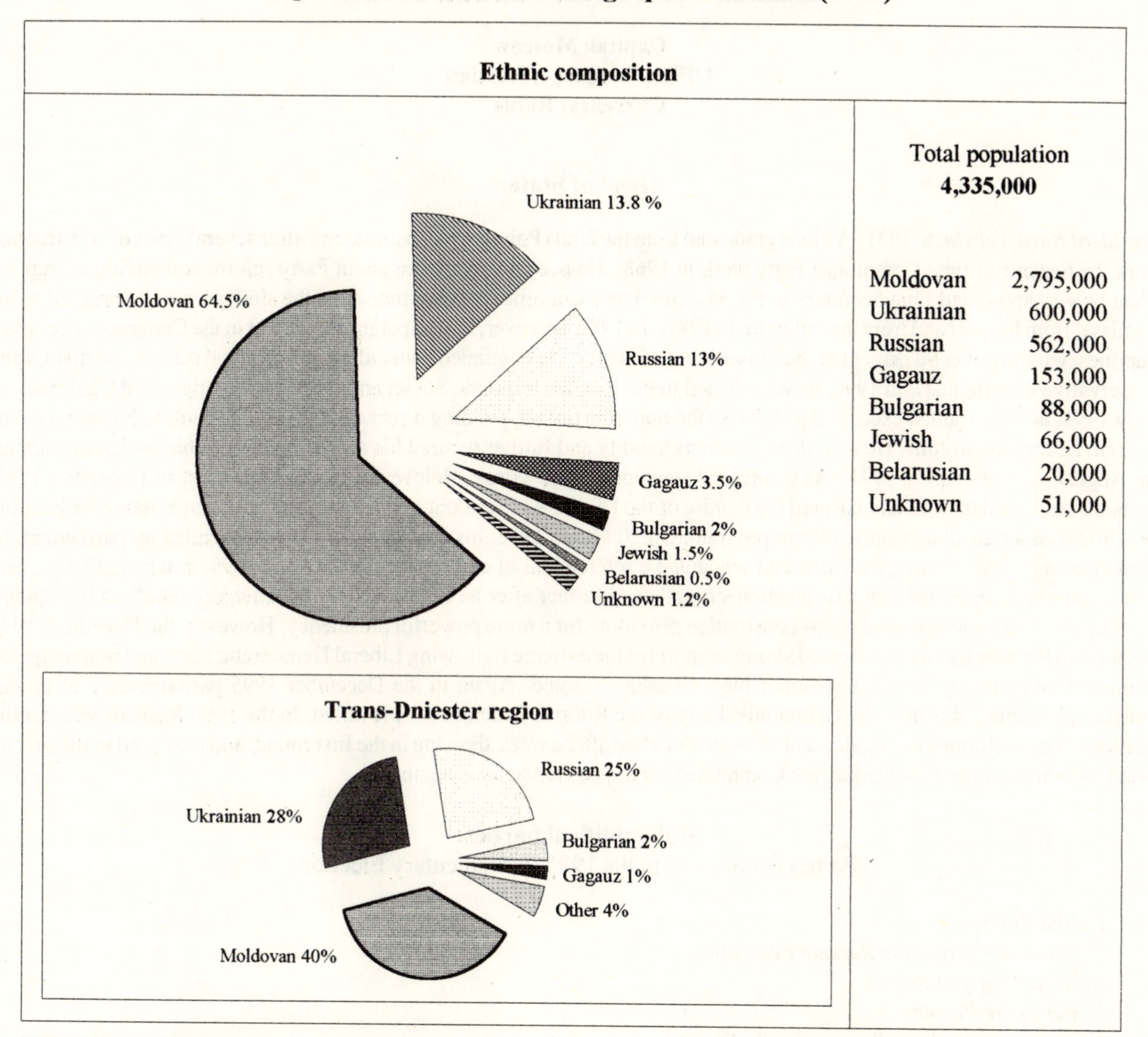


**Total population**
**4,335,000**

| | |
|---|---|
| Moldovan | 2,795,000 |
| Ukrainian | 600,000 |
| Russian | 562,000 |
| Gagauz | 153,000 |
| Bulgarian | 88,000 |
| Jewish | 66,000 |
| Belarusian | 20,000 |
| Unknown | 51,000 |

### Religious composition

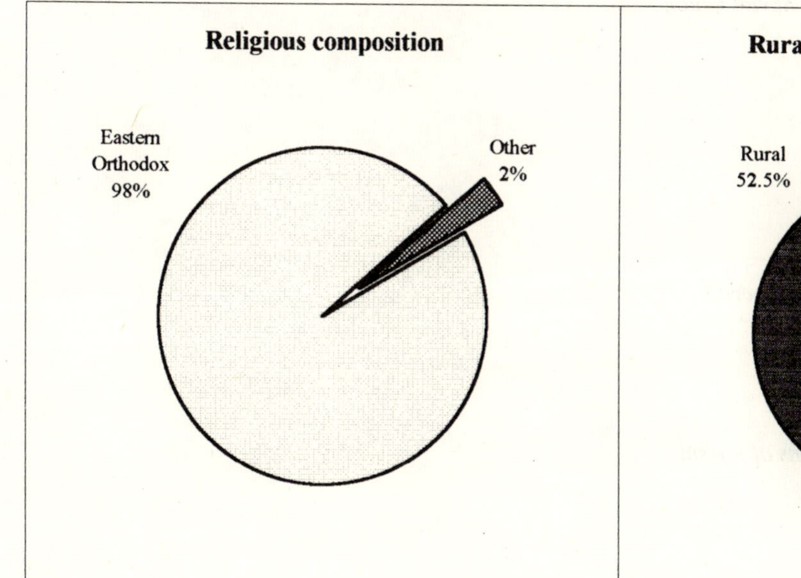


### Rural / Urban population split

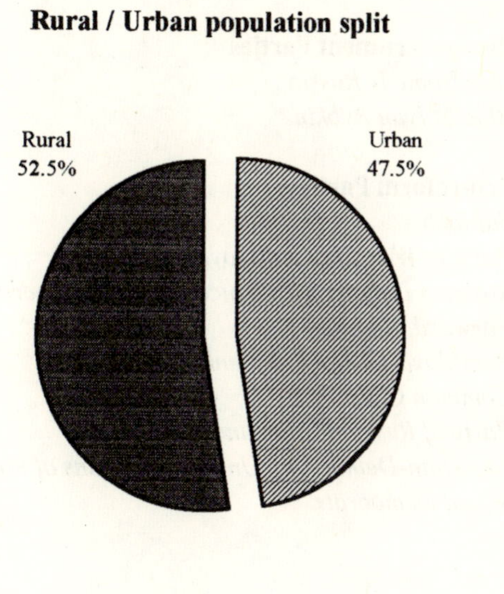


*Source:* Based on official 1989 All-Union Census.

## Russian Federation: Political Profile

**Capital: Moscow**
**Official language: Russian**
**Currency: Ruble**

### Head of State:

***President Boris Yeltsin,*** b. 1931. Yeltsin graduated from the Urals Polytechnic Institute and after several years of construction work, he began full-time Communist Party work in 1968. He became outspoken about Party reform and ending corruption when he was appointed First Secretary of the Moscow Party Committee. His criticism of the slow pace of reform led to his dismissal from his post and from the Politburo in 1987. In 1989, however, he campaigned for a seat in the Congress of People's Deputies, winning over 90 percent of the Moscow constituency. He continued demanding more radical reform and dismissing conservative Communists. In 1990, he was elected to the Russian Supreme Soviet and immediately appointed Chairman. In early 1991, he was granted executive powers by the Supreme Soviet, pending direct elections for Executive President of the Russian Federation in June. He won these elections handily and further secured his authority through his leadership against the August coup attempt in 1991. As an original signatory of the Minsk (Belovezh Forest) Agreement of December 1991, Yeltsin established the CIS and ensured the demise of the USSR. As President of the newly independent Russian Federation, he introduced a radical economic reform program. On 20 March 1993, his emergency powers suspended by parliament, he introduced a period of emergency rule and scheduled a referendum of confidence for 25 April 1995, in which 57.4 percent of the votes cast supported him. His position strengthened further after he suppressed a parliamentary rebellion in September–October 1993 and endorsed a new constitution providing for a more powerful presidency. However, the December 1993 elections to the new legislature showed strong support for the extreme right-wing Liberal Democratic Party and unanticipated softness of support for Yeltsin's reformer bloc, Russia's Choice. Again in the December 1995 parliamentary elections, Yeltsin opponents—this time the Communist Party of the Russian Federation—prevailed. In the 1996 Russian presidential elections, Yeltsin formed an alliance with Aleksandr Lebed after a weak showing in the first round, and prevailed in the second round over his strongest challenger, the Communist candidate Gennadiy Zyuganov.

### Main political parties:

(Parties Competing in the 1995 Parliamentary Election)

**Leftist Parties**
*Communist Party of the Russian Federation*
*Agrarian Party of Russia*
*Power to the People*
*Communists-Workers' Russia–For the Soviet Union*

**Pro-government Parties**
*Our Home Is Russia*
*Bloc of Ivan Rybkin*

**Pro-reform Parties**
*Yabloko*
*Party of Workers' Self-Government*
*Russia's Democratic Choice–United Democrats*
*Forward, Russia!*
*Pamfilova-Gurov-V. Lysenko*
*Common Cause*
*Party of Russian Unity and Concord*
*Christian-Democratic Union–Christians of Russia*
*Social Democrats*

*Party of Economic Freedom*
*Federal Democratic Movement*
*Bloc 89*

**Nationalist Parties**
*Liberal Democratic Party of Russia*
*Congress of Russian Communities*
*Derzhava [Power]*
*My Fatherland*
*For the Motherland*
*National-Republican Party of Russia*
*Russian All-National Movement*

**Groups**
*Women of Russia*
*Trade Unions and Industrialists–Union of Labor*
*Interethnic Union*
*Ecological Party Kedr [Cedar]*

## National legislature and distribution of seats:

### Bicameral assembly

The Federation Council (upper house): 178 deputies; 2 from each of 89 republics and regions

The State Duma (lower house): 450 deputies elected on the basis of party lists plus single-member districts:

| | |
|---|---|
| Communist Party of the Russian Federation | 34 percent |
| Our Home Is Russia | 12 percent |
| Liberal Democratic Party of Russia | 11 percent |
| Yabloko | 10 percent |
| Independents | 17 percent |
| Other parties | 14 percent |

## Russian Federation

Total Area: 6,592,772 sq. miles (17,075,200 sq. km.)

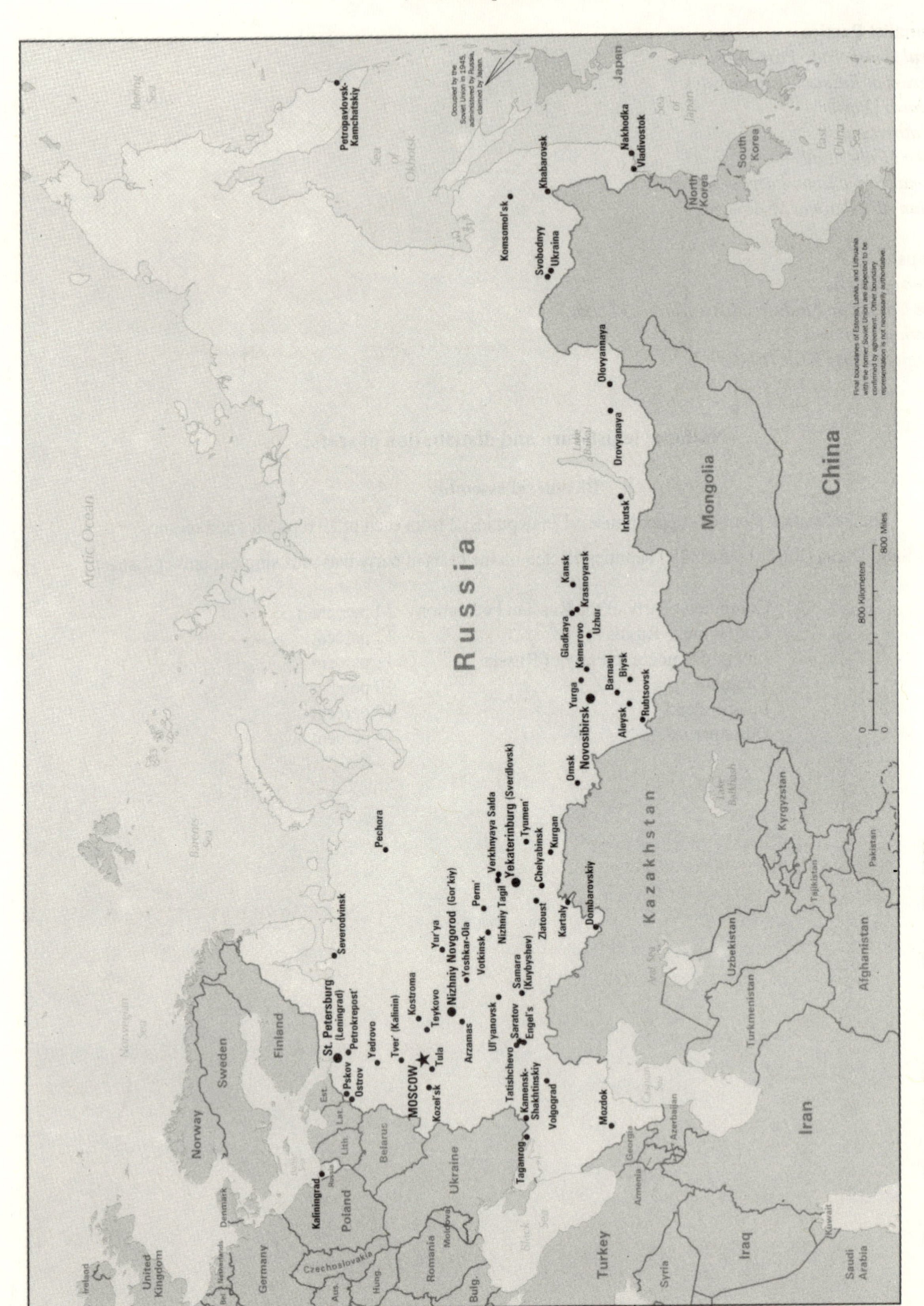


*Source*: *States of the Former Soviet Union*, CIA 1992.

# Russian Federation: Economic Statistics

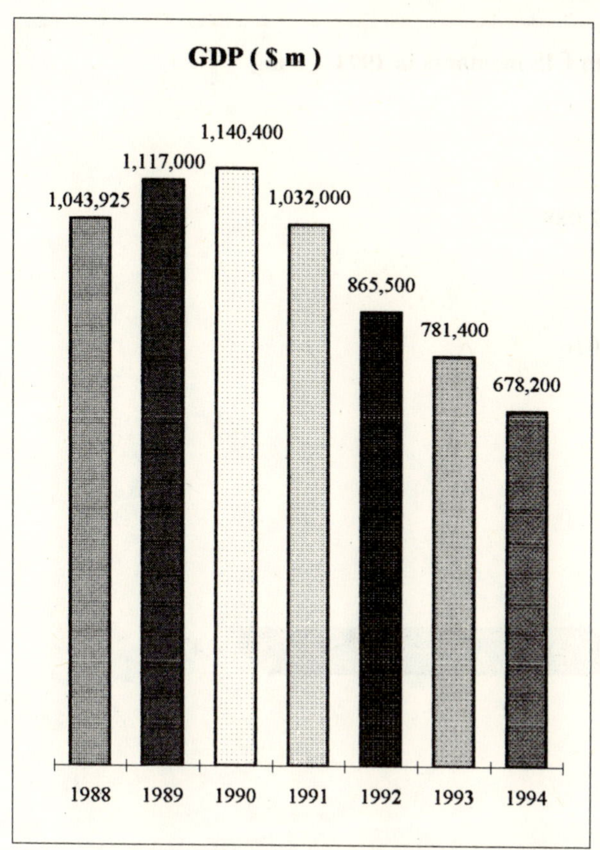


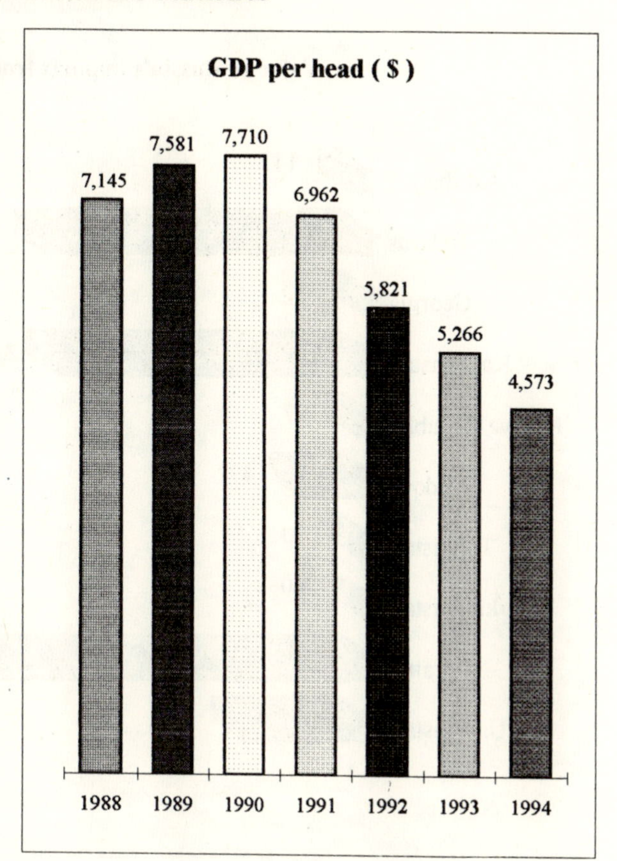


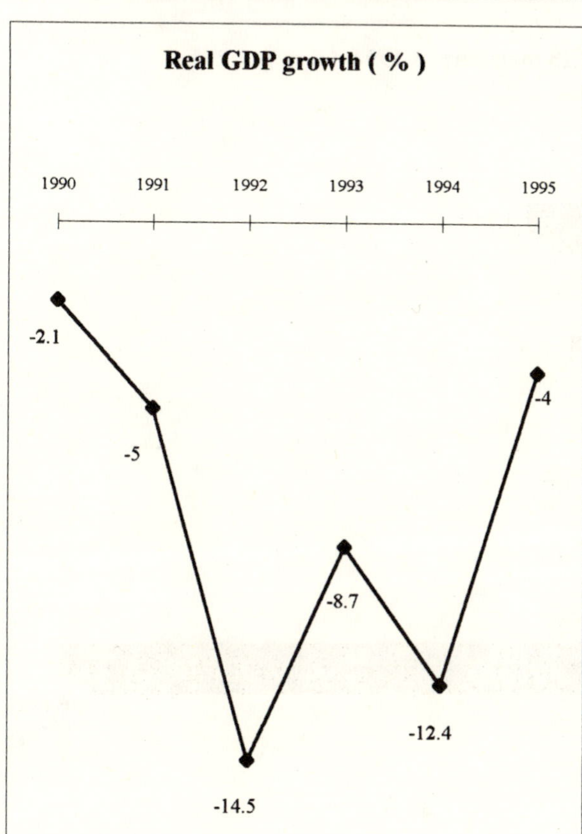


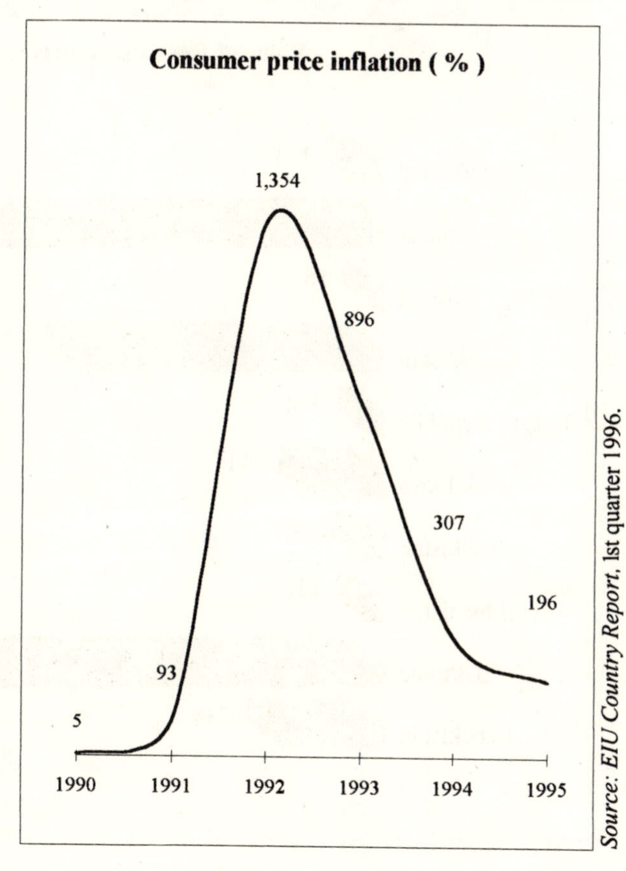


*Source: EIU Country Report*, 1st quarter 1996.

## Russian Federation: Economic Statistics

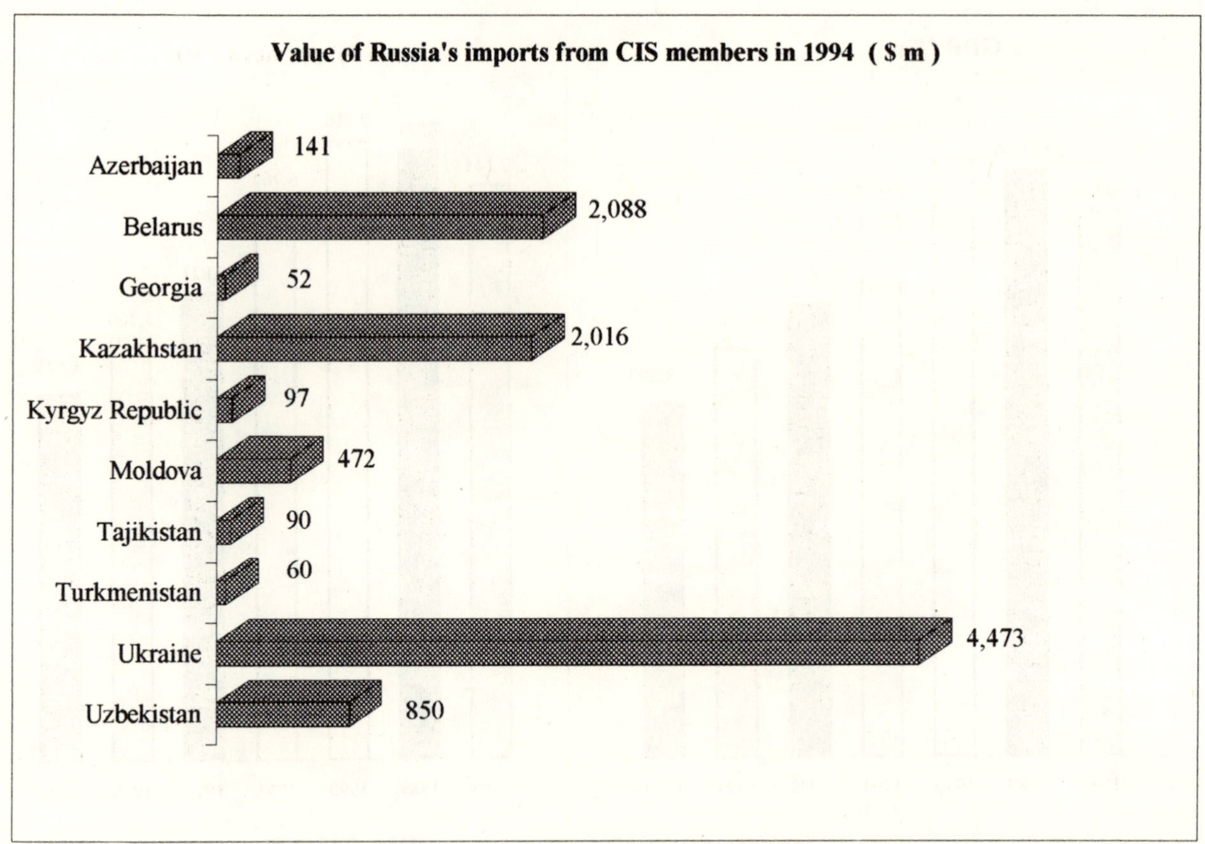


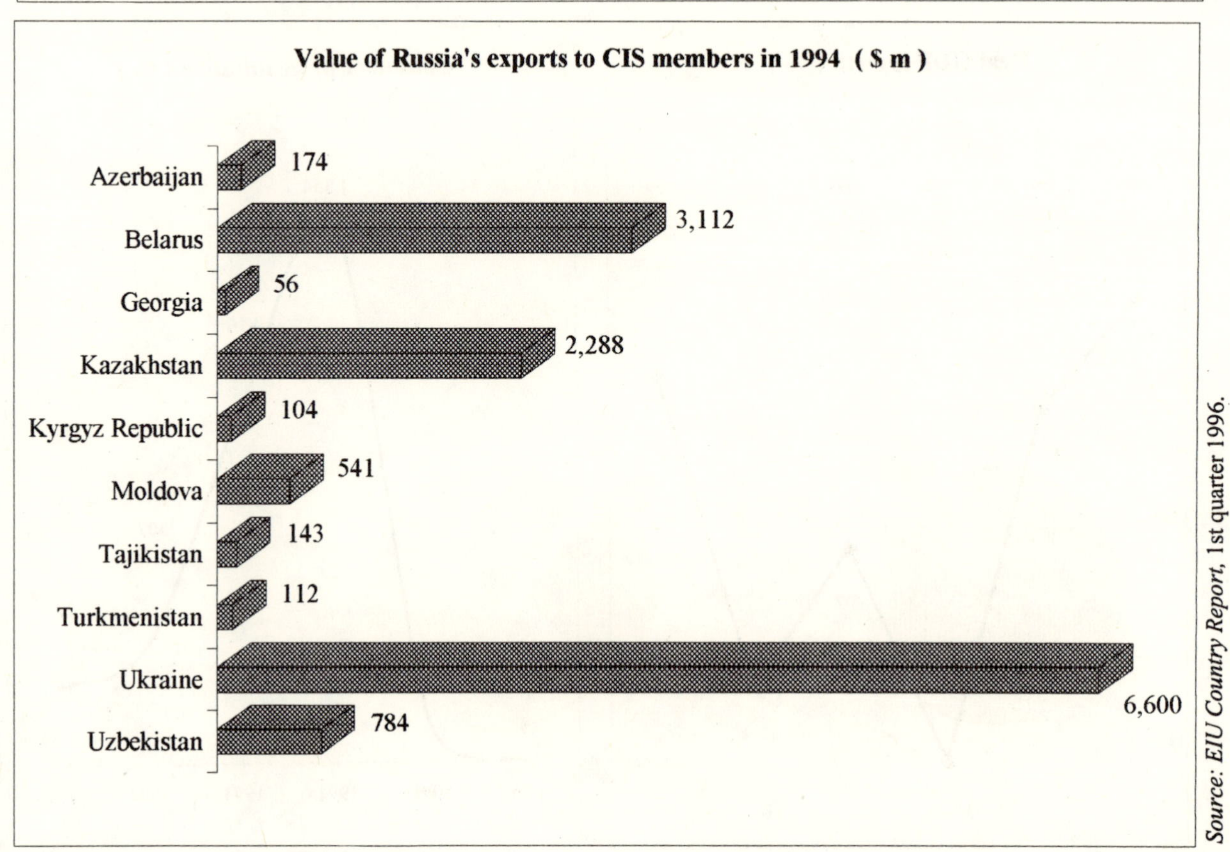


*Source: EIU Country Report*, 1st quarter 1996.

# Russian Federation: Demographic Statistics (1989)

## Ethnic composition

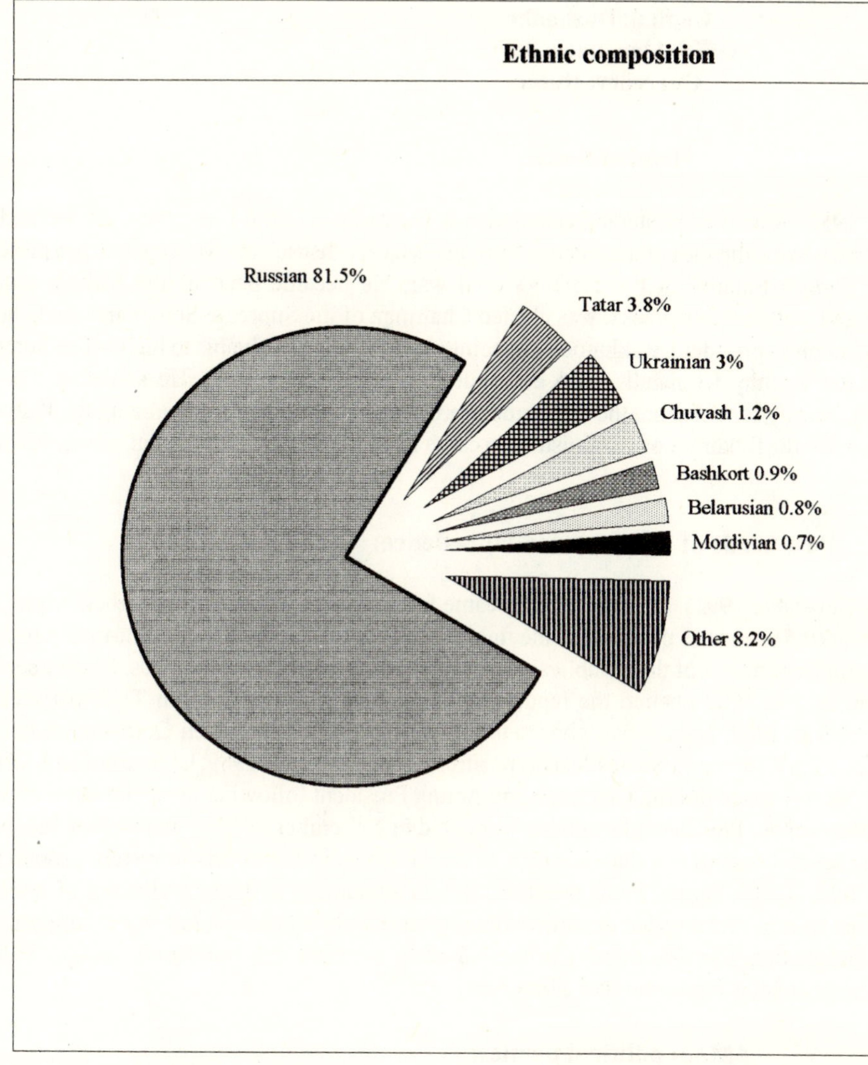


Total population
**147,022,000**

| | |
|---|---|
| Russian | 119,866,000 |
| Tatar | 5,522,000 |
| Ukrainian | 4,363,000 |
| Chuvash | 1,774,000 |
| Bashkort | 1,345,000 |
| Belarusian | 1,206,000 |
| Mordivian | 1,073,000 |
| Other: | |
| Chechen | 899,000 |
| German | 842,000 |
| Udmurt | 715,000 |
| Mari | 644,000 |
| Kazakh | 636,000 |
| Avar | 544,000 |
| Jewish | 537,000 |
| Armenian | 532,000 |
| Buriat | 417,000 |
| Ossetian | 402,000 |
| Kabard | 386,000 |
| Iakut | 380,000 |
| Dargin | 353,000 |
| Komi | 336,000 |
| Azerbaijani | 336,000 |
| Kumyk | 277,000 |
| Lezghin | 257,000 |
| Ingush | 215,000 |
| Tuvinian | 206,000 |
| Moldovan | 173,000 |
| Kalmyk | 166,000 |
| Gypsy | 153,000 |
| Karachai | 150,000 |
| Komi | 147,000 |
| Karelian | 125,000 |
| Adygei | 123,000 |
| Korean | 107,000 |
| Lak | 106,000 |
| Polish | 95,000 |
| Unknown | 1,614,000 |

## Religious composition



## Rural / Urban population split



*Source:* Based on official 1989 All-Union Census.

# Republic of Tajikistan: Political Profile

**Capital: Dushanbe**
**Official language: Tajik**
**Currency: Ruble**

## Head of State:

***President Emomali Rakhmonov,*** b. 1953. Rakhmonov studied economics at Dushanbe's Lenin University. He worked in various positions ranging from electrician to the director of a collective farm in his native district. He was reportedly a protégé of the main Kulyabi militia leader Sangak Safarov in the 1992–93 civil war. He became head of the Kulyab region administration on 2 November 1992 and, within a few weeks, was elected Chairman of the Supreme Soviet and made head of state. In November 1994 he won election as president. Rakhmonov appointed many fellow Kulyabis to high office but was careful not the alienate completely the wealthy Khojandis, who demanded disarming the militia. He achieved this by incorporating the militia into national security services, but the Khojandis were dubious about institutionalizing the Kulyabi military advantage. Military assistance for the Rakhmonov administration came from Russia and Uzbekistan, which favored a conservative communist regime.

## Past heads of state since independence:

***Rakhmon Nabiev*** (November 1991–November 1992), b. 1930. Nabiev came from peasants of reportedly Uzbek origin. He studied agricultural mechanization, working in various positions in the field until 1961, when he went into full-time work for the Communist Party at the Agricultural Department of the republican Central Committee. In the early 1970s, Nabiev served as Minister of Agriculture and from 1973 to 1982 chaired the republican Council of Ministers. When Tajik party chief Dzhabbar Rasulov died of a heart attack in 1982, Nabiev was chosen to replace him. However, when Gorbachev came to power, he began sweeping out Brezhnev-era leaders and Nabiev lost his position as party chief. In 1989, he returned to politics, being elected to the Supreme Soviet. He was made interim Chairman and Acting President following the resignation of then Supreme Soviet Chairman Kakhar Mahkamov. Presidential elections were held in November of 1991 and Nabiev received 57 percent of the vote. Nabiev made several enemies in the aftermath of the elections in his efforts to restore ground the Communists had lost. In March of 1992, Nabiev caused much political strife in attempting to dismiss Minister of Internal Affairs Mamadaez Navzhuvanov, who in turn orchestrated demonstrations against Nabiev. Nabiev lost much support and there were increasing calls for his resignation. Clashes erupted between Nabiev loyalists and opposition groups, and in September 1992 he was forced to flee, launching Tajikistan into a bitter civil war.

## Main political parties:

*Communist Party of Tajikistan,* f. 1924. Only registered party until 1991 and during 1993.
*Democratic Party of Tajikistan,* f. 1990. Secular nationalist and pro-West.
*Islamic Renaissance Party,* f. 1991.
*Lale Badakhshon,* f. 1991. Sought greater autonomy for Gorno-Badakhshon and its peoples, the Pamiri.
*Party of Economic Freedom,* f. 1993, represents interest of northern Tajikistan.
*Party of Popular Unity and Justice,* f. 1994, opposition.
*People's Party of Tajikistan,* f. 1993. Pro-communist.
*People's Democratic Party,* f. 1993, seeks to represent northern economic interests.
*Rastokhez (Rebirth),* f. 1990. Nationalist-religious party favored by intellectuals.

## National legislature and distribution of seats:

Unicameral assembly

Opposition parties widely boycotted the 1995 parliamentary elections which resulted in an overwhelming victory for the Communist Party of Tajikistan.

## Republic of Tajikistan: Political Profile

Total Area: 55,251 sq. miles (143,100 sq. km.)

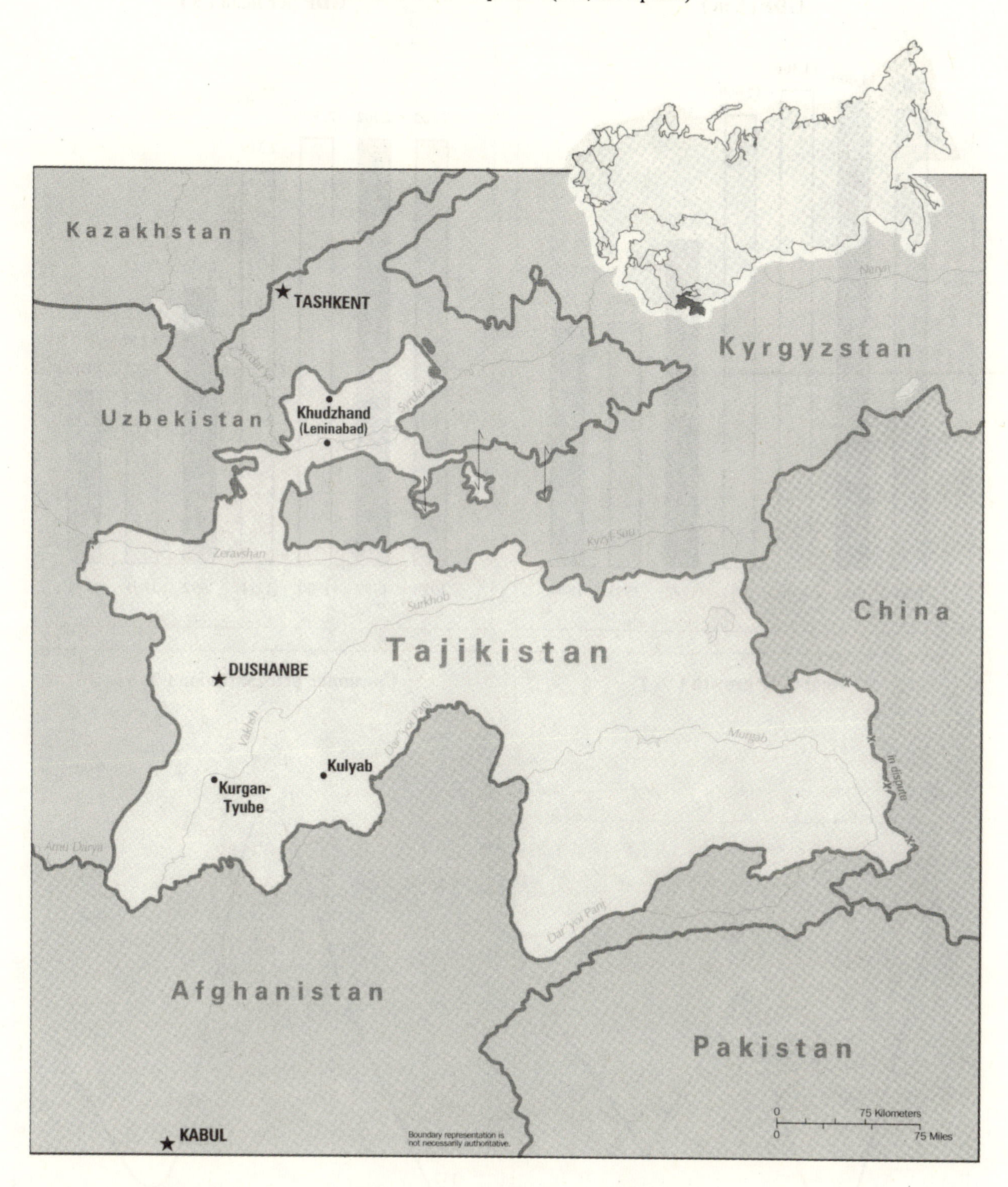


*Source: States of the Former Soviet Union,* CIA 1992.

## Republic of Tajikistan: Economic Statistics

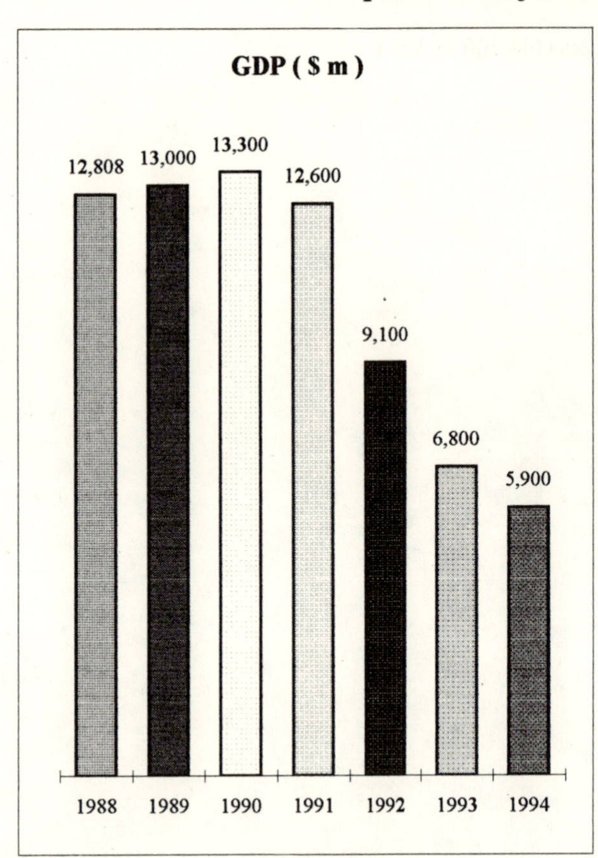


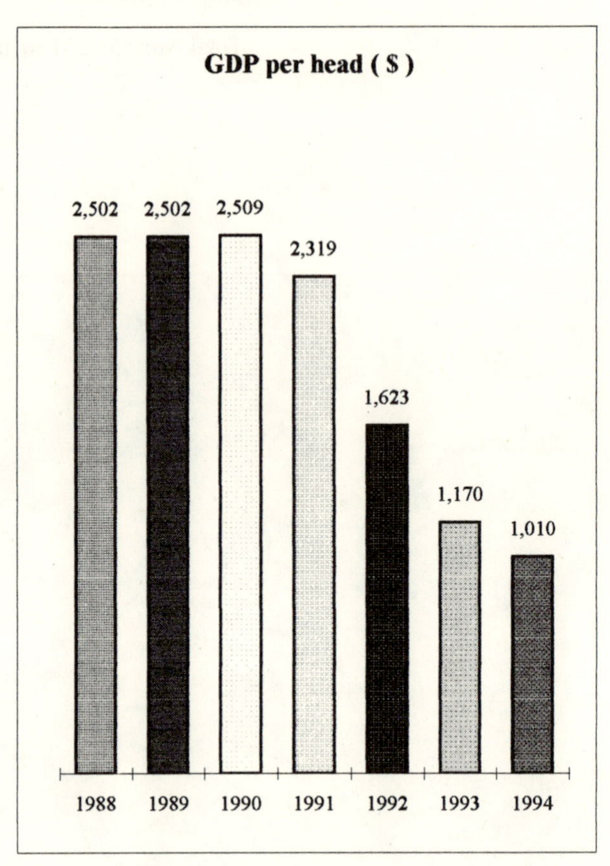


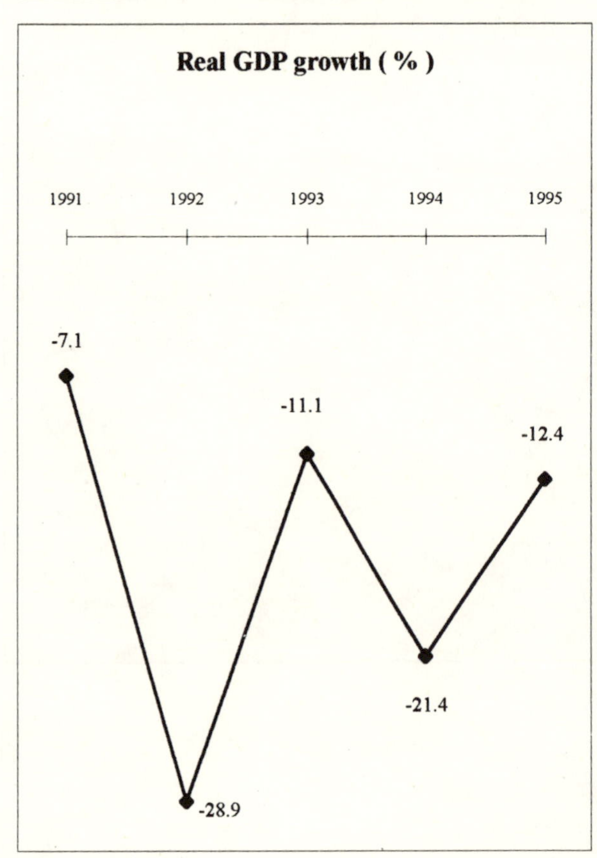


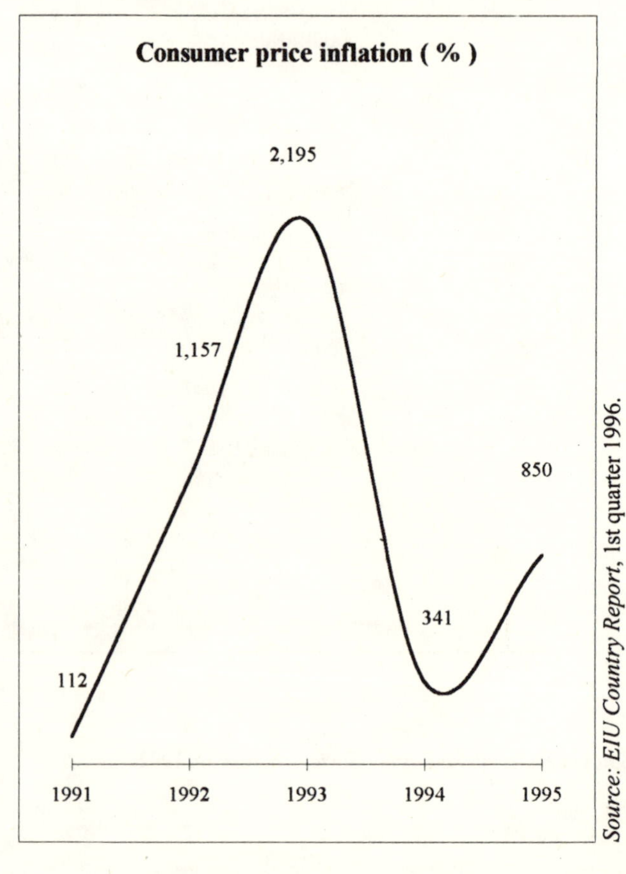


Source: *EIU Country Report*, 1st quarter 1996.

## Republic of Tajikistan: Demographic Statistics (1989)

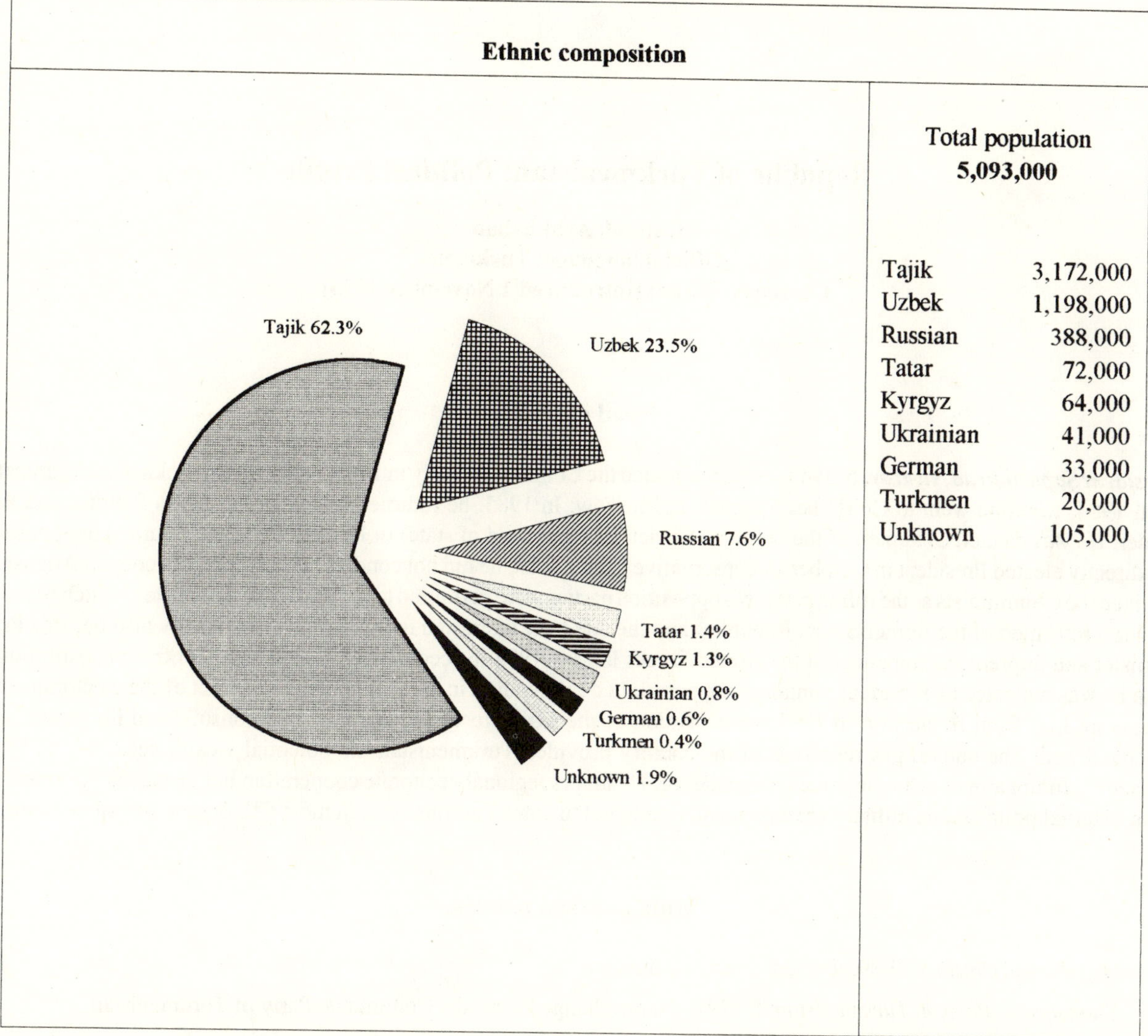


Total population
**5,093,000**

| | |
|---|---|
| Tajik | 3,172,000 |
| Uzbek | 1,198,000 |
| Russian | 388,000 |
| Tatar | 72,000 |
| Kyrgyz | 64,000 |
| Ukrainian | 41,000 |
| German | 33,000 |
| Turkmen | 20,000 |
| Unknown | 105,000 |

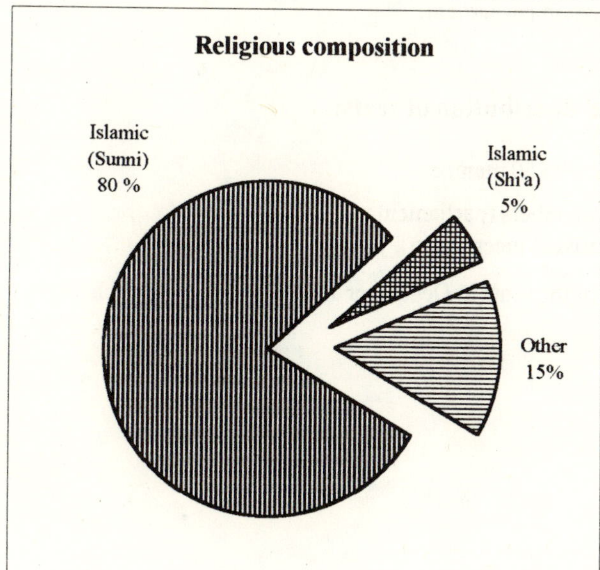


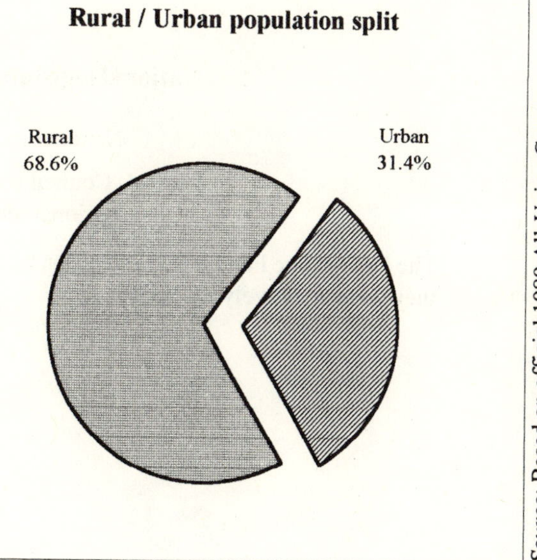


*Source:* Based on official 1989 All-Union Census.

## Republic of Turkmenistan: Political Profile

**Capital: Ashkhabad**
**Official language: Turkmen**
**Currency: Manat (introduced 1 November 1993)**

### Head of State:

***President Saparmurad Niyazov,*** b. 1940. Niyazov joined the Communist Party in 1962, heading the Ashkhabad organization until 1984, when he went to CPSU headquarters in Moscow. In 1985, he returned to Turkmenistan as Premier and Party leader. He was elected Chairman of the Supreme Soviet (de facto head of state) in January 1990 and returned unopposed as the directly elected President in October. A conservative Communist, he did not condone or condemn the coup in August and retained the Communists as the ruling party. No opposition parties were permitted to register in Turkmenistan, which remained the least reformed of the former Soviet Republics and the one least interested in independence. Niyazov also became Prime Minister and Supreme Commander of the Armed Forces in May 1992 in accordance with the new Turkmen constitution. In June he was reelected as President, running unopposed. In a referendum in January 1994, 99 percent of the electorate voted to exempt him from re-election in 1997, an indication of the extent to which Niyazov had consolidated his power since independence. The natural gas resources of the country provided Turkmenistan the potential wealth necessary to secure Niyazov's future aim of achieving true independence. He favors regional economic cooperation but distances Turkmenistan from external political and military entanglement, opting not to enter Turkmenistan in most CIS organs and agreements.

### Main political parties:

*Agzybirlik* (*Unity*), f. 1989. Popular front organization.

*Democratic Party of Turkmenistan,* f. 1991. Name changed from the Communist Party of Turkmenistan.

*Peasants' Party,* f. 1993 by deputies of the agrarian faction in parliament.

### National legislature and distribution of seats:

National Assembly with 50 members

People's Council with 100 members (parliament plus elected regional representatives) meets once a year.

The Democratic Party of Turkmenistan won most of the seats in December 1994 elections, in which they were effectively unopposed.

## Republic of Turkmenistan

Total Area: 188,456 sq. miles (488,100 sq. km.)

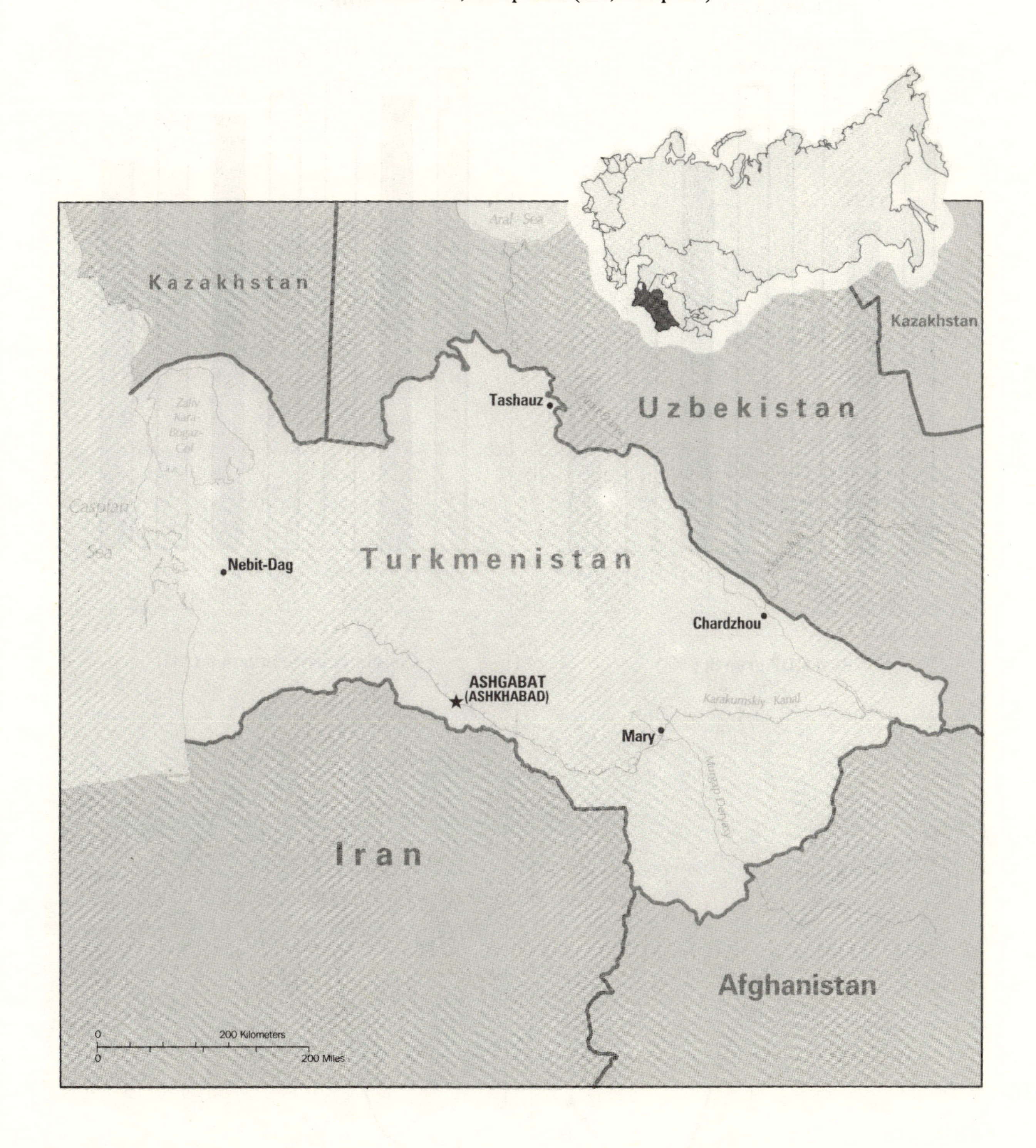


*Source: States of the Former Soviet Union,* CIA 1992.

## Republic of Turkmenistan: Economic Statistics

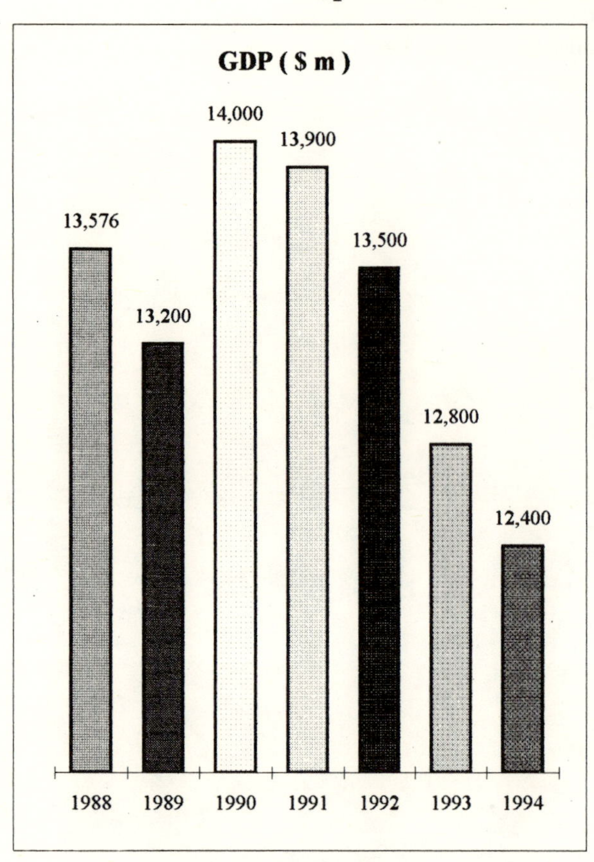


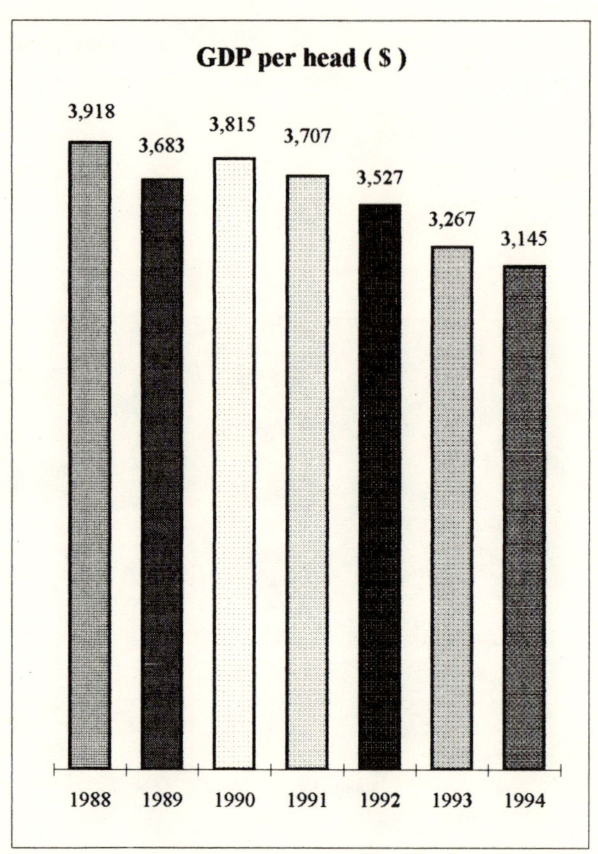


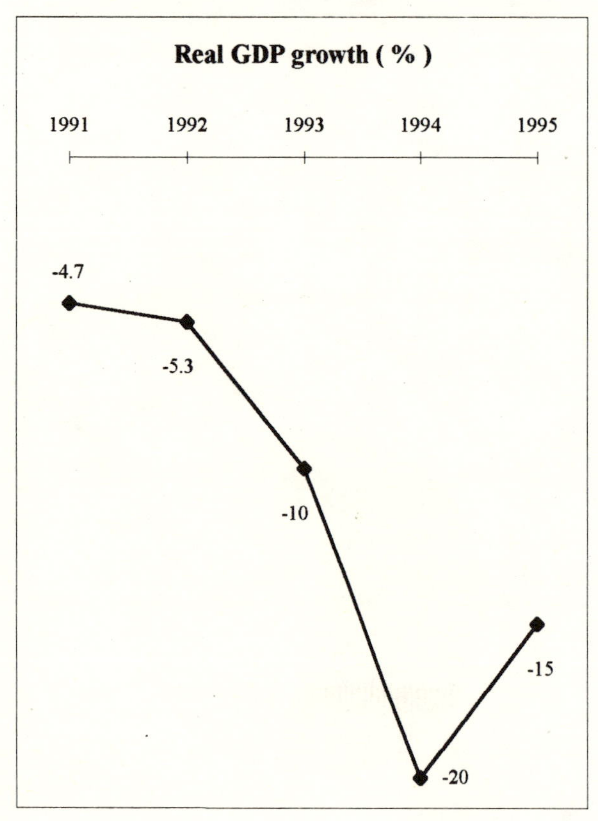


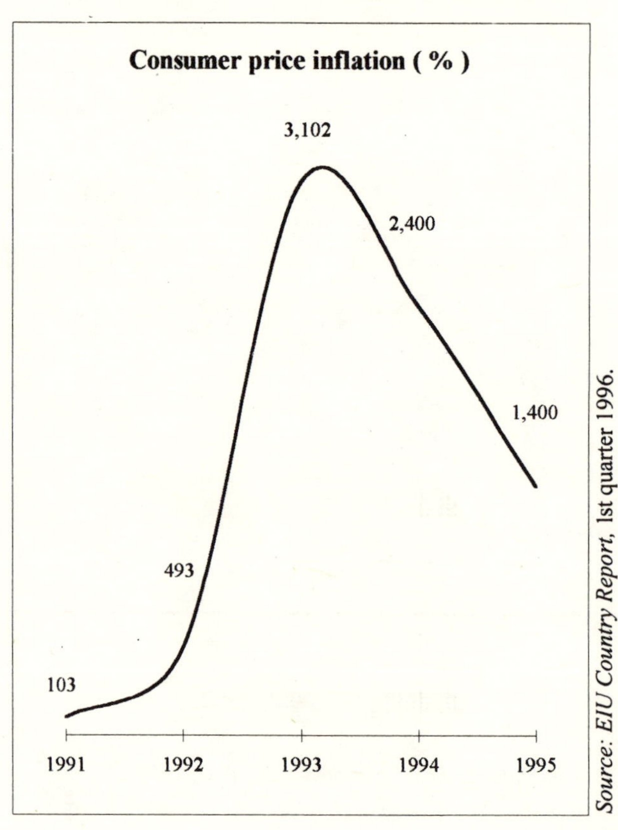


*Source: EIU Country Report,* 1st quarter 1996.

## Republic of Turkmenistan: Demographic Statistics (1989)

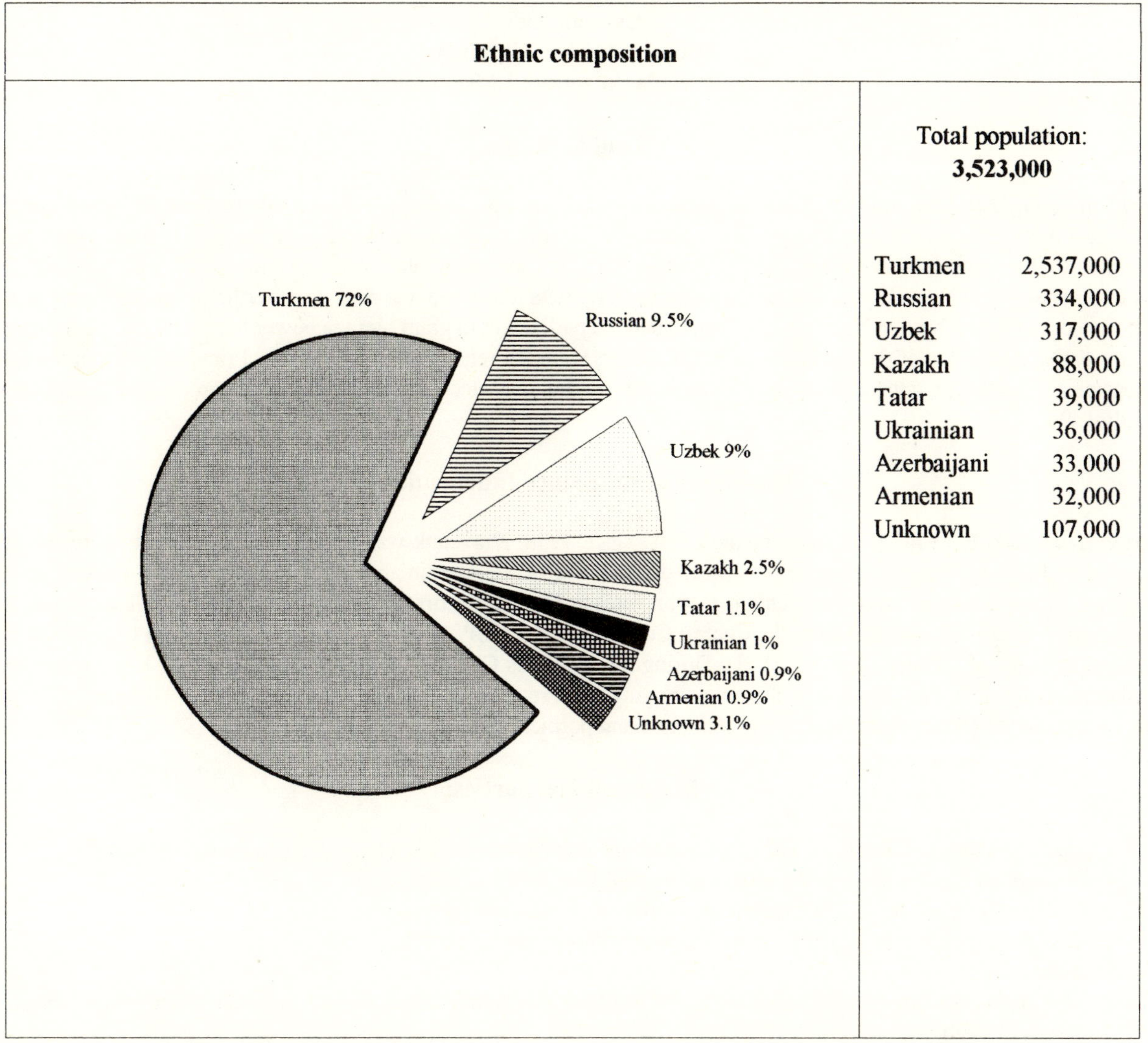


Total population:
**3,523,000**

| | |
|---|---:|
| Turkmen | 2,537,000 |
| Russian | 334,000 |
| Uzbek | 317,000 |
| Kazakh | 88,000 |
| Tatar | 39,000 |
| Ukrainian | 36,000 |
| Azerbaijani | 33,000 |
| Armenian | 32,000 |
| Unknown | 107,000 |

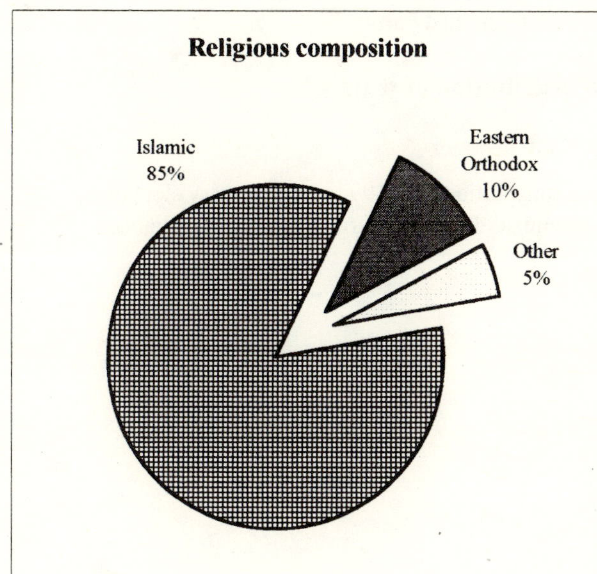


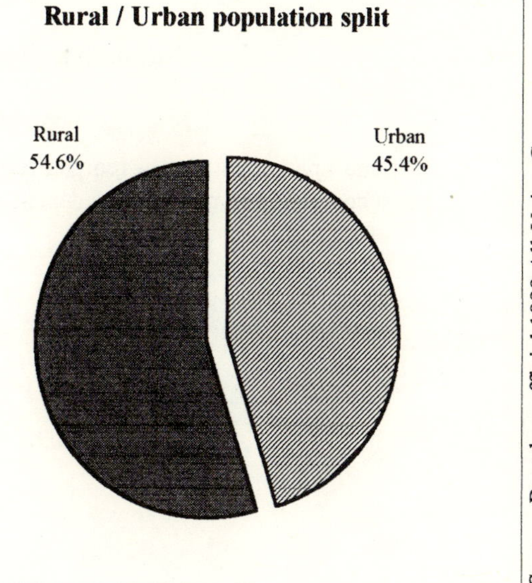


*Source*: Based on official 1989 All-Union Census.

# Ukraine: Political Profile

**Capital: Kiev**
**Official language: Ukrainian**
**Currency: Hryvnia (introduced 2 September 1996)**

## Head of State:

***President Leonid Kuchma,*** b. 1938. Kuchma graduated from Dnepropetrovsk State University. In 1960, he joined Yuzmash, the largest missile factory in the country, and eventually became manager. A member of the CPSU from 1960, he was appointed to the Central Committee of the Communist Party of Ukraine in 1981. On 13 October 1992, his nomination for Prime Minister was approved by the Ukrainian parliament. Reluctant to accept the post, he nevertheless energetically pursued his policy of market reform. In November, he persuaded the parliament to grant him the power to rule by decree for a period of six months. At the end of that time, however, his powers were not renewed. He resigned as Prime Minister in the face of increasing parliamentary opposition to his economic reform, but was elected President with 52 percent of the popular vote 10 July 1994.

## Past heads of state since Independence:

***Leonid Kravchuk,*** b. 1934. A member of the CPSU since 1958, Kravchuk was serving as the Second Secretary of the Communist Party of Ukraine when he was elected Chairman of the Ukrainian Supreme Soviet. On 1 December 1991, Kravchuk garnered 62 percent of presidential election votes to become the first President of independent Ukraine. Just a few months prior, Kravchuk resigned from the Central Committee of the CPSU, the Politburo of the Central Committee of the Communist Party of Ukraine, and the CPSU. During his tenure as Chairman of the Supreme Soviet, Kravchuk initiated legislation resulting in a Declaration of State Sovereignty, the Proclamation of State Independence of Ukraine, and various laws implementing Ukraine's sovereignty in economic and political spheres.

## Main political parties:

*Communist Party of Ukraine,* f. 1993 by Russian-speakers in eastern Ukraine.
*Inter-regional Reform Bloc* is the core of support of President Kuchma.
*Labor Party of Ukraine,* f. 1993, represents managers of state enterprises.
*Liberal Party of Ukraine,* f. 1992, represents nomenklatura entrepreneurs.
*Peasant Party.*
*Rukh,* f. 1990, the nationalist-democratic umbrella organization that united a broad spectrum of anti-communist forces in 1990–91.
*Socialist Party of Ukraine,* f. 1991, as successor to the ruling Communist Party.

## National legislature and distribution of seats:

Supreme Council of 450 deputies

Ukraine's parliament will be reorganized under a new constitution. The 1994 parliamentary elections produced a body dominated by independents and communists and characterized by shifting alliances.

# Ukraine

Total Area: 233,090 sq. miles (603,700 sq. km.)

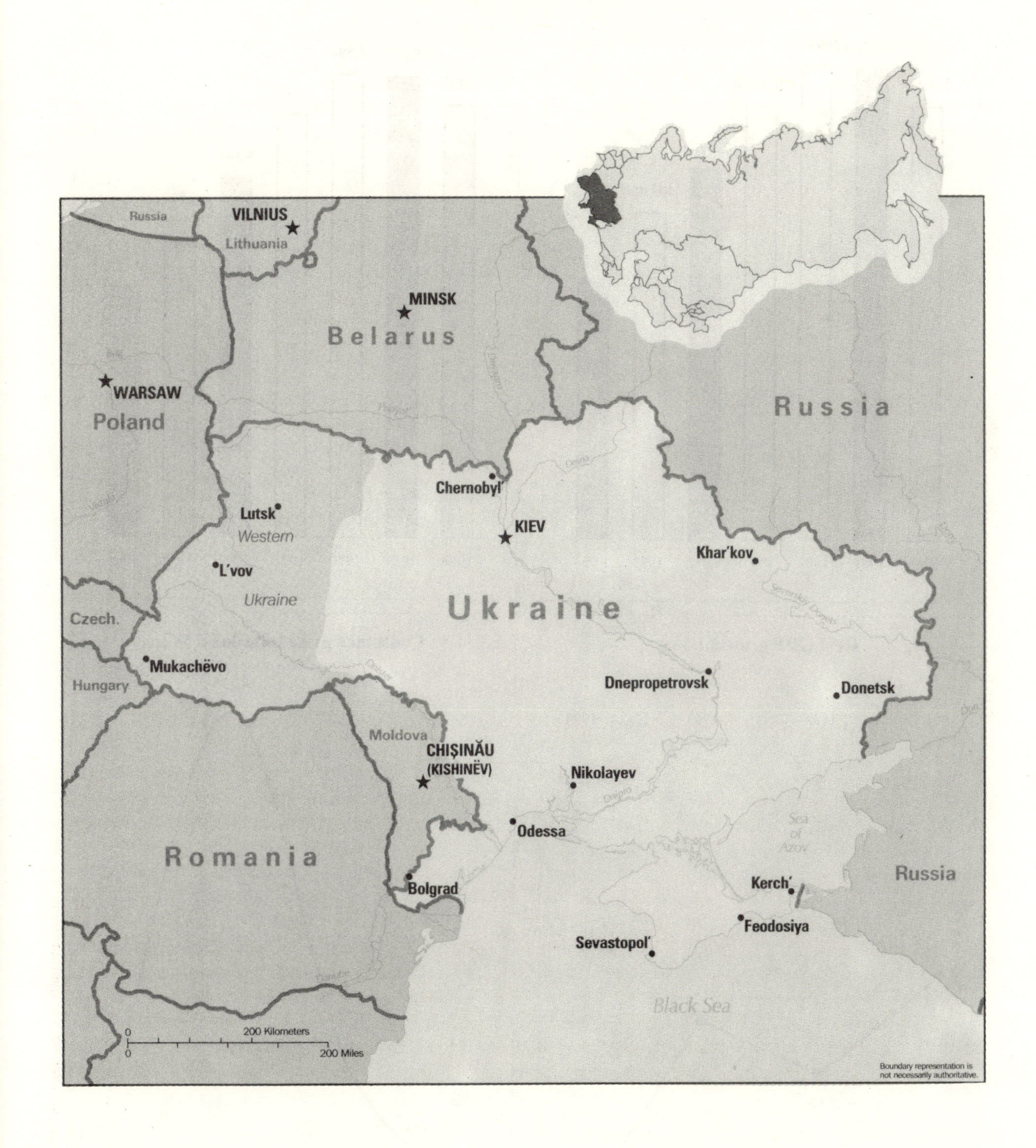


*Source: States of the Former Soviet Union,* CIA 1992.

## Ukraine: Economic Statistics

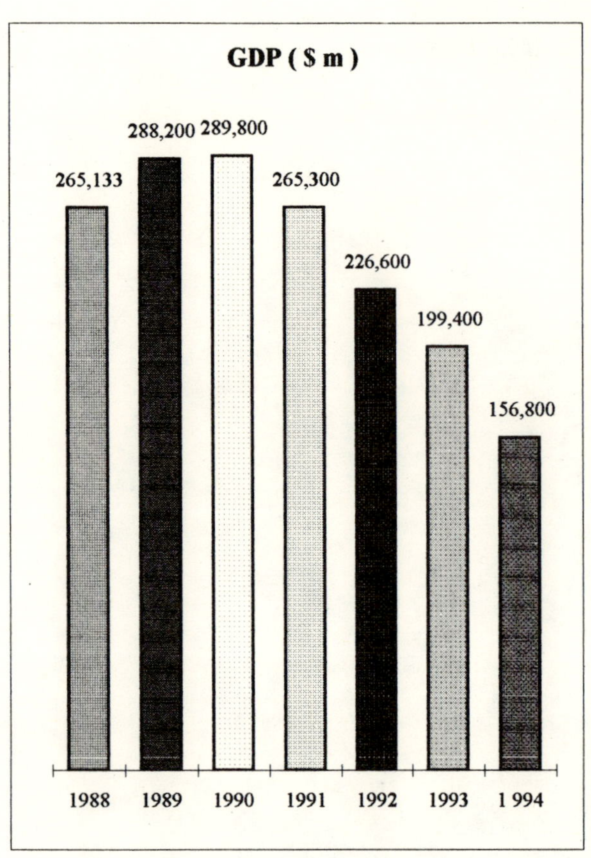


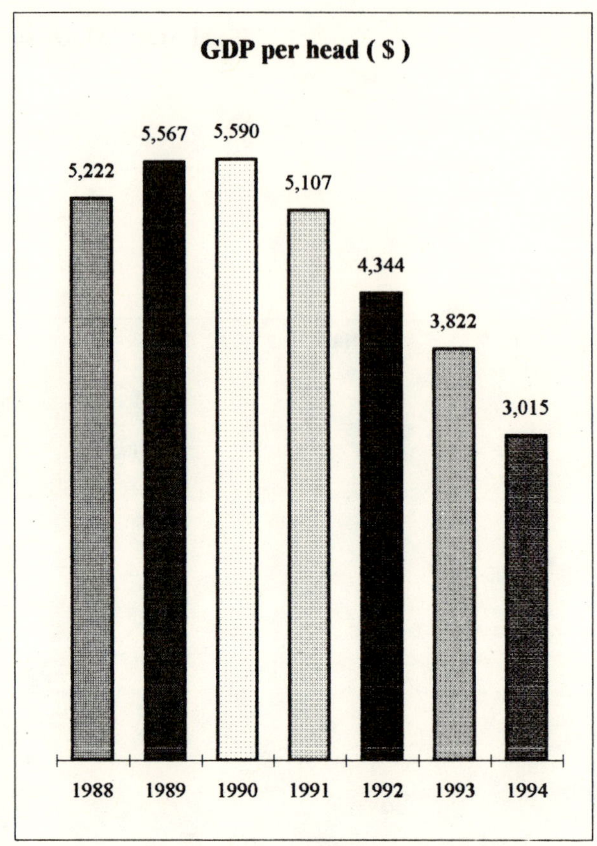


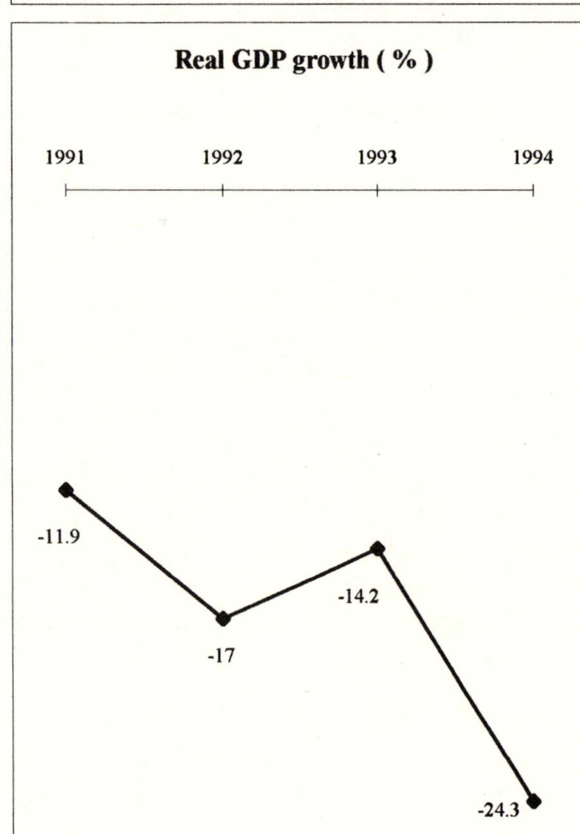


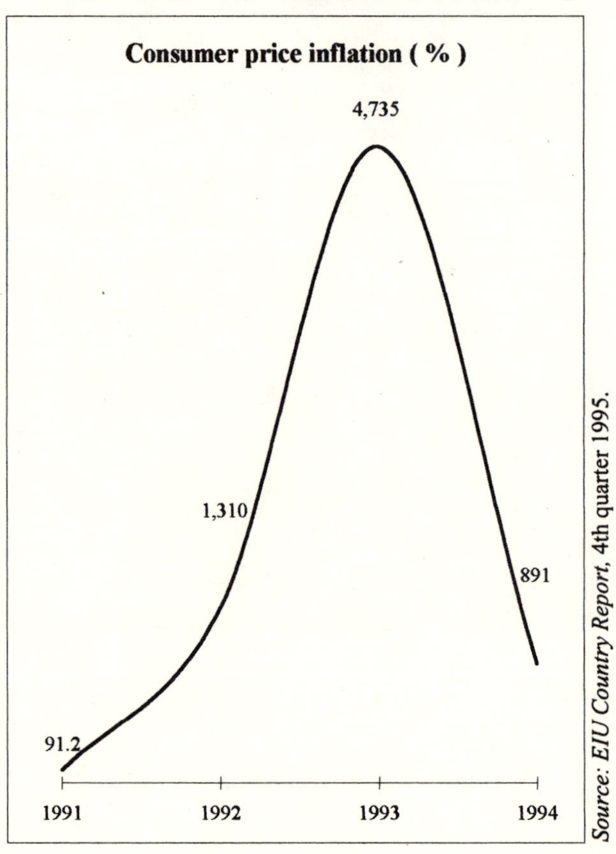


*Source: EIU Country Report*, 4th quarter 1995.

# Ukraine: Demographic Statistics (1989)

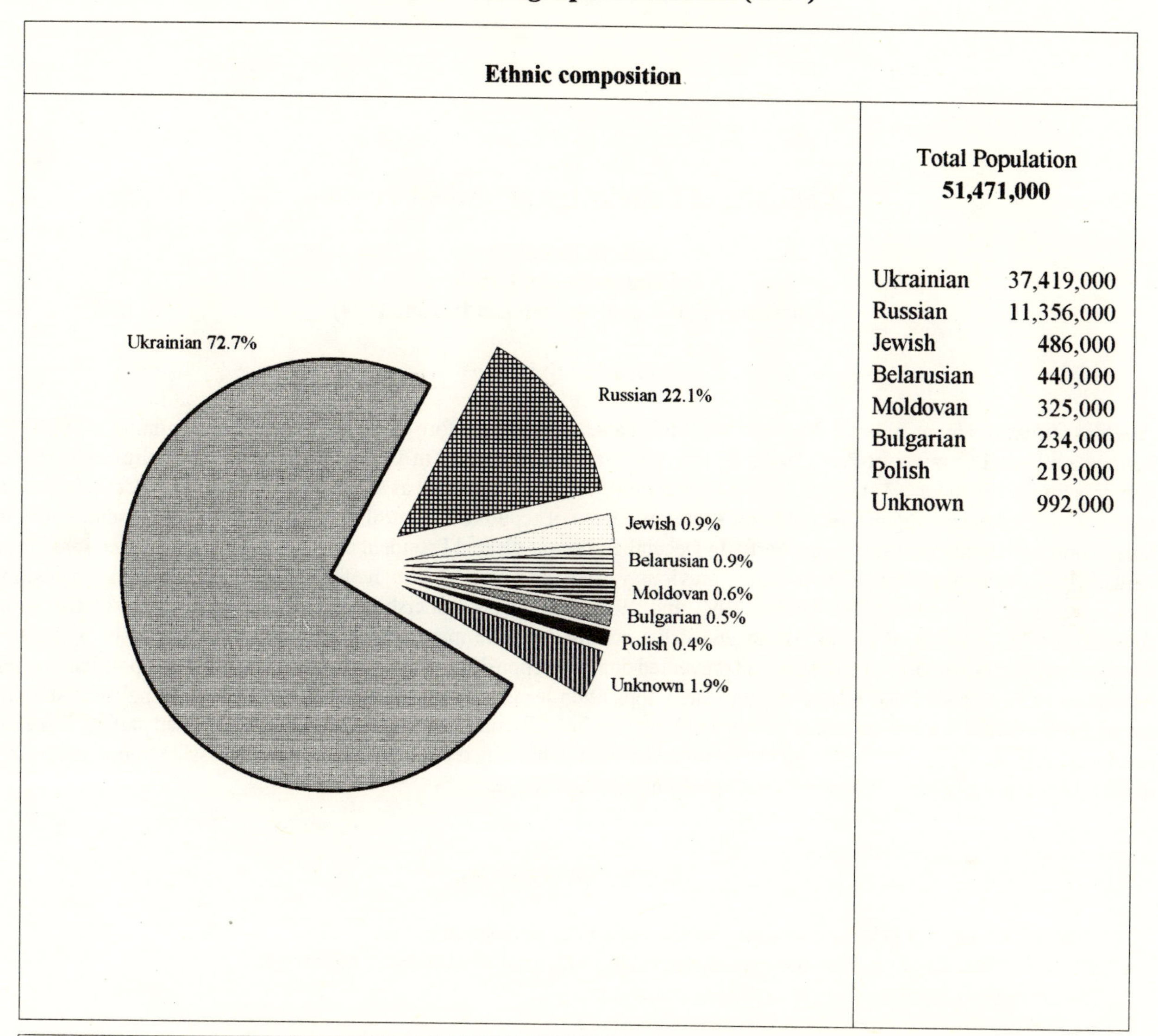


Total Population
**51,471,000**

| | |
|---|---|
| Ukrainian | 37,419,000 |
| Russian | 11,356,000 |
| Jewish | 486,000 |
| Belarusian | 440,000 |
| Moldovan | 325,000 |
| Bulgarian | 234,000 |
| Polish | 219,000 |
| Unknown | 992,000 |

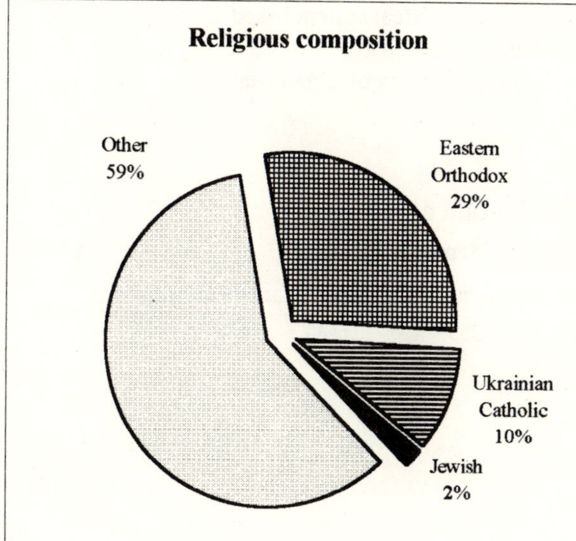


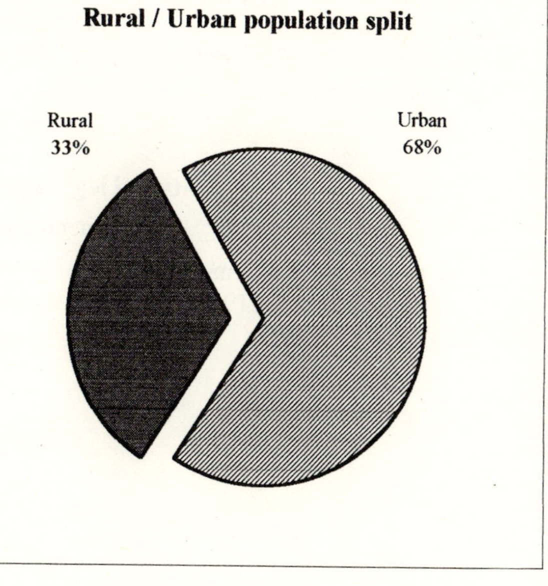


*Source:* Based on official 1989 All-Union Census.

## Republic of Uzbekistan: Political Profile

**Capital: Tashkent**
**Official language: Uzbek**
**Currency: Uzbek som (introduced 15 June 1994)**

### Head of State:

***President Islam Karimov,*** b. 1938. Karimov was a mechanical engineer before moving into economic planning in 1966. He became a regional Communist Party leader in 1986 and a republican leader in 1989. He assumed the chairmanship of the Supreme Soviet (Presidency) of Uzbekistan in March 1990. Although regarded as an old-style, conservative Communist and architect of a repressive regime, he did favor a greater degree of republican control over the central government. After the failed coup, he banned the Communist Party. In December, he was elected President of the Republic in free elections. Until January 1992, he also performed the functions previously executed by the Chairman of the Council of Ministers—a position he abolished in November 1990. During 1992 and 1993, Karimov's leadership became increasingly absolutist as he consolidated his position of power. He extended his control over the mass media, precluded the spread of opposition movements from neighboring Tajikistan, and prevented domestic opposition movements from registering as political parties. Allegations also surfaced that Karimov intimidated opposition leaders. His stated concern was ensuring political stability during the transition to a free market economy and he particularly discouraged religious or ethnically based parties. Karimov had a large role in inciting the Tajik civil war by invading the neighboring republic with Uzbek and Russian troops, ostensibly to prevent the rising Islamic movement from spreading into Uzbekistan.

### Main political parties:

*Adolat* (*Justice*), f. 1995. A pro-Islamic Social Democratic opposition party.
*Birlik,* f. 1989. Leading opposition group banned in 1992; registered as social movement.
*Erk (Freedom),* f. 1990. Only registered opposition party.
*Fatherland Progress Party*. A pro-government business party.
*Islamic Renaissance Party.* Banned in 1991; advocates introduction of political system based on tenets of Islam.
*National Revival Democratic Party,* f. 1995. An intelligentsia-based centrist party.
*People's Democratic Party of Uzbekistan,* f. 1991 to replace Communist Party of Uzbekistan.
*People's Unity Party,* f. 1995.

### National legislature and distribution of seats:

Unicameral Assembly with 250 Members

Only the People's Democratic Party of Uzbekistan and the Fatherland Progress Party were allowed to participate in the 1994/5 parliamentary elections.

# Republic of Uzbekistan

Total Area: 172,742 sq. miles (447,400 sq. km.)

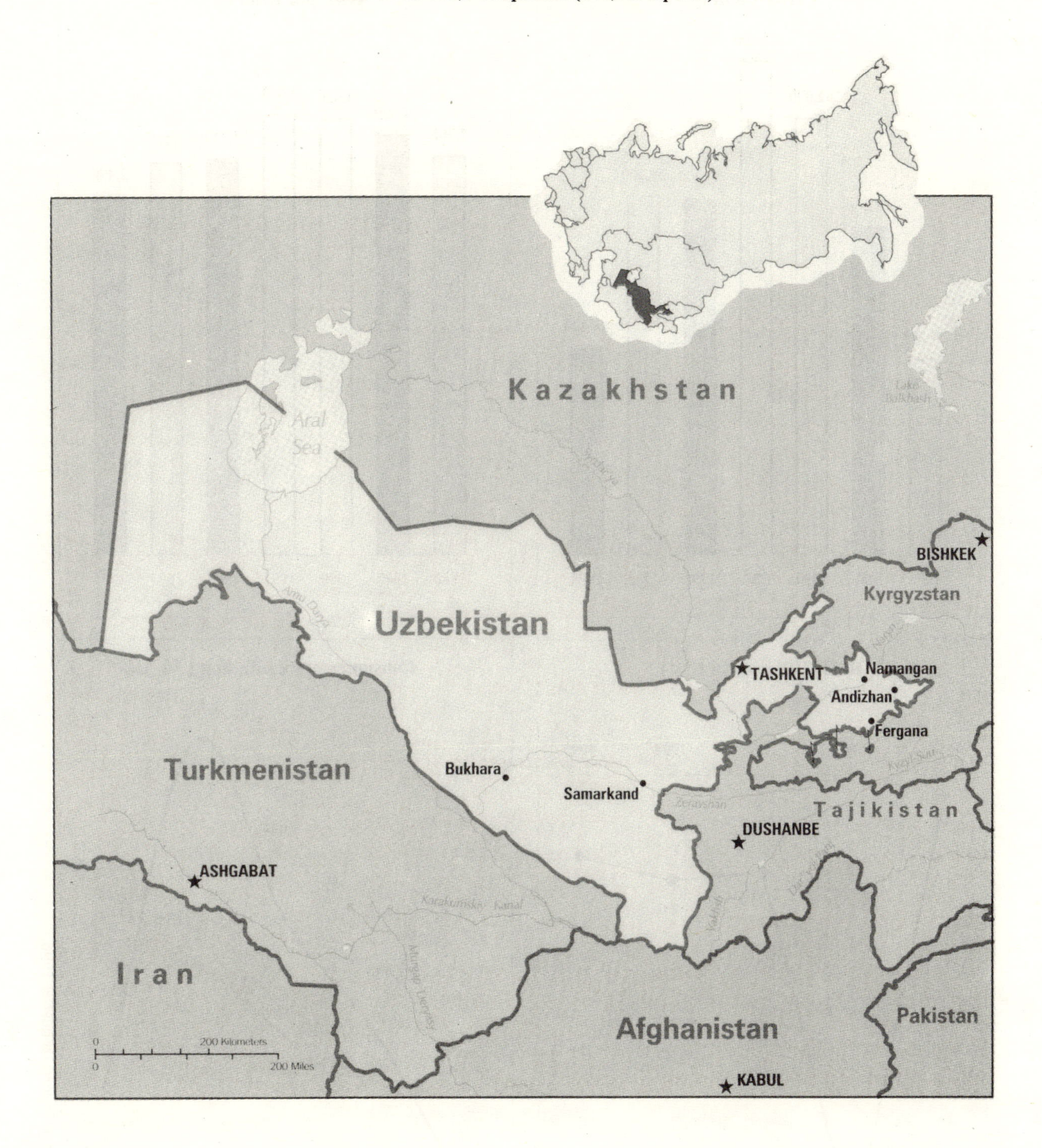


*Source: States of the Former Soviet Union,* CIA 1992.

## Republic of Uzbekistan: Economic Statistics

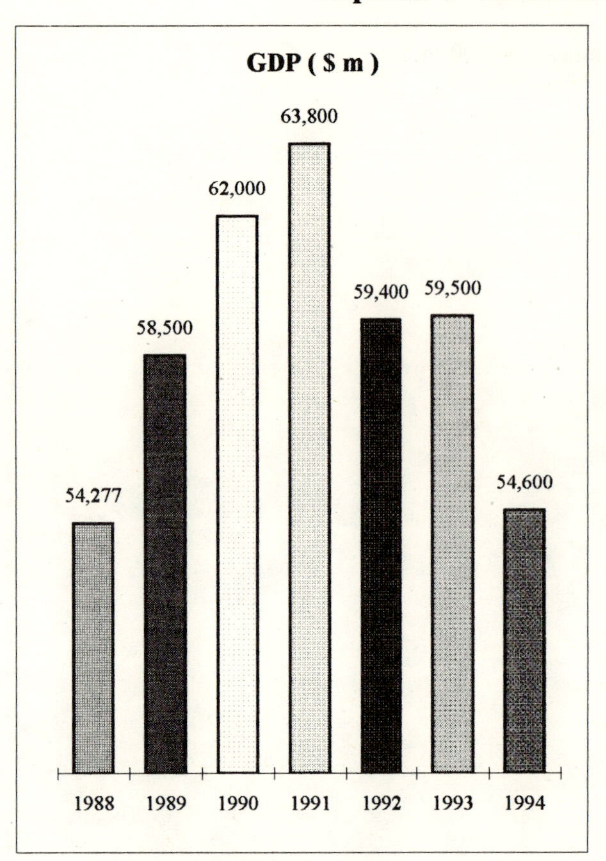


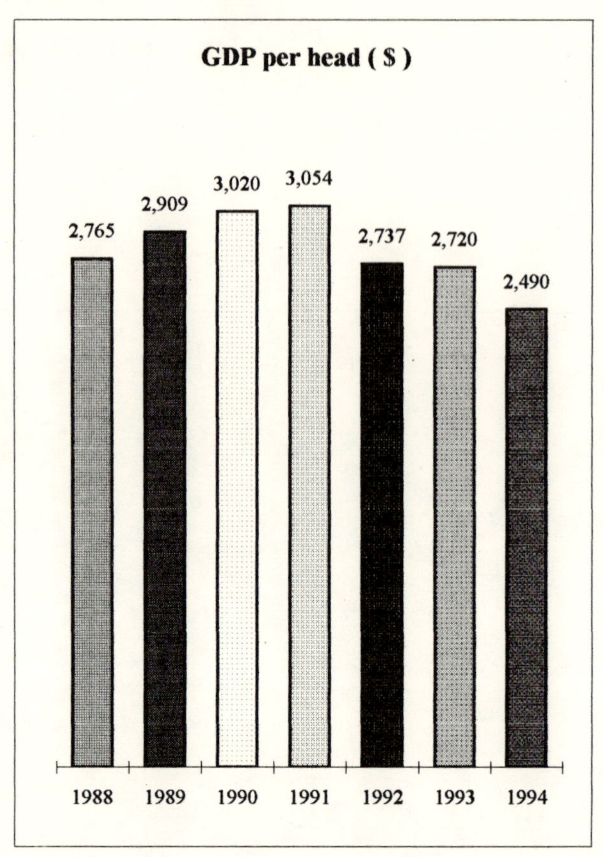


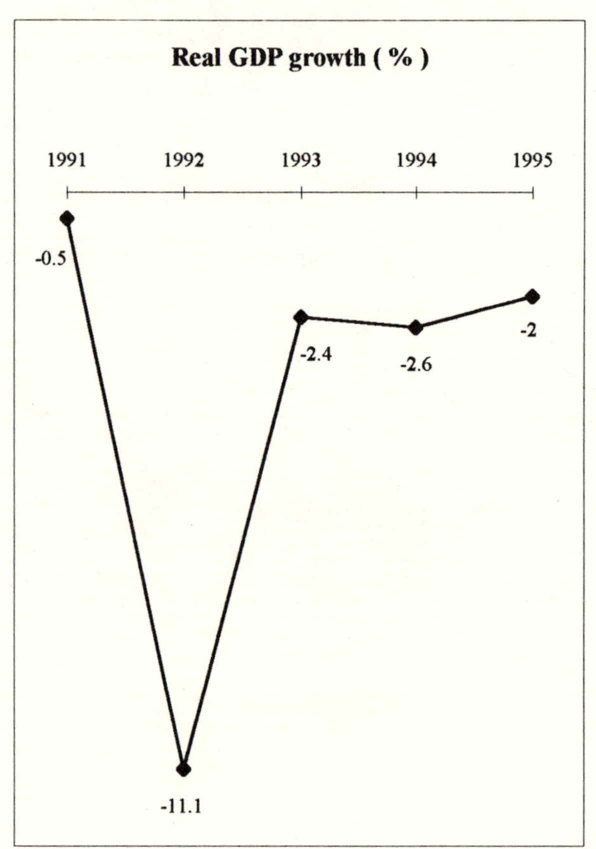


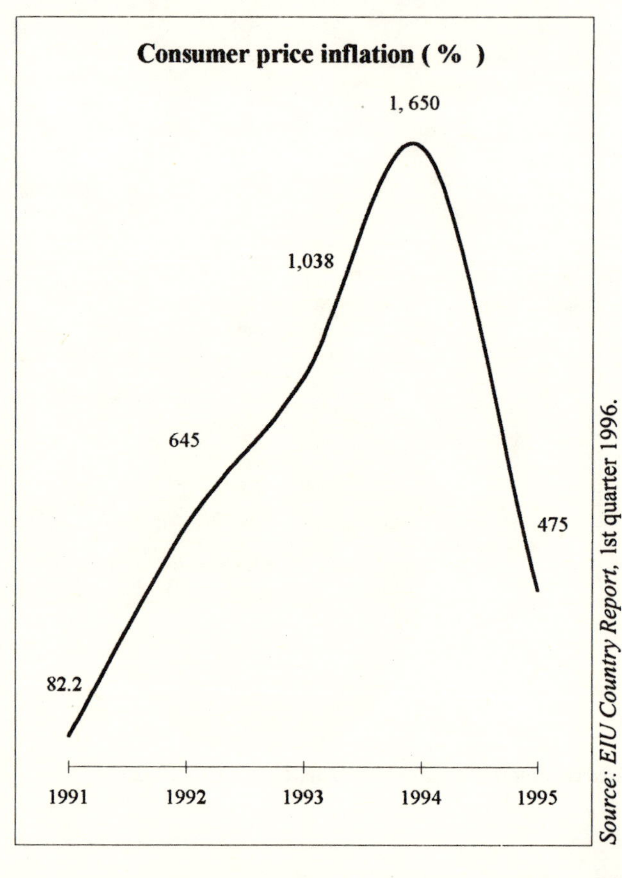


*Source: EIU Country Report,* 1st quarter 1996.

## Republic of Uzbekistan: Demographic Statistics (1989)

### Ethnic composition

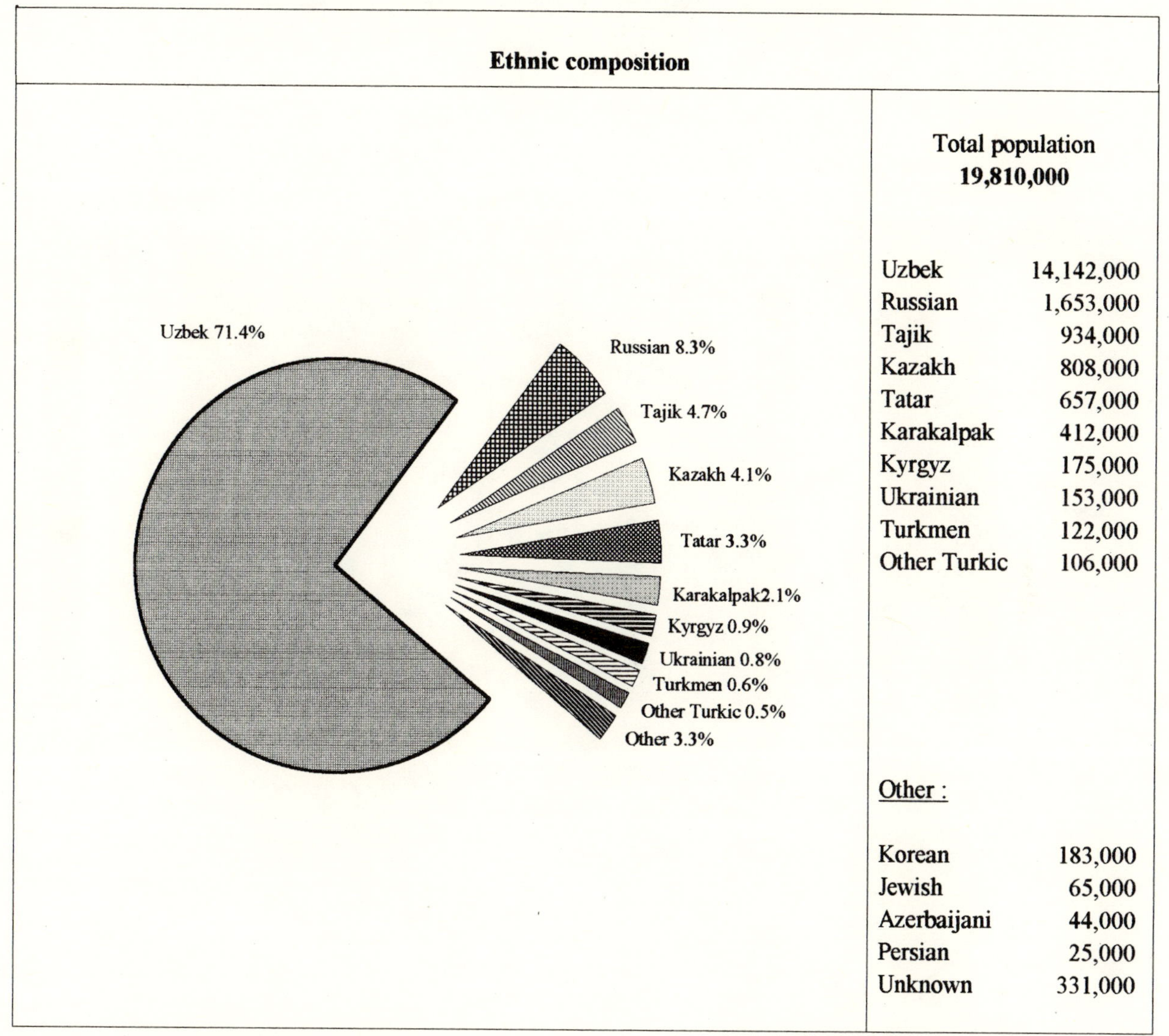


Total population
**19,810,000**

| | |
|---|---|
| Uzbek | 14,142,000 |
| Russian | 1,653,000 |
| Tajik | 934,000 |
| Kazakh | 808,000 |
| Tatar | 657,000 |
| Karakalpak | 412,000 |
| Kyrgyz | 175,000 |
| Ukrainian | 153,000 |
| Turkmen | 122,000 |
| Other Turkic | 106,000 |

Other :

| | |
|---|---|
| Korean | 183,000 |
| Jewish | 65,000 |
| Azerbaijani | 44,000 |
| Persian | 25,000 |
| Unknown | 331,000 |

### Religious composition

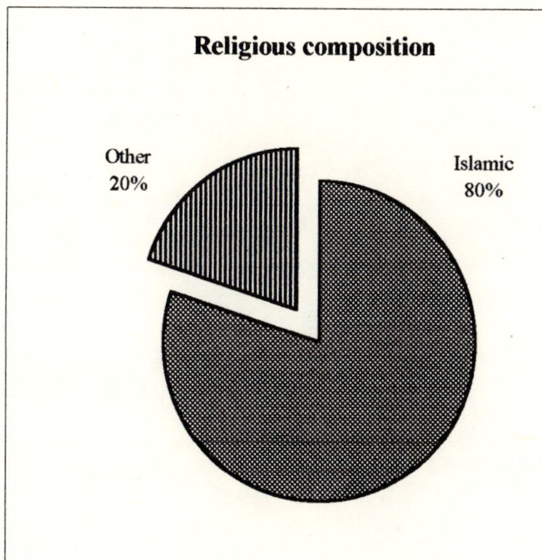


### Rural / Urban population split

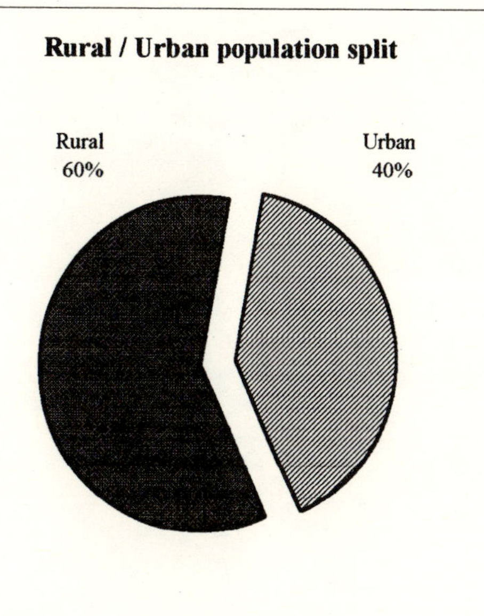


*Source:* Based on official 1989 All-Union Census.

## Appendix E

# Useful Addresses, Telephone and Fax Numbers

## Addresses, Telephone and Fax Numbers of Ministries

### ARMENIA

Office of the Prime Minister, Hrant Bagratyan
1 Government Bldg., Republic Square
375010 Yerevan, Armenia
Tel: (374–2) 520–360
Fax: (374–2) 151–036

Ministry of Foreign Affairs, Vahan Papazyan
10 Marshal Baghramyan Street
375012 Yerevan, Armenia
Tel: (374–2) 523–531
Fax: (374–2) 151–042

Ministry of Defense, Vazgen Sarkissyan
Ashtarak Shoose
Yerevan, Armenia
Tel: (374–2) 357–881
Fax: (374–2) 527–537

Ministry of Economy, Vahram Avanesyan
1 Government Bldg., Republic Square
375010 Yerevan, Armenia
Tel: (374–2)527–342
Fax: (374–2)151–069

### AZERBAIJAN

Office of the Prime Minister, Faud Quliyev
68 Lermontov Street
370066 Baku, Azerbaijan
Tel: (9–9412) 98–00–08
Fax: (9–9412) 92–91–79

Ministry of Foreign Affairs, Hasan Hasanov
3 Gandjilar Square
370016 Baku, Azerbaijan
Tel: (9–9412) 92–56–06
Fax: (9–9412) 65–10–38

Ministry of Defense, Safar Abiyev
3 Azizbekov Avenue
370073 Baku, Azerbaijan
Tel: (9–9412) 38–93–90

Ministry of External Economic Relations
Nịcat Quliyev (acting)
68 Lermontov Street
370066 Baku, Azerbaijan
Tel: (9–9412) 92–93–90
Fax: (9–9412) 92–91–79

## BELARUS

Office of the Prime Minister, Mikhail Chigir
House of Government
10 Prospekt Skoryni
220010 Minsk, Belarus
Tel: (172) 226-105
Fax: (172) 226-665

Ministry of Foreign Affairs, Vladimir Senko
Ul. Lenina, 19
220030 Minsk, Belarus
Tel: (172) 272-922
Fax: (172) 274–521

Ministry of Defense, Leonid Maltsev
Ul. Kommunisticheskaya, 1
220003 Minsk, Belarus
Tel: (172) 330–352
Fax: (172) 331-234

Ministry of Foreign Economic Relations, Mikhail Marinich
House of Government
220010 Minsk, Belarus
Sovetskaya Ul., 10
Tel. (172) 202-635
Fax: (172) 296–335

## GEORGIA

Office of the State Chancellor, Nikoloz Lekishvili
7 Ingorokva Street
380034 Tbilisi, Georgia
Tel: (995–32) 98–97–93
Fax: (995–32) 99–86–90

Ministry of Foreign Affairs, Irakli Menagharashvili
4 Chitadze
380018 Tbilisi, Georgia
Tel: (995–32) 98–93–77
Fax: (995–32) 98–72–49

Ministry of Defense, Vardiko Nadibaidze
2 Universiteti Street
380043 Tbilisi, Georgia
Tel: (995–32) 38–39–26
Fax: (995–32) 39–35–88

Ministry of Economy, Vladimir Papava
12 Chanturia Street
380062 Tbilisi, Georgia
Tel: (995–32) 23–09–25
Fax: (995–32) 29–00–63

## KAZAKHSTAN

Office of the President, Nursultan Nazarbaiev
Republic Square, 4
480091 Almaty, Kazakhstan
Tel: (7–3272) 623–016
Fax: (7–3272) 639–595

Office of the Prime Minister,
Akezhan Magzhan-uly Kazhegeldin
Government House
480091 Almaty, Kazakhstan
Tel: (7–3272) 627–966
Fax: (7–3272) 637–633

Ministry of Foreign Affairs,
Kasymzhomart K Tokayev
Aiteke Bi, 65
480091 Almaty, Kazakhstan
Tel: (7–3272) 623–538
Fax: (7–3272) 631–387

Ministry of Defense, Lt. Gen. Alibek A Kasymov
Dzhandosov Street, 53
480091 Almaty, Kazakhstan
Tel: (7–3272) 214–735
Fax: (7–3272) 282–346

Minister of Economy, Umirzak Shukeev
Zheltoksan Street, 115
480091 Almaty, Kazakhstan
Tel: (7–3272) 626–500
Fax: (7–3272) 636–605

## KYRGYZSTAN

Office of the Prime Minister, Apas Jumagulov
Government House
720000 Bishkek, Kyrgyzstan
Tel: (3312) 22–56–56
Fax: (3312) 21–86–27

Ministry of Foreign Affairs, Roza Otunbayeva
Abdumomunova 205
720003 Bishkek, Kyrgyzstan
Tel: (3312) 22–05–45
Fax: (3312) 22–57–35

Ministry of Defense, Myrzakan Subanov
Logvinenko 25
720001 Bishkek, Kyrgyzstan
Tel: (3312) 22–78–79
Fax: (3312) 22–86–48

Ministry of Industry and Trade, Andrei Iordan
Chui Prospect 106
72000 Bishkek, Kyrgyzstan
Tel: (3312) 22–38–66
Fax: (3312) 22–97–93

## MOLDOVA

Office of the Prime Minister, Andrei Sangheli
Piata Marii Adunari Nationale 1
277012 Chisinau, Moldova
Tel: (2) 23–40–30
Fax: (2) 22–22–64

Ministry of Foreign Affairs, Mihai Popov
Piata Marii Adunari Nationale 1
277033 Chisinau, Moldova
Tel: (2) 23–39–40
Fax: (2) 23–23–02

Ministry of Defense, Pavel Creanga
Hincesti Street 84
277048 Chisinau, Moldova
Tel: (2) 23–26–31
Fax: (2) 23–45–35

Ministry of Economy, Valeriu Bobutac
Piata Marii Adunari Nationale 1
277033 Chisinau, Moldova
Tel: (2) 23–31–35
Fax: (2) 23–40–64

## RUSSIA

Office of the Prime Minister, Victor Chernomyrdin
Strataya Square, 4
103132 Moscow, Russia
Tel: (095) 206–3328
Fax: (095) 260–4722

Ministry of Foreign Affairs, Yvgeny Primakov
Smolenskaya-Sennaya Square 32/34
121200 Moscow, Russia
Tel: (095) 244–4021
Fax: (095) 244–2203

Ministry of Defense, Igor Rodinov
Ul. Myasnitskaya 37
103160 Moscow, Russia
Tel: (095) 296–89–00
Fax: (095) 293–3313

Ministry of Foreign Economic Relations, Oleg Davydov
Smolenskaya-Sennaya Square, 32/34
121200 Moscow, Russia
Tel: (095) 244–1046
Fax: (095) 244 3068

## TAJIKISTAN

Office of the Prime Minister, Yakhyo Jamshed Azimov
80 Rudaki Street
734023 Dushanbe, Tajikistan
Tel: (3772) 21–06–24

Ministry of Foreign Affairs, Talbak Nazarov
42 Rudaki Street
734051 Dushanbe, Tajikistan
Tel: (3772) 21–18–08
Fax: (3772) 23–29–64

Ministry of Defense, Maj. Gen. Sherali Khayrullayev
Bokhtar Street 59
734002 Dushanbe, Tajikistan
Tel: (3372) 24–33–33

Ministry of Economics, Tukhtaboy Gafarov
42 Rudaki Street
734025 Dushanbe, Tajikistan
Tel: (3772) 23–29–44

## TURKMENISTAN

Ministry of Foreign Affairs, Boris Shikmuradov
83 Magtumguly
Ashgabat, Turkmenistan
Tel: (7–3632) 33–16–66
Fax: (7–3632) 51–14–30

Ministry of Defense, Lt-Gen. Danatar Kopekov
15 Nurberdi Pomma
Ashgabat, Turkmenistan
Tel: (7–3632) 25–73–82
Fax: (7–3632) 29–77–45

Ministry of Economy & Finance, Valeriy Otchertsov
15 Nurberdi Pomma
Ashgabat, Turkmenistan
Tel: (7–3632) 25–16–53
Fax: (7–3632) 25–65–11

## UKRAINE

Office of the Prime Minister, Pavalo Lazarenko
12/2 Krushevskiy Street
252008 Kiev, Ukraine
Tel: (380–44) 266–3263
Fax: (380–44) 293–2093

Ministry of Foreign Affairs, Hennadiy Udovenko
1 Myhailivska Square
252018 Kiev, Ukraine
Tel: (380–44) 226–3379
Fax: (380–44) 266–3169

Ministry of Defense, Valeriy Shmarov
6 Bankivska Street
252009 Kiev, Ukraine
Tel: (380–44) 226–2637
Fax: (380–44) 226–2015

Ministry of External Economic Relations & Trade, Serhiy Osyka
8 Lviv Square
252053 Kiev, Ukraine
Tel: (380–44) 226–5233
Fax: (380–44) 226–2629

## UZBEKISTAN

Office of the Prime Minister, Utkur Sultanov
House of Government
700008 Tashkent, Uzbekistan
Tel: (3712) 39–82–95
Fax: (3712) 39–86–01

Ministry of Foreign Affairs, Abdulaziz Kamilov
87 Gogol Street
700047 Tashkent, Uzbekistan
Tel: (3712) 33–64–75
Fax: (3712) 39–43–48

Ministry of Defense, Gen. Rustam Akhmedov
100 Academician Adbdullaev
700000 Tashkent, Uzbekistan
Tel: (3712) 33–66–77

Ministry of Foreign Economic Relations, Utkur Sultanov
Ul. Bajuk Ipak Yuli, 75
700077 Tashkent, Uzbekistan
Tel: (3712) 68–92–56
Fax: (3712) 68–72–31

# Addresses, Telephone and Fax Numbers of CIS Embassies and Consulates in the United States

## ARMENIA

Embassy of the Republic of Armenia
2225 R Street NW
Washington, DC 20008
Tel: (202) 319–1976
Fax: (202) 319–1982

Armenian National Committee of America
888 17th Street NW, Suite 904
Washington, DC 20006
Tel: (202) 775–1918
Fax: (202) 775–5648

## AZERBAIJAN

Embassy of the Republic of Azerbaijan
927 15th Street NW, Suite 700
Washington, DC 20005
Tel: (202) 842–0001
Fax: (202) 842–0004

U.S.–Azerbaijani Council
1030 15th Street NW, Suite 444
Washington, DC 20005
Tel: (202) 371–2288, 371–2289
Fax: (202) 371–2299

## BELARUS

Embassy of the Republic of Belarus
1619 New Hampshire Avenue NW
Washington, DC 20009
Tel: (202) 986–1604
Fax: (202) 986–1805

## GEORGIA

Embassy of the Republic of Georgia
1511 K Street NW, Suite 424
Washington, DC 20005
Tel: (202) 393–5959
Fax: (202) 393–6060

## KAZAKHSTAN

The Embassy of Kazakhstan
3421 Massachusets Avenue NW
Washington, DC 20007
Tel: (202) 333–4504
Fax: (202) 333–4509

The U.S.–Kazakhstan Council
2000 L. Street NW, Suite 200
Washington, DC 20006
Tel: (202) 416–1624
Fax: (202) 416–1865

## KYRGYZSTAN

Embassy of the Kyrgyz Republic
1511 K Street NW, Suite 706
Washington, DC 20005
Tel: (202) 347–3732
Fax: (202) 347–3718

U.S.–Kyrgyz Business Council, Inc.
700 13 Street NW, Suite 950
Washington, DC 20005
Tel: (202) 347–6540

## MOLDOVA

Embassy of Moldova
1511 K Street NW, Suites 329 and 333
Washington, DC 20005
Tel: (202) 783–3012
Fax: (202) 783–3342

Moldovan American Chamber of Commerce
2 Wisconsin Circle, Suite 510
Washington, DC 20815
Tel: (301) 656–9022
Fax: (301) 656–9008

## RUSSIA

Embassy of the Russian Federation
2650 Wisconsin Avenue NW
Washington, DC
Tel: (202) 298–5700, 5701, 5702, 5703
Fax: (202) 298–5749, 5735

Trade Representative of Russia in the U.S.A.
2001 Connecticut Avenue NW
Washington, DC 20008
Tel: (202) 232–0975
Fax: (202) 232–2917

Consulate General of the Russian Federation
2790 Green Street
San Francisco, CA 94123
Tel: (415) 202–9800
Fax: (415) 929–0306

Consulate General of the Russian Federation
2323 Westing Building
2001 6th Avenue
Seattle, WA 98121
Tel: (206) 728–1910
Fax: (206) 728–1871

Consulate General of the Russian Federation
9 East 91st Street
New York, NY 10128
Tel: (212) 348–0926
Fax: (212) 831–9162

U.S.–Russia Business Council
1701 Pennsylvania Avenue NW, Suite 650
Washington, DC 20006
Tel: (202) 739–9180
Fax: (202) 659–5920

Russian–American Chamber of Commerce
The Market Place, Tower II, Suite 735
3025 South Parker Road
Aurora, CO 80014
Tel: (303) 745–0757
Fax: (303) 745–0776

Foundation for Russian–American Economic Cooperation
1932 First Avenue, Suite 803
Seattle, WA 98101
Tel: (206) 443–1935
Fax: (206) 443–0954

## TAJIKISTAN

Tajikistan Mission to the United Nations
(c/o Russian Mission to the United Nations)
136 East 67th Street
New York, NY 10021
Tel: (212) 472–7645
Fax: (212) 628–0252

## TURKMENISTAN

Embassy of Turkmenistan
2207 Massachusets Avenue NW
Washington, DC 20008
Tel: (202) 588–1500
Fax: (202) 588–0697

## UKRAINE

Embassy of Ukraine
3350 M Street NW
Washington, DC 20007
Tel: (202) 333–0606
Fax: (202) 333–0817

The Ukraine Working Group
c/o U.S. Chamber of Commerce
1615 H Street NW
Washington, DC 20062
Tel: (202) 463–5482
Fax: (202) 463–3114

## UZBEKISTAN

Embassy of Uzbekistan
1511 K Street NW, Suites 619 and 623
Washington, DC 20005
Tel: (202) 638–4266
Fax: (202) 638–4268

American Business Club/Services
American–Uzbek Chamber of Commerce
1225 I Street, Suite 520
Washington, DC 20005
Tel: (202) 682–4718
Fax: (202) 789–1056

Uzbek–American Business Center
237 Park Avenue, Suite 20, First Floor
New York, NY 10017
Tel: (212) 580–0800
Fax: (212) 580–0010

The Eurasia Foundation
1527 New Hampshire Avenue NW
Washington, DC 20036
Tel: (202) 234–7370
Fax: (202) 234–7377
E-Mail: eurasia@eurasia.org

# Appendix F

# Bibliography


Stephen Batalden and Sandra Batalden, *The Newly Independent States of Eurasia: Handbook of Former Soviet Republics* (Phoenix, AZ: Oryx, 1993).

Business Information Service for the Newly Independent States, *Commercial Overview of Armenia* (Washington, DC: U.S. Department of Commerce, January 1995).

Business Information Service for the Newly Independent States, *Commercial Overview of Belarus* (Washington, DC: U.S. Department of Commerce, June 1994).

Business Information Service for the Newly Independent States, *Commercial Overview of Kyrgyz Republic* (Washington, DC: U.S. Department of Commerce, 1994).

Business Information Service for the Newly Independent States, *Commercial Overview of Moldova* (Washington, DC: U.S. Department of Commerce, February 1995).

Business Information Service for the Newly Independent States, *Commercial Overview of Tajikistan* (Washington, DC: U.S. Department of Commerce, October 1994).

Business Information Service for the Newly Independent States, *Commercial Overview of Turkmenistan* (Washington, DC: U.S. Department of Commerce, 1994).

Business Information Service for the Newly Independent States, *Commercial Overview of Ukraine* (Washington, DC: U.S. Department of Commerce, 1994).

Business Information Service for the Newly Independent States, *Commercial Overview of Uzbekistan* (Washington, DC: U.S. Department of Commerce, December 1994).

Central Intelligence Agency, *The World Factbook* (annual).

*C.I.S. and Eastern Europe on File* (Facts on File, Inc., 1993).

J. Denis Derbyshire and Ian Derbyshire, *Political Systems of the World* (Oxford: Helicon, 1996).

*Eastern Europe and the Commonwealth of Independent States 1994* (London: Europa Publications Limited, 1994).

*EIU Country Profile: Armenia,* 4th quarter, 1994 (Economic Intelligence Unit, Limited, 1994).

*EIU Country Report: Armenia,* 4th quarter, 1994 (Economic Intelligence Unit, Limited, 1994).

*EIU Country Profile: Azerbaijan*, 4th quarter, 1994 (Economic Intelligence Unit, Limited, 1994).

*EIU Country Report: Azerbaijan,* 4th quarter, 1994 (Economic Intelligence Unit, Limited, 1994).

*EIU Country Profile: Belarus,* 4th quarter, 1994 (Economic Intelligence Unit, Limited, 1994).

*EIU Country Report: Belarus,* 4th quarter, 1994 (Economic Intelligence Unit, Limited, 1994).

*EIU Country Profile: Georgia,* 4th quarter, 1994 (Economic Intelligence Unit, Limited, 1994).

*EIU Country Report: Georgia,* 4th quarter, 1994 (Economic Intelligence Unit, Limited, 1994).

*EIU Country Profile: Kazakhstan,* 4th quarter, 1994 (Economic Intelligence Unit, Limited, 1994).

*EIU Country Report: Kazakhstan,* 4th quarter, 1994 (Economic Intelligence Unit, Limited, 1994).

*EIU Country Profile: Kyrgystan,* 4th quarter, 1994 (Economic Intelligence Unit, Limited, 1994).

*EIU Country Report: Kyrgystan,* 4th quarter, 1994 (Economic Intelligence Unit, Limited, 1994).

*EIU Country Profile: Moldova,* 4th quarter, 1994 (Economic Intelligence Unit, Limited, 1994).

*EIU Country Report: Moldova,* 4th quarter, 1994 (Economic Intelligence Unit, Limited, 1994).

*EIU Country Profile: Russia,* 4th quarter, 1994 (Economic Intelligence Unit, Limited, 1994).

*EIU Country Report: Russia,* 4th quarter, 1994 (Economic Intelligence Unit, Limited, 1994).

*EIU Country Profile: Tajikistan,* 4th quarter, 1994 (Economic Intelligence Unit, Limited, 1994).

*EIU Country Report: Tajikistan,* 4th quarter, 1994 (Economic Intelligence Unit, Limited, 1994).

*EIU Country Profile: Turkmenistan,* 4th quarter, 1994 (Economic Intelligence Unit, Limited, 1994).

*EIU Country Report: Turkmenistan,* 4th quarter, 1994 (Economic Intelligence Unit, Limited, 1994).

*EIU Country Profile: Ukraine,* 4th quarter, 1994 (Economic Intelligence Unit, Limited, 1994).

*EIU Country Report: Ukraine,* 4th quarter, 1994 (Economic Intelligence Unit, Limited, 1994).

*EIU Country Profile: Uzbekistan,* 4th quarter, 1994 (Economic Intelligence Unit, Limited, 1994).

*EIU Country Report: Uzbekistan,* 4th quarter, 1994 (Economic Intelligence Unit, Limited, 1994).

Foreign Broadcast Information Service, *Central Eurasia: Daily Report* (Washington, DC: Foreign Broadcast Information Service).

Theodore W. Karasik, editor, *Russia and Eurasia: Facts and Figures Annual*, Volume 19 (Gulf Breeze, FL: Academic International Press, 1994).

*Keesing's Record of World Events,* 1991, vol. 37; 1992, vol. 38; 1993, vol. 39; 1994, vol. 40; 1995, vol. 41 (Longman Group, UK Limited: Harlow).

*Leksykon Panstw Swiata '94/95* (Warsaw: Kronika, 1994).

Open Media Research Institute, *The OMRI Annual Survey of Eastern Europe and the Former Soviet Union*, 1995: Building Democracy (Armonk, NY: M.E. Sharpe, 1996).

*The Statesman's Year-Book* (London: Macmillan, annual).

*Transcaucasus: A Chronology,* 1992, vols. 1 & 2 1993, vol. 3; 1994, vol. 4; 1995, vol. 4, No. 1–9. (Armenian National Committee of America: Washington, DC).

*World Reference Atlas* (New York: Dorling Kindersley Publishing, 1994).

# General Subject Index

# Name Index

# Geographic Index